Comprehensive Organometallic Chemistry II

A Review of the Literature 1982–1994

Comprehensive Organometallic Chemistry II

A Review of the Literature 1982–1994

Editors-in-Chief

Edward W. Abel
University of Exeter, UK

F. Gordon A. Stone
Baylor University, Waco, TX, USA

Geoffrey Wilkinson
Imperial College of Science, Technology and Medicine, London, UK

Volume 9

NICKEL, PALLADIUM, AND PLATINUM

Volume Editor

Richard J. Puddephatt
University of Western Ontario, London, ON, Canada

PERGAMON

UK	Elsevier Science Ltd., The Boulevard, Langford Lane, Kidlington, Oxford OX5 1GB, UK
USA	Elsevier Science Inc., 660 White Plains Road, Tarrytown, New York 10591-5153, USA
JAPAN	Elsevier Science Japan, Tsunashima Building Annex, 3-20-12 Yushima, Bunkyo-ku, Tokyo 113, Japan

First edition 1995

Library of Congress Cataloging in Publication Data
Comprehensive organometallic chemistry II : a review of the literature 1982–1994 / editors-in-chief, Edward W. Abel, F. Gordon A. Stone, Geoffrey Wilkinson
p. cm.
Includes indexes.
1. Organometallic chemistry. I. Abel, Edward W. II. Stone, F. Gordon A. III. Wilkinson, Geoffrey.
QD411.C652 1995
547'.05—dc20 95–7030

British Library Cataloguing in Publication Data
A catalogue record for this book is available from the British Library.

ISBN 0–08–040608–4 (set : alk. paper)
ISBN 0–08–042316–7 (Volume 9)

Important note
For safety reasons, readers should always consult the list of abbreviations on p. xi before making use of the experimental details provided.

∞™ The paper used in this publication meets the minimum requirements of the American National Standard for Information Sciences—Permanence of Paper for Printed Library Materials, ANSI Z39.48–1984.

Chemical structures drawn by Synopsys Scientific Systems Ltd., Leeds, UK.

Printed and bound in Great Britain by BPC Wheatons Ltd., Exeter, UK.

Contents

Preface		vii
Preface to 'Comprehensive Organometallic Chemistry'		vii
Contributors to Volume 9		ix
Abbreviations		xi
Contents of All Volumes		xv
1	Nickel Complexes with Carbonyl, Isocyanide, and Carbene Ligands C. P. KUBIAK, *Purdue University, West Lafayette, IN, USA*	1
2	Nickel–Carbon σ-Bonded Complexes A. K. SMITH, *University of Liverpool, UK*	29
3	Nickel–Carbon π-Bonded Complexes M. J. CHETCUTI, *University of Notre Dame, IN, USA*	107
4	Palladium Complexes with Carbonyl, Isocyanide and Carbene Ligands K. R. DIXON and A. C. DIXON, *University of Victoria, BC, Canada*	193
5	Palladium–Carbon σ-Bonded Complexes A. J. CANTY, *University of Tasmania, Hobart, Tas, Australia*	225
6	Palladium–Carbon π-Bonded Complexes J. A. DAVIES, *University of Toledo, OH, USA*	291
7	Platinum Complexes with Carbonyl, Isocyanide and Carbene Ligands R. J. CROSS, *University of Glasgow, UK*	391
8	Platinum–Carbon σ-Bonded Complexes G. K. ANDERSON, *University of Missouri–St. Louis, MO, USA*	431
9	Platinum–Carbon π-Bonded Complexes G. B. YOUNG, *Imperial College of Science, Technology and Medicine, London, UK*	533
Author Index		589
Subject Index		639

Preface

'Comprehensive Organometallic Chemistry', published in 1982, was well received and remains very highly cited in the primary journal literature. Since its publication, studies on the chemistry of molecules with carbon–metal bonds have continued to expand rapidly. This is due to many factors, ranging from the sheer intellectual challenge and excitement provided by the continuing production of novel results, which demand new ideas, through to the successful application of organometallic species in organic syntheses, the generation of living catalysts for polymerization, and the synthesis of precursors for materials employed in the electronic and ceramic industries. For many reasons, therefore, we judged it timely to update 'Comprehensive Organometallic Chemistry' with a new work.

Due to the scope and depth of this area of chemistry, to have merely revised each of the original nine volumes did not seem the most user-friendly or cost-effective procedure to follow. As a consequence of the sheer bulk of the literature of the subject, a revised edition would necessarily require either the elimination of much chemistry of archival value but which is still important, or the production of a set of volumes significantly larger in number than the original nine. Accordingly, we decided it would be best to use the original work as a basis for new volumes focusing on organometallic chemistry reported since 1982, with reference back to the original work when necessary. For ease of use the new volumes maintain the same general structure as employed previously but reflect the changes in substance and direction the field has undergone in the last ten years. Thus it is not surprising that the largest volume in the new work concerns the role of the transition elements in metal-mediated organic syntheses.

The expansion of organometallic chemistry since the early 1980s also led us to decide that an updating of 'Comprehensive Organometallic Chemistry' would be more effectively accomplished if each volume had one or two editors who would be responsible both for recruiting experts for the Herculean task of writing the many chapters of each volume and for overseeing the content. We are deeply indebted to the volume editors and their authors for the time and effort they have given to the project.

As with the original 'Comprehensive Organometallic Chemistry', published some thirteen years ago, we hope this new version will serve as a pivotal reference point for new work and will function to generate new ideas and perceptions for the continued advance of what will surely continue as a vibrant area of chemistry.

Edward W. Abel
Exeter, UK

F. Gordon A. Stone
Waco, Texas, USA

Geoffrey Wilkinson
London, UK

Preface to 'Comprehensive Organometallic Chemistry'

Although the discovery of the platinum complex that we now know to be the first π-alkene complex, $K[PtCl_3(C_2H_4)]$, by Zeise in 1827 preceded Frankland's discovery (1849) of diethylzinc, it was the latter that initiated the rapidly developing interest during the latter half of the nineteenth century in compounds with organic groups bound to the elements. This era may be considered to have reached its apex in the discovery by Grignard of the magnesium reagents which occupy a special place because of their ease of synthesis and reactivity. With the exception of trimethylplatinum chloride discovered by Pope, Peachy and Gibson in 1907 by use of the Grignard reagent, attempts to make stable transition metal alkyls and aryls corresponding to those of main group elements met with little success, although it is worth recalling that even in 1919 Hein and his co-workers were describing the 'polyphenyl-chromium' compounds now known to be arene complexes.

The other major area of organometallic compounds, namely metal compounds of carbon monoxide, originated in the work starting in 1868 of Schützenberger and later of Mond and his co-workers and was subsequently developed especially by Hieber and his students. During the first half of this century, aided by the use of magnesium and, later, lithium reagents the development of main group organo chemistry was quite rapid, while from about 1920 metal carbonyl chemistry and catalytic reactions of carbon monoxide began to assume importance.

In 1937 Krause and von Grosse published their classic book 'Die Chemie der Metallorganischen Verbindungen'. Almost 1000 pages in length, it listed scores of compounds, mostly involving metals of the main groups of the periodic table. Compounds of the transition elements could be dismissed in 40 pages. Indeed, even in 1956 the stimulating 197-page monograph 'Organometallic Compounds' by Coates adequately reviewed organo transition metal complexes within 27 pages.

Although exceedingly important industrial processes in which transition metals were used for catalysis of organic reactions were developed in the 1930s, mainly in Germany by Reppe, Koch, Roelen, Fischer and Tropsch and others, the most dramatic growth in our knowledge of organometallic chemistry, particularly of transition metals, has stemmed from discoveries made in the middle years of this century. The introduction in the same period of physical methods of structure determination (infrared, nuclear magnetic resonance, and especially single-crystal X-ray diffraction) as routine techniques to be used by preparative chemists allowed increasingly sophisticated exploitation of discoveries. Following the recognition of the structure of ferrocene, other major advances quickly followed, including the isolation of a host of related π-complexes, the synthesis of a plethora of organometallic compounds containing metal–metal bonds, the characterization of low-valent metal species in which hydrocarbons are the only ligands, and the recognition from dynamic NMR spectra that ligand site exchange and tautomerism were common features in organometallic and metal carbonyl chemistry. The discovery of alkene polymerization using aluminium alkyl–titanium chloride systems by Ziegler and Natta and of the Wacker palladium–copper catalysed ethylene oxidation led to enormous developments in these areas.

In the last two decades, organometallic chemistry has grown more rapidly in scope than have the classical divisions of chemistry, leading to publications in journals of all national chemical societies, the appearance of primary journals specifically concerned with the topic, and the growth of annual review volumes designed to assist researchers to keep abreast of accelerating developments.

Organometallic chemistry has become a mature area of science which will obviously continue to grow. We believe that this is an appropriate time to produce a comprehensive review of the subject, treating organo derivatives in the widest sense of both main group and transition elements. Although advances in transition metal chemistry have appeared to dominate progress in recent years, spectacular progress has, nevertheless, also been made in our knowledge of organo compounds of main group elements such as aluminium, boron, lithium and silicon.

In these Volumes we have assembled a compendium of knowledge covering contemporary organometallic and carbon monoxide chemistry. In addition to reviewing the chemistry of the elements individually, two Volumes survey the use of organometallic species in organic synthesis and in catalysis, especially of industrial utility. Within the other Volumes are sections devoted to such diverse topics as the nature of carbon–metal bonds, the dynamic behaviour of organometallic compounds in solution, heteronuclear metal–metal bonded compounds, and the impact of organometallic compounds on the environment. The Volumes provide a unique record, especially of the intensive studies conducted during the past 25 years. The last Volume of indexes of various kinds will assist readers seeking information on the properties and synthesis of compounds and on earlier reviews.

As Editors, we are deeply indebted to all those who have given their time and effort to this project. Our Contributors are among the most active research workers in those areas of the subject that they have reviewed and they have well justified international reputations for their scholarship. We thank them sincerely for their cooperation.

Finally, we believe that 'Comprehensive Organometallic Chemistry', as well as providing a lasting source of information, will provide the stimulus for many new discoveries since we do not believe it possible to read any of the articles without generating ideas for further research.

E. W. ABEL
Exeter

F. G. A. STONE
Bristol

G. WILKINSON
London

Contributors to Volume 9

Professor G. K. Anderson
Department of Chemistry, University of Missouri–St. Louis, 8001 Natural Bridge Road, St. Louis, MO 63121, USA

Professor A. J. Canty
Chemistry Department, University of Tasmania, Hobart, Tas 7001, Australia

Professor M. J. Chetcuti
Department of Chemistry and Biochemistry, University of Notre Dame, Notre Dame, IN 46556, USA

Dr. R. J. Cross
Department of Chemistry, University of Glasgow, Glasgow, G12 8QQ, UK

Professor J. A. Davies
Department of Chemistry, University of Toledo, 2801 West Bancroft Street, Toledo, OH 43606, USA

Dr. A. C. Dixon
Department of Chemistry, University of Victoria, PO Box 3055, Victoria, British Columbia, V8W 3P6, Canada

Professor K. R. Dixon
Department of Chemistry, University of Victoria, PO Box 3055, Victoria, British Columbia, V8W 3P6, Canada

Professor C. P. Kubiak
Department of Chemistry, Purdue University, 1393 Brown Laboratory, West Lafayette, IN 47907, USA

Dr. A. K. Smith
Department of Chemistry, University of Liverpool, PO Box 147, Liverpool, L69 3BX, UK

Dr. G. B. Young
Department of Chemistry, Imperial College of Science, Technology and Medicine, South Kensington, London, SW7 2AY, UK

Abbreviations

The abbreviations used throughout 'Comprehensive Organometallic Chemistry II' are consistent with those used in 'Comprehensive Organometallic Chemistry' and with other standard texts in this area. The abbreviations in some instances may differ from those commonly used in other branches of chemistry.

Ac	acetyl
acac	acetylacetonate
AIBN	2,2'-azobisisobutyronitrile
Ar	aryl
arphos	1-(diphenylphosphino)-2-(diphenylarsino)ethane
Azb	azobenzene
9-BBN	9-borabicyclo[3.3.1]nonyl
9-BBN-H	9-borabicyclo[3.3.1]nonane
BHT	2,6-di-*t*-butyl-4-methylphenol (butylated hydroxytoluene)
bipy	2,2'-bipyridyl
t-BOC	*t*-butoxycarbonyl
bsa	*N*,*O*-bis(trimethylsilyl)acetamide
bstfa	*N*,*O*-bis(trimethylsilyl)trifluoroacetamide
btaf	benzyltrimethylammonium fluoride
Bz	benzyl
can	ceric ammonium nitrate
cbd	cyclobutadiene
1,5,9-cdt	cyclododeca-1,5,9-triene
chd	cyclohexadiene
chpt	cycloheptatriene
[Co]	cobalamin
(Co)	cobaloxime $[Co(DMG)_2]$ derivative
cod	1,5-cyclooctadiene
cot	cyclooctatetraene
Cp	η^5-cyclopentadienyl
Cp*	pentamethylcyclopentadienyl
18-crown-6	1,4,7,10,13,16-hexaoxacyclooctadecane
CSA	camphorsulfonic acid
csi	chlorosulfonyl isocyanate
Cy	cyclohexyl
dabco	1,4-diazabicyclo[2.2.2]octane
dba	dibenzylideneacetone
dbn	1,5-diazabicyclo[4.3.0]non-5-ene
dbu	1,8-diazabicyclo[5.4.0]undec-7-ene
dcc	dicyclohexylcarbodiimide
dcpe	1,2-bis(dicyclohexylphosphino)ethane
ddq	2,3-dichloro-5,6-dicyano-1,4-benzoquinone
deac	diethylaluminum chloride
dead	diethyl azodicarboxylate
depe	1,2-bis(diethylphosphino)ethane
depm	1,2-bis(diethylphosphino)methane
det	diethyl tartrate (+ or –)

DHP	dihydropyran
diars	1,2-bis(dimethylarsino)benzene
dibal-H	diisobutylaluminum hydride
dien	diethylenetriamine
DIGLYME	bis(2-methoxyethyl)ether
diop	2,3-*O*-isopropylidene-2,3-dihydroxy-1,4-bis(diphenylphosphino)butane
dipt	diisopropyl tartrate (+ or –)
dma	dimethylacetamide
dmac	dimethylaluminum chloride
DMAD	dimethyl acetylenedicarboxylate
dmap	4-dimethylaminopyridine
DME	dimethoxyethane
DMF	*N,N'*-dimethylformamide
DMG	dimethylglyoximate
DMI	*N,N'*-dimethylimidazalone
dmpe	1,2-bis(dimethylphosphino)ethane
dmpm	bis(dimethylphosphino)methane
DMSO	dimethyl sulfoxide
dmtsf	dimethyl(methylthio)sulfonium fluoroborate
dpam	bis(diphenylarsino)methane
dppb	1,4-bis(diphenylphosphino)butane
dppe	1,2-bis(diphenylphosphino)ethane
dppf	1,1'-bis(diphenylphosphino)ferrocene
dpph	1,6-bis(diphenylphosphino)hexane
dppm	bis(diphenylphosphino)methane
dppp	1,3-bis(diphenylphosphino)propane
eadc	ethylaluminum dichloride
edta	ethylenediaminetetraacetate
eedq	*N*-ethoxycarbonyl-2-ethoxy-1,2-dihydroquinoline
en	ethylene-1,2-diamine
Et_2O	diethyl ether
F_6 acac	hexafluoroacetylacetonate
Fc	ferrocenyl
Fp	$Fe(CO)_2Cp$
HFA	hexafluoroacetone
hfacac	hexafluoroacetylacetonate
hfb	hexafluorobut-2-yne
HMPA	hexamethylphosphoramide
hobt	hydroxybenzotriazole
$IpcBH_2$	isopinocampheylborane
Ipc_2BH	diisopinocampheylborane
kapa	potassium 3-aminopropylamide
K-selectride	potassium tri-*s*-butylborohydride
LAH	lithium aluminum hydride
LDA	lithium diisopropylamide
LICA	lithium isopropylcyclohexylamide
LITMP	lithium tetramethylpiperidide
L-selectride	lithium tri-*s*-butylborohydride
LTA	lead tetraacetate
mcpba	*m*-chloroperbenzoic acid
MeCN	acetonitrile
MEM	methoxyethoxymethyl
MEM-Cl	β-methoxyethoxymethyl chloride

Mes	mesityl
mma	methyl methacrylate
mmc	methylmagnesium carbonate
MOM	methoxymethyl
Ms	methanesulfonyl
MSA	methanesulfonic acid
MsCl	methanesulfonyl chloride
nap	1-naphthyl
nbd	norbornadiene
NBS	*N*-bromosuccinimide
NCS	*N*-chlorosuccinimide
nmo	*N*-methylmorpholine *N*-oxide
NMP	*N*-methyl-2-pyrrolidone
Nu^-	nucleophile
ox	oxalate
pcc	pyridinium chlorochromate
pdc	pyridinium dichromate
phen	1,10-phenanthroline
phth	phthaloyl
ppa	polyphosphoric acid
ppe	polyphosphate ester
$[PPN]^+$	$[(Ph_3P)_2N]^+$
ppts	pyridinium *p*-toluenesulfonate
py	pyridine
pz	pyrazolyl
Red-Al	sodium bis(2-methoxyethoxy)aluminum dihydride
sal	salicylaldehyde
salen	*N,N'*-bis(salicylaldehydo)ethylenediamine
SEM	β-trimethylsilylethoxymethyl
tas	tris(diethylamino)sulfonium
tasf	tris(diethylamino)sulfonium difluorotrimethylsilicate
tbaf	tetra-*n*-butylammonium fluoride
TBDMS	*t*-butyldimethylsilyl
TBDMS-Cl	*t*-butyldimethylsilyl chloride
TBDPS	*t*-butyldiphenylsilyl
tbhp	*t*-butyl hydroperoxide
TCE	2,2,2-trichloroethanol
TCNE	tetracyanoethene
TCNQ	7,7,8,8-tetracyanoquinodimethane
terpy	2,2':6',2"-terpyridyl
tes	triethylsilyl
Tf	triflyl (trifluoromethanesulfonyl)
TFA	trifluoracetic acid
TFAA	trifluoroacetic anhydride
tfacac	trifluoroacetylacetonate
THF	tetrahydrofuran
THP	tetrahydropyranyl
tipbs-Cl	2,4,6-triisopropylbenzenesulfonyl chloride
tips-Cl	1,3-dichloro-1,1,3,3-tetraisopropyldisiloxane
TMEDA	tetramethylethylenediamine [1,2-bis(dimethylamino)ethane]
TMS	trimethylsilyl
TMS-Cl	trimethylsilyl chloride
TMS-CN	trimethylsilyl cyanide
Tol	tolyl
tpp	*meso*-tetraphenylporphyrin

Tr	trityl (triphenylmethyl)
tren	2,2',2''-triaminotriethylamine
trien	triethylenetetraamine
triphos	1,1,1-tris(diphenylphosphinomethyl)ethane
Ts	tosyl
TsMIC	tosylmethyl isocyanide
ttfa	thallium trifluoroacetate

Contents of All Volumes

Volume 1 Lithium, Beryllium, and Boron Groups
1 Alkali Metals
2 Beryllium
3 Magnesium, Calcium, Strontium and Barium
4 Compounds with Three- or Four-coordinate Boron, Emphasizing Cyclic Systems
5 Boron Rings Ligated to Metals
6 Polyhedral Carbaboranes
7 Main-group Heteroboranes
8 Metallaboranes
9 Transition Metal Metallacarbaboranes
10 Aluminum
11 Gallium, Indium and Thallium, Excluding Transition Metal Derivatives
12 Transition Metal Complexes of Aluminum, Gallium, Indium and Thallium
Author Index
Subject Index

Volume 2 Silicon Group, Arsenic, Antimony, and Bismuth
1 Organosilanes
2 Carbacyclic Silanes
3 Organopolysilanes
4 Silicones
5 Germanium
6 Tin
7 Lead
8 Arsenic, Antimony and Bismuth
Author Index
Subject Index

Volume 3 Copper and Zinc Groups
1 Gold
2 Copper and Silver
3 Mercury
4 Cadmium and Zinc
Author Index
Subject Index

Volume 4 Scandium, Yttrium, Lanthanides and Actinides, and Titanium Group
1 Zero Oxidation State Complexes of Scandium, Yttrium and the Lanthanide Elements
2 Scandium, Yttrium, and the Lanthanide and Actinide Elements, Excluding their Zero Oxidation State Complexes
3 Titanium Complexes in Oxidation States Zero and Below
4 Titanium Complexes in Oxidation States +2 and +3
5 Titanium Complexes in Oxidation State +4
6 Zirconium and Hafnium Complexes in Oxidation States Zero and Below
7 Metallocene(II) Complexes of Zirconium and Hafnium
8 Zirconium and Hafnium Compounds in Oxidation State +3
9 Bis(cyclopentadienyl)zirconium and -hafnium Halide Complexes in Oxidation State +4
10 Bis(cyclopentadienyl) Metal(IV) Compounds with Si, Ge, Sn, N, P, As, Sb, O, S, Se, Te or Transition Metal-centred Ligands
11 Zirconium and Hafnium Complexes in Oxidation State +4

12 Cationic Organozirconium and Organohafnium Complexes
13 Cyclooctatetraene Complexes of Zirconium and Hafnium
Author Index
Subject Index

Volume 5 Vanadium and Chromium Groups
1 Vanadium
2 Niobium and Tantalum
3 Hexacarbonyls and Carbonyl Complexes of Carbon σ-Bonded Ligands of Chromium, Molybdenum and Tungsten
4 Carbonyl Complexes of Noncarbon σ-Bonded Ligands of Chromium, Molybdenum and Tungsten
5 Organometallic Complexes of Chromium, Molybdenum and Tungsten without Carbonyl Ligands
6 π-Complexes of Chromium, Molybdenum and Tungsten, Excluding those of Cyclopentadienyls and Arenes
7 Cyclopentadienyl Complexes of Chromium, Molybdenum and Tungsten
8 Arene and Heteroarene Complexes of Chromium, Molybdenum and Tungsten
Author Index
Subject Index

Volume 6 Manganese Group
1 Manganese Carbonyls and Manganese Carbonyl Halides
2 Manganese Alkyls and Hydrides
3 Manganese Complexes Containing Nonmetallic Elements
4 Manganese Hydrocarbon Complexes Excluding Cyclopentadienyl
5 Cyclopentadienyl Manganese Complexes
6 Manganese Nitrosyl and Isonitrile Complexes
7 High-valent Organomanganese Compounds
8 Technetium
9 Low-valent Organorhenium Compounds
10 High-valent Organorhenium Compounds
Author Index
Subject Index

Volume 7 Iron, Ruthenium, and Osmium
1 Iron Compounds without Hydrocarbon Ligands
2 Mononuclear Iron Compounds with η^1–η^6 Hydrocarbon Ligands
3 Dinuclear Iron Compounds with Hydrocarbon Ligands
4 Polynuclear Iron Compounds with Hydrocarbon Ligands
5 Introduction to Organoruthenium and Organoosmium Chemistry
6 Mononuclear Complexes of Ruthenium and Osmium Containing η^1 Carbon Ligands
7 Complexes of Ruthenium and Osmium Containing η^2–η^6 Hydrocarbon Ligands: (i) Complexes not Containing Cyclobutadiene, Cyclopentadienyl or η-Arene Coligands
8 Complexes of Ruthenium and Osmium Containing η^2–η^6 Hydrocarbon Ligands: (ii) Complexes Containing Four- and Five-membered Rings (Including MCp(arene) Complexes)
9 Complexes of Ruthenium and Osmium Containing η^2–η^6 Hydrocarbon Ligands: (iii) Complexes Containing Six-, Seven- and Eight-membered Rings
10 Ruthenocenes and Osmocenes
11 Binuclear Complexes of Ruthenium and Osmium Containing Metal–Metal Bonds
12 Trinuclear Clusters of Ruthenium and Osmium: (i) Introduction and Simple Neutral, Anionic and Hydrido Clusters
13 Trinuclear Clusters of Ruthenium and Osmium: (ii) Hydrocarbon Ligands on Metal Clusters
14 Trinuclear Clusters of Ruthenium and Osmium: (iii) Clusters with Metal–Carbon Bonds to Heteroatom Ligands
15 Tetranuclear Clusters of Ruthenium and Osmium
16 Medium- and High-nuclearity Clusters of Ruthenium and Osmium
Author Index
Subject Index

Volume 8 Cobalt, Rhodium, and Iridium
1 Cobalt
2 Rhodium
3 Iridium
4 Cluster Complexes of Cobalt, Rhodium, and Iridium
Author Index
Subject Index

Volume 9 Nickel, Palladium, and Platinum
1 Nickel Complexes with Carbonyl, Isocyanide, and Carbene Ligands
2 Nickel–Carbon σ-Bonded Complexes
3 Nickel–Carbon π-Bonded Complexes
4 Palladium Complexes with Carbonyl, Isocyanide and Carbene Ligands
5 Palladium–Carbon σ-Bonded Complexes
6 Palladium–Carbon π-Bonded Complexes
7 Platinum Complexes with Carbonyl, Isocyanide and Carbene Ligands
8 Platinum–Carbon σ-Bonded Complexes
9 Platinum–Carbon π-Bonded Complexes
Author Index
Subject Index

Volume 10 Heteronuclear Metal–Metal Bonds
1 Synthesis of Compounds Containing Heteronuclear Metal–Metal Bonds
2 Heterodinuclear Compounds
3 Heteronuclear Clusters Containing C_1, C_2, C_3, ..., C_n Acyclic Hydrocarbyl Ligands
4 Binary Carbonyls, Carbonyls plus Hydrides, Carbonyls plus Phosphines, Cyclic Hydrocarbyls and Main-group Ligands without Acyclic Hydrocarbyls
5 Cluster Complexes with Bonds Between Transition Elements and Copper, Silver and Gold
6 Cluster Complexes with Bonds Between Transition Elements and Zinc, Cadmium, and Mercury
7 Catalysis and Related Reactions with Compounds Containing Heteronuclear Metal–Metal Bonds
Author Index
Subject Index

Volume 11 Main-group Metal Organometallics in Organic Synthesis
1 Lithium
2 Sodium and Potassium
3 Magnesium
4 Zinc and Cadmium
5 Boron
6 Aluminum
7 Silicon
8 Tin
9 Mercury
10 Thallium
11 Lead
12 Antimony and Bismuth
13 Selenium
14 Tellurium
Author Index
Subject Index

Volume 12 Transition Metal Organometallics in Organic Synthesis
1 Introduction and Fundamentals
2 Transition Metal Hydrides: Hydrocarboxylation, Hydroformylation, and Asymmetric Hydrogenation
3.1 Transition Metal Alkyl Complexes from Hydrometallation
3.2 Transition Metal Alkyl Complexes from RLi and CuX
3.3 Transition Metal Alkyl Complexes: Main-group Transmetallation and Insertion Chemistry
3.4 Transition Metal Alkyl Complexes: Oxidative Addition and Transmetallation

3.5 Transition Metal Alkyl Complexes: Oxidative Addition and Insertion
3.6 Transition Metal Alkyl Complexes: Multiple Insertion Cascades
3.7 Transition Metal Alkyl Complexes: Reductive Dimerization of Alkenes and Alkynes
4 Transition Metal Carbonyl Complexes
5.1 Transition Metal Carbene Complexes: Cyclopropanation
5.2 Transition Metal Carbene Complexes: Diazodecomposition, Ylide, and Insertion
5.3 Transition Metal Carbene Complexes: Alkyne and Vinyl Ketene Chemistry
5.4 Transition Metal Carbene Complexes: Photochemical Reactions of Carbene Complexes
5.5 Transition Metal Carbene Complexes: Tebbe's Reagent and Related Nucleophilic Alkylidenes
6.1 Transition Metal Alkene, Diene, and Dienyl Complexes: Nucleophilic Attack on Alkene Complexes
6.2 Transition Metal Alkene, Diene, and Dienyl Complexes: Complexation of Dienes for Protection
6.3 Transition Metal Alkene, Diene, and Dienyl Complexes: Nucleophilic Attack on Diene and Dienyl Complexes
7.1 Transition Metal Alkyne Complexes: Transition Metal-stabilized Propargyl Systems
7.2 Transition Metal Alkyne Complexes: Pauson–Khand Reaction
7.3 Transition Metal Alkyne Complexes: Transition Metal-catalyzed Cyclotrimerization
7.4 Transition Metal Alkyne Complexes: Zirconium–Benzyne Complexes
8.1 Transition Metal Allyl Complexes: Telomerization of Dienes
8.2 Transition Metal Allyl Complexes: Pd, W, Mo-assisted Nucleophilic Attack
8.3 Transition Metal Allyl Complexes: Intramolecular Alkene and Alkyne Insertions
8.4 Transition Metal Allyl Complexes: Trimethylene Methane Complexes
8.5 Transition Metal Allyl Complexes: π-Allylnickel Halides and Other π-Allyl Complexes Excluding Palladium
9.1 Transition Metal Arene Complexes: Nucleophilic Addition
9.2 Transition Metal Arene Complexes: Ring Lithiation
9.3 Transition Metal Arene Complexes: Side-chain Activation and Control of Stereochemistry
10 Synthetically Useful Coupling Reactions Promoted by Ti, V, Nb, W, Mo Reagents
11.1 Transition Metal-catalyzed Oxidations: Asymmetric Epoxidation
11.2 Transition Metal-catalyzed Oxidations: Asymmetric Hydroxylation
11.3 Transition Metal-catalyzed Oxidations: Other Oxidations
12.1 Transition Metals in Polymer Synthesis: Ziegler–Natta Reaction
12.2 Transition Metals in Polymer Synthesis: Ring-opening Metathesis Polymerization and Other Transition Metal Polymerization Techniques
Author Index
Subject Index

Volume 13 Structure Index

Structures of Organometallic Compounds Determined by Diffraction Methods

Volume 14 Cumulative Indexes

Cumulative Formula Index
Cumulative Subject Index

1
Nickel Complexes with Carbonyl, Isocyanide, and Carbene Ligands

CLIFFORD P. KUBIAK
Purdue University, West Lafayette, IN, USA

1.1 INTRODUCTION 1
1.2 MONONUCLEAR NICKEL COMPLEXES 2
1.2.1 Complexes with Carbonyl Ligands 2
1.2.2 Complexes with Carbon Dioxide and Related Ligands 5
1.2.3 Complexes with Isocyanide Ligands 10
1.2.4 Complexes with Carbene Ligands 12
1.2.5 Related Electron Pair Donor Ligands 14
1.3 DINUCLEAR NICKEL COMPLEXES 15
1.3.1 Complexes with Carbonyl Ligands 15
1.3.2 Complexes with Isocyanide Ligands 17
1.3.3 Complexes with Carbene Ligands 18
1.4 HIGHER CLUSTER COMPLEXES 19
1.4.1 Clusters with Carbonyl Ligands 19
1.4.2 Clusters with Isocyanide Ligands 22
1.4.3 Clusters with Carbene and Carbyne Ligands 23
1.5 SUMMARY 24
1.6 REFERENCES 24

1.1 INTRODUCTION

The organometallic chemistry of nickel has a long history, dating back to the preparation in 1890 of nickel carbonyl [$Ni(CO)_4$], and its use in the Mond process for refining nickel metal.[1] In general, the chemistry of nickel is dominated by the divalent Ni^{2+} and zero-valent Ni^0 oxidation states, although examples of compounds of Ni^+, Ni^{3+}, and even Ni^{4+} are known.[2] In the case of Ni^+, there is a tendency towards disproportionation, as shown in Equation (1). The alkyls and hydrides of nickel are predominantly compounds of Ni^{2+}. The presence of π-acceptor ligands such as carbonyls, isocyanides, and carbenes tends to stabilize lower oxidation states and these ligands are often encountered in zero-valent nickel chemistry. This chapter reviews the organometallic chemistry of nickel complexes with carbonyl, isocyanide, and carbene ligands. Emphasis has been placed on literature reported since publication of *COMC-I*. However, where it is appropriate for the purposes of historical perspective, reference is made to earlier work. Nickel complexes with carbonyl, isocyanide, and carbene ligands are of considerable importance as catalysts for processes including alkyl and aryl cross-coupling, alkene polymerization, hydrocyanation of alkenes, and alkene isomerization. The reader is directed to the appropriate sections of *COMC-II* (see Chapters 4, 5.1, 5.3, and 5.5, Volume 12), where the catalytic applications of these classes of compounds are discussed in more detail.

$$2[NiL_n]^+ \longrightarrow [NiL_n]^{2+} + [NiL_n] \quad (1)$$

1.2 MONONUCLEAR NICKEL COMPLEXES

1.2.1 Complexes with Carbonyl Ligands

The field of nickel organometallic chemistry began with the dramatic discovery in 1890 by Mond, Langer, and Quincke of nickel tetracarbonyl.[1] Nickel is the only metal that reacts at a significant rate with CO at STP (Equation (2)). The significance of the facile formation of nickel carbonyl by the direct combination of nickel metal and carbon monoxide was quickly realized by Mond. The formation of nickel carbonyl, which has a reported boiling point of 42 °C at 760 mm Hg, and its subsequent decomposition at higher temperatures to nickel metal and carbon monoxide is the basis for the Mond process for the purification of nickel metal, which is still used today. The carbonyl $[Ni(CO)_4]$ is a d^{10}, 18-valence-electron complex with a tetrahedral structure. Molecular orbital calculations suggest that a partial net positive charge resides on the nickel atom, an indication of the importance of nickel to CO ligand backbonding.

$$Ni + 4\,CO \xrightarrow[1\ atm]{25\ ^\circ C} [Ni(CO)_4] \quad (2)$$

The substitution of the carbonyl groups of $[Ni(CO)_4]$ by other two-electron-donor ligands, for example, phosphine, isocyanide, and pyridyl-type ligands provides the basis for the preparation of an extensive series of related d^{10}, Ni^0 organometallic compounds. In view of the 18-electron configuration of $[Ni(CO)_4]$, the substitution reactions generally proceed by a CO dissociation mechanism (Equation (3)).[3] Over 300 phosphine and phosphite ligands as well as the related group 15 arsine, AsR_3, and stibine, SbR_3, ligands have been used in the preparation of monosubstituted $[Ni(CO)_3L]$ compounds. Most of these are tabulated in *COMC-I*. These nickel compounds include those with nonstandard types of electron pair donor ligands. For example, the first phosphaallene ($L = \eta^1$-RP=C=CPh$_2$) complex of the type $[Ni(CO)_3L]$ was reported and structurally characterized.[4] Typical of the metal carbonyls, photosubstitutions are reasonably efficient. Photolysis of $[Ni(CO)_4]$ in liquid krypton in the presence of nitrogen gave as an unstable transient species $[Ni(N_2)(CO)_3]$, which was characterized by infrared spectroscopy.[5] The Ni–N_2 bond energy was estimated to be only 42 kJ mol^{-1}.[6] Numerous examples of the disubstituted $[Ni(CO)_2L_2]$ and trisubstituted $[Ni(CO)L_3]$ compounds are also known. The disubstitution of $[Ni(CO)_4]$ by diphosphines, L–L, such as dppe to form chelating diphosphine complexes of the type $[Ni(CO)_2(L\text{–}L)]$ is rather general. The complete substitution to $[NiL_4]$ complexes, however, is seldom achieved since as more electron-donating ligands, L, are added, the remaining CO ligands are stabilized by backbonding. Steric considerations of phosphine cone angles are also important.[7,8] However, of the fully substituted $[NiL_4]$ compounds prepared from $[Ni(CO)_4]$, most have poor electron donor ligands such as $L = PF_3$, PCl_3, or $P(OR)_3$. The fully substituted trialkyl- and triarylphosphine complexes, $[Ni(PR_3)_4]$, are more conveniently prepared from Ni^0 cyclic alkene compounds such as the extensively used starting material, $[Ni(cod)_2]$.

$$[Ni(CO)_4] \underset{+CO}{\overset{-CO}{\rightleftharpoons}} [Ni(CO)_3] \xrightarrow[fast]{+L} [Ni(CO)_3L] \quad (3)$$

The Ni^0 dicarbonyl phosphine $[Ni(CO)_2L_2]$ complexes may also be prepared by reduction of Ni^{II} phosphine halide complexes $[NiX_2L_2]$ (X = Cl, Br, I) in the presence of CO. Reductive elimination of dialkyl Ni^{II} complexes in the presence of CO is also an effective route to $[Ni(CO)_2L_2]$ (Equation (4)).[9]

$$[NiMe_2(PEt_3)_2] \xrightarrow{CO} [Ni(CO)_2(PEt_3)_2] + Me_2CO \quad (4)$$

The reaction of $[Ni(SiCl_3)_2(\eta\text{-}C_6H_5Me)]$ with CO affords the five-coordinate complex $[Ni(SiCl_3)_2(CO)_3]$.[10] This complex is a rare example of a Ni^{II} carbonyl complex. The structure of this complex was established by x-ray diffraction to be trigonal bipyramidal with axial $SiCl_3$ ligands.[11]

Many nitrogen donor ligand carbonyl complexes of Ni^0 are known. The additions of amine and pyridine-type ligands generally lead to unstable species. However, the use of chelating nitrogen donor ligands such as 1,10-phenanthroline or 2,2'-bipyridine (bipy) leads to stable complexes (Equation (5)).

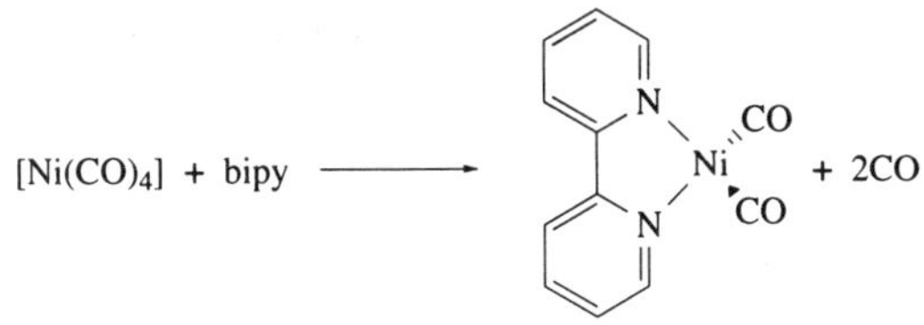

(5)

Phosphorus ylides react with $[Ni(CO)_4]$ to give monosubstituted ylide complexes with a direct Ni–C bond, as shown in Equation (6).[12]

$$[Ni(CO)_4] + MeCH{=}PCy_3 \longrightarrow [(CO)_3NiCHMePCy_3] + CO \qquad (6)$$

Nickel phosphine carbonyl and related complexes are readily oxidized to five-coordinate 18-electron nickel(II) dihalide complexes, as shown in Equation (7). Nickel carbonyl does not form a stable protonated species, in contrast with the isoelectronic $[Co(CO)_4]^-$ anion which protonates to form the acidic hydride $[HCo(CO)_4]$.

$$[Ni(CO)_2(PMe_3)_2] + X_2 \longrightarrow [NiX_2(CO)(PMe_3)_2] + CO \qquad (7)$$

The reaction of the functionalized P=C double bonds of methylenephosphanes, X–P=C(R)TMS (R = TMS, X = Cl, Cp*; R = Ph, X = Br, Cp*), with alkali-metal carbonylmetallates or carbonyl compounds {M} affords metal-substituted phosphaalkenes [{M}–P=C(R)TMS], in which the phosphorus is formally a one-electron donor to the metal center.[13]

The insertion of Ni^0 into maleic anhydride leads initially to the cyclic insertion product, but subsequently eliminates CO resulting in a nickel "metallolactone" product (Equation (8)).[14]

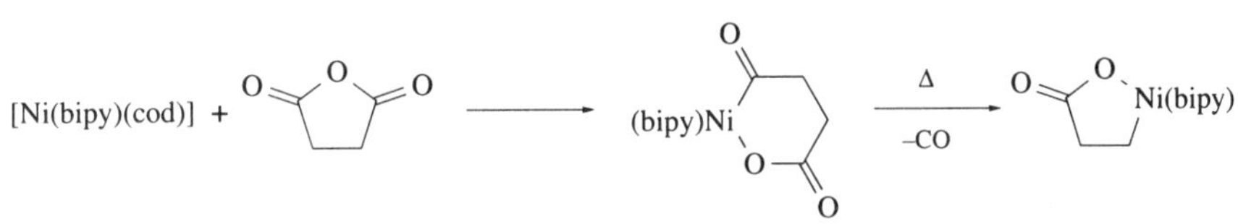

(8)

The carbonylation of nickel alkyls to afford acyls is a well-established reaction type. The nickel alkyls $[NiMe(PMe_3)_2(dmpe)][BF_4]$ and $[NiMe(dmpe)_2][BF_4]$ react with CO to give stable acyl species $[Ni\{C(O)Me\}(PMe_3)_2(dmpe)][BF_4]$ and $[Ni\{C(O)Me\}(dmpe)_2][BF_4]$, respectively. However, the dppe acyl complex $[Ni\{C(O)Me\}(PMe_3)_2(dppe)][BF_4]$ could only be observed at low temperature.[15] Oxidative addition of $ClCH_2CO_2Et$, $ClCH_2CN$, or $BrCH_2OMe$ to $[Ni(cod)_2]$ in the presence of two equivalents of PMe_3 affords the β-functionalized alkyl Ni^{II} complexes $[Ni(CH_2R)X(PMe_3)_2]$ (R = CO_2Et, X = Cl; R = CN, X = Cl; R = OMe, X = Br). The carbonylation of the methoxymethyl complex leads to a stable acyl complex, $[Ni\{C(O)CH_2OMe\}Br(PMe_3)_2]$. The complex $[Ni(CH_2CO_2Et)Cl(PMe_3)_2]$ forms the corresponding acyl complex reversibly; however, the acyl is only stable under an atmosphere of CO. The cyanomethyl complex does not form a stable acyl.[16] The insertion of phenylacetylene into the nickel–acyl carbon bonds of the complexes *trans*-$[Ni(COR)Cl(PMe_3)_2]$ yields the corresponding *trans*-(*Z*)-alkenyl ketone complex. The alkenyl ketone complexes react with CO with elimination of dimers of γ-but-2-enolactones (Equation (9)).[17] The dimeric α,β-butenolides can be produced in a one-pot synthesis, starting with the alkyl nickel complex $[Ni(R)Cl(PMe_3)_2]$, PhC≡CH, and CO. The crystal structure of the alkenyl ketone complex formed by insertion of phenylacetylene into the nickel–acyl carbon bond of the complex $[Ni(COCH_2TMS)Cl(PMe_3)_2]$ was determined to be distorted square pyramidal with an axial nickel–acyl oxygen interaction (**1**).[17]

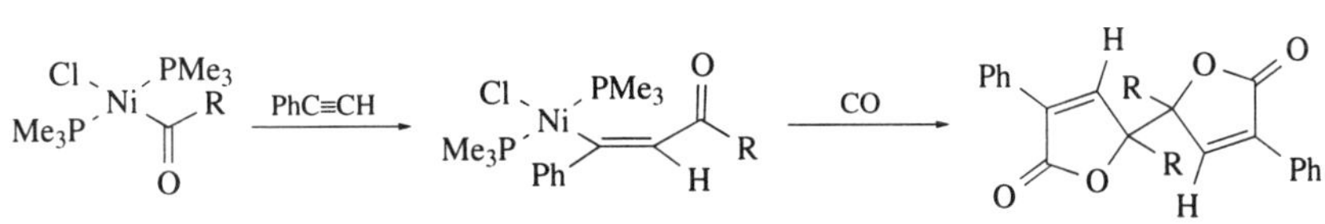

(9)

The carbonylation of nickel alkoxide complexes has also been investigated. The reactivity of a series of *trans*-$[NiX_2L_2]$ (X = Cl, Br, I; L = PMe_3, PEt_3, PEt_2Ph) complexes towards the alkoxides NaOR (R = Me, Et, Ph, CH_2Ph) under CO was studied. Only the methoxide complexes gave isolable methoxycarbonyls, *trans*-$[Ni\{C(O)OMe\}XL_2]$. The other alkoxides reacted to give a complex mixture

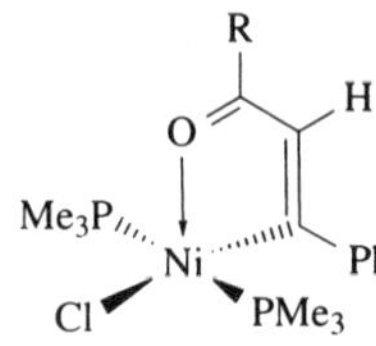

(1)

of organic and inorganic decomposition products.[18] A series of ethoxycarbonyl complexes, *trans*-[Ni{C(O)OEt}ClL$_2$] (L = PPh$_3$, PMe$_2$Ph, PCy$_3$, L$_2$ = bipy), have been prepared by the addition of ethyl chloroformate to a mixture of [Ni(cod)$_2$] and phosphine ligand.[19] The partitioning between insertions of CO into nickel–alkyl carbon vs. nickel–alkoxide oxygen bonds has been explored in connection with the carbonylation of nickel alkyl(alkoxide) complexes to give esters (Scheme 1).

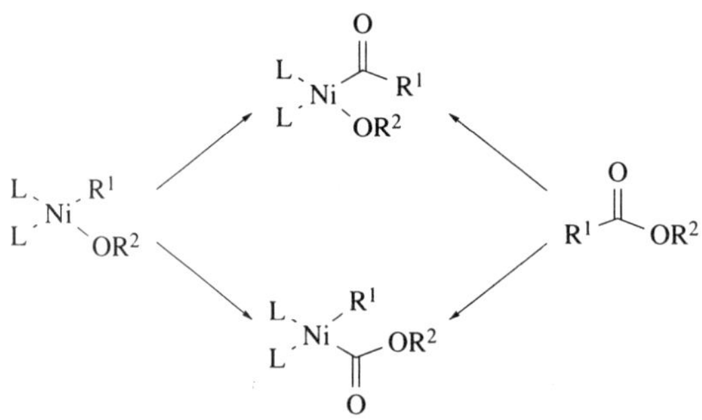

Scheme 1

Studies of the reactions of methyl(aryloxide) nickel(II) and palladium(II) complexes showed that CO insertion occurred preferentially into the metal–alkyl carbon bond to give acetyl(aryloxy) nickel and palladium complexes. Aryl acetates were thus formed by reductive elimination of the acetyl and aryloxy groups. Reductive elimination was promoted by CO and other π-acceptor ligands for the nickel complexes, while the corresponding palladium complexes did not show a promotion effect.[20,21] In platinum alkyl(alkoxide) complexes, however, CO insertion is generally observed into the Pt–O bond of the alkoxide instead of the Pt–C bond of the alkyl.[22–4] The platinum complexes are also relatively stable to reductive elimination. The monoalkyl(alkoxide) nickel complexes [NiMe(OR)(bipy)] and [NiEt(OR)(bipy)] react with CO at 1 atm pressure to give carboxylic esters in high yields. The reductive elimination of the esters is so rapid that it could not be determined whether these products were derived from an acyl(alkoxide) complex or an alkyl(alkoxycarbonyl) complex.[25]

The coupling and carbonylation of organic substrates often involve the use of [Ni(CO)$_4$]. Several examples of this chemistry are summarized below. A total synthesis of methylenomycin B was achieved in four steps starting with a cyclocondensation reaction of allyl chloride and 2-butyn-l-ol (Equation (10)).[26]

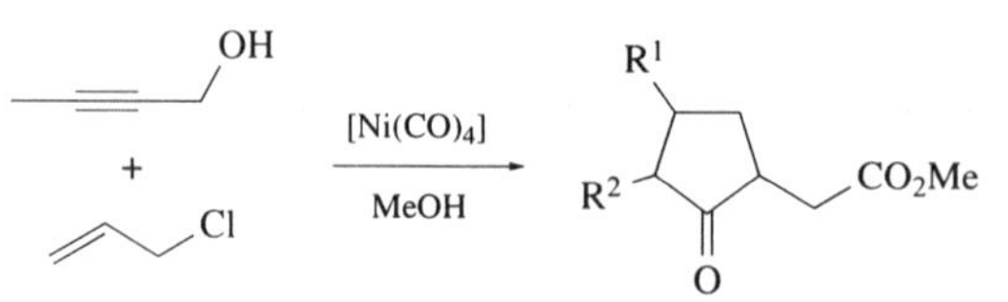

R^1 = Me; R^2 = CH$_2$OH, 80%
R^1 = CH$_2$OH; R^2 = Me, 20%

(10)

Various *gem*-dibromocyclopropanes bearing chloromethyl or mesyloxymethyl groups undergo [Ni(CO)$_4$] induced ring opening–carbonylation with alcohols or amines to afford γ,δ-unsaturated carboxylic acid derivatives (Equation (11)).[27] The silyl compounds *N,N'*-dialkyltrimethylsilylamine, 1-(trimethylsilyl)imidazole, and phenylthiotrimethylsilane have also been used in the carbonylation reactions of *gem*-dibromocyclopropanes.[28]

The [Ni(CO)$_4$] promoted additions of diphenylcyclopropenone to isothiocyanates and to CO$_2$ lead to heterocyclic spirans (Equation (12)).[29]

The anion [Ni(η-C$_5$H$_5$)(CO)]$^-$ is obtained by reduction of the dinuclear complex [Ni(η-C$_5$H$_5$)(CO)]$_2$. Potassium naphthalenide reduction gives approximately a 70% yield of K[Ni(η-C$_5$H$_5$)(CO)]. A recent

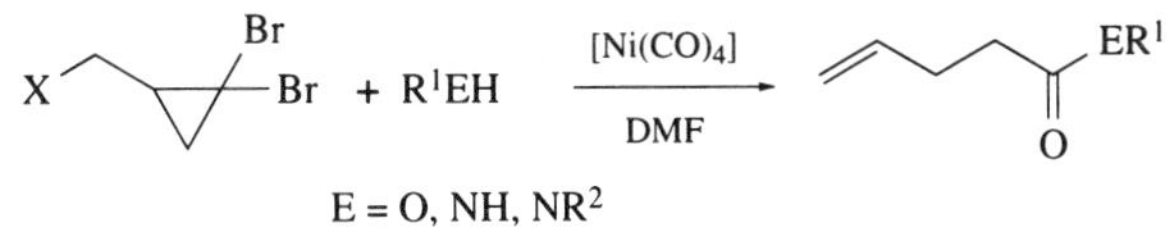

(11)

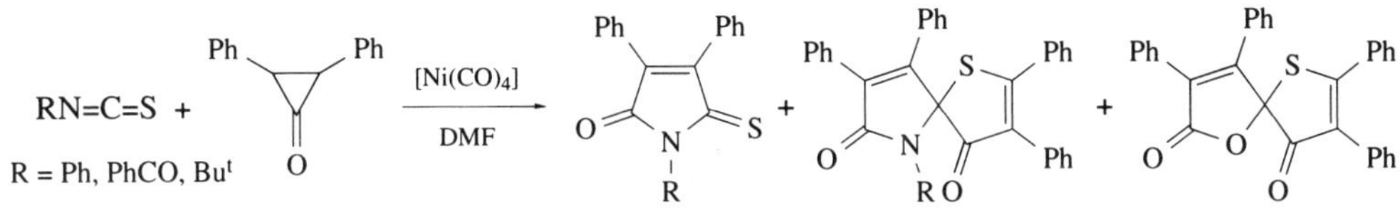

(12)

report of the use of C_8K, an alkali metal–graphite complex reducing agent,[30] claimed significantly higher (>95%) yields in the preparation of K[Ni(η-C_5H_5)(CO)].[31] The first examples of direct Ni–Ga and Ni–In bonds were brought about by the reaction of K[Ni(η-C_5H_5)(CO)] with R_2EX (E = Ga, In), as shown in Equation (13).[31]

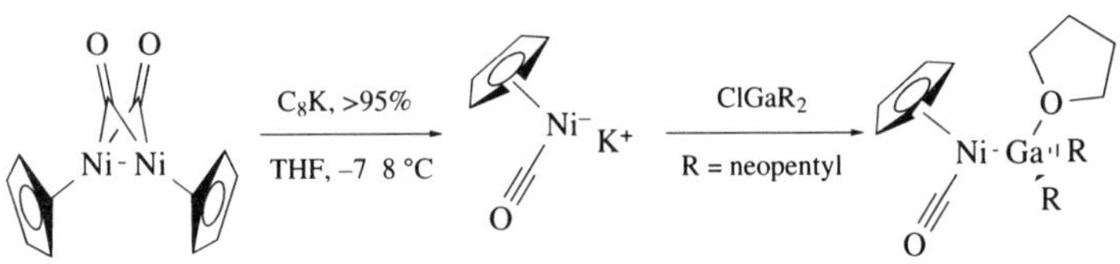

(13)

The absorption and thermal decomposition of [Ni(CO)$_4$] within the supercage of zeolite Y were studied spectroscopically. The [Ni(CO)$_4$] molecules are absorbed intact in cation-free zeolite Y and NaY, but subsequently decompose thermally with loss of CO. EXAFS data suggested that larger nickel aggregates are formed during thermal decomposition.[32,33]

1.2.2 Complexes with Carbon Dioxide and Related Ligands

Heteroatom alkenes including ketones, aldehydes, imines, oxygen, diazenes, and heterocumulenes including carbon dioxide, ketenes, isocyanates, and isothiocyanates form stable complexes with Ni^0. One of the first structurally characterized carbon dioxide complexes was prepared by the direct addition of CO_2 to [Ni(PCy$_3$)$_2$] (Equation (14)).[34] More recently, the toluene solvate of the bis(tricyclohexylphosphine) complex was obtained and structurally characterized after the reaction of *trans*-cyclododecatrienenickel(0) with CO_2 in the presence of the phosphine ligand.[35]

(14)

A novel azadienyl propylene diamine-type ligand has been used to form a reversible CO_2 carrier complex. The nitrogen atoms of the azadienyl ligands play an important role in the stabilization of the dimeric bis(CO_2) adduct (Equation (15)).[36]

The cocyclization of CO_2 with alkynes and alkenes has been a highly productive area of research. The nickel(0) trialkylphosphine complex catalyzed one-step bicyclic α-pyrone synthesis from diynes and CO_2 has been described.[37] The synthesis of silyl-substituted bicyclic α-pyrones by nickel(0) trialkylphosphine complex catalyzed cycloadditions of terminally silyl-substituted diynes with CO_2 illustrates this approach (Equation (16)).[38]

The nickel(0) bis(diphenylphosphino)butane complex catalyzed co-oligomerization of three molecules of ethoxy(trimethylsilyl)ethyne with one CO_2 led to a 90% yield of a vinyl-substituted pyrone (Equation (17)).[39] The copolymerization of CO_2 with acyclic and cyclic diynes that do not readily undergo intramolecular cyclization has been achieved. For example, when 3,11-tetradecadiyne was treated with CO_2 under pressure in a THF–MeCN solvent mixture at 110 °C in the presence of a tertiary phosphine nickel catalyst, [NiL$_2$], a CO_2–diyne alternating copolymer was obtained (Equation (18)).[40]

The reaction of either Ni^0/trialkylphosphine/acetonitrile[41] or [Ni(TMEDA)][42] with CO_2 and hex-3-yne results in the catalytic formation of tetraethyl-2-pyrone. In the case of the TMEDA complex the metallacycle formed initially can be observed to condense with a second equivalent of alkyne before reductive elimination of tetraethyl-2-pyrone (Equations (19) and (20)).[42]

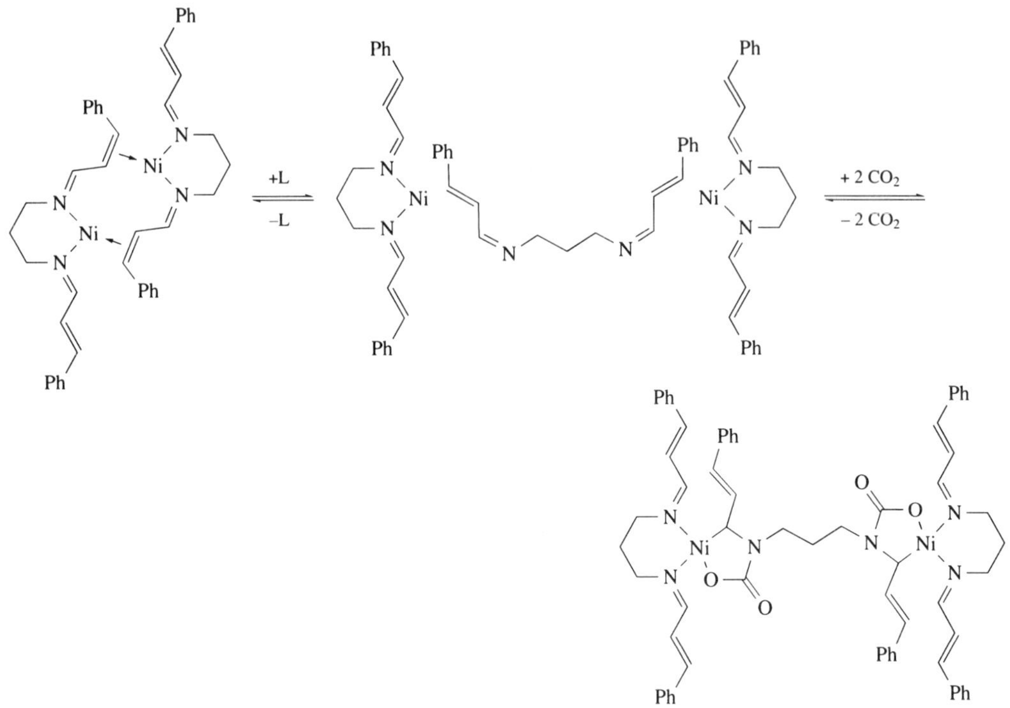

(15)

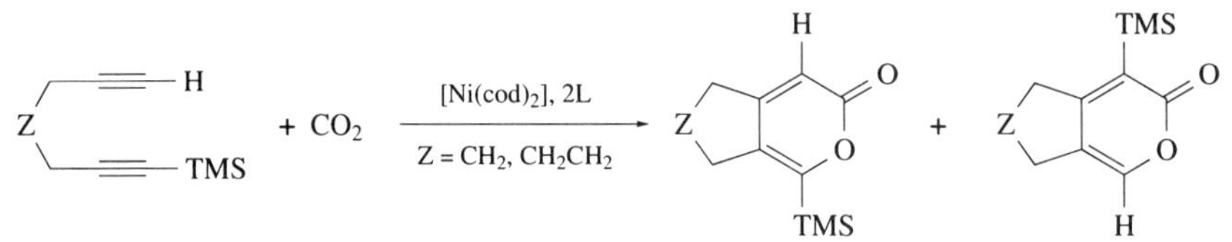

(16)

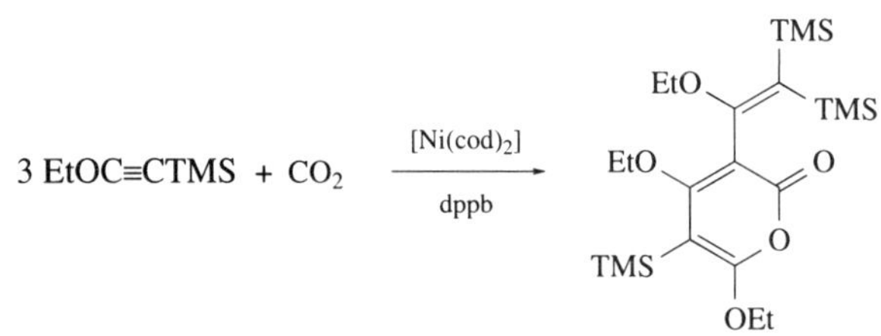

(17)

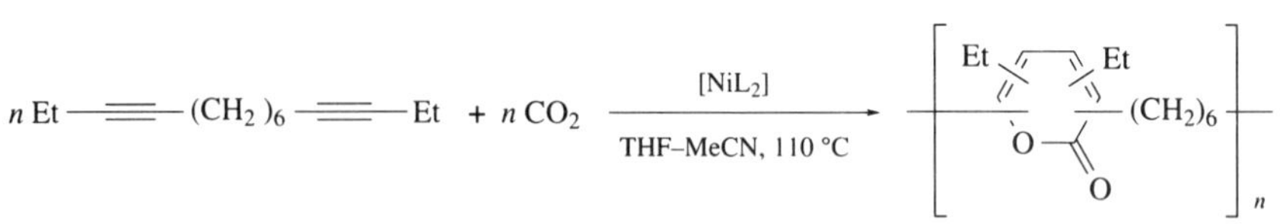

(18)

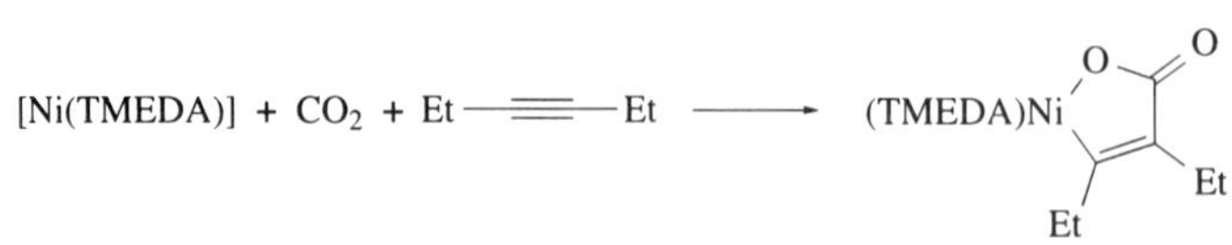

(19)

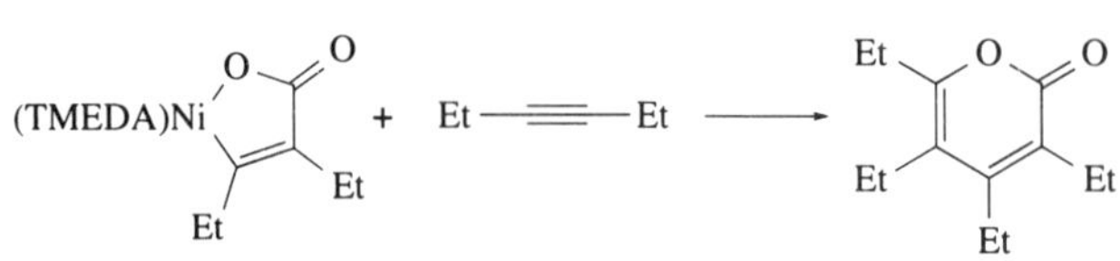

(20)

The formation of nickelacycles from CO_2 and alkenes has become the basis of an extensive organonickel-based chemistry for the carboxylation and dicarboxylation of alkenes (Equation (21)). For example, ethene undergoes coupling with CO_2 in the presence of Ni^0 to give both five- and seven-membered nickelacycles.[43] Depending on workup conditions, different carboxylic acids can be obtained. These include pentenoic acid which arises via β-elimination from the seven-membered nickelacycle, prior to hydrolysis. The dicyclohexylphosphinoethyl-2-pyridyl ligand has been used to prepare a five-membered nickelacycle from CO_2 and cyclopentene. Again, depending on workup conditions, an extensive range of functionalized cyclopentene products can be obtained, as shown in Scheme 2.[44]

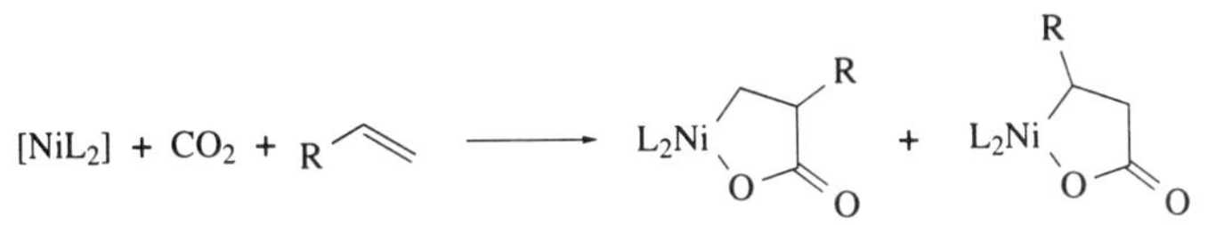

(21)

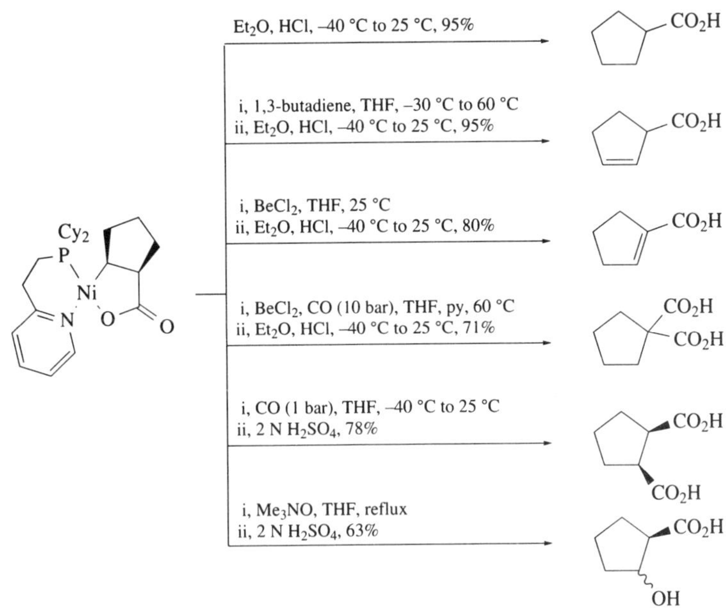

Scheme 2

Cyclooctene reacts with CO_2 under similar circumstances with the dicyclohexylphosphinoethyl-2-pyridylnickel system to give a five-membered nickelacycle. Upon protonolysis (HCl–Et_2O) cyclooctyl carboxylic acid is obtained, while treatment with dilute inorganic acids results in the formation of the β,γ-unsaturated cyclooctenyl carboxylic acid.[45] The five-membered nickelacycles formed from acyclic alkenes can be induced to undergo a ring contraction reaction to a four-membered nickelacycle (Equation (22)). This provides the basis for several new functionalization routes, including dicarboxylation and addition of ethene.[46]

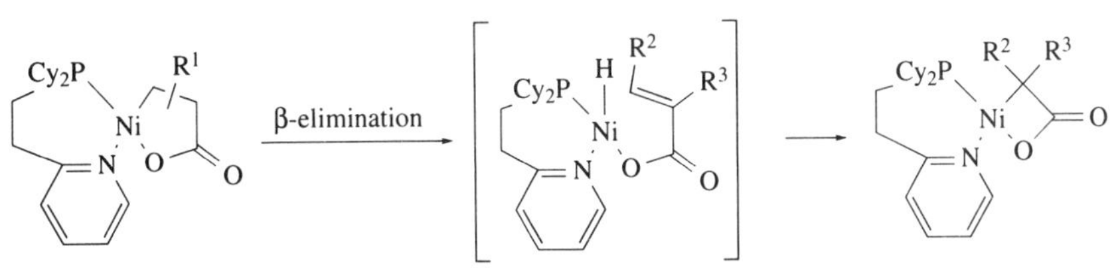

(22)

The five-membered nickelacycles derived from the coupling of an unsaturated hydrocarbon and CO_2 have also been employed in cross-coupling reactions to produce carboxylic acids by treatment of the nickelacycle with an alkyl halide (Equation (23)).[47]

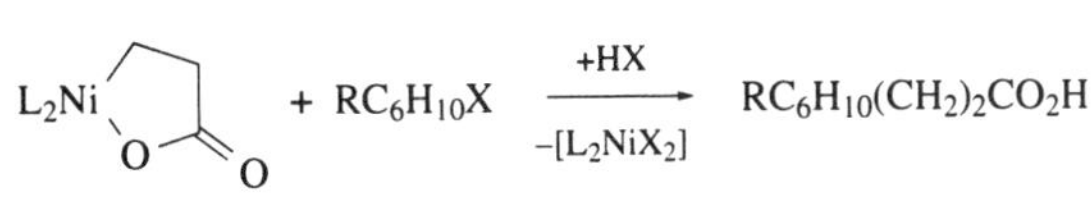

(23)

Isocyanates also have been investigated as substrates for alkene functionalization. Vinyl-substituted heterocycles such as 2-vinylfuran and 2-vinylpyridine react with phenyl isocyanate in the presence of dicyclohexylphosphinoethyl-2-pyridylnickel to form the five-membered 5-azanickelacyclopentan-4-one derivatives. Workup of the azanickelacyclopentanones affords normal and α,β-unsaturated amides.[48] The carbon–carbon bond-forming reactions between 1,3-butadiene and phenyl isocyanate can be conducted stoichiometrically or catalytically, depending on the phosphine ligands bound to nickel. The basic phosphines PEt_3 and PCy_3 induce stoichiometric coupling and azanickelacycles (Equation (24)).[49] Hydrolysis of the azanickelacycles affords the corresponding unsaturated amides, stoichiometrically. The use of the less basic triphenylphosphine results in a catalytic coupling of two 1,3-butadiene molecules and phenyl isocyanate to yield five-membered carbocyclic amides (**2**) and (**3**).[49]

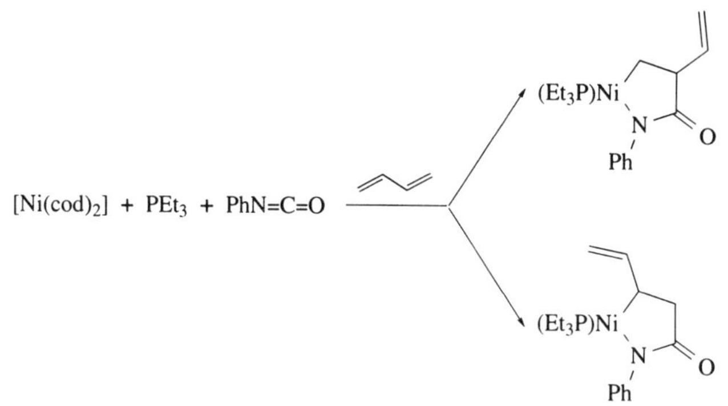

(24)

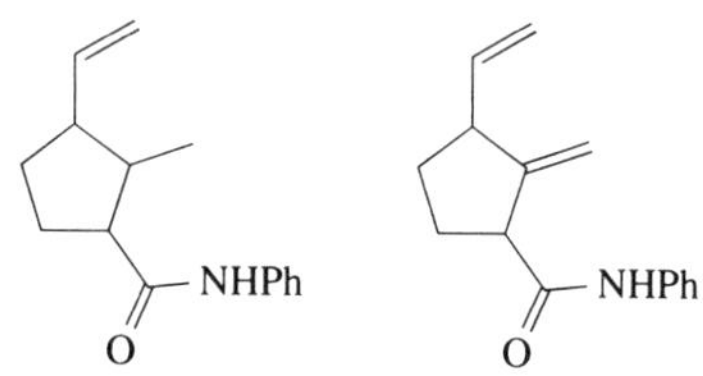

(**2**) (**3**)

The regioselectivity of the couplings of isocyanates and alkenes can be improved significantly with substituted alkenes. The reaction of styrene and phenyl isocyanate in the presence of $[Ni(cod)_2]$ and PCy_3 results in the regioselective formation of one azanickelacycle (Equation (25)).[50] The β-elimination of the azanickelacycle followed by reductive elimination of cinnamanilide constitute the basis for a nickel-catalyzed system for the conversion of styrene and isocyanates to cinnamanilides (Equation (26)).[50] Similar regioselective carbon–carbon bond-forming reactions between phenyl isocyanate and the monosubstituted alkenes $RCH{=}CH_2$ (R = OEt, SPh, CO_2Me) have also been reported.[51] The reaction of 1,1-difluoroethene with phenyl isocyanate in the presence of $[Ni(cod)_2]$ and PCy_3 results in the selective formation of tricyclohexylphosphine-5-azadifluoronickelacyclopentan-4-one (Equation (27)).[52] However, the decreased reactivity of the resulting fluorine-substituted azanickelacyclopentanones toward acid hydrolysis and electrophiles is quite different from the azanickelacyclopentanones derived from ordinary alkenes.

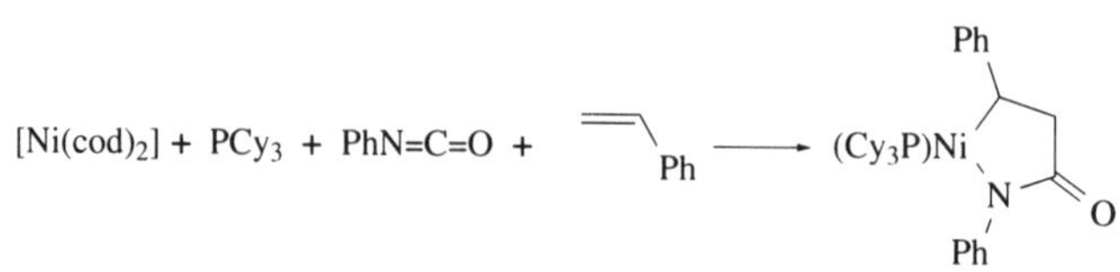

(25)

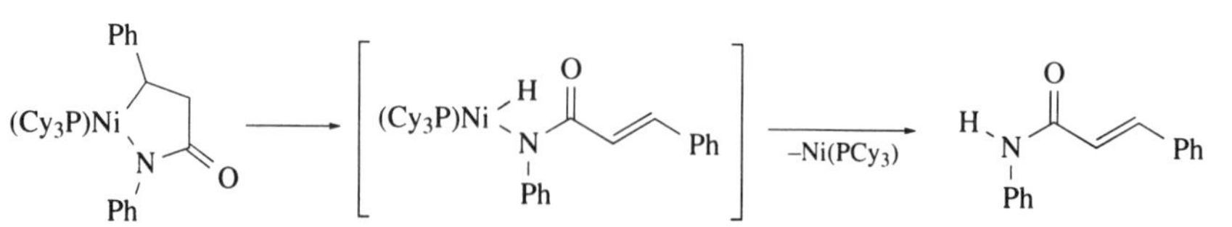

(26)

Chelating β-phosphinoenolate complexes of metals of the nickel triad have been investigated in the context of carbon–carbon bond-forming reactions with CO_2 and isocyanates. The chelating phosphinoenolate complex *cis*-$[Ni\{Ph_2PCH{=}C(O)Ph\}_2]$ reacts with aryl isocyanates with insertion of one isocyanate into each of the C–H bonds of the two coordinated enolates, namely *cis*-$[Ni\{Ph_2PC[C(O)Ph][{=}C(O)NH(Ar)]\}_2]$ (Equation (28)).[53]

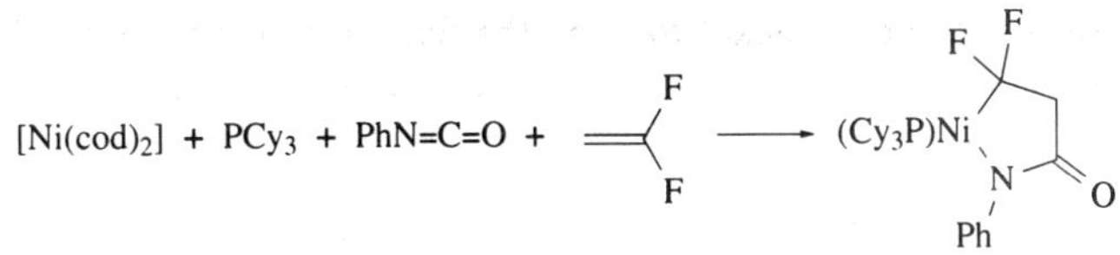

(27)

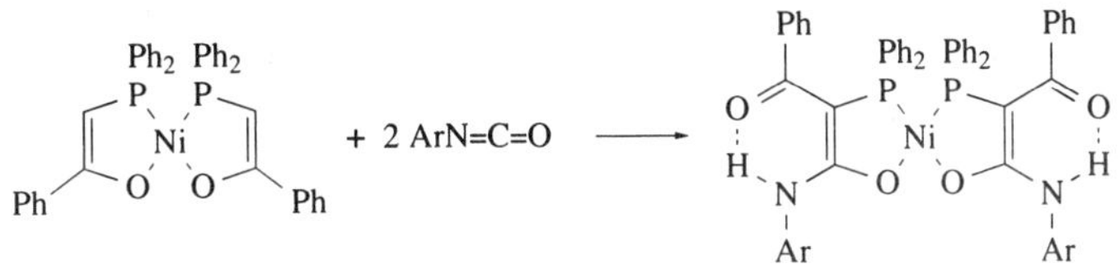

(28)

Complexes of nickel can also activate nitrous oxide toward oxygen atom transfer. The nickelacyclopentane bipyridyl complex $[Ni(C_4H_8)(bipy)]$ reacts with nitrous oxide in benzene solutions to effect oxygen atom insertion into a Ni–C bond giving the ring-expanded oxometallacycle (Equation (29)).[54]

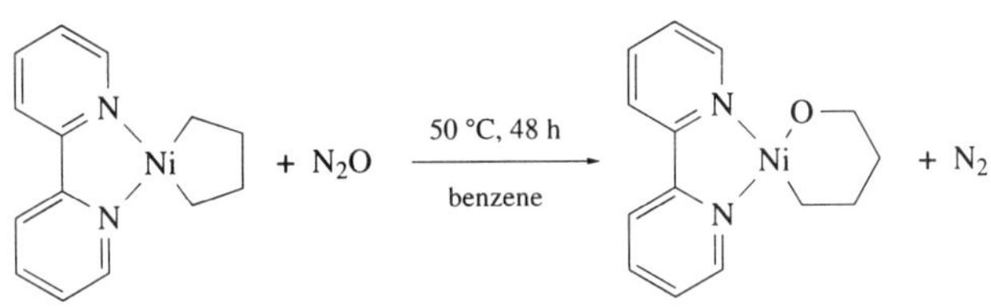

(29)

The simple ketenenickel(0) complex $[Ni(CH_2{=}C{=}O)(PPh_3)_2]$ has been reported to possess an η^2-(C,C)-bonded ketene molecule (**4**).[55] However, the first structurally characterized η^2-(C,O)-bonded nickel(0) ketene complex has now been described, and on the basis of spectroscopic data and theoretical studies it has been suggested that many of the early nickel ketene complexes may possess η^2-(C,O) structures (**5**).[56] A series of alkyl phenyl ketene complexes have been prepared of the type $[NiL_2\{Ph(R)C{=}C{=}O\}]$, and on the basis of spectroscopic data it was suggested that these too have η^2-(C,O) structures. For example, the reaction of $[Ni(cod)_2]$ with methyl phenyl ketene in the presence of triphenylphosphine leads to the formation of $[Ni(PPh_3)_2\{Ph(Me)C{=}C{=}O\}]$ which, on the basis of spectroscopic data, is assigned an η^2-(C,O) structure.[57] NMR data for these complexes, however, were interpreted in terms of a coordination mode switching between the limiting η^2-(C,O) and η^2-(C,C) structures via an associative process, presumed to involve dinuclear μ-ketene intermediates.[57–9]

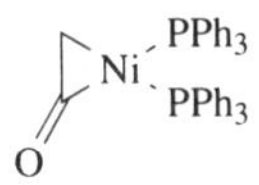

η^2-(C,C)
(**4**)

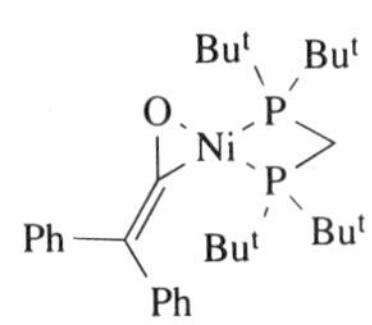

η^2-(C,O)
(**5**)

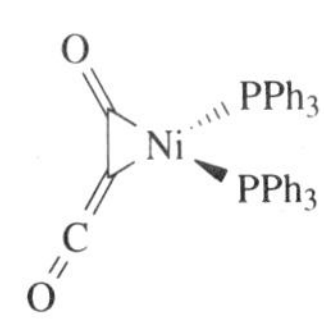

(**6**)

Carbon suboxide, O=C=C=C=O, has been shown to react with $[Ni(PPh_3)_2(cod)]$ to give the [(C,C'-η^2-C_3O_2)$Ni(PPh_3)_2$] complex (**6**).[60] Several complexes of nickel with CO_2 can be formed by apparent oxygen atom transfer to coordinated CO. For example, the complex $[L_2(NO_2)Ni{-}NO_2]$ is reported to slowly and reversibly form a CO adduct, $[L_2(NO_2)(CO)Ni{-}NO_2]$. This adduct is rapidly converted to the nitrosyl CO_2 complex $[L_2(NO_2)(CO_2)Ni{-}NO]$, which is reported to reversibly lose CO_2 with formation of the nitrosyl complex $[L_2(NO_2)Ni{-}NO]$.[61]

The electrocatalytic reduction of CO_2 to CO by $[Ni(cyclam)]^{2+}$ complexes has been reported. The intermediate species $[Ni(CO)(cyclam)]^+$ was observed by UV–visible and EPR spectroscopies. The role of cyclam N–H bonds in hydrogen bonding to CO_2 is discussed as a potentially key element of this electrocatalyst system.[62] A proposed intermediate in the electrocatalytic reduction of CO_2 by $[Ni(cyclam)]^{2+}$ is a $Ni^I{-}CO_2$ species. An *ab initio* molecular orbital study has suggested that a $Ni^I{-}CO_2$ intermediate is viable when a sixth axial ligand is present to facilitate Ni(d_{z^2}) to CO_2 charge transfer.[63]

The coupling of unsaturated hydrocarbons and carbon dioxide can be achieved under catalytic conditions in an electrochemical cell. A key component of the electrocatalytic carboxylation of unsaturated hydrocarbons is the use of a sacrificial anode (Mg, Al, or Zn). Thus, the electrocarboxylation of 4-octyne by the catalyst $[Ni(bipy)_2][BF_4]_2$ in DMF in the presence of a magnesium anode and CO_2 (1 atm, 20 °C) yields 2-propyl-2-hexenoic acid upon hydrolysis in 80% yield (E:Z = 85:15). The carboxylation of terminal alkynes under similar conditions leads to selective carboxylation in the 2-position of the alkyne.[64] The metal cations liberated by consumption of the sacrificial anode play an important role in the electrocatalysis.[65] Magnesium cations were shown to open the five-membered nickelacycle formed by coupling of the alkyne and CO_2 (Equation (30)). The electrochemical reaction stops after passage of 2 F mol^{-1} of electronic charge in the absence of Mg^{2+}. Nonconjugated diynes react with CO_2 to give products which depend on the ligands attached to nickel. A nickel–PMDTA (pentamethyldiethylenetriamine) system yields mainly linear monocarboxylic acids (Equation (31)), while the bipy complex gives primarily cyclic adducts. Diynes which possess both terminal and internal triple bonds are exclusively carboxylated at the terminal alkyne group.[66] In conjugated diyne systems, selective carboxylation occurs with the Ni–PMDTA system yielding (E)-2-vinyldiene-2-alkynyl carboxylic acids (Equation (32)).[67] Nonconjugated enynes selectively incorporate CO_2 at the triple bond, while conjugated enynes give the conjugated dienoic acids.[68] Alkenes and dienes are less reactive but can also be electrocatalytically carboxylated.[69]

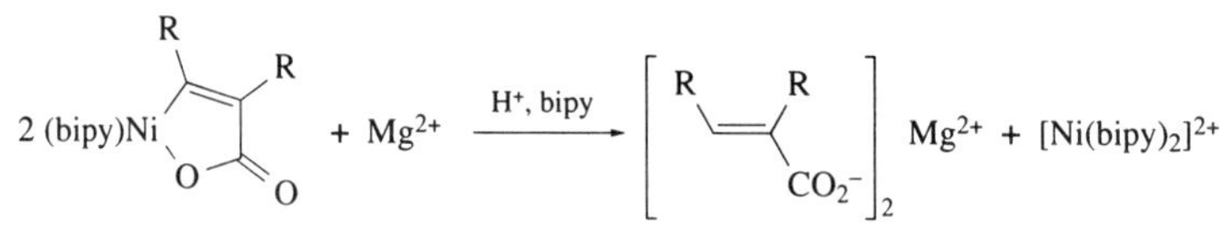

(30)

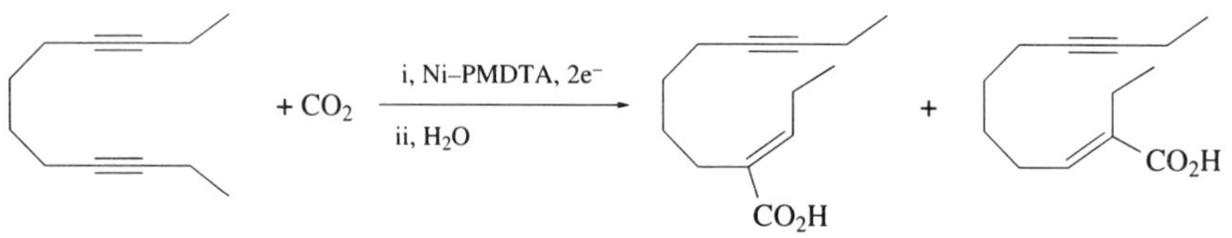

(31)

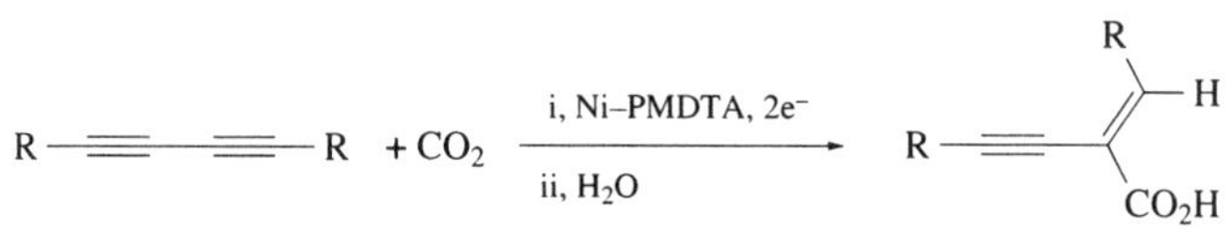

(32)

The catalytic reduction of CO_2, particularly to methane, has been an area of considerable research activity. The nickel-catalyzed reduction of CO_2 to methane has been reported to involve Ni(CO) intermediates.[70]

1.2.3 Complexes with Isocyanide Ligands

The reaction of $[Ni(CO)_4]$ with alkyl isocyanides favors the formation of the disubstituted $[Ni(CNR)_2(CO)_2]$ species owing to the greater σ-donor properties of the isocyanides and greater π-acceptor properties of CO. The variously substituted complexes, $[Ni(CNR)_n(CO)_{4-n}]$, have all been prepared, although the homoleptic isocyanides, $[Ni(CNR)_4]$, are difficult to prepare from $[Ni(CO)_4]$.[71] Only strongly π-accepting isocyanides give complete conversion of $[Ni(CO)_4]$ to the homoleptic isocyanides. For example, the reaction of $[Ni(CO)_4]$ with CF_3NC gave $[Ni(CNCF_3)_4]$. Subsequent reaction of $[Ni(CNCF_3)_4]$ with nickelocene yielded $[Ni(\eta\text{-}C_5H_5)(\mu\text{-}CNCF_3)]_2$.[72] In the case of less strongly π-accepting isocyanides, $[Ni(cod)_2]$ is a convenient starting material for the preparation of homoleptic nickel(0) isocyanide complexes. The first homoleptic isocyanosilane complex [Ni(TMS-NC)$_4$] has been prepared by the reaction of $[Ni(cod)_2]$ with trimethylcyanosilane. The reaction represents an efficient isomerization of a nitrile to an isocyanide.[73] The homoleptic isocyanide complexes can also be conveniently prepared in reasonable yields by the reaction of nickelocene with excess alkyl or aryl isocyanide (Equation (33)).[74,75]

$$[Ni(C_5H_5)_2] + CNR\ (excess) \longrightarrow [Ni(CNR)_4] + H_5C_5\text{–}C_5H_5 \qquad (33)$$

The diaryl isocyanide (**7**) was reacted with $[Ni(CO)_4]$ to produce apparently mononuclear $[Ni(CO)_2(L\text{–}L)]$ and $[Ni(L\text{–}L)_2]$ complexes.[76]

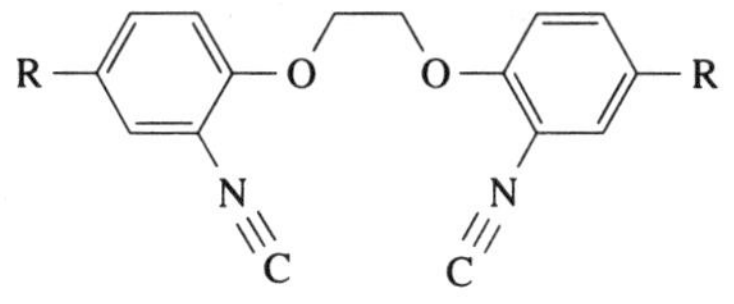

R = H, Me
(7)

The synthesis of $CN(CH_2)_nOH$ has been described. A stable open-chain isocyanoethanolnickel complex, $[Ni(CNCH_2CH_2OH)_2(PPh_3)_2][PF_6]_2$, was reported.[77] However, these ligands also readily undergo intramolecular cyclization to give carbenes (Section 1.2.4).[78]

Most Ni^0 complexes are moderately to very air sensitive and give intractable oxidation products in the presence of oxygen. However, several stable O_2 adducts have been reported. The product of the reaction of the isocyanide complex $[Ni(CNBu^t)_4]$ with oxygen has been structurally characterized by x-ray diffraction (Equation (34)).[79] The O_2 adducts have been investigated with respect to the oxidation of cyclohexene.[80,81] In general, the homoleptic Ni^0 isocyanide complexes, and particularly the alkyl isocyanides, appear to be more reactive toward electron deficient species than $[Ni(CO)_4]$. The reaction of $[Ni(CNR)_4]$ (R = Bu^t, Xy (xylyl)) with 3,4-diphenylcyclobut-3-ene-1,2-dione or 4,5-diphenylcyclopent-4-ene-1,2,3-trione leads to the formation of monosubstituted products (**8**) and (**9**).[82]

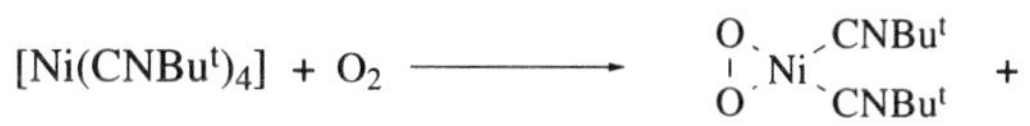

(34)

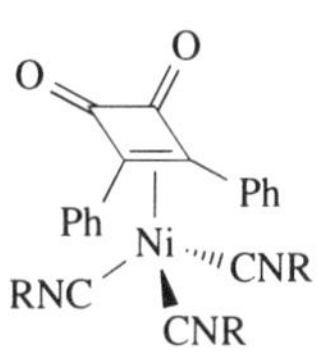

(8)

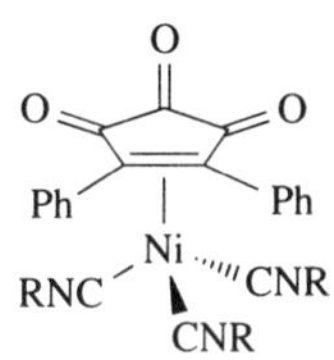

(9)

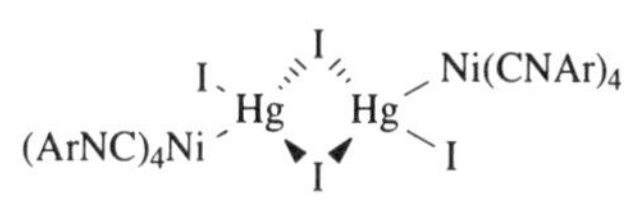

(10)

The reaction of $[NiI_2(CNR)_2]$ (R = 2,6-$Me_2C_6H_3$, 4-Br-2,6-$Me_2C_6H_2$, 2,4-Bu^t_2-6-Me-C_6H_2, 2,4,6-$Bu^t_3C_6H_2$) with CNR or PPh_3 gave $[NiI_2(CNR)_3]$ or $[NiI_2(PPh_3)(CNR)_2]$. Treatment of the complexes $[NiI_2(CNR)_2]$ with mercury in the presence of excess isocyanide results in the formation of the complexes $[Ni(CNR)_4HgI_2]_2$. The structure of the 4-Br-2,6-$Me_2C_6H_2$ isocyanide complex was determined by x-ray diffraction (**10**). These nickel–mercury complexes are viewed as donor–acceptor complexes. The Ni–Hg separation in the structurally characterized 4-Br-2,6-$Me_2C_6H_2$ isocyanide complex is 0.261 9(3) nm.[83]

Another nickel–mercury isocyanide complex, $[\{Ni_2(\mu\text{-CNMe})(CNMe)_4(dppm)\}Hg][NiCl_4]$, was found by x-ray diffraction to exist as a spirocyclic Ni_4Hg cluster with a mercury(II) center. The complex contains an $\{HgNi_4\}^{2+}$ core with direct Ni–Hg bonding.[84] The electrochemistry of $[NiX_2(CNR)_2]$ (X = Cl, Br, I; R = 2,6-$Me_2C_6H_3$, 4-Br-2,6-$Me_2C_6H_2$, 2,4-Bu^t_2-6-Me-C_6H_2, 2,4,6-$Bu^t_3C_6H_2$) complexes has been investigated at platinum and mercury electrodes. The reduction potential of the complexes $[NiX_2(CNXy)_2]$ (X = Br, I) shifted toward positive potentials by approximately 0.3 V in the presence of excess isocyanide due to the formation of the five-coordinate complex $[NiX_2(CNXy)_3]$ (X = Br, I). At a mercury electrode in the presence of excess isocyanide, the complex $[NiI_2(CNXy)_2]$ formed the species $[Ni(CNXy)_4HgI_2]_2$.[85]

The carbonylation of nickel alkyls to produce acyl species is a well-established reaction type (Section 1.2.1). Isocyanide insertion chemistry has been slower to develop, but has promising synthetic applications. In general, there are three differences between insertions of CO and insertions of isocyanides. These include: (i) the increased tendency of iminoacyl groups to bind to metals in an η^2 manner, compared with acyls; (ii) the increased tendency of isocyanides to undergo multiple insertions; and (iii) the increased stability of iminoacyl groups with respect to deinsertion.

Examples of multiple insertions of isocyanide ligands into nickel–alkyl carbon bonds are found in the reactions of *trans*-$[Ni(R)Cl(PMe_3)_2]$ (R = Me, CH_2TMS, $CH_2C_6H_4$-*o*-Me) with Bu^tNC.[86,87] The reaction of one or two equivalents of Bu^tNC with hexane solutions of *trans*-$[Ni(R)Cl(PMe_3)_2]$ (R = Me, CH_2TMS) leads to the mono- and di-insertion products (Equations (35) and (36)). The formation of the five-membered nickelacycle formed by the insertion of three equivalents of Bu^tNC was observed to proceed in the presence of an excess of Bu^tNC (Equation (37)).[87] Titration of *trans*-$[Ni(Me)Cl(PMe_3)_2]$

with varying amounts of Bu^{t}NC led to the formation of the acyclic mono-, bis-, and tris-isocyanide insertion products. The crystal structure of the $CH_2C_6H_4$-*o*-Me derivative of the cyclic tris-isocyanide insertion product was determined. In contrast with related palladium complexes, the reaction of *trans*-[Ni(R)Cl(PMe_3)$_2$] (R = Me, CH_2TMS) with cyclohexyl isocyanide gives only the mono-insertion iminoacyl products.[86]

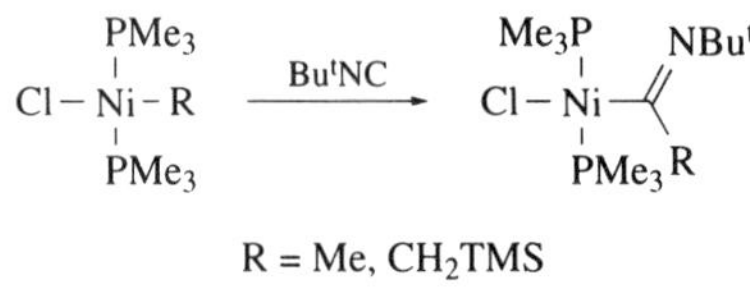
(35)

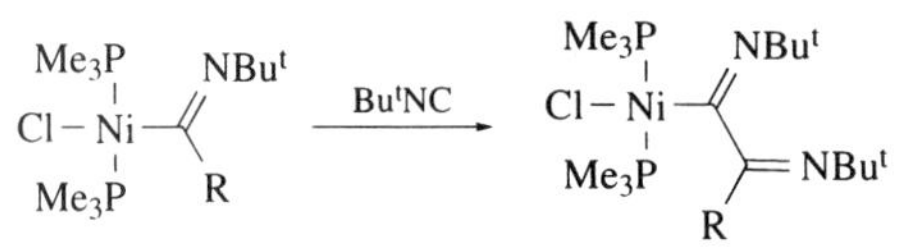
(36)

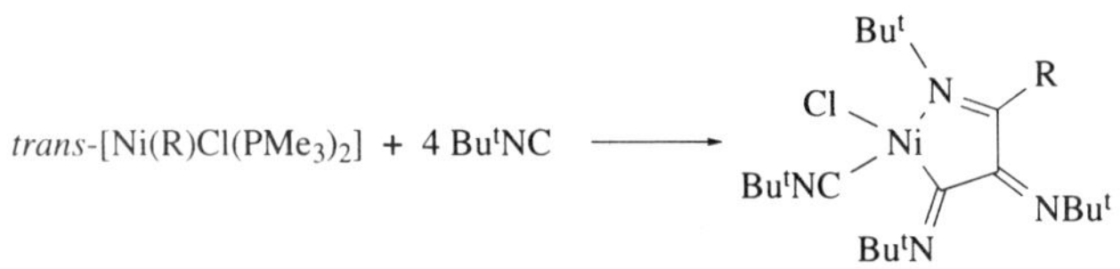
(37)

The *ortho*-η^3-benzylic dinuclear nickel complex [$Ni_2(\mu_2$-η^3,η^1-$CH_2C_6H_4$)Br(PMe_3)$_4$] reacts with isocyanides to produce an insertion product. The isocyanide insertion product can subsequently be carbonylated yielding an *N*-aryl-1-isoquinolone species (Equation (38)).[88,89]

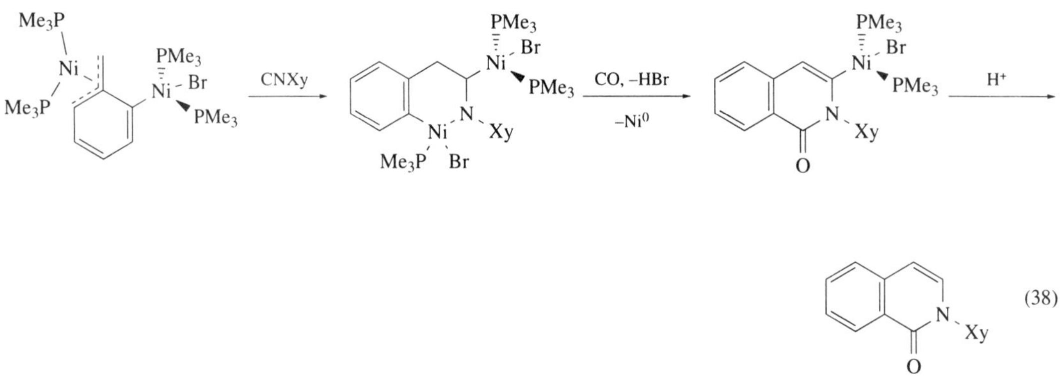
(38)

1.2.4 Complexes with Carbene Ligands

Nickel carbene complexes have been prepared both by additions to coordinated ligands like CO and by direct addition of electron-rich alkenes. Fischer-type carbenes have been prepared by the usual addition of a strong nucleophile, Nu^-, to [Ni(CO)$_4$], followed by addition of an electrophile, E^+ (Equation (39)).

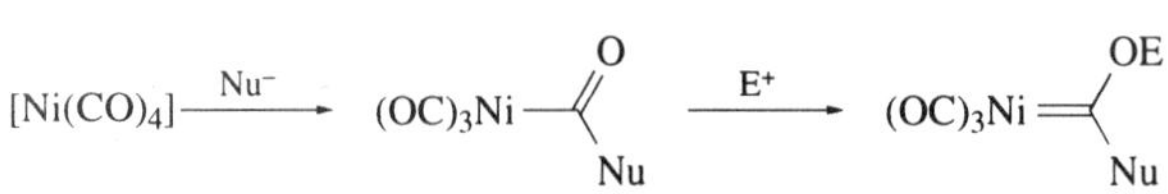
(39)

For example, the reaction of [Ni(CO)$_4$] with the nucleophile LiN($CHMe_2$)$_2$ followed by treatment with the electrophilic reagent [Et_3O][BF_4] gave the carbene complex [(CO)$_3$Ni=C(OEt)N($CHMe_2$)$_2$].[90] Fischer carbenes of this type can be converted to carbynes in the usual manner with Lewis acids (Equation (40)).[90] Similar heteroatom-stabilized carbenes have been prepared from isocyanide ligands. The addition of phenyllithium and Bu^{t}NC to tris(ethene)nickel leads to nucleophilic addition of Ph^- to the unsaturated carbon atom of the Bu^{t}NC. This species can be trapped by *N*-alkylation with MeI to obtain the aminocarbene complex (**11**).[91] Reaction of [Ni(η-C_5H_5)(CO)]$_2$ with four equivalents of

2-isocyanoethanol affords $[Ni(\eta\text{-}C_5H_5)(CNCH_2CH_2OH)_2]$ which subsequently cyclizes by intramolecular nucleophilic addition to the isocyanide carbon atoms to give a bis(carbene) complex (Equation (41)).[78]

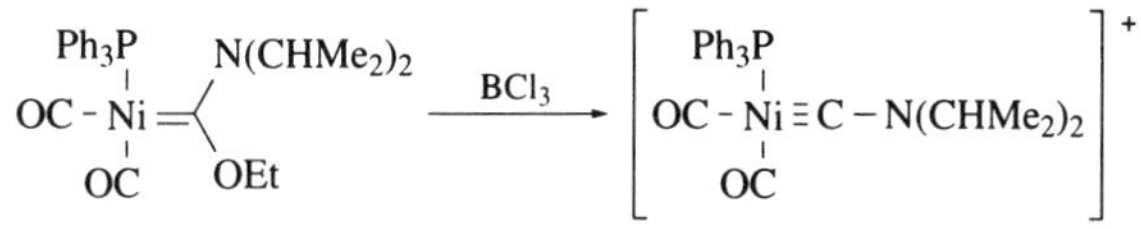

(40)

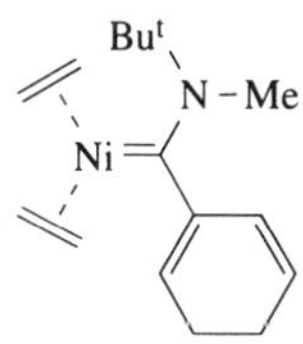

(11)

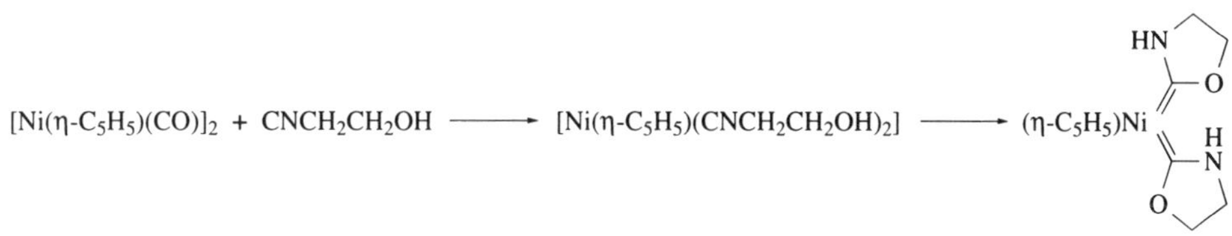

(41)

The η^2-iminoacyl complex $[Ni\{\eta^2\text{-}C(NBu^t)CH_2Bu^t\}(NC_4H_2Me_2)(PMe_3)]$ with a coordinated 2,5-dimethylpyrrolyl ligand reacts with CO_2 in the presence of water. The reaction results in cleavage of the nickel–iminoacyl nitrogen bond and protonation of the iminoacyl nitrogen atom to yield a neutral aminocarbene complex in which an η^2-carbonate ligand, derived from CO_2, is also introduced (Equation (42)). Other diprotic acids including oxalic, parabanic, and salicylic acids react with $[Ni\{\eta^2\text{-}C(NBu^t)CH_2Bu^t\}(NC_4H_2Me_2)(PMe_3)]$ to yield similar products.[92]

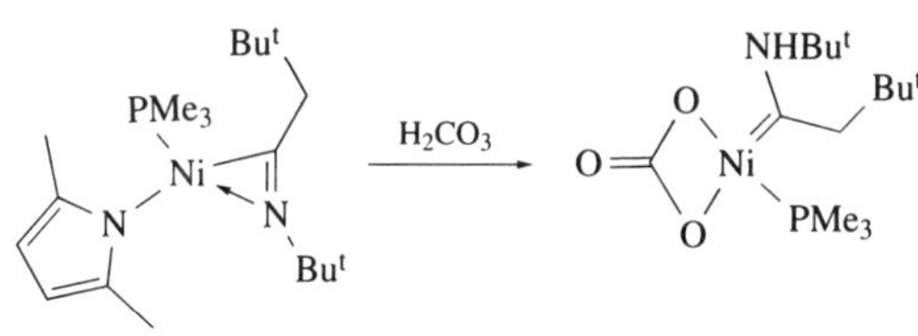

(42)

The addition of bis(1,3-diethylimidazolidin-2-ylidene) to $[Ni(CO)_4]$ affords a carbene complex directly (Equation (43)).[93]

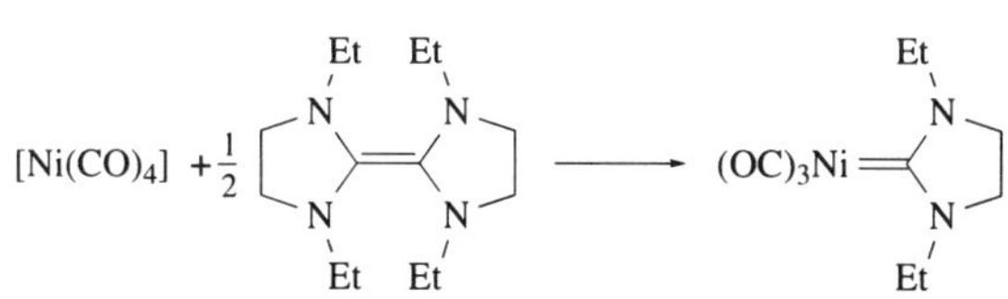

(43)

A theoretical study of the barrier to rotation about the $Ni{=}CH_2$ bond of the model nickel carbene complex $[(CO)_3Ni{=}CH_2]$ concluded that the barrier was extremely low, ~1 kJ mol^{-1}.[94]

A series of nickel carbene complexes have been prepared by the reaction of 1-lithio-3,4-dihydrofuran with $[Ni(aryl)_2L_2]$ complexes. Complexes of the type $[Ni(carbene)(aryl)L_2]$ are obtained (**12**).[95,96] The carbene-like nickel–carbon bond in the bis(trimethylphosphine)pentachlorophenyl complex is reported to be 0.007 nm shorter than a typical nickel–vinyl carbon bond. This has been cited as evidence of significant Ni=C π-bonding.[95,96] The nickel carbene complex *trans*-$[(PPhMe_2)_2(\eta^1\text{-}C_6Cl_5)Ni{=}C(OMe)Me][BF_4]$ has also been structurally characterized by the same group. The coordination geometry about the nickel is approximately square planar, and a comparatively short Ni–C(carbene) bond distance of 0.184 3(10) nm is consistent with the notion of significant π-backbonding.[97] The crystal structures of the simplest alkyl(alkoxy)carbene ligand complex *trans*-$[Ni(C_6Cl_5)(PMe_3)_2\{C(OMe)Me\}][BF_4]$ and the related methoxyvinyl complex *trans*-$[Ni(C_6Cl_5)(PMe_3)_29\{(C(OMe){=}CH_2\}]$ have been determined. Here, the Ni–C(carbene) bond length (0.1848 nm) is approximately 0.007 nm shorter than the corresponding Ni–C(vinyl) bond length.[98]

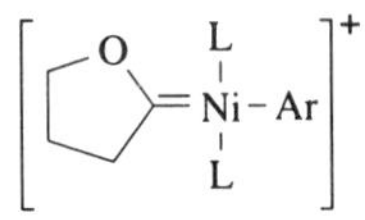

(12)

A remarkable carbene complex is formed by the reaction of $[Ni(cod)_2]$ with diphenylcarbodiimide in THF. The product was determined by x-ray diffraction to be the result of coupling three of the carbodiimides in a [3.3.0]nickelabicyclic arrangement (Equation (44)).[99]

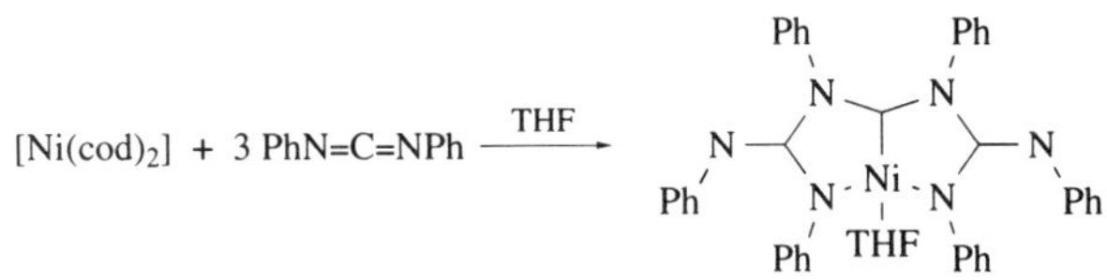

(44)

Treatment of *trans*-cyclododecatrienenickel(0) with $Me_3P{=}CH_2$ yields the stable ylide complex $[Ni(cdt)(Me_3P{=}CH_2)]$ (cdt, *trans*-cyclododecatriene). A similar ylide complex, $[Ni(CH_2{=}CH_2)_2(Me_3P{=}CH_2)]$, was formed from tris(ethene)nickel(0). These ylide complexes decomposed to give the corresponding trimethylphosphine complexes $[Ni(cdt)(PMe_3)]$ and $[Ni(CH_2{=}CH_2)_2(PMe_3)]$. Cyclopropane was observed in the presence of ethene, indicating methylene transfer from the ylide complexes to ethene.[100]

A review of complexes containing metal–carbon σ-bonds of the groups iron, cobalt, and nickel, including carbenes and carbynes, has appeared.[101]

1.2.5 Related Electron Pair Donor Ligands

Electron pair donor ligands other than carbonyls, isocyanides, and carbenes include group 14 ER_3^- (E = Si, Ge, Sn) anions, group 15 amine, arsine, and stibine ligands, and phosphorus ylides. The complex $[Ni\{Sn(NBu^t)_2(SiMe_2)\}_4]$, a stable stannylene analogue of $[Ni(CO)_4]$, was prepared from the cyclic stannylene $[Sn(NBu^t)_2(SiMe_2)]$ and $[Ni(cod)_2]$ (Equation (45)).[102]

$$[Ni(cod)_2] + 4\,[Sn(NBu^t)_2(SiMe_2)] \xrightarrow{THF} [Ni\{Sn(NBu^t)_2(SiMe_2)\}_4] + 2\,cod \qquad (45)$$

The reaction of $B_2H_4{\cdot}2\,PMe_3$ with $[Ni(CO)_4]$ results in the formation of a NiH_2B_2 chelate ring system **(13)**.[103]

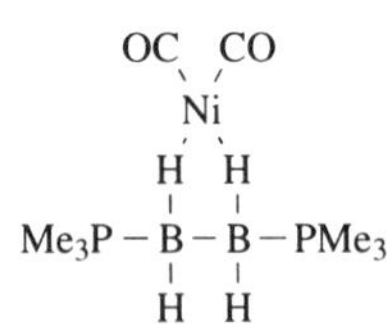

(13)

Although the organometallic chemistry of metal carbonyls is extensive, the chemistry of zero-valent metal cyanide compounds is much more limited. The nickel(0) cyanide complex $K_4[Ni(CN)_4]$ has been prepared. The complex $K_4[Ni(CN)_4]$ is insoluble in solvents with which it does not react. In the presence of diphenylacetylene and 18-crown-6, the species $[K(18\text{-crown-}6)]_2[Ni(CN)_2(\eta\text{-}PhC{\equiv}CPh)]$, the first organometallic derivative of $[Ni(CN)_4]^{4-}$, was isolated.[104] The reaction of the nickel(0) cyanide complex $[K(18\text{-crown-}6)]_2[Ni(CN)_2(\eta\text{-}PhC{\equiv}CPh)]$ with CO led to the isolation of $[K(18\text{-crown-}6)]_2$. $[Ni(CN)_2(CO)_2]$ and some $[K(18\text{-crown-}6)][Ni(CN)(CO)(\eta\text{-}PhC{\equiv}CPh)]$. The latter complex could not be isolated in pure form.[105]

Nucleophilic attack of coordinated carbonyl and isocyanide ligands is evident in the formation of Fischer-type carbene complexes of nickel (Section 1.2.4). The reaction of $[R_4N][OH]$ with the nickel(II) cyanide species $[Ni(CO)_2(CN)_2]$ results in the addition of hydroxide to a cyanide ligand and the formation of $[Ni(CO)_2(CN)\{C({=}N)OH\}]$. The formal loss of HNCO under the reaction conditions results in the formation of a nickel(0) cyanide species $[R_4N][Ni(CO)_3(CN)]$. The oxidative addition of allyl halides to this species affords $[Ni(CO)_2(CN)(\eta^1\text{-}C_3H_5)]$ and is the basis for a catalytic system for the phase transfer catalyzed carbonylation of allyl halides.[106]

1.3 DINUCLEAR NICKEL COMPLEXES

1.3.1 Complexes with Carbonyl Ligands

The organometallic chemistry of dinuclear and polynuclear nickel cluster compounds is extensive, although somewhat less systematic than the chemistry of the mononuclear complexes. Perhaps the most elementary dinuclear complex of nickel is [Ni(η-C_5H_5)(CO)]$_2$, which may be formed by the combination of [Ni(η-C_5H_5)$_2$] and [Ni(CO)$_4$]. The dinuclear structure of this material is easily cleaved by reactions with most electron pair donor ligands. The high degree of lability of [Ni(η-C_5H_5)(CO)]$_2$ may be used as the basis for the preparation of a wide variety of heterodinuclear complexes. The reaction with [$Co_2(CO)_8$], for example, leads to the heterodinuclear complex (Equation (46)). These heterodinuclear compounds may also be prepared by reaction of the metal carbonyl anions, such as Na[Co(CO)$_4$], with [Ni(η-C_5H_5)(L)(X)] complexes. Treatment of [$NiCl_2(PPh_3)_2$] with Na[M(η-C_5H_4Me)(CO)$_3$] (M = Mo, W) results in the formation of [NiM(η-C_5H_4Me)(CO)$_3$(PPh$_3$)$_2$] with two semibridging carbonyl groups. Calculations suggested that the nickel–metal interactions were weak. However, the tungsten complex can be reversibly oxidized, as observed by cyclic voltammetry, to [NiM(η-C_5H_4Me)(CO)$_3$(PPh$_3$)$_2$]$^+$.[107] The reaction of group 6 cyclopentadienyl carbonyl dimers, [M(η-C_5H_4Me)(CO)$_3$]$_2$ (M = Mo, W), with [Ni(η-C_5H_5)(CO)]$_2$ results in the formation of the heterodinuclear complexes (**14**).[108] A related methylene-bridged complex has also been reported.[109]

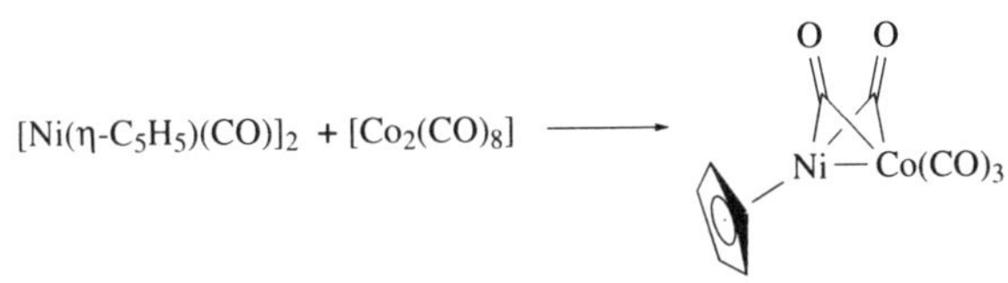

(46)

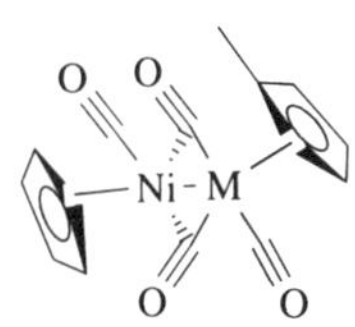

M = Mo, W
(14)

The reactions of [NiM(η-C_5H_5)(η-C_5H_4Me)(CO)$_4$] (M = Mo, W) with alkynes have been studied in some detail.[110] The nickel–tungsten complex reacts with alkynes to yield bridging alkyne complexes. The nickel–molybdenum complex reacts to form a nickelacyclobutenone complex. The tungsten–molybdenum dinuclear complexes are relatively labile with respect to dissociation into mononuclear fragments. The crossover reaction in benzene between [Ni(η-C_5H_5)(CO)W(η-C_5H_4Me)(CO)$_3$] and [Ni(η-C_5H_4Me)(CO)W(η-C_5H_5)(CO)$_3$] affords a statistical mixture of all four heterodinuclear complexes, including also [Ni(η-C_5H_5)(CO)W(η-C_5H_5)(CO)$_3$] and [Ni(η-C_5H_4Me)(CO)W(η-C_5H_4Me)(CO)$_3$]. The crossover reaction involves trace impurities. High-purity samples of the heterodinuclear complexes do not undergo crossover reactions. The dissociation of the heterodinuclear complexes, however, does not appear to play an important role in the reactions of [Ni(η-C_5H_5)(CO)W(η-C_5H_4Me)(CO)$_3$] with alkynes. Kinetic studies of the addition of alkynes to [Ni(η-C_5H_5)(CO)W(η-C_5H_4Me)(CO)$_3$] show a first-order dependence on both the concentrations of the heterodinuclear complex and the alkyne. Competition experiments between different alkynes were used to show that rate-determining alkyne addition occurs at the nickel atom of the heterodinuclear complex. It was concluded that alkyne additions occur by an associative pathway involving the heterodinuclear starting materials, not dissociated mononuclear intermediates.[110]

The reactions of [Ni(η-C_5H_5)(CO)]$_2$ and the heterodinuclear complexes with molybdenum and tungsten with allenes have been shown to result in C–C bond formation between allene fragments.[108] The direct reaction of allene or 1,1-dimethylallene with [Ni(η-C_5H_5)(CO)]$_2$ results in C–C coupling and the formation of a "tetramethyleneethane"-bridged dinuclear complex (Equation (47)).[108]

The reactions of diphosphines, distibines, and complexes of these ligands have been employed to prepare homodinuclear and heterodinuclear nickel carbonyl complexes. The reaction of [Ni(CO)$_4$] with Me_2Sb–SbMe_2 yields [(CO)$_3$Ni(μ-Me_2Sb–SbMe_2)Ni(CO)$_3$]. Similarly, the reaction of [Ni(CO)$_4$] with the pendant tetramethyldiphosphinechromium complex [(CO)$_5$Cr(Me_2PPMe_2)] affords the heterodinuclear complexes [(CO)$_3$Ni(μ-Me_2PPMe_2)Cr(CO)$_5$] and [(CO)$_2$Ni{(μ-Me_2PPMe_2)Cr(CO)$_5$}$_2$].[111–14]

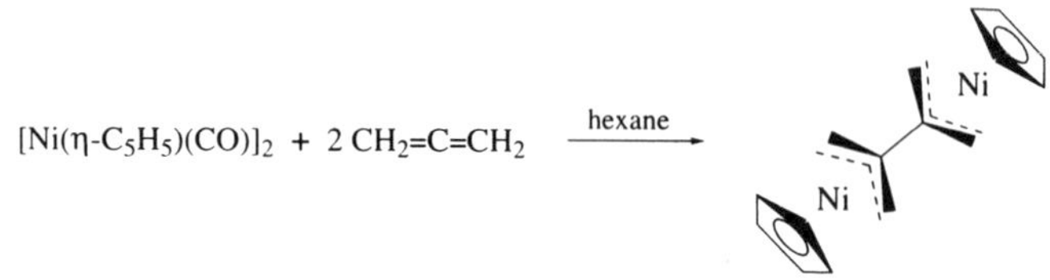

(47)

In view of the lability of many of the dinuclear compounds of nickel, ligands which are designed to bridge two metal centers have often been employed to prepare intact dinuclear complexes. The methylene-linked diphosphines bis(diphenylphosphino)methane (dppm) and bis(dimethylphosphino)methane (dmpm) in particular have been used in this regard. In contrast with the many and varied so-called "A-frame" complexes of platinum and palladium of the general type $[M_2(\mu\text{-L})X_2(dppm)_2]$ (M = Pt or Pd; X = halide or pseudohalide; L = small unsaturated molecules),[115–17] there are very few such complexes of nickel. A series of dppm-bridged d^9–d^9 dinuclear nickel complexes have been reported.[75,118–22] A rather general method for the preparation of Ni^I dinuclear complexes bridged by methylene-linked diphosphines involves the reaction of $[Ni(cod)_2]$ with the diphosphine in the presence of ABX_2 (A = O, B = S, X = Cl; A = O, B = C, X = Cl; A = PhN, B = C, X = Cl) species. The reaction with thionyl chloride, for example, yields the μ-SO "A-frame" complex (Equation (48)).[121] This μ-SO complex, the μ-vinylidene complex $[Ni_2(\mu\text{-}C{=}CH_2)(SCN)_2(dppm)_2]$,[122] and the μ-CNPh complex $[Ni_2(\mu\text{-}CNPh)Cl_2(dppm)_2]$,[121] have symmetric A-frame structures in the solid state.[75,118–20] The remainder of the diphosphine-bridged dinuclear nickel complexes exhibit a locally tetrahedral geometry at one metal center and a rectilinear geometry at the other. The structure of $[Ni_2(\mu\text{-}CO)Cl_2(dppm)_2]$ is typical of the asymmetric dinuclear complexes (**15**).[120] This is an apparent signature of internal disproportionation of two nickel(I) centers to more stable nickel(0) and nickel(II) centers. This class of complexes has thus been formulated as mixed oxidation state Ni^0–Ni^{II} species.

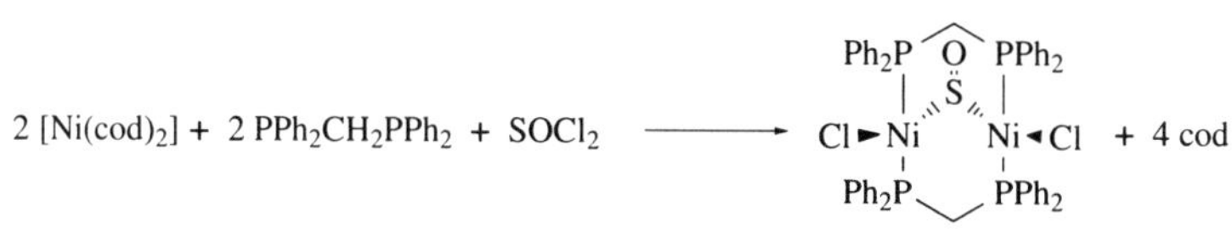

(48)

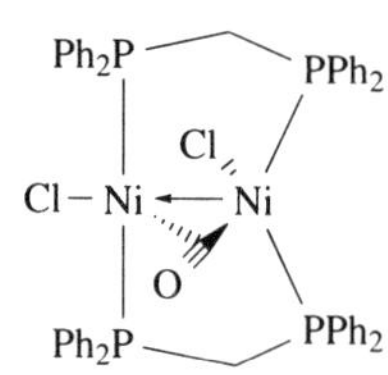

(15)

A series of dppm- and dmpm-bridged dinuclear nickel(0) complexes have been prepared from $[Ni(cod)_2]$, dppm or dmpm, and a π-acceptor ligand. The reaction of $[Ni(CO)_4]$ with dmpm was originally believed to result in the formation of the mononuclear complex $[Ni(CO)_2(dmpm)]$.[123] The product has since been shown by x-ray diffraction to be the dinuclear complex $[Ni_2(CO)_4(dmpm)_2]$.[124] The reaction of $[Ni(cod)_2]$, dppm, and CO results in the formation of $[Ni_2(\mu\text{-}CO)(\mu\text{-}dppm)_2(CO)_2]$.[118] This complex had been prepared by a less direct method involving the tripodal triphosphine $HC(PPh_2)_3$ and structurally characterized earlier (**16**).[125] A rather similar overall structure was obtained from the reaction of the cyclic tetraphosphine $(MeNPPh)_4$ with $[Ni(CO)_4]$.[126] A molecular orbital study of the related complex $[Ni_2(\mu\text{-}CO)(\mu\text{-}dmpm)_2(PMe_3)_2]$ has been undertaken.[127] Molecular orbital studies have also been described for the complexes $[Ni_2(\mu\text{-L})(\mu\text{-}dppm)_2(CNMe)_2]$, where L = CNMe, $CN(H)Me^+$, and NO^+.[128] These studies reconcile the short Ni–Ni bonding found in this class of dinuclear complexes with their "closed shell" d^{10}–d^{10} electronic configurations in terms of interactions of the $d\pi^*$ metal based fragment orbitals and μ-ligand based π^* fragment orbitals.

The organometallic photochemistry of metal–metal bonded dinuclear complexes, including $[Ni(\eta\text{-}C_5H_5)(CO)]_2$, has been reviewed.[129]

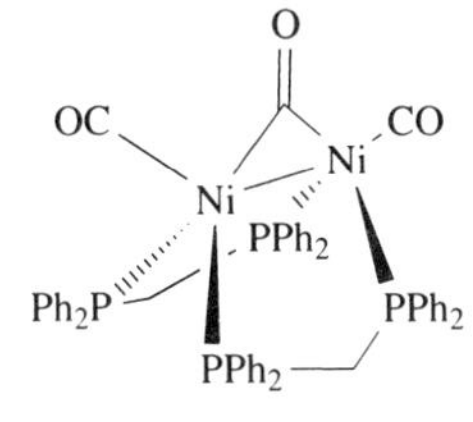

(16)

1.3.2 Complexes with Isocyanide Ligands

The preparation of Ni^{I} dinuclear complexes bridged by methylene-linked diphosphines by the reaction of $[Ni(cod)_2]$ with the diphosphine in the presence of ABX_2 species has been extended to $PhNCCl_2$. The product is a symmetric μ-phenyl isocyanide A-frame of the type $[Ni_2(\mu\text{-}CNPh)Cl_2(dppm)_2]$.[130]

Conproportionation of d^8 $[Ni^{II}(CNMe)_4][PF_6]$ and d^{10} $[Ni^0(CNMe)_4]$ results in the formation of the dinuclear complex $[Ni_2(CNMe)_8][PF_6]_2$.[75] The structure of this complex was not determined; however, the reaction of $[Ni_2(CNMe)_8][PF_6]_2$ with two equivalents of dppm yields an asymmetrically bridged complex, $[Ni_2(\mu\text{-}CNMe)(CNMe)_3(dppm)_2][PF_6]_2$ (Equation (49)). Reduction of $[Ni_2(\mu\text{-}CNMe)(CNMe)_3(dppm)_2][PF_6]_2$ with Na–Hg leads to the formation of the so-called "cradle"-type complex based on a locally tetrahedral structure about the nickel atoms in $[Ni_2(\mu\text{-}CNMe)(CNMe)_2(dppm)_2]$.[131] The μ-methyl isocyanide ligand of $[Ni_2(\mu\text{-}CNMe)(CNMe)_2(dppm)_2]$ can be protonated to form $[Ni_2\{\mu\text{-}CN(H)Me\}(CNMe)_2(dppm)_2]^+$ (Equation (50)). A comparison of the structures of the μ-CNMe and μ-CN(H)Me⁺ complexes indicated that the protonated complex had significant aminocarbyne character.[131]

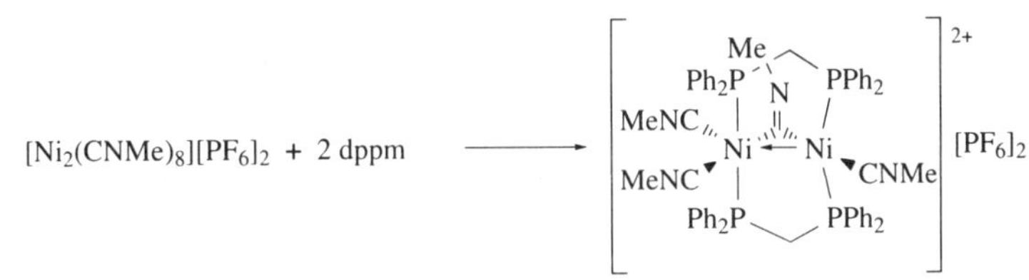

(49)

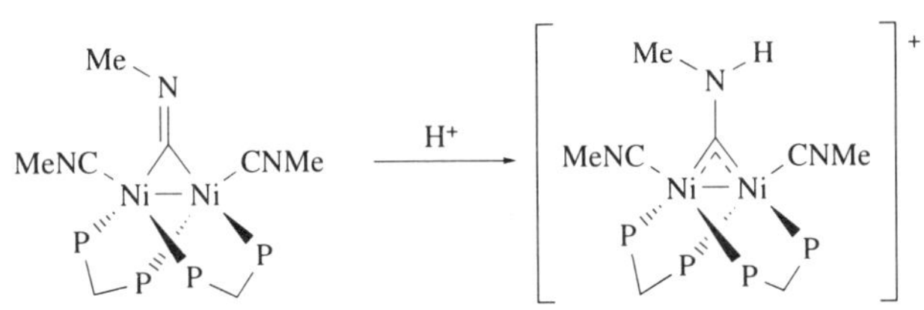

(50)

The importance of π-acceptor ligands in the stabilization of dinuclear Ni^0 complexes of this type can be seen in the comparison of the structures of the homologous series of dinuclear isocyanide complexes $[Ni_2(\mu\text{-}L)(CNMe)_2(dppm)_2]$, (L = CNMe, CN(H)Me⁺, NO⁺). Each structure consists of two approximately tetrahedrally coordinated Ni^0 centers with *cis,cis*-diphosphines. The Ni–Ni bond length of 0.247 1(2) nm for the μ-NO⁺ complex is well within the range for Ni–Ni single bonds, and significantly shorter than the corresponding distances found in the μ-CNMe and μ-CN(H)Me⁺ complexes. No *d*-orbital-based metal–metal bonding is feasible for a d^{10}–d^{10} dinuclear complex because the entire *d*-manifold is occupied. In the case of strongly π-accepting bridging ligands, however, the π*-systems of the bridging ligands generally possess the proper symmetry and are placed favorably in energy to depopulate the nickel–nickel antibonding ($d\pi^*$) molecular orbitals. This interaction can be regarded as the dinuclear complex version of metal–ligand "backbonding" (**17**).

It has the effect of partially cancelling a metal–metal antibonding interaction. The greater the π*-back-bonding ability, the stronger the M–M bonding interaction is expected to be. For the μ-isocyanide (**18**), μ-aminocarbyne (**19**), and μ-nitrosyl (**20**) complexes which have been structurally characterized, the degree of metal–metal interaction increases as the π-acidity of the bridging ligand increases. The nitrosyl ligand is generally regarded to be the strongest π-acceptor diatomic ligand, and the nitrosyl complex possesses a significantly shorter Ni–Ni distance.[128]

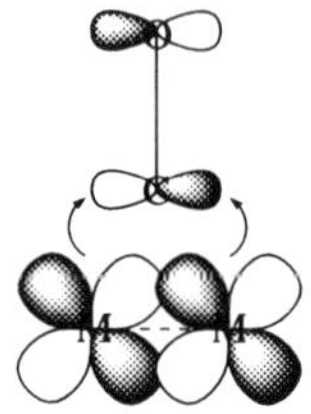

(17)

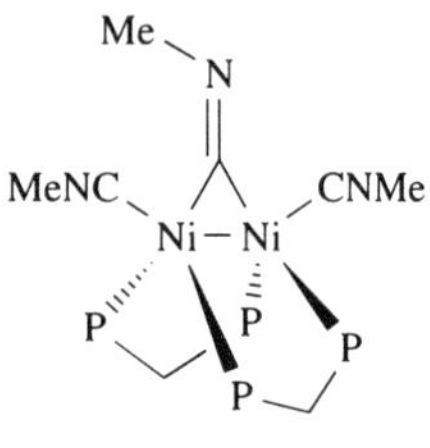

$d_{(Ni-Ni)}$ = 0.257 2(1) nm
(18)

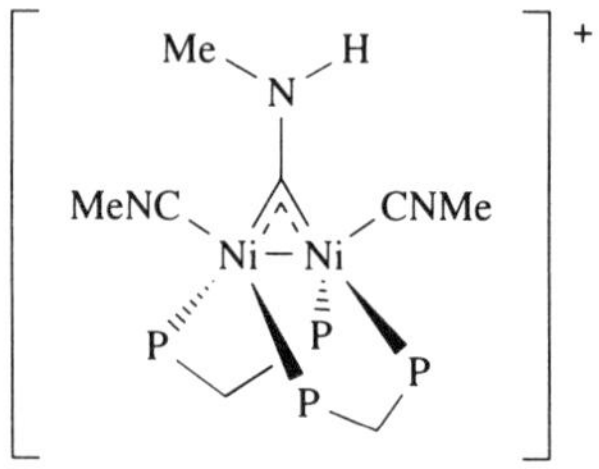

$d_{(Ni-Ni)}$ = 0.250 0(1) nm
(19)

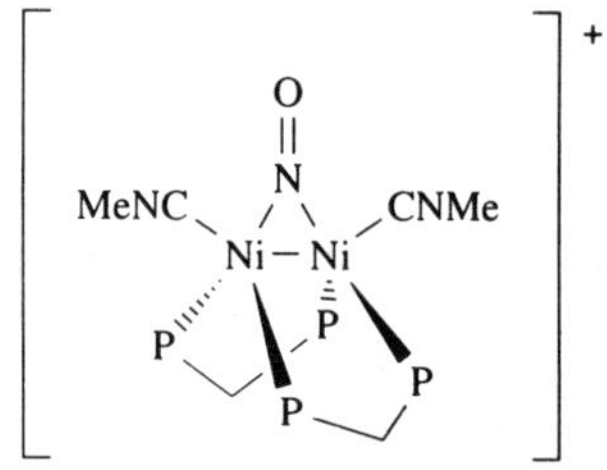

$d_{(Ni-Ni)}$ = 0.247 1(2) nm
(20)

The μ-methyl isocyanide complex $[Ni_2(\mu\text{-CNMe})(CNMe)_2(dppm)_2]$ undergoes a series of reactions with liquid CO_2 which result overall in the metathesis of the multiple bonds of methyl isocyanide and CO_2.[119,132] The reaction of the ^{13}C-substituted tris(methyl isocyanide) complex $[Ni_2(\mu\text{-*CNMe})(\text{*CNMe})_2(dppm)_2]$ with CO_2 resulted in the quantitative conversion to the completely ^{13}C-substituted tricarbonyl complex $[Ni_2(\mu\text{-*CO})(\text{*CO})_2(dppm)_2]$. The multiple bond metathesis was shown to proceed through a formal [2 + 2] cycloadduct of CO_2 with the μ-CNMe ligand of $[Ni_2(\mu\text{-CNMe})(CNMe)_2(dppm)_2]$ and a dicarbonyl complex, $[Ni_2(CNMe)(CO)_2(dppm)_2]$. The multiple bond metathesis leads to the conversion of CO_2 to *N,N'*-dimethylcarbodiimide by metathesis of the oxygen atoms of CO_2 with the methylimido groups of methyl isocyanide, as shown in Scheme 3.

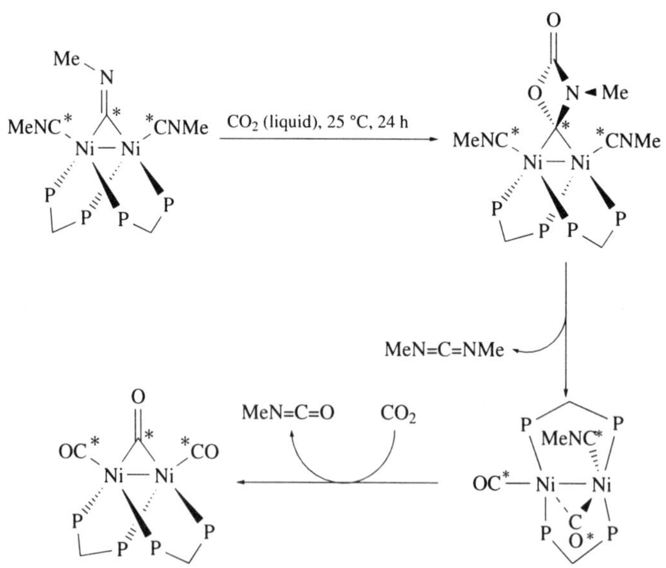

Scheme 3

1.3.3 Complexes with Carbene Ligands

A frequently employed method for the preparation of carbene complexes is the reaction of diazoalkanes, $R_2C{=}N{=}N$, with transition metal complexes. Heterodinuclear methylene-bridged complexes have also been prepared by this method. For example, the reaction of the unsaturated Ni–M (M = Mo, W) complexes $[Ni(\eta\text{-}C_5Me_5)M(CO)_3(\eta\text{-}C_5H_5)]$ with diazomethane led to the formation of the first structurally characterized nickel methylene complexes $[Ni(\eta\text{-}C_5Me_5)(\mu\text{-}CH_2)(\mu\text{-}CO)M(CO)_2(\eta\text{-}C_5H_5)]$ (Equation (51)). The further reaction of the nickel–tungsten μ-methylene complex with 2-butyne results in insertion into the Ni–CH_2 bond to give the complex $[Ni(\eta\text{-}C_5Me_5)\{\mu_2\text{-}\eta^1,\eta^3\text{-}C(Me)C(Me)CH_2\}(\mu\text{-}CO)W(CO)(\eta\text{-}C_5H_5)]$.[109,133]

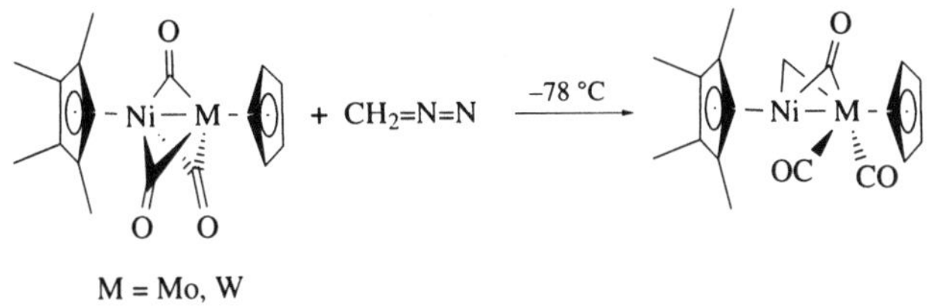

(51)

In several instances, nitrogen-coordinated diazoalkane complexes can be obtained which subsequently eliminate N_2 to form the carbene complexes. Carbon-bonded diazoalkane groups have been introduced into transition metal complexes by the reaction of $LiC(N_2)R$ or $Hg[C(N_2)R]_2$ reagents with transition metal halide complexes. The first example of a carbon-bonded nickel diazo complex, $[Ni\{C(N_2)TMS\}Cl(PMe_3)_2]$, was prepared by the reaction of one equivalent of $LiC(N_2)TMS$ with $[NiCl_2(PMe_3)_2]$ in THF below −25 °C. Photolysis of $[Ni\{C(N_2)TMS\}Cl(PMe_3)_2]$ resulted in the formation of a dinuclear complex with an Ni_2C_2 core (Equation (52)).[134] The crystal structure revealed the presence of two symmetrically bridging three-electron-donor ylide ligands and an exceedingly short Ni–Ni distance of 0.228 1(1) nm.

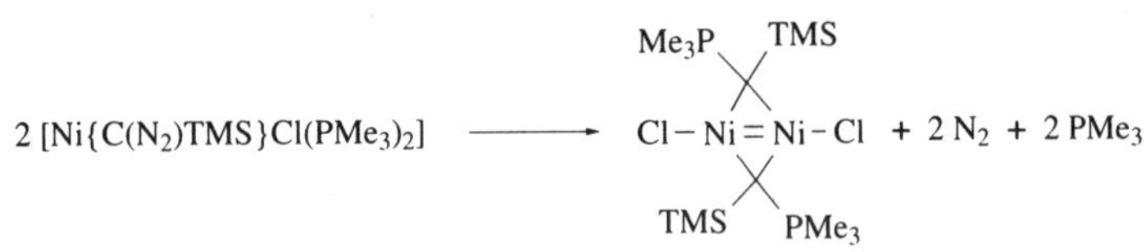

(52)

Generation of the bridging ylide complex by photoinduced elimination of N_2 from $[Ni\{C(N_2)TMS\}Cl(PMe_3)_2]$ presumably leads to a carbyne nickel intermediate which is stabilized by the addition of PMe_3 to give the observed ylide ligands. A theoretical investigation at the extended Hückel level of the bridging ylide complex has been undertaken. It was suggested that the very short nickel–nickel separation in the complex is not a consequence of strong metal–metal interactions, but rather results from the interactions between the Ni_2Cl_2 fragment and the bridging ylides.[135]

1.4 HIGHER CLUSTER COMPLEXES

1.4.1 Clusters with Carbonyl Ligands

Since the 1970s, numerous syntheses of nickel clusters from nickelocene, nickel tetracarbonyl, and nickel halides have been developed. Nickelocene and nickel carbonyl are activated toward cluster formation either by direct reduction or by reaction with nucleophiles. Alkali-metal reduction of $[Ni(CO)_4]$ produces a variety of polynuclear nickel carbonyl anions including $[Ni_5(CO)_{12}]^{2-}$, $[Ni_6(CO)_{12}]^{2-}$, and $[Ni_9(CO)_{18}]^{2-}$.[136,137] The cluster anion $[Ni_6(CO)_{12}]^{2-}$ has also been prepared by more specialized methods including the ultrasonic treatment of NiX_2 under an atmosphere of CO.[138] The $[Ni_3(CO)_3(\mu\text{-}CO)_3]$ fragment is commonly found in the nickel carbonyl clusters, including the heteronuclear cluster anions $[MNi_6(CO)_{17}]^{2-}$ (M = Cr, Mo, W).[139] For example, the oxidation of $[Ni_6(CO)_{12}]^{2-}$ with $FeCl_3$ leads to the Longoni–Chini cluster dianion, $[Ni_9(CO)_{18}]^{2-}$, quantitatively. The crystal structure of this material reveals the three $[Ni_3(CO)_3(\mu\text{-}CO)_3]$ units of the Ni_9 core to be oriented in a previously unknown nine-vertex polyhedron.[140,141] The bonding in clusters based on the $[Ni_3(CO)_3(\mu\text{-}CO)_3]$ structural unit has been examined theoretically using the fragment molecular orbital approach.[142] A new topological electron-counting theory for assigning cluster framework bonding electrons has been presented by Teo.[143] The theory is useful for predicting the electron counts for very large polyhedral clusters, including polymetallic clusters.[144] The hydrido clusters $[H_2Ni_{12}(CO)_{21}]^{2-}$ and $[H_2Ni_{12}(CO)_{21}]^{3-}$ have been obtained by treatment of $[Ni_6(CO)_{12}]^{2-}$ with water. Further deprotonation to afford the known cluster anion $[Ni_{12}(CO)_{21}]^{4-}$ could only be accomplished under forcing conditions.[145] The reaction of $[Ni_6(CO)_{12}]^{2-}$ with $AsPhCl_2$ gave $[Ni_9(AsPh)_3(CO)_{15}]^{2-}$, but the reaction with $AsPhBr_2$

gave $[Ni_{10}(AsPh)_2(CO)_{18}]^{2-}$. Both cluster anions were characterized by x-ray diffraction. The Ni_9As_3 core was viewed as analogous to a *closo*-1,2,12-trisubstituted icosahedron, while the $Ni_{10}As_2$ core was 1, 12-disubstituted.[140] The reaction of $[Ni_6(CO)_{12}]^{2-}$ with $(TMS)_2CHPCl_2$ gave a phosphinidene coupling product, $[Ni_5Cl(CO)_6\{(TMS)_2CHP{=}PCH(TMS)_2\}]^{2-}$.[146]

The chemistry of nickel carbide clusters is extensive. The reaction of $[Ni_6(CO)_{12}]^{2-}$ with $HCCl_3$ results in the formation of the carbide clusters $[Ni_{10}(CO)_{18}C]^{2-}$, $[Ni_9(CO)_{17}C]^{2-}$, and $[Ni_8(CO)_{16}C]^{2-}$.[147] Treatment of $[Ni_6(CO)_{12}]^{2-}$ with a C_2 halocarbon such as $Cl_2C{=}CCl_2$ or Cl_3CCCl_3 gives the cluster $[Ni_6(CO)_{16}C_2]^{2-}$ containing an interstitial $\{C_2\}$ group.[148] The cluster $[Ni_{10}(CO)_{16}C_2]^{2-}$ also has an interstitial $\{C_2\}$ group. Its reaction with PPh_3 induces cluster condensation, resulting in the unusual species $[Ni_{16}(CO)_{23}C_4]^{2-}$. Two interstitial dicarbide groups with short (0.138 nm) C–C separations are contained within the cluster framework.[149] The ketenylidene iron cluster $[Fe_3(CCO)(CO)_9]^{2-}$ can be converted to a carbide cluster, $[Fe_3Ni_3(CO)_{13}C]^{2-}$, by the reaction with $[Ni(CO)_4]$ or $[Ni(\eta\text{-}C_5H_5)(CO)]_2$.[150] The reaction of $[Co_3(CO)_9CCl]$ with $[Ni_6(CO)_{12}]^{2-}$ gives a complex mixture of clusters including the carbide $[Co_3Ni_9C(CO)_{20}]^{3-}$, which was structurally characterized. The new dicarbide clusters $[Co_3Ni_7C_2(CO)_{15}]^{3-}$ and $[Co_6Ni_2C_2(CO)_{16}]^{2-}$ were also obtained. The Co_3Ni_7 cluster contained interstitial carbon atoms which were separated by 0.143 nm.[151,152] The use of the nickel cluster anion $[Ni_9(CO)_{17}C]^{2-}$ instead of $[Ni_6(CO)_{12}]^{2-}$ gives another decanuclear species, $[Co_3Ni_7C_2(CO)_{16}]^{2-}$, with a C–C separation of the $\{C_2\}$ of 0.148 nm.[153]

Many of the clusters synthesized from $[Ni_6(CO)_{12}]^{2-}$ are extraordinarily large. The reaction of two equivalents of $[Ni_6(CO)_{12}]^{2-}$ with C_2Cl_6 results in a complex mixture of carbide clusters: $[Ni_{10}(CO)_{16}C_2]^{2-}$, $[Ni_{16}(CO)_{23}(C_2)_2]^{2-}$, $[HNi_{34}(CO)_{38}C_4]^{2-}$, and $[Ni_{34}(CO)_{38}C_4]^{3-}$. The reaction of $[Ni_6(CO)_{12}]^{2-}$ with dichloromethane results in the formation of some crystalline material, from which $[HNi_{34}(CO)_{38}C_4]^{2-}$ and $[Ni_{35}(CO)_{39}C_4]^{6-}$ were isolated and structurally characterized. The structures consist of cubic-close-packed Ni_{20} cores which are further elaborated. An $Ni_{28}C_4$ unit is common to both structures and consists of the carbide carbon atoms lying in four trigonal prismatic interstices.[154] The nickel carbide cluster $[HNi_{38}(CO)_{42}C_6]^{5-}$ was prepared from $[Ni_6(CO)_{12}]^{2-}$ and C_3Cl_6. It possesses an Ni_8 cube, capped on each face by the six carbide carbon atoms.[155]

Several nickel clusters with Ni_8 cubic cores have been described. A mixture of both the tetrachloro cluster $[Ni_8Cl_4(PPh_3)_4(\mu_4\text{-}PPh)_6]$ and the tetracarbonyl cluster $[Ni_8(CO)_4(PPh_3)_4(\mu_4\text{-}PPh)_6]$ is obtained by the same procedure. The highly symmetric clusters have a μ_4-PPh coordinated to each of the six faces of the Ni_8 cube.[156] Several other procedures for the preparation of $[Ni_8(CO)_4(PPh_3)_4(\mu_4\text{-}PPh)_6]$ have been developed. These involve the reduction of $[Ni_8Cl_4(PPh_3)_4(\mu_4\text{-}PPh)_6]$ under CO and the carbonylation of the unsaturated species $[Ni_8(PPh_3)_4(\mu_4\text{-}PPh)_6]$.[157] The nickel sulfido cluster $[Ni_8Cl_2(PPh_3)_6S_6]$ also could be converted to a carbonyl cluster by reduction under CO.[158]

An extensive cluster chemistry of polymetallic clusters containing nickel has developed. Several reviews of nickel clusters and nickel-containing heterometallic clusters have appeared.[159–62] The unusual polynuclear clusters prepared by Vahrenkamp are notable. For example, treatment of $[RPFeCo_2(CO)_9]$ with $[Ni(\eta\text{-}C_5H_5)(CO)]_2$ gave a mixture of products from which $[Fe_2CoNi(\eta\text{-}C_5H_5)(\mu_4\text{-}PMe)(CO)_8]$ and $[FeCoNi(\eta\text{-}C_5H_5)(\mu_3\text{-}PBu^t)(CO)_6]$ could be isolated and characterized by x-ray diffraction.[163] The nickel clusters prepared in the laboratories of Fenske and co-workers which contain group 15 (E = P, As, Sb) or group 16 (E' = S, Se, Te) heteroatoms in the cluster frameworks have also been reviewed.[164] The reaction of $[Ni(CO)_4]$ with $[Fe_3(CO)_{11}]^{2-}$ leads to $[Fe_3Ni(CO)_{12}]^{2-}$. Protonation gives the hydride $[HFe_3Ni(CO)_{12}]^-$ which was characterized by x-ray diffraction. The Fe_3Ni core is approximately tetrahedral with eight terminal and four edge-bridging carbonyls. The hydride caps an Fe_2Ni face.[165] The cluster $[NiOs_3(CO)_9(\eta^5\text{-}C_5H_5)(\mu\text{-}H)_3]$ (**21**) can be prepared in reasonable yield from $[Os_3(CO)_{12}]$ and $[Ni(\eta\text{-}C_5H_5)(CO)]_2$ in the presence of H_2.[166]

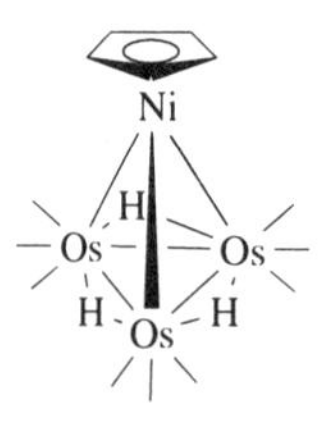

(21)

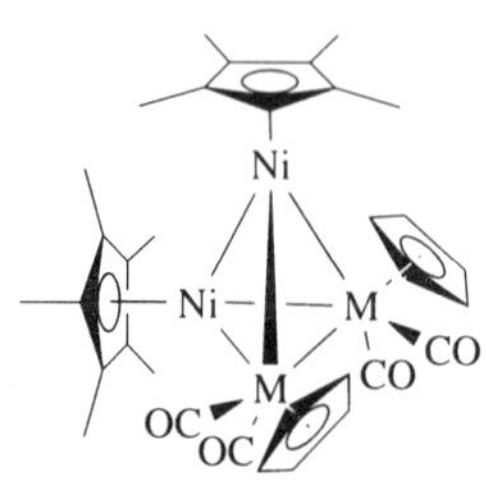

M = Mo, W
(22)

A hydride may be removed from $[NiOs_3(CO)_9(\eta\text{-}C_5H_5)(\mu\text{-}H)_3]$ by treatment with NaH, giving $[NiOs_3(CO)_9(\eta\text{-}C_5H_5)(\mu\text{-}H)_2]^-$. Electrophilic "proton equivalents" $\{M(PPh_3)\}^+$ (M = Au, Cu) may then be introduced to give trimetallic clusters of the type $[NiOs_3(CO)_9(\eta\text{-}C_5H_5)(\mu\text{-}H)_2\{\mu\text{-}M(PPh_3)\}]^-$. The gold-containing cluster was structurally characterized.[167] The corresponding triruthenium– nickel cluster $[NiRu_3(CO)_9(\eta\text{-}C_5H_5)(\mu\text{-}H)_3]$ has also been prepared in 70% yield.[166] The electrochemistry of the cluster $[CoFeNi(\mu_3\text{-}Me_3P)(\eta\text{-}C_5H_5)(CO)_8]$ has been examined. Single-electron reduction of the cluster leads only to fragmentation.[168] Pyrolysis of the heterodinuclear compounds $[Ni(\eta\text{-}C_5Me_5)(CO)M(CO)_3(\eta\text{-}C_5H_5)]$ (M = Mo, W) afforded the tetranuclear clusters $[Ni_2(\eta\text{-}C_5Me_5)_2M_2(CO)_4(\eta\text{-}C_5H_5)_2]$ (**22**).[169] This contrasts with the products obtained when the nickel cyclopentadienyl heterodinuclear complexes, $[Ni(\eta\text{-}C_5H_5)(CO)M(CO)_3(\eta\text{-}C_5H_4Me)]$ (M = Mo, W) were subjected to thermolysis, affording paramagnetic clusters with Ni_3M tetrahedral metallic cores as products.[170]

The platinum-containing clusters resulting from the reaction of $[Ni_6(CO)_{12}]^{2-}$ and $PtCl_2$ or $K_2[PtCl_4]$ include hydrides $[Ni_{38}Pt_6(CO)_{48}H_n]^{2-}$ (n = 0–3). The structures of the mono- and dihydride clusters $[Ni_{38}Pt_6(CO)_{48}H_2]^{2-}$ and $[Ni_{38}Pt_6(CO)_{48}H]^{5-}$ were determined.[171]

It is not surprising that the mechanisms of these reactions are poorly understood. Nickel tetracarbonyl and nickelocene react with each other to produce the well-known Fischer–Palm trinuclear cluster, $[Ni_3(\eta\text{-}C_5H_5)_3(\mu_3\text{-}CO)_2]$, which may be more conveniently synthesized by reduction of $[Ni(\eta\text{-}C_5H_5)(CO)]_2$ with sodium–potassium alloy in THF.[172,173] Heteronuclear clusters can also be formed as in the reaction of $[Ni(\eta\text{-}C_5H_5)(CO)]_2$ with $[Co(\eta\text{-}C_5H_5)(CO)]_2$, which produces $[Ni_2Co(\eta\text{-}C_5H_5)_3(\mu_3\text{-}CO)_2]$.[174]

The next step up in complexity from the binary nickel tetracarbonyl clusters is reached by varying the reducing agent. Reaction of nickel tetracarbonyl with $[M_2(CO)_{10}]^{2-}$ (M = Cr, Mo, W) produces dianionic clusters in which a triangular nickel core containing three bridging and three terminal carbonyls is bicapped by approximately square pyramidal $M(CO)_5$ fragments (Equation (53)).[175] Dahl has reported the synthesis of the clusters $[Ni_{10}(SnR)_2(CO)_{18}]^{2-}$ (R = Bu^n, Me) which are based on an icosahedral $Ni_{10}Sn_2$ cage, containing an additional body-centered nickel atom (Equation (54)).[176] The cluster violates the usual electron-counting rules for clusters, since it contains four more electron pairs than would be expected for this geometry.

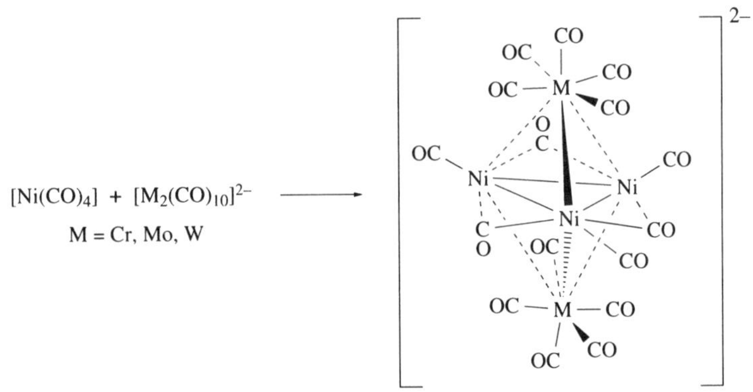

(53)

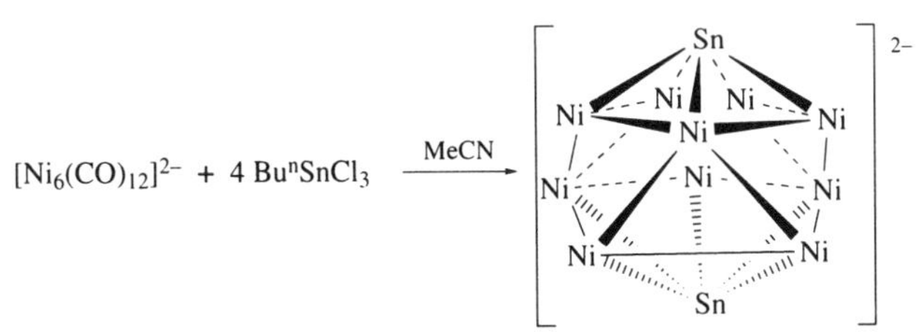

(54)

Dahl has prepared analogues of the Fischer–Palm trimer bridged by CS via reaction of $[Ni(\eta\text{-}C_5H_5)(CO)]_2$ with CS_2, and obtained a C_2S_6-bridged "dimer of trimers" side product.[177]

Nickel trimers bridged by the diphosphine dppm and two triply bridging iodide ligands have been reported (Equation (55)).[178–80] One of the capping iodides is easily substituted by π-acceptor ligands which substitute at the capping position. For example, the cluster $[Ni_3(dppm)_3(\mu_3\text{-}CO)(\mu_3\text{-}I)]I$ is prepared by the reaction of $[Ni_3(\mu_3\text{-}I)_2(dppm)_3]$ with CO (Equation (56)).

Nickel clusters capped by iodide appear to be capable of tolerating long Ni–Ni bonds. It was noted by Stone and co-workers in 1979 that nickel–nickel distances tended to fall into two narrow bands.[181] Bonding interactions led to bond distances in the range 0.235–0.245 nm; nonbonding interactions involved bond lengths in the range ca. 0.270–0.280 nm. During the 1980s a number of somewhat longer nonbonding nickel–nickel distances have been found, the longest being 0.316 nm in $[Ni_3(\mu_3\text{-}$

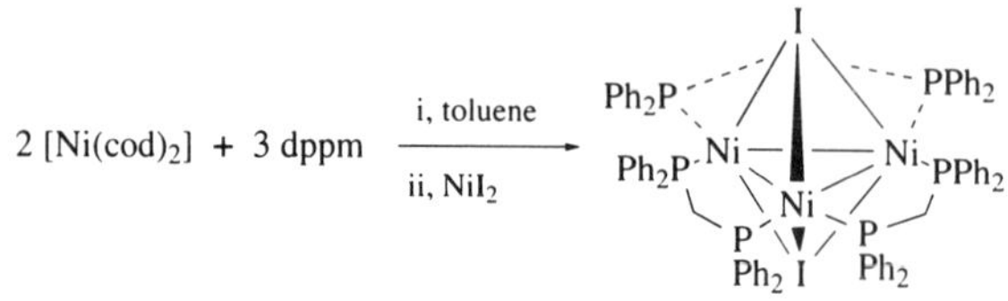

(55)

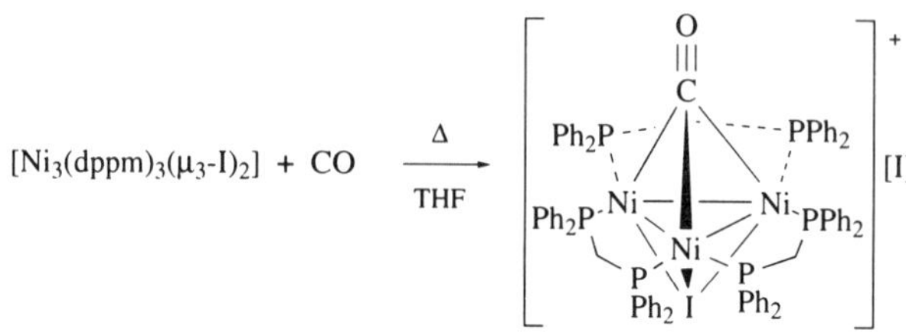

(56)

$Se)_2(PEt_3)_6]^{2+}$.[182] The significant variation in the Ni–Ni bonding distance of dimers achieved by varying the bridging ligand has already been discussed, but until recently there was no exception to the rule for clusters of higher nuclearity. However, the iodo-bridged clusters $[Ni_3(PPh_3)_3(\mu_3\text{-I})(\mu_2\text{-I})_3]$ and $[Ni_3(\mu_3\text{-I})_2(dppm)_3]^+$ have average nickel–nickel bond lengths of 0.266 nm and 0.252 nm, respectively, even though the closely related cluster $[Ni_3(\mu_3\text{-CNMe})(\mu_3\text{-I})(dppm)_3]^+$ has a "normal" average bond length of 0.242 nm.[179] Capping iodide thus appears to share with bridging dppm the ability to stabilize weak metal–metal bonds.

1.4.2 Clusters with Isocyanide Ligands

Another important class of nickel clusters which can be formed under mild conditions are the tri- and tetranuclear species formed when isocyanides or alkynes are reacted with nickel tetracarbonyl, all-*trans*-cyclododecatrienenickel(0), or bis(cyclooctadiene)nickel(0). The products contain μ_3,η^2-alkynes or isocyanide ligands bridging three metals. Interestingly, reaction of hexafluorobutyne with $[Ni(CO)_4]$ gives a tetranuclear cluster, $[Ni_4(\mu_3,\eta^2\text{-}CF_3C_2CF_3)_3(CO)_4]$, but reaction with nickel cyclooctatetraene leads to a trinuclear cluster, $[Ni_3(CO)_3(\mu_3,\eta^2\text{-}CF_3C_2CF_3)(\mu_3,\eta\text{-}C_8H_8)]$ (Equation (57)).[181] A similar reaction was reported by Muetterties in which reaction of bis(cyclooctadiene)nickel(0) with about two equivalents of Bu^tNC produced the cluster $[Ni_4(CNBu^t)_7]$ in which the isocyanides exhibit several coordination modes (Equation (58)).[183] There have been no other reactions of this type reported since 1976.

$[Ni(CO)_4] + F_3CC{\equiv}CCF_3 \longrightarrow [Ni(CO)_2(\eta^2\text{-}F_3CC{\equiv}CCF_3)] \longrightarrow [Ni_4(\mu_3,\eta^2\text{-}F_3CC{\equiv}CCF_3)_3(CO)_4] \xrightarrow{C_8H_8}$

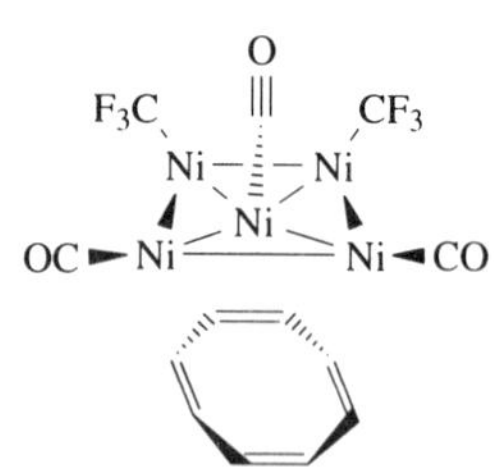

(57)

$[Ni(cod)_2] + Bu^tNC \longrightarrow [Ni_4(\mu_3,\eta^2\text{-}Bu^tNC)_3(\eta^1\text{-}Bu^tNC)_4] \xrightarrow{PhC{\equiv}CPh} [Ni_4(\mu_3,\eta^2\text{-}PhC{\equiv}CPh)_3(\eta^1\text{-}Bu^tNC)_4]$ (58)

The reaction of the 1,4-diisocyanide CN–C(Me)$_2$–(CH$_2$)$_2$–C(Me)$_2$–NC with nickel acetate produced crystals of the tetranickel cluster $[Ni_4(MeCO_2)_2(OMe)_4(1,4\text{-diisocyanide})_4][BPh_4]_2$. The crystal structure of this cluster is characterized by a cubane-like $Ni_4(OMe)_4$ core.[184]

The trinuclear isocyanide-capped clusters $[Ni_3(\mu_3,\eta^1\text{-CNMe})(\mu_3\text{-I})(CNMe)_2(dppm)_2]^+$ and $[Ni_3(\mu_3,\eta^1\text{-CNMe})(\mu_3\text{-I})(dppm)_3]^+$ were originally prepared by metallation of dinuclear Ni^0 isocyanide complexes.[178,179] Surprisingly, these clusters, reported in 1990, were the first examples of symmetrically bridging μ_3,η^1-isocyanide ligands (Equation (59)).

An unusual metallation product of $[Ni_2(\mu\text{-CNMe})(CNMe)_2(dppm)_2]$ is the mercuration product. The reaction of $[Ni_2(\mu\text{-CNMe})(CNMe)_2(dppm)_2]$ with $HgCl_2$ produces $[Ni_4Hg(\mu\text{-CNMe})_2(CNMe)_8$-

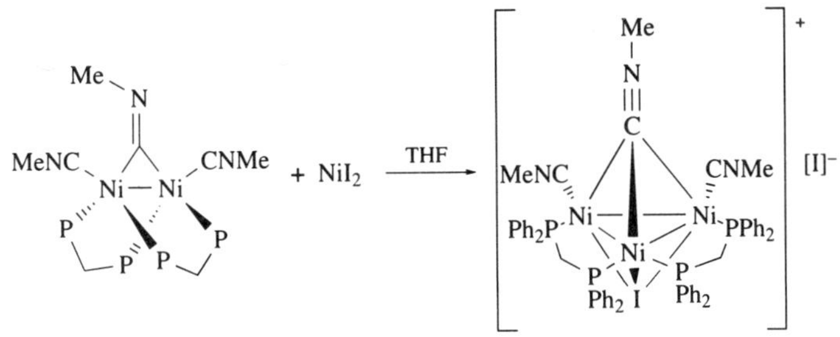

(59)

$(dppm)_2][NiCl_4]$ (Equation (60)).[84] The structure of $[Ni_4Hg(\mu\text{-}CNMe)_2(CNMe)_8(dppm)_2][NiCl_4]$ consists of a central Hg^{II} ion coordinated only to four nickel atoms, $d(\text{Ni–Hg})_{avg} = 0.265(3)$ nm. The four nickel atoms show two close pairwise Ni–Ni interactions, $d(\text{Ni–Ni})_{avg} = 0.254\,2(2)$ nm. The overall structure can be viewed as a central Hg^{II} ion coordinated by two neutral $Ni_2(\mu\text{-}CNMe)(CNMe)_4(dppm)$ fragments with a rectangular framework. The overall interaction of the two $Ni_2(\mu\text{-}CNMe)(CNMe)_4(dppm)$ fragments is similar to that of a linear bis(π-complex).

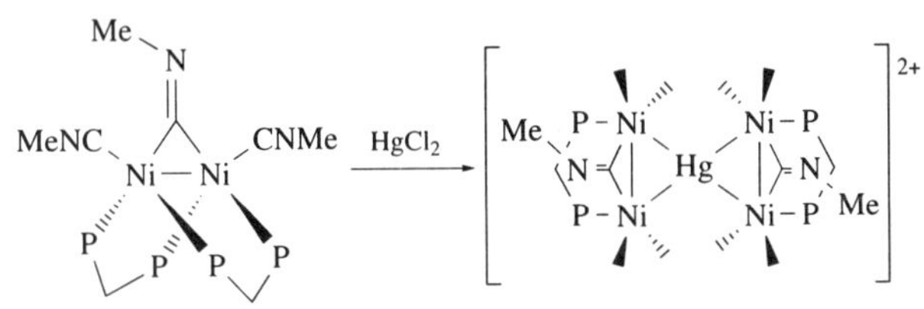

(60)

1.4.3 Clusters with Carbene and Carbyne Ligands

An extensive chemistry of di- and trinuclear complexes, especially those containing both group 6 and group 9 metals, with bridging carbene or carbyne ligands has issued forth from the laboratories of Stone.[185–7] The reaction of the mononuclear tungsten carbyne complexes of the type $[W(\equiv CR)(CO)_2(\eta^5\text{-}C_5H_5)]$ with $[Ni(cod)_2]$ gives the W_2Ni dicarbyne-bridged clusters $[NiW_2(\mu\text{-}CR)_2(CO)_4(\eta\text{-}C_5H_5)_2]$ (**23**).[187] Treatment of the trinuclear clusters with $[Pt(cod)_2]$ leads to the formation of μ_3-carbyne-capped tetranuclear clusters, $[NiPtW_2(\mu\text{-}CR)(\mu_3\text{-}CR)(CO)_4(cod)(\eta\text{-}C_5H_5)_2]$.[187] The carbyne complex $[W(\equiv CC_6H_4\text{-}4\text{-}Me)(CO)_2(\eta\text{-}C_5Me_5)]$ reacts with $[Ni(CO)_2(\eta\text{-}C_5H_5)]_2$ to give a mixture of trinuclear products, $[Ni_2W(\mu_3\text{-}CC_6H_4\text{-}4\text{-}Me)(CO)_2(\eta\text{-}C_5H_5)_2(\eta\text{-}C_5Me_5)]$ and $[NiW_2(\mu\text{-}CC_6H_4\text{-}4\text{-}Me)_2(CO)_4(\eta\text{-}C_5Me_5)_2]$.[186] Remarkable metal clusters containing chains of seven metal atoms, notably $[Ni_3W_4(\mu\text{-}CMe)_2(\mu_3\text{-}CMe)_2(CO)_8(\eta\text{-}C_5Me_5)_4]$, have also been prepared by extensions of these strategies.[185]

(23)

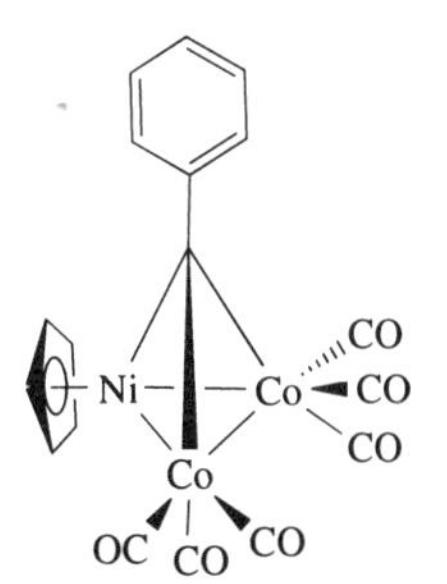

(24)

The reaction of the carbyne-capped clusters $[Co_2M(\mu_3\text{-}CR)(\eta\text{-}C_5H_5)(CO)_8]$ (M = Mo, W) with $[Ni(\eta\text{-}C_5H_5)(CO)]_2$ produced $[CoMNi(\mu_3\text{-}CR)(\eta\text{-}C_5H_5)_2(CO)_5]$. When R is a chiral group such as menthyl–C(O)O– the diastereomers of the chiral cluster can be separated by column chromatography. When R was not chiral, the chiral CoMNi clusters were separated on a modified cellulose column.[188] General syntheses of chiral cluster frameworks containing Ni–Co–Mo and Mo–Ru–Co have been reported.[189]

The carbyne cluster $[Co_2Ni(\mu_3\text{-}CPh)(\eta\text{-}C_5H_5)(CO)_6]$ has been reported and structurally characterized (**24**).[190,191] The internal rotation rates of the cyclopentadienyl and phenyl rings of this cluster have been determined by ^{13}C NMR.[192] Significantly, the net phenyl spinning rates were found to be lower than those observed in the isolobal Co_3 cluster, $[Co_3(\mu_3\text{-CPh})(CO)_9]$. This was attributed to a barrier created by the presence of nickel which splits the degeneracy of the cluster 2e molecular orbital leading to an asymmetry in the phenyl ring rotational potential energy surface.

The reaction of $[Fe_3(\mu\text{-H})(\mu\text{-COMe})(CO)_{10}]$ with $[Ni(\eta\text{-}C_5H_5)(CO)]_2$ produced the carbyne-capped cluster $[Fe_2Ni(\mu_3\text{-COMe})(\mu_3\text{-CO})(\eta\text{-}C_5H_5)_2(CO)_6]$, as well as smaller quantities of $[FeNi(\mu\text{-CO})_2(\eta\text{-}C_5H_5)_2(CO)]$ and $[Fe_3(\mu_3\text{-COMe})(\mu\text{-CO})_2(\eta\text{-}C_5H_5)(CO)_6]$.[193]

1.5 SUMMARY

A large body of new organometallic chemistry of nickel complexes with carbonyl, isocyanide, and carbene ligands has developed since the publication of *COMC-I* in 1982. Several areas are particularly noteworthy for their rapid growth. The nickel chemistry of carbon dioxide, for example, has seen enormous growth. For this reason, Section 1.2.2 has been included along with the section on complexes with carbonyl ligands (Section 1.2.1). The nickel-mediated coupling of CO_2 with alkynes has become a paradigm for the coupling of other unsaturated organic molecules, including simple alkenes, with CO_2 and with related heterocumulenes. The chemistry of nickel isocyanide complexes (Section 1.2.3) has seen significant new developments, especially in the chemistry of iminoacyl complexes and multiple insertions of isocyanides into nickel–alkyl carbon bonds. The chemistry of nickel carbene complexes (Section 1.2.4) has progressed; and our understanding of the structure and reactivity of nickel carbenes has increased significantly.

Dinuclear and higher nickel cluster compounds have been the subject of an increasing number of studies. The organometallic chemistry of nickel-containing heterodinuclear complexes has been especially active (Section 1.3.1). The differences in reactivity that occur in and are sometimes unique to dinuclear and metal cluster compounds are becoming more clear. The conversion of a μ-isocyanide ligand to a μ-carbonyl ligand via multiple-bond metathesis in the presence of CO_2 (Section 1.3.2), for example, has no known analogue in the chemistry of mononuclear complexes. Astonishing developments have occurred in the chemistry of higher nickel clusters (Section 1.4.1). The synthesis and structural characterization of clusters of very high nuclearity, such as $[HNi_{38}(CO)_{42}C_6]^{5-}$, have progressed to the point that these clusters exhibit properties which are less like those of molecules and more like those of bulk nickel metal particles. The ability to prepare heteronuclear carbyne-capped clusters which possess chiral cluster frameworks such as $[CoMoNi(\mu_3\text{-CR})(\eta\text{-}C_5H_5)_2(CO)_5]$ (Section 1.4.3) is of both fundamental and potential applied importance in the preparation of materials and catalysts.

1.6 REFERENCES

1. L. Mond, G. Langer and F. Quincke, *J. Chem. Soc.*, 1890, 749.
2. P. W. Jolly and G. Wilke, 'The Organic Chemistry of Nickel', Academic Press, New York, 1974.
3. F. Basolo, *Inorg. Chim. Acta*, 1981, **50**, 65.
4. R. Appel, F. Knoch and V. Winkhaus, *J. Organomet. Chem.*, 1986, **307**, 93.
5. W. B. Maier, M. Poliakoff, M. B. Simpson and J. J. Turner, *J. Mol. Struct.*, 1982, **80**, 83.
6. J. J. Turner, M. B. Simpson, M. Poliakoff and W. B. Maier, *J. Am. Chem. Soc.*, 1983, **105**, 3898.
7. C. A. Tolman, *Chem. Rev.*, 1977, **77**, 313.
8. T. L. Brown, *Inorg. Chem.*, 1992, **31**, 1286.
9. T. Kohara, S. Komiya, T. Yamamoto and A. Yamamoto, *Chem. Lett.*, 1979, 1513.
10. T. J. Groshens and K. J. Klabunde, *J. Organomet. Chem.*, 1983, **259**, 337.
11. S. K. Janikowski, L. J. Radonovich, T. Groshens and K. J. Klabunde, *Organometallics*, 1985, **4**, 396.
12. F. Heydenreich, *et al.*, *Isr. J. Chem.*, 1972, **10**, 293.
13. E. Niecke, *et al.*, *Chem. Ber.*, 1993, **126**, 1299.
14. E. Uhlig and B. Nestler, *Z. Chem.*, 1981, **21**, 451.
15. M. Bochmann, I. Hawkins, M. B. Hursthouse and R. L. Short, *J. Chem. Soc., Dalton Trans.*, 1990, 1213.
16. T. R. Belderrain, D. A. Knight, D. J. Irvine, M. Paneque, M. L. Poveda and E. Carmona, *J. Chem. Soc., Dalton Trans.*, 1992, 1491.
17. E. Carmona, E. Gutiérrez-Puebla, A. Monge, J. M. Marín, M. Paneque and M. L. Poveda, *Organometallics*, 1989, **8**, 967.
18. P. Giannoccaro, C. F. Nobile and M. Latronico, *Inorg. Chim. Acta*, 1990, **175**, 133.
19. W. Seidel and L. Raupach, *Z. Chem.*, 1988, **28**, 414.
20. S. Komiya, Y. Akai, K. Tanaka, T. Yamamoto and A. Yamamoto, *Organometallics*, 1985, **4**, 1130.
21. T. Yamamoto, J. Ishizu, T. Kohara and S. Komiya, *J. Am. Chem. Soc.*, 1980, **102**, 3758.
22. M. A. Bennett and T. Yoshida, *J. Am. Chem. Soc.*, 1978, **100**, 1750.
23. H. E. Bryndza, S. A. Kretchmar and T. H. Tulip, *J. Chem. Soc., Chem. Commun.*, 1985, 977.
24. H. E. Bryndza, *Organometallics*, 1985, **4**, 1686.

25. Y.-J. Kim, K. Osakada, K. Sugita, T. Yamamoto and A. Yamamoto, *Organometallics*, 1988, **7**, 2182.
26. F. Camps, J. Coll, J. M. Moretó and J. Torras, *Tetrahedron Lett.*, 1985, **26**, 6397.
27. T. Hirao, S. Nagata and T. Agawa, *Tetrahedron Lett.*, 1985, **26**, 5795.
28. T. Hirao, S. Nagata, Y. Yamana and T. Agawa, *Tetrahedron Lett.*, 1985, **26**, 5061.
29. Y. Ohshiro, *et al.*, *Tetrahedron Lett.*, 1985, **26**, 3015.
30. R. Csuk, B. I. Glänzer and A. Fürstner, *Adv. Organomet. Chem.*, 1988, **28**, 85.
31. R. A. Fischer, J. Behm, E. Herdtweck and C. Kronseder, *J. Organomet. Chem.*, 1992, **437**, C29.
32. T. Bein, S. J. McLain, G. D. Stucky, G. Woolery, D. Sayers and K. Moller, *J. Phys. Colloq.*, 1986, **1**, 277.
33. T. Bein, *et al.*, *J. Am. Chem. Soc.*, 1988, **110**, 1801.
34. M. Aresta, C. F. Nobile, V. G. Albano, E. Forni and M. Manassero, *J. Chem. Soc., Chem. Commun.*, 1975, 636.
35. A. Döhring, P. W. Jolly, C. Krüger and M. J. Romão, *Z. Naturforsch., Teil B*, 1985, **40**, 484.
36. D. Walther and V. Herzog, *Z. Chem.*, 1987, **27**, 373.
37. T. Tsuda, S. Morikawa, T. Saegusa and R. Sumiya, *J. Org. Chem.*, 1988, **53**, 3140.
38. T. Tsuda, S. Morikawa, N. Hasegawa and T. Saegusa, *J. Org. Chem.*, 1990, **55**, 2978.
39. T. Tsuda, N. Hasegawa and T. Saegusa, *J. Chem. Soc., Chem. Commun.*, 1990, 945.
40. T. Tsuda, K. Maruta and Y. Kitaike, *J. Am. Chem. Soc.*, 1992, **114**, 1498.
41. D. Walther, H. Schönberg, E. Dinjus and J. Sieler, *J. Organomet. Chem.*, 1987, **334**, 377.
42. D. Walther, G. Bräunlich, R. Kempe and J. Sieler, *J. Organomet. Chem.*, 1992, **436**, 109.
43. H. Hoberg, Y. Peres and A. Milchereit, *J. Organomet. Chem.*, 1986, **307**, C41.
44. H. Hoberg, A. Ballesteros, A. Sigan, C. Jégat and A. Milchereit, *Synthesis*, 1991, **5**, 395.
45. H. Hoberg, A. Ballesteros and A. Sigan, *J. Organomet. Chem.*, 1991, **403**, C19.
46. H. Hoberg, A. Ballesteros, A. Sigan, C. Jégat, D. Bärhausen and A. Milchereit, *J. Organomet. Chem.*, 1991, **407**, C23.
47. G. Bräunlich, R. Fischer, B. Nestler and D. Walther, *Z. Chem.*, 1989, **29**, 417.
48. H. Hoberg, D. Guhl and P. Betz, *J. Organomet. Chem.*, 1990, **387**, 233.
49. E. Hernandez and H. Hoberg, *J. Organomet. Chem.*, 1987, **327**, 429.
50. E. Hernandez and H. Hoberg, *J. Organomet. Chem.*, 1986, **315**, 245.
51. H. Hoberg and D. Guhl, *J. Organomet. Chem.*, 1989, **375**, 245.
52. H. Hoberg and D. Guhl, *J. Organomet. Chem.*, 1989, **378**, 279.
53. S. E. Bouaoud, P. Braunstein, D. Grandjean, D. Matt and D. Nobel, *Inorg. Chem.*, 1988, **27**, 2279.
54. P. T. Matsunaga, G. L. Hillhouse and A. L. Rheingold, *J. Am. Chem. Soc.*, 1993, **115**, 2075.
55. H. Hoberg and J. Korff, *J. Organomet. Chem.*, 1978, **152**, 255.
56. P. Hofmann, L. A. Perez-Moya, O. Steigelmann and J. Riede, *Organometallics*, 1992, **11**, 1167.
57. R. Sugai, A. Miyashita and H. Nohira, *Chem. Lett.*, 1988, 1403.
58. A. Miyashita, H. Shitara and H. Nohira, *J. Chem. Soc., Chem. Commun.*, 1985, 850.
59. A. Miyashita, R. Sugai and J. Yamamoto, *J. Organomet. Chem.*, 1992, **428**, 239.
60. A. K. List, M. R. Smith and G. L. Hillhouse, *Organometallics*, 1991, **10**, 361.
61. J. Kriege-Simondsen, T. D. Bailey and R. D. Feltham, *Inorg. Chem.*, 1983, **22**, 3318.
62. M. Beley, J. P. Collin, R. Ruppert and J. P. Sauvage, *J. Am. Chem. Soc.*, 1986, **108**, 7461.
63. S. Sakaki, *J. Am. Chem. Soc.*, 1990, **112**, 7813.
64. E. Duñach and J. Perichon, *J. Organomet. Chem.*, 1988, **352**, 239.
65. S. Derien, E. Duñach and J. Perichon, *J. Am. Chem. Soc.*, 1991, **113**, 8447.
66. S. Derien, J. C. Clinet, E. Duñach and J. Perichon, *J. Chem. Soc., Chem. Commun.*, 1991, 549.
67. S. Derien, J. C. Clinet, E. Duñach and J. Perichon, *J. Org. Chem.*, 1993, **58**, 2578.
68. S. Derien, J. C. Clinet, E. Duñach and J. Perichon, *J. Organomet. Chem.*, 1992, **424**, 213.
69. S. Derien, J. C. Clinet, E. Duñach and J. Perichon, *Tetrahedron*, 1992, **48**, 5235.
70. D. J. Darensbourg, C. G. Bauch and C. Ovalles, *Rev. Inorg. Chem.*, 1985, **7**, 315.
71. M. Bigorgne and A. Bouquet, *J. Organomet. Chem.*, 1963, **1**, 101.
72. D. Lentz, *Chem. Ber.*, 1985, **118**, 560.
73. E. Bessler, B. M. Barbosa, W. Hiller and J. Weidlein, *Z. Naturforsch., Teil B*, 1991, **46**, 490.
74. H. Behrens and K. Meyer, *Z. Naturforsch., Teil B*, 1966, **21**, 489.
75. D. L. DeLaet, D. R. Powell and C. P. Kubiak, *Organometallics*, 1985, **4**, 954.
76. D. T. Plummer and R. J. Angelici, *Inorg. Chem.*, 1983, **22**, 4063.
77. W. P. Fehlhammer, K. Bartel, B. Weinberger and U. Plaia, *Chem. Ber.*, 1985, **118**, 2220.
78. W. P. Fehlhammer and U. Plaia, *Z. Naturforsch., Teil B*, 1986, **41**, 1005.
79. M. Matsumoto and K. Nakatsu, *Acta Crystallogr., Sect. B*, 1975, **31**, 2711.
80. H. Arzoumanian, A. Baldy, M. Pierrot and J. F. Petrignani, *J. Organomet. Chem.*, 1985, **294**, 327.
81. H. Arzoumanian, R. Lai, J. Metzger and J. F. Petrignani, *J. Organomet. Chem.*, 1984, **267**, 207.
82. Y. Iitaka, T. Oohama, M. Shiina, T. Takizawa, Y. Yamamoto and H. Yamazaki, *J. Chem. Soc., Dalton Trans.*, 1987, 853.
83. Y. Yamamoto, K. Ehara, K. Takahashi and H. Yamazaki, *Bull. Chem. Soc. Jpn.*, 1991, **64**, 3376.
84. J. Gong, J. Huang, P. E. Fanwick and C. P. Kubiak, *Angew. Chem.*, 1990, **102**, 407.
85. Y. Yamamoto, K. Ehara and K. Takahashi, *Bull. Chem. Soc. Jpn.*, 1993, **66**, 778.
86. M. Bochmann, I. Hawkins and M. P. Sloan, *J. Organomet. Chem.*, 1987, **332**, 371.
87. E. Carmona, J. M. Marín, P. Palma and M. L. Poveda, *J. Organomet. Chem.*, 1989, **377**, 157.
88. J. Cámpora, E. Carmona, E. Gutiérrez-Puebla, P. Palma, M. L. Poveda and C. Ruiz, *Organometallics*, 1992, **11**, 11.
89. J. Cámpora, E. Gutiérrez-Puebla, A. Monge, M. L. Poveda, C. Ruiz and E. Carmona, *Organometallics*, 1993, **12**, 4025.
90. E. O. Fischer and J. R. Schneider, *J. Organomet. Chem.*, 1985, **295**, C29.
91. B. Gabor, C. Krüger, B. Marczinke, R. Mynott and G. Wilke, *Angew. Chem., Int. Ed. Engl.*, 1991, **30**, 1666.
92. E. Carmona, P. Palma and M. L. Poveda, *Polyhedron*, 1990, **9**, 1447.
93. M. F. Lappert and P. L. Pye, *J. Chem. Soc., Dalton Trans.*, 1977, 2172.
94. D. Spangler, J. J. Wendoloski, M. Dupuis, M. M. L. Chen and H. F. Schaefer, *J. Am. Chem. Soc.*, 1981, **103**, 3985.
95. M. Wada, K. Sameshima, K. Nishiwaki and Y. Kawasaki, *J. Chem. Soc., Dalton Trans.*, 1982, 793.
96. K. Miki, H. Taniguchi, Y. Kai, N. Kasai, K. Nishiwaki and M. Wada, *J. Chem. Soc., Chem. Commun.*, 1982, 1178.
97. D. J. Xu, K. Miki, Y. Kai, N. Kasai and M. Wada, *J. Organomet. Chem.*, 1985, **287**, 265.

98. D. J. Xu, K. Miki, M. Tanaka, N. Kasai, N. Yasuoka and M. Wada, *J. Organomet. Chem.*, 1989, **371**, 267.
99. H. Hoberg, G. Burkhart, C. Krüger and Y. Tsay, *J. Organomet. Chem.*, 1981, **222**, 343.
100. K. R. Pörschke, G. Wilke and R. Mynott, *Chem. Ber.*, 1985, **118**, 298.
101. M. J. Morris, *Organomet. Chem.*, 1991, **20**, 245.
102. M. Veith, L. Stahl and V. Huch, *J. Chem. Soc., Chem. Commun.*, 1990, 359.
103. S. A. Snow and G. Kodama, *Inorg. Chem.*, 1985, **24**, 795.
104. R. Del Rosario and L. S. Stuhl, *J. Am. Chem. Soc.*, 1984, **106**, 1160.
105. R. Del Rosario and L. S. Stuhl, *Organometallics*, 1986, **5**, 1260.
106. F. Joó and H. Alper, *Organometallics*, 1985, **4**, 1775.
107. L. Carlton, W. E. Lindsell, K. J. McCullough and P. N. Preston, *Organometallics*, 1985, **4**, 1138.
108. M. J. Chetcuti, P. E. Fanwick, S. R. McDonald and N. N. Rath, *Organometallics*, 1991, **10**, 1551.
109. M. J. Chetcuti, K. J. Deck, J. C. Gordon, B. E. Grant and P. E. Fanwick, *Organometallics*, 1992, **11**, 2128.
110. M. J. Chetcuti and K. A. Green, *Organometallics*, 1988, **7**, 2450.
111. A. Trenkle and H. Vahrenkamp, *Chem. Ber.*, 1981, **114**, 1366.
112. A. Trenkle and H. Vahrenkamp, *Chem. Ber.*, 1981, **114**, 1343.
113. E. Röttinger and H. Vahrenkamp, *J. Organomet. Chem.*, 1981, **213**, 1.
114. K. Fischer and H. Vahrenkamp, *Z. Anorg. Allg. Chem.*, 1981, **475**, 109.
115. K. A. Azam, A. A. Frew, B. R. Lloyd, L. Manojlovic-Muir, K. W. Muir and R. J. Puddephatt, *Organometallics*, 1985, **4**, 1400.
116. P. G. Pringle and B. L. Shaw, *J. Chem. Soc., Dalton Trans.*, 1983, 889.
117. C.-L. Lee, G. Besenyei, B. R. James, D. A. Nelson and M. A. Lilga, *J. Chem. Soc., Chem. Commun.*, 1985, 1175.
118. J. K. Gong and C. P. Kubiak, *Inorg. Chim. Acta*, 1989, **162**, 19.
119. D. L. DeLaet, R. Del Rosario, P. E. Fanwick and C. P. Kubiak, *J. Am. Chem. Soc.*, 1987, **109**, 754.
120. L. Manojlovic-Muir, K. W. Muir, W. M. Davis, H. A. Mirza and R. J. Puddephatt, *Inorg. Chem.*, 1992, **31**, 904.
121. J. K. Gong, P. E. Fanwick and C. P. Kubiak, *J. Chem. Soc., Chem. Commun.*, 1990, 1190.
122. X. L. R. Fontaine, S. J. Higgins, B. L. Shaw, M. Thornton-Pett and Y. C. Wu, *J. Chem. Soc., Dalton Trans.*, 1987, 1501.
123. R. B. King and K. S. Raghu Veer, *Inorg. Chem.*, 1984, **23**, 2482.
124. K. R. Pörschke, Y. H. Tsay and C. Krüger, *Inorg. Chem.*, 1986, **25**, 2097.
125. J. A. Osborn, G. G. Stanley and P. H. Bird, *J. Am. Chem. Soc.*, 1988, **110**, 2117.
126. K. D. Gallicano and N. L. Paddock, *Can. J. Chem.*, 1985, **63**, 314.
127. H. H. Karsch, B. Milewski-Mahrla, J. O. Besenhard, P. Hofmann, P. Stauffert and T. A. Albright, *Inorg. Chem.*, 1986, **25**, 3811.
128. K. S. Ratliff, D. L. DeLaet, J. Gao, P. E. Fanwick and C. P. Kubiak, *Inorg. Chem.*, 1990, **29**, 4022.
129. A. E. Stiegman and D. R. Tyler, *Coord. Chem. Rev.*, 1985, **63**, 217.
130. S. Hinze, J. K. Gong, P. E. Fanwick and C. P. Kubiak, *J. Organomet. Chem.*, 1993, **458**, C10.
131. D. L. DeLaet, P. E. Fanwick and C. P. Kubiak, *Organometallics*, 1986, **5**, 1807.
132. D. L. DeLaet, P. E. Fanwick and C. P. Kubiak, *J. Chem. Soc., Chem. Commun.*, 1987, 1412.
133. M. J. Chetcuti, B. E. Grant and P. E. Fanwick, *Organometallics*, 1990, **9**, 1345.
134. H. König, M. J. Menu, M. Dartiguenave, Y. Dartiguenave and H. F. Klein, *J. Am. Chem. Soc.*, 1990, **112**, 5351.
135. G. Trinquier, M. Dartiguenave, Y. Dartiguenave and M. Bénard, *Inorg. Chem.*, 1991, **30**, 4490.
136. J. C. Calabrese, L. F. Dahl, A. Cavalieri, P. Chini, G. Longoni and S. Martinengo, *J. Am. Chem. Soc.*, 1974, **96**, 2616.
137. G. Longoni, P. Chini, L. D. Lower and L. F. Dahl, *J. Am. Chem. Soc.*, 1975, **97**, 5034.
138. K. S. Suslick and R. E. Johnson, *J. Am. Chem. Soc.*, 1984, **106**, 6856.
139. T. L. Hall and J. K. Ruff, *Inorg. Chem.*, 1981, **20**, 4444.
140. D. F. Rieck, R. A. Montag, T. S. McKechnie and L. F. Dahl, *J. Am. Chem. Soc.*, 1986, **108**, 5370.
141. D. A. Nagaki, L. D. Lower, G. Longoni, P. Chini and L. F. Dahl, *Organometallics*, 1986, **5**, 1764.
142. D. G. Evans, *J. Chem. Soc., Chem. Commun.*, 1983, 675.
143. B. K. Teo, *Inorg. Chem.*, 1984, **23**, 1251.
144. B. K. Teo, G. Longoni and F. R. K. Chung, *Inorg. Chem.*, 1984, **23**, 1257.
145. A. Ceriotti, P. Chini, R. Della Pergola and G. Longoni, *Inorg. Chem.*, 1983, **22**, 1595.
146. M. M. Olmstead and P. P. Power, *J. Am. Chem. Soc.*, 1984, **106**, 1495.
147. A. Ceriotti, G. Longoni, M. Manassero, M. Perego and M. Sansoni, *Inorg. Chem.*, 1985, **24**, 117.
148. A. Ceriotti, G. Longoni, M. Manassero, N. Masciocchi, L. Resconi and M. Sansoni, *J. Chem. Soc., Chem. Commun.*, 1985, 181.
149. A. Ceriotti *et al.*, *J. Chem. Soc., Chem. Commun.*, 1985, 1402.
150. J. A. Hriljac, P. N. Swepston and D. F. Shriver, *Organometallics*, 1985, **4**, 158.
151. A. Ceriotti, R. Della Pergola, G. Longoni, M. Manassero and M. Sansoni, *J. Chem. Soc., Dalton Trans.*, 1984, 1181.
152. A. Ceriotti, *et al.*, *Chem. Abstr.*, 1984, **100**, 78 836e.
153. A. Arrigoni, A. Ceriotti, R. Della Pergola, G. Longoni, M. Manassero and M. Sansoni, *J. Organomet. Chem.*, 1985, **296**, 243.
154. A. Ceriotti *et al.*, *J. Am. Chem. Soc.*, 1986, **108**, 5370.
155. A. Ceriotti *et al.*, *J. Am. Chem. Soc.*, 1986, **108**, 8091.
156. D. Fenske, R. Basoglu, J. Hachgenei and F. Rogel, *Angew. Chem., Int. Ed. Engl.*, 1984, **23**, 160.
157. D. Fenske, J. Hachgenei and F. Rogel, *Angew. Chem., Int. Ed. Engl.*, 1984, **23**, 982.
158. D. Fenske, J. Hachgenei and J. Ohmer, *Angew. Chem., Int. Ed. Engl.*, 1985, **24**, 706.
159. G. Schmid, *Struct. Bonding (Berlin)*, 1985, **62**, 51.
160. E. Sappa and A. Tiripicchio, *New J. Chem.*, 1988, **12**, 599.
161. L. J. Farrugia, *Adv. Organomet. Chem.*, 1990, **31**, 301.
162. Y. L. Slovokhotov and Y. T. Struchkov, *Russ. Chem. Rev.*, 1985, **54**, 323.
163. M. Müller and H. Vahrenkamp, *Chem. Ber.*, 1983, **116**, 2765.
164. D. Fenske, J. Ohmer, J. Hachgenei and K. Merzweiler, *Angew. Chem., Int. Ed. Engl.*, 1988, **27**, 1277.
165. A. Ceriotti, P. Chini, A. Fumagalli, T. F. Koetzle, G. Longoni and F. Takusagawa, *Inorg. Chem.*, 1984, **23**, 1363.
166. C. Bergounhou, J. J. Bonnet, G. Lavigne and F. Papageorgiou, *Inorg. Synth.*, 1989, **26**, 360.

167. P. Braunstein, J. Rosé, A. M. Manotti-Lanfredi, A. Tiripicchio and E. Sappa, *J. Chem. Soc., Dalton Trans.*, 1984, 1843.
168. P. N. Lindsay *et al.*, *Organometallics*, 1984, **3**, 413.
169. M. J. Chetcuti, P. N. Cunningham, J. C. Gordon, B. E. Grant and J. Klaiss, *J. Organomet. Chem.*, 1990, **394**, 765.
170. M. J. Chetcuti, J. C. Huffman and S. R. McDonald, *Inorg. Chem.*, 1989, **28**, 238.
171. A. Ceriotti *et al.*, *Angew. Chem., Int. Ed. Engl.*, 1985, **24**, 697.
172. J. E. Ellis and E. A. Flom, *J. Organomet. Chem.*, 1975, **99**, 263.
173. E. O. Fischer and C. Palm, *Chem. Ber.*, 1958, **91**, 1725.
174. L. R. Byers, V. A. Uchtman and L. F. Dahl, *J. Am. Chem. Soc.*, 1981, **103**, 1942.
175. J. K. Ruff, R. P. White and L. F. Dahl, *J. Am. Chem. Soc.*, 1971, **93**, 2159.
176. J. P. Zebrowski, R. K. Hayashi and L. F. Dahl, *J. Am. Chem. Soc.*, 1993, **115**, 1142.
177. T. E. North, J. B. Thoden, A. Bjarnason and L. F. Dahl, *Organometallics*, 1992, **11**, 4338.
178. K. S. Ratliff, G. K. Broeker, P. E. Fanwick and C. P. Kubiak, *Angew. Chem. Int. Ed. Engl.*, 1990, **29**, 395.
179. K. S. Ratliff, P. E. Fanwick and C. P. Kubiak, *Polyhedron*, 1990, **9**, 1487.
180. D. A. Morgenstern, R. E. Wittrig, P. E. Fanwick and C. P. Kubiak, *J. Am. Chem. Soc.*, 1993, **115**, 6470.
181. J. L. Davidson, M. Green, F. G. A. Stone and A. J. Welch, *J. Chem. Soc., Dalton Trans.*, 1979, 506.
182. F. Cecconi, C. A. Ghilardi and S. Midollini, *Inorg. Chem.*, 1983, **22**, 3802.
183. V. W. Day, R. O. Day, J. S. Kristoff, F. J. Hirsekorn and E. L. Muetterties, *J. Am. Chem. Soc.*, 1976, **97**, 2571.
184. W. L. Gladfelter, M. W. Lynch, W. P. Schaefer, D. N. Hendrickson and H. B. Gray, *Inorg. Chem.*, 1981, **20**, 2390.
185. S. J. Davies, G. P. Elliott, J. A. K. Howard, C. M. Nunn, F. G. A. Stone and L. J. Farrugia, *J. Chem. Soc., Dalton Trans.*, 1987, 2177.
186. E. Delgado, J. Hein, J. C. Jeffery, A. L. Ratermann and F. G. A. Stone, *J. Chem. Soc., Dalton Trans.*, 1987, 1191.
187. G. P. Elliott, J. A. K. Howard, T. Mise, I. Moore, C. M. Nunn and F. G. A. Stone, *J. Chem. Soc., Dalton Trans.*, 1986, 2091.
188. R. Blumhofer and H. Vahrenkamp, *Chem. Ber.*, 1986, **119**, 683.
189. H. Vahrenkamp, *Inorg. Synth.*, 1990, **27**, 191.
190. H. Beurich, R. Blumhofer and H. Vahrenkamp, *Chem. Ber.*, 1982, **115**, 2409.
191. H. Beurich and H. Vahrenkamp, *Angew. Chem., Int. Ed. Engl.*, 1981, **20**, 98.
192. P. Yuan, M. J. Don, M. G. Richmond and M. Schwartz, *Inorg. Chem.*, 1991, **30**, 3704.
193. A. A. Aitchison and L. J. Farrugia, *Organometallics*, 1986, **5**, 1103.

2
Nickel–Carbon σ-Bonded Complexes

ANTHONY K. SMITH
University of Liverpool, UK

2.1 INTRODUCTION 29

2.2 LIGAND-FREE COMPLEXES, COMPLEXES STABILIZED BY π-BONDED LIGANDS AND NICKEL YLIDES 30

2.3 MONOLIGAND NICKEL ALKYL AND ARYL COMPLEXES 37

2.4 $[NiX(R)L_2]$ AND $[NiR_2L_n]$ COMPLEXES (n = 2–4) 40

2.4.1 Synthesis 41

2.4.1.1 Synthesis using organo–main group metal reagents 41

2.4.1.2 Oxidative addition reactions 47

2.4.1.3 Miscellaneous syntheses 53

2.4.2 Reactions 56

2.4.2.1 Reactions in which the Ni–C bond is not involved 56

2.4.2.2 Reactions in which new Ni–C bonds are generated 58

2.4.2.3 Reactions in which the alkyl or aryl group is displaced 62

2.4.3 Structure and Properties 64

2.5 η^3-ALLYLNICKEL ALKYL AND ARYL COMPLEXES 65

2.6 η-CYCLOPENTADIENYLNICKEL ALKYL AND ARYL COMPLEXES 66

2.6.1 Synthesis 66

2.6.2 Reactions and Properties 70

2.7 NICKELACARBOCYCLIC COMPLEXES 73

2.7.1 Synthesis 73

2.7.2 Reactions and Properties 78

2.8 NICKELAHETEROCYCLIC COMPLEXES 80

2.9 NICKEL COMPLEXES OF BIOLOGICAL SIGNIFICANCE 91

2.10 NICKEL CLUSTERS CONTAINING Ni–C σ-BONDS 93

2.10.1 Homometallic Clusters 93

2.10.2 Heterometallic Clusters 95

2.11 REFERENCES 99

2.1 INTRODUCTION

The subject matter in this chapter has been divided largely according to the nature of the additional ligands bonded to nickel in much the same way as in *COMC-I*.[1] Thus separate sections are devoted to 'ligand-free' nickel alkyls, complexes stabilized by group 15 donor ligands, η^3-allyl and η-cyclopentadienyl groups, and to nickelacyclic complexes in which the cycle contains an Ni–C σ-bond, including those formed in reactions with carbon dioxide. A brief account of nickel alkyl complexes of biological significance and of nickel cluster complexes containing Ni–C σ-bonds has also been included. This account refers to reactions catalysed by nickel complexes only in those cases where nickel complexes containing Ni–C σ-bonds have been isolated and characterized. The reader is referred to Volume 12 for an account of the use of nickel complexes in organic synthesis.

Interest in organonickel chemistry remains high and is maintained in particular by the role of nickel in many homogeneous catalytic reactions. Many carbon–carbon bond formation reactions such as oligomerization and polymerization of alkenes and dienes, and hydrocyanation and carbonylation reactions can be carried out by organonickel complexes. Of particular importance is the Shell Higher Olefin Process (SHOP), which is one of the largest applications of homogeneous catalysis, and has spurred the interest and activity in complexes with Ni–C σ-bonds.[2]

2.2 LIGAND-FREE COMPLEXES, COMPLEXES STABILIZED BY π-BONDED LIGANDS AND NICKEL YLIDES

This section deals with complexes of the type $[Ni(R)_2]$ or [NiX(R)]; with complexes stabilized by π-bonded ligands [Ni(R)(π-ligand)], where π-ligand is alkene or arene for example; with complexes stabilized by ligands which do not contain a group 15 donor ligand; with ionic complexes; and with ylide complexes. For convenience, the ylide complexes include those which also contain a group 15 donor ligand, but where two group 15 donors are coordinated the complexes are discussed in Section 2.4. The complexes considered are listed in Table 1.

The finely divided nickel powder prepared by the reduction of nickel halide with lithium reacts with C_6F_5I in glyme (glyme = 1,2-dimethoxyethane) to yield the solvated complex $[Ni(C_6F_5)_2(glyme)]$ and NiI_2. With C_6H_5Br, the nickel powder reacts to give $[NiBr(C_6H_5)(glyme)]$.[3] The coordinated glyme is readily displaced by phosphines, amines, sulfides, isocyanides, dienes and carbon monoxide to give many novel compounds in high yields. This represents a useful synthetic route to a number of compounds such as $[Ni(C_6F_5)_2(PPh_2H)_2]$ that are not readily prepared by other methods.

Treatment of $[Ni(acac)_2]$ or $NiBr_2$ with CaC_2 in the presence of phenylacetylene gives rise to a benzene-soluble material which could be isolated by gel-permeation chromatography. The elemental composition and relative molecular mass (3250 ± 290) of the isolated product suggest a composition of $[Ni(C_2Ph)_2]_n$ with n having a value near 13. This catalytically active material was also obtained on treatment of $[NiCl_2(glyme)]$ with 2 equiv. PhCCLi.[4] The reactivity of this material suggests that it is not simply an oligomer of nickel(II)bis(phenylacetylide); it may be related to the Reppe catalyst prepared from the reaction of nickel atoms with phenylacetylene.[5] The material is an active catalyst for the oligomerization of phenylacetylene to 1,2,4-triphenylbenzene and 1,3,5,8-tetraphenylcyclooctatetraene.

The cocondensation of nickel atoms and cyclopropane in an argon matrix at 12 K produces nickelacyclobutane.[6] Visible photolysis (500 nm) of this metallacycle results in the formation of $[Ni(CH_2)(C_2H_4)]$ which, upon UV irradiation, undergoes a hydrogen shift reaction to give the vinylnickel complex $[Ni(CH{=}CH_2)(Me)]$. The complexes were characterized by FTIR spectroscopy.

There have been numerous studies of the interaction of nickel atoms or nickel ions with hydrocarbons in the gas phase. Several excellent reviews on this chemistry have been published recently and the reader is referred to these for further information.[7–11] These studies provide valuable data on Ni–C bond enthalpies as well as mechanistic information on C–C and C–H bond cleavage and Ni–C and Ni–H bond formation processes. Table 2 gives a list of experimental and calculated (using molecular orbital calculations based on density-functional theory) enthalpies for the Ni–Me or Ni^+–Me bond.[12–17] It is usually found that a positive metal centre, Ni^+–Me, is able to stabilize the Ni–C bond to the point where it becomes stronger than the corresponding Ni^+–H bond.[12,16,17] Adsorption energies for Me on the Ni(111) surface have been calculated using a many-electron embedding theory. The values range from 164 kJ mol^{-1} for adsorption at the three-fold site to 151 kJ mol^{-1} and 143 kJ mol^{-1} for adsorption at the bridge and atop atom sites respectively.[18] Other calculations for the adsorption of Me on Ni(111) have given adsorption energy values ranging from 109 kJ mol^{-1} to 279 kJ mol^{-1}.[19–22] Of particular interest to the organometallic chemist are a theoretical study of the breaking of a C–H bond in CH_4 on both discrete transition metal complexes and on a nickel surface,[23] a study of the transition state for CH_4 elimination from [NiH(Me)], which provides some information on the charge distribution to be expected in an agostic system,[24] and the observation that the decomposition of neopentyl groups on a nickel surface proceeds via a β-methyl elimination rather than α- or γ-hydrogen elimination or coupling.[25]

There is much other useful information for the preparative chemist available from gas phase and theoretical studies. Thus, for example, it has been shown that the atomic Ni^+ ion reacts with

Table 1 Ligand-free and ionic nickel alkyl complexes.

Complex	*Reference*	*Ni–C* (nm)
Ligand-free complexes		
$[Ni(C_6F_5)_2(C_4H_{10}O_2)]$	3	
$[NiBr(C_6F_5)(C_4H_{10}O_2)]$	3	
$[Ni(C{\equiv}CPh)_2]_n$ ($n \approx 13$)	4	
$[Ni(CH_2CH_2CH_2)]$	6	
$[Ni(Me)(CH{=}CH_2)]$	6	
$[Ni\{CH_2S(O)Me_2\}(C_2H_4)_2]$	31	
$[Ni\{CH_2S(O)Me_2\}(CO)_3]$	31	
$[Ni(CH_2CH{=}CH_2)\{C(O)Bu\}(CO)_3]$	32	
$[Ni\{B(O_2C_6H_4)_2\}(\eta^1,\eta^2\text{-}C_8H_{13})]$ (**1**)	33	
$[Ni\{B(O_2C_6H_4)_2\}(\eta^2,\eta^1,\eta^2\text{-}C_9H_{15})]$ (**2**)	34	
$[Ni\{B(O_2C_6H_4)_2\}(\eta^2,\eta^1,\eta^2\text{-}C_{10}H_{17})]$ (**3**)	34	
$[Ni(C_6F_5)_2(PhCN)_2]$	43	
$[Ni(C_6F_5)_2(\eta\text{-}C_6H_5Me)]$	38,40,41	
$[Ni(C_6HF_4\text{-}2,3,5,6)_2(\eta\text{-}C_6H_5Me)]$	38,40,41	
$[Ni(C_6Cl_5)_2(\eta\text{-}C_6H_5Me)]$	38,40,41	
$[Ni(SiF_3)_2(\eta\text{-}C_6H_5Me)]$	38,40,41	
$[Ni(SiCl_3)_2(\eta\text{-}C_6H_5Me)]$	38,40,41	
$[Ni(SiCl_2Me)_2(\eta\text{-}C_6H_5Me)]$	38,40,41	
$[Ni(SiCl_3)_2(\eta\text{-}C_6H_3Me_3\text{-}1,3,5)]$	38,40,41	
$[Ni(SiCl_3)_2(\eta\text{-}C_6H_6)]$	38,40,41	
$[Ni(SiCl_3)_2(\eta^6\text{-}C_6H_5\text{–}C_6H_5)]$	39	
$[Ni(SiCl_3)_2(\eta^6\text{-}p\text{-}MeC_6H_4\text{–}C_6H_4Me)]$	39	
$[Ni(SiCl_3)_2(\eta^6\text{-}C_6H_5CH_2C_6H_5)]$	39	
$[Ni(SiCl_3)_2(\eta^6\text{-}C_6H_5CH_2CH_2C_6H_5)]$	39	
$[Ni(SiCl_3)_2\{\eta^6\text{-}C_6H_4(CH_2CH_2)_2C_6H_4\}]$	39	
$[Ni(SiCl_3)_2(\eta^6\text{-naphthalene})]$	39	
$[Ni(SiCl_3)_2(\eta^6\text{-anthracene})]$	39	
$[Ni(C_6F_5)_2(nbd)]$	43	0.191 4(2), 0.192 1(2)
$[Ni(C_6F_5)_2(cod)]$	43	0.190 6(2)
Ionic complexes		
$[Ni(\mu\text{-}F)(C_6F_5)_2]_2[NBu_4]_2$ (**4**)	44	0.190 0(3), 0.191 2(3)
$[Ni(\mu\text{-}Cl)(C_6F_5)_2]_2[NEt_4]_2$ (**5**)	44	0.188 0(3), 0.188 0(3)
$[Ni(\mu\text{-}Cl)(SiCl_3)_2]_2[2,6\text{-}Bu^t{}_2py]_2$	44	0.216 1(3),[a] 0.218 1(3)[a]
$[Ni(\mu\text{-}F)(SiCl_3)_2]_2[NBu_4]_2$	44	
$[Ni(\mu\text{-}Cl)(SiCl_3)_2]_2[NBu_4]_2$	44	
$[Ni(\mu\text{-}Br)(SiCl_3)_2]_2[NBu_4]_2$	44	
$[Ni(C_6F_5)_2(\mu\text{-}OH)]_2[NBu_4]_2$ (**6**)	45,46,47	0.186 7(8), 0.188 3(6)
$[(Ph)_2Ni(\mu\text{-}OH)(\mu\text{-}pz)Ni(Ph)_2][NBu_4]_2$	45	0.190 1(6), 0.188 7(6), 0.189 1(6), 0.189 6(6)
$[(Ph)_2Ni(\mu\text{-}OH)(\mu\text{-}3\text{-}Mepz)Ni(Ph)_2][NBu_4]_2$	45	
$[(Ph)_2Ni(\mu\text{-}OH)(\mu\text{-}3,5\text{-}Me_2pz)Ni(Ph)_2][NBu_4]_2$	45	
$[(Ph)_2Ni(\mu\text{-}OH)(\mu\text{-indazolate})Ni(Ph)_2][NBu_4]_2$	45	
$[(Ph)_2Ni(\mu\text{-}OH)(\mu\text{-triazolate})Ni(Ph)_2][NBu_4]_2$	45	
$[Ni(\mu\text{-}pz)(C_6F_5)_2]_2[NBu_4]_2$	45	
$[Ni(\mu\text{-}3\text{-}Mepz)(C_6F_5)_2]_2[NBu_4]_2$	45	
$[Ni(\mu\text{-}3\text{-indazolate})(C_6F_5)_2]_2[NBu_4]_2$	45	
$[Ni(\mu\text{-}3\text{-triazolate})(C_6F_5)_2]_2[NBu_4]_2$	45	
$[Ni(C_6F_5)_2(S_2CO)][NBu_4]_2$	46	
$[Ni(C_6F_5)_2(S_2CX)][NBu_4]$[b]	46	
$[Ni(C_6F_5)_2(\text{pyridine-2-thiolate})][NBu_4]$	47	0.189 4(6), 0.189 1(6)
$[Ni(C_6F_5)_2(acac)][NBu_4]$	47	
$[Ni(C_6F_5)_2(\text{benzoylacetonate})][NBu_4]$	47	
$[Ni(C_6F_5)_2(\text{8-hydroxyquinolinate})][NBu_4]$	47	
$[Ni(C_6F_5)_2(\text{pyrimidine-2-thiolate})][NBu_4]$	47	
$[Ni(C_6F_5)_2(\text{2,6-dimethyl-5-oxo-1,2,4-triazine-3-thiolate})][NBu_4]$	47	
$[Ni(C_6F_5)_2(N\text{-methylimidazole-2-thiolate})][NBu_4]$	47	
$[Ni(C_6F_5)_2(\text{thiozolidine-2-thiolate})][NBu_4]$	47	
$[Ni(C_6F_5)_2(\text{benzoimidazole-2-thiolate})][NBu_4]$	47	
$[Ni(C_6F_5)_2(\text{benzothiazole-2-thiolate})][NBu_4]$	47	
$[Ni(C_6F_5)_2(\mu\text{-}NCCHCN)]_2[NBu_4]_2$ (**7**)	48	
$[K_2\{Ni(C{\equiv}CC_6H_4C{\equiv}C\text{-}4)\}]_n$	49	
$[(C_2H_4)_2Ni(\mu\text{-}Me)Li(PMDTA)]$ (**8**)	50,51	0.201 2(5)
$[(C_2H_4)_2Ni(\mu\text{-}Et)Li(PMDTA)]$	50,51	
$[(C_2H_4)_2Ni(\mu\text{-}Pr^n)Li(PMDTA)]$	50,51	
$[(C_2H_4)_2Ni(\mu\text{-}Bu^n)Li(PMDTA)]$	50,51	
$[(C_2H_4)_2Ni(\mu\text{-}Pr^n)_2\{Li(PMDTA)\}_2]$	51	
$[(C_2H_4)_2Ni(\mu\text{-}Bu^n)_2\{Li(PMDTA)\}_2]$	51	

Table 1 (continued)

Complex	Reference	Ni–C (nm)
$[Li(TMEDA)_2][Ni(Pr^n)(C_2H_4)_2]$	51	
$[Li(TMEDA)_2][Ni(Bu^n)(C_2H_4)_2]$	51	
$[(C_2H_4)_2Ni(\mu\text{-Me})Mg(Me)(TMEDA)]$ (**9**)	52	0.203 1(3)
$[(C_2H_4)_2Ni(\mu\text{-Et})Mg(R)(TMEDA)]$	52	
$[(C_2H_4)_2Ni(\mu\text{-Pr}^n)Mg(R)(TMEDA)]$	52	
$[(C_2H_4)_2Ni(\mu\text{-}C_2H_4CH{=}CH_2)Mg(R)(TMEDA)]$	52	
$[(C_2H_4)_2Ni(\mu\text{-Ph})Mg(R)(TMEDA)]$	52	
$[\{(C_2H_4)_2Ni\}_2(\mu\text{-Ph})Mg(OEt_2)]$	52	
Ylide complexes		
$[Ni(CH_2PMe_3)(cdt)]$	56	
$[Ni(CH_2PMe_3)(C_2H_4)_2]$	56	
$[Ni(CH_2PPh_3)(C_2H_4)_2]$	56	
$[Ni(CH_2PMe_3)(CO)_3]$	56	
$[Ni(CH_2PPh_3)(CO)_3]$	56	
$[Ni(Ph)(CH_2PMe_3)(Ph_2PCHCHO)]$	57,61	
$[Ni(Ph)(CH_2PMe_3)(Ph_2PCHCPhO)]$	57,58,61,62	
$[Ni(Ph)(CH_2PMe_3)(Ph_2PCHCMeO)]$	57,58,61,62	
$[Ni(Ph)(CH_2PPh_3)(Ph_2PCHCMeO)]$	57,61	
$[Ni(Ph)(CH_2PPr^i_3)(Ph_2PCHCPhO)]$	58,61	
$[Ni(Ph)\{CH(Ph)PPr^i_3\}(Ph_2PCHCMeO)]$ (**11**)	57,59,61	0.189 0(4)[c], 0.196 6(4)[d]
$[Ni(Ph)\{CH(Ph)PPr^i_3\}(Ph_2PCHCPhO)]$	58,61,62	
$[Ni\{CH(Me)PPh_3\}_2(Cp)][BPh_4]$	63	
$[Ni\{CH(Me)PPh_3\}_2(Cp)][PF_6]$	63	
$[Ni\{Ph_3PCH(CH_2)_nCHPPh_3\}(Cp)][PF_6]$	63	
$[Ni\{CH_2P(Me)_2B(Me)_2P(Me)_2CH_2\}_2]$ (**12**)	64	

[a] Ni–Si. [b] X = NHEt, NHPr, NMe_2, NEt_2, NC_4H_8, NC_5H_{10}, NC_4H_8O, OMe, OEt. [c] $Ni–C_{Ph}$. [d] $Ni–C_{CH}$

Table 2 Nickel–carbon bond dissociation enthalpies or bond enthalpies.

Bond	*Substrate*	*D(Ni–C)* (kJ mol^{-1})	*Method*[a]	*Reference*
Ni–Me	NiMe	231 ± 13	GIBMS	12
Ni^+–Me	$NiMe^+$	189 ± 13	GIBMS	12
		190 ± 12	GIBMS	13
Ni^+–Me	$Ni(Me)_2^+$	>214	GIBMS	14
$Ni^+{=}CH_2$	$NiCH_2^+$	315 ± 7	GIBMS	15
Ni–Me	[Ni(Me)(CO)(Cp)]	174	MO	16
Ni^+–Me	$[Ni(Me)(CO)_4]^+$	251	MO	17

[a] GIBMS = Guided ion beam mass spectrometry; MO = Molecular orbital calculations.

n-pentylbenzene exclusively at the benzylic C–C bond with no dehydrogenation;[26] that nickel atoms react with methane under UV irradiation to generate [NiH(Me)] (characterized by FTIR) in an argon matrix;[27] and that there is a correlation between the amount of covalency in the Ni–L bond in [NiH(Me)L] (L = C_2H_4, C_2F_4, CO, PR_3, N_2, H_2O and Cl^-) and the destabilizing effect of the ligand on the Ni–Me bond.[28] It is found that the greater the covalent Ni–L bonding, the greater the destabilization of the Ni–Me bond. The 17-electron anion $[Ni(CO)_3]^-$ reacts in the gas phase with organic electrophiles by an oxidative addition process.[29] For example, $[Ni(CO)_3]^-$ reacts with bromonitrobenzenes by oxidative addition of a C–Br bond, and with benzophenone by oxidative addition of the C–C bond between a phenyl group and the carbonyl carbon. In a study of the reactions of Ni^+, $[NiCO]^+$, $[NiPF_3]^+$ and $[NiCp]^+$ with a series of aromatic compounds it was found that $[NiCp]^+$ was the most reactive of the ligated ions. This was suggested to be due to some charge transfer resulting in an increased positive charge on the metal.[30] The important conclusions from gas phase studies of reactions of alkanes with nickel ions are the facile C–C bond cleavage processes, favoured by a high Ni–C bond strength for the bare ion, and the multiple C–H and C–C bond cleavage reactions that can occur, presumably due to the high coordinative unsaturation of these ions. For nickel complexes the presence of ligands does of course restrict the activity.

The thermally unstable Ni–C σ-bonded complex $[Ni\{CH_2S(O)Me_2\}(C_2H_4)_2]$ has been obtained from the reaction of $[Ni(C_2H_4)_3]$ with dimethylmethyleneoxosulfurane in ether below 0 °C. This product decomposes violently at 0 °C, but on treatment with CO at −78 °C the carbonyl complex [Ni{$CH_2S(O)Me_2$}$(CO)_3$] is obtained which is stable to ~20 °C.[31] The acyl(alkyl)nickel carbonyl complex [Ni($CH_2CH=CH_2$)(COBu)$(CO)_3$] is formed when the pentanoylnickel acylate [Ni{=C-(Bu)OLi}$(CO)_3$] is treated with allyl bromide.[32] Treatment of [Ni$(cod)_2$] with bis(brenzcatechnicato)boric acid, $HB(O_2C_6H_4)_2$ in ether gives the η^1,η^2-cyclooctenyl complex (**1**).[33] Complex (**1**) is catalytically active for the 1,4-*trans*-polymerization of butadiene with a selectivity of 85% at 50 °C, although the activity is rather low. The analogous allyl complexes [Ni(η-C_3H_5){$(O_2C_6H_4)_2B$}] and [Ni(η^3-C_4H_7){$(O_2C_6H_4)_2B$}] react with hexa-1,5-diene by insertion into the nickel–allyl bond to give the corresponding η^2,η^1,η^2-nona- and -deca-1,8-diene-5-yl nickel complexes (**2**) and (**3**), respectively.[34] These reactions suggest that the bis(brenzcatechnicato)borate anion may have an important role in the future of organonickel chemistry.

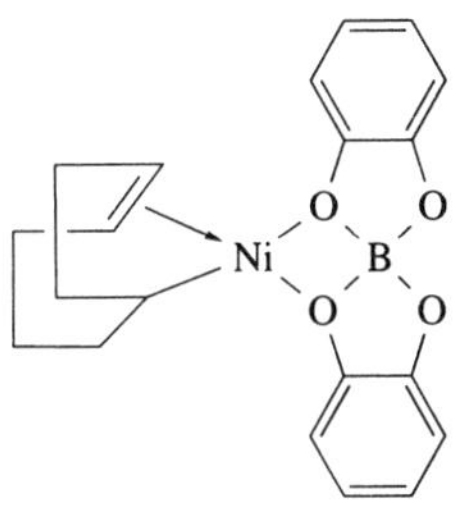

(**1**)

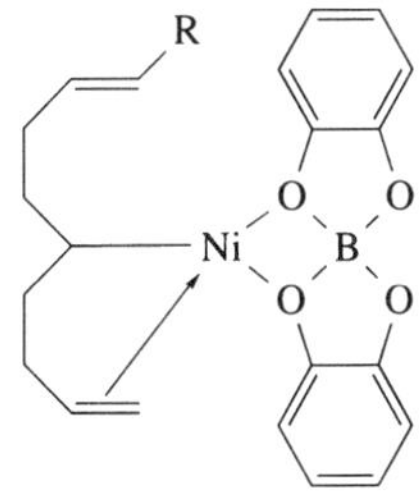

(**2**) R = H
(**3**) R = Me

Following the synthesis of the first fully characterized η^6-arenenickel complex [Ni$(C_6F_5)_2$(η-C_6H_5Me)] in 1978[35] there is now quite an extensive chemistry associated with this class of compounds. Synthetic routes to these complexes using either metal atom or solution methods have been developed, and, since the η-C_6H_5Me complexes undergo arene exchange reactions, a wide variety of complexes of the type [NiR_2(arene)] is now available.[35–9] These complexes show a very high activity for but-1-ene isomerization[40] and for the catalytic dimerization of ethene and propene.[41] For these dimerization reactions the reaction rate is dependent on both the arene (with benzene > toluene > mesitylene) and on the R group (with $SiCl_3 \gg SiF_3 > C_6F_5$). Importantly, a cocatalyst is not required in these reactions, but the addition of $AlCl_2Et$ as cocatalyst increases both the reaction rate and the catalyst lifetime. The η^6-arene ligand in the complex [Ni$(C_6F_5)_2$(η-C_6H_5Me)] can be displaced by the cyclopentadienyl ligand to give the anionic complex [Ni$(C_6F_5)_2$(Cp)]$^-$,[42] and by norbornadiene or cyclooctadiene to give [Ni$(C_6F_5)_2$(η^4-diene)] for which the crystal structures have been determined.[43] The displacement of the arene ligand in [NiR_2(η^6-arene)] complexes by treatment with tetraalkylammonium halides (Equation (1)) gives the square-planar halide-bridged alkylnickel complexes $[Ni(\mu\text{-}X)R_2]_2^{2-}$ (R = C_6F_5, X = F, Cl; R = $SiCl_3$, X = F, Cl, Br).[44] A structural comparison of the compounds $[NBu_4]_2$[{Ni(μ-F)$(C_6F_5)_2$}$_2$] (**4**) and $[NEt_4]_2$[{Ni(μ-Cl)$(C_6F_5)_2$}$_2$] (**5**) has shown that the Ni–C distances in (**4**) are slightly but significantly longer than those in (**5**). This may be attributed to slightly less π-back-bonding (Ni → C_6F_5) in the fluoride-bridged compound. In the analogous hydroxo compound, $[NBu_4]$[{Ni(μ-OH)$(C_6F_5)_2$}$_2$] (**6**)[45] the Ni–C distances are not significantly different from those in the chloro-bridged compound (**5**). The μ-hydroxo-compound (**6**) is synthesized by the reaction of the labile complex *cis*-[Ni$(C_6F_5)_2(PhCN)_2$] with aqueous tetrabutylammonium hydroxide. The hydroxo groups are highly nucleophilic, as indicated by the high field proton resonance (δ = −5.74 ppm), and react with weak protic electrophiles HL (L = pyrazolate, 3-methylpyrazolate, indazolate or triazolate) to give the corresponding complexes $[(F_5C_6)_2Ni(\mu\text{-}OH)(\mu\text{-}L)Ni(C_6F_5)_2]^{2-}$ and $[Ni(C_6F_5)_2(\mu\text{-}L)]_2^{2-}$, depending upon the Ni:HL ratio.[45] When L = 3,5-dimethylpyrazolate, only the monosubstituted derivative $[(F_5C_6)_2Ni(\mu\text{-}OH)(\mu\text{-}L)Ni(C_6F_5)_2]^{2-}$ is formed. The μ-hydroxo compound (**6**) has proved to be a useful precursor to a large number of anionic nickel complexes of the type $[Ni(C_6F_5)_2(LL)]^-$, as shown in Table 1, again by making use of the highly nucleophilic character of the μ-OH groups.[46,47] Complex (**6**) is also a catalyst for the cyclotrimerization of malononitrile, and reacts stoicheometrically with malononitrile (in the nickel:malononitrile molar ratio of 1:2) to form the unusual μ-$[CH(CN)_2]^-$ complex $[\{Ni(C_6F_5)_2(\mu\text{-}NCCHCN)\}_2]^{2-}$ (**7**).[48] The homoleptic, polymeric complex $[K_2\{Ni(C{\equiv}CC_6H_4C{\equiv}C\text{-}4)_2\}]_n$, which is very unstable towards oxidation and hydrolysis, has been isolated from the reaction between $[Ni(NCS)_2(NH_3)_4]$ and the potassium salt of *p*-diethynylbenzene in liquid ammonia.[49]

The reactions of alkenenickel(0) complexes with alkyllithium or dialkylmagnesium compounds afford a fascinating and important series of complexes in which either the alkyl groups are bridging

$$[Ni(\eta^6\text{-arene})R^1_2] + NR^2_4X \longrightarrow [NiR^1_2X]_2[NR^2_4]_2 \quad (1)$$

$X = F, Cl, Br; R^1 = C_6H_5, SiCl_3; R^2 = Et, Bu^n$

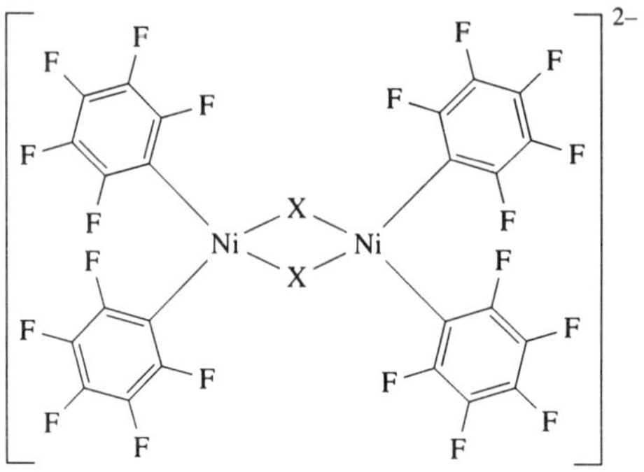

(4) X = F
(5) X = Cl
(6) X = OH

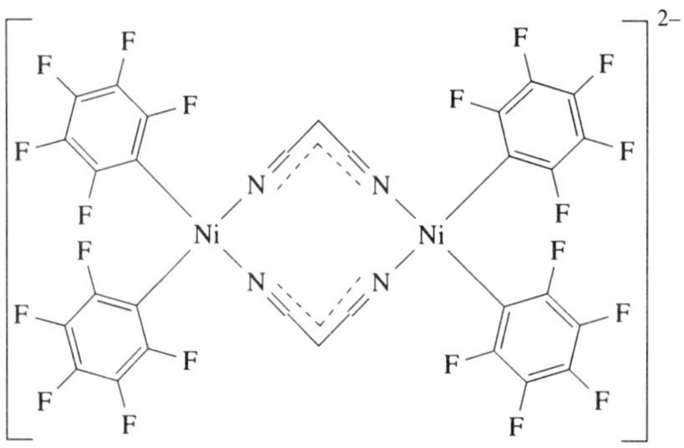

(7)

between nickel and lithium or magnesium, or in which an anionic alkylnickel complex is formed (Schemes 1 and 2).[50–3] Since an interaction between trialkylaluminum compounds and alkenenickel(0) complexes has not been detected (^{1}H NMR spectra of 1:1 mixtures of $AlEt_3$ and $[Ni(C_2H_4)_3]$ at –80 °C show only the resonances of the separate components)[54] but is assumed to play a key role in the well-known 'nickel effect' whereby ethene is dimerized to 1-butene by triethylaluminum in the presence of nickel, the study of complexes in which alkyl groups bridge between nickel and main group metals will undoubtedly provide important information on this effect. The structures of the μ-methyl complexes $[(C_2H_4)_2Ni(\mu\text{-Me})Li(PMDTA)]$ **(8)**[50] (PMDTA = pentamethyldiethylenetriamine) and $[(C_2H_4)_2Ni(\mu\text{-Me})\text{-}Mg(Me)(TMEDA)]$ **(9)**,[52] and of the anionic methylnickel complex $[Li(TMEDA)_2]\text{-}[NiMe(C_2H_4)_2]$ **(10)**[55] have been determined. The structures of **(8)** and **(9)** are characterized by the presence of a Ni–C–Li or Ni–C–Mg bridge in which the Ni–C distances (0.201 2(5) nm and 0.203 1(3) nm, respectively) are rather longer than a typical nonbridged Ni–C σ-bond as expected for a three-centre bond. The nature of the Ni–C bond in these complexes is dependent on the π-ligands (ethene, cyclododeca-1,5,9-triene (cdt) or CO) bonded to the nickel. The chemical and spectroscopic properties of the complexes indicate that the Ni–Me bond in the carbonyl complex is largely covalent whereas in the cdt or ethene derivatives it is more polar.[50] Thus, for example, the ^{13}C NMR resonances due to the methyl carbon in $[(THF)_x\text{-}Li(\mu\text{-Me})Ni(\pi\text{-ligand})]$ (where π-ligand = cdt, $(C_2H_4)_2$ or $(CO)_3$) are at –4.6 ppm, –12.2 ppm and –24.7 ppm, respectively. It is clear that the acceptor strength of the nickel is determined by the nature of the π-ligands, and increases in the order $Ni(cdt) < Ni(C_2H_4)_2 < Ni(CO)_3$. For the higher alkyls, the complexes $[(\pi\text{-donor})Ni(\mu\text{-R})Li(PMDTA)]$ (R = Et, Pr^n, Bu^n) and $[(C_2H_4)_2Ni(\mu\text{-R})_2\{Li(PMDTA)\}_2]$ (R = Pr^n, Bu^n) have been isolated as yellow crystalline solids. In addition, ionic complexes of the type $[Li(TMEDA)_2]^+[NiR(C_2H_4)_2]^-$, previously reported for R = Et,[55] have been obtained for R = Pr^n and Bu^n, but cannot be isolated without decomposition.[51] All these complexes undergo an alkyl–alkene exchange reaction with ethene in solution, providing an example of the nickel effect. Thus, in the presence of ethene, the μ-propyl or μ-butyl complexes are converted to the μ-ethyl or ionic $[Ni(Et)(C_2H_4)_2]^-$ compounds with the liberation of propene or butene (Scheme 3). They are thus important model compounds for mechanistic studies of the nickel effect in ethene dimerization. The dilithium–

nickelacyclopentane complex $[Ni(C_2H_4)(C_4H_8Li_2)(O_2C_4H_8)(OEt_2)_2]$ has been obtained by treating $[Ni(C_2H_4)_3]$ with 1,4-dilithiobutane in ether/dioxane.[51]

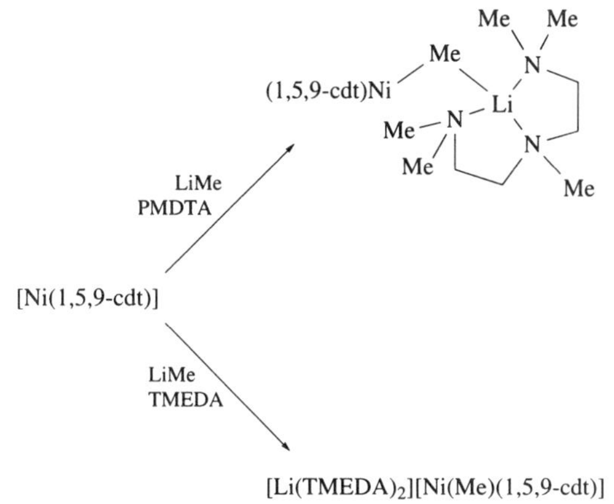

Scheme 1

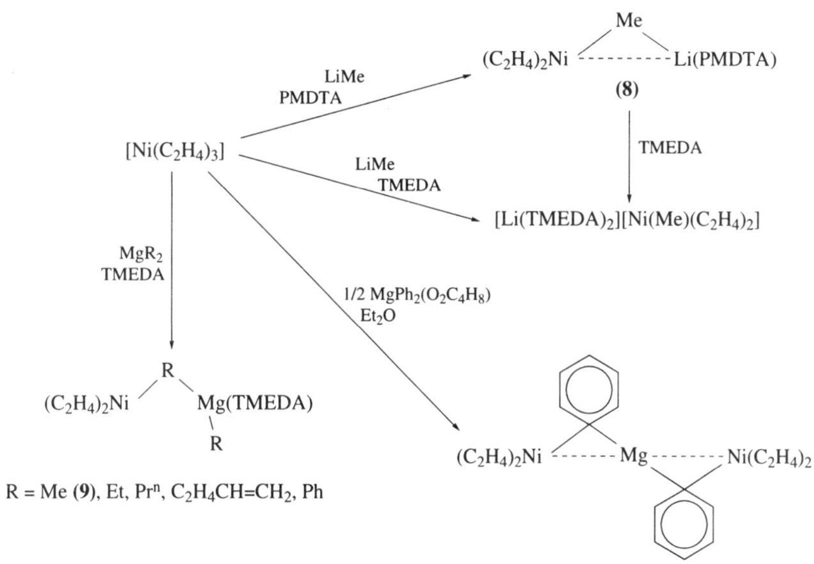

Scheme 2

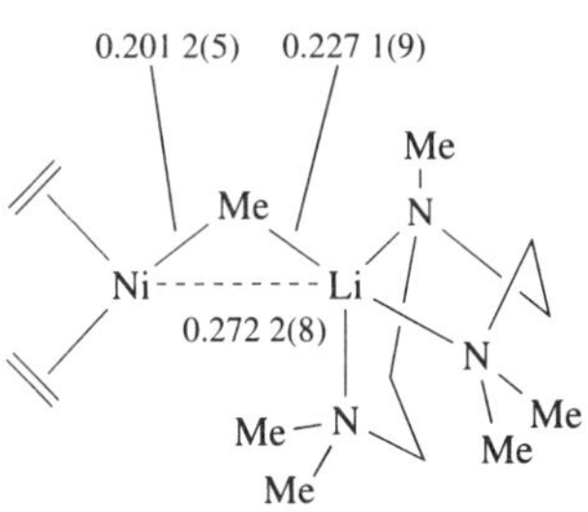

(8)

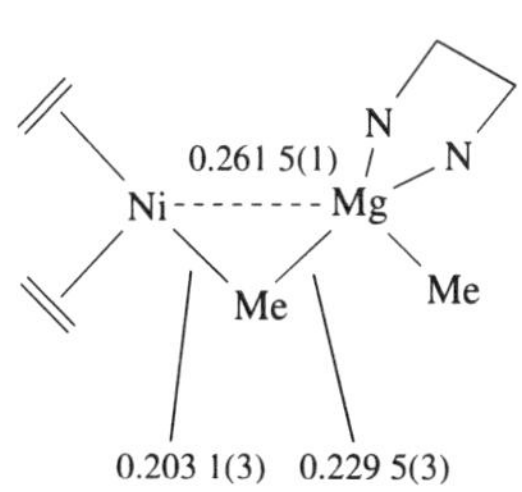

(9)

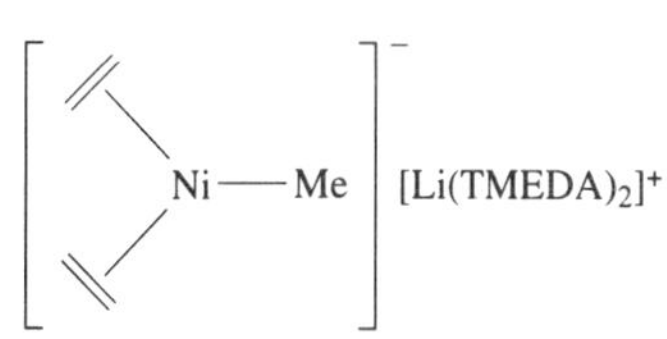

(10)

The organomagnesium complexes of nickel(0), $[(C_2H_4)_2Ni(\mu\text{-}R)MgR(donor)]$ (R = Me (**9**), Et, Pr^n, $C_2H_4CH{=}CH_2$ and Ph; donor = Et_2O, THF, dioxane or TMEDA) have been isolated from the reaction of the diorganomagnesium compound with $[Ni(C_2H_4)_3]$.[52] The complex $[\{(C_2H_4)_2Ni\}_2(\mu\text{-}Ph)_2Mg(OEt_2)]$ is formed by the reaction of $[Mg(Ph)_2(O_2C_4H_8)]$ with 2 equiv. $[Ni(C_2H_4)_3]$ in ether.[52] In solution this complex is in equilibrium with the monophenyl derivative $[(C_2H_4)_2Ni(\mu\text{-}Ph)MgPh(OEt_2)]$. ^1H and ^{13}C NMR studies of complex (**9**) and the ethyl and *n*-propyl derivatives indicate that there is no exchange in solution between the bridging and terminal organo groups at 0 °C.

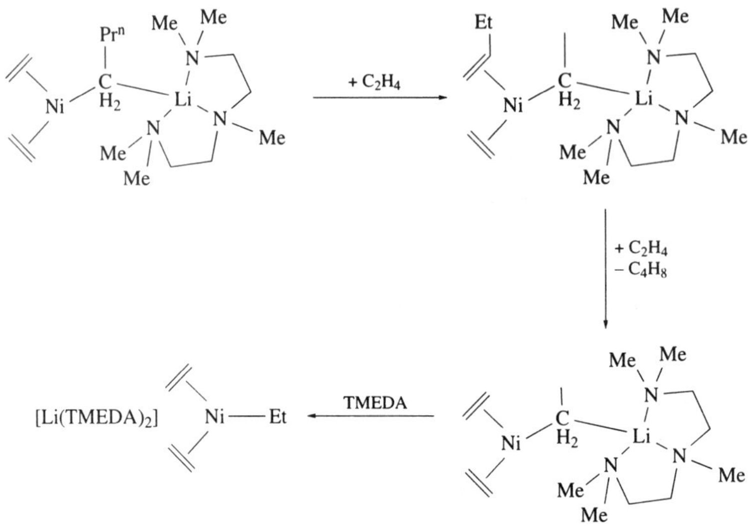

Scheme 3

The nickel ylide complexes [Ni(CH_2PMe_3)(cdt)] and [Ni(CH_2PR_3)(C_2H_4)$_2$] (R = Me or Ph) have been prepared by treating [Ni(cdt)] or [Ni(C_2H_4)$_3$] with the appropriate methylenetriorganophosphorane.[56] On treatment with CO these ylide complexes give the known tricarbonylnickel ylide [Ni(CH_2PR_3)(CO)$_3$]. Treatment of [Ni(cod)$_2$] with $Me_3P{=}CH_2$ and $Ph_3P{=}CHC(O)Ph$ yields the ylide complex [Ni(Ph)($Ph_2PCHCPhO$)(CH_2PMe_3)], which is a highly active catalyst for the polymerization of ethene.[57] A number of derivatives of this ylide complex have been synthesized (see Table 1) and it has been found that the substituents have a marked effect on the catalytic activity. Thus, on changing the chelate ligand from formyl to acetyl to benzoyl there is an increase in activity, and there is also an increase in activity with decreasing first ionization enthalpy of the ylide ligand.[57,58] The structure of the dark yellow derivative [Ni(Ph)($Ph_2PCHCMeO$)($CH_2PPr^i_3$)] (**11**) has been determined.[59] The short P–CH_2 bond length 0.176 5(4) nm) compared to the usual P–C single bond length, and the ^{13}C NMR chemical shift of −6.4 ppm for the CH_2 carbon of the ylide ligand suggests that the P^+–C^- bond polarity found in free ylides is maintained in the coordinated state. These complexes are closely related to the well-known ethene oligomerization catalysts [Ni(Ph)($Ph_2PCR^1CR^2O$)(PR_3)] (discussed in Section 2.4) used in the Shell Higher Olefin Process.[2,60] The nickel ylide complexes also catalyse the polymerization of ethyne.[61,62] Most importantly, in a stabilizing polyacrylonitrile matrix, it is possible to control the composition of the polymers; the degree of polyene conjugation is dependent on the choice of ligand. The catalysts are the most active ethyne polymerization catalysts based on nickel; in contrast, [Ni(Ph)($Ph_2PCHCPhO$)(PPh_3)] is not active under the reaction conditions used.

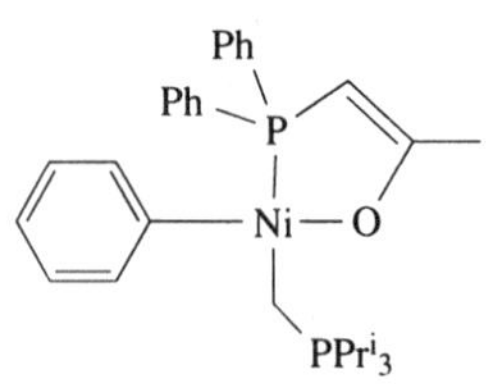

(**11**)

The ylide chelate complexes [Ni{$Ph_3PCH(CH_2)_nCHPPh_3$}(Cp)][PF_6] have been prepared by treatment of nickelocene or [NiBr(Cp)(PPh_3)] with the double ylide $Ph_3PCH(CH_2)_nCHPPh_3$ (n = 1–3).[63] The ionic ylide complexes [Ni($CHMePPh_3$)$_2$(Cp)][X] (X = BPh_4, PF_6) are obtained by treating nickelocene with ethylidenetriphenylphosphorane.[63] The bis(ylidic) nickel complex (**12**) has been prepared by treating [$NiCl_2$(PMe_3)$_2$] with lithium bis(dimethylmethylenephosphoranyl)dimethylborate.[64] The properties of (**12**) are similar to those of the previously reported[1] dihydroborate complex.

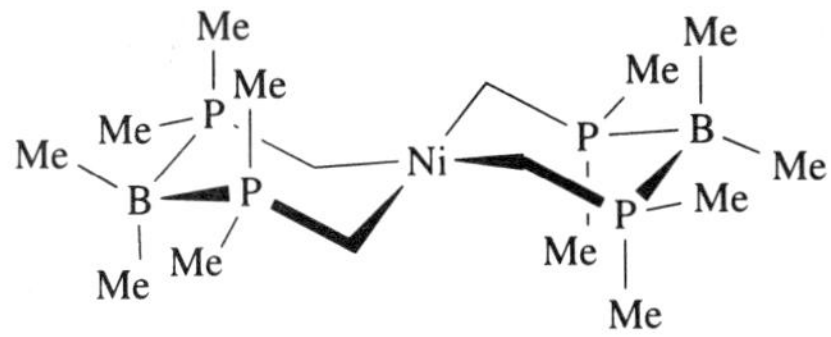

(12)

2.3 MONOLIGAND NICKEL ALKYL AND ARYL COMPLEXES

This section discusses complexes in which only one group 15 donor ligand is coordinated to the nickel atom. A list of the complexes considered is given in Table 3.

The oxidative addition of α,α'-dichloro-*p*-xylene to an equimolar mixture of $[Ni(PCy_3)_2(C_2H_4)]$ and $[Ni(cod)_2]$ gives a binuclear nickel(II) complex $[(Cy_3P)ClNi(CH_2C_6H_4CH_2\text{-}4)NiCl(PCy_3)]$ (**13**).[65] In a similar manner, the dinuclear complexes $[(Cy_3P)BrNi(CH_2C_6H_4CH_2\text{-}2)NiBr(PCy_3)]$ and $[(Ph_3P)BrNi(2,3\text{-dimethylenenaphthalene})NiBr(PPh_3)]$ are obtained from the oxidative addition of α,α'-dibromo-*o*-xylene to $[Ni(PCy_3)_2(C_2H_4)]/[Ni(cod)_2]$ and of α,α'-dibromo-2,3-dimethylnaphthalene to $[Ni(PPh_3)_2(C_2H_4)]$, respectively.[65] The dinuclear complex $[(Ph_3P)_2BrNi(p\text{-}C_6H_4)NiBr(PPh_3)_2]$ is prepared by the alternative route of treating $[NiBr_2(PPh_3)_2]$ with *p*-dilithiobenzene.[65] Treatment of $[Ni\{P(O\text{-}o\text{-}Tol)_3\}_2(C_2H_4)]$ with ethene and HCN at –40 °C leads to the quantitative formation of the nickel(II) complex $[Ni(Et)(CN)\{P(O\text{-}o\text{-}Tol)_3\}(C_2H_4)]$. This complex is an intermediate in the catalytic hydrocyanation of ethene, since treatment with $P(O\text{-}o\text{-}Tol)_3$ causes reductive elimination of propionitrile by an associative process.[66] The cyclonickelated compound $[NiCl(CH_2C_6H_4NMe_2\text{-}2)(PEt_3)]$ undergoes an insertion reaction on treatment with hexafluorobut-2-yne to yield the chloro-bridged dimer $[Ni(\mu\text{-}Cl)\{C(CF_3)=C(CF_3)CH_2C_6H_4NMe_2\text{-}2\}(PEt_3)_2]$ (**14**).[67]

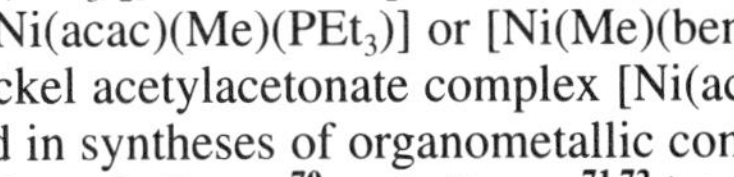

(13)

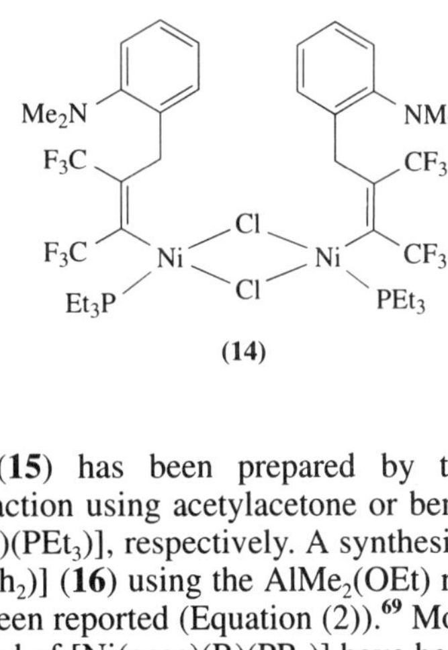

(14)

The methylnickel acetophenone oximato complex (**15**) has been prepared by treatment of $[Ni(Me)_2(PEt_3)_2]$ with acetophenone oxime.[68] A similar reaction using acetylacetone or benzoylacetone yielded $[Ni(acac)(Me)(PEt_3)]$ or $[Ni(Me)(benzoylacetonato)(PEt_3)]$, respectively. A synthesis of the new methylnickel acetylacetonate complex $[Ni(acac)(Me)(PEtPh_2)]$ (**16**) using the $AlMe_2(OEt)$ reagent often employed in syntheses of organometallic compounds has been reported (Equation (2)).[69] More details of the insertion of alkenes[70] and alkynes[71,72] into the Ni–C bond of $[Ni(acac)(R)(PR_3)]$ have been published since those reported in *COMC-I*. The reaction of $[Ni(acac)(Me)(PCy_3)]$ with ethene under pressure gives a mixture of products $[Ni(acac)(R)(PCy_3)]$ (R = Et, Pr and Bu), with the relative amounts depending upon the reaction conditions.[70] The insertion of alkynes into the Ni–C bond of $[Ni(acac)(Me)(PPh_3)]$ occurs in a regiospecific manner. Only the regioisomer resulting from methyl migration to the least-hindered alkyne carbon atom is observed (Scheme 4).[71] This requires the alkyne carbon atom with the sterically largest substituent to be bonded to the nickel atom. The reaction has been shown by isotope labelling and other studies to involve *cis*-addition, although the resulting coordinatively unsaturated vinylnickel intermediate is able to undergo isomerization of the double bond at a rate competitive with that of product formation so that usually a mixture of *cis*- and *trans*-products is observed or even, in the case of diphenylacetylene addition, exclusively the *trans*-product.[71]

The reaction of $[NiCl(Ar)(PR_3)_2]$ (Ar = Ph or α-naphthyl) with malonate ion results in the formation of the stable complex (**17**).[73] The strongly bound malonate anion in these complexes renders reductive elimination unfavourable.

The methoxy-bridged complex $[\{Ni(Me)(\mu\text{-}OMe)(PMe_3)\}_2]$ (**18**) has continued to be used as a starting material for a range of mono- and dinuclear methylnickel(II) compounds (Equations (3)–(7)).[74–7] The dinuclear methylnickel carboxylates formed in Equation (6) react reversibly with trimethylphos-

Table 3 Monoligand nickel alkyl and aryl complexes.

Complex	*Reference*	*Ni–C* (nm)
$[(Cy_3P)BrNi(\mu\text{-}CH_2C_6H_4CH_2\text{-}2)NiBr(PCy_3)]$	65	
$[(Ph_3P)BrNi$(2,3-dimethylenenaphthylene)$NiBr(PPh_3)]$	65	
$[(Cy_3P)ClNi(\mu\text{-}CH_2C_6H_4CH_2\text{-}4)NiCl(PCy_3)]$ (**13**)	65	
$[Ni(CN)(Et)\{P(o\text{-}Tol)_3\}(C_2H_4)]$	66	
$[Ni(\mu\text{-}Cl)\{C(CF_3){=}C(CF_3)CH_2C_6H_4NMe_2\text{-}2\}(PEt_3)]_2$ (**14**)	67	
$[Ni(acac)(Me)(PEt_3)]$	68	
$[Ni(acac)(Me)(PEtPh_2)]$ (**16**)	69	
$[Ni(acac)(Me)(PPh_3)]$	71	
$[Ni$(benzoylacetonate)$(Me)(PEt_3)]$	68	
$[Ni(\eta^3$-acetophenone oximate)$(Me)(PEt_3)]$ (**15**)	68	
$[Ni(acac)(Et)(PCy_3)]$	70	
$[Ni(acac)(Pr)(PCy_3)]$	70	
$[Ni(acac)(Bu)(PCy_3)]$	70	
$[Ni(acac)\{C(Ph){=}C(Ph)Me\}(PPh_3)]$	71	0.189 7
$[Ni(acac)\{C(Ph){=}C(Ph)Me\}(PMe_3)]$	72	
$[Ni(acac)\{C(Ph){=}CMe_2\}(PPh_3)]$	71	
$[Ni(acac)\{C(Ph){=}CHMe\}(PPh_3)]$	71	
$[Ni(acac)\{C(Bu^t){=}CHMe\}(PPh_3)]$	71	
$[Ni(acac)\{C(Bu^t){=}CMe_2\}(PPh_3)]$	71	
$[Ni(acac)\{C(CO_2Me){=}C(CO_2Me)Me\}(PPh_3)]$	71	
$[Ni(acac)\{C(Ph){=}C(Ph)Me\}(PMe_3)]$	72	
$[Ni(Ph)\{OC(OMe)CHC(OMe)O\}(PPh_3)]$ (**17**)	73	
$[Ni(Ph)\{OC(OMe)CHC(OMe)O\}(PCy_3)]$	73	
$[Ni(\alpha\text{-naphthyl})\{OC(OMe)CHC(OMe)O\}(PCy_3)]$	73	
$[Ni(\alpha\text{-naphthyl})\{OC(OMe)CHC(OMe)O\}(PPh_3)]$	73	
$[Ni(\alpha\text{-naphthyl})\{OC(OMe)C(CO_2Me)C(OMe)O\}(PPh_3)]$	73	
$[Ni(\alpha\text{-naphthyl})\{OC(OEt)CHC(OEt)O\}(PPh_3)]$	73	0.188 7(8)
$[Ni(C_6H_4Cl\text{-}2)\{OC(OMe)CHC(OMe)O\}(PPh_3)]$	73	
$[Ni(Me)$(8-hydroxyquinolate)$(PMe_3)]$ (**19**)	74	0.191 7(5)
$[(Me_3P)(Me)Ni(\mu\text{-}O_2CCO_2)Ni(Me)(PMe_3)]$ (**20**)	74	0.189 9(4)
$[(Me_3P)(Me)Ni\{\mu\text{-}O_2C(CH_2)CO_2\}Ni(Me)(PMe_3)]$	74	
$[(Me_3P)(Me)Ni\{\mu\text{-}O_2C(CH_2)_2CO_2\}Ni(Me)(PMe_3)]$	74	
$[(Me_3P)(Me)Ni\{\mu\text{-}O_2C(CH_2)_3CO_2\}Ni(Me)(PMe_3)]$	74	
$[(Me_3P)(Me)Ni\{\mu\text{-}O_2C(CH_2)_4CO_2\}Ni(Me)(PMe_3)]$	74	
$[\{Ni(Me)(\mu\text{-}O_2CPh)(PMe_3)\}_2]$	75	0.190 7(7)
$[\{Ni(Me)\{\mu\text{-}O_2C(Cy)\}(PMe_3)\}_2]$	75	
$[\{Ni(Me)(\mu\text{-}O_2CCHPh_2)(PMe_3)\}_2]$	75	
$[\{Ni(Me)(\mu\text{-}O_2CCPh_3)(PMe_3)\}_2]$	75	
$[\{Ni(Me)\{\mu\text{-}O_2C$(9-anthracenyl)$\}(PMe_3)\}_2]$	75	
$[\{Ni(Me)\{\mu\text{-}O_2C$(bicyclo[3.2.2]nonane-1-yl)$\}(PMe_3)\}_2]$	75	
$[\{Ni(Me)(\mu\text{-}O_2CC_6H_4NH_2\text{-}2)(PMe_3)\}_2]$	75	
$[\{Ni(Me)(\mu\text{-}O_2CC_6H_4OH\text{-}4)(PMe_3)\}_2]$	75	
$[\{Ni(Me)\{\mu\text{-}O_2C$(1-naphthyl)$\}(PMe_3)\}_2]$	75	
$[\{Ni(Me)\{\mu\text{-}O_2C$(2-naphthyl)$\}(PMe_3)\}_2]$	75	
$[\{Ni(Me)(\mu\text{-}O_2CCF_3)(PMe_3)\}_2]$	75	
$[\{Ni(Me)(\mu\text{-}O_2CCH_2Cl)(PMe_3)\}_2]$	75	
$[\{Ni(Me)(\mu\text{-}O_2CCHCl_2)(PMe_3)\}_2]$	75	
$[\{Ni(Me)(\mu\text{-}O_2CCH_2Br)(PMe_3)\}_2]$	75	
$[(Me_3P)(Me)Ni$(benzene-1,2-dioxo)$Ni(Me)(PMe_3)]$	76	
$[(Me_3P)(Me)Ni$(benzene-1,2-dioxo-4-Me)$Ni(Me)(PMe_3)]$	76	
$[(Me_3P)(Me)Ni$(benzene-1,2-dioxo-3-OMe)$Ni(Me)(PMe_3)]$	76	
$[(Me_3P)(Me)Ni\{\mu\text{-}OC_6H_3(OMe\text{-}3)O\}Ni(Me)(PMe_3)_2]$ (**21**)	76	0.192 0(7), 0.194 3(8)
$[(Me_3P)(Me)Ni\{\mu\text{-}OC_6H_4O\}Ni(Me)(PMe_3)_2]$	76	
$[(Me_3P)(Me)Ni\{\mu\text{-}OC_6H_3(Me\text{-}4)O\}Ni(Me)(PMe_3)_2]$	76	
$[(Me_3P)(Me)Ni\{\mu$-naphthalene-2,3-dioxo$\}Ni(Me)(PMe_3)_2]$	76	
$[Ni(Me)(PMe_3)_4]^+[Ni(Me)(PMe_3)$(benzene-1,2-dioxo-4-$NO_2)]^-$	76	
$[Ni(Me)(PMe_3)_4]^+[Ni(Me)(PMe_3)$(benzene-1,2-dioxo-3-$CH_2O)]^-$	76	
$[Ni(Me)(PMe_3)_4]^+[Ni(Me)(PMe_3)$(benzene-1,2-dioxo-4-$CH_2O)]^-$	76	
$[\{Ni(Me)(\mu\text{-}OH)(PEt_3)\}_2]$	77	
$[\{Ni(Me)(\mu\text{-}OMe)(PEt_3)\}_2]$	77	
$[\{Ni(Me)(\mu\text{-}OPh)(PEt_3)\}_2]$	77	
$[\{Ni(Me)(\mu\text{-}OSiPh_3)(PMe_3)\}_2]$	77	
$[\{Ni(Me)(\mu\text{-}OC_6F_5)(PMe_3)\}_2]$	77	
$[(Me_3P)(Me)Ni(\mu\text{-}OMe)(\mu\text{-}OSiPh_3)Ni(Me)(PMe_3)]$	77	
$[\{Ni(CH_2CMe_2Ph)(\mu\text{-}OH)(PMe_3)\}_2]$	78	
$[\{Ni(Bz)(\mu\text{-}OH)(PMe_3)\}_2]$	78	
$[\{Ni(CH_2C_6H_4Me\text{-}2)(\mu\text{-}OH)(PMe_3)\}_2]$ (**23**)	78,80	0.194 6(7)

Table 3 (continued)

Complex	*Reference*	*Ni–C* (nm)
$[Ni_2(Me)_2(\mu\text{-}CO_3)(PMe_3)_3]$	78	
$[Ni_2(Ph)_2(\mu\text{-}CO_3)(PMe_3)_3]$	78	
$[Ni_2(Bz)_2(\mu\text{-}CO_3)(PMe_3)_3]$	78	
$[Ni_2(CH_2\text{-}TMS)(\mu\text{-}CO_3)(PMe_3)_3]$	78	
$[(Cy_3P)(Me)Ni(O_2CO)Ni(H)(PCy_3)_2]$	79	
$[Ni_3(CH_2C_6H_4Me\text{-}2)_4(\mu^3\text{-}OH)_2(PMe_3)_2]$ (**22**)	80	0.195(1), 0.195(1), 0.194(1), 0.196(1)

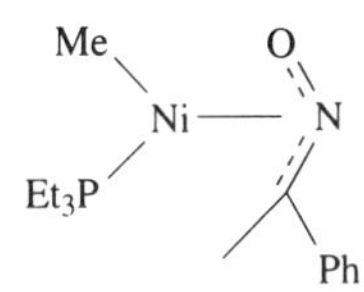

(15)

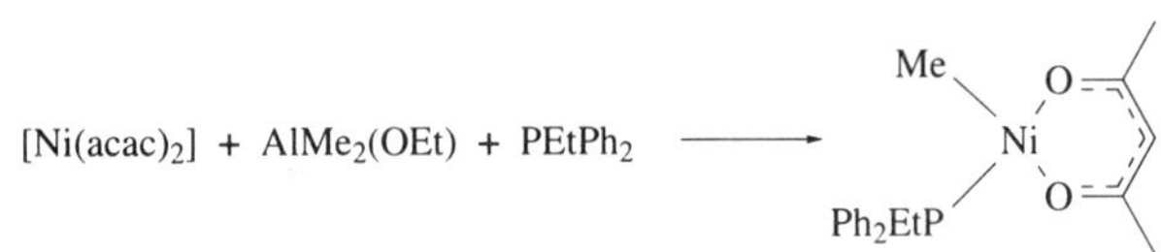

(2)

(16)

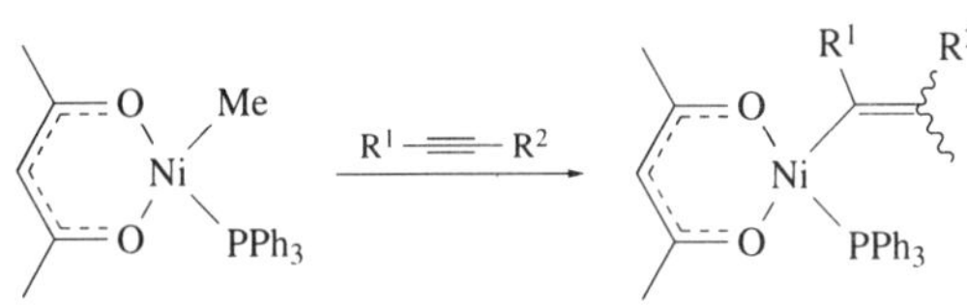

$R^1 = Ph$; $R^2 = Ph, Me, H$
$R^1 = Bu^t$; $R^2 = Me, H$
$R^1 = R^2 = CO_2Me$

Scheme 4

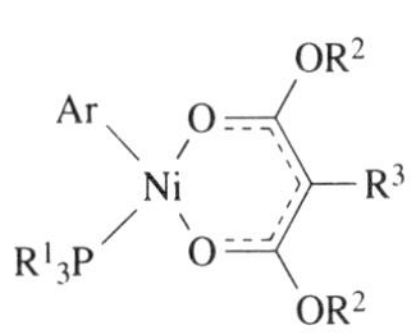

(17)

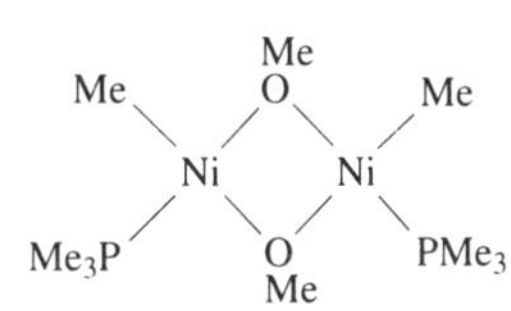

(18)

phine to give the mononuclear complexes $[Ni(Me)\{OC(O)R\}(PMe_3)_2]$ except when $R = CF_3$, where treatment with PMe_3 gives the ionic complex $[Ni(Me)(PMe_3)_4][CF_3CO_2]$.[75] If the μ-benzenedioxo complexes formed in Equation (7) have H, Me or OMe as a substituent on the benzene ring they react with trimethylphosphine to give dinuclear complexes such as (**21**).[76] However with –I substituents such as nitro or formyl as substituents on the ring, treatment with trimethylphosphine leads to the formation of ligand-rich ionic species $[Ni(Me)(PMe_3)_4]^+[(O\cdots O)Ni(Me)(PMe_3)]^-$.[76] The μ-hydroxo complexes $[\{Ni(R)(\mu\text{-}OH)(PMe_3)\}_2]$ ($R = CH_2CMe_2Ph$ or Bz) have been obtained by treating the corresponding monoalkylnickel chlorides with powdered NaOH.[77] An x-ray crystal structure analysis of the *o*-methylbenzyl derivative $[\{Ni(CH_2C_6H_4Me\text{-}2)(\mu\text{-}OH)(PMe_3)\}_2]$, as the 2,5-dimethylpyrrole adduct, shows the nickel to be in a distorted square-planar environment. The μ-hydroxo complexes react with CO_2 to form the carbonates $[Ni_2(R)_2(CO_3)(PMe_3)_3]$ ($R = Me$, CH_2TMS, Bz or Ph).[78] The carbonate $[(Cy_3P)(Me)Ni(\mu\text{-}O_2CO)NiH(PCy_3)_2]$ has been produced in low yield from the reaction between $[NiH(Me)(PCy_3)_2]$ and CO_2.[79] The *o*-methylbenzyl complex $[Ni_3(\eta^1\text{-}CH_2C_6H_4Me\text{-}2)_4(\mu_3\text{-}OH)_2(PMe_3)_2]$ (**22**) is obtained when the product of the reaction between $[NiCl_2(PMe_3)_2]$ and $[MgCl(CH_2C_6H_4Me\text{-}2)]$ is treated with a drop of water during workup. On treatment of (**22**) with PMe_3 the μ-hydroxonickel complex $[\{Ni(CH_2C_6H_4Me\text{-}2)(\mu\text{-}OH)(PMe_3)\}_2]$ (**23**) is obtained.[80] This hydroxy-bridged complex (**23**) is also formed on treatment of $[NiCl(\eta^3\text{-}CH_2C_6H_4Me\text{-}2)(PMe_3)]$ with KOH or treatment of $[Ni(CH_2C_6H_4Me\text{-}2)(NC_4H_2Me_2)(PMe_3)_2]$ with H_2O.[80]

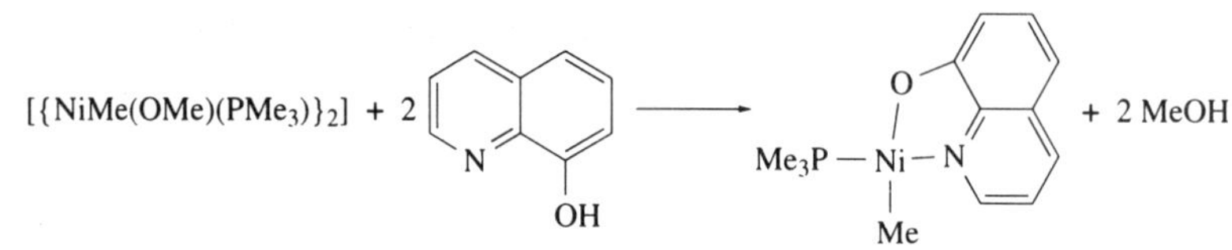
(3)

(19)

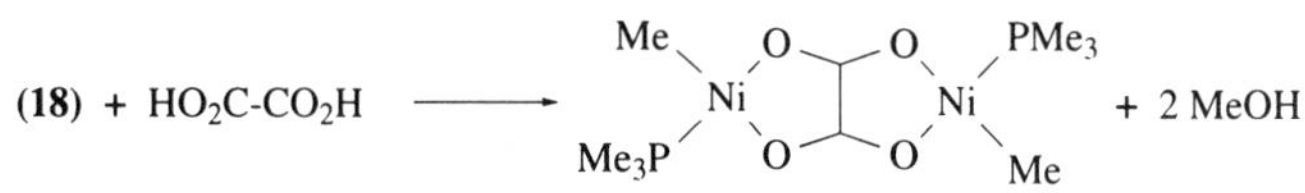
(4)

(20)

$$\textbf{(18)} + HO_2C(CH_2)_nCO_2H \longrightarrow [Ni(Me)(PMe_3)\{O_2C(CH_2)_nCO_2\}]_x \qquad (5)$$

$n = 1\text{–}4$

$$\textbf{(18)} + RCO_2H \longrightarrow [Ni(Me)(O_2CR)(PMe_3)]_2 \qquad (6)$$

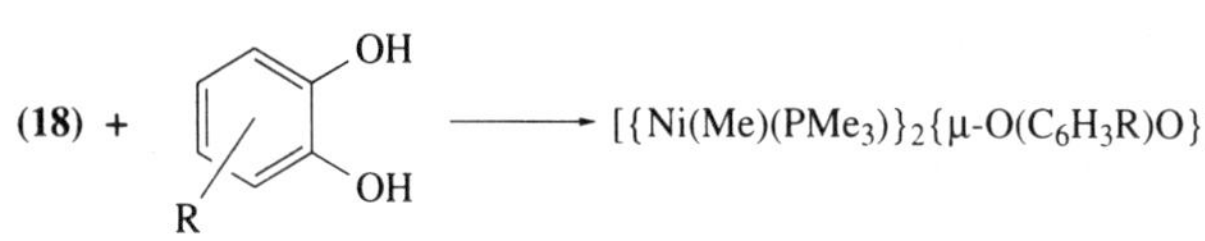
(7)

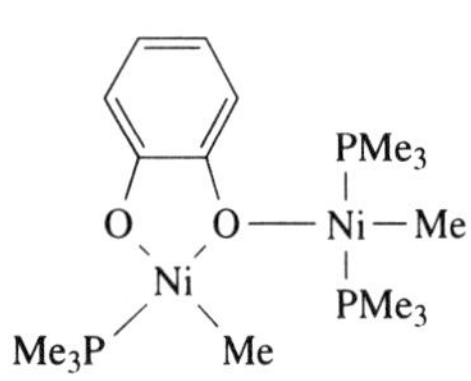

(21)

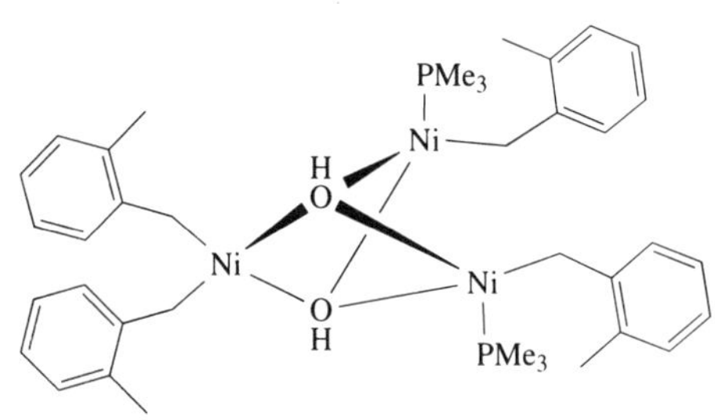

(22)

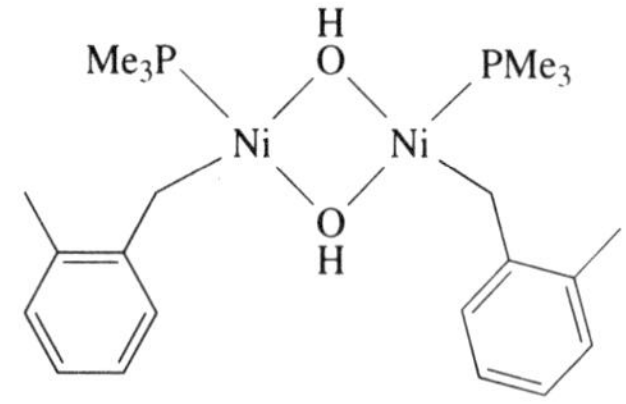

(23)

The maximum catalytic activity for ethene dimerization of the complex $[NiBr(C_6F_5)(PPh_3)_2]$ in the presence of $AgClO_4$ occurs when the mole ratio of nickel to added $AgClO_4$ is 1:2. A ^{31}P NMR study of the reaction between $[NiBr(C_6F_5)(PPh_3)_2]$ and $AgClO_4$ indicates that the species formed is $[Ni(C_6F_5)(ClO_4)(PPh_3)]$ and that this is the catalytically active species.[81] The complex $[NiCl(Me)(PH_3)(C_2H_4)]$ has been considered as a model for a polymerization catalyst. From MO symmetry considerations a rearrangement involving alkene rotation about the $Ni\text{–}C_2H_4$ bond is required. The barrier to this rotation is a function of the *trans* effect of PH_3 and steric interactions.[82]

2.4 $[NiX(R)L_2]$ AND $[NiR_2L_n]$ COMPLEXES (n = 2–4)

The great majority of organonickel complexes are stabilized by two group 15 donor ligands. Table 4 lists those complexes of general formula $[NiX(R)L_2]$ where X is an anionic ligand and L is a group

15 donor ligand. Tables 5 and 6 list the complexes of the type $[NiR_2L_n]$ and $[NiR^1R^2L_n]$, respectively. Table 7 provides a list of ionic complexes $[NiRL_n]^+[X]^-$. The lists include complexes reported in *COMC-I*[1] where new synthetic methods have been employed, where more details of the synthetic methods have been published, or where new reactions of the complexes have been investigated. Complexes in which X and R form part of a chelate ligand are not included here, but are discussed later in Section 2.8.

2.4.1 Synthesis

The two most common synthetic routes to organonickel complexes are the reaction of a nickel dihalide with an organolithium or organomagnesium reagent (for complexes of general formula $[NiX(R)L_2]$ and $[NiR_2L_2]$) and the oxidative addition or substitution of an organohalide with a nickel(0) species (for complexes of the type $[NiX(R)L_2]$). Sections 2.4.1.1 and 2.4.1.2 deal with these two principal synthetic routes. Section 2.4.1.3 covers miscellaneous methods of synthesis.

$$[NiCl_2(PMe_3)_2] + Mg(C_6H_2Me_3\text{-}2,4,6)Br \longrightarrow [NiCl(C_6H_2Me_3\text{-}2,4,6)(PMe_3)] + MgBrCl \quad (8)$$

$$[NiCl_2(PMe_3)_2] + 2\ Bu^t\!-\!\!\equiv\!\!-Li \longrightarrow [Ni(C{\equiv}CBu^t)_2(PMe_3)_2] + 2\ LiCl \quad (9)$$

$$[NiH(OAc)(PCy_3)_2] + PhMgBr \longrightarrow [NiH(Ph)(PCy_3)_2] \quad (10)$$

$$[Ni(acac)_2(TMEDA)] + MgMe_2(TMEDA) \longrightarrow [NiMe_2(TMEDA)] \quad (11)$$

$$[NiBr(Mes)(bipy)] + LiMe \longrightarrow [Ni(Me)(Mes)(bipy)] + LiBr \quad (12)$$

$$[NiBr_2(PMePh_2)_2] + 1,4\text{-}C_6F_4Li_2 \longrightarrow [\{NiBr(PMePh_2)_2\}_2(\mu\text{-}1,4\text{-}C_6F_4)] \quad (13)$$

$$[NiCl(C_6Cl_5)(PMe_2Ph)_2] + LiC{=}CH(CH_2)_nO \longrightarrow [Ni(C_6Cl_5)\{C{=}CH(CH_2)_nO\}(PMe_2Ph)_2] \quad (14)$$

$$[Ni(acac)_2] + Al(Bz)_3 + bipy \longrightarrow [Ni(Bz)_2(bipy)] \quad (15)$$

$$[Ni(acac)(TMEDA)] + AlMe_2(OEt) \longrightarrow [NiMe_2(TMEDA)] \quad (16)$$

$$[NiBr(SbPh_3)_3] + [TlBr(C_6F_5)_2] \longrightarrow [NiBr(C_6F_5)(SbPh_3)_2] \quad (17)$$

$$[Ni(C_6F_5)(CO)(PPh_3)_2] + [TlBr(C_6F_5)_2] \longrightarrow [Ni(C_6F_5)_2(PPh_3)_2] \quad (18)$$

$$[NiCl_2(bipy)] + [Yb(C_6F_5)_2] \longrightarrow [Ni(C_6F_5)_2(bipy)] \quad (19)$$

$$[NiBr_2(PEt_3)_2] + [Cd(CF_3)_2(glyme)] \longrightarrow [NiBr(CF_3)(PEt_3)_2] + [Ni(CF_3)_2(PEt_3)_2] \quad (20)$$

$$[NiCl_2(dppm\text{-}P)_2] + [Hg(C{\equiv}CPh)_2] \longrightarrow [(PhC{\equiv}C)_2Ni(\mu\text{-}dppm)_2HgCl_2] \quad (21)$$

$$[Ni(PPh_3)_4] + [HgCl(C_6F_5)] \longrightarrow [NiCl(C_6F_5)(PPh_3)_2] \quad (22)$$

$$[Ni(PPh_3)_4] + [Hg(C_6F_5)_2] \longrightarrow [Ni(C_6F_5)(HgC_6F_5)(PPh_3)_2] \quad (23)$$

2.4.1.1 Synthesis using organo–main group metal reagents

The most common organo–main group metal reagents employed are organomagnesium compounds[83,84,98,101–3,112,151,159,175,183,205] and organolithium compounds.[4,65,83,112,122,147–9,153,161,170,171,175,195,197,202,203,205] Other reagents used include compounds of organoaluminum,[69,84,183,186,189] organothallium,[36,152,175] organoytterbium,[201] organocadmium[109] and organomercury.[124,164,192] The nickel starting material used is most often a complex of general formula $[NiX_2L_2]$ where X is a halide (Equations (8) and (9) are typical examples) although other nickel complexes have been used (Equations (10) and (11)). Other reactions of interest employing organo–main group reagents are shown in Equations (12)–(25).

Table 4 $[NiX(R)L_2]$ complexes.

R	*X*	*Ligand*	*Reference*	*Ni–C* (nm)
Me	H	PCy_3	84	
	Cl	PEt_3	68,77,87	
		PPh_3	86	
		dppe	86	
		bipy	68,87	
	Br	PEt_3	77	
	OBu^t	PPh_3	85	
	OPh (**30**)	PMe_3	88	0.195(1)
		PEt_3	68	
		bipy	68	
	$OCH(CF_3)Ph$	PMe_3	88	
		dppe	89	
		bipy	89	
	OCH_2CF_3	dppe	89	
		bipy	89	
	$OCH(CF_3)_2$	dppe	89	
		bipy	89	
	$OC_6H_4NO_2$-3	PMe_3	77	
	$OC_6H_4NO_2$-4	PMe_3	77	
	OC_6H_4CHO-4	PMe_3	77	
	OC_6F_5	PMe_3	77	
	$OC_6H_4N_2Ph$-4	PMe_3	77	
	OC_6H_4CN-4	bipy	68,87,90	
		PEt_3	68,90	
	OC_6H_4Ph-4	PEt_3	68	
	8-quinolinato	PEt_3	68	
	$OC_6H_3(OMe)ONi$	PMe_3	76	
	OCOMe	PMe_3	75	
		PEt_3	91	
	OCOPh	PMe_3	75	
		PEt_3	91	
	OCOCy	PMe_3	75	
	$OCOCHPh_2$	PMe_3	75	
	OCO-9-anthracenyl	PMe_3	75	
	OCO(bicylo[3.2.2]nonane-1-yl)	PMe_3	75	
	$OCOC_6H_4NH_2$-2	PMe_3	75	
	$OCOC_6H_4OH$-4	PMe_3	75	
	OCO-1-naphthyl	PMe_3	75	
	OCO-2-naphthyl	PMe_3	75	
	$OCOCH_2Cl$	PMe_3	75	
	$OCOCHCl_2$	PMe_3	75	
	$OCOCH_2Br$	PMe_3	75	
	succinimide	PEt_3	92	
		dppe	92	
		bipy	92	
	phthalimide	PEt_3	87,92	
		dppe	92	
		bipy	92	
	diacetamide	PEt_3	92	
	imidazole	PEt_3 (**32**)	92	
	pyromellitimide	PEt_3 (**33**)	92	
	NC_4H_4	PMe_3	78	
	$NC_4H_2Me_2$-2,5	PMe_3	78	
	SPh	PEt_3	68	
		dppe	89	
	SC_6H_4Me-4	dppe	89	
	$N(SiMe_2CH_2PPh_2)_2$ (**44**)		93	
Et	Cl	bipy	68,87	
	Br	bipy	68	
	CN	$P(OC_6H_4Me\text{-}2)_3$	95	
	OEt	bipy	96	
	$OCH(CF_3)Ph$	bipy	89	
	OCH_2CF_3	bipy	89	
	$OCH(CF_3)_2$	bipy	89	
	OC_6H_4CN-4	bipy	68,87	
	OC_6H_4Ph-4	bipy	68	

Table 4 (continued)

R	*X*	*Ligand*	*Reference*	*Ni–C* (nm)
	OCOEt	bipy	68	
	OCOPh	bipy	68,91	
	$OC(CF_3)_2CH_2PPh_2$	PCy_3 (**28**)	97	
	succinimide	bipy	87,92	
	phthalimide	bipy	92	
Pr^n	$N(SiMe_2CH_2PPh_2)_2$ (**44**)	93		
$CH_2CH{=}CH_2$	$N(SiMe_2CH_2PPh_2)_2$	94		
CH_2SiMe_2Ph	Cl	PEt_3	98	
		PMe_2Ph	98	
CH_2TMS	NC_4H_4	PMe_3	78	
	$NC_4H_2Me_2$-2,5	PMe_3	78	
	S_2CNMe_2	PMe_3	99,100	
	S_2CNEt_2	PMe_3	99,100	
	$S_2CN(CHMe_2)_2$	PMe_3	99,100	
CH_2Bu^t	NC_4H_4	PMe_3	78	
	$NC_4H_2Me_2$-2,5	PMe_3	78	
	Cl	PMe_3	101	
	NCS	PMe_3	101	
CH_2CMe_2Ph	Cl	PMe_3	98	
		PMe_2Ph	98	
		PEt_3	98	
	OCO_2	PMe_3	102	
	NC_4H_4	PMe_3	78	
	$NC_4H_2Me_2$-2,5	PMe_3	78	
	S_2CNMe_2	PMe_3	99,100	
	S_2CNEt_2	PMe_3	99,100	
	$S_2CN(CHMe_2)_2$	PMe_3	99,100	
Bz	Cl	PMe_3	103	
CH_2CN	Cl	PMe_3	104	
CH_2OMe	Br	PMe_3	104	
CH_2CO_2Et	Cl	PMe_3	104	
CH_2SMe	Cl	PPh_3	105	
$CH_2C_6H_4Me$-2	Cl	PMe_3	80	
	$NC_4H_2Me_2$-2,5	PMe_3	80	
$CH_2C_6H_4Br$-2	Br	PMe_3	106	
$CH_2C_6H_4Br$-3	Br	PMe_3	107	
$CH_2C_6H_4Br$-4	Br	PMe_3	107	
$CH_2C_6H_4Ni$-2	Br	PMe_3 (**41**)	106,108	
		dmpm	106	
$CH_2C_6H_4Ni$-3	Br	PMe_3 (**42**)	107,108	
$CH_2C_6H_4Ni$-4	Br	PMe_3	107,108	0.188 1(12), 0.195 1(12)
CF_3	Br	PEt_3	109	
polyvinylchloride	Cl	bipy	110	
$CH{=}CH_2$	$N(SiMe_2CH_2PPh_2)_2$ (**44**)	93,94,138		
$CH{=}CHCl$	Cl	PPh_3	111	
$CH{=}CCl_2$	Cl	PPh_3	111	
$CCl{=}CCl_2$	Cl	PPh_3	111	
		PMe_3	112	
		PMe_2Ph	112	
		dppe	111	
$CCl{=}CCl(C_6H_4Me$-4)	Cl	PMe_3	114,115,116	0.185 5(6)
$CCl{=}CCl(C_6H_4Me$-3)	Cl	PMe_3	114	
$C(C_6H_4Me$-4)$=CCl_2$	Cl	PMe_3	114,115,116	0.191 1(5)
$C(C_6H_4Me$-3)$=CCl_2$	Cl	PMe_3	114	
$CCl{=}CCl(C_6H_4NMe_2$-4)	Cl	PMe_3	114	
$C(C_6H_4NMe_2$-4)$=CCl_2$	Cl	PMe_3	114	
$C(C_6H_4Cl$-4)$=CCl_2$	Cl	PMe_3	114,115	
$CCl{=}CCl(C_6H_4Cl$-4)	Cl	PMe_3	114,115	
$CPh{=}CPh(CH_2TMS)$	Cl	PMe_3	117	
$CPh{=}C(Me)Ph$	Cl	PMe_3 (**46**)	72,117	
	Br	PMe_3	72	
	I	PMe_3	72	
	CN	PMe_3	72	
$CPh{=}CMe_2$	Cl	PMe_3	72	
	Br	PMe_3	72	
	I	PMe_3	72	

Table 4 (continued)

R	*X*	*Ligand*	*Reference*	*Ni–C* (nm)
CMe=C(Me)Ph	Cl	PMe_3	72	
$CPh=CPh_2$	Cl	PMe_3	72	
	$OC(Ph)NPPh_2$	PPh_3 (**51**)	141	
$C(CH_2Bu^t)=NBu^t$	Cl	PMe_3 (**53**)	118	
$C(CH_2TMS)=NCy$	Cl	PMe_3	117	
	Br	PMe_3	72	
	I	PMe_3	72	
$C(CH_2TMS)=NBu^t$	Cl	PMe_3 (**53**)	117	
C(Me)=NCy	Cl	PMe_3	117	
$C(Me)=NBu^t$	Cl	PMe_3 (**53**)	117,118	
$C(CH_2CMe_2Ph)=NBu^t$	Cl	PMe_3	118	
$C=CHC_6H_4$-2-$C(O)NC_6H_3Me_2$-2,6	Br	PMe_3 (**58**)	108	
$C(C_5H_4N$-2$)=N(C_6H_4OMe$-4$)ZnCl_2$	Cl	$PMePh_2$ (**57**)	119	
C(Ph)=CHCHC(Ph)Ni	Br	PEt_3	4,120	
		depe (**35**)	120	
$C(Ph)=C(Ph)C_6H_3Cl_2$-3,5	Br	PPh_3 (**52**)	121	
		$PEtPh_2$	121	
		dppe	121	
$C(CO_2Me)=C(CO_2Me)C_6H_3Cl_2$-3,5	Br	PPh_3	121	0.189 5(7)
		dppe	121	
$C(Ph)=C(Me)C_6H_3Cl_2$-3,5	Br	PPh_3	121	
$C(Ph)=C(CO_2Et)C_6H_3Cl_2$-3,5	Br	dppe	121	
$C(Ph)=C(Ph)C(O)NEt_2$	I	$NHEt_2$	165b	
C(C≡C-TMS)=C(TMS)C≡C-TMS	Cl	PMe_3	122	
	Br	PMe_3	122	
	I	PMe_3	122	
C(C≡C-TMS)=C(Me)-TMS	Cl	PMe_3	122	
C(C≡CPh)=C(Me)Ph	Cl	PMe_3	122	
CH(Me)-8-quinoline	Br	PPh_3	124	
Ph	H	PCy_3	79,84	
	Cl	PMe_3	72,103	
		PPh_3	124,125	
	Br	PMe_3	103	
		PPh_3	125,127	
		dppe	129	
		bipy	130	
	I	PPh_3	126,127	
	CN	PCy_3	131	
		$Et_2P(CH_2)_4PEt_2$	132	
	CO_2H	PCy_3	79,84	
	OCOMe	PEt_3	91	
		bipy	91	
	OCOPh	PEt_3	91	
		bipy	91	
	$OC(Ph)CHPPh_2$	PPh_3 (**24**)	133,134, 136,141	0.189 9(15)
		PEt_3 (**27**)	136	
		$P(p\text{-}Tol)_3$	141	
	$OC(Ph)C(SO_3Na)PPh_2$	PPh_3	135	
	$OC(OMe)C(Ph)PPh_2$	PEt_3 (**27**)	136	0.186 7(6)
		PPh_3	136	
	$OC(OEt)C(Me)PPh_2$	PPh_3	141	
	$OC(Ph)CHAsPh_2$	PPh_3	133,141	
	$OC(Ph)NPPh_2$	PPh_3 (**26**)	133,141	
	OC(Me)CHC(Me)O	PPh_3	133	
	$OC_6H_3(OH)$-4-PPh_2-2	PPh_3 (**25**)	141	
	$N(SiMe_2CH_2PPh_2)_2$		93,94,138	
	$N(SiMe_2CH_2PMe_2)_2$		138	
	$HB(3\text{-}Bu^tC_3H_2N_2)_3$	PMe_3	139	
	$H_2B(3\text{-}Bu^tC_3H_2N_2)_2$	PMe_3	139	
	SH	PEt_3	140	
	SPh	PEt_3	137	
		dmpe	137	
	$S(C_6H_4Me$-4$)$	PEt_3	137	
		dmpe	137	
C_6H_4Me-2	Br	PPh_3	127	
	I	PPh_3	127	

Table 4 (continued)

R	*X*	*Ligand*	*Reference*	*Ni–C* (nm)
C_6H_4Me-4	SPh	PEt_3	137	
		dmpe	137	
	$HB(3\text{-}Bu^tC_3H_2N_2)_3$	PMe_3 (**38**)	139	0.189 1(8)
	$H_2B(3\text{-}Bu^tC_3H_2N_2)_2$	PMe_3	139	
$C_6H_4NMe_2$-4	$N(SiMe_2CH_2PPh_2)_2$		138	
C_6H_4OMe-4	SC_6H_4OMe-4	PBu^n_3	142	
C_6H_4OEt-4	$HB(3\text{-}Bu^tC_3H_2N_2)_3$	PMe_3	139	
	$H_2B(3\text{-}Bu^tC_3H_2N_2)_2$	PMe_3	139	
C_6H_4Cl-2	Cl	PPr^i_3	415	
	Br	PPh_3	143	
C_6H_4Br-4	Br	PPh_3	144	
C_6H_4Br-2	Br	PPh_3	145	
C_6H_4Ni-4	Br	PPh_3	65	
$C_6H_4CH_2Ni$	Br	PMe_3 (**43**)	106	
$C_6H_4CH_2CH_2C_6H_4Ni$	Br	PMe_3	106,107	
$C_6H_3Cl_2$-2,3	Br	PPh_3	127,128	
		dppe	146	
		bipy	146	
	I	PPh_3	127,128	
$C_6H_3Cl_2$-2,4	Br	PPh_3	128,143	
$C_6H_3Cl_2$-2,5	Br	PPh_3	128	
$C_6H_3Cl_2$-2,6	Cl	PPh_3	128	
	Br	PPh_3	127,128	
	I	PPh_3	127,128	
$C_6H_3Cl_2$-3,4	Br	PPh_3	128	
$C_6H_3Cl_2$-3,5	Br	PPh_3	127,128	
	I	PPh_3	127,128	
$C_6H_3(OMe)_2$-2,6	Cl	PMe_3	149	
		PMe_2Ph	149	
		PPh_3	149	
$C_6H_3(CH_2NMeBu^t)_2$-2,6	Br	PMe_3	330	
		PEt_3	330	
$C_6H_3(CH_2NMePh)_2$-2,6	Cl	PMe_3	330	
		PEt_3	330	
		PBu_3	330	
	Br	PMe_3	330	
		PEt_3	330	
		PBu_3	330	
	I	PMe_3	330	
		PEt_3	330	
		PBu_3	330	
$C_6H_2Me_3$-2,4,6	Br	bipy, phen	147	
		dppe	147	
		$(Ph_2P)_2NPh$	147	
		PPh_2CH_2(2-py)	146	
		$PPh_2CH_2CH_2CH{=}CH_2$	146	
		PMe_3	112	
		$PMePh_2$	112,143,150,151	
		$PEtPh_2$	143,150,151	
		PPh_3	143,150,151	
		PCy_3, PBz_3, PEt_3, $P(p\text{-}Tol)_3$	150,151	
		$P(C_6H_4Cl\text{-}4)_3$	150,151	
		$P(C_6H_4OMe\text{-}4)_3$	150,151	
		PEt_2Ph, PMe_2Ph	150,151	
		PBu_3, PBu^i_3	151	
	NCS	PR_3	150	
	NC_4H_4	PMe_3	78	
	$NC_4H_2Me_2$-2,5	PMe_3	78	
$C_6H_2Cl_3$-2,3,4	Cl	PMe_3	103	
	Br	PMe_3	103	
		$PMePh_2$	128,143,150	
		PEt_3	150	
		$PEtPh_2$	128,143,150	
		PEt_2Ph	150	
		PPh_3	128,143,150	
	NCS	PR_3	150	
$C_6H_2Cl_3$-2,4,5	Br	PPh_3	128	

Table 4 (continued)

R	X	Ligand	Reference	Ni–C (nm)
$C_6H_2Cl_3$-3,4,5	Br	PPh_3	128	
C_6HF_4-2,3,5,6	Br	$PMePh_2$	153	
C_6HCl_4-2,3,4,5	Br	PPh_3	143	
C_6HCl_4-2,3,4,6	Br	PPh_3	143	
C_6HCl_4-2,3,5,6	Br	PPh_3	143	
C_6F_5	Cl,Br,I	$SbPh_3$	152	
	Br	$PMePh_2$	150,153	
		PCy_3, PBz_3, PBu^t_3	150	
		PPh_3	124,143,150	
		PEt_2Ph, $PEtPh_2$	150	
	NCS	PR_3	150	
	SC_6F_5	PBu_3	142	
	HgC_6F_5	PPh_3	124	
	I	PEt_3	154	
C_6Cl_5	Cl, I, NCS, NCO, N_3	PCy_3, PEt_3, bipy, 2,5-lut	155	
	Cl	β-pic, γ-pic	156	
		(py)(PEt_3)	156	
		$PMePh_2$	157	
	Cl, Br, I, NCS, NCO	PBz_3	158	
	Br	$PMePh_2$	143,150	
		$PEtPh_2$	143,150	
		PPh_3	143,150	
		py	156	
		PEt_3, PCy_3, PBz_3, P(*p*-Tol)$_3$	150	
		$P(C_6H_4Cl\text{-}4)_3$	150	
		$P(C_6H_4OMe\text{-}4)_3$	150	
		$PCyPh_2$, $PBzPh_2$, PCy_2Ph	150	
		PBz_2Ph, PEt_2Ph, PMe_2Ph	150	
	NCS	PR_3	150	
		py, β-pic, γ-pic, 3,5-lut	156	
		(py)(PPh_3)	156	
	C_6Cl_5O	py, β-pic, γ-pic, 3,5-lut	156	
	NO_2	PPh_3, py	156	
C_6Br_5	Cl, Br, I, NO_2, SCN, N_3	$PRPh_2$	159,160	
C_6F_4Br-4	Br	$PMePh_2$	153	
C_6F_4Ni-4	Br	$PMePh_2$	153	
4,4'-$C_{12}F_8H$	Br	$PMePh_2$	161	0.186(1)
4,4'-$C_{12}F_8Ni$	Br	$PMePh_2$	161	
10-Cl-9-anthryl	Cl	PPh_3	162	0.187(2)
10-Br-9-anthryl	Br	PPh_3	162	
C≡C-TMS	Cl, Br	PMe_3	122	
$C{\equiv}CCH_2OH$	NCS	PR_3	123	
$C{\equiv}CCH_2Cl$	NCS	PR_3	123	
$C{\equiv}CCH_2Br$	NCS	PR_3	123	
CN(Ph)NOCO	Cl, Br	PPh_3	164	
	Cl, Br	PEt_3	164	
$C(N_2)$TMS	Cl	PMe_3	408	
C(O)CN(Ph)NOCO	Br	PEt_3	164	
C(O)Me	OCOMe	PEt_3	91	
	OCOPh	PEt_3 (**40**)	91	
	phthalimide	PEt_3	87	
	OC_6H_4CN-4	bipy	87	
C(O)Et	Cl	bipy	87	
	OCOPh	bipy	91	
	succinimide	bipy	87	
	OC_6H_4CN-4	bipy	87	
$C(O)CH_2Bu^t$	NCS	PMe_3	101	
$C(O)CH_2$TMS	S_2CNR_2	PMe_3	100	
$C(O)CH_2CMePh_2$	S_2CNR_2	PMe_3	100	
C(O)Bz	Cl	PMe_3	103	
C(O)Ph	Cl, Br	PMe_3	103	
	$OC(Ph)NPPh_2$	PPh_3	128	
	$OC(OMe)C(Ph)PPh_2$	PEt_3	136	0.185 9(2)
	$N(SiMe_2CH_2PPh_2)_2$		138	0.187 8(2)
$C(O)NEt_2$	I	PEt_3	165a	
		$NHEt_2$	165b	
3-py-6-Br	Br	PPh_3	163	

Table 4 (continued)

R	*X*	*Ligand*	*Reference*	*Ni–C* (nm)
2-py-5-Br	Br	PPh_3	163	
Ph[a]		dppe	129	
C_6F_5[a]		$(CO)_2(PPh_3)_2$	175	
C_6Cl_5[a]		$(CO)_2(PPh_3)_2$	175	
C_6HCl_4-2,3,5,6[a]		$(CO)_2(PPh_3)_2$	175	

[a] Nickel(I) complex.

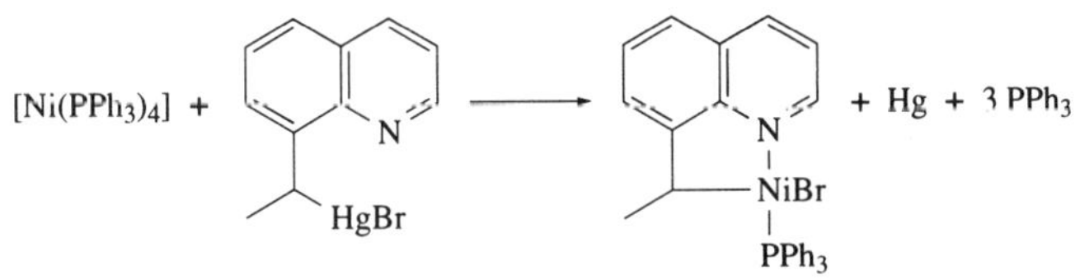

(24)

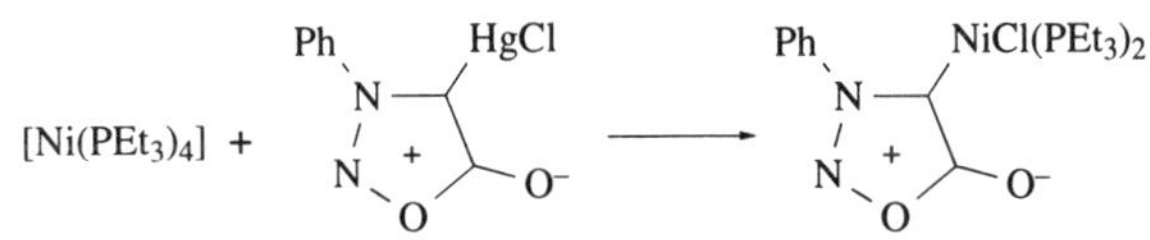

(25)

These reactions provide examples of a typical synthetic route to mixed alkyl(aryl)nickel complexes (Equation (12)),[147,148] the use of an organo dilithium compound to synthesize a dinickel complex bridged by the organo group with the potential to make an organometallic polymer (Equation (13)),[153,161] and the synthesis of an alkoxyvinylnickel complex, which is a precursor to a nickel–carbene complex (Equation (14)).[170] The use of organoaluminum reagents is illustrated in Equations (15)[186] and (16).[183] Also exemplified are the oxidation of nickel(I) species by organothallium reagents (Equations (17)[152] and (18)[175]), the use of organolanthanides as arylating reagents (Equation (19)),[201] and the use of organocadmium (Equation (20))[109] or organomercury (Equation (21))[192] reagents for halide ligand replacement. Equations (22)–(25)[124,132] provide examples of the oxidative addition of organomercury compounds to a nickel(0) complex.

In a study of the reactions of $Li(CD_3)$ and $MgBr(CD_3)$ with $NiCl_2$, nickel–carbene as well as nickel–alkyl intermediates were proposed.[83]

2.4.1.2 Oxidative addition reactions

Oxidative addition, usually of an alkyl or an aryl halide, to a nickel(0) complex is the most common route for the preparation of monoalkyl or monoaryl complexes, although the treatment of dialkylnickel complexes with protic acids has become increasingly popular (see Section 2.4.1.3). In general, the monoalkylnickel complexes are thermally more stable than the corresponding dialkylnickel complexes, partly due to the lack of reductive elimination as a decomposition route. The nickel(0) complexes used as starting materials are most often $[Ni(cod)_2]$ (Equation (26))[89,97,103,104,110,113,134,136,141,168] or $[NiL_3]$ or $[NiL_4]$ (Equation (27))[111,113,122,128,132,142,145,163,164] although $[Ni(PCy_3)(cod)]$ (Equation (28)),[65] $[Ni(PPh_3)_2(C_2H_4)]$ (Equation (29)),[65,105] electrogenerated [Ni(dppe)] (Equation (30)),[129] and activated nickel powder (Equations (31) and (32))[3,154] have also been used. It has been found that preformed freshly prepared $[Ni(PMe_3)_2(cod)]$ (from $[Ni(cod)_2]$ and PMe_3) provides a convenient starting material for oxidative addition reactions in which uncontrolled bromination would otherwise occur (Scheme 5).[106–8] Thus, for example, $[Ni(PMe_3)_4]$ reacts with $BrCH_2C_6H_4Br$-4 to give $[NiBr_2(PMe_3)_3]$, whereas with $[Ni(cod)(PMe_3)_2]$ the bromobenzylnickel complex $[NiBr(CH_2C_6H_4Br\text{-}4)(PMe_3)_2]$ is initially obtained, and further reaction can occur to give $[(Me_3P)_2BrNi(\mu\text{-}CH_2C_6H_4)NiBr(PMe_3)_2]$.[107] As expected it is the more labile bromine–benzyl bond that is involved in the first oxidative addition reaction.

$$[Ni(cod)_2] + BzCl + 2\,PMe_3 \longrightarrow [NiCl(Bz)(PMe_3)_2] + 2\,cod \qquad (26)$$

$$[Ni(PMe_3)_4] + TMS{-}C{\equiv}C{-}Cl \longrightarrow [NiCl(C{\equiv}C\text{-}TMS)(PMe_3)_2] + 2\,PMe_3 \qquad (27)$$

Table 5 [NiR_2L_2] and [NiR_2L_3] compounds.

R	*L*	*Reference*	*Ni–C* (nm)
Me	TMEDA	183	
	bipy	183,184	
	PEt_3	69,184	
	PBu_3	69	
	dmpe	183	
	dppe	69,184	
	dppp	184	
Et	bipy	184	
	phen	185	
	PEt_3	69	
	PBu_3	69	
	dppe	69	
Bz	bipy	186	
CH_2CMe_2Ph	bipy, Me_2bipy	98,187	
	Bu^t_2bipy	98	
	Ph_2phen	98	
	phen	187	
	TMEDA	101,187	
	dmpe	101	
CH_2SiMe_2Ph	bipy, Me_2bipy	98	
	Bu^t_2bipy	98	
	Me_4phen, Ph_2phen	98	
	bipym[a]	98	
	PMe_3	98	
	dmpe	188	
	dppe	98	
	depe	188	
	$Pr^i_2PCH_2CH_2PPr^i_2$ (**39**)	188	0.201 4(4)
CH_2TMS	py	101,187	0.189(1)
	bipy	187,189	
	phen, TMEDA	187	
	PMe_3	101,187	
	PMe_2Ph	187	
	dmpe	101	
	dppe	187	
$CH_2C_6H_4Me$-2	dmpe	188	
	depe	188	
	dipe[b]	188	
CF_3	$(PMe_3)_3$	190	
	PEt_3	109	
Ph	phen	185	
syn-C_6H_4Me-2	PMe_2Ph	195,196	0.193 3(16), 0.195 0(12), 0.190 3(15), 0.193 9(13)
anti-C_6H_4Me-2	PMe_2Ph	195,196	0.194 2(3)
	PMe_3	195	
$C_6H_3(OMe)_2$-2,6	PMe_3	149,197–200	0.193 3(5), 0.195 4(5)
	PMe_2Ph	149,197,199,200	
	PMe_2CH_2Li	149,197	
	$PMePhCH_2Li$	149,197	
	$PMePh_2$	149	
	$PMe_2(CH_2TMS)$	149	
	$PMePh(CH_2TMS)$	149	
C_6H_2Br-3-$(OMe)_2$-2,6	PMe_3, PMe_2Ph	200	
$C_6H_2Me_3$-2,4,6	bipy	147,148	
C_6HBr_2-3,5-$(OMe)_2$-2,6	PMe_3	198–200	0.195 8(8)
	PMe_2Ph	200	
C_6F_5	bipy	201	
	$PMePh_2$	153	
	$PHPh_2$	3	
	PPh_3	154,201	
2-C_4H_3S	PBu_3	202	
5-(2'-C_4H_3S)-2-C_4H_2S	PBu_3	202	
C≡CH	PBu_3	191	
C≡CC_6H_4C≡CH-2	PBu_3, PPh_3	193,194	
C≡CC_6H_4C≡CH-4	PBu_3, PPh_3	193,194	
C≡CPh	PBu_3	191	
	$(\mu$-dppm$)_2HgX_2$	192	

Table 5 (continued)

R	*L*	*Reference*	*Ni–C* (nm)
$C{\equiv}CC_6H_4Me$-4	$(\mu\text{-dppm})_2HgX_2$	192	
$CCl{=}CCl_2$	bipy, phen	111	
	PMe_2Ph	111	0.191(1), 0.193(3)
$C{\equiv}CTMS$	PMe_3	122	
$C{\equiv}CBu^t$	PMe_3	122	
C_6F_4H	bipy, PPh_3	201	
	$PMePh_2$	153	
C_6F_4Br	$PMePh_2$	153	
C_6F_4Ni	$PMePh_2$	153	
4,4'-$C_{12}F_8H$	$PMePh_2$	161	0.190 6(9)
4,4'-$C_{12}F_8Ni$	$PMePh_2$	161	
$(B_{10}C_2H_{10})_2SiMe_2$	bipy	203	
Ph[c]	Br(dppe)	129	

[a] bipym = 2,2'-bipyramidyl. [b] dipe = 1,2-bis(diisopropyl phosphino)ethane. [c] Nickel(III) complex.

Table 6 $[NiR^1R^2L_2]$ compounds.

R^1	R^2	*L*	*Reference*	$Ni{-}C_R{}^1$, $Ni{-}C_R{}^2$ (nm)
Me	Ph	dmpe	204	
	C_6H_4Me-2	dmpe	204	
	C_6H_4Me-4	dmpe	204	
	C_6H_4OMe-4	dmpe	204	
	C_6H_4F-4	dmpe	204	
	$C_6H_2Me_3$-2,4,6	bipy	147	
$CCl{=}CCl_2$	Ph	PMe_2Ph, PEt_3	205,206	
	C_6H_4Me-2	PMe_3	112,195	
		PMe_2Ph	195,205,206	
		$PMePh_2$	195	
		PEt_3	205,206	
	C_6H_4Me-3	PMe_3	114	
	C_6H_4Me-4	PMe_3	114–116	0.190 7(5), 0.190 8(5)
		PMe_2Ph, PEt_3	205,206	
	C_6H_4Cl-4	PMe_3	114,115	
		PEt_3	205,206	
	$C_6H_4NMe_2$-4	PMe_3	114	
	C_6H_2Br-3-$(OMe)_2$-2,6	PMe_3, PMe_2Ph	195	
	$C_6H_2Me_3$-2,4,6	PMe_2Ph	205,206	0.193 3(3), 0.193 6(3)
$C{\equiv}CBu^t$	$C(C{\equiv}C\text{-TMS}){=}C(TMS)C{\equiv}C\text{-TMS}$	PMe_3 (**49**)	122	
Ph	C_6H_4Me-2	PEt_3	205,206	
	C_6H_4Cl-2	PEt_3	205,206	
	$C_6H_2Me_3$-2,4,6	PMe_2Ph	205,206	
C_6H_4Me-2	$C_6H_2Me_3$-2,4,6	PMe_3	112	
		PMe_2Ph, PEt_3	205,206	
		bipy	205,206	
	C_6H_4Me-4	PEt_3	205,206	
	C_6H_4Cl-2	PMe_2Ph, PEt_3	205,206	
C_6H_4Me-3	$C_6H_2Me_3$-2,4,6	PMe_3, PMe_2Ph	112	
C_6H_4Me-4	C_6H_4Cl-2	PEt_3	205,206	
C_6H_4Cl-2	$C_6H_2Me_3$-2,4,6	PMe_2Ph	205,206	
C_6F_4H-4	C_6F_4Ni	$PMePh_2$	153	
$C_6H_2Me_3$-2,4,6	$C{=}CHCH_2CH_2O$	PMe_3, PMe_2Ph	171	
	$C{=}CHCH_2CH_2CH_2O$	PMe_3, PMe_2Ph	171	
C_6Cl_5	$C{=}CHCH_2CH_2O$	PMe_3 (**60**)	171,172	0.193 0(3), 0.190 9(3)
		PMe_2Ph	170,171	
	$C{=}CHCH_2CH_2CH_2O$	PMe_3, PMe_2Ph	170,171	
	$C(OMe){=}CH_2$	PMe_3 (**62**)	173a	0.193 0(5), 0.191 6(5)
4,4'-$C_{12}F_8H$	4,4'$C_{12}F_8Ni$	$PMePh_2$	161	

Oxidative addition reactions to nickel(I) complexes have also been reported (Equation (33)).[127,177] Iododichlorobenzenes give $[NiI(Ar)(PPh_3)_2]$ on reaction with $[NiBr(PPh_3)_3]$, whereas no reaction occurs with aryl chlorides.[127] A comparison of the rates of reaction of the nickel(0) complex $[Ni(PEt_3)_4]$ and the

Table 7 $[NiRL_n]^+[X]^-$ complexes.

R	n*L*	*X*	*Reference*	*Ni–C* (nm)
Me	$4PMe_3$	BPh_4	166	
		BF_4	167	
		CF_3CO_2	75	
	$(PMe_3)_2$(bipy)	BF_4	174	
	$(PMe_3)_2$(phen)	BF_4	174	
	$(PMe_3)_2$(dppe)	BF_4	174	
	$(PMe_3)_2$(dmpe)	BF_4	174	
	(PMe_3)(dmpe)	BF_4	174	
	2dmpe	BF_4	174	
Et	$Bu^t_2P(CH_2)_2PBu^t_2$ (**34**)	BF_4	168	0.194 0(5) 0.208 1(5)
	$Bu^t_2P(CH_2)_3PBu^t_2$	BF_4	168	
	o-$(Bu^t_2PCH_2)_2C_6H_4$	BF_4	168	
	$3PBu_3$, $3PPh_3$	$AlEt_4$	169	
	$3P(OEt)_3$, $3P(OPh)_3$	$AlEt_4$	169	
	$3P(OCy)_3$, $3P(OBu^s)_3$	$AlEt_4$	169	
	$3P(OPr^i)_3$	$AlEt_4$	169	
Bu^i	$3PBu_3$, $3PPh_3$	$AlBu^i_4$	169	
	$3P(OEt)_3$, $3P(OPh)_3$	$AlBu^i_4$	169	
	$3P(OCy)_3$, $3P(OBu^i)_3$	$AlBu^i_4$	169	
	$3P(OPr^i)_3$	$AlBu^i_4$	169	
CH_2TMS	$3PMe_3$	BF_4	174	0.192 0(22)
$CCl{=}CCl_2$	$(PMe_3)(NC_5H_3Me_2$-3,5)	ClO_4	112	
	$(PMe_3)(NC_5H_4Me$-3)	ClO_4	112	
	$(PMe_3)(NC_5H_4Me$-4)	ClO_4	112	
	$(PMe_2Ph)(NC_5H_4Me$-3)	ClO_4	112	
	$(PMe_2Ph)(NC_5H_3Me_2$-3,5)	ClO_4	112	
	$(PMePh_2)(NC_5H_3Me_2$-3,5)	ClO_4	112	
C(Ph)=CHCH=C(Ph)Ni	$(PEt_3)_2$(THF) (**37**)	PF_6	120	
$C(Ph){=}C(Ph)C_6H_3Cl_2$-3,5	(γ-pic)(dppe)	ClO_4	121	
$C(CO_2Me){=}C(CO_2Me)C_6H_3Cl_2$-3,5	(γ-pic)(dppe)	ClO_4	121	
$C_6H_2Me_3$-2,4,6	$(PMe_3)(NC_5H_4Me$-2)	ClO_4	112	
	$(PMe)_3(NC_5H_4Me$-3)	ClO_4	112	
	$(PMe_3)(NC_5H_4CO_2Me$-3)	ClO_4	112	
	$(PMe_3)(N_2C_3H_3Me$-2)	ClO_4	112	
	$(PMePh_2)(N_2C_3H_3Me$-1)	ClO_4	112	
	$(MeCN)(PEt_3)_2$	BF_4	151	
	$(MeCN)(PBu_3)_2$	BF_4	151	
	$(MeCN)(PBu^t_3)_2$	BF_4	151	
	$(MeCN)(PBz_3)_2$	BF_4	151	
	$(MeCN)(PMe_2Ph)_2$	BF_4	151	
	$(py)(PBu_3)_2$	BF_4	151	
	$(THF)(PBu_3)_2$	BF_4	151	
	$\{=C(CH_2)_3O\}(PMe_3)_2$	ClO_4	171	
	$\{=C(CH_2)_3O\}(PMe_2Ph)_2$	ClO_4	171	
	$\{=C(CH_2)_4O\}(PMe_3)_2$	ClO_4	171	
C_6Cl_5	3py	ClO_4	156	
	3β-pic	ClO_4	156	
	3γ-pic	ClO_4	156	
	$(py)_2(PEt_3)$	ClO_4	156	
	$(py)_2(PPh_3)$	ClO_4	156	
	$(PMe_2Ph)(NC_5H_4Me$-3)	ClO_4	112	
	$(PMe_2Ph)(NC_5H_4CO_2Me$-3)	ClO_4	112	
	$(PMe_2Ph)(NC_5H_3Me_2$-3,4)	ClO_4	112	
	$(PMe_2Ph)(N_2C_3H_3Me$-1)	ClO_4	112	
	$(PMe_2Ph)(N_2C_3H_3Me$-2)	ClO_4	112	
	$(PMe_2Ph)(N_2C_3H_3Et$-1)	ClO_4	112	
	$\{=C(CH_2)_3O\}(PMe_3)_2$	ClO_4	171	
	$\{=C(CH_2)_3O\}(PMe_3)_2$	BF_4	172	0.191 2(9) 0.183 7(7)
	$\{=C(CH_2)_3O\}(PMe_2Ph)_2$	ClO_4	170	
	$\{=C(CH_2)_4O\}(PMe_3)_2$	ClO_4	171	
	$\{=C(CH_2)_4O\}(PMe_2Ph)_2$	ClO_4	170,171	
	$\{=C(OMe)Me\}(PMe_3)_2$	BF_4	173a	
	$\{=C(OMe)Me\}(PMe_2Ph)_2$	BF_4	173b	
$(\mu_2$-η^3:η^1-CH_2-*o*-$C_6H_4)$Ni	(PMe_3)(dppm)	Br	106	
COMe	$3PMe_3$	BF_4	167	
	$(PMe_3)_2(CO)$	BF_4	167	
	(PMe_3)(dmpe)	BF_4	174	
	2dmpe	BF_4	174	

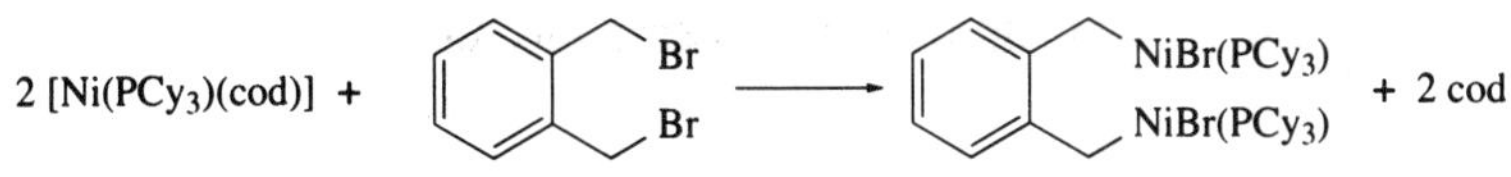

(28)

$$[Ni(PPh_3)_2(C_2H_4)] + ClCH_2SMe \longrightarrow [NiCl(CH_2SMe)(PPh_3)_2] + C_2H_4 \quad (29)$$

$$[Ni(dppe)] + PhBr \longrightarrow [NiBr(Ph)(dppe)] \quad (30)$$

$$Ni + 2\,C_6F_5I + 2\,PPh_3 \longrightarrow [Ni(C_6F_5)_2(PPh_3)_2] + I_2 \quad (31)$$

$$Ni + C_6F_5I + 2\,PEt_3 \longrightarrow [Ni(C_6F_5)(PEt_3)_2] \quad (32)$$

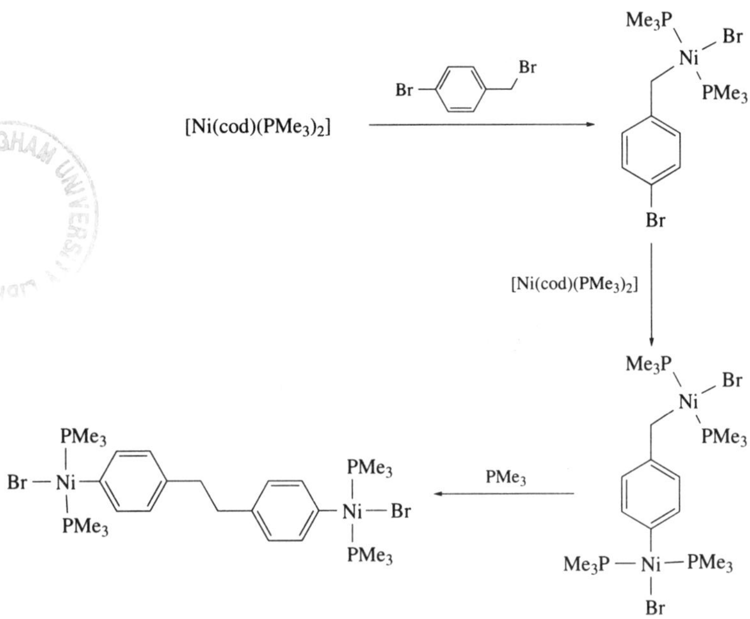

Scheme 5

nickel(I) complex $[NiX(PEt_3)_3]$ (X = Br,I) with MeBr, MeI or EtI, has shown that the reactivity of the nickel(I) complex is 10–20 times less than that of the nickel(0) complex.[177]

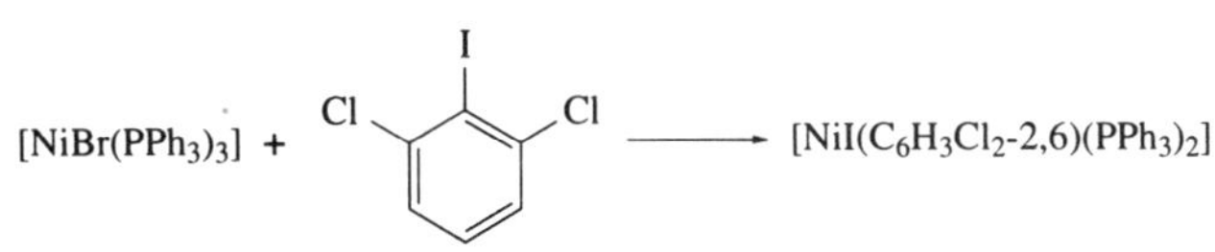

(33)

Oxidative addition with concomitant reductive elimination from a dialkylnickel(II) complex has been observed (Equation (34)).[110]

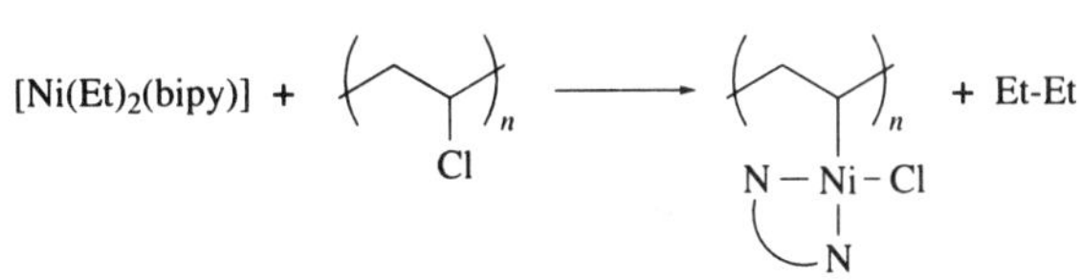

(34)

Organic compounds other than simple alkyl or aryl halides used in oxidative addition reactions include functionalized alkyl halides (Equation (35)),[104] polyvinylchloride (Equations (34) and (36)),[110] 6-iodo-3,4-methylenedioxytoluene (Equation (37)),[113] chloroalkynes (Equation (27)),[122] cyano compounds (Equation (38)),[131,132] diaryl sulfides (Equation (39)),[137,142] 9,10-dihaloanthracenes (Equation (40)),[167] and 4-bromo-3-phenylsydnone (Equation (41)).[164]

A high yield synthesis of $[NiCl(Ph)(PMe_3)_2]$ from $[Ni(PMe_3)_4]$ and C_6H_5COCl has been reported (Equation (42)).[72]

$$[Ni(cod)_2] + RCH_2Cl + 2\,PMe_3 \longrightarrow [NiCl(CH_2R)(PMe_3)_2] + 2\,cod \qquad (35)$$

$R = CO_2Et, CN, OMe$

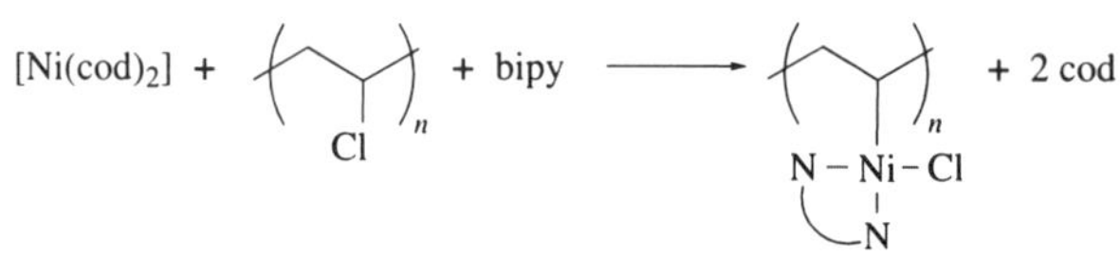

(36)

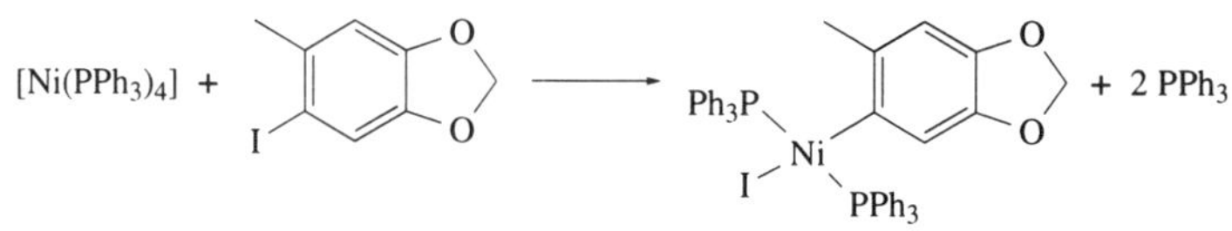

(37)

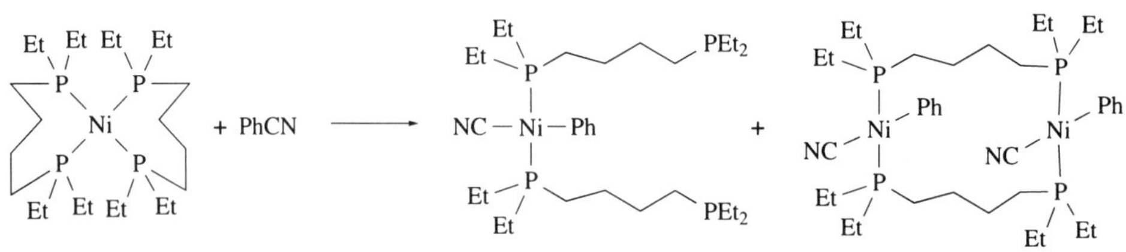

(38)

$$[Ni(PBu_3)_3] + C_6F_5SC_6F_5 \longrightarrow [Ni(C_6F_5)(SC_6F_5)(PBu_3)_2] + PBu_3 \qquad (39)$$

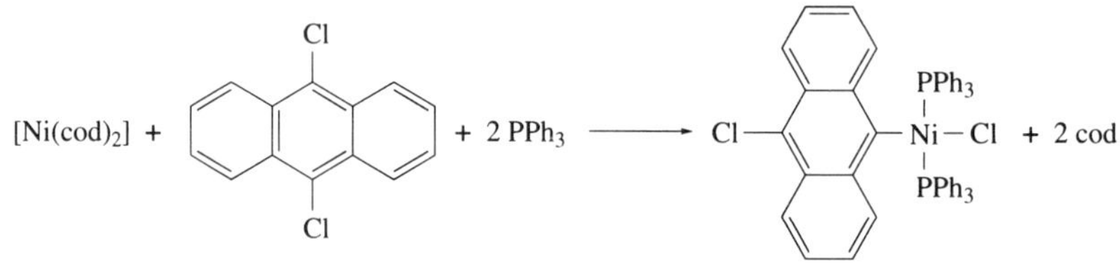

(40)

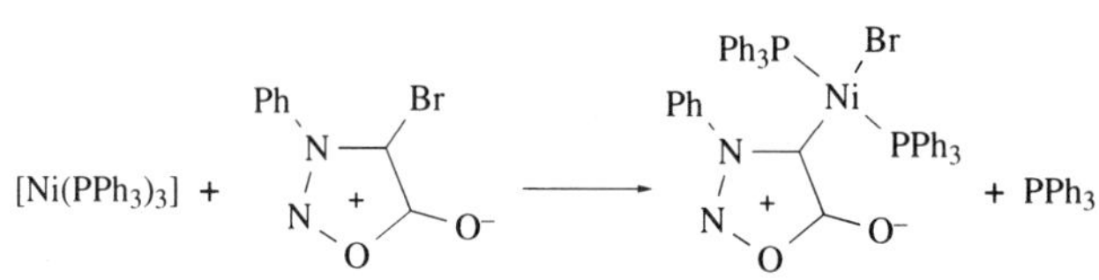

(41)

$$[Ni(PMe_3)_4] + PhCOCl \longrightarrow [NiCl(Ph)(PMe_3)_2] + CO + 2\,PMe_3 \qquad (42)$$

An important class of compound synthesized by oxidative addition reactions are organonickel complexes containing a *P,O*-chelate ligand. The importance of such complexes as catalysts or catalyst precursors for ethene polymerization[2,176] has encouraged a number of studies. A representative selection of syntheses are given in Equations (43)–(46), and in Scheme 6. The formation of the ethylnickel complex (**28**) (Scheme 6) provides an important model reaction for the Shell Higher Olefin Process. On warming (**28**) to room temperature, β-hydrogen elimination occurs to give the hydridonickel complex (**29**), which is formed directly from [Ni(cod)$_2$] and $Ph_2PCH_2C(CF_3)_2OH$ in the presence of tricyclohexylphosphine, but in the absence of ethene. These results provide support for the nickel hydride mechanism for the formation of α-alkenes from ethene rather than a metallocycloalkane mechanism.[97] The complexes (**24**) and (**27**) have been structurally characterized and show a planar chelate ring and a delocalization of the π-system as had been observed previously.[60]

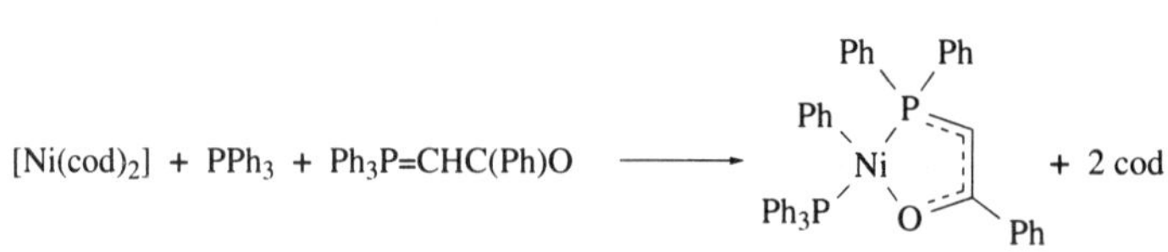

(**24**)

(43)

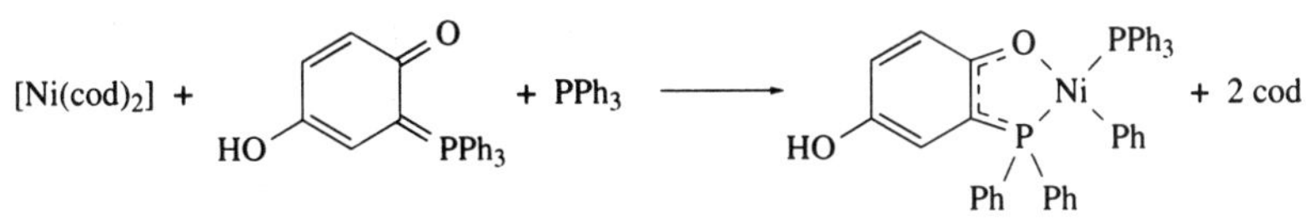

(25) (44)

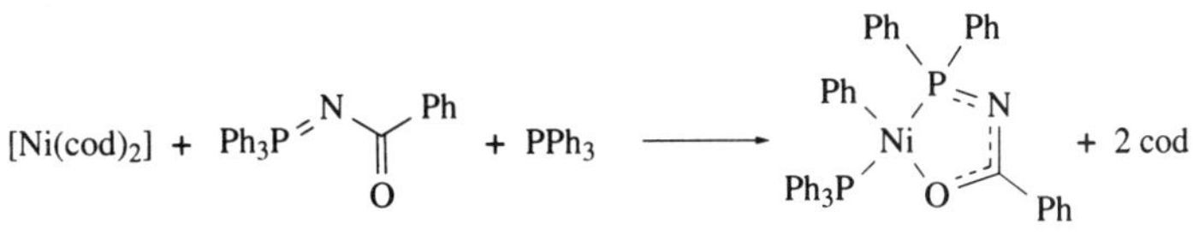

(26) (45)

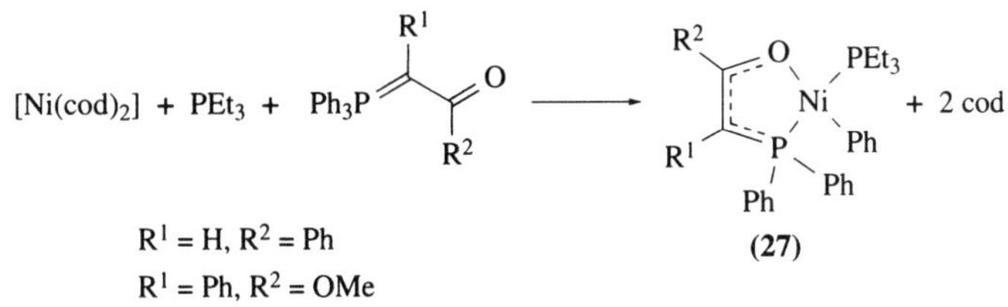

(27) (46)

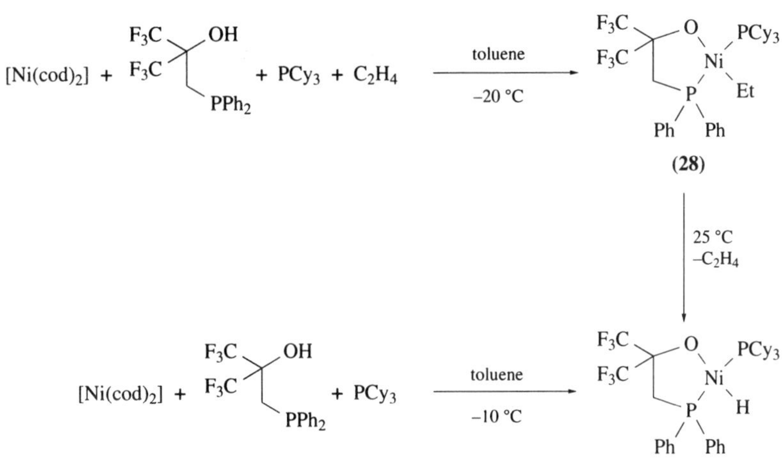

(28)

(29)

Scheme 6

Ab initio calculations on the system $Ni(PH_3)_2 + CH_4$ have indicated that the oxidative addition is exothermic by 29.4 kJ mol^{-1}, and that the planar *trans*-product is the most stable, being lower in energy than the *cis*-isomer by 14.3 kJ mol^{-1}. However, the *cis*-isomer is kinetically favoured with an activation energy for the reaction path of 74.3 kJ mol^{-1}, compared to a value of 201 kJ mol^{-1} for the *trans*-isomer.[178]

Oxidative addition reactions leading to nickelacyclic complexes in which the ring includes a Ni–C bond are considered in Sections 2.7 and 2.8.

2.4.1.3 Miscellaneous syntheses

The protonolysis of $[NiR_2L_2]$ complexes by alcohols (Equations (47)–(50)),[68,85,88,89] thiols (Equation (51)),[68,69] H_2S (Equation (52)),[140] carboxylic acids (Equation (53))[68,91] and compounds having N–H bonds (Equation (54))[92] continues to be an important synthetic route to monoalkyl- or arylnickel complexes.

$$[Ni(Me)_2(PPh_3)_2] + Bu^tOH \longrightarrow [Ni(Me)(OBu^t)(PPh_3)_2] + CH_4 \quad (47)$$

$$[Ni(Et)_2(bipy)] + (CF_3)_2CHOH \longrightarrow [Ni(Et)\{OCH(CF_3)_2\}(bipy)] + C_2H_6 \quad (48)$$

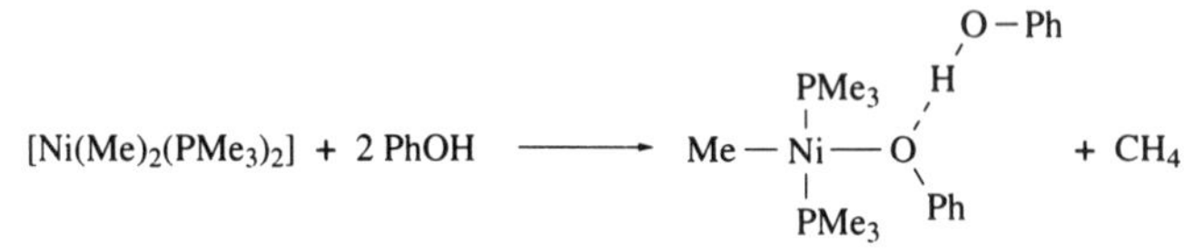

(49)

(30)

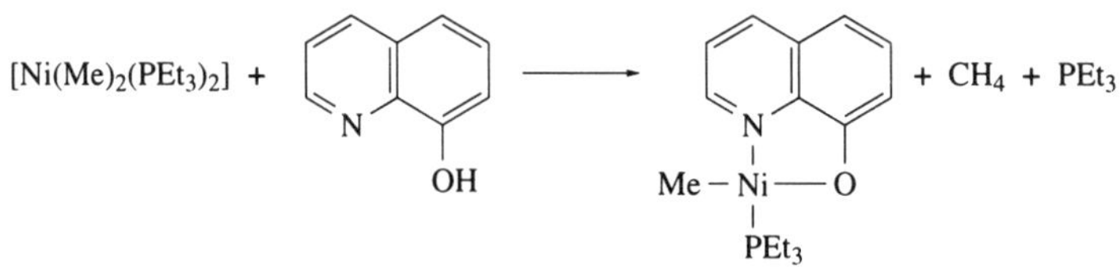

(50)

$$[Ni(Me)_2(dppe)] + PhSH \longrightarrow [Ni(Me)(SPh)(dppe)] + CH_4 \qquad (51)$$

$$[Ni(Ph)(Me)(PEt_3)_2] + H_2S \longrightarrow [Ni(Ph)(SH)(PEt_3)_2] + CH_4 \qquad (52)$$

$$[Ni(Me)_2(PEt_3)_2] + PhCO_2H \longrightarrow [Ni(Me)(OCOPh)(PEt_3)_2] + CH_4 \qquad (53)$$

$$[Ni(Me)_2(PEt_3)_2] + HN(COR)_2 \longrightarrow [Ni(Me)\{N(COR)_2\}(PEt_3)_2] + CH_4 \qquad (54)$$

(31)

It has been found that in complex (**30**) and the related $[Ni(Me)\{OCH(CF_3)Ph\}(PMe_3)_2]$ there is strong hydrogen bonding between the phenoxide or alkoxide ligand and the alcohol, both in the solid state and in solution.[88] Among the complexes of type (**31**) synthesized by a reaction of the type shown in Equation (54) are the polymeric imidozolatonickel complex (**32**) and the (pyromellitimido)dinickel complex (**33**).[92]

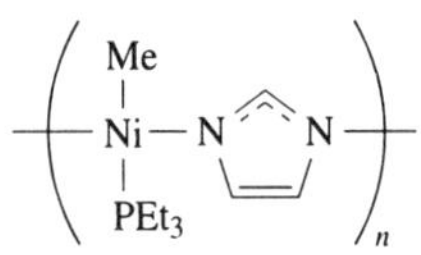

(32)

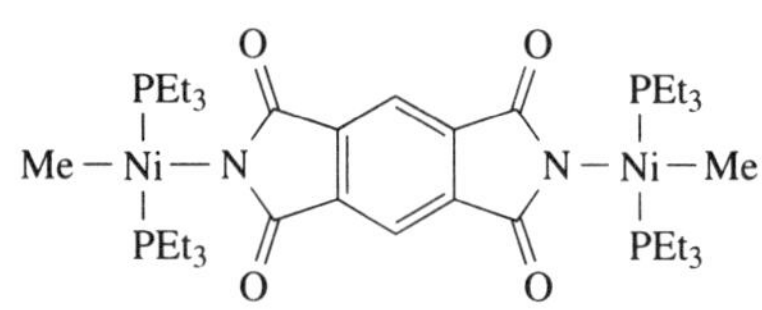

(33)

The protonation of the coordinated ethene in $[Ni(Bu^t{}_2PCH_2CH_2PBu^t{}_2)(\eta\text{-}C_2H_4)]$ with HBF_4 gives the ethylnickel complex (**34**) in which, for the first time in nickel chemistry, an agostic interaction with a hydrogen on a β-carbon atom is observed. Evidence for this $Ni\cdots H\cdots C_\beta$ interaction is provided by NMR spectroscopy (a broad resonance at –5.8 ppm is assigned to the agostic hydrogen at 173 K) and an x-ray crystal structure determination.[168] Protonation of the coordinated ethene in $[Ni\{P(OC_6H_4Me\text{-}2)_3\}_2(C_2H_4)]$ by HCN at low temperature gives the ethylnickel complex $[Ni(CN)(Et)\{P(OC_6H_4Me\text{-}2)_3\}_2]$.[95]

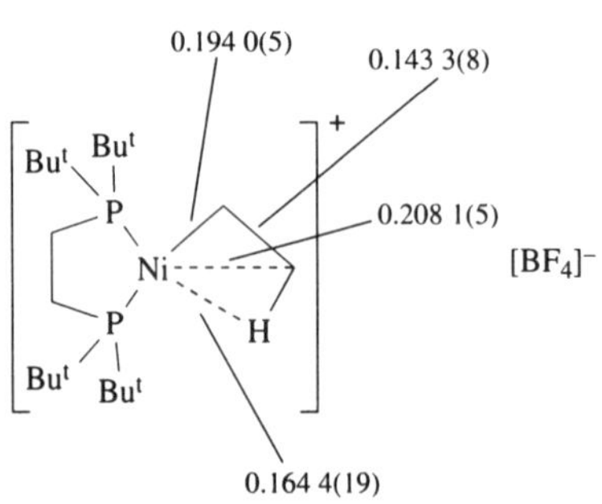

(34)

The oxidation of a Ni–C bond in the diethylnickel complex $[Ni(Et)_2(bipy)]$ by N_2O has been reported (Equation (55)).[96] It seems likely that this will provide a useful general synthetic route to alkyl(alkoxide)nickel complexes.

$$[Ni(Et)_2(bipy)] + N_2O \longrightarrow [Ni(Et)(OEt)(bipy)] + N_2 \quad (55)$$

Base-assisted dehydrohalogenation has proved to be a useful route to alkynylnickel complexes (Equation (56)).[184]

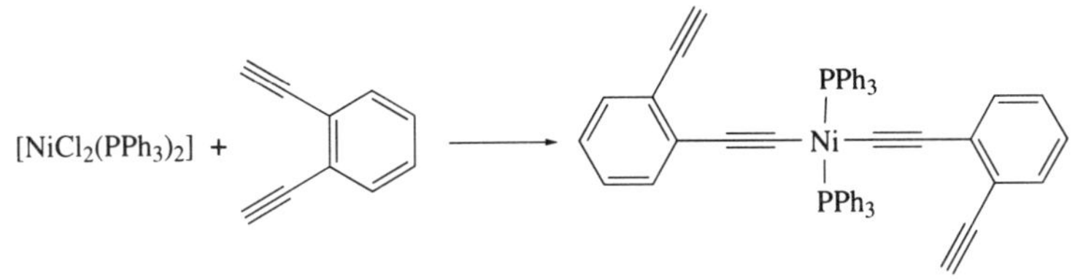

(56)

The electrochemical reduction of nickel(II) complexes in the presence of an alkyl or aryl halide can be used as a method of synthesis of both $[NiX(R)L_2]$ and $[NiR_2L_2]$ complexes (see also Equation (30)). Thus, the reduction of $[Ni(phen)_3]^{2+}$ in the presence of ethyl, octyl or phenyl bromide leads to the formation of $[NiR_2(phen)]$,[185] while the reduction of NiX_2 (X = Cl, Br) in the presence of PPh_3 and phenyl bromide or phenyl chloride gives $[NiX(Ph)(PPh_3)_2]$.[125] Nickel(II) complexes can also be chemically reduced in the presence of RX to produce $[NiX(R)L_2]$ complexes (Equation (57)).[112]

$$[NiCl_2(PMe_3)_2] + Cl_2C{=}CCl_2 \xrightarrow{NaBH_4} [NiCl(CCl{=}CCl_2)(PMe_3)_2] \quad (57)$$

Novel alkylating or arylating reagents are the silicon methylpentafluoride or phenylpentafluoride dianions $[SiRF_5]^{2-}$. This reagent has been shown to react with $[NiCl_2(PPh_3)_2]$ in the presence of 2% aqueous NH_4F to give both $[NiCl(R)(PPh_3)_2]$ and $[NiR_2(PPh_3)_2]$.[86] Treatment of the μ-1,4-diphenylbutadienylnickel complex (**35**) with $TlPF_6$ causes C–C bond cleavage to give the μ-2,3-diphenylbutadienylnickel complex (**36**), whereas a similar reaction with the bis(triethylphosphine) derivative of (**35**) gives the dication (**37**) (Scheme 7).[120] The conversion of (**35**) into (**36**) may be of importance in the catalytic cyclodimerization of phenylacetylene.

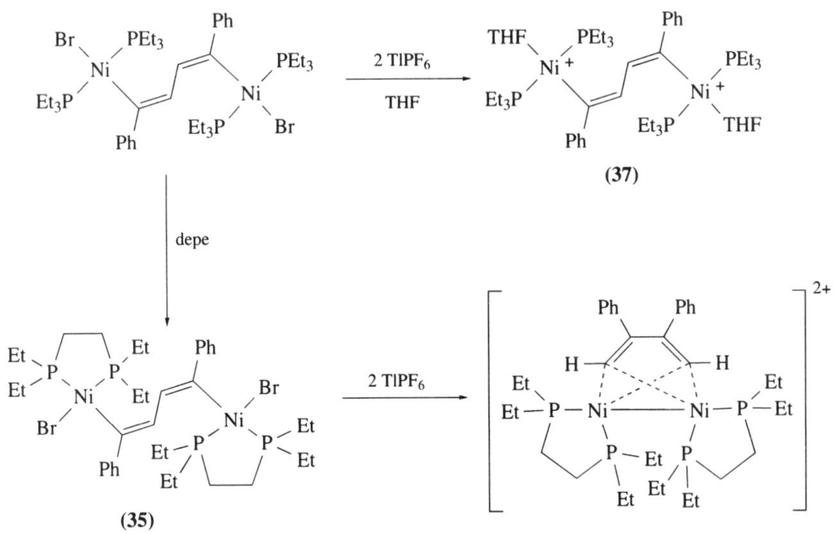

Scheme 7

The treatment of $[NiCl(C_6Cl_5)(PPh_3)_2]$ with $NaClO_4$ in the presence of excess N-donor ligands, L, such as pyridine leads to the formation of the ionic compounds $[Ni(C_6Cl_5)L_3][ClO_4]$. The action of potassium salts KX on these ionic compounds gives $[NiX(C_6Cl_5)L_2]$ (X = Cl, Br, NO_2, C_6Cl_5O or NCS).[156]

The transformation of chloroalkenyl(aryl)nickel complexes upon thermolysis[114,116] or photolysis[115,116] has been reported. The resulting nickel complexes are isomeric chloro(alkenyl) species (Equation (58)) probably formed via reductive elimination followed by oxidative addition.

$$[Ni(CCl{=}CCl_2)(C_6H_4Me\text{-}4)(PMe_3)_2] \xrightarrow{\text{heat or } h\nu} [NiCl\{CCl{=}CCl(C_6H_4Me\text{-}4)\}(PMe_3)_2] + [NiCl\{C(C_6H_4Me\text{-}4){=}CCl_2\}(PMe_3)_2] \quad (58)$$

The cocondensation of nickel with CF_3• radicals gives the unstable $[Ni(CF_3)_2]$, which may be stabilized by the addition of trimethylphosphine at low temperature to give $[Ni(CF_3)_2(PMe_3)_3]$.[190]

2.4.2 Reactions

Reviews on the chemistry of organonickel complexes $[NiR_2L_2]$ relevant to catalysis,[179,180] on the influence of heteroatoms in α- and β-functionalized alkyl transition metal compounds,[181] on the mechanisms of decomposition of organometallic compounds,[182] and an electrochemical approach to the rates and mechanisms of electron-transfer/nickel-catalysed homocoupling and carboxylation reactions,[207] have been published.

In accordance with *COMC-I*,[1] the reactions of organonickel complexes have been divided into three types: those in which the Ni–C bond is not involved, those in which new Ni–C bonds are formed, and those in which the alkyl or aryl groups are displaced.

2.4.2.1 Reactions in which the Ni–C bond is not involved

Reactions in this class are largely concerned with anionic ligand exchange in $[NiX(R)L_2]$[75–8,100,139,143,150,158,159] or neutral ligand exchange in $[NiX(R)L_2]$[111,147,155,157,205] or $[NiR_2L_2]$,[101,183,187,188,200] and have been used to synthesize a wide range of complexes. A number of kinetic studies of these reactions have been carried out although such studies are rare compared to studies on platinum(II) complexes. It has been observed that the rate of substitution of bromide ion in $[NiBr(C_6Br_5)(PRPh_2)_2]$ by N_3^-, Cl^-, I^-, NO_2^- or SCN^- was slower than for the pentachlorophenyl derivative.[159] The substitution of bromide for NCS^- in the complexes $[NiBr(Ar)(L)_2]$ (where Ar is a substituted phenyl group and L is a tertiary phosphine) appears to be governed solely by steric factors; in particular the *ortho*-substitutents on the aryl ligand have a dominant role.[143] The effect of the mutually *trans* phosphine ligands, which are *cis* to the Br^- leaving group, is also almost entirely steric.[150] Steric effects rather than amine basicity also control the rate of substitution of chloride with substituted pyridines in $[NiCl(C_6Cl_5)(PMePh_2)_2]$.[157]

Treatment of $[NiCl(R)(PMe_3)_2]$ with sodium pyrrolyl salts $[Na][NC_4H_2X_2]$ gives derivatives of the type $[Ni(R)(NC_4H_2X_2)(PMe_3)_2]$.[78] The reaction of $[NiBr(C_6H_4Me\text{-}4)(PMe_3)_2]$ with the thallium salt of the $HB(3\text{-}Bu^tC_3H_2N_2)_3$ anion gives the η^1-tris(pyrazolyl)borate nickel complex (**38**). Complex (**38**) does not react with CO, but the corresponding aroyl derivative can be prepared from $[NiBr\{C(O)C_6H_4Me\text{-}4\}(PMe_3)_2]$.[139] Dialkyldithiocarbamate derivatives $[Ni(CH_2R^1)(S_2CNR^2_2)(PMe_3)]$ and $[Ni(COR^1)(S_2CNR^2_2)(PMe_3)]$ have been synthesized by halide exchange with $[R^2_2NCS_2]^-$.[99,100] The acyl derivative is also formed on treatment of the alkyl derivative with CO.[100] The exchange reaction of $[NiX(C_6Cl_5)(PCy_3)_2]$ with PPh_3 gives $[NiX(C_6Cl_5)(PPh_3)_2]$ for X = Cl, I, NCO and N_3, but for X = NCS, $[Ni(C_6Cl_5)(NCS)(PCy_3)(PPh_3)]$ is obtained.[155] Treatment of *cis*-$[Ni(CH_2TMS)_2(py)_2]$ with monodentate or bidentate ligands yields $[Ni(CH_2TMS)_2L_2]$ for L = PMe_3 and PMe_2Ph and L_2 = dppe, bipy, phen and TMEDA, but for L = $PMePh_2$ or PPh_3 reductive elimination takes place to give $[NiL_4]$.[187]

The structure of the complex $[Ni(CH_2SiMe_2Ph)_2(Pr^i_2PCH_2CH_2PPr^i_2)]$ (**39**) formed by ligand exchange with the TMEDA derivative has been determined.[188] This complex does not undergo δ-hydrogen abstraction to form a nickelacyclic compound, unlike the analogous $[Ni(CH_2CMe_2Ph)(PMe_3)_2]$ which forms the nickelacyclic complex $[\overline{Ni\{2\text{-}CH_2C(Me)_2C}_6H_4\}(PMe_3)_2]$.[102] The irradiation of a mixture of the diorganonickel complexes $[Ni\{C_6H_3(OMe)_2\text{-}2,6\}_2(PMe_3)_2]$ and $[Ni\{C_6H_3(OMe)_2\text{-}2,6\}_2(PMe_2Ph)_2]$ gives the mixed ligand complex $[Ni\{C_6H_3(OMe)_2\text{-}2,6\}_2(PMe_3)(PMe_2Ph)]$. When the *anti*-isomer of $[Ni\{C_6H_2Br\text{-}3\text{-}(OMe)_2\text{-}2,6\}_2(PMe_2Ph)_2]$ is irradiated it partially isomerizes to give a mixture of *syn*- and *anti*-isomers in the ratio 32:68, through rotation about the Ni–C bond.[200] This steady state ratio is different from that resulting from thermal isomerization which gives a 50:50 mixture.[195]

The complexes $[Ni(Me)(PMe_3)_2(L\text{–}L)][BF_4]$ can be prepared by ligand displacement from $[Ni(Me)(PMe_3)_4][BF_4]$ (L–L = bipy, phen, dppe). With dmpe, the product $[Ni(Me)(dmpe)_2][BF_4]$ can be formed. With the more bulky alkyl ligand, CH_2TMS, the tris(trimethylphosphine) derivative $[Ni(CH_2TMS)(PMe_3)_3][BF_4]$ is formed which has a distorted square planar structure.[174] At least two strongly electron-donating phosphine ligands are required for the stabilization of cationic alkylnickel complexes, and their preferred coordination number is five in a trigonal bipyramidal structure.

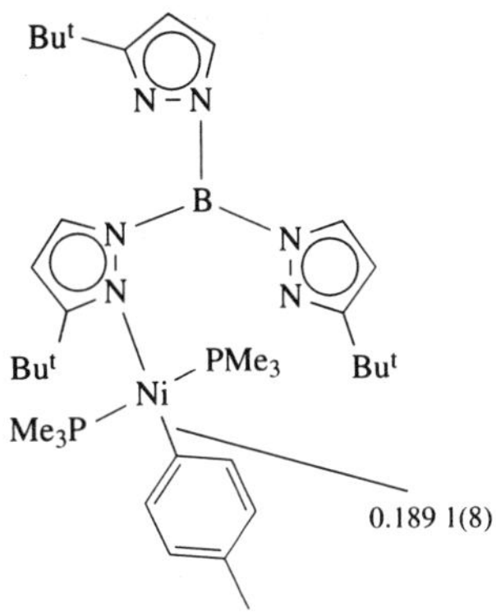

(38)

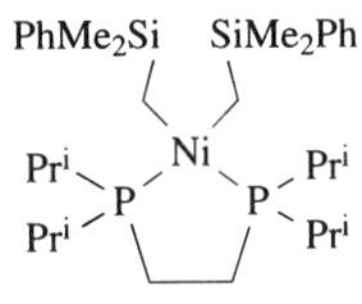

(39)

Reactions on coordinated ligands include the bromination of a dimethoxyphenylnickel complex (Equation (59)),[195] the synthesis of [Ni(Ph){OC(Ph)C(SO_3Na)PPh_2}(PPh_3)],[135] and the lithiation of tertiary methylphosphines (Scheme 8).[149,197]

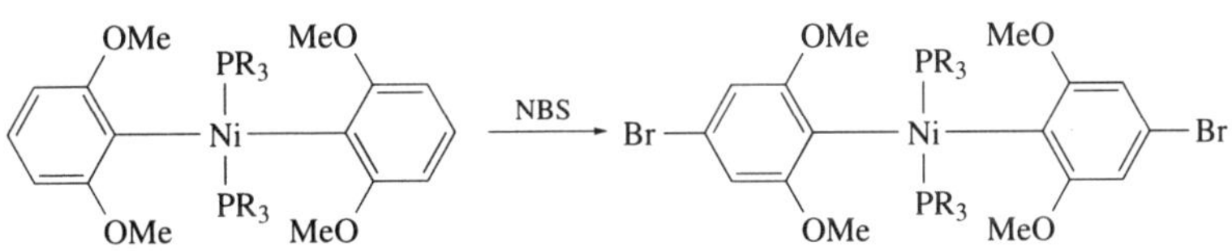

(59)

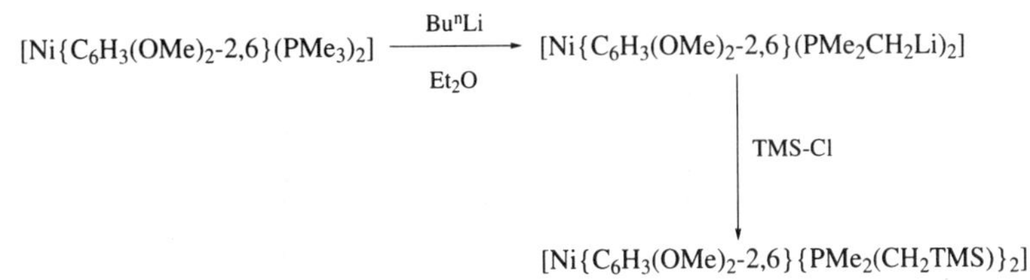

Scheme 8

The reaction of [NiH(Ph)(PCy_3)$_2$] with CO_2 proceeds by insertion at the hydride to produce the formate complex [Ni(Ph)(O_2CH)(PCy_3)$_2$]; the analogous hydridomethyl complex [NiH(Me)(PCy_3)$_2$] reacts with CO_2 to produce the hydridoformate [NiH(O_2CH)(PCy_3)$_2$] as the only isolable complex, with CH_4 and C_2H_6 being evolved. If this latter reaction is carried out above 0 °C, another low-yield product is the dimer [(Cy_3P)(Me)Ni(O_2CO)NiH(PCy_3)$_2$]. Labelling studies have shown that both the hydride and formate hydrogen in [NiH(O_2CH)(PCy_3)$_2$] originate exclusively from the hydride in [NiH(Me)(PCy_3)$_2$].[79,84]

Cyclopentadienylnickel complexes have been synthesized by treating [NiX(R)L_2] complexes with NaCp. Thus, [NiCl(CH_2SMe)(PPh_3)$_2$] reacts with NaCp to give [Ni(CH_2SMe)(PPh_3)$_2$(Cp)],[105] the complexes [NiX(CH_2R)(PMe_3)$_2$] (X = Cl, R = CO_2Et, CN; X = Br, R = OMe) react with NaCp to give [Ni(CH_2R)(PMe_3)(Cp)],[104] and [NiX($C_6H_2Me_3$-2,4,6)(PMe_3)$_2$] gives [Ni($C_6H_2Me_3$-2,4,6)(PMe_3)(Cp)] in a similar reaction.[103]

2.4.2.2 Reactions in which new Ni–C bonds are generated

The vast majority of the reactions in this class involve insertion of CO (Equations (60)–(66))[87,89–91,93,101,103,104,106–8,110,136,138,141,164,167,174,198,199,205] or alkynes[72,117,121,122,174] into a Ni–C bond.

The acylnickel complex (**40**) formed by CO insertion (Equation (60))[91] is also formed by the oxidative addition of $PhCO_2COMe$ to $[Ni(cod)_2]$ in the presence of PEt_3.[91]

$$[Ni(Me)(OCOPh)(PEt_3)_2] + CO \xrightarrow{-78\ ^\circ C} \underset{\textbf{(40)}}{[Ni(COMe)(OCOPh)(PEt_3)_2]} \tag{60}$$

$$[Ni(Me)(OC_6H_4CN\text{-}4)(PEt_3)_2] + CO \longrightarrow [Ni(COMe)(OC_6H_4CN\text{-}4)(PEt_3)_2] \tag{61}$$

A number of studies have shown that alkyl and aryl complexes of nickel(II) containing PMe_3 ligands are particularly suitable for forming stable acyl complexes on reaction with CO. The presence of two different types of Ni–C bond in complex (**41**) has allowed the selectivity of CO, insertion to be investigated. Insertion occurs first into the nickel–benzyl bond (Equation (62)). In the presence of excess CO, insertion into the nickel–aryl bond also occurs to give the unstable bis(acyl)nickel complex $[(Me_3P)_2BrNi\{C(O)CH_2C_6H_4\text{-}2\text{-}C(O)\}NiBr(PMe_3)_2]$.[106] A similar bis(acyl)nickel complex is formed on carbonylation of the binuclear nickel complex (**42**) (Equation (63)).[107] In contrast, carbonylation of the η^3-benzylnickel complex (**43**) occurs at the nickel–aryl bond (Equation (64)).[106]

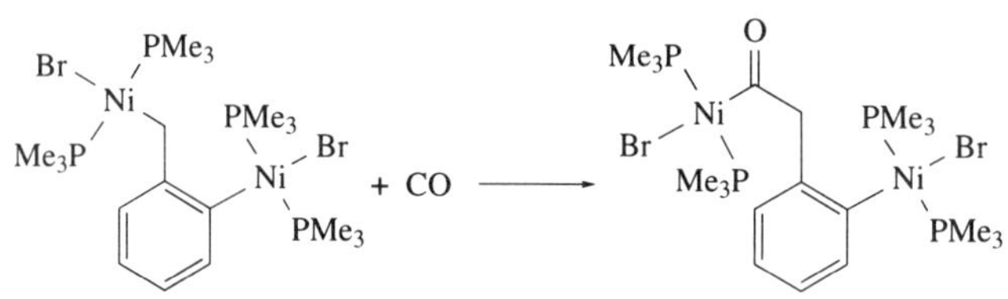

(**41**) (62)

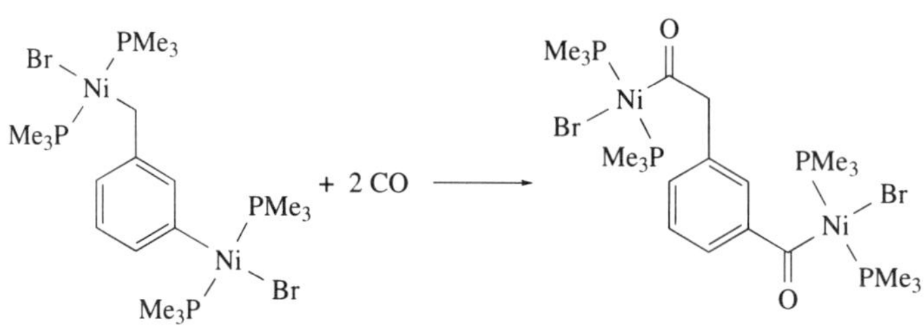

(**42**) (63)

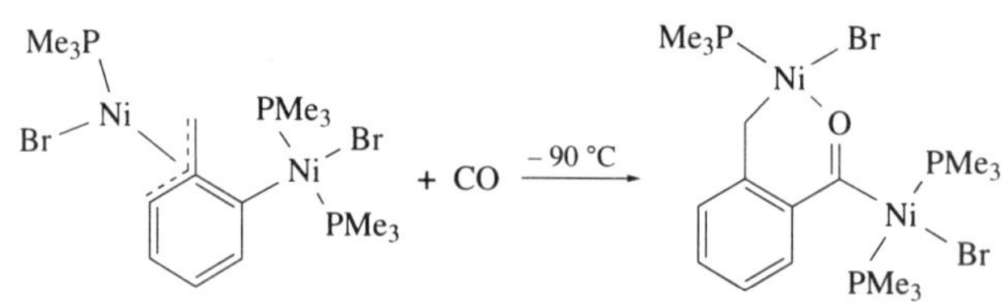

(**43**) (64)

The carbonylation of $[NiX(CH_2R)(PMe_3)_2]$ is dependent upon the substituent R. Thus, when R = OMe and X = Br, an acylnickel complex $[NiBr\{C(O)CH_2OMe\}(PMe_3)_2]$ is the stable product of the carbonylation reaction. However, when R = CN and X = Cl, a stable acyl complex is not obtained, and when R = CO_2Et and X = Cl, the corresponding acyl complex is formed reversibly and is only stable in solution under a CO atmosphere.[104] The mesityl complexes $[NiX(C_6H_2Me_3\text{-}2,4,6)(PMe_3)_2]$ (X = Cl, Br) cannot be carbonylated, probably due to the steric hindrance of the *ortho*-methyl groups.[103] The steric effect of the *ortho*-methoxy groups in $[Ni\{C_6H_3(OMe)_2\text{-}2,6\}(PMe_3)_2]$ and $[Ni\{C_6H(OMe)_2\text{-}2,6\text{-}Br_2\text{-}3,5\}(PMe_3)_2]$ is dependent on the *meta*-substituent. In the brominated derivative the conformation of the substituted phenyl groups is such that no reaction with CO is possible whereas the $C_6H_3(OMe)_2$-2,6 derivative has some flexibility to allow a CO molecule room to insert into the Ni–C bond.[198,199]

The amidonickel(II) complexes $[Ni(R^1)\{N(SiMe_2CH_2PR^2{}_2)_2\}]$ (**44**) (R^1 = Me, Pr^n, CH=CH_2, Ph) undergo an initial carbonyl insertion into the Ni–C bond (labelling studies have shown that insertion into the nickel–amide bond does not occur), but the acyl species formed undergoes reductive elimination and rearrangement to give (**45**) (Scheme 9).[93,138]

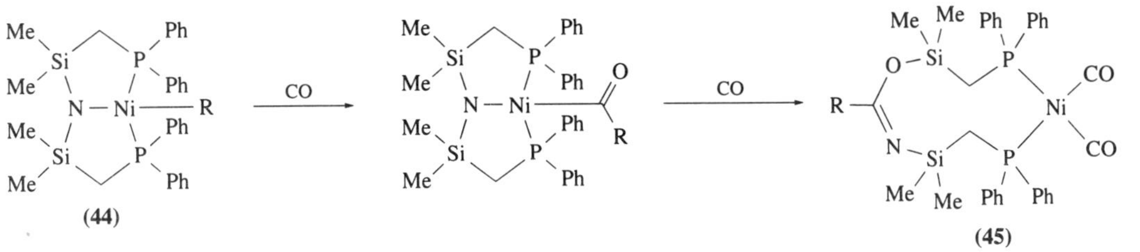

Scheme 9

The *P,O*-chelate complexes (**26**) and (**27**) react with CO to give the corresponding acyl complexes (Equations (65)[141] and (66)[136]). If the substituents on the chelating ligand in (**27**) are changed, as in the complex [Ni(Ph){OC(Ph)CHPPh$_2$}(PEt$_3$)], the initial benzoyl complex obtained on reaction with CO undergoes an elimination reaction to give [Ni(CO)$_3$(PEt$_3$)] and the ester Ph$_2$PCH=C(Ph)OCOPh, formed by coupling of the chelate ring with the benzoyl group.[136]

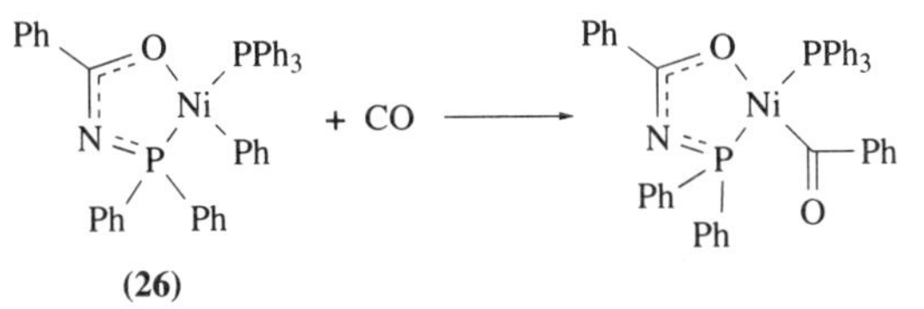

(65)

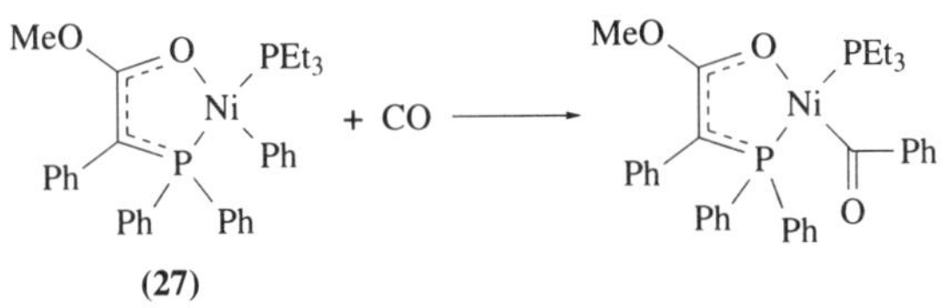

(66)

Reactions of cationic alkylnickel complexes have been little studied. The carbonylation of the ionic compound [Ni(Me)(PMe$_3$)$_4$][BF$_4$] gives the cationic acylnickel complex [Ni(COMe)(PMe$_3$)$_3$][BF$_4$] which reacts further with CO to give [Ni(COMe)(CO)(PMe$_3$)$_2$][BF$_4$], although this latter species could not be isolated.[167] The expected reduced metal–acyl back-bonding in the cationic acyl complexes is evidenced by the CO stretching frequency of 1700 cm^{-1} in [Ni(COMe)(PMe$_3$)$_3$]$^+$, a shift of 65 cm^{-1} to higher wavenumber compared to [NiCl(COMe)(PMe$_3$)$_2$]. The cationic derivative [Ni(Me)(dppe)-(PMe$_3$)$_2$][BF$_4$] reacts with CO to yield [Ni(CO)$_2$(dppe)] and PMe$_4^+$ via a reductive elimination process in which the methyl group migrates from nickel to the PMe$_3$ ligand.[174]

The σ-carbamoylnickel complex [NiI{C(O)NEt$_2$}(NHEt$_2$)$_2$] is formed when a Et$_2$NH/THF/Et$_2$O solution of [NiI$_2$(NHEt$_2$)$_2$] is treated with 1 mol equiv. CO. The carbamoyl complex readily reacts with excess CO to give Et$_2$NC(O)C(O)NEt$_2$ via an intramolecular C–C bond forming reaction.[165b] The phosphine derivative [NiI{C(O)NEt$_2$}(PEt$_3$)$_2$] is inactive under these conditions.[165a]

Alkyl- or arylnickel complexes react with alkynes to give vinylnickel species. Treatment of [NiCl(Me)(PMe$_3$)$_2$] with diphenylacetylene in methanol gives the (*Z*)-vinylnickel complex (**46**) which, on heating, equilibrates with the (*E*)-isomer (**47**) (Scheme 10).[72,117] If the reaction is carried out in diethylether a precursor complex [NiCl(Me)(PMe$_3$)$_2$(PhC≡CPh)$_{0.5}$] can be isolated.[117] The alkynes 1-phenylethyne and 1-phenylpropyne selectively produce 1-phenylvinylnickel compounds (Equation (67)).[72] The action of LiR (R = Me, C≡CBu^t) on (**46**) followed by reductive elimination gives the alkenes R(Ph)C=C(Ph)Me, while mineral acids or iodine cleave the Ni–C bond in (**46**) to afford PhCH=C(Ph)Me or Ph(I)C=C(Ph)Me, respectively.[72]

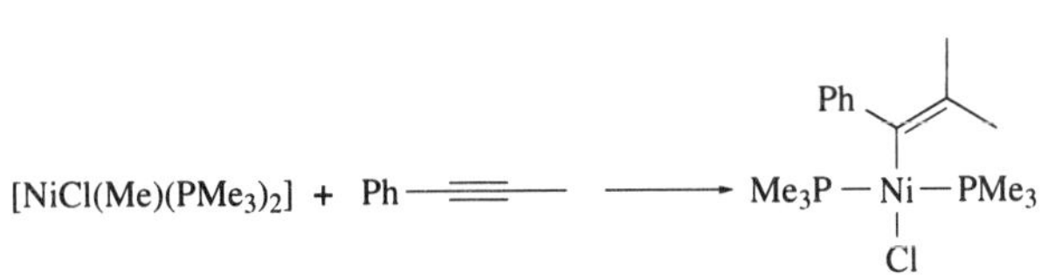

(67)

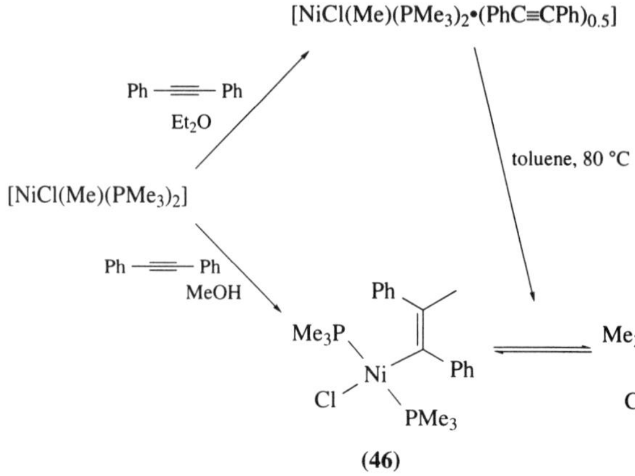

Scheme 10

Alkynyl groups can oxidatively couple at a nickel centre in the complexes [NiX(C≡C-TMS)(PMe_3)$_2$] (X = Cl, Br, I), either by heating in the solid state (for X = Cl or Br) or spontaneously in solution (for X = I) to give (**48**) (Equation (68)).[122] When the complex (**48**) (X = I) is treated with LiC≡CBu^t, (**49**) is obtained, which undergoes a CO-induced reductive elimination to give [Ni(CO)$_2$(PMe_3)$_2$] and (**50**).[122] Diphenylacetylene inserts into the Ni–Ph bond of (**26**) to give the vinylnickel complex (**51**).[141]

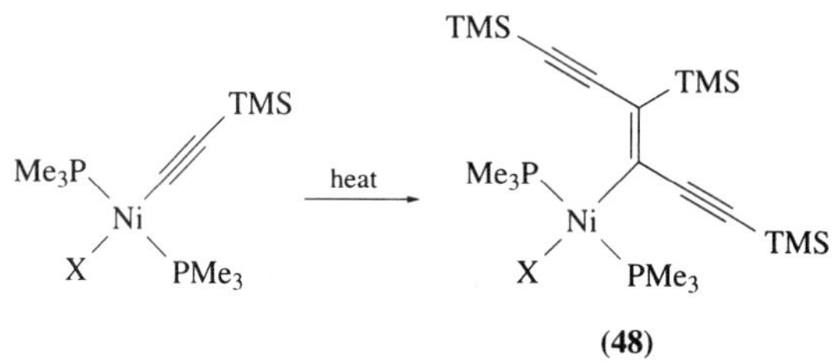

(68)

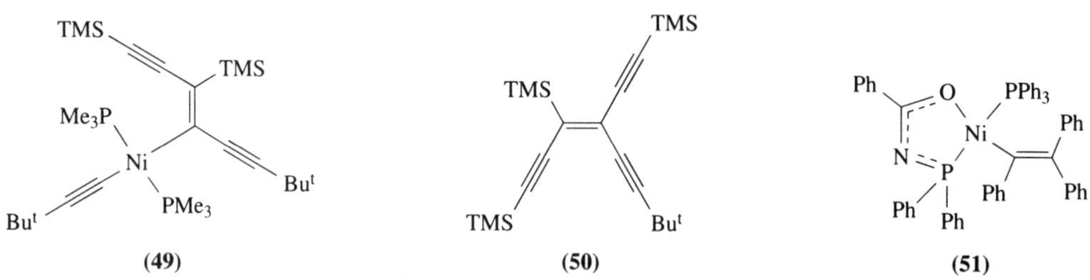

Alkyne insertion into (substituted-aryl)nickel compounds does not occur if the aryl group has *ortho*-substituents. Thus, for example, PhC≡CPh inserts into the Ni–C bond in [NiBr($C_6H_3Cl_2$-3,5)(PPh_3)$_2$], but no reaction occurs with [NiBr($C_6H_2Me_3$-2,4,6)(PPh_3)$_2$] or [NiBr($C_6H_3Cl_2$-2,5)(PPh_3)$_2$].[121] The insertion is also inhibited by the presence of free phosphine. The insertion of $MeO_2CC{\equiv}CCO_2Me$ into the Ni–C bond of [NiBr($C_6H_3Cl_2$-3,5)(PPh_3)$_2$] gives the *cis*-isomer (**52**). With $PhC{\equiv}CCO_2Et$, the only isomer obtained is the *cis*-isomer in which the sterically largest substituent is on the carbon atom bound to nickel.[121]

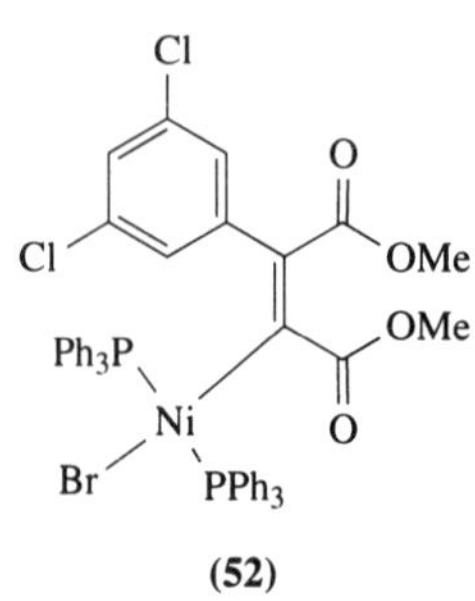

The cationic nickel complex $[Ni(Me)(PMe_3)_4][BF_4]$ does not give an insertion product on reaction with diphenylacetylene, instead reductive elimination yielding $(PMe_3)_4^+$ and $[Ni(PMe_3)_2(PhC{\equiv}CPh)]$ occurs.[174]

Migratory insertion of isocyanides into Ni–C bonds is mentioned only briefly here. The complexes $[NiCl(R)(PMe_3)_2]$ (R = Me or CH_2TMS) react with Bu^tNC to give, successively, the products of mono- and diinsertion into the Ni–C bond (Equation (69)).[117,118] With cyclohexylisocyanide this reaction gives the monoinsertion product only, while benzylisocyanide is polymerized.[117] It has also been found that the monoinsertion product (**53**) (R = CH_2TMS or CH_2Bu^t) undergoes a facile η^1–η^2 interconversion of the alkaneimidoyl ligand by the removal or addition of PMe_3 (Equation (70)).[118] Insertion of *p*-$MeOC_6H_4NC$ into the Ni–C bond of the 2-pyridylnickel complex (**55**) gives an imino(2-pyridyl)methyl complex (**56**) which could not be isolated but which was trapped as the $ZnCl_2$ adduct (**57**) (Equation (71)).[119]

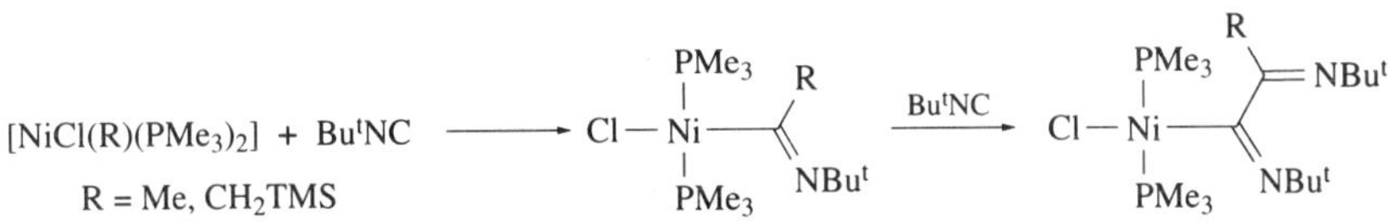

(**53**) (**54**) (69)

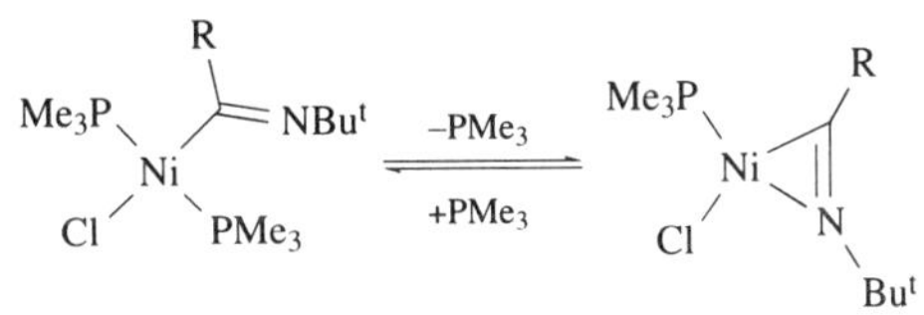

R = CH_2TMS, CH_2Bu^t

(**53**) (70)

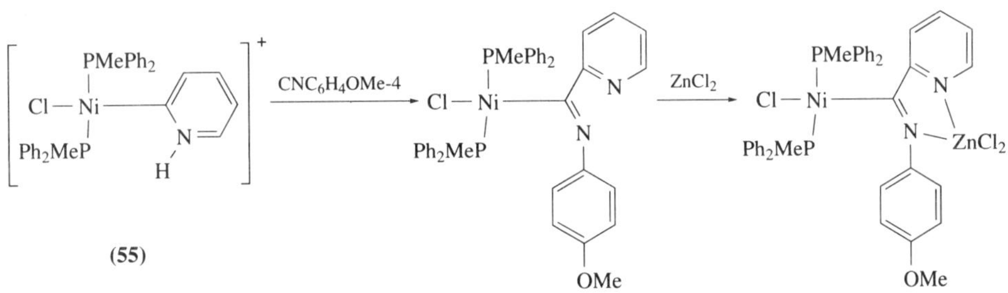

(**55**) (**56**) (**57**) (71)

The reversible reductive elimination and oxidative addition of diarylsulfides leads to aryl ligand exchange between nickel and sulfur in the complexes *trans*-$[Ni(Ar^1)(SAr^2)(PEt_3)_2]$ and *cis*-$[Ni(Ar^1)(SAr^2)(dmpe)]$ (Ar^1 = Ph, Ar^2 = C_6H_4Me-4).[137]

When the complex $[NiCl(CCl{=}CCl_2)(PPh_3)_2]$ is treated with bidentate N-donor ligands L_2 (L_2 = bipy or phen) a symmetrization reaction occurs forming $[NiCl_2(L_2)]$ and $[Ni(CCl{=}CCl_2)_2(L_2)]$.[111] The thermolysis or photolysis of $[Ni(CCl{=}CCl_2)(C_6H_4Y)(PMe_3)_2]$ (Y = NMe_2-4, Me-4, Me-3 or Cl-4) results in the formation of the monoorganonickel complexes $[Ni\{C_2Cl_2(C_6H_4Y)\}(PMe_3)_2]$, the products expected from reductive elimination followed by oxidative addition.[114,115] Reductive elimination and oxidative addition also takes place on treatment of dialkylnickel complexes with alkyl or acyl halides. The reactions with alkyl halides can be classified into three types depending on the products of the reaction (Equations (72)–(74)).[184]

$$[NiR^1{}_2L_2] + R^2X \longrightarrow [NiX(R^2)L_2] + R^1\text{-}R^1 \qquad (72)$$

$$[NiR^1{}_2L_2] + R^2X \longrightarrow \text{alkane } (R^1H) + \text{alkene } (R^2\,(-H)) \qquad (73)$$

$$[NiR^1{}_2L_2] + R^2X \longrightarrow [NiX(R^1)L_2] + R^1\text{-}R^2 \qquad (74)$$

cis-$[Ni(Me)_2(bipy)]$ and *cis*-$[Ni(Et)_2(bipy)]$ react with alkyl halides according to Equation (72); *trans*-$[Ni(Me)_2(PEt_3)_2]$ reacts with ethyl bromide mainly by Equation (73), and with phenyl bromide by Equation (72). *cis*-$[Ni(Me)_2(dppe)]$ reacts with phenyl chloride to give toluene as the major product

(Equation (74)).[184] The *cis* or *trans* nature of the diorganonickel complex is thus of prime importance in determining the course of the reaction, with the *cis*-configuration favouring reductive elimination of R–R. Theoretical studies of reductive elimination of d^8-organotransition metal complexes have been published.[208,209]

The reaction of R^2COY (Y = Cl, Br, OPh or OCOPh) with $[NiR^1{}_2L_2]$ can be classified into two types (Equations (75) and (76)).[184]

$$[NiR^1{}_2L_2] + R^2COY \longrightarrow [NiY(R^1)L_2] + R^1COR^2 \qquad (75)$$

$$[NiR^1{}_2L_2] + R^2COY \longrightarrow [NiY(R^2)L_2] + R^1COR^1 \qquad (76)$$

The nickel(I) complex $[Ni(C_6F_5)(CO)(PPh_3)_2]$ undergoes a disproportionation reaction on treatment with $P(OMe)_3$ to give $[Ni(C_6F_5)_2\{P(OMe)_3\}_2]$.[175]

Treatment of the cyclic alkoxyvinylnickel(II) complexes $[Ni(R)(\overline{C{=}CHCH_2CH_2O})L_2]$ ($R = C_6H_2Me_3$-2,4,6 or C_6Cl_5) with perchloric acid gives the corresponding cationic cyclic carbene complexes (Equation (77)).[170,171]

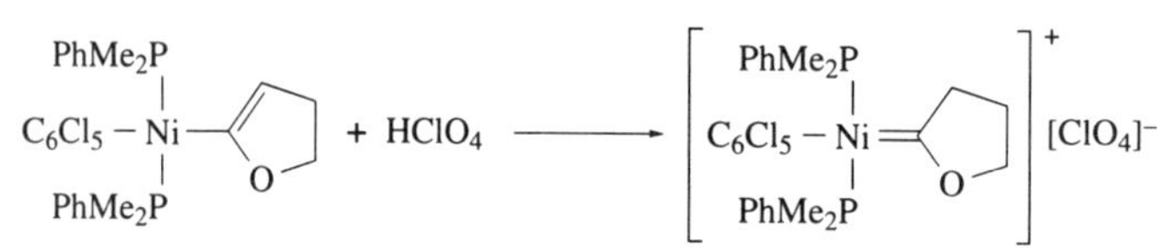

(77)

2.4.2.3 *Reactions in which the alkyl or aryl group is displaced*

The reactions in this section are mainly concerned with thermolysis,[85,92,113,128,140,166] protonolysis,[108,140,155] and reductive elimination induced by phosphines,[107,130,187,204] carbon monoxide[89,91,102,174] or alkenes.[133,146] The thermolysis of dialkyl- or diarylnickel(II) complexes shows two important pathways, reductive elimination with C–C bond formation and β-hydrogen elimination with the formation of alkenes. Protonolysis generally leads to the elimination of RH from an organonickel complex. The reaction of $[Ni(Me)_2(PPh_3)_2]$ with Bu^tOH leads to the liberation of methane and the formation of $[Ni(Me)(OBu^t)(PPh_3)_2]$. The thermolysis of the dimethylnickel complex probably proceeds by a metallacyclic process to give methane, ethane, 2,2-dimethyloxirane, 2-methylpropene and 2-methyl-2-butene.[85] The thermolysis of the ethylnickel complexes $[Ni(Et)(NR^1R^2)(bipy)]$ liberates ethane, butane and ethene, whereas exposure of the complex to air results in the formation of ethene as the main gaseous decomposition product.[92] The stability of the monoarylnickel complexes $[NiBr(Ar)(PPh_3)_2]$ ($Ar = C_6H_nCl_{5-n}$) depends largely on steric effects. Where Ar contains two *ortho*-chlorine atoms the compounds are more stable and give only RH on decomposition. However if an *ortho*-hydrogen is present, R–R is obtained on decomposition,[128] the usual decomposition pathway for arylnickel halide complexes.[113,130] The decomposition of vinylnickel halides occurs by coupling with isomerization of the double bond, except for 3-haloacrylates in which retention of geometry is observed in the coupling reaction.[113] Treatment of $[Ni(Me)(Ph)(PEt_3)_2]$ with H_2S gives $[Ni(Ph)(SH)(PEt_3)_2]$ and methane, and the thermolysis of the phenylmercaptonickel complex leads to the formation of benzene, Et_3PS and a small amount of PhSPh.[140]

The cationic complex $[Ni(Me)(PMe_3)_4]^+$ decomposes thermally to give methane as the major product (methane:ethane ratio 4:1), but photochemical decomposition gives mainly ethane (methane:ethane ratio 1:3), suggesting a different pathway.[166]

Reductive elimination from *cis*-dialkylnickel complexes induced by CO, PR_3 or alkenes occurs via an associative mechanism (Equation (78)).[179,204,213] The corresponding *trans*-dialkylnickel complexes are more stable to reductive elimination since the five-coordinate intermediates (Equation (79)) either have *trans*-alkyl groups or the alkyl groups occupy equatorial sites in the trigonal bipyramidal structure when reductive elimination is symmetry forbidden.[210]

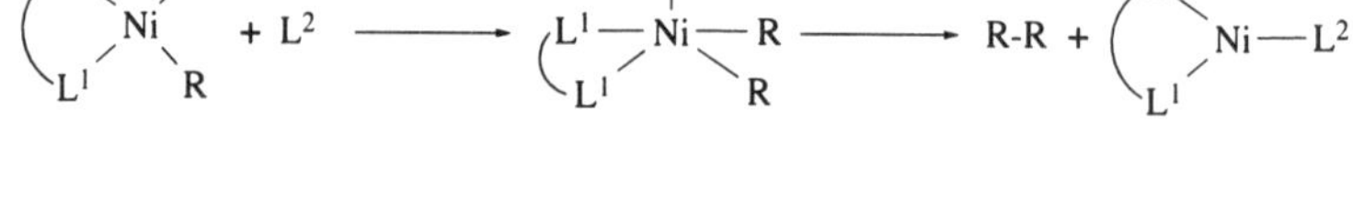

(78)

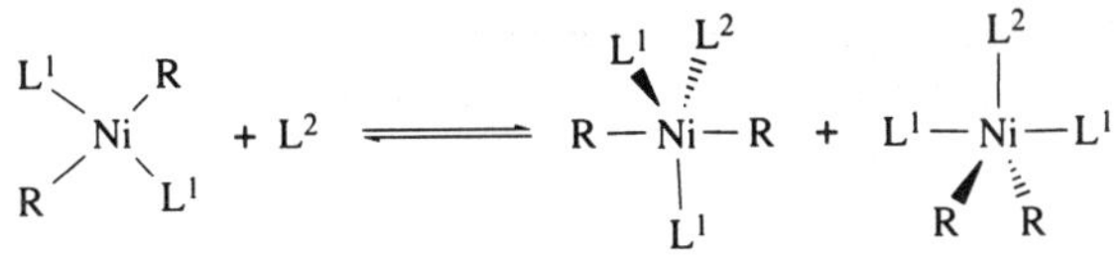

(79)

The phosphine-induced reductive elimination of a binuclear nickel complex has been mentioned earlier (Scheme 5).[107]

Reductive elimination of $[Ni(Me)_2(TMEDA)]$ is induced by strong π-acceptor molecules such as methyl acrylate, methyl vinyl ketone, acrylonitrile, tetracyanoethene, tetrafluoroethene or maleic anhydride, to give ethane and [Ni(TMEDA)(π-ligand)$_n$].[183]

Carbon monoxide induced reductive elimination of ketones (Equation (80)),[102] carboxylic esters (Equation (81)),[89] or carboxylic anhydrides (Equation (82))[91] has been observed. In general, the products of the reaction of $[NiR_2L_2]$ with CO depend upon reaction conditions, geometry, and the reactivity of the Ni–C bonds involved.

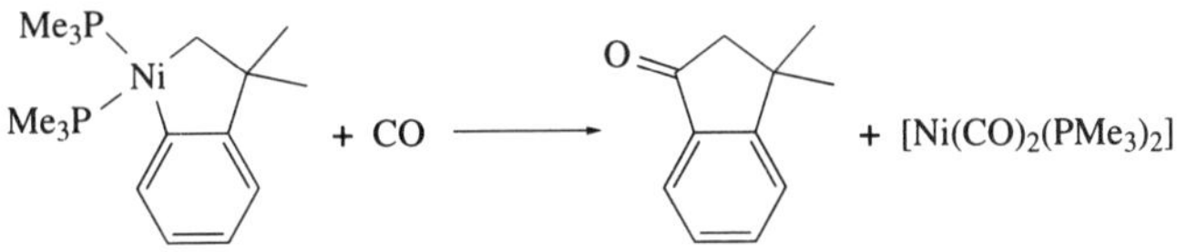

(80)

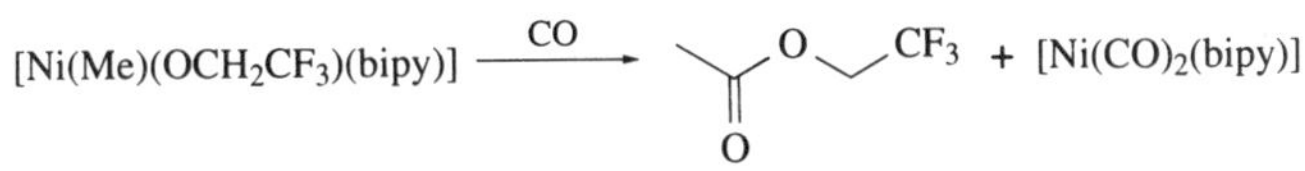

(81)

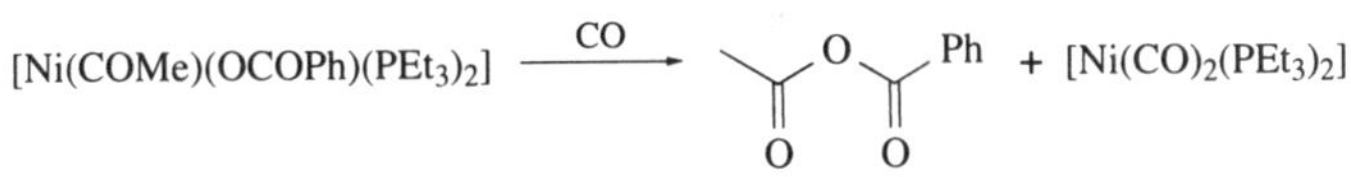

(82)

When $[Ni(Me)_2(bipy)]$ is treated with hexafluoroacetone or 2,2,2-trifluoroacetophenone, ethane is eliminated and a nickel complex with a η^2-ketone ligand is formed (Equation (83)).[211]

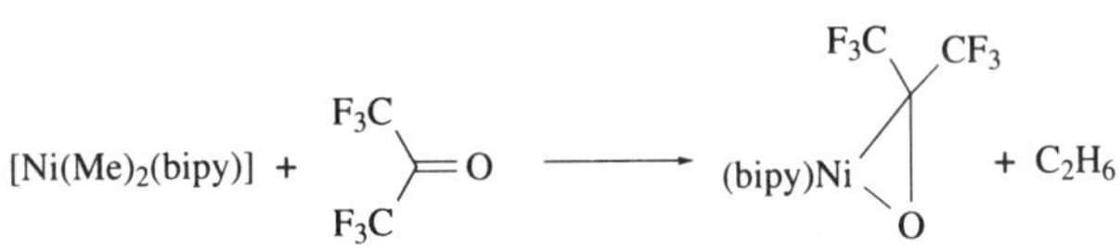

(83)

Nickel(II) complexes are effective catalysts for the oligomerization of alkenes,[176] but the factors affecting the activity and selectivity of the active catalyst towards a given alkene are still poorly defined. A study of the reaction of ethene with complexes of the type $[NiX(R)L_2]$ with R = Mes, $C_6H_3Cl_2$-3,5, $C_6H_3Cl_2$-2,3, C_6Cl_5 or C_6H_4Me-2, has been reported. Under ethene pressure migratory insertion and β-elimination of vinyl–R or butenyl–R products was observed.[146] The phenylnickel complexes (**24**) and (**26**) react with mixtures of CO_2 and ethene via insertion of CO_2 and/or ethene into the Ni–Ph bond to give, after esterification with methanol, methyl cinnamate, methyl benzoate, styrene, ethylbenzene and butylbenzene.[133]

The complexes $[NiX(C_6Cl_5)(PCy_3)_2]$ (X = Cl, I, NCS, NCO or N_3) react with HCl to give C_6Cl_5H.[155] The *N*-aryl-1-isoquinolone (**59**) is obtained by the hydrolytic cleavage of the nickel–vinyl bond in (**58**) (Scheme 11).[108]

The Ni–C bond is also cleaved on reaction with halogens (Equation (84))[148,206] or by electrochemical methods.[125,129,185,206] Thus, for example, electrochemical reduction of $[Ni(Et)_2(phen)]$ causes the catalytic reductive coupling of excess ethyl bromide in solution.[185] The electrogenerated nickel(I) complex [Ni(Ph)(dppe)] undergoes an oxidative addition reaction with phenyl bromide to give the nickel(III) complex $[NiBr(Ph)_2(dppe)]$ which reductively eliminates biphenyl to produce [NiBr(dppe)].[129]

$$[Ni(Me)(Mes)(bipy)] + I_2 \longrightarrow [NiI_2(bipy)] + C_6H_2Me_4 \quad (84)$$

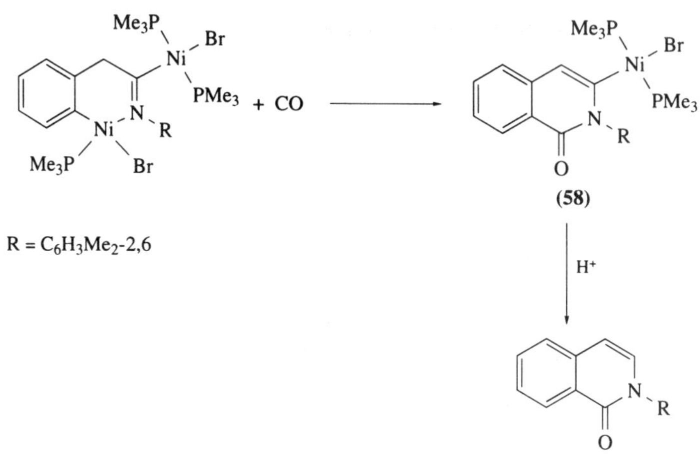

(58)

(59)

Scheme 11

2.4.3 Structure and Properties

This section deals with aspects of the structure and properties of organonickel compounds that have not been dealt with in previous sections.

Perfluorophenyl groups tend to be distorted at the carbon atom bonded to a metal, so that if a rigid, regular hexagon is used in x-ray structure refinement, misleading values for the M–C bond length can result.[212] For *cis*- and *trans*-$[MX(Y)(PR_3)_2]$ planar complexes (M = Ni, Pd or Pt), deviations in the P–M–P bond angles from ideal correlate linearly with the average PR_3 tilt angle.[213] The NMR parameters (^1H, ^{13}C, ^{31}P) of alkynyl complexes $[M(C{\equiv}CR^1)_2(PR^2{}_3)_2]$ (M = Ni, Pd, Pt) indicate π-back-bonding from metal to the π* orbitals of the alkynyl ligands and that this increases in the order Pt < Pd < Ni.[191] The ^1H NMR parameters for the *m*-tolylnickel complexes *trans*-$[NiR(C_6H_4Me\text{-}3)(PMe_2Ph)_2]$ and *trans*-$[NiR(C_6H_4Me\text{-}3)(PMe_3)_2]$ indicate that the *m*-tolyl group is orientated perpendicularly to the nickel coordination plane in the PMe_2Ph derivatives but rotate freely in the PMe_3 derivatives.[112]

The crystal structures of the *syn*- and *anti*-isomers of $[Ni(C_6H_4Me\text{-}2)_2(PMe_2Ph)_2]$ have been determined. It appears that there are negligible intramolecular repulsive forces due to the adjacent tolyl methyl groups in the *syn*-isomer.[196] A comparison between the Ni–C bond lengths in the vinylnickel complexes (**60**) and (**62**) and the corresponding carbenenickel complexes (**61**) and (**63**) has established a shortening of 0.007 nm in the carbene complexes indicative of the multiple bond character of the nickel–carbene bond.[172,173]

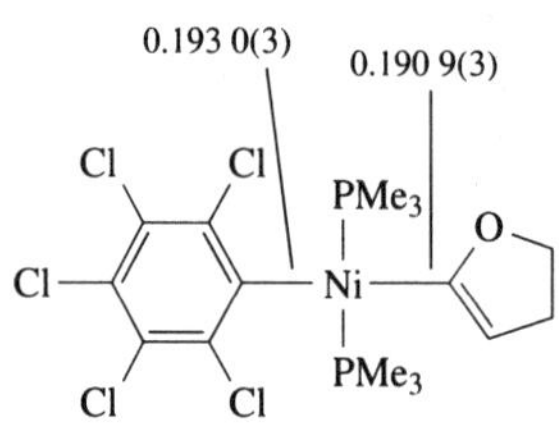

(60)

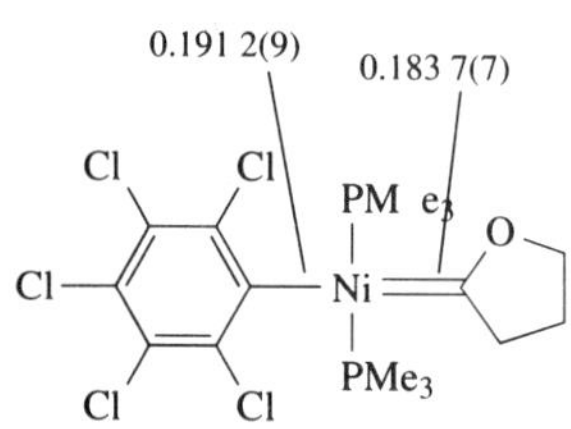

(61)

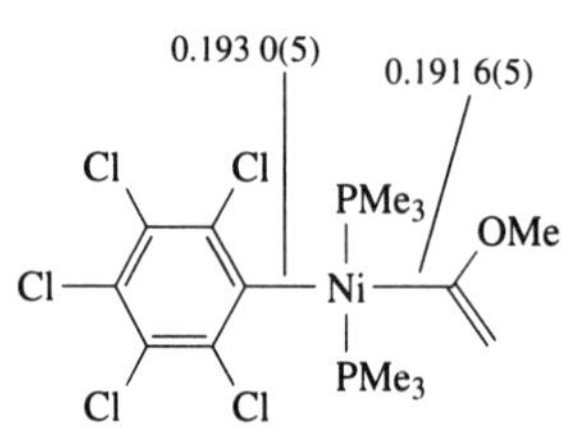

(62)

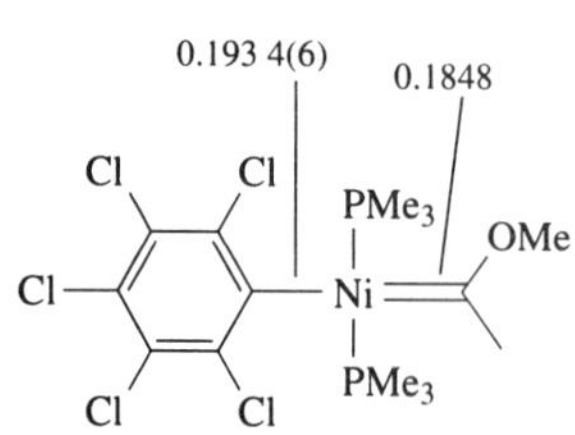

(63)

Nitrosodurene reacts with σ-alkyl- and σ-arylnickel compounds to form nitrosoduryl radicals, whereas alkenenickel complexes do not react by formation of radicals. Thus this spin-trapping method can be used as an indicator for the determination of the presence of σ-organo ligands in nickel compounds and for the determination of the relative reactivity of σ- and π-bonded organo-groups.[214,215]

Poly-2,5- and poly-2,6-pyridine coatings on glassy carbon electrodes can be produced from the nickel complexes [NiBr(2-py-5-Br)(PPh_3)$_2$] and [NiBr(3-py-6-Br)(PPh_3)$_2$], respectively.[163] The electrochemical reduction of [NiBr(C_6H_4Br-4)(PPh_3)$_2$] in acetonitrile gave a polymer coating of nickel containing poly-*para*-phenylene, in which nickel–aryl units are present. The polymer coating undergoes a reversible two-electron reduction and a reversible one-electron oxidation.[144]

2.5 η^3-ALLYLNICKEL ALKYL AND ARYL COMPLEXES

The compounds considered in this section are listed in Table 8.

Table 8 [NiR(η^3-allyl)] and [NiR(L)(η^3-allyl)] complexes.

R	*η^3-allyl*	*Ligand (L)*	*Reference*	*Ni–C* (nm)
Me	MeCHCHCHMe		216	
		CO	216	
		$(CO)_2$	216	
		PR_3[a]	217–219	
CH_2Bu^t	$CH_2C(Me)CH_2$	PMe_3	220	
CH_2TMS	$CH_2C(Me)CH_2$	PMe_3	220	
$CH_2C_6H_4$Me-2	$CH_2C(Me)CH_2$	PMe_3	220	
Ph	CH_2CHCH_2	PPh_3	221	
$C_6H_3Cl_2$-2,5	CH_2CHCH_2	PPh_3	221,222	
C_6F_5	$CH_2C(Me)CH_2$	PPh_3	221,222	
		dppe	221,222	
		dppen[b]	221,222	
	$CH_2CHCHMe$	PPh_3	221,222	
		dppen	221,222	
η^3-$CH_2C(Me)C(Me)CH(Ph)$-η^1-CHCH=NPh (**64**)			223	0.205

[a] 28 different ligands. [b] dppen = 1,2-bis(diphenylphosphino)ethene.

Allylnickel complexes are key intermediates in a number of catalytic and stoichiometric reactions and several investigations of the effect of supporting ligands on the reactivity of such complexes have been reported. In an investigation of optical induction, it has been shown that the complex [Ni(Me)(η^3-MeCHCHCHMe(L)], where L is a chiral P-ligand, reacts with CO to give the optically active 3-methyl-(*E*)-4-hexene-2-one. The extent and direction of optical induction depends on the concentration and type of P-ligand.[216] This reaction was too fast for any intermediates to be observed. However, in the absence of the chiral P-ligand, CO reacts with [{Ni(Me)(η^3-MeCHCHCHMe)}$_2$] to give [Ni(Me)(η^3-MeCHCHCHMe)$(CO)_2$] which was identified by ^{13}C NMR spectroscopy at −125 °C. On warming to −80 °C, the monocarbonyl derivative [Ni(Me)(η^3-MeCHCHCHMe)(CO)] is observed.[216] The influence of phosphine ligands on CO insertion into the allyl–nickel bond[217,218] and thermal stability[218,219] in [Ni(Me)(η^3-MeCHCHCHMe)(L)] complexes has been investigated. The electronic and steric effects of a wide range of ligands L could be separated using a multilinear regression analysis. An increase in donor ability of the phosphine ligand favours CO insertion and an increase in steric hindrance leads to a decrease in ketone formation.[217]

The alkyl(allyl)nickel complexes [Ni(R){η^3-$CH_2C(Me)CH_2$}(PMe_3)] (R = CH_2Bu^t, CH_2TMS, or $CH_2C_6H_4$Me-2) have been synthesized by treating the corresponding allylnickel chloride complex with the appropriate Grignard reagent.[220]

The phosphine-induced reductive elimination observed for dialkylnickel complexes (Section 2.4.2.3) also occurs in alkyl(allyl)nickel complexes.[221,222] Thus, while complexes of the type [Ni(C_6F_5){η^3-$CH_2C(Me)CH_2$}(PPh_3)] undergo reductive elimination at a rate 26 times faster than the corresponding palladium complex, in the presence of dppe, when the 18-electron complex [Ni(C_6F_5){η^3-$CH_2C(Me)CH_2$}(dppe)] is formed (this complex could not be isolated but was characterized spectroscopically), a rate enhancement of 10^8 was observed. The activation energy for reductive elimination from the 18-electron complex was found to be 59 ± 4 kJ mol^{-1} compared to a value of 122 ± 10 kJ mol^{-1} for the 16-electron complex.

An oxidative coupling reaction between cinnamaldehydanil and 1,3-dienes occurs on reaction with $[Ni(cod)_2]$ to give the binuclear complex (**64**).[223]

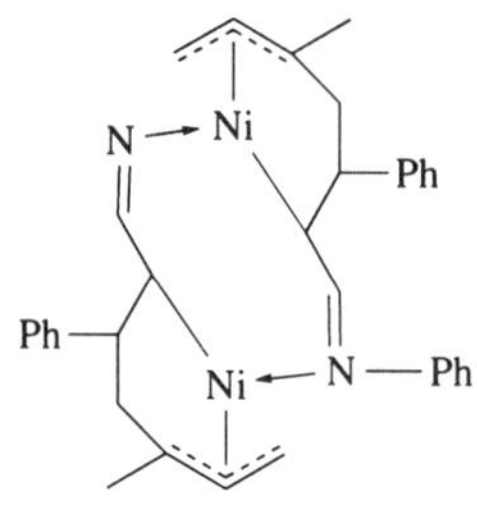

(64)

2.6 η-CYCLOPENTADIENYLNICKEL ALKYL AND ARYL COMPLEXES

The compounds considered in this section fall mainly into two categories, compounds of the type [NiR(alkene)(Cp)] and $[NiR^1(PR^2_3)(Cp)]$. These compounds are listed in Table 9.

2.6.1 Synthesis

Compounds of these types are usually prepared by the action of organolithium or organomagnesium reagents on either [NiCl(L)(Cp)] (Equations (85)–(88))[98,101,239–41,244] or on nickelocene in the presence of an alkene (Equations (89)–(92))[224,226,227,229,232,233,236,237,251,252] or an alkyne (Equation (93)).[234,235,238] A number of other preparative routes have been employed, these include the reaction of NaCp with $[NiX(R)(L)_2]$ (Equation (94)),[101,104] the reaction of Cp*H with $[Ni(C_2H_4)_3]$ (Equation (95)),[227] treatment of $NiCl_2$ with $[Li\{C_5H_4(CH_2)_3CH{=}CH_2\}]$ followed by LiR (Scheme 12),[227] and the cyclodimerization of an alkyne with $[Ni(\eta^2\text{-}S{=}PR_2)(Cp)]$ (Equation (96)).[243]

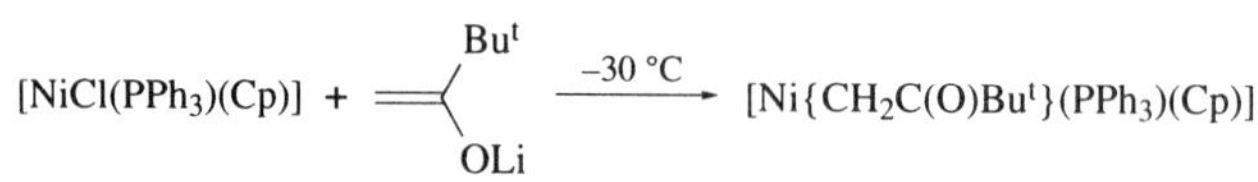

(85)

$$[NiCl(PPh_3)(Cp)] + PhSCH_2Li \longrightarrow [Ni(CH_2SPh)(PPh_3)(Cp)] \quad (86)$$

$$[NiCl(PPh_3)(Cp)] + PhSO_2CH_2Li \longrightarrow [Ni(CH_2SO_2Ph)(PPh_3)(Cp)] \quad (87)$$

$$[NiCl(PPh_3)(Cp)] + CH_2{=}CHLi \longrightarrow [(Cp)(PPh_3)Ni(CH_2CH{=}CHCH_2)Ni(PPh_3)(Cp)] \quad (88)$$

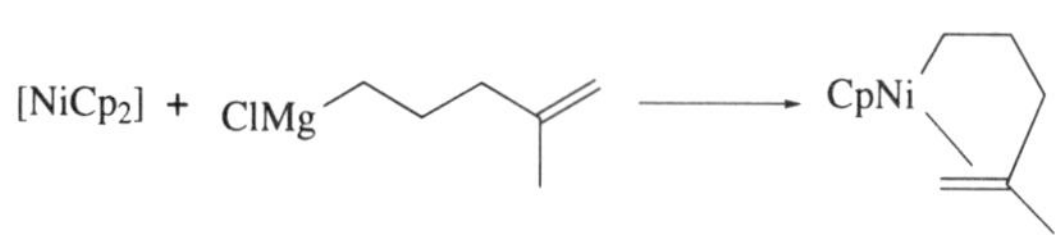

(89)

(65)

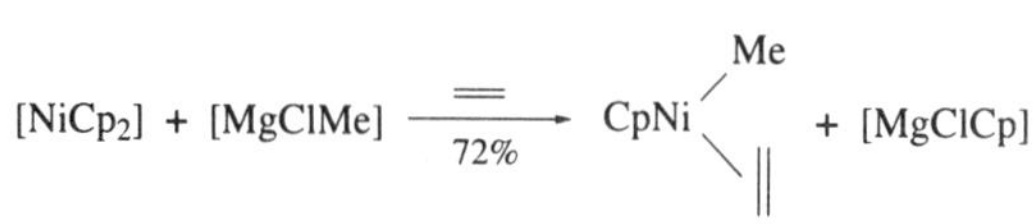

(90)

(66)

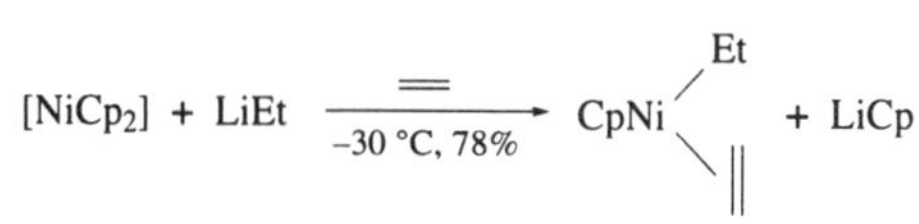

(91)

An improved synthesis of $[Ni(C{\equiv}CPh)(PPh_3)(Cp)]$, giving a 72% yield, is the CuI-catalysed reaction of $[NiCl(PPh_3)(Cp)]$ with PhC≡CH in triethylamine.[245] A new route, giving a 50% yield, to

Table 9 [NiR(L)(η-Cp)] complexes.

R	*L*	*Reference*	*Ni–C* (nm)
Me	η-C_2H_4 (**66**)	224–227	
	η-C_2H_4[a]	227	
	η-C_2H_4[b]	227	
	η²-CH_2=CHMe	226,228–230	
	η²-CH_2=$\overline{CCH_2CH_2}$ (**73**)	231	
	η²-CH_2=CHEt	232	
	cis-η²-MeCH=CHMe	232	
	trans-η²-MeCH=CHMe	232	
	η²-CH_2=CHPh	226	
	η²-CH_2=CH-TMS	226	
	η²-CH_2=CHBu[t]	226	
	η²-CH_2=CMe_2	226	
	η²-CH_2=$\overline{CCH_2CH_2CH_2}$	226	
	η²-CH_2=$\overline{CCH_2CH_2CH_2CH_2}$	226	
	η²-CH_2=CH$(CH_2)_3C_5H_4$	227	
	η²-CH_2=CH$(CH_2)_3C_5Me_4$	227	
	η²-cyclohexene	232	
	η²-CH_2=CHCH=CH_2	232,233	
	η²-CH_2=CHCH=CH_2[a]	227	
	η²-CH_2=CMeCH=CH_2	232,233	
	η²-CH_2=CHC(Me)=CH_2	226	
	η²-CH_2=CMeC(Me)=CH_2	226,232,233	
	η²-CH_2=CMeCH=CMe_2 (**70**)	226	
	cis-η²-CH_2=CHCH=CHMe	226	
	trans-η²-CH_2=CHCH=CHMe	226	
	η²-cod	232	
	η²-1,3-cyclohexadiene	232	
	η²-1,4-cyclohexadiene	232	
	η²-CH_2=CHCH_2CH_2CH=CH_2	232	
	η²-cdt	232	
	MeC≡CMe	234,235	
	μ-MeC≡CMe (**76**)	234	
	TMS-C≡C-TMS	234,235	
	PhC≡CPh	235	
	MeCN	234	
CD_3	η²-CH_2=CHMe	226,228–230	
Et	η-C_2H_4	224–226,236	
	η-C_2H_4[a]	227	
	η-C_2H_4[c]	226	
	η-C_2H_4[d]	226	
	η²-CH_2=CHMe	226,228–230	
	η²-CH_2=CH$(CH_2)_3C_5H_4$	227	
	η²-CH_2=CH$(CH_2)_3$Cp*	227	
Pr	η-C_2H_4	224–226	
	η²-CH_2=CHMe	226,228–230	
$CH(Me)_2$	η-C_2H_4	224,226,236	
	η²-CH_2=CHMe	226,229,236	
	η²-CH_2=CH$(CH_2)_3C_5H_4$	227	
Bu[n]	η-C_2H_4	225,226,236	
Bu[i]	η-C_2H_4	226,236	
	η²-CH_2=CHMe	226	
CH_2TMS	η-C_2H_4	224,226	
	η²-CH_2=CHMe	226,228–230	
CH_2Bu[t]	PMe_3	101	
$CH_2Si(Me)_2Ph$	PCy_3	98	
$CH_2C(Me)_2Ph$	PPh_3	98,101	
CH=CH_2	η-C_2H_4[a] (**67**)	237a	
$\overline{CHCH_2CH_2}$	η-C_2H_4	226,231,236	
	η²-CH_2=CHMe	226	
$CH_2CH_2\overline{CHCH_2CH_2}$	η-C_2H_4	226,231	
CH_2CH_2Ph	η-C_2H_4	225,226,231	
$CH_2CH_2C_6H_4$(TMS)-4	η-C_2H_4	225	
	η-C_2D_4	225	
$CH_2CH_2CH_2CH_2Ni$	η-C_2H_4	226	
C(Me)=CMe_2	η²-Me_2C=C(Me)Ni (**79**)	238	
CH_2CO_2Et	PMe_3	104	

Table 9 (continued)

R	*L*	*Reference*	*Ni–C* (nm)
CH_2CN	PMe_3	104	
CH_2OMe	PMe_3	104	
$CH_2C(O)Bu^t$	PPh_3	239	
$CH_2C(O)Ph$	PPh_3	239	
$CH(Me)C(O)Bu^t$	PPh_3	239	
$CH_2C(O)OBu^t$	PPh_3	239	
$CH_2C(O)OBu^t$	PPh_3[a]	239	
CH_2SPh	PPh_3	240	
CH_2SO_2Ph	PPh_3	241	0.195 6(4), 0.195 9(4)
$CH(Me)SO_2Ph$	PPh_3	242	
$CH(Et)SO_2Ph$	PPh_3	241	
$CH(CH{=}CMe_2)SO_2Ph$	PPh_3	241	0.198 1(6)
$CH(C_6H_{13})SO_2Ph$	PPh_3	241	
$CH(C_{11}H_{23})SO_2Ph$	PPh_3 (**82**)	241	
$C(CO_2Me){=}C(CO_2Me)P(Mes)_2S$	S	243	
$C(CO_2Me){=}C(CO_2Me)P(xylyl)_2S$	S	243	
$CH_2CH{=}CHCH_2Ni$	PPh_3	244	
C≡CPh	PPh_3 (**81**)	245	
$C\{{=}C(CN)_2\}C(Ph){=}C(CN)_2$	PPh_3	246	
$C(Me){=}CMe_2$		235	
C(Ph)=C(Ph)Me		235	
C(TMS)=C(TMS)Me		235	
Ph	η-C_2H_4	225,226,231	
	η-C_2H_4[a]	225	0.191 3(2)
	η$_2$-$CH_2{=}CHCH{=}CH_2$	225	
	PPh_3	247	
C_6H_4Me-4	η-C_2H_4, η-C_2H_4[a]	225	
	η^2-$CH_2{=}CHCH{=}CH_2$	225	
$C_6H_4Bu^t$-4	η-C_2H_4	225	
	η^2-$CH_2{=}CHCH{=}CH_2$	225	
$C_6H_4Bu^t$-2	PPh_3	98	
C_6H_4TMS-4	η-C_2H_4	225	
C_6H_4OMe-4	η-C_2H_4	225	
$C_6H_4NMe_2$-4	η-C_2H_4	225	
C_6H_4F-4	η-C_2H_4	225	
C_6H_4Cl-4	η-C_2H_4	225	
C_6H_4Br-4	η-C_2H_4	225	
$C_6H_4CH_2NMe_2$-2 (**80**)		248	
C_6H_4-2-N=N(Ph)		249	
$C_6H_4N{=}NC_5H_4Fe(Cp)$ (**68**)		250	
$C_6H_2Me_3$-2,4,6	PMe_3	103	
η^1,η^2-$CH_2CH_2CH{=}CH_2$		237	
η^1,η^2-$CH_2CH_2CH{=}CH_2$[a] (**71**)		237a	
η^1,η^2-$CH_2CH_2CH_2CH{=}CHMe$ (**65**)		251	
η^1,η^2-$CH_2C(Me)_2CH{=}CH_2$		251	
η^1,η^2-$CH(Me)CH_2CH_2CH{=}CH_2$		251	
η^1,η^2-$CMe_2CH_2CH_2C(Me){=}CH_2$		251	
η^1,η^2-$CH_2CH_2C(Me){=}CH_2$ (**74**)		231,251	
η^1,η^2-$CH_2CH(Me)C(Me)_2CH{=}CH_2$		251	
η^1,η^2-$CH_2CH_2CH_2C(Me){=}CH_2$		251	
η^1,η^2-$CH_2CH_2CH_2CH{=}CH_2$		251	
η^1,η^2-$CH_2CH_2CH(Me)CH{=}CH_2$		251	
η^1,η^2-$CH_2CH_2CMe_2C(Me){=}CH_2$		251	
η^1,η^2-$CHCMe_2CCH_2CH{=}CH_2$ (**69**)		252	
η^1,η^2-$CHCMe_2CHCH_2C(Me){=}CH_2$		252	
η^1,η^2-$CHCMe_2CHCH(Me)CH{=}CH_2$		252	
η^1,η^2-$CHCMe_2CHC(Me)_2CH{=}CH_2$		252	
η^1,η^2-$CHCMe_2CHCH(Me)CH{=}CHMe$		252	
η^1,η^2-$CHCMe_2CHCH(Ph)CH{=}CH_2$		252	0.193 4(3)
η^1,η^2-$CHCMe_2CHCH(CH{=}CH_2)CH{=}CH_2$		252	
η^1-$CHC(Me)_2CHCH_2CH{=}CH_2$	PMe_3	252	
η^1-$CHC(Me)_2CHCH(Me)CH{=}CH_2$	PMe_3	252	
η^1-$CHC(Me)_2CHC(Me)_2CH{=}CH_2$	PMe_3	252	
η^1-$CHC(Me)_2CHCH_2C(Me){=}CH_2$	PMe_3	252	
η^1-$CHC(Me)_2CHCH(Ph)CH{=}CH_2$	PMe_3	252	

Table 9 (continued)

R	*L*	*Reference*	*Ni–C* (nm)
(4,5,8-η;4',5',8'-η)-bi-1,1'-cyclooctenyl[a]		253	
$C{=}CHC(O)CPh_2$	PPh_3 (**83**)	254	
$C{=}CHC(O)CH_2$	PPh_3 (**83**)	254	
$C{=}C(Ph)C(O)CPh_2$	PPh_3 (**83**)	254	
$C{=}C(Ph)C(O)CH_2$	PPh_3 (**83**)	254	
$C{=}C(C_6H_4Me\text{-}4)C(O)CPh_2$	PPh_3 (**83**)	254	
$C{=}C(C_6H_4Br\text{-}4)C(O)CPh_2$	PPh_3 (**83**)	254	
$C{=}C(CH{=}CH_2)C(O)CPh_2$	PPh_3 (**83**)	254	
$C{=}C(Me)C(O)CPh_2$	PPh_3 (**83**)	254	
$C{=}CHC(O)C(Ph)Me$	PPh_3 (**83**)	254	
$C{=}C(Ph)C(O)C(Ph)H$	PPh_3 (**83**)	254	
$C{=}C(Ph)C(O)C(Me)H$	PPh_3 (**83**)	254	
CH_2PPh_2Ni	PPh_2CH_2Ni	255	0.193 5(6), 0.195 2(6)
$CH_2CH_2CH_2PPh_2$		255	
$CH_2CH_2CH_2CH_2PPh_2$		255	
$\eta^1,\eta^2\text{-}C_7H_8PMe_3$ (**84**)		256	
$\eta^1,\eta^2\text{-}C_{12}Cl_4(OMe)_2(CO_2Me)_2$ (**86**)		257	0.197 2(2)
$C(O)CH_2Bu^t$	PMe_3	101	
$C(O)CH_2OMe$	PMe_3	104	
$C(O)CH_2CO_2Et$	PMe_3	104	
$C(O)CH_2Fe(CO)_2(Cp)$	CO	427	

[a] η-Cp* complex. [b] $\eta\text{-}C_5Ph_5$ complex. [c] $\eta^5\text{-}C_5H_4Bu^t$ complex. [d] $\eta^5\text{-}C_5H_4Et$ complex.

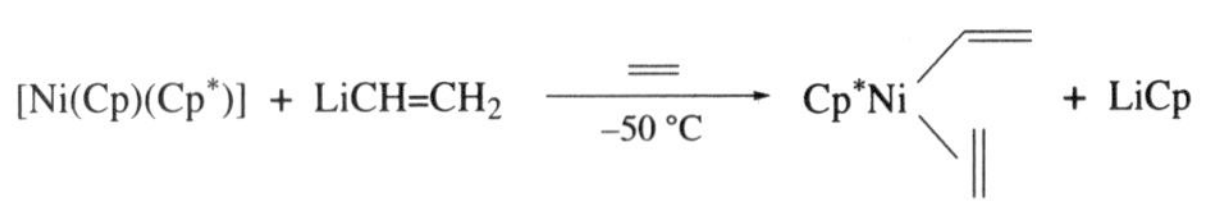

(92)

(**67**)

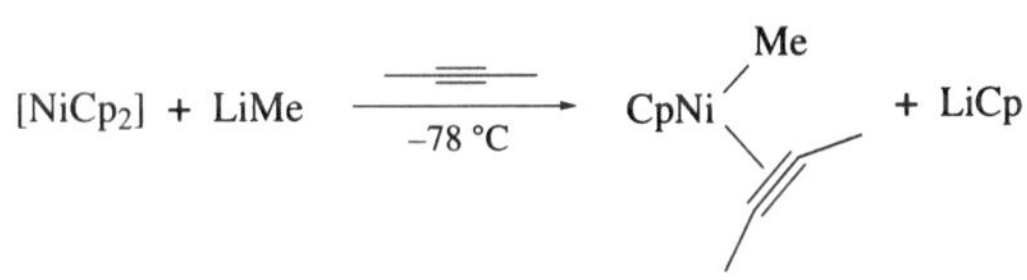

(93)

$$[NiCl(CH_2CN)(PMe_3)_2] + NaCp \longrightarrow [Ni(CH_2CN)(PMe_3)(Cp)] + NaCl \quad (94)$$

$$[Ni(C_2H_4)_3] + Cp^*H \longrightarrow [Ni(Et)(C_2H_4)(Cp^*)] \quad (95)$$

$$[NiCl_2] + [Li\{C_5H_4(CH_2)_3CH{=}CH_2\}] \longrightarrow [Ni\{\eta\text{-}C_5H_4(CH_2)_3CH{=}CH_2\}_2]$$

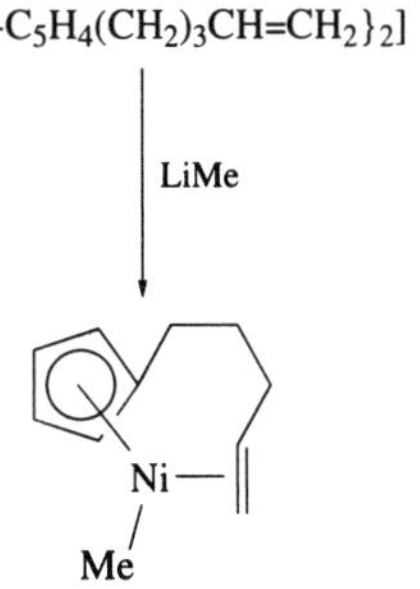

Scheme 12

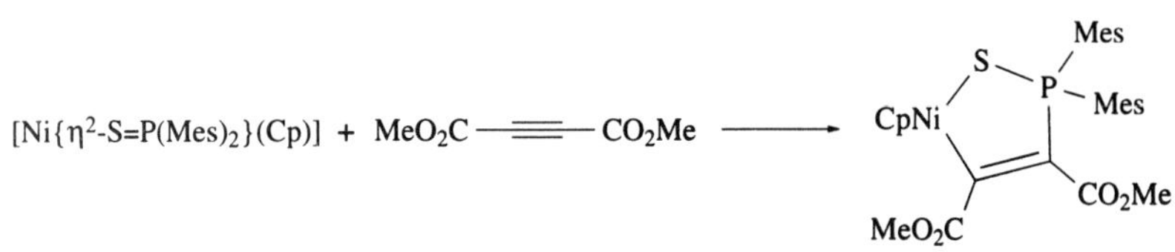

(96)

[Ni(Ph)(PPh_3)(Cp)] is to treat $NiCl_2$ with MgBrPh in the presence of PPh_3 at −60 °C, followed by the addition of cyclopentadiene.[247] The substitution of CO in the complex [NiI(CO)(Cp)] by the functional phosphine $Ph_2PCH_2CH_2CH_2Cl$ gives [NiI($Ph_2PCH_2CH_2CH_2Cl$)(Cp)] which, on treatment with sodium amalgam, forms the cyclic product [Ni($CH_2CH_2CH_2PPh_2$)(Cp)].[255] An unusual reaction is the formation of (**68**) from the treatment of nickelocene with phenylazoferrocene (Equation (97)).[250]

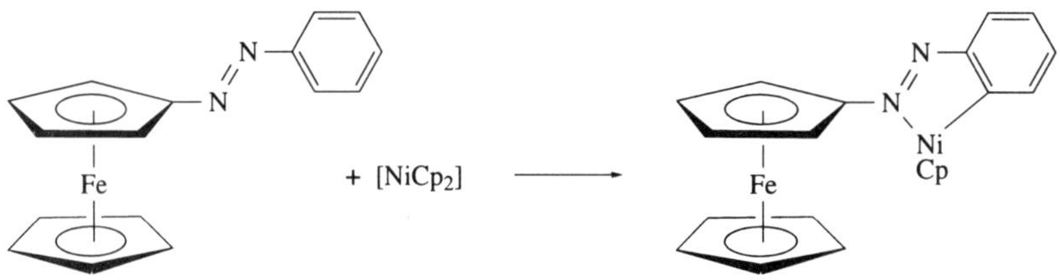

(97)

(68)

2.6.2 Reactions and Properties

Some of the chemistry of complexes of the type [Ni(R)(alkene)(Cp)] and [Ni(η^1,η^2-alkenyl)(Cp)] has been reviewed.[258,259] The ^{1}H NMR spectra of a series of η^1,η^2-alkenyl- and η^3-cycloalkenyl(cyclopentadienyl)nickel complexes have been analysed.[260]

Among the compounds of the type [Ni(R)(alkene)(Cp)], those compounds in which the alkyl group is linked to the alkene, η^1,η^2-alkenylnickel compounds such as (**65**), are much more thermally stable than compounds in which the alkyl and alkene are separate. The η^1,η^2-alkenyl compounds are thermally stable up to 70 °C and can be distilled under vacuum as red oils. The coordination geometry is pseudo-trigonal planar with a rigid conformation of the alkenyl group. This prevents a β-elimination reaction unless the Ni–C bond is broken. This is a major factor in the increased thermal stability of these compounds. At higher temperatures, when alkene decomplexation occurs, the result is the formation of an η^3-allylnickel complex via a series of β-elimination and alkene addition reactions (Scheme 13).[258] If the α- and β-carbon atoms of the 4-alkenyl group form part of a cyclopropyl group, complex (**69**), then stability with respect to β-elimination is further increased.[252,258] With 5-alkenylnickel complexes isomerization to the more stable 4-alkenylnickel species occurs (Equation (98)).[251]

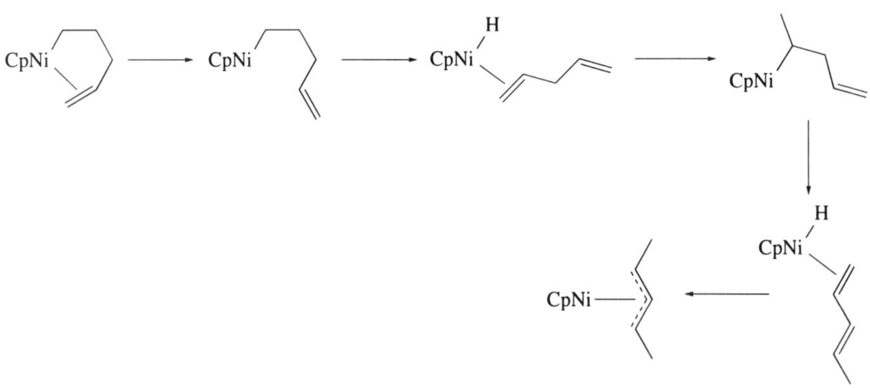

Scheme 13

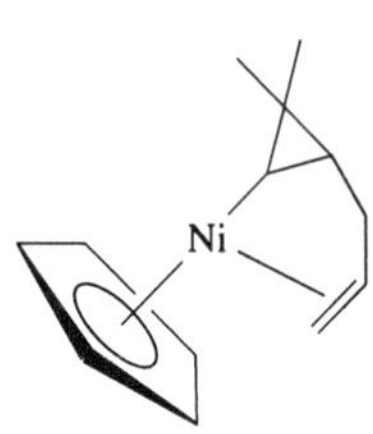

(69)

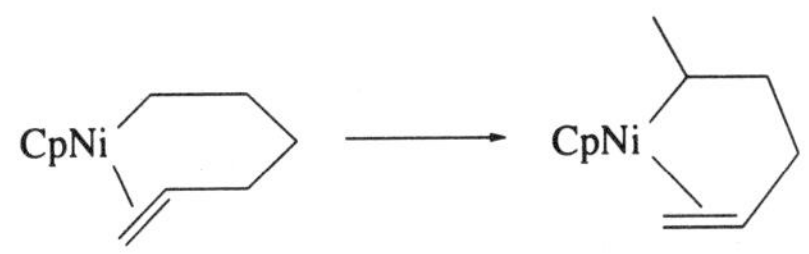

(98)

Complexes of the type [Ni(R)(η^2-alkene)(Cp)] (**66**) are stable up to ~20 °C. The complexes are trigonal planar with the alkene coordinated perpendicular to the trigonal plane. In solution there is rotation of the alkene around the coordination axis. With unsymmetrical alkenes, CH_2=CHR, two rotamers are observed below −60 °C.[226] In complexes with 1,3-dienes, such as (**70**), it is the least-substituted C=C bond that is coordinated to the nickel atom.[226] While the complex [Ni-(CH=CH_2)(CH_2=CH_2)(Cp)] cannot be isolated since it reacts even at −80 °C by insertion of the complexed ethene to give the 3-butenyl compound (Equation (99)),[237b] the corresponding Cp* derivative (**67**) can be isolated (Equation (92)) and only converts to the 3-butenyl complex (**71**) and the vinyl-coupled product (**72**) above −10 °C (Equation (100)).[237a] Other alkyl- or aryl(η^2-ethene)(cyclopentadienyl)nickel complexes react by migratory insertion of the ethene into the alkyl–nickel or aryl–nickel bond to give the primary alkyl Ni–CH_2CH_2R moiety which does not react further and is stabilized by coordination of ethene present in excess.[225,226] Deuterium-labelling studies using [Ni(Ph)(η^2-C_2H_4)(Cp)] and C_2D_4 have shown that it is the coordinated ethene that inserts into the Ni–Ph bond (Equation (101)).[225] In the case of the methyl(methylenecyclopropane)nickel complex (**73**) the initial alkene insertion product undergoes a ring-opening reaction at 20 °C to give the η^1,η^2-3-methyl-3-butenylnickel complex (**74**) which further isomerizes at 40 °C to the η^3-allyl complex (**75**) (Scheme 14).[231]

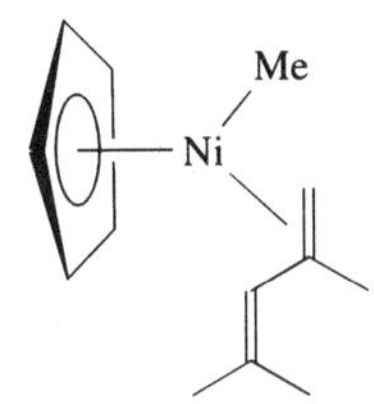

(70)

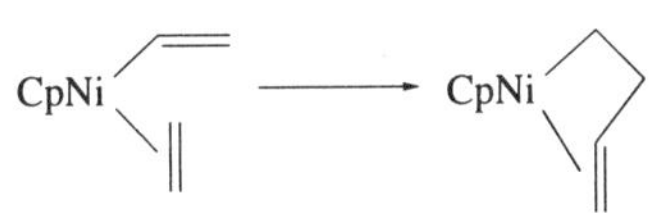

(99)

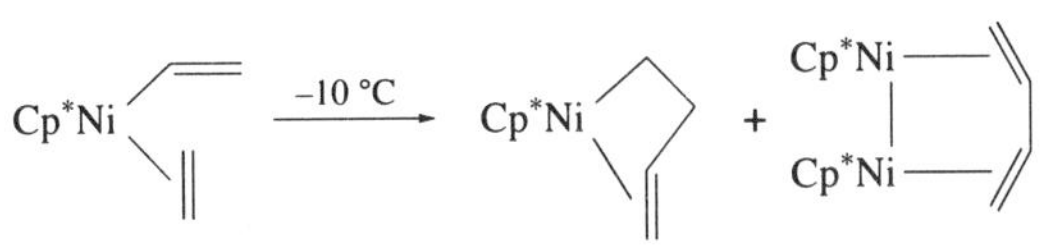

(100)

(67) **(71)** **(72)**

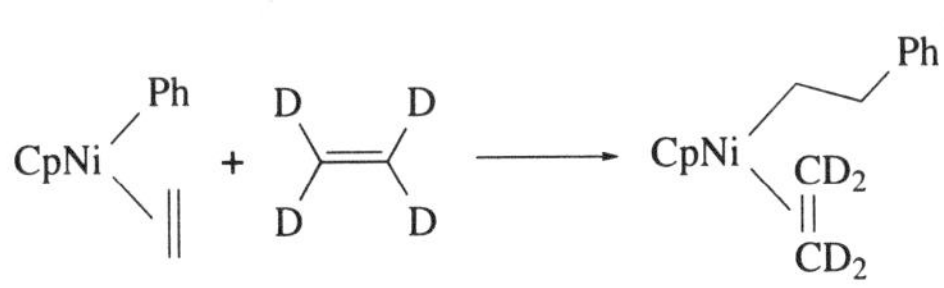

(101)

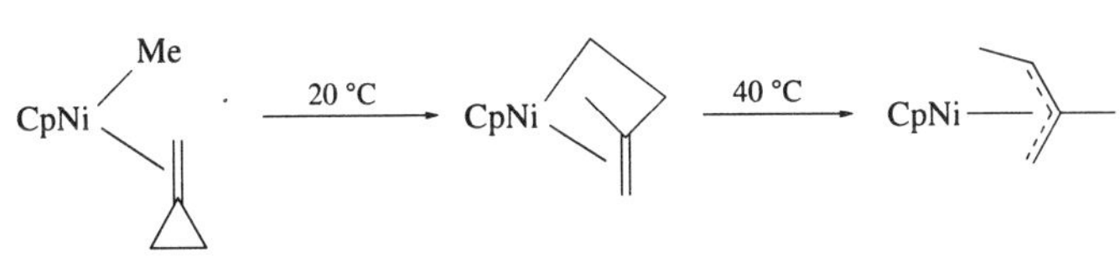

(73) **(74)** **(75)**

Scheme 14

The ethylnickel complex [Ni(Et)(η^2-C_2H_4)(Cp)] catalyses the oligomerization of ethene to a mixture of butenes, hexenes and octenes, with 1-butene as the main product.[226]

The thermal decomposition of [Ni(R)(η^2-propene)(Cp)] depends largely on the nature of the R group. If R = Me the decomposition products are methane, ethane and [$(NiCp)_3CH$]. Deuterium-labelling studies have shown that hydrogen exchange between the methyl group and the complexed alkene takes place. If β-hydrogen atoms are present β-H elimination is the main decomposition reaction.[228,230]

The alkyne complexes [Ni(Me)(η^2-RC≡CR)(Cp)] (R = Me or TMS) are also thermolabile. The 2-butyne derivative decomposes above −30 °C to liberate 2-butyne and ethane, and the μ-alkyne complex (**77**) is formed via (**76**) (Scheme 15).[234] The TMS-C≡C-TMS derivative is stable up to 0 °C but then decomposes to give the bis(trimethylsilyl) derivative of (**77**) together with the η^3-allylnickel complex (**78**) in which alkyl–alkyne coupling and rearrangement has occurred. The alkyne can be displaced from [Ni(Me)(η^2-RC≡CR)(Cp)] in an equilibrium reaction with acetonitrile (Equation (102)).[234]

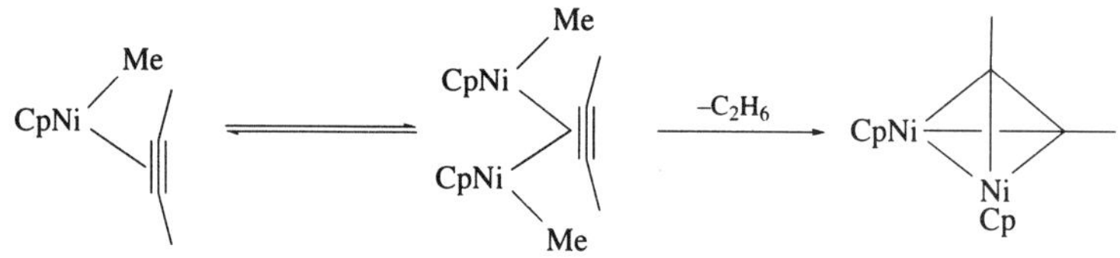

(76) **(77)**

Scheme 15

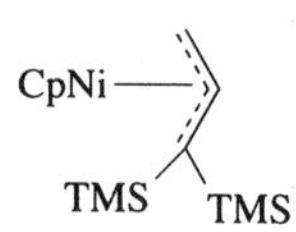

(78)

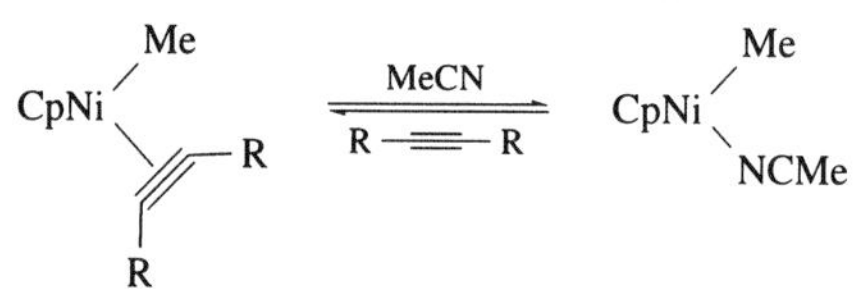

(102)

The tetrameric complex (**79**) has recently been isolated from the reaction of nickelocene with LiMe in the presence of excess 2-butyne.[238] When nickelocene is treated with phenyllithium or tolylmagnesium bromide the unstable compounds [$Ni(C_6H_4R)(Cp)$] are formed. With the $C_6H_4CH_2NMe_2$-2 derivative the complex is stabilized by the coordination of the NMe_2 group (**80**).[248]

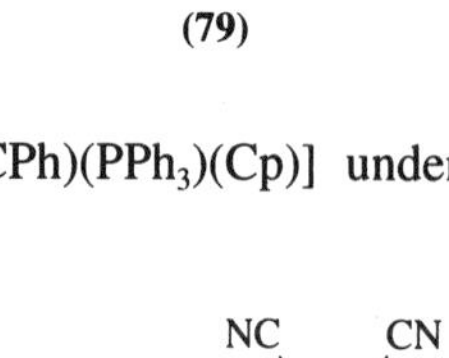

(79)

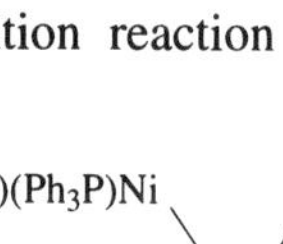

(80)

The complex [$Ni(C{\equiv}CPh)(PPh_3)(Cp)$] undergoes an addition reaction with TCNE to give (**81**) (Equation (103)).[246]

(81)

(103)

The reaction of oxygen with [$Ni\{CH(R)SO_2Ph\}(PPh_3)(Cp)$] (R = n-$C_{10}H_{21}$) (**82**) provides a new method of preparing the aldehyde RCH_2CHO.[242] Deprotonation of (**82**) with LiBu^n gives an anionic species that can be alkylated by iodomethane to give [$Ni\{CH(Me)SO_2Ph\}(PPh_3)(Cp)$].[242]

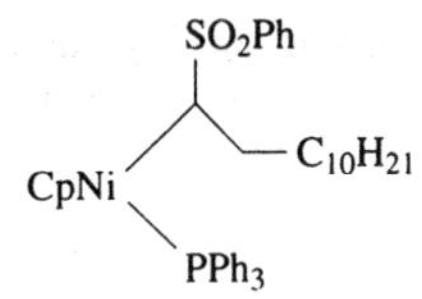

(82)

The alkynylnickel complexes [Ni(C≡CR)(PPh_3)(Cp)] react with ketenes to afford σ-cyclobut-1-en-3-onylnickel complexes (**83**) (Equation (104)).[254]

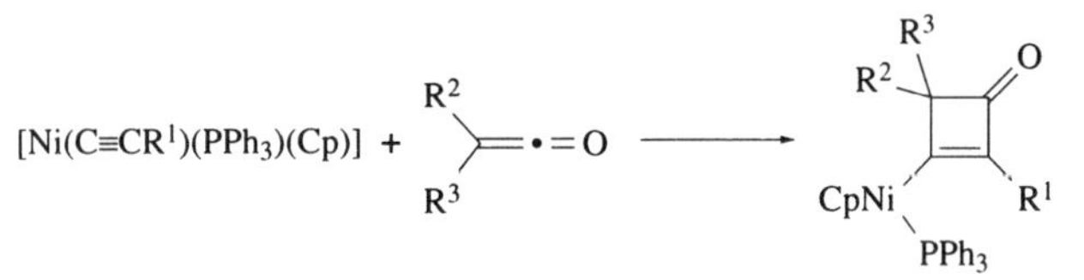

(83)

(104)

Trimethylphosphine adds to [Ni(nbd)(Cp)]$^+$ to give the η^1,η^2-nbd derivative (**84**) (Equation (105)).[256] The η^1,η^2-norbornadienenickel complex (**85**) undergoes a Diels–Alder reaction with 1,2,3,4-tetrachloro-5,5-dimethoxycyclopentadiene to give (**86**) (Equation (106)).[257]

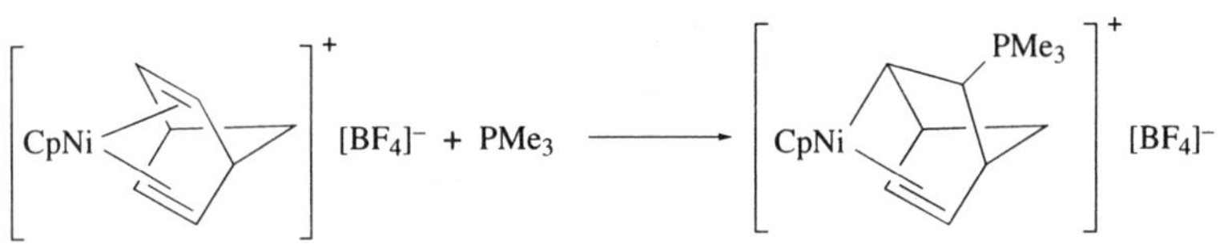

(84)

(105)

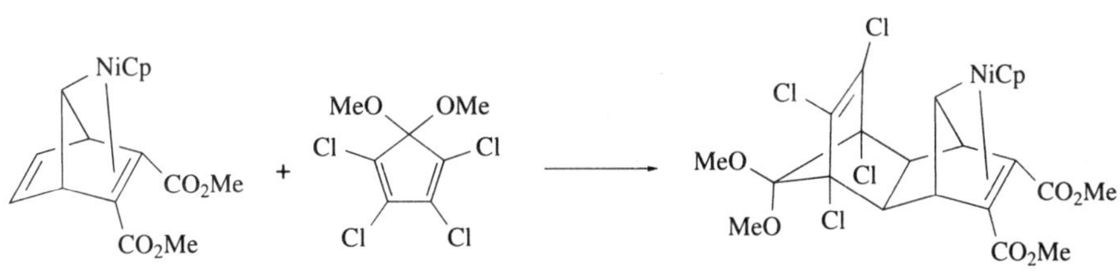

(85) **(86)**

(106)

2.7 NICKELACARBOCYCLIC COMPLEXES

In this section complexes in which nickel is bonded to two carbon atoms in a nickelacycle are discussed. The compounds are listed in Table 10.

In the gas phase, stable nickelacyclopentane ions have been produced, whereas propenenickel ions are formed rather than nickelacyclobutane ions.[261] Theoretical studies on nickelacyclopentane complexes have shown that if reductive elimination leading to cyclobutane is symmetry allowed then the formation of ethene in a C–C bond cleavage reaction will usually be symmetry forbidden and vice versa.[262] A possible low-energy pathway, not involving Ni–P bond cleavage, has also been found for the interconversion of the square-planar and tetrahedral forms of [$\overline{Ni(CH_2CH_2CH_2CH_2})(PH_3)_2$].[263]

2.7.1 Synthesis

The principal methods of synthesis of nickelacarbocyclic complexes are the addition of alkenes[273,281,283,292] or alkynes[274–8,292] to η^2-alkene or η^2-alkyne nickel complexes (Equations (107)–(110)), the oxidative addition of a strained ring organic compound to a nickel(0) complex (Equations (111)–(116)),[270,282,290,291,297–301] or the reaction of nickel(II) complexes with organolithium, organomagnesium or other carbanionic reagents (Equations (117)–(120)).[102,266,267,271,272,284,285,293]

Table 10 Nickelacarbocyclic complexes.

Complex	*Reference*	*Ni–C* (nm)
[Ni{CH$_2$C(O)}(PPh$_3$)$_2$] (**111**)	264	
[Ni{CH$_2$S(O)(Me)CH$_2$}(SacSac)]	265	0.202 1(10), 0.201 1(9)
[Ni{CH(Ph)S(O)$_2$CHPh}(PMe$_3$)$_2$]	266,267	
[Ni{CH(Ph)S(O)$_2$CHPh}(dppe)]	266,267	
[NiCl{C(Ph)(PMe$_3$)CHC(O)CH$_2$CMe$_2$Ph}(PMe$_3$)] (**104**)	268	0.194(1), 0.197(1)
[NiCH$_2$CH$_2$CH$_2$]	6	
[Ni(CH$_2$CMe$_2$CH$_2$)(PPh$_3$)$_2$] (**101**)	269	
[Ni(CH$_2$CMe$_2$CH$_2$)(dppe)]	269	
[Ni{C(Ph)=C(TMS)Si(TMS)(Mes)}(PEt$_3$)$_2$]	270	
[Ni(CH$_2$CH$_2$CH$_2$CH$_2$)(bipy)]	96	
Li$_2$[Ni(CH$_2$CH$_2$CH$_2$CH$_2$)$_2$(OEt$_2$)]	271,272	
Li$_2$[Ni(CH$_2$CH$_2$CH$_2$CH$_2$)$_2$(TMEDA)$_2$] (**96**)	272	0.199 5(3)
Li$_2$[Ni(CH$_2$CH$_2$CH$_2$CH$_2$)$_2$(THF)$_4$]	272	
[Ni(CH$_2$CH$_2$CF$_2$CF$_2$){(2,6-Pr$^i{}_2$Ph)$_2$-dad}][a]	273	
[Ni(CH=CHCF$_2$CF$_2$)(TMEDA)]	274	
[{[(Pr$^i{}_2$Ph)$_2$dad]\$\$Ni}$_2$(CH=CHCH=C\$\$H)][a] (**112**)	275,276	0.188 7(4), 0.185 5(4), 0.189 1(3), 0.185 6(3)
[{[(Me$_2$Ph)$_2$dad]Ni}$_2$(CH=CHCH=CH)][a]	275,277	0.188 2(4), 0.185 6(5)
[{[(Pr$^i{}_2$Ph)$_2$dad]Ni}$_2$(CH=CHCH=CH)Ni(CO)$_2$][a]	276	0.189 7(3), 0.190 7(3)
[Ni(CH=CHCH=CH)(Pr$^i{}_2$PCH$_2$CH$_2$PPr$^i{}_2$)(μ-HC≡CH)Ni(Pr$^i{}_2$PCH$_2$CH$_2$PPr$^i{}_2$)]	278	
[Ni{C(O)C(Ph)=C(Ph)C(O)}(bipy)] (**102**)	279	
[Ni{CH$_2$C(=CMe$_2$)C(=CMe$_2$)CH$_2$}(PPh$_3$)$_2$] (**99**)	280	
[Ni{CH$_2$CH$_2$C(=CH$_2$)C(=CH$_2$)CH$_2$CH$_2$}(bipy)] (**90**)	281	
[Ni(CH$_2$CCH$_2$CH$_2$CH$_2$CCH$_2$CH$_2$)(bipy)] (**87**)	281	
[Ni{C(=CH$_2$)CH$_2$CH$_2$CCH$_2$CH$_2$CH$_2$}(bipy)] (**88**)	281	
[Ni{CH$_2$CCH$_2$CH$_2$C(=CH$_2$)CH$_2$CH$_2$}(bipy)] (**89**)	281	
[Ni{CHCMe$_2$CHCHCMe$_2$CH}(PPh$_3$)$_2$]	282	
[Ni{CH(CO$_2$Me)CH$_2$CHC(Me)$_2$CH}(dmpe)]	283	
[Ni{CH(CO$_2$Me)CH$_2$CHC(Me)$_2$CH}(PPh$_3$)] (**91**)	283	
[Ni{C(Ph)=C(Ph)C(Ph)=CPh} (PEt$_3$)$_2$] (**97**)	284	
[Ni{C(Ph)=C(Ph)C(Ph)=CPh}(PPh$_3$)$_2$]	284,285	
[Ni{C(Ph)=C(Ph)C(Ph)=CPh}(dppe)]	284,285	
[Ni{C(Ph)=C(Ph)C(Ph)=CPh}(dmpe)]	284	
[Ni{C(Ph)=C(Ph)C(Ph)=C\$\$Ph}(depe)]	284	
[Ni{C(=CMe$_2$)C(=CMe$_2$)C(=CMe$_2$)C(=CMe$_2$)}(bipy)] (**100**)	286	
[Ni{perfluorobicyclo[3.3.0]octa-2,7-diene-4,6-diyl}(CNBu^t)$_2$] (**103**)	288	
[Ni{perfluorobicyclo[3.3.0]octa-2,7-diene-4,6-diyl}(CNBu^t)$_2$(PMe$_3$)]	288,289	0.195 5(3)
[Ni{perfluorobicyclo[3.3.0]octa-2,7-diene-4,6-diyl}(CNBu^t)$_2$(PPh$_3$)]	288,289	
[Ni{perfluorobicyclo[3.3.0]octa-2,7-diene-4,6-diyl}(CNBu^t)$_3$]	288,289	
[Ni(CH$_2$C$_6$H$_4$-2)(PEt$_3$)$_2$]	290	
[Ni(CH$_2$C$_6$H$_4$-2)(PBu$_3$)$_2$] (**93**)	290	
[Ni(CH$_2$C$_6$H$_4$-2)(PPh$_3$)$_2$]	290	
[Ni(CH$_2$C$_6$H$_4$-2)(TMEDA)]	290	
[Ni{C(TMS)$_2$C$_6$H$_4$-2}(TMEDA)]	291	0.207 5(4), 0.189 7(4)
[Ni{C(TMS)$_2$C$_6$H$_4$-2}(TEEDA)][b]	291	
[Ni{C(TMS)$_2$C$_6$H$_4$-2}(PMDTA)][c]	291	
[Ni{C(TMS)$_2$C$_6$H$_4$-2}(bipy)]	291	
[Ni{C(TMS)$_2$C$_6$H$_4$-2}(PMe$_3$)$_2$]	291	
[Ni{C(TMS)$_2$C$_6$H$_4$-2}(dppe)]	291	
[Ni{C(TMS)$_2$C$_6$H$_4$-2}(Cy$_2$PCH$_2$CH$_2$PCy$_2$)]	291	
[Ni(CH$_2$CH$_2$C$_6$H$_4$-2)(Cy$_2$PCH$_2$CH$_2$PCy$_2$)]	292	0.198 8(12), 0.195 1(12)
[Ni{C(CO$_2$Me)=C(CO$_2$Me)C$_6$H$_4$-2}(Cy$_2$PCH$_2$CH$_2$PCy$_2$)]	292	
[Ni{CH$_2$C(Me)$_2$C$_6$H$_4$-2}(PMe$_3$)$_2$] (**98**)	102,293,294	0.197(1), 0.193(1)
[Ni{CH$_2$C(Me)$_2$C$_6$H$_4$-2}(PMe$_3$)(py)] (**105**)	295	0.193 5(5), 0.191 7(5)
[Ni{CH$_2$C(Me)$_2$C$_6$H$_4$-2}(PMe$_2$Ph)$_2$]	295	
[Ni{CH$_2$C(Me)$_2$C$_6$H$_4$-2}(dmpe)]	294,295	
[Ni$_2$(μ-Cl){CHC(Me)$_2$C$_6$H$_4$-2}(PMe$_3$)$_2$]$_2$ (**109**)	296	0.191(2), 0.204(2)
[Ni$_2${CHC(Me)$_2$C$_6$H$_4$-2}(PMe$_3$)$_2$(Cp)]	296	
[Ni{CH$_2$CH$_2$C(CF$_2$)C(CH)$_4$C}(TEEDA)][b]	297	
[Ni{C(CF$_2$)CH=CHCH=CHC}(PMe$_3$)$_2$]	298	

Table 10 (continued)

Complex	Reference	Ni–C (nm)
$[Ni\{C(CF_2)CH{=}CHCH{=}CHC\}(PEt_3)_2]$	298	0.194 0(6), 0.194 5(6)
$[Ni\{C(CF_2)CH{=}CHCH{=}CHC\}(PPh_3)_2]$	298	
$[Ni\{C(CF_2)CH{=}CHCH{=}CHC\}(Cy_2PCH_2CH_2PCy_2)]$	298	
$[Ni\{C(CF_2)CH{=}CHCH{=}CHC\}(TMEDA)]$	298	
$[Ni\{C(CF_2)CH{=}CHCH{=}CHC\}(bipy)]$	298	
$[Ni(2,2'\text{-}C_6H_4\text{–}C_6H_4)(PEt_3)_2]$ (**95**)	299	
$[Ni_2(C_{24}H_{14})(PEt_3)_2]$	299	0.191(2), 0.197(2)
$[Ni\{C{=}CHCH{=}CHCH{=}C(C{=}CHCH{=}CHCH{=}CCH_2)CH_2\}(PMe_3)_2]$ (**92**)	300	
$[Ni\{C{=}CHC(CH_2CH_2){=}CCH{=}C(C{=}CHC(CH_2CH_2){=}CCHCCH_2)CH_2\}(PMe_3)_2]$ (**94**)	301	

[a] dad = 1,4-diazabutadiene. [b] TEEDA = tetraethylethylenediamine. [c] PMDTA = pentamethyldiethylenetriamine.

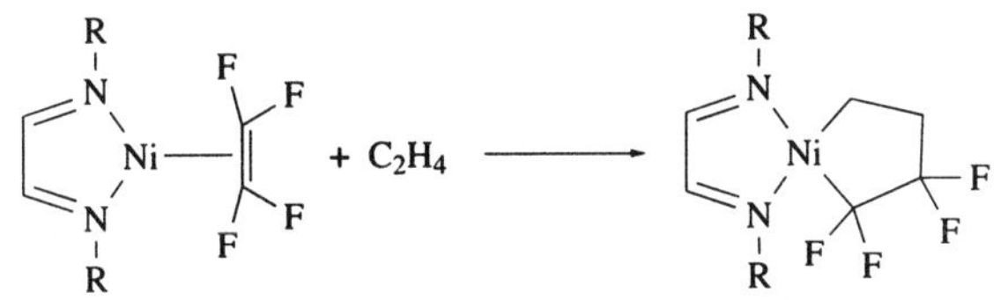

$R = C_6H_3Pr^i{}_2\text{-}2,6$ (107)

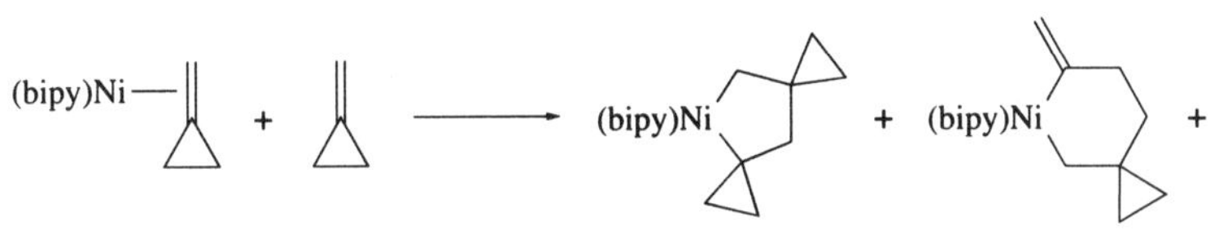

(**87**) (**88**)

(**89**) (**90**) (108)

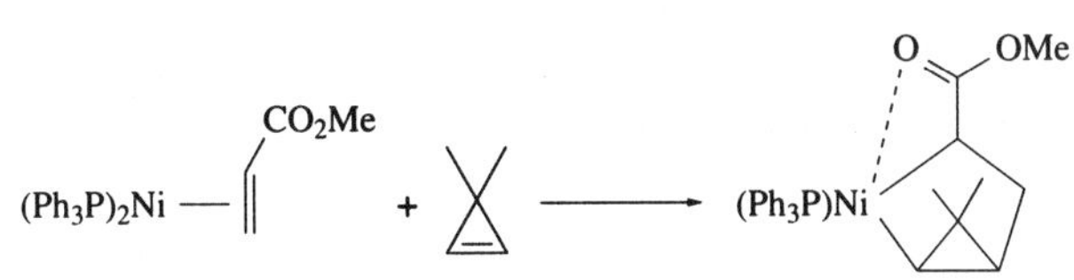

(**91**) (109)

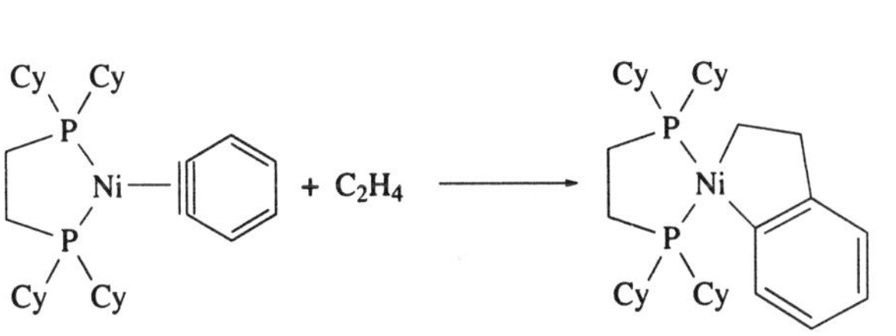

(110)

The presence of traces of iodide ion in the reaction of $[NiCl_2(PMe_3)_2]$ with $MgCl(CH_2CMe_2Ph)$ (Equation (120)) is important since in the absence of I^- only the monoalkyl complex $[NiCl(CH_2CMe_2Ph)(PMe_3)_2]$ is formed.[293] Other reactions leading to nickelacarbocycles are the reaction of allenes[280,287] or a triene[286] with nickel(0) complexes (Equations (121) and (122)), the elimination of CMe_4 by δ-hydrogen abstraction from a bis(neopentyl)nickel complex (Equation (123)),[269] ring opening

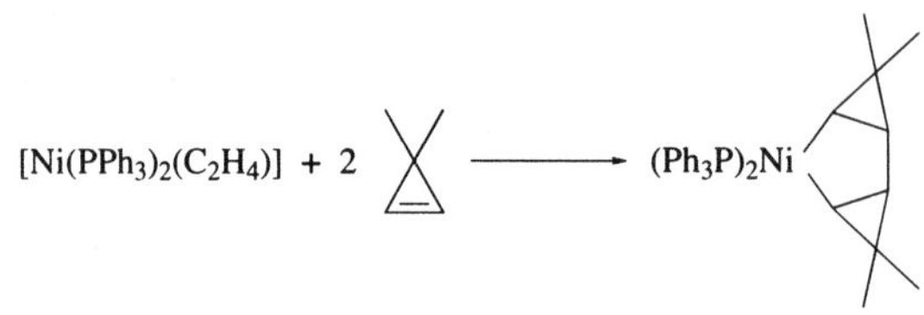

(111)

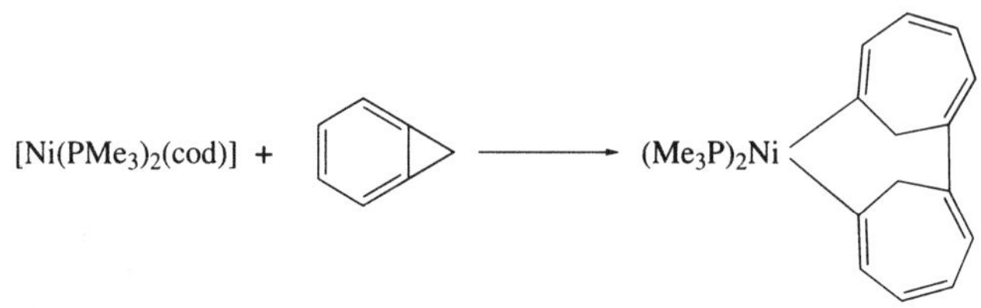

(112)

(92)

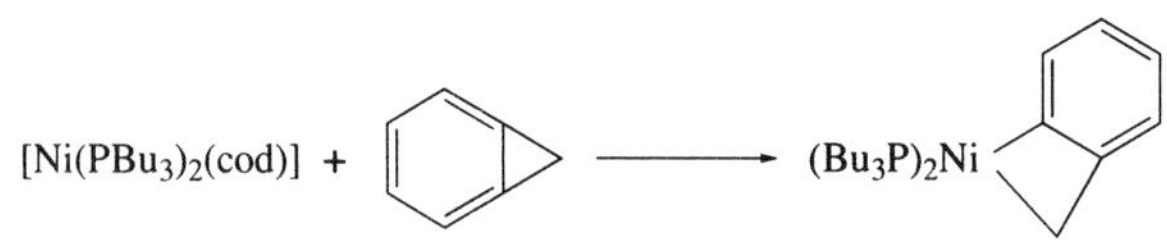

(113)

(93)

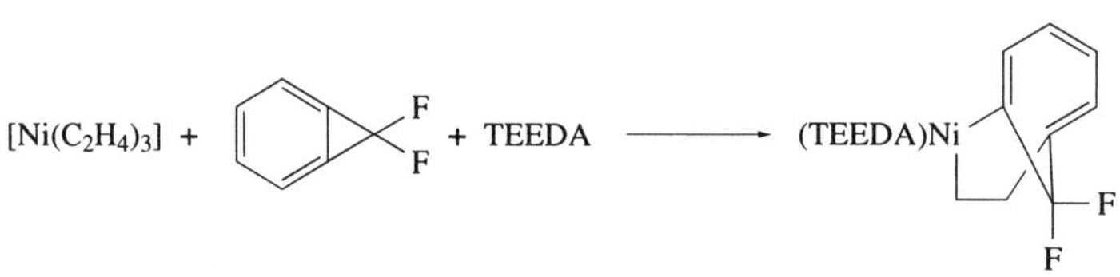

(114)

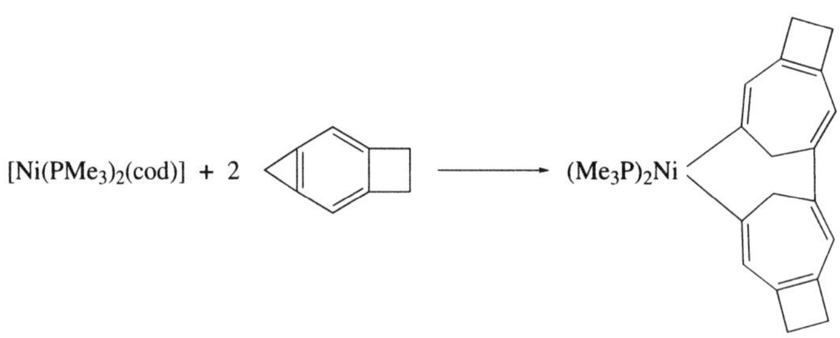

(115)

(94)

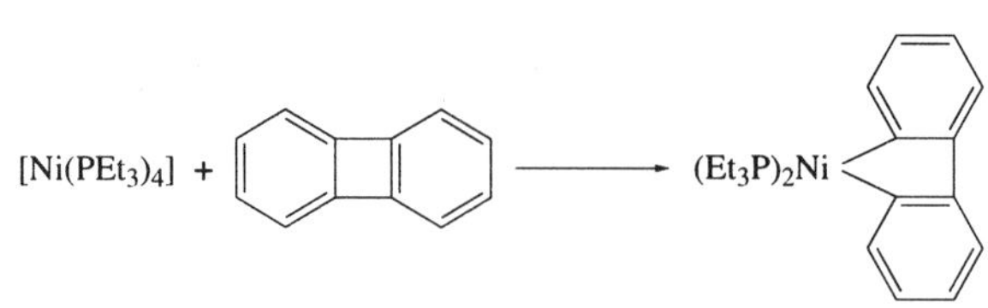

(116)

(95)

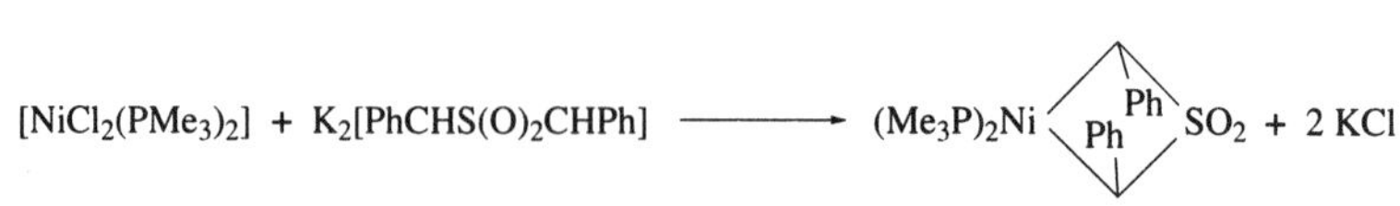

(117)

of a cyclobutenedione nickel complex (Equation (124))[279] or a cyclobutadiene complex (Equation (125)),[285] and oxidative addition to nickel(0) complexes (Equation (126)).[288,289]

In the reactions of nickel(0) complexes with strained ring systems the reaction path is dependent on the other ligands coordinated to nickel. Thus, for example, cyclopropabenzene reacts with $[Ni(PMe_3)_2(cod)]$ to give (**92**) (Equation (112)),[300] but with $[Ni(PBu_3)_2(cod)]$ the complex (**93**) is

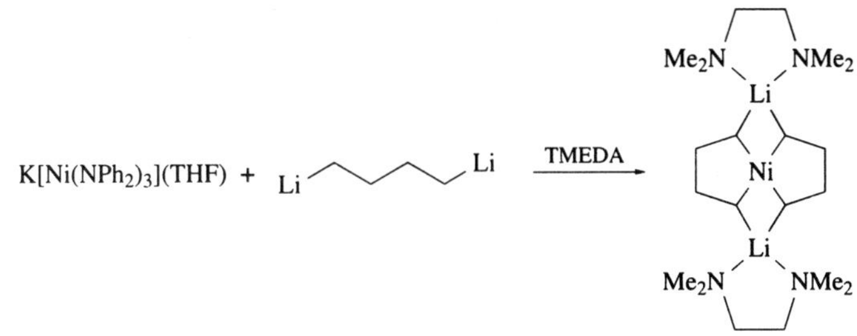

(96)

(118)

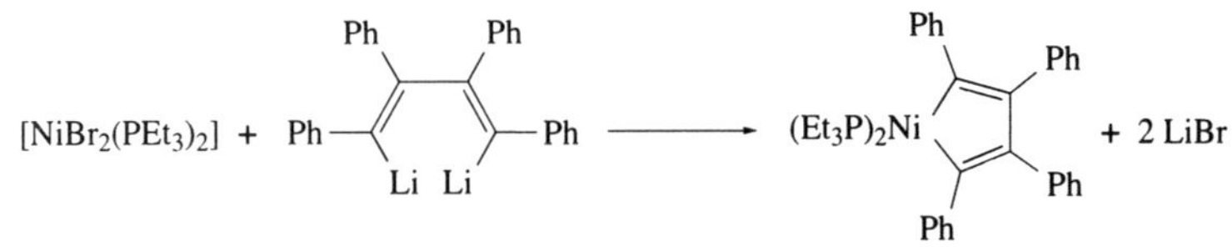

(97)

(119)

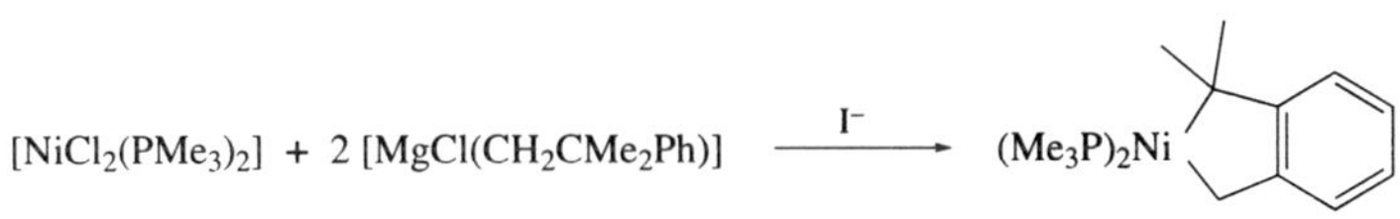

(98)

(120)

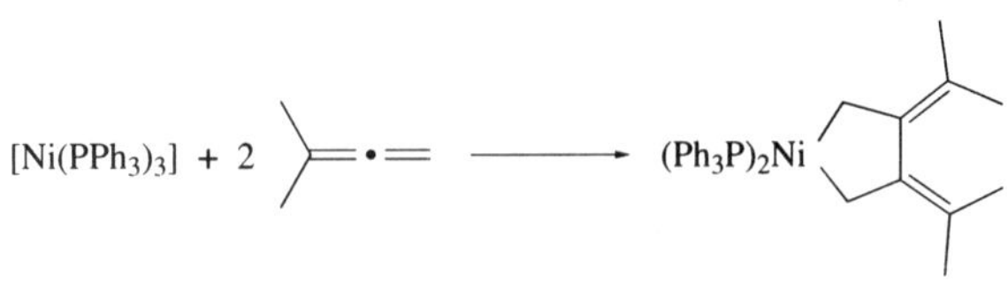

(99)

(121)

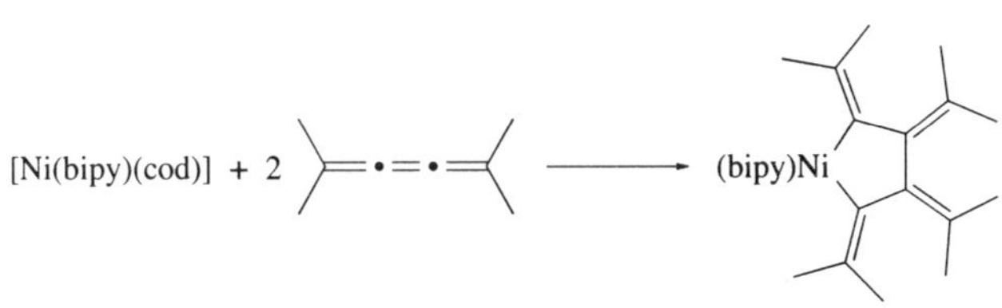

(100)

(122)

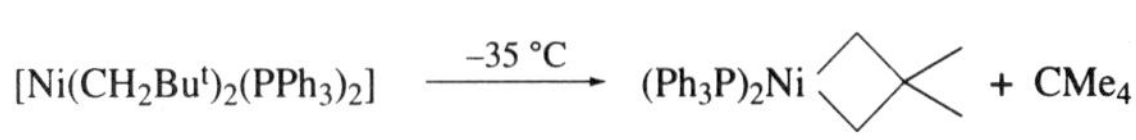

(101)

(123)

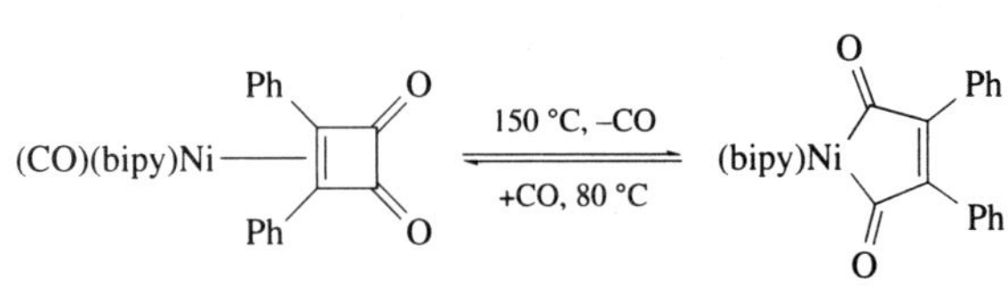

(102)

(124)

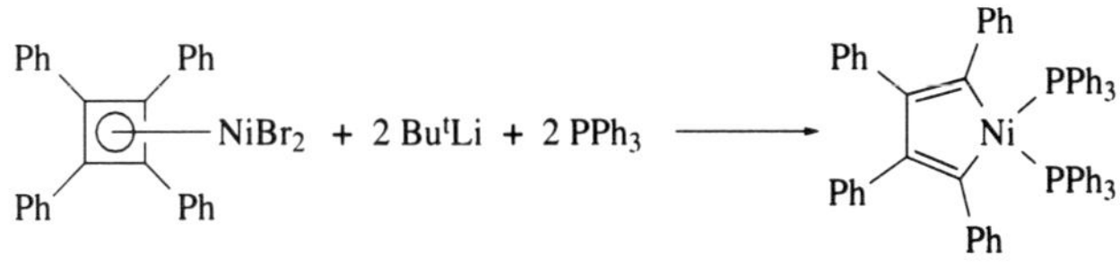

(125)

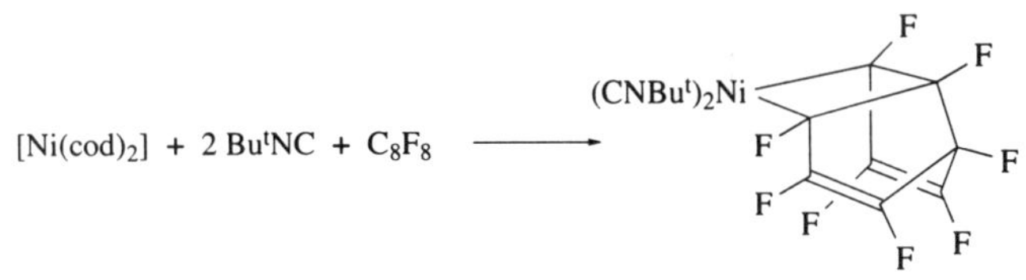

(103)

(126)

produced (Equation (113)).[290] The insertion of phenylacetylene into the Ni–C bond of the acylnickel complex $[NiCl(COR)(PMe_3)_2]$ occurs regio- and stereoselectively to give *trans*-(Z)-$[NiCl\{C(Ph){=}C(H)(COR)\}(PMe_3)_2]$.[268] When $R = CH_2CMe_2Ph$, the compound undergoes a reversible 1,2-PMe_3 shift to give the nickelacyclopropane complex (**104**) (Equation (127)).[268]

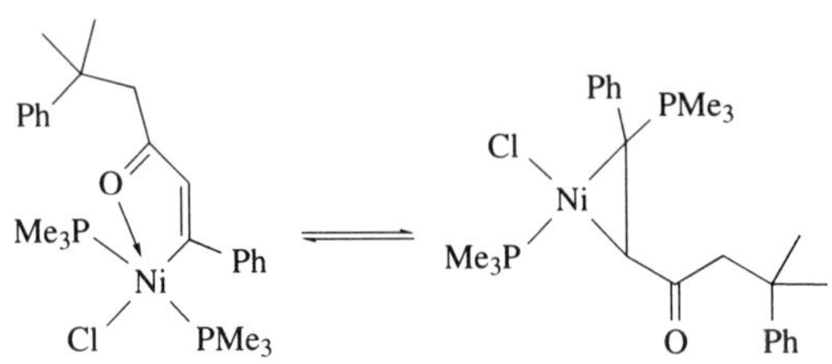

(104)

(127)

2.7.2 Reactions and Properties

The reactions of nickelacarbocyclic complexes follow a similar pattern to those observed with dialkylnickel complexes (Section 2.4.2). Thus, CO insertion with reductive elimination (Equation (128)),[102,293] alkyne insertion (Equation (129)),[102] CO_2 insertion (Equation (130)),[290] ligand exchange (Equations (131) and (132))[276,291,295] and protolysis (Equation (133))[290] are commonly observed. The ligand exchange reaction (Equation (131)) occurs exclusively at the position *trans* to the nickel–alkyl bond as expected since the alkyl group has a higher *trans* effect than the aryl ligand.[295] Complex (**98**) shows a strong tendency to undergo insertion reactions (Equations (128) and (129)), and will react with formaldehyde by insertion into the nickel–alkyl bond to give the dimeric oxanickelacycloheptene derivative (**106**),[102,293] and with CS_2 to give (**107**) (Scheme 16).[294] If the phosphine ligands in (**98**) are exchanged with dmpe, subsequent reaction with CS_2 proceeds by a different route to give (**108**) (Scheme 17).[294] Complex (**98**) also undergoes an unusual α-H abstraction when treated with $[NiCl(\eta^3\text{-}CH_2C_6H_4Me\text{-}2)(PMe_3)]$ to give the μ-alkylidenenickel complex (**109**) (Equation (134)).[296]

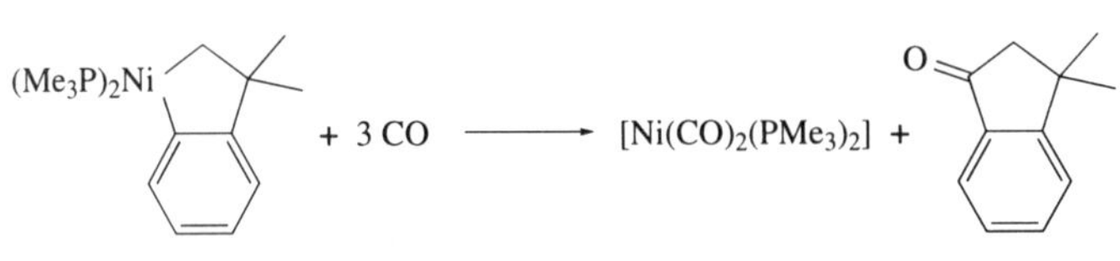

(98)

(128)

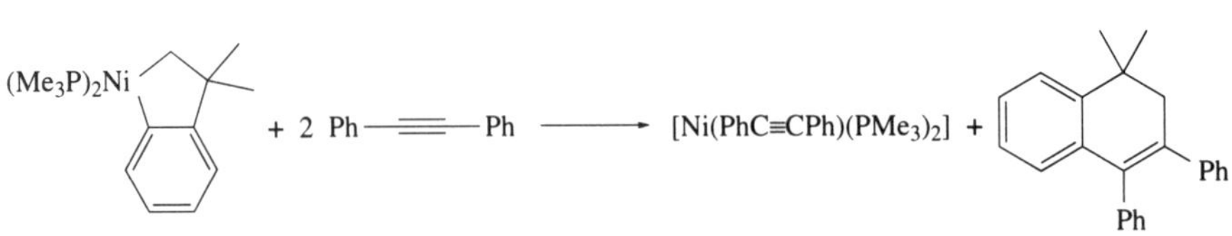

(98)

(129)

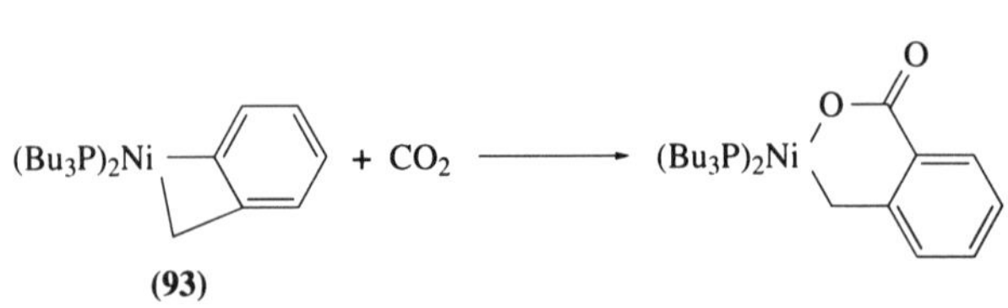

(93)

(130)

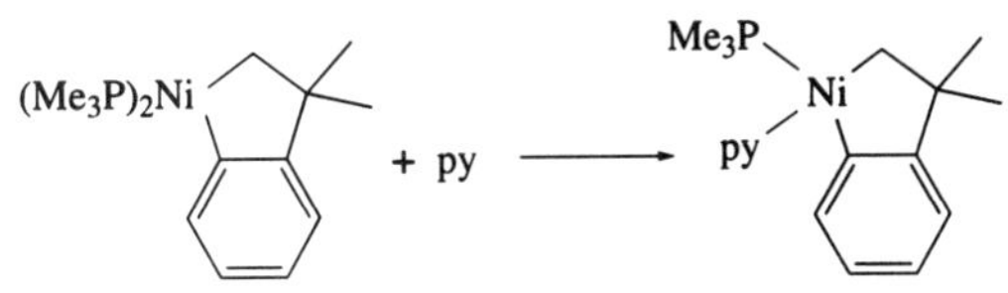

(98) (105)

(131)

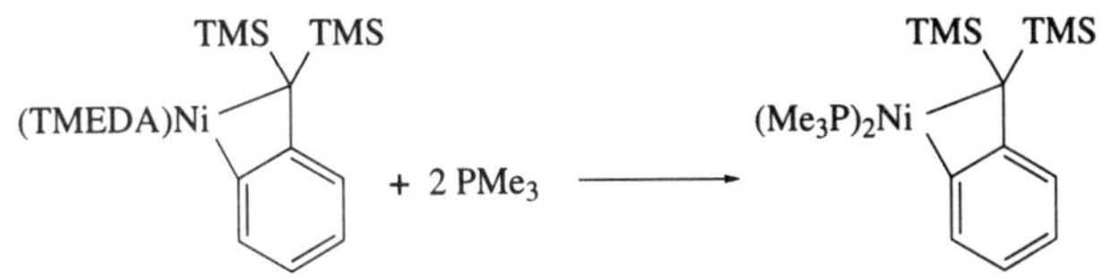

(132)

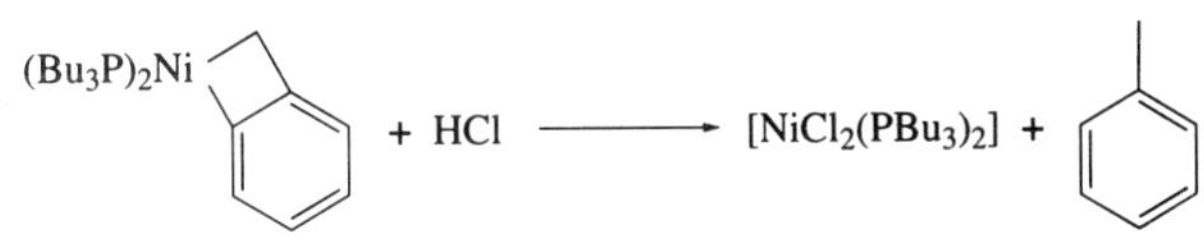

(133)

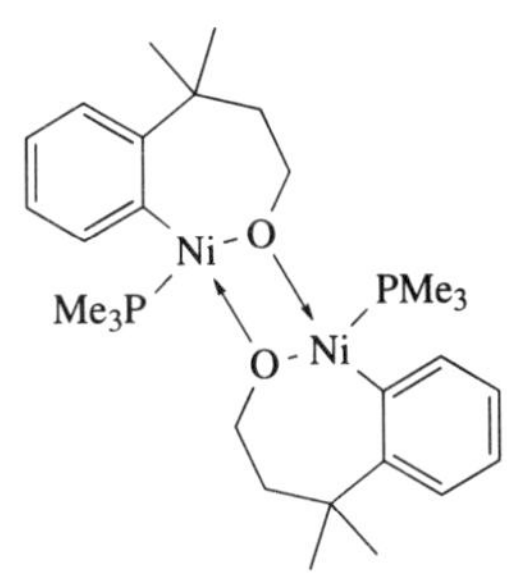

(106)

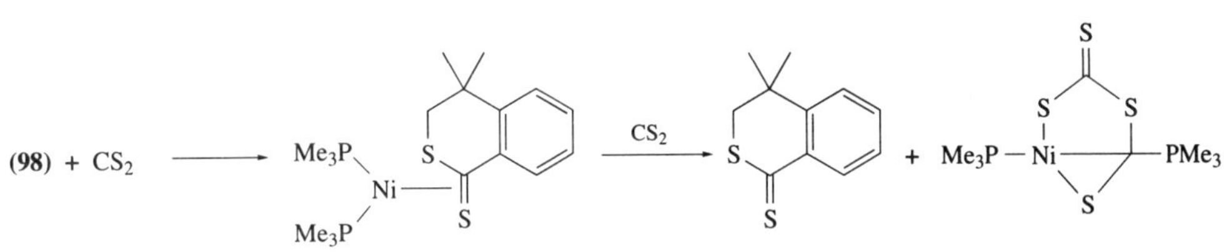

(107)

Scheme 16

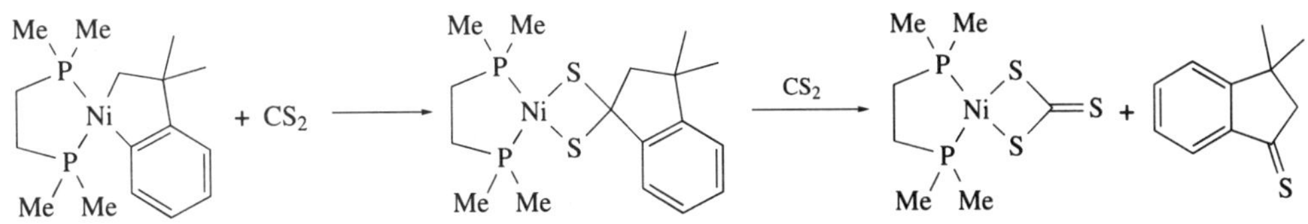

(108)

Scheme 17

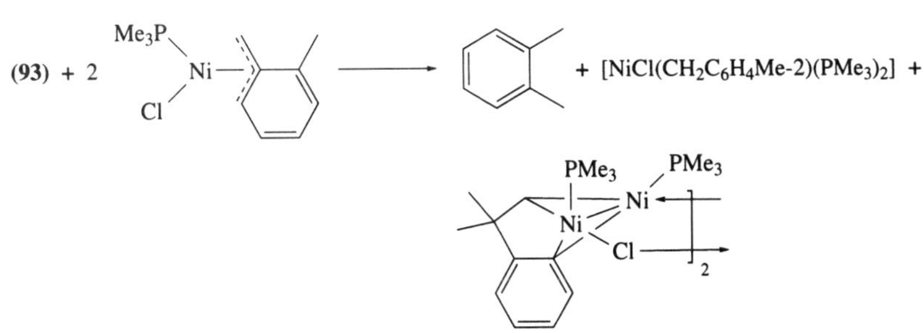

(134)

(109)

Treatment of $[\overline{Ni(CH_2CH_2CH_2CH_2})(bipy)]$ with N_2O results in the formation of the oxanickelacycle (**110**) (Equation (135)). This contrasts with the behaviour of the nickelacyclopentane complex with pyridine-*N*-oxide, which does not react, and with O_2, which results in the elimination of cyclobutane.[96]

(135)

(**110**)

The carbonylation of the nickelacyclobutane complex (**101**) yields the η^2-ketene complex (**111**) (Equation (136)).[264] Complex (**111**) can also be prepared by the reaction of $[Ni(PPh_3)_4]$ with CH_2Br_2 in the presence of zinc followed by carbonylation.[264]

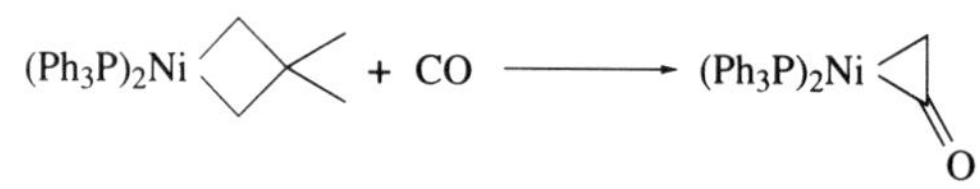

(136)

(**101**) (**111**)

Rare examples of stable, five-coordinate *cis*-dialkylnickel complexes (five-coordinate *trans*-dialkylnickel complexes are well known) are obtained by adding phosphine or isocyanide ligands to the η^1,η^1-octafluorocyclooctatetraenenickel complex (**103**).[288]

The diazabutadienenickelacyclic complex (**112**), prepared by the reaction of HC≡CH with $[Ni\{Pr^i_2Ph)_2dad\}(C_2H_4)_2]$, undergoes a dynamic process in solution which leads to the bonding situations of the nickel atoms being exchanged.[275–7] The diazabutadiene ligand on the nickel atom η^4-bonded to the C_4H_4 moiety may be exchanged with CO.[276]

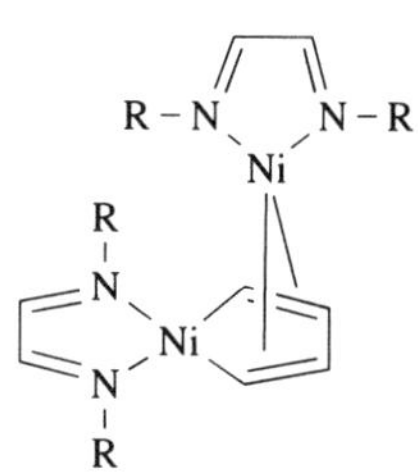

$R = C_6H_4Pr^i_2$-2,6

(**112**)

2.8 NICKELAHETEROCYCLIC COMPLEXES

This section deals with complexes in which the nickel atom is bonded to carbon and a heteroatom (N, P, O or S) in a nickelacycle. The compounds are listed in Table 11. Many of these compounds are formed by the reaction of nickel complexes with organic isocyanates or carbon dioxide, and are intermediates in the synthesis of organic compounds. This aspect of the chemistry of these complexes is discussed in Volume 12. Transition metal mediated reactions of organic isocyanates[376,377] and reactions of CO_2 with transition metal complexes[378–81] have been reviewed.

Nickel(0) complexes react with isocyanates and alkenes with C–C bond formation to produce azanickelacyclopentanones (Equations (137)[308] and (138)[316,317]). These reactions are not always regioselective. Thus, for example, with propene[308] the products formed are a mixture of (**113**) and (**114**), and an analogous mixture is obtained with 2-vinylfuran or 2-vinylpyridine.[319] With bicyclo[3.3.0]octene-2 complexes (**116**) and (**117**) are formed (Equation (139)).[322] With functional alkenes CH_2=CHR (R = OEt, SPh, CO_2Me[316] or CF_3[317]) and 4-pentenecarboxylic acid anilide[320] only one isomer, complex (**115**), is obtained. The thermodynamically less stable product is formed preferentially. With ethene, the initial five-membered ring complex (**118**) can undergo further insertion to produce the seven-membered nickelacyclic complex (**119**) (Equation (140)).[310]

Diisocyanates react with styrene in the presence of $[Ni(cod)_2]$ to give mono- or bis(azanickelacyclic) complexes depending on the ratio of the reactants (Equation (141)).[321] The coupling of isocyanates with dienes at nickel(0) yields substituted azanickelacyclopentanones (Equation (142)).[309] The isomer (**121**) was only obtained when R^1 = H and R^2 = Me.

Table 11 Nickelaheterocyclic complexes.

Complex	Reference	Ni–C (nm)
C,Sn-Heterocycle		
[Ni{CH=CHSn[CH(TMS)$_2$]$_2$}(Pr$^i_2$PCH$_2$CH$_2$PPr$^i_2$)]	302	0.193 3(4)
C,N-Heterocycles		
[Ni{C(TMS)$_2$C$_5$H$_4$N-2}$_2$] (**124**)	303	
[NiCl{CH$_2$C$_6$H$_4$NMe$_2$-2}(PMe$_3$)] (**125**)	65,304	
[NiCl{CH$_2$C$_6$H$_4$NMe$_2$-2}(PMe$_2$Ph)]	65,304	
[NiCl{CH$_2$C$_6$H$_4$NMe$_2$-2}(PEt$_3$)]	65,304	
[NiCl{CH(TMS)C$_6$H$_4$NMe$_2$-2}(PMe$_3$)]	304	
[NiCl{CH(TMS)C$_6$H$_4$NMe$_2$-2}(PMe$_2$Ph)]	304	
[NiCl{CH(TMS)C$_6$H$_4$NMe$_2$-2}(PEt$_3$)]	304	
[Ni{CH$_2$CH(Me)C(O)NH}(dppe)]	305	
[Ni{CH$_2$CH(Me)C(O)NH}(dppp)]	305	
[Ni{CH$_2$CH(Me)C(O)NH}(dppb)]	305	
[Ni{CH$_2$CH(Me)C(O)NH}(PEt$_3$)]$_4$ (**129**)	305	0.194(1), 0.196(1), 0.197(1), 0.194(1)
[Ni{CH$_2$CH(Me)C(O)NH}(PEt$_3$)$_2$]	305	
[Ni{C(Ph)=C(Ph)C(O)NPh}(TMEDA)] (**122**)	306,307	
[Ni{CH(Me)CH$_2$C(O)NPh}(PEt$_3$)] (**113**)	308	
[Ni{CH$_2$CH(Me)C(O)NPh}(PEt$_3$)] (**114**)	308	
[Ni{CH$_2$C(=CHMe)C(O)NPh}(TMEDA)] (**120**)	309	
[Ni{CH(Me)C(=CH$_2$)C(O)NPh}(TMEDA)] (**121**)	309	
[Ni{CH$_2$C(=CHMe)C(O)NPh}(bipy)]	309	
[Ni{CH$_2$CH(Me)C(O)NPh}(PCy$_3$)]	310	
[Ni{CH$_2$C(=CH$_2$)C(O)NPh}(TMEDA)]	309,311	
[Ni{CH$_2$C(=CMe$_2$)C(O)NPh}(TMEDA)] (**120**)	309,311	
[Ni{CH$_2$C(=CMe$_2$)C(O)NPh}(bipy)]	309,311	
[Ni{CH$_2$CH$_2$C(O)NPh}(PCy$_3$)] (**118**)	310,312,313	
[Ni{CH$_2$CH$_2$C(O)NPh}(bipy)]	313	
[Ni{CH$_2$CH$_2$C(O)NPh}(PPh$_3$)]	314	
[Ni{CH$_2$CH$_2$C(O)NC$_6$H$_4$Me-4}(PCy$_3$)]	310	
[Ni{CH(Me)CH$_2$C(O)NPh}(PCy$_3$)]	313	
[Ni{CH(Me)CH$_2$C(O)NPh}(bipy)]	313	
[Ni{CH(Ph)CH$_2$C(O)NPh}(PCy$_3$)]	131,314	
[Ni{CH(Ph)CH$_2$C(O)NPh}(dppe)]	314	
[Ni{CH(Ph)CH$_2$C(O)NPh}(bipy)]	314	
[Ni{CH(Ph)CH$_2$C(O)NPh}(Cy$_2$PCH$_2$CH$_2$PCy$_2$)]	314	
[Ni{CH(Ph)CH$_2$C(O)NMe}(PCy$_3$)]	314	
[Ni{CH(Ph)CH$_2$C(O)NMe}(bipy)]	314	
[Ni{CH(Ph)CH$_2$C(O)NC$_6$H$_4$Me-4}(PCy$_3$)]	314	
[Ni{CH(Ph)CH$_2$C(O)NC$_6$H$_4$Me-4}(bipy)]	314	
[Ni{CH(CH$_2$CH$_2$CHCH$_2$)CH$_2$C(O)NPh}(PCy$_3$)]	315	
[Ni{CH$_2$CH(CH$_2$CH$_2$CHCH$_2$)C(O)NPh}(PCy$_3$)]	315	
[Ni{CH(OEt)CH$_2$C(O)NPh}(PCy$_3$)] (**115**)	316	
[Ni{CH(SPh)CH$_2$C(O)NPh}(PCy$_3$)] (**115**)	316	
[Ni{CH(CO$_2$Me)CH$_2$C(O)NPh}(PCy$_3$)] (**115**)	316	
[Ni{CF$_2$CH$_2$C(O)NPh}(PCy$_3$)]	317	
[Ni{CF$_2$CH$_2$C(O)NPh}(PPh$_3$)]	317	
[Ni{CF$_2$CH$_2$C(O)NPh}(dppe)]	317	
[Ni{CF$_2$CH$_2$C(O)NPh}(bipy)]	317	
[Ni{CH(CF$_3$)CH$_2$C(O)NPh}(PCy$_3$)]	317	
[Ni{CH(CF$_3$)CH$_2$C(O)NPh}(dppe)]	317	
[Ni{CH(CF$_3$)CH$_2$C(O)NPh}(bipy)]	317	
[Ni{CH(C$_6$H$_{11}$)CH$_2$C(O)NPh}(PCy$_3$)]	318	
[Ni{CH(C$_6$H$_{11}$)CH$_2$C(O)NPh}(PEt$_3$)]	318	
[Ni{CH(furyl-2)CH$_2$C(O)NPh}(PCy$_3$)]	319	
[Ni{CH(py-2)CH$_2$C(O)NPh}(PCy$_3$)]	319	
[Ni{CH$_2$CH(furyl-2)C(O)NPh}(PCy$_3$)]	319	
[Ni{CH$_2$CH(py-2)C(O)NPh}(PCy$_3$)]	319	
[Ni{CH[CH$_2$CH$_2$C(O)NHPh]CH$_2$C(O)NPh}(PCy$_3$)]	320	
[Ni{CH(Ph)CH$_2$C(O)N(CH$_2$)$_6$NCO}(PCy$_3$)]	321	
[(Cy$_3$P)Ni{CH(Ph)CH$_2$C(O)N(CH$_2$)$_6$NC(O)CH$_2$CH(Ph)}Ni(PCy$_3$)]	321	
[Ni{CHCH$_2$CHCH$_2$CH$_2$CH$_2$CHCHC(O)NPh}(PCy$_3$)] (**116**)	322	
[Ni{CHCHCH$_2$CH$_2$CH$_2$CHCH$_2$CHC(O)NPh}(PCy$_3$)] (**117**)	322	
[Ni{CH(Ph)N(Ph)C(O)NMe}(TMEDA)]	323	
[Ni{CH(Ph)N(Ph)C(O)NBu^n}(TMEDA)]	323	
[Ni{CH(Ph)N(Ph)C(O)NBu^t}(TMEDA)]	323	
[Ni{CH(Ph)N(Ph)C(O)NPh}(TMEDA)]	323	

Table 11 (continued)

Complex	Reference	Ni–C (nm)
$[Ni\{CH(Ph)N(Ph)C(O)NC_6H_4Me\text{-}4\}(TMEDA)]$	323	
$[Ni\{CH(Ph)N(Ph)C(O)NCy\}(TMEDA)]$	323	
$[Ni\{CH(Ph)N(Ph)C(O)NC_6H_4Me\text{-}4\}(bipy)]$	323	
$[Ni\{CH(Ph)OC(O)NMe\}(TMEDA)]$	324	
$[Ni\{CH(Ph)OC(O)NPh\}(TMEDA)]$	324	
$[Ni\{CH(Ph)OC(O)NPh\}(bipy)]$	324	
$[Ni\{CH(Ph)OC(O)NPh\}(PCy_3)]$	324	
$[Ni\{CH(Ph)OC(O)NC_6H_4Me\text{-}4\}(TMEDA)]$	324	
$[Ni\{CH(Ph)OC(O)NC_6H_4Me\text{-}4\}(bipy)]$	324	
$[NiCl\{C(=NBu^t)C(=NBu^t)C(Me)=NBu^t\}(CNBu^t)]$	325	
$[NiCl\{C(=NBu^t)C(=NBu^t)C(CH_2TMS)=NBu^t\}(CNBu^t)]$	325	
$[NiCl\{C(=NBu^t)C(=NBu^t)C(CH_2C_6H_4Me\text{-}2)=NBu^t\}(CNBu^t)]$	325	0.191(1), 0.192(1)
$[Ni\{(1\text{-naphthyl-8-}NMe_2)\}_2]$ (**126**)	326	
$[NiCl\{C_6H_3(CH_2NMe_2)_2\text{-}2,6\}]$ (**130**)	327–329	
$[NiBr\{C_6H_3(CH_2NMe_2)_2\text{-}2,6\}]$ (**130**)	327–330	
$[NiI\{C_6H_3(CH_2NMe_2)_2\text{-}2,6\}]$ (**130**)	327–330	
$[NiX\{C_6H_3(CH_2NMe_2)_2\text{-}2,6\}]$ ($X = NO_2, N_3, NO_3, OH$)	329	
$[NiX\{C_6H_3(CH_2NMe_2)_2\text{-}2,6\}]$ ($X = OTf, O_2CPh, O_2CMe$)	329	
$[Ni\{OC(O)H\}\{C_6H_3(CH_2NMe_2)_2\text{-}2,6\}]$	329	0.181 4(2)
$[Ni\{C_6H_3(CH_2NMe_2)_2\text{-}2,6\}]_2[SO_4]$	329	
$[Ni\{C_6H_3(CH_2NMe_2)_2\text{-}2,6\}(H_2O)][BF_4]$	329	
$[Ni(SO_2)\{C_6H_3(CH_2NMe_2)_2\text{-}2,6\}]$	331	
$[Ni(NCS)\{C_6H_3(CH_2NMe_2)_2\text{-}2,6\}]$	332	
$[NiBr\{C_6H_3(CH_2NEt_2)_2\text{-}2,6\}]$	330	
$[NiBr\{C_6H_3(CH_2NPr^i_2)_2\text{-}2,6\}]$	330	
$[NiI\{C_6H_3(CH_2NEt_2)_2\text{-}2,6\}]$	330	
$[NiI\{C_6H_3(CH_2NPr^i_2)_2\text{-}2,6\}]$	330	
$[NiBr\{C_6H_3\{CH_2N(Me)Pr^i\}_2\text{-}2,6\}]$	330	
$[NiBr\{C_6H_3[CH_2N(Me)Bu^t]_2\text{-}2,6\}]$	330	
$[NiI\{C_6H_3[CH_2N(Me)Pr^i]_2\text{-}2,6\}]$	330	
$[NiI\{C_6H_3[CH_2N(Me)Bu^t]_2\text{-}2,6\}]$	330	
$[Ni\{C_6H_3(CH_2NEt_2)_2\text{-}2,6\}(H_2O)][CF_3SO_3]$	330	
$[Ni\{C_6H_3[CH_2N(Me)Pr^i]_2\text{-}2,6\}(H_2O)][CF_3SO_3]$	330	
$[NiCl\{C_6H_2(OMe)\text{-}p\text{-}(CH_2NMe_2)_2\text{-}2,6\}]$	328	
$[NiCl_2\{C_6H_3(CH_2NMe_2)_2\text{-}2,6\}]$ (**131**)	333,334	
$[NiBr_2\{C_6H_3(CH_2NMe_2)_2\text{-}2,6\}]$	333,334	
$[NiI_2\{C_6H_3(CH_2NMe_2)_2\text{-}2,6\}]$	333,334	0.189 8(5)
$[Ni(NO_3)_2\{C_6H_3(CH_2NMe_2)_2\text{-}2,6\}]$	332	
$[Ni(NO_2)_2\{C_6H_3(CH_2NMe_2)_2\text{-}2,6\}]$	332	
$[Ni(NCS)_2\{C_6H_3(CH_2NMe_2)_2\text{-}2,6\}(py)]$ (**132**)	332	0.190 0(9)
$[NiBr_2\{C_6H_3(CH_2NEt_2)_2\text{-}2,6\}]$	330	
$[NiBr_2\{C_6H_3[CH_2N(Me)Pr^i]_2\text{-}2,6\}]$	330	
$[Ni\{CH_2CH_2CH_2C(O)NH\}(PCy_3)]$	305,335	
$[Ni\{CH_2CH_2CH_2CH_2C(O)NPh\}(PCy_3)]$ (**119**)	310	
$[Ni\{CH_2CH_2CH_2CH_2C(O)NPh\}(bipy)]$	313	
$[NiCl\{C(CF_3)=C(CF_3)CH_2C_6H_4NMe_2\text{-}2\}(PMe_3)]$	65	
$[Ni\{C(CO_2Me)=C(CO_2Me)C(Ph)=C(Ph)C(O)NPh\}(TMEDA)]$ (**123**)	306	
C,P-Heterocycles		
$[Ni\{C(CO_2Me)=C(CO_2Me)C(COPh)PPh_2\}_2]$ (**136**)	336,337	
$[Ni\{C_6H_4PPh_2\text{-}2\}(PPh_3)_2]$ (**134**)	338	
$[NiCl\{C_6H_3(CH_2PPh_2)_2\text{-}2,6\}]$ (**133**)	339	
$[NiCl(CH_2PPh_2CHPPh_2)(PMe_3)]$	340	
$[Ni(CH_2PPh_2CHPPh_2)_2]$ (**137**)	340,341	0.201 2(6)
$[Ni(CH_2PPh_2CH_2PPh_2)_2][Br]_2$ (**138**)	341	0.201(1)
C,O Heterocycles		
$[NiCl\{C(Ph)=CHCOMe\}(PMe_3)_2]$	268	
$[NiCl\{C(Ph)=CHCOCH_2TMS\}(PMe_3)_2]$	268	0.189 6(7)
$[NiCl\{C(Ph)=CHCOCH_2Bu^t\}(PMe_3)_2]$	268	
$[NiCl\{C(Ph)=CHCOCH_2CMe_2Ph\}(PMe_3)_2]$	268	
$[NiBr\{CH_2C_6H_4C(O)NiBr(PMe_3)_2\}(PMe_3)]$	106	0.193 0(15), 0.182 4(16)
$[Ni(CH_2CH_2CH_2CH_2O)(bipy)]$ (**110**)	96	0.192 1(8)
$[Ni\{C_6H_4CMe_2CH_2CH_2O\}(PMe_3)]_2$ (**106**)	102,293	
$[Ni\{CH_2CMe_2C_6H_4C(O)O\}(PMe_3)_2]$	102,293	0.195 9(14)
$[Ni\{CH_2CH_2C(O)O\}(bipy)]$ (**147**)	342–344	
$[Ni\{CH_2CH_2C(O)O\}(dppe)]$	343,345	
$[Ni\{CH_2CH_2C(O)O\}(cod)]$	346	
$[Ni\{CH_2CH_2C(O)O\}(TMEDA)]$	335,342,347	

Table 11 (continued)

Complex	Reference	Ni–C (nm)
[Ni{$CH_2CH_2C(O)O$}($Cy_2PCH_2CH_2$-2-py)]	348	
[Ni{$CH_2CH_2C(O)O$}$(dbu)_2$] (**147**)	349	0.192 5(5)
[Ni{$CH_2CH_2C(O)O$}($Cy_2PCH_2CH_2PCy_2$)]	344	
[Ni{$CH_2CH(Me)C(O)O$}($Cy_2PCH_2CH_2$-2-py)]	348	
[Ni{$CH_2CH(Me)C(O)O$}(dppe)] (**149**)	343	
[Ni{$CH_2CH(n\text{-}C_6H_{13})C(O)O$}($Cy_2PCH_2CH_2$-2-py)]	348	
[Ni{CH(Et)CH(Et)C(O)O}($Cy_2PCH_2CH_2PCy_2$)]	344	
[Ni{$CH(Ph)CH_2C(O)O$}($Cy_2PCH_2CH_2$-2-py)]	348	
[Ni{$CH(Ph)CH_2C(O)O$}(dbu)] (**139**)	350	
[Ni{$CH(Me)CH_2C(O)O$}($Cy_2PCH_2CH_2$-2-py)]	348	
[Ni{$CH(Me)CH_2C(O)O$}(dppe)] (**138**)	335,343,351	
[Ni{$CH(Me)CH_2C(O)O$}(chiraphos)]	352	
[Ni{$CH_2CH(Ph)C(O)O$}(dbu)] (**140**)	350	
[Ni{$CH_2CH(CH{=}CHMe)C(O)O$}(cod)]	353	
[Ni{$CH_2C({=}CH_2)C(O)O$}(TMEDA)]	354	
[Ni{$CH_2C({=}CH_2)C(O)O$}(bipy)]	355	
[Ni{$CH_2C({=}CH_2)C(O)O$}($Cy_2PCH_2CH_2PCy_2$)] (**143**)	355	
[Ni{$CH_2C({=}CMe_2)C(O)O$}(bipy)]	355	
[Ni{$CH_2C({=}CHMe)C(O)O$}(bipy)]	355	
[Ni{$CH_2C({=}CHMe)C(O)O$}($Cy_2PCH_2CH_2PCy_2$)]	355	
[Ni{CH(Me)OC(O)O}(bipy)] (**145**)	356,357	0.190(1)
[Ni{CH(Et)OC(O)O}(bipy)]	356	
[Ni{$CH(n\text{-}C_6H_{13})OC(O)O$}(bipy)]	356	
[Ni{CH(Et)N(Cy)C(O)O}(TMEDA)]	358	
[Ni{CH(Ph)N(Ph)C(O)O}(TMEDA)] (**146**)	358,359	0.193 8(6)
[Ni{CH(Ph)N(Ph)C(O)O}(bipy)]	358,359	
[Ni{$CH(Pr^n)N(Cy)C(O)O$}(TMEDA)]	358	
[Ni{$CH(Pr^n)N(Cy)C(O)O$}(bipy)]	358	
[Ni{$CH(Pr^n)N(Pr^n)C(O)O$}(bipy)]	360	
[Ni{CH(CH=CHPh)N(Ph)C(O)O}(bipy)]	361	
[Ni{C(=NCy)N(Cy)C(O)O}(bipy)]	362	
[Ni{$CH_2CH_2CH_2C(O)O$}(bipy)]	342,343,345	
[Ni{$CH_2CH_2CH_2C(O)O$}(TMEDA)]	342	
[Ni{$CH_2CH_2CH_2C(O)O$}(PCy_3)]	305,335	
[Ni{$CH_2CH_2CH_2C(O)O$}(dppe)]	335	
[Ni{$CH_2CH_2CH_2C(O)O$}(chiraphos)]	352	
[Ni{$CH_2CMe_2CH_2C(O)O$}(bipy)]	342	
[Ni{$CH_2CH_2CH_2CH_2C(O)O$}(cod)]	346	
[Ni{$CH(CH_2)_6CHC(O)O$}($Cy_2PCH_2CH_2PCy_2$)]	363	
[Ni{$CH(CH_2CH_2CH{=}CHCH_2CH_2)CHC(O)O$}($Cy_2PCH_2CH_2PCy_2$)]	364	
[Ni{$CH(CH_2)_4CHC(O)O$}(TMEDA)]	342	
[Ni{$CH_2C(CH_2CH_2CH_2CH_2)CH_2C(O)O$}(bipy)]	342	
[Ni{$CH_2C(CH_2CH_2CH_2CH_2)CH_2C(O)O$}(TMEDA)]	342	
[Ni{$CH_2C(OCH_2CH_2O)CH_2C(O)O$}(bipy)]	342	
[Ni{$CH_2OCH_2C(O)O$}(bipy)]	342	
[Ni{$CH_2OCH_2C(O)O$}(TMEDA)]	342	
[Ni{$CH_2N(Ph)CH_2C(O)O$}(bipy)]	342	
[Ni{$CH_2N(Ph)CH_2C(O)O$}(TMEDA)]	342	
[Ni{$CH(CH_2CH{=}CHCH_2)CHC(O)O$}(bipy)]	365	
[Ni{$CH(CH_2CH{=}CHCH_2)CHC(O)O$}(TMEDA)]	365	
[Ni{$C(O)CH(CH_2CH{=}CHCH_2)CHC(O)O$}(bipy)]	365	
[Ni{$C(O)CH(CH_2CH{=}CHCH_2)CHC(O)O$}(TMEDA)]	365	
[Ni{$C_6H_4C(O)O$}(bipy)] (**150**)	342	
[Ni{$C_6H_4C(O)O$}(TMEDA)]	342	
[Ni{$C_6H_4C(O)O$}($Cy_2PCH_2CH_2PCy_2$)]	292	
[Ni{$C_6H_4C_6H_4C(O)O$}(bipy)]	354	
[Ni{$C_6H_4C_6H_4C(O)O$}(TMEDA)]	354	0.189 1(4)
[Ni{$CH_2C_6H_4C(O)O$}$(PBu_3)_2$]	290	
[Ni{$C(Pr^n){=}C(Pr^n)C(O)O$}(bipy)]	366	
[Ni{C(Et)=C(Et)C(O)O}(TMEDA)] (**144**)	367,368	0.194 4(7)
[Ni{C(Me)=C(Me)C(O)O}(TMEDA)]	344,369–371	
[Ni{C(Me)=C(Me)C(O)O}(bipy)]	344	
[Ni{C(Me)=C(Me)C(O)O}($Cy_2PCH_2CH_2PCy_2$)]	344	
[Ni{CH=CHC(O)O}($Cy_2PCH_2CH_2PCy_2$)]	344	
[Ni{C(Ph)=C(Ph)C(O)O}(TMEDA)]	344	
[Ni{C(Ph)=C(Ph)C(O)O}(bipy)]	344	

Table 11 (continued)

Complex	Reference	Ni–C (nm)
[Ni{C(Me)=C(Me)C(Me)=C(Me)C(O)O}(TMEDA)]	370	
[Ni{C(CF$_3$)=C(CF$_3$)C(Me)=C(Me)C(O)O}(TMEDA)]	344,370,371	
[Ni{C(CO$_2$Me)=C(CO$_2$Me)C(Me)=C(Me)C(O)O}(TMEDA)]	344,370,371	0.190 3(2)
[Ni{(dicyclopentadiene)C(O)O}(bipy)] (**141**), (**142**)	372,373	0.192 9(5)
[Ni{(cyclooctatetraene)C(O)O}(bipy)]	373	
[Ni{(norbornadiene)C(O)O}(bipy)]	344,373,374	
[Ni{(quadricyclane)C(O)O}(bipy)]	373	
C,S-Heterocycles		
[NiCl(CH$_2$PPh$_2$S)(PPh$_3$)]	375	0.199 9(8)
[Ni{SC(S)SC(PMe$_3$)S}(PMe$_3$)] (**107**)	294	

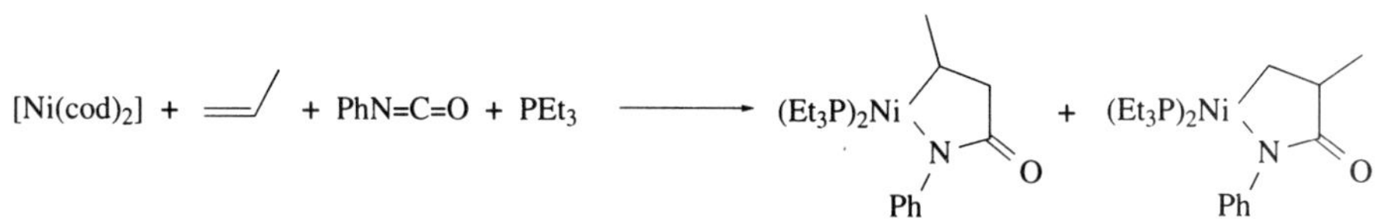

(113) **(114)** (137)

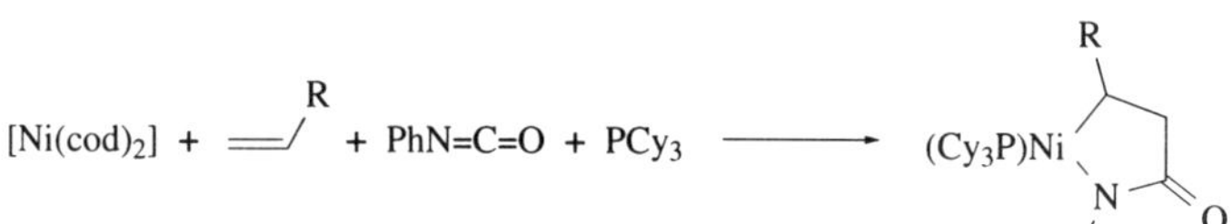

R = OEt, SPh, CO$_2$Me, CF$_3$, CH$_2$CH$_2$C(O)NHPh

(115) (138)

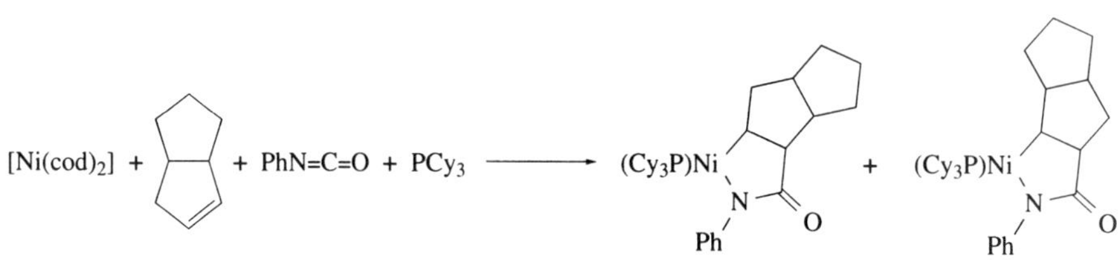

(116) **(117)** (139)

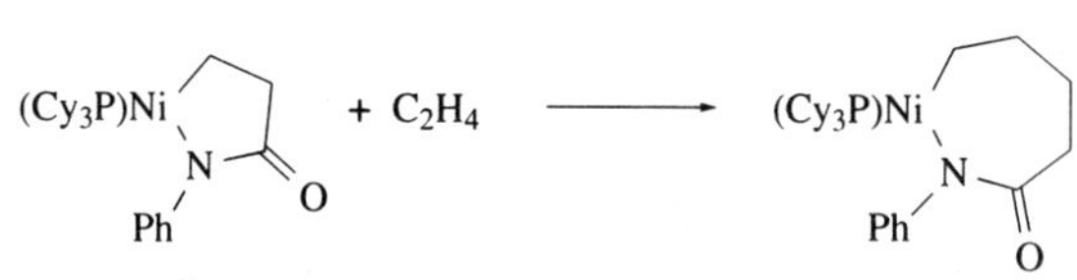

(118) **(119)** (140)

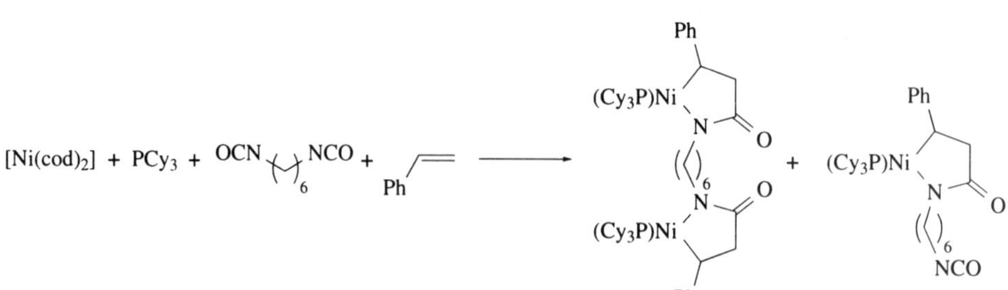

(141)

Azanickelacyclopentenes (**122**) are obtained from the cycloaddition reactions of alkynes, isocyanates and nickel(0) complexes (Equation (143)).[306] Addition of the activated alkyne MeO$_2$CC≡CCO$_2$Me to (**122**) gives the nickelaheptacyclic complex (**123**).

Coupling reactions of isocyanates with benzaldehyde (Equation (144)) and with imines (Equation (145))[323] have been reported. These reactions appear to proceed by coordination of the aldehyde or imine followed by C–O or C–N bond formation, respectively.

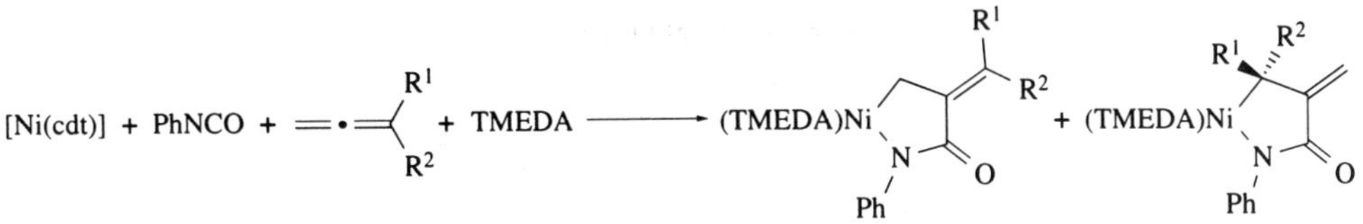

(120) (121)

(142)

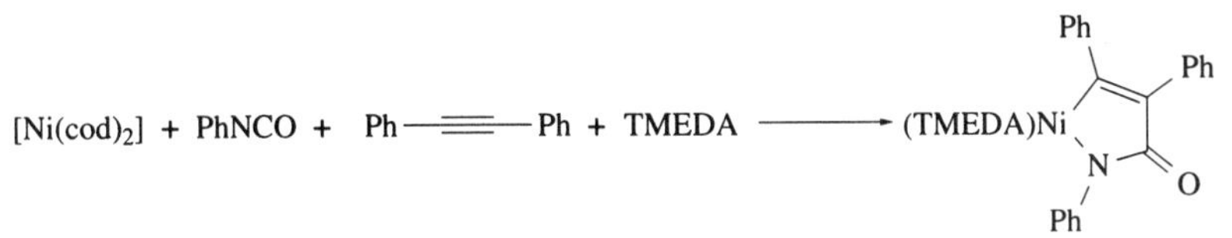

(122)

(143)

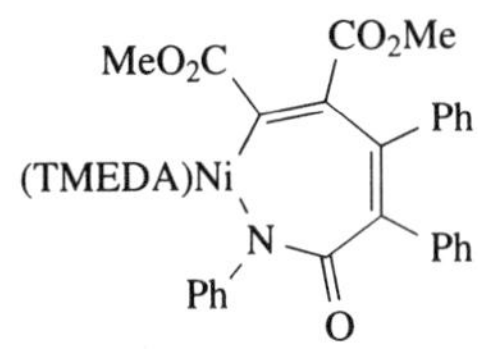

(123)

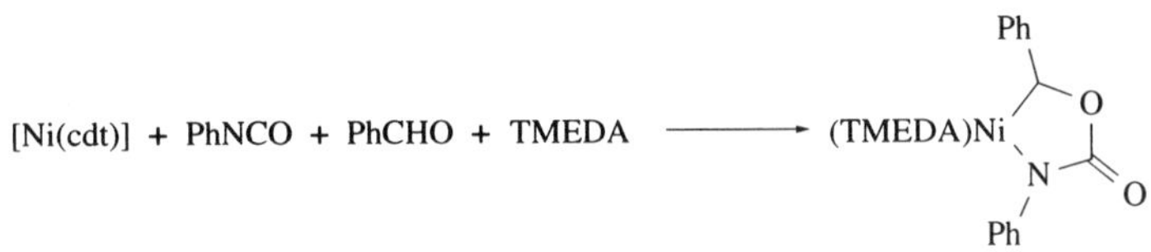

(144)

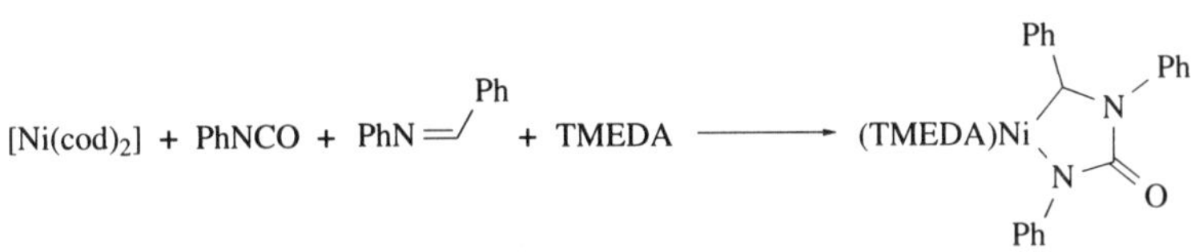

(145)

Azanickelacyclic complexes have been obtained by other routes. Nickel chloride reacts with the lithium derivative of bis(trimethylsilyl)methylpyridine in the presence of PEt_3 to give the air-stable nickel complex (**124**).[303] $[NiCl_2(PMe_3)_2]$ reacts with lithiated dimethylamino-*o*-toluene to give the nickelacycle (**125**).[304] Treatment of $NiBr_2$ with the lithium salt of the 8-dimethylaminonaphthyl-1-anion affords the homoleptic nickel complex (**126**).[326] $[Ni(cod)_2]$ reacts with α,β-unsaturated amides in the presence of PCy_3 to give the oligomeric nickelacycle (**127**) (Equation (146)).[305] The tricyclohexylphosphine ligand in (**128**) is readily replaced by PEt_3 to give (**129**), which has been shown to be tetrameric in the solid state.[305] In (**129**) each nickelacycle unit is bonded to an adjacent nickel atom in another nickelacycle through the amide carbonyl oxygen and not through the amide nitrogen as earlier assumed.[382] Treatment of (**128**) with ditertiary phosphine ligands gives monomeric complexes of the type $[\overline{Ni\{CH_2CH_2CH_2C(O)NH\}}(dppe)]$.[305] A typical reaction of these six-membered ring complexes is with CO which proceeds by insertion followed by reductive elimination to yield cyclic imides.

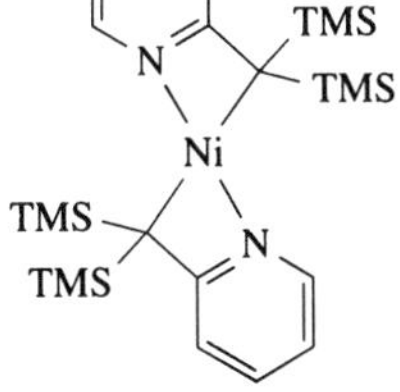

(124)

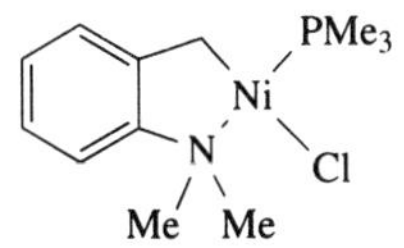

(125)

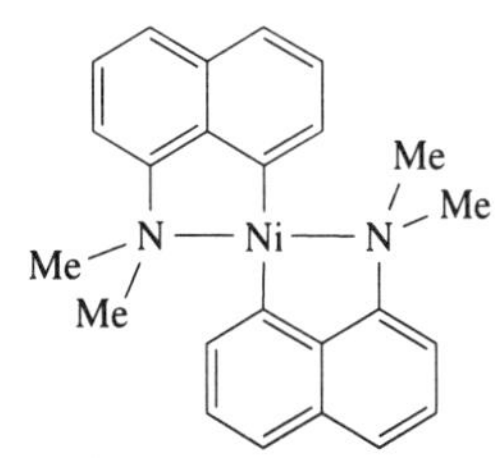

(126)

Azanickelacycles are also obtained by insertion reactions of isocyanides into nickel–alkyl bonds (Equation (147)).[325]

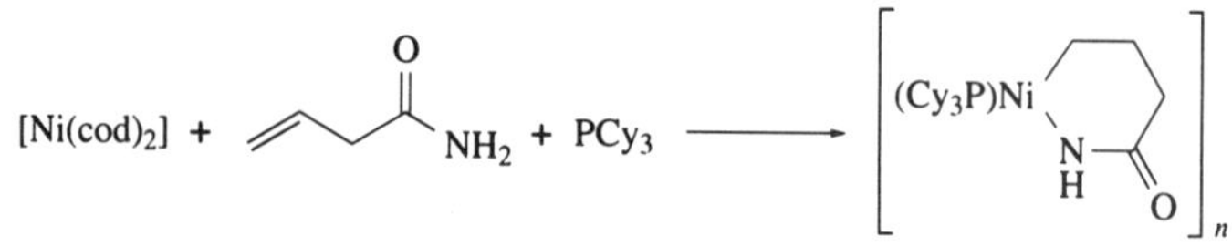

(146)

(127)

A series of compounds (**130**) containing the tridentate *N,N',C*-ligand $C_6H_3(CH_2NMe_2)_2$-2,2' has been prepared.[329,331] Photoelectron spectra show that these complexes have a very low first ionization enthalpy

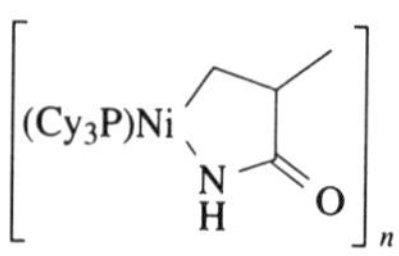

(128)

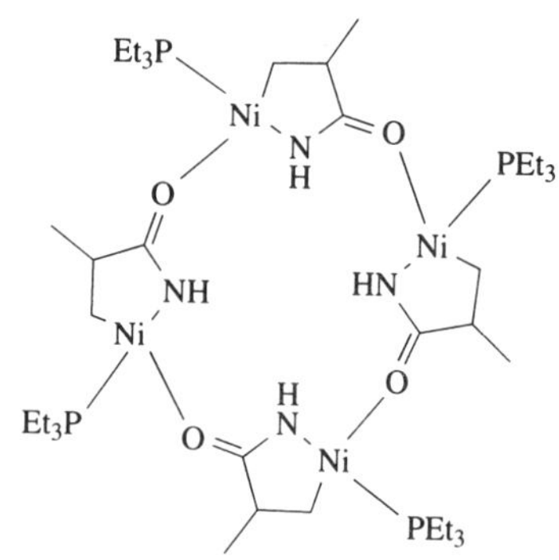

(129)

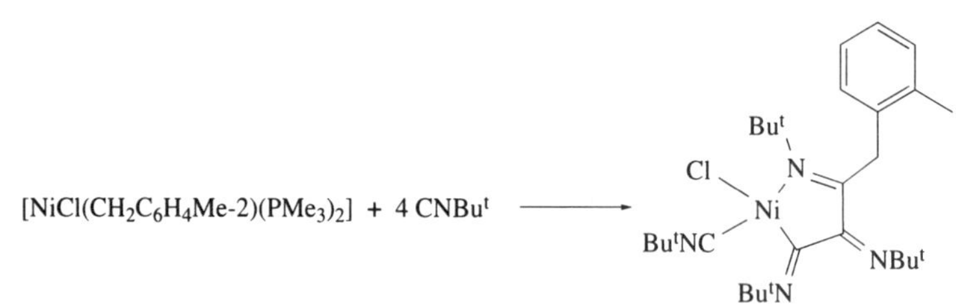

(147)

due to the presence of an occupied π-type orbital delocalized over a large part of the molecule with Ni–C antibonding character.[327] These compounds are remarkable in that they provide an easy entry into nickel(III) organometallic chemistry (Scheme 18).[332–4] The reversible one-electron oxidation of (**130**) (X = Br) to (**131**) occurs at +0.38 V. The oxidation is achieved chemically by treating (**130**) with copper(II) halides or silver(I) salts. The resulting nickel(III) complexes are generally square-pyramidal five-coordinate species although a six-coordinate nickel(III) complex $[Ni\{C_6H_3(CH_2NMe_2)_2\text{-}2,2'\}(NCS)_2(py)]$ (**132**), which has a tetragonally compressed octahedral geometry, has been isolated.[332] The crystal structures of two nickel(III) complexes have been determined but the data so far available do not allow conclusions to be made concerning the shortening or lengthening effects arising as a result of the electronic features of the bonding to nickel(III).

The presence of the readily oxidized nickel(II) centre in (**130**) has been found to be important in the use of these complexes as catalysts for the Kharasch addition of CCl_4 to methyl methacrylate.[328]

The phosphine analogue of (**130**) has been prepared (**133**).[339] The complex $[Ni(PPh_3)_2(C_2H_4)]$ reacts with lithiobenzyldiphenylphosphine to give (**134**).[338]

The nickel phosphinoenolate complex (**135**) undergoes a C–C bond-forming reaction on treatment with $MeO_2CC{\equiv}CCO_2Me$ to give the bis(alkenyl) complex (**136**) (Equation (148)).[336,337] Other alkynes were found to be unreactive under similar conditions, but the assumed inert character of the enolate moiety in polymerization catalyst precursors is now doubtful.

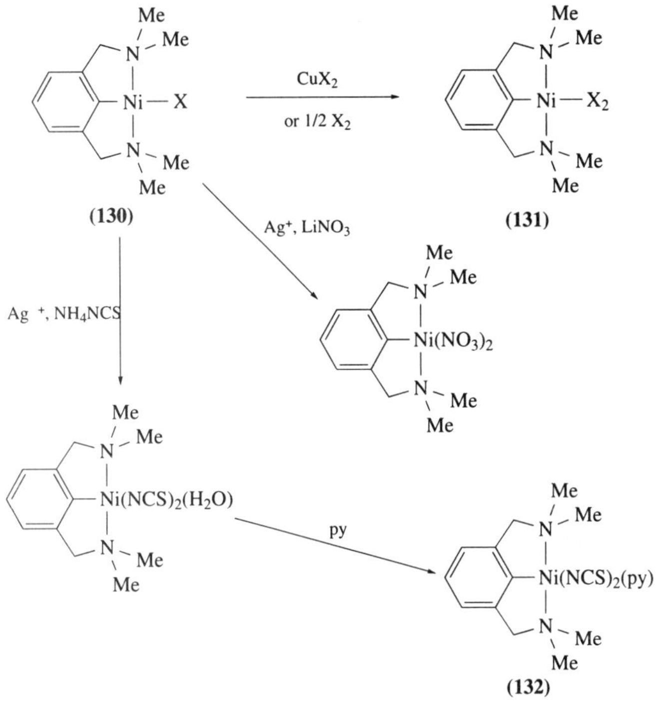

Scheme 18

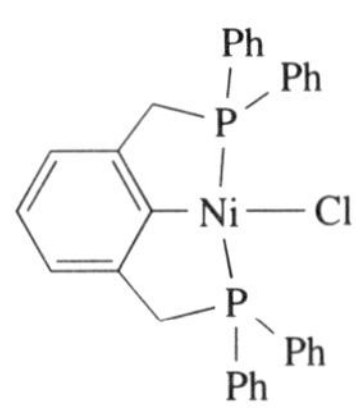

(133)

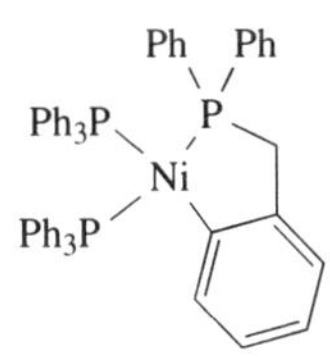

(134)

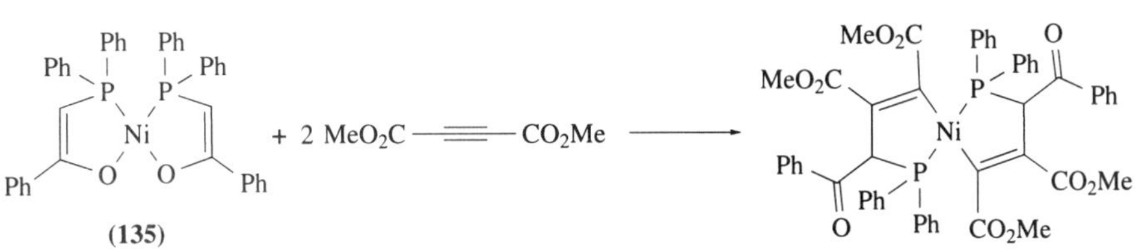

(136) (148)

Phosphorus ylide complexes forming nickelaheterocyles have been prepared (Equations (149) and (150)).[340,341] The cationic complex (**138**) can be reversibly deprotonated to yield (**137**) by treatment with $NaNH_2$. X-ray crystal structure determinations of (**137**) and (**138**) show that the delocalization in (**137**) has no apparent effect on the ylide Ni–C bond distances.

The reactions of CO_2 with nickel complexes have led to the preparation and characterization of many nickelacarboxylate complexes. The insertion of carbon dioxide into a Ni–C bond leads to C–C bond formation and proceeds via the pathway shown in Equation (151). Nickelacyclic complexes undergo a

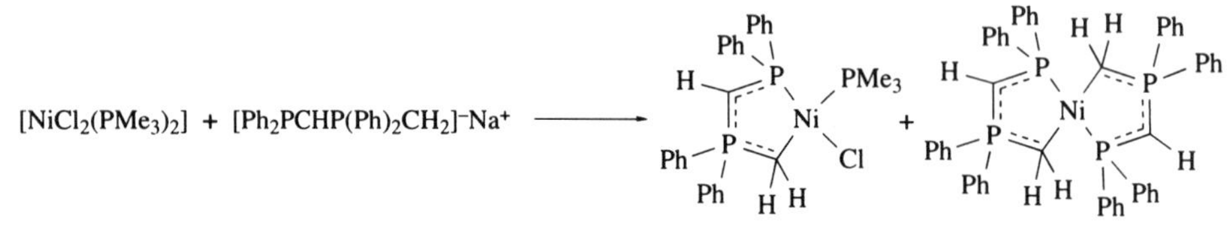

(137) (149)

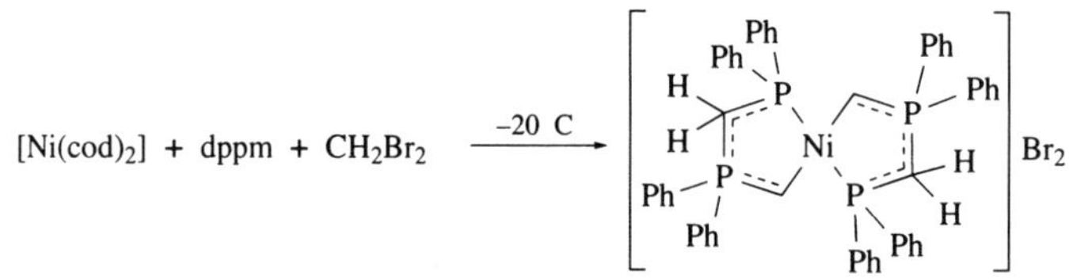

(138) (150)

similar reaction with CO_2 leading to oxanickelacycles (Equations (130), (152) and (153)). Reactions of this type have been described for many complexes of nickel and palladium, which form the most active CO_2-substrate coupling reagents.[378–81] Oxidative coupling is an important reaction in stoichiometric carbon dioxide chemistry. Coupling reactions at a nickel(0) centre between CO_2 and alkenes,[344,346,348–50] cycloalkenes,[344,363,364,374] acyclic or cyclic dienes and trienes,[353,364,372,373] allenes,[355] alkynes[344,366–71] aldehydes,[356,357] azaalkenes,[358–61] and a strained cycloalkane[373] lead to the formation of oxanickelacycles containing a Ni–C σ-bond. Some representative examples of these reactions are given in Equations (154)–(162).

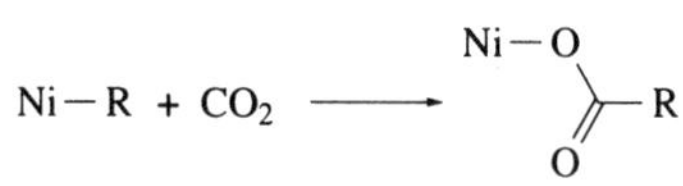

(151)

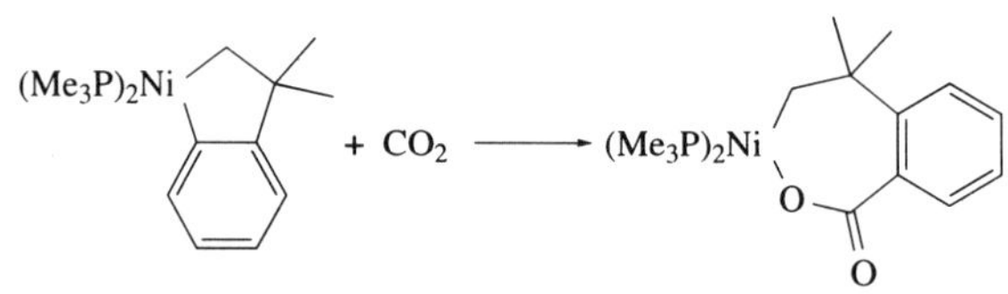

(152)

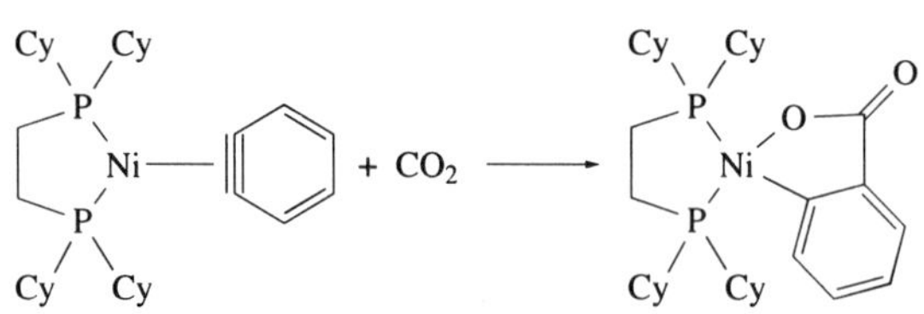

(153)

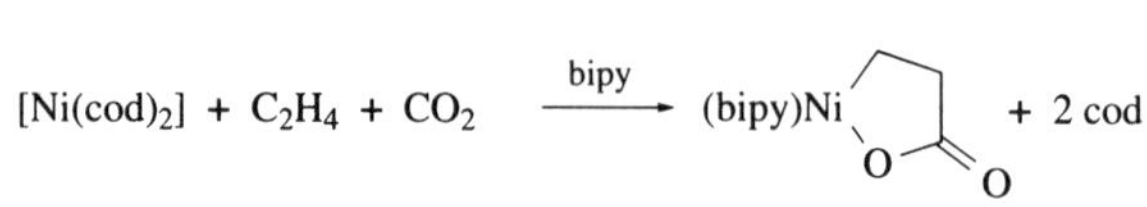

(154)

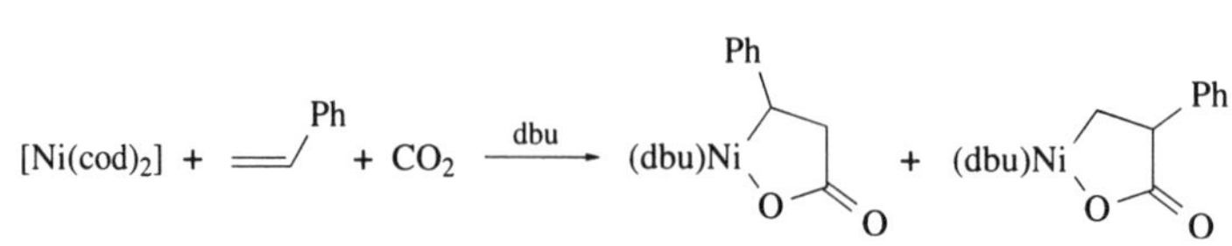

(139) (140) (155)

Monosubstituted alkenes CH_2=CHR react with CO_2 at a nickel(0) centre to give regioisomers (Equation (155)). The thermodynamically more stable product, with a Ni–CH_2 σ-bond, is obtained if the reaction mixture is heated for long periods, while at room temperature a mixture of isomers (**139**) and (**140**) is obtained, in the approximate ratio of 14:1. This suggests that the coupling reaction can proceed reversibly. With ethene and CO_2, the initially formed oxanickelapentacyclic complex will undergo further reaction with ethene to give an oxanickelaheptacyclic complex which, on treatment with acid, gives *n*-pentanoic acid.[346] In the presence of dbu (diazabicyclo[5.4.0]undec-7-ene) the coupling product

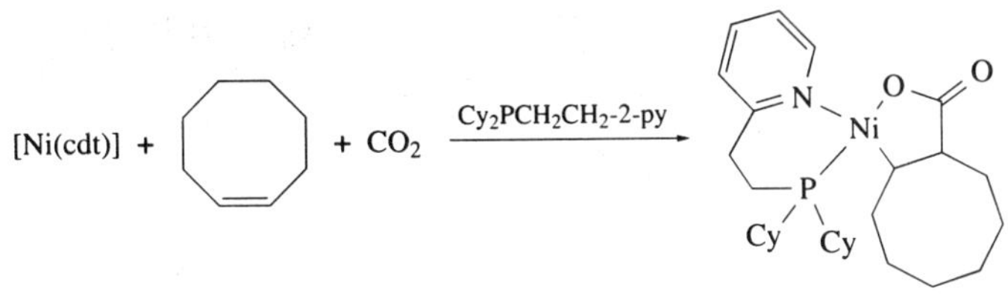

(156)

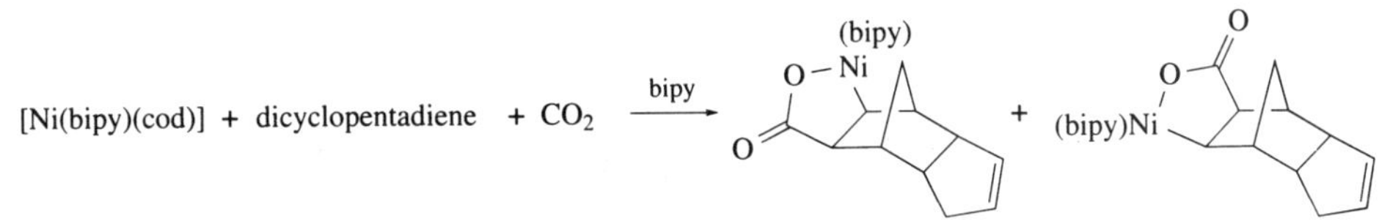

(141) **(142)**

(157)

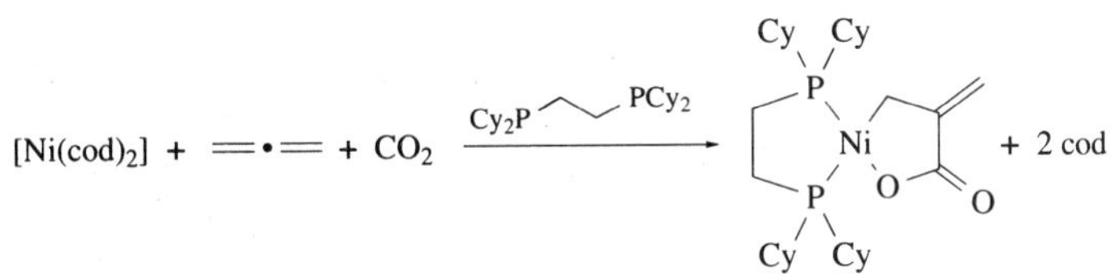

(143)

(158)

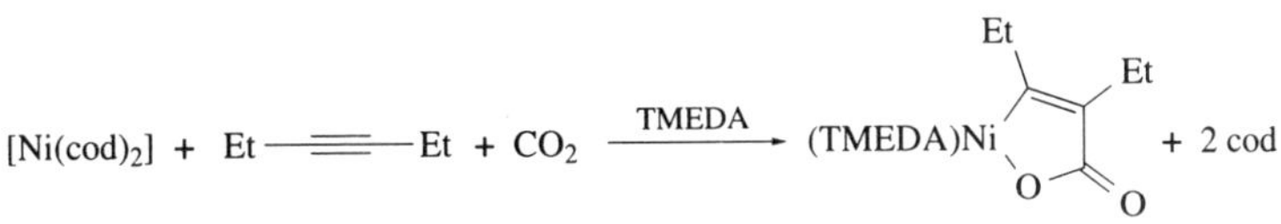

(144)

(159)

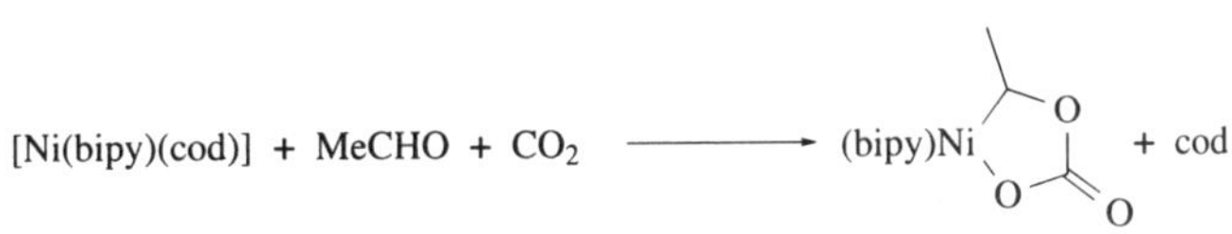

(145)

(160)

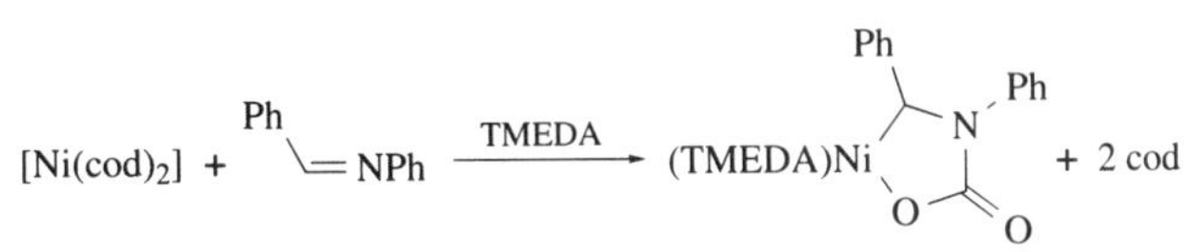

(146)

(161)

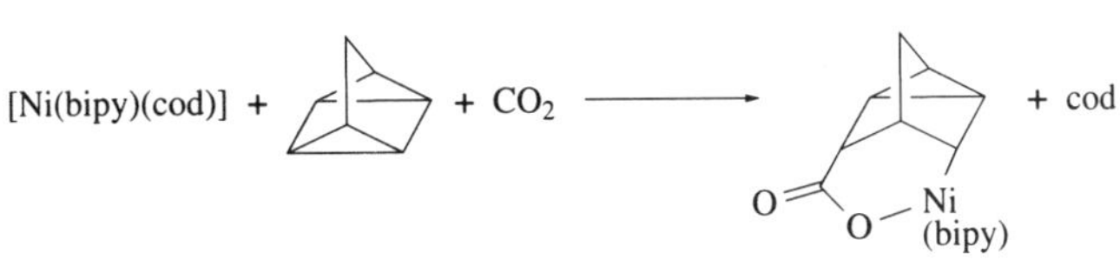

(162)

of ethene and CO_2 at nickel(0) has been isolated and structurally characterized (**147**).[349] In this case (**147**) reacts further with ethene via insertion into the Ni–C bond followed by β-elimination to give the hydridonickel complex $[NiH\{OC(O)CH_2CH_2CH{=}CH_2\}(dbu)_2]$.

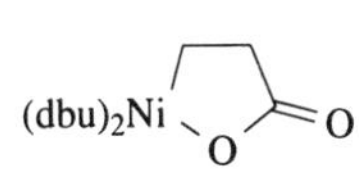

(147)

The products of the reaction between dicyclopentadiene and CO_2 at nickel(0) are the two regioisomers (**141**) and (**142**). These products are remarkably stable, decomposing to dicyclopentadiene and CO_2 at 230 °C.[372]

Allene and CO_2 couple at nickel(0) to give (**143**) (Equation (158)). If allenes of the type $CH_2=C=CHR$ or $CH_2=C=CR_2$ are used the reaction is regioselective with the only product having a Ni–CH_2 bond.[355]

Alkynes and CO_2 undergo a 1:1 coupling in the presence of nickel(0) to give nickelacyclic complexes such as (**144**) (Equation (159)).[367] A competing reaction is the reduction of CO_2 to CO, which may occur through an intermediate in which two molecules of CO_2 are coupled head-to-tail to form an oxanickelacycle.[368] Oxanickelacyclopentanones, of which (**144**) is an example, will react further with alkynes to give oxanickelacycloheptadienones (Equation (163)).[344] Complexes of this type are intermediates in the nickel-catalysed synthesis of 2-pyrones from alkynes and CO_2.[367] The electrochemical reduction of the nickel(II) complex $[Ni(bipy)_3][BF_4]_2$ generates a nickel(0) complex, which forms a nickelacyclic complex on reaction with 4-octyne and CO_2. In the presence of magnesium ions, magnesium carboxylate is eliminated and a nickel(II) complex is regenerated allowing the reaction to proceed catalytically.[366]

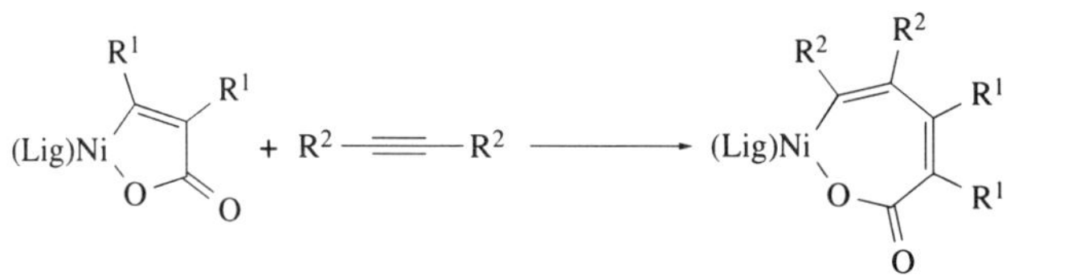

(163)

Coupling reactions of CO_2 with aldehydes (Equation (160))[356,357] and azaalkenes (Equation (161))[358] also lead to nickelacyclic products. The complex (**146**) will liberate CO_2 at 50 °C in a reversible reaction (Equation (164)).[358]

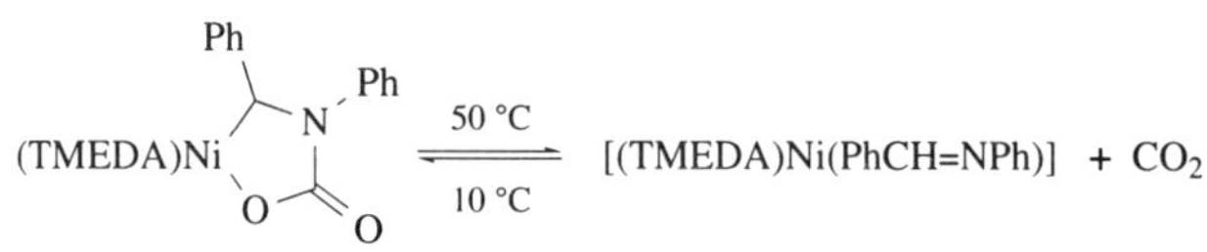

(**146**)

(164)

There are a number of other synthetic routes to oxanickelacyclic complexes which have been previously mentioned. These include the reaction of the nickelacycle (**98**) with formaldehyde to give (**106**),[102,293] the reaction of a nickelacycle with N_2O (Equation (135)),[96] and the treatment of the dinickel complex (**43**) with CO (Equation (64)).[106]

Another important route to oxanickelacyclic complexes is via the oxidative addition of cyclic carboxylic acid anhydrides (Equations (165)–(168)).[340,342,343,345,347,354,365] In these oxidative addition reactions the liberated CO is usually trapped by the nickel(0) complex leading to the formation of nickel carbonyl complexes. The oxidative addition of methylsuccinic anhydride (Equation (166)) is not regiospecific, both complexes (**148**) and (**149**) being formed. The ratio of (**148**) to (**149**) depends upon the ligand present and on the temperature of the reaction, but isomer (**149**) is generally favoured.[343] The oxanickelacyclic products may also be formed from nickel(II) complexes by treatment with triethylaluminum followed by the cyclic carboxylic acid anhydride (Equation (168)).[342]

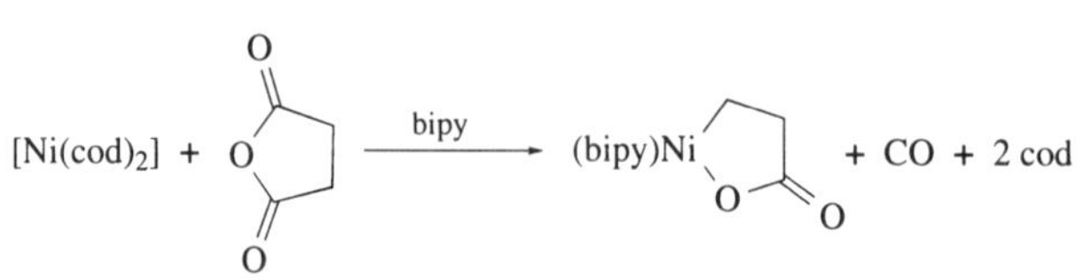

(165)

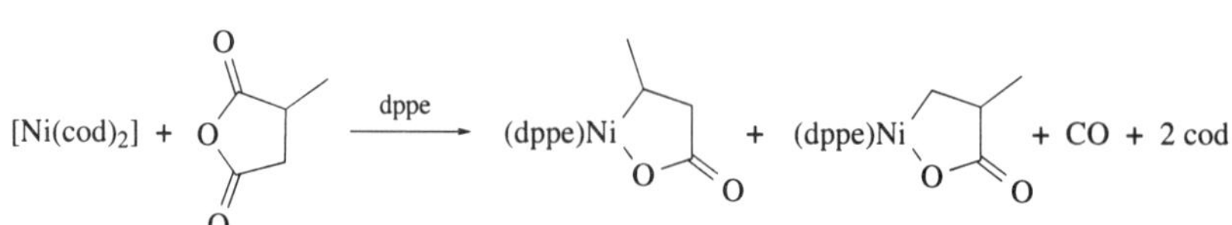

(**148**) (**149**)

(166)

When six-membered nickelacyclic products are formed, an irreversible ring contraction reaction can occur to give a five-membered nickelacycle (Equation (169)).[351,352] The ease of this reaction is strongly dependent on the ligand L, with a more sterically demanding ligand favouring ring contraction. In the

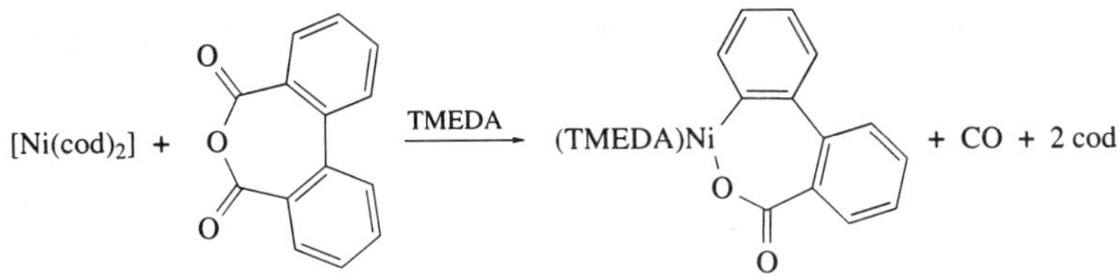

(167)

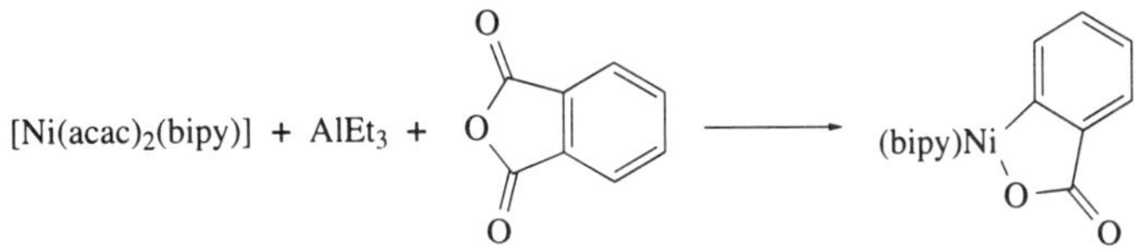

(150)

(168)

presence of dppe, ring contraction is instantaneous. The reaction produces a chiral centre and when a chiral phosphine is used to drive the reaction, asymmetry is induced in the five-membered nickelacyclic product.[351]

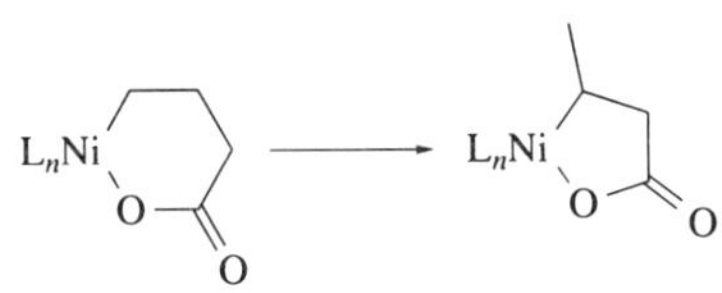

(169)

A $\overline{\text{Ni–C–P–S}}$ heterocyclic product is formed when $[NiCl_2(PPh_3)_2]$ reacts with $LiCH_2P(S)Ph_2$ (Equation (170)).[375]

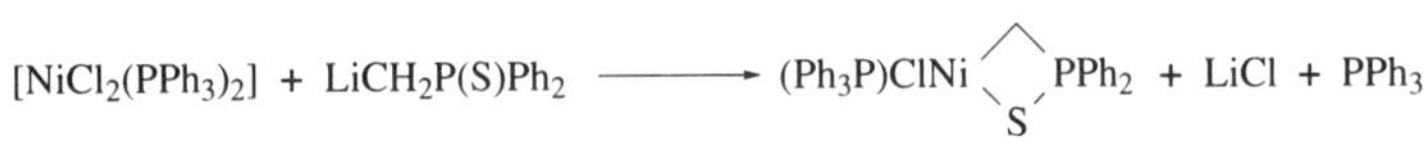

(170)

2.9 NICKEL COMPLEXES OF BIOLOGICAL SIGNIFICANCE

There are several metabolic enzymes of acetogenic and methanogenic organisms in which nickel is an essential component. Two of these enzymes, CO dehydrogenase (or acetyl coenzyme A synthase) and methyl coenzyme M reductase, catalyse reactions in which organonickel intermediates are almost certainly involved.[383–5,406,407] In CO dehydrogenase the nickel is probably present as part of a mixed nickel–iron cluster with the nickel surrounded by four sulfur donors or two sulfur and two nitrogen donors. In methyl coenzyme M reductase the nickel is surrounded by four nitrogen donors of a tetrapyrrole system. There is thus much interest in synthesizing model complexes in which organonickel is coordinated to nitrogen and/or sulfur donors only. Complexes of this type are listed in Table 12.

Significant advances have been made in recent years in the chemistry of nickel(II) and nickel(I) macrocyclic complexes.[386] Complexes containing the cyclam ligand, 1,4,8,11-tetraazacyclotetradecane, or its tetramethyl (tmc) or decamethyl (dmc) derivatives, have been the most extensively studied. The nickel(II) macrocycles can be reduced chemically, electrochemically or photochemically to generate stable nickel(I) complexes. The nickel(I) macrocycle $[Ni(tmc)]^+$ (tmc = 1,4,8,11-tetramethyl-1,4,8,11-tetraazacyclotetradecane) reacts with alkyl halides to give alkylnickel(II) complexes $[Ni(R)(tmc)]^+$ **(151)**,[387,388,392] with 1,4-dihaloalkanes to give ethene via an organonickel intermediate,[390,392] with 1,5-dihaloalkanes to give cyclopentane,[391] and with $Br(CH_2)_3OH$ to give $[Ni\{(CH_2)_3OH\}(tmc)]^+$.[392] The reactions of the more sterically crowded nickel macrocyclic complex $[Ni(dmc)]^+$ with organic halides occur some 10^4 times more slowly indicating that the reactions proceed by an inner sphere mechanism.[393] The (*R,R,S,S*)- and (*R,R,R,R*)-isomers of $[Ni(cyclam)]^{2+}$ undergo a reversible reaction with alkyl radicals, as shown in Equation (171).[394–7] It is found that the equilibrium constants for this reaction are highest for methyl radicals, less favourable for primary alkyls, and least favourable for secondary alkyls. The (*R,R,R,R*)-isomer also forms more stable complexes than the (*R,R,S,S*)-isomer.[395]

$$[Ni(cyclam)]^{2+} + R\bullet + H_2O \rightleftharpoons [NiR(cyclam)(H_2O)]^{2+} \qquad (171)$$

Table 12 Nickel complexes of biological significance.

Complex	*Reference*	*Ni–C* (nm)
$[NiMe(tmc)]^{+}$[a] (**151**)	387–395	
$[NiEt(tmc)]^{+}$[a]	387–395	
$[Ni(Pr^{n})(tmc)]^{+}$[a]	387–395	
$[Ni\{CH_2(CH_2)_3CH{=}CH_2\}(tmc)]^{+}$[a]	387–395	
$[NiMe(cyclam)(H_2O)]^{2+}$[b]	396,397	
$[NiMe(OEiBC)]$[c]	398,399	
$[NiPh(SDPDTPH)]$[d]	400	
$[NiPh(NMeTPPH)]$[e]	400	
$[NiMe(TSPP)]$[f]	401	
$[Ni(CF_3)(deuteroporphyrin)]$	402	
$[Ni(CF_3)(hematoporphyrin)]$	402	
$[NiMe\{N(CH_2CH_2SPr^{i})_3\}][BPh_4]$ (**152**)	403	0.194(2)
$[NiMe\{N(CH_2CH_2SBu^{t})_3\}][BPh_4]$	403	
$[Ni(COMe)\{N(CH_2CH_2SPr^{i})_3\}][BPh_4]$	403	0.190(2)
$[Ni(COMe)\{N(CH_2CH_2SBu^{t})_3\}][BPh_4]$	403	
$[NiMe\{Ph_2PCH_2CH_2S(CH_2)_3SCH_2CH_2PPh_2\}][BPh_4]$	404	
$[Ni(COMe)\{Ph_2PCH_2CH_2S(CH_2)_3SCH_2CH_2PPh_2\}][BPh_4]$	404	
$(NMe_4)[NiMe\{S(Me)C_6H_4S\text{-}2\}\{SC_6H_4(SMe)\text{-}2\}]$ (**153**)	405	0.195 4(7)
$F_{430M}CD_3$	406	

[a] tmc = 1,4,8,11-tetramethyl-1,4,8,11-tetraazacyclotetradecane. [b] cyclam = (*R,R,S,S*)- or (*R,R,R,R*)-1,4,8,11-tetraazacyclotetradecane. [c] OEiBC = octaethylisobacteriochlorin. [d] SDPDTPH = 5,20-diphenyl-10,15-bis(*p*-tolyl)-21-thiaporphyrin. [e] NMeTPPH = N-methyltetraphenylporphyrin. [f] TSPP = N-methyltetrakis(4-sulfonatophenyl)porphyrin.

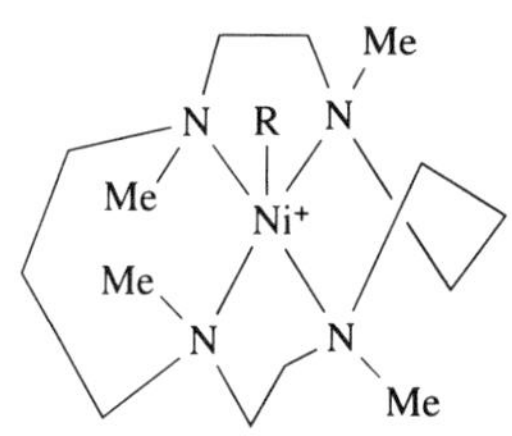

(151)

The complexes $[Ni(R)(tmc)]^{+}$ undergo hydrolysis to give RH and $[Ni(OH)(tmc)]^{+}$,[389] and react further with alkyl halides leading to the formation of coupled products R–R, alkanes and alkenes.[388]

A transient alkylnickel(III) intermediate is involved in the reaction of the nickel(I) complex $[Ni(OEiBC)]^{-}$ (OEiBC = octaethylisobacteriochlorin) with alkyl halides.[398,399] In this case the reaction proceeds by a nucleophilic S_N2-like mechanism, unlike the reaction of alkyl halides with $[Ni(tmc)]^{+}$, which reacts via a radical mechanism.

The phenylnickel(II) complexes [Ni(Ph)(SDPDTPH)] and [Ni(Ph)(NMeTPPH)] (SDPDTPH = 5,20-diphenyl-10,15-bis(*p*-tolyl)-21-thiaporphyrin; NMeTPPH = *N*-methyltetraphenylporphyrin) have been prepared by treatment of the nickel(II) monohalide porphyrin complexes with phenyl Grignard reagent.[400] *N*-methyltetrakis(4-sulfonatophenyl)porphinatonickel(II) [Ni(NMeTSPP)] undergoes a one-electron reduction to give a transient radical anion, $[Ni(NMeTSPP)]^{\cdot -}$, which can undergo a methyl migration from the pyrrole nitrogen to the nickel to give the unstable nickel(III) complex [Ni(Me)(TSPP)].[401]

The tripod ligands $N(CH_2CH_2SR)_3$ (R = Pr^{i} or Bu^{t}) have been used to prepare a series of nickel complexes of potential relevance to the nickel site in CO dehydrogenase. Thus, $[NiCl\{N(CH_2CH_2SR)_3\}]^{+}$ on treatment with MgClMe affords the methylnickel complex $[NiMe\{N(CH_2CH_2SR)_3\}]^{+}$ (**152**). This complex reacts with CO to give the corresponding acyl complex.[403] In complex (**152**) the nickel atom is in a trigonal bipyramidal ligand environment. In the complex (**153**) the coordination geometry is square planar.[405] These complexes represent rare examples of the stabilization of methylnickel complexes in the presence of nitrogen or sulfur donor atoms only. Further model compounds are provided by the S/P-donor ligand complexes (**154**) and (**155**).[404] These complexes may be chemically reduced by treatment with sodium amalgam to give red nickel(0) complexes, which have been shown to react with MeI and MeCOCl to give methyl and acyl nickel complexes (Scheme 19).[404]

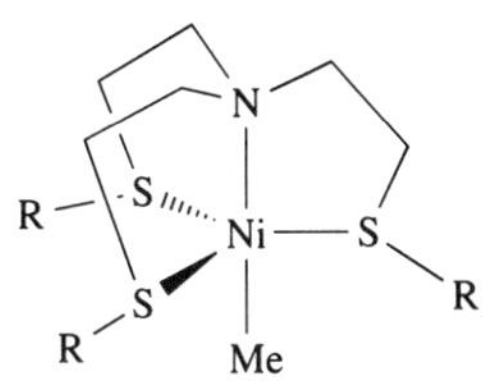

(152)

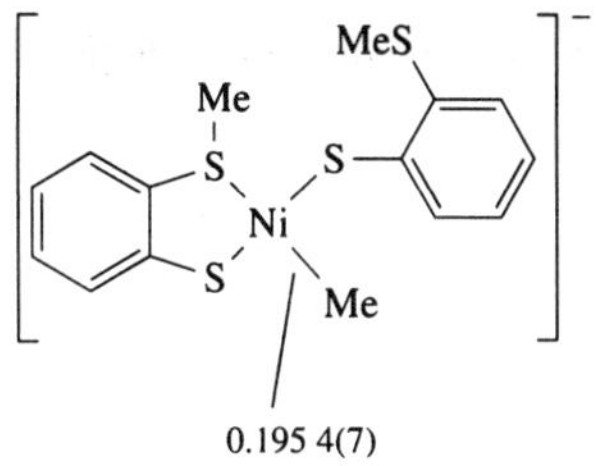

(153)

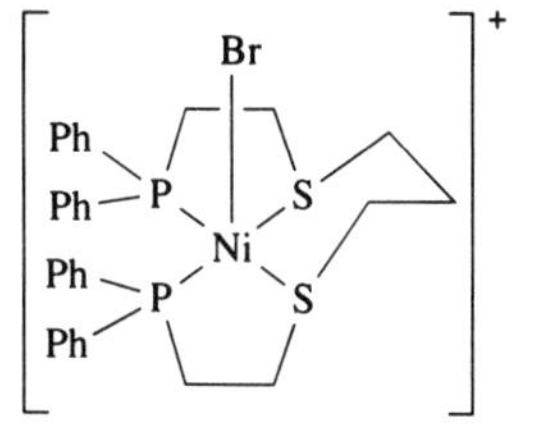

(154)

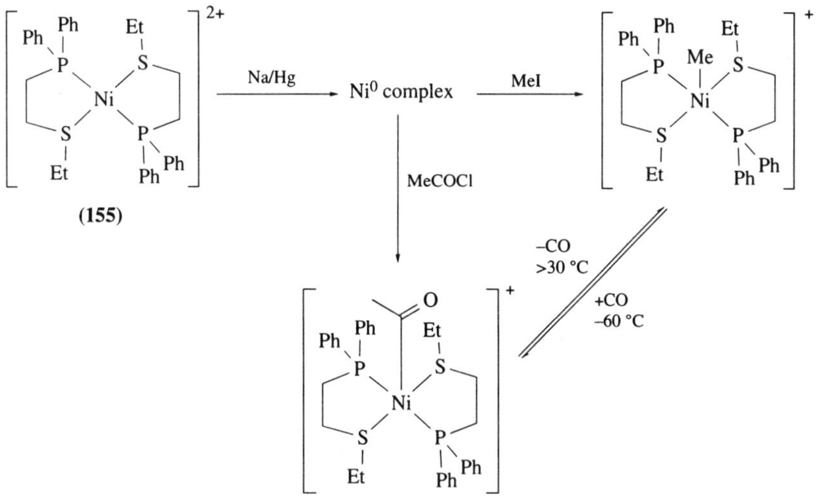

(155)

Scheme 19

2.10 NICKEL CLUSTERS CONTAINING Ni–C σ-BONDS

2.10.1 Homometallic Clusters

In this section, complexes containing at least one Ni–Ni bond and a Ni–C σ-bond are considered. The complexes are listed in Table 13. Nickel carbido cluster carbonyls are discussed elsewhere (Chapter 1, this volume) and are not considered here.

Table 13 Homonuclear nickel clusters.

Complex	*Reference*	*Ni–C* (nm)
$[Ni_2Cl_2\{\mu\text{-}C(PMe_3)TMS\}_2]$ (**156**)	408,409	0.190 6(4)
$[\{(Cp)Ni\}_2(\mu_2\text{-}HC{\equiv}CCN)]$ (**157**)	410	
$[\{(Cp)Ni\}_2(\mu_2\text{-}NCC{\equiv}CCN)]$	410	
$[\{(depe)Ni\}_2\{\mu\text{-}HC{\equiv}C(Ph)C(Ph){=}CH\}][PF]_2$ (**36**)	120	
$[\{(Cp)(\eta\text{-}C_5H_4CO_2Me)Ni_2\}\{\mu\text{-}C(Ph)CC(O)NHCH(Ph)Me\}]$ (**158**)	411	
$[\{(Cp)Ni\}_2\{\mu\text{-}C(Ph)CC(O)NHCH(Ph)Me\}]$	411	
$[\{(\eta\text{-}C_5H_4CO_2Me)Ni\}_2\{\mu\text{-}C(Ph)CC(O)NHCH(Ph)Me\}]$	411	
$[\{(\eta\text{-}C_5H_4Me)Ni\}_2\{\mu\text{-}C(Ph)CC(O)NHCH(Ph)Me\}]$	411	
$[\{(Cp)(\eta\text{-}C_5H_4Me)Ni\}_2\{\mu\text{-}C(Ph)CC(O)NHCH(Ph)Me\}]$	411	
$[\{(Cp)Ni\}_3(\mu_3\text{-}CMe)]$ (**160**)	413	0.184 0(4), 0.183 6(1)
$[\{(Cp^*)Ni\}_3(\mu_3\text{-}CMe)]$	413	
$[\{(Cp)Ni\}_3(\mu_3\text{-}CCH_2Me)]$	413	
$[\{(Cp)Ni\}_3(\mu_3\text{-}CCHMe_2)]$	413	
$[\{(Cp)Ni\}_3(\mu_3\text{-}CCO_2Me)]$	414	0.184(2)
$[\{(Pr^i{}_3P)Ni\}_3(\mu_3,\eta_2\text{-}C_6H_4)(\mu_3,\eta^2\text{-}C_6H_4C_6H_4)]$ (**161**)	415	0.1918, 0.1952, 0.2050, 0.2073, 0.1996, 0.1975, 0.1935, 0.1955

The Ni–Ni bonded dication [{(depe)Ni}$_2${μ-HC=C(Ph)C(Ph)=CH}][PF_6]$_2$ (**36**) was discussed earlier (Section 2.4.1.3).[120] The thermally unstable *C*-bonded diazoalkane nickel complex [NiCl{C(N_2)TMS}(PMe_3)$_2$] undergoes a photoinduced (300 nm) elimination of N_2 to produce the air-stable green μ-ylide dinickel complex (**156**) (Scheme 20).[408] Complex (**156**) has a very short Ni–Ni distance 0.228 1(1) nm. Electron counting would suggest a Ni=Ni double bond. However extended Hückel calculations suggest that there is little bonding interaction between the nickel atoms, but that the short Ni–Ni distance is due to the requirements of the bridging three-electron-donor ylide ligands.[409]

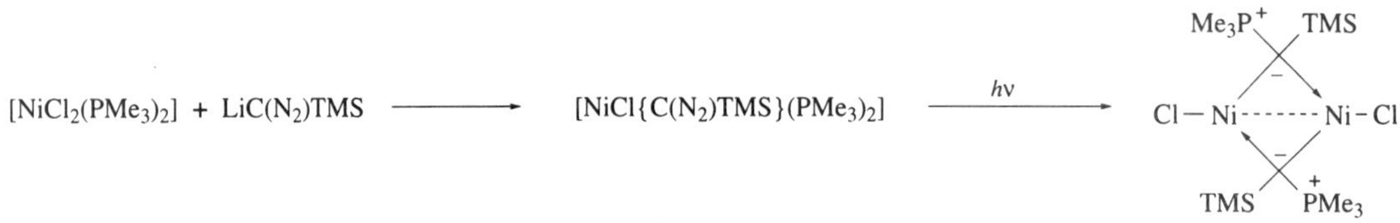

Scheme 20 **(156)**

Cyanoalkynes react slowly with nickelocene to produce a large amount of alkyne polymer and a low yield of the Ni–Ni bonded dimer (**157**).[410]

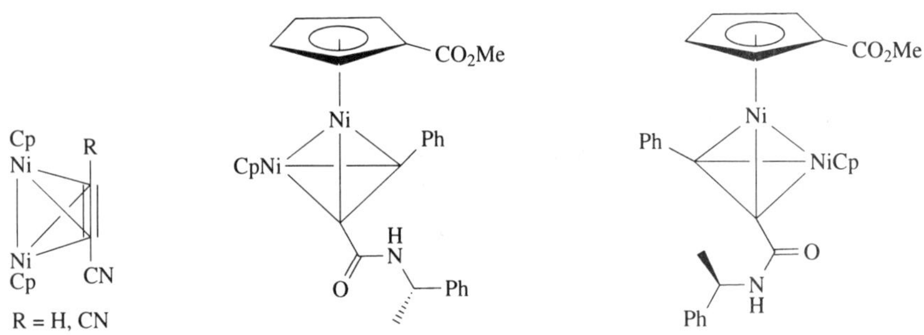

(157) **(158)** **(159)**

The chiral clusters (**158**) and (**159**) have been isolated from the reaction of a 1:1 mixture of nickelocene and [Ni($C_5H_4CO_2Me$)$_2$] with the alkyne (–)-(*R*)-PhC≡CC(O)NHCH(Me)Ph. The two diastereomers are configurationally stable with respect to a change of cluster chirality. The clusters [(CpNi)$_2${μ-PhC≡CC(O)NHCH(Me)Ph}] and [{($C_5H_4CO_2Me$)$_2$Ni}$_2${μ-PhC≡CC(O)NHCH(Me)Ph}] are also formed in this reaction.[411]

Since the first report of the synthesis of [(CpNi)$_3$($μ_3$-CPh)],[412] ($μ_3$-alkylidyne)trinickel clusters have received little attention.[1] A general method for the synthesis of these clusters is the reaction of nickelocene with vinyl or β-substituted vinyl derivatives of lithium or magnesium halide (Equation (172)).[413] The ($μ_3$-ethylidyne)trinickel cluster (**160**) ($R^1 = R^2 = H$) may also be prepared by treating nickelocene with methyllithium.[413] A metal exchange reaction (Equation (173))[414] has been used to synthesize the known cluster [(CpNi)$_3$($μ_3$-CPh)][412] and the new cluster [(CpNi)$_3$($μ_3$-CCO_2Me)].

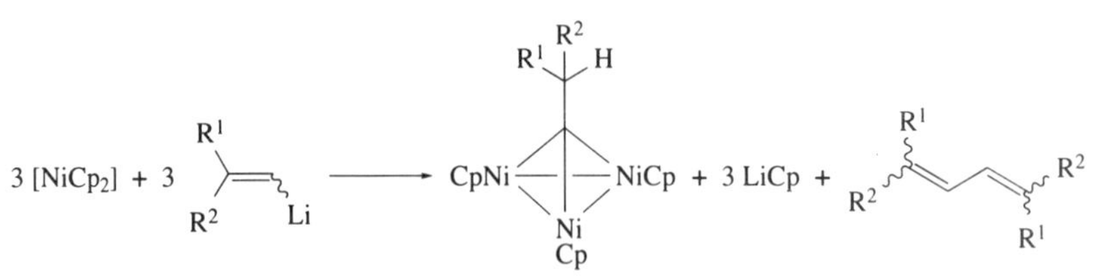

(160)

(172)

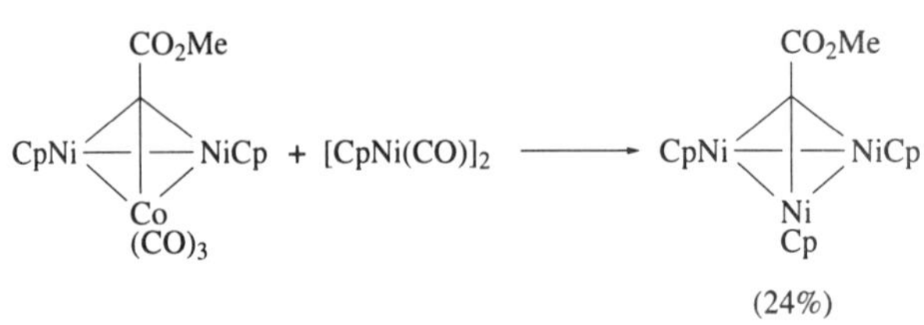

(173)

The reduction of $[NiCl(C_6H_4Cl\text{-}2)(PPr^i_3)_2]$ with sodium amalgam gives the η^2-benzyne complex $[Ni(\eta^2\text{-}C_6H_4)(PPr^i_3)_2]$ as the major product (30–50% yield), but also isolated from the product mixture (in 4–30% yield) is the unstable, black (μ_3-benzyne)(μ-2,2'-biphenylyl) dinickel complex (**161**).[415]

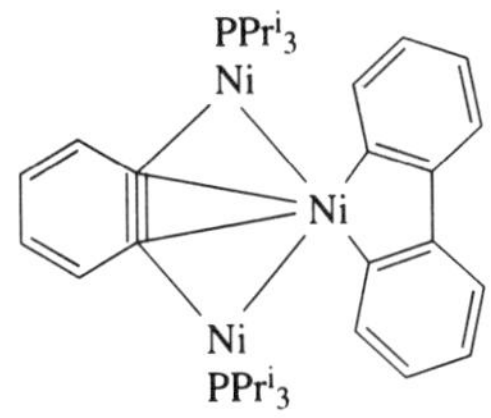

(161)

2.10.2 Heterometallic Clusters

Compounds with heteronuclear metal–metal bonds are discussed elsewhere (Volume 10), so this section will only briefly consider those heteronuclear clusters in which a Ni–C σ-bond is present. The complexes are listed in Table 14.

The heterometallic complexes $[(C_5R^1_5)NiM(CO)_4(C_5H_4R^2)]$ (**162**) (R^1 = H, Me; M = Cr, Mo, W; R^2 = H, Me) have proved to be useful starting materials for the synthesis of organonickel heterometallic complexes. The nickelacyclobutenone complexes (**163**) are obtained by the reaction between alkynes and $[NiM(CO)_4(Cp)(\eta\text{-}C_5H_4R)]$ (M = Cr, R = H; M = Mo, W, R = H, Me). These complexes slowly decarbonylate to afford the μ-alkyne species (**164**) (Equation (174)).[416–18] Complex (**162**) (M = Mo, R = Me) reacts with 1,1-dimethylallene to give the μ-allene complex (**165**), which isomerizes on silica gel to give the Ni–C σ-bonded complex (**166**).[419] Deuterium-labelling studies indicate that (**166**) is formed via a 1,4-proton shift rather than a 1,2-methyl migration process (Scheme 21).

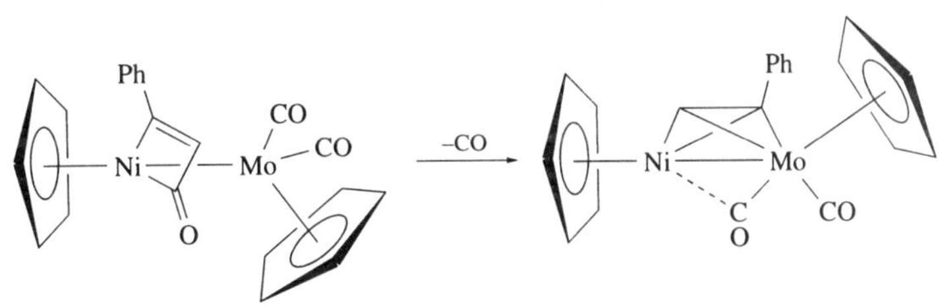

(174)

(163) **(164)**

The tricarbonyl complexes $[(Cp^*)NiM(CO)_3(Cp)]$ (M = Mo, W) react with diazomethane to give the μ-methylene complexes (**167**). The nickel–tungsten complex undergoes an insertion reaction with but-2-yne to afford (**168**), analogous to (**166**) (Scheme 22).[420] The μ-methylene complex (**167**) (M = W) undergoes a CH_2–CO coupling reaction on treatment with phenylacetylene to give (**169**). On treatment with allene, complex (**167**) gives the μ-trimethylenemethane complex (**170**) (Scheme 22).[421] Complex (**170**) is fluxional and it is not known whether it is formed by allene insertion into the W–CH_2 bond or the Ni–CH_2 bond.

The thermolysis of a mixture of $[Ni_2(PhC{\equiv}CPh)(Cp)_2]$ and $[Mo_2(CO)_6(Cp)_2]$ in refluxing *o*-xylene leads to C≡C bond cleavage and the formation of the bis(μ_3-phenylidyne) cluster (**171**).[422] $[Ni(cod)_2]$ reacts with the alkylidyne complex $[W(CC_6H_4Me\text{-}4)(CO)_2(Cp)]$ to give the trinuclear species (**172**).[423] The reaction of the phosphinide complex $[W({=}PMe)(CO)_5]$ with $[Ni_2(R^1C{\equiv}CR^2)(Cp)_2]$ affords the phosphametallacyclobutene complex (**173**).[424] The organonickel heterometallic butterfly cluster (**174**) is formed by the interaction of nickelocene with $[Re_2(\mu\text{-}H)(\mu\text{-}C{\equiv}CPh)(CO)_8]$ in refluxing *o*-xylene.[425] Nickelocene reacts with $[Fe_2(CO)(\mu\text{-}CO)\{\mu\text{-}\sigma{:}\eta^3\text{-}(CO)C_2R^1R^2\}(Cp)_2]$ (R^1 = Ph or H, R^2 = Ph) to give the heterodinuclear species (**175**).[426] One of the products of the reaction of $[Fe\{\eta^3\text{-}(MeO_2C)HC{=}C(CO_2Me)CO\}(CO)_3]^-$ with $[NiBr(PPh_3)(Cp)]$ is the heterobimetallic complex (**176**). This

Table 14 Heteronuclear nickel clusters.

Complex	*Reference*	*Ni–C* (nm)
$[NiCr\{\mu\text{-}\eta^2,\eta^2\text{-}C(Me)C(Me)\}(CO)_2(Cp)_2]$	416,417	
$[NiMo\{\mu\text{-}\eta^2,\eta^2\text{-}C(Me)C(Me)\}(CO)_2(Cp)_2]$	416	
$[NiW\{\mu\text{-}\eta^2,\eta^2\text{-}C(Me)C(Me)\}(CO)_2(Cp)_2]$	416	0.194 8(7), 0.194 5(7), 0.194 6(7), 0.195 8(6)
$[NiMo\{\mu\text{-}\eta^2,\eta^2\text{-}C(Me)C(Me)\}(CO)_2(Cp)(C_5H_4Me)]$	416	
$[NiW\{\mu\text{-}\eta^2,\eta^2\text{-}C(Me)C(Me)\}(CO)_2(Cp)(C_5H_4Me)]$	416	
$[NiCr\{\mu\text{-}\eta^2,\eta^2\text{-}C(Me)C(Et)\}(CO)_2(Cp)_2]$	416	
$[NiMo\{\mu\text{-}\eta^2,\eta^2\text{-}C(Me)C(Et)\}(CO)_2(Cp)_2]$	416	
$[NiW\{\mu\text{-}\eta^2,\eta^2\text{-}C(Me)C(Et)\}(CO)_2(Cp)_2]$	416	
$[NiMo\{\mu\text{-}\eta^2,\eta^2\text{-}C(Me)C(Et)\}(CO)_2(Cp)(C_5H_4Me)]$	416	
$[NiW\{\mu\text{-}\eta^2,\eta^2\text{-}C(Me)C(Et)\}(CO)_2(Cp)(C_5H_4Me)]$	416	
$[NiMo\{\mu\text{-}\eta^2,\eta^2\text{-}CHC(Ph)\}(CO)_2(Cp)_2]$	416	
$[NiW\{\mu\text{-}\eta^2,\eta^2\text{-}CHC(Ph)\}(CO)_2(Cp)_2]$	416	
$[NiMo\{\mu\text{-}\eta^2,\eta^2\text{-}CHC(Ph)\}(CO)_2(Cp)(C_5H_4Me)]$	416	
$[NiW\{\mu\text{-}\eta^2,\eta^2\text{-}CHC(Ph)\}(CO)_2(Cp)(C_5H_4Me)]$	416	
$[NiMo\{\mu\text{-}\eta^2,\eta^2\text{-}C(Ph)C(Ph)\}(CO)_2(Cp)(C_5H_4Me)]$	416	
$[NiW\{\mu\text{-}\eta^2,\eta^2\text{-}C(Ph)C(Ph)\}(CO)_2(Cp)(C_5H_4Me)]$	416	
$[NiW\{\mu\text{-}\eta^2,\eta^2\text{-}C(CO_2Me)C(CO_2Me)\}(CO)_2(Cp)_2]$	416	
$[NiMo\{\mu\text{-}\eta^2,\eta^2\text{-}C(Me)C(Me)C(O)\}(CO)_2(Cp)_2]$	416	
$[NiMo\{\mu\text{-}\eta^2,\eta^2\text{-}C(Me)C(Me)C(O)\}(CO)_2(Cp)(C_5H_4Me)]$	416	0.191 2(2), 0.192 7(2)
$[NiMo\{\mu\text{-}\eta^2,\eta^2\text{-}C(Me)C(Et)C(O)\}(CO)_2(Cp)_2]$	416	
$[NiMo\{\mu\text{-}\eta^2,\eta^2\text{-}C(Me)C(Et)C(O)\}(CO)_2(Cp)(C_5H_4Me)]$	416	
$[NiMo\{\mu\text{-}\eta^2,\eta^2\text{-}CHC(Ph)C(O)\}(CO)_2(Cp)_2]$	416	
$[NiMo\{\mu\text{-}\eta^2,\eta^2\text{-}CHC(Ph)C(O)\}(CO)_2(Cp)(C_5H_4Me)]$	416	
$[NiMo\{\mu\text{-}\eta^2,\eta^2\text{-}C(Ph)C(Ph)C(O)\}(CO)_2(Cp)(C_5H_4Me)]$	416	
$[NiMo\{\mu\text{-}\eta^2,\eta^2\text{-}CHC(Ph)C(O)\}(CO)_2(Cp^*)(Cp)]$ **(163)**	418	
$[NiMo\{\mu\text{-}\eta^1,\eta^3\text{-}C(Me)C(Me)CH_2\}(CO)_2(Cp)(C_5H_4Me)]$ **(166)**	419	0.191 0(4)
$[NiW\{\mu\text{-}\eta^1,\eta^3\text{-}C(Me)C(Me)CH_2\}(CO)_2(Cp^*)(Cp)]$ **(168)**	420	
$[NiMo(\mu\text{-}CH_2)(\mu\text{-}CO)(CO)_2(Cp^*)(Cp)]$ **(167)**	420,421	
$[NiW(\mu\text{-}CH_2)(\mu\text{-}CO)(CO)_2(Cp^*)(Cp)]$ **(167)**	420,421	0.190 6(5)
$[NiW\{\mu\text{-}\eta^1,\eta^3\text{-}CH_2C(CH_2)CH_2\}(\mu\text{-}CO)_2(Cp^*)(Cp)]$ **(170)**	421	0.196 2(6)
$[Ni_2Mo_2(\mu_3\text{-}CPh)_2(\mu_3\text{-}CO)_2(Cp)_4]$ **(171)**	422	0.193 7(9)
$[NiW_2(\mu\text{-}CC_6H_4Me\text{-}4)_2(CO)_4(Cp)_2]$ **(172)**	423	0.189 3(7), 0.189 5(8)
$[(Cp)Ni\{\eta^2\text{-}MePC(Ph){=}C(Ph)NiCp\}W(CO)_5]$ **(173)**	424	0.192 0(7)
$[(Cp)Ni\{\eta^2\text{-}MePC(CO_2Et){=}C(Ph)NiCp\}W(CO)_5]$	424	
$[(Cp)Ni\{\eta^2\text{-}MePCH{=}C(Ph)NiCp\}W(CO)_5]$	424	
$[(Cp)Ni\{\eta^2\text{-}MePC(Ph){=}CHNiCp\}W(CO)_5]$	424	
$[Ni_2Re_2\{\eta^1,\eta^1,\eta^2,\eta^2\text{-}C(Ph){\equiv}C(CH{=}CHPh)\}(CO)_6(Cp)_2]$ **(174)**	425	0.188(1), 0.195(1)
$[NiFe\{\mu\text{-}\sigma{:}\eta^4\text{-}C(Ph)C(Ph)C(Ph)C(Ph)\}(Cp)_2]$ **(175)**	426	0.193 2(8)
$[NiFe\{\mu\text{-}\sigma{:}\eta^4\text{-}CHCHC(Ph)C(Ph)\}(Cp)_2]$	426	
$[NiFe\{\mu\text{-}\eta^1,\eta^2\text{-}C(Ph){=}CHPh\}(CO)_3(PPh_3)(Cp)]$ **(177)**	427	
$[NiFe\{\mu\text{-}\eta^1,\eta^3\text{-}C(CO_2Me){=}CHCO_2Me\}(CO)_2(PPh_3)(Cp)]$ **(176)**	427	
$[Cp(OC)Ni\{\mu\text{-}C(O)CH_2\}Fe(CO)_2(Cp)]$ **(178)**	429	0.194 0(6)
$[Ni_3Fe(\mu_4\text{-}C{\equiv}CH)(\mu\text{-}CO)(CO)_2(Cp)_3]$	430	0.194(1)
$[(Cp)_2Ni_2Ru_2\{\mu_3\text{-}CCHC(CH_2)Me\}(CO)_6]$ **(179)**	431	0.183(3), 185(3)
$[CpNiCo_2(\mu_3\text{-}CH)(CO)_6]$	432	
$[CpNiCo_2(\mu_3\text{-}CMe)(CO)_6]$	432	
$[CpNiCo_2(\mu_3\text{-}CCO_2Me)(CO)_6]$	414	0.183(1)
$[CpNiCo_2\{\mu_3\text{-}CC(O)Ph\}(CO)_6]$	414	
$[Cp_2Ni_2Co(\mu_3\text{-}CMe)(CO)_3]$	414	
$[Cp_2Ni_2Co(\mu_3\text{-}CCO_2Me)(CO)_3]$	414	0.185(2)
$[CpNi(CpCo)_2(\mu_3\text{-}CCO_2Me)(CO)]$	414	
$[CpNiCo_2(\mu_3\text{-}CCO)(CO)_6][PF_6]$ **(181)**	433	
$[CpNiCo_2\{\mu_3\text{-}CC(O)OMe\}(CO)_6]$	433	
$[CpNiCo_2\{\mu_3\text{-}CC(O)OEt\}(CO)_6]$	433	
$[CpNiCo_2\{\mu_3\text{-}CC(O)C_4H_4N\}(CO)_6]$	433	
$[CpNiCo_2\{\mu_3\text{-}CC(O)C_8H_6N\}(CO)_6]$	433	
$[CpNiCo_2\{\mu_3\text{-}CC(O)OCHMe_2\}(CO)_6]$ **(180)**	434	
$[CpNiMoCo(\mu_3\text{-}CPh)(CO)_5(Cp)]$	435,436,438	
$[CpNiMoCo\{\mu_3\text{-}C(O)Ph\}(CO)_5(Cp)]$	435,436,438	
$[CpNiMoCo\{\mu_3\text{-}C(O)OPh\}(CO)_5(Cp)]$	435,436,438	
$[CpNiMoCo\{\mu_3\text{-}C(O)Omenthyl\}(CO)_5(Cp)]$	435,437,438	
$[CpNiCo_2Ru(\mu_4\text{-}CCMe)(CO)_9]$	439	
$[CpNiCo_2Ru(\mu_4\text{-}CCMe)(CO)_8(PPh_3)]$ **(182)**	439	0.192 1(8)
$[(Et_3P)_2(PhC{\equiv}C)Ni\{\mu\text{-}C({=}CHPh)\}Pt(PEt_3)_2][PF_6]$ **(183)**	440	

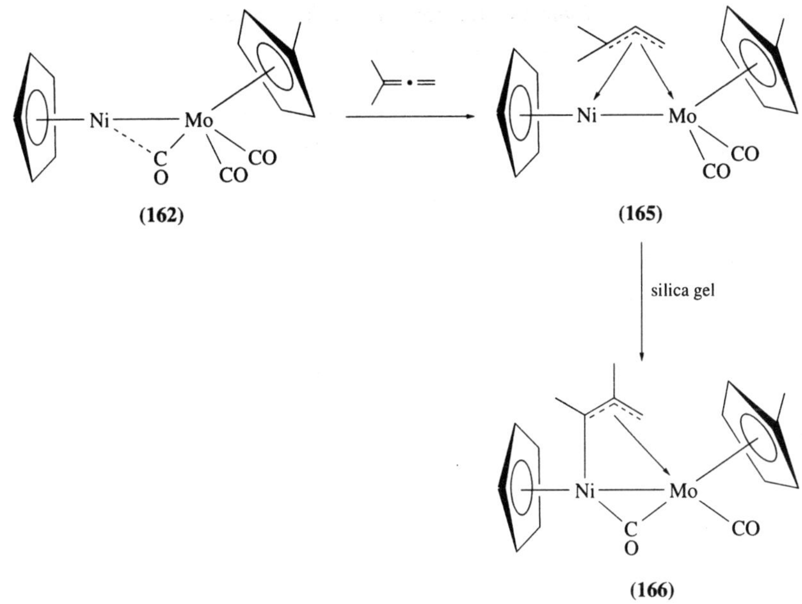

Scheme 21

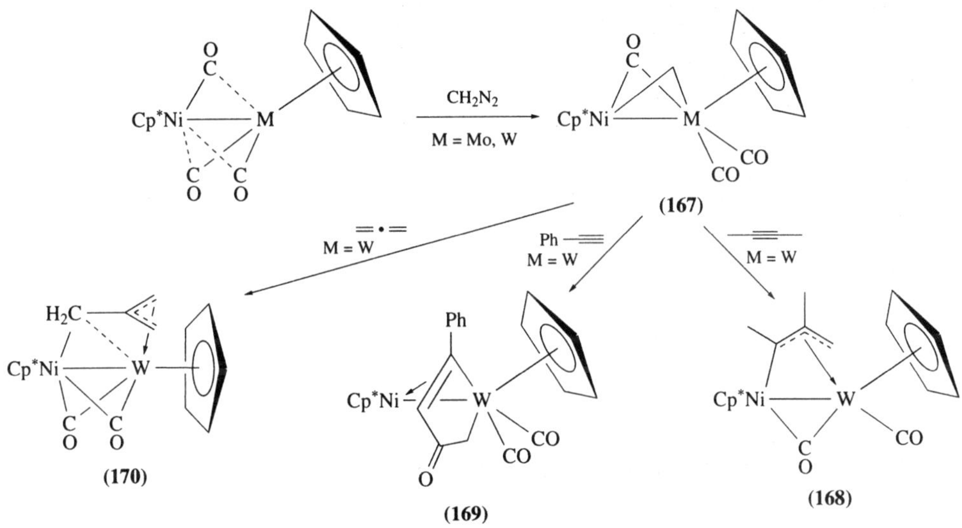

Scheme 22

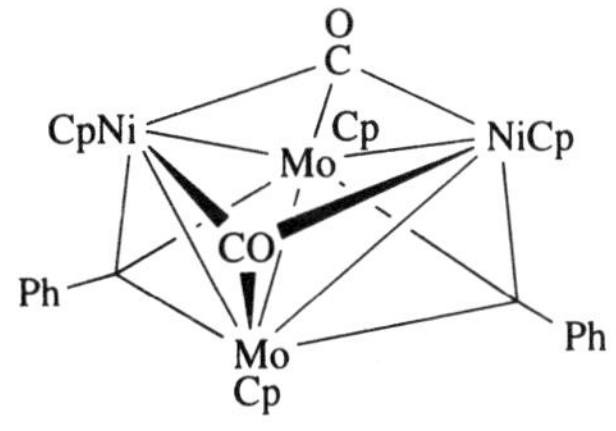

(171)

product is also obtained when a solution of $[PPh_4][FeH(CO)_4]$, $MeO_2CC{\equiv}CCO_2Me$ and $[NiBr(PPh_3)(Cp)]$ is refluxed in THF. When diphenylacetylene is used in this latter reaction the heterobimetallic nickel alkenyl complex (**177**) is formed.[427] The reaction between $[Ni(CO)(Cp)]^-$ and $[Fe\{CH_2C(O)Cl\}(CO)_2(Cp)]$ also produces a heterobimetallic product, (**178**), but in this case no Ni–Fe bond is present.[428,429]

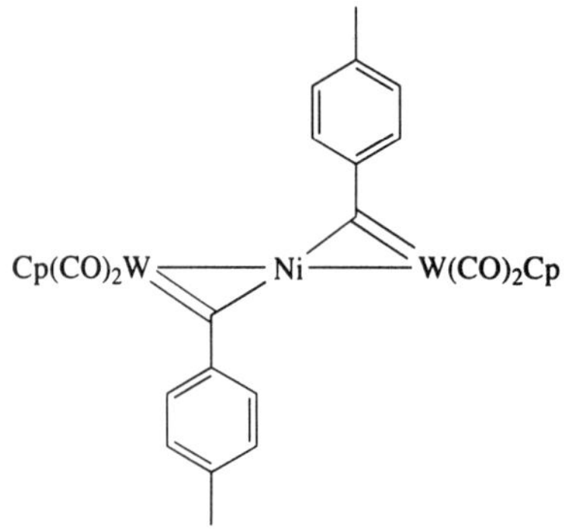

(172)

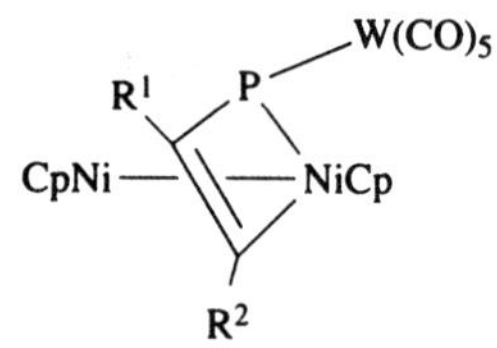

(173)

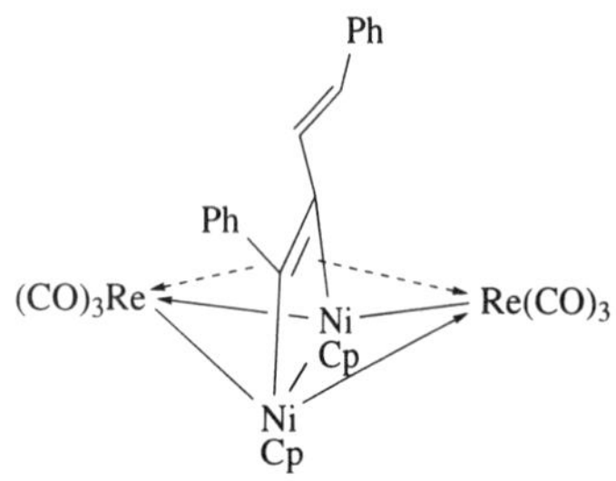

(174)

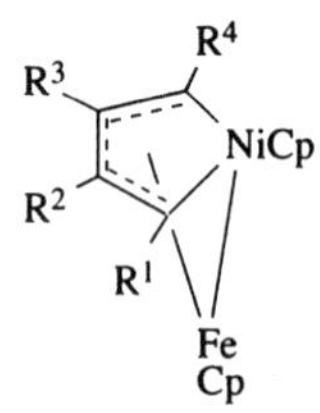

(175)

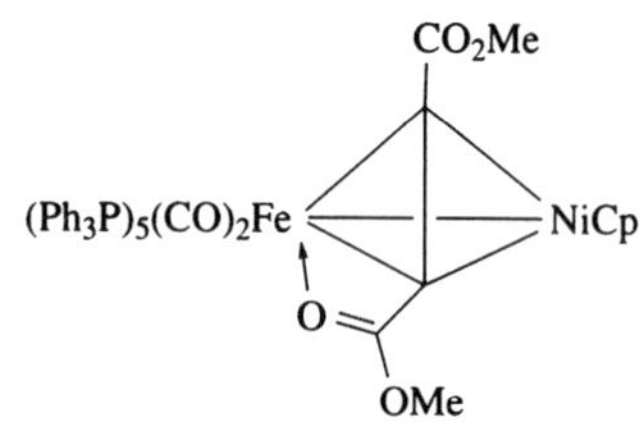

(176)

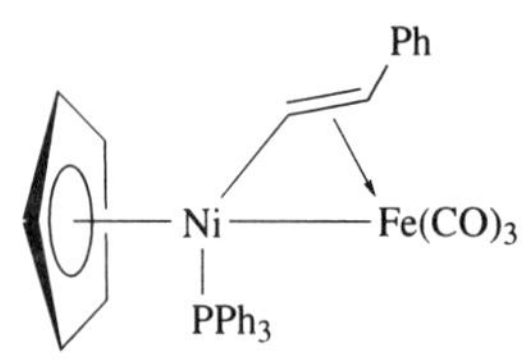

(177)

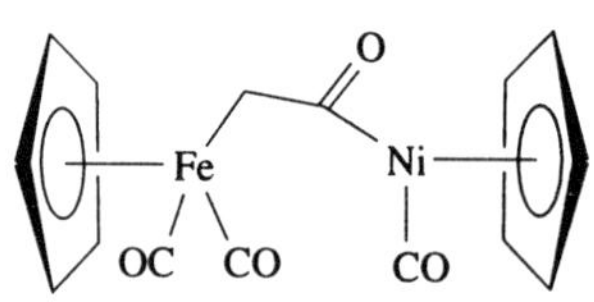

(178)

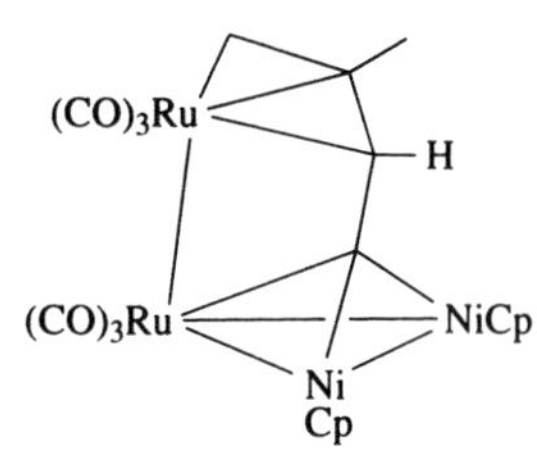

(179)

The reaction of $[Ni(CO)_2(Cp)_2]$ with $(Me)_2CHC{\equiv}CH$ followed by $[Ru_3(CO)_{12}]$ affords (**179**) as the major product.[431]

A wide range of μ_3-alkylidyne complexes containing nickel and cobalt,[414,432–4] and nickel, cobalt and molybdenum,[435–8] are known (Table 14). These complexes may be prepared by metal exchange reactions in which a $Co(CO)_3$ fragment is eliminated from the starting cluster, which is usually $[Co_3(\mu_3\text{-}CR)(CO)_9]$. This reaction proceeds in a stepwise manner allowing two different metal atoms to be incorporated into the original tricobalt framework. Treatment of the dicobaltnickel cluster (**180**) with HPF_6 gives the acylium cation (**181**). This cation reacts with alcohols to give esters and with indole or pyrrole to give Friedel–Crafts type products (Scheme 23).[433,434]

The nickel acetylide containing tetranuclear cluster (**182**) has been synthesized by deprotonation of the alkyne-bridged cluster $[Co_2Ru(CO)_9(\mu_3\text{-}RC{\equiv}CH)]$ followed by reaction with the nickel halide complex $[NiCl(PPh_3)(Cp)]$.[439]

A nickel–platinum heterobimetallic complex (**183**) has been prepared by the addition of the Pt–H bond of $[PtH(PEt_3)_2(Me_2CO)][PF_6]$ to the C≡C bond of one of the alkynyl groups in $[Ni(C{\equiv}CPh)_2(PEt_3)_2]$.[440]

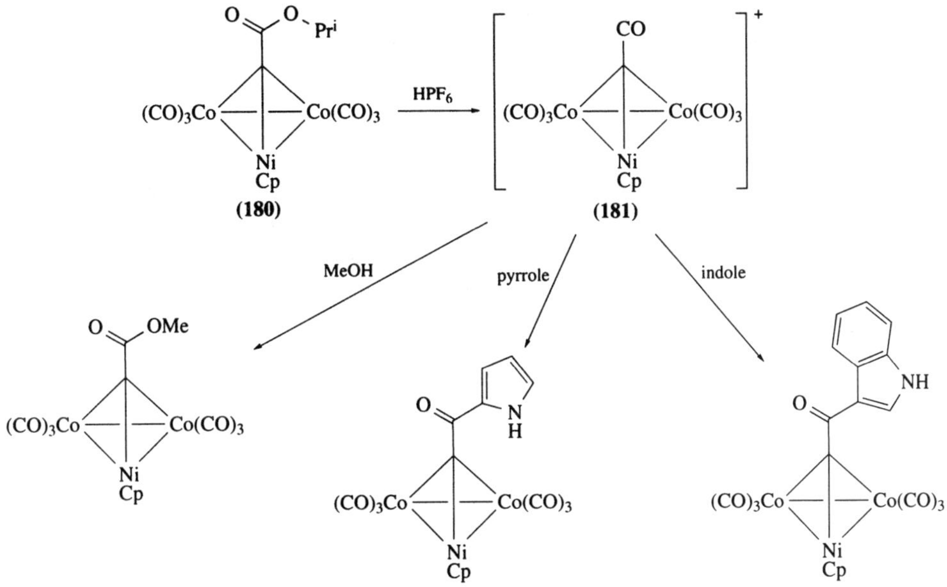

Scheme 23

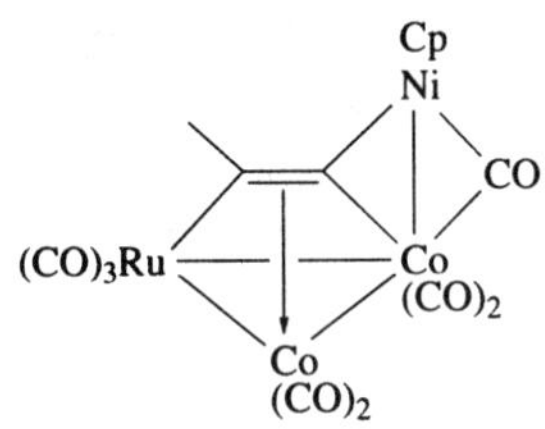

(182)

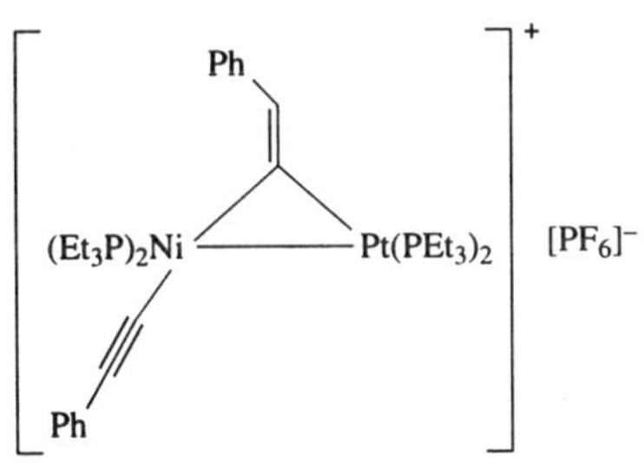

(183)

2.11 REFERENCES

1. P. W. Jolly, in 'COMC-I', vol. 6, p. 37.
2. W. Keim, *Angew. Chem., Int. Ed. Engl.*, 1990, **29**, 235.
3. A. V. Kavaliunas, A. Taylor and R. D. Rieke, *Organometallics*, 1983, **2**, 377.
4. C. J. Lawrie, K. P. Gable and B. K. Carpenter, *Organometallics*, 1989, **8**, 2274.
5. L. H. Simons and J. J. Lagowski, *Fundam. Res. Homogeneous Catal.*, 1978, **2**, 73.
6. E. S. Kline, R. H. Hauge, Z. H. Kafafi and J. L. Margrave, *Organometallics*, 1988, **7**, 1512.
7. J. Allison, *Prog. Inorg. Chem.*, 1986, **34**, 627.
8. J. A. Martinho Simões and J. L. Beauchamp, *Chem. Rev.*, 1990, **90**, 629.
9. J. C. Weisshaar, *Acc. Chem. Res.*, 1993, **26**, 213.
10. K. Eller, *Coord. Chem. Rev.*, 1993, **126**, 93.
11. K. Eller and H. Schwartz, *Chem. Rev.*, 1991, **91**, 1121.
12. R. Georgiadis, E. R. Fisher and P. B. Armentrout, *J. Am. Chem. Soc.*, 1989, **111**, 4251.
13. E. R. Fisher, L. S. Sunderlin and P. B. Armentrout, *J. Phys. Chem.*, 1989, **93**, 7375.
14. L. F. Halle, W. E. Crowe, P. B. Armentrout and J. L. Beauchamp, *Organometallics*, 1984, **3**, 1694.
15. E. R. Fisher and P. B. Armentrout, *J. Phys. Chem.*, 1990, **94**, 1674.
16. T. Zeigler, W. Cheng, E. J. Baerends and W. Ravenek, *Inorg. Chem.*, 1988, **27**, 3458.
17. T. Zeigler, V. Tschinke and A. Becke, *J. Am. Chem. Soc.*, 1987, **109**, 1351.

18. H. Yang and J. L. Whitten, *J. Am. Chem. Soc.*, 1991, **113**, 6442.
19. J. Schüle, P. Siegbahn and U. Wahlgren, *J. Chem. Phys.*, 1988, **89**, 6982.
20. E. Shustorovich, *Acc. Chem. Res.*, 1988, **21**, 189.
21. A. T. Bell and E. Shustorovich, *J. Catal.*, 1990, **121**, 1.
22. T. H. Upton, *J. Vac. Sci. Technol.*, 1982, **20**, 527.
23. J.-Y. Saillard and R. Hoffmann, *J. Am. Chem. Soc.*, 1984, **106**, 2006.
24. (a) M. R. A. Blomberg, U. Brandemark and P. E. M. Siegbahn, *J. Am. Chem. Soc.*, 1983, **105**, 5557; (b) E. Bauwe and G. Rasch, *Z. Chem.*, 1984, **24**, 304.
25. F. Zaera and S. Tjandra, *J. Am. Chem. Soc.*, 1993, **115**, 5851.
26. O. Blum, P. O'Bannon, D. Schröder and H. Schwartz, *Organometallics*, 1993, **12**, 980.
27. S.-C. Chang, R. H. Hauge, W. E. Billups, J. L. Margrave and Z. H. Kafafi, *Inorg. Chem.*, 1988, **27**, 205.
28. M. R. A. Blomberg, J. Schüle and P. E. M. Siegbahn, *J. Am. Chem. Soc.*, 1989, **111**, 6156.
29. Y. H. Pan and D. P. Ridge, *J. Am. Chem. Soc.*, 1992, **114**, 2773.
30. R. Stepnowski and J. Allison, *J. Am. Chem. Soc.*, 1989, **111**, 449.
31. K. R. Pörschke, *Chem. Ber.*, 1987, **120**, 425.
32. (a) J. L. Simunic and A. R. Pinhas, *Inorg. Chem.*, 1989, **28**, 2400; (b) A. R. Pinhas and J. W. Hershberger, *Organometallics*, 1990, **9**, 2840.
33. R. Taube, E. Weckmann, P. Böhme and J.-P. Gehrke, *Z. Anorg. Allg. Chem.*, 1989, **577**, 245.
34. R. Taube, P. Böhme and J.-P. Gehrke, *Z. Anorg. Allg. Chem.*, 1989, **578**, 89.
35. K. J. Klabunde, B. B. Anderson, M. Bader and L. J. Radonovich, *J. Am. Chem. Soc.*, 1978, **100**, 1313.
36. T. J. Groschens and K. J. Klabunde, *Organometallics*, 1982, **1**, 564.
37. S. T. Lin, T. J. Groschens and K. J. Klabunde, *Inorg. Chem.*, 1984, **23**, 1.
38. (a) S. T. Lin, R. N. Narske and K. J. Klabunde, *Organometallics*, 1985, **4**, 571; (b) S. T. Lin and K. J. Klabunde, *Organomet. Synth.*, 1986, **3**, 147; (c) S.-B. Choe and K. J. Klabunde, *ibid.*, 1986, **3**, 153; (d) S.-B. Choe and K. J. Klabunde, *ibid.*, 1986, **3**, 156; (e) M. Brezinski, K. J. Klabunde and B. B. Anderson, *ibid.*, 1986, **3**, 158.
39. S.-B. Choe and K. J. Klabunde, *J. Organomet. Chem.*, 1989, **359**, 409.
40. H. Kanai, S.-B. Choe and K. J. Klabunde, *J. Am. Chem. Soc.*, 1986, **108**, 2019.
41. S.-B. Choe, H. Kanai and K. J. Klabunde, *J. Am. Chem. Soc.*, 1989, **111**, 2875.
42. M. M. Brezinski, K. J. Klabunde, S. K. Janikowski and L. J. Radonovich, *Inorg. Chem.*, 1985, **24**, 3305.
43. M. W. Eyring and L. J. Radonovich, *Organometallics*, 1985, **4**, 1841.
44. M. M. Brezinski, J. Schneider, L. J. Radonovich and K. J. Klabunde, *Inorg. Chem.*, 1989, **28**, 2414.
45. G. López *et al.*, *Inorg. Chem.*, 1992, **31**, 1518.
46. G. López, G. Sánchez, G. García, J. García, A. Sanmartín and M. D. Santana, *Polyhedron*, 1991, **10**, 2821.
47. G. López *et al.*, *J. Organomet. Chem.*, 1992, **435**, 193.
48. G. López *et al.*, *Angew. Chem., Int. Ed. Engl.*, 1991, **30**, 716.
49. L. Ballester, M. Cano and A. Santos, *J. Organomet. Chem.*, 1982, **229**, 101.
50. K.-R. Pörschke *et al.*, *Chem. Ber.*, 1985, **118**, 275.
51. K.-R. Pörschke, K. Jonas and G. Wilke, *Chem. Ber.*, 1988, **121**, 1913.
52. W. Kaschube, K.-R. Pörschke, K. Angermund, C. Krüger and G. Wilke, *Chem. Ber.*, 1988, **121**, 1921.
53. G. Wilke, *Angew. Chem., Int. Ed. Engl.*, 1988, **27**, 185.
54. K. Fischer, K. Jonas, A. Mollbach and G. Wilke, *Z. Naturforsch., Teil. B*, 1984, **39**, 1011.
55. K. Jonas, K.-R. Pörschke, C. Krüger and Y.-H. Tsay, *Angew. Chem., Int. Ed. Engl.*, 1976, **15**, 621.
56. K.-R. Pörschke, G. Wilke and R. Mynott, *Chem. Ber.*, 1985, **118**, 298.
57. K. A. Ostoja Starzewski and J. Witte, *Angew. Chem., Int. Ed. Engl.*, 1985, **24**, 599.
58. K. A. Ostoja Starzewski and J. Witte, *Angew. Chem., Int. Ed. Engl.*, 1987, **26**, 63.
59. K. A. Ostoja Starzewski and L. Born, *Organometallics*, 1992, **11**, 2701.
60. W. Keim, F. H. Kowaldt, R. Goddard and C. Krüger, *Angew. Chem., Int. Ed. Engl.*, 1978, **17**, 466.
61. K. A. Ostoja Starzewski and J. Witte, *Angew. Chem., Int. Ed. Engl.*, 1988, **27**, 839.
62. K. A. Ostoja Starzewski and G. M. Bayer, *Angew. Chem., Int. Ed. Engl.*, 1991, **30**, 961.
63. B. L. Booth and K. G. Smith, *J. Organomet. Chem.*, 1981, **220**, 229.
64. G. Muller, D. Neugebauer, W. Geike, F. H. Köhler, J. Pebler and H. Schmidbaur, *Organometallics*, 1983, **2**, 257.
65. B. Hipler, E. Uhlig and J. Vogel, *J. Organomet. Chem.*, 1981, **218**, C1.
66. R. J. McKinney and D. C. Roe, *J. Am. Chem. Soc.*, 1985, **107**, 261.
67. C. Arlen, M. Pfeffer, J. Fischer and A. Mitschler, *J. Chem. Soc., Chem. Commun.*, 1983, 928.
68. T. Yamamoto, T. Kohara and A. Yamamoto, *Bull. Chem. Soc. Jpn.*, 1981, **54**, 2010.
69. T. Yamamoto, M. Takamatsu and A. Yamamoto, *Bull. Chem. Soc. Jpn.*, 1982, **55**, 325.
70. G. T. Crisp, S. Holle and P. W. Jolly, *Z. Naturforsch., Teil. B*, 1982, **37**, 1667.
71. J. M. Huggins and R. G. Bergman, *J. Am. Chem. Soc.*, 1981, **103**, 3002.
72. H.-F. Klein and L. Reitzel, *Chem. Ber.*, 1988, **121**, 1115.
73. G. Agnès, J. C. J. Bart, M. Calcaterra, W. Cavigiolo and C. Santini, *Transition Met. Chem.*, 1986, **11**, 246.
74. H.-F. Klein, T. Weimer, M.-J. Menu, M. Dartiguenave and Y. Dartiguenave, *Inorg. Chim. Acta*, 1988, **154**, 21.
75. H.-F. Klein, T. Weimer, M. Dartiguenave and Y. Dartiguenave, *Inorg. Chim. Acta*, 1991, **189**, 35.
76. H.-F. Klein, T. Weimer, M.-J. Menu, M. Dartiguenave and Y. Dartiguenave, *Inorg. Chim. Acta*, 1991, **189**, 45.
77. H.-F. Klein and T. Weimer, *Inorg. Chim. Acta*, 1991, **189**, 267.
78. E. Carmona, J. M. Marín, P. Palma, M. Paneque and M. L. Poveda, *Inorg. Chem.*, 1989, **28**, 1895.
79. D. J. Darensbourg, M. Y. Darensbourg, L. Y. Goh, M. Ludwig and P. Wiegreffe, *J. Am. Chem. Soc.*, 1987, **109**, 7539.
80. E. Carmona, J. M. Marín, M. Paneque and M. L. Poveda, *Organometallics*, 1987, **6**, 1757.
81. Y. Ishimura, K. Maruya, Y. Nakamura, T. Mizoroki and A. Ozaki, *Chem. Lett.*, 1981, 657.
82. B. M. Trzcinska, J. P. Fackler and A. B. Anderson, *Organometallics*, 1984, **3**, 319.
83. N. N. Kostitsyna, I. A. Oreshkin, B. A. Dolgoplosk, E. Tinyakova and A. I. Mikaya, *Dokl. Chem.*, 1984, **277**, 268.
84. M. Y. Darensbourg, M. Ludwig and C. G. Riordan, *Inorg. Chem.*, 1989, **28**, 1630.
85. A. Miyashita, J. Ishida and H. Nohira, *Tetrahedron Lett.*, 1986, **27**, 2127.
86. V. O. Efe and B. N. Ghose, *Rev. Port. Quim.*, 1983, **25**, 151.

87. T. Yamamoto, T. Kohara and A. Yamamoto, *Bull. Chem. Soc. Jpn.*, 1981, **54**, 2161.
88. Y.-J. Kim, K. Osakada, A. Takenaka and A. Yamamoto, *J. Am. Chem. Soc.*, 1990, **112**, 1096.
89. Y.-J. Kim, K. Osakada, K. Sugita, T. Yamamoto and A. Yamamoto, *Organometallics*, 1988, **7**, 2182.
90. S. Komiya, Y. Akai, K. Tanaka, T. Yamamoto and A. Yamamoto, *Organometallics*, 1985, **4**, 1130.
91. S. Komiya, A. Yamamoto and T. Yamamoto, *Chem. Lett.*, 1981, 193.
92. T. Yamamoto, T. Kohara and A. Yamamoto, *Bull. Chem. Soc. Jpn.*, 1981, **54**, 1720.
93. M. D. Fryzuk and P. A. MacNeil, *Organometallics*, 1982, **1**, 1540.
94. M. D. Fryzuk, P. A. MacNeil, S. J. Rettig, A. S. Secco and J. Trotter, *Organometallics*, 1982, **1**, 918.
95. C. A. Tolman, W. C. Seidel, J. D. Druliner and P. J. Domaille, *Organometallics*, 1984, **3**, 33.
96. P. T. Matsunaga, G. L. Hillhouse and A. L. Rheingold, *J. Am. Chem. Soc.*, 1993, **115**, 2075.
97. U. Müller, W. Keim, C. Krüger and P. Betz, *Angew. Chem., Int. Ed. Engl.*, 1989, **28**, 1011.
98. S. I. Black and G. B. Young, *Polyhedron*, 1989, **8**, 585.
99. E. Carmona, F. González, M. L. Poveda, J. M. Marín, and A. Martinez, *An. Quim., Ser. B*, 1982, **78**, 51 (*Chem. Abstr.*, 1982, 97, 144 991).
100. E. Carmona, F. González, M. L. Poveda and J. M. Marín, *Synth. React. Inorg. Metal Org. Chem.*, 1982, **12**, 185.
101. E. Carmona, M. Paneque, M. L. Poveda, R. D. Rogers and J. L. Atwood, *Polyhedron*, 1984, **3**, 317.
102. E. Carmona *et al.*, *J. Am. Chem. Soc.*, 1989, **111**, 2883.
103. E. Carmona, M. Paneque and M. L. Poveda, *Polyhedron*, 1989, **8**, 285.
104. T. R. Belderrain, D. A. Knight, D. J. Irvine, M. Paneque, M. L. Poveda and E. Carmona, *J. Chem. Soc., Dalton Trans.*, 1992, 1491.
105. J. G. Davidson, E. K. Barefield and D. G. Van Derveer, *Organometallics*, 1985, **4**, 1178.
106. J. Cámpora, E. Gutiérrez, A. Monge, M. L. Poveda and E. Carmona, *Organometallics*, 1992, **11**, 2644.
107. J. Cámpora, E. Gutiérrez, M. L. Poveda, C. Ruíz and E. Carmona, *J. Chem. Soc., Dalton Trans.*, 1992, 1769.
108. J. Cámpora, E. Carmona, E. Gutiérrez, P. Palma, M. L. Poveda and C. Ruíz, *Organometallics*, 1992, **11**, 11.
109. L. J. Krause and J. A. Morrison, *J. Chem. Soc., Chem. Commun.*, 1981, 1282.
110. T. Yamamoto, *Chem. Ind.*, 1981, 28.
111. J. Carvajal, G. Muller, J. Sales, X. Solans and C. Miravitlles, *Organometallics*, 1984, **3**, 996.
112. M. Wada and K. Sameshima, *J. Chem. Soc., Dalton Trans.*, 1981, 240.
113. M. F. Semmelhack *et al.*, *J. Am. Chem. Soc.*, 1981, **103**, 6460.
114. M. Wada, K. Nishiwaki and M. Kumazoe, *J. Chem. Soc., Chem. Commun.*, 1984, 980.
115. M. Wada, M. Kumazoe, Y. Matsuhiro and T. Erabi, *Chem. Lett.*, 1986, 1959.
116. J. Chen, Y. Kai, N. Kasai, M. Wada and M. Kumazoe, *Bull. Chem. Soc. Jpn.*, 1991, **64**, 2802.
117. M. Bochmann, I. Hawkins and M. P. Sloan, *J. Organomet. Chem.*, 1987, **332**, 371.
118. E. Carmona, P. Palma, M. Paneque and M. L. Poveda, *Organometallics*, 1990, **9**, 583.
119. B. Crociani, F. Di Bianca, A. Giovenco and A. Berton, *J. Organomet. Chem.*, 1987, **323**, 123.
120. C. J. Lawrie, H. E. Dankosh and B. K. Carpenter, *J. Organomet. Chem.*, 1991, **411**, C7.
121. J. Hernández, G. Muller, M. Rocamora, X. Solans and M. Aguiló, *J. Organomet. Chem.*, 1988, **345**, 383.
122. H.-F. Klein, H. Beck-Hemetsberger, L. Reitzel, B. Rodenhäuser and G. Cordier, *Chem. Ber.*, 1989, **122**, 43.
123. A. Furlani, M. V. Russo, P. Carusi, S. Licoccia, E. Leoni and G. Valentini, *Gazz. Chim. Ital.*, 1983, **113**, 671.
124. L. S. Isaeva, L. N. Morozova, V. V. Bashilov, P. V. Petrovskii, V. I. Sokolov and O. A. Reutov, *J. Organomet. Chem.*, 1983, **243**, 253.
125. J. C. Folest, J. Périchon, J. F. Fauvarque and A. Jutland, *J. Organomet. Chem.*, 1988, **342**, 259.
126. L. S. Isaeva, L. N. Morozova, G. I. Drogunova, V. S. Kolesov and D. N. Kravtsov, *Sov. J. Coord. Chem. (Engl. Transl.)*, 1986, **12**, 890.
127. M. Antón, N. Clos and G. Muller, *J. Organomet. Chem.*, 1984, **267**, 213.
128. M. Antón, G. Muller and J. Sales, *Transition Met. Chem.*, 1983, **8**, 79.
129. C. Amatore and A. Jutland, *Organometallics*, 1988, **7**, 2203.
130. T. Yamamoto, S. Wakabayashi and K. Osakada, *J. Organomet. Chem.*, 1992, **428**, 233.
131. W. P. Fehlhammer and M. Fritz, *Chem. Rev.*, 1993, **93**, 1243.
132. G. Favero and A. Morvillo, *J. Organomet. Chem.*, 1984, **260**, 363.
133. A. Behr, W. Keim and G. Thelen, *J. Organomet. Chem.*, 1983, **249**, C38.
134. H. Qichen, X. Minzhi, Q. Yanlong, X. Weihua, S. Meicheng and T. Youqi, *J. Organomet. Chem.*, 1985, **287**, 419.
135. Y. V. Kissin and D. L. Beach, *J. Polym. Sci. Part A: Polym. Chem.*, 1989, **27**, 147.
136. U. Klabunde, T. H. Tulip, D. C. Roe and S. D. Ittel, *J. Organomet. Chem.*, 1987, **334**, 141.
137. K. Osakada, M. Maeda, Y. Nakamura, T. Yamamoto and A. Yamamoto, *J. Chem. Soc., Chem. Commun.*, 1986, 442.
138. M. D. Fryzuk, P. A. MacNeil and S. J. Rettig, *J. Organomet. Chem.*, 1987, **332**, 345.
139. E. Gutiérrez, S. A. Hudson, A. Monge, M. C. Nicasio, M. Paneque and E. Carmona, *J. Chem. Soc., Dalton Trans.*, 1992, 2651.
140. K. Osakada, M. Maeda, Y. Nakamura, T. Yamamoto and A. Yamamoto, *Chem. Lett.*, 1986, 597.
141. W. Keim *et al.*, *Organometallics*, 1986, **5**, 2356.
142. E. Wenkert, M. E. Shepard and A. T. McPhail, *J. Chem. Soc., Chem. Commun.*, 1986, 1390.
143. M. Martinez and G. Muller, *J. Chem. Soc., Dalton Trans.*, 1989, 1669.
144. G. Schiavon, G. Zotti and G. Bontempelli, *J. Electroanal. Chem. Interfacial Electrochem.*, 1984, **161**, 323.
145. G. Schiavon, G. Bontempelli, F. Magno and S. Daniele, *J. Electroanal. Chem. Interfacial Electrochem.*, 1982, **140**, 91.
146. R. M. Ceder, J. Cubillo, G. Muller, M. Rocamora and J. Sales, *J. Organomet. Chem.*, 1992, **429**, 391.
147. W. Seidel, *Z. Chem*, 1985, **25**, 411.
148. H.-P. Abicht, H. Schmidt and K. Issleib, *Z. Chem.*, 1985, **25**, 410.
149. M. Wada, K. Nishiwaki and Y. Kawasaki, *J. Chem. Soc., Dalton Trans.*, 1982, 1443.
150. M. Martinez and G. Muller, *J. Chem. Res. (S)*, 1991, 199.
151. R. Ceder, G. Muller, J. Sales, J. Vidal, D. Niebecker and I. Tkatchenko, *J. Mol. Catal.*, 1991, **68**, 23.
152. M. Gomez and P. Royo, *An. Quim., Ser. B*, 1982, **78**, 196 (*Chem. Abstr.*, 1982, **98**, 54 159).
153. R. McDonald, K. C. Sturge, A. D. Hunter and L. Shilliday, *Organometallics*, 1992, **11**, 893.
154. H. Matsumoto, S. Inaba and R. D. Rieke, *J. Org. Chem.*, 1983, **48**, 840.

155. A. Ortiz and J. Sales, *Synth. React. Inorg. Metal Org. Chem.*, 1982, **12**, 601.
156. J. M. Coronas, C. Pold and J. Sales, *Inorg. Chim. Acta*, 1981, **48**, 87.
157. Q. Zhang, Y. Li, J. Chen, R. Chen, Z. Zhang and X. Wang, *Wuli Huaxue Xuebao*, 1985, **1**, 460 (*Chem. Abstr.*, 1986, **104**, 40 565).
158. J. M. Lopez and B. N. Spevak, *Rev. Cienc. Quim.*, 1985, **16**, 161 (*Chem. Abstr.*, 1988, **108**, 186 954).
159. Z. Zhang and H. Wang, *Gaodeng Xuexiao Huaxue Xuebao*, 1985, **6**, 910 (*Chem. Abstr.*, 1986, **105**, 226 941).
160. Z. Zhang, S. Wang, G. Yang, C. Shen and Y. Fang, *Jiegou Huaxue*, 1984, **3**, 119 (*Chem. Abstr.*, 1985, **103**, 62 902).
161. K. C. Sturge, A. D. Hunter, R. McDonald and B. D. Santarsiero, *Organometallics*, 1992, **11**, 3056.
162. Z.-H. Zhou and T. Yamamoto, *J. Organomet. Chem.*, 1991, **414**, 119.
163. G. Schiavon, G. Zotti and G. Bontempelli, *J. Electroanal. Chem. Interfacial Electrochem.*, 1985, **194**, 327.
164. L. N. Morozova, L. S. Isaeva, P. V. Petrovskii, D. N. Kravtsov, She Fan Min and V. N. Kalinin, *J. Organomet. Chem.*, 1990, **381**, 281.
165. (a) H. Hoberg, F. J. Fañanás, K. Angermund, C. Krüger and M. J. Romão, *J. Organomet. Chem.*, 1985, **281**, 379; (b) H. Hoberg and F. J. Fañanás, *Angew. Chem., Int. Ed. Engl.*, 1985, **24**, 325.
166. A. Gleizes, A. Kerkeni, M. Dartiguenave, Y. Dartiguenave and H.-F. Klein, *Inorg. Chem.*, 1981, **20**, 2372.
167. M. Bochmann, I. Hawkins, M. B. Hursthouse and R. L. Short, *J. Organomet. Chem.*, 1987, **332**, 361.
168. F. M. Conroy-Lewis, L. Mole, A. D. Redhouse, S. A. Litster and J. L. Spencer, *J. Chem. Soc., Chem. Commun.*, 1991, 1601.
169. A. A. Pozdeeva, U. M. Dzhemilev, N. R. Popod'ko, R. I. Khusnutdinov, S. I. Zhdanov and G. A. Tolstikov, *J. Organomet. Chem.*, 1989, **367**, 205.
170. M. Wada, Y. Koyama and K. Sameshima, *J. Organomet. Chem.*, 1981, **209**, 115.
171. M. Wada, K. Sameshima, K. Nishiwaki and Y. Kawasaki, *J. Chem. Soc., Dalton Trans.*, 1982, 793.
172. K. Miki, H. Taniguchi, Y. Kai, N. Kasai, K. Nishiwaki and M. Wada, *J. Chem. Soc., Chem. Commun.*, 1982, 1178.
173. (a) D. Xu, K. Miki, M. Tanaka, N. Kasai, N. Yasuoka and M. Wada, *J. Organomet. Chem.*, 1989, **371**, 267; (b) D. Xu, K. Miki, Y. Kai, N. Kasai, and M. Wada, *J. Organomet. Chem.*, 1985, **287**, 265.
174. M. Bochmann, I. Hawkins, M. B. Hursthouse and R. L. Short, *J. Chem. Soc., Chem. Commun.*, 1990, 1213.
175. E. Hernandez, I. Saez and P. Royo, *J. Organomet. Chem.*, 1985, **293**, 249.
176. J. Skupińska, *Chem. Rev.*, 1991, **91**, 613.
177. A. Morvillo and A. Turco, *J. Organomet. Chem.*, 1982, **224**, 387.
178. N. Re, M. Rosi, A. Sgamellotti, C. Floriani and M. F. Guest, *J. Chem. Soc., Dalton Trans.*, 1992, 1821.
179. A. Yamamoto, T. Yamamoto, S. Komiya and F. Ozawa, *Pure Appl. Chem.*, 1984, **56**, 1621.
180. A. Yamamoto, T. Yamamoto and F. Ozawa, *Pure Appl. Chem.*, 1985, **57**, 1799.
181. D. Steinborn, *Angew. Chem., Int. Ed. Engl.*, 1992, **31**, 401.
182. B. A. Dolgoplosk, *Russ. Chem. Rev.*, 1983, **52**, 613.
183. W. Kaschube, K.-R. Pörschke and G. Wilke, *J. Organomet. Chem.*, 1988, **355**, 525.
184. T. Yamamoto, T. Kohara, K. Osakada and A. Yamamoto, *Bull. Chem. Soc. Jpn.*, 1983, **56**, 2147.
185. W. H. Smith and Y.-M. Kuo, *J. Electroanal. Chem. Interfacial Electrochem.*, 1985, **188**, 189.
186. E. Uhlig and W. Poppitz, *Z. Anorg. Allg. Chem.*, 1981, **477**, 167.
187. E. Carmona, F. González, M. L. Poveda, J. L. Atwood and R. D. Rogers, *J. Chem. Soc., Dalton Trans.*, 1981, 777.
188. O. Boutry, M. C. Nicasio, M. Paneque, E. Carmona, E. Gutiérrez and C. Ruíz, *J. Organomet. Chem.*, 1993, **444**, 245.
189. E. Dinjus, K. H. Thiele, G. Sonnek, K. Jacob and H. Reinheckel, *Ger. (East) Pat.* 160 261 (1983) (*Chem. Abstr.*, 1983, **99**, 212 719).
190. D. W. Firsich and R. J. Lagow, *J. Chem. Soc., Chem. Commun.*, 1981, 1283.
191. A. Sebald, B. Wrackmeyer and W. Beck, *Z. Naturforsch., Teil., B*, 1983, **38**, 45.
192. X. L. R. Fontaine, S. J. Higgins, C. R. Langrick and B. L. Shaw, *J. Chem. Soc., Dalton Trans.*, 1987, 777.
193. R. Nast and A. Beyer, *J. Organomet. Chem.*, 1981, **204**, 267.
194. R. Nast, *Coord. Chem. Rev.*, 1982, **47**, 89.
195. M. Wada and M. Kumazoe, *J. Organomet. Chem.*, 1983, **259**, 245.
196. K. Miki, M. Tanaka, N. Kasai and M. Wada, *J. Organomet. Chem.*, 1988, **352**, 385.
197. M. Wada, *J. Chem. Soc., Chem. Commun.*, 1981, 680.
198. D. Xu, Y. Kai, K. Miki, N. Kasai, K. Nishiwaki and M. Wada, *Chem. Lett.*, 1983, 591.
199. D. Xu, Y. Kai, K. Miki, N. Kasai, K. Nishiwaki and M. Wada, *Bull. Chem. Soc. Jpn.*, 1984, **57**, 1459.
200. M. Wada and M. Kumazoe, *J. Chem. Soc., Chem. Commun.*, 1985, 1204.
201. G. B. Deacon, P. I. MacKinnon and T. D. Tuong, *Aust. J. Chem.*, 1983, **36**, 43.
202. S. Kotani, K. Shiina and K. Sonogashira, *J. Organomet. Chem.*, 1992, **429**, 403.
203. L. I. Zakharkin and N. F. Shemyakin, *Bull. Acad. Sci. USSR, Div. Chem. Sci.*, 1985, **33**, 2572.
204. S. Komiya, Y. Abe, A. Yamamoto and T. Yamamoto, *Organometallics*, 1983, **2**, 1466.
205. J. M. Coronas, G. Muller, M. Rocamora, C. Miravitlles and X. Solans, *J. Chem. Soc., Dalton Trans.*, 1985, 2333.
206. J. M. Coronas, G. Muller and M. Rocamora, *J. Organomet. Chem.*, 1986, **301**, 227.
207. C. Amatore and A. Jutland, *Acta Chem. Scand.*, 1990, **44**, 755.
208. A. C. Balazs, K. H. Johnson and G. M. Whitesides, *Inorg. Chem.*, 1982, **21**, 2162.
209. K. Tatsumi, R. Hoffmann, A. Yamamoto and J. K. Stille, *Bull. Chem. Soc. Jpn.*, 1981, **54**, 1857.
210. K. Tatsumi, A. Nakamura, S. Komiya, A. Yamamoto and T. Yamamoto, *J. Am. Chem. Soc.*, 1984, **106**, 8181.
211. Y.-J. Kim, K. Osakada and A. Yamamoto, *Bull. Chem. Soc. Jpn.*, 1989, **62**, 964.
212. P. G. Jones, *J. Organomet. Chem.*, 1988, **345**, 405.
213. J. Powell, *J. Chem. Soc., Chem. Commun.*, 1989, 200.
214. E. Dinjus, D. Walther, R. Kirmse and J. Stach, *Z. Anorg. Allg. Chem.*, 1983, **501**, 121.
215. E. Dinjus, D. Walther, R. Kirmse and J. Stach, *Z. Anorg. Allg. Chem.*, 1981, **481**, 71.
216. T. Bartik, I. Gerdes, P. Heimbach and H.-G. Schulte, *J. Organomet. Chem.*, 1989, **367**, 359.
217. R. Berger, H. Schenkluhn and B. Weimann, *Transition Met. Chem.*, 1981, **6**, 272.
218. H. Schenkluhn, H. Bandmann, R. Berger and E. Hübinger, *Transition Met. Chem.*, 1981, **6**, 287.
219. H. Schenkluhn, R. Berger, B. Pittel and M. Zähres, *Transition Met. Chem.*, 1981, **6**, 277.
220. E. Carmona, P. Palma and M. L. Poveda, *Polyhedron*, 1990, **9**, 757.

221. H. Kurosawa, H. Ohnishi, M. Emoto, N. Chatani, Y. Kawasaki, S. Murai and I. Ikeda, *Organometallics*, 1990, **9**, 3038.
222. H. Kurosawa, H. Ohnishi, M. Emoto, Y. Kawasaki and S. Murai, *J. Am. Chem. Soc.*, 1988, **110**, 6272.
223. D. Walther, J. Sieler and J. Kaiser, *Z. Anorg. Allg. Chem.*, 1981, **472**, 149.
224. H. Lehmkuhl, C. Naydowski, R. Benn, A. Rufińska and G. Schroth, *J. Organomet. Chem.*, 1982, **228**, C1.
225. H. Lehmkuhl *et al.*, *Chem. Ber.*, 1988, **121**, 1931.
226. H. Lehmkuhl *et al.*, *Chem. Ber.*, 1984, **117**, 3231.
227. H. Lehmkuhl *et al.*, *Chem. Ber.*, 1991, **124**, 441.
228. S. Pasynkiewicz and H. Lehmkuhl, *J. Organomet. Chem.*, 1985, **289**, 189.
229. H. Lehmkuhl, S. Pasynkiewicz, R. Benn and A. Rufińska, *J. Organomet. Chem.*, 1982, **240**, C27.
230. S. Pasynkiewicz, *J. Organomet. Chem.*, 1990, **387**, 1.
231. H. Lehmkuhl, C. Naydowski, R. Benn, A. Rufińska and G. Schroth, *J. Organomet. Chem.*, 1983, **246**, C9.
232. H. Lehmkuhl, F. Danowski, R. Benn, R. Mynott and G. Schroth, *Chem. Ber.*, 1986, **119**, 2542.
233. H. Lehmkuhl, F. Danowski, R. Benn, A. Rufińska, G. Schroth and R. Mynott, *J. Organomet. Chem.*, 1983, **254**, C11.
234. H. Lehmkuhl, F. Danowski, G. Mehler, J. Poplawska and S. Pasynkiewicz, *J. Organomet. Chem.*, 1989, **363**, 387.
235. S. Pasynkiewicz, M. Poplawska and R. Mynott, *J. Organomet. Chem.*, 1992, **429**, 135.
236. H. Lehmkuhl, C. Naydowski and M. Bellenbaum, *J. Organomet. Chem.*, 1983, **246**, C5.
237. (a) H. Lehmkuhl and T. Keil, *J. Organomet. Chem.*, 1988, **342**, C38; (b) H. Lehmkuhl and C. Naydowski, *J. Organomet. Chem.*, 1984, **277**, C18.
238. S. Pasynkiewicz, A. Pietrzykowski and M. Poplawska, *J. Organomet. Chem.*, 1993, **443**, 131.
239. E. R. Burkhardt, R. G. Bergman and C. H. Heathcock, *Organometallics*, 1990, **9**, 30.
240. R. Taube, D. Steinborn and W. Höbold, *J. Organomet. Chem.*, 1985, **284**, 385.
241. M. Julia, H. Lauron, J.-N. Verpeaux, Y. Jeannin and C. Bois, *J. Organomet. Chem.*, 1988, **358**, C11.
242. M. Julia, H. Lauron and J.-N. Verpeaux, *J. Organomet. Chem.*, 1990, **387**, 365.
243. E. Lindner, A. Nothdurft, R. Fawzi and C. Maichle, *J. Organomet. Chem.*, 1992, **435**, 213.
244. K.-H. Thiele, U. Böhme and S. Pasynkiewicz, *Z. Anorg. Allg. Chem.*, 1991, **600**, 121.
245. M. I. Bruce, M. G. Humphrey, J. G. Matisons, S. K. Roy and A. G. Swincer, *Aust. J. Chem.*, 1984, **37**, 1955.
246. M. I. Bruce, D. N. Duffy, M. J. Liddell, M. R. Snow and E. R. T. Tiekink, *J. Organomet. Chem.*, 1987, **335**, 365.
247. F. Scott, S. Cronje and H. G. Raubenheimer, *J. Organomet. Chem.*, 1987, **326**, C40.
248. A. Pietrzykowski and S. Pasynkiewicz, *J. Organomet. Chem.*, 1992, **440**, 401.
249. G. K. Anderson, R. J. Cross, K. W. Muir and L. Manojlović-Muir, *J. Organomet. Chem.*, 1989, **362**, 225.
250. G. R. Knox, P. L. Pauson and D. Willison, *J. Organomet. Chem.*, 1993, **450**, 177.
251. H. Lehmkuhl, A. Rufińska, R. Benn, G. Schroth and R. Mynott, *Liebigs Ann. Chem.*, 1981, 317.
252. H. Lehmkuhl, C. Naydowski, R. Benn, A. Rufińska, G. Schroth, R. Mynott and C. Krüger, *Chem. Ber.*, 1983, **116**, 1447.
253. A. L. Spek, J. L. de Boer, B. Fischer and J. Boersma, *Acta Crystallogr., Part C*, 1988, **44**, 1663.
254. P. Hong, K. Sonogashira and N. Hagihara, *J. Organomet. Chem.*, 1981, **219**, 363.
255. E. Lindner, F. Bouachir and W. Hiller, *Z. Naturforsch., Teil. B*, 1982, **37**, 1146.
256. N. Kuhn, U. Schwenk, M. Winter and R. Mynott, *J. Organomet. Chem.*, 1984, **260**, C43.
257. M. A. Battiste, B. G. Griggs, Jr., D. Sackett, J. M. Coxon and P. J. Steel, *J. Organomet. Chem.*, 1987, **330**, 437.
258. H. Lehmkuhl, *Pure Appl. Chem.*, 1986, **58**, 495.
259. H. Lehmkuhl, *Pure Appl. Chem.*, 1990, **62**, 731.
260. R. Benn, J. Klein, A. Rufińska and G. Schroth, *Z. Naturforsch. Teil. B*, 1981, **36**, 1595.
261. D. B. Jacobson and B. S. Freiser, *Organometallics*, 1984, **3**, 513.
262. R. J. McKinney, D. L. Thorn, R. Hoffmann and A. Stockis, *J. Am. Chem. Soc.*, 1981, **103**, 2595.
263. A. Peluso, D. R. Salahub and A. Goursot, *Inorg. Chem.*, 1990, **29**, 1544.
264. A. Miyashita, H. Shitara and H. Nohira, *J. Chem. Soc., Chem. Commun.*, 1985, 850.
265. D. S. Dudis and J. P. Fackler, *J. Organomet. Chem.*, 1983, **249**, 289.
266. K. W. Chiu, J. Fawcett, W. Henderson, R. D. W. Kemmitt and D. R. Russell, *J. Chem. Soc., Chem. Commun.*, 1986, 41.
267. K. W. Chiu, J. Fawcett, W. Henderson, R. D. W. Kemmitt and D. R. Russell, *J. Chem. Soc., Dalton Trans.*, 1987, 733.
268. E. Carmona, E. Gutiérrez-Puebla, A. Monge, J. M. Marín, M. Paneque and M. L. Poveda, *Organometallics*, 1989, **8**, 967.
269. A. Miyashita, M. Ohyoshi, H. Shitara and H. Nohira, *J. Organomet. Chem.*, 1988, **338**, 103.
270. M. Ishikawa, J. Ohshita, Y. Ito and J. Iyoda, *J. Am. Chem. Soc.*, 1986, **108**, 7417.
271. H.-O. Fröhlich, B. Hipler and B. Hofmann, *J. Organomet. Chem.*, 1992, **430**, 133.
272. H.-O. Fröhlich, R. Wyrwa and H. Görls, *J. Organomet. Chem.*, 1992, **441**, 169.
273. W. Schröder, W. Bonrath and K. R. Pörschke, *J. Organomet. Chem.*, 1991, **408**, C25.
274. W. Kaschube, W. Schröder, K. R. Pörschke, K. Angermund and C. Krüger, *J. Organomet. Chem.*, 1990, **389**, 399.
275. W. Bonrath, S. Michaelis, K. R. Pörschke, B. Gabor, R. Mynott and C. Krüger, *J. Organomet. Chem.*, 1990, **397**, 255.
276. J. C. M. Sinnema, G. H. B. Fendesak and H. tom Dieck, *J. Organomet. Chem.*, 1990, **390**, 237.
277. S. Michaelis, K. R. Pörschke, R. Mynott, R. Goddard and C. Krüger, *J. Organomet. Chem.*, 1992, **426**, 131.
278. K. R. Pörschke, *Angew. Chem., Int. Ed. Engl.*, 1987, **26**, 1288.
279. H. Hoberg and A. Herrera, *Angew. Chem., Int. Ed. Engl.*, 1981, **20**, 876.
280. D. J. Pasto and N.-Z. Huang, *Organometallics*, 1985, **4**, 1386.
281. P. Binger, M. J. Doyle and R. Benn, *Chem. Ber.*, 1983, **116**, 1.
282. T. A. Peganova *et al.*, *J. Organomet. Chem.*, 1985, **282**, 283.
283. H. M. Büch, P. Binger, R. Benn and A. Rufińska, *Organometallics*, 1987, **6**, 1130.
284. J. J. Eisch, A. M. Piotrowski, A. A. Aradi, C. Krüger and M. J. Romão, *Z. Naturforsch., Teil. B*, 1985, **40**, 624.
285. J. J. Eisch, J. E. Galle, A. A. Aradi and M. P. Boleslawski, *J. Organomet. Chem.*, 1986, **312**, 399.
286. L. Stehling and G. Wilke, *Angew. Chem., Int. Ed. Engl.*, 1985, **24**, 496.
287. D. J. Pasto, N.-Z. Huang and C. W. Eigenbrot, *J. Am. Chem. Soc.*, 1985, **107**, 3160.
288. R. T. Carl, E. W. Corcoran, Jr., R. P. Hughes and D. E. Samkoff, *Organometallics*, 1990, **9**, 838.
289. R. P. Hughes, *Adv. Organomet. Chem.*, 1990, **31**, 183.
290. R. Neidlein, A. Rufińska, H. Schwager and G. Wilke, *Angew. Chem., Int. Ed. Engl.*, 1986, **25**, 640.
291. C. Krüger, K. Laakmann, G. Schroth, H. Schwager and G. Wilke, *Chem. Ber.*, 1987, **120**, 471.
292. M. A. Bennett, T. W. Hambley, N. K. Roberts and G. B. Robertson, *Organometallics*, 1985, **4**, 1992.

293. E. Carmona, P. Palma, M. Paneque, M. L. Poveda, E. Gutiérrez-Puebla and A. Monge, *J. Am. Chem. Soc.*, 1986, **108**, 6424.
294. J. Cámpora, E. Carmona, E. Gutiérrez-Puebla, M. L. Poveda and C. Ruíz, *Organometallics*, 1988, **7**, 2577.
295. E. Carmona, M. Paneque, M. L. Poveda, E. Gutiérrez-Puebla and A. Monge, *Polyhedron*, 1989, **8**, 1069.
296. E. Carmona, E. Gutiérrez-Puebla, A. Monge, M. Paneque and M. L. Poveda, *J. Chem. Soc., Chem. Commun.*, 1991, 148.
297. R. Benn, H. Schwager and G. Wilke, *J. Organomet. Chem.*, 1986, **316**, 229.
298. H. Schwager, C. Krüger, R. Neidlein and G. Wilke, *Angew. Chem., Int. Ed. Engl.*, 1987, **26**, 65.
299. J. J. Eisch, A. M. Piotrowski, K. I. Han, C. Krüger and Y. H. Tsay, *Organometallics*, 1985, **4**, 224.
300. R. Mynott, R. Neidlein, H. Schwager and G. Wilke, *Angew. Chem., Int. Ed. Engl.*, 1986, **25**, 367.
301. D. Bläser *et al.*, *Angew. Chem., Int. Ed. Engl.*, 1989, **28**, 206.
302. C. Pluta, K. R. Pörschke, I. Ortmann and C. Krüger, *Chem. Ber.*, 1992, **125**, 103.
303. K. J. Izod and P. Thornton, *Polyhedron*, 1993, **12**, 1613.
304. C. Arlen, F. Maassarani, M. Pfeffer and J. Fischer, *Nouv. J. Chim.*, 1985, **9**, 249.
305. T. Yamamoto, K. Sano, K. Osakada, S. Komiya, A. Yamamoto, Y. Kushi and T. Tada, *Organometallics*, 1990, **9**, 2396.
306. H. Hoberg and B. W. Oster, *J. Organomet. Chem.*, 1983, **252**, 359.
307. H. Hoberg and B. W. Oster, *J. Organomet. Chem.*, 1983, **234**, C35.
308. H. Hoberg, K. Sümmermann, E. Hernandez, C. Ruppin and D. Guhl, *J. Organomet. Chem.*, 1988, **344**, C35.
309. H. Hoberg and K. Sümmermann, *J. Organomet. Chem.*, 1984, **275**, 239.
310. H. Hoberg, K. Sümmermann and A. Milchereit, *J. Organomet. Chem.*, 1985, **288**, 237.
311. H. Hoberg, E. Hernandez and K. Sümmermann, *J. Organomet. Chem.*, 1985, **295**, C21.
312. H. Hoberg, K. Sümmermann and A. Milchereit, *Angew. Chem., Int. Ed. Engl.*, 1985, **24**, 325.
313. H. Hoberg and E. Hernandez, *J. Organomet. Chem.*, 1986, **311**, 307.
314. E. Hernandez and H. Hoberg, *J. Organomet. Chem.*, 1986, **315**, 245.
315. E. Hernandez and H. Hoberg, *J. Organomet. Chem.*, 1987, **328**, 403.
316. H. Hoberg and D. Guhl, *J. Organomet. Chem.*, 1989, **375**, 245.
317. H. Hoberg and D. Guhl, *J. Organomet. Chem.*, 1989, **375**, 279.
318. H. Hoberg and D. Guhl, *Angew. Chem., Int. Ed. Engl.*, 1989, **28**, 1035.
319. H. Hoberg, D. Guhl and P. Betz, *J. Organomet. Chem.*, 1990, **387**, 233.
320. H. Hoberg and D. Bärhausen, *J. Organomet. Chem.*, 1991, **403**, 401.
321. H. Hoberg, E. Hernandez and D. Guhl, *J. Organomet. Chem.*, 1988, **339**, 213.
322. H. Hoberg and M. Nohlen, *J. Organomet. Chem.*, 1991, **412**, 225.
323. H. Hoberg and K. Sümmermann, *J. Organomet. Chem.*, 1983, **253**, 383.
324. H. Hoberg and K. Sümmermann, *J. Organomet. Chem.*, 1984, **264**, 379.
325. E. Carmona, J. M. Marín, P. Palma and M. L. Poveda, *J. Organomet. Chem.*, 1989, **377**, 157.
326. H. Drevs, *J. Organomet. Chem.*, 1992, **433**, C1.
327. J. N. Louwen, D. M. Grove, H. J. C. Ubbels, D. J. Stufkens and A. Oskam, *Z. Naturforsch., Teil. B*, 1983, **38**, 1657.
328. D. M. Grove, A. H. M. Verschuuren, G. van Koten and J. A. M. van Beek, *J. Organomet. Chem.*, 1989, **372**, C1.
329. D. M. Grove, G. van Koten, H. J. C. Ubbels, R. Zoet and A. L. Spek, *Organometallics*, 1984, **3**, 1003.
330. J. A. M. van Beek *et al.*, *Inorg. Chem.*, 1991, **30**, 3059.
331. J. Terheijden, G. van Koten, W. P. Mul, D. J. Stufkens, F. Muller and C. H. Stam, *Organometallics*, 1986, **5**, 519.
332. D. M. Grove *et al.*, *Organometallics*, 1986, **5**, 322.
333. D. M. Grove, G. van Koten, R. Zoet, N. W. Murrall and A. J. Welch, *J. Am. Chem. Soc.*, 1983, **105**, 1379.
334. D. M. Grove *et al.*, *Inorg. Chem.*, 1988, **27**, 2466.
335. K. Sano, T. Yamamoto and A. Yamamoto, *Chem. Lett.*, 1982, 695.
336. F. Balegroune, P. Braunstein, T. M. Gomes Carneiro, D. Grandjean and D. Matt, *J. Chem. Soc., Chem. Commun.*, 1989, 582.
337. P. Braunstein, T. M. Gomes Carneiro, D. Matt, F. Balegroune and D. Grandjean, *Organometallics*, 1989, **8**, 1737.
338. H. P. Abicht, K. Issleib, B. Hipler and E. Uhlig, *Synth. React. Inorg. Metal Org. Chem.*, 1982, **12**, 331.
339. H. Rimml and L. M. Venanzi, *J. Organomet. Chem.*, 1983, **259**, C6.
340. H. Schmidbaur, U. Deschler and B. Milewski-Mahrla, *Angew. Chem., Int. Ed. Engl.*, 1981, **20**, 586.
341. J. K. Gong, T. B. Peters, P. E. Fanwick and C. P. Kubiak, *Organometallics*, 1992, **11**, 1392.
342. R. Fischer, D. Walther, G. Bräunlich, B. Undeutsch, W. Ludwig and H. Bandmann, *J. Organomet. Chem.*, 1992, **427**, 395.
343. K. Sano, T. Yamamoto and A. Yamamoto, *Bull. Chem. Soc. Jpn.*, 1984, **57**, 2741.
344. H. Hoberg, D. Schaefer, G. Burkhart, C. Krüger and M. J. Romão, *J. Organomet. Chem.*, 1984, **266**, 203.
345. K. Sano, T. Yamamoto and A. Yamamoto, *Chem. Lett.*, 1983, 115.
346. H. Hoberg, Y. Peres and A. Milchereit, *J. Organomet. Chem.*, 1986, **307**, C41.
347. R. Fischer, B. Nestler and H. Schütz, *Z. Anorg. Allg. Chem.*, 1989, **577**, 111.
348. H. Hoberg, A. Ballesteros, A. Sigan, C. Jégat, D. Bärhausen and A. Milchereit, *J. Organomet. Chem.*, 1991, **407**, C23.
349. H. Hoberg, Y. Peres, C. Krüger and Y.-H. Tsay, *Angew. Chem., Int. Ed. Engl.*, 1987, **26**, 771.
350. H. Hoberg, Y. Peres and A. Milchereit, *J. Organomet. Chem.*, 1986, **307**, C38.
351. T. Yamamoto, K. Sano and A. Yamamoto, *J. Am. Chem. Soc.*, 1987, **109**, 1092.
352. K. Sano, T. Yamamoto and A. Yamamoto, *Chem. Lett.*, 1984, 941.
353. H. Hoberg and D. Schaefer, *J. Organomet. Chem.*, 1983, **255**, C15.
354. M. Döring, D. Kosemund, E. Uhlig and H. Görls, *Z. Anorg. Allg. Chem.*, 1993, **619**, 1512.
355. H. Hoberg and B. W. Oster, *J. Organomet. Chem.*, 1984, **266**, 321.
356. E. Dinjus, J. Kaiser, J. Sieler and D. Walther, *Z. Anorg. Allg. Chem.*, 1981, **483**, 63.
357. J. Kaiser, J. Sieler, U. Braun, L. Golič, E. Dinjus and D. Walther, *J. Organomet. Chem.*, 1982, **224**, 81.
358. D. Walther, E. Dinjus, J. Sieler, J. Kaiser, O. Lindqvist and L. Anderson, *J. Organomet. Chem.*, 1982, **240**, 289.
359. D. Walther, E. Dinjus and V. Herzog, *Z. Chem.*, 1984, **24**, 260.
360. D. Walther and E. Dinjus, *Z. Chem.*, 1981, **21**, 415.
361. D. Walther and E. Dinjus, *Z. Chem.*, 1984, **24**, 298.
362. D. Walther, E. Dinjus and V. Herzog, *Z. Chem.*, 1983, **23**, 188.
363. H. Hoberg, A. Ballesteros and A. Sigan, *J. Organomet. Chem.*, 1991, **403**, C19.
364. H. Hoberg and A. Ballesteros, *J. Organomet. Chem.*, 1991, **411**, C11.

365. R. Fischer, D. Walther, R. Kempe, J. Sieler and B. Schönecker, *J. Organomet. Chem.*, 1993, **447**, 131.
366. S. Dérien, E. Duñach and J. Périchon, *J. Am. Chem. Soc.*, 1991, **113**, 8447.
367. D. Walther, G. Bräunlich, R. Kempe and J. Sieler, *J. Organomet. Chem.*, 1992, **436**, 109.
368. R. Kempe, J. Sieler, D. Walther, J. Reinhold and K. Rommel, *Z. Anorg. Allg. Chem.*, 1993, **619**, 1105.
369. G. Burkhart and H. Hoberg, *Angew. Chem., Int. Ed. Engl.*, 1982, **21**, 76.
370. H. Hoberg and D. Schaefer, *J. Organomet. Chem.*, 1982, **238**, 383.
371. H. Hoberg, D. Schaefer and G. Burkhart, *J. Organomet. Chem.*, 1982, **228**, C21.
372. D. Walther, E. Dinjus, J. Sieler, L. Anderson and O. Lindqvist, *J. Organomet. Chem.*, 1984, **276**, 99.
373. A. Behr and G. Thelen, *C_1 Mol. Chem.*, 1984, **1**, 137.
374. H. Hoberg and D. Schaefer, *J. Organomet. Chem.*, 1982, **236**, C28.
375. A. M. Mazany and J. P. Fackler, *Organometallics*, 1982, **1**, 752.
376. H. Hoberg, *J. Organomet. Chem.*, 1988, **358**, 507.
377. P. Braunstein and D. Nobel, *Chem. Rev.*, 1989, **89**, 1927.
378. D. Walther, E. Dinjus and J. Sieler, *Z. Chem.*, 1983, **23**, 237.
379. A. Behr, *Angew. Chem., Int. Ed. Engl.*, 1988, **27**, 661.
380. P. Braunstein, D. Matt and D. Nobel, *Chem. Rev.*, 1988, **88**, 747.
381. D. Walther, *Coord. Chem. Rev.*, 1987, **79**, 135.
382. T. Yamamoto, K. Igarashi, S. Komiya and A. Yamamoto, *J. Am. Chem. Soc.*, 1980, **102**, 7448.
383. R. Cammack, *Adv. Inorg. Chem.*, 1988, **32**, 297.
384. R. P. Hausinger, *Microbiol. Rev.*, 1987, **51**, 22.
385. J. R. Lancaster (ed.), 'The Bioinorganic Chemistry of Nickel', VCH, New York, 1988.
386. A. G. Lappin and A. McAuley, *Adv. Inorg. Chem.*, 1988, **32**, 241.
387. A. Bakač and J. H. Espenson, *J. Am. Chem. Soc.*, 1986, **108**, 713.
388. A. Bakač and J. H. Espenson, *J. Am. Chem. Soc.*, 1986, **108**, 719.
389. M. S. Ram, J. H. Espenson and A. Bakač, *Inorg. Chem.*, 1986, **25**, 4115.
390. J. H. Espenson, M. S. Ram and A. Bakač, *J. Am. Chem. Soc.*, 1987, **109**, 6892.
391. M. S. Ram, A. Bakač and J. H. Espenson, *Inorg. Chem.*, 1988, **27**, 2011.
392. M. S. Ram, A. Bakač and J. H. Espenson, *Inorg. Chem.*, 1988, **27**, 4231.
393. N. Sadler, S. L. Scott, A. Bakač, J. H. Espenson and M. S. Ram, *Inorg. Chem.*, 1989, **28**, 3951.
394. D. G. Kelley, J. H. Espenson and A. Bakač, *J. Chem. Soc., Chem. Commun.*, 1991, 546.
395. D. G. Kelley, A. Marchaj, A. Bakač and J. H. Espenson, *J. Am. Chem. Soc.*, 1991, **113**, 7583.
396. A. Sauer, H. Cohen and D. Meyerstein, *Inorg. Chem.*, 1988, **27**, 4578.
397. R. van Eldik, H. Cohen, A. Meshulam and D. Meyerstein, *Inorg. Chem.*, 1990, **29**, 4156.
398. A. M. Stolzenberg and M. T. Stershic, *J. Am. Chem. Soc.*, 1988, **110**, 5397.
399. G. K. Lahiri, L. J. Schussel and A. M. Stolzenberg, *Inorg. Chem.*, 1992, **31**, 4991.
400. P. J. Chmielewski and L. Latos-Grazyński, *Inorg. Chem.*, 1992, **31**, 5231.
401. D. M. Guldi, P. Neta, P. Hambright and R. Rahimi, *Inorg. Chem.*, 1992, **31**, 4849.
402. D. M. Guldi, M. Kumar, P. Neta and P. Hambright, *J. Phys. Chem.*, 1992, **96**, 9576.
403. P. Stavropoulos, M. C. Muetterties, M. Carrié and R. H. Holm, *J. Am. Chem. Soc.*, 1991, **113**, 8485.
404. Y.-M. Hsiao, S. S. Chojnacki, P. Hinton, J. H. Reibenspies and M. Y. Darensbourg, *Organometallics*, 1993, **12**, 870.
405. D. Sellmann, H. Schillinger, F. Knoch and M. Moll, *Inorg. Chim. Acta*, 1992, **198–200**, 351.
406. S.-K. Lin and B. Jaun, *Helv. Chim. Acta*, 1991, **74**, 1725.
407. S. A. Raybuck *et al.*, *J. Am. Chem. Soc.*, 1987, **109**, 3171.
408. H. König, M. J. Menu, M. Dartiguenave, Y. Dartiguenave and H. F. Klein, *J. Am. Chem. Soc.*, 1990, **112**, 5351.
409. G. Trinquier, M. Dartiguenave, Y. Dartiguenave and M. Bénard, *Inorg. Chem.*, 1991, **30**, 4490.
410. R. Kergoat *et al.*, *J. Organomet. Chem.*, 1990, **389**, 71.
411. H. Brunner and M. Muschiol, *J. Organomet. Chem.*, 1983, **248**, 233.
412. T. J. Voyevodskaya, I. M. Pribytkova and Y. A. Ustynyuk, *J. Organomet. Chem.*, 1972, **37**, 187.
413. H. Lehmkuhl, C. Krüger, S. Pasynkiewicz and J. Poplawska, *Organometallics*, 1988, **7**, 2038.
414. R. Blumhofer, K. Fischer and H. Vahrenkamp, *Chem. Ber.*, 1986, **119**, 194.
415. M. A. Bennett, K. D. Griffiths, T. Okano, V. Parthasarathi and G. B. Robertson, *J. Am. Chem. Soc.*, 1990, **112**, 7047.
416. M. J. Chetcuti, C. Eigenbrot and K. A. Green, *Organometallics*, 1987, **6**, 2298.
417. M. J. Chetcuti and K. A. Green, *Organometallics*, 1988, **7**, 2450.
418. M. J. Chetcuti, B. E. Grant and P. E. Fanwick, *J. Am. Chem. Soc.*, 1989, **111**, 2743.
419. M. J. Chetcuti, S. R. McDonald and N. P. Rath, *Organometallics*, 1989, **8**, 2077.
420. M. J. Chetcuti, B. E. Grant and P. E. Fanwick, *Organometallics*, 1990, **9**, 1345.
421. M. J. Chetcuti, P. E. Fanwick and B. E. Grant, *Organometallics*, 1991, **10**, 3003.
422. A. D. Shaposhnikova *et al.*, *J. Organomet. Chem.*, 1992, **429**, 109.
423. T. V. Ashworth, M. J. Chetcuti, J. A. K. Howard, F. G. A. Stone, S. J. Wisbey and P. Woodward, *J. Chem. Soc., Dalton Trans.*, 1981, 763.
424. R. de Vaumas, A. Marinetti, F. Mathey and L. Ricard, *J. Chem. Soc., Chem. Commun.*, 1988, 1325.
425. A. D. Shaposhnikova *et al.*, *J. Organomet. Chem.*, 1991, **405**, 111.
426. S. B. Colbran, B. H. Robinson and J. Simpson, *Organometallics*, 1985, **4**, 1594.
427. I. Moldes, J. Ros, R. Yáñez, R. Mathieu, X. Solans and M. Font-Bardía, *J. Organomet. Chem.*, 1990, **395**, 305.
428. M. Akita, A. Kondoh, T. Kawahara, T. Takagi and Y. Moro-oka, *Organometallics*, 1988, **7**, 366.
429. M. Akita, A. Kondoh and Y. Moro-oka, *J. Chem. Soc., Dalton Trans.*, 1989, 1627.
430. M. Akita, M. Terada, M. Tanaka and Y. Moro-oka, *Organometallics*, 1992, **11**, 3468.
431. M. Lanfranchi, A. Tiripicchio, M. Tiripicchio Camellini, O. Gambino and E. Sappa, *Inorg. Chim. Acta.*, 1982, **64**, L269.
432. H. T. Schacht and H. Vahrenkamp, *J. Organomet. Chem.*, 1990, **381**, 261.
433. M. F. D'Agostino, M. Mlekuz and M. J. McGlinchey, *J. Organomet. Chem.*, 1988, **345**, 371.
434. M. Mlekuz, P. Bougeard, M. J. McGlinchey and G. Jaouen, *J. Organomet. Chem.*, 1983, **253**, 117.
435. H. Vahrenkamp, *J. Organomet. Chem.*, 1989, **370**, 65.
436. D. Mani and H. Vahrenkamp, *Chem. Ber.*, 1986, **119**, 3639.

437. E. Roland and H. Vahrenkamp, *Chem. Ber.*, 1984, **117**, 1039.
438. R. Blumhofer and H. Vahrenkamp, *Chem. Ber.*, 1986, **119**, 683.
439. W. Bernhardt and H. Vahrenkamp, *J. Organomet. Chem.*, 1988, **355**, 427.
440. D. Afzal and C. M. Lukehart, *Organometallics*, 1987, **6**, 546.

3
Nickel–Carbon π-Bonded Complexes

MICHAEL J. CHETCUTI
University of Notre Dame, IN, USA

3.1 INTRODUCTION 108

3.2 NICKEL COMPLEXES WITH η^2-LIGANDS 108
3.2.1 General Comments 108
3.2.2 Theoretical and Physical Studies 109
3.2.3 Nickel Ethene Complexes 110
3.2.4 Other Nickel Alkene Complexes 113
3.2.5 Heteroalkene Complexes 125
3.2.5.1 General comments 125
3.2.5.2 Nickel silylene species 125
3.2.5.3 Nickel η^2-N=C species 125
3.2.5.4 Nickel η^2-P=C species 126
3.2.5.5 Nickel η^2-C=O species 127
3.2.5.6 Nickel η^2-C=S species 129
3.2.6 Nickel η^2-Alkyne Complexes and Related Species 129
3.2.6.1 Theoretical and physical studies 129
3.2.6.2 Nickel monoalkyne complexes 132
3.2.6.3 Nickel complexes with benzyne ligands 133
3.2.6.4 Nickel complexes with di- and polyalkynes 134
3.2.6.5 Di- and polymetallic nickel alkyne complexes 135

3.3 NICKEL COMPLEXES WITH η^3-LIGANDS 136
3.3.1 General Comments 136
3.3.2 Theoretical and Physical Studies of Nickel η^3-Allyl Species 137
3.3.3 Chemistry and Reactions of Nickel η^3-Allyl Compounds 137
3.3.4 Nickel Heteroallyl Species 144
3.3.5 Nickel η^3-Benzyl Complexes 145
3.3.6 Nickel η^3-Cyclopropylium Complexes and Related Species 146

3.4 NICKEL COMPLEXES WITH η^4-LIGANDS 147
3.4.1 General Comments 147
3.4.2 Theoretical Studies 147
3.4.3 Syntheses and Reactions of Nickel Cyclic η^4-Species 148

3.5 NICKEL COMPLEXES WITH η^5-LIGANDS 149
3.5.1 General Comments and Literature 149
3.5.2 Theoretical and Physical Studies of Nickel η^5-Ligand Complexes 150
3.5.2.1 Theoretical and gas-phase studies 150
3.5.2.2 Other spectroscopic and magnetic studies 150
3.5.2.3 Electrochemical studies 152
3.5.3 The Chemistry of Nickelocene and Substituted Nickelocenes 154
3.5.3.1 $[NiCp_2]$ and monosubstituted cyclopentadienyl sandwich compounds 154
3.5.3.2 Polysubstituted cyclopentadienyl nickel sandwich compounds and fulvalene species 155
3.5.4 $[NiL^1L^2Cp]^+X^-$ and $[NiXL^1Cp]$ Complexes and their Derivatives 156
3.5.4.1 General comments 156
3.5.4.2 Chemistry of nickel $[NiL^1L^2Cp]^+ X^-$ and $[NiXL^1Cp]$ complexes 160

3.5.5 Dinuclear Nickel η^5-Ligand Complexes 167
3.5.5.1 General comments 167
3.5.5.2 Chemistry of the complex $[Ni_2(\mu\text{-}CO)_2Cp_2]$ (Ni–Ni) and its derivatives 167
3.5.5.3 Heterodinuclear complexes containing nickel η^5-ligands 169
3.5.6 Clusters with NiCp or NiCp'' Groups (Cp'' = Substituted Cp) 170
3.5.6.1 General comments 170
3.5.6.2 Homonuclear clusters with NiCp and NiCp'' groups 171
3.5.6.3 Heteronuclear clusters with NiCp and NiCp'' groups 172

3.6 NICKEL π-ARENE COMPLEXES 174
3.6.1 General Comments 174
3.6.2 Syntheses and Reactions 174

3.7 NICKEL π-BONDED POLYDECKER COMPLEXES 177
3.7.1 General Comments 177
3.7.2 Theoretical and Physical Studies 177
3.7.3 Synthetic Studies 177

3.8 BORANE AND CARBORANE COMPLEXES 180

3.9 REFERENCES 182

3.1 INTRODUCTION

There have been many developments in nickel chemistry since the publication of *COMC-I* in 1982. Various reviews summarize some aspects of this chemistry and these are detailed here. Reviews of π-complexes for the years 1981, 1982, 1983, and 1986 include sections relevant to nickelocene and nickel cyclopentadienyl complexes.[1–4] A more recent survey contains pertinent material on η^2-alkene and η^2-alkyne complexes. Organonickel chemistry, especially the dimerization and cyclooligomerization of alkenes and dienes, and their implication for homogeneous catalysis, has been surveyed.[5,6] General reviews of group 10 metal chemistry have appeared for the years 1981, 1983, 1985, and 1986[7–10] and contain much relevant material. Other reviews of interest include general surveys on the syntheses of arene complexes, syntheses by metal vapor techniques[11] and by "wet" methods.[12] Reviews of nickel alkyl complexes[13] and of their decomposition modes[14] contain many references to nickel π-complexes. A summary of nickel-catalyzed reactions has appeared.[15] Other reviews of applications of nickel complexes are listed in subsequent sections.

3.2 NICKEL COMPLEXES WITH η^2-LIGANDS

3.2.1 General Comments

Nickel complexes that contain η^2-ligands include homoleptic and heteroleptic alkene complexes, heteroalkene complexes, and alkyne complexes, among others. Nickel ethene complexes are discussed first, followed by a discussion of nickel complexes with other alkenes. Heteroalkene complexes are then covered, followed by a discussion of alkyne complexes. Some overlap is inevitable and some ethene chemistry is covered within the alkene section.

Nickel has proven to be adept at stabilizing coordinated benzyne ligands and a number of nickel η^2-benzyne complexes have been synthesized since the mid-1980s. Other interesting developments in η^2-nickel complexes are the characterization of many anionic nickel alkene complexes and the syntheses of a growing body of compounds with heteroalkene–nickel bonds.

Much of the chemistry of nickel alkene and nickel alkyne species has been developed by researchers working in Germany. The alkene complexes $[Ni(\eta\text{-}C_2H_4)_3]$ and $[Ni(cod)_2]$ and, to a lesser extent, [Ni(cyclododeca-1,5,9-triene)] ([Ni(1,5,9-cdt)]) provide the usual source of nickel(0) species as the alkene ligands are readily displaced by other ligands. The chemistry of di- and polyalkene nickel(0) compounds continues to be dominated by reactions involving $[Ni(cod)_2]$ and [Ni(1,5,9-cdt)]. Ligand substitution of the coordinated alkene bonds in these complexes, occasionally with concurrent oxidative addition of new ligands, is facile and these species remain important starting materials for new nickel(0) and nickel(II) compounds. One or more of the ligated alkene groups in these "naked-nickel" compounds can be displaced by other alkenes (including heteroalkenes), by alkynes, and by trialkylphosphine species, among others.

In general, nickel-coordinated nonconjugated dienes like cod behave in similar fashion to two separate ligated η^2-alkene functionalities; conjugated dienes such as butadiene can be η^2- or η^4-coordinated depending on the reaction conditions. Thus, the chemistry of nickel η^4-diene species is discussed jointly with that of nickel η^2-alkene complexes. However, nickel η^4-cyclobutadiene complexes and their derivatives are discussed under nickel η^4-complexes. The cationic species $[Ni(\eta^4\text{-}C_5H_6)Cp]^+$, readily produced *in situ* by protonation of nickelocene, is a useful source of the $[NiCp]^+$ fragment. Reactions of $[Ni(\eta^4\text{-}C_5H_6)Cp]$ complexes, which are common starting materials for the syntheses of $[NiL^1L^2Cp]$ or [NiXLCp] species, are covered in Section 3.5.

A large body of research has investigated the use of nickel(0) alkene complexes in coupling and cyclization reactions of organic ligands. Many of these stoichiometric reactions can be made to proceed catalytically. The active nickel(0) species are derived from $[Ni(\eta\text{-cod})_2]$ or [Ni(η-1,5,9-cdt)]. Typically, these alkene complexes are treated with trialkylphosphine ligands (often PCy_3, PPh_3, dcpe (1,2-bis(dicyclohexylphosphino)ethane)) or with chelating nitrogen ligands (bipy, TMEDA) in the presence of the organic reagents and the alkene, diene, allene, or heteroallene reagent. These reactions are not discussed unless new π-complexed nickel species were isolated and characterized.

Nickel η^2-alkene complexes are listed in Tables 1–5. Tables 1 and 2 list ethene complexes of the type $[Ni(\eta\text{-}C_2H_4)_2L]$ and $[Ni(\eta\text{-}C_2H_4)L^1L^2(L^3)]$, respectively. Alkene complexes of generic formula $[Ni(\eta\text{-alkene})L^1L^2(L^3)]$, including di- and polyalkene complexes, are tabulated in Table 3. Other mononickel alkene complexes, including those that contain tethered alkene ligands, are collected in Table 4. Di- and polynickel alkene complexes are listed in Table 5. Heteroalkene complexes are discussed later and appear in Table 6. Nickel alkene complexes have been reviewed.[16] The complex $[Ni(cod)_2]$ is a well-known cyclization catalyst for the intramolecular cyclization of various dienes, diynes, dienynes, and enynes, and its utility in these reactions has been reviewed.[17]

3.2.2 Theoretical and Physical Studies

The monoalkene species $[Ni(C_2H_4)]$ has been calculated to have a singlet ground state, with a carbon–carbon distance of 0.145 nm and a bend-back angle of 21°. A binding energy of 80 kJ mol^{-1} was calculated for this species.[18] Calculations (carried out at the CASSCF CC1 level) were performed on three different relative orientations of the C_2H_4 ligands in the bis(ethene) complex $[Ni(\eta\text{-}C_2H_4)_2]$. The calculations indicated that a molecule of D_{2d} symmetry, in which the two ethene molecules are at right angles to each other, has the lowest energy. Two simultaneous d_π–p_π bonds were believed to be present in this complex. Other alternative structures with D_{2h} and C_{2v} symmetry were calculated to have energies of 43 kJ mol^{-1} and 134 kJ mol^{-1}, respectively, higher than the D_{2d} structure.[19]

Ab initio calculations on $[Ni(\eta\text{-}C_2H_4)(PH_3)_2]$ and $[Ni(\eta\text{-}C_2H_2)(PH_3)_2]$ indicate that, as expected, the ethyne C≡C bond is stronger than the ethene C=C bond. The relative importance of σ-donation and π-back-donation was studied by analyzing the bond energies and the electronic distribution in the nickel–carbon bonds.[20] Calculations for the complexes $[M(\eta^2\text{-}L)(PH_3)_2]$ (M = Ni, Pd, Pt; L = O_2, C_2H_4, C_2H_2) indicate that the binding energy of the ligand decreases in the order $O_2 > C_2H_2 \approx C_2H_4$.[21] The mechanism of nucleophilic attack on metal alkene complexes was investigated; for $[Ni(\eta\text{-}C_2H_4)(PR_3)_2]$, slippage to an η^1-coordinated ligand was considered.[22]

An XPES (x-ray photoelectron spectra) study found a good correlation between the ionization potential of the core electrons of $[Ni(cod)_2]$ and related species, and the oxidation state of the complexes.[23] The species $[Ni(CH_2)(\eta\text{-}C_2H_4)]$ has been detected under cryogenic conditions.[24] Matrix-isolated nickel complexes with ethene, propene, and butadiene have been studied by UV–Vis, IR, and Raman spectroscopy.[25] A new model has been used for normal mode treatment of $[Ni(\eta\text{-}C_2H_4)_3]$ (as well as for $[Pt(\eta\text{-}C_2H_4)_3]$ and $[Pt(\eta\text{-}C_2H_4)Cl_3]^-$) to calculate metal–alkene force constants. For the tris(alkene) complexes, π-bonding was found to dominate the bonding. Skeletal modes were measured and assigned for these complexes.[26] The IR spectrum of $[Ni(\eta\text{-}C_2H_4)_3]$ was analyzed and assignments of the vibrational modes were made.[27]

Nickel ions dehydrogenate isomeric butenes and afford cationic $[Ni(C_4H_6)]^+$ complexes. These ions were bombarded with further molecules of butene in the mass spectrometer. They are far less reactive with alkenes than the corresponding $[Co(C_4H_6)]^+$ ions: only two separate reaction pathways, that are shown in Equations (1a) and (1b), are observed with the nickel ions.[28] Reactions of Ni^+ ions with hydrocarbons led to insertion of nickel into carbon–carbon bonds, followed by a cascade of other reactions. Cationic nickel alkene species have been observed by mass spectrometry in these reactions.[29,30]

$$[Ni]^+ + C_4H_8 \xrightarrow{-H_2} [Ni(C_4H_6)]^+ \xrightarrow{C_4H_8} \begin{cases} \xrightarrow{38\%} [Ni(C_5H_8)]^+ + C_3H_6 & (1a) \\ \xrightarrow{61\%} [Ni(C_4H_6)(C_4H_8)]^+ & (1b) \end{cases}$$

The equilibria between $[NiL_4]$ species and $[NiL_2(\eta\text{-alkene})]$ species (Equation (2)) have been studied. Values for the equilibrium constant K span a range of 10^{10} between the loosely bonded cyclohexene and the tightly bound maleic anhydride. Large differences are also seen in K when the ligand is varied. Cyanoalkenes ligate via both the alkene bond and the cyano group to the nickel atom.[31,32]

$$[NiL_4] + \text{alkene} \rightleftharpoons [Ni(\eta^2\text{-alkene})L_2] + 2\,L \qquad (2)$$

3.2.3 Nickel Ethene Complexes

A study comparing the reactions of [Ni(1,5,9-cdt)] with $[Ni(\eta\text{-}C_2H_4)_3]$ concluded that the reactions of the two species paralleled each other. However, their relative reactivities indicated that the acceptor strength of the nickel fragment decreased in the order $[Ni(CO)_3]$ > [Ni(1,5,9-cdt)] > $[Ni(\eta\text{-}C_2H_4)_3]$.[33,34]

Among the more interesting developments in the chemistry of nickel alkene complexes has been the synthesis of interesting anionic species, isolated from solutions of $[Ni(\eta\text{-}C_2H_4)_3]$ or [Ni(1,5,9-cdt)] and hydridic or carbanionic reagents. Chelating di- or triamines appear to be necessary to stabilize the cation. These complexes are isolable at low temperatures and have been characterized by extensive low-temperature ^1H and ^{13}C NMR studies. In many cases, the structures of these species have been established by x-ray diffraction studies. Tables 1 and 2 list ethene complexes of the type $[Ni(\eta\text{-}C_2H_4)_2L]$ and $[Ni(\eta\text{-}C_2H_4)L^1L^2]$, respectively.

Table 1 $[Ni(\eta\text{-}C_2H_4)_2L]$ and $[Ni(\eta\text{-}C_2H_4)_2X]^-$ species.[a]

L or X	*Cation*	*Ref.*
PMe_3		35
PPh_3		36
$Sn\{CH(TMS)_2\}_2$		37
H	**$[Na(TMEDA)_2]^+$**	38
H	**$[Li(PMPTA)]^+$** [b]	38
Me	$[Li(TMEDA)_2]^+$	39
(μ-Me)[Li(PMPTA)][b]		39
(μ-Me)MgMe(TMEDA)		40
$C(O)NMe_2$	**$[Li(PMPTA)]^+$** [b]	41
$CPh\{NBu^t(Me)\}$		42
$[CPh(NBu^t)]^-$	$[Li(TMEDA)_2]^+$	43
$N(C_2H_4)_3CH$		44
$[M^1HR_3]$ (M^1 = Al, Ga)	$(M^2)^+$ (M^2 = Li, Na)	45
N(Xyl")=CHCH=N(Xyl")[c]		46
CH_2PR_3 (R = Me, Ph)		35
(*N,N*)-dad-Ar_2[d]		47

[a] Complexes in bold were characterized by an x-ray diffraction study. [b] PMPTA = $(Me_2NC_2H_4)_2NMe$ (see structure (**1**)). [c] Xyl" = $C_6H_3Pr^i{}_2$-2,6. [d] dad = Diazadiazene; Ar = $C_6H_3Pr^i{}_2$-2,6.

A series of species of general formula $[MNiMe_mL^1{}_n(\pi\text{-}L^2)]$ (M = Li, Mg; L = chelating amine, THF; π-L = C_2H_4, 1,5,9-cdt, or CO) are formed by combination of lithium or magnesium methyl species with the appropriate organonickel reagent and the σ-donor ligand. An x-ray diffraction study of the complex $[Li(PMPTA)(\mu\text{-}Me)Ni(\eta\text{-}C_2H_4)_2]$ has been performed (structure (**1**)). The carbanionic methyl group of the organolithium reagent bridges the nickel and the lithium atoms in a three-center two-electron interaction. With TMEDA, the ionic species $[Li(TMEDA)_2]^+[NiMe(\eta\text{-}C_2H_4)_2]^-$ was isolated. This salt forms $[Ni(CO)_3Me]^-$ when exposed to CO.[39] The complex $[Ni(\eta\text{-}C_2H_4)_2(\mu\text{-}Me)Mg(Me)(TMEDA)]$ was also isolated at low temperature from solutions of $[Ni(\eta\text{-}C_2H_4)_3]$, TMEDA, and solvated $MgMe_2$. A structural study of this species revealed a nickel–magnesium distance of 0.261 5(1) nm.[40] Related syntheses of other nickel alkyls from $[Ni(\eta\text{-}C_2H_4)_3]$ and MgR_2 or LiR reagents are described in Ref. 62.

Treatment of $[Ni(\eta\text{-}C_2H_4)_3]$ with alkali metal alkyl hydrides of aluminum or gallium at low temperature affords anions such as $[Ni(\mu\text{-}HAlR_3)(\eta\text{-}C_2H_4)_2]^-$.[45] When carbon monoxide is bubbled

Table 2 $[Ni(\eta\text{-}C_2H_4)L^1L^2]$ or $[Ni(\eta\text{-}C_2H_4)L^1L^2L^3]^-$ complexes.[a]

L¹	*L² (L³)*	*Ref.*
CH_2=O	**TMEDA**	48, 49
CH=CH_2	Cp*	50
Et	η-C_5H_4R (R = Et, Bu^t)	51, 52
Et	η-C_5H_4Et	53
Et	CN, $P(OTol\text{-}2)_3$	54–6
R (R = Et, Pr^i, *c*-Pr, Bu^n, Bu^i)	Cp	53
C_2H_4R (R = Ph, *c*-C_3H_5)	Cp	57
Ph	**Cp***	50
R (R = H, Me)	Cp*	58
PMe_3	PMe_3	33, 34
PR_3 (R = Me, Ph, Cy)	dmpe	46
PR_3 (R = Cy, Et, Ph, Pr^i)	C_2H_2	59
PPh_3	PPh_3	36
$PBu^t_2C_2H_4PBu^t_2$	**H (BF_4^-)**	60
$PC_5H_2Ph_3$-2,4,6	PCy_3	61

[a] Complexes in bold were characterized by an x-ray diffraction study.

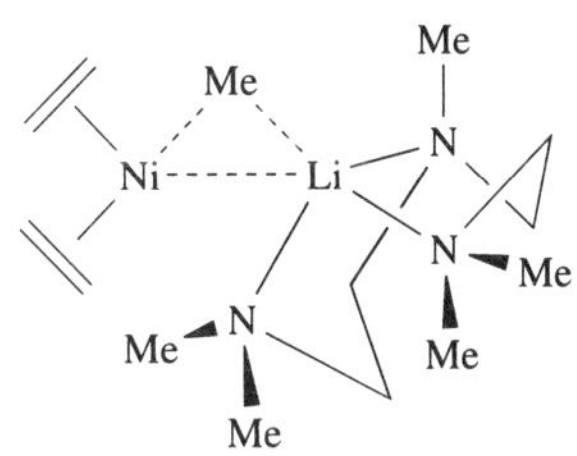

(1)

through solutions of these species, $[Ni(CO)_3H]^-$ and the free trialkylaluminum or trialkylgallium reagent are formed (Equation (3)). Similar moieties with three-center two-electron Ni–H–Al bonds are formed with [Ni(1,5,9-cdt)] and $(AlMe_2H)_2$. The complex $[NiHAlMe_2(1,5,9\text{-cdt})(C_7H_{13}N)]$ ($C_7H_{13}N$ = 1-azabicyclo[2.2.2]octane) has been structurally characterized.[63]

$$[Ni(\eta\text{-}C_2H_4)_3] \; + \; M^+[M'HR_3]^- \rightleftharpoons M^+[Ni\{(\mu\text{-}H)M'R_3\}(\eta\text{-}C_2H_4)_2]^- \qquad (3)$$

$$(M = Li, Na;\ M' = Al, Ga)$$

The ionic dinickel complexes $[Na(TMEDA)_2]^+[Ni_2(\mu\text{-}H)(\eta\text{-}C_2H_4)_4]^-$ and $[Li(PMPTA)_2]^+[Ni_2(\mu\text{-}H)(\eta\text{-}C_2H_4)_4]^-$ have been isolated. Both species contain essentially similar anions: each nickel atom is in a trigonal planar coordinate geometry, linked to two ethene molecules and to the μ-H ligand. There is a nickel–nickel interaction in both molecules (Ni···Ni = 0.259 6(1) nm in the former species and 0.258 1(1) nm in the latter); the Ni–H–Ni angles are 131° and 111°, respectively. However, the first molecule contains well-separated distinct ions in the unit cell, while the PMPTA complex is best regarded as an ion pair in the solid state, with an ion pair contact between the lithium ion moiety, a nickel atom, an ethene group, and the hydride ligand. Bonding in the anion can be explained by an interaction between an empty *p*-orbital on one metal with filled *d*-orbitals of another. The observed electron density distributions are in accord with a semiempirical MO calculation.[38]

Remarkable carbene complexes of nickel, supported by ethene ligands, have been prepared. When a solution of $[Ni(\eta\text{-}C_2H_4)_3]$ and PMPTA was treated with Bu^{t}NC, and followed by LiPh, the PMPTA complex $[Ni(\eta\text{-}C_2H_4)_2\{=CPh(NBu^t)\}Li(PMPTA)]$ was isolated (Structure (**2**)). Complex (**2**) contains a Fischer-type carbene ligand. The related anion (**3**), $[Ni(\eta\text{-}C_2H_4)_2\{=C(NBu^t)(Ph)\}]^-$ was isolated as an $[Li(TMEDA)_2]^+$ salt in a parallel reaction with TMEDA. Complex (**2**) is believed to contain a dative bond from the nitrogen atom of the carbene ligand to the lithium atom (N → Li). Methylation at the nitrogen atom of (**3**) can be achieved by using MeI and a cryptand ligand to stabilize the cation. This reaction proceeds at −30 °C to afford the carbene complex $[Ni(\eta\text{-}C_2H_4)_2\{=CPh(NMeBu^t)\}]$ (**4**) in 68% yield. These reactions are summarized in Scheme 1.

The structure of (**4**) was determined by a low-temperature x-ray diffraction study. Some pertinent bond lengths are indicated in Figure 1.[42] The ethene ligands are rotating at −30 °C and a singlet is observed at 48.6 ppm for the alkene carbon atoms in the ^{13}C NMR spectrum at this temperature. Alkene rotation can be arrested at −110 °C on the ^{13}C NMR timescale on a 400 MHz instrument.[42] One or both

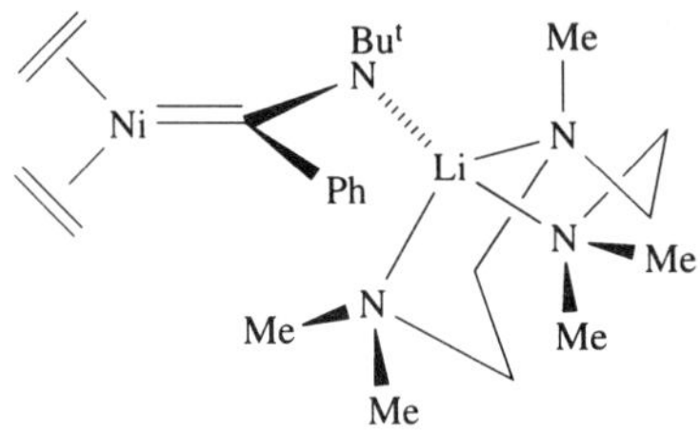

(2)

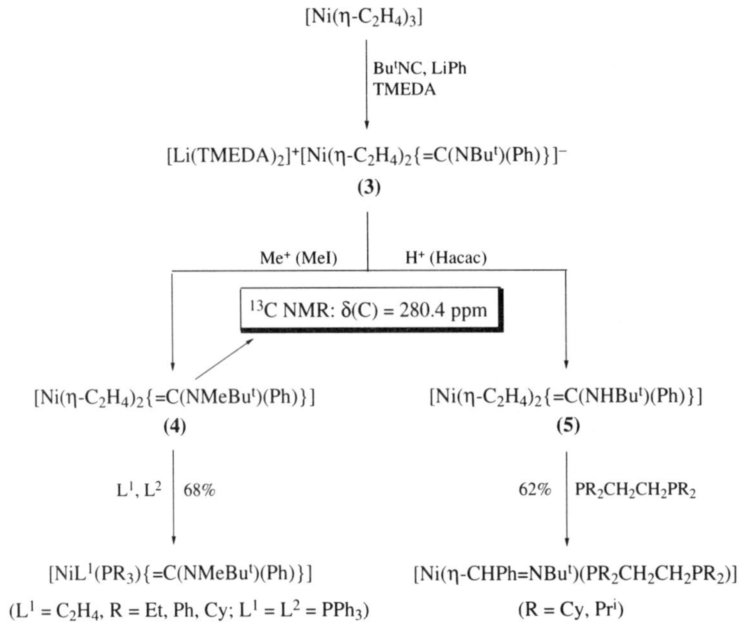

Scheme 1

of the ethene ligands in (**4**) can be replaced with trialkyl- or triarylphosphines or with CO affording the complexes $[Ni(\eta\text{-}C_2H_4)L\{=CPh(NBu^tMe)\}]$ (L = PEt_3, PCy_3, or PPh_3), $[Ni(PPh_3)_2\{=CPh(NBu^tMe)\}]$, or $[Ni(CO)_3\{=CPh(NBu^tMe)\}]$, respectively (Scheme 1).

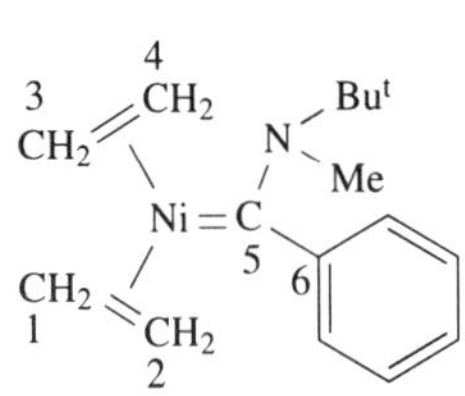

(4)

Ni–C-1 = 0.200 8(1)
Ni–C-2 = 0.200 8(1)
Ni–C-3 = 0.200 0(1)
Ni–C-4 = 0.199 9(1)

Ni–C-5 = 0.188 9(1)
N–C-5 = 0.132 0(1)

C-5–C_1(Ph) = 0.150 3(1)

Figure 1 Key bond length data (nm) for (**4**).[42]

Protonation of (**3**) with Hacac at −40 °C yields the neutral nickel carbene complex $[Ni(\eta\text{-}C_2H_4)_2\{=CPh(NHBu^t)\}]$ (**5**) in 62% yield. The carbene ligand in (**5**) isomerizes when bulky chelating phosphine ligands are added to form a π-bonded imine complex in a proton transfer reaction.[42,43]

More recently, the stannylene compound $[Ni(\eta\text{-}C_2H_4)_2(=SnR_2)]$ (R = $CH(TMS)_2$) was characterized.[37] Nickel η-alkene alkylidene complexes are implicated in the thermolysis of nickelacyclobutane species.[64]

The reaction of [Ni(1,5,9-cdt)(CO)] with $LiNMe_2$ and PMPTA followed by C_2H_4 affords $[Ni(\eta\text{-}C_2H_4)_2\{C(O)NMe_2\}(PMPTA)Li]$ whose geometry was established by an x-ray structural determination. The $[Ni(\eta\text{-}C_2H_4)_2]$ unit is σ-bonded to the planar carbamoyl unit.[41] Both $[Ni(\eta\text{-}C_2H_4)_3]$ and [Ni(1,5,9-cdt)] have been reacted with Wittig reagents. The complexes $[Ni(CH_2PR_3)(\eta\text{-}C_2H_4)_2]$ (R = Me, Ph) and $[Ni(1,5,9\text{-cdt})(CH_2PMe_3)]$, respectively, were obtained. Thermolysis of $[Ni(CH_2PMe_3)(\eta\text{-}C_2H_4)_2]$ afforded $[Ni(\eta\text{-}C_2H_4)_2(PMe_3)]$.[35]

Ethene ligand substitution in [Ni(η-C_2H_4)$_3$] leads to complexes of the type [Ni(η-C_2H_4)$_2$L], [Ni(η-C_2H_4)L_2], or [NiL_3/L_4] species depending on L. The structural parameters of some of these nickel ethene complexes have been determined. A more accurate redetermination of the structure of [Ni(η-C_2H_4)$_2$(PPh_3)] revealed that the nickel atom, the phosphorus, and the ethene carbon atoms are close to being rigorously coplanar.[36] An x-ray diffraction study of the bis(ethene) complex [Ni(η-C_2H_4)$_2$($C_7H_{13}N$)] ($C_7H_{13}N$ = 1-azabicyclo[2.2.2]octane) has also been published.[44] The rotational barrier of the ethene ligands in the complex [Ni(η-C_2H_4)$_2$(PMe_3)] was also reported.[35] Other structurally determined ethene species are shown in bold in Tables 1 and 2.

Diazabutadiene (dad-R_2) ligands displace ethene from [Ni(η-C_2H_4)$_3$] to form the violet 18-electron complex [Ni(η-C_2H_4)$_2${$η^2$-(*N,N*)-dad-Ar_2}] (Ar = $C_6H_3Pr^i_2$-2,6). This molecule is thermally stable to 20 °C under an ethene atmosphere. The diazabutadiene ligand is σ-bonded to the nickel atom via the nitrogen atoms. The ethene ligands are readily displaced with cot or with C_2H_2.[47]

The reactions of diphosphine ligands of the type $PR_2(CH_2)_nPR_2$ (n = 1, 2) with [Ni(η-C_2H_4)$_3$] may afford either mono- or dinickel complexes. The two different μ-dmpe compounds [Ni_2(η-C_2H_4)$_4$(μ-dmpe)] and [Ni_2(η-C_2H_4)$_4$(μ-dmpe)$_2$] were obtained with dmpe. Depending on the reaction conditions and the relative stoichiometry used, [Ni(η-C_2H_4)($PPr^i_2C_2H_4PPr^i_2$)] or [Ni_2(η-C_2H_4)$_4$(μ-$PPr^i_2C_2H_4PPr^i_2$)] could be isolated with bis(diisopropylphosphino)ethane. The dinuclear compound [Ni_2(η-C_2H_4)$_4$(μ-$PPr^i_2C_2H_4PPr^i_2$)] reacts with ethyne to give the structurally characterized nickel-acyclopentadiene complex [Ni_2($η^2$,$η^4$-C_4H_4)(μ-$η^2$,$η^{2'}$-C_2H_2)($PPr^i_2C_2H_4PPr^i_2$)$_2$] (in which the alkyne is the only bridging ligand). The C_4H_4 ligand is formed by the coupling of two ethyne groups. Other species are also formed in this reaction.[65] The thermally unstable compounds [Ni(η-C_2H_4)(dmpe)(PR_3)] (R = Me, Ph, Cy) all decompose above −20 °C.[46,66]

When [Ni(η-C_2H_4)$_3$] is treated with bulky chelating diphosphine ligands L_2, [Ni(η-C_2H_4)(L_2)] species can be isolated. Similar complexes with chelating ditertiary phosphine ligands have been prepared by combining [Ni(cod)$_2$] and the appropriate chelating ligand with butadiene.[67] Protonation of [Ni(η-C_2H_4)(L_2)] complexes afforded cationic ethyl species in which there are agostic hydrogen interactions. The structure of [Ni{(η-C_2H_4)(μ-H)}($PBu^t_2C_2H_4PBu^t_2$)]$^+BF_4^-$ was established by a diffraction study. Similar palladium species were also synthesized.[60]

Mixed-ligand nickel alkene alkyl complexes of the type [NiR^1(η-CH_2=CHR^2)Cp] (R^2 = H, Me) may be prepared by cyclopentadienyl ligand substitution in [NiCp_2] using a combination of an alkyl anion source (typically an alkyllithium or an organomagnesium reagent) and an alkene ligand, often ethene or propene. For example, [NiCp_2] and MgEtBr react to form [NiEt(η-C_3H_6)Cp] under a propene atmosphere. Nickel butadiene and substituted 1,3-butadiene complexes, both mono- and dinuclear, have also been obtained this way. Similar complexes are formed with ethene, and NMR evidence indicates that alkene rotation is operative in these complexes.[68]

The mechanism of this type of reaction has been inferred from ^{1}H and ^{13}C NMR studies of the intermediates formed. At −78 °C, the primary products are the nickel alkene alkyl species [NiR^1(η-CH_2=CHR^2)Cp] and LiCp. At +20 °C, however, the major complexes formed with ethene are the ethyl species [NiEt(η-C_2H_4)Cp] and [NiEt(η-C_2H_4)(η-C_5H_4Et)]. Loss of LiCp and the formation of the unsaturated alkyl species [Ni(CHR^1CR^2R^3)Cp] take place first. The reaction then proceeds via two independent pathways. Addition of CH_2=CHR to the unsaturated alkyl species affords the product derived from the predominant pathway at low temperature. The second mode of reaction proceeds via β-elimination of an alkene from the unsaturated alkyl to yield the hydride species {NiHCp}. Ethene addition and insertion into the nickel–hydrogen bond yields an organoethyl species. Further ethene insertion into a carbon–hydrogen bond of the Cp ligand may also occur to give an ethylcyclopentadienyl alkene complex. Scheme 2 summarizes these mechanistic studies.[53,57]

Related reactions with [Ni(η-C_5H_4R)$_2$] (R = Bu^t, Et), C_2H_4, and RMgX (R = Bu^t, Et) reagents afforded [NiEt(η-C_2H_4)(η-C_5H_4R)] as the final product of ethene insertion and reductive alkyl group elimination reactions.[51,52] Ethene insertion into the Ni–R bonds of [NiR(η-C_2H_4)Cp] (R = *c*-C_3H_5, Ph) occurs under ethene pressure (50 bar) followed by addition of a second molecule of ethene to afford the corresponding [Ni(C_2H_4R)(η-C_2H_4)Cp] species. The formation pathways of these products are presumed to be similar to those shown in Scheme 2.[57] Studies of the rate of insertion of the coordinated alkene into the Ni–R bonds of the complexes [NiR(η-C_2H_4)Cp] and [NiR(η-C_2H_4)Cp*] have been published.[69]

3.2.4 Other Nickel Alkene Complexes

The substitution of a Cp group from [NiCp_2] has also been used to prepare the series of $η^2$-alkene complexes [NiR^1(η-CH_2=CHR^2)Cp*] (R^1 = Me, Et; R^2 = H, CH=CH_2) in which the alkene ligands are

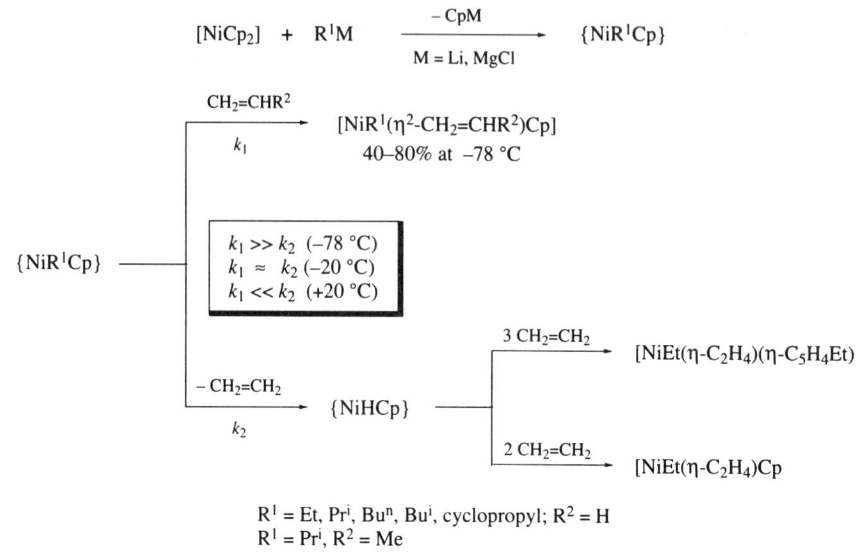

Scheme 2

ligated to NiCp*; related complexes with η-C_5Ph_5 ligands were also prepared.[58] Alkene complexes of nickel are listed in Tables 3–5.

The thermal decomposition of a series of [NiR(η-CH_2=CHMe)Cp] complexes (R = Me, Et, Pr^n, CH_2TMS) has been the subject of a labeling study investigation. Thermolysis of the methyl complex [NiMe(η-CH_2=CHMe)Cp] in toluene-d_8 proceeded as shown in Equation (4), with the products indicated being observed. No deuterated products were formed, indicating that no hydrogen–deuterium exchange with the toluene took place.

$$[\text{NiMe}(\eta\text{-CH}_2\text{=CHMe})\text{Cp}] \longrightarrow \text{CH}_4 + \text{C}_2\text{H}_4 + \text{C}_2\text{H}_6 + [\text{Ni}_3(\mu_3\text{-CH})\text{Cp}_3] \qquad (4)$$

Thermolysis of [Ni(CD_3)(η-CH_2=CHMe)Cp] did lead to hydrogen–deuterium exchange between the CD_3 group and propene. The ethyl and the *n*-propyl derivatives formed ethene and propene, respectively, via {NiHCp} formed through β-elimination processes. This transient species coupled with {NiRCp} affording the alkane and metal clusters. Higher hydrocarbons, formed as minor product, arise from alkene insertion into the Ni–R bond. The TMS derivative afforded tetramethylsilane.[72]

Nickelocene reacts with vinyllithium or the vinyl Grignard reagent in a similar fashion to that of saturated alkyllithium or Grignard reagents: a vinyl group substitutes the cyclopentadienyl ligand and MCp (M = Li or MgCl), together with the transient species {Ni(CH=CH_2)Cp}, forms. Two possible mechanistic pathways ensue. Ethene insertion into the Ni–CH=CH_2 bond followed by rearrangement of the resultant {Ni(C_2H_4CH=CH_2)Cp} species leads to the formation of the *syn*- and *anti*-isomers of [Ni(η^3-C_3H_4Me)Cp]. Dimerization of the transient complex affords the nickel–nickel bonded bridging butadiene species [Ni_2(μ-η^2,η^2-1,3-C_4H_6)Cp_2]. This dinickel complex is also formed by thermally induced reductive elimination of ethane and loss of butadiene from two molecules of [NiMe(η^2-1,3-C_4H_6)Cp].[147]

Treatment of nickelocene with methyllithium in the presence of 1,3-butadienes CH_2=CHR^1–CHR^2=CH_2 (R^1, R^2 = H, Me) affords nickel η^2-butadiene complexes in 40–75% yield. The diene with an *s-trans*-conformation is the major product; the less-substituted double bond is complexed to the nickel atom. The thermal stability of the diene complex decreases with increasing methylation of the diene.[120]

When the mixed-ligand metallocene [NiCpCp*] was treated with vinyllithium at –50 °C under an ethene atmosphere, it is the Cp group that was replaced: the fully characterized Cp* complex [Ni(CH=CH_2)(η-C_2H_4)Cp*] was formed. Ethene insertion into the nickel–vinyl bond affords [Ni(η^1,η^2-C_2H_4CH=CH_2)Cp*]. Above –10 °C, dissociation of coordinated ethene from [Ni(CH=CH_2)(η-C_2H_4)Cp*] followed by the coupling of two vinyl ligands and metal–metal bond formation afforded [Ni_2(μ-η^2,η^2-1,3-C_4H_6)$Cp*_2$].[50] These results are summarized in Scheme 3. Related work details the reactions of nickelocene and LiMe with various alkenes. The initial [NiMe(η^2-C_nH_{2n})Cp] (*n* = 2–4, 6) products could be thermolyzed to the *cis*- and *trans*-*s*-diene complexes [Ni_2(μ-η^2,η^2-CH_2=CR^1CR^2=CH_2)Cp_2] (R^1, R^2 = H, Me).[73]

Table 3 [Ni(η^2-alkene)L^1L^2] or [Ni(η^2-alkene)$L^1L^2L^3$] complexes.[a]

Alkene(s)	L^1, L^2, (L^3)	*Ref.*
C_2F_4	**TMEDA**	70
C_2H_3X (X = F, Me, CF_3)	3 PF_3	71
3 C_2H_3X (X = Cl, F, Me, CF_3)		71
C_3H_6	Cp, R (R = Me, Me-d_3, Et, Pr, TMS)	72
C_3H_6	Pr^i, Cp	53
C_nH_{2n} (n = 2–4, 6)	Me, Cp	73
$CH_2{=}C(CH_2)_2$	Me, Cp	57
trans-**CHPh=CHPh**	**$Bu^t_2PC_2H_4PBu^t_2$**	74
trans-CHPh=CHPh	2 PPh_3	74
trans-CHPh=CHPh	$PBu^t_2CH_2PBu^t_2$	75
$CH_2{=}C{=}O$	2 PPh_3	76, 77
$CH_2{=}CH(CH_2)_2CH{=}CH_2$	N(Xyl")=CHCH=N(Xyl")[b]	46
3 C_7H_{10}[c]		78
C_2H_3CHO	L_2 (L_2 = en, bipy, 2 PPh_3)	79, 80
C_2H_3CHO	**bipy**	81
C_2H_3R (R = **CO_2Me**, COMe, CN)	TMEDA	82
2 C_2H_3R (R = **CHO**, CO_2Me)	**py**	83
CHEt=C(Me)CHO	bipy	81
$C_{10}H_8$[d]	**$PPr^i_2(CH_2)_nPPr^i_2$** (n = **2**, 3)	84, 85
$C_{10}H_8$[d]	**$PBu^t_2C_2H_4PBu^t_2$**	86
$C_{14}H_{10}$[e]	2 PBu^n_3	87
$C_{14}H_{10}$[e]	PR_3 (R = Et, Bu^n)	88
η^4-$C_{14}H_8O_2$[f]	PCy_3	89
η^2-cod	bipy	90
cod	C_3H_5, (BF_4^-)	91
cod	**2C_6F_5**	92
cod	**PhC_2Ph**	93
cod	η^5-C_5H_7	94, 95
cod, η^4-C_5Ph_4O		96
cod, $C_2H_3CH{=}CHCO_2Me$		97
cod	η^5-$C_3Et_2HB_3Me_3$ (BF_4^-)	98
cod, η^4-$C_4H_2Ph_2$-2,5-$SiMe_2$		99
cod, η^4-$C_4H_2Me_2$-3,4-$SiMe_2$		100
cod	2 $PC_5H_2Ph_3$-2,4,6	61
cod, η^4-dbn[g]		101
2 η^4-dbn[g]		101
dbn, η^4-$C_2H_3C_2H_4C_2H_3$[g]		101
2 η^4-$C_4H_4SiMe_2$		102
2 η^4-$C_4H_2Ph_2$-2,5-$SiMe_2$		99
1,5,9-cdt	$PC_5H_2Ph_3$-2,4,6	61
1,5,9-cdt	**$AlMe_2H$, $N(C_2H_4)_3CH$**	63
1,5,9-cdt	CH_2PMe_3	35
c-$C_2(CO_2Me)_2CH_2CH(C_2H_3)$	2 PPh_3	103
c-$C_2(CO_2Me)_2CHMeCHMe$	2 PPh_3	104
1,2:5,6-η^4-C_8F_8	2 L (L = PMe_3, PMe_2Ph)	105, 106
1,2:5,6-η^4-$C_8H_4Ph_4$-3,4,7,8	2 PEt_3	107
$CPh(PMe_3){=}CHCOCH_2R$ (31) (R = CMe_2Ph)	**Cl, PMe_3**	108, 109
c-1,2-$C_2R_2C_2(O)_2$	bipy, CO	110, 111
***c*-3,4-$C_2Ph_2C_2(O)_2$**	**bipy, CO**	112
NR=CHCH=CHPh (R = Pr^i, **Ph**, Tol-4)	**2 PPh_3**	113, 114
2 C_4H_6	TMEDA	115
η^4-C_4H_6	**dcpe**	116
C_4H_6	Cp*, R (R = Me, Et)	58
η^4-norbornadiene	**2 C_6F_5**	92
η^4-norbornadiene	η^5-C_5H_7	94, 95
$CH_2{=}CH(CH_2)_3CH{=}CH_2$	**PPh_3**	117
η^4-C_5H_6	Cp (BF_4^-)	118
$C_2H_3CH{=}CHCO_2Me$	2 PPh_3	97
n $C_2H_3CH{=}CHCO_2Me$ (n = 1, 2)	bipy	97
$C_2H_3SiR_3$ (R = Me, OMe)	2 PPh_3	119
$(C_2H_3)_2SiMe_2$	n PPh_3 (n = 1, 2)	119
$(C_2H_3)_4Si$	PPh_3	119
$C_2H_3R^1CHR^2{=}CH_2Me$ (R^1, R^2 = H, Me)		120
C_4H_6	$PR_2C_2H_4PR_2$	67
$CH_2{=}C(CH_2)_2$	2 PPh_3	121
$C_2H_3C_6H_4(PPh_2)$-2	2 PPh_3	122
2 η^4-*c*-C_5Ph_4CO		123
$nCH_2{=}C{=}CHR$ (n = 1, 2) (R = H, Et, MeO, Ph, CN, Bu^t)	n PPh_3 (n = 1, 2)	124, 125

Table 3 (continued)

Alkene(s)	$L^1, L^2, (L^3)$	*Ref.*
$CH_2{=}C{=}CMe_2$	n PPh_3 (n = 1, 2)	124
$O{=}C{=}C{=}C{=}O$	2 PPh_3	126
η^2-C_{60}	2 PEt_3	127

[a] Bold complexes were characterized by x-ray diffraction. Unless otherwise stated, all alkenes except (η^4-)cod and (η^6-)1,5,9-cdt are η^2-ligated. Nickel-coordinated atoms are italicized. [b] Xyl'' = $C_6H_3Pr^i_2$-2,6. [c] Norbornene. [d] Naphthalene. [e] Anthracene. [f] Anthraquinone. [g] dbn = 2,3-diethyl-1,4-dimethyl-1,4-diboranaphthalene (**28**).

Table 4 Other mononickel alkene complexes.[a]

Complex	*Ref.*
$[NiBr(\eta^2,\eta^5\text{-}CH_2{=}CH(CH_2)_3C_5Me_4]$	58
$[Ni\{\eta^1,\eta^2\text{-}PPh_2[C_6H_4(C_2H_3)\text{-}2]\}]$	128
$[Ni(\eta^1,\eta^2\text{-}C_7H_8PMe_3)Cp]$ (**17**)	129
$[Ni\{\eta^1,\eta^2\text{-}C_7H_5(CO_2Me)_2\}Cp]$ (**11**)	130
$[Ni(\eta^1,\eta^2\text{-}1,4,5\text{-}C_8H_{13})(\eta^6\text{-}C_5H_5BPh)]$	131
$[Ni\{\eta^1{}_{11},\eta^2{}_{9,10}\text{-}3,4,5,6\text{-}Cl_4\text{-}12,12\text{-}(MeO)_2\text{-}9,10\text{-}(CO_2Me)_2\text{-}exo,endo\text{-}C_{10}H_{12}\}Cp]$ (13)	132
$[Ni\{\eta^1,\eta^2,\eta^2\text{-}(CH_2{=}CHCH_2CH_2)_2CH\}\{B(C_6H_4O_2\text{-}1,2)_2\}]$ (**30**)	133
$[Ni(\eta^1,\eta^2\text{-}CH_2CH_2CR^1R^2CR^3{=}CH_2)Cp]$ (R^1, R^2, R^3 = H, Me)	134, 135
$[Ni(\eta^1,\eta^2\text{-}CHCMe_2CHCR^1R^2CR^3{=}CHR^4)Cp]$	136
$[Ni(\eta^1{}_5,\eta^2{}_{1,2}\text{-}C_8H_{13})\{PPh_2CH_2C(O)O\}]$	137
$[Ni(\eta^1,\eta^2\text{-}CH_2CH_2CH{=}CH_2)Cp^*]$	50
$[Ni(\eta^1,\eta^2\text{-}CH_2CH_2CH_2CH{=}CH_2)Cp]$	138
$[NiX\{(C_2H_3C_2H_4)As(C_3H_6AsMe_2)_2\}]^+X^-$ (X = Cl, Br)	139
$[Ni(C_{60})]_n{}^+$ (n = 1, 2)	140

[a] Complexes in bold were characterized by an x-ray diffraction study. Nickel-coordinated atoms are italicized.

Table 5 Di- and polynickel alkene complexes.

Complex	*Ref.*
$[Ni_2(\eta\text{-}C_2H_4)_3(dmpe)_2]$	33,34
$[Ni_2(\eta\text{-}C_2H_4)_4(\mu\text{-}L_2)]$ (L_2 = dmpm, $PPr^i{}_2C_2H_4PPr^i{}_2$)	65
$[Ni_2(\eta\text{-}C_2H_4)_2(\mu\text{-}dmpm)_2]$	65
$[Ni_2\{\eta^1,\eta^2,\eta^{1'},\eta^{2'}\text{-}(C_8H_{12})_2\}Cp^*{}_2]$ (15)	141, 142
$[Ni_2(\eta\text{-}C_2H_4)(\eta^1,\eta^3\text{-}PEtC_5H_2Ph_3\text{-}2,4,6)_2]$	143
$[Ni_2\{\mu\text{-}\eta^2,\eta^2,\eta^{2'},\eta^{2'}\text{-}Si(CH{=}CH_2)_4\}(PPh_3)_n]$ (n = 2, 4)	119, 121
$[Ni_2(cod)_2(\eta^4,\eta^{4'}\text{-}C_{42}H_{36}O_2)]$ (16)	144
$[Ni_2(cod)_2\{\eta^4,\eta^{4'}\text{-}C_{42}H_{36}(OEt)_2\}]^{2+}(BF_4{}^-)_2$[b]	144
$[Ni_2(cod)_2\{\eta^4,\eta^{4'}\text{-}C_{42}H_{36}(OH)_2\}]^{2+}(X^-)_2$ (X = BF_4, CF_3SO_3)[b]	144
$[Ni_2(cod)_2\{\eta^4,\eta^{4'}\text{-}C_{42}H_{36}(OEt)_2\}]^{2+}(BF_4{}^-)_2$[b]	144
$[Ni_2(\mu\text{-}\eta^2,\eta^{2'}\text{-}HC_2H)(\eta\text{-}C_2H_4)_2(PEt_3)_2]$ (R = Et, Ph)	59
$[Ni_2(\mu\text{-}\eta^2,\eta^{2'}\text{-}CH_2CHCH{=}CH_2)Cp^*{}_2]$	50
$[Ni_2(s\text{-}cis,trans\text{-}\mu\text{-}\eta^2,\eta^{2'}\text{-}CH_2{=}CR^1CR^2{=}CH_2)Cp_2]$ (R^1, R^2 = H, Me)	73
$[Ni_2(\mu\text{-}\eta^2,\eta^{2'}\text{-}C_4H_6)(\eta^2\text{-}C_4H_6)_2(TMEDA)_2]$ (**7**)	115
$[Ni_2(\mu\text{-}\eta^2,\eta^{2'}\text{-}L)L_2(bipy)_2]$ (L = C_4H_6)	145
$[Ni_2(\eta\text{-}C_2H_4)\{\eta^1\text{-}(P),\eta^3\text{-}P(Et)C_5H_2Ph_3\text{-}2,4,6\}_2]$	143
$[Ni_2(\mu\text{-}\eta^1,\eta^2\text{-}CH_2{=}CHC_5H_4N)_2(PPh_3)_2]$	146
$[Ni_2(\eta^2\text{-}C_4H_6)_2(\mu\text{-}\eta^2,\eta^{2'}\text{-}C_4H_6)(bipy)_2]$	145
$[Ni_2(\mu\text{-}\eta^2,\eta^{2'}\text{-}C_4H_6)Cp_2]$	147
$[Ni_2(\mu\text{-}\eta^2,\eta^{2'}\text{-}C_4H_6)(PPh_3)_2Cp_2]$	148
$[Ni_2(\mu\text{-}\eta^1,\eta^2\text{-}NR{=}CHCH{=}CHPh)_2(PPh_3)_4]$ (R = **Pr^i**, Ph, Tol-4)	113, 114
$[Ni_2(CO)_2(NAr{=}CMeCMe{=}NAr)(\mu\text{-}\eta^2,\eta^{4'}\text{-}C_4H_4)]$[c]	149
$[Ni_2(\eta^2\text{-}NAr{=}CMeCMe{=}NAr)_2(\mu\text{-}\eta^2,\eta^{4'}\text{-}C_4H_4)]$[c]	149
$[Ni_2(\eta^2\text{-}NAr{=}CRCR{=}NAr)_2(\mu\text{-}\eta^2,\eta^{4'}\text{-}C_4H_4)]$ (R = H, Me)	150
$[Ni(\eta^2\text{-}NR{=}CHCH{=}NR)\{\eta^2,\eta^2\text{-}CH_2{=}CH(CH_2)_2CH{=}CH_2\}]$	151
$[Ni_2\{\mu\text{-}(Z,Z)\text{-}\eta^1,\eta^{2'},\eta^{1'},\eta^2\text{-}CH{=}CPhCPh{=}CH\}(depe)_2]^{2+}(PF_6{}^-)_2$	152
$[Ni_2Br_2(\mu\text{-}\eta^2,\eta^4\text{-}CPh{=}CHCH{=}CPh)(PEt_2C_2H_4PEt_2)_2]$	152
$[Ni_2Br_2(\mu\text{-}\eta^2,\eta^4\text{-}CPh{=}CHCH{=}CPh)(PEt_3)_4]$	152
$[Ni_2\{\eta^2\text{-}N(Xyl\text{-}2,6){=}CHCH{=}N(Xyl\text{-}2,6)\}_2(\mu\text{-}\eta^2,\eta^4\text{-}C_4H_4)]$	47
$[Ni_2\{PMe[W(CO)_5]C(R^1)C(R^2)\}Cp_2]$	153
$[Ni_6(\eta^{12}\text{-}C_{60})(PEt_3)_{12}]$	127

[a] Complexes in bold were characterized by an x-ray diffraction study. Nickel-coordinated atoms are italicized. [b] Derivatives of (**16**). [c] Ar = $C_6H_3Pr^i{}_2$-2,6.

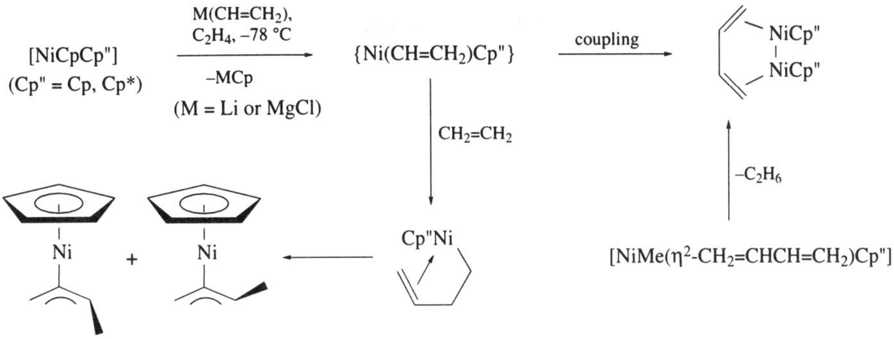

Scheme 3

Nickel 1,3-butadiene complexes with chelating ditertiary phosphine ligands have been prepared by combining [Ni(cod)$_2$] and the appropriate chelating ligand with butadiene. Initially the [Ni(η^2-C_4H_6)] species are formed, but in some cases, depending on the diphosphine ligand present, chelated η^4-butadiene compounds of the type [Ni(η^4-C_4H_6)($PR_2C_2H_4PR_2$)] (R = alkyl, aryl) can be observed at low temperature. Ligand exchange mechanisms and dynamic processes operative in these complexes were also discussed.[67] The η^4-butadiene complex [Ni(η^4-C_4H_6)(dcpe)] was characterized by an x-ray diffraction study.[116]

The reaction of alkenyl Grignard reagents with nickelocene affords tethered alkene complexes: the alkenyl group is attached to the nickel atom via a direct nickel–carbon σ-bond as well as through the alkene. Related complexes with substituents on the butenyl ligands have also been made (Scheme 4),[134] and these species are quite stable towards elimination reactions. Syntheses and substitution patterns in such complexes have been probed.[135] Related complexes (also shown) that contain a cyclopropyl group as part of the tethered chain are also stable towards β-elimination reactions: the β-hydrogen atoms are fixed in a *trans*-position on the rigid ring.[136] The solution configurations of such complexes were established by high-field NMR studies.[154] Complexes with η^1,η^2-pentenyl ligands have also been reported.[138] Other tethered alkene compounds have been made in which the alkene is attached to one of the η^5-ligands. The complex [NiBr{η^2,η^5-[CH_2=CH(CH_2)$_3C_5Me_4$]}] was made and was structurally characterized.[58]

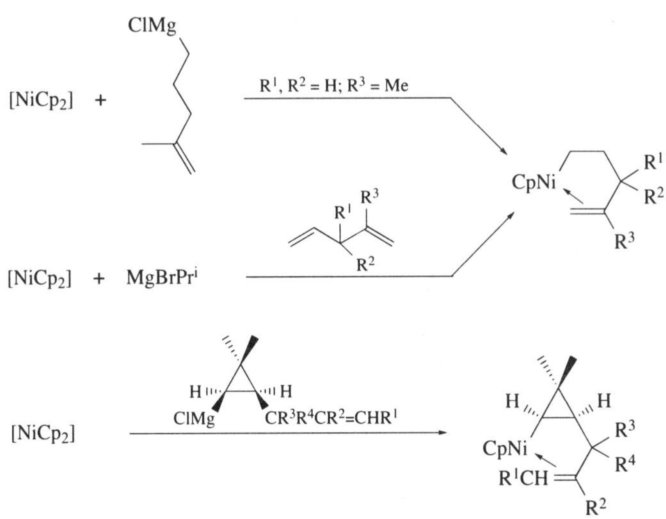

Scheme 4

Many butadiene complexes of nickel are η^2- and not η^4-coordinated. A mixture of TMEDA and 1,3-butadiene and [Ni(η-C_2H_4)$_3$] reacts below –40 °C to form [Ni(η^2-C_4H_6)$_2$(TMEDA)] (**6**) and [Ni$_2$(μ-η^2,$\eta^{2'}$-C_4H_6)(η^2-C_4H_6)$_2$(TMEDA)$_2$] (**7**): only the latter species could be crystallized from solution. Complexes (**6**) and (**7**) decompose above –40 °C, with coupling of the coordinated butadiene moieties to form the species [Ni(η^2,η^3,η^3-$C_{12}H_{18}$)] (**8**). This species has been shown to be an intermediate in the nickel-catalyzed cyclotrimerization of butadiene. Complexes (**6**) and (**7**) are the first isolated examples of nickel butadiene complexes that are catalytically active despite the presence of stabilizing ligands. Carbon dioxide reacts with (**6**) to afford [Ni(η^3-$C_3H_4CH_2CO_2$)(TMEDA)]. These and other reactions are summarized in Scheme 5.[115] The structure of the bipy analogue to (**7**), $[Ni_2(\mu\text{-}\eta^2,\eta^{2'}\text{-}C_4H_6)(\eta^2\text{-}C_4H_6)_2(bipy)_2]$, has been determined.[145]

$[Ni(\eta\text{-}C_2H_4)_3]$ + TMEDA + 1,3-C_4H_6 ⟶ (**6**) + (**7**)

(**6**) ⇌ (**7**) + C_4H_6

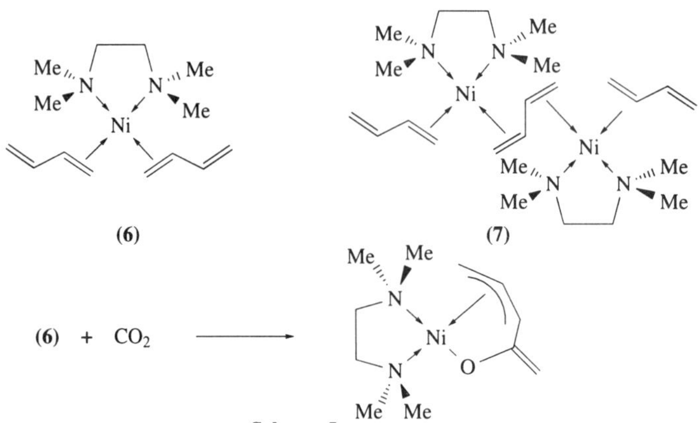

Scheme 5

When $[Ni(\eta\text{-}C_2H_4)(PPh_3)_2]$ reacts with the monoazadienes (mad) CHPh=CHCH=NR (mad-R) (R = Pr^i, Ph, Tol-4), it is the carbon–carbon double bond and not the imine carbon–nitrogen double bond that coordinates to the nickel atom. The mononuclear nickel complexes $[Ni\{\eta^2\text{-}(C,C)\text{-mad-R}\}(PPh_3)_2]$ were isolated from polymeric reaction intermediates. An x-ray study of the mad-Ph complex $[Ni(\eta^2\text{-mad-Ph})(PPh_3)_2]$ revealed that the monoazadiene had a slightly distorted *s-trans*-configuration. Further addition of mad-R resulted in the formation of the dinickel species $[Ni_2\{\mu\text{-}\eta^1(N),\eta^2(C,C)\text{-mad-R}\}_2(PPh_3)_2]$ (**9**). The NMR spectra of these compounds were recorded and compared with those of the mononickel complexes. The structure of one dinuclear species, $\{Ni_2[\mu\text{-}\eta^1(N),\eta^2(C,C)\text{-mad-Pr}^i\}_2(PPh_3)_2]$, was established crystallographically.[113,114]

The dinuclear dad-Ar$_2$ (diazadiene-Ar$_2$) species $[Ni_2\{\eta^2\text{-}(N,N)\text{-dad-Ar}_2\}_2(\mu\text{-}\eta^2,\eta^4\text{-}C_4H_4)]$ (Ar = $C_6H_3Me_2$-2,6) results from the reaction of C_2H_2 with $[Ni(\eta^2\text{-}C_2H_4)_2\{\eta^2\text{-}(N,N)\text{-NAr=CHCH=NAr}\}]$. This molecule, which is fluxional down to −80 °C, contains a nickelacyclopentadienyl ligand.[47] The nickelacyclopentadiene species $[Ni_2(\mu\text{-}\eta^2,\eta^4\text{-}C_4H_4)(CO)_2(\text{dad-Ar}_2)]$ and $[Ni_2(\mu\text{-}\eta^2,\eta^4\text{-}C_4H_4)(\text{dad-Ar}_2)]$ (Ar = $C_6H_3Pr^i_2$-2,6) are also formed from alkyne-coupling reactions; their structures were established by crystallographic studies.[149] Unlike the butadiene complexes referred to earlier, the nitrogen atoms in these diaza ligand species are σ-bonded to the nickel atom.

Substituted diazabutadiene compounds that contain 1,5-hexadiene ligands of formula $[Ni\{\eta^2\text{-}(N,N)\text{-dad-Ar}_2\}\{\eta\text{-}CH_2\text{=}CH(CH_2)_2CH\text{=}CH_2\}]$ were prepared and characterized.[46,66] Ethyne coupling leads to the dinuclear complex $[Ni_2(\eta^2,\eta^4\text{-CH=CHCH=CH})\{\eta^2\text{-}(N,N)\text{-dad-Ar}_2\}_2]$ which also contains N-bonded diazabutadiene ligands. The molecule with Ar = $C_6H_3Pr^i_2$-2,6 was structurally characterized.[150] The diazabutadiene complexes $[Ni\{\eta^2\text{-}(N,N)\text{-dad-R}_2\}\{\eta^2,\eta^2\text{-}CH_2\text{=}CH(CH_2)_2CH\text{=}CH_2\}]$, which contain η^2,η^2-1,5-hexadiene ligands, have recently been prepared.[151]

Oxidation of the nonmetal–metal bonded complex $[Ni_2Br_2(\mu\text{-}\eta^2,\eta^4\text{-CPh=CHCH=CPh})(PEt_2C_2H_4PEt_2)_2]$ with $TlPF_6$ generates the cationic butadienediyl species $[Ni_2\{\mu\text{-}\eta^2,\eta^4\text{-}(Z,Z)\text{-CH=CPhCPh=CH}\}(PEt_2C_2H_4PEt_2)_2]^{2+}$ (Ni–Ni) that contains a nickel–nickel bond and a rearranged ligand skeleton (Scheme 6).[152] A bridging phosphine ligand appears to be necessary. When the analogous reaction was attempted with *trans*-$[Ni_2Br_2(\mu\text{-}\eta^2,\eta^4\text{-CPh=CHCH=CPh})(PEt_3)_4]$ no rearrangement took place; solvent molecules replaced the coordinated bromide ligands.[152]

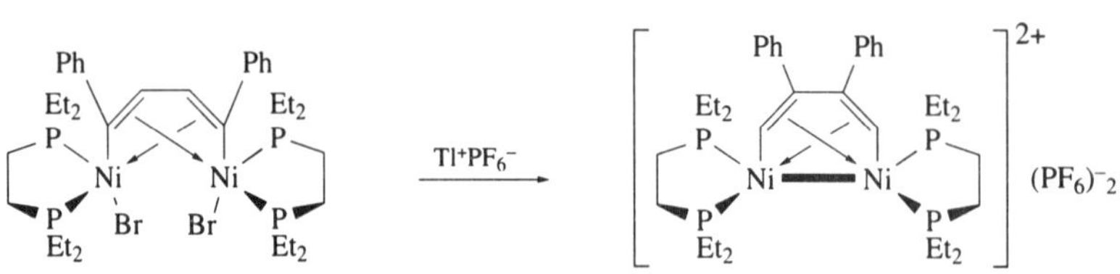

Scheme 6

A detailed procedure for the synthesis of deep violet-blue [Ni(η^2-cod)bipy] (94% yield) has been published.[91] This complex allows oxidative additions to nickel(0) that do not happen with [Ni(cod)$_2$]. A few examples include the cleavage of carbon–sulfur bonds in dibenzothiophene and the cleavage of weak carbon–carbon bonds in biphenylene.[155,156] A report that describes the preparation of [Ni(cod)$_2$] from the reduction of technical grade [Ni(acac)$_2$] with AlHBu$^t_2$, in the presence of 1,5-cod, has appeared.[157] The complex [Ni(η-2,3-norbornene)$_3$] has been structurally characterized: the three norbornene moieties lie in a trigonal plane around the central nickel atom.[78]

Some nickel η-(vinyl halide) complexes have been prepared by metal atom synthetic techniques. Nickel atoms react with substituted ethenes C_2H_3X (X = Cl, F, Me, CF_3) to afford [Ni(η^2-C_2H_3X)$_3$] species which have varying degrees of thermal stability, and decompose, depending on the nature of X, between −96 °C and −30 °C. The vinyl halide species were the least thermally stable. The 18-electron species [Ni(η^2-C_2H_3X)(PF_3)$_3$] (X = F, Me, CF_3) could be isolated when these complexes were reacted with PF_3. Their ^{1}H and ^{19}F NMR spectra were recorded at −45 °C.[71]

The methylenecyclopropane species [NiMe{η^2-CH_2=C(CH_2)$_2$}Cp] undergoes a reverse β-methyl elimination reaction followed by a ring opening of the (1-methylcyclopropyl)methyl ring to afford a 3-methyl-3-butenyl complex. At 40 °C, this species undergoes a reversible β-hydride elimination reaction, followed by a hydrogen transfer to the η^2-diene species formed, terminating the sequence of events to afford the allylic complex [NiCp(η^3-$C_3H_3Me_2$-1,2)Cp]. These results are shown in Scheme 7.[57]

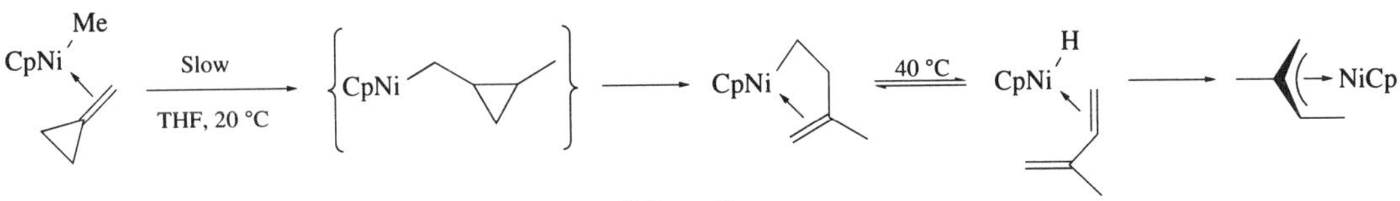

Scheme 7

Methylenecyclopropane reacts with [Ni(η-C_2H_4)(PPh_3)$_2$] to afford yellow [Ni{η^2-CH_2=C(CH_2)$_2$}(PPh_3)$_2$] in 80% yield.[121] The phosphorus atoms in this compound are not equivalent on the NMR timescale, indicating restricted rotation of the alkene ligand about the nickel atom. This species may be obtained by reductive ligation of [Ni(acac)$_2$] with [AlEt$_2$(OEt)] in the presence of the alkene in lower (44%) yield. The alkene may be substituted with other alkenes: *trans*-stilbene affords [Ni(η^2-*trans*-CHPh=CHPh)(PPh_3)$_2$], while tetravinylsilane leads to the previously reported[119] dinickel species [{Ni(PPh_3)$_2$}$_2${(η^2-CH_2=CH)$_2$Si(η^2-CH=CH_2)$_2$}].[121] Other alkene species were obtained by reacting [Ni(PPh_3)$_3$] or [Ni(η-C_2H_4)(PPh_3)$_2$] with methyl derivatives of cyclopropene or vinylcyclopropane.[158]

Trimethylsilyl- or trimethoxysilylalkene and polyalkene nickel(0) complexes have been synthesized by reducing [Ni(acac)$_2$] with aluminum alkyls in the presence of the alkene and PPh_3. The resulting 16-electron [Ni(η^2-alkene)(PPh_3)$_2$] complexes can also be prepared by substitution, using [Ni(cod)$_2$] as the nickel(0) precursor (Scheme 8).[119] Related alkene complexes can be prepared from dimethyldivinylsilane or from tetravinylsilane. These products usually contain nickel in an 18-electron configuration with both vinyl ligands coordinated to the nickel atom (or to two nickel atoms in the case of [Si(CH=CH_2)$_4$]). The 16-electron monophosphine complexes may also be isolated.[119]

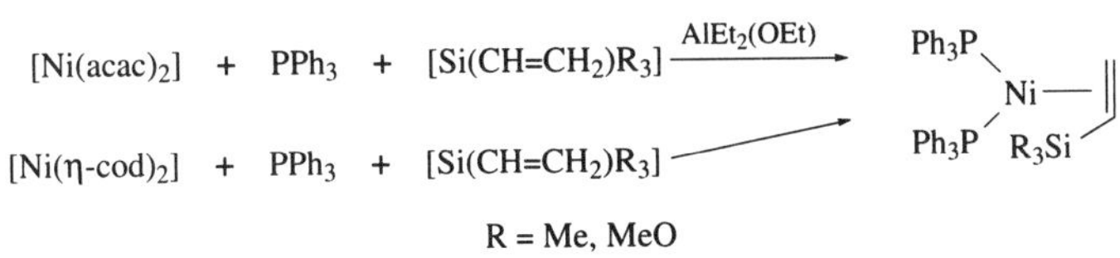

Scheme 8

The ease of addition of HC_2H and CO to [Ni(η^2-CH_2=CHCHO)L_2] was found to increase in the order L_2 = bipy < 2 PPh_3 < en.[79,80] A structural study of the C=C coordinated α,β-unsaturated aldehyde complex [Ni(η^2-CHEt=CMeCHO)(bipy)] (**10**) indicates that there is some interaction of the carbonyl carbon atom with the nickel: the aldehyde approaches η^3-coordination (Figure 2). The syntheses and chemistry of [Ni(η^2-CH_2=CHCHO)$_2$(py)] have been described and the complex was characterized by x-ray diffraction.[83] Two different syntheses were described for the compounds [Ni(η^2-CH_2=CHR)(TMEDA)] (R = CO_2Me, COMe, CN), and the methyl acrylate derivative was structurally characterized.[82]

Nickelocene reacts with dmac (dimethylaluminum chloride) to yield the σ,π-bound substituted norbornadiene complex (**11**) that stems from the Diels–Alder addition of the alkyne to a cyclopentadiene

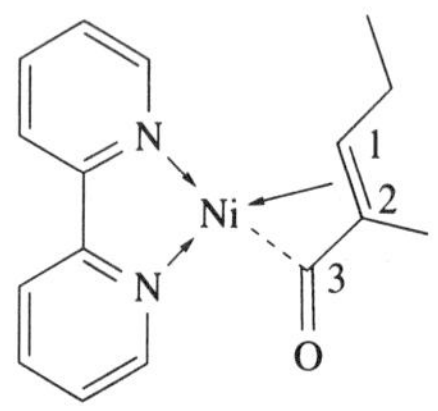

(10)

Ni–C-1 = 0.195 4(10), 0.194 5(11)
Ni–C-2 = 0.199 6(10), 0.197 6(11)
Ni–C-3 = 0.228 4(11), 0.238 9(11)

Figure 2 Bond length (nm) data for the two independent molecules of (**10**).[83]

moiety.[130] This species has been used as a precursor to seven-substituted norbornadiene derivatives. The structure of the Diels–Alder adduct of (**11**) with 2,3,4,5-tetrachloro-1,1'-dimethoxycyclopentadiene (**12**) has been studied by x-ray diffraction.[132] Bond lengths in (**13**) were compared to those of the NiCp-free organic moiety (**14**). Scheme 9 outlines these reactions and the structures of these adducts.[130,132]

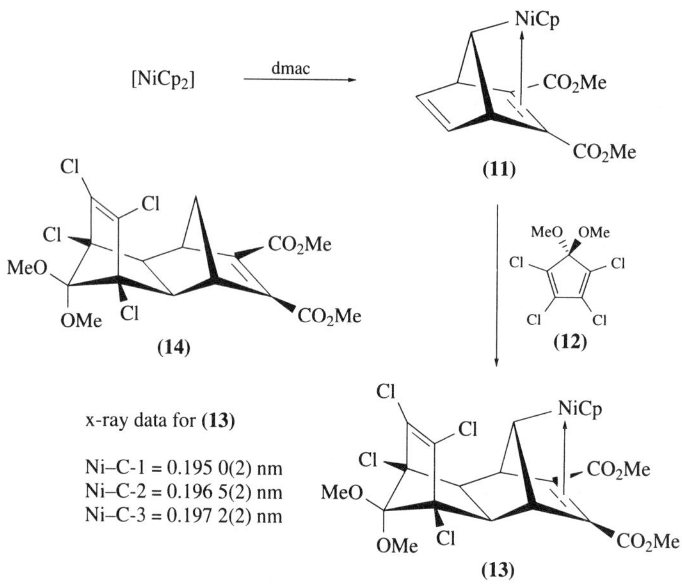

Scheme 9

A series of remarkable nickel–zinc clusters have been obtained from the reactions of nickelocene complexes with cyclopentadienylzinc species. These are discussed in Section 3.5. The (zinc-free) nickel complex $[Ni_2(\eta^2,\eta^4\text{-}C_8H_{12})_2Cp^*_2]$ (**15**) that contains a σ,π-bound, dimerized cod ligand was obtained when $[Ni(cod)_2]$ was reacted with $[ZnCp^*_2]$.[141]

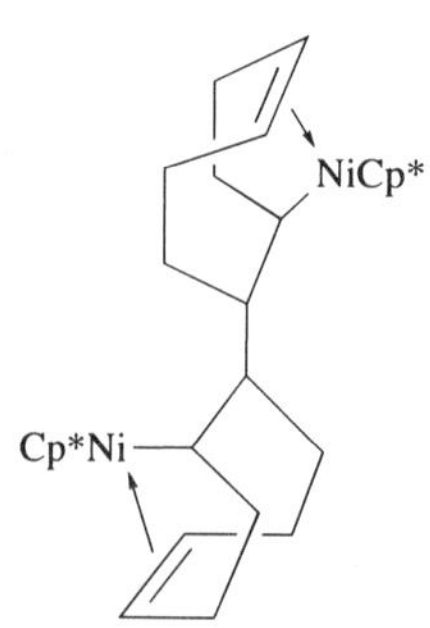

(15)

Complexes with C=C-ligated ketene ligands have been obtained by the carbonylation of NiC_3 nickelacyclobutane complexes. The complex $[Ni(\eta^2\text{-}CH_2{=}C{=}O)(PPh_3)_2]$ is also realized by treating $[Ni(PPh_3)_4]$ with a CH_2Br_2/Zn mixture under a CO atmosphere. The ketene is ligated to the nickel via both carbon atoms. It decomposes to C_2H_4 and to poorly characterized $[Ni(CO)(PPh_3)]_n$ species.[76,77]

The light-brown, air sensitive complex (**16**) was prepared by treatment of [Ni(cod)$_2$] with the free ligand. An x-ray diffraction study of (**16**) revealed a centrosymmetric structure for the molecule: the diphenylcyclopentadienone rings are almost planar. Both oxygen atoms in the complex could be protonated (with HBF_4 or CF_3SO_3H) or ethylated (with $Et_3O^+BF_4^-$) to yield dicationic derivatives in good yields. The electrochemistry of (**16**) was also investigated.[144]

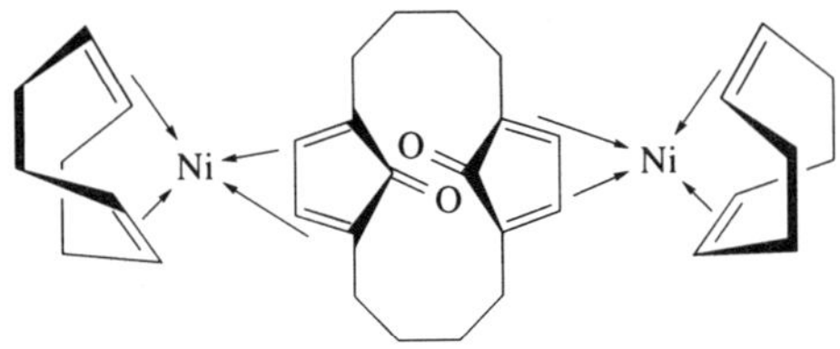

(16)

Nickel–η^2-alkene bonds in the cationic complex [Ni(η^2-nbd)(Cp)]$^+$ are not displaced by PMe_3. Instead, the PMe_3 ligand acts as a nucleophile and adds to one of the carbon–carbon double bonds of the nbd ligand, forming the η^1,η^2-phosphonium species (**17**), shown in Scheme 10.[129] Thermolysis of the product or dissolution in polar solvents liberates free nbd and leads to [Ni(PMe$_3$)$_2$Cp]$^+$.[129]

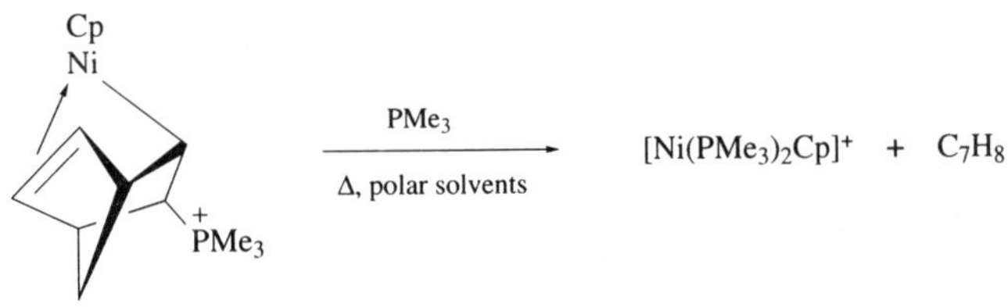

(17)

Scheme 10

The simultaneous reaction of [Ni(1,5,9-cdt)], bipy, and CH_2=CHCH=CHR (R = CO_2Me) resulted in carboxymethyl-substituted η^2-butadiene nickel complexes.[97] The initially formed species, [Ni(η^2-CH_2=CHCH=CHR)(bipy)] (**18**), coordinated as shown in Scheme 11,[97,159] reacts with excess of the diene ligand to form the bis(diene) complex [Ni(η^2-CH_2=CHCH=CHR)$_2$(bipy)] (**19**). Carboxymethylbutadiene reacts with [Ni(cod)$_2$] at 0 °C in toluene, affording [Ni(η^4-CH_2=CHCH=CHR)(cod)] (R = CO_2Me). The cod ligand can be displaced with PPh_3 or dcpe affording [Ni(η^4-CH_2=CHCH=CHR)(PPh$_3$)$_2$] or [Ni(η^4-CH_2=CHCH=CHR)(dcpe)], respectively. In the latter species, a structural study reveals marginal coordination of the *C*H=*C*HCO$_2$Me group.[97,159] (This diene also reacts with [Ni(cod)$_2$], in the presence of tertiary phosphine ligands, to give the allylic complexes (**48**), Scheme 27. This reaction is discussed in Section 3.3.)

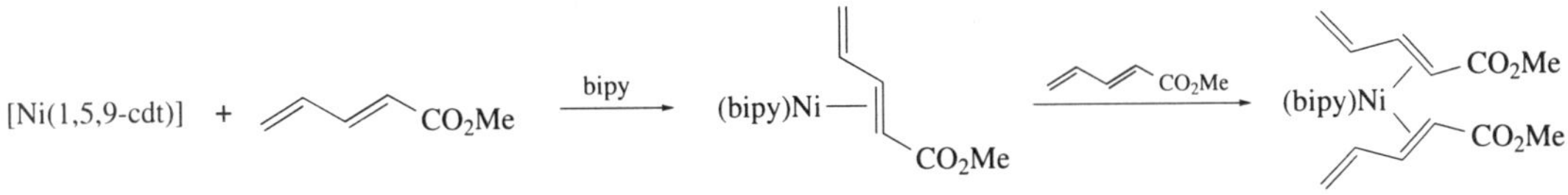

(18) **(19)**

Scheme 11

The reactions of nickel(0) complexes with allene and substituted allenes have been studied. At temperatures below –70 °C, [Ni(PPh$_3$)$_3$] reacts with a selection of substituted allenes affording mixtures of [Ni(η^2-CH_2=C=CR^{1}R^2)(PPh$_3$)$_2$] (R^1 = H, R^2 = Et, Bu^t, Ph, MeO, CN; R^1 = R^2 = Me) and *cis*- and *trans*-[Ni(η^2-CH_2=C=CR^{1}R^2)$_2$(PPh$_3$)] species. The *cis*-[Ni(η^2-CH_2=C=CR^{1}R^2)$_2$(PPh$_3$)] complexes undergo carbon–carbon coupling reactions to afford compounds which contain NiC_4 rings. The factors affecting the regio- and stereochemistry of π-complex formation and the rate of the coupling reaction were analyzed. The coupling rate increases in the order R^1, R^2 = H, Bu^t < H, Et < Me, Me ≪ H, MeO < H, Ph < H, CN.[124,125] The reactions of allene and 1,1-dimethylallene with [Ni$_2$(μ-CO)$_2$Cp$_2$] (Ni–Ni) and Ni–Mo and Ni–W species afford allyl complexes.[160,161] These are discussed in Section 3.3.

The reaction of carbon dioxide and alkynes in the presence of such nickel(0) alkene complexes and TMEDA leads to 1,1-coupling of the unsaturated organic molecules and to the formation of metallacycles of the type [{$\overline{\text{Ni–C(R}^1\text{)=C(R}^2\text{)–C(O)–O}}$}L] (L = TMEDA) that contain a nickel–oxygen bond. These complexes can be carbonylated to afford η^2-alkene bonded maleic anhydride derivatives $[Ni\{\eta^2\text{-}\overline{C(R)=C(R)}\text{-}C(O)\text{-}O\text{-}\overline{C}(O)\}L]$.[162]

The complex $[Ni(PPh_3)_2\{PPh_2C_6H_4(CH=CH_2)\text{-}2\}]$ (**20**) was prepared by alkene displacement from $[Ni(\eta\text{-}C_2H_4)(PPh_3)_2]$ or from a 1:2 mixture of $[Ni(cod)_2]$ and PPh_3 with the free (2-vinylphenyl)diphenylphosphine ligand $PPh_2C_6H_4(CH=CH_2)\text{-}2$. Initially (**20**) was not isolated but was detected by ^{31}P and ^1H NMR spectroscopy; it undergoes an intramolecular ligand exchange process above −30 °C and redistributes its ligands as shown (Equation (5)), affording $[Ni\{PPh_2C_6H_4(CH=CH_2)\text{-}2\}_2]$ (**21**). The two vinylphosphine ligands in (**21**) were believed to be bidentate.[122] Later work did result in the isolation of (**20**). The $PPh_2C_6H_4(CH=CH_2)\text{-}2$ ligand was indeed bidentate and bound to the nickel atom via the vinyl ligand and the phosphorus atom.[128]

$$\underset{(\mathbf{20})}{[NiL(PPh_3)_2]} \rightleftharpoons \underset{(\mathbf{21})}{[NiL_2]} + [Ni(PPh_3)_3] + PPh_3 \qquad (5)$$

$(L = PPh_2C_6H_4(CH=CH_2)\text{-}2)$

The polydentate arsenic ligands $As(C_2H_4CH=CH_2)_n(C_3H_6AsMe_2)_{3-n}$ (n = 1, 2) have been reacted with nickel(II) halides in ethanol. The monoalkene ligand (n = 1) affords the cationic trigonal bipyramidal complex $[Ni\{\mathit{As}(C_2H_4\mathit{C}H\text{=}\mathit{C}H_2)(C_3H_6\mathit{As}Me_2)_2\}\mathit{Br}]^+Br^-$ when reacted with $NiBr_2$ (atoms coordinated to nickel are shown in italic). The $CH=CH_2$ π-bond is equatorially coordinated to the nickel atom and a bromide ligand occupies one of the axial sites. This complex is a 1:1 electrolyte in nitromethane or dichloromethane. Nickel iodide forms the neutral compound $[Ni\mathit{I}_2\{\mathit{As}(C_2H_4CH=CH_2)(C_3H_6\mathit{As}Me_2)_2\}]$ (square-pyramidal geometry) with the same ligand. In this iodo species, and in all other complexes isolated with n = 2, coordinated alkene carbon–carbon double bonds are absent.[139]

The pentafluorophenyl complexes $[Ni(C_6F_5)_2(cod)]$ and $[Ni(C_6F_5)_2(nbd)]$ were prepared: structural studies of both species revealed square-planar coordination around the nickel atom.[92]

The carbon suboxide complex $[Ni(\eta^2\text{-}O=\mathit{C}=\mathit{C}=C=O)(PPh_3)_2]$ exhibits a similar C=C coordination mode. This species, made by reacting *in situ* solutions of $[Ni(cod)(PPh_3)_2]$ with C_3O_2, as shown in Scheme 12, extrudes CO in solution, making solution NMR studies unfeasible. The binding mode of the ligand was determined from CP–MAS (cross-polarization–magic angle spinning) ^{13}C NMR spectra of the molecule and of its ^{13}C-enriched isotopomers. An η^2-(C,C) coordination mode was indicated.[126] Carbon-13 NMR data for ligand carbon atoms are also shown in Scheme 12.

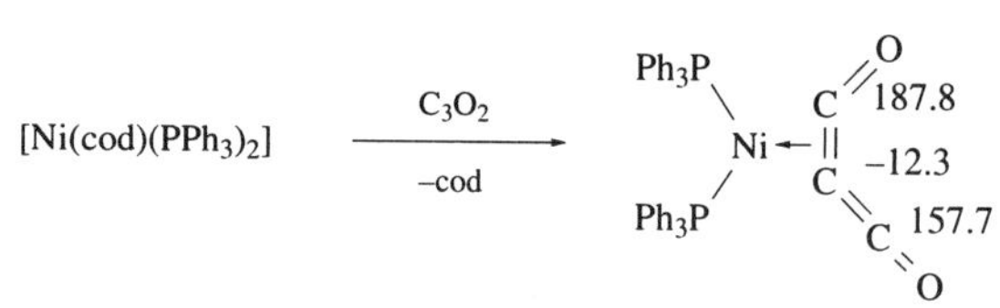

Scheme 12

Alkene complexes are often accessible from alkyne reactions. The reaction of the complex $[Ni(CO)_2bipy]$ with alkynes leads to CO–alkyne coupling reactions and to nickelacyclopentenedione species. These compounds are also accessible from free cyclobutenediones, as shown in Scheme 13.[111,112] The structure of $[Ni\{\eta^2\text{-}c\text{-}\mathit{C}Ph\mathit{C}PhC(O)C(O)\}(CO)(bipy)]$ (**23**) (R = Ph), a nickel 3,4-diphenylcyclobutenedione complex, was determined.[112] Other cyclobutenedione (and cyclopentenetrione) complexes of nickel have been reported—the ligand binds via the carbon–carbon double bond to the metal.[163]

As part of a study of the Reppe cyclotetramerization of alkynes, the tetraphenylcyclooctatetraene complex $[Ni(1,2:5,6\text{-}\eta^4\text{-}C_8H_4Ph_4\text{-}3,4,7,8)(PEt_3)_2]$ (**24**) was synthesized as shown in Scheme 14.[107] Perfluorocyclooctatetraene reacts with $[Ni(cod)_2]$ and the two-electron donor ligands L to afford the complexes $[Ni(1,2:5,6\text{-}\eta^4\text{-}C_8F_8)L_2]$ when L = PMe_3 or PMe_2Ph. Attempts with other phosphorus-donor ligands to prepare analogous species failed; when L = Bu^tNC, oxidative addition of the C_8F_8 to the nickel(0) center to afford nickel(II) complexes ensues.[105,106] A unified mechanistic scheme for the Reppe trimerization and tetramerization of alkynes was delineated. This study led to the characterization of the complex $[Ni_2\{\mu\text{-}\eta^2,\eta^2\text{-}(C_6H_4)_4\}(PEt_3)_2]$ by an x-ray diffraction study.[156]

Zerovalent nickel catalysts are well-known hydrocyanation catalysts and various studies have targeted these active species. The presence of both alkene and nitrile complexes was shown in solutions that contain nickel(0), various cyanoalkenes, and free $P(OTol\text{-}2)_3$. The identities of the various species in solution and the equilibria between the diverse complexes in solution were established by IR, ^1H

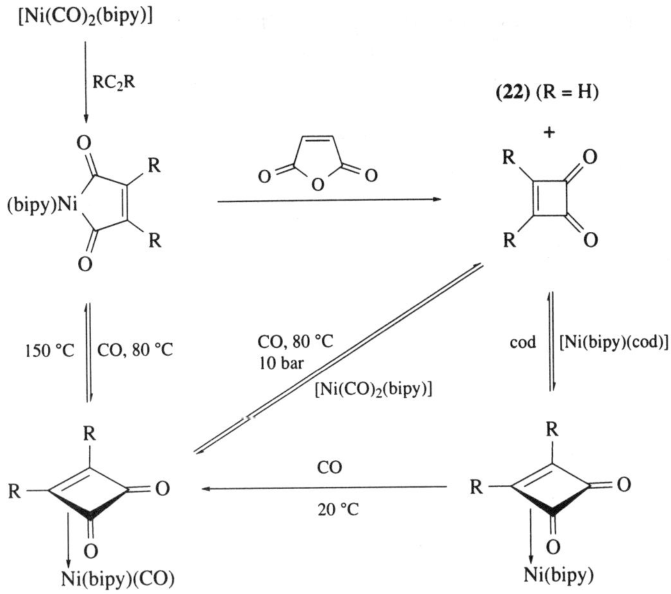

(23) (22)

Scheme 13

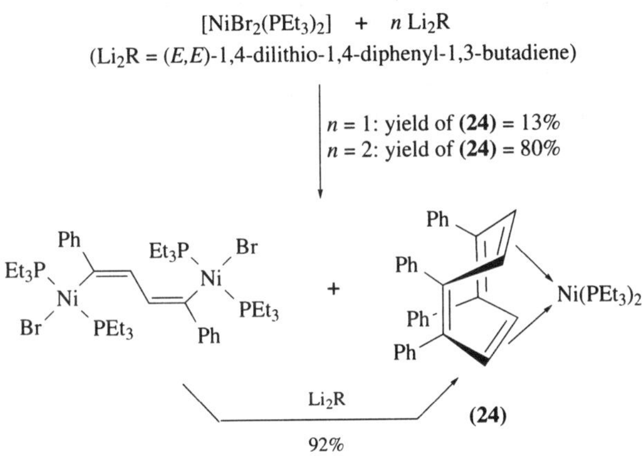

(24)

Scheme 14

NMR, spectrophotometry, and especially by ^{31}P NMR spectroscopy. Terminal alkene double bonds and conjugation with the CN group were shown to be factors that favored alkene coordination. Alkene species existing in solution included the 16-electron complexes [Ni(η^2-cyanoalkene)L_2] (L = P(OTol-2)$_3$) and 18-electron [Ni(η-cyanoalkene)$_2L_2$] compounds.[31] The complex [NiEt(CN)(η-C_2H_4){P(OR)$_3$}] (R = Tol-2) was characterized at −40 °C as part of an NMR study involving the hydrocyanation of ethene.[54,55] The use of triorganotin salts as tunable Lewis acid promoters of the nickel(0)-catalyzed hydrocyanation reactions has been studied.[164] Other studies on the mechanism of this nickel-catalyzed reaction are listed.[165–7]

The naphthalene ligand in the complex [Ni(η^2-$C_{10}H_8$)($PPr^i_2C_2H_4PPr^i_2$)] was shown to be η^2-bound to the nickel atom by an x-ray diffraction study. NMR data indicate that the molecule is fluxional in solution, and solid-state NMR (CP–MAS) studies reveal that the NiP_2 fragment undergoes shifts around the ring even in the solid state.[84,86] Spin saturation transfer solution NMR studies on the related anthracene complex [Ni(η-$C_{14}H_{10}$)(PBu^n_3)$_2$] suggest that an intramolecular dynamic process is operative that allows the NiP_2 fragment to "wander around" the fused-ring system.[87] Further work on this species and the related bis(triethylphosphine) complex indicated that the NiP_2 group migrated between the two terminal rings of the anthracene ligand. These rates are independent of the concentration of the complex, or of free added anthracene or [Ni(PR_3)$_3$] species. NMR data indicated that the anthracene ligand is η^4-coordinated to the NiP_2 group, although equilibria between two η^2-anthracene species could not be ruled out.[88] The anthraquinone ligand in [Ni(η^4-$C_{14}H_8O_2$)(PCy_3)] is also believed to be η^4-coordinated to the nickel atom.[89]

One-electron reduction of $[NiL_2(\eta^5\text{-}C_5H_7)]^+$ (L_2 = cod, norbornadiene) affords neutral radical species which decompose to $[NiCp_2]$ and other products which were not isolated. The cod complex is stable to isomerization as long as the temperature is maintained below −10 °C.[94,95] The η^1,η^2-ligated species (**25**) and an isomeric η^3-bound cyclooctenyl species (**26**) were formed from the oxidative addition of $Ph_2PCH_2CO_2H$ to $[Ni(cod)_2]$ and are in equilibrium (3:2 ratio, respectively, at ambient temperatures, Scheme 15). This mixture is an active ethene oligomerization catalyst.[137]

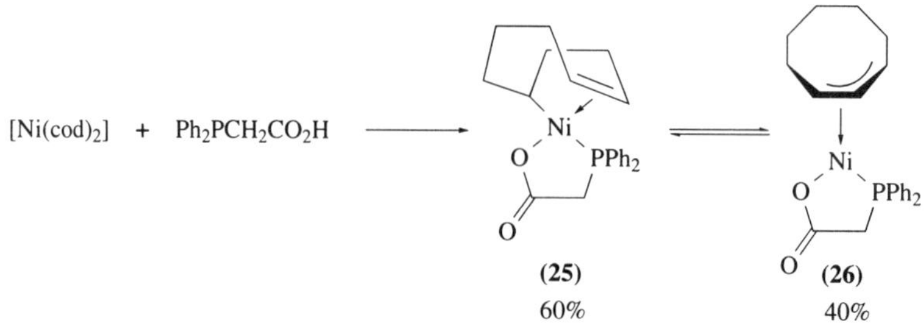

(**25**) 60% (**26**) 40%

Scheme 15

The use of 1,1-dimethylsilole and its derivatives as ligands towards nickel(0) species has been investigated, with the synthesis of hitherto-unknown η^5-silacyclopentadienyl species being the ultimate goal. The bis(η^4-1,1-dimethylsilacyclopentadiene) species $[Ni(\eta^4\text{-}C_4H_4SiMe_2\text{-}1,1)_2]$ was prepared as an impure orange oil by displacement of coordinated cod from $[Ni(cod)_2]$.[102] The mixed-ligand species $[Ni(cod)(\eta^4\text{-}C_4H_2Me_2\text{-}3,4\text{-}SiMe_2)]$ was subsequently prepared.[100] The substituted silole complexes $[Ni(\eta^4\text{-}c\text{-}CPh{=}CHCH{=}CPh\text{-}SiMe_2)_2]$ (**27**) and $[Ni(cod)(\eta^4\text{-}c\text{-}CPh{=}CH\text{-}CH{=}CPh\text{-}SiMe_2)]$ are formed from the reaction of $[Ni(acac)_2]$ and $[Ni(cod)_2]$, respectively, with c-$CHPhCH{=}CHCHPhSiMe_2$. The alkene ligands in (**27**) are tetrahedrally coordinated around the nickel atom.[99]

Displacement of one or two cod ligands from $[Ni(cod)_2]$ by dbn (**28**) (dbn = 2,3-diethyl-1,4-dimethyl-1,4-diboranaphthalene) similarly afforded the nickel species $[Ni(cod)(\eta^4\text{-dbn})]$ and $[Ni(\eta^4\text{-dbn})_2]$. Both species were structurally characterized: the nickel atom is tetrahedrally coordinated in each species. The latter complex was also obtained from the reaction of the free ligand with $[Ni(\eta^3\text{-}C_3H_5)_2]$. The complex $[Ni(\eta^4\text{-dbn})(\eta^2,\eta^2\text{-}CH_2{=}CHC_2H_4CH{=}CH_2)]$, which results from the coupling of two allyl ligands, was also isolated.[101]

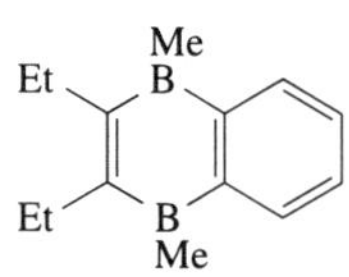

(**28**)

The catecholborane $BH(C_6H_4O_2\text{-}1,2)_2$ affords an η^1,η^2-cyclooctenyl species when treated with $[Ni(cod)_2]$. The catecholborate allyl complex $[Ni(\eta^3\text{-}C_3H_5)\{B(C_6H_4O_2\text{-}1,2)_2\}]$ (**29**) reacts with 1,5-hexadiene to afford (**30**) as shown in Scheme 16.[133]

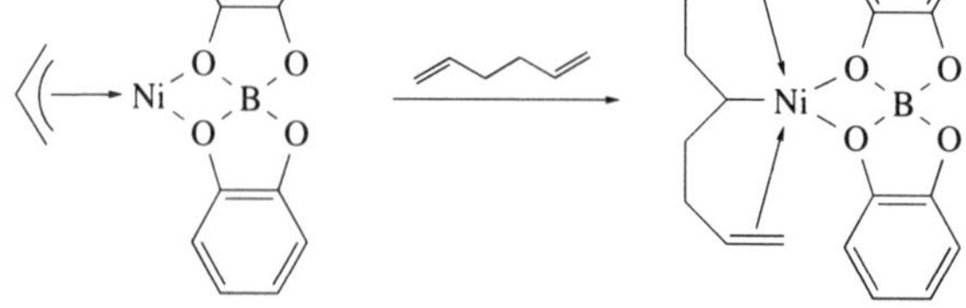

(**29**) (**30**)

Scheme 16

The nickel acyl complexes *trans*-$[NiCl(COR)(PMe_3)_2]$ (R = Me, CH_2TMS, $CH_2C_6H_4OMe$-2, CH_2CMe_2Ph) insert PhC_2H into the nickel–acyl bond, in a highly regio- and stereoselective reaction, to afford the series of σ-vinyl complexes *trans*-$[NiCl\{(E)\text{-}CPh{=}CHC(O)R\}(PMe_3)_2]$. When R = CH_2CMe_2Ph, this vinylic complex is in equilibrium with the π-bound zwitterionic species (**31**), $[NiCl\{\eta^2\text{-}CPh(PMe_3){=}CH(CH_2CMe_2Ph)\}PMe_3]$ (Scheme 17). The latter complex results from a 1,2-PMe_3 shift.[108] Structure (**31**) was subsequently determined.[109]

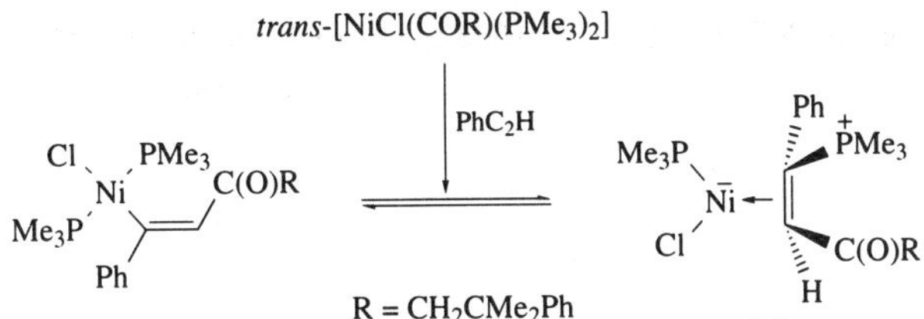

(31)

Scheme 17

Latest developments in nickel–"alkene" chemistry come from the fullerene field. The complexes $[Ni(\eta^2\text{-}C_{60})(PEt_3)_2]$[168] and $[\{Ni(PEt_3)_2\}_6(\eta^2\text{-}C_{60})]$ have been prepared and the structure of the former was established by an x-ray diffraction study. The fullerene interacts via two of its carbon atoms per nickel atom. In the Ni_6 complex, the phosphorus tips slightly out of the NiC_2 plane and the overall symmetry of the molecule approaches T_h. More recent research targeted the IR and Raman spectra of these nickel complexes and other metal fullerene compounds. The ligated metal or metals lower the icosahedral symmetry of the fullerene skeleton, allowing "silent" vibrational modes to be seen and degenerate modes to be split.[127] The complexes $[\{M(PEt_3)_2\}_n(\eta^2\text{-}C_{60})]$ (M = Ni, Pd, Pt; n = 2–4, 6) were the subject of an electrochemical study. These species were oxidized irreversibly in a metal-centered process, indicating that the bonding of C_{60} to the metal was localized to the attached carbon–carbon double bond only. Reduction was centered on the C_{60} fragment.[169]

3.2.5 Heteroalkene Complexes

3.2.5.1 General comments

A relatively large number of heteroalkene complexes have been synthesized. Many of these species are synthesized from nickel(0) alkene or cod complexes. These species include side-bound C=N, C=P, C=O, and C=S groups, and there is some evidence for transient silylene complexes. Good σ-donor ligands stabilize such nickel complexes and are always present. Not all complexes with heteroalkene ligands contain π-bound heteroalkene ligands: in many complexes of aza- or diazabutadiene or phosphaalkene ligands, the ligands bind to the nickel atom via the heteroatom only and are not η^2-coordinated. In addition, no η^2-(*C,O*)-bound C_3O_2 complexes have been isolated.[126,170] Table 6 lists heteroalkene complexes.

3.2.5.2 Nickel silylene species

The addition of aldehyde or ketone carbon–oxygen double bonds across the silicon–silicon bond of benzodisilacyclobut-3-enes is catalyzed by $[Ni(PEt_3)_4]$: nickel silylene complexes have been invoked as transient species in such reactions.[198] Species such as $[Ni\{\eta^2\text{-}(C,Si)\text{-}Si(TMS)(Mes){=}C{=}CPh\text{-}(TMS)\}(PEt_3)_2]$ and related silylene species have been proposed as intermediates in earlier work by this research group.[171,199,200]

3.2.5.3 Nickel η^2-N=C species

Nickel π-bound alkaneimidoyl complexes have been prepared by Bu^{t}NC ligand insertion into Ni–R bonds, where R = a bulky alkyl group (CH_2TMS, CH_2Bu^t) as shown in Scheme 18. These species are in equilibrium with the η^1-ligands, which are the sole products when the alkyl group is small. Related complexes (**32**) with π-bound C(R)=NBu^t ligands are formed in similar reactions. Ligands with η^1- or π-bonded η^2-alkaneimidoyls can be distinguished on the basis of their ν(CN) stretches, as shown in Scheme 18.[173]

Aryl azides react with $[Ni(cod)_2]$ or $[NiCp_2]$ to yield $[Ni(RN_4R)_2]$ species in which the nickel atom is in a distorted tetrahedral geometry. One tetraazabutadiene ligand can be displaced with Bu^{t}NC and the complex $[Ni(RN_4R)(Bu^tNC)_2]$ is formed.[201–3] The 19-electron paramagnetic intermediate $[Ni(PhN_4Ph)Cp]$, formed in the reaction with $[NiCp_2]$, was structurally characterized.[204]

Table 6

Heteroalkene	*L^1, L^2*	*Ref.*
Si(Mes)(TMS)=C=CPh(TMS)	2 PEt_3	171
2 NPh=B$(Mes)_2$		172
NBu^t=CHPh	**$PPr^i{}_2C_2H_4PPr^i{}_2$**	43
CMe=NBu^t	PMe_3, c-$NC_4H_2R_2$-2,4 (R = H, Me)	173
CR=NBu^t (R = CH_2Bu^t, CH_2TMS)	PMe_3, Cl	173
CR=NBu^t (R = CH_2Bu^t, CH_2TMS)	PMe_3, $NC_4H_2Me_2$-2,4	173, 174
2 P(Cl)=C$(TMS)_2$ (34)	**CO**	175
P[CH$(TMS)_2$]=C$(TMS)_2$ (35)	**2 PMe_3**	176, 177
P(Xyl-2,6)=CPh_2	**bipy**	178
C$(TMS)_2$=P=C$(TMS)_2$	**PEt_3, Cl**	179
CO_2	**2 PCy_3**	180–2
CO_2	$PCy_2(CH_2)_nPCy_2$ (n = 2–4)	183
Ph_2CO	bipy	184
PhCHO	**2 PCy_3**	185
H_2C=O	C_2H_4, TMEDA	46
R*C*(*O*)PO$(OMe)_2$ (36)	**2 PPh_3** (R = Me, **Et**, Ph, Tol-4, C_6H_4Cl-4)	186
PhRC=*C*=*O* (R = Me, Et)	2 L (L = PPh_3, PPh_2Tol)	187
Ph_2C=*C*=*O*	**$PBu^t{}_2CH_2PBu^t{}_2$**	75
PhMeC=*C*=*O*	2 PPh_3	188
PhN*CS*	**triphos**	189
*CS*S	triphos	190
*CS*E (E = O, S)	triphos	191
C(S)SMe	triphos (BF_4^-)	192
***SC*(S)$CH_2CMe_2C_6H_4$**	**2PMe_3**	193
SC(S)$CH_2CMe_2C_6H_4$	dmpe	193
η^1,η^2-*SC*(PMe_3)SC(=S)*S*	PMe_3	193, 194

[a] Complexes in bold were characterized by an x-ray diffraction study. Nickel-coordinated atoms are italicized.

Table 6 (continued)

Compound	*Ref.*
[$Ni_2(CO)_2${μ-η^1,η^2-P(PPh_2)=CR(TMS)}$_2$] (R = Ph, TMS)	175
[$Ni_2(CO)_2${μ-η^1,η^2-P($PClBu^t$)=C(TMS)Ph}$_2$]	195
[Ni_2(μ-η^1,η^2-CS_2)$_2$(PPh_3)$_2$]	196
[Ni{η^2-N($C_2H_4PPh_2$)*C*(=S)(SMe)(C_2H_4*P*Ph_2)$_2$}]$^+$$BPh_4^-$ (37)	191, 197

[a] Complexes in bold were characterized by an x-ray diffraction study. Nickel-coordinated atoms are italicized.

3.2.5.4 Nickel η^2-P=C species

A number of phosphaalkene nickel complexes have been prepared. Typically, these species have bulky groups attached to a multiply bonded atom which impart kinetic stability to the molecule. Either phosphorus-bonded η^1-ligand or π-bonded η^2-ligand species are obtained depending on subtle and not fully understood steric and electronic factors.

The species [Ni(bipy){η^2-P(2,6-Xyl)=CPh_2}] (**33**) has been made from [Ni(cod)(bipy)] by substitution with the free ligand as outlined below (Equation (6)). An x-ray diffraction study confirmed the η^2-coordination of the ligand (Ni–P = 0.217 7(2) nm, Ni–C = 0.198 7(6) nm, P–C = 0.183 2(6) nm).[178]

Nickel carbonyl reacts with the phosphaalkene P(Mes)=CPh_2 to afford the phosphorus-bound complexes [$Ni(CO)_3L$] and [$Ni(CO)_2L_2$] (L = P(Mes)=CPh_2).[178] Similar η^1 phosphorus-bound complexes [$Ni(CO)_3${P(Mes*)=CH_2}],[205] [$Ni(CO)_3${P(Mes*)=C=CPh_2}],[206] and dimeric [$Ni(CO)_2${η^2-(*P,P*)-P(Mes*)=CH–PCl(Mes*)}]$_2$[207] are obtained with P(Mes*)=CH_2, P(Mes*)=C=CPh_2 and P(Mes*)=CH–PCl(Mes*), respectively. However, a π-bound phosphaalkene complex is obtained by treating nickel carbonyl with P(Cl)=C$(TMS)_2$. The structure of the resulting [Ni(CO){η^2-P(Cl)=C$(TMS)_2$}$_2$] (**34**) was established by an x-ray diffraction study and some structural parameters are shown (Figure 3). Triphenylphosphine displaces the phosphaalkene from (**34**) and forms [$Ni(CO)_2(PPh_3)_2$].[175] The mononuclear π-bound complex [Ni{η^2-P[CH$(TMS)_2$]=C$(TMS)_2$}(PMe_3)$_2$] (**35**) was made by reacting [$NiCl_2(PMe_3)_2$] with the dialkylphosphido anion [P{CH$(TMS)_2$}$_2$]$^-$. An x-ray diffraction study of the

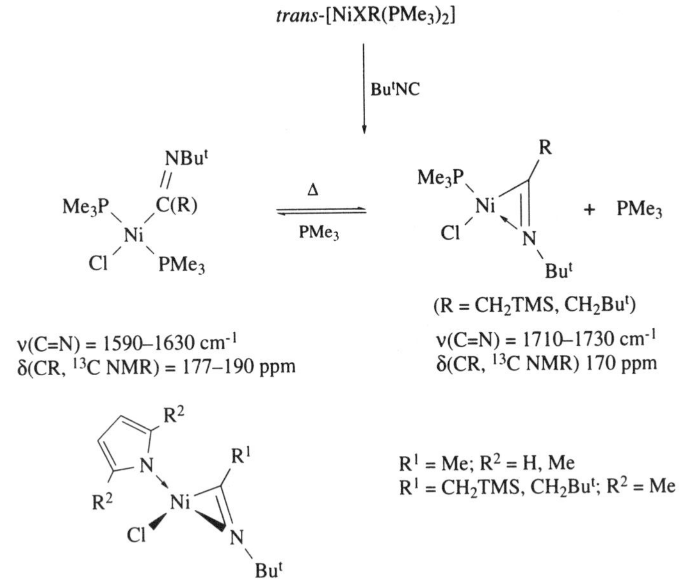

(32)

Scheme 18

complex established the η^2-coordination mode of the phosphaalkene.[176,177] Bond length data are collected in Figure 3.

[Ni(bipy)(cod)] + P(2,6-Xyl)=CPh_2 ⇌ [Ni(bipy){η^2-P(2,6-Xyl)=CPh_2}] (6)

(33)

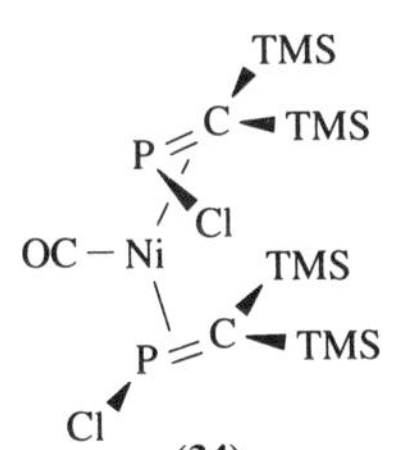

(34)

Ni–P = 0.2289
Ni–C = 0.2059
P–C = 0.1715

(mean values)

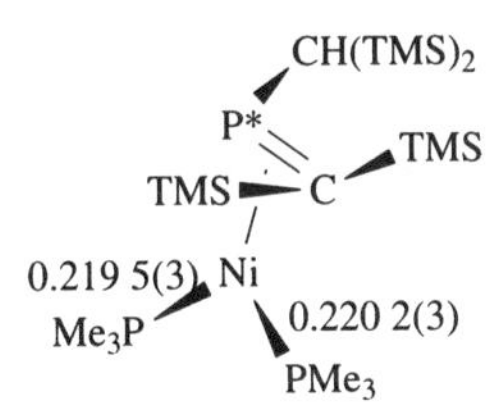

(35)

Ni–P* = 0.223 9(2)
Ni–C = 0.202 0(8)
P*–C = 0.177 3(8)
Ni–P = 0.220 2(3)
0.219 5(3)

Figure 3 Key bond lengths (nm) for the phosphaalkene complexes (**34**) and (**35**).[175–7]

The reaction of P(PPh_2)=C(TMS)(R) and [Ni$(CO)_4$] initially affords the monodentate phosphaalkene complex [Ni$(CO)_3${P(PPh_2)=C(TMS)(R)}] (R = Ph, TMS). The ligand is bonded to the nickel atom via the PPh_2 phosphorus atom. Loss of CO from this compound affords the dinickel complexes [Ni(CO){μ-η^1,η^2-P(PPh_2)=C(TMS)(R)}]$_2$, which contain a six-membered NiPPNiPP ring and π-bound phosphaalkene ligands (Scheme 19).[175] The compound [Ni(CO){μ-η^1,η^2-P[PCl(Bu^t)]=C(TMS)(Ph)}]$_2$ has a similar cyclic structure.[195] The zwitterionic species [NiCl(PEt_3){η^2-C$(TMS)_2$=P=C$(TMS)_2$}] was crystallographically characterized.[179]

3.2.5.5 Nickel η^2-C=O species

Side-bound aldehyde and ketone complexes of nickel have been prepared and some have been characterized by structural studies. The complex [Ni(η^2-Ph_2C=O)(bipy)] was prepared from the corresponding cod complex and benzophenone.[184] The structure of [Ni(η^2-PhCHO)$(PCy_3)_2$] was

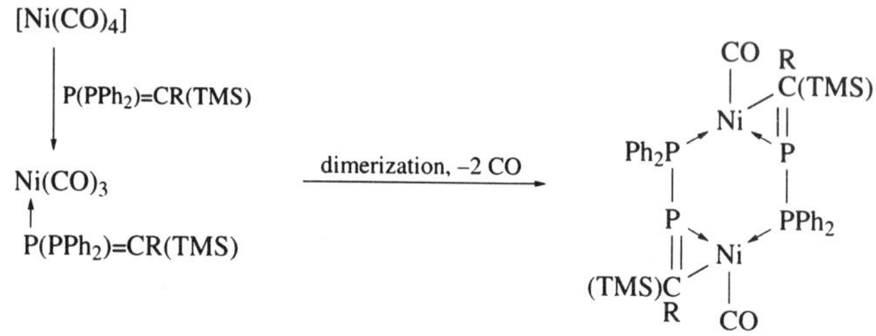

(R = TMS, Ph)

Scheme 19

established by x-ray diffraction, and shows a planar nickel geometry.[185] The formaldehyde complex $[Ni(\eta-C_2H_4)(\eta^2-H_2C{=}O)(TMEDA)]$ was prepared and its structure was also established.[48,49] Bulky trialkylphosphine or chelating diphosphine ligands impart kinetic stability and are often present as ancillary ligands. The structurally characterized η^2-CO_2 species $[Ni(\eta^2-CO_2)(PCy_3)_2]$[180] and the more recently prepared ketene complex $[Ni(\eta^2-Ph_2C{=}C{=}O)(PBu^t_2CH_2PBu^t_2)]$[75] both contain sterically demanding phosphine groups.

Extended Hückel calculations favor an η^2-(*C,O*)-ligation mode of the ketene ligands to nickel,[75] and this is borne out in most examples. In addition to the previously referred to species $[Ni(\eta^2-Ph_2C{=}C{=}O)(PBu^t_2CH_2PBu^t_2)]$, the methyl phenyl ketene ligand in the complex $[Ni(\eta^2-PhMeC{=}C{=}O)(PPh_3)_2]$ was presumed to be coordinated via the carbon–oxygen double bond of the substituted ketene ligand.[188] An η^2-(*C,O*)-ligation mode was indicated for the complexes $[Ni(\eta^2-PhRC{=}C{=}O)L_2]$ (L = PPh_3, PPh_2Tol; R = Et) but, based on IR and NMR data, the compounds $[Ni(\eta^2-CPhR{=}C{=}O)(PPh_3)_2]$ (R = H, Me) are believed to be coordinated via the carbon–carbon double bond of the ketene.[76,187]

The nickel atom in the CO_2 complex $[Ni(\eta^2-CO_2)(PCy_3)_2]$ lies in a square-planar environment.[180] The molecule was analyzed by vibrational FTIR studies at various temperatures (in the solid state and in solution), by ^{13}C and ^{31}P NMR (including CP-MAS), and by theoretical complete active space–self-consistent field (CAS–SCF) calculations. The research indicates that the CO_2 ligand in the compound probably adopts an "end-on" η^1-coordination mode in solution.[181] Solid-state and solution NMR studies performed by another group are in accord with these results. They indicate that in solution at 173 K, the CO_2 is η^2-coordinated to the nickel atom, but at ambient temperatures, the molecule is fluxional and the bonding may involve η^1-CO_2 oxygen-bonded species. *Ab initio* studies also suggest the viability of the η^1-coordination mode of CO_2 in these complexes.[182] Other Ni–CO_2 complexes have been synthesized. NMR spectroscopic studies of the species $[Ni\{PCy_2(CH_2)_nPCy_2\}(CO_2)]$ (n = 2–4) indicate that when n = 2, the CO_2 was likely to be η^1-coordinated, but the n = 3 and 4 species contained π-ligated CO_2 ligands.[183]

A series of η^2-C=O bound α-ketophosphonate complexes $[Ni\{\eta^2-RC(O)P(O)(OMe)_2\}(PPh_3)_2]$ (**36**) (R = Me, Et, Ph, Tol-4, C_6H_4Cl-4) have been prepared in good yields by standard ligand substitution methods as shown in Scheme 20. The structure of (**36**) (R = Et) was determined and confirmed the presumed η^2-C=O coordination mode of the ligand in these species. The nickel is essentially planar (Σ angles around nickel = 359.97°) and there is a weak interaction between the oxygen atom of the P=O group and the nickel atom.[186]

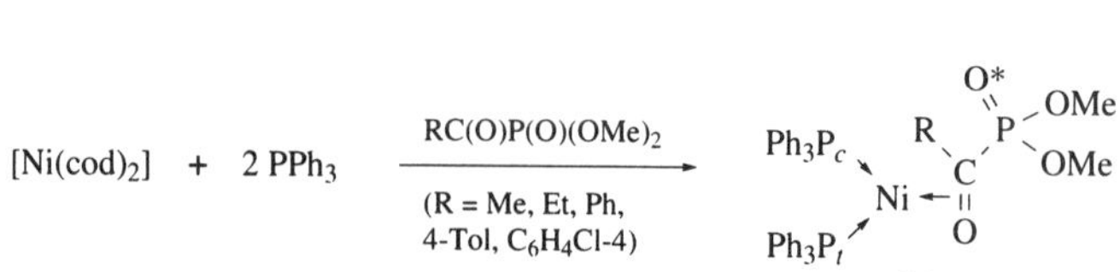

(**36**)
65–89%

Ni–C = 0.194 9(7) nm	$Ni–P_c$ = 0.221 7(2) nm	P_c–Ni–P_t = 106.12(8)°
Ni–O = 0.185 1(5) nm	$Ni–P_t$ = 0.213 6(2) nm	P_c–Ni–C = 114.2(2)°
C–O = 0.133 6(8) nm	Ni- -O* = 0.318 7(5) nm	P_t–Ni–O = 98.6(2)°

Scheme 20

3.2.5.6 Nickel η^2-C=S species

Tripodal ligands have been used to synthesize and stabilize nickel-bonded η^2-CS_2 species and the reactivity of these η^2-CS_2 complexes towards addition or displacement reactions has been investigated. Treatment of [Ni(η^3-np_3)] ($np_3 = N(C_2H_4PPh_2)_3$) with CS_2, PhNCS, or COS affords the heteroallene complexes [Ni{η^2-(*C,S*)-*CS*E}(η^3-np_3)] (E = S, PhN) and [Ni(CO)(η^3-np_3)], respectively. The cationic complex $[Ph_2PC_2H_4N(C_2H_4PPh_2)_2NiCS_2Me]^+$ **(37)**, (Figure 4) was produced unexpectedly when the η^2-CS_2 species was methylated with $MeOSO_2F$. The η^2-dithioester group formed *in situ* was attacked in an intramolecular fashion by one of the PPh_2 groups. The structure of this species (as a BPh_4^- salt) was determined by x-ray diffraction and shows that the nickel atom is in an approximately square-planar environment: two PPh_2 groups and a side-bound C=S group are ligated to the nickel atom and there is a significant Ni···N interaction.[191,197] Some pertinent bond lengths and a sketch of the structure are shown in Figure 4.

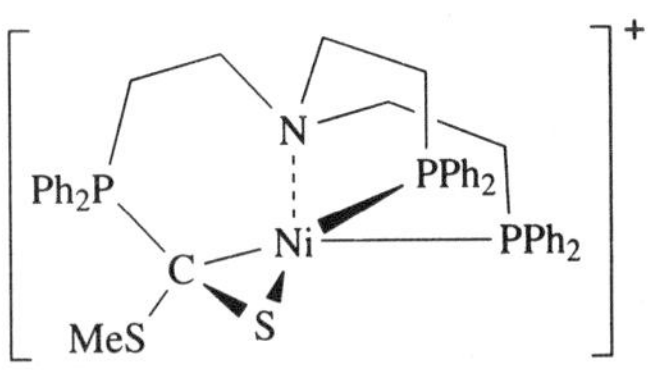

(37)

Ni–C	= 0.203(2)
Ni–S	= 0.211 7(9)
Ni–P	= 0.218(1), 0.218 8(9)
Ni- - -N	= 0.259(2)

Figure 4 Key bond lengths (nm) for **(37)**.[191,197]

The tripodal side-bound CS_2 complex [Ni{η^2-(*C,S*)-CS_2)(η^3-triphos)] has been prepared by reduction of cationic $[Ni(S_2CSMe)(triphos)]^+BPh_4^-$ with $NaBH_4$.[190] The cationic η^2-*CS*SMe complex [Ni{η^2-(*C,S*)-CS(SMe)}(η^3-triphos)]$^+$ is formed when [Ni{η^2-(*C,S*)-CS_2)(η^3-triphos)] is methylated. Protonation of the η^2-CS_2 complex with HBF_4, or the reaction of the compound with elemental oxygen, sulfur, or selenium, does not afford nickel π-complexes.[192] The CS_2 is displaced by PhNCS to afford the side-bonded PhNCS complex [Ni{η^2-(*S,C*)-PhNCS}(η^3-triphos)] in 60% yield. The structure was established by x-ray diffraction.[189] This complex and the analogous np_3 species react with hexafluoro-2-butyne affording the nickel cyclic carbene complexes [Ni{=C–S–C(CF_3)=C(CF_3)–S}L_3] (L_3 = triphos, np_3).[208] An x-ray diffraction study[196] of the complex with the empirical formula "[Ni(PPh_3)(CS_2)]" showed that the complex was dimeric. The actual formulation is [Ni(PPh_3)(η^1,η^2-CS_2)$]_2$, and each CS_2 ligand is σ-bonded to one nickel atom via a sulfur atom and π-complexed to the other nickel atom via the other carbon–sulfur double bond. A six-membered NiCSNiCS ring results.

Some nickelacyclopentanes react with CS_2 to give η^2-C=S-coordinated complexes. The structure of [Ni{η^2-*SC*(S)$CH_2CMe_2C_6H_4$}$(PMe_3)_2$] was established by an x-ray diffraction study. The nickel–carbon, nickel–sulfur, and carbon–sulfur bond lengths of the coordinated C=S group in this compound are 0.193 9(11), 0.212 9(3), and 0.176 5(11) nm, respectively. The dmpe adduct [Ni{η^2-SC(S)$CH_2CMe_2C_6H_4$}(dmpe)] was also prepared.[193]

3.2.6 Nickel η^2-Alkyne Complexes and Related Species

Table 7 lists nickel alkyne complexes of the type [Ni(η-$R^1C_2R^2$)L] and [Ni(η-$R^1C_2R^2$)L^1L^2]. Other nickel alkyne complexes are tabulated in Table 8, while Table 9 contains a short list of nickel benzyne complexes.

3.2.6.1 Theoretical and physical studies

The ethyne bonding in the compounds [Ni(C_2H_2)$(CO)_2$], [Ni(C_2H_2)$(CNH)_2$], and [Ni_2(μ-C_2H_2)$(CNH)_2$] was studied by the nonempirical self-consistent field Hartree–Fock Slater LCAO method.[255] Nickel atoms react with ethyne (with or without an inert atmosphere being present) to afford the complexes [Ni$(C_2H_2)_n$] (*n* = 1, 2). Vibrational spectroscopic data indicate that simple π-bonding links the alkyne to the nickel. The complexes decompose above 150 K.[256] Nickel atoms have also been reacted with C_2H_2 in solid argon at 12 K. The complex [Ni(η^2-C_2H_2)] was detected by FTIR. It isomerizes to the vinylidene species [Ni=C=CH_2] when exposed to visible light ($\lambda \geq 400$ nm). The

Table 7 $[Ni(\eta^2\text{-}R^1C_2R^2)L]$ and $[Ni(\eta^2\text{-}R^1C_2R^2)L^1L^2]$ complexes.[a]

R¹, R²	*L¹, L²*	*Ref.*
H, H	2 PCy_3	209
R, R (R = H, Ph, TMS)	2 L (L = PPh_3, Bu^tNC, 1/2 cod)	209
Ph, R (R = H, CO_2Et)	PPh_3, Cl	209
R, R (R = H, Me, **Ph**)	**dmpe**	210
R, R (R = H, Me, Ph)	2 PMe_3	210
R, R (R = **H**, Me, Ph)	**2 PPh_3**	33, 34
H, H	**$PPr^i_2C_2H_4PPr^i_2$**	33
Ph, H	2 PPh_3	211
Ph, Ph	2 PPh_3	212
H, H	2 L (L = PEt_3, PPh_3)	59
H, H	HC_2H, PR_3 (R = Cy, Me, Ph)	59
H, H	C_2H_4, PR_3 (R = Cy, Et, Ph, Pr^i)	59
H, H	2 PR_3 (R = Me, Et, Ph, Pr^i, OMe, OPh, OTol-2)	59
H, H	2 Bu^tNC	59
H, H	PMe_3, PCy_3	59
Ph, H	2 CO, bipy	213
Ph, Ph	dmpe	210
Ph, Ph	cod	93
H, $SnH[CH(TMS)_2]_2$	$Pr^i_2P(CH_2)_nPPr^i_2$	214
Ph, R (R = H, CO_2Et)	Cl, PPh_3	215
R, R (R = Me, TMS)	Me, Cp	216
Me, Me	Me, Cp	13
CO_2Me, CO_2Me	**2 PPh_3**	217
TMS, TMS	**2 PPh_3**	218
CH_2OH, CH_2OH	**2 PPr^i_3**	219
CH_2OH, CH_2OH	**2 PPh_3**	220
CH_2OMe, CH_2OMe	**2 PPh_3**	221
CH_2OMe, CH_2OMe	**2 PCy_3**	222
H, C≡CH	**$Pr^i_2P(CH_2)_nPPr^i_2$** (n = 2, 3)	223
R, Ph (R = H, Me, Et, Pr^i, Bu^t, Ph, $SiPh_3$)	2 PPh_3	224
R, TMS (R = Me, **Ph**)	**2 PR_3** (R = **Ph**, Cy, Tol-2)	224
Ph, Ph	2 CN $[K(18\text{-c-}6)]^{+}$ [b]	225
Ph, Ph	CN, CO $[K(18\text{-c-}6)]^{+}$ [b]	225

[a] Complexes in bold were characterized by an x-ray diffraction study. [b] 18-c-6 = 18-crown-6.

isomerization reaction is reversible on exposure to UV radiation (λ = 280–360 nm).[257] Photoelectron spectra of the $\mu\text{-}\eta^2,\eta^{2'}$-alkyne-bridged species $[Ni_2(\mu\text{-}\eta^2,\eta^{2'}\text{-}RC_2R)Cp_2]$ (Ni–Ni, R = H, CF_3) have been obtained. These data demonstrate the crucial importance of M $\rightarrow$ alkyne backbonding in these species.[258]

A series of nickel(0) alkyne complexes of general stoichiometry $[Ni(\eta^2\text{-}RC_2R)(PPh_3)_2]$ were synthesized from nickel(0) precursors as shown in Scheme 21.[259] Similar compounds $[Ni(\eta^2\text{-}XCH_2C_2CH_2X)L_2]$ have been prepared more recently. The strength of the nickel(0)–alkyne bond was studied by monitoring the ν(C≡C) stretch in the IR spectrum. For acceptor alkynes, donor ligands were best in promoting nickel(0)–alkyne bonding: the opposite is true for donor alkynes. The extent of alkyne activation towards reactions was found to be inversely proportional to the strength of complexation. The cyclooligomerization of alkynes by nickel(0) complexes (up to 10^6 turnovers per mole of catalyst) was found to be dependent on the strength of the donor ligands present. The turnover number increased with stronger donor ligands.[211]

$[Ni(\eta\text{-}C_2H_4)L_2]$ or $[NiL_4]$ or $[Ni(cod)_2]$ + 2 L $\xrightarrow{XCH_2C\equiv CCH_2X}$ $[Ni(\eta^2\text{-}XCH_2C_2CH_2X)L_2]$

L = PCy_3, PPr^i_3, PBu^n_3, PEt_3, PBu^n_2Ph, PBu^nPh_2, PPh_3, $P(o\text{-}OTol)_3$, $P(o\text{-}O\text{-}thym)_3$

X = OH, OR, OAr, O-TMS, NH_2, NHR, NR_2

Scheme 21

A linear correlation was found between the carbon–carbon triple-bond length and the coordination shift of the $(C\equiv C)_{sym}$ stretch in the series of complexes $[Ni(\eta^2\text{-}RC_2R)L_2]$ (R = H, Ph, TMS, etc.; L_2 = 2 PPh_3, 2 Bu^tNC, cod). The structures of $[Ni(\eta^2\text{-}TMSC_2TMS)(PPh_3)_2]$ and $[Ni(\eta^2\text{-}XCH_2C_2CH_2X)L_2]$ (X = CO_2Me, L = PPh_3; X = CH_2OH, L = PPh_3, PCy_3, PPr^i_3) have been deter-

Table 8 Other nickel–alkyne complexes.[a]

Complex	*Ref.*
$[Ni(\eta^2\text{-}P{\equiv}CNPr^i_2)(CO)(PCy_3)]$	226
$[Ni_2(\mu\text{-}\eta^2,\eta^2\text{-}P_2)(\eta\text{-}C_5HPr^i_4)_2]$	227
$[Ni(\eta^2,\eta^2,\eta^2\text{-}TBC)]$ [b]	228, 229
$[Ni\{\eta^2,\eta^2,\eta^2\text{-}1,4,7,10,13,16\text{-}(MeO)_6TBC\}]$ [b]	230
$[Ni_2Cl_2(\mu\text{-}\eta^2,\eta^{2'}\text{-}PhC_2H)(PBu^n_3)_4]$	215
$[Ni_2X_2(\mu\text{-}\eta^2,\eta^{2'}\text{-}PhC_2Ph)(PBu^n_3)_4]$ (X = Cl, Br)	215
$[Ni_2Cl_2(\mu\text{-}\eta^2,\eta^{2'}\text{-}PhC_2R)(PPh_3)_2]$ (R = H, CO_2Et)	215
$[Ni_2(\mu\text{-}\eta^2,\eta^{2'}\text{-}RC_2R)(dmpe)_2]$ (R = H, Me, Ph)	33, 34
$[Ni_2(HC_2H)_2(PCy_3)_2(dmpe)]$	59
$[Ni_2(\mu\text{-}\eta^2,\eta^{2'}\text{-}HC_2H)(\eta\text{-}C_2H_4)_2(PR_3)_2]$ (R = Et, Ph)	59
$[Ni_2(\mu\text{-}\eta^2,\eta^{2'}\text{-}R^1C_2R^1)(PR^2_3)_4]$ (R^1 = H, D, R^2 = Et; R^1 = H, R^2 = Me, Ph)	59
$[Ni_2(\mu\text{-}\eta^2,\eta^{2'}\text{-}HC_2H)(HC_2H)_2(PPh_3)_2]$	59
$[Ni_2(\mu\text{-}\eta^2,\eta^{2'}\text{-}HC_2H)(\eta^1,\eta^1\text{-}C_4H_4)(PPr^i_2C_2H_4PPr^i_2)_2]$	65
$[Ni_2(\eta^2,\eta^{2'}\text{-}HC{\equiv}C\text{-}C{\equiv}CH)(dippp)_2]$[c]	223
$[Ni_2\{\mu\text{-}\eta^1,\eta^2,\eta^{1'},\eta^{2'}\text{-}(C_6H_4)_4\}(PEt_3)_2]$	156
$[Ni_2\{\eta^2,\eta^{2'}\text{-}(-)\text{-}(R)\text{-}PhC_2C(O)NHCH(Me)Ph\}Cp'(\eta\text{-}C_5H_4CO_2Me)]$	231
$[Ni_2\{\eta^2,\eta^{2'}\text{-}(-)\text{-}(R)\text{-}PhC_2C(O)NHCH(Me)Ph\}Cp'_2]$	231
$[Ni_2\{\eta^2,\eta^{2'}\text{-}(-)\text{-}(R)\text{-}PhC_2C(O)NHCH(Me)Ph\}(\eta\text{-}C_5H_4CO_2Me)_2]$	231
$[Ni_2(\mu\text{-}\eta^2,\eta^{2'}\text{-}RC_2R)Cp_2]$ (R = Me, TMS)	216
$[Ni_2(\mu\text{-}\eta^2,\eta^{2'}\text{-}MeC_2Me)Cp_2]$	13
$[Ni_2(\mu\text{-}\eta^2,\eta^{2'}\text{-}PhC_2Ph)Cp_2]$	232
$[Ni_2(\mu\text{-}\eta^2,\eta^{2'}\text{-}Bu^tCP)Cp_2]$	233
$[NiCl_2(\mu\text{-}\eta^1,\eta^2\text{-}C_2TMS)_2Ti(\eta\text{-}C_5H_4TMS)_2]$	234
$[Ni(CO)(\mu\text{-}\eta^1,\eta^2\text{-}C_2TMS)_2Ti(\eta\text{-}C_5H_4TMS)_2]$	235
$[NiCo(\mu\text{-}\eta^2,\eta^{2'}\text{-}Bu^tCP)(CO)_3Cp]$	233
$[NiCo(\mu\text{-}\eta^2,\eta^{2'}\text{-}Bu^tCPML_n)(CO)_3Cp]$ ($ML_n = W(CO)_5$, $Mn(CO)_2Cp$)	233
$[NiCr(CO)_2(\mu\text{-}\eta^2,\eta^{2'}\text{-}RC_2R)Cp_2]$ (Ni–Cr; R = Me, Et)	236
$[NiMo(CO)_2(\mu\text{-}\eta^2,\eta^{2'}\text{-}R^1C_2R^2)Cp_2]$ (Ni–Mo; $R^1 = R^2$ = Me, Et, Ph)	236
$[NiMo(CO)_2(\mu\text{-}\eta^2,\eta^{2'}\text{-}R^1C_2R^2)Cp_2]$ (Ni–Mo; R^1 = Ph, R^2 = H)	236
$[NiW(CO)_2(\mu\text{-}\eta^2,\eta^{2'}\text{-}R^1C_2R^2)Cp_2]$ (Ni–W; R^1 = Ph, R^2 = H; R^1, R^2 = Me, Et, Ph)	236
$[NiML_n(\mu\text{-}\eta^2,\eta^{2'}\text{-}PhC_2CO_2Pr^i)Cp]$ (Ni–M) ($ML_n = Mn(CO)_4$, $Mo(CO)_2Cp$, $Co(CO)_3$)	237
$[NiFeMo(\mu_3\text{-}PhC_2CO_2Pr^i)(CO)_5Cp_2]$	238
$[NiCoFe(\mu_3\text{-}PhC_2R)(CO)_6Cp]$ (R = Me, Ph)	238
$[NiCp^*(\mu\text{-}PhC_2H)Ni(\mu\text{-}\eta^2,\eta^4\text{-}C_4H_2Ph_2)Mo(CO)_2Cp]$	239
$[Ni_4\{\mu\text{-}\eta^1(N),\eta^2(C,N)\text{-}MeCN\}_4(PCy_3)_4]$	240
$[Ni_2Fe_2(\mu_4\text{-}\eta^2\text{-}PhC_2Ph)(CO)_6Cp_2]$	241
$[Ni_2Fe_2(\mu\text{-}PPh_2)(\mu_4\text{-}\eta^2\text{-}C_2Ph)(CO)_5Cp_2]$	242
$[NiCp(\mu_3\text{-}\eta^2\text{-}C_2Pr^i)Fe_2(CO)_6]$	243
$[NiCp(\mu_3\text{-}\eta^2\text{-}C_2Bu^t)Fe_2(CO)_5(PPh_3)]$	243
$[Ni_2Cp_2(\mu_4\text{-}\eta^1,\eta^2\text{-}C_2H)Fe(CO)Cp''(\mu\text{-}CO)Ni(CO)]$ (Cp'' = **Cp**, Cp*)	244
$[Ni_2M_3(\mu_4\text{-}\eta^2\text{-}PhC_2Ph)(CO)_6Cp_2]$ (M = Fe, Ru)	241
$[Ni_2CpRu(\mu\text{-}PhC_2Ph)(CO)_3]$	245
$[Ni_2Cp_2Ru_2(HC_2CEt)(CO)_6]$	246
$[NiCpRu_2(\mu\text{-}PhC_2Ph)(\mu_3\text{-}CO)(CO)_3]$	247
$[NiCpRu_3(\mu\text{-}PhC_2Ph)(CO)_9]$	245
$[NiRu_4(\mu_4\text{-}C_2Pr^i)_2(\mu\text{-}PPh_2)_2(CO)_9Cp]$	248
$[NiCpRu_3(\mu_2\text{-}H)(\mu_4\text{-}\eta^2\text{-}C_2Bu^t)(CO)_9]$	249
$[NiOs_3(\mu\text{-}H)_3(Pr^iC_2PPh_2)(CO)_8Cp]$	250

[a] Species in bold were characterized by x-ray diffraction. [b] Structure (**44**), Figure 6. [c] dippp = $PBu^t_2C_3H_6PBu^t_2$.

Table 9 Nickel benzyne complexes.[a]

Compound	*Ref.*
$[Ni(\eta^2\text{-}C_6H_4)(dcpe)]$	251
$[Ni(\eta^2\text{-}C_6H_3F)(dcpe)]$	252, 253
$[Ni_2(\eta^2,\eta^{2'}\text{-}C_6H_2)(dcpe)]$	252, 253
$[Ni_3(\mu_3\text{-}\eta^2\text{-}C_6H_4)\{\mu_3\text{-}2,2'\text{-}(C_6H_4)_2\}(PPr^i_3)_3]$	254

[a] Complexes in bold characterized by x-ray diffraction.

mined.[217–19,221,222] The ^{13}C NMR chemical shifts of such alkyne complexes were reported and compared with other $[Ni(\eta^2\text{-}RC_2R)(PPh_3)_2]$ species.[212] Other research focused on the effect of increasing methylation of the alkyl group α to the alkyne carbon–carbon triple bond (Me, CH_2Me, $CHMe_2$, CMe_3).[260]

3.2.6.2 Nickel monoalkyne complexes

Alkyne complexes are accessible using synthetic methods more commonly applied in the preparation of alkene complexes—substitution of a Cp ring from $[NiCp_2]$. The alkyne complexes $[NiMe(\eta^2\text{-}RC_2R)Cp]$ (R = Me, Ph, TMS) were accessible by Cp ring displacement of one of the Cp rings at –75 °C from $[NiCp_2]$ by the alkyne and LiMe. The thermal decomposition products of two of the mononickel alkyne complexes are the dinuclear species $[Ni_2(\mu\text{-}\eta^2,\eta^{2'}\text{-}RC_2R)Cp_2]$. These dinuclear alkyne-bridged complexes, usually prepared from the reaction of $[Ni_2(\mu\text{-}CO)_2Cp_2]$ with alkynes, are thus accessible without having to utilize $[Ni(CO)_4]$ at some stage in the procedure. Allylic complexes are also obtained when R = TMS (Scheme 22).[216,260] Further studies led to the isolation of purple (**38**) containing an allylic-bound MeC_2Me/Me coupled species, as an air sensitive oil. Complex (**38**) is stable to H_2O or dilute aqueous HCl and may be distilled at high vacuum at ~100 °C. The mechanism of formation of this species was discussed.[13,261]

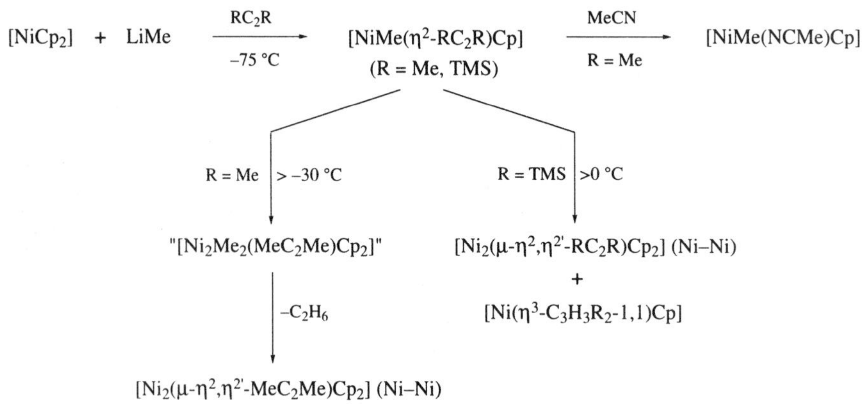

Scheme 22

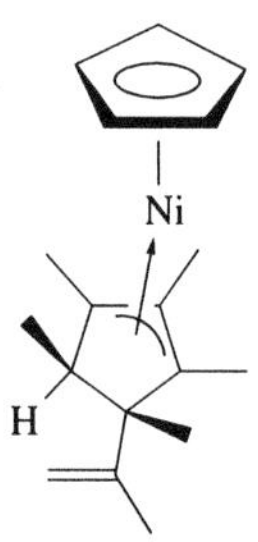

(38)

A series of alkyne complexes have been prepared from ethene complexes such as $[Ni(\eta\text{-}C_2H_4)(PMe_3)_2]$ and $[Ni_2(\eta\text{-}C_2H_4)_2(dmpe)_2]$ by alkene substitution reactions with the free alkyne. The complexes $[Ni(\eta^2\text{-}RC_2R)L_2]$ (R = H, Me, Ph; $L_2 = 2\,PMe_3$, dmpe) and $[Ni_2(\mu\text{-}RC_2R)(dmpe)_2]$ were obtained and the structure of $[Ni(\eta^2\text{-}PhC_2Ph)(dmpe)]$ (**39**) was determined. Key structural data for (**39**) are summarized in Figure 5. Chemical shift values for the ligated alkyne carbon atoms in the ^{13}C NMR spectra of these species ranged from δ = 111.4–137.9 ppm. The *ipso*-carbon of the Ph group in PhC_2Ph compounds lies in the range 135.0–140.9 ppm.[33,34,210]

Mono- and dinuclear ethyne complexes of the type $[Ni(\eta\text{-}C_2H_2)L_2]$ ($L = PR^1_3$, $P(OR^2)_3$, Bu^tNC; R^1 = Me, Et, Pr^i, Cy, Ph; R^2 = Me, Ph, Tol-2), $[Ni(\eta\text{-}C_2H_2)(\eta\text{-}C_2H_4)(PR_3)]$, $[Ni_2(\mu\text{-}\eta^2,\eta^{2'}\text{-}C_2H_2)(\eta\text{-}C_2H_4)_2(PR_3)_2]$, $[Ni_2(\mu\text{-}\eta^2,\eta^{2'}\text{-}C_2H_2)(\eta\text{-}C_2H_2)_2(PR_3)_2]$, and $[Ni(\eta\text{-}C_2H_2)_2(PR_3)]$ have been described in a paper that summarizes known nickel(0)–ethyne chemistry.[59] The structure of some of the dinuclear complexes is shown in (**40**). These complexes are made from nickel ethene species and the appropriate ethyne and two-electron donor ligand in the correct molar ratio at low temperatures (0 to –78 °C), followed by low-temperature crystallization. Polyethyne is a common side product. As stated in this manuscript, "the type of product complex formed and its thermal properties are mainly directed by both the stoichiometries employed and the steric requirements of the stabilizing ligands."[59] Detailed experimental procedures for the syntheses of these compounds are given, and their spectroscopic data are reported. The complex $[Ni(\eta\text{-}HC_2H)(PCy_3)_2]$ (reported earlier[209] but not hitherto isolated pure) is described in this manuscript. In general, as the steric bulk of the ancillary ligands increases, it becomes

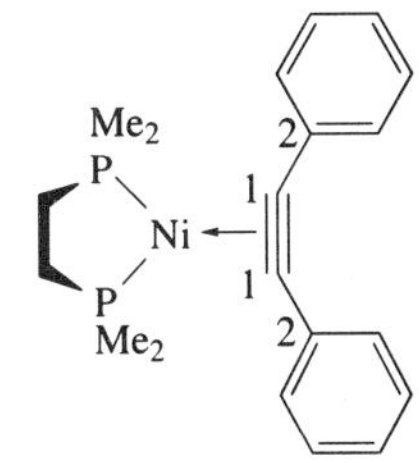

Ni–C-1 = 0.187 8(2) nm
C-1–C-1 = 0.129 9(3) nm
C-1–C-1–C-2 = 146.0(6)°

(mean values for chemically equivalent parameters)

(39)

Figure 5 Key x-ray data for [Ni(η^2-PhC$_2$Ph)(dmpe)] (**39**).[210]

more difficult to produce dinickel polyethyne species. Thus, the complexes [Ni$_2$(μ-η^2,$\eta^{2'}$-C$_2$H$_2$)(PR$_3$)$_4$] (**40**) (L = PR$_3$) are accessible when R = Ph, but not when the sterically bulkier R = Pr^i or Cy is utilized. Similarly the complex [Ni$_2$(μ-η^2,$\eta^{2'}$-HC$_2$H)(η^2-HC$_2$H)$_2$(PR$_3$)$_2$] (**40**) (L = C$_2$H$_2$) is only obtained when R = Me. Only [Ni(η-C$_2$H$_2$)L$_2$] species are obtained when L = P(OR)$_3$ or Bu^{t}NC.[59]

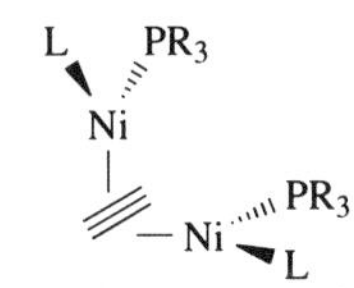

(40)
R = Ph, L = PPh$_3$, C$_2$H$_4$
R = Et, L = PEt$_3$, C$_2$H$_4$
R = Me, L = HC$_2$H

In addition to detailed experimental data for preparing various ethyne complexes, a brief structural report of [Ni(η^2-HC$_2$H)(PPh$_3$)$_2$] was presented. The nickel is close to trigonal planar (the angle between the NiC$_2$ plane and the NiP$_2$ plane is 5.2°) and the ethyne protons are bent back, away from the metal (the H–C–C angle = 148(2)°). The coordinated C≡C distance was found to be 0.128 7(7) nm.[59] The equivalent C≡C distance in the species [Ni(η^2-HC$_2$H)(PPr$^i_2$C$_2$H$_4$PPr$^i_2$)] was found to be 0.128 7(7) nm.[33]

Other papers describe syntheses of [Ni(η-HC$_2$H)(PR$_3$)$_2$] and [Ni(η^2-HC$_2$H)$_2$(PR$_3$)] species.[65,220,262,263] Alkyne complexes with triaryl phosphite ligands of the type [Ni(η-RC$_2$R){P(OC$_6$H$_4$R-2)$_3$}$_2$] have been reported.[264] The effect of varying the alkyne substituents on the complexes [Ni(η-R^{1}C$_2$R^2)(PPh$_3$)$_2$] has been investigated by using combinations of R^1 = Me, Ph and R^2 = various alkyl, TMS or SiPh$_3$ ligands. A structural study of the complex [Ni(η-PhC$_2$TMS)(PPh$_3$)$_2$] was also presented.[224]

Anionic alkyne complexes have been made. The complex [Ni(η-PhC$_2$Ph)(CN)$_2$]$^{2-}$ was isolated as a [K(18-crown-6)]$^+$ salt. The complex reacts with CO to afford [K(18-crown-6)]$^+_2$[Ni(CN)$_2$(CO)$_2$]$^{2-}$ and [K(18-crown-6)]$^+$[Ni(CN)(CO)(η-PhC$_2$Ph)]$^-$. The latter mixed CN/CO anion was not isolated pure.[225]

Nickel alkyne complexes have been obtained by reductive ligation methods. Reduction of mixtures of [NiX$_2$(PBu$^n_3$)$_2$] complexes and alkynes afforded the bridging alkyne complexes [Ni$_2$X$_2$(μ-η^2,$\eta^{2'}$-PhC$_2$R)(PBu$^n_3$)$_4$] (R = H, X = Cl; R = Ph, X = Cl, Br). When PPh$_3$ replaced the coordinated PBu$^n_3$ ligands, alkyne complexes were not formed and only the nickel(I) species [NiX(PPh$_3$)$_3$] was isolated. Alkynes do not react with [NiX(PPh$_3$)$_2$]$_n$, but when [NiCl(PPh$_3$)]$_n$ is reacted with phenyl-substituted alkynes, the paramagnetic complexes [NiCl(η^2-PhC$_2$R)(PPh$_3$)] (R = H, CO$_2$Et) are obtained along with the dinickel species [Ni$_2$Cl$_2$(μ-η^2,$\eta^{2'}$-PhC$_2$R)(PPh$_3$)$_2$] (R = H, CO$_2$Et).[215]

The nickel-catalyzed hydroboration of thioalkynes with catecholboranes is proposed to occur via η^2-alkyne intermediates.[265] When [NiL$_2$(η-C$_2$H$_4$)$_2$] (L$_2$ = PPr$^i_2$C$_2$H$_4$PPr$^i_2$) was treated with the tin(II) carbene species [Sn{CH(TMS)$_2$}$_2$], the initially formed product was the metallacycle species [L$_2$NiSnH{CH(TMS)$_2$}$_2$CH=CH]. This complex rearranges to the η^2-alkyne complex [NiL$_2${η^2-HC$_2$SnH[CH(TMS)$_2$]$_2$}].[214]

3.2.6.3 Nickel complexes with benzyne ligands

Benzyne ligands have been stabilized by ligation to nickel by using reductive ligation synthetic methods. Sodium amalgam or sodium naphthalenide reduction of [NiCl(C$_6$H$_4$Br-2)(dcpe)] yielded the mononuclear nickel benzyne complex [Ni(η^2-C$_6$H$_4$)(dcpe)] (**41**) (Scheme 23). The coordinated C≡C distance in (**41**) is 0.132 2(6) nm. Complex (**41**) undergoes insertion reactions with CO$_2$, C$_2$H$_4$, and dmac, all under mild conditions.[251] Dinuclear benzyne complexes of nickel that have been prepared include complex $[Ni_2(\eta^2,\eta^{2'}\text{-}C_6H_2)(dcpe)_2]$ (**43**) whose synthesis and that of its precursor (**42**) is shown in Scheme 24. The nickel atoms are in an essentially square planar environment in the structurally characterized complex (**43**).[253] A triply bridging benzyne ligand is present in $[Ni_3(\mu_3\text{-}\eta^2\text{-}C_6H_4)\{\mu_3\text{-}2,2'\text{-}(C_6H_4)_2\}(PPr^i{}_3)_3]$, which has also been characterized by an x-ray diffraction study.[254]

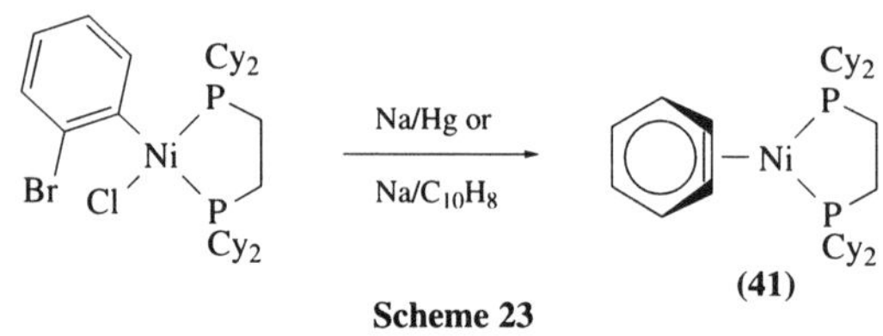

Scheme 23

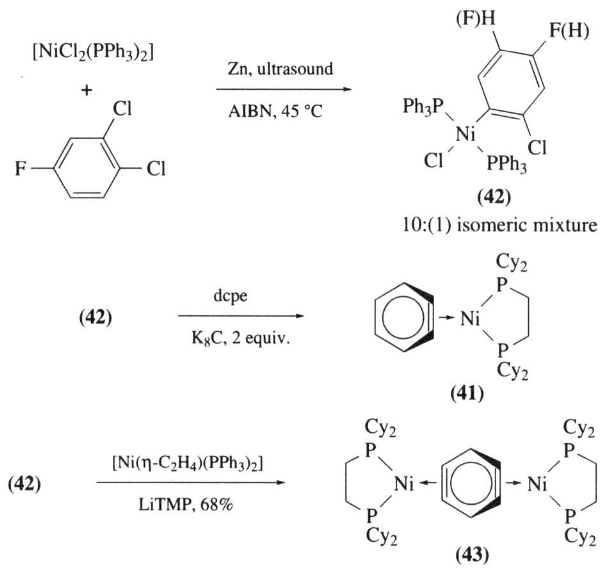

AIBN = 2,2'-azobisisobutyronitrile
TMP = 2,2,6,6-tetramethylpiperidide

Scheme 24

3.2.6.4 Nickel complexes with di- and polyalkynes

Mono- and dinickel complexes with butadiyne have been synthesized from the reaction of $[Ni(cod)(L_2)]$ ($L_2 = Pr^i{}_2P(CH_2)_nPPr^i{}_2$, $n = 2, 3$) with butadiyne, HC_4H. The complexes $[Ni(\eta^2\text{-}HC_4H)(L_2)]$ and $[Ni_2(\mu\text{-}\eta^2,\eta^{2'}\text{-}HC_4H)(L_2)]$ were prepared and characterized: representative mono- and dinickel complexes were structurally characterized.[223]

A series of cyclotrialkyne complexes of nickel which have potential as one-dimensional conductors have been prepared. Treatment of $[Ni(cod)_2]$ with the antiaromatic TBC (TBC = 1,2:5,6:9,10-tribenzocyclododeca-1,5,9-triene-3,7,11-triyne) yielded a deep-blue solution, from which needles of [Ni(TBC)] (**44**) (R = H) could be crystallized ($\nu(C{\equiv}C) = 1955\ cm^{-1}$). The structure of [Ni(TBC)] was determined by x-ray diffraction: key structural data are summarized in Figure 6 and NMR data for this complex are listed in Table 10. The molecules stack in a herringbone type (slip stacked) structure and are essentially planar (the nickel atom is 0.033 8(8) nm out of the least square-plane of the C≡C atoms). The rings are arrayed in an eclipsed conformation with an interplanar spacing of 0.329 nm. MO calculations indicate that [Ni(TBC)] has a metal-centered HOMO and a ligand-centered LUMO. The molecule undergoes a two-electron reduction (a "moderately reversible" process), and a one-electron oxidation. The purple monoanion and the blue dianion can be prepared by alkali metal reduction (M = Li, Na, K) in THF or DME with good alkali metal complexing agents (TMEDA, 18-crown-6, or 2.2.2-crypt).[228,229] Conductivity studies have been performed on [Ni(TBC)]; lithium reduction leads to a 10^4 increase in conductivity.[229,266]

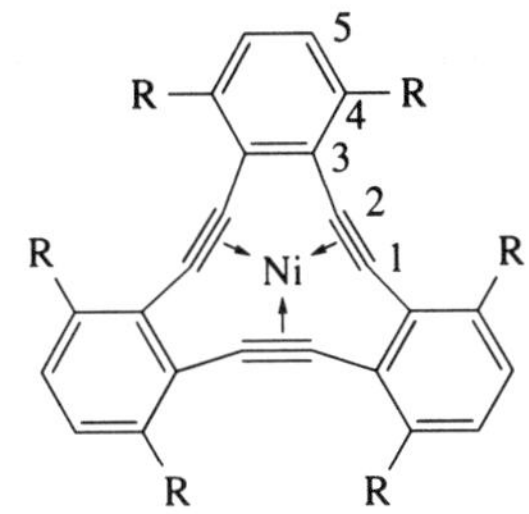

(44)

Parameter	R = H	R = MeO
Ni–C-1	0.198 5(5)	0.195 5(3)
C-1–C-2	0.124 0(10)	0.124 4(5)
C-1–C-2 –C-3	173.8(9)	

Figure 6 X-ray data for (**44**) (R = H, OMe). Distances (nm) and angles (°) are mean values of equivalent parameters.[228,230]

Table 10 ^{1}H and ^{13}C NMR data[a] for complex (**44**) (R = H).

H or C atom	*^{1}H chemical shift, δ* (ppm)	*^{13}C chemical shift, δ* (ppm)
1		101.8
3		128.2
4	7.66	130.0
5	6.80	128.1

[a] See Figure 6 for numbering scheme. Protons attached to numbered carbon atoms.

The related hexamethoxy-substituted complex [Ni{(MeO)$_6$TBC}] (**44**) (R = OMe) has also been prepared. This species stacks in layers similar to graphite with an interlayer spacing of 0.343 nm. There is an intermolecular agostic interaction in the solid state between a methoxy carbon–hydrogen bond and a nickel atom in an adjacent layer (Ni···H = 0.282 2(1) nm; Ni···H–C = 163.5(2)°), and an intraplanar, intermolecular hydrogen bond interaction between another methoxy carbon–hydrogen group and the OMe oxygen atom of an neighboring molecule (H···O = 0.255 1(2) nm; H···O–Me = 158.7(2)°).[230]

The reaction of [Ni(cod)$_2$] with PCy$_3$ and MeCN affords the unusual tetranickel complex [Ni$_4${μ-η^1(*N*),η^2(*C,N*)-MeCN}$_4$(PCy$_3$)$_4$] (**45**) in which the acetonitrile ligands are bridging and π-bound in an η^1(*N*),η^2(*C,N*) fashion. Structural parameters for the crystallographically characterized (**45**) are summarized in Figure 7.[240]

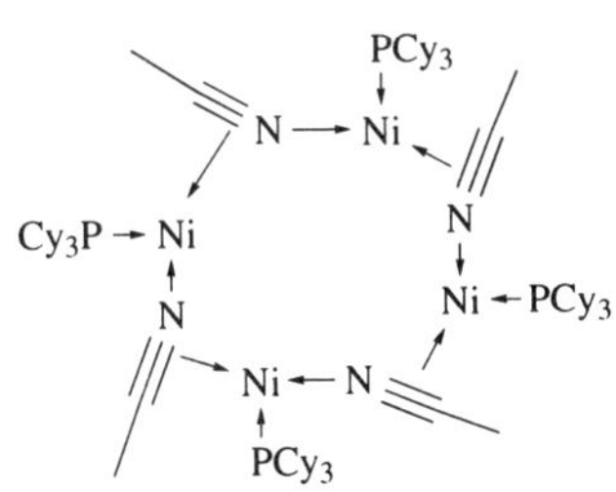

(45)

Ni–C = 0.1862
Ni–N(π) = 0.1932
Ni–N(σ) = 0.1916

N–Ni–N = 105.6(3)–111.9(3)
Ni–N–Ni = 120.4(4)–132.4(4)
N–C–C = 135.5(9)–137.0(9)

Figure 7 Key bond lengths (nm, mean values) and angles (°, range of angles) for (**45**).[240]

3.2.6.5 Di- and polymetallic nickel alkyne complexes

Members of the well-known class of μ-η^2,η$^{2'}$-alkyne complexes of the type [Ni$_2$(μ-η^2,η$^{2'}$-RC$_2$R)Cp$_2$] have been produced by the photolysis of [Ni$_2$(μ-CO)$_2$Cp$_2$] and the appropriate alkyne.[267] Chiral dinickelatetrahedrane clusters have been prepared by reacting mixtures of [NiCp$_2$] and [NiCp''$_2$] (Cp'' = Cp' or C$_5$H$_4$CO$_2$Me) with the optically active alkyne (–)-(*R*)-PhC$_2$C(O)NHCHMePh. The complexes $[Ni_2Cp_2\{\eta^2,\eta^{2'}\text{-}PhC_2C(O)NHCHMePh\}]$ (Ni–Ni), $[Ni_2Cp_2\{\mu\text{-}\eta^2,\eta^{2'}\text{-}PhC_2C(O)NHCHMePh\}]$ (Ni–Ni), and $[Ni_2CpCp'\{\eta^2,\eta^2\text{-}PhC_2C(O)NHCHMePh\}]$ (Ni–Ni) (**46**) were formed. Complex (**46**) exists as a mixture of two diastereomers that, for a given (*R*)-configuration, differ in cluster chirality. The diastereomers can be separated by chromatography and are configurationally stable. Their CD and IR spectra were reported.[231]

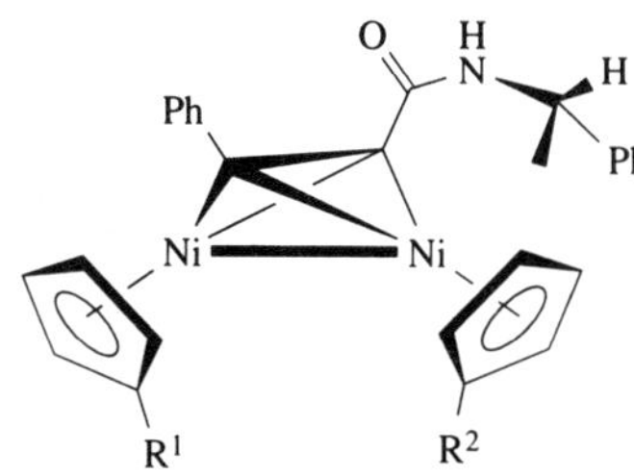

(46)
$R^1 \neq R^2$
R^1, R^2 = H, Me, CO_2Me

The reaction of $[Ni_2(\mu\text{-}\eta^2,\eta^{2'}\text{-}PhC_2Ph)Cp_2]$ (Ni–Ni) with H_2 gave no H_2 addition products below 120 °C, and above that temperature, decomposition ensued. The diphenylethyne in this complex could be exchanged with other diarylalkynes, probably via mononickel intermediates.[232]

The chiral mixed-metal clusters $[NiML_n(\mu\text{-}\eta^2,\eta^2\text{-}PhC_2CO_2Pr^i)Cp]$ (Ni–M, $ML_n = Mo(CO)_2Cp$, $Mn(CO)_4$, $Co(CO)_3$) were prepared from nickelocene, $[M(CO)L_n]_2$ and the free alkyne. The fluxional behavior of the complexes, regarded as being *nido*-t.b.p. cluster species, was investigated by NMR spectroscopy. The reaction of analogous species that contain $\mu\text{-}\eta^2,\eta^2\text{-}R^1C_2R^2$ ($R^1 = R^2$ = Ph; R^1 = Me, R^2 = Ph) ligands straddling NiCp–$Co(CO)_3$, NiCp–$Mo(CO)_2Cp$, or NiCp–NiCp units with $[Fe_2(CO)_9]$ led to trimetallic alkyne cluster species.[237,238] Heterodinuclear μ-alkyne complexes of the type $[NiM(\mu\text{-}\eta^2,\eta^2\text{-}R^1C_2R^2)(CO)_2CpCp']$ have been prepared by reaction of the metal–metal bonded complex $[Ni(CO)Cp\text{–}M(CO)_3Cp']$ (Ni–M) with the appropriate alkyne.[236,268,269] A mechanistic study of the formation of $[NiMo(CO)_2(\eta^2,\eta^2\text{-}R^1C_2R^2)CpCp']$ ($R^1C_2R^2$ = MeC_2Me, PhC_2H, PhC_2Ph) showed that the alkyne attacked the heterodinuclear units in an associative reaction: η^5- to η^3-Cp ring slippage was presumed to occur.[270]

Both phosphaalkyne and μ-P_2 complexes of nickel have been reported. The $\mu\text{-}\eta^2,\eta^{2'}$-phosphaalkyne complex $[Ni_2(\mu\text{-}\eta^2,\eta^{2'}\text{-}P{\equiv}CBu^t)Cp_2]$ (Ni–Ni) has been obtained from the reaction of $[Ni_2(\mu\text{-}CO)_2Cp_2]$ (Ni–Ni) and $P{\equiv}CBu^t$.[233] Terminal phosphaalkyne complexes also exist: the complex $[Ni(\eta^2\text{-}P{\equiv}CNPr^i_2)(CO)(PCy_3)]$ was obtained from the reaction of $[Ni(CO)_3(PCy_3)]$ with $P{\equiv}CNPr^i_2$. An x-ray diffraction study revealed $Ni\text{–}P_{P\equiv C}$ = 0.219 4(1) nm, $Ni\text{–}C_{P\equiv C}$ = 0.190 4(3) nm, and a coordinated phosphorus–carbon bond length of 0.166 5(2) nm.[226]

Bridging four-electron donor P_2 ligands that span a Ni_2 framework have been realized in the complex $[Ni_2(\mu\text{-}\eta^2,\eta^{2'}\text{-}P_2)(\eta^5\text{-}C_5HPh_4)_2]$ (Ni–Ni). The lone pair of one of the phosphorus atoms can be activated to form a phosphorus donor bond to a $W(CO)_4$ unit (Scheme 25). An oxygen atom can be chemically transferred to each phosphorus atom yielding the first known examples of μ_3-PO ligands that are coordinated to the Ni_2W core.[227,271]

3.3 NICKEL COMPLEXES WITH η^3-LIGANDS

3.3.1 General Comments

Nickel η^3-allyl complexes dominate this section. Allyl complexes often stem from reactions of nickel complexes with 1,3- or 1,2-dienes, and the vast majority of nickel allyl species have been synthesized from dienes. A growing number of η^3-benzylnickel complexes have been prepared and there has been progress in the chemistry of nickel η^3-cyclopropylium compounds since *COMC-I* was published. Allylic species are implicated in many catalytic and/or stoichiometric reactions; obtaining a clearer understanding of the mechanisms of these reactions has driven much research in this field. These catalytic reactions will be discussed selectively or not presented at all, unless new allylic species have been characterized and isolated. The use of nickel allyl species in the syntheses of polydecker complexes and some allylic species that contain boron will be discussed in Section 3.7.

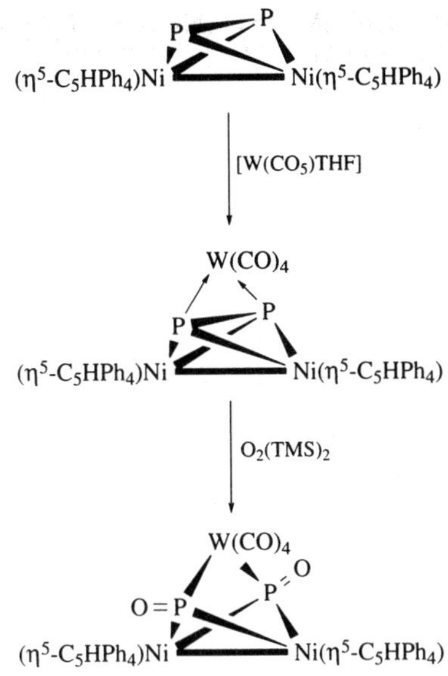

Scheme 25

Nickel allyl complexes have been reviewed by Wilke[272] and Billington.[273] The molecular basis of catalytic reactions, in which allyl complexes of group 10 metals are key intermediates, has also been reviewed.[274]

3.3.2 Theoretical and Physical Studies of Nickel η^3-Allyl Species

The electron transmission spectra of [M(η^3-C_3H_5)$_2$] have been measured in the 0–5 eV range. Three resonances associated with electron capture were noted. Multiple scattering X_α (MS-X_α) calculations on the nickel and platinum species reproduce the experimental details of these complexes.[275] The photoelectron spectra of [M(η^3-C_3H_5)$_2$] (M = Ni, Pd, Pt) were recorded.[276] The energy levels assigned by this study have been used as a starting point in MO calculations. However, a more recent study measured the photoelectron spectrum over the 20–80 eV photon energy range, using synchrotron radiation. All the transitions were fully assigned. This research demonstrated that the lowest-energy ionization (at 7.8 eV) arises from a metal *d*-orbital for [Ni(η^3-C_3H_5)$_2$], but from allyl $2p_\pi$-orbitals for the palladium and platinum complexes.[277]

The ground-state singlet electronic structure of [Ni(η^3-C_3H_5)$_2$] exhibits broken spin symmetry.[278] The structure of [Ni(η^3-C_3H_5)$_2$] was determined by a neutron diffraction study at −100 °C. This research indicated that the five allylic hydrogen atoms do not lie in the C_3 allylic plane.[279] MO calculations were carried out on [Ni(η^3-C_3H_5)$_2$].[280] The conformation of this complex (and of other transition metal compounds) was also studied by extended Hückel methods.[278]

EIMS (electron impact MS) data for the complexes [NiXL(η^3-C_3H_5)] and [Ni_2(μ-X)$_2$(η^3-C_3H_5)$_2$] (X = Cl, Br, I; L = mainly triarylphosphine or triaryl phosphite) were obtained and analyzed.[281,282] For all the [Ni_2(μ-X)$_2$(η^3-C_3H_5)$_2$] compounds except the iodo species, the most abundant peak was the parent ion: monomeric [NiI(η^3-C_3H_5)]$^+$ ions dominated the EIMS spectrum of [Ni_2(μ-I)$_2$(η^3-C_3H_5)$_2$]. Loss of NiX_2, [Ni(η^3-C_3H_5)$_2$], and the allyl group (as C_3H_5X) was observed. Co-condensation of nickel atoms with allyl halides led to the isolation of [Ni_2(μ-X)$_2$(η^3-C_3H_5)$_2$] compounds.[283]

3.3.3 Chemistry and Reactions of Nickel η^3-Allyl Compounds

Nickel allyl species and their derivatives of general formula [NiXL(η^3-allyl)] and [NiL^{1}L^2(η^3-allyl)]$^+$ are collected in Table 11. Other nickel allyl species are found in Table 12. Nickel η^3-benzyl complexes and nickel η^3-cyclopropylium species are listed in Tables 13 and 14, respectively.

Carbon dioxide couples with 1,3-pentadiene and nickel(0) species to form the allylic nickel species (**46**) and (**47**), shown in Scheme 26, which are linked to the nickel atom via a direct nickel–oxygen bond. Maleic anhydride cleaves the nickel–carbon bonds and affords sorbic acid (2,4-hexadienoic acid).[323] Related allylic species are derived from the reaction of butadiene complexes and CO_2.[357–9]

Table 11 $[NiXL^1(\eta^3\text{-ligand})]$ and $[NiL^1L^2(\eta^3\text{-ligand})]^+X^-$ complexes.[a]

η³-Ligand	*X, L¹, L²*	*Ref.*
2 C_3H_5[b]		279
2 η³-c-BMeNBu^tC_3H_3		284
C_3H_5	X (X = Cl, Br, I)	281, 282
C_3H_5	CO_2Me	285
C_3H_5	CO_2Me, PR_3	285
C_3H_5	**$GaMe_2pz_2$**	286
C_3H_5	**$BHpz_3$**	58
C_3H_5	2 L, PF_6 (2 L = cod; L = Bu^{t}NC, MeCN, PPh_3, $AsPh_3$, $SbPh_3$)	287
C_3H_5	2 L (L = $B(C_6H_4O_2\text{-}1,2)_2$)	133
C_3H_5	2 $P(OPh)_3$, X (X = Cl, Br, I, PF_6, BF_4, $AlBr_4$, $B(C_6H_4O_2\text{-}1,2)_2$, $MeC_6H_4SO_3$)	288
C_3H_5	2 L, X (X = BF_4, PF_6; 2 L = cod; L = Bu^{t}NC, $P(OPh)_3$, $P(OThym)_3$, $SbPh_3$)	289
C_3H_5	R, PPh_3 (R = Ph, $C_6H_3Cl_2$-2,5)	290
C_3H_5	Ar, PPh_3 (R = Ph, $C_6H_3Cl_2$-2,5)	291, 292
C_3H_5	Ar, L_2 (Ar = Ph, C_6F_5, $C_6H_3Cl_2$-2,5; L_2 = dppe, (*Z*)-$Ph_2PCH{=}CHPPh_2$)	291
C_3H_4Me-1 or C_3H_4Me-2	C_6F_5, L, or L_2 (L = PPh_3; L_2 = dppe, (*Z*)-$Ph_2PCH{=}CHPPh_2$)	291
C_3H_4R	2 $P(OPh)_3$	293
C_3H_4R (R = H, Me)	C_6F_5, L (L = PPh_3, dppe, dppen[c])	294
C_3H_4R-1 (R = H, Me)	C_6F_5, dppen[c]	290
C_3H_4Me-1	C_6F_5, PPh_3	290
syn- and *anti*-C_3H_4Me-1	Cp	147
C_3H_4Me-2	cod, X (X = $AlBr_4$, BF_4, PF_6)	295
C_3H_4Me-2	2 PPh_3, X	266
C_3H_4Me-2	C_6F_5, PPh_3	290
C_3H_4Me-2	C_6F_5, L_2 (L_2 = dppe, dppen[c])	290
C_3H_4Me-2	2 L (L = PMe_3; 2 L = dmpe)	296
C_3H_4Me-2	**$[C_6H_3Me(PPh_2)\text{-}2,6]_2$**[d]	297
C_3H_4Me	**cod, PF_6**	298
C_3H_4Me	**cod, OTf**	299
$C_3H_2Me_3$-1,2,3	cod, PF_6	295
$C_3H_4[C(O)NEt_2]$-2	**$NHEt_2$, $NHEt_2$, I**	300
$C_3H_4[C(O)NEt_2]$-2	**$NHEt_2$, I**	300
anti-$C_3H_3Me_2$-1,2	Cp	57
syn-$C_3H_3Me_2$-1,2	Me, L	301
syn,syn-$C_3H_3Me_2$-1,3	Me, L	302, 303
C_3H_3MePh-1,1	PPh_3, Br	304
$C_3H_3(TMS)_2$-1,1	Cp	216
2*C*$_3H_3(CMe{=}CH_2)Me$-1,2		305
c-*CHCHCH*$(CH_2)_n$ (*n* = 2–5)	Cp	306, 307
η³,η³'-C_8H_{12}	PR_3 (R = Me, Pr^i, Cy, Ph, OC_6H_4Ph-2)	308
C_8H_{13}	PPh_3, BF_4	309
η³,η³'-$C_8H_{10}(CO_2Me)_2$-1,8	**PR_3** (R = **Ph, Cy**)	97
$C_{12}H_{18}$	PMe_2(menthyl)	5,6
η³,η²,η²-$C_{12}H_{19}$[e]	PF_6 or SbF_6	310
η³,η²,η²-$C_{12}H_{19}$[e]	X (X = BF_4, OTf, $AlBr_4$, $B(C_6H_4O_2\text{-}1,2)_2$)	311
η³,η²,η²-$C_{12}H_{19}$[e,f]	**$B(Xyl^F)_4$**	312
η³,η²,η²-$C_{12}H_{19}$[e]	**PF_6**	313
η³,η²,η²-$C_{12}H_{19}$[e]	X (X = PF_6, $BF(C_6F_5)_3$, $B(Xyl^F)_4$)	314
endo-c-C_4Ph_4H	Cp	315
c-C_4Ph_4HCO	Cp	123
cis,trans-*C*$_3Me_3C_2H_2Me_2$-1,2	Cp*	316
C$_2Et_2$*C*$(CH{=}CH_2)BEtOBEt$[g]	*C*$_3Et_3BEtOBEt$	317
c-*C*$_3Me_3CHMeCMeCMe{=}CH_2$	Cp	13, 261
c-*C*$_3Me_3C_2R^1R^2$ (R^1, R^2 = CN; R^1 = H, R^2 = H, Me, CN)	Cp*	318
PMes*C(O-TMS)PMes*[h]	η³-C_8H_{13}	319, 320
P*Mes*CHPMes[h]	**Cp**	207
$P_2C_3Bu^t_3$	**η⁵-$P_3C_2Bu^t_2$**	321
2 *c*-$C_3H_3N(Bu^t)B(Me)$		284

[a] Complexes in bold were characterized by an x-ray diffraction study. Nickel-coordinated atoms are italicized. [b] Neutron diffraction study at −100 °C. [c] dppen = (*Z*)-$PPh_2CH{=}CHPPh_2$. [d] L = (*R,R*)-biphenyl-$(PPh_2)_2$-2,2'-Me_2-6,6'. [e] $C_{12}H_{19}$ = dodeca-2(*E*),6(*E*),10(*Z*)-trien-1-yl. [f] Xyl^F = $C_6H_3(CF_3)_2$-2,4. [g] Complex (**94**), Scheme 62. [h] Mes* = $C_6H_2Bu^t_3$-2,4,6.

Table 12 Other nickel η^3-ligand complexes.[a]

Compound	*Ref.*
[Ni{η^1,η^3-(C$_3$H$_4$CH$_2$CPhC$_4$H$_2$Ph$_2$)PEt(Mes$^\#$)}][b]	322
[Ni{η^1,η^3-*C*HMe*C*H*C*HCH$_2$C(O)*O*}(PR$_3$)]	323
[Ni{η^1,η^3-*C*H$_2$*C*H*C*HCHMeC(O)*O*}(PR$_3$)]	323
[Ni{η^1,η^3-*C*H$_2$*C*H*C*HCH$_2$C(O)*N*Ph}(PEt$_3$)]	324
[Ni{η^1,η^3-*C*H$_2$*C*Me*C*HCH$_2$C(O)*N*Ph}(PCy$_3$)]	324
[Ni{η^1,η^3-*C*H$_2$*C*Me*C*MeCH$_2$C(O)*O*}(PR$_3$)] (R = **Ph, Cy**)	325
[Ni(η^1,η^3-*C*$_3$H$_4$C$_2$H$_4$CH=CH*C*H$_2$)(PR$_3$)] (R = Me, Pr^i, Cy, Ph, OC$_6$H$_4$Ph-2)	308
[Ni{η^3,η^3-Ph*C*(*O*)H*C*HCH(CO$_2$Me)CH$_2$*C*H*C*H*C*H$_2$}]	326
[Ni$_2$(μ-SH)$_2$(η^3-C$_3$H$_5$)$_2$]	327
[Ni$_2$(η^3-C$_3$H$_5$)$_2$S]$_x$	328
[Ni$_2$(μ-X)$_2$(η^3-C$_3$H$_5$)$_2$] (X = Cl, Br, I)	281–3
[Ni$_2$Br$_2$(η^3-C$_3$H$_4$R-1)$_2$] (R = CH=CH$_2$, CH=CHMe, CH$_2$CH=CH$_2$)	329
[Ni$_2$Br$_2$(PPh$_3$)$_2$(η^3-C$_3$H$_3$MePh-1,1)]	304
[Ni$_2$(MeCN)$_2$(η^3-C$_3$H$_3$MePh-1,1)$_2$]$^{2+}$(BF$_4^-$)$_2$	304
[Ni$_2$(μ-Me)$_2$(η^3-C$_3$H$_3$Me$_2$-1,1)$_2$]	330
[Ni$_2$(μ-C$_3$H$_5$)(μ-X)(PR^{1}R$^2{}_2$)$_2$] (X = Cl, Br, C$_3$H$_5$; R^{1}R$^2{}_2$ = Pr$^i{}_3$, etc.)	331
[Ni$_2$(μ-Br)$_2${C$_3$H$_4$(CH$_2$TMS)-2}$_2$]	332
[Ni$_2$(η^3,η$^{3'}$-C$_4$Me$_4$C$_4$Me$_4$)Cp$_2$]	333
[Ni$_2$(η-C$_2$H$_4$)(η^1,η^3-*P*EtC$_5$H$_2$Ph$_3$-2,4,6)$_2$]	143
[Ni$_2$(*trans*-NAr=CHCH=NAr)$_2$(μ-η^3,η$^{3'}$-C$_6$H$_8$)][c]	334
[Ni$_2$(μ-Cl)$_2${η-*C*$_3$H$_3$Me(O-TMS)-1,3}$_2$]	335
[Ni$_2$(μ-Cl)$_2${η^3-*C*$_3$H$_2$Me(O-TMS)-1,1}$_2$]	335
[Ni$_2$(μ-Cl)$_2${η^3-CHR^{1}CR^{2}CH(OSiMe$_2$R^3)}$_2$] (R^1, R^2 = H, R^3 = Me, Bu^t; R^1 = Me, R^2 = H, R^3 = Bu^t; R^1 = H, R^2, R^3 = Me; R^1 = Ph, R^2 = H, R^3 = Me)	335
[Ni$_2$(μ-η^1,η$^{1'}$,η^3-*C*H$_2$*C*R^1*C*R^{2}CH$_2$CHPh*C*HCH=*N*Ph)$_2$]	336
[Ni$_2$(*p*-C$_6$H$_8$)Cp$_2$][c]	161
[NiCp(η^3,η$^{3'}$-C$_9$H$_{12}$)Mo(CO)$_2$Cp'][d]	161
[NiCp(μ-η^3,η^4-C$_4$Ph$_4$C$_5$H$_5$)CoCp]	315
[Ni$_3$(μ-S)$_2$(η^3-C$_3$H$_5$)$_3$]$^+$	337

[a] Complexes in bold were characterized by an x-ray diffraction study. Nickel-coordinated atoms are italicized.
[b] Mes$^\#$ = C$_6$H$_2$Ph$_3$-2,4,6. [c] Ar = Xyl-2,6; C$_6$H$_8$ = (*C*H$_2$)$_2$*CC*(*C*H$_2$)$_2$. [d] C$_9$H$_{12}$ = (*C*H$_2$)$_2$*C*CH$_2$C(=CH$_2$)*C*(*C*H$_2$)$_2$.

Table 13 [NiX(η^3-CH$_2$Ar)L^1] and [Ni(η^3-CH$_2$Ar)L^{1}L^2]$^+$ complexes.[a]

Ar	*L^1, L^2 or L^1, X*	*Ref.*
Ph (and **CHMePh**)[b]	PCy$_3$, CN	338
Ph	PCy$_3$, X (X = Cl, CN)	339
Ph	PMe$_3$, Cl	340
Ph	**2 PPh$_3$, PF$_6$**	341
Ph	(+)-diop	341
Tol-2	**PMe$_3$, Cl**	342
C$_6$H$_4$[*trans*-NiBr(PMe$_3$)$_2$]-*n* (*n* = 2–4)	PMe$_3$, Br	343
C$_6$H$_4$[*trans*-NiBr(PMe$_3$)$_2$]-4	PMe$_3$, Cl	344
C$_6$H$_4$[*trans*-NiBr(PMe$_3$)$_2$]-4	PMe$_3$, Br	345
Tol-2	PMe$_3$, Cl	346

[a] Species in bold characterized by x-ray diffraction. [b] η^3-CHMePh complexes also prepared.

Table 14 Cyclic η^3-nickel complexes.[a]

Complex	*Ref.*
[{Ni(μ-C$_3$Cl$_3$)(μ-Cl)(CO)$_2$}$_2$]	347
[Ni(PPh$_3$)$_2$(η^2-C$_3$Ph$_3$)]$^+$ClO$_4^-$	348, 349
[Ni(triphos)(η^3-C$_3$Ph$_3$)]$^+$ClO$_4^-$	350
[Ni(η^3-P$_3$)Cp*]	351
[Ni(η^3-P$_3$)(η-C$_5$H$_3$Bu$^t{}_2$-1,3)] (E = P, As)	352
[Ni(η^3-E$_3$)(η-C$_5$HPh$_4$)] (E = P, As)	353
[Ni$_2$(μ-Br)$_2$(CO)$_2$(η^3-C$_3$Ph$_3$)]	354
[Ni$_2$(μ-Br)$_3$(η^3-C$_3$Ph$_3$)$_2$]$^-$[C$_3$Ph$_3$]$^+$	355

[a] Complexes in bold characterized by an x-ray diffraction study.

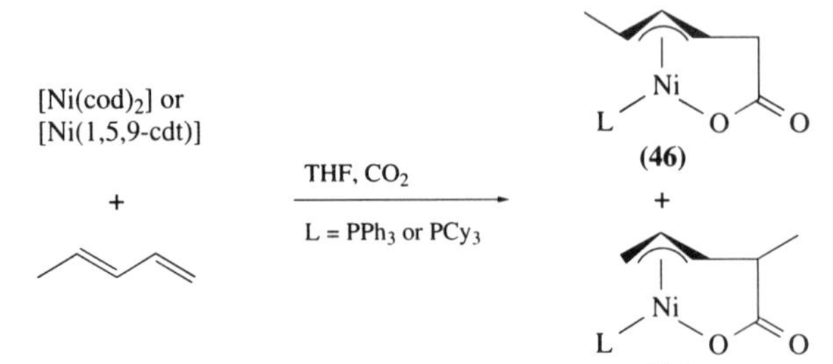

Scheme 26

Allylic species are implicated in the nickel(II)-catalyzed coupling reaction of 1,3-dienes with CO and secondary amines that leads to 2,4-dienecarboxylic amides[360] and in the coupling of butadiene with PhC(O)CH=CHCO$_2$Me to afford PhC(O)CH$_2$–CH(CO$_2$Me)CH=CHCH=CH$_2$.[326] The η^3-cyclooctenylnickel(II) species [Ni(η^3-C$_8$H$_{13}$){PPh$_2$CH$_2$C(O)O}] (**26**) referred to earlier is in equilibrium with the η^1,η^2-complex [Ni($\eta^1{}_5$,$\eta^2{}_{1,2}$-C$_8$H$_{13}$){PPh$_2$CH$_2$C(O)O}] (**25**) and is an excellent model catalyst for ethene oligomerization (Scheme 15).[137]

Nickel allyl species containing pyrazolyl gallates as ancillary ligands have been prepared. The dynamic processes occurring in these molecules have been studied by NMR spectroscopy. The structure of one of these species, the complex [Ni(η^3-C$_3$H$_5$)(GaMe$_2$pz$_2$)], was determined crystallographically,[286] as was that of the tris(pyrazolyl borate) species [Ni(η^3-C$_3$H$_5$)(HBpz$_3$)].[58] Allylic complexes have also been prepared by the oxidative addition of C$_3$H$_5$(CO$_2$Me) with [Ni(cod)$_2$], as shown in Equations (7a) and (7b). The initially formed polymeric species [Ni(η^3-C$_3$H$_5$)(CO$_2$Me)]$_n$ undergoes a redistribution reaction to form [Ni(η^3-C$_3$H$_5$)$_2$] and [Ni(CO$_2$Me)$_2$]. The complexes of general formula [Ni(η^3-C$_3$H$_5$)(CO$_2$Me)(PR$_3$)] are formed with tertiary phosphine ligands and their reactions with nucleophiles were investigated.[285]

$$[\mathrm{Ni(cod)_2}] + \mathrm{C_3H_5(CO_2Me)} \longrightarrow [\mathrm{Ni}(\eta^3\text{-}\mathrm{C_3H_5})(\mathrm{CO_2Me})]_n \qquad (7a)$$

$$[\mathrm{Ni}(\eta^3\text{-}\mathrm{C_3H_5})(\mathrm{CO_2Me})]_n \longrightarrow [\mathrm{Ni}(\eta^3\text{-}\mathrm{C_3H_5})_2] + [\mathrm{Ni(CO_2Me)_2}] \qquad (7b)$$

Reactions of dienes with nickel(0) alkene complexes often afford η^3-allyl compounds. Carboxymethylbutadiene CH$_2$=CHCH=CH(CO$_2$Me) is coupled stoichiometrically or catalytically to a mixture of products, mainly isomeric vinylcyclohexenes, when [Ni(cod)$_2$] and phosphorus donor ligands are available. High yields of the allyl complex (**48**) could be isolated from this reaction: only the *syn,syn*-isomer was obtained, as shown in Scheme 27.[97] Oxidative coupling of 2,3-dimethylbuta-1,4-diene with carbon dioxide and nickel(0) phosphine species affords the oxygen-bound nickel allyl species (**49**), shown in Scheme 28.[325] This reaction is similar to the coupling reaction depicted in Scheme 26. The structures of both complexes were determined by single-crystal x-ray diffraction studies: key bond lengths for two compounds are listed in Scheme 28. The allyl complexes [Ni$_2$Br$_2$(η^3-C$_3$H$_4$R-1)$_2$] (R = CH=CH$_2$, CH=CHMe, CH$_2$CH=CH$_2$) were prepared through oxidative addition of the appropriate bromodiene to [Ni(cod)$_2$].[329]

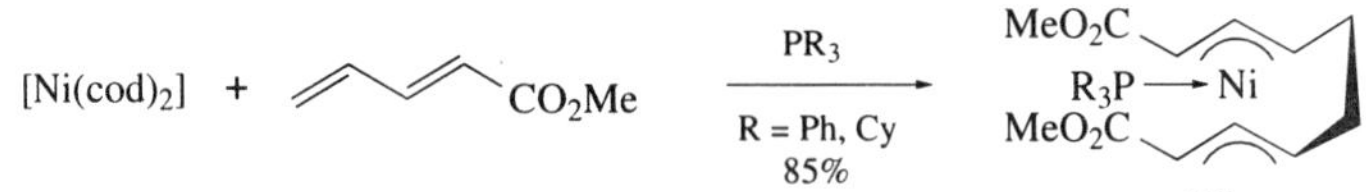

Scheme 27

The thermolysis of a series of allyl complexes [NiMeL(η^3-*syn,syn*-C$_3$H$_3$Me$_2$-1,3)] (**50**) has been investigated mechanistically (Scheme 29). The separate effects of L and [L] on the decomposition of this complex were studied. Decomposition temperature was dependent on the steric bulk (~25%) of L and more strongly affected by the electronic effect of L (~75%). The selectivity of the decomposition is equally affected by the ligand's steric and electronic effects.[302,303] *Syn*- and *anti*-isomers of the η^3-pentadienyl complexes [Ni{η^3-CH$_2$CMe$_2$CHCH=CH$_2$}Cp] were reported. These species were formed by reacting nickelocene with the cyclopropyl Grignard derivative MgCl(CHCMe$_2$CH–CH=CMe$_2$).[305]

Complexes (**51**) and (**52**) (Scheme 30) were isolated by reacting nickel(0) alkene complexes with trialkyl- or triarylphosphine ligands and 1,3-butadiene. The preparations and structures of these Ni(L) octadienediyl species and their relevance to the catalytic dimerization of dienes were probed.[308] Extensive low-temperature ^{1}H and ^{13}C NMR data and various chemical reactions were used to establish the structures of the molecules shown in Scheme 30, and of other related species.

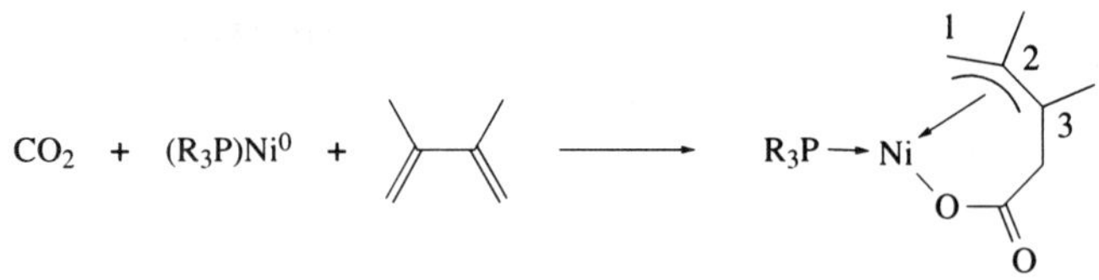

(49)

R	Ni–C-1 nm	Ni–C-2 nm	Ni–C-3 nm
Ph	0.199 4(3)	0.196 8(3)	0.202 5(3)
Cy	0.201 0(6)	0.197 0(6)	0.201 2(5)

Scheme 28

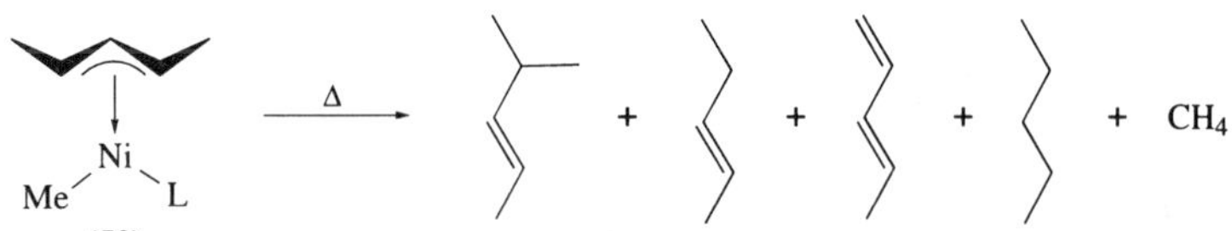

(50)

Scheme 29

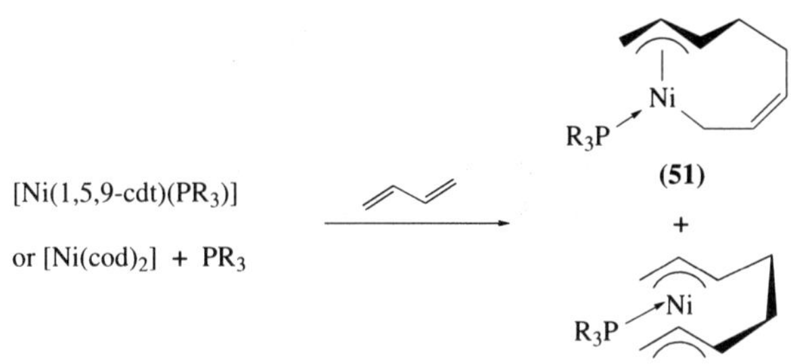

(51)

(52)

R = Me, Pr^i, Cy, Ph, OC_6H_4Ph-2

Scheme 30

Three papers describe the reactions of H_2S with [Ni(η^3-C_3H_5)Cp] (and its palladium and platinum analogues) at low temperature. The thermally unstable complex [Ni_2(μ-SH)$_2$(η^3-C_3H_5)$_2$] eliminates propene at −60 °C. Reaction of this species with [Ni(η^3-C_3H_5)$_2$] affords polymeric [Ni_2(η^3-C_3H_5)$_2$S]$_x$. The trigonal bipyramidal cationic clusters [Ni_3(μ-S)$_2$(η^3-C_3H_5)$_3$]$^+$ were obtained when the μ-hydrosulfide complex [Ni_2(μ-SH)$_2$(η^3-C_3H_5)$_2$] was deprotonated with BuLi.[327,328,337]

Various diene-coupling reactions afford η^3-allyl species. Cinnamaldehydaniline couples oxidatively with 2,3-dimethylbutadiene at a nickel(0) center to afford (**53**) (Scheme 31, $R^1 = R^2 =$ Me).[361] A series of 2,3-substituted butadienes link up with azadienes to afford other stereospecifically formed dinickel allylic complexes of type (**53**) as shown in Scheme 31.[336]

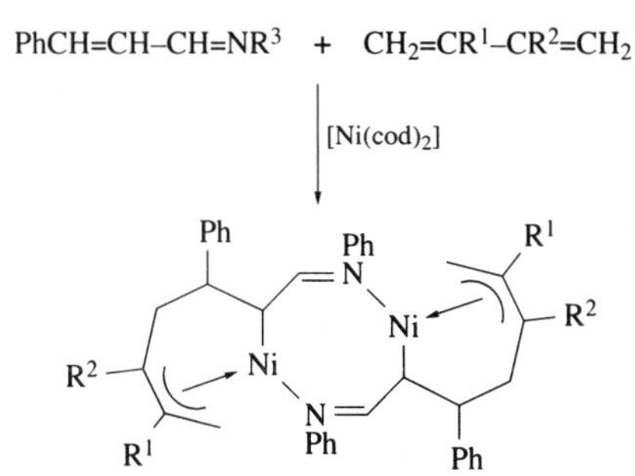

(53)

Scheme 31

Reduction of [Ni(acac)$_2$] with $AlEt_2$(OEt) and the substituted phosphapyridine species $PC_5H_2Ph_3$-2,4,6 afforded the dinuclear nickel–nickel bonded Structure (**54**), which was characterized by x-ray diffraction. The ethene is readily displaced by tertiary phosphine ligands, by other alkenes, and by alkynes. NMR studies indicate that the allyl moiety is fluxional around the six-membered ring.[143]

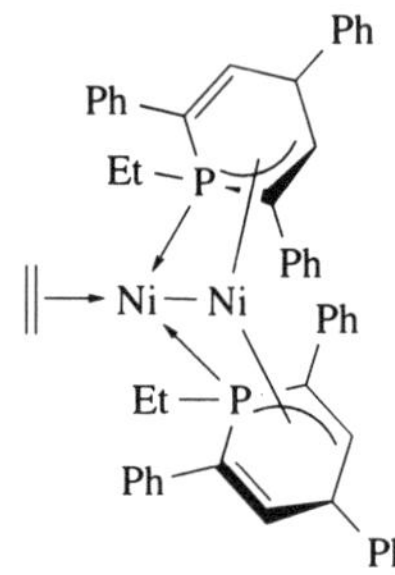

(54)

Cationic η^3-allyl complexes of the type $[Ni(\eta^3\text{-}C_3H_4R\text{-}2)L_2]^+X^-$ (L = various two-electron donor ligands, X = halide, PF_6^-, BF_4^-, etc.) are alkene and butadiene polymerization catalysts. Unraveling the mechanistic details of this reaction has led to many studies into the chemistry of this class of nickel allyl complex.

The crotyl complex $[Ni(\eta^3\text{-crotyl})(PPh_3)_2]^+$ is an effective catalyst for the oligomerization and polymerization of butadiene. The products obtained—cyclic or linear dimers, oligomers, or polymers—are dependent on the solvents and the co-catalysts used.[361] New syntheses have been described for the complexes $[Ni(cod)(\eta^3\text{-}C_3H_4R\text{-}2)]^+X^-$ ($X^- = AlBr_4^-$, BF_4^-, PF_6^-). This cationic species polymerizes 1,3-butadiene to *cis*-1,4-polybutadiene. The compound $[Ni(\eta^3\text{-}C_3H_2Me_3\text{-}1,2,3)(cod)]^+PF_6^-$ was isolated from the reaction mixture. The influence of tertiary phosphine ligands with sulfonated aryl rings on the oligomerization of C_2H_4 by $[Ni(\eta^3\text{-}C_3H_4Me\text{-}2)(cod)]^+PF_6^-$ was studied;[362] the structures of both the PF_6^- and the OTf^- salts of $[Ni(\eta^3\text{-}C_3H_4Me\text{-}2)(cod)]^+$ were determined.[298,299] Other examples of studies and applications of allyl complexes in catalysis are listed.[301]

A mechanistic study of the butadiene polymerization reaction catalyst $[Ni(\eta^3\text{-}C_3H_4R)L_2]^+PF_6^-$ has been published for L = $P(OPh)_3$. *Syn*- and *anti*-configurations of the allyl ligand were identified by analysis of the [AB] type ^{31}P NMR spectra observed for phosphorus donor ligands in these compounds. It was shown that *syn* to *anti* isomerization occurs; the more thermodynamically stable *syn*-isomer is also more active catalytically.[293] Other $[Ni(\eta^3\text{-}C_3H_5)L_2]^+PF_6^-$ species were prepared and characterized by ^1H and ^{13}C NMR spectroscopy and by elemental analysis. The effect of L on the catalytic activity of these compounds was investigated. It was shown that the catalytic activity and the *cis*-selectivity of the polymerization increased substantially with decreasing strength of the Ni–L bond in the order L = Bu^{t}NC < PPh_3 < MeCN < $AsPh_3$ < cod < $SbPh_3$. The novel $SbPh_3$ complex is a highly active, *cis*-selective catalyst.[287]

The extent of anion X^- coordination in the complexes $[Ni(\eta^3\text{-}C_3H_5)\{P(OPh)_3\}_2]^+X^-$ (X = Cl, Br, I, PF_6, BF_4, $AlBr_4$, $B(C_6H_4O_2\text{-}1,2)_2$, $MeC_6H_4SO_3$) was investigated by conductivity measurements and ^{13}C NMR spectroscopy. It was found that increasing anion coordination in the above series was accompanied by inhibition of the catalytic activity of the complexes.[288] The catalytic activity of the complexes $[Ni(\eta^3\text{-}C_3H_5)L_2]^+X^-$ (L = $P(OPh)_3$, Bu^{t}NC, $P(OThym)_3$, 1/2 cod, $SbPh_3$; $X^- = PF_6^-$, BF_4^-) was investigated as a function of L and X. The BF_4^- salts showed reduced activity and lower *cis*-selectivity in butadiene polymerization.[289]

Nickel allyl complexes of class (**55**) can be protonated to form cationic species (**56**) (Scheme 32). These monoallyl cationic species are catalytically active for the stereospecific 1,4-polymerization of butadiene.[309,363] A series of cationic complexes $[Ni(\eta^3,\eta^2,\eta^2\text{-}C_{12}H_{19})]^+X^-$ ($C_{12}H_{19}$ = dodeca-2(*E*),6(*E*),10(*Z*)-trien-1-yl; $X^- = PF_6^-$ or SbF_6^-) were prepared and their catalytic activities were investigated. Later, other salts with $X = B[C_6H_3(CF_3)_2]_4$[312] or with $X^- = BF_4^-$, $CF_3SO_3^-$, $AlBr_4^-$, $B(C_6H_4O_2\text{-}1,2)_2^-$ [313] were made. These complexes catalyze the 1,4-*cis*-polymerization of 1,3-butadiene. The structure of the PF_6^- salt of (**57**) (Figure 8) was determined by an x-ray diffraction study: geometric parameters of the compound are tabulated in Figure 8. The C_{12} chain is essentially planar and allows other molecules to coordinate to the nickel atom.[313] The complexes $(\mathbf{57}^+)X^-$ ($X^- = B[C_6H_3(CF_3)_2]_4^-$, $B(C_6F_5)_4^-$, PF_6^-) were also studied by two-dimensional NMR studies ($^{13}C/^1H$, $^1H/^1H$ COSY (correlation spectroscopy) and HMQC (heteronuclear multiquantum correlation spectroscopy)). The vicinal $^3J_{HH}$ coupling constants correlated very well with the torsion angles obtained from the x-ray data for $(\mathbf{57}^+)PF_6^-$.[314] The solid-state and solution structures of the complex are apparently the same.[314]

A series of η^3-cycloalkenyl complexes of general formula [Ni(η^3-cycloalkenyl)Cp] with cycloalkenyl ring sizes ranging between five and eight were isolated and characterized and their high-field NMR spectra were recorded. It was concluded that in the complexes with cycloalkenyl ring sizes of six to eight carbon atoms, the carbon bridge that was uninvolved in nickel bonding was bent away from the metal. These compounds were prepared by reacting a mixture of $[NiCp_2]$ and the 1,3-cycloalkadiene

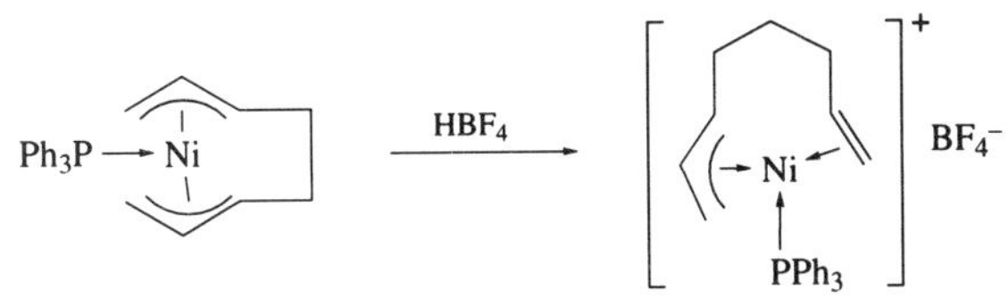

(55) (56)

Scheme 32

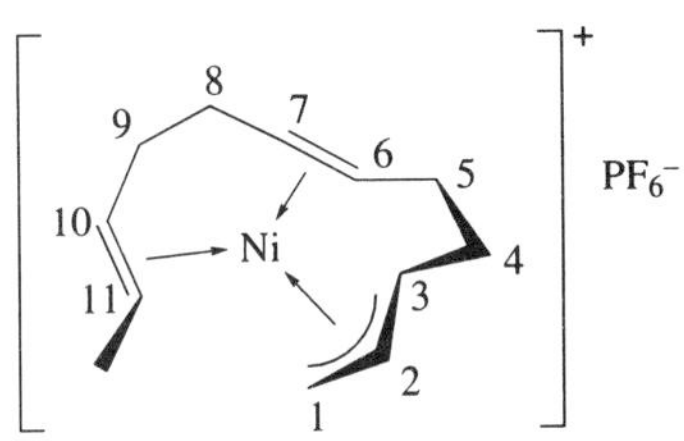

(57)

Ni–C-1 = 0.206 7(4)
Ni–C-2 = 0.211 6(4)
Ni–C-3 = 0.201 8(4)

Ni–C-6 = 0.210 9(4)
Ni–C-7 = 0.212 6(4)

Ni–C-10 = 0.213 3(4)
Ni–C-11 = 0.221 1(4)

C-1–C-2 = 0.138 7(6)
C-2–C-3 = 0.140 6(6)

C-6–C-7 = 0.136 5(5)
C-10–C-11 = 0.136 1(5)

Ni···F = 0.291 4(6)

C-1–C-2–C-3 = 118.7(4)

Figure 8 Structural parameters of the PF_6^- salt of (**57**). Bond lengths in nanometers, angles in degrees.[313]

with Pr^{i}MgBr.[306,307] High-field NMR studies were utilized to elucidate the conformers of these species in solution.[154] NMR two-dimensional studies and measurements of ^{13}C–^{13}C coupling constants were used to deconvolute the often complex spectra of [Ni(η^3-allyl)Cp] derivatives and other η^1,η^2-tethered alkene–alkyl species.[364–6]

Alkyl and aryl complexes of formula [NiRL(η^3-C_3H_5)] (R = H, alkyl, aryl; L = phosphorus donor ligands) and related allylic species have been prepared.[367] Various (alkyl)methylallylnickel complexes of the type [NiR(PMe_3)(η^3-C_3H_4Me-2)] were also synthesized along with the nickel chloroallyl species [NiClL(η^3-C_3H_4Me-2)] (L = PMe_3, dmpe).[296]

The species [MAr(PPh_3)(η^3-C_3H_5)] (M = Ni, Pd, Pt; Ar = $C_6H_3Cl_2$-2,5) undergo a carbon–carbon coupling reductive elimination reaction to yield Ar–CH_2CH=CH_2 and nickel(0) species. The relative rates of the coupling reaction were compared for the three metals. Even at 80 °C, no reductive elimination took place for the platinum complex. Rate constants at 0 °C in toluene for the palladium and nickel compounds were 2.9×10^{-3} and 5.8×10^{-2} h^{-1}, respectively. The rate of the coupling reaction increased dramatically when palladium was added to the 16-electron nickel allyl complex, but not for the palladium compound.[291] More recent work indicates that 16-electron complexes of general formula [NiAr(PPh_3)(η^3-$C_3H_4R^1R^2$-1,2)] ($R^1 = R^2$ = H; Ar = Ph, C_6H_3Cl-2,5; R^1 = H, R^2 = Me, Ar = C_6F_5; R^1 = Me, R^2 = H, Ar = C_6F_5) react with dppe or with (Z)-Ph_2PCH=$CHPPh_2$ by PPh_3 ligand substitution to afford the chelated species [NiAr(L_2)(η^3-$C_3H_3R^1R^2$-1,2)]. It is interesting that the latter 18-electron complexes exhibit an enhanced reactivity towards reductive elimination, as compared to the former 16-electron PPh_3 compounds.[290,292]

The allyl species [NiI($NHEt_2$)$_2$(η^3-C_3H_4R-2)] (R = C(O)NEt_2) has been structurally characterized and its reactions with alkenes and allene were described.[300] The zwitterionic species (**58**) was formed when [Ni(η^3-C_3H_5)$_2$] was reacted with bis(diphenylphosphino)maleic anhydride.[368] Bridging allyl complexes are present in the dinickel complexes [Ni_2(μ-C_3H_5)(μ-X)($PR^1R^2_2$)$_2$] (X = Cl, Br, C_3H_5). The allyl ligands are believed to bridge symmetrically.[331]

Nickel trimethylsiloxyallyl species have been synthesized. The structure of one of these species, the complex [Ni_2(μ-Cl)$_2${η^3-C_3H_3Me(O-TMS)-1,3}$_2$], was determined crystallographically. The structural study indicates that there is significant overlap of oxygen lone pair electron density with the allyl ligand.[335] Functionalized allylsilanes have been made by treating *in situ* solutions of [Ni_2(μ-Br)$_2${η^3-C_3H_4(CH_2TMS)-2}$_2$] with a variety of alkyl halides.[332] The complex [NiBr(PPh_3)(η^3-C_3H_3PhMe-1,1)] and the dinuclear complexes [Ni_2Br_2(PPh_3)$_2$(η^3-C_3H_3PhMe-1,1)] and [Ni_2(MeCN)$_2$(η^3-C_3H_3PhMe-1,1)$_2$]$^{2+}$ $(BF_4^-)_2$ have been synthesized.[304]

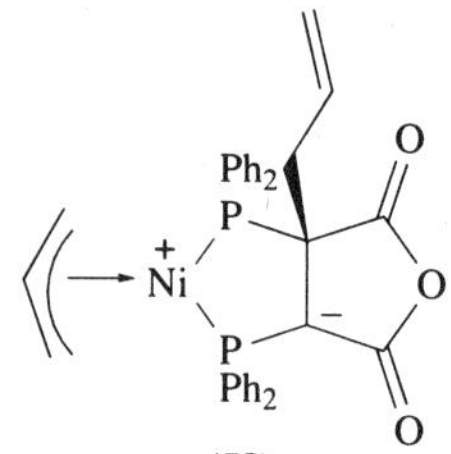

(58)

Coupled allyl ligands have been coordinated and stabilized onto nickel centers. The complex $[Ni_2\{NAr{=}C(Me)C(Me){=}NAr\}_2(\eta^3,\eta^{3'}\text{-}C_6H_8)]$ (Ar = $C_6H_3Ph_2$-2,5; C_6H_8 = 2,2'-biallyl) has been prepared and structurally characterized.[334] The previously synthesized complex $[Ni_2(\eta^3,\eta^{3'}\text{-}C_6H_8)Cp_2]$ (**59**) has been made by reacting $[Ni_2(\mu\text{-}CO)_2Cp_2]$ (Ni–Ni) with allene (Scheme 33). In contrast, the reaction of allene with the heterobimetallic species $[Ni(CO)Cp\text{–}Mo(CO)_3Cp']$ afforded (**60**), which contains a trimerized allene ligand that is η^3-coordinated to both the nickel and the molybdenum atom. Thermolysis of the complex afforded isomeric trimethylenecyclohexanes, via cyclization of the C_9H_{12} ligand.[161] The reaction of $[Ni(CO)Cp\text{–}M(CO)_3Cp']$ (M = Mo, W) with 1,1-dimethylallene proceeded differently. Initially the 2-nickel allyl complexes $[NiCp\{\mu\text{-}\eta^1(Ni),\eta^3(M)\text{-}C_3H_2Me_2\text{-}1,1\}M(CO)_2Cp']$ (Ni–M) were obtained. These complexes undergo an acid-promoted 1,4-proton shift isomerization to yield the 1-nickel allyl complexes $[NiCp\{\mu\text{-}\eta^1(Ni),\eta^3(Mo)\text{-}C_3H_2Me_2\text{-}1,2\}(\mu\text{-}CO)Mo(CO)Cp']$ (Ni–M). The structure of the latter Ni–Mo complex was established crystallographically.[160,161]

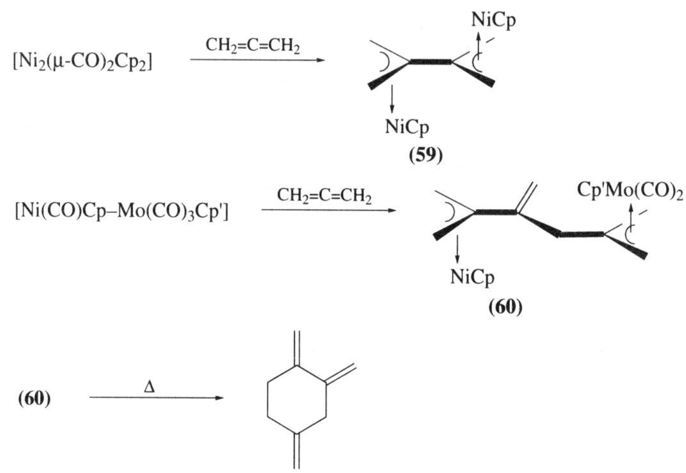

Scheme 33

Allylic species derived from Cp* ligands have been synthesized by reactions of nickel atoms with pentamethylcyclopentadiene and small organometallic molecules. When either $[Fe(CO)_5]$ or $[Co(CO)_2Cp]$ was used as the organometallic reagent, two isomers of the allylic species $[Ni(\eta^3\text{-}C_5Me_5H_2)Cp^*]$ (**61a**) and (**61b**) were obtained in 1:1 ratio, as shown in Scheme 34.[316] Mixed-metal clusters were also formed. The two isomers were identified by ^1H, ^{13}C and two-dimensional COSY NMR data. The *cis-endo*-isomer (with both methyl groups of the CH(Me)–CH(Me) group pointing towards the nickel atom) was not obtained here, although it has been identified by ^1H NMR spectroscopy in earlier work.[318] Structure (**38**), referred to earlier, also contains an η^3-allylic species derived from an η-Cp* ligand.[13,261]

3.3.4 Nickel Heteroallyl Species

In contrast with nickel η^2-heteroalkene species, few nickel η^3-heteroallyl complexes have been prepared. Nickel is present in some metalloallyl complexes such as $[NiCp\{\mu\text{-}\eta^1(Ni),\eta^3(Mo)\text{-}C_3H_2Me_2\text{-}1,2\}(\mu\text{-}CO)M(CO)Cp']$ (Ni–M),[161] but is part of the allyl ligand in such species, and is σ-bonded, not π-ligated, to the allyl group.

The phosphaalkene ligand P(Mes*)=CH[PCl(Mes*)] (Mes* = $C_6H_2Bu^t_3$-2,4,6) reacts with $K^+[Ni(CO)Cp]^-$ to produce the diphosphaallyl complex (**62a**) which is in equilibrium with (**62b**), as shown in

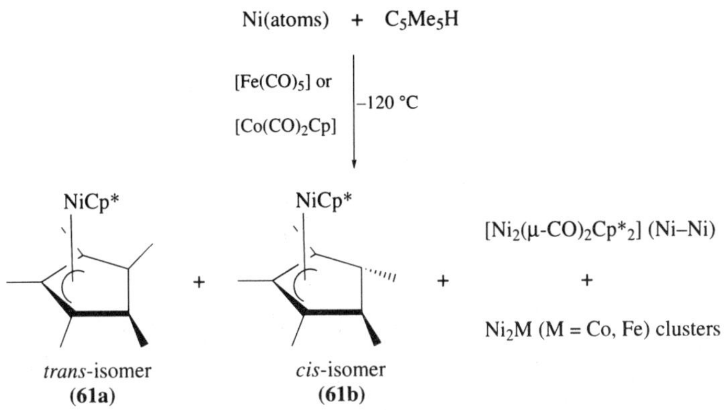

Scheme 34

Scheme 35. In solution, the *syn,syn* and *syn,anti*-conformers are in equilibrium. The *syn,anti*-conformer is obtained in the solid state and was structurally characterized.[206,207] Another diphosphaallyl species $[Ni(\eta^3\text{-}C_8H_{13})\{\eta^3\text{-}P(Mes^*)C(O\text{-}TMS)P(Mes^*)\}]$ is an active ethene polymerization catalyst, but this compound does not polymerize propene.[319,320] An azadiphosphaallyl species is believed to be an intermediate when $[NiCp_2]$ reacts with $Li^+[P(Mes^*)\text{–}P\text{–}N(Mes^*)]^-$ to afford a dinuclear complex.[369]

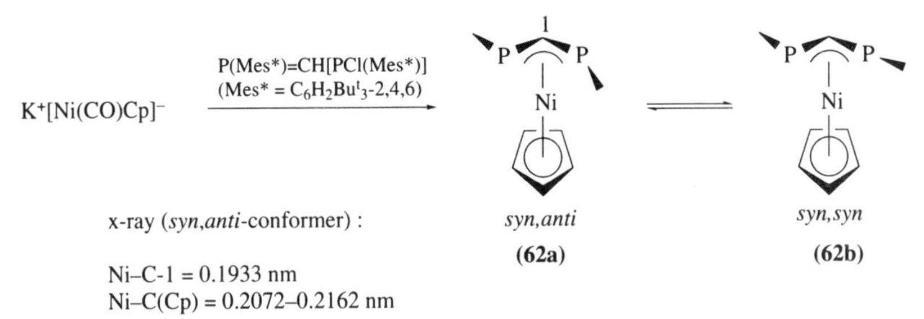

Scheme 35

The η^3-species $[Ni(\eta^3\text{-}C_4Ph_4HCO)Cp]$ (**63**) is formed when nickelocene reacts in a sealed glass tube with liquid tetraphenylcyclopentadienone at 170 °C (Scheme 36). Other products are obtained.[123] The azaborolinyl ligand $C_3H_3N(Bu^t)B(Me)$ behaves as a three-electron donor in the complex $[Ni\{\eta^3\text{-}C_3H_3N(Bu^t)B(Me)\}_2]$ (and not as a five-electron donor, as is more commonly observed).[284]

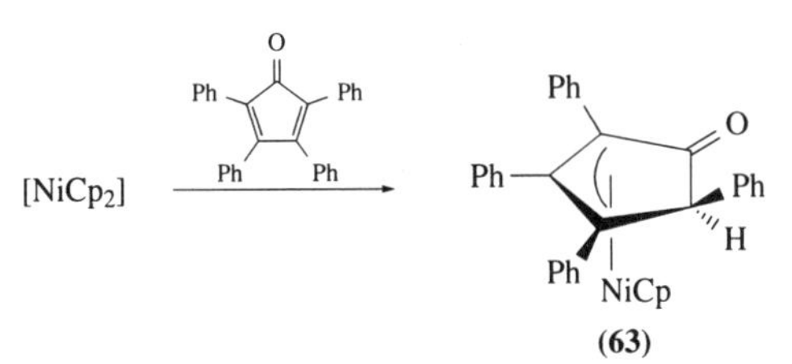

Scheme 36

3.3.5 Nickel η^3-Benzyl Complexes

Reactions that can conceivably lead to nickel η^1-benzyl complexes often afford η^3-species and an increasing number of such species have been prepared since the mid-1980s. These species are tabulated in Table 13.

The benzyl–cyanide bond of the substituted benzyl cyanide species PhCH(R)CN (R = H, Me) oxidatively adds to the Ni^0–PCy_3 species $[Ni(PCy_3)_2]$ or $[Ni_2(PCy_3)_4(N_2)]$ to afford the nickel(II) complexes $[Ni\{PhCH(R)\}(CN)(PCy_3)_n]$. This purple mixture of compounds is believed to contain both η^1- and η^3-benzylic species. The kinetics of this reaction have been studied by IR spectroscopy, and NMR, ESR, and GLC studies were also reported. The concentration of the η^3-benzyl complexes is inversely proportional to the PCy_3 concentration.[338]

Later work expanded on these reactions. Oxidative addition of $PhCH_2X$ (X = Cl, CN) to $[Ni(PCy_3)_2]$ affords the 16-electron compounds $[NiX(\eta^3\text{-}CH_2Ph)(PCy_3)]$. These purple or blue-violet complexes are believed to contain η^3-benzyl ligands: they exhibit a strong absorption band at 555 nm (18 000 cm^{-1}) and their CH_2Ph protons resonate at ~4.2 ppm in the ^1H NMR spectrum. Thermolysis produced benzaldehyde and benzyl alcohol: their relative ratios depended on X, the solvent, and the temperature of thermolysis. Isotopic labeling studies and MS data show that the extra proton in the benzyl alcohol came from the PCy_3 ligands.[339] Oxidation of $[NiX(CH_2Ph)(PCy_3)]$ (X = Cl, CN) with O_2 afforded $P(O)Cy_3$, PhCHO, and $PhCH_2OH$; it is presumed that a 1:1 O_2 adduct is formed initially.[370]

Treatment of $[Ni(cod)_2]$ with benzyl chloride and PMe_3 also afforded benzylic species (Equations (8a) and (8b)). Either η^1- or η^3-benzyl compounds can be obtained by controlling the relative equivalents of PMe_3 added to the reaction mixture.[340] Cationic π-benzyl complexes were formed from the neutral species $[NiXL_2(\eta^3\text{-}CH_2Ph)]$ (X = Cl, Br; L_2 = 2 PPh_3, (+)-diop) by halide abstraction with $TlPF_6$. The resulting $[NiL_2(\eta^3\text{-}CH_2Ph)]^+PF_6^-$ species exhibited catalytic activity towards styrene oligomerization—67% of the isostatic polymer was obtained.[341]

$$[Ni(cod)_2] + 2\,PMe_3 \xrightarrow{PhCH_2Cl} [NiCl(CH_2Ph)(PMe_3)_2] \quad (8a)$$

$$[Ni(cod)_2] + PMe_3 \xrightarrow{PhCH_2Cl} [NiCl(PMe_3)(\eta^3\text{-}CH_2Ph)] \quad (8b)$$

Nickel *o*-xylyl derivatives, analogous to some of the benzyl complexes discussed above, have been described. They are prepared by oxidative addition to $[Ni(cod)_2]$ in the presence of PMe_3. Frequently, η^1- and η^3-benzylic species are in equilibrium, and the PMe_3 ligand undergoes dissociation. The species $[NiCl(\eta^3\text{-}CH_2C_6H_4Me\text{-}2)(PMe_3)]$ was reported along with other η^1-xylyl complexes.[371] The structure of $[NiCl(\eta^3\text{-}CH_2C_6H_4Me\text{-}2)(PMe_3)]$ subsequently appeared. Nickel–carbon distances to the three benzylic carbon atoms are: $Ni\text{–}C_{CH2_2}$ = 0.193 0(8) nm; $Ni\text{–}C_{C\text{-}1\text{-}Ph}$ = 0.205 0(8) nm; $Ni\text{–}C_{C\text{-}6\text{-}Ph}$ = 0.231 8(9) nm.[342] Very recently, η^3-benzyl species such as $[NiCl(\eta^3\text{-}CH_2C_6H_4Me\text{-}2)(PMe_3)]$ have been utilized as "PMe_3 sponges." The η^3-benzyl species readily abstracts and combines with a molar equivalent of PMe_3 from a nickelacyclopentane complex to afford the corresponding η^1-species $[NiCl(\eta^1\text{-}CH_2C_6H_4Me\text{-}2)(PMe_3)_2]$. New dinickel μ-carbene species are also generated.[346] Other η^3-benzyl species, such as $[NiBr\{\eta^3\text{-}CH_2C_6H_4[\textit{trans}\text{-}NiBr(PMe_3)_2]\text{-}4\}]$, have been prepared from the reaction of $[Ni(cod)_2]$ with PMe_3 and $BrCH_2C_6H_4Br$-3 or $BrCH_2C_6H_4Br$-4 (Equation (9)).[344]

$$2\,[Ni(cod)_2] + 3\,PMe_3 \xrightarrow{BrCH_2C_6H_4Br\text{-}4} [NiBr\{\eta^3\text{-}CH_2C_6H_4[\textit{trans}\text{-}NiBr(PMe_3)_2]\text{-}4\}] \quad (9)$$

3.3.6 Nickel η^3-Cyclopropylium Complexes and Related Species

Most nickel η^3-C_3R_3 species have been prepared with R = Ph. While most triphenylcyclopropylium nickel complexes are η^3-coordinated, in a few cases, the coordination is best described as η^2-ligated. More recently, cyclic η^3-E_3 (E = P, As, Sb) species have been synthesized from reactions of various $[Ni_2(\mu\text{-}CO)_2Cp''_2]$ (Ni–Ni; Cp" = Cp or substituted Cp ligands) with E_4 species (E = P, As, Sb). These species are all collected in Table 14, but $Ni(\eta^3\text{-}E_3)$ complexes are discussed later under the chemistry of $[Ni_2(\mu\text{-}CO)_2Cp''_2]$ species.

The configuration of the C_3Cl_3 ligand in the molecule $[\{Ni(\mu\text{-}C_3Cl_3)(\mu\text{-}Cl)(CO)_2\}_2]$ has been the subject of an MO analysis.[347] The cationic species $[Ni(\eta^2\text{-}C_3Ph_3)(PPh_3)_2]^+X^-$ (X = ClO_4, PF_6) (Figure 9) has been prepared in ≥75% yield by reacting $[Ni(\eta\text{-}C_2H_4)(PPh_3)_2]$ with $[C_3Ph_3]^+X^-$. This complex decomposes in air slowly, and more rapidly in solution.[348] A structural study of $[Ni(\eta^2\text{-}C_3Ph_3)(PPh_3)_2]^+$ and its palladium and platinum analogues revealed an increasing tendency towards η^3-coordination on descending down the group from nickel to platinum. The NiP_2 plane is almost perpendicular to the C_3 plane (the angle between the C_3 plane and the NiP_2 plane = 88°) and the C_3Ph_3 ligand is best considered to be η^2-coordinated. Other structural parameters of the nickel complex are summarized in Figure 9. This compound was treated as a model for the transition state in a theoretical study of ring whizzing in these species.[349]

The η^2-C_3Ph_3 ligand is displaced with S_4N_4 affording $[\{Ni(PPh_3)_2\}_2\{\eta^1\text{-}(S), \mu\text{-}\eta^1(N)\text{-}SNSN\}_2]$. This species is assumed to be isostructural to its structurally characterized platinum analogue.[372] However, it is the PPh_3 groups that are displaced when $[Ni(\eta^2\text{-}C_3Ph_3)(PPh_3)_2]^+ClO_4^-$ is treated with triphos. The resulting ionic species $[Ni(triphos)(\eta^3\text{-}C_3Ph_3)]^+$ was crystallographically characterized. η^3-C_3Ph_3 is symmetrically coordinated to the nickel atom in this triphos cation.[350]

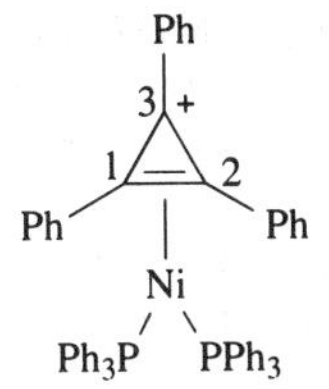

Ni–C-1 = 0.186 7(12)
Ni–C-2 = 0.200 9(12)
Ni–C-3 = 0.205 6(10)
Ni–P = 0.223 4(4), 0.224 1(4)

C-1–C-2 = 0.144 5(18)
C-1–C-3 = 0.142 5(20)
C-2–C-3 = 0.132 9(21)

Figure 9 Key bond lengths (nm) for the cationic nickel cyclopropylium species.[349]

An x-ray diffraction study of the complex $[NiBr(CO)(\eta^3\text{-}C_3Ph_3)]_2$ (Figure 10) revealed its dimeric nature. The centrosymmetric molecule contains essentially symmetrically bound C_3Ph_3 ligands: the nickel is in a distorted tetrahedral environment. Key structural parameters are shown in Figure 10.[354] The reaction of $[Ni(cod)_2]$ with $[C_3Ph_3]Br$ afforded the salt $[C_3Ph_3]^+$ $[Ni_2(\mu\text{-}Br_3)(\eta^3\text{-}C_3Ph_3)_2]^-$, whose structure was also determined. The reaction of $[NiL_2(cod)]$ (L_2 = bis(dialkylphosphine)) with $[C_3Ph_3]Br$ led to the formation of $[NiBrL_2(\eta^3\text{-}C_3Ph_3)]$.[355]

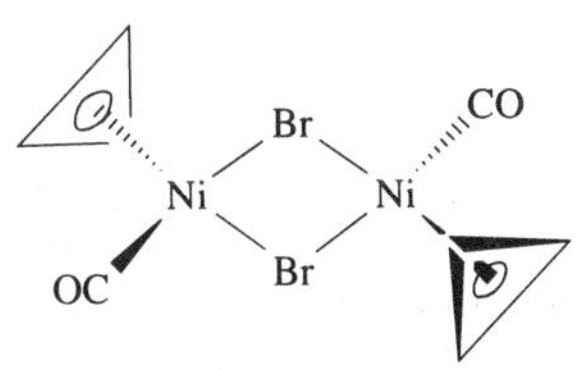

Ni–C = 0.195(1)–0.200(1)
C–C = 0.140 9(7)
Ni–Br = 0.248 2(2)
Ni–C3(centroid) = 0.180(1)
Ni···Ni = 0.329 3(3)

Figure 10 Key bond lengths (nm) for for another cationic nickel cyclopropylinm species. Carbon atoms are those of the C_3H_3 group.[354]

3.4 NICKEL COMPLEXES WITH η^4-LIGANDS

3.4.1 General Comments

This section is short as many classes of nickel η^4-ligand complexes are discussed elsewhere. Nickel η^4-cod and η^4-butadiene complexes are covered with nickel η^2-alkene species as their reactions and chemistry are intimately associated with those of nickel η^2-alkene compounds. Furthermore, nickel complexes with $\eta^4\text{-}C_5H_6$ ligands are treated with nickel cyclopentadienyl species, as $[Ni(\eta^4\text{-}C_5H_6)Cp]^+$ compounds are a useful starting material for [NiXLCp] and $[NiL_2Cp]^+$ compounds. Most complexes in this section are nickel cyclobutadiene or substituted cyclobutadiene derivatives; some examples of nickel complexes with heterocyclic η^4-ligands are included. A list of these nickel η^4-ligand complexes is presented in Table 15.

3.4.2 Theoretical Studies

The structure and bonding of $[Ni(\eta^4\text{-}C_4Ph_4)_2]$ have been modeled by a theoretical study of the cyclobutadiene parent compound $[Ni(\eta^4\text{-}C_4H_4)_2]$. Nonempirical molecular electronic structure theory (SCF) predicts that an eclipsed (D_{4h}) structure should be more stable than a staggered (D_{4d}) structure by 5.44 kJ. Equilibrium distances in the parent molecule (Figure 11) and ionization potentials were calculated.[386] Other theoretical treatments of cyclobutadiene bonding in various species have been reported.[387,388]

Table 15 Nickel complexes with cyclic η^4-ligands.[a]

Complex	*Ref.*	*Complex*	*Ref.*
$[Ni(\eta^4\text{-}C_4Me_4)\{\eta^2(S,S)\text{-}C_2S_2(CN)_2\}]$	373	$[Ni_2(\mu\text{-}Br)_2(\eta^4\text{-}C_4Ph_4)_2]$	96
$[Ni(\eta^4\text{-}C_4Ph_4)_2]$	374	$[Ni(\eta^4\text{-}C_4Ph_4)(PEt_3)_2]$	380
$[NiBr(\eta^4\text{-}C_4Ph_4)(PPh_3)]$	374	**$[Ni(\eta^4\text{-}\mathit{C}Me\mathit{C}Et\mathit{B}Et\mathit{Se}SiMe_2)_2]$**	381
$[NiBrX(\eta^4\text{-}C_4Ph_4)]$ (X = Br, I)	374	**$[Ni(\eta^4\text{-}\mathit{C}Me\mathit{C}Et\mathit{B}Et\mathit{S}SiMe_2)_2]$**	382
$[NiX_2(\eta^4\text{-}C_4R_4)]$ (R = Cy, Pr^i; X = Cl, Br)	375	**$[Ni(\eta^4\text{-}\mathit{P}Ph\mathit{B}Et\mathit{C}Et\mathit{C}MeSiMe_2)_2]$**	383
$[Ni(\eta^4\text{-}C_4R^1{}_4)(R^2N{=}CHCH{=}NR^2)]$	375	$[Ni(\eta^4\text{-}C_4R^1{}_4)(\eta\text{-}C_5R^2{}_5)]^+$ (R^1 = Me, Ph; R^2 = H, Me)	333
$[NiX_2(\eta^4\text{-}C_4H_4)]$ (X = Cl, Br, I)]	376	**$[Ni(\eta^4\text{-}\mathit{C}Bu^t\mathit{P}\mathit{C}Bu^t\mathit{P})_2]$**	384
$[Ni_2(\mu_2\text{-}\eta^4\text{-}cot)(\eta^4\text{-}C_4Ph_4)]$	377	$[NiCl_2(\eta^4\text{-}C_4Me_4)(PMe_2Ph)]$	385
$[Ni(\eta^4\text{-}C_4Me_4)(\eta^6\text{-}C_6Me_6)]^{2+}(AlCl_4^-)_2$	378	$[\{NiCl(\eta^4\text{-}C_4Me_4)\}_2WS_4]$	385
$[Ni_2(\mu\text{-}Cl)_2(\eta^4\text{-}C_4Me_4)_2]$	378	**$[Ni(\eta^4\text{-}C_4Me_4)(PMe_2Ph)(\mu\text{-}S)_2WS_2]$**	385
$[Ni_2(\mu\text{-}Br)_2(\eta^4\text{-}C_4Et_4)_2]$	379	$[Ni(\eta^4\text{-}C_4Ph_4)Cp]^{n+}$ (n = 1, 0, −1)	315

[a] Complexes in bold characterized by x-ray diffraction. Italicized atoms ligated to nickel.

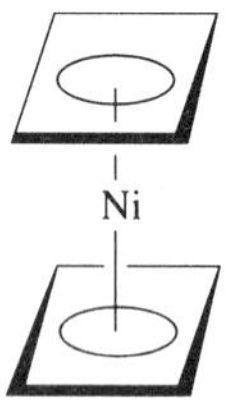

Ni–C_4, (centroid) = 0.1722 nm
C–C = 0.1453 nm
Bend-back angle of hydrogen atoms from C_4 rings = 52°

Figure 11 Calculated structural parameters for $[Ni(\eta^4\text{-}C_4H_4)_2]$.[386]

3.4.3 Syntheses and Reactions of Nickel Cyclic η^4-Species

Nickel carbonyl reacts with *cis*- or *trans*-2,3-dichlorohalobutenes and the corresponding aluminum(III) halide to give $[NiX_2(\eta^4\text{-}C_4H_4)]$ species (X = Cl, Br, I). No aluminum reagent is needed when X = Br or I.[376] A series of communications and papers in the early 1980s have described the chemistry of tetraphenylcyclobutadiene nickel complexes. The preparations of the sandwich compound $[Ni(\eta^4\text{-}C_4Ph_4)_2]$, the cot complex $[Ni_2(\eta^4\text{-}C_4Ph_4)_2(\mu\text{-}cot)]$, and the diazabutadiene species $[Ni(\eta^4\text{-}C_4Ph_4)(NR{=}CHCH{=}NR)]$ (R = Cy, Pr^i) are among some of those described in these papers (Scheme 37).[374,375,377]

Nickel(II) halides in boiling THF catalyze the trimerization of hex-3-yne (EtC_2Et) to hexaethylbenzene. The complex $[Ni_2(\mu\text{-}Br)_2(\eta^4\text{-}C_4Et_4)_2]$ was isolated from the mixture.[379] The cyclotrimerization of PhC_2Ph is catalyzed by $[Ni\{c\text{-}\overline{C(Ph)C(Ph)C(Ph)C}(Ph)\}L_2]$ ($L_2 = PPh_3$, 1/2 dppe) species and these complexes can be synthesized from the blue-black complex $[Ni_2(\mu\text{-}Br)_2(\eta^4\text{-}C_4Ph_4)_2]$, LiBu^t, and the appropriate trialkylphosphine ligand. The decarbonylation of $[Ni(\eta^4\text{-}C_4Ph_4CO)(cod)]$ was attempted in the presence of dppe, but no nickelacycles were isolated.[96] The complex $[Ni_2(\mu\text{-}Br)_2(\eta^4\text{-}C_4Ph_4)_2]$ has been used as an $\eta^4\text{-}C_4Ph_4$ transfer agent, and has been used to synthesize the complex $[Ni(\eta^4\text{-}C_4Ph_4)(PEt_3)_2]$. Detailed syntheses of both complexes have been published. Reported yields were 81% (for $[Ni_2(\mu\text{-}Br)_2(\eta^4\text{-}C_4Ph_4)_2]$) and 70%, respectively.[389,390] The chemistry of $[Ni(\eta^4\text{-}C_4Ph_4)(PEt_3)_2]$ with CO and its behavior towards oxidation, reduction, and acidolysis were studied.[380]

Reduction of cationic cyclobutadiene complexes $[Ni(\eta^4\text{-}C_4R^1{}_4)(\eta^5\text{-}C_5R^2{}_5)]^+$ (R^1 = Me, Ph; R^2 = H, Me) affords neutral radical species. The ESR of the reaction products has been interpreted in terms of a Jahn–Teller distorted structure. In some cases, depending on R^1 and R^2, the radicals may dimerize via carbon–carbon coupling at the C_4R_4 ring. One of these coupled products, (**64**), was the subject of an x-ray diffraction study.[333]

The redox chemistry of the cationic complex $[Ni(\eta^4\text{-}C_4Ph_4)Cp]^+$ was studied both electrochemically and chemically. Cyclic voltammetry studies showed that a reversible one-electron reduction led to the formation of the neutral 19-electron paramagnetic complex. Anionic $[Ni(\eta^4\text{-}C_4Ph_4)Cp]^-$ formed in an irreversible process upon further reduction. Chemical reduction of $[Ni(\eta^4\text{-}C_4Ph_4)Cp]^+$ proved to be reagent dependent. Reduction with $[Fe(\eta^6\text{-}C_6Me_6)Cp]$ or with one equivalent of cobaltocene afforded $[Ni(\eta^4\text{-}C_4Ph_4)Cp]$. Two equivalents of $[CoCp_2]$ led to formation of the mixed-metal species $[CpNi(\mu\text{-}\eta^3,\eta^4\text{-}C_4Ph_4C_5H_5)CoCp]$; a C_4Ph_4–Cp coupling reaction has taken place.[315]

The red complex $[NiCl_2(\eta^4\text{-}C_4Me_4)(PMe_2Ph)]$ was synthesized by addition of PMe_2Ph to purple $[Ni(\eta^4\text{-}C_4Me_4)Cl_2]_2$. Mixed-metal complexes which contain $Ni(\eta^4\text{-}C_4Me_4)$ groups have been synthesized

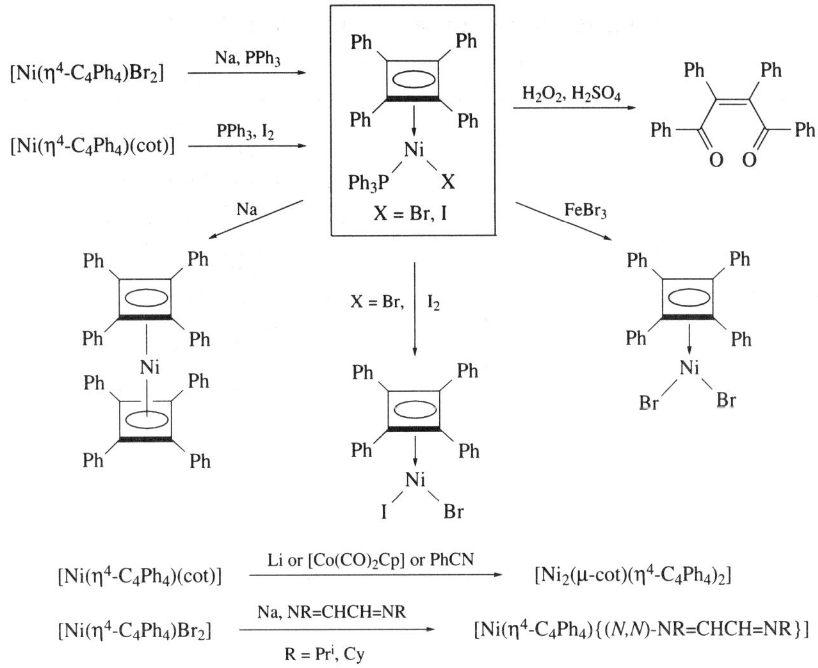

Scheme 37

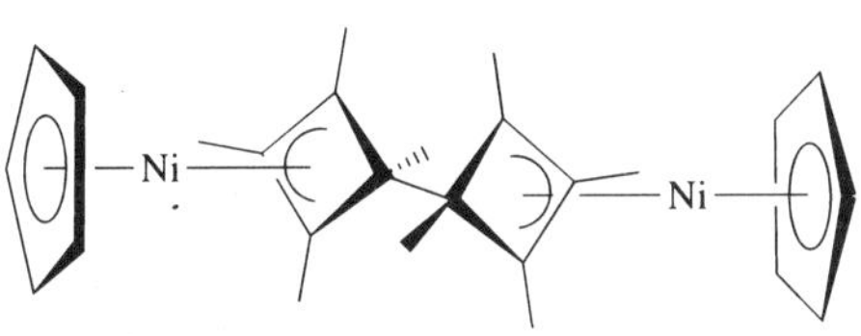

(64)

from both of these Ni(η^4-C_4Me_4) species. The Ni_2W complex [{NiCl(η^4-C_4Me_4)}$_2$$WS_4$] was prepared by reacting [Ni(η^4-C_4Me_4)Cl_2]$_2$ with WS_4^{2-}, and its thermolysis was studied. Reaction of [$NiCl_2$(η^4-C_4Me_4)(PMe_2Ph)] with WS_4^{2-} afforded the compound [Ni(η^4-C_4Me_4)(PMe_2Ph)(μ-S)$_2$$WS_2$], whose structure was established by an x-ray diffraction study. Average nickel–carbon distances are 0.205 and 0.206 nm for the two independent molecules in the unit cell.[385]

A limited number of heterocyclobutadiene species have been synthesized. The complex [Ni{η^4-PC(Bu^t)PC(Bu^t)}$_2$] was obtained from a reaction of [Ni(cod)$_2$] with the phosphaalkyne P≡CBu^t: its structure was established by an x-ray study.[384] Various η^4-thiadiborole complexes have been synthesized.[391] The syntheses and structures of a series of heterocyclic sandwich compounds that contain boron, silicon and phosphorus, sulfur, or selenium atoms in the ring systems have been described, including the species [Ni(η^4-CMeCEtBEtSeSiMe$_2$)$_2$],[381] [Ni(η^4-CMeCEtBEtSSiMe$_2$)$_2$],[382] and [Ni(η^4-PPhBEtCEtCMeSiMe$_2$)$_2$].[383]

3.5 NICKEL COMPLEXES WITH η^5-LIGANDS

3.5.1 General Comments and Literature

This section is dominated by the chemistry of nickelocene and nickel cyclopentadienyl derivatives. In addition, there has been significant progress in the chemistry of substituted cyclopentadienyl derivatives of nickel, especially that of its pentamethylcyclopentadienyl complexes. The chemistry and electrochemistry of these and of other substituted cyclopentadienylnickel complexes such as Ni(η-C_5Ph_5) and Ni(η-C_5HPh_4) have played an increasingly important role. Nickel cyclopentadienyl groups appear in many metal clusters. These species are not covered comprehensively, except when the nickel atom(s) play key or unusual roles. While the vast majority of low-valent organometallic complexes are diamagnetic, a relatively large number of organometallic paramagnetic compounds contain Ni(η-C_5R_5)

(R = H, R) groups. These species include mononickel species, dinuclear complexes, and homo- or heteronuclear metal clusters.

The utility of nickelocene and other nickel sandwich compounds in the synthesis of nickel alkene complexes and the chemistry of nickel alkene complexes of the type [NiR(η^2-C_2H_4)Cp] are not discussed in this section. Catalytic applications of nickel η^5-ligand complexes are not covered and research in this area is alluded to only if new nickel η^5-ligand species were prepared. In addition, nickel η^5-ligand species that form part of a multidecker complex or a carborane are not reviewed here. These species are covered in Sections 3.7 and 3.8.

A vast body of chemistry exists for nickelocene. Reviews of π-complexes for the years 1981–1983 and 1986 include sections relevant to nickelocene and nickel cyclopentadienyl complexes.[1–4] A survey of applications of many metallocenes (including nickelocene) as homogeneous catalysts and as reagents has appeared.[392] A recent review of bulky or supracyclopentadienyl metal derivatives contains some references to nickel-substituted cyclopentadienyl complexes.[393] Monosubstituted Cp half-sandwich compounds have also been reviewed.[394]

3.5.2 Theoretical and Physical Studies of Nickel η^5-Ligand Complexes

3.5.2.1 Theoretical and gas-phase studies

The pentadienyl complex [Ni(η^5-C_5H_7)$_2$] was the subject of a semiempirical INDO study. The magnitude of reorganization effects, Koopman's defects, and the PE spectrum of the complex were calculated.[395] Gas-phase reactions of [$NiCp_2$] (and ferrocene) with Rh^+ and Ti^+ ions were studied by ion cyclotron resonance spectroscopy.[396] The proton affinities of [$NiCp_2$] and [NiCp(NO)] have been determined. In contrast with [$Ni(CO)_4$], H^+ probably protonates the Cp ligand in these species.[397]

An experimental study of the multiphoton ionization and dissociation of [$NiCp_2$] in effusive and supersonic molecular beams was documented. Bare metal ions were formed by both resonant and nonresonant mechanisms, and were detected by MS techniques.[398,399] The gas-phase reactions of $[NiCp]^+$ and other nickel species with aromatic compounds have been studied. The $[NiCp]^+$ group exhibited the most reactivity.[400] Other research targeted the electron impact of nickelocene and other metallocenes in the gas phase. Two distinct charge separation processes of the "coulombic explosion reaction" were observed as shown in Equations (10a) and (10b). Irrespective of M (M = Ni, Co, Fe) the MCp^+/Cp^+ distance was found to be 0.50 nm and the $[Cp–Cp]^+/M^+$ distance was 0.62 nm.[401] Gas-phase reactions of $[Co(isobutene)]^+$ ions with [$NiCp_2$] were reported and analyzed.[402]

$$[NiCp_2]^{2+} \longrightarrow [NiCp]^+ + Cp^+ \qquad (10a)$$

$$[NiCp_2]^{2+} \longrightarrow [Ni]^+ + [Cp–Cp]^+ \qquad (10b)$$

A mass spectral study, set up to study doubly charged ions, revealed that the only such species detectable for nickelocene was $[NiCp_2]^{2+}$.[403] The effects of the alkyl group on the fragmentation of various substituted nickelocene complexes [Ni(η-C_5H_4R)$_2$] (R = Et, Pr^n, Pr^i, Bu^n, Bu^t) in the mass spectrometer have been analyzed. Under electron impact conditions, when the R group contained a proton on the carbon atom α to the ring, fragmentation occurred such that hydrogen migration from one ligand to another and cleavage of the bond between the metal and the hydrogen acceptor ligand ensued.[404,405] A similar mass spectroscopy study was performed for nickelocene and the substituted nickelocenes [Ni(η-C_5H_4R)$_2$] (R = Me, Et, Bu^n, Bz). The ^1H NMR spectra of these species were noted.[406,407]

3.5.2.2 Other spectroscopic and magnetic studies

Solid samples of metallocenes and other half-sandwich compounds, including nickelocene, were studied by quasi-inelastic neutron scattering, calorimetry, and vibrational spectroscopy.[408] Reorientational processes in nickelocene were also studied by incoherent quasi-inelastic neutron-scattering techniques.[409,410] Electron transmission spectroscopy has been utilized to determine the energies of low-lying unstable negative ion states for [$NiCp_2$] and other metallocenes. Calculations using multiple scattering X_α procedures imply that the observed states can be assigned to electron capture into metal

3*d*, $4e_{1g}$, or Cp π*-orbitals.[411] The first gas-phase microwave measurement of nitrogen quadrupolar coupling in a transition metal complex has been reported: the high-resolution rotational spectrum of $[Cp^{58}Ni^{14}N^{16}O]$ has been measured.[412]

Nickelocene has a large zero field splitting (ZFS) parameter (*D*) that is not easily measured by conventional ESR spectroscopy. Values for *D* have been obtained from magnetic susceptibility (χ) measurements. Ferromagnetic coupling was observed for $[NiCp_2]$ in undiluted samples, and antiferromagnetic coupling in dilute samples. By doping up to 6% nickelocene into diamagnetic samples of $[MCp_2]$ (M = Fe or Ru), a value of $D_0 = +33.6 \pm 0.3$ cm^{-1} was found for the ZFS parameter of isolated nickelocene.[413] The same paper reports an independently determined value of $D = 31.6 \pm 1.0$ cm^{-1}, obtained from inelastic neutron-scattering experiments.[413] More recently, a value of $D = 32.5 \pm 0.2$ cm^{-1} was extracted from far IR spectroscopy of $[NiCp_2]$ at 4 K under magnetic fields of 0, 1, 2, and 4 T.[414] The fulvalene complex $[Ni_2(\mu\text{-}\eta^5,\eta^5\text{-}C_5H_4C_5H_4)_2]$ exhibits significant antiferromagnetic coupling.[415]

ESR studies of the d^7 species $[NiCp_2]^+$ revealed its Jahn–Teller degeneracy. The ESR spectrum varies according to the host lattice, in accord with predictions of a continuum of energy equivalent conformations along a circular Jahn–Teller valley.[416] Detailed UV–Vis spectroscopy, magnetic measurements, and electrochemical studies have been performed on $[NiCp^*_2]$ and on the monocation $[NiCp^*_2]^+$.[417–19] The ground state of the cation $[NiCp^*_2]^+$ was found to be $^2E_{1g}$ ($e_{2g}{}^4$, $a_{1g}{}^2$, $e_{1g}{}^1$): that of $[NiCp^*_2]$ was $^3A_{2g}$ ($e_{2g}{}^4$, $a_{1g}{}^2$, $e_{1g}{}^2$). These states are the same as was found for the Cp derivatives. The UV–visible spectra of each species were reported and *d–d* transition assignments were made. Ligand field analysis led to the extraction of Δ_1, Δ_2, and the Racah electron parameter *B*. These values were found to be greater than the corresponding values for the Cp complexes. Low-temperature solid and ambient-temperature solution magnetic studies of both species were reported. (These data and magnetic data from other paramagnetic nickel complexes are collected in Table 16.) The value obtained for $[NiCp^*_2]$ is close to the expected spin-only value for a molecule where $S = 1$. Below 25 K, χ_m is independent of temperature. No ESR signal could be detected for $[NiCp^*_2]$: this was attributed to the large value of the zero field splitting parameter for the molecule.[419]

Table 16 Magnetic data for some paramagnetic nickel species.[a]

Complex	μ_{eff} (μ_B) *solid*	θ	μ_{eff} (μ_B) *solution*	μ_{eff} (BM) *solid*	*Ref.*
[Ni(bipy)Cp]			1.69		420
[Ni(dppe)Cp]			1.76		420
$[NiCp^*_2]$	2.93(1) (6–100 K)	−15	2.89(10) (304 K)[b]		419
$[NiCp^*_2]^+ BF_4^-$	1.62(1) (6–57 K)	0	1.44(10) (304 K)[b]		419
$[NiCp^*_2]^+ PF_6^-$	1.67(1) (5–75 K)	28	1.44(10) (304 K)[b]		419
$[Ni(\eta\text{-}C_5HPh_4)_2]$	2.83(3) (50–300 K)	3			421
$[Ni_2(\mu\text{-}CO)_2(\mu\text{-}\eta^2,\eta^2\text{-}LH_2)(\mu\text{-}\eta^5,\eta^5\text{-}L)]$[c,d]				2.58	422
$[Ni_2(\mu\text{-}CO)_2(\mu\text{-}\eta^2,\eta^5\text{-}LH)_2]$[c,d]				2.58	422
$[NiBr_2(\eta^5\text{-}LH)]$[d]				1.82	422
00$[Ni(\eta^5\text{-}2,3,5\text{-}C_3B_3HMe_3Et_2)_2]$ (**104**)				3.06 ± 0.02	423

[a] Magnetic moments (errors in last digit(s)), θ values obtained from the linear portion of the $\chi_m{}^{-1}$ vs. *T* curve. [b] In acetone or toluene. [c] Mixture of two isomers. $L = C_{10}H_4Bu^t{}_4$ (Scheme 51).

The effect of peralkylation of various metallocenes of nickel (as well as vanadium, chromium, and cobalt) was studied by ^{1}H and ^{13}C NMR spectroscopy. These studies indicate that the reducing power of nickelocene was enhanced with alkylation. It was inferred that there is hindered rotation of the C_5 rings in $[Ni(\eta\text{-}C_5Et_5)]_2$. The β-carbon atom of the ethyl groups is believed to be pointing away from the metal.[424] The ^{1}H NMR spectra of the substituted nickelocene species $[Ni(\eta\text{-}C_5H_4R)_2]$ (R = Me, Et, Bu^n, Bz) have been recorded.[406,407] The paramagnetic NMR spectra of $[Ni_2Cp^*_2(\mu\text{-}\eta^5,\eta^{5'}\text{-}C_5Me_4C_5Me_4)]$ was also reported.[425]

Substituted nickelocenes (and other metallocenes with vanadium, chromium, and cobalt as the central atom) of general formula $[Ni(\eta\text{-}C_5H_4EMe_3)_2]$ (E = C, Si, Ge, Sn) were reported. The paramagnetic ^{1}H ^{13}C NMR spectra of these complexes were measured. The chemistry of the substituted nickelocenes was similar to that of $[NiCp_2]$: only the complex $[Ni(\eta\text{-}C_5H_4SnMe_3)_2]$ exhibited significant differences, owing to reaction at the polar tin–carbon bond.[426] An earlier paper concerns the paramagnetic ^{29}Si and ^{119}Sn NMR spectra of $[Ni(\eta\text{-}C_5H_4TMS)_2]$, $[Ni(\eta\text{-}C_5H_4SnMe_3)_2]$, and of the vanadium, chromium, and cobalt analogues.[427] The NMR spectra of other substituted metallocenes were studied by the same research group.[428]

The ^{1}H and ^{13}C NMR spectra of $[Ni(\eta\text{-}C_5Me_4Et)_2]^+PF_6^-$ and of the metallocenium cations of other metals have been measured and from the signs of the signal shifts for the nickel cation, it was deduced that the spin density in the η-C_5Me_4Et orbitals was positive for the nickel complex. NMR data are summarized in Table 17.[429]

Table 17 ^1H and ^{13}C NMR data[a] for $[Ni(\eta\text{-}C_5Me_4Et)_2]^+PF_6^-$.

H or C atom	*^1H chemical shift*	*^{13}C chemical shift*
1		521
2, 5		575
3, 4		564
20, 50	103.7	−269.4
30, 40	102.6	−263.6
α	41.9	−185.8
β	8.7	546.1

[a] Data from Ref. 429. δ (ppm) in acetone-d_6 at 298 K. Carbon atoms 1–5 are the aromatic ring atoms, (C-1 is the C(Et) carbon). C-20–C-50 are methyl group carbon atoms attached to C-2–C-5. α and β are the ethyl carbon atoms. H numbering scheme based on C scheme.

The frequencies and normal vibrational modes of $[NiCp_2]$ and other metallocenes have been reported.[430–3] The IR and Raman spectra of $[NiCp_2\text{-}d_{10}]$ were recorded over the 30–300 K temperature range, and new assignments were proposed for some internal modes.[434]

An x-ray diffraction study of the disordered phase of $[NiCp_2]$ was reported at 295 K. The molecular packing of the metastable ordered phase at 5 K was determined by lattice energy minimization techniques.[435] An analysis of the relationships between the crystal and the molecular structures of $[MCp_2]$ (M = Ni, Fe, Ru) has been published.[436] Various differences in the crystal structures and the phase changes undergone by these metallocenes are explained.[436] Solvent and steric effects are more important than IP (ionization potential) effects of guest molecules in forming interchelates of $[NiCp_2]$ in FeOCl matrices.[437,438]

3.5.2.3 Electrochemical studies

The vast majority of electrochemical studies of nickel organometallic π-complexes have been carried out on nickelocene, on its substituted derivatives, and on other NiCp complexes. Alkylation of the cyclopentadienyl rings of $[NiCp_2]$ leads to more facile oxidation of the complex, a not unsurprising result attributed to the increased electron density of the alkylated complex. Nickelocene undergoes up to a two-electron reduction and a two-electron oxidation. The fifth redox state for $[NiCp_2]$, $[NiCp_2]^{2-}$, was established only recently. A review on the electrochemistry of mononuclear transition metal complexes contains material of interest.[439]

The electrochemistry of nickelocene was studied in neutral 1:1 molar ratios of $AlCl_3/[Bupy]^+Cl^-$ at 40 °C. Both $[NiCp_2]$ and $[NiCp_2]^+$ were found to be unstable in chloride-rich melts. Spontaneous oxidation of $[NiCp_2]$ to $[NiCp_2]^+$ occurred in acidic melts, and a stable dication formed reversibly at $E_{1/2} = +0.912$ V. For the $[NiCp_2]/[NiCp_2]^+$ couple, $E_{1/2} = -0.165$ V. The electronic spectra of these nickel(II), nickel(III), and nickel(IV) species were also recorded. The redox potentials for a series of complexes of the type $[NiL_2Cp]^+$ (L = PR_3, $P(OR)_3$; L_2 = cod, η^4-norbornadiene, dppe) were also determined by cyclic voltammetry in this medium.[440]

The cyclic voltammetry and the controlled potential electrolysis of nickelocene under a CO atmosphere were investigated. The sequence of reductions is shown in Equation (11). The electron self-exchange rate in the $[NiCp_2]/[NiCp_2]^+$ couple was investigated by NMR spectral line broadening techniques. Activation parameters were found to be as follows: $\Delta H^{\ddagger} = 6.9$ kJ mol^{-1}; $\Delta S^{\ddagger} = -63$ J mol^{-1} K^{-1}. The anionic species $[NiCp_2]^-$ could be produced electrochemically, but this anion is unstable at ambient temperatures and spontaneously decomposed to Cp^- and [NiCp].[441]

$$[NiCp_2] \xrightarrow{e^-} [NiCp_2]^- \xrightarrow[-Cp^-]{CO} [Ni(CO)Cp] \xrightarrow{e^-} [Ni(CO)Cp]^- \xrightarrow[-Cp^-]{CO} [Ni(CO)_4] \quad (11)$$

The most recent electrochemical study of nickelocene details the detection of five different oxidation states for nickel in various oxidized or reduced forms of nickelocene. For the first time, the dianion $[NiCp_2]^{2-}$ was detected. The anion $[NiCp_2]^-$ (unstable at room temperature) was found to be stable on a cyclic voltammetry timescale below −80 °C. By using cyclic voltammetry with platinum microelectrodes and $[NBu_4]^+PF_6^-$ in THF at temperatures below −60 °C, $[NiCp_2]^-$ could be reduced at very negative potentials to $[NiCp_2]^{2-}$. Based on the linear relationship between the ionization potential of a particular species and its E_0 value,[442,443] the $[NiCp_2]^-/[NiCp_2]^{2-}$ redox couple was predicted to have a value of −3.32 ± 0.1 V. Experimental values of −3.04 V (−40 °C) and −3.13 V (−60 °C) are in good agreement with the predicted E_0 value. Table 18 lists the various $[NiCp_2]^{n+}/[NiCp_2]^{(n-1)+}$ (n = −1, 0, 1, 2) redox couples for all five nickel oxidation states.[444]

Table 18 Redox potentials (V) at various temperatures and ionization potentials (eV) for $[NiCp_2]^{n+}/[NiCp_2]^{(n-1)+}$ species.

	$[M]^{2+}/[M]^{+\,a}$	$[M]^+/[M]$	$[M]/[M]^-$	$[M]^-/[M]^{2-}$
E_0 (20 °C)	0.81	0.01	−1.85	
E_0 (−40 °C)	0.81	0.01	−1.84	−3.04
E_0 (−80 °C)	0.81	0.01	−1.84	−3.13
IP $[M]^{n+}$	54.9	35.2	18.2	7.6

Source: Bard *et al.*[444]
[a] M = $NiCp_2$.

Cyclic voltammetry studies were carried out on $[Ni(PPh_3)_2Cp]^+BF_4^-$. A cathodic wave was noted at −1.25 V and a much smaller anodic wave was seen at −1.18 V ($i_p^a/i_p^c < 0.6$). Another anodic wave at −0.35 V was attributed to nickelocene. The bis(ylide) complex $[Ni(CH_2PPh_3)_2Cp]^+$ is also oxidized (E_0 = −0.21 V vs. Ag/AgCl). When at least a 10-fold excess of PPh_3 is present, the reduction of $[Ni(PPh_3)_2Cp]^+$ to the neutral nickel(I) species $[Ni(PPh_3)_2Cp]$ is fully reversible: otherwise the disproportionation equilibrium, shown in Equation (12), set up between the nickel(I) complex and the nickel(0) and nickel(II) compounds $[Ni(PPh_3)_4]$ and $[NiCp_2]$ lies to the right.[420] Electrochemical studies on other nickel cyclopentadienyl compounds of the type $[NiL_2Cp]^+$ (L = PR_3, $P(OR)_3$; L_2 = dppe, dapm, cod, nbd) have been reported.[94,95,445]

$$[Ni(PPh_3)_2Cp] \rightleftharpoons [Ni(PPh_3)_4] + [NiCp_2] \qquad (12)$$

The cyclic compounds $[Ni\{\eta^1,\eta^1\text{-}Ph_2PCH{=}C(O)R\}(\eta\text{-}C_5Ph_5)]$ (R = Ph, $(\eta\text{-}C_5H_4)FeCp$) exhibit different behavior upon electrochemical oxidation. The R = Ph complex undergoes a one-electron oxidation (at platinum electrodes) to form a nickel(III) species, which is stable for up to 6 h in solution. Reduction of the ferrocenyl complex initially involves the Fc center, followed by an intramolecular redox reaction to afford an Ni^{3+} species. This species has a lifetime of ~1 h and like the phenyl species undergoes an irreversible chemical reaction. The species in each case is believed to be a nickel(II) radical cation. The species were also investigated by thin-layer spectroelectrochemistry and ESR.[446]

The electrochemistry of the pentalene complex (**65**) was investigated. For the Ni^{II}, Ni^{II}/Ni^{II}, Ni^{III} couple, $E_{1/2}$ = −1.245 V, while for the Ni^{II}, Ni^{III}/Ni^{III}, Ni^{III} couple, $E_{1/2}$ = −0.405 V. These readings were obtained at a sweep rate of 20 mV s^{-1} in CH_2Cl_2 with $[NBu^n_4]^+[ClO_4]^-$ as the supporting electrolyte at 20 °C, relative to the Ag/AgNO$_3$ couple. A ΔE value of 0.830 V between these two couples was obtained from differential pulse voltammetry.[447]

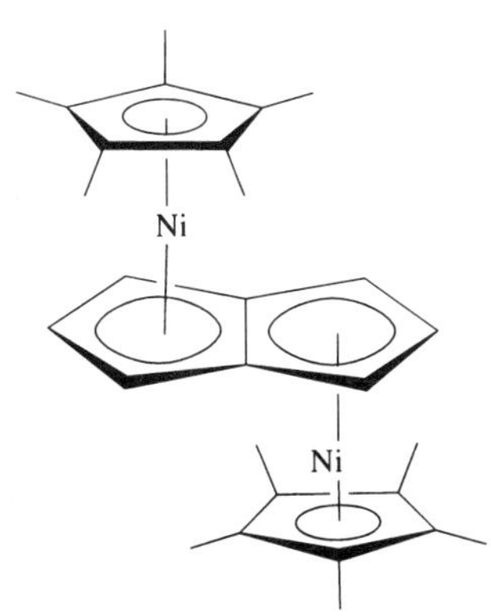

(65)

Cyclic voltammetry studies of [Ni(η-C_5HPh_4)$_2$] led to the isolation of the oxidized salt [Ni(η-C_5HPh_4)$_2$]$^+PF_6^-$. The neutral sandwich complex possesses a partly reversible reduction wave, a reversible one-electron oxidation, and a second partly reversible oxidation. The magnetic susceptibility of this complex, measured over the 50–300 K range, exhibited $\mu_{eff} = 2.83 \pm 0.03$ ($\theta = 3$) over the linear portion of the Curie–Weiss plot.[421] The complex [Ni(S_2CNMe_2)(η-C_5Ph_5)], which has no Cp analogue, was also synthesized. This species may be oxidized by a reversible electrochemical process to its one-electron oxidation product [Ni(S_2CNMe_2)(η-C_5Ph_5)]$^+BF_4^-$.[448]

The species originally formulated as [Ni_2(μ-C_2S_4)Cp*$_2$] can be oxidized to a monocation ($E_0 = 0.09$ V) and may also be reduced to a monoanion and a dianion ($E_0 = -0.92$ V and -1.43 V, respectively).[449] More recently, this complex was reformulated as [(Ni_3Cp*$_3$)$_2$(μ-C_2S_6)]. The structures of the analogous Cp and Cp' derivatives were established crystallographically.[450]

The decamethylbimetallocene complex [Ni_2Cp*$_2$(μ-η^5,η$^{5'}$-$C_5Me_4C_5Me_4$)] was prepared and its electrochemistry and paramagnetic NMR spectra were investigated.[425] The complex exhibits strong deviations from Curie behavior—all the absolute chemical shift values of the various protons in the ^{1}H NMR of the complex increase with temperature after passing through a minimum. This behavior is characteristic of strong antiferromagnetic coupling. The complex could be oxidized stepwise to give a tetracation: the various redox potentials, established by cyclic voltammetry, are listed in Table 19.[425]

Table 19 Cyclic voltammetry data for [Ni_2Cp*$_2$(μ-η^5,η$^{5'}$$C_5Me_4$-$C_5Me_4$)].[a]

Redox couple	(V)
Ni_2/Ni_2^+	0.320
Ni_2^+/Ni_2^{2+}	0.705
Ni_2^{2+}/Ni_2^{3+}	1.305[b]
Ni_2^{3+}/Ni_2^{4+}	1.495[b]

[a] Data from Ref. 425, in EtCN, 0.1 M $Bu^n_4NPF_6$, Pt disk; $E_{1/2}$ relative to internal $Cp_2Co^{0/+}$ at −21 °C, scan rate 200 mV s^{-1}, $\Delta E_p = 65$ mV. [b] Scan rate = 50 mV s^{-1}, $\Delta E_p = 125$ mV and 65 mV.

The thiolato-bridged complexes [Ni_2(μ-SR)$_2Cp_2$] (R = Ph, CH_2Ph, 2-nap, n-C_8H_{17}; $(SR)_2 = C_6H_4(CH_2S)_2$-1,2) undergo a reversible one-electron reduction; no carbon–sulfur bond cleavage occurred and E_0 values (vs. an Ag/$AgNO_3$ electrode) ranged from −1.41 V (R = Ph) to −1.63 V ($(SR)_2 = C_6H_4(CH_2S)_2$-1,2).[451,452]

3.5.3 The Chemistry of Nickelocene and Substituted Nickelocenes

3.5.3.1 [NiCp$_2$] and monosubstituted cyclopentadienyl sandwich compounds

The physical and electrochemical data for these species have been summarized in Section 3.5.2. A list of nickel η^5-ligand sandwich compounds and related species appears in Table 20. Nickelocene has been prepared by the reaction of [$NiCl_2$] and C_5H_6, in DMSO, using diethylamine as a base.[453,454]

The oxidation of various complexes, including [$NiCp_2$] with [$FeCp_2$]$^+BF_4^-$, has been reported. Nickelocene afforded high yields of [$NiCp_2$]$^+BF_4^-$.[455] Nickelocene has been prepared electrochemically in modest yields[456] and a new efficient synthesis of this sandwich compound (from $NiCl_2$ and $MgCp_2$) has been described.[457] The structure of nickelocene has been redetermined.[458]

Nickelocene affords polymeric products when treated with dithiols.[469] A mixture of [$NiCp_2$] and $LiAlH_4$ has been found to be an effective reducing and desulfurizing agent for a variety of organosulfur species.[470,471]

Good yields of the substituted nickelocene complexes [Ni(η-C_5H_4R)$_2$] have been reported even for bulky R groups, by synthesizing THF-solvated nickel(II) chloride [$NiCl_2(THF)_x$] from the elements in THF and reacting this species with $C_5H_4R^-$ ions.[460] The syntheses of [Ni(η-C_5H_4R)$_2$] (R = COMe, CO_2Me) were reported.[459] The chemistry of the substituted nickelocenes [Ni(η-$C_5H_4EMe_3$)$_2$] (E = C, Si, Ge, Sn) was similar to that of [$NiCp_2$]: only the complex [Ni(η-$C_5H_4SnMe_3$)$_2$] exhibited significant differences owing to reaction at the polar tin–carbon bond.[426]

Table 20 NiCp and substituted cyclopentadienyl sandwich compounds.[a]

Complex	*Ref.*
[NiCp$_2$]	458
[NiCp]$^{n+}$ (n = 2, 1, 0, −1, −2)	444
[NiCp]$^{n+}$ (n = 1, 0, −1)	441
[Ni(η-C$_5$H$_4$R)$_2$] (R = COMe, CO$_2$Me)	459
[Ni(η-C$_5$H$_4$R)$_2$] (R = Et, Pr, Pr^i, Bu, Bu^t)	405
[Ni(η-C$_5$H$_4$R)$_2$] (R = H, Me, Et, Bu, CH$_2$Ph)	406, 407, 460
[Ni(η-C$_5$H$_4$R)$_2$] (R = Bu^t, Et)	52
[Ni(η-C$_5$H$_4$EMe$_3$)$_2$] (E = C, Si, Ge, Sn)	426
[Ni(η-C$_5$H$_4$EMe$_3$)$_2$] (E = C, Si, Ge, Sn)	427
[NiCpCp*]	461
[Ni(η-C$_5$HPr^i$_4$)$_2$]	51, 421
[Ni(η-C$_5$HPr^i$_4$)$_2$]$^+$PF$_6^-$	421
[Ni(η^5-C$_9$H$_7$)$_2$]	462
[NiCp*$_2$]	419, 461
[NiCp*$_2$]$^+$PF$_6^-$	419, 461
[NiCp*$_2$]$^{2+}$(PF$_6^-$)$_2$	461
[Ni(η-C$_5$Me$_4$Et)$_2$]	424
[Ni(η-C$_5$Me$_4$Et)$_2$]$^+$PF$_6^-$	429
[Ni$_2$(μ-η^5,η$^{5'}$C$_5$H$_4$C$_5$H$_4$)$_2$]	415, 463
[Ni$_2$Cp$_2$(μ-η^5,η$^{5'}$C$_5$H$_4$C$_5$H$_4$)]	464
[Ni$_2$Cp*$_2$(μ-η^5,η$^{5'}$C$_5$Me$_4$C$_5$Me$_4$)]	425
[Ni$_2$(μ-η^5,η$^{5'}$-C$_5$H$_2$Me$_2$C≡CC$_5$H$_2$Me$_2$)Cp*$_2$]	465
[Ni$_2$Cp$_2$(η^5,η$^{5'}$-C$_8$H$_6$)]$^{n+}$ (n = 0, 1, 2) (**65**)[b]	447
[Ni(η^5-C$_3$Et$_2$HB$_3$Me$_3$)Cp]	98
[Ni(η^3-C$_4$Ph$_4$HCO)Cp]	123
[Ni(η^4-C$_4$R^1$_4$)(η^5-C$_5$R^2$_5$)]$^+$ (R^1 = Me, Ph; R^2 = H, Me)	333
[Ni(η^4-C$_5$Me$_5$R)Cp*]$^+$Cl$^-$ (R = H, CCl$_3$, CH$_2$Ph)	318
[Ni(η^3-C$_5$Me$_5$R^{1}R^2)Cp*] (R^1 = H, R^2 = H, Me, CN; R^1, R^2 = CN)	318
[Ni(η^4-C$_5$H$_4$CPh$_3$)Cp]$^+$BF$_4^-$	466
[Ni(η^5-C$_4$H$_4$BX)$_2$] (X = F, **Cl**, Br, I, OH)	467
[Ni(η^5-C$_4$H$_4$BNR$_2$)$_2$] (R = Me, Et, Pr^i)	467
[Ni(η^5-C$_4$H$_4$BR)$_2$] (R = H, Me, Bu, Bu^t, Ph, OH, OMe, OC$_2$Ph)	467
[Ni(η^3-P$_3$)Cp*]	351
[Ni(η^3-P$_3$)(η-C$_5$H$_3$Bu^t$_2$-1,3)] (E = P, As)	352
[Ni(η^3-E$_3$)(η-C$_5$HPh$_4$)] (E = P, As)	353
[Ni(η^3-P$_2$C$_3$Bu^t$_3$)(η^5-P$_3$C$_2$Bu^t$_2$)]	321
[Ni(η^3-C$_5$H$_4$Bu^t$_3$-1,2,4)(η-C$_5$H$_2$Bu^t$_3$-1,2,4)]	468

[a] Compounds in bold were characterized by an x-ray diffraction study. Multidecker sandwich and borane/carborane complexes are listed in Tables 27 and 28. [b] C_8H_6 = pentalene (Structure (**65**)).

3.5.3.2 Polysubstituted cyclopentadienyl nickel sandwich compounds and fulvalene species

The potassium salt of $C_5H_2Ph_4$ reacts with [Ni(NH$_3$)$_6$]Cl$_2$ to afford the tetraphenylcyclopentadienyl sandwich species [Ni(η-C$_5$HPh$_4$)$_2$]. The magnetic susceptibility, the ESR spectrum, and the x-ray structure of this molecule were reported. Average Ni–C and C_{Cp}–C_{Cp} distances in this species are 0.222(1) nm and 0.141(1) nm, respectively (cf. values for nickelocene of 0.216 4(7) nm and 0.138 1(1) nm, respectively[458]). Cyclic voltammetry studies led to the isolation of the oxidized salt [Ni(η-C$_5$HPh$_4$)$_2$]$^+$PF$_6^-$.[421] The reaction of C$_5$Ph$_5$Br with [Ni(CO)$_4$] afforded an entry into Ni(η-C$_5$Ph$_5$) group chemistry. Species such as [Ni$_2$(μ-X)$_2$(η-C$_5$Ph$_5$)$_2$] (X = Cl, Br) and [NiX(CO)(η-C$_5$Ph$_5$)] (X = Cl, Br, I), [Ni(MeCN)$_2$(η-C$_5$Ph$_5$)]$^+$BF$_4^-$ and the mixed-ligand complex [Ni(η-C$_5$Ph$_5$)Cp] were prepared.[471,472] Later the complex [Ni(S$_2$CNMe$_2$)(η-C$_5$Ph$_5$)], which has no Cp analogue, was reported. Its cationic one-electron oxidation product [Ni(S$_2$CNMe$_2$)(η-C$_5$Ph$_5$)]$^+$ BF$_4^-$ was synthesized (by a reversible electrochemical or chemical oxidation) and its magnetic susceptibility was measured. The structure of the salt was established by x-ray diffraction.[448]

Various syntheses of [NiCp*$_2$] have been described.[318,419] These are usually based on the reactions of nickel(II) salts with Cp*$^-$ anion sources. The metallocene may be oxidized chemically, using [FeCp$_2$]$^+$ X$^-$ (X = BF$_4$ or PF$_6$) as the oxidant, yielding [NiCp*$_2$]$^+$X$^-$ as brown prisms in high (90%) yields. Dicationic [NiCp*$_2$]$^{2+}$(PF$_6$)$^-_2$ was also prepared chemically by oxidation of [NiCp$_2$*] with HgCl$_2$. The orange diamagnetic complex, obtained in 38% yield, decomposes slowly (over a week) in air, under an argon atmosphere, or under vacuum. It is stable for days in an acidic solution, and its ^{1}H and ^{13}C NMR spectra (as a Cl$^-$ salt in 0.1 M HCl) were recorded.[419] The oxidation of [Ni(η-C$_5$Me$_4$Et)$_2$] has been carried out chemically, by using $AgPF_6$ as the oxidant. The cationic nickel complex $[Ni(\eta\text{-}C_5Me_4Et)_2]^+ PF_6^-$ was isolated in 64% yield as a yellow-brown solid.[429]

A series of pentamethylcyclopentadienyl metallocenes including $[NiCp^*_2]$ were prepared and characterized and their electrochemistry was studied. It was shown that $[NiCp^*_2]$ is resistant to hydrolysis and ring exchange. Cyclic voltammetry shows that the complex readily undergoes one-electron oxidation to $[NiCp^*_2]^+PF_6^-$; dicationic $[NiCp^*_2]^{2+}(PF_6^-)_2$ species could also be prepared. The reaction of $[NiCp^*_2]$ with a variety of halide sources leads to electrophilic attack on one of the Cp* rings.[461]

A comparative structural study of the η^5-indenyl complexes $[M(\eta^5\text{-}C_9H_7)_2]$ (M = Fe, Co, Ni) was undertaken to examine slip–fold distortion effects. In contrast with what is observed for the respective metallocenes, an increase in $\eta^5 \rightarrow \eta^3$ slip distortion is observed on descending the series Fe → Co → Ni (nickel is most distorted in an η^3-fashion).[462]

Various fulvalene and related compounds have been reported. The structure of the diamagnetic fulvalene complex $[Ni_2(\eta^5,\eta^{5'}\text{-}C_5H_4C_5H_4)_2]$ was determined. The complex is isomorphous with its iron analogue.[463] The complex $[Ni_2(\eta^5,\eta^{5'}\text{-}C_5H_2Me_2C{\equiv}CC_5H_2Me_2)Cp^*_2]$ was also examined in an x-ray study.[465] Dinickelocene $[Ni_2(\mu\text{-}\eta^5,\eta^{5'}\text{-}C_5H_4C_5H_4)Cp_2]$ has been synthesized.[464] Its permethylated analogue, the decamethylbimetallocene complex $[Ni_2Cp^*_2(\mu\text{-}\eta^5,\eta^{5'}\text{-}C_5Me_4C_5Me_4)]$, was prepared and its electrochemistry and paramagnetic NMR spectra were investigated.[425]

A series of interesting boracyclopentadienyl complexes, **(66)**, with a wide variety of substituents have been prepared. When complex **(66)** (X = Cl) was treated with $LiO(CH_2)_3OLi$, the linked boracyclopentadienyl species **(67)** was obtained, as shown in Scheme 38.[467] Rotational barriers in sandwich compounds of the type $[M(\eta^5\text{-}C_4H_4BR)_2]$ (M = Ni, Pd, Pt) have been measured.[474]

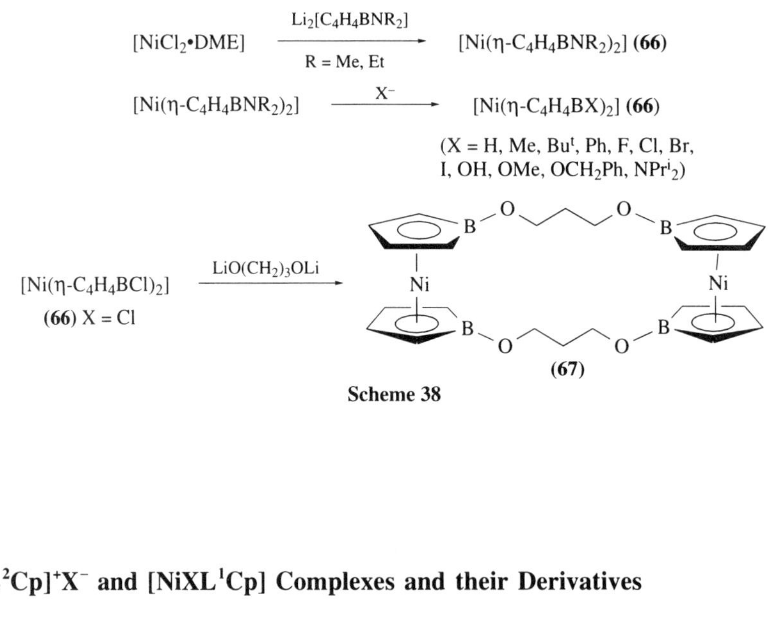

Scheme 38

3.5.4 $[NiL^1L^2Cp]^+X^-$ and $[NiXL^1Cp]$ Complexes and their Derivatives

3.5.4.1 General comments

A large body of chemistry has been developed for the nickelocene derivatives $[NiL^1L^2Cp]$ and $[NiXL^1Cp]$ (L^1 = neutral donor ligand; X = anionic ligand). Some 19-electron species of formulation $[NiL^1L^2Cp]$ are also known (L = two-electron donor ligand). There is often a delicate balance between neutral $[NiXL^1Cp]$ species and the corresponding cationic $[NiL^1L^2Cp]^+X^-$ complexes (L^1 = two-electron donor; X = halogen) that is dependent on the concentration of L^1, the solvent, and the temperature. Compounds of general formulae $[NiL^1L^2Cp]^+X^-$ and $[NiXL^1Cp]$ and their substituted Cp derivatives are collected in Table 21. Dinuclear compounds with NiCp groups are collected in Table 22.

The most commonly used synthetic entry into $[NiL^1L^2Cp]$ species is by displacement of one of the cyclopentadienyl groups in nickelocene with the appropriate ligand. Cyclopentadienyl ring displacement from nickelocene is facilitated by protonation of nickelocene with HBF_4 at low temperature: the resulting cationic species $[Ni(\eta^4\text{-}C_5H_6)Cp]^+BF_4^-$ ($(\mathbf{68}^+)BF_4^-$) is a useful source of the $[NiCp]^+$ fragment in the syntheses of $[NiL^1L^2Cp]$ or $[NiXLCp]$ complexes.[118] A large body of work describes the reactions of nickelocene and the cationic species ($\mathbf{68}^+$) with two-electron donor ligands. These ligands include

Table 21 $[NiXL^1(\eta\text{-Cp})]$ and $[NiL^1L^2(\eta\text{-Cp})]^+Y^-$ complexes.[a]

L^1,L^2	*X or Y*	*Ref.*
CO, PR_3 (R = Pr^i, Cy, Tol-2)	BF_4	475
CO	Me	476
CO	**$Ge(CH_2Bu^t)_2THF$**	476
*CO	Br	477
*CO	X (X = Br, $SnPh_3$)	478
*CO	M^+ (M = Li, Na)	478
CO^b	Cl	479
MeCN	Me	216
C_nH_{2n} (n = 2–4, 6)	Me	73
C_2H_4	R (R = Et, Pr^i, $c\text{-}C_3H_5$, Bu^n, Bu^i)	53
$C_2H_4{}^c$	Et	51–3
C_2H_4	C_2H_4R (R = Ph, $c\text{-}C_3H_5$)	57
$*C_2H_4$	**Ph**	50, 69
$*C_2H_4$	$CH{=}CH_2$	50
RC_2R (R = Me, TMS)	Me	13, 216
C_3H_6	R (R = Me, Me-d_3, Et, Pr, TMS)	72
C_3H_6	Pr^i	53
*η^4-cod	X (X = Cl, Br, PF_6)	461
$\eta^4\text{-}C_5H_6$	BF_4	118
PH_3	I	480
2 PH_3	BF_4	480
2 PMe_3	**Cl**	481
2 PMe_3	BF_4	129
PMe_3	CH_2Tol-2	342
PMe_3	$CMe{=}NBu^t$	173
PMe_3	$=CMe(NHBu^t)$	173
PMe_3	CH_2X (X = CN, OMe, CO_2Et)	483
PMe_2R (R = H, Me)	$W(CO)_3Cp'$	483
$*PR_3$	$P{=}NBu^t$	484
2 $PBu^n{}_3$	BF_4	485
2 PR_3 (R = Pr^i, Bu^t, Cy)	BF_4	475
PR_3 (R = Pr^i, Cy), $P(S)Me_3$	BF_4	475
PR_3 (R = Ph, Cy)	X (X = Cl, Br, I)	486
$P(NR_2)_3$ (R = Me, Et)	X (X = Me, Cl, Br)	487
2 $P(NR_2)_3$ (R = Me, Et)	PF_6	487
PCy_3	Cl	475
$*PHPh_2$	$W(CO)_3Cp$	483
2 PPh_2Me	BF_4	486
PPh_3	$CH_2C(O)R$ (R = Bu^t, OBu^t, Ph)	489
$*PPh_3$	**$CH_2C(O)OBu^t$**	488
PPh_3	$CHMeC(O)Bu^t$	488
PPh_3	$CHRSO_2Ph$ (R = $n\text{-}C_{11}H_{23}$)	489
PPh_3	**$CHRSO_2Ph$** (R = **H**, Et, $n\text{-}C_6H_{11}$, $n\text{-}C_{11}H_{23}$, **$CH{=}CMe_2$**)	490
PPh_3	Ph	491
PPh_3	$C{\equiv}CPh$	492
PPh_3	$C[{=}C(CN)_2]C(Ph){=}C(CN)_2$	493
PPh_3	$c\text{-}CHCR^1C({=}O)C(R^2)(R^3)$	493
PPh_3	Me	494
PPh_3	R (R = Me, Et)	495
PPh_3	$c\text{-}C{=}C(R^1)C(O)CR^2R^3$	493
PPh_3	CH_2SPh	496
PPh_3	SPh	496
PPh_3	I	497
$PPh_3{}^b$	X (X = Cl, Br, I)	479
$PPh_2(CH_2)_nCl$ (n = 1, 3, 4)	I	497
$PPh_2(CH_2)_n$ (n = 3, 4)		497
2 PPh_3	$C_5H_4PPh_3$	498
PPh_3, LCH_2 (L = Ph_3P, Ph_3As Me_2S, $Me_2S(O)$)	Br	499
$CHPPh_3(CH_2)_nCHPPh_3$ (n = 1–3)		498
'PPh_3	I	500
$*PPh_3$	X (X = Cl, Br)	461
$P(Tol\text{-}2)_3$	I	475
$P(Tol\text{-}2)_3$, SMe_2	BF_4	475
P_2Ph_4	X (X = Cl, I)	501
2 P_2Ph_4	BF_4	501
dppm or dppe	Cl or Br	502
dppe	BF_4	486

Table 21 (continued)

L¹,L²	*X or Y*	*Ref.*
η^1(*P*)-dapm	Cl	503
arphos	Cl	503
2 arphos	Cl	503
c-*P*(Cl)N(Bu^t)P(Cl)N(Bu^t)	Cl	504
2 *c*-*P*(Cl)N(Bu^t)P(Cl)N(Bu^t)	BF_4	504
2 $Ph_3PCHCOPh$ or 2 Ph_3PCHMe	Br	498
P(O)$(OR)_2$H	P(O)$(OR)_2$	505
2 R_3PSe (R = Me, Ph)	BF_4	506
$[Ph_2P(Se)]_2(CH_2)_n$ (*n* = 0–2)	BF_4	506
2 *c*-*N*HCH₂CH₂C(O)	**PF_6**	507
2 SbR_3 (R = Me, Bu^n, Ph)	BF_4	485
SbR_3 (R = Me, Bu^n, Ph)	Me	485
2 $AsPPh_4$	BF_4	508
$AsPPh_4$	Cl	508
$E^1E^2Ph_4$ (E^1, E^2 = As, Sb)	X (X = I, BF_4)	508
2 $Ph_2\mathit{Sb}CH_2SbPh_2$	BF_4	509
$Ph_2\mathit{Sb}CH_2SbPh_2$	I	509
$Ph_2\mathit{Sb}(CH_2)_n\mathit{Sb}R_2$ (*n* = 1, 3)	BF_4	509
$TeMe_2$	I	509
2 EMe_2 (E = S, Se, Te)	BF_4	475, 509
Te_2Me_2	BF_4	485
$Ph\mathit{S}(CH_2)_n\mathit{S}Ph$ (*n* = 1, 2)	BF_4	510
$Me_2\mathit{N}C_2H_4\mathit{N}Me_2$	BF_4	511
$C_6H_4(\mathit{N}Me_2)_2$-1,2	BF_4	511
2 C_6F_5[d]	**$[NEt_4]^+$**	487
2 $Ge(C_6F_5)_3$[d]	**$[Cp_3Ni_2]^+$**	512
2 PPh_2Me[e]		486
2 PR_3[e] (R = Ph, Bu^n, OMe; R_3 = $MePh_2$, Me_2Ph, $PhBu^n{}_2$, Ph_2OMe)		420
L_2[e] (L_2 = dppe, arphos, **bipy**, phen)		420
dppe[e]		486
(*S*,*S*)-S_2CNMe_2[f]	**BF_4**	448
Br[e,g]	Br	422
cod[h]	BF_4	98
L¹ and X		
*acac		513
$C_6H_4N{=}NPh$		514
$N_4(Tol\text{-}4)_2$		204
η^3-PMes*CHPMes*[i]		207
η^5-C_5H_7		305
NO		515
$\mathit{P}Ph_2CH{=}C(\mathit{O})R$[f] (R = Ph, ($\eta$-$C_5H_4$)FeCp)		446
η^3-$C_3H_3(TMS)_2$-1,1		216
η^1,η^2-$\mathit{C}HCMe_2CHCR^1R^2\mathit{C}R^3{=}\mathit{C}HR^4$		136
η^1,η^2-$\mathit{C}H_2CH_2CH_2\mathit{C}H{=}\mathit{C}H_2$		138
η^1,η^2-$\mathit{C}H_2CH_2CR^1R^2\mathit{C}R^3{=}\mathit{C}H_2$ (R^1, R^2, R^3 = H, Me)		134, 135
η^1,η^2-$C_7H_5(CO_2Me)_2$		130
η^1,η^2-$C_7H_8PMe_3$		129
η^1,η^2-$C_{16}H_{17}Cl_4O_4$		132
*η^1,η^2-$\mathit{C}H_2CH_2\mathit{C}H{=}\mathit{C}H_2$		50
η^1,η^2-$\mathit{C}(CO_2Me){=}C(CO_2Me)PR_2\mathit{S}$		516
η^1,η^2-$C_6H_4(CH_2NMe_2)$-2		517
η^2,η^5-$CH_2{=}CH(CH_2)_3C_5Me_4Br$		58

[a] Compounds in bold were characterized by an x-ray diffraction study. Italicized atoms are bonded to the nickel atom. Complexes with Cp′ or Cp* (instead of Cp) ligands are denoted (′) or (*), respectively. [b] Cp = η^5-$C_5H_2(TMS)_3$-1,2,4. [c] Cp = η-C_5H_4Et or η-$C_5H_4Bu^t$. [d] Anionic species; cation shown under the X or Y column. [e] Paramagnetic 19-electron complex. [f] Cp = η^5-C_5Ph_5. [g] Cp = η^5-$C_{10}H_4Bu^t{}_4$ (Scheme 51). [h] Cp = η^5-$C_3Et_2HB_3Me_3$. [i] Mes* = $C_6H_2Bu^t{}_3$-2,4,6.

mono- and bidentate molecules that contain group 15 (N, P, As, Sb) or group 16 (S, Se, Te) elements. Reactions of chelating ligands, with potentially coordinating atoms being selected either from within each of the above groups (e.g., P and As) or from both groups (e.g., P and Se), have been looked into.

Compounds in this class have also been prepared from $[NiX_2L_2]$ (X = halogen) complexes. These may be generated *in situ*. Thus, for example, the complex $[NiPh(PPh_3)Cp]$ may be prepared by reacting nickel(II) chloride with a PhMgX (X = halide) in the presence of cyclopentadiene and PPh_3.[490] The

Table 22 Dinuclear complexes with nickel η^5-ligands.[a]

Compound	*Ref.*
[Ni$_2$(μ-Cl)$_2${η-C$_5$H$_2$(TMS)$_3$-1,2,4}$_2$]	479
[Ni$_2$(μ-X)$_2$Cp*$_2$] (X = Cl, Br)	461, 478
[Ni$_2$(μ-CO)$_2$Cp$_2$] (Ni–Ni)	518
[Ni$_2$(μ-CO)$_2$Cp'$_2$] (Ni–Ni)	518
[Ni$_2$(μ-CO)$_2$Cp*$_2$] (Ni–Ni)	478, 519
[Ni$_2$(μ-CO)$_2$(η-C$_5$HPh$_4$)$_2$] (Ni–Ni)	353
[Ni$_2$(μ-CO)$_2$Cp"$_2$] (Ni–Ni; Cp" = Cp', Cp*, C$_5$H$_4$Bu^t, C$_5$H$_3$Bu$^t_2$-1,3)	351, 352
[Ni$_2$(μ-CO)$_2$(μ-η^2,η^5-LH)$_2$]^b	422
[Ni$_2$(μ-CO)$_2$(μ-η^2,η^2-LH$_2$)(μ-η^5,η^5-L)]^b	422
[Ni$_2$(μ-CO)$_2$(η^5-C$_2$Et$_2$BMeCHBMe)$_2$]	520
[Ni$_2$(μ-Bu^{t}CP)Cp$_2$]	233
[Ni$_2${η^1,η^2,η$^{1'}$,η$^{2'}$-(C$_8$H$_{12}$)$_2$}Cp*$_2$]	141, 142
[Ni$_2$(μ-*P*Me$_2$*S*)$_2$Cp$_2$]	521, 522
[Ni$_2$(μ-*P*R$_2$*S*)$_2$Cp$_2$] (R = Me, Ph, Cy, OMe)	523
[Ni$_2$(μ-*P*R$_2$*S*)$_2$Cp$_2$] (R = Me, OMe)	524
[Ni$_2$(μ-SPh)$_2$Cp$_2$]	496
[Ni$_2$(μ-SR)$_2$Cp$_2$] (R = **Ph**, CH$_2$Ph, C$_8$H$_{17}$, 2-naphthyl)	452
[Ni$_2${μ-C$_6$H$_4$(CH$_2$S)$_2$-1,2}Cp$_2$]	452
[Ni$_2$(μ-SePh)$_2$Cp*$_2$]	525
[Ni$_2$(μ-SMe)(CH$_2$CH$_2$)(PPh$_3$)$_2$Cp$_2$]$^+$PF$_6^-$	526
[Ni$_2$(μ-C$_2$*S*$_4$)Cp*$_2$]	449
[Ni{μ-P(O)(OMe)$_2$}$_2$Ni]	505
cis- and *trans*-[Ni$_2$(η^3,η$^{3'}$-C$_8$H$_8$)Cp*$_2$]	461
[Ni$_2$(μ-η^2,η^2-P$_2$)(η-C$_5$HPr$^i_4$)$_2$]	227
[Ni$_2$X$_2$(PPh$_3$)$_2$(μ-η^5,η^5-L)]^b (X = Br, I)	509
[Ni$_2$(μ-η^3,η^3-E$_4$)(η-C$_5$HPh$_4$)$_2$] (E = P, **As**)	353
[NiCp(μ-η^1,η^2-C$_6$H$_4$-*o*-CMe$_2$*C*H)Ni(PMe$_3$)$_2$]	482
[Ni$_2$(dppe)(PPh$_3$)$_2$Cp$_2$]$^{2+}$(BF$_4^-$)$_2$	486
[Ni$_2${μ-η^2(*C,P*)-CH$_2$PPh$_2$}$_2$Cp$_2$]	497
[Ni$_2$(μ-η^3,η$^{3'}$-C$_4$Me$_4$C$_4$Me$_4$)Cp$_2$]	333
[Ni$_2$(μ-Cl)(μ-CHCMe$_2$C$_6$H$_4$-2)Cp$_2$]	346
[Ni$_2${μ-(*P,P*)-P(C$_6$H$_2$Bu$^t_2$)-2,4-P(N=Mes)}Cp$_2$]	369
[Ni$_2$(μ-C$_{10}$Cl$_4$S$_4$)Cp$_2$]	527
[Ni$_2$(μ-PhCN$_2$S$_2$)Cp$_2$]	528
[Ni$_2$(μ-CF$_3$NC)$_2$Cp$_2$] (Ni–Ni)	529
[Ni$_2$Cl$_2${μ-c-P(Cl)N(Bu^t)P(Cl)N(Bu^t)}Cp$_2$]	504
[Ni$_2$(*cis*- and *trans*-*s*-μ-η^2,η$^{2'}$-C$_4$H$_4$R^{1}R^2-2,3)Cp$_2$] (R = H, Me)	504
[NiCp(μ-SR)$_2$TiCp$_2$]$^+$BF$_4^-$ (R = Me, Ph)	530
[NiCp(μ-SEt)$_2$NbCl$_2$Cp$_2$]	531
[NiCp*{μ-η^2,η$^{2'}$-*C*Ph*C*HC(O)*C*H$_2$}W(CO)$_2$Cp]	532
[NiCp*{μ-η^1(*Ni*),η^3(*W*)-*C*H$_2$*C*(*C*H$_2$)$_2$}W(CO)$_2$Cp]	531
[NiCp*(μ-ER)(μ-CO)M(ER)(CO)Cp'] (Ni–M; ER = **SPh**, M = Mo, **W**; ER = SePh, SMe; M = W)	525
[Ni(CO)Cp"M(CO)$_3$Cp$^\#$] (M = Mo, W; Cp" = Cp, Cp'; Cp$^\#$ = Cp', Cp)	161
[NiCp"(μ-η^1,η^3-*C*H$_2$*C*CMe$_2$)M(CO)$_2$Cp$^\#$] (M = Mo, W; Cp", Cp$^\#$ = Cp, Cp')	161
[NiCp"(μ-η^1,η^3-*C*Me*C*Me*C*H$_2$)M(CO)$_2$Cp$^\#$] (M = Mo, W; Cp", Cp$^\#$ = Cp, Cp')	161
[NiCp{μ-η^3(*Ni*),η$^{3'}$(*Mo*)-C$_9$H$_{12}$}Mo(CO)$_2$Cp']^c	161
[NiCp*M(CO)$_3$Cp'] (Ni–M; M = Mo, **W**)	535
[NiCp*(μ-CH$_2$)(μ-CO)M(CO)$_2$Cp] (Ni–M; M = Mo, **W**)	483, 534
[NiMo(μ-SR)$_2$Cp$_3$] (R = Me, Bu^t, Ph)	535
[NiCp(μ-SR)$_2$MoCp$_2$]$^+$BF$_4^-$ (R = Me, **Bu^t**, Ph; Cp" = **Cp**, Cp')	530
[NiCp{(η^2,η^4-C$_4$Ph$_4$)FeCp}]	536
[NiCp(μ-CO)$_2$Fe(CO)Cp*]	244
[NiCo(μ-η^2,η$^{2'}$-Bu^{t}Cp)(CO)$_3$Cp]	233
[NiCp(μ-SPh)$_2$Co(PMe$_3$)Cp]	530
[Ni{μ-P(O)(OMe)$_2$}$_n$M] (*n* = 2, M = Co, Cu, Zn; *n* = 3, M = Al, Cr, Fe)	505
[Ni{[μ-P(O)(OMe)$_2$]$_3$CoCp}Cp]	537
[Ni$_2${PMe[W(CO)$_5$]C(R^1)C(R^2)}Cp$_2$]	153
[Ph$_3$GeMNiCp(GePh$_3$)MNiCp(GePh$_3$)MGePh$_3$] (M = Cd, Hg)	538
[Ni$_2$Zn{μ-(*P,S*)-PR$_2$S}$_4$Cp$_2$] (R = Me, Ph, Cy, OMe)	525

[a] Compounds in bold were characterized by an x-ray diffraction study. [b] L = C$_{10}$H$_4$Bu$^t_4$ (Scheme 51).
[c] C$_9$H$_{12}$ = (*C*H$_2$)$_2$*C*CH$_2$C(=CH$_2$)*C*(*C*H$_2$)$_2$.

dimeric nickel complex [Ni$_2$(μ-CO)$_2$Cp$_2$] (Ni–Ni) also is a source of [NiL$_2$Cp] species: for example, [NiL$_2$Cp]$^+$BF$_4^-$ species are formed from [Ni$_2$(μ-CO)$_2$Cp$_2$] (Ni–Ni) and [FeCp$_2$]$^+$BF$_4^-$ in the presence of two-electron donors L.[455]

3.5.4.2 Chemistry of nickel [NiL^{1}L^{2}Cp]$^+$ X$^-$ and [NiXL^{1}Cp] complexes

Substitution reactions of [NiCp$_2$] with two-electron donor ligands L proceed via [NiL(η^1-Cp)Cp] intermediates.[539] Treatment of nickelocene with phosphonium iodide affords the phosphine complex [NiI(PH$_3$)Cp]. The bis(phosphine) species [Ni(PH$_3$)$_2$Cp]$^+$ may be obtained from the former complex by treatment with TlBF$_4$ (Equation (13)).[480]

$$[\mathrm{NiCp_2}] \xrightarrow[-\mathrm{C_5H_6}]{\mathrm{PH_4I}} [\mathrm{Ni(PH_3)(I)Cp}] \xrightarrow[-\{\mathrm{CpNi}\}^+,\ -\mathrm{TlI}]{\mathrm{TlBF_4}} [\mathrm{Ni(PH_3)_2Cp}]^+\mathrm{BF_4}^- \qquad (13)$$

The reactions of [NiCp$_2$] with the chelating ligands dppm and dppe, and with mixed phosphorus–arsenic ligands of the type Ph$_2$P(CH$_2$)$_n$AsPh$_2$ (dapm, n = 1; arphos, n = 2), were investigated. Nickelocene reacts with dppm or dppe in CCl$_4$ to yield the complexes [NiCl(η^1-L$_2$)Cp] (L = dppm, dppe). The related bromide complexes [NiBr(η^1-L$_2$)Cp] can be obtained when [NiCp$_2$] and the diphosphine ligand are reacted in a CBr$_4$/MeCN mixture.[502] With dapm in CCl$_4$, the phosphorus-bonded complex [Ni{η^1-(*P*)-dapm}Cp(Cl)] is obtained; the ionic complex [Ni(η^2-arphos)Cp]$^+$ Cl$^-$ results with arphos. Chelated ionic species are obtained with both ligands when the free ligands are reacted with the cationic complex (**68**$^+$)BF$_4^-$. These complexes react with more of the free ligand to afford [Ni{η^1-(*P*)-dapm}$_2$Cp]$^+$ or [Ni{η^1-(*P*)-arphos}$_2$Cp]$^+$. These results clearly indicate that the PPh$_2$ group has a stronger affinity for the nickel atom than the AsPh$_2$ group.[503] These reactions and others with trialkylstibines, chelating distibines, and group 16 ligands are referenced.[485,503,509,540] Some are shown in Schemes 39, 40, and 41.

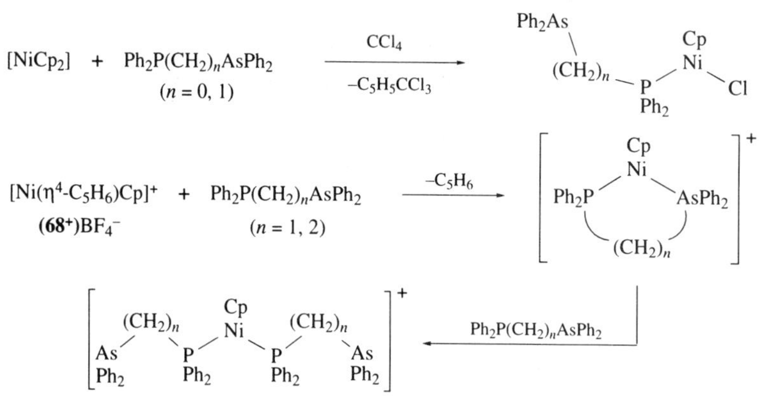

Scheme 39

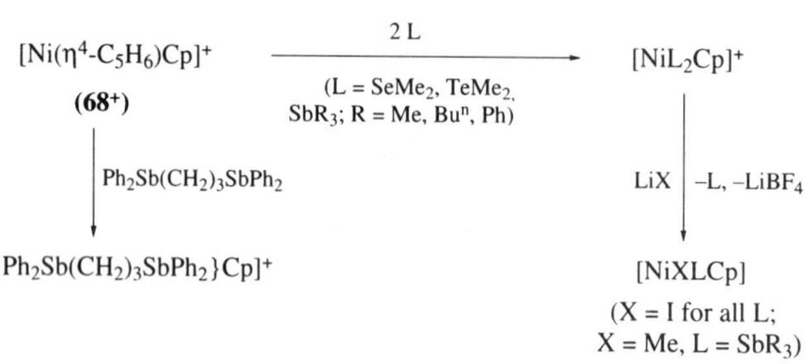

Scheme 40

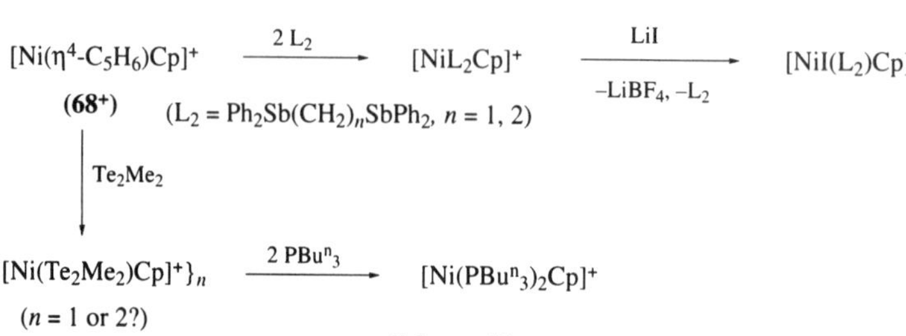

Scheme 41

Various NiCp complexes of $P(NR_2)_3$ (R = Me, Et) have been prepared by reacting the cationic norbornadiene species $[Ni(\eta^4\text{-}C_7H_8)Cp]^+$ with the free ligands. Monophosphine complexes are accessible in a ligand redistribution reaction involving nickelocene, the appropriate nickel(II) halide, and the free dialkylaminophosphine. The nickel methyl species $[NiMe\{P(NR_2)_3\}Cp]$ (R = Me, Et) have also been prepared.[487] These results are summarized in Equations (14)–(16). Another reaction that involves the cationic norbornadiene species $[Ni(\eta^4\text{-}C_7H_8)Cp]^+$ shown in Equation (16) has been referred to earlier: the phosphonium species (**16**) shown in Scheme 10 was obtained by addition of PMe_3 to $[Ni(\eta^4\text{-}C_7H_8)Cp]^+$ in nonpolar solvents. When solutions of (**16**) are warmed in polar solvents, norbornadiene and $[Ni(PMe_3)_2Cp]^+$ form.[129]

$$[NiCp_2] + NiX_2 + 2\,P(NR_2)_3 \longrightarrow [NiX\{P(NR_2)_3\}Cp] \quad (X = Cl, Br; R = Me, Et) \tag{14}$$

$$[NiCl\{P(NR_2)_3\}Cp] \xrightarrow[-LiCl]{LiMe} [NiMe\{P(NR_2)_3\}Cp] \quad (R = Me, Et) \tag{15}$$

$$[Ni(\eta^4\text{-}C_7H_8)Cp]^+ \xrightarrow[-C_7H_8]{2\,P(NR_2)_3} [Ni\{P(NR_2)_3\}_2Cp]^+ \quad (R = Me, Et) \tag{16}$$

Nickelocene reacts with P_2Ph_4 and with various combinations of the mixed ligands $E^1E^2Ph_4$ (E^1, E^2 = P, As, Sb): similar types of products are formed for all these related ligands. An equilibrium mixture of ionic $[Ni(\eta^2\text{-}P_2Ph_4)Cp]^+\,Cl^-$ and neutral $[NiCl(\eta^1\text{-}P_2Ph_4)Cp]$ resulted when HCl was added to a mixture of $[NiCp_2]$ and P_2Ph_4. Free cyclopentadiene was also liberated. The ligated η^4-cyclopentadiene in $(\mathbf{68}^+)BF_4^-$ can be replaced by two η^1-bound P_2Ph_4 moieties yielding $[Ni(\eta^1\text{-}P_2Ph_4)_2Cp]^+BF_4^-$. One of the diphosphine groups can be removed by the addition of LiI, resulting in the formation of $[NiI(\eta^1\text{-}P_2Ph_4)Cp]$.[501] The mixed-ligand systems exhibit similar behavior. However, some of the species $[Ni(\eta^2\text{-}E^1E^2{}_2Ph_4)Cp]^+$ (E = As, Sb) may be the dicationic complexes $[Ni_2(\mu\text{-}\eta^2\text{-}E^1E^2{}_2Ph_4)_2Cp_2]^{2+}$ which may contain six-membered $NiE^1E^2NiE^1E^2$ rings.[508] Nickelocene and $(\mathbf{68}^+)\,BF_4^-$ react with the cyclic species c-PCl–NBu^t–PCl–NBu^t as shown in Scheme 42. Only phosphorus-bound complexes are obtained.[504]

Tetracyanoethene reacts with $[Ni(C_2Ph)(PPh_3)Cp]$ to afford the structurally characterized complex $[Ni\{C[=C(CN)_2]C(Ph)[=C(CN)_2]\}(PPh_3)Cp]$. The carbon–carbon double bond of $C_2(CN)_4$ has effectively been cleaved and the two $C(CN)_2$ fragments have added across the alkynide carbon–carbon triple bond to afford the final product. Cyclobutenyl complexes are believed to be implicated, and have been observed for the analogous reactions of manganese and iron complexes. However, they were not seen for the nickel reaction.[492] Ketenes (even vinyl ketenes) react somewhat similarly by adding the ketene C=C across the triple bond of the alkynide to afford complexes that contain σ-bonded cyclobutenone ligands.[493]

The C_5H_6 ligand in $(\mathbf{68}^+)\,BF_4^-$ is displaced with disulfide ligands $PhS(CH_2)_nSPh$ (n = 1, 2) to afford the cationic chelates $[Ni\{PhS(CH_2)_nSPh\}Cp]^+$. These cations readily transfer an $[NiCp]^+$ fragment to nickelocene yielding the free disulfide ligand and the cationic tripledecker complex $[Ni_2(\mu\text{-}Cp)Cp_2]^+$.[510] Trialkylphosphine ligands substitute one or both dimethyl sulfide groups in the complex $[Ni(SMe_2)_2Cp]^+$ sequentially forming mixed-ligand or bis(trialkylphosphine) cyclopentadienyl complexes of nickel. The latter species was not obtained for the bulky $P(Tol\text{-}2)_3$ ligand. The cationic complexes $[Ni(PR_3)(CO)Cp]^+$ (R = Pr^i, Tol-2, Cy; L = CO, R = Pr^i, Cy; L = $PSMe_3$, R = Pr^i) and the neutral species $[NiCl(PCy_3)Cp]$ were also prepared. No 16-electron cationic complexes of formula $[Ni(PR_3)Cp]^+$ were formed, despite the relative steric bulk of the ligand.[475] Complexes with the chelating amines TMEDA and $C_6H_4(NMe_2)_2$-1,2 have also been reported, and prepared in a similar fashion from $[Ni(SMe_2)_2Cp]^+$. Some of these reactions are depicted in Scheme 43.[511]

Trialkylphosphine selenides displace the η^4-cyclopentadiene ligand in $(\mathbf{68}^+)BF_4^-$ to afford selenium-bound $[NiL_2Cp]^+BF_4^-$ (L = Me_3PSe, Ph_3PSe) species. Similar selenium-bound cyclic compounds $[Ni\{PSePh_2(CH_2)_nPSePh_2\}Cp]^+$ result from reactions with $PSePh_2(CH_2)_nPSePh_2$ (n = 0–2).[506]

One cyclopentadienyl ligand of nickelocene is displaced with $HP(O)(OR)_2$ ligands to form (**69**). This compound contains a symmetric hydrogen bond.[505] Complex (**69**) (R = Me) reacts with the metal acetylacetonates $[M(acac)_n]$ (M = Co, Cu, Zn, Ni, n = 2; M = Al, Co, Fe, n = 3) to produce $[CpNi\{(MeO)_2P(O)\}_nM]$ species.[502] The structure of the nickel–cobalt derivative $[CpNi\{(MeO)_2P(O)\}_3CoCp]$ was recently documented.[537]

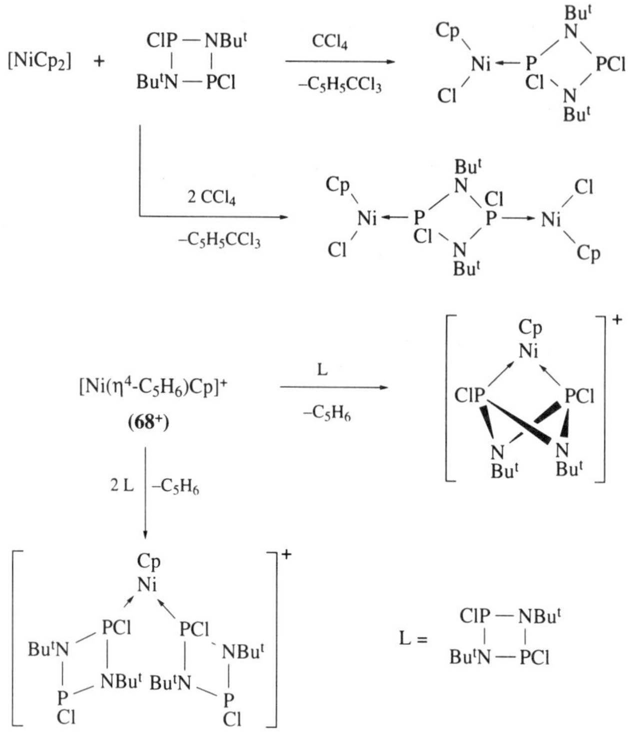

Scheme 42

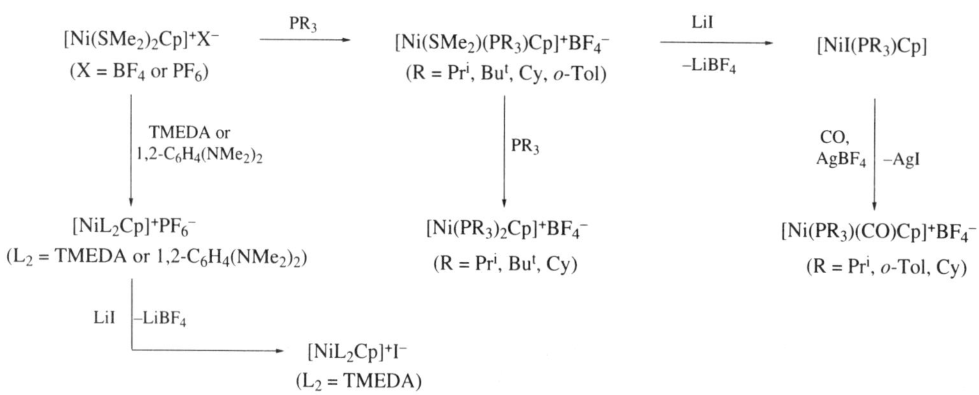

Scheme 43

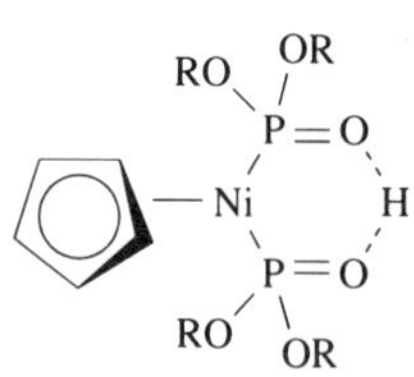

(69)

A large number of 19-electron complexes $[NiL_2Cp]$ (L = nitrogen or phosphorus donor ligands) have been prepared using six different synthetic methods. The ESR spectra of many of these species were recorded. Their stability varies with the nature of L. In general the complexes are in equilibrium with $[NiCp_2]$ and corresponding $[NiL_4]$ species; the position of the equilibrium favors the homoleptic compounds. However, dark-blue compounds [Ni(bipy)Cp] and [Ni(dppe)Cp] were especially stable. The structure of [Ni(bipy)Cp] was established by x-ray diffraction, and clearly showed that the Cp ligand was η^5- and not η^3-coordinated. The Cp ligand is planar (maximum deviation of atoms from C_5 plane = 0.0006 nm); the nickel–carbon(Cp centroid) distance is 0.1844 nm. The nickel–nitrogen distances are 0.195 5(3) nm and 0.195 8(4) nm and the NiN_2 blade is essentially perpendicular to the Cp plane (the interplanar angle between the NiP_2 and the C_5 plane = 89.1°). Cyclic voltammetry was performed on $[Ni(PPh_3)_2Cp]^+PF_6^-$.[420]

The photolysis of $[NiR(PPh_3)Cp]$ (R = Me, Et) with phosphorus-donor ligands L was studied. The radical species $[NiL_2Cp]$ that were formed were studied by ESR spectroscopy.[494,495] A carbon–carbon coupling reaction ensues when the chlorotriphenylphosphine complex $[NiCl(PPh_3)Cp]$ reacts with $Li(CH{=}CH_2)$.[148]

There are no $Ni{-}SO_2$ interactions in the complexes $[Ni\{CH(R)SO_2Ph\}(PPh_3)Cp]$ (R = H, Et, n-C_6H_{13}, n-$C_{11}H_{23}$, $CH{=}CMe_2$). Two of these species (R = H, $CH{=}CMe_2$) were structurally characterized,[490] and reactions of the n-$C_{11}H_{23}$ sulfone complex with Lewis acids and O_2 were investigated.[489]

The reaction of $[NiCp_2]$ with a variety of dialkylphosphine sulfides $R_2P(H){=}S$ (R = Me, OMe, Ph, Cy) leads to protonation of a Cp ring. The cyclic diamagnetic species $[Ni\{P(S)R_2\}Cp]_2$ (**70**) result together with free cyclopentadiene, as shown in Scheme 44. The x-ray structure of (**70**) and its reaction with dmac were documented.[521] Other reactions with Zn^{2+} are shown in Scheme 44.[522,523]

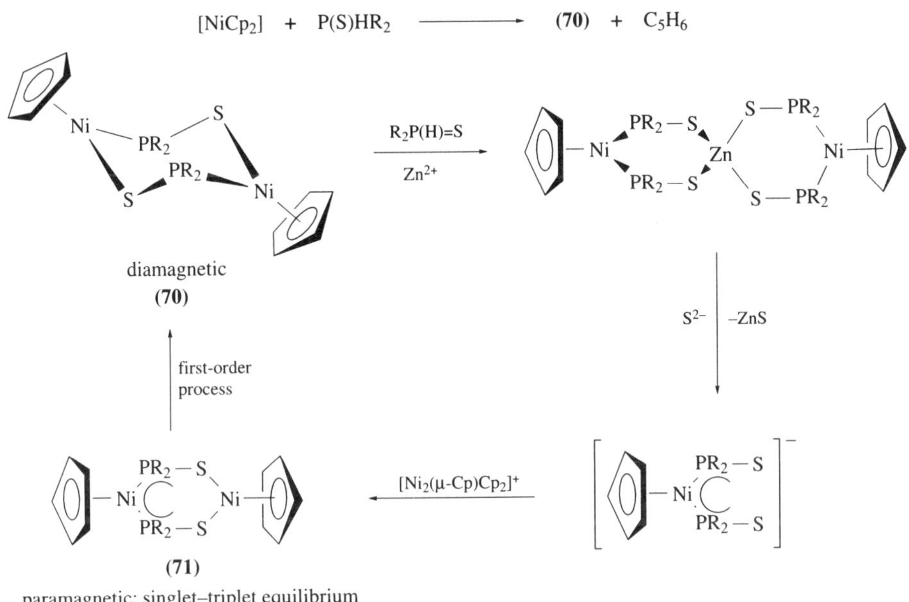

Scheme 44

Later it was shown that a NiCp fragment could be added to $[Ni\{P(S)R_2\}_2Cp]^-$ (R = Me, OMe) by treatment of this anion with the nickel tripledecker cationic species $[Ni_2(\mu\text{-}Cp)Cp_2]^+$. The resulting dinickel species $[CpNi\{\mu\text{-}(P,S)\text{-}P(S)R_2\}_2NiCp]$ (**71**) is an isomer of (**70**) and contains two $P(R_2){=}S$ bridges. Compound (**71**) exhibits unusual magnetic behavior and a temperature-dependent singlet–triplet equilibrium is operative. The Ni*Cp* signals in the ^1H NMR spectrum of this species exhibit a strong temperature dependence. Molecular orbital calculations (EHT) were performed on the model species $[P(S)H_2]^-$, $[Ni\{P(S)H_2\}_2Cp]^-$, and $[CpNi\{\mu\text{-}(P,S)\text{-}P(S)H_2\}_2NiCp]$ to better understand the bonding in this interesting molecule. By considering the species as a combination of $\{NiP_2Cp\}$ and $\{NiS_2Cp\}$ fragments, it was shown that the spin equilibrium was localized on the $\{NiS_2Cp\}$ fragment. The paramagnetic complex isomerizes in a first-order reaction to its diamagnetic isomer. The rate is independent of the solvent, and crossover studies indicate that the process is intramolecular: monomeric $[Ni\{P(S)R_2\}Cp]$ species were not involved.[524]

Spin equilibria associated with an $\{NiS_2Cp\}$ fragment have also been observed in the heterobimetallic complex $[NiCp(\mu\text{-}SR)_2MoCp_2]^+$.[535] The NMR spectra of this molecule are strongly temperature dependent. Later research described the syntheses of $[NiCp(\mu\text{-}SR)_2MoCp''_2]^+BF_4^-$ (R = Me, Bu^t, Ph; Cp'' = Cp, Cp') and the related NiTi and NiCo species $[NiCp(\mu\text{-}SR)_2TiCp''_2]^+BF_4^-$ (R = Me, Ph) and $[NiCp(\mu\text{-}SPh)_2Co(PMe_3)Cp]$. The NMR spectra of the CoNi species showed no temperature dependence. The structure of $[NiCp(\mu\text{-}SBu^t)_2MoCp_2]^+BF_4^-$ was determined by x-ray diffraction studies. The $Ni(\mu\text{-}S)_2W$ core adopts a butterfly structure with the bulky Bu^t groups located on the inside of the butterfly core. MO calculations suggest that the spin crossover results from flattening at the sulfur atom, resulting from the bulky Bu^t ligands. This prediction is borne out by the x-ray structure of the heterobimetallic complex.[530] Nickelocene also reacts with the bulkier dialkylphosphine sulfides

$R_2P(H){=}S$ (R = $C_6H_3Me_2$-2,6 or Mes). When this reaction is followed by the addition of dmac, the metallacycle [Ni{*C*(CO_2Me)=C(CO_2Me)PR_2*S*}Cp] results.[516]

The chloro complex [NiCl(PPh_3)Cp] can be alkylated with $LiCH_2SPh$ yielding the complex [Ni(CH_2SPh)(PPh_3)Cp], whose physical and spectroscopic properties (NMR, MS, magnetic measurements) were measured. Methyl phenyl sulfide is formed when the latter species is protonated; this and other related reactions are shown in Scheme 45.[497] The complex [Ni(η^2-CH_2SPh)Cp] was proposed as a possible unstable intermediate to account for these results.[496] When the chlorophosphines $PPh_2(CH_2)_nCl$ (n = 1, 3, 4) react with [NiI(PPh_3)Cp], the chlorophosphine replaces the PPh_3 ligands. Reduction of the resultant [NiI{$PPh_2(CH_2)_nCl$}Cp] with sodium amalgam gives products that depend on the value of n. When n = 1, the complex [Ni_2{μ-η^2(*C,P*)-CH_2PPh_2}$_2Cp_2$], which contains an NiCPNiCP six-membered ring system, was obtained. The compounds [Ni{$PPh_2(CH_2)_n$}Cp] result when n = 3 or 4.[497]

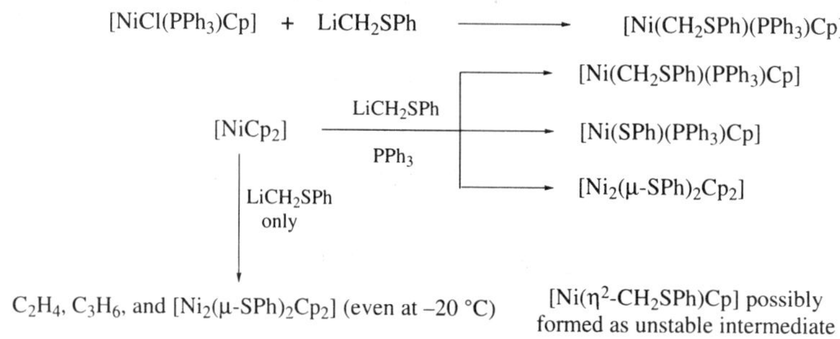

Scheme 45

The complexes [NiXLCp] (L = PPh_3, PCy_3; X = Cl, Br, I) were prepared from the corresponding nickel phosphine halide species [NiX_2L_2] and TlCp. These species could be reduced with mercury amalgam in the presence of other coordinating ligands: with CO, [$NiCp_2$], [$Ni(CO)_2L_2$], [Ni_2(μ-CO)$_2Cp_2$] (Ni–Ni), and other uncharacterized decomposition products were formed, while with alkynes, the dimetallatetrahedrane complexes [Ni_2(μ-RC_2R)Cp_2] (Ni–Ni) were formed, as shown in Scheme 46.[486] Cationic species such as [NiL_2Cp]$^+$ (L = $PMePh_2$, 1/2 dppe) and [$Ni_2(PPh_3)_2$(μ-dppe)Cp_2]$^+$ were prepared. Some of these complexes could be reduced to unstable 19-electron [NiL_2Cp] radical species with Na/Hg. The position of the equilibrium shown in Equation (17) is dependent on the nature of the PR_3 species. For PMe_2Ph, the equilibrium is shifted towards the cationic complex [Ni(PMe_2Ph)Cp]$^+$ X^-, while similar concentrations of the neutral species and the salt are present when L = PPh_3 or $PMePh_2$.[486]

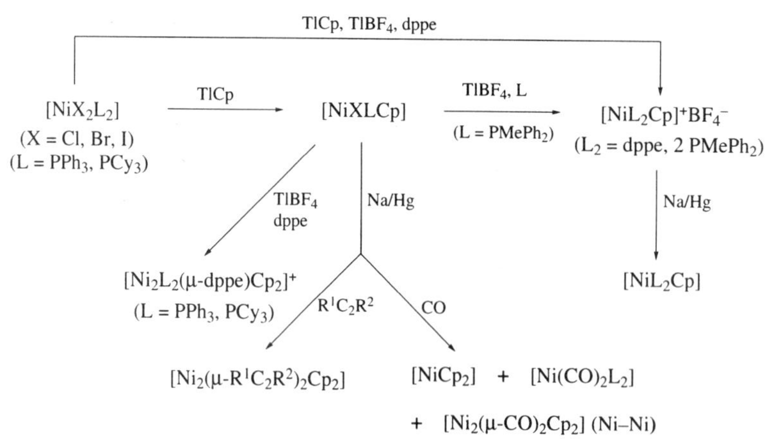

Scheme 46

(17)

A series of nickel enolates of the type [Ni{C(R^1)C(O)R^2}(PPh_3)Cp] (R^1 = H or Me; R^2 = Bu^t, OBu^t, Ph; Cp = Cp or Cp*; not all combinations made) have been prepared, and their reactions with aldehydes and other electrophilic reagents were studied. A low-temperature x-ray study was carried out on [Ni{C(H)C(O)OBu^t}(PPh_3)Cp*] (**72**). The syntheses of these compounds and structural parameters of (**72**) are summarized in Scheme 47. Treatment of [Ni{$CH_2C(O)Bu^t$}(PPh_3)Cp] with excess PPh_3 or dppe afforded the paramagnetic species [NiL_2Cp] (L_2 = 2 PPh_3, dppe; Scheme 47) whose ESR spectra were measured. A better route to these 19-electron complexes is given.[481]

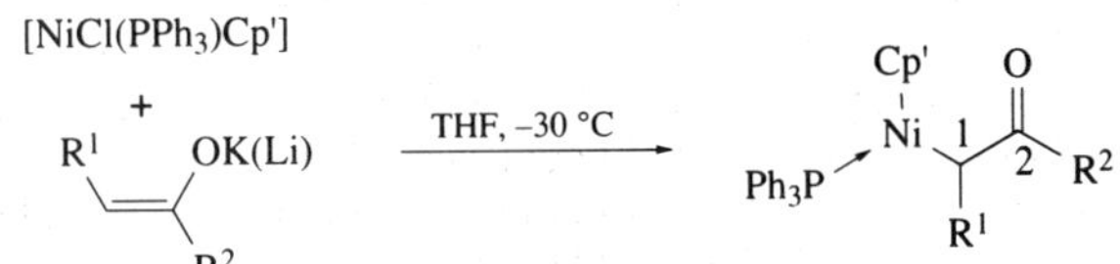

(72)

Cp" = Cp; R^1 = H; R^2 = Bu^t, OBu^t, Ph
Cp" = Cp; R^1 = Me; R^2 = Bu^t
x-ray: Cp" = Cp*; R^1 = H; R^2 = OBu^t

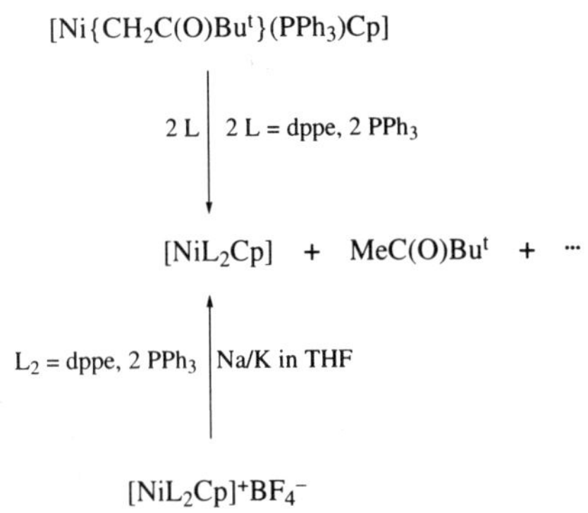

x-ray data for [Ni{$CH_2C(O)OBu^t$}(PPh_3)Cp*]

Ni–C-1	= 0.202 7(2) nm	Ni–C(Cp*)	= 0.209 8(2) nm
Ni–P	= 0.213 4(1) nm		0.215 0(2) nm
C-1–C-2	= 0.141 4(3) nm		0.213 1(2) nm
			0.213 3(2) nm
Ni–C(Cp* centroid) = 0.1761 nm			0.217 1(2) nm

Scheme 47

Various ylides have been reacted with [$NiCp_2$] and with the complex [$NiBr(PPh_3)Cp$]. Treatment of [$NiBr(PPh_3)Cp$] with the ylides Ph_3PCH_2, Ph_3AsCH_2, Me_2SCH_2, $Me_2S(O)CH_2$ affords the carbon-bonded ylidic complexes [$Ni(CH_2Y)(PPh_3)Cp]^+Br^-$ (Y = PPh_3, $AsPh_3$, SMe_2, $S(O)Me_2$). The stabilized ylide $Ph_3PCHCOPh$ affords the compound [$Ni(Ph_3PCHCOPh)_2Cp]^+Br^-$ with [$NiCp_2$]. Some reactions with these ylides are shown in Scheme 48.[498,499]

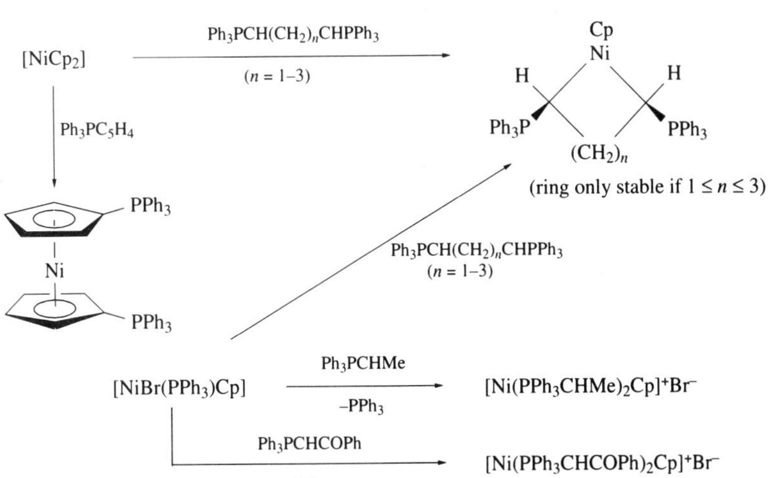

Scheme 48

The reaction of nickelocene with lithium alkyls under an alkene atmosphere is a synthetic entry into complexes of general formula [NiR(η-alkene)Cp]. These reactions have been discussed earlier. When

$[NiCp_2]$ is treated with aryllithium species in the absence of alkenes, thermally unstable {NiArCp} species result (Ar = Ph, Tol-2, Tol-3, Tol-4); their subsequent decomposition products are detailed. The thermally stable complex $[Ni\{C_6H_4(CH_2NMe_2)\text{-}2\}Cp]$ was isolated with Ar = $C_6H_4(CH_2NMe_2)$-2. This species achieves stability through electron donation from the dimethylamino group.[517]

Some chemistry of complexes that contain an NiCp" group (Cp" = η-$C_5H_2(TMS)_3$-1,2,4) has been explored. Reaction of LiCp" with $[NiCl_2]$ affords $[Ni_2(\mu\text{-}Cl)_2Cp''_2]$ or [NiCl(CO)Cp"] when CO is available. The syntheses of $[NiX(PPh_3)Cp'']$ (X = Cl, Br, I) were also detailed and the barrier to ring rotation in the half-sandwich compound $[NiCl(PPh_3)Cp'']$ was determined. A value of $\Delta G^{\ddagger} = 43.9$ kJ mol^{-1} was found for the free activation energy of Cp" rotation in this complex.[479]

The complexes $[NiXCp^*]_2$ (X = Cl, Br) have been made by reaction of LiCp* with $[NiX_2\cdot DME]$, and were used as a stepping stone in the synthesis of a series of NiCp* species. These reactions are shown in Scheme 49.[461] Pentamethylcyclopentadienylnickel species are accessible from $[Ni(acac)_2]$, as indicated in Equation (18) (carried out in THF).[513]

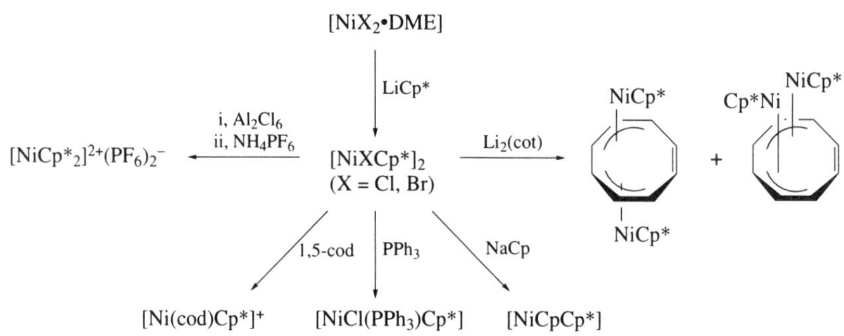

Scheme 49

$$[Ni(acac)_2] + LiCp^* \longrightarrow [Ni(acac)Cp^*] + Li(acac) \qquad (18)$$

When PCp*=NBu^t is combined with $[Ni(cod)(PR_3)_2]$ species, Cp* transfer to nickel ensues to yield the phosphorus-bonded species $[Ni(P{=}NBu^t)(PPh_3)Cp^*]$.[484] The nitrosyl complex [Ni(NO)Cp] has been prepared from polymeric $[Ni(SR)(NO)]_x$ by treatment with NaCp.[515]

The cationic complexes $[Ni(\eta^4\text{-}C_5Me_5R)Cp^*]^+$ have been synthesized by electrophilic attack of RX (R = H, CCl_3, Bz) on $[NiCp^*_2]$. These complexes in turn react with strong nucleophiles to yield neutral allylic $[Ni(\eta^3\text{-}C_5Me_5R^1R^2)Cp^*]$ species as indicated in Scheme 50.[318]

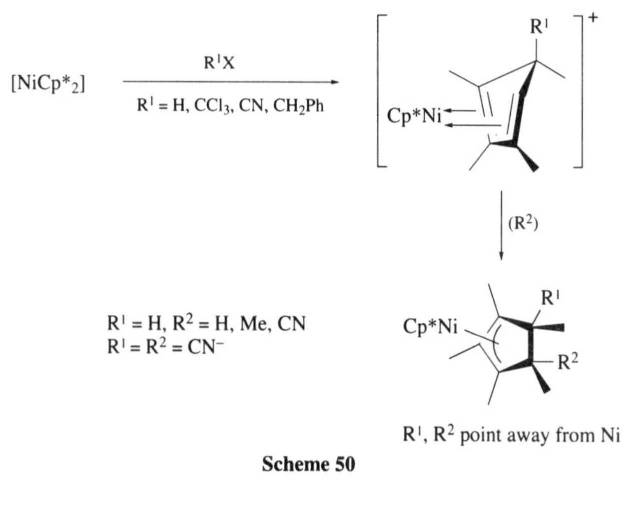

Scheme 50

3.5.5 Dinuclear Nickel η^5-Ligand Complexes

3.5.5.1 General comments

Most of the compounds in this section are derived from $[Ni_2(\mu\text{-}CO)_2Cp_2]$ (**73**) and its alkylated cyclopentadienyl derivatives. Some of the compounds in this class such as the halogen-bridged species $[Ni_2(\mu\text{-}X)_2Cp^*{}_2]$ (X = Cl, Br) have been alluded to earlier.[461] Table 22 provides a list of these dinuclear nickel η^5-ligand complexes. Applications of (**73**) and its derivatives in metal cluster synthesis are discussed in Section 3.5.6.

3.5.5.2 Chemistry of the complex $[Ni_2(\mu\text{-}CO)_2Cp_2]$ (Ni–Ni) and its derivatives

A detailed experimental synthesis of $[Ni_2(\mu\text{-}CO)_2Cp^*{}_2]$ (Ni–Ni) has been published. The process is outlined in Equations (19a) and (19b). The yields in following this procedure are typically >60% but the procedure requires the use of $[Ni(CO)_4]$.[519] A new synthesis of $[Ni_2(\mu\text{-}CO)_2Cp^*{}_2]$ (Ni–Ni) that does not require the manipulation of highly toxic $[Ni(CO)_4]$ has appeared. When $[NiBr_2\cdot DME]$ was treated with MgCp*Cl in THF, followed by sequential treatment with CO and zinc amalgam at 0 °C, the Ni_2 species $[Ni_2(\mu\text{-}CO)_2Cp^*{}_2]$ (Ni–Ni) was formed in yields that approached 60%. The preparation of the thermally unstable intermediate $[Ni_2(\mu\text{-}Br)_2Cp^*{}_2]$ and the syntheses of the compounds [NiBr(CO)Cp*], $[Ni(SnPh_3)(CO)Cp^*]$, and $[Ni(CO)Cp^*]^-M^+$ (M = Li, Na) were also described in this paper.[478]

$$[Ni(CO)_4] + LiCp^* \xrightarrow[8\ h]{THF,\ 65\ °C} [Ni(CO)Cp^*]^- \quad (19a)$$

$$[Ni(CO)Cp^*]^- \xrightarrow{Cu_2Cl_2} [Ni_2(\mu\text{-}CO)_2Cp^*{}_2]\ (Ni–Ni) \quad (19b)$$

The structures of $[Ni_2(\mu\text{-}CO)_2Cp_2]$ (Ni–Ni) (**73**) and its methylcyclopentadienyl analogue $[Ni_2(\mu\text{-}CO)_2Cp'_2]$ (Ni–Ni) were established by x-ray diffraction studies.[518] The methylcyclopentadienyl complex contains a planar centrosymmetric $Ni_2(\mu\text{-}CO)_2$ core, with inequivalent sets of Ni–CO distances. The unsubstituted Cp complex (**73**) crystallizes with two independent molecules in the unit cell: each molecule has a bent (butterfly-like) structure. Both the Cp and Cp' complexes adopt bent $Ni_2(\mu\text{-}CO)_2$ cores in solution. Key bond lengths and angles for these species are tabulated in Table 23.[518] Restricted Hartree–Fock calculations on (**73**) and related Co_2 and Fe_2 species reveal that the potential energy surface for (**73**) is very flat about the planar geometry, that is, there is little energy difference between planar and butterfly-like $Ni_2(\mu\text{-}CO)_2$ cores for this molecule. "This is entirely consistent with the experimental observations of a bent structure for (**73**) and a planar structure for $[Ni_2(\mu\text{-}CO)_2Cp'_2]$."[542] Both PES studies and Hückel MO calculations suggest that $[Ni_2(\mu\text{-}CO)_2Cp_2]$ does not contain a direct nickel–nickel bond.[542] There are many applications of (**73**) in the syntheses of heterobimetallic complexes and in metal cluster synthesis.[543,545]

Table 23 X-ray data for complex $[Ni_2(\mu\text{-}CO)_2Cp_2]$ (Ni–Ni) (**73**) and $[Ni_2(\mu\text{-}CO)_2Cp'_2]$ (Ni–Ni).

Parameter	*(73) Molecule A*	*(73) Molecule B*	*$[Ni_2(\mu\text{-}CO)_2Cp'_2]$*
Ni–Ni'	0.236 3(1)	0.235 1(1)	0.239 0(1)
Ni–C	0.186 1(5)	0.186 5(5)	0.184 0(3), 0.188 9(3)
Ni–C(Cp centroid)	0.1751 nm	0.1752 nm	0.1745 nm
Ni–C–Ni'	78.8(2)	78.1(2)	79.7(1)
C–Ni–C'	95.2(4)	93.4(4)	100.3(1)
Interplanar angles			
NiC_2–$Ni'C_2$	140.8	133.5	180.0
Ni_2C–Ni_2C'	146.0	139.2	180.0
C_5–NiC_2	98.6, 98.7	99.5, 98.0	90.0, 90.0
C_5–C'_5	22.0	29.1	0.0

Source: Byers and Dahl.[518]
Bond distances in nm; angles in degrees; esds in parentheses. C, C′ = carbonyl carbon atoms. C_5, C'_5 = Cp or Cp′ aromatic carbon atoms. Chemically equivalent values are averaged.

The dinickel species (**73**) can be reductively cleaved with C_8K quickly and cleanly, affording the anion $[Ni(CO)Cp]^-$; MeI reacts with this nickel(0) species to afford the complex [NiMe(CO)Cp].[478] The nickel dimer (**73**) reacts with 2-cyanoethanol to afford a cationic product (**74**$^+$), in which two molecules of the cyanoethanol have cyclized. The structure of (**74**$^+$)PF_6^- was determined crystallographically.[507]

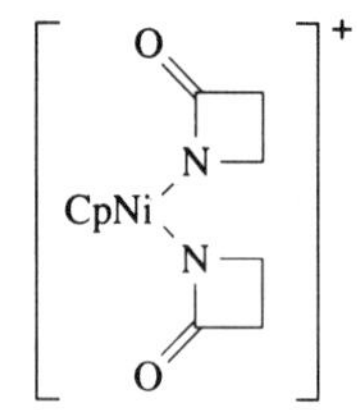

(74)

Nickelocene reacts with $[Ni(CF_3NC)_4]$ in a ligand redistribution reaction to afford the isocyanide complex $[Ni_2(\mu\text{-}CF_3NC)_2Cp_2]$.[529] The bridging germylene complex $[Ni_2\{\mu\text{-}Ge(C_6F_5)_2\}Cp_2]$ was formed from nickelocene and $[GeH_2(C_6F_5)_2]$.[512]

When 1,1',3,3'-tetra-*t*-butyl-5,5'-dihydropentafulvalene was treated with LiMe followed by $[Ni(CO)_4]$, a mixture of the dinickel complexes $[Ni_2(\mu\text{-}CO)_2(\mu\text{-}\eta^2,\eta^2\text{-}C_{10}H_6Bu^t_4)(\mu\text{-}\eta^5,\eta^5\text{-}C_{10}H_4Bu^t_4)]$ and $[Ni_2(\mu\text{-}CO)_2(\mu\text{-}\eta^2,\eta^5\text{-}C_{10}H_5Bu^t_4)_2]$ resulted as shown (Scheme 51).[422] These species are paramagnetic ($\mu = 2.58\ \mu_B$). The former species reacts with Br_2 or I_2 in the presence of PPh_3 to afford the complexes $[Ni_2X_2(PPh_3)_2(\mu\text{-}\eta^5,\eta^5\text{-}C_{10}H_4Bu^t_4)]$. However, the latter isomer $[Ni_2(\mu\text{-}CO)_2(\mu\text{-}\eta^2,\eta^5\text{-}C_{10}H_5Bu^t_4)_2]$ reacts with bromine to produce the mononuclear nickel(III) complex $[NiBr_2(\eta^5\text{-}C_5H_2Bu^t_2C_5H_3Bu^t_2)]$. This paramagnetic complex ($\mu = 1.82\ \mu_B$) was characterized crystallographically.[422]

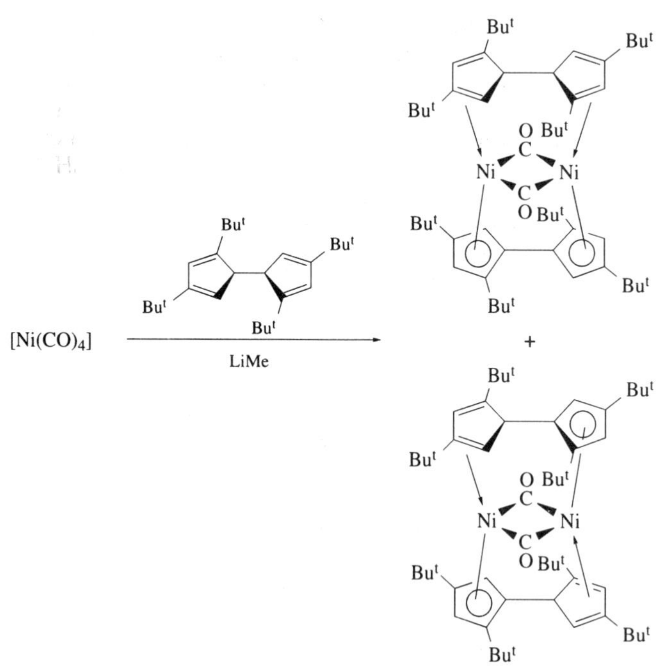

Scheme 51

The reaction of $[Ni_2(\mu\text{-}CO)_2Cp''_2]$ (Ni–Ni; Cp'' = Cp', Cp*, $C_5H_4Bu^t$, $C_5H_3Bu^t_2$-1,3) with P_4 afforded heterocubane type clusters with four or five of the vertices of the cube occupied by phosphorus atoms as shown in Scheme 52.[351,352] The stoichiometry depends on the nature of Cp'': $[Ni_3(\mu_3\text{-}P)(\eta^3\text{-}P_4)Cp^*_3]$ (**75**) was formed from $[Ni_2(\mu\text{-}CO)_2Cp^*_2]$ and P_4; mononuclear $[Ni(\eta^3\text{-}P_3)(\eta\text{-}C_5H_3Bu^t_2\text{-}1,3)]$ (**76**) resulted when Cp'' = $C_5H_3Bu^t_2$-1,3; the same reaction with nickel-ligated Cp' or $C_5H_4Bu^t$ ligands afforded compounds (**77**) which have an Ni_4P_4 heterocubane core. Similar complexes were obtained with As_4. The structures of $[Ni_3(\mu_3\text{-}E)(\eta^3\text{-}E_4)Cp^*_3]$ (E = P, As) and of $[Ni(\eta^3\text{-}P_3)Cp]$ were ascertained by x-ray diffraction studies.[351,352] More recent research studied the reactions of P_4 and As_4 with $[Ni_2(\mu\text{-}CO)_2(\eta\text{-}C_5HPh_4)_2]$ (Ni–Ni). The complexes $[Ni_2E_4Cp''_2]$ (E = P, As; Cp'' = $\eta\text{-}C_5HPh_4$) exhibit yet another structural type: a trigonal prismatic Ni_2As_4 core. Mononuclear species $[Ni(\eta^3\text{-}E)Cp'']$ were also obtained.[353]

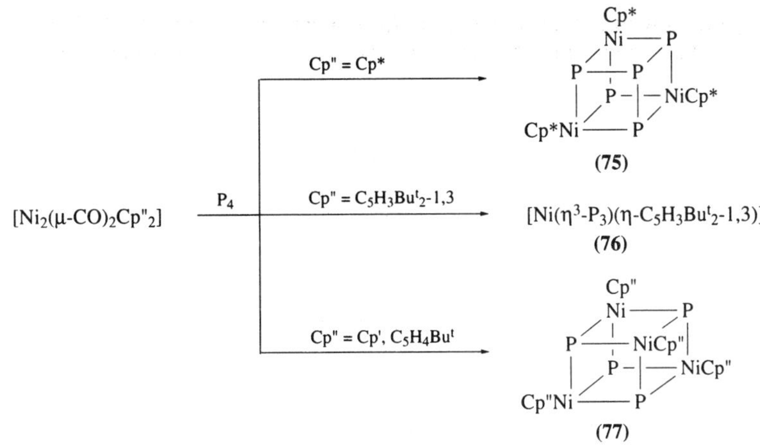

Scheme 52

3.5.5.3 Heterodinuclear complexes containing nickel η^5-ligands

Heterobimetallic complexes containing Ni(CO)Cp" or NiCp" (Cp" = Cp, Cp' or Cp*) groups linked to $M(CO)_3Cp$ (M = Cr, Mo, W) have been prepared and their chemistry researched. The species [Ni(CO)Cp"–$M(CO)_3Cp$] (**78**) are prepared by either metal–metal bond metathesis reactions between the homobimetallic species (for the Cp and Cp' species) or by treating *in situ* solutions of [NiI(CO)Cp*] with the appropriate $[M(CO)_3Cp]^-$ anion for the NiCp* compounds. Complexes (**78**) react with alkynes to afford μ-$R^1C_2R^2$ species, discussed earlier in the alkyne section.[236]

There are significant differences between the chemistry of (**78**) (Cp" = Cp or Cp') heterobimetallic complexes and that of the corresponding (**78**) (Cp" = Cp*) species. The Cp* group in NiCp* complexes is more labile and Cp* ligand transfer from the nickel atom to molybdenum or tungsten atoms has been documented in the reactions of (**78**) (Cp" = Cp*; M = Mo, W) with PhC_2H.[545] Later work led to the isolation of complex (**79**), [NiCp*(μ-PhC_2H)Ni{μ-η^2(*Mo*),η^4(*Ni*)-$C_4H_2Ph_2$}$Mo(CO)_2Cp$], which was isolated from the reaction of (**78**) (Cp" = Cp*; M = Mo) with PhC_2H. The Ni_2Mo compound contains an Ni–Ni–Mo chain and the central nickel atom was stripped of its Cp* group during the course of the reaction.[239] The Cp or Cp' ligands in the analogous (**78**) (Cp" = Cp, Cp'; M = Mo) are not labile under similar reaction conditions and Cp–Cp' ligand transfer does not take place.

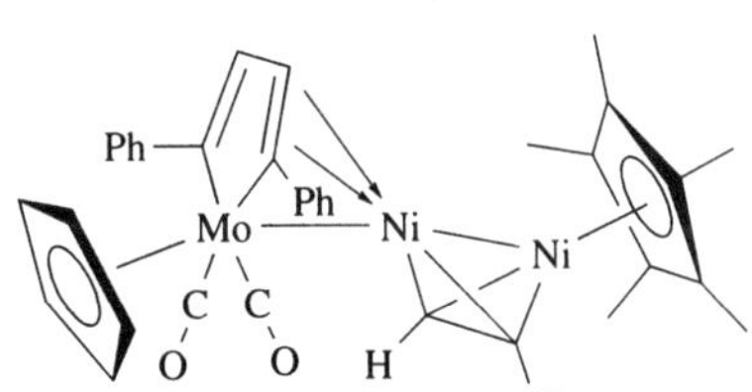

(79)

The NiCp and NiCp' species (**78**) lose CO when thermolyzed and form paramagnetic clusters with Ni_3M cores (see Section 3.5.6.3). In contrast, the NiCp* complexes (**78**) (Cp" = Cp*; M = Mo, W) readily decarbonylate when mildly heated to afford the highly reactive unsaturated paramagnetic molecules [NiCp*–$M(CO)_3Cp$] (**80**) (M = Mo, W).[533] These complexes add on various two-electron donor ligands such as trialkylphosphines and Bu^tNC yielding complexes of general formula [NiCp*(μ-$CO)_2$ML(CO)Cp]. With CH_2N_2, the complexes [NiCp*(μ-CH_2)(μ-CO)$M(CO)_2Cp$] (**81**) (Ni–M; M = Mo, W) are obtained in high yield, as shown in Scheme 53. These species are the first established examples of nickel–methylene complexes.[483,534]

The complexes [NiCp*(μ-ER)(μ-CO)M(ER)(CO)Cp'] (Ni–M; ER = SPh, M = Mo, W; ER = SePh, SMe, M = W) were prepared by reacting the appropriate [NiCp*–$M(CO)_3Cp'$] (Ni–M; M = Mo, W) species (analogous to complexes (**80**)) with the respective dialkyl disulfide or dialkyl diselenide. Compound (**82**) [$Ni_2(\mu$-$SePh)_2Cp^*_2$] (E = Se, R = Ph) was obtained as a side product in the reaction with Ph_2Se_2. The x-ray structures of [NiCp*(μ-SPh)(μ-CO)W(SPh)(CO)Cp'] and of (**82**) (E = Se, R = Ph) were determined.[525] Complexes with μ-SEt ligands spanning nickel–niobium centers have also been prepared; the structure of [NiCp(μ-SEt)$_2NbCl_2Cp_2$] was determined.[531]

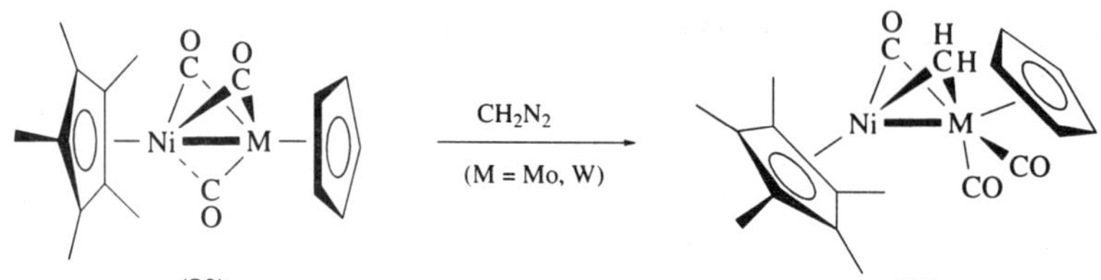

(80) (81)

Scheme 53

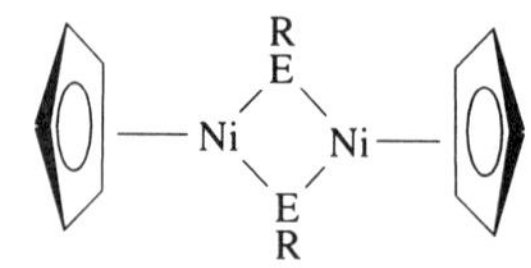

(82)

Other examples of the well-known thiolato-bridged complexes $[Ni_2(\mu\text{-}SR)_2Cp_2]$ (**82**) (E = S; R = Ph, CH_2Ph, 2-nap, $n\text{-}C_8H_{17}$; $(ER)_2 = C_6H_4(CH_2S)_2\text{-}1,2$) have been prepared by the reaction of (**73**) with various disulfides. The structures of two examples of (**82**) (E = S; R = Ph; $(ER)_2 = C_6H_4(CH_2S)_2\text{-}1,2$) were established crystallographically. Complexes (**82**) could be reduced with $LiAlH_4$ and yielded the corresponding hydrocarbon species RH in yields ranging from 55% to 80%. Intermediate $[Ni_2(\mu\text{-}SR)_2Cp_2]^-$ anions (which were not isolated) are believed to be involved. The redox chemistry of the series of complexes (**82**) was also studied: the complexes undergo a reversible one-electron reduction and no carbon–sulfur cleavage takes place.[451,452]

Other sulfur species with NiCp groups have been prepared from complex (**73**). This species reacts with 4-phenyl-1,2,3,5-dithiazole (**83**) to afford $[Ni_2(\mu\text{-}PhCN_2S_2)Cp_2]$ (Ni–Ni) (**84**) as shown in Scheme 54.[528] This unusual 19-electron red complex contains a Ni_2S_2 butterfly core and was characterized by an x-ray diffraction study. Calculations indicate that the unpaired electron density in this compound is delocalized over the nickel, the sulfur, and the carbon atoms.[528] The tetrathiolene complex $[Ni_2(\mu\text{-}C_{10}Cl_4S_4)Cp_2]$ (**85**) was prepared from the reaction of (**73**) with the free tetrathiolene ligand in 77% yield (Scheme 55). Its structure was established by a single-crystal x-ray diffraction study.[527]

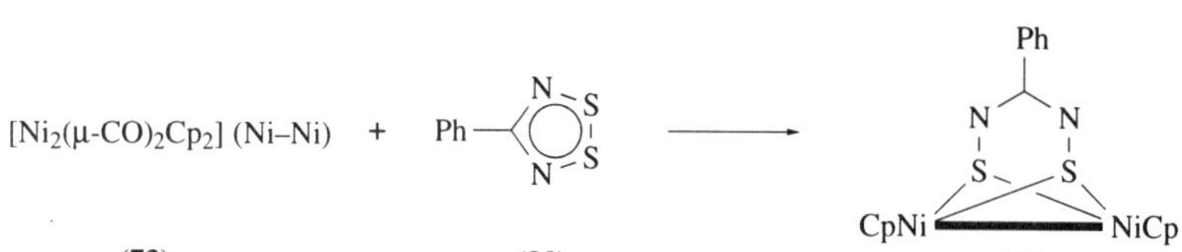

(73) (83) (84)

Scheme 54

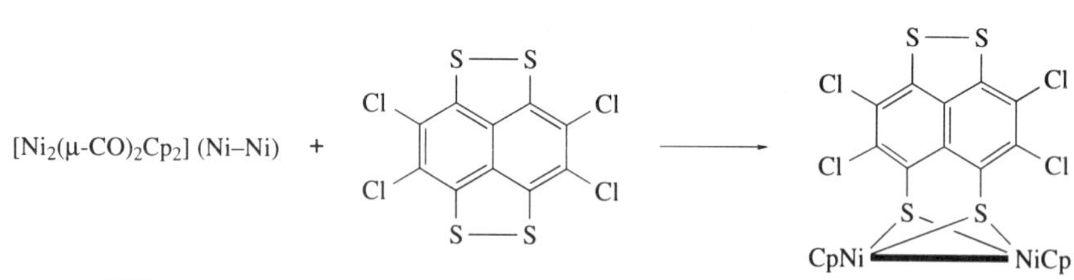

(73) (85)

Scheme 55

3.5.6 Clusters with NiCp or NiCp" Groups (Cp" = Substituted Cp)

3.5.6.1 General comments

A large number of homo- and heteronuclear metal clusters which contain NiCp and substituted NiCp groups are known. Some of these species have found catalytic applications. Many clusters which contain NiCp fragments are prepared from $[Ni_2(\mu\text{-}CO)_2Cp_2]$ (Ni–Ni) or $[NiCp_2]$ or the Cp" derivatives

(Cp" = substituted Cp) of these species. Other less common methods are metal atom syntheses or from heterometallic complexes that contain an NiCp or NiCp" fragment. The coverage, especially of mixed-metal clusters with NiCp" groups, is not comprehensive, but illustrative. Clusters are discussed when they are of unusual interest or when the nickel atom(s) play key roles. Table 24 lists homonuclear clusters that contain NiCp or NiCp" groups. Heteronuclear clusters that contain NiCp or NiCp" groups are found in Table 25. Nickel clusters, including many that contain η^5-ligated ligands, have been reviewed.[546]

Table 24 Homonuclear clusters containing nickel η^5-ligands.[a]

Cluster	*Ref.*	*Cluster*	*Ref.*
$[Ni_3(\mu_3\text{-CH})Cp_3]$	72	**$[(Ni_3Cp_3)_2(\mu\text{-}C_2S_6)]$** (Cp" = **Cp, Cp'**)	350
$[Ni_3(\mu_3\text{-CCO}_2Me)Cp_3]$	547	**$[Ni_3(\mu_3\text{-CS})_2Cp''_3]$** (Cp" = **Cp, Cp'**)	552
$[Ni_3(\mu_3\text{-CMe})Cp_nCp^*_{3-n}]$ (n = 0–3)	548	$[Ni_3(\mu_3\text{-H})_2(\eta\text{-}C_5H_5Bu^t)_3]$	553
$[Ni_3(\mu\text{-S})_2Cp_3]^+SbF_6^-$	549	$[Ni_3(\mu_3\text{-H})_2(\eta\text{-}C_5H_3Bu^t_2\text{-}1,3)_3]$	553
$[Ni_3(\mu\text{-L})_2Cp_3]$ (L = S, CO, etc.)	550	$[Ni_4H_n(\eta\text{-}C_5H_4Bu^t)_4]$	553
$[Ni_4(\mu_4\text{-Se})_2Cp_4]$	551	$[Ni_4(\mu_3\text{-P})_4Cp''_4]$ (Cp" = Cp', $C_5H_4Bu^t$)	352
$[Ni_3(\mu_3\text{-CO})_2Cp_3]$	543	**$[Ni_3(\mu_3\text{-E})(E_4)Cp^*_3]$** (E = **P, As**)	351
$[Ni_3(\mu_3\text{-CS})(\mu_3\text{-CO})Cp_3]$	552	**$[Ni_6Cp_6]^{n+}$** (**n = 0, 1**)	554

[a] Representative list. Clusters shown in bold were characterized by an x-ray diffraction study.

3.5.6.2 Homonuclear clusters with NiCp and NiCp" groups

The reaction of $[NiCp_2]$ with alkyllithium reagents and various alkenes or dienes is a valuable synthetic entry into many cyclopentadienylnickel alkene and alkyl complexes. When no alkene was introduced to the reaction mixture, tetrahedral trinickel clusters such as $[Ni_3(\mu_3\text{-CMe})Cp_3]$ (**86**) (R = Me; Scheme 56)[548] were obtained. Similar reactions with [NiCpCp*] afford mixtures of the clusters $[Ni_3(\mu_3\text{-CMe})Cp^*_{3-n}Cp_n]$ (n = 0–2); the compounds where n = 1 or 2 are inseparable by chromatography, but can be sublimed away from the Cp^*_3 cluster $[Ni_3(\mu_3\text{-CMe})Cp^*_3]$. The mechanism of formation of this cluster was probed.[548] The related cluster complex $[Ni_3(\mu\text{-CCO}_2Me)Cp_3]$ (**86**) (R = CO_2Me) has been prepared independently.[547]

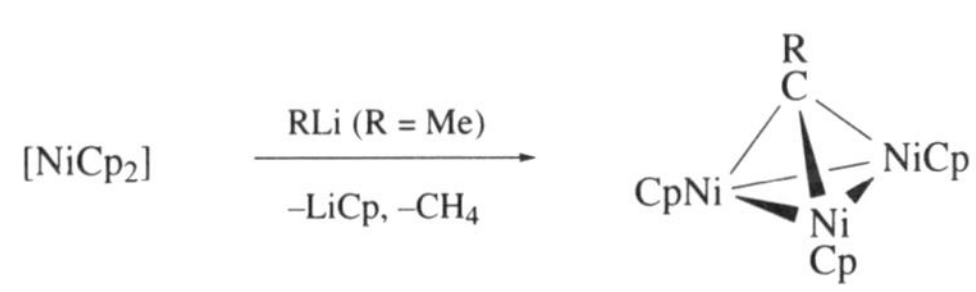

(**86**) R = Me

Scheme 56

Nickel cyclopentadienyl and substituted cyclopentadienyl groups are present in a number of paramagnetic clusters and contribute to the stability of such species. The paramagnetic cluster $[Ni_3(\mu\text{-CO})_2Cp_3]$ and its derivatives continue to attract interest. The structures of $[Ni_3(\mu_3\text{-CO})_2Cp'_3]$ and $[Ni_3(\mu_3\text{-CO})_2Cp_3]^-$ were reported and $[Ni_3(\mu\text{-CO})_2Cp^*_3]$ was also prepared. The HOMO of $[Ni_3(\mu_3\text{-CO})_2Cp'_3]$ has antibonding Ni_3 character.[568]

An x-ray diffraction study and MO calculations (Fenske–Hall) have been carried out on the sulfide-bridged Ni_3 cluster species $[Ni_3(\mu_3\text{-S})_2Cp_3]^+$ (**87**$^+$). This trinickel cluster is produced by oxidation of the neutral cluster species $[Ni_3(\mu_3\text{-S})_2Cp_3]$ (**87**) with $[FeCp_2]^+SbF_6^-$. Calculations show that the HOMO in (**87**) is antibonding in character, so a general shortening of nickel–nickel bonds is observed in the oxidized cation (**87**$^+$) compared with (**87**). The molecule has C_{2v} symmetry: nickel–nickel distances in (**87**$^+$)SbF_6^- are 0.253 6(7) nm and 0.314 5(2) nm, and the latter distance is not considered to be a nickel–nickel bonding interaction. Magnetic measurements of (**87**)$^+$ confirm the paramagnetism of the complex (μ_{eff} = 2.7 μ_B; 2.8 μ_B using Faraday's method).[549] The redox chemistry of (**87**), $[Ni_3(\mu_3\text{-CO})_2Cp_3]$, and other $[Ni_3(\mu_3\text{-L})_2Cp_3]$ species has been studied. The redox potential of the clusters was ligand dependent.[550] The structures of $[Ni_4(\mu_4\text{-Se})_2Cp_4]$ and $[Ni_4(\mu_3\text{-E})_2(PPh_3)_2Cp_4]$ (E = Se, Te) were determined crystallographically.[551] Other Ni_4 clusters have been synthesized by the same group.[575] The synthesis and structure of the Ni_3 species $[(Ni_3Cp'_3)_2(\mu\text{-}C_2S_6)]$ have been reported.[450]

Homonickel clusters have been synthesized via metal vapor techniques. Nickel atoms react with butylcyclopentadiene to yield various tri- and tetranuclear clusters including $[Ni_3(\mu_3\text{-H})_2(\eta^5\text{-}C_5H_5Bu^t)_3]$

Table 25 Heteronuclear clusters with nickel η^5-ligands.[a]

Cluster	*Ref.*
$[Ni_3Cp_3M(\mu_3\text{-}CO)_3Cp'']$ (M = Mo, W; Cp'' = Cp, **Cp'**)	555
$[Ni_2Mo\{\mu_3\text{-}As[Mo_3(\mu_3\text{-}As)(CO)_5Cp_3]\}(CO)_2Cp_3]$	556
$[NiCp^*(\mu\text{-}PhC_2H)Ni(\mu\text{-}\eta^2,\eta^4\text{-}C_4H_2Ph_2)Mo(CO)_2Cp]$	239
$[Ni_2Cp^*_2M_2Cp_2(CO)_4]$ (M = Mo, W)	533
$[Ni_2Mo_2(\mu_3\text{-}S)_2(\mu_4\text{-}CO)Cp_4]$	557
$[Ni_2Mo_2(\mu_3\text{-}CPh)(\mu_3\text{-}CO)_2Cp_4]$	558
$[NiMo_2(\mu_3\text{-}As)(CO)_4Cp_3]$	556
$[Ni_2W(\mu_3\text{-}P_2)(CO)_4(\eta\text{-}C_5HPr^i_4)_2]$	271
$[Ni_2W(\mu_3\text{-}PO)_2(CO)_4(\eta\text{-}C_5HPr^i_4)_2]$	271
$[Ni_2(\mu\text{-}\eta^3,\eta^3\text{-}P_4)(\eta\text{-}C_5HPh_4)_2\{W(CO)_5\}_2]$	353
$[NiCoMo(\mu_3\text{-}CMe)(CO)_5Cp_2]$	559
$[Ni_2Fe(\mu_3\text{-}PBu^t)(CO)_3Cp_2]$	560
$[Ni_2Fe_2(\mu_4\text{-}\eta^2\text{-}PhC_2Ph)(CO)_6Cp_2]$	241
$[Ni_2Fe_2(\mu\text{-}PPh_2)(\mu_4\text{-}\eta^2\text{-}C_2Ph)(CO)_5Cp_2]$	242
$[NiCp(\mu_3\text{-}\eta^2\text{-}C_2Pr^i)Fe_2(CO)_6]$	243
$[NiCp(\mu_3\text{-}\eta^2\text{-}C_2Bu^t)Fe_2(CO)_5(PPh_3)]$	243
$[Ni_2Cp_2(\mu_4\text{-}\eta^1,\eta^2\text{-}C_2H)Fe(CO)Cp''(\mu\text{-}CO)Ni(CO)]$ (Cp'' = **Cp**, Cp*)	244
$[Ni_2M_3(\mu_4\text{-}\eta^2\text{-}PhC_2Ph)(CO)_6Cp_2]$ (M = Fe, Ru)	241
$[Ni_2CpRu(\mu\text{-}PhC_2Ph)(CO)_3]$	245
$[Ni_2Cp_2Ru_2(HC_2CEt)(CO)_6]$	246
$[Ni_2Cp_2Ru_2\{\mu_4\text{-}C(CHCMeCH_2)\}(CO)_6]$	246
$[NiCpRu_2(\mu\text{-}PhC_2Ph)(\mu_3\text{-}CO)(CO)_3]$	247
$[NiCpRu_3(CO)_9]$	245
$[NiCpRu_3(\mu\text{-}PhC_2Ph)(CO)_9]$	245
$[NiRu_4(\mu_4\text{-}C_2Pr^i)_2(\mu\text{-}PPh_2)_2(CO)_9Cp]$	248
$[NiCpRu_3(\mu_2\text{-}H)(\mu_4\text{-}\eta^2\text{-}C_2Bu^t)(CO)_9]$	249
$[NiCpM_3(\mu\text{-}H)_3(CO)_8(PR_3)]$ (M = Ru, Os; R = Ph, Et, Cy; PR_3 = $PHPh_2$)	561
$[NiCpM_3(\mu\text{-}H)_3(CO)_7(PR_3)_2]$ (M = Ru, Os; R = Ph, Et, Cy; PR_3 = $PHPh_2$)	562
$[NiOs_3(\mu_2\text{-}H)_3(CO)_9]$	563, 564
$[Ni_3Os_3(CO)_9Cp_3]$	544, 564
$[NiOs_3(\mu\text{-}H)(\mu\text{-}C{=}CHBu^t)(CO)_9Cp]$	241, 565
$[NiOs_3(\mu\text{-}H)_3(Pr^iC_2PPh_2)(CO)_8Cp]$	250
$[NiOs_3(\mu_2\text{-}H)_2(CO)_{10}(PPh_3)_2]$	566
$[NiOs_3(\mu\text{-}H)_3(CO)_8(AsPPh_3)]$	567
$[NiOs_3(\mu\text{-}H)_3(CO)_9Cp\{\mu\text{-}(Ph_2PC_2Pr^i)Co_2(CO)_6\}]$	250
$[NiFeCo(\mu_3\text{-}PBu^t)(CO)_6Cp]$	560
$[NiFe_2Co(\mu_3\text{-}PMe)_2(CO)_8Cp]$	560
$[NiCo(\mu\text{-}\eta^2,\eta^{2'}\text{-}Bu^tCPML_n)(CO)_3Cp]$ (ML_n = $W(CO)_5$, $Mn(CO)_2Cp$)	233
$[Ni_2Co(\mu_3\text{-}CO)_2(CO)_3Cp''_2]$ (Cp'' = Cp', Cp*)	316
$[Ni_2Co(\mu_3\text{-}CO)_2Cp_3]$	543
$[Ni_2Co(\mu_3\text{-}CO)_2Cp_2Cp^*]$	568
$[K(2.2.2\text{-}crypt)]^+[Ni_2Co(\mu_3\text{-}CO)_2Cp_2Cp^*]^-$	568
$[NiCo_2(\mu_3\text{-}CCO_2Pr^i)(CO)_6Cp]$	569
$[NiCo_2(\mu_3\text{-}CCO_2Pr^i)(\mu\text{-}CO)Cp_3]$	569
$[NiCo_2(\mu_3\text{-}CMe)(CO)_6Cp]$	559
$[NiCo_2M(CO)_8CpCp']$ (M = Mo, W)	570
$[NiCoW_2(CO)_7CpCp'_2]$	570
$[\{\mu\text{-}Ni(PPh_3)Cp\}Zn_2(\mu_2\text{-}Cp)(\eta^1\text{-}Cp)_2]$	571, 572
$[Ni_2Zn_4Cp_6]$	573
$[Ni_2Cp_2Zn_4(\eta^3\text{-}Cp)_2(\eta^2\text{-}Cp^*)_2]$	574

[a] Noncomprehensive list. Clusters shown in bold were characterized by an x-ray diffraction study.

and $[Ni_4H_n(\eta^5\text{-}C_5H_5Bu^t)_4]$. The cluster $[Ni_3(\mu_3\text{-}H)_2(\eta^5\text{-}C_5H_3Bu^t_2\text{-}1,3)_3]$ was also obtained.[553] When nickelocene was reduced with the powerful reducing agent sodium naphthalenide, the cluster $[Ni_6Cp_6]$ was isolated, along with other species. This Ni_6 cluster and the cation $[Ni_6Cp_6]^+$ were structurally characterized.[554]

3.5.6.3 Heteronuclear clusters with NiCp and NiCp'' groups

The mixed-metal cluster $[Ni_2Co(\mu_3\text{-}CO)_2Cp_3]$ has been made as shown in Equation (20). The unpaired electron in this paramagnetic cluster occupies an antibonding MO.[543] The similar CoCp* cluster $[Ni_2Co(\mu_3\text{-}CO)_2Cp_2Cp^*]$ has also been studied. The nickel–nickel and nickel–cobalt distances lengthen when it undergoes a one-electron reduction with $K^+C_{10}H_8^-$ and 2.2.2-crypt. The extra electron in the anion resides in an antibonding orbital of predominately nickel character. The structures and MO schemes of these clusters were compared with those of $[Ni_3(\mu_3\text{-}CO)_2Cp'_3]$, $[Ni_3(\mu_3\text{-}CO)_2Cp_3]^-$, and $[Ni_3(\mu_3\text{-}CO)_2Cp^*_3]$.[568] The catalytic activity of $[NiCo_2(\mu_3\text{-}CMe)(CO)_6Cp]$, $[NiCoMo(\mu_3\text{-}CMe)(CO)_5Cp_2]$, and other clusters towards the photoinduced hydrosilylation of acetophenone was probed.[559]

$$[Ni_2(\mu\text{-}CO)_2Cp_2] \;+\; [Co(CO)_2Cp] \longrightarrow [Ni_2Co(\mu_3\text{-}CO)_2Cp_3] \qquad (20)$$

A series of mixed-metal Ni_3Mo and Ni_3W compounds with tetrahedral metal cores have been prepared by reacting $[Ni_2(\mu\text{-}CO)_2Cp''_2]$ (**73**) (Cp'' = Cp, Cp') with $[M_2(CO)_6Cp''_2]$ (M–M; M = Mo, W; Cp'' = Cp, Cp'). This and other synthetic pathways are shown in Scheme 57.[533,555] The resulting $[Ni_3M(\mu\text{-}CO)_3Cp''_3Cp]$ (**88**) (Cp'' = Cp, Cp') clusters have 62 valence electrons, are paramagnetic, and exhibit a temperature-dependent singlet–triplet equilibrium. The structure of the cluster (**88**) (Cp'' = Cp) was established by an x-ray diffraction study. This revealed an asymmetric Ni_3Cp_3 triangular structure, capped by a Cp'Mo unit (nickel–nickel distances were 0.248 9(2) nm, 0.251 0(2) nm, and 0.262 7(2) nm).[555] Prolonged thermolysis of the NiCp* complexes $[Ni(CO)Cp^*M(CO)_3Cp]$ (Ni–M; M = Mo, W) led to formation of the 60-electron Ni_2M_2 clusters $[Ni_2M_2(CO)_4Cp_2Cp^*_2]$. The coordinatively unsaturated species $[NiCp^*\text{–}M(CO)_3Cp]$ (**80**) Ni–M; M = Mo, W) are isolable intermediates in this decarbonylation and clustering reaction (Scheme 57).[533]

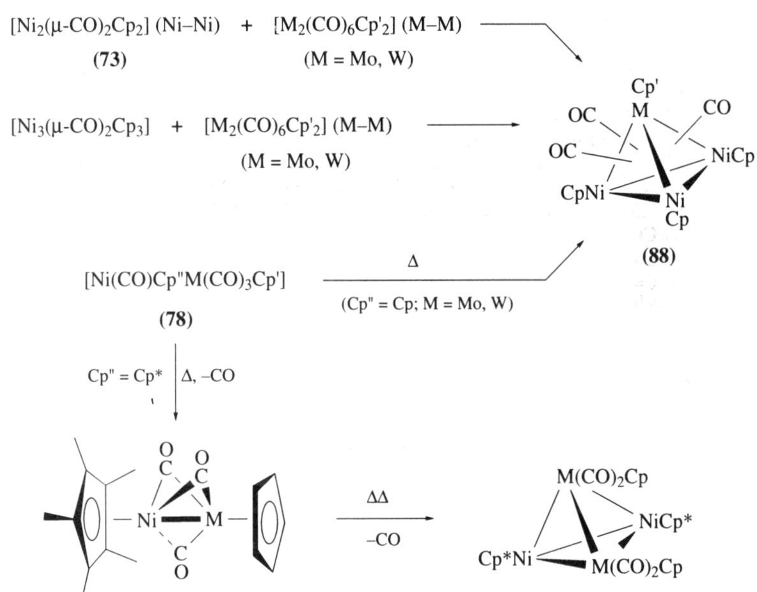

Scheme 57

Many other clusters that contain NiCp species combined with group 6 metals have emerged since the mid-1980s. These include the tetrahedral mixed-metal clusters $[NiCo_2M(CO)_8CpCp']$ (M = Mo, W) and $[NiCoW_2(CO)_7CpCp'_2]$, all of which contain NiCp fragments.[570] The reaction of $[Ni_2(\mu\text{-}PhC_2Ph)Cp_2]$ (Ni–Ni) with $[Mo_2(CO)_6Cp_2]$ afforded the tetranuclear mixed-metal alkylidyne cluster $[Ni_2Mo_2(\mu_3\text{-}CPh)(\mu_3\text{-}CO)_2Cp_4]$.[558] Other NiMo clusters containing NiCp units include $[NiMo_2(\mu_3\text{-}As)(CO)_4Cp_3]$ and $[Ni_2Mo\{\mu_3\text{-}As[Mo_3(\mu_3\text{-}As)(CO)_5Cp_3]\}(CO)_2Cp_3]$.[556] The tetranuclear butterfly cluster $[Ni_2Mo_2(\mu_3\text{-}S)_2(\mu_4\text{-}CO)Cp_4]$, prepared by refluxing $[Mo_2(CO)_2(\mu\text{-}SMe)_2Cp_2]$ with $[NiCp_2]$, contains a CO ligand that interacts with all four metal atoms.[557]

Various mixed-metal clusters that contain NiCp moieties have been prepared by reacting $Co_2Fe(\mu_3\text{-}PR)$ cluster compounds with (**73**). Species isolated depend both on the nature of R and on the reflux time with the Ni_2 species. Clusters formed included $[NiCoFe(\mu_3\text{-}PBu^t)(CO)_6Cp]$, $[Ni_2Fe(\mu_3\text{-}PBu^t)(CO)_3Cp_2]$, and $[NiCoFe_2(\mu_3\text{-}PMe)_2(CO)_8Cp]$.[560]

An interesting series of mixed-metal nickel–zinc clusters with Cp and/or Cp* ligands has been synthesized from the reactions of $[ZnCp_2]$ or $[ZnCp^*_2]$ with $[Ni(cod)_2]$. The dienyl ligands in many of these clusters are η^5-, η^3-, or even η^1-coordinated to the zinc atom, though η^5-coordination of these ligands is usually maintained to the nickel atoms. Among the clusters that have been synthesized is $[Ni_2Zn_4Cp_6]$.[573] Mixed-ligand species such as $[Ni_2Zn_4Cp_4Cp^*_2]$ have been prepared by reacting $[Ni(cod)_2]$ with mixtures of $[ZnCp_2]$ and $[ZnCp^*_2]$.[574]

Many examples of clusters that contain $NiRu_3$ and $NiOs_3$ tetrametallic cores have been reported. Some of these clusters, such as $[NiOs_3(\mu\text{-}H)_3(CO)_9]$, are catalytically active and many reports of their syntheses and applications have appeared.[241,242,245–50,544,561–3,567,576–8] Other papers are found in a review of nickel cluster species.[546]

3.6 NICKEL π-ARENE COMPLEXES

3.6.1 General Comments

A large number of nickel η^6-arene complexes have been synthesized by metal vapor techniques. Most of these species contain nickel(II). Once the existence of these species had been established, many of these complexes were prepared by "wet" synthetic methods. Some nickel arene complexes have also resulted from the reactions of metal alkynes. In many nickel(0) arene complexes such as in the naphthalene complex $[Ni(\eta^2\text{-}C_{10}H_8)(PPr^i_2C_2H_4PPr^i_2)]$,[84] the arene is η^2- and not η^6-ligated. These species have been discussed earlier. Table 26 lists nickel η^6-arene species. Nickel π-arene moieties that form part of a multidecker complex are listed together with other multidecker complexes in Table 27.

Table 26 Nickel arene complexes.[a]

Complex	*Ref.*	*Complex*	*Ref.*
$[Ni_2(\mu\text{-}\eta^1,\eta^6\text{-}PC_5H_2Ph_3\text{-}2,4,6)_2(PCy_3)_2]$	61	$\mathbf{[Ni(\eta^2\text{-}C_{10}H_8)(PBu^t_2C_2H_4PBu^t_2)]^b}$	86
$[Ni(\eta^4\text{-}C_4Me_4)(\eta^6\text{-}C_6Me_6)]^{2+}(AlCl_4^-)_2$	378	$[Ni(\eta^2\text{-}C_{14}H_{10})(PBu^n_3)_2]^c$	87
$[NiR_2(\eta\text{-}TolH)]$ (R = SiF_3, $SiCl_3$, C_6F_5)	579, 580	$[Ni(\eta^2\text{-}C_{14}H_{10})(PR_3)_2]$ (R = Et, Bu^n)^c	88
$[Ni(SiCl_3)_2(\eta\text{-}PhH)]$	581	$[Ni(\eta^4\text{-}C_{14}H_8O_2)(PCy_3)]^d$	89
$[NiR_2(\eta\text{-}ArH)]$ (R = C_6F_5, C_6F_4H, C_6Cl_5, SiX_3)	582	$[Ni_2(\mu\text{-}CO)_2(\eta^6\text{-}C_5H_5BR)_2]$ (R = Me, Ph)	584
$[Ni(SiF_3)_2(\eta\text{-}TolH)]$	583	$\mathbf{[Ni(\eta^6\text{-}C_5H_5BPh)(\eta^4\text{-}C_5H_7BPh)]}$	131
$[Ni(\eta^2\text{-}C_{10}H_8)\{PPr^i_2(CH_2)_nPPr^i_2\}]^b$ (*n* = 2, 3)	84	$[Ni(\eta^6\text{-}C_5H_5BPh)(1,4,5\text{-}\eta^3\text{-}C_8H_{13})]$	131

[a] Other nickel arene complexes are listed in Table 27 (nickel multidecker complexes). [b] $C_{10}H_8$ = naphthalene. [c] $C_{14}H_{10}$ = anthracene. [d] $C_{14}H_8O_2$ = anthraquinone. Structures determined by x-ray diffraction studies shown in bold.

3.6.2 Syntheses and Reactions

There is evidence of nickel arene species from gas-phase MS studies. The reactions of Ni^+ ions with halobenzenes have been investigated.[400] More recent studies, using FT-MS techniques, showed that $[Ni(C_6H_5X)_n]^+$ ions (*n* = 1, 2) formed with C_6H_5X (X = F, Cl) while bromo- and iodobenzene resulted in multiple dehydrohalogenation reactions and the formation of $[Ni(C_6H_4)_n]^+$ (*n* = 1–6 for PhBr; *n* = 1–5 for PhI).[604]

A series of $[NiR_2(\eta\text{-}ArH)]$ (**89**) (Ar = Ph, Tol, Mes; R = SiF_3, $SiCl_3$, C_6F_5) complexes have been prepared by metal vapor techniques and also by more conventional methods. Vaporizing nickel and co-condensing it with Si_2F_6 and toluene on to a liquid nitrogen chilled flask affords $[Ni(SiF_3)_2(\eta\text{-}TolH)]$ of ~99% purity, in yields approaching 20% (based on vaporized nickel). This green complex is highly soluble in organic solvents and decomposes when kept above 50 °C or when exposed to air.[584]

The chemistry of $[NiR_2(\eta\text{-}ArH)]$ complexes towards arene ligand substitution has been investigated. In general, metal–arene bond strengths decrease in the order MesH > TolH ≥ PhOMe > PhH. Nevertheless, benzene can substitute the toluene in the complex $[Ni(SiCl_3)_2(\eta\text{-}TolH)]$.[581,605] Carbon monoxide results in formation of the complexes $[NiR_2(CO)_3]$ when R = SiF_3 or $SiCl_3$, but reductive elimination of C_6F_5–C_6F_5 ensues when R = C_6F_5; $[Ni(CO)_4]$ is also formed in the latter case.

Derivatives of the type $[NiX_2L_2]^{2-}$ (X = halide) can be obtained from the complexes $[NiR_2(\eta\text{-}TolH)]$ (R = SiF_3, $SiCl_3$, C_6F_5).[579,606] By utilizing TlCp, the arene ligand can be replaced by an η-Cp group, and the structure of the anion $[Ni(C_6F_5)_2Cp]^-$ (as an NEt_4^+ salt) has been determined crystallographically.[607] These results[579,581,606,607] are summarized in Scheme 58.

The use of the complexes $[NiR_2(\eta\text{-}ArH)]$ (**89**) as isomerization catalysts for 1-butene has been investigated.[608] Some of these species are highly active homogeneous catalysts for the dimerization of ethene, propene, and butene. Their activity in their own right and in conjunction with alkylaluminum halides has been investigated in bromobenzene: η^4- and η^2-C_6H_5Br complexes are believed to be involved in the cycle. The catalytic activity of the compounds decreases in the order R = $SiCl_3 \gg SiF_3 > C_6F_5$.[609]

Table 27 Nickel multidecker complexes.[a]

Complex	*Ref.*
$[(NiCp)_2(\mu-\eta^5,\eta^5-Cp)]^+AlBr_4^-$	585
$[(NiCp)_2(\mu-\eta^5,\eta^5-Cp)]^+BF_4^-$	586
$[(NiCp)_2(\mu-\eta^5,\eta^5-Cp)]^+[Ni\{Ge(C_6F_5)_3\}_2Cp]^-$	512
$[Ni_2(\mu-C_4H_4BR)Cp_2]$	587
$[Ni\{(\eta^3-C_3H_4R-2)Ni(\mu-\eta^5,\eta^{5'}-BEtCEtBEtC_2Et_2)\}_2]$ (R = H, Me)	588
$[(\eta^3-C_3H_5)Ni(\mu-\eta^5,\eta^{5'}-C_2Me_2BMeCRBMe)Ni(\eta^1-C_3H_5)_2]$	579
$[Ni\{[\eta^5-C_2Et_2BMeC(H)BMe]CoCp\}_2]$	520
$[\{CpCo(\mu-\eta^5,\eta^{5'}-BMeC_2Et_2BMeCH)\}_2Ni]$	590
$[CpCo\{CMeBEtC_2Et_2BEt\}Ni(\mu-CO)]_2$	591
$[CpCo(CMeBEtC_2Et_2BEt)Ni_2(\mu-\eta^5,\eta^{5'}-CMeBEtC_2Et_2BEt)]$	591
$[Ni_2(\mu-\eta^5,\eta^{5'}-C_2B_2Et_4BMe)Cp_2]^{n-}$ (*n* = 0, 1)	592
$[Ni\{(\mu-\eta^4,\eta^5-C_2Et_2B_4H_4)Fe(\eta^6-C_9H_8)\}Cp^*]$	593
$[Ni\{CpCo(\mu-C_2Et_2B_3H_2X-5)\}_2]$ (X = **COMe**, Cl, Br, CH_2C_2Me)	594
$[NiCp\{(\mu-\eta^5,\eta^{5'}-L)Pt(\eta^5-L)\}]$ (L = $CMeC_2B_2Et_4$)	595
$[Ni_2(\mu-\eta^3,\eta^4-CH_2CHMeCCEtBEtNMeSiMe_2)_2]$	596
$[Ni_2(\mu-\eta^6,\eta^{6'}-B_2Me_2C_4H_2Me_2)Cp_2]$	597
$[Ni_2(\mu-\eta^6,\eta^{6'}-B_2Me_2C_4H_2R_2)(\eta^1-C_3H_4Me-2)_4]$	597
$[Ni_2Cp_2(\mu-\eta^5,\eta^{5'}-CMeC_2B_2Et_4)_2Pt]$	595
$[Ni_2\{\mu-\eta^5,\eta^{5'}-C_4H_4BO(CH_2)_3OBC_4H_4\}_2]$	467
$[\{CpCo(\mu-\eta^5,\eta^{5'}-L)Ni(\mu-\eta^5,\eta^{5'}-C_2Et_2BMeCHBMe)\}_2Ni]$	391
$[\{CpCo(\mu-\eta^5,\eta^{5'}-L)\}_2Ni]$ (L = $C_2Et_2BMeCHBMe$)	391
$[\{CpCo(\mu-\eta^5,\eta^{5'}-L)Ni\}_2(\mu-\eta^5,\eta^{5'}-L)]$ (L = $C_2Et_2BEtCMeBEt$)	391
$[Ni_2Cp_2(\eta^5,\eta^{5'}-C_2R^1{}_2BR^2CR^1BR^2)]^{n-}$ (*n* = 0, 1 (BF_4^-))	598
$[CpNi(\mu-\eta^5,\eta^{5'}-L)CoCp]^{n+}$ (*n* = 0, 1; L = $C_2Et_2BMeCHBMe$)	598
$[\{CpCo(\mu-\eta^5,\eta^{5'}-L)\}_2Ni]$ (L = $C_2Et_2BMeCHBMe$)	598
$[CpCo(\mu-\eta^5,\eta^{5'}-L)Ni(\mu-CO)_2FeCp]$ (L = $C_2Et_2BMeCHBMe$)	598
$[\{CpCo(\mu-\eta^5,\eta^{5'}-L^1)Ni(\mu-\eta^5,\eta^{5'}-L^2)\}_2Ni]$ (L^1 = $C_2Et_2BEtCEtBEt$; L^2 = $C_2Me_2BMeCMeBMe$)	391
$[(NiFeCp_4)(BF_4^-)_2]_n$	599
$[Ni(\mu-\eta^5,\eta^{5'}-C_2Me_2BMeCHBMe)]_n$	600
$[(\eta^3-C_3H_5)Ni(\mu-\eta^5,\eta^{5'}-C_2Me_2BMeCHBMe)(\eta^3-C_3H_5)_2]$	600
$[Ni\{(2,3-C_2Et_2B_3HXY)CoCp^*\}_2]$ (X = H, Cl, Br, I, CH_2C_2Me, COMe; Y = H, Cl)	601
$[Ni\{(C_2Et_2B_3H_2Cl)Co[C_6H_4(C_5Me_4)_2-1,4]Co(C_2Et_2B_3H_2Cl)\}_2]$	602
$[Ni\{(2,3-C_2Et_2B_3HCl_2)CoCp^*\}_2]$	603
$[Ni\{(2,3-C_2Et_2B_3Cl_2F)CoCp^*\}_2]$	603

[a] Noncomprehensive list. Structures determined by x-ray diffraction studies shown in bold.

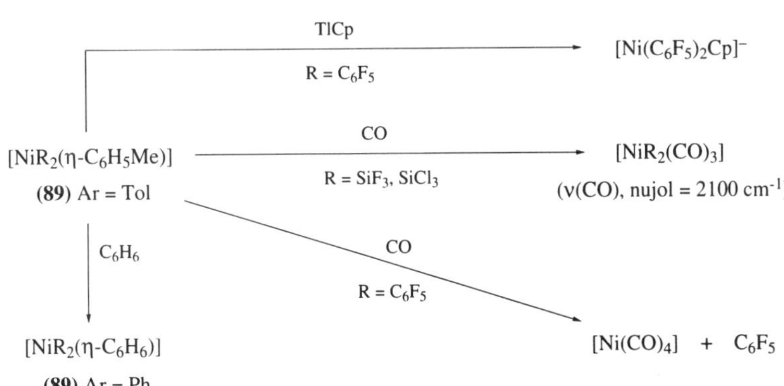

Scheme 58

Many nickel arene compounds have been prepared by more conventional methods. Treatment of nickel(II) trifluoroacetate with various lithium alkyls followed by the arene in the presence of a Lewis acid (typically BF_3) afforded complexes (**89**) (Ar = PhH) as shown in Equation (21).[582] When $[Ni(\eta^3-C_3H_4Me-2)_2]$ was treated with $SiHX_3$ in a liquid nitrogen matrix at 77 K and the solid mixture warmed to ambient temperature, the allyl ligands were displaced and the arene complex $[Ni(SiX_3)_2(\eta-TolH)]$ was obtained.[580] The syntheses of a number of nickel η^6-arene complexes have appeared in *Organometallic Syntheses*, Volume 3. These include the species $[Ni(C_6X_5)_2(\eta-TolH)]$ (X = F, Cl) formed in 86% yield from $[Ni(OAc)_2]$, toluene, and LiC_6X_5[610,611] and the complexes $[Ni(SiX_3)_2(\eta-TolH)]$ (X = F, Cl) formed in yields of 85% (X = F) and 60% (X = Cl) from $[Ni(\eta^3-C_3H_4Me-2)_2]$, $SiHX_3$, and toluene.[612,613]

The dinickel π-bound phosphapyridine complex (**90**) was prepared by thermolysis of the phosphorus-bound complex $[Ni(\eta^2-C_2H_4)(PCy_3)(PC_5H_2Ph_3-2,4,6)]$.[61] The mixed-ligand dicationic sandwich complex

$$[Ni(CO_2CF_3)_2] \xrightarrow[\text{ii, arene/BF}_3]{\text{i, RLi}} [NiR_2(\eta\text{-ArH})] \quad (21)$$

$(R = C_6F_5, C_6F_4H, C_6Cl_5, SiF_3, SiCl_3)$

$[Ni(\eta^6\text{-}C_6Me_6)(\eta^4\text{-}C_4Me_4)]^{2+}(AlCl_4^-)_2$ was isolated from a cyclooligomerization reaction of 2-butyne with aluminum(III) chloride, followed by $[Ni(CO)_4]$ in CH_2Cl_2. Other $Ni(\eta^4\text{-}C_4Me_4)$ complexes were isolated.[378,614]

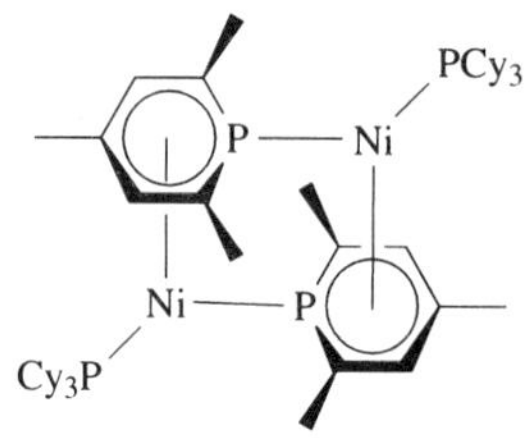

(90)

Sandwich complexes with diborolene ligands, such as (**91**), have been reported.[615] The borolene complex (**92**) $[Ni_2(\mu\text{-}CO)_2(\eta^6\text{-}C_5H_5BR)_2]$ (R = Me, Ph) has been prepared in a ligand transfer reaction from $[Co(\eta^6\text{-}C_5H_5BR)_2]$ and $[Ni(CO)_4]$ (Scheme 59).[584] No mixed-metal complex was formed. These complexes were characterized by ^1H and ^{11}B NMR, IR, and elemental analyses.[584] Other nickel borolene complexes were synthesized by reacting the free ligand with $[Ni(CO)_4]$. This led to two isomers of $[Ni(\eta^6\text{-}C_5H_5BPh)(\eta^4\text{-}C_5H_7BPh)]$ being formed, one of which was characterized by an x-ray diffraction study (Scheme 60).[131] The complex $[Ni(cod)_2]$ reacted with C_5H_6BPh to afford the cyclooctenyl species $[Ni(\eta^6\text{-}C_5H_5BPh)(\eta^1,\eta^2\text{-}1,4,5\text{-}C_8H_{13})]$ in accord with Equation (22). Strong nickel–boron interactions are present in this species.[131]

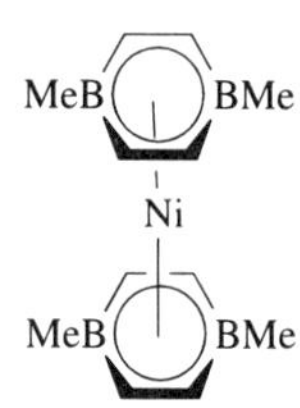

(91)

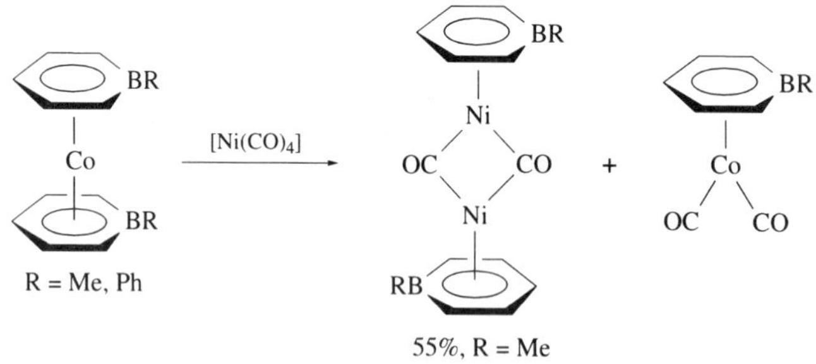

Scheme 59

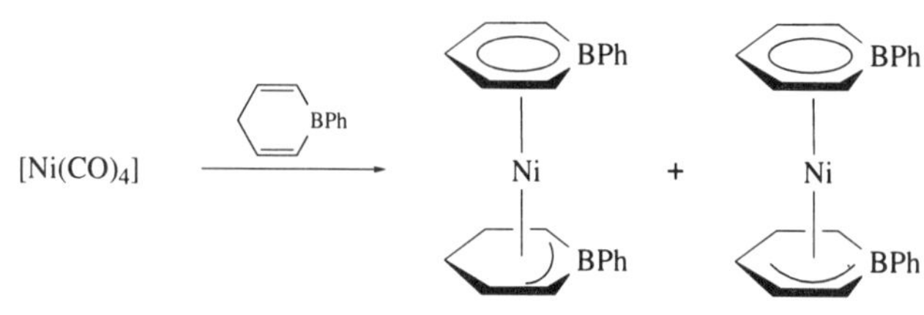

Scheme 60

$$[Ni(cod)_2] + C_5H_6BPh \longrightarrow [Ni(\eta^6\text{-}C_5H_5BPh)(\eta^1,\eta^2\text{-}1,4,5\text{-}C_8H_{13})] \quad (22)$$

3.7 NICKEL π-BONDED POLYDECKER COMPLEXES

3.7.1 General Comments

Nickel is frequently one of the metals that constitute the "filling" in many polydecker sandwich complexes. The tripledecker cationic species $[Ni_2Cp_2(\mu\text{-}Cp)]^+$ is easily prepared, but more complicated procedures are required to prepare many of the other multidecker species. Table 27 contains a selection of multidecker complexes that contain nickel π-bound ligands. A large number of such complexes have been synthesized and they are not covered comprehensively here. The majority of these contain heterocyclic rings, principally boracycles. Mixed-metal complexes, particularly those that involve cobalt, are especially common.

3.7.2 Theoretical and Physical Studies

A structural study of the tripledecker $[Ni_2(\mu\text{-}Cp)Cp_2]^+$ [586] revealed that the three Cp rings were within 3.97° of coplanarity. The structure of this tripledecker cation has been redetermined (with $[Ni\{Ge(C_6F_5)_3\}_2Cp]^-$ as the counterion).[512] The solution and solid-state IR spectra of the tripledecker complexes $[Ni_2(\mu\text{-}Cp)Cp_2]^+X^-$ ($X = BF_4$ or PF_6) have been investigated and skeletal vibrations were assigned.[616] Extended Hückel calculations were carried out on a series of tripledecker complexes that contained boron or phosphorus. Relationships between valence electron counts and stability were examined.[617]

Most multidecker diamagnetic complexes were characterized by NMR spectroscopy and MS studies. The electrochemistry of many of these species was studied; magnetic measurements and ESR spectra were recorded for most paramagnetic species.

3.7.3 Synthetic Studies

The reaction of nickelocene with Lewis acids sometimes leads to the tripledecker sandwich cationic species $[Ni_2(\mu\text{-}Cp)Cp_2]^+$. This complex is obtained when nickelocene is treated with Al_2Br_6.[585] This tripledecker species is a useful synthon and has been used synthetically to cap monocyclopentadienide species with an NiCp group.[464]

"Slipped" tripledecker complexes incorporating the fused Cp ring pentalenide dianion have been prepared for nickel and for other transition metals. Species such as (**65**) and its iron and cobalt analogues are accessible by treatment of $[M(acac)_2]$ (M = Ni, Fe, Co) with LiCp* and dilithium pentalenide. Yields are in the range 48–70% (Scheme 61).[516] The polymeric mixed-metal polydecker species $\{[NiFeCp_4]^+(BF_4^-)_2\}_n$ has been made by oxidizing $[NiCp_2]^+BF_4^-$ with $[FeCp_2]^+BF_4^-$. The polymer contains nickel(IV) and iron(II) and its stacked structure is discussed.[599]

Nickel carbonyl reacts with the diborole species (**93**) to afford a mixed-ligand π-allyl derivative (**94**) (Scheme 62). An Et group has been dehydrogenated to a vinyl moiety in this species.[317] Alkyl-2,3-dihydro-1,3-diboroles react with $[Ni(\eta^3\text{-}C_3H_5)_2]$ or $[Ni(\eta^3\text{-}C_3H_4Me\text{-}2)_2]$ to yield species such as (**95**) and other compounds.[588] Further reactions of this species with CoCp sandwich complexes yield tetradecker and pentadecker mixed-metal species such as (**96**) (Scheme 63). A similar pentadecker complex (**97**) with a different alkylation pattern to (**96**) was formed as shown in Scheme 64.[391] Allyl groups in $[Ni(\eta^3\text{-}C_3H_5)_2]$ can be displaced by tetramethyl-1,3-diborole (**98**) to afford (**99**) (Scheme 65). This complex reductively eliminates $[Ni(\eta^3\text{-}C_3H_5)_2]$ and C_6H_{10} to form the first example of a polydecker complex (**100**). Detailed experimental procedures are given for the preparation of this oxygen sensitive polymer, which is thermally stable to 500 °C and is insoluble in organic solvents.[601] Complex (**99**) is an especially labile species; tetradecker complexes were prepared from it.[589]

A series of polydecker complexes involving nickel and cobalt have been synthesized. The paramagnetic doubledecker complex $[CpNi(\mu\text{-}C_2B_2Et_4CMe)NiCp]$ has been prepared and may be reduced with potassium to the monoanion. The anion was studied by multinuclear (^1H, ^{13}C, ^{11}B) NMR. The odd electron is localized between both metals; chemical shifts span ranges of 60 (^1H), 1900 (^{13}C), and 1000 (^{11}B) ppm, respectively.[592] The tripledecker complex (**101**) and other polydecker species, both

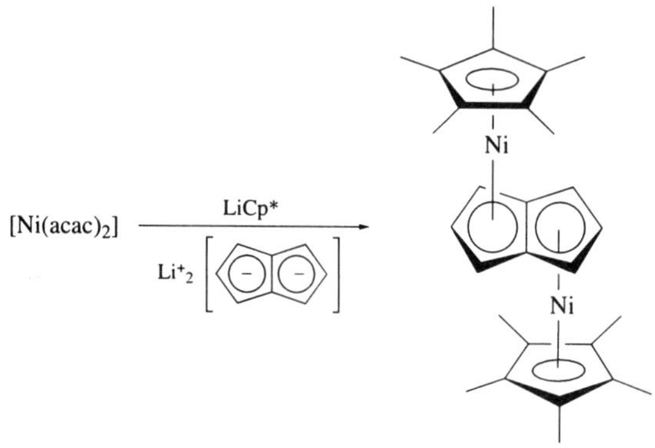

(65)
40–60%

Scheme 61

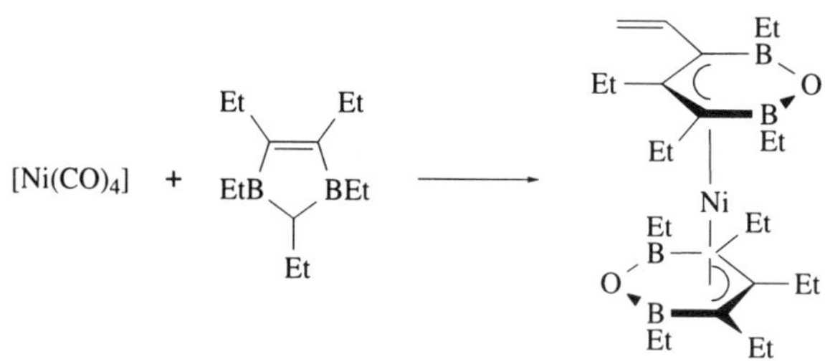

(93) **(94)**

Scheme 62

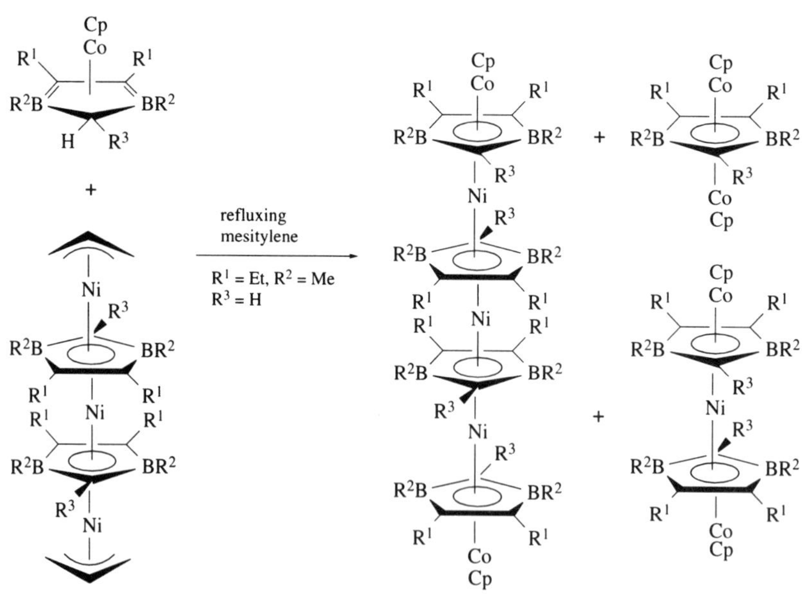

(95) **(96)**

Scheme 63

of which contain substituted 1,3-diboracyclopentadiene ligands, were prepared; complex (**101**) was structurally characterized. Nickel carbonyl has been used as a starting material for other mixed-metal polydeckers as illustrated in a typical example in Scheme 66.[590,591,618,619] Other complexes with μ-η^5,$\eta^{5'}$-$C_3B_2Me_4$ ligands involving nickel are described.[600,620,621]

A series of tripledecker complexes of general formula $[M^1M^2(\mu\text{-}C_3B_2R_5)Cp_2]$ (M^1M^2 = FeCo, Co_2, CoNi, Ni_2) and the tetradecker complex $[CpCo(\mu\text{-}C_3B_2R_5)Ni(\mu\text{-}C_3B_2R_5)CoCp]$ have been synthesized; their electronic structure was studied and their electrochemistry and ESR spectra were recorded. For the CoNi complex, the monocation was isolable and the ESR spectrum of this species exhibited ^{59}Co hyperfine splitting. There is a high degree of delocalization, and a single well-separated one-electron

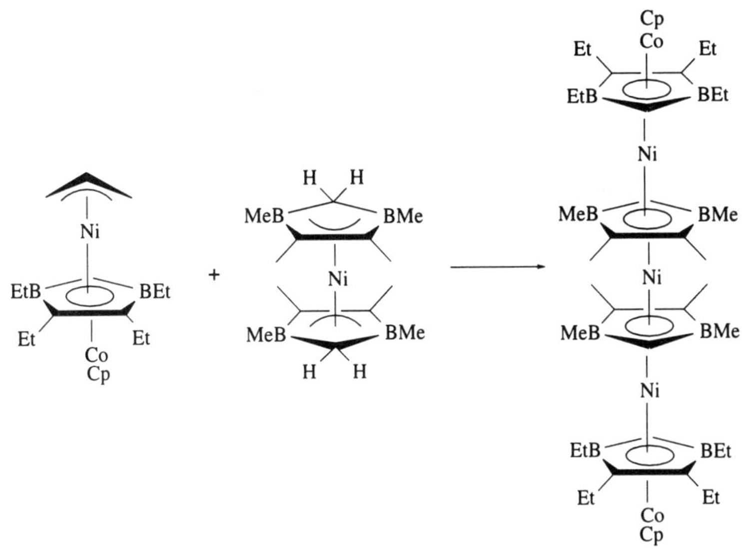

(97)

Scheme 64

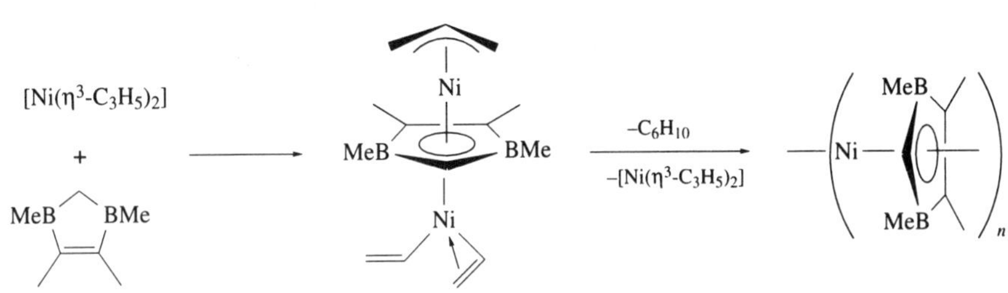

(98) **(99)** **(100)**

Scheme 65

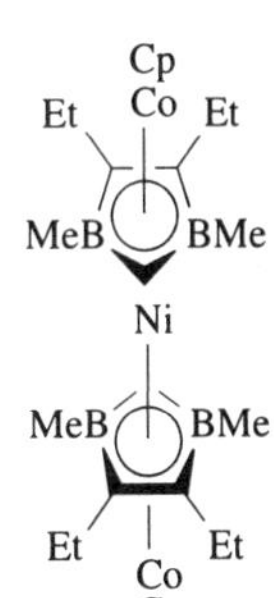

(101)

wave was seen. For the dinickel species, reduction to the 34-electron monoanion was possible. The heterogeneous charge transfer reaction, measured by a.c. polarography, indicated that very rapid electron transfer occurred.[598,622]

Tripledecker complexes of general formula $[Ni_2(\mu\text{-}\eta^5,\eta^{5'}\text{-}C_4H_4BR)Cp_2]$ have been studied.[587] Tetradecker complexes containing nickel trapped between two cobalt sandwich compounds are prepared as shown in Equation (23).[594] Other multidecker species characterized by diffraction studies include $[CpNi(\mu\text{-}\eta^5,\eta^{5'}\text{-}MeCB_2C_2Et_4)Pt(\eta^5\text{-}MeCB_2C_2Et_4)]$, $[Pt\{(\mu\text{-}\eta^5,\eta^{5'}\text{-}MeCB_2C_2Et_4)NiCp\}_2]$,[595] and $[(\eta^6\text{-}C_9H_8)Fe(\mu\text{-}\eta^4,\eta^5\text{-}Et_2C_2B_4H_4)NiCp^*]$.[593]

$$[Co(\eta^5\text{-}C_2Et_2B_3H_3X)Cp^*] \xrightarrow[\text{THF}]{[NiBr_2]} [Cp^*Co(\eta^5\text{-}C_2Et_2B_3H_3X)Ni(\eta^5\text{-}C_2Et_2B_3H_3X)CoCp^*] \quad (23)$$

$(X = COMe, Cl, Br, CH_2C{\equiv}CMe)$

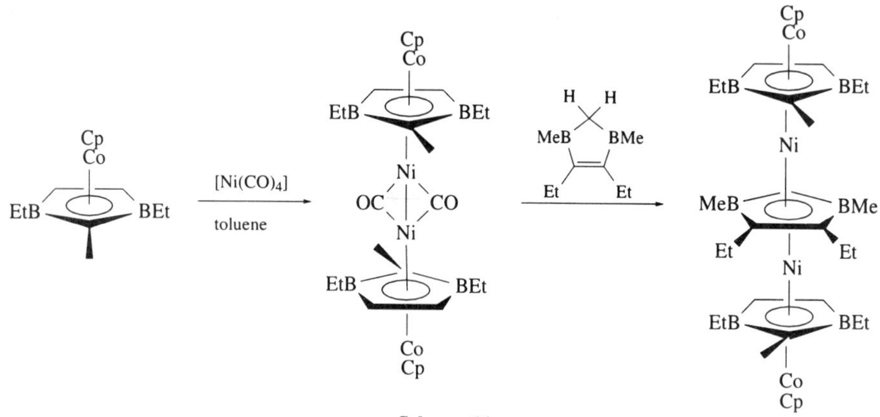

Scheme 66

A series of 2-methylallyldiborolene species of the type $[Ni_2(C_3H_4Me\text{-}2)_4(\mu\text{-}\eta^6,\eta^{6'}\text{-}B_2R^1{}_2C_4R^2{}_4)]$ have been prepared; the tripledecker complex $[Ni_2(\mu\text{-}\eta^6,\eta^{6'}\text{-}B_2Me_2C_4Me_2H_2)Cp_2]$ was structurally characterized.[597] The species $[Ni_2(\mu\text{-}\eta^3,\eta^4\text{-}CH_2CHMeCCEtBEtNMeSiMe_2)_2]$ was also synthesized and its structure was established crystallographically.[596]

3.8 BORANE AND CARBORANE COMPLEXES

Coverage in this section provides a sampling of some of the material in the literature and is not intended to be comprehensive. A selection of nickel borane and carborane species is given in Table 28.

Table 28 Nickel boranes and carboranes.[a]

Complex	*Ref.*
$[NiBr(PPh_3)(\eta^5\text{-}C_2Et_2B_4H_3PPh_3)]$	623
***closo*-$[3\text{-}(\eta^2\text{-}Ph_2Bpz_2)\text{-}3,1,2\text{-}NiC_2B_9H_{11}]^-[NMe_4]^+$**	624
$[Ni(\eta^5\text{-}2,3,5\text{-}C_3B_3HMe_3Et_2)_2]$ (104)	423
$[Ni(\eta^3\text{-}C_3H_3Me\text{-}1\text{-}Et\text{-}3)(\eta^5\text{-}2,3,5\text{-}C_3B_3HMe_3Et_2)]$	625
$[11\text{-}7,8,10\text{-}(\eta^3\text{-}C_4H_3Me_3)NiC_3B_7H_7Me_3\text{-}5,7,8]$	625
$[Ni(1,2\text{-}C_2B_9H_{11})(bipy)]$	626–8
$[(Cb_2Ni)_2Ni{\cdot}4L]$ ($Cb = \pi\text{-}(3)\text{-}1,2\text{-}C_2B_9H_{11}$; 2 L = bipy, 2 py)	629
$[Cb_2Ni]^-Na^+{\cdot}2\,L$ ($Cb = \pi\text{-}(3)\text{-}1,2\text{-}C_2B_9H_{11}$; 2 L = bipy, 2 py)	629
$[Ni(1\text{-}O_2C\text{-}1,2\text{-}C_2B_9H_{11})_2(bipy)]$	629
$[6\text{-}\eta^5\text{-}CpNi\text{-}\eta^5\text{-}1\text{-}CB_8H_9]$	630
$[10\text{-}\eta^5\text{-}CpNi\text{-}\eta^4\text{-}1\text{-}CB_8H_9]$	630
$[Ni_2Cp_2(6,7,1\text{-}CB_7H_8)]$	631, 632
$[Ni(\delta\text{-}MeO\text{-}1,2\text{-}C_2B_9H_{10})_2]$	633
$[1\text{-}Br\text{-}1,5\text{-}(PPh_3)_2\text{-}2,3\text{-}Et_2\text{-}1,2,3\text{-}NiC_2B_4H_3]$	623
$[11\text{-}(\eta^3\text{-}C_4Me_4H)\text{-}5,7,8\text{-}Me_3\text{-}11,7,8,10\text{-}NiC_3B_7H_7]$	634
closo-$[(1,1\text{-}dppe)\text{-}1,2,3\text{-}NiC_2B_4H_6]$	635
$[4,5\text{-}\mu\text{-}NiBr(dppe)\text{-}2,3\text{-}C_2B_4H_7]$	635
***commo*-$[1,1'\text{-}Ni\{2,4\text{-}(SiMe_3)_2\text{-}2,4\text{-}C_2B_4H_4\}_2]$**	636
***closo*-$[1\text{-}Ni(TMEDA)\text{-}2,4\text{-}(SiMe_3)_2\text{-}2,4\text{-}C_2B_4H_4]$**	636
$[9\text{-}CpNi\text{-}8\text{-}Me\text{-}7,8,10\text{-}C_3B_7H_9]$	637

[a] Noncomprehensive list. Compounds in bold were characterized by x-ray diffraction studies.

Extended MO calculations on $[Ni_4Cp_4(B_4H_4)]$ indicated that the polyhedral electron pair theory broke down here (and for the cobalt analogue). The nickel complex adopted a D_{2d} type structure (an elongated metal tetrahedron) based on a dodecahedron.[638]

Thermally generated nickel, iron, and cobalt atoms were co-condensed with *nido*-$2,6\text{-}C_2B_7H_{11}$ and with a variety of arenes and other functionalized organic molecules. The only characterized product for nickel (obtained with MeC_2Me and toluene) was the compound $[11\text{-}(\eta^3\text{-}C_4Me_4H)\text{-}5,7,8\text{-}Me_3\text{-}11,7,8,10\text{-}NiC_3B_7H_7]$. The nickel atom is bound via an η^3-tetramethylcyclobutenyl ligand and to a C_2B_2 section of the C_3B_7 cage which has an unusual 11-vertex cage geometry.[634]

A series of fascinating complexes are formed from the reactions of various 2,2-dihydro-1,3-diboroles. When the synthesis of (**102**) was attempted, a variety of polydecker species (**103**) were readily obtained

(Scheme 67).[639,640] The mechanism of formation of (**103**) is not known, but reactive carbon–hydrogen bonds at the 2-position appear to be required for the reaction to proceed as shown.[620,641] When [Ni(η-$C_2H_4)_3$] was reacted with excess 1,3-dimethyl-4,5-dialkyl-1,3-diboroles, the centrosymmetric cluster complex [Ni(η^5-1,4,6-Me_3-2,3-Et_2-2,3,4-$C_3B_3H)_2$] (**104**) was formed, along with the allyl complex [11,7,8,10-(η^3-$C_4H_3Me_3$)Ni$C_3B_7H_7Me_3$-5,7,8] (**105**) (Scheme 68). The structures of both (**104**) and (**105**) were established by x-ray diffraction studies. Complex (**104**) is paramagnetic and exhibits Curie–Weiss behavior, with a magnetic moment ($\mu_{eff} = 3.06 \pm 0.02$ μ_B) corresponding to two unpaired electrons. This value is similar to that of nickelocene. These results clearly indicate that the *nido*-2,3,5-tricarbahexaborate ion is isolobal with the cyclopentadienyl ion.[423,625]

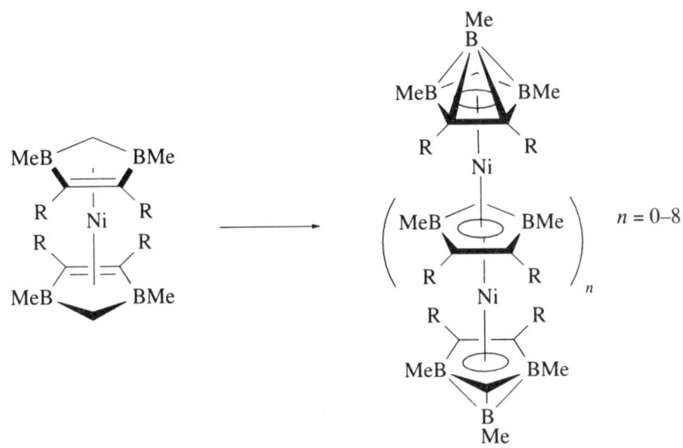

(102) **(103)**

Scheme 67

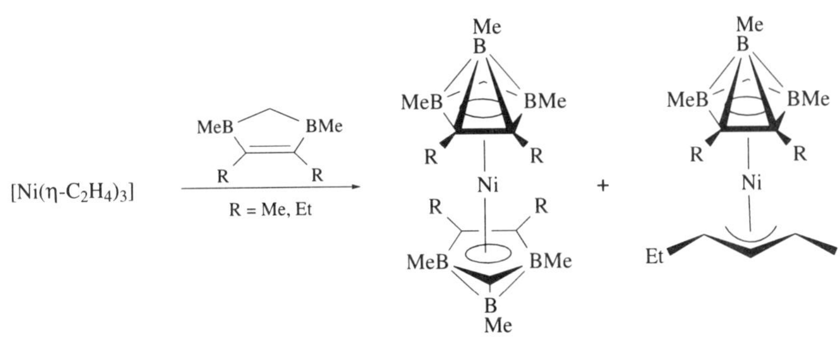

(104) **(105)**

For **(104)** (R = Et): Ni–ring center = 0.177 nm

Scheme 68

Electrochemical studies were performed on (**104**).[423,625] The electrochemistry of some polydecker NiCp sandwiches that contain boron and of [Ni(cod)(η^5-$C_3Et_2HB_3Me_3$)]$^+BF_4^-$ was investigated. The structure of the latter species and of [Ni(η^5-$C_3Et_2HB_2Me_2$)(η^5-$C_3Et_2C_3B_3Me_3$)] was established by an x-ray study.[98]

Nickelocene reacts with $Na^+(4\text{-}CB_8H_{13})^-$ to afford [6-{(η-Cp)Ni}-η^5-1-CB_8H_9]. This complex rearranges to give a 10-vertex *closo*-metallacarborane [10-(η-Cp)Ni-η^4-1-CB_8H_9].[630] Nickelaborane anions are formed when [NiCp_2] is reduced with $[NBu_4]^+{}_2[B_6H_6]^{2-}$.[642] The structure of the carborane species [Ni(δ-MeO-1,2-$C_2B_9H_{10})_2$] was established by a diffraction study.[633]

Carborane complexes whose molecular parameters were ascertained by x-ray diffraction studies include [6,7,1-$Cp_2Ni_2CB_7H_8$][631,632] and the η^5-carborane complex [NiBr(PPh_3)(η^5-$Et_2C_2B_4H_3PPh_3$)].[623] The nickel(II) species [NiX_2(dppe)] (X = Cl, Br) react with $[C_2B_4H_7]^-$ to form *closo*-[1,1-(dppe)-1,2,3-Ni$C_2B_4H_6$]. It was suggested that this species forms via the *nido*-intermediate [4,5-μ-{NiX(dppe)}-2,3-$C_2B_4H_7$]. The intermediate was observed for X = Cl by low-temperature NMR studies.[635]

The sodium salt of the nickel(II) carborane sandwich compound $Na^+(NiCb_2)^-$ reacts with pyridine or 2,2'-bipy to afford [Na(NiCb_2)·2 L] (Cb = π-(3)-1,2-$C_2B_9H_{11}$; 2 L = bipy, 2 py). The pyridyl adduct reacts with [NiCl_2] and 2,2'-bipy or pyridine to afford the mixed-valence nickel(II)/nickel(III) species [Ni(Cb_2Ni)$_2$·2 L_2] (Cb = π-(3)-1,2-$C_2B_9H_{11}$; L_2 = bipy, 90–94%; L_2 = 2py, 75%). Pyrolysis of both species at 200–230 °C generated 3-(2,2'-bipy)-*closo*-3,1,2-nickeladicarbadodecaborane-4. The same species formed when [Ni(1-O_2C-1,2-$C_2B_9H_{11}$)$_2$(bipy)] was pyrolyzed.[629] An x-ray diffraction study of the anionic nickelacarborane species $[NEt_4]^+$*closo*-[3-(η^2-Ph_2Bpz_2)-3,1,2-$NiC_2B_9H_{11}$]$^-$ revealed unexpected distortions imposed by the Ph_2Bpz_2 ligand.[624] The electrolysis of mixtures of bipy and *o*-$C_2B_{10}H_{12}$ with nickel electrodes affords [1,2-$C_2B_9H_{11}$Ni(bipy)] in yields up to 20%. A similar reaction was described with 1,7-$C_2B_{10}H_{12}$.[626] Two isomers of [Ni($C_2B_9H_{11}$)(bipy)] were later structurally characterized.[627,628]

3.9 REFERENCES

1. G. Marr and B. W. Rockett, *J. Organomet. Chem.*, 1983, **242**, 1.
2. B. W. Rockett and G. Marr, *J. Organomet. Chem.*, 1983, **257**, 275.
3. G. Marr and B. W. Rockett, *J. Organomet. Chem.*, 1984, **278**, 331.
4. B. W. Rockett and G. Marr, *J. Organomet. Chem.*, 1988, **343**, 553.
5. G. Wilke, *Angew. Chem., Int. Ed. Engl.*, 1988, **27**, 185.
6. G. Wilke, *Angew. Chem.*, 1988, **100**, 190.
7. P. A. Chaloner, *J. Organomet. Chem.*, 1987, **337**, 431.
8. P. A. Chaloner, *J. Organomet. Chem.*, 1989, **374**, 349.
9. P. A. Chaloner, *J. Organomet. Chem.*, 1992, **432**, 387.
10. P. A. Chaloner, *J. Organomet. Chem.*, 1992, **442**, 271.
11. U. Zenneck, *Angew. Chem., Int. Ed. Engl.*, 1990, **29**, 126.
12. K. Jonas, *Pure Appl. Chem.*, 1990, **62**, 1169.
13. S. Pasynkiewicz, *J. Organomet. Chem.*, 1992, **429**, 1.
14. S. Pasynkiewicz, *J. Organomet. Chem.*, 1990, **387**, 1.
15. O. N. Efimov and V. V. Strelets, *Coord. Chem. Rev.*, 1990, **99**, 15.
16. K. Jonas, *Angew. Chem., Int. Ed. Engl.*, 1985, **24**, 295.
17. K. Tamao, K. Kobayashi and Y. Ito, *Synlett*, 1992, **3**, 539.
18. P. O. Widmark, B. O. Roos and P. E. M. Siegbahn, *J. Phys. Chem.*, 1985, **89**, 2180.
19. P. E. M. Siegbahn and U. B. Brandemark, *Theor. Chim. Acta*, 1986, **69**, 119.
20. K. Kitaura, S. Sakaki and K. Morokuma, *Inorg. Chem.*, 1981, **20**, 2292.
21. T. Ziegler, *Inorg. Chem.*, 1985, **24**, 1547.
22. O. Eisenstein and R. Hoffmann, *J. Am. Chem. Soc.*, 1981, **103**, 4308.
23. A. Baldy, A. Durocher, H. Rzehak and Y. Limouzin-Marie, *Bull. Soc. Chim. Fr.*, 1981, 575.
24. E. S. Kline, R. H. Hauge, Z. H. Kafafi and J. L. Margrave, *Organometallics*, 1988, **7**, 1512.
25. T. Merle-Mejean, C. Cosse-Mertens, S. G. F. Bouchareb, J. Mascetti and M. Tranquille, *J. Phys. Chem.*, 1992, **966**, 9148.
26. J. Mink, M. Gal, P. L. Goggin and J. L. Spencer, *J. Mol. Struct.*, 1986, **142**, 467.
27. P. Csaszar, P. L. Goggin, J. Mink and J. L. Spencer, *J. Organomet. Chem.*, 1989, **379**, 337.
28. R. L. Hettich and B. S. Freiser, *Organometallics*, 1989, **8**, 2447.
29. R. Houriet, L. F. Halle and J. L. Beauchamp, *Organometallics*, 1983, **2**, 1818.
30. L. F. Halle, P. B. Armentrout and J. L. Beauchamp, *Organometallics*, 1983, **2**, 1829.
31. C. A. Tolman, *Organometallics*, 1983, **2**, 614.
32. C. A. Tolman, W. C. Seidel and L. W. Gosser, *Organometallics*, 1983, **2**, 1391.
33. K.-R. Pörschke, Y.-H. Tsay and C. Krüger, *Angew. Chem., Int. Ed. Engl.*, 1985, **24**, 323.
34. K.-R. Pörschke, Y.-H. Tsay and C. Krüger, *Angew. Chem.*, 1985, **97**, 334.
35. K.-R. Pörschke, G. Wilke and R. Mynott, *Chem. Ber.*, 1985, **118**, 298.
36. W. Dreissig and H. Dietrich, *Acta Crystallogr., Sect. B*, 1981, **37**, 931.
37. C. Pluta, K.-R. Pörschke, R. Mynott, P. Betz and C. Krüger, *Chem. Ber.*, 1991, **124**, 1321.
38. R. Goddard, C. Krüger, K. R. Pörschke and G. Wilke, *J. Organomet. Chem.*, 1986, **308**, 85.
39. K.-R. Pörschke *et al.*, *Chem. Ber.*, 1985, **118**, 275.
40. W. Kaschube, K.-R. Pörschke, K. Angermund, C. Krüger and G. Wilke, *Chem. Ber.*, 1988, **121**, 1921.
41. K. Pörschke, G. Wilke and C. Krüger, *Angew. Chem., Int. Ed. Engl.*, 1983, **22**, 547.
42. B. Gabor, C. Krüger, B. Markzinke, R. Mynott and G. Wilke, *Angew. Chem., Int. Ed. Engl.*, 1991, **30**, 1666.
43. B. Gabor, C. Krüger, B. Markzinke, R. Mynott and G. Wilke, *Angew. Chem.*, 1991, **103**, 1711.
44. W. Kaschube, K. R. Pörschke, W. Bonrath, C. Krüger and G. Wilke, *Angew. Chem., Int. Ed. Engl.*, 1989, **28**, 772.
45. K.-R. Pörschke and G. Wilke, *Chem. Ber.*, 1985, **118**, 313.
46. W. Bonrath, K. R. Pörschke and S. Michaelis, *Angew. Chem., Int. Ed. Engl.*, 1990, **29**, 298.
47. S. Michaelis, K.-R. Pörschke, R. Mynott, R. Goddard and C. Krüger, *J. Organomet. Chem.*, 1992, **426**, 131.
48. W. Schröder, K.-R. Pörschke, Y.-H. Tsay and C. Krüger, *Angew. Chem., Int. Ed. Engl.*, 1987, **26**, 919.
49. W. Schröder, K.-R. Pörschke, Y.-H. Tsay and C. Krüger, *Angew. Chem.*, 1987, **99**, 953.
50. H. Lehmkuhl and T. Keil, *J. Organomet. Chem.*, 1988, **342**, C38.
51. H. Lehmkuhl *et al.*, *Chem. Ber.*, 1984, **117**, 3231.
52. H. Lehmkuhl, C. Naydowski, R. Benn, A. Rufinska and G. Schroth, *J. Organomet. Chem.*, 1982, **228**, C1.
53. H. Lehmkuhl, C. Naydowski and M. Bellenbaum, *J. Organomet. Chem.*, 1983, **246**, C5.
54. Z. Fu, Z. Chen, Z. Cai, K. Pang, G. Zhang and H. Zhu, *Jiegou Huaxue*, 1985, **4**, 203.
55. Z. Fu, Z. Chen, Z. Cai, K. Pang, G. Zhang and H. Zhu, *Chem. Abstr.*, 1986, **105**, 217 006c.
56. R. J. McKinney and D. C. Roe, *J. Am. Chem. Soc.*, 1985, **107**, 261.
57. H. Lehmkuhl, C. Naydowski, R. Benn, A. Rufinska and G. Schroth, *J. Organomet. Chem.*, 1983, **246**, C9.
58. H. Lehmkuhl *et al.*, *Chem. Ber.*, 1991, **124**, 441.
59. K. R. Pörschke, *J. Am. Chem. Soc.*, 1989, **111**, 5691.
60. F. M. Conroy-Lewis, L. Mole, A. D. Redhouse, S. A. Litster and J. L. Spencer, *J. Chem. Soc., Chem. Commun.*, 1991, 1601.
61. H. Lehmkuhl, R. Paul and R. Mynott, *Liebigs Ann. Chem.*, 1981, 1139.

62. K.-R. Pörschke, K. Jonas and G. Wilke, *Chem. Ber.*, 1988, **121**, 1913.
63. K.-R. Pörschke, W. Kleimann, Y.-H. Tsay, C. Krüger and G. Wilke, *Chem. Ber.*, 1990, **123**, 1267.
64. A. Miyashita and R. H. Grubbs, *Tetrahedron Lett.*, 1981, **22**, 1255.
65. K.-R. Pörschke, *Angew. Chem., Int. Ed. Engl.*, 1987, **26**, 1288.
66. W. Bonrath, K. R. Pörschke and S. Michaelis, *Angew. Chem.*, 1990, **102**, 295.
67. R. Benn, P. W. Jolly, T. Joswig, R. Mynott and K.-P. Schick, *Z. Naturforsch., Teil B*, 1986, **41**, 680.
68. H. Lehmkuhl, S. Pasynkiewicz, R. Benn and A. Rufinska, *J. Organomet. Chem.*, 1982, **240**, C27.
69. H. Lehmkuhl *et al.*, *Chem. Ber.*, 1988, **121**, 1931.
70. W. Kaschube, W. Schröder, K. R. Pörschke, K. Angermund and C. Krüger, *J. Organomet. Chem.*, 1990, **389**, 399.
71. A. D. Berry, *Organometallics*, 1983, **2**, 895.
72. S. Pasynkiewicz and H. Lehmkuhl, *J. Organomet. Chem.*, 1985, **289**, 189.
73. H. Lehmkuhl, F. Danowski, R. Benn, R. Mynott and G. Schroth, *Chem. Ber.*, 1986, **119**, 2542.
74. P. Hofmann, L. A. Perez-Moya, M. E. Krause, O. Kumberger and G. Müller, *Z. Naturforsch. Teil B*, 1990, **45**, 897.
75. P. Hofmann, L. A. Perez-Moya, O. Steigelmann and J. Riede, *Organometallics*, 1992, **11**, 1167.
76. A. Miyashita, H. Shitara and H. Nohira, *J. Chem. Soc., Chem. Commun.*, 1985, 850.
77. A. Miyashita, H. Shitara and H. Nohira, *Organometallics*, 1985, **4**, 1463.
78. K. Fischer, K. Jonas, A. Mollbach and G. Wilke, *Z. Naturforsch., Teil B*, 1984, **39**, 1011.
79. Y. Ohkatsu, E. Sato and T. Osa, *Sekiyu Gakkaishi*, 1981, **24**, 270.
80. Y. Ohkatsu, E. Sato and T. Osa, *Chem. Abstr.*, 1981, **95**, 186 400v.
81. E. Dinjus *et al.*, *J. Organomet. Chem.*, 1984, **276**, 257.
82. W. Kaschube, K. R. Pörschke, K. Seevogel and C. Krüger, *J. Organomet. Chem.*, 1989, **367**, 233.
83. R. Sustmann, P. Hopp and R. Boese, *J. Organomet. Chem.*, 1989, **375**, 259.
84. R. Benn, R. Mynott, I. Topalovic and F. Scott, *Organometallics*, 1989, **8**, 2299.
85. F. Scott, C. Krüger and P. Betz, *J. Organomet. Chem.*, 1990, **387**, 113.
86. M. I. Rybinskaya, *J. Organomet. Chem.*, 1990, **383**, 113.
87. A. Stanger, *Organometallics*, 1991, **10**, 2979.
88. A. Stanger and K. P. C. Vollhardt, *Organometallics*, 1992, **11**, 317.
89. E. Uhlig, R. Fischer and R. Kirmse, *J. Organomet. Chem.*, 1982, **239**, 385.
90. E. Dinjus, D. Walther, J. Kaiser, J. Sieler and N. N. Thanh, *J. Organomet. Chem.*, 1982, **236**, 123.
91. K. I. Han, A. M. Piotrowski and J. J. Eisch, in 'Organometallic Syntheses', eds. R. B. King and J. J. Eisch, Elsevier, New York, 1986, vol. 3, pp. 112–15.
92. M. W. Eyring and L. J. Radonovich, *Organometallics*, 1985, **4**, 1841.
93. R. Goddard, B. Apotecher and H. Hoberg, *Acta Crystallogr. Sect. C*, 1987, **43**, 1290.
94. G. Lane and W. E. Geiger, *Organometallics*, 1982, **1**, 401.
95. U. Köelle and H. Werner, *J. Organomet. Chem.*, 1981, **221**, 367.
96. J. J. Eisch, J. E. Galle, A. A. Aradi and M. P. Boleslawski, *J. Organomet. Chem.*, 1986, **312**, 399.
97. H. M. Buch, G. Schroth, R. Mynott and P. Binger, *J. Organomet. Chem.*, 1983, **247**, C63.
98. J. Zwecker, T. Kuhlmann, H. Pritzkow, W. Siebert and U. Zenneck, *Organometallics*, 1988, **7**, 2316.
99. F. Carré *et al.*, *Organometallics*, 1986, **5**, 910.
100. G. T. Burns *et al.*, *Organometallics*, 1987, **6**, 1398.
101. H. Schulz, H. Pritzkow and W. Siebert, *Chem. Ber.*, 1992, **125**, 993.
102. G. T. Burns, E. Colomer and R. J. P. Corriu, *Organometallics*, 1983, **2**, 1901.
103. H. Choi, J. W. Hershberger, A. R. Pinhas and D. M. Ho, *Organometallics*, 1991, **10**, 2930.
104. H. Choi and A. R. Pinhas, *Organometallics*, 1992, **11**, 442.
105. R. P. Hughes, R. T. Carl, D. E. Samkoff, R. E. Davis and K. D. Holland, *Organometallics*, 1986, **5**, 1053.
106. R. T. Carl, E. W. Corcoran, Jr., R. P. Hughes and D. E. Samkoff, *Organometallics*, 1990, **9**, 838.
107. C. J. Lawrie, K. P. Gable and B. K. Carpenter, *Organometallics*, 1989, **8**, 2274.
108. E. Carmona, E. Gutiérrez-Puebla, A. Monge, J. M. Marín, M. Paneque and M. L. Poveda, *Organometallics*, 1989, **8**, 1744.
109. E. Carmona, E. Gutiérrez-Puebla, A. Monge, J. M. Marín, M. Paneque and M. L. Poveda, *Organometallics*, 1989, **8**, 967.
110. H. Hoberg and A. Herrera, *Angew. Chem., Int. Ed. Engl.*, 1981, **20**, 876.
111. H. Hoberg and A. Herrera, *Angew. Chem.*, 1981, **93**, 924.
112. R. Goddard and Y.-H. Tsay, *Acta Crystallogr., Sect. C*, 1988, **44**, 810.
113. C. Müller, L. Stamp and H. T. Dieck, *Z. Naturforsch., Teil B*, 1986, **41**, 519.
114. C. Müller, L. Stamp and H. T. Dieck, *J. Organomet. Chem.*, 1986, **308**, 105.
115. W. Schröder and K.-R. Pörschke, *J. Organomet. Chem.*, 1987, **322**, 385.
116. R. Benn *et al.*, *Z. Naturforsch., Teil B*, 1991, **46**, 1395.
117. B. Proft, K.-R. Pörschke, F. Lutz and C. Krüger, *Chem. Ber.*, 1991, **124**, 2667.
118. H. Kuhn and M. Winter, *Chem.-Ztg.*, 1983, **107**, 14.
119. L. S. Isaeva, T. A. Peganova, P. V. Petrovskii, F. F. Kayumov and F. G. Yusupova, *J. Organomet. Chem.*, 1983, **248**, 375.
120. H. Lehmkuhl, F. Danowski, R. Benn, A. Rufinska, G. Schroth and R. Mynott, *J. Organomet. Chem.*, 1983, **254**, C11.
121. L. S. Isaeva *et al.*, *J. Organomet. Chem.*, 1983, **258**, 367.
122. M. A. Bennett and C. Chiraratvatana, *J. Organomet. Chem.*, 1985, **296**, 255.
123. D. W. Slocum, M. Moronski, R. Gooding and S. Duraj, *J. Organomet. Chem.*, 1984, **260**, C26.
124. D. J. Pasto and N.-Z. Huang, *Organometallics*, 1985, **4**, 1386.
125. D. J. Pasto, N.-Z. Huang and C. W. Eigenbrot, *J. Am. Chem. Soc.*, 1985, **107**, 3160.
126. A. K. List, M. R. Smith, III and G. L. Hillhouse, *Organometallics*, 1991, **10**, 361.
127. P. J. Fagan, J. C. Calabrese and B. Malone, *J. Am. Chem. Soc.*, 1992, **114**, 2252.
128. M. A. Bennett, C. Chiraratvatana, G. B. Robertson and U. Tooptakong, *Organometallics*, 1988, **7**, 1394.
129. N. Kuhn, U. Schwenk, M. Winter and R. Mynott, *J. Organomet. Chem.*, 1984, **260**, C43.
130. C. Bleasdale and D. W. Jones, *J. Chem. Soc., Chem. Commun.*, 1983, 214.
131. G. E. Herberich and E. Raabe, *J. Organomet. Chem.*, 1986, **309**, 143.
132. M. A. Battiste, B. G. Griggs, Jr., D. Sackett, J. M. Coxon and P. J. Steel, *J. Organomet. Chem.*, 1987, **330**, 437.
133. R. Taube, E. Weckmann, P. Böhme and J. P. Gehrke, *Z. Anorg. Allg. Chem.*, 1989, **577**, 245.

134. E. Carmona, F. González, M. L. Poveda, J. D. Atwood and R. D. Rodgers, *J. Chem. Soc., Dalton Trans.*, 1981, 777.
135. H. Lehmkuhl, A. Rufinska, R. Benn, G. Schroth and R. Mynott, *Liebigs Ann. Chem.*, 1981, 317.
136. H. Lehmkuhl, C. Naydowski, R. Benn and A. Rufinska, *J. Organomet. Chem.*, 1981, **216**, C41.
137. M. Peuckert and W. Keim, *Organometallics*, 1983, **2**, 594.
138. H. Lehmkuhl *et al.*, *Chem. Ber.*, 1983, **116**, 2447.
139. F. M. Ashmawy, F. R. Benn, C. A. McAuliffe and D. G. Watson, *J. Organomet. Chem.*, 1985, **287**, 65.
140. Y. Huang and B. S. Freiser, *J. Am. Chem. Soc.*, 1991, **113**, 8186.
141. B. Fischer, J. Boersma, B. Kojik-Prodic and A. L. Spek, *J. Chem. Soc., Chem. Commun.*, 1985, 1237.
142. A. L. Spek, J. L. de Boer, B. Fischer and J. Boersma, *Acta Crystallogr., Sect. C*, 1988, **44**, 1663.
143. H. Lehmkuhl, R. Paul, C. Krüger, Y.-H. Tsay, R. Benn and R. Mynott, *Liebigs Ann. Chem.*, 1981, 1147.
144. P. Jutzi, U. Siemeling, A. Müller and H. Bögge, *Organometallics*, 1989, **8**, 1744.
145. W. Mayer, G. Wilke, R. Benn, R. Goddard and C. Krüger, *Monatsh. Chem.*, 1985, **116**, 879.
146. H. Hoberg, D. Guhl and P. Betz, *J. Organomet. Chem.*, 1990, **387**, 233.
147. H. Lehmkuhl and C. Naydowski, *J. Organomet. Chem.*, 1984, **277**, C18.
148. K. H. Thiele, U. Böhme and S. Pasynkiewicz, *Z. Anorg. Allg. Chem.*, 1991, **600**, 121.
149. J. C. M. Sinnema, H. tom Dieck and G. H. B. Fendesak, *J. Organomet. Chem.*, 1990, **390**, 237.
150. W. Bonrath, S. Michaelis, K. R. Pörschke, B. Gabor, R. Mynott and C. Krüger, *J. Organomet. Chem.*, 1990, **397**, 255.
151. K.-R. Pörschke, C. Pluta, B. Proft, F. Lutz and C. Krüger, *Z. Naturforsch., Teil B*, 1993, **48**, 608.
152. C. J. Lawrie, H. E. Dankosh and B. K. Carpenter, *J. Organomet. Chem.*, 1991, **411**, C7.
153. R. de Vaumas, A. Marinetti, F. Mathey and L. Ricard, *J. Chem. Soc., Chem. Commun.*, 1988, 1325.
154. R. Benn, J. Klein, A. Rufinska and G. Schroth, *Z. Naturforsch., Teil B*, 1981, **36**, 1595.
155. J. J. Eisch, L. E. Hallenbeck and K. I. Han, *J. Org. Chem.*, 1983, **48**, 2963.
156. J. J. Eisch, A. M. Piotrowski, K. I. Han, C. Krüger and Y.-H. Tsay, *Organometallics*, 1985, **4**, 224.
157. D. J. Krysan and P. B. Mackenzie, *J. Org. Chem.*, 1990, **55**, 4229.
158. L. S. Isaeva, T. A. Peganova, P. V. Petrovskii and D. N. Kravtsov, *J. Organomet. Chem.*, 1989, **376**, 141.
159. H. M. Buch, P. Binger, R. Goddard and C. Krüger, *J. Chem. Soc., Chem. Commun.*, 1983, 648.
160. M. J. Chetcuti, S. R. McDonald and N. P. Rath, *Organometallics*, 1989, **8**, 2077.
161. M. J. Chetcuti, P. E. Fanwick, S. R. McDonald and N. P. Rath, *Organometallics*, 1991, **10**, 1551.
162. H. Hoberg, D. Schaefer, G. Burkhart, C. Krüger and M. J. Romão, *J. Organomet. Chem.*, 1984, **266**, 203.
163. Y. Iitaka, T. Oohama, M. Shiina, T. Takizawa, Y. Yamamoto and H. Yamazaki, *J. Chem. Soc., Dalton Trans.*, 1987, 853.
164. R. J. McKinney and W. A. Nugent, *Organometallics*, 1989, **8**, 2871.
165. J. E. Backvall and O. S. Andell, *Organometallics*, 1986, **5**, 2350.
166. R. J. McKinney and D. C. Roe, *J. Am. Chem. Soc.*, 1986, **108**, 5167.
167. C. A. Tolman, R. J. McKinney, W. C. Seidel, J. D. Druliner and W. R. Stevens, *Adv. Catal.*, 1985, **33**, 1.
168. P. J. Fagan, J. C. Calabrese and B. Malone, *Science*, 1991, **252**, 1160.
169. S. A. Lerke, B. A. Parkinson, D. H. Evans and P. J. Fagan, *J. Am. Chem. Soc.*, 1992, **114**, 7807.
170. L. Pandolfo and G. Paiaro, *Gazz. Chim. Ital.*, 1985, **115**, 561.
171. M. Ishikawa, J. Ohshita, Y. Ito and J. Iyoda, *J. Am. Chem. Soc.*, 1986, **108**, 7417.
172. H. Chen, R. A. Bartlett, M. M. Olmstead, P. P. Power and S. C. Shoner, *J. Am. Chem. Soc.*, 1990, **112**, 1048.
173. E. Carmona, P. Palma, M. Paneque and M. L. Poveda, *Organometallics*, 1990, **9**, 583.
174. M. Bochmann, I. Hawkins and M. P. Sloan, *J. Organomet. Chem.*, 1987, **332**, 371.
175. R. Appel, C. Casser and F. Knoch, *J. Organomet. Chem.*, 1985, **297**, 21.
176. A. H. Cowley *et al.*, *J. Am. Chem. Soc.*, 1983, **105**, 3737.
177. A. H. Cowley *et al.*, *J. Am. Chem. Soc.*, 1984, **106**, 7015.
178. T. A. van der Knaap *et al.*, *J. Organomet. Chem.*, 1983, **254**, C33.
179. H.-J. Metternich, E. Niecke, J. F. Nixon, R. Bartsch, P. B. Hitchcock and M. F. Meidine, *Chem. Ber.*, 1991, **124**, 1973.
180. A. Döhring, P. W. Jolly, C. Krüger and M. J. Romão, *Z. Naturforsch., Teil B*, 1985, **40**, 484.
181. C. Jegat *et al.*, *Inorg. Chem.*, 1993, **32**, 1279.
182. M. Aresta, R. Gobetto, E. Quaranta and I. Tommasi, *Inorg. Chem.*, 1992, **31**, 4286.
183. G. Mastrolli, G. Moro, C. F. Nobile and M. Latronico, *Inorg. Chim. Acta*, 1992, **192**, 189.
184. D. Walther and E. Dinjus, *Z. Chem.*, 1981, **21**, 270.
185. J. Kaiser, J. Sieler, D. Walther, E. Dinjus and L. Golic, *Acta Crystallogr., Sect. B*, 1982, **38**, 1584.
186. H. Nakazawa, H. Nosaka, Y. Kushi and H. Yoneda, *Organometallics*, 1990, **9**, 1958.
187. A. Miyashita, R. Sugai and J. Yamamoto, *J. Organomet. Chem.*, 1992, **428**, 239.
188. R. Sugai, A. Miyashita and H. Nohira, *Chem. Lett.*, 1988, 1403.
189. C. Bianchini, D. Masi, C. Mealli and A. Meli, *J. Organomet. Chem.*, 1983, **247**, C29.
190. C. Bianchini, C. Mealli, A. Meli and G. Scapacci, *Organometallics*, 1983, **2**, 141.
191. C. Bianchini, C. A. Ghilardi, A. Meli and A. Orlandini, *J. Organomet. Chem.*, 1984, **270**, 251.
192. C. Bianchini, C. A. Ghilardi, A. Meli and A. Orlandini, *J. Organomet. Chem.*, 1985, **286**, 259.
193. J. Cápora, E. Carmona, E. Gutiérrez-Puebla, M. L. Poveda and C. Ruiz, *Organometallics*, 1988, **7**, 2577.
194. M. G. Mason, P. N. Swepston and J. A. Ibers, *Inorg. Chem.*, 1983, **22**, 411.
195. M. Nieger, G. Bruder and R. Appel, *Acta Crystallogr., Sect. C*, 1991, **47**, 655.
196. C. Bianchini, C. A. Ghilardi, A. Meli, S. Midollini and A. Orlandini, *J. Chem. Soc., Chem. Commun.*, 1983, 753.
197. C. Bianchini, C. A. Ghilardi, A. Meli and A. Orlandini, *J. Organomet. Chem.*, 1983, **246**, C13.
198. M. Ishikawa, A. Naka, S. Okazaki and H. Sakamoto, *Organometallics*, 1993, **12**, 87.
199. M. Ishikawa, S. Matsuzawa, K. Hirotsu, S. Kamitori and T. Higuchi, *Organometallics*, 1984, **3**, 1930.
200. M. Ishikawa, S. Matsuzawa, T. Higuchi, S. Kamitori and K. Hirotsu, *Organometallics*, 1985, **4**, 2040.
201. P. Overbosch, G. van Koten and O. Overbeek, *J. Am. Chem. Soc.*, 1980, **102**, 2091.
202. P. Overbosch, G. van Koten and K. Vrieze, *J. Organomet. Chem.*, 1981, **208**, C21.
203. P. Overbosch, G. van Koten and O. Overbeek, *Inorg. Chem.*, 1982, **21**, 2373.
204. P. Overbosch, G. van Koten, A. L. Spek, G. Roelofsen and A. J. M. Duisenberg, *Inorg. Chem.*, 1982, **21**, 3908.
205. R. Appel, C. Casser and F. Knoch, *J. Organomet. Chem.*, 1985, **293**, 213.
206. R. Appel, F. Knoch and V. Winkhaus, *J. Organomet. Chem.*, 1986, **307**, 93.

207. R. Appel, W. Schuhn and F. Knoch, *J. Organomet. Chem.*, 1987, **319**, 345.
208. C. Bianchini, A. Meli and G. Scapacci, *Organometallics*, 1985, **4**, 264.
209. U. Rosenthal and H. Görls, *J. Organomet. Chem.*, 1988, **348**, 135.
210. K. R. Pörschke, R. Mynott, K. Angermund and C. Krüger, *Z. Naturforsch., Teil B*, 1985, **40**, 199.
211. U. Rosenthal and W. Schulz, *J. Organomet. Chem.*, 1987, **321**, 103.
212. U. Rosenthal, G. Oehme, V. V. Burlakov, P. V. Petrovskii, V. B. Shur and M. E. Vol'pin, *J. Organomet. Chem.*, 1990, **391**, 119.
213. A. Herrera and H. Hoberg, *J. Organomet. Chem.*, 1981, **222**, 331.
214. C. Pluta, K.-R. Pörschke, I. Ortmann and C. Krüger, *Chem. Ber.*, 1992, **125**, 103.
215. E. Hernandez, I. Saez and P. Royo, *J. Organomet. Chem.*, 1985, **293**, 249.
216. H. Lehmkuhl, F. Danowski, G. Mehler, J. Poplawska and S. Pasynkiewicz, *J. Organomet. Chem.*, 1989, **363**, 387.
217. U. Rosenthal *et al.*, *J. Organomet. Chem.*, 1990, **389**, 351.
218. U. Rosenthal, W. Schulz and H. Görls, *Z. Anorg. Allg. Chem.*, 1987, **550**, 169.
219. U. Rosenthal *et al.*, *J. Organomet. Chem.*, 1991, **409**, 299.
220. H. Görls, B. Schulz, U. Rosenthal and W. Schulz, *Cryst. Res. Technol.*, 1988, **23**, 25.
221. U. Rosenthal *et al.*, *J. Organomet. Chem.*, 1990, **389**, 409.
222. U. Rosenthal *et al.*, *J. Organomet. Chem.*, 1990, **390**, 113.
223. W. Bonrath, K. R. Pörschke, G. Wilke, K. Angermund and C. Krüger, *Angew. Chem., Int. Ed. Engl.*, 1988, **27**, 833.
224. T. Bartik et al., *Organometallics*, 1992, **11**, 1235.
225. R. del Rosario and L. S. Stuhl, *Organometallics*, 1986, **5**, 1260.
226. J. Grobe, D. Le Van, M. Hegemann, B. Krebs and M. Läge, *Chem. Ber.*, 1993, **126**, 63.
227. O. J. Scherer, J. Braun, P. Walther, G. Heckmann and G. Wolmershäuser, *Angew. Chem., Int. Ed. Engl.*, 1991, **30**, 852.
228. J. D. Ferrara, C. Tessier-Youngs and W. J. Youngs, *J. Am. Chem. Soc.*, 1985, **107**, 6719.
229. J. D. Ferrara, A. A. Tanaka, C. Fierro, C. A. Tessier-Youngs and W. J. Youngs, *Organometallics*, 1989, **8**, 2089.
230. W. J. Youngs, J. D. Kinder, J. D. Bradshaw and C. A. Tessier, *Organometallics*, 1993, **12**, 2406.
231. H. Brunner and M. Muschiol, *J. Organomet. Chem.*, 1983, **248**, 233.
232. S. Slater and E. L. Muetterties, *Inorg. Chem.*, 1981, **20**, 1604.
233. R. Bartsch, J. F. Nixon and N. Sarjudeen, *J. Organomet. Chem.*, 1985, **294**, 267.
234. H. Lang, M. Herres, L. Zsolnai and W. Imhof, *J. Organomet. Chem.*, 1991, **409**, C7.
235. H. Lang and W. Imhof, *Chem. Ber.*, 1992, **124**, 1307.
236. M. J. Chetcuti, C. Eigenbrot and K. A. Green, *Organometallics*, 1987, **6**, 2298.
237. G. Jaouen, A. Marinetti, J.-Y. Saillard, B. G. Sayer and M. J. McGlinchey, *Organometallics*, 1982, **1**, 225.
238. G. Jaouen *et al.*, *Organometallics*, 1982, **1**, 753.
239. M. J. Chetcuti, L. A. DeLiberato, P. E. Fanwick and B. E. Grant, *Inorg. Chem.*, 1990, **29**, 1295.
240. D. Walther, H. Schönberg, E. Dinjus and J. Sieler, *J. Organomet. Chem.*, 1987, **334**, 377.
241. E. Sappa, A. M. Manotti-Lanfredi, G. Predieri, A. Tiripicchio and A. J. Carty, *J. Organomet. Chem.*, 1985, **288**, 365.
242. C. Weatherell, N. J. Taylor, A. J. Carty, E. Sappa and A. Tiripicchio, *J. Organomet. Chem.*, 1985, **291**, C9.
243. A. J. Carty, N. J. Taylor, E. Sappa, A. Tiripicchio and M. Tiripicchio-Camellini, *Organometallics*, 1991, **10**, 1907.
244. M. Akita, M. Terada, M. Tanaka and Y. Moro-oka, *Organometallics*, 1992, **11**, 3468.
245. E. Sappa, A. M. M. Lanfredi and A. Tiripicchio, *J. Organomet. Chem.*, 1981, **221**, 93.
246. M. Lanfranchi, A. Tiripicchio, M. Tiripicchio-Camellini, O. Gambino and E. Sappa, *Inorg. Chim. Acta*, 1982, **64**, L269.
247. E. Sappa, A. Tiripicchio and M. Tiripicchio-Camellini, *J. Organomet. Chem.*, 1981, **213**, 175.
248. M. Lanfranchi, A. Tiripicchio, E. Sappa, S. A. Maclaughlin and A. J. Carty, *J. Chem. Soc., Chem. Commun.*, 1982, 538.
249. M. Castiglioni, R. Giordano and E. Sappa, *J. Organomet. Chem.*, 1984, **275**, 119.
250. E. Sappa, G. Predieri, A. Tiripicchio and M. Tiripicchio-Camellini, *J. Organomet. Chem.*, 1985, **297**, 103.
251. M. A. Bennett, T. W. Hambley, N. K. Roberts and G. B. Robertson, *Organometallics*, 1985, **4**, 1992.
252. M. A. Bennett, J. S. Drage, K. D. Griffiths, N. K. Roberts, G. B. Robertson and W. A. Wickramasinghe, *Angew. Chem.*, 1988, **100**, 1002.
253. M. A. Bennett, J. S. Drage, K. D. Griffiths, N. K. Roberts, G. B. Robertson and W. A. Wickramasinghe, *Angew. Chem., Int. Ed. Engl.*, 1988, **27**, 941.
254. M. A. Bennett, K. D. Griffiths, T. Okano, V. Parthasarathi and G. B. Robertson, *J. Am. Chem. Soc.*, 1990, **112**, 7047.
255. P. Guerts, H. Burgers and A. Van der Avoird, *Chem. Phys.*, 1981, **54**, 397.
256. G. A. Ozin, D. F. McIntosh, W. J. Power and R. P. Messmer, *Inorg. Chem.*, 1981, **20**, 1782.
257. E. S. Kline, Z. H. Kafafi, R. H. Hauge and J. L. Margrave, *J. Am. Chem. Soc.*, 1987, **109**, 2402.
258. M. Casarin, D. Ajò, G. Granozzi, E. Tondello and S. Aime, *J. Chem. Soc., Dalton Trans.*, 1983, 869.
259. U. Rosenthal, *Z. Anorg. Allg. Chem.*, 1981, **482**, 179.
260. P. Heimbach, T. Bartik, R. Boese, H. Schenkluhn, G. Szczendzina and E. Zeppenfeld, *Z. Chem.*, 1988, **28**, 121.
261. S. Pasynkiewicz, A. Poplawska and R. Mynott, *J. Organomet. Chem.*, 1992, **429**, 135.
262. K.-R. Pörschke and R. Mynott, *Z. Naturforsch.*, 1987, **42**, 421.
263. K.-R. Pörschke, *Angew. Chem.*, 1987, **99**, 1321.
264. U. Rosenthal and W. Schulz, *Z. Chem.*, 1985, **25**, 227.
265. I. D. Gridner, N. Miyaura and A. Suzuki, *Organometallics*, 1993, **12**, 589.
266. D. Malaba, A. Djebili, L. Chen, E. A. Zarati, C. A. Tessier and W. J. Youngs, *Organometallics*, 1993, **12**, 1266.
267. E. J. Forbes and N. Iranpoor, *J. Organomet. Chem.*, 1982, **236**, 403.
268. M. C. Azar, M. J. Chetcuti, C. Eigenbrot and K. A. Green, *J. Am. Chem. Soc.*, 1985, **107**, 7209.
269. M. J. Chetcuti, J. C. Gordon, K. A. Green, P. E. Fanwick and D. Morgenstern, *Organometallics*, 1989, **8**, 1790.
270. M. J. Chetcuti and K. A. Green, *Organometallics*, 1988, **7**, 2450.
271. O. J. Scherer, J. Braun, P. Walther, G. Heckmann and G. Wolmershäuser, *Angew. Chem.*, 1991, **103**, 861.
272. G. Wilke, *Pure Appl. Chem.*, 1984, **56**, 1635.
273. D. C. Billington, *Chem. Soc. Rev.*, 1985, **14**, 93.
274. P. W. Jolly, *Angew. Chem., Int. Ed. Engl.*, 1985, **24**, 283.
275. M. Guerra, D. Jones, G. Distefano, S. Torroni, A. Foffani and A. Modelli, *Organometallics*, 1993, **12**, 2203.
276. M. C. Bohm and R. Gleiter, *Angew. Chem., Int. Ed. Engl.*, 1983, **22**, 329.

277. X. Li, G. M. Bancroft, R. J. Puddephatt, Y. F. Hu and K. H. Tan, *Inorg. Chem.*, 1992, **31**, 5162.
278. M.-D. Su and S.-Y. Chu, *New J. Chem.*, 1989, **13**, 835.
279. R. Goddard, C. Krüger, F. Mark, R. Stansfield and X. Zhang, *Organometallics*, 1985, **4**, 285.
280. G. C. Hancock, N. M. Kostic and R. F. Fenske, *Organometallics*, 1983, **2**, 1089.
281. N. Stransky, R. Herzschuh, J.-P. Gehrke and R. Taube, *J. Organomet. Chem.*, 1984, **270**, 353.
282. N. Stransky, R. Herzschuh, J.-P. Gehrke and R. Taube, *J. Organomet. Chem.*, 1984, **270**, 357.
283. V. M. Akmedov and V. G. Mardanov, *Chem. Abstr.*, 1981, **95**, 98 030u.
284. G. Schmid, D. Kampmann, U. Hohner, D. Blaser and R. Boese, *Chem. Ber.*, 1984, **117**, 1052.
285. T. Yamamoto, J. Ishizu and A. Yamamoto, *J. Am. Chem. Soc.*, 1981, **103**, 6863.
286. K. S. Chong, S. J. Rettig, A. Storr and J. Trotter, *Can. J. Chem.*, 1981, **59**, 966.
287. R. Taube, J.-P. Gehrke and U. Schmidt, *J. Organomet. Chem.*, 1985, **292**, 287.
288. R. Taube and J.-P. Gehrke, *J. Organomet. Chem.*, 1987, **327**, 419.
289. R. Taube and J.-P. Gehrke, *J. Organomet. Chem.*, 1987, **328**, 393.
290. H. Kurosawa *et al.*, *Organometallics*, 1990, **9**, 3038.
291. H. Kurosawa, *J. Organomet. Chem.*, 1987, **334**, 243.
292. G. Consiglio, O. Piccolo, L. Roncetti and F. Morandini, *Tetrahedron*, 1986, **42**, 2043.
293. R. Taube, J.-P. Gehrke and R. Radeglia, *J. Organomet. Chem.*, 1985, **291**, 101.
294. H. Kurosawa, H. Ohnishi, M. Emoto, Y. Kawasaki and S. Murai, *J. Am. Chem. Soc.*, 1988, **110**, 6272.
295. R. B. A. Pardy and I. Tkatchenko, *J. Chem. Soc., Chem. Commun.*, 1981, 49.
296. E. Carmona, P. Palma and M. L. Poveda, *Polyhedron*, 1990, **9**, 757.
297. A. Knierzinger and P. Schönholzer, *Helv. Chim. Acta*, 1992, **75**, 1211.
298. R. Kempe, J. Sieler, S. Wache and R. Taube, *Z. Kristallogr.*, 1992, **201**, 287.
299. R. Kempe, J. Sieler, S. Wache and R. Taube, *Z. Kristallogr.*, 1992, **201**, 293.
300. H. Hoberg, G. Heger, C. Krüger and Y. H. Tsay, *J. Organomet. Chem.*, 1988, **348**, 261.
301. R. Berger, H. Schenkluhn and B. Weimann, *Transition Met. Chem.*, 1981, **6**, 272.
302. H. Schenkluhn, R. Berger, B. Pittel and M. Zähres, *Transition Met. Chem.*, 1981, **6**, 277.
303. H. Schenkluhn, H. Bandmann, R. Berger and E. Hübinger, *Transition Met. Chem.*, 1981, **6**, 287.
304. T. A. Peganova, L. S. Isaeva, P. V. Petrovskii and D. N. Kravtsov, *J. Organomet. Chem.*, 1990, **384**, 397.
305. H. Lehmkuhl and C. Naydowski, *J. Organomet. Chem.*, 1982, **240**, C30.
306. R. Benn, J. Klein, A. Rufinska and G. Schroth, *Z. Naturforsch., Teil B*, 1981, **36**, 1565.
307. H. Lehmkuhl, A. Rufinska, C. Naydowski and R. Mynott, *Chem. Ber.*, 1984, **117**, 376.
308. R. Benn *et al.*, *J. Organomet. Chem.*, 1985, **279**, 63.
309. R. Kempe, J. Sieler and S. Wache, *J. Organomet. Chem.*, 1993, **455**, 241.
310. R. Taube, P. Böhme and J.-P. Gehrke, *J. Organomet. Chem.*, 1990, **399**, 327.
311. R. Taube, J.-P. Gehrke, P. Böhme and K. Scherzer, *J. Organomet. Chem.*, 1991, **410**, 403.
312. R. Taube and S. Wache, *J. Organomet. Chem.*, 1992, **428**, 431.
313. R. Taube, S. Wache, S. Sieler and R. Kempe, *J. Organomet. Chem.*, 1993, **456**, 131.
314. S. Wache and R. Taube, *J. Organomet. Chem.*, 1993, **456**, 137.
315. G. E. Herberich, W. Klein, U. Kölle and D. Spiliotis, *Chem. Ber.*, 1992, **125**, 1589.
316. J. J. Schneider, R. Goddard and C. Krüger, *Organometallics*, 1991, **10**, 665.
317. J. Edwin, W. Siebert and C. Krüger, *J. Organomet. Chem.*, 1981, **215**, 255.
318. U. Kölle, F. Khouzami and H. Lueken, *Chem. Ber.*, 1982, **115**, 1178.
319. W. Keim, R. Appel, S. Gruppe and F. Knoch, *Angew. Chem., Int. Ed. Engl.*, 1987, **26**, 1012.
320. W. Keim, R. Appel, S. Gruppe and F. Knoch, *Angew. Chem.*, 1987, **99**, 1042.
321. R. Bartsch, P. B. Hitchcock and J. F. Nixon, *J. Organomet. Chem.*, 1989, **373**, C17.
322. H. Lehmkuhl *et al.*, *Z. Naturforsch., Teil B*, 1985, **40**, 171.
323. H. Hoberg and D. Schaefer, *J. Organomet. Chem.*, 1983, **255**, C15.
324. E. Hernandez and H. Hoberg, *J. Organomet. Chem.*, 1987, **327**, 429.
325. D. Walther, E. Dinjus, L. Andersen, H. Görls, J. Sieler and O. Lindqvist, *J. Organomet. Chem.*, 1985, **286**, 103.
326. G. P. Chiusoli, M. Costa, M. G. Terenghi and I. Vinay, *Transition Met. Chem.*, 1981, **6** 125.
327. B. Bogdanovic, M. Rubach and K. Seevogel, *Z. Naturforsch., Teil B*, 1983, **38**, 592.
328. B. Bogdanovic, P. Gottsch and M. Rubach, *Z. Naturforsch., Teil B*, 1983, **38B**, 599.
329. L. S. Hegedus and S. Varaprath, *Organometallics*, 1982, **1**, 259.
330. T. Bartik, I. Gerdes, P. Heimbach and H. G. Schulte, *J. Organomet. Chem.*, 1989, **367**, 359.
331. R. Hanko, *Angew. Chem., Int. Ed. Engl.*, 1985, **24**, 704.
332. G. A. Molander and D. C. Shubert, *Tetrahedron Lett.*, 1986, **27**, 787.
333. U. Kölle *et al.*, *Chem. Ber.*, 1990, **123**, 227.
334. J. Sinnema, H. tom Dieck and G. H. B. Fendesak, *J. Organomet. Chem.*, 1990, **397**, 261.
335. J. R. Johnson, P. S. Tully, P. B. Mackenzie and M. Sabat, *J. Am. Chem. Soc.*, 1991, **113**, 6172.
336. D. Walther and D. Zeigan, *J. Organomet. Chem.*, 1981, **216**, 403.
337. R. Benn, B. Bogdanovic, P. Gottsch and M. Rubach, *Z. Naturforsch., Teil B*, 1983, **38**, 604.
338. G. Favero, A. Morvillo and A. Turco, *J. Organomet. Chem.*, 1983, **241**, 251.
339. G. Favero, S. Issa, A. Turco and U. Vettori, *J. Organomet. Chem.*, 1986, **315**, 237.
340. E. Carmona, M. Paneque and M. L. Poveda, *Polyhedron*, 1989, **8**, 285.
341. J. R. Ascenso *et al.*, *Polyhedron*, 1989, **8**, 2449.
342. E. Carmona, J. M. Marín, M. Paneque and M. L. Poveda, *Organometallics*, 1987, **6**, 1757.
343. J. Cámpora, E. Carmona, E. Gutiérrez-Puebla, P. Palma, M. L. Poveda and R. Caridad, *Organometallics*, 1992, **11**, 11.
344. E. Carmona, E. Gutiérrez-Puebla and E. Al, *J. Chem. Soc., Dalton Trans.*, 1992, 1769.
345. J. Cámpora, E. Gutiérrez-Puebla, A. Monge, M. L. Poveda, C. Ruiz and E. Carmona, *Organometallics*, 1993, **12**, 4025.
346. T. R. Belderraín *et al.*, *Organometallics*, 1993, **12**, 4431.
347. E. D. Jemmis and B. V. Prasad, *J. Organomet. Chem.*, 1988, **347**, 401.
348. C. Mealli, S. Midollini, S. Moneti and L. Sacconi, *Angew. Chem.*, 1980, **92**, 967.
349. C. Mealli, S. Midollini, S. Moneti, L. Sacconi, J. Silvestre and T. A. Albright, *J. Am. Chem. Soc.*, 1982, **104**, 95.

350. C. Mealli, S. Midollini, S. Moneti and L. Sacconi, *J. Organomet. Chem.*, 1981, **205**, 273.
351. O. J. Scherer, J. Braun and G. Wolmershäuser, *Chem. Ber.*, 1990, **123**, 471.
352. O. J. Scherer, T. Dave, J. Braun and G. Wolmershäuser, *J. Organomet. Chem.*, 1988, **350**, C20.
353. O. J. Scherer, J. Braun, P. Walther and G. Wolmershäuser, *Chem. Ber.*, 1992, **125**, 2661.
354. C. A. Ghilardi, S. Midollini and A. Orlandini, *J. Organomet. Chem.*, 1985, **295**, 377.
355. F. Cecconi, C. A. Ghilardi, S. Midollini, S. Moneti and A. Orlandini, *Angew. Chem., Int. Ed. Engl.*, 1986, **25**, 833.
356. D. Walther and E. Dinjus, *Z. Chem.*, 1982, **22**, 228.
357. D. Walther and E. Dinjus, *Z. Chem.*, 1982, **22**, 237.
358. D. Walther, E. Dinjus, J. Sieler, N. N. Thanh, W. Schade and I. Leban, *Z. Naturforsch., Teil B*, 1983, **38**, 835.
359. H. J. Riegel and H. Hoberg, *J. Organomet. Chem.*, 1984, **260**, 121.
360. P. Grenouillet, D. Neibecker and I. Tkatchenko, *J. Organomet. Chem.*, 1983, **243**, 213.
361. J. Sieler, J. Kaiser, D. Walther, L. Golic and J. Siftar, *J. Organomet. Chem.*, 1982, **224**, 313.
362. J. J. Brunet, A. Sivade and I. Tkatchenko, *J. Mol. Catal.*, 1989, **50**, 291.
363. R. Taube, J.-P. Gehrke, P. Böhme and J. Kottnitz, *J. Organomet. Chem.*, 1990, **395**, 341.
364. R. Benn, *Z. Naturforsch., Teil B*, 1982, **37**, 1054.
365. R. Benn and A. Rufinska, *J. Organomet. Chem.*, 1982, **239**, C19.
366. R. Benn and A. Rufinska, *J. Organomet. Chem.*, 1982, **238**, C27.
367. R. Bertani, G. Carturan and A. Scrivanti, *Angew. Chem., Int. Ed. Engl.*, 1983, **22**, 246.
368. D. Fenske and P. Stock, *Angew. Chem., Int. Ed. Engl.*, 1982, **21**, 356.
369. E. Niecke, B. Kramer and M. Nieger, *Organometallics*, 1991, **10**, 10.
370. G. Favero, S. I. Mohamud, A. Turco and U. Vettori, *Gazz. Chim. Ital.*, 1989, **119**, 407.
371. E. Carmona, J. M. Marín, P. Palma, M. Paneque and M. L. Poveda, *Organometallics*, 1985, **4**, 2053.
372. C. A. Ghilardi, S. Midollini, S. Moneti and A. Orlandini, *J. Organomet. Chem.*, 1985, **286**, 419.
373. R. Hemmer, H. A. Brune and U. Thewalt, *Z. Naturforsch., Teil B*, 1981, **36**, 78.
374. H. Hoberg, W. Richter and C. Fröhlich, *J. Organomet. Chem.*, 1981, **213**, C49.
375. H. Hoberg and C. Fröhlich, *J. Organomet. Chem.*, 1981, **209**, C69.
376. C. Fröhlich and H. Hoberg, *J. Organomet. Chem.*, 1981, **222**, 337.
377. H. Hoberg and C. Fröhlich, *J. Organomet. Chem.*, 1981, **204**, 131.
378. H. Hoberg and H. J. Riegel, *J. Organomet. Chem.*, 1982, **229**, 85.
379. P. Mauret and P. Alphonse, *J. Organomet. Chem.*, 1984, **276**, 249.
380. J. J. Eisch, A. M. Piotrowski, A. A. Aradi, C. Krüger and M. J. Romão, *Z. Naturforsch., Teil B*, 1985, **40**, 624.
381. R. Köster, G. Seidel, R. Boese and B. Wrackmeyer, *Chem. Ber.*, 1988, **121**, 1955.
382. R. Köster, G. Seidel, R. Boese and B. Wrackmeyer, *Chem. Ber.*, 1988, **121**, 709.
383. R. Köster, G. Seidel, R. Boese and B. Wrackmeyer, *Chem. Ber.*, 1988, **121**, 1941.
384. T. Wettling, G. Wolmershäuser, P. Binger and M. Regitz, *J. Chem. Soc., Chem. Commun.*, 1990, 1541.
385. K. E. Howard, J. R. Lockemeyer, M. A. Massa, T. B. Rauchfuss, S. R. Wilson and X. G. Yang, *Inorg. Chem.*, 1990, **29**, 4385.
386. R. M. Pitzer, J. D. Goddard and H. F. Schaefer, III, *J. Am. Chem. Soc.*, 1981, **103**, 5681.
387. M. C. Boehm, *Theor. Chim. Acta*, 1981, **60**, 233.
388. M. C. Boehm, *Chem. Phys.*, 1981, **60**, 277.
389. A. A. Aradi, A. M. Piotrowski and J. J. Eisch, in 'Organometallic Syntheses', eds. R. B. King and J. J. Eisch, Elsevier, New York, 1986, vol. 3, pp. 116–18.
390. A. M. Piotrowski and J. J. Eisch, in 'Organometallic Syntheses', eds. R. B. King and J. J. Eisch, Elsevier, New York, 1986, vol. 3, pp. 119–21.
391. T. Kuhlmann and W. Siebert, *Z. Naturforsch., Teil B*, 1985, **40**, 167.
392. A. Reissova and M. Capka, *Chem. Listy*, 1982, **76**, 34.
393. C. Janiak and H. Schumann, *Adv. Organomet. Chem.*, 1991, **33**, 291.
394. N. J. Coville, K. E. du Plooy and W. Pickl, *Coord. Chem. Rev.*, 1992, **116**, 1.
395. M. C. Boehm and B. Bunsenges, *Phys. Chem.*, 1982, **86**, 56.
396. D. B. Jacobson, G. D. Byrd and B. S. Freiser, *J. Am. Chem. Soc.*, 1982, **104**, 2320.
397. A. V. Stevens and J. V. Beauchamp, *J. Am. Chem. Soc.*, 1981, **103**, 190.
398. S. Leutwyler, U. Even and J. Jortner, *Chem. Phys.*, 1981, **58**, 409.
399. S. Leutwyler, U. Even and J. Jortner, *J. Phys. Chem.*, 1981, **85**, 3026.
400. R. Stepnowski and J. Allison, *J. Am. Chem. Soc.*, 1989, **111**, 449.
401. T. Drewello, C. B. Lebrilla, H. Schwarz and T. Ast, *J. Organomet. Chem.*, 1988, **339**, 333.
402. D. B. Jacobson and B. S. Freiser, *Organometallics*, 1985, **4**, 103.
403. P. R. Nelson, J. R. Appling, E. K. Barefield and T. F. Moran, *Inorg. Chem.*, 1986, **25**, 1510.
404. Y. A. Andrianov, O. N. Druzhkov and A. S. Smirnov, *J. Gen. Chem. USSR*, 1981, **51**, 2161.
405. Y. A. Andrianov, O. N. Druzhkov and A. S. Smirnov, *Zh. Obshch. Khim.*, 1981, **51**, 2508.
406. Z. Z. Zhang, D. Wei, Y. Chen and X. K. Wang, *Kexue Tong (For. Lang. Ed.)*, 1984, **29**, 281.
407. Z. Zhang, D. Wei, Y. Chen and X. Wang, *Chem. Abstr.*, 1984, **101**, 91 184.
408. G. Lucazeau, K. Chlor and C. Pommier, *Spectrosc. Int. J.*, 1983, **2**, 79.
409. A. B. Gardner, J. Howard, T. C. Waddington, R. M. Richardson and J. Tomkinson, *J. Chem. Phys.*, 1981, **57**, 453.
410. C. Sourisseau, G. Lucazeau, A. J. Dianoux and C. Poinsignon, *Mol. Phys.*, 1983, **48**, 367.
411. J. C. Giordan, J. H. Moore, J. A. Tossell and J. Weber, *J. Am. Chem. Soc.*, 1983, **105**, 3431.
412. S. G. Kukolich, J. V. Rund, D. J. Pauley and R. E. Bumgarner, *J. Am. Chem. Soc.*, 1988, **110**, 7356.
413. P. Baltzer, A. Furrer, J. Hulliger and A. Stebler, *Inorg. Chem.*, 1988, **27**, 1543.
414. S. Li, Y. M. Hamrick, R. J. Van Zee and J. W. Weltner, *J. Am. Chem. Soc.*, 1992, **114**, 4433.
415. F. H. Köhler, K.-H. Doll, W. Prössdorf and J. Müller, *Angew. Chem., Int. Ed. Engl.*, 1982, **21**, 151.
416. J. H. Ammeter, L. Zoller, J. Bachmann, P. E. G. Baltzer, R. Bucher and E. Deiss, *Helv. Chim. Acta*, 1981, **64**, 1063.
417. J. L. Robbins, *Energy Res. Abstr.*, 1981, **6**, 20 584.
418. J. L. Robbins, *Chem. Abstr.*, 1982, **96**, 6846j.
419. J. L. Robbins, N. Edelstein, B. Spencer and J. C. Smart, *J. Am. Chem. Soc.*, 1982, **104**, 1882.

420. E. K. Barefield *et al.*, *J. Am. Chem. Soc.*, 1981, **103**, 6219.
421. M. P. Castellani, S. J. Geib, A. L. Rheingold and W. C. Trogler, *Organometallics*, 1987, **6**, 1703.
422. P. Jutzi, J. Schnittger, W. Wieland, B. Neumann and H.-G. Stammler, *J. Organomet. Chem.*, 1991, **415**, 425.
423. J. Zwecker, H. Pritzkow, U. Zenneck and W. Siebert, *Angew. Chem., Int. Ed. Engl.*, 1986, **25**, 1099.
424. F. H. Köhler, K. H. Doll and W. Prössdorf, *J. Organomet. Chem.*, 1982, **224**, 341.
425. P. Hudeczek and F. H. Köhler, *Organometallics*, 1992, **11**, 1773.
426. F. H. Köhler and W. A. Geike, *J. Organomet. Chem.*, 1987, **328**, 35.
427. F. H. Köhler and W. A. Geike, *J. Magn. Reson.*, 1983, **53**, 297.
428. F. H. Köhler and K.-H. Doll, *Z. Naturforsch., Teil B*, 1982, **37**, 144.
429. J. Blümel, N. Hebendanz, P. Hudeczek, F. H. Köhler and W. Strauss, *J. Am. Chem. Soc.*, 1992, **114**, 4223.
430. V. T. Aleksanyan, T. I. Arsen'eva, N. N. Vyshinskii, I. I. Grinval'd and A. N. Smirnov, *Bull. Acad. Sci. USSR, Div. Chem. Sci.*, 1981, **30**, 227.
431. V. T. Aleksanyan, T. I. Arsen'eva, N. N. Vyshinskii, I. I. Grinval'd and A. N. Smirnov, *Izv. Akad. Nauk. SSSR, Ser. Khim.*, 1981, 229.
432. V. T. Aleksanyan, N. N. Vyshinskii, I. I. Grinval'd and T. I. Arsen'eva, *Bull. Acad. Sci. USSR, Div. Chem. Sci.*, 1981, **30**, 230.
433. V. T. Aleksanyan, N. N. Vyshinskii, I. I. Grinval'd and T. I. Arsen'eva, *Izv. Akad. Nauk SSSR, Ser. Khim.*, 1981, 303.
434. K. Chlor, G. Lucazeau and C. Sourisseau, *J. Raman Spectrosc.*, 1981, **11**, 183.
435. G. Clec'h and G. Calvarin, *Mol. Cryst. Liq. Cryst.*, 1981, **75**, 345.
436. D. Braga and F. Grepioni, *Organometallics*, 1992, **11**, 711.
437. H. Schaefer-Stahl, *Synth. Met.*, 1981, **4**, 65.
438. H. Schaefer-Stahl, *Chem. Abstr.*, 1981, **95**, 114 235x.
439. N. G. Connelly and W. E. Geiger, *Adv. Organomet. Chem.*, 1984, **23**, 1.
440. R. J. Gale and R. Job, *Inorg. Chem.*, 1981, **20**, 40.
441. J. D. Payne and E. Elmurr, *J. Chem. Soc., Chem. Commun.*, 1984, 1137.
442. V. V. Strelets and S. V. Kukharenko, *Organomet. Chem. USSR (Engl. Trans.)*, 1990, 3.
443. V. V. Strelets, *Coord. Chem. Rev.*, 1992, **114**, 1.
444. A. J. Bard, E. Garcia, S. Kukharenko and V. V. Strelets, *Inorg. Chem.*, 1993, **32**, 3528.
445. G. Bermudez and D. Pletcher, *J. Organomet. Chem.*, 1982, **231**, 173.
446. A. Louati and M. Huhn, *Inorg. Chem.*, 1993, **32**, 3601.
447. E. E. Bunel *et al.*, *J. Am. Chem. Soc.*, 1988, **110**, 6596.
448. M. Huhn, W. Klaüi, L. Ramacher, R. Herbst-Irmer and E. Egert, *J. Organomet. Chem.*, 1990, **398**, 339.
449. J. J. Maj, A. D. Rae and L. F. Dahl, *J. Am. Chem. Soc.*, 1982, **104**, 4278.
450. T. E. North, J. B. Thoden, A. Bjarnason and L. F. Dahl, *Organometallics*, 1992, **11**, 4338.
451. N. F. Ho, T. C. W. Mak and T.-Y. Luh, *J. Organomet. Chem.*, 1986, **317**, C28.
452. N.-F. Ho, T. C. W. Mak and T.-Y. Luh, *J. Chem. Soc., Dalton Trans.*, 1990, 3591.
453. N. Kuhn, *Chem.-Ztg.*, 1982, **106**, 146.
454. N. Kuhn, *Chem. Abstr.*, 1982, **97**, 39 126.
455. H. Schumann, *J. Organomet. Chem.*, 1986, **304**, 341.
456. J. Grobe, B. H. Schneider and H. Zimmermann, *Z. Anorg. Allg. Chem.*, 1981, **481**, 107.
457. E. G. Perevalova, M. D. Reshetova and D. A. Lemenovskii, *Koord. Khim.*, 1981, **7**, 1340.
458. P. Seiler and J. D. Dunitz, *Acta Crystallogr., Sect. B*, 1980, **36**, 2255.
459. W. P. Hart, *Diss. Abstr.*, 1981, **42**, 1447.
460. F. H. Köhler, K. H. Doll, E. Fladerer and W. A. Geike, *Transition Met. Chem.*, 1981, **6**, 126.
461. U. Köelle, B. Fuss, F. Khouzami and J. Gersdorf, *J. Organomet. Chem.*, 1985, **290**, 77.
462. S. A. Westcott, A. K. Kakkar, G. Stringer, N. J. Taylor and T. B. Marder, *J. Organomet. Chem.*, 1990, **394**, 777.
463. P. R. Sharp, K. N. Raymond, J. C. Smart and R. J. McKinney, *J. Am. Chem. Soc.*, 1981, **103**, 753.
464. H. Schottenberger, G. Ingram and D. Obendorf, *J. Organomet. Chem.*, 1992, **426**, 109.
465. E. E. Bunel *et al.*, *Organometallics*, 1988, **7**, 789.
466. S. K. Brownstein, E. J. Gabe and R. C. Hynes, *Can. J. Chem.*, 1992, **70**, 1011.
467. G. E. Herberich, D. P. J. Köffer, M. Hostalek and R. Laven, *Chem. Ber.*, 1991, **124**, 17.
468. J. J. Schneider and C. Krüger, *Chem. Ber.*, 1992, **125**, 843.
469. W. Douglas, *J. Organomet. Chem.*, 1985, **287**, C39.
470. M.-C. Chan, K.-M. Cheng, M. K. Li and T.-Y. Luh, *J. Chem. Soc., Chem. Commun.*, 1985, 1610.
471. W. Klaüi and L. Ramacher, *Angew. Chem., Int. Ed. Engl.*, 1986, **25**, 97.
472. W. Klaüi and L. Ramacher, *Angew. Chem.*, 1986, **98**, 107.
473. W. Klaüi and L. Ramacher, *Angew. Chem., Int. Ed. Engl.*, 1986, **25**, 97.
474. G. E. Herberich and M. Negele, *J. Organomet. Chem.*, 1988, **350**, 81.
475. N. Kuhn, N. Heuser and M. Winter, *J. Organomet. Chem.*, 1984, **267**, 221.
476. R. A. Fischer, J. Behm, E. Herdtweck and C. Kronseder, *J. Organomet. Chem.*, 1992, **437**, C29.
477. P. Jutzi and A. Mix, *Chem. Ber.*, 1990, **123**, 1043.
478. N. M. Boag, A. J. Goodby and R. Quyoum, *Organometallics*, 1992, **11**, 3135.
479. J. Okuda, *J. Organomet. Chem.*, 1988, **353**, C1.
480. N. Kuhn and M. Winter, *J. Organomet. Chem.*, 1986, **301**, C9.
481. M. I. Bruce, J. R. Rodgers, M. R. Snow and F. S. Wong, *J. Organomet. Chem.*, 1982, **240**, 299.
482. T. R. Belderraín, D. A. Knight, D. J. Irvine, M. Paneque, M. L. Poveda and E. Carmona, *J. Chem. Soc., Dalton Trans.*, 1992, 1491.
483. M. J. Chetcuti, K. J. Deck, P. E. Fanwick, J. C. Gordon and B. E. Grant, *Organometallics*, 1992, **11**, 2128.
484. D. Gudat and E. Niecke, *J. Chem. Soc., Chem. Commun.*, 1987, 10.
485. N. Kuhn and M. Winter, *J. Organomet. Chem.*, 1983, **243**, C83.
486. E. Hernandez and P. Royo, *J. Organomet. Chem.*, 1985, **291**, 387.
487. N. Kuhn and M. Winter, *J. Organomet. Chem.*, 1983, **252**, C5.
488. E. R. Burkhardt, R. G. Bergman and C. H. Heathcock, *Organometallics*, 1990, **9**, 30.

489. M. Julia, H. Lauron and J. N. Verpeaux, *J. Organomet. Chem.*, 1990, **387**, 365.
490. M. Julia, H. Lauron, J. N. Verpeaux, Y. Jeannin and C. Bois, *J. Organomet. Chem.*, 1988, **358**, C11.
491. F. Scott, S. Cronje and H. G. Raubenheimer, *J. Organomet. Chem.*, 1987, **326**, C40.
492. M. I. Bruce, D. N. Duffy, M. J. Liddell, M. R. Snow and E. R. J. Tielink, *J. Organomet. Chem.*, 1987, **337**, 365.
493. P. Hong, K. Sonogashira and N. Hagihara, *J. Organomet. Chem.*, 1981, **219**, 363.
494. A. Becalska, J. D. Debad, H. K. Sanati and R. H. Hill, *J. Photochem. Photobiol.*, 1991, **60**, 279.
495. A. Becalska, J. D. Debad, H. K. Sanati and R. H. Hill, *Polyhedron*, 1990, **9**, 581.
496. R. Taube, D. Steinborn and W. Höbold, *J. Organomet. Chem.*, 1985, **284**, 385.
497. E. Lindner, F. Bouachir and W. Hiller, *Z. Naturforsch., Teil B*, 1982, **37**, 1146.
498. B. L. Booth and K. G. Smith, *J. Organomet. Chem.*, 1981, **220**, 229.
499. B. L. Booth and K. G. Smith, *J. Organomet. Chem.*, 1981, **220**, 219.
500. L. Ballester, S. Perez, E. Gutiérrez-Puebla, A. Monge and C. Ruiz, *J. Organomet. Chem.*, 1991, **414**, 411.
501. N. Kuhn and M. Winter, *J. Organomet. Chem.*, 1983, **254**, C15.
502. N. Kuhn and M. Winter, *Chem.-Ztg.*, 1981, **105**, 376.
503. N. Kuhn and M. Winter, *J. Organomet. Chem.*, 1983, **243**, C42.
504. N. Kuhn and M. Winter, *J. Organomet. Chem.*, 1983, **243**, C42.
505. H. Werner and T. N. Khac, *Z. Anorg. Allg. Chem.*, 1981, **475**, 241.
506. N. Kuhn and M. Winter, *J. Organomet. Chem.*, 1983, **246**, C80.
507. W. P. Fehlhammer and U. Plaia, *Z. Naturforsch., Teil B*, 1986, **41**, 1005.
508. N. Kuhn and M. Winter, *J. Organomet. Chem.*, 1983, **256**, C5.
509. N. Kuhn and M. Winter, *J. Organomet. Chem.*, 1983, **249**, C28.
510. M. Kuhn and M. Winter, *J. Organomet. Chem.*, 1982, **234**, C1.
511. N. Kuhn and M. Winter, *J. Organomet. Chem.*, 1984, **276**, C16.
512. L. V. Pankratov *et al.*, *J. Organomet. Chem.*, 1992, **429**, 13.
513. E. E. Bunel, L. Valle and J. M. Manriquez, *Organometallics*, 1985, **4**, 1680.
514. G. K. Anderson, R. J. Cross, K. W. Muir and L. Manojlovic-Muir, *J. Organomet. Chem.*, 1989, **362**, 225.
515. P. Bladon, M. Dekker, G. R. Knox, D. Willison, G. A. Jasffari and R. J. Doedens, *Organometallics*, 1993, **12**, 1725.
516. E. Lindner, A. Nothdurft, R. Fawzi and C. Maichle, *J. Organomet. Chem.*, 1992, **435**, 213.
517. A. Pietrzykowski and S. Pasynkiewicz, *J. Organomet. Chem.*, 1992, **440**, 401.
518. L. R. Byers and L. F. Dahl, *Inorg. Chem.*, 1980, **19**, 680.
519. T. Mise and H. Yamazaki, in 'Organometallic Syntheses', eds. R. B. King and J. J. Eisch, Elsevier, New York, 1986, vol. 3, pp. 55–7.
520. W. Siebert, J. Edwin, H. Wadepohl and H. Pritzkow, *Angew. Chem.*, 1982, **94**, 464.
521. E. Lindner, F. Bouachir and W. Hiller, *J. Organomet. Chem.*, 1981, **210**, C37.
522. E. Lindner, F. Bouachir and S. Hoehne, *Chem. Ber.*, 1983, **116**, 46.
523. W. Klaüi, W. Eberspach and R. Schwarz, *J. Organomet. Chem.*, 1983, **252**, 347.
524. W. Klaüi, K. Schmidt, A. Bochmann, P. Hofmann, H. R. Schmidt and P. Stauffert, *J. Organomet. Chem.*, 1985, **286**, 407.
525. A. Bartlone, M. J. Chetcuti, P. E. Fanwick and K. A. Haller, *Inorg. Chem.*, 1993, **32**, 1435.
526. J. G. Davidson, E. K. Barefield and D. G. V. Derveer, *Organometallics*, 1985, **4**, 1178.
527. B. K. Teo, V. Bakirtzis and P. A. Snyder-Robinson, *J. Am. Chem. Soc.*, 1983, **105**, 6330.
528. A. J. Banister, I. B. Gorrell, W. Clegg and K. A. Jørgensen, *J. Chem. Soc., Dalton Trans.*, 1991, 1105.
529. D. Lentz, *Chem. Ber.*, 1985, **118**, 560.
530. H. Werner, B. Ulrich, U. Schubert, P. Hofmann and B. Zimmer-Gasser, *J. Organomet. Chem.*, 1985, **297**, 27.
531. S. Woodward, M. D. Curtis, A. L. Rheingold and B. S. Haggerty, *Organometallics*, 1992, **11**, 2140.
532. M. J. Chetcuti, P. E. Fanwick and B. E. Grant, *Organometallics*, 1990, **9**, 1345.
533. M. J. Chetcuti, B. E. Grant, P. E. Fanwick, M. J. Geselbracht and A. M. Stacy, *Organometallics*, 1990, **9**, 1343.
534. M. J. Chetcuti, B. E. Grant and P. E. Fanwick, *Organometallics*, 1991, **10**, 3003.
535. J. Weiss, *Z. Anorg. Allg. Chem.*, 1983, **502**, 165.
536. S. B. Colbran, B. H. Robinson and J. Simpson, *Organometallics*, 1985, **4**, 1594.
537. W. Klaüi, M. Huhn and R. Herbst-Irmer, *J. Organomet. Chem.*, 1991, **415**, 133.
538. S. N. Titova, V. T. Bychkov, G. A. Domachev, G. A. Razuvaev and L. N. Zakharov, *Inorg. Chim. Acta*, 1981, **50**, 71.
539. N. Kuhn and M. Winter, *J. Organomet. Chem.*, 1984, **269**, C47.
540. N. Kuhn and M. Winter, *J. Organomet. Chem.*, 1983, **252**, C86.
541. A. A. Low and M. B. Hall, *Inorg. Chem.*, 1993, **32**, 3880.
542. G. Granozzi, M. Casarin, D. Ajò and D. Osella, *J. Chem. Soc., Dalton Trans.*, 1983, 2047.
543. L. R. Byers, V. A. Uchtman and L. F. Dahl, *J. Am. Chem. Soc.*, 1981, **103**, 1942.
544. E. Sappa, M. Lanfranchi, A. Tiripicchio and M. Tiripicchio-Camellini, *J. Chem. Soc., Chem. Commun.*, 1981, 995.
545. M. J. Chetcuti, B. E. Grant and P. E. Fanwick, *J. Am. Chem. Soc.*, 1989, **111**, 2743.
546. E. Sappa and A. Tiripicchio, *New J. Chem.*, 1988, **12**, 599.
547. R. Blumhofer, K. Fischer and H. Vahrenkamp, *Chem. Ber.*, 1986, **119**, 194.
548. H. Lehmkuhl, C. Krüger, S. Pasynkiewicz and J. Poplawska, *Organometallics*, 1988, **7**, 2038.
549. T. E. North, J. B. Thoden, B. Spencer and L. F. Dahl, *Organometallics*, 1993, **12**, 1299.
550. T. Madach and H. Vahrenkamp, *Chem. Ber.*, 1981, **114**, 505.
551. D. Fenske, A. Hollnagel and K. Mertzweiler, *Angew. Chem., Int. Ed. Engl.*, 1988, **27**, 965.
552. T. E. North, J. B. Thoden, B. Spencer, A. Bjarnason and L. F. Dahl, *Organometallics*, 1992, **11**, 4326.
553. J. J. Schneider, R. Goddard, C. Krüger, S. Werner and B. Metz, *Chem. Ber.*, 1991, **124**, 301.
554. M. S. Paquette and L. F. Dahl, *J. Am. Chem. Soc.*, 1980, **102**, 6621.
555. M. J. Chetcuti, S. R. McDonald and J. C. Huffman, *Inorg. Chem.*, 1989, **28**, 238.
556. M. Gorzellik, B. Nuber, T. Bohm and M. L. Ziegler, *J. Organomet. Chem.*, 1992, **429**, 173.
557. M. D. Curtis and P. Li, *J. Am. Chem. Soc.*, 1989, **111**, 8279.
558. A. D. Shaposhnikova *et al.*, *J. Organomet. Chem.*, 1992, **429**, 109.
559. C. U. Pittman, M. G. Richmond, M. Absi-Halabi, H. Beurich, F. Richter and H. Vahrenkamp, *Angew. Chem., Int. Ed. Engl.*, 1982, **21**, 786.

560. M. Müller and H. Vahrenkamp, *Chem. Ber.*, 1983, **116**, 2765.
561. M. Castiglioni, R. Giordano and E. Sappa, *J. Organomet. Chem.*, 1988, **342**, 97.
562. M. Castiglioni, R. Giordano and E. Sappa, *J. Organomet. Chem.*, 1988, **342**, 111.
563. S. G. Shore, W.-L. Hsu, C. R. Weisenberger, M. L. Caste, M. R. Churchill and C. Bueno, *Organometallics*, 1982, **1**, 1405.
564. M. Castiglioni, E. Sappa, M. Valle, M. Lanfranchi and A. Tiripicchio, *J. Organomet. Chem.*, 1983, **241**, 99.
565. E. Sappa, A. Tiripicchio and M. Tiripicchio-Camellini, *J. Organomet. Chem.*, 1983, **246**, 287.
566. L. J. Farrugia, J. A. K. Howard, P. Mitrprachachon, F. G. A. Stone and P. Woodward, *J. Chem. Soc., Dalton Trans.*, 1981, 171.
567. G. Predieri, A. Tiripicchio, C. Vignali, E. Sappa and P. Braunstein, *J. Chem. Soc., Dalton Trans.*, 1986, 1135.
568. J. J. Maj, A. D. Rae and L. F. Dahl, *J. Am. Chem. Soc.*, 1982, **104**, 3054.
569. M. Mlekuz, P. Bougeard, M. J. McGlinchey and G. Jaouen, *J. Organomet. Chem.*, 1983, **253**, 117.
570. M. J. Chetcuti, P. N. Cunningham, J. C. Gordon, B. E. Grant and J. Klaiss, *J. Organomet. Chem.*, 1990, **394**, 765.
571. P. H. M. Budzelaar, J. Boersma, G. J. M. van der Kerk, A. L. Spek and A. J. M. Duisenberg, *J. Organomet. Chem.*, 1985, **287**, C13.
572. P. H. M. Budzelaar, J. Boersma, G. J. M. van der Kerk and A. L. Spek, *Angew. Chem.*, 1983, **95**, 335.
573. P. H. M. Budzelaar, J. Boersma, G. J. M. van der Kerk, A. L. Spek and A. J. M. Duisenberg, *Organometallics*, 1985, **4**, 680.
574. B. Fischer, H. Kleijn, J. Boersma, G. van Koten and A. L. Spek, *Organometallics*, 1989, **8**, 920.
575. D. Fenske and A. Hollnagel, *Angew. Chem., Int. Ed. Engl.*, 1989, **28**, 1390.
576. S. G. Shore, W. S. Hsu, C. R. Weisenberger, M. L. Castle, M. R. Churchill and C. Bueno, *Inorg. Chem.*, 1983, **22**, 510.
577. G. Lavigne, F. Papageorgiou, C. Bergounhou and J. J. Bonnet, *Inorg. Chem.*, 1983, **22**, 2485.
578. M. Castiglioni, R. Giordano, E. Sappa, A. Tiripicchio and M. Tiripicchio-Camellini, *J. Chem. Soc., Dalton Trans.*, 1986, 23.
579. T. J. Groshens and K. J. Klabunde, *Organometallics*, 1982, **1**, 564.
580. S.-T. Lin, T. J. Groshens and K. J. Klabunde, *Inorg. Chem.*, 1984, **23**, 1.
581. T. J. Groshens and K. J. Klabunde, *J. Organomet. Chem.*, 1983, **259**, 337.
582. S.-T. Lin, R. N. Narske and K. J. Klabunde, *Organometallics*, 1985, **4**, 571.
583. T. R. Bierschenk, M. A. Guerra, T. J. Juhlke, S. B. Larson and R. J. Lagow, *J. Am. Chem. Soc.*, 1987, **109**, 4855.
584. G. E. Herberich, H. J. Becker, B. Hessner and L. Zelenka, *J. Organomet. Chem.*, 1985, **280**, 147.
585. G. Garzon, *Rev. Lationam. Quim.*, 1982, **13**, 5.
586. E. Dubler, M. Textor, H. R. Oswald and G. B. Jameson, *Acta Crystallogr., Sect. B*, 1983, **39**, 607.
587. G. E. Herberich, I. Hausmann, B. Hessner and M. Negele, *J. Organomet. Chem.*, 1989, **362**, 243.
588. T. Kuhlmann and W. Siebert, *Z. Naturforsch., Teil B*, 1984, **39**, 1046.
589. G. Brodt and W. Siebert, *Chem. Ber.*, 1989, **122**, 633.
590. W. Siebert, J. Edwin and H. Pritzkow, *Angew. Chem., Int. Ed. Engl.*, 1982, **21**, 148.
591. W. Siebert, J. Edwin, H. Wadepohl and H. Pritzkow, *Angew. Chem., Int. Ed. Engl.*, 1982, **21**, 149.
592. F. H. Köhler, U. Zenneck, J. Edwin and W. Siebert, *J. Organomet. Chem.*, 1981, **208**, 137.
593. A. Fessenbecker, M. Stephan, R. N. Grimes, H. Pritzkow, U. Zenneck and W. Siebert, *J. Am. Chem. Soc.*, 1991, **113**, 3061.
594. K. W. Piepgrass, J. H. J. Davis, M. Sabat and R. N. Grimes, *J. Am. Chem. Soc.*, 1991, **113**, 681.
595. H. Wadepohl, H. Pritzkow and W. Siebert, *Chem. Ber.*, 1985, **118**, 729.
596. R. Köster, G. Seidel, C. Krüger, G. Müller, A. Jiang and R. Boese, *Chem. Ber.*, 1989, **122**, 2075.
597. K.-F. Wörner, J.-K. Uhm, H. Pritzkow and W. Siebert, *Chem. Ber.*, 1990, **123**, 1239.
598. J. Edwin *et al.*, *J. Am. Chem. Soc.*, 1983, **105**, 2582.
599. N. Kuhn and H. Schumann, *J. Organomet. Chem.*, 1986, **314**, C39.
600. T. Kuhlmann, S. Roth, J. Rozière and W. Siebert, *Angew. Chem., Int. Ed. Engl.*, 1986, **25**, 105.
601. K. W. Piepgrass, X. Meng, M. Hölscher, M. Sabat and R. N. Grimes, *Inorg. Chem.*, 1992, **31**, 5202.
602. J. R. Pipal and R. N. Grimes, *Organometallics*, 1993, **12**, 4459.
603. J. R. Pipal and R. N. Grimes, *Organometallics*, 1993, **12**, 4452.
604. A. Bjarnason, *Organometallics*, 1991, **10**, 1244.
605. M. M. Brezinski and K. J. Klabunde, *Organometallics*, 1983, **2**, 1116.
606. M. M. Brezinski, J. Schneider, L. J. Radonovich and K. J. Klabunde, *Inorg. Chem.*, 1989, **28**, 2414.
607. M. M. Brezinski, K. J. Klabunde, S. K. Janikowski and L. J. Radonovich, *Inorg. Chem.*, 1985, **24**, 3305.
608. H. Kanai, S. B. Choe and K. J. Klabunde, *J. Am. Chem. Soc.*, 1986, **108**, 2019.
609. S.-B. Choe, H. Kanai and K. J. Klabunde, *J. Am. Chem. Soc.*, 1989, **111**, 2875.
610. S.-T. Lin and K. J. Klabunde, in 'Organometallic Syntheses', eds. R. B. King and J. J. Eisch, Elsevier, New York, 1986, vol. 3, pp. 147–9.
611. S.-T. Lin and K. J. Klabunde, in 'Organometallic Syntheses', eds. R. B. King and J. J. Eisch, Elsevier, New York, 1986, vol. 3, pp. 150–2.
612. S.-T. Lin and K. J. Klabunde, in 'Organometallic Syntheses', eds. R. B. King and J. J. Eisch, Elsevier, New York, 1986, vol. 3, pp. 156–7.
613. S.-B. Choe and K. J. Klabunde, in 'Organometallic Syntheses', eds. R. B. King and J. J. Eisch, Elsevier, New York, 1986, vol. 3, pp. 153–5.
614. H. Hoberg, H. J. Reigel and K. Seevogel, *J. Organomet. Chem.*, 1982, **229**, 281.
615. G. E. Herberich and B. Hessner, *Chem. Ber.*, 1982, **115**, 3115.
616. I. A. Garbuzova, O. G. Garkusha, B. V. Lokshin, A. R. Kudinov and M. I. Rybinskaya, *J. Organomet. Chem.*, 1991, **408**, 247.
617. E. D. Jemmis and A. C. Reddy, *Organometallics*, 1988, **7**, 1561.
618. M. W. Whiteley, H. Pritzkow, U. Zenneck and W. Siebert, *Angew. Chem.*, 1982, **94**, 464.
619. M. W. Whiteley, H. Pritzkow, U. Zenneck and W. Siebert, *Angew. Chem., Int. Ed. Engl.*, 1982, **21**, 453.
620. W. Siebert, *Angew. Chem.*, 1985, **97**, 924.
621. T. Kuhlmann, S. Roth, J. Rozière and W. Siebert, *Angew. Chem.*, 1986, **98**, 87.
622. J. Edwin *et al.*, *Organometallics*, 1983, **2**, 1666.
623. H. A. Boyter, R. G. Swisher, E. Sinn and R. N. Grimes, *Inorg. Chem.*, 1985, **4**, 3810.

624. D. M. Schubert, C. B. Knobler, S. Trofimenko and M. F. Hawthorne, *Inorg. Chem.*, 1990, **29**, 2364.
625. J. Zwecker, H. Pritzkow, U. Zenneck and W. Siebert, *Angew. Chem.*, 1986, **98**, 1129.
626. A. A. Erdman, T. V. Baranova, Z. P. Zubreichuk, N. A. Maier and Y. A. Ol'dekop, *Metalloorg. Khim.*, 1990, **3**, 199.
627. A. A. Erdman *et al.*, *Metalloorg. Khim., Engl. Ed.*, 1989, **2**, 458.
628. A. A. Erdman *et al.*, *Metallorg. Khim.*, 1989, **2**, 477.
629. N. A. Maier, A. A. Erdman, Z. P. Zubreichuk, V. P. Propkopovich and Y. A. Ol'Dekop, *J. Organomet. Chem.*, 1985, **292**, 297.
630. B. Stibr *et al.*, *Polyhedron*, 1982, **1**, 833.
631. K. A. Solntsev *et al.*, *Koord. Khim., Engl. Ed.*, 1984, **10**, 627.
632. K. A. Solntsev *et al.*, *Koord. Khim.*, 1984, **10**, 1132.
633. Z. Janousek, J. Plesek, S. Hermanek, K. Base, L. J. Todd and W. F. Wright, *Collect. Czech. Chem. Commun.*, 1981, **46**, 2818.
634. J. J. Briguglio and L. G. Sneddon, *Organometallics*, 1986, **5**, 327.
635. L. Barton and P. K. Rush, *Inorg. Chem.*, 1986, **25**, 91.
636. H. Zhang, Y. Wang, A. K. Saxena, A. R. Oki, J. A. Maguire and N. S. Hosmane, *Organometallics*, 1993, **12**, 3933.
637. C. A. Plumb, P. J. Carroll and L. G. Sneddon, *Organometallics*, 1992, **11**, 1665.
638. D. N. Cox, D. M. P. Mingosa and R. Hoffmann, *J. Chem. Soc., Dalton Trans.*, 1981, 1788.
639. T. Kuhlmann, H. Pritzkow, U. Zenneck and W. Siebert, *Angew. Chem.*, 1984, **96**, 984.
640. T. Kuhlmann, H. Pritzkow, U. Zenneck and W. Siebert, *Angew. Chem., Int. Ed. Engl.*, 1984, **23**, 965.
641. W. Siebert, *Angew. Chem., Int. Ed. Engl.*, 1985, **24**, 943.
642. D. M. Vinitskii *et al.*, *Z. Neorg. Khim.*, 1984, **29**, 1714.

4
Palladium Complexes with Carbonyl, Isocyanide and Carbene Ligands

KEITH R. DIXON and ANDREW C. DIXON
University of Victoria, BC, Canada

4.1 INTRODUCTION 194

4.2 CARBONYL COMPLEXES 194

4.2.1 Introduction and Homoleptic Carbonyls 194
4.2.2 Theoretical Studies 194
4.2.2.1 Theoretical studies related to catalysis 194
4.2.2.2 Theoretical studies related to stability 195
4.2.3 Carbonyl Halides 195
4.2.4 Dinuclear Carbonyls 196
4.2.5 Multinuclear Carbonyls 198
4.2.5.1 Introduction 198
4.2.5.2 $[Pd_3(\mu_3\text{-}CO)(dppm)_3]^{2+}$ 198
4.2.5.3 Other trinuclear carbonyls 200
4.2.5.4 Tetranuclear carbonyls 201
4.2.5.5 Pentanuclear carbonyls 202
4.2.5.6 Hexanuclear carbonyls 202
4.2.5.7 Higher nuclearity carbonyls (n = 7–23) 203
4.2.6 Palladium Carbonyls Containing Other Metals 204
4.2.7 Catalysis by Palladium Carbonyls and Reactions of the CO Ligand 207

4.3 ISOCYANIDE COMPLEXES 208

4.3.1 Introduction 208
4.3.2 Palladium(0) Isocyanides 208
4.3.3 $[Pd_3(\mu_3\text{-}CNR)(dppm)_3]^{2+}$ 209
4.3.4 Palladium(I) Isocyanides 210
4.3.5 Palladium(II) Isocyanides 211
4.3.5.1 Introduction 211
4.3.5.2 Insertions of isocyanides into palladium–carbon bonds 211
4.3.5.3 Insertions of isocyanides into palladium–element bonds other than carbon 214
4.3.6 Structures 215

4.4 CARBENE COMPLEXES 216

4.4.1 Introduction 216
4.4.2 Synthesis 216
4.4.2.1 From coordinated isocyanides 216
4.4.2.2 Other synthetic methods 218
4.4.3 Structures 218
4.4.4 Novel Palladium Carbenes 219

4.5 REFERENCES 220

4.1 INTRODUCTION

Palladium has often been, to some extent, the forgotten sibling of the nickel triad, especially with respect to platinum chemistry. However, increasing interest in the role of palladium in automotive exhaust catalysts, and in a wide variety of organic processes promoted or catalysed by palladium and its complexes, is gradually remedying this neglect. This is especially true in carbonyl chemistry where extensive recent developments, especially in cluster chemistry, are reported below. In the areas of isocyanide and carbene chemistry, recent work has been more developmental in nature, extending the framework established by early studies.

The first edition of this book reported the literature in the area of palladium carbonyl, isocyanide and carbene complexes to about 1979.[1] This coverage is now extended to approximately the spring of 1993. Work described in the earlier edition is not repeated in detail, but an effort is made to direct the reader to its most important features. It has been necessary to exclude from the coverage large numbers of studies where palladium or its compounds are utilized as catalysts, but the primary focus is on organic chemistry, with little characterization of palladium-containing intermediates.

4.2 CARBONYL COMPLEXES

4.2.1 Introduction and Homoleptic Carbonyls

In the nickel, palladium, platinum triad, a dramatic change in the stability of carbonyl derivatives occurs between nickel and palladium. The tetracarbonyls, $[M(CO)_4]$, M = Pd or Pt, are observed only at low temperatures (<80 K) in noble gas or CO matrices; in contrast to the ambient temperature stability of $[Ni(CO)_4]$. Other homoleptic carbonyls, $[Pd(CO)_{1-3}]$, exist under matrix conditions, and infrared spectroscopy shows that all have the expected structures for a d^{10} metal: $n = 2$, linear; $n = 3$, trigonal planar; and $n = 4$, tetrahedral.[1] The electronic absorption spectrum of $[Pd(CO)_4]$ exhibits strong absorption at 45 870 cm^{-1} with a pronounced shoulder at 40 000 cm^{-1}, assigned to d–π^* and $t_2(d)$–$t_2(\pi^*)$ transitions, respectively. These energies are intermediate between the slightly larger values for $[Ni(CO)_4]$ and slightly smaller values for $[Pt(CO)_4]$.[2,3] Interestingly, when palladium atoms are co-condensed with CO and SiO moieties in an argon matrix, there is infrared evidence for silicon analogues of the palladium carbonyl species.[3]

4.2.2 Theoretical Studies

4.2.2.1 Theoretical studies related to catalysis

Theoretical studies of the interaction of CO with palladium atoms are becoming increasingly common and sophisticated, in part because of the crucial role these interactions may play in catalysis on surfaces, especially in automobile exhaust catalysts. An *ab initio* study of interactions of NO and CO with palladium and platinum is typical, and provides references to much earlier literature.[4] Carbon monoxide binds to palladium surfaces in both linear (atop) and bridging geometries, with interaction energies of 92–96 kJ mol^{-1} and 142–150 kJ mol^{-1}, respectively, and corresponding ν(CO) frequencies, 2100 cm^{-1} and 1820 cm^{-1}, as compared with free CO at 2143 cm^{-1}. The calculations for the PdCO moiety confirm the familiar σ/π, forward/back donation bonding mechanism, leading to a linear geometry. The $^1\Sigma^+$ ground state is predicted to have a relatively short Pd–C distance of 0.196 nm, and is bound by 113 kJ mol^{-1} compared to 77 kJ mol^{-1} for the platinum analogue.[4] *Ab initio* SCF calculations predict shifts for ν(CO) chemisorbed on palladium surfaces (~ −72 cm^{-1} relative to free CO), which are in good agreement with values observed in krypton matrices (−95 cm^{-1}) and on palladium hydrosol surfaces (−78 cm^{-1}).[5] All electron coupled functional pair calculations have been applied to both linear, Pd–CO, and bridging, $Pd_2(\mu\text{-}CO)$, situations, yielding binding energies of 138 kJ mol^{-1} and 230 kJ mol^{-1}, respectively.[6] All recent authors emphasize the necessity of including relativistic contributions in the calculations, noting for example that bridging CO is bound by only 13 kJ mol^{-1} when these contributions are neglected.[6] Earlier studies which do not include relativistic contributions may be flawed. Carbon monoxide chemisorbed on Pd(100) surfaces has also been studied by valence photoemission spectroscopy, and *ab initio* calculations are in good agreement with the experimental spectra,[7] as are similar many body calculations of the core–hole spectra.[8]

A model for a different aspect of catalytic behaviour, namely promoted catalysts, is provided by SCF–CI calculations for [PdCONa] and $[PdCONa]^+$. The former is bound by only 25 kJ mol^{-1}, but in the ionic species the electrostatic effect enhances both σ and π bonding and produces greater stability (63–92 kJ mol^{-1}).[9] A wide ranging study of proton affinities by Hartree–Fock–Slater methods is also of possible relevance to catalysis in that it defines the nucleophilicity of the metal species. For the tetracarbonyls, $[M(CO)_4]$, M = Ni, Pd and Pt, total protonation energies ($E[M(CO)_5{}^{n-}]$–$E[HM(CO)_5{}^{(1-n)+}]$) are calculated to be 580.8 kJ mol^{-1}, 602.2 kJ mol^{-1} and 646.4 kJ mol^{-1}, respectively.[10]

Some other theoretical results concerning CO insertion into metal–hydride or metal–alkyl bonds are described in Section 4.2.7.

4.2.2.2 Theoretical studies related to stability

Theoretical studies related to more synthetic aspects of palladium carbonyl chemistry, for example the relative stabilities of $[M(CO)_4]$, M = Ni, Pd, Pt, complexes, include extensive molecular orbital calculations based on density functional theory. A wide range of $M(CO)_n$, $n = 4$ or 6, complexes are included in the study and, as noted above, inclusion of relativistic contributions is important. The familiar σ/π bonding mechanism is again confirmed with π-bonding making the biggest contribution to the stability. Results for both intrinsic mean bond energy (Ni 748 kJ mol^{-1}, Pd 184 kJ mol^{-1}, Pt 247 kJ mol^{-1}) and the first CO bond dissociation energy (Ni 443 kJ mol^{-1}, Pd 113 kJ mol^{-1}, Pt 159 kJ mol^{-1}) show that palladium carbonyls are the least stable of the three; a property shared with other second row metals and in agreement with synthetic results. The rise in stability for platinum is considered to be due to relativistic effects.[11] Effective core potential studies of [PdCO] and $[Pd(CO)_4]$ show a similar picture,[12] and dissociation energies for $[Pd(CO)]^+$ (131.7 kJ mol^{-1}) and $[Pd(CO)_2]^+$ (247.0 kJ mol^{-1}) have been calculated as part of a detailed study of corresponding cations for first and second row transition metals. The bonding is primarily electrostatic and for comparison, corresponding values for nickel cations are 153.4 kJ mol^{-1} and 298.5 kJ mol^{-1}, respectively.[13]

4.2.3 Carbonyl Halides

Carbonyl halides of palladium were first reported in 1898. Carbonylation of $PdCl_2$ and $[Pd_2X_4L_2]$ derivatives was attempted subsequently by various workers but, as is evident from the discussions in the first edition of this book, well-characterized compounds were rare. In particular, a number of materials originally formulated as mononuclear carbonyl halides had anomalously low CO stretching frequencies and probably contained bridging carbonyls, possibly polymers, $[PdX(CO)]_n$, in many cases.[1] One well-characterized compound was the Pd^I dianion, (**1**), obtained from carbonylation of $[PdCl_4]^{2-}$.[14] The CO stretches (1906 cm^{-1}, 1966 cm^{-1} in CH_2Cl_2) are higher than normal for CO bridges involving other transition metals, but the structure has been confirmed by x-ray diffraction study of the $[Bu^n{}_4N]^+$ salt. The short Pd–Pd distance (0.2697(3) nm) suggests the presence of a metal–metal bond.[15] The polymeric materials $[PdX(CO)]_n$ probably have the related structure (**2**) and a high yield synthesis of the chloride polymer has been reported by carbonylation of $PdCl_2$ or $[Pd_2Cl_4(CO)_2]$ in acetic anhydride.[16,17]

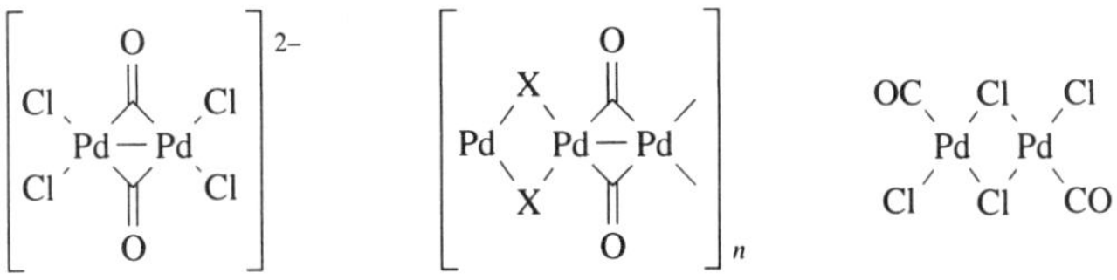

(**1**) (**2**) (**3**)

High pressure carbonylation of $PdCl_2$ in anhydrous thionyl chloride gives (**3**), a close relative of (**1**) except that it is now a Pd^{II} complex. Depending on the medium, ν(CO) is in the range 2159–2167 cm^{-1}, clearly indicating terminal CO and most probably a *trans* structure.[16,17] The bromide analogue of (**3**) is known but attempts to prepare the iodide have failed.[18] Detailed study of the equilibrium represented by Equation (1) gives the activation parameters $\Delta H^\ddagger$, −56.8 ± 0.8 kJ mol^{-1}, and $\Delta S^\ddagger$, 130 ± 3 J K^{-1} mol^{-1}, and leads to an estimate of 100 kJ mol^{-1} for the Pd–CO bond strength.[17] This low bond energy may explain some interesting contrasts between platinum and palladium carbonyl chemistry. Thus, $[Pd_2Cl_4(CO)_2]$ does not react with CO whereas the platinum analogue is rapidly cleaved to *cis*-

$[PtCl_2(CO)_2]$. By contrast, $[PdCl(CO)_3]^-$ reacts rapidly with Cl^- to displace CO whereas $[PtCl(CO)_3]^-$ is much less reactive. Other comparisons suggest that these differences are due to the low Pd–CO bond energy rather than originating from simply kinetic effects.[19]

$$2\,PdCl_{2(s)} + 2\,CO_{(SOCl2)} \xrightarrow{SOCl_2} Pd_2(CO)_2Cl_{4(SOCl2)} \qquad (1)$$

The contrast between the chloride-bridged structure of (**3**) and the carbonyl-bridged structure of (**1**) is of some interest, especially since the latter infringes the normal rule that metal carbonyl halides are 'invariably' halide bridged! Molecular orbital calculations show that the difference is critically dependent on the relative energy of a π^* bridge orbital. When the bridge is CO, the crucial orbital is bonding in character relative to its component palladium and CO orbitals. When the bridge is chlorine, the orbital is antibonding. Thus, for a Pd^I compound with CO bridges, the orbital can be filled (HOMO) and contributes a stabilizing interaction, whereas if the compound were chlorine bridged the HOMO would be destabilizing. Conversely, for a Pd^{II} compound with chlorine bridges, the orbital can be empty (LUMO) whereas filling it with the additional two electrons in Pd^I dimer would be destabilizing. It is also interesting that, despite the relatively short Pd–Pd distance, the calculations indicate that direct metal–metal bonding is almost negligible. Occupation of both bonding and antibonding components gives a net bond order close to zero and the diamagnetism and short distance are both due to coupling through the strong bridge bonds.[20]

Mononuclear palladium(II) carbonyl halides are also well established. The anions, $[PdX_3(CO)]^-$, X = Cl, Br, I, are synthesized either by carbonylation of $[Pd_2X_6]^{2-}$, X = Cl or Br,[21] or by carbonylation of PdI_2. The iodide is relatively unstable, decarbonylating rapidly both in solution and in the solid state, but both the chloride and bromide have been characterized by x-ray diffraction. The anions are closely planar, with Pd–C distances of 0.187(1) nm, only slightly longer than that found in platinum analogues (0.182 nm).[18]

Organopalladium carbonyls are relatively rare, despite the widespread use of palladium and its complexes in organic reactions, often involving carbonylation or decarbonylation processes. Interest tends to focus on the organic chemistry and palladium intermediates are generally not isolated. Carbonylation of labile solvent complexes containing perhalophenyl groups does yield organopalladium carbonyls but the products are stable only to about –30 °C. Thus, $[Pd(C_6F_5)(acetone)(PR_3)_2]^+$ and *cis*-$[Pd(C_6X_5)_2(THF)_2]$, X = F or Cl, give $[Pd(C_6F_5)(CO)(PR_3)_2]^+$ and *cis*-$[Pd(C_6X_5)_2(CO)_2]$, respectively.[1,22] The latter complexes have very high carbonyl stretching frequencies (2186 cm^{-1} and 2173 cm^{-1}) indicating that back-bonding is virtually nonexistent, consistent with the low stability. Platinum analogues are much more stable.[22] Use of nitro-substituted aryls gives somewhat more stable complexes and the $[PPh_3Bz]^+$ salt of $[PdCl_2(CO)(2\text{-methyl-6-nitrophenyl})]^-$ is the first example to be crystallographically characterized. The Pd–C length (0.185(1) nm) is short enough for significant π-bonding and $\nu(CO)$, 2100 cm^{-1}, is lower than in the related cationic and neutral complexes above.[23]

4.2.4 Dinuclear Carbonyls

The carbonyl halides, (**1**)–(**3**), discussed in Section 4.2.3, are examples of dinuclear palladium carbonyls and illustrate the basic division into complexes with bridging CO and those with only terminal CO. Other than Structure (**3**) and its other halide analogues, complexes without bridging CO are rare, although $[Pd_2Cl_2(CO)_2(2\text{-diphenylphosphinopyridine})_2]$ may be an example.[24] Bridging CO is much more common, the complexes falling into five structural types as illustrated by Structures (**4**)–(**8**); with (**4**) being an obvious analogue of (**1**). These complexes are all formally Pd^I, with $\nu(CO)$ in the range 1780–1870 cm^{-1}, and except for (**8**), have relatively short Pd–Pd distances 0.265–0.277 nm.

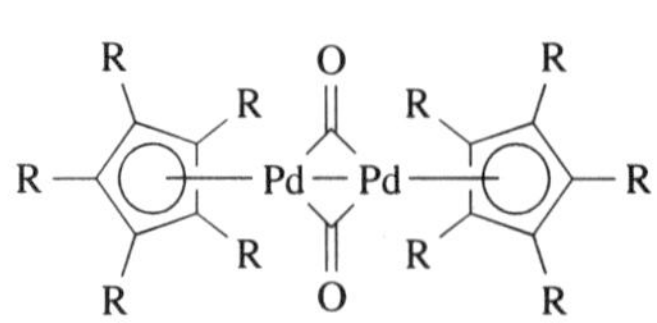

(4)

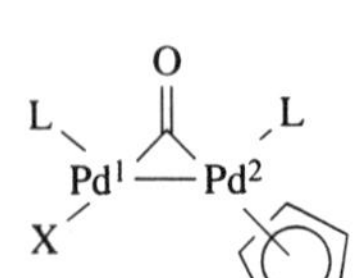

(5)

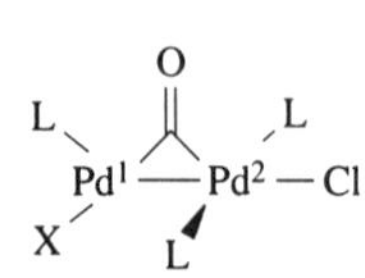

(6)

Complex (**4**), R = Ph, is formed by reductive carbonylation of $[Pd_2(\mu\text{-Cl})_2(\eta^5\text{-}C_5Ph_5)_2]$. The reducing agent is zinc dust and, interestingly, when this is omitted carbonylation yields only the mononuclear

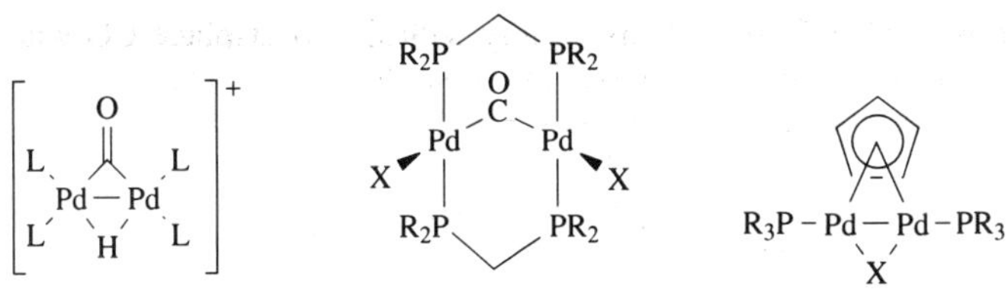

(7) (8) (9)

carbonyl [PdCl(CO)(η^5-C_5Ph_5)].[25] Similar complexes (**4**), R = Me, result from reaction of $[PdCl(CO)]_n$ polymer with $(Me_5C_5)MgCl{\cdot}THF$.[26] The analogy between Structures (**1**) and (**4**) raises some questions about electron counting in the latter complexes. In (**1**), assignment of 16 electrons (e) to each palladium is straightforward; whereas in (**4**), an η^5-Cp would result in a less probable 18e count at palladium. Unfortunately, there seems to be no structural study which might reveal slippage of the Cp rings towards an η^3 configuration.

Carbonylation of (**9**) displaces the bridging Cp ring to a terminal position and yields the air stable complexes, (**5**), X = Cl or Br, L = PEt_3 or PPr^i_3, with bridging CO absorptions at ~1800 cm^{-1}.[27] Closely related complexes result from carbonylation of $[Pd(NO_2)_2(PEt_2Ph)_2]$ in dichloromethane solution at 298 K to yield (**6**), L = PEt_2Ph,[28] or from the combination of palladium(0) in $[Pd(CO)(PPh_3)_3]$ with palladium(II) in $[PdCl_2(PhCN)_2]$ to give the triphenylphosphine analogue.[29] In the former route, the solvent appears to be the source of the chloride. The latter route can also be applied to the synthesis of mixed metal derivatives, $[PdClL(\mu\text{-}CO)PtClL_2]$ and $[PtClL(\mu\text{-}CO)PdClL_2]$, L = PPh_3,[29] and similar complexes have been obtained by reactions of $[M_4(CO)_5(PPh_3)_4]$ with $[MCl_2(PPh_3)_2]$.[30] The comparison between (**5**) and (**6**) presents an interesting problem in electron counting. If we ignore the Pd–Pd bond in (**5**) then the Pd^1 and Pd^2 centres are 14e and 18e respectively, suggesting the possibility of a donor Pd^2 to Pd^1 bond. In the solid-state structure (X = Cl, L = Pr^i_3), the Pd–Pd bond is short (0.2675(1) nm), and the electronic asymmetry is reflected in the CO bridge, which is more strongly bound to Pd^1 (0.1899(1) nm vs. 0.2082(1) nm) and is more linear at Pd^1 (Pd–C–O, 146.2(2)° vs. 129.4(2)°).[27] In (**6**), L = PEt_2Ph, the palladium centres would be 14e and 16e without the metal–metal bond (0.26521(3) nm) and the asymmetry of the CO bridge is slightly greater (0.1874(3) nm vs. 0.2110(3) nm). The authors suggest that the CO electrons should be associated mainly with Pd^1 to relieve the electronic asymmetry.[28]

Several other complexes of general formula $[(PdXL)_2CO]$ have been reported and are possibly closely related to (**5**). However, structures involving simple replacement of cyclopentadienyl in (**5**) by a one-electron donor such as halide would create 15e counts at palladium unless a metal–metal double bond is present, and unfortunately none of these complexes are especially well characterized. Thus, reaction of $[Pd_2H_2Cl_4(CO)_2]$ with L yields $[(PdClL)_2CO]$, L = PPh_3, $AsPh_3$, $SbPh_3$, dppm, or dpam;[31] reaction of $[Pd_2Cl_4(CO)_2]$ with $HCl/SnCl_2/PPh_3$ gives $[\{Pd(SnCl_3)(PPh_3)\}_2CO]$.[32] A kinetic study of the disproportionation reactions of some of these complexes with arylphosphines has also been reported.[33]

In contrast to the above examples, the complexes, (**7**), have 16e counts at both palladium centres. When L_2 is a chelating bis(diphenylphosphino)propane ligand (dppp), (**7**) is formed by methanol reduction of [PdCl(Ph)(dppp)] in the presence of triethylamine; a reaction which probably involves electrophilic attack of a cationic Pd–H intermediate on a palladium(0) carbonyl. The Pd–Pd bond in the product is 0.2767(4) nm, and there is a dihedral twist of 16° between the two palladium coordination planes. ν(CO) is 1789 cm^{-1}.[34] The triphenylphosphine analogue, (**7**), L = PPh_3, results from protonation of $[Pd(PPh_3)_4]$ or $[Pd(CO)(PPh_3)_3]$ by $CH_2(SO_2CF_3)_2$ in a carbon monoxide atmosphere[35] or from carbonylation of $[PdH(PPh_3)_3]^+$ in trifluoroacetic acid.[36] This cation displays interesting fluxional behaviour in both ^{13}C and ^1H NMR spectra. For example, in the latter spectrum the bridging hydride is a 40 Hz quintet at – 60 °C but resolves to a triplet (93 Hz) of triplets (–12 Hz) at –110 °C. The activation parameters are $\Delta H^\ddagger$ 39.7 ± 1.7 kJ mol^{-1} and $\Delta S^\ddagger$ 33 ± 17 J K^{-1} mol^{-1}, and the process probably involves cleavage of either bridge followed by rotation.[35]

The final type of dinuclear complexes are those involving CO bridges between metals which have no metal–metal bond. Examples are relatively rare, so the so-called 'A-frame' complexes typified by (**8**) are important. As is frequently the case, other metals, especially platinum, have been studied more extensively than palladium and the subject has been reviewed.[37] Many small molecules, CO included, insert into the Pd–Pd bond of (**10**) to give 'A-frame' complexes, of which (**8**), X = Cl, is typical. A detailed synthesis using $[PdCl_2(PhCN)_2]$ to prepare (**10**), R = Ph, and hence (**8**) has been reported.[38] The metal centres in (**8**) are palladium(I) and the same product results from essentially oxidative addition of $COCl_2$ to the palladium(0) complex, $[Pd_2(dppm)_3]$.[37] Analogues of (**8**), R = Me, are also stable with the less sterically demanding ligand, bis(dimethylphosphino)methane (dmpm). Reaction with $[PdCl(CO)]_n$ polymer at –75 °C yields $[Pd_2Cl_2(dmpm)_2]$, which readily adds CO to form $[Pd_2Cl_2(\mu\text{-}CO)(dmpm)_2]$.

The bridging CO stretch is at 1710 cm^{-1} in the infrared and the Pd–Pd distance is 0.3169(1) nm. Interestingly, $[Pd_2Cl_2(dmpm)_2]$ is relatively water soluble and is readily hydrolysed to $[Pd_2(OH)_2(dmpm)_2]$, which in turn carbonylates to $[Pd_2(OH)_2(\mu\text{-}CO)(dmpm)_2]$.[39] Other examples of these types of complex include (**8**), R = Ph, where X is a trifluoroacetate group rather than a halide,[40] and a complex, $[Pd_2Cl_4(CO)(Ph_2PCH_2NEt_2)_2]$, which may have structure (**11**); conceptually the result of oxidative addition of chlorine to an analogue of (**8**), oxidizing both metals to palladium(II) with concomitant formation of a metal–metal bond.[41]

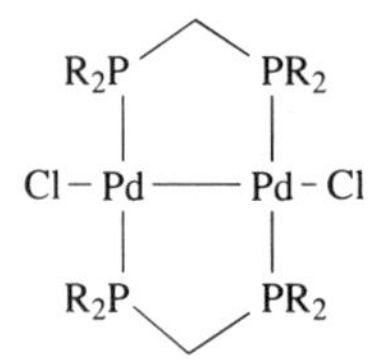

(10)

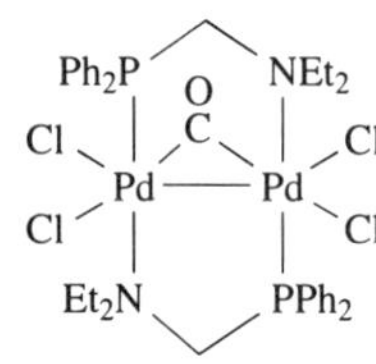

(11)

4.2.5 Multinuclear Carbonyls

4.2.5.1 Introduction

The first edition of this book reported only three types of palladium carbonyls with a nuclearity of three or greater: $[Pd_3(CO)_3(PPh_3)_{3\text{ or }4}]$, a Pd_3 triangle, with terminal phosphines and each edge bridged by CO; $[Pd_4(OAc)_4(CO)_4]$, a Pd_4 rectangle with two long edges double bridged by acetate, and two short edges double bridged by CO; and $[Pd_4(CO)_5(PPh_3)_4]$, a 'butterfly' arrangement of four palladium atoms, terminal phosphines, CO bridges on each Pd–Pd bond, and the extra CO bridging 'wingtip to wingtip' on a nonbonded Pd···Pd edge.[1] However, since the mid-1980s, this field has been rapidly extended, especially by the work of Puddephatt in Canada, Mednikov in the USSR, and their co-workers. An unusual trinuclear cation, $[Pd_3(\mu_3\text{-}CO)(dppm)_3]^{2+}$, exhibits a formal oxidation state of 2/3 and has an extensive chemistry, and in palladium(0) chemistry, complexes $[Pd_n(CO)_x(PR_3)_y]$ are now well characterized for nuclearities n = 3–8, 10, 16 and 23. The synthetic methods are serendipitous rather than logical, generally involving either reductive carbonylation of palladium(II) salts in the presence of added tertiary phosphine, or aggregation of palladium(0) compounds under CO. A particularly well-studied system uses palladium(II) acetate, with added CO, tertiary phosphines and often trifluoroacetic or acetic acid. As detailed in the succeeding sections, many different clusters have been isolated from this system, the nature of the products being very sensitive to the exact conditions, solvents and molar ratios of reactants.

4.2.5.2 $[Pd_3(\mu_3\text{-}CO)(dppm)_3]^{2+}$

(i) Synthesis and structure

An especially unusual and interesting palladium carbonyl results from reaction of palladium(II) acetate with dppm (in numbered structures throughout this article the phenyl groups of dppm have been omitted for clarity) and carbon monoxide in aqueous acetone containing trifluoroacetic acid. The reaction proceeds in 90% yield at ambient temperature according to Equation (2).

$$3\,Pd(OAc)_2 + 3\,dppm + 3\,CO + 2\,H_2O + 2\,CF_3CO_2H \longrightarrow [Pd_3(O_2CCF_3)(CO)(dppm)_3]^+ + CF_3CO_2^- + 2\,CO_2 + 6\,AcOH \quad (2)$$

The tripalladium product contains a weakly coordinated trifluoroacetate anion, but this is easily displaced by treatment with ammonium hexafluorophosphate to give a more highly conducting salt containing the parent cation $[Pd_3(\mu_3\text{-}CO)(dppm)_3]^{2+}$ (Structure (**12**)).[42]

This complex cation may be considered as a model for CO on a palladium surface, and, in this context, the role of the bridging dppm ligands is vital in preventing fragmentation during reaction chemistry. As is so often the case, the platinum analogue has been extensively studied, and the chemistry

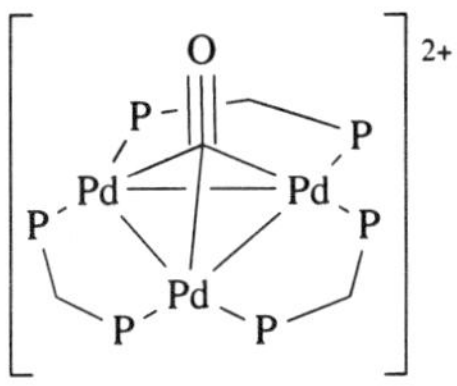

(12)

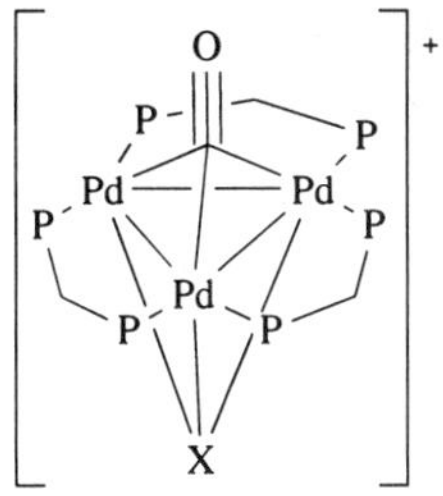

(13)

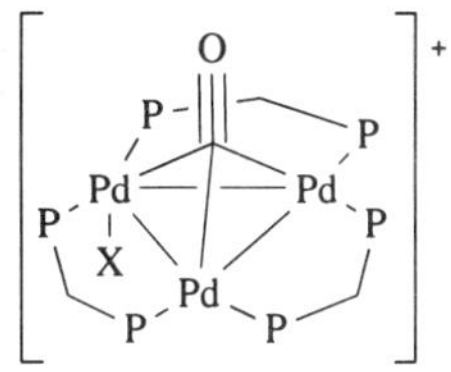

(14)

of both species has been reviewed recently.[43] The mechanism of Equation (2) is thought to involve initial formation of [PdX_2(dppm)], X = CF_3CO_2, which is reduced to a short-lived Pd^0(dppm) intermediate (probably also containing coordinated CO). This intermediate adds to [PdX_2(dppm)] to give a dinuclear framework, [$Pd_2X_2(dppm)_2$], which inserts CO to the A-frame complex, [Pd_2X_2(μ-CO)$(dppm)_2$], which in turn adds the palladium(0) intermediate to yield the final product.[42]

Although several adducts of (**12**), [$Pd_3(\mu_3$-X)(μ_3-CO)$(dppm)_3]^+$, X = CF_3CO_2, Cl, I, have been studied by x-ray diffraction (see Section 4.2.5.2(ii)), the parent cation itself has not been structurally characterized. However, the platinum analogue has been studied as a PF_6^- salt. It has an approximately planar Pt_3P_6 skeleton and short Pt–Pt distances (0.263 nm) indicative of M–M single bonds, and the adduct structures below suggest that [$Pd_3(\mu_3$-CO)$(dppm)_3]^{2+}$ is similar. One face of the Pt_3 triangle is symmetrically capped by the CO.[43] In the infrared spectrum of (**12**), PF_6^- salt, ν(CO) is at 1860 cm^{-1},[44] and Pd–Pd stretches have been identified in the Raman ($CF_3CO_2^-$ salt) at 143(*e*) and 205(a_1) cm^{-1}.[45] Carbonyl stretches in the halide adducts are slightly lower, 1815–1820 cm^{-1}.[43] The nature of the lowest energy excited states of (**12**) has been examined by a combination of spectroscopic and computational techniques. A low intensity feature at 570 nm is attributed to singlet–triplet absorption and a more intense one at 482 nm to singlet–singlet. The cluster is luminescent at 77 K but not at ambient temperature.[46]

(ii) Coordination chemistry

Complex (**12**) is formally a 42e cluster, with palladium atoms in a $\frac{2}{3}$ oxidation state. Empty $5p_z$ orbitals of a_1 and *e* symmetry on the palladiums should confer coordinatively unsaturated character, with a correspondingly extensive coordination chemistry; potentially adding up to 6e donors, although 2e or 4e are probably more likely. As noted above, the trifluoroacetate salt has a weakly coordinated anion which caps the other (remote from CO) side of the Pd_3 triangle at very long Pd–O distances (0.28–0.31 nm). One oxygen of the trifluoroacetate is bound to a single palladium and the other bridges the remaining pair of palladiums. The Pd–Pd distances are 0.2576(1)–0.2607(2) (av. 0.2598) nm. Halide ions displace the trifluoroacetate to give symmetrical caps completing a trigonal bipyramidal $Pd_3(\mu_3$-X)(μ_3-CO) unit, (**13**), but again the Pd–X distances are very long (0.295–0.308 nm for X = I).[42,43] Overall, the metal–metal distances in these various coordination adducts vary rather little (0.259–0.260 nm for palladium, 0.261–0.264 nm for platinum), confirming that essentially nonbonding metal 5*p* and 6*p* orbitals are utilized. The principal effect of the coordination is a 'slippage' of the μ_3-CO group towards asymmetrical bridging. For example, in the platinum analogue of (**14**), X = CN, Pt–CO is 0.194 nm to the platinum carrying CN, and 0.225/0.244 nm to the other platinums. The degree of 'slippage' apparently increases with the donor ability of the ligand.[43,44]

The weakly bound halide ligands in their μ_3 capping mode can, perhaps, be regarded as 6e donors, bringing the overall cluster count to coordinative saturation at 48 electrons. With 2e donor ligands there is also evidence of terminal attachment, frequently accompanied by fluxional processes which may proceed through μ_3 intermediates. For example, addition of CO to (**12**) is rapid and reversible. The product shows infrared evidence of both terminal and bridging CO but on the NMR timescale is fluxional even at −90 °C, suggesting rapid interchange between (**13**) and (**14**), X = CO.[43] Similarly, addition of cyanide to (**12**) gives (**14**), X = CN, which exhibits rapid cyanide migration even at −60 °C via a mechanism which may involve either μ_2 or μ_3 intermediates.[44]

(iii) Oxidative addition reactions

In contrast to the simple coordination chemistry of (**12**), oxidative addition does cause dramatic changes in the Pd–Pd bond lengths, thereby illustrating the importance of the dppm ligands in preventing fragmentation. For example, the initial product of addition of thiocyanate contains a terminal SCN^- ligand, but oxidative addition of the C=S bond can be promoted either thermally or photochemically to give (**15**), X = CN. The palladium atoms form an isosceles triangle (0.3508(1) nm and 0.2579(1) nm) capped by sulfur, 0.1311 nm above the Pd_3 plane and equidistant (0.2297(3) nm) from all three palladium atoms.[47] Oxidative addition of hydrogen sulfide to (**12**), studied as a possible model for poisoning of palladium catalysts, gives a similar product, (**15**), X = H,[48] and both complexes (**15**) show labile inversion at the sulfur atom.

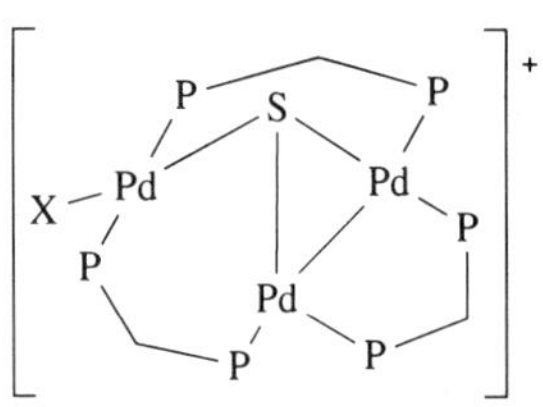

(15)

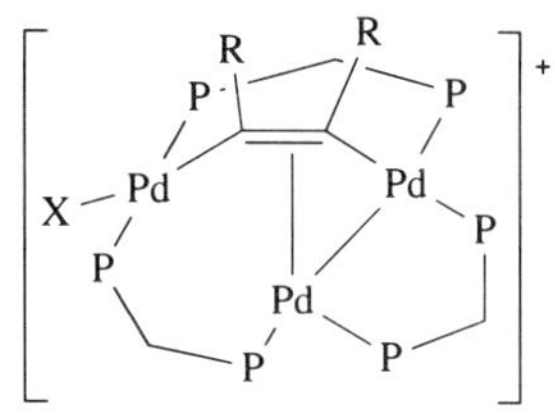

(16)

(iv) Other reactions

Chemistry of the platinum analogue of (**12**) has been more extensively developed, including: coordination chemistry of phosphines and phosphites, isocyanides, SnX_3^-, hydride and biphosphines and some other bidentate ligands; cluster expansion using reactions with electrophiles such as $(Ph_3P)Au^+$ and nucleophiles such as Pt^0 and Hg^0 complexes; and reactions with alkynes; and catalysis of the water gas shift reaction.[43] Further development of this interesting chemistry with palladium analogues may be anticipated and, for example, a very recent report deals with reaction of (**12**) with alkynes, RCCR, R = CO_2Me or CO_2Et to form (**16**), X = CF_3CO_2.[49] Reactions of (**12**) with isocyanides are described in Section 4.3.3.

4.2.5.3 Other trinuclear carbonyls

As noted above in Section 4.2.4, the reaction of $[PdCl(CO)]_n$ polymer with $(Me_5C_5)MgCl{\cdot}THF$ results in a dinuclear complex, $[Pd_2(\mu\text{-}CO)_2(\eta^5\text{-}Me_5C_5)_2]$ (**4**). Treatment of this product with strong acid, HBF_4 or CF_3SO_3H gives further aggregation to a trinuclear cation, (**17**), which is a close relative of (**12**). The cluster has a full complement of 48 electrons, the Pd–Pd distances average 0.263 nm, and the CO ligands cap both faces of the metal triangle.[26]

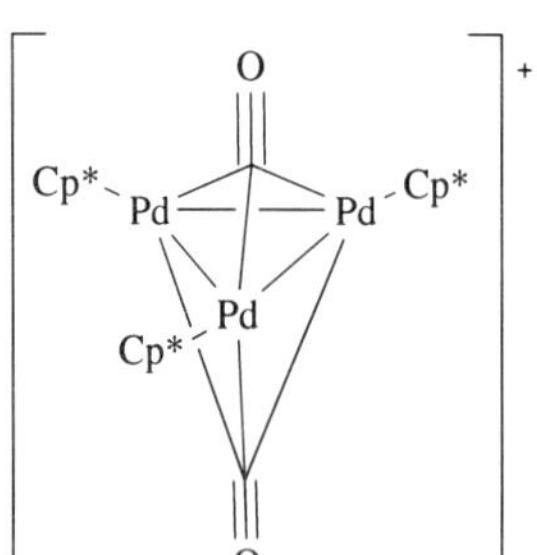

(17)

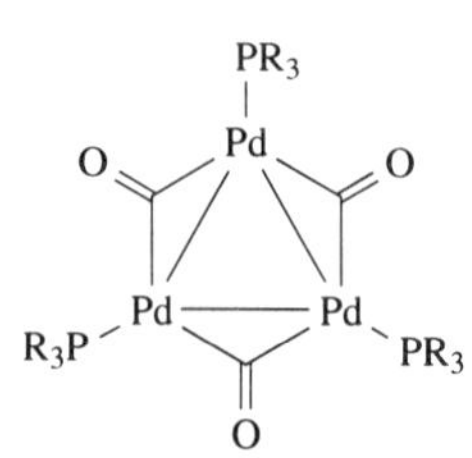

(18)

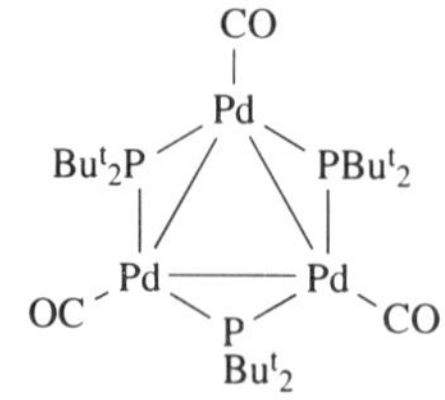

(19)

The trinuclear cluster, $[Pd_3(CO)_3(PPh_3)_3]$ (**18**), is readily prepared by reductions of palladium(II), either using $[Pd(acac)_2]$ with Al_2Et_6 in the presence of carbon monoxide and triphenylphosphine, or $[PdCl_2(PPh_3)_2]$ with amine and CO in methanol solution. Although no crystal structure is available, the complex is assumed isostructural with the well-established platinum analogue. The carbonyl stretching frequency for (**18**) is 1850 cm^{-1}. As is suggested by theoretical studies of metal triangles, these 42e clusters readily add a terminal phosphine ligand to give the 44e, $[Pd_3(CO)_3(PPh_3)_4]$.[1] Other analogues of (**18**) containing bulky phosphines, $R_3 = Bu^t_3$, Bu^t_2Ph or Cy_3, have been synthesized by a very direct procedure using carbonylation of the corresponding palladium(0) complexes, $[Pd(PR_3)_2]$. These products react with carbon disulfide to give unusual dimeric species, $[Pd_2(\mu\text{-}CS_2)_2(PR_3)_2]$.[50] An alternative approach using strongly bridging phosphido ligands derived from $Li[PBu^t_2]$ reacting with $[PdCl(CO)]_n$ polymer in THF at −78 °C gives an example, (**19**), with terminal CO groups absorbing at 2040 cm^{-1} in the infrared.[51]

4.2.5.4 Tetranuclear carbonyls

Tetranuclear palladium carbonyls are basically of two types: based respectively on a Pd_4 tetrahedron, usually with one edge opened, but sometimes symmetrical; or on Pd_4 rectangles. The most effective preparation of the former type seems to be reductive carbonylation of $[Pd(NO_2)_2L_2]$, L = PPh_3, $PMePh_2$, PMe_2Ph or PEt_3, to give a series of clusters, $[Pd_4(CO)_5L_4]$ (**20**).[28] Similar complexes can be obtained by carbonylation of $[Pd(OAc)_2]$ in the presence of tertiary phosphine.[52] The best characterized example has L = PPh_3, and x-ray diffraction shows a distorted tetrahedron of palladium atoms with one edge open to give a *nido*, 'butterfly' structure. Each palladium carries a terminal phosphine, with Pd–Pd edges bridged by CO, (**20**). The open Pd…Pd distance is 0.3209(1) nm and the bonded Pd–Pd contacts average 0.276 nm.[28] These $[Pd_4(CO)_5L_4]$ clusters have valence electron counts of 58, two less than the usual 60e for symmetrical tetrahedral clusters such as $[Co_4(CO)_{12}]$ and $[Fe_4(CO)_4Cp_4]$. In this context, it is interesting that the 60e complex, $[Pd_4(\mu\text{-}CO)_6(PBu^n_3)_4]$, formally derived by addition of a single CO to $[Pd_4(CO)_5L_4]$, does have a regular tetrahedral structure (**21**), Pd–Pd 0.2778–0.2817 nm, with terminal phosphines on each palladium and each edge bridged by CO. The 60e derivative is best isolated from carbonylation of palladium(II) acetate in the presence of PBu^n_3 in acetic acid, but can also be obtained by very slow crystallization of $[Pd_4(CO)_5(PBu^n_3)_4]$ in a CO atmosphere.[53] A similar transformation has been proposed between $[Pd_4(\mu\text{-}I)_2(\mu\text{-}CO)_2(PPh_3)_4]$ and $[Pd_4(\mu\text{-}I)_2(\mu\text{-}CO)_3(PPh_3)_4]$, derived from reaction of $[PdI_2(PPh_3)_2]$ with $[Pd_{10}(CO)_{12}(PPh_3)_6]$.[54]

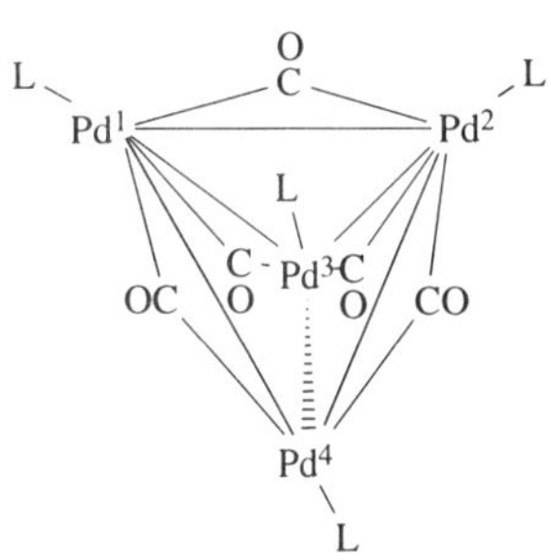

(**20**)

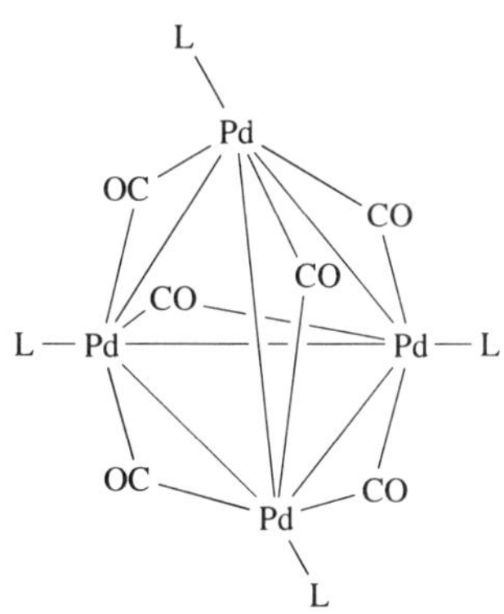

(**21**)

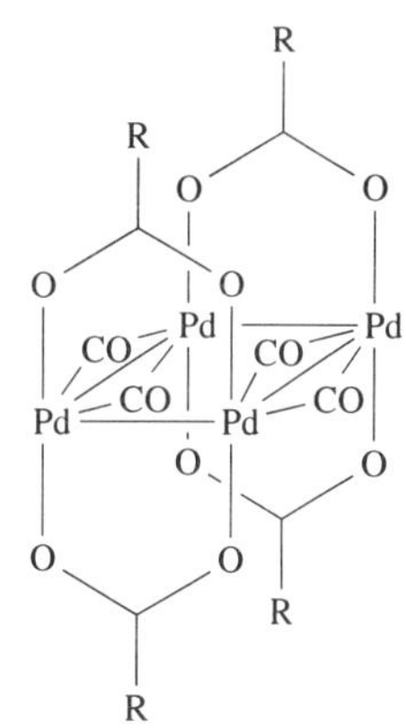

(**22**)

The synthesis of the other type of tetranuclear cluster, $[Pd_4(RCO_2)_4(CO)_4]$ (**22**), is also a reductive carbonylation but starting from palladium(II) acetate in a carboxylic acid (sometimes in benzene solution). Derivatives are known for R = Me, Et, Ph, CH_2Cl, CCl_3 and CF_3.[55,56] The mechanism of the synthesis is thought to involve a stepwise aggregation with gradual reduction in overall oxidation state. Thus, for R = Me, a coordinated CO inserts into a Pd–OAc bond and CO_2 is eliminated to give a Pd–COMe moiety. This moiety then combines with Pd–OAc to eliminate Ac_2O and form a Pd–Pd bond.[56] The structure of (**22**), R = Me, consists of two $Pd^I(\mu_2\text{-}CO)_2Pd^I$ units with a strong Pd–Pd bond, 0.2663 nm, analogous to $[Pd_2Cl_4(CO)_2]^{2-}$ and $[PdCl(CO)]_n$, (**1**) and (**2**). These two units are then joined by long Pd–Pd interactions, 0.2909 nm, which are supported by double acetate bridges, resulting overall in a Pd_4 rectangle.[55] The cluster has 60 bonding electrons, and extended Huckel calculations suggest a 'vertical' interaction on the long sides, analogous to the stacking of d^8 monomers in many 'one-dimensional' chain structures.[57]

Some reactions of $[Pd_4(RCO_2)_4(CO)_4]$ clusters involving aggregation to larger clusters have been reported (see below) and the reaction of $[Pd_4(OAc)_4(CO)_4]$ with $Na[Mo(CO)_3Cp]$ to form $Na_2[Pd_4\{Mo(CO)_3Cp\}_4]$ is described in Section 4.2.6. However, the majority of $[Pd_4(RCO_2)_4(CO)_4]$ chemistry involves degradation to smaller units. The thermolysis is interesting, resulting in carboxylic acids formed by CO_2 insertion into the C–H bonds of benzene or toluene used as solvents. The extra oxygen is apparently derived from the acetate.[58] The kinetics of reductive decomposition of $[Pd_4(OAc)_4(CO)_4]$, to Pd^0, CO, CO_2 and Ac_2O, have been described,[59] as have several reactions of $[Pd_4(OAc)_4(CO)_4]$ with donor ligands resulting in conversion to dinuclear complexes: $[Pd_2H(OAc)_2(CO)(phen)_2]$, $[Pd_2(OAc)_2(CO)(phen)_2]$, $[Pd_2(OAc)_2(CO)(dppe)_2]$, $[Pd_2(CO)(dppe)_4](OAc)_2$ and $[Pd_2(OAc)_2(CO)(PPh_3)_2]$.[60] Reactions with phenanthroline have also been briefly reported to yield some tetranuclear species: $[Pd_4(OAc)_4(CO)_2(phen)_4]$, $[Pd_4H_2(CO)_2(phen)_4](OAc)_2$ and $[Pd_4H_2(CO)_3(phen)_3](OAc)_2$; but none appear to be fully structurally characterized.[61]

4.2.5.5 Pentanuclear carbonyls

Synthesis of the only known pentanuclear palladium carbonyl clusters requires a preformed Pd_5 unit, a contrast to the aggregation reactions used for the other large palladium carbonyls. Reaction of $[Pd_2(dba)_3]$ with tertiary phosphine, ($L = PPh_3$, $P(p\text{-}C_6H_4OMe)_3$ or PMe_2Ph) under an atmosphere of sulfur dioxide gives pentanuclear clusters, $[Pd_5(SO_2)_4L_5]$, which in turn react with CO to displace SO_2, yielding $[Pd_5(\mu\text{-}CO)_2(\mu_3\text{-}SO_2)_2L_5]$ (**23**), $L = PPh_3$ or $P(p\text{-}C_6H_4OMe)_3$, or $[Pd_5(\mu\text{-}CO)(\mu_3\text{-}CO)(\mu\text{-}SO_2)_3(PMe_2Ph)_5]$ (**24**). The latter complex has been structurally characterized and the palladium framework consists of the edge-bridged tetrahedron shown in (**24**) with Pd–Pd distances 0.2736(1)–0.2968(1) nm. Each palladium carries a terminal phosphine; the SO_2 ligands bridge the Pd–Pd edges, Pd(1,3), Pd(2,5) and Pd(4,5), and the CO ligands cap faces of the tetrahedron, Pd(1,2,4) and Pd(2,3,4). Both CO caps are highly asymmetric with one short contact (0.20 nm) and two longer ones (0.230–0.273 nm). Asymmetric caps are common in palladium clusters, but this is the most asymmetric known, accounting for the unusual intermediate CO stretch at 1908 cm^{-1} in the infrared.[62]

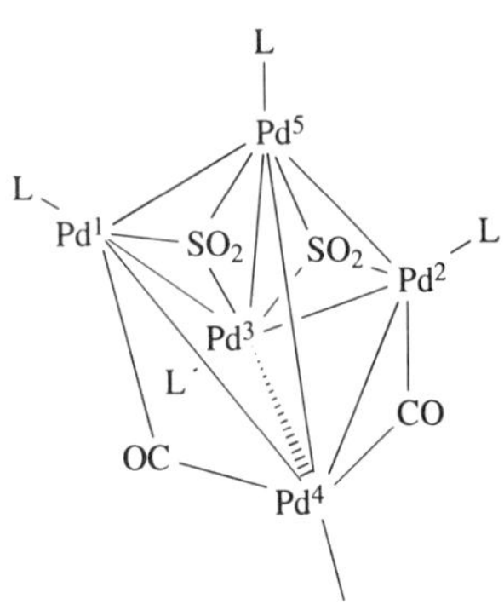

(23)

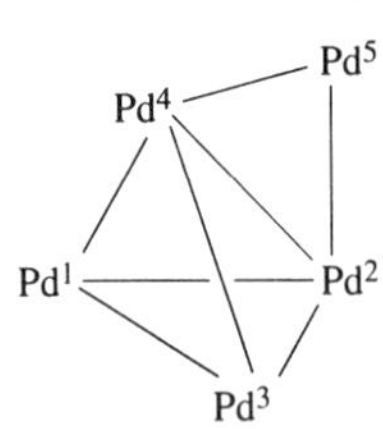

(24)

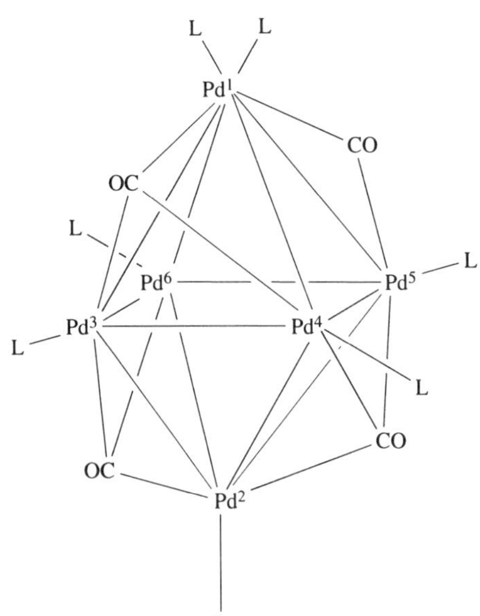

(25)

4.2.5.6 Hexanuclear carbonyls

The palladium(0) complex $[Pd(PMe_3)_4]$ in toluene solution under CO (1 atm) at 50 °C aggregates spontaneously to air sensitive red crystals of $[Pd_6(CO)_4(PMe_3)_7]$ (**25**). The palladium core forms a slightly distorted octahedron (Pd–Pd 0.2705–0.2806 nm) with single terminal phosphines on all palladiums except for one which has two phosphines, and four face-bridging CO in an alternating sequence. The stability of this cluster with only 82 valence electrons, as compared with the more usual octahedral count of 86e, has been discussed in terms of increasing *p–d* orbital separations at the right-hand end of the transition series.[63] Another hexanuclear cluster, $[Pd_6Cl_2(CO)_4(PPh_3)_4]$, also has 82 electrons but consists of two triangular fragments joined by bridging chlorine atoms.[64]

4.2.5.7 Higher nuclearity carbonyls (n = 7–23)

The series of clusters having 7, 8 and 10 palladium atoms in the cluster are derived from capping of a basic octahedral unit. Idealized structures are shown in (**26**), (**27**) and (**28**), but, as is common for d^{10} metals, the caps are asymmetric, often sufficiently so that they may be more correctly described as edge bridging. Compare for example with the edge-bridging palladium in (**24**) and with the opened tetrahedron of (**20**). The latter Pd_4 framework is compared with the caps in Pd_7, Pd_8 and Pd_{10} clusters in Structures (**29**)–(**30**), where Pd–Pd distances are in nanometers. Carbonylation of $[Pd(\eta^1,\eta^3\text{-}C_8H_{12})(PMe_3)]$ in toluene at –78 °C gives dark red crystals of $[Pd_7(CO)_7(PMe_3)_7]$. The palladium atoms adopt a monocapped octahedral arrangement (**26**) with terminal phosphine ligands on each palladium. Four CO groups are face bridging on the octahedron, spanning faces Pd(1,3,6), Pd(1,4,5), Pd(2,3,4) and Pd(2,5,6), respectively; the cap consists of a $Pd(CO)_3(PMe_3)$ fragment with all CO edge bridging to palladium atoms in the octahedron (i.e., edges Pd(7,2), Pd(7,3) and Pd(7,6)).[65] The Pd(7) cap is markedly asymmetric, with short bonds to Pd(2) and Pd(6), 0.273 nm, and a much longer distance to Pd(3), 0.318 nm, as detailed in (**30**). This Pd_4 fragment may be compared with the 'butterfly', opened tetrahedral structure of $[Pd_4(CO)_5(PPh_3)_4]$, (**29**), and with similar asymmetric caps, (**31**) and (**32**), found in the Pd_8 and Pd_{10} clusters described below. As noted above, the large *p–d* orbital separation at the right-hand end of the transition series leads in general to palladium clusters with less than the expected full complement of valence electrons, and the Pd_n units, $n = 4,6,7,8,10$, have 58, 82, 98, 110 and 136 valence electrons respectively, compared with the expected counts for symmetrical *closo* structures of 60, 86, 100, 114 and 142. However, it is interesting to compare with the clusters, $[Rh_6(CO)_{16}]$, $[Os_7(CO)_{21}]$, $[Os_8(CO)_{22}]^{2-}$ and $[Os_{10}C(CO)_{24}]^{2-}$. These form a structural series, progressing by successive capping of an octahedron similar to (**25**)–(**28**), and except for the last example, have the same total electron counts (86, 98, 110, 134) as the palladium clusters.

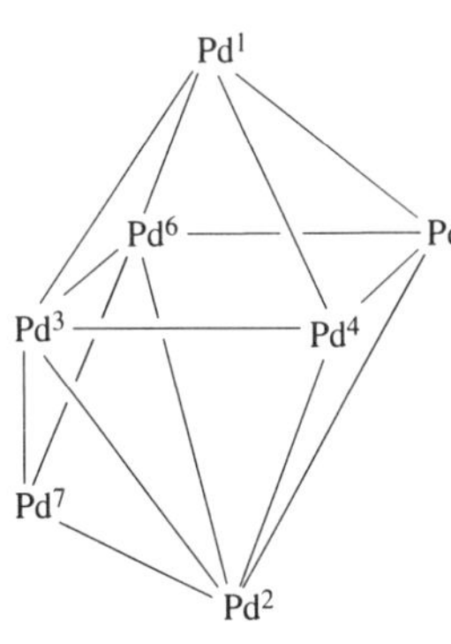

(**26**)

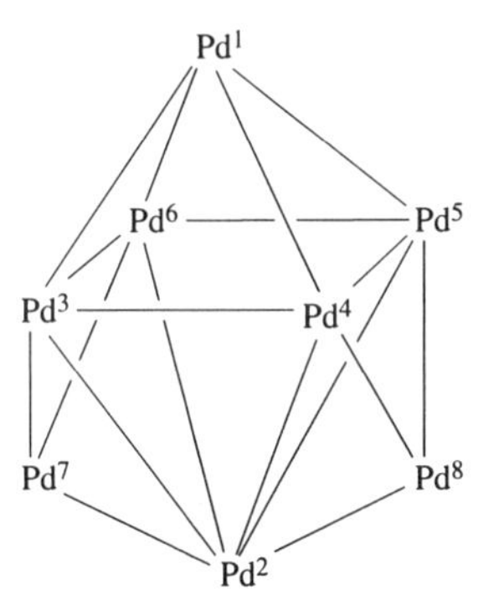

(**27**)

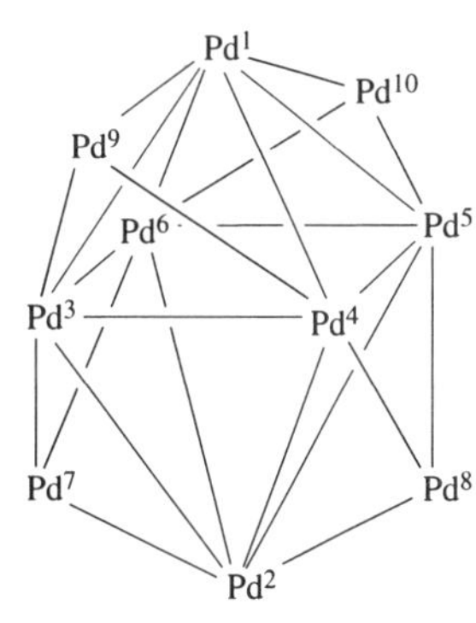

(**28**)

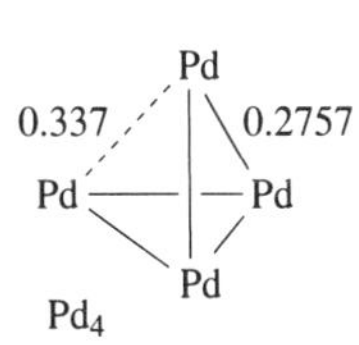

(**29**)

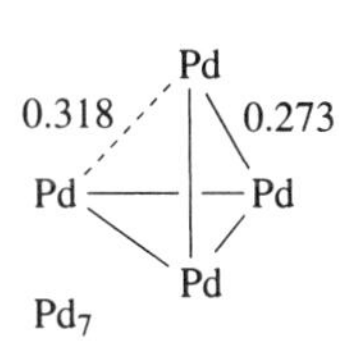

(**30**)

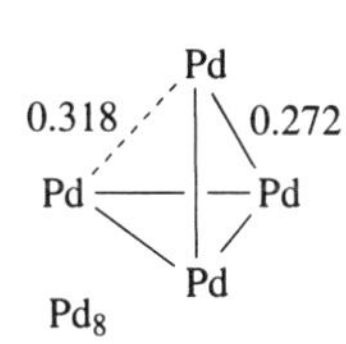

(**31**)

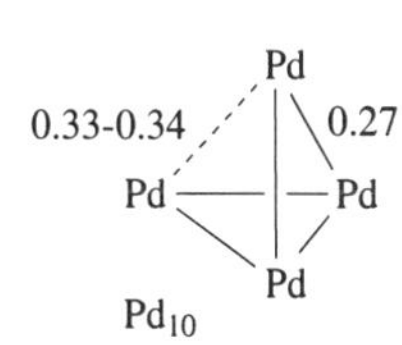

(**32**)

An octanuclear cluster results when $[Pd(dpa)_2]$ is reacted with CO and one equivalent of PMe_3 in dichloromethane solution at ambient temperature. The product, $[Pd_8(\mu_3\text{-}CO)_2(\mu_2\text{-}CO)_6(PMe_3)_7]$, is similar to the Pd_7 derivatives in being based on a capped octahedron as shown in the idealized Structure (**27**), but in this case there are two asymmetric palladium caps, (**31**). The face-bridging carbonyls span faces Pd(2,3,4) and Pd(2,5,6), and the edge-bridging carbonyls span edges Pd(1,3), Pd(1,5), Pd(7,2), Pd(7,6), Pd(8,2) and Pd(8,4).[66] An octanuclear cluster can also be obtained from reaction of $[Pd_4(CO)_5(PPh_3)_4]$ with palladium(II) acetate in toluene to yield $[Pd_8(CO)_{10}(PPh_3)_4]$. Infrared evidence is consistent with a structure similar to (**27**).[67]

As noted above (see Section 4.2.5.1), the nature of isolated products in solutions of palladium(II) acetate, with added CO and tertiary phosphines, is very sensitive to the exact conditions, solvents and molar ratios of reactants. A number of papers have appeared in the Russian literature dealing with preparations and interconversions of carbonyl phosphine clusters in the range Pd_3 to Pd_{10}.[68] The best characterized decanuclear clusters are obtained from palladium(II) acetate with CO, PEt_3 or $PBu^n{}_3$, and trifluoroacetic acid in dioxane solution, yielding $[Pd_{10}(CO)_{12}(PR_3)_6]$ and $[Pd_{10}(CO)_{14}(PR_3)_4]$. The Pd_{10} core of $[Pd_{10}(CO)_{12}(PBu_3)_6]$, (**28**), consists of a tetracapped octahedron with the caps alternating in a C_{2v} pattern. Within the octahedron the Pd–Pd distances are closely similar, 0.2825 nm, but as usual the caps are asymmetric, (**32**), with short (0.2709 nm) and long (0.3300–0.3422 nm) Pd–Pd lengths. The tertiary phosphine ligands are terminal on Pd(1), Pd(2), Pd(7), Pd(8), Pd(9) and Pd(10); carbonyls are edge bridging on Pd(7,3), Pd(7,6), Pd(8,4), Pd(8,5), Pd(9,3), Pd(9,4), Pd(10,5) and Pd(10,6), and face bridging on Pd(1,4,5), Pd(1,3,6), Pd(2,3,4) and Pd(2,5,6).[69,53] The structure of $[Pd_{10}(CO)_{14}(PBu^n{}_3)_4]$ is closely similar except for the replacement of two terminal tributylphosphine ligands by CO,[70] and this complex undergoes various substitution reactions with triphenylphosphine to give mixed $PPh_3/PBu^n{}_3/CO$ derivatives whilst maintaining the Pd_{10} nuclearity.[71]

The search for ever larger clusters is continuing. Some Pd_{12} derivatives, $[Pd_{12}(CO)_{15\text{ or }17}(PR_3)_{7\text{ or }5}]$, R = Et or Bu^n, have been briefly reported as dark-red powders stable under a CO atmosphere, but none are especially well characterized.[72] Reaction of $[Pd_4(CO)_5(PEt_3)_4]$ or $[Pd_{10}(CO)_{12}(PEt_3)_6]$ with Me_3NO serves to increase the equivalent palladium content by oxidizing some phosphine, producing $[Pd_{16}(\mu_3\text{-}CO)_7(PEt_3)_6]$, which is described as a centred icosahedral structure with three additional $Pd(\mu\text{-}CO)_2(PEt_3)$ bridging groups giving a metal core of overall D_3 symmetry.[73] The best characterized large cluster, however, is $[Pd_{23}(CO)_{22}(PEt_3)_{10}]$, obtained by aggregation of $[Pd_{10}(CO)_{12}(PEt_3)_6]$ with palladium(II) acetate under an argon atmosphere at ambient temperature. The product is obtained as black crystals containing an essentially close-packed Pd_{19} core, consisting of a body-centered and edge-bridged octahedron, with four nonclosed $[Pd(\mu\text{-}CO)_2(PEt_3)]$ capping fragments. The caps are similar to the asymmetric caps described above for smaller clusters. All vertices of the Pd_{19}, 'large octahedron', and all capping atoms are coordinated by PEt_3 and there are 14 edge-bridging and eight face-capping CO ligands.[74]

4.2.6 Palladium Carbonyls Containing Other Metals

The most general method of accessing palladium carbonyls containing other metal atoms is addition of carbonyl metallate anions to simple palladium(I) or palladium(II) starting complexes. These methods have been developed primarily by the research of Braunstein and co-workers and ultimately provide access to complexes containing up to four different types of transition metal atom. The simplest compounds are those with two different metal atoms, (**33**), synthesized by reaction of $[PdCl_2(PR_3)_2]$ with $Na[M(CO)_3Cp]$, M = Cr, Mo or W, $R_3 = Me_3$, Et_3, $Bu^n{}_3$, Ph_3 or Me_2Ph, and the mechanisms of these reactions have been discussed in terms of radical pathways.[75] The same complexes are also available in better yields (up to 90%), by the alternative approach of preforming a mixed metal complex and then adding the phosphine ligands. Thus, $[PdCl_2(PhCN)_2]$ reacts with $Na[M(CO)_3Cp]$ at −78 °C to form *in situ* a linear M–Pd–M complex, *trans*-$[\{Cp(CO)_3M\}Pd(PhCN)_2\{M(CO)_3Cp\}]$, which is subsequently converted to $[Pd_2M_2(CO)_6(PR_3)_2Cp_2]$, (**33**), by addition of tertiary phosphine.[75] Complete details of the synthesis of $[Pd_2Mo_2(CO)_6(PPh_3)Cp_2]$ by this latter route have been reported.[76] A closely related complex, $[Pd_2Mo_2(CO)_6(PBzPh_2)_2Cp_2]$ has been obtained and structurally characterized from reaction of $Na[Mo(CO)_3Cp]$ with $[Pd_2(\mu\text{-}Cl)_2\{2\text{-phenyl(diphenylphosphino)methane}\}_2]$.[77] Structural studies of (**33**), M = Cr, Mo, W, show that these clusters contain planar, triangulated M_4 cores, with Pd–Pd bonds among the shortest reported, averaging 0.258 nm. Each Pd_2M triangle is capped by a μ_3-CO (on opposite sides of the M_4 plane), and each Pd–M edge is bridged by a single μ_2-CO. However, both the caps and the bridges are very asymmetric with the CO always more strongly bound to M rather than to palladium. Overall, the complexes may be regarded as 18 electron, $[M(CO)_3Cp]^-$ anions, combining with a 22e, $[L\text{–}Pd^I\text{–}Pd^I\text{–}L]^{2+}$, unit to produce a 58 electron cluster.[75] When allowance is made for the metal–metal bond, each palladium in $[L\text{–}Pd^I\text{–}Pd^I\text{–}L]^{2+}$ is 12e, so that the $[M(CO)_3Cp]^-$ anions each require to donate 4e to complete formal 16e counts at the palladiums. The Pd–M bonds would account for two of these donated electrons, leaving only 2e to be provided by the three CO bridges, and accounting for the relatively weak interaction between these bridges and the palladium atoms.

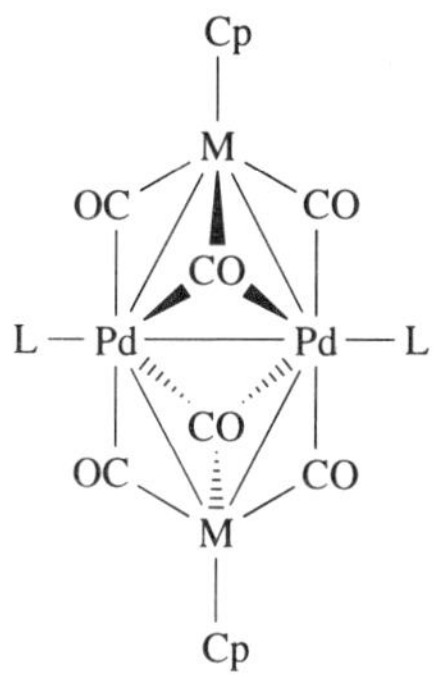

(33)

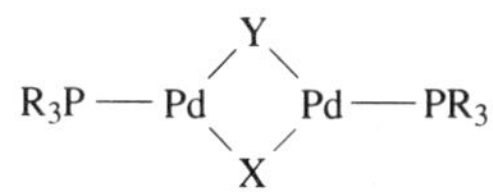

(34)

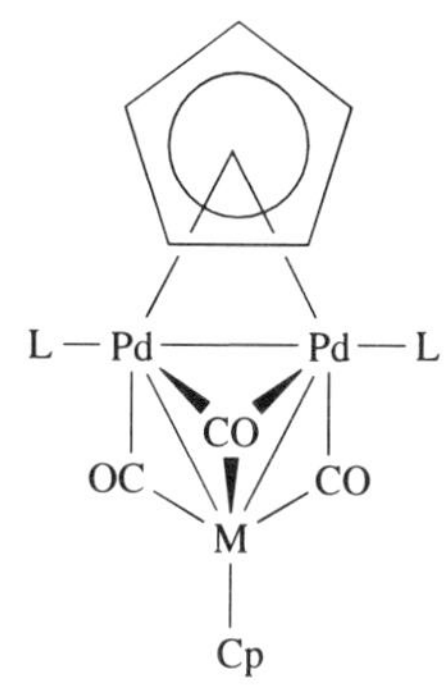

(35)

The electrochemistry of the clusters, (**33**), has been shown to involve a relatively unusual two-electron reduction to $[PdM(CO)_3(PPh_3)_2Cp]^{2-}$, rather than the one-electron steps which are more usual for clusters. Oxidation occurs in one-electron steps, and both oxidation and reduction result ultimately in rupture to mononuclear fragments.[78] $[Pd_2Mo_2(CO)_6(PPh_3)_2Cp_2]$ has also been used as a moderately active catalyst for hydrogenation and hydroformylation reactions;[79] both this complex and $[Pd_2Fe(CO)_4(dppm)_2]$ can be precursors for the deposition of mixed metal catalysts.[80]

As noted above, the Pd_2M_2 complexes, (**33**), have been regarded as $[M(CO)_3Cp]^-$ anions acting as 4e donors towards a $Pd_2L_2^{2+}$ fragment. In this context, there is an interesting analogy with the complexes, (**34**), X = halide or carboxylate, Y = η^3-allyl or η^5-Cp, prepared by Werner *et al.*[1] A similar picture for the latter complexes would require that X^- and Y^- act as 4e donors towards the $Pd_2L_2^{2+}$ moiety, to complete formal 16e counts at each palladium. As we might then expect, the X^- group can be replaced by $[M(CO)_3Cp]^-$ to yield (**35**).[81,82] Related reactions using $[V(CO)_6]^-$ or $[Co(CO)_4]^-$ lead to complexes analogous to (**35**) but having $V(CO)_6$ and $Co(CO)_4$ bridging fragments, respectively.[81,83] In these products, there are no capping CO groups as in (**35**), only two μ_2-CO bridging the M–Pd edges, with the remaining carbonyls terminal on M. Similar complexes have also been obtained using 2-methylallyl as the bridging unit instead of cyclopentadienyl.[84]

The elaboration of mixed metal palladium carbonyls into species containing more than two different metals uses a route slightly different to the reactions leading to (**33**) and (**35**). Formally the reactions involve insertion into a Pd^I–P bond as shown in Scheme 1.[85] Clearly, complex (**36**) in Scheme 1 is a very close relative of (**33**) above, with an $[L\text{–}Pd\text{–}Pd\text{–}X]^+$ fragment receiving donation of 4e from a dppm bridge and 4e from $[LCo(CO)_3]^-$; the electron excess in this latter fragment is relieved by formation of the CO caps. As in (**33**), the CO is always more strongly bound to cobalt than to palladium.[85] Close relatives of (**36**) result from other carbonyl metallate anions, $[M(CO)_3Cp]^-$, M = Mo or W,[86] $[Fe(CO)_3NO]^-$ or $[Mn(CO)_5]^-$.[87]

Reactions of (**37**), X = halide, with further carbonyl metallate anions, $[Mn(CO)_5]^-$ or $[Fe(NO)(CO)_3]^-$, result in substitution of halide to give corresponding complexes, (**37**), X = carbonyl metallate, with three different metals in the product;[88] and if the original starting material in Scheme 1 is $[PtPdCl_2(dppm)_2]$ instead of $[Pd_2Cl_2(dppm)_2]$, then clusters, (**38**), containing four different metals are accessible.[88,89] It is noteworthy that, in reactions of $[PtPdCl_2(dppm)_2]$, insertion of the carbonyl metallate (analogous to the first reaction in Scheme 1) occurs regioselectively at the more labile Pd–P bond, rather than at platinum. The insertion of carbonyl metallate fragments into the Pd–P bond has also been investigated for mononuclear palladium(II) complexes. Thus, addition of $[Co(CO)_4]^-$ to $[PdCl_2(dppm)]$ at −30 °C, gives *in situ* formation of (**39**) which is ultimately converted to (**36**). Introduction of extra dppm into this reaction 'traps' the intermediate as complex (**40**).[90] This last complex also results from combination of $[Pd(dppm)_2]^{2+}$ with $[Co(CO)_4]^-$. When the carbonyl metallate addendum is changed to $[M(CO)_3Cp]^-$, M = Mo or W, reaction with $[PdCl_2(dppm)]$ is similar to the cobalt case, yielding (**41**), but use of $[Pd(dppm)_2]^{2+}$ gives dinuclear 32 electron cations, (**42**), with a donor–acceptor metal–metal bond from M to palladium, Pd–Mo 0.2799(1) nm.[91] The cations, (**42**), are also available using the trinuclear chain complexes, *trans*-$[Pd(PhCN)_2\{M(CO)_3Cp\}_2]$, as starting materials. Reaction with two equivalents of dppm in THF at −80 °C gives (**42**) together with an analogue of (**37**), but with Co(CO) replaced by MoCp. However, use of other metals in the initial trinuclear chain leads to different products: *trans*-

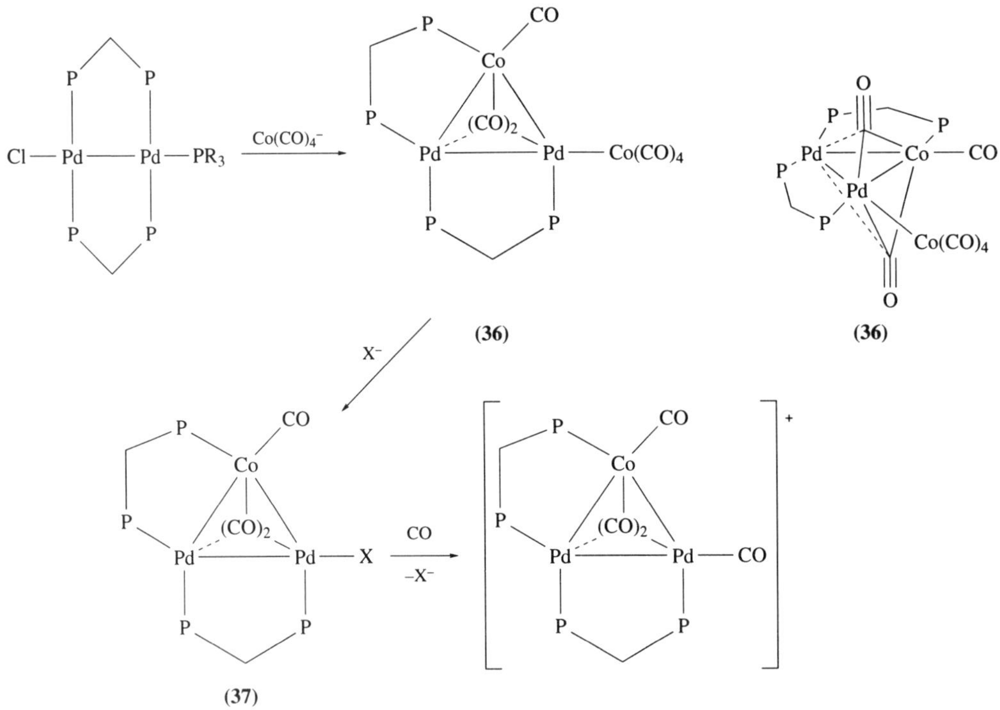

Scheme 1

$[Pd(PhCN)_2\{Mn(CO)_5\}_2]$ gives $[PdCl(\mu\text{-dppm})_2Mn(CO)_3]$ which can be carbonylated to (**43**); and *trans*-$[Pd(PhCN)_2\{Co(CO)_4\}_2]$ yields the trinuclear cluster, (**40**). Overall, the reactions are in contrast to corresponding reactions of platinum complexes, *trans*-$[Pt(PhCN)_2M_2]$, M = $Mo(Cp)(CO)_3$, $W(Cp)(CO)_3$, $Mn(CO)_5$ or $Co(CO)_4$, which are more selective, leading exclusively to $[(\eta^2\text{-dppm})Pt(\mu\text{-dppm})M]^+$ cations.[92]

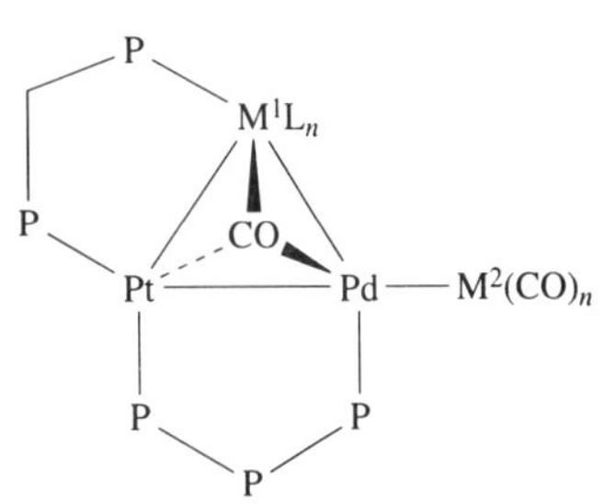

(38)

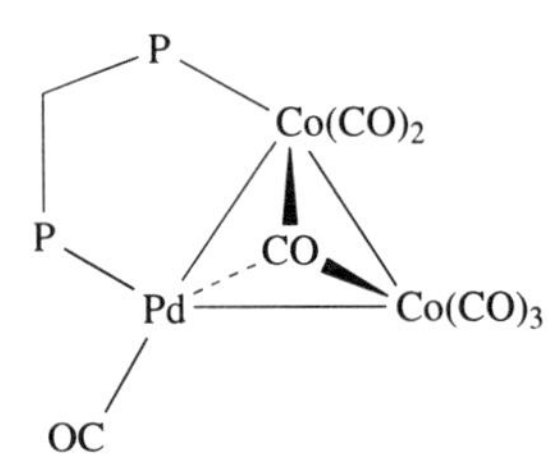

(39)

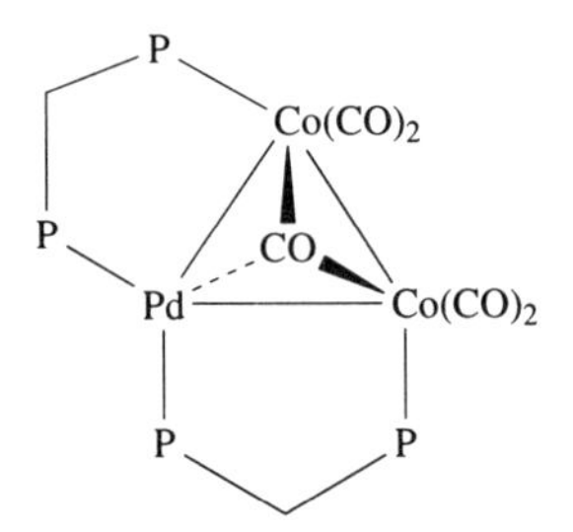

(40)

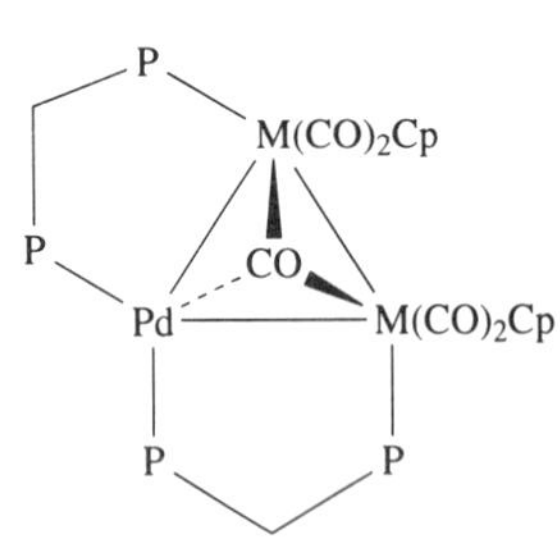

(41)

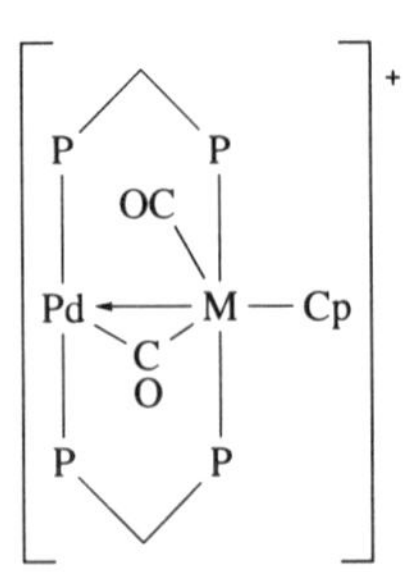

(42)

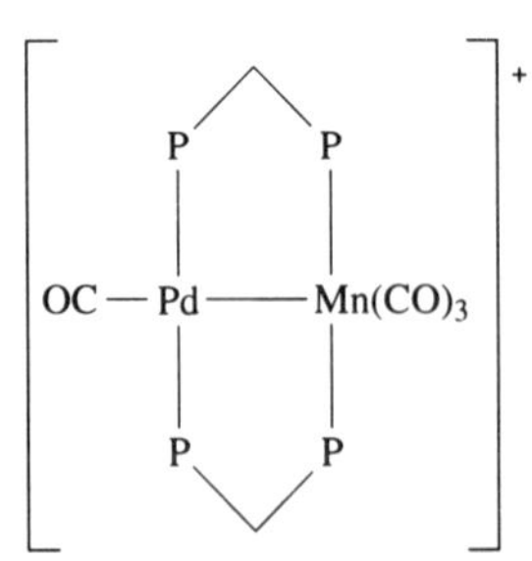

(43)

Several reactions have been reported in which other metal atoms add to a tetranuclear framework of palladium atoms. One of these is closely analogous to the carbonyl metallate additions to dinuclear species described above. Thus, $[Pd_4(OAc)_4(CO)_4]$, (**22**), reacts with $[Mo(CO)_3Cp]^-$ to give (**44**), a dianion containing a Pd_4 square edge-bridged by $Mo(CO)_3Cp$ fragments in a manner exactly analogous to complex (**33**) and (**35**).[93] However, when the carbonyl metallate is changed to $[Mn(CN)(CO)_2(\eta^5\text{-}C_5H_4Me)]^-$, reaction with $[Pd_4(OAc)_4(CO)_4]$ proceeds quite differently. Four equivalents of the metallate react quantitatively at −40 °C to yield a deep violet solution of the air stable cluster, $[Pd(CO)(\mu\text{-}NC)Mn(CO)_2(\eta^5\text{-}C_5H_4Me)]_4$, (**45**), containing an orthogonal arrangement of novel helical units.[94] The butterfly cluster, $[Pd_4(CO)_5(PEt_3)_4]$, (**20**), also adds additional metal atoms, reacting for example with excess 8-(α-bromomercuryethyl)quinoline to yield $[Pd_4Hg_2Br_2(CO)_4(PEt_3)_4]$, a cluster in which two HgBr caps have been added to the wings of the original butterfly structure to produce the metal framework (**46**). As in the original structure, (**20**), CO bridges the Pd–Pd edges and the triethylphosphine ligands are terminal at each palladium.[95] A related cluster, $[Pd_6Hg(CO)_6(PEt_3)_6]$, probably consists of a mercury atom sandwiched between two Pd_3 triangles;[96] and similar reactions with $[Sn(acac)_2]$ lead to $[Pd_3Sn_2(acac)_4(CO)_2(PEt_3)_3]$, a triangular palladium cluster having the two tin atoms bridging on the same Pd–Pd edge.[97]

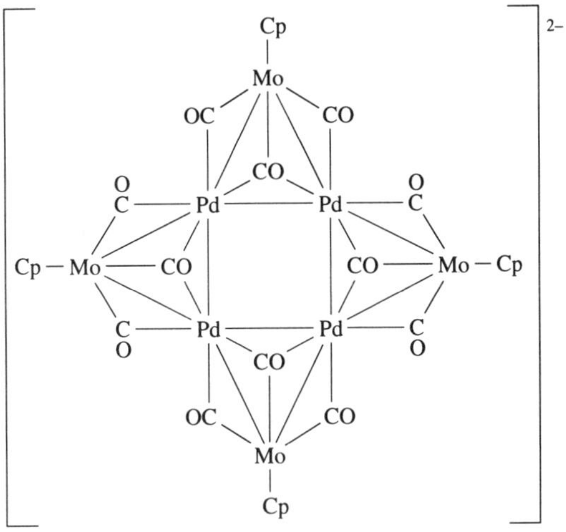

(**44**)

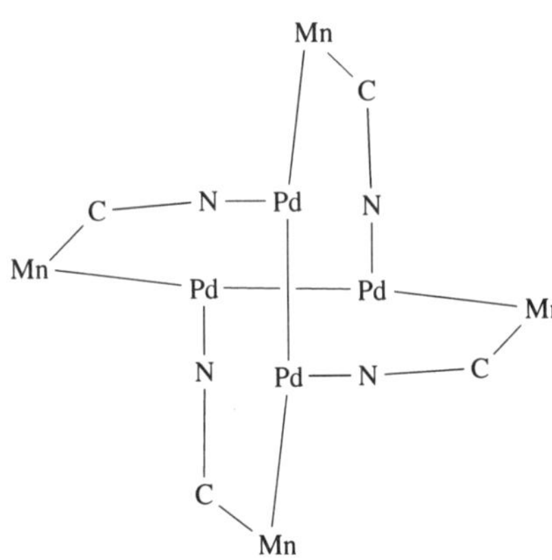

(**45**)

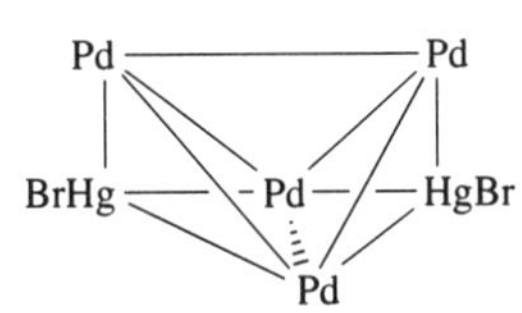

(**46**)

4.2.7 Catalysis by Palladium Carbonyls and Reactions of the CO Ligand

Palladium, both the metal and some of its complexes, is widely used in the catalysis of organic reactions, and carbonylations are especially numerous.[98] Some theoretical results relevant to catalytic processes have already been discussed in Section 4.2.2.1. In most experimental studies, the palladium intermediates are not characterized, and the interest is focused on the organic chemistry, which is beyond the scope of this book. It is clear, however, that the main reactions of the CO ligand involve nucleophilic attack at carbon. The attacking group can be an external nucleophile, or sometimes an internal nucleophile such as coordinated hydride or alkyl, and in the latter case, products are formally the result of CO insertion into the Pd–R bond. It is, of course, possible that attack by external nucleophiles also occurs via initial coordination. A well-studied recent example of nucleophilic attack is the reaction of *trans*-[Pd(COPh)(CO)$(PMe_3)_2$] with pyrrolidine. The ultimate product is *trans*-[Pd(COPh)-$(CO\overline{N(CH_2)_3C}H_2)(PMe_3)_2$], but the reaction has been shown to proceed via an intermediate containing an O-protonated carbamoyl group.[99]

Carbonyl insertion into a metal–hydride or metal–alkyl bond is a key elementary step in many catalytic processes. In the particular case of palladium complexes, several groups have modelled this process by calculations of the conversion of [PdH(R)(CO)(PH_3)] to [PdH(COR)(PH_3)], where R = H or Me. *Ab initio* calculations for R = Me provide optimized geometries for both reactant and product, and indicate a three-centred transition state involving methyl rather than CO migration. The mechanism is basically similar to that for the well-studied insertion of CO into [Mn(Me)$(CO)_5$]. Relative to platinum, weaker Pd–C bonds make the insertion reaction less endothermic and with a lower activation energy. Changes in the Pd–C bond energy also account for lower activation energies when the R group has electron-releasing substituents or when a strong *trans*-influence ligand is present, and increased activation energies for electron-withdrawing substituents.[100] An *ab initio* calculation based on the shift operator estimates ΔE as 79.8 kJ mol^{-1} and $\Delta E^{\ddagger}$ as 107.8 kJ mol^{-1} for methyl insertion.[101] Other authors have discussed the importance of electron correlation effects in calculating profiles for this reaction.[102] Comparison of methyl and hydride insertion, using second-order Moeller–Plesset perturbation calculations starting from restricted Hartree–Fock theory (MP2/RHF), confirms this mechanism and shows that hydride is more endothermic but has a lower activation energy. This reflects a greater strength for the Pd–H bond relative to Pd–Me, and also the fact that the spherical 1*s* orbital of hydrogen is more suitable for the three-centre transition state.[103] Also, in the hydride case, intramolecular hydrogen migration is much more favourable for the $[PdH(CO)]^+$ cation than in the neutral species, [PdH(CO)], which may explain why palladium catalysts are more active in methanol synthesis when cationic palladium species are also present.[104]

The rate of insertion of CO into methyl–palladium complexes containing chelate ligands, [PdClMe(P–P)] and $[PdMeL(P–P)]^+$, P–P = dppe, dppp, dppb or dppf, L = MeCN or PPh_3, increases with the flexibility of the chelate backbone. The cations react faster than the neutral complexes.[105]

Palladium is also used to catalyse carbonylation of allylic compounds. High CO pressures are normally required, suggesting that insertion into the palladium–allyl bond is a difficult process. In this context, Equation (3), R^1 = H, R^2 = Me, X = Cl, R^1 = H, R^2 = H, X = Br, R^1 = Ph, R^2 = H, X = Br, is claimed as the first example of insertion under mild conditions into an isolated palladium–allyl complex.[106]

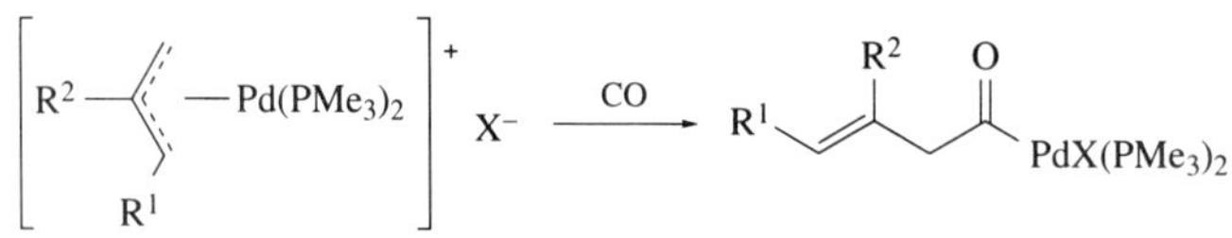

(3)

4.3 ISOCYANIDE COMPLEXES

4.3.1 Introduction

Isocyanides are effective ligands for palladium, being good σ-donors and relatively poor π-acceptors relative to the isoelectronic carbon monoxide ligand. In consequence, there are many published studies where isocyanides are present merely as 'incidental' ligands on palladium. This is especially true of palladium(II) complexes, and space precludes coverage of such incidental usage of isocyanides in this book. Also, many organic reactions involving isocyanides utilize palladium metal or palladium compounds (e.g., $[Pd(PPh_3)_4]$, $[PdCl_2(PPh_3)_2]$, PdO, etc.) as catalysts. In these reactions the primary focus is on the organic chemistry and the palladium intermediates are rarely characterized. These studies also are excluded from the present coverage, which will focus on examples where the primary interest is the isocyanide chemistry.

4.3.2 Palladium(0) Isocyanides

Early work on isocyanide complexes of palladium(0) yielded only rather poorly defined, possibly polymeric materials. Thus, reduction of aryl isocyanide complexes, $[PdI_2(CNR)_2]$, in strongly alkaline solution gave $[Pd(CNR)_2]_n$ and reaction of [Pd(η^3-cyclohexenyl)(η^5-Cp)] with CNR, R = Pr^i or Cy, gave $[Pd(CNR)_2]$.[1] The first, and still the only, type of palladium(0) isocyanide to be well characterized was obtained by metal vapour techniques, depositing palladium atoms into a solution of cyclohexyl isocyanide in THF at 160 K. The product, $[Pd_3(CNCy)_6]$ (**47**), is air sensitive in solution but quite stable in the solid state, where the structure is similar to that of known platinum analogues. Very short Pd–Pd bonds, 0.2651(2) nm, are bridged by isocyanides and each palladium also carries a terminal isocyanide.[107] Further details of this structure are given in Table 2 and Section 4.3.6. An analogous product has been prepared from $[Pd(dba)_2]$ (dba = dibenzylideneacetone) and xylyl isocyanide.[108] The only other palladium(0) isocyanides appear to be clusters which are also stabilized by SO_2 bridges. Thus, the trinuclear cluster, (**48**), $L^1 = L^2 = Bu^tNC$, derived from the reaction of $[PdL_2]_n$ with sulfur dioxide, has been known for a long time;[1] and close relatives $L^1 = PPh_3$, L^2 = xylyl isocyanide (CNXy) and $L^1 = L^2$ = CNXy have been prepared more recently. The former results from reaction of CNXy with a pentanuclear cluster, $[Pd_5(SO_2)_4(PPh_3)_5]$,[62] and the latter from reduction of $[PdCl_2(CNXy)_2]$ by sodium amalgam.[109] Isocyanides containing pentanuclear clusters are also accessible. Reaction of $[Pd_2(dba)_3]$ with xylyl isocyanide under an atmosphere of sulfur dioxide gives $[Pd_5(\mu\text{-}SO_2)_3(\mu\text{-}CNXy)_2(CNXy)_5]$, (**49**), a structure based on an edge-bridged tetrahedron of palladium atoms with three edges bridged by SO_2 and two edges bridged by isocyanide. Each palladium also carries a terminal isocyanide. The normal ranges for ν(CN) stretches in palladium isocyanides are 2080–2221 cm^{-1} for terminal ligands and 1589–1911 cm^{-1} for bridging ligands, so the values of 1990 cm^{-1}, 2047 cm^{-1} and 2142 cm^{-1} in (**49**) are somewhat unusual. This is explained by the structural parameters which show very asymmetric isocyanide bridges, more strongly bound to one of the palladiums (~0.205 vs. 0.228 nm) and with relatively linear C–N–C ligands (~170°).[110]

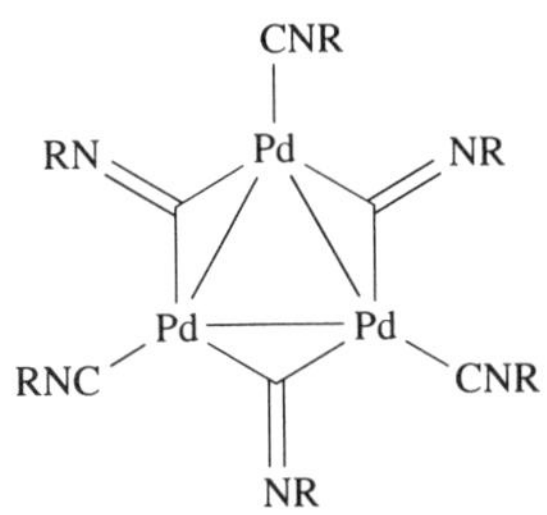

(**47**)

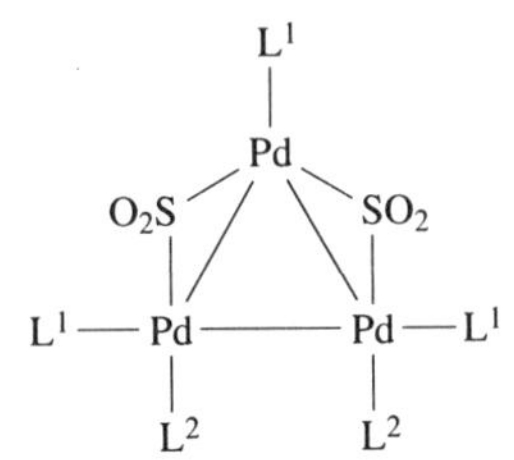

(**48**)

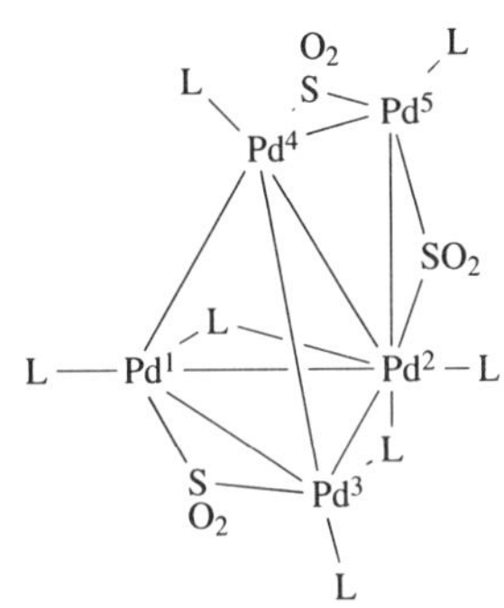

(**49**)

4.3.3 $[Pd_3(\mu_3\text{-}CNR)(dppm)_3]^{2+}$

The chemistry of the $[Pd_3(\mu_3\text{-}CO)(dppm)_3]^{2+}$ cation (**12**) is described above in Section 4.2.5.2, and the potential for analogous isocyanide chemistry is obvious. Addition of one equivalent of xylyl isocyanide to (**12**) under ambient conditions in dichloromethane yields a direct analogue (**50**), the only example of a palladium isocyanide in the very unusual $\mu_3\text{–}\eta^1$ bonding mode. The formal palladium oxidation state is $\frac{2}{3}$, ν(CN) 1972 cm^{-1}, and in the structurally characterized hexafluorophosphate salt, the metal–metal lengths are short, 0.2593(1)–0.2651(1) nm, the Pd–C bonds approximately equal, 0.215(1)–0.218(1) nm, and the isocyanide approximately linear, C–N–C 167(1)°. In common with many 42e, platinum group clusters, (**50**) adds an extra 2e donor, in this case a reversible addition of xylyl isocyanide to form (**51**). This chemistry is substantially different from corresponding platinum reactions, which form only 44e clusters, $[Pt_3(\mu_3\text{-}CO)(CNR)(dppm)_3]^{2+}$ and $[Pt_3(CNR)_2(dppm)_3]^{2+}$, containing only terminal isocyanide ligands. Reaction of (**12**) with the diisocyanide, 1,4-$C_6Me_4(NC)_2$ gives the complex cation, $[\{Pd_3(\mu\text{-}dppm)_3\}_2(CNC_6Me_4NC)]^{4+}$, in which two palladium triangles are linked by the diisocyanide ligand forming a $\mu_3\text{–}\eta^1$ attachment to each one. Addition of further diisocyanide results in oligomeric chain complexes containing similar linkages.[111] When excess xylyl isocyanide is used in the reaction with (**12**), the reaction proceeds beyond (**51**), cleaving the cluster to the palladium(I) complex cation, $[Pd_2(CNXy)_2(\mu\text{-}dppm)_2]^{2+}$, and an unidentified palladium(0) species. These two species react further to give formal insertion of a $Pd(CNXy)_2$ fragment into the Pd–Pd bond, thus 'reassembling' a trinuclear complex, (**52**).[112]

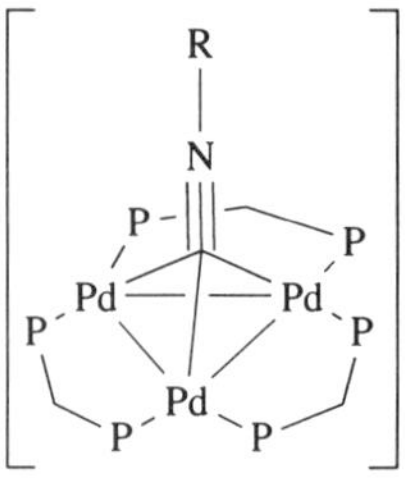

(50)

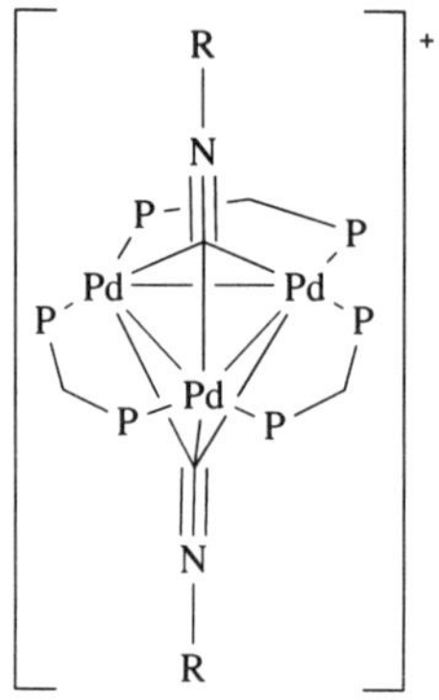

(51)

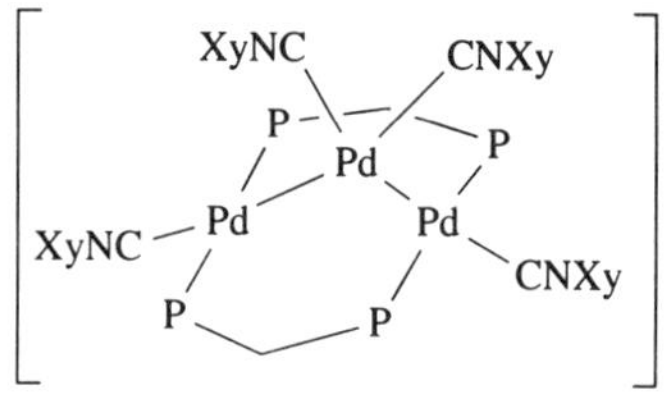

(52)

4.3.4 Palladium(I) Isocyanides

Isocyanide complexes are among the most important and accessible examples of palladium(I). In general, the complexes are dinuclear with unsupported Pd–Pd bonds, prepared either by reaction of palladium(0) with palladium(II) complexes, or by reduction of palladium(II), as illustrated by Equations (4)–(7).[1,113] The cationic examples have usually been isolated as hexafluorophosphate salts. Equation (7) is especially interesting in that it involves a spontaneous internal reduction of palladium, probably via oxidation and hydrolysis of methyl isocyanide.[1] Reduction of $[PdCl_2(CNR)_2]$ to $[Pd_2Cl_2(CNR)_2]$ (cf., Equation (4)) can also be accomplished electrochemically at a platinum electrode.[114]

$$[Pd(CNR)_2] + [PdX_2(CNR)_2] \longrightarrow [Pd_2X_2(CNR)_4] \quad (4)$$

$$[Pd_2(dba)_3] + [Pd(CNR)_4]^{2+} \longrightarrow [Pd_2(CNR)_6]^{2+} \quad (5)$$

$$[Pd(dba)_2] + trans\text{-}[PdCl_2(PhCN)_2] \xrightarrow{Bu^tNC} [Pd_2Cl_2(CNBu^t)_4]^{2+} \quad (6)$$

$$[PdCl_4]^{2-} + MeNC \longrightarrow [Pd(CNMe)_4]^{2+} \longrightarrow [Pd_2(CNMe)_6]^{2+} \quad (7)$$

The structure of $[Pd_2(CNMe)_6]^{2+}$ (**53**) has no bridging isocyanides, two PdL_3 fragments being joined by a very short (0.2531(1) nm) metal–metal bond.[1] However, the halide derivatives, $[Pd_2X_2(CNR)_4]$, were initially thought to be halide bridged on the basis of low vibrational frequencies for the Pd–X bonds. More recent work has shown that low frequencies are not unusual for terminal halides and x-ray diffraction study of $[Pd_2I_2(CNMe)_4]$ (**54**) confirms a structure similar to the homoleptic cation with an almost identical Pd–Pd length (0.2533(1) nm). Coordination about each palladium is planar with the two planes approximately perpendicular to each other (dihedral angle 85.3°).[115] X-ray diffraction study of $[Pd_2Cl_2(Bu^tNC)_4]$ shows a similar structure,[114] and an analysis of vibrational spectra for $[PdPt(MeNC)_6]^{2+}$ has shown a stretching absorption for the metal–metal bond at 160 cm^{-1}, and M–C stretches at 300–350 cm^{-1}.[116] Bridging isocyanide structures supporting a Pd–Pd bond are also possible. Thus, reaction of pyridine with $[Pd_2Cl_2(CNR)_4]$, R = xylyl, gives (**55**), an obvious close relative of the bridging carbonyl structure $[Pd_2Cl_4(\mu\text{-}CO)_2]^{2-}$ (see Section 4.2.3, Structure (**1**)), and a structure which further illustrates the problems of using infrared assignments. The ν(CN) frequencies of 2056 cm^{-1} and 1976 cm^{-1} suggest terminal isocyanides but are actually due to bridges formed by unusually linear C–N–C linkages.[117] Another bridging isocyanide complex, (**56**), is also an analogue of a carbonyl complex (see Section 4.2.4, complex (**5**)) and is prepared in a similar way.[27]

$[Pd_2(CNMe)_6]^{2+}$ is relatively unreactive,[1] although substitution reactions with other isocyanides, tertiary phosphines and halides have been described.[115] However, some interesting photochemistry has been reported with halogen atom donors such as CX_4, X = Cl or Br. The principal product is $[PdX(CNMe)_3]^+$ and the results strongly suggested involvement of radical cations, $[Pd(CNMe)_3]^+$ as the main photoproduct.[118] Radical cations have been confirmed in flash photolysis experiments, $[Pd(CNMe)_3]^+$ having a recombination rate of 10^9 mol^{-1} s^{-1}, approximately diffusion controlled in acetonitrile solution.[119]

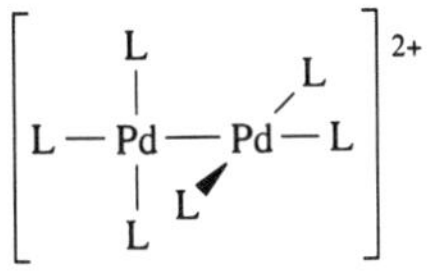

(53)

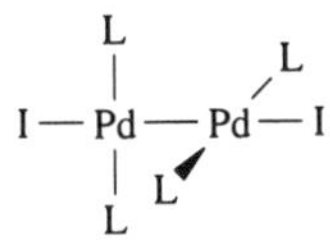

(54)

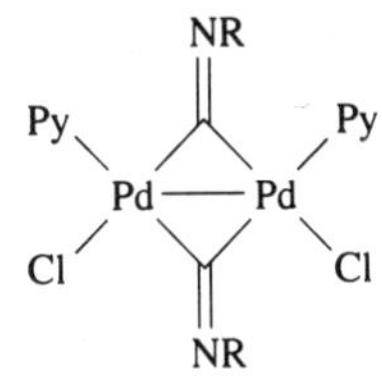

(55)

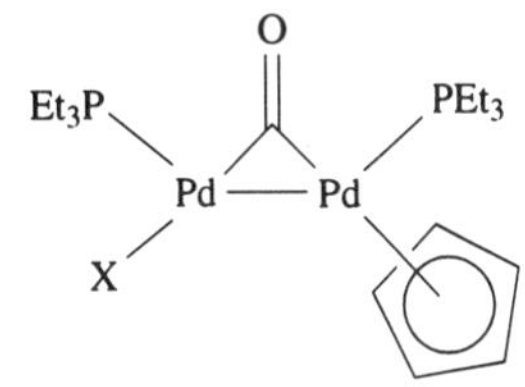

(56)

4.3.5 Palladium(II) Isocyanides

4.3.5.1 Introduction

As noted above, isocyanides are good ligands for palladium, especially palladium(II). The resulting complexes are much more stable than corresponding carbonyl derivatives, owing to the greater σ-donating ability of isocyanides relative to CO. In consequence a large variety of simple isocyanide complexes are accessible, using the direct routes common for all donor ligands in the platinum metal group. Thus, direct reactions with palladium(II) halides or tetrahalopalladate(II) anions lead to $[PdX_2(CNR)_2]$ derivatives, which are easily elaborated into product types such as $[PdCl_2(CNR)L]$, $[Pd_2X_4(CNR)_2]$, $[PdX_3(CNR)]^-$ and $[PdX(CNR)L_2]^+$.[1] Also, isocyanides are frequently present, essentially as spectator ligands in similar roles to nitrogen or phosphorus donors, in palladium complexes where the interest focuses on other aspects of the chemistry. Space precludes treatment of all these examples in the present book. More complex chemistry arises when compounds contain isocyanide together with an organic group on the palladium, so that the isocyanide has the possibility of insertion into a Pd–C bond. There are also occasional examples of isocyanide insertions into other palladium–element bonds. The insertion chemistry is described in detail in the following section. The other important reactivity pattern of coordinated isocyanides is their reaction with protonic addenda to produce carbene complexes. This chemistry is described under syntheses of carbene complexes in Section 4.4.2.1.

As noted above, syntheses of isocyanide complexes usually employ direct reactions of the isocyanide. This method can be used to attach palladium(II) acetate or $[PdCl_2(PPh_3)_2]$ to an isocyanide functionalized polymer, to generate a heterogeneous hydrogenation catalyst.[120] A more unusual synthesis is the addition of α,β-unsaturated ketones to coordinated cyanide, converting *trans*-$[Pd(CN)_2(PR_3)_2]$ to *trans*-$[Pd\{CNCR^2R^3CHR^4C(O)R^5\}_2(PR^1{}_3)_2]$.[121] Palladium isonitriles have also been investigated as organometallic liquid crystals. Complexes of the long chain isocyanides, CN-*p*-$C_6H_4OC(O)$-*p*-$C_6H_4OC_nH_{2n+1}$, $n = 6$–8, exhibit mesomorphic properties over a wide temperature range.[122]

4.3.5.2 Insertions of isocyanides into palladium–carbon bonds

(i) Single insertions

The basic reaction of insertion of isocyanides is analogous to the insertion of carbonyl into metal–carbon bonds, the archetypical example being the addition of iodomethane to $[Pd(CNBu^t)_2]$, as shown in Equation (8). The product is an iminoacyl derivative (imidoyl) resulting from intramolecular insertion of Bu^{t}NC into the initially formed Pd–Me bonds. More generally, the reaction proceeds according to Equation (9), yielding an iminoacyl complex, and several other examples were described in the first edition of this book.[1] Some more recent examples are summarized in Table 1.[123–30] All of the reactions in Table 1 follow Equation (9) to yield insertion products, (**57**), except that in the first example, there is a subsequent proton shift to yield *trans*-[PdCl{C(NHR)(CHCOMe)}].[123]

A kinetic investigation of Equation (10), insertion of CNR, R = Cy or Bu^t, into the palladium–carbon bond of (**58**), X = Cl, shows a mechanism involving initial coordination of CNR, similar to that for insertion of CO. Disappearance of the intermediate (**58**), X = CNR, is first order and probably proceeds via solvent-assisted migration of the pyridyl group on to the coordinated isocyanide, as illustrated in (**59**), to yield (**60**) as the ultimate product. As in the carbonyl case, excess 'free' isocyanide promotes the reaction.[127]

$[Pd(CNBu^t)_2]$ + MeI ⟶ *trans*-$[PdIMe(CNBu^t)_2]$

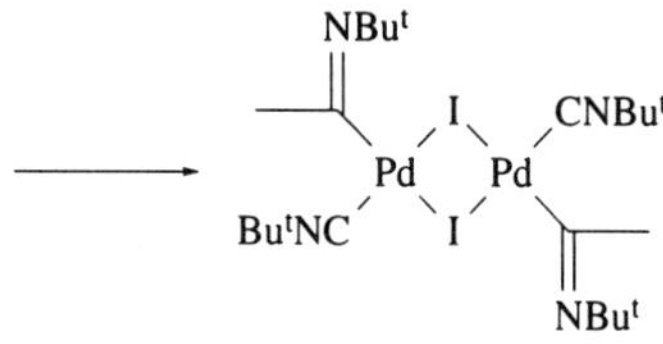

(8)

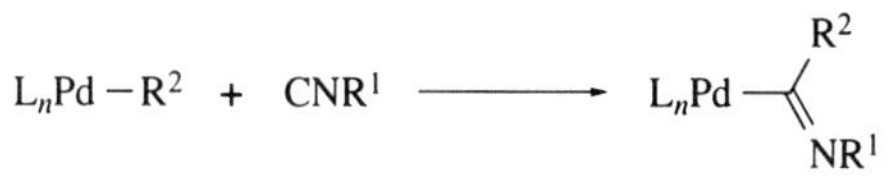

(57)

(9)

Table 1 Summary of selected insertion reactions proceeding according to Equation (9).

L_n	R^2	R^1	*References*
trans-$Cl(PPh_3)_2$	CH_2COMe	Me, *p*-C_6H_4OMe	123
trans-$Cl(CNMe)_2$	C_6Cl_5	Me	124
trans-$Br(PMePh_2)_2$	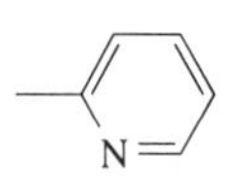	Me, Cy, Bu^t, *p*-C_6H_4O	125, 126
cis-Cl(dppe)	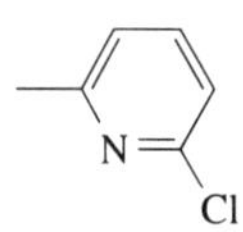	Cy, Bu^t	127
trans-$Cl(PPh_3)_2$	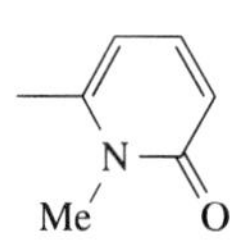	Cy, *p*-C_6H_4OMe	128
trans-$Cl(PPh_3)_2$	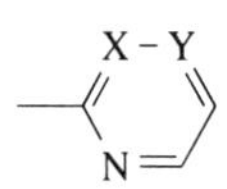 [a]	Bu^t	129
trans-$Br(PPh_3)_2$	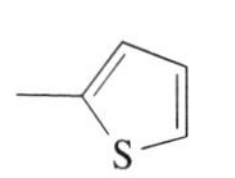	Cy, Bu^t, *p*-C_6H_4OMe	130

[a] X = Y = CH; X = CH, Y = N; or X = N, Y = CH

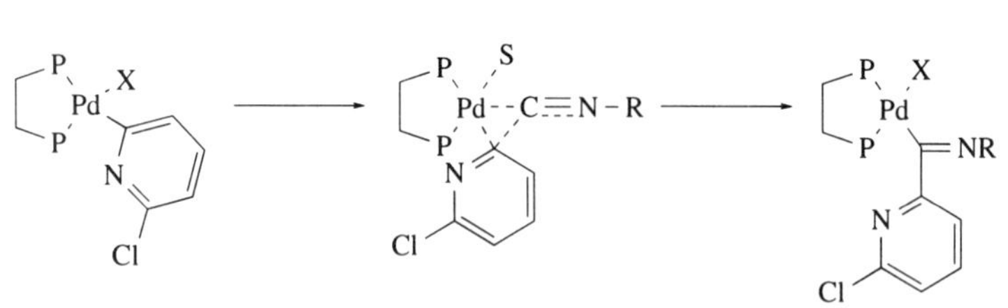

(58) (59) (60)

(10)

Insertion of isocyanide into Pd-(η-hydrocarbyls) also occurs. Thus, reactions with allylpalladium chloride dimer, $[Pd_2Cl_2(\eta^3\text{-}C_3H_5)_2]$, give (**61**),[1] and the thietane-3,3-dioxide complex, (**62**), yields regiospecific insertion into the Pd–CH(CO_2Me) bond producing the zwitterionic ring complex, (**63**).[131]

An unusual cleavage of the C≡N bond of xylyl isocyanide occurs during the reduction of $[PdCl_2(CNXy)_2]$ with sodium amalgam. There are two products, $[Pd_3(CNXy)_6]$ and (**64**), the latter arising from bond cleavage followed by insertion of isocyanide into a Pd–C bond. The fate of the carbon from the cleaved isocyanide is unknown.[109]

The presence of a basic nitrogen atom in the aminoacyl group makes it an obvious candidate for a bridging role, and in consequence, insertions of isocyanides often lead to dinuclear derivatives. For example, heating *trans*-$[PdCl(C_6F_5)(CNMe)_2]$ under reflux in benzene leads to intramolecular insertion

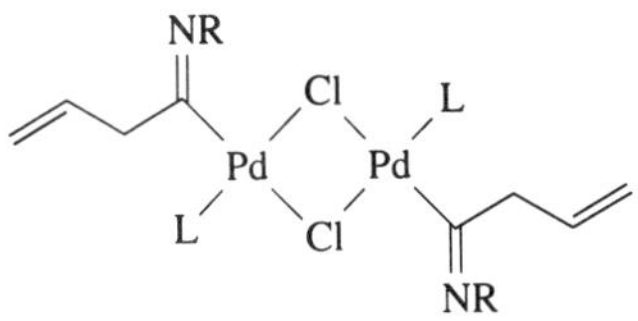

(61)

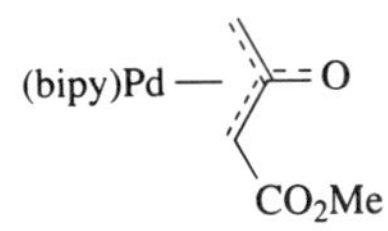

(62)

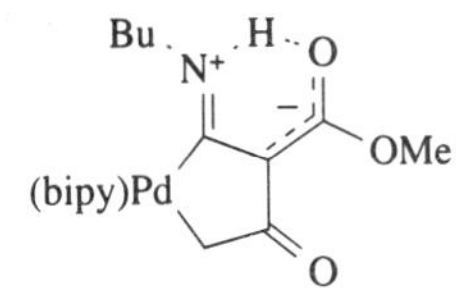

(63)

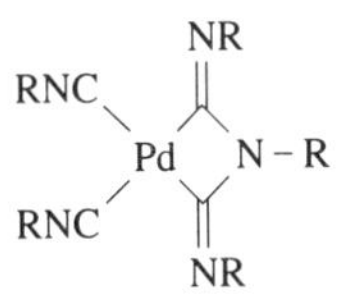

(64)

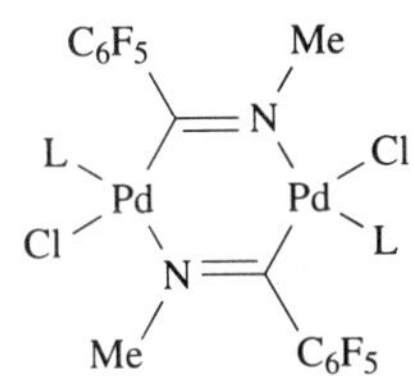

(65)

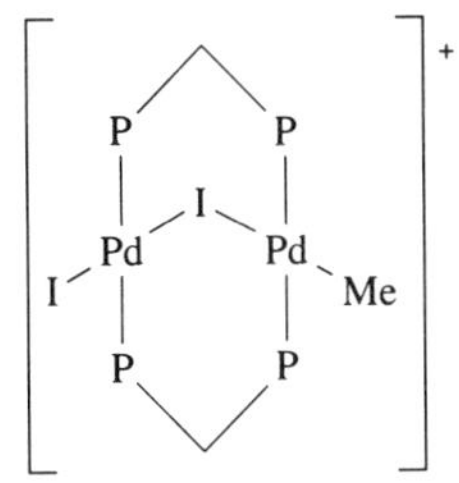

(66)

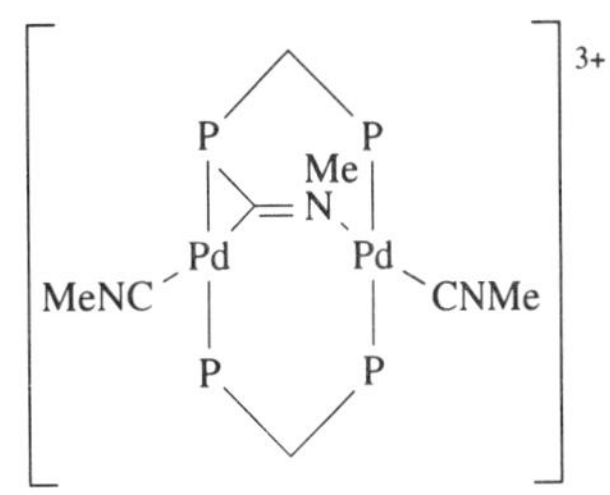

(67)

to the aminoacyl-bridged derivative, (**65**), L = CNMe. Similar products are accessible from $[Pd_2(\mu\text{-}Cl)_2(C_6F_5)_2(CNMe)_2]$ which initially yields a polymer with alternating $(\mu\text{-}Cl)_2$ and $(\mu\text{-aminoacyl})_2$ bridges. Subsequent depolymerization with a variety of donor ligands gives dinuclear products, (**65**), L = CNMe, NMe_3, py, 4Me-py, $4Bu^t$-py, PPh_3 and SC_4H_8, and the polymer also reacts with Tl(acac) or silver acetate to dinuclear complexes similar to (**65**), but with acac or acetate replacing the terminal Cl,L ligand pair.[132] Similar reactions of C_6Cl_5 complexes have also been reported.[124] These reactions are interesting since metal–perfluorocarbon bonds are normally inert to insertion reactions. Use of a series of isocyanides shows that reaction to analogues of (**65**) proceeds only with CNMe or CN(*p*-tolyl), but not with CNCy or $CNBu^t$, consistent with the idea that ease of insertion increases with nucleophilicity of the isocyanide.[133] In the context of difficulty of insertion into metal–perfluorocarbon bonds, it is also interesting that the complexes (**65**) provide a rare example of elimination of isocyanide. Thus, heating (**65**), L = PPh_3, under reflux in benzene with added triphenylphosphine for 18 h, eliminates methyl isocyanide to yield *trans*-$[PdCl(C_6F_5)(PPh_3)_2]$.[134] Formation of a bridging aminoacyl in an 'A-frame' complex is illustrated by conversion of (**66**) to (**67**) by reaction with methylisocyanide.[135]

More complex bridges may result from insertion into Pd–C bonds which are part of a chelating ligand system. Thus, insertion reactions of isocyanides with (**68**), Y = SMe or NMe_2, lead to both mononuclear products, (**69**), and to aminoacyl bridges, (**70**).[136,137] Direct palladation of various thioethers leads to a series of complexes related to (**68**), and comparison of their insertion reactions with those of NMe_2 analogues always shows greater reactivity for the sulfur derivatives.[137]

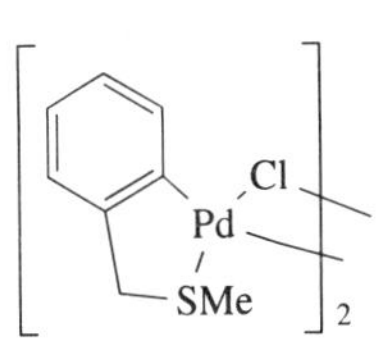

(68)

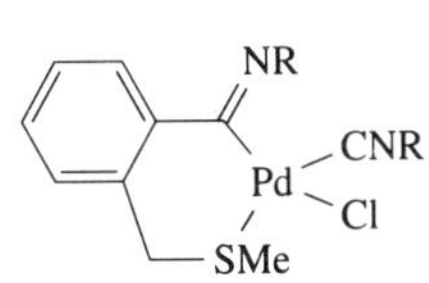

(69)

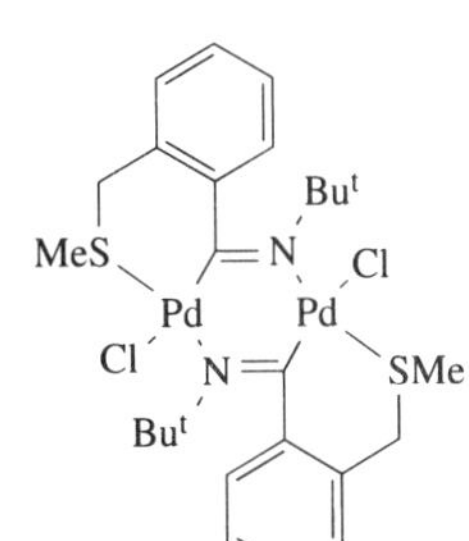

(70)

(ii) Multiple insertions

Since the products of isocyanide insertion still contain Pd–C bonds, it is evident that further insertion reactions are possible. Some typical examples of double insertion are shown in Equation (11), Ar = Ph or *p*-tolyl,[138] Equation (12), R = Me, Cy, *p*-C_6H_4OMe[125,126] and Equation (13), $n = 1$, R^1 = Et, R^2 = Ph or *p*-$C_6H_4NO_2$, $R^1 = Bu^n$, R^2 = Ph.[139] Equation (13), R^1 = Et, R^2 = Ph, *p*-tolyl or *p*-$C_6H_4NO_2$, can also yield multiple insertions up to approximately n = 100, essentially an example of living polymerization.[140]

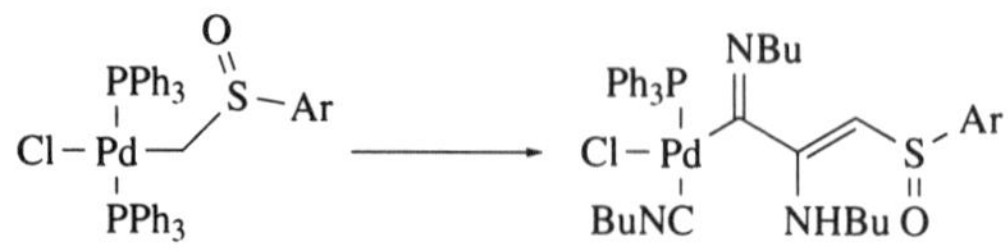

(11)

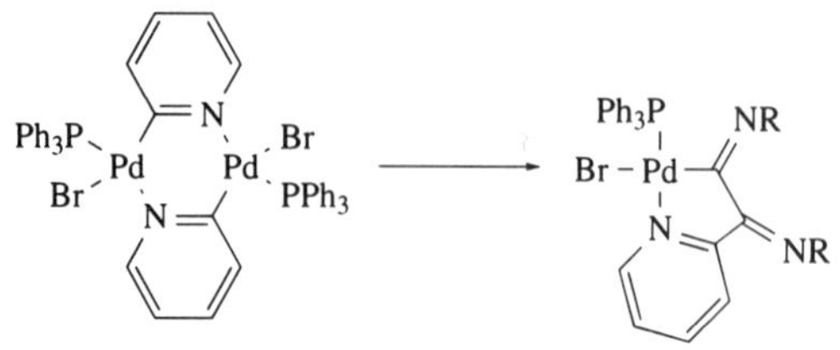

(12)

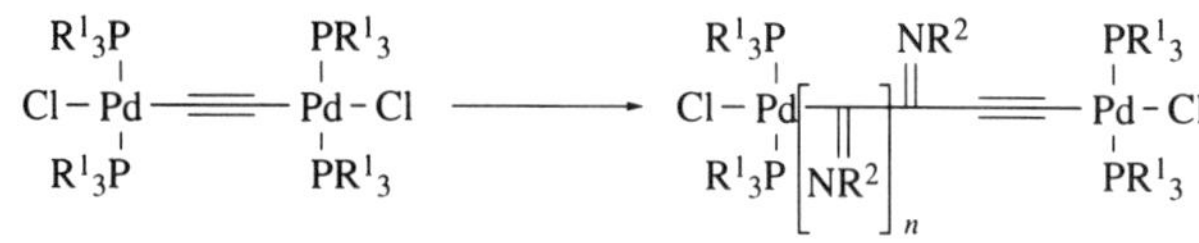

(13)

Multiple insertions can sometimes lead to unusual heterocyclic products, as in the synthesis of (**71**) as the ultimate product of insertion of $CNBu^t$ into a cyclopalladated benzylidine–aniline Schiff base complex, $[Pd_2(\mu\text{-}OAc)_2\{5'\text{-}MeOC_6H_3CH{=}NC_6H_4Me\text{-}4\}_2]$.[141]

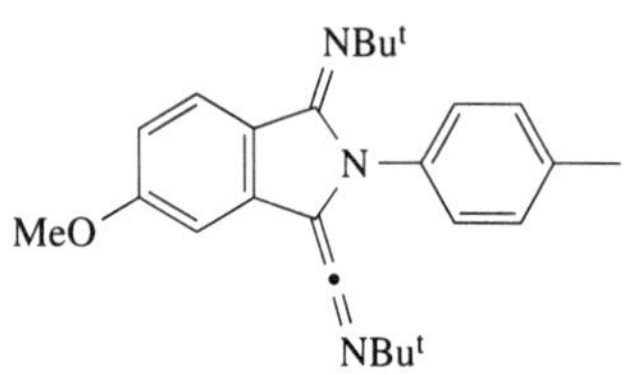

(**71**)

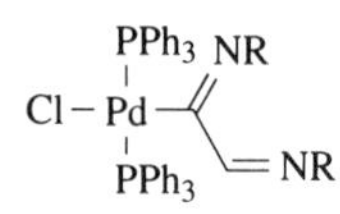

(**72**)

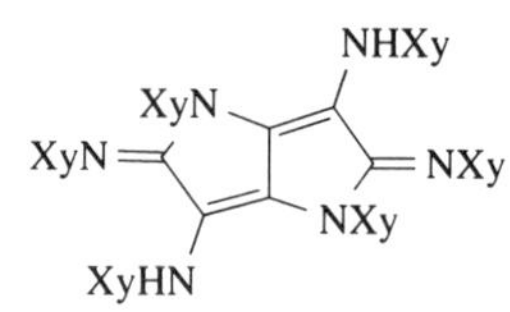

(**73**)

4.3.5.3 Insertions of isocyanides into palladium–element bonds other than carbon

The low temperature addition (–70 °C) of HCl to a solution of $[Pd(PPh_3)_4]$ containing CN(*p*-C_6H_4OMe) results in (**72**). The reaction presumably proceeds by initial oxidative addition of HCl, followed by a double insertion of isocyanide into the Pd–H bond.[142] The heterocyclic, (**73**), generated from reaction of $[PdCl_2(PPh_3)_2]$ with methylphenylsilane and xylyl isocyanide, is also a product of multiple insertion into a Pd–H bond, since the initial reaction is *in situ* generation of $[PdHCl(PPh_3)_2]$.[143]

Insertion of isocyanide into a Pd–N bond occurs in the conversion of (**74**) to (**75**), R = *p*-tolyl, by treatment with CN(*p*-tolyl),[144] and the reactions of $[Pd_2Cl_2(\mu\text{-dppm})_2]$ and its analogues to form the A-frame complexes (**76**) are examples of insertion into the Pd–Pd bond.[37,145] Formation of A-frame complexes by insertion into $[Pd_2Cl_2(\mu\text{-dppm})_2]$ has also been reported for 1,4-diisocyanobenzene, but in this case the difunctionality results in linking of two $Pd_2Cl_2(\mu\text{-dppm})_2$ fragments.[146] Infrared evidence suggests that the Pd–Pd bond of $[Pd_2Cl_2(\mu\text{-}Ph_2PNHPPh_2)_2]$ may also insert CN(*p*-tolyl) although the only well-characterized products contain terminal isocyanide.[147] Insertion into a Pd–Pt bond occurs in the conversion of $[PdPt(C_6F_5)_2(\mu\text{-dppm})_2]$ to $[PdPt(C_6F_5)_2(\mu\text{-dppm})_2(CNR)]$ by reaction with *p*-tolyl isocyanide.[148]

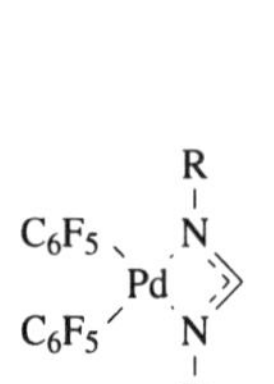

(**74**)

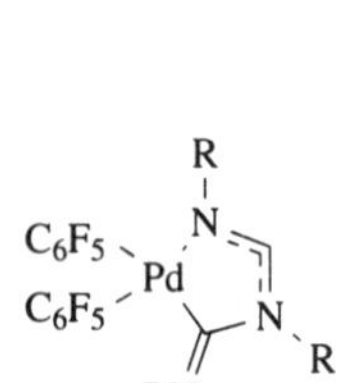

(**75**)

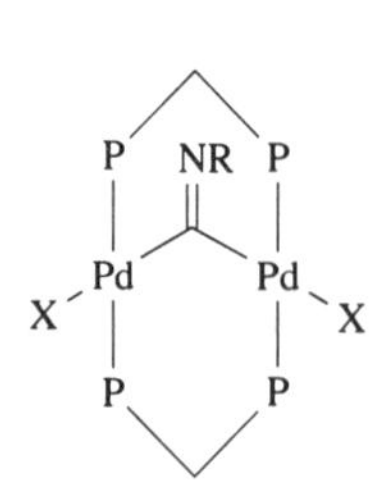

(**76**)

4.3.6 Structures

The key structural point in isocyanide complexes concerns the relative contributions of the resonance forms (**77**) and (**78**). Extensive contribution by (**78**) should of course lead to lowering of the C≡N stretching frequency and corresponding geometry changes in the ligand. Early work on palladium(II) complexes showed rather high ν(C≡N) frequencies, 2200–2225 cm^{-1} compared to 2124 cm^{-1} for free phenylisocyanide. This indicated relatively little contribution by (**78**), a view which was confirmed by x-ray results showing almost linear isocyanides with C–N–C angles 173–179°.[1] Table 2 summarizes some more recent structural results, and in general shows the same trends, except that bridging isocyanides occur in several of the complexes. The most definitive bridged example is perhaps the trinuclear cluster discussed in more detail above (see Section 4.3.2), $[Pd_3(CNCy)_6]$ (**47**), giving the distances and angles to be expected for structure (**79**), or if (**78**) were to play a role in terminal complexes. However, rather than bending of terminal isocyanides, it is noticeable that several of the bridging examples, for example (**49**) and (**55**), contain unusually linear C–N–C ligand backbones. The C≡N bond length is slightly longer in bridging isocyanides, but the change is very small, less than expected if (**79**) were a true representation, and more similar to bridging carbonyls. In (**49**) there is also considerable asymmetry of the bridges, with stronger binding to one of the two palladium atoms involved in each bridge. In combination with earlier work,[1] Table 2 suggests that overall ranges for ν(C≡N) are about 2080–2230 cm^{-1} for terminal isocyanides on palladium, and about 1580–1920 cm^{-1} for bridges. An x-ray photoelectron study of $[Pd_2Cl_2(\mu\text{-dppm})_2(\mu\text{-CNMe})]$ and $[Pd_2(CNMe)_2(\mu\text{-dppm})_2(\mu\text{-CNMe})]$ has shown that nitrogen 1*s* binding energies are ~1 eV lower for CNMe bridges relative to terminal methyl isocyanide.[149]

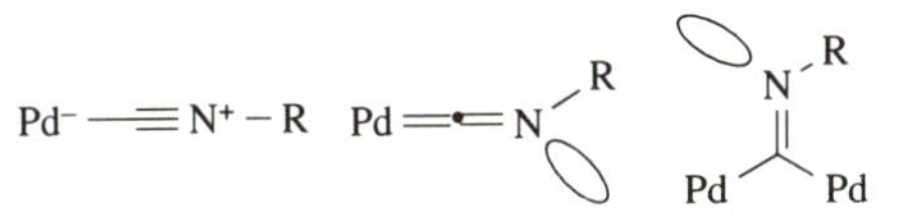

(**77**) (**78**) (**79**)

Table 2 Selected infrared and structural data for representative palladium isocyanides.[a]

Complex	ν(C≡N) (cm^{-1})	Pd–C (nm)	C≡N (nm)	N–C (nm)	Pd–C≡N (°)	C≡N–C (°)	*References*
(**47**), R = Cy	2111	0.200(1)	0.115(1)	0.146(1)	169(1)	170(1)	107
	1733	0.207(1)	0.122(2)	0.146(1)	141(1)	129(1)	
(**48**), $L^1 = PPh_3$, L^2 = CNXy	2119	0.2048(4)	0.1160(5)	0.1394(5)	164.6(3)	175.6(4)	62
	2092	0.2050(4)	0.1159(5)	0.1404(5)	160.6(4)	174.0(4)	
(**49**), L = CNXy	2142	0.200[b]	0.1155	0.1399	172[c]	174[d]	110
	2047	0.2055[e]	0.116	0.138	149[e]	170	
	1990	0.2285[e]			129[e]		
(**54**), L = CNMe	2165	0.197	0.113[f]	0.146	177.7	177.4	115
(**55**), R = Xy	2056	0.206(1)	0.118(2)	0.140(1)	143	175(1)	117
	1976	0.196(1)	0.118(2)	0.143(1)	133	158(1)	
cis-$[PdCl_2(CNCy)_2]$	2130	0.193(1)	0.113	0.145	177	174	150
trans-$[Pd(CNBu^t)_2(dppm\text{-}P)_2][BPh_4]_2$	2233	0.197(1)	0.113(2)	0.150	179(1)	178(1)	151
$[Pd_2Cl_2(\mu\text{-dppm})_2(\mu\text{-CNPh})]$	1624[g]	0.198(1)	0.128(1)	0.139(1)	126	129(1)	152
$[Pd_2(\mu\text{-SPPh}_2)_2(CNMe)_2]$	2200	0.2021(4)	0.1135(6)	0.1454(6)	177.1(4)	177.7	153

[a] Structural parameters lacking standard deviations are averages. If the range is significantly wide, then it is given in a footnote. It is often not possible to be sure which infrared frequencies are to be associated with which structural parameters. [b] Range 0.1973–0.2024 nm. [c] Range 169.3–175.4°. [d] Range 170.1–177.6°. [e] These parameters refer to the two Pd–C bonds of an asymmetric bridge. [f] Range 0.1115–0.1142 nm. [g] There is some uncertainty in this assignment since there is another absorption at 1574 cm^{-1}.

Other x-ray structure determinations providing basically similar parameters to those in Table 2 are: $[Pd_2Cl_2(CNBu^t)_4]$,[114] $[Pd(CNBu^t)\{CH(COMe)(Ph_3P)\}(\eta^3\text{-2-MeC}_3H_4)]BF_4$,[154] $[Pd_2(\mu\text{-}\eta^3\text{-in-denyl})_2(CNXy)_2]$,[155] $[Cp(CNBu^t)Rh(\mu\text{-Ph}_2PPy)PdCl(CNBu^t)]PF_6$[156] and $[Pd_2Cl_2(\mu\text{-dppm})_2(\mu\text{-CNC}_5H_5\text{-CHCH}_2CO_2Me)]$.[157] The xylyl isocyanide complex, *cis*-$[PdCl_2(CNR)_2)]$, has been shown to have a quasi-one-dimensional linear chain structure with Pd···Pd interactions ~0.41 nm and Pd···Pd···Pd angles of 170.2°.[158] $[Pd\{1,2\text{-}S_2C_2(CN)_2\}_2(CNMe)_2]$ is also a linear chain structure.[159]

4.4 CARBENE COMPLEXES

4.4.1 Introduction

Despite the many studies of carbenes since their discovery in 1964, palladium carbene chemistry remains relatively limited and recent work has tended to focus more on the early transition metals. The general framework of palladium–carbene chemistry was fairly well established when the first edition of this book was published, and much of the same ground is also covered in a 1983 book on carbenes.[1,160] Most complexes are basically palladium(II), with stoichiometry suggesting that the carbene functionality simply replaces other 2e donor ligands. Typical complex types are $[PdX_2L(carbene)]$, $[PdX_2(carbene_2)]$, $[PdXL_2(carbene)]^+$ and $[Pd(carbene)_4]^{2+}$, while anionic examples are apparently rare.[1] There has been little systematic development of a reaction chemistry for palladium carbenes beyond that described in the first edition. Consequently, with the exception of the few novel examples reported in Section 4.4.4, the primary interest in this chemistry focuses on synthetic methods, especially syntheses using addition of protonic compounds to isocyanide ligands. As in previous sections of this chapter, there is no attempt to cover organic chemistry of carbenes where palladium complexes are often used as a catalysts.

4.4.2 Synthesis

4.4.2.1 From coordinated isocyanides

The most convenient and widely used synthesis of palladium carbenes is by the addition of protonic compounds to isocyanide coordinated to palladium(II). In general, this reaction is favoured for isocyanide complexes where there is little back-bonding as indicated by a high $\nu(C{\equiv}N)$ frequency, making palladium complexes ideal candidates. The elementary reaction is shown in Equation (14).

(14)

Early work established many examples of this reaction, including one of the simplest, namely use of *cis*-$[PdCl_2(PR^1_3)(CNR^2)]$ to yield *cis*-$[PdCl_2(PR^1_3)\{C(NHR^2)Y\}]$, where R^2 = aryl and HY is an alcohol or amine.[1] The mechanism of this reaction has been studied for aryl amines and the primary step is direct nucleophilic attack of the amine nitrogen on the isocyanide carbon to give an intermediate, (**80**), which is a protonated iminoacyl complex. Subsequent steps to the final carbene involve proton transfer, which may occur either intramolecularly via a four-membered cyclic transition state, or by the intervention of a second molecule of amine acting as a catalyst. The rate of the primary step is affected by the donor ability of the entering amine, by the electrophilic character of the isocyanide carbon, and by steric crowding. The solvent also plays a role.[161] The intermediacy of (**80**) in the reaction suggests that iminoacyl complexes should be readily protonated to carbenes, and this is confirmed by the conversion by perchloric acid of $[PdCl_2(PPh_3)\{C(R^1){=}NR^2\}]$, R^1 = Ph, R^2 = *p*-Tol, or R^1 = *p*-Tol, R^2 = *p*-$C_6H_4NO_2$ or *p*-$CH_2C_6H_4NO_2$, to $[PdCl_2(PPh_3)\{C(R^1)NHR^2\}]^+$. Triethyloxonium fluoroborate accomplishes a similar transformation, adding Et^+ to yield $[PdCl_2(PPh_3)\{C(R^1)NEtR^2\}]^+$.[162] Other, more recent, examples of the simple formation of monocarbene complexes from isocyanides include reaction of *cis*-$[PdCl_2(CNR)L]$, L = tertiary phosphine, R = Bu^t or *p*-$MeOC_6H_4$, with azetidine to give (**81**),[163] reaction of $[Pd(C_6F_5)(CNR^1)L_2]^+$, L_2 = bis(tertiary phosphine), 2,2'-bipyridine, en, TMEDA or dppe, R^1 = Bu^t, Cy or *p*-Tol, with R^2H, R^2 = *p*-$MeC_6H_4NH_2$, $PhCH_2NH_2$, Me_3CNH_2, MeOH or EtOH, giving $[Pd(C_6F_5)\{C(R^2)NHR^1\}L_2]^+$,[164] and reaction of $[PdCl\{CN(p\text{-}C_6H_4X)\}(\eta^3\text{-allyl})]$, X = Cl, Me, OMe, with secondary amines, NHR^2_2, R^2 = Me or Et, giving (**82**).[165]

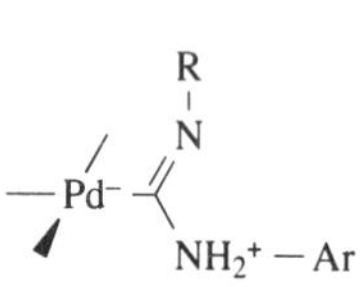

(**80**)

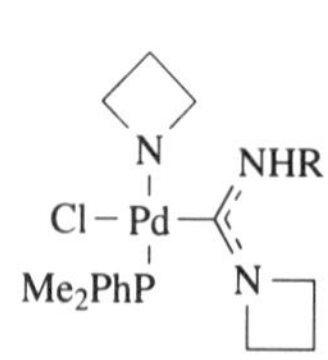

(**81**)

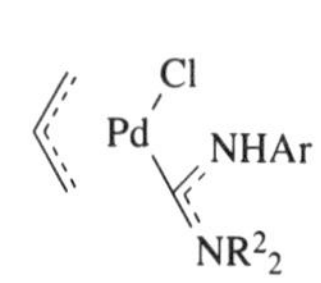

(**82**)

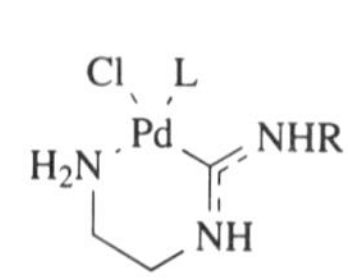

(**83**)

Complexes containing more than one carbene ligand are also readily accessible by protonation of isocyanides. Early work established reactions of primary amines with *cis*-$[PdCl_2(CNR)_2]$ to form mono- and bis-carbene complexes, and also the reaction of $[PdCl_4]^{2-}$ with methyl isocyanide and methylamine to form the $[Pd\{C(NHMe)_2\}_4]^{2+}$ cation which can be isolated as its hexafluorophosphate salt.[1] More recent examples include other homoleptic carbene cations, $[Pd\{C(NHR)[NHCH_2CH(OEt)_2]\}_4]^{2+}$,[166] and organocarbenes, *trans*-$[Pd(C_6F_5)_2\{C(NHR)(NHCH_2Ph)\}_2]$. Although these neutral organocarbene complexes form easily from *trans*-$[Pd(C_6F_5)_2(CNR)_2]$, the corresponding anions, $[Pd(C_6F_5)_3(CNR)]^-$ are unreactive towards $BzNH_2$.[167] This is interesting in that, whereas cationic carbenes are fairly common, the first edition of this book noted only a single anionic carbene, $[PdCl_3\{C(NHPh)(OMe)\}]^-$.[1] This still appears to be the only example.

The use of bifunctional amines as the protonic addendum in Equation (14) can result in chelate complexes such as (**83**), derived from reactions of *cis*-$[PdCl_2(CNR)L]$ with ethylenediamine. Other bifunctional amines such as *o*-phenylenediamine and 2-aminopyridine give similar complexes.[1] Equation (15) represents an interesting intermolecular example of such bifunctionality.[168]

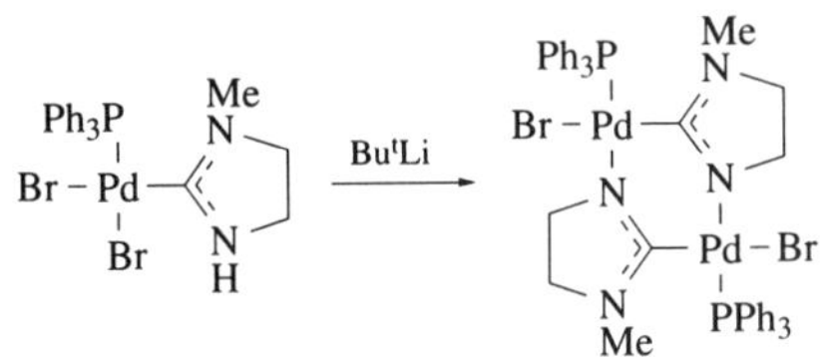

(15)

Perhaps the most interesting examples of Equation (14) are those used to synthesize heterocyclic carbenes.[169] Several different strategies have been employed but all utilize the basic idea of providing a nucleophilic group, capable of attack at the isocyanide carbon, as illustrated in Equations (16)–(19).

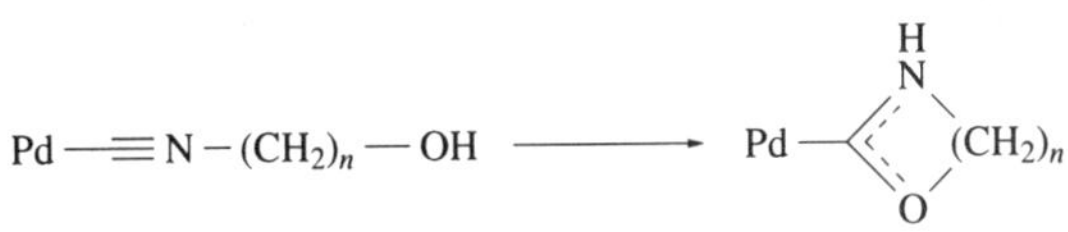

(16)

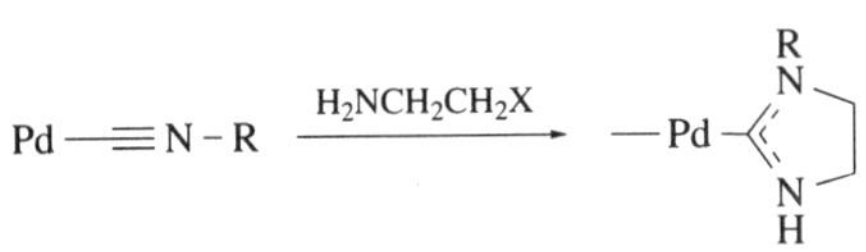

(17)

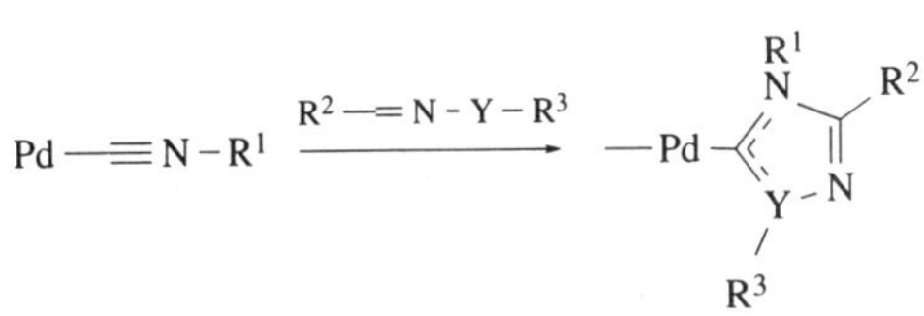

(18)

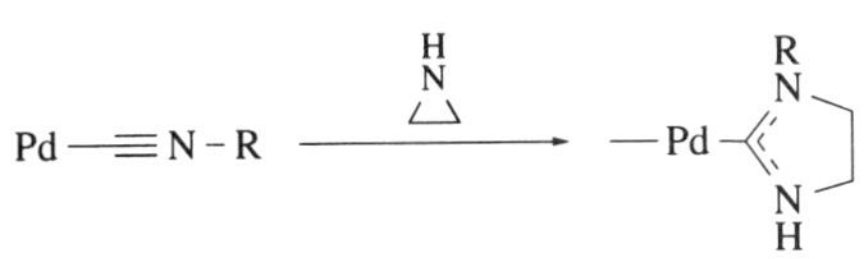

(19)

In Equation (16), the nucleophilic group is already part of the isocyanide and reactions of $CN(CH_2)_3OH$ or $CNCHR^1CHR^2OH$; $R^1 = R^2 = H$; $R^1 = Et$, $R^2 = H$; $R^1 = H$, $R^2 = Me$; with palladium(II) halides or halide complexes, $[PdX_2L_2]$, L = MeCN, PhCN, PPh_3, lead to spontaneous ring closure.[170]

In the other examples (Equations (17)–(19)), in addition to the nucleophilic group it is necessary to provide an electrophilic centre in the addendum. Ring closure is then completed by reaction of this centre with the nitrogen of an aminoacyl intermediate. Thus, in Equation (17), 2-bromoethylamine (generated *in situ* from $[BrCH_2CH_2NH_3]Br$ and $LiBu^n$) reacts with *trans*-$[PdCl(PPh_3)_2(CNR)]$, R = Me, *p*-$MeOC_6H_4$, *p*-Tol, to form air stable cyclic carbenes,[171] via an intermediate which can be represented as (**84**). Similar reactions using 2-haloalcohols were successful for platinum complexes but failed for palladium even at −50 °C. This failure is attributed to preferential alkoxide attack at palladium with subsequent decomposition by β-hydrogen elimination.[169] The strategy in Equation (18) provides both nucleophilic and electrophilic centres in a 1,3-dipolar addendum such as a nitrilimine ($R^1–C^+=N–N^-–R^2$) or nitrilylide ($R^1–C^+=N–CH^-–R^2$), where R^1 and R^2 are aryl groups. These addenda react with *cis*-[$PdCl_2(PR_3)(CNTol)$] complexes.[1,172] Finally, Equation (19) provides the electrophilic centre in the form of the strained three-membered ring of aziridine, leading to reactions with *cis*-[$PdCl_2(PPh_3)(CNR)$], R = Bu^t, *p*-MeOC$_6$H$_4$, or propylacrylate, to form cylcic aminocarbenes. Presumably the reaction proceeds by initial attachment of aziridine nitrogen to the isocyanide carbon, followed by intramolecular attack of the nucleophilic imino nitrogen on an adjacent aziridine methylene group (compare intermediate (**84**)). Use of *cis*-[$PdCl_2(CNR)_2$] complexes in these reactions leads to bis(cyclic carbene) complexes. Reactions of thiirane are similar but oxirane reacts only with platinum isocyanides, not palladium which again leads to decomposition similar to that observed in the haloalkoxide reactions above.[169,173] Increase of the ring size from aziridine to azetidine, $H\overline{N(CH_2)_2C}H_2$, gives only acyclic carbenes, (**81**), with no subsequent ring closure.[163]

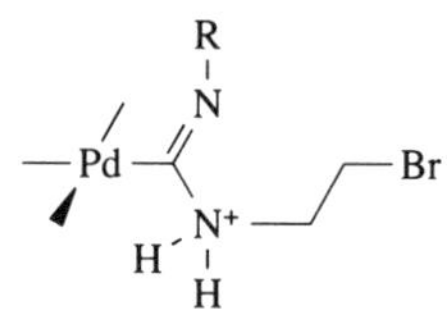

(84)

4.4.2.2 *Other synthetic methods*

The first edition of this book described several syntheses of palladium carbenes from organic carbene precursors, such as *gem*-dichloropropenes and electron-rich alkenes. However, little further work has been reported except for more detailed accounts of the reactions of the electron-rich alkene (**85**).[174] These reactions were described in preliminary form in the first edition.[1]

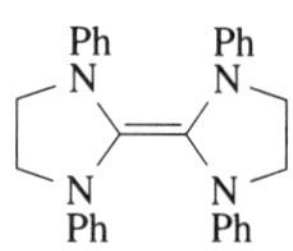

(85)

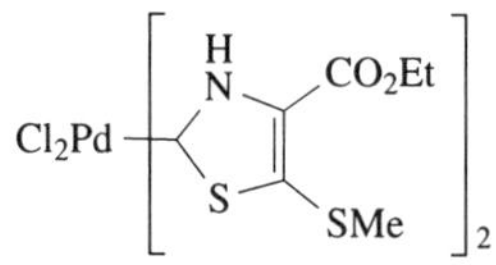

(86)

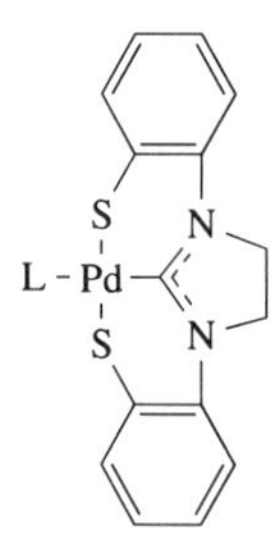

(87)

A palladium carbene, (**86**), has been synthesized by direct carbene transfer using the corresponding chromium carbene complex, [Cr{$\overline{NHC(CO_2Et)C(SMe)S}$}$(CO)_5$], reacting with [$PdCl_2$(cod)].[175] Another possible route utilizes organolithium reagents, Li[C(OMe)=CH_2] or Li[$\overline{C{=}CH(CH_2)_nO}$], $n = 2$ or 3, reacting with *trans*-[$PdCl(C_6Cl_5)(PMe_2Ph)_2$] to yield palladium allyl derivatives which can be subsequently protonated with perchloric acid to the cationic carbene complexes *trans*-[Pd{C(OMe)Me}$(C_6Cl_5)(PMe_2Ph)_2$]$^+$ and *trans*-[Pd{$\overline{CCH_2(CH_2)_nO}$}$(C_6Cl_5)(PMe_2Ph)_2$]$^+$.[176] Finally, an unusual template synthesis uses a 'one pot' reaction of [$PdCl_2$(cod)] with $CH(OEt)_3$ and 1,2-ethanediamine-*N,N*'-bis(2-benzenethiol). The initial product is a sulfur-bridged dimer which can be subsequently cleaved by pyridine or tertiary phosphine to (**87**).[177]

4.4.3 Structures

In palladium(II) carbenes, the palladium atom, carbenoid carbon and the other two atoms directly bound to the carbenoid carbon, are generally coplanar, and except where constrained by chelation, this plane lies approximately orthogonal to the palladium coordination plane. Early crystal structures showed a range of Pd–C bond lengths from 0.1948–0.2074 nm, and comparison of these distances with a calculated Pd–C(sp^2) of 0.225 nm suggests that most complexes have at least a degree of π-bonding. The carbene ligand has a high *trans*-influence.[1] More recent x-ray determinations with corresponding Pd–C lengths are: (**81**) 0.1990(6) nm,[163] the iodide salt of (**88**),[166] chloride salt of (**89**) 0.202 nm[170] and (**90**) 0.1943(9) nm.[173] These structures confirm the early picture and also include the first examples of cyclic palladium carbenes to be structurally characterized. Theoretical calculations suggest a bond length of 0.186 nm for a Pd=CH_2 model fragment, and a bond energy of 213 kJ mol^{-1}. This energy is similar to that calculated for the Pd–O bond (199 kJ mol^{-1}), whereas for most earlier transition metals the oxides are much more strongly bound.[178] Some structures of more unusual palladium carbenes are included in Section 4.4.4.

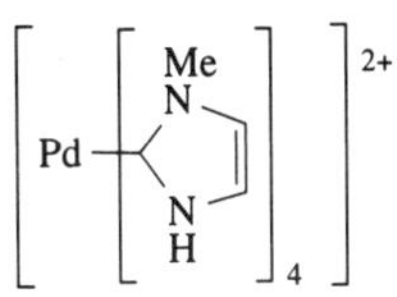

(88)

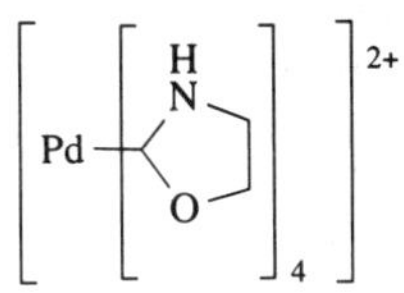

(89)

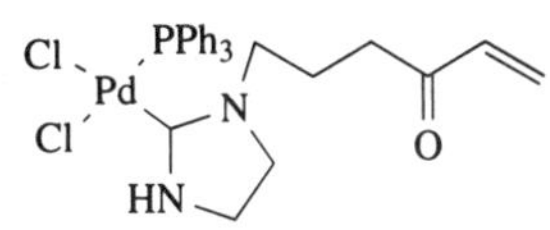

(90)

4.4.4 Novel Palladium Carbenes

A palladium cluster carbene has been prepared by reaction of the tetranuclear, $[Pd_4(OAc)_4(CO)_4]$, with diphenyldiazomethane. An extended x-ray absorption fine structure (EXAFS) study of the product, $[Pd_4(OAc)_4(CPh_2)_4]$ (**91**), shows that each palladium is surrounded by two palladium atoms at 0.27 nm, a third palladium at 0.38 nm, and pairs of light atoms at 0.19 nm, 0.214 nm and 0.346 nm. This suggests that the starting carbonyl structure (Section 4.2.5.4, Structure (**22**)) is considerably modified during the replacement of carbonyl by carbene, resulting in a Pd_4 square, with each edge bridged by one diphenyl carbene and one acetate group, (**91**).[179] One may perhaps quibble as to whether these bridges should be regarded as carbenes, or as diphenylmethylene groups (forming dimetallocyclopropane rings) but the analogy with the carbonyl cluster does support the carbene view, as does the reactivity. For example, reaction with 1-hexene gives cyclopropanation to $Me(CH_2)_3\overline{CHCH_2C}Ph_2$, a typical carbene reactivity pattern. Another interesting observation is that in this reaction, and in hydrogenation of (**91**) to diphenylmethane, some benzophenone is produced. This also occurs on thermolysis of (**91**). The source of the oxygen appears to be the acetate group, similar to thermolysis of the corresponding carbonyl cluster, (**22**), where the product is carbon dioxide.[58,179]

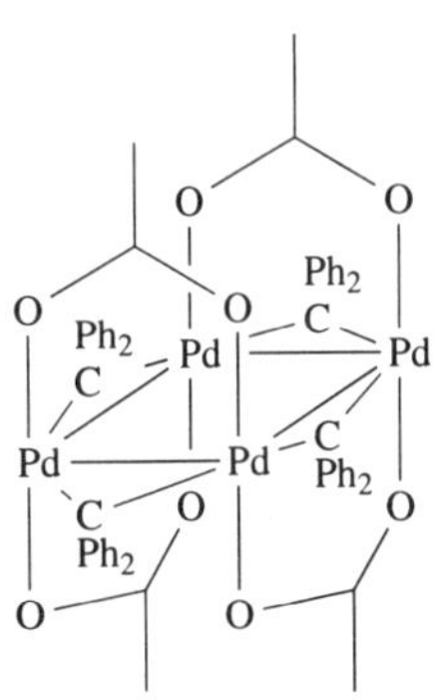

(91)

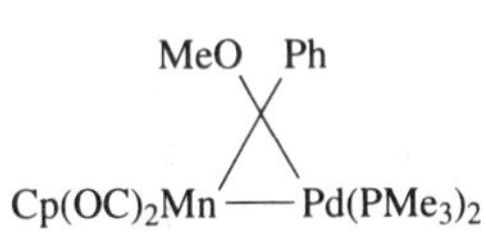

(92)

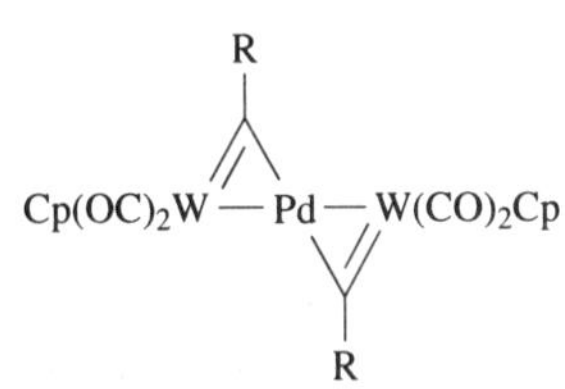

(93)

Other examples of bridging carbenes are to be found in a general synthesis of heteronuclear dimetal complexes using reactions of M^0 complexes, M = Ni, Pd or Pt, with mononuclear metal carbene complexes. In particular, $[Pd(dba)_2]$ reacts with $[Mn\{C(OMe)Ph\}(CO)_2Cp]$ and trimethylphosphine to form (**92**).[180] A close relative, (**93**), R = *p*-Tol, is formed by reaction of [Pd(bicyclo[2.2.1]heptene)$_3$] with $[W(\equiv CR)(CO)_2Cp]$, but an x-ray diffraction study of the platinum analogue shows the carbon much closer to tungsten than to platinum. Thus, these should probably be regarded as tungsten, rather than palladium, carbenes.[181]

Another novel palladium carbene is the infinite chain, mixed metal, polymer (**94**) produced by reaction of $[PdCl_2(PhCN)_2]$ with $[Pt(dppm)_2]$, in benzene at 25 °C. The coordination units of platinum and palladium are coplanar and the Pd–C bond is remarkably short, 0.163(5) nm, suggesting extensive double bond character compared with the more usual value of about 0.195 nm for other palladium carbenes.[182] The authors suggest that the polymer be regarded as three-coordinate palladium(0) and four-coordinate platinum(II). Although this is somewhat at odds with the method of preparation, it may perhaps explain the short Pd–C bond since all other well-determined distances are for palladium(II) carbenes.

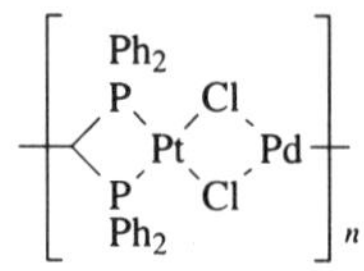

(94)

4.5 REFERENCES

1. P. M. Maitlis, P. Espinet and M. J. H. Russell, in 'COMC-I', vol. 6, pp. 243–349.
2. (a) G. A. Ozin and A. J. L. Hanlan, *Inorg. Chem.*, 1979, **18**, 2091; (b) A. B. P. Lever, G. A. Ozin, A. J. L. Hanlan, W. J. Power and H. B. Gray, *ibid.*, 1979, **18**, 2088.
3. T. Mehner, R. Koeppe and H. Schnoeckel, *Angew. Chem.*, 1992, **104**, 653.
4. G. W. Smith and E. A. Carter, *J. Phys. Chem.*, 1991, **95**, 2327; 10828.
5. P. Schwerdtfeger, J. S. McFeaters, J. J. Moore, D. M. McPherson, R. P. Cooney, G. A. Bowmaker, M. Dolg and D. Andrae, *Langmuir*, 1991, **7**, 116.
6. M. R. A. Blomberg, C. B. Lebrilla and P. E. M. Siegbahn, *Chem. Phys. Lett.*, 1988, **150**, 522.
7. (a) M. Ohno and W. Von Niessen, *Surf. Sci.*, 1992, **269–270(Pt. A)**, 258; (b) M. Ohno and W. Von Niessen, *Phys. Rev. B: Cond. Matter*, 1992, **45**, 9382.
8. P. Decleva and M. Ohno, *Chem. Phys.*, 1992, **160**, 353.
9. V. Bonacic-Koutecky, J. Koutecky, P. Fantucci and V. Ponec, *J. Catal.*, 1988, **111**, 409.
10. T. Ziegler, *Organometallics*, 1985, **4**, 675.
11. T. Ziegler, V. Tschinke and C. Ursenbach, *J. Am. Chem. Soc.*, 1987, **109**, 4825.
12. (a) C. M. Rohlfing and P. J. Hay, *J. Chem. Phys.*, 1985, **83**, 4641; (b) J. Koutecky, G. Pacchioni and P. Fantucci, *Chem. Phys.*, 1985, **99**, 87.
13. L. A. Barnes, M. Rosi and C. W. J. Bauschlicher, *J. Chem. Phys.*, 1990, **93**, 609.
14. P. L. Goggin and J. Mink, *J. Chem. Soc., Dalton Trans.*, 1974, 534.
15. P. L. Goggin, R. J. Goodfellow, I. R. Herbert and A. G. Orpen, *J. Chem. Soc., Chem. Commun.*, 1981, 1077.
16. F. Calderazzo and D. B. Dell'Amico, *Inorg. Chem.*, 1981, **20**, 1310.
17. D. B. Dell'Amico, F. Calderazzo and N. Zandona, *Inorg. Chem.*, 1984, **23**, 137.
18. B. P. Andreini, D. B. Dell'Amico, F. Calderazzo, M. G. Venturi and G. Pelizzi, *J. Organomet. Chem.*, 1988, **354**, 369.
19. (a) F. Calderazzo and D. B. Dell'Amico, *Pure Appl. Chem.*, 1986, **58**, 561; (b) D. B. Dell'Amico, F. Calderazzo, C. A. Veracini and N. Zandona, *Inorg. Chem.*, 1984, **23**, 3030; 4781.
20. N. M. Kostic and R. F. Fenske, *Inorg. Chem.*, 1983, **22**, 666.
21. J. Browning, P. L. Goggin, R. J. Goodfellow, M. G. Norton, A. J. M. Rattray, B. F. Taylor and J. Mink, *J. Chem. Soc., Dalton Trans.*, 1977, 2061.
22. R. Uson, J. Fornies, M. Tomas and B. Menjon, *Organometallics*, 1985, **4**, 1912.
23. J. Vicente, A. Arcas, M. V. Borrachero, A. Tiripicchio and M. Tiripicchio-Camellini, *Organometallics*, 1991, **10**, 3873.
24. A. Maisonnat, J. P. Farr and A. L. Balch, *Inorg. Chim. Acta*, 1981, **53**, L217.
25. J. Powell and N. I. Dowling, *Organometallics*, 1983, **2**, 1742.
26. N. M. Boag, D. Boucher, J. A. Davies, R. W. Miller, A. A. Pinkerton and R. Syed, *Organometallics*, 1988, **7**, 791.
27. (a) K. Zenkeert, P. Thometzek and H. Werner, *Angew. Chem.*, 1985, **97**, 520; (b) H. Werner, P. Thometzek, K. Zenkert, R. Goddard and H. J. Kraus, *Chem. Ber.*, 1987, **120**, 365.
28. R. D. Feltham, G. Elbaze, R. Ortega, C. Eck and J. Dubrawski, *Inorg. Chem.*, 1985, **24**, 1503.
29. R. Bender, P. Braunstein, A. Tiripicchio and M. Tiripicchio-Camellini, *J. Chem. Soc., Chem. Commun.*, 1984, 42.
30. V. V. Bashilov, E. G. Mednikov, S. S. Kurasov, P. V. Petrovskii, N. K. Eremenko and V. I. Sokolov, *Izv. Akad. Nauk SSSR, Ser. Khim.*, 1987, **1987**, 2360 (*Chem. Abstr.*, **109**, 110 622f).
31. (a) A. A. Lamberov, V. K. Polovnyak and N. S. Akhmetov, *Zh. Obshch. Khim.*, 1987, **57**, 630. (*Chem. Abstr.*, **108**, 75 594d); (b) V. P. Linev, O. N. Temkin, V. K. Polovnyak, L. G. Bruk and G. V. Romanov, *Zh. Obshch. Khim.*, 1981, **51**, 2536 (*Chem. Abstr.*, **96**, 143 050j).
32. M. E. Lamberova, A. S. Khramov, V. K. Polovnyak and N. S. Akhmetov, *Zh. Obshch. Khim.*, 1990, **60**, 2069 (*Chem. Abstr.*, **114**, 74 085a).
33. I. V. Loginova and V. K. Polovnyak, *Zh. Obshch. Khim.*, 1990, **60**, 271 (*Chem. Abstr.*, **113**, 78 658s).
34. M. Portnoy, F. Frolow and D. Milstein, *Organometallics*, 1991, **10**, 3960.
35. A. R. Siedle, R. A. Newmark and W. B. Gleason, *Inorg. Chem.*, 1991, **30**, 2005.
36. V. N. Zudin, V. D. Chinakov, V. M. Nekipelov, V. A. Likholobov and Y. I. Ermakov, *J. Organomet. Chem.*, 1985, **289**, 425.
37. R. J. Puddephatt, *Chem. Soc. Rev.*, 1983, **12**, 99.
38. A. L. Balch and L. S. Benner, *Inorg. Synth.*, 1982, **21**, 47.
39. (a) M. L. Kullberg, F. R. Lemke, D. R. Powell and C. P. Kubiak, *Inorg. Chem.*, 1985, **24**, 3589; (b) M. L. Kullberg and C. P. Kubiak, *Organometallics*, 1984, **3**, 632.
40. T. E. Krafft, C. I. Hejna and J. S. Smith, *Inorg. Chem.*, 1990, **29**, 2682.
41. P. Dagnac, R. Turpin and R. Poilblanc, *J. Organomet. Chem.*, 1983, **253**, 123.
42. (a) B. R. Lloyd, L. Manojlovic-Muir, K. W. Muir and R. J. Puddephatt, *Organometallics*, 1993, **12**, 1231; (b) L. Manojlovic-Muir, K. W. Muir, B. R. Lloyd and R. J. Puddephatt, *J. Chem. Soc., Chem. Commun.*, 1985, 536.
43. R. J. Puddephatt, L. Manojlovic-Muir and K. W. Muir, *Polyhedron*, 1990, **9**, 2767.
44. M. C. Jennings, R. J. Puddephatt, L. Manojlovic-Muir, K. W. Muir and B. N. Mwariri, *Organometallics*, 1992, **11**, 4164.
45. P. D. Harvey, M. Drouin, A. Michel and D. Perreault, *J. Chem. Soc., Dalton Trans.*, 1993, 1365.
46. P. D. Harvey and R. Provencher, *Inorg. Chem.*, 1993, **32**, 61.
47. G. Ferguson, B. R. Lloyd, L. Manojlovic-Muir, K. W. Muir and R. J. Puddephatt, *Inorg. Chem.*, 1986, **25**, 4190.
48. M. C. Jennings, N. C. Payne and R. J. Puddephatt, *Inorg. Chem.*, 1987, **26**, 3776.

49. M. Rashidi, G. Schoettel, J. J. Vittal and R. J. Puddephatt, *Organometallics*, 1992, **11**, 2224.
50. (a) D. H. Farrar, R. R. Gukathasan and K. Won, *J. Organomet. Chem.*, 1984, **275**, 263; (b) D. H. Farrar, R. R. Gukathasan and S. A. Morris, *Inorg. Chem.*, 1984, **23**, 3258; (c) S. Otsuka and T. Yoshida, *J. Am. Chem. Soc.*, 1977, **99**, 2134.
51. A. M. Arif, D. E. Heaton, R. A. Jones and C. M. Nunn, *Inorg. Chem.*, 1987, **26**, 4228.
52. E. G. Mednikov and N. K. Eremenko, *Izv. Akad. Nauk SSSR, Ser. Khim.*, 1981, **1981**, 2400.
53. E. G. Mednikov, N. K. Eremenko, S. P. Gubin, Y. L. Slovokhotov and Y. T. Struchkov, *J. Organomet. Chem.*, 1982, **239**, 401.
54. (a) E. G. Mednikov, *Metalloorg. Khim.*, 1989, **2**, 317 (*Chem. Abstr.*, **111**, 125 664m); (b) E. G. Mednikov, *Organomet. Chem. USSR*, 1989, **2**, 148.
55. I. I. Moiseev, T. A. Stromnova, M. N. Vargaftik, G. J. Mazo, L. G. Kuz'mina and Y. T. Struchkov, *J. Chem. Soc., Chem. Commun.*, 1978, 27.
56. T. A. Stromnova, M. N. Vargaftik and I. I. Moiseev, *J. Organomet. Chem.*, 1983, **252**, 113.
57. C. Mealli and D. M. Proserpio, *J. Am. Chem. Soc.*, 1990, **112**, 5484.
58. T. A. Stromnova, I. N. Busygina, N. Y. Tikhonova and I. I. Moiseev, *Mendeleev Commun.*, 1991, **1991**, 58 (*Chem. Abstr.*, **115**, 208 194s).
59. T. A. Stromnova, M. N. Vargaftik and I. I. Moiseev, *Izv. Akad. Nauk SSSR, Ser. Khim.*, 1983, **1983**, 31 (*Chem. Abstr.*, **98**, 9638f).
60. (a) I. P. Stolyarov, *et al.*, *Izv. Akad. Nauk SSSR, Ser. Khim.*, 1986, **1986**, 942 (*Chem. Abstr.*, **105**, 107 227t); (b) T. A. Stromnova and M. N. Vargaftik, *ibid.*, 1980, **1980**, 1411 (*Chem. Abstr.*, **93**, 106 186a); (c) T. A. Stromnova, V. S. Sergienko, A. V. Kisin, M. A. Porai-Koshits and I. I. Moiseev, *ibid.*, 1987, **1987**, 894 (*Chem. Abstr.*, **108**, 186 946q).
61. (a) I. P. Stolyarov, E. V. Evdokimova and I. I. Moiseev, *Koord. Khim.*, 1989, **15**, 1545 (*Chem. Abstr.*, **112**, 170 940e); (b) M. N. Vargaftik, T. A. Stromnova, T. S. Khodashova, M. A. Porai-Koshits and I. I. Moiseev, *ibid.*, 1981, **7**, 132 (*Chem. Abstr.*, **94**, 113 637x).
62. A. D. Burrows, J. C. Machell, D. M. P. Mingos and H. R. Powell, *J. Chem. Soc., Dalton Trans.*, 1992, 1521.
63. H. F. Klein, M. Mager, U. Floerke and H. J. Haupt, *Organometallics*, 1992, **11**, 3915.
64. (a) E. G. Mednikov, N. K. Eremenko, Y. L. Slovokhotov and Y. T. Struchkov, *Metalloorg. Khim.*, 1989, **2**, 1289 (*Chem. Abstr.*, **112**, 244 837a); (b) E. G. Mednikov, E. A. Grebenshchikova and N. K. Eremenko, *Izv. Akad. Nauk SSSR, Ser. Khim.*, 1986, **1986**, 2155 (*Chem. Abstr.*, **108**, 22 026p).
65. R. Goddard, P. W. Jolly, C. Krueger, K. P. Schick and G. Wilke, *Organometallics*, 1982, **1**, 1709.
66. M. Bochmann, I. Hawkins, M. B. Hursthouse and R. L. Short, *Polyhedron*, 1987, **6**, 1987.
67. E. G. Mednikov, N. K. Eremenko and G. A. Khivintseva, *Koord. Khim.*, 1987, **13**, 106 (*Chem. Abstr.*, **106**, 206 628b).
68. (a) E. G. Mednikov, N. K. Eremenko and S. P. Gubin, *Koord. Khim.*, 1984, **10**, 711. (*Chem. Abstr.*, **102**, 142 160a); (b) E. G. Mednikov and N. K. Eremenko, *Izv. Akad. Nauk SSSR, Ser. Khim.*, 1983, **9**, 243 (*Chem. Abstr.*, **98**, 171 880q); (c) E. G. Mednikov, N. K. Eremenko, S. P. Gubin, Y. L. Slovokhotov, Y. T. Struchkov and V. A. Mikhailov, *Dokl. Akad. Nauk SSSR*, 1981, **259**, 121 (*Chem. Abstr.*, **95**, 161 184q); (d) V. V. Bashilov, A. A. Musaev, V. I. Sokolov and O. A. Reutov, *ibid.*, 1987, **294**, 368 (*Chem. Abstr.*, **107**, 125 862s; (e) E. G. Mednikov and N. K. Eremenko, *Izv. Akad. Nauk SSSR, Ser. Khim.*, 1984, **1984**, 2781 (*Chem. Abstr.*, **102**, 178 045f); (f) E. G. Mednikov and N. K. Eremenko, *ibid.*, 1982, **1982**, 2540 (*Chem. Abstr.*, **98**, 89 636f).
69. E. G. Mednikov, N. K. Eremenko, V. A. Mikhailov, S. P. Gubin, Y. L. Slovokhotov and Y. T. Struchkov, *J. Chem. Soc., Chem. Commun.*, 1981, 989.
70. E. G. Mednikov, N. K. Eremenko, Y. L. Slovokhotov, Y. T. Struchkov and S. P. Gubin, *J. Organomet. Chem.*, 1983, **258**, 247.
71. E. G. Mednikov and N. K. Eremenko, *Koord. Khim.*, 1986, **12**, 802 (*Chem. Abstr.*, **105**, 144 957b).
72. E. G. Mednikov, N. K. Eremenko, Y. L. Slovokhotov and S. P. Gubin, *J. Organomet. Chem.*, 1980, **202**, C102.
73. (a) E. G. Mednikov, *Metalloorg. Khim.*, 1991, **4**, 1237 (*Chem. Abstr.*, **116**, 50 346b); (b) E. G. Mednikov, Y. L. Slovokhotov and Y. T. Struchkov, *ibid.*, 1991, **4**, 123 (*Chem. Abstr.*, **115**, 29 589c).
74. (a) E. G. Mednikov, N. K. Eremenko, Y. L. Slovokhotov and Y. T. Struchkov, *J. Organomet. Chem.*, 1986, **301**, C35; (b) E. G. Mednikov, *Metalloorg. Khim.*, 1991, **4**, 885 (*Chem. Abstr.*, **115**, 208 230a); (c) E. G. Mednikov, N. K. Eremenko, Y. L. Slovokhotov and Y. T. Struchkov, *Zh. Vses. Khim. Ova.*, 1987, **32**, 101 (*Chem. Abstr.*, **107**, 248 869g).
75. (a) R. Bender, P. Braunstein, Y. Dusausoy and J. Protas, *Angew. Chem.*, 1978, **90**, 637; (b) R. Bender, P. Braunstein, J. M. Jud and Y. Dusausoy, *Inorg. Chem.*, 1983, **22**, 3394.
76. P. Braunstein, R. Bender and J. Jud, *Inorg. Synth.*, 1989, **26**, 341.
77. H. Abicht, R. Barth, K. Peters, E. M. Peters and H. G. von Schnering, *Z. Chem.*, 1986, **26**, 409.
78. (a) R. Jund, P. Lemoine, M. Gross, R. Bender and P. Braunstein, *J. Chem. Soc., Dalton Trans.*, 1985, 711; (b) R. Jund, P. Lemoine, M. Gross, R. Bender and P. Braunstein, *J. Chem. Soc., Chem. Commun.*, 1983, 86.
79. C. U. J. Pittman, W. Honnick, M. Absi-Halabi, M. G. Richmond, R. Bender and P. Braunstein, *J. Mol. Catal.*, 1985, **32**, 177.
80. (a) P. Braunstein, R. Bender and J. Kervennal, *Organometallics*, 1982, **1**, 1236; (b) P. Braunstein *et al.*, *Angew. Chem.*, 1988, **100**, 972.
81. H. Werner, P. Thometzek, C. Krueger and H. J. Kraus, *Chem. Ber.*, 1986, **119**, 2777.
82. H. Werner, H. J. Kraus and P. Thometzek, *Chem. Ber.*, 1982, **115**, 2914.
83. H. Werner and P. Thometzek, *Angew. Chem.*, 1982, **94**, 707.
84. P. Thometzek and H. Werner, *J. Organomet. Chem.*, 1983, **252**, C29.
85. (a) P. Braunstein, C. De Meric de Bellefon, M. Ries, J. Fischer, S. E. Bouaoud and D. Grandjean, *Inorg. Chem.*, 1988, **27**, 1327. (b) P. Braunstein, C. De Meric de Bellefon, M. Ries, J. Fischer, S. E. Bouaoud and D. Grandjean, *Inorg. Chem.*, 1989, **28**, 1422.
86. P. Braunstein, J. M. Jud, Y. Dusausoy and J. Fischer, *Organometallics*, 1983, **2**, 180.
87. (a) P. Braunstein, J. M. Jud and J. Fischer, *J. Chem. Soc., Chem. Commun.*, 1983, 5; (b) P. Braunstein, C. De Meric de Bellefon, M. Ries and J. Fischer, *Organometallics*, 1988, **7**, 332.
88. (a) P. Braunstein, C. De Meric de Bellefon and M. Ries, *J. Organomet. Chem.*, 1984, **262**, C14; (b) P. Braunstein, C. De Meric de Bellefon and M. Ries, *Inorg. Chem.*, 1990, **29**, 1181.
89. (a) P. Braunstein, M. Ries, C. De Meric de Bellefon, Y. Dusausoy and J. P. Mangeot, *J. Organomet. Chem.*, 1988, **355**, 533; (b) P. Braunstein, C. De Meric de Bellefon and M. Ries, *Inorg. Chem.*, 1988, **27**, 1338.

90. P. Braunstein, C. De Meric de Bellefon and B. Oswald, *Inorg. Chem.*, 1993, **32**, 1649.
91. P. Braunstein, B. Oswald, A. DeCian and J. Fischer, *J. Chem. Soc., Dalton Trans.*, 1991, 2685.
92. P. Braunstein, C. De Meric de Bellefon, B. Oswald, M. Ries, M. Lanfranchi and A. Tiripicchio, *Inorg. Chem.*, 1993, **32**, 1638.
93. (a) T. A. Stromnova, I. N. Busygina, S. B. Katser, A. S. Antsyshkina, M. A. Porai-Koshits and I. I. Moiseev, *J. Chem. Soc., Chem. Commun.*, 1988, 114; (b) T. A. Stromnova, I. N. Busygina, S. B. Katser, A. S. Antsyshkina, M. A. Porai-Koshits and I. I. Moiseev, *Dokl. Akad. Nauk SSSR*, 1987, **295**, 890 (*Chem. Abstr.*, **109**, 190 534s); (c) T. A. Stromnova, I. N. Busygina, S. B. Katser, A. S. Antsyshkina, M. A. Porai-Koshits and I. I. Moiseev, *Izv. Akad. Nauk SSSR, Ser. Khim.*, 1987, **1987**, 1435 (*Chem. Abstr.*, **108**, 149 863q).
94. P. Braunstein, B. Oswald, A. Tiripicchio and M. Tiripicchio-Camellini, *Angew. Chem.*, 1990, **102**, 1206.
95. E. G. Mednikov, V. V. Bashilov, V. I. Sokolov, Y. L. Slovokhotov and Y. T. Struchkov, *Polyhedron*, 1983, **2**, 141.
96. E. G. Mednikov, N. K. Eremenko, V. V. Bashilov and V. I. Sokolov, *Inorg. Chim. Acta*, 1983, **76**, L31.
97. (a) V. V. Bashilov, V. I. Sokolov, Y. L. Slovokhotov, Y. T. Struchkov, E. G. Mednikov and N. K. Eremenko, *J. Organomet. Chem.*, 1987, **327**, 285; (b) V. V. Bashilov, V. I. Sokolov, Y. L. Slovokhotov, Y. T. Struchkov, E. G. Mednikov and N. K. Eremenko, *Dokl. Akad. Nauk SSSR*, 1987, **292**, 863 (*Chem. Abstr.*, **106**, 197 841x).
98. (a) A. Yamamoto *et al.*, *Pure Appl. Chem.*, 1991, **63**, 687; (b) L. Huang, F. Ozawa and A. Yamamoto, *Organometallics*, 1990, **9**, 2603.
99. (a) L. Huang, F. Ozawa, K. Osakada and A. Yamamoto, *J. Organomet. Chem.*, 1990, **383**, 587; (b) L. Huang, F. Ozawa, K. Osakada and A. Yamamoto, *Organometallics*, 1989, **8**, 2065; (c) L. Huang, F. Ozawa and A. Yamamoto, *Organometallics*, 1990, **9**, 2603.
100. (a) N. Koga and K. Morokuma, *J. Am. Chem. Soc.*, 1986, **108**, 6136; (b) N. Koga and K. Morokuma, *ibid*, 1985, **107**, 7230.
101. N. Koga and K. Morokuma, *Chem. Phys. Lett.*, 1990, **172**, 243.
102. A. Dedieu, S. Sakaki, A. Strich and P. E. M. Siegbahn, *Chem. Phys. Lett.*, 1987, **133**, 317.
103. N. Koga and K. Morokuma, *New J. Chem.*, 1991, **15**, 749.
104. G. Pacchioni, P. Fantucci, J. Koutecky and V. Ponec, *J. Catal.*, 1988, **112**, 34.
105. G. P. C. M. Dekker, C. J. Elsevier, K. Vrieze and P. W. N. M. Van Leeuwen, *Organometallics*, 1992, **11**, 1598.
106. F. Ozawa, T. Son, K. Osakada and A. Yamamoto, *J. Chem. Soc., Chem. Commun.*, 1989, 1067.
107. C. G. Francis, S. I. Khan and P. R. Morton, *Inorg. Chem.*, 1984, **23**, 3680.
108. A. Christofides, *J. Organomet. Chem.*, 1983, **259**, 355.
109. Y. Yamamoto and H. Yamazaki, *J. Chem. Soc., Dalton Trans.*, 1989, 2161.
110. A. D. Burrows, H. Fleischer and D. M. P. Mingos, *J. Organomet. Chem.*, 1992, **433**, 311.
111. M. Rashidi, E. Kristof, J. J. Vittal and R. J. Puddephatt, *Inorg. Chem.*, 1994, **33**, 1497.
112. M. Rashidi, J. J. Vittal and R. J. Puddephatt, *J. Chem. Soc., Dalton Trans.*, 1994, 1283.
113. M. F. Rettig and P. M. Maitlis, *Inorg. Synth.*, 1990, **28**, 110.
114. (a) Y. Yamamoto, K. Takahashi and H. Yamazaki, *Bull. Chem. Soc. Jpn.*, 1987, **60**, 2665; (b) Y. Yamamoto and H. Yamazaki, *ibid.*, 1985, **58**, 1843.
115. N. M. Rutherford, M. M. Olmstead and A. L. Balch, *Inorg. Chem.*, 1984, **23**, 2833.
116. C. Sourisseau, J. Hervieu and M. C. Vitorge, *Spectrochim. Acta, Part A*, 1980, **36**, 153.
117. Y. Yamamoto and H. Yamazaki, *Inorg. Chem.*, 1986, **25**, 3327.
118. T. D. Miller, M. A. St. Clair, M. K. Reinking and C. P. Kubiak, *Organometallics*, 1983, **2**, 767.
119. (a) P. A. Metcalf and C. P. Kubiak, *J. Am. Chem. Soc.*, 1986, **108**, 4682; (b) M. K. Reinking, M. L. Kullberg, A. R. Cutler and C. P. Kubiak, *ibid.*, 1985, **107**, 3517.
120. B. Corain, M. Basato, P. Main, M. Zecca, R. De Pieri and R. Arshady, *J. Organomet. Chem.*, 1987, **326**, C43.
121. W. Weigand, U. Nagel and W. Beck, *Chem. Ber.*, 1990, **123**, 439.
122. T. Kaharu and S. Takahashi, *Chem. Lett.*, 1992, **1992**, 1515.
123. R. Bertani, C. B. Castellani and B. Crociani, *J. Organomet. Chem.*, 1984, **269**, C15.
124. R. Uson, J. Fornies, P. Espinet, L. Pueyo and E. Lalinde, *J. Organomet. Chem.*, 1986, **299**, 251.
125. A. Mantovani and B. Crociani, *J. Organomet. Chem.*, 1982, **236**, C37.
126. B. Crociani, F. Di Bianca, R. Bertani and C. B. Castellani, *Inorg. Chim. Acta*, 1985, **101**, 161.
127. (a) A. Campagnaro, A. Mantovani and P. Uguagliati, *Inorg. Chim. Acta*, 1985, **99**, L15; (b) A. Mantovani, *Monatsh. Chem.*, 1983, **114**, 1045.
128. B. Crociani, F. Di Bianca, A. Giovenco and A. Scrivanti, *J. Organomet. Chem.*, 1984, **269**, 295.
129. R. Bertani, A. Berton, F. Di Bianca and B. Crociani, *J. Organomet. Chem.*, 1986, **303**, 283.
130. A. Mantovani, L. Calligaro and A. Pasquetto, *Inorg. Chim. Acta*, 1983, **76**, L145.
131. W. Henderson, R. D. W. Kemmitt, L. J. S. Prouse and D. R. Russell, *J. Chem. Soc., Dalton Trans.*, 1990, 781.
132. R. Uson, J. Fornies, P. Espinet, E. Lalinde, P. G. Jones and G. M. Sheldrick, *J. Chem. Soc., Dalton Trans.*, 1982, 2389.
133. R. Uson, J. Fornies, P. Espinet and E. Lalinde, *J. Organomet. Chem.*, 1983, **254**, 371.
134. (a) R. Uson, J. Fornies, P. Espinet, E. Lalinde, P. G. Jones and G. M. Sheldrick, *J. Organomet. Chem.*, 1985, **288**, 249; (b) R. Uson, J. Fornies, P. Espinet, E. Lalinde, P. G. Jones and G. M. Sheldrick, *ibid.*, 1983, **253**, C47.
135. C. L. Lee, C. T. Hunt and A. L. Balch, *Organometallics*, 1982, **1**, 824.
136. Y. Yamamoto and H. Yamazaki, *Inorg. Chim. Acta*, 1980, **41**, 229.
137. (a) J. Dupont, M. Pfeffer, J. C. Daran and Y. Jeannin, *Organometallics*, 1987, **6**, 899; (b) J. Dupont and M. Pfeffer, *J. Chem. Soc., Dalton Trans.*, 1990, 3193.
138. P. Veya, C. Floriani, A. Chiesi-Villa and C. Guastini, *Organometallics*, 1993, **12**, 253.
139. (a) H. Ogawa, T. Joh and S. Takahashi, *J. Chem. Soc., Chem. Commun.*, 1988, 561; (b) K. Onitsuka, H. Ogawa, T. Joh, S. Takahashi, Y. Yamamoto and H. Yamazaki, *J. Chem. Soc., Dalton Trans.*, 1991, 1531.
140. K. Onitsuka, T. Joh and S. Takahashi, *Angew. Chem.*, 1992, **104**, 893.
141. A. Albinati, P. S. Pregosin and R. Ruedi, *Helv. Chim. Acta*, 1985, **68**, 2046.
142. (a) P. L. Sandrini, A. Mantovani and B. Crociani, *J. Organomet. Chem.*, 1980, **185**, C13; (b) B. Crociani, A. Mantovani and A. Scrivanti, *ibid.*, 1982, **233**, 387.
143. T. Tanase, T. Ohizumi, K. Kobayashi and Y. Yamamoto, *J. Chem. Soc., Chem. Commun.*, 1992, 707.
144. R. Uson, J. Fornies, P. Espinet and E. Lalinde, *J. Organomet. Chem.*, 1987, **334**, 399.

145. (a) R. Uson, J. Fornies, P. Espinet, F. Martinez, C. Fortuno and B. Menjon, *J. Organomet. Chem.*, 1983, **256**, 365; (b) M. L. Kullberg and C. P. Kubiak, *Inorg. Chem.*, 1986, **25**, 26.
146. R. Uson, J. Fornies, P. Espinet and C. Fortuno, *Inorg. Chim. Acta*, 1984, **87**, 207.
147. R. Uson, J. Fornies, R. Navarro, M. Tomas, C. Fortuno, J. I. Cebollada and A. J. Welch, *Polyhedron*, 1989, **8**, 1045.
148. J. Fornies, F. Martinez, R. Navarro, A. Redondo, M. Tomas and A. J. Welch, *J. Organomet. Chem.*, 1986, **316**, 351.
149. P. Brant, L. S. Benner and A. L. Balch, *Inorg. Chem.*, 1979, **18**, 3422.
150. Y. Kitano and T. Hori, *Acta Crystallogr., Sect. B*, 1981, **37**, 1919.
151. M. M. Olmstead, C. L. Lee and A. L. Balch, *Inorg. Chem.*, 1982, **21**, 2712.
152. M. A. Khan and A. J. McAlees, *Inorg. Chim. Acta*, 1985, **104**, 109.
153. B. Messbauer, H. Meyer, B. Walther, M. J. Heeg, A. F. M. Rahman and J. P. Oliver, *Inorg. Chem.*, 1983, **22**, 272.
154. G. Facchin, R. Bertani, L. Zanotto, M. Calligaris and G. Nardin, *J. Organomet. Chem.*, 1989, **366**, 409.
155. T. Tanase, T. Nomura, Y. Yamamoto and K. Kobayashi, *J. Organomet. Chem.*, 1991, **410**, C25.
156. S. Lo Schiavo, E. Rotondo, G. Bruno and F. Faraone, *Organometallics*, 1991, **10**, 1613.
157. A. W. Hanson, A. J. McAlees and A. Taylor, *J. Chem. Soc., Perkin Trans. 1*, 1985, 441.
158. M. Drouin, D. Perreault, P. D. Harvey and A. Michel, *Acta Crystallogr., Sect. C: Cryst. Struct. Commun.*, 1991, **C47**, 752.
159. N. G. Connelly, J. G. Crossley and A. G. Orpen, *J. Chem. Soc., Chem. Commun.*, 1992, 1568.
160. K. H. Datz, H. Fischer, P. Hoffmann, F. R. Freissl, U. Schubert and K. Weiss, in 'Transition Metal Carbene Complexes', Verlag Chemie, Weinheim, 1983.
161. (a) U. Belluco, B. Crociani, R. Michelin and P. Uguagliati, *Pure Appl. Chem.*, 1983, **55**, 47; (b) U. Belluco, R. A. Michelin, P. Uguagliati and B. Crociani, *J. Orgmet. Chem.*, 1983, **250**, 565.
162. K. Hiraki, Y. Fuchita and T. Masumoto, *Bull. Chem. Soc. Jpn.*, 1980, **53**, 1171.
163. R. Bertani, M. Mozzon, F. Benetollo, G. Bombieri and R. A. Michelin, *J. Chem. Soc., Dalton Trans.*, 1990, 1197.
164. R. Uson, J. Fornies, P. Espinet, A. Garcia and A. Sanau, *Transition Met. Chem. (Weinheim, Ger.)*, 1983, **8**, 11.
165. A. Scrivanti, G. Carturan and B. Crociani, *Organometallics*, 1983, **2**, 1612.
166. W. P. Fehlhammer, T. Bliss, J. Fuchs and G. Holzmann, *Z. Naturforsch., Teil B*, 1992, **47**, 79.
167. R. Uson, J. Fornies, P. Espinet, R. Navarro and E. Lalinde, *Transition Met. Chem. (Weinheim, Ger.)* 1984, **9**, 277.
168. R. Bertani, M. Mozzon, R. A. Michelin, F. Benetollo, G. Bombieri, T. J. Castilho and A. J. L. Pombeiro, *Inorg. Chim. Acta*, 1991, **189**, 175.
169. U. Belluco *et al.*, *Inorg. Chim. Acta*, 1992, **198–200**, 883.
170. W. P. Fehlhammer, K. Bartel, U. Plaia, A. Voelkl and A. T. Liu, *Chem. Ber.*, 1985, **118**, 2235.
171. R. A. Michelin, L. Zanotto, D. Braga, P. Sabatino and R. J. Angelici, *Inorg. Chem.*, 1988, **27**, 93.
172. Y. Fuchita, K. Hidaka, S. Morinaga and K. Hiraki, *Bull. Chem. Soc. Jpn.*, 1981, **54**, 800.
173. (a) R. A. Michelin, M. Mozzon, M. Zecca, B. Corain, O. Piazzi and G. Zanotto, *Inorg. Chim. Acta*, 1990, **174**, 3; (b) R. Bertani, M. Mozzon and R. A. Michelin, *Inorg. Chem.*, 1988, **27**, 2809; (c) R. Bertani, M. Mozzon, L. Zanotto and R. A. Michelin, *NATO ASI Ser., Ser. C, 269 (Adv. Met. Carben)* 1989, 67 (*Chem. Abstr.*, **112**, 15 863b).
174. (a) K. Hiraki, K. Sugino and M. Onishi, *Bull. Chem. Soc. Jpn.*, 1980, **53**, 1976; (b) K. Hiraki and K. Sugino, *J. Organomet. Chem.*, 1980, **201**, 469.
175. W. P. Fehlhammer, D. Achatz, U. Plaia and A. Voelkl, *Z. Naturforsch., Teil B*, 1987, **42**, 720.
176. (a) M. Wada, Y. Koyama and K. Sameshima, *J. Organomet. Chem.*, 1981, **209**, 115; (b) M. Wada and Y. Koyama, *ibid.*, 1980, **201**, 477.
177. D. Sellmann, W. Prechtel, F. Knoch and M. Moll, *Inorg. Chem.*, 1993, **32**, 538.
178. P. E. M. Siegbahn, *Chem. Phys. Lett.*, 1993, **201**, 15.
179. (a) T. A. Stromnova, I. N. Busygina, D. I. Kochubei and I. I. Moiseev, *J. Organomet. Chem.*, 1991, **417**, 193; (b) I. N. Busygina, T. A. Stromnova, D. I. Kochubei, M. Y. Botnikov and I. I. Moiseev, *Metalloorg. Khim.*, 1991, **4**, 69. (*Chem. Abstr.*, **114**, 207 479f); (c) T. A. Stromnova, I. N. Busygina, D. I. Kochubei and I. I. Moiseev, *Mendeleev Commun.*, 1991, 1 (*Chem. Abstr.*, **115**, 71 871p).
180. M. Berry, J. A. K. Howard and F. G. A. Stone, *J. Chem. Soc., Dalton Trans.*, 1980, 1601.
181. T. V. Ashworth, M. J. Chetcuti, J. A. K. Howard, F. G. A. Stone, S. J. Wisbey and P. Woodward, *J. Chem. Soc., Dalton Trans.*, 1981, 775.
182. S. I. Al-Resayes, P. B. Hitchcock and J. F. Nixon, *J. Chem. Soc., Chem. Commun.*, 1986, 1710.

5
Palladium–Carbon σ-Bonded Complexes

ALLAN J. CANTY
University of Tasmania, Hobart, Tas, Australia

5.1 INTRODUCTION 226

5.2 PALLADIUM(I) COMPLEXES 226

5.3 PALLADIUM(II) YLIDES 227
- *5.3.1 Complexes of Phosphorus Ylides* 227
- *5.3.2 Complexes of Nitrogen, Arsenic and Sulfur Ylides* 229
- *5.3.3 Structures* 230

5.4 PALLADIUM ALKYNYL COMPLEXES 230
- *5.4.1 Synthetic Methods* 230
- *5.4.2 Reactions and Structures* 231

5.5 PALLADIUM(II) ALKYL, ARYL AND RELATED COMPLEXES 233
- *5.5.1 Formation of Pd–C σ-bonds* 233
  - *5.5.1.1 From palladium metal or palladium(0) complexes* 233
  - *5.5.1.2 From palladium(II) complexes and organometallic reagents* 238
  - *5.5.1.3 From cyclopalladation reactions* 242
  - *5.5.1.4 From palladium(II) complexes and alkenes* 248
  - *5.5.1.5 From palladium complexes and alkynes* 251
  - *5.5.1.6 From palladium complexes and carbon monoxide* 255
  - *5.5.1.7 From palladium complexes and isocyanides* 257
  - *5.5.1.8 From other synthetic methods* 259
- *5.5.2 Reactions of Organopalladium(II) Complexes* 261
  - *5.5.2.1 Metathesis and related reactions* 261
  - *5.5.2.2 Isomerization and decomposition of organopalladium(II) complexes* 265
  - *5.5.2.3 Reactions with acids, including alcohols, thiols, amines, peracids and peroxides* 267
  - *5.5.2.4 Reactions with halogens and organohalides* 268
  - *5.5.2.5 Other reactions* 270
- *5.5.3 Structures* 271

5.6 RADICAL COMPLEXES 272

5.7 PALLADIUM(IV) COMPLEXES 272
- *5.7.1 Complexes of Bidentate Nitrogen Donor Ligands and Intramolecular Coordination Complexes* 273
- *5.7.2 Complexes of Tripodal Nitrogen or Sulfur Donor Ligands* 275
- *5.7.3 Structures and Reactions of Complexes* 276

5.8 REFERENCES 276

5.1 INTRODUCTION

The chemistry of organopalladium complexes is classified here according to the formal oxidation state of palladium, and for oxidation state +2 the chemistry of ylide and alkynyl complexes is treated separately from other complexes containing Pd–C σ-bonds. Reference to the corresponding chapter in *COMC-I*[1] and to the monograph by Maitlis[2] is assumed for all sections, and selected references to the early literature are also given. Examples of most classes of complex and reaction system are discussed, and both the topics covered and references to the original literature appearing from 1981 to early 1993 are intended to be comprehensive.

5.2 PALLADIUM(I) COMPLEXES

The development of hydrocarbylpalladium(I) chemistry commenced with the report that $[Al_2Me_6]$ reacts with $[PdCl(\mu\text{-dppm})]_2$ at −78 °C to give a solution containing the palladium(I) species (**1**). Complex (**1**) disproportionates at ~10 °C to form the palladium(II) complex (**2**) and a palladium(0) species, and may be protonated by ethanol to form the palladium(II) complex (**3**) (Scheme 1).[3] Reaction of $[PdCl(\mu\text{-dppm})]_2$ with excess alkylating agent results in the formation of the dimethyl analogue of (**1**), $[Pd(Me)(\mu\text{-dppm})]_2$, which reacts with ethanol at −40 °C to form the hydrido-bridged complex $[(PdMe)_2(\mu\text{-dppm})_2(\mu\text{-H})]^+$.

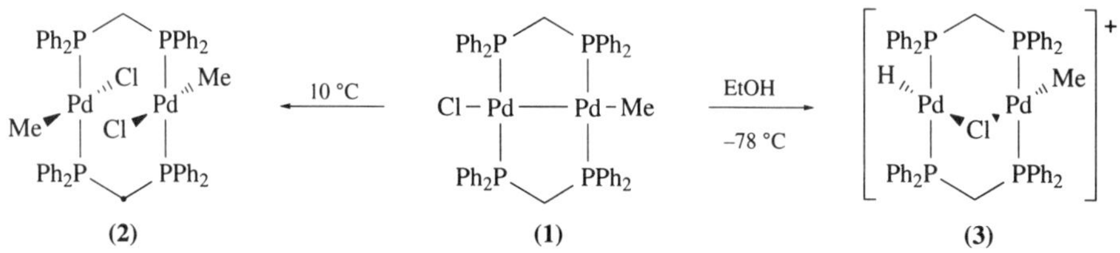

Scheme 1

In contrast to the low stability of methylpalladium(I) species, an extensive series of stable metal–metal-bonded complexes containing pentafluorophenyl[4–9] or pentachlorophenyl[4,10] groups bonded to palladium(I) has been obtained on the reaction of metal(0) reagents with organometal(II) complexes of η^1-diphosphines, as illustrated in Scheme 2. For closely related reactions forming the mixed metal species $[XPt(\mu\text{-dppm})_2Pd(C_6F_5)]$ (X = Cl, Br, C_6F_5)[8] and $[(C_6F_5)Pt(\mu\text{-dppm})_2PdX]$ (X = Cl, Br),[9] employing platinum(0) and palladium(0) reagents, respectively, a preference occurs for halogen over C_6F_5 transfer from *trans*-$[MX(C_6F_5)(\eta^1\text{-dppm})_2]$ to the metal(0) reagent.

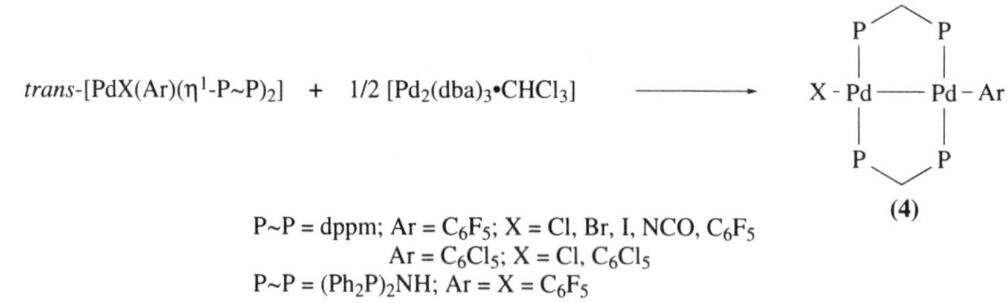

Scheme 2

X-ray structural studies have been reported for $[\{Pd(C_6Cl_5)\}_2(\mu\text{-dppm})_2]$[10] and $[ClPt(\mu\text{-dppm})_2Pd(C_6F_5)]$,[8] showing approximately square planar coordination with Pd–Pd 0.267 04(21) nm, Pd–Pt 0.264 3(1) nm, and Pd–C distances of 0.212 7(12) nm and 0.207 6(14) nm, respectively.

Insertion chemistry has been developed for electrophiles such as ArN_2^+, SO_2, CO, RNC and dimethyl acetylenedicarboxylate (DMAD). Addition of aryldiazonium salts to $[\{Pd(C_6F_5)\}_2\{\mu\text{-}(Ph_2P)_2NH\}_2]$, $[XPt(\mu\text{-dppm})_2Pd(C_6F_5)]$ (X = Cl, C_6F_5), $[(C_6F_5)Pt(\mu\text{-dppm})_2PdCl]$ or $[ClPd(\mu\text{-dppm})_2Pd(C_6F_5)]$ gives cations of structure (**5**).[7–10] Insertion of SO_2 to form $[ClPd(\mu\text{-dppm})_2(SO_2\text{-}\mu\text{-}S)Pd(C_6Cl_5)]$ is reversible,[10] but $[Pd(C_6Cl_5)(\mu\text{-dppm})]_2$ does not react with ArN_2^+, SO_2 or CO. Carbon monoxide also does not react with $[\{Pd(C_6F_5)\}_2\{\mu\text{-}(Ph_2P)_2NH\}_2]$ or other dimers that contain a C_6F_5 group at each metal centre, but unstable complexes are detectable by IR spectroscopy on reaction with [Pd–Pd] or [Pt–Pd] dimers containing only one C_6F_5 group bonded to a palladium centre.[8] DMAD inserts into Pt–Pd bonds of dppm-bridged complexes to give (**6**),[8,9] and isocyanides form a complex chemistry based on a range of complexes of general formulation (**7**)–(**9**);[5,6,8–10] for example, addition of excess RNC favours the formation of (**9**), and reversible conversion between (**7**) and (**8**) is possible for some complexes. Complexes involving diisocyanide bridging between two [Pd–Pd] centres have also been described.[6,10]

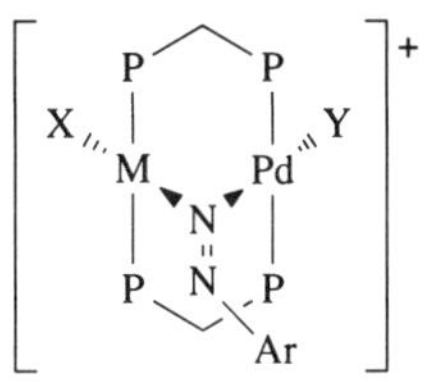

(**5**)

M = Pd; X = Y = C_6F_5; X = Cl, Y = C_6Cl_5
M = Pt; X = C_6F_5, Y = C_6F_5 or Cl
X = Cl, Y = C_6F_5

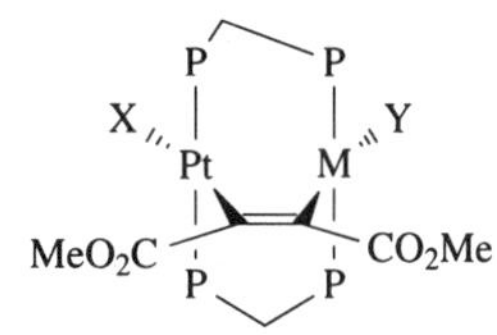

(**6**)

X = Cl, Y = C_6F_5
X = C_6F_5, Y = Cl
X = Y = C_6F_5

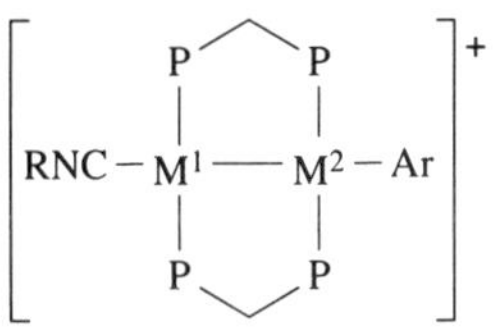

(**7**)

$M^1 = M^2$ = Pd
M^1 = Pd, M^2 = Pt
M^1 = Pt, M^2 = Pd

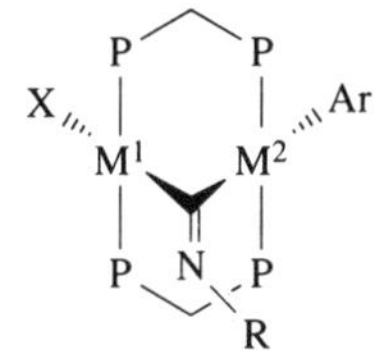

(**8**)

$M^1 = M^2$ = Pd; M^1 = Pd, M^2 = Pt
M^1 = Pt, M^2 = Pd; X = Cl or Ar

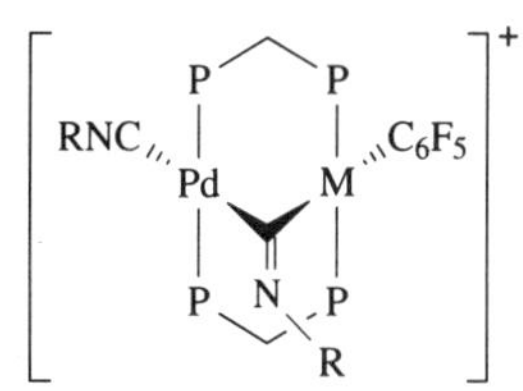

(**9**) M = Pd, Pt

5.3 PALLADIUM(II) YLIDES

5.3.1 Complexes of Phosphorus Ylides

The first ylide complexes of palladium[11–14] were obtained on addition of ylides to $PdCl_2$ or related simple species such as $[PdCl_4]^{2-}$ or *trans*-$[PdCl_2(NCPh)_2]$. A recent structural study of an analogue (**10**)* of the first reported complex[11] has shown it to have a *trans*-configuration,[15] and a more recent example of a complex of this stoichiometry is $[PdCl_2\{CH_2P(NMe_2)_3\}_2]$.[16] Formal charges for ylides and their complexes, illustrated in $Bu_3P^+–C^-HCOPh$ and (**10**), have been omitted in the remaining formulae and structures. Several complexes of stoichiometry $[Pd_2Cl_4(ylide)_2]$ have been obtained by HX abstraction from phosphonium salts such as $[Ph_3PCH_2COMe]_2[Pd_2Cl_6]$ with bases, although they may also be formed from $PdCl_2$; for example, (**11**)* was obtained from a 1:1.72 ratio of $PdCl_2$ and $Bu_3PCHCOPh$ in refluxing benzene.[17] Complex (**11**) forms (**10**) on addition of excess ylide in acetonitrile. Recent work indicates that longer reaction times or use of higher temperatures in the syntheses of complexes of type (**10**) or (**11**) may result in cyclometallation reactions, illustrated by the contrast between the reaction of $PdCl_2$ with Ph_3PCHCO_2Me in 1:2 ratio in acetonitrile at ambient temperature to form $[PdCl_2\{CH(CO_2Me)PPh_3\}_2]$[18] and the reactions in acetonitrile at reflux shown in Scheme 3.[18,19]

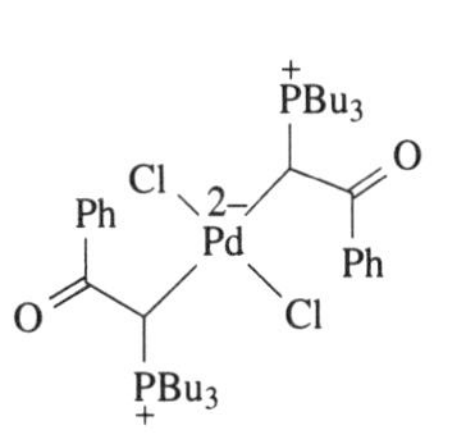

(**10**)

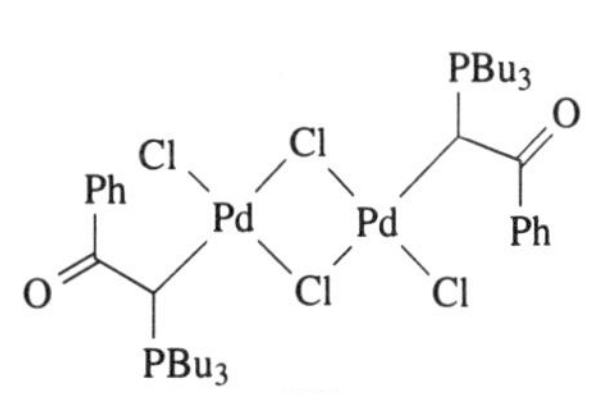

(**11**)

Complex (**13**) (R = Me) forms mononuclear species by bridge-splitting reactions,[19] and simple monomeric 1:1 complexes are obtained in the same way[20] and from direct syntheses using substrates containing strong donor ligands that are retained; for example, $[PdCl\{CH(COPh)PPh_2R\}(\eta\text{-}ONBu^t_2)]$

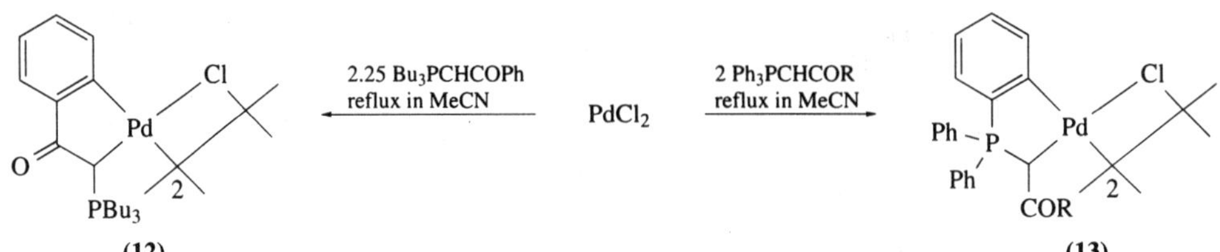

(12) **(13)**

Scheme 3

(R = Me, Ph) from $[PdCl_2(ONBu^t_2)]_2$,[21] *trans*-$[PdI\{CH_2PO(OR)_2\}(PPh_3)_2]$ (R = Me, Et) on the oxidative addition of iodophosphonates to $[Pd(PPh_3)_4]$[22] and (**14**) on quaternization occurring in the reaction of *cis*-$[PdI(CH_2I)(PPh_3)_2]$ with PEt_3.[23]

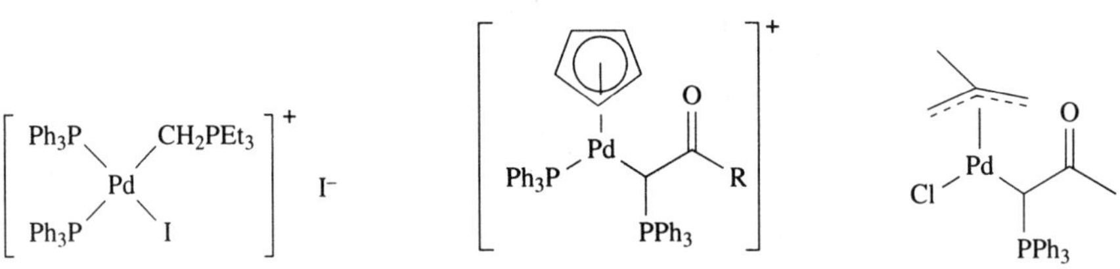

(14) **(15)** **(16)**

Organopalladium(II) complexes also react with ylides; for example, $[Pd(C_6F_5)(THT)(\mu\text{-}Cl)]_2$ (THT = tetrahydrothiophene) with $Ph_3PCHCOMe$ to form $[PdCl(C_6F_5)\{CH(COMe)PPh_3\}(THT)]$,[24] $[PdCl(Cp)(PPh_3)]$ with $Ph_3PCHCOR$ (R = Me, Ph) in the presence of $AgClO_4$ to form (**15**)[25] and $[Pd(\eta\text{-}2\text{-}MeC_3H_4)(\mu\text{-}Cl)]_2$ with $Ph_3PCHCOMe$ to form (**16**)*.[26,27] Complexes (**16**) are in equilibrium with the reagents in solution, and some of these complexes have been shown to react with $AgBF_4$ and isocyanides to give ionic complexes $[Pd(\eta\text{-}2\text{-}MeC_3H_4)\{CH(COMe)PPh_3\}(CNR)][BF_4]$.[28]

Ylides containing donor groups readily form chelate ylide complexes, as illustrated in Scheme 4 where the smaller chelate ring favours the monoylide complex,[29] and the reaction of $[PdAr_2(\mu\text{-}X)]_2^{2-}$ or $[PdAr(THT)(\mu\text{-}Cl)]_2$ with 2 equiv. $Ph_2P(CH_2)_nPPh_2CHCOPh$ to form (**19**) where the major isomer in a mixture is shown for X = Cl.[30] Simple dibromo analogues of (**19**) (X = Ar = Br; *n* = 1, 2) and related complexes have been obtained efficiently by phase transfer methods in which $Na_2[PdCl_4]$ or *trans*-$[PdCl_2(NCPh)_2]$ and $[Ph_2P(CH_2)_nPPh_2CH_2COR][Br]$ are reacted in an aqueous–organic solvent mixture containing KOH and 18-crown-6.[31] Under these conditions it appears that OH^- transfer to the organic phase results in the formation and stabilization of the ylide.[31,32] Complexes containing sulfur bis(ylides) and chelate phosphorus ylides are obtained similarly, for example, by the reaction of $[Pd\{(CH_2)_2SOMe\text{-}C,C\}(\mu\text{-}I)]_2$ to form (**20**)*; in the absence of phase transfer methods reaction for *n* = 2 ceases at the intermediate $[PdI\{(CH_2)_2SOMe\text{-}C,C\}\{PPh_2(CH_2)_2PPh_2CH_2COR\text{-}P\}][I]$.[31]

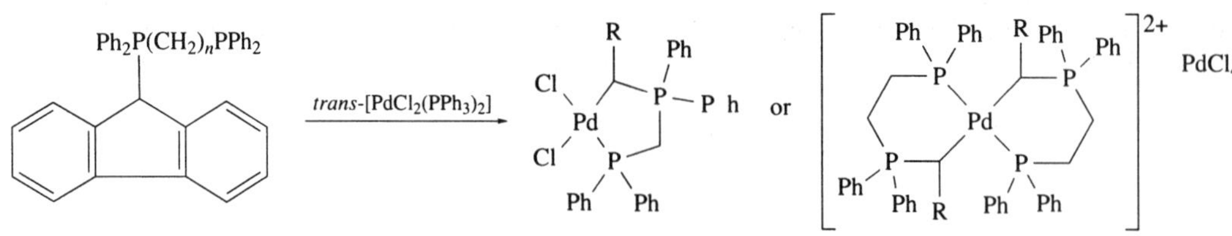

(**17**) *n* = 1 (R = $C_{12}H_8$) (**18**) *n* = 2 (R = $C_{12}H_8$)

Scheme 4

Complexes of phosphorus bis(ylides) have been extensively developed.[12–14] For example, $[PR_2(CH_2)_2]^-$ reacts with palladium(II) precursors to form bidentate (R = Me, Et, Bu^t; $R_2 = (CH_2)_4$, $(CH_2)_5$) or bridging bidentate (R = Me; $R_2 = (CH_2)_4$, $(CH_2)_5$) complexes, and $[BH_2(PMe_2CH_2)_2]^-$ forms (**21**).[33] Silver bis(ylide) reagents are useful in synthesis, as shown in Scheme 5, where bridge-splitting reactions similar to those giving (**24**) also allow the synthesis of [PdCl{bis(ylide)}(L)] (L = py, PPh_3) and, from TlX, [PdX{bis(ylide)}] (X = acac, Cp).[34]

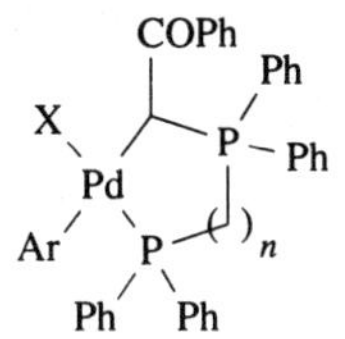

(19)

$n = 1$: Ar = X = C_6F_5, C_6Cl_5
Ar = C_6F_5 or C_6Cl_5, X = Cl
$n = 2$: Ar = X = C_6F_5; Ar = C_6Cl_5, X = Cl

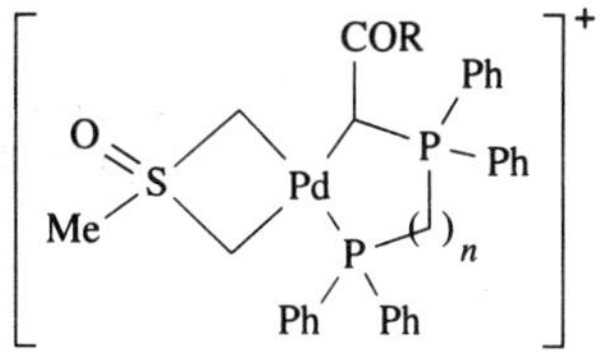

(20)

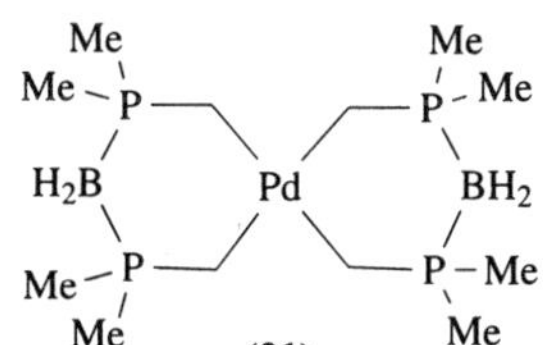

(21)

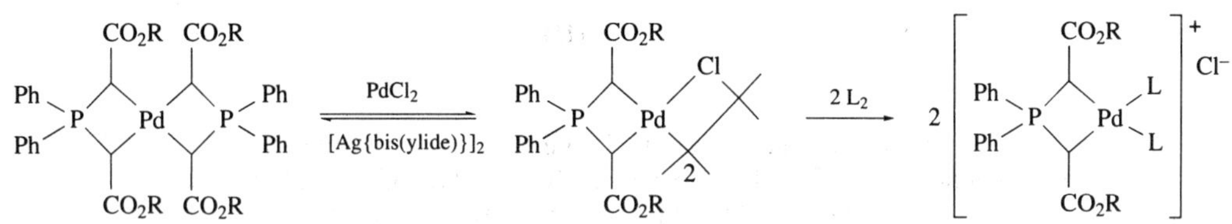

(22) (23) (24)

R = Me, Et; L_2 = 2 py, bipy, dppe

Scheme 5

5.3.2 Complexes of Nitrogen, Arsenic and Sulfur Ylides

Complexes of stoichiometry $[PdX_2(N\text{-, As- or S-ylide})_2]$, $[PdX(S\text{-ylide})(\mu\text{-}X)]_2$, $[PdX_2(\text{As- or S-ylide})(L)]$, $[Pd\{S\text{-bis(ylide)}\}(\mu\text{-}X)]_2$,[35] $[Pd\{CH_2(COPh)SMe_2\}(PPh_3)(Cp)][ClO_4]$,[25] $[PdCl(\eta\text{-ONBu}^t)(\text{As- or S-ylide})]$,[21] the arsenic analogue of the chelate ylide complex (**13**) (R = Me)[19] and chelate ylides of nitrogen[36] have been obtained by methods similar to those used for the phosphine complexes.[37] For example, reaction of the 2-pyridylcarbonylmethylide $4\text{-MeC}_5H_4NCHC(O)(2\text{-}C_5H_4N)$ with $[PdCl_2(SMe_2)_2]$ gives the complex $[PdCl_2\{4\text{-MeC}_5H_4NCHC(O)(2\text{-}C_5H_4N)\text{-}C,N\}]$.[36] The development of phase transfer methods has led to the synthesis of a series of sulfur ylide complexes in water (Scheme 6),[38] and bridge-splitting reactions for (**25**)* with sulfur, nitrogen and phosphorus donor ligands to give mononuclear complexes have been reported.[39] Similar mononuclear complexes have been obtained directly by inclusion of phosphine ligands in phase transfer syntheses,[32] for example, in the reaction of *trans*-$[PdCl_2(PPh_3)_2]$ with $[Me_3SO][I]$ to form $[PdI\{(CH_2)_2SOMe\text{-}C,C\}(PPh_3)]$, and in the presence of PPh_3 to form $[Pd\{(CH_2)_2SOMe\text{-}C,C\}(PPh_3)_2][I]$.

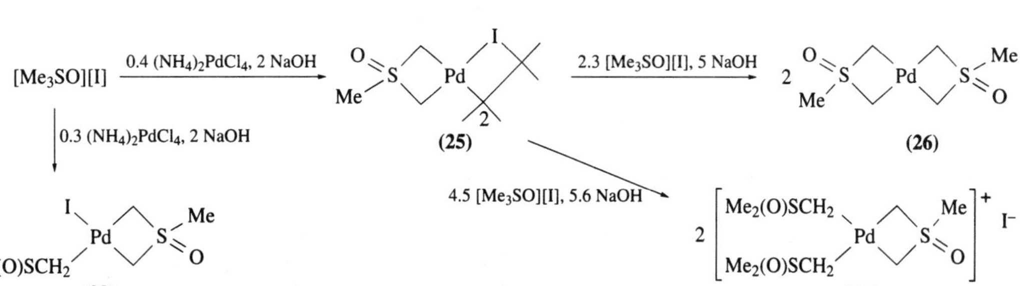

Scheme 6

5.3.3 Structures

Structures have been reported for the complexes (**10**),[15] (**11**),[17] (**16**) (R = Ph),[26] (**19**) (Ar = X = Cl, $n = 2$),[40] (**20**) ($n = 1$, R = Ph),[31] (**22**) (R = Et),[34] (**25**),[39] (**28**),[38] [PdCl{CH(TMS)PMe_2Ph}(cod)],[20] [PdCl{$(CHCO_2Et)_2PPh_2$-*C,C*}(L)] (L = py, PPh_3),[34] [PdI{$(CH_2)_2$SOMe-*C,C*}$(PPh_3)_2$], [Pd{$(CH_2)_2$SOMe-*C,C*}$(PPh_3)_2$][I][32] and [Pd{CH(COMe)PPh_3}(CNBu^t)(η-2-MeC_3H_4)][BF_4].[28] The Pd–C(ylide) distances fall within the range 0.201(1)–0.219 3(3) nm.

5.4 PALLADIUM ALKYNYL COMPLEXES

5.4.1 Synthetic Methods

Alkynylpalladium(II) complexes are synthesized by the reaction of an alkynyl anion or a covalent main group alkynyl with a palladium(II) precursor, or by oxidative addition of a terminal alkyne or a chloroalkyne to a palladium(0) precursor. Homoleptic alkynyls $[Pd(C{\equiv}CR)_4]^{2-}$ and polymeric species $[Pd(C{\equiv}C\text{-}p\text{-}C_6H_4C{\equiv}C)_2]_n^{2n-}$ may be obtained on the reaction of $[Pd(CN)_2(en)]$ with potassium alkynyls in liquid ammonia.[41,42] The anion $[Pd(C{\equiv}CBu^t)_4]^{2-}$ has also been detected as a product of addition of excess [Li(C≡CBu^t)] to *trans*-$[PdCl_2(PPh_3)_2]$ in THF.[43] However, alkynyl complexes are generally stabilized by phosphines (L); complexes of the type $[PdX(C{\equiv}CR)L_2]$ and $[Pd(C{\equiv}CR)_2L_2]$ are usually *trans*, and the same is presumably also true for the complex ion $[Pd(C{\equiv}C\text{-}p\text{-}C_6H_4C{\equiv}C)_2(PPh_3)_2]^{2-}$.[42] The presence of bidentate ligands allows the isolation of *cis*-dialkynyl complexes[44,45] and $[Pd(C_6F_5)(C{\equiv}CR)(dppe)]$ (R = Bu^t, Ph),[46] although *trans*-[Pd(C≡CPh)$_2$(η^1-dppm)$_2$] has been isolated by extrusion of $HgCl_2$ from the heteronuclear complex [{*trans*-$(PhC{\equiv}C)_2$Pd}(μ-dppm)$_2$($HgCl_2$)] on addition of sulfide.[47]

Several complexes have been obtained using $[Hg(C{\equiv}CPh)_2]$[47] or Grignard reagents,[48] the reaction of palladium(0) reagents with α-hydroxyalkynes[49] and the reaction of $[Pd(PPh_3)_4]$ with ClC≡CCl to form a phosphoniumalkynyl complex *cis*-$[Pd^-Cl_2(C{\equiv}CP^+Ph_3)(PPh_3)]$*.[50] Other strategies employing alkynylsilver(I) reagents,[46] platinum(II) reagents[51] or redistribution reactions are illustrated in Scheme 7.[52,53]

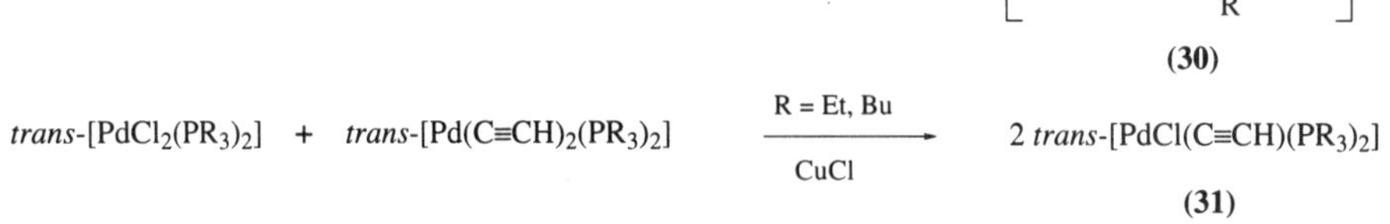

Scheme 7

The best-developed synthetic methods involve oxidative addition reactions of $[Pd(PPh_3)_4]$ or $[Pd(P{\sim}N)_2]$ with RC≡CX to form a wide range of complexes *trans*-$[PdX(C{\equiv}CR)(PPh_3)_2]$ (X = Cl, Br, I)[50,54–6] and (**32**)*,[57] respectively, or reactions of palladium(II) substrates with alkali metal alkynides,[41–3,56,58–71] with free alkynes[62] usually in a base as solvent to assist deprotonation[41] (NH_3,[44] NEt_3,[56,63] $NHEt_2$[52,64–9] or piperidine[66,67]) and often with copper(I) as a catalyst,[52,64,66,68,69] and with alkynyltin(IV) reagents.[45,70–2] Recent syntheses using alkynyllithium reagents include those of (**33**) ($R^1 = R^2$ = Me),[59] (**34**),[73] *trans*-[Pd(Ph)(C≡CBu^t)$(PEt_3)_2$],[60] carboranylalkynyl complexes[74] and the stepwise syntheses of *trans*-$[PdCl(C{\equiv}CR)(PMe_3)_2]$ and *trans*-$[Pd(C{\equiv}CR)_2(PMe_3)_2]$ from *trans*-$[PdCl_2(PMe_3)_2]$ and [Li(C≡CR)] (R = Pr, hexyl, TMS).[56] Complexes synthesized from free alkynes in basic conditions include *trans*-$[PdCl(C{\equiv}CR)(PMe_3)_2]$ (R = Pr, hexyl, TMS), *trans*-$[Pd(C{\equiv}CR)_2(PEt_3)_2]$ (R = H, Me, CH_2OH, 2-$C_6H_4C{\equiv}CH$, 4-$C_6H_4C{\equiv}CH$),[63,67] [Pd(C≡CR){$PPh_2CH_2CH_2NC$(Me)=CHCMe(O)-*P,N,O*}] (R = Bu, Ph, CO_2Me)[68] and (**35**).[75] The hydrido complex *trans*-$[PdH(C{\equiv}CH)(PCy_3)_2]$ has been detected by NMR spectroscopy on the reaction of *trans*-$[PdH(NHPh)(PCy_3)_2]$ with ethyne,[76] a reaction closely related to the free alkyne route in which the 'base' is now coordinated to palladium.

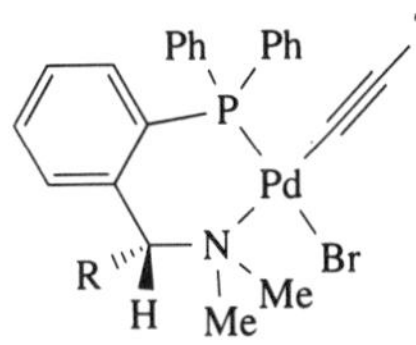

(32) R = H, Me

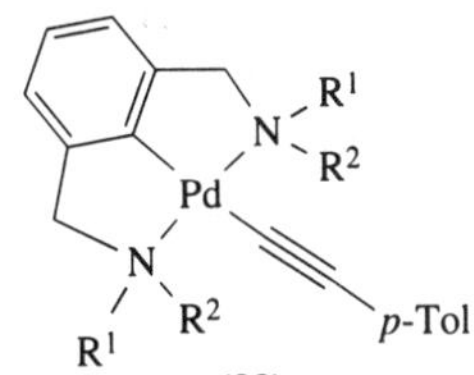

(33)

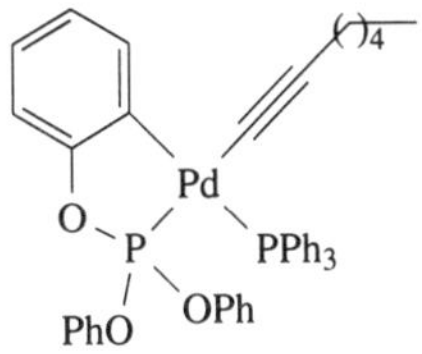

(34)

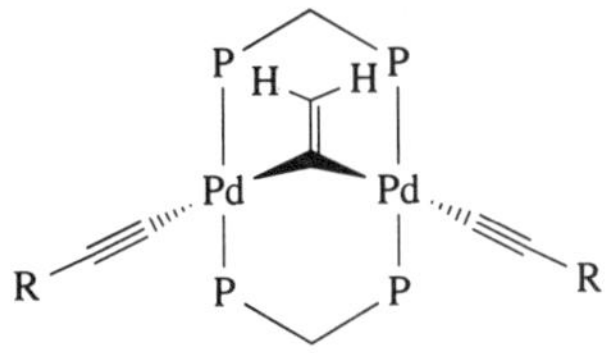

(35)
P~P = dppm; R = CMe_2OH, Ph, *p*-Tol

Intramolecular coordination complexes of cyclopalladated *N*-nitrosoanilines such as [Pd{2-[N(O)NMe]C_6H_4-*N,C*}(μ-Cl)]$_2$ react with [Sn(Me)$_3$(C≡CPh)] or [Sn(Bu)$_3${C≡C(CH_2)$_4$Me}] to form the organic products 2-[N(O)NMe]C_6H_4C≡CR (R = Ph or (CH_2)$_4$Me),[73] but the potential for applications of tin(IV) reagents in the synthesis of complexes is illustrated by the reactions in Scheme 8, together with isolation of the bimetallic complexes [Pd(C≡CR)($SnMe_2Cl$)(dppe)] (R = H, Me, Bu^t) on reaction of [PdCl($SnMe_2Cl$)(dppe)] with [Sn(Me)$_2$(C≡CR)$_2$].[45]

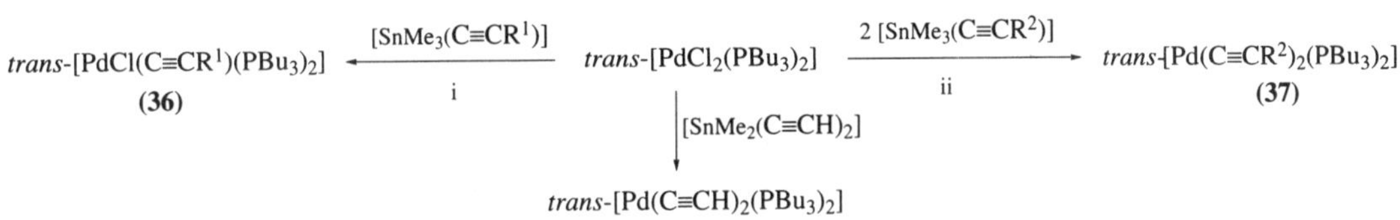

i, R^1 = Me, Bu^t,[70] *p*-$C_6H_4C_6H_4$C≡CMn(CO)$_5$;[71] ii, R^2 = *p*-C_6H_4C≡C$SnMe_3$[72]

Scheme 8

Dinuclear complexes such as (**39**) have been prepared by reactions of dialkynes with palladium(II) complexes in the presence of a base,[67,69] by procedures directly related to this method ((**40**)[77] and (**41**)[53] in Scheme 9) and by the reaction of *trans*-[$PdCl_2(PBu_3)_2$] with [1,4-($SnMe_3$C≡C)$_2C_6H_4$] to give (**39**) (X = Cl, R^1 = Bu, R^2 = H).[72] The mixed metal di- and trinuclear complexes *trans,trans*-[{($R_3P)_2ClM^1$}C≡C{$M^2Cl(PR_3)_2$}] (R = Me, Et, Bu; $M^1 = M^2$ = Pd;[53] M^1 = Pt, M^2 = Pd[52]) and *trans,trans,trans*-[Pt{C≡C–C≡CPdCl(PBu_3)$_2$}$_2$(PBu_3)$_2$][77] are obtained by strategies similar to those giving (**41**); for example, the trinuclear complex is formed on the reaction of *trans*-[$PdCl_2(PBu_3)_2$] with *trans*-[Pt(C≡C–C≡CH)$_2$(PBu_3)$_2$] in 2:1 ratio. An extensive series of polymers containing bridging units such as those depicted in (**39**) and (**40**) have been obtained by similar procedures, with interest in the field partly stimulated by the occurrence of interesting physical properties such as nonlinear optical effects.[65,67,72,77–9]

5.4.2 Reactions and Structures

The species [Pd(C≡C-*p*-C_6H_4C≡C)$_2$]$^{2-}$ is reduced by potassium in liquid ammonia to form [Pd(C≡C-*p*-C_6H_4C≡C)$_2$]$^{4-}$ which may be isolated as the potassium salt, but the electronic structure of the alkynyl present precludes the straightforward assignment of formal oxidation state for the palladium centre in the product.[42]

Polymerization reactions are discussed above, and heating of the complexes *trans*-[PdX(C≡CR)(PMe_3)$_2$] (X = Br, R = Bu^t, TMS; X = I, R = TMS) results in oxidative trimerization of the alkynyl group to give *trans*-[PdX{C(C≡CR)=C(R)C≡CR}(PMe_3)$_2$]*.[56] Metathesis reactions for anionic ligands *trans* to alkynyl groups have been reported,[53,67,80,81] and this type of reaction is useful in generating polymers, for example, (**42**) from the reaction of (**40**) (R = Bu) with 4,4'-bipyridyl in the presence of KPF_6.[80] The palladium–alkynyl bond appears to be preferentially cleaved by acids in

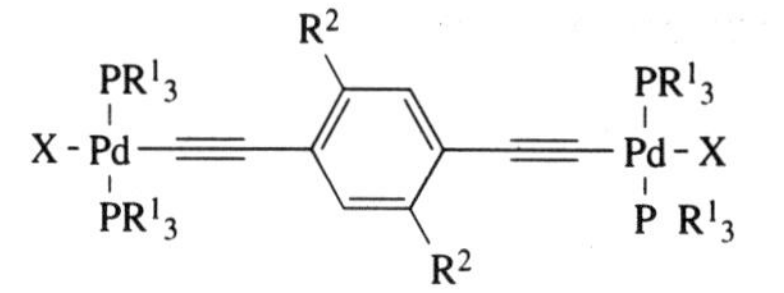

(39)

X = Cl: R^1 = Et, R^2 = H; R^1 = Bu, R^2 = H, Me, Et
X = Br: R^1 = Et, R^2 = H

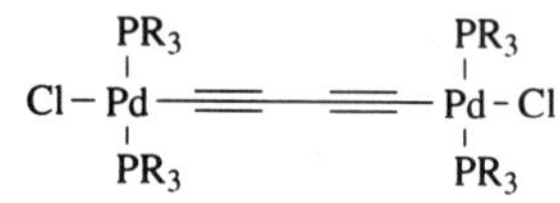

(40)

R = Et, Bu

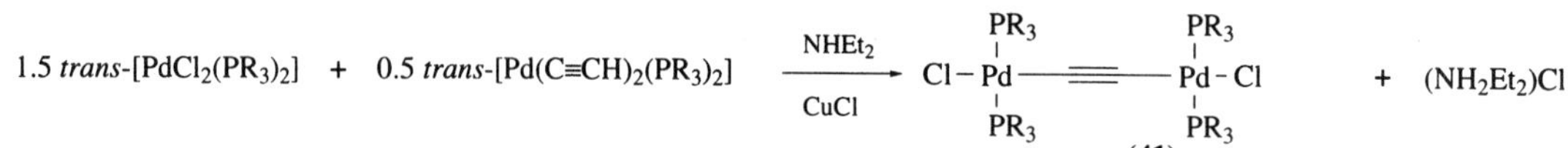

(41)

R = Me, Et, Bu: shorter reaction times give **(31)**

Scheme 9

complexes that also contain Pd–C_6F_5, Pd–C_6Cl_5 or Pd–{C(CO_2Me)=CH(CO_2Me)} bonds; for example, *trans*-[Pd(C≡CH)(C_6Cl_5)(PMe_2Ph)$_2$] reacts with $HClO_4$ to give *trans*-[Pd(ClO_4)(C_6Cl_5)(PMe_2Ph)$_2$].[63]

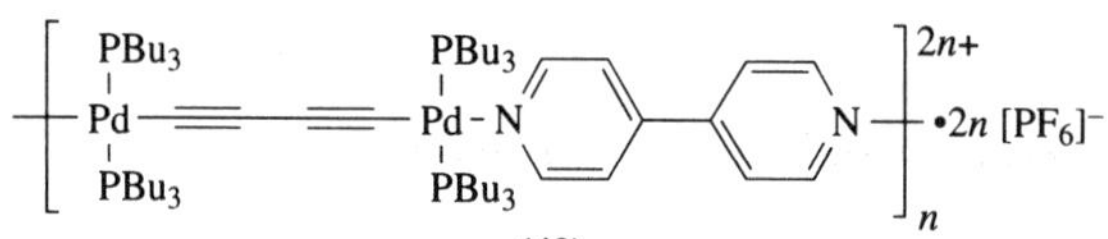

(42)

Single aryl isocyanide insertion occurs at ambient temperature for *trans*-[PdCl(C≡CR)(PEt_3)$_2$] (R = Ph, 4-MeOC_6H_4) and at one palladium centre in (**39**) (X = Cl; R = Et, Bu) and (**40**) (R = Bu), but double insertion occurs for μ-ethynediyl complexes in refluxing THF.[81] The cation *trans,trans*-[{(Et_3P)$_2$(PhNC)Pd}$_2$(μ-C≡C)]$^{2+}$ is an intermediate in the formation of (**43**)* from [{(Et_3P)$_2$ClPd}$_2$(μ-C≡C)]. The presence of a platinum centre in μ-ethynediyl complexes allows multiple insertion at the palladium centre, for example, to give the soluble block cooligomer (**46**) starting from (**44**) (Scheme 10).[82]

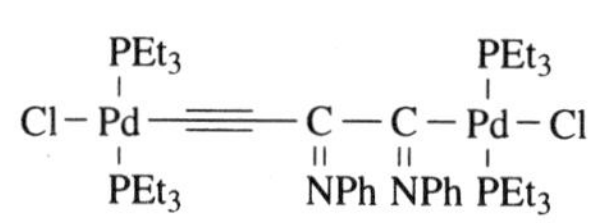

(43)

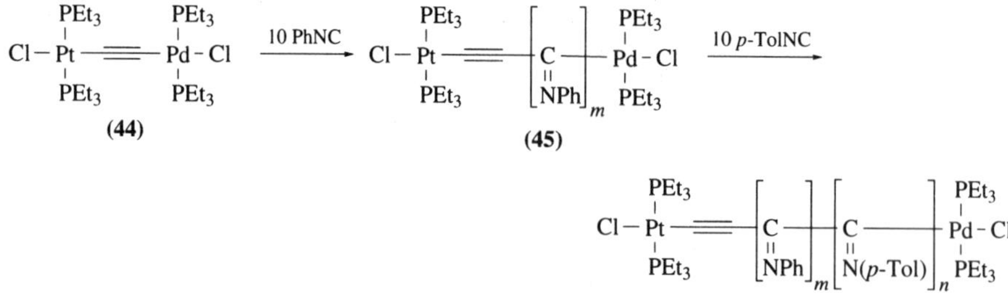

(44) **(45)** **(46)**

m = *n* = 10

Scheme 10

Six structural studies including those for the complexes (**32**) (R = H) and (**43**) give Pd–C(alkynyl) distances within the range 0.191(1)–0.204(2) nm[50,55,57,81,83,84] and, in complexes where Pd–C(*sp*) and Pd–C(sp^2) distances may be compared in the same molecule, little (**43**)[81] or no variation[84] has been detectable.

5.5 PALLADIUM(II) ALKYL, ARYL AND RELATED COMPLEXES

Procedures for the formation of Pd–C σ-bonds are presented in this section, followed by accounts of the reactions of organopalladium(II) complexes. Overlap between these themes also occurs; for example, reactions of alkynes with inorganic substrates to form Pd–C σ-bonds and insertion of alkynes into preexisting Pd–C σ-bonds are treated together in Section 5.5.1.5. Presentation of important synthetic reactions is not limited to Section 5.5.1; for example, reactions discussed elsewhere include the formation of monoorganopalladium(II) complexes via the acid cleavage of a Pd–C σ-bond in diorganopalladium(II) complexes, isomerization processes for the Pd–Alk group and aspects of the reactivity of palladium(I) complexes, palladium(II) ylides and palladium(II) alkynyls.

5.5.1 Formation of Pd–C σ-bonds

Synthetic methods for formation of Pd–C σ-bonds are classified according to the reagent classes used. Where more than one of the major reagents is used, the chemistry is discussed in the later section; for example, syntheses employing both alkenes and organomercury(II) reagents are presented in Section 5.5.1.4 rather than 5.5.1.2, and those employing both alkenes and carbon monoxide in Section 5.5.1.6 rather than 5.5.1.4.

5.5.1.1 From palladium metal or palladium(0) complexes

(i) Reactions of organohalides and pseudohalides

The oxidative addition reaction of palladium(0) complexes with organic halides, as in Equation (1), has become one of the most general synthetic methods in organopalladium(II) chemistry,[85] and the related direct reaction of palladium metal has interesting applications but is less widely applicable. Reaction of palladium vapour with organic halides at low temperature gives 'PdXR' (RX = CF_3I, C_2F_5I, $C_8F_{17}Br$) in low to moderate yield,[86,87] and the vapour reacts similarly with $CF_3\cdot$ radicals to form $[Pd(CF_3)_2]$.[88] Addition of ligands to 'PdXR' generates complexes, and *trans*-$[PdCl(Bz)(PEt_3)_2]$ has been formed in this way.[89] Higher yields of complexes *trans*-$[PdX(R)(PEt_3)_2]$ are obtained from reactions employing palladium powders as slurries formed by potassium reduction of *trans*-$[PdCl_2(PEt_3)_2]$ followed by addition of RX (RX = PhX (Cl, I, CN), C_6F_5X (X = Br, I)). An extensive series of $[PdI(C_6F_5)]$ complexes and [PdIPh(bipy)] have been formed similarly by reduction of $PdCl_2$, oxidative addition and addition of ligands.[90] Electrochemical oxidation of a palladium anode in the presence of C_6F_5Br and ligands also generates $[PdBr(C_6F_5)(L_2)]$ (L_2 = PEt_3, 2 py, bipy).[91]

$$[PdL_m] + RX \longrightarrow [PdXRL_n] + (m-n)L \qquad (1)$$

Palladium(0) reagents for synthesis generally contain dibenzylideneacetone (dba) or monodentate or bidentate phosphines as ligands, although a small number of nitrogen donor and isocyanide systems have been reported. Reagents are often generated *in situ* by reduction of palladium(II) complexes or addition of ligands to palladium(0) species such as $[Pd(dba)_2]$. Organic halide reagents include alkyl, benzyl, allyl, alkenyl, aryl and other halides, discussed in this order following initial comments on mechanisms.

Precise mechanisms of Equation (1) are yet to be established, although several aspects are well understood.[85,92] For reactions of reagents containing monodentate phosphines several studies suggest that coordinatively unsaturated complexes are the reactive intermediate(s), and recent electrochemical studies indicate that $[Pd(PPh_3)_2]$ is an intermediate in oxidative addition of aryl halides to $[Pd(PPh_3)_4]$,[93] or of aryl halides and 1,2-dichloroethene to electrochemically reduced *trans*-$[PdCl_2(PPh_3)_2]$.[94,95] Several phosphine reagents $[PdL_2]$ have been isolated and they react with organohalides, for example, to form *trans*-$[PdCl(COPh)(PR_3)_2]$ (R = Ph, OPh)[96] and *trans*-$[PdIPh(PCy_3)_2]$.[97] The reactivity of aryl halides toward palladium(0) reagents follows the order ArI > ArBr > ArCl > ArCN, illustrated by the yields of products in Scheme 11.[98–103] Electron-withdrawing substituents on the aryl group favour the oxidative addition reaction; for example, π-coordination of chlorobenzene to $[FeCp]^+$ *[104] or $[Cr(CO)_3]$[105] permits activation of the Ar–Cl bond as illustrated in Scheme 12. Activation of chlorides and cyanides is also facilitated by strongly donating phosphines; for example, bromobenzene[106] and 2-amino-5-nitrobromobenzene[107] react with $[Pd(PPh_3)_4]$ to form *trans*-$[PdBrAr(PPh_3)_2]$, and chlorobenzene is unreactive toward $[Pd(PPh_3)_4]$ but reacts readily with $[Pd(PCy_3)_2(dba)]$ to form *trans*-$[PdCl(Ph)(PCy_3)_2]$*.[108]

$$[Pd(dba)_2] \xrightarrow{ArX,\ L} [PdX(Ar)(L\text{-}N,N')]$$

L = TMEDA: ArX = PhI (95%), PhBr (8%), 1-NapI (96%), *p*-TolI (88%), *o*-TolI (100%)
2-$(HOCH_2CH_2)C_6H_4Br$ (58%)
L = bipy: ArX = PhI (70%), PhBr (5%); L = $NMe(CH_2CH_2NMe_2)_2$: ArX = PhI (84%)

Scheme 11

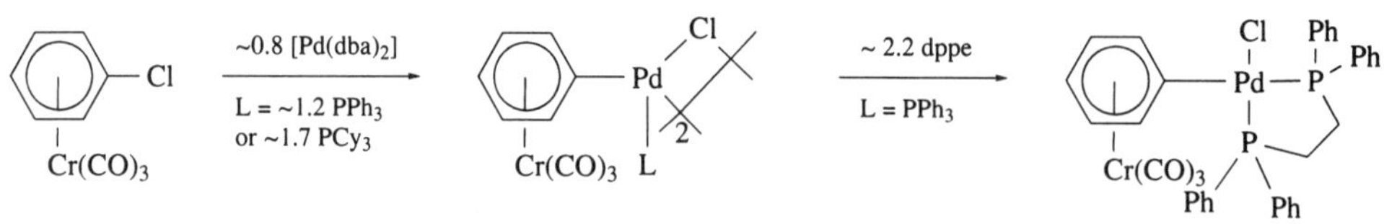

(47) **(48)**

Scheme 12

Inversion of configuration has been observed for several systems, in particular for benzyl halides,[85,109–11] which is consistent with occurrence of the S_N2 mechanism (Scheme 13). The partial loss of inversion for reactions of $[Pd(PPh_3)_4]$ with benzyl halides was initially explained as resulting from exchange of a benzyl group via an intermediate of structure $[(Ph_3P)_nPd\cdots Bz\cdots PdCl(PPh_3)_2]^{\ddagger}$,[109] but detailed studies of analogous reactions of $[Pd(PEt_3)_3]$ indicate that at least for this system a mechanism involving radicals probably occurs.[111]

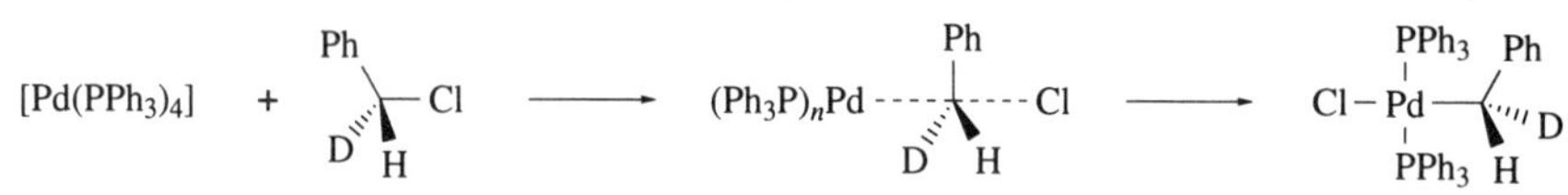

(49)

Scheme 13

Stereochemical studies of the reactions of allenyl halides (Scheme 14) are consistent with interaction of palladium(0) with the double bond adjacent to the Bu^t group to give an ionic species (**50**) followed by a 1,3-shift of palladium to give (**51**).[112] Allenylpalladium(II) complexes are also accessible by oxidative addition of allenyl bromides and, in contrast to the chloro complexes, the *trans*-complexes undergo a slow isomerization to the *cis*-isomers in CH_2Cl_2 or $CHCl_3$.[113] Oxidative addition reactions of propynyl chlorides, $R^1C{\equiv}C{-}CR^2R^3Cl$, also result in the formation of allenylpalladium(II) complexes *trans*-$[PdCl(CR^1HC{=}C{=}CR^2R^3)(PPh_3)_2]$, except for bulky R^1 (Bu^t, TMS) with $R^2 = R^3 = H$ when propynyl complexes are formed.[114]

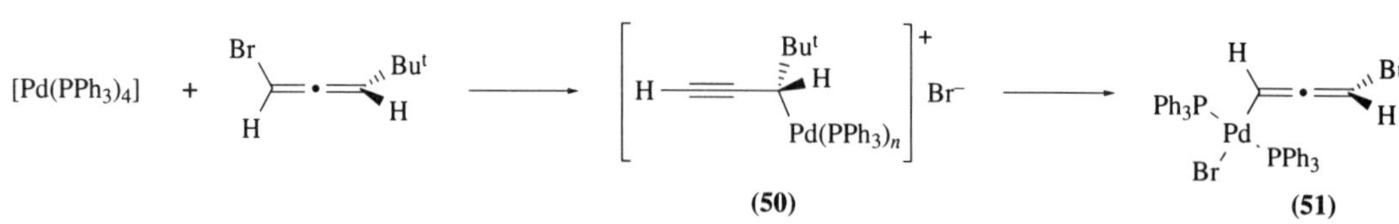

(50) **(51)**

Scheme 14

Evidence for one-electron processes has been presented for the reaction of Pr^{i}I with $[Pd(PEt_3)_3]$[115] but appears to be lacking for other systems,[116] and it has been demonstrated that [Pd(dppf)(cot)] reacts with alkenyl halides via coordination (Scheme 15).[117]

The simple halides MeI and/or MeBr react readily with palladium(0) phosphine complexes,[118,119] including $[Pd(PPh_3)_2]$,[96] and related reagents containing more complex phosphine and nitrogen donor ligands. Iodomethane reacts with $[Pd_2(dba)_3{\cdot}CHCl_3]$/bipy[120] and $[Pd(2,9\text{-}Me_2phen)(CH_2{=}CHCN)]$[121] to form [PdIMe(bipy)] (79% yield) or $[PdIMe(2,9\text{-}Me_2phen)]$ in equilibrium with (**52**) (X = I, L = $CH_2{=}CHCN$). The complexes [PdXR(P~N)] (P~N as in (**32**), RX = MeI, MeBr*),[57] (**53**)[122,123] and (**54**)[108] are formed from the reagents $[Pd(P{\sim}N\text{-}P,N)_2]$, $[Pd(P_3{\sim}N\text{-}P,P,P)]$ and $[Pd(PCy_3)_2(dba)]$, respectively. Benzyl[85,109–11,124] and related halides[125,126] have been widely studied, as illustrated in Scheme

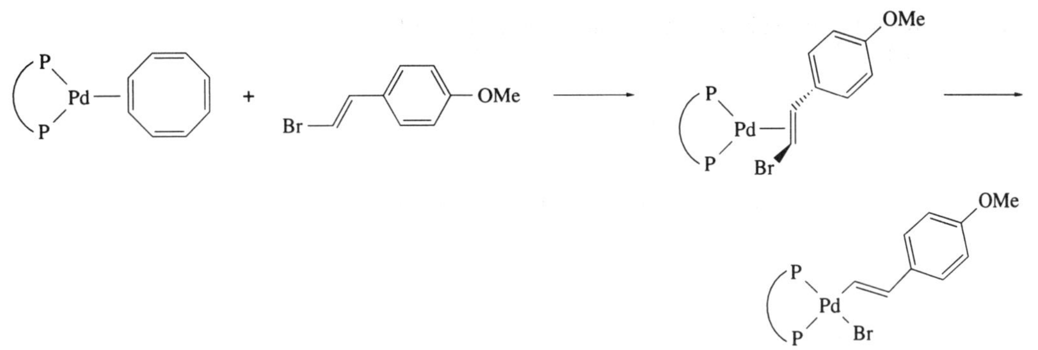

Scheme 15

13 and (**53**), the reaction of BzCl with *trans*-[$PdCl_2(PCy_3)_2$] after reduction with potassium to form (**54**) (X = Cl, Y = Ph)[127] and the reaction of [$Pd_2(dba)_3{\cdot}CHCl_3$]/4.2 PPh_3 with [Cr{CN(2-ICH_2)C_6H_4}$(CO)_5$] to form *trans*-[PdI{($CH_2C_6H_4NC$)Cr$(CO)_5$}$(PPh_3)_2$].[128] Halomethyl ketones react with [PdL_4] to form *trans*-[PdCl{$CH_2C(O)Me$}$(PPh_3)_2$][129] and *trans*-[PdBr{$CH_2C(O)Bu^t$}$(PR_3)_2$] (R = Ph, *p*-Tol),[130] and oxidative addition of CF_3Br,[131] $NCCH_2I$[132] and $RSCH_2Cl$ (R = Me, Ph)[133] occurs readily to [$Pd(PPh_3)_4$]. Simple styryl and allyl halides undergo oxidative addition, including (*E*)- and (*Z*)-styryl bromides with [PdL_4] to give the corresponding (*E*)- and (*Z*)-complexes *trans*-[PdBr(CH=CHPh)L_2] (L = PPh_3, $PMePh_2$),[85,134] and propen-2-yl chloride with [Pd(diop)(CH_2=CH_2)] to form [PdCl(CH_2CH=CH_2)(diop)][135] (diop = 2,3-*O*-isopropylidene-2,3-dihydroxy-1,4-bis(diphenylphosphino)butane). Alkenyl halides containing electron-withdrawing halogen or other substituents have been widely studied, including CXF=CF_2 (X = Cl, Br)[136] and XCH=$CHCO_2R$ (X = Cl, I),[137] and a range of simple complexes have been obtained from reactions with [$Pd(PMePh_2)_2$(PhCH=CH_2)] generated *in situ*, for example, *trans*-[PdBr(CR^1=CR^2R^3)$(PMePh_2)_2$] ($R^1 = R^2 = R^3$ = H; $R^1 = R^2$ = H, R^3 = Me or Ph; R^1 = H, $R^2 = R^3$ = Ph).[138]

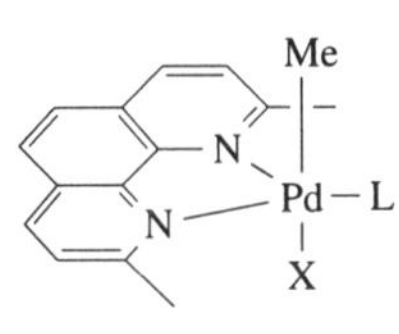

(52)

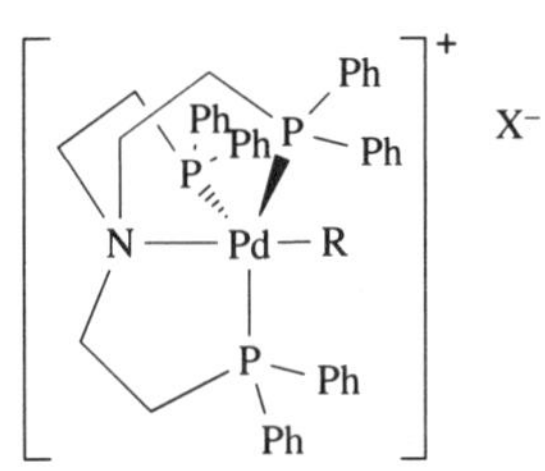

(53)
RX = MeI, EtI
X = BPh_4 for RY = PrI
BuI, BzI, CH_2Cl_2

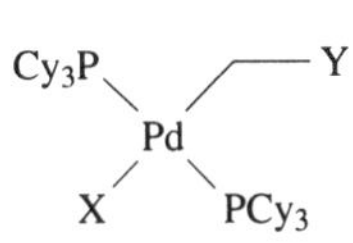

(54)
X = Cl, Y = Ph

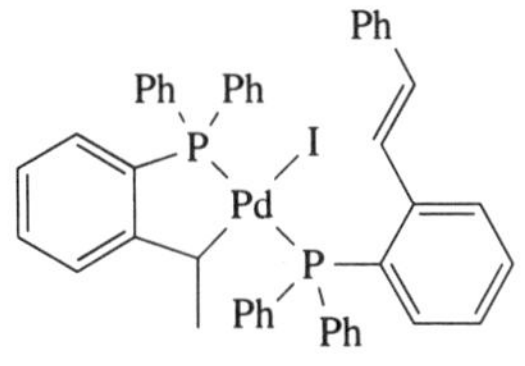

(55)

Aryl halides react readily with palladium(0) complexes of nitrogen donor ligands (Scheme 11) and [$Pd(PPh_3)_4$].[139–41] Iodobenzene reacts with [$Pd(diop)_2$] to form *trans*-[PdIPh(diop)] which forms an equilibrium in solution with polymeric species formulated as [PdIPh(μ-diop)]$_n$,[142] and with bis[(2-vinylphenyl)diphenylphosphine]palladium(0) to form (**55**), presumably via oxidative addition followed by rearrangements.[143] A wide range of mononuclear complexes have been obtained from R^1ClC=N(R^2)[144–6] and RClC=S=O reagents,[147] C_6F_5Br[148] and halogenated heterocycles such as 2,6-dichloropyridine and related pyridines,[149–52] halogenated pyrimidines,[153,154] and related reagents including 4-iodo-3*H*-phenothiazin-3-one,[155] 4-bromo-3-phenylsydnone[156] and 1,3-dialkyl-5-fluorouracils.[157]

Complexes *trans*-[PdX(COR)L_2] have been obtained from the reactions of palladium(0) substrates with a range of reagents RCOCl (R = Me, Cy, Bz, Ph, *p*-Tol, C_6F_5, [Fe(Cp)(η-C_5H_4)], OMe, OEt, $OCH_2CH_2C{\equiv}CR^1$ (R^1 = Me, TMS), $(CH_2)_nC{\equiv}CMe$ (*n* = 2, 3), $(CH_2)_nC{\equiv}CH$ (*n* = 2, 3)),[158–65] and similar approaches result in the formation of [Pd–COCOR] complexes (R = Me, Ph, OMe, OEt, $OCH_2C{\equiv}CMe$, $O(CH_2)_nCH{=}CH_2$ (*n* = 2–4)).[166–8]

Intramolecular coordination complexes have been obtained on the reaction of $[Pd(dba)_2]$ or $[Pd(PPh_3)_4]$ with organic halides,[169–80] including complexes from $[Pd(dba)_2]$ and alkyl or acyl chlorides (**56**)[169,170] or aryl bromides to give (**57**)–(**60**) and the complex [PdBr(N~C~N)] (N~C~N as in (**33**)).[177] Aryl bromides containing intramolecular donor groups react with $[Pd(dba)_2]$ to give products in higher yield than obtained with simple aryl bromides in the presence of nitrogen donor ligands; for example, 2-(MeN=CH)C_6H_4Br reacts with $[Pd(dba)_2]$ to give [Pd{2-(MeN=CH)C_6H_4-*C,N*}(μ-Br)$]_2$ in 98% yield,[171] but PhBr reacts with $[Pd(dba)_2]$ in the presence of bipy to give [PdBrPh(bipy)] in only 5% yield.[100] The reaction of $[Pd(PPh_3)_4]$ with $pyCH_2Cl$ results in bridging by the picoline group rather than halogen ((**61**) (R = H, L = PPh_3*);[181–3] halopyridines also react and, for substitution in the 2-position of pyridines and related heterocycles, dinuclear complexes (**62**) are formed. A complex with the same general structure as (**61**), but with R as a 2-pyridyl group intramolecularly coordinated in place of L, [PdCl(py_2PhC)$]_2$*, has been obtained from the reaction of $[Pd(dba)_2]$ with py_2PhCCl.[184]

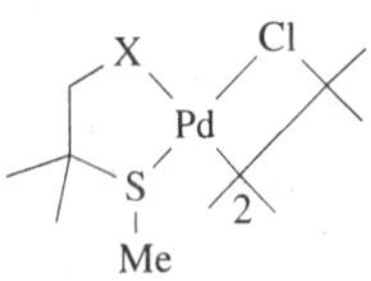

(56)
X = CH(OMe), C=O

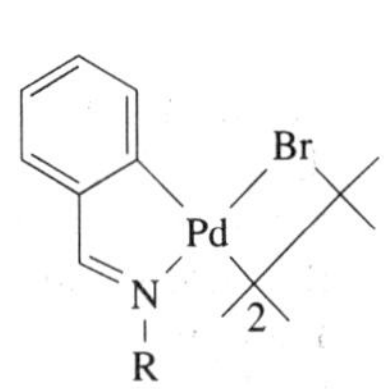

(57)[171,172]

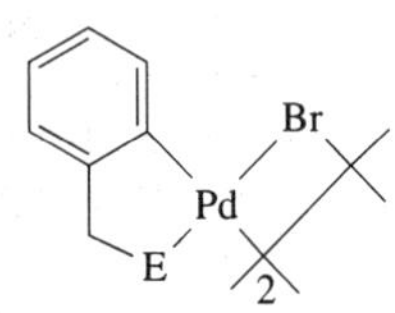

(58)
E = NR^1R^2,[173] PR^1R^2, AsR^1_2[174]
2,6-$Cl_2C_6H_3$CH=N[172]

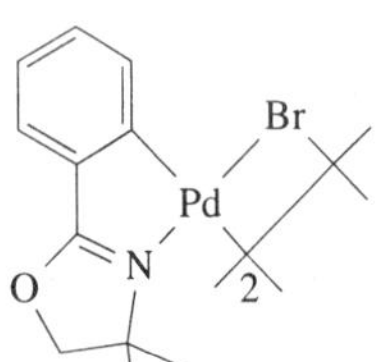

(59)[175]

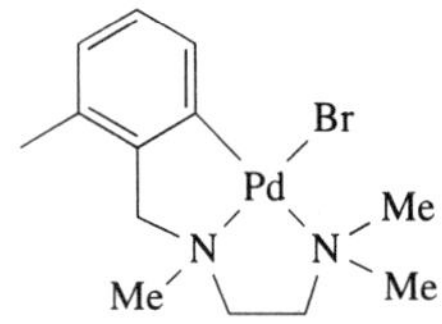

(60)[176]

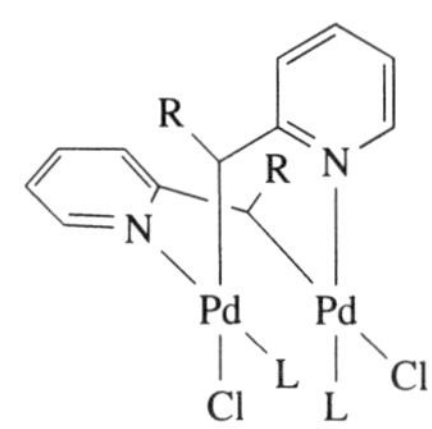

(61)

Methyl halides react with $[Pd_2(dppm)_3]$ to form dinuclear A-frame complexes (**2**) (X = Br, I), and in a similar way with CH_2X_2, $CHMeX_2$ (X = Cl, Br, I) or 1,2-$C_6H_4I_2$ to form neutral complexes with bridging CH_2, CHR or C_6H_4 groups,[119] where the C_6H_4 group bridges as in (**6**). Complex (**2**) (X = Br, I) forms an equilibrium with $[(MePd)_2(\mu\text{-dppm})_2(\mu\text{-X})]^+$.[119] Oxidative addition reactions employing the C–Br or C–Cl bonds of *trans*-[NiBr(2-BrC_6H_4)$(PPh_3)_4$], $CH_2{=}CX_2$ or *trans*-[MCl(CCl=CCl_2)$(PPh_3)_2$] give similar dinuclear complexes,[75,185] including (**63**) (M = Ni*) from the reaction of *trans*-[NiCl(CCl=CCl_2)$(PPh_3)_2$] with $[Pd(PPh_3)_4]$ in the presence of dppm.[185]

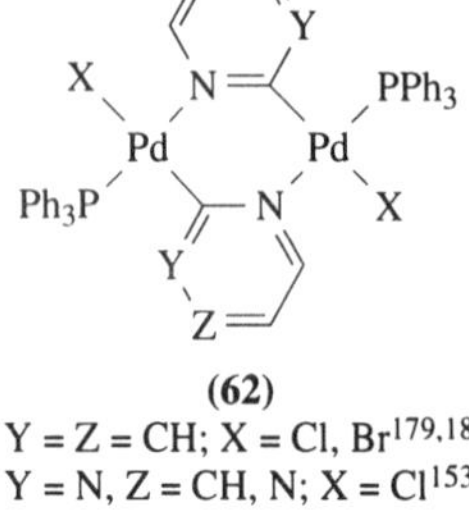

(62)
Y = Z = CH; X = Cl, Br[179,180]
Y = N, Z = CH, N; X = Cl[153]

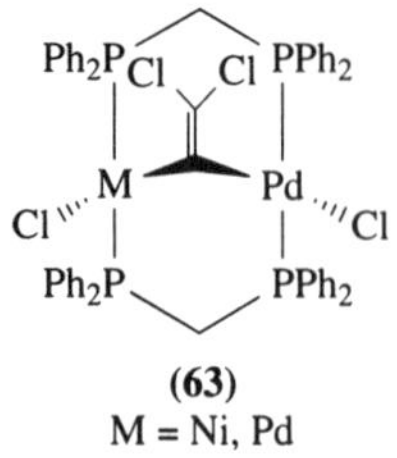

(63)
M = Ni, Pd

(ii) Reactions of other reagents

Palladium(0) phosphine complexes condense with $(CF_3)_2CO$ and $(CF_3)_2C{=}C(CN)_2$ to form cyclic complexes,[136] and $[Pd(PPh_3)_4]$ oxidatively inserts into the C-1–C-2 bond of 1,1,2,2-tetracyanocyclopropane to form a palladacyclobutane complex,[186] and into a CH_2–C(aryl) bond of 1*H*-cyclopropa[*b*]naphthalene to form an unstable complex tentatively assigned as a palladacyclobutarene.[187] Octafluorocyclooctatetraene undergoes a rearrangement on reaction with $[Pd(PPh_3)_4]$ or $[Pd_2(dba)_3]$/Bu^{t}NC to form the palladacyclic complexes (**64**),[188] and $[Pd(PMe_3)_2(PhCH{=}CH_2)]$ and $[Pd(PCy_3)_2]$ react with 3-butenoic acid to form complexes (**65**)* and (**66**).[189,190] Complex (**65**) forms an equilibrium with a small proportion of the PMe_3 analogue of (**66**).

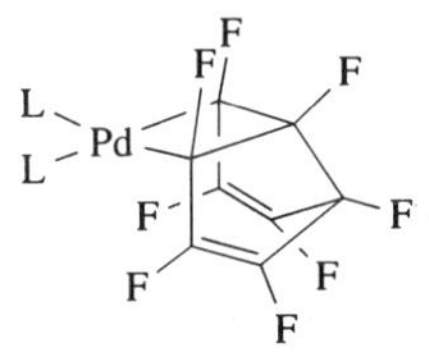

(**64**) L = PPh_3, Bu^{t}NC

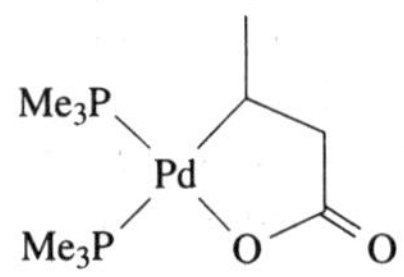

(**65**)

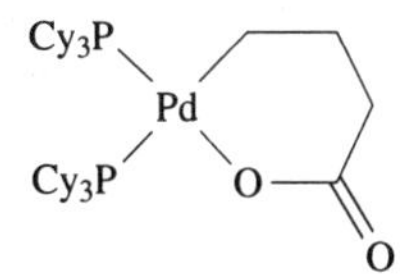

(**66**)

Palladium(0) also activates P–C and As–C bonds to form complexes; for example, $[Pd(PEt_3)_3]$ reacts with $PPh_2(C_6F_5)$ to give impure *trans*-$[Pd(C_6F_5)(PPh_2)(PEt_3)_2]$,[148] and cyanopalladium(0) complex ions react with $[MPh_4]^+$ (M = P, As) in liquid ammonia to form $[Pd(CN)_3Ph]^{2-}$.[191] The complexes $[Pd(PEt_3)_2(CH_2{=}CH_2)]$ and $[Pd(dppe)(CH_2{=}CH_2)]$ insert into the strained P–C bonds of $[W(CO)_5\{P(Ph)C(Ph){=}CPh\}]$ and its unsaturated analogue.[192,193] Retention of configuration occurs on insertion into (*E*)- and (*Z*)-isomers, as illustrated for the formation of (**68**)* from (*Z*)-(**67**), consistent with formation of a transition state involving η^2-coordination of the P–C bond to palladium. Several P–C bond activation processes involve palladium(II) complexes as reactants, but some appear to proceed via palladium(0) intermediates,[118] for example, the reaction of $[Pd(OAc)_2(PPh_3)_2]$ with formic acid to give $[Pd(Ph)(PPh_3)(\mu\text{-}O_2PPh_2\text{-}O,O)]_2$*,[194] and the reaction of *trans*-$[PdEt_2(PMe_3)_2]$ with $R^1C(O)P(O)(OR^2)_2$ to give aroyl(phosphonato)palladium(II) complexes *trans*-$[Pd(COR^1)\{P(O)(OR^2)_2\}(PMe_3)_2]$ (R^1 = Ph, *p*-Tol; R^2 = Me, Et).[195]

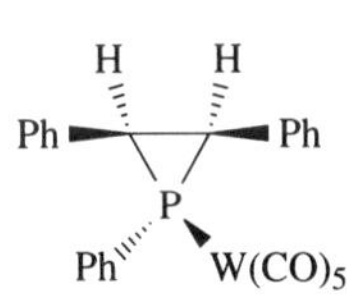

(*Z*)-(**67**)

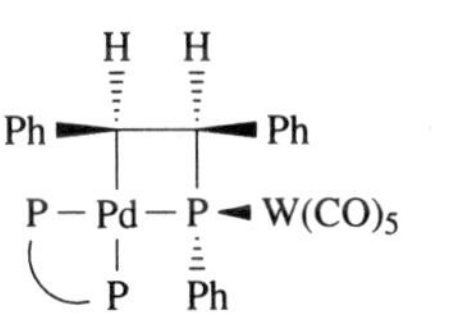

(**68**)

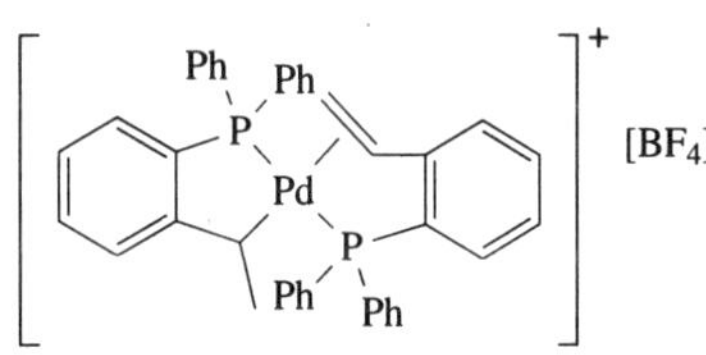

(**69**)

Aryldiazonium electrophiles react with $[Pd(PPh_3)_4]$ to form aryldiazenido complexes that readily eliminate nitrogen to form $[PdAr(PPh_3)_3]^+$,[196] and HBF_4 reacts with bis[(2-vinylphenyl)diphenylphosphine]palladium(0) to form (**69**),[197] and with $[Pd(P{\sim}P)(CH_2{=}CH_2)]^+$ (P~P = $Bu^t_2P(CH_2)_nPBu^t_2$ (*n* = 2, 3), 1,2-$(Bu^t_2P)_2C_6H_4$) to form [Pd(Et)(P~P)] where the ethyl group forms a strong agostic interaction.[198]

Organomercury(II) halides react with palladium(0) reagents in the presence of neutral ligands to give metallic mercury and complexes in high yields,[199–203] for example, [PdCl(*p*-Tol)(phen)],[199] *trans*-$[PdCl(COR)(PPh_3)_2]$ (R = OMe, NEt_2),[200] and the intramolecular coordination complexes (**70**) (R = H)[202] and $[PdBr(CHMeCH_2CH_2NMe_2\text{-}C,N)(py)]$.[203] The last complex has the same sign of optical rotation as its chiral organomercury(II) precursor. Diorganomercury(II) compounds and $[TlBr(C_6F_5)_2]$ transfer two organic groups in similar redox reactions to give, for example, *trans*-$[Pd\{C({=}N_2)CO_2Et\}_2(PBu_3)_2]$*,[204] *cis*-[Pd(C~N)$_2$] (C~N as in (**58**) (E = NMe_2))[205] and *cis*-$[Pd(C_6F_5)_2(PPh_3)_2]$.[4]

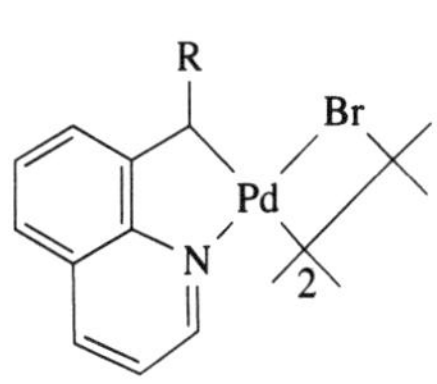

(**70**)

5.5.1.2 From palladium(II) complexes and organometallic reagents

Mercury, aluminum, lithium and magnesium reagents are of particular importance in organopalladium synthesis, and are dealt with after a brief survey of other reagents.

Tetramethyltin reacts with *trans*-$[PdBrBz(PPh_3)_2]$ to form *trans*-$[PdBrMe(PPh_3)_2]$,[110] and a series of complexes, mainly intramolecular coordination complexes, are obtained on the reaction of *trans*-$[PdCl_2(NCPh)_2]$ with organotin and -silicon reagents $[M(R^1)_3(CH_2R^2)]$ (M = Sn, Si),[206–8] such as *trans*-$[Pd(CH_2CH_2CH_2NMe_2$-*C,N*)(μ-Cl)$]_2$* and the chloro analogue of (**70**) (R = H).[207] Silyloxy reagents such as $TMS[OC(=CH_2)CH_2CH_2CH=CH_2]$ react with *trans*-$[PdCl_2(NCPh)_2]$ to form isolable intramolecular coordination complexes involving alkene group coordination, $[Pd\{\eta^2\text{-}CH_2{=}CHCH_2CH_2C(O)CH_2\}$-(μ-Cl)$]_2$, and on reflux in acetonitrile this complex and substituted analogues form products resulting from cyclization of the organo group, $[Pd\{CH_2CHCH_2C(O)CH_2CH_2\}(NCMe)Cl]_n$.[209] The complexes *trans*-$[PdBr(CF_3)(PEt_3)_2]$ and *trans*-$[Pd(CF_3)_2(PEt_3)_2]$ are obtained on the reaction of $[Cd(CF_3)_2(MeOCH_2CH_2OMe)]$ with *trans*-$[PdBr_2(PEt_3)_2]$.[210] Early demonstrations that methyl groups can be transferred from platinum(II) to palladium(II),[211,212] and from $Li[CuMe_2]$ to $[PdCl_2(cod)]$,[213] do not appear to have been developed further except for a recent report that 1:1 $LiBu^i$/CuCN reacts with $[PdCl(S_2CNMe_2)(PEt_3)]$ to form the iso-butyl complex $[PdBu^i(S_2CNMe_2)(PEt_3)]$.[214] The related complex $[Pd(CH_2CH_2CN)(S_2CNMe_2)(PEt_3)]$ may be obtained using $[ZnI(CH_2CH_2CN)]$,[214] and a series of aryl(allyl)palladium(II) complexes are obtained from [ZnClAr] and $[PdCl(allyl)(L)]_2$, for example, $[PdPh(\eta\text{-}2\text{-}MeC_3H_4)(PPh_3)]$.[215] Similar reactions of $[Tl(2,3,5,6\text{-}C_6Cl_4H)_3]$ give related complexes.[216] The reaction of $[TlCl(C_6F_5)_2]$ with *trans*-$[PdCl_2(L)_2]$ in 1:1 mol ratio results in the formation of $[Pd(C_6F_5)(L)(\mu\text{-}Cl)]_2$ (e.g., L = PPh_3, $PMePh_2$)[4] and [TlCp] reacts with $[PdCl(L)(\mu\text{-}O_2CMe)]_2$ to form fluxional complexes $[Pd(\eta^1\text{-}Cp)(\eta\text{-}Cp)(L)]$.[217]

Diphenylmercury(II) reacts with *trans*-$[PdCl_2(PEt_3)_2]$ to form *trans*-$[PdClPh(PEt_3)_2]$ and [HgClPh],[218] and related reactions give diazomethyl complexes such as *trans*-$[PdCl\{C(=N_2)R\}(PPh_3)_2]$ (R = CO_2Et, COMe).[204] The alkene complexes *cis*-$[Pd(CH_2{=}CHCMe_2CH_2E)Cl(CH_2Cl)]$ (E = NMe_2, SMe) are formed on reaction of the dichloropalladium(II) analogues with $[Hg(CH_2Cl)_2]$.[219] Diorganomercury(II) compounds react with $[PdCl(PMePh_2)(\mu\text{-}Cl)]_2$ to give $[PdR(PMePh_2)(\mu\text{-}Cl)]_2$ (R = Ph, Bz),[220] and related anions $[Pd_2Cl_6]^{2-}$ also react with retention of halogen bridging for a range of substituted aryl groups to form (**71**) and (**72**).[221–4] One of these mercurials, forming (**72**), also reacts with *trans*-$[PdCl_2(NCPh)_2]$ to form (**73**), involving an unusual rearrangement of arene substituents.[224] Intramolecular coordination occurs on the reaction of appropriate diorganomercurials to give monomeric or dimeric complexes, such as (**74**)*[225] and (**75**).[226]

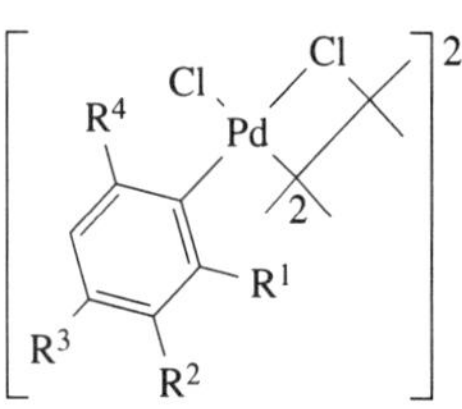

(**71**)

R^1 = Me, R^2 = R^3 = H, R^4 = NO_2
R^1 = R^3 = R^4 = NO_2, R^2 = H

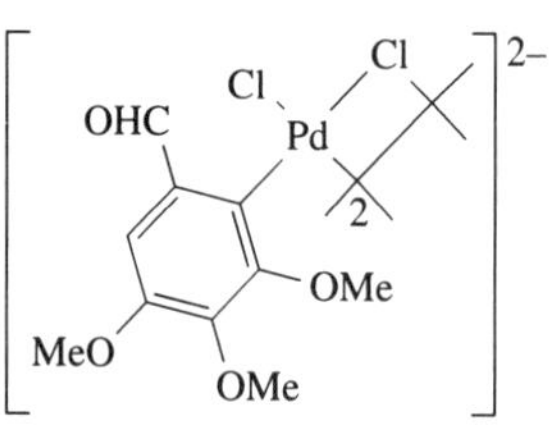

(**72**)

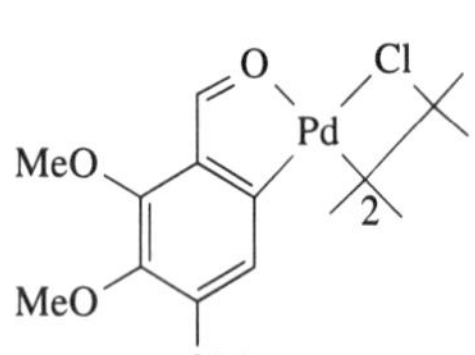

(**73**)

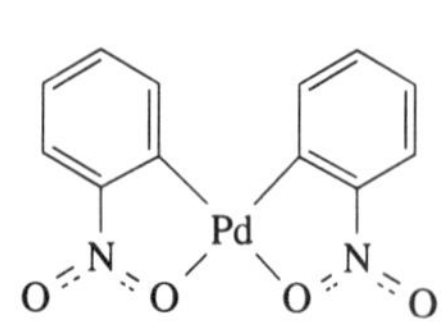

(**74**)

Monoorganomercury(II) compounds also react with palladium(II) substrates,[227] for example, $[HgCl(CO_2Me)]$ with *trans*-$[PdClBz(PMe_3)_2]$ to form *trans*-$[PdBz(CO_2Me)(PMe_3)_2]$,[228] and [HgClAr] with palladium(II) acetate to form (**76**)[229] and $[Pd(N{\sim}C{\sim}N)(O_2CMe)]$ (N~C~N as in (**33**) (R^1 = R^2 = Me)).[205]

The reactions of $[Al(OEt)R_2]_2$, or preferably $[Al_2(OEt)_3R_3]$, with $[Pd(acac)_2]$ in the presence of phosphines is an ideal route to *trans*-$[PdR_2L_2]$ complexes, including *trans*-$[PdBu_2(PMe_2Ph)_2]$ and methyl, ethyl and propyl complexes with a range of phosphines.[230–4] Complexes of bidentate ligands are

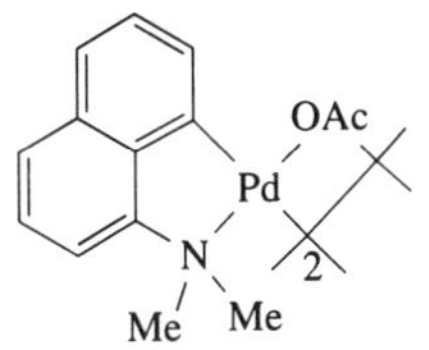

(75)

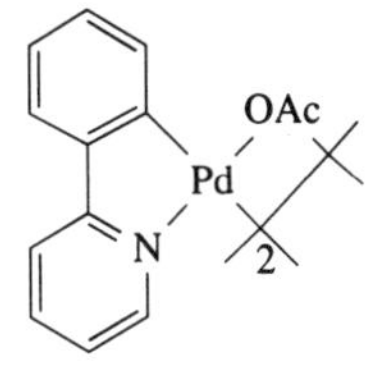

(76)

also accessible, in particular [PdR_2(dppe)] (R = Me, Et, Pr)[230] and the nitrogen donor complexes [$PdEt_2(L_2)$] (L_2 = bipy, 4,4'-bipy, phen) and [$PdPr_2$(bipy)].[235,236] The importance of supporting ligands is illustrated by the formation of palladium metal on reaction of [$Pd(acac)_2$] with [Al_2Et_6],[237] contrasting with formation of [PdMe(L)(μ-Cl)]$_2$ (L = PEt_3, PBu_3, PMe_2Ph) in high yield on reaction of [PdCl(L)(μ-Cl)]$_2$ with [Al_2Me_6].[238]

Transmetallation using lithium reagents is the most widely used route to simple *cis*-dialkylpalladium(II) complexes of monodentate phosphine and arsine ligands, with alkyl groups including Me, Et, CH_2CMe_3, CH_2TMS and $CH_2C(CH_2CH_2)_3CH$. Two equivalents or a slight excess of alkyllithium reagent is generally used in reactions with [$PdCl_2L_2$] at low temperature to give *cis*-[PdR_2L_2],[48,60,118,232,239–43] for example, *cis*-[$PdMe_2(PEt_3)_2$].[60] Methyllithium also reacts with [Pd(η-C_3H_5)(PPh_3)$_2$]$^+$ to form *cis*-[$PdMe_2(PPh_3)_2$][244] and with [Pd{2-(Pr^{i}N=CH)C_6H_4-*C*}(PMe_3)$_3$]$^+$ to form *trans*-[PdMe{2-(Pr^{i}N=CH)C_6H_4-*C*}(PMe_3)$_2$].[245]

It has been demonstrated that complexes such as *trans*-[$PdMe_2(PEt_2Ph)_2$], prepared independently from [$Pd(acac)_2$] and [$Al_2(OEt)_2Me_4$] in the presence of PEt_2Ph, isomerize to the *cis*-isomer on reaction with [LiMe].[232,246–8] This process involves the formation of palladates such as [$PdMe_3(L)$]$^-$ and [$PdMe_4$]$^{2-}$, and these and related species have been detected in NMR studies.[246] Isomerization of *trans*- to *cis*-complexes also occurs via formation of bridging species (**77**)[232] such as '[L(Me)Pd(μ-Ar)$_2$PdIL$_2$]' for the formation of *cis*-[PdMeAr(PEt_2Ph)$_2$] (Ar = *p*-XC_6H_4 for X = H, Me, F, Cl, CF_3) from *trans*-[PdBrAr(PEt_2Ph)$_2$] and excess [LiMe].[249] Similar species (**78**) have been proposed as intermediates in *cis–trans* equilibria for [$PdMe_2L_2$] catalysed by [$MgMe_2$].[250] Isomerization may also be assisted by added phosphine.[118] However, isolation of *trans*-isomers is possible for syntheses using alkyllithium reagents, for example, *trans*-[$PdR_2(PEt_3)_2$] (R = Ph, (*E*)-Bu^{t}CH=CH) from *trans*-[$PdCl_2(PEt_3)_2$], *trans*-[PdPhR(PEt_3)$_2$] (R = Me, Ph, *p*-Tol, (*E*)-Me(CH_2)$_5$CH=CH) from *trans*-[PdIPh(PEt_3)$_2$][60] and *trans*-[PdMeAr(PEt_2Ph)$_2$] (Ar = Ph, *o*-Tol) from *trans*-[PdBrAr(PEt_2Ph)$_2$].[249]

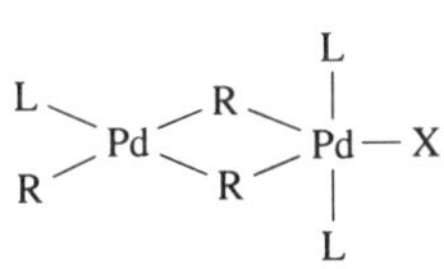

(77)

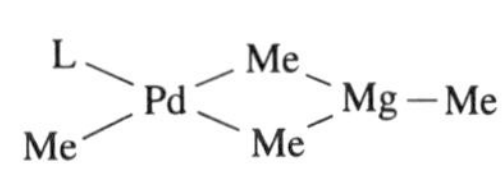

(78)

The few monoalkylpalladium(II) complexes which have been isolated using alkyllithium reagents include *cis*-[PdBrMe(P~N)] (P~N as in (**32**))[57] and a series of dialkyldithiocarbamate complexes *cis*-[PdR(S_2CNMe_2)(PEt_3)] (R = Me, Pr^n, Pr^i, Bu^n, Bu^s) and [PdR{$S_2CN(CH_2)_4$}(PEt_3)] (R = Pr^n, Pr^i)*.[251] However, the simple [PdMe] complexes (**79**) are readily obtained (Scheme 16)[252] and are excellent precursors for substitution reactions.[120,253–6]

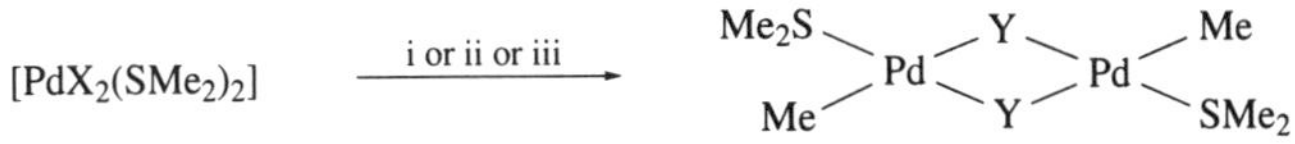

(79)

i, [MeLi] (X = Y = Cl, Br); ii, [MeLi]/LiY (X = Cl, Y = Br, I); iii, 2 [MeLi]/MeY (X = Cl, Y = Br, I)

Scheme 16

Complexes [$PdX_2(L_2)$] readily give *cis*-[PdR_2] complexes for L_2 as neutral bidentate phosphorus,[48,257,258] arsenic,[48] the mixed P~N donor shown in (**32**) (R = Me)[57] and other mixed donor ligands,[48,243] and nitrogen donor ligands including bipy[48,257,259] and TMEDA,[260,261] for example, [PdMePh(TMEDA)] from [PdIPh(TMEDA)] and [LiMe].[98,100,262] The *trans*-spanning ligand transphos forms *trans*-[$PdMe_2$(transphos)] from [$PdCl_2$(transphos)] and [LiMe] (transphos = 2,11-bis(diphenylphosphinomethyl)benzo[*c*]phenanthrene).[118] Complexes of bipy, and weak donor ligands such as pyridazine, (CH_2SMe)$_2$[120] and bidentate poly(pyrazolyl)alkanes[120,253,263] may be obtained by generating dimethylpalladium(II) species from [$PdCl_2(SMe_2)_2$] and [LiMe] at low temperature followed by addition of ligand. Similarly, two of the isolable complexes of weak donor ligands, [$PdMe_2$(TMEDA)][261] and [$PdMe_2$(pyridazine)]$_n$ (where *n* is probably 2),[120] are valuable reagents for the preparation of other complexes by displacement of TMEDA or pyridazine, in particular by phosphines, bidentate nitrogen donors, 1,4,7-trithiacyclononane (9S3)[264] and ligands containing groups that are sensitive to [LiMe]. The TMEDA complex [PdBrBz(TMEDA)] reacts with [LiMe] to form a 4:1 ratio of [PdMeBz(TMEDA)] and [$PdMe_2$(TMEDA)],[265] and intramolecular coordination complexes react also with [LiMe], for example, [PdCl(C~N~N)] to form (**80**)[176] and [PdCl(P~C~P)] to form (**81**) (R = H).[266]

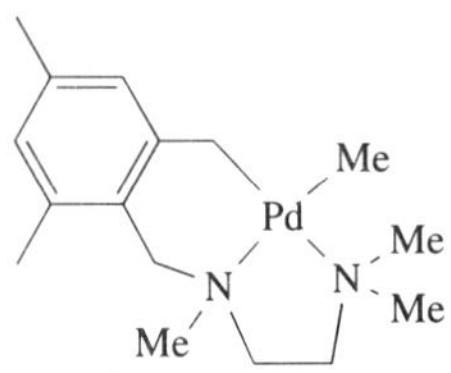

(**80**)

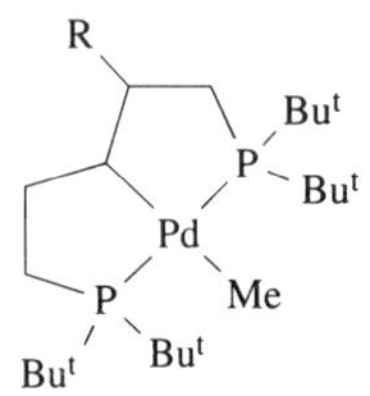

(**81**)

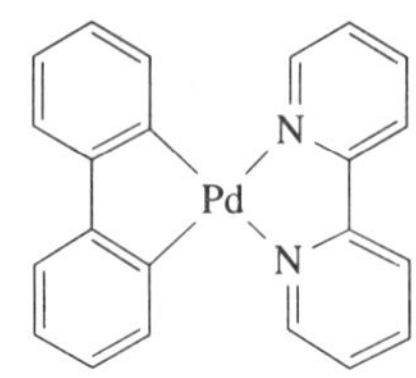

(**82**)

Palladacycloalkane complexes isolated from organolithium reagents are limited to five-membered rings in [$Pd(CH_2CH_2CH_2CH_2)(L_2)$], obtained from [$Li_2(CH_2)_4$] and [$PdCl_2(L_2)$] ($L_2$ = 2 PMe_2Ph, TMEDA, dppe, dppb, dcpe)[267–71] (dcpe = 1,2,-bis(dicyclohexylphosphino)ethane), and related complexes from [$Li_2(CH_2CMe_2CMe_2CH_2)$] and [$PdCl_2L_2$] (L_2 = TMEDA, dppe).[257,269]

Several simple *trans*-diarylpalladium(II) complexes have been isolated using [LiAr] reagents, such as *trans*-[$PdAr_2(PEt_3)_2$] (Ar = Ph, *p*-$NMe_2C_6H_4$, *o*-Tol),[48,239] *trans*-[$PdAr_2(PEt_2Ph)_2$] (Ar = Ph, *m*-Tol)[246,272] and a mixture of pentafluorophenylpalladium(II) species on the reaction of *trans*-[$PdCl_2(PEt_3)_2$] with [$Li(C_6F_5)$].[273] Intramolecular coordination complexes also react with [LiAr] reagents, for example, to form [Pd(*o*-Tol)(N~C~N)] (N~C~N as in (**33**) with $R^1 = R^2$ = Me)[274] and *cis*-[PdPh(C~P)(PPh_3)] (C~P as in (**34**)).[73] Complex (**82**) has been obtained from 2,2'-dilithiobiphenyl and [$PdCl_2(SEt_2)_2$] followed by addition of bipy.[275] The synthesis of monoaryl complexes is illustrated by the formation of [$PdCl(C_{10}H_6OMe$-*C*)$(SEt_2)_2$] from [$PdCl_2(SEt_2)_2$] and 8-methoxynaphthyllithium,[226] and complexes containing [Pd{C(Ar)=CMe_2}],[276] [Pd{C(=N_2)CO_2Et}][204] and [Pd{CH($POPh_2$)}][277] groups have also been prepared from organolithium reagents.

The less powerfully alkylating [MgRX] reagents have been used to synthesize a range of complexes including *trans*-[PdH(Me)$(PR_3)_2$] (R = Pr^i, Cy),[278] [$PdR(S_2CNMe_2)(PEt_3)$] (R = Pr^n, Pr^i, Bu^n, Bu^s),[214] [PdR{N($SiMe_2CH_2PPh_2$)$_2$-*P,N,P*}] (R = Me, CH_2CH=CH_2)[279] and complexes resulting from partial alkylation, for example, *trans*-[$PdBrMe(PEt_3)_2$][48] and [PdXR(dppe)].[271] Monoaryl complexes have also been obtained from [MgArBr], including complexes *trans*-[$PdBrPh(XEt_2)_2$] (X = Se, Te),[280] *trans*-[PdBr(*o*-Tol)$(PMe_2Ph)_2$][281] and *trans*-[$PdBrAr(PEt_3)_2$] (Ar = 2- or 4-vinylphenyl).[282] However, dialkylation is also possible; for example, [$MgBr(CH_2Bu^t)$] reacts with [$PdCl_2$(bipy)] to form [$Pd(CH_2Bu^t)_2$(bipy)],[257] [$MgCl(CH_2TMS)$] reacts similarly with [$PdCl_2(L_2)$] (L_2 = TMEDA,[264] dppe[283]), addition of ligands to a solution of [MgClMe] and *trans*-[$PdCl_2(NCPh)_2$] at −70 °C in THF gives [$PdMe_2$] complexes of PPh_3, $P(OPh)_3$ and dppe,[284] and addition of [MgR_2] to [$Pd(OAc)_2$] and excess L_2 at −78 °C gives *cis*-[PdR_2L_2] (R = Me, Bz, CH_2CMe_2Ph; L_2 = 2 PMe_3, dmpe) and [$Pd(CH_2TMS)_2$(dmpe)].[285]

An extensive halocarbon chemistry has been developed, including a perfluorinated nbd complex,[286] but principally involving C_6F_5, C_6Cl_5 and related complexes.[4,273,287] As for hydrocarbyl systems, [LiR] reagents are more reactive than [MgRX] and complexes of weak donor ligands are excellent substrates for subsequent metathesis reactions, examples being [$Pd(C_6F_5)_2(THT)_2$] and [$Pd(C_6F_5)_3(THT)]^-$. Complex ions such as [$Pd(C_6F_5)_2(\mu\text{-}X)]_2^{2-}$ (X = Cl, Br) and [$PdR_4]^{2-}$ (R = C_6F_5, C_6Cl_5) have been isolated, and a generally higher stability of C_6F_5 complexes noted.

Stable bromide-bridged anions [$PdR_2(\mu\text{-Br})]_2^{2-}$ (R = CH_2SO_2-*p*-Tol) have also been isolated on the reaction of [LiR] with [$NBu_4]_2[Pd_2Br_6$].[288]

Cyclopentadienyl and allyl groups may be present as ancillary ligands; for example, [PdCl(η-C_3H_5)(PCy_3)] reacts with [LiMe] to form [PdMe(η-C_3H_5)(PCy_3)], [Pd(η-C_3H_5)(μ-Cl)]$_2$ reacts at −78 °C to give an unstable species formulated as [Pd(η-C_3H_5)(μ-Me)]$_2$,[289] and [PdX(η-Cp)(PPh_3)] reacts with the potassium enolates of pinacolone and *t*-butyl acetate to form [Pd(CH_2COR)(η-Cp)(PPh_3)] (R = Bu^t, OBu^t).[290] Low-temperature NMR studies indicate that the black pyrophoric product obtained on reaction of [$MgEt_2$] with [Pd(η-C_3H_5)$_2$] in THF is formed via [PdEt(η-C_3H_5)(η^1-C_3H_5)]$^-$ and [$PdEt_2$(η-C_3H_5)]$^-$.[291]

Intramolecular coordination is a strategy widely used to stabilize metal–carbon bonds, and alkali metal derivatives [MR] of the anions (**83**)–(**87**) and the ligand systems shown in (**58**), (**70**) (R = H) and

(**75**) have been employed in the synthesis of palladium(II) complexes. Homoleptic complexes *cis*-[Pd(C~E)$_2$] have been obtained directly from [M(C~E)],[292–6] for example, C~E = (**83**)–(**86**) and as in (**58**) (E = PR$_2$, AsPh$_2$) and (**75**), and from reactions of [M(C~E)] with cyclometallated complexes [Pd(C~E)(μ-Cl)]$_2$, for example, C~N as in (**58**) (E = NMe$_2$). Mixed ligand complexes have been obtained by the latter method,[292–7] for example, [Pd{2-(NMe$_2$CH$_2$)C$_6$H$_4$-*C,N*}(C~E)] (C~E = (**84**), (**85**) and as in (**58**) (E = PR$_2$) and (**75**)), [Pd(L)(C~E)] (L = (**70**) (R = H), C~E = (**86**)* and as in (**58**) (E = NMe$_2$) and (**75**)), [Pd(L)(C~N)] (L = (**86**), C~N as in (**75**)), and complexes obtained from cyclopalladated azobenzene [Pd{2-(PhN$_2$)C$_6$H$_4$-*C,N*}(μ-Cl)]$_2$, although homoleptic species may also be obtained in mixtures of products.[297] Organolithium reactions on the coordinated ligands are also possible, illustrated by the reaction of *cis*-[Pd{2-(PPh$_2$CH$_2$)C$_6$H$_4$-*C,P*}$_2$] with [BuLi] followed by MeI to give methyl substitution on the ligand skeleton, *cis*-[Pd{2-(PPh$_2$CHMe)C$_6$H$_4$-*C,P*}]$_2$.[298] Complex (**88**)* is formed on the reaction of [Pd(NCMe)$_4$]$^{2+}$ with the dipotassium salt of 10,10'-dihydroxy-9,9'-biphenanthryl.[299]

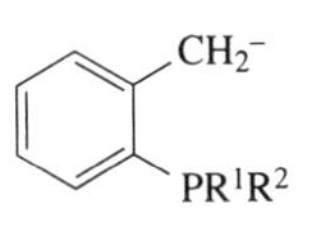

(**83**)

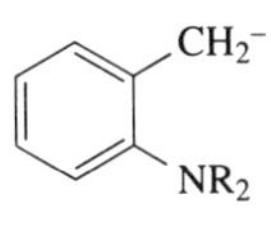

(**84**)

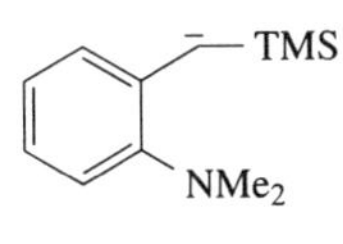

(**85**)

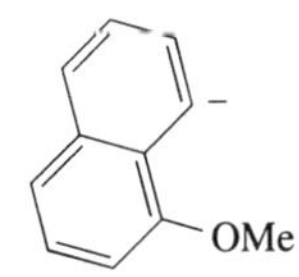

(**86**)

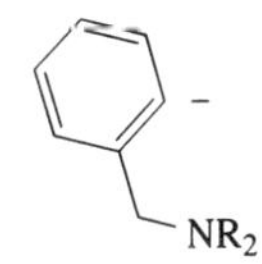

(**87**)

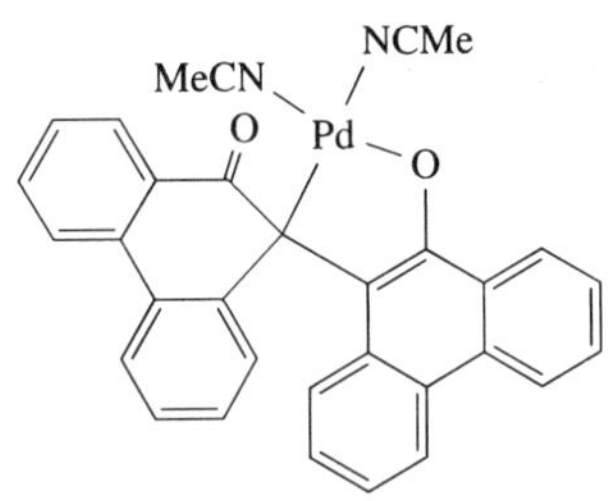

(**88**)

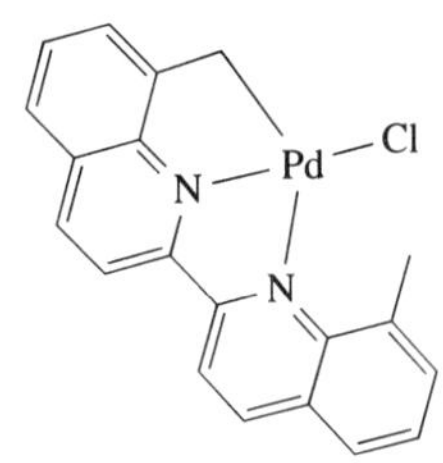

(**89**)

Complexes of the [N~C~N]$^-$ ligand shown in (**33**) are obtained on the reaction of [Li(N~C~N)] with [PdCl$_2$(cod)] in the presence of Br$^-$ to give [PdBr(N~C~N)] (R^1 = R^2 = Me, Et; R^1 = Me, R^2 = Bu^t).[300,301] Related complexes (**89**),[302] (**90**)[303] and (**91**)[304] are obtained by similar routes, with the formation of (**90**) accompanied by the presence of *cis*-[Pd(C~N)]$_2$. The complexes [Pd(C~N)(μ-Cl)]$_2$ (C~N as in (**58**) (E = NMe$_2$) and (**70**) (R = H)) react with [Li(CHRPPh$_2$)] (R = CN, CO$_2$Et) in 1:1 mol ratio to form dinuclear complexes,[305,306] such as (**92**)*, and a dinuclear complex (**61**)* (R = Ph, L = 4-Mepy) is also formed on the reaction of [PdCl$_2$(4-Mepy)$_2$] with [Li{2-(CHPh)C$_5$H$_4$N}].[184]

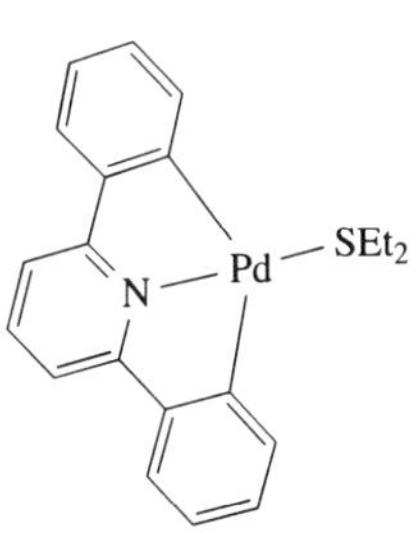

(**90**)

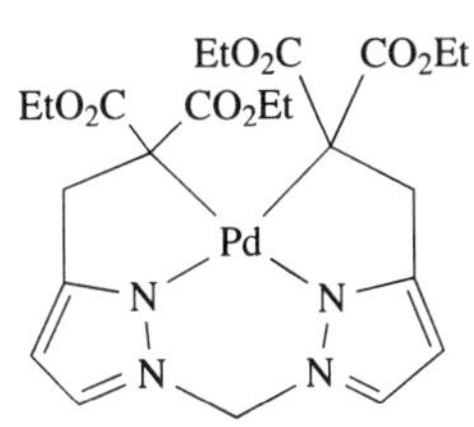

(**91**)

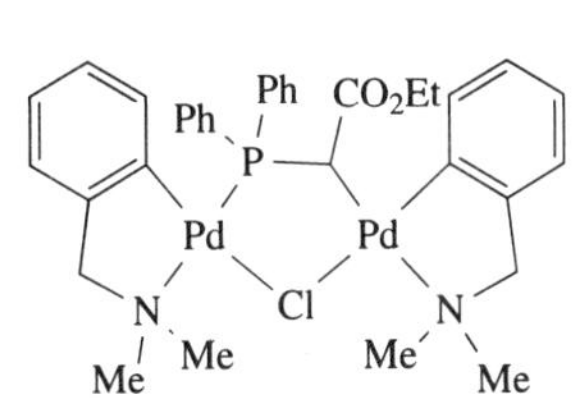

(**92**)

Anions of dialkyl malonates, and of related reagents containing acidic methylene groups, are formed in aqueous or nonaqueous solutions on addition of bases, and they react readily with palladium(II) salts to form complexes,[307–12] for example, Na$_2$[Pd{CH(CO$_2$Et)$_2$}$_2$(μ-Cl)]$_2$,[311] *trans*-[Pd{CH(COMe)(CO$_2$Et)}$_2$(2-Mepy)$_2$][307] and *cis*-[Pd{CH(CO$_2$Me)$_2$}$_2$(bipy)].[312] This strategy has resulted in the synthesis of a series of intramolecular coordination complexes for reagents containing electron-withdrawing groups,[313–21] representative examples being (**93**), (**94**) and (**95**).

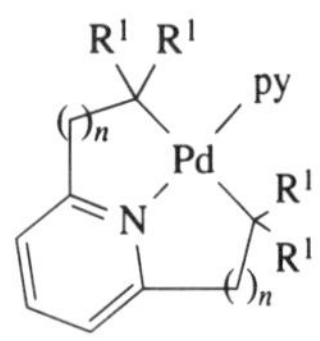
(93)

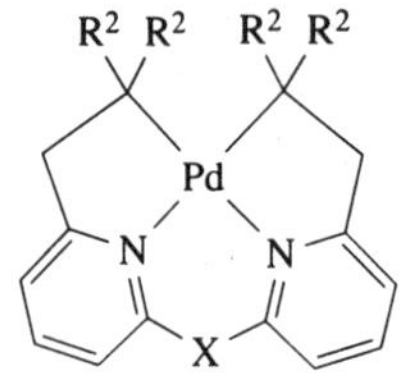
(94)

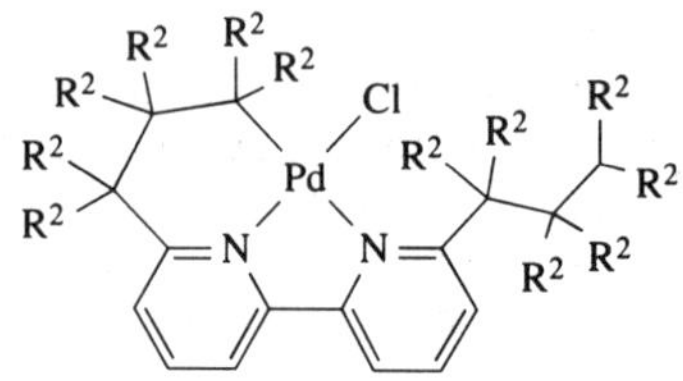
(95)

$R^1 = CO_2Et$, $n = 1$; $R^1 = CO_2Me$, $n = 2$; $R^2 = CO_2Me$, X = —, CH_2CH_2, CH=CH, C=O, CHMe(CN)

5.5.1.3 From cyclopalladation reactions

The stability of organometallic compounds is enhanced when the organic group participates in intramolecular coordination, and the ability of palladium(II) to undergo facile cyclopalladation reactions (Equation (2)) has led to a rapid and continuing development of organopalladium chemistry based on this strategy. Recognition that intramolecular coordination complexes have applications in organic synthesis, resolution of enantiomers, catalysis, photochemistry, bioinorganic chemistry and ordered mesophases has been important in guiding the development of cyclopalladation chemistry. Extensive reviews have been published,[263,320,322–35] including recent detailed discussions of mechanisms.[176,322,323] Diagrams in this section are, in general, limited to showing the basic ligand skeleton with omission of substituents.

PdX_2 + HC~E ⟶ 1/2 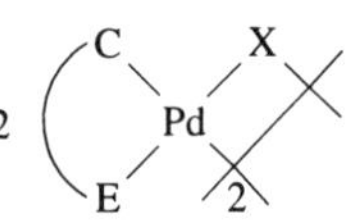 + HX (2)

The most widely used synthetic procedures include reaction of the organic substrate with (i) $[\{Pd(OAc)_2\}_3]$ in an organic solvent or acetic acid; (ii) $[PdCl_4]^{2-}$ in methanol and/or water; and (iii) $[PdCl_4]^{2-}$ with NaOAc in methanol or acetic acid. Reaction mixtures, which are often heterogeneous, may require heating. In some cases trinuclear complexes of formulation $[\{Pd(C\text{~}N)(\mu\text{-}OAc)_2\}_2Pd]$ are obtained.[336–9] Less widely used reagents include [Pd(β-diketonate)$_2$],[340–2] and directly from coordination complexes $[PdX_2(E\text{~}CH\text{-}E)_2]$ which may be isolated under milder (or different) conditions or as by-products of cyclopalladation. Complexes may also be generated by exchange of ligands in acidic media,[343–9] with representative examples shown in Equations (3)[347] and (4).[346] The forward reaction favours ligands with electron-withdrawing groups, and the mechanism appears to involve dissociation of N~CH to give coordinatively unsaturated palladium(II) species that are active toward cyclopalladation.[344]

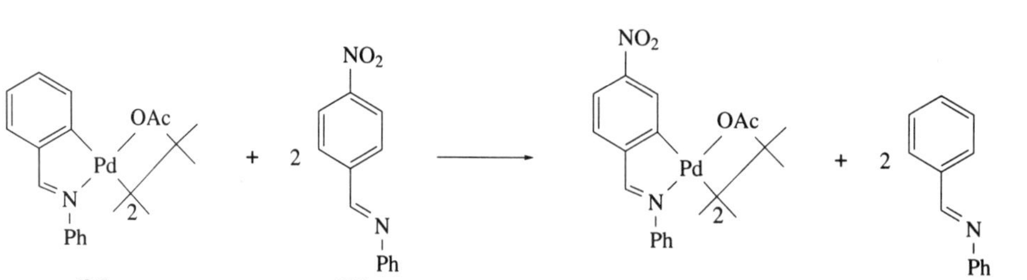
(96) **(97)** **(98)** (3)

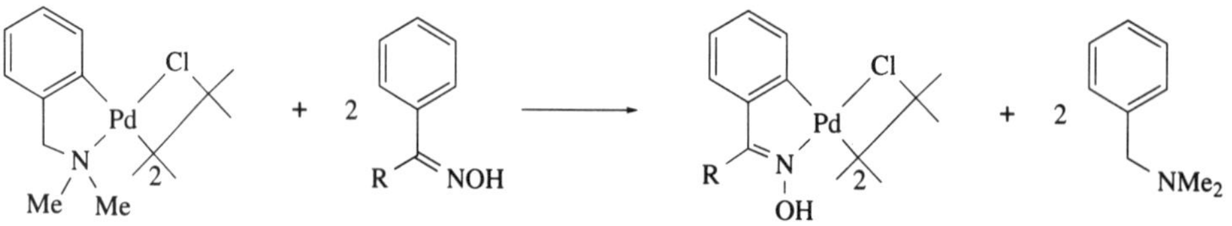
(99) **(100)** (4)

Mechanistic and synthetic studies of the reaction shown in Equation (2) have been dominated by nitrogen donor systems. In general, an electrophilic mechanism appears to operate, probably via a 14-electron intermediate,[330,350,351] such as '$[Pd(HC\text{~}E\text{-}E)(OAc)_2]$', where the Pd···HC interaction occurs in the coordination plane. For cyclopalladation of aryl groups, intermediate(s) may resemble those shown in model complexes (**101**)* (η^1-arene),[352] (**102**)* (η^2-arene)[353] and (**103**) (arenonium).[354] Consistent with an electrophilic mechanism, cyclopalladation generally occurs on the aryl ring with more electron-donating groups,[355–60] for example (**104**) ($R^1 = NMe_2$, $R^2 = NO_2$, X = OAc)[357] and (**105**) (R^1 = Me, $R^2 = NO_2$),[358] although cyclopalladation of the ring with an electron-withdrawing group is favoured in the exchange reaction between (**96**) and (**97**).

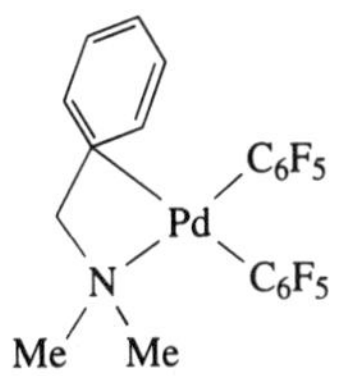

(**101**)

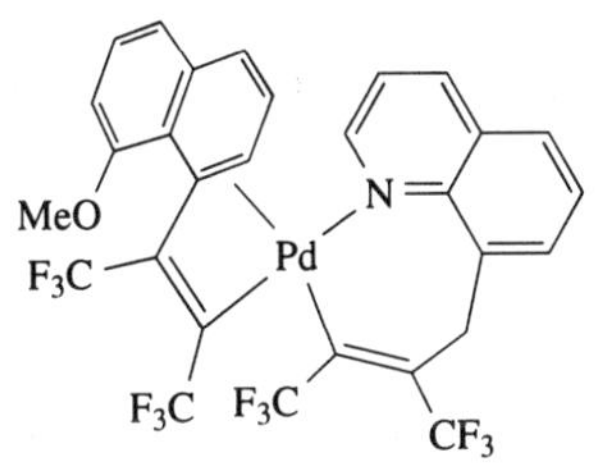

(**102**)

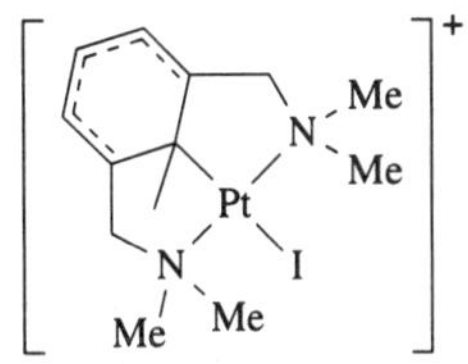

(**103**)

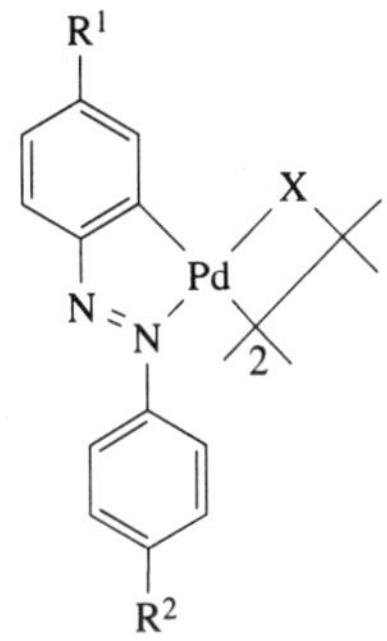

(**104**)

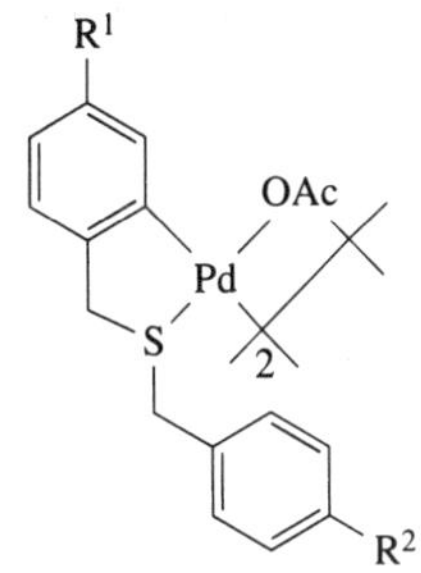

(**105**)

Five-membered chelate ring formation is generally favoured over four- and six-membered rings, and palladation of sp^2 is favoured over sp^3 carbon atoms, but there is a complex interplay of factors operating in these selectivities that renders general principles applicable to all ligand systems difficult to establish. Thus, although very few four-membered rings have been synthesized, requiring bulky *t*-butylphosphines with the chelate skeleton shown in (**106**), an increasing number of six-membered ring complexes, such as in (**80**)[176] and (**107**)–(**121**), and Pd–C(sp^3) systems, for example, (**70**) (R = H,[351,395] Me,[351] TMS[207,208]), (**80**), (**81**) (R = H,[266,396] Me[397]), (**84**),[339,398] (**106**), (**109**), (**112**), (**116**), (**122**)–(**136**) and (**137**), (**142**) and (**145**) in Scheme 17, have been obtained by cyclopalladation. Changes in reagent and/or solvent conditions affect the course of palladation for some ligands, for example, selectivity in Ar^1 or Ar^2 palladation for $NMe(CH_2Ar^1)(CH_2Ar^2)$,[344] formation of (**124**) (MeOH, R = MePh) or (**126**) (benzene, R^1R^2 = MePh), and the reactions shown in Scheme 17.[351,414,417,418]

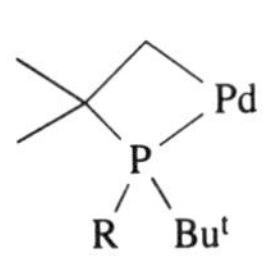

(**106**)
R = Bu^t,[361,362] Ph[363]

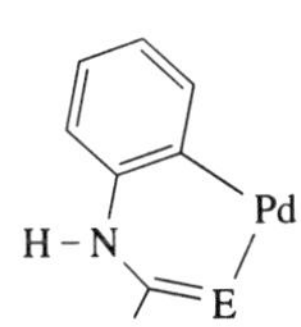

(**107**)
E = O,[364–369] NR2[365,370,371]

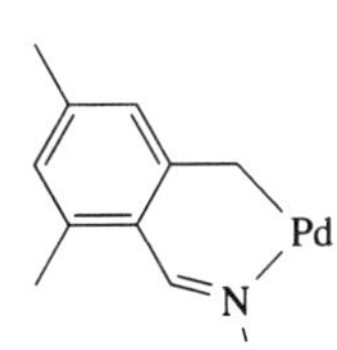

(**108**)[372,373]

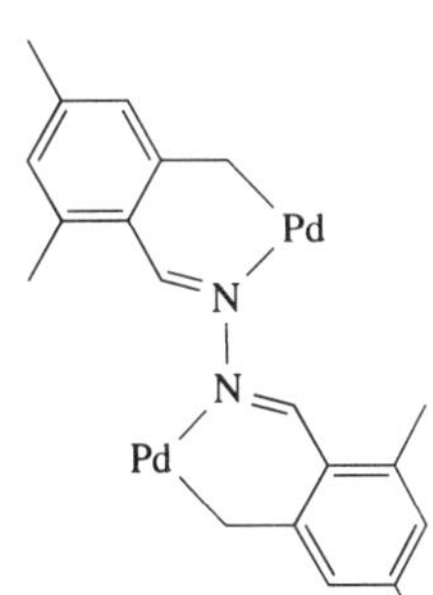

(**109**)[374]

Selectivity in the position of palladation is of major interest, as illustrated by (**98**), (**104**), (**105**), (**137**) and (**139**), (**140**) and (**142**), and (**144**) and (**145**). Both symmetric and unsymmetric diarylazo compounds have been extensively studied,[341,347,357,419–25] and for the 1-arylazonaphthalenes (**111**), palladation occurs at C-2 ($R^1 = R^2 = R^3 = R^4$ = H), C-1 (R^2 = OMe, R^3 = H, R^4 = Me), or as in (**111**)

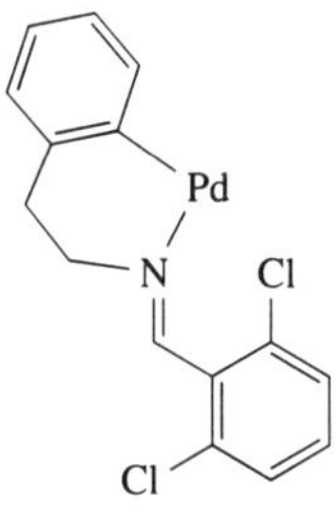

(110)[375]

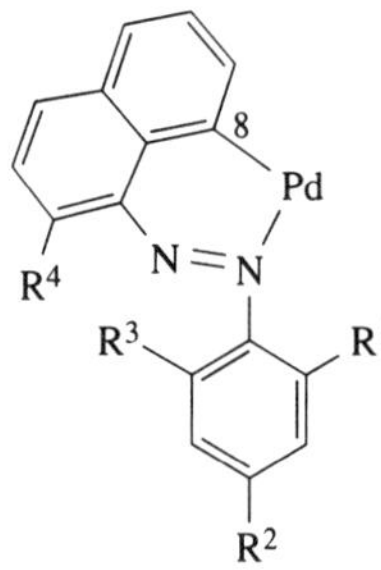

(111)[376,377]

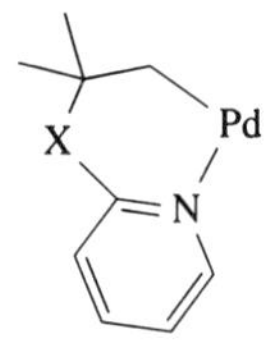

(112)
X = CO,[378] CH_2[379]

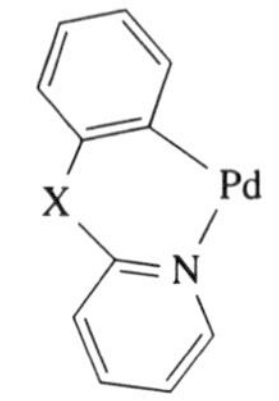

(113)
X = CO,[380] NH[381]
CH_2,[380–384] CHPh[385]

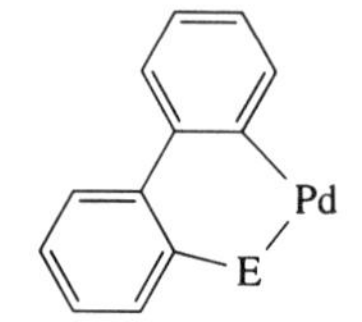

(114)
E = NMe_2,[386,387] SMe[349]

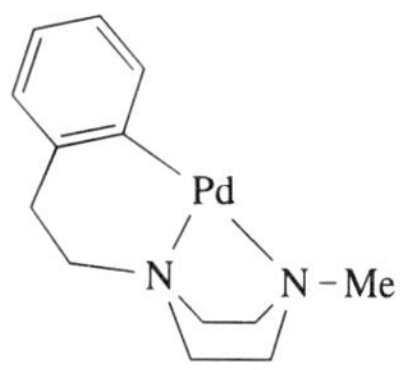

(115)[176]

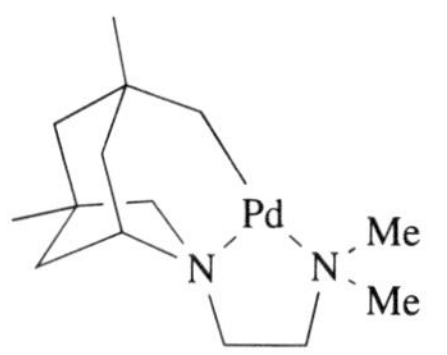

(116)[176]

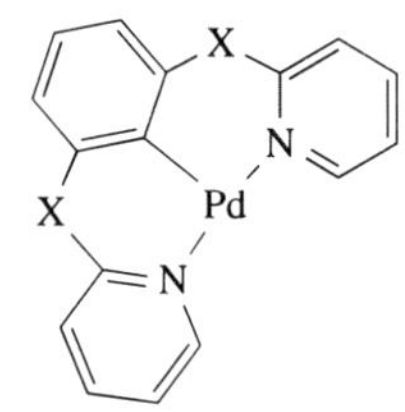

(117) X = NH,[388] CO, CHMe[389]

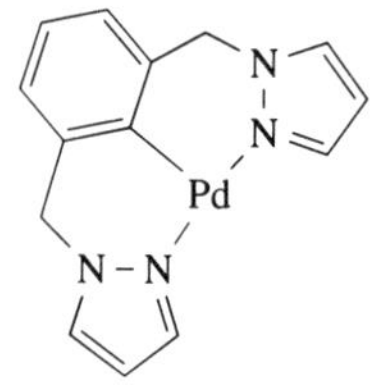

(118)[390]

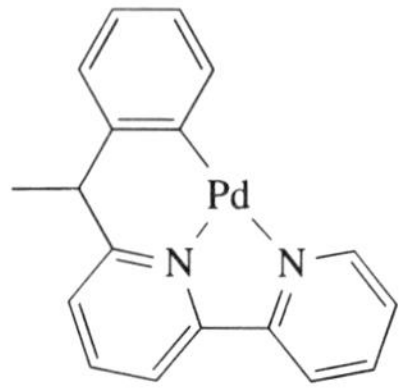

(119)[391]

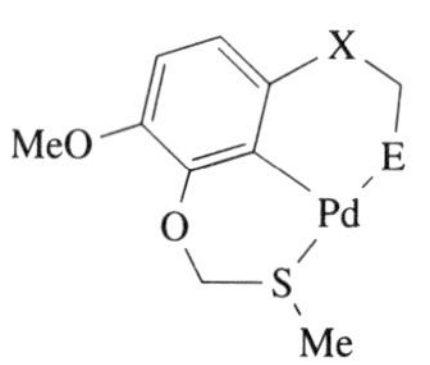

(120)
E = SMe; X = O, CH[392]
E = N(R)Me; X = —[393]

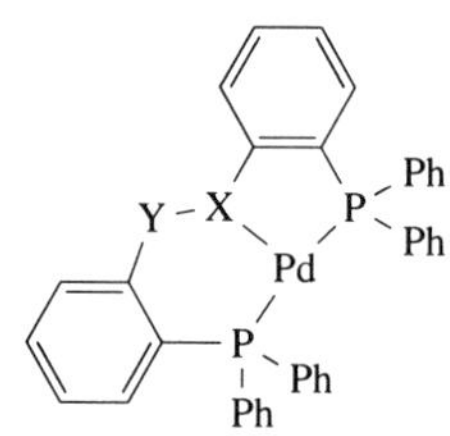

(121)[394]
X–Y = CH–CH_2, C=CH

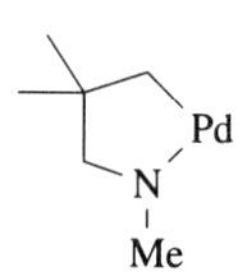

(122)[336]

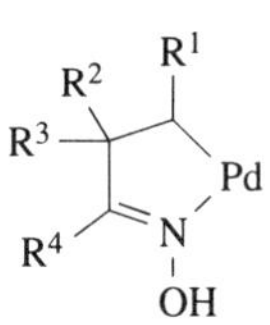

(123)[399–403]

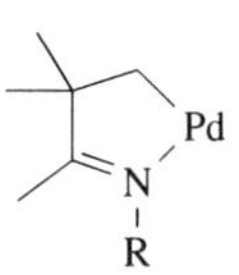

(124)
R = N=CBu^{t}Me,[399] NMePh[404]

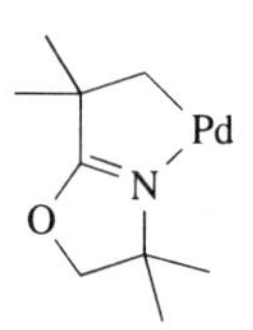

(125)[337]

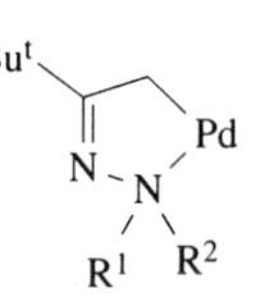

(126)[404,405]

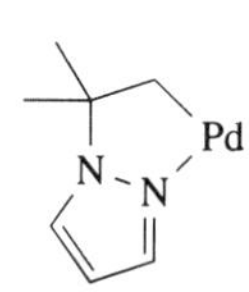

(127)[338]

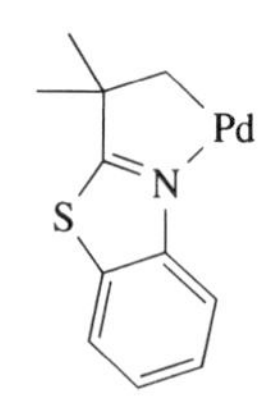

(128)[406]

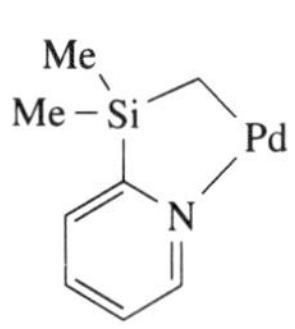

(129)[407]

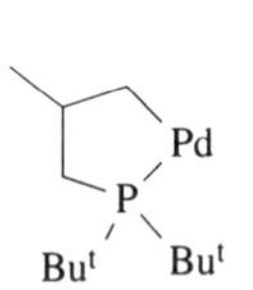

(130)[408]

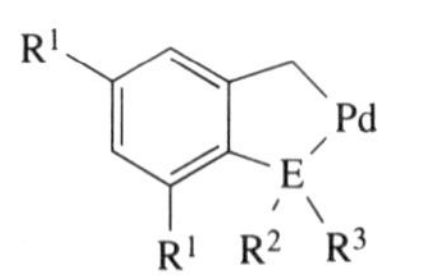

(131)
E = P: R^1 = H, R^2 = R^3 = *o*-Tol,[409]
R^2 = *o*-Tol, R^3 = Bu^t
E = As or P: R^1 = Me, R^2 = R^3 = Mes[410–412]

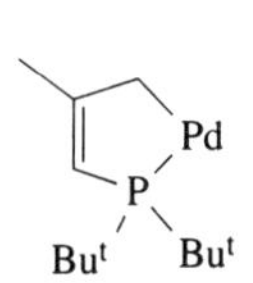

(132)[413]

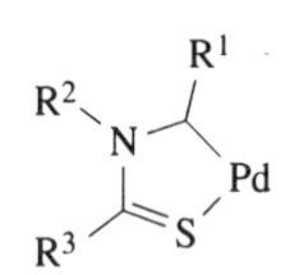

(133)[349,414,415]

(R^1 = R^3 = R^4 = Me, R^2 = OH), depending on which position is blocked sterically, establishing the preferred sites for C(sp^2) activation as C-4 > C-1 > C-8.[377]

Tertiary benzylamines react readily to form palladacycles,[340,356,395,426–38] with applications in organic synthesis[139,333] and in the resolution of enantiomers,[439–44] such as (**146**) and (**147**), from which the optically active arsines were isolated.[444] Cyclopalladation of primary benzylamines involves unusual conditions, for example, addition of $AgBF_4$ to *trans*-[$PdI_2(NH_2CH_2Ph)_2$] in ethyl acetate to give [Pd{2-

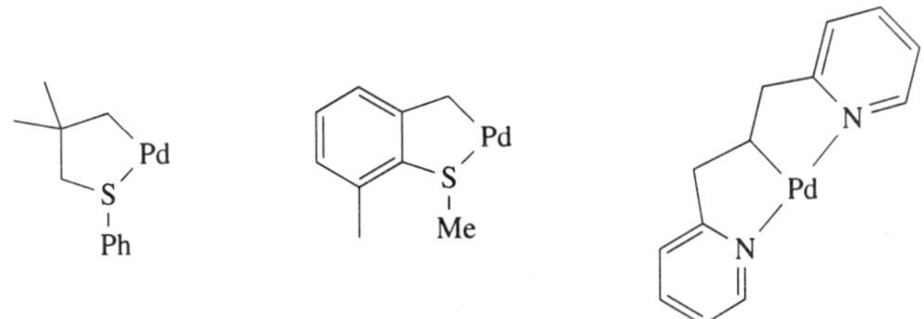

(134)[349] **(135)**[349] **(136)**[416]

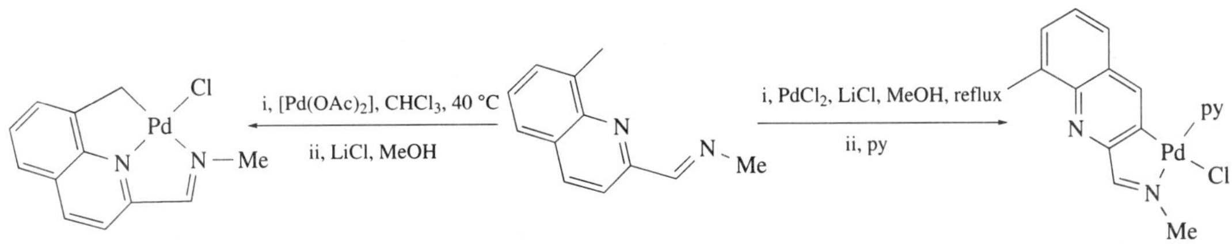

(137) **(138)** **(139)**

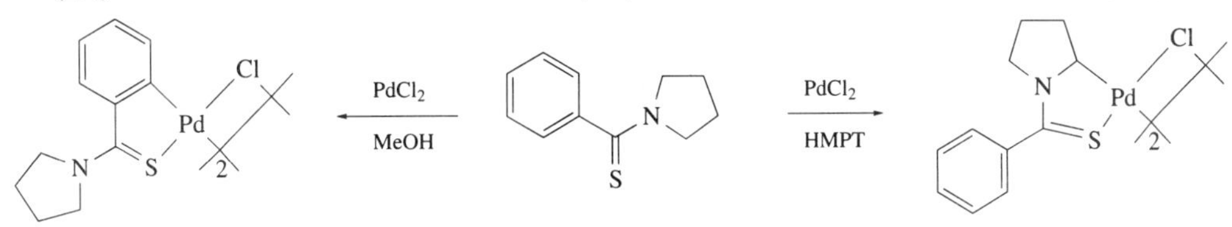

(140) **(141)** **(142)**

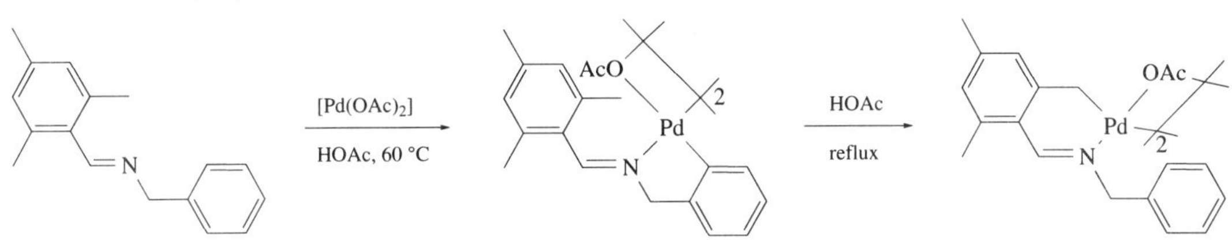

(143) **(144)** **(145)**

Scheme 17

(NH$_2$CH$_2$)C$_6$H$_4$-*C,N*}(μ-I)]$_2$.[445] For secondary benzylamines, cyclopalladation by [PdCl$_4$]$^{2-}$ requires bulky R^1 and a substituent R^2 in PhCHR^{2}NHR^1,[436,446] but cyclopalladation by more reactive [Pd(OAc)$_2$] occurs readily for BzNHMe on heating at 50 °C in benzene for 1 d.[447] Oximes are also important in organic synthesis, with cyclopalladated intermediates involving alkyl (**123**) or aryl groups (**100**).[341,346,448–51] Studies of cyclopalladated metallocenes[331,452–7] mainly involve ferrocene derivatives, such as (**148**) and (**149**). The formation of optically active (**148**) by Na$_2$[PdCl$_4$] in the presence of the sodium salt of *N*-acetyl-D-leucine is consistent with carboxylate coordination during the reaction,[331] as is the assistance that sodium acetate provides for cyclometallation reactions using [PdCl$_4$]$^{2-}$. Arylimines Ar^{1}C=N(Ar^2)[70,458–68] and arylalkylimines ArC=NAlk[139,245,433,458,461,468–76] have been investigated in considerable detail, including studies of selectivity in palladation position for unsymmetrically substituted Ar^1 and Ar^2 groups, selectivity for aryl palladation when Alk = Bz[477,478] and CH$_2$CH$_2$Ar,[479] and the synthesis of diazepam and prazepam derivatives.[480–2] Cyclopalladation of aldehydes[483–5] and 2-vinylpyridines[486] has been reported, for example, to form [Pd{2-NMe$_2$C$_6$H$_4$C(O)-*C,N*}(μ-Cl)]$_2$ and [Pd{2-[HC=C(CH$_2$OH)]C$_5$H$_4$N-*C,N*}(μ-Cl)]$_2$, and the aryl group palladated systems (**59**)[487] and (**150**)–(**152**) have also been studied.

An extensive series of Ar^1–Ar^2 species have been cyclopalladated, illustrated by the groups in (**76**),[340,341,494–8] (**149**) and (**153**)–(**167**), including a catenand complex exhibiting motif (**167**).[524] Cyclopalladation to form tridentate intramolecular coordination complexes is well established ((**81**) (R = H, Me),[266,396] (**115**)–(**121**), (**136**), (**137**), (**149**) (E = 6-bipy) and (**164**)–(**179**)), and a few tetradentate complexes have also been isolated ((**180**) and (**181**)) including a nitrogen analogue of (**180**).[546]

The initial report of dipalladation of *N,N,N',N'*-tetraethylxylenediamines[547] has been reexamined and extended to include other donor groups and ring systems.[548–52] Polynuclear complexes may be readily characterized by bridge-splitting reactions typical of acetato- and halo-bridged complexes, for example, to form acetylacetonato complexes (**182**)–(**186**). Dipalladated complexes in which the palladium atoms are bonded to different rings have been reported,[507,553–8] for example, (**109**), (**173**) and (**185**)–(**188**),

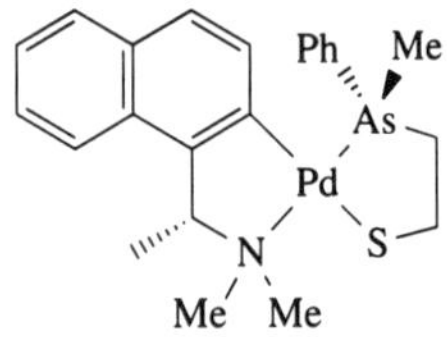

(146)

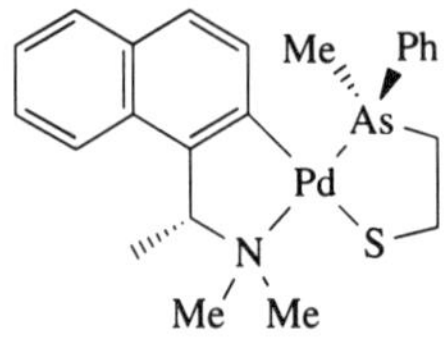

(147)

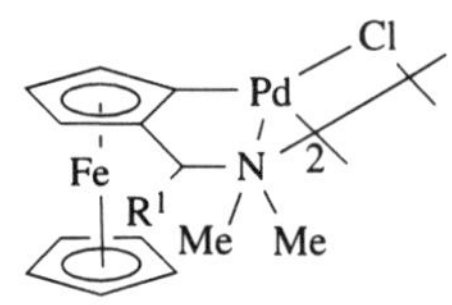

(148) $R^1 = (CH_2)_5CO_2R^2$

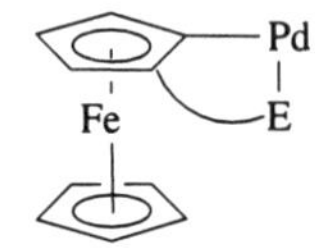

(149)[456]
M = Fe, Ru, E = 2-py (C~N)
6-bipy(C~N~N)

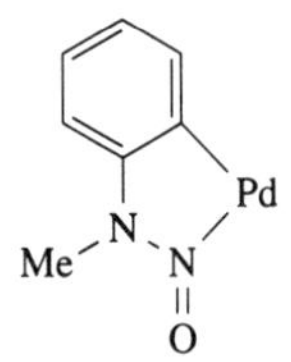

(150)[369,432,488,489]

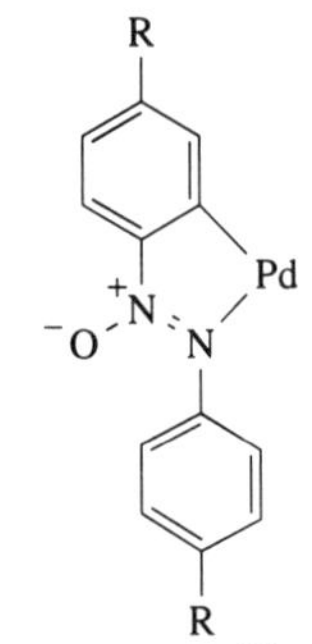

(151)[490]

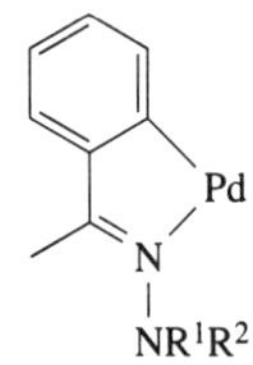

(152)[491–493]

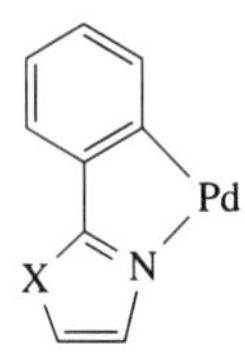

(153)
X = NR,[499,500] O,[501] S[501–503]

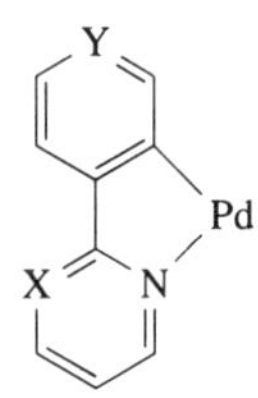

(154)
X = CH, Y = N[504]
X = N, Y = CH[425,505–507]

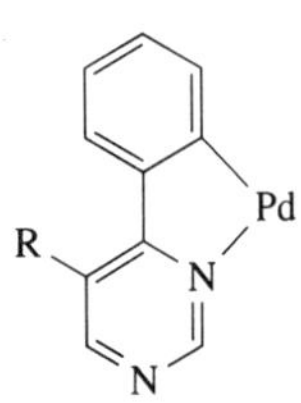

(155)
R = H, Me, Ph, p-$NO_2C_6H_4$[507]

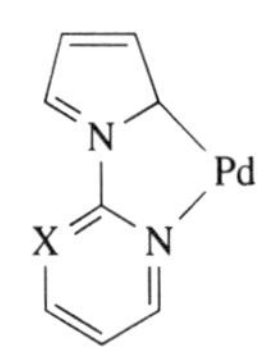

(156)
X = N,[508] CH[509]

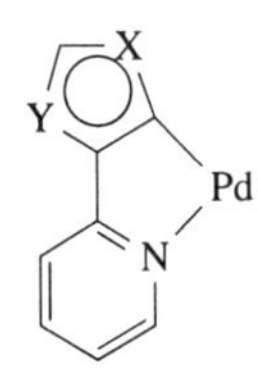

(157)
X = S, Y = CH[510]
X = CH, Y = S[510,511]

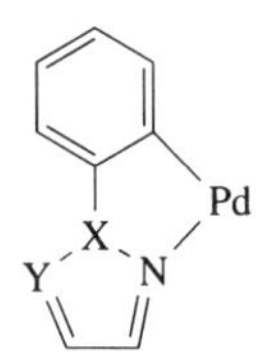

(158)
X = Y = N[512]
X = N, Y = CH[513,514]

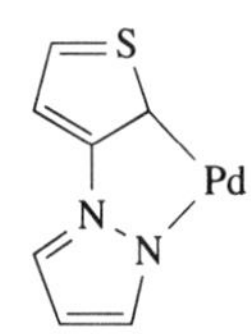

(159)[515]

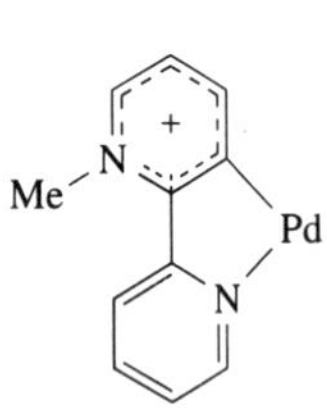

(160)[516,517]

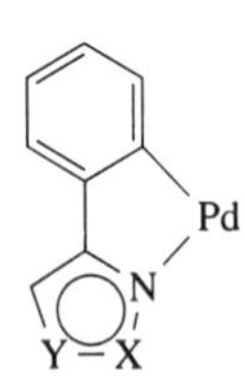

(161)
X = NMe, Y = CH[178]
X = CH, Y = NMe[518]

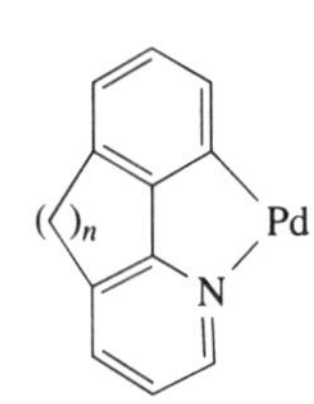

(162) n = 1–4[342,497]

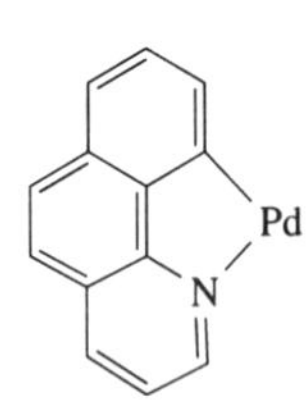

(163)[497]

where (**188**) shows the product of reaction of PEt_3 with the initial insoluble product of cyclopalladation. Monopalladated products for the ligands shown in (**183**), (**185**), (**186**) and (**188**) have also been isolated.[507,552,553,557,559,560] Acetato-bridged complexes of dinucleating ligands react with a range of nitrogen donor ligands to form complexes with the latter present as a $[C{\sim}N]^-$ bridge;[561–4] for example, 2-vinylpyridine reacts in MeOH as solvent to form a $[2\text{-}C(CH_2OMe)C_5H_4N\text{-}C,N]^-$ bridge and N-propylimidazole forms (**189**) with the same dinucleating system.[564]

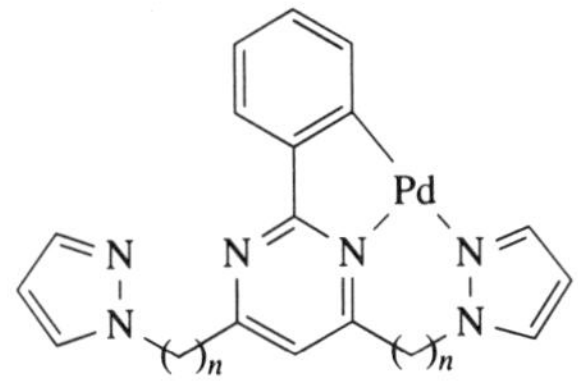

(164)[519] n = 0, 1

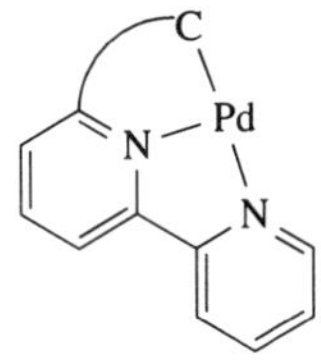

(165)
C = 2-thienyl,[520] C_6H_4[521,522]

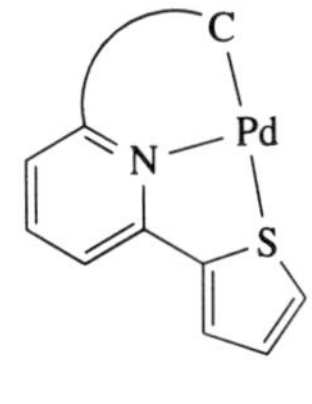

(166)
C = 2-thienyl, C_6H_4[523]

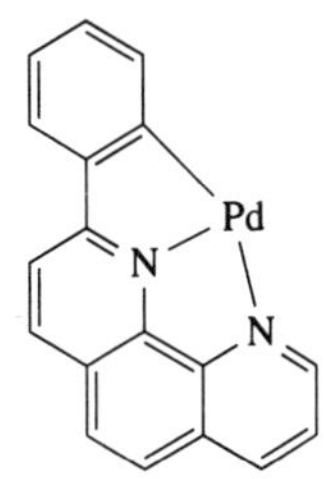

(167)[524,525]

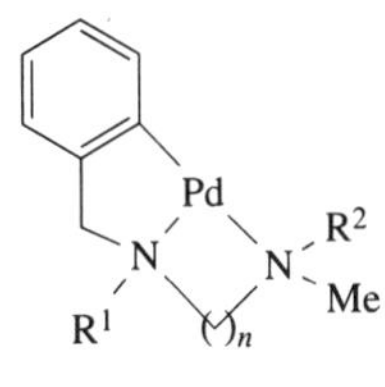

(168)
n = 2: $R^1 = R^2$ = Me[176]
n = 3: R^1 = Me, R^2 = Bz[526]

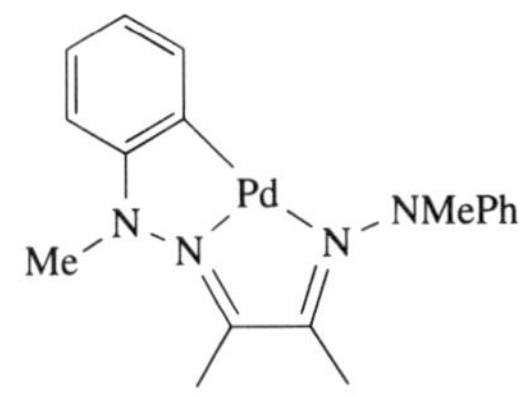

(169)[527]

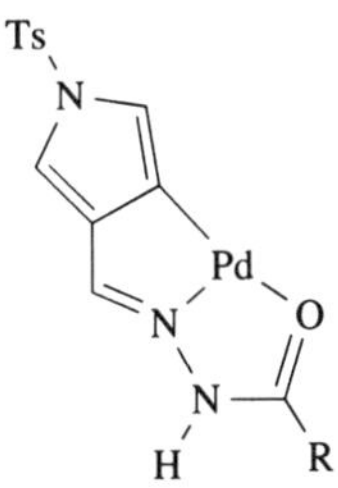

(170)[528]

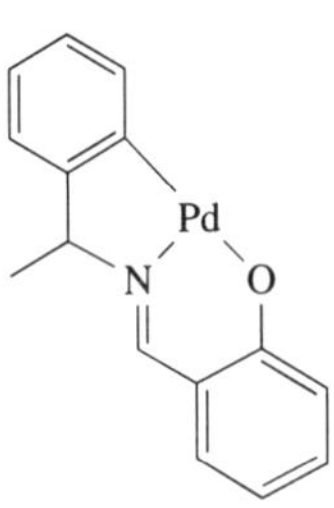

(171)[529]

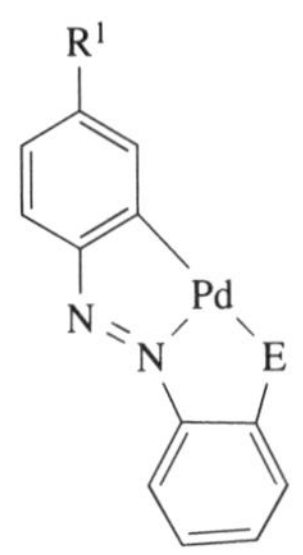

(172)
E = O,[530,531] SR^2,[532–534] $S(O)R^2$[535]

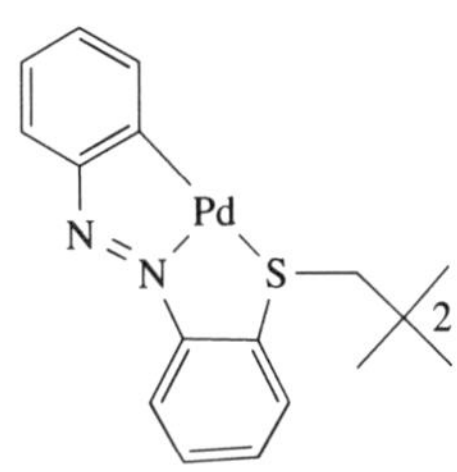

(173)[536]

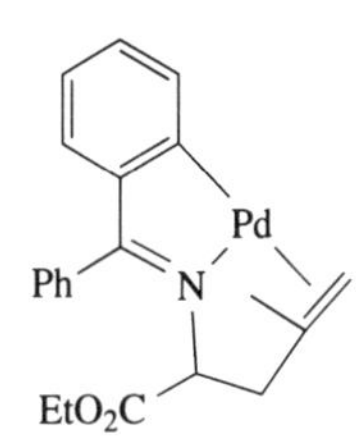

(174)[537]

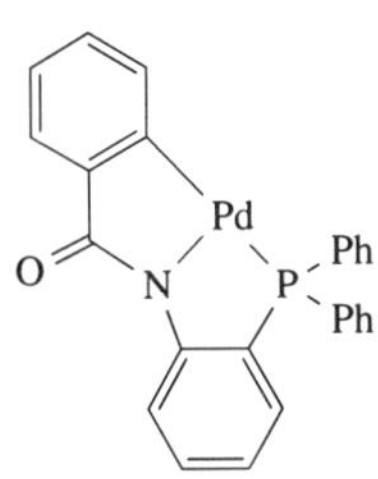

(175)[538]

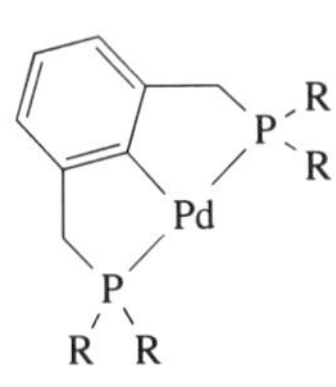

(176)
R = Bu^t,[539] Ph[540]

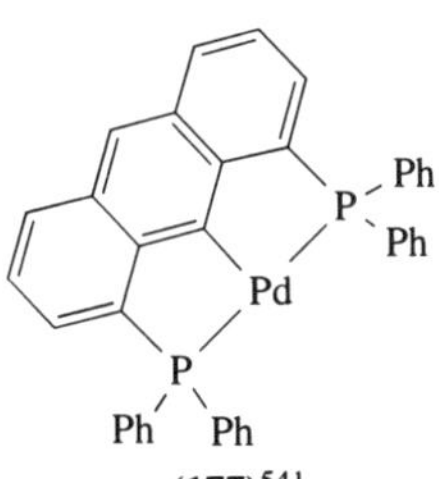

(177)[541]

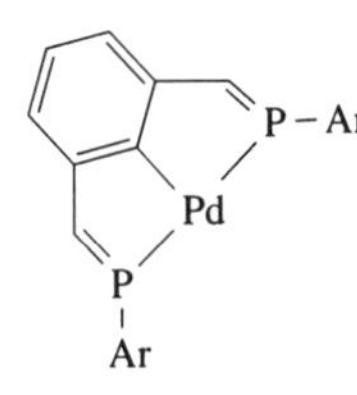

(178)[542]
Ar = 2,4,6-$Bu^t_3C_6H_2$

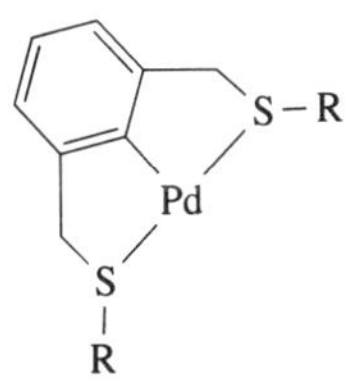

(179)
R = Me,[349] Bu^t[543]

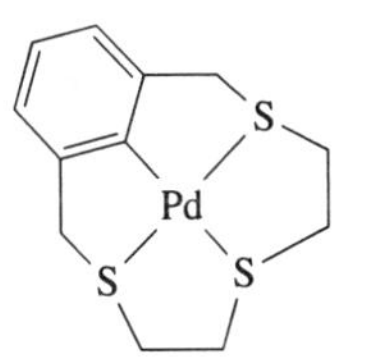

(180)[544]

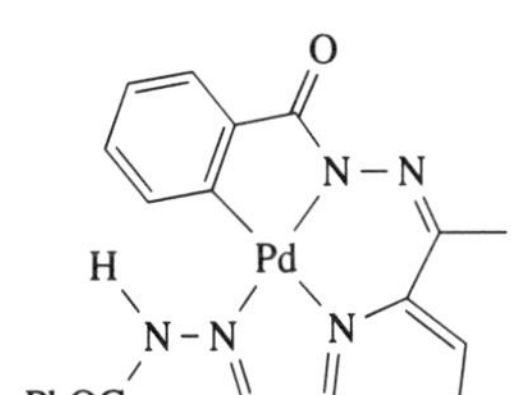

(181)[545]

Complexes of triaryl phosphites and $P(OAr)(OEt)_2$ undergo cyclopalladation in high-boiling solvents; for example, *cis*-$[PdI_2\{P(OPh)_3\}]$ reacts with PdI_2 to form $[Pd\{2\text{-}[P(OPh)_2O]C_6H_4\text{-}C,P\}(\mu\text{-}I)]_2$[565] in which the ligand skeleton is identical to that of (**34**). The presence of bulky groups encourages

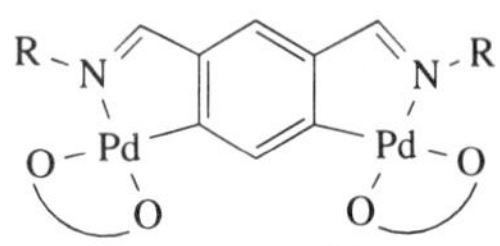

(182)[551]

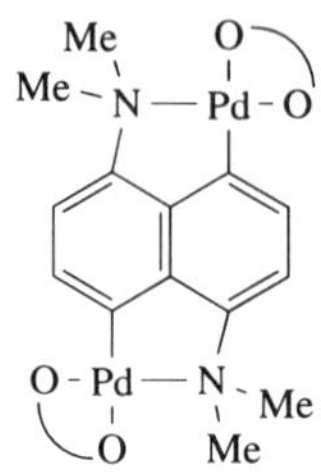

(183)[552]

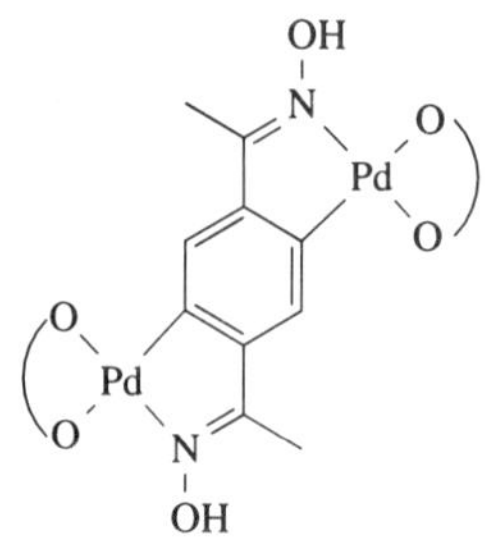

(184)[552]

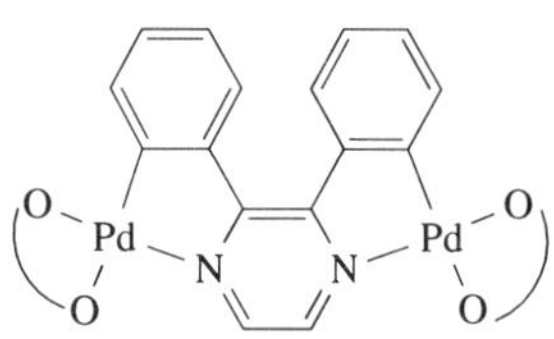

(185)[553]

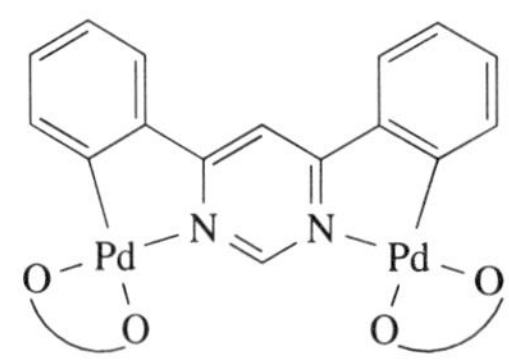

(186)[507]

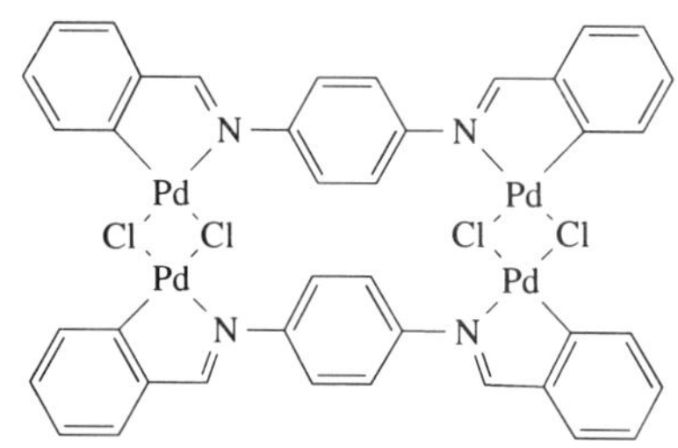

(187)[554]

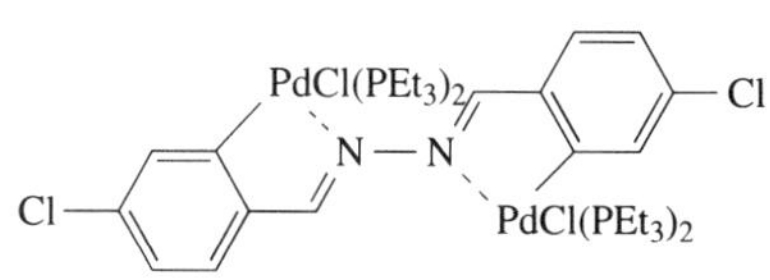

(188)[555]

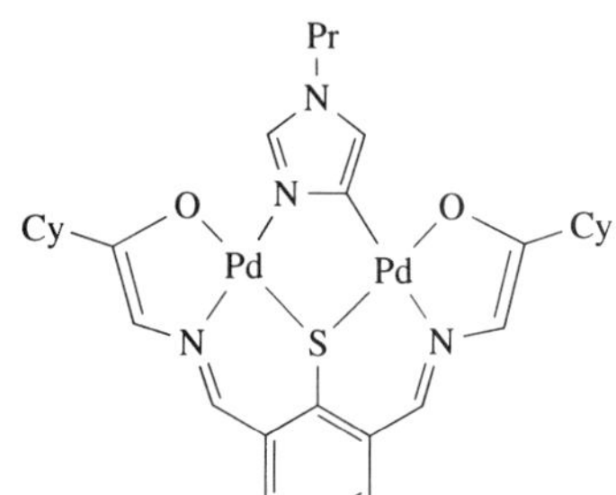

(189)

cyclopalladation of alkyl- and arylphosphines,[566–9] illustrated by (**130**), (**131**) and the facile cyclopalladation of PBu^t_2Ph and PBu^t_3 to form (**106**). Cyclopalladation of $P(CH_2Ph)Ph_2$ occurs with palladium(II) acetate at the *ortho*-position of the benzyl group.[566]

Complexes containing oxygen donor groups are shown in (**107**), (**120**) and (**170**)–(**172**), and the acetanilide system (**107**) has applications in organic synthesis. Cyclopalladation of sulfur donors is illustrated in (**105**),[358] (**114**), (**120**), (**133**)–(**135**), (**140**), (**142**), (**166**), (**172**), (**173**), (**179**), (**180**) and in the cyclopalladation of 1-thiomethylnaphthalene[349] and dibenzyl sulfoxides.[570] Palladation of the phenyl ring of 2-$NO_2C_6H_4SN(H)Ph$,[571] aryl rings of thiones[341,349] and related molecules,[572] the furan and thiophene rings of 2-$[Me_2NC(E)]C_4H_3X$ (X = O, E = S, Se; X = S, E = Se) and 3-$[Me_2NC(E)]C_4H_3X$ (X = O, S; E = Se),[573–5] and the substituted cyclopentadienyl ring of $[Fe(\eta\text{-}Cp)\{\eta\text{-}C_5H_4C(S)R\}]$ (R = Bu^t, NMe_2)[576] occurs to form five-membered rings. The thioethers $PhCH_2SR$ (R = Me, Bu^i, Bu^t, Ph) undergo similar cyclopalladation reactions for the benzyl group,[577,578] and $PhSCH_2Bu^t$ also undergoes palladation at a methyl group to form a five-membered ring as illustrated in (**134**).

Intramolecular coordination complexes are readily characterized by NMR spectroscopy, with recent examples including ^{13}C,[432,579] ^{15}N,[432] 2D NOESY[433] and studies of fluxionality;[384,407,420] applications of FAB[481] and electrospray[580] mass spectrometry for characterization have also been reported.

5.5.1.4 From palladium(II) complexes and alkenes

Addition of a nucleophile to an alkene catalysed by palladium(II) is an important procedure in organic synthesis, and Pd–C σ-bonded species are involved as intermediates in these processes.[329,581,582] Two closely related general mechanisms are established, involving either internal attack by a coordinated nucleophile on a coordinated alkene, also regarded as alkene insertion reactions, or external attack by a nucleophile on a coordinated alkene. Nucleophiles known to attack externally include hydroxide, alkoxide, carboxylate, malonate, β-diketonate, azide, chloride, *t*-butyl peroxide[583] and amines. External attack is expected to lead to *trans*-addition, as established for the reaction of Equation (5),[584–6] and the reactions of $[CH(CO_2Me)_2]^-$ and OH^- with $[Pd(\eta\text{-Cp})(PPh_3)(\textit{trans}\text{-CHD=CHD})]^+$ and $[PdCl_2(cod)]$ to give (**193**) and (**194**) (R = H), respectively.[587,588] In studies of low-temperature reactions related to that giving (**192**)* via (**191**), with ethene as the alkene and Et_2NH as the nucleophile, it has been shown that an intermediate zwitterion *trans*-$[Pd^-Cl_2(CH_2CH_2N^+Et_2H)(NEt_2H)]$ is formed prior to chelate ring formation.[589]

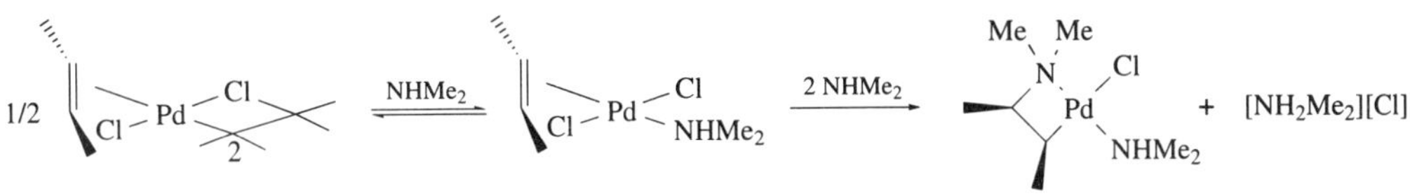

(5)

(**190**) (**191**) (**192**)

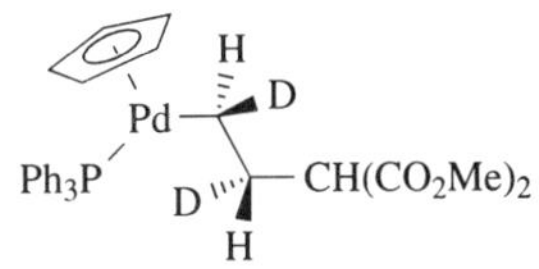

(**193**)

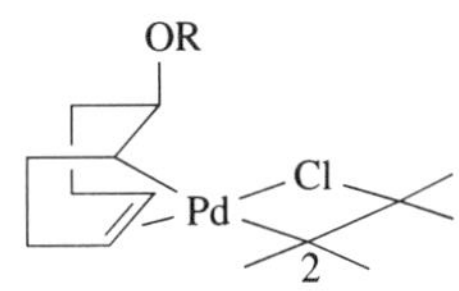

(**194**)

A range of coordinated diene complexes have been studied,[588,590–5] including the reaction of $[PdCl_2(\textit{cis}\text{-1,3-divinylcyclohexane})]$ with $[Ag(O_2CMe)]$ to give (**195**),[594] and methoxide addition to $[PdCl_2(1,4\text{-cod})]$ to form (**196**).[592] The double bond of the norbornene ring in $[PdCl_2(\text{dicyclopentadiene})]$ is unsymmetrically coordinated[596] and is the preferred site for nucleophilic attack, consistent with theoretical studies indicating that unsymmetrical binding assists with activation of the double bond[597] in addition to the effect of binding to an electrophilic palladium(II) centre.[598] Theoretical analyses are also consistent with the observed nucleophilic attack at the terminal carbon of conjugated dienes.[599] Alkenes containing a cyclopropane ring react with *trans*-$[PdCl_2(NCPh)_2]$ to form σ,π-complexes in which the ring has been cleaved,[600–2] as illustrated in Equation (6).[602]

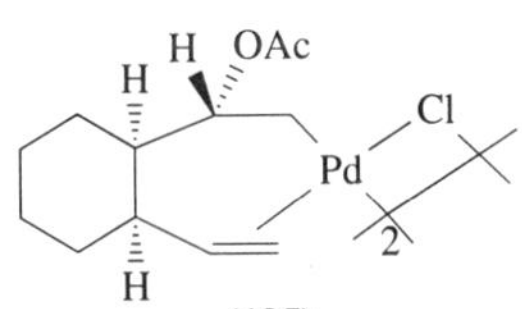

(**195**)

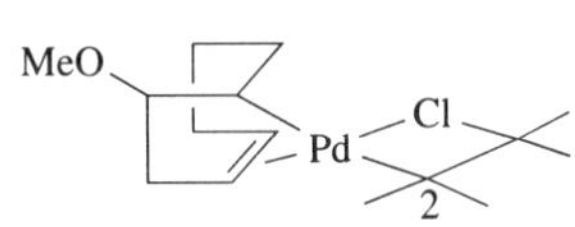

(**196**)

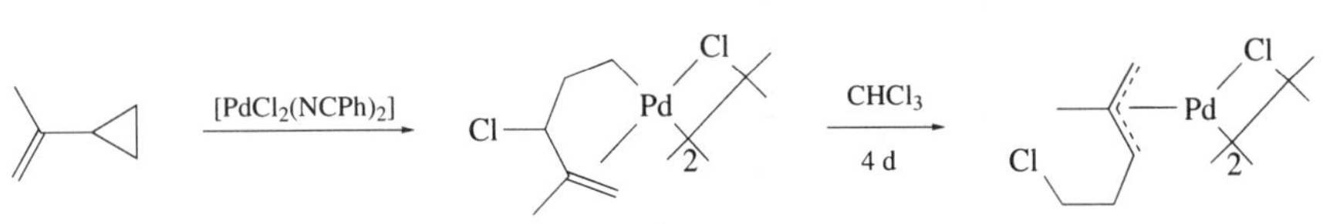

(6)

(**197**)

Intramolecular coordination complexes formed by nucleophilic attack on coordinated alkenes are illustrated in (**192**), and in (**198**) for attack on coordinated $CH_2{=}CHCH_2E$ (R^1 = Nu)[603,604] or $\mathbf{CH_2}{=}CHCMe_2CH_2E$ (R^3 = $\mathbf{CH_2}$Nu).[605,606] This approach has been extended to include the reaction of $CH_2{=}CHCH_2CMe_2CH_2NMe_2$ with $[PdX_2(NCPh)_2]$ to give seven-membered ring complexes (**199**),[607] and intramolecular attack[608,609] by a deprotonated malonate group of coordinated $ECH_2CH{=}CHCH_2CH_2CH(CO_2Me)_2$ to form (**200**).[609] Palladium(II) acetate reacts with $CH_2{=}CHCN$ to form $[Pd\{C(CN){=}CHOC(O)Me\}(\mu\text{-}O_2CMe)]_2$* in which the nitrile groups are coordinated to palladium centres in adjacent dimers.[610]

Addition of a coordinated nucleophile (insertion) occurs in a *cis*-fashion, and for the synthesis of stable Pd–C-bonded complexes is generally confined to electrophilic or highly strained alkenes. Examples include the conversion of $[PdCl(C_6F_5)(1,5\text{-cod})]$ to (**201**),[611] and addition of the nitro group in $[PdCl(NO_2)(\text{alkene})]$ to form complexes such as (**202**)* obtained from the dicyclopentadiene complex.[612] Tetracyanoethene reacts with the enolate group of (**203**) to form (**204**)*.[613] Insertion of alkenes into the Pd–C bond of cyclopalladated complexes generally results in decomposition and the formation of

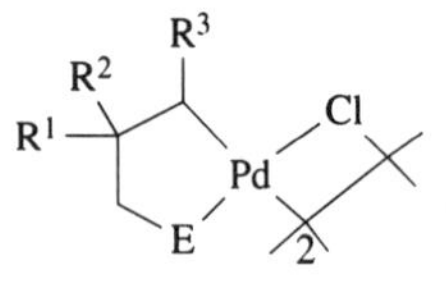

(198)
E = NMe_2, SR^4
R^1 = Nu, $R^2 = R^3$ = H
$R^1 = R^2$ = Me, $R^3 = CH_2Nu$

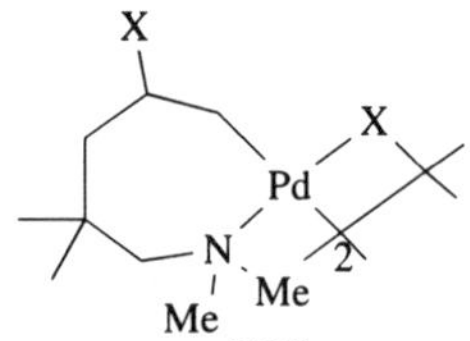

(199)
X = Cl, Br

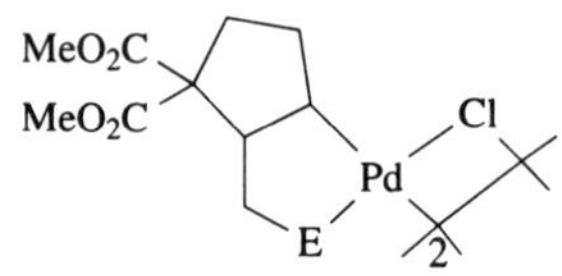

(200)
E = NMe_2, SMe

organic products resulting from the insertion reaction;[102,364,457,614–18] for example, reaction of the *N,N*-dimethylbenzylamine complexes $[Pd(C{\sim}N)(\mu\text{-}X)]_2$ (X = Cl, O_2CMe) with styrene is catalysed by H^+ to give palladium metal and N~C–CH=CHPh.[615,616] However, the reaction of the chloro dimer with activated cyclopropenes allows isolation of complexes which react with pyridine to give (**205**).[619] Ring opening also occurs for R^1 = H, $R^2 = CO_2Me$ to give an allylpalladium(II) complex as a by-product.[619]

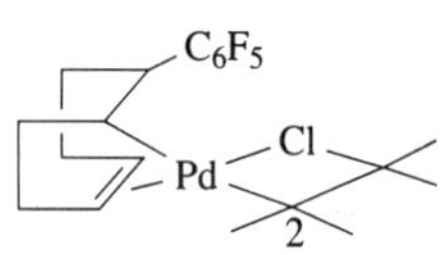

(201)

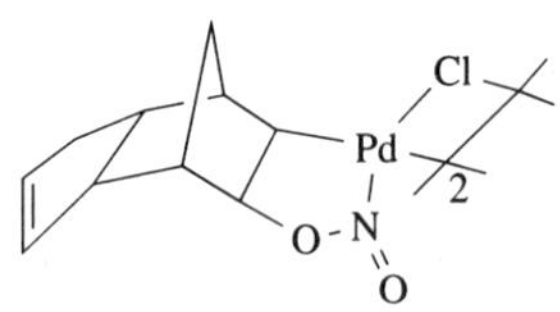

(202)

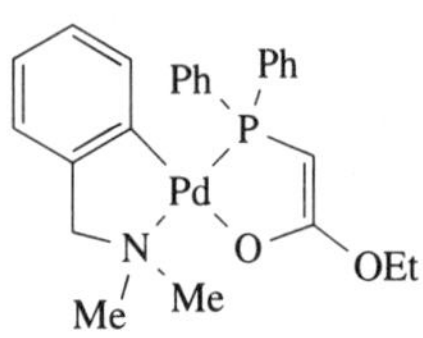

(203)

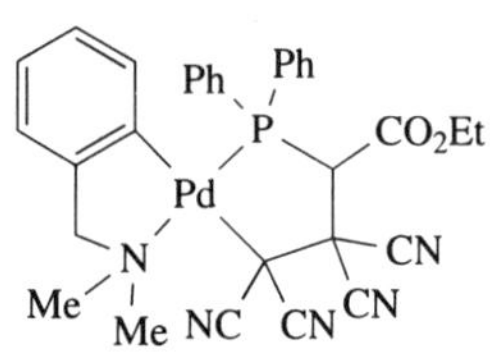

(204)

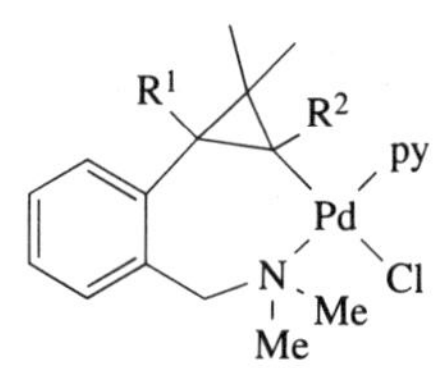

(205)
$R^1 = R^2 = CO_2Me$
R^1 = H, $R^2 = CO_2Me$

Norbornene,[620] nbd,[621,622] butadiene[623,624] and isoprene[623] react with allylpalladium(II) complexes to form Pd–C σ-bonded complexes, and $[Pd(\eta\text{-}Cp)(\eta\text{-}C_3H_5)]$ reacts with 3,3-dimethylcyclopropene in the presence of phosphines to give palladacycles such as (**206**)* and (**207**)*.[624,625] Acrylonitrile reacts with $[PdH(S_2CNMe_2)(PEt_3)]$ to form $[Pd\{CH(CN)Me\}(S_2CNMe_2)(PEt_3)]$,[626] and 1-hexene reacts similarly to form an isomeric mixture of 1-, 2- and 3-hexyl complexes.[214] Maleic anhydride reacts with $[Pd(C_6F_5)(CH_2CH{=}CH_2)(dppen)]$ via cycloaddition to form (**208**)*,[627] and 1,5-hexadiene reacts with $[PdBr(C_6F_5)(NCMe)_2]$ to form the alkene complex $[Pd\{\eta^2\text{-}CH_2{=}CH(CH_2)_2CHCH_2C_6F_5\}(\mu\text{-}Br)]_2$ prior to slow isomerization to give an allylpalladium(II) complex (dppen = bis(diphenylphosphino)ethene).[628] Norbornene inserts into the palladium–alkenyl bond of $[PdBr(CH{=}CH\text{-}p\text{-}MeOC_6H_4)(dppf)]$,[117] and into the Pd–COR bond of *trans*-$[Pd(COR)(PPh_3)_2(NCMe)]^+$ to form cation (**209**)* with the *cis,exo*-geometry.[629] Complex (**209**) is stabilized toward decomposition by the presence of intramolecular coordination. Similar cationic complexes are formed with nbd and dicyclopentadiene, and the reactions are assumed to proceed via alkene displacement of MeCN to give a square planar geometry with the alkene bond coplanar with the coordination plane, as indicated from kinetic studies of this[629] and related reactions[163] and theoretical analyses.[630] Analogous insertion reactions into palladium–acetyl bonds have been observed for complexes of bidentate ligands to give (**210**),[631] for the reactions of $[Pd(COMe)(dppp)(NCMe)]^+$ or $[PdCl(COMe)(dppp)]$ with nbd,[632] on addition of $[Ag(O_3SCF_3)]$ and norbornene or dicyclopentadiene to nitrogen donor complexes such as [PdI(COMe)(bipy)], for example, to form (**211**)*,[633] and for related reactions of $[Pd(COMe)(bipy)(CO)]^+$ [634] and [PdCl(COMe){bis(p-methoxyphenylimino)acenaphthene}].[635] Alkene exchange reactions have been reported for a range of palladium(II) substrates and alkenes, for example, the reaction of *trans*-$[Pd(COPr^i)(PPh_3)_2(NCMe)]^+$ with $CH_2{=}CH_2$ to form the Pd–COEt analogue.[164]

The addition of aryl groups to alkenes using arylmercury(II) reagents with palladium complexes as catalysts has developed into an important synthetic strategy in organic synthesis, and Equation (7) illustrates the stabilization of an organopalladium(II) intermediate by intramolecular coordination.[636] An insertion reaction occurs on addition of $[HgPh_2]$ to $[PdCl_2(nbd)]$, and similar reactions occur for

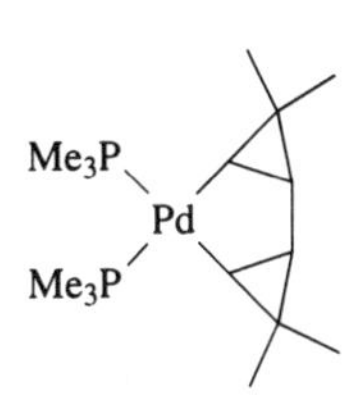

(206)

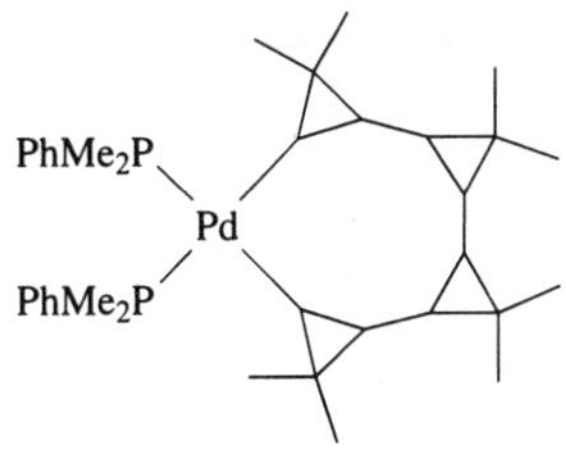

(207)

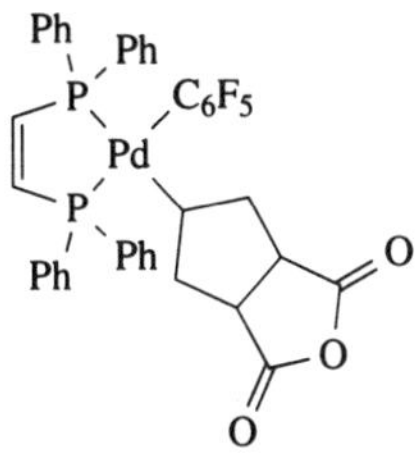

(208)

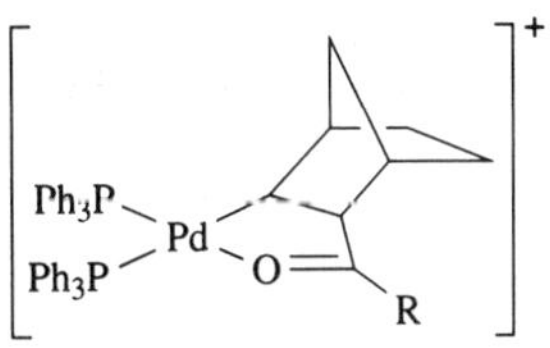

(209)
R = Me, Et, Ph (unstable)

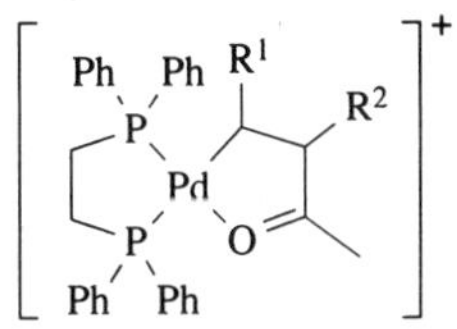

(210)
$R^1 = CO_2Me$, $R^2 = H$
$R^1{\sim}R^2 = (CH_2)_3$, $(CH_2)_5$

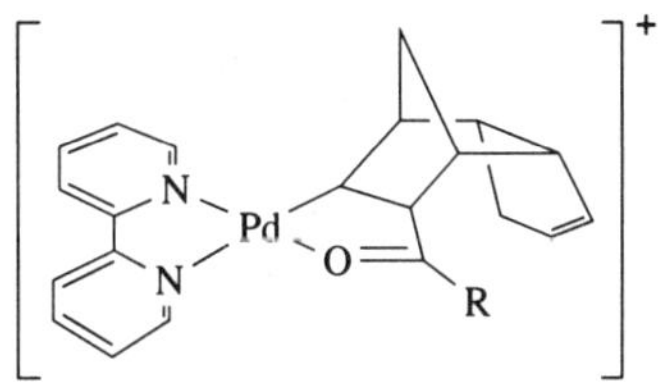

(211)

$Li_2[PdCl_4]$/norbornene with vinylmercury(II) halides[637] or [HgClPh].[638] When the latter reaction is followed by addition of phen and a base (OPh^-), cyclization occurs to form (**212**).[639,640] Similar complexes may be generated by oxidative addition of aryl bromides to $[Pd(PPh_3)_4]$ in the presence of norbornene, followed by OPh^-.[641] The reactions of nbd and norbornene with *trans*-$[PdIAr(PPh_3)_2]$[642] and of nbd with $[Pd(PPh_3)_4]$/3-iodo-2-cyclohexen-1-one[643] give complexes involving η^2-coordination of the Ar or R group, such as (**213**)*, indicating that other alkene insertion reactions may also involve η^2-coordination of unsaturated R groups. The cation $[PdPh\{NMe(CH_2CH_2NMe_2)_2\text{-}N,N',N''\}]^+$ reacts with norbornene to give (**214**)*, presumably involving formation of '$[PdPh(alkene)\{NMe(CH_2CH_2NMe_2)_2\text{-}N,N'\}]^+$' followed by insertion to give an intermediate similar to (**213**) and a subsequent rearrangement to give (**214**) that may involve a palladium(IV) intermediate.[103]

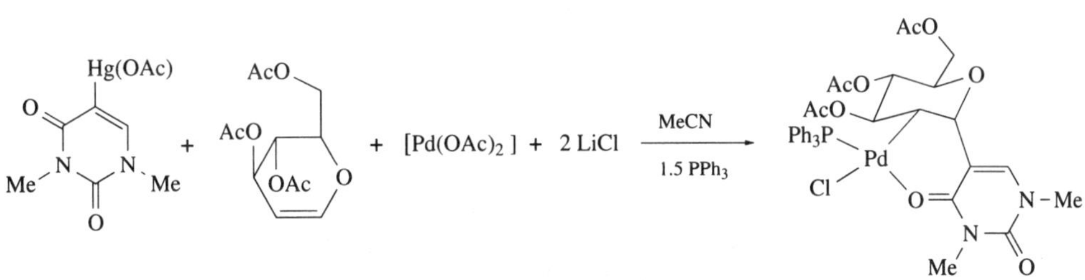

(7)

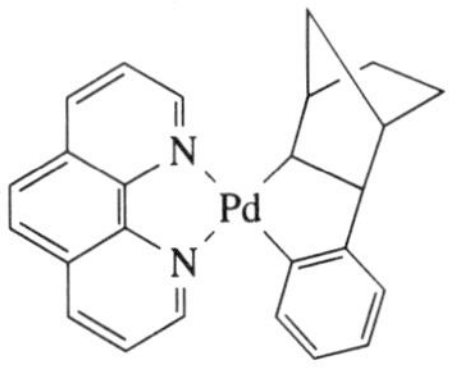

(212)

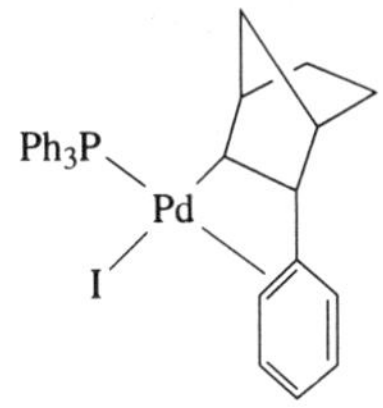

(213)

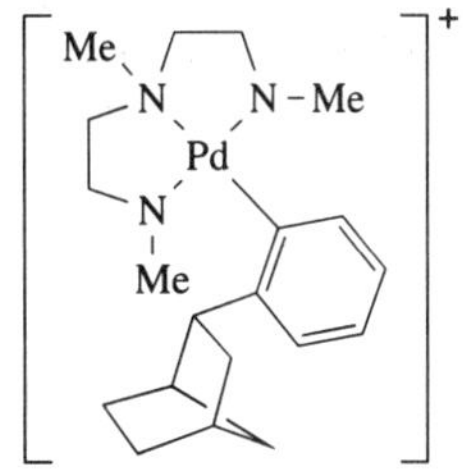

(214)

5.5.1.5 From palladium complexes and alkynes

A wide range of structural and bonding systems have been obtained using alkynes as reagents, in addition to alkynyl complexes discussed in Section 5.4. The early literature has been reviewed extensively, in particular for reactions resulting in oligomerization of alkynes and the isolation of complexes related to this process.[581,644,645] In this section reactions of alkynes with palladium(I) substrates are reviewed, followed by reactions with Pd^{II}–X bonds and organopalladium(II) complexes.

Alkynes insert into Pt^{I}–Pd^{I} bonds to form (**6**),[8,9] and into the Pd^{I}–Pd^{I} bond of $[Pd(\mu\text{-}E^1\sim E^2)(\mu\text{-}X)]_2$ where the bridging ligands $E^1\sim E^2$ have one carbon atom linking the donor atoms, to give palladium(II) complexes (**215**). At least one electron-withdrawing group in the alkyne is generally required; for example, complexes have been obtained for $R^1 = R^2 = CO_2Me$ for all of the ancillary ligands studied. However, the presence of H^+ as a catalyst has allowed the isolation of (**215**) from $[Pd(\mu\text{-dppm})(\mu\text{-Cl})]_2$ and RC≡CR (R = H, Ph, *p*-Tol), and for these reactions the insertion is reversible.[647] DMAD reacts with retention of the Pd–Pd bond in $[\{(R_3P)Pd\}_2(\mu\text{-Cp})(\mu\text{-Br})]$ to form $[\{(R_3P)BrPd\}\{\mu\text{-C(CO}_2\text{Me)=C(CO}_2\text{Me)}\}\{Pd(\eta\text{-Cp})(PR_3)\}]$ (R = Et, Pr^i).[652]

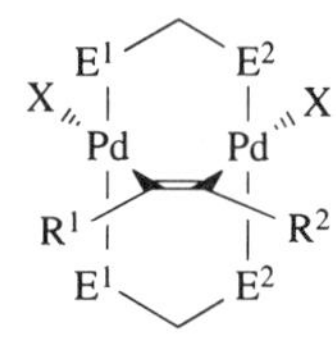

(**215**)
$E^1\sim E^2$ = dppm, X = Cl,[646,647] Br, I[646]
$E^1\sim E^2$ = PPh_2py,[648] Ppy_3,[649] X = Cl
$E^1\sim E^2$ = $PPh_2(CH_2SBz)$, X = Cl[650]
$E^1\sim E^2$ = $CH(SBz)_2$, X = Cl[651]

Insertion of alkynes, usually electrophilic, into Pd–H bonds has been examined for several systems, for example, insertion of $PhC{\equiv}CCF_3$ to form *trans*-$[Pd(ONO_2)\{C(Ph)=CH(CF_3)\}(PCy_3)_2]$,[653] and insertion of HC≡CPr or $MeC{\equiv}CCO_2Me$ into $[PdH(S_2CNMe_2)(PEt_3)]$ to form isomeric mixtures containing $[Pd\{C(Pr)=CH_2\}]$ and (*E*)-[Pd(CH=CHPr)] groups, or both (*E*)- and (*Z*)-isomers containing $[Pd\{C(CO_2Me)=CHMe\}]$ and $[Pd\{C(Me)=CHCO_2Me\}]$ groups.[626] Insertion of DMAD into a Pd–NHPh bond has been observed, resulting in the formation of *trans*-$[PdPh\{C(CO_2Me)=C(NHPh)(CO_2Me)\}(PMe_3)_2]$*.[654]

Cyclopalladated imine complexes such as $[Pd\{5\text{-(MeO)}C_6H_3CH=N(p\text{-Tol})\text{-}C,N\}(\mu\text{-}O_2CMe)]_2$ react with DMAD in MeOH in a manner closely related to that observed for nucleophilic attack by **MeO**$^-$ on coordinated alkenes to give $[Pd\{5\text{-(MeO)}C_6H_3CH=N(p\text{-Tol})\text{-}C,N\}\{C(CO_2Me)=C(\mathbf{OMe})C(O)OMe\text{-}C,O\}]$.[618] A similar reaction occurs between $HC{\equiv}CCMe_2NMe_2$ and $Li_2[PdCl_4]$ to give $[Pd\{CH=C(Cl)CMe_2NMe_2\text{-}C,N\}(\mu\text{-Cl})]_2$,[655] but closely related $HC{\equiv}CCMe_2CH_2SMe$ reacts with *trans*-$[PdCl_2(NCPh)_2]$ in air to give several products including (**216**)*.[656] Simple alkynes undergo complex reactions with chloropalladium(II) substrates;[581,644,645] for example, MeC≡CMe and $Bu^tC{\equiv}CH$ trimerize to give symmetrically substituted benzenes via several intermediates, with isolated complexes such as (**217**) and (**218**) regarded as suitable models for some of the intermediates that contain Pd–C σ-bonds. Reactions of dialkyl acetylenedicarboxylates with $[Pd_2(dba)_3]$ or $[Pd(dba)_2]$ in the presence of ligands give butadiene complexes of type (**219**), for example, $L_2 = 2\,PPh_3$, bipy,[657] cod, nbd*[658] and *N*-aryldiazadienes (dad);[659] $[Pd\{C_4(CO_2Me)_4\}]_n$ also reacts with ligands to form $[Pd\{C_4(CO_2Me)_4\text{-}C,C\}L]_2$, for example, L = 2,6-$Me_2py$ where the dimeric structure involves carbonyl group coordination from the neighbouring palladium centre.[660] The butadiene complexes exhibit catalytic properties;[658,661–3] for example, (**219**) (L = dad) acts as a catalyst precursor for the cycloaddition of methoxyallene and DMAD.[662] Complex (**220**) has been isolated from this reaction, but it is not an intermediate, and (**221**) has been obtained from a related reaction with 1-hexene and DMAD.[663]

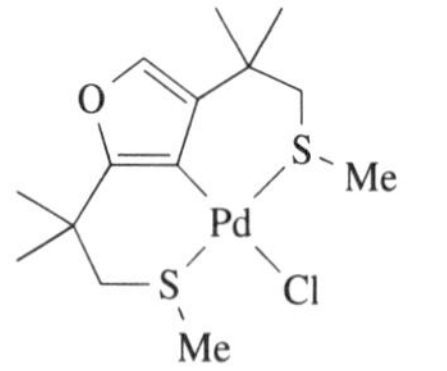

(**216**)

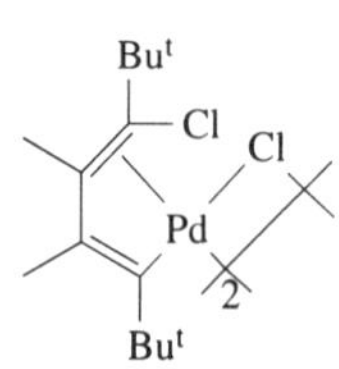

(**217**)

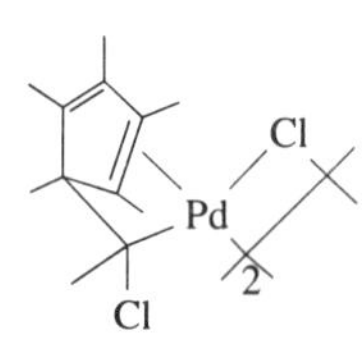

(**218**)

Electrophilic alkynes react with acetylacetonate[664] and related phosphinoenolate complexes to give alkenyl complexes via C–C bond formation, as illustrated in Scheme 18.[613,665] Complex (**223**)* isomerizes on heating to form (**224**), and the closely related *cis*-$[Pd(C\sim P)_2]$* analogue of (**223**) isomerizes on heating to form the *trans*-isomer* where the ligand retains the same coordination mode.[665,666] Alkenyl complexes are also formed on reaction of $CF_3C{\equiv}CCF_3$ with allylpalladium(II) complexes,[667] on insertion of DMAD into Pd–C≡CR bonds,[668] and in a condensation reaction between

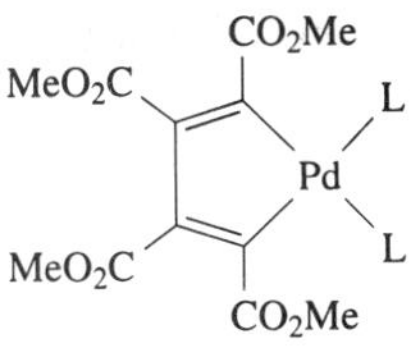

(219)

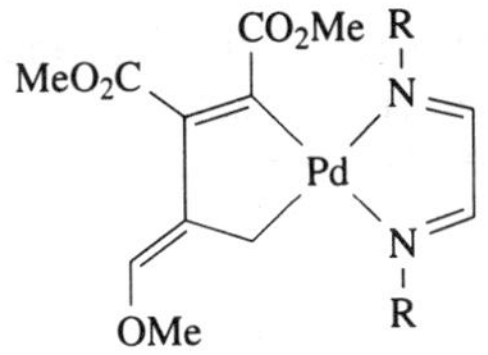

(220)
R = 2,6-(Pr^i)$_2$$C_6H_3$

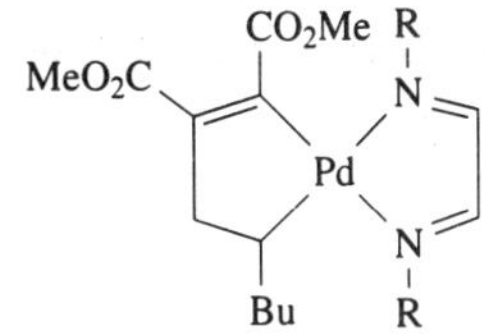

(221)
R = 2,6-(Pr^i)$_2$$C_6H_3$

DMAD and *trans*-[PdCl{C(Me)=N(*p*-Tol)}(PEt_3)$_2$] to give (**225**)*.[669] Intramolecular insertion occurs on heating the complexes *trans*-[PdCl{C(O)O(CH_2)$_n$C≡CR^1}(PR^2_3)$_2$] to give (**226**),[159,163] and both kinetic and ^{31}P NMR studies of these reactions[163] offer strong support for theoretical analyses indicating that alkyne insertion occurs from a planar '[PdClPC(C≡C)]' group.[630]

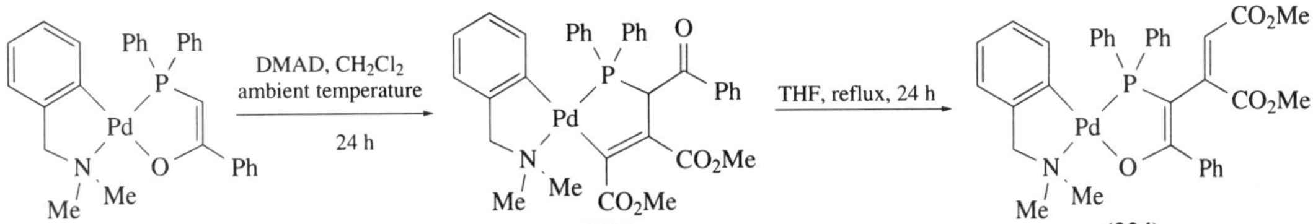

(222) **(223)** **(224)**

Scheme 18

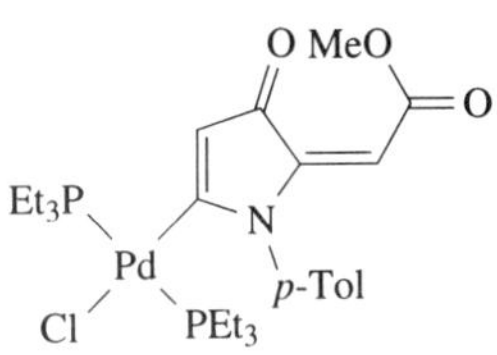

(225)

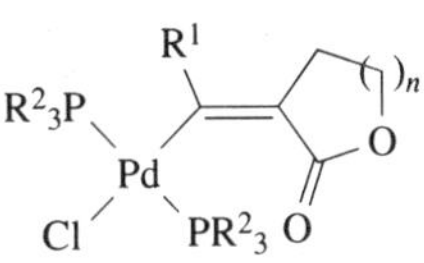

(226)

n = 1: R^1 = H, R^2 = Ph; R^1 = Me, R^2 = Ph, *p*-Tol
n = 2: R^1 = Me, R^2 = Ph

Electrophilic alkynes have been shown to insert into Pd–C bonds of Pd–Me,[670,671] Pd–Bz,[672] and Pd–COMe complexes.[626] An additional insertion of DMAD occurs for [PdCl{C(CO_2Me)=C-Me(CO_2Me)}(dppe)] to give the σ-butadienyl complex [PdCl{C(CO_2Me)=C(CO_2Me)C-(CO_2Me)=CMe(CO_2Me)}(dppe)],[671] and (**227**) is obtained as a by-product from the reaction of *trans*-[PdCl(Bz)(PPh_3)$_2$] with DMAD.[672]

Insertion of alkynes into Pd–C(sp^3) and Pd–C(sp^2) bonds of intramolecular coordination systems[335,673–5] generally gives products involving incorporation of one (**228**),[208,292,353,365,385,434,577,675–8] two (**229**)[208,365,434,474,577,674,675,677–81] or three alkyne units (**230**)–(**232**)[365,678,682,683] into the Pd–C bond. The weakly coordinated $[BF_4]^-$ group in (**230**) has Pd···F 0.235 5(5) nm. There are also close relationships between reactions resulting in (**229**)–(**232**) and the earlier work giving (**217**) and (**218**). Organic products of interest in organic synthesis are often obtained rather than complexes, in particular for reactions of alkynes with iodo-bridged reagents [Pd(C~N)(μ-I)]$_2$ and cations such as [Pd(C~N)(NCMe)$_2$]$^+$. Reactions that give isolable complexes usually employ [Pd(C~E)$_2$], [Pd(C~E^1)(C'~E^2)] or chloro-bridged dimers [Pd(C~E)(μ-Cl)]$_2$ as reagents. Ligand skeletons that react with alkynes to form monoinsertion complexes (**228**) include those derived from PhCH_2SMe,[577] benzo[*h*]quinoline (**162**),[353,675] acetanilide and amidine (**107**),[365] 2-benzylpyridine,[678] *N,N*-dimethylbenzylamine,[292,674–6] those illustrated in (**114**)[387] and those that give complex (**102**).[353] Skeletons that give diinsertion products (**229**) include cyclopalladated PhCH_2SMe,[577] *N,N*'-dimethylaminomethylferrocene ((**148**) (R^1 = H)),[674,680,681,684] 8-methylquinoline,[208,437,675,677,679] amidine[365] and *N,N*'-dimethylbenzylamine.[434,437,674,677]

The regioselectivity of insertion reactions for unsymmetrical alkynes PhC≡CCO_2Et generally favours the phenyl group occurring adjacent to palladium (R^2 = Ph in (**228**), (**229**) and (**233**)),[577,677,678] although a mixture of isomers is found for Structure (**228**) on insertion into chloro-bridged dimers of cyclopalladated *N,N*-dimethylbenzylamine[677] and 2-benzylpyridine.[678] Reactions to form diinsertion products (**229**) are favoured by electron-rich alkynes, for example, PhC≡CPh and MeC≡CMe, and the *trans,cis*-butadiene geometry is assumed to result from isomerization of a cyclic complex with the *cis,cis*-geometry shown in (**217**). Kinetic studies indicate that the insertion reactions are first order in

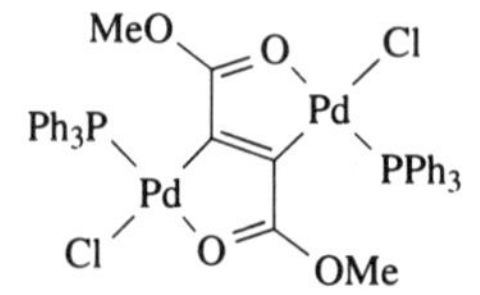

(227)

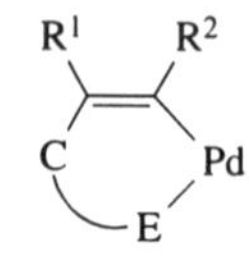

(228)

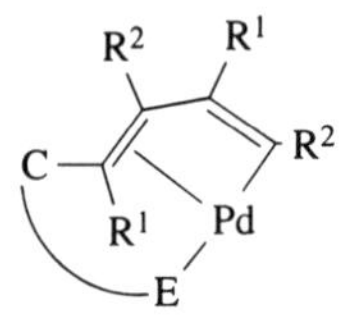

(229)

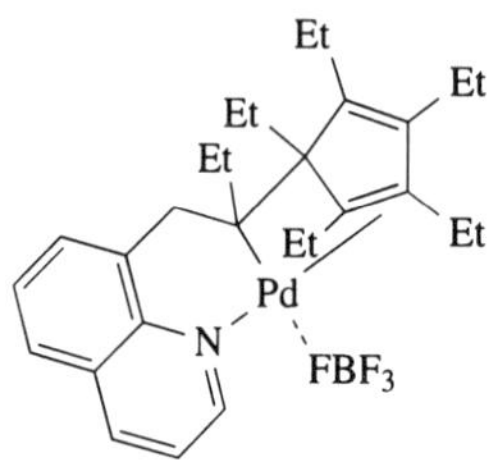

(230)[682,683]

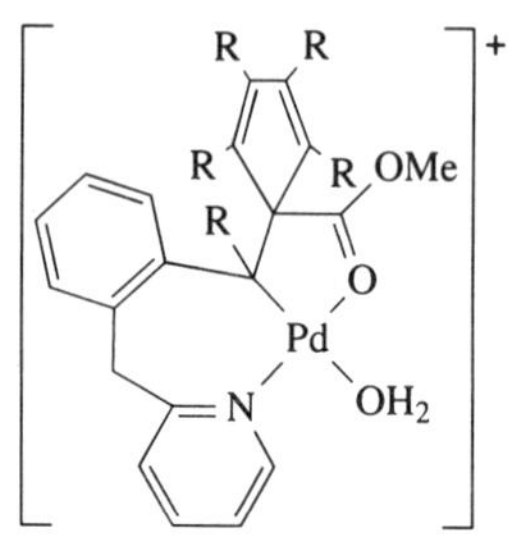

(231)[678]
R = CO_2Me

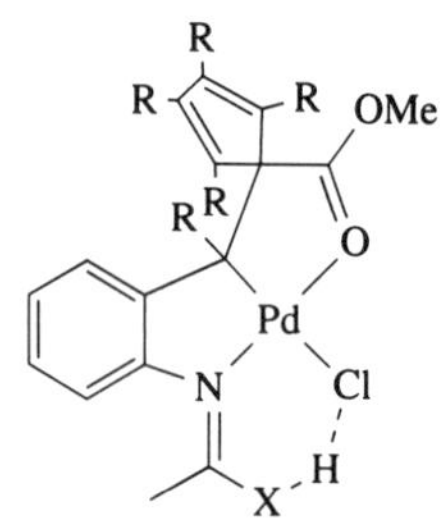

(232)[365]
X = O, NPh; R = CO_2Me

each reagent, the monoinsertion step is rate determining, electron-donating groups on the palladium substrate assist the monoinsertion step, and for monoinsertion suggest that the reaction occurs from the intact $[Pd(C{\sim}N)(\mu\text{-}Cl)]_2$ dimer.[437] A mechanism involving cleavage of one Pd–Cl bond to create a site for alkyne interaction at one palladium centre is suggested.

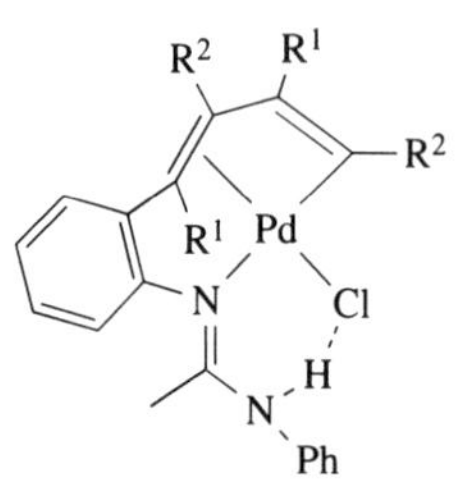

(233)[365]

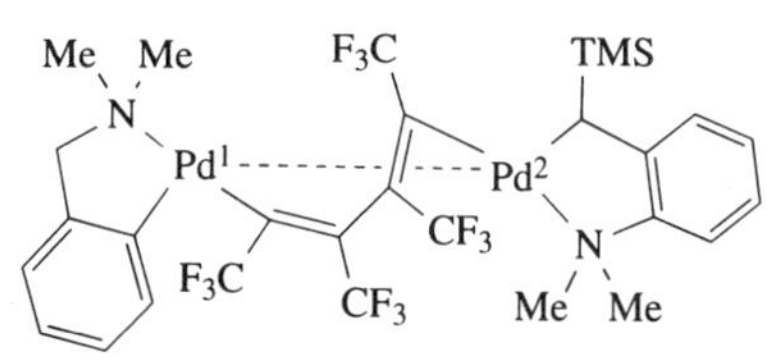

(234)

Complexes of structure (**228**) incorporating one alkyne unit generally occur as seven- or eight-membered palladocyclic rings, although excess $CF_3C{\equiv}CCF_3$ reacts with the chloro-bridged dimer of cyclopalladated acetanilide (**107**) (E = O, R = Me) to form an eight-membered ring complex which rearranges on addition of pyridine to form a six-membered ring complex with *C,N*-coordination, $[PdCl\{2\text{-}[OC(Me)NH]C_6H_4C(CF_3){=}C(CF_3)\text{-}C,N\}(py)]$.[365] Structure (**229**) occurs with eight- or nine-membered palladocyclic rings $[Pd(CR^2{\sim}E)]$, including eight-membered (**233**) which results from the reaction of excess $R^1C{\equiv}CR^2$ ($R^1 = R^2$ = Ph; $R^1 = CO_2Et$, R^2 = Ph) with the chloro-bridged dimer containing (**107**) (E = NPh). A similar flexibility in coordination mode of the amidine ligand is also shown in the formation of (**232**) (X = NPh), and by the cyclopalladated acetanilide group (**107**) (E = O) forming (**232**) (X = O) involving nitrogen atom coordination. Complexes with bulky substituents, or ligand geometries precluding adoption of the preferred ring sizes for (**228**) and (**229**), form different structures, for example, allyl complexes[208,365,678,685] or (**234**) on reaction of $[Pd(C{\sim}N^1)(C'{\sim}N^2)]$ with excess $CF_3C{\equiv}CCF_3$.[686] The palladium atoms of (**234**) have a T-shaped 'PdC_2N' geometry, with the nature of the remaining bonding unresolved although it may involve weak $Pd{\cdots}\eta^2$-alkene interaction(s) and a $Pd^2 \rightarrow Pd^1$ interaction (0.275 6(1) nm).

Metallacarbynes form monoinsertion products closely related to (**228**) in which the metallacarbyne unit is η^2-bonded to palladium, and for one system η^1-intermediates have been isolated on the reaction of $[M\{{\equiv}C(p\text{-}Tol)\}(CO)_2(\eta\text{-}Cp)]$ (M = Mo, W) with the $[Pd(C{\sim}N)(\mu\text{-}Cl)]_2$ precursor, as illustrated in (**235**)* for tungsten.[687] The complex has Pd–C 0.208 4(7) nm and the Pd–C–W angle is 89.26(26)°.

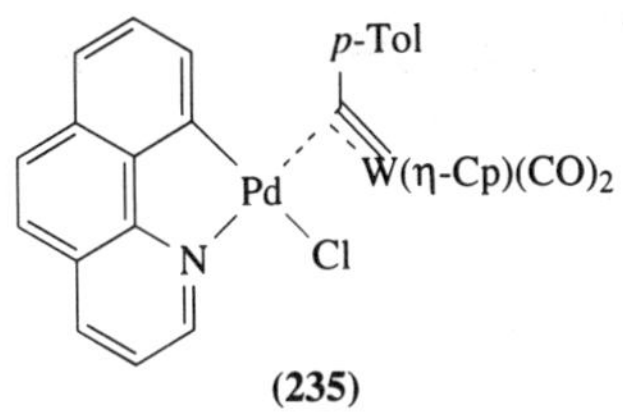

(235)

5.5.1.6 From palladium complexes and carbon monoxide

(i) Complexes of CO and insertion reactions

Carbonyl complexes for several organopalladium(II) systems form at 1 atm of CO, for example, *trans*-$[Pd(COPh)(PMe_3)_2(CO)]^+$ on addition of $AgBF_4$ to the chloro analogue in acetone followed by addition of CO,[688] *cis*-$[PdCl_2(Ar)(CO)]^-$ on addition of CO to $[PdCl(2\text{-}Me,6\text{-}NO_2C_6H_3)(\mu\text{-}Cl)]_2^{2-}$ (**71**),[689] and on addition of CO to pentahalophenyl complexes.[4,690,691] Several intramolecular coordination complexes form CO complexes, including [Pd(C~N-*C*)(C~O-*C*,η^2-aryl)(CO)] from (**102**),[353] $[Pd(C\text{\textasciitilde}N\text{\textasciitilde}N)(CO)]^+$ from $[Pd(C\text{\textasciitilde}N\text{\textasciitilde}N)(acetone)]^+$ (C~N~N as in (**119**)),[694] and complexes formed by bridge splitting of $[Pd(C\text{\textasciitilde}N)(\mu\text{-}Cl)]_2$ (C~N as in (**75**))[692,693] and (**198**) ($R^1 = CH(CO_2Et)_2$, $R^2 = R^3 = H$, $E = NMe_2$).[695]

Reactions of intramolecular coordination complexes with CO lead more commonly to products of decomposition after insertion of CO into the Pd–C bond,[139,450,696] for example, $PhCH_2NMe_2$ and $[2\text{-}(NMe_2CH_2)C_6H_4]_2CO$ in 1:2 ratio from the reaction of *cis*-[Pd{2-$(NMe_2CH_2)C_6H_4$-*C,N*}$_2$],[297] and heterocyclic compounds from cyclopalladated diazepam and prazepam.[482] Acyl groups as part of new intramolecular systems have also been obtained, for example, [Pd{2-$(SMeCH_2)C_6H_4C(O)$-*C,S*}(μ-Cl)]$_2$,[693] and the analogous products of CO insertion into the chloro derivatives of (**119**)[694] and (**229**) ($R^1 = R^2 = Et$, C~E = 2-$(NMe_2CH_2)C_6H_4$).[674] In closely related nitrogen and sulfur donor intramolecular coordination systems, insertion into the sulfur donor complexes is more facile and may be reversible, as demonstrated for [Pd{2-$(SMeCH_2)C_6H_4$-*C,S*}(μ-Cl)]$_2$.[692,693]

Palladacyclic $[\overline{Pd(CH_2CH_2CH_2CH_2})L_2]$[257] and dialkylpalladium(II) complexes $[PdR_2L_2]$ react with CO to give ketones and, in some cases, other decomposition products for a range of alkyl groups and ligands such as monodentate and bidentate phosphines[230,257] and TMEDA.[261] However, for monoorganopalladium(II) substrates, many Pd–COR complexes have been formed under mild conditions, for example, for palladium bonded to η-allyl,[697] Me,[3,238,260,626,633–5,698–708] Et,[709] Pr^i,[626] Bz,[109,220,228,710] picolyl,[183] $[Cr(CO)_5(\eta\text{-}C_6H_5)]$,[105] pyridyl,[149,151,152,711] other aryl[97,99,106,108,220,700,707,712] and alkenyl groups,[138] and for donor ligands including monodentate[97,105,106,108,138,151,152,183,220,228,238,697,700,701,703,709,711,712] and bidentate phosphines,[699,705] TMEDA,[260,261] bipy,[633,634] terpy[704] and other bidentate[635,706,710] and tridentate nitrogen donors,[99,704] P~N donors,[698] thiolates,[709] dimethyldithiocarbamate,[626] acac and monothioacetylacetonate.[707,708] For allylpalladium(II) complexes, CO insertion products are isolable when the product is stable toward dissociation of ligands, in particular for cationic complexes as reagents; for example, $[Pd(\eta\text{-}C_3H_5)(PMe_3)_2][Br]$ forms *trans*-$[PdBr(COCH_2CH{=}CH_2)(PMe_3)_2]$.[697] Double carbonylation occurs for the allenyl complex *trans*-$[PdBr(CH{=}CH{=}CH_2)(PPh_3)_2]$, resulting in formation of a 2(3*H*)-furan-5-yl complex *trans*-$[PdBr\{\overline{C{=}C(CMe{=}CH_2)CH_2C(O)O}\}(PPh_3)_2]$.[713]

Insertion of CO into palladium–alkyl bonds occurs with retention of configuration at the alkyl carbon atom,[92] and is believed to involve the alkyl group as a nucleophile in an alkyl migration mechanism.[247,248,714,715] Early kinetic studies are consistent with initial CO coordination to give a five-coordinate intermediate from which insertion occurs, or from which dissociation of ligand occurs to give a four-coordinate complex prior to insertion.[716] Subsequent studies have been interpreted generally in terms of either mechanism, depending upon the ease of dissociation of donor groups from the palladium(II) substrate.[99,698,699,717] The different behaviour of simple *cis*- and *trans*-$[PdAlk_2L_2]$ (L = phosphine) complexes toward CO, for example, *cis*-$[PdEt_2L_2]$ to give EtCHO and $CH_2{=}CH_2$, *cis*-$[PdMe_2L_2]$ to give Me_2CO and MeCOCOMe, and *trans*-$[PdR_2L_2]$ to give R_2CO (R = Me, Et, Pr, Bu), may be explained by the requirement for square planar geometry and migration of the alkyl group to the coordinated CO ligand.[248] For bidentate phosphines as ligands in [PdClMe(P~P)], kinetic studies indicate that the rate of CO insertion increases with ring size and flexibility of the ligand,[699] which is consistent with theoretical calculations.[718,719] Low-temperature IR spectra permit detection of an intermediate $[PdMe(PAN)(CO)]^+$ (PAN = 1-dimethylamino-8-diphenylphosphinonaphthalene-*P,N*) on the reaction of CO with [PdXMe(PAN)] or [PdMe(PAN)(NCMe)][X] (X = O_3SCF_3, BF_4) in CH_2Cl_2 to form acetyl complexes,[698] and the cation $[Pd(COMe)(bipy)(CO)]^+$ has been detected by NMR at low temperature on demethylation of $[PdMe_2(bipy)]$ with 1 equiv. $[Et_2OH]^+$ in CD_2Cl_2 followed by addition of CO.[634] Insertion occurs also for cationic complexes such as $[PdMe(L_3)]^+$ (L_3 = terpy,[704] 2,6-$(Me_2NCH_2)_2C_5H_3N$[99]) and $[PdMe(L_2)(NCMe)]^+$ (L_2 = dppe,[699,705] dppf,[699] $Ph_2P(CH_2)_3NMe_2$[698]), and complexes related to (**211**) where the bidentate nitrogen donor ligand is bis(*p*-methoxyphenylimino)acenaphthene.[635]

Acyl complexes may be obtained on oxidative addition of organohalides to $[Pd(PPh_3)_3(CO)]$ under mild conditions,[92,720] although CO at 20 atm was used in the reaction with *trans*-PhCH=CHCH$_2$Br to give *trans*-[PdBr{(*E*)-PhCH=CHCH$_2$CO}(PPh$_3$)$_2$] in equilibrium with the *trans*-(*E*)-PhCH$_2$CH=CHCO isomer in solution.[721]

Insertion of CO is reversible for some complexes, with decarbonylation occurring on gentle heating and/or the absence of CO gas. Examples include [PdCl(COMe)(2,9-Me$_2$phen)],[706] *trans*-[PdCl(COPh)(PCy$_3$)$_2$],[108] and intramolecular coordination complexes[693,694] such as [Pd(COAr){2,6-(Me$_2$NCH$_2$)$_2$C$_5$H$_3$N-*N,N',N''*}][O$_3$SCF$_3$]*.[99]

Carbon monoxide inserts into the Pd–C bond of *trans*-[Pd(OAr)Me(PEt$_3$)$_2$] and *trans*-[Pd(SPh)Et(PMe$_3$)$_2$]*,[709] but the closely related alkoxides [Pd{OCH(CF$_3$)$_2$}Me(dppe)][722] and [Pd(OMe)Me(P~P)] (P~P = (*S,S*)-(PPh$_2$CHMe)$_2$CH$_2$)[723] react differently, undergoing insertion into the Pd–O bond as illustrated in Equations (8) and (9).

$$trans\text{-}[Pd(SPh)Et(PMe_3)_2] + CO \longrightarrow trans\text{-}[Pd(SPh)(COEt)(PMe_3)_2] \longrightarrow PhSCOEt \quad (8)$$

$$[Pd(OR)Me(P{\sim}P)] + CO \longrightarrow [Pd(CO_2R)Me(P{\sim}P)] \longrightarrow MeCO_2R \quad (9)$$

Dimeric complexes [PdR(L)(μ-Cl)]$_2$ (R = Alk, Ar, including [Cr(CO)$_5$(η-C$_6$H$_5$)])[105,220,238] and Pd–Me A-frame complexes undergo insertion, for example, $[HPd(\mu\text{-dppm})_2(\mu\text{-Cl})PdMe]^+$,[3] and for both Pd–C bonds in $[\{PdMe(\mu\text{-dppm})\}_2(\mu\text{-X})]^+$ (X = Br, I,[702] H[3]).

(ii) Syntheses involving CO and other reagents

In addition to insertion into cyclopalladated complexes, and insertion into a 2-picolyl complex to give [PdCl{2-(COCH$_2$)py-*C,N*}(PPh$_3$)],[183] palladocyclic complexes have also been obtained by carrying out reactions of coordinated alkenes with amines in the presence of CO. Thus, reactions similar to that shown in Equation (5) yield complexes such as [PdCl(COCH$_2$CHRNEt$_2$-*C,N*)(NEt$_2$H)] (R = H, Me) from ethene and propene,[724] and a mixture of isomers [PdCl(COCHR^{1}CHR^{2}NEt$_2$-*C,N*)(NEt$_2$H)] (R^1 = H, R^2 = Et; R^1 = Et, R^2 = H) from 1-butene.[725] Acyl complexes are also obtained on reactions of alkenes, CO and *trans*-[PdCl$_2$(PPh$_3$)$_2$] in ethanol,[726–8] and low-temperature NMR studies of copolymerization reactions show that $[Pd(COMe)(bipy)(CO)]^+$ (**236**) reacts with *p*-Bu^{t}C$_6$H$_4$CH=CH$_2$ to give species (**237**) and (**238**) as an equilibrium mixture that reacts further with CO to give (**239**) (Scheme 19).[634]

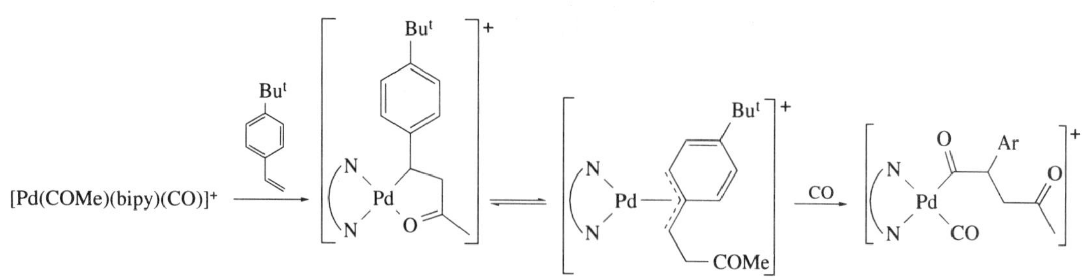

(236) **(237)** **(238)** **(239)**

Scheme 19

Alkoxides and amines also act as nucleophiles toward coordinated CO,[729] allowing the synthesis of *trans*-[PdX{CO$_2$Me}L$_2$][730–2] and *trans*-[PdX{CONR^{1}R^2}L$_2$] (R^1 = R^2 = Alk;[733–5] R^1 = H, R^2 = Alk[733] or Ar,[736] CHMeCO$_2$Me, CH$_2$CO$_2$Et,[737] L = phosphine donor). Unstable complexes [Pd(O$_3$SCF$_3$)(CO$_2$Me)(P~P)] (P~P = dppe, dppp, dppb) have been detected,[738] and a series of benzoyl(carbamoyl) complexes (**242**) (NR$_2$H = NMe$_2$H, piperidine, pyrrolidine) and an *O*-ethylated complex (**243**) have been isolated as part of detailed NMR studies establishing the reaction chemistry and rapid equilibria between (**241**) and (**242**) illustrated in Scheme 20.[734,735] Complex (**242**)

(NR_2 = piperidine = pip) reacts with pip/[pipH][BF_4]/CO to form PhCOCOpip.[735] *trans*-[PdCl(COMe)-($PMePh_2)_2$],[166] intramolecular coordination complexes such as [PdCl{COC(Et)HCH_2NEt_2-*C,N*}($NHEt_2$)],[725] and a complex of cyclopalladated quinoline-8-carbaldehyde[485] react similarly, indicating the occurrence of a second carbonyl insertion reaction. Complexes of the tridentate ligand 2,6-($PPh_2CH_2)_2C_6H_3N$ (pnp) have also been obtained, including $[Pd(COR)(pnp)]^+$ (R = OMe, OPh, NHAlk, NHAr, $NAlk_2$)[736] and a carbomethoxy complex of the deprotonated ligand (**244**).[739]

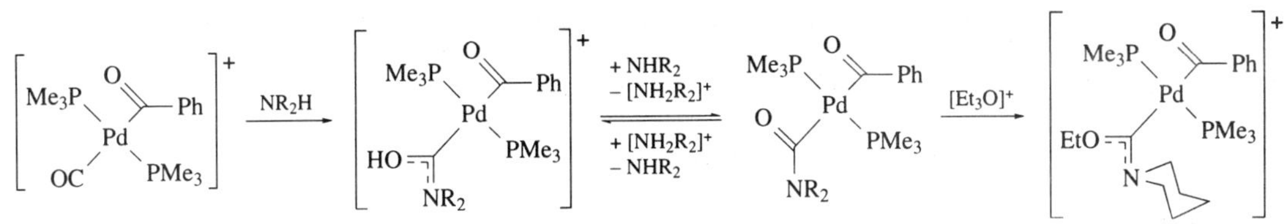

(**240**) (**241**) (**242**) (**243**)

Scheme 20

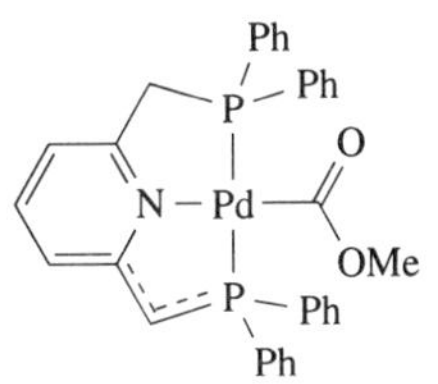

(**244**)

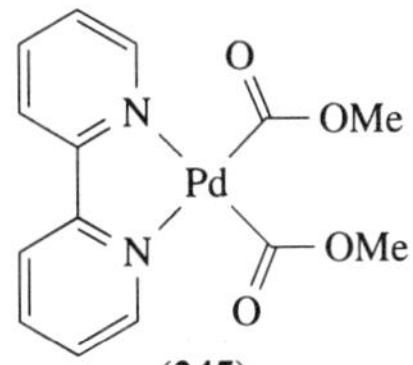

(**245**)

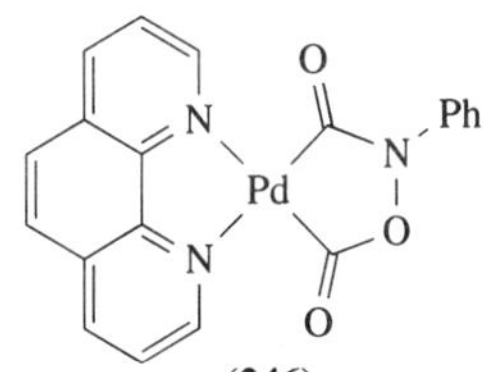

(**246**)

The complex *trans*-[$PdCl_2(PPh_3)_2$] reacts with CO/NEt_3 in MeOH at 40–80 atm (50 °C) to form *trans*-[PdCl(CO_2Me)($PPh_3)_2$],[740] but in a similar reaction under much milder conditions [$PdCl_2$(bipy)] reacts with CO/[Li(OMe)] in MeOH/CH_2Cl_2 at 1 atm to form the bis(carbomethoxy) complex (**245**)*.[741] In other reactions relevant to catalytic systems, mixtures of [$Pd(OAc)_2$]/3 phen/40 $PhNO_2$ in EtOH and [$PdCl_2(NCPh)_2$]/2 PPh_3 in BuONO react under CO at 30 atm (80 °C) and 1 atm, respectively, to form (**246**)*[742] and [$PdCl_2(CO_2Bu)(NO)(PPh_3)$].[743]

5.5.1.7 *From palladium complexes and isocyanides*

Isocyanides (R^1NC) react with alkylpalladium(II) and some allylpalladium(II) complexes to form iminoacyl complexes, [Pd–C(=NR^1)R^2] (R^2 = alkyl, allyl). Additional insertion is possible to form [Pd–C(=NR^1)C(=NR^1)R^2] and [Pd–C(=NR^1)C(=NR^1)C(=NR^1)R^2] units. Structural and spectroscopic studies indicate that the diimine unit of [Pd–C(=NR^1)C(=NR^1)R^2] adopts the (*E*)-*trans*-conformation (**247**)[744–6] expected from theoretical analyses.[746] Isocyanides react with *trans*-[PdCl(CH_2COMe)($PPh_3)_2$] to form the β-ketoamine complexes *trans*-[PdCl{C(NHR)=CHCOMe}($PPh_3)_2$] (R = Me, *p*-$MeOC_6H_4$) exhibiting intramolecular H···O hydrogen bonding.[129]

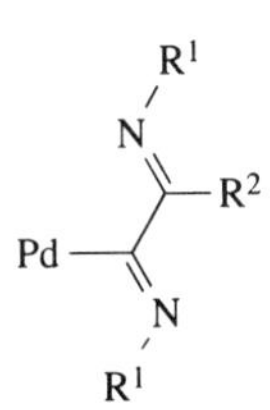

(**247**)

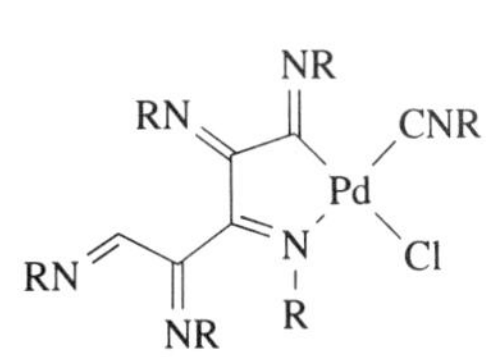

(**248**) R = 2,6-Xyl

Single, double or triple insertion products similar to those above are formed on the reaction of *p*-$MeOC_6H_4NC$ with [$Pd(PPh_3)_4$] in the presence of HCl, presumably via oxidation of palladium(0) by HCl and insertion into the Pd–H bond.[747] Insertion into a Pd–H bond is also believed to occur in the complex reaction of *trans*-[$PdCl_2(PPh_3)_2$] with excess SiH_2MePh and 2,6-XylNC to form (**248**)*.[748] Reaction of *p*-TolNC with the *N,N'*-di-*p*-tolylformamidinato complex ion [$Pd(C_6F_5)_2${η^3-N(*p*-Tol)CHN(*p*-Tol)}]$^-$ gives the five-membered ring intramolecular coordination complex ion [Pd{C(=N-*p*-Tol)N(*p*-Tol)CHN(*p*-Tol)}($C_6F_5)_2$]$^-$.[749]

Insertion occurs into Pd–Ar bonds, with a well-developed chemistry reported for Ar as phenyl, thienyl,[750] 2-pyridyl[711,745,751–4] and related heterocycles,[150,711,754] C_6F_5 and C_6Cl_5,[4] and several intramolecular coordination systems. Methyl isocyanide inserts into the Pd–C_6F_5 bond of $[Pd(C_6F_5)(CNMe)(\mu\text{-}Cl)]_2$ and the Pd–C_6Cl_5 bond of *trans*-$[PdCl(C_6Cl_5)(CNMe)_2]$ on reflux in benzene to form complexes containing *syn*-iminoacyl bridges, such as in (**249**).[755] Several derivatives have been characterized crystallographically, including $[PdCl\{\mu\text{-}C(C_6F_5)\text{=}NMe\text{-}C,N\}(thiophene)]_2$,[756] $[Cl_2Pd\{\mu\text{-}C(C_6F_5)\text{=}NMe\text{-}C,N\}_2Pd(TMEDA)]$*,[757,758] a mononuclear complex with an *anti*-iminoacyl group (**250**)* formed on addition of MeNC to a dimer,[759] and the cyclic tetranuclear complex (**249**)* formed on addition of [Ag(OAc)] to the corresponding chloro tetramer.[755] Pentachlorophenyl chemistry similar to that leading to the C_6F_5 derivative (**250**) has also been developed.[760] The A-frame complex $[MePd(\mu\text{-}dppm)_2(\mu\text{-}I)PdI]^+$ reacts with MeNC to form $[\{(MeNC)Pd\}_2(\mu\text{-}dppm)_2\{\mu\text{-}CMe\text{=}NMe\}]^{3+}$ with the bridging iminoacyl present as in (**249**).[702]

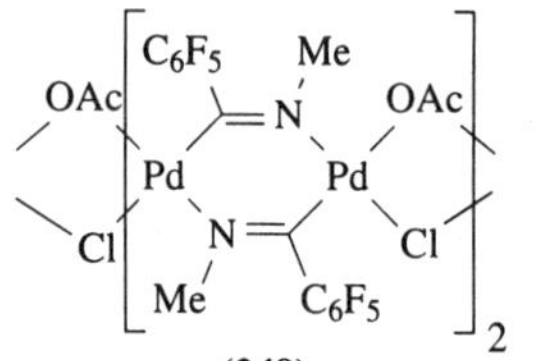

(**249**)

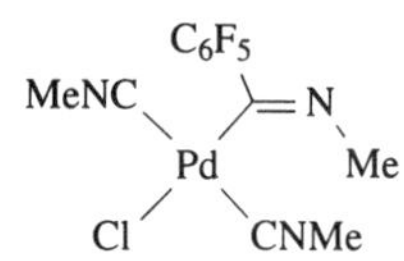

(**250**)

Reaction of the allenyl complex *trans*-$[PdBr(CH\text{=}C\text{=}CMe_2)(PPh_3)_2]$ with Bu^{t}NC gives an unstable vinyl ketenimine complex *trans*-$[PdBr\{C(\text{=}CMe_2)CH\text{=}C\text{=}NBu^t\}(PPh_3)_2]$.[761]

Single insertion into the Pd–C bond occurs for the chloro derivative of (**179**) (R = Me) to form (**251**), and into other Pd–C(sp^3) or Pd–C(sp^2) intramolecular coordination complexes $[Pd(C\sim E)(\mu\text{-}Cl)]_2$ to form similar complexes, where C~E is cyclopalladated 1-thiomethylnaphthalene,[693] *N,N*-dimethylbenzylamine[762] or as shown in (**75**), (**114**) (E = SMe), (**134**) or (**135**).[693] The ligands containing thiomethyl coordination are generally more reactive towards insertion than those containing NMe_2 donor groups. The complex where C~E is cyclopalladated 1-dimethylaminonaphthalene reacts with BzNC to give (**252**), and $PhCH_2SMe$ forms complexes analogous to (**251**) or (**252**), depending upon reagent stoichiometry.[692,693] Single or double insertion for $[Pd(C\sim N)(\mu\text{-}Cl)]_2$ may also occur with loss of intramolecular coordination, for example, reaction of the complex of cyclopalladated *N*-methyl-*N*-nitrosoaniline (**150**) with Bu^{t}NC to form *trans*-$[PdCl\{2\text{-}(O\text{=}NNMe)C_6H_4C(\text{=}NBu^t)\text{-}C\}(CNBu^t)_2]$.[489]

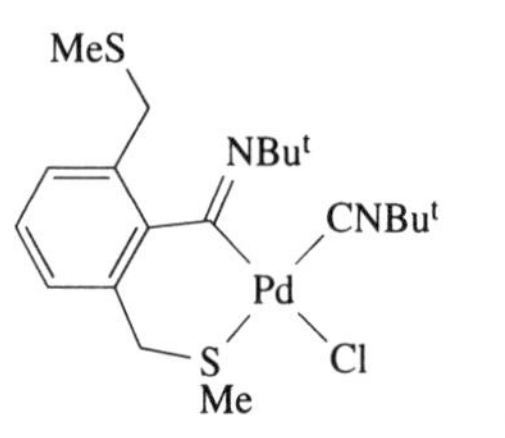

(**251**)

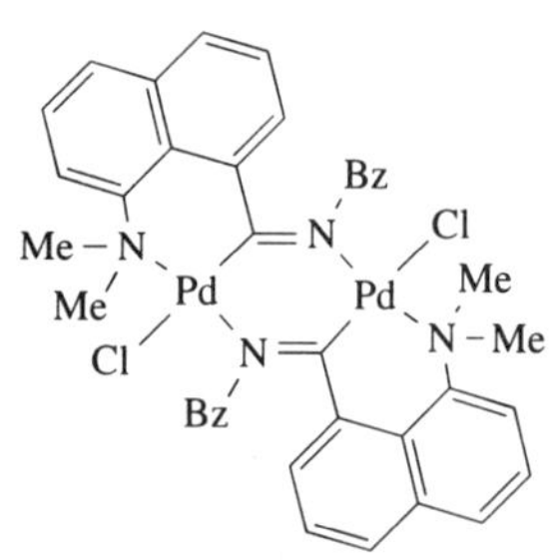

(**252**)

Diisocyanoarenes react with *trans*-$[PdBrMeL_2]$ to form oligomers of structure (**253**), two of which (n = 1*, 3*) have been structurally characterized.[763] A related reaction, for which R^1 = H, R^2 = *p*-Tol and L = bis[(*S*)-2-methylbutyl]phenylphosphine, resulted in the synthesis and separation of isomers with opposite helical configurations for n = 5.[764]

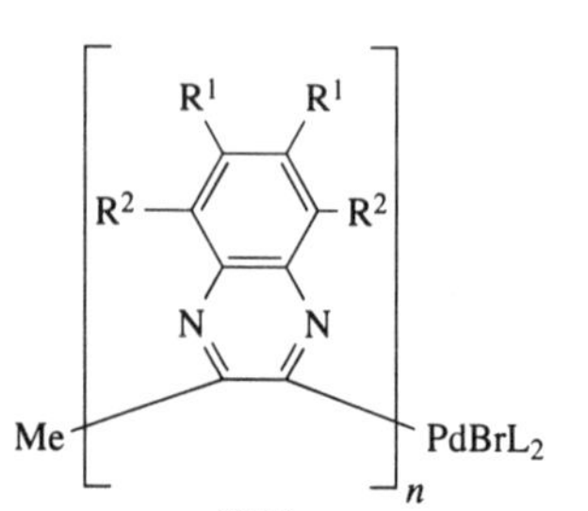

(**253**)

L = $PMePh_2$, n = 1–8, $R^1 = R^2$ = Me

5.5.1.8 From other synthetic methods

Several reaction systems that cannot be readily classified into Sections 5.5.1.1–7 are mentioned here, commencing with insertion and elimination reactions, followed by the interaction of palladium complexes with arenes and reagents containing activated $C(sp^3)$–H bonds, conversion of polyhapto complexes to η^1-complexes, and the chemistry of acetylacetonates. The synthesis of monoorganopalladium(II) complexes from diorganopalladium(II) complexes using a wide range of reactants is discussed also in Sections 5.5.2.3 and 5.5.2.4.

Diazoacetonitrile[765] and $(CF_3)_2CN_2$[766] react with Pd–Cl bonds to form complexes containing Pd–CH(Cl)CN and Pd–$C(CF_3)_2$Cl groups, and diazomethane reacts with the intramolecular coordination complex [$PdCl_2(\eta^2$-CH_2=$CH_2CMe_2CH_2NMe_2$)] in Et_2O/EtOH to form [PdCl(CH_2R)(η^2-CH_2=$CH_2CMe_2CH_2NMe_2$)] (R = H, Cl*, OMe, OEt).[767,768] The product distribution for this reaction depends upon solvent conditions,[219,767] and other methylene insertion products obtained include [PdX(CH_2X)(cod)] (X = Cl, Br) and [PdCl(CH_2Cl)(L_2)] (L_2 = CH_2=$CHCMe_2CH_2SMe$, TMEDA, $(Me_2NCH_2)_2CMe_2$, $(MeSCH_2)_2CMe_2$, dppe, dppp). Insertion into the Pd–Cl bond is more facile than insertion into the Pd–Br bond, and all of the complexes show some tendency to revert to the starting dichloro complexes.

Addition of PPh_3 to [PdCl(PPh_3){μ-C(C_6F_5)=NMe-*C,N*}]$_2$, a dimer with bridges identical to that for (**249**), results in elimination of MeNC to form *trans*-[PdCl(C_6F_5)$(PPh_3)_2$];[758] *trans*-[PdCl{$COCO_2(CH_2)_2C$≡CMe}$(PPh_3)_2$] eliminates CO on reflux in toluene to form the Pd–$CO_2(CH_2)_2C$≡CMe analogue,[168] and *trans*-[PdCl(COCOR)L_2] eliminate CO at ambient temperature in solution to form *trans*-[PdCl(COR)L_2] (R = Me, L = $PMePh_2$; R = Me, Ph, L = PPh_3).[166,167] Thallium(I) polyfluorobenzoates react with [$PdX_2(L_2)$] (X = Cl, Br; L_2 = 2 py, 2 PPh_3, bipy, phen, dppe) to provide a general route to a wide range of polyfluoroarylpalladium(II) complexes via decarboxylation of carboxylatopalladium(II) intermediates, although fluorine atoms in both *ortho*-positions are required for elimination of CO_2 to occur.[769,770]

Activation of P–C and As–C bonds by palladium(0) species is discussed in Section 5.5.1.1(ii), and activation by palladium(II) reagents also occurs in reactions whose mechanisms are not well understood;[771] for example, [$Pd(OAc)_2$] reacts with PPh_3 in MeOH at reflux to form *trans*-[PdPh(PPh_3)(μ-OAc)]$_2$ in 41% yield.[772]

Activation of C–H bonds of benzene solvent occurs at 70 °C on addition of [$Pd(OAc)_2$] and thioethers to form the diphenyltripalladium(II) complexes [{PdPh(SR^1R^2)(μ-OAc)$_2$}$_2$Pd] in 9–37% yield.[773] The complexes form as mixtures of geometrical isomers, based on a *cis*- or *trans*-Pd–Ph disposition, with the *trans*-isomer shown as (**254**) (E = S). For some thioethers, benzene activation has been shown to occur directly from the complexes [$Pd(OAc)_2(SR^1R^2)_2$] in benzene at 70 °C, and activation also occurs for a mixture of [$Pd(OAc)_2$] and $SePr_2$ in benzene at reflux to form (**254**) (E = Se).[774] Several thioether complexes of type (**255**) form at higher temperatures (reflux in benzene) and they are also obtained directly from [$Pd(OAc)_2$] and the thioether under these conditions.[775] Activation of methyl-substituted arenes usually occurs at more than one site but favours the more electron-rich and less hindered sites.[774,776] Coupling of arene activation with metathesis procedures is suitable for the synthesis of some complexes that have wider applications in organopalladium chemistry, for example, [PdCl(Ar)(bipy)] (Ar = Ph (74% yield), 2,5-xylene (64%)).

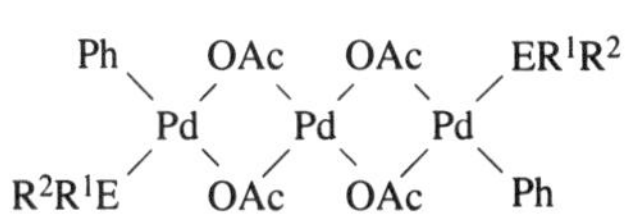

(254)
E = S: $R^1 = R^2$ = Et, Pr, Pr^i, Bu^i
$R^1 = Bu^t$, $R^2 = CH_2Bu^t$
E = Se: $R^1 = R^2$ = Pr

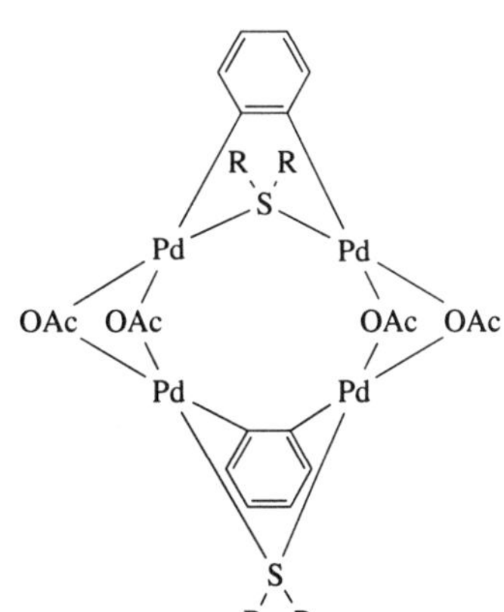

(**255**) R = Et, Pr, Pr^i

In reactions related to those forming (**254**), and to cyclopalladation chemistry, $[PdCl_2(dppp)]$ reacts with $MeCONHC(=CH_2)CO_2H$ to form (**256**)[777] and $[Pd(OAc)_2]$ reacts with $PhCOCO_2H$ in the presence of SBu_2 and Na_2CO_3 to form the oxalato-bridged complex $[\{PdPh(SBu_2)\}_2(\mu\text{-}C_2O_4)]$.[778]

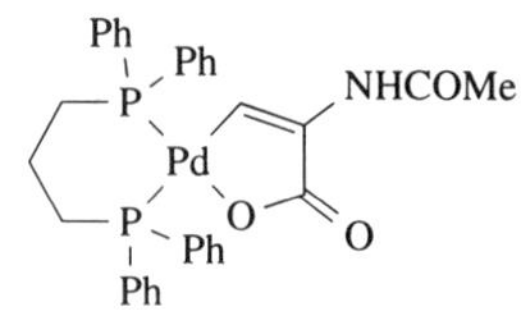

(256)

In ring-opening reactions resembling those discussed in Section 5.5.1.1(ii), cyclopropabenzene reacts with equimolar quantities of $[Pd(\eta\text{-}Cp)(\eta\text{-}C_3H_5)]$ and PMe_3 to form $[PdBz(\eta^5\text{-}C_5H_4CH_2CH{=}CH_2)(PMe_3)]$.[779] A complex ring-opening reaction also occurs on addition of $[Pd(Me)(PMe_3)(\mu\text{-}Cl)]_2$ to the titanacyclic complex $[\overline{Ti(CH_2CMe_2C}H_2)(\eta\text{-}Cp)_2]$, forming (**257**) and isobutene.[780] The palladium(I) dimer $[PdCl(\mu\text{-}dppm)]_2$ reacts with nitromethane to form $[PdCl(CH_2NO_2)(\mu\text{-}dppm)]_2$,[781] and Pd–C bond formation, probably involving free-radical reactions, occurs on the interaction of *trans*-$[PdCl_2(NCPh)_2]$ with tertiary amines;[782] for example, $NEtPr^i_2$ forms a mixture of (**258**) and (**259**). A structural analysis of the enamine complex (**258**) shows a distorted square planar geometry and indicates the presence of a weak Pd···C-2 interaction (0.270 5(9) nm, Pd–C-1–C-2 = 101.0(6)°), and thus a similar interaction is shown for (**259**).

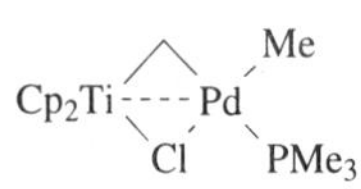

(257)

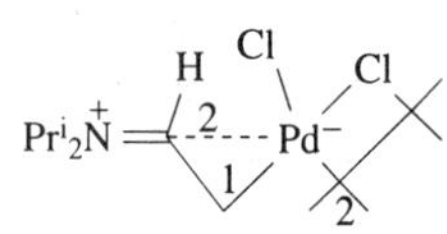

(258)

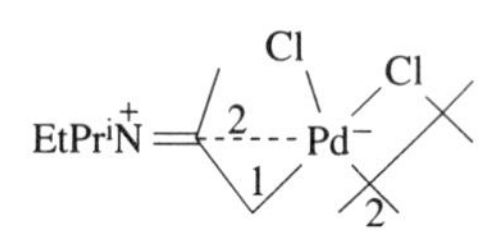

(259)

Allyl and cyclopentadienyl bonding may be converted to Pd–C σ-bonding on the reaction of π-complexes with a range of ligands,[627,783–90] and similar reactions occur for benzallyl complexes; for example, $[Pd(\eta^3\text{-}Bz)(PEt_3)_2]^+$ reacts with Cl^- to form *trans*-$[PdClBz(PEt_3)_2]$.[791] Bis(allyl)palladium(II) reacts with PMe_3, tetracyanoethene or fumaronitrile (L) to form $[Pd(\eta^1\text{-}CH_2CH{=}CH_2)(\eta\text{-}C_3H_5)L]$,[789] and with $PR_2CH_2CH_2PR_2$ ($R = Pr^i$, Bu^t) at –40 °C to form $[Pd(\eta^1\text{-}CH_2CH{=}CH_2)_2(PR_2CH_2CH_2PR_2)]$.[790] The diallyl-bridged complex $[\{Pd(hfacac)(\eta\text{-}C_3H_4\text{–}CH_2)\}_2]$ (hfacac = hexafluoroacetylacetonate) reacts with 2 equiv. PPr^i_3 to form $[\{Pd(hfacac)(PPr^i_3)(\eta^1\text{-}CH_2CH{=}CHCH_2)\}_2]$.[788] The aryl(allyl)palladium(II) complexes $[Pd(Ar)(\eta\text{-}C_3H_5)(PPh_3)]$ ($Ar = C_6F_5$, $p\text{-}HC_6Cl_4$) react with dppe and $(Ph_2PCH)_2$ to form $[PdAr(\eta^1\text{-}CH_2CH{=}CH_2)(L_2)]$.[627] For the cyclopentadienyl(allyl)palladium(II) complexes $[Pd(\eta\text{-}Cp^*)(\eta\text{-}C_3H_5)]$ and $[Pd(\eta\text{-}Cp)(\eta\text{-}2\text{-}RC_3H_4)]$ ($R = Me$, Bu^t), reaction with PMe_3 and a range of phosphines, respectively, leads to the formation of $[Pd(\eta\text{-}Cp^*)(\eta^1\text{-}CH_2CH{=}CH_2)(PMe_3)]$ and $[Pd(\eta^1\text{-}Cp)(PR_3)(\eta\text{-}2\text{-}RC_3H_4)]$.[783,786,792] When R = Me the latter complexes undergo rapid exchange with the η^1-2-methylpropenyl(cyclopentadienyl)palladium(II) isomer, for example, $[Pd(\eta\text{-}Cp)(\eta^1\text{-}CH_2CN{=}CH_2)(PMe_3)]$. Allyl groups may also be removed completely from the coordination sphere; for example, the cyclic η-allyl complexes $[\overline{Pd\{\eta\text{-}C_3H_4(CH_2)_2CH{=}CHC}H_2\}(PR_3)]$ (R = Me, Cy) react with acetylacetone to form $[Pd\{\eta^1\text{-}CH_2CH{=}CH(CH_2)_3CH{=}CH_2\}(acac\text{-}O,O')(PR_3)]$.[793]

Low-temperature NMR studies indicate that PBu_3 reacts with the cyclopalladated azobenzene complex $[Pd\{2\text{-}(PhN_2)C_6H_4\text{-}C,N\}(\eta\text{-}Cp)]$ in two series of reversible equilibria.[794] One series involves monodentate azobenzene complexes in equilibrium, $[Pd(C{\sim}N\text{-}C)(\eta\text{-}Cp)(PBu_3)]$ with *trans*-$[Pd(C{\sim}N\text{-}C)(\eta^1\text{-}Cp)(PBu_3)_2]$, and the other involves equilibria between *cis*-$[Pd(C{\sim}N\text{-}C,N)(\eta^1\text{-}Cp)(PBu_3)]$, *cis*-$[Pd(C{\sim}N\text{-}C)(\eta^1\text{-}Cp)(PBu_3)_2]$ and $[Pd(C{\sim}N\text{-}C)(PBu_3)_3]^+[Cp]^-$. At ambient temperature irreversible reactions occur to give *trans*-$[PdCl(C{\sim}N\text{-}C)(PBu_3)_2]$.

Complexes containing the allylic group $[Pd\{\eta^3\text{-}2\text{-}(OH)C_3H_3COR\}]$ (R = Me, OEt) undergo reactions with monodentate or bidentate ligands to form $Pd\text{–}CH_2COCH_2COR$ complexes,[795–7] such as $[Pd(CH_2COCH_2COEt)(acac\text{-}O,O')(PPh_3)]$. This complex exhibits two binding modes for β-diketonate ligands, and an extensive chemistry has also been developed for complexes in which palladium(II) is bonded to the central carbon atom;[1,795–804] for example, $[Pd(acac)_2]$ reacts with an equimolar amount of PPh_3,[798] PEt_3, PPr^i_3,[799] $NHMe_2$, $NHEt_2$,[340] py and picolines[804] to form $[Pd\{CH(COMe)_2\text{-}C\}(acac\text{-}O,O')(L)]$. Addition of more than 1 equiv. of a donor ligand frequently leads to displacement of $[acac]^-$ and related ligands from palladium. This process, and the formation of monodentate *O*-coordination, is favoured by the presence of electron-withdrawing substituents on the β-diketonate,[340,800] and occurs commonly for $[tfacac]^-$ (tfacac = trifluoroacetylacetonate); for example, addition of 1 equiv. PCy_3 to

$[Pd(tfacac)_2]$ gives [Pd(tfacac-*O,O*'){*O*-C(Me)=CHCOCF$_3$-*O*}(PCy$_3$)],[800] and NMR studies show that $[Pd(hfacac)_2]$ reacts with 1 equiv. pyridine to form the isomers [Pd(hfacac-*C*)(hfacac-*O,O*')(py)] and [Pd(hfacac-*O,O*')(hfacac-*O*)(py)].[805] On further addition of pyridine sequential reactions occur to give [Pd(hfacac-*O,O*')(py)$_2$]$^+$, [Pd(hfacac-*O*)(py)$_3$]$^+$ and $[Pd(py)_4]^{2+}$. The complex $[Pd(tfacac)_2]$ reacts with a mixture of excess 2,6-Me$_2$py and PPh$_3$ to give [Pd(*C*H$_2$COCHC*O*CF$_3$-*C,O*)(2,6-Me$_2$py)(PPh$_3$)]*,[806] and [PdCl(C~N~N)] reacts with [Na(acac)] to form [Pd{CH(COMe)$_2$-*C*}(C~N~N)] (C~N~N as in (**165**) with C = 2-thienyl).[520]

5.5.2 Reactions of Organopalladium(II) Complexes

Reactions of organopalladium(II) complexes may be classified as involving either retention or modification of the organic group bonded to palladium, or loss of Pd–C bond(s). Examples of both categories are dealt with in earlier sections, in particular reactions involving alkenes, alkynes, carbon monoxide and isocyanides. In this section, other reactions involving retention of the Pd–C bond are discussed (Section 5.5.2.1) followed by reactions involving mainly Pd–C bond cleavage.

5.5.2.1 Metathesis and related reactions

Exchange reactions between free and cyclopalladated groups are illustrated in Equations (3) and (4), and intramolecular exchange of Ar groups between palladium and phosphorus centres in [PdIAr(PPh$_3$)$_2$] (Equation (10)) and related complexes occurs readily in THF.[807] A mechanism involving oxidative addition of the P–C bond to palladium(II) has been proposed. Proton NMR spectra indicate that further exchange to form complexes containing PPh(*p*-Tol)$_2$, and intermolecular exchange of phosphines, also occurs.

$$\textit{trans}\text{-[PdI(}p\text{-Tol)(PPh}_3)_2] \rightleftharpoons \textit{trans}\text{-[PdIPh\{PPh}_2(p\text{-Tol)\}(PPh}_3)] \quad (10)$$

Reactions involving exchange of donor ligands are numerous, and they are crucial in affecting the stability and reactivity of Pd–C σ-bonds. Representative examples are only presented here and extensive treatments are given elsewhere, for example, for pentahalophenyl and related complexes[4,808–15] and the chemistry of poly(pyrazolyl)alkanes and borates as ligands.[263] Exchange reactions at square planar palladium(II) centres are much faster than for analogous platinum(II) centres, and exchange often occurs via the axial interaction of an incoming ligand or an uncoordinated donor group of a polydentate ligand. Examples of this phenomenon include the rapid exchange of coordinated and uncoordinated thioether groups in [PdMe$_2$(1,4,7-trithiacyclononane-*S,S*')],[264] and of pyrazole and pyridine groups in tripod ligands in [PdMe$_2$(L$_3$-*N,N*')] and [PdIMe(L$_3$-*N,N*')], for example, for L$_3$ = (pz)$_3$CH, (py)$_3$CH and (pz)$_2$(py)CH[253,263] (pz = pyrazolyl). Spectroscopic studies indicate that exchange between free and coordinated MeCN occurs for *trans*-[Pd(COMe)(PPh$_3$)$_2$(NCMe)][BF$_4$] in CDCl$_3$,[629] and between free and coordinated phosphine for [PdPr^i(S$_2$CNMe$_2$)(PEt$_3$)].[214] Exchange occurs for *trans*-[PdClBzL$_2$] (L = PPh$_3$, PMePh$_2$), but not for closely related *trans*-[PdClBz(PBu$_3$)$_2$], *trans*-[PdClPhL$_2$] (L = PPh$_3$, PMePh$_2$, PBu$_3$) or *trans*-[PdCl(COBz)(PMePh$_2$)$_2$], although the last complex does undergo exchange in the presence of excess phosphine.[220]

Exchange of phosphines between complexes is illustrated in Equation (11) (R = Pr^i,[164] Ph,[629] OCH$_2$CH$_2$C≡CMe[163]), and exchange of phosphines between neutral complexes also occurs for [PdPr(S$_2$CNMe$_2$)(PMe$_3$)] and [PdBu(S$_2$CNMe$_2$)(PEt$_3$)],[214] and between cations *trans*-[Pd(COR)(L)$_2$(NCMe)]$^+$ (R = Me,[164] Ph[629]) with different L. Exchange of Cl$^-$ and MeCN groups occurs between *trans*-[PdCl(COMe)(PPh$_3$)$_2$] and *trans*-[Pd(COPh)(NCMe)(PPh$_3$)$_2$]$^+$.[164]

$$\textit{trans}\text{-[PdCl(COR)(PPh}_3)_2] + \textit{trans}\text{-[PdCl(COR)\{P(}p\text{-Tol)}_3\}_2] \rightleftharpoons 2\,\textit{trans}\text{-[PdCl(COR)(PPh}_3)\{P(p\text{-Tol)}_3\}] \quad (11)$$

In another exchange process a solution of the dimers [PdPh(PBu$_3$)(μ-Cl)]$_2$ and [PdBz(PMePh$_2$)(μ-Cl)]$_2$ in CDCl$_3$ forms a mixture of the reactants together with [(Bu$_3$P)PhPd(μ-Cl)$_2$PdBz(PMePh$_2$)] without phosphine exchange between PdPh and PdBz centres.[220] The dimer [PdPh(PMePh$_2$)(μ-Cl)]$_2$ reacts with PPh$_3$ to give an almost statistical mixture of *trans*-[PdClPh(PMePh$_2$)$_2$], *trans*-[PdClPh(PPh$_3$)$_2$] and *trans*-[PdClPh(PMePh$_2$)(PPh$_3$)], again indicating the presence of an exchange process in the presence of free phosphine.

The replacement of a weak neutral donor ligand by a more stabilizing stronger donor ligand is a major aspect of the synthesis of complexes. Important systems include addition of neutral ligands to

[PdMe$_n$(SMe$_2$)$_m$]$^{2-n}$ at low temperature for the synthesis of PdMe and PdMe$_2$ complexes,[120,253] addition at low temperature to [PdMe$_2$(NCPh)$_2$],[284] and at a range of temperatures to the isolable complexes [PdMe$_2$(L)] (L = pyridazine (probably a dimer),[120,253,254,263] TMEDA[260,261,264,808]), [PdMe(Ph)-(TMEDA)],[262] [PdMe(SMe$_2$)(μ-X)]$_2$,[120,253–6,816] [PdMe(bipy)(NCMe)]$^+$,[255] [PdIMe(TMEDA)],[633] [PdClMe(cod)],[698,699,704,723,817] [Pd(CH$_2$CH$_2$CH$_2$CH$_2$)(TMEDA)],[267–9] [PdIPh(TMEDA)][98] and other TMEDA complexes,[260,264] [PdCl{2,4,6-(NO$_2$)$_3$C$_6$H$_2$}(L$_2$)] (L$_2$ = 2-tetrahydrothiophene, cod),[818] [PdI(C$_6$F$_5$)(MeOCH$_2$CH$_2$OMe)],[90] [PdBr(C$_6$F$_5$)(NCMe)$_2$][819] and [Pd(C$_6$X$_5$)$_2$L$_2$] (X = F, Cl),[4,352,691,820,821] for example, the formation of (**101**) from *cis*-[Pd(C$_6$F$_5$)$_2$(THF)$_2$] and PhCH$_2$NMe$_2$.[352] Strong donor ligands may also be replaced by stronger ligands, for example, PPh$_3$ by phen,[161,310] PMePh$_2$ or PMe$_2$Ph by PEt$_3$,[97,151–3,232,822] and monodentate by bidentate phosphines.[131,149,233,249,310,627] However, weaker ligands such as PPh$_3$ will displace PEt$_3$ when present in excess,[555] and PEt$_3$ will displace bidentate dppe.[556]

Anionic ligands may also replace weak neutral ligands; for example, dioxane in [Pd(C$_6$F$_5$)$_2$(dioxane)] is replaced by PhCN to give *cis*-[Pd(C$_6$F$_5$)$_2$(NCPh)$_2$][823] which reacts with OH$^-$ to give [Pd(C$_6$F$_5$)$_2$(μ-OH)]$_2$$^{2-}$.[824] Deprotonation of the uncoordinated nitrogen of [PdCl{2-[NH(Ph)N=CR]C$_6$H$_4$-*C,N*}L] by OMe$^-$ generates a strong anion donor site leading to the formation of dimers (**260**).[492,825]

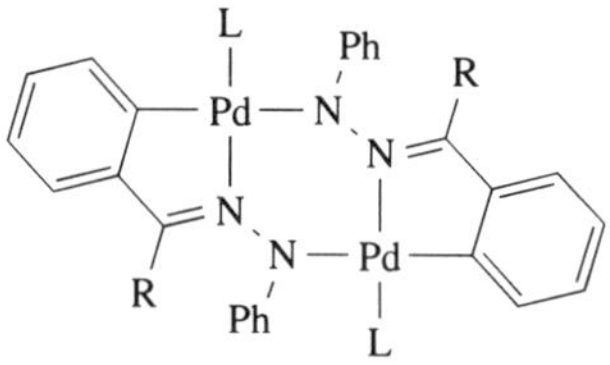

(**260**) R = Me, Ph; L = 4-Mepy, P(OMe)$_3$

Removal of halogen groups is usually effected by addition of a silver(I) or thallium(I) salt,[99,102,103,125,131,164,167,176,224,255,288,384,389,556,629,677,678,694,698,699,816,819,826–9] for example, the reaction of [PdR$_2$(μ-Br)]$_2$$^{2-}$ (R = CH$_2$SO$_2$-*p*-Tol) with AgClO$_4$ in MeCN to give *cis*-[PdR$_2$(NCMe)$_2$] followed by addition of bidentate ligands,[288] the reaction of [PdIAr(TMEDA)] with [Ag(O$_3$SCF$_3$)] followed by addition of a ligand to give [PdAr(L$_3$-*N,N',N''*)]$^+$ (L$_3$ = NMe(CH$_2$CH$_2$NMe$_2$)$_2$, 2,6-(NMe$_2$CH$_2$)$_2$C$_5$H$_3$N),[99,103] addition of AgClO$_4$ to [PdBrR(μ-Br)]$_2$$^{2-}$ (R = C$_6$F$_5$, C$_6$Cl$_5$) to form 'PdBrR' which reacts with neutral ligands to give complexes,[830] and the isolation of [PdBr(C$_6$F$_5$)(NCMe)$_2$] from the dibromo anion dimer in MeCN.[819] The complex [PdBr(C$_6$F$_5$)(NCMe)$_2$] forms the dimer [Pd(C$_6$F$_5$)(NCMe)(μ-Br)]$_2$ on dissolution in chloroform. Reactions employing silver(I) or thallium(I) salts are often carried out in a donor solvent with precipitation of MX, for example, addition of AgBF$_4$ to [PdIMe(bipy)][255] or *trans*-[PdClR(PPh$_3$)$_2$] (R = Me, Bz)[829] in acetonitrile to give [PdR(L$_2$)(NCMe)]$^+$. Coordinated solvent, or the anion of the M^I salt, may be subsequently replaced by stronger donor neutral ligands[102,167,176,224,255,823,829] or anions[101,107,131,490,707,708,816] illustrated above and by the synthesis of acetylacetonate and related complexes[490] such as [PdPh(acac)(PPh$_3$)] and [PdPh(PPh$_3$)(sacac)],[707,708] and the reaction of PhO$^-$ with [Pd(CH$_2$TMS)(dppe)(NCMe)]$^+$ to form [Pd(CH$_2$TMS)(dppe)(OPh)] (sacac = monothioacetylacetonate).[831] Examples from intramolecular coordination chemistry[827,828] include: addition of PhO$^-$ [177] or *N,N'*-disubstituted formamidines[832] to [Pd{2,6-(NMe$_2$CH$_2$)$_2$C$_6$H$_3$-*N,C,N*}(OH$_2$)]$^+$, addition of bipyrimidine to [Pd{2-(NMe$_2$CH$_2$)C$_6$H$_4$-*C,N*}(NCMe)$_2$]$^+$ to give mononuclear and dinuclear species,[828] addition of PPh$_3$ to [Pd(C$_6$F$_5$){2-(NMe$_2$CH$_2$)C$_6$H$_4$-*C,N*}(NCMe)]$^+$,[837] addition of PMe$_3$ to [Pd{2-(Pr^{i}N=CH)C$_6$H$_4$-*C,N*}(NCMe)$_2$] to give [Pd{2-(Pr^{i}N=CH)C$_6$H$_4$-*C*}(PMe$_3$)$_3$]$^+$ followed by NHPh$^-$ to give *trans*-[Pd{2-(Pr^{i}N=CH)C$_6$H$_4$-*C*}(NHPh)(PMe$_3$)$_2$],[245] and addition of phen or adenine to cationic cyclopalladation complexes to give (**261**)*[833] and (**262**)*.[834]

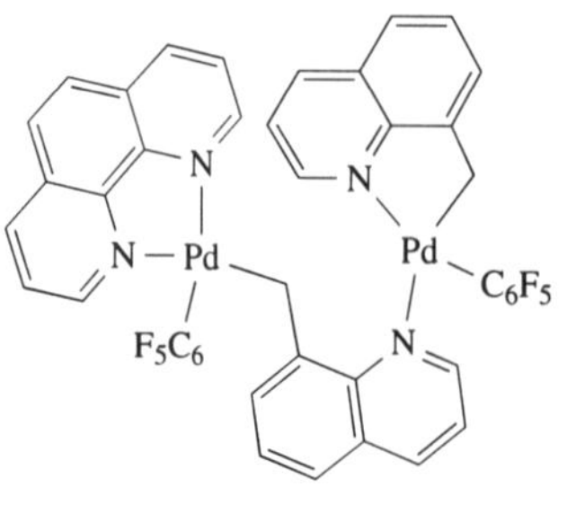

(**261**)

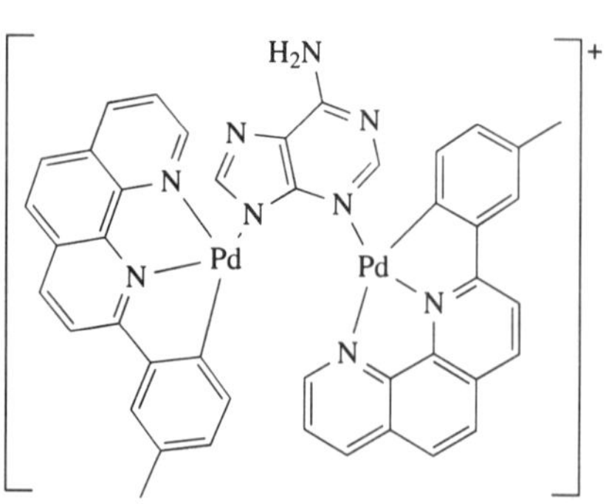

(**262**)

Removal of coordinated anions as insoluble salts encourages polydentate ligands acting as monodentates in complexes to adopt polydentate coordination, for example, addition of $[\mathrm{Ag(O_3SCF_3)}]$ to $[\mathrm{PdCl\{2\text{-}(NMe_2CH_2)C_6H_4\text{-}}C,N\mathrm{\}(Ph_2PCH_2CO\text{-}}P)]$ to give a cation with bidentate $\mathrm{Ph_2PCH_2COPh\text{-}}P,O$,[835] addition of $\mathrm{AgNO_3}$ to $[\mathrm{PdBr\{2\text{-}(HOCH_2CH_2)C_6H_4\text{-}}C\mathrm{\}(TMEDA)}]$ to give $[\mathrm{Pd\{2\text{-}(HOCH_2CH_2)C_6H_4\text{-}}C,O\mathrm{\}(TMEDA)}]^+$,[102] and the reaction of the bromo-bridged cyclopalladated 8-methylquinoline analogue of (**92**) with $\mathrm{AgPF_6}$ in $\mathrm{CH_2Cl_2}$ to form a trinuclear complex containing $\mu_3\text{-}\mathrm{OH}^-$ and $\mu_3\text{-}[\mathrm{PPh_2CHCO_2Et\text{-}}P,C,O]^-$ groups.[836]

The anion of a silver(I) or thallium(I) salt may also act as a ligand,[107,414,517,585,586,698,823,837–9] for example, on formation of $[\mathrm{PdPh\{}O\mathrm{\text{-}N(Ph)NO\text{-}}O\mathrm{\}(PPh_3)_2}]$ from the chloro analogue,[838] and the reaction of $[\mathrm{Tl(SC_6F_5)}]$ with $[\mathrm{Pd(C_6F_5)_2(\mu\text{-}Br)}]_2^{2-}$ to form $[\mathrm{Pd(C_6F_5)_2(\mu\text{-}SC_6F_5)}]_2^{2-}$.[839] The dichloro complex of the cyclopalladation system (**160**) reacts with $\mathrm{AgNO_3}$ to give $[\mathrm{Pd(C{\sim}N)(ONO_2)_2}]$, for which careful hydrolysis and addition of $\mathrm{NaClO_4}$ results in formation of $[\mathrm{Pd(C{\sim}N)(ONO_2)(OH_2)}][\mathrm{ClO_4}]\cdot\mathrm{H_2O}$.[517,840] Addition of $\mathrm{AgNO_3}$ to *trans*-$[\mathrm{PdArBr(PPh_3)_2}]$ gives *trans*-$[\mathrm{Pd(ONO_2)Ar(PPh_3)_2}]$, from which *trans*-$[\mathrm{PdXAr(PPh_3)_2}]$ ($\mathrm{Ar = 2\text{-}NH_2\text{-}5\text{-}NO_2C_6H_3}$) are obtained on reaction with $\mathrm{Cl^-}$ or $\mathrm{I^-}$.[107] Complex (**192**) reacts similarly with $[\mathrm{Ag(O_3SCF_3)}]$ to give $[\mathrm{Pd(C{\sim}N)(NMe_2H)(O_3SCF_3)}]$ from which cationic complexes such as $[\mathrm{Pd(C{\sim}N)\{NH_2(CH_2)}_n\mathrm{NH_2\}}]^+$ ($n = 2, 3$) have been obtained.[585] Thallium salts of polydentate ligands also readily exchange ligands, for example, $[\mathrm{Tl(acac)}]$,[125,819,841,842] and the bulky poly(pyrazolyl)borate anion in $[\mathrm{Tl\{BH(3\text{-}Bu^tpz)_3\}}]$ reacts with *trans*-$[\mathrm{PdBrPh(PMe_3)_2}]$ to form a complex with a monodentate poly(pyrazolyl)borate group, $[\mathrm{Pd\{BH(3\text{-}Bu^tpz)_3\text{-}}N\mathrm{\}Ph(PMe_3)_2}]$.[843]

Replacement of one anionic ligand by another is often achieved by direct reaction without the driving force of highly insoluble AgX or TlX precipitation,[133,151,310,760,844] in particular when the anion is a strong donor ligand and/or a polydentate ligand, for example, poly(pyrazolyl)borates[125,161,263,310] and dialkyldithiocarbamates,[149,310,747,845] zwitterionic $\mathrm{Ph_3P^+CMe_2CS^-_2}$[846] and $\mathrm{R_3P^+CS^-_2}$ (R = Et, Cy),[847] succinimidate[848] and methoxide.[723] The oxygen atom of alkoxide groups in *trans*-$[\mathrm{Pd(OR^1)Me(L)_2}]$ (L = phosphine donor) forms a hydrogen bond with alcohols ($\mathrm{R^2OH}$); exchange of the alkoxide group with $\mathrm{R^2OH}$ may also occur, and alkoxide complexes may also react with aryl esters as illustrated in Equation (12).[233]

$$[\mathrm{Pd\{OCH(CF_3)Ph\}(Me)(dppe)}] + \mathrm{RCOXAr} \longrightarrow [\mathrm{Pd(XAr)(Me)(dppe)}] + \mathrm{RCO_2CH(CF_3)Ph} \qquad (12)$$

R = Me, Ph; X = O, S

Addition of monodentate anions is a procedure frequently employed to exchange bridging groups in dimeric complexes, for example, for conversion of $[\mathrm{PdMe(SMe_2)(\mu\text{-}Cl)}]_2$[252] and $[\mathrm{Pd\{2,4,6\text{-}(NO_2)_3C_6H_2\}Cl(\mu\text{-}Cl)}]_2^{2-}$ [222] to the corresponding bromo and iodo species. This approach is particularly useful in intramolecular coordination chemistry,[178,337,339,358,364,375,408,460,473,476,487,506,514,548,554,559,560,566,756,849] mainly involving conversion of acetate to halide bridges.

The dimeric anions $[\mathrm{Pd(2\text{-}Me\text{-}6\text{-}NO_2C_6H_3)Cl(\mu\text{-}Cl)}]_2^{2-}$ (**71**) and (**72**) react with bipy to form $[\mathrm{PdClAr(bipy)}]$, and (**71**) reacts with DMSO to give $[\mathrm{PdCl_2Ar(DMSO)}]^-$.[222,224] However, with the donors $\mathrm{PPh_3}$ and $\mathrm{AsPh_3}$ added in 2:1 ratio to palladium, the anions (**71**) and $[\mathrm{PdCl\{2,4,6\text{-}(NO_2)_3C_6H_2\}(\mu\text{-}Cl)}]_2^{2-}$ form *trans*-$[\mathrm{PdClAr(PPh_3)_2}]$ and *trans*-$[\mathrm{PdClAr(AsPh_3)_2}]$, respectively.[221,818] Neutral ligands such as $\mathrm{PPh_3}$ and CO also cleave the bridging $\mathrm{C_6F_5}$ groups in $[\mathrm{(C_6F_5)_2Pd(\mu\text{-}C_6F_5)_2Pt(C_6F_5)_2}]^{2-}$ to give $[\mathrm{M(C_6F_5)_3(CO)}]^-$ (M = Pd, Pt).[690]

Bridge-splitting reactions by neutral monodentate[173,225,238,306,337,358,364,365,370,375,378,379,383,385,395,401,403,407,412,420,450,454,455,460,472,476–8,483,486,489,492,502,504,505,508,509,531,548,549,551,555–7,565,566,574,677,678,693,695,827,849–52] and bidentate ligands[225,337,441,443,455,473,489,505,549,695,849,853,854] are widely used in intramolecular coordination chemistry, as illustrated in Scheme 12. Bidentate ligands used for this purpose include en,[827] bipy,[225,505,549,849,853] TMEDA,[101] dppm,[225,460,473,828,853,854] dppe[225,337,455,487,489,556,695] and 1,5-cod.[225] In some cases bidentate ligands react with retention of a bridge,[460,473,695,853–5] for example, to form (**263**)[695] and (**264**)*[473] on addition of the diphosphines in a dimer:ligand ratio of 1:1.

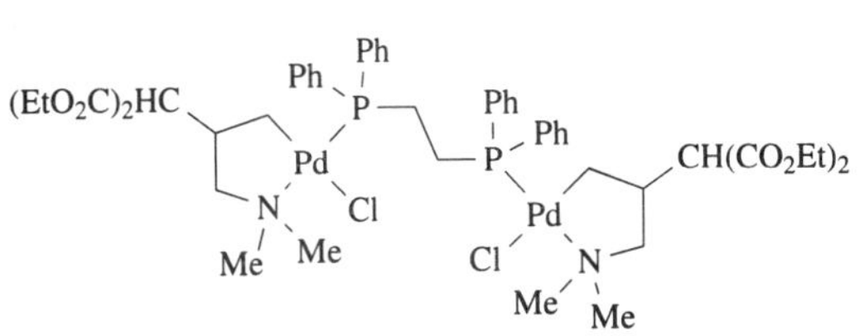

(263)

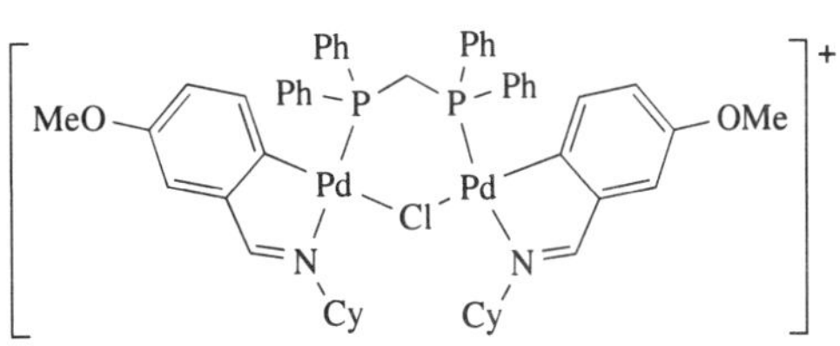

(264)

The anion $[PdR_2(\mu\text{-Br})]_2^{2-}$ reacts with dppm to form $[(RPd)_2(\mu\text{-dppm})]^{2-}$ (R = CH_2SO_2-*p*-Tol) in which the diphosphine bridges in the same manner as dppe in (**263**).[288]

Monodentate[432] and bidentate anions[201] are also applied in bridge-splitting reactions, in particular β-diketonates,[178,225,370,378,385,406,414,425,464,476,505,507,508,514,518,548,550–3] dithiocarbamates[495,504] and poly(pyrazolyl)-borates;[263] for example, addition of $K[B(pz)_4]$ to $[PdMe(SMe_2)(\mu\text{-I})]_2$ followed by PPh_3 gives $[PdMe\{B(pz)_4\text{-}N,N'\}(PPh_3)]$. Other reagents include $[PhCOCHCSPh]^-$,[548] $[Ph_2PCH_2CO_2]^-$,[856] $[MePhAsCH_2CH_2S]^-$ to give (**146**) and (**147**),[444] oxine,[505] $[2\text{-}(CHO)C_6H_4O]^-$,[551] $[2\text{-}(RN{=}CH)C_6H_4O]^-$,[549,551] $[PhN{=}N\text{-}C(R){=}NO]^-$,[857] and the 2-$pyS^-$ group which forms polynuclear species.[858] Cyclopentadienyl and substituted cyclopentadienyl,[363,370,408,420,555] allyl[370] and $[\eta^3\text{-MeC(N-}p\text{-Tol)}_2]^-$ groups[370] are also useful reagents for bridge splitting.

Monodentate phosphines cleave the Pd–N bonds of (**61**) (R = H, L = PPh_3)[181,183] and (**62**) (Y = Z = CH, X = Cl)[822] and, although dppe and $[BH(pz)_3]^-$ cleave the Pd–N bonds of (**61**) (R = H, L = PPh_3) to give mononuclear species,[183,859] for example, $[Pd\{BH(pz)_3\text{-}N,N'\}(2\text{-}CH_2C_5H_4N\text{-}C)(PPh_3)]$, the closely related ligands $[BH_2(pz)_2]^-$ and $[B(pz)_4]^-$ displace the chloro and PPh_3 groups with retention of the dimeric structure, for example, $[Pd\{BH_2(pz)_2\text{-}N,N'\}(\mu\text{-}2\text{-}CH_2C_5H_4N\text{-}C,N)]_2$.[859]

The addition of neutral ligands to intramolecular coordination complexes containing the [Pd(C~E-*C,E*)] group may result in displacement of the neutral donor group E,[340,676] for example, addition of py to $[Pd\{2\text{-}(HOCH_2CH_2)C_6H_4\text{-}C,O\}(TMEDA)]^+$ to form $[Pd\{2\text{-}(HOCH_2CH_2)C_6H_4\text{-}C\}(TMEDA)(py)]^+$,[102] addition of phosphines to $[Pd(\eta\text{-Cp})\{2\text{-}(PhN_2)C_6H_4\text{-}C,N\}]$ to form $[Pd(\eta\text{-Cp})\{2\text{-}(PhN_2)C_6H_4\text{-}C\}(L)]$,[420,860] and for the last system further addition of phosphine gives $[PdL_4]$.[860] Displacement of the neutral donor may also occur in bridge-splitting reactions on addition of excess anion[432,548] or neutral ligand,[375,460,477,555,557,827] for example, addition of excess CN^- to the chloro-bridged complex of a nitrosamine (**150**) to give $[Pd(CN)_3\{2\text{-}(O{=}NNMe)C_6H_4\text{-}C\}]^-$, although a similar reaction for cyclopalladated $P(OPh)_3$ ceases with formation of $[Pd(CN)_2\{2\text{-}[P(OPh)_2O]C_6H_4\text{-}C,P\}]^-$.[432]

Dimeric complexes may be assembled by removal of a ligand from a mononuclear complex or addition of a palladium(II) reagent to generate a stoichiometry expected to lead to dimerization via ligand bridging. For example, the dimeric complexes (**62**) (Y = N, Z = N or CH, X = Cl)[153,711] and (**62**) (X = Cl, bridging group as 6-chloropyridyl)[149,151] may be formed on H_2O_2 oxidation of a PPh_3 group in the monomers *trans*-$[PdCl(R)(PPh_3)_2]$, and *cis*-$[Pd\{2\text{-}(AsPh_2CH_2)C_6H_4\text{-}C,As\}_2]$ reacts with $[PdCl_2(SEt_2)_2]$ to form $[Pd\{2\text{-}(AsPh_2CH_2)C_6H_4\text{-}C,As\}(\mu\text{-Cl})]_2$.[295] Elemental sulfur[629] or *trans*-$[PdCl_2(NCPh)_2]$ may be used to remove phosphines; for example, *trans*-$[PdCl(COPr^i)(PPh_3)_2]$ reacts with the latter reagent to give $[Pd(COPr^i)(PPh_3)(\mu\text{-Cl})]_2$.[164] Hydrogen peroxide and Bu^tO_2H react similarly with *trans*-$[PdCl(COEt)(PPh_3)_2]$ to form $[Pd(COEt)(PPh_3)(\mu\text{-Cl})]_2$,[861] but merely displace the hydroxy group of $[Pd(OH)(CF_3)(dppe)]$ to form $[Pd(OOR)(CF_3)(dppe)]$ (R = H, Bu^t).[131] Removal of a ligand does not always result in dimerization; for example, although *trans*-$[PdClBz(PPh_3)_2]$ reacts with *trans*-$[PdCl_2(NCPh)_2]$ to form $[PdBz(PPh_3)(\mu\text{-Cl})]_2$,[829] closely related *trans*-$[PdClBz(PCy_3)_2]$ reacts with $K_2[PdCl_4]$ to form the neutral benzallyl complex $[PdCl(\eta\text{-Bz})(PCy_3)]$.[127]

The synthesis of polynuclear complexes from the reaction of organopalladium(II) complexes with inorganic or organometallic complexes has been developed for several systems, including addition of the α-diimine bidentate donor *trans*-$[PdCl\{C[{=}N(p\text{-MeOC}_6H_4)](2\text{-}C_5H_4N)\}(PPh_3)_2]$ to a $PdCl_2$ acceptor.[862] The complex *cis*-$[Pd(C_6F_5)_2(NCPh)_2]$ reacts with $[MX_2L_2]$ to form $[(C_6F_5)_2Pd(\mu\text{-X})_2ML_2]$, for example, $[MX_2L_2]$ = $[NiCl_2(dppe)]$ or $[Pt(SCN)_2(PEt_3)_2]$*.[863] The anions $[PdR_2(\mu\text{-Br})]_2^{2-}$ undergo aryl group transfer reactions with $[Pd_2Br_6]^{2-}$ to form $[PdBrR(\mu\text{-Br})]_2^{2-}$ (R = C_6F_5, C_6Cl_5).[864] The weakly donating THF is displaced from *cis*-$[Pd(C_6F_5)_2(THF)_2]$ by $[MCl_4]^{2-}$ to form $[(C_6F_5)_2Pd(\mu\text{-Cl})_2M(\mu\text{-Cl})_2Pd(C_6F_5)_2]^{2-}$ (M = Pd, Pt), and a similar strategy allows the synthesis of $[(C_6F_5)_2Pd(\mu\text{-Br})_2Pt(\mu\text{-Br})_2Pt(\mu\text{-Br})_2Pd(C_6F_5)_2]^{2-}$ from $[Pt_2Br_6]^{2-}$.[865] Complex ions containing μ-aryl groups, $[(C_6F_5)_2Pd(\mu\text{-}C_6F_5)_2M(C_6F_5)_2]^{2-}$ (M = Pd, Pt), are obtained on the reaction of *cis*-$[Pd(C_6F_5)_2(THF)_2]$ with $[M(C_6F_5)_4]^{2-}$,[690] and polynuclear pentafluorophenyl complexes with single-bridged pseudohalide or halide groups have been isolated,[866,867] for example, on the reaction of $[Pd(SCN)(C_6F_5)(bipy)]$ with *trans*-$[Pd(OClO_3)(C_6F_5)(PPh_3)_2]$ to give $[(bipy)(C_6F_5)Pd(\mu\text{-SCN})Pd(C_6F_5)(PPh_3)_2][ClO_4]$.[867]

Dimers or polynuclear complexes are also accessible from mononuclear intramolecular coordination complexes; for example, *cis*-$[Pd\{2\text{-}(NO_2)C_6H_4\text{-}C,O\}_2]$ (**74**)[225] and *cis*-$[Pd\{2\text{-}(NMe_2CH_2)C_6H_4\text{-}C,N\}_2]$[205] react with MCl_2 (M = Pd, Hg) and $[M(OAc)_2]$ (M = Pd, Hg), respectively, to form $[Pd(C{\sim}E\text{-}C,E)(\mu\text{-X})]_2$, and the 8-methoxynaphthyl complex *cis*-$[Pd(C_{10}H_6OMe\text{-}C,O)_2]$ reacts with $[Hg(OAc)_2]$ to form $[Pd(C_{10}H_6OMe\text{-}C,O)(\mu\text{-OAc})]_2$*.[226] The complex $[Pd\{2\text{-}(NMe_2CH_2)C_6H_4\text{-}C,N\}\{2\text{-}(NMe_2)C_6H_4CHTMS\text{-}C,N\}]$ reacts with $[PdCl_2(SMe_2)_2]$ to form chloro-bridged dimers for each intramolecular coordination group,[685] but reaction with closely related $[PdCl_2(SEt_2)_2]$ gives these products together with the μ-benzylidene complex (**265**).[868] Single halide bridged complexes are obtained on the reaction of $[M^1X(N{\sim}C{\sim}N)]$ with $[M^2(N{\sim}C{\sim}N)(OH_2)][BF_4]$ to give (**266**) (E = NMe_2),[869] and of $[PdX(C_6F_5)(C{\sim}N)]^-$ with $[Pd(C_6F_5)(C{\sim}N)(NCMe)]$ to give (**267**).[837] A hydrido-bridged complex (**266**) (E = PPh_2) has been obtained on addition of sodium formate to $[Pd(P\sim C\sim P)(OCMe_2)][O_3SCF_3]$ in MeOH,[826] and the reaction of $[Pt(cod)_2]$/2 PPh_3 with $[Pd(N\sim C\sim N)(OH_2)][BF_4]$ and hydrogen gives the dihydrido-bridged complex $[(Ph_3P)_2Pt(\mu\text{-}H)_2Pd(N\sim C\sim N)][BF_4]$ (N~C~N as in (**266**)).[869] A Pt → Pd interaction is assumed to be present in the product of reaction of *cis*-$[PtMe_2(dppm\text{-}P)_2]$ with [PdCl(Me)(cod)], namely $[(Me)_2Pt(\mu\text{-}dppm)_2Pd(Me)][Cl]$.[817]

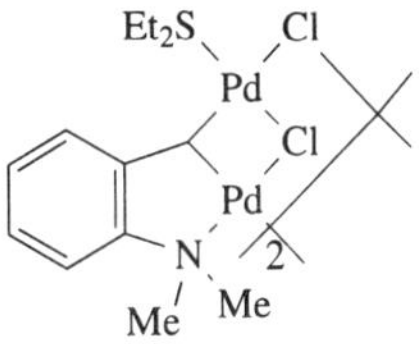

(**265**)

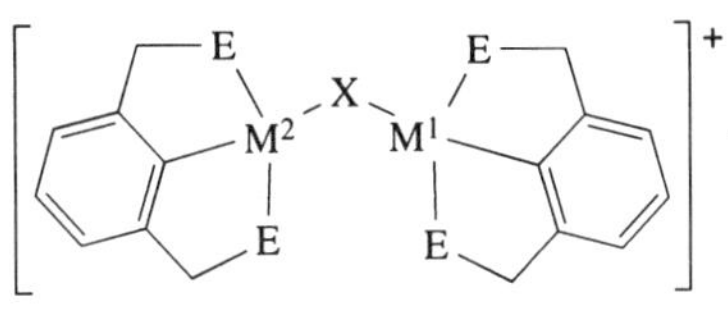

(**266**)
E = NMe_2: $M^1 = M^2$ = Pd, X = Cl, Br, I, CN, Y = BF_4
M^1 = Pt, M^2 = Pd, X = Br, Y = BF_4
E = PPh_2: $M^1 = M^2$ = Pd, X = H, Y = O_3SCF_3

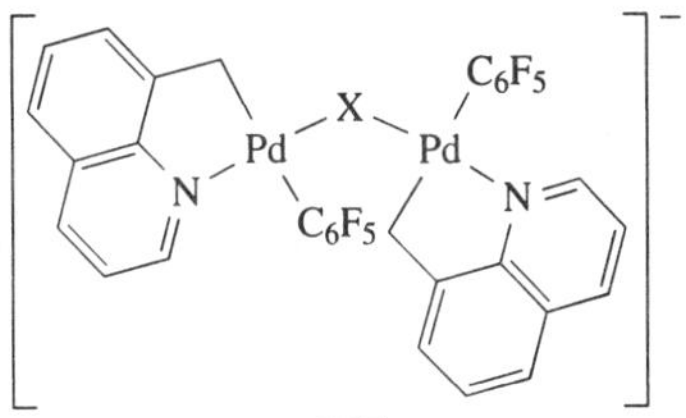

(**267**)
X = Cl, Br

5.5.2.2 Isomerization and decomposition of organopalladium(II) complexes

Isomerization and exchange reactions are discussed elsewhere, in particular regarding complexes (**65**) and (**223**), intermediates (**77**) and (**78**) in *cis–trans* equilibria, the equilibria illustrated in Schemes 19 and 20 and Equations (10) and (11), isomerization occurring in synthesis using $LiBu^t/CuCN$ and in alkene and alkyne insertion reactions, reversible CO insertion reactions, alkene exchange reactions exhibited by *trans*-$[Pd(COR)(NCMe)(PPh_3)_2]^+$, and exchange processes discussed in the introductory paragraphs of Section 5.5.2.1. Other aspects are mentioned here when relevant to considerations of decomposition.

(i) Monoorganopalladium(II) complexes

The stability of monoalkylpalladium(II) complexes is enhanced by several factors common to organotransition metal chemistry, in particular the presence of bidentate ligands, the formation of intramolecular coordination systems, and the absence of β-hydrogen atoms; for example, [PdIMe(TMEDA)] is stable but [PdIEt(TMEDA)] decomposes over 1 d in $CDCl_3$ to form palladium metal.[265] Five-membered rings are preferred for intramolecular coordination complexes, as discussed in Section 5.5.1.3, and illustrated by the facile rearrangement of (**192**) in benzene to form *trans*-$[PdCl(CH_2CH_2CHMeNMe_2\text{-}C,N)(NHMe_2)]$.[586] This isomerization reaction is also favoured by a release of steric repulsion and the increased Pd–C bond strength expected on replacement of a Pd–$CHAlk_2$ group by a Pd–CH_2Alk group. The latter effect has been estimated as 6.7 kJ mol^{-1} from studies of the isomerization of a series of remarkably stable dimethyldithiocarbamate complexes $[PdR(S_2CNMe_2)(PEt_3)]$; for example, the propyl and butyl isomers are preferred to the $CHMe_2$ and CHMeEt isomers, respectively, by ratios of 10:1 in equilibria at 75 °C in toluene.[214] Similar isomerizations occur for equilibria involving the acyl complexes *trans*-$[Pd(COR)(NCMe)(PPh_3)_2][BF_4]$ in $CDCl_3$, for example, for R = Pr^n and Pr^i, $K_{eq} = 5.1 \pm 0.3$ in favour of the Pr^n complex, and kinetic studies are consistent with the involvement of an intermediate $[PdH(alkene)(CO)(PPh_3)_2]^+$ in the latter equilibrium.[164]

Decomposition of monoorganopalladium(II) compounds is usually a complex process; for example, *trans*-$[PdXMe(PEt_3)_2]$ (X = Br, I, CN) decompose in solution at 120–60 °C to give methane, ethane, ethene and *trans*-$[PdX_2(PEt_3)_2]$.[844] The complex *trans*-$[Pd\{OCH(CF_3)Ph\}Me(PMe_3)_2\cdot\{HOCH(CF_3)Ph\}]$, in which the alcohol solvate is hydrogen bonded to the coordinated oxygen atom, decomposes to give mainly methane and alcohol;[233] $[Pd\{OCH(CF_3)Ph\}Me(dppe)]$ behaves similarly to give the alcohol together with lesser amounts of methane, ethane, ethene and CF_3COPh.[722]

The cation *trans*-$[PdMe(NCCD_3)(PPh_3)_2]^+$ forms $[PMePh_3]^+$ in $CDCl_3$ at 50 °C, but *trans*-$[PdBz(NCCD_3)(PPh_3)_2]^+$ forms BzH and Bz_2 in $CDCl_3/C_6D_6$ at 65 °C in a process involving benzyl radicals.[829] Addition of $AgBF_4$ to *trans*-$[PdClBz(PPh_3)_2]$ in $C_6D_6/CDCl_3$, to generate a cation in the absence of a stabilizing neutral donor such as CD_3CN, results in the formation of C_6D_5Bz at 85 °C and this product may result from an intermediate with coordinated C_6D_6.[829] A complex decomposition process also occurs for the 8-methoxynaphthyl complex $[PdCl(C_{10}H_6OMe\text{-}C)(SEt_2)_2]$ which forms the C–C coupled product $(C_{10}H_6OMe)_2$ at 0 °C in solution.[226]

The complexes *trans*-[$Pd(COR^1)\{PO(OR^2)_2\}(PMe_3)_2$] ($R^1$ = Ph, *p*-Tol; R^2 = Me, Et)[195] and the aryloxide complex *trans*-[$Pd(COMe)\{4\text{-}(NC)C_6H_4O\}(PEt_3)_2$][703] decompose in solution to form $R^1PO(OR^2)_2$ and 4-($MeCO_2$)C_6H_4CN, respectively, and kinetic studies for the latter complex are consistent with the presence of both direct elimination and elimination via dissociation of PEt_3.

The dimer [$PdBz(PPh_3)(\mu\text{-}Cl)]_2$ decomposes at 65 °C in $CDCl_3$ to form BzCl,[829] and the A-frame complex (**3**) eliminates methane at –20 °C.[3] The acetyl analogue of (**3**) and [$(PdCOMe)_2(\mu\text{-dppm})_2(\mu\text{-}H)$] eliminate MeCHO at 25 °C in CD_2Cl_2.

(ii) Diorganopalladium(II) complexes

Decomposition of diorganopalladium(II) complexes has been studied extensively,[85,247,248,870] including from a theoretical standpoint[85,871–4] with an emphasis on reactions giving C–C bond formation as a fundamental step in catalysis and organic synthesis mediated by palladium complexes.

The stability of diorganopalladium(II) complexes is enhanced by the presence of intramolecular coordination, electron-withdrawing organo groups, strong donor ligands and the *trans*-[PdR_2] configuration; for example, *cis*-[$Pd(C_6F_5)_2L_2$] complexes are well known but the analogous *cis*-[$PdPh_2L_2$] complexes have not been reported,[48] *trans*-[$PdPh_2(PEt_3)_2$] has been isolated but not the PPh_3 analogue,[60] and *trans*-[$PdMe_2$(transphos)] is stable to 80 °C but *cis*-[$PdMe_2L_2$] (L = PPh_3, $PMePh_2$) rapidly eliminate ethane at 45 °C.[118] Bidentate nitrogen donor ligands give particularly stable *cis*-[$PdMe_2(L_2)$][120,261] and *cis*-[$PdMePh(L_2)$] complexes.[100,262]

Thermal decomposition of *cis*-[PdR_2L_2] (L = monodentate phosphine) generally results in the formation of R–R and palladium(0) species, although *cis*-[$Pd(CH_2Bu^t)_2(PPh_3)_2$] eliminates Bu^tMe with formation of palladacyclic [$\overline{Pd(CH_2CMe_2CH_2)}(PPh_3)_2$].[241] Theoretical studies[871] are consistent with indications from kinetic and spectroscopic studies that reductive elimination is an intramolecular process preceded by dissociation of L to give a three-coordinate intermediate.[118,232,875] Addition of phosphine, suppressing formation of a three-coordinate intermediate, leads to the formation of products other than R–R from reductive elimination; for example, *cis*-[$Pd(CD_3)_2(PMePh_2)_2$] in C_6D_6 at 60 °C in the presence of $PMePh_2$ forms methane, of which CD_3H comprises 98% and CD_4 2%.[232] Hydrogen atom abstraction via cyclopalladated $PMePh_2$ is suggested to explain the formation of CD_3H as the dominant product. The complex *cis*-[$Pd(CH_2CD_3)_2(PMe_2Ph)_2$] behaves similarly; for example, in C_6D_6 at 70 °C it forms $CD_3CH_2CH_2CD_3$, but with added PMe_2Ph forms CD_3Me and CD_2=CH_2.[876]

Reductive elimination occurs from the *cis*-bidentate ligand complexes [$PdMe_2$(dppe)],[118,232] [$PdAr(CH_2CH{=}CH_2)(PPh_2CH{=}CHPPh_2)$] (Ar = C_6F_5, 2,5-$Cl_2C_6H_3$),[877] [$Pd(CH_2CH{=}CH_2)_2(R_2PCH_2CH_2PR_2)$] (R = Pr^i, Bu^t)[790] and [$PdMe(CO_2Me)\{(Ph_2PCHMe)_2CH_2\}$],[723] but α-elimination together with β-elimination occurs for [$PdMe_2$(1,4,7-trithiacyclononane)][264] and [$PdMe_2(L_2)$] (L_2 = TMEDA, TMPDA, dmpe) to liberate methane, ethane and ethene.[261] For the latter complexes, dissociation of one arm of the ligand may occur, and deuteration studies indicate that hydrogen abstraction to form methane occurs from the second methyl group and from the solvent benzene. Decomposition occurs via both reductive elimination and β-elimination for *cis*-[$PdEt_2$(dppe)],[232] and β-elimination to form alkane and alkene occurs for *cis*-[PdR_2(bipy)] (R = Et, Pr) in Ph_2CH_2 at 80 °C (Scheme 21).[235,236] Addition of electron-deficient alkenes to *cis*-[PdR_2(bipy)] results in the occurrence of reductive elimination only, suggesting that β-hydrogen elimination is prevented by alkene coordination, and thus allowing the application of [Pd(bipy)(fumaronitrile)] as a catalyst for the coupling of tetraalkyltin(IV) reagents with organohalides containing β-hydrogen atoms.[878] Alkenes such as dimethyl maleate or diphenylethyne are often added to solutions in studies of decomposition of complexes. Although they do not appear to coordinate to palladium(II) phosphine complexes, and thus do not participate in the decomposition process, they do coordinate to palladium(0) product(s) to form [PdL_2(alkene)] and thus prevent further decomposition to palladium metal or involvement of palladium(0) species in the decomposition of palladium(II) complexes.

$$\text{Et–Et} + \text{Pd} + \text{bipy} \xleftarrow{\text{CH}_2\text{=CHCO}_2\text{Me}} [\text{PdEt}_2(\text{bipy})] \longrightarrow \text{Me–Me} + \text{CH}_2\text{=CH}_2 + \text{Pd} + \text{bipy}$$

Scheme 21

The complex *trans*-[$PdMe_2(PMePh_2)_2$] eliminates Me_2 on thermolysis, and in the presence of free phosphine eliminates mainly MeH, a behaviour exhibited also by the *cis*-isomer.[232] Calculations indicate that reductive elimination from *trans*-complexes via phosphine dissociation and subsequent isomerization to place methyl groups adjacent for reductive elimination is a high-energy process.[871] Isomerization may occur via a bimolecular intermediate (**268**), similar to (**77**), promoted by the presence of *cis*-$[PdMe_2L_2]$, since reductive elimination from a mixture of *trans*-$[PdMe_2(PMePh_2)_2]$ and *cis*-$[Pd(CD_3)_2(PMePh_2)_2]$ in the presence of $PMePh_2$ at 30 °C in CH_2Cl_2 gives $(CD_3)_2$, $MeCD_3$ and Me_2, consistent with exchange of Me and CD_3 groups between palladium(II) centres.[232] Structure (**77**) for an intermediate, rather than (**268**), allows exchange of Ar^1 and Ar^2 groups on reductive elimination of a mixture of *trans*-$[PdAr^1{}_2L_2]$ and *trans*-$[PdIAr^2L_2]$ as shown in Scheme 22.[879] The complexes *trans*-$[PdR^1(CH{=}CHR^2)L_2]$ (R^1 = Ar,[60] Alk;[134] L = monodentate phosphine) and related complexes[654] form C–C bonds on decomposition, but *trans*-$[Pd(CH_2CD_3)_2L_2]$ forms CH_2DCD_3 and $CH_2{=}CD_2$ in a clean reaction for which kinetic studies are consistent with direct β-elimination without phosphine dissociation.[231]

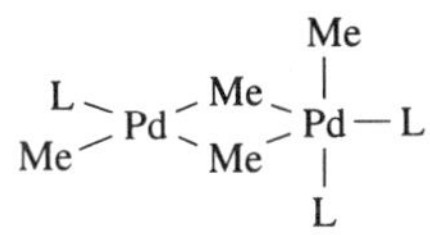

(**268**)

trans-$[PdAr^1{}_2L_2]$ + *trans*-$[PdIAr^2L_2]$ —— dimethyl maleate (dmm), toluene, 60 °C ——→

Ar^1–Ar^1 + Ar^1–Ar^2 + Ar^2–Ar^2 + *trans*-$[PdIAr^1L_2]$ + *trans*-$[PdAr^2L_2]$ + $[PdL_2(alkene)]$

Ar^1 = Ph, Ar^2 = *o*-Tol; alkene = dmm or its product of isomerization (dimethyl fumarate); stoichiometry not indicated

Scheme 22

Both *cis*- and *trans*-$[PdEt_2(PMe_2Ph)_2]$ form Et_2, $CH_2{=}CH_2$ and EtH in 1:2:2 ratio on photolysis in benzene,[270] in contrast to the thermal decomposition of the *cis*- (giving Et_2) and *trans*-isomers (giving $CH_2{=}CH_2$ and EtH).[232] Addition of $PMePh_2$ slows the photochemical decomposition and favours elimination of $CH_2{=}CH_2$ and EtH, and deuteration studies indicate the occurrence of intramolecular elimination, consistent with intramolecular isomerization by the four-coordinate complex to a *cis*-geometry followed by elimination.

Thermal decomposition in toluene for $[\overline{Pd(CH_2CMe_2CH_2})(PPh_3)_2]$ gives mainly 1,1-dimethylcyclopropane,[241] and the palladacyclopentane complexes $[\overline{Pd(CH_2CH_2CH_2CH_2})(L_2)]$ (L_2 = bipy, TMEDA, dppe) give mainly a mixture of 1- and 2-butenes, with 1-butene dominant for dppe, and together with *n*-butane for the nitrogen donor ligands.[269] The monodentate phosphine complexes $[\overline{Pd(CH_2CH_2CH_2CH_2})L_2]$ (L = PPh_3, PMe_2Ph) give mainly 1-butene,[269,270] but photochemical decomposition of the PMe_2Ph complex gives mainly ethane (75–80%) and 1-butene (~20%).

Complexes of structure [PdAr(η-allyl)(L)] decompose to form Ar–allyl,[215] but [PdMe(η-allyl)(L)] forms Me_2 as the major product, probably via a dinuclear species since $[Pd(\eta\text{-allyl})(\mu\text{-Me})]_2$ also forms mainly Me_2 at 0 °C.[289] Reductive elimination of 1,5-hexadiene occurs on addition of PPh_3 to $[Pd(CH_2CH{=}CH_2)(\eta\text{-}C_3H_5)(TCNE)]$,[880] and $[Pd(CH_2CH_2X)(\eta\text{-Cp})(PPh_3)]$ (X = OMe, OPr^i, $CH(COMe)_2$) decompose to form CpH and $CH_2{=}CHX$.[587]

5.5.2.3 Reactions with acids, including alcohols, thiols, amines, peracids and peroxides

The Pd–C bond generally reacts slowly in hydrolysis reactions, and reactivity towards H^+ appears varied; for example, the nitrogen atom of 2-pyridylpalladium(II)[822,881] and related complexes[153] is protonated by $HClO_4$, but many hydrocarbylpalladium(II) complexes are cleaved by H^+ to give a hydrocarbon (Equation (13)),[170,787,882,883] for example, on addition of HCl to *trans*-$[PdMe_2(PEt_2Ph)_2]$[712] or $[PdMe_2(Ph_2PCH_2CH_2SMe)]$[243] to form [PdClMe] complexes, addition of HI to *trans*-$[PdR_2(PEt_2Ph)_2]$ to form *trans*-$[PdIR(PEt_2Ph)_2]$ (R = Me, *m*-Tol),[272] addition of $[Et_2OH][BAr_4]$ to $[PdMe_2(bipy)]$ in MeCN to form $[PdMe(bipy)(NCMe)][BAr_4]$[634] and on reaction of carboxylic acids with $[PdMe_2(PEt_3)_2]$ to form *trans*-$[Pd(O_2CR)Me(PEt_3)_2]$.[701] Preferential cleavage of the Pd–(η^1-C_3H_5) bond occurs for $[Pd(C_6F_5)(\eta^1\text{-}C_3H_5)(dppe)]$ on addition of HCl to form $[PdCl(C_6F_5)(dppe)]$, but for $[Pd(C_6F_5)(\eta\text{-}C_3H_5)(PPh_3)]$ cleavage of the Pd–C_6F_5 bond occurs[627] and a similar selectivity for retention of π-allyl coordination is observed in the reactions of other complexes, for example $[Pd(\eta^1\text{-Cp})(\eta\text{-Cp})(PPh_3)]$ with CF_3CO_2H to form $[Pd(O_2CCF_3)(\eta\text{-Cp})(PPh_3)]$.[217] Excess HCl reacts with $[PdMe_2(TMEDA)]$ to form methane and $[PdCl_2(TMEDA)]$,[261] and with $[\overline{Pd(CH_2CH_2CH_2CH_2})(L_2)]$ (L_2 = TMEDA, bipy, dppe, 2 PPh_3) to form mainly butane, except for L_2 = dppe where substantial quantities of butenes are also formed.[267]

$$[PdMe_2] \xrightarrow[-MeH]{H^+} [PdMe]^+ \xrightarrow[-MeH]{H^+} Pd^{2+} \qquad (13)$$

Intramolecular coordination complexes such as $[Pd(C{\sim}N)(\mu\text{-}X)]_2$ undergo Pd–C bond cleavage with strong mineral acids,[333,695] although careful addition of HY can result in replacement of X by Y.[560] Acetic acid in $CDCl_3$ reacts with $[Pd\{2\text{-}(NMe_2CH_2)C_6H_4\text{-}C,N\}\{2\text{-}(EPh_2CH_2)C_6H_4\text{-}C,E\}]$ (E = P, As) and $[Pd\{2\text{-}(AsPh_2CH_2)C_6H_4\text{-}C,As\}\{2\text{-}(PPh_2CH_2)C_6H_4\text{-}C,P\}]$ to liberate the nitrogen and arsenic donor ligands, respectively, to form $[Pd(C{\sim}E)(\mu\text{-}O_2CMe)]_2$ which are stable to acetic acid.[884] A series of complexes $[PdCl(C{\sim}N)(PPh_3)]$, for example, $[C{\sim}N]^-$ as in cyclopalladated tertiary benzylamines, 8-methylquinoline ((**70**) (R = H)), azobenzene ((**104**) ($R^1 = R^2$ = H)), 2-benzylpyridine ((**113**) (X = CH_2)) and (**148**) (R^1 = H), decompose in acetic acid at 80 °C to give a brown-black precipitate, but these reactions are more rapid on addition of LiCl to give HC~N and $[PdCl_2(PPh_3)_2]$ cleanly.[850] The reactions are assumed to occur via an intermediate with the nitrogen donor atom uncoordinated and protonated, $[Pd^-Cl_2(C{\sim}N^+H\text{-}C)(PPh_3)]$, and models for this type of intermediate include $[PdCl_2\{2,4,6\text{-}(NO_2)_3C_6H_3\text{-}C\}(DMSO)]^-$ [222] and $[Pd\{2\text{-}(NBz_2CH_2)C_6H_4\text{-}C\}(hfacac)(PPh_3)]$.[340]

With alcohols in the presence of a base decomposition occurs to form palladium metal, and alcohols will also react directly with Pd–C bonds; for example, MeOH and EtOH react with *trans*-$[PdCl(COMe)(PPh_3)_2]$ to form $MeCO_2R$ (R = Me, Et),[885] and heating solutions of $[PdPh(Bu^t_2acac)(PPh_3)]$ or $[Pd\{2\text{-}(NMe_2CH_2)C_6H_4\text{-}C,N\}(Bu^t_2acac)]$ in MeOH in the presence of excess PPh_3 results in the elimination of arene.[886] The complexes *trans*-$[PdR^1_2(PMe_3)_2]$ (R^1 = Me, Et) react under mild conditions with an equimolar quantity of a range of alcohols ArOH or $R^2(CF_3)CHOH$ (R^2 = Ph, CF_3), resulting in cleavage of one Pd–C bond to form *trans*-$[Pd(OR^3)R^1(PMe_3)_2]$ (R^3 = Ar or $R^2(CF_3)CH$).[233,703,712] The complexes $[PdMe_2(L_2)]$ (L_2 = bipy,[703] dmpe, dppe[722]) and $[Pd(CH_2TMS)_2(dmpe)]$[831] undergo analogous reactions, and the dppe complex together with *trans*-$[PdEt_2(PMe_3)_2]$[887] reacts with PhSH in a similar way. The thiols *p*-RC_6H_4SH (R = Me, Br) cleave the Pd–C bonds of $[Pd(dppe)\{2\text{-}(NC)C_6H_4CH_2\text{-}\mu\text{-}C,N\}]_2^{2+}$ to form the dimeric coordination complexes $[Pd(dppe)(\mu\text{-}SAr)]_2^{2+}$.[888]

Acidic amines such as NH_2R (R = SO_2CF_3, $COCF_3$, COPh) react with *cis*-$[PdMe_2(PEt_3)_2]$ or $[PdMe_2(L_2)]$ (L_2 = dmpe, dcpe) to give methane and [Pd(NHR)Me] complexes.[258] Protonated amines cause similar cleavage reactions to occur; for example, $[PdMe_2(dppe)]$ reacts with $[NH_nEt_{4-n}]^+$ (*n* = 1–4) to give $[PdMe(dppe)(NH_mEt_{3-m})]^+$ (*m* = 0–3),[889] $[PdMe_2(dmpe)]$ reacts with $[PhNH_3]^+$ to give $[PdMe(dmpe)(NH_2Ph)]^+$,[283] and [PdMe(P~C~P)] (P~C~P as in (**81**) (R = H)) reacts with $[NH_2Et_2]^+$ in the presence of water to give $[Pd(P{\sim}C{\sim}P)(OH_2)]^+$.[266]

The reaction of *m*-chloroperbenzoic acid, *m*-$ClC_6H_4CO_3H$, with *trans*-$[PdClBz(PPh_3)_2]$ results in oxidation of Ph_3P to Ph_3PO and release of the benzyl group as *m*-$ClC_6H_4CO_3Bz$,[890] but reaction with a range of intramolecular coordination complexes such as $[Pd\{2\text{-}(PhN_2)C_6H_4\text{-}C,N\}(\mu\text{-}Cl)]_2$[891] and complexes containing (**172**)[531,533,535] and (**173**)[536] results in insertion of oxygen to form complexes in which an aryloxy group is coordinated to palladium(II). The complex $[Pd\{2\text{-}(NMe_2CH_2)C_6H_4\text{-}C,N\}_2]$ is oxidized by Bu^tO_2H to $[Pd\{2\text{-}(NMe_2CH_2)C_6H_4O\text{-}O,N\}\{2\text{-}(NMe_2CH_2)C_6H_4\text{-}C,N\}]$, but oxidation of the second Pd–C bond requires the presence of $[VO(acac)_2]$ as a catalyst, and the catalyst is also required for oxidation of $[Pd\{2\text{-}(NMe_2CH_2)C_6H_4\text{-}C,N\}(\mu\text{-}Cl)]_2$.[892] This reactivity pattern shows that the more nucleophilic palladium centres are most susceptible to oxidation, illustrated further by the inertness of the Pd–C bonds of *trans*-$[PdCl(COEt)(PPh_3)_2]$,[861] $[Pd(OH)(CF_3)(dppe)]$[131] and cationic $[Pd\{2\text{-}(NMe_2CH_2)C_6H_4\text{-}C,N\}(NCMe)_2]^+$ [892] toward oxidation by Bu^tO_2H, and consistent with an enhancement of rate in *m*-$ClC_6H_4CO_3H$ oxidation for complexes containing electron-rich aryl rings bonded to palladium.[536] Reactions of both Bu^tO_2H and *m*-$ClC_6H_4CO_3H$ in oxidations are believed to occur via associative mechanisms.

Molybdenum peroxide reacts with a wide range of arylpalladium(II) complexes in alcohols (ROH) to form ROAr, for example, $[Pd(ONO_2)Ar(TMEDA)]$ (Ar = Ph, Nap), and the alkylpalladium(II) complex $[Pd(O_2CMe)(C{\sim}N{\sim}N)]$ (C~N~N as in (**116**)) reacts in a similar way.[101,893] For reactions carried out in CH_2Cl_2, analogous chloro derivatives are formed, and for both alcohols and CH_2Cl_2 as solvents a palladium(IV) intermediate involving a '$Pd(\mu\text{-}O)_2Mo$' group may occur.

5.5.2.4 Reactions with halogens and organohalides

Reactions with halogens usually result in the cleavage of Pd–C bonds and, although organic halides also cleave Pd–C bonds, they are often unreactive towards monoorganopalladium(II) complexes. Reaction mechanisms are not fully established for most of the reactions discussed, although oxidative addition of X_2 or R^1X to $[Pd^{II}R^2R^3]$ complexes to form a palladium(IV) intermediate followed by reductive elimination to form the palladium(II) product and other products, such as R^2X or R^1R^2, is often canvassed. Reactions for which palladium(IV) intermediates have been detected spectroscopically are presented in Section 5.7, even though compelling evidence for palladium(IV) intermediates is available for some reactions mentioned below. Section 5.5.2.4(ii) should be considered in conjunction with Section 5.7.

(i) Halogens

The complexes $[PdMe_2(bipy)]$, *trans*-$[PdX(CH_2SMe)(PPh_3)_2]$ (X = Cl, SCN, NO_3)[133] and *trans*-$[PdCl(COR)(PPh_3)_2]$ (R = Me, Ph, Bz)[165] are readily cleaved by the least reactive halogen, iodine, to form organoiodides, and the A-frame complex $[\{Pd(COMe)\}_2(\mu\text{-dppm})_2(\mu\text{-I})][I]$ reacts with X_2 (X = Br, I) to form $[PdX_2(dppm)]$ and MeCOX.[702] Palladium–carbon bond cleavage involving I_2 as the reagent is a complex process for $[PdAlk_2(phosphine)_2]$,[209] but cleavage of (**194**) by Cl_2 is a clean process occurring with retention of configuration at carbon.[894] The palladacycles $[Pd(CH_2CH_2CH_2CH_2)(L_2)]$ (L_2 = TMEDA, bipy, dppe, 2 PPh_3) react with Br_2 to form $Br(CH_2)_4Br$.[267] Excess Cl_2 cleaves the Pd–Ar bond of the pyridin-2-yl complex *trans*-$[PdCl(6\text{-}ClC_5H_3N)(PPh_3)_2]$, but stoichiometric addition of I_2 results in replacement of only the chloro ligand,[151] and stoichiometric addition of Br_2 to $[Pd(C_6F_5)(\eta^1\text{-}C_3H_5)(dppe)]$ or $[Pd(C_6F_5)(\eta\text{-}C_3H_5)(PPh_3)]$ results in selective cleavage of the η^1-allyl or C_6F_5 groups, respectively.[627]

The Pd–C bond of intramolecular coordination complexes,[333,373,534] for example, $[Pd(C\sim E)(\mu\text{-}X)]_2$ with C~E as in (**107**) (E = O, R^1 = Me)[364] and (**148**) (R^1 = H),[855] is cleaved by halogens, and the complexes [PdX(C~N~N)] (C~N~N as in (**168**) (*n* = 2), X = Cl, I) react with X_2 to give the coordination complexes $[PdX_2(2\text{-}XC_6H_4CH_2NMeCH_2CH_2NMe_2\text{-}N,N')]$.[176] Stoichiometric reaction of $[Pd(C\sim E)_2]$ is also possible with retention of one Pd–C bond, for example, addition of Cl_2 to (**74**) to form $[Pd\{2\text{-}(NO_2)C_6H_4\text{-}C,O\}(\mu\text{-}Cl)]_2$.[225]

(ii) Organohalides

The successful synthesis of many monoorganopalladium(II) complexes by oxidative addition of excess organohalides to palladium(0) substrates is an indication of their low reactivity toward organohalides. The uncoordinated nitrogen atom in 2-pyridylpalladium(II) complexes may be alkylated on reaction with organohalides,[845] and MeI reacts with *trans*-$[PdX(CH_2SMe)(PPh_3)_2]$, $[PdX(CH_2SMe\text{-}C,S)(PPh_3)]$ and related complexes to exchange X for I.[895] Related exchanges occur on the reaction of *trans*-$[Pd(SPh)Et(PMe_3)_2]$ with a range of organohalides (Scheme 23),[879] on reaction of *trans*-$[Pd(OPh)Et(PMe_3)_2\cdot HOPh]$ with allyl chloride to form *trans*-$[PdClEt(PMe_3)_2]$ and $CH_2{=}CHCH_2OPh$,[887] and on reaction of *trans*-$[Pd(NHPh)\{2\text{-}(Pr^iN{=}CH)C_6H_4\text{-}C\}(PMe_3)_2]$ with MeI to form *trans*-$[PdI\{2\text{-}(Pr^iN{=}CH)C_6H_4\text{-}C\}(PMe_3)_2]$ and NHMePh.[245]

$$trans\text{-}[Pd(SPh)Et(PMe_3)_2] + RX \longrightarrow trans\text{-}[PdXEt(PMe_3)_2] + RSPh$$

RX = MeI, BzBr, $CH_2{=}CHCH_2Cl$, MeCOCl

Scheme 23

Dimethylpalladium(II) species obtained at low temperature on the reaction of [LiMe] with $[PdCl_2(SMe_2)_2]$ in diethyl ether react readily with MeY (Y = Br, I) to form $[PdMe(SMe_2)(\mu\text{-}Y)]_2$ (Scheme 16), and although the *cis*-$[PdR_2]$ complexes *cis*-$[PdMe_2L_2]$ (L_2 = 2 PPh_3, 2 $PMePh_2$,[875] dmpe,[261] P~N (**32**)[57]) react cleanly with MeI to form $[PdIMe(L_2)]$ and ethane, *cis*-$[PdMe_2(dppe)]$ forms $[PdI_2(dppe)]$ and ethane.[209] The related complexes $[PdR_2(dppe)]$ (R = Et, Pr) react with MeI to form $[PdI_2(dppe)]$ and a range of products including RMe, RI, MeH and alkenes,[209] but $[Pd(C_6F_5)(CH_2CH{=}CHR^1)(dppe)]$ (R^1 = H, Me) react with CR^2Cl_3 (R^2 = H, Cl) to form $CH_2{=}CHCR^1CR^2Cl_2$ and $[PdCl(C_6F_5)(dppe)]$.[627]

The complex $[Pd(CH_2Bu^t)_2(bipy)]$ reacts with BzBr to form $[PdBr(CH_2Bu^t)(bipy)]$ and $BzCH_2Bu^t$;[257] $[PdMe_2(TMEDA)]$ reacts with MeCOCl to form [PdClMe(TMEDA)] and Me_2CO, and with EtI to form [PdIR(TMEDA)] (R = Me, Et), Me_2 and EtMe in ratios consistent with statistical reductive elimination from an undetected palladium(IV) intermediate '$[PdIMe_2Et(TMEDA)]$'.[265] Such an intermediate has been detected by NMR in the analogous reaction of $[PdMe_2(bipy)]$ with EtI.[896] A closely related reaction occurs on the addition of MeI to (**82**) to form $[PdI\{2\text{-}(o\text{-}Tol)C_6H_4\text{-}C\}(bipy)]$ and, although (**82**) is stable in CH_2Cl_2 it reacts cleanly under UV irradiation to form [PdI_2(bipy)] and [2-(CH_2Cl)C_6H_4]$_2$.[275] Palladium(IV) species may occur in these and related reactions but the involvement of radicals is possible in some reactions, as indicated by the formation of bibenzyl in the reactions illustrated in Scheme 24.[57,265]

[Pd(Me)$_2$(L_2)] + BzBr ⟶ *a* [PdBr(Me)(L_2)] + *b* Bz–Bz + *c* Bz–Me + *d* Me–Me

L_2 = 2 PPh_3, P~N as in (**32**); *a* approx. 0.95, b:c:d approx. 2:3:1

Scheme 24

A mixture of *cis*- and *trans*-[PdMe{(*E*)-CH=CHPh}(PPh_3)$_2$] reacts with CD_3I to form (*E*)-CD_3CH=CHPh and (*E*)-MeCH=CHPh,[134] and [$PdMe_2$(transphos)] forms $MeCD_3$ only, consistent with formation of a transient palladium(IV) intermediate [$PdMe_2$(CD_3)(transphos)]$^+$ containing a *trans*-[$PdMe_2$] moiety and adjacent Me and CD_3 groups.[118]

trans-[$PdEt_2$(PMe_3)$_2$] reacts with neat iodobenzene to form *trans*-[PdIPh(PEt_3)$_2$], presumably via reductive elimination and subsequent oxidative addition of PhI to [Pd(PEt_3)$_2$].[735] The complexes *trans*-[$PdAr_2$($PEtPh_2$)$_2$] (Ar = Ph,[879] *m*-Tol[272]) in toluene react with ArI to form Ar_2 and *trans*-[PdIAr($PEtPh_2$)$_2$]. *trans*-[$PdPh_2$($PEtPh_2$)$_2$] reacts with *m*-TolI to form Ph_2, *m*-TolPh and (*m*-Tol)$_2$,[879] and *trans*-[PdPh(*p*-Tol)(PEt_3)$_2$] reacts similarly with PhI to form Ph_2 and *p*-TolPh.[60] The formation of a range of coupling products in the latter and related reactions[60,110,272,879] has been attributed, from detailed kinetic studies including the effect of added *trans*-[$PdXRL_2$], to the occurrence of organic group exchange reactions involving intermediates (**77**) and (**268**),[60,272,879] rather than to palladium(IV) intermediates, in particular for processes involving aryl groups and monodentate ligands.

A range of intramolecular coordination complexes [Pd(C~E)(μ-X)]$_2$, for example, for [C~E]$^-$ as [2-(NR_2CH_2)C_6H_4]$^-$ [428] and [2-(NR=CH)C_6H_4]$^-$,[897] and as in (**59**),[898] (**107**) (E = O, R^1 = Me)[366] and (**198**) (E = NMe_2, R^1 = R^2 = Me, R^3 = CHO),[849] react with organohalides (in particular RCOCl) to form coupling[139,333,366,428,897,898] and other products.[849,897] However, the complex [Pd{2-(NMe_2)C_6H_4CH(TMS)-*C,N*}(μ-Br)]$_2$ is resistant to attack by BzBr following its formation on reaction of [Pd{2-(NMe_2)C_6H_4CH(TMS)-*C,N*}{2-(NMe_2CH_2)C_6H_4-*C,N*}] with BzBr.[293] The complex [Pd(C~N~N)Me] (**80**) reacts with MeCOCl to give [PdCl(C~N~N)] and acetone.[176]

The complex [Pd(η^1-Cp)(η-Cp)(PPr^i_3)] reacts with CHX_3 (X = Cl, Br, I) to form [PdX(η-Cp)(PPr^i_3)],[217] and the kinetics and product distribution of Ar–allyl^1 and Ar–allyl^2 formed during the decomposition of [Pd(2,5-$Cl_2C_6H_3$)(η-allyl^1)(PPh_3)] in the presence of added allyl^2–Cl indicate that palladium(IV) intermediates may be formed in addition to the role of allyl^2–Cl as an η^2-alkene ligand.[899]

5.5.2.5 Other reactions

Organopalladium(II) complexes are readily reduced by $NaBH_4$ or LAH to palladium metal and hydrocarbons;[333,401] for example, (**212**) reacts with $NaBD_4$ in MeOD to form C_6H_4D–$C_7H_{10}D$,[639] but Pd–C bonds may be stabilized toward reduction by appropriate ligands; for example, [PdMe(P~C~P)] (P~C~P as in (**81**) (R = H)) reacts with LAH in THF to form [PdH(P~C~P)].[266] Reduction reactions using hydrogen are less frequently reported, for example, reaction with [$PdMe_2$(TMEDA)],[261] [Pd{2-(NMe_2CH_2)C_6H_4-*C,N*}(Bu^t_2acac)][886] and [Pd(C~N)(μ-Cl)]$_2$ (C~N as in (**263**))[695] to liberate alkanes or arene.

The reaction of *trans*-[PdCl{(*S*)-CHDPh}(PEt_3)$_2$] with Na[BPh_4] gives the benzallyl cation [Pd(η-CHDPh)(PEt_3)$_2$]$^+$, which reacts with Cl^- to give the neutral chloro complex with 94% retention of configuration.[791] In a similar reaction involving loss of Pd–C σ-bonding, a benzallyl complex is formed on the removal of one PCy_3 group from *trans*-[PdClBz(PCy_3)$_2$].[127]

The trityl cation reacts in the usual way to abstract H^- from [$\overline{Pd(CH_2CH_2CH_2C}H_2)$($L_2$)] ($L_2$ = bipy, dppe, 2 PPh_3) to form [Pd(η-C_3H_4Me)(L_2)]$^+$,[267,271] but cleaves the Pd–C bonds of [Pd(CH_2Bu^t)$_2$(dppe)] to form $Ph_3CCH_2Bu^t$.[257]

The 2,5-dioxo-3,6-dichloro-1,4-benzoquinone dianion (chloroanilate) forms [$PdCl_2$($C_6Cl_2O_4$-C^3,C^6)]$^{2-}$, and reaction of this complex ion with a series of phosphines results in loss of Pd–C bonding to give [Pd($C_6Cl_2O_4$-O^1,O^2)L_2].[900]

Silver(I) is often used to remove halide ions from complexes, but [Ag(ClO_4)(L)] will also remove a pentafluorophenyl group from [Pd(C_6F_5)$_3$(THT)]$^-$ to form *cis*-[Pd(C_6F_5)$_2$(PPh_3)(THT)] (L = PPh_3) and from [Pd(C_6F_5)$_4$]$^{2-}$ to form [Pd(C_6F_5)$_3$(THT)]$^-$ (L = THT).[901]

Oxygen gas under pressure reacts with $[\{Pd(COMe)\}_2(\mu\text{-}dppm)_2(\mu\text{-}I)][I]$ to form $(MeCO)_2O$ and $[PdI(dppm)]_2$,[702] and the complexes $[PdR_2(dmpe)]$ (R = Me, CH_2TMS) and [PdMe(P~C~P)] (P~C~P as in (**81**) (R = H)) in MeCN undergo one-electron oxidative cleavage electrochemically, or chemically using $[FeCp_2]^+$, to generate $[PdR(dmpe)(NCMe)]^+$ and $[Pd(P{\sim}C{\sim}P)(NCMe)]^+$, respectively.[831] A similar reaction occurs on addition of acrylonitrile to a mixture of $[Pd(CH_2TMS)_2(dmpe)]$ and $Ag[BPh_4]$ in the dark to give $[PdMe(dmpe)(\eta^2\text{-}CH_2{=}CHCN)]^+$.[283]

Aryl allyl thioethers such as $PhSCH_2CH{=}CH_2$ react with *trans*-$[PdR_2(PMe_3)_2]$ (R = Me, Et) to give *trans*-$[Pd(SPh)R(PMe_3)_2]$.[887] Palladium–methyl bonds of *cis*-$[PdMe_2(PMePh_2)_2]$ are cleaved by silanes such as $SiHCl_3$, $SiHMeCl_2$, $SiCl_4$, SiH_2Ph_2, SiH_2MePh and $TMS(CH_2CH_2PPh_2)$; for example, $[Pd\{SiMe_2(CH_2CH_2PPh_2)\text{-}Si,P\}_2]$ could be isolated and *cis*-$[Pd(SiHPh_2)_2(PMePh_2)_2]$ detected prior to Si–Si coupling.[242] The complex $[Pd\{2\text{-}(NMe_2CH_2)C_6H_4\text{-}C,N\}\{Ph_2PCH{=}C(Ph)O\text{-}P,O\}]$ and its cyclopalladated 8-methylquinoline analogue react with PCl_2Ph/py in THF to give coordination complexes resulting from complex P–C coupling reactions, such as $[PdCl_2\{Ph_2PCH{=}C(Ph)OP(Ph)C_6H_4(2\text{-}CH_2NMe_2H)\text{-}P,P\}][Cl]$.[902]

Carbon dioxide reacts with $[PdR_2L_2]$ (R = Me, Et; L = PEt_3, $PMePh_2$) and $[PdMe_2(PPh_3)_2]$ in the presence of traces of water to form *trans*-$[Pd(OCO_2H)RL_2]$ and methane.[903,904] A similar reaction occurs with dry CO_2 in the presence of an amine R^1R^2NH ($R^1 = R^2$ = Et, H; R^1 = H for R^2 = Bu, Ph, Bz) to form carbamato complexes *trans*-$[Pd(O_2CNR^1R^2)MeL_2]$ (L = PEt_3, $PMePh_2$, PPh_3), although for $NHPh_2$ as the amine, *trans*-$[Pd(NPh_2)MeL_2]$ is formed.[904]

Carbon disulfide inserts into the Pd–C bond of *trans*-$[PdIMe(PPh_3)_2]$ to form *trans*-$[PdI(S_2CMe)(PMe_3)_2]$,[905] but the Pd–C bond is retained on reaction with $[Pd\{OCH(CF_3)Ph\}Me(dppe)]$ to form $[Pd\{S_2COCH(CF_3)Ph\}Me(dppe)]$,[722] and on reaction with *trans*-$[PdI(COMe)(PMe_3)_2]$ and $[PdMe(PMe_3)_3][BPh_4]$ to form complexes containing zwitterionic $Me_3P^+CS^-_2$ in place of one phosphine.[905]

Sulfur dioxide inserts into both Pd–C bonds of $[PdMe_2(TMEDA)]$,[906] $[PdR_2(dppe)]$ (R = Me, Bu)[267] and $[\overline{Pd(CH_2CH_2CH_2CH_2})(L_2)]$ (L_2 = TMEDA,[906] dppe[267]) to give disulfinato complexes, for example, $[Pd(SO_2Me\text{-}S)_2(TMEDA)]$. Both CO_2 and SO_2 insert into the Pd–C bond of η^1-allyl complexes.[907]

Nitric oxide does not react with $[PdI(p\text{-}NO_2C_6H_4)(PPh_3)_2]$ or [PdClPh(dppe)], but NO_2 forms *trans*-$[Pd(ONO_2)_2(PPh_3)(OPPh_3)]$ and $[PdCl(ONO_2)(dppe)]$, respectively.[908]

5.5.3 Structures

This section is restricted to a consideration of Pd–C bond lengths, structures that do not exhibit the square planar geometry characteristic of palladium(II), and comparisons between palladium(II) and platinum(II) complexes.

Palladium–carbon distances are generally shorter for Pd–C(sp^2) than for Pd–C(sp^3) bonds, although comparisons require caution owing to variations in bond length exhibited by closely related complexes. For example, Pd–C(sp^2) lengths occur within the range 0.189 0(5)–0.200 4(5) nm for $[PdCl\{2\text{-}(HON{=}CMe)C_6H_4\text{-}C,N\}(L)]$ (L = py and substituted pyridines),[346] Pd–C(sp^3) occur within the range 0.208 1(4)–0.213 0(8) nm for [PdMe(dmpe)L][X] (L = NH_3, NH_2Et, $NHEt_2$, NEt_3),[889] and the complexes $[PdMe_2(L_2)]$ have Pd–C(sp^3) 0.202 6(3) nm and 0.202 9(3) nm (L_2 = TMEDA) and 0.208 7(4) nm (L_2 = dmpe).[261] For complexes where both the Ar and Alk groups have identical *trans*-donor groups, shorter Pd–C(sp^2) bonds also occur, for example, [PdMePh(bipy)] with Pd–Ph 0.198 5(3) nm and Pd–Me 0.202 0(5) nm, and [PdMePh(TMEDA)] with 0.198 5(3) nm and 0.210 7(3) nm, respectively.[100,262] The ion *cis*-$[PdCl_2(2\text{-}NO_2\text{-}6\text{-}MeC_6H_3\text{-}C)(CO)]^-$ has Pd–CO 0.185 3(10) nm and Pd–C(sp^2) 0.200 0(5) nm.[689]

The η^1-phenyl group in complex (**101**) has Pd–C 0.233 5(7) nm,[352] and the η^2-phenyl groups in complexes (**102**)[353] and (**213**)[642] exhibit Pd–C(*ipso*) distances 0.243 3(3) nm and 0.243 0(5) nm and Pd–C(*ortho*) distances 0.257 1(4) nm and 0.259 1(6) nm, respectively.

Deviations from the regular square planar geometry are mentioned earlier for complexes (**234**)[686] and (**258**),[782] and for $[PdClMe(2,9\text{-}Me_2phen)]$ a dihedral angle of 39.3° between the phen and coordination planes, and a twist angle of 15.3° between the 'PdC(Cl)' and 'PdN_2' planes are attributed to steric effects dominated by the PdMe···Me interaction.[706]

Pseudo square pyramidal geometry occurs in several 2-iminophenyl and related complexes involving weak 'axial' Pd···N interactions of 0.255 9(1)–0.2860 nm,[245,347,493,524,555] for example, 0.277 9(3) nm for *trans*-$[Pd\{2\text{-}(Pr^iN{=}C)C_6H_4\text{-}C\}(PMe_3)_3][BF_4]$[245] and 0.274 3(5) nm for (**188**).[555] A cyclopalladated catenand complex exhibiting the 'Pd(C~N~N)' group shown in (**167**) has an additional phen group coordinated monodentate, but the distorted square planar palladium centre has the second nitrogen atom of this phen group at 0.255 9(1) nm from palladium.[524] Trigonal bipyramidal geometry occurs for (**53**) (R = Me, X = I) and for (**52**) (X = Cl, L = maleic anhydride).[256]

The dinuclear complex (**2**) has Pd···Pd at the nonbonding distance of 0.338 1(1) nm, but a shorter distance of 0.303 1(1) nm in the related chloro-bridged complex (**3**) has been interpreted as indicating the presence of a weak Pd···Pd interaction.[3]

Each pair of isostructural palladium(II) and platinum(II) complexes *trans*-$[MCl(COPr)(PPh_3)_2]$[726] and [MCl(C~N~S)] (C~N~S as in (**172**) (R^1 = Me, E = SMe))[532] has M–C within 2σ, although data collected at −162 °C for the isostructural pair *cis*-$[MMe_2(PMePh_2)_2]$ show that the Pd–C bonds (0.208 9(3) nm, 0.209 0(3) nm) are ~0.003 0(4) nm shorter than Pt–C bonds but the Pd–P bonds are ~0.003 9(1) nm longer than Pt–P bonds.[909] The isostructural intramolecular coordination complexes $[M\{2\text{-}(PhN_2)C_6H_4\text{-}C,N\}(\eta\text{-}Cp)]$ have identical bond lengths (0.196 0(3) nm (Pd), 0.196 2(6) nm (Pt)), and the nonisostructural nickel analogue has Ni–C 0.186 3(4) nm and 0.184 7(4) nm in two independent molecules.[910]

5.6 RADICAL COMPLEXES

Monomeric coordination complexes with the formal oxidation state +3 for palladium are paramagnetic and, except for a few complexes of tripod ligands, the complexes isolated have been shown by EPR studies to be palladium(II) complexes where the paramagnetism is associated with the ligand skeleton.[911,912] Organopalladium complexes with ligand-centred radicals[913] include $[PdCl(3,5\text{-}Bu^t_2\text{-}4\text{-}OC_6H_2\text{-}C)(PPh_3)_2]$ obtained on oxidation ($-e^-$, $-H^+$) of (**269**) by $[Pb(O_2CMe)_4]$,[201] and the complexes [Pd(C~E)(L-*O,O'*)] obtained on addition of (**270**) to cyclopalladated azobenzene complexes (M = Na),[914] or to the σ,π-bonded complex (**194**) (R = Me) (M = Tl).[915] The palladium(II) complex $[Pd(\eta\text{-}C_5Ph_5)(\sigma,\pi\text{-}L)]$ (σ,π-L as in (**194**) with R = OCOPh) is obtained on oxidation of $[Pd(\eta\text{-}C_5Ph_5)(cod)]$ with benzoyl peroxide, and this and related complexes undergo reversible electrochemical oxidation to form 'palladium(III)' complexes.[595] Complex (**260**) (R = Me, L = $P(OMe)_3$) is oxidized by $[FeCp_2]^+$ to form isolable tetrapalladia cations, such as (**271**), and (**271**) may be reduced by $[CoCp_2]$ to form the neutral analogue.[825] The cations (+1, +2, +3, +4) and the neutral complex are formulated as palladium(II) complexes and undergo fully reversible electron transfer. One-electron reduction of [PdCl(P~C~P)] (P~C~P as in (**178**)) gives a product exhibiting an EPR spectrum indicating that reduction has taken place on the ligand.[542]

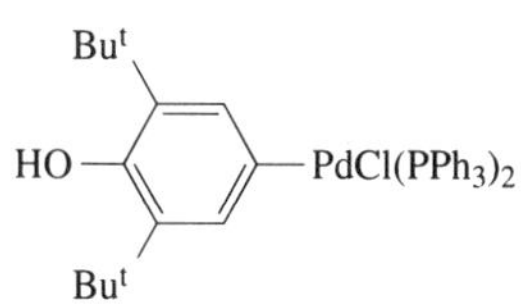

(**269**)

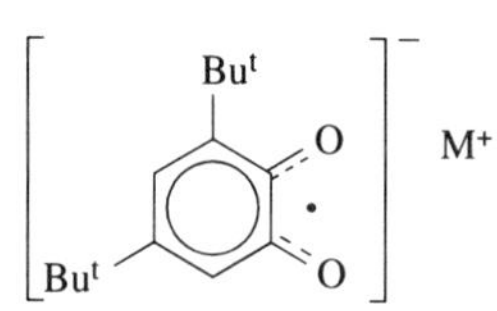

(**270**)

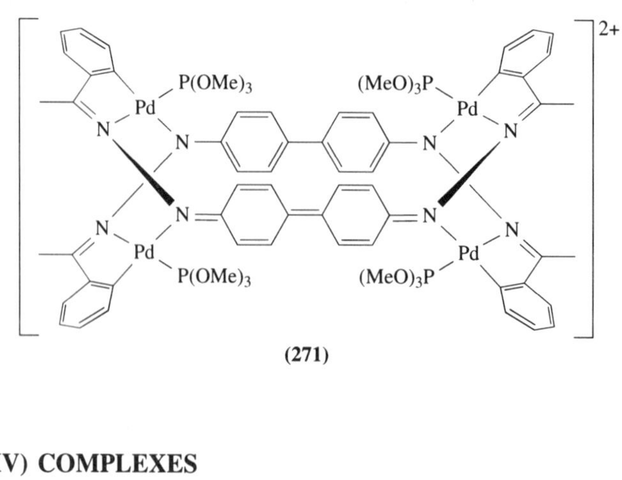

(**271**)

5.7 PALLADIUM(IV) COMPLEXES

Hydrocarbylpalladium(IV) complexes were first isolated and structurally characterized in 1986,[100,261,263–5,639,916–19] commencing with the report of $[PdIMe_3(bipy)]$*,[916] or well characterized in solution without isolation in 1988[176,260,920] following earlier syntheses of pentafluorophenylpalladium(IV) complexes,[921,922] ^{1}H NMR studies indicating the possible presence of a $Pd^{IV}Me$ complex prior to decomposition,[695] and an extensive series of reactions of organopalladium(II) complexes that appeared to be consistent with the formation of undetected palladium(IV) intermediates.[2,85,110,116,118,133,134,230,257,366,875,894,895,923,924] Some of these reactions, and more recently reported reactions for which undetected palladium(IV) species have been discussed as possible intermediates,[57,101,103,164,165,176,260,261,263,265,275,293,342,489,639,807,899,917,918,925,926] have been mentioned in previous sections, in particular for processes involving P–C bond activation (Equation (10)),[807] oxidation using molybdenum peroxide (Section 5.5.2.3),[101] reactions of organopalladium(II) complexes with halogens[85,165,176,230,894] or organohalides (Section 5.5.2.4)[2,57,85,110,116,118,133,134,164,176,230,257,260,261,265,275,293,366,875,895,899,923,926] and the synthesis of (**214**).[103]

Organopalladium(IV) intermediates may occur in several organic synthesis and catalysis procedures[917,918] involving oxidative addition of carbon–halogen bonds,[85,164,293,366,639,920,926–30] C–C bonds,[931] C–P bonds[932] or N–S bonds,[933] and other processes including cycloaddition.[925,934–42] Well-characterized complexes to date are formed by oxidative addition of halogens or organohalides ($C(sp^3)$–X) (X = Br or I except for one report for $CH_2=CHCH_2Cl$[920]) to palladium(II) complexes of polydentate nitrogen or sulfur donor ligands. All isolated triorganopalladium(IV) complexes have the octahedral *fac*-$[PdC_3]$ configuration, and the ligands are such that they allow straightforward assignment of formal oxidation state as +4. The complex $[PdCl_2(CO_2Bu)(NO)(PPh_3)]$, formed on the reaction of $[PdCl_2(NCPh)_2]$, PPh_3, CO and butyl nitrite, may be regarded formally as containing $[Pd^{II}(NO^+)] \leftrightarrow [Pd^{IV}(NO^-)]$.[743] Reductive elimination via C–C bond formation is the main decomposition route reported for most complexes, except for some pallada(IV)cyclopentane complexes which give butenes.[919] Other processes may occur for exceptionally stable complexes of some tripod ligands whose decomposition has yet to be explored.

5.7.1 Complexes of Bidentate Nitrogen Donor Ligands and Intramolecular Coordination Complexes

The complexes *fac*-$[PdXMe_3(L_2)]$ have been isolated by oxidative addition of MeI to $[PdMe_2(L_2)]$ (L_2 = bipy,[916,943] phen[943] and the related diimine ligands (**272**) and (**273**)[944]), and of MeX (X = Br, I) to $[PdMe_2(TMEDA)]$,[261] and detected in solution at −30 °C to 0 °C on addition of MeI to $[PdMe_2(L_2)]$ where L_2 are planar bidentate ligands similar to bipy[943] or more flexible and bulky ligands such as $(py)_2CHMe$, $(pz)_2CH_2$[945] or $(pz)_2CHMe$.[946] Complexes of ligands (**272**) and (**273**) appear to be more stable than those of bipy or phen, and complexes of the flexible ligands are least stable. Kinetic studies are consistent with the classical S_N2 mechanism for oxidative addition of MeI or BzBr to $[PdMe_2(L_2)]$ (L_2 = bipy, phen), and the more nucleophilic $[PtMe_2(L_2)]$ complexes react faster than the palladium analogues but with similar activation parameters.[947,948] S_N2 reactions are sensitive to steric effects, and consistent with occurrence of this mechanism $[PdMe_2(bipy)]$ reacts more slowly with EtI than MeI, resulting in NMR detection of a palladium(IV) species that could not be isolated since for EtI, but not MeI, oxidative addition is too slow compared with the subsequent reductive elimination reaction.[896] Methyl iodide reacts with *cis*-$[PdMe_2(PMePh_2)_2]$ ~6600 times more slowly than with $[PdMe_2(bipy)]$ under similar conditions,[875,948] at least partly accounting for the isolation of $[PdIMe_3(bipy)]$ as a crystalline solid but an inability to detect formation of the possible palladium(IV) intermediate in the reaction of *cis*-$[PdMe_2(PMePh_2)_2]$ to form *trans*-$[PdIMe(PMePh_2)_2]$ and ethane.

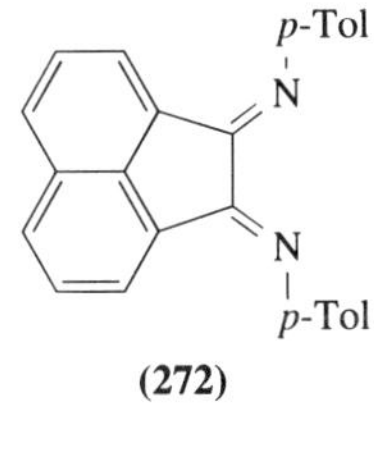

(**272**)

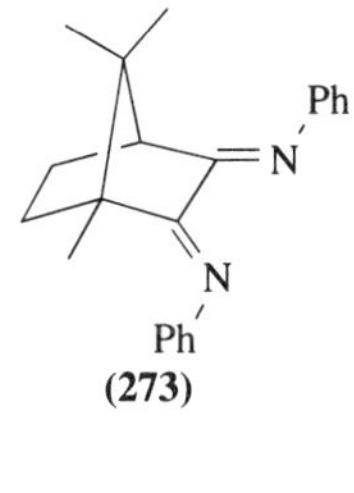

(**273**)

Of particular importance to the establishment of an S_N2 mechanism, cationic intermediate(s) such as $[PdMe_2(CD_3)(bipy)\{(CD_3)_2CO\}]^+$ are detected in NMR studies at −50 °C,[943,949,950] and methyl triflate reacts with $[PdMe_2(TMEDA)]$ in $(CD_3)_2CO$ or CD_3CN to form $[PdMe_3(TMEDA)(solvent)]^+$.[261] Thus, the formation of palladium(IV) complexes from simple organohalides is generally assumed to occur as illustrated in Equation (14), with the possible involvement of a five-coordinate cation prior to and/or following solvent coordination. In contrast to *trans*-addition of CD_3I to $[PtMe_2(bipy)]$,[951] the Me and

CD_3 groups of $[PdMe_2(CD_3)(bipy)\{(CD_3)_2CO\}]^+$ are scrambled between positions *trans* to bipy and $(CD_3)_2CO$ at −50 °C.[950] The fluxionality of palladium(IV) is also illustrated by NMR studies of $[PdMe_3(TMEDA)(CD_3CN)]^+$ allowing determination of $\Delta G^\ddagger = \sim 53$ kJ mol^{-1} for the scrambling process of methyl groups,[261] and scrambling may be assumed to occur via five-coordinate species.

$$[PdMe_2(L_2)] + MeI \longrightarrow fac\text{-}[PdMe_3(L_2)(solvent)][I] \longrightarrow fac\text{-}[PdIMe_3(L_2)] \quad (14)$$

Kinetic studies indicate that reductive elimination of ethane from $[PdIMe_3(bipy)]$ is a first-order process in solution and occurs mainly via I^- dissociation. Carbon–carbon bond formation is a concerted process since the energy of activation, ~65 kJ mol^{-1}, is much less than the Pd–C bond energy.[947] NMR studies show that reductive elimination from $[PdIMe_3(L_2)]$ (L_2 as in (**272**) and (**273**)) is also a first-order process, yielding an activation energy of ~60 kJ mol^{-1}, similar to that obtained from kinetic studies for $[PdIMe_3(bipy)]$.[947] The facile reductive elimination from complexes of flexible bidentate ligands, for example, TMEDA or $(pz)_2CHMe$, compared with more rigid ligands such as bipy may reflect the presence of lower activation energies for the formation of five-coordinate (or solvated octahedral) cation intermediates(s) or for C–C bond formation resulting from the wider range of coordination geometries possible for the flexible ligands. Estimates of Pd–Me and Pd–Bz bond energies, ~130 kJ mol^{-1} and ~123 kJ mol^{-1}, respectively, obtained from DSC studies of the reactions in Scheme 25, appear to be the only reported attempts to determine experimentally the strengths of Pd–C σ-bonds.[948,952]

$$fac\text{-}[PdIMe_3(L_2)](s.) \xrightarrow{80\text{–}120\ °C} [PdIMe(L_2)](s.) + Me\text{–}Me(g.)$$

$$fac\text{-}[PdBrMe_2Bz(L_2)](s.) \xrightarrow{85\text{–}145\ °C} a\,\{[PdBrBz(L_2)](s.) + Me\text{–}Me(g.)\} + b\,\{[PdBrMe(L_2)](s.) + Bz\text{–}Me(g.)\}$$

L_2 = bipy, $a = 0.8$, $b = 0.2$; L_2 = phen, $a = b = 0.5$

Scheme 25

Dissociation of X^-, as demonstrated for reductive elimination, is also a key factor in other aspects of reactivity; for example, iodo complexes of planar ligands such as bipy form equilibria with $[PdMe_3(L_2)(NCCD_3)]^+$ (L_2 = bidentate ligand) in CD_3CN[943] but iodide does not react with $[PdMe_3(TMEDA)(NCCD_3)]^+$ in this solvent, presumably partly reflecting the steric effects of NMe groups.[261] The steric demands of TMEDA are also illustrated by the structures adopted by $[PdXMe_2R(L_2)]$, which for L_2 = bipy (R = CH_2Ar (Ar = *p*-YC_6H_4, C_6F_5, 1-Nap, 2-Nap),[816] $CH_2CH{=}CH_2$, $CH_2CH{=}CHPh$,[896] CH_2COAr[949,950]) have R *trans* to X, but for L_2 = TMEDA, *trans* to TMEDA (R = Bz, $CH_2CH{=}CH_2$) to reduce NMe···Bz or NMe···$CH_2CH{=}CH_2$ interactions.[265,953] For (**272**) and (**273**) as ligands, the complexes $[PdBrMe_2Bz(L_2)]$ occur as a mixture of isomers.

Steric effects of TMEDA may also be responsible for the slower reaction of [PdMePh(TMEDA)] than [PdMePh(bipy)] with MeI at 0 °C in $(CD_3)_2CO$, and the inability to detect a palladium(IV) species for the former reaction which leads directly to the formation of [PdI(Ph)(TMEDA)] and ethane.[262,953]

The relatively high stability of $[PdBrMe_2(CH_2Ar)(bipy)]$ compared to other $[PdXR_3(L_2)]$ complexes has allowed the synthesis of hydrocarbyl-bridged complexes by reaction of $[PdMe_2(bipy)]$ with 1,3-$(CH_2Br)_2C_6H_4$ followed by reaction with $[MMe_2(bipy)]$ to form $[(bipy)Me_2BrPd\{CH_2C_6H_4CH_2\}MBrMe_2(bipy)]$ (M = Pd, Pt).[816]

Reductive elimination from $[PdXMe_2R(L_2)]$ (L_2 = bipy, phen, TMEDA) under mild conditions (0–40 °C) in solution is more selective than for the solid-state reactions illustrated in Scheme 25, for example, to give $[PdBr(CH_2Ar)(bipy)]$ and Me_2 only,[816] $[PdX(CH_2Ph)(TMEDA)]$ (X = Br, I) and Me_2 only,[265] with less selectivity but still favouring Me_2 for the phen analogues.[816] However, $[PdBrMe_2Bz(L_2)]$ (L_2 = (**272**)) undergoes reductive elimination at ~50 °C in $CDCl_3$ by a first-order process to give ~60% Me_2 and ~40% MeBz.[944] Elimination of MeI also occurs as a minor pathway for $[PdIMe_3(L_2)]$ (L_2 = (**272**)),[944] and MeBr has also been observed together with ethane (major product) from $[PdBrMe_2(CH_2COPh)(L_2)]$ (L_2 = bipy, phen).[949] Allylpalladium(II) products and ethane are obtained from $[PdXMe_2(CH_2CH{=}CH_2)(L_2)]$ (L_2 = bipy, phen,[896] TMEDA[265]).

The phenylpalladium(IV) complex $[PdIMe_2Ph(bipy)]$ (**274**), formed on oxidative addition of MeI to [PdMePh(bipy)] at 0 °C followed by isolation at −60 °C, undergoes facile reductive elimination at 0 °C (Equation (15)).[262,953] Complex (**274**) occurs as a mixture of isomers in 1:1 ratio, but [PdXMePhBz(bipy)] (X = Br, I) occur as one isomeric form with Bz *trans* to X and undergo highly selective reductive elimination of toluene at 0 °C in solution (Scheme 26). Complexes (**275**) are formed in similar oxidative addition reactions[639,920] and, although the configuration of one of these complexes (RX = MeI) is uncertain, it is known to undergo facile intramolecular reductive elimination of the aryl and methyl groups to form $[PdI(2\text{-}MeC_6H_4C_7H_{10}\text{-}C)(phen)]$.[639] The palladacyclopentane complexes

(**276**) decompose very slowly and cleanly in chloroform at 40 °C to give 1- and 2-butenes as the volatile products and crystalline $[PdBr_2(bipy)]$.[919] This is the only well-characterized system to date for which decomposition is not via simple reductive elimination.

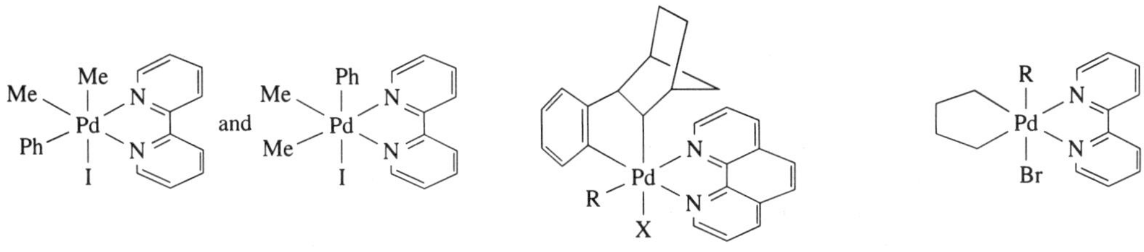

(274) **(275)** **(276)**

RX = MeI, CH_2=$CHCH_2Cl$, CH_2=$CHCH_2Br$ R = Bz, CH_2COPh, CH_2CH=CHPh

$$fac\text{-}[PdIMe_2Ph(bipy)] \xrightarrow{0\,°C} 0.8\,\{[PdIPh(bipy)] + Me\text{–}Me\} + 0.2\,\{[PdIMe(bipy)] + Ph\text{–}Me\} \quad (15)$$

$$fac\text{-}[PdXMePhBz(bipy)] \xrightarrow[X = Br,\ I]{0\,°C} [PdXBz(bipy)] + Me\text{–}Ph$$

Scheme 26

Chlorine oxidizes pentafluorophenylpalladium(II) complexes of nitrogen donor ligands to form palladium(IV) complexes $[PdCl_3(C_6F_5)(L_2)]$ (L_2 = bipy, phen, TMEDA) and $[PdCl_2(C_6F_5)_2(L_2)]$ (L_2 = bipy, phen, en, pn).[921,922] The configuration and decomposition chemistry of these unstable octahedral complexes have not been reported. Dimethylpalladium(II) complexes of (**272**) and (**273**) react with iodine in $CDCl_3$ to form unstable complexes $[PdI_2Me_2(L_2)]$, for which ^1H NMR spectra indicate a *trans*-configuration for L_2 = (**272**), but in the presence of an additional minor isomer for L_2 = (**273**).

An early indication that unstable palladium(IV) intramolecular coordination complexes may be accessible[695] has been realized recently by the reaction of MeI with [PdMe(C~N~N)] (**80**) at −3 °C to give $[PdIMe_2(C\text{\textasciitilde}N\text{\textasciitilde}N)]$ prior to reductive elimination of ethane, and chlorine oxidation of [PdCl(C~N~N)] (C~N~N as in (**168**) (n = 2)) to form predominantly $[PdCl_3(C\text{\textasciitilde}N\text{\textasciitilde}N)]$.[176]

5.7.2 Complexes of Tripodal Nitrogen or Sulfur Donor Ligands

Cationic complexes $[PdMe_3(L_3)][I]$ (L_3 = 1,4,7-trithiacyclononane (9S3),[264] or $(pz)_3CH$* and related ligands containing pz, py and *N*-methylimidazol-2-yl (mim) groups[943,954–6]) may be isolated on oxidative addition of MeI to palladium(II) precursors. Neutral poly(pyrazolyl)borate complexes may be obtained according to Scheme 27,[263,917–19] where oxidative addition occurs to the intermediate anion $[PdMe_2\{BR(pz)_3\}]^-$.

$$1/n\ [PdMe_2(pyridazine)]_n \xrightarrow[\text{ii, MeI}]{\text{i, K[BR(pz)}_3\text{] (R = H, pz)}} [PdMe_3\{BR(pz)_3\}]$$

$$[Pd(\overline{CH_2CH_2CH_2CH_2})(TMEDA)] \xrightarrow[\text{ii, MeI or BzBr}]{\text{i, K[BH(pz)}_3\text{]}} [Pd(\overline{CH_2CH_2CH_2CH_2})R\{BH(pz)_3\}]$$

(R = Me or Bz)

Scheme 27

The complexes $[PdMe_2R\{(pz)_3CH\}][X]$ (RX = MeI, BzBr, CH_2=$CHCH_2Br$) undergo facile reductive elimination of ethane, as illustrated in Equation (16) for an unstable allyl complex; the bromo analogue gives $(pz)_3CH$, Me_2 and the double salt $[Pd(\eta\text{-}C_3H_5)\{(pz)_3CH\}][PdBr_2(\eta\text{-}C_3H_5)]$.[956] However, the poly(pyrazolyl)borate, 9S3 (**277**)[264] and $(py)(mim)_2CH$[956] complexes (**278**) are stable at ambient temperature and in solution at 60 °C, and complexes (**278**) occur as a mixture of isomers with R *trans* to py or mim groups. The high stability reflects the higher basicity of $[(pz)_3BR]^-$ and $(py)(mim)_2CH$, together with an expected reduced tendency for ease of formation of five-coordinate intermediates; for example, free 9S3 has almost the same conformation as that expected in the complex.[957,958] Reductive elimination from the stable cations $[PdMe_2R\{(py)(mim)_2CH\}]^+$ (R = Me, Et, CH_2CH=CH_2) may be conveniently studied by electrospray mass spectrometry.[580] The parent ions together with the palladium(II) cations $[PdMe\{(py)(mim)_2CH\}]^+$ and $[PdR\{(py)(mim)_2CH\}]^+$ are observed at low ion source energy. Increasing the ion source energy causes increased reductive elimination and further fragmentation resulting from collisionally activated decomposition with solvent molecules within the ion source.

$$[PdMe_2(CH_2CH{=}CH_2)\{(pz)_3CH\}][I] \longrightarrow 1/2\,[PdI(\eta\text{-}C_3H_5)]_2 + (pz)_3CH + Me{-}Me \qquad (16)$$

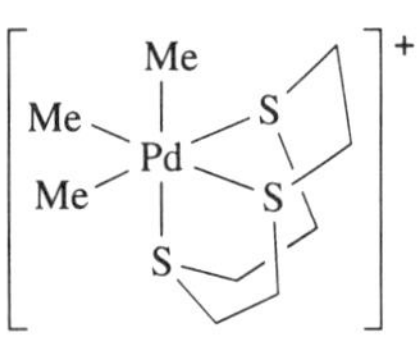

(277)

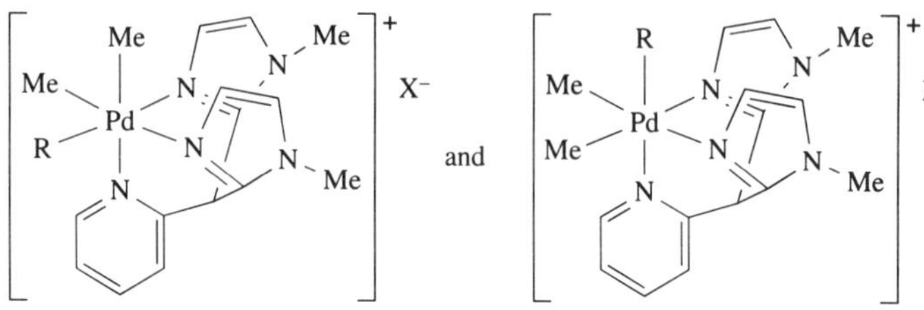

(278) RX = EtI, PrI, BzBr, $CH_2{=}CHCH_2Br$, $CH_2{=}CHCH_2I$

5.7.3 Structures and Reactions of Complexes

The complexes $[PdIMe_3(bipy)]$,[916,943] $[PdBrMe_2(CH_2\text{-}p\text{-}BrC_6H_4)(phen)]$,[816] $[PdBrMe_2(CH_2COPh)(bipy)]$,[949,950] $[PdMe_3\{(pz)_3CH\}][I]$[943] and $[PdMe_3\{(pz)_3BR\}]$ (R = H, pz)[263,917,918] have the expected octahedral geometry with *fac*-$[PdC_3]$ units, and the x-ray structural studies do not reveal differences from bond distances typical of palladium(II) complexes. The complex $[PdIMe_3(bipy)]$ has Pd–I 0.283 4(1) nm, compared with Pt–I 0.284 3(1) nm in *fac*-$[PtIMe_3\{(3,5\text{-}Me_2pz)_2CH_2\}]$,[959] and the isostructural complexes $[MMe_3\{(pz)_3CH\}][I]$ exhibit Pd–N ~0.004 nm longer than Pt–N, consistent with similar trends in M^0–P and M^{II}–P bond distances in other isostructural pairs.[909,960]

Anion exchange is readily effected on reaction of $[PdMe_2R\{(pz)_3CH\}][I]$ (R = Me, Bz)[943,956] or $[PdMe_3(9S3)][I]$[264] with $AgBF_4$ or $AgNO_3$, respectively, and provided reactions are carried out quickly at 0 °C the bromo group in $[PdBrMe_2Bz(bipy)]$ may be exchanged for pseudohalo, carboxylato and other halo groups.[961]

Reaction of organopalladium(IV) complexes of bidentate ligands with $[Pt^{II}Me_2]$ complexes,[948] or with $[Pd^{II}Me_2]$ complexes that are known to form palladium(IV) complexes that are more stable than the palladium(IV) reactant,[262,953] results in alkyl halide transfer from palladium(IV) to platinum(II) or palladium(II), as illustrated in Equations (17) and (18). Proton NMR studies of these redox reactions, and UV–visible kinetic studies of the reaction in Equation (17), suggest that the major pathway for reaction involves halide dissociation to form a cation which is activated toward nucleophilic attack by platinum(II) or palladium(II). The probable intermediate for alkyl transfer, $[(L^1{}_2)Me_2M\cdots R^1\cdots PdR^2R^3(L^2{}_2)]^{+\ddagger}$, is closely related to those proposed for S_N2 oxidative addition of alkyl halides to palladium(0) substrates, such as (**49**) in Scheme 13. The mechanism proposed is consistent with the observed preference for organic group transfer $PhCH_2 > Me$, since nucleophilic attack on BzX occurs faster than on MeX by factors as high as 500,[962] and this would be expected to favour benzyl halide transfer.

$$[PdBrMe_2Bz(phen)] + [PtMe_2(phen)] \longrightarrow \qquad (17)$$
$$0.85\,\{[PdMe_2(phen)] + [PtBrMe_2Bz(phen)]\} + 0.15\,\{[PdMeBz(phen)] + [PtBrMe_3(phen)]\}$$

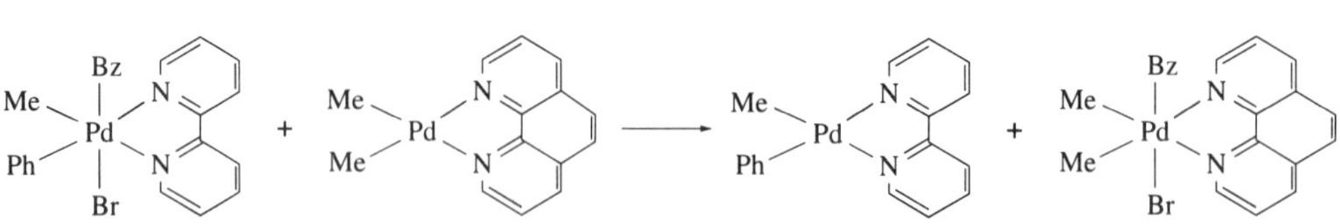

(18)

5.8 REFERENCES

1. P. M. Maitlis, P. Espinet and M. J. H. Russell, in 'COMC-I', vol. 6, p. 279.
2. P. M. Maitlis, 'The Organic Chemistry of Palladium', Academic Press, New York, 1971, vol. 1.
3. S. J. Young *et al.*, *J. Am. Chem. Soc.*, 1988, **110**, 5744.

4. R. Uson and J. Fornies, *Adv. Organomet. Chem.*, 1988, **28**, 219.
5. R. Uson, J. Fornies, P. Espinet, F. Martinez, C. Fortuno and B. Menjon, *J. Organomet. Chem.*, 1983, **256**, 365.
6. R. Uson, J. Fornies, P. Espinet and C. Fortuno, *Inorg. Chim. Acta*, 1984, **87**, 207.
7. R. Uson, J. Fornies, R. Navarro and J. I. Cebollada, *J. Organomet. Chem.*, 1986, **304**, 381.
8. J. Fornies, F. Martinez, R. Navarro, A. Redondo, M. Tomas and A. J. Welch, *J. Organomet. Chem.*, 1986, **316**, 351.
9. R. Uson, J. Fornies, F. Martinez, R. Navarro and M. C. Frias, *Inorg. Chim. Acta*, 1987, **132**, 217.
10. P. Espinet *et al.*, *J. Organomet. Chem.*, 1986, **317**, 105.
11. P. A. Arnup and M. C. Baird, *Inorg. Nucl. Chem. Lett.*, 1969, **5**, 65.
12. L. Weber, in 'The Chemistry of the Metal–Carbon Bond', eds. F. R. Hartley and S. Patai, Wiley, New York, 1982, vol. 1, p. 91.
13. W. C. Kaska, *Coord. Chem. Rev.*, 1983, **48**, 1.
14. H. Schmidbaur, *Angew. Chem., Int. Ed. Engl.*, 1983, **22**, 907.
15. J. A. Albanese, A. L. Rheingold and J. L. Burmeister, *Inorg. Chim. Acta*, 1988, **150**, 213.
16. Y. Yamamoto, *Bull. Chem. Soc. Jpn.*, 1984, **57**, 43.
17. J. A. Albanese, D. L. Staley, A. L. Rheingold and J. L. Burmeister, *J. Organomet. Chem.*, 1989, **375**, 265.
18. J. Vicente, M. T. Chicote and F. Fernandez-Baeza, *J. Organomet. Chem.*, 1989, **364**, 407.
19. J. A. Teagle and J. L. Burmeister, *Inorg. Chim. Acta*, 1986, **118**, 65.
20. R. M. Buchanan and C. G. Pierpont, *Inorg. Chem.*, 1979, **18**, 3608.
21. M. Okunaka, G. Matsubayashi and T. Tanaka, *Inorg. Nucl. Chem. Lett.*, 1976, **12**, 813.
22. I. J. B. Lin, L. T. C. Kao, F. J. Wu, G. H. Lee and Y. Wang, *J. Organomet. Chem.*, 1986, **309**, 225.
23. N. J. Kermode, M. F. Lappert, B. W. Skelton, A. H. White and J. Holton, *J. Organomet. Chem.*, 1982, **228**, C71.
24. R. Uson, J. Fornies, R. Navarro, P. Espinet and C. Mendivil, *J. Organomet. Chem.*, 1985, **290**, 125.
25. M. Onishi, Y. Ohama, K. Hiraki and H. Shintani, *Polyhedron*, 1982, **1**, 539.
26. G. Facchin, R. Bertani, M. Calligaris, G. Nardin and M. Mari, *J. Chem. Soc., Dalton Trans.*, 1987, 1381.
27. P. Uguagliati, L. Canovese, G. Facchin and L. Zanotto, *Inorg. Chim. Acta*, 1992, **192**, 283.
28. G. Facchin, R. Bertani, L. Zanotto, M. Calligaris and G. Nardin, *J. Organomet. Chem.*, 1989, **366**, 409.
29. N. Holy, U. Deschler and H. Schmidbaur, *Chem. Ber.*, 1982, **115**, 1379.
30. R. Uson, J. Fornies, R. Navarro and A. M. Ortega, *J. Organomet. Chem.*, 1987, **334**, 389.
31. I. J. B. Lin, H. C. Shy, C. W. Liu, L.-K. Liu and S.-K. Yeh, *J. Chem. Soc., Dalton Trans.*, 1990, 2509.
32. R. F. Wu, I. J. B. Lin, G. H. Lee, M. C. Cheng and Y. Wang, *Organometallics*, 1990, **9**, 126.
33. G. Muller, U. Schubert, O. Orama and H. Schmidbaur, *Chem. Ber.*, 1979, **112**, 3302.
34. J. Vicente, M.-T. Chicote, I. Saura-Llamas, M.-J. Lopez-Munoz and P. G. Jones, *J. Chem. Soc., Dalton Trans.*, 1990, 3683.
35. R. K. Singhal, A. K. Awasthi and N. K. Mishra, *Indian J. Chem., Sect. A*, 1984, **23**, 1046.
36. G.-E. Matsubayashi and Y. Kondo, *J. Organomet. Chem.*, 1981, **219**, 269.
37. L. Weber, *Angew. Chem., Int. Ed. Engl.*, 1983, **22**, 516.
38. I. J. B. Lin, L. Hwan, H. C. Shy, M. C. Chen and Y. Wang, *J. Organomet. Chem.*, 1986, **315**, 135.
39. M. C. Cheng, S. M. Peng, I. J. B. Lin, B. H. H. Meng and C. H. Liu, *J. Organomet. Chem.*, 1987, **327**, 275.
40. H. Takahashi, Y. Oosawa, A. Kobayashi, T. Saito and Y. Sasaki, *Bull. Chem. Soc. Jpn.*, 1977, **50**, 1771.
41. R. Nast, *Coord. Chem. Rev.*, 1982, **47**, 89.
42. L. Ballester, M. Cano and A. Santos, *J. Organomet. Chem.*, 1982, **229**, 101.
43. E.-I. Negishi, K. Akiyoshi and T. Takahashi, *J. Chem. Soc., Chem. Commun.*, 1987, 477.
44. R. Nast, H.-P. Muller and V. Pank, *Chem. Ber.*, 1978, **111**, 1627.
45. C. Stader and B. Wrackmeyer, *J. Organomet. Chem.*, 1985, **295**, C11.
46. P. Espinet *et al.*, *J. Organomet. Chem.*, 1991, **403**, 253.
47. C. R. Langrick, P. G. Pringle and B. L. Shaw, *Inorg. Chim. Acta*, 1983, **76**, L263.
48. G. Calvin and G. E. Coates, *J. Chem. Soc.*, 1960, 2008.
49. J. H. Nelson, A. W. Verstuyft, J. D. Kelly and H. B. Jonassen, *Inorg. Chem.*, 1974, **13**, 27.
50. K. Sunkel, *J. Organomet. Chem.*, 1992, **436**, 101.
51. J. Fornies, M. A. Gomez-Saso, E. Lalinde, F. Martinez and M. T. Moreno, *Organometallics*, 1992, **11**, 2873.
52. K. Onitsuka, T. Joh and S. Takahashi, *Bull. Chem. Soc. Jpn.*, 1992, **65**, 1179.
53. H. Ogawa, K. Onitsuka, T. Joh, S. Takahashi, Y. Yamamoto and H. Yamazaki, *Organometallics*, 1988, **7**, 2257.
54. J. Burgess, M. E. Howden, R. D. W. Kemmitt and N. S. Sridhara, *J. Chem. Soc., Dalton Trans.*, 1978, 1577.
55. E. van der Voort, A. L. Spek and W. de Graaf, *Acta Crystallogr., Part C*, 1987, **43**, 2311.
56. H.-F. Klein, B. Zettel, U. Florke and H.-J. Haupt, *Chem. Ber.*, 1992, **125**, 9.
57. W. de Graaf, S. Harder, J. Boersma, G. van Koten and J. A. Kanters, *J. Organomet. Chem.*, 1988, **358**, 545.
58. H. Masai, K. Sonogashira and N. Hagihara, *Bull. Chem. Soc. Jpn.*, 1971, **44**, 2226.
59. J. Terheijden *et al.*, *J. Organomet. Chem.*, 1986, **315**, 401.
60. E.-I. Negishi, T. Takahashi and K. Akiyoshi, *J. Organomet. Chem.*, 1987, **334**, 181.
61. W. J. Blau, H. J. Byrne, D. J. Cardin and A. P. Davey, *J. Mater. Chem.*, 1991, **1**, 245.
62. S. D. Perera and B. L. Shaw, *J. Organomet. Chem.*, 1991, **402**, 133.
63. M. Wada and Y. Koyama, *J. Organomet. Chem.*, 1980, **201**, 477.
64. K. Sonogashira, T. Yatake, Y. Tohda, S. Takahashi and N. Hagihara, *J. Chem. Soc., Chem. Commun.*, 1977, 291.
65. S. Takahashi, E. Murata, K. Sonogashira and N. Hagihara, *J. Polym. Sci. Polym. Chem. Ed.*, 1980, **18**, 661.
66. R. Nast and A. Beyer, *Z. Naturforsch., Teil B*, 1980, **35**, 924.
67. R. Nast and A. Beyer, *J. Organomet. Chem.*, 1980, **194**, 125.
68. K. Hiraki, T. Masumoto, Y. Fuchita and Y. Zegi, *Bull. Chem. Soc. Jpn.*, 1981, **54**, 1044.
69. S. Takahashi, H. Morimoto, E. Murata, S. Kataoka, K. Sonogashira and N. Hagihara, *J. Polym. Sci., Polym. Chem. Ed.*, 1982, **20**, 565.
70. A. Sebald, C. Stader, B. Wrackmeyer and W. Bensch, *J. Organomet. Chem.*, 1986, **311**, 233.
71. S. J. Davies, B. F. G. Johnson, J. Lewis and M. S. Khan, *J. Organomet. Chem.*, 1991, **401**, C43.
72. M. S. Khan *et al.*, *J. Organomet. Chem.*, 1992, **424**, 87.
73. A. Albinati, S. Affolter and P. S. Pregosin, *J. Organomet. Chem.*, 1990, **395**, 231.
74. L. I. Zakharkin, A. I. Kovredov and V. A. Ol'shevskaya, *Izv. Akad. Nauk SSSR, Ser. Khim.*, 1982, 673.

75. X. L. R. Fontaine, S. J. Higgins and B. L. Shaw, *J. Chem. Soc., Dalton Trans.*, 1988, 1179.
76. A. L. Seligson, R. L. Cowan and W. C. Trogler, *Inorg. Chem.*, 1991, **30**, 3371.
77. K. Sonogashira, S. Kataoka, S. Takahashi and N. Hagihara, *J. Organomet. Chem.*, 1978, **160**, 319.
78. S. Takahashi, Y. Takai, H. Morimoto and K. Sonogashira, *J. Chem. Soc., Chem. Commun.*, 1984, 3.
79. P. L. Porter, S. Guha, K. Kang and C. C. Frazier, *Polymer*, 1991, **32**, 1756.
80. K. Onitsuka, H. Ogawa, T. Joh and S. Takahashi, *Chem. Lett.*, 1988, 1855.
81. K. Onitsuka, H. Ogawa, T. Joh, S. Takahashi, Y. Yamamoto and H. Yamazaki, *J. Chem. Soc., Dalton Trans.*, 1991, 1531.
82. K. Onitsuka, T. Joh and S. Takahashi, *Angew. Chem., Int. Ed. Engl.*, 1992, **31**, 851.
83. U. Behrens and K. Hoffmann, *J. Organomet. Chem.*, 1977, **129**, 273.
84. T. Yasuda, Y. Kai, N. Yasuoka and N. Kasai, *Bull. Chem. Soc. Jpn.*, 1971, **50**, 2888.
85. J. K. Stille, in 'The Chemistry of the Metal–Carbon Bond', eds. F. R. Hartley and S. Patai, Wiley, New York, 1985, vol. 2, p. 625.
86. K. J. Klabunde, B. B. Anderson and K. Neuenschwander, *Inorg. Chem.*, 1980, **19**, 3719.
87. K. J. Klabunde and R. Campostrini, *J. Fluorine Chem.*, 1989, **42**, 93.
88. D. W. Firsich and R. J. Lagow, *J. Chem. Soc., Chem. Commun.*, 1981, 1283.
89. J. S. Roberts and K. J. Klabunde, *J. Am. Chem. Soc.*, 1977, **99**, 2509.
90. A. V. Kavaliunas, A. Taylor and R. D. Rieke, *Organometallics*, 1983, **2**, 377.
91. J. J. Habeeb and D. G. Tuck, *J. Organomet. Chem.*, 1977, **139**, C17.
92. J. K. Stille and K. S. Y. Lau, *Acc. Chem. Res.*, 1977, **10**, 434.
93. C. Amatore and F. Pfluger, *Organometallics*, 1990, **9**, 2276.
94. C. Amatore, M. Azzabi and A. Jutand, *J. Am. Chem. Soc.*, 1991, **113**, 8375.
95. C. Amatore, M. Azzabi and A. Jutand, *J. Am. Chem. Soc.*, 1991, **113**, 1670.
96. H. Urata, H. Suzuki, Y. Moro-oka and T. Ikawa, *J. Organomet. Chem.*, 1989, **364**, 235.
97. F. Ozawa, N. Kawasaki, H. Okamoto, T. Yamamoto and A. Yamamoto, *Organometallics*, 1987, **6**, 1640.
98. W. de Graaf, J. van Wegen, J. Boersma, A. L. Spek and G. van Koten, *Recl. Trav. Chim. Pays-Bas*, 1989, **108**, 275.
99. B. A. Markies, P. Wijkens, J. Boersma, A. L. Spek and G. van Koten, *Recl. Trav. Chim. Pays-Bas*, 1991, **110**, 133.
100. B. A. Markies *et al.*, *J. Organomet. Chem.*, 1994, **482**, 191.
101. P. L. Alsters, J. Boersma and G. van Koten, *Organometallics*, 1993, **12**, 1629.
102. P. L. Alsters, J. Boersma, W. J. J. Smeets, A. L. Spek and G. van Koten, *Organometallics*, 1993, **12**, 1639.
103. B. A. Markies, P. Wijkens, H. Kooijman, A. L. Spek, J. Boersma and G. van Koten, *J. Chem. Soc., Chem. Commun.*, 1992, 1420.
104. T. Aoki, Y. Ishii, Y. Mizobe and M. Hidai, *Chem. Lett.*, 1991, 615.
105. V. Dufaud, J. Thivolle-Cazat, J.-M. Basset, R. Mathieu, J. Jaud and J. Waissermann, *Organometallics*, 1991, **10**, 4005.
106. W. R. Moser, A. W. Wang and N. K. Kildahl, *J. Am. Chem. Soc.*, 1988, **110**, 2816.
107. J. Vicente, J. A. Abad and J. A. Sanchez, *J. Organomet. Chem.*, 1988, **352**, 257.
108. M. Huser, M.-T. Youinou and J. A. Osborn, *Angew. Chem., Int. Ed. Engl.*, 1989, **28**, 1386.
109. K. S. Y. Lau, P. K. Wong and J. K. Stille, *J. Am. Chem. Soc.*, 1976, **98**, 5832.
110. D. Milstein and J. K. Stille, *J. Am. Chem. Soc.*, 1979, **101**, 4981.
111. Y. Becker and J. K. Stille, *J. Am. Chem. Soc.*, 1978, **100**, 838.
112. C. J. Elsevier and P. Vermeer, *J. Org. Chem.*, 1985, **50**, 3042.
113. J. M. A. Wouters, R. A. Klein, C. J. Elsevier, L. Haming and C. H. Stam, *Organometallics*, 1994, **13**, 4568.
114. C. J. Elsevier, H. Kleijn, J. Boersma and P. Vermeer, *Organometallics*, 1986, **5**, 716.
115. A. V. Kramer and J. A. Osborn, *J. Am. Chem. Soc.*, 1974, **96**, 7832.
116. J. K. Stille and K. S. Y. Lau, *J. Am. Chem. Soc.*, 1976, **98**, 5841.
117. J. M. Brown and N. A. Cooley, *Organometallics*, 1990, **9**, 353.
118. A. Gillie and J. K. Stille, *J. Am. Chem. Soc.*, 1980, **102**, 4933.
119. A. L. Balch, C. T. Hunt, C.-L. Lee, M. M. Olmstead and J. P. Farr, *J. Am. Chem. Soc.*, 1981, **103**, 3764.
120. P. K. Byers and A. J. Canty, *Organometallics*, 1990, **9**, 210.
121. M. E. Cucciolito, A. Panunzi, F. Ruffo and V. De Felice, *Gazz. Chim. Ital.*, 1989, **119**, 461.
122. F. Cecconi, C. A. Ghilardi, S. Midollini, S. Moneti, A. Orlandini and G. Scapacci, *J. Chem. Soc., Dalton Trans.*, 1989, 211.
123. C. A. Ghilardi, S. Midollini, S. Moneti, A. Orlandini and J. A. Ramirez, *J. Chem. Soc., Chem. Commun.*, 1989, 304.
124. H. Werner and W. Bertleff, *Chem. Ber.*, 1983, **116**, 823.
125. M. Onishi, T. Ito and K. Hiraki, *J. Organomet. Chem.*, 1981, **209**, 123.
126. P. Zanello, R. Seeber, A. Cinquantini and R. Ros, *J. Inorg. Nucl. Chem.*, 1981, **43**, 1095.
127. F. Bergamin, F. Morandini and A. Turco, *Gazz. Chim. Ital.*, 1990, **120**, 57.
128. R. A. Michelin, G. Facchin and P. Uguagliati, *Inorg. Chem.*, 1984, **23**, 961.
129. R. Bertani, C. B. Castellani and B. Crociani, *J. Organomet. Chem.*, 1984, **269**, C15.
130. R. A. Wanat and D. B. Collum, *Organometallics*, 1986, **5**, 120.
131. G. Strukul, R. Ros and R. A. Michelin, *Inorg. Chem.*, 1982, **21**, 495.
132. R. McCrindle, G. Ferguson, A. J. McAlees, M. Parvez and P. J. Roberts, *J. Chem. Soc., Dalton Trans.*, 1982, 1699.
133. H. D. McPherson and J. L. Wardell, *Inorg. Chim. Acta*, 1983, **75**, 37.
134. M. K. S. Loar and J. K. Stille, *J. Am. Chem. Soc.*, 1981, **103**, 4174.
135. M. Hodgson, D. Parker, R. J. Taylor and G. Ferguson, *J. Chem. Soc., Chem. Commun.*, 1987, 1309.
136. V. A. Mukhedkar and J. A. Mukhedkar, *J. Inorg. Nucl. Chem.*, 1981, **43**, 2801.
137. L. V. Rybin *et al.*, *J. Organomet. Chem.*, 1985, **288**, 119.
138. T. I. Son, H. Yanagihara, F. Ozawa and A. Yamamoto, *Bull. Chem. Soc. Jpn.*, 1988, **61**, 1251.
139. R. F. Heck, 'Palladium Reagents in Organic Synthesis', Academic Press, 1985.
140. N. A. Bumagin, N. P. Andryukhova and I. P. Beletskaya, *Organomet. Chem. USSR*, 1989, **2**, 467.
141. E. van der Voort, P. van der Sluis and A. L. Spek, *Acta Crystallogr., Part C*, 1987, **43**, 461.
142. W. de Graaf, J. Boersma, G. van Koten and C. J. Elsevier, *J. Organomet. Chem.*, 1989, **378**, 115.
143. M. A. Bennett and P. N. Kapoor, *J. Organomet. Chem.*, 1987, **336**, 257.
144. K. Hiraki, Y. Fuchita and T. Masumoto, *Bull. Chem. Soc. Jpn.*, 1980, **53**, 1171.
145. F. Porta, S. Cenini, P. Del Buttero and S. Maiorana, *J. Organomet. Chem.*, 1980, **194**, 211.

146. A. Mantovani, G. Facchin, T. Boschi and B. Crociani, *J. Organomet. Chem.*, 1981, **206**, C11.
147. J. W. Gosselink, G. van Koten and K. Vrieze, *J. Organomet. Chem.*, 1981, **222**, 131.
148. D. R. Fahey and J. E. Mahan, *J. Am. Chem. Soc.*, 1976, **98**, 4499.
149. A. Mantovani, *J. Organomet. Chem.*, 1983, **255**, 385.
150. B. Crociani, F. Di Bianca, A. Giovenco and A. Scrivanti, *J. Organomet. Chem.*, 1984, **269**, 295.
151. K. Isobe, K. Nanjo, Y. Nakamura and S. Kawaguchi, *Bull. Chem. Soc. Jpn.*, 1986, **59**, 2141.
152. K. Isobe, Y. Nakamura, T. Miwa and S. Kawaguchi, *Bull. Chem. Soc. Jpn.*, 1987, **60**, 149.
153. B. Crociani, F. Dibianca, A. Giovenco and A. Scrivanti, *J. Organomet. Chem.*, 1985, **291**, 259.
154. T. Benneche, *Acta Chem. Scand.*, 1990, **44**, 927.
155. H. Konishi, C. Matsumura, T. Okano and Y. Kiji, *J. Organomet. Chem.*, 1989, **364**, 245.
156. V. N. Kalinin, S. F. Min and P. V. Petrovskii, *J. Organomet. Chem.*, 1989, **379**, 195.
157. H. Urata, M. Tanaka and T. Fuchikami, *Chem. Lett.*, 1987, 751.
158. D. Milstein and J. K. Stille, *J. Org. Chem.*, 1979, **44**, 1613.
159. T. F. Murray and J. R. Norton, *J. Am. Chem. Soc.*, 1979, **101**, 4107.
160. Y. Yamamoto and A. Yanagi, *Chem. Pharm. Bull.*, 1982, **30**, 2003.
161. M. Onishi, K. Hiraki, H. Konda, Y. Ishida, Y. Ohama and Y. Uchibori, *Bull. Chem. Soc. Jpn.*, 1986, **59**, 201.
162. W. Keim, J. Becker and A. M. Trzeciak, *J. Organomet. Chem.*, 1989, **372**, 447.
163. E. G. Samsel and J. R. Norton, *J. Am. Chem. Soc.*, 1984, **106**, 5505.
164. J. S. Brumbaugh and A. Sen, *J. Am. Chem. Soc.*, 1988, **110**, 803.
165. M. Kubota, S. C. Boegeman, R. N. Keil and C. G. Webb, *Organometallics*, 1989, **8**, 1616.
166. F. Ozawa, T. Sugimoto, T. Yamamoto and A. Yamamoto, *Organometallics*, 1984, **3**, 692.
167. A. Sen, J.-T. Chen, W. M. Vetter and R. R. Whittle, *J. Am. Chem. Soc.*, 1987, **109**, 148.
168. W. M. Vetter and A. Sen, *J. Organomet. Chem.*, 1989, **378**, 485.
169. R. McCrindle and A. J. McAlees, *J. Chem. Soc., Dalton Trans.*, 1983, 127.
170. R. McCrindle, D. K. Stephenson, A. J. McAlees and J. M. Willson, *J. Chem. Soc., Dalton Trans.*, 1986, 641.
171. P. W. Clark and S. F. Dyke, *J. Organomet. Chem.*, 1984, **276**, 421.
172. J. Albert, J. Barro and J. Granell, *J. Organomet. Chem.*, 1991, **408**, 115.
173. P. W. Clark and S. F. Dyke, *J. Organomet. Chem.*, 1985, **281**, 389.
174. H.-P. Abicht, *J. Organomet. Chem.*, 1986, **311**, 57.
175. G. Balavoine, J. C. Clinet, P. Zerbib and K. Boubekeur, *J. Organomet. Chem.*, 1990, **389**, 259.
176. P. L. Alsters, P. F. Engel, M. P. Hogerheide, M. Copijn, A. L. Spek and G. van Koten, *Organometallics*, 1993, **12**, 1831.
177. P. L. Alsters *et al.*, *Organometallics*, 1992, **11**, 4124.
178. M. T. Alonso, O. Juanes, J. de Mendoza and J. C. Rodriguez-Ubis, *J. Organomet. Chem.*, 1992, **430**, 335.
179. T. A. Anderson, R. J. Barton, B. E. Robertson and K. Venkatasubramanian, *Acta Crystallogr., Part C*, 1985, **41**, 1171.
180. K. Nakatsu, K. Kinoshita, H. Kanda, K. Isobe, Y. Nakamura and S. Kawaguchi, *Chem. Lett.*, 1980, 913.
181. M. Onishi, K. Hiraki, K. Maeda and T. Itoh, *J. Organomet. Chem.*, 1980, **188**, 245.
182. K. Nakatsu, K. Kafuku, H. Yamaoka, K. Isobe, Y. Nakamura and S. Kawaguchi, *Inorg. Chim. Acta*, 1981, **54**, L69.
183. K. Isobe, Y. Nakamura and S. Kawaguchi, *Bull. Chem. Soc. Jpn.*, 1989, **62**, 1802.
184. A. J. Canty, N. J. Minchin, L. M. Engelhardt, B. W. Skelton and A. H. White, *Aust. J. Chem.*, 1988, **41**, 651.
185. X. L. R. Fontaine, S. J. Higgins, B. L. Shaw, M. Thornton-Pett and Y. Wu, *J. Chem. Soc., Dalton Trans.*, 1987, 1501.
186. M. Lenarda, R. Ros, M. Graziani and U. Belluco, *J. Organomet. Chem.*, 1974, **65**, 407.
187. P. J. Stang, L. Song and B. Halton, *J. Organomet. Chem.*, 1990, **388**, 215.
188. R. T. Carl, E. W. Corcoran, Jr., R. P. Hughes and D. E. Samkoff, *Organometallics*, 1990, **9**, 838.
189. K. Osakada, M.-K. Doh, F. Ozawa and A. Yamamoto, *Organometallics*, 1990, **9**, 2197.
190. T. Yamamoto *et al.*, *Organometallics*, 1990, **9**, 2396.
191. R. Nast, J. Bulck and R. Kramolowsky, *Chem. Ber.*, 1975, **108**, 3461.
192. D. Carmichael, P. B. Hitchcock, J. F. Nixon, F. Mathey and A. Pidcock, *J. Chem. Soc., Chem. Commun.*, 1986, 762.
193. D. Carmichael, P. B. Hitchcock, J. F. Nixon, F. Mathey and L. Ricard, *J. Chem. Soc., Chem. Commun.*, 1989, 1389.
194. N. Y. Kozitsyna, A. M. Ellern, M. Y. Antipin, Y. T. Struchkov and I. I. Moiseev, *Mendeleev Commun.*, 1991, 92.
195. H. Nakazawa, Y. Matsuoka, I. Nakagawa and K. Miyoshi, *Organometallics*, 1992, **11**, 1385.
196. R. Yamashita, K. Kikukawa, F. Wada and T. Matsuda, *J. Organomet. Chem.*, 1980, **201**, 463.
197. M. A. Bennett, C. Chiraratvatana, G. B. Robertson and U. Tooptakong, *Organometallics*, 1988, **7**, 1403.
198. F. M. Conroy-Lewis, L. Mole, A. D. Redhouse, S. A. Litster and J. L. Spencer, *J. Chem. Soc., Chem. Commun.*, 1991, 1601.
199. V. I. Sokolov, V. V. Bashilov and O. A. Reutov, *J. Organomet. Chem.*, 1975, **97**, 299.
200. V. I. Sokolov, G. Z. Suleimanov, A. A. Musaev and O. A. Reutov, *Izv. Akad. Nauk SSSR, Ser. Khim.*, 1980, 2093.
201. E. R. Milaeva, A. Z. Rubezhov, A. I. Prokoph'ev, A. G. Milaev and O. Y. Okhlobystin, *J. Organomet. Chem.*, 1980, **188**, C43.
202. V. I. Sokolov, V. V. Bashilov, A. A. Musaev and O. A. Reutov, *J. Organomet. Chem.*, 1982, **225**, 57.
203. V. V. Bashilov, E. V. Maskaeva, P. V. Petrovskii and V. I. Sokolov, *J. Organomet. Chem.*, 1985, **292**, 89.
204. S.-I. Murahashi *et al.*, *Organometallics*, 1986, **5**, 356.
205. A. F. M. J. van der Ploeg, G. van Koten and K. Vrieze, *J. Organomet. Chem.*, 1981, **222**, 155.
206. H. Nishiyama, M. Matsumoto, T. Matsukura, R. Miura and K. Itoh, *Organometallics*, 1985, **4**, 1911.
207. J. W. Suggs and K. S. Lee, *J. Organomet. Chem.*, 1986, **299**, 297.
208. M. T. Pereira, M. Pfeffer and M. A. Rotteveel, *J. Organomet. Chem.*, 1989, **375**, 139.
209. Y. Ito, H. Aoyama and T. Saegusa, *J. Am. Chem. Soc.*, 1980, **102**, 4519.
210. L. J. Krause and J. A. Morrison, *J. Chem. Soc., Chem. Commun.*, 1981, 1282.
211. J. P. Visser, W. W. Jager and C. Masters, *Recl. Trav. Chim. Pays-Bas*, 1975, **94**, 70.
212. R. J. Puddephatt and P. J. Thompson, *J. Chem. Soc., Dalton Trans.*, 1975, 1810.
213. M. Rudler-Chauvin and H. Rudler, *J. Organomet. Chem.*, 1977, **134**, 115.
214. D. L. Reger, D. Garza and L. Lebioda, *Organometallics*, 1992, **11**, 4285.
215. H. Kurosawa *et al.*, *J. Am. Chem. Soc.*, 1987, **109**, 6333.
216. S. Numata and H. Kurosawa, *J. Organomet. Chem.*, 1977, **131**, 301.

217. H. Werner, H.-J. Kraus, U. Schubert, K. Ackerman and P. Hofmann, *J. Organomet. Chem.*, 1983, **250**, 517.
218. R. J. Cross and R. Wardle, *J. Chem. Soc. (A)*, 1970, 840.
219. R. McCrindle, G. J. Arsenault, R. Farwaha, A. J. McAlees and D. W. Sneddon, *J. Chem. Soc., Dalton Trans.*, 1989, 761.
220. G. K. Anderson, *Organometallics*, 1983, **2**, 665.
221. J. Vicente, A. Arcas, M. V. Borrachero and M. B. Hursthouse, *J. Chem. Soc., Dalton Trans.*, 1987, 1655.
222. J. Vicente, A. Arcas, M. V. Borrachero, E. Molins and C. Miravitlles, *J. Organomet. Chem.*, 1989, **359**, 127.
223. J. Vicente, A. Arcas, M. V. Borrachero, E. Molins and C. Miravitlles, *J. Organomet. Chem.*, 1992, **441**, 487.
224. J. Vicente, J.-A. Abad and P. G. Jones, *Organometallics*, 1992, **11**, 3512.
225. J. Vicente *et al.*, *J. Chem. Soc., Dalton Trans.*, 1988, 141.
226. E. Wehman, G. van Koten, J. T. B. H. Jastrzebski, H. Ossor and M. Pfeffer, *J. Chem. Soc., Dalton Trans.*, 1988, 2975.
227. L.-Y. Chia and W. R. McWhinnie, *J. Organomet. Chem.*, 1980, **188**, 121.
228. D. Milstein, *J. Chem. Soc., Chem. Commun.*, 1986, 817.
229. E. C. Constable, A. M. W. Cargill Thompson, T. A. Leese, D. G. F. Reese and D. A. Tocher, *Inorg. Chim. Acta*, 1991, **182**, 93.
230. T. Ito, H. Tsuchiya and A. Yamamoto, *Bull. Chem. Soc. Jpn.*, 1977, **50**, 1319.
231. F. Ozawa, T. Ito, Y. Nakamura and A. Yamamoto, *J. Am. Chem. Soc.*, 1980, **102**, 6457.
232. F. Ozawa, T. Ito, Y. Nakamura and A. Yamamoto, *Bull. Chem. Soc. Jpn.*, 1981, **54**, 1868.
233. Y.-J. Kim, K. Osakada, A. Takenaka and A. Yamamoto, *J. Am. Chem. Soc.*, 1990, **112**, 1096.
234. F. Ozawa, T.-I. Son, S. Ebina, K. Osakada and A. Yamamoto, *Organometallics*, 1992, **11**, 171.
235. R. Sustmann, J. Lau and M. Zipp, *Recl. Trav. Chim. Pays-Bas*, 1986, **105**, 356.
236. R. Sustmann and J. Lau, *Chem. Ber.*, 1986, **119**, 2531.
237. F. K. Schmidt, G. V. Ratovskii, T. V. Dmitrieva, I. N. Ivleva and Y. G. Borodko, *J. Organomet. Chem.*, 1983, **256**, 309.
238. Y. Hayashi, K. Isobe, Y. Nakamura and S. Okeya, *J. Organomet. Chem.*, 1986, **310**, 127.
239. F. Basolo, J. Chatt, H. B. Gray, R. G. Pearson and B. L. Shaw, *J. Chem. Soc.*, 1961, 2207.
240. B. Wozniak, J. D. Ruddick and G. Wilkinson, *J. Chem. Soc. (A)*, 1971, 3116.
241. A. Miyashita, M. Ohyoshi, H. Shitara and H. Nohira, *J. Organomet. Chem.*, 1988, **338**, 103.
242. U. Schubert and C. Muller, *J. Organomet. Chem.*, 1989, **373**, 165.
243. A. M. Saleem and H. A. Hodali, *Transition Met. Chem.*, 1991, **16**, 458.
244. H. Urata, H. Suzuki, Y. Moro-oka and T. Ikawa, *J. Organomet. Chem.*, 1989, **364**, 235.
245. L. A. Villanueva, K. Abboud and J. M. Boncella, *Organometallics*, 1991, **10**, 2969.
246. H. Nakazawa, F. Ozawa and A. Yamamoto, *Organometallics*, 1983, **2**, 241.
247. A. Yamamoto, T. Yamamoto, S. Komiya and F. Ozawa, *Pure Appl. Chem.*, 1984, **56**, 1621.
248. A. Yamamoto, *J. Organomet. Chem.*, 1986, **300**, 347.
249. F. Ozawa, K. Kurihara, M. Fujimori, T. Hidaka, T. Toyoshima and A. Yamamoto, *Organometallics*, 1989, **8**, 180.
250. F. Ozawa, K. Kurihara, T. Yamamoto and A. Yamamoto, *J. Organomet. Chem.*, 1985, **279**, 233.
251. D. L. Reger, D. G. Garza and L. Lebioda, *Organometallics*, 1991, **10**, 902.
252. P. K. Byers, A. J. Canty, L. M. Engelhardt and A. H. White, *J. Chem. Soc., Dalton Trans.*, 1986, 1731.
253. P. K. Byers, A. J. Canty and R. T. Honeyman, *J. Organomet. Chem.*, 1990, **385**, 417.
254. P. K. Byers, A. J. Canty, B. W. Skelton and A. H. White, *J. Organomet. Chem.*, 1987, **336**, C55.
255. P. K. Byers, A. J. Canty, B. W. Skelton and A. H. White, *J. Organomet. Chem.*, 1990, **393**, 299.
256. V. G. Albano, C. Castellari, M. E. Cucciolito, A. Panunzi and A. Vitagliano, *Organometallics*, 1990, **9**, 1269.
257. P. Diversi, D. Fasce and R. Santini, *J. Organomet. Chem.*, 1984, **269**, 285.
258. D. R. Schaad and C. R. Landis, *Organometallics*, 1992, **11**, 2024.
259. M. Bochmann, G. Wilkinson and G. B. Young, *J. Chem. Soc., Dalton Trans.*, 1980, 1879.
260. W. de Graaf, J. Boersma, D. M. Grove, A. L. Spek and G. van Koten, *Recl. Trav. Chim. Pays-Bas*, 1988, **107**, 299.
261. W. de Graaf, J. Boersma, W. J. J. Smeets, A. L. Spek and G. van Koten, *Organometallics*, 1989, **8**, 2907.
262. B. A. Markies, A. J. Canty, M. D. Janssen, A. L. Spek, J. Boersma and G. van Koten, *Recl. Trav. Chim. Pays-Bas*, 1991, **110**, 477.
263. P. K. Byers, A. J. Canty and R. T. Honeyman, *Adv. Organomet. Chem.*, 1992, **34**, 1.
264. M. A. Bennett, A. J. Canty, J. K. Felixberger, L. M. Rendina, C. Sutherland and A. C. Willis, *Inorg. Chem.*, 1993, **32**, 1951.
265. W. de Graaf, J. Boersma and G. van Koten, *Organometallics*, 1990, **9**, 1479.
266. A. L. Seligson and W. C. Trogler, *Organometallics*, 1993, **12**, 738.
267. P. Diversi, G. Ingrosso, A. Lucherini and S. Murtas, *J. Chem. Soc., Dalton Trans.*, 1980, 1633.
268. P. Diversi, G. Ingrosso and A. Lucherini, *Inorg. Synth.*, 1983, **22**, 167.
269. P. Diversi *et al.*, *J. Chem. Soc., Dalton Trans.*, 1988, 133.
270. F. Ozawa, A. Yamamoto, T. Ikariya and R. H. Grubbs, *Organometallics*, 1982, **1**, 1481.
271. P. Barabotti, P. Diversi, G. Ingrosso, A. Lucherini and F. Nuti, *J. Chem. Soc., Dalton Trans.*, 1984, 2517.
272. F. Ozawa, M. Fujimori, T. Yamamoto and A. Yamamoto, *Organometallics*, 1986, **5**, 2144.
273. F. J. Hopton, A. J. Rest, D. T. Rosevear and F. G. A. Stone, *J. Chem. Soc. (A)*, 1966, 1326.
274. J. Terheijden *et al.*, *J. Organomet. Chem.*, 1986, **315**, 401.
275. C. Cornioley-Deuschel and A. von Zelewsky, *Inorg. Chem.*, 1987, **26**, 3354.
276. C. J. Cardin, D. J. Cardin, H. E. Parge and A. C. Sullivan, *J. Chem. Soc., Dalton Trans.*, 1986, 2315.
277. J. Browning, G. W. Bushnell and K. R. Dixon, *Inorg. Chem.*, 1981, **20**, 3912.
278. L. Abis, R. Santi and J. Halpern, *J. Organomet. Chem.*, 1981, **215**, 263.
279. M. D. Fryzuk, P. A. MacNeil, S. J. Rettig, A. S. Secco and J. Trotter, *Organometallics*, 1982, **1**, 918.
280. S. Sergi, F. Faraone, L. Silvesto and R. Pietropaolo, *J. Organomet. Chem.*, 1971, **33**, 403.
281. G. K. Anderson, R. J. Cross, S. A. Leaman, F. J. Robertson, D. S. Rycroft and M. Rocamora, *J. Organomet. Chem.*, 1990, **388**, 221.
282. R. G. Miller, R. D. Stauffer, D. R. Fahey and D. R. Parnell, *J. Am. Chem. Soc.*, 1970, **92**, 1511.
283. A. L. Seligson and W. C. Trogler, *Organometallics*, 1993, **12**, 744.
284. N. Garty and M. Michman, *J. Organomet. Chem.*, 1972, **36**, 391.
285. R. Tooze, K. W. Chiu and G. Wilkinson, *Polyhedron*, 1984, **3**, 1025.
286. P. K. P. Claire, C. J. Jones, J. A. McCleverty, P. L. Coe and M. G. B. Drew, *J. Organomet. Chem.*, 1992, **424**, 105.

287. G. Lopez *et al.*, *J. Chem. Soc., Dalton Trans.*, 1990, 1621.
288. P. Espinet, J. M. Martinez-Ilarduya and C. Perez-Briso, *J. Organomet. Chem.*, 1993, **447**, 145.
289. Y. Hayashi, K. Matsumoto, Y. Nakamura and K. Isobe, *J. Chem. Soc., Dalton Trans.*, 1989, 1519.
290. E. R. Burkhardt, R. G. Bergman and C. H. Heathcock, *Organometallics*, 1990, **9**, 30.
291. B. Bogdanovic, S. C. Huckett, U. Wilczok and A. Rufinska, *Angew. Chem., Int. Ed. Engl.*, 1988, **27**, 1513.
292. C. Arlen, M. Pfeffer, O. Bars and D. Grandjean, *J. Chem. Soc., Dalton Trans.*, 1983, 1535.
293. F. Maassarani, M. Pfeffer, G. Le Borgne, J. T. B. H. Jastrzebski and G. van Koten, *Organometallics*, 1987, **6**, 1111.
294. H.-P. Abicht and K. Issleib, *J. Organomet. Chem.*, 1980, **185**, 265.
295. H.-P. Abicht and K. Issleib, *J. Organomet. Chem.*, 1985, **289**, 201.
296. J. Dehand, A. Mauro, H. Ossor, M. Pfeffer, R. H. De A. Santos and J. R. Lechat, *J. Organomet. Chem.*, 1983, **250**, 537.
297. T. Janecki, J. A. D. Jeffreys, P. L. Pauson, A. Pietrzykowski and K. J. McCullough, *Organometallics*, 1987, **6**, 1553.
298. H.-P. Abicht, P. Lehniger and K. Issleib, *J. Organomet. Chem.*, 1983, **250**, 609.
299. S. H. Bergens, P.-H. Leung, B. Bosnich and A. L. Rheingold, *Organometallics*, 1990, **9**, 2406.
300. D. M. Grove, G. van Koten, J. N. Louwen, J. G. Noltes, A. L. Spek and H. J. C. Ubbels, *J. Am. Chem. Soc.*, 1982, **104**, 6609.
301. J. A. M. van Beek, G. van Koten, G. P. C. M. Dekker, E. Wissing, M. C. Zoutberg and C. H. Stam, *J. Organomet. Chem.*, 1990, **394**, 659.
302. G. R. Newkome and D. W. Evans, *Organometallics*, 1987, **6**, 2592.
303. C. Cornioley-Deuschel, T. Ward and A. von Zelewsky, *Helv. Chim. Acta*, 1988, **71**, 130.
304. O. Juanes, J. de Mendoza and J. C. Rodriguez-Ubis, *J. Organomet. Chem.*, 1989, **363**, 393.
305. P. Braunstein, D. Matt, J. Fischer, L. Ricard and A. Mitschler, *Nouv. J. Chim.*, 1980, **4**, 493.
306. P. Braunstein, D. Matt, Y. Dusausoy, J. Fischer, A. Mitschler and L. Ricard, *J. Am. Chem. Soc.*, 1981, **103**, 5115.
307. S. Okeya, S. Kawaguchi, N. Yasuoka, Y. Kai and N. Kasai, *Chem. Lett.*, 1976, 53.
308. S. Okeya and S. Kawaguchi, *Inorg. Chem.*, 1977, **16**, 1730.
309. K. Hiraki, M. Onishi and H. Matsuo, *J. Organomet. Chem.*, 1980, **185**, 111.
310. M. Onishi, H. Yamamoto and K. Hiraki, *Bull. Chem. Soc. Jpn.*, 1980, **53**, 2540.
311. T. Kawato, T. Uechi, H. Koyama, H. Kanatomi and Y. Kawanami, *Inorg. Chem.*, 1984, **23**, 764.
312. G. R. Newkome, V. K. Gupta, H. C. R. Taylor and F. R. Fronczek, *Organometallics*, 1984, **3**, 1549.
313. G. R. Newkome *et al.*, *J. Am. Chem. Soc.*, 1981, **103**, 3423.
314. G. R. Newkome, V. K. Gupta and F. R. Fronczek, *Organometallics*, 1982, **1**, 907.
315. G. R. Newkome, D. K. Kohli and F. R. Fronczek, *J. Am. Chem. Soc.*, 1982, **104**, 994.
316. G. R. Newkome, W. E. Puckett, V. K. Gupta and F. R. Fronczek, *Organometallics*, 1983, **2**, 1247.
317. G. R. Newkome *et al.*, *Inorg. Chem.*, 1985, **24**, 811.
318. G. R. Newkome, G. E. Kiefer, Y. A. Frere, M. Onishi, V. K. Gupta and F. R. Fronczek, *Organometallics*, 1986, **5**, 348.
319. G. R. Newkome, D. W. Evans, G. E. Kiefer and K. J. Theriot, *Organometallics*, 1988, **7**, 2537.
320. G. R. Newkome, W. E. Puckett, V. K. Gupta and G. E. Kiefer, *Chem. Rev.*, 1986, **86**, 451.
321. A. Yoneda, G. R. Newkome and K. J. Theriot, *J. Organomet. Chem.*, 1991, **401**, 217.
322. A. D. Ryabov, *Chem. Rev.*, 1990, **90**, 403.
323. V. V. Dunina, O. A. Zalevskaya and V. M. Potapov, *Russ. Chem. Rev.*, 1988, **57**, 250 (*Usp. Khim.*, 1988, **57**, 434).
324. I. Omae, *Chem. Rev.*, 1979, **79**, 287.
325. I. Omae, *Coord. Chem. Rev.*, 1979, **28**, 97.
326. I. Omae, *Coord. Chem. Rev.*, 1980, **32**, 235.
327. I. Omae, *Coord. Chem. Rev.*, 1982, **42**, 245.
328. I. Omae, *Coord. Chem. Rev.*, 1988, **83**, 137.
329. I. Omae, 'Organometallic Intramolecular Coordination Compounds', Elsevier, Amsterdam, 1986.
330. A. J. Deeming and I. P. Rothwell, *Pure Appl. Chem.*, 1980, **52**, 649.
331. V. I. Sokolov, *Pure Appl. Chem.*, 1983, **55**, 1837.
332. E. C. Constable, *Polyhedron*, 1984, **3**, 1037.
333. A. D. Ryabov, *Synthesis*, 1985, 233.
334. D. W. Evans, G. R. Baker and G. R. Newkome, *Coord. Chem. Rev.*, 1989, **93**, 155.
335. M. Pfeffer, *Recl. Trav. Chim. Pays-Bas*, 1990, **109**, 567.
336. Y. Fuchita, K. Hiraki and Y. Matsumoto, *J. Organomet. Chem.*, 1985, **280**, C51.
337. G. Balavoine and J. C. Clinet, *J. Organomet. Chem.*, 1990, **390**, C84.
338. M. T. Alonso, O. Juanes, J. de Mendoza and J. C. Rodriguez-Ubis, *J. Organomet. Chem.*, 1992, **430**, 349.
339. M. Pfeffer, E. Wehman and G. van Koten, *J. Organomet. Chem.*, 1985, **282**, 127.
340. S. Okeya *et al.*, *Bull. Chem. Soc. Jpn.*, 1981, **54**, 1978.
341. A. R. Siedle, *J. Organomet. Chem.*, 1981, **208**, 115.
342. R. P. Thummel and Y. Jahng, *J. Org. Chem.*, 1987, **52**, 73.
343. A. D. Ryabov and G. M. Kazankov, *J. Organomet. Chem.*, 1984, **268**, 85.
344. A. D. Ryabov, *Inorg. Chem.*, 1987, **26**, 1252.
345. A. D. Ryabov, A. V. Eliseev, E. S. Sergeyenko, A. V. Usatov, L. I. Zakharkin and V. N. Kalinin, *Polyhedron*, 1989, **8**, 1485.
346. A. D. Ryabov *et al.*, *Inorg. Chem.*, 1992, **31**, 3083.
347. J. Granell, D. Sainz, J. Sales, X. Solans and M. Font-Altaba, *J. Chem. Soc., Dalton Trans.*, 1986, 1785.
348. R. M. Ceder, M. Gomez and J. Sales, *J. Organomet. Chem.*, 1989, **361**, 391.
349. J. Dupont, N. Beydoun and M. Pfeffer, *J. Chem. Soc., Dalton Trans.*, 1989, 1715.
350. A. D. Ryabov, I. K. Sakodinskaya and A. K. Yatsimirsky, *J. Chem. Soc., Dalton Trans.*, 1985, 2629.
351. A. J. Deeming and I. P. Rothwell, *J. Organomet. Chem.*, 1981, **205**, 117.
352. L. R. Falvello, J. Fornies, R. Navarro, V. Sicilia and M. Tomas, *Angew. Chem., Int. Ed. Engl.*, 1990, **29**, 891.
353. H. Ossor, M. Pfeffer, J. T. B. H. Jastrzebski and C. H. Stam, *Inorg. Chem.*, 1987, **26**, 1169.
354. J. Terheijden, G. van Koten, I. C. Vinke and A. L. Spek, *J. Am. Chem. Soc.*, 1985, **107**, 2891.
355. M. I. Bruce, B. L. Goodall and F. G. A. Stone, *J. Chem. Soc., Dalton Trans.*, 1978, 687.
356. B. J. Brisdon, P. Nair and S. F. Dyke, *Tetrahedron*, 1981, **37**, 173.
357. Y. Wakatsuki, H. Yamazaki, P. A. Grutsch, M. Santhanam and C. Kutal, *J. Am. Chem. Soc.*, 1985, **107**, 8153.

358. K. Hiraki, Y. Fuchita and Y. Kage, *J. Chem. Soc., Dalton Trans.*, 1984, 99.
359. J. M. Vila, A. Suarez, M. T. Pereira, E. Gayoso and M. Gayoso, *Polyhedron*, 1987, **6**, 1003.
360. M. Ghedini, S. Armentano, F. Neve and S. Licoccia, *J. Chem. Soc., Dalton Trans.*, 1988, 1565.
361. A. B. Goel, S. Goel and D. Vanderweer, *Inorg. Chim. Acta*, 1981, **54**, L267.
362. R. G. Goel and W. O. Ogini, *Organometallics*, 1982, **1**, 654.
363. H. Werner and H. J. Kraus, *J. Organomet. Chem.*, 1981, **204**, 415.
364. H. Horino and N. Inoue, *J. Org. Chem.*, 1981, **46**, 4416.
365. J. Dupont, M. Pfeffer, J.-C. Daran and J. Gouteron, *J. Chem. Soc., Dalton Trans.*, 1988, 2421.
366. S. J. Tremont and H. U. Rahman, *J. Am. Chem. Soc.*, 1984, **106**, 5759.
367. V. S. Goncharov, V. S. Raida and E. E. Sirotkina, *Izv. Sib. Otd. Akad. Nauk SSSR, Ser. Khim.*, 1988, 112 (*Chem. Abstr.*, 1989, **111**, 57 197j).
368. A. D. Ryabov, I. K. Sakodinskaya and A. K. Yatsimirsky, *J. Organomet. Chem.*, 1991, **406**, 309.
369. W. Mossi, A. J. Klaus and P. Rys, *Helv. Chim. Acta*, 1992, **75**, 2531.
370. J. Barker, N. D. Cameron, M. Kilner, M. M. Mahmoud and S. C. Wallwork, *J. Chem. Soc., Dalton Trans.*, 1991, 3435.
371. V. A. Dorokhov, K. L. Cherkasova and A. I. Lutsenko, *Izv. Akad. Nauk SSSR, Ser. Khim.*, 1987, 2351.
372. J. Albert, J. Granell, J. Sales, X. Solans and M. Font-Altaba, *Organometallics*, 1986, **5**, 2567.
373. J. Albert, J. Granell and J. Sales, *Polyhedron*, 1989, **8**, 2725.
374. R. M. Ceder, J. Granell and J. Sales, *J. Organomet. Chem.*, 1986, **307**, C44.
375. J. Albert, M. Gomez, J. Granell, J. Sales and X. Solans, *Organometallics*, 1990, **9**, 1405.
376. K. Gehrig, M. Hugentobler, A. J. Klaus and P. Rys, *Inorg. Chem.*, 1982, **21**, 2493.
377. M. Hugentobler, A. J. Klaus, H. Mettler, P. Rys and G. Wehrle, *Helv. Chim. Acta*, 1982, **65**, 1202.
378. K. Hiraki, M. Nakashima, T. Uchiyama and Y. Fuchita, *J. Organomet. Chem.*, 1992, **428**, 249.
379. Y. Fuchita, K. Hiraki and T. Uchiyama, *J. Chem. Soc., Dalton Trans.*, 1983, 897.
380. Y. Fuchita, K. Hiraki and Y. Kage, *Bull. Chem. Soc. Jpn.*, 1982, **55**, 955.
381. M. Nonoyama, *Transition Met. Chem.*, 1982, **7**, 281.
382. K. Hiraki, Y. Fuchita and K. Takechi, *Inorg. Chem.*, 1981, **20**, 4316.
383. K. Hiraki and Y. Fuchita, *Inorg. Synth.*, 1989, **26**, 208.
384. V. A. Polyakov and A. D. Ryabov, *J. Chem. Soc., Dalton Trans.*, 1986, 589.
385. A. J. Canty, N. J. Minchin, B. W. Skelton and A. H. White, *J. Chem. Soc., Dalton Trans.*, 1986, 2205.
386. J. Dupont, M. Pfeffer, M. A. Rotteveel, A. De Cian and J. Fischer, *Organometallics*, 1989, **8**, 1116.
387. J. Dupont, M. Pfeffer, L. Theurel, M. A. Rotteveel, A. De Cian and J. Fischer, *New J. Chem.*, 1990, **15**, 551.
388. M. Nonoyama, *Polyhedron*, 1985, **4**, 765.
389. A. J. Canty, N. J. Minchin, B. W. Skelton and A. H. White, *J. Chem. Soc., Dalton Trans.*, 1987, 1477.
390. A. J. Canty, R. T. Honeyman, B. W. Skelton and A. H. White, *J. Organomet. Chem.*, 1990, **389**, 277.
391. G. Minghetti, M. A. Cinellu, G. Chelucci, S. Gladiali, F. Demartin and M. Manassero, *J. Organomet. Chem.*, 1986, **307**, 107.
392. R. A. Holton and R. V. Nelson, *J. Organomet. Chem.*, 1980, **201**, C35.
393. R. A. Holton, M. P. Sibi and W. S. Murphy, *J. Am. Chem. Soc.*, 1988, **110**, 314.
394. M. A. Bennett and P. W. Clark, *J. Organomet. Chem.*, 1976, **110**, 367.
395. M. Pfeffer, *Inorg. Synth.*, 1989, **26**, 211.
396. N. A. Al-Salem, H. D. Empsall, R. Markham, B. L. Shaw and B. Weeks, *J. Chem. Soc., Dalton Trans.*, 1979, 1972.
397. R. J. Errington, W. S. McDonald and B. L. Shaw, *J. Chem. Soc., Dalton Trans.*, 1982, 1829.
398. J. Dehand, C. Mutet and M. Pfeffer, *J. Organomet. Chem.*, 1981, **209**, 255.
399. A. G. Constable, W. S. McDonald, L. C. Sawkins and B. L. Shaw, *J. Chem. Soc., Dalton Trans.*, 1980, 1992.
400. V. Rocherolle, J. C. Lopez, A. Olesker and G. Lukacs, *J. Chem. Soc., Chem. Commun.*, 1988, 513.
401. K. Carr, H. M. Saxton and J. K. Sutherland, *J. Chem. Soc., Perkin Trans. 1*, 1988, 1599.
402. J. E. Baldwin, C. Najera and M. Yus, *J. Chem. Soc., Chem. Commun.*, 1985, 126.
403. A. P. Wells and W. Kitching, *Organometallics*, 1992, **11**, 2750.
404. B. Galli *et al.*, *J. Chem. Soc., Dalton Trans.*, 1985, 1155.
405. B. Galli, F. Gasparrini, L. Maresca, G. Natile and G. Palmieri, *J. Chem. Soc., Dalton Trans.*, 1983, 1483.
406. K. Hiraki, Y. Fuchita, M. Nakashima and H. Hiraki, *Bull. Chem. Soc. Jpn.*, 1986, **59**, 3073.
407. Y. Fuchita, M. Nakashima, K. Hiraki, M. Kawatani and K. Ohnuma, *J. Chem. Soc., Dalton Trans.*, 1988, 785.
408. A. R. H. Bottomley, C. Crocker and B. L. Shaw, *J. Organomet. Chem.*, 1983, **250**, 617.
409. A. L. Rheingold and W. C. Fultz, *Organometallics*, 1984, **3**, 1414.
410. E. C. Alyea and J. Malito, *J. Organomet. Chem.*, 1988, **340**, 119.
411. E. C. Alyea, G. Ferguson, J. Malito and B. L. Ruhl, *Organometallics*, 1989, **8**, 1188.
412. E. C. Alyea, G. Ferguson, J. Malito and B. L. Ruhl, *Can. J. Chem.*, 1988, **66**, 3162.
413. W. J. Youngs *et al.*, *Organometallics*, 1983, **2**, 917.
414. Y. Tamaru, M. Kagotani and Z.-I. Yoshida, *Angew. Chem., Int. Ed. Engl.*, 1981, **20**, 980.
415. K. Miki, N. Tanaka and N. Kasai, *Acta Crystallogr., Part B*, 1981, **37**, 447.
416. K. Hiraki, Y. Fuchita and Y. Matsumoto, *Chem. Lett.*, 1984, 1947.
417. A. J. Deeming, I. P. Rothwell, M. B. Hursthouse and K. M. Abdul Malik, *J. Chem. Soc., Dalton Trans.*, 1979, 1899.
418. J. Albert, R. M. Ceder, M. Gomez, J. Granell and J. Sales, *Organometallics*, 1992, **11**, 1536.
419. A. M. M. Lanfreddi, F. Ugozzoli, M. Ghedini and S. Licoccia, *Inorg. Chim. Acta*, 1984, **86**, 165.
420. G. K. Anderson, R. J. Cross, L. Manojlovic-Muir and M. Rocamora, *Organometallics*, 1988, **7**, 1520.
421. G. K. Anderson, R. J. Cross, K. W. Muir and L. Manojlovic-Muir, *J. Organomet. Chem.*, 1989, **362**, 225.
422. M. I. Bruce, M. J. Liddell and G. N. Pain, *Inorg. Synth.*, 1989, **26**, 171.
423. S. Armentano, A. Crispini, G. de Munno, M. Ghedini and F. Neve, *Acta Crystallogr., Part C*, 1991, **47**, 966.
424. S. Armentano, A. Crispini, G. de Munno, M. Ghedini and F. Neve, *Acta Crystallogr., Part C*, 1991, **47**, 2545.
425. M. Ghedini, F. Neve, D. Pucci, E. Cesarotti and M. Grassi, *J. Organomet. Chem.*, 1992, **438**, 343.
426. A. D. Ryabov, V. A. Polyakov and A. Yatsimirsky, *Inorg. Chim. Acta*, 1984, **91**, 59.
427. M. E. K. Cartoon and G. W. H. Cheeseman, *J. Organomet. Chem.*, 1982, **234**, 123.
428. P. W. Clark, H. J. Dyke, S. F. Dyke and G. Perry, *J. Organomet. Chem.*, 1983, **253**, 399.

429. N. Barr, S. F. Dyke, G. Smith, C. H. L. Kennard and V. McKee, *J. Organomet. Chem.*, 1985, **288**, 109.
430. T. C. Jones, A. J. Nielson and C. E. Rickard, *Aust. J. Chem.*, 1984, **37**, 2179.
431. A. J. Deeming, M. Nafees Meah, P. A. Bates and M. B. Hursthouse, *J. Chem. Soc., Dalton Trans.*, 1988, 2193.
432. S. Affolter and P. S. Pregosin, *J. Organomet. Chem.*, 1990, **398**, 197.
433. P. S. Pregosin and F. Wombacher, *Magn. Res. Chem.*, 1991, **29**, S106.
434. N. Beydoun, M. Pfeffer, A. De Cian and J. Fischer, *Organometallics*, 1991, **10**, 3693.
435. O. A. Zalevskaya, V. V. Dunina, V. M. Potapov, L. G. Kuz'mina and Y. T. Struchkov, *Zh. Obshch. Khim.*, 1985, **55**, 1332.
436. V. V. Dunina, O. A. Zalevskaya, I. P. Smolyakova and V. M. Potapov, *Zh. Obshch. Khim.*, 1986, **56**, 674.
437. A. D. Ryabov, R. van Eldik, G. Le Borgne and M. Pfeffer, *Organometallics*, 1993, **12**, 1386.
438. G. Wu, A. L. Rheingold, S. J. Geib and R. F. Heck, *Organometallics*, 1987, **6**, 1941.
439. K. Tani *et al.*, *J. Am. Chem. Soc.*, 1977, **99**, 7876.
440. D. G. Allen, G. M. McLaughlin, G. B. Robertson, W. L. Steffen, G. Salem and S. B. Wild, *Inorg. Chem.*, 1982, **21**, 1007.
441. J. W. L. Martin, J. A. L. Palmer and S. B. Wild, *Inorg. Chem.*, 1984, **23**, 2664.
442. X. C. Wang, Y. X. Cui, T. C. W. Mak and H. N. C. Wong, *J. Chem. Soc., Chem. Commun.*, 1990, 167.
443. S. Y. M. Chooi, T. S. A. Hor, P.-H. Leung and K. F. Mok, *Inorg. Chem.*, 1992, **31**, 1494.
444. P. H. Leung, G. M. McLaughlin, J. W. L. Martin and S. B. Wild, *Inorg. Chem.*, 1986, **25**, 3392.
445. A. Avshu, R. D. O'Sullivan, A. W. Parkins, N. W. Alcock and R. M. Countryman, *J. Chem. Soc., Dalton Trans.*, 1983, 1619.
446. L. G. Kuz'mina, O. Y. Burtseva, M. A. Porai-Koshits, V. V. Dunina, O. A. Zalevskaya and V. M. Potapov, *Zh. Obshch. Khim.*, 1989, **59**, 2525.
447. Y. Fuchita and H. Tsuchiya, *Inorg. Chim. Acta*, 1993, **209**, 229.
448. W. Kitching, in 'Comprehensive Organic Chemistry', eds. B. M. Trost, I. Fleming and S. V. Ley, Pergamon, Oxford, 1991, vol. 7, pp. 630–1.
449. B. L. Shaw and I. Shepherd, *J. Chem. Soc., Dalton Trans.*, 1979, 1634.
450. A. J. Nielson, *J. Chem. Soc., Dalton Trans.*, 1981, 205.
451. A. K. Yatsimirsky, G. M. Kazankov and A. D. Ryabov, *J. Chem. Soc., Perkin Trans. 2*, 1992, 1295.
452. M. Nonoyama and K. Hamamura, *J. Organomet. Chem.*, 1991, **407**, 271.
453. J. C. Gaunt and B. L. Shaw, *J. Organomet. Chem.*, 1975, **102**, 511.
454. T. Izumi, M. Maemura, K. Endoh, T. Oikawa, S. Zakosi and A. Kasahara, *Bull. Chem. Soc. Jpn.*, 1981, **54**, 836.
455. C. Lopez, J. Sales, X. Solans and R. Zquiak, *J. Chem. Soc., Dalton Trans.*, 1992, 2321.
456. I. R. Butler, *Organometallics*, 1992, **11**, 74.
457. V. I. Sokolov, L. L. Troitskaya and N. S. Kruschova, *J. Organomet. Chem.*, 1983, **250**, 439.
458. J. M. Vila, M. T. Pereira, E. Gayoso and M. Gayoso, *Transition Met. Chem.*, 1986, **11**, 342.
459. J. M. Vila, A. Suarez, M. T. Pereira, E. Gayoso and M. Gayoso, *Polyhedron*, 1987, **6**, 1003.
460. M. T. Pereira, J. M. Vila, A. Suarez, E. Gayoso and M. Gayoso, *Gazz. Chim. Ital.*, 1988, **118**, 783.
461. G. Z. Wu, S. J. Geib, A. L. Rheingold and R. F. Heck, *J. Org. Chem.*, 1988, **53**, 3238.
462. M. C. Navarro-Ranninger, A. Alvarez-Valdes, M. J. Camazon, J. Roman and R. Lozano, *J. Organomet. Chem.*, 1987, **331**, 107.
463. J. L. Garcia-Ruano, I. Lopez-Solera, J. R. Masaguer, C. Navarro-Ranninger, J. H. Rodriguez and S. Martinez-Carrera, *Organometallics*, 1992, **11**, 3013.
464. M. J. Baena, P. Espinet, M. B. Ros and J. L. Serrano, *Angew. Chem., Int. Ed. Engl.*, 1991, **30**, 711.
465. A. Crispini, G. de Munno, M. Ghedini and F. Neve, *J. Organomet. Chem.*, 1992, **427**, 409.
466. J. M. Vila *et al.*, *Polyhedron*, 1993, **12**, 171.
467. P. S. Pregosin and R. Ruedi, *J. Organomet. Chem.*, 1984, **273**, 401.
468. J. M. Vila, A. Suarez, M. T. Pereira, E. Gayoso and M. Gayoso, *Synth. React. Inorg. Metal-Org. Chem.*, 1988, **18**, 47.
469. S. F. Dyke and S. N. Quessy, *Transition Met. Chem.*, 1982, **7**, 233.
470. R. Grigg and J. Devlin, *J. Chem. Soc., Chem. Commun.*, 1986, 631.
471. A. L. Rheingold and D. L. Staley, *Acta Crystallogr., Part C*, 1988, **44**, 572.
472. J. M. Vila, M. Gayoso, M. T. Pereira, A. Romar, J. J. Fernandez and M. Thornton-Pett, *J. Organomet. Chem.*, 1991, **401**, 385.
473. J. M. Vila *et al.*, *Inorg. Chim. Acta*, 1991, **179**, 171.
474. J. Albert, J. Granell, J. Sales and X. Solans, *J. Organomet. Chem.*, 1989, **379**, 177.
475. D. A. Widdowson and I. R. Girling, *Tetrahedron Lett.*, 1982, **23**, 4281.
476. J. M. Vila, M. Gayoso, M. T. Pereira, M. Lopez, G. Alonso and J. J. Fernandez, *J. Organomet. Chem.*, 1993, **445**, 287.
477. J. Albert, J. Granell and J. Sales, *J. Organomet. Chem.*, 1984, **273**, 393.
478. P. W. Clark, S. F. Dyke, G. Smith and C. H. L. Kennard, *J. Organomet. Chem.*, 1987, **330**, 447.
479. J. Albert, J. Granell and J. Sales, *Synth. React. Inorg. Metal-Org. Chem.*, 1989, **19**, 1009.
480. M. A. Cinellu, M. L. Ganadu, G. Minghetti, F. Cariati, F. Demartin and M. Manassero, *Inorg. Chim. Acta*, 1988, **143**, 197.
481. M. A. Cinellu, G. Minghetti, G. Banditelli, A. L. Bandini, B. Pelli and P. Traldi, *Inorg. Chim. Acta*, 1989, **161**, 57.
482. M. A. Cinellu, S. Gladiali, G. Minghetti, S. Stoccoro and F. Demartin, *J. Organomet. Chem.*, 1991, **401**, 371.
483. C. G. Anklin and P. S. Pregosin, *J. Organomet. Chem.*, 1983, **243**, 101.
484. A. Albinati, C. G. Anklin, F. Ganazzoli, H. Ruegg and P. S. Pregosin, *Inorg. Chem.*, 1987, **26**, 503.
485. F. Ozawa, I. Yamagami, M. Nakano, F. Fujisawa and A. Yamamoto, *Chem. Lett.*, 1989, 125.
486. G. R. Newkome, K. J. Theriot, B. K. Cheskin, D. W. Evans and G. R. Baker, *Organometallics*, 1990, **9**, 1375.
487. G. Balavoine, J. C. Clinet, P. Zerbib and K. Boubekeur, *J. Organomet. Chem.*, 1990, **389**, 259.
488. A. G. Constable, W. S. McDonald and B. L. Shaw, *J. Chem. Soc., Dalton Trans.*, 1980, 2282.
489. A. Albinati, S. Affolter and P. S. Pregosin, *J. Organomet. Chem.*, 1990, **395**, 231.
490. M. Ghedini, S. Morrone, G. De Munno and A. Crispini, *J. Organomet. Chem.*, 1991, **415**, 281.
491. C. A. O'Mahoney, I. P. Parkin, D. J. Williams and J. D. Woollins, *J. Chem. Soc., Dalton Trans.*, 1989, 1179.
492. P. Espinet, G. Garcia, F. J. Herrero, Y. Jeannin and M. Philoche-Levisalles, *Inorg. Chem.*, 1989, **28**, 4207.
493. J. Granell, R. Moragas, J. Sales, M. Font-Bardia and X. Solans, *J. Organomet. Chem.*, 1992, **431**, 359.
494. M. A. Gutierrez, G. R. Newkome and J. Selbin, *J. Organomet. Chem.*, 1980, **202**, 341.
495. J. Selbin and M. A. Gutierrez, *J. Organomet. Chem.*, 1981, **214**, 253.

496. J. Selbin, K. Abboud, S. F. Watkins, M. A. Gutierrez and F. R. Fronczek, *J. Organomet. Chem.*, 1983, **241**, 259.
497. J. Selbin and M. A. Gutierrez, *J. Organomet. Chem.*, 1983, **246**, 95.
498. C. A. Craig and R. J. Watts, *Inorg. Chem.*, 1989, **28**, 309.
499. K. Hiraki, Y. Fuchita, H. Nakaya and S. Takakura, *Bull. Chem. Soc. Jpn.*, 1979, **52**, 2531.
500. W. Hiller, A. Castineiras, O. Gonzalez, M. Gayoso and J. M. Vila, *Acta Crystallogr., Part C*, 1989, **45**, 1431.
501. M. R. Churchill, H. J. Wasserman and G. J. Young, *Inorg. Chem.*, 1980, **19**, 762.
502. K. Hiraki, Y. Fuchita and S. Takakura, *J. Organomet. Chem.*, 1981, **210**, 273.
503. K. Hiraki, Y. Fuchita, M. Nakashima and H. Hiraki, *Bull. Chem. Soc. Jpn.*, 1986, **59**, 3073.
504. E. C. Constable, *J. Chem. Soc., Dalton Trans.*, 1985, 1719.
505. M. Ghedini and D. Pucci, *J. Organomet. Chem.*, 1990, **395**, 105.
506. M. Ghedini, D. Pucci, G. De Munno, D. Viterbo, F. Neve and S. Armentano, *Chem. Mater.*, 1991, **3**, 65.
507. G. B. Caygill, R. M. Hartshorn and P. J. Steel, *J. Organomet. Chem.*, 1990, **382**, 455.
508. M. Nonoyama, *Transition Met. Chem.*, 1987, **12**, 1.
509. M. Nonoyama, *J. Organomet. Chem.*, 1984, **262**, 407.
510. M. Nonoyama and S. Kajita, *Transition Met. Chem.*, 1981, **6**, 163.
511. E. C. Constable and L. R. Sousa, *J. Organomet. Chem.*, 1992, **427**, 125.
512. J. A. C. Allison, H. S. El Khadem and C. A. Wilson, *J. Heterocycl. Chem.*, 1975, **12**, 1275.
513. A. A. Watson, D. A. House and P. J. Steel, *J. Organomet. Chem.*, 1986, **311**, 387.
514. G. B. Caygill and P. J. Steel, *J. Organomet. Chem.*, 1987, **327**, 115.
515. M. Nonoyama, *J. Organomet. Chem.*, 1982, **229**, 287.
516. F. L. Wimmer and S. Wimmer, *Transition Met. Chem.*, 1985, **10**, 238.
517. P. Castan, J. Jaud, S. Wimmer and F. L. Wimmer, *J. Chem. Soc., Dalton Trans.*, 1991, 1155.
518. A. Suarez *et al.*, *Z. Anorg. Allg. Chem.*, 1986, **535**, 213.
519. G. B. Caygill and P. J. Steel, *J. Organomet. Chem.*, 1990, **395**, 375.
520. E. C. Constable, R. P. G. Henney, P. R. Raithby and L. R. Sousa, *J. Chem. Soc., Dalton Trans.*, 1992, 2251.
521. E. C. Constable, R. P. G. Henney, T. A. Leese and D. A. Tocher, *J. Chem. Soc., Dalton Trans.*, 1990, 443.
522. E. C. Constable, R. P. G. Henney, T. A. Leese and D. A. Tocher, *J. Chem. Soc., Chem. Commun.*, 1990, 513.
523. E. C. Constable, R. P. G. Henney and D. A. Tocher, *J. Chem. Soc., Dalton Trans.*, 1992, 2467.
524. A. J. Blake, C. O. Dietrich-Buchecker, T. I. Hyde, J.-P. Sauvage and M. Schroder, *J. Chem. Soc., Chem. Commun.*, 1989, 1663.
525. R. A. Berthon, S. B. Colbran and D. C. Craig, *Polyhedron*, 1992, **11**, 243.
526. M. G. Clerici, B. L. Shaw and B. Weeks, *J. Chem. Soc., Chem. Commun.*, 1973, 516.
527. G. Bombieri *et al.*, *Chem. Commun.*, 1971, 1415.
528. M. Nonoyama, *Inorg. Chim. Acta*, 1988, **145**, 53.
529. H. Yang, M. A. Khan and K. M. Nicholas, *J. Chem. Soc., Chem. Commun.*, 1992, 210.
530. E. Steiner and F. A. L'Eplatteria, *Helv. Chim. Acta*, 1978, **61**, 2264.
531. A. K. Mahapatra, D. Bandyopadhyay, P. Bandyopadhyay and A. Chakravorty, *Inorg. Chem.*, 1986, **25**, 2214.
532. S. Chattopadhyay, C. Sinha, P. Basu and A. Chakravorty, *Organometallics*, 1991, **10**, 1135.
533. C. K. Pal, S. Chattopadhyay, C. Sinha and A. Chakravorty, *J. Organomet. Chem.*, 1992, **439**, 91.
534. S. Chattopadhyay, C. Sinha, P. Basu and A. Chakravorty, *J. Organomet. Chem.*, 1991, **414**, 421.
535. C. Sinha, D. Bandyopadhyay and A. Chakravorty, *Inorg. Chem.*, 1988, **27**, 1173.
536. S. Chattopadhyay, C. Sinha, S. B. Choudhury and A. Chakravorty, *J. Organomet. Chem.*, 1992, **427**, 111.
537. N. Steiner, U. Nagel and W. Beck, *Chem. Ber.*, 1988, **121**, 1759.
538. D. Hedden, D. M. Roundhill, W. C. Fultz and A. L. Rheingold, *Organometallics*, 1986, **5**, 336.
539. C. J. Moulton and B. L. Shaw, *J. Chem. Soc., Dalton Trans.*, 1976, 1020.
540. H. Rimml and L. M. Venanzi, *J. Organomet. Chem.*, 1983, **259**, C6.
541. M. W. Haenel, D. Jakubik, C. Kruger and P. Betz, *Chem. Ber.*, 1991, **124**, 333.
542. A. Jouaiti, M. Geoffroy, G. Terron and G. Bernardinelli, *J. Chem. Soc., Chem. Commun.*, 1992, 155.
543. J. Errington, W. S. McDonald and B. L. Shaw, *J. Chem. Soc., Dalton Trans.*, 1980, 2312.
544. G. R. Giesbrecht, G. S. Hanan, J. E. Kickham and S. J. Loeb, *Inorg. Chem.*, 1992, **31**, 3286.
545. G. J. Palenik and T. J. Giordano, *J. Chem. Soc., Dalton Trans.*, 1987, 1175.
546. K. Hiraki, J. Tsutsumida and Y. Fuchita, *Chem. Lett.*, 1986, 337.
547. S. Trofimenko, *J. Am. Chem. Soc.*, 1971, **93**, 1808.
548. S. Chakladar *et al.*, *J. Chem. Soc., Dalton Trans.*, 1992, 3119.
549. S. Chakladar, P. Paul and K. Nag, *Polyhedron*, 1991, **10**, 1513.
550. K. K. Nanda, K. Nag, K. Venkatsubramanian and P. Paul, *Inorg. Chim. Acta*, 1992, **196**, 195.
551. S. Chakladar, P. Paul, K. Venkatsubramanian and K. Nag, *J. Chem. Soc., Dalton Trans.*, 1991, 2669.
552. I. G. Phillips and P. J. Steel, *J. Organomet. Chem.*, 1991, **410**, 247.
553. P. J. Steel and G. B. Caygill, *J. Organomet. Chem.*, 1990, **395**, 359.
554. J. M. Vila, M. Gayoso, M. T. Pereira, M. C. Rodriguez, J. M. Ortigueira and M. Thornton-Pett, *J. Organomet. Chem.*, 1992, **426**, 267.
555. J. Granell *et al.*, *J. Chem. Soc., Dalton Trans.*, 1983, 2441.
556. J. Granell, J. Sales and J. Vilarrasa, *Transition Met. Chem.*, 1984, **9**, 203.
557. R. M. Ceder, J. Sales, X. Solans and M. Font-Altaba, *J. Chem. Soc., Dalton Trans.*, 1986, 1351.
558. J. L. Casas, E. Gayoso, J. M. Vila, M. T. Pereira and M. Gayoso, *Synth. React. Inorg. Metal-Org. Chem.*, 1991, **21**, 263.
559. P. Espinet, J. Etxebarria, M. Marcos, J. Perez, A. Remon and J. L. Serrano, *Angew. Chem., Int. Ed. Engl.*, 1989, **28**, 1065.
560. P. Espinet, E. Lalinde, M. Marcos, J. Perez and J. L. Serrano, *Organometallics*, 1990, **9**, 555.
561. J. G. Hughes and R. Robson, *Inorg. Chim. Acta*, 1979, **35**, 87.
562. R. Robson, *Inorg. Chim. Acta*, 1982, **57**, 71.
563. T. E. Crossley, P. Davies, T. N. Huckerby, M. Louey and R. Robson, *Inorg. Chim. Acta*, 1984, **85**, 199.
564. B. F. Hoskins and I. A. S. McDonald, *Aust. J. Chem.*, 1984, **37**, 1395.
565. A. Albinati, S. Affolter and P. S. Pregosin, *Organometallics*, 1990, **9**, 379.
566. K. Hiraki, Y. Fuchita and T. Uchiyama, *Inorg. Chim. Acta*, 1983, **69**, 187.

567. B. L. Shaw, *J. Organomet. Chem.*, 1980, **200**, 307.
568. S. Hietkamp, D. J. Stufkens and K. Vrieze, *J. Organomet. Chem.*, 1979, **168**, 351.
569. A. J. Deeming, I. P. Rothwell, M. B. Hursthouse and K. M. Abdul Malik, *J. Chem. Soc., Dalton Trans.*, 1980, 1974.
570. R. Ruger, W. Rittner, P. G. Jones, W. Isenberg and G. M. Sheldrick, *Angew. Chem., Int. Ed. Engl.*, 1981, **20**, 382.
571. A. Guarnieri and A. W. Parkins, *J. Organomet. Chem.*, 1989, **361**, 399.
572. B. Bogdanovic, C. Kruger and P. Locatelli, *Angew. Chem., Int. Ed. Engl.*, 1979, **18**, 684.
573. H. Mizuno and M. Nonoyama, *Polyhedron*, 1990, **9**, 1287.
574. M. Nonoyama and K. Nonoyama, *Polyhedron*, 1991, **10**, 2265.
575. H. Mizuno, M. Kita, J. Fujita and M. Nonoyama, *Inorg. Chim. Acta*, 1992, **202**, 183.
576. M. Nonoyama and K. Hamamura, *J. Organomet. Chem.*, 1991, **407**, 271.
577. J. Dupont and M. Pfeffer, *J. Organomet. Chem.*, 1987, **321**, C13.
578. Y. Fuchita, K. Hiraki, T. Yamaguchi and T. Maruta, *J. Chem. Soc., Dalton Trans.*, 1981, 2405.
579. P. J. Steel and G. B. Caygill, *J. Organomet. Chem.*, 1987, **327**, 101.
580. A. J. Canty, P. R. Traill, R. Colton and I. M. Thomas, *Inorg. Chim. Acta*, 1993, **210**, 91.
581. P. M. Maitlis, P. Espinet and M. J. H. Russell, in 'COMC-I', vol. 6, p. 351; 363.
582. J.-E. Backvall, *Acc. Chem. Res.*, 1983, **16**, 335.
583. N. Oshima, Y. Hamatani, H. Fukui, H. Suzuki and Y. Moro-oka, *J. Organomet. Chem.*, 1986, **303**, C21.
584. B. Akermark and K. Zetterberg, *J. Am. Chem. Soc.*, 1984, **106**, 5560.
585. R. Arnek and K. Zetterberg, *Organometallics*, 1987, **6**, 1230.
586. L. Zhang and K. Zetterberg, *Organometallics*, 1991, **10**, 3806.
587. H. Kurosawa, T. Majima and N. Asada, *J. Am. Chem. Soc.*, 1980, **102**, 6996.
588. J. K. Stille and D. E. James, *J. Organomet. Chem.*, 1976, **108**, 401.
589. L. S. Hegedus, B. Akermark, K. Zetterberg and L. F. Olsson, *J. Am. Chem. Soc.*, 1984, **106**, 7122.
590. M. Akbarzadeh and C. B. Anderson, *J. Organomet. Chem.*, 1980, **197**, C5.
591. E. Rotondo, F. C. Priolo, A. Donato and R. Pietropaolo, *J. Organomet. Chem.*, 1983, **251**, 273.
592. G. R. Wiger, S. S. Tomita, M. F. Rettig and R. M. Wing, *Organometallics*, 1985, **4**, 1157.
593. D. J. Evans and L. A. P. Kane-Maguire, *J. Organomet. Chem.*, 1986, **312**, C24.
594. C. Moberg, L. Sutin and A. Heumann, *Acta Chem. Scand.*, 1991, **45**, 77.
595. G. A. Lane, W. E. Geiger and N. G. Connelly, *J. Am. Chem. Soc.*, 1987, **109**, 402.
596. L. L. Wright, R. M. Wing and M. F. Rettig, *J. Am. Chem. Soc.*, 1982, **104**, 610.
597. O. Eisenstein and R. Hoffmann, *J. Am. Chem. Soc.*, 1981, **103**, 4308.
598. A. D. Cameron, V. H. Smith, Jr. and M. C. Baird, *J. Chem. Soc., Dalton Trans.*, 1988, 1037.
599. P. Sautet, O. Eisenstein and K. M. Nicholas, *Organometallics*, 1987, **6**, 1845.
600. M. F. Rettig, D. E. Wilcox and R. S. Fleischer, *J. Organomet. Chem.*, 1981, **214**, 261.
601. M. Parra-Hake, M. F. Rettig, R. M. Wing and J. C. Woolcock, *Organometallics*, 1982, **1**, 1478.
602. M. Parra-Hake, M. F. Rettig, J. L. Williams and R. M. Wing, *Organometallics*, 1986, **5**, 1032.
603. A. C. Cope, J. M. Kliegman and E. C. Friedrich, *J. Am. Chem. Soc.*, 1967, **89**, 287.
604. R. A. Holton and R. A. Kjonaas, *J. Am. Chem. Soc.*, 1977, **99**, 4177.
605. R. McCrindle, G. Ferguson, M. A. Khan, A. J. McAlees and B. L. Ruhl, *J. Chem. Soc., Dalton Trans.*, 1981, 986.
606. R. McCrindle and A. J. McAlees, *J. Chem. Soc., Dalton Trans.*, 1983, 127.
607. R. McCrindle, E. C. Alyea, G. Ferguson, S. A. Dias, A. J. McAlees and M. Parvez, *J. Chem. Soc., Dalton Trans.*, 1980, 137.
608. K. Isomura, N. Okada, M. Saruwatari, H. Yamasaki and H. Taniguchi, *Chem. Lett.*, 1985, 385.
609. R. A. Holton and J. R. Zoeller, *J. Am. Chem. Soc.*, 1985, **107**, 2124.
610. M. Lenarda, G. Nardin, G. Pellizer, E. Braye and M. Graziani, *J. Chem. Soc., Chem. Commun.*, 1985, 1536.
611. A. C. Albeniz, P. Espinet, Y. Jeannin, M. Philoche-Levisalles and B. E. Mann, *J. Am. Chem. Soc.*, 1990, **112**, 6594.
612. M. A. Andrews, T. C.-T. Chang, C.-W. F. Cheng, T. J. Emge, K. P. Kelly and T. F. Koetzle, *J. Am. Chem. Soc.*, 1984, **106**, 5913.
613. F. Balegroune, D. Grandjean, D. Lakkis and D. Matt, *J. Chem. Soc., Chem. Commun.*, 1992, 1084.
614. C. H. Chao, D. W. Hart, R. Bau and R. Heck, *J. Organomet. Chem.*, 1979, **179**, 301.
615. A. D. Ryabov, I. K. Sakodinskaya and A. K. Yatsimirsky, *J. Chem. Soc., Perkin Trans. 2*, 1983, 1511.
616. A. D. Ryabov, I. K. Sakodinskaya and A. K. Yatsimirsky, *J. Organomet. Chem.*, 1991, **406**, 309.
617. N. Barr, J. P. Bartley, P. W. Clark, P. Dunstan and S. F. Dyke, *J. Organomet. Chem.*, 1986, **302**, 117.
618. A. Albinati, P. S. Pregosin and R. Ruedi, *Helv. Chim. Acta*, 1985, **68**, 2046.
619. F. Maassarani, M. Pfeffer, G. Le Borgne and D. Grandjean, *Organometallics*, 1986, **5**, 1511.
620. M. Zocchi, G. Tieghi and A. Albinati, *J. Organomet. Chem.*, 1971, **33**, C47.
621. R. P. Hughes and J. Powell, *J. Organomet. Chem.*, 1971, **30**, C45.
622. J. A. Sadownick and S. J. Lippard, *Inorg. Chem.*, 1973, **12**, 2659.
623. R. Benn *et al.*, *Organometallics*, 1985, **4**, 1945.
624. P. Binger, H. M. Buch, R. Benn and R. Mynott, *Angew. Chem., Int. Ed. Engl.*, 1982, **21**, 62.
625. H. M. Buch and C. Kruger, *Acta Crystallogr., Part C*, 1984, **40**, 28.
626. D. L. Reger and D. G. Garza, *Organometallics*, 1993, **12**, 554.
627. H. Kurosawa, A. Urabe, K. Miki and N. Kasai, *Organometallics*, 1986, **5**, 2002.
628. A. C. Albeniz and P. Espinet, *Organometallics*, 1991, **10**, 2987.
629. J. S. Brumbaugh, R. R. Whittle, M. Parvez and A. Sen, *Organometallics*, 1990, **9**, 1735.
630. D. L. Thorn and R. Hoffmann, *J. Am. Chem. Soc.*, 1978, **100**, 2079.
631. F. Ozawa, T. Hayashi, H. Koide and A. Yamamoto, *J. Chem. Soc., Chem. Commun.*, 1991, 1469.
632. G. P. C. M. Dekker, C. J. Elsevier, K. Vrieze, P. W. N. M. van Leeuwen and C. F. Roobeek, *J. Organomet. Chem.*, 1992, **430**, 357.
633. B. A. Markies, M. H. P. Rietveld, J. Boersma, A. L. Spek and G. van Koten, *J. Organomet. Chem.*, 1992, **424**, C12.
634. M. Brookhart, F. C. Rix, J. M. De Simone and J. C. Barborak, *J. Am. Chem. Soc.*, 1992, **114**, 5894.
635. R. van Asselt, E. E. C. G. Gielens, R. E. Rulke and C. J. Elsevier, *J. Chem. Soc., Chem. Commun.*, 1993, 1203.
636. U. Hacksell, H. T. Kalinkoski, D. F. Barofsky and G. D. Daves, Jr., *Acta Chem. Scand., Ser. B*, 1985, **39**, 469.

637. R. C. Larock, S. S. Hershberger, K. Takagi and M. A. Mitchell, *J. Org. Chem.*, 1986, **51**, 2450.
638. H. Horino, M. Arai and N. Inoue, *Tetrahedron Lett.*, 1974, 647.
639. M. Catellani and G. P. Chiusoli, *J. Organomet. Chem.*, 1988, **346**, C27.
640. M. Catellani and G. P. Chiusoli, *J. Organomet. Chem.*, 1992, **425**, 151.
641. M. Catellani and G. P. Chiusoli, *J. Organomet. Chem.*, 1992, **437**, 369.
642. C.-S. Li, C.-H. Cheng, F.-L. Liao and S.-L. Wang, *J. Chem. Soc., Chem. Commun.*, 1991, 710.
643. C.-S. Li, D.-C. Jou, C.-H. Cheng, F.-L. Liao and S.-L. Wang, *Organometallics*, 1993, **12**, 3553.
644. P. M. Maitlis, P. Espinet and M. J. H. Russell, in 'COMC-I', vol. 6, p. 456.
645. P. M. Maitlis, *J. Organomet. Chem.*, 1980, **200**, 161.
646. C.-L. Lee, C. T. Hunt and A. L. Balch, *Inorg. Chem.*, 1981, **20**, 2498.
647. S. J. Higgins and B. L. Shaw, *J. Chem. Soc., Dalton Trans.*, 1988, 457.
648. J. T. Mague, *Organometallics*, 1986, **5**, 918.
649. Y. Xie, C.-L. Lee, Y. P. Yang, S. J. Rettig and B. R. James, *Can. J. Chem.*, 1992, **70**, 751.
650. Y. Fuchita, K. I. Hardcastle, K. Hiraki and M. Kawatani, *Bull. Chem. Soc. Jpn.*, 1990, **63**, 1961.
651. Y. Fuchita, H. Maruyama, M. Kawatani and K. Hiraki, *Polyhedron*, 1991, **10**, 561.
652. H. Werner, P. Thometzek, K. Zenkert, R. Goddard and H.-J. Kraus, *Chem. Ber.*, 1987, **120**, 365.
653. H. C. Clark and C. R. Milne, *J. Organomet. Chem.*, 1978, **161**, 51.
654. L. A. Villanueva, K. A. Abboud and J. M. Boncella, *Organometallics*, 1992, **11**, 2963.
655. T. Yukawa and S. Tsutsumi, *Inorg. Chem.*, 1968, **7**, 1458.
656. R. McCrindle, G. Ferguson, G. J. Arsenault, A. J. McAlees and M. Parvez, *J. Organomet. Chem.*, 1983, **246**, C19.
657. T. Ito, S. Hasegawa, Y. Takahashi and Y. Ishii, *J. Organomet. Chem.*, 1974, **73**, 401.
658. H. Suzuki, K. Itoh, Y. Ishii, K. Simon and J. A. Ibers, *J. Am. Chem. Soc.*, 1976, **98**, 8494.
659. H. tom Dieck, C. Munz and C. Muller, *J. Organomet. Chem.*, 1990, **384**, 243.
660. L. D. Brown, K. Itoh, H. Suzuki, K. Hirai and J. A. Ibers, *J. Am. Chem. Soc.*, 1978, **100**, 8232.
661. K. Itoh, K. Hirai, M. Sasaki, Y. Nakamura and H. Nishiyama, *Chem. Lett.*, 1981, 865.
662. C. Munz, C. Stephan and H. tom Dieck, *J. Organomet. Chem.*, 1990, **395**, C42.
663. C. Munz, C. Stephan and H. tom Dieck, *J. Organomet. Chem.*, 1991, **407**, 413.
664. D. R. Russell and P. A. Tucker, *J. Chem. Soc., Dalton Trans.*, 1975, 1743.
665. F. Balegroune, P. Braunstein, T. M. G. Carneiro, D. Grandjean and D. Matt, *J. Chem. Soc., Chem. Commun.*, 1989, 582.
666. P. Braunstein, T. M. G. Carneiro, D. Matt, F. Balegroune and D. Grandjean, *Organometallics*, 1989, **8**, 1737.
667. T. G. Appleton, H. C. Clark, R. C. Poller and R. J. Puddephatt, *J. Organomet. Chem.*, 1972, **39**, C13.
668. Y. Tohda, K. Sonogashira and N. Hagihara, *J. Chem. Soc., Chem. Commun.*, 1975, 54.
669. H. C. Clark, C. R. C. Milne and N. C. Payne, *J. Am. Chem. Soc.*, 1978, **100**, 1164.
670. N. Chaudhury, M. G. Kekre and R. J. Puddephatt, *J. Organomet. Chem.*, 1974, **73**, C17.
671. H. C. Clark, C. R. C. Milne and C. S. Wong, *J. Organomet. Chem.*, 1977, **136**, 265.
672. K. Hiraki, T. Itoh, K. Eguchi and M. Onishi, *J. Organomet. Chem.*, 1983, **241**, C16.
673. M. Pfeffer, *Pure Appl. Chem.*, 1992, **64**, 335.
674. W. Tao, L. J. Silverberg, A. L. Rheingold and R. F. Heck, *Organometallics*, 1989, **8**, 2550.
675. A. Bahsoun, J. Dehand, M. Pfeffer, M. Zinsius, S.-E. Bouaoud and G. Le Borgne, *J. Chem. Soc., Dalton Trans.*, 1979, 547.
676. C. Arlen, M. Pfeffer, O. Bars and G. Le Borgne, *J. Chem. Soc., Dalton Trans.*, 1986, 359.
677. F. Maassarani, M. Pfeffer and G. Le Borgne, *Organometallics*, 1987, **6**, 2029.
678. F. Maassarani, M. Pfeffer and G. Le Borgne, *Organometallics*, 1987, **6**, 2043.
679. M. Pfeffer, M. A. Rotteveel, G. Le Borgne and J. Fischer, *J. Org. Chem.*, 1992, **57**, 2147.
680. M. Pfeffer, M. A. Rotteveel, J.-P. Sutter, A. De Cian and J. Fischer, *J. Organomet. Chem.*, 1989, **371**, C21.
681. J.-P. Sutter, M. Pfeffer, A. De Cian and J. Fischer, *Organometallics*, 1992, **11**, 386.
682. G. Wu, A. L. Rheingold and R. F. Heck, *Organometallics*, 1986, **5**, 1922.
683. A. L. Rheingold, G. Wu and R. F. Heck, *Inorg. Chim. Acta*, 1987, **131**, 147.
684. M. Pfeffer, J.-P. Sutter, A. De Cian and J. Fischer, *Organometallics*, 1993, **12**, 1167.
685. F. Maassarani, M. Pfeffer and G. van Koten, *Organometallics*, 1989, **8**, 871.
686. F. Maassarani, M. Pfeffer, G. Le Borgne, E. Wehman and G. van Koten, *J. Am. Chem. Soc.*, 1984, **106**, 8002.
687. P. F. Engel, M. Pfeffer, J. Fischer and A. Dedieu, *J. Chem. Soc., Chem. Commun.*, 1991, 1274.
688. L. Huang, F. Ozawa, K. Osakada and A. Yamamoto, *Organometallics*, 1989, **8**, 2065.
689. J. Vicente, A. Arcas, M. V. Borrachero, A. Tiripicchio and M. T. Camellini, *Organometallics*, 1991, **10**, 3873.
690. R. Uson, J. Fornies, M. Tomas, J. M. Casas and R. Navarro, *J. Chem. Soc., Dalton Trans.*, 1989, 169.
691. R. Uson, J. Fornies, M. Tomas and B. Menjon, *Organometallics*, 1985, **4**, 1912.
692. J. Dupont, M. Pfeffer, J. C. Daran and Y. Jeannin, *Organometallics*, 1987, **6**, 899.
693. J. Dupont and M. Pfeffer, *J. Chem. Soc., Dalton Trans.*, 1990, 3193.
694. M. A. Cinellu, S. Gladiali and G. Minghetti, *J. Organomet. Chem.*, 1989, **363**, 401.
695. E. L. Weinberg, B. K. Hunter and M. C. Baird, *J. Organomet. Chem.*, 1982, **240**, 95.
696. R. D. O'Sullivan and A. W. Parkins, *J. Chem. Soc., Chem. Commun.*, 1984, 1165.
697. F. Ozawa, T.-I. Son, K. Osakada and A. Yamamoto, *J. Chem. Soc., Chem. Commun.*, 1989, 1067.
698. G. P. C. M. Dekker *et al.*, *Organometallics*, 1992, **11**, 1937.
699. G. P. C. M. Dekker, C. J. Elsevier, K. Vrieze and P. W. N. M. van Leeuwen, *Organometallics*, 1992, **11**, 1598.
700. H. Werner and W. Bertleff, *J. Chem. Res. (S)*, 1978, 201.
701. S. Komiya, A. Yamamoto and T. Yamamoto, *Chem. Lett.*, 1981, 193.
702. C.-L. Lee, C. T. Hunt and A. L. Balch, *Organometallics*, 1982, **1**, 824.
703. S. Komiya, Y. Akai, K. Tanaka, T. Yamamoto and A. Yamamoto, *Organometallics*, 1985, **4**, 1130.
704. R. E. Rulke *et al.*, *Inorg. Chim. Acta*, 1990, **169**, 5.
705. F. Ozawa, T. Hayashi, H. Koide and A. Yamamoto, *J. Chem. Soc., Chem. Commun.*, 1991, 1469.
706. V. De Felice, V. G. Albano, C. Castellari, M. E. Cucciolito and A. De Renzi, *J. Organomet. Chem.*, 1991, **403**, 269.
707. K. J. Cavell, H. Jin, B. W. Skelton and A. H. White, *J. Chem. Soc., Dalton Trans.*, 1992, 2923.
708. K. J. Cavell, H. Jin, B. W. Skelton and A. H. White, *J. Chem. Soc., Dalton Trans.*, 1993, 1973.
709. K. Osakada, Y. Ozawa and A. Yamamoto, *Bull. Chem. Soc. Jpn.*, 1991, **64**, 2002.

710. R. van Asselt and C. J. Elsevier, *Organometallics*, 1992, **11**, 1999.
711. R. Bertani, A. Berton, F. Di Bianca and B. Crociani, *J. Organomet. Chem.*, 1986, **303**, 283.
712. F. Ozawa, T. Sugimoto, Y. Yuasa, M. Santra, T. Yamamoto and A. Yamamoto, *Organometallics*, 1984, **3**, 683.
713. J. M. A. Wouters, M. W. Avis, C. J. Elsevier, C. E. Kyriakidis and C. H. Stam, *Organometallics*, 1990, **9**, 2203.
714. F. Ozawa and A. Yamamoto, *Chem. Lett.*, 1981, 289.
715. N. Koga and K. Morokuma, *J. Am. Chem. Soc.*, 1986, **108**, 6136.
716. P. E. Garrou and R. F. Heck, *J. Am. Chem. Soc.*, 1976, **98**, 4115.
717. G. K. Anderson and R. J. Cross, *Acc. Chem. Res.*, 1984, **17**, 67.
718. H. Berke and R. Hoffmann, *J. Am. Chem. Soc.*, 1978, **100**, 7224.
719. S. Sasaki, K. Kitaura, K. Morokuma and K. Ohkubo, *J. Am. Chem. Soc.*, 1983, **105**, 2280.
720. K. Kudo, M. Sato, M. Hidai and Y. Uchida, *Bull. Chem. Soc. Jpn.*, 1973, **46**, 2820.
721. H. Matsuzaka, Y. Hiroe, M. Iwasaki, Y. Ishii, Y. Koyasu and M. Hidai, *J. Org. Chem.*, 1988, **53**, 3832.
722. Y.-J. Kim, K. Osakada, K. Sugita, T. Yamamoto and A. Yamamoto, *Organometallics*, 1988, **7**, 2182.
723. I. Toth and C. J. Elsevier, *J. Chem. Soc., Chem. Commun.*, 1993, 529.
724. L. S. Hegedus *et al.*, *Inorg. Chem.*, 1977, **16**, 1887.
725. F. Ozawa, M. Nakano, I. Aoyama, T. Yamamoto and A. Yamamoto, *J. Chem. Soc., Chem. Commun.*, 1986, 382.
726. R. Bardi, A. M. Piazzesi, G. Cavinato, P. Cavoli and L. Toniolo, *J. Organomet. Chem.*, 1982, **224**, 407.
727. R. Bardi, A. M. Piazzesi, A. Del Pra, G. Cavinato and L. Toniolo, *Inorg. Chim. Acta*, 1985, **102**, 99.
728. G. Cavinato and L. Toniolo, *J. Organomet. Chem.*, 1990, **398**, 187.
729. A. Yamamoto *et al.*, *Pure Appl. Chem.*, 1991, **63**, 687.
730. F. Rivetti and U. Romano, *J. Organomet. Chem.*, 1978, **154**, 323.
731. G. Del Piero and M. Cesari, *Acta Crystallogr., Part B*, 1979, **35**, 2411.
732. M. I. Terekhova, N. E. Petrova, R. R. Shifrina and E. S. Petrov, *Zh. Obshch. Khim.*, 1988, **58**, 658.
733. C. R. Green and R. J. Angelici, *Inorg. Chem.*, 1972, **11**, 2095.
734. L. Huang, F. Ozawa, K. Osakada and A. Yamamoto, *J. Organomet. Chem.*, 1990, **383**, 587.
735. L. Huang, F. Ozawa and A. Yamamoto, *Organometallics*, 1990, **9**, 2603.
736. G. Vasapollo, C. F. Nobile and A. Sacco, *J. Organomet. Chem.*, 1985, **296**, 435.
737. W. Beck and B. Purucker, *J. Organomet. Chem.*, 1976, **112**, 361.
738. G. P. C. M. Dekker, C. J. Elsevier, K. Vrieze, P. W. N. M. van Leeuwen and C. F. Roobeek, *J. Organomet. Chem.*, 1992, **430**, 357.
739. A. Sacco, G. Vasapollo, C. F. Nobile, A. Piergiovanni, M. A. Pellinghelli and M. Lanfranchi, *J. Organomet. Chem.*, 1988, **356**, 397.
740. G. Cavinato and L. Toniolo, *J. Organomet. Chem.*, 1993, **444**, C65.
741. G. D. Smith, B. E. Hanson, J. S. Merola and F. J. Waller, *Organometallics*, 1993, **12**, 568.
742. P. Leconte *et al.*, *J. Chem. Soc., Chem. Commun.*, 1990, 1616.
743. P. Brechot, Y. Chauvin, D. Commereuc and L. Saussine, *Organometallics*, 1990, **9**, 26.
744. B. Crociani, G. Bandoli and D. A. Clemente, *J. Organomet. Chem.*, 1980, **184**, 269.
745. B. Crociani, M. Sala, A. Polo and G. Bombieri, *Organometallics*, 1986, **5**, 1369.
746. B. Crociani and G. Granozzi, *Inorg. Chim. Acta*, 1987, **132**, 197.
747. B. Crociani, A. Mantovani and A. Scrivanti, *J. Organomet. Chem.*, 1982, **233**, 387.
748. T. Tanase, O. Ohizumi, K. Kobayashi and Y. Yamamoto, *J. Chem. Soc., Chem. Commun.*, 1992, 707.
749. R. Uson, J. Fornies, P. Espinet and E. Lalinde, *J. Organomet. Chem.*, 1987, **334**, 399.
750. A. Mantovani, L. Calligaro and A. Pasquetto, *Inorg. Chim. Acta*, 1983, **76**, L145.
751. A. Mantovani and B. Crociani, *J. Organomet. Chem.*, 1982, **236**, C37.
752. A. Mantovani, *Monatsh. Chem.*, 1983, **114**, 1045.
753. A. Campagnaro, A. Mantovani and P. Uguagliati, *Inorg. Chim. Acta*, 1985, **99**, L15.
754. B. Crociani, F. Dibianca, R. Bertani and C. B. Castellani, *Inorg. Chim. Acta*, 1985, **101**, 161.
755. R. Uson *et al.*, *J. Chem. Soc., Dalton Trans.*, 1986, 259.
756. R. Uson, J. Fornies, P. Espinet, E. Lalinde, P. G. Jones and G. M. Sheldrick, *J. Chem. Soc., Dalton Trans.*, 1982, 2389.
757. R. Uson, J. Fornies, P. Espinet, E. Lalinde, P. G. Jones and G. M. Sheldrick, *J. Organomet. Chem.*, 1983, **253**, C47.
758. R. Uson, J. Fornies, P. Espinet and E. Lalinde, *J. Organomet. Chem.*, 1988, **349**, 269.
759. R. Uson, J. Fornies, P. Espinet, E. Lalinde, P. G. Jones and G. M. Sheldrick, *J. Organomet. Chem.*, 1985, **288**, 249.
760. R. Uson, J. Fornies, P. Espinet, L. Pueyo and E. Lalinde, *J. Organomet. Chem.*, 1986, **299**, 251.
761. J. M. A. Wouters, R. A. Klein, C. J. Elsevier, M. C. Zoutberg and C. H. Stam *Organometallics*, 1993, **12**, 3864.
762. Y. Yamamoto and H. Yamazaki, *Inorg. Chim. Acta*, 1980, **41**, 229.
763. Y. Ito, E. Ihara, M. Murakami and M. Shiro, *J. Am. Chem. Soc.*, 1990, **112**, 6446.
764. Y. Ito, E. Ihara and M. Murakami, *Angew. Chem., Int. Ed. Engl.*, 1992, **31**, 1509.
765. K. Matsumoto, Y. Odaira and S. Tsutsumi, *Chem. Commun.*, 1968, 832.
766. J. Ashley-Smith, J. Clemens, M. Green and F. G. A. Stone, *J. Organomet. Chem.*, 1969, **17**, P23.
767. R. McCrindle and D. W. Sneddon, *J. Organomet. Chem.*, 1985, **282**, 413.
768. G. Ferguson and B. L. Ruhl, *Acta Crystallogr., Part C*, 1984, **40**, 2020.
769. G. B. Deacon and I. L. Grayson, *Transition Met. Chem.*, 1983, **8**, 131.
770. G. B. Deacon, S. J. Faulks and G. N. Pain, *Adv. Organomet. Chem.*, 1986, **25**, 237.
771. K. Kikukawa and T. Matsuda, *J. Organomet. Chem.*, 1982, **235**, 243.
772. M. B. Hursthouse, O. D. Sloan, P. Thornton and N. P. C. Walker, *Polyhedron*, 1986, **5**, 1475.
773. Y. Fuchita, K. Hiraki, Y. Kamogawa, M. Suenaga, K. Tohgoh and Y. Fujiwara, *Bull. Chem. Soc. Jpn.*, 1989, **62**, 1081.
774. Y. Fuchita, H. Oka and M. Okamura, *Inorg. Chim. Acta*, 1992, **194**, 213.
775. Y. Fuchita, M. Akiyama, Y. Arimoto, N. Matsumoto and H. Okawa, *Inorg. Chim. Acta*, 1993, **205**, 185.
776. Y. Fuchita, M. Kawakami and K. Shimoke, *Polyhedron*, 1991, **10**, 2037.
777. R. D. W. Kemmitt, S. Mason and D. R. Russell, *J. Organomet. Chem.*, 1991, **415**, C9.
778. R. Kramer, K. Polborn and W. Beck, *J. Organomet. Chem.*, 1992, **441**, 333.
779. H. Schwager, R. Benn and G. Wilke, *Angew. Chem., Int. Ed. Engl.*, 1987, **26**, 67.

780. F. Ozawa, J. W. Park, P. B. Mackenzie, W. P. Schaefer, L. M. Henling and R. H. Grubbs, *J. Am. Chem. Soc.*, 1989, **111**, 1319.
781. J. A. Davies, S. Dutremez, A. A. Pinkerton and M. Vilmer, *Organometallics*, 1991, **10**, 2956.
782. R. McCrindle, G. Ferguson, G. J. Arsenault, A. J. McAlees and D. K. Stephenson, *J. Chem. Res. (S)*, 1984, 360.
783. H. Werner and A. Kuhn, *Angew. Chem., Int. Ed. Engl.*, 1979, **18**, 416.
784. B. Henc *et al.*, *J. Organomet. Chem.*, 1980, **191**, 449.
785. S. Lin and P. Boudjouk, *J. Organomet. Chem.*, 1980, **187**, C11.
786. H. Werner, G. T. Crisp, P. W. Jolly, H.-J. Kraus and C. Kruger, *Organometallics*, 1983, **2**, 1369.
787. R. Benn, P. W. Jolly, R. Mynott and G. Schenker, *Organometallics*, 1985, **4**, 1136.
788. A. Behr, G. v. Ilsemann, W. Keim, C. Kruger and Y.-H. Tsay, *Organometallics*, 1986, **5**, 514.
789. R. Bertani, A. Berton, G. Carturan and R. Campostrini, *J. Organomet. Chem.*, 1988, **349**, 263.
790. J. Krause, W. Bonrath and K. R. Porschke, *Organometallics*, 1992, **11**, 1158.
791. Y. Becker and J. K. Stille, *J. Am. Chem. Soc.*, 1978, **100**, 845.
792. H. Werner, A. Kuhn and C. Burschka, *Chem. Ber.*, 1980, **113**, 2291.
793. P. W. Jolly, R. Mynott, B. Raspel and K.-P. Schick, *Organometallics*, 1986, **5**, 473.
794. G. K. Anderson, R. J. Cross, S. Fallis and M. Rocamora, *Organometallics*, 1987, **6**, 1440.
795. S. Kawaguchi, *Coord. Chem. Rev.*, 1986, **70**, 51.
796. N. Yanase, Y. Nakamura and S. Kawaguchi, *Inorg. Chem.*, 1980, **19**, 1575.
797. K. Yamada, S. Baba, Y. Nakamura and S. Kawaguchi, *Bull. Chem. Soc. Jpn.*, 1983, **56**, 1393.
798. S. Baba, T. Ogura and S. Kawaguchi, *Inorg. Nucl. Chem. Lett.*, 1971, **7**, 1195.
799. H. Werner and H. J. Kraus, *Chem. Ber.*, 1980, **113**, 1072.
800. S. Okeya, Y. Nakamura and S. Kawaguchi, *Bull. Chem. Soc. Jpn.*, 1981, **54**, 3396.
801. A. R. Siedle and L. H. Pignolet, *Inorg. Chem.*, 1981, **20**, 1849.
802. T. Kurokawa, K. Miki, N. Tanaka and N. Kasai, *Bull. Chem. Soc. Jpn.*, 1982, **55**, 45.
803. S. Okeya, H. Yoshimatsu, Y. Nakamura and S. Kawaguchi, *Bull. Chem. Soc. Jpn.*, 1982, **55**, 483.
804. B. K. Sahu and B. K. Mohapatra, *J. Indian Chem. Soc.*, 1982, **59**, 732.
805. A. R. Siedle, R. A. Newmark and L. H. Pignolet, *Inorg. Chem.*, 1983, **22**, 2281.
806. S. Okeya *et al.*, *Bull. Chem. Soc. Jpn.*, 1982, **55**, 2134.
807. K.-C. Kong and C.-H. Cheng, *J. Am. Chem. Soc.*, 1991, **113**, 6313.
808. J. Fornies, C. Fortuno, R. Navarro, F. Martinez and A. J. Welch, *J. Organomet. Chem.*, 1990, **394**, 643.
809. J. Fornies, R. Navarro and E. P. Urriolabeitia, *J. Organomet. Chem.*, 1990, **390**, 257.
810. G. Lopez, G. Garcia, G. Sanchez, M. D. Santana, J. Ruiz and J. Garcia, *Inorg. Chim. Acta*, 1991, **188**, 195.
811. G. Lopez, J. Ruiz, G. Garcia, J. M. Marti, G. Sanchez and J. Garcia, *J. Organomet. Chem.*, 1991, **412**, 435.
812. G. Lopez *et al.*, *Organometallics*, 1992, **11**, 4090.
813. J. Fornies, F. Martinez, R. Navarro and E. P. Urriolabeitia, *Polyhedron*, 1990, **9**, 2181.
814. J. Fornies, R. Navarro, M. Tomas and E. P. Urriolabeitia, *Organometallics*, 1993, **12**, 940.
815. R. Uson, J. Fornies, M. A. Uson and S. Herrero, *J. Organomet. Chem.*, 1993, **447**, 137.
816. P. K. Byers, A. J. Canty, B. W. Skelton, P. R. Traill, A. A. Watson and A. H. White, *Organometallics*, 1990, **9**, 3080.
817. A. T. Hutton, P. G. Pringle and B. L. Shaw, *J. Chem. Soc., Dalton Trans.*, 1985, 1677.
818. J. Vicente, A. Arcas, M. V. Borrachero, M. L. de Goicoechea, M. Lanfredi and A. Tiripicchio, *Inorg. Chim. Acta*, 1990, **177**, 247.
819. A. C. Albeniz, P. Espinet, C. Foces-Foces and F. H. Cano, *Organometallics*, 1990, **9**, 1079.
820. R. Uson, J. Fornies, M. Tomas, B. Menjon, R. Navarro and J. Carnicer, *Inorg. Chim. Acta*, 1989, **162**, 33.
821. A. C. Albeniz, J. C. Cuevas, P. Espinet, J. de Mendoza and P. Prados, *J. Organomet. Chem.*, 1991, **410**, 257.
822. F. Benetollo, G. Bombieri, F. Di Bianca and B. Crociani, *Inorg. Chim. Acta*, 1991, **188**, 51.
823. C. de Haro, G. Garcia, G. Sanchez and G. Lopez, *J. Chem. Res. (S)*, 1986, 119.
824. G. Lopez *et al.*, *Inorg. Chem.*, 1991, **30**, 2605.
825. P. Espinet, M. Y. Alonso, G. Garcia-Herbosa, J. M. Ramos, Y. Jeannin and M. Philoche-Levisalles, *Inorg. Chem.*, 1992, **31**, 2501.
826. H. Rimml and L. M. Venanzi, *J. Organomet. Chem.*, 1984, **260**, C52.
827. H. Adams, N. A. Bailey, T. N. Briggs, J. A. McCleverty and H. M. Colquhoun, *J. Chem. Soc., Dalton Trans.*, 1982, 1521.
828. J. Fornies, R. Navarro and V. Sicilia, *Polyhedron*, 1988, **7**, 2659.
829. E. Gretz and A. Sen, *J. Am. Chem. Soc.*, 1986, **108**, 6038.
830. R. Uson, J. Fornies, J. A. Nalda, M. J. Lozano, P. Espinet and A. C. Albeniz, *Inorg. Chim. Acta*, 1989, **156**, 251.
831. A. L. Seligson and W. C. Trogler, *J. Am. Chem. Soc.*, 1992, **114**, 7085.
832. D. M. Grove, G. van Koten, H. J. C. Ubbels, K. Vrieze, L. C. Niemann and C. H. Stam, *J. Chem. Soc., Dalton Trans.*, 1986, 717.
833. J. Fornies, R. Navarro, V. Sicilia and M. Tomas, *Organometallics*, 1990, **9**, 2422.
834. J. W. Suggs, M. J. Dube and M. Nichols, *J. Chem. Soc., Chem. Commun.*, 1993, 307.
835. S.-E. Bouaoud, P. Braunstein, D. Grandjean, D. Matt and D. Nobel, *Inorg. Chem.*, 1986, **25**, 3765.
836. P. Braunstein, J. Fischer, D. Matt and M. Pfeffer, *J. Am. Chem. Soc.*, 1984, **106**, 410.
837. J. Fornies, R. Navarro, V. Sicilia, F. Martinez and A. J. Welch, *J. Organomet. Chem.*, 1991, **408**, 425.
838. M. Ahmed, A. J. Edwards, C. J. Jones, J. A. McCleverty, A. S. Rothin and J. P. Tate, *J. Chem. Soc., Dalton Trans.*, 1988, 257.
839. R. Uson, J. Fornies, M. A. Uson and J. A. Apaolaza, *Inorg. Chim. Acta*, 1991, **187**, 175.
840. F. L. Wimmer, S. Wimmer and P. Castan, *J. Organomet. Chem.*, 1992, **424**, 99.
841. N. Ishio, Y. Nakamura and S. Okeya, *Bull. Chem. Soc. Jpn.*, 1984, **57**, 1012.
842. H. Tanaka, K. Isobe, S. Kawaguchi and S. Okeya, *Bull. Chem. Soc. Jpn.*, 1984, **57**, 456.
843. E. Gutierrez, S. A. Hudson, A. Monge, M. C. Nicasio, M. Paneque and E. Carmona, *J. Chem. Soc., Dalton Trans.*, 1992, 2651.
844. A. Morvillo and A. Turco, *J. Organomet. Chem.*, 1984, **276**, 431.
845. B. Crociani, F. Di Bianca, A. Fontana and R. Bertani, *J. Organomet. Chem.*, 1992, **425**, 155.
846. R. Uson, J. Fornies, M. A. Uson and M. M. Carranza, *Inorg. Chim. Acta*, 1989, **155**, 71.

847. R. Uson, J. Fornies, M. A. Uson, J. F. Yague, P. G. Jones and K. Meyer-Base, *J. Chem. Soc., Dalton Trans.*, 1986, 947.
848. H. Adams, N. A. Bailey, T. N. Briggs, J. A. McCleverty, H. M. Colquhoun and D. J. Williams, *J. Chem. Soc., Dalton Trans.*, 1986, 813.
849. R. McCrindle and A. J. McAlees, *J. Organomet. Chem.*, 1983, **244**, 97.
850. A. D. Ryabov, *J. Organomet. Chem.*, 1984, **268**, 91.
851. H. van der Poel and G. van Koten, *J. Organomet. Chem.*, 1981, **217**, 129.
852. G. Ferguson, A. J. McAlees, R. McCrindle and B. L. Ruhl, *Acta Crystallogr., Part B*, 1982, **38**, 2253.
853. A. B. Goel and S. Goel, *Inorg. Chim. Acta*, 1985, **98**, 67.
854. J. M. Vila, M. Gayoso, J. J. Fernandez, J. M. Ortigueira and A. Suarez, *Polyhedron*, 1990, **9**, 2741.
855. M. Onishi, K. Hiraki and A. Iwamoto, *J. Organomet. Chem.*, 1984, **262**, C11.
856. P. Braunstein, D. Matt, D. Nobel, S.-E. Bouaoud and D. Grandjean, *J. Organomet. Chem.*, 1986, **301**, 401.
857. P. Bandyopadhyay, P. K. Mascharak and A. Chakravorty, *J. Chem. Soc., Dalton Trans.*, 1982, 675.
858. A. J. Deeming, M. N. Meah, P. A. Bates and M. B. Hursthouse, *J. Chem. Soc., Dalton Trans.*, 1988, 2193.
859. M. Onishi, K. Hiraki, T. Itoh and Y. Ohama, *J. Organomet. Chem.*, 1983, **254**, 381.
860. G. K. Anderson, S. E. Saum, R. J. Cross and S. A. Morris, *Organometallics*, 1983, **2**, 780.
861. J. Collin, D. Commereuc and Y. Chauvin, *J. Mol. Catal.*, 1982, **14**, 133.
862. R. Bertani, A. Berton, F. Di Bianca and B. Crociani, *J. Organomet. Chem.*, 1988, **348**, 411.
863. G. Lopez *et al.*, *J. Chem. Soc., Dalton Trans.*, 1991, 3311.
864. R. Uson, J. Fornies, J. A. Nalda, M. J. Lozano, P. Espinet and A. C. Albeniz, *Inorg. Chim. Acta*, 1989, **156**, 251.
865. R. Uson, J. Fornies, M. Tomas, B. Menjon, J. Carnicer and A. J. Welch, *J. Chem. Soc., Dalton Trans.*, 1990, 151.
866. R. Uson, J. Fornies, M. A. Uson and M. L. Mas, *Inorg. Chim. Acta*, 1990, **168**, 59.
867. R. Uson, J. Fornies, P. Espinet and R. Navarro, *Inorg. Chim. Acta*, 1984, **82**, 215.
868. F. Maassarani, M. Pfeffer, A. L. Spek, A. M. M. Schreurs and G. van Koten, *J. Am. Chem. Soc.*, 1986, **108**, 4222.
869. J. Terheijden, G. van Koten, D. M. Grove, K. Vrieze and A. L. Spek, *J. Chem. Soc., Dalton Trans.*, 1987, 1359.
870. J. M. Brown and N. A. Cooley, *Chem. Rev.*, 1988, **88**, 1031.
871. K. Tatsumi, R. Hoffman, A. Yamamoto and J. K. Stille, *Bull. Chem. Soc. Jpn.*, 1981, **54**, 1857.
872. M. J. Calhorda, J. M. Brown and N. A. Cooley, *Organometallics*, 1991, **10**, 1431.
873. J. J. Low and W. A. Goddard, III, *J. Am. Chem. Soc.*, 1986, **108**, 6115.
874. J. J. Low and W. A. Goddard, III, *Organometallics*, 1986, **5**, 609.
875. A. Moravskiy and J. K. Stille, *J. Am. Chem. Soc.*, 1981, **103**, 4182.
876. F. Ozawa, K. Kurihara, T. Yamamoto and A. Yamamoto, *Bull. Chem. Soc. Jpn.*, 1985, **58**, 399.
877. H. Kurosawa, H. Ohnishi, M. Emoto, Y. Kawasaki and S. Murai, *J. Am. Chem. Soc.*, 1988, **110**, 6272.
878. R. Sustmann, J. Lau and M. Zipp, *Tetrahedron Lett.*, 1986, **27**, 5207.
879. F. Ozawa, T. Hidaka, T. Yamamoto and A. Yamamoto, *J. Organomet. Chem.*, 1987, **330**, 253.
880. R. Bertani, A. Berton, G. Carturan and R. Campostrini, *J. Organomet. Chem.*, 1988, **349**, 263.
881. B. Crociani, F. Di Bianca, A. Giovenco and A. Scrivanti, *J. Organomet. Chem.*, 1983, **251**, 393.
882. E. Rotondo, F. P. Cusmano, G. Neri, A. Donato and R. Pietropaolo, *J. Organomet. Chem.*, 1985, **292**, 429.
883. H. D. McPherson and J. L. Wardell, *Inorg. Chim. Acta*, 1984, **81**, 33.
884. A. D. Ryabov, A. K. Yatsimirsky and H.-B. Abicht, *Polyhedron*, 1987, **6**, 1619.
885. T.-W. Lai and A. Sen, *Organometallics*, 1984, **3**, 866.
886. D. A. White and G. W. Parshall, *Inorg. Chem.*, 1970, **9**, 2358.
887. K. Osakada, Y. Ozawa and A. Yamamoto, *J. Chem. Soc., Dalton Trans.*, 1991, 759.
888. R. Ros, R. A. Michelin, T. Boschi and R. Roulet, *Inorg. Chim. Acta*, 1979, **35**, 43.
889. A. L. Seligson and W. C. Trogler, *J. Am. Chem. Soc.*, 1991, **113**, 2520.
890. I. J. Harvie and F. J. McQuillin, *J. Chem. Soc., Chem. Commun.*, 1976, 369.
891. A. K. Mahapatra, D. Bandyopadhyay, P. Bandyopadhyay and A. Chakravorty, *J. Chem. Soc., Chem. Commun.*, 1984, 999.
892. P. L. Alsters, H. T. Teunissen, J. Boersma and G. van Koten, *Recl. Trav. Chim. Pays-Bas*, 1990, **109**, 487.
893. P. L. Alsters, J. Boersma and G. van Koten, *Tetrahedron Lett.*, 1991, **32**, 675.
894. P. K. Wong and J. K. Stille, *J. Organomet. Chem.*, 1974, **70**, 121.
895. H. D. McPherson and J. L. Wardell, *Inorg. Chim. Acta*, 1984, **81**, 39.
896. P. K. Byers, A. J. Canty, P. R. Traill and A. A. Watson, *J. Organomet. Chem.*, 1990, **390**, 399.
897. J. S. McCallum, J. R. Gasdaska, L. S. Liebeskind and S. J. Tremont, *Tetrahedron Lett.*, 1989, **30**, 4085.
898. J. C. Clinet and G. Balavoine, *J. Organomet. Chem.*, 1991, **405**, C29.
899. H. Kurosawa, M. Emoto and Y. Kawasaki, *J. Organomet. Chem.*, 1988, **346**, 137.
900. R. F. Johnston, P. K. Sen Gupta, M. B. Ossain, R. van der Helm, W.-Y. Jeong and R. A. Holwerda, *Acta Crystallogr., Part C*, 1990, **46**, 1796.
901. R. Uson, J. Fornies, M. Tomas, I. Ara, J. M. Casas and A. Martin, *J. Chem. Soc., Dalton Trans.*, 1991, 2253.
902. P. Braunstein, D. Matt, D. Nobel and J. Fischer, *J. Chem. Soc., Chem. Commun.*, 1987, 1530.
903. R. J. Crutchley, J. Powell, R. Faggiani and C. J. L. Lock, *Inorg. Chim. Acta*, 1977, **24**, L15.
904. F. Ozawa, T. Ito and A. Yamamoto, *Chem. Lett.*, 1979, 735.
905. H. Werner and W. Bertleff, *Chem. Ber.*, 1980, **113**, 267.
906. P. Diversi *et al.*, *J. Chem. Soc., Dalton Trans.*, 1988, 461.
907. P. W. Jolly, *Angew. Chem., Int. Ed. Engl.*, 1985, **24**, 283.
908. C. J. Jones, J. A. McCleverty, A. S. Rothin, H. Adams and N. A. Bailey, *J. Chem. Soc., Dalton Trans.*, 1986, 2055.
909. J. M. Wisner, T. J. Bartczak and J. A. Ibers, *Organometallics*, 1986, **5**, 2044.
910. G. K. Anderson, R. J. Cross, K. W. Muir and L. Manojlovic-Muir, *J. Organomet. Chem.*, 1989, **362**, 225.
911. A. L. Balch, *Comments Inorg. Chem.*, 1984, **3**, 51.
912. K. K. Pandey, *Coord. Chem. Rev.*, 1992, **121**, 1.
913. E. R. Milaeva, A. Z. Rubezhov, A. I. Prokoph'ev and O. Yu. Okhlobystin, *J. Organomet. Chem.*, 1980, **193**, 135.
914. G. A. Razuvaev, G. A. Abakumov, I. A. Teplova, K. G. Shal'nova, L. G. Abakumova and V. K. Cherkasov, *Izv. Akad. Nauk SSSR, Ser. Khim.*, 1980, 2484.
915. G. A. Abakumov, I. A. Teplova, V. K. Cherkasov, K. G. Shal'nova and G. A. Razuvaev, *Izv. Akad. Nauk SSSR, Ser. Khim.*, 1984, 1402.

916. P. K. Byers, A. J. Canty, B. W. Skelton and A. H. White, *J. Chem. Soc., Chem. Commun.*, 1986, 1722.
917. A. J. Canty, *Acc. Chem. Res.*, 1992, **25**, 83.
918. A. J. Canty, *Platinum Metals Rev.*, 1993, **37**, 2.
919. A. J. Canty and P. R. Traill, *J. Organomet. Chem.*, 1992, **435**, C8.
920. M. Catellani and B. E. Mann, *J. Organomet. Chem.*, 1990, **390**, 251.
921. R. Uson, J. Fornies and R. Navarro, *J. Organomet. Chem.*, 1975, **69**, 307.
922. R. Uson, J. Fornies and R. Navarro, *Synth. React. Inorg. Metal-Org. Chem.*, 1977, **7**, 235.
923. D. Milstein and J. K. Stille, *J. Am. Chem. Soc.*, 1979, **101**, 4992.
924. R. A. Holton and K. J. Natalie, *Tetrahedron Lett.*, 1981, **22**, 267.
925. A. D. Ryabov, A. V. Eliseev and A. K. Yatsimirsky, *Appl. Organomet. Chem.*, 1988, **2**, 101.
926. J. S. McCallum, J. R. Gasdaska, L. S. Liebeskind and S. J. Tremont, *Tetrahedron Lett.*, 1989, **30**, 4085.
927. N. A. Bumagin, I. G. Bumagina and I. P. Beletskaya, *Zh. Org. Khim.*, 1984, **20**, 457.
928. M. Catellani, G. P. Chiusoli and M. Costa, *Pure Appl. Chem.*, 1990, **62**, 623.
929. M. Catellani, G. P. Chiusoli and C. Castagnoli, *J. Organomet. Chem.*, 1991, **407**, C30.
930. O. Reiser, M. Weber and A. de Meijere, *Angew. Chem., Int. Ed. Engl.*, 1989, **28**, 1037.
931. H. Alper and M. Saldana-Maldonado, *Organometallics*, 1989, **8**, 1124.
932. H. Nakazawa, Y. Matsuoka, I. Nakagawa and K. Miyoshi, *Organometallics*, 1992, **11**, 1385.
933. N. Kamigata, M. Satoh and M. Yoshida, *J. Organomet. Chem.*, 1991, **401**, C26.
934. A. D. Ketley, L. P. Fisher, A. J. Berlin, C. R. Morgan, E. H. Gorman and T. R. Steadman, *Inorg. Chem.*, 1967, **6**, 657.
935. F. T. T. Ng and P. M. Henry, *J. Org. Chem.*, 1973, **38**, 3338.
936. G. Oehme, *J. Prakt. Chem.*, 1984, **326**, 779.
937. D. E. Rudisill and J. K. Stille, *J. Org. Chem.*, 1989, **54**, 5856.
938. B. M. Trost, *Acc. Chem. Res.*, 1990, **23**, 34.
939. B. M. Trost and A. S. K. Hashmi, *Angew. Chem., Int. Ed. Engl.*, 1993, **32**, 1085.
940. S. K. Meegalla, N. J. Taylor and R. Rodrigo, *J. Org. Chem.*, 1992, **57**, 2422.
941. I. Guibert, D. Neibecker and I. Tkatchenko, *J. Chem. Soc., Chem. Commun.*, 1989, 1850.
942. H. tom Dieck, C. Muller, E. T. K. Haupt and D. Bolze-Kuhrt, *Z. Naturforsch.*, 1987, **42**, 853.
943. P. K. Byers, A. J. Canty, B. W. Skelton and A. H. White, *Organometallics*, 1990, **9**, 826.
944. R. van Asselt, E. Rijnberg and C. J. Elsevier, *Organometallics*, 1994, **13**, 706.
945. P. K. Byers, A. J. Canty, R. T. Honeyman and A. A. Watson, *J. Organomet. Chem.*, 1990, **385**, 429.
946. P. K. Byers, A. J. Canty, R. T. Honeyman, B. W. Skelton and A. H. White, *J. Organomet. Chem.*, 1992, **433**, 223.
947. P. K. Byers, A. J. Canty, M. Crespo, R. J. Puddephatt and J. D. Scott, *Organometallics*, 1988, **7**, 1363.
948. K.-T. Aye, A. J. Canty, M. Crespo, R. J. Puddephatt, J. D. Scott and A. A. Watson, *Organometallics*, 1989, **8**, 1518.
949. A. J. Canty, A. A. Watson, B. W. Skelton and A. H. White, *J. Organomet. Chem.*, 1989, **367**, C25.
950. P. K. Byers, A. J. Canty, B. W. Skelton, P. R. Traill, A. A. Watson and A. H. White, *Organometallics*, 1992, **11**, 3085.
951. M. Crespo and R. J. Puddephatt, *Organometallics*, 1987, **6**, 2548.
952. J. A. Martinho Simoes and J. L. Beauchamp, *Chem. Rev.*, 1990, **90**, 629.
953. B. A. Markies, A. J. Canty, J. Boersma and G. van Koten, *Organometallics*, 1994, **13**, 2053.
954. P. K. Byers, A. J. Canty, B. W. Skelton and A. H. White, *J. Chem. Soc., Chem. Commun.*, 1987, 1093.
955. P. K. Byers and A. J. Canty, *J. Chem. Soc., Chem. Commun.*, 1988, 639.
956. D. G. Brown, P. K. Byers and A. J. Canty, *Organometallics*, 1990, **9**, 1231.
957. S. R. Cooper and S. C. Rawle, *Struct. Bonding*, 1990, **72**, 1.
958. A. J. Blake and M. Schroder, *Adv. Inorg. Chem.*, 1990, **35**, 1.
959. H. C. Clark, G. Ferguson, V. K. Jain and M. Parvez, *J. Organomet. Chem.*, 1984, **270**, 365.
960. S. Otsuka, T. Yoshida, M. Matsumoto and K. Nakatsu, *J. Am. Chem. Soc.*, 1976, **98**, 5850.
961. A. J. Canty, P. R. Traill, B. W. Skelton and A. H. White, *J. Organomet. Chem.*, 1992, **433**, 213.
962. G. N. Schrauzer and E. Deutsch, *J. Am. Chem. Soc.*, 1969, **91**, 3341.

6
Palladium–Carbon π-Bonded Complexes

JULIAN A. DAVIES
University of Toledo, OH, USA

6.1 INTRODUCTION 292

6.2 π-COMPLEXES OF ALKENES 293

6.2.1 Interactions of Alkenes with Metallic Palladium 293
6.2.2 Complexation of Alkenes to Palladium: Structure and Bonding 294
6.2.2.1 Theoretical investigations 294
6.2.2.2 Electronic spectroscopy 296
6.2.2.3 Vibrational spectroscopy 296
6.2.2.4 Nuclear magnetic resonance spectroscopy 297
6.2.2.5 Electrochemistry 300
6.2.2.6 Thermodynamic (equilibrium constant) studies 300
6.2.3 Synthetic Methods 302
6.2.3.1 Synthesis of palladium(0)–monoene complexes 302
6.2.3.2 Synthesis of palladium(II)–monoene complexes 304
6.2.3.3 Synthesis of palladium(0)–diene complexes 307
6.2.3.4 Synthesis of palladium(I)–diene complexes 307
6.2.3.5 Synthesis of palladium(II)–diene complexes 308
6.2.3.6 Synthesis of palladium(II)–cyclobutadiene complexes 309
6.2.4 Reactions of Palladium–Alkene Complexes 310

6.3 π-COMPLEXES OF ALKYNES 316

6.3.1 Interactions of Alkynes with Metallic Palladium 316
6.3.2 Complexation of Alkynes to Palladium: Structure and Bonding 316
6.3.2.1 Theoretical and structural investigations 316
6.3.3 Synthetic Methods 317
6.3.3.1 Synthesis of palladium(0)–alkyne complexes 317
6.3.3.2 Synthesis of palladium(I)–alkyne complexes 319
6.3.3.3 Synthesis of palladium(II)–alkyne complexes 320
6.3.3.4 Synthesis of palladium–alkyne cluster compounds 320
6.3.4 Reactions of Alkynes with Palladium Complexes 320

6.4 π-ALLYLIC COMPLEXES 323

6.4.1 Complexation of Allylic Ligands to Palladium: Structure and Bonding 323
6.4.1.1 Theoretical and structural investigations 323
6.4.1.2 Photoelectron spectroscopy and allied techniques 324
6.4.1.3 Nuclear magnetic resonance spectroscopy 325
6.4.2 Synthetic Methods 328
6.4.2.1 Reactions of monoenes with palladium salts and complexes 331
6.4.2.2 Reactions of 1,3-, 1,4-, 1,5-, 1,6-, and 1,7-dienes with palladium salts and complexes 335
6.4.2.3 Reactions of 1,2-dienes (allenes) with palladium salts and complexes 341
6.4.2.4 Reactions involving ring opening of cycloalkanes by palladium salts and complexes 341
6.4.2.5 Reactions involving oxidative addition of allylic substrates to palladium(0) 343
6.4.2.6 Miscellaneous reactions leading to π-allylic complexes 347
6.4.3 Reactions of π-Allylic Complexes 349
6.4.3.1 Nucleophilic attack at palladium 350
6.4.3.2 Halide exchange in $[Pd_2(\mu\text{-}Cl)_2(\eta^3\text{-allyl})_2]$ and related reactions 353

6.4.3.3 *Thermal and photochemical decomposition* 354
6.4.3.4 *Oxidation to aldehydes and ketones* 355
6.4.3.5 *Halogenation* 356
6.4.3.6 *Reduction and charge reversal* 357
6.4.3.7 *Reactions with carbon monoxide* 358
6.4.3.8 *Nucleophilic attack at carbon* 360
6.4.4 *Diene Oligomerization and Telomerization* 371
6.4.5 *Cycloaddition Reactions Implicating Trimethylenemethanepalladium Intermediates* 373

6.5 CYCLOPENTADIENYL COMPLEXES 375
6.5.1 *Synthetic Methods* 375
6.5.1.1 *Synthesis of terminal cyclopentadienyl complexes* 375
6.5.1.2 *Synthesis of bridged cyclopentadienyl complexes* 377
6.5.2 *Reactions of Palladium–Cyclopentadienyl Complexes* 378

6.6 ARENE COMPLEXES 381
6.6.1 *Interactions of Arenes with Metallic Palladium* 381
6.6.2 *Palladium–Arene Complexes* 381

6.7 REFERENCES 383

6.1 INTRODUCTION

In the first edition of *COMC* work on π-complexes of palladium was discussed in several different chapters. Useful introductory material on the chemistry of palladium(0), palladium(I), palladium(II), and palladium(IV) was detailed in *Palladium: Introduction and General Principles*[1] and information on the preparation and reactions of palladium(0)–monoene and –diene complexes appeared in *Complexes of Palladium(0)*.[2] Work on di-, tri-, and tetranuclear complexes, including those of palladium(I) with alkyne, allylic, arene, and cyclopentadienyl ligands, was covered in *Palladium(I) and Cluster Compounds*.[3] The chemistry of alkenes and alkynes with palladium was described in three chapters: *Monoolefin and Acetylene Complexes of Palladium*,[4] *Diene Complexes of Palladium*,[5] and *Palladium Complexes Derived from Reactions with Acetylenes*.[6] Palladium complexes with allylic ligands were treated separately in *Allylic Complexes of Palladium(II)*,[7] while cyclopentadienyl and arene complexation was detailed in *Cyclopentadienyl and Arene Complexes of Palladium(II)*.[8] Much closely related chemistry was covered in the chapter *Compounds with Palladium–Carbon σ-Bonds*.[9] The first volume in the collection entitled *The Chemistry of the Metal–Carbon Bond* contains chapters on the synthesis of transition metal alkene and alkyne complexes,[10] π-allyl complexes,[11] η^4-butadiene and cyclobutadiene complexes,[12] η^5-bonded ligands,[13] and η^6-, η^7-, and η^8-bonded ligands.[14] Subsequent volumes in the collection include many chapters of relevance to the chemistry of palladium π-complexes, such as that covering carbon–carbon bond formation using π-allylpalladium complexes.[15] The chapter *Organometallic and Homogeneous Catalytic Chemistry of Palladium and Platinum*[16] in *Chemistry of the Platinum Group Metals: Recent Developments* updated coverage of much of the chemistry of palladium π-complexes from earlier collections. Annual surveys of the chemistry of nickel, palladium, and platinum, including sections relating to π-complexes, have appeared covering some of the period since *COMC-I* was published (1981–1986).[17–21] The developing area of organopalladium(IV) chemistry has been reviewed.[22]

In recent years organometallic chemistry has been impacted upon by the increasing availability of x-ray crystallography, by new developments in NMR spectroscopy (e.g., multinuclear methods,[23] two-dimensional techniques,[24] and high-resolution solid-state measurements[25]), by increasingly advanced theoretical and computational treatments of structure and bonding,[26] and by rapid advances in other techniques. These developments have contributed to an expanded understanding of the chemistry of palladium π-complexes.

In this chapter the focus is on developments that have followed the publication of *COMC-I*, and descriptions of earlier work can be found in the compilations cited above. This chapter is divided into sections concerning the chemistry of π-complexes of alkenes, alkynes, allylic ligands, cyclopentadienyls, and arenes with palladium. Overlap between the somewhat artificial divisions is inevitable and cross-references to other relevant sections are provided throughout. The application of homogeneous palladium catalysts to transformations of organic substrates is not covered comprehensively in this chapter, although illustrative examples are included where appropriate, since full treatment of this topic is available in the material cited above and elsewhere within these volumes.

6.2 π-COMPLEXES OF ALKENES

6.2.1 Interactions of Alkenes with Metallic Palladium

The major driving forces behind the study of organopalladium chemistry, particularly that involving unsaturated organic compounds, have been the importance of catalytic reactions of hydrocarbons that involve palladium and the relationship between heterogeneous and homogeneous catalytic reactions.[27] Although not a major focus of this chapter, a few introductory comments on recent developments in the study of heterogeneous palladium catalysts and model systems are in order. Palladium metal promotes a variety of transformations of hydrocarbons including hydrogenolysis, isomerization, cyclization, hydrogen–deuterium exchange, aromatization, hydrogenation, and so on. In general, the catalytic properties of palladium mirror those of platinum but with a number of small, yet significant, differences. Skeletal reactions (hydrogenolysis, isomerization, etc.) of a family of C_6 alkanes have been compared in cases where palladium black and platinum black were employed as catalysts.[28] It was found that palladium black gave rise to a higher selectivity for alkene formation and for aromatization than did platinum black, and this difference was attributed to competing reaction pathways. Palladium was said to favor formation of surface π-complexes, while reactions on platinum favored σ-type intermediates.

Extended Hückel molecular orbital calculations,[29] designed to model the low-temperature adsorption of ethene on platinum and palladium surfaces, indicated that, for Pt(111), π-complexation was less favorable than formation of a di-σ-intermediate. On Pd(111) the two forms were found to be energetically comparable. An experimental study of the bonding of ethene on Pd(111) at low temperature with the use of near-edge x-ray absorption fine structure (NEXAFS) and photoelectron spectroscopy (PES) showed that ethene was π-bonded in a flat-lying geometry. On annealing to 300 K, the chemisorbed ethene was found to desorb reversibly with essentially no detectable conversion to other species.[30] The low-temperature (175 K) phase formed between Pd(111) and ethene has also been characterized by angle-resolved UV–PES and molecular beam measurements.[31] These studies indicated the formation of a π-bound species with the C–C axis parallel to the palladium surface. On raising the temperature, dehydrogenation, self-hydrogenation, and the formation of C_4 and C_6 products were found to occur. A comparative study of the chemisorption of ethene on Ni(111), Pd(111), and Pt(111) with the use of extended Hückel calculations within a tight-binding formalism has been presented. There appeared to be a slight preference for the twofold site (di-σ-bonded) over the on-top site (π-bonded) in the case of Ni(111) and Pt(111), whereas the opposite was found for Pd(111).[32] The interaction of ethene with Pd(111) has been studied experimentally by temperature-programmed reaction spectroscopy and high-resolution electron energy loss spectroscopy (HREELS).[33] Both di-σ-bonded and π-bonded forms were found to be stable at 80 K. The π-bonded form was found to desorb between 100 K and 300 K, whereas the di-σ-bonded form underwent dehydrogenation to generate coadsorbed atomic hydrogen and a surface-bound intermediate, believed to be a vinyl species.

The vibrational spectra of ethene complexes of nickel and platinum have been examined by IR and Raman methods (see also Section 6.2.2.3) and treated by normal coordinate analysis.[34] The results for $[M(C_2H_4)_3]$ (M = Ni, Pt) have been compared with data obtained by electron energy loss spectroscopy (EELS) for ethene adsorbed on Ni(111), Pd(111), and Pt(111) at low temperature. Analysis of force constants suggested that chemisorbed ethene was largely rehybridized to $C(sp^3)$ on Ni(111) and Pt(111), but remained alkenic on Pd(111) where it was less tightly bound.

A study[35] of the interaction of ethene with the Pd(110) surface by HREELS, low-energy electron diffraction (LEED), and desorption techniques has demonstrated that, at 90 K, ethene was π-bonded to the surface. Heating caused some desorption of intact ethene and some rearrangement until, above 300 K, dehydrogenation to surface ethynyl species, hydrogen adatoms, and other unstable, possibly vinylic, surface groups occurred. Ethene desorption occurred by recombination of the unstable surface groups and hydrogen adatoms; hydrogen was also lost by adatom recombination. The surface ethynyl groups decomposed to produce hydrogen and carbon adatoms which were left on the palladium surface. In other studies of ethene of Pd(110), no stable species intermediate between π-bonded ethene and surface-bound ethynyl (e.g., ethylidyne or vinyl) were observed[36] (cf. also Pd(111), above, where species believed to be surface-bound vinyls have been detected).

Studies of supported palladium catalysts have generated results that are more difficult to interpret unambiguously than those that have arisen from studies of clean metal surfaces. An investigation of the interaction of ethene with palladium-on-silica by IR spectroscopy has shown[37] that different procedures for catalyst preparation give rise to materials with different IR spectra and, hence, different surface-bound species. Of the two preparation methods investigated, both resulted in some π-bound ethene, but one also showed evidence for a di-σ-bonded form while the other showed IR absorptions tentatively attributed to a species that involved π-bonding of a dehydrogenated hydrocarbon to two surface palladium atoms. With alkenes other than ethene, binding to supported palladium catalysts has the potential to be even more complex. Palladium-on-alumina and palladium-on-Li_2O-doped-alumina have been used as styrene hydrogenation catalysts and IR spectra of styrene adsorbed on these materials have been reported.[38] Doping of the alumina support affected the IR spectrum of the styrene adsorbed on the dispersed palladium. Bands attributed to π-bound styrene, surface vinyl groups, and styrene bound by π-bonding of the aromatic ring to palladium have been observed for the undoped catalyst, while the IR spectrum of the doped material indicated an even more complex array of surface species. In relation to π-bonding of styrene to palladium through its aromatic ring, studies of palladium-on-carbon as a hydrogenation catalyst for alkenes with aromatic substituents have suggested that such interactions may be important in stereocontrol.[39]

The electronic properties of palladium deposited on carbon supports are known to vary with cluster size,[40] which is a function of the method used to prepare the catalyst. The wide variety of methods for the preparation of palladium-on-carbon catalysts have been reviewed.[41] It has been observed[3] that the palladium in 5% palladium-on-carbon (graphite) is likely to consist of small, irregularly shaped clusters that are only weakly bound to the support. Since small palladium clusters have only modest binding energies, it has been suggested that palladium(0)–alkene complexes might be formed that could act as substrates in subsequent hydrogenation steps. As triethylamine is a common solvent for alkene hydrogenations, it has been noted that the formation of homogeneous palladium–alkene complexes, stabilized by amine ligands, might occur and could be important in controlling product distributions, activities, and so on (interestingly, it has been reported that the interaction of palladium π-allyl complexes with higher aliphatic amines produced species that catalyzed the selective hydrogenation of dienes to monoenes[42]). This area has been investigated theoretically through all-electron-Hartree–Fock gradient calculations[43] and geometry optimizations on systems involving up to three palladium atoms and a number of relevant organic fragments, including C_2H_4, NH_3, and so on. The results of these very large calculations support the idea that it might be favorable for palladium atoms to become detached from the surface of a heterogeneous catalyst and form π-alkene complexes. In other words, the binding of ethene to palladium is competitive with the binding of palladium to small palladium clusters. It was further speculated that such homogeneous species might be important intermediates, as yet unrecognized experimentally, in alkene hydrogenation.

6.2.2 Complexation of Alkenes to Palladium: Structure and Bonding

Bonding in palladium–alkene complexes is usefully described by the Dewar–Chatt–Duncanson model,[44,45] as discussed in *COMC-I*.

6.2.2.1 Theoretical investigations

Since 1980, *ab initio* calculations on complexes of second- and third-row transition metal complexes, including those with alkenes, have been reported. Such calculations had earlier been problematic because of the very large number of basis functions on heavy metals needed to treat all of the electrons explicitly. Relativistic shifts in the valence electrons of platinum and, to a lesser degree, palladium suggest that nonrelativistic treatments may lead to serious errors. In a study of $[MCl_3(C_2H_4)]^-$ (M = Pd, Pt), *ab initio* relativistic effective core potentials have been employed in order to replace the chemically inert core, and hence reduce the problem to one of valence electrons, and also to incorporate relativistic terms into the potential.[46] Unlike many other *ab initio* and approximate calculations, such methods allow for the reliable estimation of geometric parameters for ligands coordinated to transition metals. The results on geometry from the calculations on Zeise's salt (M = Pt) were compared with those obtained earlier by a neutron diffraction study[47] and were found to be in good agreement. Thus, the bending back of the methylene groups upon coordination of ethene (12.6° from the calculation vs. 16.3° from the neutron structure), the lengthening of the carbon–carbon bond (0.0053 nm calculated vs. 0.0037 nm experimental), and the metal–alkene bond distance (0.211 nm calculated, changing to 0.206 nm when 3*d* functions were added to the carbon atoms, vs. 0.202 nm experimental) were all reasonably determined in the *ab initio* calculations. The more stable, upright coordination of ethene to the $[PtCl_3]^-$ fragment was calculated to be bound by ~117 kJ mol^{-1} and the barrier to rotation about the platinum–alkene axis was calculated to be ~63 kJ mol^{-1}, in agreement with experimental values (~42–59 kJ mol^{-1}) and implying rotation without dissociation. For the palladium analogue, where no neutron structure has been reported, the upright binding of ethene to the $[PdCl_3]^-$ fragment was again found to be favored, but here by only ~29 kJ mol^{-1} compared with the in-plane form and by just ~50 kJ mol^{-1} compared with dissociation. The weaker binding of ethene to palladium was manifested in smaller calculated distortions of the alkene upon binding. It was suggested that this weaker binding of ethene to palladium in comparison with platinum may explain the propensity of palladium complexes to act as alkene oxidation catalysts, that is, although the carbon–carbon bond was calculated to be weaker in the platinum complex, the tighter binding might result in lower overall reactivity. Orbital population analysis was discussed in qualitative terms through the Dewar–Chatt–Duncanson model of metal–alkene binding. With the MCl_3 fragment in the *xy* plane (Cl atoms lying along $\pm x$ and $-y$) and the upright ethene along $+y$, the "filled" $d(xy)$-orbital is of appropriate symmetry to interact with the ethene π^*-orbital, and the "empty" $d(x^2 - y^2)$-, *s*-, and $p(y)$-orbitals are able to interact with the occupied π-orbital of ethene. The population analysis showed that the cooperative σ-donor, π-acceptor mode of binding was of decreased magnitude in the palladium system compared with platinum (i.e., for platinum, 0.24 electron was transferred from the alkene to the platinum σ-orbitals and 0.22 electron was returned by "back donation" from the platinum $5d(xy)$-orbital, whereas for palladium, 0.12 electron was donated to the palladium σ-orbitals and 0.07 electron was transferred to the alkene by π-donation). The lower level of cooperation in the σ-donor, π-acceptor mode of bonding for palladium was manifested in the overall decrease in calculated binding energy of the alkene.

A semiempirical, self-consistent field (SCF) molecular orbital method for the calculation of energies and geometries of transition metal organometallic compounds, CNDO-S^2, has been presented with parameterization including H, C, N, O, Cl, Fe, Ni, and Pd atoms. The method has been utilized to examine the anion $[PdCl_3(C_2H_4)]^-$.[48] The binding energy for ethene to the $[PdCl_3]^-$ fragment was calculated to be ~42 kJ mol^{-1} (cf. ~50 kJ mol^{-1} by the *ab initio* method described above) and the barrier to rotation about the palladium–alkene axis was calculated as ~25 kJ mol^{-1} (cf. ~29 kJ mol^{-1} by the *ab initio* method). The bending back of the methylene groups upon coordination of ethene, the lengthening of the carbon–carbon bond, and the metal–alkene bond distance were all calculated to within reasonable agreement of the values found by the *ab initio* method. This less time-consuming calculational approach has been employed to model proposed steps in the oxidation of alkenes by a palladium–nitrito complex, a catalytic reaction for which experimental data are available.[49]

Extended Hückel molecular orbital calculations that compared $[PdCl_3(C_2H_4)]^-$ with its ethylidene isomer $[PdCl_3(CHMe)]^-$ have been reported.[50] The geometry chosen for the palladium–alkene complex was based upon the *ab initio* calculations, described earlier, while the geometry of the ethylidene isomer was based on x-ray structural studies of related compounds. The calculations supported a low binding energy for ethene to the $[PdCl_3]^-$ fragment (~42 kJ mol^{-1}) and a low level of charge transfer accompanying bond formation. The binding of ethylidene to $[PdCl_3]^-$ (both singlets) was found to be very favorable, and the isomerization energy for conversion of $[PdCl_3(C_2H_4)]^-$ to its ethylidene isomer was calculated to be high. The absolute values of these energy terms may not be meaningful, given the approximate method of calculation.

Calculations comparing the binding of ethene and ethyne to palladium are described in Section 6.3.2.1.

The effect of coordination on the conformation of diene ligands (see also Section 6.2.2.4) has been investigated[51] by comparison of x-ray crystallographic data for the complexes $[PdCl_2(diene)]$ (diene = cod, cycloocta-1,4-diene, cyclonona-1,5-diene) with the low-energy conformations of the free dienes as calculated by molecular mechanics. The analogous complex of cyclohepta-1,4-diene, which was characterized spectroscopically, also formed part of the study. It was concluded that cod and cyclohepta-1,4-diene undergo little conformational change on coordination but that the conformations of cycloocta-1,4-diene and cyclonona-1,5-diene are significantly affected. The former is coordinated in a high-energy boat–boat conformation while the latter diene changes from a calculated low-energy chair form in the free state to a twist-boat–chair conformation upon coordination. Such processes are likely to have significant activation energy barriers. It was postulated that the conformational energy profile for free cyclonona-1,5-diene served as an indicator of the conformational energy "stored" in the coordinated form. Coordination of 3-methyl-1,4-cyclooctadiene (3-Me-1,4-cod) to palladium(II) has been shown[52] to give rise to the boat–chair conformer of $[PdCl_2(3\text{-Me-}1,4\text{-cod})]$, and molecular mechanics calculations have indicated that the boat–chair conformation of free 3-Me-1,4-cod may be the preferred, low-energy form. Conformational effects are manifested in the kinetics and thermodynamics of diene complexation to palladium(II).[53]

The migration of an ML_n unit within the periphery of a cyclic polyene ("ring-whizzing," the organometallic analogue of a sigmatropic rearrangement) has been examined both theoretically and experimentally[54] in the case of $[M(PPh_3)_2(C_3Ph_3)]X$ (M = Ni, Pd, Pt; X = ClO_4, PF_6). The reaction of $[M(PPh_3)_2(C_2H_4)]$ with triphenylcyclopropenium salts allowed isolation of the nickel and palladium complexes and these were characterized crystallographically in three cases: $[Ni(PPh_3)_2(C_3Ph_3)][PF_6]$, $[Pd(PPh_3)_2(C_3Ph_3)][PF_6]{\cdot}C_6H_6$, and $[Pd(PPh_3)_2(C_3Ph_3)][ClO_4]$. Along with structural data already known for M = Pt, X = PF_6, the complexes comprised a series showing progressive movement of the ML_2 fragment over the face of the cyclopropenium cation. It was proposed that the structures chart the pathway for movement of the ML_2 fragment from one η^2-position to an equivalent η^2-geometry. The proposed pathway for fluxionality is shown in Scheme 1.

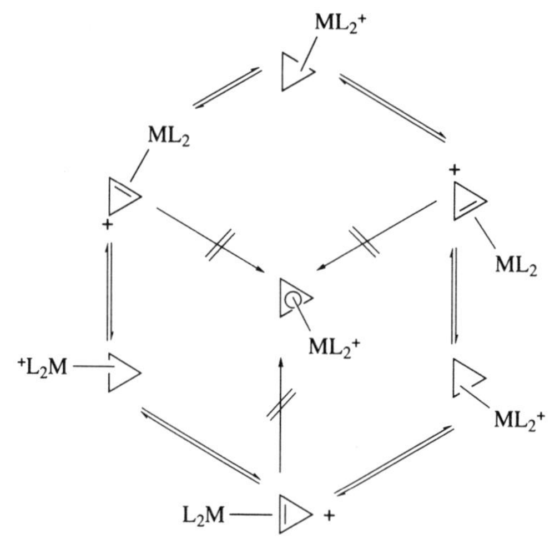

Scheme 1

For a hypothetical, ground-state η^2-structure with the metal fragment coordinated to and lying above the double bond, three possible geometries exist (Scheme 1). Interconversion via an η^3-transition state would imply that three channels exist from the transition state to the ground-state structure, but this is forbidden by the McIver–Stanton rules[55] concerning symmetry of potential surfaces. Three points of lower energy than that of the hypothetical η^3-structure must exist that serve as transition states for the fluxional process. Extended Hückel molecular orbital calculations on a model $[Ni(PH_3)_2(cyclopropenium)]^+$ system were performed and confirmed that the hypothetical η^3-structure was at higher energy than the three crossover structures depicted in Scheme 1. Of the four compounds characterized crystallographically, two were found to resemble ground-state structures, one resembled a crossover transition-state structure, and one structure lay in between. None of the structures resembled the hypothetical η^3-intermediate. Fluxionality in metal–polyene and –polyenyl complexes, including examples from palladium chemistry, has been reviewed.[56]

6.2.2.2 Electronic spectroscopy

The UV–visible spectrum of $[PdCl_3(C_2H_4)]^-$, generated in dry THF solution by treatment of $[Pd_2Cl_4(C_2H_4)_2]$ with LiCl(THF), has been described.[57] Although the electronic spectrum of $[PtCl_3(C_2H_4)]^-$ in aqueous solution was described some years ago,[58] no analogous spectrum of the palladium complex had been reported, presumably because of the rapid redox reaction that occurs between $[PdCl_3(C_2H_4)]^-$ and water. It was found that solutions of $[PdCl_3(C_2H_4)]^-$ in dry THF, saturated with ethene, were sufficiently stable below 273 K to allow spectroscopic study. The assignments were based on the assumption that $[PdCl_3(C_2H_4)]^-$ has the same structure as its platinum analogue, which was also studied under the same conditions. The one-electron schematic energy-level diagram and the associated transitions are shown in Figure 1.

6.2.2.3 Vibrational spectroscopy

As a consequence of the Dewar–Chatt–Duncanson model of metal–alkene binding, ν(C=C) is expected to decrease on coordination of an alkene. In *COMC-I* it was shown that $\Delta\nu$(C=C) of 100 cm^{-1} was typical for palladium, whereas $\Delta\nu$(C=C) of ~200 cm^{-1} was observed for platinum.

The IR and Raman spectra of the cod complexes $[Rh_2Cl_2(cod)_2]$, $[PtCl_2(cod)]$, and $[PdCl_2(cod)]$ have been analyzed[59] and assignments given for bands I and II (the ν(C=C) and in-plane alkene C–H wagging

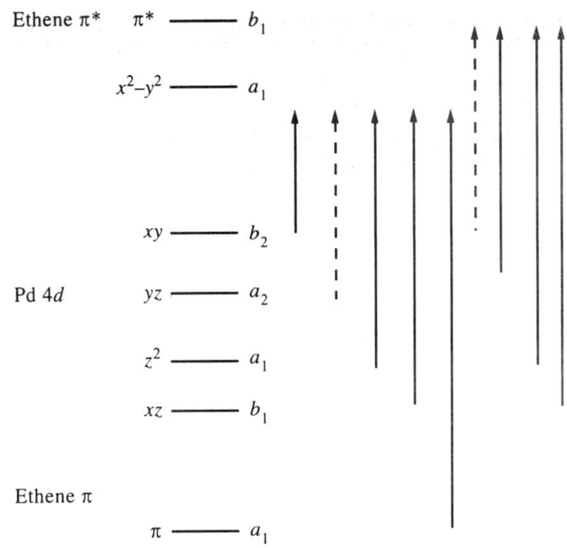

Figure 1 Schematic molecular orbital diagram for $[PdCl_3(C_2H_4)]^-$. The chlorine 3s- and 3p-orbitals have been excluded. Dashed lines indicate dipole-forbidden transitions. The two shown here are vibronically allowed (reproduced by permission of Munksgaard from *Acta Chem. Scand.*, 1989, **43**, 938).

mode, which are often very strongly coupled), the alkene C–H wagging and rocking modes, and for the metal–alkene stretching modes. The combined shifts in frequencies of bands I and II were found to lie in the order Rh > Pt > Pd, which was taken to indicate that the total interaction between the metals and diene ligands followed the same order. The frequencies of the in-phase and out-of-phase symmetric metal–alkene stretching modes were also found to follow the same order. The increase in frequencies of the out-of-plane alkene C–H modes was observed to follow the order Pt > Rh > Pd, as were the frequencies of the bands attributed to symmetric metal–alkene stretching modes, implying that the extent of the π-component to metal–alkene bonding was Pt > Rh > Pd, that is, metal–alkene π-backbonding was found to be more important in the platinum–diene complex than in the palladium analogue, although the contribution of backbonding was suggested to be significant in both cases. This situation contrasts with the case of monoene coordination where an analysis of cyclopentene complexes of platinum(II) and palladium(II) suggested that π-backbonding was very much more significant for platinum than for palladium.

A closely related study of the vibrational spectra of the norbornadiene complexes $[Rh_2Cl_2(nbd)_2]$, $[PtCl_2(nbd)]$, and $[PdCl_2(nbd)]$, which included a reinvestigation of the IR and Raman spectra of norbornadiene, has also been reported.[60] Analysis of the shifts in frequencies of bands I and II and the out-of-plane alkene C–H wagging modes, and the relative energies of the bands assigned to metal–alkene motions, was consistent with an overall order for the metal–alkene interaction of Rh > Pt > Pd and an overall order for the extent of π-bonding of Rh ≈ Pt > Pd.

6.2.2.4 Nuclear magnetic resonance spectroscopy

The complexes $[Pd(\eta\text{-}C_5H_5)(alkene)(PPh_3)]^+$ (alkene = $CH_2{=}CHPh$ (**1**), $CH_2{=}CH_2$ (**2**)), $[Pt(\eta\text{-}C_5H_5)(CH_2{=}CH_2)(PPh_3)]^+$ (**3**), and $[Pt(\eta^3\text{-}2\text{-}MeC_3H_4)(alkene)(PPh_3)]^+$ (alkene = $CH_2{=}CHPh$ (**4**), $CH_2{=}CH_2$ (**5**), (*E*)-CHMe=CHMe (**6**)) have formed the basis of an NMR study[61] designed to evaluate the importance of steric and electronic effects in determining barriers to rotation about the M–alkene axes and also to compare the relative π-donor abilities of $[Pd(\eta\text{-}C_5H_5)(PPh_3)]^+$ and $[Pt(\eta^3\text{-}2\text{-}MeC_3H_4)(PPh_3)]^+$ structural fragments.

In (**1**), (**2**), and (**3**) the alkene lies perpendicular to the plane defined by palladium, phosphorus, and the Cp centroid (confirmed crystallographically for (**1**)), that is, the compounds are best regarded as 18-electron species. In compound (**4**), an x-ray crystallographic structure determination showed that the alkene double bond was parallel to the coordination plane of platinum, while for (**6**) preliminary x-ray data gave an angle of ~67° between the carbon–carbon double-bond axis and the coordination plane, that is, compounds of this type are best regarded as 16-electron species. Barriers to rotation, $\Delta G^{\neq}_{rot}$, of ~54 kJ mol^{-1} at 273 K for (**3**), ~37 kJ mol^{-1} at 183 K for (**5**), and ~46 kJ mol^{-1} at 218 K for (**6**) were determined. Thus, the barrier to rotation for the 16-electron complex (**5**), which was estimated and may

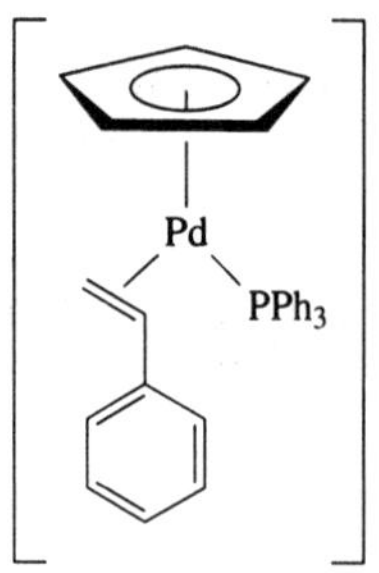

(1)

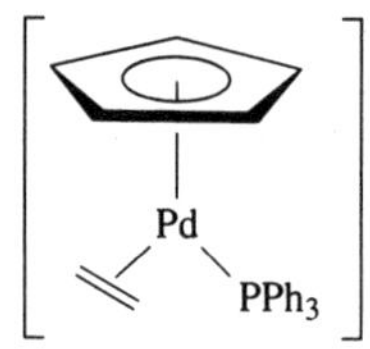

(2)

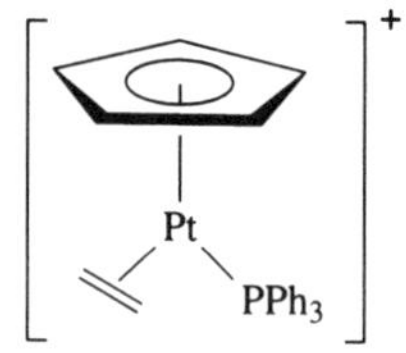

(3)

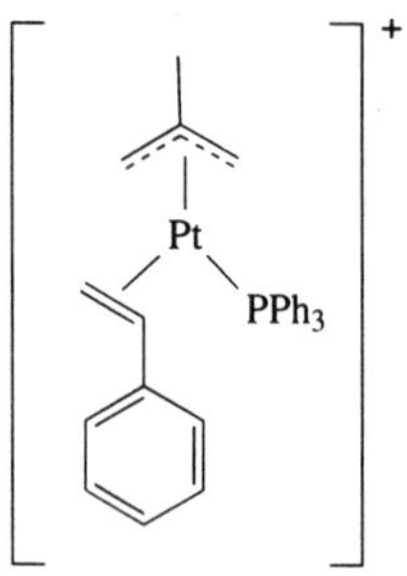

(4)

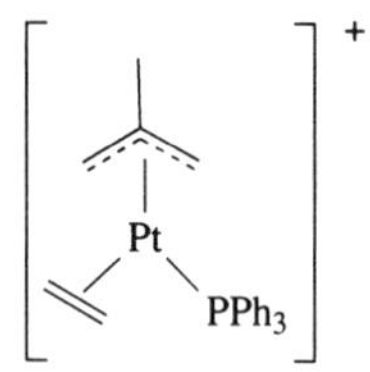

(5)

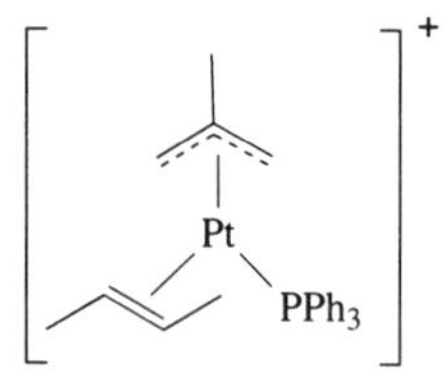

(6)

represent an upper limit, is seen to be smaller than that measured for the 18-electron complex (**3**), although both contain platinum–ethene moieties. The origin of this difference in rotational barriers was ascribed to differences in platinum–alkene bonding in the electronically different complexes, that is, the alkene is more tightly bound in (**3**) than in (**5**). Metal–alkene binding was further investigated through determination of equilibrium constants for alkene complexation according to Equation (1).

$$[M]\text{–NCPh} + \text{alkene} \rightleftharpoons [M]\text{–alkene} + \text{PhCN} \qquad (1)$$

$$[M] = [Pd(\eta\text{-Cp})(PPh_3)]^+,\ [Pt(\eta^3\text{-2-MeC}_3H_4)(PPh_3)]^+$$

Values of the equilibrium constants determined by NMR for Equation (1) are shown in Table 1. The extraordinary feature of these data is that the values for binding of $[Pd(\eta\text{-}C_5H_5)(PPh_3)]^+$ to alkenes are ~10 times larger than the values for binding of $[Pt(\eta^3\text{-2-MeC}_3H_4)(PPh_3)]^+$, with the exception of the values for the alkene (*E*)-2-butene, which are closer in magnitude.

Table 1 Equilibrium constants for Equation (1) (in $CDCl_3$ at 298 K).

Alkene	*$[M] = [Pd(\eta\text{-}C_5H_5)(PPh_3)]^+$*	*$[M] = [Pt(\eta^3\text{-2-MeC}_3H_4)(PPh_3)]^+$*
$CH_2{=}CH_2$	15 ± 3	1.3 ± 0.3
$CH_2{=}CHMe$	0.81 ± 0.09	0.12 ± 0.05
$CH_2{=}CHEt$	0.87 ± 0.09	0.10 ± 0.05
(*E*)-CHMe=CHMe	0.03 ± 0.01	0.020 ± 0.005
$CH_2{=}CHPh$	0.26 ± 0.05	0.035 ± 0.006

Source: Kurosawa *et al.*[61]

Typical equilibrium constants for binding of alkenes to platinum and palladium, for example, those found for Equation (2), are ~100 times larger for platinum than for palladium.

$$[MCl_4]^{2-} + \text{alkene} \rightleftharpoons [MCl_3(\text{alkene})]^- + Cl^- \qquad (2)$$

The results of this study were interpreted in terms of the effect of the cyclopentadienyl anion on the binding of alkenes to palladium(II), that is, that a palladium(II) fragment bearing an η-cyclopentadienyl fragment is a more effective π-donor towards alkenes than a platinum(II) fragment containing an η^3-allyl ligand. The atypical behavior of (*E*)-2-butene in these systems was tentatively related to the structural distortion of (**6**) where the alkene assumes a position intermediate between in-plane and out-of-plane coordination (see the discussion above of structural data). An x-ray crystal structure determination has also been reported for the complex $[Pd(\eta\text{-}C_5H_5)(CH_2{=}CHPh)(PEt_3)][BF_4]$ where the alkene carbon–carbon bond is similarly inclined (at an angle of 77.3°) to the plane defined by palladium, phosphorus, and the Cp centroid.[62]

An NMR study[63] of the mononuclear palladium(0) and platinum(0) complexes of dibenzylideneacetone (dba) has demonstrated that the coordinated alkene groups of dba are static in the *s-trans* conformation, while the uncoordinated alkene groups are fluxional between *s-cis* and *s-trans* conformations (Equation (3)).

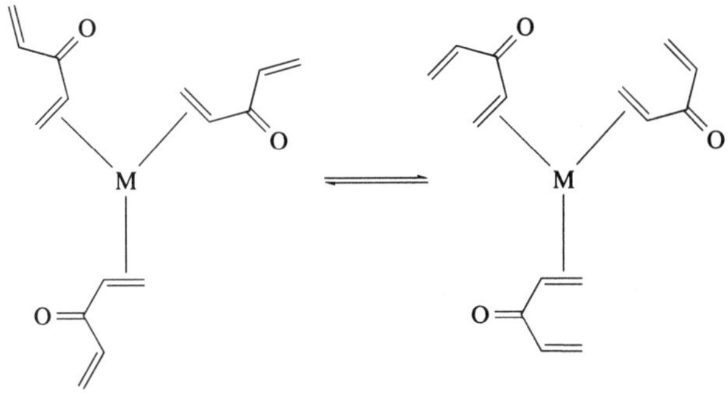

(3)

The variable-temperature ^1H NMR study of $[Pd(dba)_3]$ and $[Pt(dba)_3]$, which included investigations of selectively deuterated forms, showed that solutions of the palladium complex were predominantly in the *s-trans* (bound), *s-cis* (free) form, while solutions of the platinum complex contained significant amounts of the *s-trans* (bound), *s-trans* (free) conformer. The origin of this difference was attributed to differences in metal–alkene π-backbonding in the palladium and platinum complexes. Thus, the *s-trans*, *s-trans* conformer of free dba is disfavored with respect to other conformations, since repulsive interactions between the alkene β-hydrogens arise. Upon coordination, out-of-plane bending of the alkene methylene groups should reduce this unfavorable interaction and, accordingly, the greater the extent of metal–alkene π-backbonding, the more effective this distortion should be in reducing these unfavorable interactions. Platinum was thus said to be better able to support the *s-trans* (bound), *s-trans* (free) conformation of the dba ligand than palladium.

The 1,3-butadiene complexes $[Pd(R_2PCH_2CH_2PR_2)(diene)]$ (R = Pr^i (**7**), Bu^t (**8**), Cy (**9**)) have been examined by solid-state ^{13}C and ^{31}P NMR spectroscopies employing cross-polarization (CP) and magic angle spinning (MAS) techniques.[64]

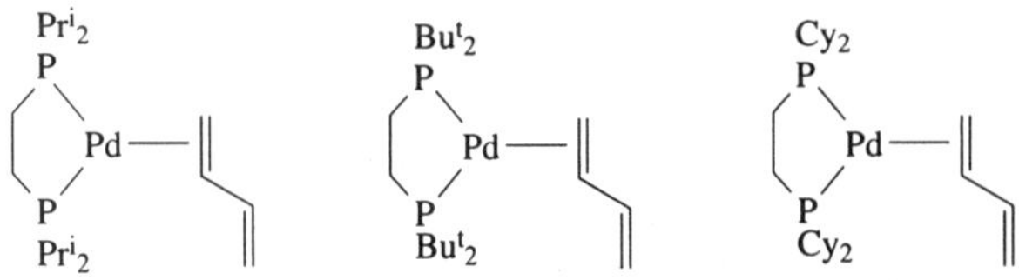

(7) **(8)** **(9)**

The spectra confirmed η^2-binding of 1,3-butadiene to palladium in the solid state, as found in solution, and this was demonstrated by an x-ray crystal structure determination of (**9**). The ^{13}C CP–MAS NMR spectrum of (**8**) was found to be temperature dependent, suggesting exchange between the free and coordinated double bonds of 1,3-butadiene, as observed in solution. This was confirmed by a two-dimensional ^{13}C CP–MAS magnetization transfer experiment which gave cross-peaks as a result of exchange between the inner and outer carbon atoms, as shown in Equation (4).

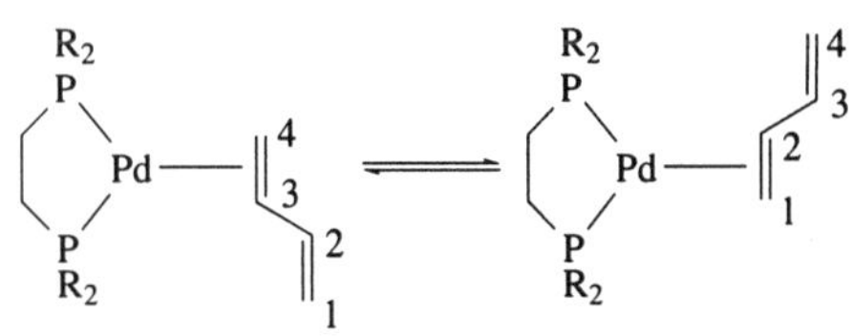

(4)

The measurement of the intensities of the diagonal and cross-peaks and knowledge of the mixing time allowed the value of $\Delta G^{\neq}$ for the exchange process to be estimated as 51 kJ mol^{-1}. The ^{31}P CP–MAS NMR spectrum of (**8**) remained as an AB pattern through the variable-temperature study, indicating that no exchange of phosphorus sites occurred. A mechanism for the process represented by Equation (4) has been proposed that involves conversion of the coordinated *s-trans* η^2-diene, via the *s-cis* η^2-form, to an η^4-intermediate. The collapse of the η^4-intermediate thus leads to exchange of the coordinated and free double bonds. No such process was observed for **(7)** at 300 K.

6.2.2.5 Electrochemistry

The dinuclear palladium(0) and platinum(0) complexes of dba, and a number of derivatives with substituted dba ligands, have been investigated by cyclic voltammetry.[65] Measurements were performed on DME solutions, 0.2 M in tetra-*n*-butylammonium perchlorate, at 243 K and 296 K employing a 0.1 V s^{-1} scan rate and an unspecified working electrode material. The cyclic voltammograms of the parent dba complexes were interpreted in terms of two reversible couples corresponding to $[M_2(dba)_3]^{0/-}$ and $[M_2(dba)_3]^{-/2-}$, a reversible couple corresponding to $(dba)^{-/2-}$, and two additional, irreversible steps of unspecified origin. With the 4,4'-dichloro derivative of dba as the ligand a reversible couple attributed to $(ligand)^{0/-}$ was identified, which suggested ligand dissociation from $[M_2(ligand)_3]$ in solution and which, in the platinum case, masked the $[M_2(ligand)_3]^{-/2-}$ couple. Oxidation occurred in a single irreversible step for all complexes examined, except for those of the 4,4'-dichloro derivative of dba where oxidation peaks were not observed. The half-wave potentials determined for $[M_2(ligand)_3]^{0/-}$ were found to be 0.29–0.45 V more positive than those for $(ligand)^{0/-}$, and this was interpreted in terms of a LUMO for $[M_2(ligand)_3]$ that is more lower lying than the π*-orbital of the free ligand. The half-wave potentials determined for $[M_2(ligand)_3]^{0/-}$ and $[M_2(ligand)_3]^{-/2-}$ were found to be largely independent of the nature of M and linearly related to the half-wave potentials for $(ligand)^{0/-}$ with slopes greater than unity. Although these data might suggest a LUMO for $[M_2(ligand)_3]$ that is predominantly of ligand π*-character, the observation of just two reduction steps (and not three, one for each ligand) was interpreted in terms of a "delocalized" LUMO of twofold degeneracy. The slopes of >1 observed in the plots of half-wave potentials were taken as an indication that the energies of the LUMOs of $[M_2(ligand)_3]$ were more sensitive to substituent effects than the energies of the ligand π*-orbitals.

A related study of the cyclic voltammetry of palladium(0) and platinum(0) complexes containing both alkene ligands and neutral, bidentate nitrogen-donor ligands (e.g., phen, bipy, and substituted derivatives) has been performed.[66] Complexes of the type [M(alkene)(bidentate)] (M = Pd, Pt; alkene = TCNE, maleic anhydride, (*E*)-2-butenedinitrile, dimethyl fumarate, diethyl fumarate, dimethyl maleate) were examined as DME solutions, 0.2 M in tetra-*n*-butylammonium perchlorate, at 243 K employing a 0.1 V s^{-1} scan rate and an unspecified working electrode material. Complexes containing 4,7-diphenyl-1,10-phenanthroline as the bidentate nitrogen donor exhibited four reversible one-electron steps in reductive sweeps while all other complexes exhibited two such steps. A representative pair of cyclic voltammograms is shown in Figure 2.

The results of this study are summarized as follows: (i) the half-wave potentials ($E_{1/2}$) for $(complex)^{0/-}$ were more positive than the $E_{1/2}$ values for $(bidentate)^{0/-}$ by 0.4–0.9 V and more negative than the $E_{1/2}$ values for $(alkene)^{0/-}$ by 0.1–1.5 V; (ii) the $E_{1/2}$ values for $(complex)^{0/-}$ and $(complex)^{-/2-}$ were more positive by ~0.1 V for platinum than for palladium; (iii) the $E_{1/2}$ values for $(complex)^{2-/3-}$ and $(complex)^{3-/4-}$ for the complexes containing 4,7-diphenyl-1,10-phenanthroline were essentially independent of the nature of M; and (iv) the differences in $E_{1/2}$ values between $(complex)^{0/-}$ and $(complex)^{-/2-}$ were nearly identical to the differences in $E_{1/2}$ values between $(bidentate)^{0/-}$ and $(bidentate)^{-/2-}$. These results were interpreted in terms of a LUMO for [M(alkene)(bidentate)] that is dominated by the π*-orbital of the bidentate ligand. The third and fourth reduction steps observed for complexes of 4,7-diphenyl-1,10-phenanthroline were tentatively attributed to processes involving a redox orbital largely localized on the bidentate ligand. Since the π*-orbitals of the bidentate ligands are lower lying than the π*-orbitals of the free alkene ligands (e.g., the $E_{1/2}$ value for $(TCNE)^{0/-}$ is 2.4 V more positive than the $E_{1/2}$ value for $(bipy)^{0/-}$), this interpretation suggests that coordination significantly stabilizes the π*-orbitals of the bidentate ligands and destabilizes the π*-orbitals of the alkene ligands.

6.2.2.6 Thermodynamic (equilibrium constant) studies

Equilibrium constants for Equation (5) (see also Section 6.2.2.4 and Equation (1)) have been determined by UV–visible spectroscopy.[67]

The advantage of the UV–visible determination over earlier, closely related, equilibrium constant determinations by NMR methods (Equation (1) and Table 1) is that more dilute solutions can be

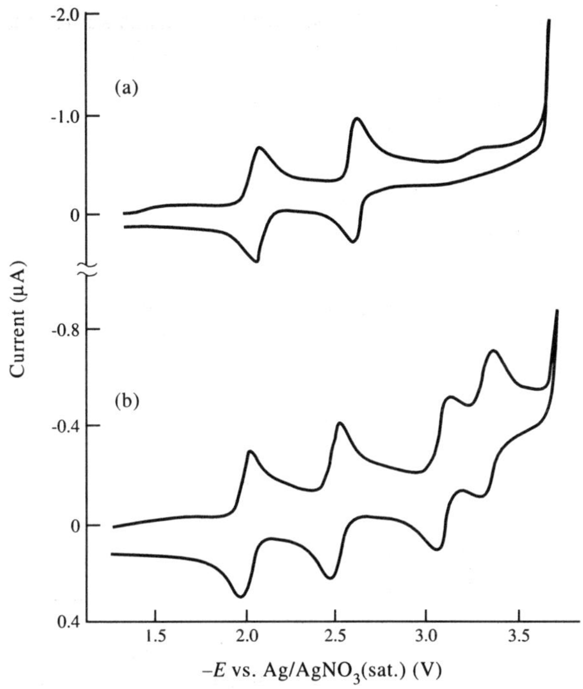

Figure 2 Cyclic voltammograms of (a) [Pd(bipy)(alkene)] and (b) [Pd(bidentate)(alkene)] (bidentate = 4,7-diphenyl-1,10-phenanthroline; alkene = (*E*)-2-butenedinitrile) as 0.3 mM solutions in DME (0.2 M in tetra-*n*-butylammonium perchlorate) at 243 K with a scan rate of 0.1 V s^{-1} (reproduced by permission of the Chemical Society of Japan from *Bull. Chem. Soc. Jpn.*, 1985, **58**, 2323).

$$[\{M\}\text{–NCR}^1][X] + \text{alkene} \rightleftharpoons [\{M\}\text{–alkene}][X] + R^1CN \qquad (5)$$

$\{M\} = [Pd(\eta\text{-Cp})(PR^2_3)]^+$; $R^1 = o\text{-MeC}_6H_4$; alkene = $p\text{-YC}_6H_4CH{=}CH_2$

Y = H, Cl, Me, OMe; X = ClO_4, BF_4 when R^2 = Ph; X = ClO_4 when R^2 = Et, Bu^n

employed, thus minimizing possible effects due to ion-pairing phenomena. The results of this study showed that smaller ΔG^0 values were obtained with styrenes bearing electron-donating substituents and that variations in ΔG^0 were dominated by the contribution of ΔS^0, with little substituent dependence observed for ΔH^0. In order to probe further the effects governing palladium–alkene bonding in these systems, x-ray structure determinations of [(**1**)][PF_6], [Pd(η-C_5H_5)(PPh_3)(*p*-$MeOC_6H_4CH{=}CH_2$)][BF_4] (**10**), and [Pd(η-C_5H_5)(PPh_3)(*p*-$ClC_6H_4CH{=}CH_2$)][BF_4] (**11**) were performed.

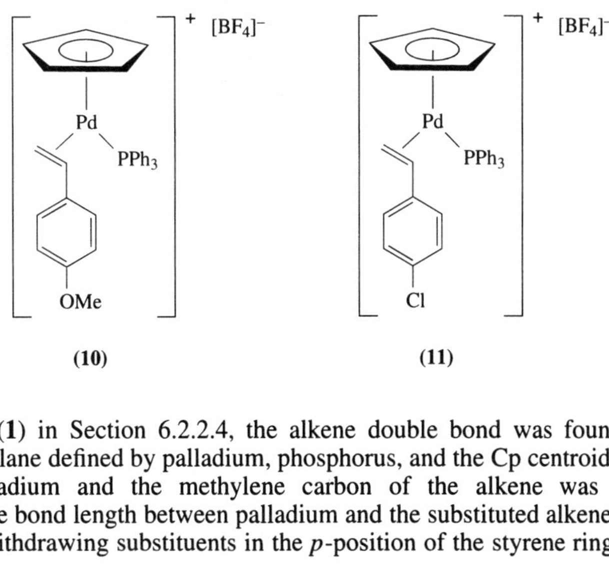

(10) **(11)**

As described for (**1**) in Section 6.2.2.4, the alkene double bond was found to lie approximately perpendicular to the plane defined by palladium, phosphorus, and the Cp centroid in each case. The bond length between palladium and the methylene carbon of the alkene was essentially substituent independent, while the bond length between palladium and the substituted alkene carbon became shorter with more electron-withdrawing substituents in the *p*-position of the styrene ring. The latter trend is the opposite to that observed earlier[68] from x-ray crystallographic studies of 16-electron platinum(II) complexes of the type *trans*-[$PtCl_2$(*p*-MeC_5H_4N)(*p*-$YC_6H_4CH{=}CH_2$)] (Y = H, NMe_2) where, in addition, measurements of $^1J(^{195}Pt, ^{13}C)$ have confirmed that this nonsymmetrical bonding remains in solution. The bonding in the 18-electron palladium(II) complexes has been interpreted in terms of binding of a d^8 ML_4 fragment to the alkene to produce an "octahedral" geometry, where the alkene π*-orbital interacts well with a filled metal orbital (b_2) in the "axial" plane, that is, π-backbonding is unusually important in these types of palladium(II)–alkene complexes.

6.2.3 Synthetic Methods

6.2.3.1 Synthesis of palladium(0)–monoene complexes

In *COMC-I* several convenient methods for the synthesis of palladium(0) complexes of monoenes were described, namely: (i) displacement reactions of [$Pd(cod)_2$], for example, with ethene to produce [$Pd(C_2H_4)_3$]; (ii) related displacement reactions of [$Pd_2(dba)_3$] or [$Pd(dba)_3$]; (iii) reductive elimination reactions of palladium(II)–dialkyl complexes in the presence of alkenes; and (iv) metal atom vapor techniques. In this section methods are discussed for the synthesis of palladium(0)–monoene complexes including those compounds where a diene functions as a ligand by coordination of just one double bond to palladium.

Perfluoronorbornadiene[69] and related ligands[70] have been found to react with [$Pd(PPh_3)_4$] to generate [$Pd(PPh_3)_2(\eta^2\text{-}C_7F_8)$] and analogues. Similar reactions of platinum(0) were also reported and [$Pt(PPh_3)_2(\eta^2\text{-}C_7F_8)$] was found to react with a further equivalent of [$Pt(PPh_3)_4$] to generate [$Pt_2(PPh_3)_4(\mu\text{-}\eta^2\text{-}\eta^2\text{-}C_7F_8)$]. No dinuclear complex of palladium(0) was reported, however.

The reduction of (*R*)-[$PdCl_2$(diop)] (where diop is 2,3-*O*-isopropylidene-2,3-dihydroxy-1,4-bis-(diphenylphosphino)butane) with sodium borohydride in ethene-saturated dichloromethane–ethanol at 243 K followed by warming to room temperature has allowed the isolation of [$Pd(C_2H_4)$(diop)] in 70% yield.[71] The complex was characterized crystallographically (and so became the first structurally characterized palladium complex of ethene) which revealed only a small torsion angle between the PdP_2 and PdC_2 planes and a short carbon–carbon bond length, indicative of a weak palladium–alkene interaction. The ethene ligand was found to be readily displaced by either strained or electron-poor alkene donors, such as norbornene and TCNE, which thus allowed access to a range of [Pd(alkene)(diop)] complexes.

Photolysis of [$Pd(C_2O_4)L_2$] (L = PBu^n_3, L_2 = dppe) in acetonitrile or dimethyl sulfoxide solutions has been found[72] to lead to liberation of CO_2 and generation of putative [PdL_2] equivalents which could be trapped with C_2H_4 or C_2F_4 to generate unstable solutions of [$Pd(alkene)L_2$]. The unisolable complexes were characterized by NMR methods. The corresponding chemistry of platinum oxalates has been more fully developed and useful synthetic routes have evolved.[73]

The β-substituted, σ-vinylpalladium(II) complexes, [$PdX(CH{=}CHCO_2R)(PPh_3)_2$] (X = Cl, I; R = Me, Et), which are readily available through oxidative addition of the (*E*)- and (*Z*)-β-haloacrylates to [$Pd(PPh_3)_4$], undergo thermal rearrangement[74] into the isomeric η^2-alkene-ylide complexes [$PdX\{CH(CO_2R){=}CHPPh_3\}(PPh_3)$] (Equation (6)).

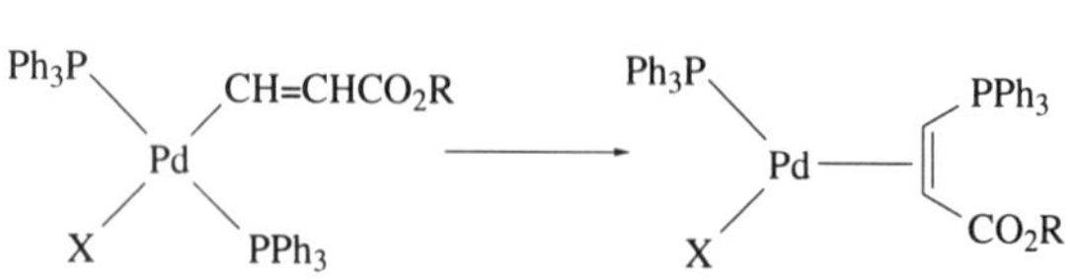

(6)

The x-ray crystal structure where X = I and R = Me was reported and showed coordination of the 1-triphenylphosphonium-2-carbomethoxyethene ligand to palladium with the double bond essentially in the coordination plane. The reaction is unusual since the reverse process, Equation (7), is more commonly observed.

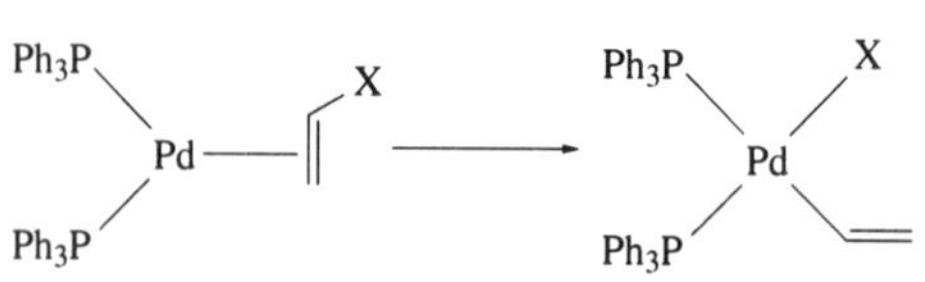

(7)

The cationic palladium(II)–allyl complexes [Pd(η^3-2-RC_3H_4)(bidentate)]$^+$ (bidentate = α-diimine; R = H, Me) have been shown[75] to react with tetraphenylborate anion in the presence of activated alkenes (alkene = fumaronitrile, dimethyl fumarate, maleic anhydride) to produce [Pd(η^2-alkene)(bidentate)] with liberation of $PhCH_2CR{=}CH_2$ and triphenylboron. The reaction was proposed to occur via ion pairing between the cationic complex and BPh_4^-, followed by rate-determining transfer of a phenyl group to palladium and rapid reductive elimination of the allylbenzene (see Section 6.4.3.8). [Pd(η^2-alkene)(bidentate)] complexes have previously been prepared[76] by the more standard method of dba displacement from palladium(0) precursors.

Since palladium(0) complexes of simple alkenes exhibit generally low thermal stabilities, alkenes containing a remote donor atom have been employed in synthesis in order to offer greater stabilization. The ligand *o*-$CH_2{=}CHC_6H_4PPh_2$ (*o*-styryldiphenylphosphine) has been shown[77] to react with [Pd(C_5H_5)(η^3-C_3H_5)] to generate [Pd(*o*-$CH_2{=}CHC_6H_4PPh_2$)$_2$]. The x-ray crystal structure of this complex showed that one ligand was coordinated to palladium in a bidentate fashion and the other ligand was coordinated through only the phosphorus, that is, the complex was found to be three-coordinate and approximately trigonal planar in geometry. The structure contrasts with those of the nickel and platinum analogues where four-coordinate, tetrahedral complexes are formed. The 16-electron palladium complex (cf. 18-electron nickel and platinum) compensates partially for its coordinative unsaturation through a weak agostic interaction with the α-hydrogen of the uncoordinated vinyl group.

[Pd(C_5Me_5)(η^3-C_3H_5)] and analogues with substituted allyl ligands have been shown[78,79] to react with tertiary phosphines and phosphites (L) in a stepwise fashion with initial formation of [Pd(η^1-C_5Me_5)(η^1-C_3H_5)(L)]. Reaction with a second molecule of L led to hydrogen transfer from a methyl group on the C_5Me_5 ring to the allyl ligand to generate the tetramethylfulvenepalladium(0) complex [Pd(η^2-$CH_2{=}C_5Me_4$)L_2] (elimination of isobutene was confirmed with the 2-methylallyl complex). The x-ray crystal structure of the complex where L = PMe_3 confirmed η^2-binding of the fulvene to palladium. The reaction is not limited to [Pd(C_5Me_5)(η^3-C_3H_5)] and also proceeds with the methylcyclopentadienyl analogue.[80] This reaction contrasts with similar reactions of [Pd(C_5H_5)(η^3-C_3H_5)] with bidentate phosphine ligands in the presence of norborn-2-ene where [Pd(bidentate)(η^2-norbornene)] complexes were formed.[81]

The bis(allyl)palladium(II) complex [Pd(η^3-C_3H_5)$_2$] has been shown to be a useful precursor for the generation of palladium(0)–alkene complexes supported by the bidentate phosphine ligands $Pr^i_2PCH_2CH_2PPr^i_2$ and $Bu^t_2PCH_2CH_2PBu^t_2$.[82] The preparative routes developed are outlined in Scheme 2.

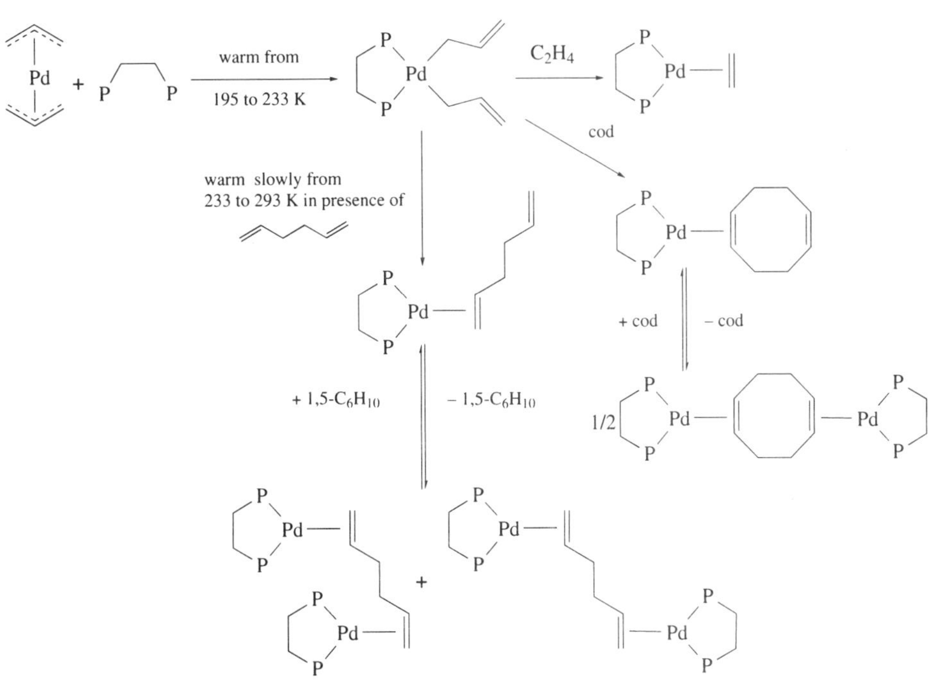

Scheme 2

6.2.3.2 Synthesis of palladium(II)–monoene complexes

In *COMC-I* two general methods for the synthesis of palladium(II)–alkene complexes were described, namely the direct reaction of $PdCl_2$ with an alkene and the displacement of weakly bound ligands (L) from $[PdCl_2L_2]$ or $[Pd_2Cl_4L_2]$. The former method is typically unsatisfactory for routine synthetic use since it is dependent on the history of the $PdCl_2$ employed and is frequently accompanied by formation of π-allyl complexes and by competing alkene oxidation. The latter method, for example, displacement of PhCN from $[PdCl_2(NCPh)_2]$, has been used more extensively. Nucleophilic attack on Pd^{II}–diene complexes (see Section 6.2.4) may also lead to monoene complex formation.

(i) Neutral, four-coordinate palladium(II)–monoene complexes

The reactions of some unusual, electron-rich 1,3-diaza-2-sila-4-cyclopentenes (**12**) with palladium(II) halides have been investigated and the x-ray crystal structure of a complex of the type $[Pd_2Cl_4(alkene)_2]$, prepared from $[PdCl_2(NCPh)_2]$, reported.[83]

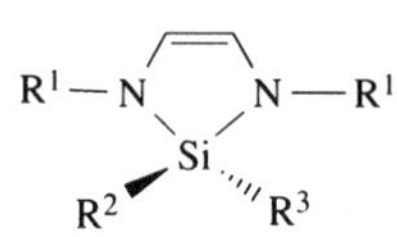

(**12**)

The syntheses are somewhat unusual since electron-rich alkenes more typically react with palladium(II) compounds to generate products derived from cleavage of the alkene into carbene fragments.[84]

Simple palladium(II) precursors such as $[Na]_2[PdCl_4]$ and $PdCl_2$ (as the DMF complex) have been shown to react with the alkene *C*-allylglycine and its substituted derivatives to generate products whose nature was found to depend on the starting material employed, reaction conditions (especially pH), and the substituents on the *C*-allylglycine ligand.[85] Thus *C*-allylglycine reacted with $[Na]_2[PdCl_4]$ to generate the simple bis(*N,O*-chelate) complex (Equation (8)).

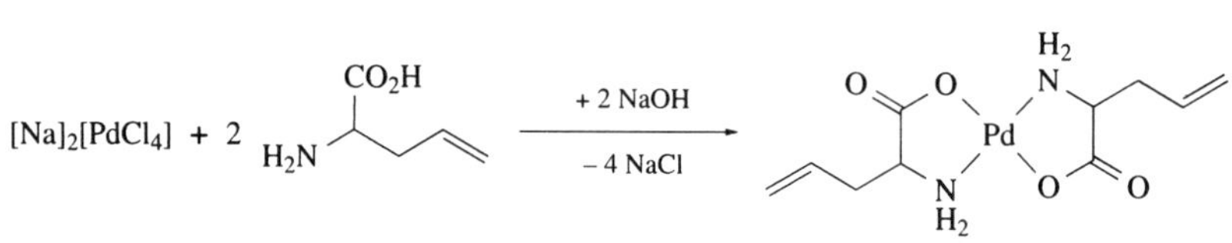

(8)

$[PdCl_2(DMF)_2]$ was found to react with *C*-allylglycine and a number of substituted derivatives to produce *N*,η^2-alkene chelate complexes (Equation (9)).

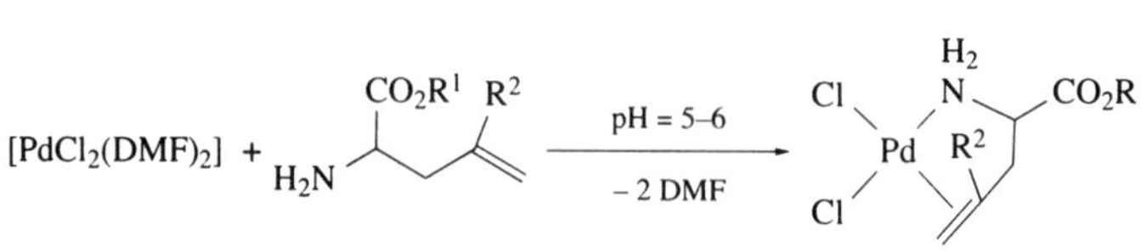

(9)

$R^1 = R^2 = H$; $R^1 = H$, $R^2 = Me$; $R^1 = Me$, $R^2 = H$; $R^1 = Et$, $R^2 = Me$

The parent complex with *C*-allylglycine bound as an *N*,η^2-alkene chelate was characterized by x-ray crystallography. *N*-(diphenylmethylene)-*C*-allylglycine was shown to react with $[Na]_2[PdCl_4]$ to yield a complex in which the ligand had undergone orthometallation (Equation (10)).

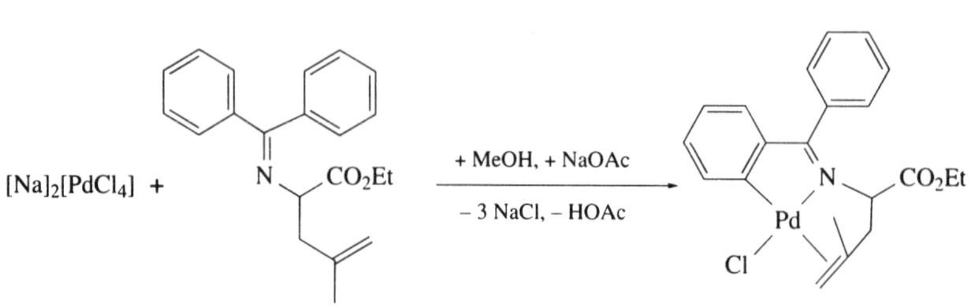

(10)

The structure of a closely related platinum(II) complex orthometallated in the same fashion was reported. A novel η^2-vinyl alcohol complex of palladium(II) has been prepared[86] by treatment of (**13**) with hydrogen chloride in acetone to effect a keto–enol tautomerism that generated (**14**).

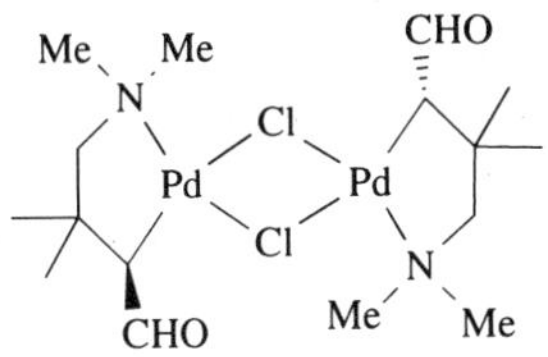

(**13**)

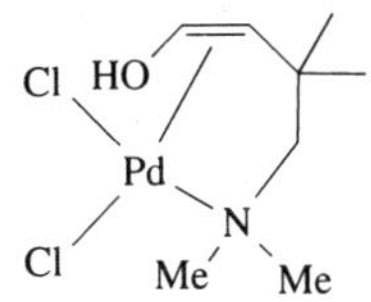

(**14**)

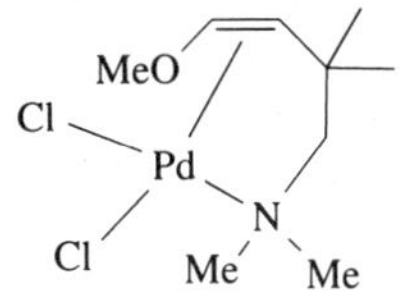

(**15**)

The x-ray structure of (**14**) has been determined and indicated that the alkene is bound at ~79° to the coordination plane. The alkene carbon bearing the hydroxy group was found to form the longer bond to palladium (see Section 6.2.2.6). The x-ray crystal structure of the closely related vinyl ether complex (**15**) has also been reported.[87]

Complexes (**13**), (**14**), and (**15**) are of interest since treatment of (**16**) with methanol in the presence of base leads to formation of both a reduced product, (**17**), and oxidized products, (**13**) and (**18**), through a mechanism in which (**14**) and (**15**) have been suggested as intermediates.

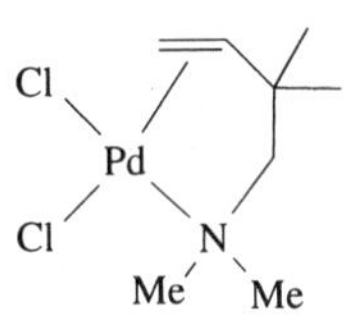

(**16**)

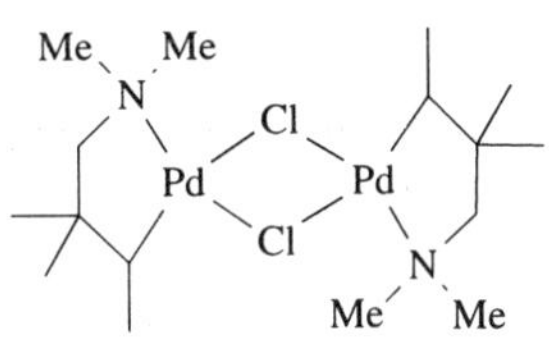

(**17**)

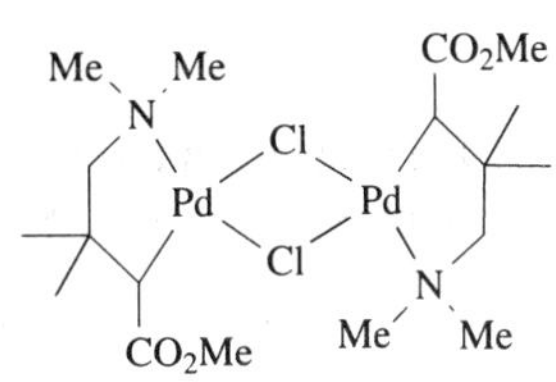

(**18**)

Complex (**16**) is a useful synthetic intermediate since it has been shown to react with diazomethane in diethyl ether–ethanol to generate the product of carbene insertion into the palladium–chlorine bond, which has been characterized crystallographically,[88] thus allowing access to complexes with *cis*-Pd–R and Pd–alkene functionalities through further manipulation of the Pd–CH_2Cl group.

In (**16**), coordination of the alkene double bond is supported through the Pd–N interaction with the remote amine group. This general approach has allowed the synthesis of a variety of Pd^{II}–alkene complexes where the alkene forms part of a chelating bidentate ligand. For example, the ligand (**19**) is reported to react with $[Na]_2[PdCl_4]$ to produce a chelate complex with palladium bound to both the alkene and thioether groups.[89]

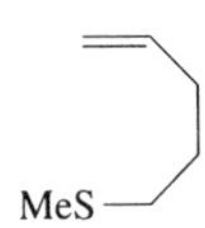

(**19**)

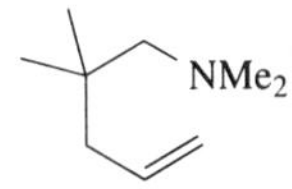

(**20**)

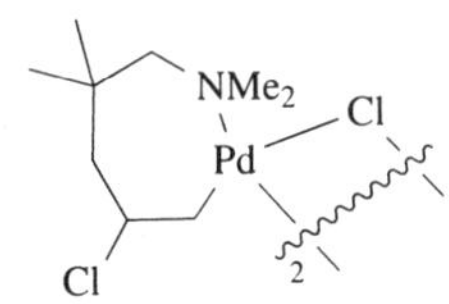

(**21**)

In contrast, the reaction of (**20**) with $[PdCl_2(NCPh)_2]$ led to the formation of (**21**), in which the alkene group has undergone insertion into a palladium–chlorine bond.[90]

The multiple insertion of alkynes into the palladium–carbon bonds of cyclopalladated five- and six-membered ring compounds, typically derived from nitrogen-donor ligands, may give rise to alkene complexes where only one of several double bonds is coordinated to palladium. The formation of cyclopalladated compounds is outside the scope of this chapter but the subject has been reviewed.[91,92]

The cyclopalladated derivative of *N*-benzylidenebenzylamine was reported[93] to react with two equivalents of diphenylethyne according to Equation (11).

The nine-membered ring product was characterized by x-ray crystallography. Closely related triple-insertion reactions of cyclopalladated 8-methylquinoline have also been reported.[94] The insertion of alkynes into the palladium–carbon bonds of cyclopalladated compounds is more typically performed under conditions where palladium is extruded and organic heterocycles are formed; such processes are further discussed in Section 6.3.4.

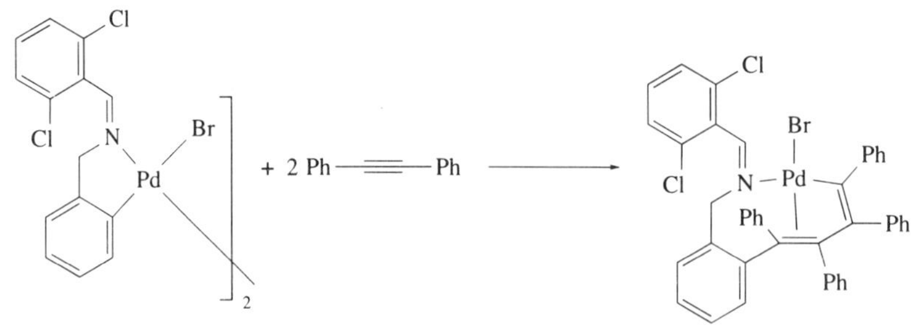

(11)

(ii) Neutral, five-coordinate palladium(II)–monoene complexes

Despite the fact that a large number of five-coordinate platinum(II)–alkene complexes have been prepared and characterized,[95] relatively few good synthetic methods have been reported for the preparation of analogous palladium(II) complexes. For example, although $[Pt_2Cl_4(C_2H_4)_2]$ has been used as a precursor to five-coordinate platinum(II)–alkene complexes through reactions with bidentate ligands in the presence of alkenes, similar reactions with the palladium analogue of Zeise's dimer yield only $[PdCl_2(bidentate)]$. However, the reactions of $[Pd_2Cl_2(Me)_2(SMe_2)_2]$ with planar, sterically demanding nitrogen-donor ligands in the presence of selected alkenes have been shown to generate five-coordinate palladium(II)–alkene complexes of the type [PdClMe(bidentate)(alkene)] in reactions directly analogous to those of platinum(II).[96] The x-ray crystal structure of the palladium complex where bidentate = 2,9-dimethyl-1,10-phenanthroline and alkene = maleic anhydride was determined. The complex was found to be essentially trigonal bipyramidal with the alkene in the equatorial plane and the chloride and methyl groups in axial positions.

Qualitative NMR measurements showed that the palladium–alkene bonds in these five-coordinate complexes were less labile with respect to alkene dissociation than the palladium–alkene bonds in typical four-coordinate complexes and that the barriers to rotation were higher (see also Section 6.2.2.4).

(iii) Cationic, five-coordinate palladium(II)–monoene complexes

Cationic monoene complexes of the type $[Pd(\eta\text{-}C_5H_5)(PR_3)(alkene)][X]$, which may be regarded formally as five-coordinate, have been prepared[97] as isolable solids by halide abstraction from neutral precursors with silver salts of poorly coordinating anions (Equation (12)).

$$[Pd(\eta\text{-}Cp)Br(PR^1_3)] + AgX + CH_2{=}CHR^2 \longrightarrow [Pd(\eta\text{-}Cp)(PR^1_3)(CH_2{=}CHR^2)][X] + AgBr \quad (12)$$

X = ClO_4, R^1 = Ph, R^2 = H, Me, Ph, *p*-YC_6H_4 (Y = Cl, Me, OMe); X = ClO_4, R^1 = Et, R^2 = Ph; X = ClO_4, R^1 = Bu^n, R^2 = Ph, *p*-YC_6H_4 (Y = Cl, Me, OMe, NO_2, MeCO); X = BF_4, R^1 = Bu^n, R^2 = Ph, *p*-YC_6H_4 (Y = Cl, OMe)

Additional complexes of this type were generated in solution without isolation by displacement of nitrile ligands from cationic precursors (Equation (13)).

$$[Pd(\eta\text{-}Cp)(PR^1_3)(L)][X] + alkene \rightleftharpoons [Pd(\eta\text{-}Cp)(PR^1_3)(alkene)][X] + L \quad (13)$$

L = PhCN, *o*-MeC_6H_4CN; alkene = $CH_2{=}CHR^2$ (R^2 = various substituents)

The cationic alkene complexes were typically stable towards decomposition in acetone or chloroform solutions at ambient temperature over several days but decomposed rapidly at temperatures above 323 K. Decomposition of the styrene complex [(**1**)][ClO_4] led to identification of $[Pd(\eta\text{-}C_5H_5)(PPh_3)_2][ClO_4]$ and the styrene dimer *trans*-1,3-diphenyl-1-butene in solution.

The cationic alkene complexes were found to react with selected nucleophiles by attack on the coordinated alkene to generate neutral alkyl complexes (Equation (14)).

$$[(\mathbf{2})][ClO_4] + NaY \longrightarrow [Pd(\eta\text{-}Cp)(PPh_3)(CH_2CH_2Y)] + NaClO_4 \quad (14)$$

Y = OMe, OPr^i, $CH(COMe)_2$

6.2.3.3 Synthesis of palladium(0)–diene complexes

Ligand displacement from palladium(0) precursors continues to be employed in the synthesis of palladium(0)–diene complexes. Thus, [Pd(quinone)(cod)] complexes of type (**22**) have been synthesized by the reaction of $[Pd_2(dba)_3]$ with quinones and cod.[98]

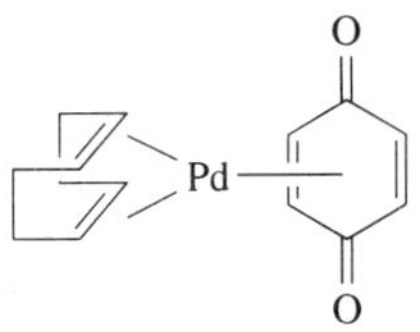

(**22**)

The structural assignment of η^4-quinone coordination was based on NMR and IR spectroscopic data. The cod ligand in complexes of type (**22**) could be displaced by ligands (L) such as PPh_3, thus allowing access to complexes of the formula $[Pd(quinone)L_2]$, where the hapticity of the quinone ligand was unspecified. In closely related platinum(0) systems it is known from x-ray structural studies that η^2-quinone coordination is commonly seen.[99] Similarly, the reactions of $[Pd(PPh_3)_4]$ with 5,8-dihydro-1,4-naphthoquinone and 5,8,9,10-tetrahydro-1,4-naphthoquinone are reported[100] to produce the η^2-quinone complexes (**23**) and (**24**), respectively.

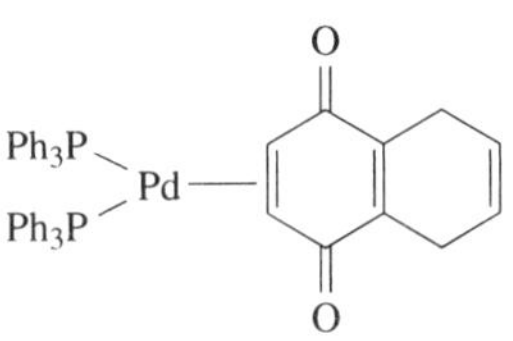

(**23**)

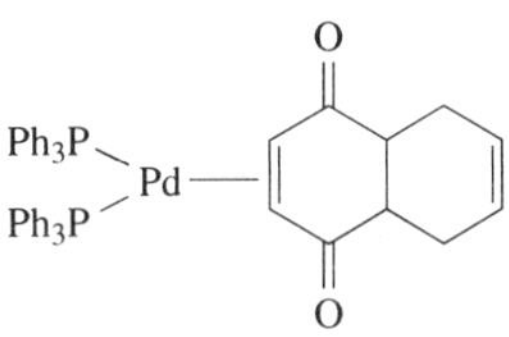

(**24**)

The reaction of $[Pd_2(dba)_3]\cdot CHCl_3$ with 1,4-naphthoquinone alone was found[98] to produce $[Pd(1,4\text{-naphthoquinone})_2]$. A similar reaction with *p*-benzoquinone generated an insoluble, possibly polymeric, compound.

6.2.3.4 Synthesis of palladium(I)–diene complexes

The vast majority of palladium(I) complexes exist as diamagnetic dimers; mononuclear paramagnetic species are exceedingly uncommon and typically unisolable. However, neutral palladium(I) radicals have been obtained by the electrochemical, one-electron reduction of the palladium(II) cations $[Pd(\eta\text{-}C_5Ph_5)(\eta^4\text{-diene})]^+$ (diene = cod, nbd, dibenzocyclooctatetraene) and the dibenzocyclooctatetraene complex proved to be isolable.[101] Initial investigations of $[Pd(\eta\text{-}C_5H_5)(cod)]^+$ by cyclic voltammetry showed that the reduction to palladium(I) was reversible only at sweep rates above several volts per second at ambient temperature. At 0.5 V s^{-1} and 263 K, the ratio of anodic to cathodic currents was 0.8, allowing a room temperature half-life of ~0.1 s to be estimated for the neutral radical. With pentaphenylcyclopentadienyl as the supporting ligand, greater stabilities (longer half-lives) were found. $[Pd(\eta\text{-}C_5Ph_5)(cod)]^+$ gave rise to a diffusion-controlled, chemically reversible reduction wave ($E^0 = -0.47$ V) with both mercury and platinum electrodes. Bulk electrolysis in dichloromethane with a platinum electrode produced a solution whose ESR spectrum was consistent with formation of the neutral radical $[Pd(\eta\text{-}C_5Ph_5)(cod)]$. Thus, the ESR spectrum in dichloromethane at 233 K gave a *g* value of 2.0706 with the central signal flanked by satellites due to ^{105}Pd with $\langle a\rangle_{Pd} = 25 \times 10^{-4}$ T. $[Pd(\eta\text{-}C_5Ph_5)(\eta^4\text{-diene})]^+$ (diene = dibenzocyclooctatetraene) was reduced at a higher potential ($E^0 = -0.22$ V) than $[Pd(\eta\text{-}C_5Ph_5)(cod)]^+$ and bulk electrolysis allowed generation of the neutral radical which was found to be stable for several hours at 300 K. The radical was isolated from the electrolyzed solution, although separation of all the background electrolyte (tetra-*n*-butylammonium hexafluorophosphate) proved problematic. The palladium(I)–dibenzocyclooctatetraene complex has been the subject of a detailed ESR study.[102]

6.2.3.5 Synthesis of palladium(II)–diene complexes

A wide variety of useful synthetic methods for the preparation of palladium(II) complexes of 1,4- and 1,5-dienes have been reported and *COMC-I* details many of these. Two methods of general use are as follows: (i) the direct reaction of certain dienes with $[Na]_2[PdCl_4]$ in a suitable solvent, often acetone or water, which is reported to lead to formation of $[PdCl_2(diene)]$—the use of nucleophilic solvents, for example, ethanol, for such reactions may lead to σ-bonded derivatives in certain cases through nucleophilic attack on a coordinated diene; and (ii) the displacement of nitrile ligands from $[PdCl_2(NCR)_2]$ (R is typically Ph), monoene ligands from $[Pd_2Cl_4(monoene)_2]$, or CO from "[PdCl(CO)]" which may be employed to generate $[PdCl_2(diene)]$. In these reactions isomerization of an alkene substrate, for example, isomerization of cycloocta-1,3-diene to the 1,5-isomer, may occur. 1,3nbhyph;Dienes more typically react with $[Na]_2[PdCl_4]$ or $[PdCl_2(NCR)_2]$ to generate π-allyl complexes (see Section 6.4.2.2), although alkene intermediates have been isolated in certain cases. Cyclobutadiene complexes are often accessible through the reactions of R^1CCR^2 with $[PdCl_2(NCR)_2]$ (see Section 6.2.3.6).

The synthesis of $[PdCl_2(cod)]$ from $PdCl_2$ and cod, and its reaction with methoxide anion, has been detailed as an undergraduate laboratory experiment.[103] Aqueous $Pd(ClO_4)_2$ has been reported[104] to react with excess cod in THF to produce $[Pd(1,3\text{-}\eta\text{-}C_8H_{13})(cod)][ClO_4]$, a compound isolated previously[105] as the tetrafluoroborate salt from the reaction of $[PdCl_2(cod)]$ with $AgBF_4$ in the presence of cod. It has been suggested that $[Pd(cod)_2]^+$ is an intermediate in the reaction that abstracts hydride from the excess cod to generate a $1\text{-}\sigma\text{-}4,5\text{-}\eta\text{-}C_8H_{13}$ complex that subsequently rearranges to $[Pd(1,3\text{-}\eta\text{-}C_8H_{13})(cod)]^+$. The cod ligand in $[Pd(1,3\text{-}\eta\text{-}C_8H_{13})(cod)]^+$ is reportedly displaced by a range of neutral ligands (PPh_3, $AsPh_3$, bidentate nitrogen donors) and by halide ions.

The complex $[Pd(C_6F_5)Cl(cod)]$ has been prepared[106] by *trans*-arylation between $[Pd(C_6F_5)_2(solvent)_2]$ (solvent = THF, diethyl ether) and $[PdCl_2(NCPh)_2]$, followed by treatment with cod (Equations (15) and (16)).

$$[Pd(C_6F_5)_2(solvent)_2] + [PdCl_2(NCPh)_2] \xrightarrow[\text{solvent = THF, diethyl ether}]{} [Pd_2(\mu\text{-}Cl)_2(C_6F_5)_2(NCPh)_2] + 2\ \text{solvent} \quad (15)$$

$$[Pd_2(\mu\text{-}Cl)_2(C_6F_5)_2(NCPh)_2] + 2\ \text{cod} \longrightarrow 2\ [Pd(C_6F_5)Cl(cod)] \quad (16)$$

The complex was characterized crystallographically and the x-ray structure indicated that the high *trans*-influence of the C_6F_5 group weakened the palladium–alkene interaction of the double bond coordinated *trans* to this group. In solution the complex was found to undergo an intramolecular double-bond insertion into the Pd–C_6F_5 bond to generate the allyl complex $[Pd_2(\mu\text{-}Cl)_2(6\text{-}C_6F_5\text{-}1,3\text{-}\eta^3\text{-}C_8H_{12})_2]$ and the σ,π-complex $[Pd_2(\mu\text{-}Cl)_2(8\text{-}C_6F_5\text{-}1:4,5\text{-}\eta^3\text{-}C_8H_{12})_2]$ (Equation (17)).

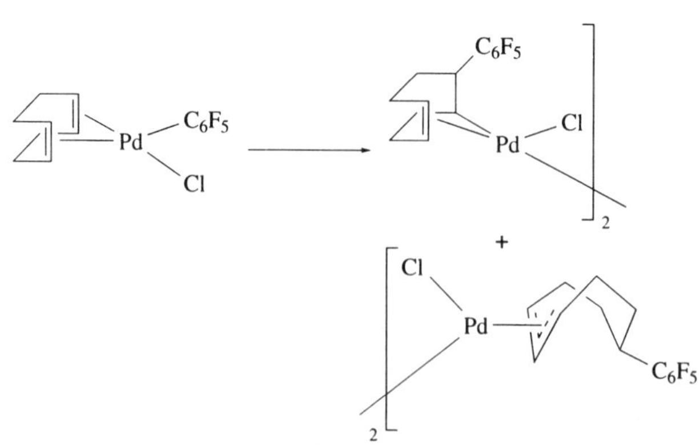

(17)

The process illustrated in Equation (17) is of interest since it represents isomerization, by *endo*-insertion, of a complex containing mutually *cis* Pd–R and Pd–alkene functionalities, a reaction commonly invoked in catalytic cycles that had previously been directly observable only for palladacyclic complexes. In contrast with the process represented by Equation (17), where both an allyl complex and a σ,π-complex are formed from a common precursor, the reaction of 1,5-hexadiene with $[Pd(C_6F_5)Br(NCMe)_2]$ forms a σ,π-complex in a fast reaction that is followed by slow conversion to a π-allyl complex[107] (Scheme 3).

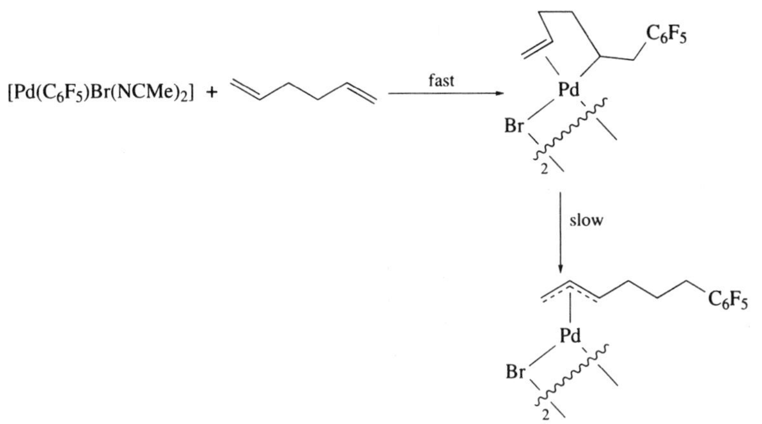

Scheme 3

The isomerization of a σ,π-complex into a π-allyl complex has a precedent in the chemistry of cyclooctenylpalladium systems, where it has been shown that palladium migration occurs with face retention.[108]

The reaction between *cis*-$[M^1(C_6X^1{}_5)_2(THF)_2]$ (M^1 = Pd, Pt; X^1 = F, Cl) and $[M^2X^2{}_2(cod)]$ (M^2 = Pd, Pt; X^2 = F, Cl, I) has been shown[109] to produce homo- and heteronuclear complexes of the type $[(C_6X^1{}_5)_2M^1(\mu\text{-}X^2)_2M^2(cod)]_n$. Interestingly, the complexes have been found to be dinuclear ($n = 1$) in solution but tetranuclear ($n = 2$) in the solid state. The x-ray crystal structure of one example demonstrated that the crystalline complexes existed as eight-membered rings consisting of alternating "$M^1(C_6X^1{}_5)_2$" and "M^2(cod)" units held together by single μ-X^2 bridges.

Halide abstraction from $[Pd_2(\mu\text{-Cl})_2(\eta^3\text{-}C_3H_5)_2]$ or $[Pd_2(\mu\text{-Cl})_2(\eta^3\text{-2-MeC}_3H_4)_2]$ in acetone followed by addition of $[Co(CO)_3(\eta^4\text{-cot})]$ has been reported to generate compounds of the type (**25**), described as "pseudo-triple-decker."[110]

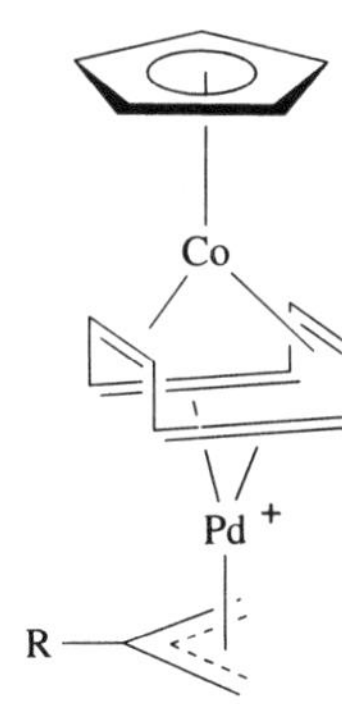

R = H, Me

(25)

6.2.3.6 Synthesis of palladium(II)–cyclobutadiene complexes

Additional examples of the synthesis of substituted cyclobutadiene complexes of palladium(II) from disubstituted alkynes and palladium(II) precursors have been reported since *COMC-I* was published and more x-ray structural data on these compounds have been published.

The reaction of bis-*p*-methoxyphenylethyne (AnCCAn) with $[Na]_2[PdCl_4]$ in ethanol has been shown[111] to generate the salt $[Pd_2(\mu\text{-Cl})_3(\eta^4\text{-}C_4An_4)_2]_2[Pd_2Cl_6]$ (**26**), which, on treatment with HCl, formed $[Pd_2(\mu\text{-Cl})_2Cl_2(\eta^4\text{-}C_4An_4)_2]$ (**27**). Compound (**27**) was characterized by x-ray crystallography.

Compounds analogous to (**26**) and (**27**) have been isolated previously, for example from PhCCPh, but (**27**) represented only the second cyclobutadiene complex of palladium to be characterized crystallographically. There were significant structural differences between (**27**) and the compound structurally characterized previously[112] (**28**).

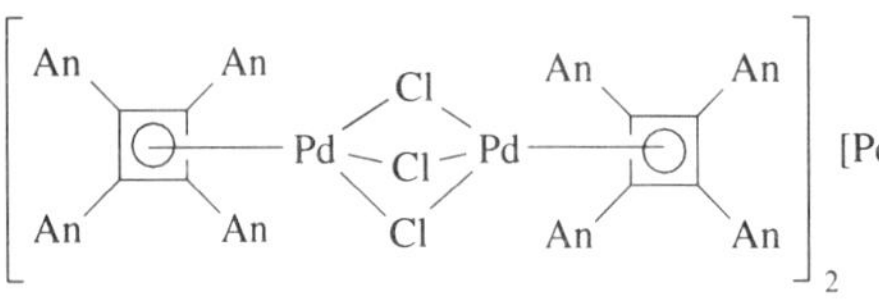

(26)

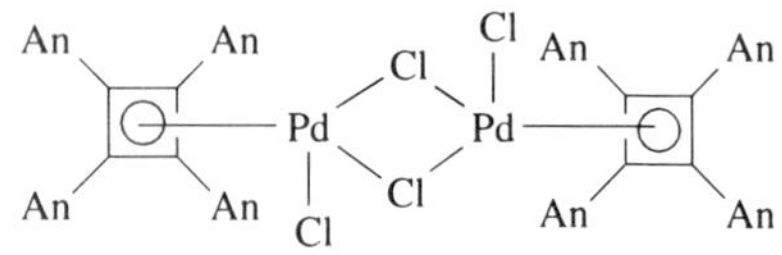

(27)

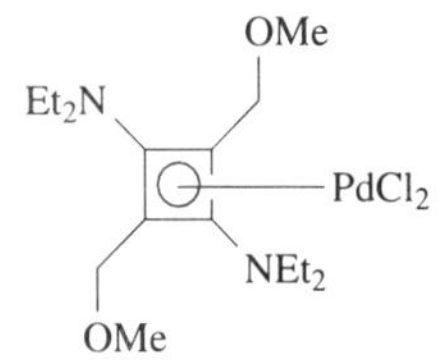

(28)

Thus, in (**27**) the C_4 ring was essentially square planar, whereas in (**28**) the ring was folded about the diagonal $MeOCH_2C\cdots CCH_2OMe$ axis at an angle of 155°, with the palladium attached to the convex side.

The reaction of $[AlCl_3(C_4Me_4)]$, formed by cyclodimerization of 2-butyne with $AlCl_3$, with $[PdCl_2(NCPh)_2]$ was reported to produce $[PdCl_2(\eta^4\text{-}C_4Me_4)]$, which was characterized spectroscopically and through degradation studies.[113]

6.2.4 Reactions of Palladium–Alkene Complexes

Three basic classes of reactions for palladium–alkene complexes were described in *COMC-I*: (i) substitution of L in *trans*-$[PdCl_2(L)(\text{alkene})]$, an uncommon reaction that is seen extensively with the platinum(II) analogues; (ii) substitution of the alkene in palladium(0)– and palladium(II)–alkene complexes, a reaction type that forms the basis for many synthetic methods; and (iii) nucleophilic attack on coordinated alkenes. The last class of reaction is of importance as it forms the basis of many stoichiometric and catalytic transformations of alkenes based on palladium chemistry.

Palladium(0)– and palladium(II)–alkene complexes continue to be exploited as valuable starting materials through substitution reactions and many descriptions of useful syntheses are extant in the literature. Two illustrative examples are cited here and others are found in Section 6.2.3.3. The potentially chelating phosphine–amide ligand *o*-$Ph_2PC_6H_4NHC(O)Me$ was reported[114] to react with $[Pd_2(dba)_3]\cdot dba$ in toluene solution to produce [Pd(dba){*o*-$Ph_2PC_6H_4NHC(O)Me$}$_2$], in which the dba ligand was coordinated through one double bond, a monodentate phosphine–amide ligand was coordinated through phosphorus, and the second phosphine–amide ligand was chelated through phosphorus and nitrogen. Formation of the *P,N*-chelate, rather than the possible *P,O*-chelate, was rationalized on the basis of preferred chelate ring size. The reaction of $[PdCl_2(cod)]$ with the potentially dinucleating macrocyclic ligand (**29**) produced the complex $[PdCl_2(\mathbf{29})]$ as the *anti*-isomer, with the ligand bound through the two phosphorus donors and providing a cavity potentially capable of binding other metals.[115]

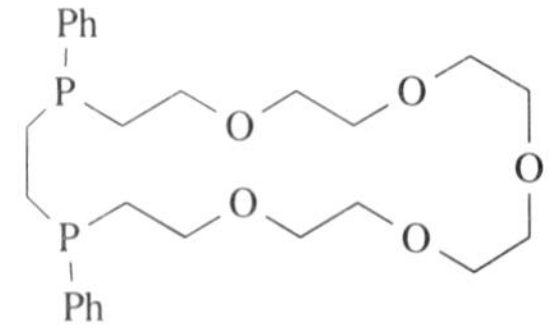

(29)

The origin of activation towards nucleophilic attack by coordination of an alkene to a metal has been examined theoretically.[116,117] It was suggested that "slippage" of a symmetrically bound η^2-alkene gives rise to a species that is activated towards attack by nucleophiles (Scheme 4).

To test this hypothesis, a conformationally constrained complex $[PdCl_2(\textit{endo}\text{-}\eta^4\text{-dicyclopentadiene})]$, which shows selectivity in reactions with nucleophiles, was examined by x-ray crystallography.[118] While the midpoints of both double bonds were found to be at normal distances from palladium, one was found to be slipped with respect to the $PdCl_2$ plane such that its midpoint was 0.034 nm from the plane and

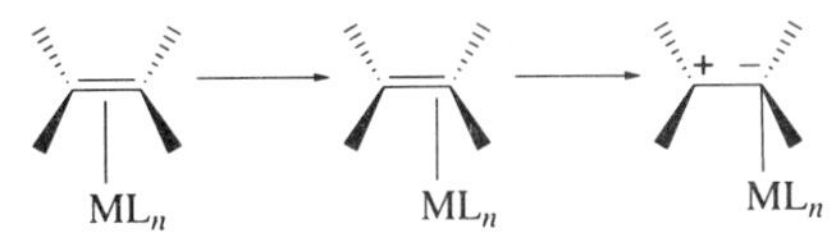

Scheme 4

one carbon was brought closer to the plane. However, this slippage did not result in the second carbon moving further from palladium because the alkene was tilted by 13° from its normal 90° angle with respect to the $PdCl_2$ plane. The result of these two distortions was that the second carbon was brought closer to palladium than the first. The other coordinated double bond exhibited no such distortions. The predicted site of nucleophilic attack, that is, the carbon atom of the distorted double bond that was farthest from palladium, was that found in experimental studies.

The dienes *cis*- and *trans*-1,2-divinylcyclohexane form π-complexes with palladium without conformational restraints. In solution [$PdCl_2$(η^4-*trans*-1,2-divinylcyclohexane)] has been found[119] to adopt a square-planar geometry with the alkene double bonds nearly parallel and perpendicular to the $PdCl_2$ plane and the cyclohexane ring in a chair conformation. The complex of *cis*-1,2-divinylcyclohexane was found to be fluxional between two chair conformers. The solid-state structures, determined by x-ray crystallography, closely resembled the solution structures with two crystallographically independent molecules of [$PdCl_2$(η^4-*cis*-1,2-divinylcyclohexane)] in the unit cell. Oxidative cyclizations of *cis*- and *trans*-1,2-divinylcyclohexane with $Pd(OAc)_2$–MnO_2–*p*-benzoquinone–HOAc were found to be highly stereoselective, producing (**30**) and (**31**), respectively.

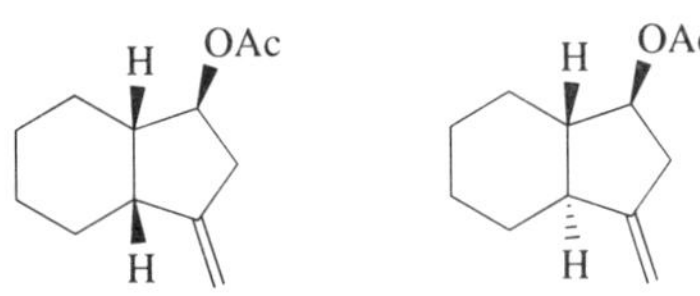

(**30**) (**31**)

Nucleophilic attack *trans* to palladium at either of the double bonds of [$PdCl_2$(η^4-*cis*-1,2-divinylcyclohexane)] would lead to (**30**), but studies of the oxidative cyclization of a diene substituted to allow differentiation of sites indicated that only the coordinated double bond in the equatorial position was attacked in the process that led to formation of products. The suggestion from these results is that one chair conformer of [$PdCl_2$(η^4-*cis*-1,2-divinylcyclohexane)] led to formation of the (1*R*,6*S*,7*S*)-product while the other conformer led to the product with the opposite absolute configuration. However, (**31**) was isolated as the stereoisomer of (1*S**,6*S**,7*S**)-configuration only, implying that only one double bond was attacked. The oxidative cyclization of [$PdCl_2$(η^4-*cis*-1,2-divinylcyclohexane)] and [$PdCl_2$(η^4-*trans*-1,2-divinylcyclohexane)] has been studied in detail[120,121] and the results suggest that modest structural differences, even when remote from the site of attack, may be important in controlling product formation.

Conclusions based on regiochemical control of nucleophilic attack must be treated with caution since the possibility of equilibria between σ- and π-complexes exists. Thus, amination of 1-hexene by diethylamine in the presence of [$PdCl_2(NCPh)_2$] and the ligand (**32**) has been examined,[122] where the intermediate σ-complexes were cleaved with lead acetate to yield the oxidized products (**33**) and (**34**).

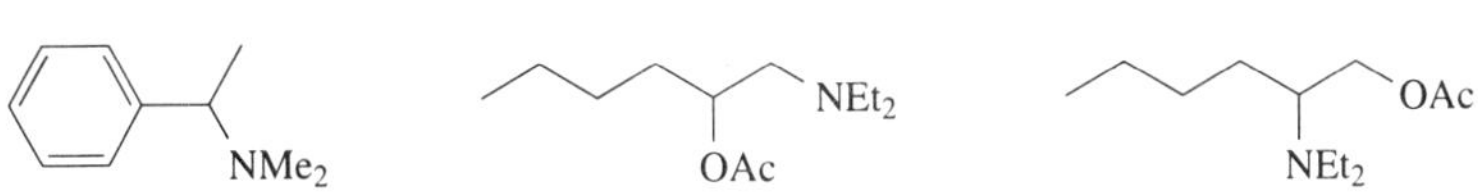

(**32**) (**33**) (**34**)

At short reaction times (~1 min) the ratio of (**33**) to (**34**) was ca. 2:1, whereas after 30 min a 1:1 ratio was obtained, that is, although the predicted product is that of Markownikoff addition, the kinetic product results from anti-Markownikoff addition. These results were explained in terms of an equilibration between the two intermediate σ-complexes via the π-complex of 1-hexene. Each of the σ-complexes was cleaved in an essentially irreversible step (Scheme 5).

Nucleophilic attack on the coordinated diene ligands in [MX_2(cod)] (M = Pd, Pt; X = Cl, Br) by cyclohexylamine has been the subject of kinetic studies.[123] It was found that, when X = Br, the palladium complex was about 70 times more reactive than the platinum analogue, and this was attributed to poorer π-backbonding in the palladium complex. The bromo complex was found to be more reactive than the chloro compound.

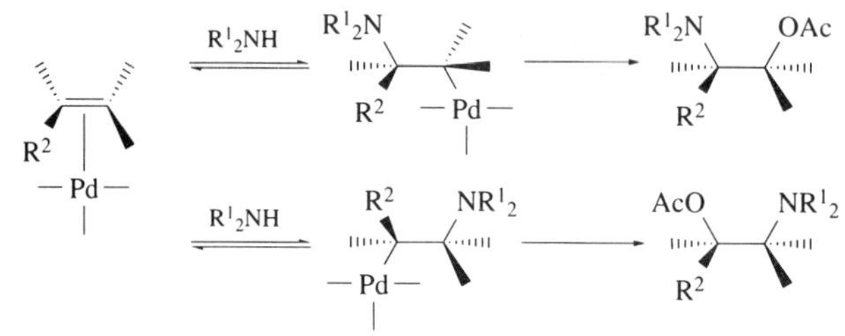

Scheme 5

The intimate mechanism of nucleophilic attack on coordinated alkenes has been a subject of debate for many years. Two limiting pathways can be envisaged (Scheme 6).

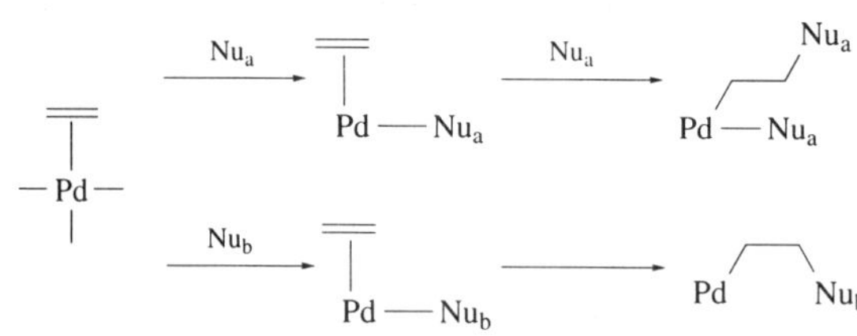

Scheme 6

One group of nucleophiles (Nu_a in Scheme 6) generally add by external *trans*-attack (e.g., water–hydroxide, carboxylates, alkoxides, and amines) while a second group (Nu_b in Scheme 6) typically add by *cis*-attack through intramolecular migration (e.g., hydride, alkyl, and aryl). The issue of prior coordination is not necessarily a major factor in controlling the pathway, since coordination of a nucleophile might be followed by *trans*-attack of a second equivalent. An important issue, therefore, is the relative reactivity of different coordinated nucleophiles towards intramolecular *cis*-migration and whether or not such a process is competitive with external *trans*-attack. This situation has been examined theoretically with the use of Hartree–Fock SCF MO calculations employing an effective core potential approximation.[124,125] The complexes *trans*-$[Pd(Nu)_2(H_2O)(C_2H_4)]$ (Nu = H, Me, OH, F) were studied with the assumption that the reacting complexes had square-planar geometries. Pd–Nu distances were optimized while internal ligand geometries and other palladium–ligand distances were fixed. In order to assess the possibility of ligand migration, frontier orbitals were examined. The HOMO was the orbital describing the Pd–Nu bond and the LUMO was the π*-orbital, which was approximated as the π*-orbital of free ethene since SCF calculations do not describe the energies of empty orbitals very well. The HOMO–LUMO separations were found to be large for Pd–OH and Pd–F complexes and small for Pd–H and Pd–Me complexes, suggesting that the latter two systems may be susceptible to migration. The requirement of a low-energy HOMO for Pd–Nu to undergo *cis*-migration correlates with a high-energy HOMO for the free nucleophile, that is, the nucleophile should be soft. The calculations agree with experimental results where classically soft nucleophiles typically undergo *cis*-migration and classically hard nucleophiles typically undergo external *trans*-attack. The situation for nucleophiles of intermediate hardness–softness remains poorly defined.

A theoretical study[126] of the *cis*-migration of hydride in *cis*-$[PdH_2(PH_3)(CH_2CX_2)]$ (X = H, F) by the *ab initio* restricted Hartree–Fock method with energy gradient technique indicated that the transition state involved a four-centered interaction and that the palladium–hydrogen distance in the transition state was not greatly changed from that in the reactant, while the carbon–carbon distance was not greatly altered from that of free ethene (see above discussion). These results indicated that the transition state is reached early, while shortening of the palladium–carbon distances indicated that the transition state is "tight." Alkene insertion into a metal–hydrogen bond has also been studied theoretically for the entire second row of the *d*-block.[127] The SCF calculations on model $[MH(C_2H_4)]$ units were supplemented by examination of systems with one or two extra hydride ligands in order to begin probing the effects of supporting ligands on the insertion process. The major differences in the energetics of insertion between the different metals, and between systems with and without additional hydride ligands, could be explained by a dominant role of repulsion between nonbonding metal electrons and alkene electrons. No relationships between barrier heights and the initial metal–alkene or metal–hydride bond strengths were found.

The general applicability of theoretical and experimental work on model systems to the discussion of working stoichiometric and catalytic organic syntheses is always open to debate, and this has certainly been the case in work aimed at elucidation of the mechanism of the palladium-catalyzed oxidation[128] of simple alkenes to aldehydes and ketones. For the stoichiometric alkene oxidation reaction[129] shown in Equation (18), the rate expression shown in Equation (19) is followed over a given range of conditions ([Pd] = 0.005–0.04 M; $[Cl^-]$ = 0.1–1.0 M; $[H^+]$ = 0.04–1.0 M).

$$[PdCl_4]^{2-} + C_2H_4 + H_2O \longrightarrow Pd + C_2H_4O + 2\,HCl + 2\,Cl^- \quad (18)$$

$$\frac{-d[\text{alkene}]}{dt} = \frac{k_1[PdCl_4^{2-}][\text{alkene}]}{[H^+][Cl^-]^2} \quad (19)$$

It is widely agreed that the $[Cl^-]^2$ term in Equation (19) arises because two chloride ligands are displaced from palladium by water and the alkene. The origin of the proton inhibition is more widely debated and has its origins in the hydroxypalladation step. The kinetics are consistent with either *cis*-addition of coordinated hydroxy in the slow step or *trans*-attack of external water in an equilibrium step followed by rate-determining decomposition of the hydroxyalkyl complex. External attack of hydroxide has been eliminated[130] in the case of allyl alcohol oxidation (which follows a rate law of the form given in Equation (19)), since kinetic studies show that such a mode of attack would give rise to a rate constant for the slow step that is ~100 times higher than rate constants for diffusion-controlled processes in aqueous solution. With allyl alcohol, *trans*-attack of external water in an equilibrium step could also be eliminated through the use of selectively deuterated substrates to monitor possible isomerization through reversible hydroxypalladation; that is, with this substrate both possible modes of *trans*-attack could be excluded.

Further light on the complexities of palladium-catalyzed alkene oxidation has been shed through studies of the isomerization of (**35**) in aqueous solution into an equilibrium mixture of (**35**) and (**36**) (the latter shown with isotope labeling that resulted from reaction in ^{18}O-enriched water) under typical Wacker oxidation conditions (i.e., $[Cl^-] < 1.0$ M; $[H^+] < 0.5$ M).[131]

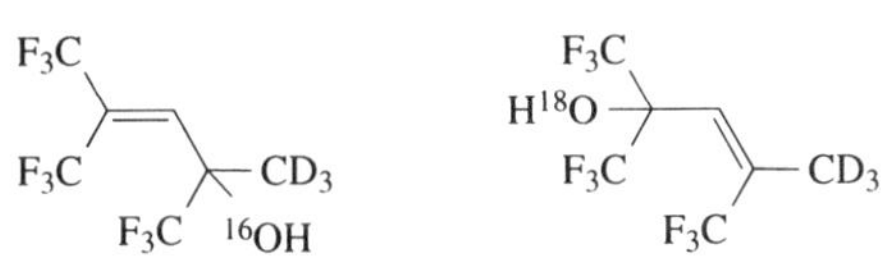

(**35**) (**36**)

The rate expression was found to be that shown in Equation (20) (cf. Equation (19)).

$$\frac{-d[\text{substrate}]}{dt} = \frac{k_1[PdCl_4^{2-}][\text{substrate}]}{[H^+][Cl^-]^2} \quad (20)$$

The rate of the exchange reaction of (**35**) with ^{18}O-enriched water was found to be the same as the rate of isomerization, which requires that isomerization and exchange both occur through hydroxypalladation. The kinetic results were found to be consistent with proton loss in a preequilibrium step followed by *cis*-hydroxypalladation. Use of chiral (*E*)-(**35**) gave (**36**) with the opposite configuration, which was also interpreted in terms of *cis*-hydroxypalladation. A study[132] of the same process under conditions of high chloride ion concentration ($[Cl^-] > 2.0$ M) gave the rate expression shown in Equation (21).

$$\frac{-d[\text{substrate}]}{dt} = \frac{k_1[PdCl_4^{2-}][\text{substrate}]}{[H^+][Cl^-]} \quad (21)$$

The single-chloride dependency is consistent with water attack on an intermediate π-complex from outside the coordination sphere. Use of chiral (*E*)-(**35**) gave (**36**) with the same configuration, but as the (*Z*)-isomer, which was also taken as indicative of *trans*-attack. The results indicated that stereochemical studies, often taken to indicate mechanism, are sensitive to chloride ion concentration, and that studies performed under one set of conditions may not relate to studies performed under different conditions.

The kinetics of alkene oxidation have also been shown to be significantly substrate dependent. Thus, 2-cyclohexenol gave a rate expression that was different from that found for acyclic alkenes under low chloride ion concentrations.[133]

In addition to detailed studies of the mechanism of nucleophilic attack on coordinated alkenes, there are a large number of reports of the application of palladium catalysts in organic synthesis where reactions are postulated to proceed through alkene complexation and nucleophilic attack. For example, 2-styrylbenzamide is converted into 1-hydroxy-3-phenylisoquinoline in 62% yield by treatment with $[Li]_2[PdCl_4]$ in the presence of triethylamine. The mechanism shown in Scheme 7 was presented.

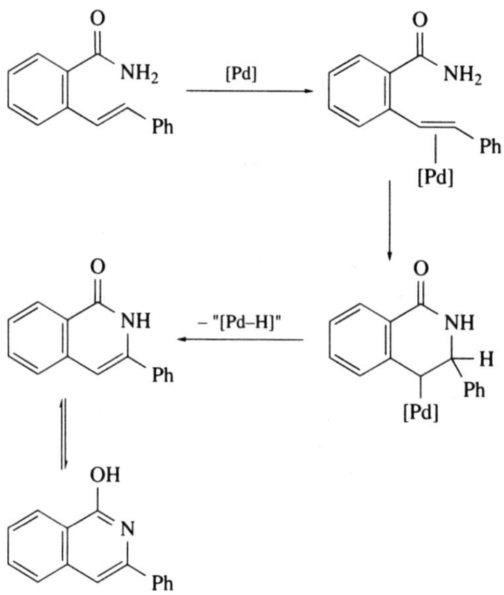

Scheme 7

The cross-coupling of an unsaturated organic halide and a main group organometallic, catalyzed by complexes of nickel and palladium (commonly referred to as the Stille cross-coupling when organostannanes are employed), is important in synthetic organic chemistry and has wide-ranging applications, for example, in the synthesis of heterocyclic species.[134] Examples of applications in organic[135] and organometallic[136] synthesis with palladium complexes as catalysts are shown in Scheme 8.

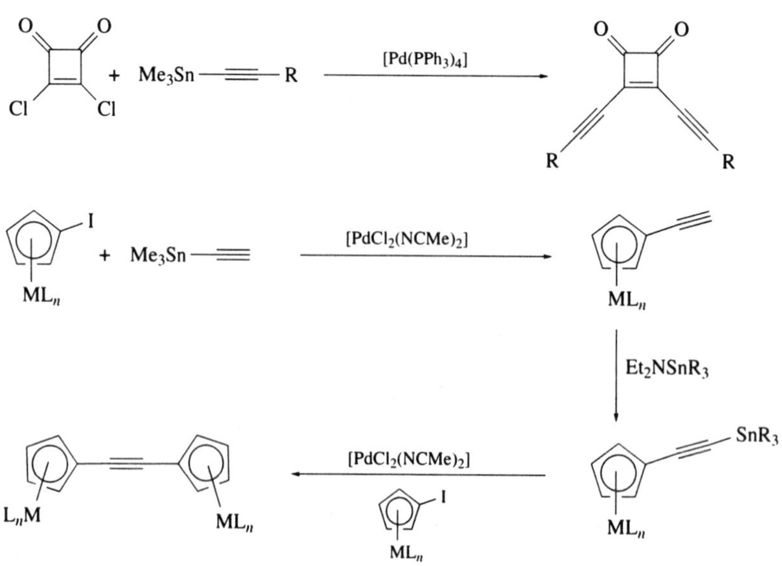

Scheme 8

Palladium(II)–alkene complexes are often invoked as intermediates in palladium-catalyzed cross-coupling reactions of vinylic substrates. For example, in a report[137] on the reaction shown in Equation (22), (**37**) was suggested as an intermediate.

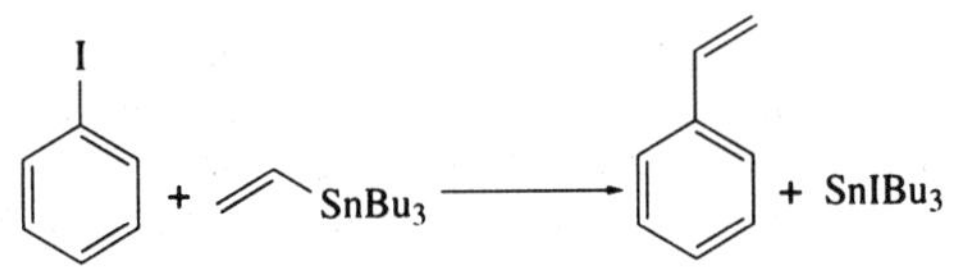

(22)

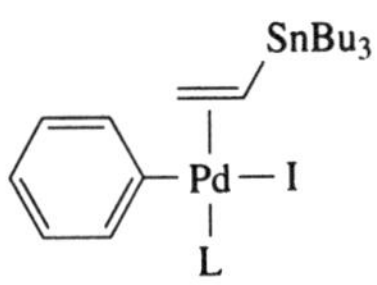

(**37**)

On the basis of NMR studies and isotope-partitioning experiments, a mechanism for the vinylic cross-coupling shown in Equation (23) has been constructed in which there is good evidence for the proposed intermediates.[138] The mechanism, which excludes the participation of palladium(IV) intermediates that had been postulated previously, is outlined in Scheme 9.

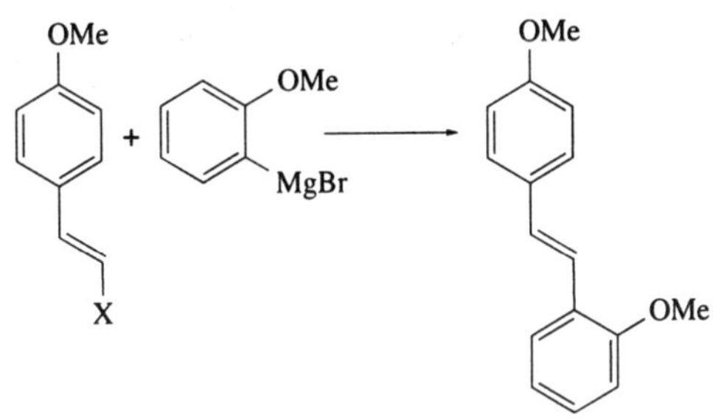

(23)

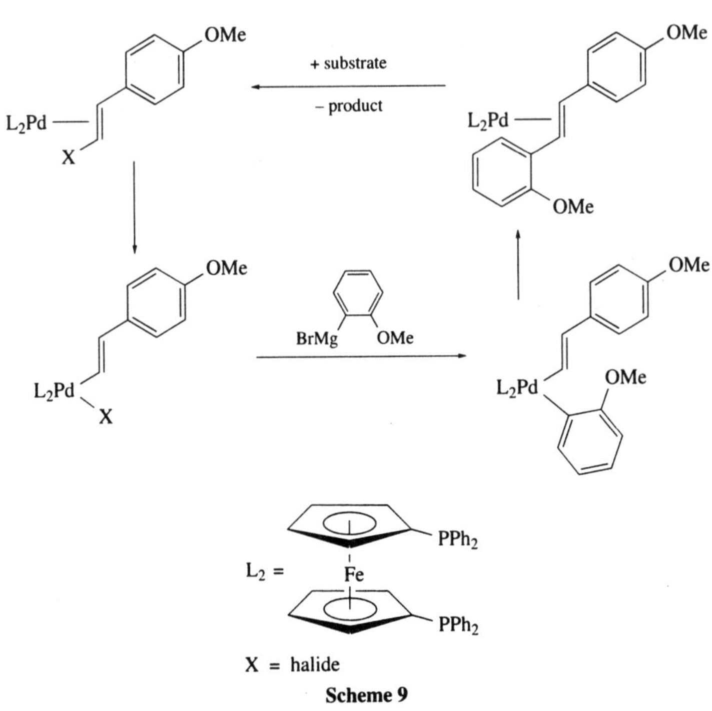

Scheme 9

The dimeric palladium(II) complex of 1,2-di-*t*-butyl-3,4-dimethylcyclobutadiene, $[Pd_2(\mu\text{-}Cl)_2Cl_2(C_4Me_2Bu^t{}_2)_2]$, has been found[139] to react with bases by abstraction of a methyl proton to produce an η^3-methylenecyclobutenyl complex and not the exocyclic η^3-allyl product originally proposed (Scheme 10).

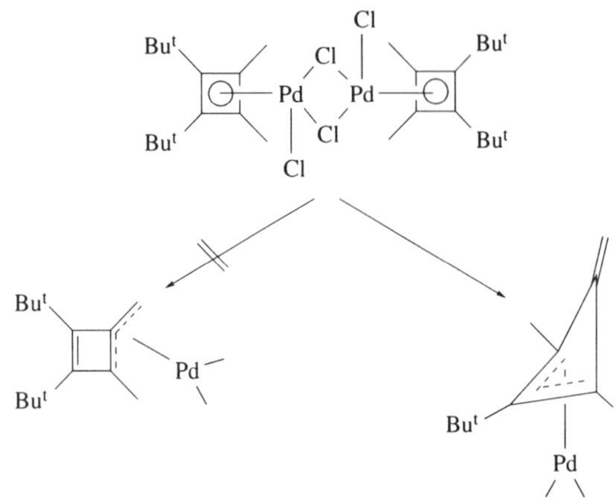

Scheme 10

An x-ray crystal structure determination on the acetylacetonato derivative [Pd{C_4:CH_2(Me)Bu^t_2}(acac)] confirmed η^3-binding at the 1,3-positions with an uncoordinated methylene group at C-4.

6.3 π-COMPLEXES OF ALKYNES

6.3.1 Interactions of Alkynes with Metallic Palladium

The interaction of ethyne with Ni(111) and Pt(111) surfaces has been studied extensively, both theoretically[140–2] and experimentally,[143] and a fairly clear picture of the interactions has evolved. Investigations of ethyne interactions with palladium surfaces, however, have been less definitive in resolving ambiguities. The temperature-dependent behavior of ethyne on Pd(111) over the temperature range 150–300 K has been studied by HREELS.[144] The low-temperature chemisorbed form, postulated to be close to a threefold site with the C–C axis parallel to the surface, was found to transform to a new species above 200 K, believed to be vinylidene (CCH_2), and, if surface hydrogen was present, ethylidyne (CMe) formation was dominant above 250 K. Above 300 K transformation to surface-bound CH groups was observed and this cleavage product appeared to result from the low-temperature alkyne form rather than from ethylidyne. Studies of ethyne on Pd(111) and Pd(110) with HREELS indicated that thermal processing to ~400 K caused formation of surface alkynyl (CCH, alkynlide) groups on Pd(110) and that these species were also formed, along with ethylidyne, following adsorption of ethyne on Pd(111) at 300 K.[145] The EEL spectra indicated that both carbon atoms of the surface alkynyl species were involved in bonding to palladium. For comparison purposes, the EEL spectra of adsorbed benzene and its deuterated forms were recorded. No evidence of benzene formation from adsorbed ethyne was found, although other studies have suggested that this oligomerization does occur on the low-Miller-index planes of palladium.[146]

6.3.2 Complexation of Alkynes to Palladium: Structure and Bonding

The complexation of alkynes to palladium in organometallic complexes is conveniently described in terms of the Dewar–Chatt–Duncanson model,[44,45] as discussed in *COMC-I*.

6.3.2.1 Theoretical and structural investigations

SCF calculations on the coordination of ethyne to atomic palladium have been described and compared with results of CAS–CI (complete active space–configuration interaction) calculations.[147] In common with results on the interaction of π-ligands with other closed-shell metal atoms, it was found that most of the interaction was due to dispersion forces and that the contribution of the Dewar–Chatt–Duncanson model was negligible. Thus, Mulliken population analysis at the equilibrium SCF distance showed very little σ-donation (0.008 electron) and almost no π-backbonding (0.004

electron). The binding energy for the ethyne–palladium system was found to lie in the range 4.6–8.8 kJ mol^{-1}. Optimization of the geometry of the ethyne fragment showed a very small increase in the carbon–carbon bond length (0.0004 nm) and a modest bending of the ethyne (~2°). Closely related CAS–SCF and contracted CI calculations on the ethyne–palladium system confirmed coordination to atomic palladium with minimal distortion of ethyne at the overall energy minimum.[148]

The situation for alkynes coordinated to palladium(0) in complexes is very different to that calculated for atomic palladium, and this is exemplified in the x-ray crystal structures of a number of simple [PdL_2(alkyne)] complexes. The x-ray crystal structure[149] of [$Pd(PPh_3)_2(MeO_2CCCCO_2Me)$], (**38**), revealed that the bend-back angles of the alkyne substituents were 33.6(1)° and 35.1(7)°, consistent with rehybridization on bonding to palladium(0), as expected from the Dewar–Chatt–Duncanson model. The structures of [$Pd(PCy_3)_2(F_3CCCCF_3)$] (**39**)[150] and its platinum analogue [$Pt(PCy_3)_2(F_3CCCCF_3)$] (**40**)[151] revealed little change in geometry for the alkyne fragment on changing from palladium(0) to platinum(0).

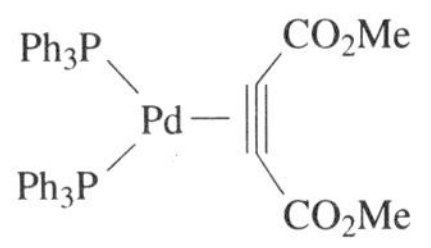

(38)

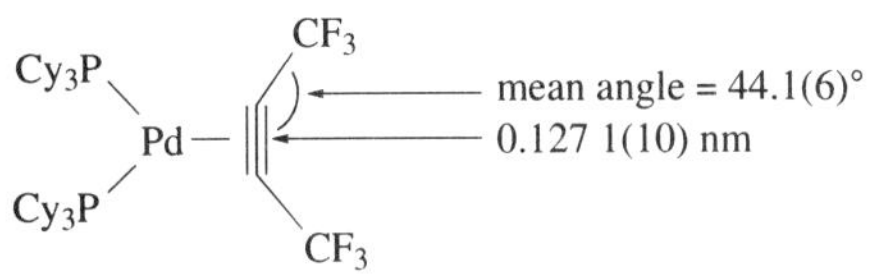

(39)

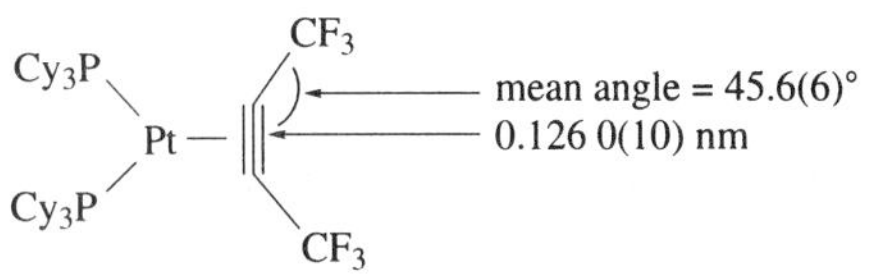

(40)

Hartree–Fock–Slater calculations on the binding of [$M(PH_3)_2$] (M = Ni, Pd, Pt) to ethene and ethyne have been described as part of a broader study.[152] Within the limits of the approximate method, the results showed that ligand-to-metal π-backbonding is more important than metal-to-ligand σ-bonding for stability. Ethene and ethyne gave rise to similar binding energies to palladium and, for a given unsaturated ligand, stability was found to follow the order Ni > Pt > Pd.

Palladium(II) complexes of simple alkynes are rare and little is known about structure and bonding in these systems. The complexes *cis*-[$M(C_6F_5)_2(PhCCPh)_2$] (M = Pd, Pt) have been isolated from the reactions of *cis*-[$M(C_6F_5)_2(THF)_2$] with two equivalents of PhCCPh in dichloromethane and the x-ray crystal structure of the platinum(II) complex determined.[153] The α-carbons of the pentafluorophenyl groups, the platinum, and the midpoints of the alkyne triple bonds were found to lie in a square plane with the carbon–carbon triple bonds inclined slightly with respect to the plane. The carbon–carbon multiple-bond lengths were found to be unchanged from that found in free PhCCPh and the coordinated alkynes were close to linear, suggesting that there is little backbonding in these M^{II} systems.

6.3.3 Synthetic Methods

6.3.3.1 Synthesis of palladium(0)–alkyne complexes

In *COMC-I* the preparation of palladium(0)–alkyne complexes, [PdL_2(alkyne)], by ligand displacement from [PdL_4] (L = tertiary phosphine) with disubstituted alkynes (RCCR) was described. Similar reactions with terminal alkynes (RCCH) were reported to lead to palladium(II)–alkynylhydride complexes via C–H oxidative addition, although palladium(0)–alkyne complexes may be involved as intermediates. Alkynes bearing electron-withdrawing substituents, such as trifluoromethyl groups, are known to undergo complex reactions with palladium(0) precursors that lead to alkyne oligomerization.

Two new methods for the preparation of palladium(0)–alkyne complexes have been reported that may prove to be of general use. The bis(silyl)palladium(II) complex [$Pd(SiHMe_2)_2$(dcpe)] (dcpe = $Cy_2PCH_2CH_2PCy_2$) has been shown[154] to react with MeO_2CCCCO_2Me according to Equation (24).

The silylation of the alkyne occurred stereospecifically and presumably proceeded via insertion of the alkyne into a Pd–Si bond followed by reductive elimination to generate the organic product and a nonlinear [Pd(dcpe)] fragment, which was trapped by a second equivalent of the alkyne.

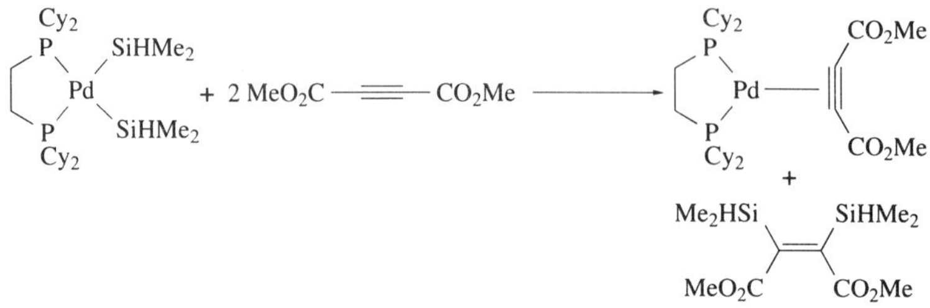

(24)

[Pd(dcpe)] fragments have also been generated by 254 nm photolysis of the oxalate $[Pd(C_2O_4)(dcpe)]$.[155] Interestingly, dimerization of the fragment occurred, producing $[Pd_2(\mu\text{-dcpe})_2]$ which could be isolated and which was crystallographically characterized. In solution an equilibrium between the mononuclear and dinuclear species, with the latter as the major component present, was proposed (Equation (25)).

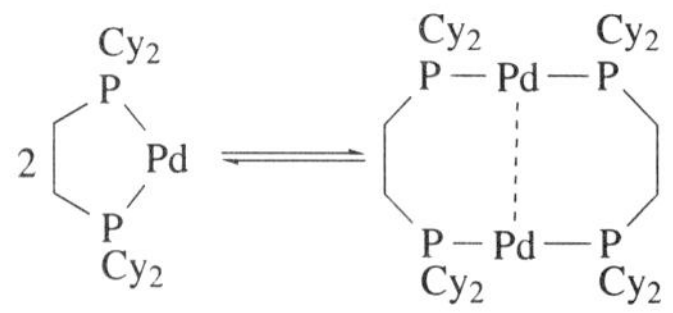

(25)

Reactions of the dinuclear complex with MeO_2CCCCO_2Me and PhCCPh each produced mononuclear complexes, [Pd(dcpe)(alkyne)], but also different dinuclear species, as shown in Scheme 11.

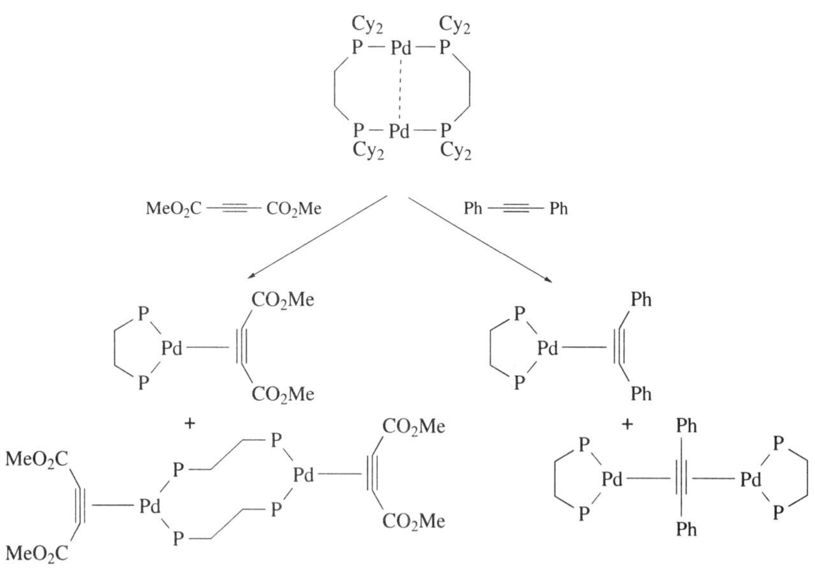

Scheme 11

The 10-membered ring framework of the original dinuclear species was retained in the major product of the reaction with MeO_2CCCCO_2Me, although the palladium–palladium distance increased from 0.276 11(5) nm to ~0.67 nm (an unreported x-ray structure of the novel alkyne complex was cited).

Despite conventional wisdom that terminal alkynes undergo C–H oxidative addition with palladium(0) to generate alkynylhydrides, novel complexes of alkynes supported by the bidentate phosphine ligands $Pr^i_2PCH_2CH_2PPr^i_2$ and $Bu^t_2PCH_2CH_2PBu^t_2$ have been prepared[82] from transformations of $[Pd(\eta^3\text{-}C_3H_5)_2]$, as shown in Scheme 12 (see Scheme 2 also).

A review has been published[156] that includes discussion of matrix isolation studies on the interactions of palladium with alkynes and cocondensation reactions of alkynes with atomic palladium. Useful synthetic procedures may evolve from these studies.

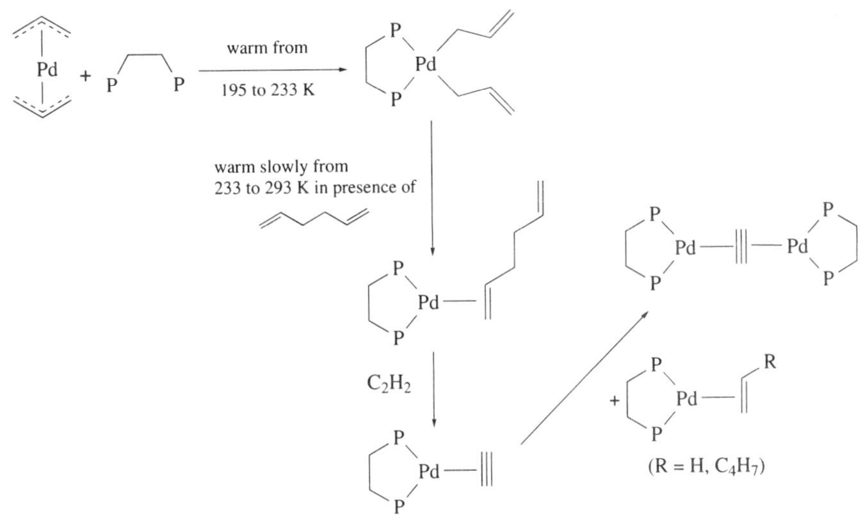

Scheme 12

6.3.3.2 Synthesis of palladium(I)–alkyne complexes

The preparation of palladium(I)–alkyne complexes of two different types was described in *COMC-I*: (i) dinuclear compounds of the type $[Pd_2Cl_2(\mu\text{-dppe})_2]$ were reported to react with disubstituted alkynes bearing electron-withdrawing groups (i.e., CF_3 or CO_2Me) to produce "A-frame" complexes by addition of the alkyne across the palladium–palladium bond; and (ii) compounds with two $[Pd(\eta\text{-}C_5R_5)]$ fragments bound to a single disubstituted alkyne were reported to be formed from reactions such as that between RCCR and $[Pd_3(OAc)_6]$ (where R = Ph).

The practical application of the reaction between $[Pd_2X_2(\mu\text{-dppe})_2]$ (X = Cl, Br, I) and alkynes has been extended to include alkynes without electron-withdrawing substituents (RCCH where R = H, Ph, $p\text{-}C_6H_4Me$) through acid catalysis of the addition reaction by $HBF_4{\cdot}Et_2O$.[157] It is likely that protonation of the palladium–palladium bond activates $[Pd_2X_2(\mu\text{-dppe})_2]$ towards alkyne addition. The reactions of $[Pd_2Cl_2(\mu\text{-dppe})_2]$ and $[Pd_2Cl_2(\mu\text{-dppm})_2]$ with the heteroatom-substituted alkynes ICCI, ClCCCl, PhCCCl, MeSCCSMe, and MeSCCMe have also been investigated. ICCI reacted with $[Pd_2Cl_2(\mu\text{-dppe})_2]$ to generate products with a bridging dichlorovinylidene group, $[Pd_2XY(\mu\text{-}CCCl_2)(\mu\text{-dppe})_2]$ (X = Y = Cl; X = Y = I; X = Cl, Y = I), which resulted from a 1,2-halide shift reaction accompanied by Pd–Cl/C–I bond metathesis.[158] $[PdCl_2(dppe)]$ was also formed in the reaction. The complex with X = Y = Cl was isolated and characterized crystallographically. ICCI was found to react with $[Pd_2Cl_2(\mu\text{-dppm})_2]$ in the dark to generate $[Pd_2I_2(\mu\text{-}CCCl_2)(\mu\text{-dppm})_2]$ and, when reactions were performed in the light, formation of the vinylidene complex was accompanied by generation of the unusual trigonal-bipyramidal palladium(II) complex $[Pd_2I_2(\mu\text{-}I_2)(\mu\text{-dppm})_2]$, which was crystallographically characterized. $[Pd_2Cl_2(\mu\text{-dppe})_2]$ was found to react with ClCCCl to generate $[Pd_2Cl_2(\mu\text{-}CCCl_2)(\mu\text{-dppe})_2]$ exclusively, while $[Pd_2Cl_2(\mu\text{-dppm})_2]$ reacted with ClCCCl to produce a mixture of $[Pd_2Cl_2(\mu\text{-}CCCl_2)(\mu\text{-dppm})_2]$ and the isomeric, bridged-alkyne complex $[Pd_2Cl_2(\mu\text{-ClCCCl})(\mu\text{-dppm})_2]$, which was crystallographically characterized. The isomeric bridged-vinylidene and bridged-alkyne complexes were found to equilibrate in solution.[159] The monosubstituted haloalkyne PhCCCl was found to react with $[Pd_2Cl_2(\mu\text{-dppe})_2]$ to generate a vinylidene-bridged complex where crystallographic disorder prevented an unambiguous structural assignment.[160] The thioalkyne MeSCCSMe was found to react with both $[Pd_2Cl_2(\mu\text{-dppe})_2]$ and $[Pd_2Cl_2(\mu\text{-dppm})_2]$ to generate bridged-alkyne complexes $[Pd_2Cl_2(\mu\text{-MeSCCSMe})(\mu\text{-diphosphine})_2]$, which were crystallographically characterized. The unsymmetrical thioalkyne MeSCCMe, however, was found to react with $[Pd_2Cl_2(\mu\text{-dppm})_2]$ to yield the vinylidene complex $[Pd_2Cl_2\{\mu\text{-CC(SMe)Me}\}(\mu\text{-dppm})_2]$, which was also characterized crystallographically, via a 1,2-sulfur shift reaction.[161] The coordination of alkynes to dinuclear palladium(I) complexes and the palladium(I)-mediated 1,2-group shift reaction display a dependence on the nature of the alkyne substituents that has yet to be completely unraveled.

6.3.3.3 Synthesis of palladium(II)–alkyne complexes

The synthesis of the di(*t*-butyl)alkyne complex $[Pd_2Cl_4(alkyne)_2]$ from the reaction of two equivalents of di(*t*-butyl)alkyne with $[Pd_2Cl_4(C_2H_4)_2]$ was the only known synthesis of a palladium(II)–alkyne complex when *COMC-I* was published. The complexes *cis*-$[M(C_6F_5)_2(PhCCPh)_2]$ (M = Pd, Pt) have now been isolated from the reactions of *cis*-$[M(C_6F_5)_2(THF)_2]$ with two equivalents of PhCCPh (see Section 6.3.2.1), but no other potentially general synthetic methods appear to have been described.

6.3.3.4 Synthesis of palladium–alkyne cluster compounds

In addition to the syntheses of the mononuclear and dinuclear palladium–alkyne complexes discussed in Sections 6.3.3.1, 6.3.3.2, and 6.3.3.3, the synthesis of the first trinuclear palladium complex containing a triply bridging alkyne has been reported.[162] It was found that reaction of $[Pd_3(\mu\text{-}CO)_3(\mu\text{-}dppm)_3][CF_3CO_2]_2$ with alkynes bearing electron-withdrawing substituents, RCCR (where R = CO_2Me or CO_2Et), followed by treatment with ammonium hexafluorophosphate, allowed isolation of the compounds $[Pd_3(\mu_3\text{-}\eta^2\text{-}RCCR)(CF_3CO_2)(\mu\text{-}dppm)_3][PF_6]$ in >90% yields. Subsequent reactions with chloride ion produced $[Pd_3(\mu_3\text{-}\eta^2\text{-}RCCR)Cl(\mu\text{-}dppm)_3][PF_6]$. The compound $[Pd_3(\mu_3\text{-}\eta^2\text{-}RCCR)(CF_3CO_2)(\mu\text{-}dppm)_3][PF_6]\cdot MeOH\cdot H_2O$ (R = CO_2Me), shown as (**41**), was crystallographically characterized.

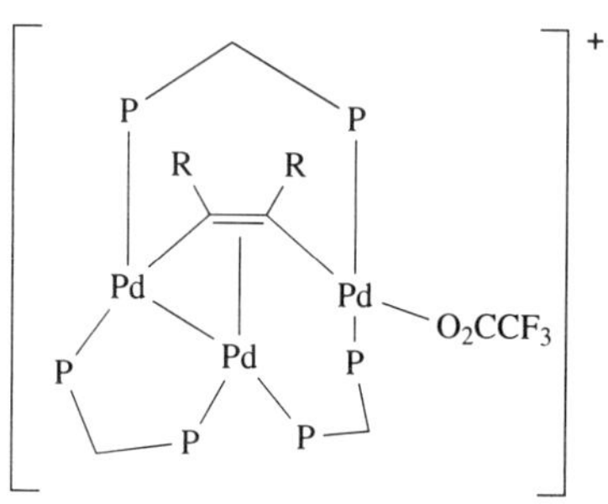

(41)

6.3.4 Reactions of Alkynes with Palladium Complexes

Alkynes typically react with palladium(II) precursors to generate products of alkyne dimerization, trimerization, or tetramerization and similar reactions, although less studied, occur with palladium(0). Useful discussion of this area is included in *COMC-I*. The concept of including alkyne oligomerization steps into organic syntheses based on transformations of cyclopalladated precursors[163] has more recently been developed (see also Section 6.2.3.2(i)). For example, cyclopalladated 2-biphenylyl sulfide is reported[164] to react with diphenylethyne to produce a complex incorporating two PhCCPh units into each cyclopalladated group. The complex was characterized by x-ray crystallography and shown to be a novel dipalladium species with arene moieties coordinated in an η^4-fashion (Scheme 13, where L = NCMe).

On standing in solution at room temperature, the dipalladium complex was found to extrude palladium(0) and generate the organic product shown in Scheme 13. Cyclopalladated *N,N*-dimethyl-2-biphenylamine was found to react differently, with generation of a spirocyclic cyclohexadienyl complex (Scheme 14, where L = THF or H_2O).

The spirocyclic compound was similarly found to extrude palladium(0) on standing in solution, with formation of the annelated product shown in Scheme 14. The structural assignment of the spirocyclic compound as a cyclohexadienyl complex of palladium was based on comparison of spectroscopic data with those of a similar compound formed from the reaction of palladated 2-benzylpyridine with diphenylethyne, which was crystallographically characterized.[165] In this case sequential insertion of the two alkyne units was observed. A mechanism for the reaction was proposed (Scheme 15).

In order to rationalize the formation of the spirocyclic compound it was proposed that generation of an unstable 10-membered ring intermediate, where palladium interacts with the benzylic ring, would be followed by heterolytic palladium–carbon cleavage and ring closure. A similar mechanism has been proposed for the conversion of (**42**) to (**43**).[166]

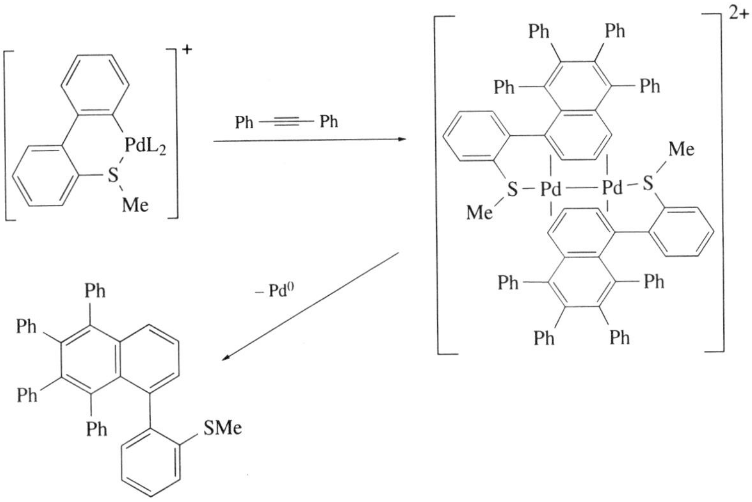

Scheme 13

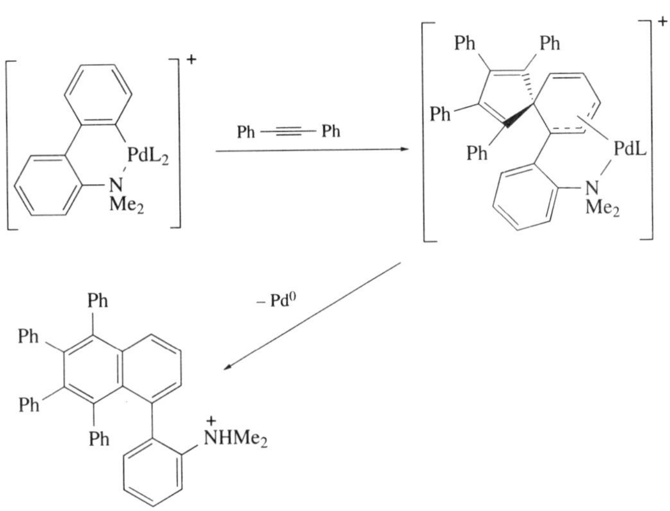

Scheme 14

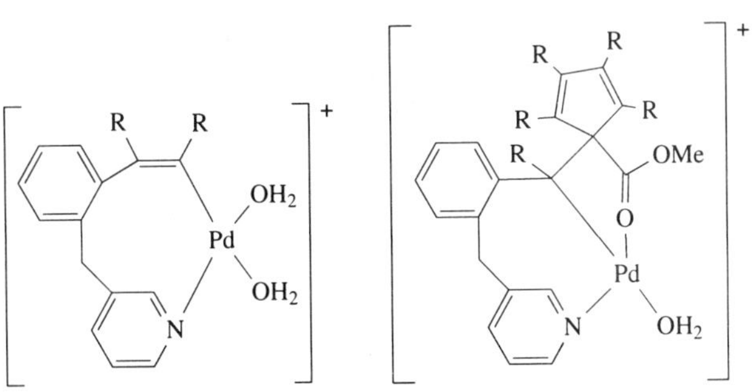

(42) (43)

Compound (**43**), characterized crystallographically (R = CO_2Me) as the triflate salt, formally results from the insertion of three alkyne units into the palladium–carbon bond of cyclopalladated 2-benzylpyridine.

The insertion of alkynes into the palladium–carbon bonds of cyclopalladated compounds appears to be controlled by three important variables. First, the nature of the donor atom in the palladacyclic substrate affects the outcome of the reaction, and this has been studied[167] through investigations into the reactivity of complexes (**44**) and (**45**), where the donors compared are =O and =NPh.

Compound (**44**) reacted with diphenylethyne to generate (**46**), whereas compound (**45**) gave (**47**).

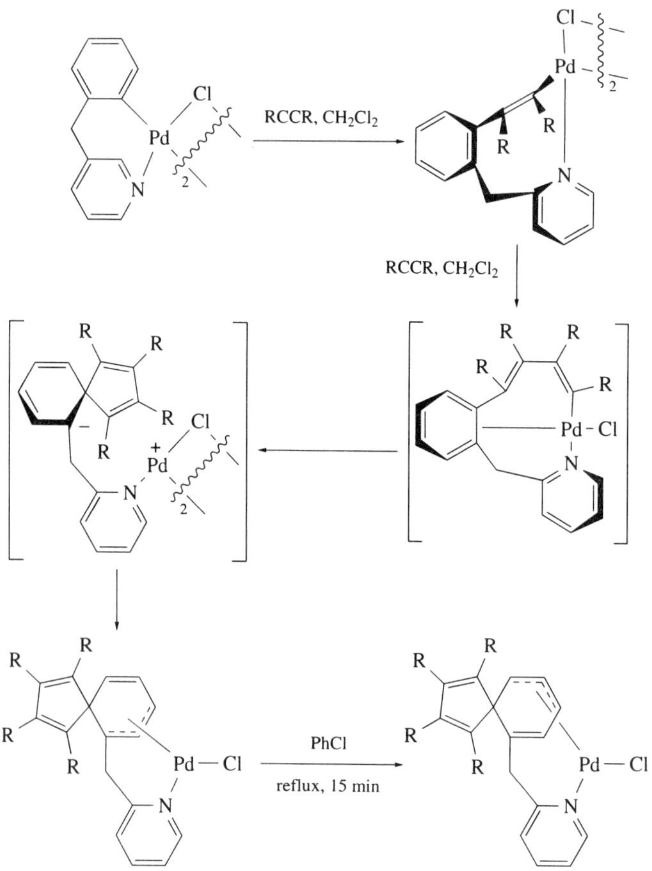

Scheme 15

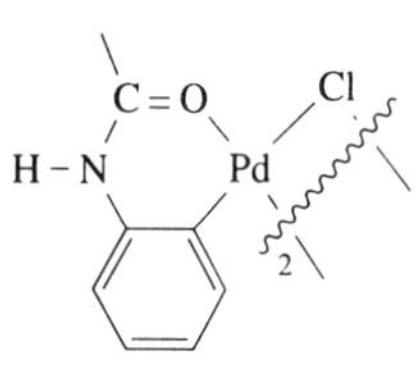

(44)

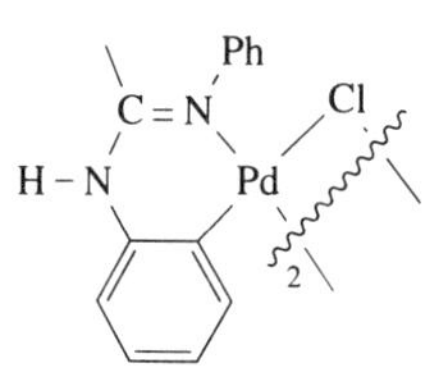

(45)

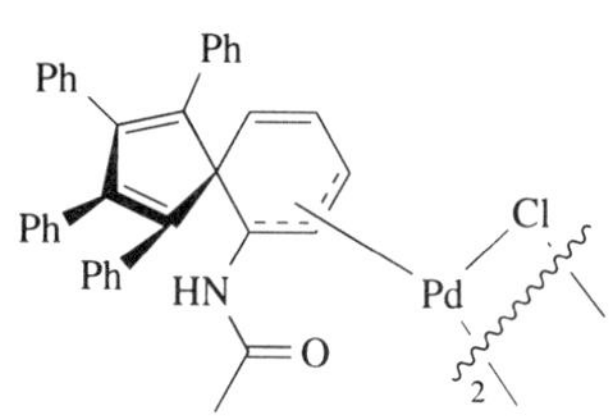

(46)

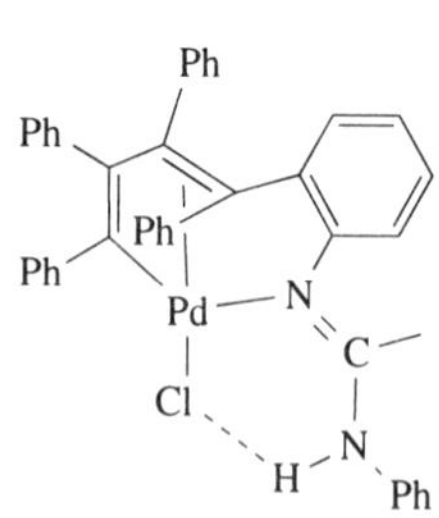

(47)

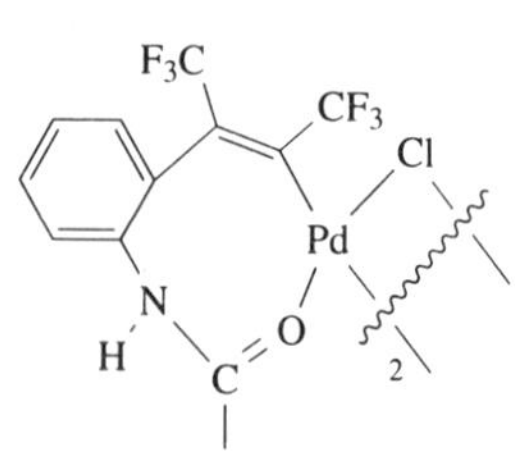

(48)

Second, the nature of the substituents on the alkyne also plays a significant role in governing the outcome of alkyne insertion reactions. For example, although **(44)** reacted with diphenylethyne to produce **(46)**, reaction with hexafluorobut-2-yne produced **(48)**.

Similar differences in reactivity between diphenylethyne and 3-hexyne have been noted in reactions with cyclopalladated *N,N*-dimethylbenzylamine.[168]

Third, the size of the chelate ring in the cyclopalladated substrate and the presence or absence of additional heteroatoms within the chelate ring are also important considerations. Compound **(45)** contains a six-membered ring with an additional heteroatom and reacts with diphenylethyne to generate

(**47**). Equation (11) illustrates the reaction of diphenylethyne with a substrate containing a five-membered ring with no additional heteroatom. Both factors appear to be important in governing the outcome. Additional complications may occur during depalladation when this alkyne insertion–oligomerization strategy is applied in organic syntheses. Thus, although cyclopalladated dimethylaminomethylferrocene (**49**) reacted with diphenylethyne at room temperature in dichloromethane in a fashion analogous to the transformation shown in Equation (11), reaction in refluxing chlorobenzene for 3 h produced the depalladated compound (**50**) in 15% yield, in which the C–NMe_2 bond of the substrate had been cleaved.[169]

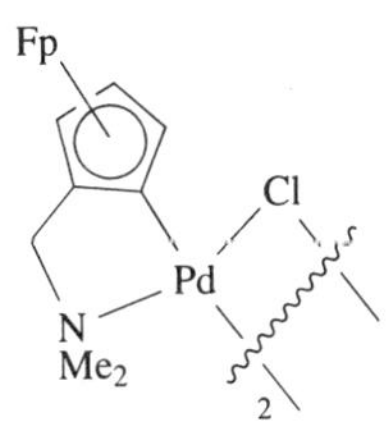

(49)

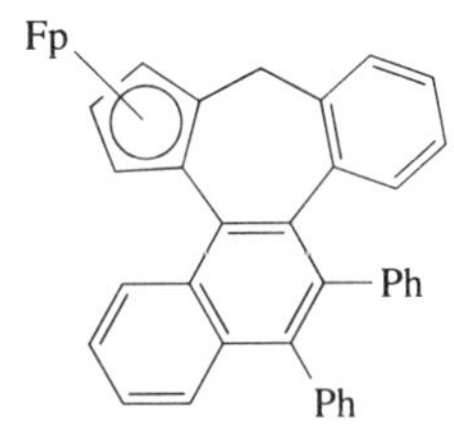

(50)

6.4 π-ALLYLIC COMPLEXES

6.4.1 Complexation of Allylic Ligands to Palladium: Structure and Bonding

6.4.1.1 Theoretical and structural investigations

Bonding of π-allylic ligands to palladium is conveniently rationalized for the parent allyl group in terms of the η^3-allyl ligand acting as a bidentate four-electron donor. Many π-allylic complexes of palladium can be regarded as d^8 square-planar systems containing a bound allylic anion. The allyl group may also function as an η^1-ligand, σ-bonded to the metal center, but specific consideration of complexes containing such ligands is outside the scope of this chapter. In *COMC-I* structural data obtained for π-allylic complexes of palladium were reviewed. In the complex $[Pd_2(\mu\text{-Cl})_2(\eta^3\text{-}C_3H_5)_2]$ the η^3-allyl ligands were found to be symmetrically bound to palladium, that is, the three carbon atoms were shown to be approximately equidistant from palladium, the carbon–carbon bond lengths were essentially equal, and the C-1–C-2–C-3 angle of approximately 120° indicated that the central carbon may be considered as sp^2 hybridized. The central Pd_2Cl_2 unit was shown to be essentially planar while the allyl groups were inclined at an angle of $111.5 \pm 0.9°$ to that plane. Hydrogen atoms have typically not been located in x-ray structure determinations of π-allylpalladium complexes and no neutron structures appear to have been reported. Calculation of hydrogen positions has indicated that *anti*-protons are closer to palladium than *syn*-protons and, as such, are typically more shielded in ^{1}H NMR spectra. Substitution of a methyl group in the C-2 position or addition of four methyl groups on the terminal carbons (1,1,3,3-tetramethylallyl) gives rise to compounds where the methyl carbons are not in the same plane as the allyl carbon backbone. In the complex $[Pd_2(\mu\text{-Cl})_2\{\eta^3\text{-}1,3\text{-}(Me)_2C_3H_3\}_2]$ the methyl carbons were similarly found to be noncoplanar with the allylic carbon backbone and, in this case, the central Pd_2Cl_2 unit was bent along the chlorine–chlorine axis. Unsymmetrical π-allylpalladium complexes result from either modification of the allyl group, for example, substitution in the C-1 position, or from bridge cleavage of a symmetrical complex such as $[Pd_2(\mu\text{-Cl})_2(\eta^3\text{-}C_3H_5)_2]$, for example, with a neutral ligand such as PPh_3 to generate $[PdCl(\eta^3\text{-}C_3H_5)(PPh_3)]$. In the former group of complexes rationalization of structural parameters can be complicated, but in the latter class of compound the palladium–carbon distances may be understood in terms of the relative *trans*-influences of the different ligands. A full discussion of this topic is included in *COMC-I*.

Molecular mechanics parameters have now been developed[170] to allow calculations on η^3-allylpalladium systems within the MM2 force field. In general, application of the MM2 force field to calculations on organotransition metal complexes is problematic since a classical valence bond model must be employed. Thus, for an η^3-allylpalladium system the interactions of the three allyl carbon atoms with palladium are modeled through three single palladium–carbon bonds. When two supporting ligands are included on palladium, the palladium atom is modeled with five single bonds. This is not handled well by present MM2 programs where a maximum of four bonds per atom and only a single value for a given angle type are typically allowed. These problems were overcome by inclusion of "dummy atoms" to extend the force field. The parameterization was based upon x-ray structural data and on results of *ab initio* calculations with electron correlation of all valence electrons for different geometries of an unsubstituted cationic η^3-allylpalladium fragment. The *ab initio* calculations on $[Pd(C_3H_5)]^+$ allow the bonding to be discussed in terms of the interactions between a palladium cation and a neutral allyl radical. Thus, in the allyl radical the unpaired electron was found to reside in the π-orbital having a nodal plane through the central carbon, as expected. The ground state of the palladium cation ($^2D(4d^9)$) has one unpaired electron, which could be placed in a *d*-orbital pointing towards the two terminal allyl carbon atoms and of the appropriate symmetry to interact with the allyl orbital containing the unpaired electron. Although the traditional model suggests transfer of the unpaired electron from the palladium cation to the allyl radical to generate an allyl anion and divalent palladium, population analysis instead suggested formation of an essentially covalent bond between palladium and the terminal allyl carbon atoms. The doubly occupied π-orbital of the allyl group interacts in a σ-sense with palladium, an interaction commonly viewed as backbonding from the allyl group into the empty 5*s*- and 5*p*-orbitals on palladium. Population analysis, however, indicated 1.0 electron on the terminal carbons, 0.7 electron on the central carbon, and only 0.2 electron in the 5*s*- and 5*p*- orbitals of palladium, although there was some uncertainty in the latter value. Calculations on $[Pd(C_3H_5)]^+$ fragments bearing either two chlorine or two ammonia supporting ligands suggested that the essential features of palladium–allyl bonding remained unchanged and that the $4d^9$ configuration for palladium was dominant. Molecular mechanics calculations were performed to estimate the relative energies of the *syn*- and *anti*-forms of η^3-crotylpalladium complexes with a range of 2,9-disubstituted phenanthroline ligands, known from experiment to shift the *syn–anti* equilibrium to favor the *anti*-form. Comparisons of experimental data on the percentage of *anti*-isomer present at equilibrium with calculated results from the MM2 program showed reasonable agreement, at least in trends if not in absolute values.

6.4.1.2 Photoelectron spectroscopy and allied techniques

Photoelectron spectra and molecular orbital calculations have been utilized to assign the ordering of molecular orbitals in π-allyl complexes of palladium. The assignment of peaks observed in photoelectron spectra of π-allyl complexes has been a subject of much controversy. In particular, the assignment of the low-energy ionization peak (peak I) of bis(allyl)metal complexes, an indicator of the nature of the HOMO in such compounds, has been debated. Early work on He^I and He^{II} PES of $[Pd(\eta^3\text{-}C_3H_5)_2]$, $[Pd(\eta^3\text{-}2\text{-}MeC_3H_4)_2]$, and $[Pt(\eta^3\text{-}C_3H_5)_2]$ led to assignment of peak I to ionization of the ligand $7a_u$-orbital, the in-phase linear combination of the two nonbonding allylic π-orbitals, in all three complexes.[171] However, a He^I, He^{II}, and variable-energy photoelectron study[172] showed that the relative intensities of peak I in the spectra of $[Pd(\eta^3\text{-}C_3H_5)_2]$ and $[Ni(\eta^3\text{-}C_3H_5)_2]$ changed differently as a function of photon energy: in the nickel complex peak I increased in intensity from 32 eV to 70 eV, while in the palladium complex peak I decreased. For both compounds the relative intensities changed smoothly from 20 eV to 80 eV, suggesting that there are no strong resonance effects for the molecular orbitals of the nickel complex to which the unusual intensity change might be attributed. The conclusion, therefore, was that peak I assignments must be different for the nickel and palladium complexes and that, while ionization from the ligand-based $7a_u$-orbital is reasonable for $[Pd(\eta^3\text{-}C_3H_5)_2]$, in the nickel analogue 3*d* ionization must give rise to peak I.

Palladium $3d_{5/2}$ core binding energies have been measured by electron spectroscopy for chemical analysis (ESCA) for a series of palladium(0) and palladium(II) compounds, including a number of π-allyl complexes.[173] The results are illustrated in Figure 3, where it is seen that the binding energies of the π-allyl complexes are nearly constant, suggesting that the initial charge distribution and dynamic polarization behavior around palladium are very similar in each of the compounds. Palladium thus appears to mediate charge effects among ligands in π-allyl complexes without significant alteration in its own charge. A series of *ab initio* SCF calculations on "$[Pd(\eta^3\text{-}C_3H_5)]$," $[Pd(\eta^3\text{-}C_3H_5)_2]$, and "$[PdCl_2(\eta^3\text{-}C_3H_5)]$" species revealed little change in the gross atomic charge on palladium through the series, while charges on the allyl subunits varied significantly. These results were interpreted in terms of a donor–acceptor bonding model which allows palladium to bind the allyl subunit without significant perturbation of its charge. Similar experimental results were seen in a study of $[Pd_2(\mu\text{-}X)_2(\eta^3\text{-}C_3H_5)_2]$ (where X = Cl, Br, I), where the palladium $3d_{5/2}$ core binding energies (337.6 eV, 337.7 eV, and 337.9 eV, respectively) were shown to be almost independent of the nature of the bridging halide.[174]

Further light on the ionization energy sequence of $[M(\eta^3\text{-}C_3H_5)_2]$ (M = Ni, Pd, Pt) has been shed by measurement of electron affinities by means of electron transmission spectroscopy (ETS).[175] The electron attachment (AE) energies are the negative values of the electron affinities (EA) and so can be

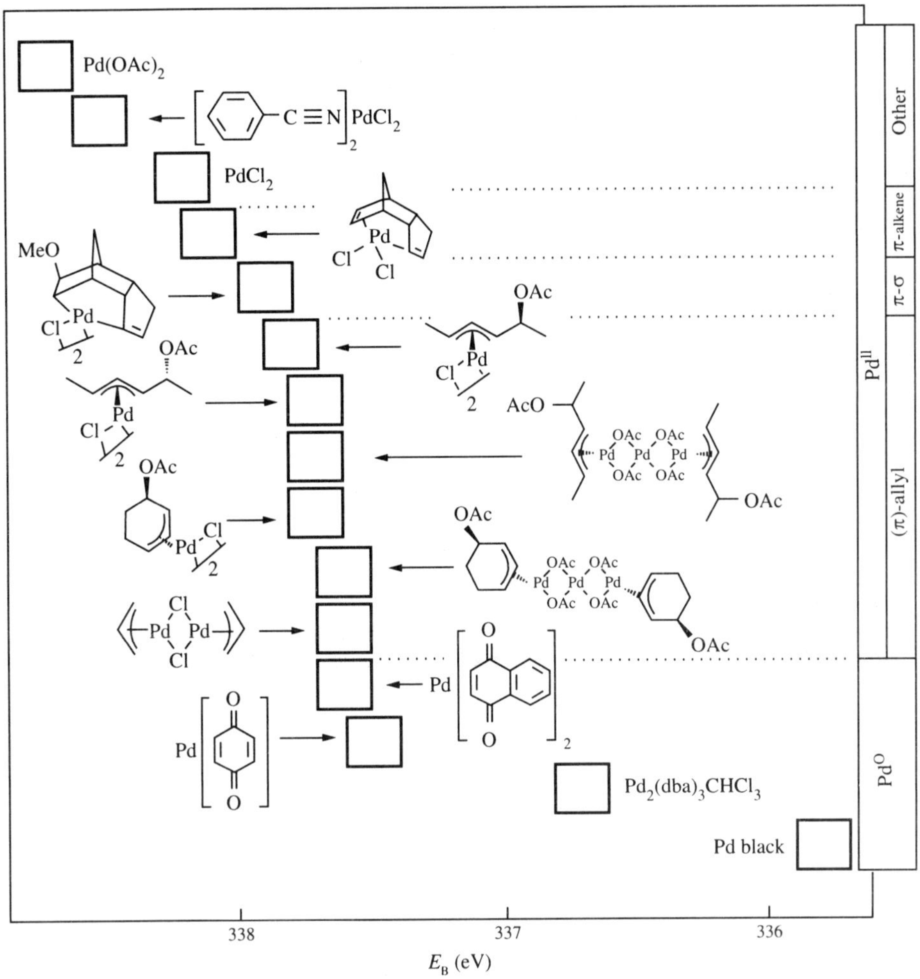

Figure 3 Palladium $3d_{5/2}$ core binding energies of palladium complexes and salts measured by ESCA (reproduced by permission of the American Chemical Society from *Organometallics*, 1992, **11**, 1784).

associated with the energies of empty molecular orbitals. Calculation of AE and IE (ionization energy) values by MS–Xα methods allowed comparison of experimental data with calculated energies for the nickel and palladium complexes. The measured spectra in the 0–5 eV energy range each exhibited three resonances due to electron capture by three empty ligand π*-orbitals. The AE values were well reproduced by the MS–Xα calculations on the nickel and palladium complexes. The IE values from the MS–Xα calculations supported the assignment of peak I in the photoelectron spectrum of $[Ni(\eta^3\text{-}C_3H_5)_2]$ to ionization from a ligand-based molecular orbital (and not from the nickel 3*d* orbital, *vide supra*). The experimental AE values for $[Pd(\eta^3\text{-}C_3H_5)_2]$ were in good agreement with those calculated, but the calculations failed to reproduce the IE values of the localized *d*-orbitals, possibly because the assumed palladium-ligand geometry was not that really adopted by the complex.

6.4.1.3 Nuclear magnetic resonance spectroscopy

The dynamic behavior of π-allylpalladium complexes has been a subject of considerable interest for a number of years. In *COMC-I* several fundamentally different dynamic processes were discussed. The NMR spectra of complexes of type **(51)** are illustrative.

It has been shown that in the absence of excess $[Pd_2(\mu\text{-}Cl)_2(\eta^3\text{-allyl})_2]$, or excess PR^2_3, **(51)** (where $R^1 = H$; $PR^2_3 = PPh^2_3$) exhibits a ^1H NMR spectrum that varies with temperature. As the temperature increased, signals due to H_a, H_b, H_c, and H_d coalesced into two signals located at the averaged positions of $H_a + H_d$, and $H_b + H_c$. Further coalescence at high temperature led to a single resonance at the averaged positions of the H_a, H_b, H_c, and H_d signals, indicative of a dynamic η^1-allyl. Addition of phosphine to complexes of type **(51)** may lead to a number of processes that involve nucleophilic attack

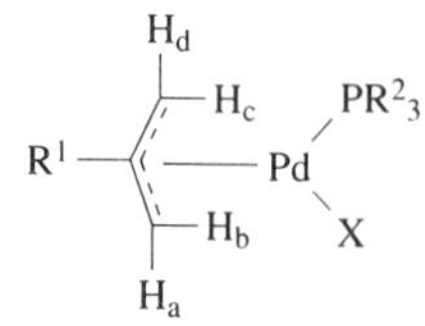

(51)

on palladium (see Section 6.4.3.1). Phosphine exchange via a five-coordinate intermediate, chloride displacement to generate $[Pd(\eta^3\text{-allyl})(PR_3)_2]^+$, and equilibria involving $[Pd(\eta^1\text{-allyl})(PR_3)_3]^+$ may occur, depending on the nature and amount of the tertiary phosphine present and the temperature. These processes may lead to dynamic exchange of *syn-* and *anti-*groups of the allyl ligand, which is governed by the steric and electronic properties of the substituents, the steric and electronic properties of the phosphine, and the temperature. Complete discussion of this area is provided in *COMC-I* and reports of new work on the solution dynamics of $[Pd_2(\mu\text{-Cl})_2(\eta^3\text{-allyl})_2]$ and $[Pd(\eta^3\text{-allyl})L_2]^+$ complexes that illustrate these features continue to be published.[176–9]

Facile exchange of allyl groups between two different palladium centers has now been observed[180] for an unusual ion pair generated by the reaction of $[Pd_2(\mu\text{-Cl})_2(\eta^3\text{-allyl})_2]$ with TMEDA, and characterized crystallographically (Equation (26)).

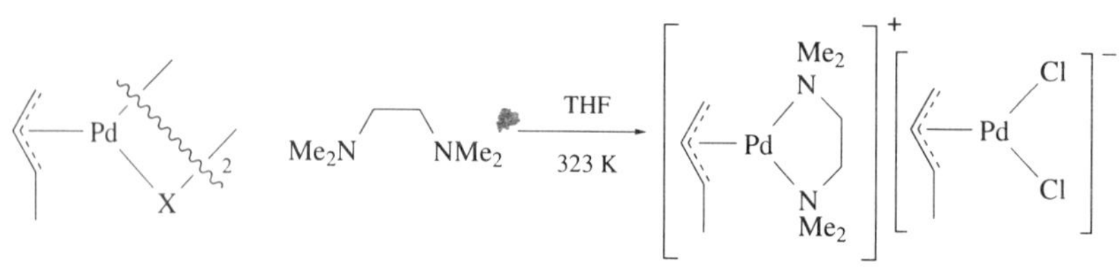

(26)

Interestingly, separate complexes containing the anionic and cationic parts of the ion pair could also be prepared (Scheme 16).

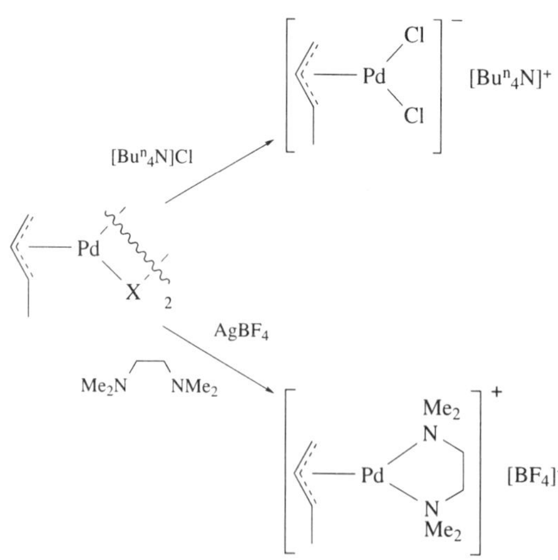

Scheme 16

The ^1H NMR spectrum of the ion pair at room temperature consisted of a series of broad peaks of indeterminate shape. At 248 K two distinct sets of peaks due to two different allyl groups could be observed. Assignments were possible since the anionic and cationic parts of the ion pair could be examined independently as a result of the syntheses shown in Scheme 16. At 320 K the signals due to the two different allyl groups of the ion pair coalesced at the averaged shift value. Kinetic parameters for the site exchange were determined by line-broadening techniques, by measurement of the broadening of the ^{13}C resonances due to the central allylic carbons in both the anion and the cation, which gave rise to two independent sets of data since the exchange rate could be measured for each ion. Kinetic parameters are listed in Table 2.

Table 2 Kinetic parameters for π-allyl ligand exchange.

	Cation	*Anion*
E_a(kJ)	70.7 ± 6.7	68.2 ± 3.5
$\Delta H^{\ddagger}$(kJ)	68.2 ± 6.7	66.1 ± 3.5
$\Delta S^{\ddagger}$(eu)	8 ± 6	6.0 ± 3
Correlation	0.991 529	0.997 352

Source: Hegedus *et al.*[180]

The data in Table 2 indicate that the entropy change in the transition state is close to zero, suggesting exchange in a tight ion pair. The observation that the rate of exchange was independent of concentration further supports the involvement of a tight ion pair. Evidence was presented that the exchange involved migration of allyl groups between palladium centers and not exchange of chloride and phosphine. Thus, *syn–anti* exchange was observed for the ion pair, but not for the independently prepared cation, that is, *syn–anti* exchange accompanies allyl exchange between metal centers. The x-ray crystal structure of the closely related cationic allylic complex $[Pd(\eta^3\text{-}1\text{-}PhC_3H_4)(TMEDA)][BF_4]$ has been reported.[181]

Dynamic processes have been investigated by two-dimensional NMR techniques and many examples of the application of such methods are extant in the literature. Ligands such as (**52**), (**53**), (**54**), and (**55**) form η^3-allylpalladium chloride complexes that undergo diastereomerization in solution[182,183] (Equation (27)).

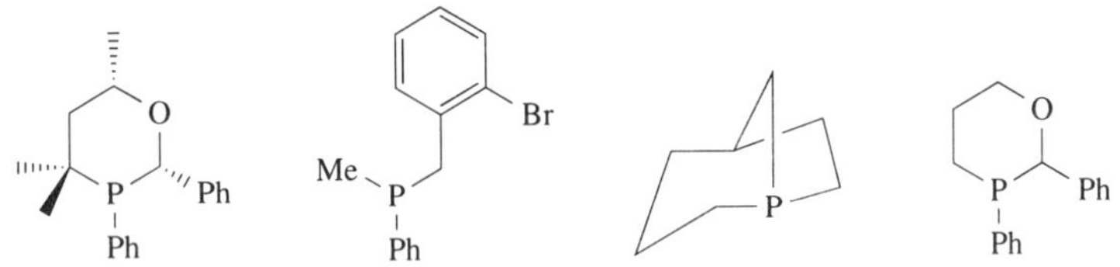

(**52**) (**53**) (**54**) (**55**)

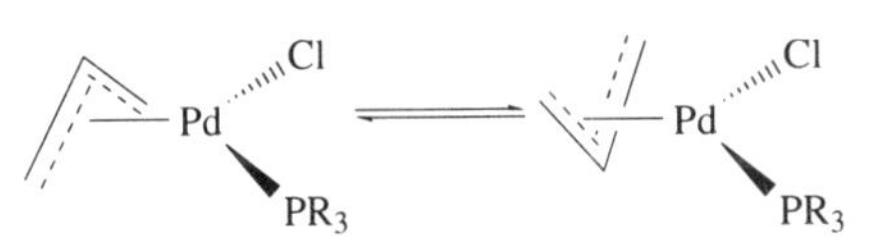

(27)

Proton two-dimensional exchange NMR spectra of the η^3-allylpalladium chloride complexes displayed two types of cross-peak network: one at room temperature or short mixing times and one at higher temperatures or longer mixing times, depending on the ligands. The first network of cross-peaks has been attributed to a fast η^3–η^1 conversion combined with a preferred rotation about a carbon–carbon bond and the second to a slower η^3–η^1 conversion combined with both carbon–carbon and palladium–carbon rotations. Studies with range of phosphorus donors demonstrated that both processes could be retarded by sterically demanding ligands.

The diastereomerization of the chiral cationic complex $[Pd(\eta^3\text{-}2\text{-}MeC_3H_4)L_2]^+$ (where L_2 = (*S*)-(*N*-diphenylphosphino)(2-diphenylphosphinoxymethyl)pyrrolidine) has been investigated.[184] One diastereomer was isolated by crystallization and characterized by x-ray crystallography. Dissolution of the isolated diastereomer in chloroform-d led to slow isomerization as the sample temperature was raised. At 295 K the thermodynamic isomer distribution was obtained. The isomers were found to undergo slow exchange (i.e., they appeared static on the timescale of the measurement) and ^{31}P magnetization transfer measurements suggested an upper limit of the rate constant for exchange of 0.5 s^{-1} at 295 K. In the presence of added ligand the isolable diastereomer underwent more rapid isomerization. Although the ^{31}P NMR spectrum displayed no signals due to intermediate species or added ligand, a DANTE (delays alternating with nutations for tailored excitation) spin saturation experiment demonstrated that a dynamic process was taking place. In the presence of added ligand, the rate constant for exchange was determined to be 1.5 s^{-1} at 295 K. In the presence of a large excess of added ligand, signals due to free ligand were detected as broad peaks in the ^{31}P NMR spectrum. A DANTE spin saturation experiment showed that excitation of one of the signals due to free ligand led to transfer to one of the signals due to complexed ligand, suggesting that isomerization of the complex involved coordination of the added ligand. In order to probe whether or not added ligand facilitated isomerization by nucleophilic attack at palladium, leading to η^3–η^1 conversion of the allyl group, a phase-sensitive NOESY (nuclear

Overhauser enhancement spectroscopy) spectrum was recorded with ligand added to the sample. Analysis of the spectrum allowed interchanging pairs of allylic protons to be determined and the results were found to be consistent with isomerization by η^3–η^1–η^3 interconversion of the allyl group. Studies of other closely related complexes using these NMR techniques have also been reported.[185]

A combination of two-dimensional NOESY (nuclear Overhauser enhancement spectroscopy), COSY (correlated spectroscopy), and ^{31}P, ^{1}H correlation measurements on the optically active β-pineneallyl–(*R*)-(+)-BINAP complex of palladium (where BINAP = (**56**)) allowed the assignment of the resonances due to the *o*-protons of the four nonequivalent aromatic rings and those due to the 15 protons of the allylic ligand.[186]

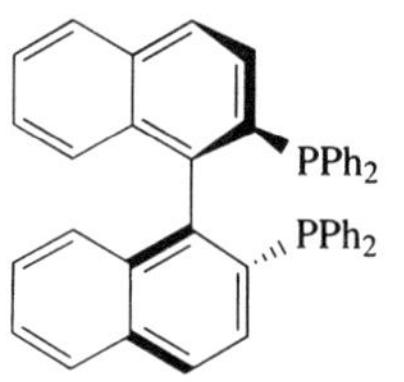

(56)

A combination of two-dimensional NOE data, calculations, and molecular graphics allowed a three-dimensional solution structure of the complex to be deduced. Such studies point the way for future solution structure determinations by NMR methods. In order to provide useful NOE data for mapping proton positions, specifically designed reporter ligands have been utilized that contain protons or methyl groups positioned in such a way that they can interact across a metal center with protons of a coordinated allyl group. The concept has been put into practice for a series of [Pd(η^3-allyl)(nitrogen chelate)]$^+$ complexes where the nitrogen chelate functions as a reporter ligand. Three-dimensional solution structures could be deduced in favorable cases.[187–9]

The phosphine ligand (**57**) functions as a paramagnetic shift reagent and has been utilized to elucidate the solution structures of [Pd(η^3-allyl)Cl(**57**)] complexes.[190]

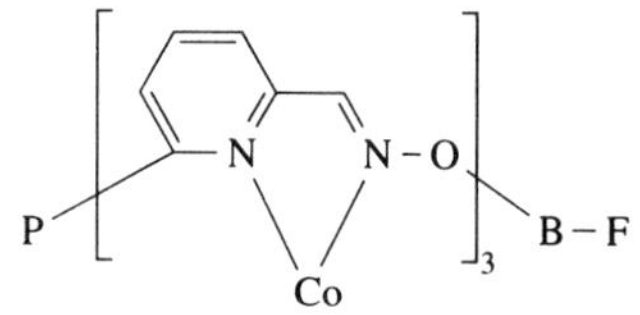

(57)

Resonances were assigned based on estimated relative distances from the paramagnetic cobalt(II) center in the phosphine ligand (**57**) and by comparison of spectra for complexes with methyl substituents in different positions on the allyl group. The method allowed for ready identification of isomers. For example, in the 1-methylallyl and 1,1-dimethylallyl complexes the spectra clearly indicated that the methyl-substituted termini of the allyl groups were located *trans* to (**57**). From geometrical parameters based on model compounds, deuterium quadrupole splittings and isotropic shifts were predicted and compared with measured values for compounds with selectively deuterated allyl ligands. Fitting required that the terminal protons of the 2-methylallyl complex be twisted out of the plane of the allyl carbon backbone, and this was confirmed for the model complex [Pd(η^3-2-MeC$_3$H$_4$)Cl(PPh$_3$)] by x-ray crystallography (including refinement of hydrogen atom positions).

6.4.2 Synthetic Methods

The major synthetic routes for the production of π-allylic complexes of palladium were discussed in detail in *COMC-I*. The methods described included the following general types.

(i) The reactions of palladium salts with monoenes (substituted propenes), typically performed with the use of protic solvents, were reported to lead to π-allyl complexes under suitable conditions. This general method had been employed with simple monoenes, cyclic monoenes, and functionalized monoenes. Representative examples, discussed in *COMC-I*, are illustrated in Scheme 17.

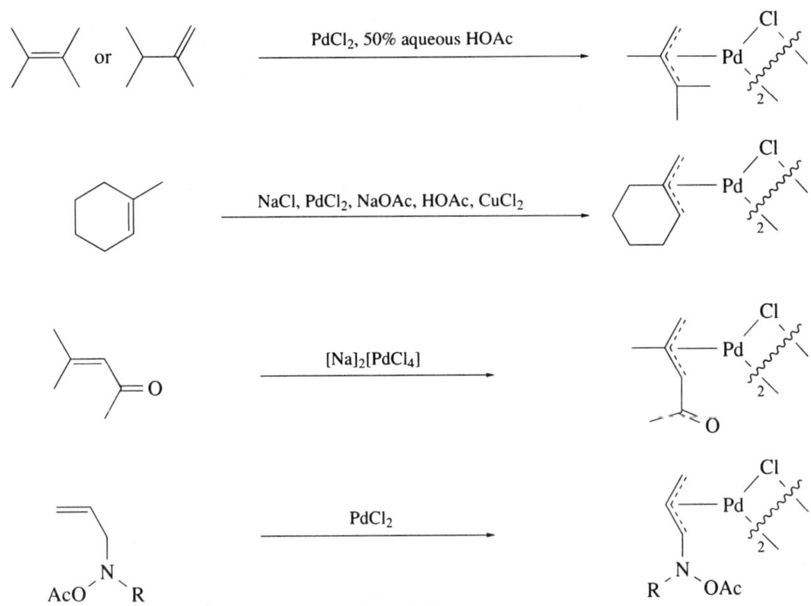

Scheme 17

(ii) The reactions of palladium salts with dienes, typically performed with 1,3-dienes in nucleophilic solvents, were described as routes to π-allyl complexes. In general it appeared that such reactions occurred via initial coordination of the diene by one double bond to palladium, followed by addition of palladium and X across the double bond to generate the π-allylic complex. Complications were known to arise since the loss of X^- and H^+ from the carbonium ion may occur and the carbonium ion may also rearrange. Examples, discussed in detail in *COMC-I*, are shown in Scheme 18, including the case of *in situ* generation of a palladium–phenyl complex from PhHgCl and [Li][$PdCl_3$] and subsequent addition to a 1,3-diene. A related reaction of a preformed palladium–1,5-diene complex containing a palladium–aryl group is illustrated in Equation (17) (Section 6.2.3.5).

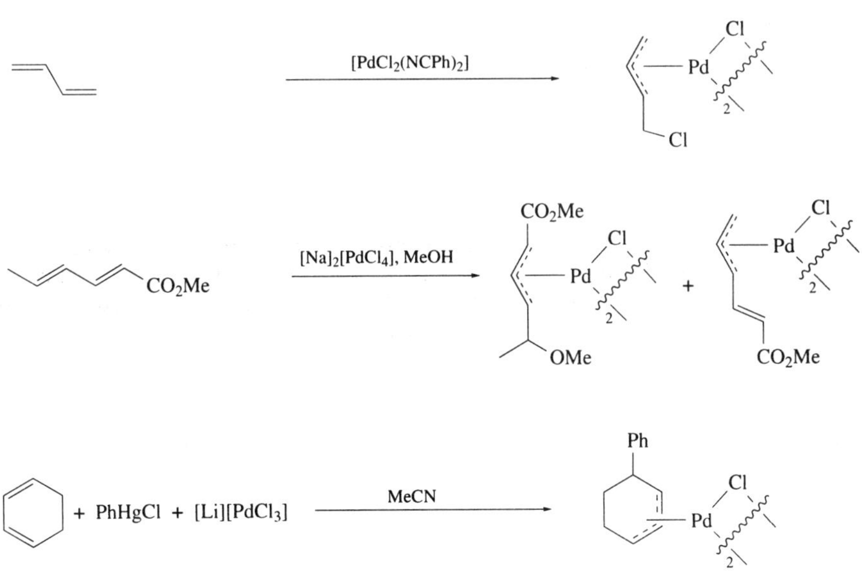

Scheme 18

1,2-Dienes (allenes) were reported to react with palladium salts to generate π-allyl complexes. The probable mechanism was described as Pd–Cl addition to a double bond to generate an η^1-allyl or η^1-vinyl intermediate, followed by rearrangement of the η^1-allyl to an η^3-allyl complex or reaction of the η^1-vinyl intermediate with a second equivalent of allene. Typical reactions, discussed further in *COMC-I*, are illustrated in Scheme 19.

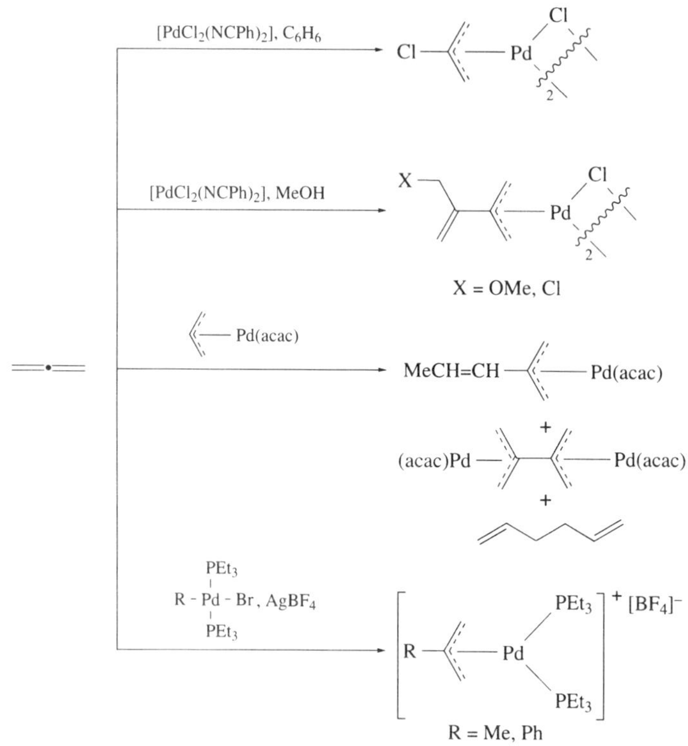

Scheme 19

(iii) Reactions of palladium salts or labile palladium(II) precursors with substituted cyclopropanes and cyclopropenes were reported to occur typically through ring opening to generate π-allyl complexes. Examples, as detailed in *COMC-I*, are shown in Scheme 20.

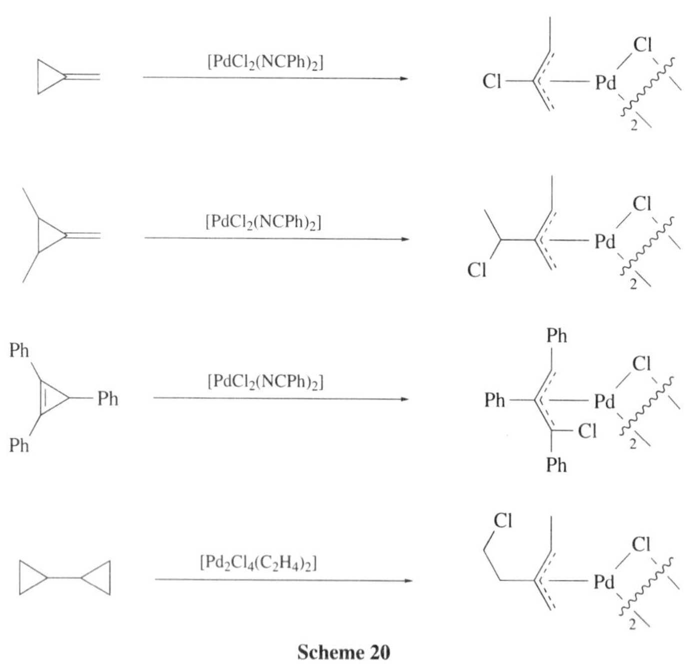

Scheme 20

(iv) Allyl halides, allyl alcohol, and related substrates were reported to yield π-allyl complexes when treated with palladium salts under certain reaction conditions. Scheme 21 shows some examples, described in detail in *COMC-I*.

Allyl halides were also reported to undergo oxidative addition to palladium(0) precursors to generate π-allyl complexes, while allylic Grignard reagents and related organometallics were said to react with palladium halide salts and complexes to generate π-allyl complexes.

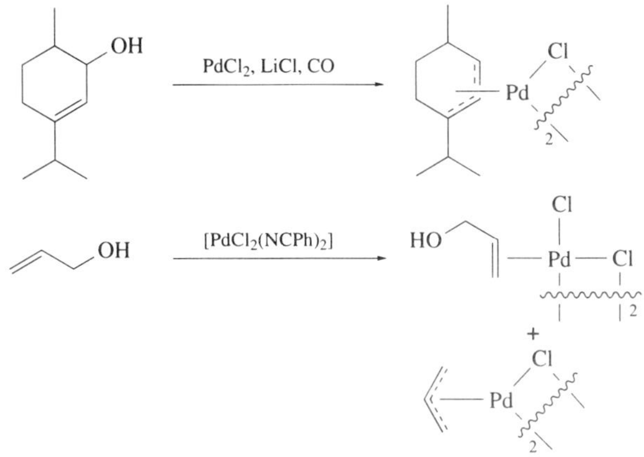

Scheme 21

(v) Alkyne dimerization may also provide a route to π-allyl complexes. An example is shown in Scheme 10 (Section 6.2.4) and the topic is discussed fully in *COMC-I*.

Many additional examples of the synthesis of π-allyl complexes by these general methods have been reported since the publication of *COMC-I*. A review article that focuses on synthesis and reactivity appeared in 1985.[191]

6.4.2.1 Reactions of monoenes with palladium salts and complexes

The mechanism of the reaction between methylenecyclohexane and palladium dichloride to produce a π-allyl complex has been investigated.[192,193] While it is generally agreed that the first step in such a reaction is formation of a palladium–alkene complex, the subsequent step has several possible pathways. Thus, oxidative addition of a C–H bond to palladium, followed by elimination of HCl, removal of a proton by an internal base (e.g., a chloride ligand), or deprotonation by an external base (e.g., a solvent molecule) are possible routes. Isotope effects were measured for the reaction shown in Equation (28).

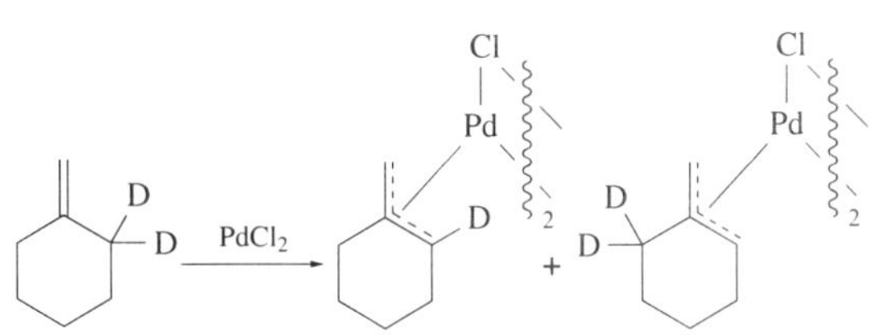
(28)

Reactions in acetic acid, both with and without added sodium acetate, gave $k_H/k_D = 3.5 \pm 0.1$ at 333 K, consistent with proton removal by a coordinated chloride ligand with a hydrogen transfer angle of 120–140°. With DMF, or with benzene containing two equivalents of DMF, as the solvent, isotope effects of 4.55 and 4.32, respectively, at 333 K were obtained and values of 3.66 and 5.19 were found at 294 K and 359 K, respectively. These data were interpreted in terms of proton abstraction by an external base. It was suggested that DMF might displace chloride that could then function as the external base. The palladium-catalyzed isomerization of methylenecyclohexane to 1-methylcyclohexene was found to proceed with $k_H/k_D = 1.8 \pm 0.5$ at 333 K, thus indicating significantly different isotope effects for allylic palladation and isomerization.

The specificity of mechanistic investigations to a particular reaction system is illustrated by comparison of the work described above with studies of the reactions of palladium trifluoroacetate with alkenes.[194] The trifluoroacetate anion is nonbasic and only poorly nucleophilic and, as such, may not participate in reactions that might involve proton abstraction (i.e., π-allyl formation) or nucleophilic attack on coordinated alkenes (i.e., alkene oxidation). It was found that reaction of palladium trifluoroacetate with β-pinene in acetone-d_6 led to identification by NMR of a π-allyl complex, believed to be dimeric by analogy with the related acetates, which could be converted to the chloride-bridged dimer in 83% overall yield by treatment with tetra-*n*-butylammonium chloride (Scheme 22).

A wide range of acyclic alkenes and alkenes exocyclic to a ring system were found to behave normally and reacted in a fashion analogous to the reaction of β-pinene. The chemoselectivity of the

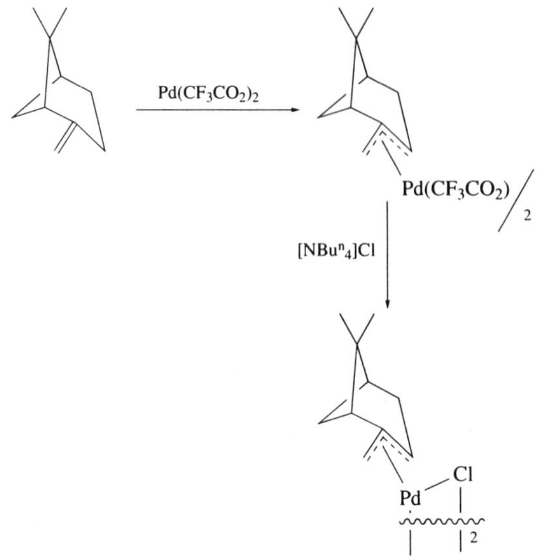

Scheme 22

method was found to be higher than that of previously reported syntheses. For example, geranylacetone produced only the terminal allyl complex in 54% yield (83% based on recovered starting material) (Equation (29)), whereas a 1:1 ratio of isomers was obtained by the more standard route based on reactions with palladium dichloride.

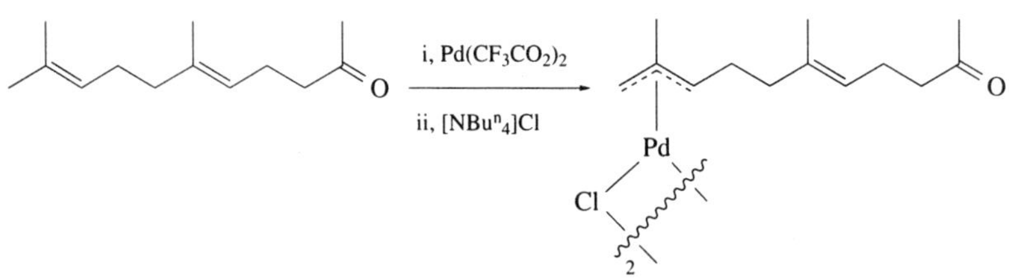

(29)

The stereochemistry of the substrate was found not to be a determining factor in predicting the stereochemistry of the allyl complex since, in all cases, *syn*-complexes were produced. Steric bulk of neighboring substituents and kinetic deactivation by alkyl substituents were found to be important factors in controlling reactivity, possibly influencing an initial equilibrium step of palladium–alkene complexation, since compound (**58**) was recovered unchanged, whereas (**59**) reacted to produce (**60**) in 92% yield.

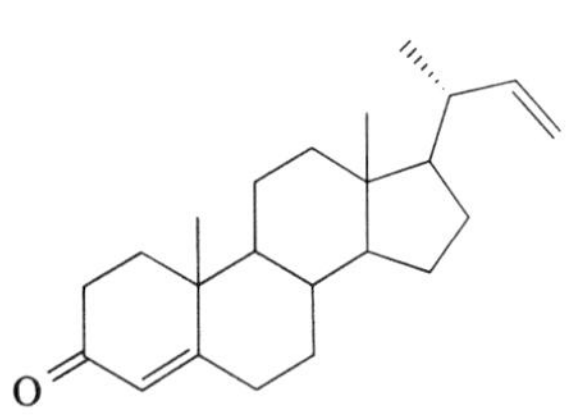

(58)

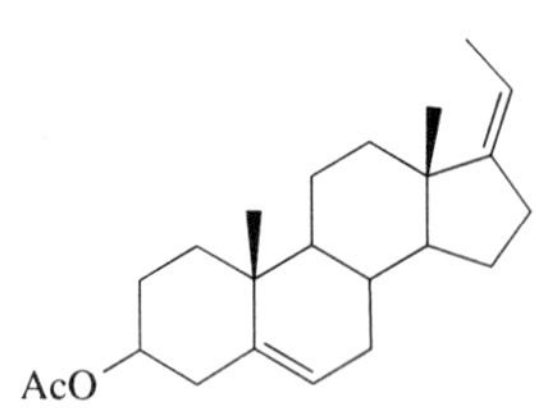

(59)

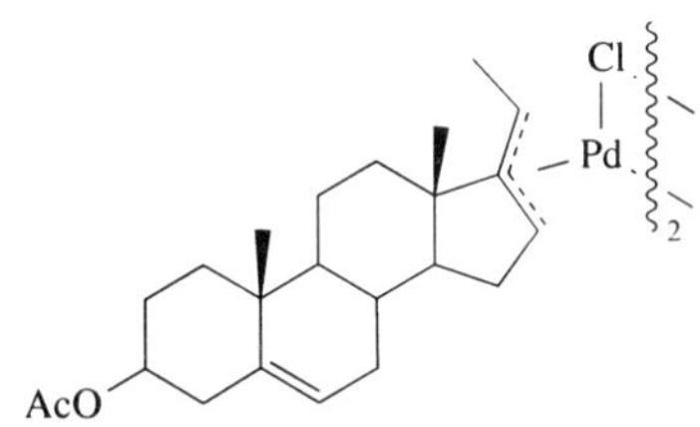

(60)

Endocyclic alkenes and simple methylenecycloalkanes were found to behave in a different fashion. Thus, cyclohexene disproportionated to benzene and cyclohexane in a reaction found to be catalytic in palladium trifluoroacetate. 4-*t*-Butylcyclohexene behaved similarly, while methylenecyclohexane and 1-methylcyclohexene produced largely toluene. The reactions are summarized in Scheme 23.

The observation of disproportionation reactions was interpreted in terms of a mechanism for π-allyl formation involving an intermediate palladium hydride. Thus it seems likely that in the absence of external bases a hydride route for deprotonation of the substrate might be dominant. Kinetic studies of the $[Li]_2[Pd_2Cl_6]$-catalyzed allylic hydrogen–deuterium exchange reaction of α-methylstyrene in acetic acid-d_4 have similarly suggested an intermediate π-allyl species formed via a hydride route.[195]

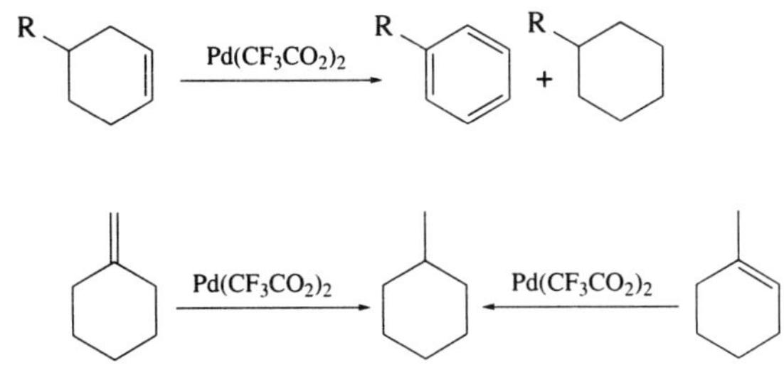

R = H, Bu^t

Scheme 23

The stereochemistry of hydrogen abstraction during π-allyl formation has been investigated in the case of steroid 4-en-3-ones.[196] Compounds (**61**) and (**62**) were treated with $[Na]_2[PdCl_4]$ in THF under reflux for 40 h to generate (**63**) (where X = H or D).

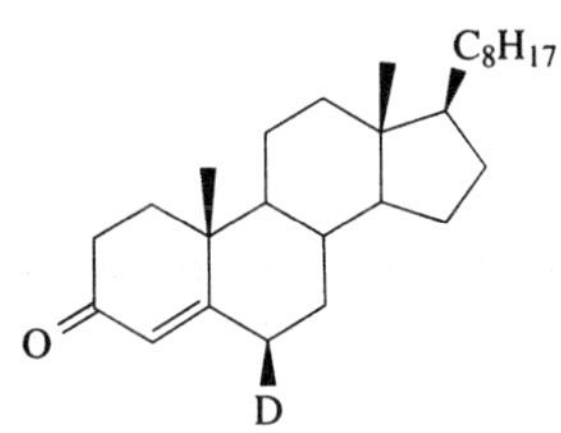

(61)

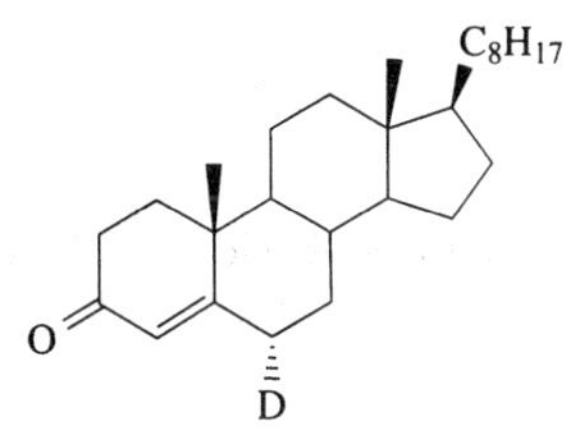

(62)

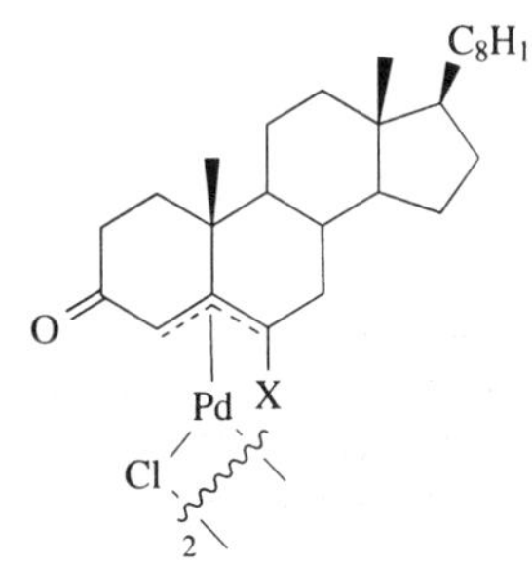

(63)

The proportion of deuterium in the C-6 position (X) of the palladium complexes (**63**) was estimated by integration of ^{1}H NMR spectra, and it was found that (**61**) produced (**63**) with only 15–18% retention of deuterium, whereas (**62**) produced (**63**) with 85–88% deuterium retention. The process is complicated by evolution of hydrogen chloride and by traces of water in the dried solvents which could be involved in hydrogen–deuterium exchange processes. Nonetheless, the results indicated that π-allyl formation occurred with highly stereoselective loss of the 6β-axial proton, producing the complex with palladium bound to the α-face of the steroid. The α-stereochemistry was confirmed in the case of the [α-(4-6η^3)Pd(acac)] derivative of progesterone by x-ray crystallography.[197] The dimeric, chloride-bridged π-allyl complex derived from testosterone has similarly been characterized crystallographically and was shown to have palladium coordinated to the α-face of the steroid.[198]

In contrast, reactions of the steroids calciferol ((**64**) where R = C_9H_{17}) or cholecalciferol ((**64**) where R = C_8H_{17}) with $[PdCl_2(NCPh)_2]$ generated the complexes (**65**) where ^{1}H and ^{13}C NMR studies[199] suggested complexation of palladium to both the α- and β-faces of the steroids.

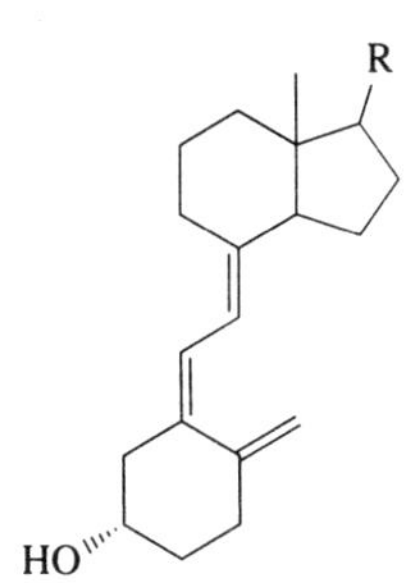

(64)

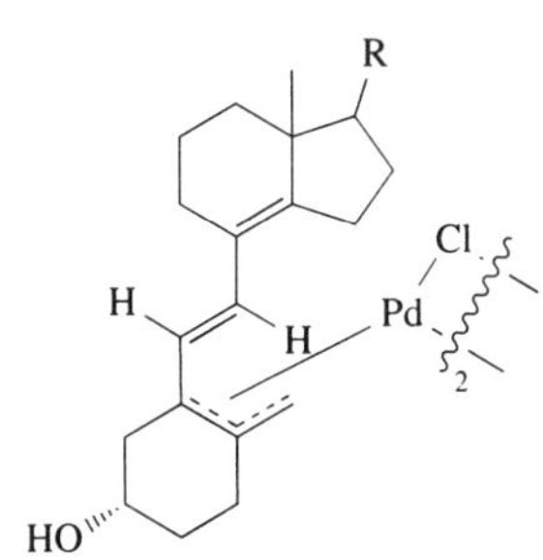

(65)

The ^{1}H NMR spectra of palladium–π-allyl complexes derived from steroids are known to exhibit considerable solvent effects which may complicate stereochemical assignments based on this technique.[200]

The formation of π-allyl complexes from other natural products, such as the alkaloid brucine,[201] has been investigated and π-allyl complexes continue to play an important role in the synthesis of such materials.[15]

The use of a detergentless microemulsion for the preparation of π-allyl complexes of palladium has been described.[202] Hexane, 2-propanol, and water were used to prepare a microemulsion by titration of aqueous $[H]_2[PdCl_4]$ and hexane to clarity with 2-propanol. The resultant medium was translucent, golden-brown, and surface active. Addition of 2,4,4-trimethyl-1-pentene followed by reflux (313 K) for 24 h allowed isolation of (**66**) in 52% yield. The compound was characterized crystallographically as (**66**)·2 $CHCl_3$.

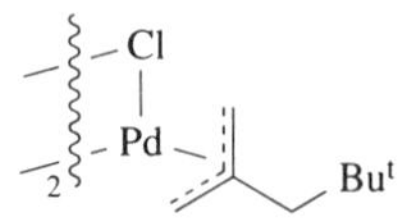

(66)

The reaction of vinylmercurials with $[Li]_2[PdCl_4]$ and acyclic alkenes is reported[203] to produce π-allyl complexes regiospecifically (cf. Scheme 18 for a related diene reaction and Section 6.4.2.2). The original mechanism proposed involved *in situ* generation of an organopalladium complex and *cis*-addition to the alkene, followed by a *cis*-β-elimination of palladium and the allylic hydrogen to produce a π-alkene complex which collapsed via a σ-allyl intermediate to form the π-allyl complex. However, it was found that simple monocyclic alkenes reacted in the same fashion as acyclic alkenes and an organopalladium adduct derived from a cyclic alkene cannot undergo *cis*-β-elimination. Accordingly, a revised mechanism was proposed to account for the reactions of cyclic alkenes. Steps following the *cis*-addition to the alkene are illustrated in Scheme 24 for the case of cyclopentene as the substrate.

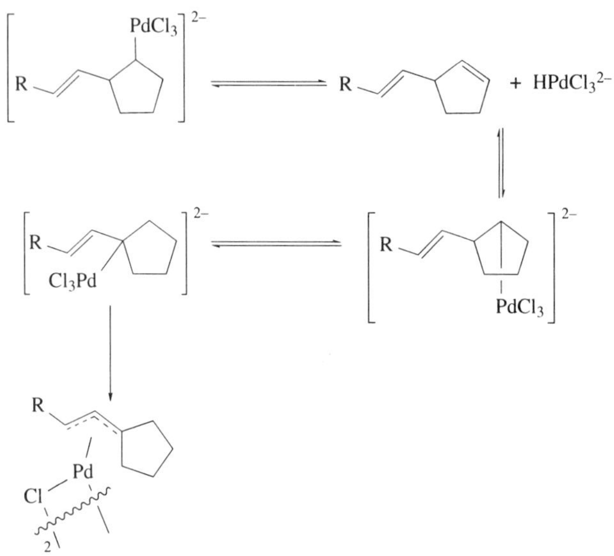

Scheme 24

Consistent with the proposed mechanism was the observation that 1,4-diene formation accompanied generation of the π-allyl complexes. When the reaction was performed in the presence of triethylamine, formation of 1,4-dienes was the dominant pathway, thus providing a useful synthetic route to these compounds.[204] A variety of substituted vinylmercurials and cyclic alkenes of differing ring size were found to be useful reagents. Bicyclic alkenes were found to react differently, producing *cis-exo*-alkylpalladium complexes. Equation (30) shows one such example.

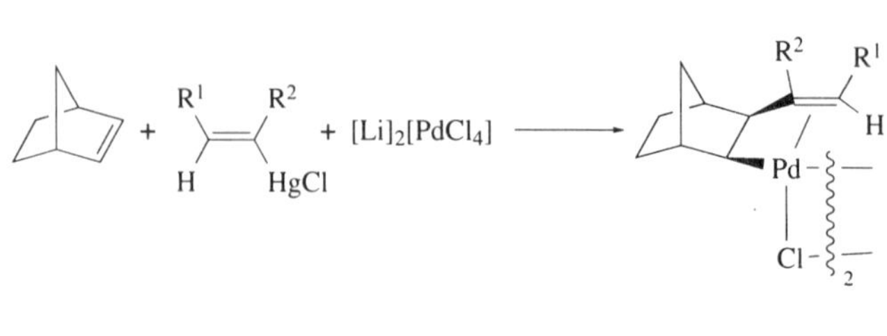

(30)

The stability of these alkylpalladium complexes was attributed to the absence of a β-hydrogen *cis* to palladium, which prevented elimination, and the coordination of the neighboring double bond.

The *in situ* generation of [RPdX] equivalents from organomercurials and palladium salts and subsequent addition to unsaturated substrates to generate intermediate π-allyl compounds which are subject to attack by external or internal nucleophiles (see Section 6.4.3.8) have been utilized in a number of important organic syntheses. Some aspects of the generation and reactivity of [RPdX] equivalents have been discussed in a 1989 review.[205] The synthesis of alkenyllactones from unsaturated carboxylic acids is representative.[206] The use of organomercurials may be avoided in certain cases by generation of the [RPdX] equivalents through reaction of organohalides or triflates (e.g., vinyl chloride or triflate) with palladium salts. Addition of a palladium–halogen (triflate) bond across the carbon–carbon double bond followed by elimination of HX is known to generate the organopalladium species which then may react further via π-allyl formation. Alkenyllactone synthesis by this route has also been described.[207]

The addition of an organopalladium intermediate, generated *in situ*, to an alkene with resultant formation of a π-allyl complex has been encountered during investigations of the reactions of vinylsilanes with palladium salts.[208] Thus, trimethylvinylsilane was found to react with palladium dichloride in DME in the presence of methanol to produce a chloride-bridged palladium dimer, containing the η^3-1-methyl-1-(trimethylsilyl)allyl ligand, as the *syn*-isomer. Heating this compound at 60 °C in chloroform solution generated the *anti*-isomer, which was crystallographically characterized. The proposed mechanism involved initial reaction of trimethylvinylsilane with palladium dichloride to generate a vinylpalladium chloride intermediate by elimination of chlorotrimethylsilane. The organopalladium intermediate was proposed to add to trimethylvinylsilane and π-allyl formation then occurred in the normal manner (Scheme 25).

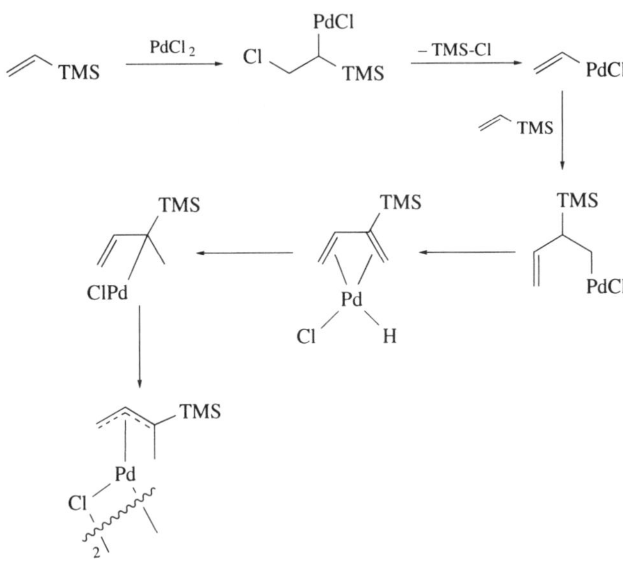

Scheme 25

Optically active allylsilanes have been utilized in reactions with lithium chloropalladate to generate optically active π-allyl complexes.[209] The mechanism was believed to involve addition of a palladium–chlorine bond across the double bond of the allylsilane, followed by elimination of the chlorosilane (cf. Scheme 25) and formation of the π-allyl complex.

6.4.2.2 Reactions of 1,3-, 1,4-, 1,5-, 1,6-, and 1,7-dienes with palladium salts and complexes

The reactions of 1,3-dienes with palladium salts in protic solvents continue to be utilized for the synthesis of π-allyl complexes. Thus, 1,4-dimethyl-, 1-isopropyl-4-methyl-, and 1-*t*-butyl-4-methylcyclohexadiene have been shown[210] to react with $[Na]_2[PdCl_4]$ in methanol to generate a single isomer of the corresponding dimeric π-allyl complex. Two stereoisomers were obtained from the reaction with 2-isopropyl-5-methylcyclohexa-1,5-diene. It was found that a 2.5- to threefold excess of diene was required in order to obtain good yields of the complexes and it was shown that the hydrogen atom which is incorporated stereo- and regioselectively into the newly formed π-allyl ligand (i.e., onto the terminal carbon of the diene from the same face as the coordinated palladium) came from the excess diene and not the solvent. The overall reaction is illustrated in Equation (31).

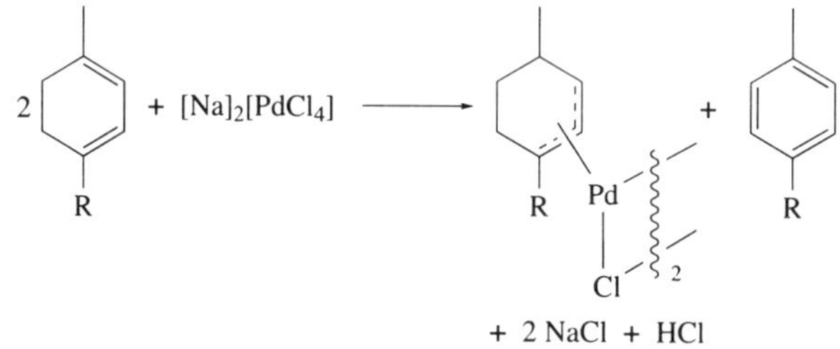

(31)

Dehydro-β-pinene has been shown[211] to react with palladium salts in nucleophilic solvents to generate π-allyl complexes via a *cis*-oxypalladation (Scheme 26), where structural assignments were based upon ^{1}H NMR spectra. The reactions contrast with the more typical *trans*-attack of oxygen nucleophiles on diene complexes (Section 6.2.4).

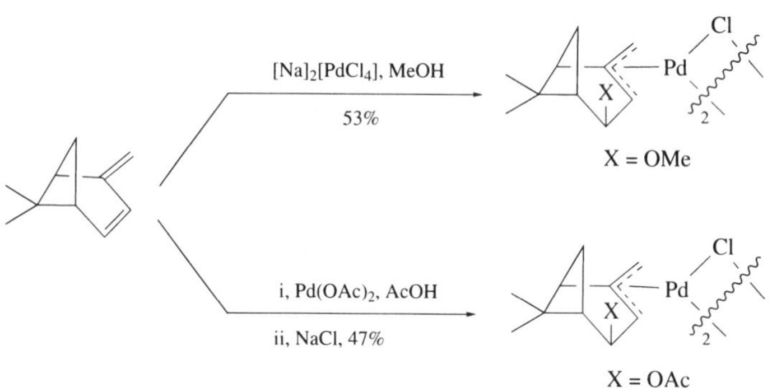

Scheme 26

Sulfonylpalladation of 1,3-dienes by treatment of palladium dichloride with 1.5–3.0 equivalents of diene and two equivalents of sodium alkylsulfinate in acetic acid at 343–353 K has been reported.[212] Highly substituted 1,3-dienes were found to be unreactive, or produced intractable products, and the reaction was found to be sensitive to the nature of the alkyl group present in the sodium alkylsulfinate. Certain 1-substituted and 1,2-disubstituted acyclic dienes generated regioisomeric mixtures of π-allyl complexes.

1-Silyl dienol ethers have been shown to react with palladium(II) salts and complexes to generate a variety of types of π-allylpalladium complexes. The reactions are summarized in Scheme 27, where it is seen that simple transmetallation products, complexes resulting from formal decarbonylation, and products of reaction with the solvent may result, depending upon reaction conditions.[213]

A mechanism for the decarbonylation process was presented that involved isomerization of the initially formed π-allyl complex, via a π-bound ketene intermediate, to a σ-acylpalladium species that underwent decarbonylation.

Novel π-allyl complexes have been isolated during investigations of the palladium-catalyzed telomerization of butadiene with acetic acid to yield acetoxyoctadienes.[214] A mechanism for the telomerization that involves dinuclear palladium–palladium-bonded complexes was proposed (Scheme 28).

The dinuclear complex initially formed was isolated from the catalytically active reaction mixture and characterized crystallographically. The palladium–palladium distance was found to be 0.29 nm, suggesting metal–metal bonding. This compound was found to react with butadiene in the absence of acetic acid to generate the known complex (**67**), while the reaction of bis(hexafluoroacetylacetonato)-palladium(II) ([Pd(F_6acac)$_2$]) with butadiene in methanol produced (**68**), in which the F_6acac ligand is bidentate. Compound (**68**) was characterized crystallographically.

Complex (**68**) could also be prepared by treatment of [Pd_2(μ-OAc)$_2$(μ-1-3-η:6-8-η-C_8H_{12})] with hexafluoroacetylacetone and was found to react with tertiary phosphines to produce palladium–allyl and palladium–alkyl complexes (Scheme 29).

The sequence of reactions described during this study is strongly suggestive of the involvement of dinuclear palladium compounds in the catalytic cycle for butadiene telomerization in the presence of nucleophiles.

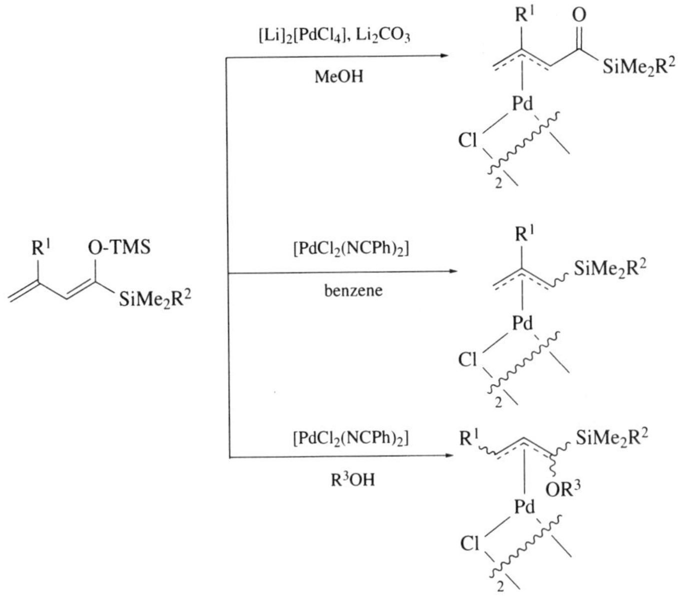

Scheme 27

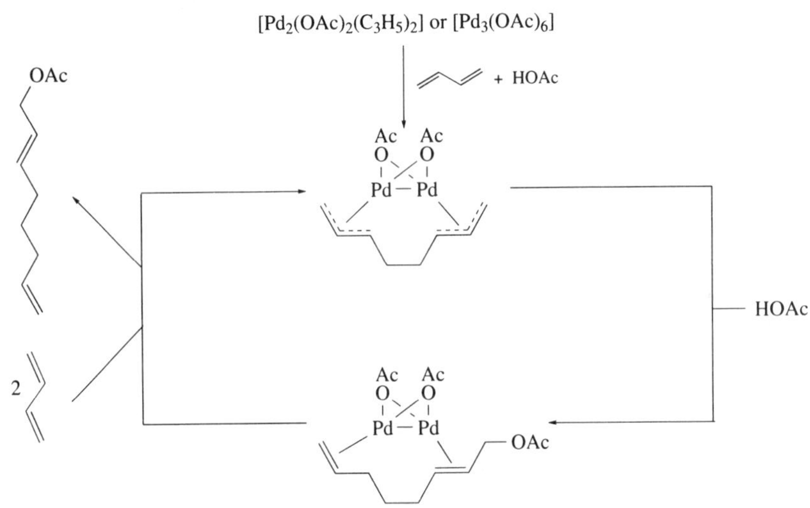

Scheme 28

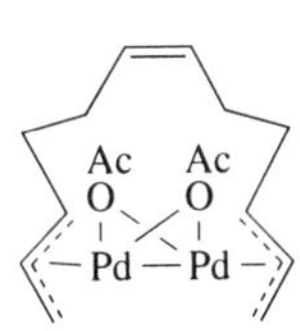

(67)

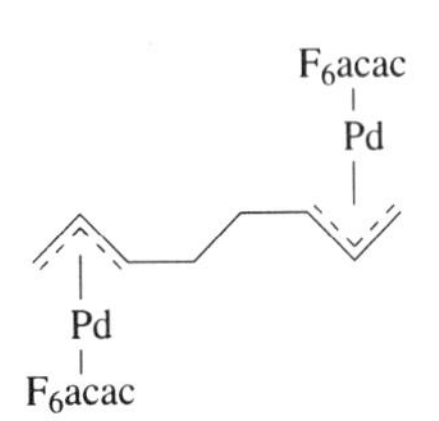

(68)

Polynuclear palladium–allyl complexes have also been prepared[215] by the reaction of $[Pd_3(OAc)_6]$ with 2,6-disubstituted pyrylium salts in aqueous acetic acid–sodium acetate (Equation (32)).

The complex with R = *t*-butyl was characterized crystallographically.

The *in situ* generation of organopalladium species has been widely employed in reactions with 1,3-dienes that lead to π-allyl complexes. Aryl-, hydrido-, and carbomethoxypalladium chloride intermediates have been found to add to conjugated dienes to generate dimeric, chloride-bridged π-allyl complexes.[216] Equation (33) illustrates the general reaction for phenylmercury(II) chloride.

Phenyl group addition to different types of double bonds within the 1,3-diene systems was examined. It was found that the order of reactivity towards "phenylpalladium chloride" was

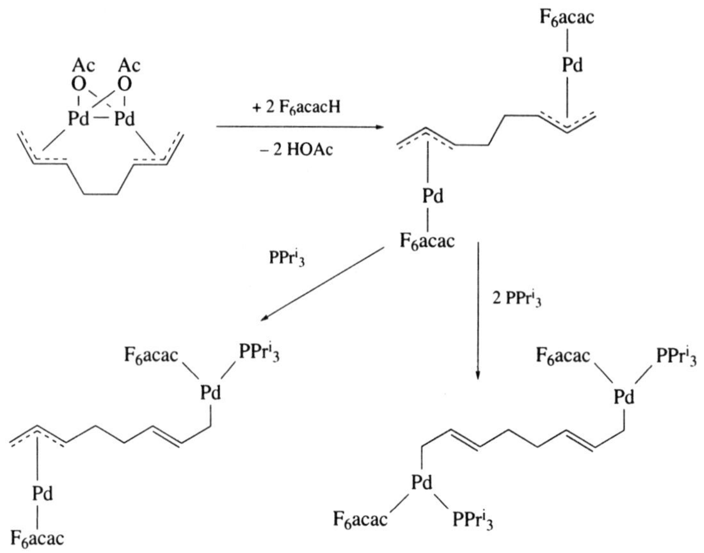

Scheme 29

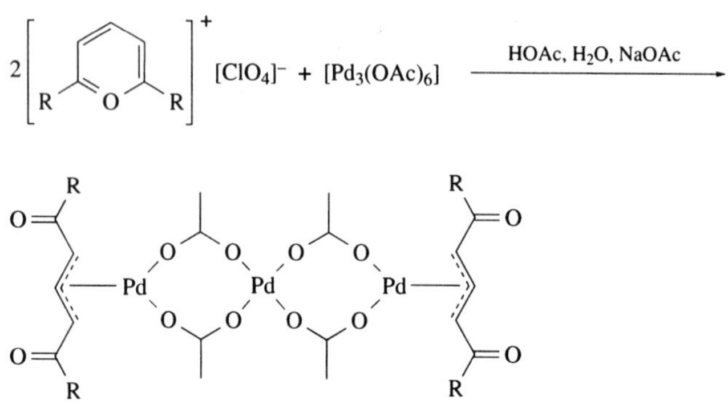

(32)

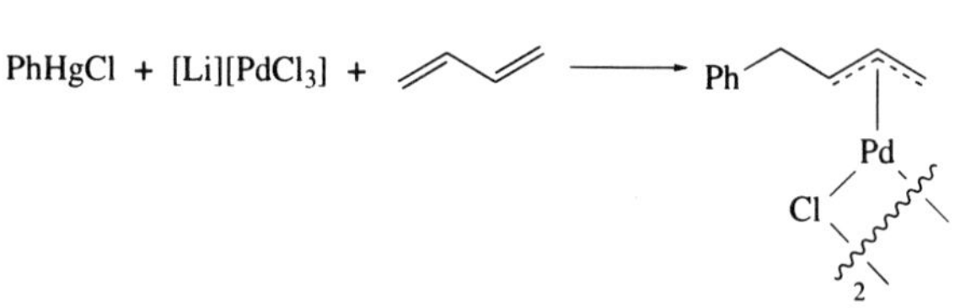

(33)

CH_2=CHR > CH_2=CR_2 > RCH=CHR, in keeping with the ability of these double-bond types to coordinate to palladium. With mercurials containing at least one sp^3-bonded hydrogen β to mercury, the intermediate organopalladium species apparently underwent facile β-hydride transfer, with liberation of alkene and formation of a palladium hydride species which reacted with the added diene (Scheme 30).

The *in situ* generation of organopalladium species from organomercurials and palladium salts and their addition to monoenes and 1,3-dienes have been extended to include 1,4-, 1,5-, 1,6-, and 1,7-dienes as substrates. Remote palladium migration is reported to occur with generation of π-allylpalladium complexes in good yields.[217,218] Scheme 31 shows typical examples.

Addition to 1,4-dienes was found to be highly regioselective, with attack occurring exclusively on the least-substituted double bond. The reaction of [PhPdCl] equivalents with 1,4-cyclohexadiene did not produce a π-allylpalladium complex but rather the allylic chloride (**69**) derived from elimination.

With 1,5-, 1,6-, and 1,7-dienes the palladium is required to migrate past increasing numbers of sp^3 carbon centers in order to generate a π-allylpalladium complex. With 1,5-dienes the addition of [RPdCl] equivalents was followed by palladium hydride elimination and subsequent addition of [HPdCl] equivalents to the substrate, and this became a complicating factor. Use of ethylmercury(II) chloride as a source of [HPdCl] equivalents (Scheme 31) via β-hydride elimination from [EtPdCl] allowed the hydride-derived products to be obtained cleanly. 1,6-Heptadiene was found to react smoothly with [HPdCl] equivalents but 1,7-octadiene generated two isomeric π-allylpalladium complexes (Scheme 31). The reason for this difference between the 1,6- and 1,7-diene systems is not clear. A wide variety of

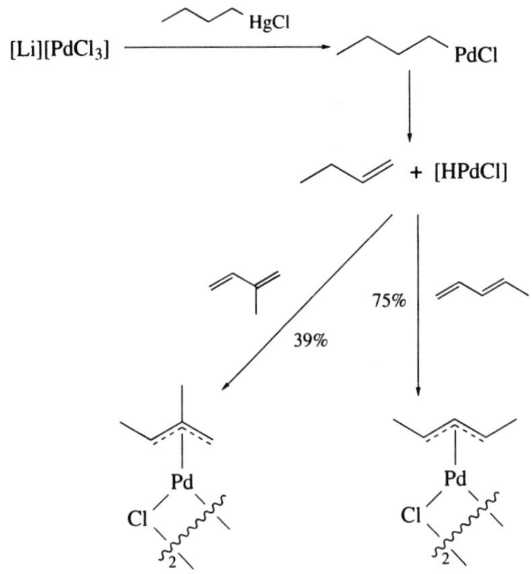

Scheme 30

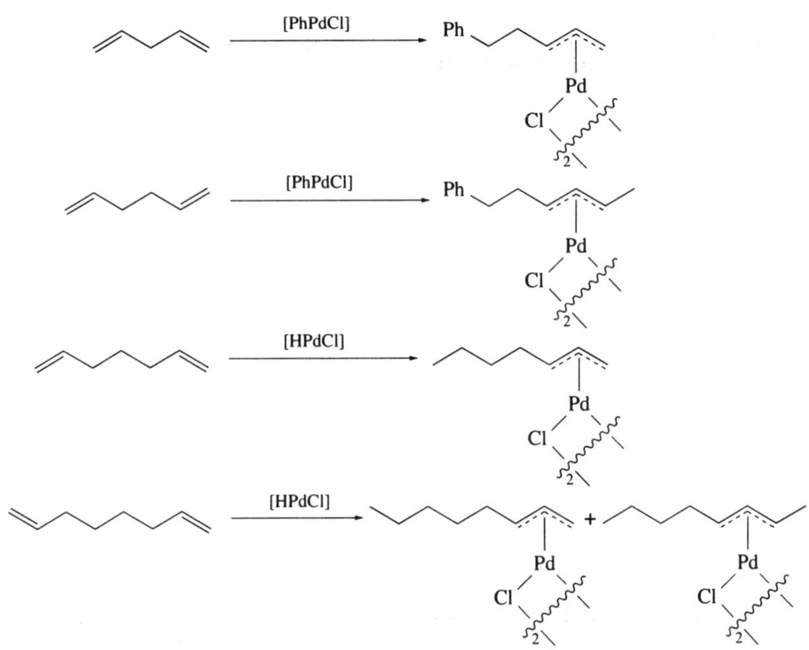

Scheme 31

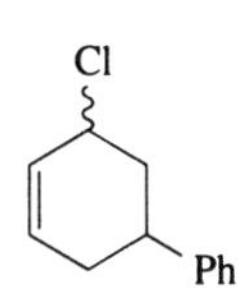

(69)

organomercurials, including those with carboxylic acid, phenol, alcohol, and amide functionalities, are reported[219] to react with conjugated and nonconjugated dienes (also vinylcyclopropanes, see Section 6.4.2.4) and [Li][$PdCl_3$] to generate π-allylpalladium complexes. Intramolecular displacement reactions provide access to oxygen and nitrogen heterocycles (see Section 6.4.3.8).

Much of the work with organomercurials is predicated on *in situ* generation of [RPdCl] equivalents which, in certain cases (e.g., R = Et), may lead to the formation of reactive [HPdCl] equivalents by β-hydride elimination. Isolable [RPdX] synthons in the form of "[(C_6F_5)PdBr]," stabilized as [Pd(C_6F_5)Br(NCMe)$_2$], have now been developed.[220] Acyclic and cyclic 1,3-dienes (and one example of a cyclic 1,4-diene) reacted by insertion into the Pd–C_6F_5 bond to generate π-allylpalladium complexes. The substrates investigated and the resultant products are illustrated in Scheme 32 (where the pentafluorophenyl group is abbreviated as Pf).

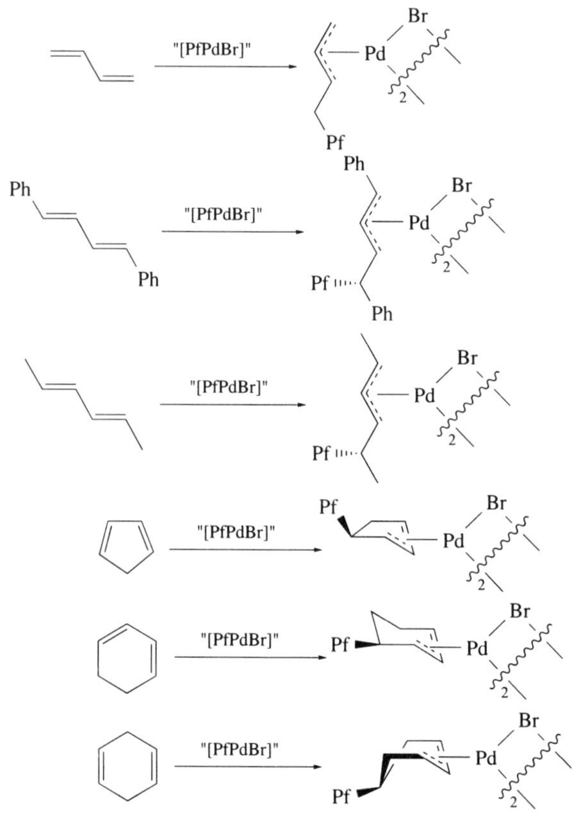

Scheme 32

Analysis by ^1H and ^{19}F NMR spectroscopies indicated the presence of four different stereoisomers for the dimeric complexes with allyl groups lacking a plane of symmetry and two stereoisomers for the dimeric complexes containing allyl groups with planes of symmetry. The relevance of the reactions to alkene arylation (the Heck reaction) was demonstrated by treatment of styrene with [PfPdBr] equivalents which produced *trans*-PhCH=CH(C_6F_5) along with palladium metal. The x-ray crystal structure of one isomer of the dimeric complex derived from reaction with 1,3-cyclohexadiene was determined and showed a *trans*-arrangement of the allyl groups about the central Pd_2Br_2 core. A *cis*-arrangement of palladium and pentafluorophenyl groups about the cyclohexenyl ring was observed, consistent with that expected from *endo*-attack.

Organopalladium addition to nonconjugated dienes to generate π-allyl complexes by remote palladium migration has been effected without the use of organomercurials.[221] Thus, aryl iodides, nonconjugated dienes, and carbon nucleophiles have been shown to react in the presence of palladium catalysts to generate coupled products. The reported mechanism involved initial generation of arylpalladium equivalents, addition to the least-substituted double bond of the diene, palladium migration to generate a π-allyl complex, and displacement of the coupled product by attack of the carbon nucleophile on the π-allyl substituent.

Dimethyl-1,4-cyclohexadienes have been shown[222] to react with [$PdCl_2(NCMe)_2$] in methanol to produce (5-methoxy-1-3-η^3-cyclohexenyl)palladium complexes. The reactions presumably proceed by initial diene coordination, followed by nucleophilic attack to generate the methoxy-substituted σ,π-complex, and rearrangement to the π-allyl complex. The presence of methyl substituents (e.g., with 1,4-dimethyl-1,4-cyclohexadiene as the substrate) on the initially formed diene complex leads to competition between attack at the most-substituted carbon atom and formation of the least-hindered σ,π-complex. The competition between pathways was found to be dependent on both temperature and the presence or absence of base.

6.4.2.3 Reactions of 1,2-dienes (allenes) with palladium salts and complexes

Scheme 19 illustrates the typical reactions of allene with palladium precursors that lead to formation of π-allyl complexes. Such compounds continue to be of interest in terms of their reactions with nucleophiles (see Section 6.4.3.8(iv)) and halogens (see Section 6.4.3.5).

The reaction of (**70**) with allene is reported[223] to occur through insertion of allene into the palladium–palladium bond to generate (**71**), characterized crystallographically, in which the two palladium centers are bridged by an η^1,η^3-allyl group, that is, the allyl group was found to be π-bonded to one palladium and σ-bonded in the C-2 position to the second palladium.

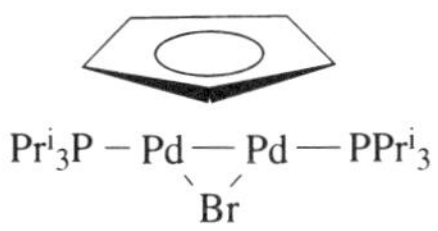

(**70**)

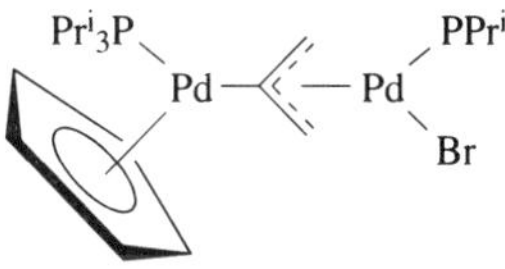

(**71**)

Compound (**71**) represents the first example in palladium chemistry of an allyl group functioning as both a π-allyl ligand and a σ-alkyl ligand.

6.4.2.4 Reactions involving ring opening of cycloalkanes by palladium salts and complexes

Cyclopropane activation by palladium compounds continues to be studied, particularly with regard to the intimate details of ring activation and the resultant stereochemistry of palladation. The reaction of (+)-2-carene with $[PdCl_2(NCMe)_2]$ in nonnucleophilic solvents, such as chloroform or benzene, to produce six-membered and seven-membered π-allyl complexes (Equation (34)) has been investigated.[224–6]

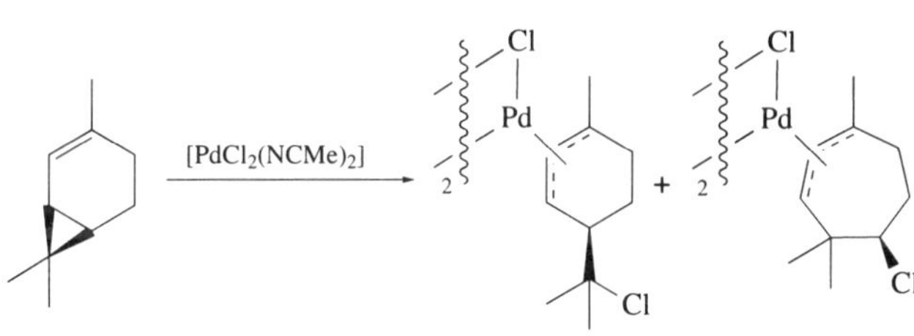

(34)

Analogous reactions performed in nucleophilic solvents, such as methanol or acetic acid, allowed isolation of the corresponding methoxy- and acetoxypalladation products. The chloropalladation reaction exhibited an unusual solvent dependence. Thus, reaction in chloroform–2% ethanol gave largely the six-membered ring product, whereas reaction in benzene produced the seven-membered ring complex as the major product. Formation of the six-membered ring product was found to occur by ring opening with inversion, whereas the seven-membered ring product was formed by ring opening with retention, resulting in overall *trans*-chloropalladation in the latter case. The available evidence did not allow a definitive mechanism to be proposed.

Chloropalladation of *exo*-9-methylbicyclo[6.1.0]non-4-ene (**72**) with $[PdCl_2(NCPh)_2]$ in chloroform-d has been found[227] to proceed with inversion of configuration at both of the newly formed asymmetric centers to generate the cyclooctenyl complex (**73**).

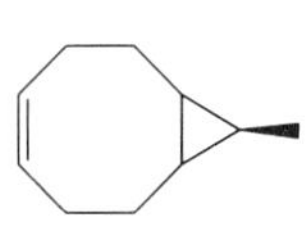

(**72**)

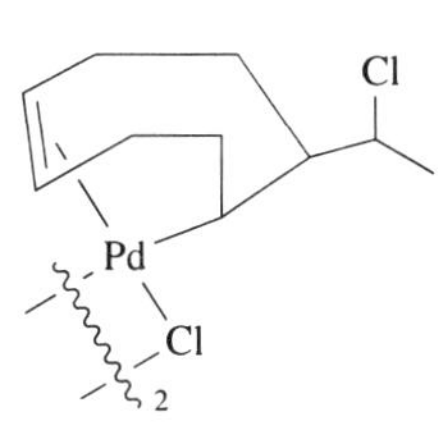

(**73**)

The x-ray crystal structure of (**73**) confirmed the 1,4,5-η^3-(σ,π)-interaction of the cyclooctenyl ligand with palladium and established the *anti*-orientation of palladium and the α-chloroethyl group, thus confirming inversion at C-1 upon electrophilic attack by palladium. The sequence ordering of atoms on C-9 established inversion upon nucleophilic attack at that carbon by chloride. The results suggest an intermediate corner-palladated cyclopropane which would give rise to an inverted product at C-1 following nucleophilic attack at C-9 by chloride.

The reaction of *cis*-9-methylenebicyclo[6.1.0]nonane with [$PdCl_2(NCPh)_2$] at 293 K was found[228] to yield a single isomer of the corresponding π-allyl complex, whereas *trans*-9-methylenebicyclo[6.1.0]nonane generated a pair of isomers. Interconversion of the products was not observed.[229] Scheme 33 illustrates the reactions.

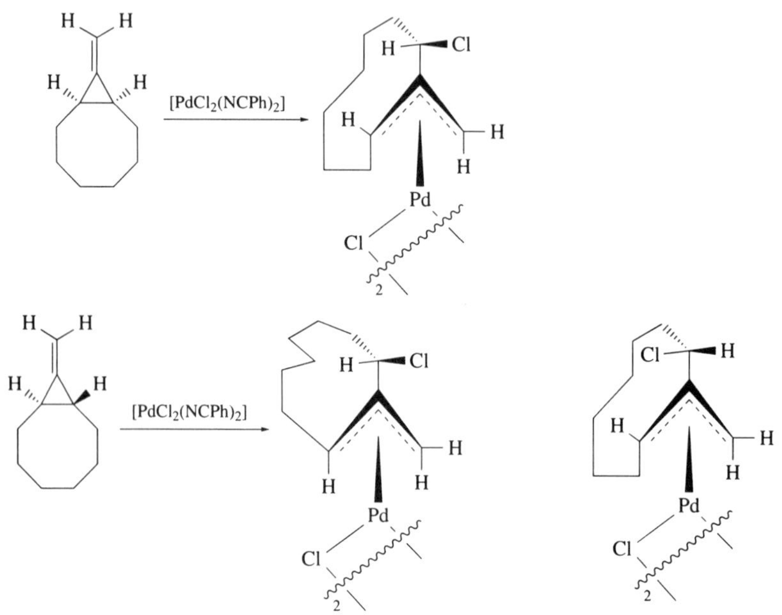

Scheme 33

The results were attributed to a stereospecific disrotatory ring opening during chloropalladation. Extended Hückel molecular orbital calculations indicated that disrotatory opening of an η^2-methylenecyclopropane coordinated in a square-planar d^8 complex is orbital-symmetry allowed providing the alkene lies perpendicular to the coordination plane; in contrast, conrotatory opening is allowed if the alkene lies in the square plane. The experimental observations thus appeared to support the theoretical predictions.

Vinylcyclopropane and derivatives with substituted vinyl groups have been found[230] to react with [$PdCl_2(NCPh)_2$] in chloroform-d through rapid chloropalladation generating 1,2,5-η^3-σ,π-complexes. In the case of vinylcyclopropane, the precursor to chloropalladation was observed by ^1H and ^{13}C NMR spectroscopies and tentatively identified as a π-complex that might involve a palladium–cyclopropane interaction. Over time the 1,2,5-η^3-σ,π-complexes rearranged in solution to 1-3-η^3-allyl complexes. In the case of the 1,2,5-η^3-σ,π-complex derived from vinylcyclopropane, allyl formation was found to involve a hydrogen shift. An example of each type of complex, the 1,2,5-η^3-σ,π- and 1-3-η^3-allyl, was characterized crystallographically.

The reaction of organopalladium intermediates, generated *in situ* from organomercurials and palladium salts, with alkenyl- or methylenecyclopropanes and alkenyl- or methylenecyclobutanes has been reported to lead to π-allyl complexes of palladium.[231,232] The general reaction is illustrated for the case of alkenylcyclobutanes in Equation (35).

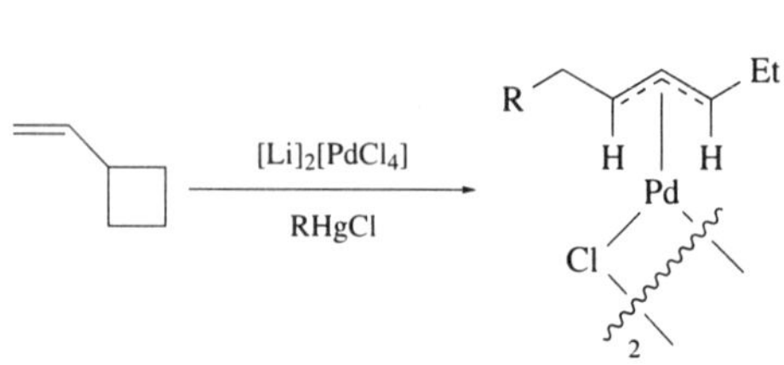

(35)

Aryl, methyl, vinyl, and heterocyclic organomercurials could all be employed in these reactions, while organomercurials containing hydrogens β to mercury gave rise to products resulting from the conversion of [RPdCl] into [HPdCl] equivalents and subsequent reaction with the substrate. The reaction of [PhPdCl] equivalents with methylenecyclopropane produced a π-allyl complex derived from initial addition of the phenyl group to the internal alkene carbon atom, whereas methylenecyclobutane gave rise to a π-allyl complex via initial addition of palladium to the internal carbon. The essential features of these reactions are illustrated in Scheme 34.

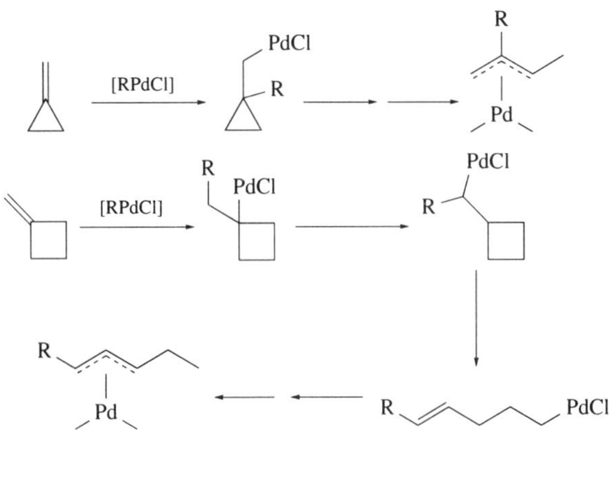

Scheme 34

The reaction with methylenecyclobutane is remarkable since palladium migration by β-hydride elimination and readdition occurred without alkene formation. Palladium-catalyzed reactions of methylenecycloalkanes based on *in situ* formation of [RPdX] and intermediate π-allyl generation have been described.[233]

The reactions of organopalladium intermediates with vinylcyclopropanes have been investigated in detail,[234] and it was shown that vinylcyclopropane, carbomethoxymercury(II) acetate, and lithium chloropalladate reacted to produce methyl sorbate. The same product was found when cyclopropylmercury(II) chloride, lithium chloropalladite, and methyl acrylate were reacted together, thus supporting the idea of a π-allylic intermediate. Scheme 35 outlines the reactions.

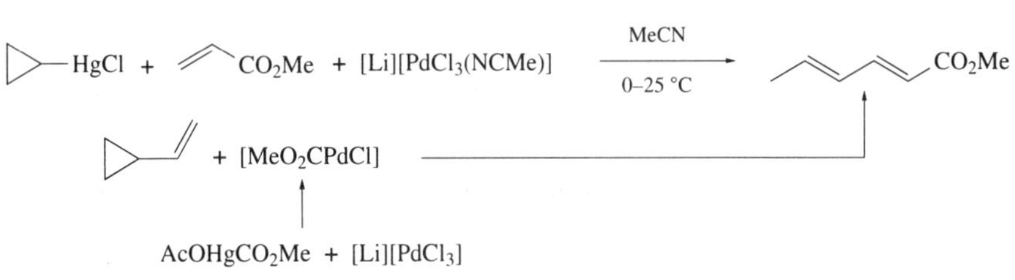

Scheme 35

Opening of substituted vinylcyclopropanes by palladium complexes to produce π-allyl species has been incorporated into a process for ring-opening polymerization.[235]

Enantiomerically pure π-allyl complexes of palladium have been synthesized by ring opening of bicyclic alkenes.[236] (+)-2-Carene (>97% purity) underwent ring opening with [HPdCl] equivalents, generated *in situ*, to produce the allyl complex (**74**) in 19% yield. The same complex was obtained from (+)-3-carene (>99% purity) in 23% yield. Compound (**74**) was characterized by x-ray crystallography (cf. Equation (34) where the products of chloropalladation of 2-carene are shown).

Related reactions of other optically pure bicyclic alkenes generated π-allyl complexes which appeared to be optically pure from comparison of the rotations measured for (*R*)- and (*S*)-enantiomers.

6.4.2.5 Reactions involving oxidative addition of allylic substrates to palladium(0)

An additional method for the preparation of π-allylpalladium complexes directly from palladium metal has been reported.[237] It was found that commercial palladium black reacted with allylic bromides

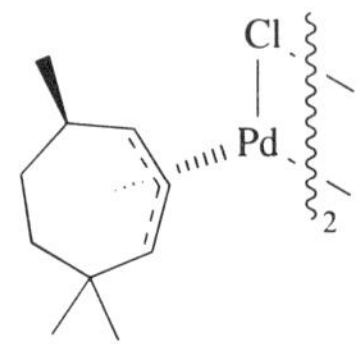

(74)

and iodides to generate dinuclear, halide-bridged π-allyl complexes in yields ranging from poor to excellent. Ultrasound increased the yields in some cases. Allyl chloride has been employed to trap palladium(0) generated by photolysis of the azide, $[Pd_2(N_3)_6]^{2-}$, in acetonitrile solution.[238] The formation of $[Pd_2(\mu\text{-}Cl)_2(\eta^3\text{-}C_3H_5)_2]$ was monitored spectrophotometrically. Similarly, the thermally unstable complex (**75**) is reported to undergo reductive elimination to generate $[TiCl(Et)(\eta\text{-}C_5H_5)_2]$ and $[Pd(PMe_3)]$ equivalents which could be trapped by 3-chloro-2-methylpropene to generate $[PdCl(\eta^3\text{-}2\text{-}MeC_3H_4)(PMe_3)]$.[239]

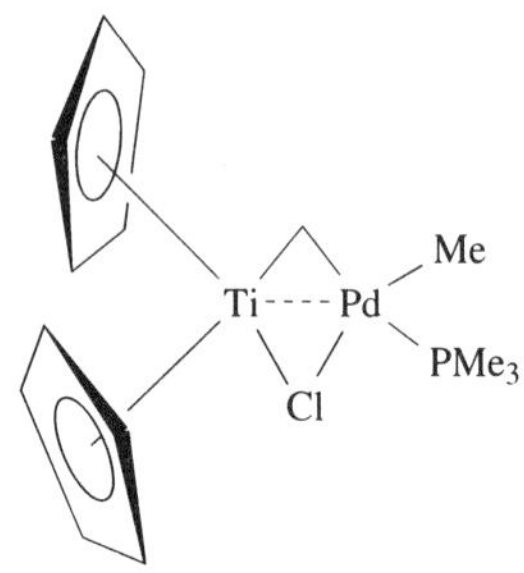

(75)

A π-allylpalladium intermediate has been proposed, but neither identified nor isolated, during toluene oxidation in a fuel cell containing a palladium-black anode.[240]

The reactions of allylic substrates with $[PdL_2(\text{styrene})]$ (L = tertiary phosphine, e.g., PPh_3), generated by treatment of *trans*-$[PdEt_2L_2]$ with styrene, have been investigated. Reactions of allylic formates[241] and allylic carbonates[242] yielded bis(phosphine)palladium–allyl cations, whereas reaction of allyl phenyl sulfide[243] generated a bridged palladium(I) species. The reactions are illustrated in Scheme 36.

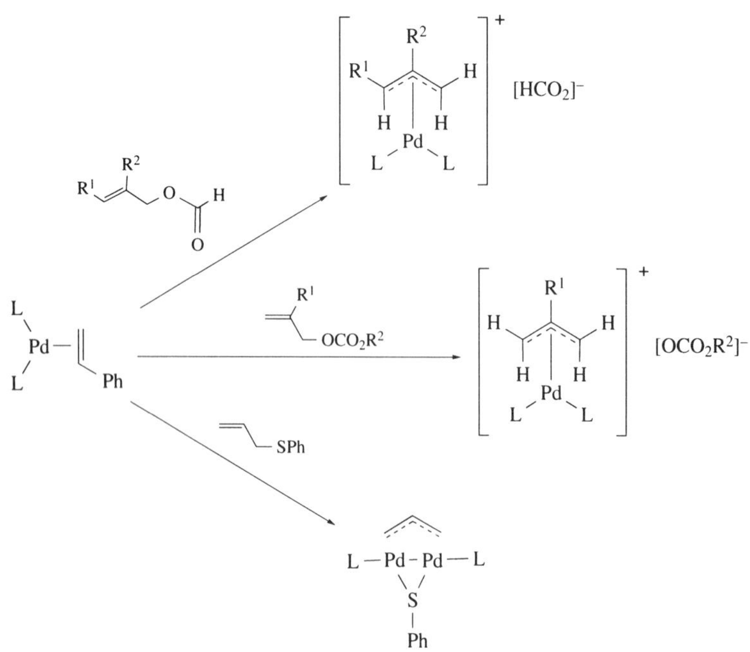

Scheme 36

The oxidative addition of an optically active allylic acetate to a palladium(0) species, generated *in situ* by reduction of $[PdCl_2(dppe)]$ with diisobutylaluminum hydride (dibal-H) in the presence of triphenylphosphine (i.e., presumably $[Pd(PPh_3)(dppe)]$), has been shown[244] to occur with inversion of configuration, producing a cationic palladium(II)–allyl complex in 44% yield (Equation (36)).

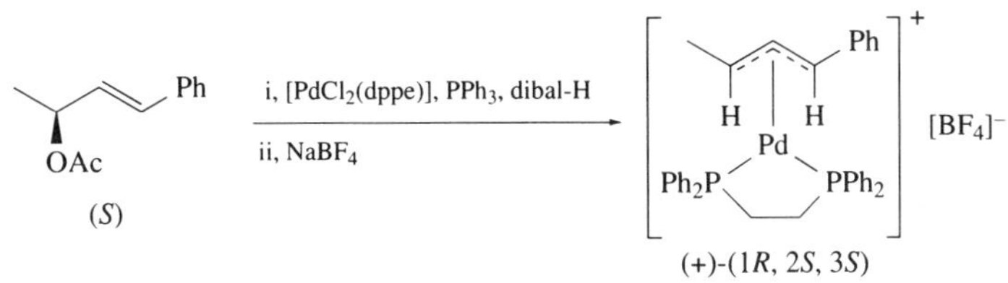

(36)

The enantiomeric purity of the complex was confirmed by comparison of its optical rotation with that of an authentic sample generated from an optically active allylsilane (see Section 6.4.2.1).

Reactions of palladium(0) complexes containing sterically demanding tertiary phosphines with allylic substrates have also been investigated.[245,246] $[Pd(PCy_3)_2]$ was found to react with allyl acetate and allyl aryl ethers to generate π-allyl complexes containing a single tertiary phosphine ligand. The second equivalent of phosphine was eliminated as a phosphonium salt. *N*-allyldiethylamine did not react with $[Pd(PCy_3)_2]$, whereas *N*-allyltriethylammonium bromide reacted by elimination of triethylamine to generate a π-allyl complex. Reactions are summarized in Scheme 37.

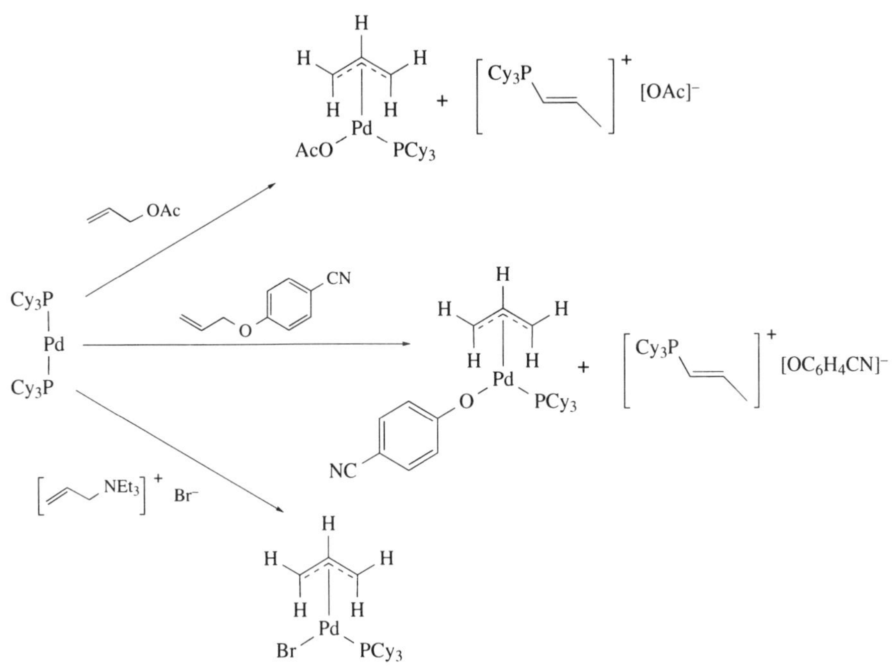

Scheme 37

Allyl phenyl sulfide and selenide were found to produce dinuclear complexes with structures analogous to that of the dinuclear compound shown in Scheme 36. Allyl alcohol and 1-methylallyl alcohol were found to react with $[Pd(PCy_3)_2]$ to generate not π-allyl complexes but $[Pd(\text{diallyl ether})(PCy_3)]$ (**76**) and $[Pd\{meso\text{-bis(methylallyl)ether}\}(PCy_3)]$, respectively, along with mixtures of diallylic ethers.

The complex $[Pd(OAc)(\eta^3\text{-}C_3H_5)(PCy_3)]$ was found to react with PCy_3 by nucleophilic attack on the allyl ligand to generate propenyltri(cyclohexyl)phosphonium acetate. The reaction thus produced "$[Pd(PCy_3)]$" equivalents which reacted with additional $[Pd(OAc)(\eta^3\text{-}C_3H_5)(PCy_3)]$ to generate $[Pd_2(\mu\text{-}OAc)(\mu\text{-}C_3H_5)(PCy_3)_2]$, thus providing an additional route to these dinuclear complexes.

Although the reactions of palladium(0) complexes with allylic esters typically occur through an *anti*-mechanism, as does subsequent nucleophilic attack by stabilized carbon nucleophiles, *syn*-complexation may be promoted by precoordination of palladium to the allylic leaving group, for example, in the form of a phosphinoacetate moiety.[247,248] Scheme 38 outlines possible pathways.

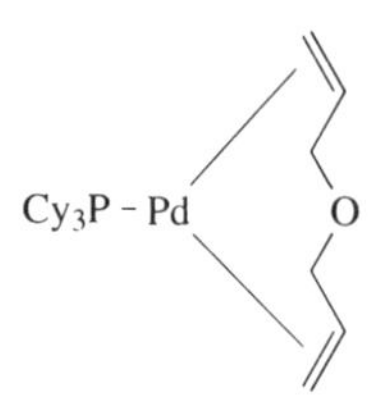

(76)

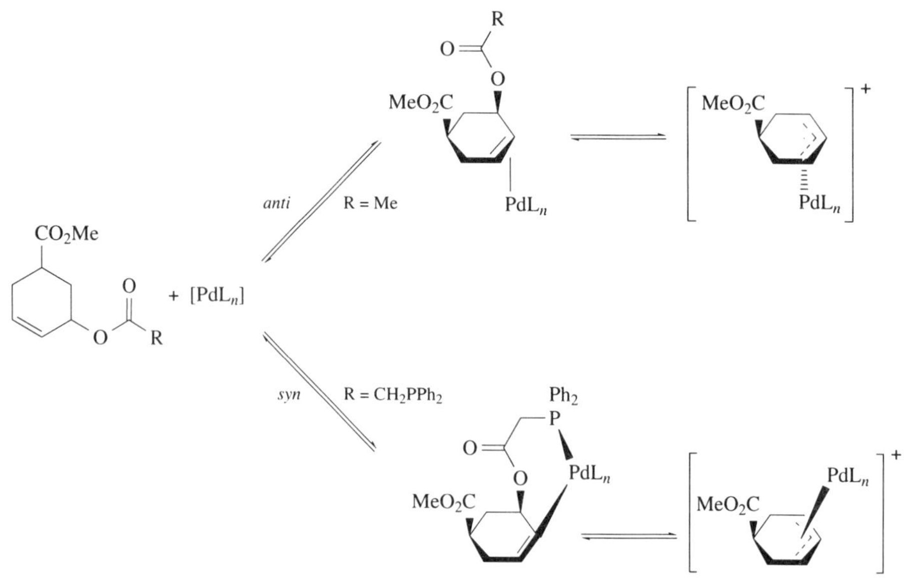

Scheme 38

Nucleophilic attack on the π-complexes shown in Scheme 38, which were generated catalytically *in situ* without identification, by $[Li][CH(CO_2Et)_2]$ was found to lead to the corresponding *cis*- and *trans*-disubstituted cyclohexene derivatives, with generation of a 3:2 product ratio in the case of the phosphinoacetate as substrate. Work with a range of substrates suggested that the *syn*-mechanism may be enforced by precoordination to specifically designed leaving groups, thus opening new possibilities for organic syntheses via palladium-catalyzed transformations of allylic substrates.

Predominant *syn*-oxidative addition of allylic halides to palladium(0)–alkene complexes has also been observed in cases where the alkene ligands bear electron-withdrawing groups. It was suggested that an electron-deficient palladium center may interact with the halide substituent of the allylic substrate, thus promoting *syn*-oxidative addition.[249]

Allylic trifluoroacetates are reported[250] to react with $[Pd(dba)_2]$ to produce dimeric, trifluoroacetate-bridged π-allylpalladium complexes, thus providing additional routes to these compounds (see Section 6.4.2.1). Subsequent reactions with neutral ligands allowed access to cationic complexes of the type $[Pd(\eta^3\text{-allyl})L_2]^+$, and use of the sterically encumbered 2,9-dimethyl-1,10-phenanthroline ligand was found to affect the *syn–anti* stereochemistry.

Functional group differentiation has been observed in reactions of allylic substrates with palladium(0). 2-Haloallyl acetates have been found[251,252] to react with alkynes under palladium(0)-catalyzed conditions via pathways that involved carbon–halogen oxidative addition and not π-allyl formation. Scheme 39 outlines the possible reactions, of which only path B was observed.

Although no palladium complexes were isolated during the study, the work suggests limits on the range of functionalities that might be tolerated during the preparation of π-allyl complexes from allylic substrates and palladium(0) precursors.

Similar limits on the range of allylic substrates suitable for preparation of π-allyl complexes are indicated by work on palladium-catalyzed transformations of allylic compounds with oxygen functional groups. Allyloxycarbonyl derivatives of alcohols are reported[253] to undergo conversion into allyl ethers in the presence of $[Pd(PPh_3)_4]$. The suggested mechanism involves oxidative addition to palladium(0), with generation of a cationic π-allyl intermediate and an alkoxycarbonyl anion, followed by loss of carbon dioxide from the anion and attack of the liberated alkoxide anion on the π-allyl cation (Scheme 40).

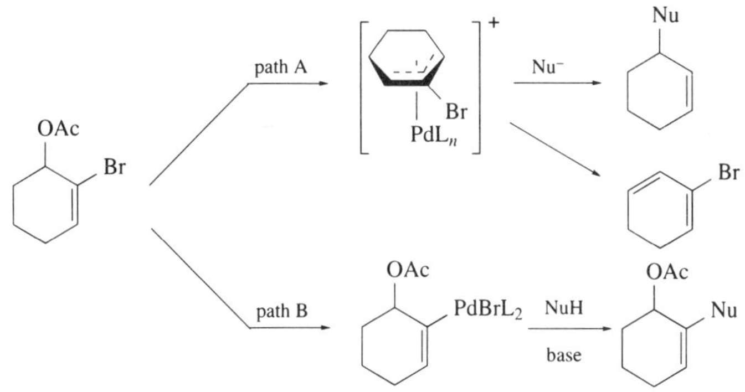

Scheme 39

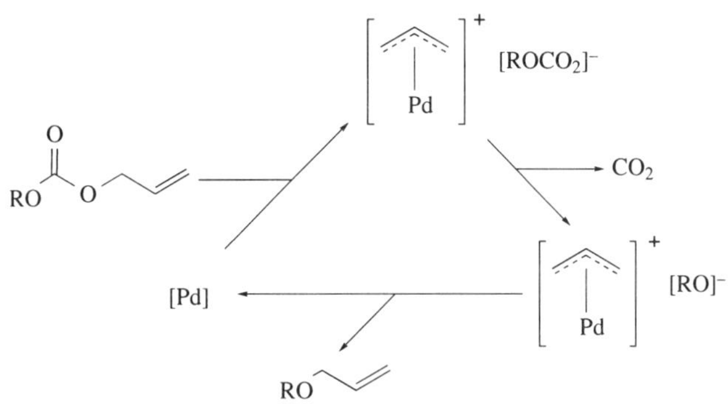

Scheme 40

Allylic β-ketocarboxylates similarly undergo decarboxylation in reactions catalyzed by palladium(0).[254–7] Scheme 41 illustrates a possible mechanism.

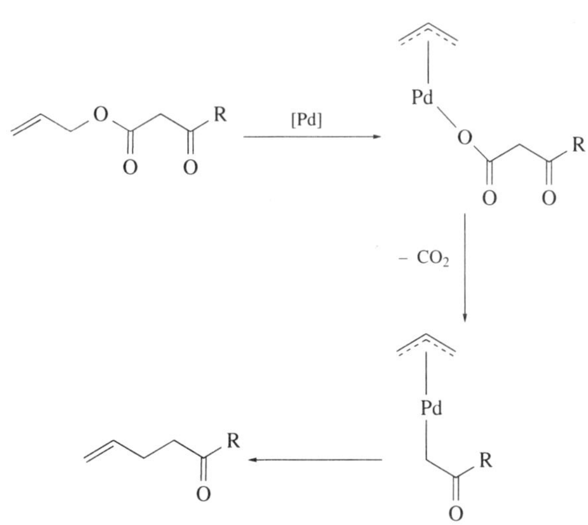

Scheme 41

6.4.2.6 Miscellaneous reactions leading to π-allylic complexes

Nucleophilic attack on a coordinated π-allyl group typically leads to elimination of an organic product (see Section 6.4.3.8) and the liberated palladium fragment may be trapped by a new, different allylic substrate. Allyl exchange under catalytic allylation conditions has been employed to transform $[Pd(\eta^3\text{-}C_3H_5)L_2]^+$ (L_2 = diphosphine) into $[Pd(\eta^3\text{-}1,3\text{-}Ph_2C_3H_3)L_2]^+$ and related substituted-allyl complexes, by treatment with 1.2 equivalents of sodium malonate followed by two equivalents of 1,3-diphenyl-2-propene or other allylic substrates.[258]

Complexes of the type [Pd(allyl)(allyl')] have been synthesized by reaction of $[Pd_2(\mu\text{-}Cl)_2(\eta^3\text{-allyl})_2]$ with Grignard reagents (allyl'MgX). Typically, reaction of allyl'MgX with $[Pd_2(\mu\text{-}Cl)_2(\eta^3\text{-allyl})_2]$ in diethyl ether results in formation of homo- and cross-coupled 1,5-dienes upon elimination, but if dioxane is introduced into the ether solvent in order to precipitate magnesium salts, only cross-coupled 1,5-dienes result from elimination. Accordingly, use of dioxane–ether allowed generation of [Pd(allyl)(allyl')] complexes from Grignard reagents and the complexes could be isolated as thermally sensitive solids and characterized by low-temperature NMR methods.[259,260]

Insertion reactions of cyclopalladated compounds with alkynes have typically been found to occur via insertion of one, two, or three alkyne units to generate products with bound alkene groups (see Section 6.2.3.2(i)) or bound arene moieties (see Section 6.6.2). However, the reaction of the dimeric, cyclopalladated complex $[Pd_2(\mu\text{-}Cl)_2\{CH(TMS)C_6H_4NMe_2\text{-}2\}_2]$ with alkynes bearing at least one electron-withdrawing group has been found to produce π-allyl complexes regioselectively.[261] The reaction, illustrated in Equation (37), was proposed to occur via insertion of the alkyne into the palladium–carbon bond followed by a 1,3-shift of the trimethylsilyl group onto the newly palladated carbon atom.

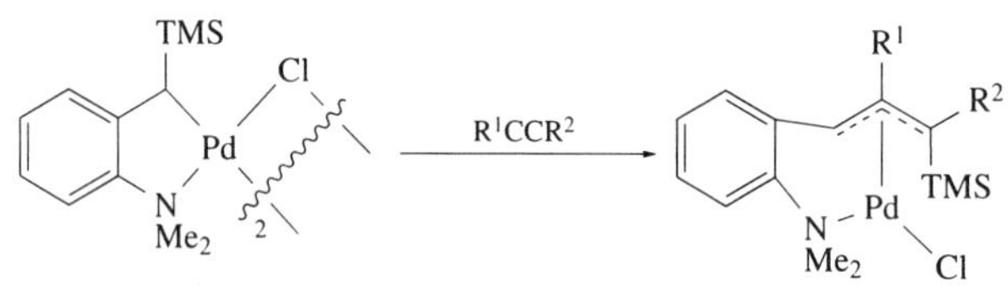

(37)

The π-allyl complex, shown in Equation (37) with the trimethylsilyl group in the *anti*-position, was found to isomerize slowly in solution to the *syn*-form.

The reaction of $[Pd_2(dba)_3]$ with 3,5-di-*t*-butyl-1,2-benzoquinone is reported[262,263] to generate, in addition to the monomeric, square-planar complex $[Pd(dbsq)_2]$, a by-product identified by x-ray crystallography as $[Pd_2\{Pd(dbsq)_2\}_2]$ (dbsq is a semiquinone derived from 3,5-di-*t*-butyl-1,2-benzoquinone). The latter was found to consist of two planar *cis*-$[Pd(dbsq)_2]$ units that were bridged by two palladium atoms π-bonded to the allylic regions of the dbsq ligands. The allylic portion of the structure is represented in (**77**).

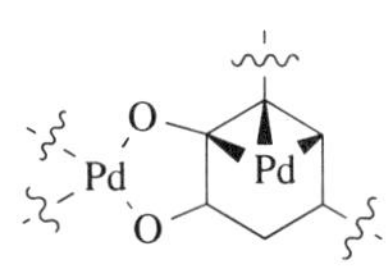

(77)

The bis(η^3-allyl)palladium portions of the dimer were found to be oriented in a *trans*-fashion about the central unit.

A number of π-allylpalladium complexes containing hindered phenoxide substituents in the C-2 position of the allyl group, for example (**78**), have been prepared by standard methods (for example, (**78**) was synthesized by reaction of 2,6-di-*t*-butyl-4-isopropenylphenol with palladium dichloride in glacial acetic acid) and then transformed by oxidation with lead(IV) oxide into the corresponding phenoxyl radicals.[264]

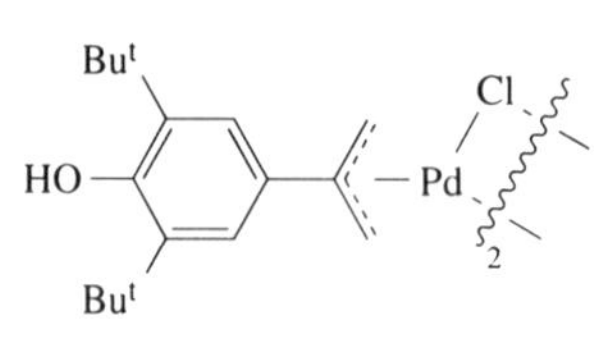

(78)

The radical species were characterized by ESR spectroscopy and the spectra were found to be essentially unchanged with use of a number of different solvents and at temperatures over the range 213–313 K, indicating significant stability for the paramagnetic compounds. The spectrum of (**78**) exhibited the expected coupling to ^{105}Pd and indicated equivalence of the four allylic protons, implying *syn–anti* exchange.

6.4.3 Reactions of π-Allylic Complexes

In *COMC-I* the reactions of π-allylpalladium complexes were discussed in a number of separate categories.

(i) *Nucleophilic attack at the metal*. The reactions considered under this category included many examples of the bridge cleavage of $[Pd_2(\mu\text{-}X)_2(\eta^3\text{-allyl})_2]$ and related compounds. NMR spectroscopy has played an important role in understanding such processes; selected examples have been discussed in Section 6.4.1.3 and additional examples are presented in Section 6.4.3.1.

(ii) *Halide exchange in $[Pd_2(\mu\text{-}Cl)_2(\eta^3\text{-allyl})_2]$ and related reactions*. The reactions of $[Pd_2(\mu\text{-}Cl)_2(\eta^3\text{-allyl})_2]$ with NaX (X = Br, I, N_3, etc.) were shown to lead to halide exchange, while treatment of $[Pd_2(\mu\text{-}X)_2(\eta^3\text{-allyl})_2]$ with silver salts of classically noncoordinating anions (Y) in the presence of neutral ligands (L) allowed isolation of the salts $[Pd(\eta^3\text{-allyl})L_2][Y]$. The reactions of thallium reagents, such as TlC_5H_5, were described as leading to products of the type $[Pd(C_5H_5)(\eta^3\text{-allyl})]$.

(iii) *Reactions in which the η^3-allyl ligand is transferred to another metal*. The lability of the η^3-allyl ligand in palladium complexes has allowed for the development of a number of allyl transfer processes. For example, $[Pd_2(\mu\text{-}Cl)_2(\eta^3\text{-allyl})_2]$ was shown to react with $[Na][Co(CO)_4]$ in the presence of triphenylphosphine to generate $[Co(\eta^3\text{-allyl})(CO)_2(PPh_3)]$.

(iv) *Thermal decomposition*. The thermal decomposition of $[Pd_2(\mu\text{-}Cl)_2(\eta^3\text{-allyl})_2]$ was shown to generate allyl chloride and palladium metal. 2-Substituted π-allylpalladium dimers typically produced low yields of hydrocarbons in addition to allylic chlorides and palladium metal.

(v) *Oxidation of allylic complexes to aldehydes and ketones*. Three basic categories were discussed: (a) the oxidation of alkenes under conditions where π-allylpalladium complexes are likely intermediates (e.g., oxidation with palladium dichloride in aqueous acetic acid) typically produced carbonyl compounds in complex, pH-dependent reactions; (b) inorganic oxidants such as manganese(IV) oxide and dichromate ion were shown to react with π-allylpalladium complexes to produce unsaturated organic carbonyls; and (c) the hydrolysis of $[Pd_2(\mu\text{-}Cl)_2(\eta^3\text{-}C_3H_5)_2]$ with alkaline boiling water was reported to produce acetone as the oxygenated product.

(vi) *Halogenation*. The majority of halogenation reactions of π-allylpalladium complexes were shown to yield halogenated organic products. For example, $[Pd_2(\mu\text{-}Cl)_2(\eta^3\text{-}C_3H_5)_2]$ was reported to react with bromine in methanol in the presence of sodium bromide to liberate allyl bromide or, if excess bromine was used, 1,2,3-tribromopropane.

(vii) *Reduction*. Hydrogenation, reduction with hydride sources, and reduction with alcoholic bases were among the reduction reactions described. Hydrogenation typically produced monoene or saturated hydrocarbon products and reactions with hydride sources proceeded similarly. Monoenes were typically found upon treatment of $[Pd_2(\mu\text{-}Cl)_2(\eta^3\text{-}C_3H_5)_2]$ and related compounds with alcoholic base.

(viii) *Reactions with acids*. π-Allylpalladium complexes are sensitive to protic acids. Methanol was found to be sufficiently acidic to cause decomposition of $[Pd(\eta^3\text{-}C_3H_5)_2]$ to palladium metal with liberation of $MeCH{=}CH_2$ and $MeOCH_2CH{=}CH_2$.

(ix) *Reactions with carbon monoxide*. Carbon monoxide was shown to react stoichiometrically with π-allylpalladium complexes, such as $[Pd_2(\mu\text{-}Cl)_2(\eta^3\text{-}C_3H_5)_2]$, in much the same way as other neutral two-electron donors to generate cleavage products that showed dynamic NMR behavior. With excess carbon monoxide, carbonylation of the allylic group was shown to result in loss of the corresponding butenoyl halide from $[Pd_2(\mu\text{-}X)_2(\eta^3\text{-allyl})_2]$. Such reactions are known to be exceptionally sensitive to variations in the nature of the ligands, temperature, pressure, and so on.

(x) *Nucleophilic attack at carbon*. Although nucleophilic attack at the allyl ligand of π-allylpalladium complexes typically results in liberation of a substituted, unsaturated organic product, substitution reactions on intact complexes were shown to be possible where the allyl ligand bears a leaving group in a suitable position (e.g., substitution of chloride by alkoxide in a chloromethyl group attached to C-1 or C-3 of the allyl ligand). The role of π-allylpalladium complexes as electrophilic reagents for allylation of nucleophiles is of increasing importance in synthetic organic chemistry and is treated in Chapter 8.2, Volume 12.

(xi) *Reactions with dienes*. Dienes are reported to react with π-allylpalladium complexes to generate coupled products. It seems likely that diene coordination is followed by insertion, possibly involving an η^1-allyl intermediate, and rearrangement. Butadiene was shown to undergo a variety of reactions with π-allylpalladium precursors, typically leading to dimerization or trimerization (see Scheme 28 for a mechanistic proposal for the telomerization of butadiene with acetic acid catalyzed by palladium acetate complexes).

Full discussion of each of the areas briefly mentioned above is given in *COMC-I* and details of new work in certain areas are presented in the following sections.

6.4.3.1 Nucleophilic attack at palladium

Additional examples of the bridge cleavage reactions of $[Pd_2(\mu\text{-}Cl)_2(\eta^3\text{-}C_3H_5)_2]$ by neutral ligands (L) to produce $[PdCl(\eta^3\text{-}C_3H_5)L]$ have been reported. Cleavage by purines, pyrimidine bases, and their nucleosides (i.e., adenine, adenosine, hypoxanthine, inosine, cytosine, and cytidine) has been described.[265] The $[PdCl(\eta^3\text{-}C_3H_5)L]$ complexes were formed with binding to N-7 in the case of purine bases and their nucleosides, and with binding to N-3 in the case of cytosine and cytidine.

$[Pd_2(\mu\text{-}Cl)_2(\eta^3\text{-}C_3H_5)_2]$ and $[Pd_2(\mu\text{-}Cl)_2(\eta^3\text{-}2\text{-}MeC_3H_4)_2]$ are reported[266] to undergo bridge cleavage upon treatment with the keto-stabilized ylides $Ph_3PC(H)COMe$ and $Ph_3PC(H)COPh$ in dichloromethane solution. The complexes $[PdCl(\eta^3\text{-}2\text{-}XC_3H_4)\{Ph_3PC(H)COR\}]$ (X = H, Me; R = Me, Ph) were obtained in high yield and characterized crystallographically (for X = Me, R = Me). NMR evidence indicated the presence of two diastereomers in solution, since coordination of palladium to the ylide carbon atom generated an unsymmetric center in the unsymmetrical complex, although the x-ray structure indicated only a single diastereomer in the solid state. The ylide ligands were found to be readily displaced by PPh_3 or chloride ion. Halide abstraction (see Section 6.4.3.2) from $[PdCl(\eta^3\text{-}2\text{-}XC_3H_4)\{Ph_3PC(H)COMe\}]$ with silver tetrafluoroborate, followed by addition of isonitrile (CNR; R = alkyl, aryl), allowed isolation of $[Pd(\eta^3\text{-}2\text{-}XC_3H_4)(CNR)\{Ph_3PC(H)COMe\}][BF_4]$ complexes without displacement of the coordinated ylide. The compound where X = Me, R = Bu^t was characterized crystallographically.[267]

Isolation of allylic palladium–alkene complexes by tin(II) chloride facilitated bridge cleavage of $[Pd_2(\mu\text{-}Cl)_2(\eta^3\text{-}2\text{-}MeC_3H_4)_2]$ and its platinum analogue in the presence of ethene, styrene, or norbornene has been described.[268] The x-ray structures of the complexes $[M(SnCl_3)(\eta^3\text{-}2\text{-}MeC_3H_4)(styrene)]$ (M = Pd, Pt) were determined and complete multinuclear magnetic resonance studies, including two-dimensional ^1H NOESY and ^{13}C–^1H correlation studies, were reported. The crystallography indicated that, in both compounds, the alkene double bonds were almost parallel to the coordination planes, while both the NMR and x-ray structural data indicated a moderate *trans*-influence for both the trichlorostannate and alkene ligands with the former being larger. Comparison of data for the palladium and platinum compounds suggested that the extent of π-backbonding to the alkenes was larger for platinum than for palladium (see also Section 6.2.2).

The reaction of $[Pd_2(\mu\text{-}Cl)_2(\eta^3\text{-}C_3H_5)_2]$ with PPh_3, generating $[PdCl(\eta^3\text{-}C_3H_5)(PPh_3)]$, has been shown to proceed differently in the presence of sodium methoxide. In this case, the bridged dipalladium(I) complex (**79**) was obtained. The compound was characterized crystallographically.[269]

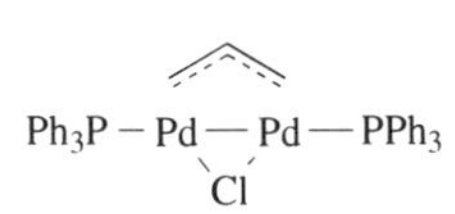

(79)

A number of routes to ligand-bridged homo- and heteronuclear complexes have been developed based on bridge cleavage reactions of $[Pd_2(\mu\text{-}Cl)_2(\eta^3\text{-}C_3H_5)_2]$. 1,2-Bis(imino)alkylpalladium(II) complexes have been shown[270] to react with $[Pd_2(\mu\text{-}Cl)_2(\eta^3\text{-}C_3H_5)_2]$, or the 2-methylallyl analogue, as shown in Scheme 42 (where L = triarylphosphine or arsine; R = H, alkyl, aryl).

Sequential treatment of $[Pd_2(\mu\text{-}Cl)_2(\eta^3\text{-}C_3H_5)_2]$ with 1,2,4-triazole and $[Rh(acac)(CO)_2]$ has been reported[271] to lead to formation of (**80**), which was characterized crystallographically.

In a related process, the complex $[PdCl(\eta^3\text{-}C_3H_5)(Hpz)]$ (Hpz = pyrazole) has been shown[272] to react with $[Rh(acac)(CO)_2]$ in the presence of sodium azide in methanol to generate (**81**) in 80% yield.

Similarly, $[RuCl_2(\eta^6\text{-}p\text{-cymene})(Hpz)]$ has been shown[273] to react with $[Pd(acac)(\eta^3\text{-}C_3H_5)]$ to generate $[(\eta^6\text{-}p\text{-cymene})ClRu(\mu\text{-}Cl)(\mu\text{-}pz)Pd(\eta^3\text{-}C_3H_5)]$ (where pz = pyrazolyl). Similar reactions with other $[Pd(acac)(\eta^3\text{-allyl})]$ complexes were also reported and the x-ray crystal structure of $[(\eta^6\text{-}p\text{-cymene})ClRu(\mu\text{-}Cl)(\mu\text{-}pz)Pd(\eta^3\text{-}C_8H_{11})]$ was determined.

The reaction of the salt $[K]_2[(pz)_3B\text{–}B(pz)_3]$ with $[Pd_2(\mu\text{-}Cl)_2(\eta^3\text{-}C_3H_5)_2]$ has allowed isolation and characterization by x-ray crystallography of the bridged complex $[(\eta^3\text{-}C_3H_5)Pd\{(\mu\text{-}pz)_2(pz)B\text{–}B(pz)(\mu\text{-}pz)_2\}Pd(\eta^3\text{-}C_3H_5)]$.[274] The complex contained two pz groups bridging between boron and palladium with an additional terminal pz group on each boron.

Treatment of $[Pd_2(\mu\text{-}Cl)_2(\eta^3\text{-}2\text{-}MeC_3H_4)_2]$ with two equivalents of $[K][Fe\{Si(OMe)_3\}(CO)_3(\eta^1\text{-}dppm)]$ in THF produced the heterodinuclear complex (**82**) in 84% yield.[275]

The trianionic cluster $[Re_7C(CO)_{21}]^{3-}$ exhibits similarities to $[C_5H_5]^-$ and $[C_2B_9H_{11}]^{2-}$ (dicarbollide anion) in its reaction chemistry. Treatment of $[Pd_2(\mu\text{-}Cl)_2(\eta^3\text{-}1\text{-}PhC_3H_4)_2]$ with $[NEt_4]_3[Re_7C(CO)_{21}]$ allowed isolation of $[Re_7C(CO)_{21}Pd(\eta^3\text{-}1\text{-}PhC_3H_4)]$, which was characterized crystallographically.[276] The

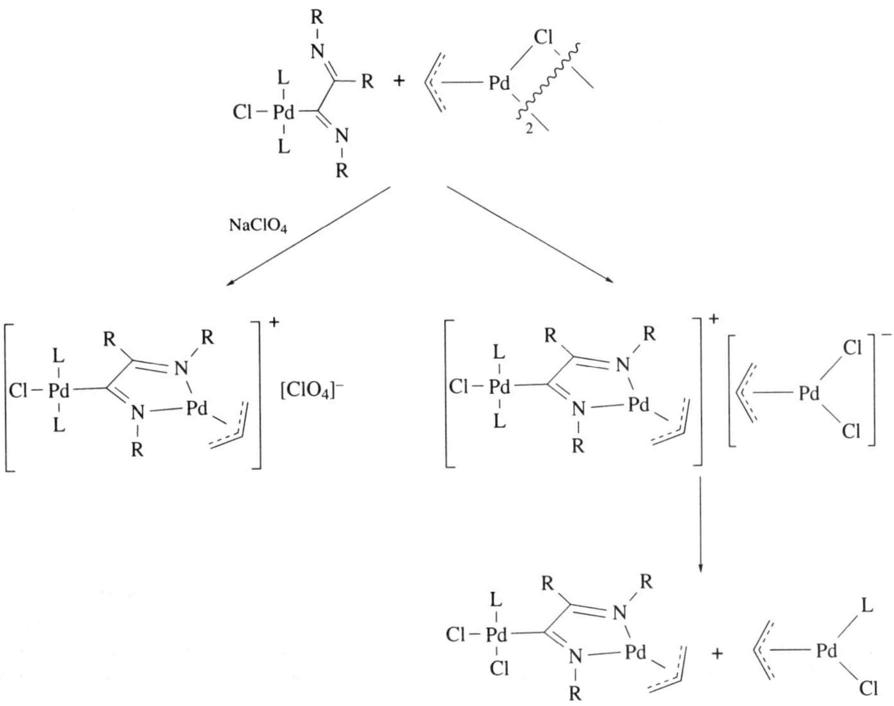

Scheme 42

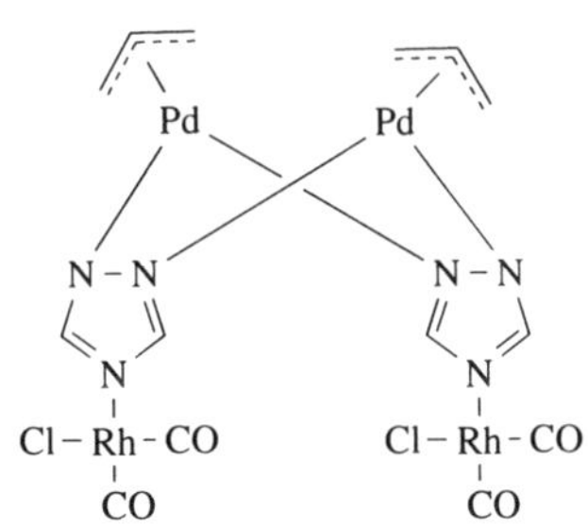

(80)

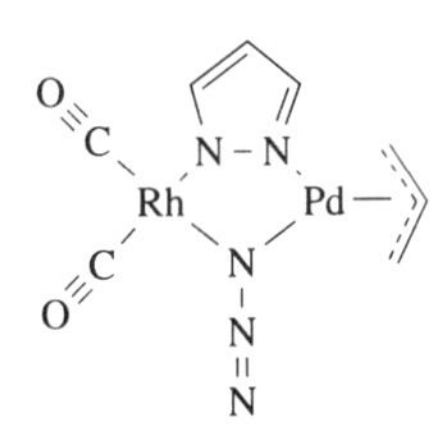

(81)

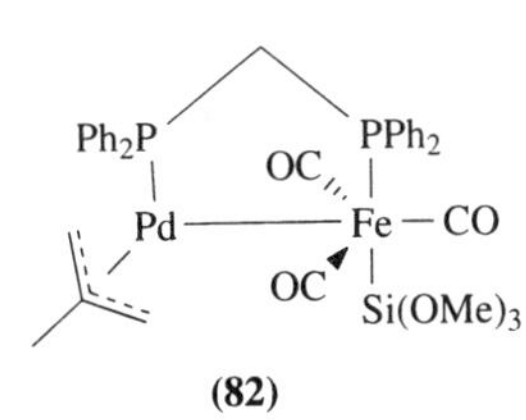

(82)

structure was found to consist of an octahedral arrangement of six rhenium atoms, capped in 1,4-fashion by $[Re(CO)_3]$ and $[Pd(\eta^3\text{-}1\text{-}PhC_3H_4)]$ units, giving rise to an approximate octahedral geometry for palladium. Unlike the analogue $[Pd(C_5H_5)(\eta^3\text{-}1\text{-}PhC_3H_4)]$, where the palladium is slipped with respect to the cyclopentadienyl ring giving rise to "allylic" bonding, in the cluster there was no clear evidence of slippage of palladium towards two rheniums and away from the third that made up a triangular face. The geometry of the cluster was thus described as the *closo*, unslipped type.

$[Pd_2(\mu\text{-}Cl)_2(\eta^3\text{-}2\text{-}MeC_3H_4)_2]$ has been shown to react with $Se(TMS)_2$ to generate the hexanuclear cluster $[Pd_6(\mu_4\text{-}Se)_3(\eta^3\text{-}2\text{-}MeC_3H_4)_6]$.[277] X-ray crystallographic characterization showed that the cluster exhibited a trigonal-prismatic geometry for the Pd_6 core with three faces of the prism bridged by μ_4-Se ligands.

Chiral ferrocenylphosphine bidentate ligands modified by monoaza or diaza crown ethers (L_2) have been shown[278] to react with $[Pd_2(\mu\text{-}Cl)_2(\eta^3\text{-}C_3H_5)_2]$ to produce $[Pd(\eta^3\text{-}C_3H_5)L_2]^+$ in which the bidentate ligand provides a pendant crown ether conveniently located to interact with an incoming nucleophile. Such compounds are of interest as potential catalytic intermediates in asymmetric allylation reactions.

Reactions of $[Pd_2(\mu\text{-}Cl)_2(\eta^3\text{-allyl})_2]$ with the sterically demanding bidentate ligand 2,9-dimethyl-1,10-phenanthroline (L_2) in the presence of silver tetrafluoroborate have been employed[279] to prepare $[Pd(\eta^3\text{-allyl})L_2]^+$ complexes in which the ligand sterically interferes with *syn*-allylic substituents. The *syn–anti* ratios could be affected by this approach and this is important in terms of stereocontrol in allylic substitution reactions (see Section 6.4.3.8(iii)).

The reaction of $[Pd(\eta^3\text{-}C_3H_5)(PPh_3)_2][PF_6]$ with potassium *t*-butylperoxide in dichloromethane has been shown[280] to produce $[Pd(OOBu^t)(\eta^3\text{-}C_3H_5)(PPh_3)]$ by nucleophilic attack on palladium. Such compounds are of interest as possible intermediates in metal-catalyzed alkene oxidation by peroxides.

Reactions of $[Pd(\eta^3\text{-}C_3H_5)_2]$, $[Pd(C_5H_5)(\eta^3\text{-}C_3H_5)]$, and their substituted analogue with nucleophiles, especially tertiary phosphines, continue to be of interest. $[Pd(\eta^3\text{-}C_3H_5)_2]$ has been shown[281] to react with triphenylphosphine at 183 K to give a 1:1 adduct in which one allyl ligand was converted to the η^1-form, that is, $[Pd(\eta^3\text{-}C_3H_5)(\eta^1\text{-}C_3H_5)(PPh_3)]$. At 273 K in toluene solution the compound was found to react further to liberate 1,5-hexadiene and form $[Pd_2(\mu\text{-}\eta^3\text{-}C_3H_5)_2(PPh_3)_2]$ (**83**).

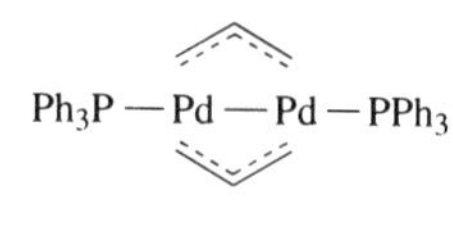

(83)

Compound (**83**) was characterized crystallographically. The allyl ligands were shown to be neither internally symmetrical nor symmetrically bonded to the two palladium atoms, indicating significant palladium–carbon σ-character and partial localization of the carbon–carbon double bond in each allylic ligand. $[Pd(\eta^3\text{-}2\text{-}MeC_3H_4)_2]$ has been shown[282] to react with PCy_3 to produce a 1:1 adduct shown to be fluxional in benzene solution at room temperature. In contrast, the reaction of $[Pt(\eta^3\text{-}2\text{-}MeC_3H_4)_2]$ with PPr^i_3 generated $[Pt(\eta^3\text{-}2\text{-}MeC_3H_4)(\eta^1\text{-}2\text{-}MeC_3H_4)(Pr^i_3P)]$, which was rigid in toluene solution up to a temperature of 363 K. $[Pd(\eta^3\text{-}C_3H_5)_2]$ has been found to react with the aminobis(imino)phosphorane (**84**) to generate a 1:1 adduct, characterized crystallographically as (**85**).[283]

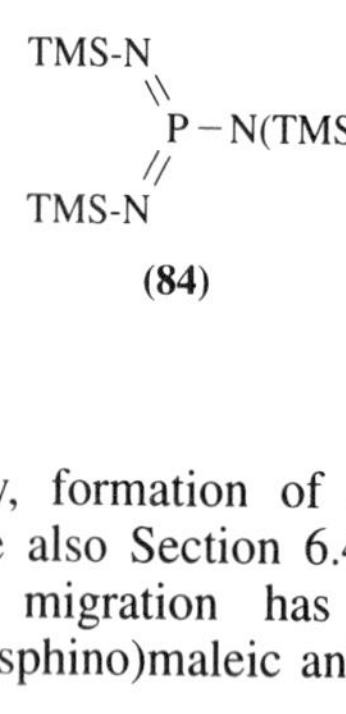

(84)

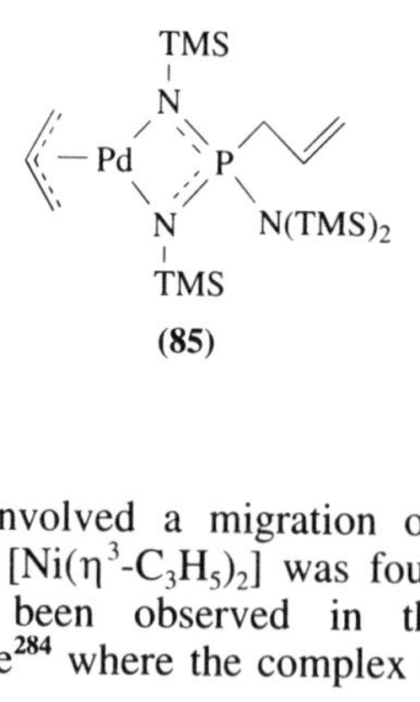

(85)

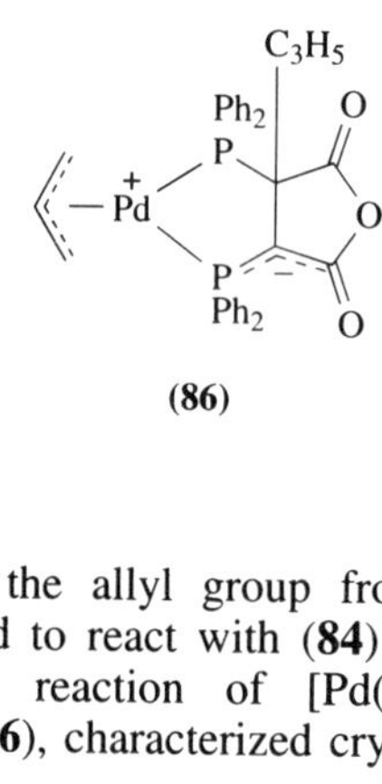

(86)

Unexpectedly, formation of (**85**) involved a migration of the allyl group from palladium to phosphorus (see also Section 6.4.3.3). $[Ni(\eta^3\text{-}C_3H_5)_2]$ was found to react with (**84**) in an analogous fashion. Allyl migration has also been observed in the reaction of $[Pd(\eta^3\text{-}C_3H_5)_2]$ with bis(diphenylphosphino)maleic anhydride[284] where the complex (**86**), characterized crystallographically, was produced.

$[Pd(C_5H_5)(\eta^3\text{-}2\text{-}MeC_3H_4)]$ and $[Pd(C_5H_5)(\eta^3\text{-}2\text{-}Bu^tC_3H_4)]$ have been shown to react with bulky tertiary phosphines (PPr^i_3, PBu^i_3, PPh_2Bu^t, PCy_3, $PPhCy_2$) to produce 1:1 adducts. NMR studies of the $[Pd(C_5H_5)(2\text{-}Bu^tC_3H_4)(PR_3)]$ complexes over the temperature range 213–333 K indicated η^3-bonding of the allylic ligand and η^1-bonding of the cyclopentadienyl group, and this was confirmed crystallographically for $[Pd(\eta^1\text{-}C_5H_5)(\eta^3\text{-}2\text{-}Bu^tC_3H_4)(PPr^i_3)]$, whereas investigations of the $[Pd(C_5H_5)(2\text{-}MeC_3H_4)(PR_3)]$ complexes revealed the presence of both $[Pd(\eta^5\text{-}C_5H_5)(\eta^1\text{-}2\text{-}MeC_3H_4)(PR_3)]$ and $[Pd(\eta^1\text{-}C_5H_5)(\eta^3\text{-}2\text{-}MeC_3H_4)(PR_3)]$ structural forms in solution.[285]

The carboxylate-bridged dimers $[Pd_2(\mu\text{-}O_2CR)_2(\eta^3\text{-}2\text{-}XC_3H_4)_2]$ (R = Me, Ph, CF_3; X = H, Me, Pr^i, Bu^t) have been shown to react with equimolar amounts of PPr^i_3 in hexane to produce the dinuclear complexes (**87**).[286]

The reaction of η^3-allylhydrogensulfidopalladium with $[Pd(\eta^3\text{-}C_3H_5)_2]$ or $[Pd(C_5H_5)(\eta^3\text{-}C_3H_5)]$ has been found[287] to produce the known cluster $[Pd_2(\eta^3\text{-}C_3H_5)_2S]_x$, by elimination of propene or cyclopentadiene, which reacted with triphenylphosphine to generate (**88**), characterized crystallographically.[288]

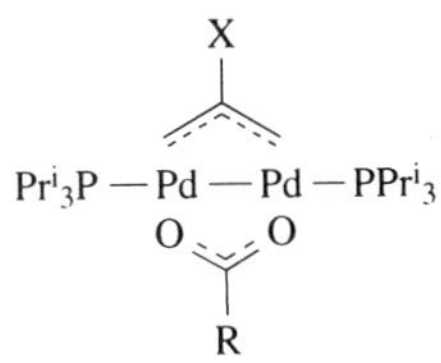
(87)

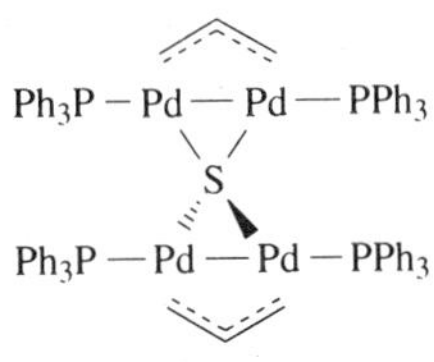
(88)

6.4.3.2 Halide exchange in [$Pd_2(\mu\text{-}Cl)_2(\eta^3\text{-allyl})_2$] and related reactions

It has been shown[289] that [$Pd_2(\mu\text{-}X^1)_2(\eta^3$-allylnorcamphene$)_2$] ($X^1$ = Cl, Br) is unaffected by treatment with MX^2_n (M = Na, K, Li, Hg, Fe; X^2 = Cl, Br, OAc) in hot acetic acid or a buffered acetate medium. Treatment with CuX^2_2, however, was found to lead to rapid decomposition (Scheme 43).

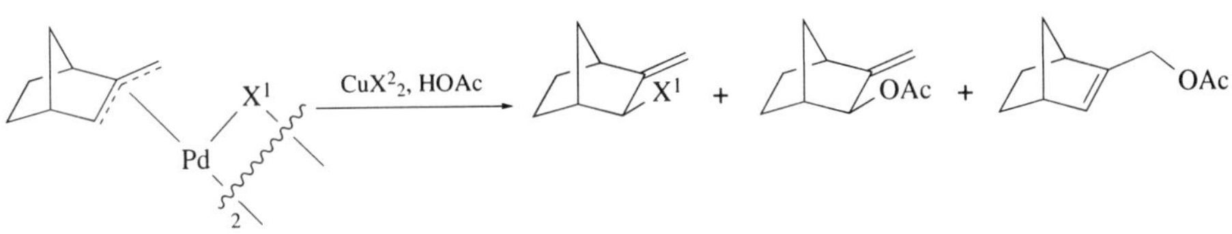
Scheme 43

The results were interpreted in terms of metathetical exchange of the bridging halide in [$Pd_2(\mu\text{-}X^1)_2(\eta^3$-allylnorcamphene$)_2$] by X^2, thus producing CuX^1_2, prior to decomposition.

Reactions of silver salts of classically noncoordinating anions with [$Pd_2(\mu\text{-}Cl)_2(\eta^3\text{-allyl})_2$] and mononuclear π-allylpalladium complexes continue to be widely used in synthesis. Reactions of [$Pd_2(\mu\text{-}Cl)_2(\eta^3\text{-allyl})_2$] with $AgBF_4$ and TMEDA have led to the isolation of [$Pd(\eta^3$-allyl)(TMEDA)][BF_4] complexes[290] (see also Section 6.4.1.3). The complexes were compared with [$Pd(C_5H_5)(\eta^3\text{-}1\text{-}EtO_2CC_3H_4)$] and [$RuCl(\eta^3\text{-}1\text{-}EtO_2CC_3H_4)(\eta\text{-}C_6H_6)$], thus allowing comparison of four-, five-, and six-coordinate complexes containing the $\eta^3\text{-}1\text{-}EtO_2CC_3H_4$ ligand. The x-ray crystal structures of [$Pd(\eta^3\text{-}1\text{-}EtO_2CC_3H_4)$(TMEDA)][$BF_4$] and the precursor [$Pd_2(\mu\text{-}Cl)_2(\eta^3\text{-}1\text{-}EtO_2CC_3H_4)_2$] were determined. The palladium–allyl geometry in the TMEDA complex was found to be quite unsymmetrical, with the substituted carbon significantly closer to palladium than the unsubstituted carbon.

Treatment of [$Pd_2(\mu\text{-}Cl)_2(\eta^3\text{-}C_3H_5)_2$] and its 2-methylallyl analogue with silver perchlorate followed by H_2L (where H_2L = 2,2'-biimidazole, H_2bim; 2,2'-bibenzimidazole, H_2bzim) allowed isolation of [$Pd(\eta^3$-allyl)(H_2L)$]^+$ as the perchlorate salts.[291] The neutral complexes [$Pd(\eta^3$-allyl)(HL)] were obtained by treatment of [$Pd_2(\mu\text{-}Cl)_2(\eta^3\text{-allyl})_2$] with two equivalents of H_2L in the presence of KOH, or by reaction of [Pd(acac)(η^3-allyl)] with one equivalent of H_2L. Tetranuclear complexes [$Pd_4(\mu\text{-}L)_2(\eta^3\text{-allyl})_4$] were formed by treatment of [$Pd_2(\mu\text{-}Cl)_2(\eta^3\text{-allyl})_2$] with one equivalent of H_2L in the presence of excess methanolic KOH, or by reaction of [Pd(acac)(η^3-allyl)] with H_2L (2:1 stoichiometry) in methanol. The x-ray crystal structure of the tetranuclear complex [$Pd_4(\mu\text{-bim})_2(\eta^3\text{-}C_3H_5)_4$]·$CH_2Cl_2$ was determined. The essential features of the structure are shown as (**89**).

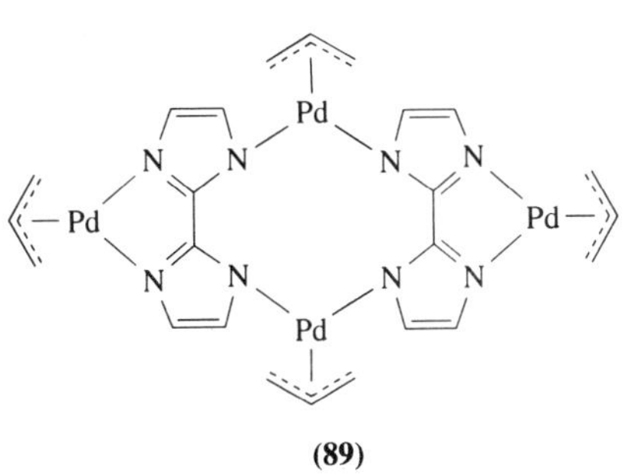
(89)

Heteronuclear complexes, structurally related to (**89**), were prepared by reaction of [Pd(acac)(η^3-allyl)] with [Rh(Hbim)(cod)] or by reaction of [Pd(Hbzim)(η^3-allyl)] with [Rh(acac)(cod)].

Treatment of $[Pd_2(\mu\text{-}Cl)_2(\eta^3\text{-allyl})_2]$ complexes containing pendant alkene substituents with silver tetrafluoroborate has led to the isolation of compounds containing both coordinated allyl and alkene groups.[292] Equation (38) illustrates the reaction.

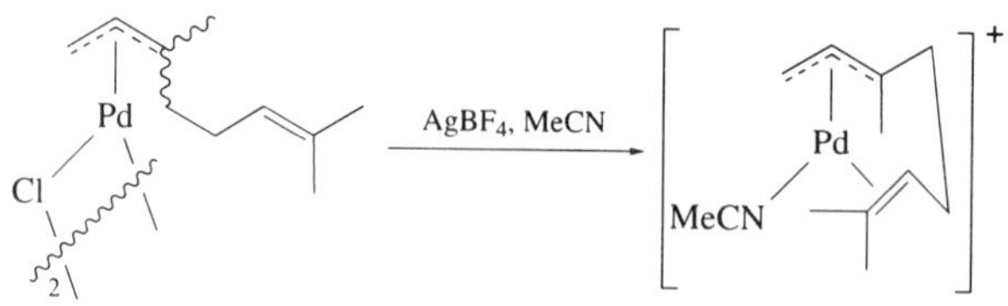

(38)

X-ray crystallography indicated an almost in-plane coordination of the double bond, which formed an angle of 26° with respect to the plane defined by the palladium and nitrogen atoms, the center of gravity of the allylic group, and the midpoint of the double bond.

The reactions of $[Pd_2(\mu\text{-}Cl)_2(\eta^3\text{-}C_3H_5)_2]$ with silver salts of *closo*-borate ions have been investigated.[293,294] Thus, treatment of the chloride-bridged dimer with $Ag_2B_{10}Br_{10}$ in acetonitrile or benzonitrile allowed isolation of $[Pd(\eta^3\text{-}C_3H_5)(NCR)_2]_2[B_{10}Br_{10}]$. The x-ray structure of the acetonitrile complex as a benzene solvate was determined.

Reactions of $[Pd(\eta^3\text{-}C_3H_5)(solvent)_2]^+$ intermediates, typically generated *in situ* by halide abstraction from $[Pd_2(\mu\text{-}Cl)_2(\eta^3\text{-}C_3H_5)_2]$ by silver salts in poorly coordinating solvents, with a variety of donors have been described. For example, treatment of $[Pd(\eta^3\text{-}C_3H_5)(solvent)_2][BF_4]$ (solvent = water or acetone) with the anion derived from treatment of (**90**) with triethylamine led to isolation of the trinuclear complex $[Rh(pyS)_2(\mu_3\text{-}pyS)_2\{Pd(\eta^3\text{-}C_3H_5)\}_2][BF_4]$ (pyS = pyridine-2-thionato).[295]

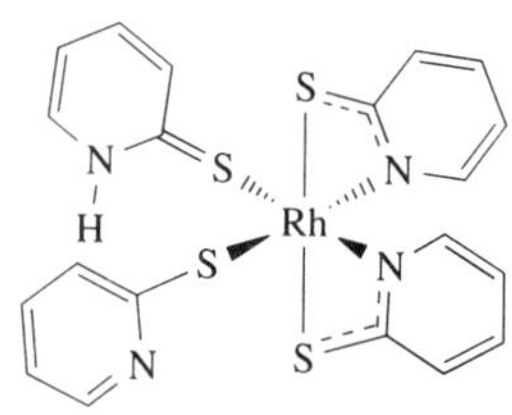

(90)

X-ray crystallographic characterization of the trinuclear compound revealed that the structure consisted of a cyclic, bidentate $[Pd_2(\mu\text{-}pyS)_2(\eta^3\text{-}C_3H_5)_2]$ unit coordinated to a $[Rh(pyS)_2]^+$ fragment through two sulfur atoms.

6.4.3.3 Thermal and photochemical decomposition

The utility of π-allylpalladium complexes as precursors for the metal–organic chemical vapor deposition (MOCVD) of palladium films has been explored.[296] Palladium films are of interest since they may serve as cost-effective replacements for gold in the construction of contacts to integrated circuit boards. Palladium films of thicknesses up to 2 μm were grown on substrates of glass, steel, copper, and aluminum by MOCVD at 532 K and 10^{-4} torr with $[Pd(\eta^3\text{-}C_3H_5)_2]$, $[Pd(\eta^3\text{-}2\text{-}MeC_3H_4)_2]$, and $[Pd(C_5H_5)(\eta^3\text{-}C_3H_5)]$ as precursors. The first two precursors led to films of >99% palladium and <1% carbon, while the last produced a film with ~5% residual carbon. $[Pd(\eta^3\text{-}C_3H_5)_2]$ produced propene (~59 mol.%) and various hexadienes (~30 mol.%) as the major organic products, whereas $[Pd(C_5H_5)(\eta^3\text{-}C_3H_5)]$ gave rise to cyclopentadiene (~43 mol.%), propene (~38 mol.%), and traces of hexadienes (<1 mol.%). The product distribution was interpreted in terms of a radical mechanism for the thermal decomposition.

The thermal decomposition of π-allylpalladium complexes impregnated into silica has been utilized as a method for the preparation of palladium-on-silica catalysts, subsequently employed in the dehydrogenation of 2-propanol and the hydrogenation of 1-hexene. The nature of the organic products formed by thermal decomposition of the impregnated complexes was not described.[297]

Thermolysis of the complex [Pd(η^3-C_3H_5)(PPh_3)$_2$][BF_4] at >403 K in toluene solution in a sealed reactor under an inert atmosphere was reported[298] to produce allyldiphenylphosphine (~10%), biphenyl (35%), and benzene (~7%). Similar thermolyses of [Pd_2(μ-Cl)$_2$(η^3-C_3H_5)$_2$] in the presence of triarylphosphines were said to produce about 5% allyldiphenylphosphine.

Photodeposition of palladium on polyimide substrates has been investigated with [Pd(C_5H_5)(η^3-C_3H_5)] as the precursor.[299] Photolysis of [Pd(C_5H_5)(η^3-C_3H_5)] vapor at 337 nm with a nitrogen laser that delivered 15 mJ per pulse with a flux density of <1 J cm^{-2} allowed deposition of palladium films up to 50 nm thick. Purity was estimated at >99%. No investigations of the nature of the organic products formed during such vapor-phase photolyses were described.

6.4.3.4 Oxidation to aldehydes and ketones

Treatment of π-allylpalladium complexes with typical oxidants routinely gives rise to aldehydes and ketones. In rare instances remote functional groups present in an allylic ligand can be oxidized without decomposition of the complex. For example, oxidation of the testosterone π-allylpalladium dimer (**91**) with pyridinium chlorochromate produced the androstenedione complex (**92**), along with the corresponding mononuclear compounds (tentatively identified as a mixture of *cis*- and *trans*-isomers) resulting from bridge cleavage of (**92**) by pyridine.[300]

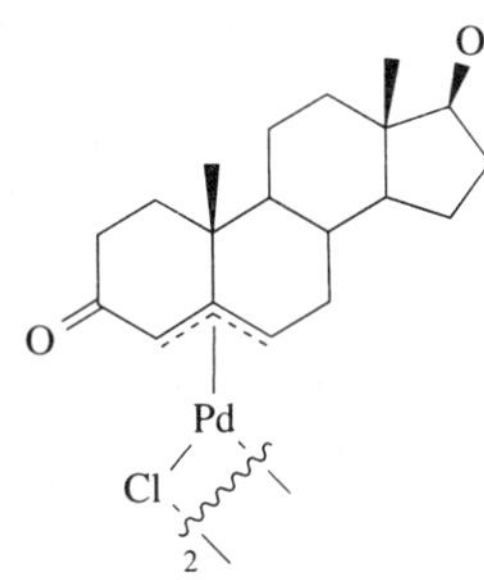

(91)

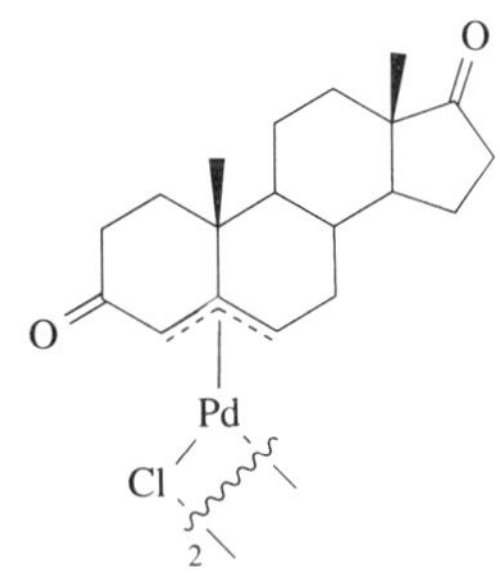

(92)

More typically, oxidation of steroidal π-allylpalladium complexes (e.g., with sodium dichromate in sulfuric acid–diethyl ether or manganese dioxide in sulfuric acid–aqueous acetic acid) generates α,β-unsaturated ketones in low yield along with by-products such as carboxylic acids. Chromium(VI) oxide in DMF containing sufficient diethyl ether to dissolve the palladium complex and a trace of sulfuric acid has been reported to oxidize steroidal complexes such as (**93**) to α,β-unsaturated ketones of type (**94**) in fair to good yields.[301]

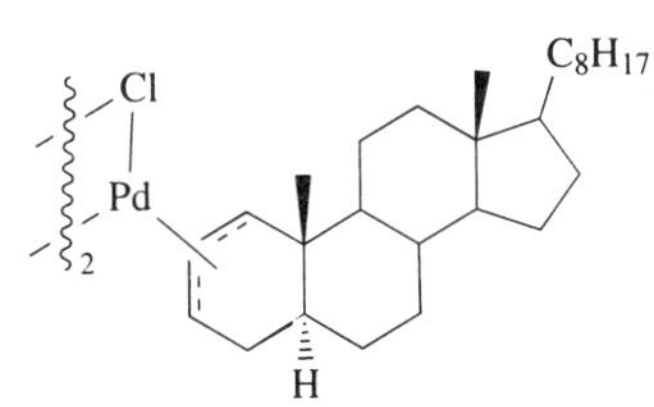

(93)

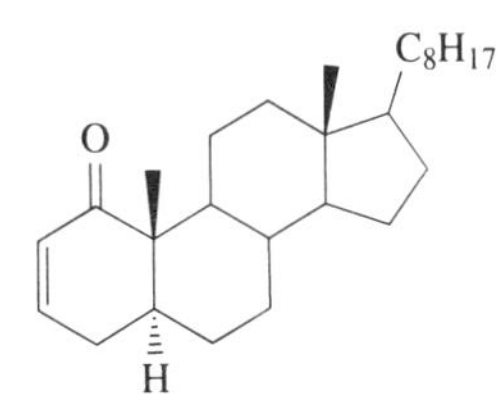

(94)

The reaction of π-allylpalladium complexes with [MoO_2(acac)$_2$]–Bu^tO_2H–pyridine is reported[302] to occur regioselectively to generate products of hydroxylation at the internal allylic position along with α,β-unsaturated ketones. One example is shown in Equation (39).

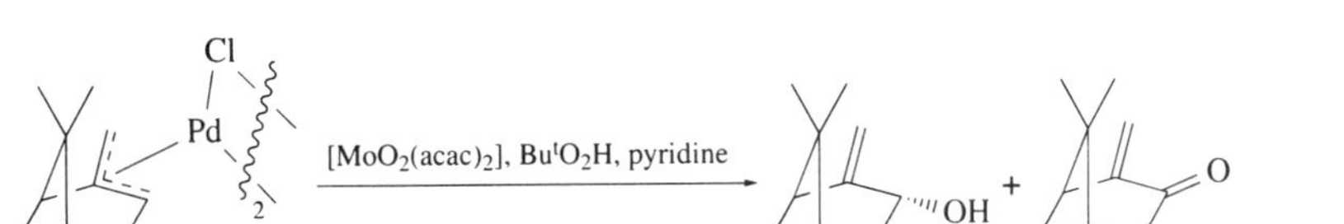

(39)

The reaction was found to display a remarkable solvent effect, with chlorocarbons (dichloromethane, carbon tetrachloride, etc.) favoring formation of the hydroxylation products while benzene, THF, and hexane promoted formation of the α,β-unsaturated ketones.

Several detailed studies of the oxidation of allylic substrates by palladium(II) salts have been reported. Allyl alcohol was oxidized by aqueous $[PdCl_4]^{2-}$ to produce β-hydroxypropanal, hydroxyacetone, acrolein, propanal, propene, and traces of acetone. Through labeling studies based on the use of selectively deuterated allyl alcohol, it was shown that β-hydroxypropanal and hydroxyacetone arose via a hydroxypalladation–hydride-shift mechanism and acrolein was generated by a direct hydride abstraction from the alcohol carbon. Reduction of acrolein in a secondary step yielded propanal, while propene arose from decomposition of the π-allylpalladium intermediate. Oxidation of propene by $[PdCl_4]^{2-}$ was found to be a likely pathway for acetone formation.[303] The oxidation of α-methylstyrene by palladium(II) acetate in acetic acid has been studied and the effects of reaction time, temperature, and reagent concentration on product distribution investigated.[304] At 353 K, eight products, (**95**)–(**101**) and a trace product identified as an oxidative dimer (*m*/*e* = 234) of unknown structure, were noted.

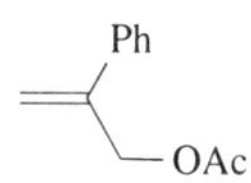

(**95**)

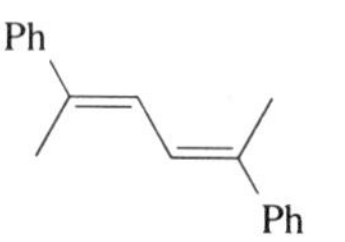

(**96**)

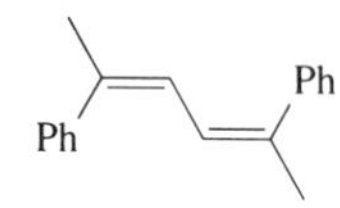

(**97**)

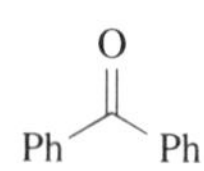

(**98**)

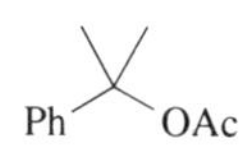

(**99**)

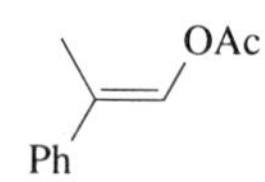

(**100**)

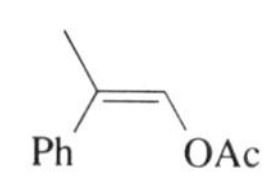

(**101**)

Compound (**95**) was the major product under these conditions, while (**96**) and (**97**) were formed in moderate yields. Compounds (**98**)–(**101**) and the unidentified dimer were minor products, with the first two of these also being produced in the absence of palladium acetate. Their formation was attributed to autoxidation and acid-catalyzed addition, respectively. Addition of increasing amounts of sodium acetate had only a marginal effect on product distribution but the ratio of substrate to palladium was noted to alter the distribution markedly, with the yields of dimers increasing while formation of the enolic acetate was suppressed. Increasing the reaction temperature favored allylic acetate formation at the expense of the oxidative dimers. At long reaction times it was found that, once the oxidant was consumed and formation of oxidative dimers and enolic products was complete, allylic acetate formation continued. The results indicated that the allylic acetate and oxidative dimers did not arise from a common intermediate. Independently synthesized samples of the π-allylpalladium acetate dimer of α-methylstyrene were found to be remarkably resistant to solvolysis under typical reaction conditions but, in the presence of added α-methylstyrene, (**95**) was generated as the sole product. It was demonstrated that the π-allylpalladium complex did not promote oxidation of the added alkene by use of *trans*-β-methylstyrene as an additive. No products derived from *trans*-β-methylstyrene were formed and only (**95**) was produced, that is, it appeared that added alkene catalyzed decomposition of the π-allylpalladium acetate dimer of α-methylstyrene to generate the allylic acetate (**95**). Enolic acetates and oxidative dimers therefore appeared to be formed by a second pathway, for which the nature of the organopalladium intermediates was not identified. The role of homogeneous and heterogeneous palladium catalysts in alkene oxidation and the involvement of π-allylpalladium intermediates have been the subject of several review articles.[305–7]

6.4.3.5 Halogenation

The reaction of allene with $[PdBr_2(NCPh)_2]$ allows isolation of complex (**102**) by a process similar to that in Scheme 19. The reaction of (**102**) with two equivalents of bromine produced 2,3-bis(bromomethyl)-1,3-butadiene (**103**) in 92% yield, while reaction with four equivalents of bromine generated tetrakis(bromomethyl)ethene (**104**).[308]

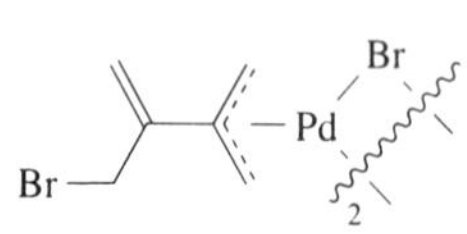

(**102**)

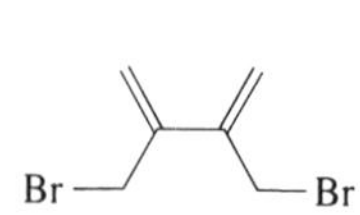

(**103**)

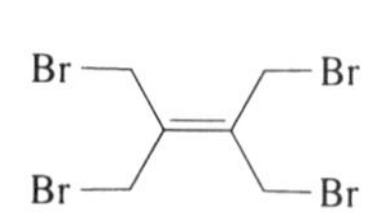

(**104**)

A mechanism for the formation of (**103**) from (**102**) that involved oxidative addition of bromine to generate a palladium(IV) intermediate was proposed. The oxidative cleavage of (**102**), generated *in situ*, by copper(II) bromide to generate (**103**) in 32% yield has also been reported, although in this case 2,3-dibromopropane (5%) and (**104**) (2%) were also formed.[309] The closely related reactions of (**102**), and its analogue prepared from allene and $[PdCl_2(NCPh)_2]$, with ICl and IBr have also been reported.[310]

6.4.3.6 Reduction and charge reversal

Hydride-capture steps have been incorporated into a number of transformations of allylic substrates catalyzed by palladium compounds. Palladium-containing species have typically not been isolated during such studies and the existence of π-allylpalladium intermediate complexes that are subject to attack by hydride is inferred from the nature of the organic products obtained. A wide variety of hydride sources have been employed in such palladium-catalyzed organic syntheses (e.g., sodium hydride,[311] tri(*n*-butyl)tin hydride,[312] diisobutylaluminum hydride,[313] formic acid, etc.). Since applications of palladium catalysts to synthetic organic chemistry generally lie outside the scope of this chapter, only the example of formic acid–formate as a hydride source is considered here.

The selective hydrogenolysis of alkenyloxiranes to homoallylic alcohols has been accomplished with palladium phosphine catalysts and formic acid as the hydride source.[314] The hydride derived from formic acid was found to attack the allyl group of a proposed π-allylpalladium intermediate from the palladium side, giving rise to stereoselectivity in the nucleophilic attack (see Section 6.4.3.8(iii)). The basic mechanism, illustrating the delivery of hydride from formic acid, via palladium, to a coordinated allyl group, is shown in Scheme 44.

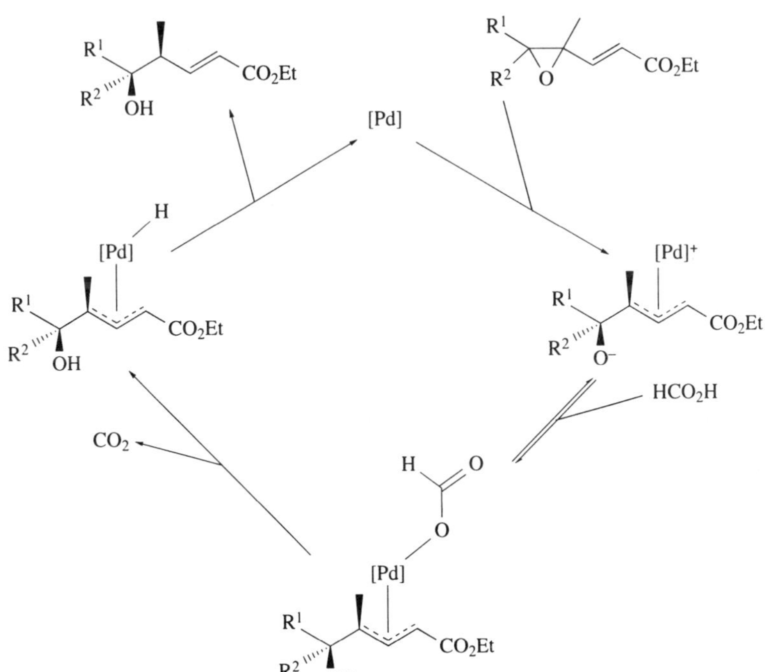

Scheme 44

Palladium-catalyzed tandem cyclization–hydride-capture processes have been developed[315] where oxidative addition of a C–X (X = Br, I, OTf (where Tf represents trifluoromethanesulfonyl), etc.) bond in a suitable substrate to palladium(0) leads to an organopalladium species with a proximate double bond or 1,3-diene system capable of cyclization to generate, respectively, an alkylpalladium or π-allylpalladium intermediate. Exchange of Pd–X with hydride, for example, from formate, may then lead to reductive elimination of a new carbon–hydrogen bond. Anions other than hydride may also be employed to capture the cyclized product.[316–18] Scheme 45 outlines the proposed pathways in such cyclizations.[319]

An example of the application of this methodology in synthesis is the conversion of (**105**) to (**106**) in 90% yield with a 1.3:1.0 (*E*):(*Z*) ratio.[320]

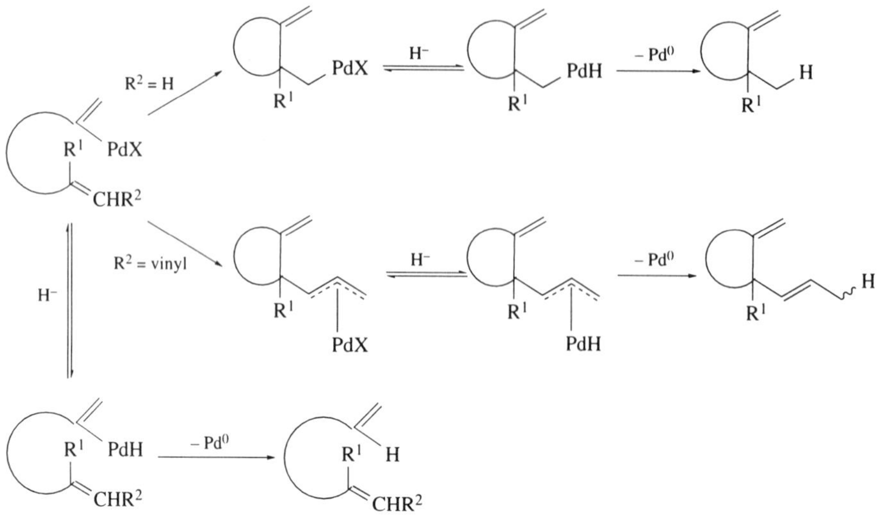

Scheme 45

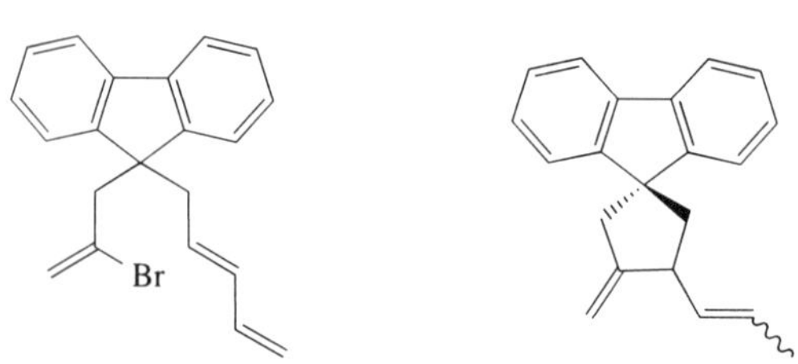

(105) **(106)**

Closely related to the hydride-capture processes described earlier are the so-called charge reversal (umpolung) reactions of π-allylpalladium complexes. In such reactions a π-allylpalladium complex, often unidentified and presumed to be an intermediate, is treated with a reducing agent to effect transformation of the normally electrophilic π-allyl group into a nucleophilic allyl anion equivalent. This allyl anion equivalent may be captured by electrophiles, including H^+, to generate isolable products. For example, treatment of allylic acetates with palladium(0) catalysts, where π-allylpalladium complexes are presumably formed, along with tin(II) as a reducing agent and an aldehyde as an electrophile, led to formation of the alcohol resulting from formal nucleophilic attack of an allyl anion equivalent on the carbonyl group of the electrophile.[321,322] The allyl anion equivalent is known to be transferred via an allyltin(IV) intermediate and such allyltin species have actually been generated by charge reversal of π-allylpalladium intermediates, formed from allylic acetates and palladium(0) catalysts, with samarium(II) iodide in the presence of triorganotin chlorides.[323] Zinc has also been employed to effect charge reversal, leading to carbonyl group allylation of electrophilic substrates via π-allylpalladium intermediates.[324] Other catalytic reactions have employed a range of reducing agents to effect charge reversal, including $P(OPr^i)_3$, dimethylglyoxime, and alkoxide anions.[325–7]

Treatment of allylic acetates bearing terminal trifluoromethyl groups with samarium(II) iodide in the presence of $[Pd(PPh_3)_4]$ as a catalyst led to formation of difluorodienes along with coupled, dimeric products. The former were attributed to fluoride ion expulsion from allylic anions, generated through charge reversal, while the latter were believed to be formed from coupling of allylic radicals, which might be formed prior to anion generation .[328]

6.4.3.7 Reactions with carbon monoxide

Palladium-promoted transformations of allylic substrates under carbonylation conditions continue to be of interest. Many such reactions appear to proceed through pathways in which the critical carbonylation step occurs through transformations of σ-bonded, rather than π-bonded, intermediates. For example, the reaction of π-allylpalladium hexafluoroacetylacetonate with norbornene[329] generated the expected addition product (cf. Equation (30) for a closely related reaction) which reacted with carbon monoxide in refluxing methanol to generate carbonylation products (Equation (40)).

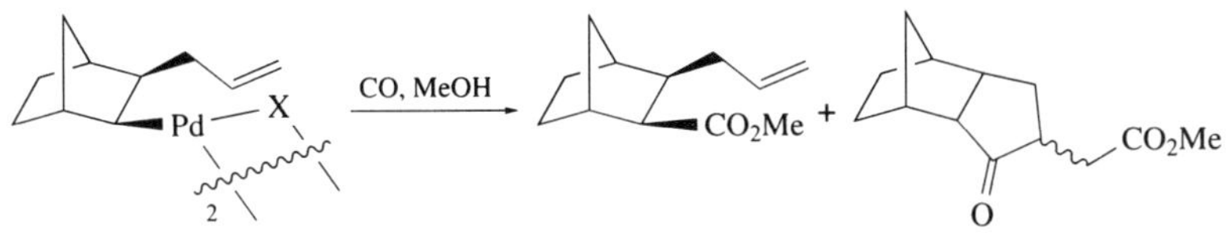

(40)

Related reactions involving intramolecular alkene insertion[330] or alkyne insertion[331] into π-allylpalladium complexes, followed by carbonylation of σ-bonded intermediates, have been described.

The low-pressure carbonylation of π-allylpalladium complexes in the presence of carboxylic acid salts has been reported.[332] Typically, carbonylation of halide-bridged π-allylpalladium complexes has required extreme conditions of temperature and pressure (ca. 325–350 K, 9.65×10^6 Pa–19.31×10^6 Pa CO, 5 h), but in the presence of carboxylic acid salts the reaction was found to proceed at 298 K and low CO pressure (3.45×10^5 Pa initial CO pressure) (Equation (41)).

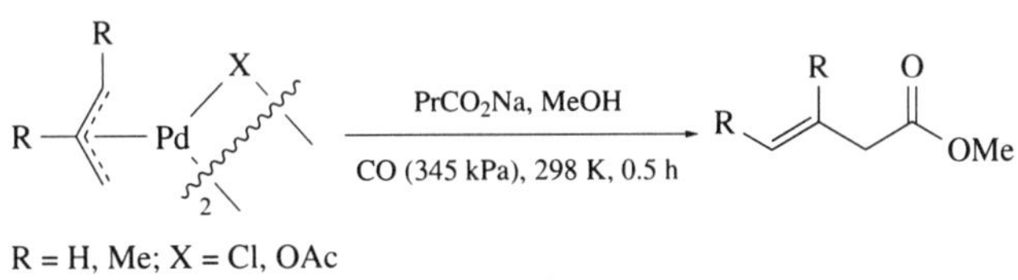

(41)

The proposed mechanism involved initial formation of a carbomethoxy(π-allyl)palladium complex, promoted by the carboxylic acid anions, which underwent reductive elimination, possibly via a σ-allyl intermediate. A carbomethoxy(π-allyl)palladium complex $[Pd(CO_2Me)(\eta^3\text{-}2\text{-}MeC_3H_4)(PPh_3)]$ has been isolated from the reaction of $[Pd_2(\mu\text{-}Cl)_2(\eta^3\text{-}2\text{-}MeC_3H_4)_2]$ with methyl formate and sodium methoxide in the presence of triphenylphosphine and carbon monoxide. Although these reaction conditions might seem far removed from those of π-allylpalladium-mediated carbonylations, methyl formate reacts much as carbon monoxide–methanol in such carbonylation reactions.[333]

Although many π-allylpalladium complexes react with carbon monoxide in largely unspecified ways (e.g., $[Pd(\eta^3\text{-}C_3H_5)_2]$ is reported to produce 1,5-hexadiene and an uncharacterized palladium–carbonyl species upon treatment with carbon monoxide in toluene[281]), some well-defined examples of carbon monoxide reactions have been described. Treatment of $[Pd_2(\mu\text{-}Cl)_2(\eta^3\text{-}2\text{-}MeC_3H_4)_2]$ with tin(II) chloride in toluene, followed by bubbling with carbon monoxide, has allowed isolation of $[Pd(SnCl_3)(\eta^3\text{-}2\text{-}MeC_3H_4)(CO)]$. The platinum analogue was prepared similarly. NMR spectroscopy indicated dynamic behavior in solution through dissociation of carbon monoxide, while x-ray crystallography showed that the palladium and platinum complexes were isostructural. The x-ray structure of $[Pd(SnCl_3)(\eta^3\text{-}2\text{-}MeC_3H_4)(CO)]$ represented the first structural characterization of a mononuclear palladium–carbonyl complex.[334] Such work is of particular interest since tin(II) chloride has been employed as an additive in the hydrocarboxylation of alkenes catalyzed by palladium(II)–phosphine complexes and π-allylpalladium intermediates have been proposed.[335]

The insertion of carbon monoxide into a palladium–allyl bond in an isolated π-allylpalladium complex has now been reported.[336] It was found that $[Pd(\eta^3\text{-allyl})(PMe_3)_2][X]$ (X = Cl, Br) complexes in dichloromethane solution reacted with carbon monoxide at room temperature to yield 3-butenoylpalladium products, *trans*-$[Pd(3\text{-butenoyl})X(PMe_3)_2]$. The parent allyl complex (X = Cl) and the 2-methylallyl analogue (X = Br) reacted over several hours at atmospheric pressure while the 1-phenylallyl complex (X = Br) required 10 atm CO for complete reaction over 4 h. It was found that $[PdCl(\eta^3\text{-}2\text{-}MeC_3H_4)(PMe_3)]$ underwent no reaction, even under pressurized conditions, while $[Pd(\eta^3\text{-}C_3H_5)(PMePh_2)_2][Br]$ generated *trans*-$[Pd(COCH_2CH{=}CH_2)Br(PMePh_2)_2]$ upon treatment with CO at 10 atm. Interestingly, the tetrafluoroborate salt of $[Pd(\eta^3\text{-}C_3H_5)(PMe_3)_2]^+$ did *not* undergo carbonyl insertion, even under 20 atm CO over 24 h, indicating a significant counterion dependence for the carbonylation process. Two possible pathways for the insertion reaction were outlined: (i) nucleophilic attack at palladium by the counterion to generate a σ-allyl complex, which then bound CO to form a five-coordinate intermediate from which insertion occurred; and (ii) nucleophilic attack at palladium by carbon monoxide to generate a σ-allyl complex which then underwent insertion to generate a three-coordinate intermediate that was attacked by the counterion. The reaction of *trans*-$[Pd(COCH_2CH{=}CH_2)Br(PMe_3)_2]$ with piperidine under 50 atm CO was found to produce both amide (**107**) and α-ketoamide (**108**), the latter resulting from double carbonylation of the original allylic substrate.

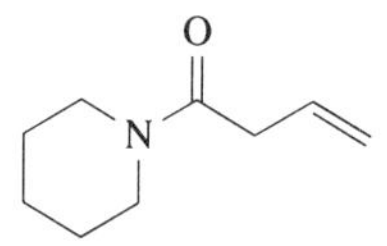

(107)

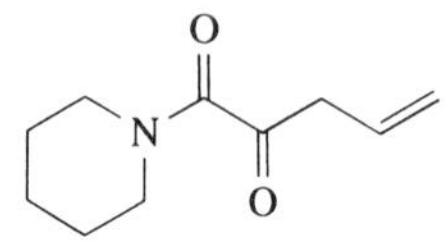

(108)

6.4.3.8 Nucleophilic attack at carbon

Palladium-catalyzed reactions of allylic substrates with nucleophiles have assumed an important place in organic synthesis and, accordingly, there is much interest in the reactions of π-allylpalladium complexes, invoked as intermediates in such reactions, with nucleophiles. Nucleophiles typically attack π-allylpalladium complexes at C-1 or C-3 of the allyl ligand, although there is an increasing body of evidence that attack at C-2, the central allylic carbon, may be more common than once thought (see Section 6.4.3.8(i)). The questions that arise during nucleophilic attack concern the intimate mechanism, since this controls the regiochemistry and stereochemistry of the derived products. Thus, assuming that attack occurs exclusively at the terminal allylic carbons, different products may result when C-1 and C-3 of the allyl ligand are different by virtue of bearing different substituents or because the palladium complex contains different ligands *trans* to C-1 and C-3. Similarly, different products may result when nucleophilic attack occurs from an external nucleophile (*trans*-attack with respect to palladium) and when it occurs with prior coordination of the nucleophile to palladium (*cis*-attack). Deduction of a mechanistic pathway by examination of product distributions in palladium-catalyzed allylation reactions is not a routine approach since reactions may occur under either kinetic or thermodynamic control and there exist pathways for isomerization of π-allylpalladium intermediates. Despite these complications, much progress has been made in the elucidation of reaction mechanisms and this has allowed for progress in areas such as asymmetric allylation with chiral palladium–phosphine catalysts.[337–48] The general area of enantioselective homogeneous catalysis involving transition metal–allyl intermediates was reviewed in 1989.[349]

(i) Attack at C-1 and C-3 vs. attack at C-2

The origin of activation of allylic ligands towards nucleophilic attack on coordination to palladium has been examined theoretically by CNDO-type semiempirical SCF–MO calculations.[350] The reaction considered was nucleophilic attack by hydroxide ion on the complexes $[PdCl_2(C_3H_5)]^-$, $[PdCl(C_3H_5)(CO)]$, $[PdCl(C_3H_5)(PH_3)]$, $[Pd_2(\mu\text{-}Cl)_2(C_3H_5)_2]$, and $[Pd(C_3H_5)(PH_3)_2]^+$. The hydroxide ion was employed as a model "hard" nucleophile, although it was recognized that attack by hydroxide on a coordinated allyl ligand to produce allyl alcohol is not a known reaction. Several features concerning the electronic structures of the π-allylpalladium complexes were noted: (i) coordination significantly reduced the electron densities found on C-1 and C-3 of the allyl ligand with respect to the parent allyl anion; (ii) significant electron donation from the $n\pi$-orbital of the allyl ligand to palladium, augmented by some donation from the π-orbital, occurred; and (iii) the LUMO was mainly composed of Pd $4d(xy)$, which accepted electron density from the allyl $n\pi$-orbital. The attack of hydroxide was modeled as a *trans*-attack with the oxygen lone pair approaching perpendicular to the face of the allyl ligand. It was found that the total energy for attack at C-2 was higher than the total energies for attack at C-1 and C-3, and so attack at the central allylic carbon was not considered further. The total energies for attack at C-1 and C-3 differed for unsymmetrical complexes but the differences were considered too small to be meaningful. A comparison of total energy changes during hydroxide attack suggested that attack was favored by neutral ligands over anionic ligands and that neutral π-acceptor ligands promoted the reaction more than neutral ligands without π-acceptor abilities. The latter conclusion was based on comparisons of carbonyl and phosphine systems with and without Pd $4d\pi$–CO π^* (or Pd $4d\pi$–P $3d\pi$) interactions. Not surprisingly, the anionic complex $[PdCl_2(C_3H_5)]^-$ was calculated to be less reactive towards hydroxide than the cationic complex $[Pd(C_3H_5)(PH_3)_2]^+$. The calculated ligand effects were found to be in agreement with empirical observations that neutral ligands, such as triphenylphosphine, promote palladium-catalyzed allylation reactions. Changes in electronic structure during hydroxide attack were generally found to be consistent with conversion of palladium(II) to palladium(0), as the allyl ligand converted to an η^2-alkene prior to dissociation.

Certain experimental observations now suggest that the theoretical model, which explains some aspects of nucleophilic attack at C-1 and C-3, must be modified to account for attack of selected nucleophiles at C-2. Thus, although there is a large body of evidence that supports external attack of, for example, stabilized carbon nucleophiles at C-1 and C-3, it has been demonstrated[351] that branched ester enolates react with $[Pd_2(\mu\text{-}Cl)_2(C_3H_5)_2]$ in the presence of triethylamine and hexamethylphosphoramide (HMPA) to produce alkylated cyclopropanes by attack at C-2. Scheme 46 outlines the possible pathways involved in attack of a stabilized carbon nucleophile $[CHE_2]^-$ ($E = CO_2R$, SO_2R, etc.) at C-1 and C-3, and for a branched ester enolate at C-2.

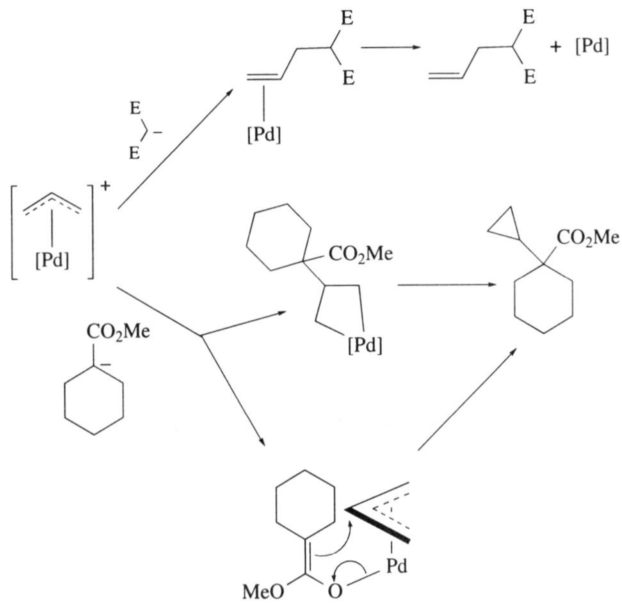

Scheme 46

Palladium(0) complexes of chelating diphosphines have been found[352] to catalyze the coupling of allylic acetates and ketene silyl acetals to yield allylated carboxylic acid esters by attack of the silyl enolate at C-1 and C-3, and, in certain cases, cyclopropane derivatives through attack at C-2. Product analysis by NMR spectroscopy indicated that the attack at C-2 occurred from the side opposite to palladium (Scheme 46). Other related work on stoichiometric and catalytic reactions involving π-allylpalladium complexes similarly implicates attack at C-2.[353–5] These experimental results have led to new theoretical work concerning nucleophilic attack on π-allylpalladium complexes, which has been supplemented by observations on the reactions of π-allylplatinum compounds where metallacyclobutanes, of the type invoked in Scheme 46, are accessible through nucleophilic attack at C-2.[356] Theoretical analysis of $[Pd(C_3H_5)L_2]^+$ suggested that the C-2-centered molecular orbital is not as destabilized as commonly thought and that it may be competitive with the orbital centered on C-1, C-3, and palladium as the LUMO in such systems. It was observed that the initial molecular orbital picture of $[Pd(C_3H_5)L_2]^+$ is merely an indicator of regioselectivity in nucleophilic attack and that it says nothing about possible attractive–repulsive interactions encountered during the process. Despite the relatively low number of atoms in $[Pd(C_3H_5)L_2]^+$, application of rigorous computational methods is problematic since the least-motion pathway involved in nucleophilic attack requires nontrivial structural rearrangements. Within the limitations of the extended Hückel molecular orbital method, it could be observed that the reaction leading to metallacyclobutane formation involved a barrier attributable to a four-electron repulsion between C-2 and the incoming nucleophile early in the pathway. Stronger nucleophiles gave rise to smaller barriers, but very strong nucleophiles—which could perhaps eliminate the barrier—directly attacked the metal center. Thus, attack at C-2 might be anticipated with certain metal centers and selected nucleophiles under kinetically controlled conditions. Factors responsible for delineating the reaction pathway have yet to be fully elucidated.

(ii) Regiochemistry of nucleophilic attack at C-1 and C-3

Control of regiochemistry during nucleophilic attack on π-allylpalladium complexes is a complex issue in which steric and electronic effects, often working in tandem, are important considerations.

Nucleophilic attack on a π-allylpalladium complex in which C-1 and C-3 bear sterically inequivalent substituents might be envisaged as proceeding by two different pathways, one in which the least-hindered carbon center is attacked (path A in Scheme 47) or one in which the least-hindered η^2-alkene complex is produced (path B in Scheme 47).

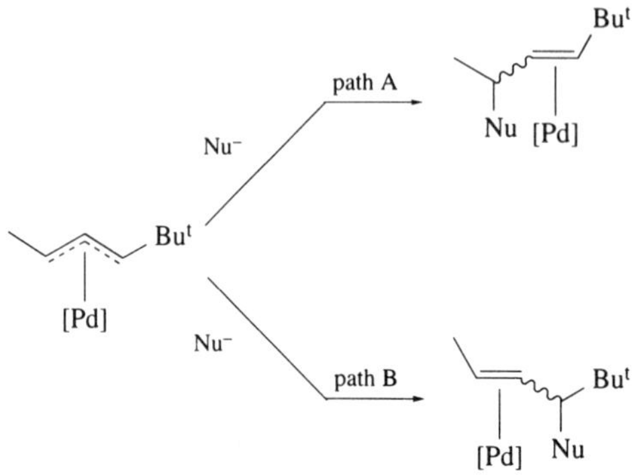

Scheme 47

The reactions of the steroidal π-allyl complexes (**109**) and (**110**) with potassium acetate in DMF at 313–318 K for 8 h are illustrative.[357] Compound (**109**) produced (**111**), while (**110**) produced (**112**). In each case the stereochemistry of nucleophilic attack (see Section 6.4.3.8(iii)) was that due to *trans*-attack with respect to palladium, while the regiochemistry resulted from attack at C-3 due to steric hindrance by the methyl substituent.

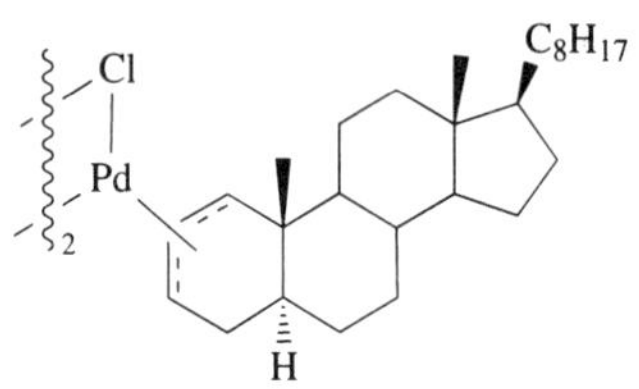

(**109**)

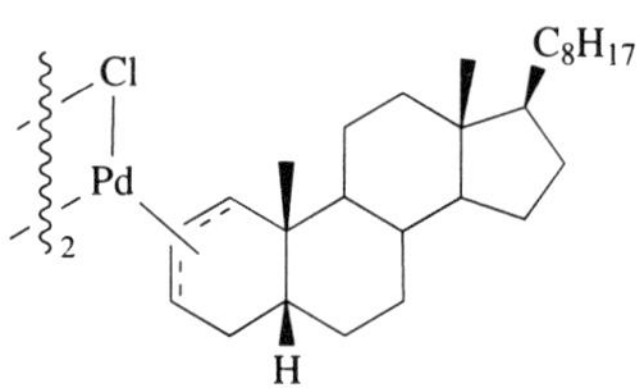

(**110**)

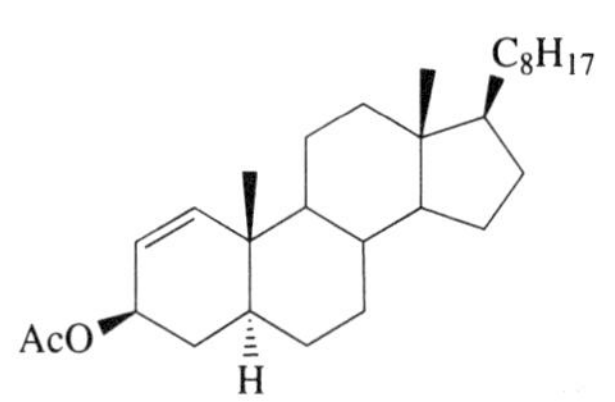

(**111**)

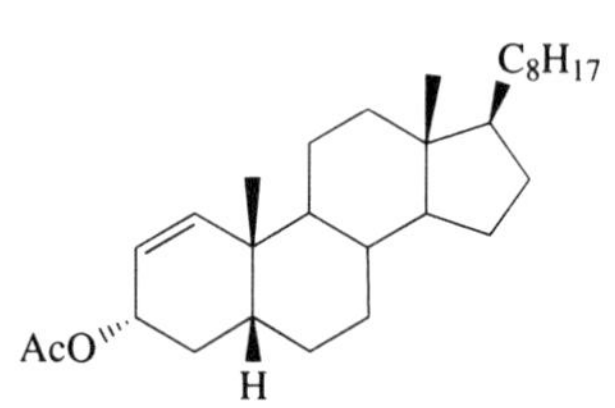

(**112**)

A series of allylic acetates of type (**113**), where R = Pr^n, Bu^n, Bu^i, Pr^i, isopentyl, and CD_3, and the allylic acetate (**114**) were reacted with nucleophiles (**115**), (**116**), [Na][CH(CO_2Me)$_2$], SnBu$^n_3$OPh, and PhZnCl in the presence of 5–10% [Pd(PPh_3)$_4$] in THF at room temperature.[358]

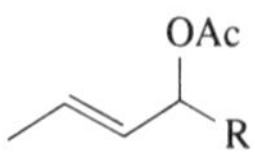

(**113**)

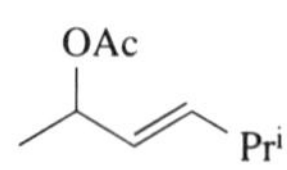

(**114**)

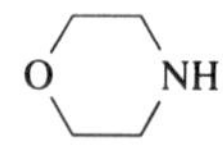

(**115**)

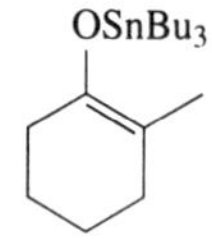

(**116**)

Product analyses indicated a remarkable regioselectivity, with the nucleophiles (**115**), (**116**), [Na][CH(CO_2Me)$_2$], and SnBu$^n_3$OPh attacking the least-hindered carbon atom in each substrate with selectivities of 80–100%, while PhZnCl attacked the most-hindered carbon atom in each case with selectivities >99%. The "symmetric" allylic substrate (**113**), where R = CD_3, reacted nonregioselectively with each nucleophile. Palladium-catalyzed equilibration of the regioisomeric products showed that the thermodynamic ratios were close to 1:1, as might be expected for simple disubstituted alkenes. Based on other studies, PhZnCl was anticipated to react via initial coordination of the phenyl group to palladium, followed by reductive elimination, while the remaining nucleophiles were anticipated to react be direct external attack. The possibility was raised that the high regioselectivity of PhZnCl might be due to the intermediacy of species such as (**117**) and (**118**), where (**117**) was favored over (**118**) on steric grounds and so reductive elimination led to regioselective phenyl transfer to the most-hindered carbon of the allyl group.

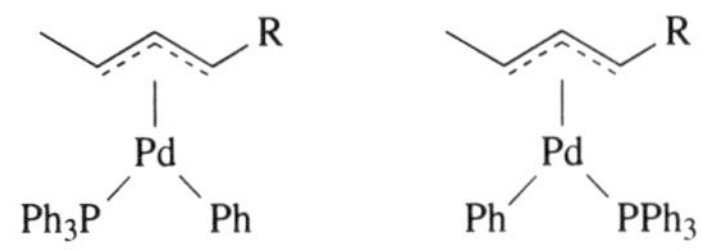

(**117**) (**118**)

Phenyl transfer from PhZnCl to the most-hindered carbon atom of proposed π-allylic intermediates has been observed in other catalytic systems,[359] and attack of amine nucleophiles[360] and stabilized carbon nucleophiles[361] at the least-hindered carbon has similarly been found for other substrates. In contrast with the reactions of PhZnCl, palladium complexes of 1,1'-bis(diphenylphosphino)ferrocene have been shown to catalyze the cross-coupling of allylic ethers with aryl-Grignard reagents, such that aryl transfer to the least-substituted position of a proposed allylic intermediate was found. Stoichiometric reactions of a preformed η^3-1-methylallylpalladium complex with PhMgBr in the presence of phosphine ligands also proceeded with regioselective phenyl transfer to the least-substituted carbon of the allyl group.[362] Phenyl transfer from sodium tetraphenylborate to 3-chloro-2-methylenecycloalkylpalladium–π-allyl complexes has been investigated and found to involve attack at the least-substituted carbon atom, although effects of differing ring sizes in the allylic complexes were recognized as factors that may play a role in controlling regioselectivity.[363] Certain other organometallic nucleophiles react as PhZnCl, with regioselective transfer of the nucleophile to the most-substituted carbon center of the allyl group. For example, the acylnickel anion $[Li][Ni(Bu^nCO)(CO)_n]$ has been found to react with $[Pd(\eta^3\text{-}1\text{-}MeC_3H_4)(PPh_3)_2][BF_4]$ to generate largely $CH_2{=}CHCH(Me)COBu^n$ (11% of the other regioisomer was obtained).[364] PhZnCl and stabilized carbon nucleophiles, for example, $[Na][CH(CO_2Me)_2]$, have been shown to react differently with 1-alkenylcyclopropyl esters and related substrates under palladium(0)-catalyzed conditions.[365] Scheme 48 outlines the possible reaction pathways.

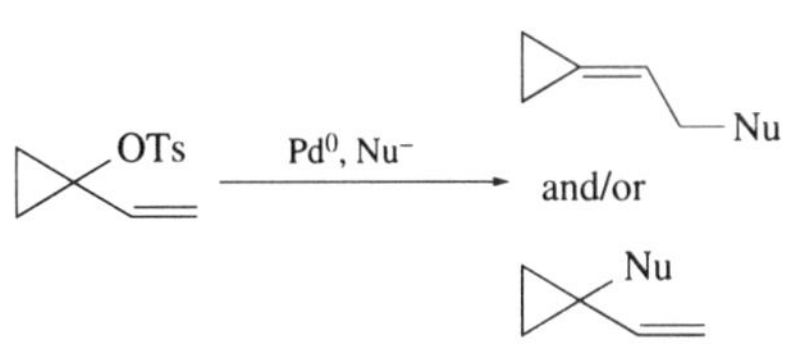

Scheme 48

PhZnCl gave rise to phenyl transfer exclusively to the cyclopropyl carbon, while stabilized carbon nucleophiles gave rise to exclusive attack at the primary carbon. The existence of intermediate unsymmetrical π-allylpalladium compounds subject to external nucleophilic attack by stabilized carbon nucleophiles and attack following coordination to palladium for unstabilized carbon nucleophiles (see Section 6.4.3.8(iii)) was invoked to explain the regioselectivities observed.

Phenyl transfer processes have been examined for both spontaneous and ligand-promoted reductive elimination of allylbenzenes from $[Pd(\eta^3\text{-allyl})(Ar)(L)]$ (where L = phosphine or arsine). Kinetic evidence was presented that neither η^1-allylpalladium nor 14-electron fragments resulting from ligand dissociation during carbon–carbon bond formation were involved in the reaction.[366,367]

Regioselectivity in palladium-catalyzed reactions of allylic acetates with stabilized carbon nucleophiles has been compared with that found in stoichiometric reactions of preformed neutral and cationic π-allylpalladium complexes with the same nucleophiles.[368] Scheme 49 outlines the general approach.

It was found that for the 3-methylbutenyl system ($R^1 = R^2 = Me$ in Scheme 49) dimethyl malonate reacted predominantly at the most-substituted position, while the bulkier nucleophile diethyl malonate reacted predominantly at the least-substituted position. In this system catalytic reactions, and reactions of both the neutral complex and the cationic complex, proceeded similarly, providing further evidence for the involvement of η^3-allylpalladium intermediates in the catalytic reactions. NMR evidence was

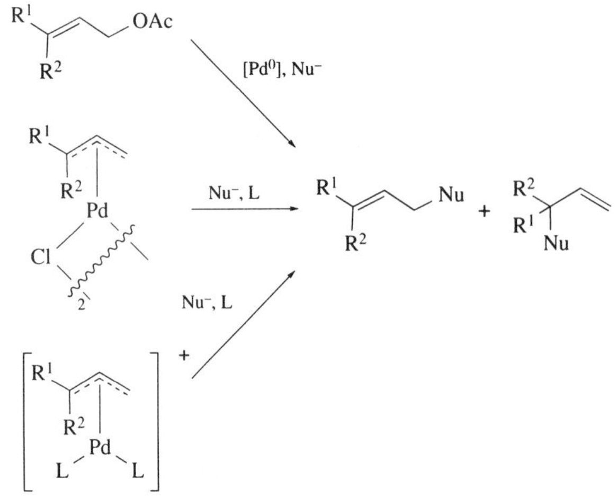

Scheme 49

used to argue against the intermediacy of η^1-allylpalladium species in the catalytic reactions, although such species have been implicated in other transformations of η^3-allylpalladium complexes. For example, reaction of (**119**) with diethylamine and carbon monoxide in THF at 298 K produced (**120**) with the regioisomer of (**120**) undetected, suggestive of an η^1-allylpalladium intermediate.[369]

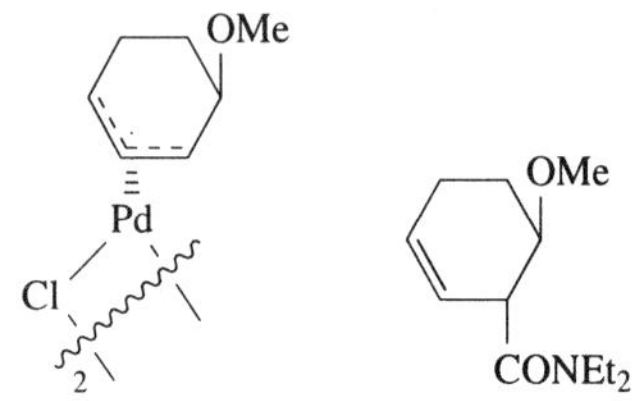

(119) **(120)**

Steric effects of ancillary ligands on the regiochemistry of nucleophilic attack have been investigated during work on the reactions of alkenylzirconium reagents with steroidal π-allylpalladium complexes.[370,371] Scheme 50 illustrates the synthesis and reactions of a steroidal π-allylpalladium complex with a $[ZrCl(\eta\text{-}C_5H_5)_2(alkenyl)]$ reagent, showing the two regioisomers that were formed.

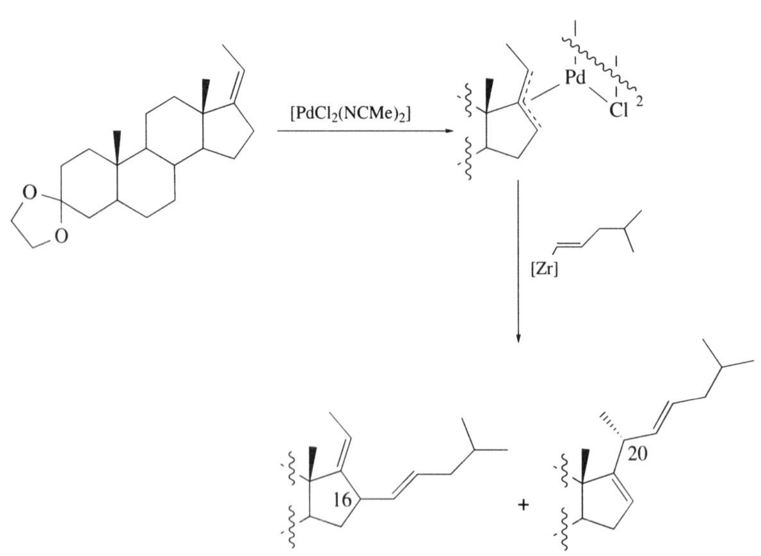

Scheme 50

Reaction of the steroidal complex with the alkenylzirconium reagent led to a mixture of regioisomers corresponding to nucleophilic attack at C-20 and C-16 (59% total yield, 2:3 ratio), along with formation of the original steroidal alkene. Effects of phosphine ligands were examined through the addition of varying amounts of triaryl- and alkyldiarylphosphines of differing steric bulk. No control of regiochemistry was observed in any fashion that could be correlated with steric effects due to the phosphine ligands. Variation in solvent similarly had little effect on regiochemistry. Equilibration of the π-allyl complex with maleic anhydride prior to addition of the alkenylzirconium reagent gave rise to C-20 and C-16 coupled products in a ratio of >7:1. The control of regiochemistry was attributed to formation of an unsymmetrical π-allyl complex of the type [Pd(allyl)Cl(L)]. Reaction with the alkenylzirconium compound then generated [Pd(allyl)(alkenyl)(L)], which reductively eliminated the coupled product.

The existence of intermediate unsymmetrical π-allyl complexes in reactions where regioselectivity is observed is supported by NMR studies,[372] which have shown that π-acceptor ligands, such as tertiary phosphines, induce large, downfield shifts for the carbon resonances of the most-substituted termini of unsymmetrical π-allyl complexes. The observed shifts suggest significant charge differences between the termini of the allyl groups in such compounds and this may be a controlling factor in regioselective nucleophilic attack.

Product structure has been found to be an important consideration in understanding regiocontrol of nucleophilic attack. For example, Scheme 51 shows a palladium-catalyzed reaction where attack of a pendant nucleophile on a π-allyl intermediate might yield either a spirocycle or a cycloalkenyl product.

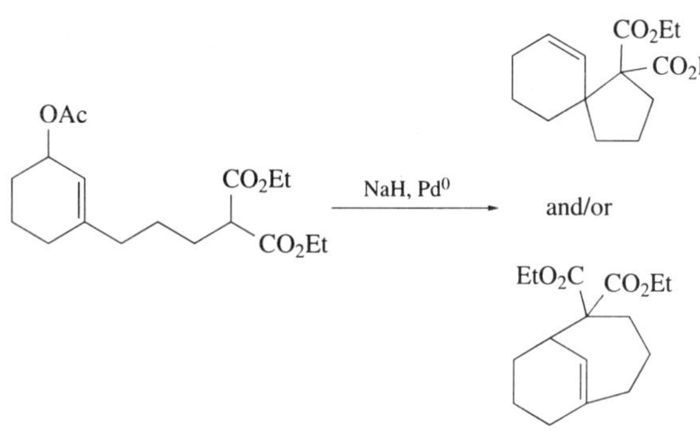

Scheme 51

The spirocycle was the only product obtained since the regioisomer is an anti-Bredt alkene.[373] The general methodology shown in Scheme 51 could be applied to the synthesis of a number of spirocyclic compounds with [5.5], [5.4], and [4.4] ring systems, since the formation of a bridgehead alkene is disfavored for these ring sizes. Similar processes involving simple ring closures typically proceed with regiochemistries influenced by the ring sizes of the possible isomeric products.[374,375]

In order to probe electronic effects on regioselectivity while maintaining approximately constant steric effects, palladium-catalyzed nucleophilic substitution reactions of the type illustrated in Equation (42) have been investigated.

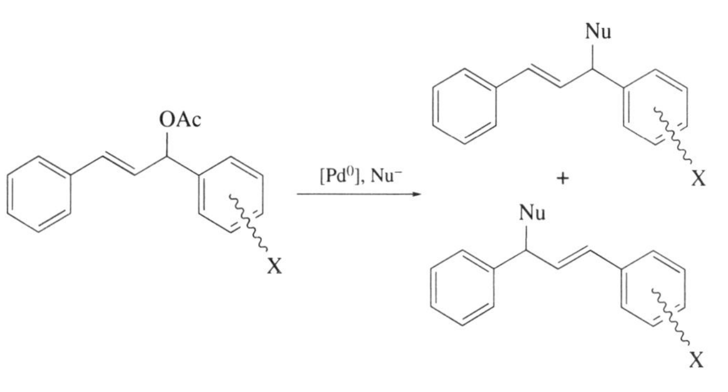

(42)

Substrates with 4-fluoro, 4-bromo, and 4-methyl substituents were reacted with the nonstabilized carbon nucleophile allenyltributyltin,[376] while substrates with 3-methyl, 4-chloro, and 4-methyl substituents were reacted with the stabilized carbon nucleophile sodium acetylacetonate.[377] In each case, the product distribution varied little from 50:50, suggesting that these remote electronic effects play virtually no role in the control of regiochemistry. In related work, substrates analogous to those in Equation (42), but bearing a 4-nitro substituent on one ring and 4-methoxy or 4-chloro substituents on the other, showed considerable regioselectivity in palladium-catalyzed reactions with sodium acetylacetonate and the sodium salt of triacetic acid lactone.[378] In all cases, regioselective attack on the allylic position remote from the electron-withdrawing nitro group was observed. The observations are in accord with electron donation to palladium giving rise to unequal carbocation characteristics for the terminal allylic carbons. The donor ability of the allylic carbons was thought to be reduced by electron-withdrawing substituents and enhanced by electron-donating substituents, thus giving rise to an unsymmetrical π-allyl intermediate.

Stabilized carbon nucleophiles and nitrogen nucleophiles have been shown to attack π-allylpalladium complexes bearing an alkyl group at one allylic terminus and an ester functionality at the other in a regioselective manner. Attack at the allyl carbon atom remote from the ester functionality was found to be favored in accord with the proposal that an asymmetrical π-allyl intermediate was involved.[379]

The presence of polar functional groups on an allylic complex may play an important role in control of regioselectivity. For example, opening of cyclopentadiene monoepoxide by palladium(0) appears to produce a π-allylic intermediate bearing a polar hydroxy functionality, which directs incoming nucleophiles to the distal end of the allyl group (Scheme 52).[380,381]

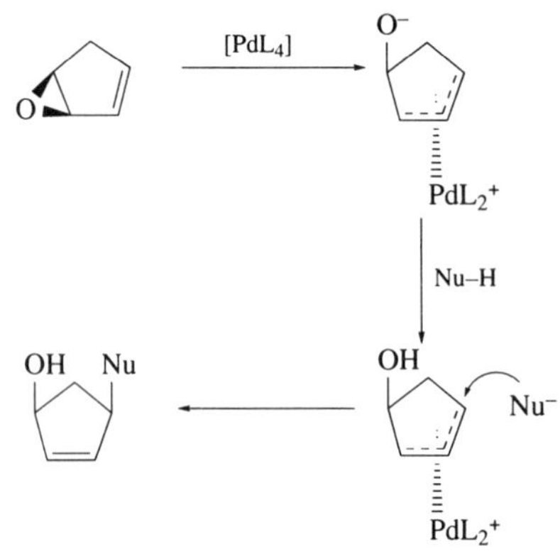

Scheme 52

Both steric and electronic effects have been shown to be important in regiochemical control during palladium-catalyzed reactions of *cis*-3,4-diacetoxycyclopent-1-ene with the stabilized carbon nucleophile [Na][CH(CO_2Et)$_2$].[382] Scheme 53 shows that, in principle, three products could arise.

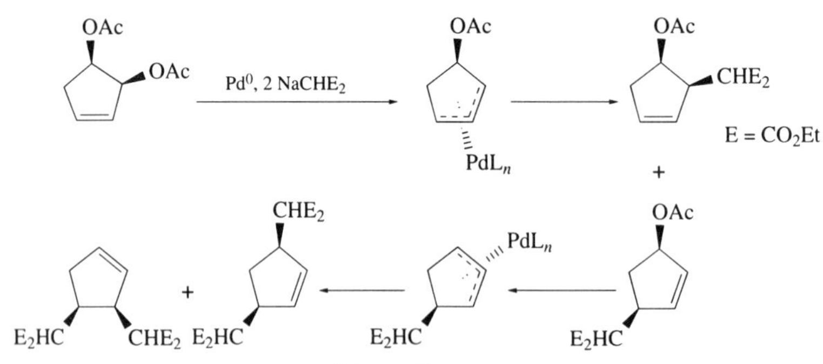

Scheme 53

However, a catalytic reaction employing [Pd(PPh_3)$_4$] produced only the 3,5-dimalonated compound with none of the 3,4-regioisomer. The regiospecificity was attributed to steric hindrance in the second nucleophilic attack by the diethyl malonate substituent introduced in the first step, and to the electronic effect of this electron-withdrawing substituent, promoting attack at the remote allylic terminus.

One problem that arises in determinations of regioselectivity is that certain nucleophiles, such as amines, form allylation products in which the amine substituent may act as a leaving group. Thus, in the simple picture of nucleophilic attack (Scheme 54), the displacement step may become essentially reversible.

Such reversibility leads to thermodynamic control of regioselectivity and this may be usefully exploited to accomplish isomerization of allylic compounds to their thermodynamically favored regioisomers.[383] Reactions involving carbon nucleophiles lead to products via carbon–carbon bond formation and so such processes have typically been regarded as irreversible and kinetically controlled.

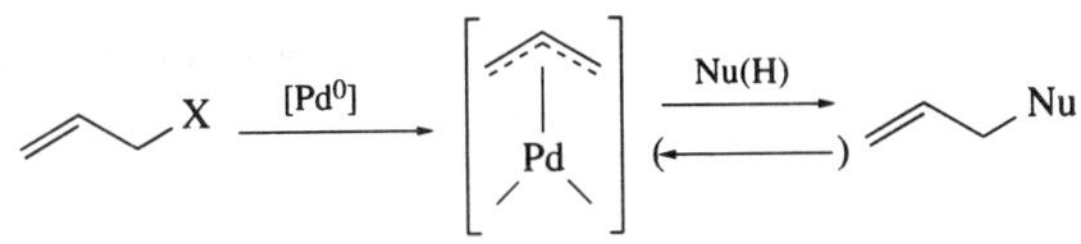

Scheme 54

Studies of the palladium(0)-catalyzed reactions of dienyl acetates with dialkyl malonates suggest that this latter assumption may be erroneous.[384] It was shown that regiochemistry was very dependent on reaction conditions and that at short reaction times and low temperatures kinetically controlled product formation occurred, while at longer reaction times and higher temperatures thermodynamic control of regiochemistry was evident. It was found that at higher temperatures and longer reaction times the kinetic product rearranged to the thermodynamic regioisomer in the presence of the palladium(0) catalyst. These observations strongly suggest that carbon–carbon bond cleavage in the reaction product, with the dialkyl malonate anion as the leaving group, is possible under palladium(0)-catalyzed reaction conditions.

(iii) Stereochemistry of nucleophilic attack at C-1 and C-3

Investigations of the stereochemistry of nucleophilic attack on π-allylpalladium complexes have shown that two basic pathways can be followed: (i) external attack (*trans* to palladium) leading to inversion; and (ii) prior coordination to palladium followed by reductive elimination (*cis*-attack) leading to retention. σ-Allylpalladium intermediates may also play a role in controlling the stereochemistry of nucleophilic attack on π-allylpalladium complexes.[385] Although many studies have made use of diastereomeric π-allylpalladium complexes in reactions with nucleophiles in order to probe stereochemistry, it can be argued that control of inversion vs. retention might be governed by stereochemical factors in the diastereomeric systems and may actually be independent of the inherent nature of the nucleophile employed. Accordingly, enantiomeric π-allylpalladium complexes have been utilized[386] to determine stereochemistry in some reactions with nucleophiles. Reaction of (–)-(1*S*,2*R*,3*R*)-[$Pd_2(\mu\text{-}Cl)_2(1\text{-}Me\text{-}3\text{-}PhC_3H_3)_2$] (82% *ee*) with the sodium salt of dimethyl malonate (two equivalents) in the presence of triphenylphosphine (two equivalents) in benzene solution for 1 h generated the (*R*)-isomer of the allylated product in ca. 79% *ee* (86% yield) along with its regioisomer (5% yield), as shown in Scheme 55.

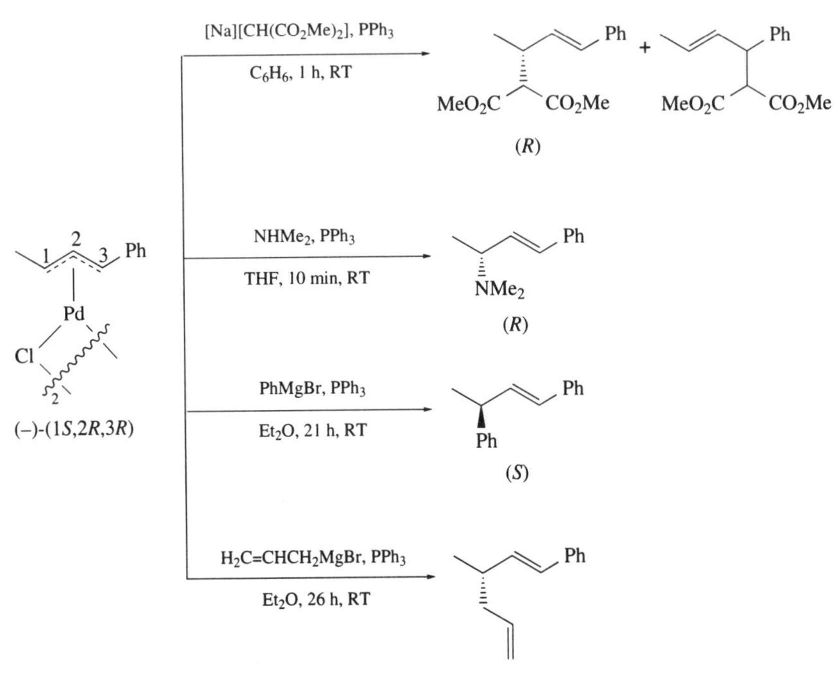

Scheme 55

The results thus indicated that nucleophilic attack of the stabilized carbon nucleophile occurred with inversion at C-1 (see Scheme 55 for atom numbering), implying external attack. Inversion at C-1 was similarly found with dimethylamine as the nucleophile (Scheme 55). Both phenylmagnesium bromide and allylmagnesium bromide were found to react with retention at C-1, implying prior coordination to palladium with subsequent reductive elimination of the product.

Reaction of (−)-(1*S*,2*R*,3*R*)-$[Pd_2(\mu\text{-}Cl)_2(1\text{-}Me\text{-}3\text{-}PhC_3H_3)_2]$ (83% *ee*) with sodium acetylacetonate in THF at 273 K has been shown to produce the corresponding acetylacetonato complex, which should have the same configuration and enantiomeric purity as the dimeric starting material. Subsequent treatment with excess triphenylphosphine in THF at room temperature generated the product of nucleophilic attack of the acetylacetonate anion on the π-allyl ligand, (*R*)-[1-{(*E*)-styryl}ethyl]acetylacetone, in 94% yield (87% *ee*) along with $[Pd(PPh_3)_4]$. The result demonstrates that nucleophilic attack occurred with inversion at C-1, implying displacement of acetylacetonate anion from palladium by triphenylphosphine and subsequent external nucleophilic attack. Crossover experiments demonstrated the intermolecular nature of the nucleophilic attack.[387] Related low-temperature NMR studies on $[Pd(\eta^3\text{-}2\text{-}MeC_3H_4)(acac)]$ in the presence of triphenylphosphine support initial formation of the ion pair $[Pd(\eta^3\text{-}2\text{-}MeC_3H_4)(PPh_3)_2]^+[acac]^-$ which, on warming, was transformed into the coupled organic product and palladium(0).[388]

Stereochemical studies of the attack of ketone enolate anions on π-allylpalladium complexes have shown that these nucleophiles undergo external *trans*-attack with no evidence for prior coordination.[389]

The reactivity of 5-carbomethoxy-(1,2,3-η^3)-cyclohexenylpalladium complexes towards acetate has demonstrated that the stereochemistry of nucleophilic attack is dependent on subtle structural factors. Scheme 56 illustrates the processes investigated.[390]

In the absence of external nucleophiles, both the *cis*- and *trans*-complexes reacted only by internal migration, with the *trans*-complex reacting much faster than the *cis*-isomer. The addition of lithium acetate had no effect on the stereochemistry of the product formed from the *trans*-complex, but the *cis*-complex yielded increasing amounts of product derived from external attack. Consideration of the structural differences between the *cis*- and *trans*-complexes, based on x-ray crystallographic and NMR data, suggested that the stereochemistry of nucleophilic attack was influenced by unfavorable 1,3-diaxial interactions that could occur along the possible reaction pathways and that such interactions kinetically controlled the stereochemical outcome. The dual stereoselectivity of carboxylate nucleophiles towards π-allylpalladium complexes has been observed in other, related systems.[391,392]

Loss of stereospecificity in palladium-catalyzed nucleophilic substitution reactions of allylic substrates has been noted in many reactions. For example, the reaction shown in Equation (43) (where R = CO_2Me, CH_2OAc) proceeded to yield a 2:1 mixture of *cis*- and *trans*-isomers.[393]

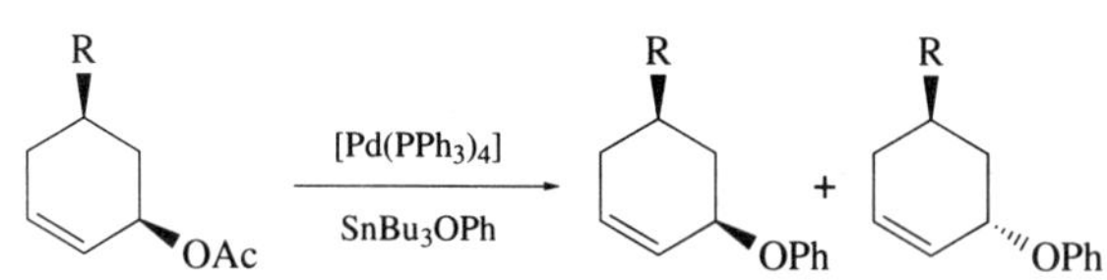

(43)

Mechanisms that lead to this loss of stereospecificity have been investigated.[394,395] Three basic pathways for the loss of stereospecificity have been suggested. One possibility is competition between external attack and attack with prior coordination; another route might involve isomerization of the allylic substrate through a palladium-catalyzed *cis–trans* isomerization process; and a third possibility is isomerization of a π-allylpalladium intermediate through nucleophilic attack of palladium(0) on a coordinated allyl group. Scheme 57 outlines the possible pathways with an allylic acetate as the substrate.

It has been shown that in amination reactions of allylic acetates the loss of stereospecificity arises from both isomerization of the substrate and isomerization of the π-allylpalladium intermediate through nucleophilic attack of palladium(0). With the stabilized carbon nucleophile $[Na][CH(SO_2Ph)_2]$, only the isomerization of the π-allylpalladium intermediate was detected. Isomerization of preformed π-allylpalladium complexes in the presence of $[Pd(PPh_3)_4]$ has been shown to involve attack of triphenylphosphine on the coordinated allyl group to produce allylic phosphonium salts with inversion of configuration, implicating this pathway as a mechanism for loss of stereospecificity in palladium-catalyzed nucleophilic substitution reactions. In the case of alkoxide anions as nucleophiles, it has been suggested[396] that loss of stereospecificity in nucleophilic substitution reactions might be attributed to a dual pathway involving both external attack and coordination followed by internal attack.

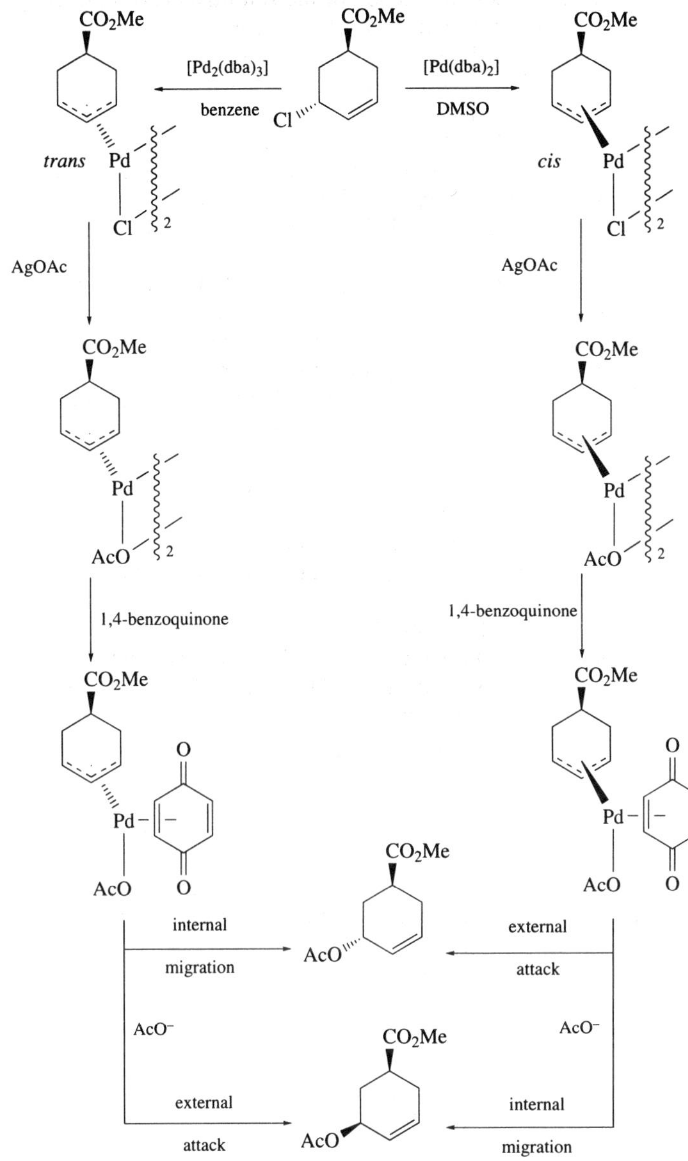

Scheme 56

(iv) Nucleophilic attack on π-allyl complexes derived from allene

The reaction of allene with $[PdCl_2(NCPh)_2]$ can be performed under conditions that give rise to a π-allylpalladium complex in which two allene units are joined together via their central carbon atoms (see Scheme 19). Reactions of such complexes with nucleophiles are of particular interest since there exist two different types of site for attack, that is, C-1 and C-3 of the π-allyl ligand, or the allylic $-CH_2Cl$ substituent. Scheme 58 illustrates that these reactions may be performed sequentially in a controlled fashion. Reaction of the dimeric, chloride-bridged π-allyl complex with the anion of dimethyl malonate resulted in substitution at the allylic $-CH_2Cl$ group and then addition of a phosphine ligand, which presumably caused bridge cleavage and so led to enhanced reactivity of the π-allyl group, and a base to cause deprotonation, led to cyclization, producing the exocyclic diene in over 50% yield, and varying amounts of the bis-adduct.[397]

The reaction sequence was found to be heavily dependent on the nature of the nucleophile, with the dianion of phenyl cyanomethyl sulfone producing the corresponding exocyclic diene exclusively while the dianion of nitromethane decomposed the complex, but without diene formation.

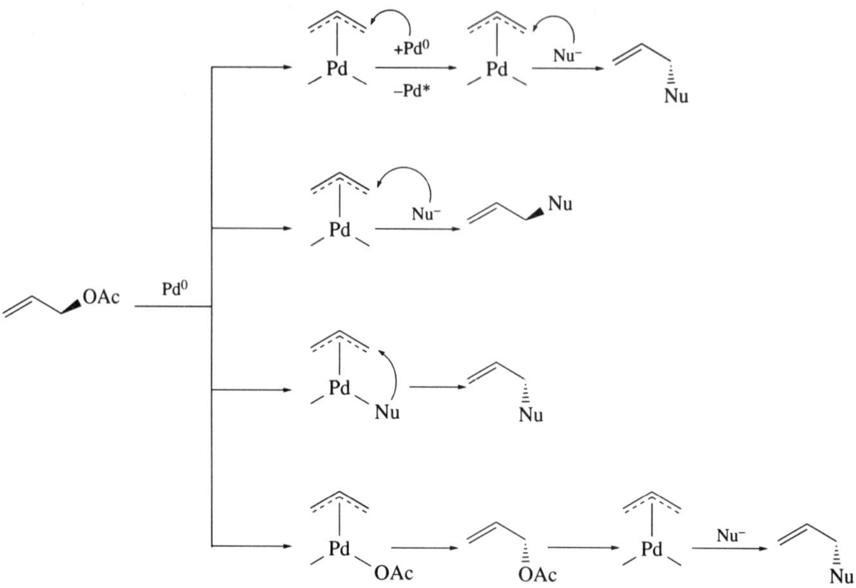

Scheme 57

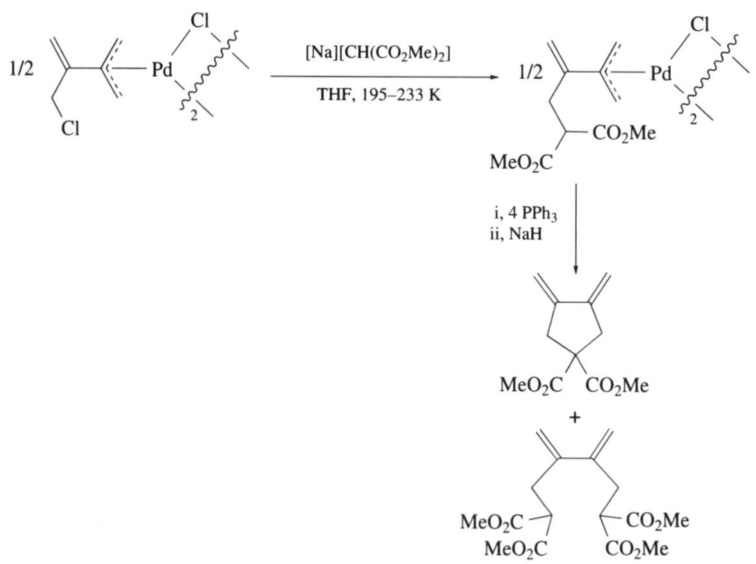

Scheme 58

The addition of [RPdX] equivalents to allene (see Scheme 19 for a well-defined example), followed by nucleophilic attack on the resultant π-allyl complex, has allowed reactions to be developed that are catalytic in palladium, and organic syntheses based on this carbopalladation strategy have been reported.[398–402]

(v) 1,4-Addition reactions of conjugated dienes via π-allyl complexes

Palladium-catalyzed 1,4-addition reactions of conjugated dienes proceed through a series of reactions of interest in this chapter. Coordination of the diene substrate to palladium (see Section 6.2.3.5), followed by nucleophilic attack (see Section 6.2.4), produces a π-allylpalladium complex which is then attacked by a second nucleophile to generate the 1,4-addition product. Much of the work in this area has been based on the use of catalytic amounts of palladium, where intermediates were not detected and mechanisms were deduced through examination of the regiochemistry and stereochemistry of the products formed.

The reactions of 1,3-cyclohexadiene in acetic acid–lithium acetate, catalyzed by palladium(II) acetate, with benzoquinone as the oxidant, are illustrative. Under halide-free conditions the reaction is reported to generate the diacetoxylation product with overall *trans*-stereochemistry (>90%). In contrast, catalytic amounts of lithium chloride led to formation of the *cis*-diacetoxylation product (>93%), while reaction in the presence of two equivalents of lithium chloride generated the 1-acetoxy-4-chloro-2-alkene with *cis*-stereochemistry (>98%).[403–5] The proposed mechanism for the formation of these products via internal or external nucleophilic attack on π-allylpalladium intermediates is shown in Scheme 59.

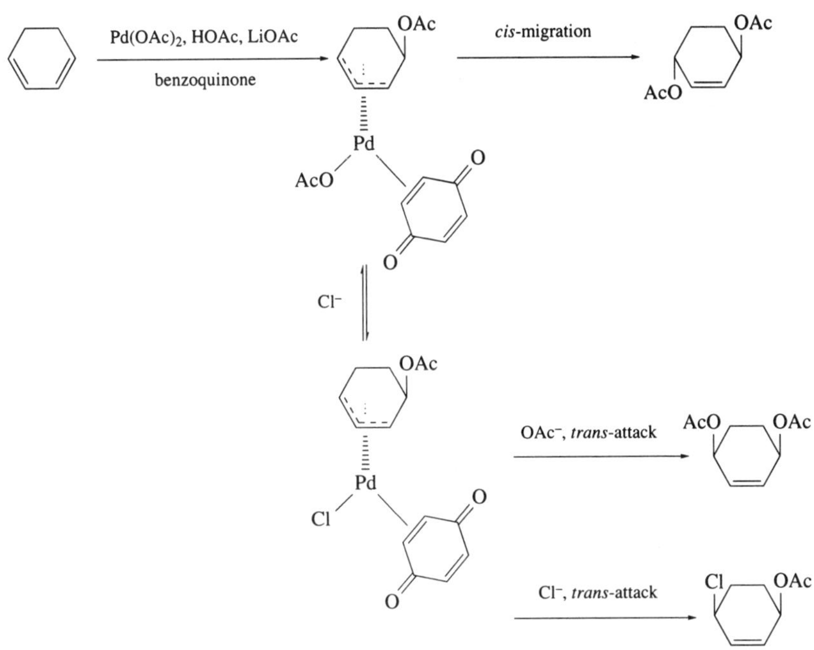

Scheme 59

NMR evidence for quinone coordination has been presented[406] and the observation of acid catalysis in related dialkoxylation reactions of 1,3-dienes[407] was interpreted in terms of protonation of a coordinated quinone, aiding the formation of hydroquinone during the oxidation step. Reoxidation of the hydroquinone may be promoted by certain metal–macrocycle complexes which effect electron transfer from molecular oxygen.[408] Binding of the oxidant to palladium has been shown to be particularly effective when quinones bearing sulfoxide substituents are employed in 1,4-addition reactions.[409] The palladium-catalyzed 1,4-addition methodology outlined in Scheme 59 has been expanded to incorporate addition of internal oxygen[410] and nitrogen[411] nucleophiles, leading to fused ring systems. Work on such intramolecular palladium-catalyzed 1,4-additions of nucleophiles to conjugated dienes has been reviewed.[412] Interestingly, the $Pd(OAc)_2$–HOAc–benzoquinone–manganese(IV) oxide system catalyzes the oxidation of simple alkenes, such as cyclohexene, to allylic acetates,[413] although in these cases the mechanism of oxidation is less clear.

6.4.4 Diene Oligomerization and Telomerization

The oligomerization of butadiene and its telomerization in the presence of nucleophiles are catalyzed by palladium compounds in reactions that involve π-allylic intermediates. The palladium acetate-catalyzed telomerization of butadiene with acetic acid was discussed in Section 6.4.2.2 and a mechanistic proposal involving dinuclear intermediates outlined in Scheme 28. It has, however, been shown[414] that $[Pd(\eta^3\text{-}2\text{-}MeC_3H_4)_2]$ reacts with tertiary phosphines (L) and either butadiene or isoprene in a stepwise fashion through intermediates of the types $[Pd(\eta^1,\eta^3\text{-allyl})_2(L)]$ and $[Pd_2(\mu\text{-}\eta^3\text{-allyl})_2(L)_2]$ to form $[Pd(\eta^1,\eta^3\text{-}C_8H_{12})(L)]$ and $[Pd(\eta^1,\eta^3\text{-}Me_2C_8H_{10})(L)]$. The reactions thus allow isolation of the elusive mononuclear η^1,η^3-octadienediylpalladium complexes earlier suggested as intermediates in diene oligomerization–telomerization reactions.

Telomerization of butadiene in the presence of ethanol and carbon monoxide with a palladium acetate–triphenylphosphine (1:4) catalyst has been shown[415] to produce ethyl nonadienoate (**121**) as a mixture of *cis*- and *trans*-isomers.

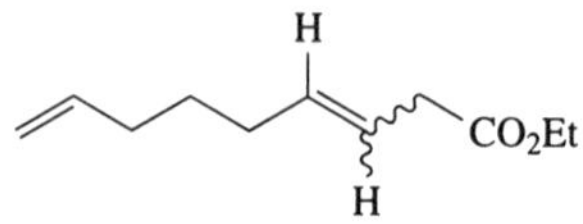

(121)

The formation of *cis*- and *trans*-isomers was attributed to initial formation of both *syn*- and *anti*-isomers of a bis(π-allyl)palladium intermediate, with the former leading to the *trans*-product and the latter to the *cis*-product. The pathways are outlined in Scheme 60.

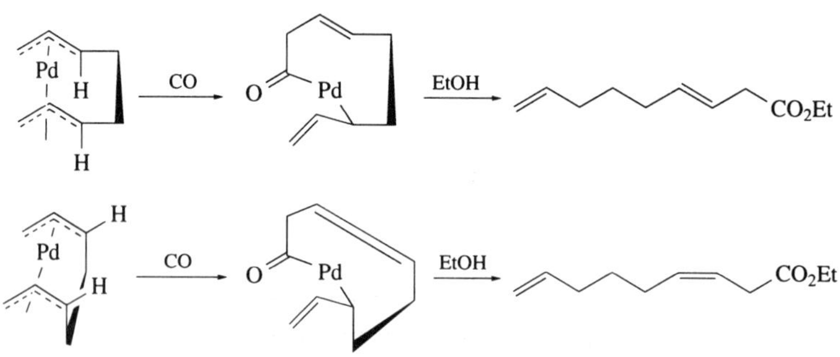

Scheme 60

Reduction of (**121**) to the corresponding alcohols, tosylation, and reduction with lithium aluminum hydride then generated 1,6-nonadiene as a mixture of *cis*- and *trans*-isomers (20:80 ratio), thus allowing access to C_9 dienes from the C_4 precursor.

Higher telomers have been generated from butadiene and methanol with cationic π-allylpalladium complexes as catalysts.[416] Thus, treatment of $[Pd_2(dba)_3]{\cdot}CHCl_3$ with $[\{CH_2{=}C(Me)CH_2O\}P(NMe_2)_3]$-$[PF_6]$ followed by addition of toluene, filtration to remove metallic palladium, addition of methanol and butadiene, and heating in an autoclave at 353 K for 20 h effected a 94% conversion of the butadiene to a mixture of $H(C_4H_6)_nOMe$ telomers of types (**122**) and (**123**).

(122) **(123)**

Product analysis indicated that telomers with even values of *n* predominated and that (**123**) with $n = 4$ was preferentially formed. The precatalyst in the system was believed to be of the type $[Pd(\eta^3\text{-}2\text{-}MeC_3H_4)L_n]^+$ (L = dba, HMPA). Similar reactions with preformed $[Pd(\eta^3\text{-}2\text{-}MeC_3H_4)(cod)][PF_6]$ as the catalyst precursor led to the same product distribution as that obtained with the catalyst generated *in situ*, whereas $[Pd(\eta^3\text{-}2\text{-}MeC_3H_4)(PR_3)_2][PF_6]$ (R = Ph, Bu^n) complexes led to mixtures of oligomers and telomers with C_8 compounds predominant. $[Pd(\eta^3\text{-}2\text{-}MeC_3H_4)(cod)][PF_6]$ has been found to be inactive towards butadiene alone but $[Pd(\eta^3\text{-}2\text{-}MeC_3H_4)L_2][BF_4]$ (L_2 = bidentate aminophosphinite (**124**), where R = H, Me) complexes generated C_8, C_{12}, and C_{16} oligomers.[417]

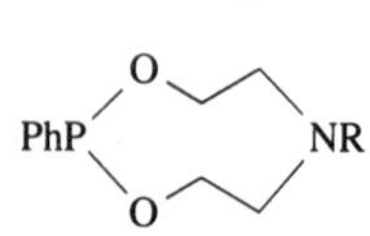

(124)

The C_8 and C_{16} oligomers were mainly linear, whereas the C_{12} oligomers were largely branched, suggesting that C_{16} oligomers arose through dimerization of C_8 compounds, whereas C_{12} oligomers were formed by butadiene insertions into an intermediate η^3-2-methylallylpalladium species.

The telomerization of butadiene or isoprene with tertiary allylic amines catalyzed by $[PdCl_2(PPh_3)_2]$–$AlEt_3$ and a polystyrene-supported analogue has been reported.[418] The reaction of butadiene with allyldimethylamine produced (**125**) with an (*E*):(*Z*) ratio of ca. 14:1 along with a small amount of (**126**). With a large excess of butadiene, (**125**) reacted further to produce (**127**).

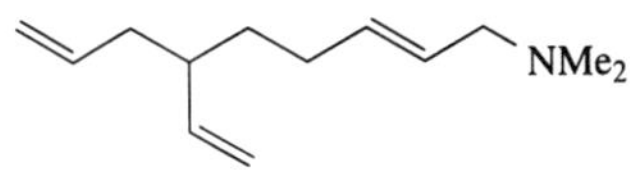

(125)

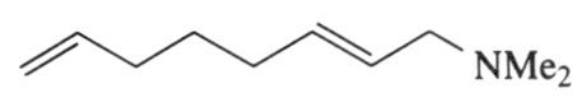

(126)

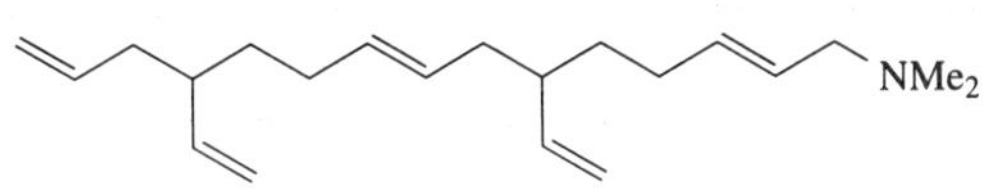

(127)

Compound (**125**) resulted from an unusual carbon–nitrogen bond cleavage of the allylic amine and addition of the amine function and the allyl residue regioselectively to the 1- and 6-positions of a butadiene dimer. The mechanism of the carbon–nitrogen bond cleavage process was presumed to involve reaction with a palladium(0) species generated by reduction of the palladium(II) precatalyst by triethylaluminum.

The selective hydrodimerization of butadiene to 1,7-octadiene has been accomplished with a catalyst system consisting of palladium acetate (one equivalent), triethylamine (two equivalents), a tertiary phosphine (4.88 equivalents), and formic acid (61.8 equivalents) as a hydride source, dissolved in DMF containing butadiene (56 equivalents).[419] With triphenylphosphine as the ligand a 96% conversion with 87% selectivity for the 1,7-isomer was achieved at 343 K over 1.5 h. Use of formic acid-d_2 led to 1,7-octadiene-d_2 with no deuterium incorporated into the terminal positions. This result was interpreted in terms of a mechanism involving a bis(π-allyl)palladium hydride intermediate which transferred hydride selectively to C-3 during reductive elimination when phosphine ligands were present. In the absence of added phosphine, the catalyst system gave rise to 1,6-octadiene-d_2 with one terminal deuterium, that is, in the absence of phosphine reductive elimination occurred at C-1.

The reaction of $[Pd(C_5H_5)(\eta^3\text{-}C_3H_5)]$ with trimethylphosphine and allylidenecyclopropane at 243 K with warming to room temperature led to the isolation of an octadienediyl complex,[420] characterized crystallographically as (**128**). Treatment of (**128**) with one equivalent of $Me_2P(CH_2)_2PMe_2$ in diethyl ether at 273 K led to formation of (**129**) which was found to rearrange quantitatively and irreversibly to (**130**), characterized crystallographically, within a few hours at room temperature.

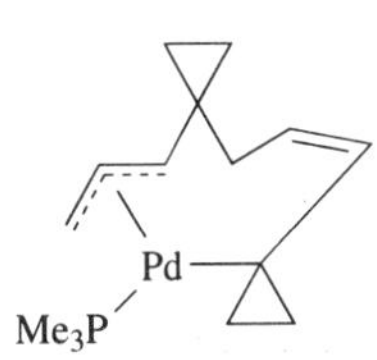

(128)

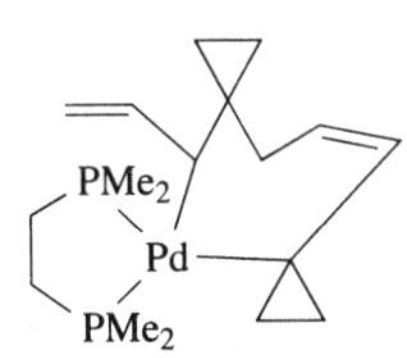

(129)

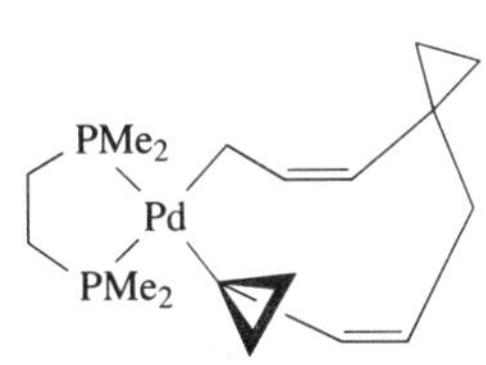

(130)

These reactions model those anticipated in the latter stages of 1,3-diene oligomerization. Thus, (**129**) is structurally related to the intermediate proposed in Scheme 60 for butadiene telomerization, that is, both are eight-membered palladacyclic systems, each with an internal, uncoordinated double bond and an exocyclic vinyl group.

6.4.5 Cycloaddition Reactions Implicating Trimethylenemethanepalladium Intermediates

Early work suggested[421] that an unisolable trimethylenemethanepalladium complex could be generated by treatment of $[Pd_2(\mu\text{-Cl})_2\{2\text{-}(CH_2Cl)C_3H_4\}_2]$ with antimony pentafluoride at low temperature, as shown in Equation (44).

Such trimethylenemethanepalladium complexes are of interest[422] since cycloaddition reactions of the type illustrated schematically in Equation (45) may be accomplished with palladium catalysts.

Examples of such reactions include those illustrated in Equations (46)[423] and (47).[424]

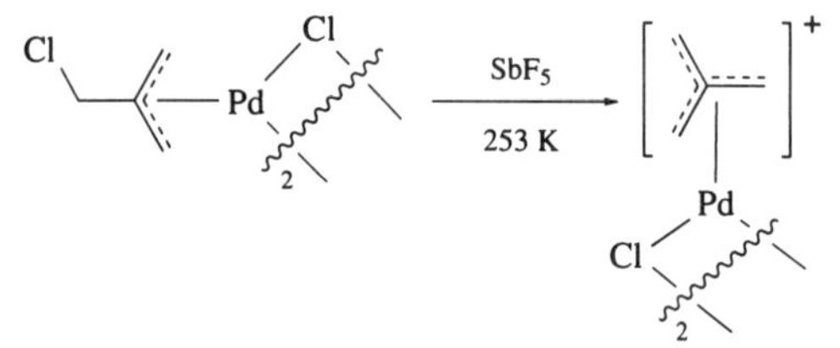

(44)

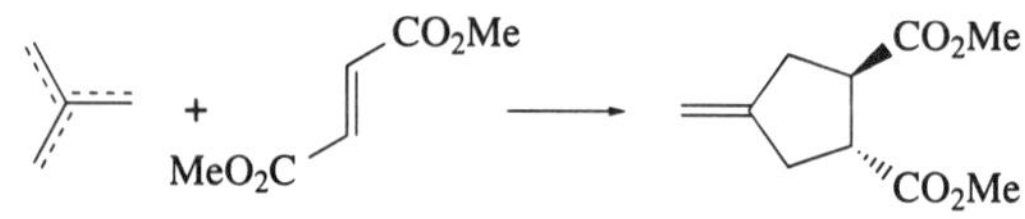

(45)

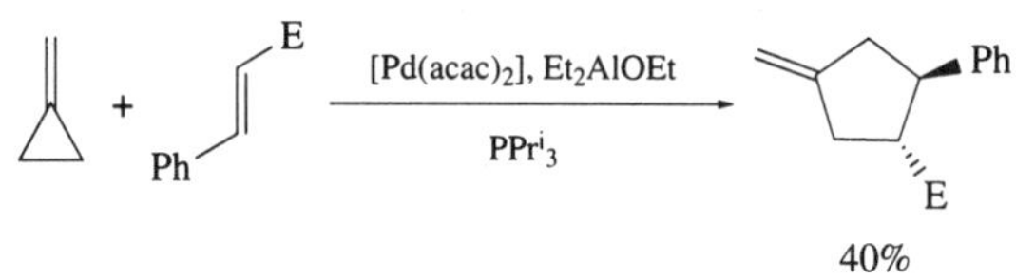

(46)

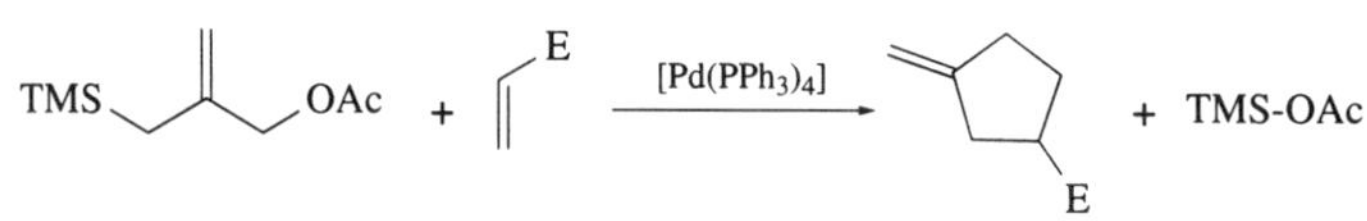

(47)

The proposed mechanism for the class of transformation illustrated in Equation (47), which implicates generation of a trimethylenemethanepalladium intermediate and subsequent reaction with an alkene, is shown in Scheme 61.

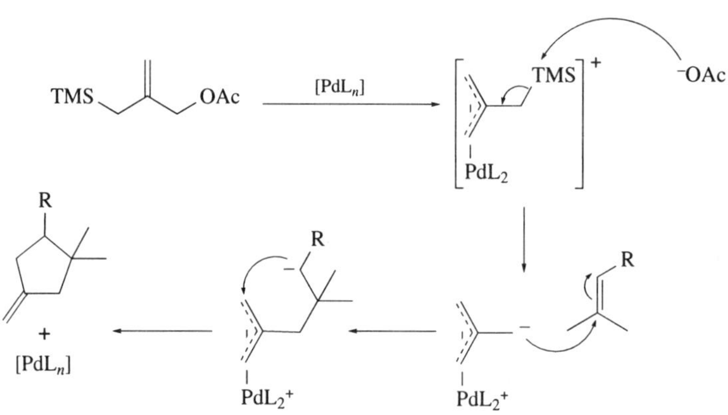

Scheme 61

There exists considerable indirect evidence for a pathway of the type shown in Scheme 61. Isolable trimethylenemethane complexes of osmium and iridium have been synthesized from [(2-(acetoxymethyl)-3-allyl]trimethylsilane, the precursor shown in Scheme 61, and, in the absence of silylophilic reagents such as acetate anion, the complex $[Pd\{\eta^3\text{-}2\text{-}(CH_2TMS)C_3H_4\}(PPh_3)_2]^+$ has been isolated as the hexafluorophosphate salt from a closely related reaction.[425] Trapping through alkylation supports the existence of an intermediate electrophilic π-allyl species, while deuteration studies suggest that a nucleophilic trimethylenemethane complex is involved in the catalytic reactions.[426] In Scheme 61 the reaction of the proposed trimethylenemethanepalladium intermediate with the alkene is depicted as a distal attack. In certain systems there is evidence of prior coordination of the alkene, again suggesting that more than one pathway for attack on an organopalladium intermediate exists and that stereocontrol may be achieved.[427] A number of theoretical studies have probed the possible structure of trimethylenemethanepalladium species of the type shown in Scheme 61. Fenske–Hall calculations[428] on a trimethylenemethane moiety bound to a $[Pd(PH_3)_2]$ fragment have suggested that the dihedral angle between the PdP_2 plane and the plane of the trimethylenemethane group is not 90° but rather ca. 96°, thus suggesting a close analogy to related π-allyl systems.

In addition to cycloaddition reactions with alkenes, palladium-mediated cyclizations that implicate trimethylenemethanepalladium intermediates have been reported with other acceptors. Conjugated dienes bearing electron-withdrawing substituents have been employed in reactions where either five- or seven-membered ring compounds were formed,[429] while imines with electron-withdrawing substituents on either carbon or nitrogen have also been found[430] to be effective acceptors. With enones as acceptors the normal 1,4-mode of addition could be switched to 1,2-addition by co-catalysis with In^{3+} which presumably binds to the enone oxygen and so promotes nucleophilic attack at the carbonyl carbon.[431] The application of palladium-catalyzed cycloaddition reactions that implicate trimethylenemethanepalladium intermediates in the total synthesis of tetracyclic diterpene derivatives has been reported.[432]

6.5 CYCLOPENTADIENYL COMPLEXES

6.5.1 Synthetic Methods

6.5.1.1 Synthesis of terminal cyclopentadienyl complexes

A significant number of palladium(II) complexes containing terminal η-cyclopentadienyl ligands are known and the common synthetic routes, discussed in *COMC-I*, have been based on reactions of palladium halide complexes with sources of cyclopentadienyl anion, such as NaC_5H_5, TlC_5H_5, or $[Fe(C_5H_5)Br(CO)_2]$. Indenyl complexes have similarly been prepared from the reactions of palladium halide complexes with sodium indenide. An alternative strategy described in *COMC-I* involved the generation of a substituted cyclopentadienyl moiety by oligomerization of disubstituted alkynes. For example, $[Pd_2(\eta\text{-}C_5Ph_5)_2(\mu\text{-PhCCPh})]$ resulted from the reaction of diphenylethyne with bis(acetato)-palladium(II).

Reactions of palladium halide complexes with sources of cyclopentadienyl anion continue to be explored as routes to palladium cyclopentadienyl complexes. For example, the reaction of TlC_5H_5 with the ortho-metallated complex (**131**) allowed isolation of $[Pd(\eta\text{-}C_5H_5)\{P(OPh)_2\}(OC_6H_4)]$, which has been characterized crystallographically.[433]

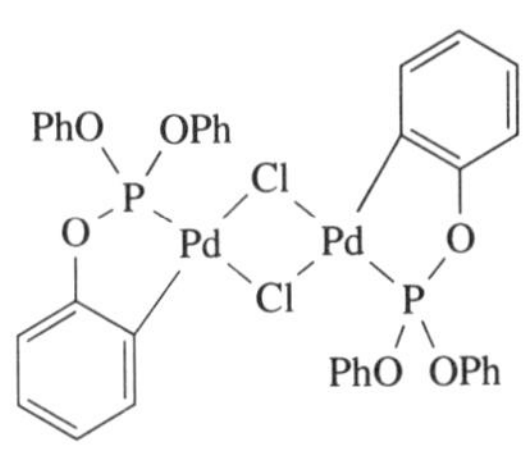

(131)

The direct reaction of a source of cyclopentadienyl anion with a palladium halide complex has been extended to include substituted cyclopentadienyls. Thus, pentaphenylcyclopentadienylsodium was reported[434] to react with $[Pd_2Cl_2(\eta^3\text{-all})_2]$ (where all = allyl, 2-chloroallyl, 2-methylallyl, 1-methylallyl, 1,1-dimethylallyl, 1-carbomethoxyallyl) to produce $[Pd(\eta\text{-}C_5Ph_5)(\eta^3\text{-all})]$. Subsequent reactions of $[Pd(\eta\text{-}C_5Ph_5)(\eta^3\text{-2-chloroallyl})]$ with tertiary phosphines allowed isolation of a series of compounds of type $[PdCl(\eta\text{-}C_5Ph_5)(PR_3)]$. Similar treatment of $[Pd(\eta\text{-}C_5Ph_5)(\eta^3\text{-allyl})]$ with PR_3 resulted in the formation of zerovalent palladium complexes and coupled organic products. Pyrolysis of $[Pd(\eta\text{-}C_5Ph_5)(\eta^3\text{-2-chloroallyl})]$ under controlled conditions produced the halide-bridged complex $[Pd_2Cl_2(\eta\text{-}C_5Ph_5)_2]$ with elimination of allene. The same complex was also generated through the reaction of $[PdCl_2(NCPh)_2]$ with NaC_5Ph_5 in THF solution. The reaction of the dimer $[Pd_2Cl_2(\eta\text{-}C_5Ph_5)_2]$ with the π-acids carbon monoxide, ethene, and allene led to equilibrium formation of kinetically labile, bridge cleavage products in solution. However, treatment of $[Pd_2Cl_2(\eta\text{-}C_5Ph_5)_2]$ with alkynes, or with CO, in the presence of activated zinc dust generated bridged palladium(I) complexes (Scheme 62).

The alkyne complexes were isolated in 30–75% overall yields and characterized by NMR, UV–visible, IR, and mass spectroscopies. The carbonyl complex was found to be less stable and the structure was assigned largely on the basis of IR data ($\nu(CO) = 1872\ cm^{-1}$).

The pentamethylcyclopentadienyl complex $[Pd_2(\mu\text{-}CO)_2(\eta\text{-}C_5Me_5)_2]$ has been synthesized through the reaction of $[PdCl(CO)]_n$ with $C_5Me_5MgCl{\cdot}THF$.[435] A single band in the IR spectrum at $1839\ cm^{-1}$ was taken to indicate that the bridging carbonyls had coplanar C–O vectors, in contrast with related nickel systems where the central core is puckered. Protonation of the dimer with HBF_4 or CF_3SO_3H led to formation of the trinuclear cluster $[Pd_3(\eta\text{-}C_5Me_5)_3(\mu\text{-}CO)_2]^+$, a member of the Fischer–Palm family, which was characterized crystallographically as the triflate salt. Cyclic voltammetry on the 48-valence-

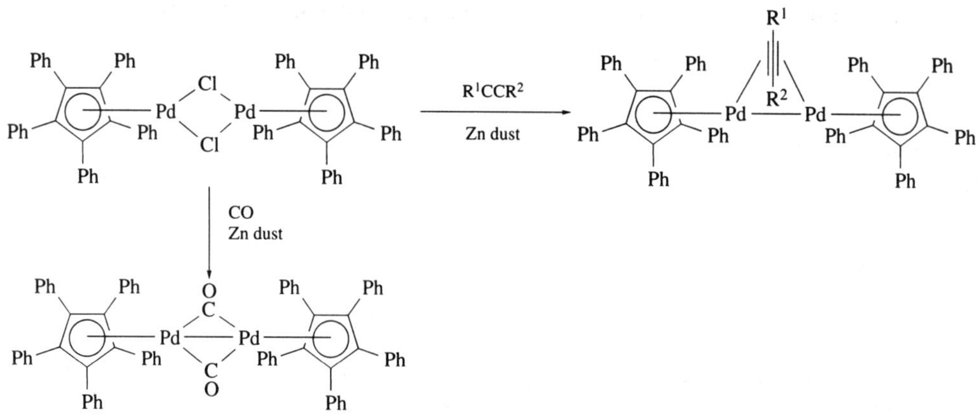

$R^1 = R^2 = Ph$; $R^1 = R^2 = H$; $R^1 = R^2 = Et$; $R^1 = R^2 =$ *p*-tolyl; $R^1 = Ph$, $R^2 = H$; $R^1 = R^2 = CO_2Me$

Scheme 62

electron cluster showed two discrete redox processes which indicated a reversible reduction to the neutral 49-electron cluster and a less reversible reduction to the 50-electron anionic cluster.

Functionally substituted cyclopentadienyl complexes of palladium have been prepared[436] through reactions of thallium or sodium salts of substituted cyclopentadienyl anions with palladium halide complexes (Scheme 63).

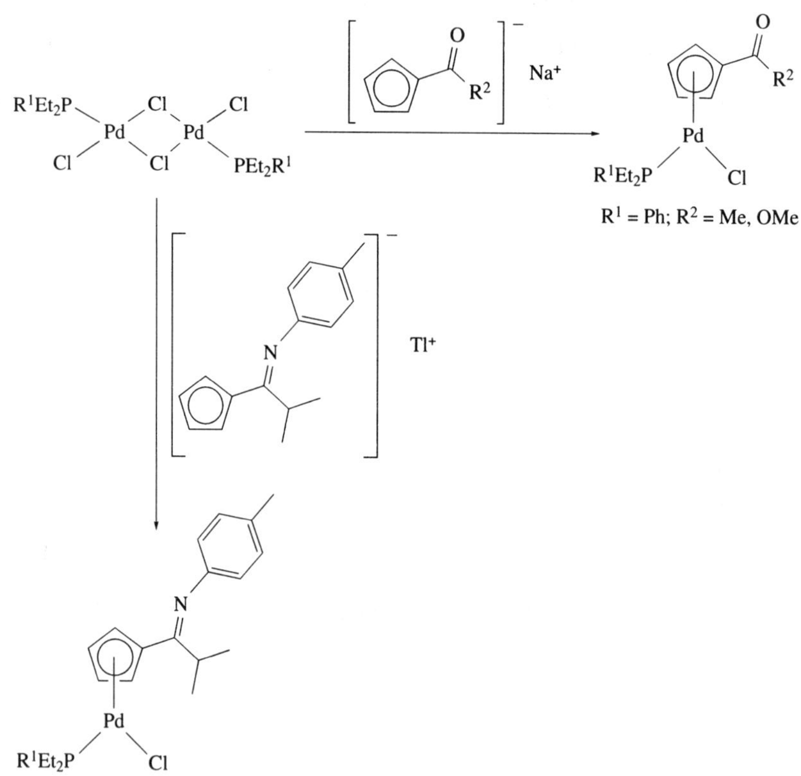

$R^1 = Ph$; $R^2 = Me, OMe$

$R^1 = Ph, Et$

Scheme 63

The complexes were characterized by NMR and IR spectroscopies and by elemental analyses. Triphenylphosphonium cyclopentadienylide has been shown[437] to react with $[Pd(diene)(acetone)_2]^{2+}$ (diene = cod, nbd), generated *in situ* from $[PdCl_2(diene)]$ and $AgPF_6$, to produce $[Pd(\eta\text{-}C_5H_4PPh_3)(diene)][PF_6]_2$. The platinum(II) analogues were also synthesized. Complexes of the type $[PdL_2(acetone)_2][PF_6]_2$ (L = monodentate phosphine, arsine, stibine; L_2 = bidentate diphosphine, diarsine), similarly generated *in situ*, were found[438] to react directly with cyclopentadiene to generate $[Pd(\eta\text{-}C_5H_5)L_2][PF_6]$ (Equation (48)).

$$[PdL_2(Me_2CO)_2][PF_6]_2 + C_5H_6 \longrightarrow [Pd(\eta\text{-}Cp)L_2][PF_6] + HPF_6 + 2\,Me_2CO \qquad (48)$$

The complex $[Pd(\eta\text{-}C_5H_5)(SbPh_3)_2][PF_6]\cdot CH_2Cl_2$ was characterized crystallographically. Dinuclear complexes of the type $[Pd_2(\mu\text{-}OH)_2Ph_2(PR_3)_2]$ (R = Ph, Cy) have similarly been shown[439] to react directly with cyclopentadiene or methylcyclopentadiene to generate the corresponding $[Pd(\eta\text{-}C_5H_4X)Ph(PR_3)]$ (X = H, Me) complexes in yields from 65% to 95%. These latter methods obviate the need for toxic reagents such as TlC_5H_5 in certain syntheses.

An unusual example of a transmetallation reaction between $[PdCl_2(dppe)]$ and $[ZrCl(CH_2CH_2CMe_3)(C_5H_5)_2]$ in the presence of $AgCF_3SO_3$ to generate $[Pd(\eta\text{-}C_5H_5)(dppe)][CF_3SO_3]$ has been described. The product was characterized crystallographically.[440]

The reaction of $[Pd_2(\mu\text{-}O_2CMe)_2Cl_2(PR_3)_2]$ with TlC_5H_5 (see also Section 6.5.1.2) has been reported to initially produce $[Pd(\eta\text{-}C_5H_5)(\eta^1\text{-}C_5H_5)(PR_3)]$.[441] The compounds were found to be fluxional in solution. A high-energy process, involving η^1–η^5 exchange of the two cyclopentadienyl rings, and a lower-energy process, presumed to be a metallotropic rearrangement, were identified from a ^1H and ^{13}C NMR study. The temperature of coalescence for the η^1–η^5 exchange was found to vary with the size of the phosphine ligand. An x-ray crystal structure of $[Pd(\eta^5\text{-}C_5H_5)(\eta^1\text{-}C_5H_5)(PEt_3)]$ confirmed η^1- and η^5-bonding of the two cyclopentadienyl rings, and extended Hückel molecular orbital calculations on a model $[Pd(\eta\text{-}C_5H_5)Me(PH_3)]$ system indicated that stronger bonding of palladium to one carbon of the C_5H_5 ring should be anticipated in this and analogous systems.

6.5.1.2 Synthesis of bridged cyclopentadienyl complexes

Several routes to palladium complexes with bridging cyclopentadienyl ligands were described in *COMC-I*. $[Pd(C_5H_5)(\eta^3\text{-all})]$ (all = a range of allylic ligands) complexes were known to react with two equivalents of L (L = phosphine or phosphite) to generate dinuclear compounds of type (**132**) which could also be generated through reaction of $[Pd(C_5H_5)(\eta^3\text{-all})]$ with $[PdL_2]$ (L = PPr^i_3 or PCy_3).

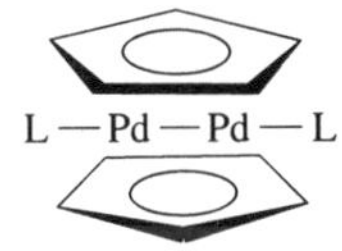

(**132**)

X-ray structural studies on complexes of type (**132**) have shown that the bridging cyclopentadienyl ligands, found in an *anti*-arrangement, might best be considered as "allyl-ene" systems with trihapto coordination. The reduction of $[Pd(\eta\text{-}C_5H_5)(X)(PR_3)_2]$ was reported to produce related complexes of the type $[Pd_2(\mu\text{-}\eta^3\text{-}C_5H_5)(\mu\text{-}X)(PR_3)_2]$.

Molecular orbital calculations have shed light on the unusual "allyl-ene" bonding of the cyclopentadienyl ligands in complexes such as (**132**). Examination of 14 examples of $[Pd_2(\mu\text{-}A)(\mu\text{-}B)L_2]$ systems (L = PH_3, Cl; A and B = allyl, butadiene, cyclopentadienyl, benzene) by the iterative Fenske–Hall method showed that the $[Pd_2(\mu\text{-}A)L_2]$ fragment has only two low-lying empty orbitals and thus functions as a four-electron acceptor. For ligands such as cyclopentadienyl and benzene, normally six-electron donors, the result is a partial localization of the π-electrons. In the case of bridging cyclopentadienyl ligands this localization results in the "allyl-ene" mode of coordination.[442]

Complexes of type (**132**) (where L = tertiary phosphine) have now been synthesized through reduction of $[Pd(C_5H_5)(O_2CMe)(PR_3)]$ with $NaK_{2.8}$ and through reaction of $[Pd_2(\mu\text{-}O_2CMe)_2Cl_2(PR_3)_2]$ with four equivalents of TlC_5H_5.[443] The latter reaction passes through an intermediate $[Pd(\eta\text{-}C_5H_5)(\eta^1\text{-}C_5H_5)(PR_3)]$ complex. Compound (**132**) (L = $P(OPh)_3$) was prepared by the reaction of $[Pd_2Cl_4\{P(OPh)_3\}_2]$ with TlC_5H_5. The reactions are summarized in Scheme 64.

An x-ray structure determination (L = PEt_3) confirmed an almost linear arrangement of L–Pd–Pd–L with "allyl-ene" coordination of the *anti*-cyclopentadienyl groups. The reaction of $[Pd_2(\mu\text{-}O_2CMe)_2Cl_2(PR_3)_2]$ with TlC_5Me_5 proceeds in a fashion different from that with TlC_5H_5 shown in Scheme 64. Thus, treatment of $[Pd_2(\mu\text{-}O_2CMe)_2X_2(PR_3)_2]$ (X = Cl; R = Ph, Me, Pr^i; X = Br; R = Pr^i) with TlC_5Me_5 generated $[Pd(\eta\text{-}C_5Me_5)(X)(PR_3)]$ and attempts to replace the halide ligand with either C_5H_5 or C_5Me_5 failed.[444] An x-ray crystal structure (X = Cl; R = Pr^i) showed slippage of the palladium across the face of the pentamethylcyclopentadienyl ring, interpreted as a tendency towards η^1-coordination in such systems.

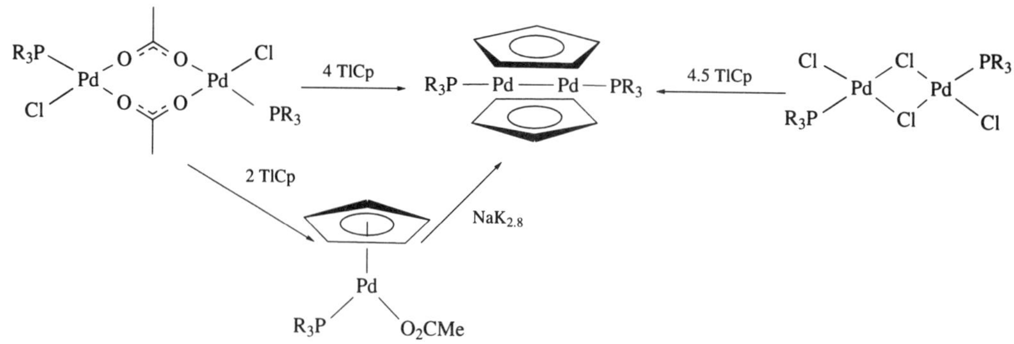

Scheme 64

Novel dinuclear complexes with bridging indenyl ligands have been prepared[445] through the reaction of $[PdCl_2(CNR)_2]$ with indenyllithium. An x-ray structure determination (where R = 2,6-dimethylphenyl) showed that an RNC–Pd–Pd–CNR unit was sandwiched between μ-η^3-indenyl groups in a *syn*-arrangement. Attempts to prepare analogous complexes from $[PdCl_2(PPh_3)_2]$ or $[PdCl_3(CO)]^-$ failed.

6.5.2 Reactions of Palladium–Cyclopentadienyl Complexes

Reactions of palladium–cyclopentadienyl complexes described in *COMC-I* fall into two basic categories: (i) those in which the η-C_5H_5 ligand functions as a spectator group, maintaining its integrity throughout the reactions; and (ii) those in which the C_5H_5 ligand is cleaved from palladium or changes its mode of binding. Additional examples of both types of reaction have been reported.

The electrochemistry of the dipalladium(I) complex $[Pd_2(\mu\text{-PhCCPh})(\eta\text{-}C_5Ph_5)_2]$ has been probed by cyclic voltammetry.[446] The compound exhibited two diffusion-controlled, reversible one-electron oxidations and a single one-electron reduction. Measurements in dichloromethane (0.1 M in $[NBu^n{}_4][PF_6]$) at a platinum electrode showed that the two oxidations (E^0 = + 0.52 V and + 1.13 V vs. SCE) were reversible with peak separations of ca. 70 mV down to sweep rates as slow as ca. 50 mV s^{-1}. The reduction was only fully reversible at sweep rates over 200 mV s^{-1}. The radical cation, formed electrochemically by one-electron oxidation, was also generated chemically by oxidation with silver ion. The paramagnetic compound was isolated as a microcrystalline solid and characterized by elemental analysis, cyclic voltammetry, and ESR spectroscopy. The ESR spectrum of a frozen dichloromethane solution exhibited a single line centered at $g = 2.044$. The radical cation was found to react with neutral donor ligands, L (L = PPh_3; L_2 = dppe, cod), according to Equation (49).

$$2\,[Pd_2(\mu\text{-PhCCPh})(\eta\text{-}C_5Ph_5)_2]^+ + 4\,L \longrightarrow$$
$$2\,[Pd(\eta\text{-}C_5Ph_5)L_2]^+ + [Pd_2(\mu\text{-PhCCPh})(\eta\text{-}C_5Ph_5)_2] + PhCCPh \quad (49)$$

The reactions described by Equation (49) illustrate that oxidation resulted in enhanced reactivity since the diamagnetic neutral compound $[Pd_2(\mu\text{-PhCCPh})(\eta\text{-}C_5Ph_5)_2]$ is generally inert to nucleophiles of the type employed. The palladium(II) complexes $[Pd(\eta\text{-}C_5Ph_5)L_2]^+$ were found to undergo generally reversible one-electron reductions and oxidations, thus opening the possibility of new routes to palladium(I)– and palladium(III)–cyclopentadienyl complexes (see Section 6.2.3.4). Despite the commonly observed lability of the palladium–cyclopentadienyl bond, the unusual processes described occur without cleavage. A mechanism for the cleavage process illustrated by Equation (49) has been proposed[447] and work in this general area summarized.[448]

Although addition reactions of small molecules with compounds containing palladium–palladium bonds are commonly accompanied by cleavage of the bond and bridge formation, addition reactions of $[Pd_2(\mu\text{-Br})(\mu\text{-}C_5H_5)(PR_3)_2]$ with carbon monoxide, sulfur dioxide, and methyl isocyanide are reported[449] to occur without cleavage of the palladium–palladium bond and with a change of bonding mode of the cyclopentadienyl ligand from bridging to terminal. Scheme 65 illustrates the reported reactions.

The x-ray crystal structure of the sulfur dioxide complex (R = Et) showed considerable asymmetry in binding of the bridging unit to the two electronically dissimilar palladium fragments. A subsequent study[450] expanded the number of bridging groups to include other isocyanides and also MeO_2CCCCO_2Me. An x-ray crystal structure of $[(\eta\text{-}C_5H_5)(PPr^i{}_3)Pd(\mu\text{-CO})PdBr(PPr^i{}_3)]$ was reported which indicated asymmetry of the bridging group.

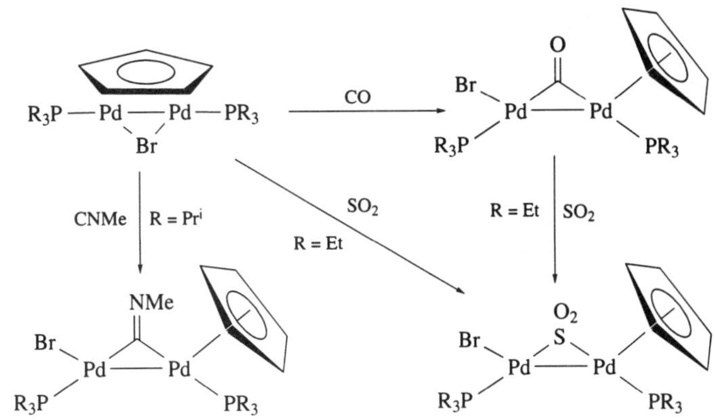

Scheme 65

The reaction of $[Pd(C_5H_5)(\eta^3\text{-}C_3H_5)]$ with bis(diphenylphosphino)maleic anhydride produced a novel complex, characterized crystallographically, in which the C_5H_5 ligand had undergone insertion into a phosphorus–carbon bond of the phosphine ligand.[451] Subsequent treatment with iodine led to generation of a complex containing a new type of phosphine ligand.[452] The compound was characterized by x-ray crystallography. Scheme 66 illustrates the reactions.

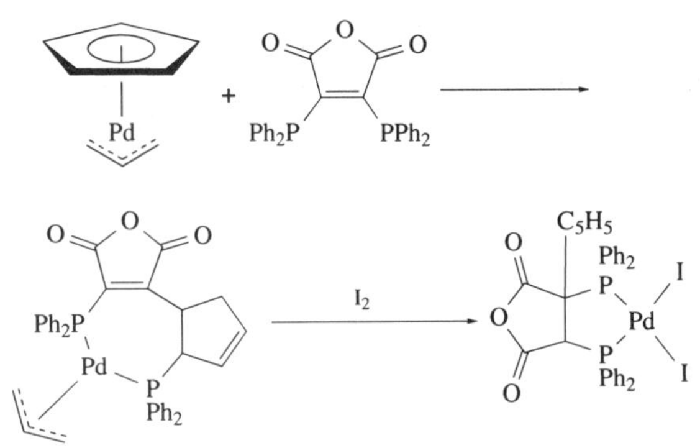

Scheme 66

The lability of the cyclopentadienyl group is further exemplified by the reaction of $[Pd(\eta\text{-}C_5H_5)(\eta^4\text{-}cod)][BF_4]$ with $[Tl][TlC_2B_9H_{11}]$ to generate, along with other products, a complex in which the cyclopentadienyl ligand was replaced by the dicarbadodecaborane unit. The complex was characterized crystallographically.[453]

The cyclopalladated compound (**133**) has been shown[454] to react with tertiary phosphines (PEt_3, $PBu^n{}_3$) by an initial, reversible displacement of the coordinated nitrogen.

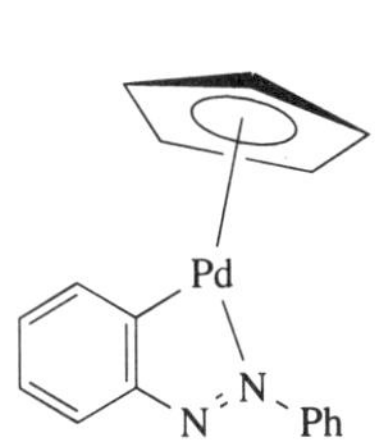

(133)

The reversibility of this process led to attack of liberated phosphine on $[Pd(C_6H_4\text{-}2\text{-}NNPh)(\eta\text{-}C_5H_5)(PR_3)]$ to produce *trans*-$[Pd(C_6H_4\text{-}2\text{-}NNPh)(\eta^1\text{-}C_5H_5)(PR_3)_2]$. The second step in this process was found to be dependent on the size of the tertiary phosphine employed and could be prevented by use of PCy_3.[455] Thus, the intermediate complex $[Pd(C_6H_4\text{-}2\text{-}NNPh)(C_5H_5)(PCy_3)]$ could be isolated and was characterized crystallographically. Compound (**133**) and its nickel and platinum analogues have been characterized by x-ray crystallography.[456] The palladium and platinum complexes showed unusual geometries for the M–C_5H_5 units, which may relate to the facile η^5–η^1 conversions observed during reactions with nucleophiles.

The reactions of $[Pd(C_5H_5)(\eta^3\text{-}C_3H_5)]$ with 3,3-dimethylcyclopropene in the presence of tertiary phosphines have been shown[457] to produce palladacycloalkanes through oxidative coupling of the alkene. Scheme 67 illustrates the reactions described.

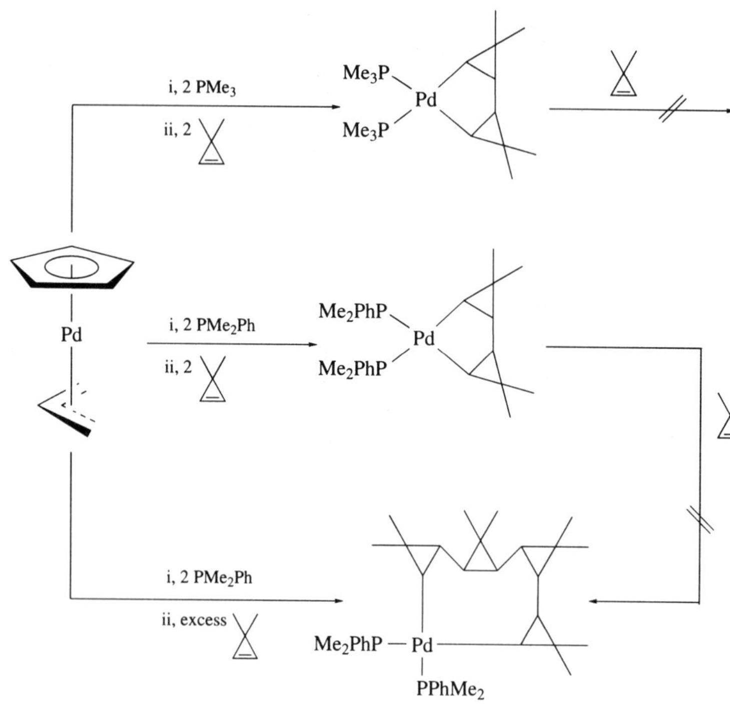

Scheme 67

Cyclopropabenzenes have been shown[458] to react with $[Pd(C_5H_5)(\eta^3\text{-}C_3H_5)]$ in the presence of tertiary phosphines to form products in which the cyclopentadienyl and allyl ligands were coupled, or products in which both groups were displaced, as shown in Scheme 68.

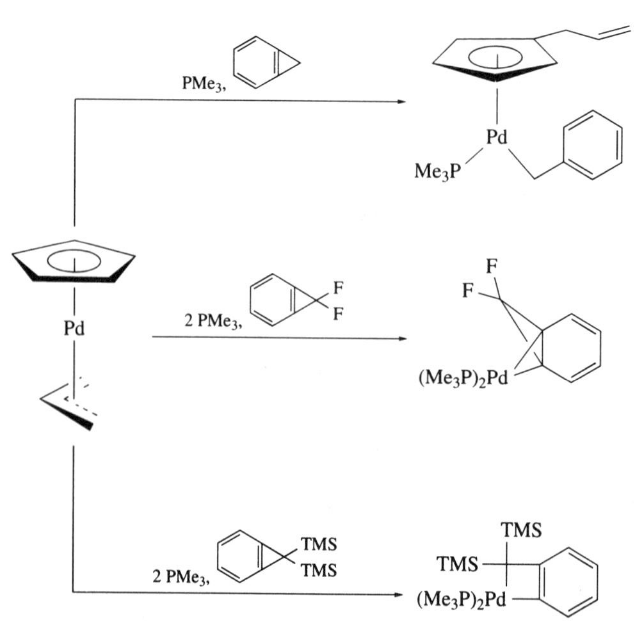

Scheme 68

6.6 ARENE COMPLEXES

6.6.1 Interactions of Arenes with Metallic Palladium

The adsorption of benzene, *m*-xylene, pyridine, and 2,6-dimethylpyridine on Pd(111) at room temperature has been studied by angle-resolved PES and electron energy loss spectroscopy (EELS).[459] Surface-bound benzene was found to be adequately described as an absorbate complex of C_{6v} symmetry, that is, one where benzene was bound parallel to the palladium surface. Both *m*-xylene and 2,6-dimethylpyridine were similarly found to be bound parallel to the surface but pyridine was adsorbed in an inclined geometry. The results for benzene on Pd(111) differ from those obtained by an angle-resolved PES study[460] of benzene on Pt(111) where the parallel-bound absorbate complex was found to be best described as C_{3v}. The strength of the metal–benzene interaction was proposed to be greater for Pt(111) than for Pd(111).

6.6.2 Palladium–Arene Complexes

Palladium–arene complexes have been invoked as intermediates in catalytic and stoichiometric C–H bond activation reactions of benzene and other aromatics that involve palladium salts and complexes. For example, a kinetic study[461] of the arylation of alkenes catalyzed by palladium(II) acetate (Equation (50)) led to a proposed mechanism that involved π-arene intermediates of unspecified hapticity (Scheme 69).

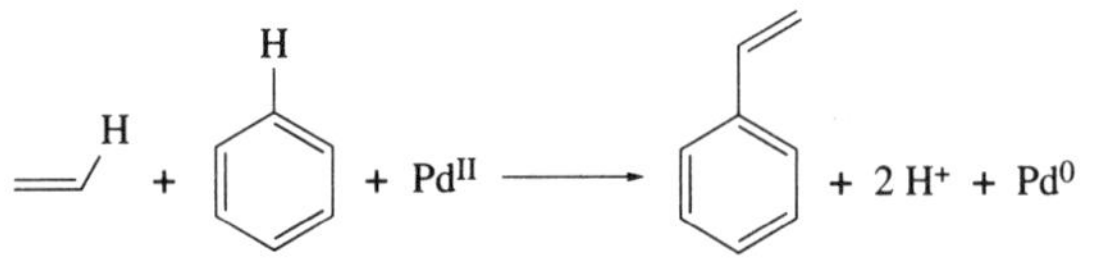

(50)

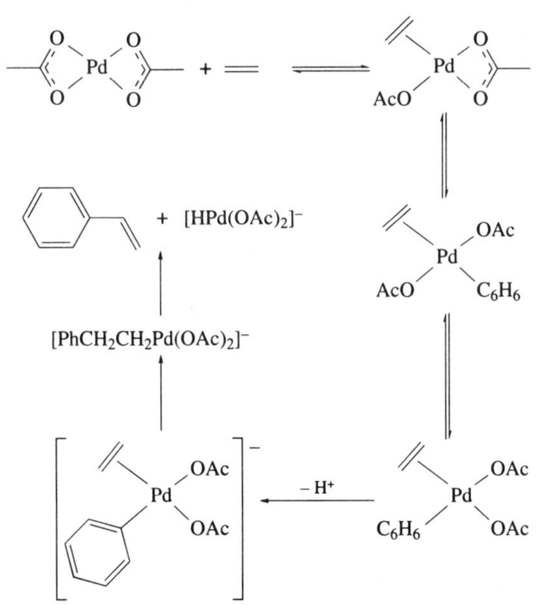

Scheme 69

Other C–H activation reactions of arenes based upon palladium chemistry are known and the area has been discussed in a review.[462]

In *COMC-I* two unusual palladium(I)–benzene complexes were described of structure (**134**) where X = Cl or $AlCl_4$.

The compounds were formed during the reaction of palladium dichloride, aluminum, aluminum trichloride, and benzene and were characterized crystallographically.[463] In both compounds the Cl–Pd–Pd–Cl unit was found to be essentially linear but the metal–arene interactions were different in the two examples. In the case where X = Cl, the palladium atoms were found to lie over one edge of

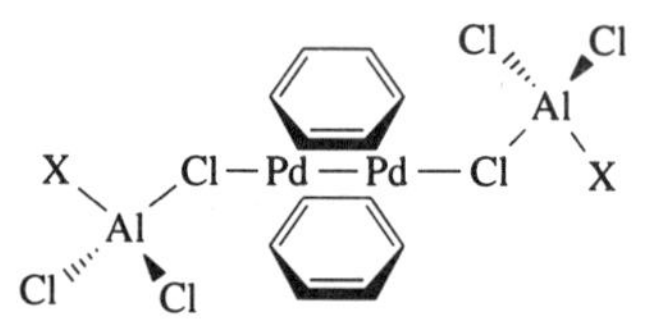

(134)

both benzene rings, whereas where X = $AlCl_4$ the Pd–Pd axis was found to bisect the benzene rings from corner to corner. In both cases it seemed that four carbon atoms of each ring were within bonding distances of palladium while two were significantly further away. The bonding situation was described in *COMC-I* as unparalleled among transition metal–arene complexes and little is known about such compounds.[464] The complex with X = Cl has since been the subject of a solid-state NMR study.[465] Proton resonance linewidths and longitudinal relaxation times were determined as a function of temperature. It was found that the coordinated benzene ligands underwent reorientational motion around their sixfold symmetry axes both at room temperature and at liquid nitrogen temperature. The activation energy for this motion was calculated to be ca. 5.9 kJ mol^{-1}, a value similar to those found in other, simpler complexes containing two arene groups. The results of the NMR study were interpreted in terms of extremely delocalized bonding between the palladium atoms and the benzene ligands. It was further suggested that the arrangement of the arene units with respect to the dipalladium core was mainly due to crystal packing effects.

Examples of both η^2- and η^4-arene coordination to palladium have been described. Complexes containing intramolecular η^2-arene interactions have been isolated during investigations of the catalytic reactions of norbornadiene and norbornene with aryl iodides in the presence of $[PdCl_2(PPh_3)_2]$ and zinc metal,[466] shown for the case with norbornadiene in Equation (51).

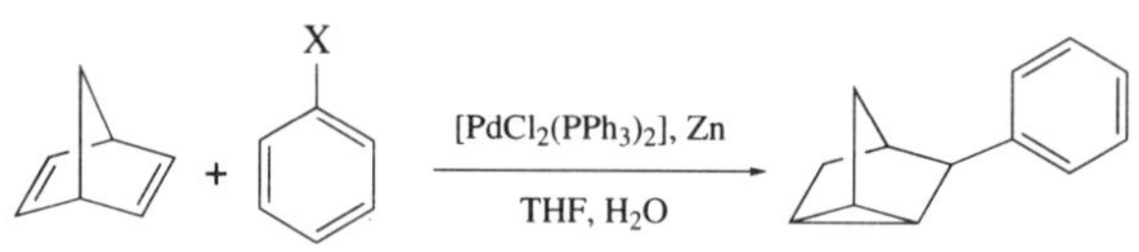

(51)

It was found that an η^2-arene complex could be isolated from the catalytic system or prepared independently from the reaction of $[PdI(Ar)(PPh_3)_2]$ with excess norbornadiene or the reaction of $[Pd(PPh_3)_4]$ with norbornadiene and aryl iodide. The latter reaction is illustrated by Equation (52).

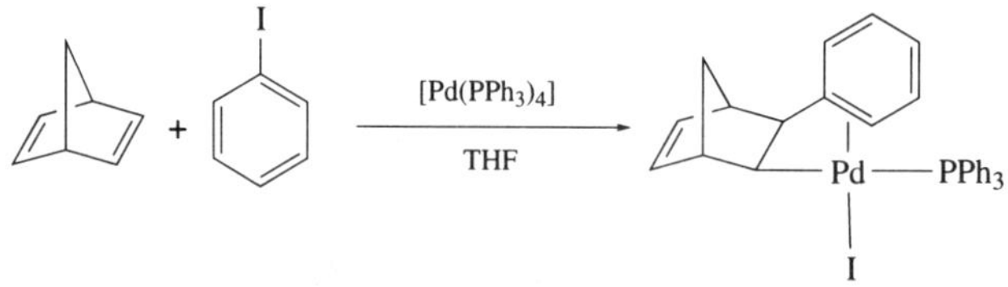

(52)

An x-ray structure of the product of Equation (52) showed η^2-coordination of the aryl group attached to the norbornenyl moiety. A number of substituted aryl iodides gave analogous products. Interestingly, addition of nucleophiles such as PPh_3, dppe, or pyridine did not lead to displacement of the bound arene group from palladium and the compounds were found to be quite stable in air.

The insertion of disubstituted alkynes in the palladium–carbon bonds of certain cyclopalladated compounds has led to the formation of products with either η^2- or η^4-coordination of aryl groups. One such example involving η^4-coordination is illustrated in Scheme 13. The reactions of hexafluorobut-2-yne with cyclopalladated compounds of the 1-methoxynaphth-8-yl chelate, which forms relatively labile palladium–oxygen bonds, have been reported[467] to lead to the products (**135**), (**136**), and (**137**), each of which involves η^2-binding of an aryl group.

Compound (**135**) was characterized by x-ray crystallography which confirmed the presence of an η^2-aryl interaction with palladium. Compound (**137**) was found to react with carbon monoxide by displacement of the coordinated amine group, and not by displacement of the η^2-arene, to generate (**138**).

The reaction of benzyldimethylamine with *cis*-$[Pd(C_6F_5)_2(THF)_2]$ in chloroform has been shown to produce *cis*-$[Pd(C_6F_5)_2(PhCH_2NMe_2)]$ in 74% yield. The x-ray crystal structure of the complex[468] revealed a most unusual η^1-arene interaction between palladium and the *ipso*-carbon of the benzyldimethylamine ligand. Thus, the *cis*-$Pd(C_6F_5)_2$ moiety was coordinated to the nitrogen atom of

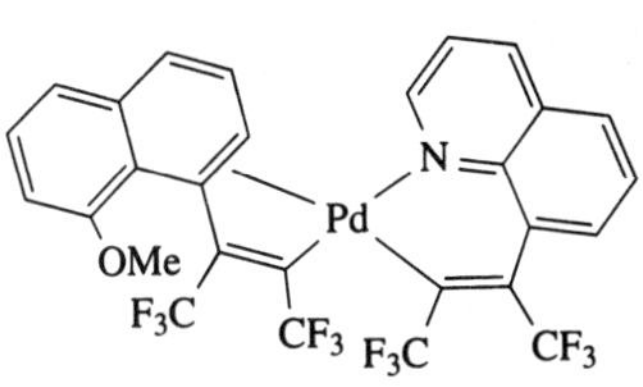

(135)

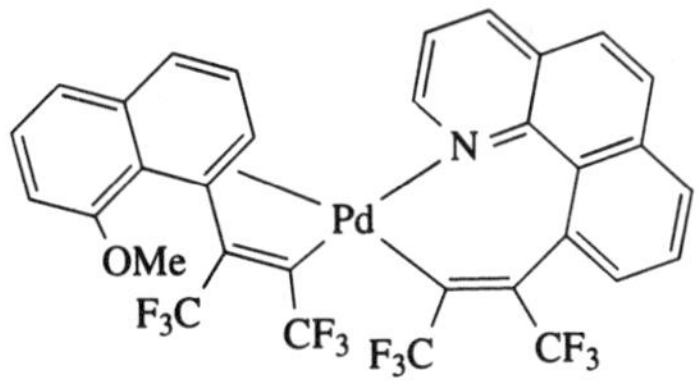

(136)

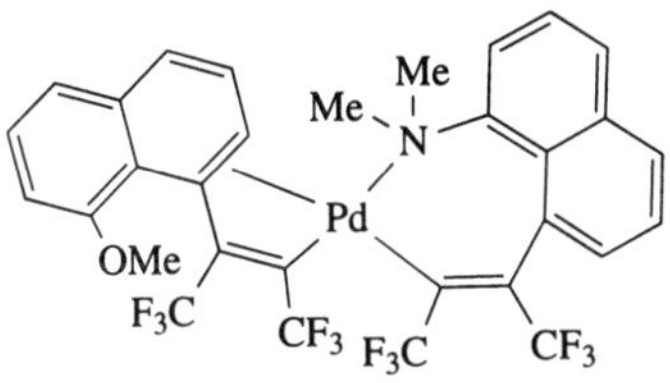

(137)

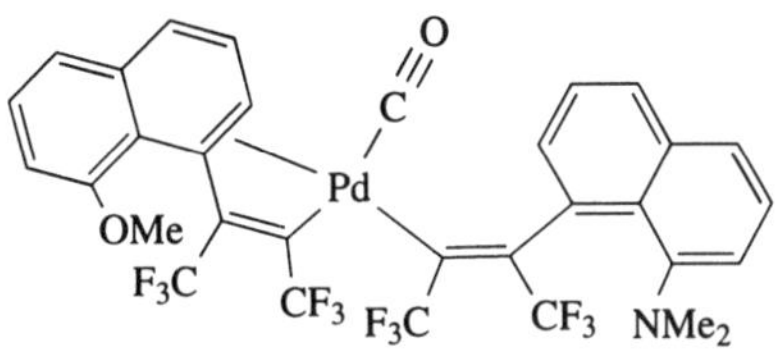

(138)

benzyldimethylamine in the normal fashion but the phenyl ring of the ligand was found to lie at ca. 92° to the least-squares plane containing palladium, nitrogen, and the two *ipso*-carbon atoms of the pentafluorophenyl groups. The palladium atom was found to lie over the phenyl group *ipso*-carbon atom of the benzyldimethylamine ligand and outside bonding distances to the adjacent carbon atoms, that is, η^2-coordination was excluded. With the palladium atom lying at right angles to the phenyl ring, directly over one carbon atom, the suggestion was made that palladium must interact with the $p(\pi)$-orbital of that carbon atom.

The reaction of phenol with $[Pd(PBu^t_3)_2]$ in hexane in normal light (but not in the dark) has been reported[469] to produce (**139**) in 8% yield.

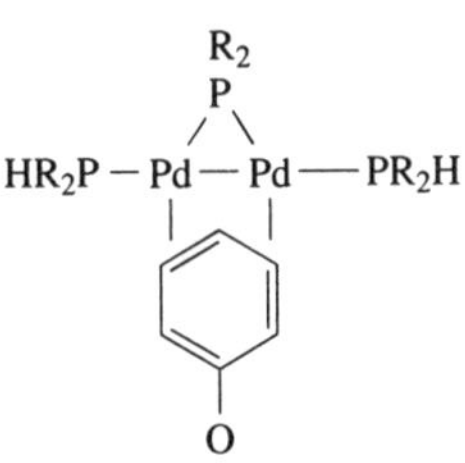

R = Bu^t

(139)

The dinuclear, zwitterionic compound $[Pd_2(\mu\text{-}PBu^t_2)(\mu\text{-}\eta^2\text{:}\eta^2\text{-}C_6H_5O)(PBu^t_2H)_2]\cdot 3\,PhOH$ was characterized by x-ray crystallography and found to contain a bound arene moiety in the form of a μ-η^2:η^2-phenoxo group. Not shown in (**139**) is the extended hydrogen-bonding network that involved the additional phenol molecules shown in the molecular formula. Such hydrogen-bonding features are common to many crystalline aryloxide complexes.

6.7 REFERENCES

1. P. M. Maitlis, P. Espinet and M. J. H. Russell, in 'COMC-I', vol. 6, p. 233.
2. P. M. Maitlis, P. Espinet and M. J. H. Russell, in 'COMC-I', vol. 6, p. 243.
3. P. M. Maitlis, P. Espinet and M. J. H. Russell, in 'COMC-I', vol. 6, p. 265.
4. P. M. Maitlis, P. Espinet and M. J. H. Russell, in 'COMC-I', vol. 6, p. 351.
5. P. M. Maitlis, P. Espinet and M. J. H. Russell, in 'COMC-I', vol. 6, p. 363.
6. P. M. Maitlis, P. Espinet and M. J. H. Russell, in 'COMC-I', vol. 6, p. 455.
7. P. M. Maitlis, P. Espinet and M. J. H. Russell, in 'COMC-I', vol. 6, p. 385.
8. P. M. Maitlis, P. Espinet and M. J. H. Russell, in 'COMC-I', vol. 6, p. 447.
9. P. M. Maitlis, P. Espinet and M. J. H. Russell, in 'COMC-I', vol. 6, p. 279.
10. G. S. Lewandos, in 'The Chemistry of the Metal–Carbon Bond', eds. F. R. Hartley and S. Patai, Wiley, Chichester, 1982, vol. 1, p. 287.
11. P. Powell, in 'The Chemistry of the Metal–Carbon Bond', eds. F. R. Hartley and S. Patai, Wiley, Chichester, 1982, vol. 1, p. 325.

12. G. Marr and B. W. Rockett, in 'The Chemistry of the Metal–Carbon Bond', eds. F. R. Hartley and S. Patai, Wiley, Chichester, 1982, vol. 1, p. 367.
13. G. Marr and B. W. Rockett, in 'The Chemistry of the Metal–Carbon Bond', eds. F. R. Hartley and S. Patai, Wiley, Chichester, 1982, vol. 1, p. 443.
14. G. Marr and B. W. Rockett, in 'The Chemistry of the Metal–Carbon Bond', eds. F. R. Hartley and S. Patai, Wiley, Chichester, 1982, vol. 1, p. 463.
15. J. Tsuji, in 'The Chemistry of the Metal–Carbon Bond', eds. F. R. Hartley and S. Patai, Wiley, Chichester, 1985, vol. 3, p. 163.
16. G. K. Anderson, in 'Chemistry of the Platinum Group Metals: Recent Developments', ed. F. R. Hartley, Elsevier, Amsterdam, 1991 (vol. 11 in Studies in Inorganic Chemistry), p. 338.
17. P. A. Chaloner, *J. Organomet. Chem.*, 1992, **442**, 271.
18. P. A. Chaloner, *J. Organomet. Chem.*, 1992, **432**, 387.
19. P. A. Chaloner, *J. Organomet. Chem.*, 1989, **374**, 349.
20. P. A. Chaloner, *J. Organomet. Chem.*, 1988, **357**, 51.
21. P. A. Chaloner, *J. Organomet. Chem.*, 1987, **337**, 431.
22. A. J. Canty, *Acc. Chem. Res.*, 1992, **25**, 83.
23. J. A. Davies, in 'The Chemistry of the Metal–Carbon Bond', eds. F. R. Hartley and S. Patai, Wiley, Chichester, 1982, vol. 1, p. 813.
24. A. E. Derome, 'Modern NMR Techniques for Chemistry Research', Pergamon, Oxford, 1987.
25. C. A. Fyfe, 'Solid State NMR for Chemists', C. F. C. Press, Guelph, 1983.
26. D. R. Salahub and M. C. Zerner (eds.), 'The Challenge of *d* and *f* Electrons: Theory and Computation', American Chemical Society Symposium Series, Washington, 1989, vol. 394.
27. M. R. Albert and J. T. Yates, Jr., 'The Surface Scientist's Guide to Organometallic Chemistry', American Chemical Society, Washington, 1987.
28. Z. Paàl and P. Tètènyi, *Appl. Catal.*, 1981, **1**, 9.
29. P. Sautet and J.-F. Paul, *Catal. Lett.*, 1991, **9**, 245.
30. L. P. Wang, W. T. Tysoe, R. M. Ormerod, R. M. Lambert, H. Hoffmann and F. Zaera, *J. Phys. Chem.*, 1990, **94**, 4236.
31. W. T. Tysoe, G. L. Nyberg and R. M. Lambert, *J. Phys. Chem.*, 1984, **88**, 1960.
32. Y.-T. Wong and R. Hoffmann, *J. Chem. Soc., Faraday Trans.*, 1990, **86**, 4083.
33. E. M. Stuve and R. J. Madix, *J. Phys. Chem.*, 1985, **89**, 105.
34. J. Mink, M. Gal, P. L. Goggin and J. L. Spencer, *J. Mol. Struct.*, 1986, **142**, 467.
35. M. Nishijima, J. Yoshinobu, T. Sekitani and M. Onchi, *J. Chem. Phys.*, 1989, **90**, 5114.
36. J. Yoshinobu, T. Sekitani, M. Onchi and M. Nishijima, *J. Electron Spectrosc. Relat. Phenom.*, 1990, **54–55**, 697.
37. D. I. James and N. Sheppard, *J. Mol. Struct.*, 1982, **80**, 175.
38. M. Udrea, C. Capat, L. Frunza and D. Crisan, *Geterog. Katal.*, 1983, **5**(1), 297 ('Proceedings of the Vth International Symposium on Heterogeneous Catalysis, Varna, 1983', Part 1, p. 297).
39. Y. Yamamoto, S. Nishii and T. Ibuka, *J. Chem. Soc., Perkin Trans. 1*, 1989, 1703.
40. M. De Crescenzi *et al.*, *Solid State Commun.*, 1990, **73**, 251.
41. V. A. Semikolenov, *Russ. Chem. Rev.*, 1992, **61**, 320.
42. O. P. Parenago, G. M. Cherkashin, G. N. Bondarenko, L. P. Shuikina and V. M. Frolov, *Kinet. Katal.*, 1990, **31**, 825.
43. H. Sellers, *J. Comput. Chem.*, 1990, **11**, 754.
44. M. J. S. Dewar, *Bull. Soc. Chim. Fr.*, 1951, **18**, C79.
45. J. Chatt and L. A. Duncanson, *J. Chem. Soc.*, 1953, 2339.
46. P. J. Hay, *J. Am. Chem. Soc.*, 1981, **103**, 1390.
47. R. A. Love, T. F. Koetzle, G. J. B. Williams, L. C. Andrews and R. Bau, *Inorg. Chem.*, 1975, **14**, 2653.
48. M. J. Filatov, O. V. Gritsenko and G. M. Zhidomirov, *J. Mol. Catal.*, 1989, **54**, 462.
49. M. A. Andrews, T. C.-T. Chang, C.-W. F. Cheng, T. J. Emge, K. P. Kelly and T. F. Koetzle, *J. Am. Chem. Soc.*, 1984, **106**, 5913.
50. O. V. Gritsenko, M. I. Mitkov, A. A. Bagatur'yants, G. L. Kamalov and V. B. Kazanskii, *Russ. J. Phys. Chem.*, 1986, **60**, 703.
51. M. F. Rettig, R. M. Wing and G. R. Wiger, *J. Am. Chem. Soc.*, 1981, **103**, 2980.
52. G. R. Wiger, S. S. Tomita, M. F. Rettig and R. M. Wing, *Organometallics*, 1985, **4**, 1157.
53. J. K. K. Sarhan, S.-W. Foong Murray, H. M. Asfour, M. Green, R. M. Wing and M. Parra-Hake, *Inorg. Chem.*, 1986, **25**, 243.
54. C. Mealli, S. Midollini, S. Moneti, L. Sacconi, J. Silvestre and T. A. Albright, *J. Am. Chem. Soc.*, 1982, **104**, 95.
55. R. E. Stanton and J. W. McIver, Jr., *Acc. Chem. Res.*, 1974, **7**, 72.
56. B. E. Mann, *Chem. Soc. Rev.*, 1986, **15**, 167.
57. L.-F. Olsson and A. Olsson, *Acta Chem. Scand.*, 1989, **43**, 938.
58. R. G. Denning, F. R. Hartley and L. M. Venanzi, *J. Chem. Soc. (A)*, 1967, 1322.
59. D. W. Wertz and M. A. Moseley, *Inorg. Chem.*, 1980, **19**, 705.
60. D. W. Wertz and M. A. Moseley, *Spectrochim. Acta, Part A*, 1980, **36**, 467.
61. H. Kurosawa, N. Asada, A. Urabe and M. Emoto, *J. Organomet. Chem.*, 1984, **272**, 321.
62. K. Miki, M. Yama, Y. Kai and N. Kasai, *J. Organomet. Chem.*, 1982, **239**, 417.
63. H. Tanaka and H. Kawazura, *Bull. Chem. Soc. Jpn.*, 1980, **53**, 1743.
64. R. Benn *et al.*, *Z. Naturforsch., Teil B*, 1991, **46**, 1395.
65. N. Ito, T. Saji and S. Aoyagui, *J. Electroanal. Chem.*, 1983, **144**, 153.
66. N. Ito, T. Saji and S. Aoyagui, *Bull. Chem. Soc. Jpn.*, 1985, **58**, 2323.
67. K. Miki, O. Shiotani, Y. Kai, N. Kasai, H. Kanatani and H. Kurosawa, *Organometallics*, 1983, **2**, 585.
68. S. C. Nyburg, K. Simpson and W. Wong-Ng, *J. Chem. Soc., Dalton Trans.*, 1976, 1865.
69. B. L. Booth, S. Casey and R. N. Haszeldine, *J. Organomet. Chem.*, 1982, **226**, 289.
70. B. L. Booth, S. Casey, R. P. Critchley and R. N. Haszeldine, *J. Organomet. Chem.*, 1982, **226**, 301.
71. M. Hodgson, D. Parker, R. J. Taylor and G. Ferguson, *J. Chem. Soc., Chem. Commun.*, 1987, 1309.
72. R. S. Paonessa, A. L. Prignano and W. C. Trogler, *Organometallics*, 1985, **4**, 647.

73. W. C. Trogler, in 'Excited States and Reactive Intermediates', American Chemical Society Symposium Series, Washington, 1986, vol. 307, p. 177.
74. L. V. Rybin *et al.*, *J. Organomet. Chem.*, 1985, **288**, 119.
75. B. Crociani, F. Di Bianca, P. Uguagliati, L. Canovese and A. Berton, *J. Chem. Soc., Dalton Trans.*, 1991, 71.
76. K. J. Cavell, D. J. Stufkens and K. Vrieze, *Inorg. Chim. Acta*, 1980, **47**, 67.
77. M. A. Bennett, C. Chiraratvatana, G. B. Robertson and U. Tooptakong, *Organometallics*, 1988, **7**, 1403.
78. H. Werner, G. T. Crisp, P. W. Jolly, H.-J. Kraus and C. Krüger, *Organometallics*, 1983, **2**, 1369.
79. H.-J. Kraus and H. Werner, *Angew. Chem.*, 1982, **94**, 871.
80. G. T. Crisp, S. Holle, P. W. Jolly, R. Mynott and H. Werner, *Z. Naturforsch., Teil B*, 1984, **39**, 990.
81. N. Carr, B. J. Dunne, L. Mole, A. G. Orpen and J. L. Spencer, *J. Chem. Soc., Dalton Trans.*, 1991, 863.
82. J. Krause, W. Bonrath and K. R. Pörschke, *Organometallics*, 1992, **11**, 1158.
83. M. Zettlitzer, H. tom Dieck and L. Stamp, *Z. Naturforsch., Teil B*, 1986, **41**, 1230.
84. K. Hiraki, K. Sugino and M. Onishi, *Bull. Chem. Soc. Jpn.*, 1980, **53**, 1976.
85. N. Steiner, U. Nagel and W. Beck, *Chem. Ber.*, 1988, **121**, 1759.
86. R. McCrindle, G. Ferguson, A. J. McAlees and B. L. Ruhl, *J. Organomet. Chem.*, 1981, **204**, 273.
87. R. McCrindle, G. Ferguson, M. A. Khan, A. J. McAlees and B. L. Ruhl, *J. Chem. Soc., Dalton Trans.*, 1981, 986.
88. G. Ferguson and B. L. Ruhl, *Acta Crystallogr., Part C*, 1984, **40**, 2020.
89. E. W. Abel, D. G. Evans, J. R. Koe, V. Sik, P. A. Bates and M. B. Hursthouse, *J. Chem. Soc., Dalton Trans.*, 1989, 985.
90. R. McCrindle, E. C. Alyea, G. Ferguson, A. A. Dias, A. J. McAlees and M. Parvez, *J. Chem. Soc., Dalton Trans.*, 1980, 137.
91. G. R. Newkome, W. E. Puckett, V. K. Gupta and G. E. Kiefer, *Chem. Rev.*, 1986, **86**, 451.
92. D. W. Evans, G. R. Baker and G. R. Newkome, *Coord. Chem. Rev.*, 1989, **93**, 155.
93. J. Albert, J. Granell, J. Sales and X. Solans, *J. Organomet. Chem.*, 1989, **379**, 177.
94. A. L. Rheingold, G. Wu and R. F. Heck, *Inorg. Chim. Acta*, 1987, **131**, 147.
95. V. G. Albano, F. Demartin, A. De Renzi, G. Morelli and A. Saporito, *Inorg. Chem.*, 1985, **24**, 2032.
96. V. G. Albano, C. Castellari, M. E. Cucciolito, A. Panunzi and A. Vitagliano, *Organometallics*, 1990, **9**, 1269.
97. H. Kurosawa, T. Majima and N. Asada, *J. Am. Chem. Soc.*, 1980, **102**, 6996.
98. M. Hiramatsu, K. Shiozaki, T. Fujinami and S. Sakai, *J. Organomet. Chem.*, 1983, **246**, 203.
99. M. J. Chetcuti *et al.*, *J. Chem. Soc., Dalton Trans.*, 1981, 284.
100. M. Hiramatsu, T. Fujinami and S. Sakai, *J. Organomet. Chem.*, 1981, **218**, 409.
101. G. A. Lane, W. E. Geiger and N. G. Connelly, *J. Am. Chem. Soc.*, 1987, **109**, 402.
102. J. A. DeGray, W. E. Geiger, G. A. Lane and P. H. Rieger, *J. Am. Chem. Soc.*, 1991, **30**, 4100.
103. C. T. Bailey and G. C. Lisensky, *J. Chem. Educ.*, 1985, **62**, 896.
104. K. R. Nagasundara, N. M. N. Gowda and G. K. N. Reddy, *Indian J. Chem., Sect. A*, 1980, **19**, 1194.
105. M. Green, D. M. Grove, J. L. Spencer and F. G. A. Stone, *J. Chem. Soc., Dalton Trans.*, 1977, 2228.
106. A. C. Albéniz, P. Espinet, Y. Jeannin, M. Philoche-Levisalles and B. E. Mann, *J. Am. Chem. Soc.*, 1990, **112**, 6594.
107. A. C. Albéniz and P. Espinet, *Organometallics*, 1991, **10**, 2987.
108. M. Parra-Hake, M. F. Rettig and R. M. Wing, *Organometallics*, 1983, **2**, 1013.
109. R. Usón, J. Forniés, M. Tomás, B. Menjón and A. J. Welch, *Organometallics*, 1988, **7**, 1318.
110. A. Salzer, T. Egolf and W. von Philipsborn, *J. Organomet. Chem.*, 1981, **221**, 351.
111. P. J. Ridgwell, P. M. Bailey and P. M. Maitlis, *J. Organomet. Chem.*, 1982, **233**, 373.
112. J. D'Angelo, J. Ficini, S. Martinon, C. Riche and A. Sevin, *J. Organomet. Chem.*, 1977, **177**, 265.
113. H. Hoberg, H. J. Riegel and K. Seevogel, *J. Organomet. Chem.*, 1982, **229**, 281.
114. S. Park, D. Hedden, A. L. Rheingold and D. M. Roundhill, *Organometallics*, 1986, **5**, 1305.
115. L. Wei, A. Bell, S. Warner, I. D. Williams and S. J. Lippard, *J. Am. Chem. Soc.*, 1986, **108**, 8302.
116. O. Eisenstein and R. Hoffmann, *J. Am. Chem. Soc.*, 1980, **102**, 6148.
117. O. Eisenstein and R. Hoffmann, *J. Am. Chem. Soc.*, 1981, **103**, 4308.
118. L. L. Wright, R. M. Wing and M. F. Rettig, *J. Am. Chem. Soc.*, 1982, **104**, 610.
119. C. Moberg, L. Sutin, I. Csöregh and A. Heumann, *Organometallics*, 1990, **9**, 974.
120. C. Moberg, L. Sutin and A. Heumann, *Acta Chem. Scand.*, 1991, **45**, 77.
121. C. Moberg and L. Sutin, *Acta Chem. Scand.*, 1992, **46**, 1000.
122. J.-E. Bäckvall and E. E. Björkman, *Acta Chem. Scand., Ser. B*, 1984, **38**, 91.
123. D. J. Evans and L. A. P. Kane-Maguire, *J. Organomet. Chem.*, 1986, **312**, C24.
124. J.-E. Bäckvall, E. E. Björkman, L. Pettersson and P. Siegbahn, *J. Am. Chem. Soc.*, 1984, **106**, 4369.
125. J.-E. Bäckvall, E. E. Björkman, L. Pettersson and P. Siegbahn, *J. Am. Chem. Soc.*, 1985, **107**, 7265.
126. N. Koga and K. Morokuma, in 'Quantum Chemistry: The Challenge of Transition Metals and Coordination Chemistry', ed. A. Veillard, NATO ASI Series, Series C, North Atlantic Treaty Organization, Brussels, 1986, p. 351.
127. P. E. M. Siegbahn, *J. Am. Chem. Soc.*, 1993, **115**, 5803.
128. P. M. Henry, 'Palladium Catalyzed Oxidation of Hydrocarbons', Reidel, Dordrecht, 1980.
129. N. Gregor, K. Zaw and P. M. Henry, *Organometallics*, 1984, **3**, 1251.
130. W. K. Wan, K. Zaw and P. M. Henry, *Organometallics*, 1988, **7**, 1677.
131. J. W. Francis and P. M. Henry, *Organometallics*, 1991, **10**, 3498.
132. J. W. Francis and P. M. Henry, *Organometallics*, 1992, **11**, 2832.
133. K. Zaw and P. M. Henry, *Organometallics*, 1992, **11**, 2008.
134. K. Undheim and T. Benneche, *Heterocycles*, 1990, **30**, 1155.
135. Y. Rubin, C. B. Knobler and F. Diederich, *J. Am. Chem. Soc.*, 1990, **112**, 1607.
136. C. L. Sterzo and J. K. Stille, *Organometallics*, 1990, **9**, 687.
137. V. Farina and B. Krishnan, *J. Am. Chem. Soc.*, 1991, **113**, 9585.
138. J. M. Brown and N. A. Cooley, *J. Chem. Soc., Chem. Commun.*, 1988, 1345.
139. P. J. Ridgwell, P. M. Bailey, S. N. Wetherell, E. A. Kelley and P. M. Maitlis, *J. Chem. Soc., Dalton Trans.*, 1982, 999.
140. A. Gavezzotti, E. Ortoleva and M. Simonetta, *J. Chem. Soc., Faraday Trans. 1*, 1982, **78**, 425.
141. A. Gavezzotti, E. Ortoleva and M. Simonetta, *Nouv. J. Chim.*, 1983, **7**, 137.
142. M. Simonetta and A. Gavezzotti, *J. Mol. Struct. (Theochem.)*, 1984, **107**, 75.

143. R. J. Koestner, M. A. Van Hove and G. A. Somorjai, *J. Phys. Chem.*, 1983, **87**, 203.
144. J. A. Gates and L. L. Kesmodel, *Surf. Sci.*, 1983, **124**, 68.
145. L. L. Kesmodel, G. D. Waddill and J. A. Gates, *Surf. Sci.*, 1984, **138**, 464.
146. T. M. Gentle and E. L. Muetterties, *J. Phys. Chem.*, 1983, **87**, 2469.
147. I. Garcia-Cuesta, J. Miralles-Sabater, A. M. Sànchez-de-Meràs, M. Merchàn and I. Nebot-Gil, *Mol. Phys.*, 1989, **66**, 659.
148. P.-O. Widmark, G. J. Sexton and B. O. Roos, *J. Mol. Struct. (Theochem.)*, 1986, **135**, 235.
149. J. A. McGinnety, *J. Chem. Soc., Dalton Trans.*, 1974, 1038.
150. D. H. Farrar and N. C. Payne, *J. Organomet. Chem.*, 1981, **220**, 239.
151. B. W. Davies and N. C. Payne, *Inorg. Chem.*, 1974, **13**, 1848.
152. T. Ziegler, *Inorg. Chem.*, 1985, **24**, 1547.
153. R. Usón, J. Forniés, M. Tomás, B. Menjón and A. J. Welch, *J. Organomet. Chem.*, 1986, **304**, C24.
154. Y. Pan, J. T. Mague and M. J. Fink, *Organometallics*, 1992, **11**, 3495.
155. Y. Pan, J. T. Mague and M. J. Fink, *J. Am. Chem. Soc.*, 1993, **115**, 3842.
156. R. W. Zoellner and K. J. Klabunde, *Chem. Rev.*, 1984, **84**, 545.
157. S. J. Higgins and B. L. Shaw, *J. Chem. Soc., Dalton Trans.*, 1988, 457.
158. J. A. Davies, A. A. Pinkerton, M. Vilmer and R. Syed, *J. Chem. Soc., Chem. Commun.*, 1988, 47.
159. M. P. Vilmer, Ph.D. Thesis, University of Toledo, 1991.
160. S. Zheng, M.Sc. Thesis, University of Toledo, 1993.
161. J. A. Davies, K. Kirschbaum and C. Kluwe, *Organometallics*, 1994, **13**, 3664.
162. M. Rashidi, G. Schoettel, J. J. Vittal and R. J. Puddephatt, *Organometallics*, 1992, **11**, 2224.
163. A. D. Ryabov, *Russ. Chem. Rev.*, 1985, **54**, 153.
164. J. Dupont, M. Pfeffer, M. A. Rotteveel, A. De Cian and J. Fischer, *Organometallics*, 1989, **8**, 1116.
165. F. Maassarani, M. Pfeffer and G. Le Borgne, *J. Chem. Soc., Chem. Commun.*, 1986, 488.
166. F. Maassarani, M. Pfeffer and G. Le Borgne, *Organometallics*, 1987, **6**, 2043.
167. J. Dupont, M. Pfeffer, J.-C. Daran and J. Gouteron, *J. Chem. Soc., Dalton Trans.*, 1988, 2421.
168. W. Tao, L. J. Silverberg, A. L. Rheingold and R. F. Heck, *Organometallics*, 1989, **8**, 2550.
169. M. Pfeffer, M. A. Rotteveel, J.-P. Sutter, A. De Cian and J. Fischer, *J. Organomet. Chem.*, 1989, **371**, C21.
170. P.-O. Norrby, B. Åkermark, F. Haeffner, S. Hansson and M. Blomberg, *J. Am. Chem. Soc.*, 1993, **115**, 4859.
171. M. C. Böhm, R. Gleiter and C. D. Batich, *Helv. Chim. Acta*, 1980, **63**, 990.
172. X. Li, G. M. Bancroft, R. J. Puddephatt, Y. F. Hu, Z. Liu and K. H. Tan, *Inorg. Chem.*, 1992, **31**, 5162.
173. F. Bökman, A. Gogoll, L. G. M. Pettersson, O. Bohman and H. O. G. Siegbahn, *Organometallics*, 1992, **11**, 1784.
174. G. B. Petrov, L. M. Markovich, A. V. Ryabtsev and A. P. Belov, *Zh. Strukt. Khim.*, 1983, **24**, 113.
175. M. Guerra, D. Jones, G. Distefano, S. Torroni, A. Foffani and A. Modelli, *Organometallics*, 1993, **12**, 2203.
176. A. P. Belov and M. K. Andari, *Dokl. Akad. Nauk SSSR*, 1984, **275**, 629.
177. M. K. Andari, A. V. Krylov and A. P. Belov, *Zh. Strukt. Khim.*, 1984, **25**, 167.
178. S. Liu, C. R. Lucas, M. J. Newlands and J.-P. Charland, *Inorg. Chem.*, 1990, **29**, 4380.
179. S. A. Godleski, K. B. Gundlach, H. Y. Ho, E. Keinan and F. Frolow, *Organometallics*, 1984, **3**, 21.
180. L. S. Hegedus, B. Åkermark, D. J. Olsen, O. P. Anderson and K. Zetterberg, *J. Am. Chem. Soc.*, 1982, **104**, 697.
181. N. W. Murrall and A. J. Welch, *J. Organomet. Chem.*, 1986, **301**, 109.
182. V. A. Zschunke, H. Meyer and I. Nehls, *Z. Anorg. Allg. Chem.*, 1982, **494**, 189.
183. H. Meyer and A. Zschunke, *J. Organomet. Chem.*, 1984, **269**, 209.
184. E. Cesarotti, M. Grassi, L. Prati and F. Demartin, *J. Organomet. Chem.*, 1989, **370**, 407.
185. E. Cesarotti, M. Grassi, L. Prati and F. Demartin, *J. Chem. Soc., Dalton Trans.*, 1991, 2073.
186. H. Rüegger, R. W. Kunz, C. J. Ammann and P. S. Pregosin, *Magn. Reson. Chem.*, 1991, **29**, 197.
187. A. Albinati, C. Ammann, P. S. Pregosin and H. Rüegger, *Organometallics*, 1990, **9**, 1826.
188. A. Albinati, R. W. Kunz, C. J. Ammann and P. S. Pregosin, *Organometallics*, 1991, **10**, 1800.
189. A. Togni, G. Rihs, P. S. Pregosin and C. Ammann, *Helv. Chim. Acta*, 1990, **73**, 723.
190. J. W. Faller, C. Blankenship, B. Whitmore and S. Sena, *Inorg. Chem.*, 1985, **24**, 4483.
191. P. W. Jolly, *Angew. Chem., Int. Ed. Engl.*, 1985, **24**, 283.
192. D. R. Chrisope and P. Beak, *J. Am. Chem. Soc.*, 1986, **108**, 334.
193. D. R. Chrisope, P. Beak and W. H. Saunders, Jr., *J. Am. Chem. Soc.*, 1988, **110**, 230.
194. B. M. Trost and P. J. Metzner, *J. Am. Chem. Soc.*, 1980, **102**, 3572.
195. A. D. Ryabov, A. V. Eliseev and A. K. Yatsimirsky, *Appl. Organomet. Chem.*, 1988, **2**, 101.
196. D. J. Collins, W. R. Jackson and R. N. Timms, *Aust. J. Chem.*, 1980, **33**, 2663.
197. D. J. Collins, B. K. M. Gatehouse, W. R. Jackson, G. A. Kakos and R. N. Timms, *J. Chem. Soc., Chem. Commun.*, 1980, 138.
198. P. Hütter, T. Butters, W. Winter, D. Handschuh and W. Voelter, *Liebigs Ann. Chem.*, 1982, 1111.
199. C. Mahé, H. Patin, M.-T. Van Hulle and D. H. R. Barton, *J. Chem. Soc., Perkin Trans. 1*, 1981, 2504.
200. C. A. Horiuchi, T. Takayama, Y. Koike, M. Tanaka and J. Y. Satoh, *Chem. Lett.*, 1989, 277.
201. A. A. Danopoulos and S. M. Paraskewas, *Inorg. Chim. Acta*, 1983, **71**, 259.
202. J. B. Murphy, S. L. Holt, Jr. and E. M. Holt, *Inorg. Chim. Acta*, 1981, **48**, 29.
203. R. C. Larock, K. Takagi, S. S. Hershberger and M. A. Mitchell, *Tetrahedron Lett.*, 1981, **22**, 5231.
204. R. C. Larock and K. Takagi, *J. Org. Chem.*, 1988, **53**, 4329.
205. G. D. Davies, Jr. and A. Hallberg, *Chem. Rev.*, 1989, **89**, 1433.
206. R. C. Larock, D. J. Leuck and L. W. Harrison, *Tetrahedron Lett.*, 1988, **29**, 6399.
207. R. C. Larock (Iowa State University Research Foundation, Inc.), *US Pat.* 4 948 905 A (1990) (*Chem. Abstr.*, 1991, **114**, 61 927b).
208. T. Ohta, T. Hosokawa, S.-I. Murahashi, K. Miki and N. Kasai, *Organometallics*, 1985, **4**, 2080.
209. T. Hayashi, M. Konishi and M. Kumada, *J. Chem. Soc., Chem. Commun.*, 1983, 736.
210. S. Imaizumi, T. Matsuhisa and V. Senda, *J. Organomet. Chem.*, 1985, **280**, 441.
211. T. Hosokawa, Y. Imada and S.-I. Murahashi, *Tetrahedron Lett.*, 1982, **23**, 3373.
212. Y. Tamaru *et al.*, *J. Org. Chem.*, 1983, **48**, 4669.
213. S. Ogoshi, K. Ohe, N. Chatani, H. Kurosawa and S. Murai, *Organometallics*, 1991, **10**, 3813.

214. A. Behr, G. v. Ilsemann, W. Keim, C. Krüger and Y.-H. Tsay, *Organometallics*, 1986, **5**, 514.
215. L. Yu. Ukhin, N. A. Dolgopolova, L. G. Kuz'mina and Yu. T. Struchkov, *J. Organomet. Chem.*, 1981, **210**, 263.
216. F. G. Stakem and R. F. Heck, *J. Org. Chem.*, 1980, **45**, 3584.
217. R. C. Larock and K. Takagi, *Tetrahedron Lett.*, 1983, **24**, 3457.
218. R. C. Larock and K. Takagi, *J. Org. Chem.*, 1984, **49**, 2701.
219. R. C. Larock, L. W. Harrison and M. H. Hsu, *J. Org. Chem.*, 1984, **49**, 3664.
220. A. C. Albéniz, P. Espinet, C. Foces-Foces and F. Cano, *Organometallics*, 1990, **9**, 1079.
221. R. C. Larock, Y. de Lu, A. C. Bain and C. E. Russell, *J. Org. Chem.*, 1991, **56**, 4589.
222. S. S. Hall and B. Åkermark, *Organometallics*, 1984, **3**, 1745.
223. A. Kühn, Ch. Burschka and H. Werner, *Organometallics*, 1982, **1**, 496.
224. J.-E. Bäckvall and E. E. Björkman, *J. Chem. Soc., Chem. Commun.*, 1982, 693.
225. E. E. Björkman and J.-E. Bäckvall, *Acta Chem. Scand. Ser. A*, 1983, **37**, 503.
226. D. Wilhelm, J.-E. Bäckvall, R. E. Nordberg and T. Norin, *Organometallics*, 1985, **4**, 1296.
227. M. Parra-Hake, M. F. Rettig, R. M. Wing and J. C. Woolcock, *Organometallics*, 1982, **1**, 1478.
228. P. R. Clemens, R. P. Hughes and L. D. Margerum, *J. Am. Chem. Soc.*, 1981, **103**, 2428.
229. R. P. Hughes and C. S. Day, *Organometallics*, 1982, **1**, 1221.
230. M. Parra-Hake, M. F. Rettig, J. L. Williams and R. M. Wing, *Organometallics*, 1986, **5**, 1032.
231. R. C. Larock and S. Varaprath, *J. Org. Chem.*, 1984, **49**, 3432.
232. R. C. Larock and S. Varaprath (Iowa State University Research Foundation, Inc.), *US Pat.* 4 632 996 (1986) (*Chem. Abstr.*, 1987, **106**, 138 638c).
233. G. Balme, G. Fournet and J. Gore, *Tetrahedron Lett.*, 1986, **27**, 3855.
234. W. Fischetti and R. F. Heck, *J. Organomet. Chem.*, 1985, **293**, 391.
235. M. Suzuki, S. Sawada and T. Saegusa, *Macromolecules*, 1989, **22**, 1505.
236. R. C. Larock, H. Song, S. Kim and R. A. Jacobson, *J. Chem. Soc., Chem. Commun.*, 1987, 834.
237. Y. Inoue, J. Yamashita and H. Hashimoto, *Synthesis*, 1984, 244.
238. A. Vogler, C. Quett and H. Kunkely, *Ber. Bunsen ges. Phys. Chem., 1988, 92, 1486.*
239. F. Ozawa, J. W. Park, P. B. Mackenzie, W. P. Schaefer, L. M. Henling and R. H. Grubbs, *J. Am. Chem. Soc.*, 1989, **111**, 1319.
240. K. Otsuka, K. Ishizuka, I. Yamanaka and M. Hatano, *J. Electrochem. Soc.*, 1991, **138**, 3176.
241. M. Oshima, I. Shimizu, A. Yamamoto and F. Ozawa, *Organometallics*, 1991, **10**, 1221.
242. F. Ozawa, T.-I. Son, S. Ebina, K. Osakada and A. Yamamoto, *Organometallics*, 1992, **11**, 171.
243. K. Osakada, Y. Ozawa and A. Yamamoto, *J. Organomet. Chem.*, 1990, **399**, 341.
244. T. Hayashi, T. Hagihara, M. Konishi and M. Kumada, *J. Am. Chem. Soc.*, 1983, **105**, 7767.
245. T. Yamamoto, O. Saito and A. Yamamoto, *J. Am. Chem. Soc.*, 1981, **103**, 5600.
246. T. Yamamoto, M. Akimoto, O. Saito and A. Yamamoto, *Organometallics*, 1986, **5**, 1559.
247. I. Starý and P. Kocovský, *J. Am. Chem. Soc.*, 1989, **111**, 4981.
248. I. Starý, J. Zajíček and P. Kocovský, *Tetrahedron*, 1992, **48**, 7229.
249. H. Kurosawa *et al.*, *J. Am. Chem. Soc.*, 1992, **114**, 8417.
250. A. Vitagliano, B. Åkermark and S. Hansson, *Organometallics*, 1991, **10**, 2592.
251. G. C. Nwokogu, *Tetrahedron Lett.*, 1984, **25**, 3263.
252. G. C. Nwokogu, *J. Org. Chem.*, 1985, **50**, 3900.
253. F. Guibé and Y. Saint M'Leux, *Tetrahedron Lett.*, 1981, **22**, 3591.
254. T. Tsuda, Y. Chujo, S. Hishi, K. Tawara and T. Saegusa, *J. Am. Chem. Soc.*, 1980, **102**, 6384.
255. T. Tsuda, M. Okada, S. Nishi and T. Saegusa, *J. Org. Chem.*, 1986, **51**, 421.
256. J. C. Fiaud and L. Aribi-Zouioueche, *Tetrahedron Lett.*, 1982, **23**, 5279.
257. J. Nokami, H. Watanabe, T. Mandai, M. Kawada and J. Tsuji, *Tetrahedron Lett.*, 1989, **30**, 4829.
258. R. L. Halterman and H. L. Nimmons, *Organometallics*, 1990, **9**, 273.
259. A. Goliaszewski and J. Schwartz, *J. Am. Chem. Soc.*, 1984, **106**, 5028.
260. A. Goliaszewski and J. Schwartz, *Tetrahedron*, 1985, **41**, 5779.
261. F. Maassarani, M. Pfeffer and G. van Koten, *Organometallics*, 1989, **8**, 871.
262. G. A. Fox and C. G. Pierpont, *J. Chem. Soc., Chem. Commun.*, 1988, 806.
263. G. A. Fox and C. G. Pierpont, *Inorg. Chem.*, 1992, **31**, 3718.
264. E. R. Milaeva, A. Z. Rubezhov, A. I. Prokof'ev and O. Y. Okhlobystin, *J. Organomet. Chem.*, 1980, **193**, 135.
265. B. Taqui Khan, K. Murali Mohan and S. M. Zakeeruddin, *Organomet. Chem. USSR*, 1990, **3**, 372.
266. G. Facchin, R. Bertani, M. Calligaris, G. Nardin and M. Mari, *J. Chem. Soc., Dalton Trans.*, 1987, 1381.
267. G. Facchin, R. Bertani, L. Zanotto, M. Calligaris and G. Nardin, *J. Organomet. Chem.*, 1989, **366**, 409.
268. A. Musco *et al.*, *Organometallics*, 1988, **7**, 2130.
269. J. Sieler, M. Helms, W. Gaube, A. Svensson and O. Lindqvist, *J. Organomet. Chem.*, 1987, **320**, 129.
270. B. Crociani, R. Bertani, T. Boschi and G. Bandoli, *J. Chem. Soc., Dalton Trans.*, 1982, 1715.
271. A. Tiripicchio, F. J. Lahoz, L. A. Oro, M. T. Pinillos and C. Tejel, *Inorg. Chim. Acta*, 1985, **100**, L5.
272. F. H. Cano, C. Foces-Foces, L. A. Oro, M. T. Pinillos and C. Tejel, *Inorg. Chim. Acta*, 1987, **128**, 75.
273. M. P. Garcia, A. Portilla, L. A. Oro, C. Foces-Foces and F. H. Cano, *J. Organomet. Chem.*, 1987, **322**, 111.
274. C. P. Brock, M. K. Das, R. P. Minton and K. Niedenzu, *J. Am. Chem. Soc.*, 1988, **110**, 817.
275. P. Braunstein, M. Knorr, H. Piana and U. Schubert, *Organometallics*, 1991, **10**, 828.
276. T. J. Henly, S. R. Wilson and J. R. Shapley, *Inorg. Chem.*, 1988, **27**, 2551.
277. D. Fenske, A. Hollnagel and K. Merzweiler, *Z. Naturforsch., Teil B*, 1988, **43**, 634.
278. M. Sawamura, H. Nagata, H. Sakamoto and Y. Ito, *J. Am. Chem. Soc.*, 1992, **114**, 2586.
279. B. Åkermark, S. Hansson and A. Vitagliano, *J. Am. Chem. Soc.*, 1990, **112**, 4587.
280. N. Oshima, Y. Hamatani, H. Fukui, H. Suzuki and Y. Moro-Oka, *J. Organomet. Chem.*, 1986, **303**, C21.
281. P. W. Jolly, C. Krüger, K.-P. Schick and G. Wilke, *Z. Naturforsch., Teil B*, 1980, **35**, 926.
282. A. Kühn and H. Werner, *Chem. Ber.*, 1980, **113**, 2308.
283. W. Keim, R. Appel, A. Storeck, C. Krüger and R. Goddard, *Angew. Chem., Int. Ed. Engl.*, 1981, **20**, 116.
284. D. Fenske and P. Stock, *Angew. Chem., Int. Ed. Engl.*, 1982, **21**, 356.

285. H. Werner, A. Kühn and Ch. Burschka, *Chem. Ber.*, 1980, **113**, 2291.
286. H. Werner and H.-J. Kraus, *Chem. Ber.*, 1980, **113**, 1072.
287. B. Bogdanović, P. Göttsch and M. Rubach, *Z. Naturforsch., Teil B*, 1983, **38**, 599.
288. B. Bogdanović, R. Goddard and M. Rubach, *Acta Crystallogr., Part C*, 1989, **45**, 1511.
289. Y. Castanet and F. Petit, *J. Chem. Res. (S)*, 1982, 238.
290. D. P. Grant, N. W. Murrall and A. J. Welch, *J. Organomet. Chem.*, 1987, **333**, 403.
291. R. Usón *et al.*, *J. Chem. Soc., Dalton Trans.*, 1983, 1729.
292. R. Ciajolo, M. A. Jama, A. Tuzi and A. Vitagliano, *J. Organomet. Chem.*, 1985, **295**, 233.
293. G. A. Kukina, V. S. Sergienko, Yu. L. Gaft, I. A. Zakharova and M. A. Porai-Koshits, *Inorg. Chim. Acta*, 1980, **45**, L257.
294. I. A. Zakharova *et al.*, *Inorg. Chim. Acta*, 1981, **47**, 181.
295. A. J. Deeming, M. N. Meah, P. A. Bates and M. B. Hursthouse, *J. Chem. Soc., Dalton Trans.*, 1988, 235.
296. J. E. Gozum, D. M. Pollina, J. A. Jensen and G. S. Girolami, *J. Am. Chem. Soc.*, 1988, **110**, 2688.
297. G. A. Domrachev, V. A. Varyukhin and B. A. Nesterov, *React. Kinet. Catal. Lett.*, 1983, **22**, 281.
298. A. B. Goel, *Tetrahedron Lett.*, 1984, **25**, 4599.
299. Y.-G. Kim, S. Bialy, R. W. Miller, J. T. Spencer, P. A. Dowben and S. Datta, *Mater. Res. Soc. Symp. Proc.*, 1990, **158**, 103.
300. D. J. Collins, W. R. Jackson and R. N. Timms, *Aust. J. Chem.*, 1980, **33**, 2761.
301. J. Y. Satoh and C. A. Horiuchi, *Bull. Chem. Soc. Jpn.*, 1981, **54**, 625.
302. K. Jitsukawa, K. Kaneda and S. Teranishi, *J. Org. Chem.*, 1983, **48**, 389.
303. K. Zaw, M. Lautens and P. M. Henry, *Organometallics*, 1985, **4**, 1286.
304. R. O. C. Norman, C. B. Thomas and G. Watson, *J. Chem. Soc., Perkin Trans. 2*, 1980, 1099.
305. J. M. Davidson, P. C. Mitchell and N. S. Raghavan, *Front. Chem. React. Eng.*, 1984, **1**, 300.
306. J. E. Lyons, G. Suld and C.-Y. Hsu, in 'Homogeneous Heterogeneous Catalysis, Proceedings of the 5th International Symposium on Relatively Homogeneous Heterogeneous Catalysis', eds. Yu. I. Ermakov and V. A. Likholobov, VNU Science Press, Utrecht, 1986, p. 117.
307. J. E. Lyons, *Catalysis Today*, 1988, **3**, 245.
308. S. Mashhood Ali, S. Tanimoto and T. Okamoto, *J. Org. Chem.*, 1988, **53**, 3639.
309. L. S. Hegedus, N. Kambe, Y. Ishii and A. Mori, *J. Org. Chem.*, 1985, **50**, 2240.
310. S. Mashhood Ali, S. Tanimoto and T. Okamoto, *Bull. Inst. Chem. Res., Kyoto Univ.*, 1989, **67**, 89.
311. T. Takemoto, Y. Nishikimi, M. Sodeoka and M. Shibasaki, *Tetrahedron Lett.*, 1992, **33**, 3531.
312. F. Guibé, A.-M. Zigna and G. Balavoine, *J. Organomet. Chem.*, 1986, **306**, 257.
313. M. W. Hutzinger and A. C. Oehlschlager, *J. Org. Chem.*, 1991, **56**, 2918.
314. M. Oshima, H. Yamazaki, I. Shimizu, M. Nisar and J. Tsuji, *J. Am. Chem. Soc.*, 1989, **111**, 6280.
315. R. Grigg, S. Sukirthalingam and V. Sridharan, *Tetrahedron Lett.*, 1991, **32**, 2545.
316. B. Burns *et al.*, *Tetrahedron Lett.*, 1988, **29**, 5565.
317. B. Burns, R. Grigg, V. Sridharan, P. Stevenson, S. Sukirthalingam and T. Worakun, *Tetrahedron Lett.*, 1989, **30**, 1135.
318. R. Grigg, V. Sridharan, S. Sukirthalingam and T. Worakun, *Tetrahedron Lett.*, 1989, **30**, 1139.
319. B. Burns, R. Grigg, V. Santhakumar, V. Sridharan, P. Stevenson and T. Worakun, *Tetrahedron*, 1992, **48**, 7297.
320. B. Burns, R. Grigg, P. Ratananukul, V. Sridharan, P. Stevenson and T. Worakun, *Tetrahedron Lett.*, 1988, **29**, 4329.
321. Y. Masuyama, J. P. Takahara and Y. Kurusu, *J. Am. Chem. Soc.*, 1988, **110**, 4473.
322. J. P. Takahara, Y. Masuyama and Y. Kurusu, *J. Am. Chem. Soc.*, 1992, **114**, 2577.
323. T. Tabuchi, J. Inanaga and M. Yamaguchi, *Tetrahedron Lett.*, 1987, **28**, 215.
324. Y. Masuyama, N. Kinugawa and Y. Kurusu, *J. Org. Chem.*, 1987, **52**, 3702.
325. B. M. Trost and G. B. Tometzki, *J. Org. Chem.*, 1988, **53**, 915.
326. W. A. Donaldson, *Tetrahedron*, 1987, **43**, 2901.
327. Y. Tamaru, M. Kagotani, R. Suzuki and Z. Yoshida, *J. Org. Chem.*, 1981, **46**, 3376.
328. Y. Hanzawa, S. Ishizawa, Y. Kobayashi and T. Tabuchi, *Chem. Pharm. Bull.*, 1990, **38**, 1104.
329. R. C. Larock, K. Takagi, J. P. Burkhart and S. S. Hershberger, *Tetrahedron*, 1986, **42**, 3759.
330. K. Yamamoto *et al.*, *Chem. Lett.*, 1989, 955.
331. N. C. Ihle and C. H. Heathcock, *J. Org. Chem.*, 1993, **58**, 560.
332. D. Milstein, *Organometallics*, 1982, **1**, 888.
333. W. Keim, J. Becker, P. Kraneburg and R. Greven, *J. Mol. Catal.*, 1989, **54**, 37.
334. M. Grassi, S. V. Meille, A. Musco, R. Pontellini and A. Sironi, *J. Chem. Soc., Dalton Trans.*, 1989, 615.
335. Yu. G. Noskov, N. A. Novikov, M. I. Terekhova and É. S. Petrov, *Kinet. Katal.*, 1991, **32**, 331.
336. F. Ozawa, T.-I. Son, K. Osakada and A. Yamamoto, *J. Chem. Soc., Chem. Commun.*, 1989, 1067.
337. T. Hayashi, K. Kanehira, H. Tsuchiya and M. Kumada, *J. Chem. Soc., Chem. Commun.*, 1982, 1162.
338. P. R. Auburn, P. B. Mackenzie and B. Bosnich, *J. Am. Chem. Soc.*, 1985, **107**, 2033.
339. P. B. Mackenzie, J. Whelan and B. Bosnich, *J. Am. Chem. Soc.*, 1985, **107**, 2046.
340. D. H. Farrar and N. C. Payne, *J. Am. Chem. Soc.*, 1985, **107**, 2054.
341. Y. Ito, M. Sawamura, M. Matsuoka, Y. Matsumoto and T. Hayashi, *Tetrahedron Lett.*, 1987, **28**, 4849.
342. T. Hayashi, K. Kanehira, T. Hagihara and M. Kumada, *J. Org. Chem.*, 1988, **53**, 113.
343. T. Hayashi, A. Yamamoto, Y. Ito, E. Nishioka, H. Miura and K. Yanagi, *J. Am. Chem. Soc.*, 1989, **111**, 6301.
344. T. Hayashi, K. Kishi, A. Yamamoto and Y. Ito, *Tetrahedron Lett.*, 1990, **31**, 1743.
345. M. Yamaguchi, T. Shima, T. Yamagishi and M. Hida, *Tetrahedron Lett.*, 1990, **31**, 5049.
346. K. Hiroi and J. Abe, *Chem. Pharm. Bull.*, 1991, **39**, 616.
347. O. Reiser, *Angew. Chem., Int. Ed. Engl.*, 1993, **32**, 547.
348. P. von Matt and A. Pfaltz, *Angew. Chem., Int. Ed. Engl.*, 1993, **32**, 566.
349. G. Consiglio and R. M. Waymouth, *Chem. Rev.*, 1989, **89**, 257.
350. S. Sakaki, M. Nishikawa and A. Ohyoshi, *J. Am. Chem. Soc.*, 1980, **102**, 4062.
351. L. S. Hegedus, W. H. Darlington and C. E. Russell, *J. Org. Chem.*, 1980, **45**, 5193.
352. C. Carfagna, L. Mariani, A. Musco, G. Sallese and R. Santi, *J. Org. Chem.*, 1991, **56**, 3924.
353. H. M. R. Hoffmann, A. R. Otte and A. Wilde, *Angew. Chem., Int. Ed. Engl.*, 1992, **31**, 234.
354. M. Formica, A. Musco, R. Pontellini, K. Linn and C. Mealli, *J. Organomet. Chem.*, 1993, **448**, C6.
355. C. Carfagna, R. Galarini, A. Musco and R. Santi, *J. Mol. Catal.*, 1992, **72**, 19.

356. C. Carfagna, R. Galarini, K. Linn, J. A. López, C. Mealli and A. Musco, *Organometallics*, 1993, **12**, 3019.
357. C. A. Horiuchi and J. Y. Satoh, *J. Chem. Soc., Perkin Trans. 1*, 1982, 2595.
358. E. Keinan and M. Sahai, *J. Chem. Soc., Chem. Commun.*, 1984, 648.
359. A. Stolle, J. Salaün and A. de Meijere, *Synlett*, 1991, 327.
360. A. Arcadi, E. Bernocchi, S. Cacchi, L. Caglioti and F. Marinelli, *Gazz. Chim. Ital.*, 1991, **121**, 369.
361. W. A. Donaldson, D. J. Stepuszek and J. A. Gruetzmacher, *Tetrahedron*, 1990, **46**, 2273.
362. T. Hayashi, M. Konishi, K.-I. Yokota and M. Kumada, *J. Organomet. Chem.*, 1985, **285**, 359.
363. W. A. Donaldson and J. Wang, *J. Organomet. Chem.*, 1990, **395**, 113.
364. L. S. Hegedus and R. Tamura, *Organometallics*, 1982, **1**, 1188.
365. A. Stolle, J. Ollivier, P. P. Piras, J. Salaün and A. de Meijere, *J. Am. Chem. Soc.*, 1992, **114**, 4051.
366. H. Kurosawa, M. Emoto, A. Urabe, K. Miki and N. Kasai, *J. Am. Chem. Soc.*, 1985, **107**, 8253.
367. H. Kurosawa *et al.*, *J. Am. Chem. Soc.*, 1987, **109**, 6333.
368. B. Åkermark, S. Hansson, B. Krakenberger, A. Vitagliano and K. Zetterberg, *Organometallics*, 1984, **3**, 679.
369. J.-E. Bäckvall, R. E. Nordberg, K. Zetterberg and B. Åkermark, *Organometallics*, 1983, **2**, 1625.
370. J. S. Temple and J. Schwartz, *J. Am. Chem. Soc.*, 1980, **102**, 7382.
371. J. S. Temple, M. Riediker and J. Schwartz, *J. Am. Chem. Soc.*, 1982, **104**, 1310.
372. B. Åkermark, B. Krakenberger, S. Hansson and A. Vitagliano, *Organometallics*, 1987, **6**, 620.
373. S. A. Godleski and R. S. Valpey, *J. Org. Chem.*, 1982, **47**, 381.
374. G. Fournet, G. Balme, J. J. Barieux and J. Gore, *Tetrahedron*, 1988, **44**, 5821.
375. G. Fournet, G. Balme and J. Gore, *Tetrahedron Lett.*, 1987, **28**, 4533.
376. E. Keinan and M. Peretz, *Tetrahedron*, 1983, **48**, 5302.
377. T. Hayashi, A. Yamamoto and Y. Ito, *Chem. Lett.*, 1987, 177.
378. M. Prat, J. Ribas and M. Moreno-Mañas, *Tetrahedron*, 1992, **48**, 1695.
379. R. Tanikaga, T. X. Jun and A. Kaji, *J. Chem. Soc., Perkin Trans. 1*, 1990, 1185.
380. D. R. Deardorff, D. C. Myles and K. D. Macferrin, *Tetrahedron Lett.*, 1985, **26**, 5615.
381. D. R. Deardorff, R. G. Linde, II, A. M. Martin and M. J. Shulman, *J. Org. Chem.*, 1989, **54**, 2759.
382. R. S. Valpey, D. J. Miller, J. M. Estes and S. A. Godleski, *J. Org. Chem.*, 1982, **47**, 4717.
383. Y. Tamaru, Z. Yoshida, Y. Yamada, K. Mukai and H. Yoshioka, *J. Org. Chem.*, 1983, **48**, 1293.
384. Y. I. M. Nilsson, P. G. Andersson and J.-E. Bäckvall, *J. Am. Chem. Soc.*, 1993, **115**, 6609.
385. J. C. Fiaud and J. L. Malleron, *Tetrahedron Lett.*, 1981, **22**, 1399.
386. T. Hayashi, M. Konishi and M. Kumada, *J. Chem. Soc., Chem. Commun.*, 1984, 107.
387. T. Hayashi, A. Yamamoto and Y. Ito, *J. Organomet. Chem.*, 1988, **338**, 261.
388. H. Kurosawa, K. Ishii, Y. Kawasaki and S. Murai, *Organometallics*, 1989, **8**, 1756.
389. B. Åkermark and A. Jutand, *J. Organomet. Chem.*, 1981, **217**, C41.
390. H. Grennberg, V. Langer and J.-E. Bäckvall, *J. Chem. Soc., Chem. Commun.*, 1991, 1190.
391. J.-E. Bäckvall, R. E. Nordberg, E. E. Björkman and C. Moberg, *J. Chem. Soc., Chem. Commun.*, 1980, 943.
392. J.-E. Bäckvall, R. E. Nordberg and D. Wilhelm, *J. Am. Chem. Soc.*, 1985, **107**, 6892.
393. E. Keinan, M. Sahai, Z. Roth, A. Nudelman and J. Herzig, *J. Org. Chem.*, 1985, **50**, 3558.
394. J.-E. Bäckvall, K. L. Granberg and A. Heumann, *Isr. J. Chem.*, 1991, **31**, 17.
395. K. L. Granberg and J.-E. Bäckvall, *J. Am. Chem. Soc.*, 1992, **114**, 6858 (see correction: *J. Am. Chem. Soc.*, 1994, **116**, 10 853)
396. S. A. Stanton, S. W. Felman, C. S. Parkhurst and S. A. Godleski, *J. Am. Chem. Soc.*, 1983, **105**, 1964.
397. L. S. Hegedus, N. Kambe, R. Tamura and P. D. Woodgate, *Organometallics*, 1983, **2**, 1658.
398. M. Ahmar, J.-J. Barieux, B. Cazes and J. Gore, *Tetrahedron*, 1987, **43**, 513.
399. M. Ahmar, B. Cazes and J. Gore, *Tetrahedron*, 1987, **43**, 3453.
400. B. Cazes, V. Colovray and J. Gore, *Tetrahedron Lett.*, 1988, **29**, 627.
401. N. Chaptal, V. Colovray-Gotteland, C. Grandjean, B. Cazes and J. Gore, *Tetrahedron Lett.*, 1991, **32**, 1795.
402. B. Friess, B. Cazes and J. Gore, *Bull. Soc. Chim. Fr.*, 1992, **129**, 273.
403. J.-E. Bäckvall and R. E. Nordberg, *J. Am. Chem. Soc.*, 1981, **103**, 4959.
404. J.-E. Bäckvall, R. E. Nordberg and J.-E. Nyström, *Tetrahedron Lett.*, 1982, **23**, 1617.
405. J.-E. Bäckvall, S. E. Byström and R. E. Nordberg, *J. Org. Chem.*, 1984, **49**, 4619.
406. J.-E. Bäckvall and A. Gogoll, *Tetrahedron Lett.*, 1988, **29**, 2243.
407. J.-E. Bäckvall and J. O. Vågberg, *J. Org. Chem.*, 1988, **53**, 5695.
408. J.-E. Bäckvall, R. B. Hopkins, H. Grennberg, M. M. Mader and A. K. Awasthi, *J. Am. Chem. Soc.*, 1990, **112**, 5160.
409. H. Grennberg, A. Gogoll and J.-E. Bäckvall, *J. Org. Chem.*, 1991, **56**, 5808.
410. J.-E. Bäckvall and P. G. Andersson, *J. Am. Chem. Soc.*, 1992, **114**, 6374.
411. J.-E. Bäckvall and P. G. Andersson, *J. Am. Chem. Soc.*, 1990, **112**, 3683.
412. J.-E. Bäckvall, *Pure Appl. Chem.*, 1992, **64**, 429.
413. A. Heumann and B. Åkermark, *Angew. Chem., Int. Ed. Engl.*, 1984, **23**, 453.
414. R. Benn *et al.*, *Organometallics*, 1985, **4**, 1945.
415. N. T. Byrom, R. Grigg, B. Kongkathip, G. Reimer and A. R. Wade, *J. Chem. Soc., Perkin Trans. 1*, 1984, 1643.
416. P. Grenouillet, D. Neibecker, J. Poirier and I. Tkatchenko, *Angew. Chem., Int. Ed. Engl.*, 1982, **21**, 767.
417. S. Agbossou, M. C. Bonnet and I. Tkatchenko, *Nouv. J. Chim.*, 1985, **9**, 311.
418. T. Antonsson and C. Moberg, *Organometallics*, 1985, **4**, 1083.
419. C. U. Pittman, Jr., R. M. Hanes and J. J. Yang, *J. Mol. Catal.*, 1982, **15**, 377.
420. H. M. Büch, P. Binger, R. Benn, C. Krüger and A. Rufinska, *Angew. Chem., Int. Ed. Engl.*, 1983, **22**, 774.
421. J. Lukas and P. A. Kramer, *J. Organomet. Chem.*, 1971, **31**, 111.
422. M. D. Jones and R. D. W. Kemmitt, *Adv. Organomet. Chem.*, 1987, **27**, 279.
423. P. Binger and A. Germer, *Chem. Ber.*, 1981, **114**, 3325.
424. B. M. Trost and D. M. T. Chan, *J. Am. Chem. Soc.*, 1983, **105**, 2315.
425. M. D. Jones and R. D. W. Kemmitt, *J. Chem. Soc., Chem. Commun.*, 1985, 811.
426. B. M. Trost and D. M. T. Chan, *J. Am. Chem. Soc.*, 1983, **105**, 2326.
427. B. M. Trost and T. N. Nanninga, *J. Am. Chem. Soc.*, 1985, **107**, 1075.

428. D. J. Gordon, R. F. Fenske, T. N. Nanninga and B. M. Trost, *J. Am. Chem. Soc.*, 1981, **103**, 5974.
429. B. M. Trost, T. N. Nanninga and D. M. T. Chan, *Organometallics*, 1982, **1**, 1543.
430. B. M. Trost and C. M. Marrs, *J. Am. Chem. Soc.*, 1993, **115**, 6636.
431. B. M. Trost, S. Sharma and T. Schmidt, *J. Am. Chem. Soc.*, 1992, **114**, 7903.
432. L. A. Paquette, D. R. Sauer, D. G. Cleary, M. A. Kinsella, C. M. Blackwell and L. G. Anderson, *J. Am. Chem. Soc.*, 1992, **114**, 7375.
433. A. Albinati, S. Affolter and P. S. Pregosin, *Organometallics*, 1990, **9**, 379.
434. J. Powell and N. I. Dowling, *Organometallics*, 1983, **2**, 1742.
435. N. M. Boag, D. Boucher, J. A. Davies, R. W. Miller, A. A. Pinkerton and R. Syed, *Organometallics*, 1988, **7**, 791.
436. M. Onishi, K. Hiraki, Y. Ohama and A. Kurosaki, *J. Organomet. Chem.*, 1985, **284**, 403.
437. G. Tresoldi, F. Faraone, P. Piraino and F. A. Bottino, *J. Organomet. Chem.*, 1982, **231**, 265.
438. N. K. Roberts, B. W. Skelton, A. H. White and S. B. Wild, *J. Chem. Soc., Dalton Trans.*, 1982, 2093.
439. V. V. Grushin, C. Bensimon and H. Alper, *Organometallics*, 1993, **12**, 2737.
440. F. Bachechi, R. Lehmann and L. M. Venanzi, *J. Crystallogr. Spectrosc. Res.*, 1988, **18**, 721.
441. H. Werner, H.-J. Kraus, U. Schubert, K. Ackermann and P. Hofmann, *J. Organomet. Chem.*, 1983, **250**, 517.
442. L. Zhu and N. M. Kostić, *Organometallics*, 1988, **7**, 665.
443. H. Werner, H.-J. Kraus, U. Schubert and K. Ackermann, *Chem. Ber.*, 1982, **115**, 2905.
444. H.-J. Kraus, H. Werner and C. Krüger, *Z. Naturforsch., Teil B*, 1983, **38**, 733.
445. T. Tanase, T. Nomura, Y. Yamamoto and K. Kobayashi, *J. Organomet. Chem.*, 1991, **410**, C25.
446. K. Broadley, G. A. Lane, N. G. Connelly and W. E. Geiger, *J. Am. Chem. Soc.*, 1983, **105**, 2486.
447. K. Broadley, N. G. Connelly, G. A. Lane and W. E. Geiger, *J. Chem. Soc., Dalton Trans.*, 1986, 373.
448. N. G. Connelly, in 'Molecular Electrochemistry of Inorganic, Bioinorganic and Organometallic Compounds', eds. A. J. L. Pombeiro and J. A. McCleverty, NATO ASI Series, Series C, Kluwer Academic Publishers, Dordrecht, 1993, vol. 385, p. 317.
449. P. Thometzek, K. Zenkert and H. Werner, *Angew. Chem., Int. Ed. Engl.*, 1985, **24**, 516.
450. H. Werner, P. Thometzek, K. Zenkert, R. Goddard and H.-J. Kraus, *Chem. Ber.*, 1987, **120**, 365.
451. D. Fenske and A. Christidis, *Z. Naturforsch., Teil B*, 1981, **36**, 518.
452. D. Fenske and A. Christidis, *Z. Naturforsch., Teil B*, 1983, **38**, 1295.
453. D. E. Smith and A. J. Welch, *Acta Crystallogr., Part C*, 1986, **42**, 1717.
454. G. K. Anderson, R. J. Cross, S. Fallis and M. Rocamora, *Organometallics*, 1987, **6**, 1440.
455. G. K. Anderson, R. J. Cross, L. Manojlovic-Muir, K. W. Muir and M. Rocamora, *Organometallics*, 1988, **7**, 1520.
456. G. K. Anderson, R. J. Cross, K. W. Muir and L. Manojlovic-Muir, *J. Organomet. Chem.*, 1989, **362**, 225.
457. P. Binger, H. M. Büch, R. Benn and R. Mynott, *Angew. Chem., Int. Ed. Engl.*, 1982, **21**, 62.
458. H. Schwager, R. Benn and G. Wilke, *Angew. Chem., Int. Ed. Engl.*, 1987, **26**, 67.
459. F. P. Netzer and J. U. Mack, *J. Chem. Phys.*, 1983, **79**, 1017.
460. J. S. Somers, M. E. Bridge and D. R. Lloyd, *Spectrochim. Acta*, 1987, **43A**, 1549.
461. I. V. Kozhevnikov and S. Ts. Sharapova, *React. Kinet. Catal. Lett.*, 1980, **15**, 49.
462. W. D. Jones, in 'Selective Hydrocarbon Activation: Principles and Progress', eds. J. A. Davies, P. L. Watson, J. F. Liebman and A. Greenberg, VCH, New York, 1990, p. 113.
463. G. Allegra, G. T. Casagrande, A. Immirzi, L. Porri and G. Vitulli, *J. Am. Chem. Soc.*, 1970, **92**, 289.
464. E. L. Muetterties, J. R. Bleeke, E. J. Wucherer and T. A. Albright, *Chem. Rev.*, 1982, **82**, 499.
465. E. R. Mognaschi and A. Chierico, *J. Phys. C, Solid-State Phys.*, 1985, **18**, 3815.
466. C.-S. Li, C.-H. Cheng, F.-L. Liao and S.-L. Wang, *J. Chem. Soc., Chem. Commun.*, 1991, 710.
467. H. Ossor, M. Pfeffer, J. T. B. H. Jastrzebski and C. S. Stam, *Inorg. Chem.*, 1987, **26**, 1169.
468. L. R. Falvello, J. Forniés, R. Navarro, V. Sicilia and M. Tomás, *Angew. Chem., Int. Ed. Engl.*, 1990, **29**, 891.
469. M. Sommovigo, M. Pasquali, P. Leoni, D. Braga and P. Sabatino, *Chem. Ber.*, 1991, **124**, 97.

7
Platinum Complexes with Carbonyl, Isocyanide and Carbene Ligands

RONALD J. CROSS
University of Glasgow, UK

7.1 PLATINUM CARBONYL COMPLEXES 392
7.1.1 Monomeric Platinum(0) Carbonyls 392
7.1.2 Polynuclear Platinum(0) Carbonyls 392
7.1.2.1 Syntheses 392
7.1.2.2 Structures 393
7.1.2.3 Spectral studies 394
7.1.2.4 Reactions 395
7.1.3 Polynuclear Platinum Carbonyl Anions 397
7.1.3.1 Spectral studies 398
7.1.3.2 Reactions 399
7.1.4 Platinum(I) Carbonyls 399
7.1.5 Triangular dppm-bridged Clusters 401
7.1.6 Platinum(II) Carbonyls 404
7.1.6.1 Monocarbonyl complexes 405
7.1.6.2 Platinum(II) dicarbonyls 409
7.1.6.3 18-Electron platinum(II) carbonyls 410
7.1.6.4 Binuclear platinum(II) carbonyls 410
7.1.7 Platinum(IV) Carbonyls 411

7.2 PLATINUM ISOCYANIDE COMPLEXES 411
7.2.1 Mononuclear Platinum(0) Isocyanides 411
7.2.2 Polynuclear Platinum(0) Isocyanides 411
7.2.3 Platinum(I) Isocyanide Complexes 413
7.2.4 Platinum(II) Isocyanide Complexes 415
7.2.4.1 Monoisocyanide compounds 416
7.2.4.2 Diisocyanide compounds 417
7.2.4.3 Tri- and tetraisocyanide compounds 419
7.2.5 Platinum(IV) Isocyanide Complexes 419

7.3 PLATINUM CARBENE COMPLEXES 419
7.3.1 New Preparative Routes 420
7.3.1.1 Preparation from nitrogen ylides 420
7.3.1.2 Electrophilic attack on coordinated CF_3 420
7.3.1.3 Template condensation of thiolate amine chelating carbene precursors 421
7.3.2 C,N-Carbene Complexes 421
7.3.3 C,O-Carbene Complexes 421
7.3.4 N,N-Carbene Complexes 423
7.3.5 N,O-Carbene Complexes 425
7.3.6 O,O- and S,S-Carbene Complexes 426

7.4 REFERENCES 426

7.1 PLATINUM CARBONYL COMPLEXES

Growth in platinum carbonyl complexes has been steady since the publication of *COMC-I*, but most of the new work extends and adds detail to the structure outlined previously. For this reason the layout of this chapter mimics that of *COMC-I*, to facilitate cross-referencing. An exception is the family of trinuclear complexes bridged by dppm ligands, developed by Puddephatt and co-workers. These compounds have no counterpart in *COMC-I* and are discussed after platinum(I) carbonyls.

A review is available that describes preparative routes, reactions, insertions and catalyses.[1]

7.1.1 Monomeric Platinum(0) Carbonyls

Compounds of platinum and carbon monoxide alone remain accessible only by matrix isolation methods. 'PtCO' itself has been the subject of several *ab initio* calculations.[2,3] The stability of $[Pt(CO)_4]$ has been calculated to lie between that of $[Ni(CO)_4]$ and $[Pd(CO)_4]$, in agreement with experimental findings.[3]

The 16-electron complex $[Pt(PPh_3)_3]$ readily adds CO at −80 °C to form $[Pt(CO)(PPh_3)_3]$ (^{31}P NMR δ 12.3; J_{Pt-P} 3537 Hz), but at −65 °C $[Pt(CO)_2(PPh_3)_2]$ is produced (δ 7.7; J_{Pt-P} 3232 Hz).[4] The higher temperature required by the second process presumably reflects a change from an associative to a dissociative mechanism. Dihydrogen from $[PtH_2(PEt_3)_2]$, which displays rapid *cis–trans* isomerism in solution, is displaced by π-acceptors such as CO (Equation (1)).[5] The more-or-less tetrahedral ligand arrangements in $[Pt(CO)(PR_3)_3]$ and $[Pt(CO)_2(PR_3)_2]$ are well established, so it is of interest that *ab initio* MO calculations on the carbene–carbonyl species $[Pt(CH_2)(CO)(PH_3)_2]$ found a *cis*-planar geometry to be the most stable.[6]

$$[PtH_2(PEt_3)_2] \; + \; 2\,CO \; \rightleftharpoons \; [Pt(CO)_2(PEt_3)_2] \; + \; H_2 \qquad (1)$$

7.1.2 Polynuclear Platinum(0) Carbonyls

A review of homonuclear platinum clusters up to 1985 includes carbonyl, isocyanide and sulfur dioxide complexes.[7] The IR data on 41 carbonyl clusters are tabulated.

Most activity relates to complexes of the types $[Pt_3(CO)_3(PR_3)_3]$, $[Pt_3(CO)_3(PR_3)_4]$, $[Pt_4(CO)_5(PR_3)_4]$ and $[Pt_5(CO)_6(PR_3)_4]$, all described in *COMC-I*. There are still no binuclear complexes in this class. Several high-yield preparative routes have been reported, but the relationship between the complexes of different nuclearities is still not understood.

7.1.2.1 Syntheses

The reductions of *cis*-$[PtCl_2(CO)(PR_3)]$ under a carbon monoxide atmosphere by either zinc dust in THF or acetone, or $Na[BH_4]$ in methanol, led to good yields of $[Pt_3(\mu\text{-}CO)_3(PBu^n{}_3)_3]$, $[Pt_3(\mu\text{-}CO)_3\{P(C_2H_4CN)_3\}_3]$, $[Pt_3(\mu\text{-}CO)_3(PCy_3)_3]$, $[Pt_3(\mu\text{-}CO)_3(PPh_3)_3]$, $[Pt_4(\mu\text{-}CO)_5(PMe_2Ph)_4]$ or $[Pt_5(CO)(\mu\text{-}CO)_5(PPh_3)_4]$. The nuclearity of the products was clearly dependent on the electronic as well as the steric properties of PR_3, and earlier assertions that more bulky phosphines lead to low-nuclearity products are inadequate.[8] In the presence of free phosphine, reductions of these types are known to lead to mononuclear products and it is quite conceivable that such species might act as intermediates in the cluster syntheses. Among the triphenylphosphine derivatives, where $[Pt_3(CO)_3(PPh_3)_4]$ is known to be more stable than $[Pt_3(CO)_3(PPh_3)_3]$, the presence of free PPh_3 in solution dramatically effects the outcome of the syntheses. The reduction of *cis*-$[PtCl_2(CO)(PPh_3)]$ by zinc under carbon monoxide produces $[Pt_5(CO)_6(PPh_3)_4]$ in 76% yield, but the presence of 1/3 mol. equiv. of PPh_3 in the reaction mixture results in an 80% yield of $[Pt_3(CO)_3(PPh_3)_4]$ instead.

A variation of the above procedure starts from solutions of $[PtCl_2(MeCN)_2]$ or $[PtCl_2(cod)]$ in acetonitrile under carbon monoxide.[9] The complexes $[Pt_3(CO)_3L_3]$ (L = $PPr^i{}_3$, $PPr^i{}_2Ph$, PCy_2Ph or PCy_3) or $[Pt_4(CO)_5L_4]$ (L = PEt_3, PBz_3 or PMe_2Ph) were obtained in good yield by reduction with $Na[BH_4]$ in the presence of free phosphine. When the phosphine was PPh_3 or $PMePh_2$, however, mixtures of both the above types along with $[Pt_3(CO)_3L_4]$ were obtained. This uncertainty over the final products when L is PPh_3 was further compounded by a study of the reduction of $[PtCl_2(PPh_3)_2]$ in the presence of carbon monoxide and sodium hydroxide.[10] Mixtures of $[Pt_3(CO)_3L_3]$, $[Pt_3(CO)_3L_4]$, $[Pt_4(CO)_5L_4]$ and $[Pt_5(CO)_6L_4]$, along with $[PtCl(CO_2Me)L_2]_2$, were reported.

Equations (2) and (3) represent a more systematic approach to cluster synthesis using platinum dibenzylideneacetone ([Pt(dba)$_2$]) (**1**) to increase cluster nuclearity.[11] The method does not appear to proceed beyond Pt_5 clusters, however, and the equilibrium of Equation (4) demonstrates the importance of reaction conditions since the tetranuclear complexes, instead of adding an extra platinum unit, can fragment.[12] Thermolysis of [$Pt_5(CO)_6(PEt_3)_4$] in decane produced 14% of [$Pt_{17}(CO)_{12}(PEt_3)_8$].[11] Treatment of [$Pt(PBu^t_3)_2$] with carbon monoxide produces [$Pt_3(CO)_3(PBu^t_3)_3$],[13] as does the reaction of *trans*-[$PtH_2(PBu^t_3)_2$] with carbon monoxide,[14] in contrast to the simplicity of Equation (1).

$$[Pt_3(CO)_3(PEt_3)_4] + [Pt(dba)_2] \longrightarrow [Pt_4(CO)_5(PEt_3)_4] \qquad (2)$$

$$[Pt_4(CO)_5(PEt_3)_4] + [Pt(dba)_2] \longrightarrow [Pt_5(CO)_6(PEt_3)_4] \qquad (3)$$

$$2\,[Pt_3(CO)_3(PEt_3)_4] + 3\,CO \rightleftharpoons [Pt_4(CO)_5(PEt_3)_4] + 2\,[Pt(CO)_2(PEt_3)_2] \qquad (4)$$

The room-temperature reactions of [$PtCl_2L_2$] with K[$Fe(CO)_3(NO)$] led unexpectedly to [$Pt_5(CO)_6L_4$] (L = PPh_3, $PMePh_2$, $P(OMe)Ph_2$ or $P(OMe)_2Ph$), often in high yields.[15] Similarly, *trans*-[$PtCl_2(PEt_3)_2$] reacted with (NEt_4)[$CrH(CO)_5$] to produce 80% of [$Pt_4(CO)_5(PEt_3)_4$].[16]

PtL_3 (L is the aminoiminophosphine Bu^t(TMS)NP=NBu^t reacts with carbon monoxide to produce [$Pt_3(CO)_3L_3$]; IR and NMR spectroscopic analyses indicate that in this case the carbonyls are terminal and phosphorus atoms bridge, as in (**2**).[17] The carbene-bridged tungsten–platinum compound [$(OC)_5W\{\mu$-C(OMe)(*p*-Tol)$\}$Pt(cod)] reacts with carbon monoxide to produce two isomers of [$Pt_3(CO)_3\{\mu$-C(OMe)(*p*-Tol)$\}_3$], also with terminal carbonyls.[18] The molecular structure of one of them, (**3**), has been reported, along with its ^{1}H and ^{13}C NMR parameters and ν(CO) bands at 2031 cm^{-1} and 2029 cm^{-1}.

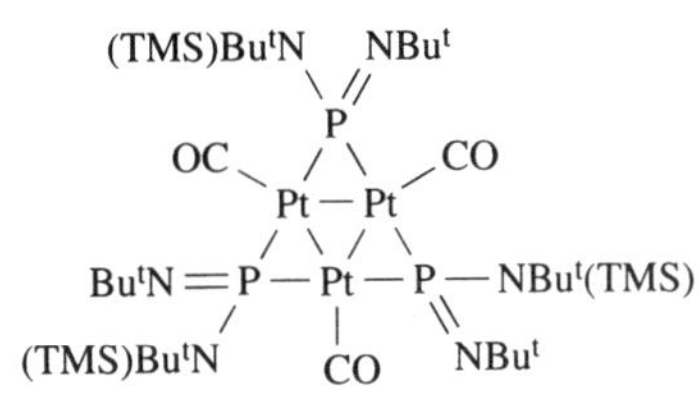

(**2**)

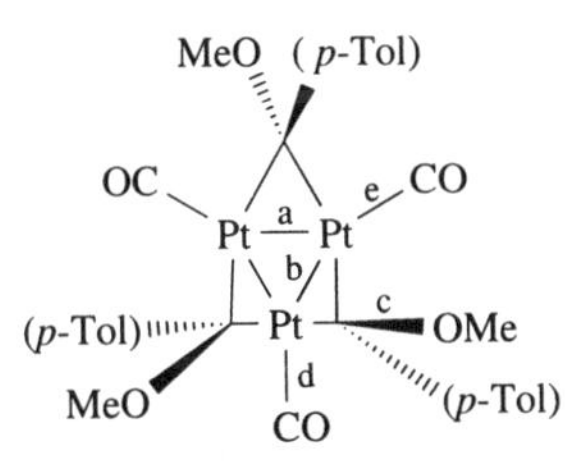

(**3**)

a, 0.262 1(1) nm
b, 0.261 9(1) nm
c, 0.213(3) nm
d, 0.211(3) nm
e, 0.182(4) nm

7.1.2.2 Structures

Several new x-ray crystal structures on phosphine–carbonyl clusters complement those reported previously. The salient features of [$Pt_3(CO)_3(PPh_3)_3$] are shown in (**4**).[10,19] The structure of [$Pt_3(CO)_3(PBu^t_2Ph)_3$] (**5**) has the CO molecules bent upwards out of the Pt_3 plane by 7°.[20] This contrasts with the planar skeleton of [$Pt_3(CO)_3(PCy_3)_3$], reported previously, with less sterically demanding phosphines. Calculated and observed phosphine conformations in molecules of this type led to a size ordering of the phosphine ligands as $PBu^t_3 \gg PBu^t_2Pr^i > PBu^t_2Ph > PCy_3 \gg PCy_2Ph$.

Crystal structures of the tetranuclear clusters [$Pt_4(CO)_5(PMe_2Ph)_4$] (**6**)[21] and [$Pt_4(CO)_5(PEt_3)_4$] (**7**)[22] both reveal open tetrahedral ('butterfly') arrangements of platinum atoms. The latter contains two independent molecules in its unit cell, each of broadly similar shape. Two determinations of the structure of [$Pt_5(CO)_6(PPh_3)_4$] (**8**) show this 70-electron pentanuclear cluster to have an edge-capped tetrahedral arrangement of metal atoms with two elongated tetrahedral edges.[15,23]

Molecules of [$Pt_4(\mu$-CO)$_2(\mu$-dppm)$_3\{Ph_2$*P*$CH_2P(O)Ph_2\}$], isolated from solutions of the water–gas shift catalyst [$Pt_2(\mu$-H)(H)$_2(\mu$-dppm)$_2$]PF_6 and carbon monoxide after exposure to air and characterized

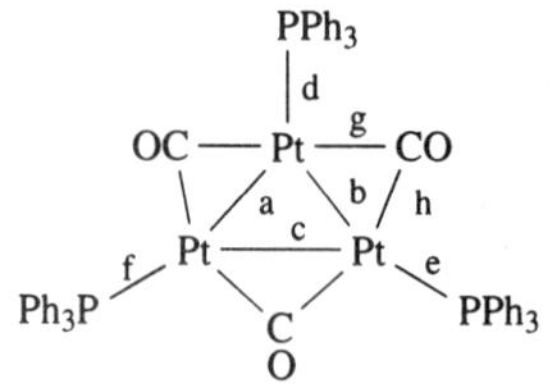

(4)

a, 0.2665 nm
b, 0.2657 nm
c, 0.2667 nm
d, 0.2261 nm
e, 0.2257 nm
f, 0.2258 nm
g, 0.2039 nm
h, 0.2078 nm

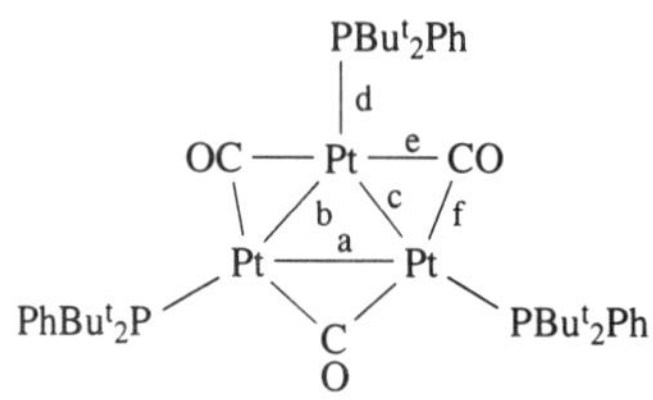

(5)

a, 0.267 3(1) nm; 0.268 3(1) nm
b, 0.267 1(1) nm; 0.268 4(1) nm
c, 0.267 7(1) nm; 0.267 3(1) nm
d, 0.229 2(5) nm; 0.230 2(4) nm
e, 0.199 7(20) nm; 0.202 1(17) nm
f, 0.203 2(18) nm; 0.201 8(17) nm

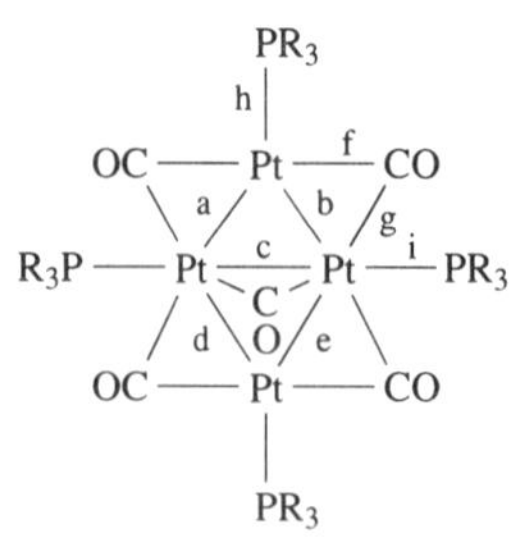

(6) $PR_3 = PMe_2Ph$	**(7)** $PR_3 = PEt_3$
a, 0.272 80(14) nm	a, 0.2691 nm; 0.2704 nm
b, 0.270 64(15) nm	b, 0.270 91 nm; 0.2705 nm
c, 0.273 12(14) nm	c, 0.2728 nm; 0.2737 nm
d, 0.275 37(15) nm	d, 0.2766 nm; 0.2745 nm
e, 0.275 37(13) nm	e, 0.2762 nm; 0.2744 nm
f, 0.205 7(24) nm	f, 0.205 nm; 0.209 nm
g, 0.214 5(24) nm	g, 0.224 nm; 0.188 nm
h, 0.223 3(6) nm	h, 0.2222 nm; 0.2206 nm
i, 0.226 1(6) nm	i, 0.2279 nm; 0.2258 nm

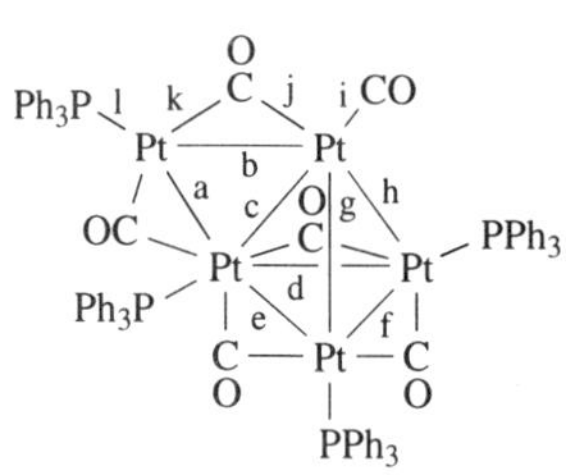

(8)

a, 0.270 83(5) nm	g, 0.291 82(5) nm
b, 0.269 21(5) nm	h, 0.286 06(5) nm
c, 0.279 10(5) nm	i, 0.186(1) nm
d, 0.274 74(5) nm	j, 0.200 4(9) nm
e, 0.275 31(5) nm	k, 0.208 1(9) nm
f, 0.267 19(5) nm	l, 0.226 0(2) nm

crystallographically, bear several similarities to the 'butterfly' Pt_4 clusters just described, and also to the triangular dppm-bridged Pt_3 cations described in Section 7.1.5. Cluster (**9**) shows salient features.[24] The x-ray structure determination of the large cluster obtained from the thermolysis of $[Pt_5(CO)_6(PEt_3)_4]$ found it to be $[Pt_{17}(\mu\text{-}CO)_4(CO)_8(PEt_3)_8]$,[25] a distorted, centred icosahedron with four additional platinum caps.

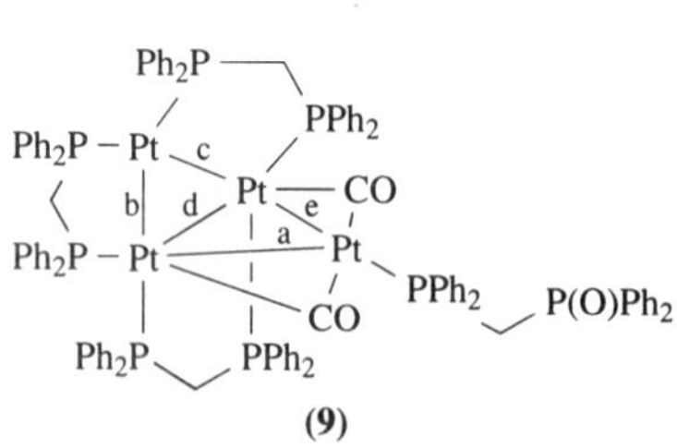

(9)

a, 0.273 9(1) nm
b, 0.261 1(2) nm
c, 0.261 7(2) nm
d, 0.270 0(1) nm
e, 0.270 8(2) nm

7.1.2.3 Spectral studies

Other than to detect bridge and terminal CO groups, IR spectroscopy is relatively unhelpful in determining the structures of these clusters. For example, the trinuclear complexes $[Pt_3(\mu\text{-}CO)_3(PR_3)_3]$ can exhibit from two to five absorptions in the 1730–1865 cm^{-1} region.[13,26] On the other hand, ^{31}P and

^{195}Pt NMR spectroscopy can provide a simple means of distinguishing $[Pt_3(CO)_3(PR_3)_3]$ from $[Pt_3(CO)_3(PR_3)_4]$. In the former, the three platinum and three phosphorus atoms are equivalent, and their spectra reveal the expected pattern for the four isotopomers of Pt_3.[13,26] Interestingly, the addition of free PBu^t_3 to solutions of $[Pt_3(CO)_3(PBu^t_3)_3]$ does not alter the appearance of its NMR spectra, indicating that neither ligand addition nor rapid exchange occurs with this bulky group.[13] The complexes $[Pt_3(CO)_3(PR_3)_4]$ of type (**10**) reveal two types of phosphorus and platinum in their NMR spectra, clearly indicating that no rapid phosphine exchange can be operating. Values of $^1J_{Pt\text{-}Pt}$ lie in the range 1808–1840 Hz for the base-to-base platinum atoms of (**10**), but only 385–586 Hz for base to apex.[27]

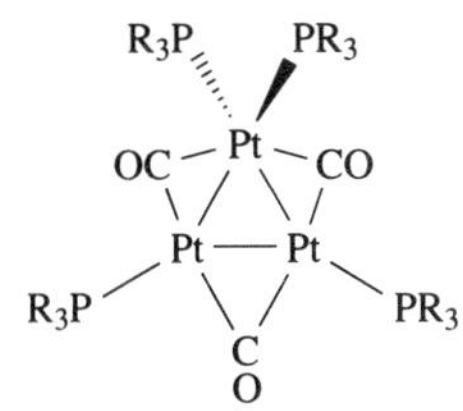

(10)

The ^{31}P and ^{195}Pt NMR spectra of tetranuclear clusters of types (**6**) and (**7**) are indicative of equivalent phosphorus atoms, with the expected five Pt_4 isotopomers,[21,28] and are clearly at variance with the x-ray-determined 'butterfly' structures. Spectra of the PEt_3 complex retained their integrity down to 180 K. The fluxional process leading to this observed simplification of the spectra must be intramolecular since couplings are retained at the high-temperature limits. One candidate for the process responsible is the folding motion shown in Scheme 1.

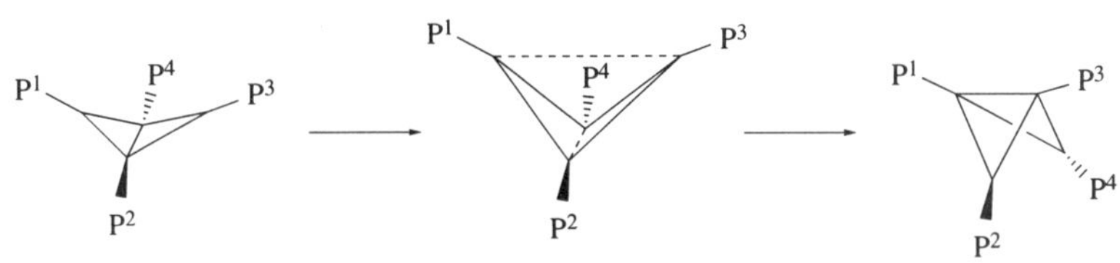

Scheme 1

$[Pt_3(CO)_3L_4]$, $[Pt_4(CO)_5L_4]$ (L = PPh_3) and $[Pt_5(CO)_6L_4]$ (L = PEt_3) are thermally labile and slowly solvolysed by THF or acetone. They could not be analysed by electron impact mass spectroscopy. Field desorption methods, however, allowed observation of $M^{+\cdot}$ as well as $(M\text{–}xCO)^{+\cdot}$.[29] Electronic and magnetic CD spectra of $[Pt_3(CO)_3(PBu^t_3)_3]$ in MeCN have been interpreted using MO calculations. The transitions in the UV–visible range are to E' or A_2'' D_{3h} spin-orbit states of excited configurations that are primarily metal-to-ligand charge transfer in character.[30]

7.1.2.4 Reactions

In solution, $[Pt_3(CO)_3(PPh_3)_4]$ can convert to $[Pt_5(CO)_6(PPh_3)_4]$ when contacted by various inorganic supports, including magnesium oxide, titanium(IV) oxide, Kieselgel, Aerosil and some aluminas.[31] The mechanism is uncertain, but protonation of the clusters by the support could be involved. This discovery has implications for the preparation of catalysts by impregnation, and column chromatography can no longer be regarded as an innocent separation technique for clusters of this type. Some zerovalent platinum carbonyl clusters are known to react with carbonyl complexes of other metals to produce heteronuclear clusters.[15] This class of compounds is described elsewhere.

As would be expected for zerovalent compounds, the clusters readily undergo oxidation reactions. Scheme 2 illustrates this for $[Pt_3(CO)_3L_3]$ clusters (L = PBu^t_3 or PBu^t_2Ph), which are oxidized by HCl or H_2S as well as by Cl_2 or I_2.[32–4] They also undergo a variety of ligand exchange reactions, some of which can alter the nuclearity of the cluster. Isocyanides replace some or all of both the CO and phosphine ligands. These are described in Section 7.2. Reactions of $[Pt_3(CO)_3(PR_3)_3]$ with other phosphines are somewhat dependent on the sizes of the phosphines involved, but can rapidly lead to replacement of all three PR_3 groups by an associative mechanism, the Pt_3 triangles remaining intact.[35] In this way coordinated PBu^t_3 has been replaced by PBu^t_2Ph; PBu^t_2Ph by PBu^tPh_2, PCy_3 or PPr^i_3; and PCy_3 by PPr^i_3. Mixtures of $[Pt_3(CO)_3(PBu^t_2Ph)_3]$ with $[Pt_3(^{13}CO)_3(PBu^t_2Ph)_3]$ or $[Pt_3(CO)_3(PCy_3)_3]$ lead to scrambling of the CO or PR_3, respectively, over about 1 h at room temperature. Treatment of $[Pt_3(CO)_3(PCy_3)_3]$ by dppe or dppp (biL) led to the 44-electron clusters $[Pt_3(CO)_3(PCy_3)_2(biL)]$,[36] an interesting result in view of the failure of many $[Pt_3(CO)_3(PR_3)_3]$ clusters to add another PR_3.

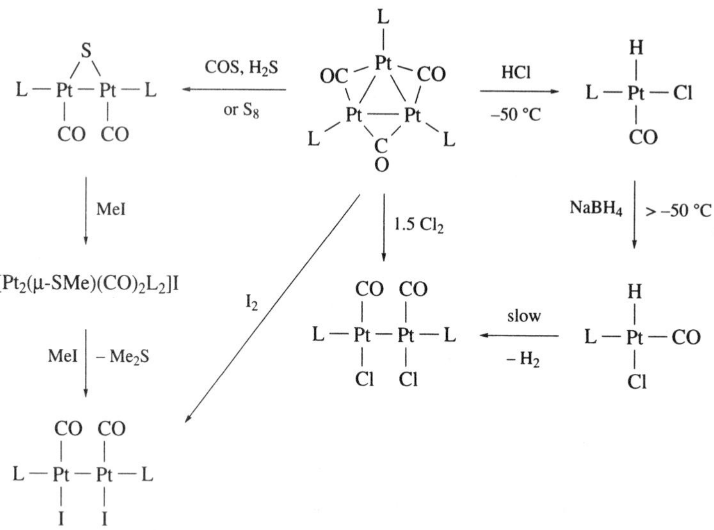

Scheme 2

Replacement of CO by SO_2 in these clusters is a straightforward process, although its larger steric bulk can make it difficult to substitute all the CO ligands. Nevertheless sulfur dioxide treatment of $[Pt_3(CO)_3L_3]$ produces $[Pt_3(SO_2)_3L_3]$ when L is PCy_3,[37,38] $P(C_2H_4CN)_3$, PPh_3 or PBu^n_3.[39] When L is the more bulky PBu^t_3, PBu^t_2Ph or $PBu^t_2Pr^i$, $[Pt_3(CO)_3L_3]$ reacts initially via cluster degradation to $[Pt_2(\mu\text{-}SO_2)(CO)_2L_2]$; these can then be converted to $[Pt_3(\mu\text{-}SO_2)(\mu\text{-}CO)_2L_3]$ on heating.[20,35] Some reactions also produce $[Pt_3(\mu\text{-}SO_2)_2(\mu\text{-}CO)L_3]$. With L = PCy_3 or PCy_2Ph the tendency to preserve cluster nuclearity is greater,[20,40,41] but with careful control all the intermediates can be isolated (Scheme 3, where L = PCy_3).[40,41]

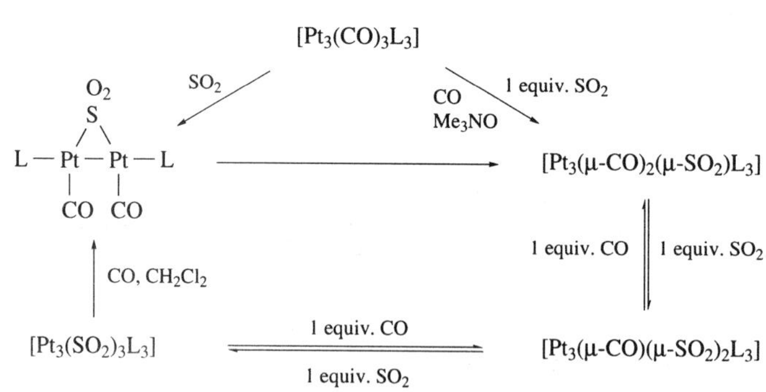

Scheme 3

$[Pt_3(\mu\text{-}SO_2)_3(PCy_3)_3]$ reacts sequentially with carbon monoxide to produce $[Pt_3(CO)(SO_2)_2(PCy_3)_3]$ then $[Pt_3(CO)_2(SO_2)(PCy_3)_3]$,[8] but only with the assistance of Me_3NO (which abstracts SO_2 as Me_3NSO_3) can the tricarbonyl derivative be generated.[40,41] The crystal structure of $[Pt_3(\mu\text{-}CO)(\mu\text{-}SO_2)_2(PCy_3)_3]$ (**11**) shows a close similarity to the tris(μ-carbonyl) analogues with little difference in Pt–Pt bond lengths.

The reaction of $[Pt_3(CO)_3(PPh_3)_4]$ with sulfur dioxide involves loss of a phosphine to form $[Pt_3(SO_2)_3(PPh_3)_3]$, and $[Pt_4(CO)_5(PMe_2Ph)_4]$ reacts by cluster degradation to form $[Pt_3(SO_2)_3(PMe_2Ph)_3]$.[39] The reverse of these reactions leads to pentanuclear complexes in both cases (Equation (5)), so sequential treatment of these initial carbonyl complexes by sulfur dioxide then carbon monoxide would seem to offer a route for specifically increasing cluster nuclearity. $[Pt_5(CO)_6(PPh_3)_4]$ and sulfur dioxide gave the complex shown in (**12**).[39]

$$[Pt_3(SO_2)_3L_3] + CO \longrightarrow [Pt_5(CO)_6L_4] \qquad (5)$$

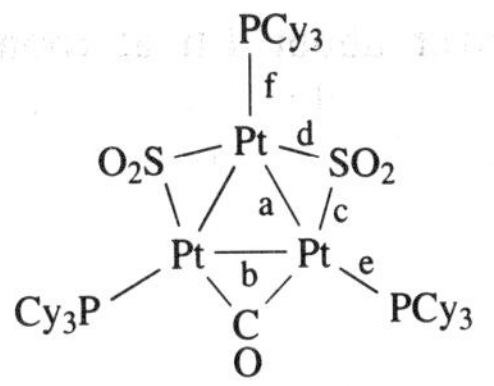

a,	0.271 0(1) nm
b,	0.267 8(2) nm
c,	0.226 4(3) nm
d,	0.228 0(3) nm
e,	0.230 8(2) nm
f,	0.230 0(3) nm

(11)

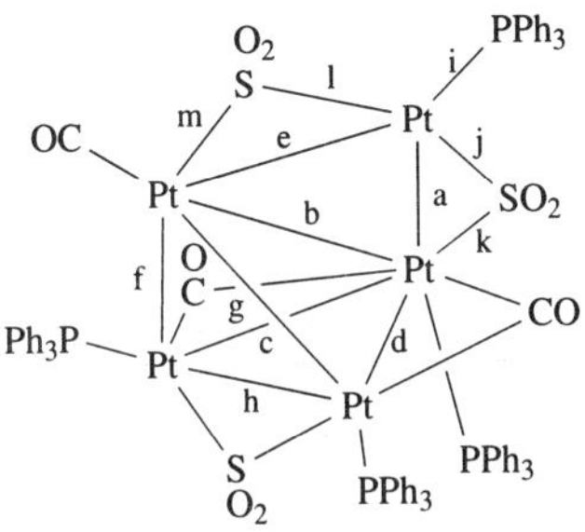

a,	0.282 6(1) nm	h,	0.279 3(1) nm
b,	0.275 1(1) nm	i,	0.227 4(6) nm
c,	0.276 3(1) nm	j,	0.224 4(6) nm
d,	0.276 6(1) nm	k,	0.231 3(6) nm
e,	0.278 4(1) nm	l,	0.227 4(7) nm
f,	0.287 7(1) nm	m,	0.222 7(7) nm
g,	0.280 8(1) nm		

(12)

Scheme 4 (L = PBu^t_2Ph) shows the degradative reactions of $[Pt_3(CO)_3(PBu^t_2Ph)_3]$ with CS_2 and COS.[35] $[Pt_3(CO)_3(PPh_3)_4]$ reacts with COS to give the analogous product, but CS_2 behaves only as a phosphine abstractor with this complex, producing $[Pt_5(CO)_6(PPh_3)_4]$.[8]

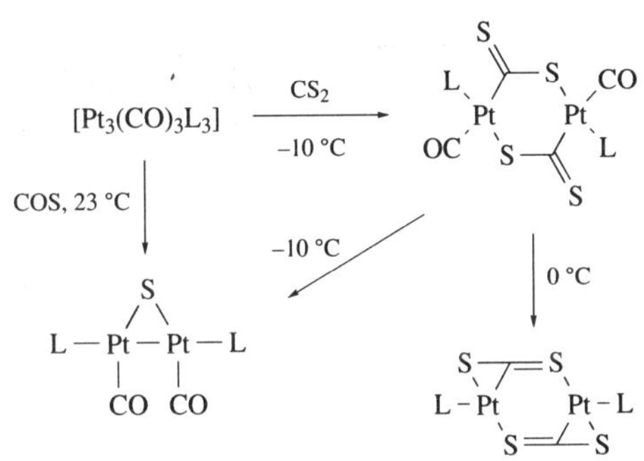

Scheme 4

One other property of these trinuclear carbonyl clusters is worthy of note. They can behave as Lewis bases towards a variety of metal ion acceptors,[42] including copper(I), silver(I), gold(I), zinc(II) and mercury(II). The heteronuclear products are dealt with elsewhere.

7.1.3 Polynuclear Platinum Carbonyl Anions

These complexes, $[Pt_3(\mu\text{-}CO)_3(CO)_3]_n^{2-}$, where n = 2–6 and 10, have considerable potential as catalysts and catalyst precursors. Many clusters of this type have been described in a review.[43] They are usually made by reductive carbonylation of platinum halides, and systematic high-yield preparations of the $[NBu^n_4]^+$ salts have been outlined in *Inorganic Syntheses*.[44] $Na_2[Pt_{30}(CO)_{60}]$ is first isolated since it is only sparingly soluble, and this can be degraded to $[Pt_3(CO)_6]_2^{2-}$ by sodium hydroxide under carbon monoxide. $[Pt_3(CO)_6]_5^{2-}$ can readily be isolated from the alkaline carbonylation of $[PtCl_6]^{2-}$, and disproportionation reactions (Equations (6) and (7)) can then provide $[Pt_3(CO)_6]_4^{2-}$ and $[Pt_3(CO)_6]_3^{2-}$.

$$2\,[Pt_3(CO)_6]_5^{2-} + [Pt_3(CO)_6]_2^{2-} \longrightarrow 3\,[Pt_3(CO)_6]_4^{2-} \quad (6)$$

$$[Pt_3(CO)_6]_5^{2-} + 2\,[Pt_3(CO)_6]_2^{2-} \longrightarrow 3\,[Pt_3(CO)_6]_3^{2-} \quad (7)$$

Room-temperature treatment of $[Pt(C_3H_5)_2]$ impregnated on magnesium oxide with carbon monoxide or its mixture with hydrogen produces $[Pt_3(CO)_6]_n^{2-}$ (n = 3 or 4) on the magnesium oxide surface, and the clusters can be extracted into solution by cation metathesis using $[NBu^n{}_4]Br$.[45] Since the nuclearity can be effectively controlled by the magnesium oxide basicity or the gas mixture, this method has been proposed as an efficient synthetic route, although the starting material is a little inconvenient. $Na_2[PtCl_6]\cdot 6H_2O$ can be used instead of bis(allyl)platinum, but higher reaction temperatures are required and product mixtures are unavoidable.

The structures of all the aforementioned anions consist of stacked columns of the triangular units, and were described in *COMC-I*. Some of their reactions, however, produce high-nuclearity platinum clusters of more complex shapes.[46] Refluxing the yellow-green solutions of $(AsPh_4)_2[Pt_3(CO)_6]_5$ under nitrogen produced $(AsPh_4)_2[Pt_{24}(CO)_{22}(\mu\text{-}CO)_8]$ and $(AsPh_4)_2[Pt_{26}(CO)_{23}(\mu\text{-}CO)_9]$. The core of the former has c.c.p. platinum atoms as found in bulk platinum metal, with two mirror-related faces of 10 coplanar platinum atoms, one 12-coordinate 'bulk' platinum atom, and six nine-coordinate 'terracelike' surface atoms. The latter cluster has a core of h.c.p. *a*:*b*:*a* layers with seven, 12 and seven platinum atoms of pseudo-D_{3h} symmetry. Refluxing $(AsPh_4)_2[Pt_3(CO)_6]_3$, followed by treatment with CF_3CO_2H, leads to yet another anionic cluster, $(AsPh_4)_2[Pt_{38}(CO)_{44}]$.[46] The core of this cluster has pseudooctahedral symmetry and a c.c.p. arrangement of platinum atoms. Six platinum atoms in an octahedron are totally encapsulated 12-coordinate, while eight others have a nine-coordinate 'surface' environment. A crystal disorder prevented clarification of the CO environments.

The reasons for the columnar stacking (and their distortions) and the requirement for the −2 charge on the triangular stacked clusters have been approached from extended Hückel theory.[47] A topological approach has also been applied to describe and rationalize these clusters.[48] $[Pt_3(CO)_6]_n{}^{2-}$ (n = 2–5) are regarded as being built from edge-localized bonds with additional Möbius delocalizations on the top and bottom triangles.

7.1.3.1 Spectral studies

Pressure-induced shifts in the first three absorption bands of the visible spectra of $[Pt_3(CO)_6]_n{}^{2-}$ (n = 2–5) have been applied to derive possible assignments.[49] The interpretation is compatible with the extended Hückel calculations.[47] An alternative MO-based explanation has been advanced, however, describing the absorptions as envelopes of many electronic transitions to a high density of excited states.[50]

The ^{195}Pt NMR spectra of $[Pt_3(CO)_6]_n{}^{2-}$ (n = 1–5), along with ^{13}C NMR data for n = 2–4, and over a range of temperatures, provided evidence for intramolecular rotation of the Pt_3 triangles about the main threefold axes.[51] Interexchange of the triangular units was also evident, but there was no exchange of bridge and terminal carbonyl groups within each $[Pt_3(CO)_3(\mu\text{-}CO)_3]$ unit.

Magnetic moment measurements on the $[Bu_4N]^+$ salts of $[Pt_3(CO)_6]_n{}^{2-}$ (n = 2 or 3) and $[Pt_{19}(CO)_{22}]^{4-}$, and on $(PPN)_2[Pt_{38}(CO)_{44}H_x]$, found each to be diamagnetic at room temperature but weakly paramagnetic at low temperatures.[52] This anomalous paramagnetism obeyed a simple Curie law. The value of X_p increased with cluster size, which was attributed to the decreasing energy gap between HOMO and LUMO as the cluster size increased. Overall μ_{eff} fell in the narrow range of 0.35–0.55 μ_B per cluster, and the effective moment per platinum atom of $(0.1 \pm 0.02)\mu_B$ is very close to that observed for platinum crystallites with about 400 atoms.

A number of very high-nuclearity platinum carbonyl anions have been analysed by ^{252}Cf desorption mass spectrometry, which shows promise as a characterization technique for compounds containing such species.[53] For example, the formation of gas-phase $[Pt_{26}(CO)_x]_n$ monocharged ion oligomers (n = 1–20) was observed from solid $[Pt_{26}(CO)_{32}]^{2-}$ compounds.

The clusters $[Pt_{24}(CO)_{30}]^n$, $[Pt_{26}(CO)_{32}]^n$ and $[Pt_{38}(CO)_{44}]^n$ exhibit unusually reversible voltammetric and spectroelectrochemical behaviour, with the Pt_{24} cluster showing a sequence of redox steps from $n = 0$ to $n = -6$, the Pt_{26} cluster transforming in two-electron steps over $n = 0$ to $n = -10$, and the Pt_{38} cluster ranging from $n = 0$ to $n = -7$ with the even charges being the more stable.[46,54] The CO stretching frequencies, obtained by FTIR spectroscopy during thin-layer cyclic voltammetry, show an almost linear progression at both terminal and bridge CO, diminishing by 10–20 cm^{-1} per electron added. Comparisons with the CO surface layers on platinum electrodes establishes a clear link between these large, ionizable clusters and chargeable metal surfaces. The effects and rates of substitution of ^{13}CO for ^{12}CO in $[Pt_{24}(CO)_{30}]^n$ and $[Pt_{26}(CO)_{32}]^n$ (n = 0 to n = −6) have also been followed by FTIR spectroelectrochemical methods, and again compared with ^{13}CO–^{12}CO adlayers on Pt(1.1.1) surfaces.[55] Exchange is slow, incomplete and nonrandom, with many deviations from first-order kinetics. The results suggest the operation of active substitution centres for CO exchange, followed by migrations from these to some other sites.

7.1.3.2 Reactions

Triphenylphosphine reacts with the anion $[Pt_3(CO)_6]_5^{2-}$ to sequentially degrade the cluster (Equations (8) and (9)).[12] Mixing solutions of the anions $[Pt_3(CO)_6]_2^{2-}$ and $[Ni_3(CO)_6]_2^{2-}$ produces[51] the heteronuclear cluster $[Ni_3Pt_3(CO)_{12}]^{2-}$. The anions $[Pt_3(CO)_6]_n^{2-}$ (n = 3 or 5) have been introduced by ion exchange methods into graphite discs pretreated with quaternary ammonium salts in order to prepare metal cluster attached electrodes.[56] The electrode treated with the Pt_{15} cluster was highly selective for the anodic oxidation of methanol. Catalytic reactions of these clusters include the use of $[Pt_3(CO)_6]_5^{2-}$ in the hydrogenation of aldehydes, ketones, cyanides and alkenes, and $[Pt_3(CO)_6]_4^{2-}$, which oxidizes water to oxygen and catalyses the reduction of benzoquinone by hydrogen or water.[57]

$$[Pt_3(CO)_6]_5^{2-} + 3\,PPh_3 \longrightarrow [Pt_3(CO)_6]_4^{2-} + [Pt_3(CO)_3(PPh_3)_3] \qquad (8)$$

$$[Pt_3(CO)_6]_4^{2-} + 3\,PPh_3 \longrightarrow [Pt_3(CO)_6]_3^{2-} + [Pt_3(CO)_3(PPh_3)_3] \qquad (9)$$

7.1.4 Platinum(I) Carbonyls

The compounds which adhere strictly to this description are Pt–Pt-bonded binuclear species. Their characteristics are very dependent on the substituent ligands employed, so each compound type is dealt with separately instead of drawing together preparations, structures and then properties.

A convenient route to the ions $[Pt_2X_4(CO)_2]^{2-}$ (X = Cl, Br or I) is the treatment of THF solutions of $[PtX_3(CO)]^-$ by controlled amounts of formic acid and $NPr^n{}_3$.[58] Their NMR spectra show J_{Pt-Pt} at 5250 Hz (X = Cl) or 4770 Hz (X = Br). Mixed halide analogues are readily made and the heavy halides show a preference for the sites collinear with the Pt–Pt bonds. Above 300 K, halide site exchange occurs at a given platinum atom, but is synchronized between both platinum atoms of the dianions. Scheme 5, involving CO transfer via μ-CO, explains this.

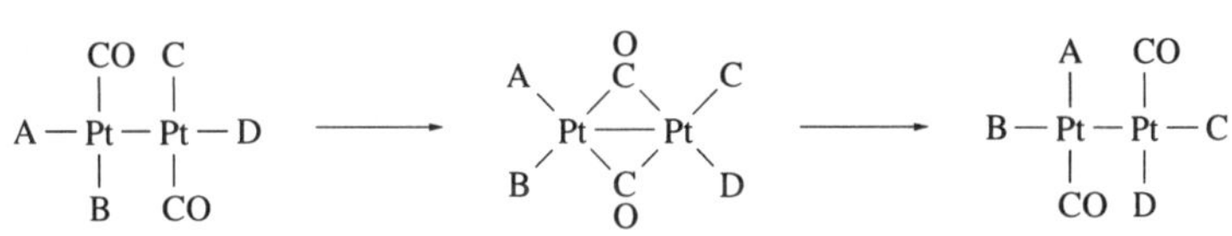

Scheme 5

The reactions of *cis*-$[PtR_2(CO)_2]$ (R = C_6F_5 or C_6Cl_5) with $[Pt(C_2H_4)(PPh_3)_2]$ produce (**13**).[59] The structure of the pentafluorophenyl derivative shows that it adopts the expected staggered conformation about the Pt–Pt bond of 0.259 9(1) nm. The carbonyls of (**13**) are readily displaced by isocyanides, and I_2 cleaves the metal–metal bond oxidatively, although MeI does not react. $[Pt(CO)_2(PPh_3)_2]$ reacts with the alkyne $C_2(CO_2Me)_2$ to produce (**14**), the structure of which is compatible with a formal one-electron oxidation at platinum.[60] MO calculations on this used valence *s*-orbital polarizability to obtain a good fit with the observed NMR coupling constants.[61]

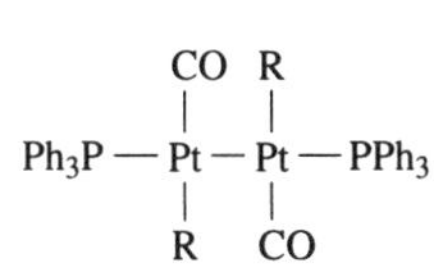

(13)

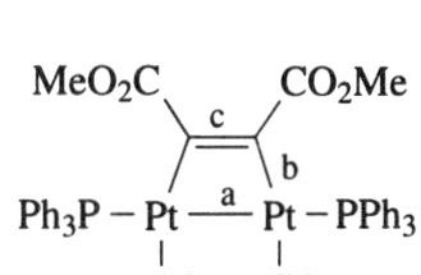

a, 0.263 54(8) nm
b, 0.207 0(11) nm
c, 0.134 1(22) nm

(14)

$MgCp_2$ and $(NBu^n_4)_2[Pt_2Cl_4(CO)_2]$ lead to good yields of $[Pt_2(CO)_2(\eta\text{-}Cp)_2]$ (**15**) along with some $[Pt(CO)(\eta^1\text{-}Cp)(\eta\text{-}Cp)]$.[62] The NMR spectra of (**15**) (^{13}CO-enriched) reveal rapid CO site exchange, presumably via μ-CO. (**15**) reacts with C_2Ph_2 in the presence of Me_3NO to produce (**16**). The anion $[PtCl_3(CO)]^-$ reacts with two equivalents of bis(pentamethylcyclopentadienyl)magnesium ($Mg(Cp^*)_2$) to produce (**17**) in yields of over 80%.[63] With only one $Mg(Cp^*)_2$, (**17**) is still obtained, along with unreacted anion and some [PtCl(CO)Cp*], suggesting that this latter compound is a reaction intermediate. Like the unsubstituted cyclopentadienyl analogue, the variable-temperature NMR spectra of (**17**) are consistent with rapid CO exchange across the Pt–Pt bond (0.263 6(1) nm). UV photolysis of (**15**) or (**17**) in matrices of argon, nitrogen or carbon monoxide at 20 K produced symmetrically bridged, coloured $[Pt_2(\mu\text{-}CO)Cp_2]$ or $[Pt_2(\mu\text{-}CO)Cp^*_2]$, respectively.[64] When these matrices were further irradiated at the wavelength responsible for their colours, the ejected CO recombined. The same unsaturated monocarbonyl fragments could be observed at room temperature using time-resolved IR spectroscopy. They have lifetimes over 2 ms, even in the presence of CO or MeCN. In CO matrices, prolonged irradiation of either compound leads ultimately, via Pt–Pt scissions, to $[Pt(CO)_4]$.

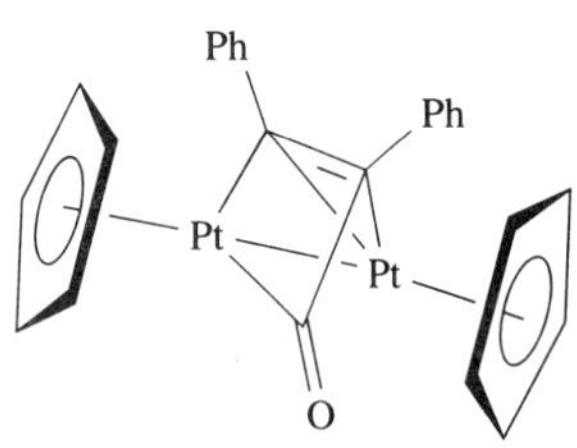

(16)

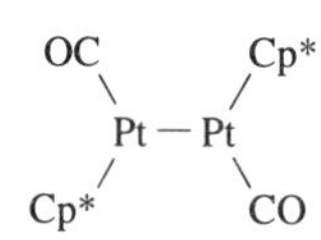

(17)

A number of compounds with bridging CO have been isolated. $[Pt(CO)(PPh_3)_3]$ reacts with $[PtCl_2(PhCN)_2]$ to produce (**18**) (X = Cl). The crystal structures of this and its bromo analogue reveal distinctly asymmetric CO bridges.[65] The hydrido binuclear platinum(II) complex cations $[Pt_2H_2(\mu\text{-}H)(biL)_2]^+$ (biL = dppe, dppp or dppb) react with carbon monoxide to eliminate H_2 and form highly coloured (**19**) with both H and CO bridges.[66] Values of $\nu(CO)$ fall in the range 1720–1750 cm^{-1}.

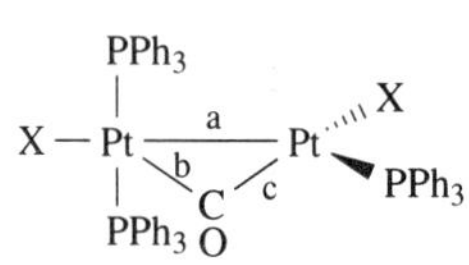

(18)

X = Cl	X = Br
a, 0.264 3(3) nm	a, 0.265 4(1) nm
b, 0.210(3) nm	b, 0.228 1(13) nm
c, 0.188(3) nm	c, 0.190 1(13) nm

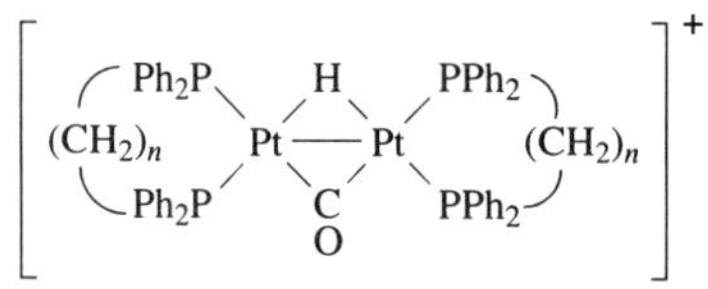

(19)

$n = 2$ ($[BF_4]^-$ salt)

a, 0.271 6(1) nm
b, 0.155(4) nm
c, 0.201 8(1) nm

Binuclear platinum(I) complexes with bridging sulfur are quite common. An example is $[Pt_2(\mu\text{-}S)(CO)(PPh_3)_3]$ (**20**), a probable intermediate in the reaction of $[Pt(PPh_3)_3]$ with COS producing, ultimately, $[Pt(CO)(PPh_3)_3]$, $[Pt(COS)(PPh_3)_2]$ and $[Pt(COS)_2(PPh_3)_2]$.[67] The sulfur atom of (**20**) is a strong nucleophile, forming complexes with copper(I), silver(I) and gold(I).[68] Reactions of (**5**) with COS, H_2S or S_8 produce $[Pt_2(\mu\text{-}S)(CO)_2(PR_3)_2]$, the sulfur of which can be methylated by MeI (Scheme 2),[33] or oxidized by *m*-chloroperbenzoic acid to form $[Pt_2(\mu\text{-}SO_2)(CO)_2(PR_3)_2]$ (**21**). Treatment of (**21**) with Me_3NO in the presence of carbon monoxide regenerates (**5**).

The insertion of carbon monoxide into Pt–Pt of $[Pt_2Cl_2(\mu\text{-}dppm)_2]$ was described in *COMC-I*. The same behaviour has been found in the mixed phosphine–arsine complex $[Pt_2Cl_2(\mu\text{-}Ph_2PCH_2AsPh_2)_2]$, which exists as 'head-to-head' and 'head-to-tail' isomers.[69] The production of (**22**) and (**23**) is accompanied by some isomerizations, compatible with an associative process. The crystal structure of (**24**) reveals a Pt–Pt bond of 0.262 0(1) nm.[70]

The elimination of hydrogen from di- and trihydrides of dppm-bridged binuclear platinum complexes affords another synthetic route to carbonyls of this type, and contrasts with the reactions of the dppe,

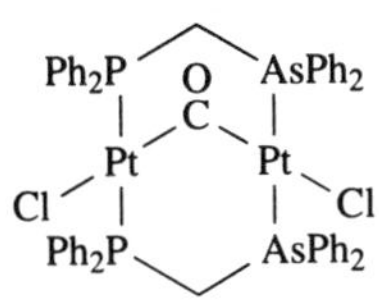

(22)

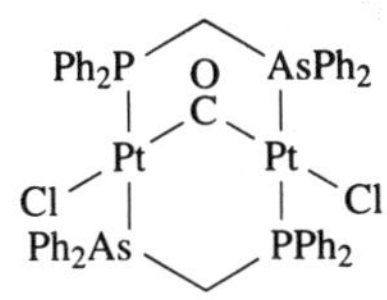

(23)

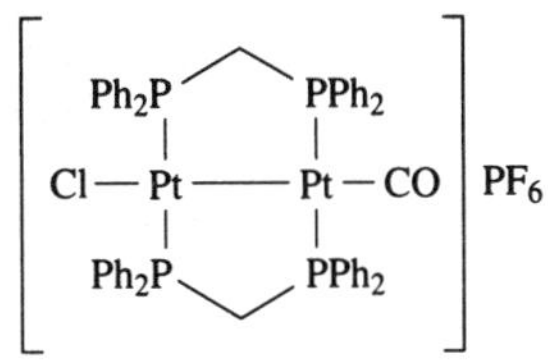

(24)

dppp and dppb analogues which produce (**19**). Scheme 6 and the reaction in Equation (10) illustrate this.[71] The ions $[Pt_2H(CO)(\mu\text{-biL})_2]^+$ (biL = dppm or $Et_2PCH_2PEt_2$, depm) rapidly and reversibly add one or two CO groups to form $[Pt_2H(CO)_2(\mu\text{-biL})_2]^+$ and $[Pt_2H(CO)_3(\mu\text{-biL})_2]^+$. Scheme 7 shows these reactions.[72] They are rapid on the NMR timescale even at room temperature and low carbon monoxide pressures. The overall reaction depicts a reversible change from formal diplatinum(I) to a Pt^0–Pt^{II} combination, and throws light on both the mechanism of ligand substitution at Pt^I and the fission of Pt–Pt bonds in such compounds.

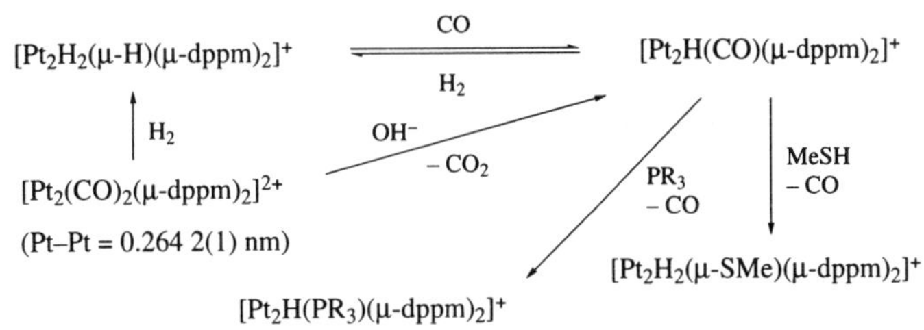

Scheme 6

$$[Pt_2H_2(\mu\text{-Cl})(\mu\text{-dppm})_2]^+ + CO \longrightarrow \textbf{(24)} + H_2 \qquad (10)$$

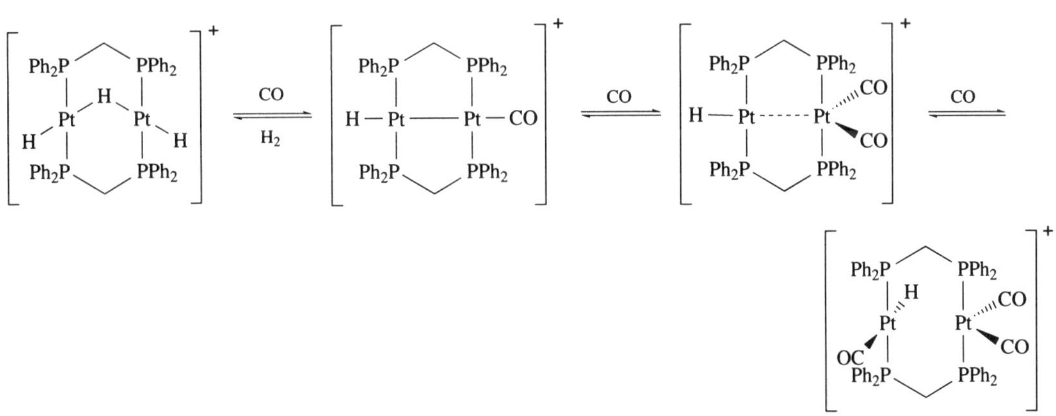

Scheme 7

Finally, the reaction of $[Pt_2Cl_2(\mu\text{-dppm})_2]$ with $NOBF_4$ to form $[Pt_2Cl_2(\mu\text{-NO})(\mu\text{-dppm})_2]BF_4$ is worthy of note, as nitrosyl complexes of platinum remain quite rare. This one is the first reported NO bridgehead 'A-frame' complex.[73]

7.1.5 Triangular dppm-bridged Clusters

This field of chemistry has been developed almost exclusively by Puddephatt and co-workers. It is based on the cation of (**25**), in which the platinum atoms have a formal oxidation number of 2/3. As the bridging dppm ligands hold the cluster together, it retains its structural integrity throughout many reactions, unlike the clusters described in Section 7.1.2. The behaviour of the three platinum atoms might thus be compared with a platinum crystallite surface, where the metal atoms have a similar local environment. The chemistry of this family of compounds was reviewed in 1990.[74]

Synthesis of the cation in (**25**), which contains a rare triply bridging CO, is achieved by CO reduction of $[PtX_2(dppm)]$ ($X = CF_3CO_2^-$) in aqueous acetone (Equation (11)).[75] Scheme 8 ($X = CF_3CO_2^-$), which

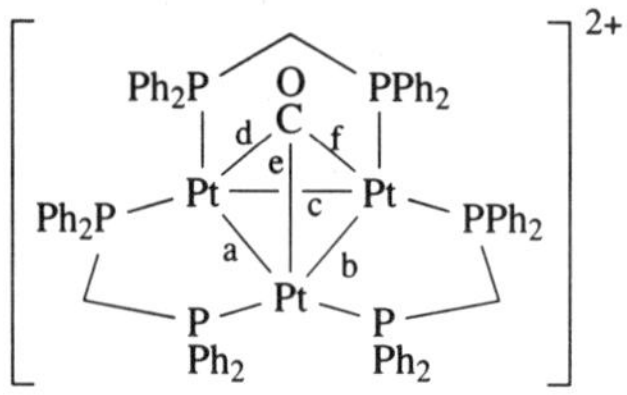

(25)

is supported by the detection in solution of several of the intermediates, describes the likely course of the reaction. The sequence of reductions can proceed beyond (**25**) to $[Pt_4(\mu\text{-}H)(\mu\text{-}CO)_2(\mu\text{-dppm})_3(\text{dppm-}P)]^+$,[76] which in turn can oxidize to $[Pt_4(\mu\text{-}H)(\mu\text{-}CO)_2(\mu\text{-dppm})_3\{Ph_2PCH_2P(O)Ph_2\}]^+$ or $[Pt_4(\mu\text{-}CO)(\mu\text{-dppm})_3\{Ph_2PCH_2P(O)Ph_2\}]^+$,[24] a step suppressed or reversed by free HX.

$$3\,[PtX_2(dppm)] + 3\,CO + 2\,H_2O \xrightarrow{100\,°C} [Pt_3(\mu_3\text{-}CO)(\mu\text{-dppm})_3]X_2 + 4\,HX + 2\,CO_2 \quad (11)$$

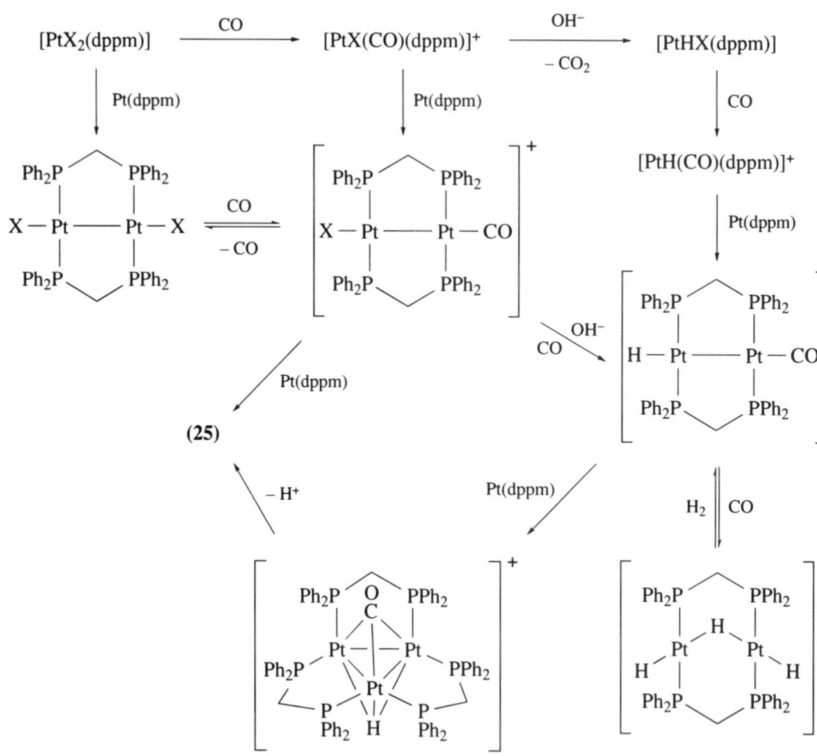

Scheme 8

The six phosphorus atoms of (**25**) lie close to the Pt_3 plane, and the 12 phenyl rings essentially shield and protect the sites above and below the triangle.[75] The crowded nature of the Pt_3 unit in these compounds is illustrated by a crystal structure of an SCN^- adduct, $[Pt_3(\mu_3\text{-}CO)(SCN)(\mu\text{-dppm})_3]PF_6$.[77] Bonding calculations based on Pt_3L_6 and CO^{2+} indicate the HOMO to be a filled Pt–Pt bonding orbital, with the LUMO an antibonding Pt–Pt orbital or nonbonding combinations of the platinum $6p_z$ orbitals.[78] Like many other clusters, (**25**) is luminescent at low temperatures and the nature of its lowest level excited state has been related to its UV–visible, emission and polarized excitation emission spectra.

The ν(CO) of (**25**) falls at 1750 cm^{-1} with anion $CF_3CO_2^-$, or 1765 cm^{-1} with PF_6^-. Its ^1H NMR spectrum reveals the expected AB quartet for the dppm CH_2 units. The effects of the four Pt_3 isotopomers are apparent in its ^{31}P and ^{195}Pt spectra.[77] A large (ca. 150 Hz) P–P coupling constant is assigned to the near-*trans* P–Pt–Pt–P units, and the value for $J_{Pt\text{-}Pt}$ (ca. 540 Hz) is close to those for other triangular clusters.

Cation (**25**) readily exchanges CO in solution at ambient temperatures, and unstable adduct (**26**) (Scheme 9) is reversibly produced at 1 atm pressure.[79] IR bands at 1760 cm^{-1} and 2075 cm^{-1} clearly indicate the presence of both μ_3-carbonyl and terminal carbonyl in (**26**), but its NMR spectra fit an ion of D_{3h} symmetry even at −90 °C. A very rapid fluxional process interconverts (**26**) and (**26a**), possibly via $[Pt_3(\mu_3\text{-}CO)_2(\mu\text{-dppm})_3]^{2+}$, which also allows the migration of the terminal CO. At higher carbon monoxide pressures, (**27**) can be observed in solution (Scheme 9). The low energies associated with the fluxional processes of Scheme 9 have great significance for the mobility of carbon monoxide on platinum crystallite surfaces.

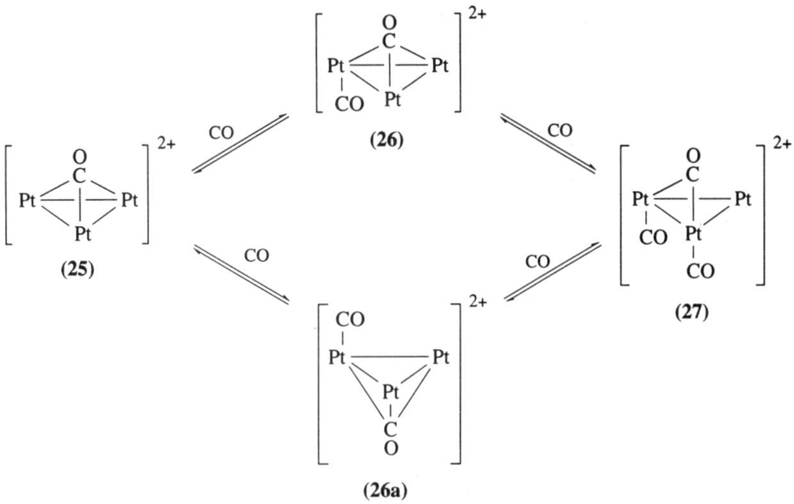

Scheme 9

Several other ligands behave like CO and produce adducts analogous to (**26**). Cyanide adds to (**25**) to yield (**28**).[80] Its x-ray crystallographic characterization was insufficiently accurate to unambiguously differentiate between CO and CN, but ^{13}CO and ^{13}CN labelling established the relative positions. This ion, too, is highly fluxional even down to −90 °C, and motions via μ_3-CN are possible. An interesting structural feature is that the triply bridging carbonyl is displaced significantly towards the platinum of highest coordination number. Tertiary phosphines or phosphites also readily attach to (**25**), and the crystal structure of one of them, (**29**), reveals this same feature.[81] These adducts also are fluxional with the PR_3 groups migrating around the Pt_3 triangles. The motion of the PF_3 adduct could not be frozen out even at −90 °C, leading to the speculation that the PF_3 may, in fact, be triply bridging. MO calculations on $[Pt_3(\mu_3\text{-}CO)(PH_3)(\mu\text{-}H_2PCH_2PH_2)_3]^{2+}$ enable comparisons to be made with $[Pt_3(\mu_3\text{-}H)(PH_3)(\mu\text{-}H_2PCH_2PH_2)_3]^{+}$.[82]

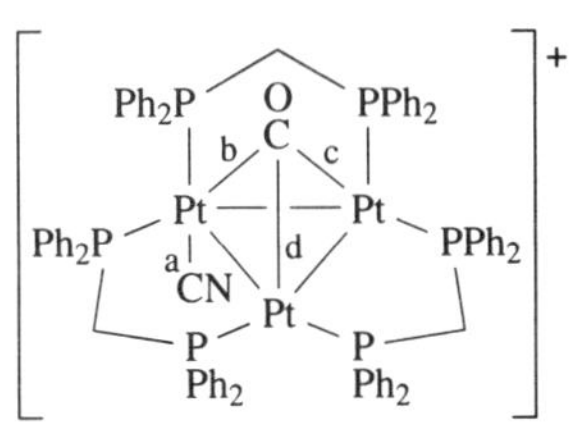

a, 0.207(2) nm
b, 0.194(2) nm
c, 0.244(2) nm
d, 0.225(2) nm

(28)

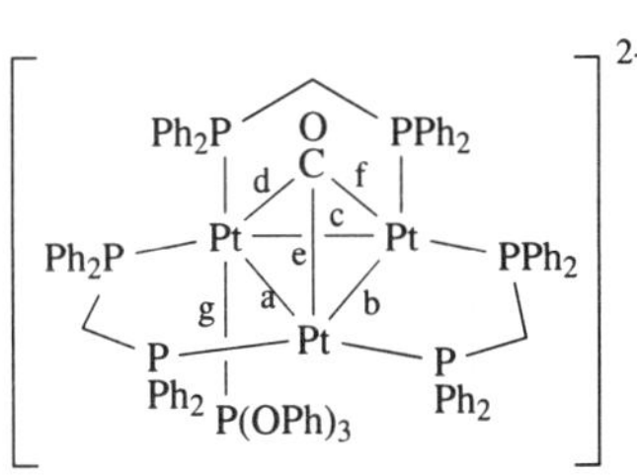

a, 0.265 6(2) nm
b, 0.263 6(2) nm
c, 0.262 6(2) nm
d, 0.193(3) nm
e, 0.216(3) nm
f, 0.227(3) nm
g, 0.229 3(8) nm

(29)

Bidentate ligands $R_2NCS_2^-$, R_3PCS_2, dppm and dmpm also add reversibly to (**25**), producing fluxional μ-carbonyl complexes (**30**), in which the bidentate ligands LL and CO migrate around the Pt_3 triangles.[83] The crystal structure of (**30**) (LL = $Me_2NCS_2^-$) confirms the geometry. Halide, X^-, adducts of (**25**) have the symmetrical structures $[Pt_3(\mu_3\text{-}X)(\mu_3\text{-}CO)(\mu\text{-dppm})_3]^+$, but thiocyanate is sulfur-bonded to one platinum only.[77,84] The $[Pt_3(SCN)(\mu_3\text{-}CO)(\mu\text{-dppm})_3]^+$ so formed reacts photochemically to produce $[Pt_3(\mu_3\text{-}S)(CN)(\mu\text{-dppm})_3]^+$, a reaction retarded by free CO. CS_2 and COS form[84] sulfur-bonded analogues of the SCN^- adduct of (**25**), but H_2S and H_2Se produce $[Pt_3H(\mu_3\text{-}E)(\mu\text{-dppm})_3]^+$ (E = sulfur or selenium).[85] Bu^{t}SH also produces $[Pt_3H(\mu_3\text{-}S)(\mu\text{-dppm})_3]^+$ with (**25**), but most other thiols lead instead[86] to $[Pt_3H(\mu_3\text{-}SR)(\mu\text{-dppm})_3]^{2+}$. A rapid preequilibrium between the reactants and $[Pt_3(\mu_3\text{-}CO)(RSH)(\mu\text{-dppm})_3]^{2+}$ can be detected spectroscopically below 0 °C.

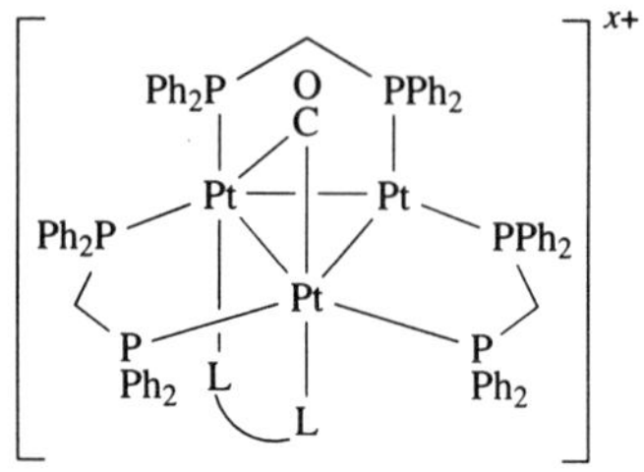

(30)

Hydride from $Na[BH_4]$ reacts with (**25**) via unstable adduct $[Pt_3(\mu_3\text{-}H)(CO)(\mu\text{-dppm})_3]^+$ to eliminate the (now terminal) CO and produce $[Pt_3(\mu_3\text{-}H)(\mu\text{-dppm})_3]^+$. This $Pt_3(\mu_3\text{-}H)$ unit undergoes ready inversion and may well represent the mechanism whereby surface hydrogen migrates into bulk metal.[87] Cation (**25**) is known to catalyse the water–gas shift reaction and the triangular hydrido species is involved in this process (Scheme 10).[24,74] Alkynes also displace the CO from (**25**).[88] Mercury atoms or SnX_3^- groups add to (**25**) to triply bridge the open Pt_3 face. A second SnX_3^- displaces the carbon monoxide molecule.[89]

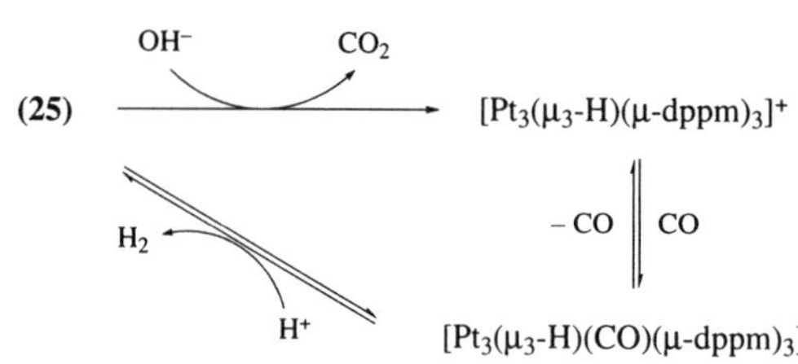

Scheme 10

A number of other clusters related to (**25**) but with dppm replaced by other phosphines are known. $[Pt_3(\mu_3\text{-}CO)(\mu\text{-}R_2PCH_2PR_2)_3]^{2+}$ (R = *p*-Tol or 3,5-$Me_2C_6H_3$) are made in the same way as (**25**),[74] and $[Pt_2(dmpm)_3(PPh_3)]$ reacts with carbon monoxide in aqueous methanol to produce the 46-electron cluster (**31**), which clearly resembles its dppm analogue.[90] Carbon-13 readily replaces CO in (**31**), but SO_2, CS_2 and C_2Ph_2 do not react.

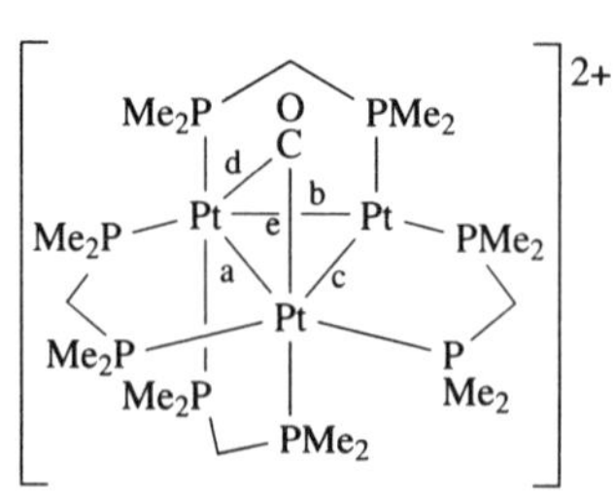

a, 0.264 8(1) nm
b, 0.262 8(1) nm
c, 0.262 0(1) nm
d, 0.204 9(8) nm
e, 0.206 1(9) nm

(31)

7.1.6 Platinum(II) Carbonyls

Most platinum(II) carbonyls are mononuclear 16-electron square-planar species, but the broad range of other ligands attached to the platinum centre means that their properties vary widely. They can be anionic, neutral or cationic.

7.1.6.1 Monocarbonyl complexes

(i) Syntheses

The syntheses of monocarbonyl complexes commonly fall into one of three categories:

(a) replacement of another ligand by CO;

(b) cleavage by CO of a ligand-bridged binuclear complex (which can be seen as a special case of (a)); and

(c) conversion of one carbonyl complex into another by replacement of an ancillary ligand. Table 1 lists examples.

Table 1 Syntheses of platinum(II) monocarbonyls.

Substrate	*Reaction type and reactants*[a]	*Products*	*Notes*	*Ref.*
$[PtCl_2(PPh_3)_2]$	(a) CO + NaX	$[PtCl(CO)(PPh_3)_2]X$	X = BF_4, ClO_4, PF_6, Ts	91
[PtCl(triL)]	(a) CO + $NaBPh_4$	$[Pt(CO)(triL)]BPh_4$	triL = $Bu^t_2PC_2H_4CHC_2H_4PBu^t_2$	92
trans-$[PtHX(PBu^t_3)_2]$	(a) CO	*trans*-$[PtH(CO)(PBu^t_3)_2]X$	X = Cl, Br, I, CF_3CO_2, NO_3, CN	14
trans-$[PtX_2L(py)]$	(a) CO	*cis*-$[PtX_2(CO)L]$	X = Cl, Br; L = CH(py)Et	93
$[PtCl_3(R_2SO)]^-$	(a) CO	$[PtCl_3(CO)]^-$	By-products R_2S and CO_2[b]	94
$[Pt(CR^1{=}CR^2R^3)L_3]^+$	(a) CO	$[Pt(CR^1{=}CR^2R^3)(CO)L_2]^+$	L = PPh_3; various R^1, R^2, R^3	95
trans-[Pt(OTf)-($CR^1{=}CR^2R^3)L_2$]	(a) CO	$[Pt(CR^1{=}CR^2R^3)(CO)L_2]^+$	L = PPh_3; various R^1, R^2, R^3	95
(**32**)	(a) CO	$[Pt(C_6F_5)(CO)(C_6H_4$-2-CPh=*N*N=$CPh_2)]$	η^2-Arene displaced	96
$[Pt_2(\mu\text{-}Cl)_2Me_2(C_2H_4)_2]$	(a) CO	$[Pt_2(\mu\text{-}Cl)_2Me_2(CO)_2]$	Cl bridges remain intact	97
$[Pt_2(\mu\text{-}I)_2(C_6H_4$-2-$CH_2$*N*$H_2)_2]$	(b) CO	$[PtI(CO)(C_6H_4$-2-$CH_2NH_2)]$		98
$[Pt_2(\mu\text{-}Cl)_2(C_6H_4$-2-N=*N*Ph$)_2]$	(b) CO	$[PtCl(CO)(C_6H_4$-2-N=*N*Ph$)]$		99
$[Pt_2(\mu\text{-}X)_2X_2L_2]$	(b) CO	*trans*-, then *cis*-$[PtX_2(CO)L]$	X = Cl, Br, I; L = PR_3 or AsR_3	100
$[Pt_2(\mu\text{-}Cl)_2Cl_2(PCy_3)_2]$	(b) CO	*trans*-$[PtCl_2(CO)(PCy_3)]$	Very slow isomerization to *cis*	101
$[Pt_2(\mu\text{-}Cl)_2Cl_2(AsBu^t_3)_2]$	(b) CO	*cis*-$[PtCl_2(CO)(AsBu^t_3)]$		102
cis-$[PtCl_2(CO)L]$	(c) TlCp	$[PtCl(CO)(\eta^1\text{-}Cp)L]$	L = PR_3; Cp is *trans* to L	103
$[PtClR^1(CO)L]$	(c) TlCp	$[PtR^1(CO)(\eta^1\text{-}Cp)L]$	R^1 = Ph or C≡CPh; L = PR^2_3; CO is *trans* to R^1	104
cis-$[PtCl_2(CO)L]$	(c) HgR^1_2	$[PtClR^1(CO)L]$	R^1 is *ortho*-aryl; L = PR^2_3	105
cis-$[PtCl_2(CO)(AsBu^t_3)]$	(c) L	*cis*-$[PtCl_2(CO)L]$	L = PBu^t_3, PCy_3, P(*o*-Tol)$_3$ PPh_3	106
$[PtCl_3(CO)]^-$	(c) $SnCl_2$	*cis*-$[PtCl_2(SnCl_3)(CO)]^-$, $[Pt(SnCl_3)_4(CO)]^{2-}$ and *trans*-$[PtCl(SnCl_3)_2(CO)]^-$	CO atmosphere needed	107
$[Pt_2(\mu\text{-}Cl)_2Cl_2(CO)_2]$	(c) L	*cis*-$[PtCl_2(CO)L]$	L = Me_2SO, Et_2SO, C_6H_{10}[c]	108
cis-$[PtCl_2(CO)(Et_2SO)]$	(c) en	$[Pt(CO)(Et_2SO)(en)]Cl_2$		108a
cis-$[Pt(C_6F_5)_2(CO)_2]$	(c) L	$[Pt(C_6F_5)_2(CO)L]$	L = $Ph_2C{=}N{-}N{=}CPh_2$	96

[a] See text. [b] CO is displaced by excess R_2SO, making its reduction to R_2S catalytic. [c] HCl converts the coordinated cyclohexene to chlorocyclohexane, regenerating $[Pt_2Cl_4(CO)_2]$.

The isomerization of *trans*-$[PtX_2(CO)L]$ (obtained from carbon monoxide cleavage of $[Pt_2X_4L_2]$) to its *cis*-isomer is catalysed by free carbon monoxide, so the *trans*-isomer can only be isolated in its absence.[100] Even so, the *trans*-form tends to lose CO in solution, so its isomerization becomes autocatalytic. The associative part of this process may well involve pseudorotation of a five-coordinate intermediate formed from *trans*-$[PtX_2(CO)L]$ and carbon monoxide. Free carbon monoxide also catalyses *cis–trans* isomerizations of $[PtCl_2(PR_3)_2]$, and related addition–elimination reactions are involved here, too.[109]

Cis–trans isomerizations of $[PtX_2(CO)L]$ are accelerated by UV irradiation.[100,110] This appears to take place from a singlet excited state of relatively long lifetime, with Pt–CO bond breaking followed by catalysis by the freed CO (Equations (12)–(14)).[110]

$$cis\text{-}[PtCl_2(CO)L] \xrightarrow{h\nu} cis\text{-}[PtCl_2(CO)L]^* \qquad (12)$$

$$cis\text{-}[PtCl_2(CO)L]^* \longrightarrow [PtCl_2L] + CO \qquad (13)$$

$$cis\text{-}[PtCl_2(CO)L]^* \xrightarrow{CO} trans\text{-}[PtCl_2(CO)L] \qquad (14)$$

Equation (15) (R = Me or Ph), representing a CO complex formation accompanied by η-Cp to η^1-Cp conversion, is a variation of preparative method (a).[111]

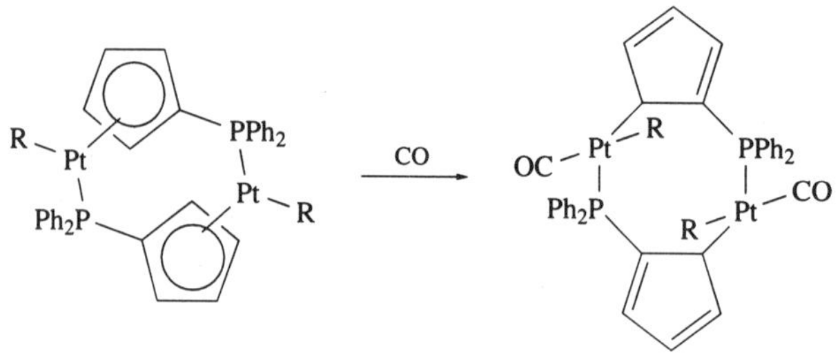

(15)

(ii) Structures and spectra

Crystal structures have been reported for *trans*-$[PtCl(CO)(PPh_3)_2]BF_3Cl$,[112] *cis*-$[Pt(C_6F_5)_2(CO)\{SC(S)PCy_3\}]$[113] and (**32**)–(**35**).[98,114,115] Only when CO is *trans* to the high *trans*-influence C_6F_5 is Pt–CO substantially longer (0.190 1(9) nm) than in the other cases (ca. 0.185 nm). (**35**) exhibits fluxional behaviour in its NMR spectra, reversibly losing CO.

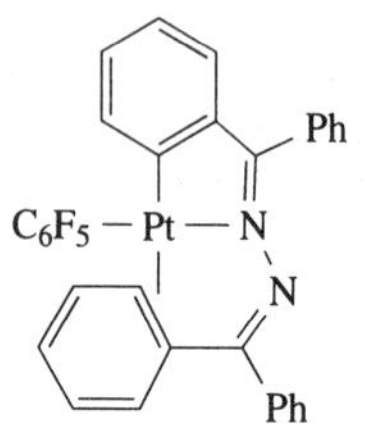

(32)

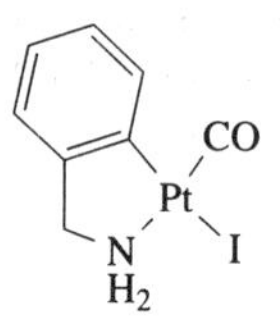

(33)

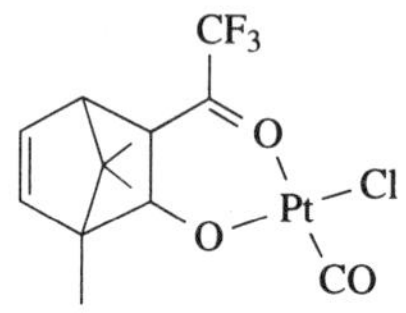

(34)

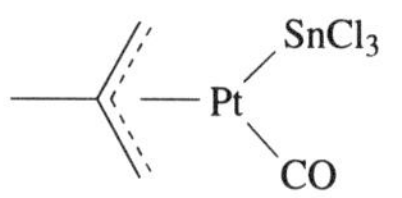

(35)

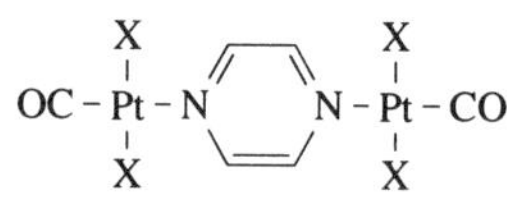

(36)

Comparisons of the solution IR spectra of $[PtCl_3(CO)]^-$ and $[Pt_2Cl_4(CO)_2]$ with those of their palladium analogues indicate that π-backbonding is significantly greater in the platinum complexes, a factor which might contribute to their lower reactivity.[116] The IR, ^1H NMR and UV spectra of *trans*-$[PtX_2(CO)L]$ (X = Cl or Br; L = $ArNH_2$, NH_3, py, pyO (pyridine oxide) or imidazole) and (**36**) show that the shielding of the aromatic protons is affected by CO, and ν(Pt–C) and ν(Pt–N) are increased by electron-releasing substituents when L is a substituted aniline, showing that the electronic effects are transmitted to these bonds.[117] The usual red shift is observed in the $5d \rightarrow \pi^*$ transition when chlorine is replaced by bromine.

The three isomers of [PtClPh(CO)L] (L = $PMePh_2$), (**37**)–(**39**), have been examined. (**37**) was made by treating *cis*-$[PtCl_2(CO)L]$ with $HgPh_2$, and (**38**) and (**39**) by the action of carbon monoxide on $[Pt_2(\mu\text{-}Cl)_2Ph_2L_2]$, which exists in solution as a rapidly interconverting *cis–trans* mixture.[118] Table 2 lists salient IR and NMR parameters, along with those of *cis*-$[PtCl_2(CO)L]$ for comparison.[118–20] The effects of the large *trans* influence of Ph on $^1J_{Pt\text{-}P}$ and $^1J_{Pt\text{-}C}$ are evident, allowing unambiguous identification of isomers of this type. Examination of ^{13}C NMR data on 28 neutral monocarbonylplatinum(II) complexes revealed a much larger dependence of δC and $^1J_{Pt\text{-}C}$ on the nature of the *cis* (to ^{13}CO) ligands than had previously been suspected.[119] Selective irradiation revealed that $^1J_{Pt\text{-}C}$ are positive.[120] The two-bond couplings between ^{13}C and phosphine P are large and positive for *trans* coupling, and small and negative for *cis*.

The electronic spectra and luminescence of complexes *cis*-$[PtPh_2(CO)L]$ (L = ring-*N*-bonded 5-aminoquinoline or amine-bonded 9-aminoacridine) indicate that the $[PtPh_2(CO)]$ moiety behaves as an electron withdrawer on the aromatic rings and affects the $\pi \rightarrow \pi^*$ transitions of the ligands.[121] Their luminescence properties (in nanosecond timescale) are compatible with the emissions being from singlet π^* excited states.

Table 2 Spectroscopic parameters of some platinum carbonyls.

	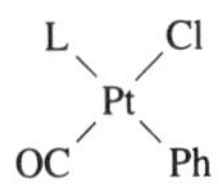 (37)	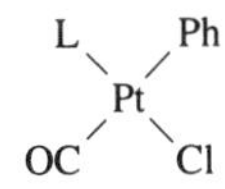 (38)	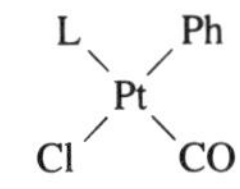 (39)	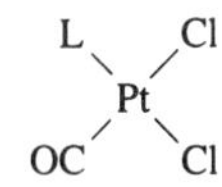
ν(CO) (cm^{-1})	2087	2106	2116	2115
δP (ppm)	6.76	−1.72	5.71	−0.21
($^1J_{Pt-P}$) (Hz)	(1402)	(3920)	(3481)	(2946)
δC (CO) (ppm)	162.1	177.4	173.3	156.3
($^1J_{Pt-C}$) (Hz)	(1947)	(906)	(1427)	(1764)
$^2J_{Pt-C}$ (Hz)	8.2	6.1	157.8	6.4
δ^{195}Pt	−515.5	−732.2	−649.7	

(iii) Ligand replacement reactions

CO is readily replaced by other ligands and the stereoretentive nature of the reactions enables them to be exploited to synthesize otherwise inaccessible isomers. Thus, *cis*-[Pt(C≡CPh)$_2$L$_2$] can be made from *cis*-[Pt(C≡CPh)$_2$(CO)L] (L = tertiary phosphine).[122] The volatility of the eliminated CO is probably responsible for its frequent displacement, as there is evidence from many reactions in which other ligands are replaced that CO is quite strongly bonded to Pt^{II}. For example, SEt$_2$ is displaced by pyridine from [PtPh$_2$(CO)(SEt$_2$)], and a variety of tertiary phosphines preferentially displace 5-aminoquinoline (L) from *cis*-[PtPh$_2$(CO)L].[123] Kinetic evidence supports the operation of the usual associative displacement mechanism in each case.

The establishment of *cis*-[PtCl$_2$(CO)L] (L = PR$_3$ or AsR$_3$) along with SnCl$_2$ as hydroformylation catalysts[124] has ensured careful spectroscopic examination of such systems,[125] and here, too, the Pt–CO often remain intact. Intermediates detected in solutions of such catalytic mixtures by ^{1}H, ^{13}C, ^{31}P, ^{119}Sn and ^{195}Pt NMR investigations include *trans*-[PtCl(CO)L$_2$]$^+$, *trans*-[PtCl(SnCl$_3$)$_2$(CO)]$^-$, [Pt(SnCl$_3$)$_5$]$^{3-}$ and [PtCl(SnCl$_3$)$_2$L]$^-$, but few could be isolated.

Cationic *trans*-[PtCl(CO)(PEt$_3$)$_2$]$^+$ reacts with water to produce *trans*-[PtCl(CO$_2$H)(PEt$_3$)$_2$], which can eliminate carbon dioxide to form *trans*-[PtHCl (PEt$_3$)$_2$].[126] *Trans*-[PtX(CO)L$_2$]BF$_4$ (X = Cl or Br; L = PPh$_3$ or AsPh$_3$) reacts with Na[ON(CF$_3$)$_2$] to produce (**40**).[127] Cations [PtCl(CO)(SbPh$_3$)$_2$]$^+$ and [PtI(CO)(PPh$_3$)$_2$]$^+$ do not undergo this reaction, however.

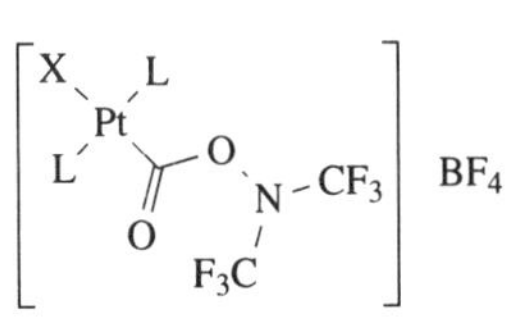

(40)

(iv) Migratory insertion reactions

Most of these reactions occur at platinum(II) monocarbonyls. Since the publication of *COMC-I* there have been a number of developments, and some results have been reviewed.[128] Of the three isomers (**37**)–(**39**) Table 2, only complexes with the same geometry as (**37**) undergo migratory insertions, equilibrating in solution with acyl or aroyl dimers (e.g., Equation (16)).[118] Replacement of chlorine by bromine or iodine displaces the equilibrium to the right, a consequence of stronger halide bridges.[129] Phosphine ligands, L, of increasing *trans* influence promote the CO insertion, a factor compatible with R migration to CO being the operating pathway. Beyond a critical size, however, the bulk of L prevents the migration proceeding. In most cases addition of free L to the equilibrium of Equation (16) causes complete conversion to *trans*-[PtX(COR)L$_2$], even when the equilibrium lies well to the left. With vanishingly small equilibrium constants, however, CO is displaced by L instead. Scheme 11 summarizes this.

$$2\,(\mathbf{37}) \rightleftharpoons [\mathrm{Pt_2(\mu\text{-}Cl)_2(COR)_2L_2}] \qquad (16)$$

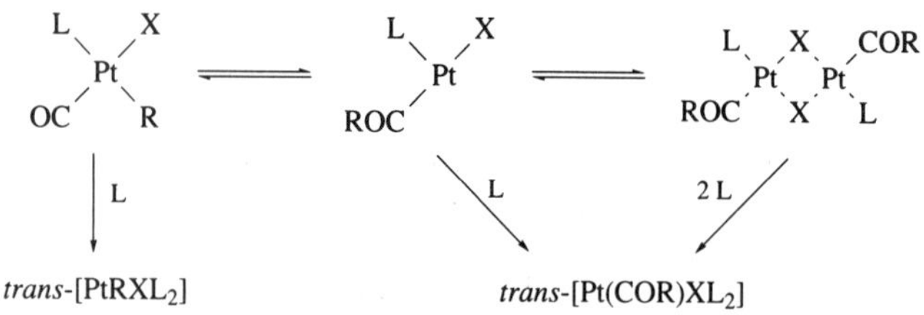

Scheme 11

The nature of the organic group is also of critical importance to equilibrium (16) or Scheme 11. The tendency towards R migration diminishes along the series Et > Ph > Me > Bz, C≡CR, $CCl{=}CCl_2$, η^1-Cp. Electron-donating *meta-* and *para*-aryl substituents promote insertion, but *ortho* substituents prevent it, presumably due to steric congestion.[130]

In agreement with experimental findings, MO calculations on $[PtMeF(CO)(PH_3)]$ determined the easiest migration to be in the isomer with PH_3 *trans* to Me, and the result was a T-shaped acyl fragment $[Pt(COMe)F(PH_3)]$.[131] With $[PtHMe(CO)(PH_3)]$, on the other hand, calculations suggest that migration of Me *trans* to H is the lowest-energy process.[132] As yet there is no experimental confirmation of this.

Two reaction pathways operate in the carbonylation of *trans*-$[PtRXL_2]$ to produce *trans*-$[Pt(COR)XL_2]$.[133] Migratory insertion from the five-coordinate intermediate $[PtX(CO)RL_2]$ leads to the product in one route. The other involves reversible replacement of L by CO and a series of isomerizations until the isomer of [PtX(CO)RL] with R *trans* to L (tertiary phosphines) results and undergoes migratory insertion as in Scheme 11. In a similar manner, the introduction of CO into solutions of *trans*-$[PtPh(SnCl_3)L_2]$ (L = $PMePh_2$ or PPh_3) results in $SnCl_3^-$ displacement to form *trans*-$[PtPh(CO)L_2]SnCl_3$, followed by its reattack leading, eventually, to *trans*-$[Pt(COPh)(SnCl_3)L_2]$.[134] Carbonylations of the allyl complexes *trans*-$[PtCl(\eta^1\text{-allyl})L_2]$ (L = PPh_3 or PCy_3) likewise lead to *trans*-$[PtCl(COC_3H_5)L_2]$, but when L_2 is $Ph_2PCH{=}CHPPh_2$, the resulting *cis* complex does not react with CO.[135] CO treatment of [PtXPh(biL)] (biL = dppe, dppm or $Ph_2PC_2H_4AsPh_2$) also fails to produce any insertion, perhaps due to steric constraint or the inability of CO to displace a chelating ligand.[136] It is interesting in this context to note that the 1-cyclohexenyl complexes $[Pt(C_6H_9)(CO_2H)(biL)]$ and $[Pt(C_6H_9)(CO_2Me)(biL)]$ do undergo carbonyl insertion when biL is the flexible dppp or dppb, but not with biL as dppe.[137] CO insertion into Pt–Me links of [PtMeX(biL)] (X = BF_4 or SO_3CF_3; biL = 1-dimethylamino-8-diphenylphosphinonaphthalene) also proceeds readily, by reversible X replacement.[138]

The vinyl complexes *trans*-$[PtBr(CR^1{=}CR^2{}_2)(PEt_3)_2]$ (R^1 = adamantyl, R^2 = H; R^1 = Ph or naphthyl, R^2 = Me) have longer than expected Pt–C bonds and readily undergo CO insertion at 1 atm pressure.[139] The analogous reaction of *trans*-$[PtI(CH{=}CMe_2)(PPh_3)_2]$ has been followed by NMR spectroscopy and found to proceed by CO replacement of PPh_3, followed by migratory insertion.[140] With the related *trans*-$[Pt(OTf)(CMe{=}CH_2)(PPh_3)_2]$, however, CO displaces the triflate but does not insert into Pt–C.[95c] The insertions of CO into the Pt–C bonds of $[PtR(\beta\text{-dik})(PPh_3)]$ (R = Me or Ph; β-dik are various β-diketonates) proceed by first-order kinetics, and are sensitive to the nature of β-dik.[141]

The reverse process to all the above insertions, decarbonylation, has often been employed to produce CO complexes. It usually requires a vacant coordination site, provided by the reaction in Equation (17) by Ag^+ removal of chloride.[142] The rearrangement reaction shown in Equation (18) might proceed by a decarbonylation then insertion.[143]

$$trans\text{-}[\mathrm{PtCl(COCOPh)(PPh_3)_2}] + \mathrm{AgBF_4} \longrightarrow trans\text{-}[\mathrm{Pt(COPh)(CO)(PPh_3)_2}]\mathrm{BF_4} + \mathrm{AgCl} \qquad (17)$$

$$trans\text{-}[\mathrm{PtEt(COCOPh)(PPh_3)_2}] \longrightarrow trans\text{-}[\mathrm{Pt(COEt)(COPh)(PPh_3)_2}] \qquad (18)$$

Nucleophilic attack at coordinated carbonyl leads to products like those of CO insertion. Equations (19) and (20) may be examples,[144,145] although there is evidence that at least some proceed by attack at platinum followed by migration.[144]

$$cis\text{-}[\mathrm{Pt(COPh)(CO)(PPh_3)_2}]^+ + \mathrm{RLi} \longrightarrow cis\text{-}[\mathrm{Pt(COPh)(COR)(PPh_3)_2}] \qquad (19)$$

R = Me or Ph

$$trans\text{-}[Pt(COPh)(CO)(PPh_3)_2]^+ + HNR_2 \longrightarrow trans\text{-}[Pt(COPh)(CONR_2)(PPh_3)_2] \quad (20)$$

7.1.6.2 Platinum(II) dicarbonyls

Platinum(II) dicarbonyls tend to be less accessible than monocarbonyls due to ready loss or replacement of CO. In 1990 work on dihalide derivatives was reviewed.[146] An earlier misconception over a material which was analysed as $[Pt_2Cl_4(CO)_3]$ has been cleared by x-ray diffraction studies, which showed it to be a mixture of $[PtCl_2(CO)_2]$ and $[Pt_2Cl_4(CO)_2]$.[147] *cis*-$[PtCl_2(CO)_2]$ can be synthesized by the cocondensation of platinum atoms and oxalyl chloride,[148] though it is more conventionally prepared from the action of CO on $[Pt_2Cl_4(CO)_2]$. When this latter reaction is performed in thionyl chloride at low temperatures and in the dark, *trans*-$[PtCl_2(CO)_2]$ results and can be precipitated by addition of hexane.[149] This readily loses CO to revert to the halide-bridged complex, or isomerizes to the more stable *cis* form.

The dibromo and diiodo platinum dicarbonyls become progressively less stable compared with $[PtCl_2(CO)_2]$.[150] They can be prepared by the action of dry HX on *cis*-$[PtCl_2(CO)_2]$, by the action of CO on $[Pt_2Br_4(CO)_2]$ (this passes through the *trans* form en route to the more isolable *cis* material) or by the action of CO on PtI_2 (which gives predominantly the *trans* isomer). $[PtI_2(CO)_2]$ maintains a rapid *cis–trans* equilibrium under a carbon monoxide atmosphere. On lowering the carbon monoxide pressure over either solid or solution, it converts to $[Pt_2I_4(CO)_2]$. Values of J_{Pt-C} decrease along the series Cl > Br > I in these complexes, indicative of a decreasing σ-content of the Pt–CO bonds.

White, solid organoderivatives *cis*-$[PtR_2(CO)_2]$ (R = C_6F_5 or C_6Cl_5) have been obtained by displacement of THF from $[PtR_2(THF)_2]$ by carbon monoxide at 1 atm pressure.[151] Equation (21) produces the *trans* isomer. Addition of PPh_3, py or halide to any of these displaces only one CO ligand. The action of CO on $[Pt_2Me_4(SMe_2)_2]$ reversibly replaces SMe_2 to yield first *cis*-$[PtMe_2(CO)(SMe_2)]$ then *cis*-$[PtMe_2(CO)_2]$.[97] *cis*-$[PtCl_2(CO)_2]$ reacts with alkenes to form *cis*-$[PtCl_2(CO)(alkene)]$.[108b]

$$trans\text{-}[PtR_2Cl(CO)]^- + CO + Ag^+ \longrightarrow trans\text{-}[PtR_2(CO)_2] + AgCl \quad (21)$$

The reactions of *cis*-$[PtCl_2(CO)_2]$ with alkynes have received extensive study: the products are dependent both on reaction conditions and on the alkyne (Scheme 12).[152–5] In the presence of water and excess CO, *cis*-$[PtCl_2(CO)_2]$ generates the triangular clusters $[Pt_3(CO)_6]_n^{2-}$ along with $(H_3O)[PtCl_3(CO)]$. These in turn react with $R^1C{\equiv}CCO(R^2)$ to produce complexes of the type shown in (**41**) and (**42**) (R^1, R^2 = Me, OMe; Ph, Me; or Pr, Et).[156]

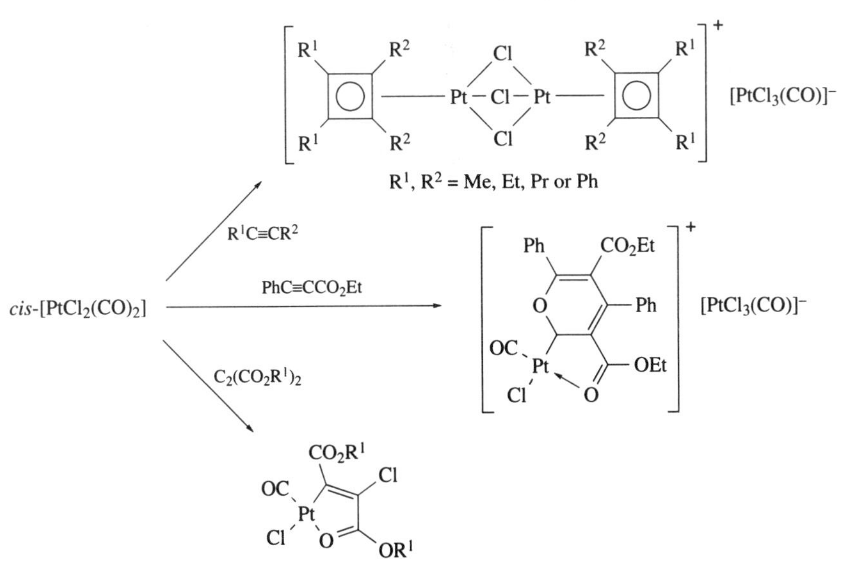

Scheme 12

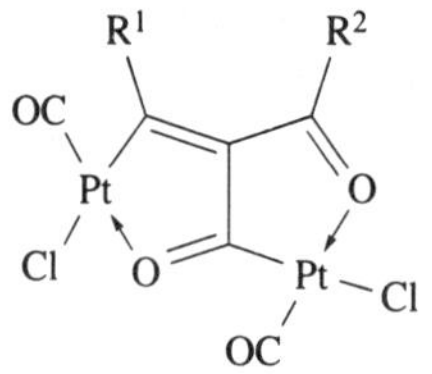

(41)

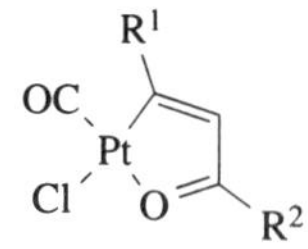

(42)

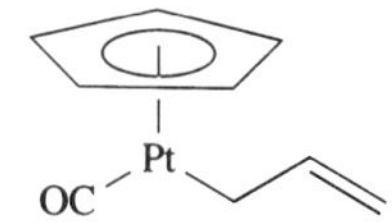

(43)

7.1.6.3 18-Electron platinum(II) carbonyls

18-Electron platinum(II) carbonyls are quite rare, most platinum(II) carbonyls being square-planar 16-electron species. [Pt(η-C_3H_5)(η-Cp)] reacts with CO to produce (**43**) as a yellow oil.[157] However, 16-electron [Pt(η-C_3H_5)(PPh_3)(CO)]BF_4 is produced when [Pt(η-C_3H_5)(cod)]BF_4 is sequentially treated with PPh_3 then CO.[158] [Pt(η-Cp)(η^1-Cp)(CO)] (along with the Pt^I dimer (**15**)) is produced by the reaction of $[Pt_2Cl_4(CO)_2]^{2-}$ with $MgCp_2$ at 0 °C,[62] or by displacing C_2H_4 from [Pt(η-Cp)(η^1-Cp)(C_2H_4)] by CO. It isomerizes to the two vinylic η^1-cyclopentadienyls in solution (Scheme 13), a behaviour which resembles several main-group cyclopentadienyls.

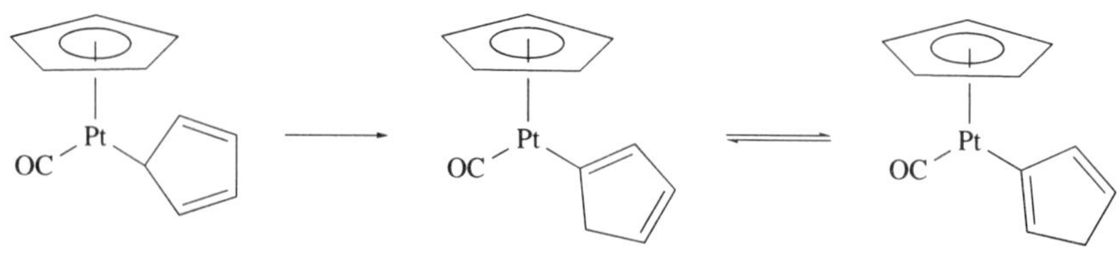

Scheme 13

[Pt(η-Cp*)X(CO)] (X = Cl or Br) are among the products from the reactions of $[PtX_3(CO)]^-$ and $Mg(Cp^*)_2$.[63,159] These react with tertiary phosphines, L, to produce 16-electron [PtX(η^1-Cp*)(CO)L], which in turn eliminate either CO or X^- to form [Pt(η-Cp*)XL] or [Pt(η-Cp*)(CO)L]X, respectively.

7.1.6.4 Binuclear platinum(II) carbonyls

Salient structural details of centrosymmetric [$Pt_2(\mu$-I$)_2I_2(CO)_2$] are shown in (**44**).[150b] [$Pt_2(\mu$-Cl$)_2Me_2(CO)_2$], obtained from [$Pt_2(\mu$-Cl$)_2Me_2(C_2H_4)_2$] and CO, is a white solid, unstable above 0 °C, with ν(CO) at 2040 cm^{-1} in CCl_2H_2.[97] Cocondensation of platinum atoms with CF_3COCl produces unstable [PtCl(CF_3CO)], which rearranges to (**45**).[160] The anion (**46**) was produced from $SnCl_2$ insertion into the Pt–Pt bond of $[Pt_2Cl_4(CO)_2]^{2-}$. This and its bromo analogue were isolated as their $[Pr_4N]^+$ salts and characterized by their ^{119}Sn and ^{195}Pt NMR spectra.[161]

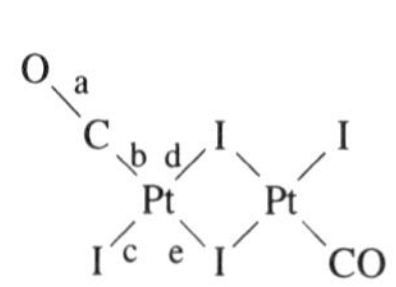

a, 0.106(4) nm
b, 0.188(3) nm
c, 0.259 6(3) nm
d, 0.263 0(2) nm
e, 0.262 2(3) nm

(44)

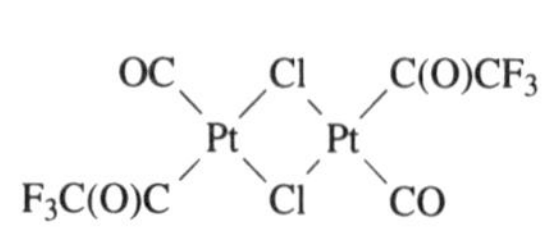

(45)

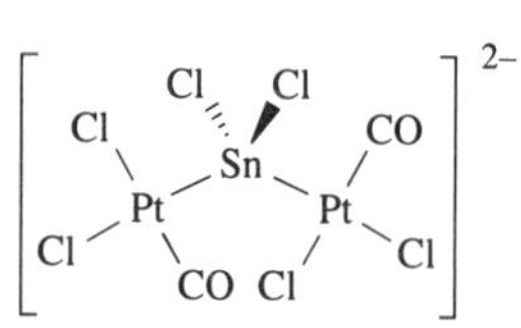

(46)

The bidentate ligand $Ph_2PCH_2NEt_2$ (ddpa) displaces CO from $[Pt_2Cl_4(CO)_2]^{2-}$ to produce [$Pt_2Cl_4(\mu$-ddpa$)_2]^{2-}$, which retains the Pt–Pt bond.[162] CO then cleaves this bond to form [$Pt_2Cl_4(\mu$-CO$)(\mu$-ddpa$)_2]^{2-}$.

A different reaction sequence is followed when CO reacts with neutral $[Pt_2Cl_4(\mu\text{-ddpa})_2]$, which does not have a Pt–Pt link. One chloride is first displaced, and this then reattacks to produce Pt–Pt-bonded $[Pt_2Cl_4(\mu\text{-CO})(\mu\text{-ddpa})_2]$. In none of these complexes was the relative orientation of the asymmetric ddpa ligands established.

7.1.7 Platinum(IV) Carbonyls

Platinum(IV) carbonyls remain uncommon. Chlorine reacts cleanly with $[PtCl_3(CO)]^-$ in thionyl chloride to produce $[PtCl_5(CO)]^-$, which was isolated as its $[NH_2Pr^i_2]^+$ salt.[163] The anion is stable in solution but reacts readily with water to produce carbon dioxide. Its CO stretching frequency of 2191 cm^{-1} is exceptionally high and, as might be expected, its value of $J_{Pt\text{-}C}$ (1267 Hz) is much lower than for platinum(II) carbonyls.[150] The platinum(IV) pyridine ylide complexes $[PtX_2IMe(py)\{CH(py)Et\}]$ react with CO in benzene to produce (**47**) (X = Cl or Br).[93] The methylplatinum(IV) carbonyl (**48**) was produced from *fac*-$[PtMe_3I(SMe_2)_2]$ and CO or from *cis*-$[PtMe_2(CO)(SMe_2)]$ and MeI.[97]

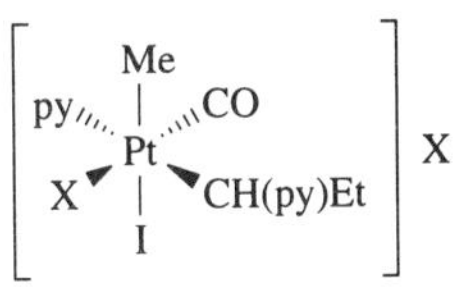

(47)

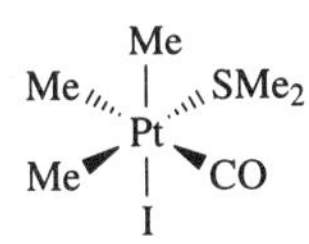

(48)

7.2 PLATINUM ISOCYANIDE COMPLEXES

7.2.1 Mononuclear Platinum(0) Isocyanides

Mononuclear platinum(0) isocyanides are quite uncommon. The aryltetraazadiene complexes (**49**) were isolated from the reactions of $[Ni(Ar_2N_4)_2]$ and $[Pt(cod)_2]$ in the presence of Bu^{t}NC.[164] The tetraazadienes only transfer in the presence of free isocyanide ligand, via intermediates $[PtNi(Ar_2N_4)_2(CNBu^t)_2]$ which probably incorporate bridging ligands. Reduction of $[PtCl_2L_2]$ (L = 2,6-xylyl isocyanide) by sodium amalgam in THF in the presence of free L produces (**50**) and (**51**) as well as clusters $[Pt_3L_6]$ and $[Pt_6HgL_{12}]$.[165] These probably proceed through $[PtL_2]$, which readily trimerize. Further reduction in the presence of L then leads to (**50**) and (**51**).

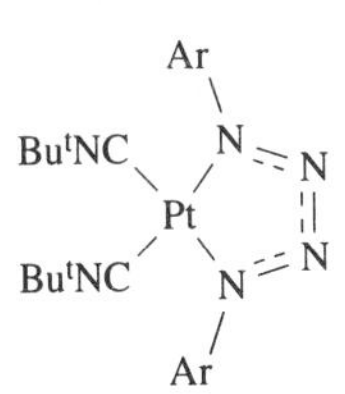

(49)

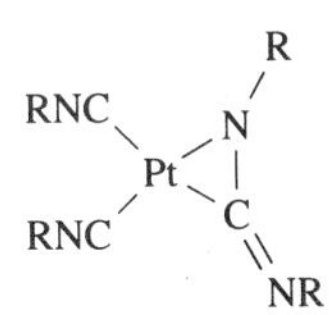

(50)

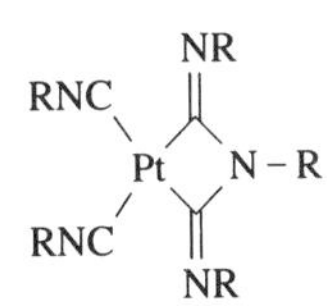

(51)

One reason for the rarity of mononuclear isocyanide complexes of Pt^0 is indicated from studies of alkene displacement from $[Pt(alkene)(PPh_3)_2]$ by RNC.[166] Although $[Pt(CNR)_2(PPh_3)_2]$ (R = 2,6-xylyl) was isolated as a yellow, air-sensitive, crystalline solid (ν(CN) at 2043 cm^{-1} and 1991 cm^{-1}), the complex is labile in solution and decomposes to high-nuclearity products.

7.2.2 Polynuclear Platinum(0) Isocyanides

As with platinum carbonyls, triangular Pt_3 clusters are very common. The isocyanide ligands readily adopt terminal or bridging positions, and in the former they rival (and can displace) tertiary phosphines. Thus, the family of neutral complexes $[Pt_3(\mu\text{-CNR})_3(CNR)_3]$ are analogous to $[Pt_3(\mu\text{-CO})_3(PR_3)_3]$ and are not prone to stacking. $[Pt(PBu^t_3)_2]$ reacts with $CNBu^t$ by displacement of both phosphines to form the orange-red solid $[Pt_3(\mu\text{-CNBu}^t)_3(CNBu^t)_3]$.[13] The same complex was produced by the action of $CNBu^t$ on *trans*-$[PtH_2(PBu^t_3)_2]$, possibly by the same route after H_2 elimination.[14] Interestingly, $[Pt(PBu^t_3)_2]$ and only one equivalent of $CNBu^t$ still lead only to $[Pt_3(CNBu^t)_6]$, whereas even an excess of CO displaces only one phosphine on each platinum.[13] $[Pt(cod)_2]$ and Pr^iNC yield only $[Pt_3(CNPr^i)_6]$, rather than mononuclear complexes.[167]

Electrochemical reductions of *cis*-$[PtCl_2(CNR)_2]$ also produce clusters.[168,169] Potentiometric reductions in MeCN at mercury pool electrodes appeared to yield either $[Pt_2Cl_2(CNR)_4]$, $[Pt_3(CNR)_6]$ or $[Pt_6Hg(CNR)_{12}]$, depending on the charge consumed and the reduction potential (R = 2,4-Bu^t_2-6-MeC_6H_2, 2,4,6-$Bu^t_3C_6H_2$ or 2,6-xylyl).[168] The main reduction waves on the d.c. polarograms related to the process of Equation (22).[169] The platinum(I) species resulted from the disproportionation reaction in Equation (23).

$$3\,[PtCl_2(CNR)_2] + 6\,e^- \longrightarrow [Pt_3(CNR)_6] + 6\,Cl^- \qquad (22)$$

$$[Pt_3(CNR)_6] + 3\,[PtCl_2(CNR)_2] \longrightarrow 3\,[Pt_2Cl_2(CNR)_4] \qquad (23)$$

The Pt–Pt bonds in the clusters are readily broken. $[Pt_3(CNR)_6]$ reacts with $[Ni(N_4Ar_2)_2]$ to produce **(49)**[164] (R = Bu^t) and with electron-withdrawing alkenes to form $[Pt(\eta^2\text{-}C_2X_4)(CNR)_2]$ (R = 2,6-xylyl).[167] With $SiHMePh_2$ or diphenylcyclopropenone, $[Pt_3(CNR)_6]$ reacts to form **(52)** or **(53)**, respectively,[167] and with sulfur or CS_2 to produce **(54)** or **(55)**, respectively.[170]

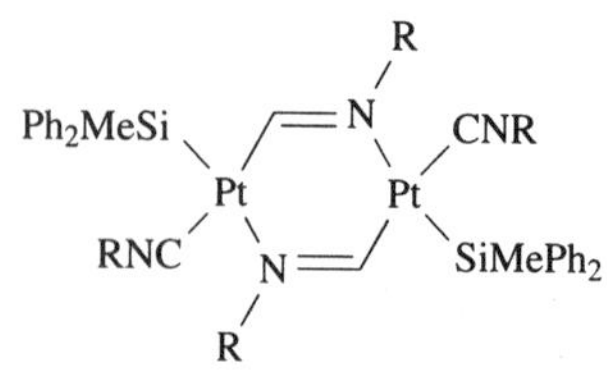

(52)

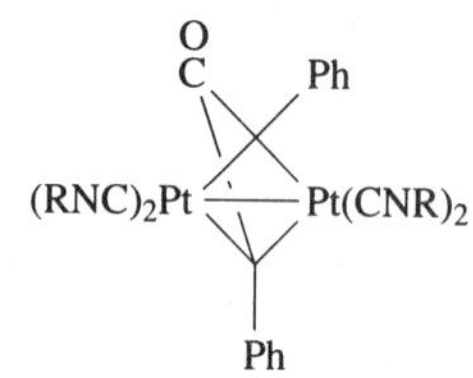

(53)

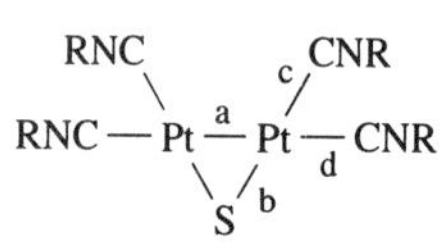

a, 0.260 4(2) nm
b, 0.228 6(10) nm
c, 0.190(3) nm
d, 0.195(2) nm

(54)

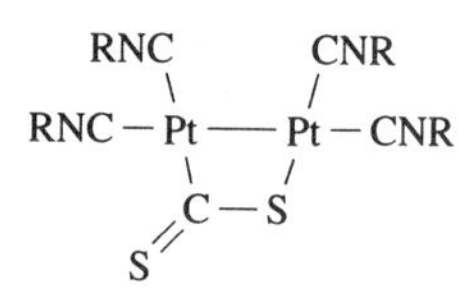

(55)

One platinum isocyanide cluster of higher nuclearity has been described.[171] $[Pt_7(CNR)_{12}]$ (R = 2,6-xylyl) was obtained as a dark-brown solid from the reduction of *cis*-$[PtCl_2(CNR)_2]$ by sodium amalgam in THF. Its structure is best described as a Pt_5 trigonal bipyramid with two edges bridged by platinum atoms. Six of the platinum atoms have terminal isocyanides (L in **(56)**) and there are five normal μ_2-bridging ligands. The 12th isocyanide spans three platinum atoms using both carbon and nitrogen.

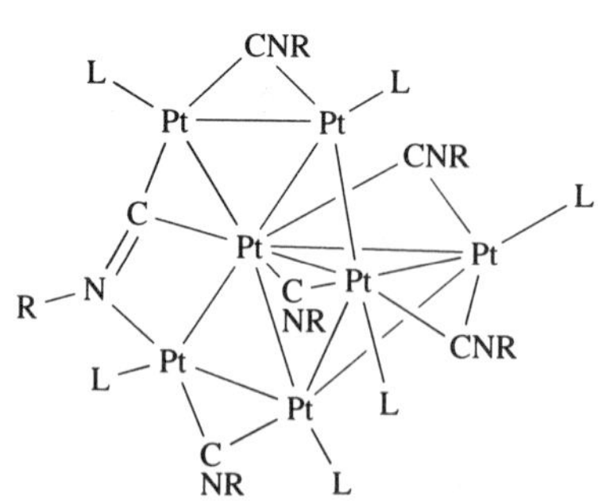

(56)

Many triangular Pt_3 isocyanide clusters containing other ligands have also been described. 2,6-Xylyl isocyanide with $[Pt_3(\mu\text{-}CO)_3(PCy_3)_3]$ produces $[Pt_3(\mu\text{-}CO)(\mu\text{-}CNR)_2(CNR)(PCy_3)_2]$ **(57)** and/or $[Pt_3(\mu\text{-}CNR)_3(CNR)_2(PCy_3)]$ **(58)** (both confirmed crystallographically), irrespective of the ratio of reactants used.[172] The bulk of a bridging isocyanide is too great to allow it to span a Cy_3P–Pt–Pt–PCy_3 unit, a factor which must contribute to the stoichiometries observed.

Treatment of the phosphazine compounds PtL_3 (L = R^1R^2N–P=NR^1) by Bu^tNC leads to **(59)**.[17] The dimensions of the compound with R^1 = Bu^t and R^2 = TMS are very similar to those previously reported

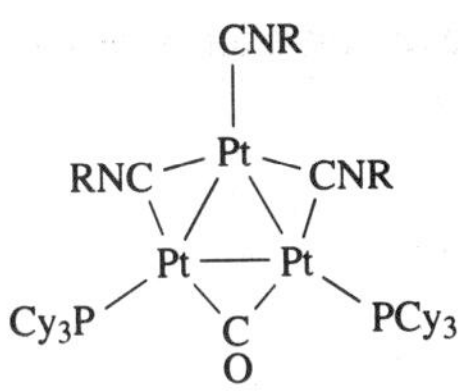

(57)

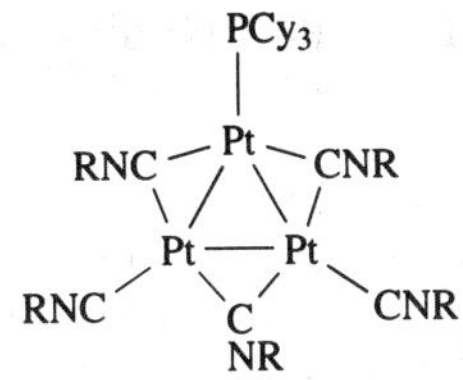

(58)

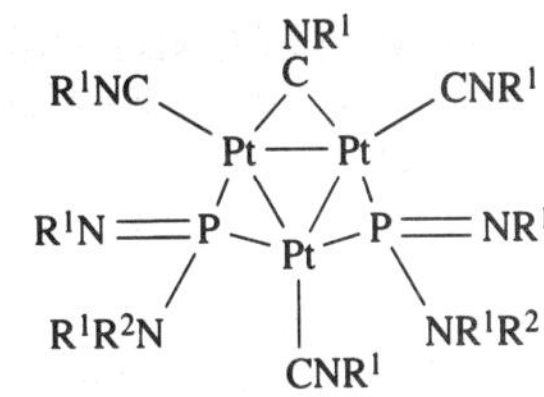

(59)

for $[Pt_3(\mu\text{-}CNR)_3(CNR)_3]$. The reaction of Bu^{t}NC with $[(OC)_5W\{\mu\text{-}C(OMe)$*p*-Tol$\}Pt(cod)]$ leads to (**60**).[18]

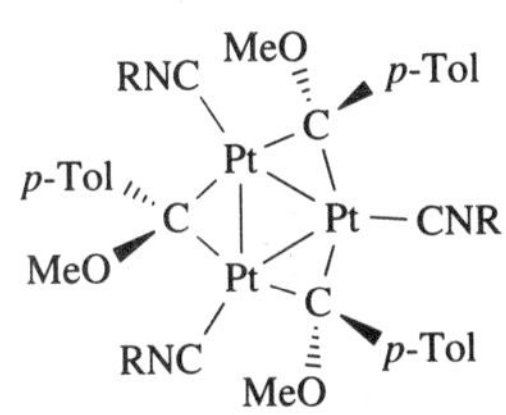

(**60**)

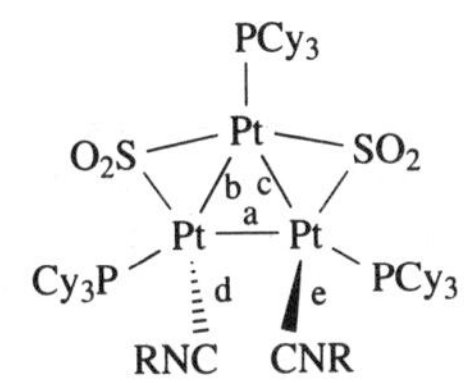

(**61**)

a, 0.284 22(9) nm
b, 0.272 82(9) nm
c, 0.272 54(9) nm
d, 0.198(2) nm
e, 0.196(2) nm

Isocyanide has been found to displace bridging sulfur dioxide molecules from some triangular clusters. $[Pt_3(\mu\text{-}SO_2)_3(PCy_3)_3]$ reacts with two RNC (R = 2,6-xylyl) to form the 44-electron cluster (**61**).[173] Treatment of (**61**) with more RNC causes cluster breakdown to produce (**62**). Mixed carbonyl–SO_2 triangular clusters react in the same way with RNC and lead to the same products (Scheme 14),[41] illustrating again the fine balance between substitution and cluster fragmentation in these reactions. The bridging carbonyls of the 'butterfly' tetranuclear cluster $[Pt_4(\mu\text{-}H)(\mu\text{-}CO)_2(\mu\text{-}dppm)_3(dppm\text{-}P)]^+$ are readily displaced by MeNC to produce $[Pt_4(\mu\text{-}H)(\mu\text{-}CNMe)_2(\mu\text{-}dppm)_3(dppm\text{-}P)]^+$.[76]

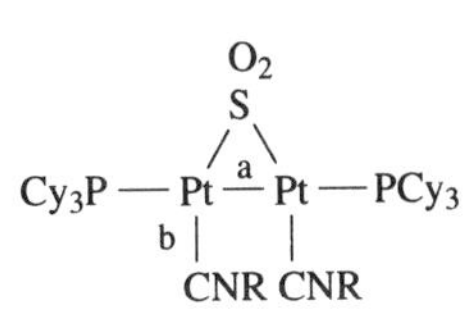

a, 0.269 34(5) nm
b, 0.196(1) nm

(**62**)

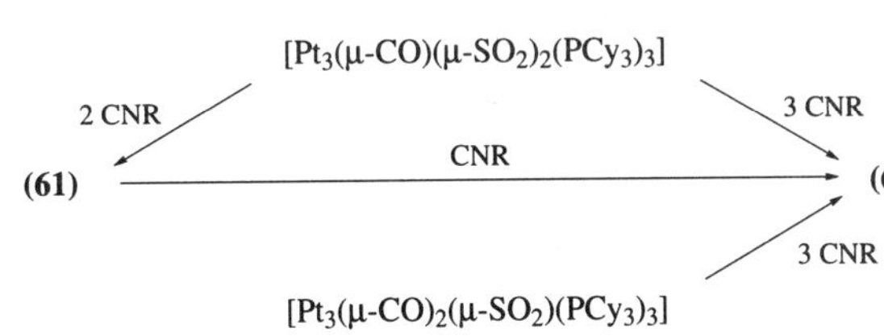

Scheme 14

7.2.3 Platinum(I) Isocyanide Complexes

Carbonyl compounds of structure type (**13**) can be converted into compounds of structure type (**63**) (where $R^1 = C_6F_5$ or C_6Cl_5; R^2 = *p*-Tol, Cy or Bu^t) by R^2NC.[59] $[Pt(C_2H_4)(PPh_3)_2]$ reacts with *cis*-$[PtR^1_2(CNR^2)_2]$ to produce (**63**) directly. The formation of $[Pt_2Cl_2(CNR)_4]$ from *cis*-$[PtCl_2(CNR)_2]$ and $[Pt_3(CNR)_6]$ has been described (Equation (23)), and is part of the electrochemical generation of this bimolecular platinum(I) compound from *cis*-$[PtCl_2(CNR)_2]$.[168] The chlorides of this platinum(I) complex can be replaced by more CNR in the presence of silver salts to yield $[Pt_2(CNR)_6]^{2+}$. In all of these complexes, an absorption at 320 nm can be assigned to a $\sigma \rightarrow \sigma^*$ transition of the Pt–Pt bonds. Hückel MO calculations confirm that the staggered conformation of complexes like (**63**) is more stable than eclipsed, planar ones.[168]

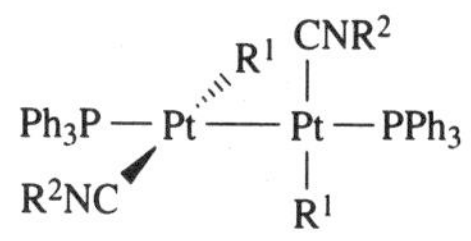

(63)

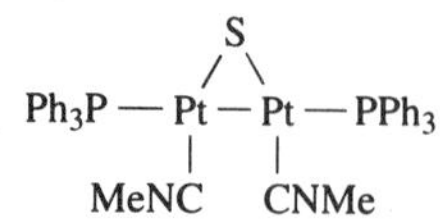

(64)

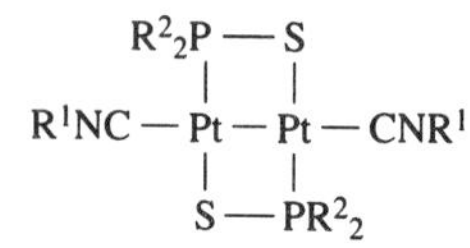

(65)

MeCN displaces CO and one PPh_3 from $[Pt_2(\mu\text{-}S)(CO)(PPh_3)_3]$ to yield **(64)** (similar to **(54)** and **(62)**), which has $\nu(CN)$ at 2160 cm^{-1} in its IR spectrum.[174] The complexes $[Pt_2(\mu\text{-}SPR^2{}_2)_2(PR^1{}_3)_2]$ (R^1 and R^2 are alkyl or aryl) exchange both terminal phosphines for CNMe or $CNBu^t$ to produce **(65)**.[175]

Scheme 15, depicting the reactions of $[Pt_2X_2(\mu\text{-dppm})_2]$ (X = Cl or I) and *p*-TolNC, illustrates the fine balance between retention of the Pt–Pt bonds and cleavage by insertion.[176] All the compounds were characterized by IR, ^1H NMR and ^{31}P NMR spectra: those with bridging isocyanides readily undergo *N*-protonation or -alkylation. $[Pt_2H_3(biL)_2]X$ (biL = dppe, dppp or dppb; X = Cl or BF_4) react with RNC (R = Me, *p*-Tol or Bu^t) to eliminate H_2 and form $[Pt_2H(CNR)(biL)_2]X$.[66] When R is Me or *p*-Tol, the hydrogen and isocyanide are bridging ($\nu(CN)$ about 1645 cm^{-1}), with structures resembling **(19)**. When R is Bu^t, however, both H and $CNBu^t$ are terminal ($\nu(PtH)$ at 1980 cm^{-1}; $\nu(CN)$ at 2160 cm^{-1}).

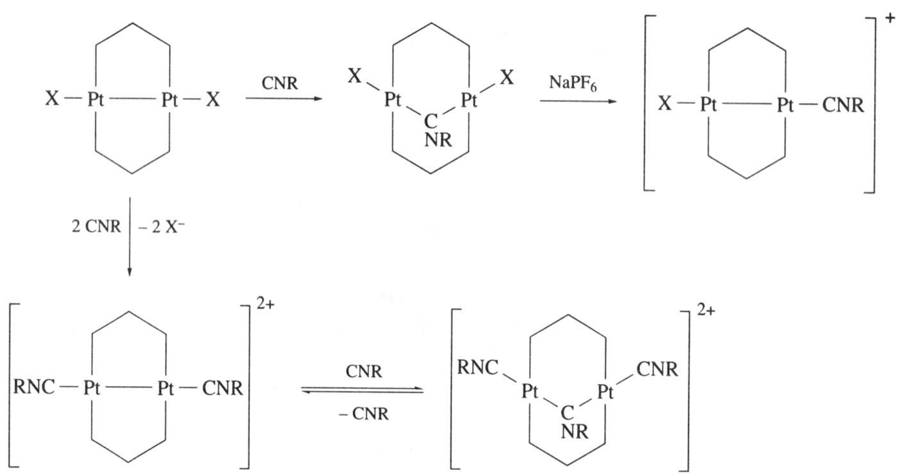

Scheme 15

As with platinum carbonyl complexes, a number of isocyanide compounds have fractional (or mixed) oxidation numbers. Two types of trinuclear complex, each with formal oxidation numbers of 2/3, are described here. $[Pt(C_2H_4)(PPh_3)_2]$ reacts with $[Pt(CNR)_4]^{2+}$ or $[PtCl_2(CNR)_2]$ and CNR (R = 2,6-xylyl) to produce a linear complex ion **(66)**, as well as binuclear **(67)**.[166] Potentiostatic electrolysis of $[Pt(CNR)_4]^{2+}$ leads to $[Pt_3(CNR)_8]^{2+}$ and $[Pt_2(CNR)_6]^{2+}$, related to **(66)** and **(67)**, respectively, and both can be isolated as yellow PF_6^- salts.[177] The reaction of this trinuclear compound with PPh_3 led to **(66)**, which was also characterized crystallographically. $[Pt_3(CNR)_8]^{2+}$ and dppm, on the other hand, produced the bent trinuclear cation **(68)**. The Pt–Pt bonds in this average at 0.2593 nm, and the Pt–Pt–Pt angle is 80 °.[177] Cation **(66)** has ν(CN) at 2144 cm and 2126 cm^{-1}, whereas **(68)** has these bands at 2174 cm^{-1}, 2118 cm^{-1} and 2104 cm^{-1}.

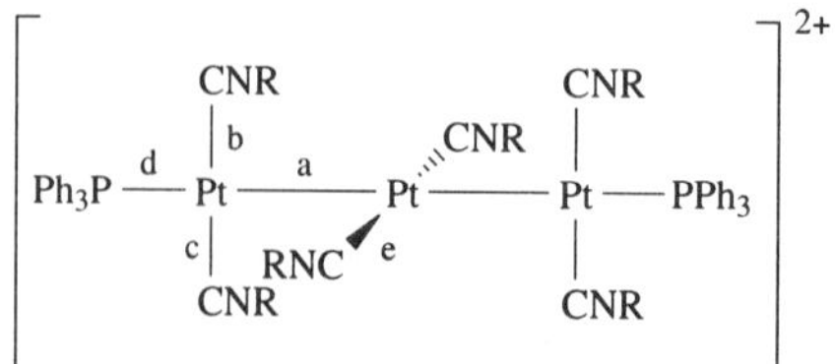

a, 0.264 37(4) nm
b, 0.193(1) nm
c, 0.191(1) nm
d, 0.233 4(1) nm
e, 0.195(1) nm

(66)

The second family of trinuclear complexes of oxidation number 2/3 consists of triangular clusters related to, and often derived from, carbonyl complex **(25)**. MeNC reacts with **(25)** to produce $[Pt_3(CNMe)(\mu_3\text{-}CO)(\mu\text{-dppm})_3]^{2+}$ via intermediate $[Pt_3(CNMe)(\mu_2\text{-}CO)(\mu\text{-dppm})_3]^{2+}$.[178] A second mole of

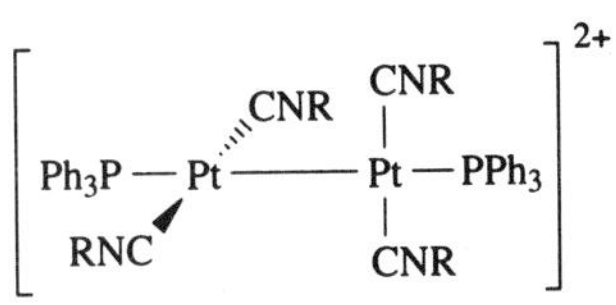

(67)

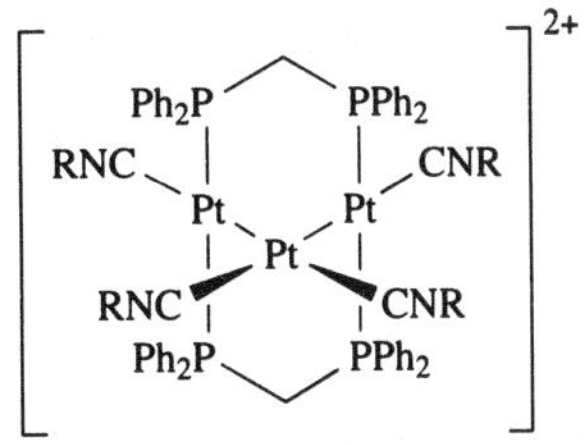

(68)

MeNC removes the CO to leave the 44-electron cluster $[Pt_3(CNMe)_2(\mu\text{-dppm})_3]^{2+}$ (**69**). The two terminal isocyanides on the same platinum have Pt–C distances of 0.197(2) nm.[179] Both $[Pt_3(CNMe)(\mu_3\text{-CO})(\mu\text{-dppm})_3]^{2+}$ and (**69**) are fluxional in solution, with the terminal isocyanides showing apparent threefold symmetry by NMR spectroscopy. This mode of bonding and fluxionality might resemble that of RCN on Pt(1.1.1) surfaces.[178]

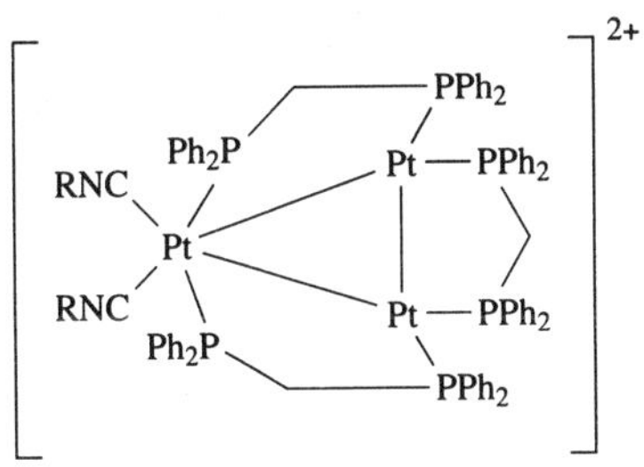

(69)

The carbonyl complex (**31**) also slowly reacts with MeNC to produce $[Pt_3(\mu\text{-CNMe})(\mu\text{-dppm})_4]^{2+}$.[90]

Excess MeNC reacts with any of the triangular dppm compounds above to cleave some of the Pt–Pt bonds. The products are first $[Pt_2(CNMe)_2(\mu\text{-dppm})_2]^{2+}$, then $[Pt_2(CNMe)_2(\mu\text{-CNMe})(\mu\text{-dppm})_2]^{2+}$ and $[Pt_2(CNMe)_3(\mu\text{-dppm})_2]^{2+}$ (Scheme 16).[178] The last two cations readily interconvert in solution, and easily lose CNMe. Complex (**69**) (R = 2,6-xylyl) also reacts by Pt–Pt bond cleavage with excess isocyanide, in this instance to produce (**68**) (with iodide counterions).[179] This compound can be produced directly from (**25**) with RNC and iodide.

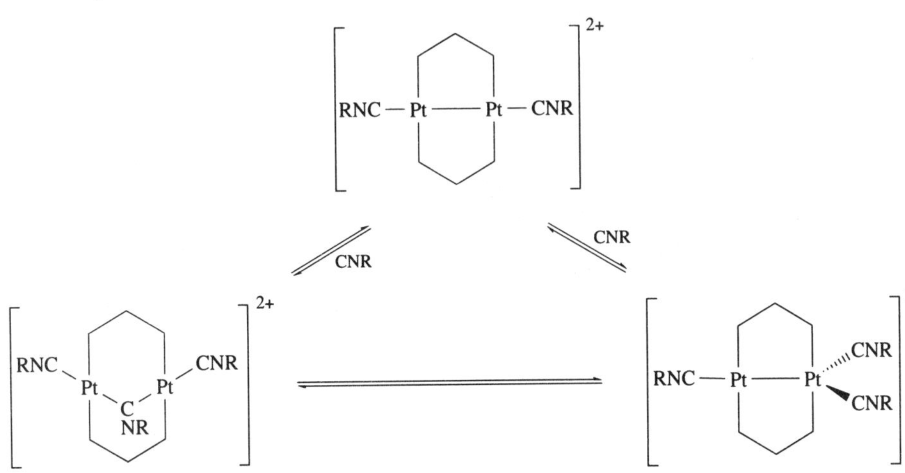

Scheme 16

7.2.4 Platinum(II) Isocyanide Complexes

There are fewer results to report in this section than in the analogous platinum(II) carbonyl section, despite the versatility of RNC as ligands and the variety available. The effects of variations in the R groups available have not been systematically investigated, though examples with quite exotic R groups are known. Unlike the carbonyls, however, tri- and tetraisocyanide complexes are known as well as mono- and diisocyanide compounds. We will examine in turn the mononuclear complexes with one, two, and then three and four isocyanides.

7.2.4.1 Monoisocyanide compounds

One convenient synthesis of monoisocyanide compounds is by cleavage of halide-bridged binuclear complexes. Like their carbonyl analogues, it appears that the *trans*-isomers are produced first in these reactions.[180] At low temperatures, and in the absence of free isocyanide which catalyses isomerizations, these isomers can be isolated as pale-yellow crystalline compounds. At ambient temperatures, however, isomerizations to the more common *cis* forms proceed (Scheme 17). With bulky PR^1_3, this isomerization process is very slow.[101,180] The geometry of such phosphine–isocyanide compounds can be readily determined from their ^{31}P NMR spectra. In particular, the *trans* complexes reveal a characteristic triplet structure from coupling to ^{14}N, with $^3J_{P-N}$ values of about 10 Hz. $^1J_{Pt-P}$ values are near 2900 Hz for the *cis* isomers, about 300 Hz larger than for *trans*. This order is the opposite to that found with the analogous carbonyls, and probably results from the stronger σ-donor bond characteristic of RNC, producing a greater *trans* influence.[180] The sterically cluttered *cis*-$[PtCl_2(CN\text{-}p\text{-}Tol)(PBu^t_2Pr)]$ has been observed to exist in solution as two rotamers about Pt–P by ^{31}P NMR spectroscopy at or below −30 °C.

$$[Pt_2Cl_4(PR^1_3)_2] + 2\,R^2NC \xrightarrow{\text{low temperature}} trans\text{-}[PtCl_2(CNR^2)(PR^1_3)] \longrightarrow cis\text{-}[PtCl_2(CNR^2)(PR^1_3)]$$

Scheme 17

The formation of monoisocyanide complexes by displacement of other ligands is quite common, as might be expected as a consequence of their ligand strength. There is evidence that R^3NC can displace X^- from *trans*-$[PtX_2(CNR^1)(PR^2_3)]$, formed as above: the reactions of $[Pt_2I_4(PBu^t_2Pr)_2]$ with 2:1 mole equivalents of *p*-TolNC at −78 °C led to *cis*-$[PtI(CN\text{-}p\text{-}Tol)_2(PBu^t_2Pr)]I$, with ν(CN) at 2190 cm^{-1} and 2170 cm^{-1}.[180] Weak ligands such as THF[181] or cod[158] are readily displaced by RNC (Equations (24) and (25)). Other examples of halide displacement include Equation (26).[92] From *cis*-$[PtCl_2(PR^1_3)_2]$, R^2NC can replace either Cl^- or PR^1_3,[109] although the reactions are reversible and slow polymerization of the isocyanide will eventually result in regenerating the starting material. In order to exploit these reactions, therefore, they should be carried out in the presence either of silver salts to remove halide or sulfur to oxidize PR^1_3.[109] In a similar fashion, a variety of aminovinyl isocyanides, L, react with *cis*-$[PtCl_2(PPh_3)_2]$ to produce *trans*-$[PtClL(PPh_3)_2]Cl$, but unless the ionic chloride is replaced by noncoordinating PF_6^-, it displaces PPh_3 to give *cis*-$[PtCl_2L(PPh_3)]$.[182]

$$[Pt(C_6F_5)_3(THF)]^- + CNR \longrightarrow [Pt(C_6F_5)_3(CNR)]^- + THF \qquad (24)$$

$$[Pt(\eta\text{-}C_3H_5)(cod)]^+ + PPh_3 + Bu^tNC \longrightarrow [Pt(\eta\text{-}C_3H_5)(CNBu^t)(PPh_3)]^+ + cod \qquad (25)$$

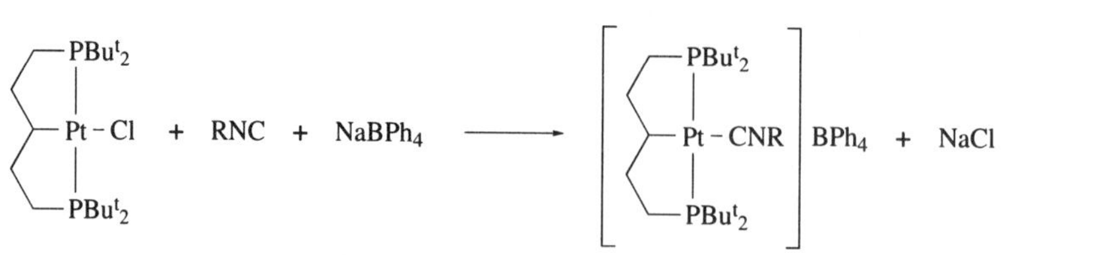

(26)

PMe_3 is displaced from *cis*-$[PtMe_2(PMe_3)_2]$ by Bu^tNC to yield *cis*-$[PtMe_2(CNBu^t)(PMe_3)]$ (J_{Pt-P} = 1686 Hz; $^2J_{Pt-N}$ = 62 Hz), with no insertion into Pt–C.[183] Bu^tNC also displaces one PPh_3 or $AsPh_3$ (L) from platinacyclobutanones (**70**),[184] or a variety of phosphorus ligands from (**71**).[185] One product has structure type (**72**). The ligands of (**73**) (L^1, X = Cl; L^2, X = I) displace one PR_3 from $[PtCl_2(PR_3)_2]$ to produce *cis*-$[PtCl_2(L^1)(PR_3)]$, or cod from $[PtCl_2(cod)]$ to produce *cis*-$[PtCl_2(L^1)_2]$.[186] When L^2 reacts with $[PtCl_2(PMePh_2)_2]$, however, halide exchange is encountered and $[PtClI(L^1)\text{-}(PMePh_2)]$ is produced as a major species along with *cis*-$[PtCl_2(L^1)(PMePh_2)]$ and $[PtI_2(L^2)(PMePh_2)]$. L^2 reacts with $[Pt(C_2H_4)(PPh_3)_2]$ by oxidative addition to form (**74**).

One of the most important reactions of isocyanide complexes is their conversion to carbene complexes following nucleophilic attack at the coordinated carbon. These are discussed in Section 7.3. A second important class of reaction is isocyanide insertion. Many more of these have been reported for palladium complexes than for platinum, but insertion of Bu^tNC into the four-membered rings depicted in (**71**) and (**72**) produces ring-expanded zwitterionic compounds of type (**75**) and (**76**),

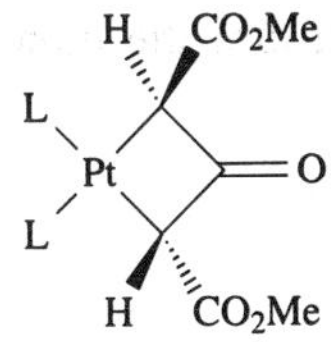

(70)

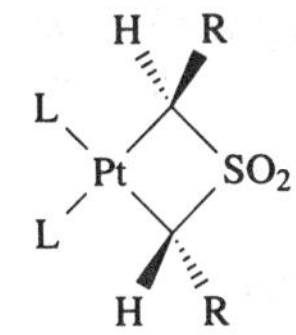

(71)

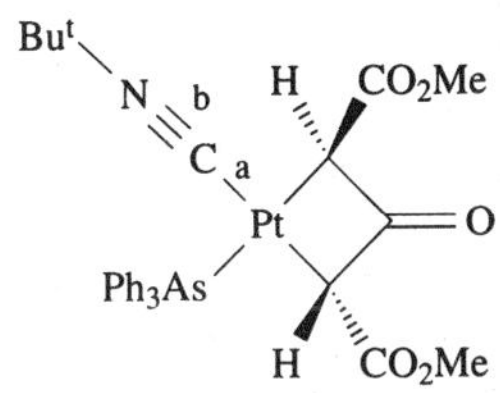

a, 0.193 3(11) nm
b, 0.114 3(14) nm

(72)

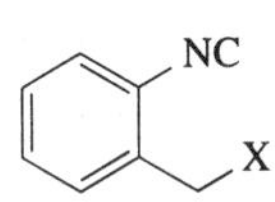

(73)

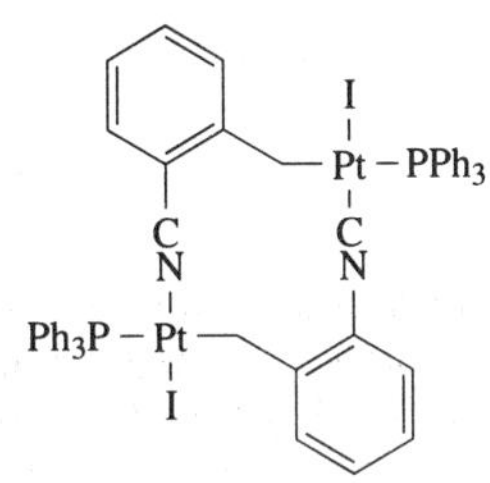

(74)

respectively.[184,185] The electrochemical behaviour of *cis*-[$PtCl_2$(CNAr)($PMePh_2$)] and *trans*-[PtX(CNAr)($(PR_3)_2$]ClO_4 (Ar = *p*-ClC_6H_4 or *p*-$MeOC_6H_4$) has been examined, and is dominated by two-electron processes.[187]

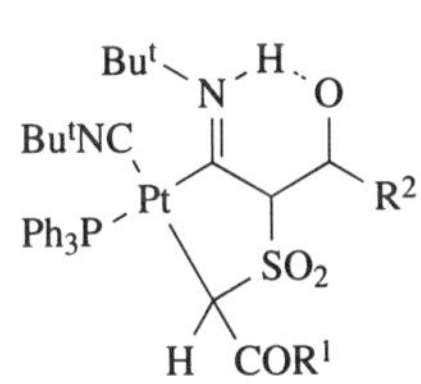

(75)

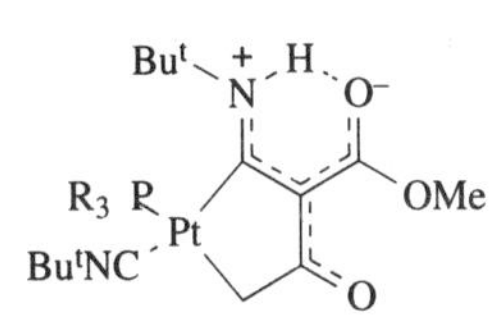

(76)

7.2.4.2 Diisocyanide compounds

The stronger ligand characteristics of RCN compared with CO mean that diisocyanide complexes are both more stable and more numerous. It is inevitable that RCN are sometimes relegated into the category of spectator ligands, employed to occupy coordination sites while the main thrust of the investigation is elsewhere in the molecule. Thus, the reaction of octafluorocyclooctatetraene and [$Pt(cod)_2$] in the presence of Bu^{t}NC yields (**77**),[188] and *cis*- and *trans*-[$PtR^1_2(THT)_2$] (R^1 = C_6F_5; THT = tetrahydrothiophene) yield the corresponding isomers of [$PtR^1_2(CNR^2)_2$] (R^2 = Me, Cy, Bu^t or *p*-Tol).[181] [Pt(η-Cp)(η-C_3H_5)] reacts with excess Bu^{t}NC to form *trans*-[Pt(η^1-Cp)(η^1-C_3H_5)($CNBu^t)_2$], with ν(CN) at 2188 cm^{-1}.[157] In solution, this slowly isomerizes to an equilibrium with the *cis* isomer (ν(CN) at 2205 cm^{-1} and 2174 cm^{-1}). Cationic [Pt(η-C_3H_5)(cod)]$^+$ reacts with one PPh_3 and excess Bu^{t}NC to form [Pt(η^1-C_3H_5)($CNBu^t)_2$(PPh_3)]$^+$, confirmed as *trans* by its single ν(CN) at 2208 cm^{-1} and the low value of J_{Pt-P} (1476 Hz) in its ^{31}P NMR spectrum.[158]

The Ph_2P(η-C_5H_4)-bridged complexes of Equation (15) (R = Me or Ph) react with Bu^{t}NC to displace the cyclopentadienyls, yielding *trans*-[PtR($CNBu^t)_2$($PPh_2C_5H_4$)].[111] Treatment of the bisphosphine or arsine complexes (**70**) by 3-Bu^{t}NC leads both to neutral ligand replacement and isocyanide insertion producing (**78**).[184] [$PtCl_2$(dppm)] reacts with Bu^{t}NC to give binuclear (**79**) by Cl^- elimination: this was isolated as its PF_6^- salt.[189] [$PtCl_2$(dppm-*P*)$_2$], on the other hand, reacts with 2 RNC (R = Me or Bu^t) to

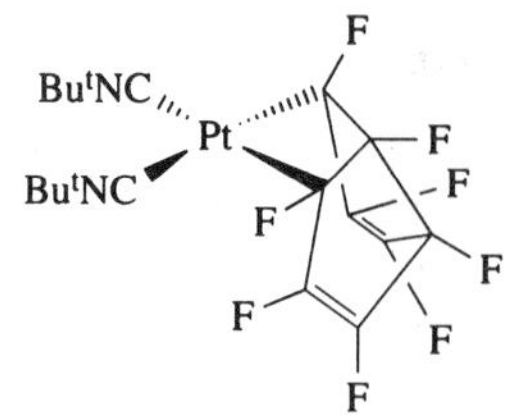

(77)

form $[Pt(CNR)_2(dppm\text{-}P)_2]^{2+}$, also isolated as PF_6^- salts. These molecules are fluxional in solution, exchanging free and coordinated phosphorus atoms. The free phosphorus atoms are also available to bridge to other metals to create heterobimetallic compounds.

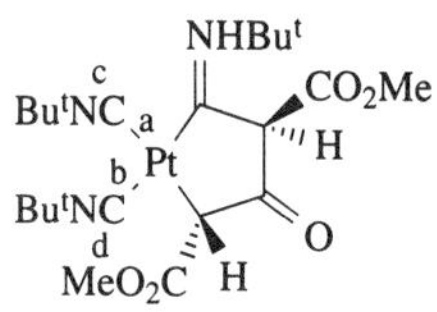

a, 0.198 3(7) nm
b, 0.199 3(8) nm
c, 0.114 4(9) nm
d, 0.115 4(9) nm

(78)

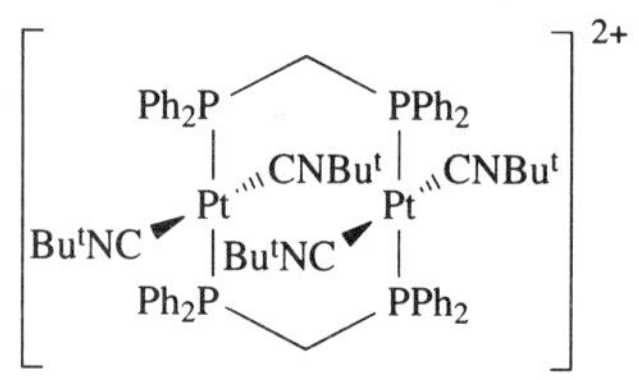

(79)

The reactions of *trans*-$[Pt(CN)_2(PR^5{}_3)_2]$ (R = Et or Ph) with a variety of α,β-unsaturated ketones in the presence of HBF_4/Et_2O leads to a series of substituted isocyanide complexes.[190] The reactions probably proceed by protonation of the coordinated cyanides to give *trans*-$[Pt(CNH)_2(PR^5{}_3)_2]^{2+}$, followed by addition to the alkene (Equation (27)).

$$trans\text{-}[Pt(CNH)_2(PR_3)_2]^{2+} + 2\,R^1R^2C{=}CR^3C(O)R^4 \longrightarrow trans\text{-}[Pt(CNCR^1R^2CHR^3C(O)R^4)_2(PR_3)_2]^{2+} \quad (27)$$

The platinum(III) anionic species $[Pt_2(HPO_4)_4(py)_2]^{2-}$ reacts readily with Bu^{t}NC in methanol to form $[Pt(CN)_2(CNBu^t)_2]$ (ν(CN) at 2256 cm^{-1}, 2235 cm^{-1} and 2164 cm^{-1}) by reductive cleavage.[191] The cyanide source is presumably methanol-hydrolysed Bu^{t}NC. The same cyano species has been produced from the reflux of Bu^{t}NC and $[PtCl_4]^{2-}$ in CH_2Cl_2. Chelating bisisocyanide ligands are relatively uncommon, but the reactions of two such species, (**80**) (R = H or Bu^t), with platinum complexes have been examined.[192] When R = Bu^t, $K_2[PtCl_4]$ reacts with (**80**) to produce $[PtCl_2(\mathbf{80})]$ as a cream powder (ν(CN) at 2238 cm^{-1} and 2212 cm^{-1}), but when R = H, (**80**) reacts with $(Et_4N)_2[PtCl_4]$ producing $[Pt(\mathbf{80})_2]$ $[PtCl_4]$ instead (ν(CN) at 2286 cm^{-1} (sh) and 2238 cm^{-1}).

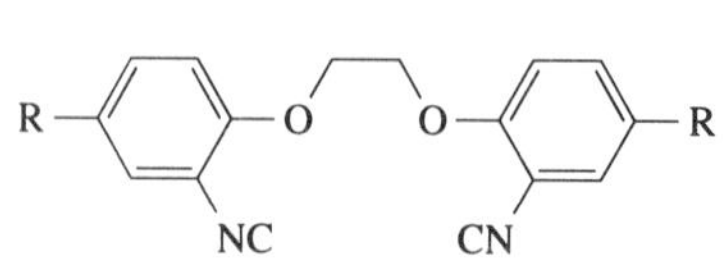

(80)

Electrochemical reduction studies of *cis*- and *trans*-$[PtX_2(CNR)_2]$ (X = Cl, Br or I; R = Cy or Bu^t) on gold or platinum electrodes in propene carbonate or acetonitrile solution revealed a one-electron irreversible step, very dependent on X.[193] Potentiostatic reductive coulometry in the cavity of an ESR spectrometer gave signals attributable to platinum(I) paramagnetic complexes, liberating X^-. A second one-electron reductive step led to decomposition, depositing a solid platinum-containing residue on the electrode.

7.2.4.3 Tri- and tetraisocyanide compounds

The $L_{(III)}$ absorption edge extended x-ray absorption fine structure (EXAFS) spectra of $[PtEt(CNMe)_3]PF_6$ as the solid and in acetone solution provide corroborating evidence for weak Pt···Pt interactions in the solid, measured at 0.357 9(2) nm by x-ray crystallographic studies.[194] The interpretation of the solution spectrum in terms of Pt···Pt interactions was ambiguous. The formation of $[Pt(CNR)_4]^{2+}$ (R = 2,6-xylyl; ν(CN) at 2209 cm^{-1}), isolated as its PF_6^- salt, was achieved by treating $[PtCl_2(CNR)_2]$ with CNR.[166] Thermal analysis of the solid-phase transformations of $[Pt(CNCy)_4]$ $[Pt(CN)_4]$ revealed almost complete loss of CNCy on heating.[195]

7.2.5 Platinum(IV) Isocyanide Complexes

The weakly bridging Me_2S ligands of $[Pt_2Me_8(\mu\text{-}SMe_2)_2]$ are readily displaced by RNC (R = Me or 2,6-xylyl) to produce colourless, crystalline *cis*-$[PtMe_4(CNR)_2]$ (ν(CN) at 2218 cm^{-1} (Me) or 2156 cm^{-1} (xylyl)).[196] Both of these complexes decompose cleanly on heating, either as solids or in solution, to C_2H_6 and *cis*-$[PtMe_2(CNR)_2]$. (These latter two complexes can also be prepared by the action of RNC on $[Pt_2Me_4(\mu\text{-}SMe_2)_2]$.) Kinetic studies on the reductive elimination reactions showed dissociation of RNC to precede ethane elimination in both cases, and the positive kinetic isotope effect detected with deuteromethyl analogues led to speculation that an agostic Pt···HC interaction employing the vacant coordination site is a feature of these processes. Since one of the platinum(II) products of this reaction, the volatile *cis*-$[PtMe_2(CNMe)_2]$, has some potential for chemical vapour deposition reactions,[197] it might be presumed that the platinum(IV) precursor also has.

7.3 PLATINUM CARBENE COMPLEXES

Platinum carbene compounds are dominated by platinum(II). The vast majority of them contain at least one heteroatom attached to the carbenoid carbon, and π-delocalization to these atoms affords some stabilization to the system. The preparation of a carbene complex without such heteroatoms has been described (Equation (28)) by Lu *et al.*[198] Extensive delocalization around the cycloheptatrienylidene ring of this complex undoubtedly compensates for the lack of heteroatoms and adds to its stability. The carbene carbon resonates at δ 210.5 ppm in its ^{13}C NMR spectrum, and $J_{Pt\text{-}C}$ is 969 Hz. The cycloheptatrienylidene complex was isolated as a yellow-green solid, which decomposes on melting at 263 °C. It is of interest that the same ligand attached to platinum(0) is bound not through a single carbon atom as a carbenoid species, but through two carbon atoms as an η^2-allene complex.[198]

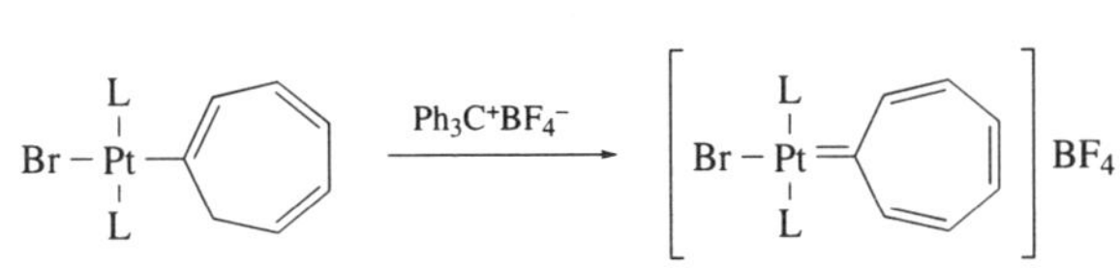

(28)

Preparative methods for the numerous complexes with heteroatoms were categorized and discussed in *COMC-I*. Most common were those routes in which the Pt–C were already present, and include

(i) nucleophilic attack of alcohols, amines or thiols on isocyanide complexes,
(ii) nucleophilic attack on alkyne complexes, and
(iii) protonation of C=O, C=S or C=NR in complexes in which the carbon is attached to platinum.

Other routes include isomerizations or rearrangements within platinum carbene compounds, transfer of the carbene moiety from other metals to platinum, and formation from carbene precursors. This section will first examine some new preparative routes to add to the above compendium, and then consider advances of the early 1990s classified by the various carbene heteroatoms. A survey of preparative methods leading primarily to cyclic carbene complexes has been published.[199]

7.3.1 New Preparative Routes

7.3.1.1 Preparation from nitrogen ylides

The reaction of $[Me_2N{=}CH_2]Cl$ with $[Pt(PPh_3)_3]$ produces (**81**) (Scheme 18), described as a bidentate ylide complex, which is converted by heat to the carbene complex (**82**).[200] The IR spectrum of (**82**) has an absorption at 1612 cm^{-1}, assigned to ν(CN). Its crystal structure reveals quite short Pt–C (0.196(1) nm) and C–N (0.125(1) nm), consistent with substantial delocalization of the multiple bonds. The carbene plane lies at 84° to the platinum(II) coordination plane. The conversion of (**81**) to (**82**) requires a coordinating anion: the BF_4^- analogue of (**81**) is stable. Scheme 18 outlines the likely reaction route.

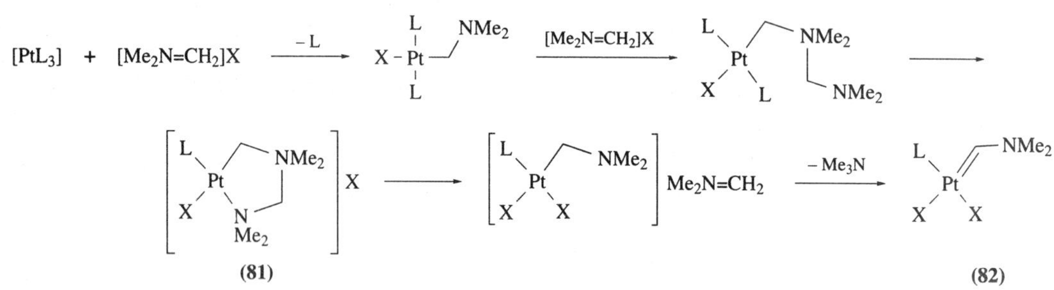

Scheme 18

7.3.1.2 Electrophilic attack on coordinated CF_3

The CF_3 groups of the complexes *trans*-$[PtX(CF_3)(PR_3)_2]$ (X = H or Cl) are strongly bonded to platinum, weakening the C–F bonds and rendering them susceptible to electrophilic attack by HBF_4 in ether.[201] The resulting cationic CF_2 complexes are very reactive, and are rapidly attacked by alcohols or thiols to produce stable complexes of carbenes with two oxygen or sulfur heteroatoms (Scheme 19). The reactive difluorocarbene intermediates were not isolated, but were detected in solution. Thus, *trans*-$[PtH(CF_2)(PPh_3)_2]BF_4$ was fully characterized by its low-temperature ^1H, ^{19}F and ^{31}P NMR spectra.[201b] The crystal structure of *trans*-$[PtH(\overline{COC_2H_4O})(PPh_3)_2]BF_4$ revealed Pt–C to be 0.201(2) nm, longer than for carbenes *trans* to chlorine, and presumably due to the higher *trans* influence of hydrogen. The (carbene) C–O values of 0.130(1) nm indicate substantial π-bonding; Pt–H is 0.172(8) nm. The cyclic carbene ligand itself is almost planar and at a dihedral angle of 61.3° to the platinum coordination plane.

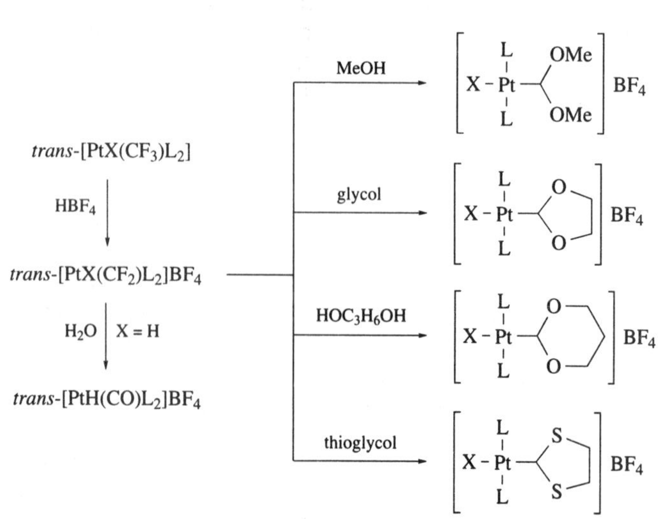

Scheme 19

7.3.1.3 Template condensation of thiolate amine chelating carbene precursors

The reaction of (**83**) with $CH(OEt)_3$ in the presence of $[PtCl_2(cod)]$ leads to the binuclear bridging chelate carbene complex (**84**).[202] The condensation mechanism is not yet known. Prior carbene formation is possible, as is an insertion into a Pt–N bond of a platinum complex of (**83**). The bridges of (**84**) are readily cleaved by pyridine or PMe_3 to form mononuclear complexes. All these complexes are remarkably stable, not reacting even with concentrated HCl. This might be attributed to the tridentate nature of the carbene ligand, and strong π-backbonding from platinum to carbon.

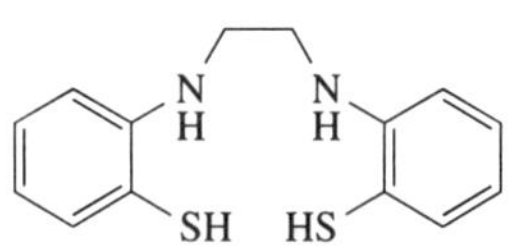

(**83**)

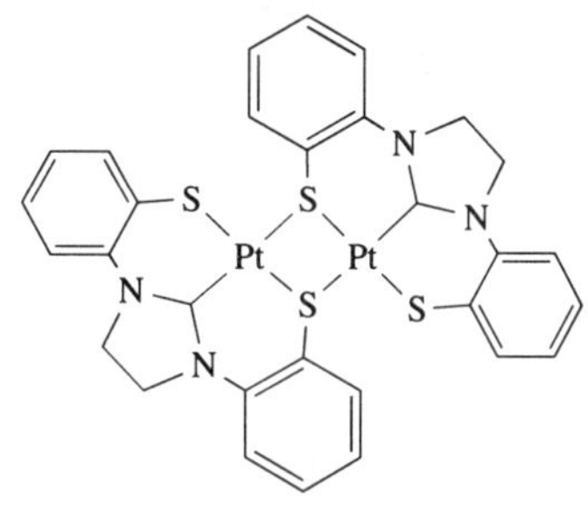

(**84**)

7.3.2 C,N-Carbene Complexes

PhC≡CH reacts slowly with *trans*-$[PtCl_2(NHR^1R^2)L]$ (L = PMe_2Ph or PBu_3) to produce *cis*-$[PtCl_2\{C(NR^1R^2)CH_2Ph\}L]$ (R^1, R^2 = H, Ph; H, C_6H_4-4-NO_2; H, C_6H_4-3-OMe; or Ph_2).[203] These reacted with more L to produce *trans*-$[PtCl\{C(NR^1R^2)CH_2Ph\}L_2]Cl$. Both the neutral and cationic aminocarbene complexes are markedly less reactive than their alkoxycarbene analogues, and show no tendency to undergo hydrogen–deuterium exchange with CD_3OD or to react with Cl^-. An adaptation of this reaction route used 2-aminophenylethyne in place of separate amines and ethyne, and produced cyclic C,N-carbene complex (**85**).

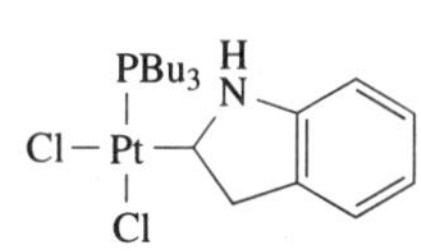

(**85**)

Cyclic C,N-carbene complexes have also been made from phosphonium-substituted isocyanide complexes (Scheme 20).[204,205] Treatment with NEt_3 forms highly reactive ylide functions, which then form the cyclic ylide–carbenes. *Trans*-complexes were similarly made, as were cationic complexes of type (**86**). Extensive delocalizations around the C_4N rings are indicated by the longer-than-usual C–N distances. The carbene ring planes lie perpendicular to the coordination planes. Complex (**86b**) (R_3 = Ph_2Bz; L = PPh_3) is readily deprotonated by potassium hydroxide at the benzyl group, which then displaces a PPh_3 by a further cyclization (Scheme 21).

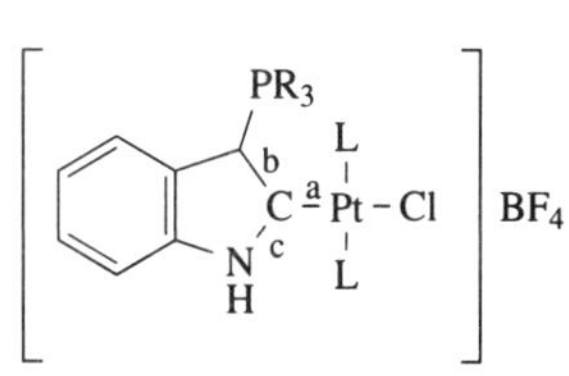

(**86**)

(**a**) L = PPh_3; R = Me
a, 0.199(1) nm
b, 0.141(1) nm
c, 0.138(1) nm

(**b**) L = PPh_3; R_3 = Ph_2Bz
a, 0.203 6(8) nm
b, 0.137(1) nm
c, 0.134(1) nm

7.3.3 C,O-Carbene Complexes

The preparative method of choice, nucleophilic attack by alcohols on ethyne complexes, has been further exploited by with additional detail. Normally only monosubstituted alkynes, $R^1C≡CH$, are

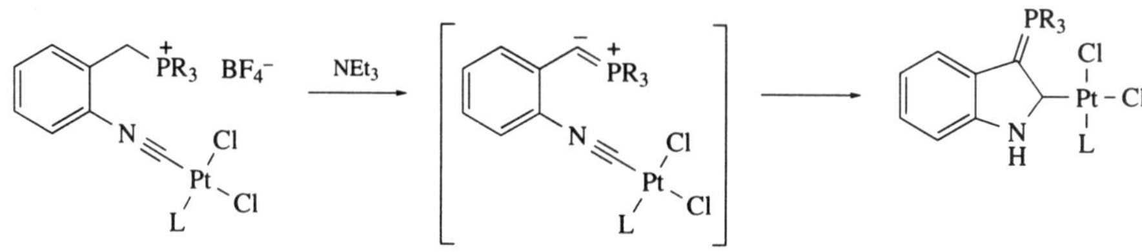

Scheme 20

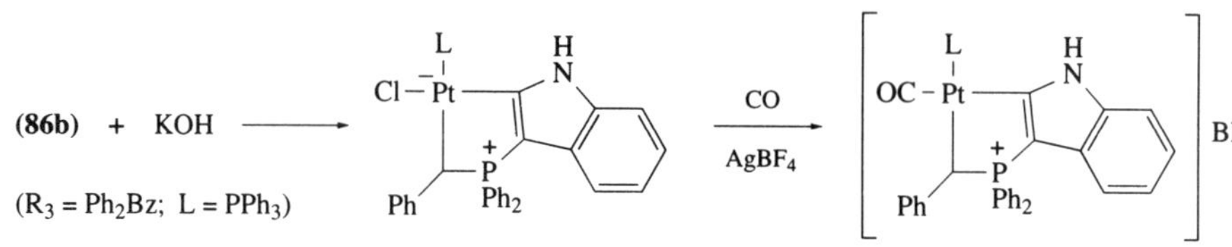

Scheme 21

converted into alkoxycarbenes by this route, but TMS-C≡CR[1] have been used instead in a reaction sequence involving Si–C cleavage (Scheme 22 (L = PMe_3 or PEt_3; X = H or Me; R^2 = Me or Et)).[206] The mechanism probably involves replacement of coordinated R^2OH by TMS-C≡CR[1], followed by alcohol reattack to leave coordinated R^1C≡CH. In these cationic carbene complexes, the chemical shifts of the carbene carbon (ca. 320 ppm) are indicative of substantial cationic character. The reduction of $[PtCl_6]^{2-}$ by Pr^iOH in the presence of Bu^tC≡CH or TMS-C≡CH reveals an interesting effect of replacing the carbon substituent by silicon, which can be cleaved.[207] Complex (**87**) was isolated with the former reagent, but (**88**) from the latter. As usual both carbene ligands of each complex lie approximately perpendicular to the platinum coordination planes, and bond lengths suggest extensive Pt–C–O delocalization.

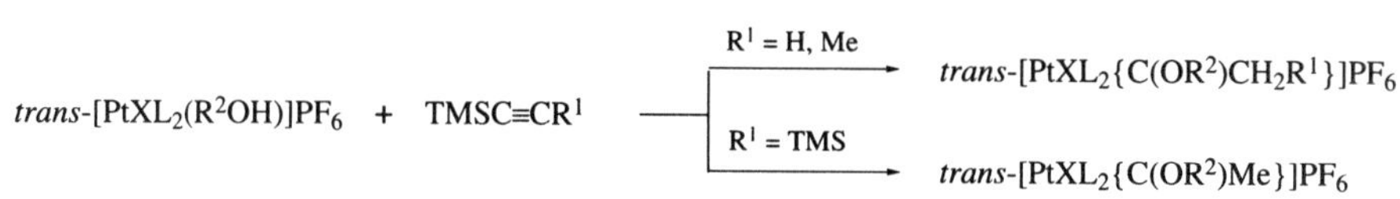

Scheme 22

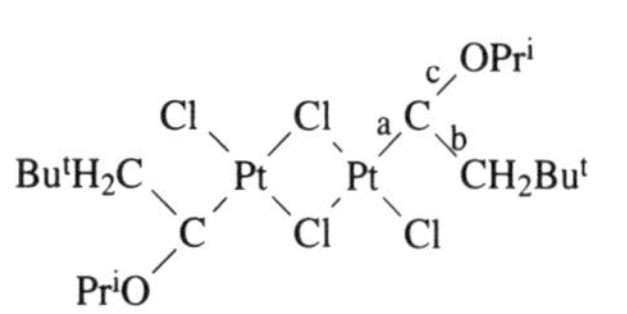

a, 0.183(5) nm
b, 0.151(8) nm
c, 0.131(7) nm

(**87**)

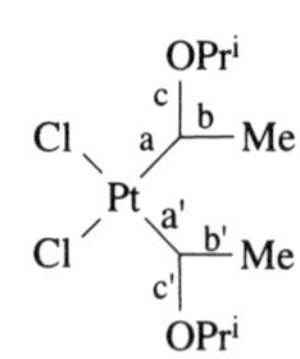

a, 0.192(2) nm a', 0.197(2) nm
b, 0.158(3) nm b', 0.151(3) nm
c, 0.124(2) nm c', 0.126(2) nm

(**88**)

$[PtCl_2(CO)_2]$ reacts with PhC≡CCO_2Et in MeCN to produce the cyclic carbene complex (**89**).[154] $\delta^{13}C$ is 193.1 ppm; the remaining CO is easily replaced by PPh_3. Dimensions of the carbene ligand (perpendicular to the plane) are given alongside (**89**).

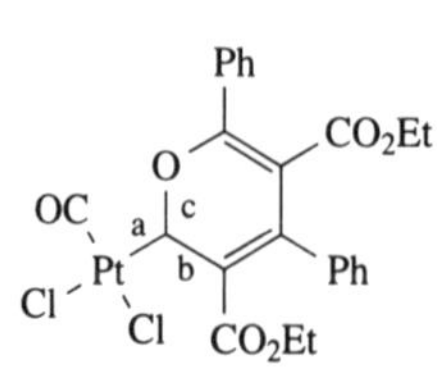

a, 0.193 3(8) nm
b, 0.141 9(9) nm
c, 0.136 1(8) nm

(**89**)

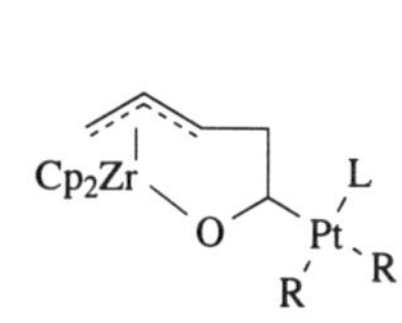

(**90**)

Cationic *trans*-$[Pt(C_6Cl_5)\{C(OMe)Me\}L_2]ClO_4$ (L = PMe_2Ph) can be made by protonating either the α-methoxyvinyl complex *trans*-$[Pt(C_6Cl_5)\{C(OMe){=}CH_2\}L_2]$ or the ethynyl *trans*-$[Pt(C_6Cl_5)$-$(C{\equiv}CH)L_2]$ in methanol.[208] Cyclic carbene complexes *trans*-$[Pt(C_6Cl_5)\{\overline{C(CH_2)_nO}\}L_2]ClO_4$ (n = 3 or 4) are similarly made from cyclic alkoxyvinyl precursors.[209] The carbonium ion character of the carbene carbons is again evident from the $\delta^{13}C$ values of about 300 ppm. The relative Brønsted acidities of carbenes in these compounds were established by determining the equilibrium constants of reactions such as Equation (29), and found to be $=C(CH_2)_3O < =C(OMe)Me < =C(CH_2)_4O$.

$$\{Pt\}C(OMe)Me \;+\; \{M\}C(OMe){=}CH_2 \;\rightleftharpoons\; \{Pt\}C(OMe){=}CH_2 \;+\; \{M\}C(OMe)Me \qquad (29)$$

The unusual bimetallic carbene complexes (**90**) were isolated from *cis*-$[PtR_2(CO)L]$ and $[Zr(\eta\text{-}Cp)_2(\eta\text{-}C_4H_6)]$ ($R = C_6F_5$; L = CO or PPr^i_3).[210] When L is CO, UV irradiation causes a geometry change of (**90**) to *trans*, presumably via reversible CO elimination. Heating reconverts to (**90**).

The reaction of *trans*-$[PtCl(COMe)(PPh_3)_2]$ with $SnCl_2$ has been found to produce *trans*-$[PtCl\{C(OSnCl_2)Me\}(PPh_3)_2]$, as well as $[Pt(SnCl_3)(COMe)(PPh_3)_2]$.[211] $\delta^{13}C$ for the platinum-bound carbon is 246.2 ppm, and one canonical structure is carbenoid in character (Equation (30)).

(30)

7.3.4 N,N-Carbene Complexes

The preparations of these remain dominated by nucleophilic attack of amines at coordinated isocyanides. The first report of the isolation of complexes $[PtCl_2(CNR^1)(PR^2_3)]$ of *trans* geometry also recounted their conversion to carbene complexes.[180] $NH_2(p\text{-Tol})$ reacts with *trans*-$[PtCl_2\{CN(p\text{-Tol})\}(PR^1_3)]$ to form *trans*-$[PtCl_2\{C(NH\text{-}p\text{-Tol})_2\}(PR^1_3)]$. The attack of secondary amines on *cis*-$[PtCl_2L(PPh_3)]$ (L = various aminovinyl isocyanides) both converts the coordinated isocyanides to carbenes and replaces Cl^- by amine.[182] The resulting $[PtCl(PPh_3)(NHR^1R^2)\{C(NR^1R^2)(NHR^3)\}]Cl$ (with Cl^- *trans* to PPh_3) can be isolated as their PF_6^- salts, but heating the chloride complexes drives off amine to produce *cis*-$[PtCl_2(PPh_3)\{C(NR^1R^2)(NHR^3)\}]$.

The reaction mechanism of this class of reactions has been the subject of a kinetic investigation.[212] The main conclusions drawn were that

(i) the first step is direct attack of amine nitrogen at the coordinated isocyanide carbon;

(ii) the rate of this step is greatly influenced by the donor ability of the amine, the electrophilic character of the isocyanide, steric crowding and the solvent, leading overall to an enormous range of reaction rates; and

(iii) the conversion of the primary adduct to the final carbene is often assisted by the entering amine itself, acting as a bifunctional catalyst as in (**91**).[212]

With such large variations in rate possible, it is perhaps not surprising that neither benzylamine nor methanol reacts with the anions $[Pt(C_6F_5)_3(CNR)]^-$ (R = Bu^t, Cy, Me or *p*-Tol).[181] Benzylamine does, however, react with the neutral *p*-tolyl isocyanide complexes *cis*- and *trans*-$[Pt(C_6F_5)_2(CNR)_2]$ to form the two isomers of $[Pt(C_6F_5)_2\{C(NH\text{-}p\text{-Tol})(NHCH_2Ph)\}_2]$.

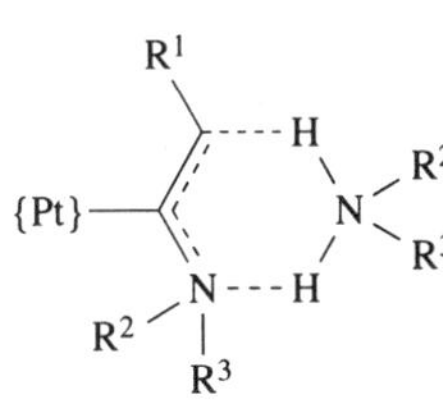
(**91**)

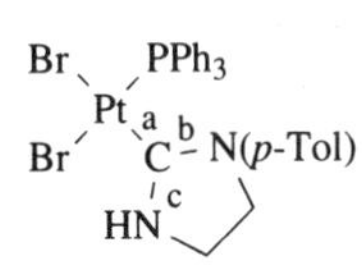
a, 0.182(1) nm
b, 0.134(1) nm
c, 0.136(1) nm

(**92**)

2-Bromoethylamine hydrochloride and BuLi react with the isocyanide complexes *trans*-$[PtX(CNR)L_2]BF_4$ (X = halide or Me; R = Me, *p*-Tol or *p*-anisyl; L = PPh_3 or $PMePh_2$) to produce cyclic carbene complexes *trans*-$[PtX\{\overline{CN(R)C_2H_4NH}\}L_2]BF_4$.[213] A minor product from one reaction, (**92**), was characterized crystallographically and reveals the usual extensive delocalization in the carbene ligand and disecting planes at 102°. Similar compounds were synthesized from $BrC_2H_4NH_2$ and *cis*-$[PtCl_2(CNR)(PPh_3)]$: Scheme 23 outlines the likely mechanism. Cyclic five-membered carbene compounds of this type have also been prepared from the action of aziridine on the same isocyanide complexes. Equation (31) shows the probable pathway.[214]

The amine hydrogen of these cyclic carbene ligands can be removed by BuLi. The resulting imino intermediate will react with electrophiles such as allyl bromide to afford new, functionalized carbene complexes,[213] or dimerize (Equation (32); R = *p*-anisyl; L = PPh_3).[215] The related complexes *cis*-

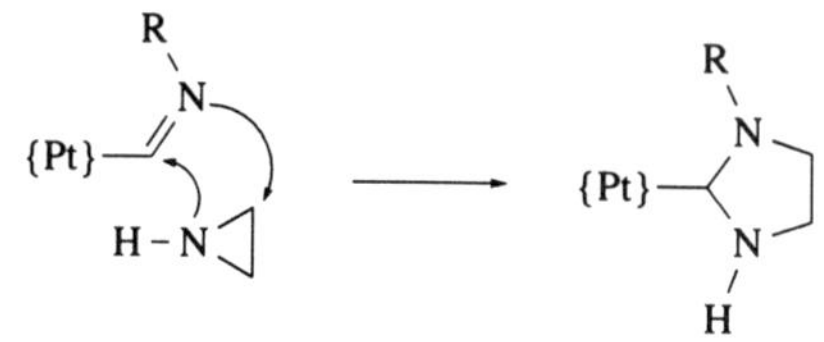

(31)

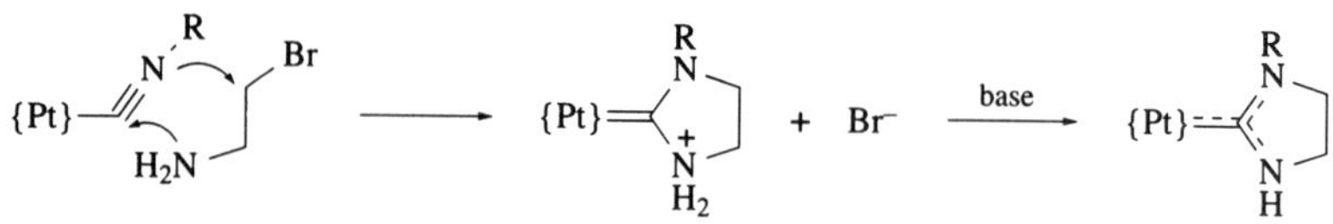

Scheme 23

$[PtCl_2\{\overline{CN(R)C_2H_4N}H\}_2]$ react with *cis*-$Ph_2PCH{=}CHPPh_2$ (diphoe) to liberate the chlorides and produce $[Pt\{\overline{CN(R)C_2H_4N}H\}_2(diphoe)]Cl_2$.[216]

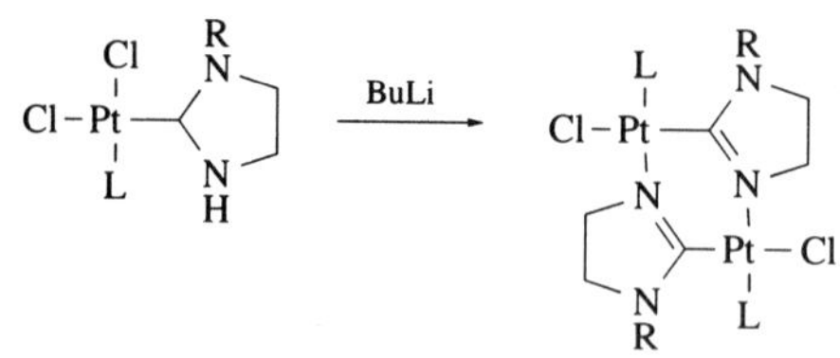

(32)

It is interesting to note that the less reactive heterocycle azetidine does not react with platinum isocyanides by ring opening, but nevertheless produces carbene complexes (Equations (33) and (34)).[217] Piperidine reacts in the same way, and the complexes $[PtHR(CN\text{-}p\text{-anisyl})(PPh_3)]$ (R = CH_2CN, CH_2CF_3 or CF_3) react in analogous fashion with azetidine or piperidine.[218] Complex (**93**), with planes intersecting at 73.7°, is one example.

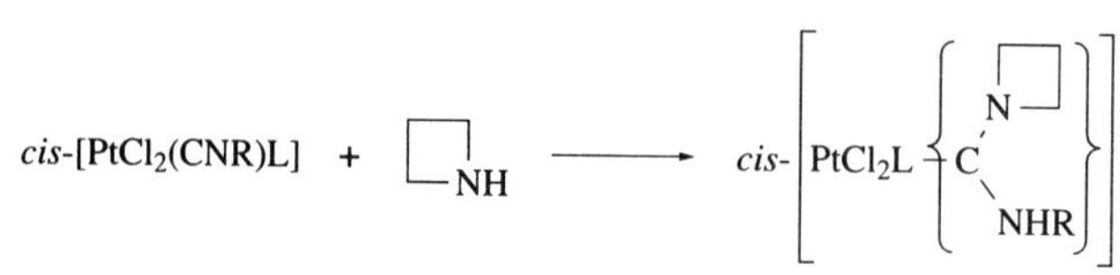

(33)

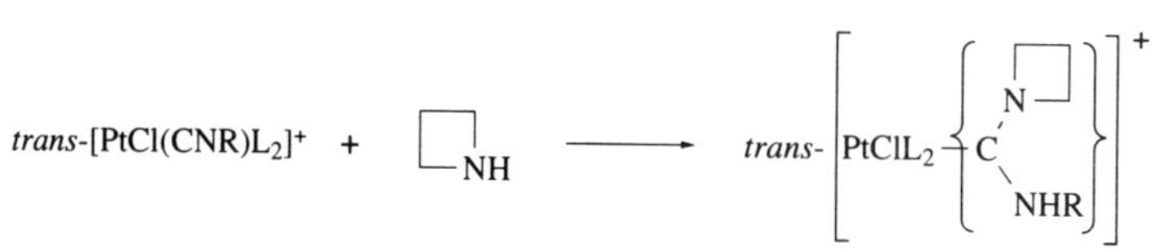

(34)

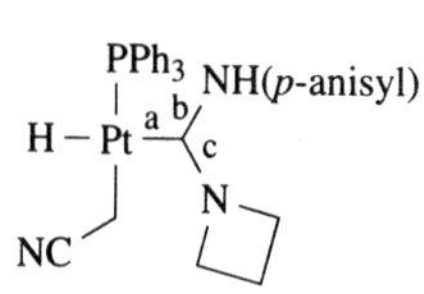

a, 0.206 9(4) nm
b, 0.135 7(5) nm
c, 0.131 3(5) nm

(93)

Complex (**93**) and its CH_2CF_3 analogue react with PPh_3 to produce (**94**). This, in turn, is protonated by HBF_4–Et_2O to produce *trans*-$[PtH(PPh_3)\{C(NH\text{-}p\text{-anisyl})\overline{NC_2H_4C}H_2\}]BF_4$.[218] The CF_3 analogue of (**93**), on the other hand, reacts with dppe or diphoe to liberate the formamide (**95**) (or (**96**) from the piperidine analogue), giving $[PtH(CF_3)(biL)]$.

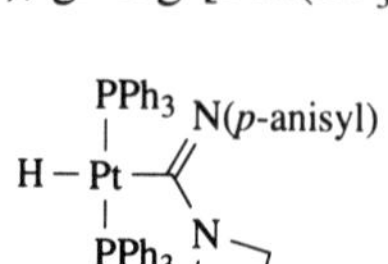

(94)

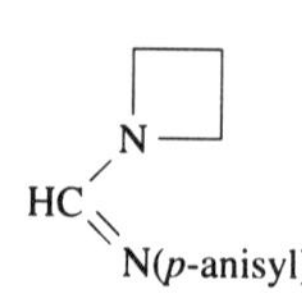

(95)

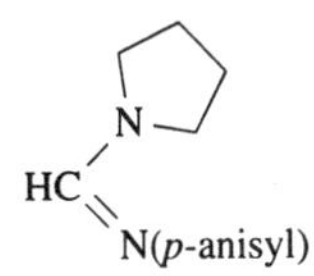

(96)

7.3.5 N,O-Carbene Complexes

cis-$[Pt(CONEt_2)(COCOPh)(PPh_3)_2]$ reacts with iodine to yield *cis*- and *trans*-$[PtI_2(PPh_3)\{C(OCOPh)(NEt_2)\}]$ and *cis*-$[PtI(PPh_3)_2\{C(OCOPh)(NEt_2)\}]I$, of which (**97**) is the major product (90%).[219] Other N,O-carbene complexes have been formed from alcohol attack on coordinated isocyanides, including cyclic complexes such as (**98**), made from *trans*-$[PtCl(CN\text{-}p\text{-}Tol)(PPh_3)_2]BF_4$, BrC_2H_4OH and BuLi.[220] The CN and CO bond lengths in (**97**) and (**98**) are again typical of extensive π-delocalization. Several analogues of (**98**) have been described, and their formation reactions are sensitive to steric crowding. Thus, when (CN-*p*-Tol) is replaced by CNCy, the reaction proceeds only to about 25%, and complexes containing $CNBu^t$ or PCy_3 did not react at all.[220] Neutral dicarbene complexes *cis*-$[PtBr_2(\overline{COC_2H_4N}R)_2]$ were made from *cis*-$[PtCl_2(CNR)_2]$ and BrC_2H_4OH.

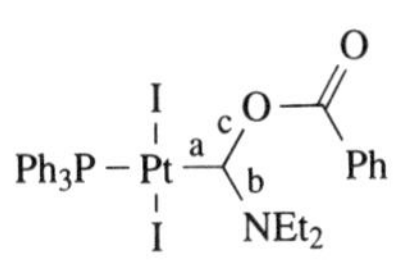

a, 0.196 9(7) nm
b, 0.132 4(8) nm
c, 0.140 3(8) nm
dihedral 101°

(97)

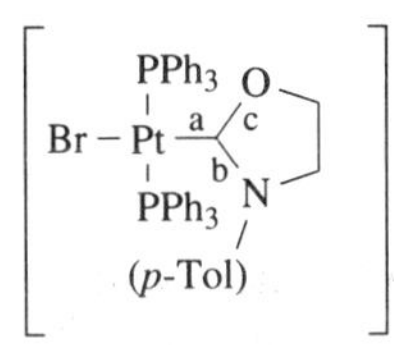

a, 0.191(1) nm
b, 0.130(1) nm
c, 0.133(1) nm
dihedral 93°

(98)

Compared with azetidine, oxirane alone was unreactive towards isocyanide complexes, but in the presence of NaCl or ClC_2H_4OH it reacted to form *cis*-$[PtCl_2(PPh_3)(\overline{COC_2H_4N}R)]$ from *cis*-$[PtCl_2(PPh_3)(CNR)]$ (R = *p*-Tol or *p*-anisyl) and *trans*-$[PtCl(PPh_3)_2(\overline{COC_2H_4N}R)]BF_4$ from *trans*-$[PtCl(PPh_3)_2(CNR)]BF_4$.[214] *cis*-$[PtCl_2(CN\text{-}p\text{-}anisyl)_2]$ reacted with oxirane to form the mixed isocyanide–aminooxycarbene complex *cis*-$[PtCl_2(CN\text{-}p\text{-}anisyl)(\overline{COC_2H_4N}\text{-}p\text{-}anisyl)]$. Similar cyclic carbene complexes are also preparable from the reactions of 2-chloroethanol and BuLi on isocyanide complexes.[215]

Cyclic five-membered N,O-carbene complexes have also been synthesized from the more reactive azetidine (along with $[NH_3C_2H_4X]X$) and carbonyl complexes (Equations (35) and (36); L = PPh_3 and X = Cl or CF_3).[221] These NH-containing carbenes are readily deprotonated by BuLi, and can then react with alkyl halides to form substituted carbenes, or dimerize to complexes like (**99**).

$$trans\text{-}[PtX(CO)L_2]BF_4 + H_2\overline{CCH_2NH} \longrightarrow trans\text{-}[PtXL_2(\overline{COC_2H_4NH})]BF_4 \qquad (35)$$

$$cis\text{-}[PtCl_2(CO)L] + H_2\overline{CCH_2NH} \longrightarrow cis\text{-}[PtCl_2L(\overline{COC_2H_4NH})] \qquad (36)$$

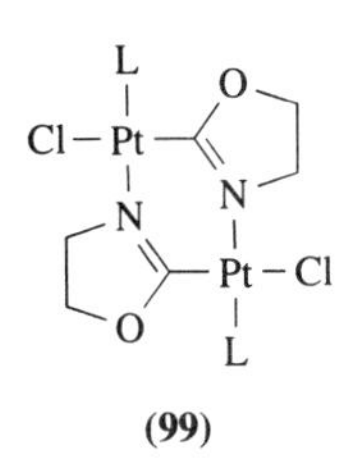

(99)

Both five- and six-membered aminooxycarbene complexes have been prepared from the action of $CNCHR^1CHR^2OH$ ($R^1 = R^2 = H$; $R^1 = Et$, $R^2 = H$; $R^1 = H$, $R^2 = Me$) or $CNCH_2CH_2CH_2OH$ on $[PtX_2L_2]$ or PtX_2 (X^- is halide or ClO_4^-; L = MeCN, PhCN or PPh_3).[222] While the triphenylphosphine complex $[Pt(\overline{CNHC_2H_4O})_2(PPh_3)_2](BF_4)_2$ is of a similar type to many already described, the reactions of the non-phosphine-containing materials give rise to a series of stable, water-soluble tetracarbene compounds $[Pt(\overline{CNHCHR^1CHR^2O})_4]X_2$ and $[Pt(\overline{CNHC_3H_6O})_4]X_2$ which have no parallel. The structure of a palladium analogue suggests that all the carbene ligands lie perpendicular to the coordination plane.[222]

Scheme 24 (L = PEt_3) illustrates two reactions of aminoalkoxycarbene complexes whereby they can be converted to their carbonyl or isocyanide 'precursors'.[223] Fast atom bombardment (FAB) mass spectra of some aminoalkoxycarbene complexes have been described.[224]

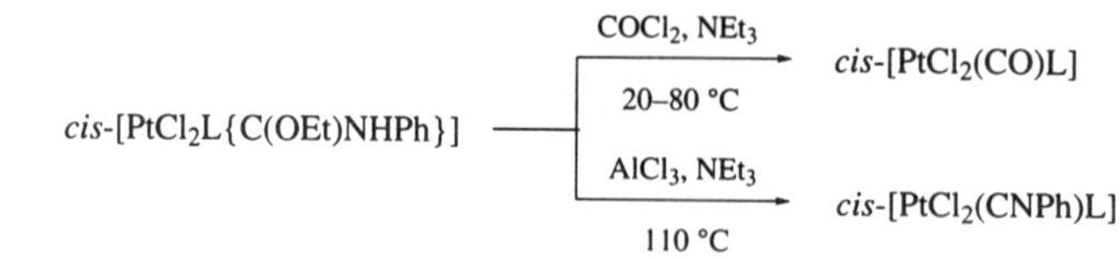

Scheme 24

7.3.6 O,O- and S,S-Carbene Complexes

The syntheses of cyclic examples such as *trans*-$[PtH(\overline{COC_2H_4O})(PPh_3)_2]^+$ have been described in Section 7.3.1.2.[201] The reactions of some of these with diphoe are in dramatic contrast to those of complexes such as *cis*-$[PtCl_2\{\overline{CN(R)C_2H_4NH}\}_2]$.[216] Whereas the latter react by Cl^- displacement, the O,O-carbene species are degraded to CO_2 and C_2H_4. Chloride reacts with these cationic species to break open the ring and form acyl derivatives, along with a variety of other products. Scheme 25 (L = PPh_3) shows these reactions.[216] Thermal rearrangement of the S,S-carbene complexes in acetonitrile involves hydride migration to the carbene ligand (Equation (37)).[225] FAB mass spectra on these O,O- and S,S-cyclic complexes have been described.[226]

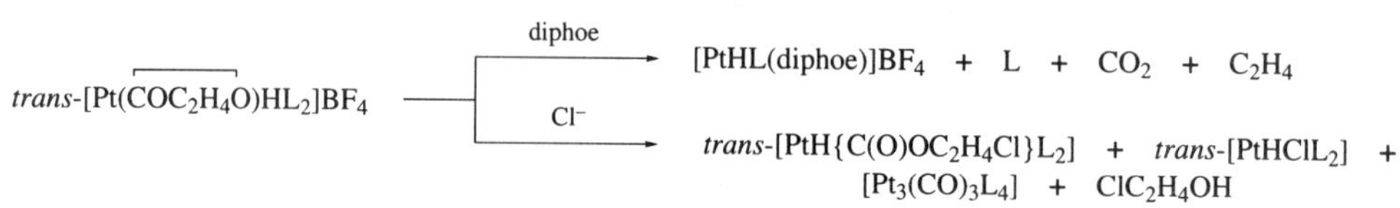

Scheme 25

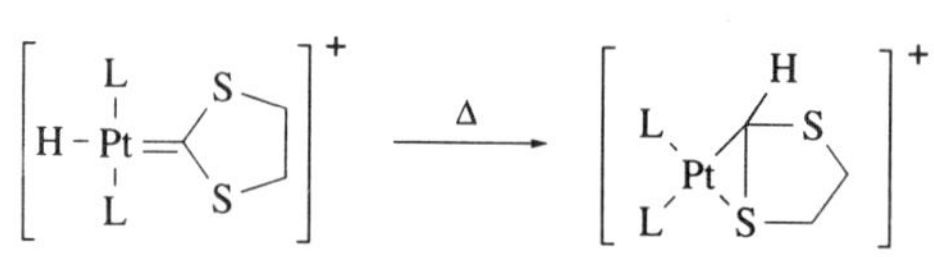

(37)

Dppe reacts with the noncyclic S,S-carbene complex $[PtI(PPh_3)\{C(SMe)_2\}_2]I$ to displace PPh_3 and one carbene to afford $[PtI(dppe)\{C(SMe)_2\}]I$. The carbene ligands in either $[PtI(PPh_3)\{C(SMe)_2\}_2]I$ or $[PtI_2(PPh_3)\{C(SMe)_2\}]$ are broken down on reactions with $[Pt(C_2H_4)(PPh_3)_2]$ to produce binuclear $[(Ph_3P)IPt(\mu\text{-}SMe)(\mu\text{-}CSMe)PtI(PPh_3)]$.[227] $[PtI(dppe)\{C(SMe)_2\}]I$ reacts with $[Pt(\eta^2\text{-}CS_2)(dppe)]$ to produce (**100**).[228] This latter binuclear complex, which can be regarded as an S,S-platinacyclic carbene complex, can also be synthesized from the action of MeI on $[Pt(CS_2)(dppe)]$. The two carbenoid C–S bond lengths are compatible with extensive π-delocalization.

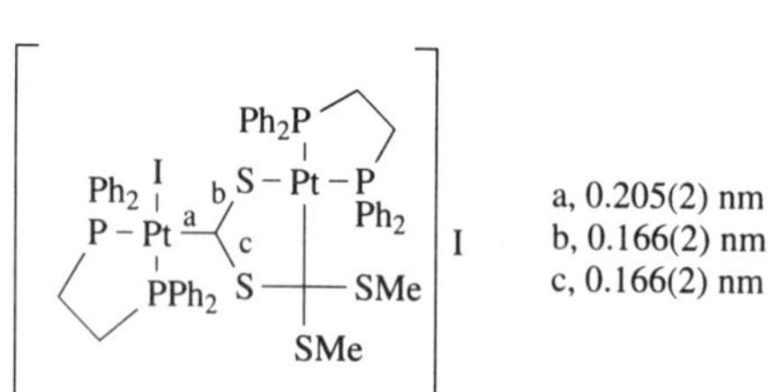

a, 0.205(2) nm
b, 0.166(2) nm
c, 0.166(2) nm

(100)

7.4 REFERENCES

1. H. C. Clark and V. K. Jain, *Coord. Chem. Rev.*, 1984, **55**, 151.
2. (a) H. Basch and D. Cohen, *J. Am. Chem. Soc.*, 1983, **105**, 3856; (b) H. Basch, *Chem. Phys. Lett.*, 1985, **116**, 58; (c) A. Gavezzotti, G. F. Tantardini and M. Simonetta, *ibid.*, 1986, **129**, 577; (d) G. W. Smith and E. A. Carter, *J. Phys. Chem.*, 1991, **95**, 2327.

3. C. M. Rohlfing and P. J. Hay, *J. Chem. Phys.*, 1985, **83**, 4641.
4. A. Sen and J. Halpern, *Inorg. Chem.*, 1980, **19**, 1073.
5. R. S. Paonessa and W. C. Trogler, *J. Am. Chem. Soc.*, 1982, **104**, 1138.
6. S. Nakamura and K. Morokuma, *Organometallics*, 1988, **7**, 1904.
7. D. M. P. Mingos and R. W. M. Wardle, *Transition Met. Chem.*, 1985, **10**, 441.
8. D. G. Evans, M. F. Hallam, D. M. P. Mingos and R. W. M. Wardle, *J. Chem. Soc., Dalton Trans.*, 1987, 1889.
9. K.-H. Dahmen, A. Moor, R. Naegeli and L. M. Venenzi, *Inorg. Chem.*, 1991, **30**, 4285.
10. W. Zhai and Z. Zhao, *Youji Huaxue*, 1986, 134 (*Chem. Abstr.*, 1986, **107**, 115 724j).
11. N. K. Eremenko and S. P. Gubin, *Pure Appl. Chem.*, 1990, **62**, 1179.
12. N. K. Eremenko, E. G. Mednikov and S. P. Gubin, *Koord. Khim.*, 1984, **10**, 617.
13. R. G. Goel, W. O. Ogini and R. C. Srivastava, *J. Organomet. Chem.*, 1981, **214**, 405.
14. R. G. Goel, W. O. Ogini and R. C. Srivastava, *Organometallics*, 1982, **1**, 819.
15. R. Bender, P. Braunstein, J. Fischer, L. Ricard and A. Mitschler, *Nouv. J. Chim.*, 1981, **5**, 81.
16. P. Braunstein and B. Oswald, *J. Organomet. Chem.*, 1987, **328**, 229.
17. O. J. Scherer, R. Konrad, E. Guggolz and M. L. Ziegler, *Angew. Chem., Int. Ed. Engl.*, 1982, **21**, 297.
18. J. C. Jeffery, I. Moore, M. Murray and F. G. A. Stone, *J. Chem. Soc., Dalton Trans.*, 1982, 1741.
19. A. Jiang and Q. Cong, *Jiegon Huaxue*, 1985, **4**, 96 (*Chem. Abstr.*, 1985, **105**, 124 739d).
20. R. A. Burrow, D. H. Farrer and J. J. Irwin, *Inorg. Chim. Acta*, 1991, **181**, 65.
21. A. Moor, P. S. Pregosin, L. M. Venenzi and A. J. Welch, *Inorg. Chim. Acta*, 1984, **85**, 103.
22. R. F. Klevtsova, E. N. Yurchenko, L. A. Glinskaya, E. B. Burgina, N. K. Eremenko and V. V. Bakakin, *J. Struct. Chem.*, 1985, **26**, 216.
23. Yu. L. Stovokhotov and Yu. T. Struchkov, *Koord. Khim.*, 1986, **12**, 553.
24. A. A. Frew, R. H. Hill, L. Manojlovic-Muir, K. W. Muir and R. J. Puddephatt, *J. Chem. Soc., Chem. Commun.*, 1982, 198.
25. S. S. Kurasov, N. K. Eremenko, Yu. L. Slovokhotov and Yu. T. Struchkov, *J. Organomet. Chem.*, 1989, **361**, 405.
26. A. Moor, P. S. Pregosin and L. M. Venenzi, *Inorg. Chim. Acta*, 1981, **48**, 153.
27. A. Moor, P. S. Pregosin and L. M. Venenzi, *Inorg. Chim. Acta*, 1982, **61**, 135.
28. V. F. Yudanov, N. K. Eremenko, E. G. Mednikov and S. P. Gubin, *J. Struct. Chem.*, 1984, **25**, 41.
29. Th. Wurminghausen, H. J. Reinecke and P. Braunstein, *Org. Mass Spectrom.*, 1980, **15**, 38.
30. H.-R. C. Jaw and W. R. Mason, *Inorg. Chem.*, 1990, **29**, 3452.
31. R. Bender and P. Braunstein, *J. Chem. Soc., Chem. Commun.*, 1983, 334.
32. C. Couture, D. H. Farrar, D. S. Fisher and R. R. Gukathasan, *Organometallics*, 1987, **6**, 532.
33. C. S. Browning and D. H. Farrar, *Organometallics*, 1989, **8**, 813.
34. A. B. Goel and S. Goel, *Inorg. Nucl. Chem. Lett.*, 1980, **16**, 379.
35. C. S. Browning, D. H. Farrar, R. R. Gukathasan and S. A. Morris, *Organometallics*, 1985, **4**, 1750.
36. M. F. Hallam, N. D. Howells, D. M. P. Mingos and R. W. M. Wardle, *J. Chem. Soc., Dalton Trans.*, 1985, 845.
37. J. M. Ritchey and D. C. Moody, *Inorg. Chim. Acta*, 1983, **74**, 271.
38. A. B. Goel and S. Goel, *Inorg. Chim. Acta*, 1982, **65**, L77.
39. C. E. Briant, D. G. Evans and D. M. P. Mingos, *J. Chem. Soc., Dalton Trans.*, 1986, 1535.
40. M. F. Hallam and D. M. P. Mingos, *J. Organomet. Chem.*, 1986, **315**, C35.
41. S. G. Bott, A. D. Burrows, O. J. Ezomo, M. F. Hallam, J. G. Jeffrey and D. M. P. Mingos, *J. Chem. Soc., Dalton Trans.*, 1990, 3335.
42. (a) A. Stockhammer, K.-H. Dahmen, T. Gerfin, L. M. Venenzi, V. Gramlich and W. Petter, *Helv. Chim. Acta*, 1991, **74**, 989; (b) A. Albinati *et al.*, *Inorg. Chem.*, 1992, **31**, 2223.
43. K. C. C. Kharas and L. F. Dahl, *Adv. Chem. Phys.*, (part 2) 1988, **70**, 1.
44. (a) A. Ceriotti, G. Longoni and M. Marchionna, *Inorg. Synth.*, 1989, **26**, 316; (b) M. J. D'Anielbo, C. J. Carr and M. G. Zammit, *ibid.*, 1989, **26**, 319.
45. J. Puga, R. Patrini, K. M. Sanchez and B. C. Gates, *Inorg. Chem.*, 1991, **30**, 2479.
46. J. D. Roth, G. J. Lewis, L. K. Safford, X. D. Jiang, L. F. Dahl and M. J. Weaver, *J. Am. Chem. Soc.*, 1992, **114**, 6159.
47. D. J. Underwood, R. Hoffmann, K. Tatsumi, A. Nakamura and Y. Yamamoto, *J. Am. Chem. Soc.*, 1985, **107**, 5968.
48. R. B. King, *Inorg. Chim. Acta*, 1986, **116**, 119.
49. R. T. Roginski, J. R. Shapley, H. G. Drickamer and M. J. D'Aniello, *Chem. Phys. Lett.*, 1987, **135**, 525.
50. R. G. Woolley, *Chem. Phys. Lett.*, 1988, **143**, 145.
51. C. Brown, B. T. Heaton, A. D. Towl, P. Chini, A. Fumagalli and G. Longoni, *J. Organomet. Chem.*, 1979, **181**, 233.
52. B. K. Teo, F. J. DiSalvo, J. V. Waszczak, G. Longoni and A. Ceriotti, *Inorg. Chem.*, 1986, **25**, 2262.
53. C. J. McNeal, J. M. Hughes, G. J. Lewis and L. F. Dahl, *J. Am. Chem. Soc.*, 1991, **113**, 372.
54. G. J. Lewis *et al.*, *J. Am. Chem. Soc.*, 1990, **112**, 2831.
55. J. D. Roth, G. J. Lewis, X. D. Jiang, L. F. Dahl and M. J. Weaver, *J. Phys. Chem.*, 1992, **96**, 7219.
56. K. Machida, A. Fukuoka, M. Ichikawa and M. Enyo, *J. Electrochem. Soc.*, 1991, **138**, 1958.
57. (a) S. Bhaduri and K. R. Sharma, *J. Chem. Soc., Dalton Trans.*, 1982, 727; (b) A. Basu, S. Bhaduri and K. R. Sharma, *J. Chem. Soc., Dalton Trans.*, 1984, 2315.
58. N. M. Boag, P. L. Goggin, R. J. Goodfellow and I. R. Herbert, *J. Chem. Soc., Dalton Trans.*, 1983, 1101.
59. R. Uson, J. Fornies, P. Espinet, C. Fortuno, M. Tomas and A. J. Welch, *J. Chem. Soc., Dalton Trans.*, 1989, 1583.
60. Y. Koie, S. Shinoda, Y. Saito, B. J. Fitzgerald and C. G. Pierpont, *Inorg. Chem.*, 1980, **19**, 770.
61. Y. Koie, S. Shinoda and Y. Saito, *Inorg. Chem.*, 1981, **20**, 4408.
62. N. M. Boag, R. J. Goodfellow, M. Green, B. Hessner, J. A. K. Howard and F. G. A. Stone, *J. Chem. Soc., Dalton Trans.*, 1983, 2585.
63. N. M. Boag, *Organometallics*, 1988, **7**, 1446.
64. A. J. Dixon, S. Firth, A. Haynes, M. Poliakoff, J. J. Turner and N. M. Boag, *J. Chem. Soc., Dalton Trans.*, 1988, 1501.
65. (a) R. J. Goodfellow, I. R. Herbert and A. G. Orpen, *J. Chem. Soc., Chem. Commun.*, 1983, 1386; (b) R. Bender, P. Braunstein, A. Tiripicchio and M. Tiripicchio-Camellini, *J. Chem. Soc., Chem. Commun.*, 1984, 42.
66. G. Minghetti *et al.*, *Inorg. Chem.*, 1983, **22**, 2332.
67. T. R. Gaffney and J. A. Ibers, *Inorg. Chem.*, 1982, **21**, 2860.
68. M. F. Hallam, M. A. Luke, D. M. P. Mingos and I. D. Williams, *J. Organomet. Chem.*, 1987, **325**, 271.

69. A. L. Balch, R. R. Guimerani, J. Lineham and F. E. Wood, *Inorg. Chem.*, 1985, **24**, 2021.
70. L. Manojlovic-Muir, K. W. Muir and T. Solomun, *J. Organomet. Chem.*, 1979, **179**, 479.
71. J. R. Fisher *et al.*, *Organometallics*, 1982, **1**, 1421.
72. A. J. McLennan and R. J. Puddephatt, *Organometallics*, 1986, **5**, 811.
73. M. Ghedini, F. Neve, C. Mealli, A. Tiripicchio and F. Ugozzoli, *Inorg. Chim. Acta*, 1990, **178**, 5.
74. R. J. Puddephatt, L. Manojlovic-Muir and K. W. Muir, *Polyhedron*, 1990, **9**, 2767.
75. G. Ferguson, B. R. Lloyd and R. J. Puddephatt, *Organometallics*, 1986, **5**, 344.
76. G. Douglas *et al.*, *Organometallics*, 1991, **10**, 3927.
77. G. Ferguson, B. R. Lloyd, L. Manojlovic-Muir, K. W. Muir and R. J. Puddephatt, *Inorg. Chem.*, 1986, **25**, 4190.
78. (a) C. Mealli, *J. Am. Chem. Soc.*, 1985, **107**, 2245; (b) D. G. Evans, *J. Organomet. Chem.*, 1988, **352**, 397; (c) P. D. Harvey and R. Provencher, *Inorg. Chem.*, 1993, **332**, 61.
79. B. R. Lloyd, A. Bradford and R. J. Puddephatt, *Organometallics*, 1987, **6**, 424.
80. M. C. Jennings, R. J. Puddephatt, L. Manojlovic-Muir, K. W. Muir and B. N. Mwariri, *Organometallics*, 1992, **11**, 4164.
81. A. M. Bradford, G. Douglas, L. Manojlovic-Muir, K. W. Muir and R. J. Puddephatt, *Organometallics*, 1990, **9**, 409.
82. R. Ramachandran, D.-S. Yang, N. C. Payne and R. J. Puddephatt, *Inorg. Chem.*, 1992, **31**, 4236.
83. A. M. Bradford, R. J. Puddephatt, G. Douglas, L. Manojlovic-Muir and K. W. Muir, *Organometallics*, 1990, **9**, 1579.
84. A. M. Bradford, M. C. Jennings and R. J. Puddephatt, *Organometallics*, 1989, **8**, 2367.
85. M. C. Jennings, N. C. Payne and R. J. Puddephatt, *Inorg. Chem.*, 1987, **26**, 3776.
86. M. C. Jennings and R. J. Puddephatt, *Inorg. Chem.*, 1988, **27**, 4280.
87. B. R. Lloyd and R. J. Puddephatt, *J. Am. Chem. Soc.*, 1985, **107**, 7785.
88. G. Douglas, L. Manojlovic-Muir, K. W. Muir, M. Rashidi, C. M. Anderson and R. J. Puddephatt, *J. Am. Chem. Soc.*, 1987, **109**, 6527.
89. (a) G. Schoettel, J. J. Vittal and R. J. Puddephatt, *J. Am. Chem. Soc.*, 1990, **112**, 6400; (b) G. Douglas, M. C. Jennings, L. Manojlovic-Muir, K. W. Muir and R. J. Puddephatt, *J. Chem. Soc., Chem. Commun.*, 1989, 159.
90. S. S. M. Ling, N. Hadj-Bagheri, L. Manojlovic-Muir, K. W. Muir and R. J. Puddephatt, *Inorg. Chem.*, 1987, **26**, 231.
91. M. Kubota and F. S. Rosenberg, *Synth. React. Inorg. Metal-Org. Chem.*, 1985, **15**, 1037.
92. J. R. Briggs, A. G. Constable, W. S. McDonald and B. L. Shaw, *J. Chem. Soc., Dalton Trans.*, 1982, 1225.
93. R. D. Gillard, M. F. Pilbrow, S. Leonard, M. C. Rendle and C. F. H. Tipper, *Polyhedron*, 1982, **1**, 689.
94. V. Yu. Kukushkin and A. I. Moiseev, *J. Gen. Chem. USSR (Engl. Transl.)*, 1990, **60**, 607.
95. (a) P. J. Stang, Z. Zhong and M. H. Kowalski, *Organometallics*, 1990, **9**, 833; (b) Z. Zhong, P. J. Stang and A. M. Arif, *ibid.*, 1990, **9**, 1703; (c) P. J. Stang and Z. Zhong, *ibid.*, 1992, **11**, 1026.
96. J. Fornies, B. Menjon, N. Gomez and M. Tomas, *Organometallics*, 1992, **11**, 1187.
97. J. D. Scott and R. J. Puddephatt, *Organometallics*, 1986, **5**, 1253.
98. A. Avshu, R. D. O'Sullivan, A. W. Parkins, N. W. Alcock and R. M. Countryman, *J. Chem. Soc., Dalton Trans.*, 1983, 1619.
99. R. J. Cross, N. H. Tennent and G. M. Warden, *J. Chem. Soc., Dalton Trans.*, 1986, 1059.
100. G. K. Anderson and R. J. Cross, *J. Chem. Soc., Dalton Trans.*, 1980, 1988.
101. G. K. Anderson, H. C. Clark and J. A. Davies, *Inorg. Chem.*, 1981, **20**, 944.
102. R. G. Goel, W. O. Ogini and R. C. Srivastava, *Inorg. Chem.*, 1981, **20**, 3611.
103. R. J. Cross and A. J. McLennan, *J. Chem. Soc., Dalton Trans.*, 1983, 359.
104. R. J. Cross and A. J. McLennan, *J. Organomet. Chem.*, 1983, **255**, 113.
105. R. J. Cross and J. Gemmill, *J. Chem. Res. (S)*, 1984, 30.
106. R. C. Srivastava, *J. Chem. Res. (S)*, 1985, 330.
107. I. R. Herbert, P. S. Pregosin and H. Ruegg, *Inorg. Chim. Acta*, 1986, **112**, 29.
108. (a) I. V. Pakhomova, Yu. N. Kukushkin, Yu. N. Martinov, N. A. Gudova and A. M. Bespozvannyl, *J. Gen. Chem. USSR (Engl. Transl.)*, 1981, **51**, 2292; (b) H. Alper, Y. J. Huang, D. B. Dell'Amico, F. Calderazzo, N. Pasqualetti and C. A. Veracini, *Organometallics*, 1991, **10**, 1665.
109. R. J. Cross and I. G. Phillips, *J. Chem. Soc., Dalton Trans.*, 1982, 2261.
110. C. Y. Mok, S. G. Tan and G. C. Chan, *Inorg. Chim. Acta*, 1990, **176**, 43.
111. G. K. Anderson, M. R. Lin and M. Y. Chiang, *Organometallics*, 1990, **9**, 288.
112. S. L. Rusakov, T. V. Lysyak, G. M. Apal'kova, A. I. Gusev, Yu. Ya. Kharitonov and I. S. Kolomnikov, *Koord. Khim.*, 1988, **14**, 229.
113. R. Uson, J. Fornies, M. A. Uson, J. F. Yague, P. G. Jones and K. Meyer-Base, *J. Chem. Soc., Dalton Trans.*, 1986, 947.
114. P. Scheer, V. Schurig, L. Walz and H. G. von Schnering, *Acta Crystallogr., Sect. C*, 1991, **C47**, 743.
115. M. Grassi, S. V. Meille, A. Musco, R. Pontellini and A. Sironi, *J. Chem. Soc., Dalton Trans.*, 1989, 615.
116. F. Calderazzo and D. B. Dell'Amico, *Inorg. Chem.*, 1981, **20**, 1310.
117. (a) G. A. Foulds, P. S. Hall and D. A. Thornton, *J. Mol. Struct.*, 1984, **117**, 95; (b) P. S. Hall, D. A. Thornton and G. A. Foulds, *Polyhedron*, 1987, **6**, 85.
118. G. K. Anderson and R. J. Cross, *J. Chem. Soc., Dalton Trans.*, 1979, 1246.
119. G. K. Anderson, R. J. Cross and D. S. Rycroft, *J. Chem. Res. (S)*, 1979, 120.
120. G. K. Anderson, R. J. Cross and D. S. Rycroft, *J. Chem. Res. (S)*, 1980, 240.
121. L. M. Scolaro, G. Alibrandi, R. Romeo, V. Ricevuto and S. Campagna, *Inorg. Chem.*, 1992, **31**, 2074.
122. R. J. Cross and M. F. Davidson, *J. Chem. Soc., Dalton Trans.*, 1986, 1987.
123. (a) R. Romeo, A. Grassi and L. M. Scolaro, *Inorg. Chem.*, 1992, **31**, 4383; (b) R. Romeo, G. Arena and L. M. Scolaro, *Inorg. Chem.*, 1992, **31**, 4879.
124. H. C. Clark and J. A. Davies, *J. Organomet. Chem.*, 1981, **213**, 503.
125. (a) G. K. Anderson, H. C. Clark and J. A. Davies, *Inorg. Chem.*, 1983, **22**, 427, 434; (b) G. K. Anderson, C. Billard, H. C. Clark, J. A. Davies and C. S. Wong, *Inorg. Chem.*, 1983, **22**, 439; (c) G. K. Anderson, H. C. Clark, J. A. Davies, G. Ferguson and M. Parvez, *J. Crystallogr. Spectrosc. Res.*, 1982, **12**, 449.
126. M. Catellani and J. Halpern, *Inorg. Chem.*, 1980, **19**, 516.
127. B. L. Booth, R. N. Haszeldine and R. G. G. Holmes, *J. Chem. Soc., Dalton Trans.*, 1982, 523.
128. G. K. Anderson and R. J. Cross, *Acc. Chem. Res.*, 1984, **17**, 67.
129. G. K. Anderson and R. J. Cross, *J. Chem. Soc., Dalton Trans.*, 1980, 712.

130. R. J. Cross and J. Gemmill, *J. Chem. Soc., Dalton Trans.*, 1981, 2317.
131. S. Sakaki, K. Kitaura, K. Morokuma and K. Ohkubo, *J. Am. Chem. Soc.*, 1983, **105**, 2280.
132. N. Koga and K. Morokuma, *J. Am. Chem. Soc.*, 1986, **108**, 6136.
133. G. K. Anderson and R. J. Cross, *J. Chem. Soc., Dalton Trans.*, 1980, 1434.
134. G. K. Anderson, H. C. Clark and J. A. Davies, *Organometallics*, 1982, **1**, 64.
135. S. Dire, R. Campostrini, G. Carturan, M. Calligaris and G. Nardini, *J. Organomet. Chem.*, 1990, **390**, 267.
136. G. K. Anderson, H. C. Clark and J. A. Davies, *Inorg. Chem.*, 1981, **20**, 3607.
137. (a) M. A. Bennett and A. Rokicki, *Organometallics*, 1985, **4**, 180; (b) M. A. Bennett and A. Rokicki, *J. Organomet. Chem.*, 1983, **244**, C31.
138. G. P. C. M. Dekker *et al.*, *Organometallics*, 1992, **11**, 1937.
139. C. J. Cardin, D. J. Cardin, H. E. Parge and A. C. Sullivan, *J. Chem. Soc., Dalton Trans.*, 1986, 2315.
140. P. J. Stang, Z. D. Zhong and A. M. Arif, *Organometallics*, 1992, **11**, 1017.
141. K. J. Cavell, H. Jin, B. W. Skelton and A. H. White, *J. Chem. Soc., Dalton Trans.*, 1992, 2923.
142. (a) A. Sen, J.-T. Chen, W. M. Vetter and R. R. Whittle, *J. Am. Chem. Soc.*, 1987, **109**, 148; (b) T. M. Huang *et al.*, *Organometallics*, 1991, **10**, 1020.
143. J.-T. Chen, T.-M. Huang, M.-C. Cheng and Y. Wang, *Organometallics*, 1991, **10**, 2838.
144. J.-T. Chen, T.-M. Huang, M.-C. Cheng and Y. Wang, *Organometallics*, 1990, **9**, 539.
145. L. Huang, F. Ozawa and A. Yamamoto, *Organometallics*, 1990, **9**, 2612.
146. F. Calderazzo, *J. Organomet. Chem.*, 1990, **400**, 303.
147. L. D. Polyachenok, S. V. Yasinetskaya, V. P. Bochin and O. G. Polyachenok, *Russ. J. Inorg. Chem. (Engl. Transl.)*, 1983, **28**, 1221.
148. P. R. Brown, F. G. N. Cloke and M. L. H. Green, *Polyhedron*, 1985, **4**, 869.
149. D. B. Dell'Amico, F. Calderazzo, C. A. Veracini and N. Zandona, *Inorg. Chem.*, 1984, **23**, 3030.
150. (a) F. Calderazzo and D. B. Dell'Amico, *Pure Appl. Chem.*, 1986, **58**, 561; (b) B. P. Andreini, D. B. Dell'Amico, F. Calderazzo, M. G. Venturi, G. Pelizzi and A. Segre, *J. Organomet. Chem.*, 1988, **354**, 357; (c) B. Andreini, D. B. Dell'Amico, F. Calderazzo and N. Pasqualetti, *Inorg. Synth.*, 1992, **29**, 188.
151. (a) R. Uson, J. Fornies, M. Tomas and B. Menjon, *Organometallics*, 1985, **4**, 1912; (b) R. Uson, J. Fornies, M. Tomas and B. Menjon, *ibid.*, 1986, **5**, 1581.
152. F. Canziani, F. Galimberti, L. Garlaschelli, M. C. Malatesta and A. Albinati, *Gazz. Chim. Ital.*, 1982, **112**, 323.
153. F. Canziani, C. Allevi, L. Garlaschelli, M. C. Malatesta and F. Ganazzoli, *J. Chem. Soc., Dalton Trans.*, 1984, 2637.
154. F. Canziani, F. Galimberti, L. Garlaschelli, M. C. Malatesta, A. Albinati and F. Ganazzoli, *J. Chem. Soc., Dalton Trans.*, 1983, 827.
155. M. Bonfa, L. Garlaschelli and M. C. Malatesta, *Gazz. Chim. Ital.*, 1986, **116**, 317.
156. (a) C. Allevi, L. Garlaschelli, M. C. Malatesta and F. Ganazzoli, *J. Chem. Soc., Dalton Trans.*, 1988, 1949; (b) C. Allevi, M. Bonfa, L. Garlaschelli and M. C. Malatesta, *Gazz. Chim. Ital.*, 1990, **120**, 711; (c) C. Allevi, L. Garlaschelli, M. C. Malatesta and F. Ganazzoli, *Organometallics*, 1990, **9**, 1383; (d) C. Allevi, L. Garlaschelli, M. C. Malatesta, F. Ganazzoli and A. Albinati, *Gazz. Chim. Ital.*, 1992, **122**, 215.
157. N. M. Boag, M. Green, J. L. Spencer and F. G. A. Stone, *J. Chem. Soc., Dalton Trans.*, 1980, 1220.
158. N. M. Boag, M. Green, J. L. Spencer and F. G. A. Stone, *J. Chem. Soc., Dalton Trans.*, 1980, 1208.
159. N. M. Boag, R. Quyoum and K. M. Rao, *J. Chem. Soc., Chem. Commun.*, 1992, 114.
160. S. T. Lin and K. J. Klabunde, *Inorg. Chem.*, 1985, **24**, 1961.
161. R. J. Goodfellow and I. R. Herbert, *Inorg. Chim. Acta*, 1982, **65**, L161.
162. P. Dagnac, R. Turpin and R. Poilblanc, *J. Organomet. Chem.*, 1983, **253**, 123.
163. D. B. Dell'Amico, F. Calderazzo, F. Marchetti and S. Merlino, *J. Chem. Soc., Dalton Trans.*, 1982, 2257.
164. P. Overbosch and G. van Koten, *J. Organomet. Chem.*, 1982, **229**, 193.
165. Y. Yamamoto and H. Yamazaki, *J. Chem. Soc., Dalton Trans.*, 1989, 2161.
166. C. E. Briant, D. I. Gilmour and D. M. P. Mingos, *J. Organomet. Chem.*, 1986, **308**, 381.
167. C. Aristides, *J. Organomet. Chem.*, 1983, **259**, 355.
168. Y. Yamamoto, K. Takahashi, K. Matsuda and H. Yamazaki, *J. Chem. Soc., Dalton Trans.*, 1987, 1833.
169. K. Takahashi, Y. Yamamoto, K. Matsuda and H. Yamazaki, *Electrochim. Acta*, 1988, **33**, 1489.
170. Y. Yamamoto and H. Yamazaki, *J. Chem. Soc., Dalton Trans.*, 1986, 677.
171. Y. Yamamoto, K. Aoki and H. Yamazaki, *Organometallics*, 1983, **2**, 1377.
172. C. E. Briant, D. I. Gilmour, D. M. P. Mingos and R. W. M. Wardle, *J. Chem. Soc., Dalton Trans.*, 1985, 1693.
173. D. M. P. Mingos, I. D. Williams and M. J. Watson, *J. Chem. Soc., Dalton Trans.*, 1988, 1509.
174. C. T. Hunt, G. B. Matson and A. L. Balch, *Inorg. Chem.*, 1981, **20**, 2270.
175. B. Messbauer, H. Meyer, B. Walther, M. J. Heeg, A. F. M. Maqsudar Rahman and J. P. Oliver, *Inorg. Chem.*, 1983, **22**, 272.
176. K. R. Grundy and K. N. Robertson, *Organometallics*, 1983, **2**, 1736.
177. (a) Y. Yamamoto, K. Takahashi and H. Yamazaki, *J. Am. Chem. Soc.*, 1986, **108**, 2458; (b) Y. Yamamoto and H. Yamazaki, *Organometallics*, 1993, **12**, 933.
178. R. J. Puddephatt, M. Rashidi and J. J. Vittal, *J. Chem. Soc., Dalton Trans.*, 1991, 2835.
179. A. M. Bradford, N. C. Payne, R. J. Puddephatt, D.-S. Yang and T. B. Marder, *J. Chem. Soc., Chem. Commun.*, 1990, 1462.
180. J. R. Briggs, C. Crocker and B. L. Shaw, *Inorg. Chim. Acta*, 1981, **51**, 15.
181. R. Uson, J. Fornies, P. Espinet, R. Navarro and E. Lalinde, *Transition Met. Chem.*, 1984, **9**, 277.
182. C. Herdeis and W. Beck, *Chem. Ber.*, 1983, **116**, 3205.
183. K. W. Chin, C. G. Howard, G. Wilkinson, A. M. R. Gates and M. B. Hursthouse, *Polyhedron*, 1982, **1**, 803.
184. R. D. W. Kemmitt, P. McKenna, D. R. Russell and L. J. S. Prouse, *J. Chem. Soc., Dalton Trans.*, 1989, 345.
185. W. Henderson, R. D. W. Kemmitt, L. J. S. Prouse and D. R. Russell, *J. Chem. Soc., Dalton Trans.*, 1990, 781.
186. R. A. Michelin, G. Facchin and P. Uguagliati, *Inorg. Chem.*, 1984, **23**, 961.
187. (a) P. Zanello, R. Seeber, A. Cinquantini and B. Crociani, *Transition Met. Chem.*, 1980, **5**, 226; (b) P. Zanello, R. Seeber, B. Crociani and M. Nicolini, *ibid.*, 1980, **5**, 45.
188. R. T. Carl, E. W. Corcoran, R. P. Hughes and D. E. Samkoff, *Organometallics*, 1990, **9**, 838.
189. C. R. Langrick, P. G. Pringle and B. L. Shaw, *J. Chem. Soc., Dalton Trans.*, 1984, 1233.

190. (a) M. Schaal, W. Weigand, U. Nagel and W. Beck, *Chem. Ber.*, 1985, **118**, 2186; (b) W. Weigand, U. Nagel and W. Beck, *ibid.*, 1990, **123**, 439.
191. H. L. Conder, F. A. Cotton, L. R. Falvello, S. Han and R. A. Walton, *Inorg. Chem.*, 1983, **22**, 1887.
192. M. L. Winzenburg, J. A. Kargol and R. J. Angelici, *J. Organomet. Chem.*, 1983, **249**, 415.
193. P. Lemoine, A. Giraudeau, M. Gross and P. Braunstein, *J. Organomet. Chem.*, 1980, **202**, 447.
194. N. Carr *et al.*, *J. Chem. Soc., Chem. Commun.*, 1990, 1369.
195. Yu. N. Kukushkin, L. V. Vreblevskaya, S. I. Bakhireva and E. N. Kalyukova, *Russ. J. Inorg. Chem. (Engl. Transl.)*, 1980, **25**, 723.
196. S. Roy, R. J. Puddephatt and J. D. Scott, *J. Chem. Soc., Dalton Trans.*, 1989, 2121.
197. R. Kumar, S. Roy, M. Rashidi and R. J. Puddephatt, *Polyhedron*, 1989, **8**, 551.
198. Z. Lu, W. M. Jones and W. R. Winchester, *Organometallics*, 1993, **12**, 1344.
199. U. Belluco *et al.*, *Inorg. Chim. Acta*, 1992, **200**, 883.
200. E. K. Barefield, A. M. Carrier, D. J. Sepelak and D. G. Van Derveer, *Organometallics*, 1982, **1**, 103.
201. (a) R. A. Michelin, G. Facchin and R. Ros, *J. Organomet. Chem.*, 1985, **279**, C25; (b) R. A. Michelin, R. Ros, G. Guadalupi, G. Bombieri, F. Benetollo and G. Chapuis, *Inorg. Chem.*, 1989, **28**, 840.
202. D. Sellmann, W. Prechtel, F. Knoch and M. Moll, *Inorg. Chem.*, 1993, **32**, 538.
203. R. J. Cross, M. F. Davidson and M. Rocamora, *J. Chem. Soc., Dalton Trans.*, 1988, 1147.
204. (a) G. Facchin, R. Campostrini and R. A. Michelin, *J. Organomet. Chem.*, 1985, **294**, C21; (b) R. A. Michelin, G. Facchin, D. Braga and P. Sabatino, *Organometallics*, 1986, **5**, 2265.
205. R. A. Michelin, M. Mozzon, G. Facchin, D. Braga and P. Sabatino, *J. Chem. Soc., Dalton Trans.*, 1988, 1803.
206. H. C. Clark, V. K. Jain and G. S. Rao, *J. Organomet. Chem.*, 1983, **259**, 275.
207. V. B. Pukhnarevich, Yu. T. Struchkov, G. G. Aleksandrov, S. P. Suschinskaya, E. O. Tsetlina and M. G. Voronkov, *Sov. J. Coord. Chem. (Engl. Transl.)*, 1979, **5**, 1195.
208. M. Wada and Y. Koyama, *J. Organomet. Chem.*, 1980, **201**, 477.
209. M. Wada, Y. Koyama and K. Sameshima, *J. Organomet. Chem.*, 1981, **209**, 115.
210. G. Erker and B. Menjon, *Chem. Ber.*, 1990, **123**, 1327.
211. H. J. Ruegg, P. S. Pregosin, A. Scrivanti, L. Toniolo and C. Botteghi, *J. Organomet. Chem.*, 1986, **316**, 233.
212. U. Belluco, B. Crociani, R. Michelin and P. Uguagliati, *Pure Appl. Chem.*, 1983, **55**, 47.
213. R. A. Michelin, L. Zanotto, D. Braga, P. Sabatino and R. J. Angelici, *Inorg. Chem.*, 1988, **27**, 93.
214. R. Bertani, M. Mozzon and R. A. Michelin, *Inorg. Chem.*, 1988, **27**, 2809.
215. R. Bertani *et al.*, *Inorg. Chim. Acta*, 1991, **189**, 175.
216. R. Bertani, M. Mozzon, R. A. Michelin, T. J. Castilho and A. J. L. Pombeiro, *J. Organomet. Chem.*, 1992, **431**, 117.
217. R. Bertani, M. Mozzon, F. Benetollo, G. Bombieri and R. A. Michelin, *J. Chem. Soc., Dalton Trans.*, 1990, 1197.
218. R. A. Michelin, R. Bertani, M. Mozzon, L. Zanotto, F. Benetollo and G. Bombieri, *Organometallics*, 1990, **9**, 1449.
219. J.-T. Chen, W.-H. Tzeng, F.-Y. Tsai, M.-C. Cheng and Y. Wang, *Organometallics*, 1991, **10**, 3954.
220. R. A. Michelin, L. Zanotto, D. Braga, P. Sabatino and R. J. Angelici, *Inorg. Chem.*, 1988, **27**, 85.
221. L. Zanotto, R. Bertani and R. A. Michelin, *Inorg. Chem.*, 1990, **29**, 3265.
222. W. P. Fehlhammer, K. Bartel, U. Plaia, A. Volkl and A. T. Liu, *Chem. Ber.*, 1985, **118**, 2235.
223. W. P. Fehlhammer, A. Mayr and G. Christian, *J. Organomet. Chem.*, 1981, **209**, 57.
224. G. Cetini, L. Operti, G. A. Vaglio, A. L. Bandini, G. Banditelli and G. Minghetti, *Org. Mass Spectrom.*, 1989, **24**, 479.
225. R. A. Michelin *et al.*, *Organometallics*, 1993, **12**, 2372.
226. R. Bertani, M. Mozzon, R. A. Michelin, R. Seraglia and P. Traldi, *Org. Mass Spectrom.*, 1992, **27**, 1187.
227. E. Ma, G. Semelhago, A. Walker, D. H. Farrar and R. R. Gukathasan, *J. Chem. Soc., Dalton Trans.*, 1985, 2595.
228. S. Myrvold, O. A. Nassif, G. Semelhago, A. Walker and D. H. Farrar, *Inorg. Chim. Acta*, 1986, **117**, 17.

8

Platinum–Carbon σ-Bonded Complexes

GORDON K. ANDERSON
University of Missouri–St. Louis, MO, USA

8.1 INTRODUCTION 432

8.2 PLATINUM(I) COMPOUNDS 432

8.3 PLATINUM(II) COMPOUNDS 433

8.3.1 Alkyls and Aryls 433
8.3.1.1 Preparation 433
8.3.1.2 Structure and properties 445
8.3.1.3 Reactions 453
8.3.2 Haloalkyls and Haloaryls 467
8.3.2.1 Preparation 467
8.3.2.2 Structure and properties 472
8.3.2.3 Reactions 473
8.3.3 Platinacycloalkanes 477
8.3.3.1 Preparation 478
8.3.3.2 Structure and properties 481
8.3.3.3 Reactions 481
8.3.4 Alkenyls 483
8.3.4.1 Preparation 483
8.3.4.2 Structure and properties 487
8.3.4.3 Reactions 487
8.3.5 Alkynyls 490
8.3.5.1 Preparation 490
8.3.5.2 Structure and properties 491
8.3.5.3 Reactions 493
8.3.6 Ylides 495
8.3.6.1 Preparation 495
8.3.6.2 Structure and properties 495
8.3.6.3 Reactions 496

8.4 PLATINUM(III) COMPOUNDS 496

8.5 PLATINUM(IV) COMPOUNDS 497

8.5.1 Preparation 497
8.5.1.1 Oxidative addition to platinum(II) complexes 497
8.5.1.2 Platinacyclobutanes 497
8.5.1.3 Reaction of $[PtMe_3X]_4$ 498
8.5.2 Structure and Properties 500
8.5.3 Reactions 504

8.6 COMPOUNDS CONTAINING METALLATED LIGANDS 507

8.6.1 Preparation 507
8.6.1.1 Compounds containing metallated phosphorus or arsenic ligands 507
8.6.1.2 Compounds containing metallated nitrogen ligands 510
8.6.1.3 Compounds with other metallated ligands 512
8.6.2 Structure and Properties 512
8.6.3 Reactions 514

8.6.3.1 Compounds containing metallated phosphorus or arsenic ligands 514
8.6.3.2 Compounds containing metallated nitrogen ligands 515

8.7 CARBON–HYDROGEN BOND ACTIVATION 517

8.8 REFERENCES 518

8.1 INTRODUCTION

The organometallic chemistry of platinum has grown enormously since the publication in the early 1970s of the first definitive texts dealing with the chemistry of platinum.[1] Advances in this area up to 1979 were reviewed in *COMC-I*,[2] and a more recent treatment was published in 1991.[3] Other, more specialized, reviews have appeared, and these will be referenced in the appropriate sections.

Although platinum is less widely used in catalysis than its second-row congener palladium, the organometallic chemistry of platinum is far-reaching. This includes the existence of many stable complexes with highly unusual structures and properties, as well as a significant number of successful homogeneous catalytic systems. The relative inertness of platinum, compared with palladium, often allows the chemistry of catalytic systems to be modeled using platinum complexes.

Organoplatinum complexes exist for each of the five oxidation states from platinum(0) to platinum(IV). For platinum(0) the organic ligands are neutral, two-electron donors, such as alkenes or alkynes (which are discussed in Chapter 9, this volume), or carbonyls, isonitriles or carbenes (Chapter 7, this volume). Compounds containing platinum–carbon σ-bonds are thus restricted to the oxidation states platinum(I)–(IV). The largest number of organoplatinum compounds involve the metal in its oxidation state 2+, but there is also a significant chemistry of platinum(IV). By contrast, there are relatively few stable organoplatinum(I) or -(III) complexes, and these are usually of a dimeric nature.

8.2 PLATINUM(I) COMPOUNDS

Platinum(I) complexes are usually dimeric and hence diamagnetic, containing a Pt–Pt bond. There are few such complexes containing unsupported metal–metal bonds. The carbonyl anions $[Pt_2X_4(CO)_2]^{2-}$ (X = Cl, Br, I) have been known for some time.[4] The neutral species $[Pt(C_6X_5)(CO)(PPh_3)]_2$ (X = F, Cl), in which the phosphine ligands occupy the terminal positions *trans* to the Pt–Pt bond, have been prepared by the disproportionation reaction of *cis*-$[Pt(C_6X_5)_2L_2]$ (L = CO, CNR) and $[Pt(C_2H_4)(PPh_3)_2]$.[5] The pentafluorophenyl derivative has been characterized by an x-ray diffraction study, which reveals a Pt–Pt distance of 0.2599(1) nm.

There is an extensive chemistry of platinum(I) based on the dppm-bridged dimer $[Pt_2Cl_2(\mu\text{-dppm})_2]$,[6] which has been summarized previously.[7] In contrast related organoplatinum(I) species are rare. The analogous phenylalkynyl complex $[Pt_2(C{\equiv}CPh)_2(\mu\text{-dppm})_2]$ has been isolated from the reaction of $[Pt_2Cl_2(\mu\text{-dppm})_2]$ with PhC≡CH in the presence of hydrazine hydrate or by deprotonation of $[Pt_2(C{\equiv}CPh)_2(\mu\text{-H})(\mu\text{-dppm})_2]^+$ by NaOPr^i,[8] but a general route to organoplatinum(I) species of the type $[Pt_2R_2(\mu\text{-dppm})_2]$ has not yet been developed. Diplatinum and mixed platinum–palladium complexes containing one pentafluorophenyl ligand have been prepared.[9] The diplatinum species is prepared according to Equation (1), but such a method does not lend itself to the synthesis of diorgano compounds since it requires the transfer of the chloride ligand between platinum centers.

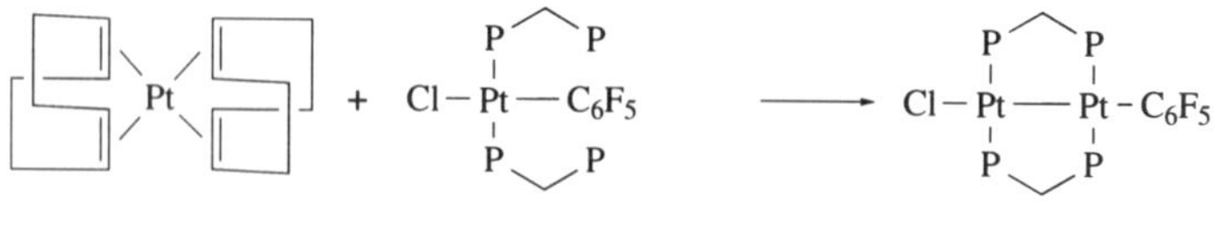

(1)

Other approaches to dppm-bridged organoplatinum(I) complexes include oxidative addition to $[Pt_2(\mu\text{-dppm})_3]$ and reductive elimination from platinum(II) dimers. Thus, addition of methyl iodide to $[Pt_2(\mu\text{-dppm})_3]$ in the presence of KPF_6 leads initially to $[MePt(\mu\text{-dppm})_2Pt(dppm\text{-}P)]PF_6$,[10] although further oxidation may occur (see below). This route has been demonstrated to be quite general, leading to complexes of the type $[RPt(\mu\text{-dppm})_2Pt(dppm\text{-}P)]PF_6$ (R = Et, CH_2CD_3, Pr, Bu).[11] Alternatively, the complexes $[MePt(\mu\text{-dppm})_2PtL]^+$ (L = dppm, PPh_3, PMe_2Ph) may be obtained by reductive elimination of H_2 from $[MePt(\mu\text{-H})(\mu\text{-dppm})_2PtH]^+$ in the presence of the appropriate ligand.[10] Solvent-stabilized species, $[MePt(\mu\text{-dppm})_2Pt(S)]^+$, have been prepared by the photochemically induced reductive elimination of ethane from $[Pt_2Me_3(\mu\text{-dppm})_2]^+$ in acetone, acetonitrile, or dichloromethane solution.[12]

Dppm-bridged platinum(I) complexes of the type $[Pt_2XY(\mu\text{-dppm})_2]$ are characterized by a virtually planar Pt_2P_4XY framework, and a linear X–Pt–Pt–Y unit. It has been pointed out that within this X–Pt–Pt–Y framework of dppm- or dmpm-bridged platinum(I) species the *trans* influence of X is felt primarily at the Pt–Pt bond, although ligands of very high *trans* influence, such as H or Me, may exert a weaker long-range *trans* influence on the Pt–Y bond.[13] In complexes of the type $[XPt(\mu\text{-dppm})_2Pt(PR_3)]^{n+}$ (n = 1 or 2), the value of ^1J(Pt,P) observed for the terminal phosphine is insensitive to the nature of X,[14] an observation that is paralleled by the similar Pt–P bond lengths in $[ClPt(\mu\text{-dppm})_2Pt(PPh_3)]^+$ (0.2333(8) nm) and $[HPt(\mu\text{-dppm})_2Pt(dppm\text{-}P)]^+$ (0.2347(4) nm). In contrast, the Pt–Pt distances in these two complexes differ considerably (0.2665(2) nm and 0.2770(2) nm), and it might be expected that the ^2J(Pt,P) value for coupling to the terminal phosphorus might be more sensitive to the *trans* influence of X. Indeed, this is the case, the values of ^2J(Pt,P) in $[XPt(\mu\text{-dppm})_2Pt(PPh_3)]^{n+}$ decreasing in the order X = I (1290 Hz), Cl (1232 Hz) > PPh_3 (1008 Hz) > H (596 Hz) > Me (490 Hz).[14]

The platinum(I) dimer $[MePt(\mu\text{-dppm})_2Pt(dppm\text{-}P)]PF_6$ undergoes further oxidative addition of methyl iodide to generate the platinum(II) A-frame complex, $[Pt_2Me_2(\mu\text{-I})(\mu\text{-dppm})_2]PF_6$.[10] Although $[MePt(\mu\text{-dppm})_2Pt(dppm\text{-}P)]PF_6$ is thermally stable, the corresponding ethyl complex undergoes thermally or photochemically induced β-elimination to yield ethene and the hydridoplatinum cation $[HPt(\mu\text{-dppm})_2Pt(dppm\text{-}P)]^+$. The rate-determining step apparently involves cleavage of the Pt–Pt bond, or dissociation of one end of a bridging dppm ligand, with the β-elimination step taking place subsequently.[11]

The perfluorophenyl complex $[ClPt(\mu\text{-dppm})_2Pt(C_6F_5)]$ reacts with a range of ligands, either by substitution of the chloride or by insertion into the Pt–Pt bond (Scheme 1). With cyclohexyl or 4-tolyl isocyanide, substitution and/or insertion occurs, depending on the reaction conditions.[9]

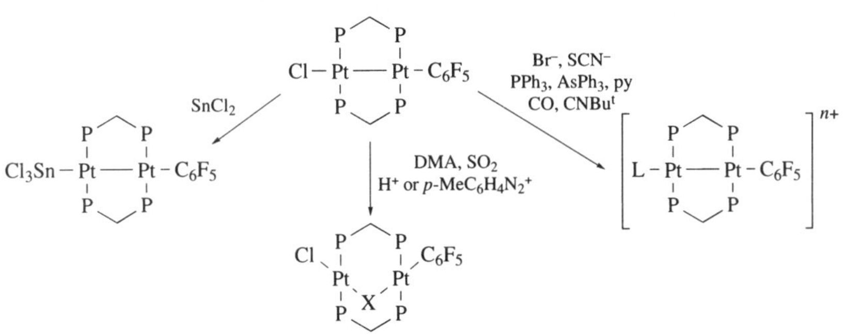

Scheme 1

8.3 PLATINUM(II) COMPOUNDS

σ-Bonded organoplatinum(II) compounds are more numerous than those of the other oxidation states, and they provide the greatest variety in terms of the types of structures available. There are large numbers of alkyl and aryl species and, since the late 1970s, the chemistry of their haloaryl analogues has been developed. These, along with alkenyl and alkynyl complexes, will be discussed in this section, in addition to the chemistry of ylide complexes and platinacycloalkanes.

8.3.1 Alkyls and Aryls

8.3.1.1 Preparation

There are several general methods suitable for the preparation of alkyl- or arylplatinum(II) complexes. One of the simplest, and most general, is by oxidative addition of an organic halide to a platinum(0) precursor, such as $[Pt(PPh_3)_4]$ or $[Pt(C_2H_4)(PPh_3)_2]$. Other commonly used methods include transfer of an organic group from another metal center, and further modification of an organometallic complex by substitution of the supporting ligands. These, and a number of less general methods, will be discussed here.

(i) Oxidative addition reactions

Many examples of oxidative addition to platinum(0) complexes have been described and their mechanisms have been discussed previously.[2] Many recent examples concern functionalized organic halides or additions to platinum(0) species containing less common ligands. Thus, for example, addition of $ClCH_2SR$ (R = Me, Ph, C_6H_4Me-4) to $[Pt(PPh_3)_4]$ in refluxing benzene produces *trans*-$[PtCl(CH_2SR)(PPh_3)_2]$, which undergoes halide exchange with MeI, EtI, or EtBr.[15] In contrast, with the more basic dimethylphenylphosphine, reaction with $ClCH_2SMe$ does not generate an organometallic compound but yields *cis*-$[PtCl_2(PMe_2Ph)_2]$ only.[15] Oxidative addition of $BrCH_2CH_2OPh$ to $[Pt(PPh_3)_4]$ produces *trans*-$[PtBr(CH_2CH_2OPh)(PPh_3)_2]$.[16]

Aryl halides may also be employed. Treatment of $[Pt(PPh_3)_4]$ with 2,6-dichloropyridine generates *trans*-$[PtCl(2\text{-}NC_5H_3Cl)(PPh_3)_2]$, in which the chloride and/or phosphine ligands may be substituted.[17] Treatment with dimethyl sulfate in the presence of $NaClO_4$ results in methylation of the pyridine nitrogen (Scheme 2).[18] Addition of methyl iodide to zerovalent complexes containing the ambidentate ligands tris(2-pyridyl)phosphine or 2-(diphenylphosphino)pyridine yields *trans*-$[PtIMeL_2]$, but without methylation of the pyridine nitrogens.[19] Oxidative addition of 4-nitrohalobenzenes to $[M(PEt_3)_4]$ (M = Pd or Pt) leads to derivatives of the type *trans*-$[MX(C_6H_4NO_2\text{-}4)(PEt_3)_2]$, materials which are suitable for second-harmonic generation. The molecular structure of *trans*-$[PtI(C_6H_4NO_2\text{-}4)(PEt_3)_2]$ has been determined.[20]

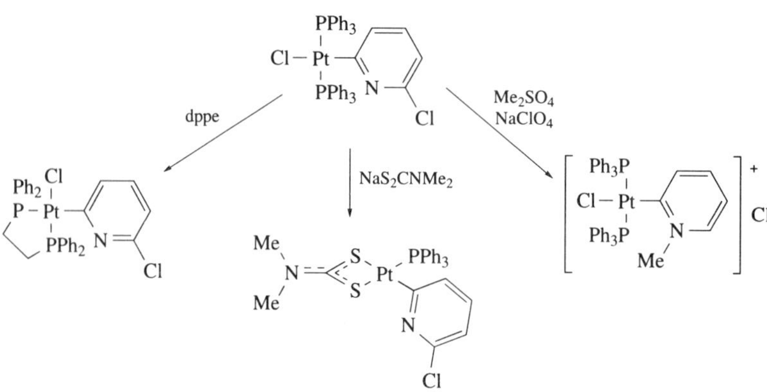

Scheme 2

Treatment of $[Pt(PPh_3)_4]$ with 2-nitrosoaniline produces an intermediate of the form $[Pt(PPh_3)_2L]$, which subsequently undergoes oxidative addition of MeI or EtCl.[21] With benzyl or benzoyl chloride, however, the dichloro complex is obtained. Two-electron reduction of *cis*-$[PtCl_2(PPh_3)_2]$ generates a coordinatively unsaturated bis(phosphine)platinum(0) species, which readily produces organo-platinum(II) complexes by oxidative addition as shown in Equation (2).[22]

$$cis\text{-}PtCl_2(PPh_3)_2 \xrightarrow[-2Cl^-]{+2e^-} \text{"}Pt(PPh_3)_2\text{"} \xrightarrow{RX} trans\text{-}PtXR(PPh_3)_2 \qquad (2)$$

RX = MeI, PhCOCl

Reduction of the five-coordinate $[PtH(pp_3)]^+$ cation (pp_3 = tris(2-diphenylphosphinoethyl)phosphine) with sodium amalgam yields a dimeric platinum(0) species which, on treatment with methyl iodide, generates a trigonal bipyramidal cation, $[PtMe(pp_3)]^+$, in which the methyl group occupies one of the axial sites.[23]

Protonation of the norbornene complexes $[Pt(C_7H_{10})L_2]$ ($L_2 = R_2P(CH_2)_nPR_2$, R = Cy, Bu^t, n = 2, 3, or L_2 = 1,4-$C_6H_4(CH_2PBu^t_2)_2$ leads to the norbornylplatinum(II) cation, which contains a β-C–H–Pt agostic interaction. The dbpe (1,2-bis(di-*t*-butylphosphino)ethane) derivative (**1**) has been characterized by x-ray diffraction methods, in which the Pt–C_α and Pt–C_β distances are 0.2096(4) nm and 0.2309(5) nm, respectively.[24] The Pt–H coupling constant ranges from 79 Hz when L_2 is $C_6H_4(CH_2PBu^t_2)_2$ to 324 Hz in the dcpe (1,2-bis(dicyclohexylphosphinoethane)) derivative.

α-Ketoacylplatinum(II) complexes have been prepared by the oxidative addition reaction shown in Equation (3).[25,26] The structure of *trans*-$[PtCl(COCOPh)(PPh_3)_2]$ in the solid state has been determined,[25] and it exhibits a planar, transoid arrangement of the C_2O_2 unit. In contrast, *trans*-$[Pt(COCOOMe)(H_2O)(PPh_3)_2]OTf$, obtained on attempted crystallization of the neutral triflate complex, contains a planar *s-cis* structure.[27]

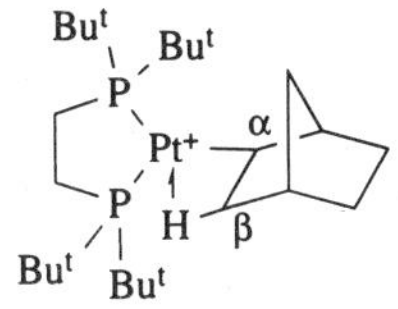

(1)

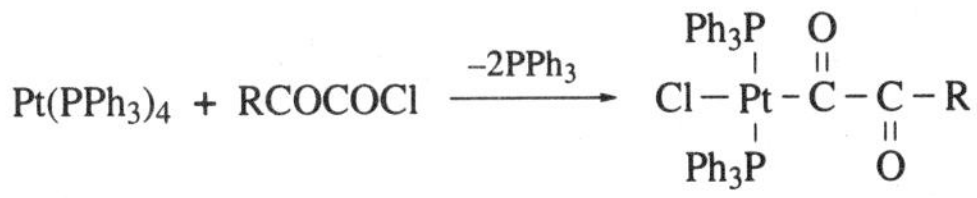

(3)

R = Me, Ph, C_6H_4F-4, C_6H_4Cl-4, C_6H_4Me-4, $OCH_2CH_2C{\equiv}CMe$, $O(CH_2)_nCH{=}CH_2$ (*n* = 2–4)

Oxidative addition to $[Pt(C_2H_4)(PPh_3)_2]$ of the activated C–H bond in $CH_2(SO_2CF_3)_2$ produces first *trans*-$[PtH\{CH(SO_2CF_3)_2\}(PPh_3)_2]$, which exhibits restricted rotation about the Pt–C bond, but on standing this isomerizes to the corresponding *cis* form by a first-order process.[28] In the presence of a coordinating solvent, such as MeCN, DMF, H_2O, MeOH, or THF, displacement of the $CH(SO_2CF_3)_2^-$ anion takes place, as shown in Equation (4). In methanol or THF solution, cleavage of a P–Ph bond in one of the PPh_3 ligands occurs to produce (**2**), the structure of which has been confirmed by an x-ray diffraction study.[28] In fact, it has been shown that refluxing an acetone solution of $[Pt(C_2H_4)(PPh_3)_2]$ itself results in P–Ph bond rupture, but in this case the trimeric complex (**3**) is produced, two forms of which, with different Pt–Pt distances and Pt–P–Pt angles, have been characterized by x-ray crystallography.[29]

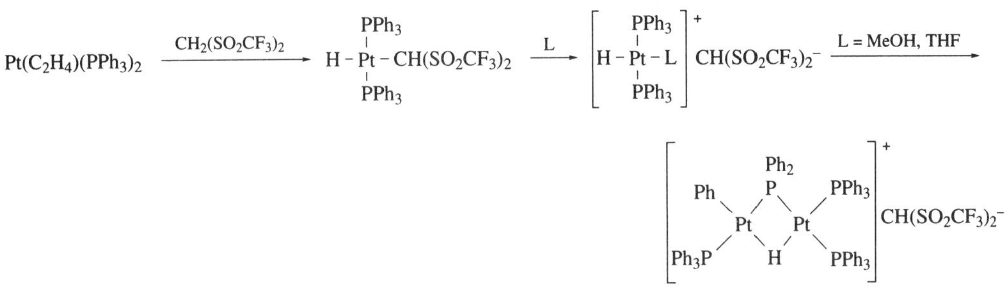

(2) (4)

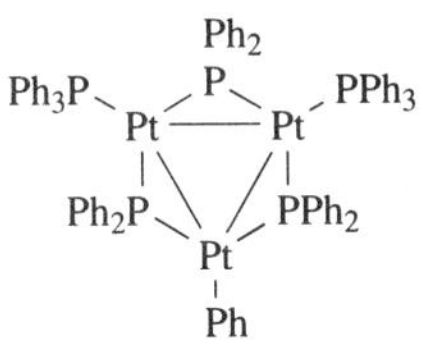

(3)

Activation of P–Ph bonds has also been found in the reactions of $[Pt(C_2H_4)(PPh_3)_2]$ or $[Pt(PPh_3)_3]$ with a series of nitrile oxides. These are illustrated in Equations (5) and (6), the structures of (**4a**) and (**4b**) having been confirmed by x-ray crystallography.[30] A theoretical study of aryl or alkyl migration from phosphorus to a 16-electron, d^8 metal center has shown that this is an allowed process involving a relatively low energy barrier,[31] suggesting that such processes may be more widespread in platinum(II) chemistry than is currently recognized.

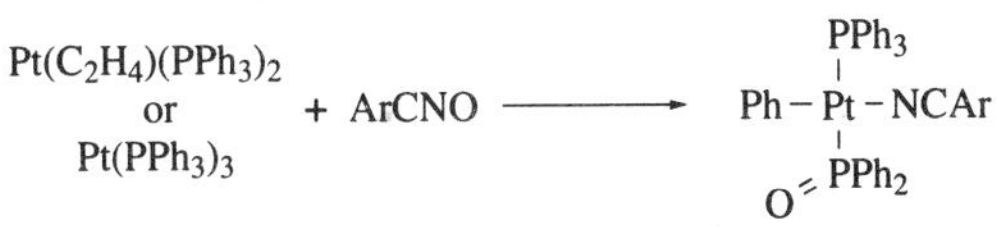

(5)

(**4a**) Ar = 3,5-Cl_2-2,4,6-Me_3C_6
(**4b**) Ar = 9-anthracenyl
(**4c**) Ar = 2,4,6-$(MeO)_3C_6H_2$
(**4d**) Ar = 2,4,6-$Me_3C_6H_2$

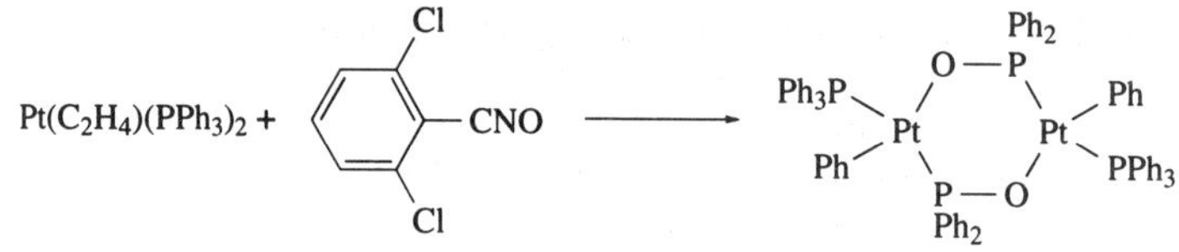

(6)

Reaction of $[Pt(PPh_3)_4]$ with 4-$MeC_6H_4CH_2SO_2H$ occurs by oxidative addition of the S–H bond, whereas with CH_2=$CH(CH_2)_nSO_2H$ (n = 2, 3) insertion of the double bond into the Pt–H bond thus formed yields the metalacyclic compounds (**5**) and (**6**). On heating to 70 °C, (**6**) undergoes a ring contraction to give (**7**).[32]

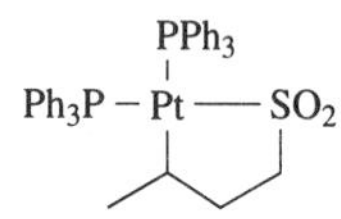

(**5**)

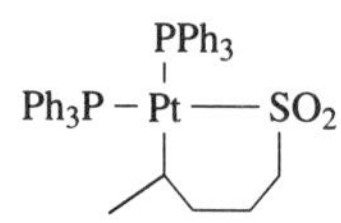

(**6**)

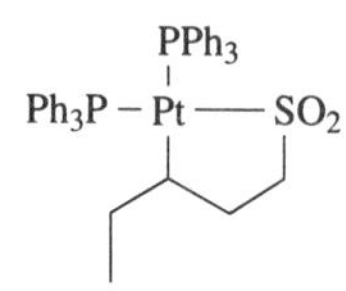

(**7**)

When bis(1,5-cyclooctadiene)platinum(0) is treated with succinic anhydride in the presence of 1 or 2 equiv. of PCy_3, insertion of platinum into one of the C–O bonds occurs to generate species of the form $[Pt(COCH_2CH_2COO)(PCy_3)_n]$ (n = 1, 2). The PCy_3 ligand(s) may be displaced by dppe to generate (**8**), from which the anhydride may be recovered by addition of bromine.[33]

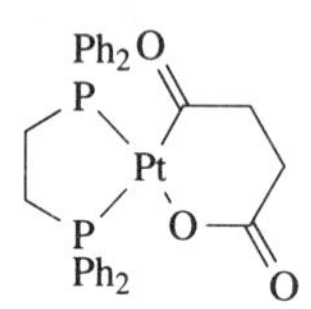

(**8**)

Reactions of platinum(0) complexes with organic derivatives of the group 14 metals also lead to the formation of platinum–carbon bonds. With mixed alkyl(aryl)stannanes, arylplatinum compounds are produced preferentially. Thus, when $[Pt(C_2H_4)(PPh_3)_2]$ is treated with $Me_{4-n}SnAr_n$ (n = 1–3) the products are *cis*-$[PtAr(SnMe_{4-n}Ar_{n-1})(PPh_3)_2]$,[34] although the reactions are complicated by the fact that the platinum complexes catalyze scrambling of the organic substituents in the stannanes.[35] These reactions are likely to occur via platinum(IV) intermediates, involving an oxidative addition–reductive elimination sequence. The preference for insertion of platinum(0) into phenyl–tin bonds is illustrated further by the reaction of $[Pt(C_2H_4)(PPh_3)_2]$ with the five-membered ring compound $Ph_2\overline{Sn(CH_2)_3C}H_2$ in benzene solution. Here the ring-expansion product is not formed, and again a species of the type *cis*-$[PtPh(SnR_3)(PPh_3)_2]$ is obtained.[36] Surprisingly, the reaction of $[Pt(C_2H_4)(PPh_3)_2]$ with trimethyltin chloride results in cleavage of a tin–methyl bond, rather than the Sn–Cl bond, to generate *cis*-$[PtMe(SnClMe_2)(PPh_3)_2]$.[37] Reactions of zerovalent platinum complexes with organolead compounds also show a preference for insertion into Pb–Ph bonds. Treatment of $[Pt(C_2H_4)(PPh_3)_2]$ with $PbPh_4$, Pb_2Ph_6, $PbPhMe_3$, or $PbPh_3Br$ results in insertion of the metal into a Pb–Ph bond. In this case, however, the reaction with $PbMe_3Cl$ produces $[PtCl(PbMe_3)(PPh_3)_2]$ by cleavage of the Pb–Cl bond.[38]

The mixed group 14–group 16 species $Ph_3SnTeAr$ (Ar = Ph, C_6H_4Me-4, C_6H_4OMe-4) react with $[Pt(PPh_3)_4]$ or $[Pt(C_2H_4)(PPh_3)_2]$ to generate first *cis*-$[Pt(SnPh_3)(TeAr)(PPh_3)_2]$ which, on reaction with further $Ph_3SnTeAr$, is converted into *cis*-$[PtPh(SnPh_2TeAr)(PPh_3)_2]$.[39] The rearrangement again probably occurs via a platinum(IV) intermediate.

Reactions of $[Pt(PPh_3)_4]$ with R^1HgCl have been shown to yield *trans*-$[PtClR^1(PPh_3)_2]$, which, on treatment with $NaSR^2$, may be converted into *trans*-$[PtR^1(SR^2)(PPh_3)_2]$. Addition of $PhHgSC_6H_4F$-4 to the zerovalent precursor, however, allows *trans*-$[PtPh(SC_6H_4F\text{-}4)(PPh_3)_2]$ to be generated directly.[40]

(ii) Reactions with main group organometallics

The most widely used sources of alkyl and aryl groups for the preparation of organoplatinum(II) compounds are organolithium (RLi) and Grignard (RMgX) reagents, which react, in general, by exchanging R for a halide attached to platinum. These are highly reactive species, and they often react indiscriminately with all halides available, producing mixtures of products unless an excess of the organometallic is employed. The reactions are exothermic and are usually performed at low temperatures to prevent decomposition.

Organoplatinum(II) complexes containing readily displaced ligands, such as [PtR_2(diene)] and [PtClR(diene)],[41] or *cis*-[$PtR_2(DMSO)_2$] and *cis*-[$PtClR(DMSO)_2$],[42] are particularly useful as they may serve as precursors to a wide range of organometallic compounds by substitution of the weakly bound ligands. Another valuable precursor is the thioether-bridged dimer [$Pt_2Me_4(\mu\text{-}SMe_2)_2$], obtained by treatment of *cis*-[$PtCl_2(SMe_2)_2$] with MeLi.[43] It has been employed in an improved synthesis of [$PtMe_2$(cod)],[44] and its versatility is illustrated by the reactions in Scheme 3.[44,45]

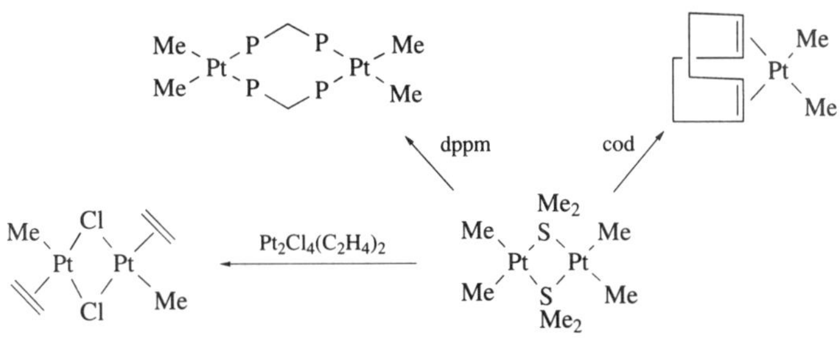

Scheme 3

A majority of alkyl- and arylplatinum(II) complexes are prepared by substitution of 1,5-cyclooctadiene from [PtR_2(cod)] or [PtClR(cod)] by neutral ligands, or by direct reaction of complexes of the type [$PtXYL_2$] (L = PR_3, AsR_3; L_2 = bidentate ligand) with RLi or RMgX. Thus, reaction of [$PtCl_2$(cod)] with 2 equiv. of 2-$MeC_6H_4CH_2MgBr$, followed by 2 equiv. of PEt_3, leads to *cis*-[$Pt(CH_2C_6H_4Me\text{-}2)_2(PEt_3)_2$],[46] and treatment of *trans*-[$PtClH(PR_3)_2$] with MeMgBr produces *trans*-[$PtHMe(PR_3)_2$] (R = Cy, Pr^i, Et).[47] Addition of excess Grignard reagent to the partially fluorinated diphosphine complexes [$PtCl_2\{(C_6F_5)_2PCH_2\}_2$] and [$PtCl_2\{(C_2F_5)_2PCH_2\}_2$] generates the corresponding dialkyl or diaryl derivatives.[48]

Brune and co-workers have synthesized an extensive array of symmetrical and mixed diaryl complexes by the general reaction sequences shown in Scheme 4. These methods have been used successfully to prepare complexes containing 3- or 4-substituted aryls containing electron-releasing[49] or electron-withdrawing[50] groups. With *trans*-[$PtCl_2(PPh_3)_2$], however, reaction with PhMgBr produces a mixture of *cis*- and *trans*-[$PtPh_2(PPh_3)_2$].[51] Further variations on this theme have included the preparation of bis(tributylphosphine) complexes,[52] complexes with chelating ligands,[53] and species with variously substituted triarylphosphines.[54]

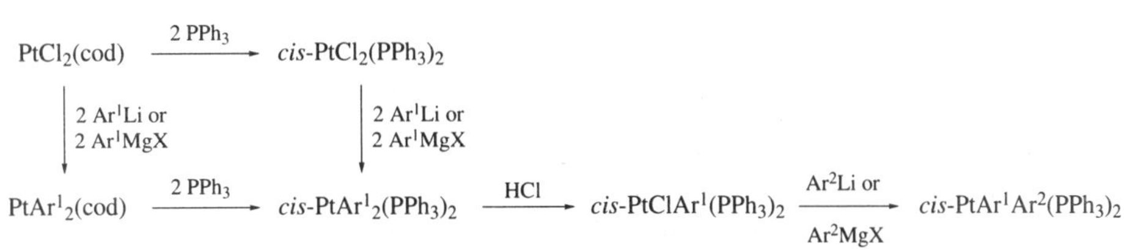

Scheme 4

With 2-substituted aryls the situation is more complex, the nature of the product depending on both the substituent and the supporting ligands. Reaction of *cis*-[$PtCl_2(PPh_3)_2$] with excess 2-$MeOC_6H_4Li$[55] or 7-methylnaphthyllithium[56] generates the diaryl complex *cis*-[$PtAr_2(PPh_3)_2$], whereas with 2-$Me_2NC_6H_4Li$ or 2-$CF_3C_6H_4Li$ only one chloride is replaced.[57] A second, less bulky aryl group may be introduced to yield a mixed aryl complex, however.[57] In contrast, [$PtCl_2$(cod)] reacts in each instance to produce a diaryl complex.[57,58]

2,6-Substituted aryls push the reactivity limits still further, as shown in Equations (7)–(9). Again, disubstitution occurs in the cyclooctadiene complex,[59–61] but with PPh_3 ligands the steric bulk of the aryl group may prevent replacement of the second chloride,[59] allowing the formation of only *trans*-[$PtAr_2(PPh_3)_2$],[59] or even preclude substitution of the diene from [$PtAr_2$(cod)].[60] With mesityllithium, only the monosubstituted products *trans*-[$PtCl(C_6H_2Me_3\text{-}2,4,6)(SEt_2)_2$][62] and [$PtBr(C_6H_2Me_3$2,4,6)-(cod)][63] are obtained even with relatively small supporting ligands.

Treatment of the dithiocarbamate complexes [$Pt(S_2CNR^1_2)Cl(PEt_3)$] (R^1 = Me, Et) with R^2Li or R^2MgBr produces a range of alkylplatinum complexes, without cleavage of the Pt–S bonds.[64] Reaction

$$\mathrm{PtCl_2(PPh_3)_2} + n\mathrm{ArLi} \longrightarrow cis\text{-}\mathrm{PtClAr(PPh_3)_2} \qquad (7)$$

$$\mathrm{Ar} = 2,6\text{-}\mathrm{(MeO)_2C_6H_3},\ 2,4,6\text{-}\mathrm{(MeO)_3C_6H_2}$$

$$\mathrm{PtCl_2(cod)} + n\mathrm{ArLi} \longrightarrow \mathrm{PtAr_2(cod)} \xrightarrow{\mathrm{2PPh_3}} trans\text{-}\mathrm{PtAr_2(PPh_3)_2} \qquad (8)$$

$$\mathrm{Ar} = 2,4,6\text{-}\mathrm{(MeO)_3C_6H_2}$$

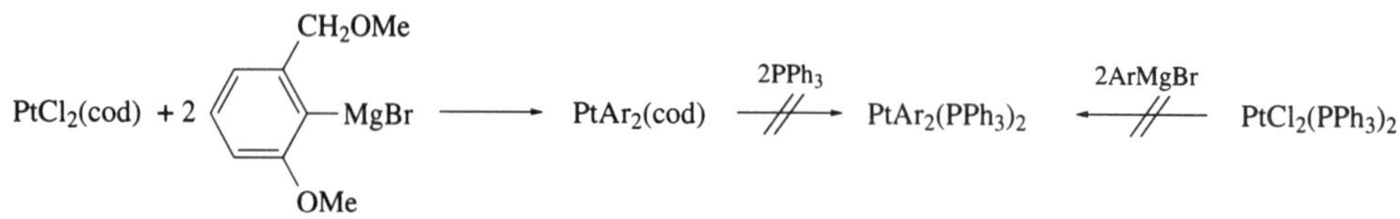

(9)

of the chloro-bridged dimer $[Pt_2(\mu\text{-}Cl)_2Cl_2(PEt_3)_2]$ with $LiCH\{P(S)Ph_2\}_2$ proceeds according to Equation (10), further treatment with HCl cleaving the Pt–C bond.[65]

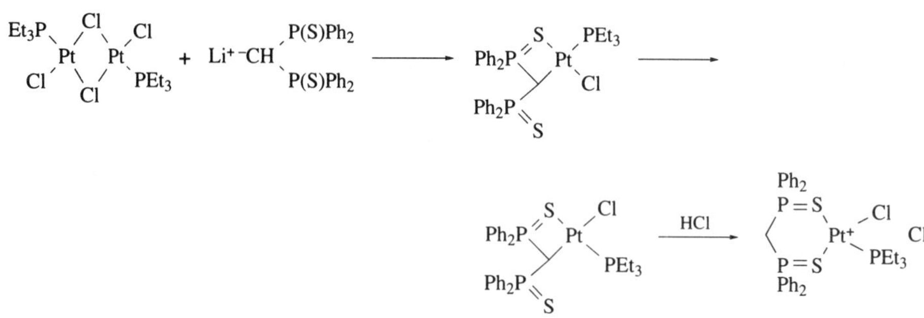

(10)

Deprotonation of the activated methyl compounds MeX (X = NO_2, CN, COMe, COPh, $COC_6H_4NO_2$-4, COC_6H_4Me-2) with potassium hydroxide in the presence of 18-crown-6 generates the corresponding carbanions, which react with complexes of the type $[PtCl_2L_2]$ to generate either $[Pt(CH_2X)_2L_2]$ or $[PtCl(CH_2X)L_2]$.[66] In the case of dimethyl malonate, deprotonation occurs with K_2CO_3, and complexes of the type $[Pt\{CH(CO_2Me)_2\}_2(N\text{-}N)]$ (N-N = bipy, *o*-phen) are formed readily.[67]

Whereas organolithium and Grignard reagents are powerful alkylating and arylating agents, the use of organotin compounds often provides a milder, and more selective, route to organoplatinum(II) complexes. This may be illustrated by two syntheses of the mixed ligand complex $[PtClMe(PMe_2Ph)(AsPh_3)]$ (Scheme 5). With MeLi both chlorides are replaced, necessitating the removal of one with HCl, but this extra step may be obviated by use of the stannane.[68] Similarly, reactions of $[PtCl_2(cod)]$ with aryllithium reagents have been shown to produce $[PtAr_2(cod)]$ (see above), but the corresponding $ArSnMe_3$ produces [PtClAr(cod)] (Ar = C_6H_4X-4; X = NO_2, NMe_2, OMe, Me, CF_3) cleanly.[69]

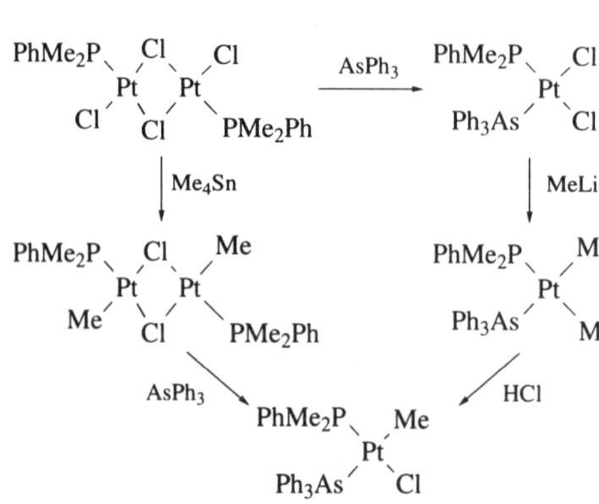

Scheme 5

Organotin reagents have been used to prepare a series of substituted naphthylplatinum complexes of the types [PtClAr(diene)] and $[PtAr_2(diene)]$ (diene = cod, nbd, 1,5-hexadiene),[70] disubstitution occurring with the more reactive aryltributylstannanes. The use of 1,4-bis(trimethylstannyl)benzene allows the synthesis of phenyl-bridged diplatinum species, as shown in Equation (11).[71] These reactions also illustrate the tendency of mixed alkyl(aryl)stannanes to act as arylating agents. Stannanes may be employed as alkylating agents, however, as in the reaction of *cis*-[$PtCl_2(DMSO)_2$] with tetramethyltin to give *cis*-[$PtMe_2(DMSO)_2$].[72]

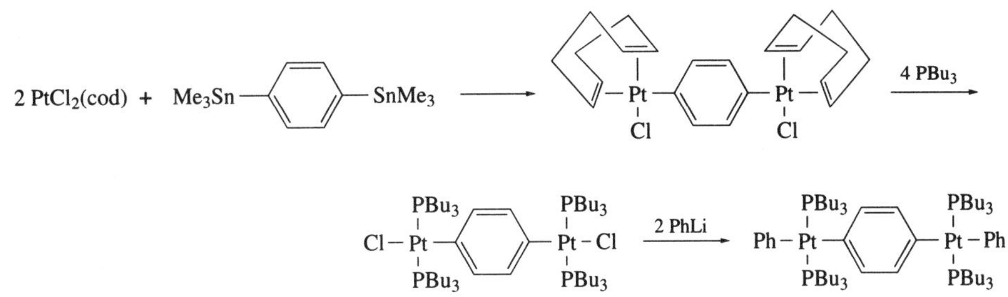

(11)

Arylplatinum derivatives have also been obtained by organic group transfer from diarylmercury[73] or arylboron compounds.[74]

(iii) Displacement of weakly bound ligands

As mentioned in the previous section, organoplatinum(II) complexes containing readily displaced ligands, such as dienes or thioethers, are excellent precursors for the preparation of organometallic species with more nucleophilic ligands. The latter are formed, usually at ambient temperature, simply by addition of the appropriate ligand(s) to a solution of the precursor. In particular, such reactions have been used extensively in the preparation of complexes containing nitrogen or phosphorus ligands, including chelating and bridging bidentate moieties.

Treatment of [$PtMe_2$(nbd)] with pyridine or NH_3 results in displacement of the diene. Norbornadiene is displaced more readily than 1,5-cyclooctadiene from the analogous complex.[75] Similarly, the reaction of *cis*-[$PtPh_2(SEt_2)_2$] with pyridine yields *cis*-[$PtPh_2(py)_2$].[76] With bidentate nitrogen ligands similar displacements occur. Addition of $R_2NCH_2CH_2NR_2$ (R = H, Me), bipy, 3,3'-bipyridazine, 2,2'-bipyrazine, or 3,6-bis(2'-pyridyl)pyridazine to [PtR_2(diene)] (R = Me, Ph; diene = cod, nbd) produces a species of the type [PtR_2(N-N)].[75,77] With the bulky alkyl groups neo-pentyl, adamantyl, 1-adamantylmethyl, CH_2SiMe_2Ph or $CH_2SiMe_2CH{=}CH_2$, a series of complexes of the type (**9**) is formed.[78–81] Reaction of [PtClMe(cod)] with bidentate nitrogen ligands yields species of the type [PtClMe(N-N)].[82] Thioether complexes *cis*-[$PtPh_2(SR_2)_2$] and [$Pt_2Me_4(\mu\text{-}SR_2)_2$] also give chelate complexes with a range of bidentate nitrogen ligands, including 2,2'-bipyridine, 2,2'-bipyrimidine and 2,2'-bithiazoline.[83–6]

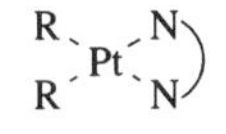

(**9**)

R = Np, Ad, CH_2Ad, CH_2SiMe_2Ph, $CH_2SiMe_2CH{=}CH_2$

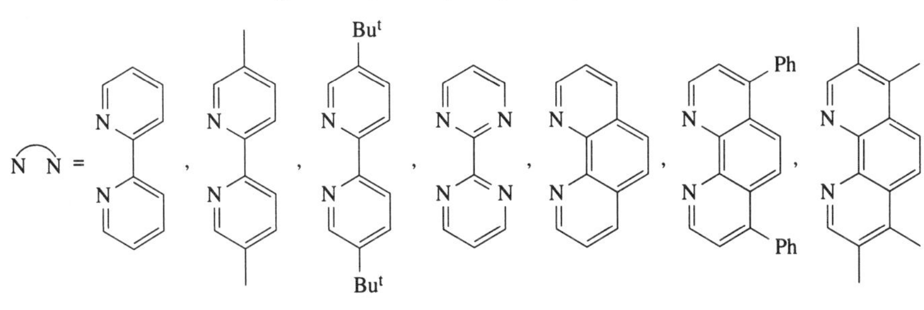

Ligands such as 2,2'-bipyrimidine may function as simple chelates or as bridging ligands. Thus, in reactions with organoplatinum precursors, with a ligand to platinum ratio of 1:1, chelate complexes are formed, but with excess platinum the bridged species is obtained.[87,88] Indeed, a second, different metal may also be introduced,[89] as shown in Equation (12). Reaction of [$PtMe_2$(cod)] with pyrazine (pyz) and 1 equiv. of $P(C_6H_4Me\text{-}4)_3$ yields *cis*-[$PtMe_2$(pyz){$P(C_6H_4Me\text{-}4)_3$}], which is in equilibrium in solution with a pyrazine-bridged dimer (Equation (13)).[90]

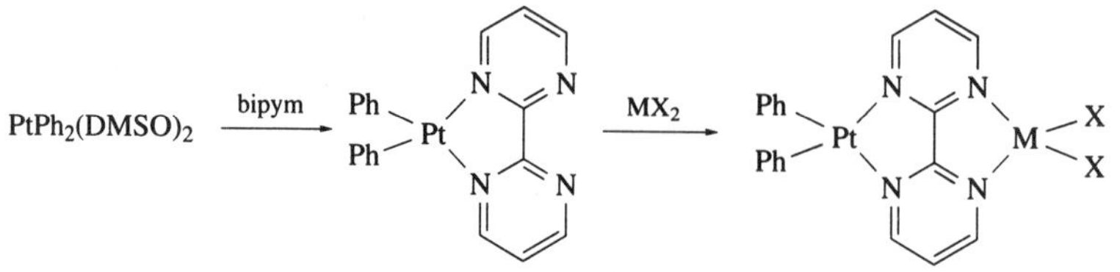

(12)

$MX_2 = PtPh_2, HgCl_2, MnCl_2$

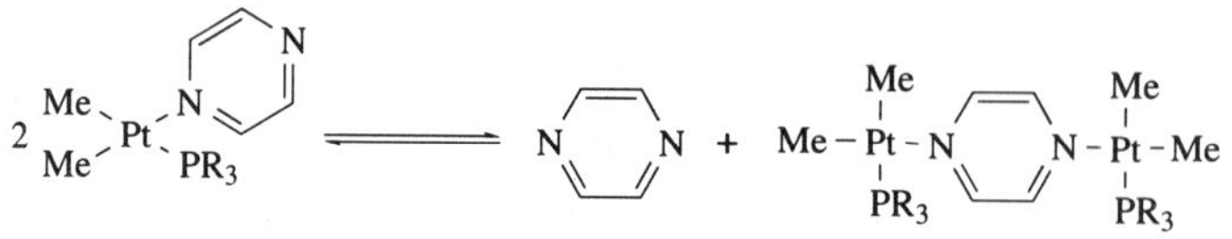

(13)

Complexes containing anionic tris(pz)borate (pz = pyrazolyl) ligands have been described previously.[2] Organoplatinum compounds of these, and of the neutral carbon analogue, tris(pz)methane, have been reviewed.[91] Treatment of [PtClMe(cod)] with $AgPF_6$, followed by $HC(C_3H_3N_2)_3$ and a π-acceptor ligand, yields five-coordinate species of the type $[PtMe\{(C_3H_3N_2)_3CH\}L]PF_6$ (L = CO, C_2H_4, MeC≡CPh, C_4F_6, DMAD (dimethyl acetylenedicarboxylate)), in which all three pz rings are coordinated to the metal.[92] In contrast, when $[PtMe_2(cod)]$ is treated with $HC(C_3H_3N_2)_3$ a complex of the form $[PtMe_2\{(C_3H_3N_2)_3CH\}]$ is obtained initially, but on heating in the presence of py or triphenylphosphine metallation of one of the pz rings takes place (Scheme 6).[93] The bis(pz)dimethylborate and -gallate anions react with [PtClMe(cod)] to produce $[PtMe\{(C_3H_3N_2)_2MMe_2\}L]$ (M = B, L = C_2Ph_2; M = Ga, L = CO, PPh_3, C_2Ph_2), in which both rings are bound to platinum.[94]

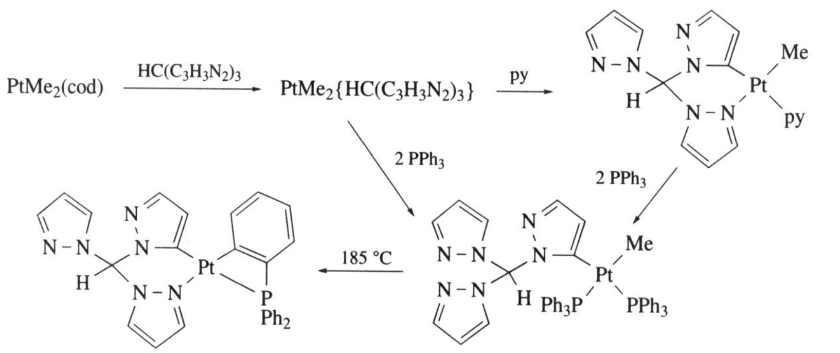

Scheme 6

Platinum–diene complexes also serve as valuable precursors to a wide range of phosphine and diphosphine species. These include the displacement of cyclooctadiene from [PtClMe(cod)] by the water-soluble $P(CH_2OH)_3$,[95] two isomers of the indenylphosphine $Ph_2PC_9H_7$,[96] and the bidentate ligands dppe and appe 1-diphenylarsino-2-diphenylphosphinoethane,[97] and from [PtClPh(cod)] by several bulky phosphines and bidentate ligands.[98] $[PtMe_2(cod)]$ reacts with monodentate[95,96,99,100] and bidentate[101–3] phosphorus ligands according to Equations (14) and (15). Analogous reactions of $[PtR_2(cod)]$ (R = $CH_2(TMS)$, $CH_2SiMe_2CH{=}CH_2$, CH_2SiMe_2Ph) with 2 equiv. of PR_3 or with $Ph_2P(CH_2)_nPPh_2$ also result in substitution of the diene.[104–6] The reaction of $[Pt(CH_2SiMe_2CH{=}CH_2)_2(cod)]$ with the bulky PCy_3 ligand generates (**10**).[105] Aryl complexes of the type $[Pt(C_6H_4X\text{-}4)_2L_2]$ (X = CF_3, TMS, OMe) are formed by displacement of cyclooctadiene by (–)-2,3-*O*-isopropylidene-2,3-dihydroxy-1,4-bis(diphenylphosphino)butane ((–)-diop) or its butyl analogue,[107] or by monodentate phosphines.[54]

$$PtMe_2(cod) + 2\,L \longrightarrow PtMe_2L_2 \qquad (14)$$

L = $P(CH_2OH)_3$, $Ph_2PC_9H_7$, $Me_2PCH_2CH_2SiX_nMe_{3-n}$, $2\text{-}Me_2PC_6H_4SiFMe_2$,

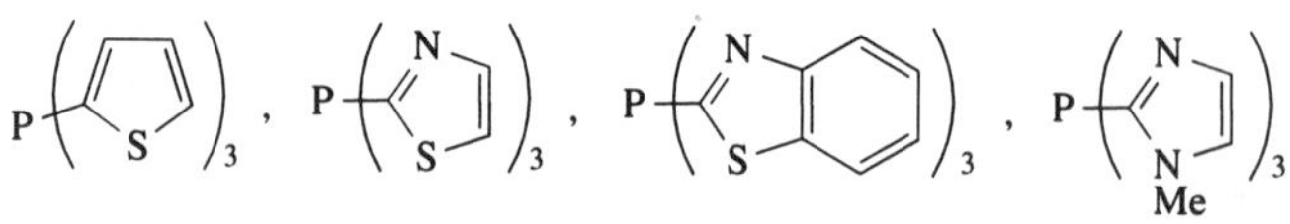

$$PtMe_2(cod) + L_2 \longrightarrow PtMe_2L_2 \quad (15)$$

L = $Ph_2P(CH_2)_3PHPh$, $Cy_2P(CH_2)_3PHPh$, $Ph_2PCH_2PPhCH_2PPh_2$, $(Ph_2P)_2C{=}CH_2$, $(Ph_2PCH_2)_2C{=}CH_2$

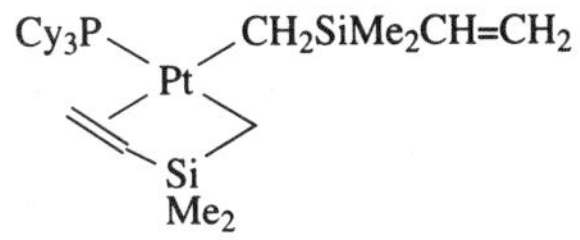

(10)

Complexes containing unsymmetrical ligands have also been prepared by displacement of cyclooctadiene, as shown in Equations (16)–(18).[108–11] Diplatinum compounds are obtained from the reactions of $[PtMe_2(SMe_2)_2]$ with 0.5 equiv. of $E(CH_2CH_2CH{=}CH_2)_2$ (E = S, Se),[112] and from the reactions of [PtClR(cod)] (R = Me, Et, Np, Bz, Ph) with $TlC_5H_4PPh_2$.[113] The latter exist in two forms, which equilibrate in solution, as depicted in Equation (19).

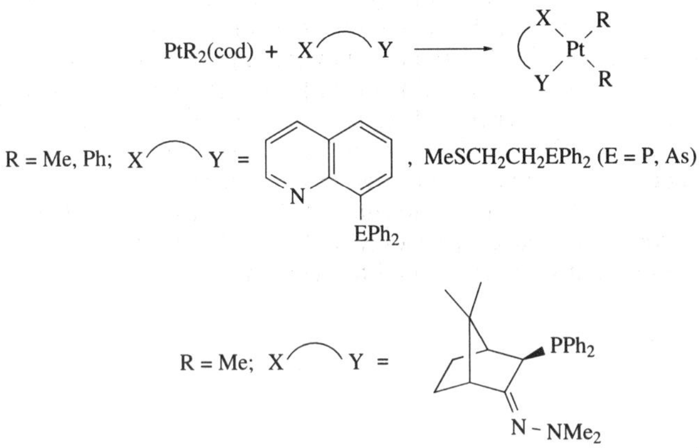

(16)

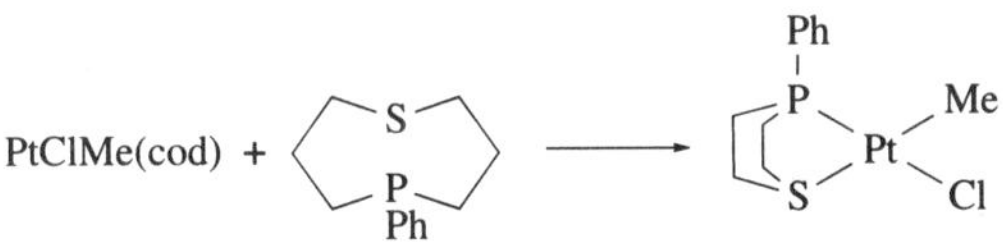

(17)

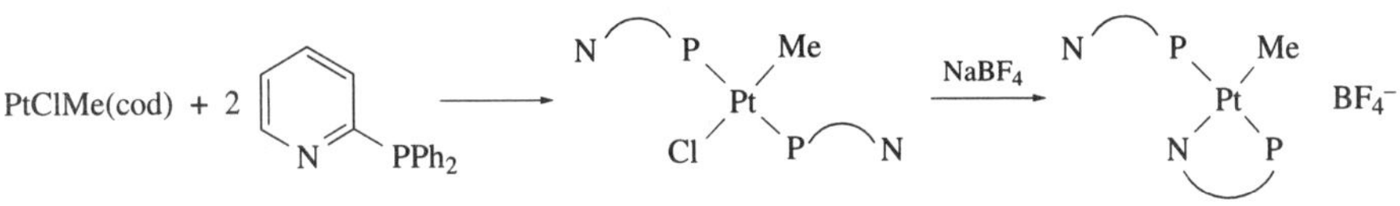

(18)

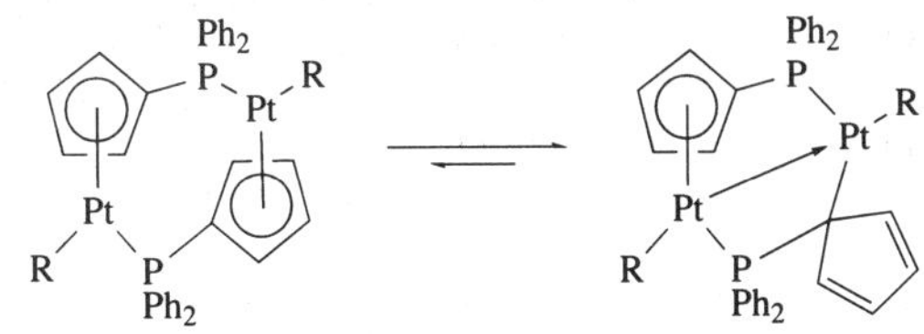

(19)

(iv) Complexes containing dppm and related ligands

The chemistry of complexes containing dppm or related ligands of the type $R_2PCH_2PR_2$ has developed tremendously since the late 1970s. Owing to the small bite size of these ligands, they tend to form ligand-bridged complexes rather than chelate compounds. A number of articles reviewing this area have appeared,[7,114,115] mostly dealing with dinuclear systems, but a report describing the chemistry of triplatinum complexes has also appeared.[116]

There are a number of approaches to organoplatinum(II) species with dppm ligands. Treatment of $[PtCl_2(dppm)]$ with excess organolithium or Grignard reagents generates the dialkyl or diaryl species, $[PtR_2(dppm)]$, in which the dppm ligand is chelating.[117,118] Alternatively, the same products may be obtained by displacement of cod from [PtR_2(cod)]. With [PtClR(cod)] (R = Me, Ph) a mixture of [PtClR(dppm)] and the A-frame complex [Pt_2R_2(μ-Cl)(μ-dppm)$_2$]Cl is obtained, whereas when R = COPh the A-frame complex is the sole product.[119] The same mixture of products is formed when [$PtMe_2$(dppm)] is treated with 1 equiv. of HCl.[117]

Addition of dppm or related ligands to [PtR_2(cod)] or [Pt_2Me_4(μ-SMe_2)$_2$] yields monomeric or dimeric products, depending on the nature of the ligand. Monomeric complexes are found when bis(diisopropylphosphino)methane (dippm) is employed,[120] but with dmpm, depm,[120] bis(diphenylarsino)methane (dpam),[121] $Pr^iHPCH_2PHPr^i$,[122] or $(RO)_2PCH_2P(OR)_2$ (R = Me, Et, Ph, 4-MeC_6H_4)[123] dimers are produced. A mixture of compounds (**11**) and (**12**) is obtained when [Pt_2Me_4(μ-SMe_2)$_2$] is treated with dppm.[120] With excess dppm,[118] complexes of the type *cis*-[PtR_2(dppm-*P*)$_2$], containing pendant ligands, are formed. Analogous complexes are obtained with dmpm[124] or dpam.[121] Cations containing one pendant dppm ligand, [PtR(dppm-*PP*)(dppm-*P*)]$^+$, are formed by the reaction of [PtClR(cod)] with 2 equiv. of dppm.[125]

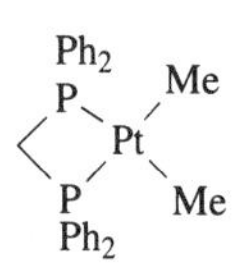
(11)

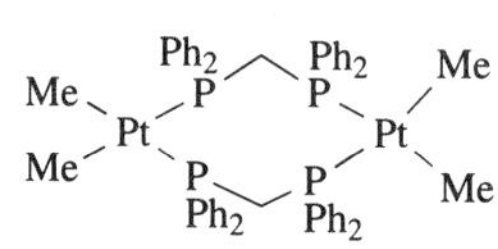
(12)

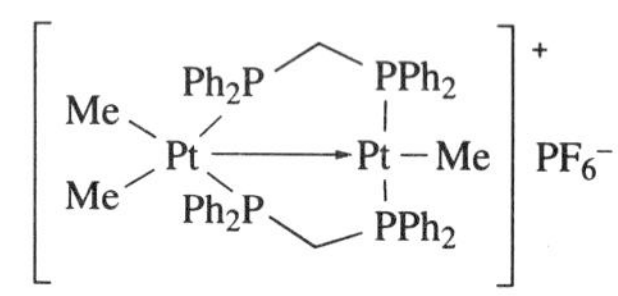
(13)

The dimeric complex (**13**), which contains a Pt → Pt dative bond, was first prepared by the reaction of [$PtMe_2$(dppm)] with [$PtCl_2$(dppm)] or [PtClMe(dppm)] in the presence of KPF_6, or by treatment of the dimethylplatinum complex with HPF_6.[126] It, and related species, have been prepared subsequently by treatment of *cis,cis*-[Pt_2Me_4(μ-dppm)$_2$] with HCl,[120] or by addition of [PtR^1_2(dppm-*P*)$_2$] to [$PtClR^2$(cod)],[118] or [$PtPh_2$(dmpm-*P*)$_2$] to [PtClR(SMe_2)$_2$] (R = Me, Ph).[124] The latter approach allows the synthesis of compounds containing different organic groups. Similarly, reactions of [PtR^1(dppm-*PP*)(dppm-*P*)]PF_6 with [$PtClR^2$(cod)] allow unsymmetrical A-frame complexes to be prepared (Equation (20)).[125]

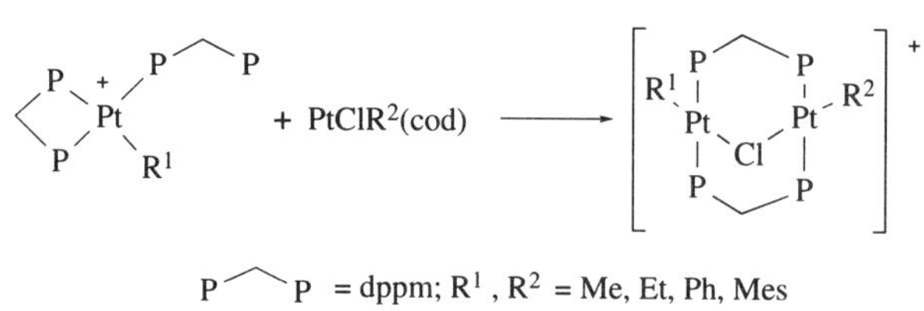

(20)

Organometallic complexes may also be prepared by reaction of dppm-bridged platinum(I) complexes with diazomethane. The reaction of [Pt_2Cl_2(μ-dppm)$_2$] proceeds according to Equation (21). The reaction is first order in both the platinum complex and CH_2N_2, and the rate-determining step is believed to be transfer of an electron pair from the Pt–Pt bond to the CH_2 group.[127] The analogous reaction of [$Pt_2(PPh_3)_2$(μ-dppm)$_2$][PF_6]$_2$ with CH_2N_2 also leads to an A-frame product.[128] The reaction of the [HPt(μ-dppm)$_2$PtL]$^+$ cations with excess diazomethane are less straightforward (Scheme 7).[129]

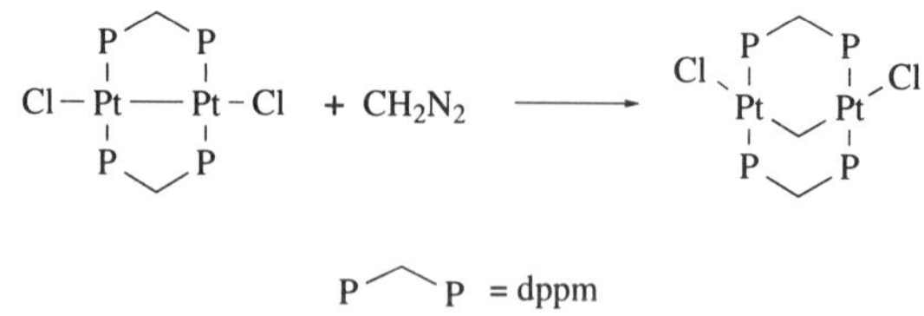

(21)

(v) Insertion reactions

There are a number of insertion or migration reactions which generate or modify already existing Pt–C σ-bonds. These include CO insertions into Pt–C or Pt–O bonds (which are discussed in Chapter 7, this volume), alkene insertions into Pt–H bonds, and insertions into platinum–halogen bonds (which will be presented in Section 8.3.2). Here we shall focus on platinum hydrides as precursors to organometallic complexes. Such reactions have been described previously.[2]

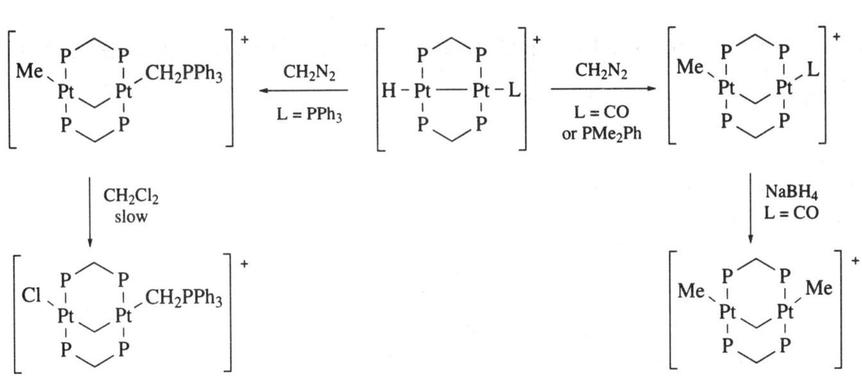

Scheme 7

Treatment of *trans*-[PtClHL$_2$] (L = PMe(OMe)$_2$, P(OMe)$_3$) with methyl acrylate at ambient temperature generates *cis*-[PtCl(CHMeCO$_2$Me)L$_2$], which does not rearrange to the expected *trans*-isomer.[130] A second surprising feature is the formation of the branched alkyl only; insertion of monosubstituted alkenes usually results in a mixture of alkyl isomers, the less branched one predominating. No reaction occurs between *trans*-[PtClH(PEt$_3$)$_2$] and CH$_2$=CHCO$_2$Me under similar conditions.[130] The hydridoplatinum cation *trans*-[PtH(PPh$_3$)$_2$(acetone)]$^+$ reacts with a disubstituted methylenecyclopropane to give products arising from alkene insertion or ring opening (Equation (22)),[131] whereas the triethylphosphine analogue reacts with thiophene as shown in Equation (23).[132]

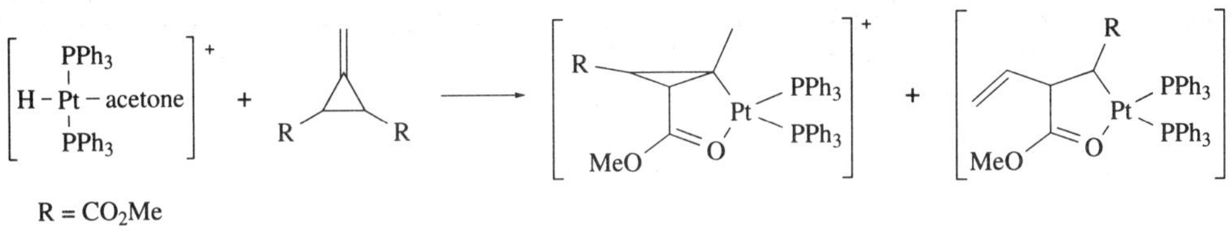

(22)

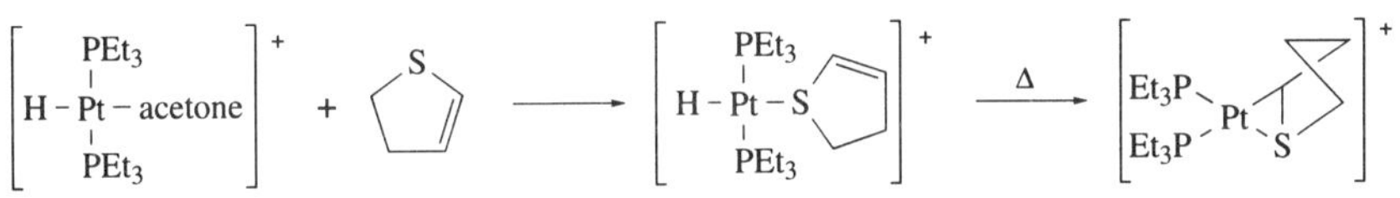

(23)

The reaction of *trans*-[PtH(NHPh)(PEt$_3$)$_2$] with acrylonitrile occurs by insertion into the Pt–N bond, in preference to the Pt–H bond.[133] In contrast, when [PtHMe(dmpe)] is allowed to react with CNC$_6$H$_4$Cl-4 at −60 °C the insertion of isocyanide is into the Pt–H bond. The dimethyl analogue does undergo insertion, but only at temperatures in excess of 100 °C, indicating that insertion into the Pt–H bond is about 10^8 times faster.[134]

Chloroplatinum complexes, in the presence of SnCl$_2$, are well known to catalyze the hydrogenation and hydroformylation of alkenes.[135] Model studies have led to the isolation of organoplatinum species which are potential intermediates in these catalytic reactions. Treatment of *cis*-[PtCl$_2$(PPh$_3$)$_2$] with 1-hexene and CO in ethanol at 100 °C leads to *trans*-[PtCl(COC$_6$H$_{13}$)(PPh$_3$)$_2$], which, in the presence of SnCl$_2$, is effective in the catalytic hydroformylation of 1-hexene.[136] Similar reactions with propene or styrene yield *trans*-[PtCl(COR)(PPh$_3$)$_2$] (R = Pr, CH$_2$CH$_2$Ph).[137,138] The hydride complexes *trans*-[PtH(SnCl$_3$)L$_2$] (L = PPh$_3$, PMePh$_2$, PEt$_3$) react with ethene or propene at −90 °C to produce *cis*-[PtR(SnCl$_3$)L$_2$] (R = Et, Pr), which isomerizes on warming.[139] The results of a similar reaction of *trans*-[PtH(SnCl$_3$)(PPh$_3$)$_2$] with 1-pentene and CO are shown in Equation (24).[140]

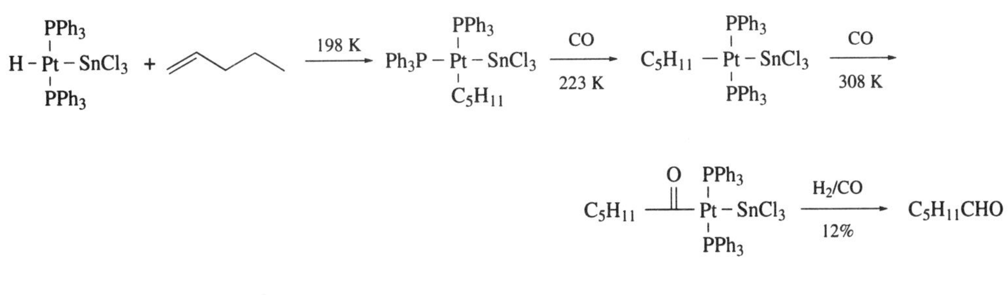

(24)

(vi) Bimetallic compounds

Hydride-bridged complexes may be prepared by reaction of a metal hydride with a solvent-stabilized organoplatinum cation. Treatment of $[Pt(cod)_2]$ with PR^1_3 and dihydrogen forms $[PtH_2(PR^1_3)_2]$ (R^1 = Et, Cy, Ph) *in situ*, which reacts with *trans*-$[PtX(S)(PR^2_3)_2]$ (X = H, Ph; R^2 = Et, Cy, Ph; S = solvent) to form the dimeric cation (**14**), in which one platinum is four-coordinate and the other is five-coordinate.[141] $[Pt_2H(\mu\text{-}H)_2(PEt_3)_4]^+$ is also formed by $NaBH_4$ reduction of *trans*-$[PtClH(PEt_3)_2]$. The reaction of $[PtH_2(PPh_3)_2]$ with $[PtPh(acetone)(PPh_3)_2]^+$ is complicated by P–Ph bond cleavage, and the $[Pt_2Ph(\mu\text{-}H)(\mu\text{-}PPh_2)(PPh_3)_3]^+$ cation has been characterized by x-ray diffraction.[141] Addition of sodium formate to *trans*-$[PtAr(MeOH)(PEt_3)_2]BF_4$, after anion exchange, yields $[Pt_2Ar_2(\mu\text{-}H)(PEt_3)_4]BPh_4$ (Ar = Ph, 4-MeC_6H_4, 2,4-$Me_2C_6H_3$),[142] whereas treatment of *trans*-$[PtH(MeOH)(PEt_3)_2]OTf$ with $NaBPh_4$ produces (**15**), by transfer of a phenyl group from the tetraphenylborate anion.[143]

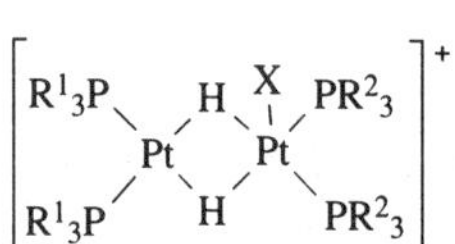

(14)

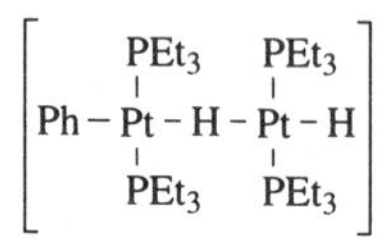

(15)

Mixed metal, hydride-bridged species may be obtained similarly. Addition of $[IrH_5(PEt_3)_2]$ to $[PtR(MeOH)(PEt_3)_2]^+$ proceeds according to Equation (25).[144] The reaction of $[PtR(S)(PEt_3)_2]^+$ (R = H, Ph) with $[WH_2Cp_2]$ yields two hydride-bridged species, (**16**) and (**17**), the phenyl derivatives having been characterized by x-ray crystallography.[145]

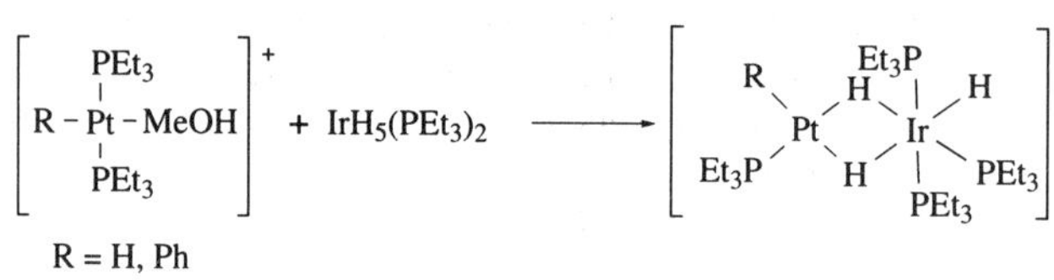
(25)

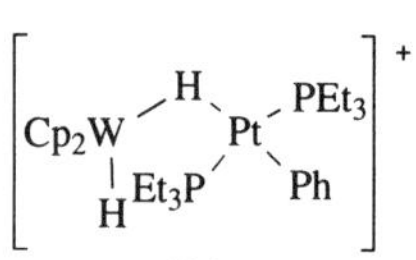

(16)

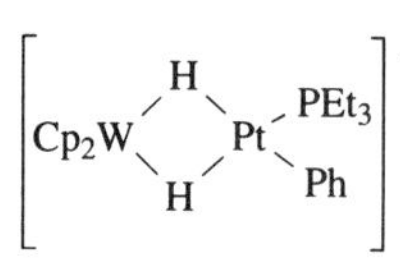

(17)

Complexes containing Pt–Hg bonds have been obtained by addition of HgR_2 (R = Me, Et, chloroaryls) to $[Pt(PPh_3)_3]$. The products, $[(Ph_3P)_2RPtHgR]$, undergo loss of mercury on heating to yield $[PtR_2(PPh_3)_2]$.[146] When R is a chloroaryl group, reaction with $[Pt(C_2H_4)(PPh_3)_2]$ results in insertion of the platinum into the second Hg–R bond to produce $[\{(Ph_3P)_2RPt\}_2Hg]$.[147] A complex containing a Pt–W bond has been obtained by reaction of $[PtClMe(cod)]$ with $Na[WCp(CO)_3]$.[148]

Complexes in which organoplatinum moieties are linked to zirconium or ytterbium by means of diphenylphosphinocyclopentadienyl (dppc) ligands have been prepared. These reactions involve displacement of cyclooctadiene by free phosphino groups. Treatment of $[PtClPh(cod)]$ or $[PtPh_2(cod)]$ with $[ZrCl_2(\eta\text{-}C_5H_4PPh_2)_2]$ yields $[Cl_2Zr(\mu\text{-}\eta\text{-}C_5H_4PPh_2)_2PtXPh]$,[149] whereas reaction of $[Yb(THF)(\eta\text{-}C_5H_4PPh_2)_2]$ with $[PtMe_2(cod)]$ in THF solution generates $[(THF)_2Yb(\mu\text{-}\eta\text{-}C_5H_4PPh_2)_2PtMe_2]$.[150] Another approach to early–late transition metal compounds is shown in Equation (26). Here the hydroxymethyl groups of the coordinated phosphines are condensed with two of the zirconium methyls.[151] A platinum–tantalum complex containing a bridging CH_2 group (**18**) has been obtained by the reverse process, namely attachment of two alkoxyphosphines to tantalum followed by reaction of the phosphino groups with $[PtMe_2(cod)]$.[152] Methylene-bridged platinum–titanium species have been prepared by reaction of $[Ti(CH_2CMe_2CH_2)Cp_2]$ with $[PtClMe(SMe_2)_2]$.[153]

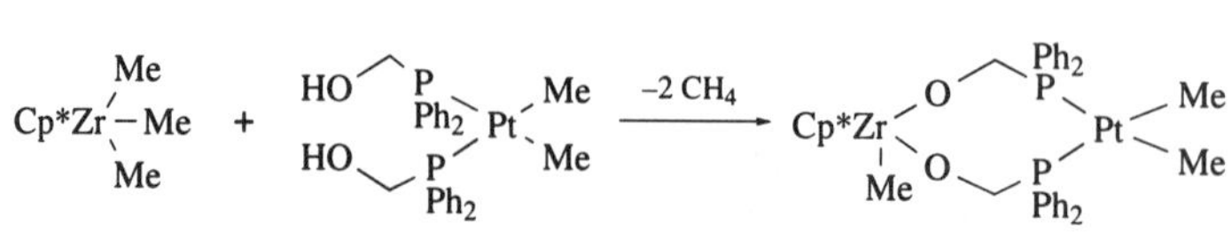
(26)

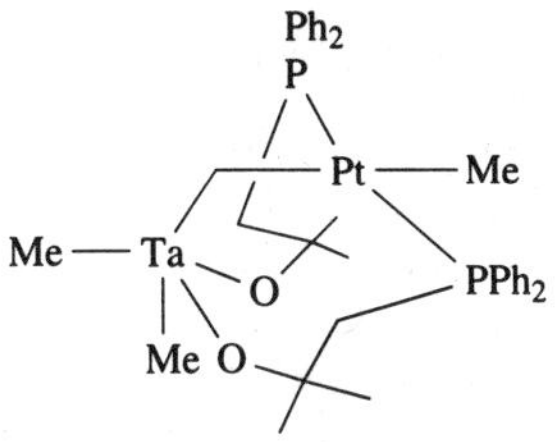

(18)

8.3.1.2 Structure and properties

(i) X-ray diffraction studies

X-ray diffraction data for a significant number of platinum(II) alkyl and aryl compounds have been tabulated previously.[2,154] It has been pointed out that typical Pt–C single bond lengths should be 0.208 nm and 0.198 nm for sp^3 and sp^2 carbons, respectively.[2] In this section, selected data which have been reported since 1980 are gathered in Tables 1–4. The complexes adopt approximately square-planar geometry at platinum. The Pt–Me bond lengths in Tables 1 and 2 span a range from 0.202 nm to 0.220 nm, the shortest distances occurring when the methyl group lies *trans* to a relatively low *trans* influence ligand, such as chloride or a nitrogen or oxygen donor. The longest Pt–C distances are found in *cis,cis*-$[Pt_2Me_4(\mu\text{-dmpm})_2]$, where the methyl groups lie *trans* to the dmpm phosphorus atoms. In *cis,cis*-$[Me_2Pt(\mu\text{-dppm})_2Pt(C_6H_4Me\text{-}2)_2]$[118] the Pt–alkyl and Pt–aryl bond lengths are essentially identical.

The phenyl complexes listed in Table 3 exhibit Pt–C bond distances between 0.199 nm and 0.210 nm, all longer than that predicted for a Pt–C(sp^2) single bond, suggesting no significant π-overlap. The shortest distances again involve chloride or a nitrogen ligand in the *trans* position, whereas the longest Pt–C bonds arise when the *trans* ligand is hydride or phenyl. In Table 4 data are collected for substituted phenyl complexes. These cover a similar range of Pt–C bond distances, and it is not possible to assign the small changes in bond length to the presence of electron-withdrawing or electron-releasing substituents on the phenyl rings. In general, the aryl rings lie approximately perpendicular to the square plane around platinum.

(ii) Stability of Pt–C bonds

Determination of the equilibrium constants for the reactions shown in Equation (27) allows the following order of decreasing platinum–ligand bond strengths to be obtained:

$$\text{Pt–C}(sp) > \text{Pt–O} > \text{Pt–H} > \text{Pt–C}(sp^3) > \text{Pt–N}$$

There is a linear correlation between the Pt–X and H–X bond dissociation energies.[182] Similarly, microcalorimetric measurements of the enthalpies of thermal decomposition of several compounds leads[183] to the following order of bond enthalpies:

$$\text{Pt–C}(sp^2),\ \text{Pt–Cl} > \text{Pt–C}(sp^3) > \text{Pt–S}$$

The cleavage of Pt–C bonds by electrophiles will be discussed in Section 8.3.1.3.

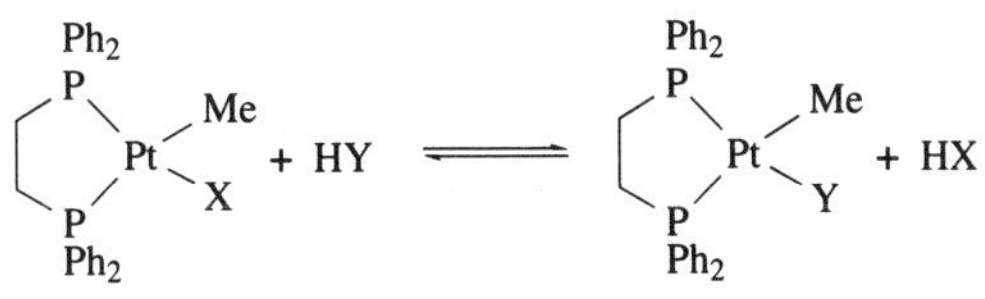

(27)

X, Y = H, R, C≡CR, OH, OR, NH_2, SH, CN

(iii) Spectroscopic studies

The applications of several spectroscopic techniques to organoplatinum compounds have been outlined previously.[2] Here we shall concentrate on NMR spectroscopy and electronic spectroscopy.

Table 1 X-ray diffraction data for monomeric methylplatinum(II) complexes.

Complex	*Pt–C* (nm)	*Other bond lengths* (nm)	*Ref.*
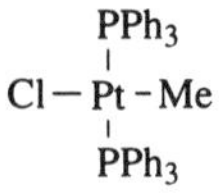	0.208(1)	Pt–Cl 0.2431(3) Pt–P 0.2295(3) 0.2298(3)	155
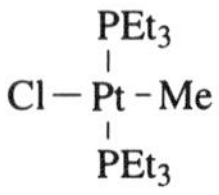	[0.2018][a]	Pt–Cl [0.2346][a] Pt–P 0.2293(2)	155
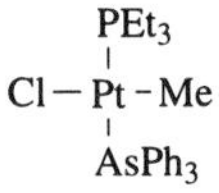	0.2076(16)	Pt–Cl 0.2397(5) Pt–P 0.2271(6) Pt–As 0.2423(2)	156
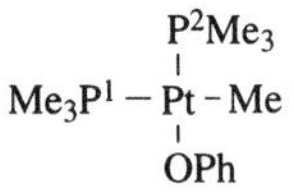	0.208(1)	Pt–O 0.2153(9) Pt–P^1 0.2362(4) Pt–P^2 0.2128(4)	157
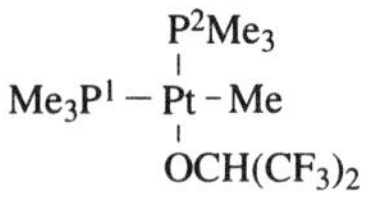	0.213(2)	Pt–O 0.207(1) Pt–P^1 0.2317(4) Pt–P^2 0.2181(4)	157
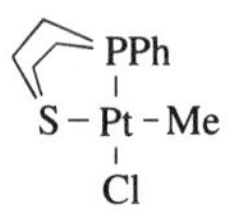	0.2084(7)	Pt–Cl 0.2365(2) Pt–P 0.2191(2) Pt–S 0.2374(2)	110
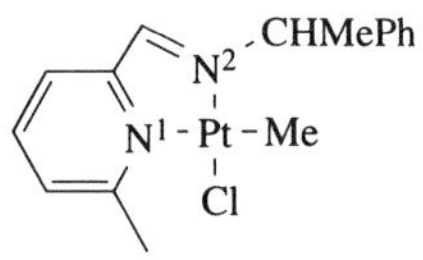	0.2044(8)	Pt–Cl 0.2290(2) Pt–N^1 0.2213(5) Pt–N^2 0.1993(5)	82
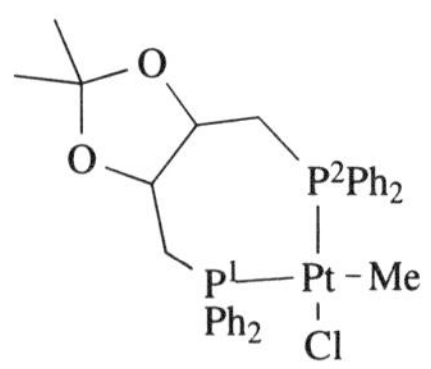	0.217(1)	Pt–Cl 0.2323(5) Pt–P^1 0.2322(4) Pt–P^2 0.2233(4)	158
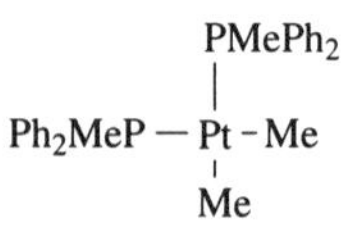	0.2089(3) 0.2090(3)	Pt–P 0.2326(1) 0.2321(1)	159
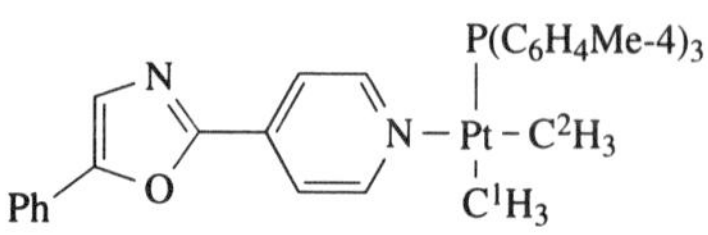	Pt–C^1 0.2082(7) Pt–C^2 0.2059(8)	Pt–P 0.2286(2) Pt–N 0.2122(6)	90
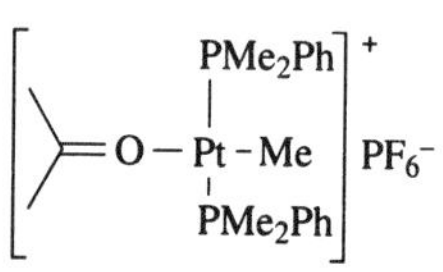	0.2037(7)	Pt–O 0.2167(4) Pt–P 0.2302(2) 0.2313(2)	160

Table 1 (continued)

Complex	*Pt–C* (nm)	*Other bond lengths* (nm)	*Ref.*
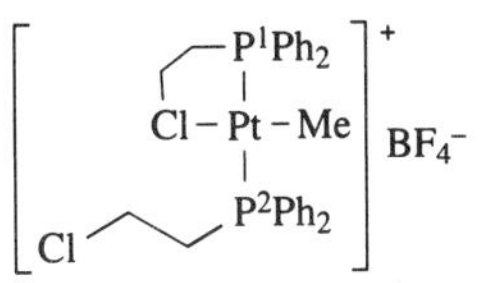	0.2050(7)	Pt–Cl 0.2425(2) Pt–P[1] 0.2275(2) Pt–P[2] 0.2309(2)	161
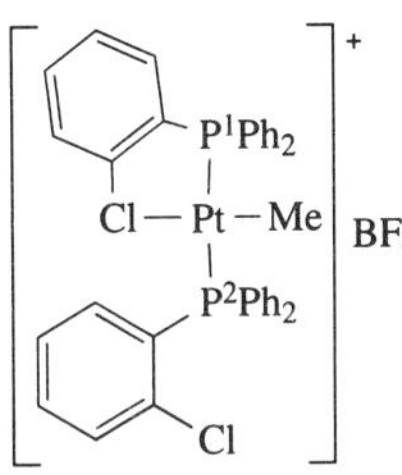	0.2053(6)	Pt–Cl 0.2462(1) Pt–P[1] 0.2278(1) Pt–P[2] 0.2310(1)	161
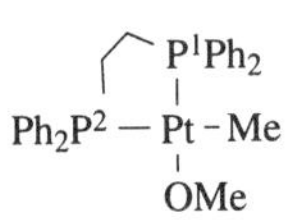	0.2145(17)	Pt–O 0.1990(10) Pt–P[1] 0.2211(3) Pt–P[2] 0.2281(3)	162
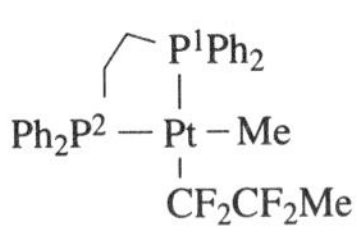	Pt–Me 0.2106(9) Pt–CF_2 0.2074(9)	Pt–P[1] 0.2258(2) Pt–P[2] 0.2298(2)	162
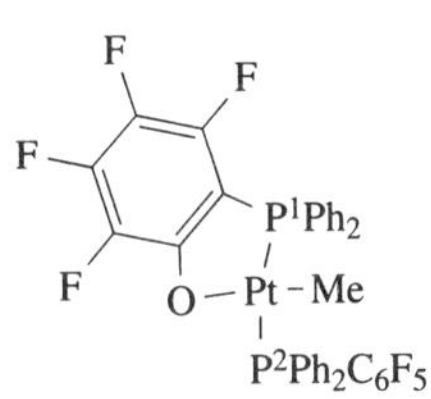	0.208(1)	Pt–O 0.212(1) Pt–P[1] 0.2259(2) Pt–P[2] 0.2273(2)	163

[a] Calculated, positions unrefined

Several nuclei are used widely in the characterization of organoplatinum compounds. Principal among these are ^{1}H, ^{13}C, and ^{31}P, although ^{19}F and ^{195}Pt have also found application. The use of ^{195}Pt NMR spectroscopy is not widespread, although some data have been reported,[184] owing to the very large chemical shift range observed for that nucleus. Couplings of other nuclei to platinum are more easily obtained, and these are usually diagnostic of the stereochemistry of square-planar platinum complexes. This phenomenon has been used to effect for a series of diarylbis(phosphine)platinum species.[185] In complexes of the type $[MX(Ar)(PEt_3)_2]$, platinum is found to induce a smaller paramagnetic shift of the *ipso*-carbon than either nickel or palladium.[186]

The magnitudes of couplings to platinum, particularly ^{1}J(Pt,P), may be used to evaluate the NMR *trans* influence of the *trans* ligand. For example, in the complexes *trans*-$[PtX(C_6H_4Y)(PEt_3)_2]$ and $[Pt(C_6H_4Y)(PEt_3)_3]BF_4$, the magnitude of ^{1}J(Pt,P) *trans* to the aryl group increases as Y becomes more electron withdrawing, consistent with a lowering of the *trans* influence of the aryl group, whereas the *cis* coupling decreases.[187] In analogous complexes of the type *cis*-$[Pt(CH_2Bu^t)_2(PR_3)_2]$ and $[Pt(CH_2(TMS))_2(PR_3)_2]$, the TMS moiety is found to exert a weaker NMR *trans* influence than neopentyl.[104] The subject of NMR *trans* influence has been discussed in detail previously.[188,189]

NMR spectroscopy has also been used to investigate numerous fluxional processes. These include hindered rotations about platinum–carbon bonds, particularly in 2-substituted arylplatinum complexes, where the barrier to rotation increases with the size of the substituent,[190] and in η^1-cyclopentadienyls.[191] Extensive ^{1}H, ^{31}P, and ^{195}Pt NMR data have been reported for dppm-bridged A-frame complexes.[192] Those containing chloride bridges are static on the NMR timescale, but the hydride-bridged complexes of the type $[Pt_2X_2(\mu\text{-}H)(\mu\text{-dppm})_2]^+$ undergo inversion, which may be detected through the coalescence

Table 2 X-ray diffraction data for dimeric methylplatinum(II) complexes.

Complex		*Pt–C* (nm)	*Other bond lengths* (nm)		*Ref.*
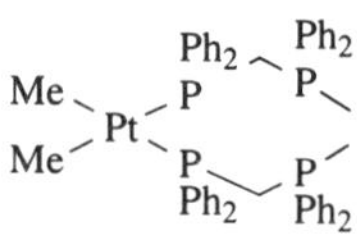		0.209(4)	Pt–P	0.2299(9)	120
		0.208(4)		0.2294(8)	
		0.214(6)		0.2300(8)	
		0.205(4)		0.2304(8)	
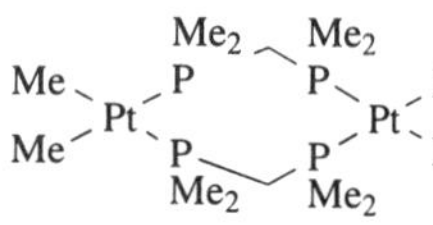		0.2124(12)	Pt–P	0.2280(3)	120
		0.2161(12)		0.2280(3)	
		0.2133(12)		0.2264(3)	
		0.2202(9)		0.2275(4)	
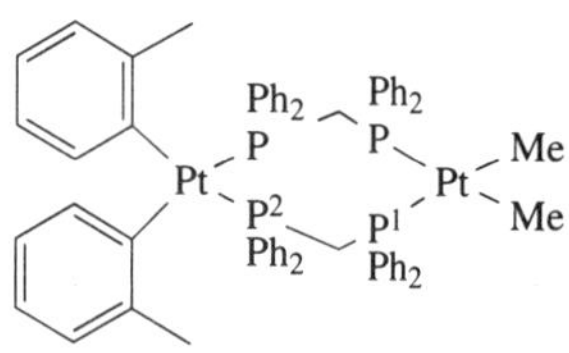	Pt–Me	0.211(2)	Pt–P[1]	0.2279(5)	118
		0.214(2)		0.2292(5)	
	Pt–aryl	0.209(1)	Pt–P[2]	0.2337(5)	
		0.211(1)		0.2322(4)	
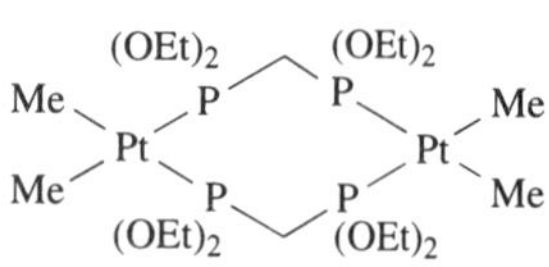		0.215(3)	Pt–P	0.2233(6)	123
		0.210(3)		0.2242(6)	
		0.217(3)		0.2230(5)	
		0.207(3)		0.2243(6)	
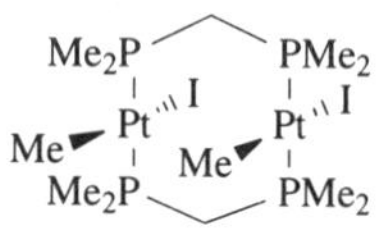		0.210(1)	Pt–I	0.2696(1)	164
			Pt–P	0.2274(3)	
				0.2278(3)	
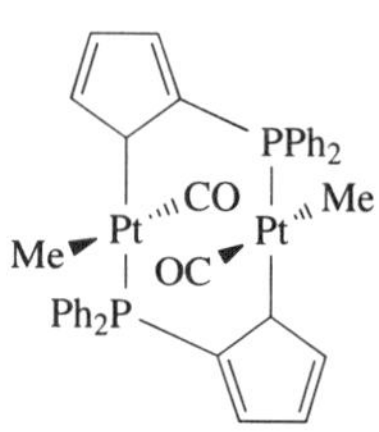		0.2088(11)	Pt–CO	0.1897(12)	165
			Pt–Cp	0.2160(11)	
			Pt–P	0.2298(3)	
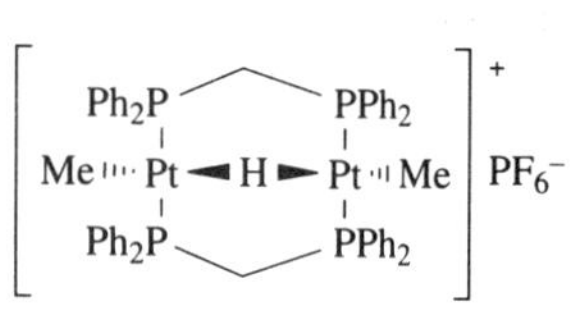		0.213(2)	Pt–P	0.2283(3)	166
		0.207(2)		0.2274(3)	
				0.2295(3)	
			Pt–Pt	0.2932(1)	
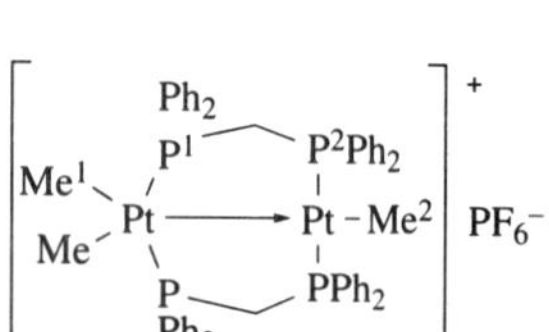	Pt–Me[1]	0.208(3)	Pt–P[1]	0.2314(7)	126
		0.210(3)		0.2338(7)	
	Pt–Me[2]	0.206(4)	Pt–P[2]	0.2263(7)	
				0.2293(7)	
			Pt–Pt	0.2769(1)	
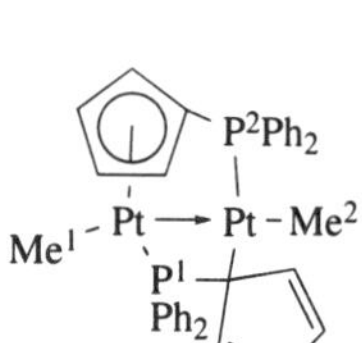	Pt–Me[1]	0.2069(12)	Pt–P[1]	0.2202(2)	113
	Pt–Me[2]	0.2118(13)	Pt–P[2]	0.2189(2)	
			Pt–Pt	0.2723(1)	

Table 3 X-ray diffraction data for phenylplatinum(II) complexes.

Complex	*Pt–C* (nm)	*Other bond lengths* (nm)	*Ref.*
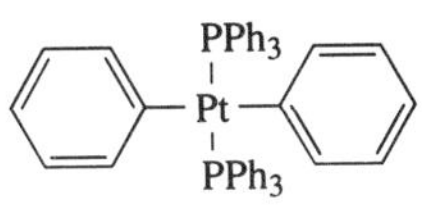	0.2080(8)	Pt–P 0.2299(2)	51
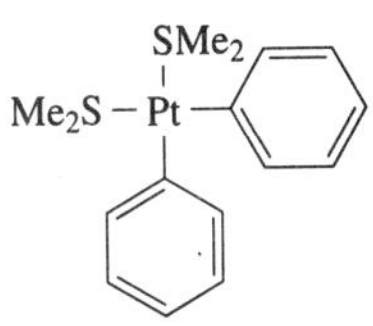	0.2036(7) 0.2010(7)	Pt–S 0.2370(2) 0.2389(2)	167
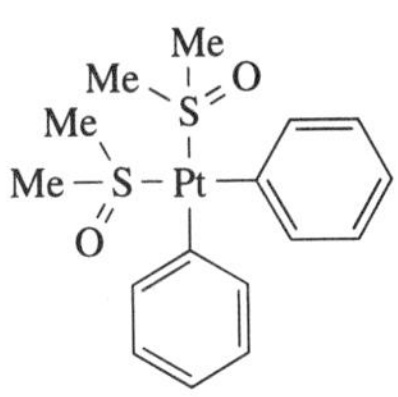	0.2043(5) 0.2049(6)	Pt–S 0.2315(2) 0.2324(2)	168
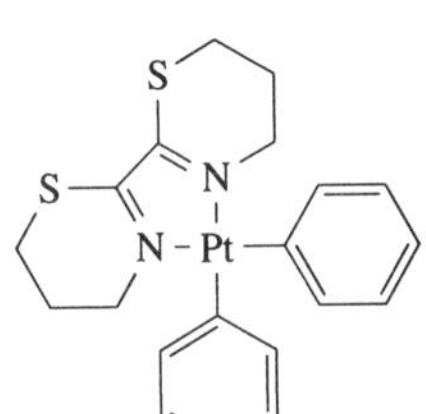	0.2003(8) 0.2013(7)	Pt–N 0.2103(6) 0.2075(5)	169
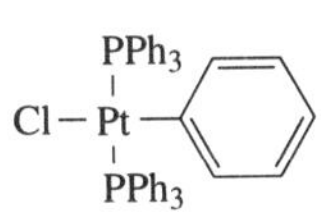	0.200(2)	Pt–Cl 0.2408(5) Pt–P 0.2303(6) 0.2297(6)	170
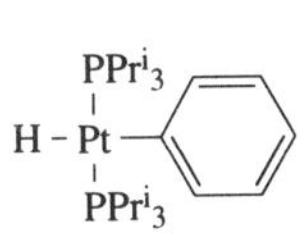	Molecule 1 0.210(2) Molecule 2 0.205(2)	Pt–P 0.2272(4) 0.2271(4) Pt–P 0.2279(4) 0.2265(5)	171
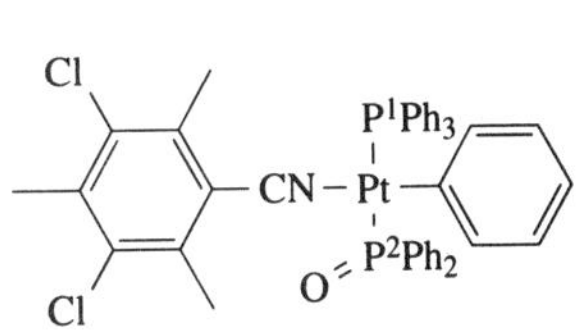	0.203(2)	Pt–N 0.198(2) Pt–P[1] 0.2347(6) Pt–P[2] 0.2316(5)	30
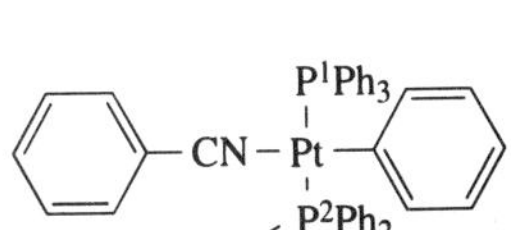	0.205(1)	Pt–N 0.201(1) Pt–P[1] 0.2354(2) Pt–P[2] 0.2324(2)	30
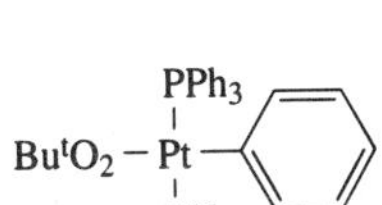	0.193(1)	Pt–O 0.199(1) Pt–P 0.2304(5) 0.2312(5)	172

Table 3 (continued)

Complex		Pt–C (nm)	Other bond lengths (nm)		Ref.
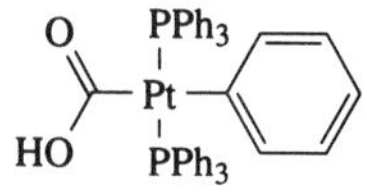	Pt–Ph Pt–CO₂H	0.2071(8) 0.2050(9)	Pt–P	0.2279(3) 0.2283(3)	173
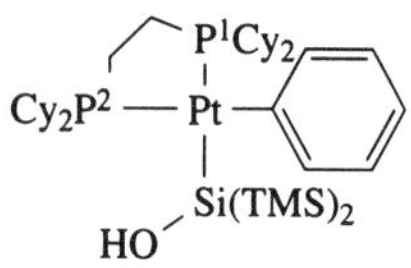		0.2069(7)	Pt–Si Pt–P^1 Pt–P^2	0.2360(2) 0.2333(1) 0.2282(1)	174
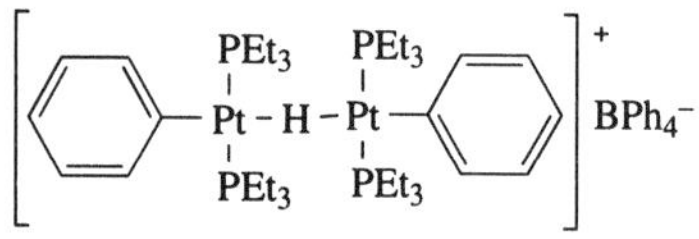		0.205(2) 0.205(2)	Pt–P	0.2306(6) 0.2312(6) 0.2303(7) 0.2318(7)	142
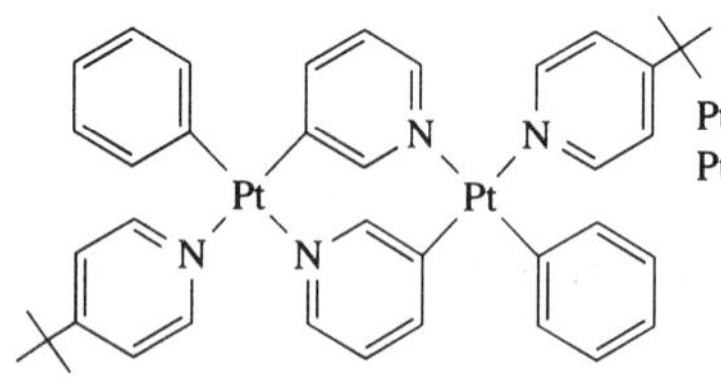	Pt–Ph Pt–C_5H_4N	0.1993(10) 0.1996(9)	P–N_{ter} P–N_{br}	0.2107(7) 0.2142(9)	175

of the CH_2 hydrogens of the dppm ligands.[193] The inversion is proposed to occur through a linear X–Pt–H–Pt–X intermediate.

Two-dimensional NMR exchange spectroscopy (EXSY) has been developed as a method for evaluating rate constants in exchanging systems,[194] the intensity of the off-diagonal peaks being related to the rate of exchange between the species giving rise to those signals. The method has been employed successfully for many organoplatinum(IV) compounds, which will be discussed in Section 8.5.2. ^{31}P–^{31}P 2D NMR exchange spectroscopy has also been used qualitatively to detect slow exchange in systems which appear static by one-dimensional techniques. This approach has been applied to platinum–trichlorostannate complexes, for example.[195,196] Insertion of $SnCl_2$ into the Pt–Cl bond of [PtCl(COEt)(dppb)] yields the corresponding trichlorostannate complex (Equation (28)). Complex (**19**) exhibits sharp resonances in its $^{31}P\{^1H\}$ NMR spectrum, but the 2D exchange spectrum indicates there is exchange between P^1 and P^2. Moreover, this has been shown to proceed by an intramolecular process, involving a five-coordinate transition state.[196]

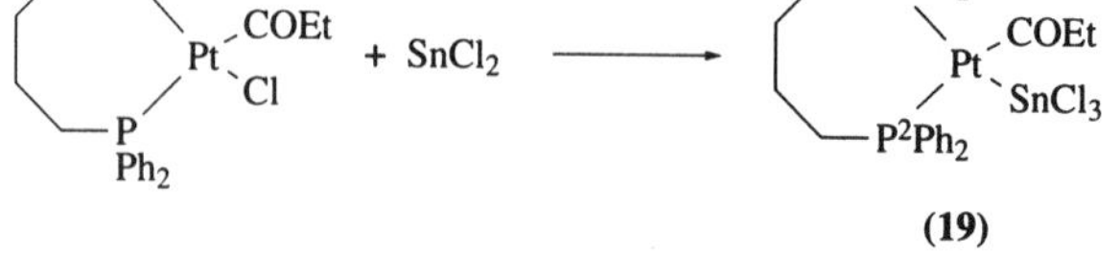

(**19**) (28)

Extensive studies have been made of the electronic spectra of platinum complexes containing metallated ligands. These will be presented in Section 8.6.2 Fewer studies have been carried out for simple alkyl and aryl complexes. Alkyl compounds of the type *cis*- or *trans*-[PtXY$(PEt_3)_2$] (X, Y = Cl, Br, Me, Et) all exhibit strong absorptions in the ultraviolet region, which have been assigned as Pt(5*d*) → P(3*d*) metal-to-ligand charge-transfer (MLCT) bands.[197] The phenyl compounds [$PtPh_2L_2$] (L_2 = bipy, Me_2bipy, bipym) also display two MLCT bands in the region 310–460 nm. They undergo two one-electron reductions, the first being a reversible reduction of the ligand and the second an irreversible, ligand-based reaction.[198] In contrast, the long-wavelength absorptions observed for aryl

Table 4 X-ray diffraction data for arylplatinum(II) complexes.

Complex	*Pt–C* (nm)	*Other bond lengths* (nm)		*Ref.*
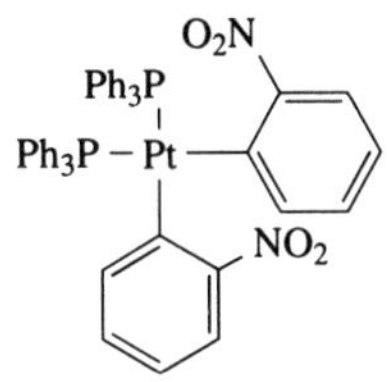	0.2060(6) 0.2061(6)	Pt–P	0.2313(2) 0.2332(1)	57
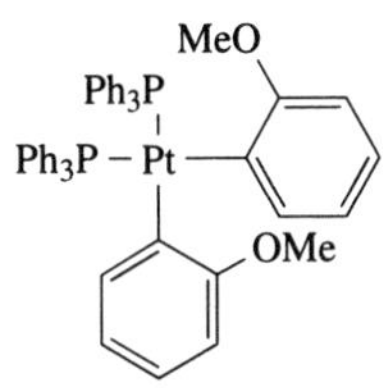	0.2033(12) 0.2034(13)	Pt–P	0.2309(2) 0.2313(3)	55
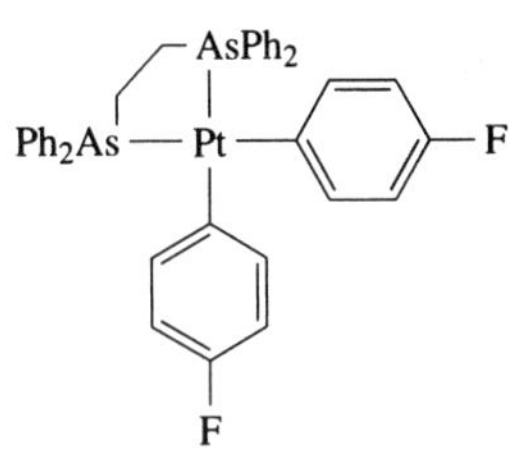	0.2034(6) 0.2035(8)	Pt–As	0.2407(1) 0.2409(1)	176
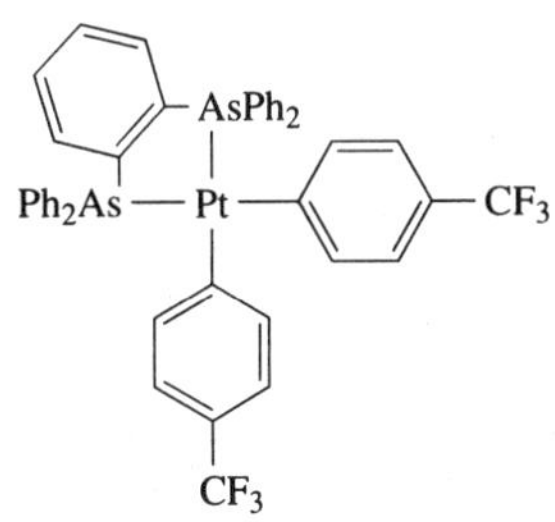	0.2040(5) 0.2042(6)	Pt–As	0.2396(1) 0.2396(1)	176
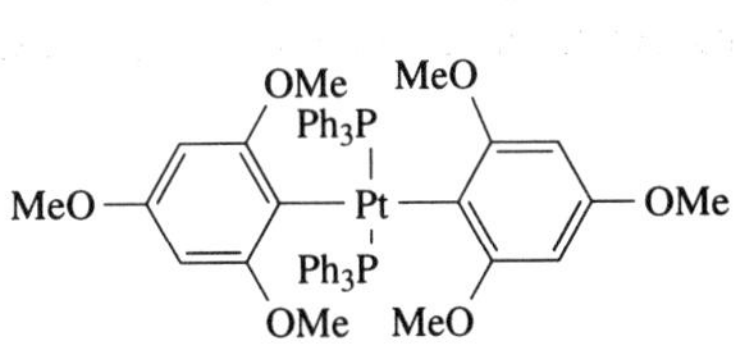	0.2104(5)	Pt–P	0.2295(1)	59
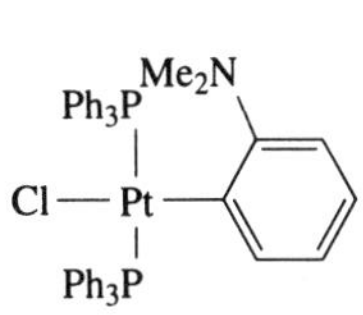	0.2030(12)	Pt–Cl Pt–P	0.2416(3) 0.2310(4) 0.2319(3)	57
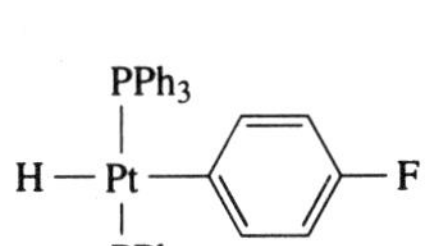	0.2066(23)	Pt–P	0.2256(6) 0.2264(5)	177

Table 4 (continued)

Complex		*Pt–C* (nm)	*Other bond lengths* (nm)	*Ref.*
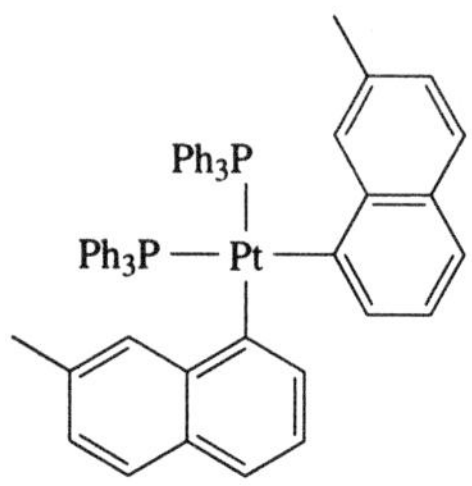		0.208(2) 0.208(2)	Pt–P 0.232(1) 0.234(1)	178
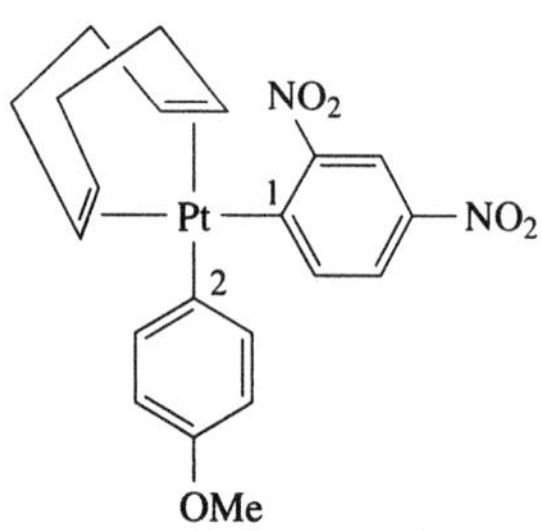	Pt–C[1] Pt–C[2]	0.2023(9) 0.2048(7)		179
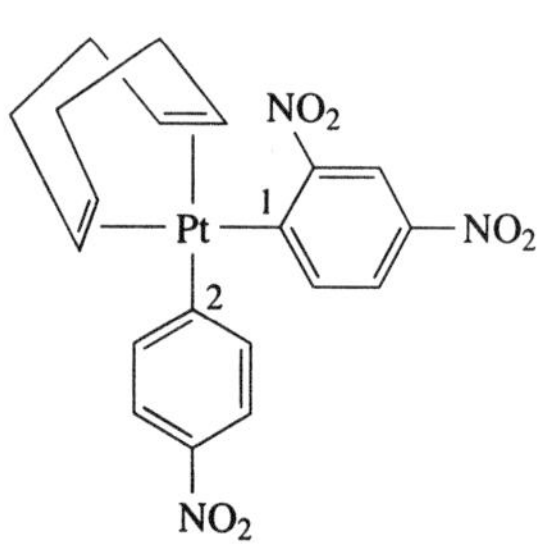	Pt–C[1] Pt–C[2]	0.2021(7) 0.2032(8)		179
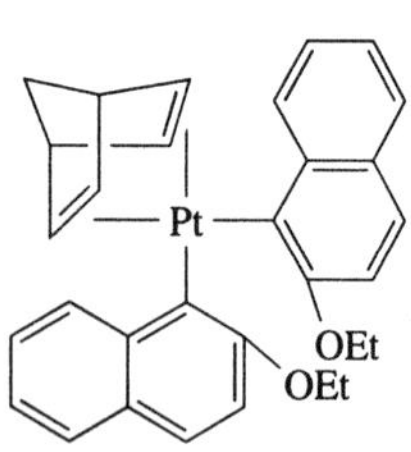		0.201(2) 0.208(2)		180
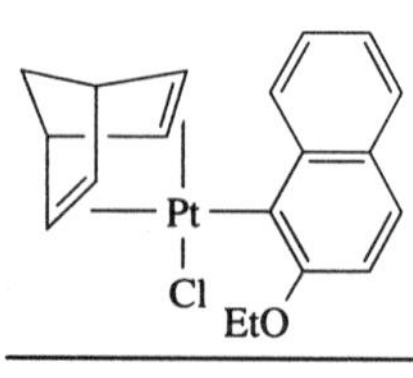		0.203(1)	Pt–Cl 0.2309(3)	181

complexes of the type *cis*- and *trans*-[$PtAr_2(PBu_3)_2$] and *cis*-[$PtAr_2\{P(C_6H_4NMe_2\text{-}4)_3\}_2$], which are influenced by the nature of the substituents on the phenyl rings, have been assigned to aryl-to-metal charge transfers.[199]

The ultraviolet photoelectron spectra of a series of *cis*-[$PtMe_2L_2$] complexes, containing phosphorus, arsenic, nitrogen, or carbon ligands, have been investigated.[200] These have revealed that the two molecular orbitals with the lowest ionization energies are the Pt–L σ-orbitals, the next four having mostly nonbonding platinum $5d$ character. The σ-donor ability of L, which is reflected in the first ionization energy, decreases in the following order:

$$P(NMe_2)_3 > TMEDA > PMePh_2, \quad PMe_2Ph > PEt_3 > AsMe_3$$
$$PMe_3 > PMe(OEt)_2 > P(OEt)_3 > P(OMe)_3 > cod > nbd$$

The π-acceptor ability of the ligands, reflected in the energy separation of the platinum d_{xy} and d_{xz} orbitals, lies in the order

$$PMePh_2, PMe_2Ph, P(OMe)_3, P(OEt)_3, PMe(OEt)_2, P(NMe_2)_3 >$$
$$AsMe_3, PMe_3, PEt_3, cod, nbd > TMEDA$$

The absorption and luminescence spectra of [$PtPh_2$(CO)L] and [$PtPh_2L_2$] (L = 5-aminoquinoline) have been studied. The $PtPh_2$(CO) or $PtPh_2$ fragments are found to act as electron-withdrawing substituents for the aromatic systems.[201]

8.3.1.3 Reactions

(i) Thermal decomposition reactions

Oxidative addition to zerovalent platinum and reductive elimination from platinum(II) complexes of H–H, C–H, or C–C bonds have been the subjects of a number of theoretical studies, and the following observations have been made.[202–7] The barrier to reductive elimination from square-planar complexes is lower when the leaving groups are strong σ-donors, but the reaction is retarded by strong σ-donors *trans* to the leaving groups. Elimination from *trans*-[PtR_2L_2], which proceeds by dissociation of L, is inhibited by a significant barrier to isomerization of the T-shaped [PtR_2L] intermediate. If the potential leaving groups are poor donors, however, the rate of isomerization may exceed that of elimination.[202] The kinetic barrier to reductive elimination of methane from *cis*-[PtHMe(PMe_3)$_2$] is small,[204] and carbon–hydrogen coupling of *cis* groups is facile in platinum(II) complexes.[207] In contrast, carbon–carbon coupling of alkyl fragments occurs much more readily for palladium(II) than for platinum(II) species.[207]

Carbon–hydrogen coupling has been reported with alkyl groups containing electron-withdrawing substituents. Elimination of $MeCF_3$ from *cis*-[PtH(CH_2CF_3)(PPh_3)$_2$] is thought to occur by a concerted, unimolecular process, without prior dissociation of PPh_3.[208] Similarly, photochemically induced reductive elimination from *trans*-[PtH{(CH_2)$_n$CN}(PPh_3)$_2$] (n = 1–3) proceeds by initial isomerization, followed by an intramolecular elimination step.[209]

Reductive elimination also occurs from dppm-bridged A-frame complexes containing alkyl groups. Elimination of H_2 (but not CH_4) from [Pt_2HMe(μ-H)(μ-dppm)$_2$]$^+$ is promoted by PPh_3, PMe_2Ph, dppm, or py (Equation (29)).[10] The reaction is second order, and the intermediate was observed by low-temperature NMR spectroscopy when L = PPh_3.[10] Methane elimination from [Pt_2Me_2(μ-H)(μ-dppm)$_2$]$^+$ is induced by addition of PMe_2Ph, but it is much slower than H_2 elimination in the previous case.[166] Both of these eliminations may also be induced photochemically.[210] Photolysis of [Pt_2Me_3(μ-dppm)$_2$]$^+$ in acetonitrile, acetone, or dichloromethane solution results in ethane formation, primarily by an intramolecular process, but with pyridine cleavage of the dimer occurs.[12,210]

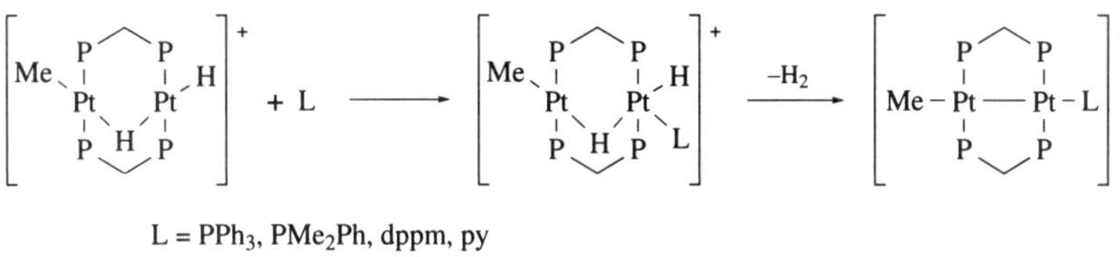

(29)

The largest number of reductive elimination reactions, however, involve the elimination of two aryl groups. This occurs by a concerted mechanism, and may be induced thermally or photochemically. It has been pointed out previously[2] that reductive elimination from [Pt(C_6H_4X-4)$_2$(PPh_3)$_2$] occurs stereospecifically to produce 4-X$C_6H_4C_6H_4$X-4. Similar results have been obtained in reductive elimination from [Pt(C_6H_4Me-4)$_2${(C_6F_5)$_2PCH_2$}$_2$].[48]

Brune and co-workers have made extensive studies of reductive elimination from biaryl complexes. They used differential-scanning calorimetry to determine the elimination temperatures for several series of compounds. For complexes of the type *cis*-$[Pt(C_6H_4X\text{-}4)_2(PPh_3)_2]$,[50] which contain identical aryl groups, the activation energy for reductive elimination decreases in the order

$$X = CF_3 > MeO > F > H > NMe_2 > Cl$$

For the analogous 3-substituted aryl complexes,[49,50] a slightly different order was obtained:

$$X = CF_3 > F > Cl > H, Me, OMe, NMe_2$$

With mixed biaryl complexes, the highest activation barriers were again found for those containing two electron-withdrawing substituents.[49,50] Alternatively, attempts to prepare mixed biaryl complexes containing one electron donor and one electron acceptor led to spontaneous reductive elimination, supporting the hypothesis that there is a donor–acceptor interaction in the transition state in the reductive elimination reaction.[61,69]

Reductive elimination of biaryls also takes place under photochemical conditions. When irradiated with long-wavelength UV radiation (288 nm or 312 nm, but not 254 nm), *cis*-$[Pt(C_6H_4X\text{-}4)_2(PBu_3)_2]$ ($X = CF_3$, Me, Bu^t, OMe) undergoes elimination by means of a regiospecific pericyclic reaction at platinum.[211] Analogous eliminations occur from $[PtAr_2L_2]$, as shown in Equation (30),[107,212] and by introducing 2- or 3-substituted aryl rings it has been shown that photochemical biaryl elimination occurs only when both rings are oriented perpendicular to the platinum square plane.[213]

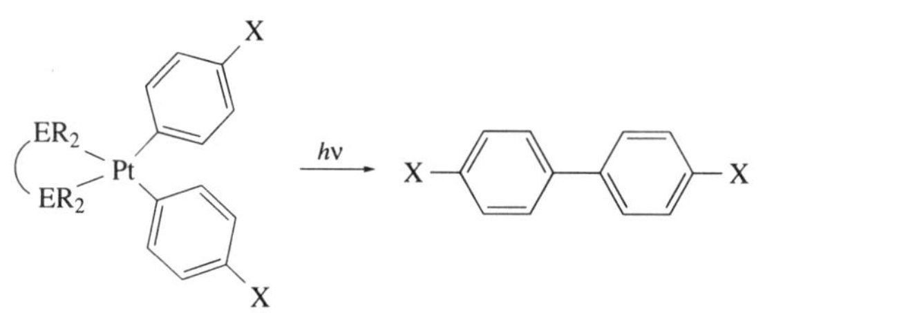

(30)

X = CF_3, F, Br, Me, Pr^i, Bu^t, TMS

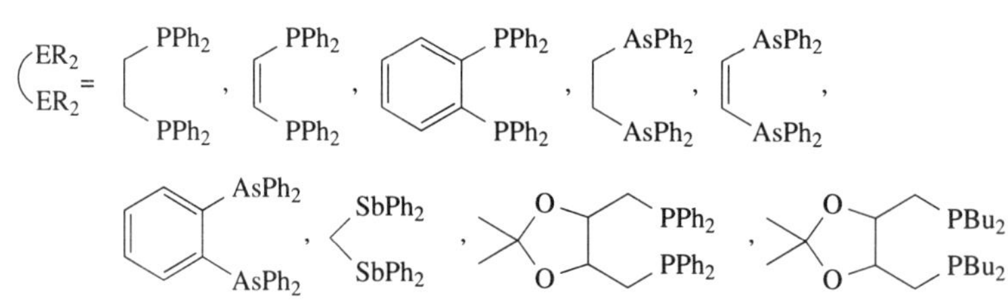

Heating solutions of $[PtAr_2(bipy)]$ does not result in reductive elimination of biaryl but, rather, metallation of bipyridine takes place along with loss of arene (Equation (31)).[214] In the analogous pyridine system, it has been shown that there is competition between biaryl elimination and loss of arene by metallation. In the case of *cis*-$[Pt(C_6H_4CF_3\text{-}4)_2(py)_2]$, reductive elimination takes place cleanly without dissociation of pyridine. With *cis*-$[Pt(C_6H_4X\text{-}4)_2(py)_2]$ (X = H, Bu^t), however, loss of C_6H_5X is a temperature-dependent competitive reaction, which is suppressed by deuteration of the aryl group or pyridine. The mechanism of the latter process may involve reversible hydrogen transfer from the aryl group to platinum, which would provide a rare example of aryl group isomerization prior to elimination.[215]

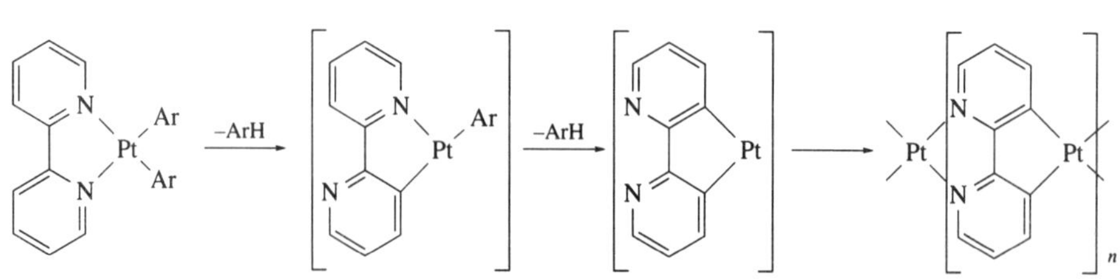

(31)

Thermal decomposition reactions of complexes of the type *cis*-$[PtR^1R^2L_2]$ (R^1,R^2 = alkyl) have been studied extensively. In general, the reactions proceed by ligand dissociation, followed by C–H oxidative addition to platinum and, finally, reductive elimination of alkane.[216] When the alkyl groups contain β-hydrogens, β-elimination is the predominant pathway, and when $R^1 \neq R^2$ the rate of β-elimination is proportional to the number of β-hydrogens on each group.[217]

The simplest case in which β-hydrogens are involved is the diethylplatinum system. Heating *cis*-$[PtEt_2(PEt_3)_2]$ in cyclohexane at 118 °C results in elimination of ethane and the observation of $[Pt(C_2H_4)(PEt_3)_2]$ as the only platinum-containing product.[218] For the decomposition of $[Pt(C_2H_5)(C_2D_5)(PEt_3)_2]$, no isotope effect is found by comparison with $[Pt(C_2H_5)_2(PEt_3)_2]$ and $[Pt(C_2D_5)_2(PEt_3)_2]$. This suggests that in the T-shaped intermediate $[PtEt_2(PEt_3)]$, formed by PEt_3 dissociation, each ethyl group is equally likely to undergo β-elimination, and some fluxional motion must occur which is faster than reductive elimination.[219] In the absence of free ligand the rate-determining step is dissociation of PEt_3. As the PEt_3 concentration is increased the rate-limiting process becomes the reductive elimination of ethane from $[PtEtH(C_2H_4)(PEt_3)]$, and at high PEt_3 concentrations the slowest step is probably loss of ethane from $[PtEtH(C_2H_4)(PEt_3)_2]$.[220]

Studies of β-elimination from $[Pt(^{13}CH_2Me)X(cod)]$ (X = $^{13}CH_2Me$, Cl, I) using quantitative ^{13}C NMR spectroscopy at 100–110 °C resulted in changes in the concentrations of $[Pt(^{13}CH_2Me)X(cod)]$ and $[Pt(CH_2{}^{13}Me)X(cod)]$ with time. This indicates that the β-elimination step is reversible, and a mechanism which accounts for these observations is shown in Scheme 8.[221] Protonation of $[PtEt_2\{Bu^t{}_2P(CH_2)_nPBu^t{}_2\}]$ (n = 2, 3) results in loss of ethane and formation of the agostic complex **(20)**, in which the extent of hydrogen transfer from the β-carbon to platinum is controlled by the size of the chelate ring, the smaller ring favoring transfer to the metal.[222]

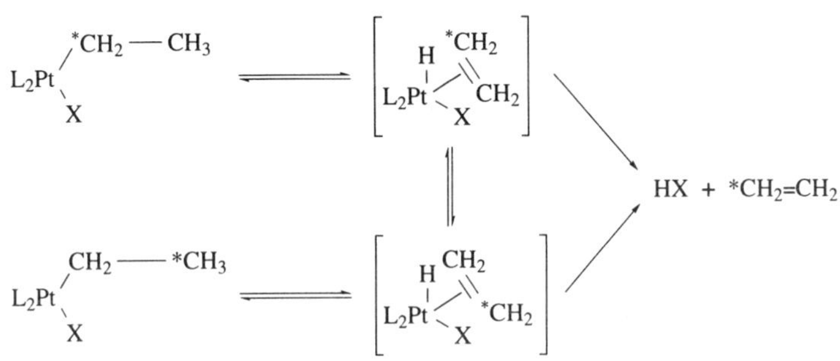

Scheme 8

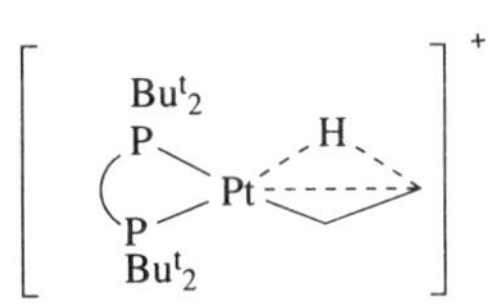

(20)

Elimination of ethene from $[Pt_2Et_3(\mu\text{-dppm})_2]SbF_6$ represents a rare example of β-elimination from a dimeric complex (Equation (32)). The reaction is intramolecular, and the β-elimination step is not rate determining. Dissociation of a PPh_2 group does not occur, and reductive elimination of ethane from the product does not take place under the reaction conditions.[223]

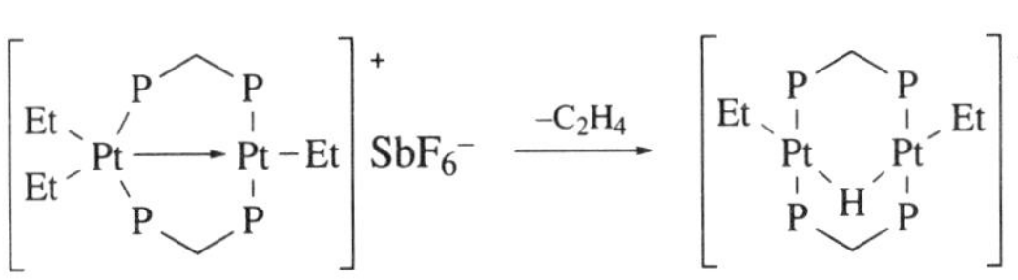

(32)

Thermal decomposition of bis(cyclobutyl)- or bis(cyclopentyl)bis(triethylphosphine)platinum complexes at 35–40 °C proceeds by β-elimination and reductive elimination to yield the corresponding cycloalkane and the cycloalkene–platinum(0) complex, by a mechanism analogous to that for the diethyl species.[224] In contrast, the bis(cyclopropyl) derivative decomposes at 148 °C to give cyclopropane and propene in a 2:1 ratio and $[Pt(PEt_3)_3]$, by a complex mechanism.[224]

Thermolysis of the aryloxyalkyl complex *trans*-$[PtBr(CH_2CH_2OPh)(PPh_3)_2]$ proceeds by β-aryloxy elimination and β-hydrogen elimination to give $CH_2{=}CH_2$ and $CH_2{=}CHOPh$ in a 2:1 ratio. In the presence of PPh_3 or $P(OEt)_3$ this ratio increases, and with 10 equiv. of $P(OEt)_3$ a product ratio of 5.7:1 is obtained.[16]

In dialkyl complexes without β-hydrogens, the major products are formed by intramolecular C–H activation, followed by reductive elimination, but only at elevated temperatures. At 157 °C in cyclohexane, *cis*-$[Pt(CH_2Bu^t)_2(PEt_3)_2]$ decomposes to generate neopentane and the platinacyclobutane complex $[Pt(CH_2CMe_2CH_2)(PEt_3)_2]$.[225] The reactions of the corresponding 2,2-dimethybutyl, 2,2-dimethylpentyl and 2,2,4,4-tetramethylpentyl complexes are shown in Scheme 9. Complexes (**21**) and (**23**) decompose 10^4 times faster, and (**22**) 50 times faster, than the neopentyl derivative.[226] In contrast to its neopentyl analogue, thermal decomposition of *cis*-$[Pt(CH_2(TMS))_2L_2]$ (L = PEt_3, $PMePh_2$, PPh_3) does not lead to a cyclic derivative. Instead, β-Si–Me scission occurs to produce *cis*-$[PtMe(CH_2SiMe_2CH_2(TMS))L_2]$, probably by way of an η^2-silylene intermediate.[227]

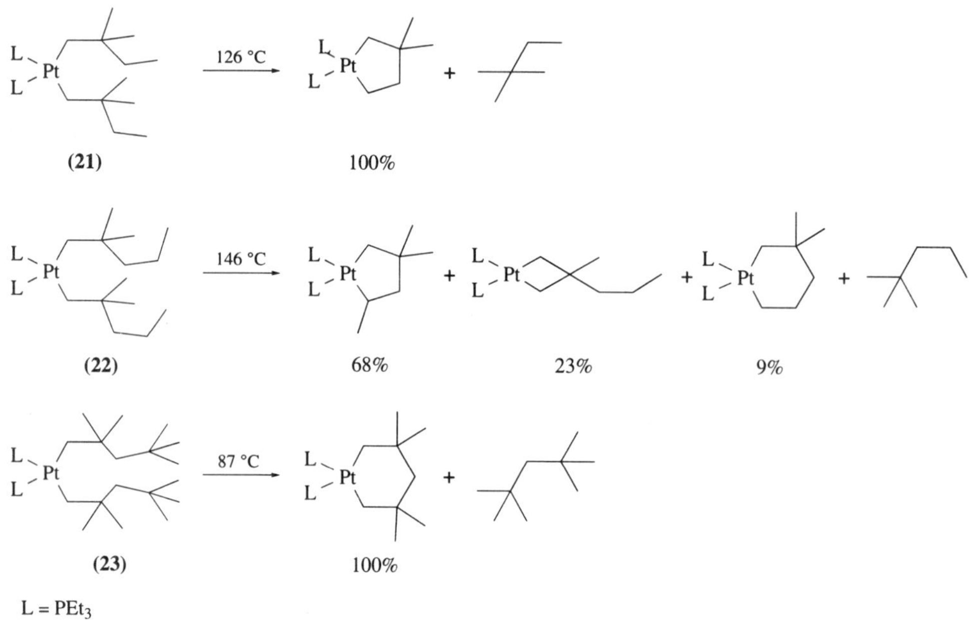

Scheme 9

Thermal decomposition of complexes of the form *cis*-$[Pt(CH_2CMe_2Ph)_2L_2]$ (L = PEt_3, PPh_3; L_2 = cod, bipy, 2,2'-bipyrimidyl (bipym), phen, 4,7-Ph_2phen, 3,4,7,8-Me_4phen, dppe) proceeds by δ-C–H activation of the aryl ring, as shown in Equation (33).[228–30] The reaction is fastest for the bulky, monodentate ligand PPh_3. Ligand dissociation is involved, but C–H addition to platinum is apparently the rate-limiting step. Complex (**25**) is also formed from *cis*-$[Pt(CH_2CMe_2Ph)(C_6H_4Bu^t\text{-}2)(PEt_3)_2]$ (**26**), where aromatic C–H activation is favored over alkyl δ-C–H scission, but the reaction is slower than the formation of (**25**) from (**24**), thereby eliminating (**26**) as a possible intermediate in that reaction.[231] Alternatively, refluxing *cis*-$[Pt(CH_2C_6H_4Me\text{-}2)_2L_2]$ (L = PEt_3, PPh_3) in xylene solution results in alkyl δ-C–H activation to form a five-membered ring, rather than aromatic C–H addition to produce a four-membered ring. When L = $PPh_2(C_6H_4Me\text{-}2)$, metallation of the phosphine occurs preferentially (Equation (34)).[46]

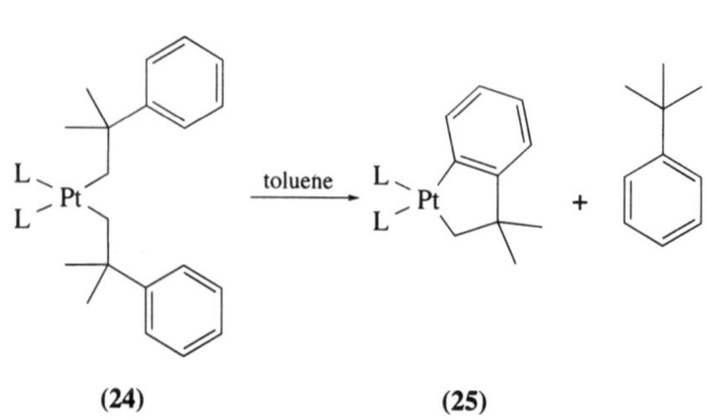

(33)

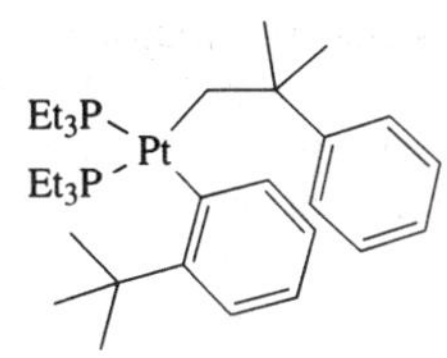

(26)

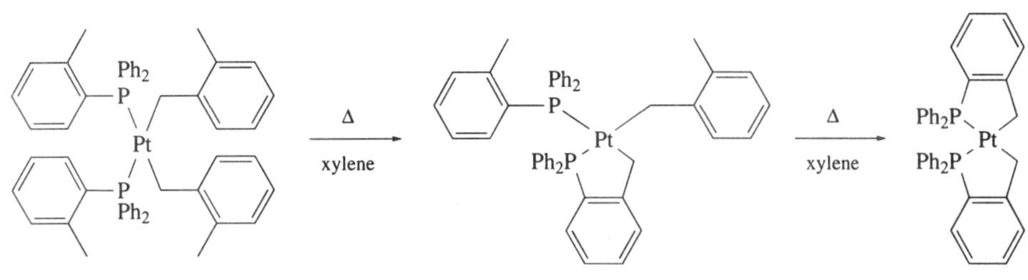

(34)

β-Elimination from $[PtXR(PEt_3)_2]$ (R = *n*-alkyl) yields *trans*-$[PtXH(PEt_3)_2]$ and the appropriate alkene. Ethene elimination from *trans*-$[PtClEt(PEt_3)_2]$ occurs at 158 °C in cyclohexane, the rate-determining step probably being dissociation of C_2H_4 from a five-coordinate intermediate.[232] In a kinetic study of β-elimination from *cis*-$[PtClR(PEt_3)_2]$ (R = Pr, Bu) it was found that the reaction is first order, and it is unaffected by the presence of chloride ion up to a concentration of 0.2 M. No intermediates could be detected but, in the absence of free chloride, isomerization competes with the elimination.[233] The propyl and butyl complexes *cis*-$[PtXR(PEt_3)_2]$ undergo β-elimination 10 times faster than the corresponding ethyl complexes.[234] The alkene distribution is dependent on X, and the rate of elimination decreases in the order

$$X = I > NCSe > Br > NCS > Cl > NO_2 > N_3$$

Thermal decomposition of *trans*-$[PtCl(CH_2Bu^t)(PCy_3)_2]$ produces *trans*-$[PtClH(PCy_3)_2]$ and 1,1-dimethylcyclopropane. This is proposed to proceed by oxidative addition of a δ-C–H bond to give a platinacyclobutane intermediate, as shown in Scheme 10.[235] Thermolysis of the (1-methylcyclobutyl)-methyl complexes *trans*-$[PtClR(PMe_3)_2]$ or [PtClR(dmpe)] at 140 °C, which do not possess β-hydrogens, leads to C–C bond activation. Ring opening occurs to give 2-methyl-1,5-hexadiene, with *trans*-$[PtClH(PMe_3)_2]$ being identified among the platinum-containing products in the case of the PMe_3 system.[236]

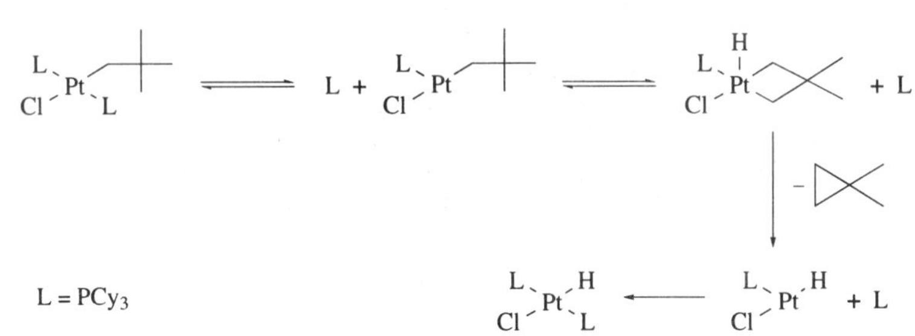

Scheme 10

In the case of methylplatinum species of the type *trans*-$[PtXMe(PR_3)_2]$, where the decomposition pathways discussed above are unavailable, the predominant process is homolytic cleavage of the Pt–C bond. The methyl radicals then abstract hydrogen from the solvent or the tertiary phosphine to give methane.[237] Heating decalin solutions of *trans*-$[PtXMe(PEt_3)_2]$ (X = Cl, Br, I, CN) to 170–200 °C produces CH_4, $[PtX_2(PEt_3)_2]$, and Pt metal. With $[PtX(CD_3)(PEt_3)_2]$ in decalin-d_{18}, CD_3H and CD_4 are formed in ratios which are independent of X and the temperature employed.[237]

Methylplatinum(II) complexes have been used as precursors for chemical vapor deposition (CVD) of platinum films.[238] The process may be accelerated by addition of H_2 (which also reduces the carbon deposits) or by photolysis. Decomposition of *cis*-$[PtMe_2(CNMe)_2]$ produces methane as the volatile product, and when D_2 is used, MeD is formed. The major part of the CVD process is believed to be of a heterogeneous nature.[238]

The bis(methoxy)platinum complex [Pt(OMe)$_2$(dppe)] decomposes at ambient temperature to produce methanol, formaldehyde, and small amounts of carbon monoxide, via β-hydrogen elimination. The platinum-containing products include the metallated phosphine complex [Pt$_2$(μ-C$_6$H$_4$PPhCH$_2$CH$_2$PPh$_2$)$_2$], [Pt(OMe)(CO$_2$Me)(dppe)], and [Pt(CO$_2$Me)$_2$(dppe)].[162] The corresponding diethyl complex releases ethane and ethene only slowly at 160 °C. The mixed complex [Pt(CH$_2$Me)(OMe)(dppe)] decomposes at an intermediate temperature of 100 °C to yield C_2H_4, C_2H_6, MeOH and (HCHO)$_n$, β-elimination from the ethyl group being favored by 1.3 kJ mol^{-1}.[162] Decomposition of *trans*-[PtMe(OMe)(PPh$_3$)$_2$] leads to *trans*-[PtMeH(PPh$_3$)$_2$] and formaldehyde.[239]

(ii) Electrophilic cleavage of platinum–carbon bonds

A number of reagents have been used in the electrophilic cleavage of platinum–carbon bonds. Treatment of *cis*-[PtPh$_2$(PEtPh$_2$)$_2$] or *trans*-[PtXR(PEtPh$_2$)$_2$] (X = Cl, Br; R = Me, Et, Pr, Bu, Bz) with HX in MeOH–H$_2$O (9:1) results in stereospecific cleavage of one Pt–C bond to generate *cis*-[PtXPh(PEtPh$_2$)$_2$] or *trans*-[PtX$_2$(PEtPh$_2$)$_2$].[240] The proposed mechanism of these reactions, which involves two possible pathways, is shown in Scheme 11. In the case of *trans*-[PtX(CH$_2$Bu^t)(PEt$_3$)$_2$] the bulky neopentyl group inhibits the k_{Cl} pathway.[241]

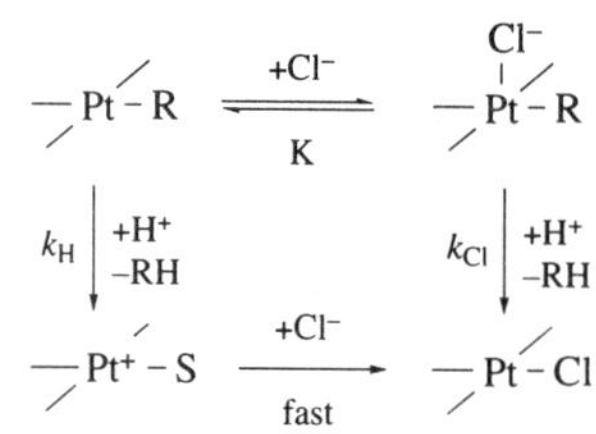

Scheme 11

In the reaction of *cis*-[PtMePh(PMePh$_2$)$_2$] with HCl, cleavage of the Pt–Me bond occurs preferentially.[242] This is in contrast to the reactions of [PtMe(C$_6$H$_4$Me-4)(cod)][243] and [PtMe(C$_6$H$_2$Me$_3$-2,4,6)(cod)],[63] in which platinum–aryl bond cleavage takes place. The reactions of [PtMePh(PMePh$_2$)$_2$] and [PtMe(C$_6$H$_4$Me-4)(cod)] with HgCl$_2$ or [PtI$_2$(PMe$_2$Ph)$_2$] proceed analogously to those with HCl.[243] Platinum–aryl bond cleavage is proposed to occur by an S$_E$2 mechanism, whereas Pt–Me rupture probably takes place by an oxidative addition–reductive elimination sequence.[243]

Similar electrophilic bond-cleavage processes are involved in the transfer of cyclopentadienyl, phenyl, alkynyl, or methyl groups between platinum(II) centers. Treatment of *cis*-[PtR$_2$(CO)(PMePh$_2$)] (R = η^1-Cp, Ph, C≡CPh) with *cis*-[PtCl$_2$(CO)(PMePh$_2$)] results in R for chloride exchange to give species of the form [PtClR(CO)(PMePh$_2$)]. These reactions are also likely to take place by means of an S$_E$2 (cyclic) mechanism.[244] Reaction of *cis*-[PtMe$_2$(SMe$_2$)$_2$] with *cis*-[PtI$_2$(PMePh$_2$)$_2$] produces [PtIMe(SMe$_2$)(PMePh$_2$)], where exchange of neutral ligands occurs in addition to methyl for iodide exchange, but when *cis*-[PtMe$_2$(PMe$_2$Ph)$_2$] is treated with [PtI$_2$(SMe$_2$)$_2$] exchange of methyl and iodide is much faster than any neutral ligand substitution and [PtIMe(PMe$_2$Ph)$_2$] and [PtIMe(SMe$_2$)$_2$] are formed.[245]

(iii) Substitution and isomerization reactions

Several studies of ligand substitution reactions in organoplatinum(II) compounds have been carried out. Such reactions involve the observed rate laws

$$k_{obs} = k_1 + k_2[Y]$$

the relative importance of the two terms being dependent on the π-acceptor ability of the incoming ligand Y.[2] Substitution reactions involving tertiary phosphines have been studied most extensively and associative mechanisms have been postulated. Traditionally, such mechanisms have been assumed to operate in other cases also, but there is increasing evidence that dissociative mechanisms are implicated in many situations. This may be illustrated by a series of studies of substitution in methyl- and phenylplatinum(II) complexes of dimethyl sulfide or dimethyl sulfoxide.[76,167,246–9]

The substitution of DMSO from *cis*-$[PtPh_2(DMSO)_2]$ by bipyridine has been shown to involve a preequilibrium in which one DMSO ligand dissociates to generate a T-shaped intermediate.[246] There is a single observable step, the kinetics being consistent with a competition between DMSO and bipy for the $[PtPh_2(DMSO)]$ intermediate.[247] Similarly, reactions of *cis*-$[PtPh_2(SMe_2)_2]$ with bidentate nitrogen, sulfur, or phosphorus ligands proceed by dissociation of a thioether ligand, induced by the high *trans* influence of the phenyl groups.[167] The reactions of the corresponding methyl complexes, *cis*-$[PtMe_2L_2]$ (L = DMSO, SMe_2), with bidentate ligands also take place predominantly by a dissociative pathway, indicating that this mechanism is not peculiar to the phenyl derivatives.[248] Ligand exchange in *cis*-$[PtPh_2(SMe_2)_2]$, *cis*-$[PtPh_2(DMSO)_2]$, and *cis*-$[PtMe_2(DMSO)_2]$ has been studied as a function of pressure. In each case a positive activation volume is found (4.7 ± 0.5, 5.5 ± 0.8, and 4.9 ± 0.5 cm^3 mol^{-1}, respectively), which is supportive of a dissociative mechanism.[249] A comparison has been made between the reactions of *cis*-$[PtPh_2(SEt_2)_2]$ and *cis*-$[PtPh_2(CO)(SEt_2)]$ with pyridine and other ligands. The former reacts by the dissociative mechanism discussed previously, whereas a bimolecular substitution pathway in which there is discrimination between different nucleophiles is involved for the carbonyl complex. This has been explained in terms of the LUMO lying perpendicular to the square plane in the latter, which renders it more suited to nucleophilic attack.[76]

Cyanide exchange in the $[PtMe(CN)_3]^{2-}$ anion has been investigated.[250] Exchange is much slower than in the tetracyano ion, substitution of the *cis* and *trans* CN^- occurring at 10^{-3} and 10^{-7} times, respectively, the rate found in $[Pt(CN)_4]^{2-}$. The *cis* effect is believed to be primarily of an entropic nature, whereas the *trans* effect is enthalpy controlled.[250]

Isomerization of square-planar platinum(II) complexes has been investigated extensively.[251] The traditionally accepted mechanism of such reactions was one of consecutive displacement, an associative process involving an ionic intermediate. A pseudorotation mechanism, involving rearrangement of a neutral, five-coordinate intermediate, was also postulated to account for certain reactions, particularly in nonpolar solvents which would not support ionic intermediates.[251] An alternative mechanism for the spontaneous isomerization of complexes of the type *cis*-$[PtXR(PEt_3)_2]$, namely one involving halide dissociation, had also been proposed.[252] This last proposal prompted considerable debate on the relative merits of dissociative and solvent-assisted pathways,[251] which have proved difficult to distinguish.

The high-pressure kinetics of the *cis–trans* isomerization of $[PtXR(PEt_3)_2]$ reveal an activation volume of -12 cm^3 mol^{-1}, consistent with an associative first step. This is proposed to be an initial solvolysis step, the rate of which is greater than that of the subsequent isomerization step (for which $\Delta V^{\ddagger}$ is positive), which may involve solvent dissociation or an intramolecular rearrangement.[253] Similarly, for the isomerization of *cis*-$[PtXPh(PEt_3)_2]$ in methanol, the activation volumes were determined to be 6.4 ± 0.4, 5.2 ± 0.3, and 7.2 ± 0.8 cm^3 mol^{-1} for the chloride, bromide, and iodide complexes, respectively.[254] The reaction rates were found to be independent of the concentration of free halide, and the rate-determining step was proposed to be the isomerization of the solvent-stabilized cation, $[PtPh(MeOH)(PEt_3)_2]^+$.[254]

In contrast, extensive kinetic studies of the spontaneous isomerization of *cis*-$[PtXR(PEt_3)_2]$ (R = alkyl, aryl) have provided support for the dissociative mechanism.[255–9] Studies of the *cis–trans* isomerization of $[PtCl(C_6H_4Me\text{-}3)(PEt_3)_2]$ in hydroxylic solvents suggested that the role of the solvent is to promote Pt–Cl bond cleavage in the rate-determining step, the degree of solvation of the leaving chloride by hydrogen bonding decreasing in the order

$$\text{MeOH} > \text{EtOH} > \text{PrOH} > \text{BuOH} > \text{Pr}^i\text{OH}$$

Thus, isomerization shows a marked dependence on the nature of the solvent, whereas displacement of chloride by thiourea, which occurs by an associative process, does not.[255,256] Initial state–transition state analysis of the isomerization of *cis*-$[PtCl(C_6H_4F\text{-}4)(PEt_3)_2]$ is consistent with a dissociative reaction, the reactivity trends being dominated by solvation of the leaving chloride.[257]

The spontaneous *cis–trans* isomerizations of the alkyl complexes $[PtClR(PEt_3)_2]$ (R = Me, Et, Pr, Bu, $CH_2(TMS)$) in Pr^iOH solution are also inhibited by free Cl^-, and are proposed to occur via three-coordinate, T-shaped intermediates.[258] The complexes containing β-hydrogens exhibit higher reactivity, and it is suggested that there may be an interaction between platinum and a β-hydrogen in the T-shaped intermediate. The situation is complicated, however, by the fact that β-elimination and isomerization are competing processes when R = Et, Pr or Bu.[258] In the case of *cis*-$[PtCl(CH_2Bu^t)(PEt_3)_2]$ in Pr^iOH solution further complications arise: two pathways are observed, one proceeding through $[PtNp(S)(PEt_3)_2]^+$, whereas the second involves phosphine dissociation.[259]

Cleavage of the Pt–Me bond in *cis*-$[PtMeR(PEt_3)_2]$ (R = Me, Ph, mesityl) by HA (A = ClO_4, BF_4, OSO_2CF_3) in the presence of methanol generates *cis*-$[PtR(MeOH)(PEt_3)_2]^+$. Isomerization of these cations is retarded by free methanol, suggesting that isomerization of the solvent-stabilized cation also occurs by a dissociative pathway and the interconversion of T-shaped intermediates.[260]

The photoisomerization of *cis*- and *trans*-[$PtClPh(PEt_3)_2$] in acetonitrile solution occurs by a mechanism different from the thermal reaction. The *cis* to *trans* and *trans* to *cis* isomerizations also take place by distinctly different mechanisms. The former is proposed to occur by intramolecular twisting from a low-lying ligand field state, whereas the latter involves a dissociative path from a charge-transfer state.[261]

Alkyl group isomerization has been investigated in a series of dithiocarbamate complexes of the form [$PtR^1(S_2CNR^2_2)(PR^3_3)$] ($R^2$ = Me, Et; R^3 = Et, Ph).[262] Alkyl isomerization takes place at 120 °C. The *t*-butyl derivative isomerizes completely to its isobutyl analogue at this temperature, whereas an equilibrium is established between the *n*- and isopropyl and *n*- and *s*-butyl, complexes, the *n*-isomer being favored by a 9:1 ratio in each case. The free energy difference of 7.1 kJ mol^{-1} represents the Pt–C bond energy difference between the linear and branched forms, assuming that steric constraints imposed by the other ligands are minimal.[262]

(iv) *Insertion reactions*

Formal insertion of carbon monoxide, isocyanides, sulfur dioxide, alkenes, or alkynes into platinum–carbon bonds has been described.[2] Most of the examples described recently have involved carbon monoxide. Many of these will be discussed in Chapter 7, this volume, but they will also be covered briefly here. A number of reviews of carbonyl insertion reactions, including those of organoplatinum compounds, have appeared.[263,264] Theoretical treatments of the insertion of CO into Pt–C bonds have also been reported.[265] The term insertion is really a misnomer and, although rare examples exist where a true insertion is apparently involved,[266] the reactions are generally believed to proceed by organic group migration to coordinated CO.

The platinum–diene complex [PtClPh(cod)] reacts readily with carbon monoxide (1 atm) at ambient temperature to produce [PtCl(COPh)(cod)], whereas the corresponding methyl compound does not react under these conditions. The diphenyl and dimethyl complexes, alternatively, do not undergo CO insertion but displacement of the diene occurs instead.[267] Reactions of *cis*-[$PtCl_2(CO)L$] (L = PEt_3, PMe_2Ph, $PMePh_2$, PPh_3, PCy_3) with organomercurials lead to one isomer of [PtClR(CO)L] (R *trans* to L) which is in equilibrium with the chloro-bridged acyl or aroyl complex [$Pt_2(\mu\text{-}Cl)_2(COR)_2L_2$]. In contrast, [$Pt_2(\mu\text{-}Cl)_2R_2L_2$] reacts with CO to produce the other two isomers of [PtClR(CO)L], which do not undergo CO insertion readily. This has been explained in terms of the requirement for a *cis* arrangement of the R and CO ligands, and the high *trans* influence of the tertiary phosphine weakening the Pt–C bond in the first isomer.[268] The equilibrium constant for the reaction shown in Equation (35) has been determined for a range of aryl groups. Electron-donating 3- or 4-substituents favor the aroylplatinum dimer, and a correlation with Hammett σ values was found. The enthalpy of reaction becomes less negative with electron-withdrawing substituents, and this may reflect a strengthening of the platinum–aryl bonds. 2-Substituted aryl groups do not undergo insertion, presumably for steric reasons.[269]

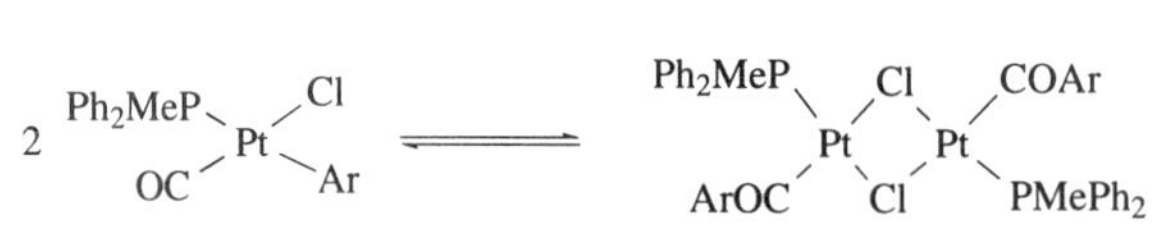

(35)

Carbonylation of the bidentate ligand complexes [$PtClPhL_2$] proceeds at ambient temperature when L_2 = dppp or $Ph_2P(CH_2)_nNMe_2$ (n = 2, 3), but not when L_2 = dppe.[270] Similarly, the rate of CO insertion into the Pt–Me bond of [$PtClMeL_2$] depends on the donor atoms and the size of the chelate ring,[271] and decreases in the order

bipy > 1-diphenylphosphino-8-dimethylaminonaphthalene > $Ph_2P(CH_2)_3NMe_2$ > dppp > dppe

The cross-coupling of alkyl iodides with $NaBPh_4$ is catalyzed by [$Pt(PPh_3)_4$].[272] A key step in the catalytic cycle is the insertion of CO into the Pt–C bond of [$PtIR(PPh_3)_2$], a reaction which has been studied in some detail.[273] Reaction of *trans*-[$PtCl(COC_6H_{13})(PPh_3)_2$] with $NaBPh_4$ has been shown to give the coupling product, $C_6H_{13}COPh$.[272]

Chloroplatinum complexes, when employed in conjunction with tin(II) chloride, catalyze homogeneous hydrogenation and hydroformylation reactions.[3] In the latter, CO insertion into a Pt–C bond is a vital step. The role of the trichlorostannate ligand, formed by insertion of $SnCl_2$ into the Pt–Cl bond, in promoting such reactions has been discussed at length. This ability has been attributed at various times to the *trans* effect or *trans* influence of the $SnCl_3$ ligand, and to its stabilization of a five-coordinate species. In the carbonylation of *trans*-[$PtPh(SnCl_3)L_2$] (L = $PMePh_2$, PPh_3), the primary function of the $SnCl_3$ group appears to be to serve as a suitable leaving group.[274]

The *cis*-$[PtCl_2(PPh_3)_2]$–$SnCl_2{\cdot}2H_2O$ system has been investigated extensively. Treatment of *cis*-$[PtCl_2(PPh_3)_2]$ with 1-hexene and CO in ethanol solution at 100 °C yields *trans*-$[PtCl(COC_6H_{13})(PPh_3)_2]$ which, with $SnCl_2$ as cocatalyst, effects regioselective hydroformylation of 1-hexene. In general, when the $[PtCl_2(PPh_3)_2]$–$SnCl_2$ system is used to hydroformylate α-alkenes, the catalyst is recovered as *trans*-$[PtCl(COR)(PPh_3)_2]$ and $[PtH(CO)(PPh_3)_2]SnCl_3$.[136,138,275] Low-temperature NMR studies of *trans*-$[PtCl(COC_6H_{13})(PPh_3)_2]$–$SnCl_2$ and related systems reveal the formation of *trans*-$[Pt(SnCl_3)(COR)(PPh_3)_2]$ and a species assigned tentatively as *trans*-$[PtCl\{C(OSnCl_2)R\}(PPh_3)_2]$. Addition of $SnCl_2$ and dihydrogen produces the aldehyde and *trans*-$[PtH(SnCl_3)(PPh_3)_2]$.[276] Addition of $SnCl_2$ to a CD_2Cl_2 solution of *trans*-$[PtClH(PPh_3)_2]$ at −80 °C generates several hydridoplatinum species, but only *trans*-$[PtH(SnCl_3)(PPh_3)_2]$ inserts ethene at this temperature to produce *cis*-$[PtEt(SnCl_3)(PPh_3)_2]$. Carbon monoxide promotes the isomerization of this complex even at −80 °C, and this is accompanied by displacement of $SnCl_3^-$ by CO. At 20 °C, CO insertion takes place.[277] These reactions are summarized in Scheme 12. Addition of dihydrogen generates propanal. The diphosphine complexes $[PtClEtL_2]$ (L_2 = dppp, dppb) have been used as models for alkene hydroformylation. Addition of CO produces the corresponding propionyl complex, and reaction with $SnCl_2$ yields $[PtEt(SnCl_3)L_2]$, which is unstable in the absence of ethene. A simple catalytic cycle which accounts for the hydroformylation process is shown in Scheme 13.[278]

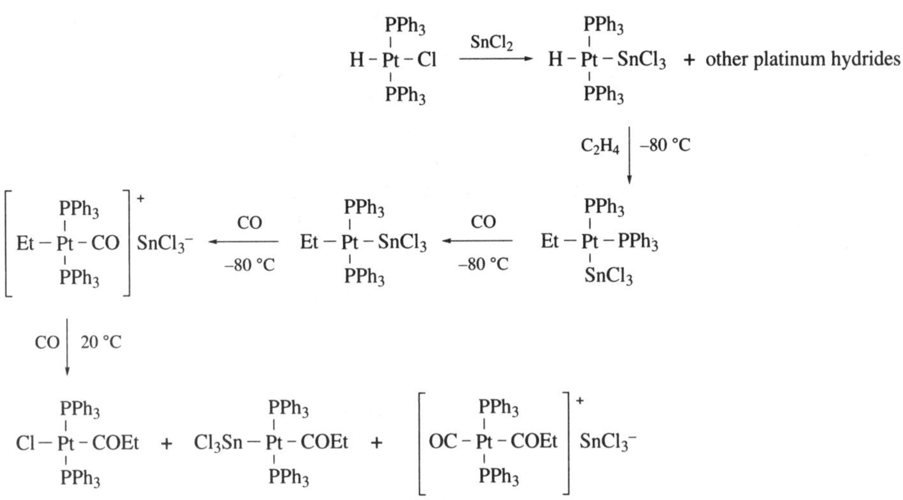

Scheme 12

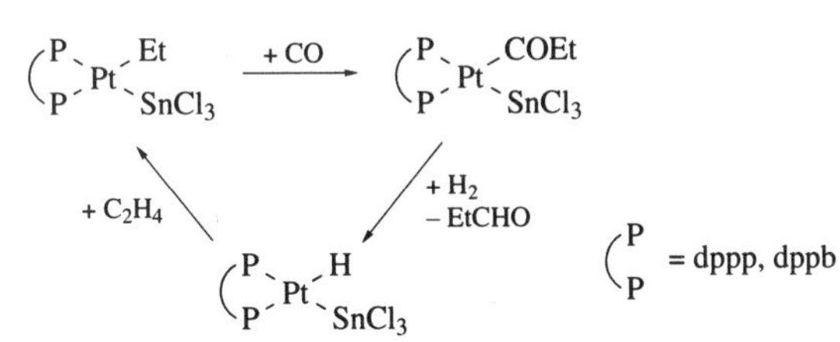

Scheme 13

In diphosphine complexes containing alkoxy or amido ligands, insertion of CO takes place into the Pt–O or Pt–N bond in preference to the Pt–C bond. Thus, carbonylation of [PtMe(OMe)(dppe)] leads to the alkoxycarbonyl complex [PtMe(COOMe)(dppe)] as the sole product in a reaction that shows a first-order dependence on both the complex and CO. Low-temperature ^{13}C NMR measurements suggest that a five-coordinate intermediate is involved. $[Pt(OMe)_2(dppe)]$ reacts sequentially with 2 equiv. of CO to give $[Pt(COOMe)_2(dppe)]$.[279] With the amido complex $[PtMe(NHCH_2Ph)(dppe)]$, reaction with CO yields $[PtMe(CONHCH_2Ph)(dppe)]$.[280] Similarly, *cis*-$[PtMe\{OCH(CF_3)_2\}(PMe_3)_2]$ reacts with CO to produce *trans*-$[PtMe\{COOCH(CF_3)_2\}(PMe_3)_2]$, but the corresponding phenoxy complex is resistant to carbonyl insertion.[281] Alkoxycarbonyl complexes have also been obtained in the reactions of $[PtCl_2(dppf)]$ with NaOMe and CO in methanol[282] and by treatment of the *trans*-$[PtCl(CO)(PPh_3)_2]^+$ cation with a primary alcohol.[283]

The chemistry of hydroxyplatinum(II) complexes has been studied extensively by Bennett and co-workers, and the syntheses of such compounds have been well documented.[284] On carbonylation they produce metalacarboxylic acids, which are implicated as intermediates in the oxidation of CO by metal ions in solution, the water gas shift reaction, and the formation of metal hydrides.[285] Treatment of [PtMe(OH)L_2] (L_2 = dppe, dppp) with CO produces [PtMe(COOH)L_2], insertion into the Pt–O bond again being preferred over the Pt–C bond, but above –20 °C decarboxylation occurs to produce the unstable [PtMeHL_2].[286] In the case of *trans*-[PtPh(OH)$(PEt_3)_2$], carbonylation produces *trans*-[PtPh(COOH)$(PEt_3)_2$], which has been characterized crystallographically as a hydrogen-bonded dimer.[173] In formamide or methylformamide the complex ionizes as *trans*-[PtPh(CO)$(PEt_3)_2$]OH. On heating to 100 °C the platinacarboxylic acid loses 0.5 equiv. of CO to generate the CO_2-bridged complex, [Pt_2Ph_2(μ-CO_2)$(PEt_3)_4$].[173] The metalacarboxylic acids [PtR(COOH)L_2] (R = Me, Ph; L = PPh_3, $PMePh_2$, PCy_3; L_2 = dppe, dppp, 1,2-bis(diphenylphosphino)ethene (diphoe)) may be converted into the corresponding methoxycarbonyl complex by addition of methanol in the presence of triethylamine.[287]

A number of "double carbonylation" reactions involving palladium and, to a lesser degree, platinum have been reported. The key mechanistic question is whether CO inserts into a metal–acyl bond, or whether the product is obtained by reductive elimination of two acyl fragments. The latter appears to be the case. α-Ketoacyl complexes have been prepared by oxidative addition to platinum(0),[25,26] and *trans*-[PtCl(COCOPh)$(PPh_3)_2$] has been characterized in the solid state. The ketoacyl ligand exhibits a planar C_2O_2 unit, in which the two C=O bonds are *anti* to one another. The compounds lose carbon monoxide readily on heating.[25]

Bis(acyl)platinum complexes have been prepared by alkylation or amidation of *cis*-[Pt(COPh)(CO)$(PPh_3)_2$]BF_4. The first crystallographically characterized neutral bis(acyl)platinum complex, *cis*-[Pt$(COPh)_2(PPh_3)_2$],[288] and mixed carbamoyl–acyl complex, *cis*-[Pt($CONEt_2$)(COPh)-$(PPh_3)_2$],[289] have been reported. Carbamoyl–acyl complexes have also been obtained by treatment of *trans*-[PtCl(COPh)$(PPh_3)_2$] with a secondary amine under 10 atm of carbon monoxide[290] and bis(acyl) species by reaction with dialkylzinc and CO.[291] When Et_2Zn is added to the *trans*-[Pt(COCOPh)-(THF)$(PPh_3)_2$]$^+$ cation an ethylplatinum species is formed initially, but in a noncoordinating solvent this rearranges to a bis(acyl) derivative (Equation (36)).[292]

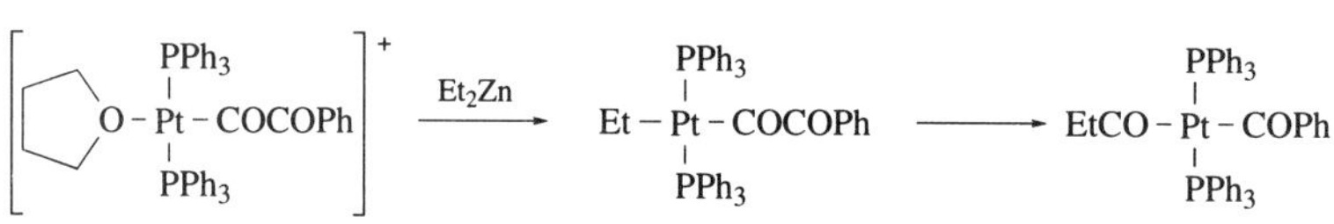

(36)

The carbonylation of [PtMe(OMe)(dppe)] was mentioned above. Reaction with C_2F_4 proceeds analogously by insertion into the Pt–O bond to give [PtMe(CF_2CF_2OMe)(dppe)]. The product has been characterized by x-ray crystallography, the Pt–CF_2 bond (0.2074(9) nm) being slightly shorter than the Pt–Me bond (0.2106(9) nm).[162] Another example of a selective insertion is the reaction of *cis*-[PtMePh$(PMePh_2)_2$] with sulfur dioxide. Here insertion occurs into the platinum–methyl bond only.[242]

A rare example of a reversible, formal β-alkyl insertion is observed for the chelated Pt(alkene)(alkyl) species (**27**) obtained as shown in Equation (37). This complex is stable in nitromethane at 125 °C. Reversible alkene insertion has been demonstrated for the deuterium-labeled complex (**28**), which equilibrates with (**30**) through the intermediacy of the insertion product (**29**) (Equation (38)).[293]

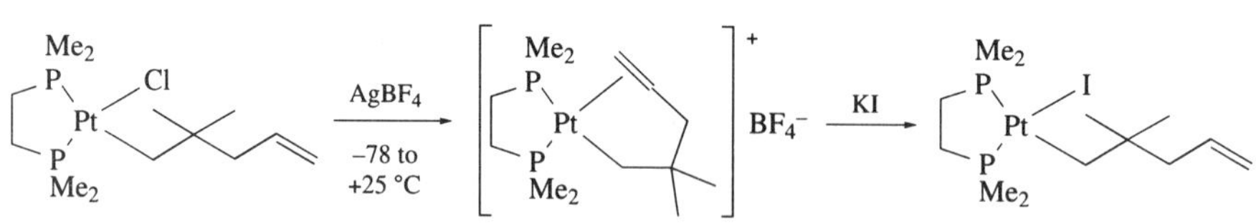

(37)

(**27**)

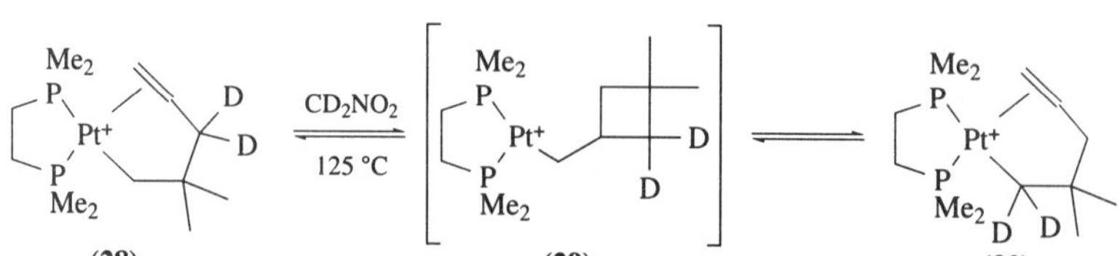

(38)

(**28**) (**29**) (**30**)

Finally, [Pt$(PPh_3)_4$] catalyzes the regioselective methylation of 1,2-epoxy-2-methylpropane with methyl iodide. The reaction has been shown to proceed by oxidative addition of MeI to give *trans*-[PtIMe$(PPh_3)_2$], followed by a ring opening–insertion step to yield *trans*-[PtI(OCH_2Bu^t)$(PPh_3)_2$]. Cleavage of the Pt–O bond with HCl generates 2,2-dimethylpropanol.[294]

(v) Oxidative addition reactions

Oxidative addition of organic halides to organoplatinum(II) complexes produces platinum(IV) species, many of which are thermodynamically stable. Although there are other routes to organoplatinum(IV) compounds, which will be discussed in Section 8.5.1, a major approach is by means of such oxidative addition processes.

A convenient organoplatinum(IV) precursor, $[Pt_2Me_8(\mu\text{-}SMe_2)_2]$, is obtained by treatment of *cis*-$[PtCl_2(SMe_2)_2]$ with excess MeLi and iodomethane. The bridging thioether ligands may be displaced by a range of neutral ligands to generate species of the type *cis*-$[PtMe_4L_2]$ (L = PMe_2Ph, $PMePh_2$; L_2 = dppm, dppe, bipy, phen).[295] Displacement of the dimethyl sulfide ligands from $[Pt_2R_4(\mu\text{-}SMe_2)_2]$ (R = Me, Ph) first, followed by oxidative addition of MeI, yields $[PtIR_2MeL_2]$. Reaction of $[Pt_2Me_4(\mu\text{-}SMe_2)_2]$ with methyl iodide and dppm gives the dppm-bridged complex $[Pt_2Me_6(\mu\text{-}I)_2(\mu\text{-dppm})]$.[296] Two platinum centers have also been linked by hydrocarbyl bridges through double oxidative addition processes.[297–9] Reaction of *cis*-$[PtMe_2L_2]$ (L = $AsMe_2Ph$, $AsMe_2(C_6H_4Me\text{-}2)$, SMe_2) with α,α-dibromo-*o*-xylene proceeds according to Equation (39). Displacement of L from complex (**31**) by dppm gives the corresponding dppm-bridged derivative. α,α-Dibromo-*p*-xylene reacts analogously, and alkyl-bridged complexes are available by double oxidative addition of $I(CH_2)_nI$.[299]

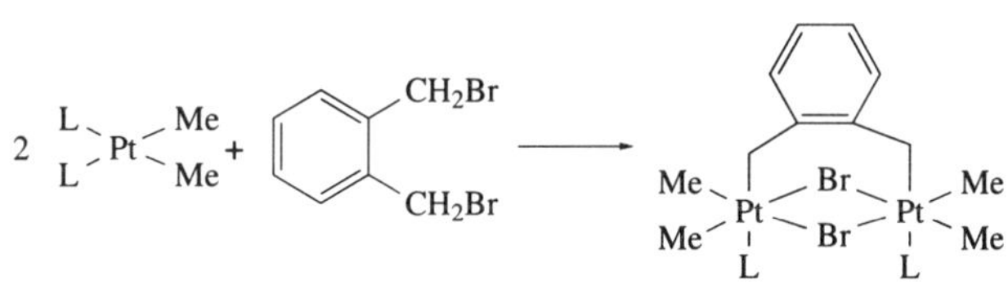

(39)

(**31**)

Oxidative addition of iodomethane to the dmpm-bridged complex *cis,cis*-$[Pt_2Me_4(\mu\text{-dmpm})_2]$ occurs first at one platinum to give a platinum(II)–platinum(IV) complex, and with a second equivalent an iodo-bridged species is formed (Equation (40)).[300,301] The dppm analogue does not react with MeI. With the unsymmetrical complex $[Ph_2Pt(\mu\text{-dmpm})_2PtMe_2]$, addition of iodomethane occurs exclusively at the dimethylplatinum center.[124] Addition of MeI to *cis*-$[PtR_2(\text{dppm-}P)_2]$ generates $[PtIMeR_2(\text{dppm-}P)_2]$ by stereospecific *trans* oxidative addition.[118] Halogen addition to *cis,cis*-$[Pt_2Me_4(\mu\text{-dmpm})_2]$ also produces a platinum(II)–platinum(IV) complex, in this case by methyl migration between platinum centers. In the case of iodine addition, further reaction occurs to cleave the methyl group from the platinum(II) center (Equation (41)).[300,302]

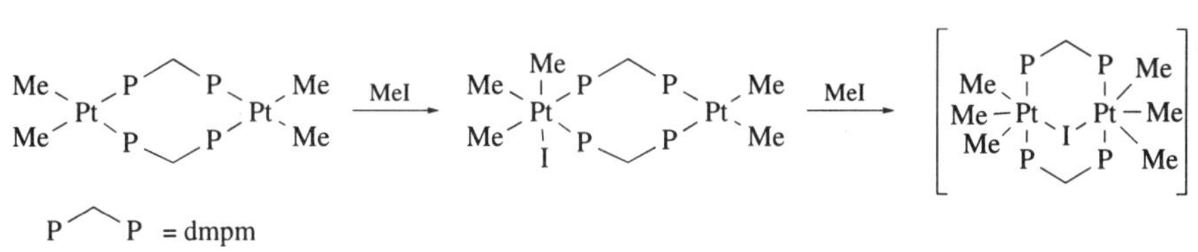

(40)

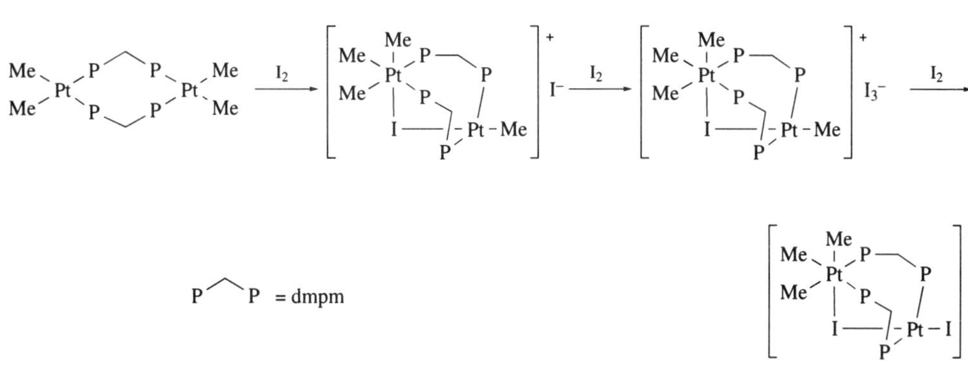

(41)

An example of neighboring group participation is found in the oxidative addition of bromomethane to $[PtMe_2L_2]$ (L = $Me_2EC_6H_3(OMe)_2$-2,4; E = P, As) to give *fac*-$[PtBrMe_3L_2]$. The phosphine and arsine complexes react 180 and 250 times faster than the analogous complexes of PMe_2Ph or $AsMe_2Ph$, respectively.[303]

A large number of oxidative addition reactions involving organoplatinum(II) complexes containing nitrogen ligands have been reported. Addition of iodomethane to $[PtMe_2L_2]$ (L = py; L_2 = bipy, en, tmen (tetramethylethylenediamine)) produces *fac*-$[PtMe_3IL_2]$, but when L = NH_3 a mixture of $[Pt_2(\mu\text{-}I)_2Me_6(NH_3)_2]$ and *fac*-$[PtMe_3(NH_3)_3]I$ is obtained.[75] Treatment of the tris(pz)methane complex $[PtMe_2\{N,N\text{-}(C_3H_3N_2)_3CH\}]$ with warm py results in metallation and loss of methane, and further addition of MeI generates $[PtMe_2\{C,N,N\text{-}(C_3H_3N_2)_3CH\}(py)]I$ (**32**), in which the tris(pz)methane serves as a tripodal ligand.[304] In contrast, addition of iodomethane to the 1,1-bis(pz)ethane complex $[PtMe_2\{N,N\text{-}(C_3H_3N_2)_2CHMe\}]$ simply produces *fac*-$[PtMe_3I\{N,N\text{-}(C_3H_3N_2)_2CHMe\}]$ (**33**), in which there is a short contact between the CH group and the iodide both in the solid state and in solution.[305] The complex $[PtMe_2L_2]$ (L_2 = 2,2'-bithiazoline) undergoes oxidative addition of MeI, CH_2Cl_2, or CH_2Br_2.[306] Organoplatinum(IV) polymers have been prepared by treatment of $[PtMe_2(bipy)]$ with $CH_2{=}CMeCO_2CH_2CH_2Br$, followed by alkene polymerization. The analogous 4,4'-bis(*t*-butyl)bipyridine complex gives rise to soluble polymers.[307]

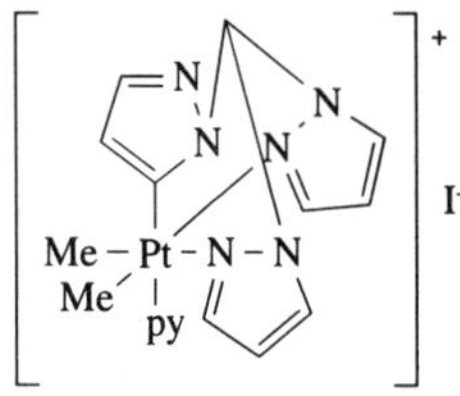

(32)

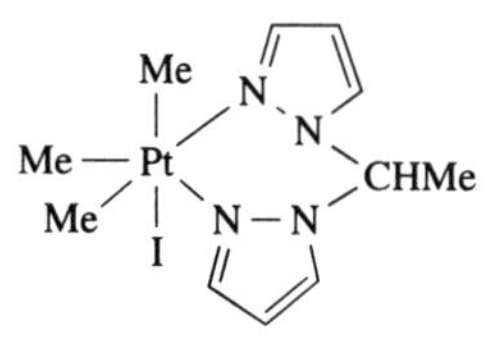

(33)

Organoplatinum(IV) complexes may also be prepared by addition of iodine to a platinum(II) precursor. Treatment of $[PtMe_2(cod)]$ or $[Pt_2R_4(\mu\text{-}SEt_2)_2]$ (R = Me, Ph) with bidentate or potentially tripodal ligands, followed by I_2, yields compounds of the type (**34**) or (**35**).[308–10]

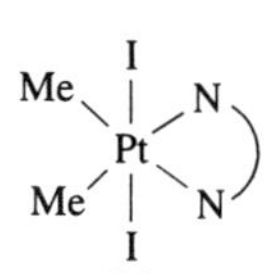

(34)

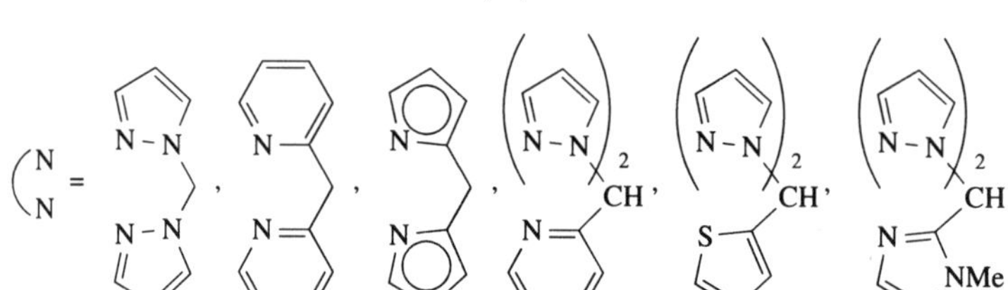

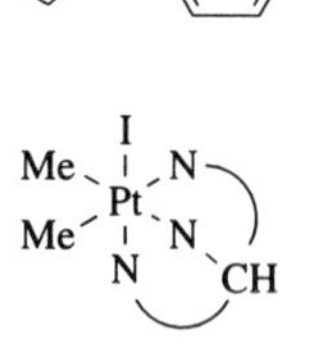

(35)

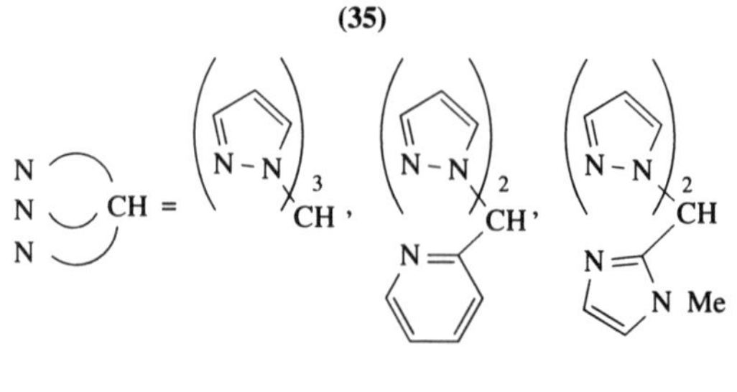

Addition of alcohols to $[PtMe_2L_2]$ (L_2 = bipy, phen) produces first $[PtMe_2(OR)(H_2O)L_2]OH$ (R = Me, Et, Pr^i), which on treatment with NH_4PF_6 yields $[PtMe_2(OR)(H_2O)L_2]PF_6$. Addition of HCl to the methoxy compound gives $[PtMe_2Cl(OMe)L_2]$.[311] Epoxide ring opening in the presence of carbon dioxide leads to coupling of these fragments and oxidative addition to $[PtMe_2(bipy)]$ or $[PtMe_2(phen)]$ (Equation (42)),[312] while opening of a β-lactone also results in oxidative addition (Equation (43)).[313]

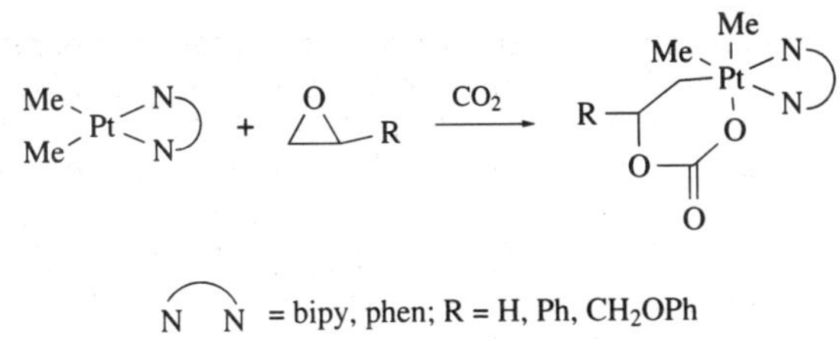

(42)

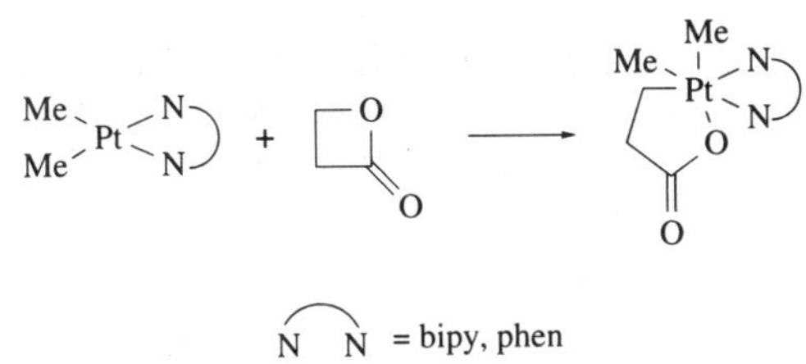

(43)

Diplatinum(IV) complexes are obtained from the double oxidative addition of diiodoalkanes to $[PtMe_2(phen)]$. *Trans* addition of $I(CH_2)_nI$ occurs to give $[PtMe_2I\{(CH_2)_nI\}(phen)]$ (n = 0–5), and with a second equivalent of $[PtMe_2(phen)]$ a bridged complex of the type $[\{PtMe_2I(phen)\}_2\{\mu\text{-}(CH_2)_n\}]$ (n = 2–5) is produced.[314] Tetranuclear platinum(II)–platinum(IV) species are formed in the case of the bipyrimidyl system (Equation (44)).[315]

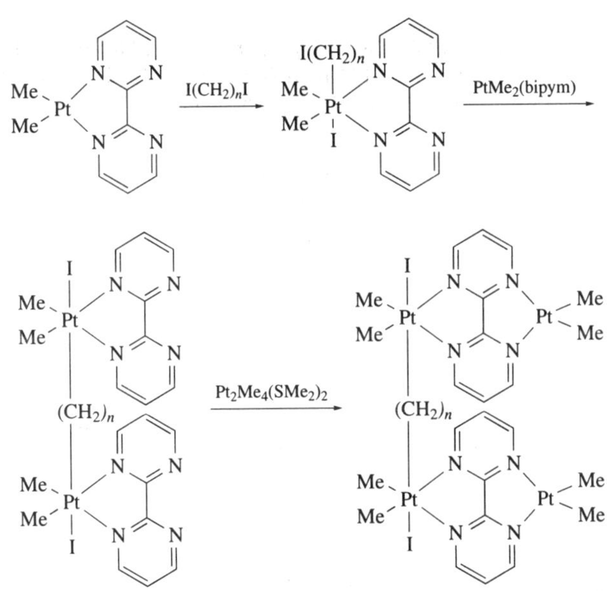

(44)

Mixed metal complexes containing methylene linkages have also been synthesized. Oxidative addition of $[Au(CH_2Cl)L]$ or $[Hg(CH_2Cl)Cl]$ to $[PtMe_2(bipy)]$ produces species of the form $[PtClMe_2(bipy)(CH_2M)]$ (M = AuL, HgCl).[316] Alkyl chlorides do not add as readily to platinum(II) as the corresponding iodides, but the presence of gold or mercury activates the C–Cl bond, and the rate of oxidative addition is enhanced by a factor of 10^3. Complexes containing direct Pt–Au or Pt–Ag bonds are formed by treatment of $[PtMe_2(bipy)]$ with $[AuX(PPh_3)]$ (X = NO_3, BF_4, PF_6) or $AgBF_4$.[317] Mixed palladium(IV)–platinum(IV) compounds are generated by treatment of $[PdMe_2(bipy)]$ with α,α-dibromo-*m*-xylene, followed by addition of $[PtMe_2(bipy)]$ (Equation (45)).[318]

Several studies have been carried out aimed at determining the mechanisms of oxidative addition of organic halides to organoplatinum(II) complexes. The photochemically initiated addition of Pr^iI to $[PtMe_2(phen)]$ in air yields both the expected oxidative addition product and $[PtMe_2I(OOPr^i)(phen)]$. The reaction proceeds by a free-radical mechanism. Irradiation of the lowest-energy MLCT band of $[PtMe_2(phen)]$ leads to a triplet excited state, which initiates the free-radical chain mechanism by iodine atom abstraction from Pr^iI. The $Pr^i\cdot$ radical acts as the chain carrier, and termination occurs by attack of $Pr^i\cdot$ on a methyl or phenanthroline ligand.[319] The reaction of $[PtMe_2(phen)]$ with Pr^iI or Bu^tI in the presence of an activated alkene proceeds by an oxidative addition–insertion sequence (Equation (46)). This again takes place by a free-radical mechanism, the chain carrier being the alkyl radical.[320]

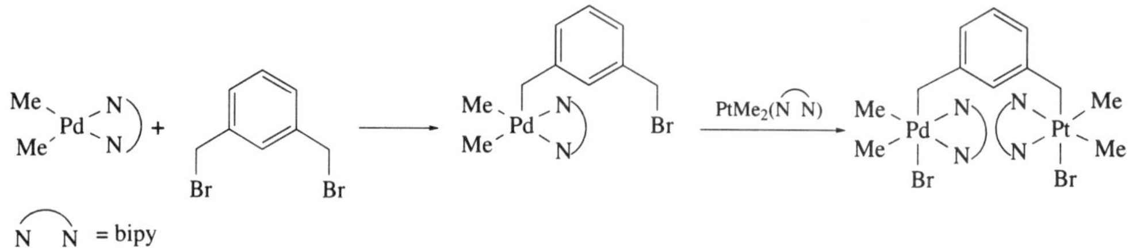

(45)

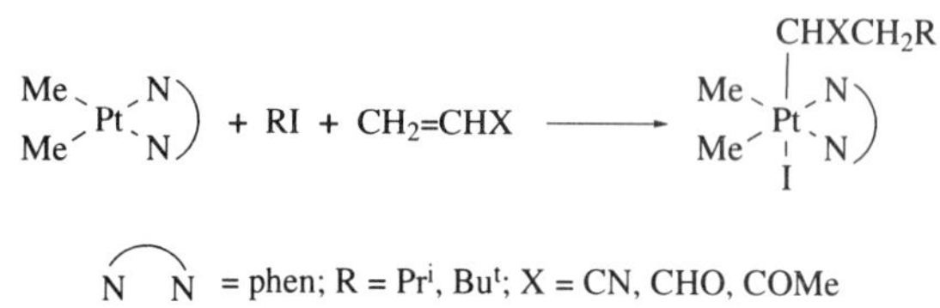

(46)

The oxidative addition of primary alkyl iodides to dimethylplatinum(II) species takes place by an S_N2 mechanism. In fact, the ionic intermediate $[PtMe_3(SMe_2)_2(CD_3CN)]I$ has been observed by NMR spectroscopy during the low-temperature reaction of *cis*-$[PtMe_2(SMe_2)_2]$ with iodomethane in CD_3CN solution.[298] Analogous ionic species have been detected at −20 °C during reactions of $[PtMe_2L_2]$ ($L = PMe_2Ph$, L_2 = bipy) with primary alkyl halides in CD_3CN solution.[321] A significant steric effect is found in these reactions. The addition of MeI to *cis*-$[PtMe_2(SMe_2)_2]$ occurs 20 times faster than the corresponding reaction with $[Pt_2Me_4(\mu\text{-}SMe_2)_2]$.[298] Iodomethane adds to $[PtMe_2(phen)]$ much more rapidly than other primary alkyl iodides, the relative rates of reaction for MeI : EtI : PrI : BuI being $10^3 : 1.0 : 0.5 : 0.6$. The effect of the metal is detected even through four bonds, in that $[PtMe_2I(phen)\{(CH_2)_3I\}]$ reacts 5.6 times faster than PrI.[322] Addition of MeI to dimethylplatinum complexes containing bis(diimine) ligands reveals that addition always occurs more rapidly with mononuclear complexes than with related dinuclear species. In dinuclear platinum(II)–platinum(IV) derivatives, the rate of addition to the platinum(II) center is dependent largely on steric effects.[323] A comparison of the rates of addition of MeI or $PhCH_2Br$ to the palladium and platinum complexes $[MMe_2L_2]$ (L_2 = bipy, phen) shows that the platinum complex reacts 7–22 times faster, and the phenanthroline complex reacts slightly faster than its bipyridine counterpart.[324]

(vi) Reactions of coordinated ligands

Where there is a reactive site on a coordinated ligand, reaction may occur there rather than at platinum. Protonation or methylation of the 2-pyridinyl complexes *trans*-$[PtBr(2\text{-}C_5H_4N)L_2]$ ($L = PEt_3$, PPh_3) with $HClO_4$ or Me_2SO_4 results in addition to the pyridinyl nitrogen to generate cationic species of the type *trans*-$[PtBr(2\text{-}C_5H_4NR)L_2]^+$, which exhibit restricted rotation about the Pt–C bond.[325] Similarly, addition of the activated alkyl chlorides RCH_2Cl (R = CN, COMe) to *trans*-$[PtCl(2\text{-}C_5H_4N)(PPh_3)_2]$ in the presence of $NaClO_4$ does not proceed by oxidative addition to platinum, but alkylation at nitrogen occurs instead. Reaction of $[PtCl(2\text{-}C_5H_4N)(dppe)]$ with $MeCOCH_2Cl$ produces $[PtCl(2\text{-}C_5H_4NCH_2COMe)(dppe)]^+$, and the dimethyldithiocarbamate complex $[Pt(S_2CNMe_2)(2\text{-}C_5H_4N)(PPh_3)]$ reacts with RCH_2Cl to give $[Pt(S_2CNMe_2)(2\text{-}C_5H_4NCH_2R)(PPh_3)]^+$, the rate of reaction decreasing in the order R = COMe > CN > Ph.[326]

The 1,1-bis(diphenylphosphino)ethene complexes $[PtR^1_2\{(Ph_2P)_2C{=}CH_2\}]$ (R^1 = Me, C_6H_4Me-2) undergo Michael addition reactions with amines or hydrazines to give $[PtR^1_2\{Ph_2PCH(CH_2X)PPh_2\}]$ ($X = NR^2R^3$, $NHNHR^2$).[103,327] The 1,1-bis(diphenylphosphinomethyl)ethene complex $[PtMe_2\text{-}\{(Ph_2PCH_2)_2C{=}CH_2\}]$ undergoes ligand isomerization in the presence of NaOH–EtOH, and deprotonation with MeLi, as shown in Equation (47).[103]

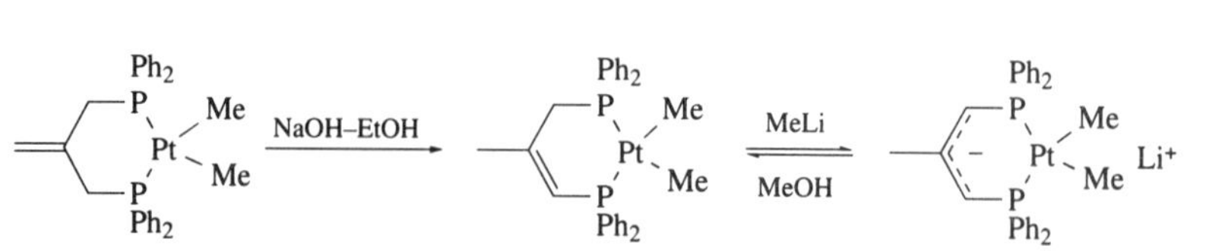

(47)

8.3.2 Haloalkyls and Haloaryls

8.3.2.1 Preparation

There has been significant growth in the chemistry of haloaryl complexes in recent years. Anionic complexes of the type $[PtAr_4]^{2-}$ and $[PtAr_3L]^-$, in particular, have been produced in increasing numbers and they are generally more stable than their phenyl counterparts. The use of such anionic compounds has been reviewed.[328a] Although the chemistry of chloromethyl and iodomethyl complexes is developing, there are relatively few trifluoromethyl complexes of platinum.[328b] The synthesis of haloalkyls and haloaryls may be accomplished by oxidative addition to platinum(0) precursors, by transfer of the organic fragment from another metal center, by modification of a complex by substitution of weakly held ligands, or, in the case of haloalkyls, by insertion reactions. Each of these methods will be discussed here. The distinction between displacement of coordinated ligands as a synthetic method or as a reaction of a haloalkyl or haloaryl complex is subjective and, for the purpose of this work, reactions that involve a second metal center in forming bimetallic complexes will be considered in Section 8.3.2.3.

(i) Oxidative addition reactions

There are few examples of oxidative addition as a method for the preparation of haloaryl- or haloalkylplatinum(II) compounds. This is in contrast to the situation with aryl and alkyl species described in Section 8.3.1.1.

Treatment of $[Pt(PPh_3)_3]$ with HgR_2 generates $[(Ph_3P)_2RPtHgR]$ ($R = C_6Cl_5$, C_6HCl_4, $C_6H_2Cl_3$) which, on addition of $[Pt(C_2H_4)(PPh_3)_2]$, produces the trimetallic species $[Hg\{PtR(PPh_3)_2\}_2]$.[147,329] Similarly, reaction of $[Pt(C_2H_4)(PPh_3)_2]$ with $[Cp(CO)_3MHgC_6Cl_5]$ (M = Mo, W) results in insertion of platinum into the Hg–C bond to give *cis*-$[Pt(C_6Cl_5)(PPh_3)_2\{HgM(CO)_3Cp\}]$.[330] In contrast to these results, the reaction of $[Pt(PPh_3)_3]$ with $[Au(C_6Cl_5)(PPh_3)]$ proceeds by insertion of platinum into one of the C–Cl bonds, rather than into the Au–C bond (Equation (48)),[331] whereas with $[AuMe(PPh_3)]$ insertion into the Au–Me bond does take place.[332]

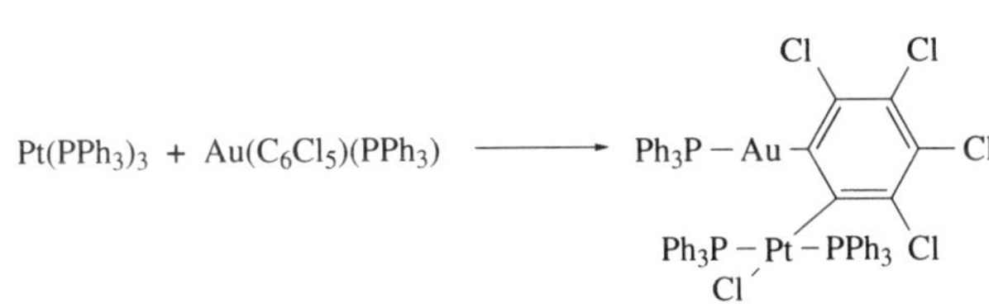

(48)

In principle, the simplest method for the generation of organoplatinum species is oxidative addition to platinum atoms. This approach has been used to prepare compounds such as *trans*-$[Pt(C_6F_5)_2(PEt_3)_2]$. Reaction of platinum atoms with C_6F_5Br produces $Pt(C_6F_5)Br$, which disproportionates to $Pt(C_6F_5)_2$ and $PtBr_2$. The bis(aryl)platinum species may then be trapped with triethylphosphine.[333]

Reaction of $[Pt(C_2H_4)L_2]$ ($L = PPh_3$, $P(C_6H_4Me\text{-}4)_3$, PEt_3, PPr^i_3, PEt_2Ph; L_2 = dppe) with CH_2X_2 (X = Cl, Br, I), either thermally or photochemically, results in the formation of *trans*-$[PtX(CH_2X)L_2]$.[334] Similarly, addition of CH_2Cl_2 to $[Pt\{Ph_2PCH_2CH_2)_3N\}(PPh_3)]$ generates $[Pt\{(Ph_2PCH_2CH_2)_3N\}(CH_2Cl)]Cl$.[335]

(ii) Reactions with main group organometallics

Few examples of such reactions have been reported recently for haloalkyl complexes, whereas an extensive chemistry of haloaryls has been demonstrated. One or two trifluoromethyl groups may be transferred from $Cd(CF_3)_2{\cdot}$diglyme to *cis*-$[PtI_2(PEt_3)_2]$ to generate $[PtI(CF_3)(PEt_3)_2]$ or $[Pt(CF_3)_2(PEt_3)_2]$ in modest yields.[336] Reaction of *cis*-$[PtCl_2(PPh_3)_2]$ with 1-lithioundecafluorobicyclo[2.2.1]heptane produces *cis*-$[PtCl(C_7F_{11})(PPh_3)_2]$. This compound is stable to HCl in ethanol, but decomposes in the presence of H_2.[337]

A range of organometallic reagents has been employed in the synthesis of haloarylplatinum complexes. Treatment of $[PtCl_2L_2]$ (L_2 = 1,2-bis(dimethylarsino)benzene) with excess C_6F_5Li yields $[Pt(C_6F_5)_2L_2]$,[338] and addition of 2 equiv. of C_6F_5MgBr to K_2PtCl_4 in the presence of dioxane gives *trans*-$[Pt(C_6F_5)_2(dioxane)_2]$, from which the dioxane ligands may be displaced by pyridine *N*-oxide, DMF or

DMSO.[339] When $PtCl_2$ or K_2PtCl_4 is allowed to react with excess C_6F_5Li or C_6F_5MgBr in the absence of coordinating ligands, the tetraaryl anions $[Pt(C_6F_5)_4]^{2-}$ are formed, which may be isolated as their Bu_4N^+ salts. Reaction of $[PtCl_2(THT)_2]$ (THT = tetrahydrothiophene) generates $Bu_4N[Pt(C_6F_5)_3(THT)]$, from which the THT may be displaced by a range of neutral ligands, whereas with $[Bu_4N]_2[Pt_2(\mu\text{-}Br)_2Br_4]$ the bromide-bridged complex $[Bu_4N]_2[Pt_2(\mu\text{-}Br)_2(C_6F_5)_4]$ is obtained. Analogous reactions take place with C_6Cl_5Li or C_6Cl_5MgCl.[340] Pentafluorophenyl derivatives have also been obtained by transfer from $Yb(C_6F_5)_2$.[341]

Partially fluorinated or chlorinated aryl complexes have also been prepared in large numbers. Treatment of [$PtCl_2$(diene)] (diene = norbornadiene, 1,5-hexadiene) with 1 equiv. of an aryltrimethyltin reagent yields [PtClR(diene)] (R = C_6F_5, C_6F_4H-4, C_6F_4OMe-4), whereas [PtR_2(nbd)] is obtained with 2 equiv. of $Me_3SnC_6F_4OMe$-4. $Bu_3SnC_6F_5$ is found to be a more selective reagent than its methyl analogue, whereas $Et_3SnC_6F_5$ is less selective.[342] Reactions of $[PtCl_2L_2]$ (L = py; L_2 = bipy) with $Ba(OSOR)_2$ (R = C_6F_5, C_6F_4H) in boiling pyridine proceed according to Equations (49) and (50).[343] Reaction of [$PtCl_2$(cod)] with TlOCOR is accompanied by decarboxylation to give [PtClR(cod)] (R = C_6F_5, C_6F_4H-4), and similar reactions of $[PtCl_2L_2]$ (L = PPh_3; L_2 = dppe, dppp) with $C_6F_4(COOTl)_2$ are shown in Scheme 14.[344] Treatment of *cis*-$[PtCl_2(PPh_3)_2]$ with $Hg(C_6Cl_nH_{5-n})_2$ (n = 2–5) generates *trans*-$[PtClR(PPh_3)_2]$.[345] Although not organometallic reagents, the fluorinated benzoic acids $C_6F_5CO_2H$ or $C_6F_4HCO_2H$ react with $[Pt(OH)_2L_2]$ (L = py; L_2 = bipy) to give $[Pt(OCOR)_2L_2]$, which undergo decarboxylation in refluxing py to give $[PtR_2L_2]$.[346]

$$PtCl_2L_2 + Ba(OSOR)_2 \longrightarrow PtR_2L_2 + BaCl_2 + SO_2 \qquad (49)$$

$$2\,PtCl_2L_2 + Ba(OSOR)_2 \longrightarrow 2\,PtClRL_2 + BaCl_2 + SO_2 \qquad (50)$$

$$L = py;\ L_2 = bipy;\ R = C_6F_5,\ C_6F_4H$$

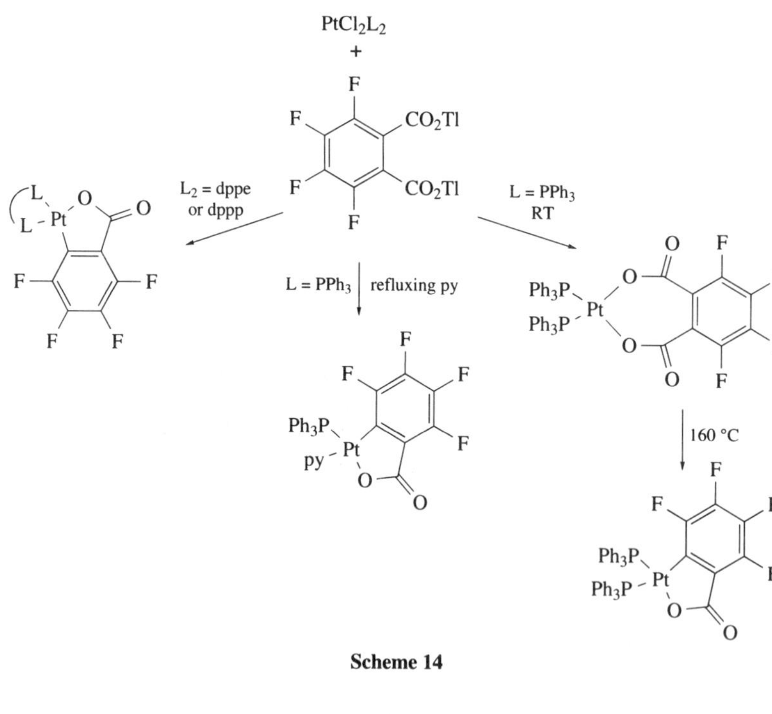

Scheme 14

(iii) Displacement of weakly bound ligands

The halide-bridged complexes $[Bu_4N]_2[Pt_2(\mu\text{-}X^2)_2(C_6X^1{}_5)_4]$ (X^1 = F, Cl; X^2 = Cl, Br) may be cleaved by a range of neutral ligands,[340,347] or the bridging halide may be replaced by PPh_2[348] or SC_6F_5.[349] Chloride removal from $[Bu_4N]_2[PtCl(C_6F_5)_3]$ with silver perchlorate generates the aryl-bridged complex $[Bu_4N]_2[Pt_2(\mu\text{-}C_6F_5)_2(C_6F_5)_4]$,[350] which is a useful precursor to many other compounds. Treatment with 2,2'-bipyridine generates $[Bu_4N]_2[Pt_2(C_6F_5)_6(\mu\text{-bipy})]$, whereas with 1,10-phenanthroline cleavage occurs to give *cis*-$[Pt(C_6F_5)_2(phen)]$ and $[Bu_4N]_2[Pt(C_6F_5)_4]$.[351] Addition of further $AgClO_4$ to $[Bu_4N]_2[Pt_2(\mu\text{-}C_6F_5)_2(C_6F_5)_4]$ produces $[Bu_4N][Pt_2Ag(\mu\text{-}C_6F_5)_2(C_6F_5)_4]\cdot OEt_2$ (**36**).[350]

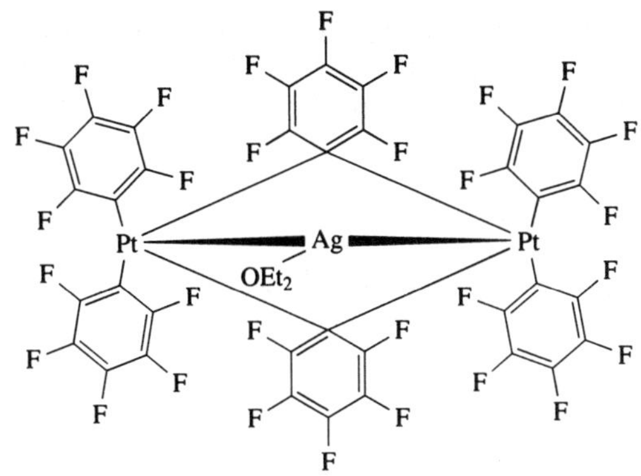

(36)

When the chloride-bridged complex $[Bu_4N]_2[Pt_2(\mu\text{-}Cl)_2(C_6F_5)_4]$ is allowed to react with tetrabutylammonium hydroxide, the corresponding hydroxide-bridged dimer is obtained. The bridging OH exhibits a ^1H NMR resonance at −1.21 ppm, indicative of its highly nucleophilic nature.[352] This is illustrated by the reactions shown in Scheme 15.[353–5]

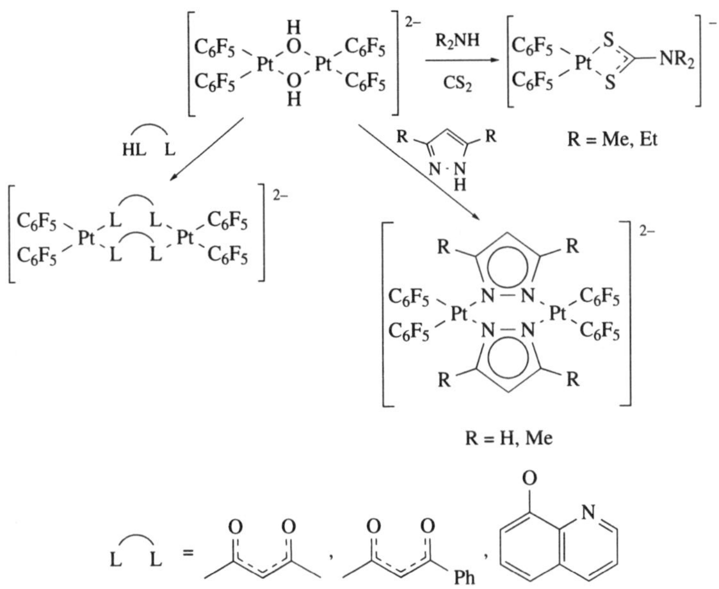

Scheme 15

Neutral ligands may also be displaced readily. The diene may be substituted in $[PtR_2(diene)]$ ($R = C_6F_5$, C_6F_4H; diene = nbd, cod, dicyclopentadiene, 1,5-hexadiene)[343] or $[Pt(CF_3)_2(nbd)]$[356] by a variety of N, O, P, or S ligands, and addition of halide to $[Pt(CF_3)_2(nbd)]$ generates the bridged anion $[Pt_2(\mu\text{-}X)_2(CF_3)_4]^{2-}$,[356] but the great majority of haloaryl complexes have been prepared by substitution of coordinated THF or THT. Introduction of carbon monoxide to *cis*-$[Pt(C_6X_5)_2(THF)_2]$ (X = F, Cl) yields *cis*-$[Pt(C_6X_5)_2(CO)_2]$, for which the high ν(CO) values in their IR spectra are indicative of negligible Pt → CO π-backbonding.[357] Similarly, THF may be displaced by SMe_2, THT, or $OPPh_3$,[358] and tetrahydrothiophene may be displaced from $Bu_4N[Pt(C_6F_5)_3(THT)]$.[338,359]

Addition of $PCy_3{\cdot}CS_2$ to *cis*-$[Pt(C_6X_5)_2(THF)_2]$ (X = F, Cl) does not produce the expected tricyclohexylphosphine species, but a chelate complex is formed instead (Equation (51)).[360] The reaction of *cis*-$[Pt(C_6F_5)_2(THF)_2]$ with $Ph_3PCMe_2CS_2$ proceeds analogously.[361] Treatment of *cis*-$[Pt(C_6F_5)_2(THF)_2]$ with PhC≡CPh produces the bis(alkyne) complex,[362] whereas with *cis*-$[Pt(C{\equiv}CR)_2L_2]$ (R = Ph, Bu^t; L = PPh_3; L_2 = dppe, cod) complexes of type (**37**) are obtained.[363] With $[Pt(C_6F_5)R^2(C{\equiv}CR^1)_2]^{2-}$ ($R^2 = C_6F_5$, $C{\equiv}CR^1$; R^1 = Ph, Bu^t) alkynyl-bridged anions (**38**) are produced.[363] Some reactions of $[Pt(C_6F_5)_2L_2]$ (L = THF, THT) with dppm[364] or bis(diphenylphosphino)amine (dppa)[365] are shown in Scheme 16.

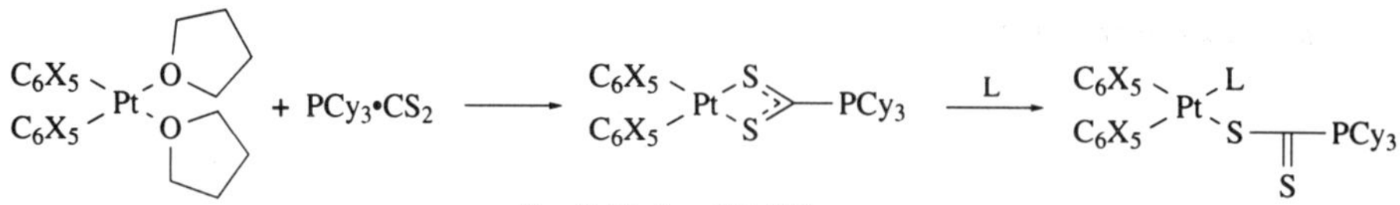

X = F, Cl; L = CO, PPh₃, py (51)

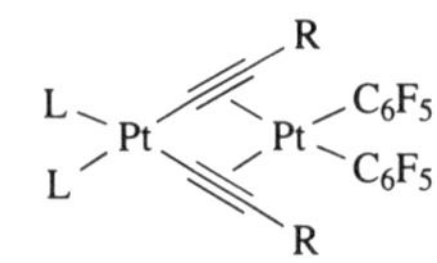

(37)

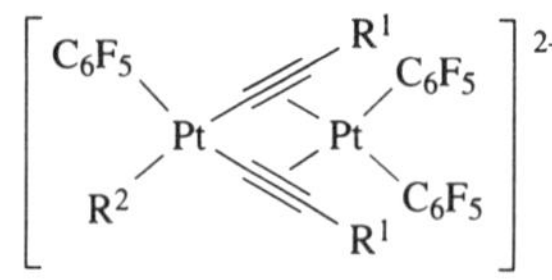

(38)

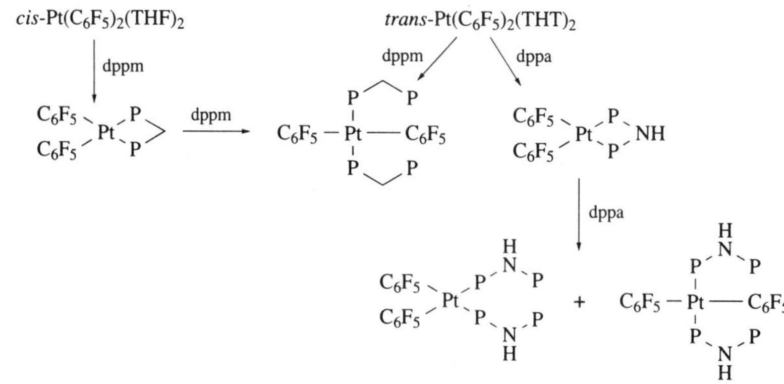

Scheme 16

The aryl-bridged anions $[Pt_2(\mu\text{-}C_6F_5)_2(C_6F_5)_4]^{2-}$ mentioned above may be prepared alternatively by reaction of *cis*-$[Pt(C_6F_5)_2(THF)_2]$ with $[Bu_4N]_2[Pt(C_6F_5)_4]$. This method also lends itself to the preparation of mixed palladium–platinum species.[366] The reactivity of these anions is illustrated in Scheme 17. Tri- and tetranuclear complexes have been obtained from the reactions of *cis*-$[Pt(C_6F_5)_2(THF)_2]$ with the $[PtX_4]^{2-}$ or $[Pt_2(\mu\text{-}X)_2X_4]^{2-}$ (X = Cl, Br) anions. Again, this strategy may be applied to the synthesis of mixed metal complexes (Scheme 18).[367]

Bridged compounds of the type $[Pt_2(\mu\text{-}L)_2(C_6X_5)_4]$ (X = F, Cl; L = SMe_2, THT, $SPPh_3$, 4-methoxypyridine *N*-oxide) may be generated by reaction of *cis*-$[Pt(C_6X_5)_2(THF)_2]$ with 1 equiv. of the appropriate ligand.[358] Reaction of *trans*-$[PtX_2(THT)_2]$ (X = Cl, Br, I, C_6F_5) with the $[Pt(C_6F_5)_3(acetone)]^-$ anion generates the triplatinum species *trans*-$[PtX_2\{(\mu\text{-}THT)Pt(C_6F_5)_3\}_2]$.[368] With $[PtCl_2(cod)]$, the THF complex *cis*-$[Pt(C_6X_5)_2(THF)_2]$ (X = F, Cl) gives $[(C_6X_5)_2Pt(\mu\text{-}Cl)_2Pt(cod)]$. This approach may also be employed to prepare palladium–platinum species, and a tetranuclear complex, $[\{(C_6F_5)_2Pt\}_2(\mu\text{-}Br)_4\{Pd(cod)\}_2]$, has been isolated and structurally characterized.[369]

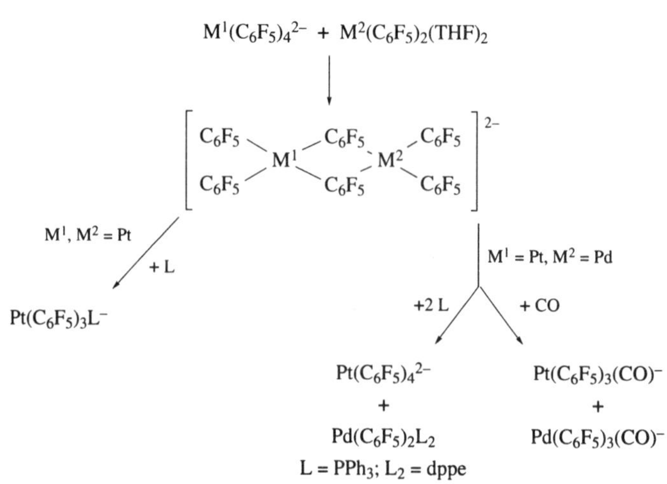

Scheme 17

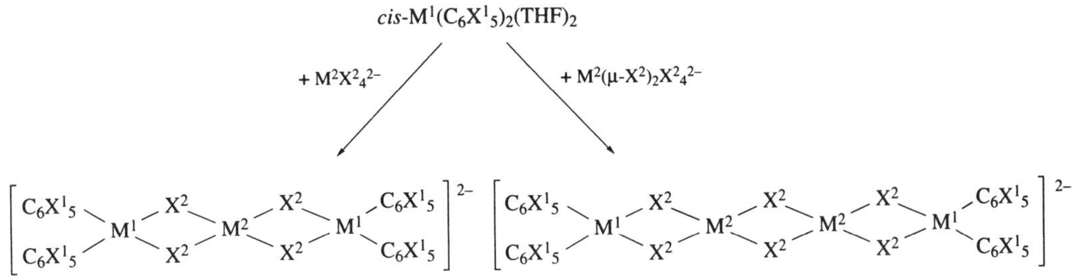

M^1, M^2 = Pt, Pd; X^1 = F, Cl; X^2 = Cl, Br

Scheme 18

(iv) Insertion reactions

The preparation of the fluoroalkyl complex [PtMe(CF_2CF_2OMe)(dppe)] by insertion of C_2F_4 into the Pt–O bond of the alkoxy precursor provides one example of insertion of a halogenated molecule,[370] but haloalkyls have been generated in significant numbers by formal carbene insertion, using diazomethane or a substituted analogue, into a platinum–halide bond. Treatment of [PtX_2(cod)] (X = Cl, I) with excess diazomethane produces the bis(chloromethyl)- or bis(iodomethyl)platinum complex,[371] from which the diene may be displaced readily (Scheme 19).[371,372] The analogous reaction of [$PtCl_2$(cod)] with $(MeO)_2P(O)CHN_2$ proceeds by sequential insertion of $(MeO)_2P(O)CH$ fragments to yield [Pt{CHClP(O)$(OMe)_2$}$_2$(cod)], which exists in solution as a mixture of diastereomers. The molecular structure of the *R,R*-form has been determined; spontaneous resolution of the diastereomers occurs on crystallization.[373] The complexes [PtClR(cod)] (R = Me, Et, Ph) react with CH_2N_2 to yield [Pt(CH_2Cl)R(cod)] which, on treatment with 10% $(CF_3)_2CHOH$ in $CDCl_3$ solution, rearrange to [PtCl(CH_2R)(cod)].[374] This represents a model for the polymerization of diazomethane, catalyzed by transition metal species.

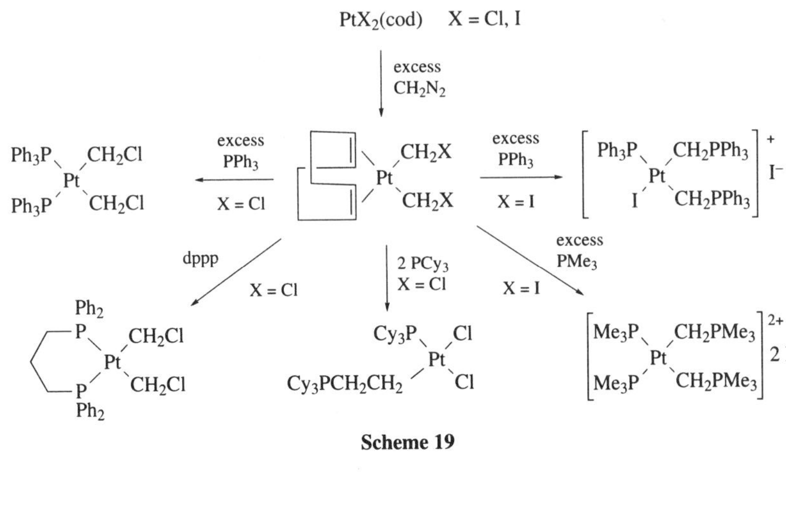

Scheme 19

The ambidentate ligand complexes [$PtCl_2$(CH_2=$CHCMe_2CH_2E$)] (E = NMe_2, SMe) and [$PtCl_2$(CH_2=$CHSiMe_2CH_2SMe$)] react with CH_2N_2 to insert CH_2 into one Pt–Cl bond.[375] Treatment of *trans*-[PtClH$(PPh_3)_2$] with CH_2N_2 yields *trans*-[Pt(CH_2Cl)H$(PPh_3)_2$], which rearranges to *trans*-[PtCl(Me)$(PPh_3)_2$]. It is observed that insertion occurs only when the Pt–Cl bond lies *trans* to a ligand of high *trans* influence.[376] With [PtX_2(dppm)] (X = Cl, Br, I), addition of CH_2N_2 generates [Pt$(CH_2X)_2$(dppm)], some reactions of which are shown in Scheme 20.[377] Reaction of [PtX_2(dppm)] or [$PtCl_2L_2$] (L = PPh_3; L_2 = cod, $Ph_2P(CH_2)_nPPh_2$, n = 1–4) with $EtOCOCHN_2$ produces a racemic mixture of enantiomers of [PtX(CHXCOOEt)L_2], but when L_2 = (*R,R*)-diop, (*S,S*)-chiraphos ((2*S*,3*S*)-bis(diphenylphosphino)butane) or (*S,S*)-skewphos ((2*S*,4*S*)-bis(diphenylphosphino)pentane) some diastereoselectivity is observed.[378,379]

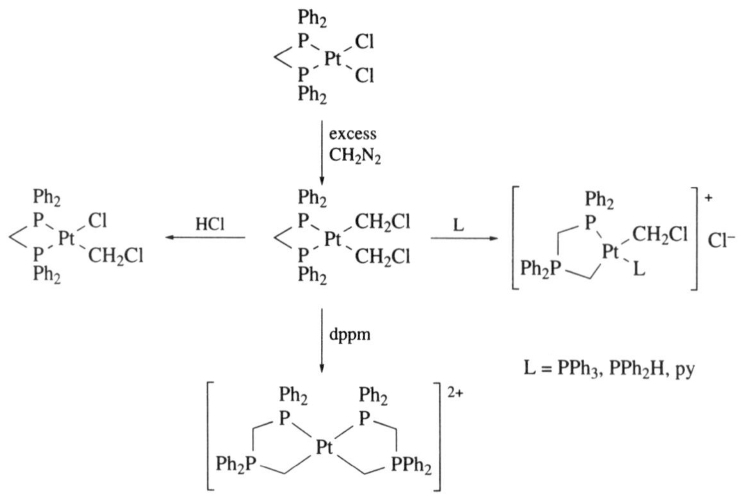

Scheme 20

8.3.2.2 Structure and properties

(i) X-ray diffraction studies

X-ray diffraction data for fluoroalkyl and chloromethyl complexes are given in Tables 5 and 6. For the trifluoromethyl complexes the Pt–C bond distances range from 0.201 nm (*trans* to H) to 0.219 nm (*trans* to P), and in the chloromethyl compounds the Pt–C distances span the range 0.202 nm (*trans* to N) to 0.211 nm (*trans* to P). There are very few examples where direct comparisons can be made between the bond distances in these and their alkyl analogues. In *trans*-[$PtClR(PPh_3)_2$] (R = Me, CF_3)[155,383] the Pt–C bond lengths are equal, whereas in [$Pt(Me)(CF_2CF_2Me)(dppe)$] (Table 1) the Pt–CF_2 distance is 0.003 nm shorter.[162] In the acyl complexes, *trans*-[$PtCl(COR)(PMePh_2)_2$] (R = Me, CF_3), the Pt–$COCF_3$ bond is 0.007 nm shorter, and presumably stronger, but it exerts a weaker *trans* influence on the Pt–Cl bond.[381]

Solid-state structural data for haloarylplatinum(II) complexes are given in Tables 7 and 8. In neutral complexes the Pt–C distances lie between 0.201 nm (*trans* to N) and 0.210 nm (*trans* to P), whereas the anionic complexes cover a wider range (0.196 nm to 0.212 nm). Again, there are few opportunities to compare directly bond lengths in aryl and haloaryl compounds.

(ii) Stability of Pt–C bonds

As mentioned previously, perhaloaryl complexes often exhibit thermodynamic stability beyond that of their aryl counterparts. Trifluoromethyl complexes, too, contain strong platinum–carbon bonds. The high stability of fluorocarbon compounds, which is probably largely kinetic in origin, has been attributed variously to the ionic nature of the Pt–C bond, the high electronegativity of the fluorocarbon group which results in contraction of the metal *d* orbitals, and hence decreased repulsions, and better Pt–C σ-overlap, and the possibility of π-back-donation from the filled platinum *d*-orbitals into the σ*-orbitals of the fluorocarbon.[2]

The kinetic stability of trifluoromethyl complexes is illustrated by the relative ease of displacement of the diene from [PtR_2(diene)]. Norbornadiene is more readily displaced than cyclooctadiene, and it is more easily displaced from [$Pt(CF_3)_2$(nbd)] than from [$Pt(Me)_2$(nbd)].[75] The stabilities of the chloromethyl complexes [$Pt(CH_2Cl)_2L_2$] and [$PtCl(CH_2Cl)L_2$] in $CDCl_3$ solution depend strongly on the nature of L. With cyclooctadiene or chelating arylphosphines the complexes are stable indefinitely, whereas the triarylphosphine complexes decompose slowly, and the monodentate or bidentate alkylphosphine compounds decompose rapidly.[391]

Table 5 X-ray diffraction data for fluoroalkylplatinum(II) complexes.

Complex	*Pt–C* (nm)	*Other bond lengths* (nm)		*Ref.*
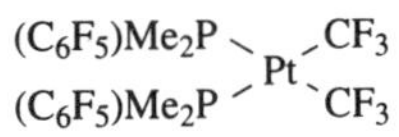	0.2058(4)	Pt–P	0.2326(1)	380
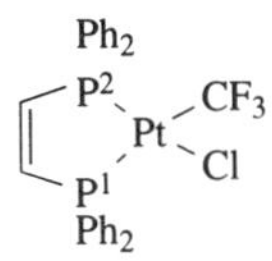	0.2188(8)	Pt–P Pt–P[1] Pt–P[2]	0.2346(4) 0.2275(4) 0.2236(4)	381
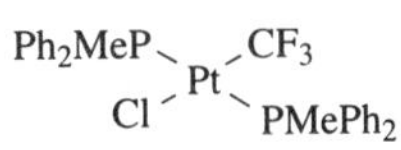	[0.209][a]	Pt–Cl Pt–P	[0.242][a] 0.2320(2)	382
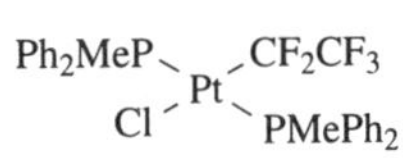	Molecule 1 0.2013(12)	Pt–Cl Pt–P	0.2361(3) 0.2329(3) 0.2341(3)	382
	Molecule 2 0.1990(12)	Pt–Cl Pt–P	0.2365(4) 0.2326(3) 0.2340(3)	
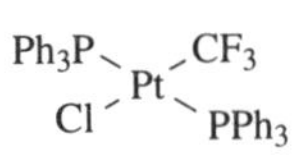	0.2080[b]	Pt–Cl Pt–P	0.2400[b] 0.2328(2)	383
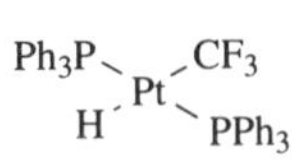	0.2009(8)	Pt–H Pt–P	0.172(9) 0.2273(3) 0.2274(2)	384

[a] Calculated, positions unrefined. [b] Disordered.

(iii) Spectroscopic studies

Photoelectron spectra of *cis*-$[Pt(CF_3)_2L_2]$ complexes have been measured (L = PEt_3, $AsMe_3$; L_2 = cod, TMEDA). These reveal that the highest occupied molecular orbitals are associated with the M–L σ-bonds, and the next four are platinum 5*d* orbitals,[392] in agreement with the situation found for the corresponding methylplatinum compounds. In contrast to Pt–Me σ-bonds, however, the Pt–CF_3 σ-bonds have significant C 2*s* and C–F antibonding character.[392]

The trifluoromethyl group is found to exhibit a high NMR *trans* influence.[393] For the complexes *trans*-$[PtHR(PPh_3)_2]$, the *trans* influence of R, as measured by ^1J(PtH), decreases in the order

$$CF_3 > C_6H_5 > C_6H_9 > (CH_2)_2CN > (CH_2)_3CN > Me > CH_2CN > Cl$$

In contrast, infrared measurements of ν(PtH) provide a different result:

$$C_6H_9 > (CH_2)_3CN > C_6H_5, Me > CF_3 > Cl$$

Determination of the NMR *cis* influence by measuring ^1J(PtP) provides a moderate value for CF_3:

$$Cl > CH_2CN > CF_3, Me, C_6H_5, (CH_2)_2CN > (CH_2)_3CN$$

8.3.2.3 Reactions

(i) Reactions with other metal complexes

Reactions of platinum(II)–haloaryl complexes with silver(I) reagents have been alluded to briefly in Section 8.3.2.1 as a means of preparing new haloaryl species. Treatment of the neutral complexes *trans*-

Table 6 X-ray diffraction data for chloromethylplatinum(II) complexes.

Complex	*Pt–C* (nm)	*Other bond lengths* (nm)	*Ref.*
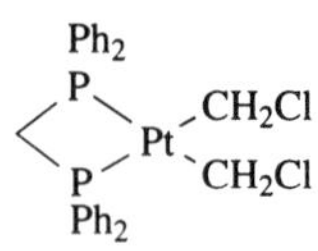	0.2069(8) 0.2082(9)	Pt–P 0.2286(3) 0.2296(2)	377
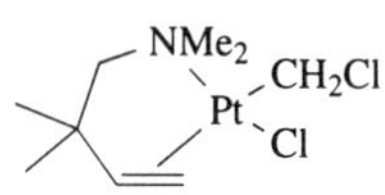	0.2057(13)	Pt–Cl 0.2301(3) Pt–N 0.2086(9)	375
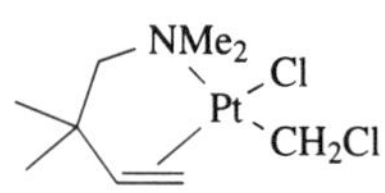	Molecule 1 0.2016(11) Molecule 2 0.203(2)	Pt–Cl 0.2310(3) Pt–N 0.2176(8) Pt–Cl 0.2318(5) Pt–N 0.218(1)	375
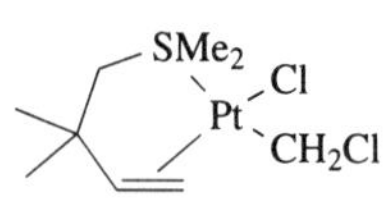	0.2032(7)	Pt–Cl 0.2314(2) Pt–S 0.2375(2)	375
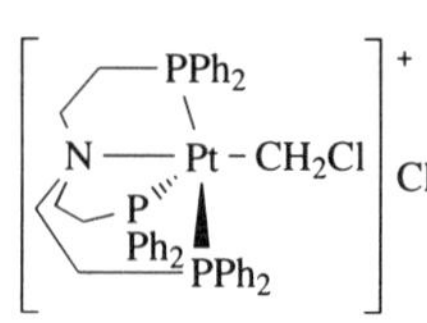	0.206(4)	Pt–N 0.218(2) Pt–P 0.2348(9) 0.2399(9) 0.2343(9)	335
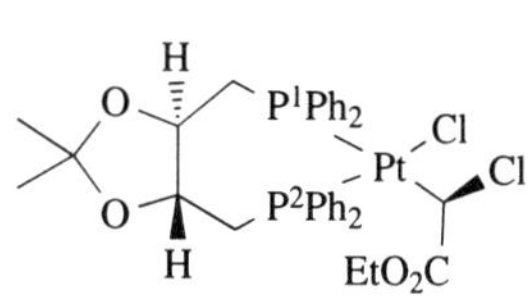	0.2114(13)	Pt–Cl 0.2362(3) Pt–P^1 0.2320(3) Pt–P^2 0.2233(3)	379
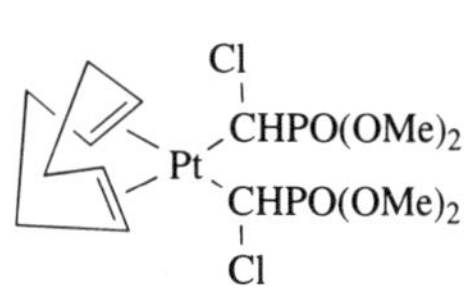	0.2088(11)		373

[PtH(C_6X_5)L^1_2] (X = F, Cl; L^1 = PEt_3, PMe_3, PPr^i_3) with [ML^2(solvent)]$^+$ (M = Ag, Au; L^2 = PPh_3, PEt_3, PPr^i_3, PCy_3) yields species of the type *trans*-[PtRL^1_2(μ-H)ML^2]$^+$.[394,395] In *trans*-[Pt(C_6F_5)(PEt_3)$_2$(μ-H)Au(PPh_3)]$^+$ the platinum–gold distance is 0.2714(1) nm, indicating the presence of a Pt–Au bond.[395] Further elaboration of [Pt(C_6X_5)(PEt_3)$_2$(μ-H)Ag(PPh_3)]$^+$ with *trans*-[PtH(C_6X_5)(PEt_3)$_2$] generates the trimetallic complex [Ag{(μ-H)Pt(C_6X_5)(PEt_3)$_2$}$_2$]$^+$.[396] In other cases, the silver salt acts simply to remove coordinated halide[397] or, in the case of [Ag(C≡CR)]$_n$, to transfer an organic group to platinum.[398] When *trans*-[PtH(C_6Cl_5)L_2] (L = PEt_3, PMe_3) is treated with silver triflate in the presence of RHgI at −80 °C, a hydride-bridged platinum–mercury compound is obtained.[399]

The reactions of the anionic platinum complexes Bu_4N[Pt(C_6F_5)$_n$(C_6Cl_5)$_{3-n}$L] (*n* = 0–3; L = PPh_3, py, THT) with [Ag(ClO_4)(PPh_3)] produce adducts, which are formulated as containing Pt → Ag dative bonds.[400,401] With *trans*-[PtBr_2(C_6Cl_5)(PPh_3)]$^-$, a tetrametallic complex, [(Ph_3P)Ag(μ-Br)PtBr(C_6Cl_5)(PPh_3)]$_2$, containing weak and strong silver–bromine interactions, is isolated.[402] The platinum dianion *trans*-[PtCl_2(C_6Cl_5)$_2$]$^{2-}$ reacts with silver nitrate to produce an oligomeric complex, [(C_6Cl_5)$_2$Pt(μ-Cl)$_2$Ag]$_n^{n-}$, composed of PtCl_2(C_6Cl_5)$_2$ units linked by silver atoms. The oligomer may be cleaved by a range of nucleophiles.[403] There is no suitable bridging group in the case of [Pt(C_6X_5)$_4$]$^{2-}$ (X = F, Cl), so with [Ag(ClO_4)L] a dative-bonded platinum–silver complex is obtained.[401] Reaction of lead(II) nitrate with [Bu_4N]$_2$[Pt(C_6F_5)$_4$] gives [Bu_4N]$_2$[Pb{Pt(C_6F_5)$_4$}$_2$], containing a linear Pt–Pb–Pt unit.[404]

Table 7 X-ray diffraction data for monomeric pentahalophenylplatinum(II) complexes.

Complex		*Pt–C* (nm)	*Other bond lengths* (nm)	*Ref.*
$[Bu_4N]_2[Pt(C_6Cl_5)_4]$		0.2086(14)		386
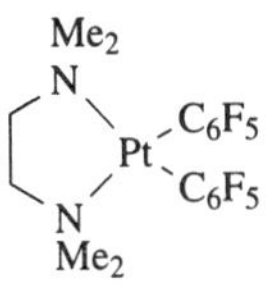		0.2006(6)	Pt–N 0.2141(5)	387
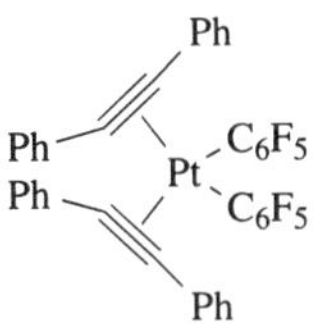		0.2048(3)		362
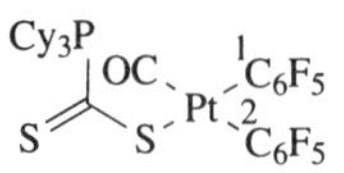	Pt–C[1] Pt–C[2]	0.2056(6) 0.2045(7)	Pt–CO 0.1910(9) Pt–S 0.2348(2)	360
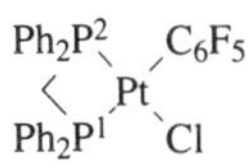		0.2078(4)	Pt–Cl 0.2360(1) Pt–P[1] 0.2295(1) Pt–P[2] 0.2224(1)	388
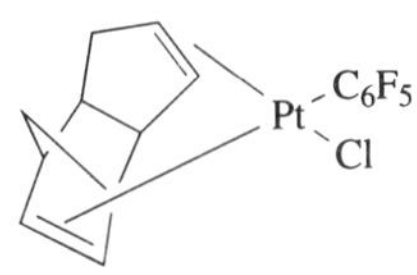		0.202(1)	Pt–Cl 0.2307(4)	342

Reaction of the dianion $[Pt_2(\mu\text{-}X)_2(C_6F_5)_4]^{2-}$ (X = Cl, Br) with silver perchlorate in diethyl ether, or with $[Ag(ClO_4)L]$ at −30 °C, produces $[Pt_2(\mu\text{-}AgL)(\mu\text{-}X)_2(C_6F_5)_4]^-$ (L = OEt_2, PPh_3, $PMePh_2$).[405] In contrast, the analogous reaction of the pentachlorophenyl complex produces a hexanuclear species, $[Bu_4N]_2[\{Pt(C_6Cl_5)_2\}_4(\mu\text{-}Cl)_4(\mu\text{-}Ag)_2]$.[406] Reaction of $[Bu_4N]_2[Pt_2(\mu\text{-}Cl)_2(C_6F_5)_4]$ with tin(II) chloride and $AgClO_4$ produces the trigonal pyramidal species $[Pt_3(\mu\text{-}Cl)_3(\mu_3\text{-}Sn)(C_6F_5)_6]$, which is formulated as containing three $Pt^{II} \rightarrow Sn^{II}$ dative bonds.[407] The trimeric platinum complex $[PtX_2\{(\mu\text{-}THT)Pt(C_6F_5)_3\}_2]$ reacts with $AgClO_4$ in CH_2Cl_2–toluene solution to yield $[PtX_2\{(\mu\text{-}THT)(C_6F_5)_3PtAg(\eta^2\text{-}C_6H_5Me)\}_2]$.[408]

(ii) Catalytic reactions of trifluoromethyl compounds

Cationic trifluoromethylplatinum(II) complexes have been found to catalyze the oxidation of a variety of unsaturated organic compounds, using H_2O_2 or Bu^tOOH as oxidant. Reaction of *trans*-$[Pt(CF_3)(OH)L_2]$ (L = PPh_3, $PMePh_2$, PMe_2Ph) with Bu^tOOH generates *trans*-$[Pt(CF_3)(OOBu^t)L_2]$,[409] which is active in the oxidation of 1-octene to 2-octanone.[410] Diphosphine complexes are found to be more active in the ketonization or epoxidation of alkenes,[411,412] and those which form rigid, five-membered chelate rings, such as diphoe, have been found to be most effective.[413] The kinetics of the epoxidation of 1-octene using $[Pt(CF_3)(OH)(diphoe)]$ or $[Pt(CF_3)(CH_2Cl_2)(diphoe)]^+$ and H_2O_2 suggest that the actual oxidant is a PtOOH species, which interacts with a platinum–alkene complex in the oxygen-transfer step.[414] Polybutadienes may also be converted into polyketones using $[Pt(CF_3)(CH_2Cl_2)(diphoe)]BF_4$ and Bu^tOOH.[415] Related complexes containing chiral diphosphine ligands, such as chiraphos or (*R*)-1,2-bis(diphenylphosphino)propane (prophos), epoxidize 1-octene in the presence of H_2O_2 to give products with modest enantiomeric excesses of up to 41%.[416]

The $[Pt(CF_3)(CH_2Cl_2)(dppe)]ClO_4$–$H_2O_2$ system also catalyzes the oxidation of secondary alcohols to ketones,[417] and $[Pt(CF_3)XL_2]$ (X = OH, OPh, $CH_2Cl_2^+$; L_2 = dppe, diphoe, 1,2-$(Ph_2P)_2C_6H_4$) is effective in promoting hydroxylation of aromatic compounds with H_2O_2.[418] In the latter, the reactivity increases

Table 8 X-ray diffraction data for dimeric and trimeric halophenylplatinum(II) complexes.

Complex	*Pt–C* (nm)	*Other bond lengths* (nm)		*Ref.*
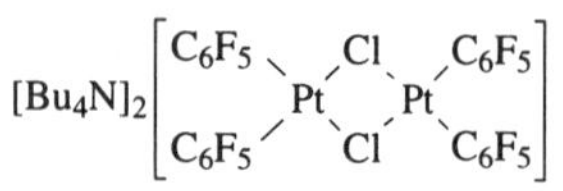	0.1977(10) 0.1991(10)	Pt–Cl	0.2394(3) 0.2387(3)	389
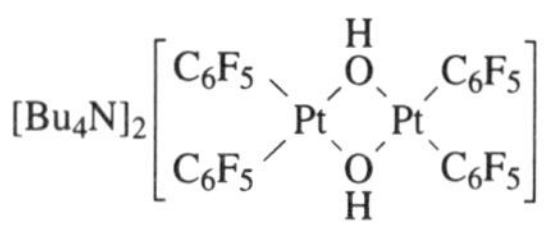	0.197(1) 0.197(2)	Pt–O	0.207(2)	353
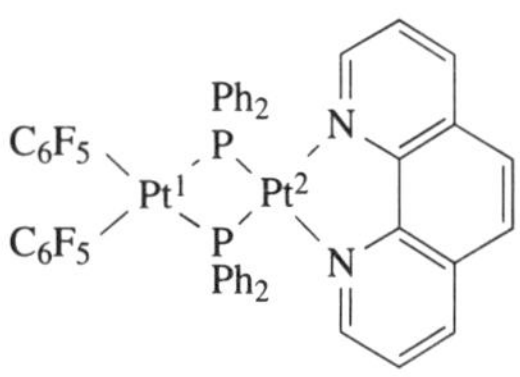	0.2099(6) 0.2069(5)	Pt¹–P Pt²–P Pt–N	0.2294(1) 0.2288(1) 0.2269(1) 0.2279(1) 0.2127(5) 0.2135(5)	348
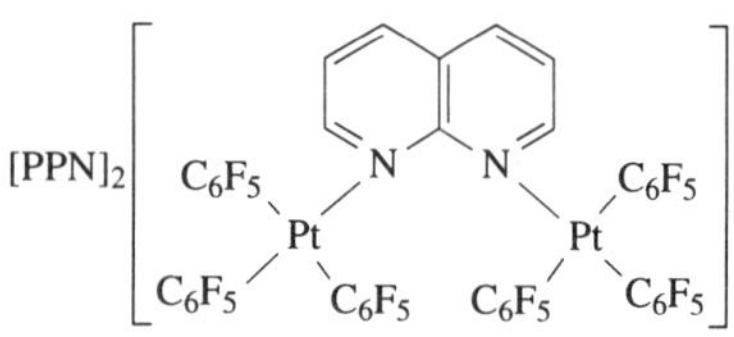	0.2117(7) 0.2107(6) 0.2039(4)	Pt–N	0.2115(7)	351
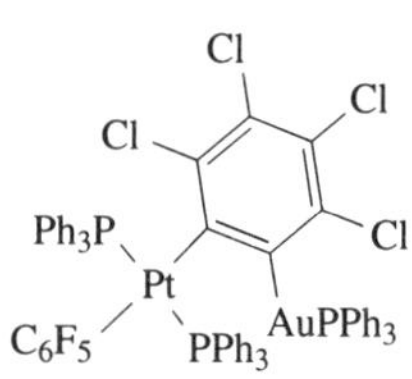	0.2013(8)	Pt–Cl Pt–P	0.2406(2) 0.2310(3) 0.2330(3)	390
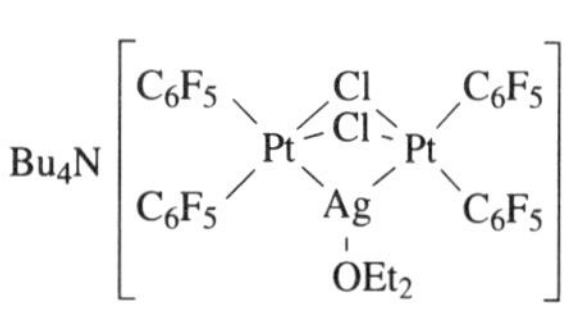	0.1960(13) 0.1987(13) 0.1975(13) 0.1956(15)	Pt–Cl Pt–Ag Pt–Pt	0.2401(3) 0.2406(3) 0.2404(3) 0.2414(3) 0.2782(1) 0.2759(1) 0.3261(1)	389
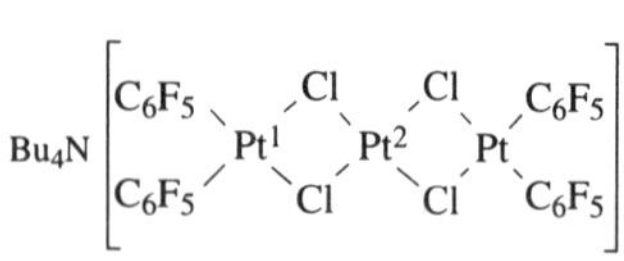	0.2017(5) 0.2017(5)	Pt¹–Cl Pt²–Cl	0.2441(2) 0.2430(2) 0.2326(3) 0.2313(2)	367
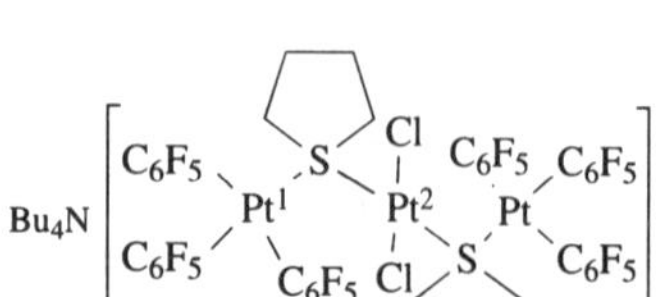	0.2089(10) 0.2097(9) 0.2023(9)	Pt¹–S Pt²–S Pt–Cl	0.2303(4) 0.2295(5) 0.2285(5)	368

with the presence of electron-withdrawing groups on the aromatic ring, and the reaction is proposed to occur by electrophilic metallation of the aromatic ring to give an arylplatinum species, followed by oxygen transfer from a PtOOH moiety.[418]

(iii) Other reactions

Haloarylplatinum complexes exhibit considerable thermal stability but, like their aryl counterparts, are susceptible to cleavage by acids. Treatment of $[Bu_4N]_2[Pt(C_6F_5)_4]$ with 1 equiv. of HCl yields the $[PtCl(C_6F_5)_3]^{2-}$ anion, whereas when the reaction is carried out in a 1:2 ratio the dimeric species $[Pt_2(\mu\text{-}Cl)_2(C_6F_5)_4]^{2-}$ is formed.[419] The chloride in $[PtCl(C_6F_5)_3]^{2-}$ may be substituted by reaction with KX (X = Br, I, CN, SCN), or by a neutral ligand.[419] The pentachlorophenyl complex $[Bu_4N]_2[Pt(C_6Cl_5)_4]$ behaves similarly. Treatment with HCl in the presence of tetrahydrothiophene produces $Bu_4N[Pt(C_6Cl_5)_3(THT)]$. $[Bu_4N]_2[PtCl_2(C_6Cl_5)_2]$ also reacts with HCl by cleavage of a Pt–C_6Cl_5 bond.[420] Reaction of *trans*-$[PtX(CF_3)(PPh_3)_2]$ (X = H, Cl) with HBF_4–Et_2O and an alcohol or thiol, however, generates a carbene complex.[421]

The bis(trifluoromethyl)platinum(II) complexes, *cis*-$[Pt(CF_3)_2L_2]$ (L = py; L_2 = bipy, TMEDA), undergo oxidative addition of Cl_2 or Br_2 in dry CH_2Cl_2 to give the product of *trans* addition, but this isomerizes slowly (Equation (52)). Oxidative addition of MeI, $PhCH_2Br$, or CF_3I occurs also, only one isomer being formed with CF_3I. Mixtures of isomers are observed in the other cases, the geometry being dependent on the nitrogen ligands.[422]

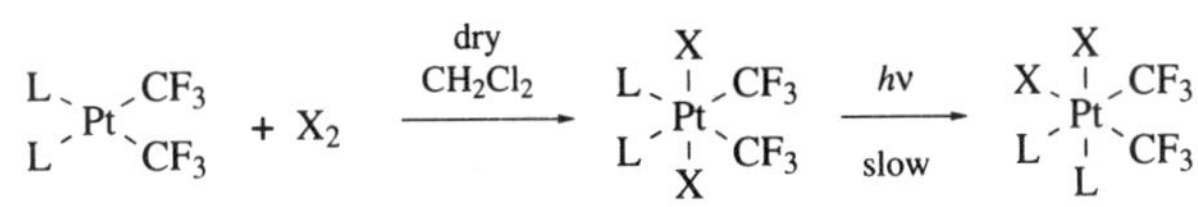

(52)

X = Cl, Br; L = py or L_2 = bipy, TMEDA

The chloromethyl or iodomethyl complexes *trans*-$[PtX^1(CH_2X^2)(PPh_3)_2]$ (X^1 = Cl, Br, I; X^2 = Cl, I) react with $AgBF_4$ in the presence of a tertiary phosphine according to Equation (53). This is believed to occur by migration of X^2 from carbon to platinum to leave a Pt=CH_2 species, which undergoes electrophilic attack on one of the phenyl rings of a PPh_3 ligand to form a five-membered ring.[423] Treatment of $[Pt(CH_2I)_2(cod)]$ with silver trifluoroacetate and methanol produces the first example of a bis(hydroxymethyl)platinum complex. This undergoes intramolecular cyclization to form a 3-oxaplatinacyclobutane (Scheme 21).[424]

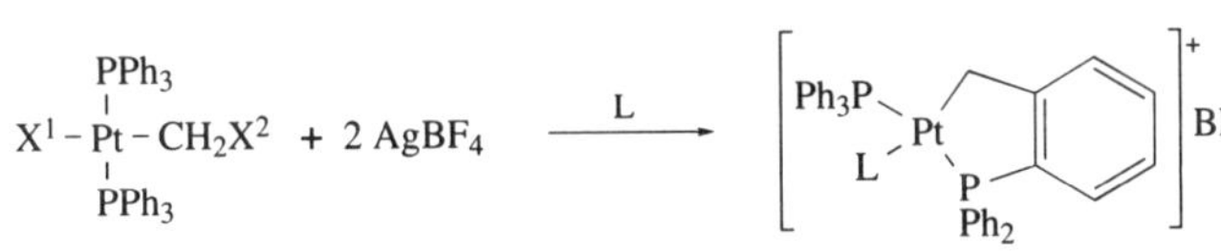
(53)

X^1 = Cl, Br,I; X^2 = Cl, I; L = PPh_3, $P(C_6H_4Me\text{-}4)_3$

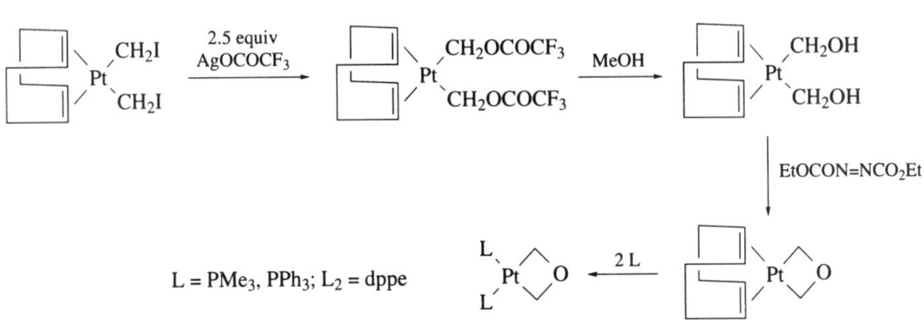

Scheme 21

8.3.3 Platinacycloalkanes

Platinacyclobutane complexes are of interest owing to their role as models for alkene metathesis and transition metal-catalyzed rearrangements of small organic rings. Synthetic routes to such complexes, particularly those of platinum(IV), their structures, and reactions have been reviewed previously.[2]

8.3.3.1 Preparation

Platinacycloalkanes and related compounds have been prepared by a number of methods. These include the action of diorganolithium or di-Grignard reagents on platinum(II) halides. Treatment of $[PtCl_2(dppm)]$ with $Li(CH_2)_4Li$ yields a platinacyclopentane derivative,[425] and reaction of $[PtCl_2(cod)]$ with $Me_2C(CH_2MgBr)_2$ proceeds according to Equation (54).[426] Similarly, addition of $CR^1R^2(CH_2CH_2MgBr)_2$ (R^1, R^2 = H, Me) followed by 2 equiv. of PPh_3 or PMe_2Ph generates the platinacyclohexanes, $[\overline{Pt(CH_2CH_2CR^1R^2CH_2C}H_2)L_2]$.[427] Reaction of $[PtI_2(cod)]$ with a xylyl Grignard reagent produces a five-membered ring (Equation (55)).[428] Complex (**39**) (L = PEt_3), prepared according to Equation (54), may be obtained alternatively by thermal decomposition of the bis(neopentyl) complex *cis*-$[Pt(CH_2Bu^t)_2(PEt_3)_2]$.[225]

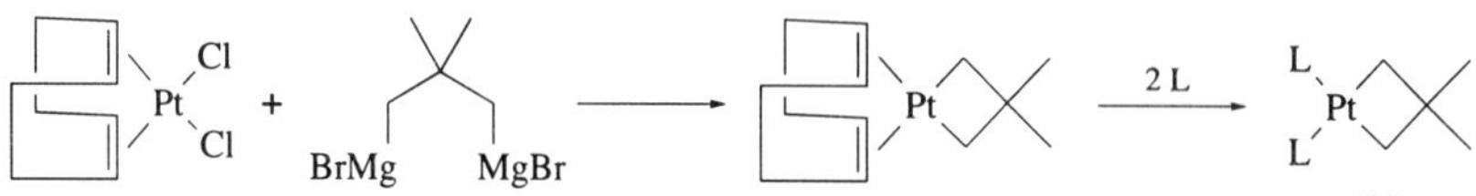

(54)

(**39**)

L = PEt_3, PMe_3, PPh_3, PBu_3

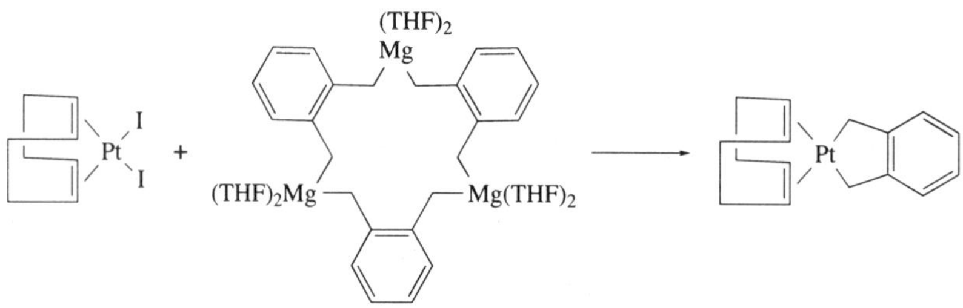

(55)

Other, less general, routes are also available. Treatment of *cis*-$[PtCl\{CH(TMS)_2\}(PMe_2Ph)_2]$ with Bu^tLi in toluene solution proceeds by deprotonation of one of the silylmethyl groups, followed by displacement of chloride and ring closure to give $[\overline{Pt(CH_2SiMe_2C}H(TMS))(PMe_2Ph)_2]$.[429] When $[Pt(CH_2Cl)_2(dppe)]$ is treated with excess Pr^iNH_2 in chlorobenzene solution, a 3-platinaazetidine is generated, whereas with $MeOH–Me_2NCH(OMe)_2$ an azetidinium salt is obtained (Scheme 22).[430]

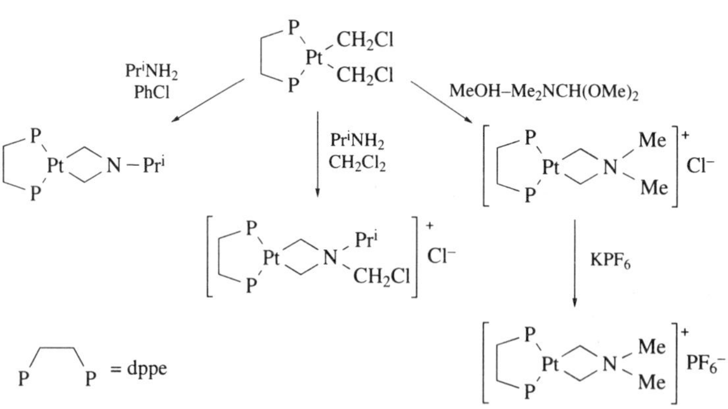

Scheme 22

Reaction of $[Pt(cod)_2]$ with 2,4-pentadienoic acid methyl ester, followed by PMe_3, generates the platinacyclopentane complex (**40**),[431] whereas alkene and allyl complexes are formed with $[Ni(cod)_2]$. Treatment of $[Pt(dba)_2]$ (dba = dibenzylideneacetone) with 1,1-dimethylallene and a diimine ligand results in a species of type (**41**).[432] Reaction of $[Pt(\eta\text{-}C_3H_5)L_2]OAc$ (L = PPh_3, PCy_3; L_2 = dppb) with the activated alkene $Me_2C{=}C(OMe)(O\text{-}TMS)$ proceeds by attack at the central carbon of the allyl group to generate a platinacyclobutane (Equation (56)), and $[PtCl(\eta\text{-}C_3H_5)]_4–PCy_3$ catalyzes the reaction shown in Equation (57).[433]

Aromatic platinacyclic compounds have also been prepared. The reaction of $[PtCl_2(cod)]$ with 1,2-dihydro-1-magnesacyclobutabenzene proceeds according to Equation (58).[434] Complexes of the biphenyl dianion have been generated by interaction of dichloroplatinum complexes with $Li_2C_{12}H_8$,[435] and substituted analogues have also been prepared.[436] Reaction of *cis*-$[PtCl_2L_2]$ with 2,2'-dilithiodiphenyl-

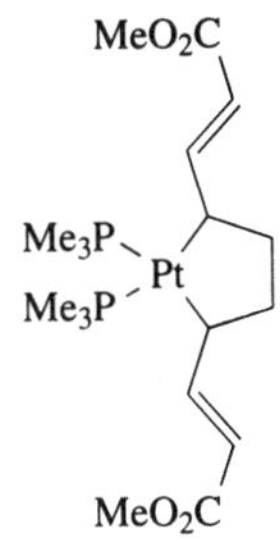

(40)

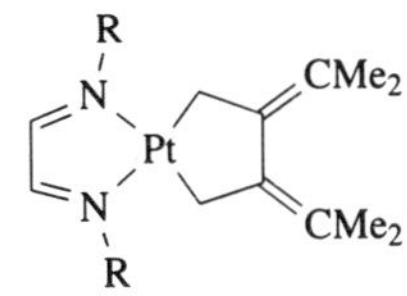

$R = 2,6\text{-}Me_2C_6H_3$, $2,6\text{-}Bu^t{}_2C_6H_3$

(41)

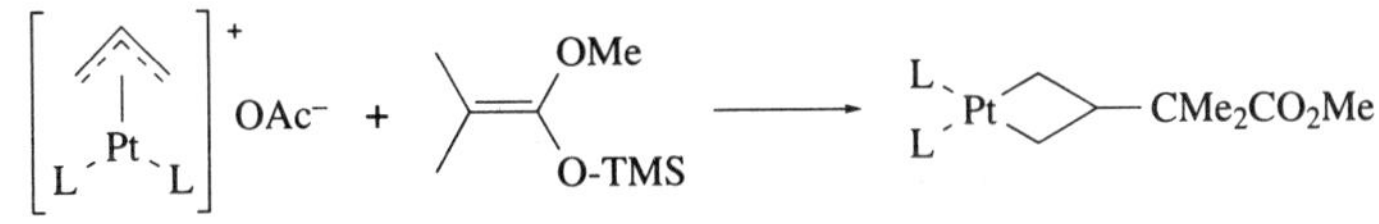

$L = PPh_3$, PCy_3; $L_2 =$ dppb

(56)

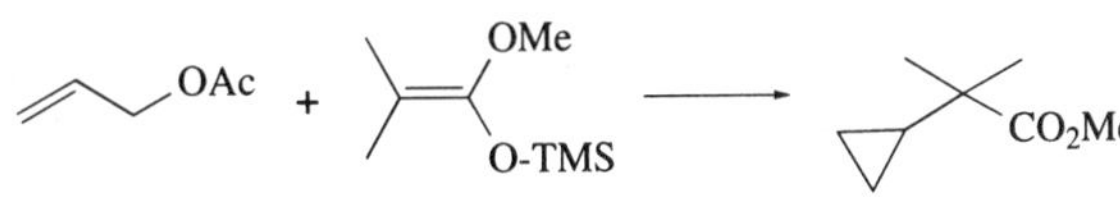

(57)

methane produces six-membered platinacycles (Equation (59)). The platinacycle adopts a boat conformation in the solid state ($L = PEt_3$), and each is rigid on the NMR timescale up to a temperature of 100 °C.[437] Platinacycles have also been obtained by reactions of platinum(0) complexes with compounds containing strained rings. Treatment of $[Pt(PPh_3)_3]$ with cycloheptatetraene yields (**42**),[438] and reactions with cyclopropa[*b*]naphthalenes produce (**43**)[439] and (**44**).[440]

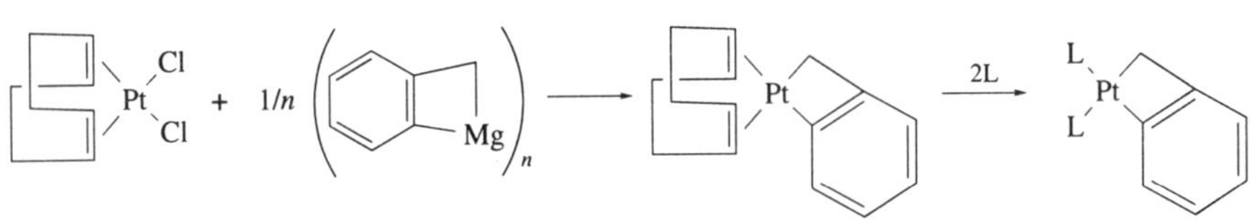

$L = PMe_3$, PEt_3, PBu_3, PMe_2Ph, $PMePh_2$

(58)

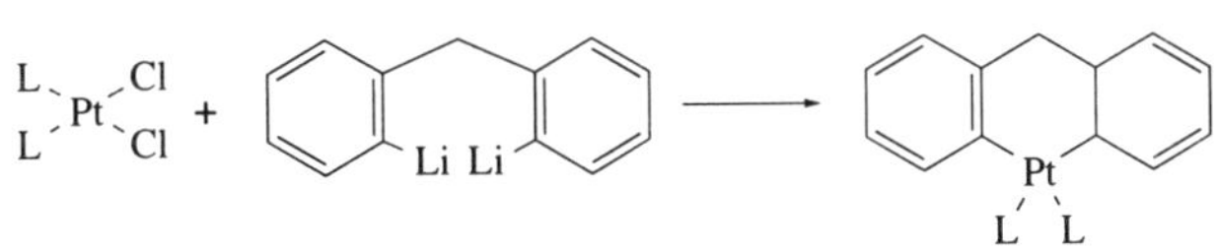

$L = PMe_3$, PEt_3, PMe_2Ph, $PMePh_2$, PPh_3; $L_2 =$ dppm, dppe, cod

(59)

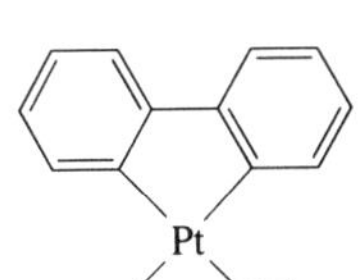

(42)

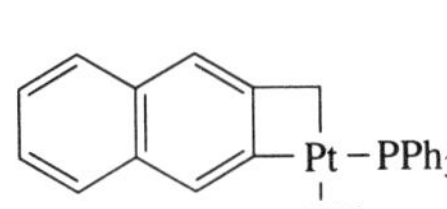

(43)

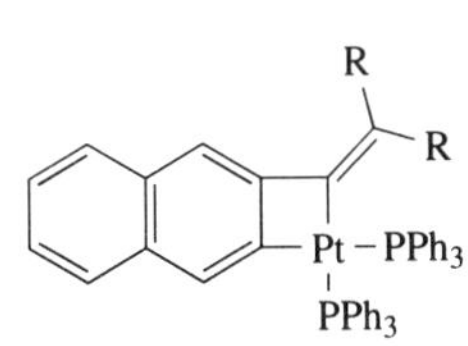

$R = Ph$, $C_6H_4Me\text{-}4$

(44)

A series of 2-platinacyclobutanones were prepared by Kemmitt and co-workers. Reaction of $[Pt(CO_3)L_2]$ ($L = PPh_3$, $AsPh_3$, $PMePh_2$, PMe_2Ph; $L_2 =$ dppe) or $[Pt(PPh_3)_4]$ with RCH_2COCH_2R ($R = CO_2Me$, CO_2Et, CO_2Pr) yields $[\overline{Pt(CHRCOCHR})L_2]$ (**45**), and the same product ($R = CO_2Me$, $L = PPh_3$) is obtained starting from $[Pt(PhCH{=}CHPh)(PPh_3)_2]$, $[Pt(CO_4)(PPh_3)_2]$, or $[Pt(OCOPh)_2(PPh_3)_2]$.[441] The carbonate complexes $[Pt(CO_3)L_2]$ ($L = PPh_3$, $AsPh_3$) also react with $(MeCOCH_2)_2CO$ in warm ethanol to give $[\overline{Pt\{CH(COMe)COCH}(COMe)\}L_2]$.[442] Reaction of *cis*-$[PtCl_2L_2]$ with $K_2[PhCHCOCHPh]$ takes place according to Equation (60), the products having a puckered four-membered ring.[443] $[Pt(PhCH{=}CHPh)(PPh_3)_2]$ catalyzes the isomerization of $(TMS)CH_2COCH_2Cl$ to $CH_2{=}C(O\text{-}TMS)CH_2Cl$, which reacts with the platinum(0) complex to generate the platinacyclobutanone, $[\overline{Pt(CH_2COCH_2})(PPh_3)_2]$.[444]

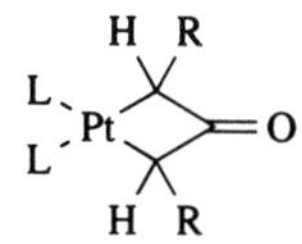

L = PPh_3, $AsPh_3$, $PMePh_2$, PMe_2Ph or L_2 = dppe
R = CO_2Me, CO_2Et, CO_2Pr

(45)

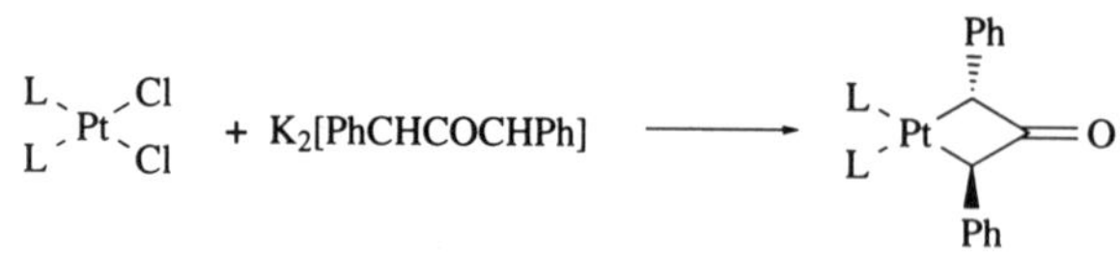

L = PPh_3. $AsPh_3$; L_2 = dppe, cod

(60)

Related platinacyclobutane sulfones and sulfoxides have been synthesized by the action of $Li_2[PhCHSO_2CHPh]$ on *cis*-$[PtCl_2L_2]$ (L = PPh_3, $AsPh_3$, SEt_2),[445] or by the reaction of *cis*-$[PtCl_2(PPh_3)_2]$ with $(RCH_2)_2SO_n$ (n = 1, R = COPh; n = 2, R = COPh, CO_2Me) in the presence of silver(I) oxide.[446] Phosphine oxide compounds have also been produced by the reaction shown in Equation (61).[447]

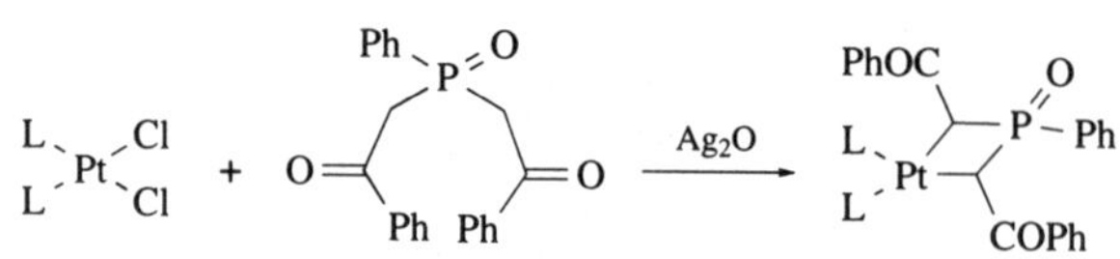

L = PPh_3. $PMePh_2$; L_2 = dppe, cod

(61)

Platinacyclopentadienes have been prepared by reaction of $[Pt_3(CNBu^t)_6]$ with activated alkynes,[448] and other platinacycloalkenes have been produced by opening of small, strained rings. Reactions of a platinum(0) precursor with substituted cyclopropenes are depicted in Scheme 23.[449–52] The reaction of $[Pt(PhCH{=}CHPh)(PPh_3)_2]$ with benzocyclobutenedione is also shown in Scheme 23.[450] Treatment of $Hg[Pt_3(CNC_6H_3Me_2\text{-}2,6)_6]$ with diphenylbutanedione generates a platinacyclopentanedione derivative.[453]

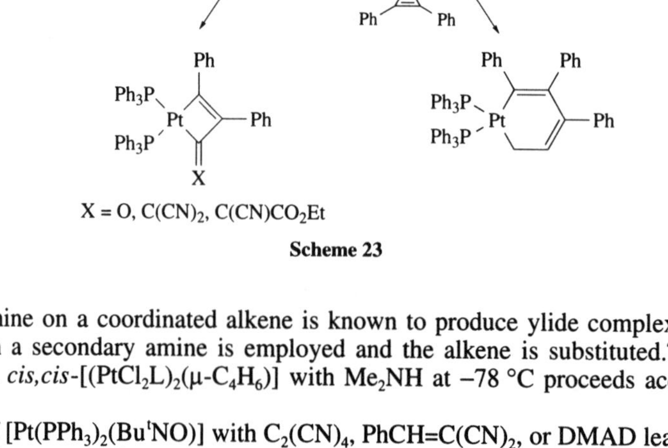

X = O, $C(CN)_2$, $C(CN)CO_2Et$

Scheme 23

Attack of an amine on a coordinated alkene is known to produce ylide complexes. Cyclic products are preferred when a secondary amine is employed and the alkene is substituted.[454] Treatment of the butadiene complex *cis,cis*-$[(PtCl_2L)_2(\mu\text{-}C_4H_6)]$ with Me_2NH at −78 °C proceeds according to Equation (62).[455]

The reactions of $[Pt(PPh_3)_2(Bu^tNO)]$ with $C_2(CN)_4$, $PhCH{=}C(CN)_2$, or DMAD lead to cyclic products (**46**) and (**47**).[456] Treatment of *cis*-$[PtCl_2L_2]$ (L = PPh_3, $PMePh_2$, PMe_2Ph, $PBzPh_2$; L_2 = dppm, dppe, dppp, dppb) with α-acetamidocinnamic acid or 2-acetamidoacrylic acid and silver(I) oxide results in five-membered ring complexes.[457]

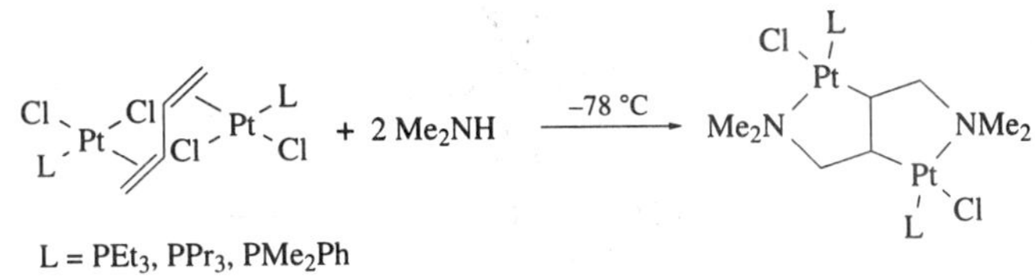 (62)

L = PEt_3, PPr_3, PMe_2Ph

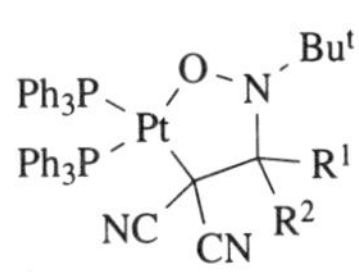

(46)

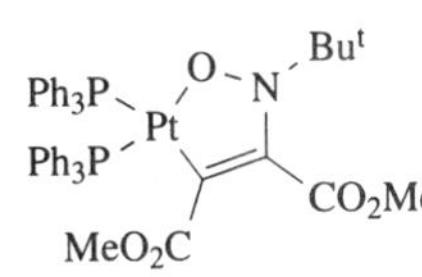

(47)

8.3.3.2 *Structure and properties*

X-ray diffraction data for platinacycloalkanes and related complexes are presented in Table 9. There are no significant differences between the Pt–C bond lengths in platinacycloalkanes and those in acyclic alkyl complexes, and the same observation holds for aryl species also. The four-membered ring is puckered in platinacyclobutanes and related species, giving a relatively short Pt···C or Pt···S transannular distance, and ring inversion in $[Pt\{CH(COMe)COCH(COMe)\}(PPh_3)_2]$, for example, takes place on the NMR timescale through a planar transition state.[442]

Platinum–carbon distances in platinacycloalkanes range from 0.203 nm (*trans* to N) to 0.219 nm (*trans* to P) and, for the few structures available, show little dependence on ring size. The Pt–C bond lengths in the platinacyclobutanones $[Pt(CHRCOCHR)(PPh_3)_2]$ (R = COMe, CO_2Me)[441,442] are about 0.008 nm longer than those in the related platinacyclobutane $[Pt(CH_2CMe_2CH_2)(PEt_3)_2]$.[225]

It has been pointed out that there is little ring strain in $[Pt(CH_2CMe_2CH_2)(PEt_3)_2]$. In fact, there is considerably less strain than in cyclobutane itself. This observation has been used to rationalize why metalacyclobutanes are more important intermediates in organometallic reactions than C_4 rings are in organic chemistry.[461]

Whereas diarylplatinum(II) complexes undergo reductive elimination both thermally and photochemically (Section 8.3.1.3), complexes containing substituted biphenyl ligands are inert towards elimination. This is explained in terms of the C–C bond forcing the two aromatic rings to remain coplanar, whereas a perpendicular arrangement of the two rings is required for elimination.[436,462]

8.3.3.3 *Reactions*

The thermal decomposition of platinacyclobutanes and -pentanes has been investigated. Heating a cyclohexane solution of $[Pt(CH_2CMe_2CH_2)(PEt_3)_2]$, from which β-hydrogen elimination cannot occur, results in a heterogeneous, autocatalytic reaction to produce neopentane, 1,1-dimethylcyclopropane, ethene, and platinum metal. In the presence of mercury(0) the reaction becomes homogeneous and 1,1-dimethylcyclopropane is the major organic product.[463] Decomposition of the corresponding PPr^i_3 and PCy_3 complexes is homogeneous, 1,1-dimethylcyclopropane and $[Pt(PR_3)_2]$ being formed. The reaction is inhibited by the presence of free phosphine, suggesting the operation of a dissociative reaction pathway.[463] Similarly, decomposition of $[Pt(CH_2CMe_2CMe_2CH_2)(PEt_3)_2]$ gives 2,2,3,3-tetramethylbutane, formed by a heterogeneous process which is inhibited by addition of mercury(0), and *t*-butylmethylcyclopropane, which is generated homogeneously.[464] In the case of the unsubstituted platinacyclopentane, $[Pt(CH_2CH_2CH_2CH_2)\{P(c\text{-}C_5H_9)_3\}_2]$, thermal decomposition yields 1-butene and $[Pt\{P(c\text{-}C_5H_9)_3\}_2]$, with trace amounts of 2-butenes, butadiene, butane, and cyclopentane. The reaction is first order up to a concentration of 15 mM, and two mechanistic pathways operate, namely, intermolecular hydrogen atom transfer and phosphine metallation with hydrogen transfer into the butene. β-Elimination does not occur.[465]

Addition of iodomethane to $[Pt(CH_2CH_2CH_2CH_2)(dppm)]$ results in oxidative addition to give two equilibrating isomers with a facial arrangement of the carbon ligands.[466] Similarly, oxidative addition of MeI to the 1-oxa-3-platinacyclobutane $[Pt(CH_2OCH_2)(PMe_3)_2]$ generates *fac*-$[Pt(CH_2OCH_2)IMe(PMe_3)_2]$.[467] Treatment of $[Pt(CH_2OCH_2)(PMe_3)_2]$ with tetracyanoethene results in elimination of the oxirane ring, whereas with acetyl chloride ring opening occurs (Scheme 24).[467]

Table 9 X-ray diffraction data for platinacycloalkanes and related complexes.

Complex	Pt–C (nm)	Other bond lengths (nm)	Ref.
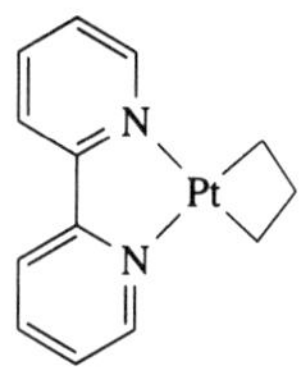	0.2030(10) 0.2037(10)	Pt–N 0.2103(4) 0.2107(8)	458
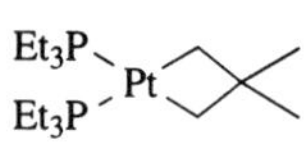	0.2080(6) 0.2086(6)	Pt–P 0.2282(2) 0.2287(1)	225
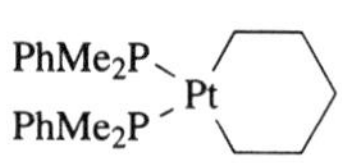	0.2097(34) 0.2195(16)	Pt–P 0.2268(4) 0.2289(3)	427
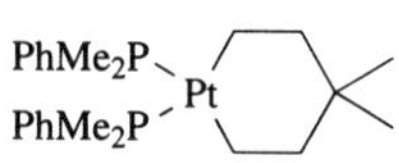	0.2089(13) 0.2102(13)	Pt–P 0.2273(3) 0.2290(3)	427
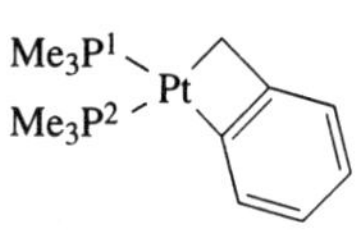	Pt–C_{alkyl} 0.2112(7) Pt–C_{aryl} 0.2060(8)	P–P^1 0.2282(3) Pt–P^2 0.2295(2)	434
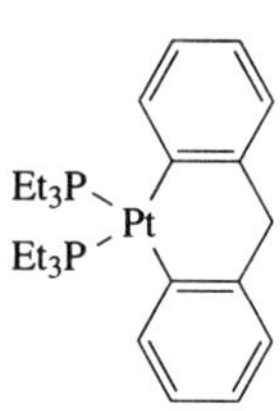	0.2062(11) 0.2038(11)	Pt–P 0.2323(3) 0.2303(3)	437
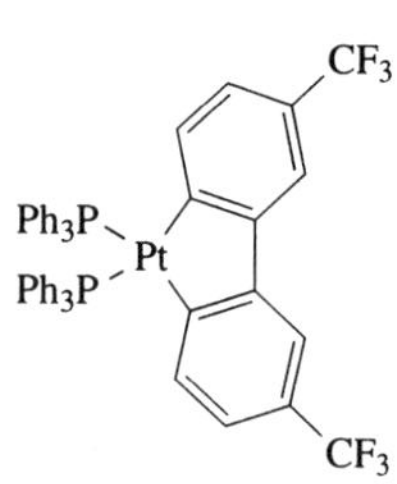	0.2058(7) 0.2065(7)	Pt–P 0.2328(2) 0.2346(2)	459
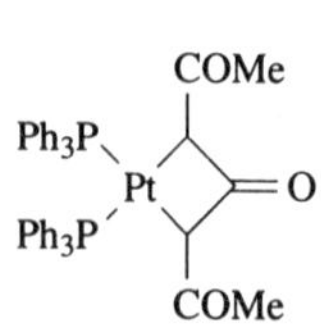	0.2153(5) 0.2183(7)	Pt–P 0.2295(2) 0.2309(1)	442
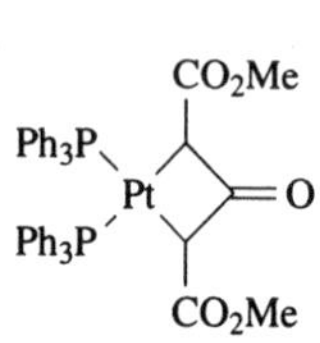	0.2133(5) 0.2155(6)	Pt–P 0.2272(2) 0.2300(1)	441

Reaction of [Pt(CHRCOCHR)L_2] (R = CO_2Me) with $CNBu^t$ leads first to ligand substitution, then to insertion–hydrogen atom migration to give a species of type (**48**).[468] Reaction of [Pt(CHRSO_2CHR)L_2] proceeds similarly.[469] Treatment of the 1-oxa-2-platinacyclobutane [Pt{OC$(CN)_2$C$(CN)_2$}$(AsPh_3)_2$] with

Table 9 (continued)

Complex		*Pt–C* (nm)	*Other bond lengths* (nm)	*Ref.*
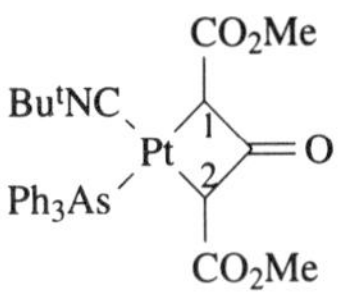	Pt–C[1] Pt–C[2]	0.2121(10) 0.2057(14)	Pt–As 0.2398(1) Pt–CNBu^t 0.1933(11)	460
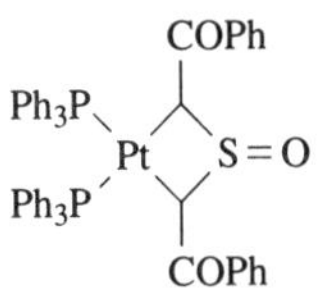		0.2123(6) 0.2163(7)	Pt–P 0.2290(2) 0.2319(2)	446
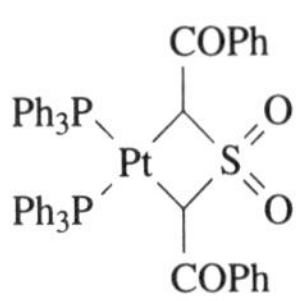		0.2106(10) 0.2125(11)	Pt–P 0.2285(3) 0.2327(3)	446
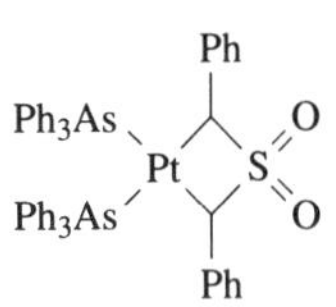		0.2063(21) 0.2082(21)	Pt–As 0.2422(2) 0.2403(2)	445
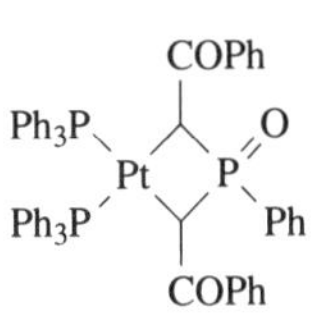		0.2137(7) 0.2162(7)	Pt–P 0.2283(2) 0.2301(2)	447

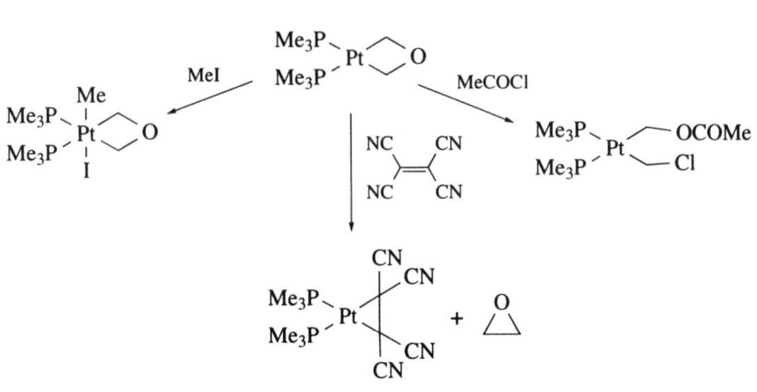

Scheme 24

a tertiary phosphine results in ligand substitution, ring opening, and formation of a cyanoplatinum species, *trans*-[Pt(CN){OC(CN)=C(CN)$_2$}L$_2$] (L = PEt$_3$, PMe$_2$Ph, PEt$_2$Ph, PMePh$_2$).[470]

8.3.4 Alkenyls

8.3.4.1 Preparation

Alkenylplatinum(II) complexes may be prepared by a number of methods. These include oxidative addition to platinum(0) precursors, transfer of an alkenyl group from another metal, and insertion of an alkyne into a platinum–element bond.

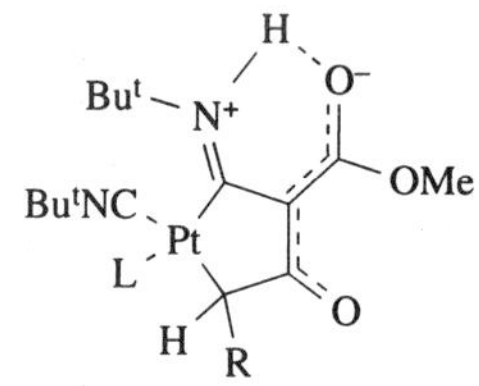

R = CO_2Me
(48)

Stang and co-workers have developed the oxidative addition of alkenyl triflates to platinum(0) as a method for the preparation of stable alkenylplatinum(II) complexes. Reaction of $[Pt(PPh_3)_4]$ with a series of alkenyl triflates proceeds to give $[Pt(CR^1{=}CR^2R^3)(PPh_3)_3]OTf$, which on treatment with halide gives *trans*-$[PtX(CR^1{=}CR^2R^3)(PPh_3)_2]$ (Equation (63)).[471] With $[Pt(PPh_3)_2(C_2H_4)]$ as precursor, *trans*-$[Pt(OTf)(CR^1{=}CR^2R^3)(PPh_3)_2]$ is obtained.[472] The reaction of $[Pt(PPh_3)_4]$ has been shown to occur by rate-determining formation of a platinum–alkene complex, followed by oxidative addition of the carbon–triflate bond, and subsequent displacement of the triflate by PPh_3.[473] Further elaboration of this chemistry includes displacement of the PPh_3 ligands by $PMePh_2$ or $Ph_2PCH_2CH_2PPhCH_2CH_2PPh_2$.[474] Reaction of $[Pt(PPh_3)_4]$ with an alkynylvinyl triflate gives either a σ-enynyl complex, the expected oxidative addition product, or a σ-butatrienyl species, depending on the size of the substituents (Equation (64)).[475]

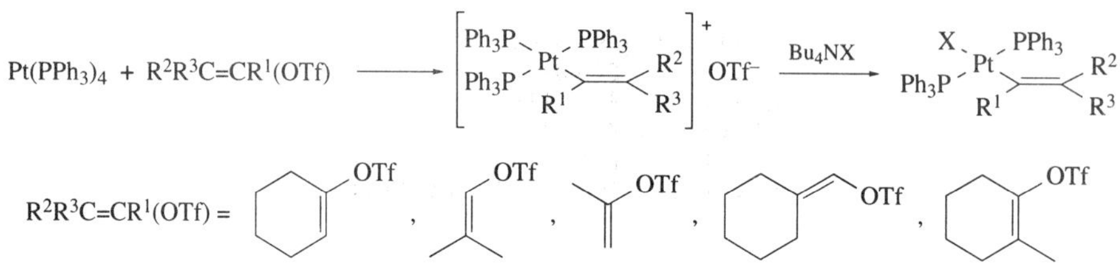

(63)

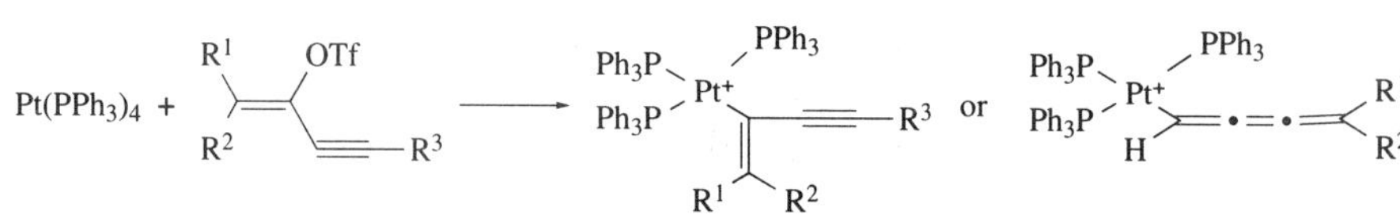

(64)

$R^1 = R^2 = Me$
$R^3 = H, Me, Bu^t, TMS, Ph$

$R^1 = R^2 = Me, Ph$ or Ph, Ph
$R^3 = H$

Oxidative addition of alkenyl halides may also occur. Reaction of $[Pt(PPh_3)_3]$ with 1,2-dibromonorbornene generates an alkene complex initially, but heating causes a rearrangement to the alkenyl complex (**49**).[476] Treatment of $[Pt(PPh_3)_2(C_2H_4)]$ with tetrachlorocyclopropene yields the ring-opened product (**50**).[477]

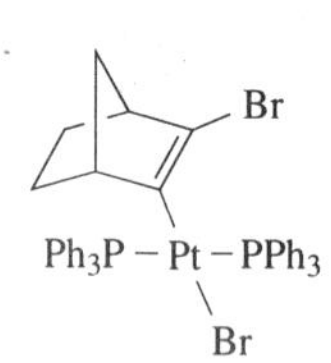

(49)

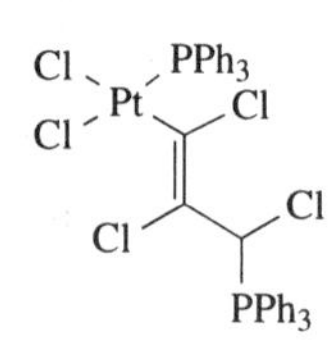

(50)

2-Thienyl complexes of the type *trans*-$[PtX(2\text{-}C_4H_3S)(PPh_3)_2]$ have been prepared by oxidative addition of $2\text{-}BrC_4H_3S$ or by reaction with $2\text{-}ClHgC_4H_3S$.[478] Alternatively, the thienyl group may be introduced by transfer from an organotin or organolithium reagent, and such groups have been used to link two or more platinum centers (Equation (65)).[479] Sterically hindered alkenyl complexes, containing naphthyl or adamantyl substituents have been prepared by reaction of *cis*-$[PtCl_2(PEt_3)_2]$ with the appropriate organolithium or Grignard reagent.[480]

Alkenyl complexes may also be prepared by insertion of an alkyne into a platinum–hydrogen bond. Treatment of the bulky phosphine complexes *trans*-$[PtH_2L_2]$ ($L = PCy_3, PPr^i_3, PBu^t_2Bu, PBu^t_2Me$) with an activated alkyne $R^1C{\equiv}CR^2$ generates *trans*-$[PtH(CR^1{=}CHR^2)L_2]$, the product being the result of *trans* addition to the triple bond (Equation (66)). The rate-determining step of the reaction is bimolecular, and

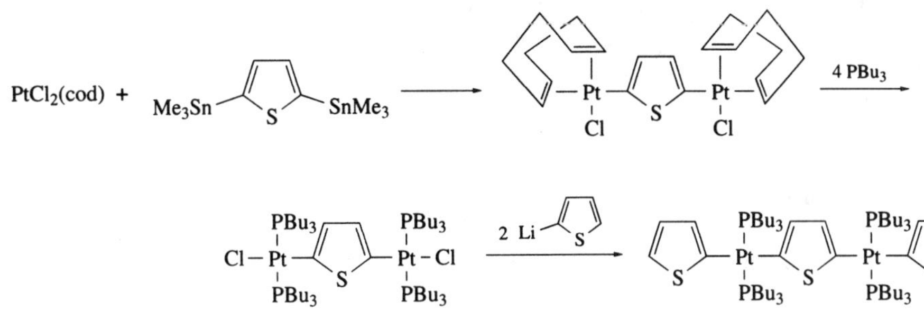

(65)

an electron-transfer mechanism is involved.[481] The dppm-bridged complex $[Pt_2H_2(\mu\text{-}H)(\mu\text{-dppm})_2]PF_6$ reacts with $CF_3C{\equiv}CCF_3$ or DMAD by reductive elimination and insertion to give a terminal alkenyl complex, which also contains a bridging alkyne, as shown in Scheme 25.[482] Addition of excess ethyne to the triplatinum cation $[Pt_3(\mu_3\text{-}H)(\mu\text{-dppm})_3]^+$ produces a species containing an ethenyl moiety, in addition to a triply bridging ethyne.[483]

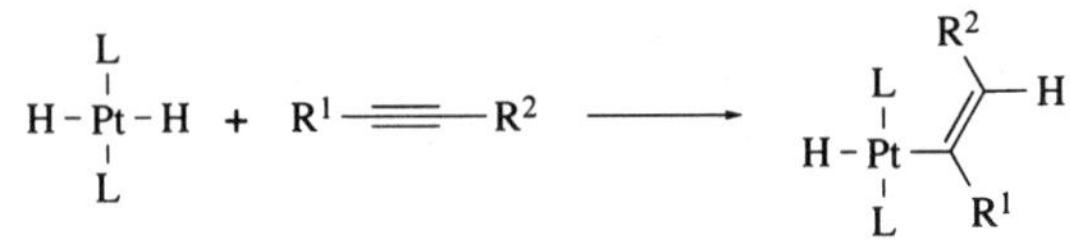

(66)

L = PCy3, PPri3, PBut2Bu, PBut2Me
R1 = R2 = CF3, CO2Me
R1 = CO2Me, R2 = H, Me, Ph
R1 = CN, R2 = H, Me, Ph
R1 = CF3, R2 = H, Ph
R1 = CO2CH2CH=CH2, CO2CH2C≡CH, R2 =H

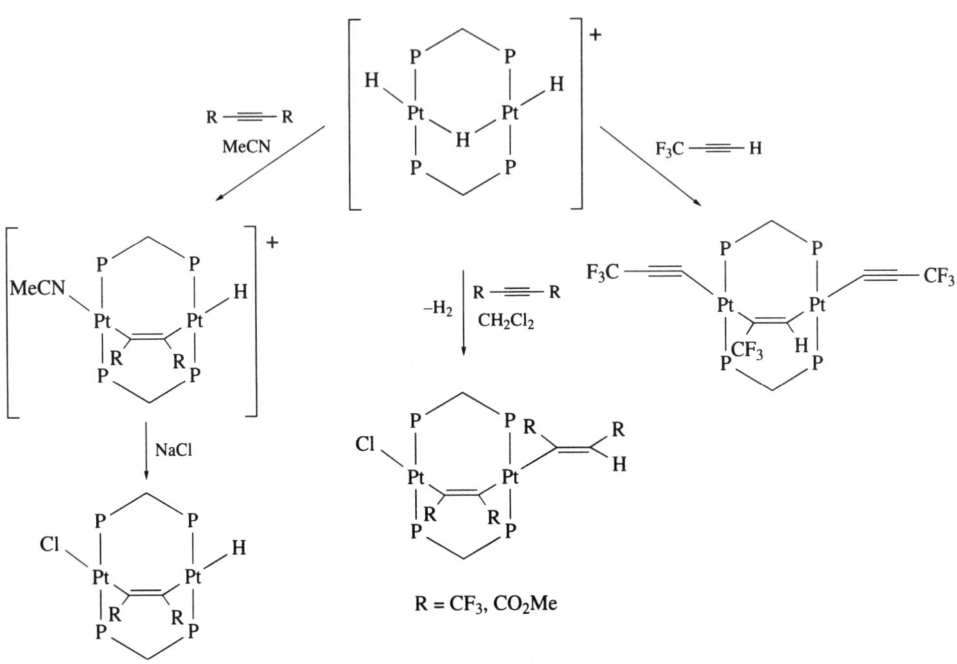

R = CF3, CO2Me

Scheme 25

Insertion of functionalized alkynes into the Pt–H bond of *trans*-$[PtH(S)(PEt_3)_2]^+$ (S = THF, acetone) leads to alkenylplatinum species. Reactions with compounds of the type $R^1{}_n(R^2C{\equiv}C)_{3-n}P{=}O$ (n = 0–2) result in ring compounds,[484,485] and reactions with alkynylplatinum complexes produce bridging alkenyl species (Scheme 26).[486] Insertion of an ethynyl group of $[Pt(C{\equiv}CH)_2(PEt_3)_2]$ into a Pt–Me bond proceeds similarly (Equation (67)).[486]

Treatment of the cyclohexyne complexes $[Pt(\eta^2\text{-}C_6H_8)L_2]$ ($L = PMePh_2$; L_2 = dppe, dppp, dppb, diphoe) with HX generates cyclohexenyl species (Equation (68)),[487] and further treatment with CO results in hydroxy- or methoxycarbonyl complexes when X = OH or OMe.[487,488] Reaction of

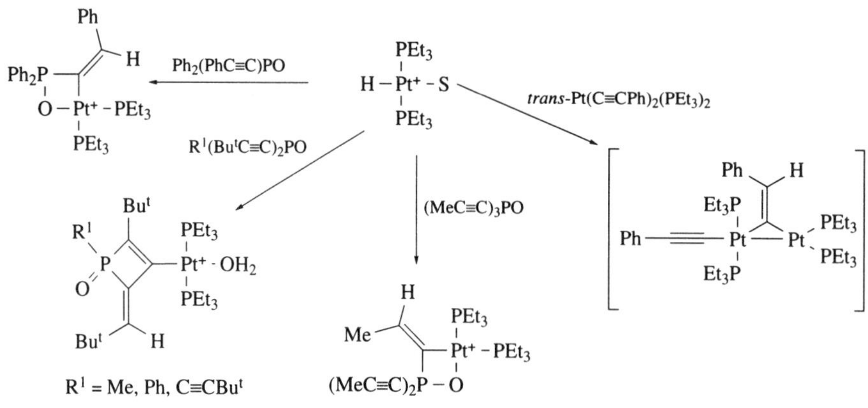

Scheme 26

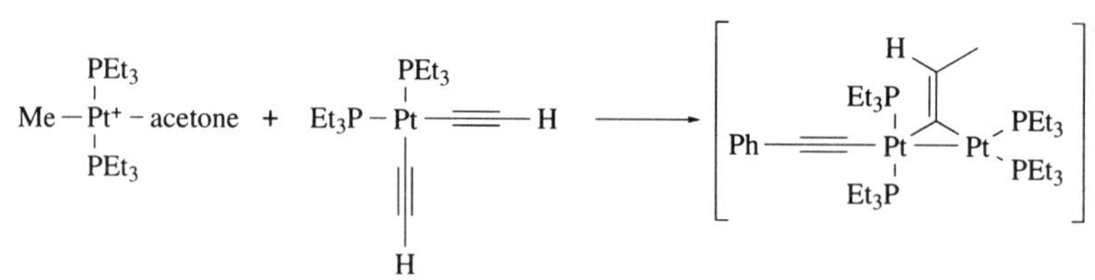

(67)

$[Pt(PPh_3)_2(PhC{\equiv}CPh)]$ with $CH_2(SO_2CF_3)_2$ produces first *trans*-$[Pt(CPh{=}CHPh)\{CH(SO_2CF_3)_2\}(PPh_3)_2]$, which exhibits restricted rotation about the Pt–$CH(SO_2CF_3)_2$ bond in solution, but on standing in moist air this is converted in to *trans*-$[Pt(CPh{=}CHPh)(H_2O)(PPh_3)_2][CH(SO_2CF_3)_2]$.[489] When $[PtCl_2(dppe)]$ is treated with $AgPF_6$ in CH_2Cl_2–MeOH solution, the bis(methanol) cation is formed which, on addition of DMAD, generates the bis(alkenyl)platinum complex, $[Pt\{C(CO_2Me){=}C(OMe)(CO_2Me)\}_2(dppe)]$.[490] The bis-chelate complex *cis*-$[Pt(OCPh{=}CHPPh_2)_2]$ reacts with DMAD by cleavage of the Pt–O linkages and insertion as shown in Equation (69).[491]

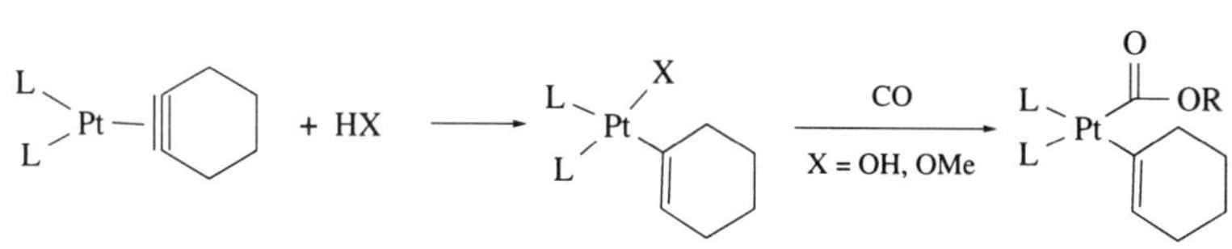

(68)

$L = PMePh_2$; L_2 = dppe, dppp, dppb, diphoe; X = Cl, SPh, OCOMe, CH_2CN, OH, OMe

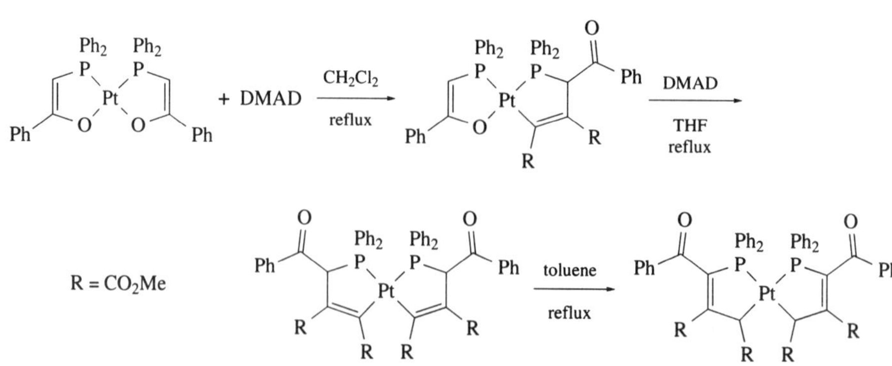

(69)

An unexpected insertion has been reported for the reaction of $[Pt_2(\mu\text{-}Cl)_2(SnCl_3)_2(PEt_3)_2]$ with hexafluorobutyne. Instead of acting simply as a leaving group, the trichlorostannate ligand exhibits some nucleophilicity in attacking the coordinated alkyne to produce $[Pt_2(\mu\text{-}Cl)_2\{C(CF_3){=}C(CF_3)(SnCl_3)\}_2(PEt_3)_2]$.[492]

8.3.4.2 Structure and properties

A number of alkenylplatinum(II) complexes have been structurally characterized, and data on bond distances are collected in Table 10. The complexes exhibit Pt–C distances in the range 0.199–0.214 nm, at the shorter end of the range found for alkylplatinum compounds (0.202–0.220 nm). All of the Pt–C bond distances are greater than the sum of the covalent radii (0.198 nm), providing little evidence for any π-component to the Pt–C(sp^2) bond. As expected, the shortest distances are observed when the alkenyl group lies *trans* to a ligand of weak *trans* influence (H_2O, OAc),[489,494] and the longest are those where the Pt–C bond lies *trans* to H or C≡CR,[472,481] or when the alkenyl group is part of a very bulky ligand.[480] In the few cases where comparisons can be made, the alkenyl ligand appears to exert a *trans* influence similar to that of an alkyl group.

8.3.4.3 Reactions

Many of the reactions reported for alkenylplatinum(II) complexes involve carbonylation. Treatment of *trans*-[Pt(CR^1=CR$^2_2$)(OTf)(PPh_3)$_2$] (R^1 = Me, R^2 = H; R^1 = H, R^2 = Me) with CO, followed by NaOMe or NaOEt, results in reductive elimination of an unsaturated ester (Equation (70)).[496] Carbonylation of the corresponding chloride complex proceeds by CO insertion, and further treatment with silver triflate and ethanol gives the ester (Equation (71)).[497]

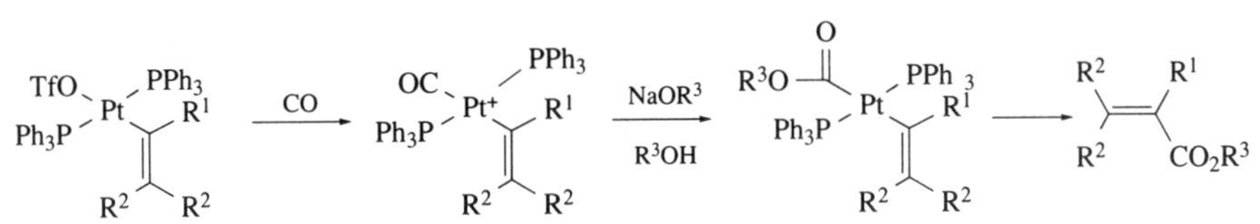

(70)

R^1 = Me, R^2 = H, R^3 = Me, Et
R^1 = H, R^2 = Me, R^3 = Me, Et

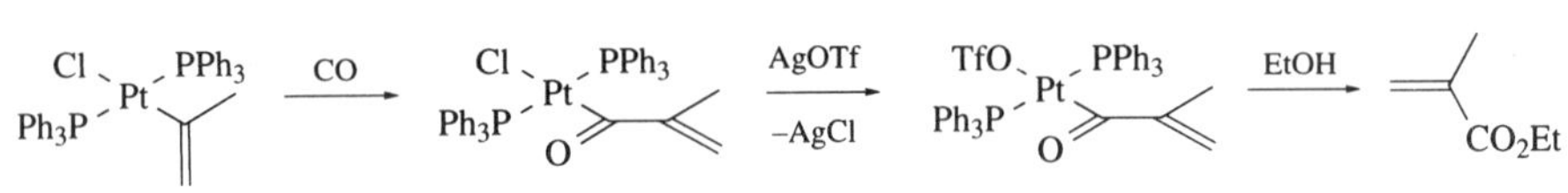

(71)

Carbonylation of the cyclohexenyl complex [Pt(C_6H_9)(OMe)(dppp)] results in insertion into the Pt–O bond and, on further reaction, insertion into the Pt–C bond. In contrast, the analogous dppe and diphoe complexes do not undergo insertion into the cyclohexenyl–platinum bond.[498] The allenyl complex *trans*-[PtBr(CH=C=CMe$_2$)(PPh_3)$_2$] undergoes double carbonylation and cyclization, as shown in Equation (72).[499]

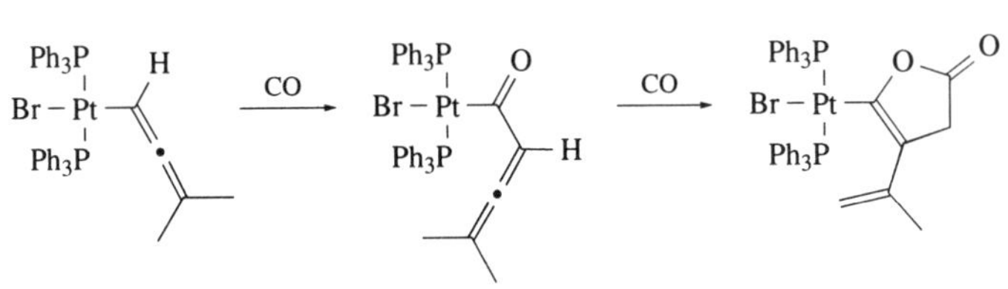

(72)

Reaction of *trans*-[Pt(CR^1=CR$^2_2$)(OTf)(PPh_3)$_2$] (R^1 = Me, Bu^t, R^2 = H; R^1 = H, R^2 = Me) with ethene produces the η-allyl complex [Pt(η-CHMeCR^{1}CR$^2_2$)(PPh_3)$_2$]OTf by insertion of ethene into the Pt–C bond.[500] Addition of K[ReHCp(CO)$_2$] to the [Pt(CMe=CHMe)(PPh_3)$_3$]$^+$ cation results in a 2-butene complex of rhenium, presumably via the intermediacy of a bridging hydride and reductive elimination of the hydride and alkenyl groups from platinum.[501] Treatment of *trans*-[Pt(CMe=CH$_2$)(OTf)(PPh_3)$_2$] with Bu^{t}C≡CLi gives *cis*-[Pt(CMe=CH$_2$)(C≡CBu^t)(PPh_3)$_2$] as the kinetic product, which then rearranges to its thermodynamically favored *trans* isomer. Both isomers undergo reductive elimination, but the *cis* isomer eliminates at much lower temperatures. Addition of free PPh_3 inhibits elimination from the *trans* isomer, but has little effect on the *cis* isomer. Concerted elimination from the *cis* isomer is proposed, whereas elimination from the *trans* form requires PPh_3 dissociation, followed by isomerization, then concerted reductive elimination from the *cis* isomer.[502]

Table 10 X-ray diffraction data for alkenylplatinum(II) complexes.

Complex	*Pt–C* (nm)	*Other bond lengths* (nm)	*Ref.*
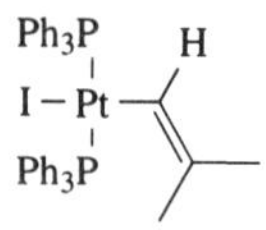	0.2032(7)	Pt–I 0.2709(4) Pt–P 0.2304(1) 0.2309(1)	474
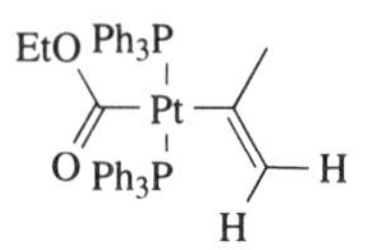	0.2071(6)	Pt–CO_2Et 0.2092(6) Pt–O 0.2292(1) 0.2299(1)	493
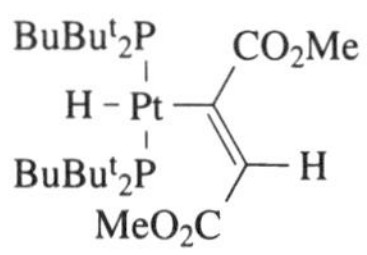	0.2086(12)	Pt–P 0.2293(4) 0.2311(4)	481
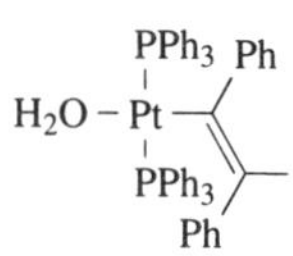	0.199(2)	Pt–O 0.221(1) Pt–P 0.2311(4) 0.2322(4)	489
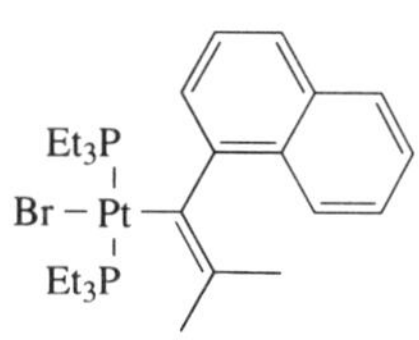	0.2086(18)	Pt–Br 0.2537(2) Pt–P 0.2297(5) 0.2320(5)	480
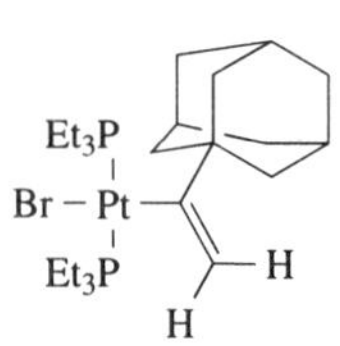	0.214(2)	Pt–Br 0.2549(2) Pt–P 0.2301(5) 0.2310(5)	480
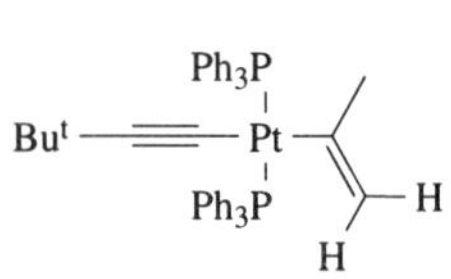	0.210(1)	Pt–C≡C 0.204(1) Pt–P 0.2293(3) 0.2296(3)	472
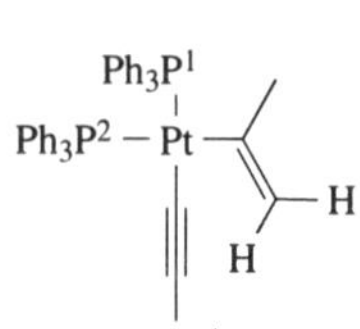	0.205(2)	Pt–C≡C 0.197(2) Pt–P^1 0.2286(4) Pt–P^2 0.2340(4)	472
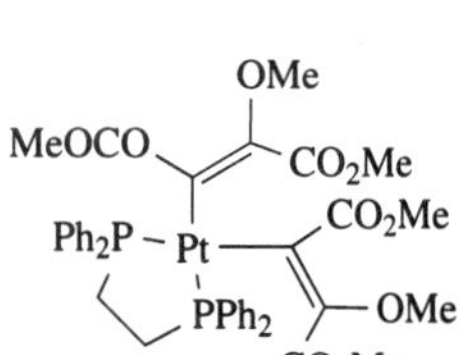	0.2074(4)	Pt–P 0.2286(1)	490

Table 10 (continued)

Complex	*Pt–C* (nm)	*Other bond lengths* (nm)	*Ref.*
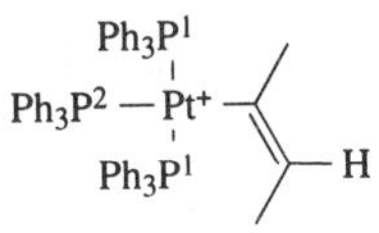	0.207(1)	Pt–P^1 0.2333(3) 0.2376(3) Pt–P^2 0.2372(3)	473
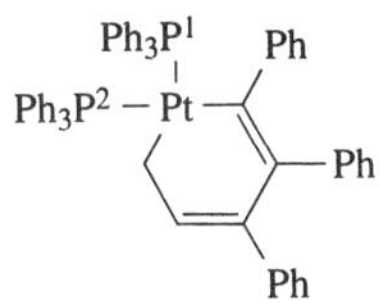	0.2088(9)	Pt–CH_2 0.2123(10) Pt–P^1 0.2313(3) Pt–P^2 0.2306(3)	451
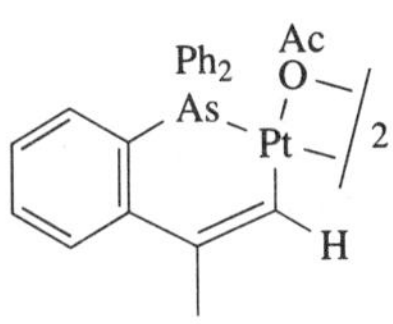	0.199(2)	Pt–As 0.2267(2) Pt–O 0.216(2) 0.210(1)	494
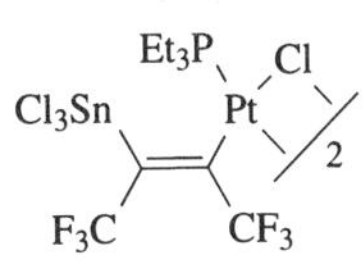	0.1982(6)	Pt–P 0.2230(2) Pt–Cl 0.2396(2) 0.2415(2)	492
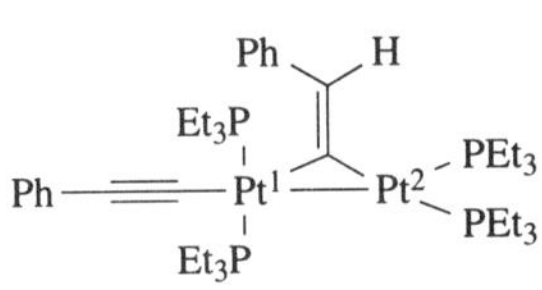	Pt^1–C 0.210(2) Pt^2–C 0.193(2)	Pt^1–C≡C 0.217(2) Pt^1–Pt^2 0.2750(2) Pt^1–P 0.2325(5) 0.2329(5) Pt^2–P 0.2228(8) 0.2351(8)	495

The bridging vinylidene complex $[(Et_3P)_2Pt(\mu\text{-}C{=}CHPh)Pt(C{\equiv}CPh)(PEt_3)_2]BF_4$ reacts as shown in Scheme 27.[503,504] The chloride complex $[(Et_3P)ClPt(\mu\text{-}C{=}CHPh)Pt(C{\equiv}CPh)(PEt_3)_2]$ emits orange-red luminescence at 696 nm when it is excited at 410 nm at 77 K in an ethanol–methanol matrix.[503]

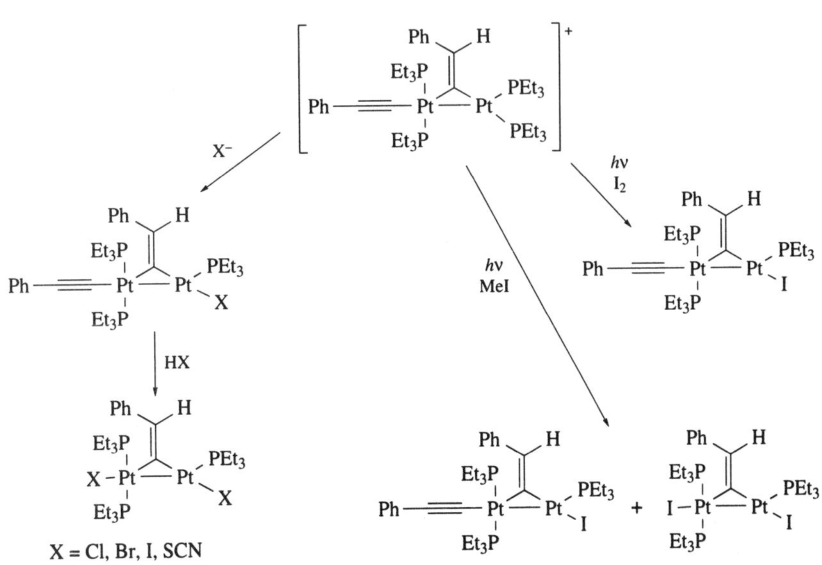

Scheme 27

8.3.5 Alkynyls

8.3.5.1 Preparation

A number of methodologies are available for the synthesis of alkynylplatinum(II) complexes. Most involve transfer of an alkynyl group from another metal to a platinum halide complex, others require addition of the alkyne itself in the presence of a base, and there are a few examples of oxidative addition to platinum(0) precursors.

The reactions of $[PtCl_2L_2]$ (L_2 = dppe, dmpe) with $Me_3SnC{\equiv}CR$ or $Me_2Sn(C{\equiv}CR)_2$ (R = H, Me, TMS, Ph) in refluxing THF give $[Pt(C{\equiv}CR)_2L_2]$,[505] and *trans*-$[PtCl_2(PEt_3)_2]$ reacts similarly.[506] Mixed bis(alkynyl)platinum complexes can be obtained by treatment of $[PtCl_2(PEt_3)_2]$ with 1 equiv. of $Me_3SnC{\equiv}CR^1$, followed by $Me_3SnC{\equiv}CR^2$.[506] Alkynyl groups have also been transferred to platinum(II) from silver,[398,507] gold,[508] or mercury.[509,510]

The reactions of $[PtCl_2(dppm)]$ with alkynylating agents are less straightforward. Reaction with $Me_2Sn(C{\equiv}CR)_2$ (R = H, Me, Bu^t, Ph) is reported to give $[Pt(C{\equiv}CR)_2(dppm)]$ in greater than 90% yield,[505,511] but a trace of free phosphine catalyzes rearrangement to the face-to-face dimer, *trans,trans*-$[Pt_2(C{\equiv}CR)_4(\mu\text{-dppm})_2]$ (**51**).[511] Reactions with $LiC{\equiv}CR$ (R = Me, CF_3, Ph, C_6H_4Me-4) also give face-to-face dimers.[512,513] Treatment of $[PtCl_2(dppm)]$ with $LiC{\equiv}CBu^t$ in THF solution has been reported to yield the fluxional, bridging alkynyl complex $[Pt_2(C{\equiv}CBu^t)_2(\mu\text{-}\eta^1,\eta^2\text{-}C{\equiv}CBu^t)(\mu\text{-dppm})_2]Cl$.[514] Reaction of $[Pt(dppm)_2]Cl_2$ with 2 equiv. of $LiC{\equiv}CPh$, or treatment of $[PtCl_2(dppm)]$ with $NaC{\equiv}CH$ and dppm in liquid ammonia, leads to species of the type *trans*-$[Pt(C{\equiv}CR)_2(dppm\text{-}P)_2]$, which have been used in the preparation of heterobimetallic complexes with silver, gold, rhodium, iridium, and tungsten.[515,516] Treatment of *cis*-$[PtCl_2(SMe_2)_2]$ with $AuC{\equiv}CBu^t$, followed by dppm,[517] or addition of $AuC{\equiv}CBu^t$ to $[Pt_2(\mu\text{-dppm})_3]$,[518] also produces heterobimetallic alkynylplatinum species.

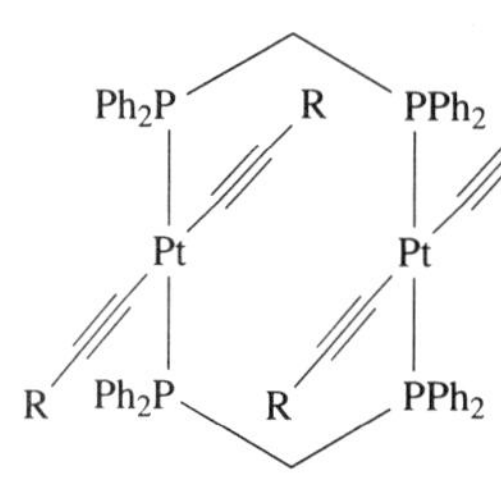

(51)

Oligomeric or polymeric alkynylplatinum species have been prepared from *trans*-$[PtCl_2(PBu_3)_2]$ according to Equation (73). Adjustment of the stoichiometry leads to di- or trimetallic species, or higher oligomers. Where $X = C_6H_4$, average molecular masses of 100 000 have been obtained. The tributylarsine precursor reacts similarly.[519] Oligomeric alkynyl-bridged species have also been prepared by reaction of terminal ethynyl complexes, and these will be discussed in Section 8.3.5.3.

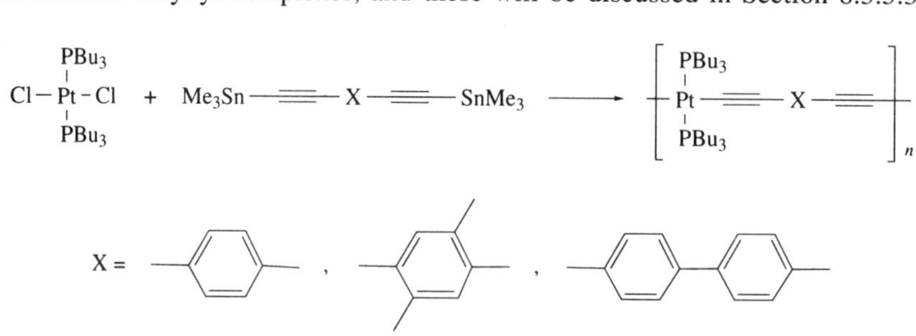

(73)

Bis(alkynyl)platinum complexes have been generated by various means starting from the alkyne itself. Treatment of *cis*-$[PtCl_2(PMePh_2)_2]$ with $PhC{\equiv}CH$ in the presence of sodium ethoxide yields *cis*-$[Pt(C{\equiv}CPh)_2(PMePh_2)_2]$, whereas the analogous reaction using Et_2NH–CuI as base gives the *trans* isomer.[520] The products of the reactions of $[PtCl_2L_2]$ (L_2 = dppm, depm, dippm) or $[Pt_2Cl_4(\mu\text{-dmpm})_2]$ with propyne in the presence of sodium alkoxide depend on the diphosphine employed (Scheme 28).[521] Treatment of *cis*-$[PtCl_2(PPh_3)_2]$ with $RC{\equiv}CH$ (R = CMe_2OH, CMeEtOH, CHPhOH, Ph) in 30% ammonia solution yields *cis*-$[Pt(C{\equiv}CR)_2(PPh_3)_2]$, whereas with Et_2NH in chloroform or dichloromethane only monosubstitution takes place.[522] Analogous reactions of *trans*-$[PtClH(PPh_3)_2]$ with $RC{\equiv}CH$–Et_2NH generate *trans*-$[PtH(C{\equiv}CR)(PPh_3)_2]$.[523]

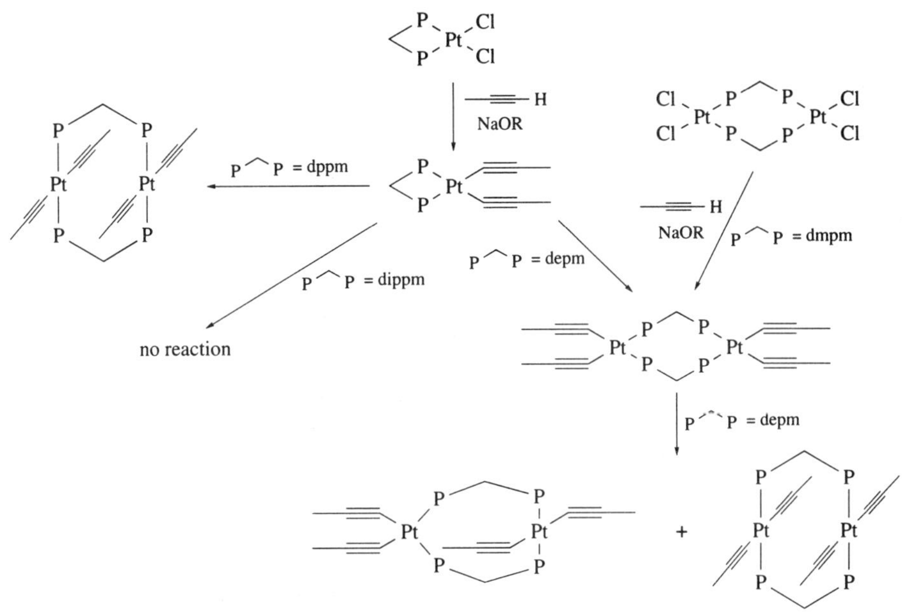

Scheme 28

Acid–base reactions of $[Pt(C_2O_4)L_2]$ (L = PEt_3, PPh_3; L_2 = dppe, dmpe, dppm) with phenylethyne also result in bis(alkynyl)platinum complexes. The dppm complex rearranges, however, in the presence of the oxalic acid formed to *cis,cis*-$[Pt_2(C{\equiv}CPh)_4(\mu\text{-dppm})_2]$. In contrast, UV irradiation or heating with a trace of free dppm converts $[Pt(C{\equiv}CPh)_2(dppm)]$ into the *trans,trans* isomer.[524] Other mild approaches to alkynyl complexes include the reactions of *trans*-$[PtCl_2(EBu_3)_2]$ (E = P, As) with 1,3,5-$C_6H_3(C{\equiv}CH)_3$ to generate trimetallic or oligomeric species,[525] and the reactions of *trans*-$[PtClH(PPh_3)_2]$ with an alkyne in the presence of an alcohol as shown in Equation (74).[526]

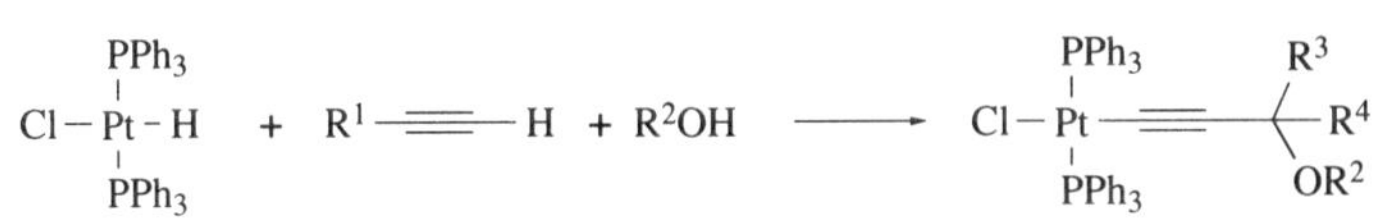

(74)

R^1 = CMe_2OH, CMeEtOH, $C_6H_{10}OH$; R^2 = Me, Et, Bu, $CH_2CHMeCH_2Me$

R^3, R^4 = Me, Me; Me, Et; $-CH(CH_2)_4CH-$

A few examples of the preparation of alkynylplatinum(II) complexes by oxidative addition to platinum(0) precursors have been reported. Reactions of $[Pt(PPh_3)_2(C_2H_4)]$ with PhC≡CX (X = $SnEt_3$, $SnPh_3$), TMS-C≡CX (X = Cl, Br), $Et_3SiC{\equiv}CI$, or MeC≡CI leads to *cis*- or *trans*-$[PtX(C{\equiv}CR)(PPh_3)_2]$. With $PhC{\equiv}CSnEt_3$ or TMS-C≡CX the intermediate alkyne complex could be detected.[527] The reaction of $[Pt(PPh_3)_2(C_2H_4)]$ with FC≡CH yields the alkyne–platinum(0) complex, whereas with ClC≡CH oxidative addition occurs to produce *trans*-$[PtCl(C{\equiv}CH)(PPh_3)_2]$.[528] Irradiation of the alkyne complex $[Pt(PPh_3)_2(DMAD)]$ generates *cis*-$[Pt(CO_2Me)(C{\equiv}CCO_2Me)(PPh_3)_2]$, but oxidative addition does not occur with RC≡CR (R = Ph, CH_2OH, CMe_2OH) or RC≡CH (R = CH_2OH, CO_2H, CH_2NH_2).[529]

8.3.5.2 Structure and properties

Structural data for a number of alkynylplatinum(II) complexes are presented in Table 11. The Pt–C distances are typically shorter than those found for alkyl or alkenyl compounds, lying within the range 0.190–0.206 nm. All but one lie above 0.191 nm, the value predicted for a Pt–C(*sp*) single bond,[2] indicating little evidence for significant π-character in the Pt–C bond. The C≡C distances may be longer or shorter than that found for ethyne itself, ranging from 0.115 nm to 0.125 nm, with the exception of those in *trans*-$[Pt(C{\equiv}CMe)_2(PEt_3)_2]$. In this case the bond length is reported to be very short at 0.104 nm, although there is some disorder in the structure.[506]

The longest Pt–C distances are found *trans* to other carbon ligands,[472,506] the shortest being *trans* to chloride[522,528] and, surprisingly, hydride.[523] Alkynyl ligands appear to exert a slightly weaker *trans*

Table 11 X-ray diffraction data for alkynylplatinum(II) complexes.

Complex	*Pt–C* (nm)	*Other bond lengths* (nm)	*Ref.*
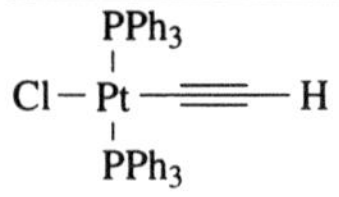	0.1933(12)	Pt–Cl 0.2373(3) C≡C 0.1154(19)	528
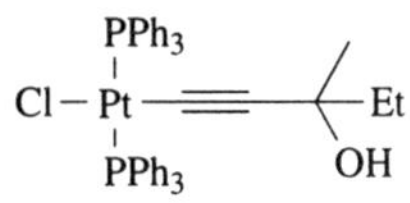	0.1988(9)	Pt–Cl 0.2336(4) Pt–P 0.2321(4) 0.2320(4) C≡C 0.115(2)	522
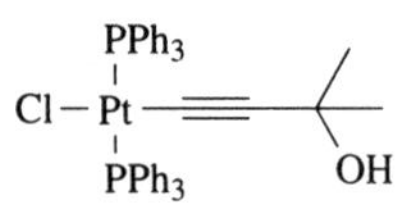	0.195(2)	Pt–Cl 0.2356(5) Pt–P 0.2311(4) 0.2320(4) C≡C 0.118(2)	522
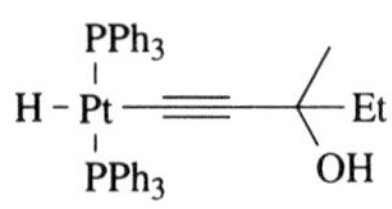	0.190(1)	Pt–P 0.2267(5) 0.2279(4) C≡C 0.125(2)	523
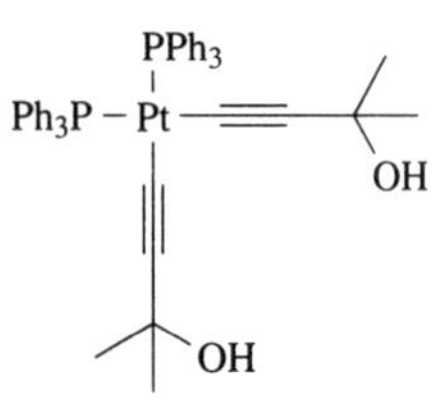	0.199(2) 0.202(2)	Pt–P 0.2307(5) 0.2331(5) C≡C 0.117(2) 0.119(2)	522
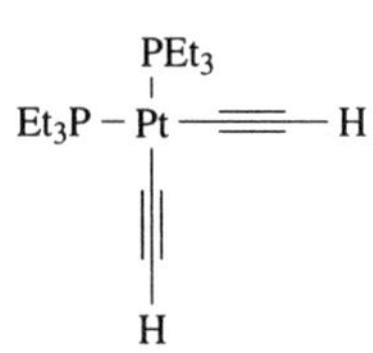	0.1949(16) 0.1987(15)	Pt–P 0.2308(5) 0.2312(4) C≡C 0.1205(25) 0.1241(30)	530
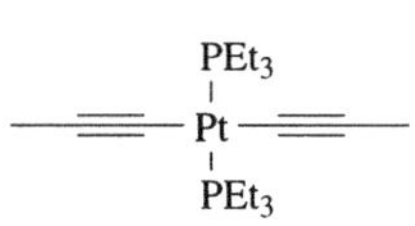	0.2062(8)	Pt–P 0.2284(2) C≡C 0.1039(11)	506
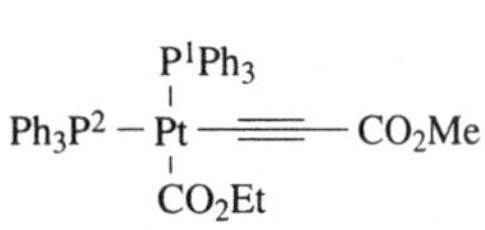	0.1991(8)	Pt–CO_2Et 0.2024(8) Pt–P^1 0.2347(2) Pt–P^2 0.2298(2) C≡C 0.1180(11)	529
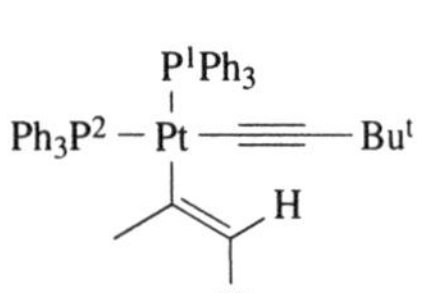	0.197(2)	Pt–C=C 0.205(2) Pt–P^1 0.2340(4) Pt–P^2 0.2286(4) C≡C 0.122(2)	472
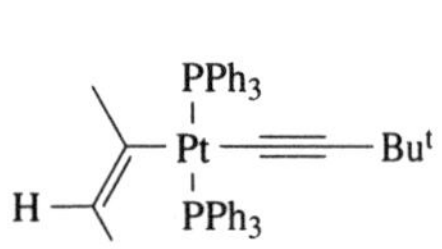	0.204(1)	Pt–C=C 0.210(2) Pt–P 0.2293(3) 0.2296(3) C≡C 0.121(2)	472
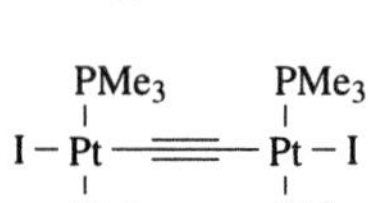	0.1973(30) 0.1980(38)	Pt–I 0.2646(4) 0.2652(3) Pt–P 0.2280(11) 0.2307(12) 0.2304(10) 0.2308(10) C≡C 0.1179(48)	531

influence than alkyl or alkenyl groups. Comparisons of the Pt–Cl distances in *trans*-[PtCl(C≡CH)(PPh_3)$_2$] (0.237 nm)[528] and *trans*-[PtCl(Me)(PPh_3)$_2$] (0.243 nm),[155] and of the Pt–P separations in *cis*-[Pt(C≡CH)$_2$(PEt_3)$_2$] (0.231 nm),[530] *cis*-[Pt(C≡CCMe_2OH)$_2$(PPh_3)$_2$] (average 0.219 nm)[522] and *cis*-[Pt(Me)$_2$($PMePh_2$)$_2$] (0.232 nm),[159] indicate that the bonds *trans* to alkynyl ligands are marginally shorter. Perhaps the best comparisons are for *cis*-[Pt(CO_2Et)(C≡CCO_2Me)(PPh_3)$_2$], where the Pt–P separations are 0.235 nm and 0.230 nm *trans* to CO_2Et and C≡CCO_2Me, respectively,[529] and for *cis*-[Pt(CMe=CH_2)(C≡CBu^t)(PPh_3)$_2$], in which the Pt–P distances are 0.234 nm and 0.229 nm,[472] the shorter distance, and presumably the stronger bond, again lying *trans* to the alkynyl group.

In contrast to the conclusions drawn from x-ray diffraction data, NMR chemical shifts (δC and δPt) for *cis*- and *trans*-[Pt(C≡CR^1)$_2$(PR^2_3)$_2$] complexes have been interpreted as indicating the existence of some π-backbonding from platinum into the C≡C π*-orbitals, although the degree of backbonding is less for platinum than for palladium or nickel.[532] Two-dimensional NMR techniques have been used with the complexes [Pt(C≡CH)$_2$(depe)] and *trans*-[Pt(C≡CH)$_2$(PBu_3)$_2$] to determine the signs of various coupling constants. The couplings $^2J_{PC}$ and $^4J_{PH}$ are found to be of opposite sign in the *cis* and *trans* species, whereas $^3J_{PC}$ and $^3J_{PtH}$ are positive, irrespective of the geometry.[533]

Studies have been made of site-selective luminescence spectroscopy of *trans*-[PtClHL_2] and *trans*-[Pt(C≡CR)$_2L_2$] (L = PEt_3; R = H, Ph). These give rise to intense emission spectra at 77 K, and the differences between the spectra have been explained in terms of changes in the charge-transfer character of the excited states.[534]

8.3.5.3 Reactions

The bis(alkynyl)platinum complexes [Pt(C≡CPh)$_2L_2$] (L = $PMePh_2$, PPh_3) undergo *cis–trans* isomerization, catalyzed by mercury(II), copper(I), or tin(II) halides. This is proposed to take place by reversible transfer of an alkynyl group, isomerization of [PtCl(C≡CPh)L_2], and transfer of C≡CPh back to platinum. This is shown for the case of $HgCl_2$ in Equations (75)–(77).[520] Treatment of the bis(alkynyl) complexes *cis*-[Pt(C≡CR)$_2$(CO)($PMePh_2$)] with the corresponding dichloro complex results in alkynyl–chloride exchange, and this may occur by a S_E2(cyclic) mechanism.[535] The mixed alkynyl–cyclopentadienyl complexes [Pt(C≡CR)(η^1-C_5H_5)(CO)L] (L = $PMePh_2$; C≡CR *trans* to CO) also react stereospecifically with *cis*-[PtCl_2(CO)L] to generate [PtCl(C≡CR)(CO)L] and [PtCl(η^1-C_5H_5)(CO)L]. Mercury(II) chloride cleaves the platinum–cyclopentadienyl linkage preferentially.[536]

$$cis\text{-Pt(C≡CPh)}_2L_2 + HgCl_2 \longrightarrow cis\text{-PtCl(C≡CPh)}L_2 + \text{PhC≡CHgCl} \quad (75)$$

$$cis\text{-PtCl(C≡CPh)}L_2 \longrightarrow trans\text{-PtCl(C≡CPh)}L_2 \quad (76)$$

$$trans\text{-PtCl(C≡CPh)}L_2 + \text{PhC≡CHgCl} \longrightarrow trans\text{-Pt(C≡CPh)}_2L_2 + HgCl_2 \quad (77)$$

Alkynylplatinum complexes are converted into alkenyl species by reaction with trialkylboranes. With 1 equiv. of BMe_3 or BEt_3, *trans*-[Pt(C≡CMe)$_2$(PEt_3)$_2$] yields *trans*-[Pt(C≡CMe){(*E*)-CMe=CR(BR_2)}(PEt_3)$_2$] (R = Me, Et), whereas the bis(ethynyl)platinum species reacts with 2 equiv. of BR_3 at 25 °C to produce *trans*-[Pt{(*E*)-CH=CR(BR_2)}$_2$(PEt_3)$_2$].[537] With the diphosphine complexes [Pt(C≡CH)$_2L_2$] (L = dppe, depe), reaction with BR_3 yields a platinacyclopentadiene derivative [Pt{CH=CRC(BR_2)=CH}L_2], which rearranges to [Pt(η-$C_4BH_2Me_3$)L_2] when R = Me.[538]

Reaction of *trans*-[Pt(C≡CR^1)$_2$(PEt_3)$_2$] (R^1 = Me, Bu, Ph) with R^2Li (R^2 = Bu, Bu^t, Ph) in a hydrocarbon solvent produces a lithium-bridged dimer, which may be solvated in diethyl ether, THF, or TMEDA to give Li[Pt(C≡CR^1)$_2R^2$(PEt_3)]. In the case of Li[Pt(C≡CPh)$_2$Bu(PEt_3)], treatment with further BuLi results in alkynyl–alkyl exchange to give Li[Pt(C≡CPh)Bu_2(PEt_3)] and PhC≡CLi.[539]

Compounds containing bridging alkynyl groups have been prepared from ethynyl complexes according to Equation (78). An amine is required to effect removal of HCl, and the reactions are catalyzed by copper(I) chloride. The terminal chlorides may be replaced by metathesis with KI.[540] Mixed platinum–nickel[541] and platinum–palladium[542] complexes have also been isolated. The mixed platinum–palladium species are best prepared by reaction of *trans*-[PtCl(C≡CH)L_2] with *trans*-[PdCl_2L_2] (L = PMe_3, PEt_3, PBu_3), since mixtures of products are obtained when an ethynylpalladium precursor is employed.[542] The oligomeric species [–Pd(PBu_3)$_2$C≡CC≡CPt(PBu_3)$_2$C≡CC≡C–]$_n$ function as lyotropic liquid crystals in trichloroethene or THF solution.[543]

The alkynyl-bridged complex *trans,trans*-[PdPtCl_2(μ-C≡C)(PEt_3)$_4$] reacts with aryl isocyanides by insertion into the Pd–C bond exclusively. A complex resulting from double insertion has been isolated (Equation (79)), but further reaction occurs with excess isocyanide, and this represents an example of

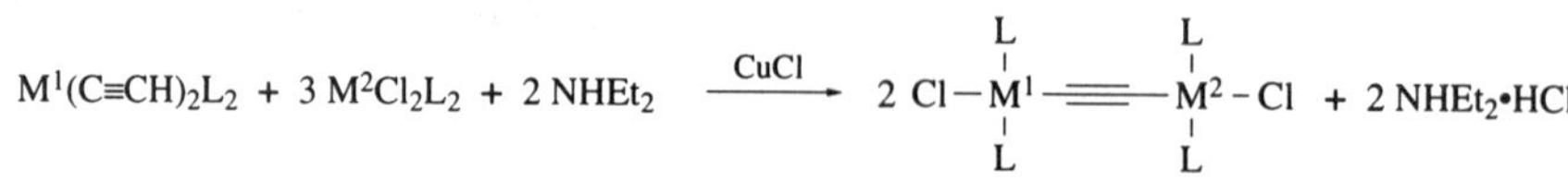

(78)

M^1, M^2 = Pd, Pt; L = PMe_3, PEt_3, PBu_3

living polymerization.[544] The phenylethynyl complex [Pt(C≡CPh)$_2$(PPh$_3$)$_2$] catalyzes phenylethyne polymerization,[545] and [Pt(C≡CCH$_2$NHCH$_2$Ph)$_2$(PPh$_3$)$_2$] is moderately effective in catalyzing the polymerization of *N*-benzylpropargylamine.[546]

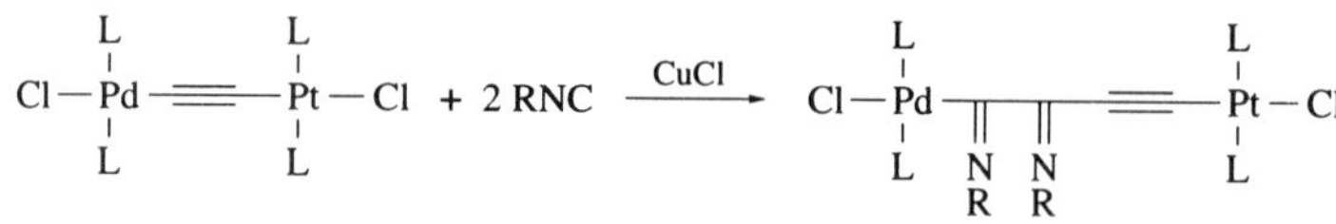

(79)

L = PEt_3; R = Ph, $C_6H_4NO_2$-4, C_6H_4Me-4

As mentioned in Section 8.3.4.1, alkynylplatinum complexes react with hydridoplatinum cations to generate alkenyl-bridged species,[486] and the reactions of the alkenyl-bridged alkynylplatinum dimers have been described in Scheme 27.[503] Treatment with excess HX results in cleavage of the platinum–alkynyl linkage.[503]

The monodentate dppm complexes *trans*-[Pt(C≡CR)$_2$(dppm-*P*)$_2$] have been employed as precursors to bimetallic dppm-bridged complexes containing copper, silver, gold, mercury, rhodium, iridium, molybdenum, tungsten, or a second platinum (Scheme 29).[8,547–52] The alkynyl-bridged platinum–rhodium species [Pt(C≡CR)(μ-η^1,η^2-C≡CR)(μ-dppm)$_2$Rh(CO)]PF$_6$ (R = Me, Ph, C$_6$H$_4$Me-4, CH$_2$CH$_2$Ph, CMe=CH$_2$) are fluxional at ambient temperature, owing to interchange of the pseudoaxial and pseudoequatorial hydrogens of the dppm ligands and of the terminal and bridging alkynyl groups.[549] A platinum–molybdenum species has also been obtained by reaction of [Mo(CO)$_3$(dppm-*P,P*)(dppm-*P*)] with *trans*-[PtH(C≡CPh)(PPh$_3$)$_2$].[553] Platinum–tungsten complexes *trans*-[Pt{C$_2$W$_2$(OBu^t)$_5$}$_2$(PMe$_2$Ph)$_2$] have been prepared by elimination of *t*-butyl alcohol from *trans*-[Pt(C≡CH)$_2$(PMe$_2$Ph)$_2$] and the [W$_2$(OBu^t)$_6$] dimer.[554]

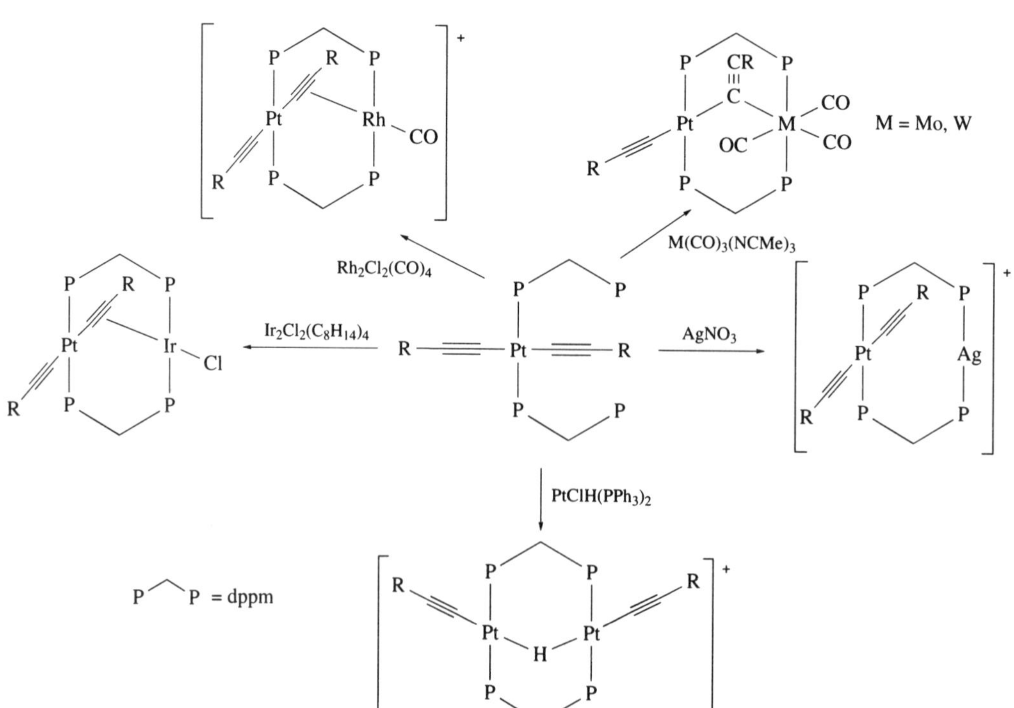

Scheme 29

8.3.6 Ylides

8.3.6.1 *Preparation*

The majority of platinum(II) ylide complexes that have been prepared are phosphorus compounds. They have been prepared in a number of ways, including addition of a preformed ylide species, and several variations on the theme of oxidative addition of organic dihalides to zerovalent platinum–phosphine complexes and attack of a phosphine on a halomethyl moiety.

The phosphorus ylide $Bu^t_2MeP{=}CH_2$ reacts with $[PtMe_2(cod)]$ to give *cis*-$[PtMe_2(CH_2PMeBu^t_2)_2]$, whereas with a dichloroplatinum complex the reaction proceeds as shown in Equation (80).[555] In the reaction of $PtCl_2$ with $Ph_3P{=}CHCOMe$, metallation of one of the phenyl rings occurs to generate the dimeric complex (**52**).[556] Addition of dimethylmethyleniminium chloride to $[Pt(PPh_3)_3]$ yields the bidentate ylide complex $[\overline{Pt(CH_2NMe_2CH_2}NMe_2)Cl(PPh_3)]Cl$ which, on heating in acetonitrile, is converted into the carbene complex *cis*-$[PtCl_2(CHNMe_2)(PPh_3)]$.[557]

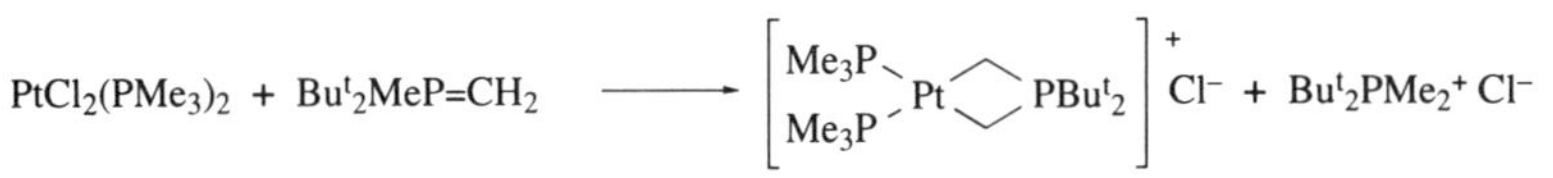

(80)

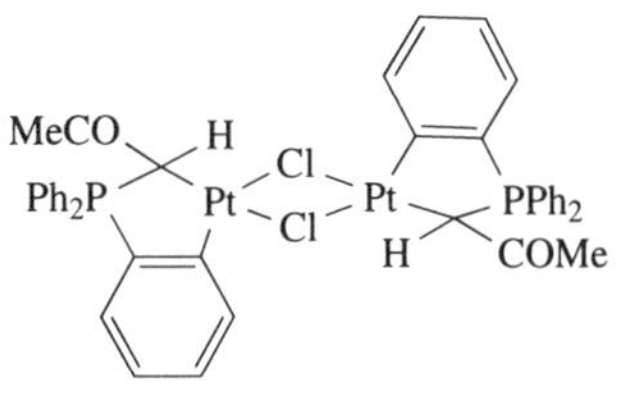

(**52**)

Oxidative addition of $(RO)_2P(O)CH_2I$ (R = Me, Et) to $[Pt(PPh_3)_4]$ proceeds to give *cis*-$[PtI\{CH_2P(O)(OR)_2\}(PPh_3)_2]$, which isomerizes slowly to its *trans* form.[558] Whereas addition of $CH_2X^1X^2$ (X^1, X^2 = Cl, Br, I, except $X^1 = X^2$ = Cl) or $CHBr_3$ to $[Pt(PPh_3)_2(C_2H_4)]$ leads to simple oxidative addition,[559] the reaction of $[Pt(PEt_3)_3]$ with diiodomethane gives *trans*-$[PtI(CH_2PEt_3)(PEt_3)_2]I$. This has been proposed to occur via the intermediacy of *cis*-$[PtI(CH_2I)(PEt_3)_2]$, which is susceptible to attack by the nucleophilic PEt_3. Similar ylide complexes are formed by attack of a tertiary phosphine on *cis*- or *trans*-$[PtX(CH_2X)(PPh_3)_2]$,[559] and the reactions of $[Pt(PPh_3)_2(C_2H_4)]$ with CH_2X_2 in the presence of added triphenylphosphine also lead to ylide complexes.[560]

Treatment of $[Pt(CH_2I)_2(cod)]$ with excess triphenylphosphine generates *cis*-$[PtI(CH_2PPh_3)_2(PPh_3)]I$, whereas the corresponding reaction of $[Pt(CH_2Cl)_2(cod)]$ yields *cis*-$[Pt(CH_2Cl)_2(PPh_3)_2]$.[372,561] The reaction of $[Pt(CH_2I)_2(cod)]$ with 5 equiv. of PMe_3 produces *cis*-$[Pt(CH_2PMe_3)_2(PMe_3)_2]I_2$.[372] The chloromethyl complex $[Pt(CH_2Cl)_2(cod)]$ reacts with tricyclohexylphosphine to give the unexpected complex *cis*-$[PtCl_2(CH_2CH_2PCy_3)(PCy_3)]$.[371]

Treatment of the dppm-bridged platinum(I) cation $[Pt_2H(PPh_3)(\mu\text{-dppm})_2]^+$ with excess diazomethane leads to the methylene-bridged ylide complex $[Pt_2Me(CH_2PPh_3)(\mu\text{-}CH_2)(\mu\text{-dppm})_2]^+$ which, in the presence of dichloromethane, may be converted into $[Pt_2Cl(CH_2PPh_3)(\mu\text{-}CH_2)(\mu\text{-dppm})_2]^+$.[129] No insertion occurs into the Pt–P bond of the corresponding PMe_2Ph complex.

Ylide complexes may alternatively be prepared by formal insertion of a methylene equivalent into the Pt–N bond of a pyridine complex. Thus, cleavage of Zeise's dimer with pyridine yields *trans*-$[PtCl_2(C_2H_4)(NC_5H_5)]$, which reacts with CH_2N_2 or $MeCHN_2$ in the presence of excess pyridine to give *trans*-$[PtCl_2(CHRNC_5H_5)(NC_5H_5)]$ (R = H, Me). The reaction with $EtOCOCHN_2$ at −30 °C generates a related product, but on warming to ambient temperature this rearranges to *cis*-$[PtCl\{CH(CO_2Et)Cl\}(NC_5H_5)_2]$.[562]

8.3.6.2 *Structure and properties*

Bond distances obtained from x-ray diffraction data for some phosphorus ylide complexes are presented in Table 12. The Pt–C distances lie in the range 0.208–0.214 nm, which corresponds to the middle of the envelope found for Pt–Me bond lengths. The ylide ligand exerts a *trans* influence comparable to that of the methyl group.

Table 12 X-ray diffraction data for platinum(II) ylide complexes.

Complex		Pt–C (nm)	Other bond lengths (nm)	Ref.
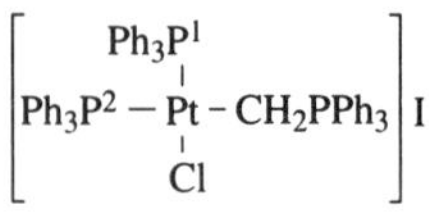		0.212(2)	Pt–Cl 0.240(1) Pt–P[1] 0.225(1) P–P[2] 0.233(1)	560
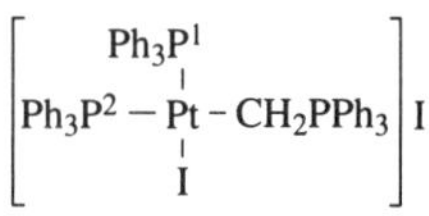		0.212(2)	Pt–I 0.266(1) Pt–P[1] 0.226(1) Pt–P[2] 0.233(1)	560
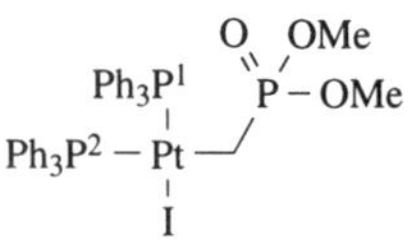		0.2124(6)	Pt–P[1] 0.2257(2) Pt–P[2] 0.2331(2) Pt–I 0.26561(7)	558
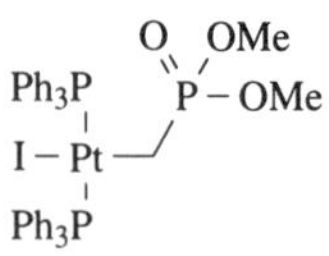		0.2080(8)	Pt–P 0.2305(2) 0.2312(2) Pt–I 0.26774(8)	558
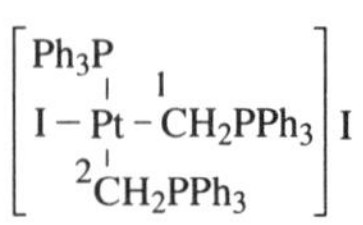	Pt–C[1] Pt–C[2]	0.2095(21) 0.2084(22)	Pt–P 0.2296(6) Pt–I 0.2679(2)	372
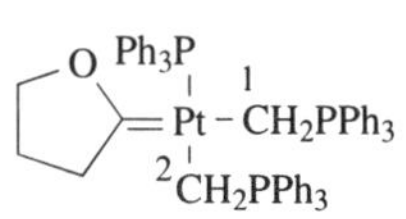	Pt–C[1] Pt–C[2]	0.2122(21) 0.2092(22)	$Pt–C_{carbene}$ 0.1990(22) Pt–P 0.2299(6)	561
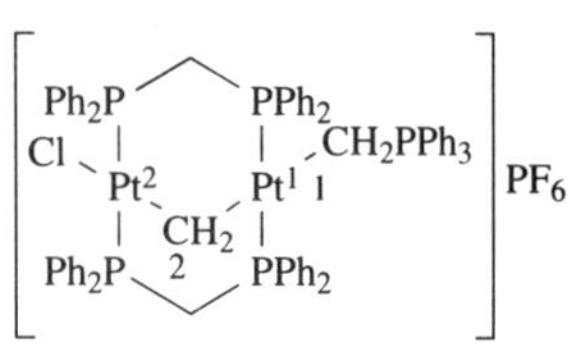	Pt[1]–C[1] Pt[1]–C[2] Pt[2]–C[2]	0.2145(21) 0.2131(26) 0.2130(25)	Pt[1]–P 0.2283(7) 0.2265(7) Pt[2]–P 0.2258(7) 0.2276(7) Pt[2]–Cl 0.2429(8)	129

8.3.6.3 Reactions

The ylide complex *cis*-$[PtCl(CH_2PPh_3)(PPh_3)_2]Cl$, formed by reaction of *cis*- or *trans*-$[PtCl(CH_2Cl)(PPh_3)_2]$ with triphenylphosphine, undergoes loss of PPh_3 in refluxing benzene to yield *cis*-$[PtCl_2(CH_2PPh_3)(PPh_3)]$. This reacts irreversibly with dppe to yield $[PtCl(CH_2PPh_3)(dppe)]Cl$, and reversibly with PPh_3, $AsPh_3$, or $P(NMe_2)_3$. Photolysis of *cis*-$[PtCl_2(CH_2PPh_3)(PPh_3)]$ results in CH_2 extrusion to give *cis*-$[PtCl_2(PPh_3)_2]$.[563] The substitution reactions are enhanced by the presence of aluminum(III) chloride. Addition of $AlCl_3$ to a CH_2Cl_2 solution of *cis*-$[PtCl_2(CH_2PPh_3)(PPh_3)]$ is proposed to give the three-coordinate cation $[PtCl(CH_2PPh_3)(PPh_3)]^+$, which reacts with $AsPh_3$ or SMe_2 to produce $[PtCl(CH_2PPh_3)(PPh_3)L]$.[564]

8.4 PLATINUM(III) COMPOUNDS

Organometallic complexes of platinum(III) are rare. Prior to 1980 they were limited exclusively to carboxylate-bridged species.[2] Since then, a small number of related complexes with other bridging ligands have been prepared and the first mononuclear platinum(III) compound has been isolated.

The platinum(II) complex $[Bu_4N]_2[Pt(C_6Cl_5)_4]$ has been oxidized with Cl_2, Br_2, I_2, or $TlCl_3$, or electrochemically to produce the first known mononuclear platinum(III) compound, $Bu_4N[Pt(C_6Cl_5)_4]$.[386]

It is blue and stable to air and moisture, with a melting point of 210 °C. The analogous pentafluorophenyl complex could not be prepared, however. The molecular structure of $Bu_4N[Pt(C_6Cl_5)_4]$ has been determined. The anion exhibits Pt–C bond distances in the range 0.2090(8) nm to 0.2097(7) nm, which are identical, within experimental error, with those found for the corresponding platinum(II) anion.[386]

A number of dimeric platinum(III) complexes bridged by 2-hydroxypyridinate ligands have been prepared. By analogy to the preparation of $[Pt_2Me_4(\mu\text{-}O_2CMe)_2(NC_5H_5)_2]$ from $[Pt_2Me_4(\mu\text{-}SEt_2)_2]$ and silver(I) acetate, reaction of $[Pt_2Me_4(\mu\text{-}SEt_2)_2]$ with $Ag(2\text{-}OC_5H_4N)$ (Aghp) or related salts in the presence of pyridine proceeds according to Equation (81).[565] Complexes containing only one pyridine in a terminal site have also been isolated.[566] Without added pyridine, complexes which retain one diethyl sulfide ligand in a terminal site have been obtained. In the complexes $[Pt_2Me_4(\mu\text{-}Xhp)_2(SEt_2)]$ (X = Me, H, F) and $[Pt_2Me_4(\mu\text{-}Xhp)_2(py)_n]$ (n = 1, 2; X = H, F, Cl, Br, Me) the Pt–Pt bond lengths lie between 0.2543(1) nm and 0.2571(1) nm, indicative of a strong metal–metal interaction. The Pt–Me distances also lie within a narrow range (0.202–0.206 nm).[567] The bis(pyridine) complexes are found as the head-to-tail isomer in the solid state, whereas the mono(pyridine) species are obtained in the head-to-head form.[565,566] NMR studies reveal that the head-to-head species undergo slow exchange of the terminal ligand(s) with free ligand in solution.[568]

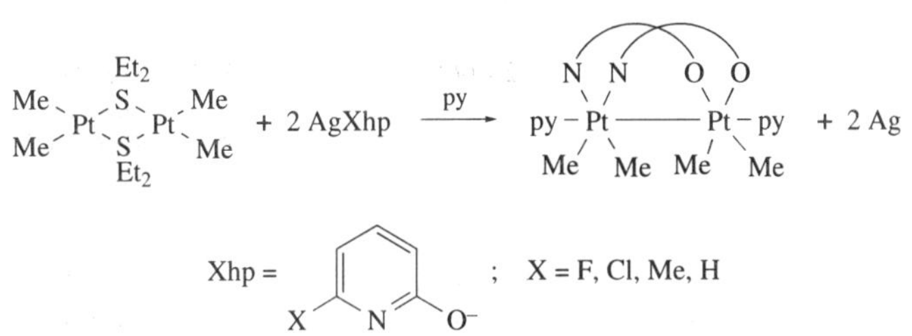

(81)

8.5 PLATINUM(IV) COMPOUNDS

8.5.1 Preparation

The number of organometallic compounds of platinum(IV) has grown tremendously since 1980. A number of methods for their preparation have been developed, the major ones being oxidative addition to platinum(II) precursors, insertion into strained cyclopropane rings to give platinacyclobutanes, and cleavage of the tetrameric species $[PtMe_3X]_4$ by neutral ligands.

8.5.1.1 Oxidative addition to platinum(II) complexes

The scope and mechanisms of these reactions were discussed in some detail in Section 8.3.1.3. There are many examples of oxidative addition of alkyl halides to platinum(II) complexes containing nitrogen or phosphorus ligands and of addition of X_2 to organoplatinum precursors. Trifluoromethylplatinum(IV) complexes may be obtained by addition of X_2 or RX to $[Pt(CF_3)_2L_2]$ (L_2 = bipy, TMEDA),[422] or by oxidative addition of CF_3I to $[PtMe_2L_2]$.[75] In some cases, double oxidative addition to platinum(0) leads to a platinum(IV) complex. This is illustrated by the reactions of $[Pt(PPh_3)_2(C_2H_4)]$ with $Ph_2P(CH_2)_nSnR_3$ (n = 2, 3; R = Me, Ph). When R = Ph, the expected platinum(II) product is obtained, but with R = Me the organoplatinum(IV) complex $[PtMe_2\{SnMe_2(CH_2)_nPPh_2\}_2]$ is produced.[569]

8.5.1.2 Platinacyclobutanes

The general method for the synthesis of platina(IV)cyclobutanes is the reaction of Zeise's dimer, $[Pt_2Cl_2(\mu\text{-}Cl)_2(C_2H_4)_2]$, with cyclopropane derivatives in the presence of a neutral ligand, usually pyridine. The "simple" cyclopropanes employed include methylcyclopropane,[570] $\overline{CH_2CHR^1CHR^2}$ ($R^1 = R^2$ = Me, Ph, C_6H_4Me-4; R^1 = Ph, R^2 = Me, D),[571] and also those bearing a remote alcohol function[572] or an electron-withdrawing keto group.[573] Platinacyclobutanes have also been prepared from bicyclic cyclopropanes, including bicyclo[6.1.0]nonane[574] and bicyclo[4.1.0]heptane.[575] A series of complexes has been obtained from tricyclic precursors based on the norbornane or norbornene fragment.

Treatment of *exo*-tricyclo[3.2.1.0[2,4]]oct-6-ene or *exo,exo*-tetracyclo[3.3.1.0[2,4].0[6,8]]nonane with Zeise's dimer and pyridine yields (**53**) or (**54**) respectively.[576] With *endo*-tricyclo[3.2.1.0[2,4]]oct-6-ene, however, a platinacyclobutane is not obtained.[577] In the case of *exo*-tricyclo[3.2.1.0[2,4]]octan-6-one, two isomers are obtained in a 2.3:1 ratio, the major isomer being formed by insertion into the cyclopropyl bond closer to the ketone functionality.[578]

(53)

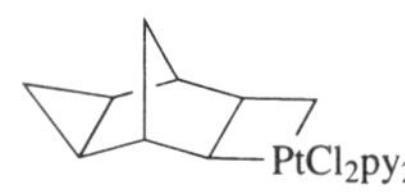

(54)

Jennings and co-workers[579] achieved the first diastereoselective synthesis of 2,3,4-trisubstituted platina(IV)cyclobutanes. Reaction of norbornene with ethyl diazoacetate, followed by reduction with LAH, yields an epimeric mixture of alcohols in a 1.7:1 ratio. Addition of Zeise's dimer results in preferential reaction with the minor, *syn* isomer, and addition of the appropriate amount results in quantitative precipitation of the platinacyclobutane derivative (Equation (82)). Subsequent addition of $[Pt_2Cl_2(\mu\text{-}Cl)_2(C_2H_4)_2]$ to the remaining *anti* isomer generates the other diastereomer. Each diastereomer may be converted to its bis(pyridine) complex.

(82)

8.5.1.3 Reaction of $[PtMe_3X]_4$

Addition of methanethiol to $[PtMe_3X]_4$ (X = Cl, Br, I) results in tetrameric species of the type $[(PtMe_3SMe)_2(PtMe_3X)_2]$ (X = Br, I) or $[(PtMe_3SMe)_3(PtMe_3Cl)]$.[580] Similarly, reaction of the hydroxy complex $[PtMe_3OH]_4$ with RSH (R = Me, Et, Pr^i, Bz, Ph) yields $[PtMe_3SR]_4$, which is unreactive toward cleavage by donor ligands.[581]

Most reactions of the trimethylplatinum(IV) tetramers, $[PtMe_3X]_4$, with donor ligands result in rupture of the tetrameric unit, and the majority of reports involve cleavage by multidentate ligands. Abel and co-workers have made extensive studies of such reactions, particularly those involving bidentate group 16 donor ligands, where inversion at the donor atoms may be studied conveniently by NMR spectroscopy. Reactions of $[PtMe_3X]_4$ with species of the type $MeE(CH_2)_nEMe$ (E = S, Se; n = 2, 3) yield *fac*-$[PtMe_3X\{MeE(CH_2)_nEMe\}]$,[582] which have been characterized by ^{13}C, ^{77}Se, and ^{195}Pt NMR spectroscopy,[583] whereas the corresponding reactions with MeECHREMe (R = H, Me) or MeEEMe produce dimeric complexes. Analogous reactions with MeSeCH=CHSeMe,[584] $RTe(CH_2)_3TeR$ (R = Me, Ph), 1,2-$(MeTe)_2C_6H_4$,[585] or $[Fe(C_5H_4EMe)_2]$ (E = S, Se)[586] also generate complexes of the type *fac*-$[PtMe_3XL_2]$. The potentially terdentate ligand $S(CH_2SMe)_2$ reacts to give species of the form (**55**),[587] and $(MeS)_2CHCH(SMe)_2$ also gives chelate complexes of type (**56**).[588] Treatment of (**56**) with further $[PtMe_3X]_4$ yields a dimeric, bis(chelate) complex,[589] whereas addition of $(MeS)_2C{=}C(SMe)_2$ to $[PtMe_3Br]_4$ produces the dimeric complex $[(PtMe_3Br)_2\{\mu\text{-}(MeS)_2C{=}C(SMe)_2\}]$ directly.[590] The platinum(II) complex $[Pt(SMe)_2(dppe)]$ also serves as a chelating ligand in its reaction with $[PtXMe_3]_4$ to yield *fac*-$[PtMe_3X(\mu\text{-}SMe)_2Pt(dppe)]$.[591]

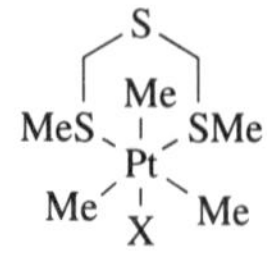

(55)

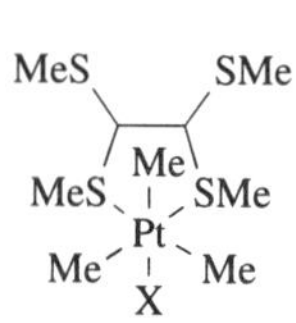

(56)

Reactions of $[PtMe_3Cl]_4$ with NaS_2AsMe_2,[592] or of $[PtMe_3I]_4$ with NaS_2PPh_2[593] or $NaS_2P(OR)_2$,[594] produce dimeric complexes, which may be cleaved by excess pyridine to give monomeric complexes, as shown in Equation (83).

The bidentate bis(pyrazolyl)methane $H_2C(C_3HR_2N_2)_2$ (R = H, Me) and bis(2-pyridinyl)methane ligands react with $[PtMe_3X]_4$ (X = Cl, I, OAc, NO_3) to give *fac*-$[PtMe_3XL_2]$.[595] With terpyridine, *fac*-$[PtMe_3Cl(terpy)]$ is formed, in which one of the terminal nitrogens is uncoordinated, and the ligand

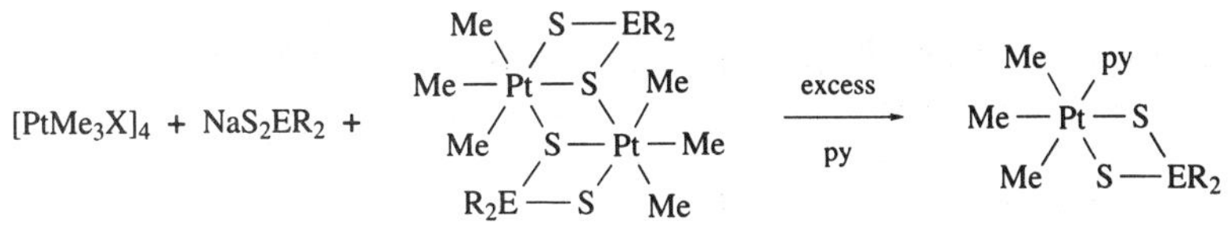

(83)

$ER_2 = AsMe_2, PPh_2, P(OR)_2$

oscillates between bidentate forms in solution.[596] With several potentially terdentate Schiff bases, ionic *fac*-trimethylplatinum(IV) cations are formed,[597] and the diastereomeric products from the reactions of $[PtMe_3X]_4$ with the optically active, bidentate ligands $PhMeCHN{=}CR\text{-}2\text{-}C_5H_4N$ (R = H, Me) could be separated, but they isomerize in solution.[598]

Cationic *fac*-$[PtMe_3L_3]^+$ complexes are formed from the reactions of $[PtMe_3X]_4$ with the terdentate ligands $MeE^1(CH_2)_nE^2(CH_2)_nE^1Me$ (n = 3, $E^1 = E^2$ = S; n = 2, E^1 = S, Se, E^2 = O, S, Se, SS),[599] or with bis(3-aminopropyl)phenylphosphine.[600] With the macrocyclic sulfur ligands 1,4,7-trithiacyclononane, 1,4,7,10-tetrathiacyclododecane, or 10-oxa-1,4,7-trithiacyclododecane, complexes of the type *fac*-$[PtMe_3L_3]$ are formed, in which the oxygen or one of the sulfur atoms is uncoordinated in the latter two cases. In the tetrathia derivative, exchange between coordinated and free sulfur atoms occurs.[601]

Treatment of $[PtMe_3X]_4$ with the tripodal ligands $HC(C_3H_3N_2)_3$, $HC(C_5H_4N)_3$, or $CH(C_3HMe_2N_2)_2(C_5H_4N)$ in the presence of $AgPF_6$ generates *fac*-$[PtMe_3L_3]PF_6$.[595] With anionic tripodal ligands, neutral species of the form *fac*-$[PtMe_3L_3]$ are obtained. These include the tris(3,5-dimethylpyrazolyl)borate derivative *fac*-$[PtMe_3\{(C_3HMe_2N_2)_3BH\}]$,[602] and the heterobimetallic species formed according to Equations (84)[603] and (85).[604]

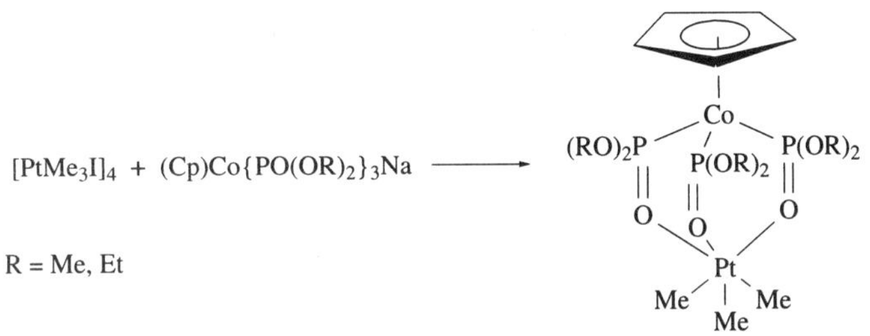

(84)

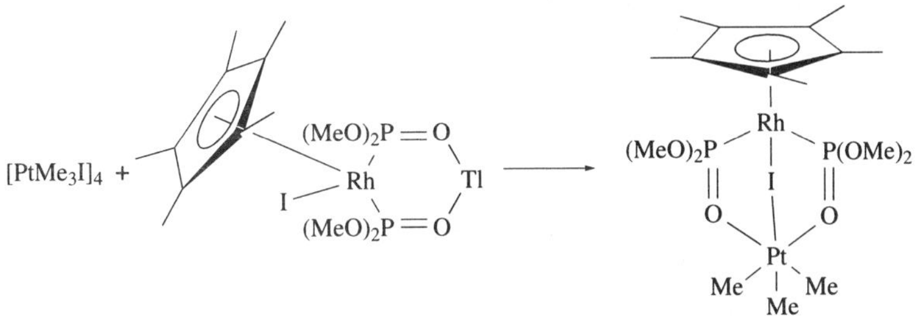

(85)

Interactions of $[PtMe_3X]_4$ with bidentate ligands in which the donor atoms are bonded directly together or are separated by only one spacer atom lead to dimeric products, in which the platinum centers are bridged by the ligand and also the two halides. Thus, reactions with MeEEMe (E = S, Se)[605,606] or PhTeTePh[607] occur as shown in Equation (86). The platinum centers still adopt facial configurations. The cyclic ligands $Me_2\overline{CCH_2EECH_2}$ (E = S, Se) react similarly.[608]

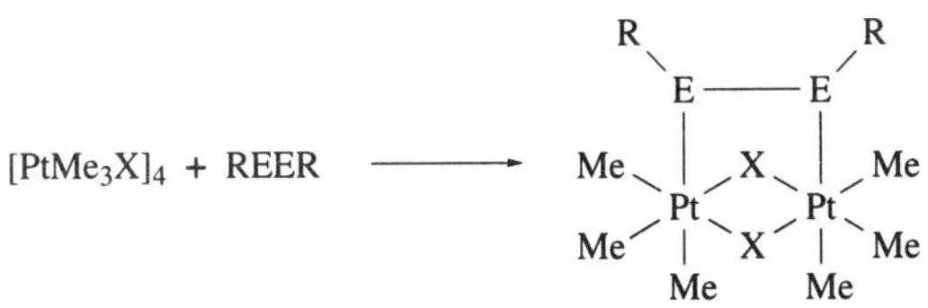

(86)

ER = SMe, SeMe, TePh; X = Br, I

When the donor atoms are separated by one atom, as in MeECHREMe (E = S, Se; R = H, Me), the products adopt structure (**57**).[609] In solution these complexes undergo four distinct dynamic processes, namely, six-membered ring reversal, sulfur or selenium inversion, sulfur or selenium switching between platinum atoms, and Pt–X cleavage leading to methyl group scrambling.[609] When 2 equiv. of $Me_2AsSAsMe_2$ are added to $[PtMe_3X]_4$, the dimeric complex $[\{PtMe_3(\mu\text{-}X)\}_2(\mu\text{-}Me_2AsSAsMe_2)]$ is formed, but a bis(ligand) complex, in which coordination is through one arsenic only, is formed when 8 equiv. are added.[610] Although the donor atoms are separated by two carbons, the diaryldichalcogenides (**58**) also react with $[PtMe_3X]_4$ (X = Cl, Br) to generate bridged complexes.[611]

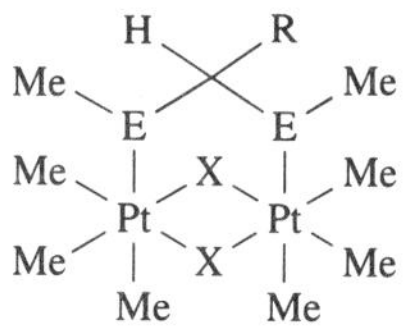

E = S, Se; R = H, Me; X = Cl, Br, I
(57)

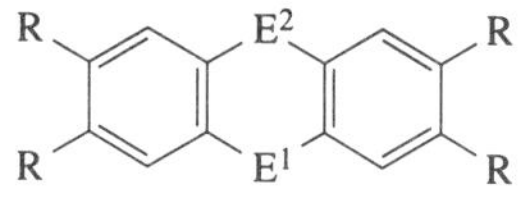

E^1 = S, E^2 = S, Se
E^1 = E^2 = Se
E^1 = E^2 = Te
(58)

The potentially terdentate ligands 1,3,5-trithiacyclohexane[612] and tris(methylthio)methane[613] also function as bidentate, bridging ligands ((**59**) and (**60**)). Intramolecular rearrangement between pairs of coordinated sulfur atoms takes place in solution. In the dimeric complexes of 1,3,5,7-tetrathiacyclooctane (**61**), the platinum atoms undergo a series of 90° pivots between adjacent pairs of coordinated sulfur atoms.[614]

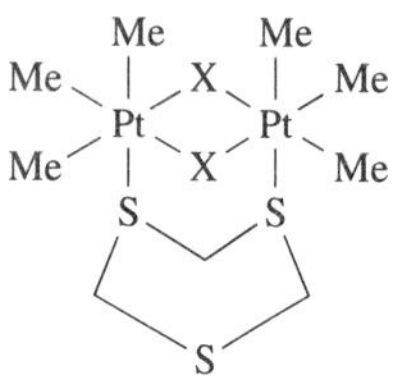

(59)

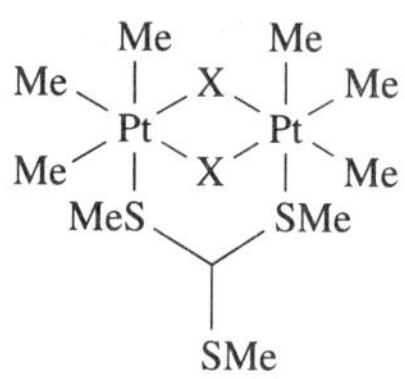

(60)

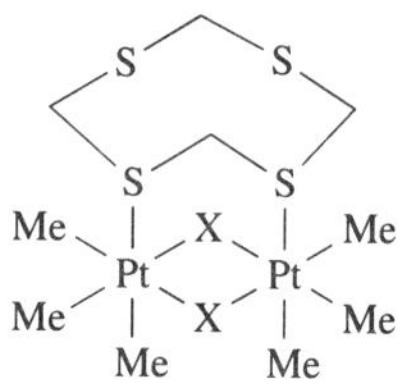

X = Cl, Br, I
(61)

8.5.2 Structure and Properties

X-ray diffraction data for a number of organoplatinum(IV) complexes are presented in Table 13. In each case, the platinum adopts approximately octahedral geometry, and multidentate ligands are involved in most cases. For methylplatinum(IV) species the Pt–C distances lie between 0.201 nm and 0.215 nm. This range is similar to that found for platinum(II) compounds, indicating that the oxidation state of the metal has little effect on the Pt–C bond distance. The structures of the tetrameric species $[PtMe_3X]_4$ (X = Br, I, SCN, SPh) have also been determined.[615] For the three platinacycloalkanes listed in Table 13, the Pt–C distances are also similar to those found for related platinum(II) derivatives. No alkenyl complexes and only one example of an alkynylplatinum(IV) complex have been structurally characterized.

UV photoelectron spectra have been obtained for $[PtMe_3(acac)]_2$, $[PtMe_3(MeCOCHCOCF_3)L]$ (L = H_2O, SMe_2), and $[PtMe_3(CF_3COCHCOCF_3)(H_2O)]$. These illustrate that the platinum 5*d* electrons are extensively delocalized.[625] The electronic spectra of $[PtMe_3I]_4$ and $[PtMe_3I(bipy)]$ have also been discussed.[626] ^{195}Pt NMR studies of the tetramers $[PtMe_3X]$ reveal that the ^{195}Pt resonance frequency decreases in the order X = OH > SPh > SMe > N_3, Cl > Br > SCF_3 > I.[627] From ^1H–^{195}Pt cross-polarization experiments with $[PtMe_3(acac)]_2$ and $[(PtMe_3)_2SO_4]\cdot 4H_2O$, it has been shown that the chemical shift anisotropy of the ^{195}Pt chemical shielding is of the order of 1000 ppm, and hence of such a magnitude that a significant contribution of chemical shift anisotropy to the ^{195}Pt relaxation parameters T_1 and T_2 is evident in solution at higher magnetic fields.[628] Such effects had been observed previously in platinum(II) complexes also.[629]

Abel and co-workers have made extensive studies of several fluxional processes in trimethylplatinum(IV) systems. In particular, they have reported the use of 2D EXSY (NOESY) NMR spectroscopy as a tool for determining rate constants in complex exchange networks. They have illustrated this by using the ^{195}Pt 2D NMR technique to study sulfur inversion in $[PtMe_3X(MeSCH_2CH_2SR)]$ (R = Me, Et),[194] and ^1H 2D EXSY to monitor three-site methyl exchange.[630]

Table 13 X-ray diffraction data for organoplatinum(IV) complexes.

Complex		Pt–C (nm)	Other bond lengths (nm)		Ref.
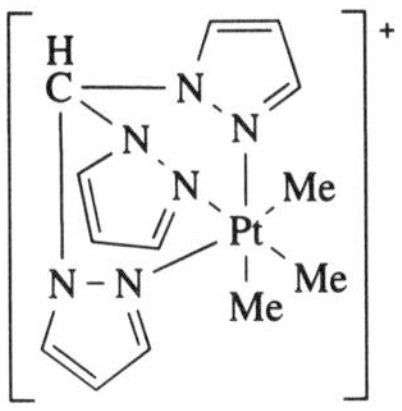		0.2031(8) 0.2056(7) 0.2056(7)	Pt–N	0.2156(6) 0.2156(5) 0.2189(5)	615
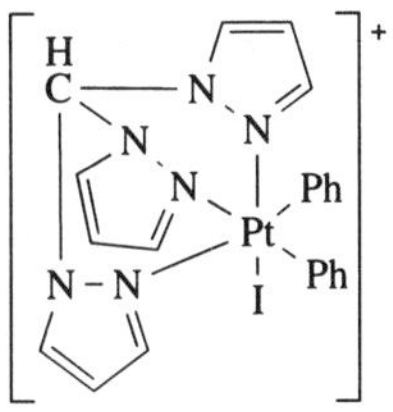		0.205(3) 0.209(3)	Pt–N Pt–I	0.220(2) 0.215(2) 0.201(2) *trans* to I 0.2601(3)	310
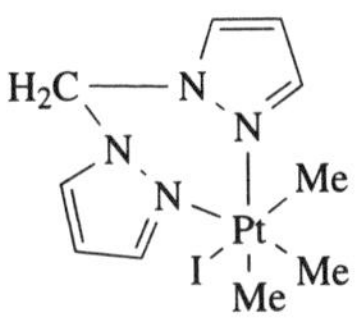		0.2077(6) *trans* to I 0.2032(5) *cis* to I	Pt–N Pt–I	0.2236(4) 0.2843(1)	595
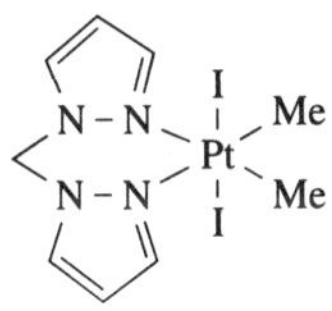		0.2081(8)	Pt–N Pt–I	0.2183(7) 0.2651(1) 0.2647(1)	308
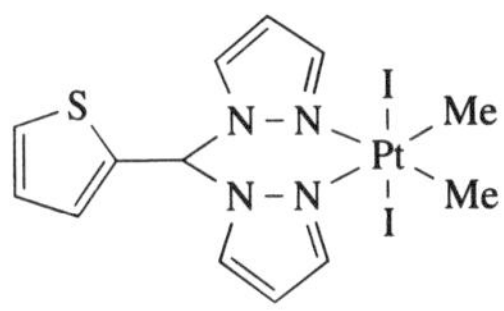		0.2070(7) 0.2097(9)	Pt–N Pt–I	0.2181(5) 0.2192(6) 0.26649(6) 0.26199(7)	309
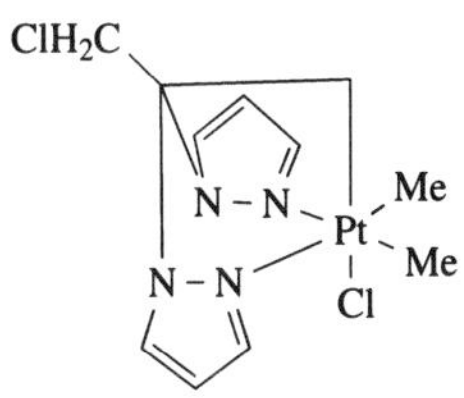	Pt–Me Pt–CH_2	0.2043(7) 0.2035(9) 0.2030(6)	Pt–N Pt–Cl	0.2166(4) 0.2163(6) 0.2443(2)	616
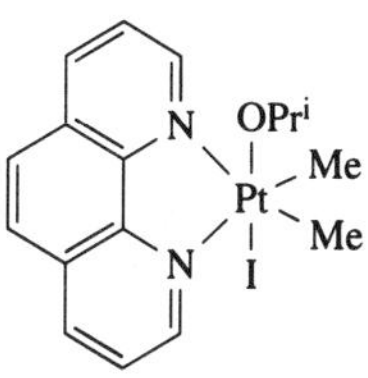		0.2067(8) 0.2048(9)	Pt–N Pt–O Pt–I	0.2164(7) 0.2175(7) 0.2032(6) 0.2624(1)	319

Table 13 (continued)

Complex	Pt–C (nm)	Other bond lengths (nm)		Ref.
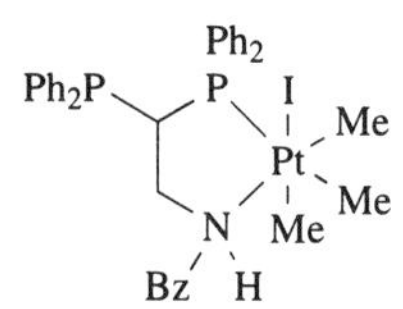	*Molecule 1*			
	0.213(3) *trans* to P	Pt–P	0.2406(9)	103
	0.201(3) *trans* to N	Pt–N	0.225(2)	
	0.207(3) *trans* to I	Pt–I	0.2853(6)	
	Molecule 2			
	0.211(3) *trans* to P	Pt–P	0.2359(9)	
	0.211(3) *trans* to N	Pt–N	0.222(2)	
	0.215(3) *trans* to I	Pt–I	0.2825(6)	
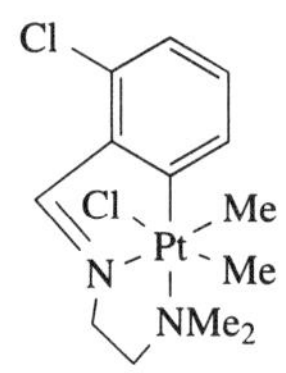	0.2036(13) *trans* to Cl	Pt–N	0.2254(11) *trans* to aryl	617
	0.2068(15) *trans* to N		0.2045(13) *trans* to Me	
	0.1999(12) aryl	Pt–Cl	0.2456(3)	
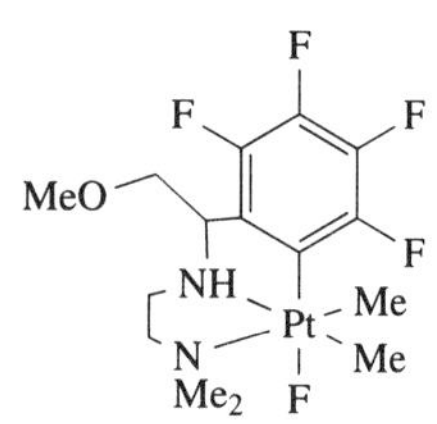	0.2060(14) *trans* to NH	Pt–N	0.2422(10) NMe_2	618
	0.2038(14) *trans* to NMe_2		0.2173(9) NH	
	0.1978(9) aryl	Pt–F	0.2070(5)	
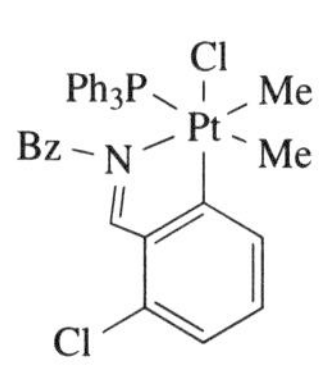	0.2097(10) *trans* to P	Pt–N	0.2152(7)	619
	0.2048(9) *trans* to N	Pt–P	0.2431(2)	
	0.2027(10) aryl	Pt–Cl	0.2422(2)	
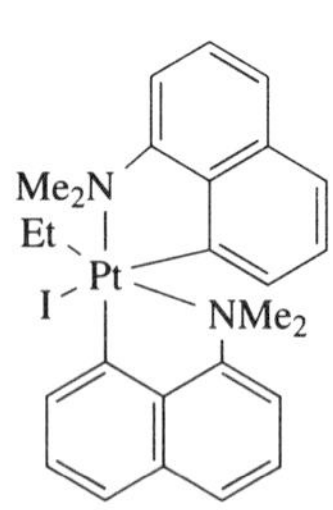	0.2098(5) Pt–Et	Pt–N	0.2406(4) *trans* to Et	620
	0.2009(6) *trans* to I		0.2266(4) *trans* to aryl	
	0.2005(4) *trans* to N	Pt–I	0.27239(5)	
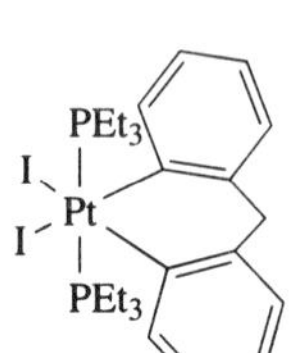	0.2093(15)	Pt–P	0.2381(3)	437
	0.2078(14)		0.2391(3)	
		Pt–I	0.2755(1)	
			0.2814(1)	
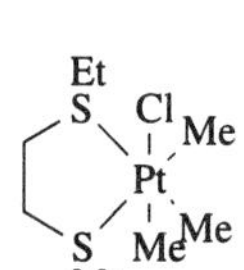	0.2048(12) *trans* to Cl	Pt–Cl	0.2474(5)	621
	0.2055(11) *trans* to SMe	Pt–SMe	0.2443(5)	
	0.2061(14) *trans* to SEt	Pt–SEt	0.2442(5)	

Table 13 (continued)

Complex	*Pt–C* (nm)	*Other bond lengths* (nm)	*Ref.*
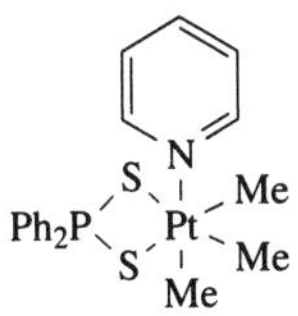	0.210(1) *trans* to N 0.206(2) 0.203(1)	Pt–N 0.2239(9) Pt–S 0.2530(3) 0.2532(3)	593
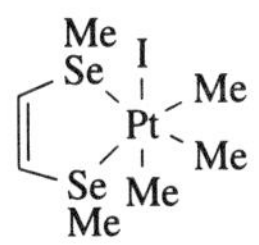	0.208(2) *trans* to I 0.210(2) 0.212(2)	Pt–Se 0.2531(2) 0.2535(3) Pt–I 0.2791(2)	584
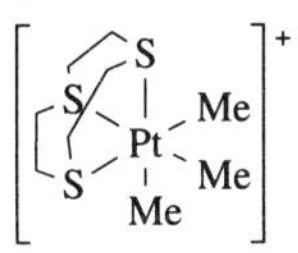	0.2073(7) 0.2087(7) 0.2075(7)	Pt–S 0.2411(3) 0.2405(3) 0.2405(3)	601
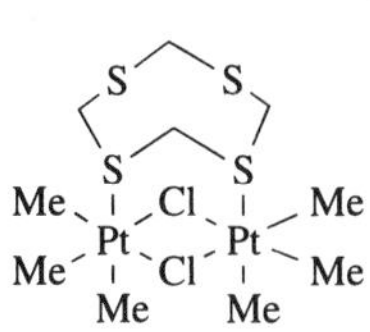	0.2057(26) *trans* to Cl 0.2054(25) 0.2038(21) 0.2049(26) 0.2068(28) *trans* to S 0.2043(24)	Pt–S 0.2472(5) 0.2478(5) Pt–Cl 0.2493(5) 0.2505(5) 0.2506(5) 0.2507(5)	614
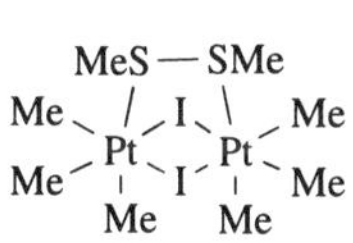	0.209(3) *trans* to Cl 0.206(3) 0.202(3) 0.211(3) 0.206(3) *trans* to Se 0.208(3)	Pt–Se 0.2564(3) 0.2563(3) Pt–I 0.2798(2) 0.2778(2) 0.2790(2) 0.2780(2)	606
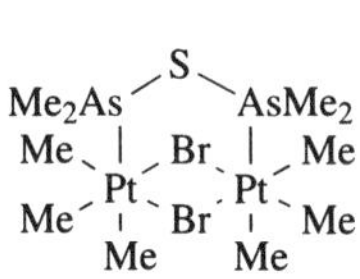	0.2009(21) *trans* to Br 0.2047(22) 0.1985(21) 0.1990(23) 0.2084(21) *trans* to As 0.2037(21)	Pt–As 0.2493(4) 0.2493(4) Pt–Br 0.2611(5) 0.2638(5) 0.2613(5) 0.2653(5)	610
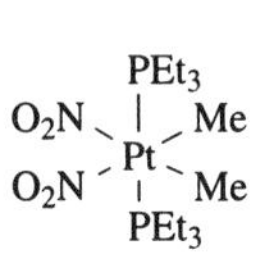	0.2111(8) 0.2077(9)	Pt–N 0.2184(6) 0.2198(8) Pt–P 0.2430(3) 0.2337(3)	622
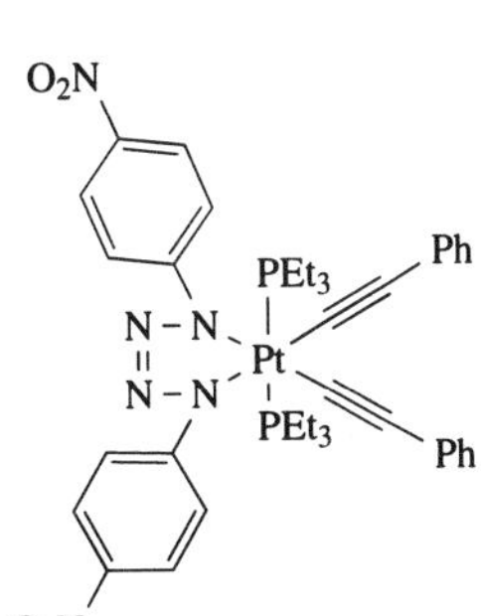	0.199(1) 0.200(1)	Pt–N 0.2119(9) 0.211(1) Pt–P 0.2401(3) 0.2396(3) C≡C 0.120(2) 0.119(2)	623

Table 13 (continued)

Complex	*Pt–C* (nm)	*Other bond lengths* (nm)	*Ref.*
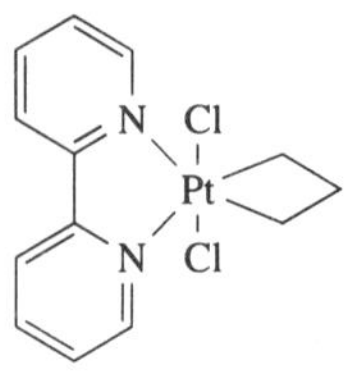	0.207(4)	Pt–N 0.220(3) Pt–Cl 0.223(2)	458
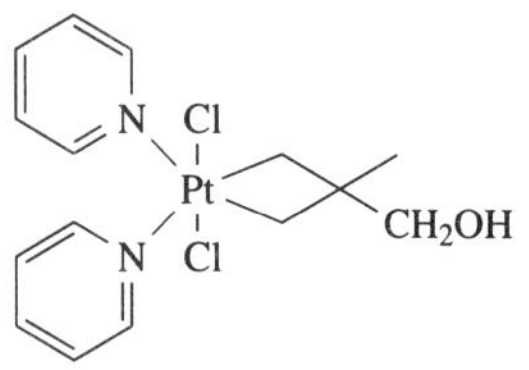	0.2039(6) 0.2042(7)	Pt–N 0.2207(5) 0.2222(5) Pt–Cl 0.2310(2) 0.2318(2)	572
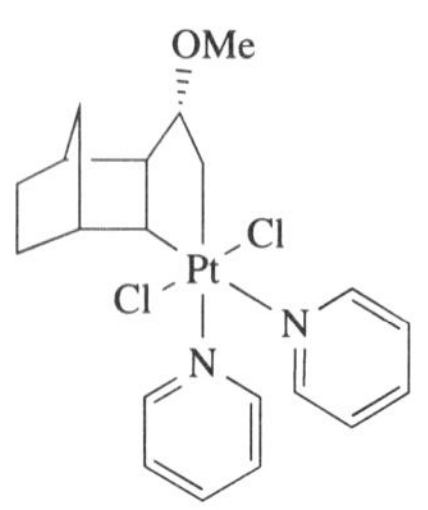	0.204(1) CH_2 0.208(1) CH	Pt–N 0.224(1) *trans* to CH_2 0.226(1) *trans* to CH Pt–Cl 0.2319(3) 0.2325(4)	579

In monomeric trimethylplatinum(IV) complexes containing chelating group 16 ligands, two major fluxional processes take place. The lower energy process is pyramidal inversion at sulfur, selenium or tellurium. Examples of $\Delta G^{\ddagger}$ values are 48 kJ mol^{-1} for $[PtMe_3Br(SMe_2)_2]$[631] and 56–58 kJ mol^{-1} for the chelate complexes $[PtMe_3X(MeSCH_2SCH_2SMe)]$.[587] Higher values of $\Delta G^{\ddagger}$ are found for selenium[584] or tellurium[585] inversion. In the complexes $[PtMe_3X\{MeE^1(CH_2)_nE^2Me\}]$ (E^1, E^2 = S, Se; n = 2, 3), the lowest barrier to inversion is found with $MeS(CH_2)_3SMe$ and the highest with $MeSe(CH_2)_2SeMe$.[582,632] A second, higher-energy process is intramolecular scrambling of the methyl groups on platinum. This has $\Delta G^{\ddagger}$ values in the range 80–90 kJ mol^{-1}, depending on the nature of the other ligands.[584,587,588]

In dimeric systems, inversion and methyl scrambling still occur, but the inversion is a synchronous process involving chalcogen pairs. In $[(PtMe_3)_2(\mu\text{-}I)_2(\mu\text{-}\overline{EECH_2CMe_2CH_2})]$ (E = S, Se), for example, the barrier to inversion is about 20 kJ mol^{-1} higher than for single site inversion.[608] In the case of $[(PtMe_3)_2(\mu\text{-}X)_2(\mu\text{-}MeSCH_2SeMe)]$, inversion of both sulfur and selenium occurs, and also ring reversal, ligand switching between platinum atoms, and scrambling of the platinum methyls.[633] When 1,3,5-trithiahexane acts as a bridging ligand (**59**), a series of 60° pivots about the individual Pt–S bonds average the ligand methylene and the equatorial platinum methyl environments, whereas a higher energy ligand dissociation–recombination leads to scrambling of all the methyl groups.[612,634] In (**60**), sulfur inversion, a 1,3-metal pivoting motion, and scrambling of the platinum methyls by $PtMe_3$ rotation all occur.[613] Similarly, in (**61**), a series of 90° 1,5-metal pivots and 120° rotations of the $PtMe_3$ groups account for the variable-temperature NMR observations.[614]

8.5.3 Reactions

The most extensively studied reactions of organoplatinum(IV) compounds are those involving platinacyclobutanes. The ready, but stereospecific, skeletal rearrangements of platinacyclobutanes have been discussed in terms of the isolobal analogy. It has been suggested that $[\overline{Pt(CH_2CH_2CH_2)}Cl_2L]$, formed by loss of L from the six-coordinate complex, is analogous to $C_4H_7^+$, a nonclassical carbonium ion, and the $PtCl_2L$ fragment rearranges to a pyramidal geometry and migrates across the face of the C_3

unit.[635] The rearrangement has also been described in terms of permutational isomerism of Jahn–Teller wheels.[636]

Thermolysis or photolysis of the platinacyclobutanes [Pt(C_3H_5Me)Cl_2L_2] results in ylide formation when L = py or 2-Mepy, but when L = MeCN the 1-butene complex *trans*-[PtCl_2(CH_2=CHCH_2Me)(MeCN)] is obtained.[570,637–40] The alkene could be produced by means of an α- or a β-elimination. The mechanism of alkene formation was investigated for the complexes [Pt(*trans*-CHMeCHMeCD_2)Cl_2]$_4$ and [Pt(CHMeCMe_2CD_2)Cl_2]$_4$; reaction with py or 2-Mepy was followed by release of the free alkene by treatment with PPh_3 (Equation (87)). The observed products were consistent with the α-elimination mechanism.[641] The skeletal isomerization of platinacyclobutanes may occur faster than decomposition, however, and this makes interpretation of product distributions difficult.[642] In the case of [Pt(CHPhCH_2CH_2)Cl_2(py)$_2$] (**62**), the complex could be obtained in optically pure form by kinetic resolution, and the optical activity was lost during the isomerization to (**63**) depicted in Equation (88).[643]

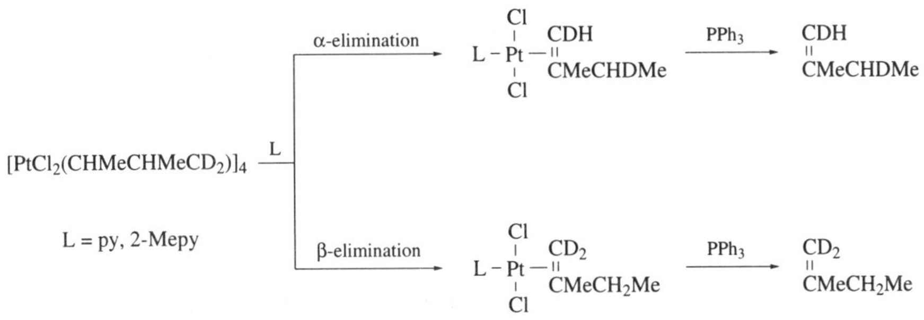

(87)

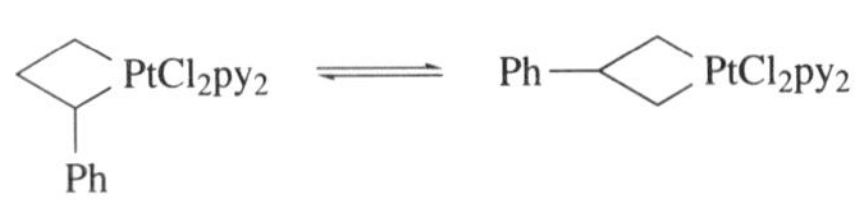

(**62**) (**63**) (88)

Treatment of [Pt($CH_2CR_2CH_2$)X_2L_2] (X = Cl, Br; R = H, Me; L = py, 4-Mepy) with iodide or thiocyanate in acetonitrile solution results in cyclopropane elimination. The reaction is first order in platinum but zero order in the nucleophile, and a mechanism involving rate-determining ligand loss has been proposed.[644] Following loss of L, attack of I^- or SCN^- on the coordinatively unsaturated complex would occur, before reductive elimination of the C_3 ring from [Pt($CH_2CR_2CH_2$)X_2YL]$^-$. In a related reaction, attack of an alkene on [Pt(CH_2CHRCH_2)X_2L_2] (X = Cl, Br; R = H, Me; L_2 = phen) results in fission of the C_3 moiety; ethene is produced when R = H, but both C_2H_4 and C_3H_6 are detected when R = Me.[645] When [Pt($CH_2CH_2CH_2$)X_2L_2] (X = Cl, Br; L_2 = bipy, phen, TMEDA) is treated with PPh_3, PPh_2H or dppm, the major products are ethene, propene, and cyclopropane. In acetonitrile or DMSO solvent, ethene is obtained in >80% yield, and this has been proposed to be formed via an alkene–ylide intermediate.[646]

Treatment of the platinacyclobutane derived from bicyclo[6.1.0]nonane (**64**) with KCN in the absence of a coordinating solvent regenerates bicyclo[6.1.0]nonane, in addition to 1-methylcyclooctene, the *exo*-methylene derivative, and cyclononene.[574] Thermal decompositions and the reaction with diazomethane of the platinacyclobutane complex (**53**) are shown in Scheme 30.[647] The trisubstituted platinacyclobutanes react stereospecifically with CH_2N_2, as shown in Equation (89).[579] Electrochemical reduction of [Pt($CH_2CH_2CH_2$)Cl_2(bipy)] does not rupture the platinacyclobutane ring, but produces chloride ions and the platina(II)cyclobutane complex [Pt($CH_2CH_2CH_2$)(bipy)].[648]

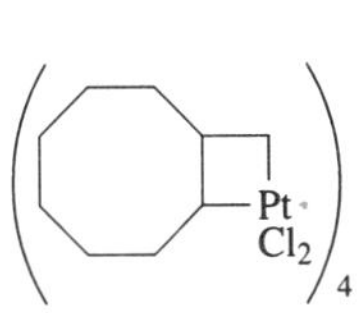

(**64**)

Platinacyclobutane derivatives of the type [Pt{CH_2CR^1(CHR^2OR^3)CH_2}Cl_2(py)$_2$] undergo ring expansion in aqueous acetone, by a first-order process, to yield the platinacyclopentane species [Pt{CHR^2CR^1(OH)CH_2CH_2}Cl_2(py)$_2$].[649]

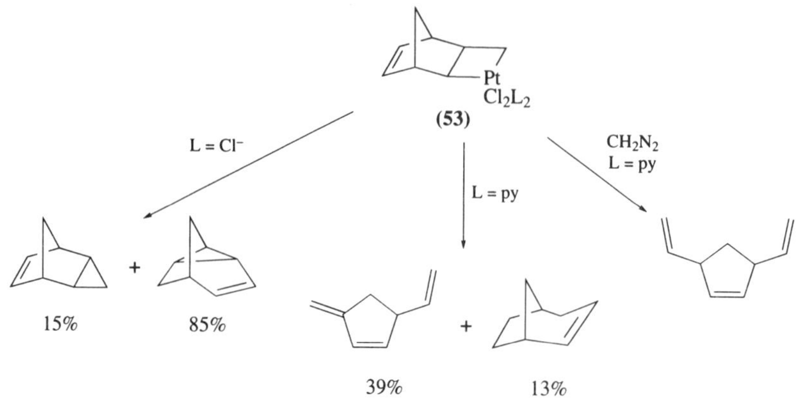

Scheme 30

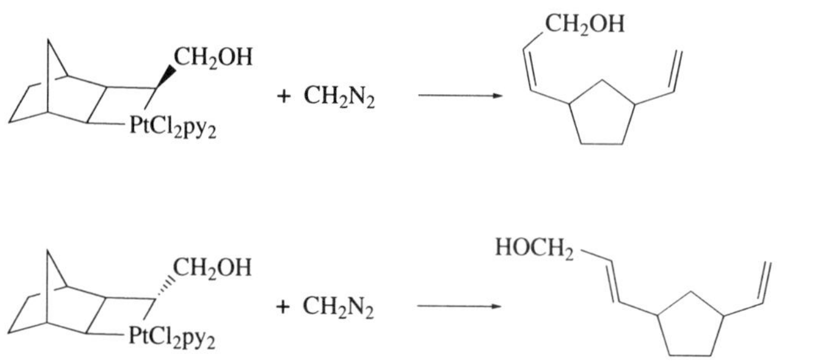

(89)

Photochemical decomposition of the platinacyclopentanes $[\overline{Pt(CH_2CH_2CH_2C}H_2)I_2L_2]$ (L = PMe_2Ph, PPh_3) in CH_2Cl_2, CH_2Br_2, or DMSO solution leads to ethene and 1-butene, whereas with $[\overline{Pt(CH_2CH_2CH_2C}H_2)IMeL_2]$, methane, *n*-pentane and traces of *n*-butane are also obtained.[650]

Thermal decomposition of the tetramethylplatinum(IV) species *cis*-$[PtMe_4(CNR)_2]$ (R = Me, $C_6H_3Me_2$-2,6) leads to *cis*-$[PtMe_2(CNR)_2]$ and ethane. Deuterium-labeling studies indicate that the elimination is an intramolecular process. There is a kinetic isotope effect of 1.30 ± 0.04, and the reaction is proposed to occur via loss of isocyanide to generate a five-coordinate species in which there is a platinum–hydrogen agostic interaction. Differential scanning calorimetric measurements lead to a bond dissociation energy of 133 ± 5 kJ mol^{-1} for the mutually *trans* Pt–Me bonds.[651]

Photolysis of $[PtMe_4(bipy)]$ using 436 nm or 473 nm radiation generates the $[PtMe_3(bipy)]\cdot$ radical, which may be trapped by CH_2Cl_2, MeOH, RC≡CH, or RCO_2Me as *fac*-$[PtMe_3X(bipy)]$ (X = Cl, OMe, C≡CR, OCOR).[652] Thermal reactions of $[PtMe_4L_2]$ (L_2 = bipy, phen) with HX (X = Cl, $OCOCF_3$, OAc, OPh, SPh, OMe) or $HgCl_2$ also lead to species of the type *fac*-$[PtMe_3XL_2]$. These reactions occur by initial electrophilic attack at one of the mutually *trans* Pt–Me bonds. Reaction with C_2X_4 (X = F, CN) also results in Pt–Me cleavage, forming $[\{PtMe_3L_2\}_2(\mu\text{-}C_2X_4)]$. With sulfur dioxide, insertion into one of the mutually *trans* Pt–Me bonds occurs to give *fac*-$[PtMe_3(SO_2Me)L_2]$.[653]

The bis(trifluoromethyl)platinum(IV) complex $[PtIMe(CF_3)_2(py)_2]$ slowly loses iodomethane in acetone solution, a process that is accelerated by light or the presence of free pyridine.[422] The dimethyl analogue, $[PtIMe_2(CF_3)(py)_2]$, formed by oxidative addition of CF_3I to *cis*-$[PtMe_2(py)_2]$, reacts with $AgClO_4$, followed by perchloric acid, to generate *fac*-$[PtMe_2(CF_3)(H_2O)_3]ClO_4$, without cleavage of the Pt–C bonds.[75]

The polymeric $[PtMe_2Br(OH)]_n$ reacts with hydroxide ion to generate the inert dimeric anion $[PtMe_2Br(OH)(\mu\text{-}OH)]_2^{2-}$, but with methanol the neutral dimer $[PtMe_2Br(MeOH)(\mu\text{-}OH)]_2$ is formed.[654] In the reactions of $[PtMe_2Br_2]_n$, $[PtMe_2(OH)_2(H_2O)_{1.5}]_n$ and *fac*-$[PtMe_2Br(H_2O)_3]^+$ with cyanide ion, substitution of the groups *trans* to methyl occurs much more rapidly than for those *cis* to methyl. The substitution reactions compete with reductive elimination.[655]

Glycinatoplatinum(IV) complexes $[\overline{Pt(OCOCH_2N}H_2)Me_2X(H_2O)]$ (X = Br, OH) have been prepared by treatment of the $[PtMe_2X(H_2O)_3]^+$ cation with sodium glycinate. The three possible isomers with the two methyl and X groups in a facial arrangement are obtained in solution. Reactions with excess $NaOCOCH_2NH_2$ yield $[Pt(OCOCH_2NH_2)_2Me_2X(OH)]^{2-}$, containing two monodentate glycinate ligands, or $[\overline{Pt(OCOCH_2N}H_2)_2Me_2X]^{2-}$, in which one ligand is chelating.[656] Heating a basic solution of $Na[\overline{Pt(OCOCH_2N}H_2)Me_2Br(OH)]$ gives the amido-bridged anion (**65**).[657] Exchange of the labile D_2O ligand (which lies *trans* to methyl in each case) in $[\overline{Pt(OCOCH_2N}H_2)Me_2Br(D_2O)]$ and

$[Pt(OCOCH_2NH_2)Me_3(D_2O)]$ with solvent water, with an associated nitrogen or oxygen migration, occurs by both associative and dissociative pathways.[658]

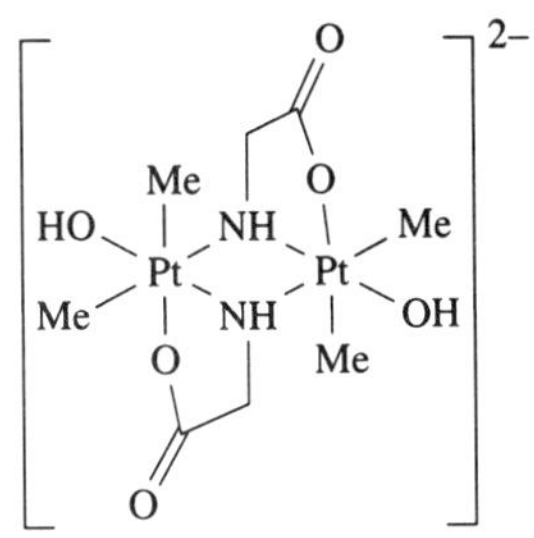

(65)

Treatment of *cis*-$[PtMe_2(OH)_4]^{2-}$ with potassium iminodiacetate or methyliminodiacetate yields first the $[PtMe_2(OH)_2L]^{2-}$ dianion, in which the ligand functions in a bidentate manner, and subsequently gives $[PtMe_2(H_2O)L]^-$, in which the ligand is terdentate. The terdentate complex undergoes a slow isomerization, which is enhanced by light (Equation (90)).[659]

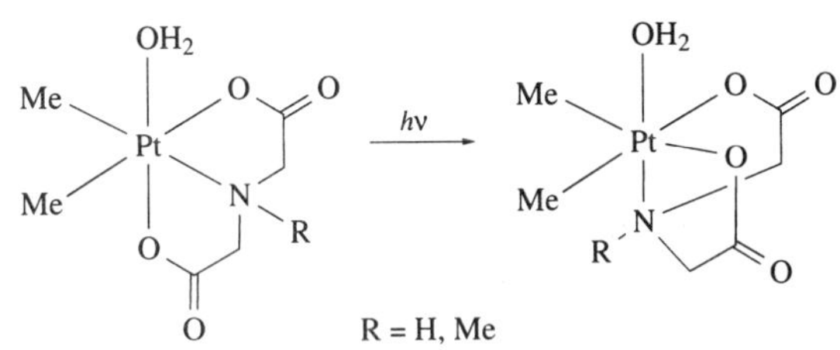

(90)

8.6 COMPOUNDS CONTAINING METALLATED LIGANDS

8.6.1 Preparation

8.6.1.1 Compounds containing metallated phosphorus or arsenic ligands

Cyclometallation of tertiary phosphines and phosphites is well known, especially for those containing aryl substituents or particularly bulky groups. Often these reactions produce stable compounds containing five-membered chelate rings. The driving force may alternatively be the relief of steric crowding in the precursor.

Treatment of $PtCl_2$, K_2PtCl_4 or $[PtCl_2(NCPh)_2]$ with tri-*t*-butylphosphine proceeds according to Equation (91).[660] The dimer may be cleaved by numerous ligands. The bis(phosphine)platinum(0) complex, $[Pt(PBu^t_3)_2]$, reacts with HCl at ambient temperature to give *trans*-$[PtClH(PBu^t_3)_2]$, but on standing this undergoes loss of H_2 to generate $[Pt(CH_2CMe_2PBu^t_2)Cl(PBu^t_3)]$.[661] Refluxing $[Pt(PBu^t_3)_2]$ in chloroform solution yields the metallated dimer $[Pt_2(CH_2CMe_2PBu^t_2)_2(\mu\text{-}Cl)_2]$, whereas addition of iodomethane gives first $[Pt_2(\mu\text{-}I)_2Me_2(PBu^t_3)_2]$, but this undergoes elimination of methane to produce $[Pt_2(CH_2CMe_2PBu^t_2)_2(\mu\text{-}I)_2]$.[662] In contrast to the reactions of PBu^t_3, metallation of the arsenic analogue does not occur.[663] Trimesitylphosphine, $P(C_6H_2Me_3\text{-}2,4,6)_3$, also undergoes metallation with $PtCl_2$, $[PtCl_2(cod)]$ or $[PtCl_2(NCR)_2]$.[664] With $PBu^t_2Bu^i$, metallation of the isobutyl group occurs preferentially.[665]

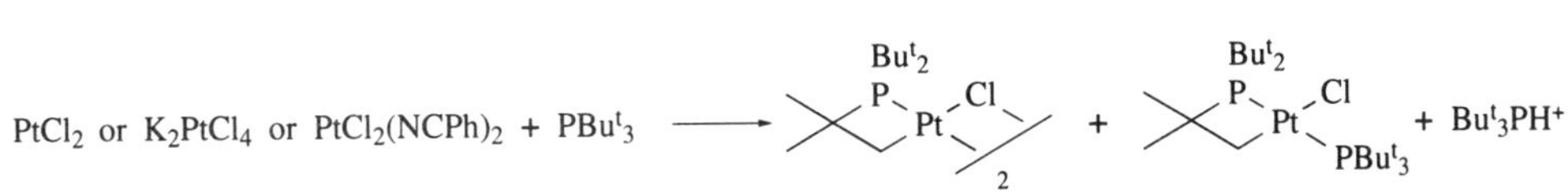

(91)

Examples of the formation of five-membered chelate rings are shown in Equations (92)[666] and (93).[667] In the former, the aromatic rings are held by the bridging oxygen in an ideal position for metallation. Treatment of K_2PtCl_4 with the (amidophenyl)phosphine, $Ph_2P(C_6H_4NHCOPh\text{-}2)$, leads initially to $[PtCl_2L_2]$, which is deprotonated in the presence of base, and on heating undergoes metallation (Equation (94)).[668] Reaction of $PtCl_2$ with 1,3-bis(diphenylphosphinomethyl)benzene leads to a metallated complex containing two five-membered rings,[669] and addition of $(TMS)_3CPPh_2$ to $[PtCl_2(NCPh)_2]$ gives (**66**),[670] the result of formal loss of TMS-Cl.

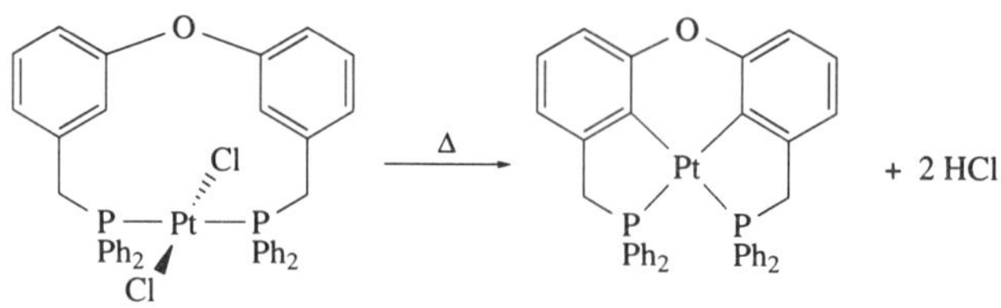

(92)

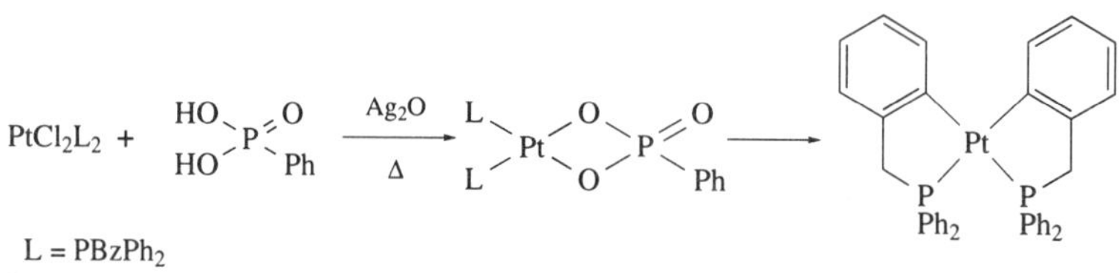

(93)

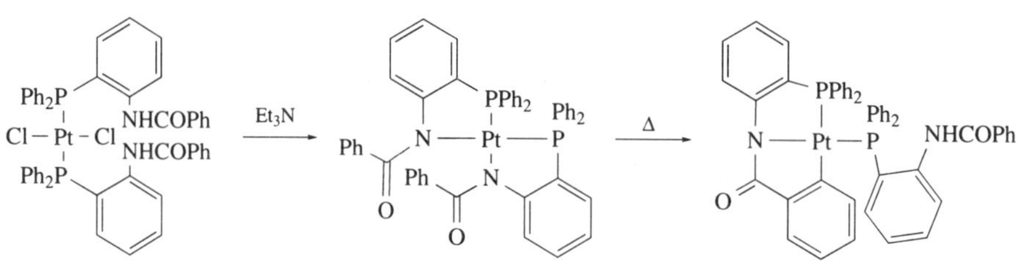

(94)

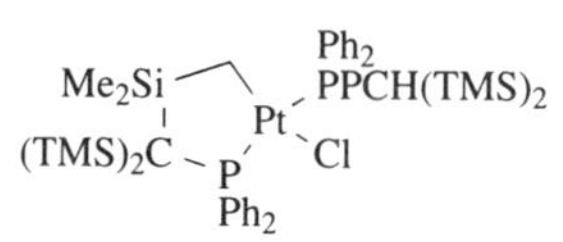

(**66**)

Metallation may also occur by opening of strained rings. Reaction of $PtCl_2$ or $[PtCl_2(NCPh)_2]$ with the (cyclopropylmethyl)phosphine $Bu^t_2PCH_2\overline{CHCH_2CH_2}$ proceeds according to Equation (95).[671] With the (cyclopropyl)phosphine $Bu^t_2P\overline{CHCH_2CH_2}$, activation of a *t*-butyl group occurs preferentially,[672] and in the case of $Bu^t_2PCH_2\overline{CHCH_2CH_2CH_2}$, activation of a C–H bond of the cyclobutyl ring occurs.[673] Heating the (adamantylmethyl)phosphine complexes $[PtMe_2L_2]$ ($L = PMe_2(CH_2C_9H_{13})$, $PMe(CH_2C_9H_{13})_2$) results in platination of one adamantyl group and elimination of methane.[674] Ring opening of a coordinated phosphacyclopropene occurs to give a four-membered ring, as shown in Equation (96).[675]

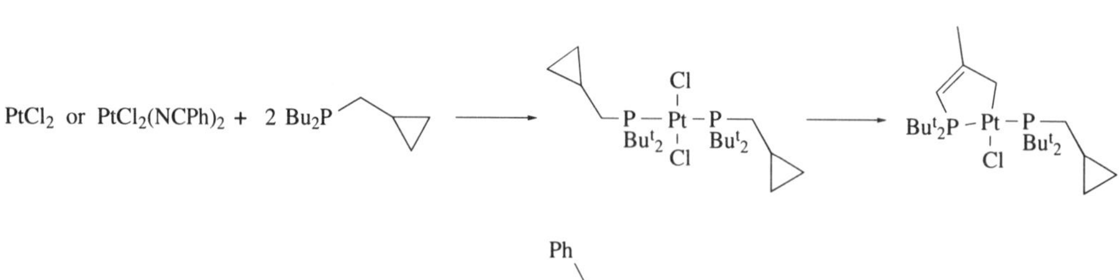

(95)

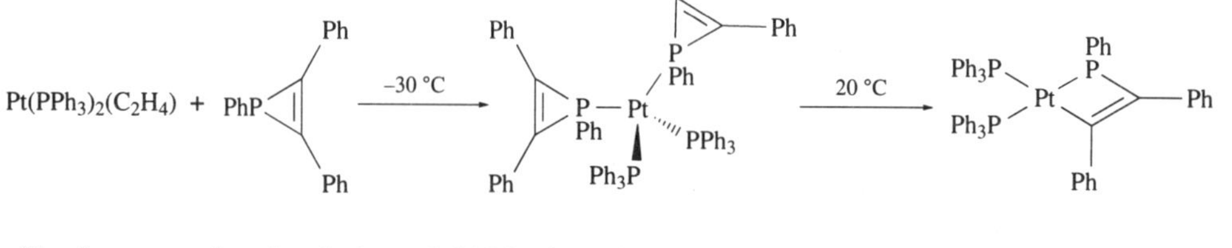

(96)

Heating a methanol solution of [PtMe(OH)(dppe)] results in reduction to platinum(I) and *ortho*-metallation of one of the phenyl groups of the dppe ligand to give the dimeric complex (**67**).[676] A related complex is formed in low yield by sodium tetrahydroborate reduction of $[PtCl_2(dppp)]$.[677] A more rational approach to *ortho*-metallated dimers involves reaction of $[PtCl_2(SEt_2)_2]$ with $LiC_6H_4PPh_2$, followed by addition of $[Pt(PPh_3)_3]$, to generate complex (**68**).[678] Treatment of $[PtCl_2(SEt_2)_2]$ with 2 equiv. of $LiC_6H_4CH_2ER_2$ leads to monomeric complexes of the type *cis*-$[Pt(C_6H_4CH_2ER_2\text{-}2)_2]$ (E = As, R = Me; E = P, R = Me, Bu^t, Ph).[679,680]

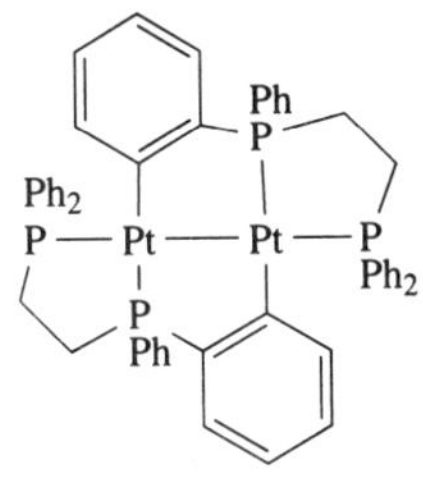

(**67**)

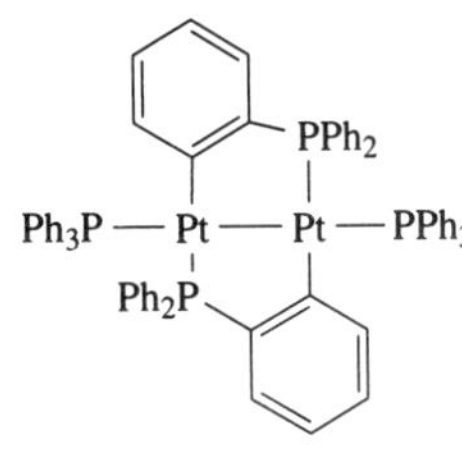

(**68**)

Protonation of the platinum(0) complex $[Pt(CH_2{=}CHC_6H_4PPh_2)_2]$, in which each ligand is coordinated through the phosphorus and the C=C double bond, yields complex (**69**),[681] whereas deprotonation of the platinum(II) complex *cis*-$[PtCl_2(CH_2{=}CMeC_6H_4AsPh_2)]$ also generates metallated products (Scheme 31).[682] Reaction of dimethylhydrazine with *cis*-$[PtCl_2(PPh_2CH{=}CH_2)_2]$ occurs by successive conjugate additions to the vinyl groups, including attack of one phosphorus on a vinyl group of an adjacent ligand, to yield (**70**).[683]

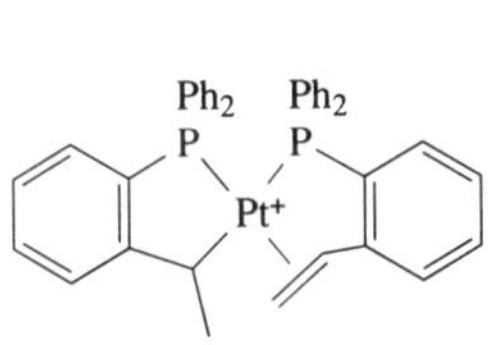

(**69**)

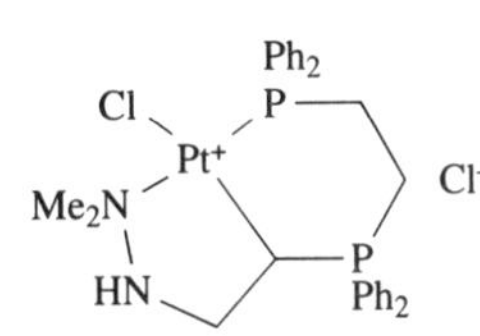

(**70**)

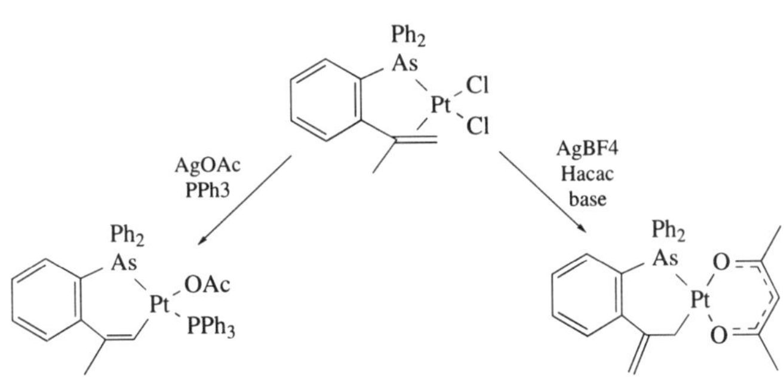

Scheme 31

Cyclometallated complexes may also be obtained by oxidative addition to platinum(0) precursors. Addition of $2\text{-}BrC_6H_4CH_2ER_2$ (E = As, R = Me; E = P, R = Bu^t, Ph) to $[Pt(dba)_2]$ yields the bromo-bridged species $[Pt_2(C_6H_4CH_2ER_2)_2(\mu\text{-}Br)_2]$, which may be cleaved by other ligands.[684] Similarly, addition of 2 equiv. of $Cy_2P(CH_2)_3Cl$ to $[Pt(cod)_2]$ produces $[Pt\{(CH_2)_3PCy_2\}Cl\{Cy_2P(CH_2)_3Cl\}]$.[685] Oxidative addition of activated C–H bonds also occurs, and some examples are illustrated in Scheme 32.[686,687]

Insertion of a vinyl group of a coordinated phosphine ligand into a Pt–H bond in the same molecule leads to a cyclometallated product (Equation (97)).[688] In the case of $[Pt(OCPh{=}CHPPh_2)_2]$, addition of excess DMAD leads to rearrangement and insertion, and again produces cyclometallated species, as shown previously in Equation (69).[491,689] Finally, insertion of CO or $CNC_6H_3Me_2\text{-}3,5$ into the Pt–O bond of $[Pt(OCMe_2CH_2PPh_2)Cl(PPh_2CH_2CMe_2OH)]$ gives a metallated complex with a six-membered chelate ring.[690]

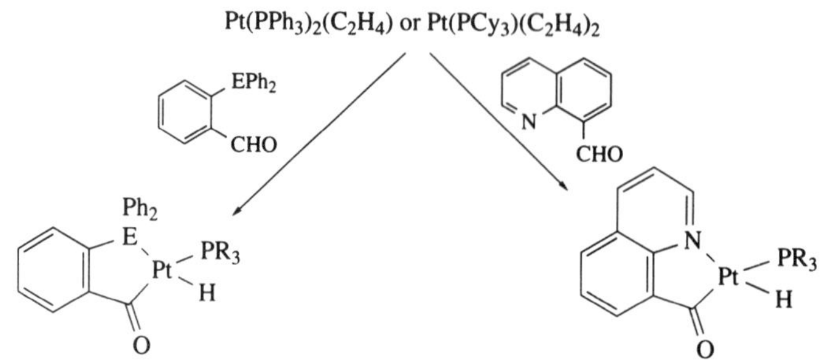

R = Ph, Cy; E = P, As

R = Cy

Scheme 32

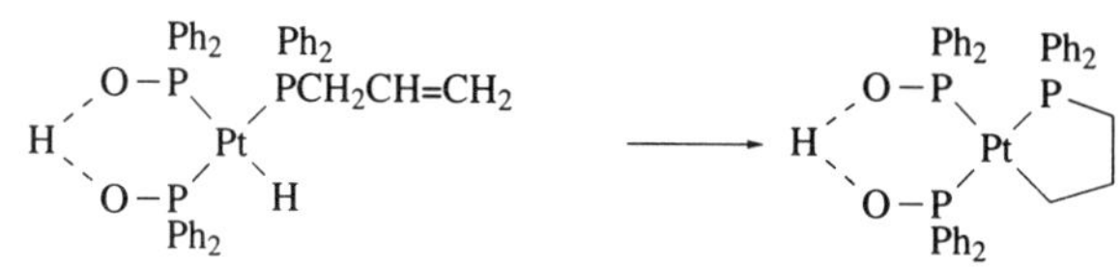

(97)

8.6.1.2 Compounds containing metallated nitrogen ligands

Metallation reactions involving nitrogen ligands such as azobenzene, benzylamines, benzoquinoline, and 8-methylquinoline have been used widely for many years. Reviews of metallation reactions with the platinum metals and nitrogen-donor ligands have appeared.[691] Four types of metallation process were described, namely anionic displacements, C–H or C–C insertions, oxidative additions of C–H, C–C, C–X, or C–M bonds to zerovalent metals, and insertion into an existing metalacycle.[691]

Many metallation reactions occur when a platinum salt is treated with a nitrogen ligand. Addition of 6-substituted bipyridines to K_2PtCl_4 leads to metallation to give species of types (**71**) and (**72**).[692–4] Reaction of K_2PtCl_4 with *N*-methylbipyridyl nitrate yields, after loss of HCl, a product in which the methylated ring is metallated.[695] Also, metallation occurs with *N*-alkyl-*N*-nitrosoanilines[696] and arylamidines.[697] Oligomeric diplatinum or mixed palladium–platinum dimetallated complexes (**73**) are obtained with $(4\text{-}ClC_6H_4CH{=}N)_2$.[698]

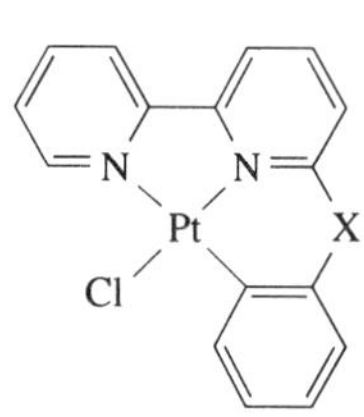

X = –, CHMe
(71)

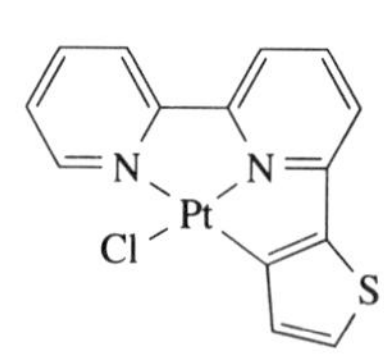

(72)

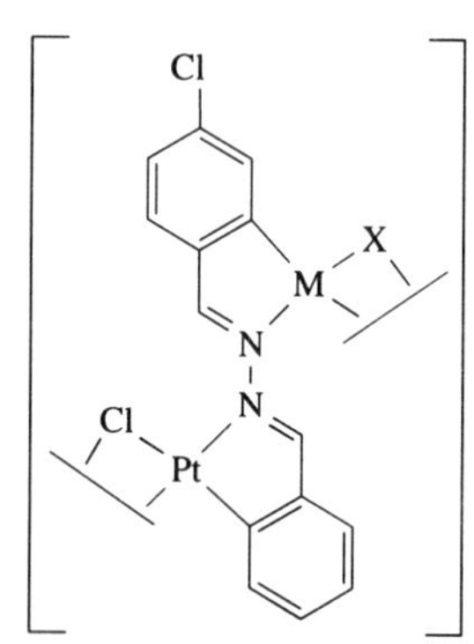

MX = PtCl, Pd(OAc)
(73)

Treatment of the platinum(0) precursor $[Pt(PCy_3)(C_2H_4)_2]$ with quinoline-8-carboxaldehyde results in oxidative addition, leading to a cyclometallated product (Scheme 32).[687] Metallation of the aldehyde function also takes place in platinum(II) complexes (Equation (98)).[699] The corresponding bromide complexes react about 70 times faster than the chlorides, and addition of $SnCl_2$ or AgOTf catalyzes the reactions, indicating the importance of cationic intermediates.[699] Treatment of $[Pt_2Me_4(\mu\text{-}SMe_2)_2]$ with $C_6F_5CH{=}NCH_2CH_2NMe_2$[618] or $C_6F_5CH{=}NCH_2C_6H_4X\text{-}2$ (X = H, Cl, Br)[700] results in intramolecular C–F activation, even in the presence of the weaker C–H, C–Cl, or C–Br bond in the latter. With $2\text{-}XC_6H_4CH{=}NCH_{2CH}CH_2NMe_2$, oxidative addition of a C–H, O–H, or C–X bond occurs, depending on the nature of X (Scheme 33), and C–F activation does not take place.[617] With *N*-benzylidene-benzylamines, C–H or C–X activation can again occur.[619] Cyclometallation has also been observed in tris(pyrazolyl)methane and related derivatives.[616,701]

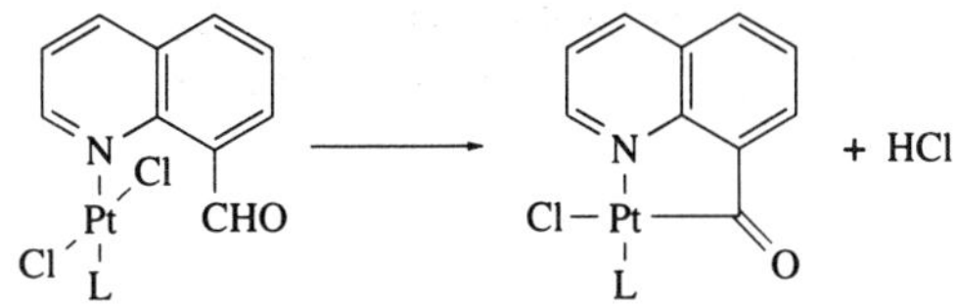

(98)

L = PEt_3, PBu_3, PPh_3, $P(C_6H_4Me)_3$, $AsEt_3$, $AsPr^i_3$, $As(C_6H_4Me)_3$

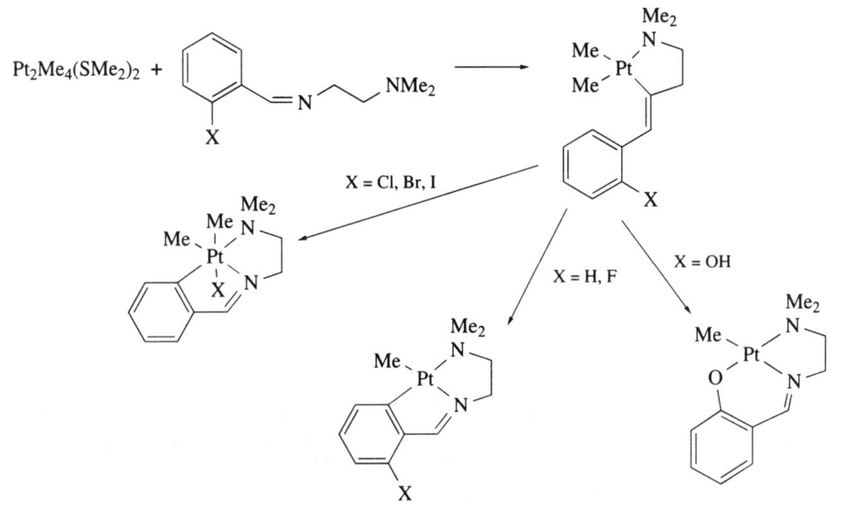

Scheme 33

Reaction of $[Pt_2(\mu\text{-}Cl)_2(\eta\text{-}C_3H_5)_2]$ with benzo[*c*]quinoline, 8-methylquinoline, or Schiff's base tetralones leads to chloro-bridged metallated complexes. Cleavage of $[Pt_2(\mu\text{-}Cl)_2Cl_2(PEt_3)_2]$ with 8-methylquinoline (L) first gives *trans*-$[PtCl_2(PEt_3)L]$, in which the quinoline is N-bonded and the methyl group forms a weak interaction with the platinum center. Further treatment with TlOTf, however, leads to metallation (Equation (99)).[702]

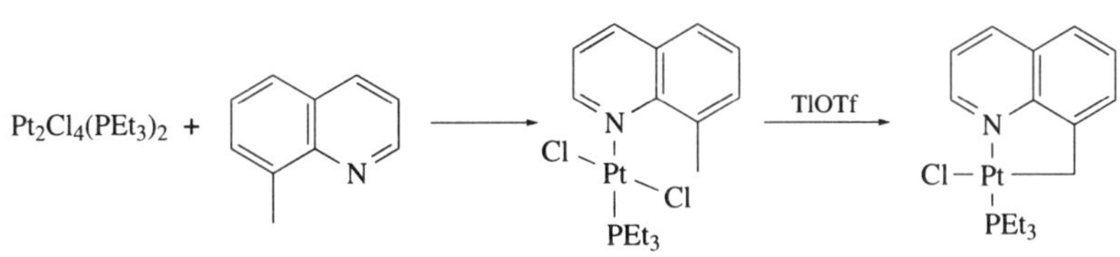

(99)

Metallated complexes may alternatively be prepared by transfer of a nitrogen-substituted organic fragment from another metal. This approach has been employed in the syntheses of the terdentate complexes of type (**74**). Complexes containing the 2,6-bis(dimethylaminomethyl)phenyl ligand have been studied extensively by van Koten and co-workers. Variations on this type of compound have been prepared by reaction of $[PtCl_2(SEt_2)_2]$ with $LiC_6H_3(CH_2NR^1R^2)_2$ ($R^1 = R^2 = Et$; $R^1 = Me$, $R^2 = Bu^t$, Ph)[703] or with a cyclic ligand to produce complex (**75**).[704] Similarly, reaction of $[PtCl_2(SEt_2)_2]$ or $[PtCl_2(cod)]$ with 2 equiv. of $LiC_{10}H_6X$-8 (X = NMe_2, OMe) generates species containing two metallated ligands.[705] Metallation also occurs with $Hg(C_6H_4NO_2\text{-}2)_2$[706] or $Hg(C_6H_4CH_2NMe_2\text{-}2)_2$.[707] An unusual example occurs in the reaction of $[Pt(PPh_3)_3]$ shown in Equation (100).[708]

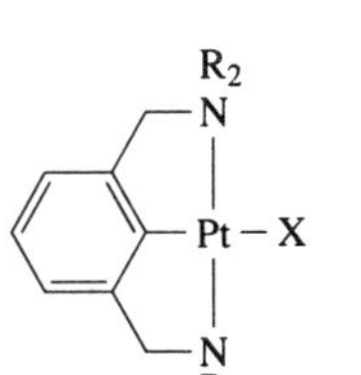

(74)

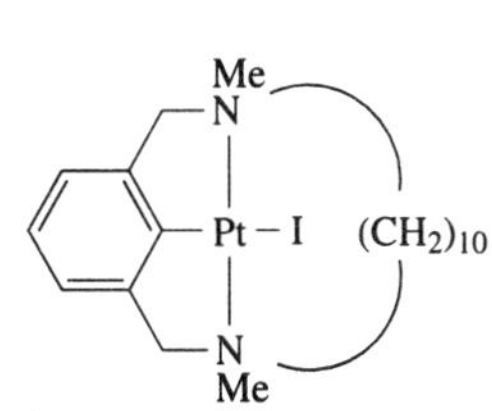

(75)

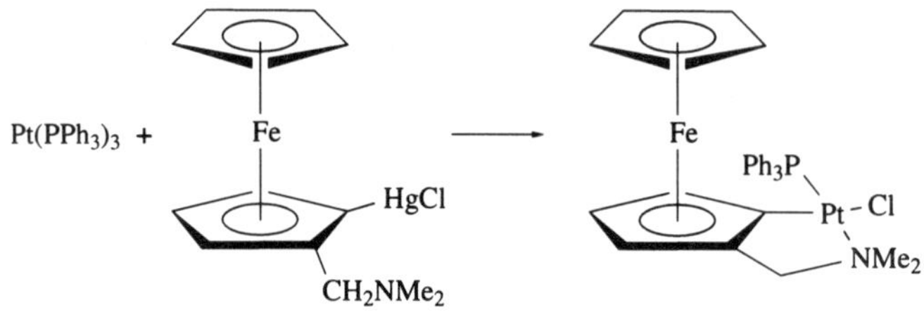

(100)

8.6.1.3 Compounds with other metallated ligands

Complexes containing metallated oxygen, sulfur, or silicon ligands have also been reported. Treatment of K_2PtCl_4 with salicylaldehyde in the presence of potassium carbonate and a tertiary phosphine, phosphite, or arsine produces the acylplatinum complexes $[\overline{Pt(COC_6H_4O})L_2]$.[709] Reactions with 2-hydroxynaphthaldehyde or 2-hydroxy-3-methoxybenzaldehyde generate analogous products.[710] Treatment of $[\overline{Pt(COC_6H_4O})L_2]$ with HCl results in cleavage of the Pt–O bond to give $[PtCl(COC_6H_4OH)L_2]$.[710] Addition of the 2- or 3-substituted *N,N*-dimethylfurancarbothioamide to K_2PtCl_4 results in loss of HCl and formation of the chloro-bridged complexes (**76**) and (**77**), which may be cleaved by neutral ligands.[711]

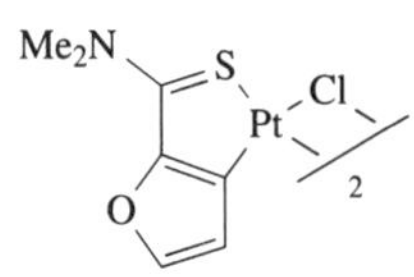

(76)

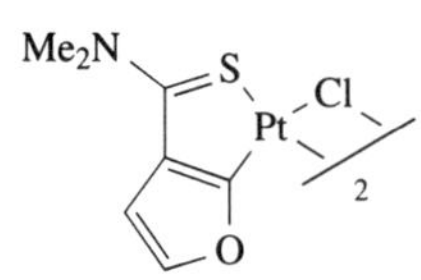

(77)

Treatment of $[PtCl_2(cod)]$ with a crown thioether in the presence of $AgBF_4$ results in metallation of the aryl ring (Equation (101)). The product is susceptible to oxidative addition with I_2, and with $PMePh_2$ undergoes substitution of the sulfur *trans* to the aryl ring.[712] The first example of the metallation of an arylsilyl ligand is illustrated in Equation (102). Elimination of HCl generates the metallated species which, in the presence of water, undergoes addition across the Si–C bond to give the final product.[713]

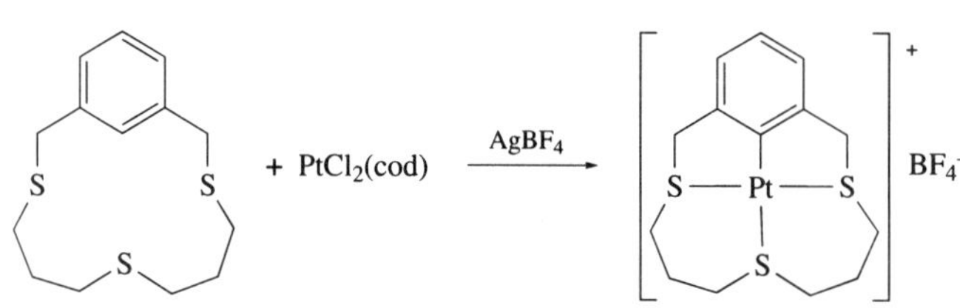

(101)

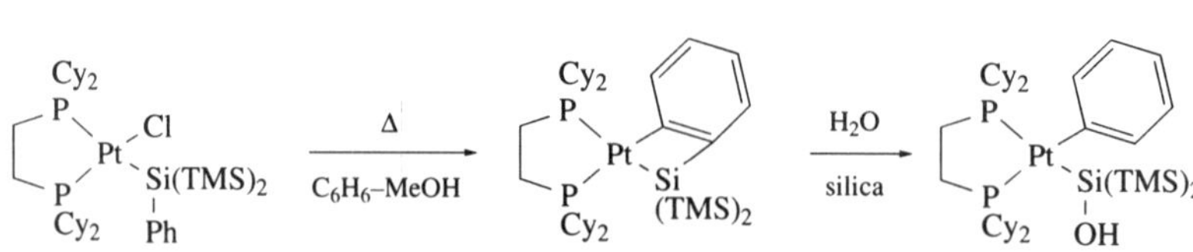

(102)

8.6.2 Structure and Properties

X-ray diffraction data for a number of metallated phosphorus ligand compounds are given in Table 14. The Pt–C bond distances are normal for alkyl or aryl groups, ranging from 0.204 nm to 0.213 nm.

It was noted in Section 8.6.1 that heating a methanol solution of [Pt(OH)Me(dppe)] results in metallation at an *ortho*-position in one of the phenyl rings of dppe, giving the metal–metal-bonded platinum(I) complex $[Pt_2\{\mu\text{-}C_6H_4PPh(CH_2)_2PPh_2\}_2]$ (**67**).[676] The Pt–Pt bond distance in (**67**) is 0.2628(1) nm. The analogous complex containing metallated dppp ligands, $[Pt_2\{\mu\text{-}C_6H_4PPh(CH_2)_3PPh_2\}_2]$, reacts with $[Au(PPh_3)BF_4]$, $Ag(O_2CCF_3)$, $HgCl_2$, $Hg(O_2CCF_3)_2$, or I_2 to generate species of the form $[Pt_2(\mu\text{-}Z)\{\mu\text{-}C_6H_4PPh(CH_2)_3PPh_2\}_2]^{n+}$ ($n = 0, 1$). The species with $Z = Au(PPh_3)^+$ and $Hg(O_2CCF_3)_2$ have been

Table 14 X-ray diffraction data for complexes containing metallated phosphorus ligands.

Complex	*Pt–C* (nm)	*Other bond lengths* (nm)	*Ref.*
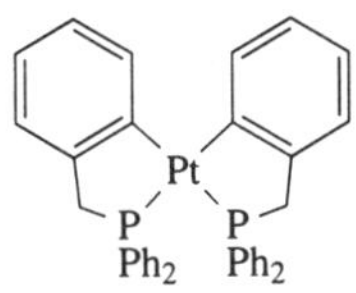	0.2064(4) 667	Pt–P 0.2282(1)	
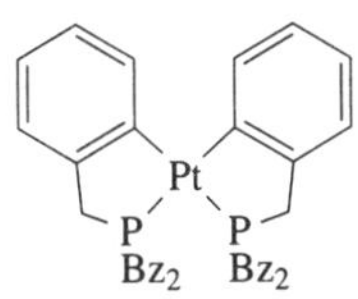	0.2062(7) 0.2071(7)	Pt–P 0.2276(2) 0.2278(2)	714
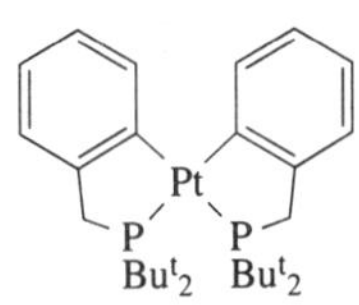	0.209(6) 0.213(6)	Pt–P 0.235(2) 0.238(2)	715
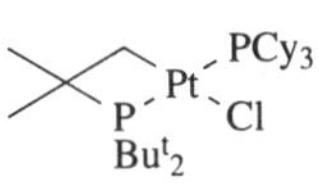	0.2043(9)	Pt–PCy_3 0.2297(2) Pt–PBu^t_2 0.2269(2) Pt–Cl 0.2409(2)	716
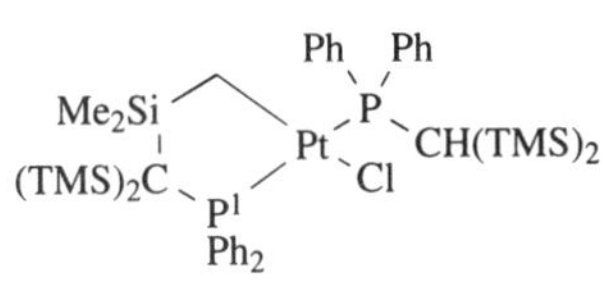	0.2087(10)	Pt–P 0.2299(3) Pt–P^1 0.2310(3) Pt–Cl 0.2402(3)	670
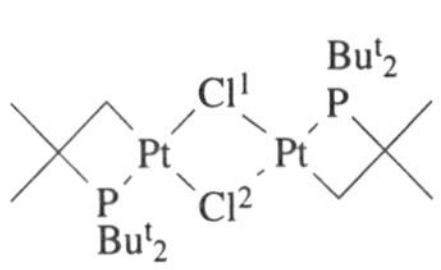	0.2062(6)	Pt–P 0.2187(2) Pt–Cl^1 0.2409(2) Pt–Cl^2 0.2455(2)	717
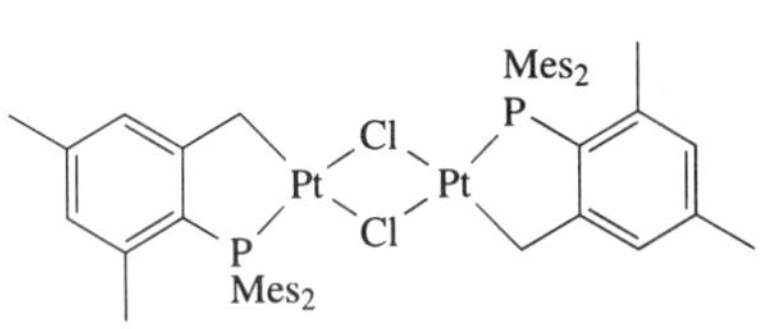	0.2035(1)	Pt–P 0.2210(1)	718

characterized by x-ray crystallography, and their Pt–Pt bond distances are 0.2703(1) nm and 0.27608(7) nm, respectively.[677,719] These distances indicate that the Pt–Pt bond is not broken on addition of these bridging moieties to $[Pt_2\{\mu\text{-}C_6H_4PPh(CH_2)_3PPh_2\}_2]$. The longer Pt–Pt separation in $[Pt_2\{\mu\text{-}Hg(O_2CCF_3)_2\}\{\mu\text{-}C_6H_4PPh(CH_2)_3PPh_2\}_2]$ is indicative of a weaker bond in this case. There are good correlations between the Pt–Pt bond distance and the ^1J(Pt,P) coupling *trans* to the Pt–Pt bond and the ^2J(Pt,P) couplings (Table 15). The two-bond couplings are small and the one-bond coupling is large in the iodide-bridged complex in which there is no formal Pt–Pt bond.[677]

The Pt–Pt bond distance in $[Pt_2(\mu\text{-}C_6H_4PPh_2)_2(PPh_3)_2]$ (**68**), in which the PPh_3 ligands occupy terminal sites, is 0.2630(1) nm. Oxidative addition of iodine generates the A-frame complex $[Pt_2(\mu\text{-}I)(\mu\text{-}C_6H_4PPh_2)_2(PPh_3)_2]I$, with a concomitant lengthening of the Pt–Pt distance to 0.2931(2) nm.[678]

The crystal structure of the bis(dimethylaminomethyl)phenyl complex $[Pt\{C_6H_3(CH_2NMe_2)_2\}I]$ has been reported, for which the Pt–C bond is relatively short at 0.1933(6) nm,[720] and the UV photoelectron spectra of $[Pt\{C_6H_3(CH_2NMe_2)_2\}X]$ (X = Cl, Br, I) provide evidence for a π-interaction of the phenyl group with the metal *d*-orbitals.[721]

The spectroscopic properties of platinum complexes containing metallated nitrogen ligands have been studied extensively by von Zelewsky and others, and reviews have appeared.[722,723] Complexes containing

Table 15 $^{195}Pt–^{31}P$ coupling constants in complexes of the type shown below.

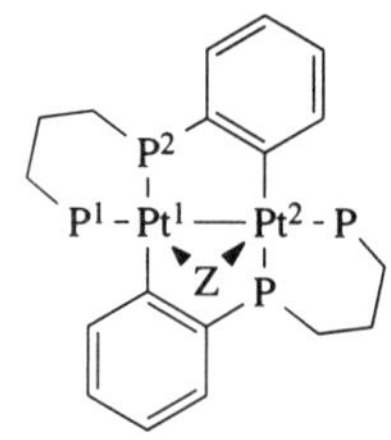

Z	$^1J(Pt^1, P^1)$ (Hz)	$^2J(Pt^2, P^1)$ (Hz)	$^2J(Pt^2, P^2)$ (Hz)
	1775	996	−143
$Ag(O_2CCF_3)$	2209	767	−139
$Au(PPh_3)^+$	2586	538	−126
$HgCl_2$	3066	312	−174
$Hg(O_2CCF_3)_2$	3353	239	−126
I^+	3719	73	0

two metallated 2-phenylpyridine or 2-(2-thienyl)pyridine ligands have been prepared,[724,725] in addition to the mixed bis(chelate) complex.[726] The crystal structure of *cis*-bis(2,6-diphenylpyridinato-*N,C*$^{2'}$)platinum(II) provides the first example of "helical chirality" in a square-planar complex with nonchiral ligands.[727] The bis(chelate)platinum(II) complexes with phenylpyridine, thienylpyridine, or related ligands typically exhibit strong luminescence from MLCT excited states in the 500–600 nm range.[728–35] Photochemically induced oxidative addition of alkyl halides to these complexes produces one isomer only, whereas the thermal reactions give mixtures of isomers.[736]

8.6.3 Reactions

8.6.3.1 Compounds containing metallated phosphorus or arsenic ligands

The cyclometallated dimers $[Pt_2(CH_2CMe_2PBu^t_2)_2(\mu\text{-}Cl)_2]$ undergo metathesis reactions with NaSR (R = Et, Pr, Pr^i, Bu, Bu^i),[737] NaSePh,[738] or NaTeR (R = C_6H_4Me, C_6H_4OMe).[739] The chloro-bridged dimers are cleaved by neutral ligands, including tertiary phosphines.[660,740] Pyrazolyl-bridged species are obtained when $[Pt_2(CH_2CMe_2PBu^t_2)_2(\mu\text{-}Cl)_2]$ is treated with pyrazole or dimethylpyrazole in the presence of NaOH.[741] The metallated trimesitylphosphine and -arsine dimers may also be cleaved with pyrazole in the presence of base.[742] With sodium tetrahydroborate, the monomeric chlorides $[Pt(CH_2CMe_2PBu^t_2)ClL]$ (L = PR_3, $P(OR)_3$) are converted into the corresponding hydrides, but when L = CO the triplatinum species $[Pt_3(\mu\text{-}CO)_3(PBu^t_3)_3]$ are formed.[743] Tin(II) chloride inserts into the Pt–Cl bonds of $[Pt(CH_2CMe_2PBu^t_2)ClL]$.[744]

Bromination of $[Pt(C_6H_4CH_2PR_2)_2]$ (R = Me, Ph) proceeds according to Equation (103), oxidative addition being the preferred reaction, whereas with palladium cleavage of the Pd–C bonds predominates.[745] Treatment of $[Pt(C_6H_4CH_2PR_2)_2]$ with 2 equiv. of BuLi results in deprotonation of the two methylene groups, and subsequent addition of MeI, TMS-Cl, Ph_3PAuCl, or Ph_2PCl generates $[Pt(C_6H_4CHXPR_2)_2]$.[746] The metallated complex $[Pt(Bu^t_2PCH_2CH_2CHCH_2CH_2PBu^t_2)Cl]$, obtained by reaction of $[PtCl_2(CNBu^t)_2]$ with $Bu^t_2P(CH_2)_5PBu^t_2$, undergoes chloride substitution or deprotonation to give an alkene complex (Scheme 34).[747]

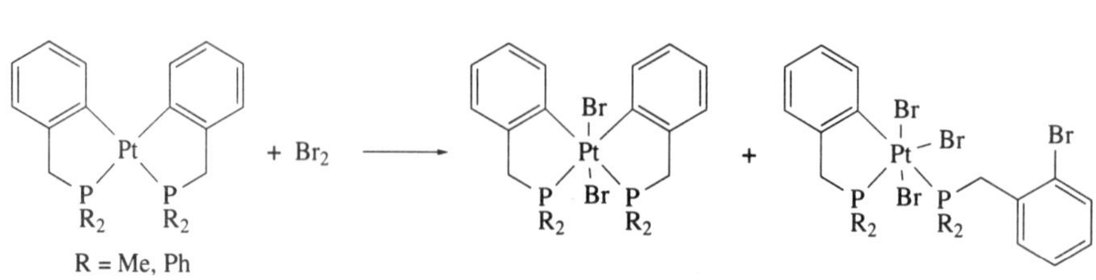

(103)

Oxidative addition of iodomethane to the platinum(I) dimer $[Pt_2\{\mu\text{-}C_6H_4PPh(CH_2)_2PPh_2\}_2]$ (**67**) generates a fluxional complex, in which the methyl group is either bridging or migrating rapidly between

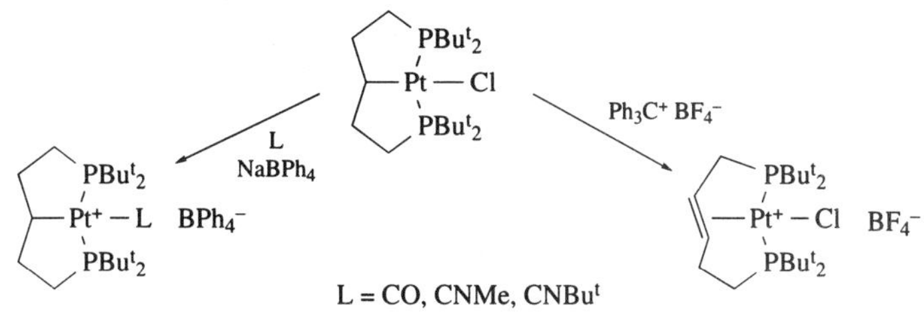

Scheme 34

platinum atoms. This reacts with neutral ligands or with methyllithium, or may be deprotonated by NaOMe (Scheme 35).[748,749] The bridging methylene complex may be methylated to give a fluxional ethyl complex.[749] Addition of $AuPPh_3^+$ to (**67**), its dppp analogue, or to $[Pt_2(\mu\text{-}C_6H_4PPh_2)_2(PPh_3)_2]$ (**68**) generates a bridging gold complex,[719] and addition of iodine to (**68**) produces an iodide-bridged cation.[678]

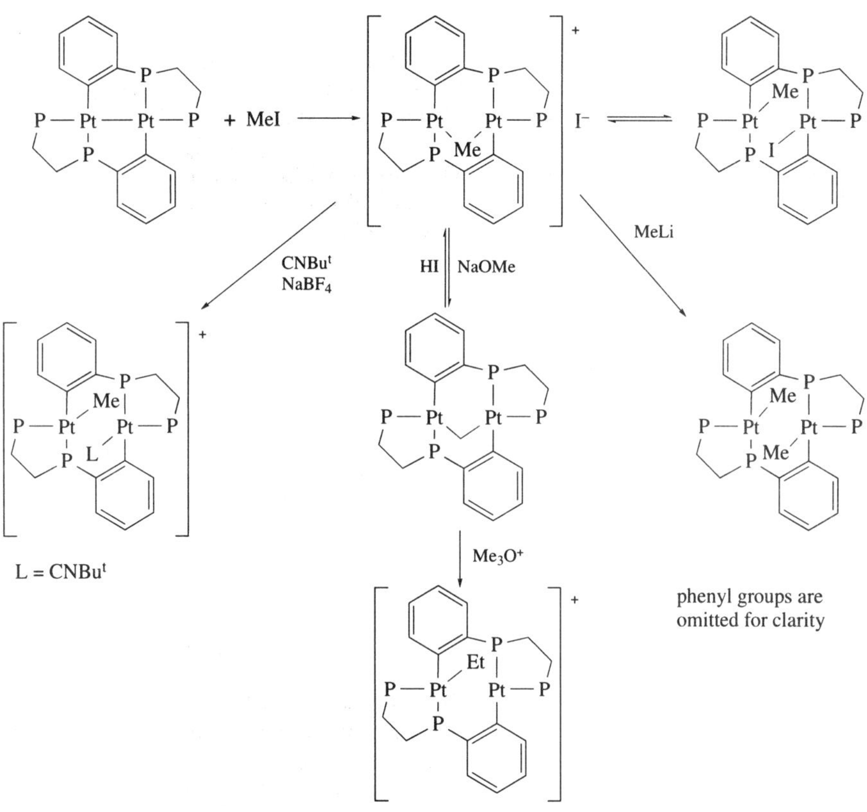

Scheme 35

8.6.3.2 *Compounds containing metallated nitrogen ligands*

Platinum(II) complexes of the terdentate 2,6-bis(dimethylaminomethyl)phenyl ligand have been introduced in Section 8.6.1. Treatment of $[Pt\{C_6H_3(CH_2NMe_2)_2\}Br]$ with water in the presence of $AgBF_4$ yields the aquo cation, from which the water molecule may be displaced by CO or PPh_3.[750] Addition of sodium formate or sodium phenoxide to $[Pt\{C_6H_3(CH_2NMe_2)_2\}(H_2O)]^+$ also results in displacement of H_2O,[751,752] whereas reactions with organolithium reagents generate $[Pt\{C_6H_3(CH_2NMe_2)_2\}R]$ (R = Ph, C_6H_4Me, C≡CPh, $C{\equiv}CC_6H_4Me$).[753] Addition of iodomethane to $[Pt\{C_6H_3(CH_2NMe_2)_2\}(H_2O)]BF_4$ produces the metal-substituted carbonium ion (**78**). The latter reacts under aqueous conditions to regenerate the aquo complex, but at low temperatures and in suitable solvents bases such as PhC≡C$^-$, $CH(CO_2Me)_2^-$, OMe^-, or OH^- produce 2,4- or 2,5-cyclohexadiene complexes by nucleophilic addition to the C_6H_3 ring.[754] The addition of the methyl group to the aromatic ring is believed to involve initial methylation at platinum, followed by a 1,2-methyl shift to give the metal-substituted carbonium ion, or reductive elimination and coordination of H_2O to produce the aquo complex (Scheme 36).[755] In contrast to the reaction with iodomethane, treatment of the aquo cation with RI (R = Et, Pr^i, Bz, CH_2=$CHCH_2$) in moist acetone gives the platinum(IV) cation $[Pt_2(\mu\text{-}I)_2I_2\{C_6H_3(CH_2NHMe_2)_2\}_2]^{2+}$.[756]

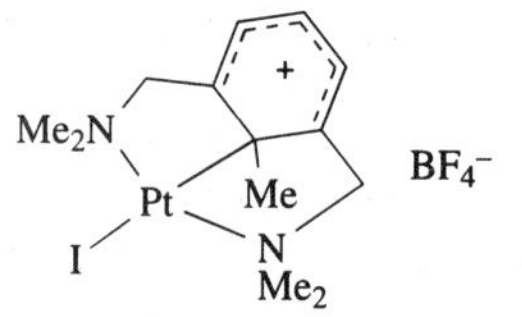

(78)

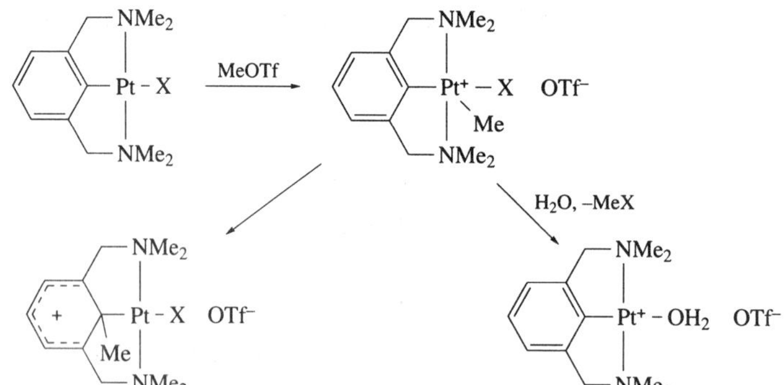

X = Cl, Br, I, C_6H_4Me

Scheme 36

Oxidative addition of halogens to $[Pt\{C_6H_3(CH_2NMe_2)_2\}X]$ (X = Cl, Br, I) yields *mer*-$[Pt\{C_6H_3(CH_2NMe_2)_2\}X_3]$,[757,758] which is also obtained with 2 equiv. of CuX_2.[757] The addition of I_2 to $[Pt\{C_6H_3(CH_2NMe_2)_2\}I]$ at ambient temperature permitted the isolation of an iodine complex, containing a Pt–I–I linkage, which has been structurally characterized.[759] Reaction of $[Pt\{C_6H_3(CH_2NMe_2)_2\}X]$ with sulfur dioxide does not result in insertion, but reversible formation of a five-coordinate SO_2 complex takes place.[760] The aquo complex $[Pt\{C_6H_3(CH_2NMe_2)_2\}(H_2O)]^+$ reacts with *N,N'*-di-4-tolylformamidine (Hdptf) by displacement of water, and further treatment with NaOH yields $[Pt\{C_6H_3(CH_2NMe_2)_2\}(dptf)]$ (Equation (104)). The latter species is fluxional at ambient temperature, the intramolecular exchange process being consistent with the existence of a chelated formamidinate intermediate.[761] Dinuclear complexes have been prepared according to Equation (105).[762]

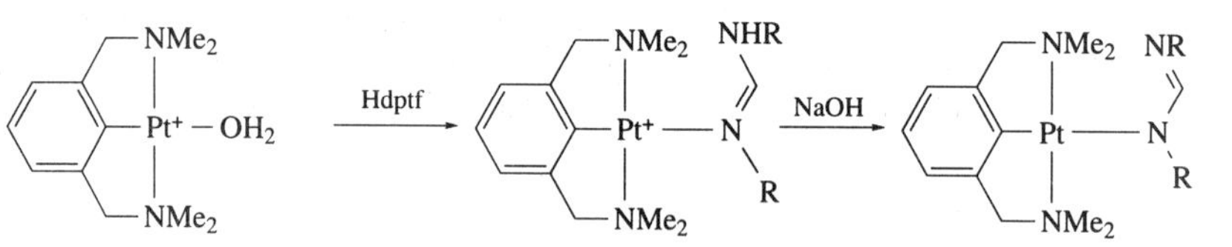

(104)

R = C_6H_4Me-4

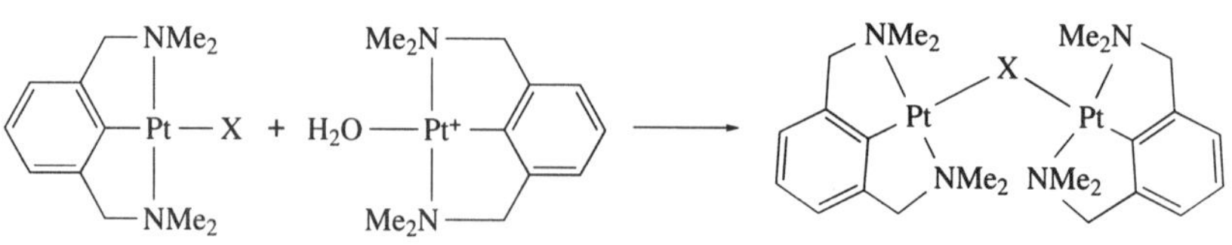

(105)

X = Cl, Br, I, CN

$[Pt\{C_6H_3(CH_2NMe_2)_2\}Br]$ undergoes oxidative addition with mercury(II) acetate, but the reaction is complicated by the formation of platinum–mercury-bonded species.[763] Reactions with Hg(4-MeC_6H_4NXNR)Cl (X = CH, N; R = Me, Et, Pr^i) generate five-coordinate species, in which the mercury sits above the square plane.[764] Several of these have been characterized by x-ray diffraction.[764,765] With $[Ag(4\text{-}MeC_6H_4NXNR)]_n$, five-coordinate complexes with Pt–Ag bonds are formed, but with silver(I) carboxylates substitution of the bromide takes place.[766]

The complexes *cis*-[$Pt(C_6H_4CH_2NMe_2)_2$] and *cis*-[$Pt(CH_2C_6H_4NMe_2)_2$] react with mercury(II) carboxylates to generate Pt–Hg-bonded species,[767] but the former, when treated with $Tl(OCOR)_3$, produces the platinum(IV) complex [$Pt(C_6H_4CH_2NMe_2)_2(OCOR)_2$].[768] Both *cis*-[$Pt(C_6H_4CH_2NMe_2)_2$] and the dimethylaminonaphthyl derivative *cis*-[$Pt(C_{10}H_6NMe_2)_2$] react with X_2 (X = Br, I) or CuX_2 (X = Cl, Br) to generate platinum(IV) complexes.[769] Reaction of *cis*-[$Pt(C_{10}H_6NMe_2)_2$] with Me_2SnBr_2 in benzene or CH_2Cl_2 solution gives a hydride complex, [$Pt(C_{10}H_6NMe_2)_2HBr$], which exhibits a ^1H NMR resonance at –20.35 ppm, with $^1J_{PtH}$ = 1540 Hz. In contrast, reactions with R_2SnX_2 (R = Me, Ph; X = Cl, Br) in CH_2Cl_2–MeOH solution produce species containing Pt–H–N agostic interactions. When X = Br, the ^1H NMR resonance for the agostic hydrogen appears at 15.8 ppm, with $^1J_{PtH}$ = 180 Hz.[770]

Metallated azobenzene complexes have been used as precursors to cyclopentadienyl derivatives.[771,772] The terdentate azobenzene complexes [$Pt(RC_6H_3N{=}NC_6H_4E)Cl$] (R = H, E = SMe, SBz; R = Me, E = SMe, SOBz) undergo oxidative addition with chlorine to give *mer*-[$Pt(RC_6H_3N = NC_6H_4E)Cl_3$].[773] The 6-(1-methylbenzyl)-2,2'-bipyridine derivative (**79**) reacts with CO at 80–90 atm, or at 1 atm in the presence of $AgBF_4$, to give the carbonyl cation,[774] and with sodium tetrahydroborate to produce the corresponding hydride.[775]

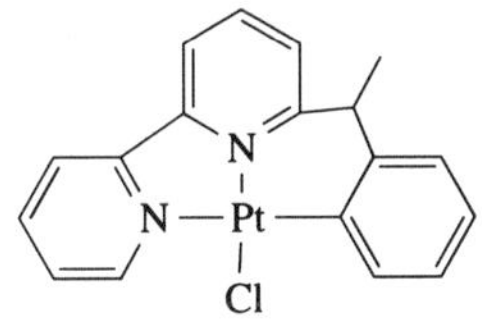

(79)

8.7 CARBON–HYDROGEN BOND ACTIVATION

A number of examples of C–H bond activation have been described already, particularly in the formation of metallated ligand complexes, but these generally involved C–H bonds that were activated in some way. The addition of unactivated C–H bonds, in alkanes or arenes, to transition metal centers has long been a goal in organometallic chemistry, and it is one that has been achieved since the early 1980s.[776]

Saillard and Hoffmann[777] predicted that a bent d^{10} ML_2 fragment should be effective in promoting oxidative addition of C–H bonds. Whitesides and co-workers demonstrated that the [Pt(dcpe)] moiety, generated by reductive elimination of neopentane from [$PtH(CH_2Bu^t)$(dcpe)], does indeed activate C–H bonds. Thus, heating [$PtH(CH_2Bu^t)$(dcpe)] in the appropriate solvent results in oxidative addition of such molecules as benzene, mesitylene, cyclopentane, and tetramethylsilane. Kinetic studies of the reaction with benzene (Equation (106)) reveal that the rate-determining step is the reductive elimination of neopentane, and the dcpe ligand remains coordinated during that step. For the rate determining step, the activation free energy is $\Delta G^{\ddagger}$ = 108 kJ mol^{-1}. The intermediacy of the 14-electron fragment was substantiated by trapping it with PhC≡CPh or dcpe.[778] The [$PtH(CH_2Bu^t)$(dcpe)] system has also been employed in the activation of Al–C, Ga–C, or In–C bonds (Equation (107)).[779]

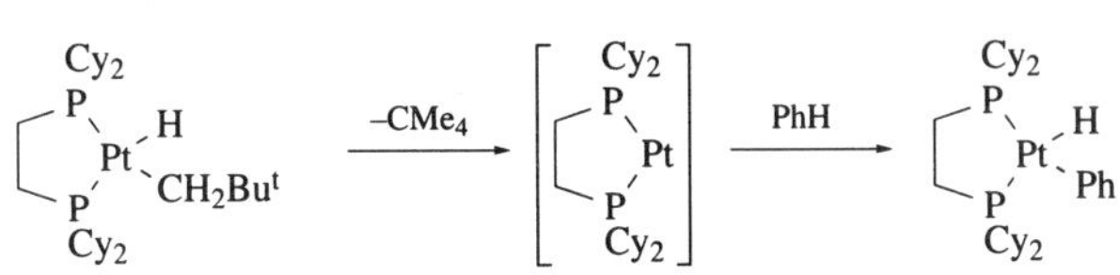

(106)

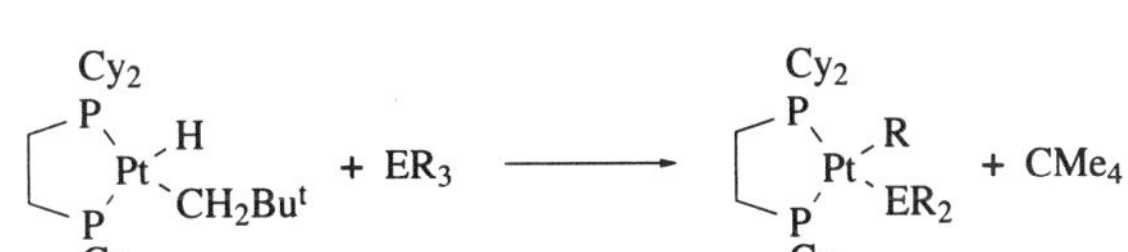

(107)

E = Al, Ga, In; R = CH_2Bu^t, CH_2(TMS)

The related complex [$PtH(CH_2Bu^t)(Bu^t{}_2PCH_2PBu^t{}_2)$] has been prepared and structurally characterized.[780] The compound loses neopentane in benzene solution at ambient temperature, but C–H bond activation does not occur. Instead a dimeric complex is formed. With tetramethylsilane, oxidative addition of the Si–Me bond occurs, whereas with the dcpe complex C–H activation was found. Loss of Me_4Si from [$PtMe(TMS)(Bu^t{}_2PCH_2PBu^t{}_2)$] produces the dimer.[780] The [$Pt(Bu^t{}_2PCH_2PBu^t{}_2)$] moiety does insert into a C–F bond of hexafluorobenzene.[781] These reactions are summarized in Scheme 37.

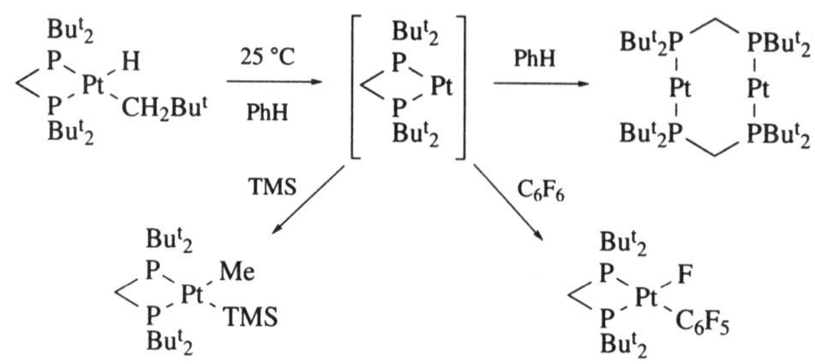

Scheme 37

The bis(neopentyl)platinum complex [$Pt(CH_2Bu^t)_2$(S-prephos)] (S-prephos = (+)-(1*S*)-Ph_2PCH-$PhCH_2PPh_2$), has been prepared and structurally characterized. Further treatment with HCl, then $NaBH(OMe)_3$, produces [$PtH(CH_2Bu^t)$(S-prephos)]. This undergoes loss of neopentane, and the 14-electron fragment inserts into the C–H bond of PhC≡CH. With $HC{\equiv}CCO_2Bu^t$, however, the alkyne complex is obtained, and in benzene alone decomposition occurs.[782]

A detailed study has been made of the reaction of *trans*-[$Pt(CH_2Bu^t)(OTf)(PMe_3)_2$] with C_6D_6 at 133 °C, which produces Bu^tCH_2D and *trans*-[$Pt(C_6D_5)(OTf)(PMe_3)_2$]. The reaction rate is reduced by addition of $Bu_4N^+OTf^-$ and is increased by $Bu_4N^+BF_4^-$. These observations suggest that triflate dissociation is involved, and electrophilic substitution or oxidative addition reactions are then possible C–D bond-activation mechanisms.[783]

Electrochemical reduction of platinum(II) complexes to generate coordinatively unsaturated PtL_2 fragments has also been studied. Two-electron reduction of *cis*-[$PtCl_2(PR_3)_2$] produces [$Pt(PR_3)_2$], which may be trapped with PhC≡CPh or undergo oxidative addition of MeI, PhCOCl, or PhX.[784] In the case of the dcpp complex (dcpp = 1,3-bis(dicyclohexylphosphino)propane), reduction in acetonitrile–benzene solution produces [$PtH(C_6H_5)$(dcpp)], which undergoes exchange with C_6D_6 at ambient temperature to give [$PtD(C_6D_5)$(dcpp)].[785]

Finally, Shul'pin and co-workers have made extensive studies of C–H activation by platinum(IV) salts. Reactions of H_2PtCl_6 with a range of aromatic substrates in the presence of $CF_3CO_2H/SiO_2{\cdot}NH_3$ yield σ-bonded arylplatinum(IV) species of the form $NH_4[PtCl_4Ar(NH_3)]$.[786] The products have been characterized by ^{195}Pt NMR spectroscopy.[787] The reactions are first order in platinum and in the arene.[788] Irradiation of a $PtCl_6^{2-}$–acetone solution produces the $[PtCl_5(CH_2COMe)]^{2-}$ anion, which eliminates $MeCOCH_2Cl$ on prolonged irradiation.[789] Irradiation of $PtCl_6^{2-}$ solutions containing organotin, -lead, -mercury, or -boron compounds also leads to anionic organoplatinum(IV) species.[790]

8.8 REFERENCES

1. F. R. Hartley, 'The Chemistry of Platinum and Palladium', Applied Science, London, 1973; U. Belluco, 'The Organometallic and Coordination Chemistry of Platinum', Academic Press, London, 1974.
2. F. R. Hartley, in 'COMC-I', vol. 6, p. 471.
3. G. K. Anderson, in 'Chemistry of the Platinum Group Metals. Recent Developments', ed. F. R. Hartley, Elsevier, Amsterdam, 1991, p. 338.
4. P. L. Goggin and R. J. Goodfellow, *J. Chem. Soc., Dalton Trans.*, 1973, 2355.
5. R. Uson, J. Fornies, P. Espinet, C. Fortuno, M. Tomas and A. J. Welch, *J. Chem. Soc., Dalton Trans.*, 1989, 1583.
6. M. P. Brown, R. J. Puddephatt, M. Rashidi and K. R. Seddon, *J. Chem. Soc., Dalton Trans.*, 1977, 951.
7. R. J. Puddephatt, *Chem. Soc. Rev.*, 1983, **12**, 99; A. L. Balch, in 'Homogeneous Catalysis with Metal Phosphine Complexes', ed. L. H. Pignolet, Plenum, New York, 1983, p. 167; A. L. Balch, *Comments Inorg. Chem.*, 1984, **3**, 51.
8. P. J. M. Ssebuwufu, *Inorg. Chim. Acta*, 1987, **134**, 185; C. R. Langrick, P. G. Pringle and B. L. Shaw, *J. Chem. Soc., Dalton Trans.*, 1985, 1015.
9. R. Uson, J. Fornies, P. Espinet and C. Fortuno, *J. Chem. Soc., Dalton Trans.*, 1986, 1849; R. Uson, J. Fornies, F. Martinez, R. Navarro and M. C. Frias, *Inorg. Chim. Acta*, 1987, **132**, 217.
10. K. A. Azam, R. J. Puddephatt, M. P. Brown and A. Yavari, *J. Organomet. Chem.*, 1982, **234**, C31; K. A. Azam, M. P. Brown, R. H. Hill, R. J. Puddephatt and A. Yavari, *Organometallics*, 1984, **3**, 697.
11. M. P. Brown, A. Yavari, R. H. Hill and R. J. Puddephatt, *J. Chem. Soc., Dalton Trans.*, 1985, 2421.
12. R. H. Hill and R. J. Puddephatt, *Organometallics*, 1983, **2**, 1472.
13. S. S. M. Ling and R. J. Puddephatt, *Polyhedron*, 1986, **5**, 1423.
14. R. J. Blau and J. H. Espenson, *Inorg. Chem.*, 1986, **25**, 878.
15. H. D. McPherson and J. L. Wardell, *J. Organomet. Chem.*, 1983, **254**, 261.
16. S. Komiya and T. Shindo, *J. Chem. Soc., Chem. Commun.*, 1984, 1672.
17. A. Mantovani, *J. Organomet. Chem.*, 1983, **255**, 385.
18. B. Crociani, F. Di Bianca, A. Giovenco and A. Scrivanti, *J. Organomet. Chem.*, 1984, **269**, 295.
19. Y. Xie and B. R. James, *J. Organomet. Chem.*, 1991, **417**, 277.
20. W. Tam and J. C. Calabrese, *Chem. Phys. Lett.*, 1988, **144**, 79.

21. S. Al-Ahmad, H. Hudali and M. H. Zaghal, *Inorg. Chim. Acta*, 1988, **147**, 1.
22. J. A. Davies, C. T. Eagle, D. E. Otis and U. Venkataraman, *Organometallics*, 1986, **5**, 1264.
23. M. Peter, M. Probst, P. Peringer and E. P. Muller, *J. Organomet. Chem.*, 1991, **410**, C29.
24. N. Carr, B. J. Dunne, A. G. Orpen and J. L. Spencer, *J. Chem. Soc., Chem. Commun.*, 1988, 926.
25. A. Sen, J.-T. Chen, W. M. Vetter and R. R. Whittle, *J. Am. Chem. Soc.*, 1987, **109**, 148.
26. W. M. Vetter and A. Sen, *J. Organomet. Chem.*, 1989, **378**, 485.
27. J.-T. Chen, Y.-S. Yeh, G.-H. Lee and Y. Wang, *J. Organomet. Chem.*, 1991, **414**, C64.
28. A. R. Siedle, R. A. Newmark and W. B. Gleason, *J. Am. Chem. Soc.*, 1986, **108**, 767.
29. R. Bender, P. Braunstein, A. Tiripicchio and M. Tiripicchio-Camellini, *Angew. Chem., Int. Ed. Engl.*, 1985, **24**, 861.
30. W. Beck *et al.*, *J. Chem. Soc., Chem. Commun.*, 1981, 446.
31. J. V. Ortiz, Z. Havlas and R. Hoffmann, *Helv. Chim. Acta*, 1984, **67**, 1.
32. J. S. Hallock, A. S. Galiano-Roth and D. B. Collum, *Organometallics*, 1988, **7**, 2486.
33. K. Sano, T. Yamamoto and A. Yamamoto, *Chem. Lett.*, 1983, 115; *Bull. Chem. Soc. Jpn.*, 1984, **57**, 2741.
34. C. Eaborn, K. Kundu and A. Pidcock, *J. Chem. Soc., Dalton Trans.*, 1981, 1223.
35. T. A. K. Al-Allaf, C. Eaborn, K. Kundu and A. Pidcock, *J. Chem. Soc., Chem. Commun.*, 1981, 55.
36. J. D. Koola and U. Kunze, *Inorg. Chim. Acta*, 1983, **76**, L283.
37. G. Butler, C. Eaborn and A. Pidcock, *J. Organomet. Chem.*, 1980, **185**, 367.
38. T. A. K. Al-Allaf, G. Butler, C. Eaborn and A. Pidcock, *J. Organomet. Chem.*, 1980, **188**, 335.
39. B. L. Khandelwal, S. K. Gupta and K. Kundu, *Inorg. Chim. Acta*, 1990, **178**, 35.
40. S. I. Pombrik, A. A. Bezrukova, L. S. Golovchenko, A. S. Peregudov, A. Z. Rubezhov and D. N. Kravtsov, *J. Organomet. Chem.*, 1985, **293**, C1.
41. H. C. Clark and L. E. Manzer, *J. Organomet. Chem.*, 1973, **59**, 411.
42. C. Eaborn, K. Kundu and A. Pidcock, *J. Chem. Soc., Dalton Trans.*, 1981, 933.
43. R. J. Puddephatt, M. A. Thomson, L. Manojlovic-Muir, K. W. Muir, A. A. Frew and M. P. Brown, *J. Chem. Soc., Chem. Commun.*, 1981, 805.
44. R. Bassan, K. H. Bryars, L. Judd, A. W. G. Platt and P. G. Pringle, *Inorg. Chim. Acta*, 1986, **121**, L41.
45. J. D. Scott and R. J. Puddephatt, *J. Chem. Soc., Chem. Commun.*, 1984, 193; *Organometallics*, 1986, **5**, 1253.
46. S. D. Chappell and D. J. Cole-Hamilton, *J. Chem. Soc., Chem. Commun.*, 1980, 238; *J. Chem. Soc., Dalton Trans.*, 1983, 1051.
47. L. Abis, R. Santi and J. Halpern, *J. Organomet. Chem.*, 1981, **215**, 263.
48. R. K. Merwin, R. C. Schnabel, J. D. Koola and D. M. Roddick, *Organometallics*, 1992, **11**, 2972.
49. H. A. Brune and J. Ertl, *Liebigs Ann. Chem.*, 1980, 928; J. Ertl, D. Grafl and H. A. Brune, *Z. Naturforsch., Teil B*, 1982, **37**, 1082; H. A. Brune, J. Ertl, D. Grafl and G. Schmidtberg, *Chem. Ber.*, 1982, **115**, 1141.
50. U. Bayer and H. A. Brune, *Z. Naturforsch., Teil B*, 1983, **38**, 226; 621; H. A. Brune, R. Hess and G. Schmidtberg, *Z. Naturforsch., Teil B*, 1984, **39**, 1772; *J. Organomet. Chem.*, 1986, **303**, 429.
51. J. Ertl, T. Debaerdemaeker and H. A. Brune, *Chem. Ber.*, 1982, **115**, 3860.
52. W.-D. Muller and H. A. Brune, *Chem. Ber.*, 1985, **118**, 4347; W.-D. Muller, G. Schmidtberg and H. A. Brune, *Chem. Ber.*, 1985, **118**, 4653.
53. H. A. Brune, R. Hess and G. Schmidtberg, *Chem. Ber.*, 1985, **118**, 2011.
54. H. A. Brune, M. Falck, R. Hemmer and H. G. Alt, *Chem. Ber.*, 1984, **117**, 2803.
55. T. Debaerdemaeker, H.-P. Klein, M. Wiege and H. A. Brune, *Z. Naturforsch., Teil B*, 1981, **36**, 958.
56. H. A. Brune, W. Schafer, K.-H. Spohn and C. Weisemann, *J. Organomet. Chem.*, 1986, **316**, 367.
57. M. Wiege, H. A. Brune, H.-P. Klein and U. Thewalt, *Z. Naturforsch., Teil B*, 1982, **37**, 718; H. A. Brune, M. Wiege and T. Debaerdemaeker, *Z. Naturforsch., Teil B*, 1984, **39**, 359; H. A. Brune, G. Schmidtberg and M. Wiege, *Z. Naturforsch., Teil B*, 1984, **39**, 517.
58. B. Stapp, G. Schmidtberg and H. A. Brune, *Z. Naturforsch., Teil B*, 1986, **41**, 514.
59. H. A. Brune, M. Wiege and T. Debaerdemaeker, *Z. Naturforsch., Teil B*, 1984, **39**, 907; R. Baumgartner, G. Laube, G. Schmidtberg and H. A. Brune, *J. Organomet. Chem.*, 1987, **332**, 221.
60. R. Baumgartner and H. A. Brune, *J. Organomet. Chem.*, 1987, **332**, 379.
61. H. A. Brune, B. Stapp and G. Schmidtberg, *Chem. Ber.*, 1986, **119**, 1845.
62. H. Tobita, H. Habazaki and H. Ogino, *Bull. Chem. Soc. Jpn.*, 1987, **60**, 797.
63. K. A. Fallis, G. K. Anderson and N. P. Rath, *Organometallics*, 1993, **12**, 2435.
64. D. L. Reger, J. C. Baxter and D. G. Garza, *Organometallics*, 1990, **9**, 16.
65. J. Browning, G. W. Bushnell, K. R. Dixon and A. Pidcock, *Inorg. Chem.*, 1983, **22**, 2226.
66. J. F. Almeida and A. Pidcock, *J. Organomet. Chem.*, 1981, **209**, 415.
67. G. R. Newkome, K. J. Theriot, F. R. Fronczek and B. Villar, *Organometallics*, 1989, **8**, 2513.
68. H. C. Clark, A. B. Goel, V. K. Jain, K. G. Tyers and C. S. Wong, *J. Organomet. Chem.*, 1987, **321**, 123.
69. H. A. Brune, B. Stapp and G. Schmidtberg, *J. Organomet. Chem.*, 1986, **307**, 129.
70. C. Weisemann and H. A. Brune, *J. Organomet. Chem.*, 1986, **312**, 133; H. A. Brune, G. Schmidtberg and C. Weisemann, *J. Organomet. Chem.*, 1989, **371**, 121; C. Weisemann, G. Schmidtberg and H. A. Brune, *J. Organomet. Chem.*, 1989, **365**, 403; 1989, **362**, 63.
71. W.-D. Muller and H. A. Brune, *Chem. Ber.*, 1986, **119**, 759.
72. D. Minniti and M. F. Parisi, *Inorg. Chim. Acta*, 1991, **188**, 127.
73. J. Vicente, J. A. Abad, F. Teruel and J. Garcia, *J. Organomet. Chem.*, 1988, **345**, 233.
74. K. Siegmann, P. S. Pregosin and L. M. Venanzi, *Organometallics*, 1989, **8**, 2659.
75. T. G. Appleton, J. R. Hall, D. W. Neale and M. A. Williams, *J. Organomet. Chem.*, 1984, **276**, C73; T. G. Appleton, J. R. Hall and M. A. Williams, *J. Organomet. Chem.*, 1986, **303**, 139.
76. R. Romeo, A. Grassi and L. M. Scolaro, *Inorg. Chem.*, 1992, **31**, 4383.
77. M. Ghedini, M. Longeri and F. Neve, *Inorg. Chim. Acta*, 1987, **132**, 223; M. Ghedini, F. Neve, M. Longeri and M. C. Bruno, *Inorg. Chim. Acta*, 1988, **149**, 131.
78. V. Christou and G. B. Young, *J. Organomet. Chem.*, 1989, **368**, 391.
79. M. Bochmann, G. Wilkinson and G. B. Young, *J. Chem. Soc., Dalton Trans.*, 1980, 1879.
80. B. C. Ankianiec and G. B. Young, *Polyhedron*, 1989, **8**, 57.

81. R. D. Kelly and G. B. Young, *J. Organomet. Chem.*, 1989, **361**, 123.
82. V. G. Albano, D. Braga, V. De Felice, A. Panunzi and A. Vitagliano, *Organometallics*, 1987, **6**, 517.
83. G. Alibrandi, D. Minniti, L. M. Scolaro and R. Romeo, *Inorg. Chem.*, 1989, **28**, 1939.
84. S. Lanza and E. Rotondo, *Inorg. Chim. Acta*, 1988, **146**, 205.
85. A. J. Canty and R. T. Honeyman, *J. Organomet. Chem.*, 1990, **387**, 247.
86. M. C. Aversa, P. Bonaccorsi, M. Cusumano, P. Giannetto and D. Minniti, *J. Chem. Soc., Dalton Trans.*, 1991, 3431.
87. V. F. Sutcliffe and G. B. Young, *Polyhedron*, 1984, **3**, 87.
88. J. D. Scott and R. J. Puddephatt, *Inorg. Chim. Acta*, 1984, **89**, L27.
89. S. Lanza, *Inorg. Chim. Acta*, 1983, **75**, 131.
90. D. L. Thorn and J. C. Calabrese, *J. Organomet. Chem.*, 1988, **342**, 269.
91. P. K. Byers, A. J. Canty and R. T. Honeyman, *Adv. Organomet. Chem.*, 1992, **34**, 1.
92. H. C. Clark and M. A. Mesubi, *J. Organomet. Chem.*, 1981, **215**, 131.
93. A. J. Canty and N. J. Minchin, *J. Organomet. Chem.*, 1982, **226**, C14; A. J. Canty, N. J. Minchin, J. M. Patrick and A. H. White, *J. Chem. Soc., Dalton Trans.*, 1983, 1253.
94. S. Nussbaum and A. Storr, *Can. J. Chem.*, 1985, **63**, 2550.
95. J. W. Ellis, K. N. Harrison, P. A. T. Hoye, A. G. Orpen, P. G. Pringle and M. B. Smith, *Inorg. Chem.*, 1992, **31**, 3026.
96. K. A. Fallis, G. K. Anderson and N. P. Rath, *Organometallics*, 1992, **11**, 885.
97. A. B. Goel and S. Goel, *Inorg. Chim. Acta*, 1982, **59**, 237.
98. G. K. Anderson, H. C. Clark and J. A. Davies, *Inorg. Chem.*, 1981, **20**, 3607.
99. J. Grobe and W. Hildebrandt, *Z. Anorg. Allg. Chem.*, 1990, **590**, 65.
100. S. S. Moore and G. M. Whitesides, *J. Org. Chem.*, 1982, **47**, 1489.
101. D. W. Meek, R. Waid, K. D. Tau, R. M. Kirchner and C. N. Morimoto, *Inorg. Chim. Acta*, 1982, **64**, L221.
102. A. L. Balch, R. R. Guimerans and J. Linehan, *Inorg. Chem.*, 1985, **24**, 290.
103. F. S. M. Hassan, S. J. Higgins, G. B. Jacobsen, B. L. Shaw and M. Thornton-Pett, *J. Chem. Soc., Dalton Trans.*, 1988, 3011; B. L. Shaw and J. D. Vessey, *J. Chem. Soc., Dalton Trans.*, 1992, 1929.
104. S. K. Thomson and G. B. Young, *Polyhedron*, 1988, **7**, 1953.
105. R. D. Kelly and G. B. Young, *Polyhedron*, 1989, **8**, 433.
106. B. C. Ankianiec and G. B. Young, *Polyhedron*, 1991, **10**, 1411.
107. H. A. Brune, R. Hohenadel and G. Schmidtberg, *J. Organomet. Chem.*, 1986, **311**, 27.
108. A. M. Saleem and H. A. Hodali, *Inorg. Chim. Acta*, 1990, **174**, 223; *Synth. React. Inorg. Met-Org. Chem.*, 1990, **20**, 1143.
109. S. D. Perera, B. L. Shaw and M. Thornton-Pett, *J. Chem. Soc., Dalton Trans.*, 1992, 999.
110. S. D. Toto, M. M. Olmstead, B. W. Arbuckle, P. K. Bharadwaj and W. K. Musker, *Inorg. Chem.*, 1990, **29**, 691.
111. V. K. Jain, V. S. Jakkal and R. Bohra, *J. Organomet. Chem.*, 1990, **389**, 417.
112. E. W. Abel, D. G. Evans, J. R. Koe, M. B. Hursthouse and M. Mazid, *J. Chem. Soc., Dalton Trans.*, 1992, 663.
113. G. K. Anderson, M. Lin and M. Y. Chiang, *Organometallics*, 1990, **9**, 288; M. Lin, K. A. Fallis, G. K. Anderson, N. P. Rath and M. Y. Chiang, *J. Am. Chem. Soc.*, 1992, **114**, 4687.
114. B. Chaudret, B. Delavaux and R. Poilblanc, *Coord. Chem. Rev.*, 1988, **86**, 191.
115. G. K. Anderson, *Adv. Organomet. Chem.*, 1993, **35**, 1.
116. R. J. Puddephatt, L. Manojlovic-Muir and K. W. Muir, *Polyhedron*, 1990, **9**, 2767.
117. S. J. Cooper, M. P. Brown and R. J. Puddephatt, *Inorg. Chem.*, 1981, **20**, 1374.
118. F. S. M. Hassan, D. M. McEwan, P. G. Pringle and B. L. Shaw, *J. Chem. Soc., Dalton Trans.*, 1985, 1501; A. T. Hutton, P. G. Pringle and B. L. Shaw, *J. Chem. Soc., Dalton Trans.*, 1985, 1677.
119. G. K. Anderson, H. C. Clark and J. A. Davies, *J. Organomet. Chem.*, 1981, **210**, 135.
120. S. S. M. Ling, R. J. Puddephatt, L. Manojlovic-Muir and K. W. Muir, *Inorg. Chim. Acta*, 1983, **77**, L95; L. Manojlovic-Muir, K. W. Muir, A. A. Frew, S. S. M. Ling, M. A. Thomson and R. J. Puddephatt, *Organometallics*, 1984, **3**, 1637.
121. G. B. Jacobsen and B. L. Shaw, *J. Chem. Soc., Chem. Commun.*, 1985, 692; *J. Chem. Soc., Dalton Trans.*, 1987, 151.
122. W. Radecka-Paryzek, A. J. McLennan and R. J. Puddephatt, *Inorg. Chem.*, 1986, **25**, 3097.
123. L. Manojlovic-Muir, I. R. Jobe, B. J. Maya and R. J. Puddephatt, *J. Chem. Soc., Dalton Trans.*, 1987, 2117.
124. N. Hadj-Bagheri and R. J. Puddephatt, *Polyhedron*, 1988, **7**, 2695.
125. K. A. Fallis, C. Xu and G. K. Anderson, *Organometallics*, 1993, **12**, 2243.
126. M. P. Brown *et al.*, *Inorg. Chem.*, 1981, **20**, 1500.
127. S. Muralidharan and J. H. Espenson, *Inorg. Chem.*, 1983, **22**, 2786.
128. S. Muralidharan and J. H. Espenson, *J. Am. Chem. Soc.*, 1984, **106**, 8104.
129. K. A. Azam, A. A. Frew, B. R. Lloyd, L. Manojlovic-Muir, K. W. Muir and R. J. Puddephatt, *J. Chem. Soc., Chem. Commun.*, 1982, 614; *Organometallics* 1985, **4**, 1400.
130. W. R. Meyer and L. M. Venanzi, *Angew. Chem., Int. Ed. Engl.*, 1984, **23**, 529.
131. T. G. Attig, R. J. Ziegler and C. P. Brock, *Inorg. Chem.*, 1980, **19**, 2315.
132. G. N. Glavee, L. M. Daniels and R. J. Angelici, *Organometallics*, 1989, **8**, 1856.
133. R. L. Cowan and W. C. Trogler, *Organometallics*, 1987, **6**, 2451; *J. Am. Chem. Soc.*, 1989, **111**, 4750.
134. N. W. Alcock, J. M. Brown and T. D. MacLean, *J. Chem. Soc., Chem. Commun.*, 1984, 1689.
135. B. R. James, 'Homogeneous Hydrogenation', Wiley, New York, 1973.
136. R. Bardi, A. M. Piazzesi, A. Del Pra, G. Cavinato and L. Toniolo, *J. Organomet. Chem.*, 1982, **234**, 107.
137. R. Bardi, A. M. Piazzesi, G. Cavinato, P. Cavoli and L. Toniolo, *J. Organomet. Chem.*, 1982, **224**, 407.
138. R. Graziani, G. Cavinato, U. Casellato and L. Toniolo, *J. Organomet. Chem.*, 1988, **353**, 125.
139. A. B. Permin and V. S. Petrosyan, *Appl. Organomet. Chem.*, 1990, **4**, 111.
140. M. Gomez, G. Muller, D. Sainz, J. Sales and X. Solans, *Organometallics*, 1991, **10**, 4036.
141. J. Jans, R. Naegeli, L. M. Venanzi and A. Albinati, *J. Organomet. Chem.*, 1983, **247**, C37; F. Bachechi *et al.*, *Inorg. Chem.*, 1983, **22**, 1031.
142. D. Carmona, R. Thouvenot, L. M. Venanzi, F. Bachechi and L. Zambonelli, *J. Organomet. Chem.*, 1983, **250**, 589.
143. C. Sorato and L. M. Venanzi, *Inorg. Synth.*, 1989, **26**, 134.
144. A. Immirzi, A. Musco, P. S. Pregosin and L. M. Venanzi, *Angew. Chem., Int. Ed. Engl.*, 1980, **19**, 721; P. Boron, A. Musco and L. M. Venanzi, *Inorg. Chem.*, 1982, **21**, 4192.
145. A. Albinati, R. Naegeli, A. Togni and L. M. Venanzi, *Organometallics*, 1983, **2**, 926; *J. Organomet. Chem.*, 1987, **330**, 85.

146. O. Rossell, J. Sales and M. Seco, *J. Organomet. Chem.*, 1982, **236**, 415; M. Crespo, O. Rossell, J. Sales and M. Seco, *J. Organomet. Chem.*, 1984, **273**, 415.
147. O. Rossell, M. Seco, I. Torra, X. Solans and M. Font-Altaba, *J. Organomet. Chem.*, 1984, **270**, C63.
148. S. Komiya and I. Endo, *Chem. Lett.*, 1988, 1709.
149. G. K. Anderson and M. Lin, *Inorg. Chim. Acta*, 1988, **142**, 7; *Organometallics*, 1988, **7**, 2285.
150. G. B. Deacon, A. Dietrich, C. M. Forsyth and H. Schumann, *Angew. Chem., Int. Ed. Engl.*, 1989, **28**, 1370; G. B. Deacon, C. M. Forsyth, W. C. Patalinghug, A. H. White, A. Dietrich and H. Schumann, *Aust. J. Chem.*, 1992, **45**, 567.
151. S. M. Baxter, G. S. Ferguson and P. T. Wolczanski, *J. Am. Chem. Soc.*, 1988, **110**, 4231.
152. S. M. Baxter and P. T. Wolczanski, *Organometallics*, 1990, **9**, 2498.
153. F. Ozawa, J. W. Park, P. B. Mackenzie, W. P. Schaefer, L. M. Henling and R. H. Grubbs, *J. Am. Chem. Soc.*, 1989, **111**, 1319.
154. A. G. Orpen, L. Brammer, F. H. Allen, O. Kennard, D. G. Watson and R. Taylor, *J. Chem. Soc., Dalton Trans.*, 1989, S1.
155. R. Bardi and A. M. Piazzesi, *Inorg. Chim. Acta*, 1981, **47**, 249; R. Bardi, A. M. Piazzesi and G. Bruno, *Cryst. Struct. Commun.*, 1981, **10**, 807.
156. A. B. Goel, S. Goel and D. Vanderveer, *Polyhedron*, 1984, **3**, 373.
157. K. Osakada, Y.-J. Kim and A. Yamamoto, *J. Organomet. Chem.*, 1990, **382**, 303.
158. N. C. Payne and D. W. Stephan, *J. Organomet. Chem.*, 1982, **228**, 203.
159. J. M. Wisner, T. J. Bartczak and J. A. Ibers, *Organometallics*, 1986, **5**, 2044.
160. A. G. Thayer and N. C. Payne, *Acta Crystallogr., Sect. C*, 1986, **42**, 1302.
161. T. M. Gomes-Carneiro, R. D. Jackson, J. H. Downing, A. G. Orpen and P. G. Pringle, *J. Chem. Soc., Chem. Commun.*, 1991, 317.
162. H. E. Bryndza, J. C. Calabrese and S. S. Wreford, *Organometallics*, 1984, **3**, 1603; H. E. Bryndza, J. C. Calabrese, M. Marsi, D. C. Roe, W. Tam and J. E. Bercaw, *J. Am. Chem. Soc.*, 1986, **108**, 4805.
163. S. Park, M. Pontier-Johnson and D. M. Roundhill, *Inorg. Chem.*, 1990, **29**, 2689.
164. L. Manojlovic-Muir, S. S. M. Ling and R. J. Puddephatt, *J. Chem. Soc., Dalton Trans.*, 1986, 151.
165. G. K. Anderson, M. Lin and N. P. Rath, *Organometallics*, 1990, **9**, 2880.
166. M. P. Brown, *et al.*, *J. Chem. Soc., Dalton Trans.*, 1982, 299.
167. G. Alibrandi, G. Bruno, S. Lanza, D. Minniti, R. Romeo and M. L. Tobe, *Inorg. Chem.*, 1987, **26**, 185.
168. R. Bardi, A. Del Pra, A. M. Piazzesi and M. Trozzi, *Cryst. Struct. Commun.*, 1981, **10**, 301.
169. G. Bruno, S. Lanza and F. Nicolo, *Acta Crystallogr., Sect. C*, 1990, **46**, 765.
170. W. Conzelmann, J. D. Koola, U. Kunze and J. Strahle, *Inorg. Chim. Acta*, 1984, **89**, 147.
171. G. B. Robertson and P. A. Tucker, *Acta Crystallogr., Sect. C*, 1983, **39**, 1354.
172. G. Strukul, R. A. Michelin, J. D. Orbell and L. Randaccio, *Inorg. Chem.*, 1983, **22**, 3706.
173. M. A. Bennett, G. B. Robertson, A. Rokicki and W. A. Wickramasinghe, *J. Am. Chem. Soc.*, 1988, **110**, 7098.
174. L. S. Chang, M. Pontier Johnson and M. J. Fink, *Organometallics*, 1991, **10**, 1219.
175. A. C. Skapski, V. F. Sutcliffe and G. B. Young, *J. Chem. Soc., Chem. Commun.*, 1985, 609.
176. H. A. Brune, R. Klotzbucher, K. Berhalter and T. Debaerdemaeker, *J. Organomet. Chem.*, 1989, **369**, 321.
177. M. Crespo, J. Sales, X. Solans and M. F. Altaba, *J. Chem. Soc., Dalton Trans.*, 1988, 1617.
178. T. Debaerdemaeker, K. Berhalter, C. Weisemann and H. A. Brune, *Acta Crystallogr., Sect. C*, 1987, **43**, 1253.
179. T. Debaerdemaeker, B. Stapp and H. A. Brune, *Acta Crystallogr., Sect. C*, 1987, **43**, 473.
180. T. Debaerdemaeker, C. Weisemann and H. A. Brune, *Acta Crystallogr., Sect. C*, 1987, **42**, 432.
181. T. Debaerdemaeker, C. Weisemann and H. A. Brune, *J. Organomet. Chem.*, 1988, **350**, 91.
182. H. E. Bryndza, L. K. Fong, R. A. Paciello, W. Tam and J. E. Bercaw, *J. Am. Chem. Soc.*, 1987, **109**, 1444; H. E. Bryndza, P. J. Domaille, W. Tam, L. K. Fong, R. A. Paciello and J. E. Bercaw, *Polyhedron*, 1988, **7**, 1441.
183. G. Al Takhin, H. A. Skinner and A. A. Zaki, *J. Chem. Soc., Dalton Trans.*, 1984, 371.
184. R. Benn, R-D. Reinhardt and A. Rufinska, *J. Organomet. Chem.*, 1985, **282**, 291.
185. H. A. Brune, J. Unsin, H. G. Alt, G. Schmidtberg and K.-H. Spohn, *Chem. Ber.*, 1984, **117**, 1606; H. A. Brune, W.-D. Muller and K.-H. Spohn, *Z. Naturforsch., Teil B*, 1986, **41**, 211; W.-D. Muller and H. A. Brune, *J. Organomet. Chem.*, 1986, **299**, 391; H. G. Alt, W.-D. Muller, J. Unsin and H. A. Brune, *J. Organomet. Chem.*, 1986, **307**, 121.
186. J. Granell, G. Muller, M. Rocamora and J. Vilarrasa, *Magn. Reson. Chem.*, 1986, **24**, 243.
187. D. P. Arnold and M. A. Bennett, *Inorg. Chem.*, 1984, **23**, 2117.
188. A. Pidcock, R. E. Richards and L. M. Venanzi, *J. Chem. Soc. A*, 1966, 1707.
189. T. G. Appleton, H. C. Clark and L. E. Manzer, *Coord. Chem. Rev.*, 1973, **10**, 335.
190. R. Baumgartner, G. Schmidtberg and H. A. Brune, *J. Organomet. Chem.*, 1988, **345**, 221; R. Baumgartner and H. A. Brune, *J. Organomet. Chem.*, 1988, **350**, 115.
191. G. K. Anderson, D. M. Black, R. J. Cross, F. J. Robertson, D. S. Rycroft and R. K. M. Wat, *Organometallics*, 1990, **9**, 2568.
192. M. C. Grossel, J. R. Batson, R. P. Moulding and K. R. Seddon, *J. Organomet. Chem.*, 1986, **304**, 391.
193. K. A. Azam and R. J. Puddephatt, *Organometallics*, 1983, **2**, 1396; R. J. Puddephatt *et al.*, *J. Am. Chem. Soc.*, 1983, **105**, 5642.
194. E. W. Abel, T. P. J. Coston, K. G. Orrell, V. Sik and D. Stephenson, *J. Magn. Reson.*, 1986, **70**, 34.
195. H. Ruegger and P. S. Pregosin, *Inorg. Chem.*, 1987, **26**, 2912.
196. C. Ammann, P. S. Pregosin and A. Scrivanti, *Inorg. Chim. Acta*, 1989, **155**, 217.
197. D. A. Roberts, W. R. Mason and G. L. Geoffroy, *Inorg. Chem.*, 1981, **20**, 789.
198. P. S. Braterman and J.-I. Song, *Inorg. Chim. Acta*, 1991, **183**, 131.
199. H. Palkovits, U. Ziegler, G. Schmidtberg and H. A. Brune, *J. Organomet. Chem.*, 1988, **338**, 119.
200. D.-S. Yang, G. M. Bancroft, J. D. Bozek, R. J. Puddephatt and J. S. Tse, *Inorg. Chem.*, 1989, **28**, 1; D.-S. Yang, G. M. Bancroft, L. Dignard-Bailey, R. J. Puddephatt and J. S. Tse, *Inorg. Chem.*, 1990, **29**, 2487.
201. L. M. Scolaro, G. Alibrandi, R. Romeo, V. Ricevuto and S. Campagna, *Inorg. Chem.*, 1992, **31**, 2074.
202. K. Tatsumi, R. Hoffmann, A. Yamamoto and J. K. Stille, *Bull. Chem. Soc. Jpn.*, 1981, **54**, 1857.
203. A. C. Balazs, K. H. Johnson and G. M. Whitesides, *Inorg. Chem.*, 1982, **21**, 2162.
204. A. Flores-Riveros and O. Novaro, *J. Organomet. Chem.*, 1982, **235**, 383.
205. S. Obara, K. Kitaura and K. Morokuma, *J. Am. Chem. Soc.*, 1984, **106**, 7482.
206. J. J. Low and W. A. Goddard, III, *Organometallics*, 1986, **5**, 609.

207. J. J. Low and W. A. Goddard, III, *J. Am. Chem. Soc.*, 1986, **108**, 6115.
208. R. A. Michelin, S. Faglia and P. Uguagliati, *Inorg. Chem.*, 1983, **22**, 1831.
209. S. Sostero, O. Traverso, R. Ros and R. A. Michelin, *J. Organomet. Chem.*, 1983, **246**, 325.
210. K. A. Azam, R. H. Hill and R. J. Puddephatt, *Can. J. Chem.*, 1984, **62**, 2029.
211. H. A. Brune, U. Mayr and G. Schmidtberg, *Z. Naturforsch., Teil B*, 1986, **41**, 1281.
212. R. Klotzbucher and H. A. Brune, *J. Organomet. Chem.*, 1986, **299**, 399; H. A. Brune, R. Klotzbucher and G. Schmidtberg, *J. Organomet. Chem.*, 1989, **365**, 389; *J. Organomet. Chem.*, 1989, **371**, 113.
213. H. A. Brune, E. Hupfer, G. Schmidtberg and A. Baur, *J. Organomet. Chem.*, 1992, **424**, 225.
214. A. C. Skapski, V. F. Sutcliffe and G. B. Young, *J. Chem. Soc., Chem. Commun.*, 1985, 609.
215. S. E. Himmel and G. B. Young, *Organometallics*, 1988, **7**, 2440.
216. G. M. Whitesides, *Pure Appl. Chem.*, 1981, **53**, 287.
217. S. Komiya, Y. Morimoto, A. Yamamoto and T. Yamamoto, *Organometallics*, 1982, **1**, 1528.
218. R. G. Nuzzo, T. J. McCarthy and G. M. Whitesides, *Inorg. Chem.*, 1981, **20**, 1312.
219. T. J. McCarthy, R. G. Nuzzo and G. M. Whitesides, *J. Am. Chem. Soc.*, 1981, **103**, 1676.
220. T. J. McCarthy, R. G. Nuzzo and G. M. Whitesides, *J. Am. Chem. Soc.*, 1981, **103**, 3396.
221. H. E. Bryndza, *J. Chem. Soc., Chem. Commun.*, 1985, 1696.
222. L. Mole, J. L. Spencer, N. Carr and A. G. Orpen, *Organometallics*, 1991, **10**, 49.
223. K. A. Azam, M. P. Brown, S. J. Cooper and R. J. Puddephatt, *Organometallics*, 1982, **1**, 1183.
224. R. G. Nuzzo, T. J. McCarthy and G. M. Whitesides, *J. Am. Chem. Soc.*, 1981, **103**, 3404.
225. P. Foley, R. DiCosimo and G. M. Whitesides, *J. Am. Chem. Soc.*, 1980, **102**, 6713; J. A. Ibers, R. DiCosimo and G. M. Whitesides, *Organometallics*, 1982, **1**, 13.
226. R. DiCosimo, S. S. Moore, A. F. Sowinski and G. M. Whitesides, *J. Am. Chem. Soc.*, 1982, **104**, 124.
227. S. K. Thomson and G. B. Young, *Organometallics*, 1989, **8**, 2068.
228. D. C. Griffiths and G. B. Young, *Polyhedron*, 1983, **2**, 1095.
229. D. C. Griffiths and G. B. Young, *Organometallics*, 1989, **8**, 875.
230. B. C. Ankianiec, D. T. Hardy, S. K. Thomson, W. N. Watkins and G. B. Young, *Organometallics*, 1992, **11**, 2591.
231. D. C. Griffiths, L. G. Joy, A. C. Skapski, D. J. Wilkes and G. B. Young, *Organometallics*, 1986, **5**, 1744.
232. R. L. Brainard and G. M. Whitesides, *Organometallics*, 1985, **4**, 1550.
233. G. Alibrandi, D. Minniti, R. Romeo and P. Vitarelli, *Inorg. Chim. Acta*, 1984, **81**, L23.
234. G. Alibrandi, L. M. Scolaro, D. Minniti and R. Romeo, *Inorg. Chem.*, 1990, **29**, 3467.
235. R. L. Brainard, T. M. Miller and G. M. Whitesides, *Organometallics*, 1986, **5**, 1481.
236. T. C. Flood and J. A. Statler, *Organometallics*, 1984, **3**, 1795.
237. A. Morvillo, G. Favero and A. Turco, *J. Organomet. Chem.*, 1983, **243**, 111; A. Morvillo and A. Turco, *J. Organomet. Chem.*, 1983, **258**, 383.
238. B. Nixon, P. R. Norton, E. C. Ou, R. J. Puddephatt, S. Roy and P. A. Young, *Chem. Mater.*, 1991, **3**, 222; N. H. Dryden *et al.*, *Chem. Mater.*, 1991, **3**, 677.
239. S. Park, M. Pontier-Johnson and D. M. Roundhill, *J. Am. Chem. Soc.*, 1989, **111**, 3101.
240. G. Alibrandi *et al.*, *Inorg. Chim. Acta*, 1985, **100**, 107.
241. G. Alibrandi, D. Minniti, R. Romeo, P. Uguagliati, L. Calligaro and U. Belluco, *Inorg. Chim. Acta*, 1986, **112**, L15.
242. R. J. Puddephatt and M. A. Stalteri, *J. Organomet. Chem.*, 1980, **193**, C27.
243. J. K. Jawad, R. J. Puddephatt and M. A. Stalteri, *Inorg. Chem.*, 1982, **21**, 332.
244. R. J. Cross and A. J. McLennan, *J. Chem. Soc., Dalton Trans.*, 1983, 359; *J. Chem. Soc., Dalton Trans.*, 1984, 205.
245. J. D. Scott and R. J. Puddephatt, *Organometallics*, 1983, **2**, 1643.
246. S. Lanza, D. Minniti, R. Romeo, P. Moore, J. Sachinidis and M. L. Tobe, *J. Chem. Soc., Chem. Commun.*, 1984, 542.
247. S. Lanza, D. Minniti, P. Moore, J. Sachinidis, R. Romeo and M. L. Tobe, *Inorg. Chem.*, 1984, **23**, 4428.
248. D. Minniti, G. Alibrandi, M. L. Tobe and R. Romeo, *Inorg. Chem.*, 1987, **26**, 3956.
249. U. Frey, L. Helm, A. E. Merbach and R. Romeo, *J. Am. Chem. Soc.*, 1989, **111**, 8161.
250. A. Y. C. Hung, J. C. Woolcock, M. F. Rettig and R. M. Wing, *Inorg. Chem.*, 1992, **31**, 810.
251. G. K. Anderson and R. J. Cross, *Chem. Soc. Rev.*, 1980, **9**, 185.
252. R. Romeo, P. Uguagliati and U. Belluco, *J. Mol. Catal.*, 1975, **1**, 325.
253. H. Kelm, W. J. Louw and D. A. Palmer, *Inorg. Chem.*, 1980, **19**, 843.
254. W. J. Louw, R. van Eldik and H. Kelm, *Inorg. Chem.*, 1980, **19**, 2878.
255. R. Romeo, D. Minniti and S. Lanza, *Inorg. Chem.*, 1980, **19**, 3663.
256. M. J. Blandamer, J. Burgess and R. Romeo, *Inorg. Chim. Acta*, 1982, **65**, L179.
257. M. J. Blandamer, J. Burgess, D. Minniti and R. Romeo, *Inorg. Chim. Acta*, 1985, **96**, 129.
258. G. Alibrandi, M. Cusumano, D. Minniti, L. M. Scolaro and R. Romeo, *Inorg. Chem.*, 1989, **28**, 342.
259. G. Alibrandi, L. M. Scolaro and R. Romeo, *Inorg. Chem.*, 1991, **30**, 4007.
260. G. Alibrandi, D. Minniti, L. M. Scolaro and R. Romeo, *Inorg. Chem.*, 1988, **27**, 318.
261. L. L. Costanzo, S. Giuffrida and R. Romeo, *Inorg. Chim. Acta*, 1980, **38**, 31.
262. D. L. Reger, D. G. Garza and J. C. Baxter, *Organometallics*, 1990, **9**, 873.
263. F. Calderazzo, *Angew. Chem., Int. Ed. Engl.*, 1977, **16**, 299; E. J. Kuhlmann and J. J. Alexander, *Coord. Chem. Rev.*, 1980, **33**, 195.
264. G. K. Anderson and R. J. Cross, *Acc. Chem. Res.*, 1984, **17**, 67.
265. S. Sakaki, K. Kitaura, K. Morokuma and K. Ohkubo, *J. Am. Chem. Soc.*, 1983, **105**, 2280; N. Koga and K. Morokuma, *J. Am. Chem. Soc.*, 1986, **108**, 6136.
266. H. Brunner and H. Vogt, *Angew. Chem., Int. Ed. Engl.*, 1981, **20**, 405; H. Brunner, B. Hammer, I. Bernal and M. Draux, *Organometallics*, 1983, **2**, 1595; T. C. Flood and K. D. Campbell, *J. Am. Chem. Soc.*, 1984, **106**, 2853.
267. G. K. Anderson, H. C. Clark and J. A. Davies, *Inorg. Chem.*, 1981, **20**, 1636.
268. G. K. Anderson and R. J. Cross, *J. Chem. Soc., Dalton Trans.*, 1979, 1246; *J. Chem. Soc., Dalton Trans.*, 1980, 712.
269. R. J. Cross and J. Gemmill, *J. Chem. Soc., Dalton Trans.*, 1981, 2317.
270. G. K. Anderson and G. J. Lumetta, *Organometallics*, 1985, **4**, 1542.
271. G. P. C. M. Dekker *et al.*, *Organometallics*, 1992, **11**, 1937.
272. T. Kondo, Y. Tsuji and Y. Watanabe, *J. Organomet. Chem.*, 1988, **345**, 397.

273. P. E. Garrou and R. F. Heck, *J. Am. Chem. Soc.*, 1976, **98**, 4115; G. K. Anderson and R. J. Cross, *J. Chem. Soc., Dalton Trans.*, 1980, 1434.
274. G. K. Anderson, H. C. Clark and J. A. Davies, *Organometallics*, 1982, **1**, 64.
275. G. Cavinato and L. Toniolo, *J. Organomet. Chem.*, 1983, **241**, 275; G. Moretti, C. Botteghi and L. Toniolo, *J. Mol. Catal.*, 1987, **39**, 177.
276. A. Scrivanti, G. Cavinato, L. Toniolo and C. Botteghi, *J. Organomet. Chem.*, 1985, **286**, 115; H. J. Ruegg, P. S. Pregosin, A. Scrivanti, L. Toniolo and C. Botteghi, *J. Organomet. Chem.*, 1986, **316**, 233.
277. A. Scrivanti, A. Berton, L. Toniolo and C. Botteghi, *J. Organomet. Chem.*, 1986, **314**, 369.
278. A. Scrivanti, C. Botteghi, L. Toniolo and A. Berton, *J. Organomet. Chem.*, 1988, **344**, 261.
279. H. E. Bryndza, *Organometallics*, 1985, **4**, 1686; H. E. Bryndza, S. A. Kretchmar and T. H. Tulip, *J. Chem. Soc., Chem. Commun.*, 1985, 977.
280. H. E. Bryndza, W. C. Fultz and W. Tam, *Organometallics*, 1985, **4**, 939.
281. K. Osakada, Y.-J. Kim and A. Yamamoto, *J. Organomet. Chem.*, 1990, **382**, 303.
282. A. Sacco, G. Vasapollo, C. F. Nobile, A. Piergiovanni, M. A. Pellinghelli and M. Lanfranchi, *J. Organomet. Chem.*, 1988, **356**, 397.
283. C. F. Shiba and W. H. Waddell, *J. Organomet. Chem.*, 1983, **241**, 119.
284. D. P. Arnold and M. A. Bennett, *J. Organomet. Chem.*, 1980, **199**, 119; *Inorg. Chem.*, 1984, **23**, 2110; M. A. Bennett and A. Rokicki, *Inorg. Synth.*, 1989, **25**, 100.
285. M. A. Bennett, *J. Mol. Catal.*, 1987, **41**, 1.
286. D. P. Arnold and M. A. Bennett, *J. Organomet. Chem.*, 1980, **199**, C17.
287. I. Torresan, R. A. Michelin, A. Marsella, A. Zanardo, F. Pinna and G. Strukul, *Organometallics*, 1991, **10**, 623.
288. J.-T. Chen, T.-M. Huang, M.-C. Cheng and Y. Wang, *Organometallics*, 1990, **9**, 539.
289. T.-M. Huang, J.-T. Chen, G.-H. Lee and Y. Wang, *Organometallics*, 1991, **10**, 175.
290. L. Huang, F. Ozawa and A. Yamamoto, *Organometallics*, 1990, **9**, 2612.
291. G.-L. Huang, T.-M. Huang and J.-T. Chen, *Inorg. Chem.*, 1992, **31**, 4034.
292. J.-T. Chen, T.-M. Huang, M.-C. Cheng and Y. Wang, *Organometallics*, 1991, **10**, 2838.
293. T. C. Flood and S. P. Bitler, *J. Am. Chem. Soc.*, 1984, **106**, 6076.
294. T. Hase, A. Miyashita and H. Nohira, *Chem. Lett.*, 1988, 219.
295. M. Lashanizadehgan, M. Rashidi, J. E. Hux, R. J. Puddephatt and S. S. M. Ling, *J. Organomet. Chem.*, 1984, **269**, 317.
296. M. Rashidi, Z. Fakhroeian and R. J. Puddephatt, *J. Organomet. Chem.*, 1991, **406**, 261.
297. A. T. Hutton, B. Shabanzadeh and B. L. Shaw, *J. Chem. Soc., Chem. Commun.*, 1982, 1345.
298. R. J. Puddephatt and J. D. Scott, *Organometallics*, 1985, **4**, 1221.
299. J. D. Scott, M. Crespo, C. M. Anderson and R. J. Puddephatt, *Organometallics*, 1987, **6**, 1772.
300. S. S. M. Ling, R. J. Puddephatt, L. Manojlovic-Muir and K. W. Muir, *J. Organomet. Chem.*, 1983, **255**, C11.
301. S. S. M. Ling, I. R. Jobe, L. Manojlovic-Muir, K. W. Muir and R. J. Puddephatt, *Organometallics*, 1985, **4**, 1198.
302. S. S. M. Ling, N. C. Payne and R. J. Puddephatt, *Organometallics*, 1985, **4**, 1546.
303. A. G. Constable, C. R. Langrick, B. Shabanzadeh and B. L. Shaw, *Inorg. Chim. Acta*, 1982, **65**, L151.
304. A. J. Canty, R. T. Honeyman, B. W. Skelton and A. H. White, *Inorg. Chim. Acta*, 1986, **114**, L39.
305. P. K. Byers, A. J. Canty, R. T. Honeyman, B. W. Skelton and A. H. White, *J. Organomet. Chem.*, 1992, **433**, 223.
306. E. Rotondo, A. Giannetto and S. Lanza, *J. Organomet. Chem.*, 1990, **396**, 115.
307. S. Achar, J. D. Scott and R. J. Puddephatt, *Organometallics*, 1992, **11**, 2325.
308. H. C. Clark, G. Ferguson, V. K. Jain and M. Parvez, *Organometallics*, 1983, **2**, 806.
309. A. J. Canty, R. T. Honeyman, B. W. Skelton and A. J. White, *J. Organomet. Chem.*, 1990, **396**, 105.
310. A. J. Canty, R. T. Honeyman, B. W. Skelton and A. J. White, *J. Organomet. Chem.*, 1992, **424**, 381.
311. P. K. Monaghan and R. J. Puddephatt, *Inorg. Chim. Acta*, 1982, **65**, L59; *Organometallics*, 1984, **3**, 444.
312. K.-T. Aye, G. Ferguson, A. J. Lough and R. J. Puddephatt, *Angew. Chem., Int. Ed. Engl.*, 1989, **28**, 767; K.-T. Aye, L. Gelmini, N. C. Payne, J. J. Vittal and R. J. Puddephatt, *J. Am. Chem. Soc.*, 1990, **112**, 2464.
313. K.-T. Aye, D. Colpitts, G. Ferguson and R. J. Puddephatt, *Organometallics*, 1988, **7**, 1454.
314. P. K. Monaghan and R. J. Puddephatt, *Inorg. Chim. Acta*, 1983, **76**, L237; *Organometallics*, 1985, **4**, 1406.
315. J. D. Scott and R. J. Puddephatt, *Organometallics*, 1986, **5**, 1538.
316. G. J. Arsenault, M. Crespo and R. J. Puddephatt, *Organometallics*, 1987, **6**, 2255.
317. G. J. Arsenault, C. M. Anderson and R. J. Puddephatt, *Organometallics*, 1988, **7**, 2094.
318. A. J. Canty and A. A. Watson, *Inorg. Chim. Acta*, 1988, **154**, 5; P. K. Byers, A. J. Canty, B. W. Skelton, P. R. Traill, A. A. Watson and A. H. White, *Organometallics*, 1990, **9**, 3080.
319. G. Ferguson, M. Parvez, P. K. Monaghan and R. J. Puddephatt, *J. Chem. Soc., Chem. Commun.*, 1983, 267; G. Ferguson, P. K. Monaghan, M. Parvez and R. J. Puddephatt, *Organometallics*, 1985, **4**, 1669; R. H. Hill and R. J. Puddephatt, *J. Am. Chem. Soc.*, 1985, **107**, 1218.
320. P. K. Monaghan and R. J. Puddephatt, *Organometallics*, 1983, **2**, 1698; 1986, **5**, 439.
321. M. Crespo and R. J. Puddephatt, *Organometallics*, 1987, **6**, 2548.
322. P. K. Monaghan and R. J. Puddephatt, *J. Chem. Soc., Dalton Trans.*, 1988, 595.
323. J. D. Scott and R. J. Puddephatt, *Organometallics*, 1986, **5**, 2522.
324. K.-T. Aye, A. J. Canty, M. Crespo, R. J. Puddephatt, J. D. Scott and A. A. Watson, *Organometallics*, 1989, **8**, 1518.
325. B. Crociani, F. Di Bianca, A. Giovenco and A. Scrivanti, *J. Organomet. Chem.*, 1983, **251**, 393.
326. B. Crociani, F. Di Bianca, A. Fontana and R. Bertani, *J. Organomet. Chem.*, 1992, **425**, 155.
327. X. L. R. Fontaine, F. S. M. Hassan, S. J. Higgins, G. B. Jacobsen, B. L. Shaw and M. Thornton-Pett, *J. Chem. Soc., Chem. Commun.*, 1985, 1635.
328. (a) R. Uson, J. Fornies and M. Tomas, *J. Organomet. Chem.*, 1988, **358**, 525; (b) J. A. Morrison, *Adv. Organomet. Chem.*, 1993, **35**, 211.
329. O. Rossell, M. Seco and I. Torra, *J. Chem. Soc., Dalton Trans.*, 1986, 1011.
330. P. Braunstein, O. Rossell, M. Seco, I. Torra, X. Solans and C. Miravitlles, *Organometallics*, 1986, **5**, 1113.
331. O. Rossell and J. Sales, *Inorg. Chim. Acta*, 1981, **64**, L43.
332. O. Rossell and J. Sales, *Inorg. Chim. Acta*, 1981, **53**, L1.
333. S. T. Lin and K. J. Klabunde, *Inorg. Chem.*, 1985, **24**, 1961.

334. O. J. Scherer and H. Jungmann, *J. Organomet. Chem.*, 1981, **208**, 153; R. A. Head, *J. Chem. Soc., Dalton Trans.*, 1982, 1637; C. Bartocci, A. Maldotti, S. Sostero and O. Traverso, *J. Organomet. Chem.*, 1983, **253**, 253.
335. C. A. Ghilardi, S. Midollini, S. Moneti, A. Orlandini and J. A. Ramirez, *J. Chem. Soc., Chem. Commun.*, 1989, 304; C. A. Ghilardi, S. Midollini, S. Moneti, A. Orlandini, G. Scapacci and A. Traversi, *J. Chem. Soc., Dalton Trans.*, 1990, 2293.
336. L. J. Krause and J. A. Morrison, *J. Chem. Soc., Chem. Commun.*, 1981, 1282.
337. P. P. K. Claire, C. J. Jones, J. A. McCleverty, P. L. Coe and M. G. B. Drew, *J. Organomet. Chem.*, 1992, **424**, 105.
338. R. Uson, J. Fornies, P. Espinet and G. Alfranca, *Synth. React. Inorg. Met.-Org. Chem.*, 1980, **10**, 579.
339. G. Lopez, G. Garcia, N. Cutillas and J. Ruiz, *J. Organomet. Chem.*, 1983, **241**, 269.
340. R. Uson, J. Fornies, F. Martinez and M. Tomas, *J. Chem. Soc., Dalton Trans.*, 1980, 888; R. Uson, J. Fornies, F. Martinez, M. Tomas and I. Reoyo, *Organometallics*, 1983, **2**, 1386.
341. G. B. Deacon, P. I. Mackinnon and T. D. Tuong, *Aust. J. Chem.*, 1983, **36**, 43.
342. G. B. Deacon, B. M. Gatehouse and K. T. Nelson-Reed, *J. Organomet. Chem.*, 1989, **359**, 267.
343. G. B. Deacon and I. L. Grayson, *J. Organomet. Chem.*, 1985, **292**, 1.
344. G. B. Deacon and K. T. Nelson-Reed, *J. Organomet. Chem.*, 1987, **322**, 257; D. Anastasiou, G. B. Deacon and B. M. Gatehouse, *J. Organomet. Chem.*, 1987, **329**, 267.
345. O. Rossell, J. Sales and M. Seco, *J. Organomet. Chem.*, 1981, **205**, 133.
346. G. B. Deacon and I. L. Grayson, *Trans. Met. Chem.*, 1982, **7**, 97.
347. G. Lopez *et al.*, *Organometallics*, 1992, **11**, 4090.
348. J. Fornies, C. Fortuno, R. Navarro, F. Martinez and A. J. Welch, *J. Organomet. Chem.*, 1990, **394**, 643.
349. R. Uson, J. Fornies, M. A. Uson and J. A. Apaolaza, *Inorg. Chim. Acta*, 1991, **187**, 175.
350. R. Uson, J. Fornies, M. Tomas, J. M. Casas, F. A. Cotton and L. R. Falvello, *J. Am. Chem. Soc.*, 1985, **107**, 2556; R. Uson *et al.*, *Organometallics*, 1988, **7**, 2279.
351. R. Uson, J. Fornies, M. Tomas, J. M. Casas and C. Fortuno, *Polyhedron*, 1989, **8**, 2209.
352. G. Lopez, G. Garcia, J. Ruiz, G. Sanchez, J. Garcia and C. Vicente, *J. Chem. Soc., Chem. Commun.*, 1989, 1045.
353. G. Lopez *et al.*, *J. Chem. Soc., Dalton Trans.*, 1992, 53.
354. G. Lopez, J. Ruiz, G. Garcia, C. Vicente, J. M. Marti and V. Rodriguez, *J. Organomet. Chem.*, 1992, **436**, 121.
355. G. Lopez *et al.*, *J. Chem. Soc., Dalton Trans.*, 1992, 1681.
356. T. G. Appleton, R. D. Berry, J. R. Hall and D. W. Neale, *J. Organomet. Chem.*, 1989, **364**, 249.
357. R. Uson, J. Fornies, M. Tomas and B. Menjon, *Organometallics*, 1985, **4**, 1912; 1986, **5**, 1581.
358. R. Uson, J. Fornies, M. Tomas, B. Menjon, R. Navarro and J. Carnicer, *Inorg. Chim. Acta*, 1989, **162**, 33.
359. R. Uson, J. Fornies, P. Espinet, R. Navarro and E. Lalinde, *Trans. Met. Chem.*, 1984, **9**, 277.
360. R. Uson, J. Fornies, M. A. Uson, J. F. Yague, P. G. Jones and K. Meyer-Base, *J. Chem. Soc., Dalton Trans.*, 1986, 947.
361. R. Uson, J. Fornies, M. A. Uson and M. M. Carranza, *Inorg. Chim. Acta*, 1989, **155**, 71.
362. R. Uson, J. Fornies, M. Tomas, B. Menjon and A. J. Welch, *J. Organomet. Chem.*, 1986, **304**, C24.
363. J. Fornies, M. A. Gomez-Saso, E. Lalinde, F. Martinez and M. T. Moreno, *Organometallics*, 1992, **11**, 2873.
364. R. Uson, J. Fornies, P. Espinet, R. Navarro and C. Fortuno, *J. Chem. Soc., Dalton Trans.*, 1987, 2077.
365. R. Uson, J. Fornies, R. Navarro and J. I. Cebollada, *J. Organomet. Chem.*, 1986, **304**, 381.
366. R. Uson, J. Fornies, M. Tomas, J. M. Casas and R. Navarro, *J. Chem. Soc., Dalton Trans.*, 1989, 169.
367. R. Uson, J. Fornies, M. Tomas, B. Menjon, J. Carnicer and A. J. Welch, *J. Chem. Soc., Dalton Trans.*, 1990, 151.
368. R. Uson, J. Fornies, M. Tomas and I. Ara, *J. Chem. Soc., Dalton Trans.*, 1989, 1011.
369. R. Uson, J. Fornies, M. Tomas, B. Menjon and A. J. Welch, *Organometallics*, 1988, **7**, 1318.
370. H. E. Bryndza, J. C. Calabrese and S. S. Wreford, *Organometallics*, 1984, **3**, 1603; H. E. Bryndza, *Organometallics*, 1985, **4**, 406.
371. R. McCrindle, G. J. Arsenault and R. Farwaha, *J. Organomet. Chem.*, 1985, **296**, C51; R. McCrindle, G. Ferguson, G. J. Arsenault, M. J. Hampden-Smith, A. J. McAlees and B. L. Ruhl, *J. Organomet. Chem.*, 1990, **390**, 121.
372. J. F. Hoover and J. M. Stryker, *Organometallics*, 1988, **7**, 2082.
373. G. Ferguson, J. F. Gallagher, A. J. McAlees, R. McCrindle, J. Phillips and G. J. B. Williams, *J. Organomet. Chem.*, 1992, **430**, C23.
374. R. McCrindle, G. J. Arsenault, R. Farwaha, M. J. Hampden-Smith and A. J. McAlees, *J. Chem. Soc., Chem. Commun.*, 1986, 943.
375. R. McCrindle, G. Ferguson, G. J. Arsenault, A. J. McAlees, B. L. Ruhl and D. W. Sneddon, *Organometallics*, 1986, **5**, 1171.
376. R. McCrindle, G. J. Arsenault, R. Farwaha, M. J. Hampden-Smith, R. E. Rice and A. McAlees, *J. Chem. Soc., Dalton Trans.*, 1988, 1773.
377. N. W. Alcock, P. G. Pringle, P. Bergamini, S. Sostero and O. Traverso, *J. Chem. Soc., Dalton Trans.*, 1990, 1553.
378. P. Bergamini, E. Costa, S. Sostero, A. G. Orpen and P. G. Pringle, *Organometallics*, 1991, **10**, 2989.
379. P. Bergamini, E. Costa, S. Sostero, A. G. Orpen and P. G. Pringle, *Organometallics*, 1992, **11**, 3879.
380. L. Manojlovic-Muir, K. W. Muir, T. Solomun, D. W. Meek and J. L. Peterson, *J. Organomet. Chem.*, 1978, **146**, C26.
381. A. Del Pra, G. Zanotti, A. Piazzesi, U. Belluco and R. Ros, *Trans. Met. Chem.*, 1979, **4**, 381.
382. M. A. Bennett, H.-K. Chee and G. B. Robertson, *Inorg. Chem.*, 1979, **18**, 1061.
383. R. Bardi, A. Piazzesi, R. Ros and P. Cavoli, *Cryst. Struct. Commun.*, 1981, **10**, 855.
384. R. A. Michelin, R. Ros, G. Guadalupi, G. Bombieri, F. Benetollo and G. Chapuis, *Inorg. Chem.*, 1989, **28**, 840.
385. M. A. Bennett, K.-C. Ho, J. C. Jeffery, G. M. McLaughlin and G. B. Robertson, *Aust. J. Chem.*, 1982, **35**, 1311.
386. R. Uson, J. Fornies, M. Tomas, B. Menjon, K. Sunkel and R. Bau, *J. Chem. Soc., Chem. Commun.*, 1984, 751; R. Uson *et al.*, *Organometallics*, 1986, **5**, 1576.
387. G. B. Deacon, K. T. Nelson and E. R. T. Tiekink, *Acta Crystallogr., Sect. C*, 1991, **47**, 1819.
388. G. B. Deacon, K. T. Nelson and E. R. T. Tiekink, *Acta Crystallogr., Sect. C*, 1991, **47**, 955.
389. R. Uson, J. Fornies, M. Tomas, J. M. Casas, F. A. Cotton and L. R. Falvello, *Inorg. Chem.*, 1987, **26**, 3482.
390. X. Solans, C. Miravitlles, J. M. Arrieta, G. Germain and J. P. Declercq, *Acta Crystallogr., Sect. B*, 1982, **38**, 1812.
391. R. McCrindle, G. J. Arsenault, A. Gupta, M. J. Hampden-Smith, R. E. Rice and A. J. McAlees, *J. Chem. Soc., Dalton Trans.*, 1991, 949.
392. D.-S. Yang, G. M. Bancroft, R. J. Puddephatt and J. S. Tse, *Inorg. Chem.*, 1990, **29**, 2496.
393. R. A. Michelin and R. Ros, *J. Chem. Soc., Dalton Trans.*, 1989, 1149.

394. H. Lehner, D. Matt, P. S. Pregosin, L. M. Venanzi and A. Albinati, *J. Am. Chem. Soc.*, 1982, **104**, 6825.
395. A. Albinati, H. Lehner, L. M. Venanzi and M. Wolfer, *Inorg. Chem.*, 1987, **26**, 3933.
396. A. Albinati, F. Demartin, L. M. Venanzi and M. K. Wolfer, *Angew. Chem., Int. Ed. Engl.*, 1988, **27**, 563.
397. M. Crespo and J. Sales, *J. Organomet. Chem.*, 1987, **322**, 405.
398. P. Espinet, J. Fornies, F. Martinez, M. Sotes, E. Lalinde, M. T. Moreno, A. Ruiz and A. J. Welch, *J. Organomet. Chem.*, 1991, **403**, 253.
399. B. S. McGilligan, L. M. Venanzi and M. Wolfer, *Organometallics*, 1987, **6**, 946.
400. F. A. Cotton, L. R. Falvello, R. Uson, J. Fornies, M. Tomas, J. M. Casas and I. Ara, *Inorg. Chem.*, 1987, **26**, 1366.
401. R. Uson, J. Fornies, M. Tomas, I. Ara, J. M. Casas and A. Martin, *J. Chem. Soc., Dalton Trans.*, 1991, 2253.
402. R. Uson, J. Fornies, M. Tomas and I. Ara, *Inorg. Chim. Acta*, 1991, **186**, 67.
403. R. Uson, J. Fornies, M. Tomas, J. M. Casas, F. A. Cotton and L. R. Falvello, *Polyhedron*, 1986, **5**, 901; *Inorg. Chem.*, 1986, **25**, 4519.
404. R. Uson, J. Fornies, L. R. Falvello, M. A. Uson and I. Uson, *Inorg. Chem.*, 1992, **31**, 3697.
405. R. Uson, J. Fornies, M. Tomas, J. M. Casas, F. A. Cotton and L. R. Falvello, *Inorg. Chem.*, 1987, **26**, 3482.
406. R. Uson, J. Fornies, M. Tomas and J. M. Casas, *Angew. Chem., Int. Ed. Engl.*, 1989, **28**, 748.
407. R. Uson, J. Fornies, M. Tomas and I. Uson, *Angew. Chem., Int. Ed. Engl.*, 1990, **29**, 1449.
408. R. Uson, J. Fornies, M. Tomas and I. Ara, *J. Chem. Soc., Dalton Trans.*, 1990, 3151.
409. G. Strukul, R. Ros and R. A. Michelin, *Inorg. Chem.*, 1982, **21**, 495.
410. G. Strukul, R. A. Michelin, J. D. Orbell and L. Randaccio, *Inorg. Chem.*, 1983, **22**, 3706.
411. G. Strukul and R. A. Michelin, *J. Am. Chem. Soc.*, 1985, **107**, 7563.
412. G. Strukul, R. Sinigalia, A. Zanardo, F. Pinna and R. A. Michelin, *Inorg. Chem.*, 1989, **28**, 554.
413. A. Zanardo, R. A. Michelin, F. Pinna and G. Strukul, *Inorg. Chem.*, 1989, **28**, 1648.
414. A. Zanardo, F. Pinna, R. A. Michelin and G. Strukul, *Inorg. Chem.*, 1988, **27**, 1966.
415. A. Iraqi and D. J. Cole-Hamilton, *Polyhedron*, 1991, **10**, 993; *J. Mater. Chem.*, 1992, **2**, 183.
416. R. Sinigalia, R. A. Michelin, F. Pinna and G. Strukul, *Organometallics*, 1987, **6**, 728.
417. R. Zennaro, F. Pinna, G. Strukul and H. Arzoumanian, *J. Mol. Catal.*, 1991, **70**, 269.
418. A. Marsella, S. Agapakis, F. Pinna and G. Strukul, *Organometallics*, 1992, **11**, 3578.
419. R. Uson, J. Fornies, M. Tomas and R. Fandos, *J. Organomet. Chem.*, 1984, **263**, 253.
420. R. Uson, J. Fornies, M. Tomas, I. Ara and B. Menjon, *J. Organomet. Chem.*, 1987, **336**, 129.
421. R. A. Michelin, G. Facchin and R. Ros, *J. Organomet. Chem.*, 1985, **279**, C25.
422. T. G. Appleton, R. D. Berry, J. R. Hall and D. W. Neale, *J. Organomet. Chem.*, 1988, **342**, 399.
423. Z.-Y. Yang and G. B. Young, *J. Chem. Soc., Dalton Trans.*, 1984, 2019.
424. J. F. Hoover and J. M. Stryker, *J. Am. Chem. Soc.*, 1989, **111**, 6466.
425. P. G. Pringle and B. L. Shaw, *J. Chem. Soc., Dalton Trans.*, 1984, 849.
426. H. J. R. de Boer, O. S. Akkerman and F. Bickelhaupt, *J. Organomet. Chem.*, 1987, **336**, 447.
427. N. W. Alcock, K. H. Bryars and P. G. Pringle, *J. Organomet. Chem.*, 1990, **386**, 399.
428. M. F. Lappert, T. R. Martin, C. L. Raston, B. W. Skelton and A. H. White, *J. Chem. Soc., Dalton Trans.*, 1982, 1959.
429. H. Suzuki, T. Tsukui and Y. Moro-Oka, *J. Organomet. Chem.*, 1986, **299**, C35.
430. R. McCrindle, G. Ferguson and A. J. McAlees, *J. Chem. Soc., Chem. Commun.*, 1990, 1524.
431. H. M. Buch, G. Schroth, R. Mynott and P. Binger, *J. Organomet. Chem.*, 1983, **247**, C63.
432. H. tom Dieck, G. Fendesak and C. Munz, *Polyhedron*, 1991, **10**, 255.
433. C. Carfagna, R. Galarini, A. Musco and R. Santi, *Organometallics*, 1991, **10**, 3956.
434. H. J. R. de Boer, G. Schat, O. S. Akkerman, F. Bickelhaupt, M. de Wit and C. H. Stam, *Organometallics*, 1989, **8**, 1288.
435. C. Cornioley-Deuschel and A. von Zelewsky, *Inorg. Chem.*, 1987, **26**, 3354; C. B. Blanton, Z. Murtaza, R. J. Shaver and D. P. Rillema, *Inorg. Chem.*, 1992, **31**, 3230.
436. R. Hohenadel and H. A. Brune, *J. Organomet. Chem.*, 1988, **350**, 101; H. A. Brune, H. Roth, T. Debaerdemaeker and H.-M. Schiebel, *J. Organomet. Chem.*, 1991, **402**, 435.
437. N. W. Alcock, K. H. Bryars and P. G. Pringle, *J. Chem. Soc., Dalton Trans.*, 1990, 1433.
438. W. R. Winchester, M. Gawron, G. J. Palenik and W. M. Jones, *Organometallics*, 1985, **4**, 1894.
439. P. J. Stang, L. Song and B. Halton, *J. Organomet. Chem.*, 1990, **388**, 215.
440. P. J. Stang, L. Song, Q. Lu and B. Halton, *Organometallics*, 1990, **9**, 2149.
441. R. D. W. Kemmitt, P. McKenna, D. R. Russell and L. J. S. Sherry, *J. Organomet. Chem.*, 1983, **253**, C59; D. A. Clarke *et al.*, *J. Chem. Soc., Dalton Trans.*, 1984, 1993.
442. A. Imran, R. D. W. Kemmitt, A. J. W. Markwick, P. McKenna, D. R. Russell and L. J. S. Sherry, *J. Chem. Soc., Dalton Trans.*, 1985, 549.
443. K. W. Chiu, W. Henderson, R. D. W. Kemmitt, L. J. S. Prouse and D. R. Russell, *J. Chem. Soc., Dalton Trans.*, 1988, 427.
444. M. D. Jones, R. D. W. Kemmitt, J. Fawcett and D. R. Russell, *J. Chem. Soc., Chem. Commun.*, 1986, 427.
445. K. W. Chiu, J. Fawcett, W. Henderson, R. D. W. Kemmitt and D. R. Russell, *J. Chem. Soc., Chem. Commun.*, 1986, 41; *J. Chem. Soc., Dalton Trans.*, 1987, 733.
446. W. Henderson, R. D. W. Kemmitt, J. Fawcett, L. J. S. Prouse and D. R. Russell, *J. Chem. Soc., Chem. Commun.*, 1986, 1791.
447. R. D. W. Kemmitt, S. Mason, M. R. Moore, J. Fawcett and D. R. Russell, *J. Chem. Soc., Chem. Commun.*, 1990, 1535.
448. N. M. Boag, G. H. M. Dias, M. Green, J. L. Spencer, F. G. A. Stone and J. Vicente, *J. Chem. Soc., Dalton Trans.*, 1981, 1981.
449. S. Forcolin, G. Pellizer, M. Graziani, M. Lenarda and R. Ganzerla, *J. Organomet. Chem.*, 1980, **194**, 203.
450. J. Burgess, R. D. W. Kemmitt, N. Morton, C. T. Mortimer and M. P. Wilkinson, *J. Organomet. Chem.*, 1980, **191**, 477.
451. N. A. Grabowski, R. P. Hughes, B. S. Jaynes and A. L. Rheingold, *J. Chem. Soc., Chem. Commun.*, 1986, 1694.
452. R. C. Hemond, R. P. Hughes, D. J. Robinson and A. L. Rheingold, *Organometallics*, 1988, **7**, 2239.
453. Y. Iitaka, T. Oohama, M. Shiina, T. Takizawa, Y. Yamamoto and H. Yamazaki, *J. Chem. Soc., Dalton Trans.*, 1987, 853.
454. M. Green, J. K. K. Sarhan and I. M. Al-Najjar, *Organometallics*, 1984, **3**, 520.
455. J. R. Briggs, C. Crocker, W. S. McDonald and B. L. Shaw, *J. Chem. Soc., Dalton Trans.*, 1982, 457.
456. M. Pizzotti, F. Porta, S. Cenini, F. Demartin and N. Masciocchi, *J. Organomet. Chem.*, 1987, **330**, 265.
457. R. D. W. Kemmitt, S. Mason and D. R. Russell, *J. Organomet. Chem.*, 1991, **415**, C9.

458. R. J. Klingler, J. C. Huffman and J. K. Kochi, *J. Am. Chem. Soc.*, 1982, **104**, 2147.
459. T. Debaerdemaeker, R. Hohenadel and H. A. Brune, *J. Organomet. Chem.*, 1991, **410**, 265.
460. R. D. W. Kemmitt, P. McKenna, D. R. Russell and L. J. S. Prouse, *J. Chem. Soc., Dalton Trans.*, 1989, 345.
461. S. S. Moore, R. DiCosimo, A. F. Sowinski and G. M. Whitesides, *J. Am. Chem. Soc.*, 1981, **103**, 948.
462. H. A. Brune, R. Hohenadel, G. Schmidtberg and U. Ziegler, *J. Organomet. Chem.*, 1991, **402**, 179.
463. R. DiCosimo and G. M. Whitesides, *J. Am. Chem. Soc.*, 1982, **104**, 3601.
464. G. M. Whitesides *et al.*, *Organometallics*, 1985, **4**, 1819.
465. T. M. Miller and G. M. Whitesides, *Organometallics*, 1986, **5**, 1473.
466. S. Tangestaninejad, M. Rashidi and R. J. Puddephatt, *J. Organomet. Chem.*, 1991, **412**, 445.
467. J. F. Hoover and J. M. Stryker, *Organometallics*, 1989, **8**, 2973.
468. R. D. W. Kemmitt, P. McKenna, D. R. Russell and L. J. S. Prouse, *J. Chem. Soc., Dalton Trans.*, 1989, 345.
469. W. Henderson, R. D. W. Kemmitt, L. J. S. Prouse and D. R. Russell, *J. Chem. Soc., Dalton Trans.*, 1990, 781.
470. P. Dall'Antonia, M. Graziani and M. Lenarda, *J. Organomet. Chem.*, 1980, **186**, 131.
471. M. H. Kowalski and P. J. Stang, *Organometallics*, 1986, **5**, 2392.
472. M. H. Kowalski, A. M. Arif and P. J. Stang, *Organometallics*, 1988, **7**, 1227; P. J. Stang and M. H. Kowalski, *J. Am. Chem. Soc.*, 1989, **111**, 3356.
473. P. J. Stang, M. H. Kowalski, M. D. Schiavelli and D. Longford, *J. Am. Chem. Soc.*, 1989, **111**, 3347.
474. P. J. Stang, Z. Zhong and M. H. Kowalski, *Organometallics*, 1990, **9**, 833.
475. P. J. Stang and A. K. Dutta, *Organometallics*, 1989, **8**, 1024.
476. P. G. Gassman and I. Gennick Cesa, *Organometallics*, 1984, **3**, 119.
477. E. Leidl, M. Steimann, K. Sunkel and W. Beck, *Eur. J. Solid State Inorg. Chem.*, 1991, **28**, 841.
478. L.-Y. Chia and W. R. McWhinnie, *J. Organomet. Chem.*, 1980, **188**, 121.
479. S. Kotani, K. Shiina and K. Sonogashira, *J. Organomet. Chem.*, 1992, **429**, 403.
480. C. J. Cardin, D. J. Cardin, H. E. Parge and A. C. Sullivan, *J. Chem. Soc., Dalton Trans.*, 1986, 2315.
481. H. C. Clark *et al.*, *J. Am. Chem. Soc.*, 1986, **108**, 6961.
482. R. J. Puddephatt and M. A. Thomson, *Inorg. Chim. Acta*, 1980, **45**, L281; *Inorg. Chem.*, 1982, **21**, 725.
483. M. Rashidi and R. J. Puddephatt, *Organometallics*, 1988, **7**, 1636.
484. C. M. Lukehart, A. T. McPhail, D. R. McPhail, J. B. Myers and H. K. Soni, *Organometallics*, 1989, **8**, 1007.
485. A. C. Dema, X. Li, C. M. Lukehart and M. D. Owen, *Organometallics*, 1991, **10**, 1197.
486. D. Afzal, P. G. Lenhert and C. M. Lukehart, *J. Am. Chem. Soc.*, 1984, **106**, 3050; D. Afzal and C. M. Lukehart, *Organometallics*, 1987, **6**, 546.
487. M. A. Bennett and A. Rokicki, *Aust. J. Chem.*, 1985, **38**, 1307.
488. M. A. Bennett and A. Rokicki, *Organometallics*, 1985, **4**, 180.
489. A. R. Siedle, W. B. Gleason, R. A. Newmark and L. H. Pignolet, *Organometallics*, 1986, **5**, 1969.
490. C. S. Browning, D. H. Farrar, O. A. Nassif and A. Walker, *Inorg. Chim. Acta*, 1988, **144**, 105.
491. P. Braunstein, T. M. Gomes Carneiro, D. Matt, F. Balegroune and D. Grandjean, *Organometallics*, 1989, **8**, 1737.
492. H. C. Clark, G. Ferguson, A. B. Goel and B. L. Ruhl, *Organometallics*, 1984, **3**, 15.
493. P. J. Stang and Z. Zhong, *Organometallics*, 1992, **11**, 1026.
494. M. K. Cooper, P. J. Guerney, H. J. Goodwin and M. McPartlin, *J. Chem. Soc., Dalton Trans.*, 1982, 757.
495. D. Afzal, P. G. Lenhert and C. M. Lukehart, *J. Am. Chem. Soc.*, 1984, **106**, 3050.
496. Z. Zhong, P. J. Stang and A. M. Arif, *Organometallics*, 1990, **9**, 1703; P. J. Stang and Z. Zhong, *Organometallics*, 1992, **11**, 1026.
497. P. J. Stang, Z. Zhong and A. M. Arif, *Organometallics*, 1992, **11**, 1017.
498. M. A. Bennett and A. Rokicki, *J. Organomet. Chem.*, 1983, **244**, C31.
499. J. M. A. Wouters, M. W. Avis, C. J. Elsevier, C. E. Kyriakidis and C. H. Stam, *Organometallics*, 1990, **9**, 2203.
500. Z. Zhong, R. J. Hinkle, A. M. Arif and P. J. Stang, *J. Am. Chem. Soc.*, 1991, **113**, 6196.
501. C. P. Casey and Y. Wang, *Organometallics*, 1992, **11**, 13.
502. P. J. Stang and M. H. Kowalski, *J. Am. Chem. Soc.*, 1989, **111**, 3356.
503. E. Baralt *et al.*, *Organometallics*, 1989, **8**, 2417.
504. E. Baralt and C. M. Lukehart, *Inorg. Chem.*, 1991, **30**, 319.
505. A. Sebald and B. Wrackmeyer, *Z. Naturforsch., Teil B*, 1983, **38**, 1156.
506. A. Sebald, C. Stader, B. Wrackmeyer and W. Bensch, *J. Organomet. Chem.*, 1986, **311**, 233.
507. O. M. Abu-Salah, A. R. Al-Ohaly and H. A. Al-Qahtani, *Inorg. Chim. Acta*, 1986, **117**, L29.
508. R. J. Cross, M. F. Davidson and A. J. McLennan, *J. Organomet. Chem.*, 1984, **265**, C37.
509. R. J. Cross and J. Gemmill, *J. Chem. Soc., Dalton Trans.*, 1984, 199.
510. S. J. Higgins, R. Taylor and B. L. Shaw, *J. Organomet. Chem.*, 1987, **325**, 285.
511. A. Sebald and B. Wrackmeyer, *Z. Naturforsch., Teil B*, 1985, **40**, 1481.
512. R. J. Puddephatt and M. A. Thomson, *J. Organomet. Chem.*, 1982, **238**, 231.
513. P. G. Pringle and B. L. Shaw, *J. Chem. Soc., Chem. Commun.*, 1982, 581; C. R. Langrick, D. M. McEwan, P. G. Pringle and B. L. Shaw, *J. Chem. Soc., Dalton Trans.*, 1983, 2487.
514. N. W. Alcock, T. J. Kemp, P. G. Pringle, P. Bergamini and O. Traverso, *J. Chem. Soc., Dalton Trans.*, 1987, 1659.
515. D. M. McEwan, P. G. Pringle and B. L. Shaw, *J. Chem. Soc., Chem. Commun.*, 1982, 1240.
516. S. W. Carr, P. G. Pringle and B. L. Shaw, *J. Organomet. Chem.*, 1988, **341**, 543.
517. L. Manojlovic-Muir, A. N. Henderson, I. Treurnicht and R. J. Puddephatt, *Organometallics*, 1989, **8**, 2055.
518. G. J. Arsenault, L. Manojlovic-Muir, K. W. Muir, R. J. Puddephatt and I. Treurnicht, *Angew. Chem., Int. Ed. Engl.*, 1987, **26**, 86; L. Manojlovic-Muir, K. W. Muir, I. Treurnicht and R. J. Puddephatt, *Inorg. Chem.*, 1987, **26**, 2418.
519. S. J. Davies, B. F. G. Johnson, M. S. Khan and J. Lewis, *J. Chem. Soc., Chem. Commun.*, 1991, 187; M. S. Khan *et al.*, *J. Organomet. Chem.*, 1992, **424**, 87.
520. R. J. Cross and M. F. Davidson, *Inorg. Chim. Acta*, 1985, **97**, L35; *J. Chem. Soc., Dalton Trans.*, 1986, 1987.
521. A. J. McLennan and R. J. Puddephatt, *Organometallics*, 1985, **4**, 485.
522. A. Furlani, S. Licoccia, M. V. Russo, A. Chiesi Villa and C. Guastini, *J. Chem. Soc., Dalton Trans.*, 1984, 2197.
523. A. Furlani, S. Licoccia, M. V. Russo, A. Chiesi Villa and C. Guastini, *J. Chem. Soc., Dalton Trans.*, 1982, 2449.
524. G. K. Anderson and G. J. Lumetta, *J. Organomet. Chem.*, 1985, **295**, 257.

525. M. S. Khan *et al.*, *Z. Anorg. Allg. Chem.*, 1992, **616**, 121.
526. A. Furlani, S. Licoccia, M. V. Russo and C. Guastini, *J. Chem. Soc., Dalton Trans.*, 1980, 1958.
527. G. Butler, C. Eaborn and A. Pidcock, *J. Organomet. Chem.*, 1981, **210**, 403.
528. K. Sunkel, *J. Organomet. Chem.*, 1988, **348**, C12.
529. M. Kubota, W. G. Sly, B. D. Santarsiero, M. S. Clifton and L. Kuo, *Organometallics*, 1987, **6**, 1257.
530. J. R. Phillips, G. A. Miller and W. C. Trogler, *Acta Crystallogr., Sect. C*, 1990, **46**, 1648.
531. H. Ogawa, K. Onitsuka, T. Joh, S. Takahashi, Y. Yamamoto and H. Yamazaki, *Organometallics*, 1988, **7**, 2257.
532. A. Sebald, B. Wrackmeyer and W. Beck, *Z. Naturforsch., Teil B*, 1983, **38**, 45.
533. B. Wrackmeyer, *Z. Naturforsch., Teil B*, 1991, **46**, 35.
534. L. Sacksteder, E. Baralt, B. A. DeGraff, C. M. Lukehart and J. N. Demas, *Inorg. Chem.*, 1991, **30**, 2468.
535. R. J. Cross and J. Gemmill, *J. Chem. Soc., Chem. Commun.*, 1982, 1343; 1983, 756.
536. R. J. Cross and A. J. McLennan, *J. Organomet. Chem.*, 1983, **255**, 113.
537. A. Sebald and B. Wrackmeyer, *J. Chem. Soc., Chem. Commun.*, 1983, 309.
538. A. Sebald and B. Wrackmeyer, *J. Chem. Soc., Chem. Commun.*, 1983, 1293; *J. Organomet. Chem.*, 1986, **304**, 271.
539. A. Sebald, B. Wrackmeyer, C. R. Theocharis and W. Jones, *J. Chem. Soc., Dalton Trans.*, 1984, 747.
540. H. Ogawa, T. Joh, S. Takahashi and K. Sonogashira, *J. Chem. Soc., Chem. Commun.*, 1985, 1220; H. Ogawa, K. Onitsuka, T. Joh, S. Takahashi, Y. Yamamoto and H. Yamazaki, *Organometallics*, 1988, **7**, 2257.
541. K. Sonogashira, K. Ohga, S. Takahashi and N. Hagihara, *J. Organomet. Chem.*, 1980, **188**, 237.
542. K. Onitsuka, T. Joh and S. Takahashi, *Bull. Chem. Soc. Jpn.*, 1992, **65**, 1179.
543. S. Takahashi, H. Morimoto, Y. Takai, K. Sonogashira and N. Hagihara, *Mol. Cryst. Liq. Cryst.*, 1981, **72**, 101.
544. K. Onitsuka, T. Joh and S. Takahashi, *Angew. Chem., Int. Ed. Engl.*, 1992, **31**, 851.
545. A. Furlani, S. Licoccia, M. V. Russo, A. Camus and N. Marsich, *J. Polym. Sci., Part A*, 1986, **24**, 991.
546. A. Furlani, R. Paolesse, M. V. Russo, A. Camus and N. Marsich, *Polymer*, 1987, **28**, 1221.
547. D. M. McEwan, P. G. Pringle and B. L. Shaw, *J. Chem. Soc., Chem. Commun.*, 1982, 859.
548. G. R. Cooper, A. T. Hutton, D. M. McEwan, P. G. Pringle and B. L. Shaw, *Inorg. Chim. Acta*, 1983, **76**, L267.
549. A. T. Hutton, C. R. Langrick, D. M. McEwan, P. G. Pringle and B. L. Shaw, *J. Chem. Soc., Dalton Trans.*, 1985, 2121.
550. D. M. McEwan, D. P. Markham, P. G. Pringle and B. L. Shaw, *J. Chem. Soc., Dalton Trans.*, 1986, 1809.
551. W. S. McDonald, P. G. Pringle and B. L. Shaw, *J. Chem. Soc., Chem. Commun.*, 1982, 861; G. R. Cooper, A. T. Hutton, C. R. Langrick, D. M. McEwan, P. G. Pringle and B. L. Shaw, *J. Chem. Soc., Dalton Trans.*, 1984, 855.
552. A. Blagg, A. T. Hutton, P. G. Pringle and B. L. Shaw, *Inorg. Chim. Acta*, 1983, **76**, L265; *J. Chem. Soc., Dalton Trans.*, 1984, 1815.
553. A. Blagg and B. L. Shaw, *J. Chem. Soc., Dalton Trans.*, 1987, 221.
554. R. J. Blau, M. H. Chisholm, K. Folting and R. J. Wang, *J. Chem. Soc., Chem. Commun.*, 1985, 1582; *J. Am. Chem. Soc.*, 1987, **109**, 4552.
555. H. Schmidbaur, *Pure Appl. Chem.*, 1980, **52**, 1057.
556. M. L. Illingsworth, J. A. Teagle, J. L. Burmeister, W. C. Fultz and A. L. Rheingold, *Organometallics*, 1983, **2**, 1364.
557. E. K. Barefield, A. M. Carrier, D. J. Sepelak and D. G. Vanderveer, *Organometallics*, 1982, **1**, 103.
558. I. J. B. Lin, L. T. C. Kao, F. J. Wu, G. H. Lee and Y. Wang, *J. Organomet. Chem.*, 1986, **309**, 225.
559. N. J. Kermode, M. F. Lappert, B. W. Skelton, A. H. White and J. Holton, *J. Chem. Soc., Chem. Commun.*, 1981, 698; *J. Organomet. Chem.*, 1982, **228**, C71.
560. C. Engelter, J. R. Moss, M. L. Niven, L. R. Nassimbeni and G. Reid, *J. Organomet. Chem.*, 1982, **232**, C78; C. Engelter, J. R. Moss, L. R. Nassimbeni, M. L. Niven, G. Reid and J. C. Spiers, *J. Organomet. Chem.*, 1986, **315**, 255.
561. J. F. Hoover and J. M. Stryker, *J. Am. Chem. Soc.*, 1990, **112**, 464.
562. T. W. Hanks, R. A. Ekeland, K. Emerson, R. D. Larsen and P. W. Jennings, *Organometallics*, 1987, **6**, 28.
563. O. J. Scherer and H. Jungmann, *Z. Naturforsch., Teil B*, 1981, **36**, 1663.
564. O. J. Scherer and H. Jungmann, *J. Organomet. Chem.*, 1982, **228**, C61.
565. D. P. Bancroft, F. A. Cotton, L. R. Falvello and W. Schwotzer, *Inorg. Chem.*, 1986, **25**, 763.
566. D. P. Bancroft and F. A. Cotton, *Inorg. Chem.*, 1988, **27**, 1633.
567. D. P. Bancroft and F. A. Cotton, *Inorg. Chem.*, 1988, **27**, 4022.
568. E. S. Peterson, D. P. Bancroft, D. Min, F. A. Cotton and E. H. Abbott, *Inorg. Chem.*, 1990, **29**, 229.
569. C. Muller and U. Schubert, *Chem. Ber.*, 1991, **124**, 2181.
570. R. J. Al-Essa, R. J. Puddephatt, D. C. L. Perkins, M. C. Rendle and C. F. H. Tipper, *J. Chem. Soc., Dalton Trans.*, 1981, 1738.
571. R. J. Al-Essa, R. J. Puddephatt, P. J. Thompson and C. F. H. Tipper, *J. Am. Chem. Soc.*, 1980, **102**, 7546.
572. J. T. Burton, R. J. Puddephatt, N. L. Jones and J. A. Ibers, *Organometallics*, 1983, **2**, 1487.
573. J. O. Hoberg and P. W. Jennings, *Organometallics*, 1991, **10**, 8.
574. E. J. Parsons and P. W. Jennings, *J. Am. Chem. Soc.*, 1987, **109**, 3973.
575. E. J. Parsons and P. W. Jennings, *Organometallics*, 1988, **7**, 1435.
576. M. Waddington and P. W. Jennings, *Organometallics*, 1982, **1**, 385; 1370.
577. M. D. Waddington, J. A. Campbell, P. W. Jennings and C. F. Campana, *Organometallics*, 1983, **2**, 1269.
578. R. A. Ekeland and P. W. Jennings, *J. Organomet. Chem.*, 1985, **281**, 397.
579. W. D. Neilsen, R. D. Larsen and P. W. Jennings, *J. Am. Chem. Soc.*, 1988, **110**, 3307; 1988, **110**, 8657.
580. E. W. Abel, K. G. Orrell and D. Stephenson, *J. Organomet. Chem.*, 1989, **373**, 401.
581. E. W. Abel, K. Kite and B. L. Williams, *J. Organomet. Chem.*, 1982, **231**, 271.
582. E. W. Abel, A. R. Khan, K. Kite, K. G. Orrell and V. Sik, *J. Chem. Soc., Dalton Trans.*, 1980, 1169; 1175.
583. E. W. Abel, K. G. Orrell and A. W. G. Platt, *J. Chem. Soc., Dalton Trans.*, 1983, 2345.
584. E. W. Abel, S. K. Bhargava, K. G. Orrell, A. W. G. Platt, V. Sik and T. S. Cameron, *J. Chem. Soc., Dalton Trans.*, 1985, 345.
585. E. W. Abel, K. G. Orrell, S. P. Scanlan, D. Stephenson, T. Kemmitt and W. Levason, *J. Chem. Soc., Dalton Trans.*, 1991, 591.
586. E. W. Abel, N. J. Long, K. G. Orrell, A. G. Osborne and V. Sik, *J. Organomet. Chem.*, 1989, **378**, 473.
587. E. W. Abel, M. Z. A. Chowdhury, K. G. Orrell and V. Sik, *J. Organomet. Chem.*, 1983, **258**, 109.

588. E. W. Abel, T. P. J. Coston, K. M. Higgins, K. G. Orrell, V. Sik and T. S. Cameron, *J. Chem. Soc., Dalton Trans.*, 1989, 701; E. W. Abel, T. P. J. Coston, K. G. Orrell and V. Sik, *J. Chem. Soc., Dalton Trans.*, 1989, 711.
589. E. W. Abel, T. P. J. Coston, K. G. Orrell and V. Sik, *J. Chem. Soc., Dalton Trans.*, 1990, 49.
590. E. W. Abel, K. Kite and B. L. Williams, *J. Chem. Soc., Dalton Trans.*, 1983, 1017.
591. E. W. Abel *et al.*, *Polyhedron*, 1987, **6**, 1261.
592. E. W. Abel, M. A. Beckett, P. A. Bates and M. B. Hursthouse, *Polyhedron*, 1988, **7**, 1855.
593. S. Chaudhury, V. K. Jain, V. S. Jakkal and K. Venkatasubramanian, *J. Organomet. Chem.*, 1992, **424**, 115.
594. R. Visalakshi and V. K. Jain, *Transition Met. Chem.*, 1990, **15**, 278.
595. H. C. Clark, G. Ferguson, V. K. Jain and M. Parvez, *J. Organomet. Chem.*, 1984, **270**, 365.
596. E. W. Abel, N. J. Long, K. G. Orrell, A. G. Osborne, H. M. Pain and V. Sik, *J. Chem. Soc., Chem. Commun.*, 1992, 303.
597. D. H. Goldsworthy and K. Kite, *J. Organomet. Chem.*, 1987, **319**, 257.
598. E. W. Abel, K. Kite and P. S. Perkins, *Polyhedron*, 1987, **6**, 319.
599. E. W. Abel, K. Kite and P. S. Perkins, *Polyhedron*, 1986, **5**, 1459; 1987, **6**, 549.
600. M. A. Beckett, D. P. Cassidy and A. J. Duffin, *Inorg. Chim. Acta*, 1991, **189**, 229.
601. E. W. Abel *et al.*, *J. Chem. Soc., Chem. Commun.*, 1987, 978; *J. Organomet. Chem.*, 1988, **341**, 559.
602. S. Roth, V. Ramamoorthy and P. R. Sharp, *Inorg. Chem.*, 1990, **29**, 3345.
603. A. Nettle *et al.*, *Polyhedron*, 1988, **7**, 2095.
604. M. Valderrama, M. Scotti, P. Campos, H. Werner and G. Muller, *Chem. Ber.*, 1990, **123**, 1005.
605. E. W. Abel, A. R. Khan, K. Kite, K. G. Orrell and V. Sik, *J. Chem. Soc., Dalton Trans.*, 1980, 2220.
606. E. W. Abel, A. R. Khan, K. Kite, M. B. Hursthouse, K. M. A. Malik and M. A. Mazid, *J. Organomet. Chem.*, 1982, **235**, 121.
607. E. W. Abel *et al.*, *Polyhedron*, 1988, **7**, 1169.
608. E. W. Abel, P. K. Mittal, K. G. Orrell, V. Sik and T. S. Cameron, *J. Chem. Soc., Chem. Commun.*, 1984, 1312; E. W. Abel, P. K. Mittal, K. G. Orrell and V. Sik, *J. Chem. Soc., Dalton Trans.*, 1986, 961.
609. E. W. Abel, A. R. Khan, K. Kite, K. G. Orrell and V. Sik, *J. Chem. Soc., Dalton Trans.*, 1980, 2208.
610. E. W. Abel, M. A. Beckett, P. A. Bates and M. B. Hursthouse, *J. Organomet. Chem.*, 1987, **325**, 261.
611. H. Kacholdt, P. Berges, G. Klar and W. Hinrichs, *Transition Met. Chem.*, 1987, **12**, 515.
612. E. W. Abel, M. Booth, K. G. Orrell, G. M. Pring and T. S. Cameron, *J. Chem. Soc., Chem. Commun.*, 1981, 29.
613. E. W. Abel, T. E. MacKenzie, K. G. Orrell and V. Sik, *J. Chem. Soc., Dalton Trans.*, 1986, 2173.
614. E. W. Abel, G. D. King, K. G. Orrell, V. Sik, T. S. Cameron and K. Jochem, *J. Chem. Soc., Dalton Trans.*, 1984, 2047.
615. P. K. Byers, A. J. Canty, B. W. Skelton and A. H. White, *Organometallics*, 1990, **9**, 826.
616. A. J. Canty, R. T. Honeyman, B. W. Skelton and A. H. White, *J. Organomet. Chem.*, 1990, **389**, 277.
617. C. M. Anderson, R. J. Puddephatt, G. Ferguson and A. J. Lough, *J. Chem. Soc., Chem. Commun.*, 1989, 1297; C. M. Anderson, M. Crespo, M. C. Jennings, A. J. Lough, G. Ferguson and R. J. Puddephatt, *Organometallics*, 1991, **10**, 2672.
618. C. M. Anderson, M. Crespo, G. Ferguson, A. J. Lough and R. J. Puddephatt, *Organometallics*, 1992, **11**, 1177.
619. M. Crespo, M. Martinez, J. Sales, X. Solans and M. Font-Bardia, *Organometallics*, 1992, **11**, 1288.
620. W. J. J. Smeets, A. L. Spek, J. A. M. van Beek and G. van Koten, *Acta Crystallogr., Sect. C*, 1992, **48**, 1303.
621. E. W. Abel *et al.*, *J. Chem. Soc., Dalton Trans.*, 1988, 521.
622. A. R. Middleton, G. Wilkinson, M. B. Hursthouse and N. P. Walker, *J. Chem. Soc., Dalton Trans.*, 1982, 663.
623. J. Geisenberger, U. Nagel, A. Sebald and W. Beck, *Chem. Ber.*, 1983, **116**, 911.
624. R. Allmann and D. Kucharczyk, *Z. Kristallogr.*, 1983, **165**, 227; W. Massa, G. Baum, B.-S. Seo and J. Lorberth, *J. Organomet. Chem.*, 1988, **352**, 415; B. Vance, *J. Organomet. Chem.*, 1987, **336**, 441; D. C. Craig and I. G. Dance, *Polyhedron*, 1987, **6**, 1157.
625. D.-S. Yang, G. M. Bancroft and R. J. Puddephatt, *Inorg. Chem.*, 1990, **29**, 2118.
626. H. Kunkely and A. Vogler, *Coord. Chem. Rev.*, 1991, **111**, 15.
627. T. G. Appleton and J. R. Hall, *Aust. J. Chem.*, 1980, **33**, 2387.
628. D. M. Doddrell, P. F. Barron, D. E. Clegg and C. Bowie, *J. Chem. Soc., Chem. Commun.*, 1982, 575.
629. J.-Y. Lallemand, J. Soulie and J.-C. Chottard, *J. Chem. Soc., Chem. Commun.*, 1980, 436.
630. E. W. Abel, I. Moss, K. G. Orrell, V. Sik and D. Stephenson, *J. Chem., Soc., Dalton Trans.*, 1987, 2695.
631. E. W. Abel, A. R. Khan, K. Kite, K. G. Orrell and V. Sik, *J. Organomet. Chem.*, 1982, **225**, 357.
632. E. W. Abel, S. K. Bhargava, K. Kite, K. G. Orrell, V. Sik and B. L. Williams, *J. Chem. Soc., Dalton Trans.*, 1982, 583.
633. E. W. Abel, K. Kite, K. G. Orrell, V. Sik and B. L. Williams, *J. Chem. Soc., Dalton Trans.*, 1981, 2439.
634. E. W. Abel, M. Booth, G. King, K. G. Orrell, G. M. Pring and V. Sik, *J. Chem. Soc., Dalton Trans.*, 1981, 1846.
635. C. N. Wilker and R. Hoffmann, *J. Am. Chem. Soc.*, 1983, **105**, 5285.
636. A. Ceulemans, *Polyhedron*, 1991, **10**, 1587.
637. R. J. Al-Essa and R. J. Puddephatt, *J. Chem. Soc., Chem. Commun.*, 1980, 45.
638. R. D. Gillard, M. F. Pilbrow, S. Leonard, M. C. Rendle and C. F. H. Tipper, *Polyhedron*, 1982, **1**, 689.
639. R. J. Puddephatt, M. C. Rendle and C. F. H. Tipper, *J. Organomet. Chem.*, 1984, **269**, 305.
640. B. M. Cushman and D. B. Brown, *Inorg. Chem.*, 1981, **20**, 2490.
641. S. S. M. Ling and R. J. Puddephatt, *J. Chem. Soc., Chem. Commun.*, 1982, 412.
642. R. J. Al-Essa, S. S. M. Ling and R. J. Puddephatt, *Organometallics*, 1987, **6**, 951.
643. T. H. Johnson, *Inorg. Chim. Acta*, 1981, **53**, L95.
644. G. E. Riley, C. F. H. Tipper and R. J. Puddephatt, *J. Organomet. Chem.*, 1981, **208**, 429.
645. M. C. Rendle and C. F. H. Tipper, *J. Organomet. Chem.*, 1982, **224**, 321.
646. D. C. L. Perkins, R. J. Puddephatt, M. C. Rendle and C. F. H. Tipper, *J. Organomet. Chem.*, 1980, **195**, 105.
647. P. W. Jennings, R. E. Ekeland, M. D. Waddington and T. W. Hanks, *J. Organomet. Chem.*, 1985, **285**, 429.
648. R. J. Klingler, J. C. Huffman and J. K. Kochi, *J. Organomet. Chem.*, 1981, **206**, C7.
649. J. T. Burton and R. J. Puddephatt, *J. Am. Chem. Soc.*, 1982, **104**, 4242; *Organometallics*, 1986, **5**, 1312.
650. D. C. L. Perkins, R. J. Puddephatt and C. F. H. Tipper, *J. Organomet. Chem.*, 1980, **191**, 481.
651. S. Roy, R. J. Puddephatt and J. D. Scott, *J. Chem. Soc., Dalton Trans.*, 1989, 2121.
652. J. E. Hux and R. J. Puddephatt, *J. Organomet. Chem.*, 1988, **346**, C31.
653. J. E. Hux and R. J. Puddephatt, *Inorg. Chim. Acta*, 1985, **100**, 1.
654. N. H. Agnew, T. G. Appleton and J. R. Hall, *Aust. J. Chem.*, 1982, **35**, 881.

655. T. G. Appleton, J. R. Hall and M. A. Williams, *Aust. J. Chem.*, 1987, **40**, 1565.
656. N. H. Agnew, T. G. Appleton and J. R. Hall, *Inorg. Chim. Acta*, 1980, **41**, 71; 1980, **41**, 85.
657. N. H. Agnew, T. G. Appleton and J. R. Hall, *Inorg. Chim. Acta*, 1981, **50**, 137.
658. T. G. Appleton, J. R. Hall, N. S. Ham, F. W. Hess and M. A. Williams, *Aust. J. Chem.*, 1983, **36**, 673.
659. T. G. Appleton, R. D. Berry, J. R. Hall and J. A. Sinkinson, *Inorg. Chem.*, 1991, **30**, 3860.
660. H. C. Clark, A. B. Goel, R. G. Goel and S. Goel, *Inorg. Chem.*, 1980, **19**, 3220; R. G. Goel and W. O. Ogini, *Organometallics*, 1982, **1**, 654.
661. R. G. Goel, W. O. Ogini and R. C. Srivastava, *Organometallics*, 1982, **1**, 819.
662. R. G. Goel, W. G. Ogini and R. C. Srivastava, *J. Organomet. Chem.*, 1981, **214**, 405.
663. R. G. Goel, W. G. Ogini and R. C. Srivastava, *Inorg. Chem.*, 1981, **20**, 3611.
664. E. C. Alyea and J. Malito, *J. Organomet. Chem.*, 1988, **340**, 119.
665. A. R. H. Bottomley, C. Crocker and B. L. Shaw, *J. Organomet. Chem.*, 1983, **250**, 617.
666. U. Baltensperger, J. R. Gunter, S. Kagi, G. Kahr and W. Marty, *Organometallics*, 1983, **2**, 571.
667. R. D. W. Kemmitt, S. Mason, J. Fawcett and D. R. Russell, *J. Chem. Soc., Dalton Trans.*, 1992, 851.
668. D. Hedden, D. M. Roundhill, W. C. Fultz and A. L. Rheingold, *J. Am. Chem. Soc.*, 1984, **106**, 5014; *Organometallics*, 1986, **5**, 336.
669. H. Rimml and L. M. Venanzi, *J. Organomet. Chem.*, 1983, **259**, C6; 1984, **260**, C52.
670. S. S. Al-Juaid, C. Eaborn, P. B. Hitchcock, J. D. Smith, L. Zanotto and P. N. Kapoor, *J. Organomet. Chem.*, 1990, **394**, 69.
671. W. J. Youngs and J. A. Ibers, *J. Am. Chem. Soc.*, 1983, **105**, 639; *Organometallics*, 1983, **2**, 979.
672. B. L. Simms, M. Shang, J. Lu, W. J. Youngs and J. A. Ibers, *Organometallics*, 1987, **6**, 1118.
673. B. L. Simms and J. A. Ibers, *J. Organomet. Chem.*, 1987, **330**, 279.
674. M. Hackett and G. M. Whitesides, *Organometallics*, 1987, **6**, 403.
675. S. S. Al-Juaid *et al.*, *J. Chem. Soc., Chem. Commun.*, 1988, 1156; S. S. Al-Juaid, D. Carmichael, P. B. Hitchcock, A. Marinetti, F. Mathey and J. F. Nixon, *J. Chem. Soc., Dalton Trans.*, 1991, 905.
676. D. P. Arnold, M. A. Bennett, M. S. Bilton and G. B. Robertson, *J. Chem. Soc., Chem. Commun.*, 1982, 115.
677. G. P. C. M. Dekker *et al.*, *Inorg. Chim. Acta*, 1992, **195**, 203.
678. M. A. Bennett, D. E. Berry, S. K. Bhargava, E. J. Ditzel, G. B. Robertson and A. C. Willis, *J. Chem. Soc., Chem. Commun.*, 1987, 1613.
679. H.-P. Abicht and K. Issleib, *J. Organomet. Chem.*, 1980, **185**, 265.
680. H.-P. Abicht and K. Issleib, *J. Organomet. Chem.*, 1985, **289**, 201.
681. M. A. Bennett, C. Chiraratvatana, G. B. Robertson and U. Tooptakong, *Organometallics*, 1988, **7**, 1394.
682. M. K. Cooper, P. J. Guerney, H. J. Goodwin and M. McPartlin, *J. Chem. Soc., Dalton Trans.*, 1982, 757.
683. A. M. Herring, D. P. Markham, B. L. Shaw and M. Thornton-Pett, *J. Chem. Soc., Chem. Commun.*, 1987, 643.
684. H.-P. Abicht, *J. Organomet. Chem.*, 1986, **311**, 57.
685. E. Lindner, F. Bouachir, R. Fawzi and D. Hubner, *J. Organomet. Chem.*, 1982, **235**, 345.
686. C. A. Ghilardi, S. Midollini, S. Moneti and A. Orlandini, *J. Chem. Soc., Dalton Trans.*, 1988, 1833.
687. J. J. Koh, W.-H. Lee, P. G. Williard and W. M. Risen, Jr., *J. Organomet. Chem.*, 1985, **284**, 409.
688. P. W. N. M. van Leeuwen, C. F. Roobeek, J. H. G. Frijns and A. G. Orpen, *Organometallics*, 1990, **9**, 1211.
689. F. Balegroune, P. Braunstein, T. M. Gomes-Carneiro, D. Grandjean and D. Matt, *J. Chem. Soc., Chem. Commun.*, 1989, 582.
690. N. W. Alcock, A. W. G. Platt, H. H. Powell and P. G. Pringle, *J. Organomet. Chem.*, 1989, **361**, 409.
691. G. R. Newkome, W. E. Puckett, V. K. Gupta and G. E. Kiefer, *Chem. Rev.*, 1986, **86**, 451; D. W. Evans, G. R. Baker and G. R. Newkome, *Coord. Chem. Rev.*, 1989, **93**, 155.
692. E. C. Constable, R. P. G. Henney, T. A. Leese and D. A. Tocher, *J. Chem. Soc., Chem. Commun.*, 1990, 513.
693. G. Minghetti, M. A. Cinellu, G. Chelucci, S. Gladiali, F. Desmartin and M. Manassero, *J. Organomet. Chem.*, 1986, **307**, 107.
694. E. C. Constable, R. P. G. Henney, P. R. Raithby and L. R. Sousa, *Angew. Chem., Int. Ed. Engl.*, 1991, **30**, 1363.
695. S. Dholakia, R. D. Gillard and F. L. Wimmer, *Inorg. Chim. Acta*, 1983, **69**, 179.
696. A. G. Constable, W. S. McDonald and B. L. Shaw, *J. Chem. Soc., Dalton Trans.*, 1980, 2282.
697. J. Barker, N. D. Cameron, M. Kilner, M. M. Mahmoud and S. C. Wallwork, *J. Chem. Soc., Dalton Trans.*, 1991, 3435.
698. R. M. Ceder and J. Sales, *J. Organomet. Chem.*, 1984, **276**, C31; *J. Organomet. Chem.*, 1985, **294**, 389.
699. C. G. Anklin, P. S. Pregosin, F. J. Wombacher and H. J. Ruegg, *Organometallics*, 1990, **9**, 1953.
700. M. Crespo, M. Martinez and J. Sales, *J. Chem. Soc., Chem. Commun.*, 1992, 822.
701. A. J. Canty, R. T. Honeyman, B. W. Skelton and A. H. White, *J. Organomet. Chem.*, 1992, **430**, 245.
702. P. S. Pregosin, F. Wombacher, A. Albinati and F. Lianza, *J. Organomet. Chem.*, 1992, **418**, 249.
703. J. A. M. van Beek, G. van Koten, G. P. C. M. Dekker, E. Wissing, M. C. Zoutberg and C. H. Stam, *J. Organomet. Chem.*, 1990, **394**, 659.
704. J. Terheijden *et al.*, *Organometallics*, 1987, **6**, 89.
705. E. Wehman, G. van Koten, C. T. Knaap, H. Ossor, M. Pfeffer and A. L. Spek, *Inorg. Chem.*, 1988, **27**, 4409.
706. J. Vicente, M-T. Chicote, J. Martin, P. G. Jones, C. Fittschen and G. M. Sheldrick, *J. Chem. Soc., Dalton Trans.*, 1986, 2215.
707. A. F. M. J. van der Ploeg, G. van Koten and K. Vrieze, *J. Organomet. Chem.*, 1981, **222**, 155.
708. L. L. Troitskaya and V. I. Sokolov, *J. Organomet. Chem.*, 1987, **328**, 169.
709. H. Motschi, P. S. Pregosin and H. Ruegger, *J. Organomet. Chem.*, 1980, **193**, 397.
710. C. Anklin, P. S. Pregosin, F. Bachechi, P. Mura and L. Zambonelli, *J. Organomet. Chem.*, 1981, **222**, 175.
711. M. Nonoyama, *Inorg. Chim. Acta*, 1989, **157**, 9; *Transition Met. Chem.*, 1990, **15**, 366.
712. G. S. Hanan, J. E. Kickham and S. J. Loeb, *Organometallics*, 1992, **11**, 3063.
713. L. S. Chang, M. P. Johnson and M. J. Fink, *Organometallics*, 1991, **10**, 1219.
714. W. Porzio, *Inorg. Chim. Acta*, 1980, **40**, 257.
715. H. P. Abicht *et al.*, *Z. Anorg. Allg. Chem.*, 1982, **494**, 55.
716. D. F. Mullica, E. L. Sappenfield and D. H. Leschnitzer, *Acta Crystallogr. Sect. C*, 1991, **47**, 874.
717. A. B. Goel, S. Goel and D. Vanderveer, *Inorg. Chim. Acta*, 1981, **54**, L267.

718. E. C. Alyea, G. Ferguson, J. Malito and B. L. Ruhl, *Organometallics*, 1989, **8**, 1188.
719. M. A. Bennett, D. E. Berry and K. A. Beveridge, *Inorg. Chem.*, 1990, **29**, 4148.
720. W. J. J. Smeets, A. L. Spek, A. J. M. Duisenberg, J. A. M. van Beek and G. van Koten, *Acta Crystallogr., Sect. C*, 1986, **43**, 463.
721. J. N. Louwen, D. M. Grove, H. J. C. Ubbels, D. J. Stufkens and A. Oskam, *Z. Naturforsch., Teil B*, 1983, **38**, 1657.
722. M. Maestri, C. Deuschel-Cornioley and A. von Zelewsky, *Coord. Chem. Rev.*, 1991, **111**, 117.
723. C. A. Craig, F. O. Garces, R. J. Watts, R. Palmans and A. J. Frank, *Coord. Chem. Rev.*, 1990, **97**, 193.
724. L. Chassot and A. von Zelewsky, *Helv. Chim. Acta*, 1983, **66**, 2443; L. Chassot, E. Muller and A. von Zelewsky, *Inorg. Chem.*, 1984, **23**, 4249.
725. L. Chassot and A. von Zelewsky, *Inorg. Chem.*, 1987, **26**, 2814.
726. C. Deuschel-Cornioley, R. Luond and A. von Zelewsky, *Helv. Chim. Acta*, 1989, **72**, 377.
727. C. Deuschel-Cornioley, H. Stoeckli-Evans and A. von Zelewsky, *J. Chem. Soc., Chem. Commun.*, 1990, 121.
728. M. Maestri, D. Sandrini, V. Balzani, L. Chassot, P. Jolliet and A. von Zelewsky, *Chem. Phys. Lett.*, 1985, **122**, 375.
729. S. Bonafede, M. Ciano, F. Bolletta, V. Balzani, L. Chassot and A. von Zelewsky, *J. Phys. Chem.*, 1986, **90**, 3836.
730. D. Sandrini, M. Maestri, V. Balzani, L. Chassot and A. von Zelewsky, *J. Am. Chem. Soc.*, 1987, **109**, 7720.
731. D. Sandrini *et al.*, *Gazz. Chim. Ital.*, 1988, **118**, 661.
732. M. Maestri, D. Sandrini, V. Balzani, A. von Zelewsky, C. Deuschel-Cornioley and P. Jolliet, *Helv. Chim. Acta*, 1988, **71**, 1053.
733. F. Barigelletti *et al.*, *Inorg. Chem.*, 1988, **27**, 3644.
734. R. Schwartz, G. Gliemann, P. Jolliet and A. von Zelewsky, *Inorg. Chem.*, 1989, **28**, 1053.
735. M. Maestri, D. Sandrini, A. von Zelewsky and C. Deuschel-Cornioley, *Inorg. Chem.*, 1991, **30**, 2476.
736. L. Chassot and A. von Zelewsky, *Helv. Chim. Acta*, 1986, **69**, 1855.
737. H. C. Clark, V. K. Jain and G. S. Rao, *J. Organomet. Chem.*, 1985, **279**, 181.
738. V. K. Jain and S. Kannan, *J. Organomet. Chem.*, 1991, **405**, 265.
739. V. K. Jain and S. Kannan, *J. Organomet. Chem.*, 1992, **418**, 349.
740. A. B. Goel, S. Goel and H. C. Clark, *Synth. React. Inorg. Met.-Org. Chem.*, 1981, **11**, 289.
741. A. B. Goel, S. Goel and D. Vanderveer, *Inorg. Chim. Acta*, 1984, **82**, L9.
742. E. C. Alyea, S. A. Dias and F. Bonati, *Transition Met. Chem.*, 1981, **6**, 24.
743. A. B. Goel and S. Goel, *Inorg. Chim. Acta*, 1983, **69**, 233.
744. A. B. Goel and S. Goel, *Inorg. Chim. Acta*, 1984, **82**, 41.
745. H.-P. Abicht and K. Issleib, *Z. Anorg. Allg. Chem.*, 1983, **500**, 31.
746. H.-P. Abicht, P. Lehniger and K. Issleib, *J. Organomet. Chem.*, 1983, **250**, 609.
747. J. R. Briggs, A. G. Constable, W. S. McDonald and B. L. Shaw, *J. Chem. Soc., Dalton Trans.*, 1982, 1225.
748. D. P. Arnold, M. A. Bennett, G. M. McLaughlin, G. B. Robertson and M. J. Whittaker, *J. Chem. Soc., Chem. Commun.*, 1983, 32.
749. D. P. Arnold, M. A. Bennett, G. M. McLaughlin and G. B. Robertson, *J. Chem. Soc., Chem. Commun.*, 1983, 34.
750. D. M. Grove, G. van Koten, J. N. Louwen, J. G. Noltes, A. L. Spek and H. J. C. Ubbels, *J. Am. Chem. Soc.*, 1982, **104**, 6609.
751. D. M. Grove, G. van Koten, H. J. C. Ubbels, R. Zoet and A. L. Spek, *J. Organomet. Chem.*, 1984, **263**, C10.
752. P. L. Alsters *et al.*, *Organometallics*, 1992, **11**, 4124.
753. J. Terheijden *et al.*, *J. Organomet. Chem.*, 1986, **315**, 401.
754. D. M. Grove, G. van Koten and H. J. C. Ubbels, *Organometallics*, 1982, **1**, 1366.
755. J. Terheijden, G. van Koten, I. C. Vinke and A. L. Spek, *J. Am. Chem. Soc.*, 1985, **107**, 2891.
756. M. F. Davidson, D. M. Grove, G. van Koten and A. L. Spek, *J. Chem. Soc., Chem. Commun.*, 1989, 1562.
757. J. Terheijden, G. van Koten, J. L. de Booys, H. J. C. Ubbels and C. H. Stam, *Organometallics*, 1983, **2**, 1882.
758. G. van Koten, J. Terheijden, J. A. M. van Beek, I. C. M. Wehman-Ooyevaar, F. Muller and C. H. Stam, *Organometallics*, 1990, **9**, 903.
759. J. A. M. van Beek, G. van Koten, W. J. J. Smeets and A. L. Spek, *J. Am. Chem. Soc.*, 1986, **108**, 5010.
760. J. Terheijden, G. van Koten, W. P. Mul, D. J. Stufkens, F. Muller and C. H. Stam, *Organometallics*, 1986, **5**, 519.
761. D. M. Grove, G. van Koten, H. J. C. Ubbels, K. Vrieze, L. C. Niemann and C. H. Stam, *J. Chem. Soc., Dalton Trans.*, 1986, 717.
762. D. M. Grove, G. van Koten, H. J. C. Ubbels and A. L. Spek, *J. Am. Chem. Soc.*, 1982, **104**, 4285; J. Terheijden, G. van Koten, D. M. Grove, K. Vrieze and A. L. Spek, *J. Chem. Soc., Dalton Trans.*, 1987, 1359.
763. A. F. M. J. van der Ploeg, G. van Koten, K. Vrieze, A. L. Spek and A. J. M. Duisenberg, *J. Chem. Soc., Chem. Commun.*, 1980, 469.
764. A. F. M. J. van der Ploeg, G. van Koten and K. Vrieze, *J. Organomet. Chem.*, 1982, **226**, 93; A. F. M. J. van der Ploeg, G. van Koten, K. Vrieze, A. L. Spek and A. J. M. Duisenberg, *Organometallics*, 1982, **1**, 1066.
765. A. F. M. J. van der Ploeg, G. van Koten and C. Brevard, *Inorg. Chem.*, 1982, **21**, 2878.
766. A. F. M. J. van der Ploeg, G. van Koten and K. Vrieze, *Inorg. Chem.*, 1982, **21**, 2026.
767. A. F. M. J. van der Ploeg, G. van Koten, K. Vrieze and A. L. Spek, *Inorg. Chem.*, 1982, **21**, 2014.
768. A. F. M. J. van der Ploeg, G. van Koten and K. Vrieze, *Inorg. Chim. Acta*, 1982, **58**, 35.
769. J. A. M. van Beek, G. van Koten, I. C. M. Wehman-Ooyevaar, W. J. J. Smeets, P. van der Sluis and A. L. Spek, *J. Chem. Soc., Dalton Trans.*, 1991, 883.
770. I. C. M. Wehman-Ooyevaar, D. M. Grove, P. van der Sluis, A. L. Spek and G. van Koten, *J. Chem. Soc., Chem. Commun.*, 1990, 1367.
771. G. K. Anderson, S. E. Saum, R. J. Cross, S. A. Morris, *Organometallics*, 1983, **2**, 780; G. K. Anderson and R. J. Cross, *J. Chem. Soc., Chem. Commun.*, 1986, 1502; G. K. Anderson, R. J. Cross, S. Fallis and M. Rocamora, *Organometallics*, 1987, **6**, 1440; G. K. Anderson, R. J. Cross, L. Manojlovic-Muir and K. W. Muir, *J. Organomet. Chem.*, 1989, **362**, 225.
772. R. J. Cross, N. H. Tennent and G. M. Warden, *J. Chem. Soc., Dalton Trans.*, 1986, 1059.
773. S. Chattopadhyay, C. Sinha, P. Basu and A. Chakravorty, *J. Organomet. Chem.*, 1991, **414**, 421; *Organometallics*, 1991, **10**, 1135.
774. M. A. Cinellu, S. Gladiali and G. Minghetti, *J. Organomet. Chem.*, 1989, **363**, 401.
775. G. Minghetti, M. A. Cinellu, S. Stoccoro, G. Chelucci and A. Zucca, *Inorg. Chem.*, 1990, **29**, 5137.

776. A. H. Janowicz and R. G. Bergman, *J. Am. Chem. Soc.*, 1982, **104**, 352; J. K. Hoyano and W. A. G. Graham, *J. Am. Chem. Soc.*, 1982, **104**, 3723.
777. J. Y. Saillard and R. Hoffmann, *J. Am. Chem. Soc.*, 1984, **106**, 2006.
778. M. Hackett, J. A. Ibers, P. Jernakoff and G. M. Whitesides, *J. Am. Chem. Soc.*, 1986, **108**, 8094; M. Hackett, J. A. Ibers and G. M. Whitesides, *J. Am. Chem. Soc.*, 1988, **110**, 1436; M. Hackett and G. M. Whitesides, *J. Am. Chem. Soc.*, 1988, **110**, 1449.
779. R. A. Fischer, H. D. Kaesz, S. I. Khan and H.-J. Muller, *Inorg. Chem.*, 1990, **29**, 1601; R. A. Fischer and J. Behm, *J. Organomet. Chem.*, 1991, **413**, C10.
780. P. Hofmann, H. Heiss, P. Neiteler, G. Muller and J. Lachmann, *Angew. Chem., Int. Ed. Engl.*, 1990, **29**, 880.
781. P. Hofmann and G. Unfried, *Chem. Ber.*, 1992, **125**, 659.
782. A. Saare and L. Dahlenburg, *Z. Naturforsch., Teil B*, 1992, **47**, 247.
783. R. L. Brainard, W. R. Nutt, T. R. Lee and G. M. Whitesides, *Organometallics*, 1988, **7**, 2379.
784. J. A. Davies and C. T. Eagle, *Organometallics*, 1986, **5**, 2149; J. A. Davies, C. T. Eagle, D. E. Otis and U. Venkataraman, *Organometallics*, 1989, **8**, 1080.
785. J. A. Davies and R. J. Staples, *Polyhedron*, 1991, **10**, 909.
786. G. B. Shul'pin, A. E. Shilov, A. N. Kitaigorodskii and J. V. Zeile Krevor, *J. Organomet. Chem.*, 1980, **201**, 319; G. B. Shul'pin, *J. Organomet. Chem.*, 1981, **212**, 267; G. B. Shul'pin and A. N. Kitaigorodskii, *J. Organomet. Chem.*, 1981, **212**, 275.
787. A. N. Kitaigorodskii, V. M. Nekipelov, A. T. Nikitaev and G. B. Shul'pin, *J. Organomet. Chem.*, 1984, **275**, 295.
788. G. B. Shul'pin, G. V. Nizova and A. T. Nikitaev, *J. Organomet. Chem.*, 1984, **276**, 115.
789. G. V. Nizova, M. V. Serdobov, A. T. Nikitaev and G. B. Shul'pin, *J. Organomet. Chem.*, 1984, **275**, 139.
790. G. B. Shul'pin, G. V. Nizova, A. N. Kitaigorodskii and M. V. Serdobov, *J. Organomet. Chem.*, 1984, **275**, 273; G. B. Shul'pin and G. V. Nizova, *J. Organomet. Chem.*, 1984, **276**, 109.

9
Platinum–Carbon π-Bonded Complexes

G. BRENT YOUNG
Imperial College of Science, Technology and Medicine, London, UK

9.1 INTRODUCTION 534

9.2 ALKENE AND RELATED COMPLEXES 534

9.2.1 η^2-Monoalkene Complexes (Monodentate) 534
9.2.1.1 Zerovalent derivatives 535
9.2.1.2 Divalent derivatives 541
9.2.1.3 Theoretical aspects 550
9.2.2 η^2-Monoalkene Complexes (Polydentate) 551
9.2.2.1 Zerovalent derivatives: preparation, structure and reactivity 551
9.2.2.2 Divalent derivatives: preparation, structure and reactivity 551
9.2.2.3 Tetravalent derivatives 553
9.2.3 η^n-Polyene Complexes (Unconjugated) 553
9.2.3.1 Zerovalent derivatives 553
9.2.3.2 Divalent derivatives 554
9.2.4 η^2-Allene, η^2-X=CR_2 and Related Complexes 558
9.2.4.1 η^2-Cumulene complexes: preparation, structure and reactivity 558
9.2.4.2 η^2-Ketene complexes: preparation, structure and reactivity 559
9.2.4.3 Complexes of η^2-X=CR_2: preparation, structure and reactivity 560
9.2.5 η^n-Polyalkene Complexes (Conjugated) 560
9.2.5.1 Preparation, structure and reactivity 560
9.2.6 η^2-Fullerene Complexes 561
9.2.6.1 Preparation, structure and reactivity 561
9.2.7 Bimetallic Alkene Complexes 561
9.2.7.1 Homobimetallic complexes: preparation, structure and reactivity 561
9.2.7.2 Heterobimetallic complexes: preparation, structure and reactivity 563
9.2.8 Polymetallic Alkene Complexes 563
9.2.8.1 Homometallic complexes: preparation, structure and reactivity 563
9.2.8.2 Heterometallic complexes: preparation, structure and reactivity 563

9.3 ALKYNE AND RELATED COMPLEXES 564

9.3.1 η^2-Monoalkyne Complexes 564
9.3.1.1 Zerovalent derivatives 565
9.3.1.2 Divalent derivatives 568
9.3.1.3 Theoretical aspects 569
9.3.2 η^2-Aryne and Related Complexes 569
9.3.2.1 Preparation, structure and reactivity 569
9.3.3 η^2-X≡CR Complexes 569
9.3.3.1 Preparation, structure and reactivity 569
9.3.4 Bimetallic Alkyne Complexes 570
9.3.4.1 Homobimetallic complexes: preparation, structure and reactivity 570
9.3.4.2 Heterobimetallic complexes: preparation, structure and reactivity 571
9.3.5 Polymetallic Alkyne Complexes 572
9.3.5.1 Homometallic complexes: preparation, structure and reactivity 572
9.3.5.2 Heterometallic complexes: preparation, structure and reactivity 572

9.4 ALLYL AND RELATED COMPLEXES 573

9.4.1 η^3-Allyl Complexes 573
9.4.1.1 Preparation 573
9.4.1.2 Structural aspects 575
9.4.1.3 Reactions 576
9.4.1.4 Theoretical aspects 576
9.4.2 Related η^3-Hydrocarbyl Complexes 577
9.4.2.1 η^3-Benzyl complexes: preparation, structure and reactivity 577
9.4.2.2 η^3-Butadienyl complexes: preparation, structure and reactivity 577
9.4.2.3 η^3-Propargyl/allenyl complexes: preparation, structure and reactivity 577
9.4.3 η-Trimethylenemethane and Related Complexes 578
9.4.3.1 Preparation, structure and reactivity 578

9.5 CYCLOBUTADIENE AND RELATED COMPLEXES 579
9.5.1 η^4-Cyclobutadiene Complexes 579
9.5.1.1 Preparation, structure and reactivity 579

9.6 CYCLOPENTADIENYL AND RELATED COMPLEXES 579
9.6.1 η^5-Cyclopentadienyl Complexes 579
9.6.1.1 Mono- and divalent derivatives: preparation, structure and reactivity 580
9.6.1.2 Tetravalent derivatives: preparation, structure and reactivity 581
9.6.2 η^5-Heterocyclopentadienyl Complexes 581
9.6.2.1 Preparation, structure and reactivity 581

9.7 ARENE AND RELATED COMPLEXES 582
9.7.1 η^6-Arene Complexes: Preparation, Structure and Reactivity 582

9.8 REFERENCES 582

9.1 INTRODUCTION

The discovery by Zeise[1] of the anion $[PtCl_3(\eta\text{-}C_2H_4)]^-$ in the early nineteenth century can be said to mark the advent of organometallic chemistry—of the transition elements, at least. Although not appreciated structurally at that time, this was also the first example of a π-organotransition metal, in which the electrons provided by carbon for metal–carbon bonding are drawn, formally, from π-orbitals of the ligating fragment. Appropriately, the chemistry of π-complexes of platinum continues to reflect the vigour of organometallic chemistry as a whole.

This chapter attempts to provide a view of developments in this area from 1980 to 1993, as a supplement to the excellent coverage provided by Hartley in *COMC-I*.[2] The organizational policies are broadly similar to the original. In scope, the survey is limited to the chemistry of identifiable, characterized π-organoplatinum derivatives. It does not aspire to document the manifold use of platinum π-complexes as catalyst precursors for a wide variety of organic transformations. Neither does it, in the main, discuss compounds of this type proposed as intermediates in metal-mediated synthetic schemes. Information on these topics should be sought in Volume 12.

The primary literature is the major source of references, but important reviews are cited wherever possible. Since 1979, in addition to a reference text featuring organoplatinum chemistry[3] and a review of the organic chemistry of platinum(IV),[4] regular, recurrent surveys have continued to appear, covering various aspects of interest. Most notable are the annual surveys of organoplatinum, published periodically in the *Journal of Organometallic Chemistry*[5] and those included in Specialist Periodical Reports, published annually by the Royal Society of Chemistry, on π-alkene, π-diene and π-alkyne complexes,[6] as well as π-cyclopentadienyl, π-arene and related derivatives.[7] In addition, during this period, several excellent texts have appeared on general aspects of organotransition metals, each of which summarizes salient features of π-organoplatinum chemistry.[8–11]

9.2 ALKENE AND RELATED COMPLEXES

9.2.1 η^2-Monoalkene Complexes (Monodentate)

η^2-Monoalkene monovalent complexes include ligands which bind to platinum via a single unsaturated C=C site but are attached to the metal by no other covalent interaction.

9.2.1.1 Zerovalent derivatives

In general, (η^2-monoalkene)platinum(0) complexes continue to have, as ancillary ligands, other unsaturated hydrocarbon fragments or group 15 donor ligands and most commonly favour three- (16-electron) or four-coordinate (18-electron) structures.

(i) Preparation

Preparations of zerovalent platinum derivatives directly from platinum metal by, for example, vapour co-condensation methods, have not developed apace with the corresponding chemistry of other metals. This may reflect the physical properties of the metal but, just as likely, is due to the high relative success of solution techniques for platinum. Synthetic approaches, therefore, involve ligand displacements and are of two broad types, namely, those which involve reduction of the metal and those which do not. Commercially available forms of platinum precursor—other than the metal itself—feature platinum(II) or platinum(IV) and the zerovalent state is readily available from the former via a range of reduction procedures. Many of the complexes formed by these routes are precursors for platinum(0) derivatives, by further ligand displacements. Reductive paths will, therefore, be considered first.

(a) Preparations via reduction of Pt^{II} precursors. The only homoleptic monoalkene complex accessible directly by the reduction of Pt^{II} precursors appears to be tris(η^2-norbornene)platinum(0). This is obtained by reduction of $PtCl_2$(cod), using dilithium cyclooctatetraenide, in the presence of norbornene (bicyclo[2.2.1]hept-2-ene) and is one of several versatile homoleptic alkene–platinum reagents whose syntheses have been reported definitively.[12]

Complexes of the general class $PtL_2(\eta\text{-}C_2H_4)$ (L = a group 15 donor) continue to be valued reagents for a variety of applications. A straightforward, improved preparation of $Pt(PPh_3)_2(\eta\text{-}C_2H_4)$ involves treatment of a dichloromethane/ethanol solution of $PtCl_2(PPh_3)_2$ with $NaBH_4$ in the presence of ethene, at ambient pressure. Reportedly, no inert atmosphere or solvent prepurification is necessary.[13a] The first analogues with phosphite ligands, $Pt[P(o\text{-}C_6H_4\text{-}2\text{-}OMe)_3]_2(\eta\text{-}C_2H_4)$, may be produced by this route.[13b] Similar reaction of $PtCl_2$(diop) affords optically active Pt(diop)(η-C_2H_4),[14] while $PtBr_2[1,2\text{-}(Ph_2PCH_2)_2C_6H_4]$ affords $Pt[1,2\text{-}(Ph_2PCH_2)_2C_6H_4](\eta\text{-}C_2H_4)$.[15] For other tertiary phosphine analogues, alternative reduction strategies have often proven necessary. Sodium naphthalenide is effective in the preparation of a range of $PtL_2(\eta\text{-}C_2H_4)$ (L = PEt_3, PPr^i_3, PEt_2Ph, PPh_3;[16] PCy_3;[17] L_2 = dppe,[15] dcpe (1,2-(1,3-bis(dicyclohexylphosphino)propane)bis(dicyclohexylphosphino)ethane), dcpp (1,3-bis(dicyclohexylphosphino)propane), dcpb.[16,17] Hydrazine, too, can be employed as reductant, as in the preparation of $Pt(PPh_3)_2(\eta^2$-norbornene) from $PtBr_2(PPh_3)_2$ in presence of norbornene.[18]

Another convenient route to $Pt(PEt_3)_2(\eta\text{-}C_2H_4)$ is provided by the thermolytic rearrangement of *cis*-$PtEt_2(PEt_3)_2$ via β-hydrogen transfer and reductive C–H elimination.[19] Corresponding η^2-cyclobutene and η^2-cyclopentene complexes are produced—although not on a preparative scale—via thermolyses of *cis*-$Pt(\eta^1\text{-}c\text{-}C_4H_7)_2(PEt_3)_2$ and *cis*-$Pt(\eta^1\text{-}c\text{-}C_5H_9)_2(PEt_3)_2$, respectively (the bis($\eta^1$-cyclopropyl)platinum analogue yielded no detectable η^2-cyclopropene product).[20] A related method involves thermolytic reductive elimination of neopentane from $Pt(H)(CH_2CMe_3)$(dcpe) to generate the reactive fragment [Pt^0(dcpe)], which can be trapped by a variety of unsaturated substrates, including alkenes, to generate Pt(dcpe)(η^2-alkene) (alkene = $H_2C{=}CHBu^t$, $H_2C{=}C(Pr^i)_2$ or $Me_2C{=}CMe_2$). Reductive elimination–oxidative addition of an aromatic C–H bond from/to [Pt^0(dcpe)] is proposed to involve the η^2-benzene intermediate Pt(dcpe)(η^2-C_6H_6).[21] A further example of an intramolecular, reductive process is provided by the reaction of $PtH_2(PMe_3)_2$ with alkenes to produce $Pt(PMe_3)_2(\eta$-alkene) (alkene = $H_2C{=}CH_2$, $H_2C{=}CHBu^t$ or (*E*)-PhHC=CHPh).[22]

Photochemically induced reductive elimination of CO_2 from oxalatoplatinum(II) precursors (Equation (1)) has proven a versatile new route to several $PtL_2(\eta\text{-}C_2X_4)$ derivatives (X = H, L = PEt_3, L_2 = dppe; X = F, L = PEt_3),[23] and there seems no reason that this should not continue to develop as an entirely general method.

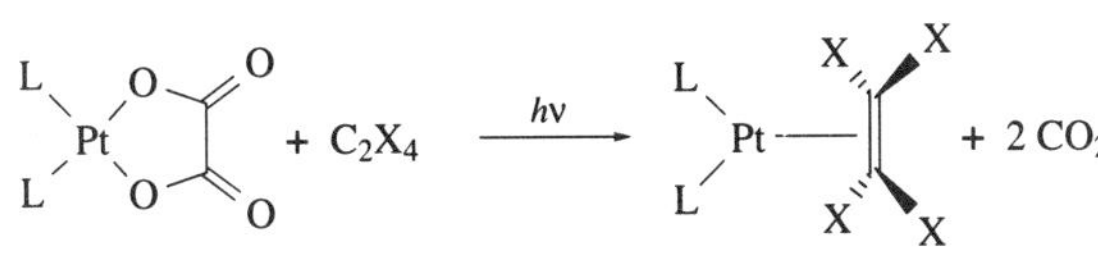

(1)

(b) Preparations via alkene displacement from Pt^0 precursors. The unconjugated diene derivative bis(η^2,η^2-cycloocta-1,5-diene)platinum(0) (see Section 9.2.3.1(i)) has continued to be an invaluable precursor to other zerovalent platinum complexes, including those with monoalkenes.[7] The synthesis of

$Pt(\eta\text{-}C_2H_4)_3$ from this reagent has been reported definitively.[24] The chemistry of these and related 'ligand-free' platinum complexes has been reviewed.[25]

$Pt(cod)_2$ can be used to generate complexes with both η^2-alkene and phosphine (or other) ligands by treatment with appropriate stoichiometric amounts of reagents as, for example, in the preparation of complexes in the class $PtL_2(\eta^2\text{-fulvene})$ with tertiary phosphine and arsine ligands ($L = PPh_3$, PCy_3 or $AsPh_3$). For 2,3,4,5-tetraphenylfulvene, where the ring is tetrasubstituted, coordination occurs via the exocyclic unsaturation. If, on the other hand, the exocyclic double bond is disubstituted, steric constraints favour binding at a ring site (Scheme 1).[26] Related complexes $PtL^1(\eta^2\text{-fulvene})(\eta\text{-}C_2H_4)$ are accessible by η^2-ethene displacement by 2,3,4,5-tetraphenylfulvene from $PtL^2(\eta\text{-}C_2H_4)_2$ ($L^2 = PPh_3$, $PMePh_2$, PMe_2Ph, $AsPh_3$ or $CNBu^t$). Bis(η^2-ethene)platinum precursors of this latter type are prepared from $Pt(cod)_2$—via $Pt(\eta\text{-}C_2H_4)_3$, formed *in situ*—by treatment with L^2.[27] The congener $Pt(PBu^t_3)(\eta^2\text{-}C_2H_4)_2$ is, however, difficult to isolate by this route and $Pt(PBu^t_3)(\eta^2\text{-norbornene})_2$ is a more practical source of the $Pt(PBu^t_3)$ fragment.[28] Both $Pt(cod)_2$ and $Pt(PCy_3)(\eta\text{-}C_2H_4)_2$ provide useful precursors to (η^2-quinone)platinum complexes.[29]

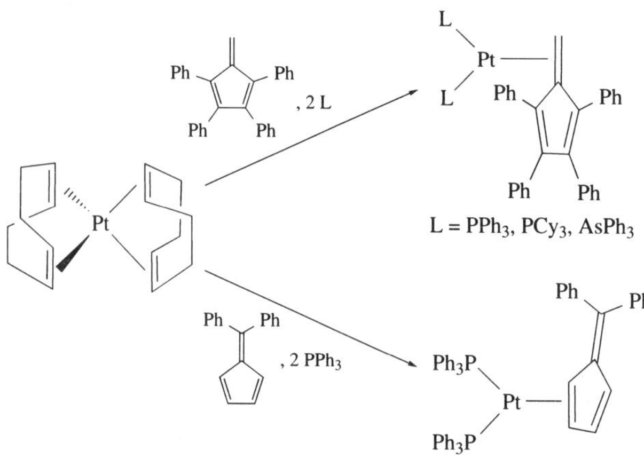

Scheme 1

A range of $PtL_2(\eta^2\text{-norbornene})$complexes with bulky bidentate phosphines, L_2, has been synthesized by reaction of $Pt(\eta^2\text{-norbornene})_3$ with the appropriate diphosphine (L_2 = dcpe, dcpp, dbpe(1,2-bis(di-*t*-butylphosphino)ethene), dbpp(1,3-bis(di-*t*-butylphosphino)propane), 1,2-bis(*t*-butylphosphinomethyl)benzene).[30] Addition of dioxygen, at low temperature, to $Pt(\eta\text{-}C_2H_4)_3$, leads to the derivative $Pt(\eta\text{-}O_2)(\eta\text{-}C_2H_4)_2$, which has been identified spectroscopically in solution but not isolated.[31]

An important new class of zerovalent platinum derivatives, $Pt(\eta\text{-}C_2H_4)(N\text{–}N')$, has been developed by reaction of $Pt(\eta\text{-}C_2H_4)_3$ with the requisite bidentate nitrogen-donor ligand, N–N' (N–N' = 2-methyl-1,10-phenanthroline or 2,9-dimethyl-1,10-phenanthroline). Subsequent reaction with other alkenes provides a general entry to a family of complexes, $Pt(\eta^2\text{-alkene})(N\text{–}N')$.[32]

Reactions of tris(dibenzylideneacetone)platinum(0) with other alkenes may also be considered in this preparative category. For example, its treatment with equimolar amounts of maleic anhydride and the unconjugated diene (see Section 9.2.3) hexa-1,5-diene yields $Pt(\eta^2\text{-}C_4H_2O_3)(\eta^2,\eta^2\text{-}C_6H_{10})$.[33]

Arguably the prototype of bis(phosphine)(η^2-alkene)platinum complexes, $Pt(PPh_3)_2(\eta\text{-}C_2H_4)$, remains a widely used precursor to other analogues in the same family. Several derivatives with unusual alkene ligands have been isolated by this approach. The strained anti-Bredt alkene bicyclo[3.3.1]non-1-ene reacts readily below ambient temperature to afford $Pt(PPh_3)_2(\eta^2\text{-}C_9H_{14})$ in high yield (Scheme 2, path (a)). No corresponding complex with a planar steric analogue, 2-methylbut-1-ene, could be prepared, indicating the unusual ligating power of this bridgehead alkene.[34] A similar complex, featuring the strained, highly pyramidalized alkene, tricyclo-[$3.3.1.0^{3,7}$]non-3(7)-ene, can be obtained by reaction of $Pt(PPh_3)_2(\eta\text{-}C_2H_4)$ with 3,7-diiodotricyclo[$3.3.1.0^{3,7}$]nonane, in the presence of sodium amalgam (Scheme 2, path (b)).[35] It is not clear in this case whether reduction occurs prior (or subsequent) to interaction of the dihaloalkane with platinum. A 3-vinylcylobutene species prepared by this method favours coordination via the ring unsaturation, despite the presence of bulky methylester groups on C-1 and C-2.[36] Synthesis of $Pt(PPh_3)_2[\eta\text{-}C_2H(CO_2Me)_3]$ employs the same approach.[37] Alkenes $R^1R^2C{=}CR^3R^4$, with a variety of other electronegative as well as electron-donor substituents (R^1 = CN; R^2, R^3, R^4 = (variously) CN, MeO, EtO, Me, Ph, 4-$O_2NC_6H_4$ or H), will also displace ethene,[38,39] as will the unusual related cycloalkenes, 1,2-dicyanocyclobutene and 1,3-di-*t*-butyl-2,5-dioxo-3-cyclopentene-1-carbonitrile.[40] Structural and synthetic aspects of $Pt(PPh_3)_2[\eta\text{-}C_2(CN)_4]$ have been reviewed.[41] The acrylonitrile complex $Pt[1,2\text{-}(Ph_2CH_2)C_6H_4](\eta\text{-}H_2C{=}CHCN)$ is prepared likewise. Ethene is readily displaced by other alkenes from the phosphite analogue $Pt[P(o\text{-}C_6H_4\text{-}2\text{-}OMe)_3]_2(\eta\text{-}C_2H_4)$ to yield $Pt[P(o\text{-}C_6H_4\text{-}2\text{-}OMe)_3]_2(\eta\text{-alkene})$ (alkene = norborene, styrene, *trans*-stilbene, $H_2C{=}CCl_2$, $H_2C{=}CH(CN)$ or $H_2C{=}CH(CHO)$).[13b]

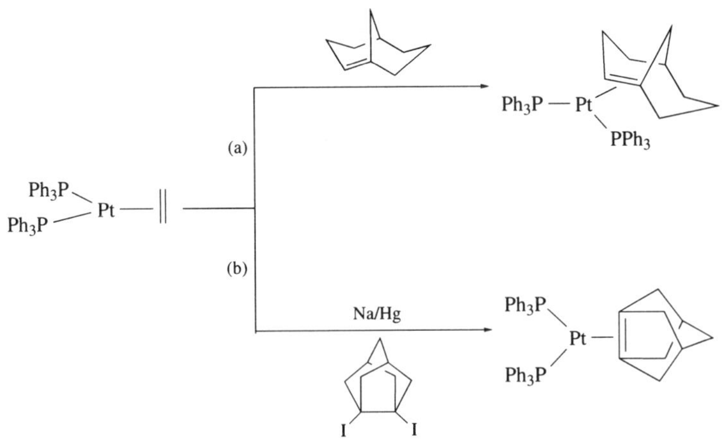

Scheme 2

(c) Preparations by other ligand displacements from Pt^0 precursors. Displacement of phosphine ligands by monoalkenes from platinum(0) complexes is often unfavourable. For example, the equilibrium constant for the reaction of ethene with $Pt(PPh_3)_3$ has been estimated as 0.12 (Equation (2)). Nevertheless, with a 2:1 platinum: ethene ratio at −100 °C, due to the thermodynamic favourability of the formation of $Pt(PPh_3)_4$, $Pt(PPh_3)_2(\eta\text{-}C_2H_4)$ is produced (Equation (3)). Similar results with but-1-ene and buta-1,3-diene—but not cyclohexene—suggest a relative binding order: ethene > but-1-ene > buta-1,3-diene > cyclohexene.[42] Complexes $Pt(PPh_3)_2[\eta^2\text{-}R^1R^2C{=}CR^3R^4]$ with electronegatively substituted alkenes[26] (see Section 9.2.1.1(ii)) may also be prepared by displacement of PPh_3 from $Pt(PPh_3)_3$, as can derivatives with 2,3-disubstituted norbornene, $Pt(PPh_3)_2(\eta^2\text{-}C_7H_8\text{-}2,3\text{-}XY)$ (X = Cl, Y = Cl, Br or H; X = Br, Y = Br, I or H).[13] Occasionally, selective replacement of one phosphine by another can be employed to generate complexes of the class $PtL^1L^2(\eta\text{-}C_2H_4)$. Reaction of $Pt(PPh_3)_2(\eta\text{-}C_2H_4)$ with PR_3 at low temperature (183–243 K) affords significant concentrations of $Pt(PPh_3)(PR_3)(\eta\text{-}C_2H_4)$ (R = Cy, CH_2Ph or $4\text{-}C_6H_4Me$).[43]

$$Pt(PPh_3)_3 + C_2H_4 \rightleftharpoons Pt(\eta^2\text{-}C_2H_4)(PPh_3)_2 + PPh_3 \qquad K \approx 0.12 \tag{2}$$

$$2\,Pt(PPh_3)_3 + C_2H_4 \rightleftharpoons Pt(\eta^2\text{-}C_2H_4)(PPh_3)_2 + Pt(PPh_3)_4 \tag{3}$$

(ii) Structural aspects

More comprehensive structural details based on diffractometric studies are given in Volume 12. Only salient features will be discussed here. Zerovalent platinum complexes of monoalkenes are seldom more than tricoordinate and, therefore, a major preoccupation has been the orientation of the coordinated alkene relative to the coordination plane of the host complex. A neutron diffraction study has confirmed a trigonal coplanar structure for the homoleptic example, $Pt(\eta\text{-}C_2H_4)_3$.[44] Similar, quasi-trigonal coplanar geometry is adopted by $Pt(PCy_3)(\eta\text{-}C_2H_4)(\eta^2\text{-}C_2F_4)$. Here the electronic differences between the two alkenes are reflected in the structure. For the perfluorinated ligand, both the C–C distance and 'bend-back' angle are larger than those for ethene, while the M–C length is smaller (Structure (**1**)).[45] These observations are consistent with a greater degree of π-acceptance by the more electronegative alkene. Infrared and Raman spectra of $Pt(\eta\text{-}C_2H_4)_3$ and $Pt(\eta\text{-}C_2D_4)_3$, as solids and solutions, allows full vibrational analysis of all skeletal modes, consistent with the trigonal coplanar assignment.[46] In contrast, on the basis of solution infrared studies, the dioxygen adduct, $Pt(\eta\text{-}O_2)(\eta\text{-}C_2H_4)_2$, is proposed to favour a structure in which both ethene ligands are mutually parallel and oriented perpendicular to the PtO_2 plane.[31] In a structurally related compound, $Pt(\eta^2\text{-}C_4H_2O_3)(\eta^2,\eta^2\text{-}C_6H_{10})$, the coordinated carbon atoms of maleic anhydride and one of the alkene termini of hexa-1,5-diene are coplanar, but the remaining η^2-

bound residue of the diene, while completing an approximate trigonal arrangement around platinum, is canted 27° out of the coordination plane. This can be ascribed to steric limitations associated with the unconjugated diene. The more electronegative alkene exhibits the smallest Pt–C distances, while the out-of-plane alkene is furthest from the metal.[33]

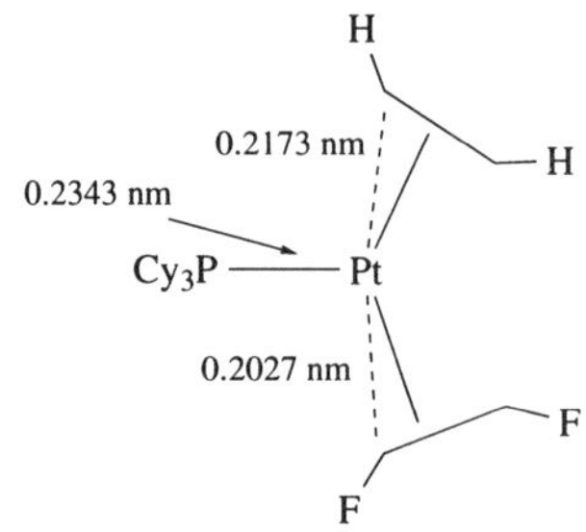

(1)

Compounds of the general type $PtL_2(\eta^2\text{-alkene})$, where L represents a phosphorus-donor ligand, remain the most plentiful among (η^2-monoalkene)platinum(0) examples. Alkene coordination is generally coplanar with the PtP_2 skeleton—or as nearly so as interligand steric interactions will allow. Some representative dimensions of the PtC_2P_2 framework are summarized in Table 1.

The complex with the strained alkene, tricyclo-[3.3.1.0^{3,7}]non-3(7)-ene, $Pt(PPh_3)_2(\eta^2\text{-}C_9H_{12})$ displays a large spin–spin coupling, $^1J_{Pt-C}$ (407 Hz), consistent with the high degree of pyramidalization at the carbon atom (cf. the corresponding ethene complex, $^1J_{Pt-C} = 194$ Hz).[35] A multinuclear NMR study of $Pt(PPh_3)_2(\eta\text{-}C_2H_{4-n}X_n)$ reveals that the value of $^1J_{Pt-P}$ is insensitive to variations in X—in contrast with analogous alkyne complexes (n = 0–4; X = CN or CO_2Me). Platinum-195 NMR shift measurements indicate that for X = CN platinum is monotonically deshielded as n increases. The dependence is less marked and nonmonotonic for X = CO_2Me.[39] A similar spectroscopic study of a related family of derivatives, $Pt(PPh_3)_2\{\eta\text{-}(NC)_2C{=}CR^1{}_2\}$, $Pt(PPh_3)_2\{\eta\text{-}(NC)_2C{=}CHR^2\}$ and $Pt(PPh_3)_2\{\eta\text{-}(NC)(R^3)C{=}CHR^2\}$ (R^1 = Me, OMe or OEt; R^2 = OMe, Ph or C_6H_4-4-NO_2; R^3 = CO_2Me) concludes that π-retrodonation is significant in complexes with strongly electron-withdrawing substituents (e.g., CN). Alkoxy groups, on the other hand, have the opposite effect and render the alkene less suitable for coordination to a d^{10} metal fragment. Where both types of substituent are present, it is suggested that an asymmetric or slipped coordination mode explains the unusually low values of $^2J_{Pt-C}$.[47]

(iii) Reactions

Many aspects of reactivity—notably alkene-displacement reactions—feature in the various preparative sections, since (η-alkene)platinum(0) complexes are common precursors to many other classes of π-organoplatinum species. Other types of reactivity may feature either the coordinated ligand or the metal centre.

(a) Alkene dissociation–displacement reactions. Ethene is readily displaced from $Pt(diop)(\eta\text{-}C_2H_4)$ by electron-poor or strained alkenes such as, for example, the chiral acrylamide, $H_2C{=}C(O)NHCHMePh$, *rac-trans*-5,6-dimethylnorbornene dicarboxylate and 5-isopropenyl-2-methylcyclohex-2-enone ((*R*)- or (*S*)-carvone). These reactions are the basis of a method for ^{31}P NMR analysis of the enantiomeric purity of chiral alkenes and allenes.[48]

(b) Protonation and related electrophilic addition. Treatment of $Pt(PPh_3)_2(\eta\text{-}C_2H_4)$ with strong protic acids HX (X = Cl or OSO_3CF_3) and H_2SO_4 liberates ethane and forms $PtX_2(PPh_3)_2$. Neither dihydrogen nor ethene is a detectable product. This occurs, proposedly, via the platinum(II) intermediate, $[Pt(Et)(PPh_3)_2]^+$.[49] Stoichiometrically controllable reactions of this type have led to some structurally unusual derivatives, which cast appreciable light on the intimate nature of such protonations. For example, treatment of the η^2-norbornene complexes, $Pt(P\text{–}P)(\eta\text{-}C_7H_{10})$ with noncoordinating protic acids (HBF_4 or $HOSO_2CF_3$) leads to cationic $[Pt(P\text{–}P)(\eta^2\text{-}C_7H_{11})]^+$, in which the norbornyl ligand is bound by both a σ-Pt–C bond and a β-agostic interaction with an *exo*-C–H (or C–D) bond (Equation (4)); P–P = dcpe, dcpp, dbpe, dbpp or 1,2-$(Ph_2P)_2C_6H_4$).[30] In a similar fashion, several related η^2-styrene complexes, $Pt(P\text{–}P)(\eta^2\text{-}C_8H_8)$, also undergo acid protonation (HBF_4), affording $[Pt(P\text{–}P)(\eta^3\text{-}MeCHC_6H_5)]^+$, which exhibit η^3-allyl-style coordination of the α-methylbenzyl ligand (P–P = dcpe, dcpp, dbpe or dbpp). These species are fluxional in solution at ambient temperature, via η^3-σ-conversion coupled with β-hydrogen migration. The activation barriers to this intramolecular rearrangement are lowest for the sterically most demanding bidentate phosphine, P–P.[50]

Table 1 Selected dimensions of the $PtC^1C^2P^1P^2$ skeleton in complexes of the type $Pt(PR_3)_2(\eta^2\text{-alkene})$.

Complex	*Bond length* (nm)					*Bond angle* (°)			
	Pt–C^1	*Pt–C*^{2}	*C*^{1}*–C*^{2}	*Pt–P*^{1}	*Pt–P*^{2}	*C*^{1}*–Pt–C*^{2}	*P*^{1}*–Pt–P*^{2}	R_w	Ref.
$Pt(PCy_3)_2(\eta^2\text{-}C_2H_4)$	0.2137(7)		0.1440(7)	0.2284(1)		39.4(2)	116.33(7)	0.039	17a
$Pt[P(O\text{-}2\text{-}C_6H_4OMe)_3]_2(\eta^2\text{-}C_2H_4)$	0.2128(5)	0.2131(4)	0.1417(7)	0.2228(2)	0.2227(2)	38.9(2)	105.8(1)	0.034	13b
$Pt(dcpp)(\eta^2\text{-}C_2H_4)$	0.214(1)	0.210(1)	0.139(20)	0.207(1)	0.210(1)	38(5)	113.0(4)	0.160	17b
$Pt(diop)(\eta^2\text{-}C_2H_4)$	0.2109(5)	0.2100(5)	0.1402(9)	0.2261(4)	0.2254(1)	38.9(3)	105.25(4)	0.026	14c[e]
$Pt[1,2\text{-}(Ph_2PCH_2)_2(C_6H_4)](\eta^2\text{-}C_2H_4)$	0.2108(18)	0.2122(17)	0.145(2)	0.2260(3)	0.2257(3)	40.2(6)	105.0(1)	0.059	16[e],17[e]
$Pt(dbpe)(\eta^2\text{-}C_7H_{10})$[a]	0.2110(7)	0.2108(8)	0.1460(11)	0.2273(2)	0.2274(2)		89.2(1)	0.059	30
$Pt(PPh_3)_2[\eta^2\text{-}(MeO_2C)HC{:}C(CO_2Me)_2]$	0.213(2)	0.210(1)	0.144(2)	0.2283(4)	0.2286(3)	39.9(6)	105.3(1)	0.069	37[e]
$Pt(PPh_3)_2[\eta^2\text{-}C_{10}H_{12}O_4]$[b]	0.2120(6)	0.2097(6)	0.1486(9)	0.2290(2)	0.2294(2)	41.2(2)	105.5(1)	0.0371	36
$Pt(PPh_3)_2(\eta^2\text{-}C_{18}H_{14})$[c]	0.215(2)	0.224(2)	0.152(3)	0.2301(6)	0.2280(6)	40.5(7)	103.8(2)	0.064	26
$Pt(PPh_3)_2[\eta^2\text{-}C_4H_4(CN)_2]$	0.2091(3)	0.2077(3)	0.1504(4)	0.2284(1)	0.2301(1)	42.3(1)	103.90(3)	0.038	40
$Pt(PPh_3)_2(\eta^2\text{-}C_{12}H_{19}O_2N)$[d]	0.2121(8)	0.2170(7)	0.1451(13)	0.2316(2)	0.2310(2)	39.5(3)	106.6(1)	0.050	40

[a] C_7H_{10} = bicyclo[2.2.1]hept-2-ene (norbornene). [b] $C_{10}H_{12}O_4$ = 1,2-di(methoxycarbonyl)-3-vinylcyclobutene. [c] $C_{18}H_{14}$ = diphenylfulvene. [d] $C_{12}H_{19}O_2N$ = 1,3-di-*t*-butyl-2,5-dioxocyclopentene-1-carbonitrile. [e] Also gives and compares structural data for earlier analogues.

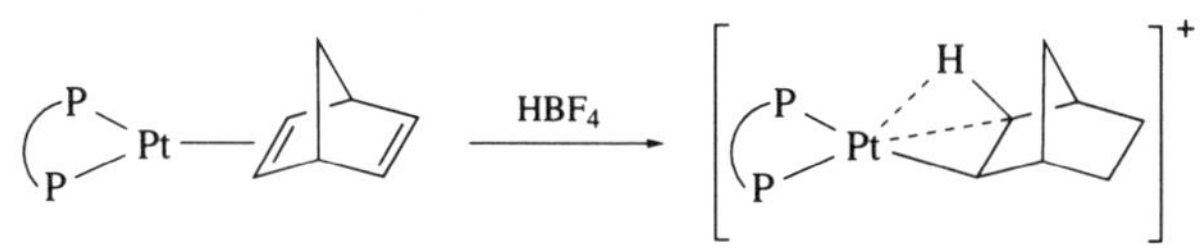

(4)

(c) Oxidative addition. Oxidative additions of C–C bonds frequently involve hydrocarbons with strained rings. An example is the addition of cyclopropa[*b*]naphthalene to $Pt(PPh_3)_2(\eta\text{-}C_2H_4)$ which generates a platinacyclobutene (Equation (5)).[51] Enthalpy measurements on oxidative addition of strained cycloalkenones to $Pt(PPh_3)_2(\eta\text{-}(E)\text{-}PhHC{=}CHPh)$ are proposed to reflect the difference in strain relief for the disparate rings (Scheme 3), leading to the conclusion that the resultant five-membered platinacycle is not appreciably strained.[52]

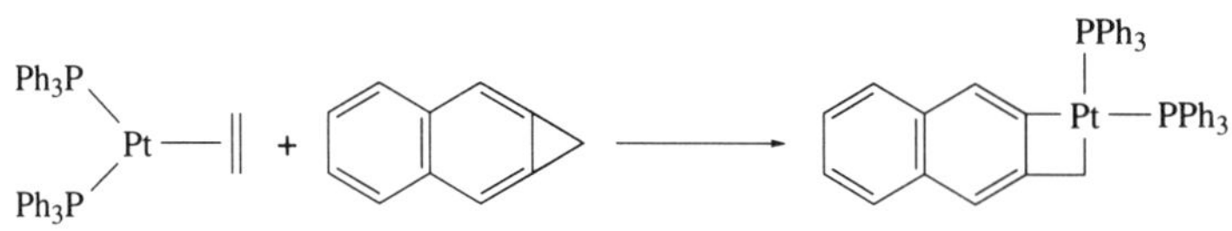

(5)

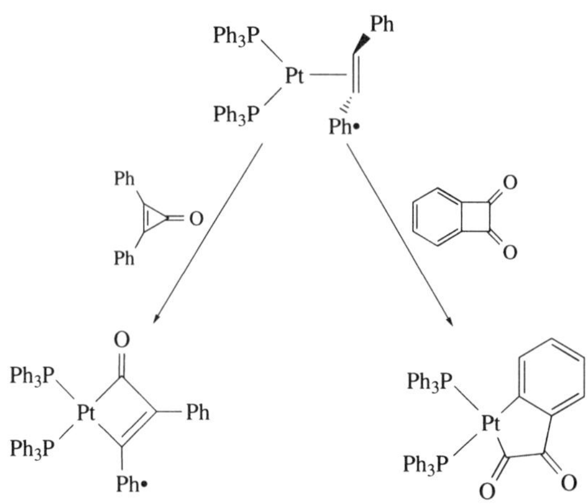

Scheme 3

More common oxidative addends are the C–X bonds of halocarbons. An interesting instance is the isomerization of derivatives of 2,3-disubstituted norbornenes, $Pt(PPh_3)_2(\eta^2\text{-}C_7H_8\text{-}2,3\text{-}XY)$ to yield *trans*-$PtX(\eta^1\text{-}2\text{-}C_7H_8\text{-}3\text{-}Y)(PPh_3)_2$ (X = Cl, Y = Cl, Br or H; X = Br, Y = Br, I or H).[12] A reaction sequence involving oxidative addition, R migration and hydrogen transfer explains the reactions of R–X with the anti-Bredt alkene complex, $Pt(PPh_3)_2(\eta^2\text{-}C_9H_{14})$ (Scheme 4; R–X = MeI or $PhCH_2Br$).[34] A variety of substrates, including R–X, X–X, H–X Si–X, Sn–X and Hg–X, add oxidatively to Pt(N–N')(η^2-alkene) to give pentacoordinate platinum(II) derivatives (N–N' = bidentate nitrogen donor; see Section 9.2.1.2(iv)).

Vinyl triflates add to zerovalent derivatives with a variety of interesting outcomes. For example, $Pt(PPh_3)_2(\eta\text{-}C_2H_4)$ reacts with $R^1{}_2C{=}CR^2(OSO_2CF_3)$ in ethene-saturated media to give cationic allyl complexes, $[Pt(PPh_3)_2(R^1{}_2CCR^2CHMe)]^+$, as a result of migratory ethene insertion reactions (see Section 9.2.1.2(iii)) of the initially formed intermediate, *trans*-$Pt(CR^2{=}CR^1{}_2)(OSO_2CF_3)(PPh_3)_2$ (R^1 = H, R^2 = Me; R^1 = Me, R^2 = H; R^1 = H, R^2 = Bu^t).[53]

A novel example in this category is the oxidative insertion of $Pt(PPh_3)_2$ fragment derived from $Pt(PPh_3)_2(\eta\text{-}C_2H_4)$ into a P–C bond in the (η^1-phosphirane)tungsten complex, $[\overline{P(Ph)C(Ph)C}\text{-}(Ph)](CO)_5$, to afford isomeric platinaphosphacyclobutanes (Equation (6)).[54] $Pt(PPh_3)_2(\eta\text{-}C_2H_4)$ also undergoes oxidative addition of the S–S bonds in cyclic thiosulfinates.[55]

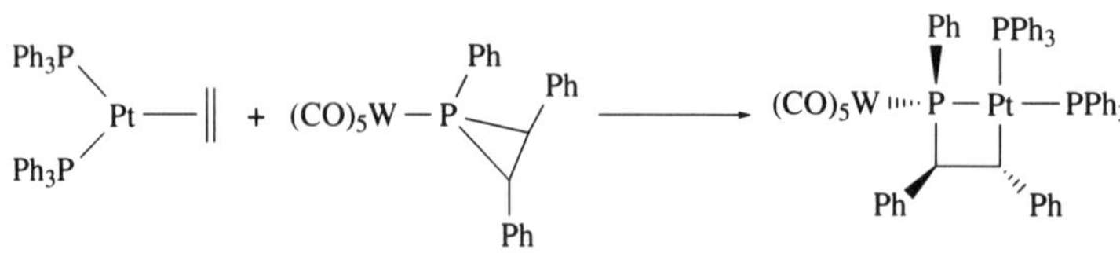

(6)

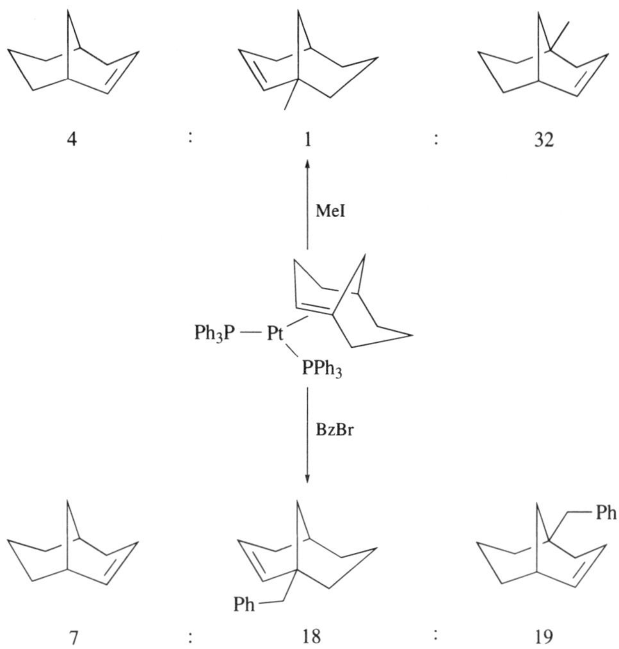

Scheme 4

9.2.1.2 Divalent derivatives

Like the Zeise anion—of which they are the lineal descendants—most complexes of monoalkenes with platinum are tetracoordinate and planar (16-electron) derivatives, although the number of known pentacoordinate (18-electron) complexes is growing (these are discussed in Section 9.2.1.2(i)(d)).

(i) Preparation

The majority of synthetic approaches continues to feature ligand displacements from platinum(II) parent compounds. Zeise's anion and the related neutral dimer, $Pt_2(\eta\text{-}C_2H_4)_2Cl_4$, are among the popular precursors.

(a) Preparations via halide displacement. Preparations via halide displacement are of two types: (i) those in which X^- is replaced by η^2-alkene; and (ii) those in which a different ancillary ligand substitutes for X^- in a precursor with an established η^2-alkene ligand. An improved preparation of Zeise's salt has been reported, in which $SnCl_2$ catalyses the reaction between C_2H_4 and K_2PtCl_4 in aqueous solution.[56] Reaction of the aqueous Zeise anion with glycine in the presence of acetate generates a mixture of *cis*-(*N*,alkene) and *trans*-(*N*,alkene) isomers of $[PtCl(N,O\text{-}gly)(\eta\text{-}C_2H_4)]^+$, in which the thermodynamic *cis* product predominates. Analogous complexes with the water-soluble prochiral alkenes allyl alcohol and allyl sulfonate, $H_2C{=}CHCH_2(OSO_3^-)$ may also be prepared.[57] This work has been extended subsequently to include a broader range both of amino acids (glycine, α-amino isobutyric acid, sarcosine and (*S*)-proline) and of prochiral alkenes (allyl alcohol, but-3-en-2-ol and 2-methylbut-3-en-2-ol).[58] The same general technique affords *trans*-(*N*,alkene)-$PtCl[OC(O)CH_2CH_2NH_2](\eta\text{-}C_2H_4)$ upon reaction of $K[PtCl_3(\eta\text{-}C_2H_4)]$ with the anion from β-alanine. Mechanistic analysis reveals an associative substitution process.[59] Reaction of dimethylsulfoxide with $K[PtCl_3(\eta\text{-}C_2H_4)]$ affords *cis*-$[PtCl_2(DMSO)(\eta\text{-}C_2H_4)]$, in which DMSO is bonded via sulfur.[60] Zeise's salt is also the precursor to the series of complexes *trans*-$[PtCl_2(QNO)(\eta\text{-}C_2H_4)]$ (QNO = quinoline *N*-oxide, 4-chloroquinoline *N*-oxide, 2-methylquinoline *N*-oxide, 4-methylquinoline *N*-oxide or isoquinoline *N*-oxide).[61]

The inclusion of silver(I) has been a frequent adjunct to halide replacement in these systems. Substituted styrene derivatives $\{Pt[\eta^3\text{-}H_2CC(Me)CH_2](PPh_3)(\eta\text{-}H_2C{=}CH\text{-}4\text{-}C_6H_4Y)\}^+$ are obtained by successive treatments of $\{PtCl[\eta^3\text{-}H_2CC(Me)CH_2](PPh_3)\}$ with $AgPF_6$ and the requisite styrene derivative (Y = H, Me, OMe, NMe_2, Cl or NO_2).[62]

(b) Preparations via alkene displacement. As was the case with zerovalent derivatives, exchange of one coordinated alkene for another can afford new complexes, especially of more unusual alkenes. The alkene replaced is commonly ethene. Thus, styrene analogues, *trans*-$PtCl_2(QNO)(\eta^2\text{-}H_2C{=}CHPh)$, are derived by displacement of ethene from the quinoline *N*-oxide complexes described above.[46] Reaction of *trans*-$PtCl_2(py)(\eta\text{-}C_2H_4)$ with the anti-Bredt alkene bicyclo[3.3.1]non-1,2-ene affords *trans*-$PtCl_2(py)(\eta^2\text{-}C_9H_{14})$ (Equation (7)). In contrast with the zerovalent case, an analogous complex with 2-methylbut-1-ene *is* accessible.[34]

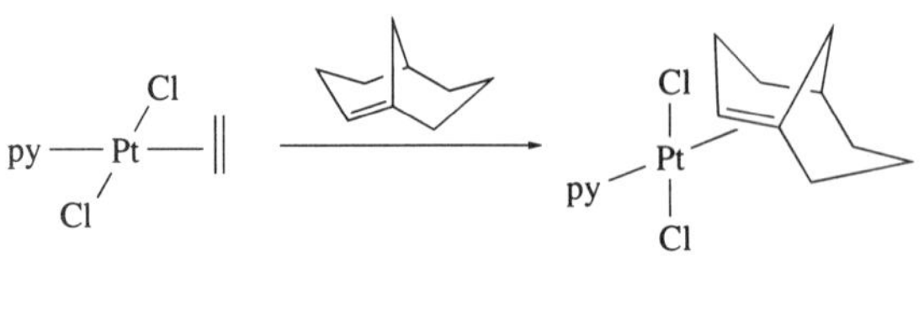

(7)

(c) Preparations via bridge cleavage. The η^2-styrene analogue of the Zeise anion is accessible by interaction of styrene and $[Bu^n{}_4N]_2[Pt_2Cl_6]$, while direct addition of styrene to $[PtCl_2]_n$ yields stereoisomers of *cis*-$PtCl_2(\eta\text{-}H_2C{=}CHPh)_2$.[63] A variety of terminal alkenes have been discovered to react with diplatinum species, $[PtCl(\mu\text{-}Cl)(PR^2{}_3)]_2$, giving $PtCl_2(PR^2{}_3)(\eta\text{-}H_2C{=}CHR^1)$ (R^1 = H, CH_2OH, CH(OH)Me, $C(OH)Me_2$ or C(O)H; $PR^2{}_3$ = PMe_3, PEt_3, $PPr^n{}_3$ or $PPhMe_2$). At low temperatures, conversion to the *trans*-isomer is virtually quantitative. Subsequent, slow isomerization affords isolable *cis*-products.[64] Treatment of the related dimer, $[Pt(CO)Cl(\mu\text{-}Cl)]_2$, with cyclohexene leads directly to *cis*-$PtCl_2(CO)(\eta^2\text{-}C_6H_8)$. Corresponding complexes of cyclopentene, propene and ethene have been characterized but not isolated.[65] Neutral complexes $PtCl[\eta^3\text{-}H_2CC(Me)CH_2](\eta\text{-}H_2C{=}CHR)$ are accessible by reaction of $\{Pt[\eta^3\text{-}H_2CC(Me)CH_2](\mu\text{-}Cl)\}_2$ with $H_2C{=}CHR$ (R = Me, Et, Bz or 4-C_6H_4Y; Y = H Me, Cl, NO_2, NMe_2 or OMe).[66] Cationic homologues, $[Pt(PPh_3)[\eta^3\text{-}H_2CC(Me)CH_2](\eta\text{-}H_2C{=}CHC_6H_4Y)]PF_6$, result from sequential treatment of $PtCl(PPh_3)[\eta^3\text{-}H_2CC(Me)CH_2]$ with $AgPF_6$ and $H_2C{=}CHC_6H_4\text{-}4\text{-}Y$ (Y = H, Me, Cl, OMe, NMe_2 or NO_2).[62] Other members of this family, $[Pt(PPh_3)(\eta^2\text{-}Z\text{-}C_4H_8)\{\eta^3\text{-}H_2CC(Me)CH_2\}]^+$, are prepared in a similar fashion (alkene = C_2H_4, (*Z*)-C_4H_8, (*E*)-C_4H_8, $H_2C{=}CHPh$, $H_2C{=}CH\text{-}2\text{-}C_6H_4Cl$, (*E*)-MeHC=CHPh).[67] The related complex $Pt(C_6F_5)(\eta^3\text{-}H_2C{=}CHPh)\{\eta^3\text{-}H_2CC(Me)CH_2\}$ arises from treatment of $\{Pt[\eta^3\text{-}H_2CC(Me)CH_2](\mu\text{-}Cl)\}_2$ with LiC_6F_5.[68] $Pt(SnCl_3)(\eta^2\text{-alkene})\{\eta^3\text{-}H_2CC(Me)CH_2\}$ is prepared by addition of alkene to a mixture of $SnCl_2$ and $[PtCl\{\eta^3\text{-}H_2CC(Me)CH_2\}]_2$ (alkene = ethene, styrene or norbornene).[69]

Alkylplatinum homologues of the Zeise dimer $[Pt(Me)(\eta\text{-}C_2H_4)(\mu\text{-}Cl)]_2$ react with donor species L to give *trans*-$PtClMe(L)(\eta\text{-}C_2H_4)$, where L may be SMe_2, py or C_2H_4.[70] Similarly, reaction of C_2H_4 with dinuclear $[PtMe_2(\mu\text{-}SMe_2)]_2$ yields *cis*-$PtMe_2(SMe_2)(\eta\text{-}C_2H_4)$.[71]

(d) Pentacoordinate complexes. Many ligand-substitution reactions of divalent platinum are proposed to involve pentacoordinate intermediates, and there has been continuing interest in detectable and isolable examples of this class of complex. One of the most significant recent advances in this area has featured monoalkene complexes of platinum(II), mainly with a variety of bidentate nitrogen-donor ligands (N–N') as ancillary ligands. These may be saturated (diamines) or unsaturated (diimines, pyridines or polypyridyls). This chemistry has been reviewed comprehensively, including full tabulation of all complexes isolated to date.[72] Earlier aspects also appear in a review of diazabutadienes as ligands.[73] Some representative syntheses are outlined here.

These derivatives mainly have the general formulations *trans*-$PtX_2(N\text{–}N')(\eta^2\text{-alkene})$ and *trans*-$PtXY(N\text{–}N')(\eta^2\text{-alkene})$ (X = Cl, Br or I; Y = H, alkyl, aryl, SnR_3, HgR, etc.). Synthetic strategies fall into four categories: (i) ligand displacement by a bidentate nitrogen-donor from tetracoordinate (mononuclear or dinuclear) platinum(II) complex with the η^2-alkene function already bound; (ii) alkene addition to a tetracoordinate (mononuclear) platinum(II) complex with the N–N' ligand already bound; (iii) oxidative addition of X–X, X–Y or R–X to tricoordinate platinum(0) precursors, $Pt(N\text{–}N')(\eta^2\text{-alkene})$; and (iv) exchange of η^2-ethene, X or Y on a prototype complex, *trans*-$PtXY(N\text{–}N')(\eta^2\text{-}C_2H_4)$. Methods (i) and (iv) are often employed consecutively. Methods (i) and (ii) may also be used in tandem, such that the alkene and the nitrogen donor are introduced simultaneously to the reaction (Scheme 5).

Much of the early work featured method (i) to *trans*-$PtX_2(N\text{–}N')(\eta^2\text{-alkene})$. Addition of diimine or diamine to $K[PtX_3(\eta^2\text{-alkene})]$ affords a range of pentacoordinate derivatives (alkene = ethene, but-2-enes (*E*) and (*Z*)), styrene, maleic anhydride, dimethylmaleate, methyl acrylate or acrolein; X = Cl or Br; N–N' = RN=CHCH=NR or $RHNCH_2CH_2NHR$ (R = Pr^i, Bu^t or $EtMe_2C$)).[74] A similar family of complexes with diastereomeric diamine ligands (*R,R*)-*N,N'*-dimethyl-*N,N'*-bis(α-methylbenzyl)-1,2-

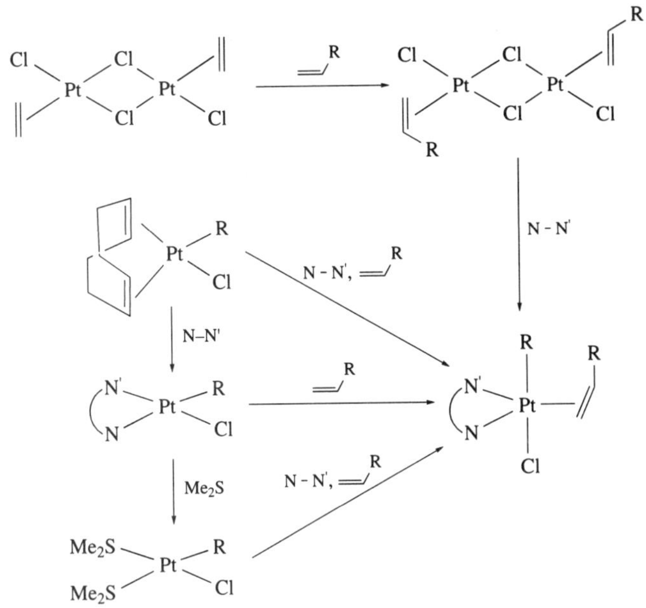

Scheme 5

ethanediamine ((*R,R*)-mben) and (*R,S*)-*N,N*'-dimethyl-*N,N*'-bis(α-methylbenzyl)-1,2-ethanediamine ((*R,S*)-mben) as well as different coordinated alkenes is accessible via both (i) and (iv), that is, by reaction of diamine with $[PtX_3(\eta\text{-}C_2H_4)]^-$ and subsequent displacement of ethene by an alternative alkene (N–N' = (*R,R*)-mben; alkene = acrolein, acrylonitrile, fumarodinitrile ((*E*)-1,2-dicyanoethene), maleic anhydride or maleimide; X = Cl or Br. N–N' = (*R,S*)-mben; alkene = ethene, acrolein, acrylonitrile, fumarodinitrile, maleic anhydride or maleimide; X = Cl or Br).[75] Treatment of binuclear precursors $[PtClMe(\eta\text{-}C_2H_4)]_2$ with diamines or diimines yields $PtClMe(N–N')(\eta\text{-}C_2H_4)$ (N–N' = TMEDA, $Pr^iHNCH_2CH_2NHPr^i$, $Bu^tN{=}CHCH{=}NBu^t$, $Me_2NN{=}CMeCMe{=}NNMe_2$, C_5H_4N-2-CH=NR, 6-Me-C_5H_4N-2-CH=NR (R = Me, Bu^t, Bz or CHMePh) or 2,9-dimethyl-1,10-phenanthroline (2,9-Me_2phen)].[76]

Reaction of PtClY(N–N') with alkenes leads to PtClY(N–N')(η^2-alkene) where Y = Me (method (ii)). The same complexes may also be obtained by the action of the alkenes on PtClY(N–N')(η-C_2H_4) (method (iv)), which is the preferred general method where Y = Cl (N–N' = C_5H_4N-2-CH=NR or 6-Me-C_5H_4N-2-CH=NR (R = Me, Pr^i, Bu^t, Bz or CHMePh)).[77] Several variations on methods (i) and (ii) may be employed to generate a further range of derivatives. Ethene can be displaced from dimeric [PtClR(η-C_2H_4)]$_2$ by other alkenes, giving [PtClR(η^2-alkene)]$_2$, which may be reacted in turn with N–N' to afford *trans*-PtClMe(N–N')(η^2-alkene). Advantage can be taken of the substitution lability of cycloocta-1,5-diene (see Section 9.2.3.2(iii)(a)) or Me_2S to produce the same range of complexes. Either PtClR(cod) or PtClR(SMe_2)$_2$ may be treated with N–N' and alkene sequentially (or simultaneously) to generate *trans*-PtClMe(N–N')(η^2-alkene) (alkene = C_2H_4, $H_2C{=}CHMe$, $H_2C{=}CHBu^t$, $H_2C{=}CMe_2$, (*E*)-MeHC=CHMe, (*Z*)-MeHC=CHMe, $H_2C{=}CH(CN)$, (*E*)-(NC)HC=CH(CN), $H_2C{=}CH(OH)$, (*E*)-ClHC=CHCl, $H_2C{=}CH(CO_2Me)$, maleic anhydride or maleimide; N–N' = TMEDA, $Bu^tHNCH_2CH_2NHBu^t$, $Bu^tN{=}CHCH{=}NBu^t$, $Me_2NN{=}CMeCMe{=}NNMe_2$, C_5H_4N-2-CH=NR, 6-Me-C_5H_4N-2-CH=NR (R = Me, Pr^i, Bu^t, Bz, CHMePh, 4-$MeOC_6H_4$ or NMe_2), quin-2-CH=NBu^t, bipy, 2,9-Me_2phen, 2-Mephen or 2,2'-biquinolyl (biquin)).[78]

Oxidative addition of X_2 or R–X to tricoordinate complexes Pt(N–N')(η^2-alkene) (method (iii)) leads to PtX_2(N–N')(η^2-alkene) or PtXR(N–N')(η^2-alkene), respectively (alkene = C_2H_4, C_3H_6 or *trans*-$C_2H_2(CO_2Me)_2$; N–N' = 2-Mephen or 2,9-Me_2phen; X_2 = Cl_2, Br_2 or I_2; R–X = Me–Br, Me–I, Et–Br, Et–I, Ph–I, $ClCH_2$–Cl, Bz–Br, Bz–I or PhMeHC–Br).[26] Analogous addition of HCl to zerovalent Pt(N–N')[η-*cis*-$C_2H_2(CO_2Me)_2$] gives the hydrido(η^2-alkene)platinum(II) complex *trans*-Pt(H)Cl(N–N')[η-*cis*-$C_2H_2(CO_2Me)_2$]. The corresponding η^2-dimethylfumarate derivative is obtained by treatment of Pt(N–N')[η-*trans*-$C_2H_2(CO_2Me)_2$] with TMS-Cl in the presence of H_2O (N–N' = 2,9-Me_2phen).[79] Various chloro(alkyl)mercury(II) derivatives HgR^2Cl add similarly to Pt(2,9-Me_2phen)[η-*cis*-$C_2H_2(CO_2R^1)_2$], generating *trans*-PtCl(HgR^2)(N–N')[η-*cis*-$C_2H_2(CO_2R^1)_2$] (R^1 = Me, Et or Bu^t; R^2 = (variously) Me, Et or Bu^t).[80] An analogous platinum(0) precursor, Pt(2,9-Me_2phen)[η-*trans*-$C_2H_2(CO_2R)_2$], undergoes reversible oxidative addition of $SnCl_2Me_2$ to give PtCl($SnClMe_2$)(2,9-Me_2phen)[η-*trans*-$C_2H_2(CO_2R)_2$].[81] A broad range of related complexes, PtCl(Y)(2,9-Me_2phen)[η^2-

alkene], has proven accessible by this route (alkene = C_2H_4, C_3H_6, *trans*-$C_2H_2(CO_2Me)_2$ or *cis*-$C_2H_2(CO_2Me)_2$; Y = $SnMe_3$, $SnClMe_2$, $SnCl_2Me$, $SnPh_3$ or $SnClPh_2$).[82]

Cationic complexes, $[PtR(L)(N–N')(\eta^2\text{-alkene})]^+$, may be prepared directly from $[PtClR(N–N')(\eta^2\text{-alkene})]$ by treatment with $AgBF_4$ or AgOTf (Tf = trifluoromethanesulfonyl) in the presence of coordinating species Y (alkene = C_2H_4, C_3H_6 or $C_2H_3(CO_2Me)$; R = Me, 4-C_6H_4OMe or 4-$C_6H_4CF_3$; N–N' = 2,9-Me_2phen or 6-Me-C_5H_4N-2-CH=NR (R = Bu^t, CHMePh or 4-$MeOC_6H_4$); L = MeCN, Bu^tCN, CF_3CN, Me_2S, MeH_2N, PhH_2N or py).[83] Corresponding derivatives $[PtCl(L)(N–N')(\eta^2\text{-alkene})]^+$ are obtained by a less direct route. Dimeric $Pt_2Cl_4L_2$ is cleaved with alkene to generate monomeric *cis*-$PtCl_2(L)(\eta^2\text{-alkene})$, which is then treated with $AgBF_4$ in the presence of acetonitrile to yield $[PtCl(NCMe)(L)(\eta^2\text{-alkene})]^+$. Displacement of MeCN with N–N' affords the cationic product (alkene = C_2H_4 or (*E*)-2-C_4H_8; N–N' = 2,9-Me_2phen, phen, bipy or 6-Mebipy; L = MeCN, Me_2HN, py, 4-Mepy or 4-Cl-py).[84]

Success in syntheses of these pentacoordinate derivatives depends primarily on: (i) the electron-withdrawing character of the coordinated alkene, (ii) strong Pt–N interaction, and (iii) sufficient steric obstruction close to the ligating nitrogen atoms of N–N' to destabilize the tetracoordinate planar product of alkene dissociation.[72]

(ii) Structural aspects

General aspects of structure (and reactivity) of monoalkene complexes have been reviewed, including coupling of unsaturated hydrocarbons at metal centres,[85] σ–π rearrangements of organometals[86] and structure, bonding and stability relationships.[87] The application of ^{13}C NMR to the characterization of alkene–metal complexes has been reviewed and features structure and dynamics of platinum derivatives.[88]

Some interesting and unusual spectroscopic studies of the Zeise anion have appeared. A single-crystal polarized electronic absorption spectrum of $[PtCl_3(\eta\text{-}C_2H_4)]^-$ comprises five major absorptions between 18 000 and 40 000 cm^{-1}. The indications are of strong σ-donation but weak π-acceptance by ethene towards platinum. Ethene, it is concluded, is a more effective σ-donor than NH_3. The photodissociation of ethene from the Zeise anion is rationalized on the basis of a photoexcitation which populates the (primarily σ-antibonding) $d_{x^2-y^2}$ orbital.[89] From measurements of chemical shift tensors obtained from ^{13}C NMR spectra of both $[PtCl_3(\eta\text{-}C_2H_4)]^-$ and $Pt_2Cl_4(\eta\text{-}C_2H_4)_2$ in the solid state, it is deduced that the C–C bond order is intermediate between one and two and that alkene rotation is not a feature of the solid-state structures.[90]

A continuing interest in this area is the phenomenon of rotation of the η^2-alkene and, while it may feature in routine spectroscopic characterizations of new (η^2-alkene)platinum complexes, several NMR studies have been devoted particularly to this facet. Rotational barriers ($\Delta G^\ddagger$ = 61–67 kJ mol^{-1}) in *trans*-PtBrCl(py-4-Y)(η-C_2H_4) are found to diminish with decreasing pK_a of substituted pyridine ligand (Y = H, Me, Cl or CN).[91] The coordinated *trans*-but-2-ene in *trans*-$PtCl_2(NHMe_2)(\eta^2\text{-}(E)\text{-}C_4H_8)$ is rotating freely (on the NMR timescale) between −40 and +25 °C.[92] The η-complex of the anti-Bredt alkene, bicyclo[3.3.1]non-1,2-ene, *trans*-$PtCl_2(py)(\eta^2\text{-}C_9H_{14})$ has been compared with its ethene analogue, *trans*-$PtCl_2(py)(\eta\text{-}C_2H_4)$. The large ^{13}C NMR shifts and large $^1J_{Pt-C}$ values suggest more metallacyclopropane character for bridgehead alkene coordination, which is at odds with the lower rotational barrier for η^2-C_9H_{14} than for η-C_2H_4.[29] Variable-temperature NMR measurements also furnish structural insight into the ligand-dependent, rotational stereoisomerism in 4-toluidine complexes, *cis*-$PtCl_2(NH_2C_6H_4\text{-4-Me})(\eta\text{-}(E)\text{-}R^1HC{=}CHR^2)$ (R^1 = H, R^2 = Et, CH_2Bu^t, OEt, OBu^t or CH_2OMe; R^1 = Me, R^2 = OEt). Conformation (**2a**) is adopted when R^2 has no hydrogen-bonding capacity and steric repulsions can dominate the conformational preference. Structure (**2b**) is favoured when the alkoxo group R^2 may hydrogen bond with the ligating nitrogen, outweighing unfavourable steric considerations. For ethylvinyl ether, for example, there is no indication of rotation of either the alkene or the amine ligand below +50 °C. Coordinated ethene, on the other hand, rotates freely above −25 °C.[93]

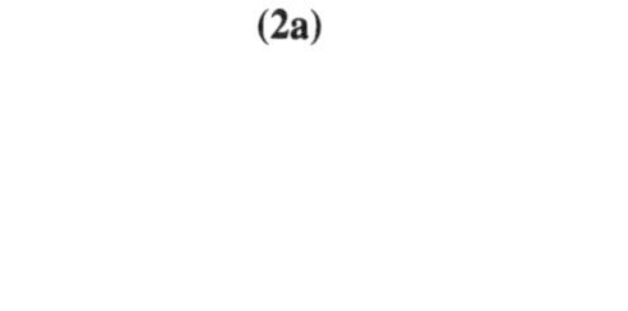

(2a)

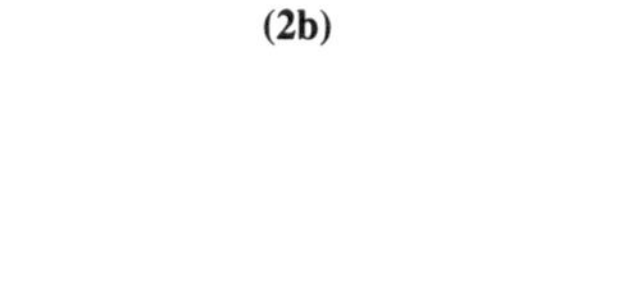

(2b)

Likewise, multinuclear NMR studies reveal the structural characteristics of imine complexes, $PtCl_2(Z)$-$R^1C=N(R^2)\{CH(CH_2)_3E\}(\eta$-$H_2C=CH_2)$ (R^1 = Me, Pr^i, Bu^i, Bu^s or Bu^t; R^2 = H, Me or Ph; E = O or S). These systems undergo (Z)–(E) imine isomerization, either spontaneously or catalysed by trace amounts of free amine. Increasing the degree of substitution slows the isomerization.[94–6] In the related system, $PtCl_2((Z)$-$RC=N(Me)\{CH(CH_2)_3NMe\})(\eta$-alkene), the observed values of $^nJ_{Pt-H}$ and $^nJ_{Pt-C}$ for H-3 and C-3 of the pyrrolyl ring are considered too large for purely five- and four-bond couplings, respectively, and are inferred to indicate an agostic Pt---H-3 interaction (alkene = $H_2C=CH_2$, $H_2C=CHMe$ or (Z)-MeHC=CHMe; R = Me, Pr^i, Bu^i, Bu^s or Bu^t).[97]

Several other alkene–platinum systems with ancillary nitrogen-donor ligands furnish interesting structural results from x-ray diffraction studies. In the pyrazole complex, *trans*-$PtCl_2(N_2C_3H_4)(\eta$-$C_2H_4)$, normal platinum–alkene parameters are observed, with the ethene coordinated perpendicular to the $PtNCl_2$ plane.[98] In the related pyrazolido derivative, $Pt_2(\mu$-$Cl)(\mu$-$N_2C_3H_3)(\eta$-$C_2H_4)_2$, the Pt_2ClN_2 ring is planar and the pyrazolide plane is tilted 5.7° off this plane. Both ethene ligands are *trans* to nitrogen and, again, perpendicular to the $PtNCl_2$ plane.[99] In a corresponding complex of 2,6-dimethylpiperidine, *cis*-$PtCl_2(NC_5H_8Me_2)(\eta$-$C_2H_4)$, the alkene lies exactly perpendicular to the plane of coordination.[100] The same is true of the O,N-chelate complex formed by the anion of β-alanine, $PtCl\{OC(O)CH_2CH_2NH_2\}(\eta$-$C_2H_4)$.[59] This is one of several structural investigations in the chemistry of alkene–platinum complexes with biologically significant ancillary ligands or their relatives. The Zeise anion reacts with glycine (gly-H) in the presence of a weak base (acetate), to form an equilibrium mixture of *cis*- and *trans*-(N,alkene) isomers of the chelating glycinate complex $PtCl_2(\eta^2$-*N,O*-gly)(η-C_2H_4). In their ^1H NMR spectra, the corresponding complexes with water-soluble, prochiral alkenes, allyl alcohol, $H_2C=CHCH_2OH$ and allyl sulfonate, $Na[H_2C=CHCH_2OSO_3]$, exhibit distributions of two rotamers for each of the geometric isomers.[57] In an extension of this work, comparison of equilibrium distributions of diastereomeric *cis*- and *trans*-(*N*,alkene) isomers of the amino acid (AA) complexes, $PtCl(\eta^2$-*N,O*-AA$)(\eta^2$-alkene), by multinuclear NMR, reveals a metal-centred effect on the stereoselectivity of prochiral alkene coordination (amino acid = glycine, α-aminoisobutyric acid, sarcosine or (*S*)-proline; alkene = allyl alcohol, but-3-en-2-ol or 2-methylbut-3-en-2-ol (2-mb)). Stereoselectivity is not significant for *trans* species but for the *cis* configurations with proline and sarcosine, stereoselectivity is substantial, especially for 2-methylbut-3-en-2-ol (>40:1 preference). X-ray diffractometric structure determinations on the preferred isomers for *cis*-$PtCl(\eta^2$-*N,O*-sarcosine)(η-2-mb) and *trans*-$PtCl(\eta^2$-*N,O*-(*S*)-proline)(η-2-mb) both exhibit alkene coordination perpendicular to the PtONCl plane.[58] The latter conformation for the alkene emerges from a similar study of *cis*-$PtCl_2(OSMe_2)(\eta$-$C_2H_4)$, in which DMSO is bonded to sulfur. This complex is one of several from whose structural characteristics the *cis* and *trans* influence of η-C_2H_4 is assessed.[60]

The barrier to ethene rotation is higher for the cyclopentadienyl derivative $[Pt(PPh_3)(\eta$-$C_2H_4)(\eta$-$C_5H_5)]^+$ than for the methallyl analogue, $[Pt(PPh_3)(\eta$-$C_2H_4)\{\eta$-$H_2CCMeCH_2\}]^+$. The structure of $[Pt(PPh_3)(\eta$-$C_2H_4)(\eta$-$C_5H_5)]^+$ in the solid state (verified by x-ray diffractometry) shows the alkene perpendicular to the plane defined by the platinum and phosphorus atoms and the Cp centroid.[101] Similarly, the structures of both isomers, $[Pt(PPh_3)(\eta^2$-(Z)-$C_4H_8)\{\eta$-$H_2CCHCHMe\}]^+$ and $[Pt(PPh_3)(\eta^2$-(E)-$C_4H_8)\{\eta,\eta$-$H_2CCMeCH_2\}]^+$, have also been confirmed by x-ray diffractometry and reveal that, for the (Z)-isomer, the alkene C=C vector lies in the Pt(allyl)P plane, but lies 67° out-of-plane in the (*E*)-isomer. Other alkenes coordinated to the same platinum fragment show wide variations in their orientation relative to the coordination plane: C_2H_4, 0°; $H_2C=CHPh$, 11.1°; $H_2C=CH$-2-C_6H_4Cl, 24.5°; (*E*)-MeHC=CHPh, 46.1°. These are ascribed primarily to steric demands of the η-alkene ligand. Extended Hückel calculations, modelling the frontier orbitals of the fragment $Pt(PH_3)(C_3H_5)^+$, suggest that bound ethene should show an energetic preference (of about 6.7 kJ mol^{-1}) for the perpendicular (relative to the in-plane) geometry.[67] The related complex, $Pt(C_6F_5)(\eta$-$H_2C=CHPh)\{\eta$-$H_2CC(Me)CH_2\}$, has an η-alkene orientation angle of 10.2°, comparable with its cationic analogue with PPh_3 as the ligand in place of perfluoroaryl.[68] The corresponding angle for $Pt(SnCl_3)(\eta$-$H_2C=CHPh)\{\eta$-$H_2CC(Me)Me\}$ is measured at 20°.[69]

The molecular structures of 16 pentacoordinate complexes of the type PtXY(N–N')(η^2-alkene) have been analysed by x-ray diffraction methods. The salient features of these are collected in Table 2. Structural and stereochemical characteristics based on these and on spectrocopic measurements have been summarized in a review.[72] Coordination geometry is generally trigonal bipyramidal (regarding the alkene as a monodentate group) in which alkene and N–N' are equatorially coplanar, with X and Y axial. Stereoisomerism may arise as a result of asymmetric substitution on the alkene and/or on N–N'. The equatorially bound alkene generates characteristically high field coordination shifts in both ^1H and ^{13}C NMR spectra of such pentacoordinate species. NMR spectroscopy also provides insight into several types of fluxionality associated with these systems, including alkene rotation and configuration inversion at the ligating nitrogen atom, where N–N' is an aliphatic, prochiral diamine.

Table 2 Selected dimensions of the $PtClYC^1C^2N^1N^2$ skeleton in complexes of the type PtCl(Y)(N–N')(η^2-alkene).[a]

Complex	Bond length (nm) *Pt–C¹*	*Pt–C²*	*C¹–C²*	*Pt–N¹*	*Pt–N²*	*Pt–Cl*	*Pt–Y*	*N¹–Pt–N²* (°)
$PtCl_2$(MeHNNCMeCMeNNHMe)(η-C_2H_4)	0.207(4)		0.146(2)	0.222(1)		0.2304(3)		72.0(4)
$PtCl_2$(Me_2NNCHCHNNMe_2)(η-C_2H_4)	0.208(1)	0.212(1)	0.150(2)	0.232(1)	0.225(1)	0.234(1)	0.225(1)	74(2)
$PtCl_2$((*R,R*)-R^1MeNCH_2CH_2NMeR^1)(η-C_2H_4)[b]	0.209(2)		0.157(4)	0.237(1)		0.2319(5)		77(1)
$PtCl_2$((*R,R*)-R^1HNCH_2CH_2NHR^1)(η-C_2H_4)[b]	0.205(2)		0.147(4)	0.227(1)		0.2286(4)		76.9(5)
$PtCl_2$((*R,S*)-R^1HNCH_2CH_2NHR^1)(η-C_2H_4)[b]	0.208(2)	0.205(2)	0.143(4)	0.228(2)	0.225(2)	0.2301(7)	0.2319(5)	78.0(6)
$PtCl_2$(2,9-Me_2phen)(η-C_2H_4)	0.2083(7)		0.141(1)	0.2236(5)		0.2311(3)		74.5(2)
PtCl(SnClPh_2)(2,9-Me_2phen)(η-C_2H_4)	0.207(1)	0.209(1)	0.141(3)	0.221(1)	0.219(1)	0.2478(3)	0.2534(2)	76.3(3)
[$PtCl_2$(2,9-Me_2phen)(η-C_2H_4)]$^+$	0.2075(6)	0.2070(6)	0.140(1)	0.2236(4)	0.2230(4)	0.2304(1)	0.2027(4)	74.8(1)
$PtCl_2$(Me_2NNCHCHNNMe_2)(η-H_2C:CHCHMeEt)	0.211(3)	0.208(3)	0.137(4)	0.225(3)	0.225(3)	0.239(2)	0.224(1)	70(3)
[$PtCl_2$(Me_2NNCHCHNNMe_2)]$_2$(η^2,η^2-C_4H_6)	0.206(1)	0.216(1)	0.144(1)	0.2245(6)	0.2188(6)	0.2295(2)	0.2309(2)	71.0(2)
$PtCl_2$(Me_2NNCHCHNNMe_2)(η-(*E*)-C_4H_6)	0.206(2)	0.209(2)	0.135(2)	0.226(1)	0.222(1)	0.2293(5)	0.2299(5)	69.7(5)
PtClMe(6-Me-py-2-2R^1)(η-C_2H_4)	0.2051(6)	0.2056(6)	0.142(1)	0.2239(5)	0.2188(4)	0.2257(2)	0.2055(6)	74.1(2)
$PtCl_2$(Bu^tNCHCHNBu^t_2)(η-H_2C:CHPh)	0.216(4)	0.220(4)	0.153(5)	0.220(3)	0.231(3)	0.2299(9)	0.2323(9)	74.7(9)
$PtCl_2$(bipy)(η-H_2C:CMeC_6H_4-2-NMe_2)	0.202(7)	0.220(8)	0.146(8)	0.225(4)	0.215(5)	0.232(2)	0.207(5)	77(2)
$PtCl_2$(R^1MeNCH_2CH_2NMeR^1){η-(*E*)-(NC)HC:CH(CN)}[b]	0.2073(5)		0.148(1)	0.2287(4)		0.2314(1)		81.2(2)
$PtCl_2$(2,9-Me_2phen){η-HC:CH(CH_2)$_2$CH(OMe)CH(CH_2)$_2$}	0.2052(7)	0.2075(7)	0.144(1)	0.2249(5)	0.2236(6)	0.2457(2)	0.206(1)	74.4(2)

[a] Data summarized from Albano *et al.*[72] [b] R^1 = CHMePh.

(iii) Reactions

(a) Alkene dissociation–displacement reactions. A kinetic evaluation (by ^{1}H NMR lineshape analysis) of ethene exchange in both the Zeise anion and the dimer indicates that the reaction is first order in free alkene. Estimates for activation parameters for $Pt_2Cl_4(\eta\text{-}C_2H_4)_2$ ($\Delta H^\ddagger = 30 \pm 4$ kJ mol^{-1}, $\Delta S^\ddagger = -23 \pm 20$ J mol^{-1} K^{-1}) and for $[PtCl_3(\eta\text{-}C_2H_4)]^-$ ($\Delta H^\ddagger = 43 \pm 5$ kJ mol^{-1}, $\Delta S^\ddagger = -43 \pm 18$ J mol^{-1} K^{-1}) emerge to explain the slower reaction of the mononuclear anion. At low temperature, *trans*-$PtCl_2(\eta\text{-}C_2H_4)_2$ is formed from $Pt_2Cl_4(\eta\text{-}C_2H_4)_2$. These reactions are largely enthalpy controlled (in contrast with corresponding complexes with acac).[102] From kinetic analyses of the interaction of $[PtCl_n(OH_2)_{4-n}]^{(2-n)-}$ and ethene ($n = 0$–4), it emerges that, for *trans*-$PtCl_2(OH_2)_2$, parallel pathways give *trans*-$PtCl_2(OH_2)(\eta\text{-}C_2H_4)$ and *trans*-$[PtCl(OH_2)_2(\eta\text{-}C_2H_4)]^+$, by rates in the ratio 1:1.5. Alkenes, it is concluded, are relatively inefficient entering groups, comparable in reactivity with DMSO.[103] Exchange of *trans*-1,2-dichloroethene (dce), 2,3-dimethylbut-2-ene (dmb) and 2-methylbut-2-ene (mbn) in optically active complexes *trans*-$PtCl_2L(\eta\text{-}(S)\text{-mbn})$ is observable by following decay of circular dichroism at ca 24 000 cm^{-1} (L = py, 4-CN-py, 4-Cl-py, 4-Me-py, 4-NH_2-py, aniline, 4-chloroaniline or 4-methylaniline). For L = py, the rate increases as the incoming alkene varies in the order dmb < dce < mbn, which is supported by both steric and electronic arguments. Rates for incoming dce, on the other hand, diminish as the +I effect of the *para* substituent on L increases. It is concluded that the σ-character of the Pt–N bond is an important contributor to the *trans* effect of L.[104] Exchange of 3,3-dimethylbut-1-ene by styrene in the related system, *trans*-$PtCl_2(py'')(\eta^2\text{-alkene})$, is catalysed by intramolecular formation of an *O,N*-chelate interaction (py" = 2-(2-hydroxyethyl)-6-methylpyridine or 2-(2-hydroxypropyl)-6-methylpyridine).[105] Dynamic NMR measurements lead to estimations of varying activation parameters ($\Delta H^\ddagger = 54$–75 kJ mol^{-1} and $\Delta S^\ddagger = -12$–38 J mol^{-1} K^{-1}) for alkene dissociation in a series of 4-substituted pyridine-*N*-oxide complexes, *trans*-$PtCl_2(ONC_5H_4\text{-4-Y})(\eta^2\text{-alkene})$ (alkene = 4-methylbut-1-ene; Y = H, Me, CO_2Me, Cl or NO_2).[106] Alkene dissociation also has a crucial role in the *trans–cis* isomerization of *trans*-$PtX_2(PR_3)(\eta^2\text{-alkene})$ (X = Cl or Br; L = PBu_3 or PMe_2Ph; alkene = ethene, propene or hept-1-ene). The *trans* isomers lose alkene in a photochemically influenced equilibrium with the corresponding dimer, from which the more stable *cis* isomers are formed. Exchange of free alkene is faster in every case than isomerization, which is retarded by the presence of free alkene for X = Cl, but slightly accelerated for X = Br. A dissociative path appears to predominate for chloro complexes, while an associative route is more important for bromo analogues.[107]

Equilibrium measurements for the reaction of $[PtL_n(NCPh)]^+$ with ethene to yield corresponding ethene complexes $[PtL_n(\eta\text{-}C_2H_4)]^+$ reveal equilibrium constants (K) which vary in the order: PtL_n = $[Pt(PPh_3)\{\eta\text{-}H_2CCH(Me)CH_2\}]^+$ > $[Pt(PPh_3)Cp]^+$. For reactions of 4-substituted styrenes, 4-Y-$C_6H_4CH{=}CH_2$, with $[Pt(PPh_3)_2Cp]^+$, the observed values of K increase with increasing electron-donating capacity of Y (Y = H, Me, OMe, NMe_2, Cl or NO_2).[101] $[Pt(PPh_3)(NCPh)\{\eta\text{-}H_2CCH(Me)CH_2\}]^+$ shows similar reactivity with the same series of 4-substituted styrenes, $H_2C{=}CH\text{-4-}C_6H_4Y$. A linear correlation of K with Hammett parameters emerges, with $\rho = -1.32$, indicating the importance to bonding of σ-donation from the alkene to platinum.[62] A related examination of equilibration of the *cis*- and *trans*-but-2-ene analogues, $[Pt(PPh_3)(\eta^2\text{-}Z\text{-}C_4H_8)\{\eta\text{-}H_2CCMeCH_2\}]^+$ and $[Pt(PPh_3)(\eta^2\text{-}(E)\text{-}C_4H_8)\{\eta\text{-}H_2CCMeCH_2\}]^+$, for which $K = 0.5$, indicates a reversal of the normal stability trend (*cis* > *trans*) for metal–alkene coordination.[69] The 2-chloro- and 2-methylstyrene derivatives in the series $[Pt(PPh_3)(\eta\text{-}H_2C{=}CHC_6H_4Y)\{\eta\text{-}H_2CCMeCH_2\}]^+$ exhibit lower solution stabilities than their 3- and 4-substituted analogues—as a result, it is proposed, of in-plane steric repulsion. A reversed stability order is observed in complexes where the coordinated alkene lies perpendicular to the coordination plane (see Section 9.2.1.2(i)(c)).[69] Stability trends in neutral complexes, $PtCl(\eta\text{-}H_2C{=}CHR)\{\eta\text{-}H_2CC(Me)CH_2\}$ (R = Me, Et, Bz or 4-C_6H_4Y; Y = H, Me, Cl, OMe, NMe_2 or NO_2) and cationic relatives, $[Pt(PPh_3)(\eta\text{-}H_2C{=}CHR)\{\eta\text{-}H_2CC(Me)CH_2\}]ClO_4$ (R = Me, Et or Bz), indicate greater electrophilic activation of a given alkene in cationic derivatives.[66] For a related family of styrene complexes, $Pt(C_6F_5)(\eta\text{-}H_2C{=}CHC_6H_4Y)\{\eta\text{-}H_2CC(Me)CH_2\}$ (Y = H, 4-Me, 4-Cl, 4-OMe or 3-NO_2), the platinum–alkene bond strength is only weakly dependent on the electronic properties of Y. For equilibration with the homologue where Y = H, K does not deviate appreciably from unity and, from a Hammett correlation, $\rho = -0.38$.[68] Similarly, the solution stabilities of styrene complexes, $PtCl_2(py)(\eta\text{-}H_2C{=}CHC_6H_4Y)$, exhibit higher stability for the more electron-donating alkenes and show good correlation with the Hammett LFER, $\rho = -0.82$. These data are in accordance with the greater contribution from σ-donation than π-retrodonation to η-alkene–platinum bonding in these systems,[108] and may be compared with those for $[Pt(PPh_3)(\eta\text{-}H_2C{=}CH\text{-4-}C_6H_4Y)\{\eta\text{-}H_2CCH(Me)CH_2\}]^+$, where $\rho = -1.32$.[62]

An overview of applications of two-dimensional NOESY NMR techniques includes an analysis of the fluxional behaviour of the styrene complex, $[Pt(SnCl_3)(\eta\text{-}C_2H_3Ph)\{\eta\text{-}H_2CC(Me)CH_2\}]$. This suggests that $SnCl_3$ and styrene exchange processes are operating, but that there is no η^1,η^3-isomerization of the methallyl ligand.[109]

Extensive studies of alkene dissociation from PtXY(N–N')(η^2-alkene) illuminate the energetics of pentacoordination (relative to tetracoordination) for platinum(II) systems of this type. For many such derivatives, alkene loss is a spontaneous and (effectively) irreversible process, while for numerous others—notably those where there is appreciable steric and/or electronic destabilization of the four-coordinate geometry—equilibria are observable. These results (including pK values) have been tabulated in a recent survey of the topic.[72] There is evidence that alkene dislocation in these species may sometimes follow a stepwise route, via preliminary loss of X^- from the inner coordination sphere to form planar tetracoordinate [PtY(N–N')(η^2-alkene)]X, which, in turn, undergoes alkene displacement by X^- to give PtXY(N–N').

(b) Nucleophilic addition. General aspects of this reactivity, towards cyclic π-hydrocarbons—including attack of amine nucleophiles on coordinated cycloocta-1,5-diene—have been surveyed comprehensively.[110,111] They have also been the focus of theoretical studies (see Section 9.2.1.3).

4-Methylpyridine (py') adds first to the coordinated ethene in *cis*-$PtCl_2$(py)(η-C_2H_4) to afford *cis*-$PtCl_2$(py)($CH_2CH_2NC_5H_4$-4-Me). This subsequently undergoes chlorine displacement by a further molecule of 4-methylpyridine to yield *cis*-[PtCl(py)(py')($CH_2CH_2NC_5H_4$-4-Me)]Cl. There is no evidence for initial formation of *cis*-[PtCl(py)(py')(η-C_2H_4)]Cl.[112] Attack by $NHMe_2$ on the alkene in *trans*-$PtCl_2$($NHMe_2$)(η-$C_2H_3R^1$) gives [$PtCl_2$($CH_2CHR^1NHMe_2$)($NHMe_2$)] by a Markownikoff pathway, unless the steric hindrance of R^1 is too great (R^1 = Me or Pr^n). Equilibrium constants for formation are sterically determined.[113] Similarly, addition of amine (am) (am = NH_2R^1 or NHR^2_2; R^1, R^2 (variously) = Me, Pr^n, Bu^n, Bu^s, Bu^t, C_5H_{11}, C_7H_{13}, CH_2Ph, etc., morpholine or piperidine) to *cis*-$PtCl_2$(Y)(η-C_2H_4) yields, sequentially, neutral acyclic *cis*-$PtCl_2$(CH_2CH_2am)(Y) (Y = Me_2SO, PPh_3 or $NHMe_2$) and cationic, acyclic [*cis*-PtCl(CH_2CH_2am)(am)(Y)]Cl (Y = Me_2SO, NH_3 or py). With bulky R^1 groups, deprotonation—by excess NHR^2_2—occurs, to afford cyclometallated amine species, *cis*-Pt($CH_2CH_2NR^2_2$)Cl(Y) (Y = Me_2SO or PPh_3; R^1 = Me).[114]

A cyclometallated product is also the outcome of reaction of the cationic substrate, [PtCl(TMEDA)(η-C_2H_4)]$^+$, with NCO^- to form Pt{CH_2CH_2NHC(=O)}Cl(NCO)(TMEDA), as a result of nucleophilic addition of cyanate to the coordinated ethene, followed by hydrogen abstraction from solvent. An x-ray diffraction analysis has confirmed this formulation. Nucleophilic attack is also observed upon addition of NO_2^- and N_3^- to give acyclic products Pt($CH_2CH_2NO_2$)(NO_2)(TMEDA) and PtCl($CH_2CH_2N_3$)(TMEDA). CN^- merely displaces Cl^-.[115,116] A related system, pentacoordinate $PtCl_2$(TMEDA)(η-C_2H_4), which is stable at low temperatures, spontaneously reverts to tetracoordination at 0 °C. The resulting system, [PtCl(TMEDA)(η-C_2H_4)]Cl, is then subject to a double nucleophilic attack by TMEDA to yield [(TMEDA)ClPt(CH_2)$_2$$NMe_2$($CH_2$)$_2$$NMe_2$($CH_2$)$_2$PtCl(TMEDA)]$^{2+}$.[117,118] A related double attack attends the reaction of [PtCl(TMEDA)(η-C_2H_4)]$^+$ with $NHEt_2$, which leads to [(TMEDA)ClPtCH_2CH_2N(Et)$_2$$CH_2CH_2$PtCl(TMEDA)]$^+$, the structure of which has been confirmed by a diffractometric study. A variety of secondary and primary amines, as well as NH_3, react with [PtCl(TMEDA)(η-C_2H_4)]$^+$ to afford di-, tri- and tetrametallated ammonium ions, [R_{4-n}N{PtCl(TMEDA)(η-C_2H_4)}$_n$]$^+$ (n = 2, 3 and 4, respectively). The enhanced electrophilicity of η-C_2H_4 in these systems and the absence of a formal negative charge on platinum in the adducts (stabilizing the Pt–C σ-bond) are regarded as crucial to the success of these syntheses.[119] The same cationic platinum substrate also reacts with 'stabilized' carbanionic nucleophiles, L^-, to give PtCl(CH_2CH_2L)(TMEDA) (HL = pentane-2,4-dione, ethylacetoacetate, diethylmalonate). Subsequent reaction with HCl does not reverse C–C formation but cleaves the Pt–C bond instead, releasing $MeCH_2L$ and $PtCl_2$(TMEDA).[120]

The Zeise anion reacts with Bu^tNO in an unusual addition, which yields Pt{CH_2CH=N(Bu^t)O}Cl(N(O)Bu^t), after elimination of HCl. Attack by free Bu^tNO is considered unlikely, since nitrosoalkanes do not normally behave as nucleophiles. More probably, reaction occurs directly between the two ligating species, both of which are activated by platinum coordination.[121]

(c) Migratory alkene insertion. Allyl complexes, [Pt(PPh_3)$_2$($R^1_2CCR^2CHMe$)]$^+$, are formed as a result of rate-limiting, migratory alkene insertion, when *trans*-Pt(CR^2=CR^1_2)(OSO_2CF_3)(PPh_3)$_2$ is exposed to ethene (R^1 = H, R^2 = Me; R^1 = Me, R^2 = H; R^1 = H, R^2 = Bu^t).[53] Likewise, treatment of *trans*-Pt(CMe=CH_2)(OSO_2CF_3)(PPh_3)$_2$ with alkylacrylates yields, ultimately, metallacyclic [Pt{CH_2CH_2C(OR)=O}(PPh_3)$_2$]$^+$, via a succession of insertion, hydrogen transfer and elimination steps (Scheme 6; R = Me or Et).[122]

An unusual example of reversible insertion in this category is provided by the thermolytic reaction of [Pt(η-H_2C=$CHCMe_2CH_2$)(dmpe)]$^+$. Thermolysis of the γ-d_2 isomer generates a 1:1 mixture of the γ-d_2 and α-d_2 isomers. This represents a β-alkyl transfer option for the 3,3-dimethylcyclobutylmethylplatinum intermediate (formed by migratory insertion) with which available β-hydrogen migration does not compete (Equation (8)). In another rare example of β-alkyl migration, [Pt(η-H_2C=$CHCH_2CH_2$)(dmpe)]$^+$ is the major thermolytic product of the corresponding cyclobutylmethyl derivative, [Pt(CH_2-c-C_4H_7)(dmpe)]$^+$.[123]

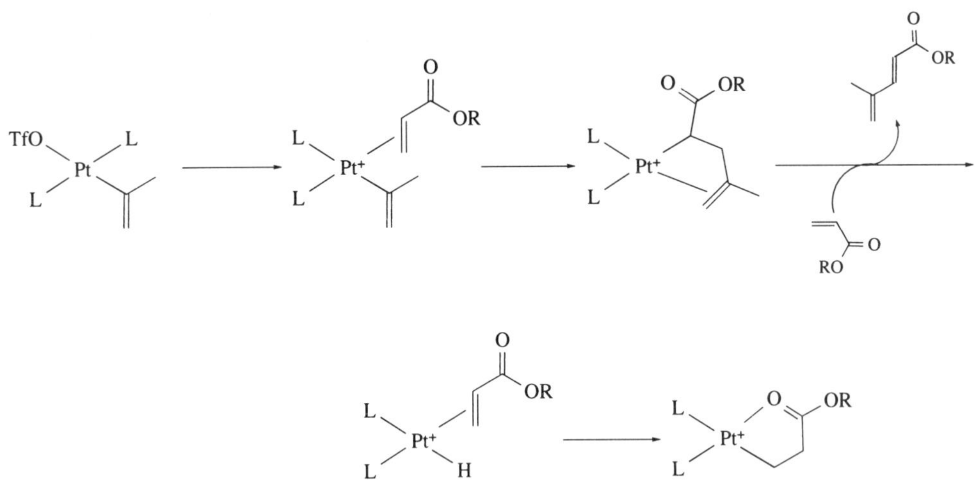

Scheme 6

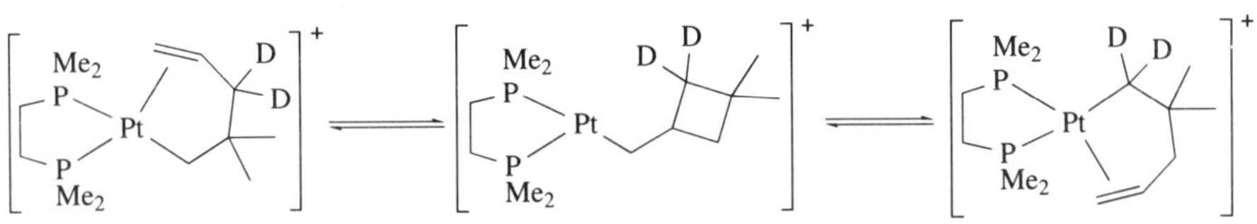

(8)

The pentacoordinate derivative PtPh(NCMe)(N–N')(η-C_2H_4) in the presence of ethene and chloride ion undergoes migratory insertion to afford Pt(CH_2CH_2Ph)(NCMe)(N–N')(η-C_2H_4). In the absence of ethene, a different insertion product, Pt(C_6H_4-2-CH_2Me)(NCMe)(N–N'), is formed instead (N–N' = 6-Me-py-2-CH=NPh).[72]

(d) Oxidative addition at platinum. As is the case for zerovalent platinum systems, oxidative C–C addition requires the impetus of a strained ring. Addition of cyclopropyl derivatives to platinum(II) centres to give platinacyclobutanes or allylplatinum complexes has been featured in a comprehensive review.[124] A series of more recent studies documents the reactions of the Zeise dimer with a range of cyclopropanes, $C_3H_3R^1R^2R^3$, ultimately to give platinacyclobutanes, Pt(η^1,η^1-$C_3H_3R^1R^2R^3$)Cl_2L_2 with interesting reactive characteristics. Where L = MeCN, for example, thermolytic rearrangement affords alkene complexes, *trans*-$PtCl_2$(NCMe)(η-$C_3H_3R^1R^2R^3$) (R^1, R^2, R^3 = (variously) H, Me or Ph).[125,126] Notable results emerge from related reactions of polycyclic species. Reaction of bicyclo[1.1.0]butane with the Zeise dimer, followed by two equivalents of pyridine, yields the platinabicyclo[1.1.1]pentane, as a consequence of oxidative insertion of platinum into the transannular C–C bond (Scheme 7).[127] Similar reaction of norbornane- and norbornene-derived cyclopropanes gives complexes in which the resulting platinacyclobutane ring remains *syn* to the methylene bridge in the hydrocarbon skeleton.[128] A similar sequence of reactions with bicyclo[4.1.0.]heptane or bicyclo[6.1.0]nonane also affords comparable metallacycles (Scheme 7). With the latter hydrocarbon, addition of only one equivalent of pyridine leads instead to the *exo*-methylenecyclooctane complex, *trans*-$PtCl_2$(py)[η^2-H_2C=CH(CH_2)CH_2}, which rearranges slowly to the corresponding η^2-1-methylcyclooctene analogue (Scheme 7).[129] An interesting variant on this general pattern of reactivity emerges from reaction of bicycloalkenes, such as bicyclo[4.1.0]hept-2-ene (Equation (9)). Here, the reaction is both regio- and stereospecific; the nucleophile arrives at C-3 *exo* to the metal to yield a transoid product. Similar results are obtained with the corresponding cyclopentene and cycloheptene. The cyclooctene gave only the dimeric η^2-complex, $Pt_2Cl_4(\eta^2$-$C_9H_{14})_2$. This, however, reacts with bipy to give a platinacyclobutane.[130] The Zeise dimer reacts with 7-methyl-1-methoxybicyclo[4.1.0.]hept-3-ene to give, after treatment with pyridine, *cis*-N,Cl-Pt(η^2,η^1-$C_8H_{10}O$)(Cl)(py) (Equation (10)). Only one diastereomer forms, and it appears that the primary process is oxidative insertion of the metal into the MeC–COMe bond of the tetrasubstituted cyclopropane, the first such insertion to be observed.[131]

Unusual examples of (effectively) oxidative addition of the C–H bond in ethene emerge when $Pt(PPh_3)_2$(η-C_2H_4) interacts with triflato derivatives of iridium and rhodium. With Ir(OSO_2CF_3)(CO)$(PPh_3)_2$, the binuclear species $(Ph_3P)_2$Pt(μ-H)(μ,η-H_2C=CH)Ir(CO)$(PPh_3)_2$ is produced. As confirmed by an x-ray diffraction structural analysis, the bridging ethenyl group is η^1-bound to platinum and η^2-bound to iridium.[132] When reacted with Rh$(OSO_2CF_3)_2$(PMe_3)Cp*, an analogously dimeric, cationic product, [$(Ph_3P)_2$Pt(μ-H)(μ,η-H_2C=CH)Rh(PMe_3)Cp*]$(OSO_2CF_3)_2$, is recovered.[133]

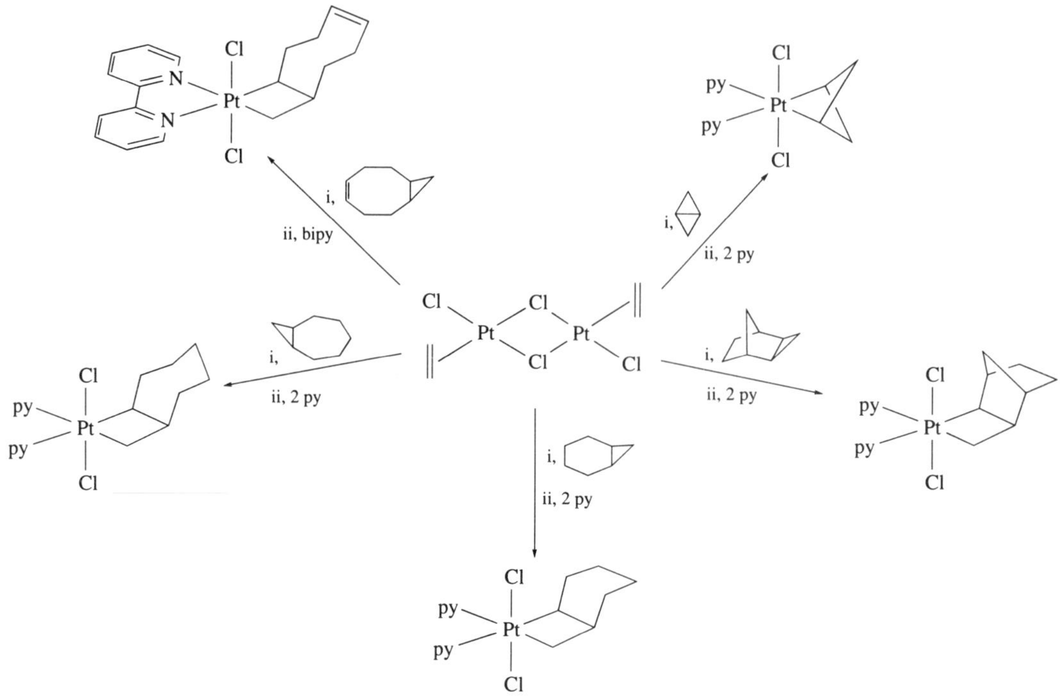

Scheme 7

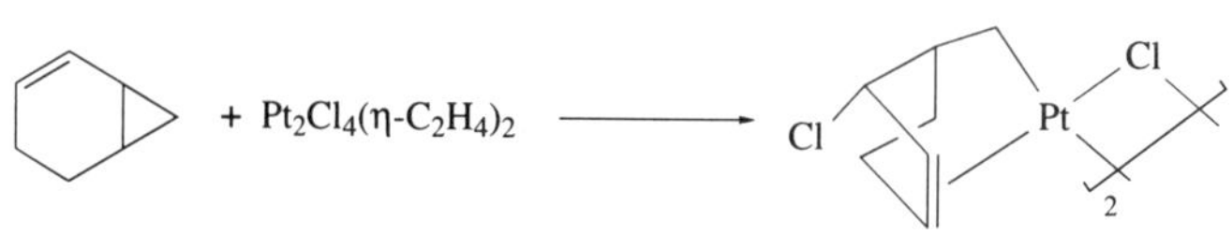

(9)

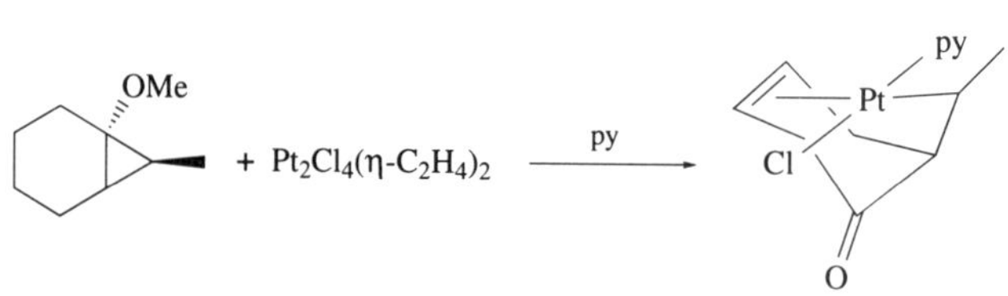

(10)

9.2.1.3 Theoretical aspects

By their nature, calculations aimed at a better theoretical understanding of both structure and reactivity of π-hydrocarbon complexes of platinum are limited, effectively, to the simplest systems—mainly η^2-ethene complexes. Hartree–Fock SCF calculations of optimized geometries for $Pt(\eta\text{-}C_2H_4)_3$ and $Pt(\eta\text{-}C_2H_4)_2(\eta\text{-}C_2F_4)$ suggest a trigonal coplanar arrangement of the alkenes, with C=C lengthening and angular modifications consistent with conventional views of bonding.[46] Comparative Hartree–Fock–Slater molecular orbital calculations on zerovalent complexes of the type $M(\eta\text{-}L)(PH_3)_2$ ($L = C_2H_4$, C_2H_2 or O_2; M = Pt, Pd or Ni) has been used to assess the relative importance of retrodative binding in the Chatt–Duncanson–Dewar model. Binding energies emerge in the orders $O_2 > C_2H_2 \approx C_2H_4$ and nickel > palladium > platinum.[134] Relativistic Hartree–Fock *ab initio* calculations on $Pt(\eta\text{-}C_2H_4)(PH_3)_2$ and the corresponding complex of the highly pyramidalized alkene, tricyclo[3.3.1.$0^{3,7}$]non-3(7)-ene[35] (see Section 9.2.1.1(i)(b)), have also been carried out. These calculations predict a higher barrier to alkene rotation as well as a higher alkene binding energy for constrained alkenes of this type, consistent with increasing π-acidity of pyramidalized alkenes.[135]

Ab initio calculations on the Zeise anion also continue to provide insight into this, the earliest example. Calculated structural parameters are in agreement with experimental observations, in that the ratio of alkene binding energy (117 kJ mol^{-1}) to the rotational barrier (63 kJ mol^{-1}) would indicate rotation without dissociation, as is observed. Rotational barriers and binding energies increase in the order palladium < platinum.[136] A later extended Hückel study of $[MCl_3(\eta\text{-}C_2H_4)]^-$ (M = Pt or Pd) explains the differing capacities of the two metals for *cis–trans* isomerization of alkenes. In this case, the C=C torsional barrier is predicted to undergo a greater reduction (31%) upon palladium(II) coordination than occurs in corresponding platinum(II) complexes.[137] Alkene activation towards nucleophilic addition as a result of palladium coordination has also been the subject of an *ab initio* Hartree–Fock molecular orbital calculation. These results indicate heightened susceptibility of cationic alkene–metal complexes, relative to neutral or anionic congeners.[138] This same reactivity, for a variety of *d*-configurations (including d^8 and d^{10}), has been analysed generally by means of extended Hückel methods.[139] These suggest that a 'slippage' mechanism is likely to play a governing role, although INDO calculations on the same systems do not require such a path.[140]

Zeise's anion and Pt(η-C_2H_4)L_2 also feature in a more broad review of molecular orbital theory in interpretation of organometallic reactions.[141] In addition, the general energy correlations between M(η-C_2H_4) and M(η^1-CH_2)$_2$—as well as those between M(η-C_2H_2) and M(η^1-CH)$_2$—have been the subject of an extended Hückel treatment.[142]

9.2.2 η^2-Monoalkene Complexes (Polydentate)

In this case, in addition to a single (η-C_2)platinum interaction, the alkenic ligand is also linked to the metal via one (or more) other ligating atoms. Complexes of this general type have been reviewed.[143]

9.2.2.1 Zerovalent derivatives: preparation, structure and reactivity

Preparations of zerovalent platinum derivatives of this type parallel those for nonchelating alkene ligands. Pt(cod)$_2$ has been used to generate complexes with both η^2-alkene and phosphine ligands. Treatment with two equivalents of 2-styryldiphenylphosphine (Scheme 8) affords Pt(η-C_2H_3-2-$C_6H_4PPh_2$)$_2$, in which both ligands are bonded via η^2-C_2- and *P*-coordination (in contrast to the palladium complex of the same stoichiometry).[144] Reaction with only one equivalent of 2-styryldiphenylphosphine, in addition to two of triphenylphosphine (Scheme 8), yields the related species Pt(η-C_2H_3-2-$C_6H_4PPh_2$)(PPh_3)$_2$.[145] When dppe is included instead of PPh_3, however, only Pt(η-C_2H_3-2-$C_6H_4PPh_2$)$_2$ and Pt(dppe)$_2$ are recovered.

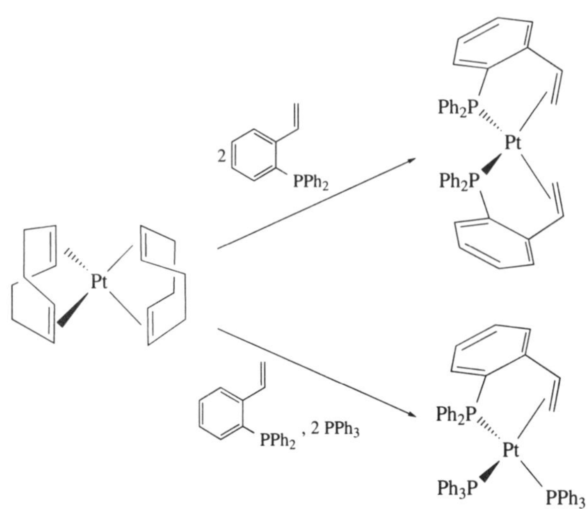

Scheme 8

9.2.2.2 Divalent derivatives: preparation, structure and reactivity

(i) Group 14 donors

In the first example of such an insertion reaction, platinum(II) (in the Zeise dimer) reacts with a tetrasubstituted cyclopropane, 7-methyl-1-methoxybicyclo[4.1.0.]hept-3-ene, to give (subsequent to addition of pyridine) *cis*-N,Cl-Pt(η^2,η^1-$C_8H_{10}O$)(Cl)(py) (Equation (10)). This structure has been confirmed in an x-ray diffraction study, showing the alkene perpendicular to the tetragonal coordination plane.[131] Related regio- and stereospecific reactivity emerges from reaction of bicycloalkenes, such as bicyclo[4.1.0.]hept-3-ene and the corresponding cyclopentene and cycloheptene (see Section 9.2.1.2(iii)(d)).[130] In a rare example of β-alkyl migration, [Pt(η-H_2C=$CHCH_2CH_2$)(dmpe)]$^+$ is the major thermolytic product of the corresponding cyclobutylmethyl derivative, [Pt($CH_2C_4H_7$)(dmpe)]$^+$. Other members of the same family—for instance, [Pt(η-H_2C=$CHCH_2CH_2$)(dmpe)]$^+$—are prepared from corresponding $PtXRL_2$, by halide abstraction with Ag^+.[123] Reaction of Pt(CH_2SiMe_2CH=CH_2)$_2$(cod) with one equivalent of a monodentate ligand, L, generates Pt(η-H_2C=$CHSiMe_2CH_2$)(CH_2SiMe_2CH=CH_2)(L), in which one vinylsilylmethyl group is η^1,η^2-bound to the metal (L = py, PMe_3, PEt_3, $PMePh_2$, PPh_3 or PCy_3).[146]

Some of the earliest derivatives in this category were the η^2,η^1-cyclooctenylplatinum complexes formed by nucleophilic attack (by methoxide, for example) on the η^2,η^2-cycloocta-1,5-diene parent species. This family still continues to provide interesting examples, as in the isomeric complexes formed on reaction of [Pt(η^2,η^1-$C_8H_{12}OMe$)(Cl)]$_2$ with HC(PPh_2)[P(S)Ph_2]$_2$ (Equation (11)).[147]

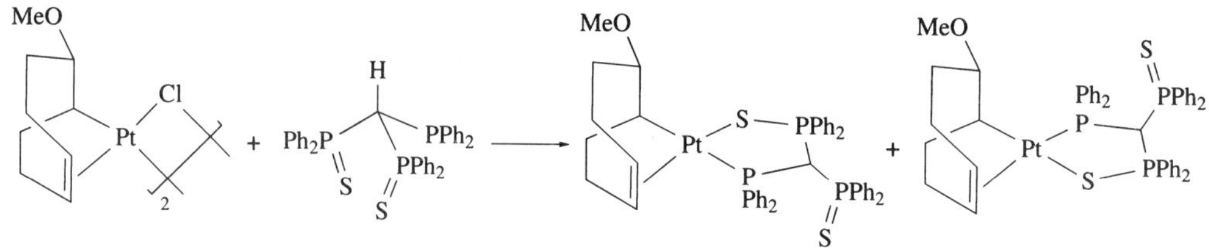

(11)

(ii) Group 15 donors

Reaction of 2-aminostyrene and related ligands with $PtCl_2$, $PtBr_2$ or (better) the Zeise anion, affords complexes of the general type Pt(η-H_2C=CR^1-2-$C_6H_4NR^2_2$)X_2 (R^1 = H, R^2 = H, Me or Ph; R^1 = Me, R^2 = H, Me or Bz; R^1 = Ph, R^2 = Me). All, except R^1 = H, R^2 = Ph, adopt the η^2,*N*-chelate configuration. Congeners Pt(η-H_2C=CR^1-2-$C_6H_4NR^2_2$)ClX are accessible by treatment of the appropriate dichloro derivative with NaX (X = Br, I, N_3 or SCN; R^2 = R^1 = Me). From measurements of ^2J(Pt–H_{alkene}), a *trans* influence series emerges: $N_3^- > NCO^- > SCN^- > Cl^- > Br \approx I^-$.[148] Related amido-bridged dimeric complexes can be obtained by deprotonation, using an anion-exchange resin. The structure of Pt_2Cl_2{η-*N*-2-C_6H_4-CH=CH_2}$_2$, solved by diffraction methods, exhibits alkene coordination not quite perpendicular to the tetragonal platinum plane, as a result of steric strain.[149]

An unusual approach to phosphorus bidentate monoalkene complexes is exemplified by the reaction of a cyclopropylphosphonium salt with K_2PtCl_4 (Equation (12)).[150]

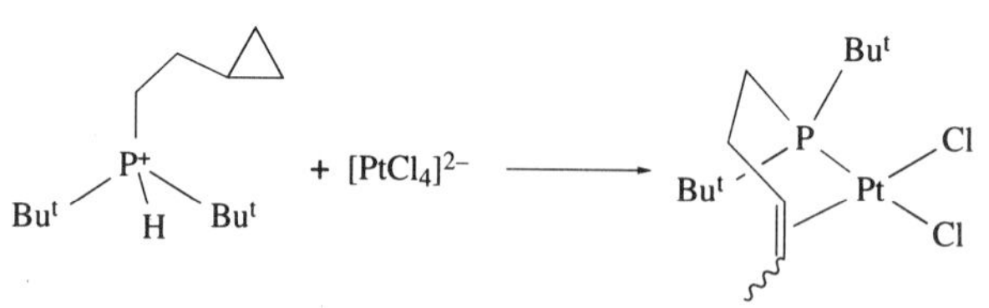

(12)

The cyclometallated η-isopropenyl group in [Pt(η^2-H_2C=CMe-2-$C_6H_4AsPh_2$)(acac)]$^+$ does not undergo the usual 1,2-addition when treated with potential nucleophiles, NEt_3, OH^- or $MeCO_2^-$. The product of these reactions is the new platinacyclic complex, Pt{CH_2C(=CH_2)-2-$C_6H_4AsPh_2$}(acac), which isomerizes when treated with $MeCO_2H$, yielding Pt{CH=C(Me)-2-$C_6H_4AsPh_2$}(acac), as verified by x-ray diffraction. Here the incumbent species act as bases rather than nucleophiles, first deprotonating the methyl substituent of the η-isopropenyl group.[151] In contrast, addition of methoxide to the cyclometallated η-vinyl substituent in a related system, $PtCl_2$(η^2-H_2C=CH-2-$C_6H_4AsPh_2$), does proceed in the conventional manner, as has been shown by a diffractometric study of the hexafluoropentane-2,4-dionato (acac") derivative, Pt{CH(CH_2OMe)-2-$C_6H_4AsPh_2$}(acac"). The structure of the parent complex has also been verified by x-ray diffraction.[152]

(iii) Group 16 donors

Treatment of $PtCl_4^{2-}$ with thiacyclooct-4-ene yields $PtCl_2(C_7H_{12}S)$, featuring both η^2-alkene and η^1-*S* coordination.[153] A range of compounds with polyfunctional alkenic ligands, of types $PtX_2[EMe(CH_2)_2CH{=}CH_2]$, $PtX_2\{E[(CH_2)_nCR{=}CR_2]_2\}$,[154] $[PtI\{S[(CH_2)_2CMe{=}CH_2]\}]_2$,[155] and $PtMe_2\{E[(CH_2)_2CH{=}CH_2]_2\}$[156] have been synthesized (E = S or Se; X = Cl, Br or I; R = H or Me; *n* = 2 or 3), which also display η^1-S or η^1-Se binding in addition to one η^2-alkene interaction with platinum. NMR studies show several types of dynamic behaviour for these compounds, including exchange of free and coordinated alkene functions and configuration inversion at sulfur and selenium. $PtX_2[EMe(CH_2)_2CH{=}CH_2]$ undergoes a remarkably facile dealkylation of E induced by addition of Br^- or I^-, leading to the sulfido- or selenido-bridged dimer, $Pt_2Cl_2\{\mu,\eta\text{-}ECH_2CH_2CMe{=}CH_2\}_2$. The x-ray structure shows a *syn* arrangement of the chelate rings and a Pt_2E_2 ring puckered about the (nonbonded) E---E vector.

9.2.2.3 Tetravalent derivatives

Alkene coordination to platinum(IV) remains relatively rare. An unusual exception is achieved by oxidative insertion of one platinum(II) centre of a binuclear analogue of the Zeise dimer into the cyclopropane function of η^2-bicyclo[6,1,0]non-4-ene (see Section 9.2.7.1).

9.2.3 η^n-Polyene Complexes (Unconjugated)

9.2.3.1 Zerovalent derivatives

Like (η^2-monoalkene)platinum(0) complexes, ancillary ligands are generally other unsaturated hydrocarbon fragments or group 15 donor ligands. Most common are three- (16-electron) or four-coordinate (18-electron) structures. The synthesis and reactivity of $Pt(cod)_2$ and related complexes have been reviewed comprehensively.[157–9]

(i) Preparation

A standard preparative procedure for $Pt(cod)_2$—via $Pt(nbd)_3$, from reduction of $PtCl_2(cod)$ by $Li_2(cot)$—has been reported.[12] This compound, as noted, is a useful precursor to other platinum(0) derivatives, usually by displacement of both cycloocta-1,5-diene ligands. An exception is its reaction with benzo-1,4-quinones, forming either Pt(cod)(η^2,η^2-quinone) (quinone = benzo-1,4-quinone, 2,3,5,6-tetramethylbenzo-1,4-quinone, 2,6-dimethylbenzo-1,4-quinone, 2,6-di(*t*-butyl)benzo-1,4-quinone, 2,5-dimethylbenzo-1,4-quinone or 2,5-diphenylbenzo-1,4-quinone) or Pt(cod)(η^2-quinone) (quinone = 2-methylbenzo-1,4-quinone or naphtho-1,4-quinone).[160]

Reaction of $Pt(PPh_3)_2(\eta\text{-}C_2H_4)$ with tetramethyldivinyldisiloxane affords $Pt(PPh_3)\{[\eta\text{-}H_2C{=}CHSi(Me)_2]_2O\}$.[161] A related, homoleptic derivative, $Pt_2\{[\eta\text{-}H_2C{=}CHSi(Me)_2]_2O\}_3$, has also been prepared by treatment of $Pt(\eta^4\text{-cod})_2$ with the same divinyldisiloxane. This, in turn, reacts with maleic anhydride to give $Pt\{[\eta\text{-}H_2C{=}CHSi(Me)_2]_2O\}(\eta^2\text{-}C_2H_2O_3)$.[162] $Pt_2\{[\eta\text{-}H_2C{=}CHSi(Me)_2]_2O\}_3$ and a variety of related zerovalent derivatives seem likely to be the major products of the reaction of $H_2PtCl_6{\cdot}xH_2O$ (Speier's hydrosilation catalyst) with the divinyldisiloxane at 50 °C. Such compounds are precursors to notably active catalysts for alkene hydrosilation. Addition of PR_3 to the product mixture yields $Pt(PR_3)\{[\eta\text{-}H_2C{=}CHSi(Me)_2]_2O\}$ (R = Bu^t or Cy).[163]

(ii) Structural aspects

The molecular structure of $Pt(cod)_2$ has been determined by x-ray diffractometry and displays a quasitetrahedral disposition of the four η^2-alkene groups.[164] A similar double-bond arrangement emerges from the x-ray diffractometric analysis of the structures of two of the series, Pt(cod)(η^2,η^2-quinone) (quinone = 2,3,5,6-tetramethylbenzo-1,4-quinone or 2,6-di(*t*-butyl)benzo-1,4-quinone).[165]

The structures of $Pt(PPh_3)\{[\eta\text{-}H_2C{=}CHSi(Me)_2]_2O\}$,[161] *rac*-$Pt(PBu^t_3)_2\{[\eta\text{-}H_2C{=}CHSi(Me)_2]_2O\}$[166] and of $Pt_2\{[\eta\text{-}H_2C{=}CHSi(Me)_2]_2O\}_3$[167] all indicate that a trigonal coplanar arrangement of double bonds around the metal is favoured, despite the steric constraints imposed upon the chelate diene.

(iii) Reactions

Many of the reactions form the basis of other preparative schemes. In a characteristic reaction the 1,5-diene $H_2C{=}CHCH_2CH{=}CHCO_2Me$ reacts with $Pt(cod)_2$ in the presence of PMe_3 to give the platinacycle, $\overline{Pt\{CH((E)\text{-}CH{=}CHCO_2Me)CH_2CH_2}CH((E)\text{-}CH{=}CHCO_2Me)(PMe_3)_2$.[168] $Pt(cod)_2$ is also a versatile precursor to triangulo-Pt_3 species, many of which are not accessible by other routes. For example, when reacted with CNR and PCy_3, it yields $Pt_3(\mu\text{-CNR})_3(PCy_3)_3$ ($R = C_6H_3\text{-}2,6\text{-}Me_2$).[169]

9.2.3.2 Divalent derivatives

Complexes of the types $PtX_2(\eta^2,\eta^2\text{-diene})$, $PtR_2(\eta^2,\eta^2\text{-diene})$ and $PtXR(\eta^2,\eta^2\text{-diene})$ (X = halide; R = hydrocarbyl or fluorocarbyl) predominate under this heading and—especially the latter two—continue to proliferate, largely because of their wide utility as precursors to other organoplatinum complexes.

(i) Preparation

Preparative routes parallel those for monoalkene derivatives and are not subdivided here. A standard synthesis has been reported for $PtCl_2(cod)$, in which hydrated H_2PtCl_4 is treated with cycloocta-1,5-diene in acetic acid. Similar procedures lead to the η^2,η^2-norbornadiene and η^2,η^2-dicyclopentadiene analogues as well as $Cl_2Pt(\mu\text{-cot})PtCl_2$.[170]

There has been considerable interest in the use of exocyclic alkenes to impose unusual forms of coordination at platinum(II) centres. By reaction of the exocyclic diene 3,7-dimethylenebicyclo[3.3.1]nonane with the Zeise dimer (or appropriate derivative), the unusual pentacoordinate complexes $PtCl_2(\eta^2\text{-alkene})(\eta^2,\eta^2\text{-}C_9H_{16})$ and $PtClMe(\eta\text{-}C_2H_4)(\eta^2,\eta^2\text{-}C_9H_{16})$ can be prepared (alkene = ethene, propene, (*Z*)-but-2-ene, (*E*)-but-2-ene or styrene).[171] The monoexocyclic dienes, 5-methylenecycloheptene (mch) and 5-methylenecyclooctene (mco) have also featured prominently. Treatment of Na_2PtCl_4 with mch affords $PtCl_2(\eta^2,\eta^2\text{-mch})$,[172] while reaction of mco with $Pt_2Cl_4(\eta\text{-}C_2H_4)_2$ yields $PtCl_2(\eta^2,\eta^2\text{-mco})$. The latter product undergoes metathetical replacement of Cl^- on reaction with LiBr, but $PtBr_2(\eta^2,\eta^2\text{-mco})$ is itself labile, dimerizing via dissociation of the exocyclic double bond to give $Pt_2Br_4(\eta^2\text{-mco})_2$.[173] Replacement of halide in these complexes by hydrocarbyl ligands can be achieved using organotin(IV) derivatives as transmetallation agents. Treatment of $PtCl_2(\eta^2,\eta^2\text{-mch})$ with one or two equivalents of $SnPhMe_3$ leads to $PtClPh(\eta^2,\eta^2\text{-mch})$ and $PtPh_2(\eta^2,\eta^2\text{-mch})$, respectively. Corresponding interaction of $PtCl_2(\eta^2,\eta^2\text{-mco})$ with $SnRMe_3$ (R = Me, Ph or 4-Cl-C_6H_4), however, yields the dimers, $Pt_2R_2(\mu\text{-Cl})_2(\eta^2\text{-mco})_2$, again through loss of exocylic coordination.[174] These compounds are notable for their anomalous structural features (see Section 9.2.3.2(ii)) and unusual reactivity, one aspect of which is the readily controllable synthesis of $PtClR(\eta^2,\eta^2\text{-diene})$ and $PtR_2(\eta^2,\eta^2\text{-diene})$ by mere variation in applied quantity of transmetallation reagent.

More generally, while complexes $PtR_2(\eta^2,\eta^2\text{-diene})$ are obtained by halide substitution in $PtX_2(\eta^2,\eta^2\text{-diene})$—usually via an excess organomagnesium or organolithium species—it is rarer that $PtXR(\eta^2,\eta^2\text{-diene})$ can be synthesized reliably by stoichiometric manipulation of the transmetallation source. Frequently, such strategies lead to complex mixtures, although there are some specific exceptions—often with the milder organotin and organomercury reagents. In view of their wide synthetic utility, a selection of compounds from these two classes, whose preparations have been reported since 1980, is collected in Table 3.[123,164,166,167,175–211]

Pentacoordination, though rare, has begun to emerge. Reaction of $PtMe(NCMe)(2,9\text{-}Me_2\text{-phen})(\eta\text{-}C_2H_4)]^+BF_4$ with cycloocta-1,5-diene affords $PtMe(NCMe)(2,9\text{-}Me_2\text{-phen})(\eta^2,\eta^2\text{-}C_8H_{12})]^+BF_4$, in which the diene spans one equatorial and one axial site in the trigonal bipyramid.[212] In contrast, the pentacoordinate complex, $PtCl_2(\eta\text{-}C_2H_4)(\eta^2,\eta^2\text{-}C_9H_{16})$, formed with the exocyclic diene 3,7-dimethylenebicyclo[3.3.1]nonane, has the three alkene substituents coplanar in the equatorial plane of an approximate trigonal bipyramid.[171]

(ii) Structural aspects

Two x-ray diffraction studies on $PtCl_2(cod)$ reveal approximately perpendicular orientations of both C=C bonds to the tetragonal coordination plane. The contention that one ligating alkene shows significant slippage—consistent with enhanced reactivity towards nucleophilic addition—seems less

Table 3 Complexes $PtR^1_2(\eta^4\text{-diene})$, $PtR^1R^2(\eta^4\text{-diene})$, $[PtLR^1(\eta^4\text{-diene})]^+$ and $PtXR^1(\eta^4\text{-diene})$.

Diene	*R^1*	*$R^2/X/L$*	*Ref.*
cod	Me		175
nbd	Me		208
cod	Me	Guanosine	176
cod	Me	Cytidine	176
cod	Me	Adenosine	176
cod	Me	MeOH	176
cod	Et		177
cod	CH_2Bu^t		178
nbd	CH_2Bu^t		209
cod	CH_2CMe_2Et		179
cod	$CH_2CMe_2Pr^n$		179
cod	$CH_2CMe_2CH_2Bu^t$		179
cod	$CH_2CH_2Bu^t$		179
cod	$CH_2C(Me)CH_2CH_2CH_2$		123
cod	CH_2-1-adamantyl		180
cod	CH_2CMe_2Ph		181
cod	CH_2TMS		182
nbd	CH_2TMS		182
cod	$CH_2SiMe_2CH{=}CH_2$		146a
nbd	$CH_2SiMe_2CH{=}CH_2$		146a
cod	$CH_2SiMe_2C_6H_4R^2$		165
nbd	$CH_2SiMe_2C_6H_4R^2$		165
nbd	Me	CF_3	167,[a]183
nbd	CF_3		167,[a]183
cod	CH_2Cl		166
cod	CH_2Br		166
cod	CH_2I		184
cod	CH_2OH		184
cod	$CH_2OC(O)CF_3$		184
cod	CH_2OCCH_2		184
cod	$CH_2CMe_2CH_2$		185
cod	$CH_2CMe_2CMe_2CH_2$		179
cod	$CH_2C_6H_4$-2-CH_2		186
cod	$CH_2CH{=}CH_2$		187
cod	$CH_2CH{=}CHPh$		187
cod	$CH_2CH{=}CHPh$	η^1-Cp	188
cod	CH_2Ph	Me	189
cod	C_6H_2-2,4,6-Me_3	Me	189
cod	Me	η^1-Cp	190[a]
cod	Me	η^1-7-$C_7H_5(CF_3)_2$	191[a]
cod	Ph	η^1-Cp	192
cod	Ph		193–4,197
cod	C_6H_4-4-Bu^t		193–4
cod	C_6H_4-4-CF_3		193–4
cod	C_6H_4-4-Me		194
cod	C_6H_4-4-NMe_2		194
cod	C_6H_4-4-F		194,197
cod	C_6H_4-4-OMe		194
cod	C_6H_4-3-F		194
cod	C_6H_3-2,6-Me_2		195
cod	C_6H_3-2,6-$(OMe)_2$		195
cod	C_6H_3-2,6-$(CF_3)_2$		195
cod	C_6H_2-2,4,6-$(OMe)_3$		195
cod	C_6H_2-2,4-Me_2-6-OMe		195–6
cod	C_6H_2-2,4-Me_2-6-OEt		195–6
cod	C_6H_2-2,4-Me_2-6-OPr^i		196
nbd	Ph		197
nbd	C_6H_4-4-F		197
Hexa-1,5-diene	Ph		197
Hexa-1,5-diene	C_6H_4-4-F		197
cod	C_6F_4-4-H		197
cod	2,2'-C_6H_4-C_6H_4		198
cod	4,4'-$(Bu^t)_2$-2,2'-C_6H_3-C_6H_3		199–200
cod	1-Naphthyl		201
nbd	1-Naphthyl		201
Hexa-1,5-diene	1-Naphthyl		201
cod	C_6H_4-4-NO_2		202[a]

Table 3 (continued)

Diene	*R[1]*	*R[2]/X/L*	*Ref.*
cod	C_6H_4-4-NO_2	C_6H_4-4-OMe	202[a]
cod	Me	Cl	203
nbd	Me	I	184
cod	Et	Cl	177,203
cod	Et	I	177
cod	CH_2Bu^t	Cl	204
cod	CH_2CMe_2Ph	Cl	181
cod	Bz	Cl	189
cod	CH_2TMS	Cl	182
nbd	CF_3	I	183
nbd	CF_2Me	I	183
cod	CH_2CHR^1=CHR^2	X	187[b]
cod	1-Indenyl	Cl	205
cod	Ph	Cl	203
cod	C_6H_4-3-NO_2	Cl	206
cod	C_6H_4-3-NH_3	Cl	206
cod	1-Naphthyl	Cl	201
nbd	1-Naphthyl	Cl	201
Hexa-1,5-diene	1-Naphthyl	Cl	201
nbd	2-methoxy-1-naphthyl	Cl	211
cod	C_6H_2-2,4,6-Me_3	Br	190
nbd	C_6F_4-4-H	Cl	210
nbd	C_6F_4-4-OMe	Cl	210
Hexa-1,5-diene	C_6F_5	Cl	210
Hexa-1,5-diene	C_6F_4-4-H	Cl	210
Hexa-1,5-diene	C_6F_4-4-OMe	Cl	210
Dicyclopentadiene	C_6H_5	Cl	210[a]
nbd	C_6F_5	Cl	210
cod		$OCH_2C_6H_4$-4-Me	207

[a] X-ray diffraction structure determination. [b] R^1 = H, Me or Cl; R^2 = H, Me or Ph; X = Cl or Br.

well grounded on the basis of the low-temperature analysis.[213] The structure of the 2,5-dimethylhexa-1,5-diene analogue, $PtCl_2(\eta^2,\eta^2\text{-}C_8H_{14})$, has been determined similarly. Here, both alkenes are again perpendicularly disposed, with each showing a slightly longer bond (0.01 nm) to the methyl substituted carbon atom.[173]

The structures of several complexes of 5-mch and 5-mco, analysed by x-ray diffractometry, all clearly demonstrate the in-plane and perpendicular coordination modes of the exocyclic and internal alkenes, respectively. The structure of $PtCl_2$(mch) also shows an unequivocal difference in the platinum–alkene distance for the exocyclic (0.221(1) nm) and internal (0.210(1) nm) double bonds. The longer Pt–C_2 interaction for the exocyclic case is attributed to steric rather than electronic factors.[172] The diphenylplatinum homologue, $PtPh_2$(mch), shows similar structural variations. The 5-mco analogues, PtX_2(mco), likewise, show disparities in their metal–exocyclic and metal–internal alkene interactions (X = Cl or Br). Monophenylation (X = Cl) results in substitution *trans* to the exocyclic C=C, and the higher *trans*-influence of phenyl relative to chloride results in an even more exaggerated increase in the Pt–C_{exo} distances.[173] A diffractometric study of the structure of the pentacoordinate complex, $PtCl_2(\eta\text{-}C_2H_4)(\eta^2,\eta^2\text{-}C_9H_{16})$, formed with the exocyclic diene 3,7-dimethylenebicyclo[3.3.1]nonane, confirms the equatorial-coplanar arrangement of all three alkene groups.[171]

A series of UV photoelectron spectroscopic studies of $PtR_2(\eta^2,\eta^2$-diene) are complemented by SCF-MS-Xα calculations on *cis*-$PtMe_2(\eta\text{-}C_2H_4)_2$ (R = Me or CF_3; diene = cod or nbd). The two most ionizable molecular orbitals are platinum–ligand σ-orbitals, followed by four essentially nonbonding platinum 5*d* levels. The σ-donor ability of the dienes varies in the order cod > nbd, although both are less capable in this respect than nitrogen donors, phosphines, arsines and phosphites. The π-acceptor capacities of cod and nbd are adjudged similar and comparable with that of PMe_3, though less than that of TMEDA.[214]

(iii) Reactions

A review of the photochemistry of alkyltransition metals contains references of relevance to this section.[215]

(a) Diene displacement. The substitution lability of diene ligands in platinum(II) derivatives is responsible for their wide use as precursors to corresponding complexes with other, most commonly nitrogen- or phosphorus-donor, ligands. For example, dimethylplatinum derivatives, $PtMe_2(\eta^2,\eta^2\text{-diene})$ (diene = nbd or cod), react with nitrogen-donor ligands, L and L_2, to give $PtMe_2L_2$ (L = py; L_2 = bipy, en or TMEDA). Significantly, nbd is more readily displaced than cod, proposedly on account of its less accommodating bite.[216] Displacement of nbd from $Pt(CF_3)(Me)(nbd)$ by bidentate dppe, bipy and TMEDA or monodentate py and DMSO occurs more easily than from $Pt(CF_3)_2(nbd)$. Addition of MeCN generates an equilibrium mixture of parent and substitution product, while DMF does not react.[183] Diene displacements by 1,2-diaminoethane, TMEDA, phen, NH_3 and PPh_3 from $PtR^1_2(\eta^2,\eta^2\text{-diene})$ is most facile for the hexadiene example ($R^1 = C_6F_5$ or C_6F_4-4-H; diene = nbd, cod, dicyclopentadiene or hexa-1,5-diene).[197] Phosphines, PR^2_3, displace cod readily from $PtCl_2(cod)$, $PtR^1_2(cod)$, PtClPh(cod) and PtCl(COPh)(cod), unless R^2 is too bulky, when alkene displacement can be inhibited entirely (R^1 = Me or Ph; $PR^2_3 = PPh_3$, $PMePh_2$, PCy_3, $P(4\text{-Tol})_3$, $P(Mes)_3$, dppm, dppe or dpae (1,2-bis(diphenylarsino)ethane)). The molecularity of products is a function of the steric bulk of PR^2_3 for monodentate cases and by ligand bite for bidentate species. Product geometry is controlled by the mutual *trans* influences of R^1, Cl and PR^2_3.[217] Relatedly, displacement of cod from $PtCl_2(cod)$ by CO and PR^2_3 is exploited to produce mixed ligand complexes *cis*-$PtCl_2(CO)(PR^2_3)$ ($PR^2_3 = PEt_3$, PBu^n_3, $PMePh_2$, PPh_3, $P(C_6H_4\text{-4-F})_3$ or PCy_3). Reaction of CO with $PtR^1_2(cod)$ affords $PtR^1_2(CO)_2$, while with $PtClR^1(cod)$ it leads to the migratory CO insertion product, $PCl(COR)^1)(cod)$ (R^1 = Me or Ph).[218] Addition of one equivalent of bulky PR^2_3 to $PtCl_2(cod)$ is reported to generate $[PtCl(PR^2_3)(cod)]Cl$, while treatment of $PtCl_2(cod)$ with PR^2_3, in the presence of $NaBH_4$, provides a useful route to good yields of *trans*-$PtClH(PR^2_3)_2$ and *trans*-$PtH_2(PR^2_3)_2$ ($PR^2_3 = PPr^i_3$, PBu^t_2Ph, PBu^t_2Me, PBu^t_3 or PCy_3).[219] Many more diene-displacement reactions may be found in the references given in Table 2.

The enhanced dissociation lability of the in-plane η-alkene element in complexes of the monoexocyclic dienes, 5-mch and 5-mco have been noted.[173,174]

(b) Nucleophilic addition. Nucleophilic addition reactions—notably those of coordinated cod to give η^1,η^2-cyclooctenyl derivatives (Scheme 9)—continue to attract interest. A kinetic investigation of addition of cyclohexylamine to $MBr_2(cod)$, to give mononuclear $MX_2\{\eta^1,\eta^2\text{-}C_8H_{12}(NHCy)\}$, indicates that the ease of reaction is highly metal dependent (M = Pd or Pt). The palladium complex reacts about 70-fold more rapidly than its platinum analogue. This is ascribed to poorer retrodonation from palladium(II) in accord with its inherently higher ionization potential.[220] Reaction of $PtCl_2(cod)$ with MeO^- and subsequent reactions of the η^1,η^2-cyclooctenyl products have been reexamined. As confirmed by an x-ray diffraction study, the primary product is dimeric, $[Pt(\mu\text{-OMe})(\eta^1,\eta^2\text{-}C_8H_{12}OMe)]_2$.[221] Reduction of this or its chloro-bridged homologue with tetrahydroborate, in presence of bulky PR_3, affords good yields of *trans*-$PtH_2(PR_3)_2$ ($PR_3 = PPr^i_3$, PBu^t_2Ph, PBu^t_2Me, $PBu^t_2Bu^n$, PBu^t_3 or PCy_3).[222] The dication in $[Pt(dppp)(cod)](OSO_2CF_3)_2$ undergoes *exo*-nucleophilic addition of PR_3 on the diene ligand (Equation (13); R = Ph or Bu^n).[223] In contrast, reaction of $PtPh(\eta^1\text{-Cp})(cod)$ with PPh_3 leads to attack at the metal, with *endo*-migration of the cyclopentadienyl group to cod (Equation (14)).[192] Addition of hexfluorobut-2-yne to a related system, $PtMe(\eta^1\text{-Cp})(cod)$, occurs on the C-5 ligand, generating a 2,3-bis(trifluoromethyl)-7-norbornadienylplatinum(II) product, $PtMe\{\eta^1\text{-}C_7H_5(CF_3)_2\}(cod)$.[190,191]

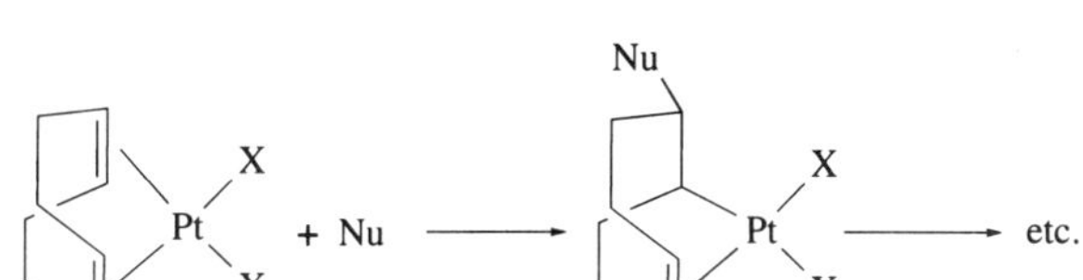

Scheme 9

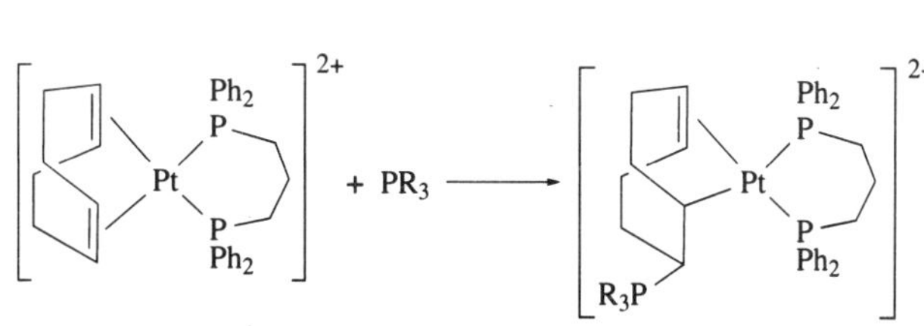

(13)

R = Ph, Bu^n

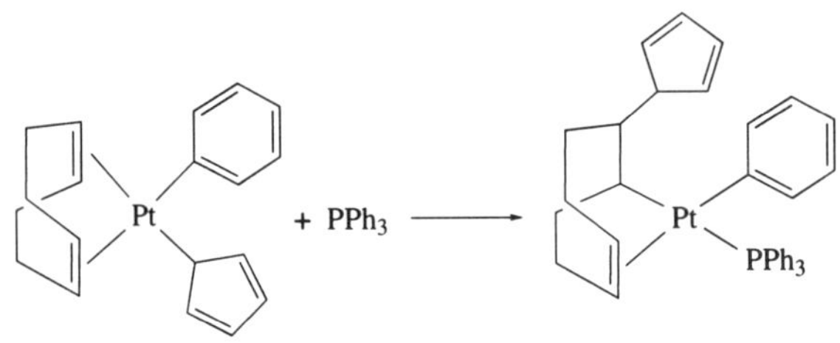

(14)

Nucleophilic attack on $PtCl_2(mco)$, by MeO^- and Ph^-, occurs at the exocyclic alkene site as confirmed by diffractometric studies.[174]

Amino acid esters, $H_2NCHR^2CO_2R^1$, add to the coordinated diene in $PtCl_2$(η-hexa-1,5-diene) to form, first, $PtCl_2(\eta^2\text{-}H_2C{=}CHCH_2CH_2CHCH_2NH_2CHR^2CO_2R^1)$. Treated with Na_2CO_3, this undergoes N-deprotonation and loss of Cl^- to afford (stereospecifically) a Pt–Pt bonded dimer (confirmed by x-ray diffraction) spanned by the aminoester (R^1 = Et, R^2 = H; R^1 = Et, R^2 = Bz; R^1 = Me, R^2 = Pr^i). Introduction of $AgBF_4$ leads to further chlorine abstraction and formation of a new dicationic dimer, in which the ester groups are bound to platinum via oxygen (R^1 = Et, R^2 = H).[224]

(c) Miscellaneous: migrations, insertions, redox reactions. A thermolytic study of ^{13}C-labelled $Pt(*CH_2Me)_2$(cod), $PtCl(*CH_2Me)$(cod) and $PtI(*CH_2Me)$(cod) reveals the variations in terms of the relative abundance of $PtY(*CH_2Me)$(cod) and $PtY(CH_2*Me)$(cod) with time. For Y = Et, the activation barrier to β-hydride transfer is estimated at 121 kJ mol^{-1}. For Y = Cl and Y = I the corresponding estimates are 122 and 112 kJ mol^{-1}, respectively. Diene dissociation is not considered a likely mechanistic contributor.[177]

PtX_2(cod) reacts with diazomethane via, effectively, carbene insertion into both Pt–X bonds, generating halomethylplatinum species, $Pt(CH_2X)_2$(cod) (X = Cl or Br).[166] In a similar fashion, dimethyldiazomethylphosphonate, $N_2CHP(O)(OMe)_2$, attacks the Pt–Cl bonds in $PtCl_2$(cod), yielding first $PtCl\{CH(Cl)P(O)(OMe)_2\}$(cod) and, finally, stereoisomers of $Pt\{CH(Cl)P(O)(OMe)_2\}_2$(cod). Stereoresolution occurs spontaneously on crystallization; x-ray diffraction shows only the (*R,R*) isomer.[225]

A series of studies of the hydrogenation of $PtR_2(\eta^2,\eta^2$-diene), heterogeneously catalysed by platinum, gave interesting mechanistic insight into alkene hydrogenation and alkyl transfers on a metal surface (R = Me, Et, Pr^i, Bu^i, CH_2Bu^t, Ph, *endo*-2-norbornyl or *exo*-2-norbornyl; diene = cyclooctadiene, norbornadiene, dicyclopentadiene or homohypostrophene). The evidence indicates that R groups are transferred to and from the bulk platinum surface regio- and stereospecifically and that—where a distinction can be made—the diene coordinates to the metal surface via its *endo* face (the face initially coordinated to mononuclear platinum). The relative rates for C–H activation in surface-bound alkyl groups is $\alpha > \beta > \delta > \gamma \approx \varepsilon$, similar to those in homogeneous Pt–R systems, except for α-hydride transfer (which is least facile in the latter case). This is ascribed to the possibility of bimetallic interactions on the metal surface. Overall, hydrogenation may be mass-transport or reaction-rate limited, depending on ambient conditions of pressure and temperature.[226]

9.2.4 η^2-Allene, η^2-X=CR_2 and Related Complexes

9.2.4.1 η^2-Cumulene complexes: preparation, structure and reactivity

As nonconjugated π-systems, cumulenes tend to coordinate to platinum in a manner analogous to monoalkenes. Most reactions also reflect those of coordinated alkenes. Displacement of ethene from Pt(diop)(η-C_2H_4) yields the corresponding η^2-allene complex, Pt(diop)(η^2-C_3H_4).[14a] Likewise, cycloheptatetraene—like its benzo and dibenzo homologues—coordinates to the zerovalent substrate $Pt(PPh_3)_2$ by η^2-coordination of the allene function. Interconversions of this structure, via the cycloheptatrienylidene tautomer, may be followed by ^{1}H NMR spectroscopy (Equation (15)). The latter isomeric form of binding—which is preferred in Pt^{II} complexes—lies at least 113 kJ mol^{-1} higher in energy for Pt^0. Preference for the allene structure is rationalized—on the basis of extended Hückel calculations—by particularly efficient retrodative bonding. Further reaction with a second molecule of cycloheptatetraene affords a platinacycle (Equation (16)).[227–9] The outcome of reaction of 1-bromocycloheptene with base in presence of $Pt(PPh_3)_3$ is crucially dependent on the reactants. With $KOBu^t$, the η^2-cyclohepta-1,2-diene complex is produced. Bromocycloheptatrienes behave similarly, to afford corresponding η^2-cycloallene complexes. This contrasts with the reactions of the same bromocycloalkene and bromocycloalkatrienes with $LiNPr^i_2$, which generates cycloalkyne complexes (see Section 9.3.1.1(i)).[230] Ethene may also be displaced from $Pt(PPh_3)_2(\eta\text{-}C_2H_4)$ by substituted butatrienes to give complexes in which the cummulene is η^2-bound at the terminal C=C bond. The reaction outcomes are mostly regiospecific; in only one case—with $Ph_2C{=}C{=}C{=}CH(GeEt_3)$—are the two possible regioisomers observed (Equation (17)).[231]

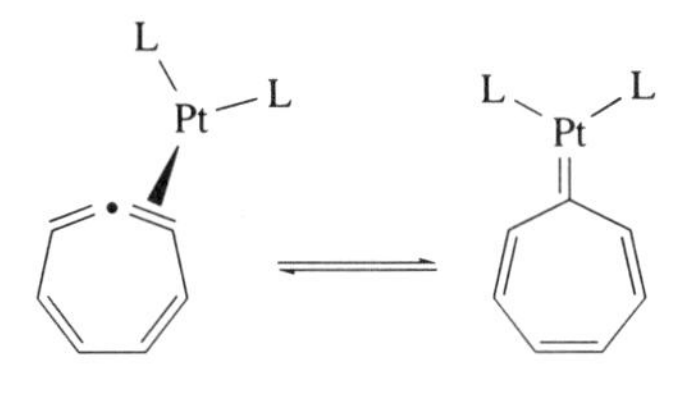

(15)

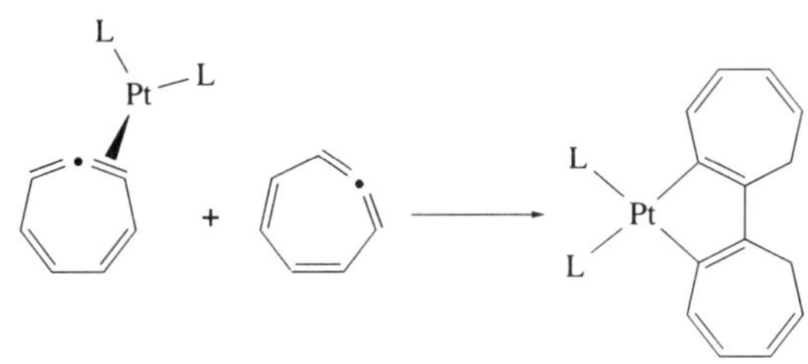

(16)

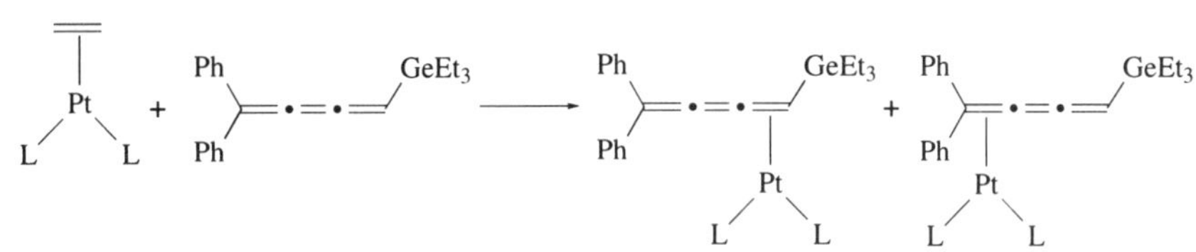

(17)

Divalent platinum derivatives with a chiral amine ligand, *trans*-$[PtCl_2\{(S)\text{-}\alpha\text{-methylbenzylamine}\}(\eta^2\text{-}R^1R^2C{=}C{=}CR^3H)]$, have been synthesized from the corresponding *trans*-(η^2-ethene)platinum derivative, by treatment with the appropriate allene ($R^1 = Et$, $R^2 = Me$, $R^3 = Bu^t$; $R^1 = Bu^t$, $R^2 = Me$, $R^3 = Bu^t$). Platinum-195 NMR spectroscopy allows enantiomeric determination of the chiral trisubstituted allenes.[232] The η^2-dimethylallene complex, *cis*-$PtCl_2(PPh_3)(\eta^2\text{-}H_2C{=}C{=}CMe_2)$, undergoes nucleophilic addition of 2-chloroaniline at the unsubstituted terminus to yield, first, the β-ammonioalkyl derivative. This reacts via self-protonolysis of the Pt–C σ-bond to afford the η^2-enamine isomer (Scheme 10). An x-ray diffractometric analysis of the molecular structure of the latter complex shows a C=C vector nearly orthogonal to the coordination plane, as might be expected.[233] Treatment of the related derivative, *cis*-$PtCl_2(PMe_2Ph)(\eta^2\text{-}H_2C{=}C{=}CH_2)$, with $HON{=}CMe_2$ leads to nucleophilic attack, followed by 1,3-hydrogen migration to give, ultimately, the metallacyclic species $Pt[CMe{=}CHN({=}CH_2)O]Cl(PMe_2Ph)$.[234]

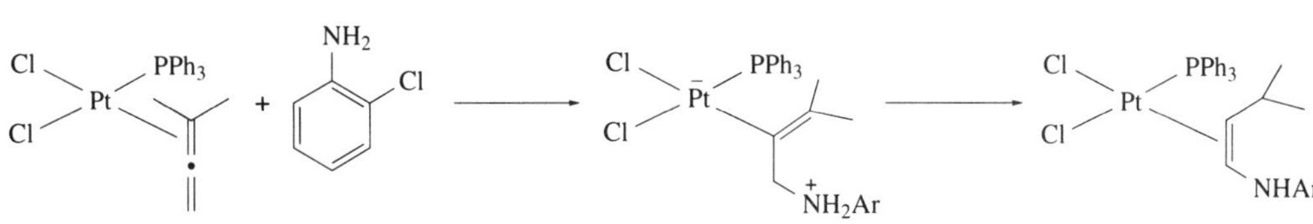

Scheme 10

9.2.4.2 η²-Ketene complexes: preparation, structure and reactivity

Treatment of $Pt(PPh_3)_4$ with CH_2Br_2 in presence of metallic zinc powder, under carbon monoxide (3 atm) leads to the ketene derivative, $Pt(PPh_3)_2(\eta^2\text{-}H_2C{=}C{=}O)$. The same complex is accessible by displacement of ethene from $Pt(PPh_3)_2(\eta\text{-}C_2H_4)$ by ketene. Relatedly, reaction of ketene with the Zeise dimer affords $PtCl_2(\eta^2\text{-}H_2C{=}C{=}O)$, which reacts in turn with PMe_2Ph to give $PtCl_2(PMe_2Ph)_2(\eta^2\text{-}H_2C{=}C{=}O)$. All these have been characterized spectroscopically as η^2-(*C,C*)-ketene complexes. Thermolytic decompositions yield mainly alkenes as well as methane, carbon monoxide and ketene.

Treatment with H_2 releases acetaldehyde, ethanol and (mainly) saturated hydrocarbons. All react with LAH to liberate ethanol in good yield.[235]

9.2.4.3 *Complexes of η^2-X=CR$_2$: preparation, structure and reactivity*

A review of η^2-acyl and related metal complexes—although it includes a few examples of group 8 metal derivatives—cites no platinum examples of η^2-X=CR$_2$ coordination.[236]

Reaction of formaldehyde with $Pt(PPh_3)_2(\eta\text{-}C_2H_4)$ results in ethene displacement to give $Pt(PPh_3)_2(\eta^2\text{-}O{=}CH_2)$.[237] Recently, there has been considerable synthetic progress featuring multiple bonding between carbon and heavier *p*-block elements. Not surprisingly, some of these fragments have emerged into the organotransition metal arena. Phosphaalkenes and their metal complexes feature in a comprehensive review.[238] Although the phosphaalkene in $Pt[Ph_2C{=}P(Mes)](PPh_3)_2$ adopts the η^1-mode of coordination in the solid state, it favours η^2-bonding in solution.[239] In contrast, when the cyclic azaphosphetine, $R^1\overline{PN(R^2)P{=}C}R^1$ displaces ethene from $Pt(PR^3_3)_2(\eta\text{-}C_2H_4)$, the resulting complex displays only η^1-ligation, via the two-coordinate phosphorus atom ($R^1 = CEt_3$, $R^2 = Bu^t$, R^3 = Et or Ph). Interestingly (although strictly outside the scope of this survey), the (η^2-diphosphine)platinum(0) complexes $Pt(PPh_3)_2(\eta\text{-}P_2Ph_2)$ and $Pt(dppe)_2(\eta\text{-}P_2Ph_2)$ have been structurally characterized. These are formed by reaction of Li(Ph)PP(Ph)Li with $PtCl_2L_2$.[240]

Several (η^2-disilene)platinum species, $PtL_2(\eta\text{-}Si_2R_4)$, have proven accessible (L_2 = dppe or dcpe; R = Me, Pr^i, Ph or Mes)[241]—some by displacement of ethene from $PtL_2(\eta\text{-}C_2H_4)$, others from platinum(II) precursors with reduction. To date, though, no (η^2-silene)platinum binding, $Pt(\eta\text{-}R^1{}_2C{=}SiR^2{}_2)$, has been observed directly.[242] Silene complexes of this type are, however, clearly implicated as reaction intermediates in the thermolytic rearrangement of $Pt(CH_2SiMe_2R)_2L_2$ to give $Pt(R)(CH_2SiMe_2CH_2SiMe_2R)_2L_2$ (R = Me or aryl; L = PMe_3, PEt_3, PMe_2Ph, $PMePh_2$ or PPh_3; L_2 = dcpe).[243]

An unusual case of η^2-coordination results from thermolysis of the cationic hydridoplatinum carbene complex *trans*-$[PtH(PPh_3)_2(=\overline{CSCH_2CH_2S})]BF_4$. Hydrogen migration to the α-carbon atom generates an η^2-*C,S*-coordinated ring (Equation (18)), the plane of which lies almost perpendicular to the PtP_2CS coordination plane (according to x-ray diffractometric evidence).[244]

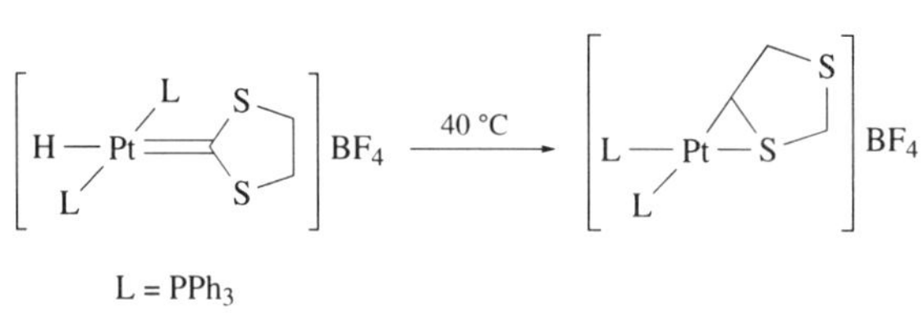

(18)

9.2.5 η^n-Polyalkene Complexes (Conjugated)

9.2.5.1 *Preparation, structure and reactivity*

As noted previously,[2] 1,3-dienes and their derivatives do not readily form isolable η^4-complexes with platinum, in either the zerovalent or the divalent states. Interaction with platinum(0) species leads, in general, to oligomerization reactions, while that with platinum(II) usually gives products with η^2-coordination.[2] More recently, little divergence from these preferences has emerged.

Interaction of $Pt(PPh_3)_3$ and buta-1,3-diene at low temperature (−100 °C) generates an equimolar mixture of $Pt(PPh_3)_4$ and $Pt(PPh_3)_2(\eta^2\text{-}C_4H_6)$. The latter complex can be synthesized more conveniently via reduction of $Pt(PPh_3)_2(\eta\text{-}O_2)$ by $NaBH_4$ in the presence of the diene.[42] Zerovalent derivatives $Pt(R_2PCH_2CH_2PR_2)(\eta^2\text{-}C_4H_6)$ are obtained by reaction of $Pt(cod)_2$ with buta-1,3-diene in the presence of $R_2PCH_2CH_2Pr_2$ (R = Pr^i or Bu^t). Molecular structure determination by x-ray diffractometry for the palladium analogue (R = Bu^t) reveals a (*S*)-*trans* conformation for η^2-coordinated butadiene and MC_2P_2 coplanarity. Multinuclear NMR studies suggest similar structures for the platinum derivatives. An (η^4-butadiene)–platinum intermediate is implicated in the η^2-η^2 exchange behaviour of these complexes.[245] Treatment of $Pt(SO_2)(PPh_3)_2$ with buta-1,3-diene or isoprene—or reaction of $Pt(PPh_3)_2(\eta^2\text{-}C_4H_5R)$ with SO_2—yields $Pt(SO_2)(PPh_3)_2(\eta^2\text{-}C_4H_5R)$ (R = H or Me). Low-temperature NMR studies indicate two rotamers in which the η^2-C–C vector is perpendicular to a planar $Pt(SO_2)(PPh_3)_2$ unit, with mutually *cis*-phosphine ligands. Subsequent thermolysis results in loss of the diene and generation of the novel triangulo cluster, $Pt_3(\mu\text{-}Ph)(\mu\text{-}PPh_2)(\mu\text{-}SO_2)(PPh_3)_3$.[246]

Reaction of buta-1,3-diene with dimeric $Pt_2Cl_4(PR_3)_2$ at low temperatures affords monomeric *trans*-$PtCl_2(PPr^n{}_3)(\eta^2\text{-}C_4H_6)$. At ambient temperature, this gives way to a dimeric structure in which butadiene bridges adjacent *trans*-$PtCl_2(PR_3)$ units with separate η^2-interactions (Equation (19)) (see Section 9.2.7.1).[247] The structure of the (η-cyclopentadiene)platinum(II) derivative, $PtCl_2(3\text{-}Pr^i\text{-}1,2,4,5,5\text{-}Me_5C_5)$, has been reported. It is an unexpected product of the reaction (Equation (20)) of 1,2,3-tri(*t*-butyl)-3-vinylcycloprop-1-ene with $Pt_2Cl_4(\eta^2\text{-}C_2H_4)_2$. Although technically classifiable as an example of η^4-coordination, the authors favour description as a tetragonal-planar η^2,η^2-diene complex, on the crystallographic evidence that none of the five bound elements (including the midpoints of the two C–C bonds) deviate from planarity by more than 0.002 nm.[248]

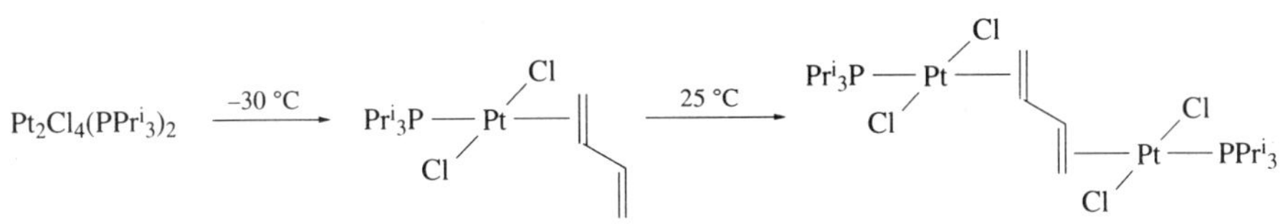

(19)

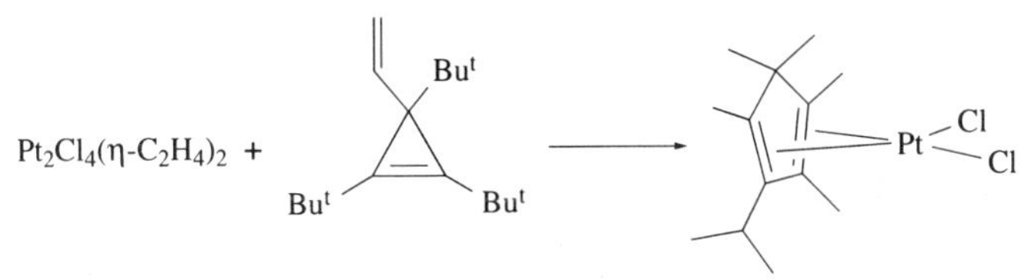

(20)

9.2.6 η^2-Fullerene Complexes

9.2.6.1 Preparation, structure and reactivity

Despite their extended π-delocalization, buckminsterfullerenes—of which C_{60} is the archetypal example—generally exhibit η^2-coordination to *d*-block metals. Complexes of this type have already been the subject of a general review.[249] More specifically, the synthesis and molecular structure of the dark-green $Pt(\eta^2\text{-}C_{60})(PPh_3)_2$ have been described.[250] Preparation is achieved by treatment of $Pt(\eta\text{-}C_2H_4)(PPh_3)_2$ with C_{60}. The notable structural feature is that the η^2-coordinated C–C bond—which is shared between two six-membered rings of the cage—lies coplanar with $Pt(PPh_3)_2$. Reaction of excess $Pt(PEt_3)_4$ with C_{60} leads to isolation of $[(\eta^2)_6\text{-}C_{60}][M(PEt_3)_2]_6$, which has also been characterized by x-ray diffractometry and multinuclear NMR studies. Again, there is η^2-coordination to C–C elements shared by adjacent six-membered rings and each C–C vector is almost coplanar with its bound $Pt(PPh_3)_2$ unit. The overall structure is highly symmetrical, with the six platinum atoms in an octahedral arrangement about the C_{60} framework, generating essentially ideal T_h symmetry. Lower homologues, $[(\eta^2)_n\text{-}C_{60}][Pt(PEt_3)_2]_n$, exist as mixtures of site-linkage isomers (n = 2–4).[251] Electrochemical studies of $Pt(\eta^2\text{-}C_{60})(PPh_3)_2$ and of $[(\eta^2)_n\text{-}C_{60}][Pt(PEt_3)_2]_n$ (n = 2–4 or 6) suggest that reduction is C_{60} centred, but that the irreversible oxidation is metal centred. There appears to be little retrodative $d\pi$-bonding interaction beyond the C–C attachment site.[252]

9.2.7 Bimetallic Alkene Complexes

9.2.7.1 Homobimetallic complexes: preparation, structure and reactivity

The simplest form of homobimetallic complexes are relatives of the Zeise dimer, $[Pt_2Cl_4(\eta\text{-}C_2H_4)_2]$. One example (see Section 9.2.3.2(i)) arises as a consequence of the dissociation-labile exocyclic double bond in $PtBr_2(\eta^2,\eta^2\text{-mco})$ which dimerizes to give $Pt_2Br_4(\eta^2\text{-mco})_2$, subsequent to treatment of $PtCl_2(\eta^2,\eta^2\text{-mco})$ with LiBr.[173]

A notable development in this area has been the evolution of alkylplatinum homologues of $[Pt_2Cl_4(\eta\text{-}C_2H_4)_2]$. Reaction of $[Pt_2Cl_4(\eta^2\text{-}C_2H_4)_2]$ with $[Pt_2Me_4(\mu\text{-}SMe_2)_2]$ generates $[Pt_2(\mu\text{-}Cl)_2Me_2(\eta\text{-}C_2H_4)_2]$. In addition, $PtMe_2(SMe_2)_2$ may also act as the methylating agent, but conventional transalkylation agents such as $SnMe_4$ are ineffective. The products readily undergo ethene displacement by other alkenes and alkynes (e.g., PhHC=CHPh and PhC≡CPh), although $F_3CC{\equiv}CCF_3$ yields a Pt–C insertion product.

Bridge cleavage, with a variety of monodentate donors, L, affords monomeric PtClMe(L)(η-C_2H_4) (L = py, SMe_2 or η[2]-C_2H_4).[253] [Pt_2(μ-Cl)$_2$$Me_2$(η-$C_2H_4$)$_2$] reacts with the Zeise dimer to produce the asymmetric analogue [Pt_2(μ-Cl)$_2$ClMe(η-C_2H_4)$_2$] (Equation (21)). Reaction of dimeric [Pt_2Me_4(μ-SMe_2)$_2$] with ethene gives *cis*-$PtMe_2$(SMe_2)(η-C_2H_4).[121] Chloro-bridged dimers, [{Pt(μ^2-Cl)($CH_2CH_2COC_6H_4$-4-R)(η-C_2H_4)}$_2$] have been synthesized by an unusual route in which [Pt_2Cl_4(η^2-C_2H_4)$_2$] reacts with a siloxycylopropane (R = H or Cl; Scheme 11). The β-platinum ketone, [{Pt(μ^2-Cl)(CH_2CH_2COPh)(η-C_2H_4)}$_2$], undergoes ethene displacement by CO to produce a transient dimeric species which rearranges to a more stable mononuclear complex in which the ketonic oxygen is metal bound (Scheme 11). An x-ray diffractometric study of molecular structure of [{Pt(μ^2-Cl)($CH_2CH_2COC_6H_4$-4-Cl)(η-C_2H_4)}$_2$] confirms the chloro-bridged structure, with η-C_2H_4 perpendicular to the $PtCl_2C$ coordination plane.[254] Asymmetric, halide-bridged binuclear complexes, [(η^1-$C_6X^1_5$)$_2$Pt(μ-X^2)$_2$Pt(cod)] are produced by interaction of Pt(η^1-$C_6X^1_5$)$_2$(THF)$_2$ and PtX^2_2(cod) (X^1 = F or Cl; X^2 = Cl, Br or I).[255]

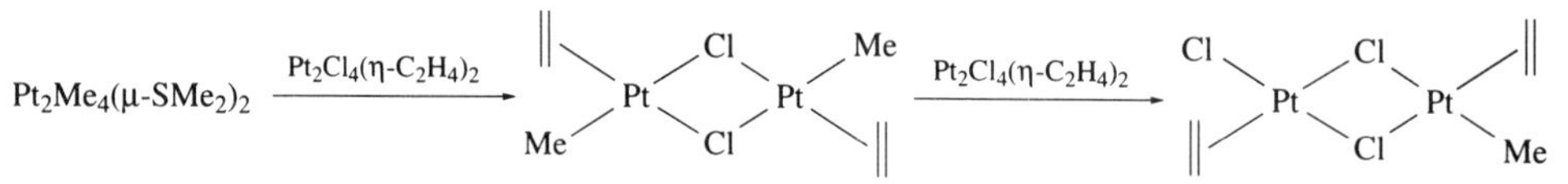

(21)

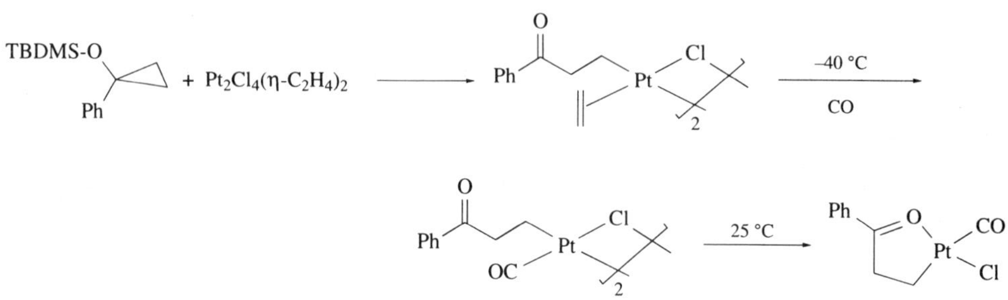

Scheme 11

An unusual, related derivative features a rare example of η^2-alkene coordination to platinum(IV). Admixture of the Zeise dimer and its analogue with η^2-bicyclo[6,1,0]non-4-ene leads to oxidative insertion of one platinum(II) centre into the cyclopropane function of η^2-bicyclo[6,1,0]non-4-ene. The product is an asymmetric, dinuclear complex in which, presumably, η^2-alkene coordination is supported by η^1,η^1-ligation within the same fragment (Equation (22)).[256]

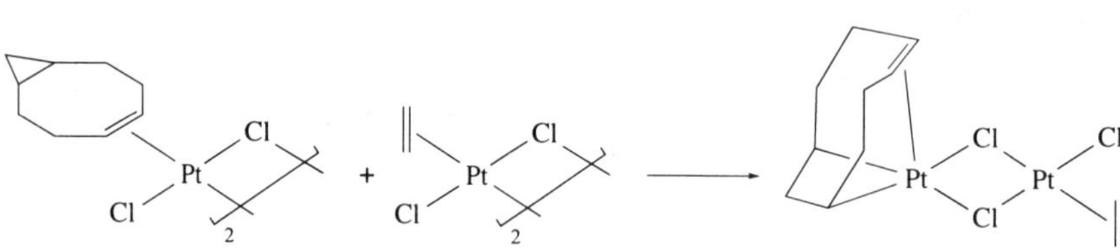

(22)

Amido-bridged dimeric species are obtained by deprotonation of the corresponding 2-aminostyrene derivatives, Pt(η-H_2C=CR^1-2-$C_6H_4NR^2_2$)X_2 (R^1 = H, R^2 = H, Me or Ph; R^1 = Me, R^2 = H, Me or Bz; R^1 = Ph, R^2 = Me), using an anion-exchange resin.[148]

Treatment of $Pt_2Cl_4L_2$ (L = PMe_2Ph, PEt_3 or PPr^n_3) with 1,3-butadiene yields *cis,cis*-[$Pt_2Cl_4L_2$(μ-C_4H_6)]. The diene is prochiral and the molecular structure (L = PMe_2Ph), from x-ray diffractometry, exhibits *meso* isomerism, with the C=C vectors perpendicular to their respective *cis*-$PtCl_2P$ coordination planes. Reaction of butadiene with [NBu^n_4]$_2$[Pt_2X_4] affords analogous cationic complexes, [NBu^n_4]$_2$[Pt_2X_6(μ-C_4H_6)] (X = Cl or Br). Several of these complexes undergo nucleophilic addition of amines to yield, ultimately, an unusual dimetallic structure in which two platinum centres are bridged by the resulting diaminodiyl ligand (Equation (23)). Treatment of *cis,cis*-[$Pt_2Cl_4L_2$(μ-C_4H_6)] with excess $NHMe_2$ gives LClPt($Me_2NCH_2CHCHCH_2NMe_2$)PtClL, with two *trans*-fused five-membered rings (L = PMe_2Ph or PEt_3), as confirmed by x-ray diffractometry (L = PEt_3). There is similar formation of (three isomers of) a diaminodiyl-bridged system by reaction of the primary amine NH_2Me with *cis,cis*-[Pt_2Cl_4(PPr^n_3)$_2$(μ-C_4H_6)]. Corresponding isoprene complexes are observable spectroscopically, but were not isolated.[247]

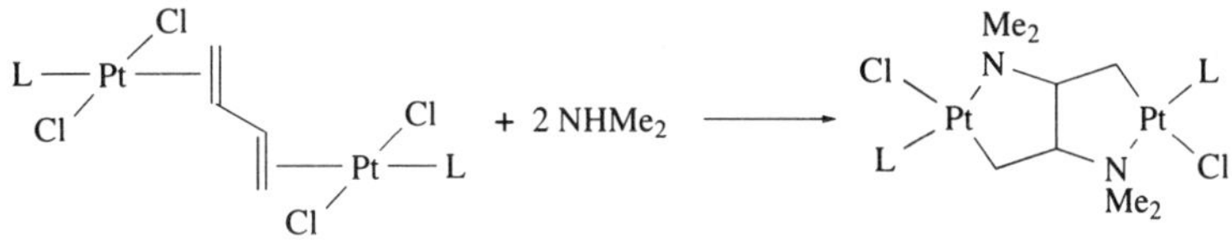

(23)

9.2.7.2 Heterobimetallic complexes: preparation, structure and reactivity

Many examples of heterobimetallic complexes exhibit a preference for 1,5-cyclooctadiene as a platinum ligand. As for the homonuclear examples (see Section 9.2.7.1), halide-bridged dimers, $[(\eta^1\text{-}C_6X^1{}_5)_2M^1(\mu\text{-}X^2)_2M^2(cod)]$ can be synthesized by interaction of $M^1(\eta^1\text{-}C_6X^1{}_5)_2(THF)_2$ (M^1 = Pd or Pt; X^1 = F or Cl) and $M^2X^2(cod)$ (M^2 = Pd or Pt; X^2 = Cl, Br or I).[255] Treatment of the phosphido-bridged species, $\{[(MeO)_3Si](OC)_3Fe(\mu\text{-}PPh_2)PtL_2\}$ with HBF_4 (L_2 = cod; L = PPh_3) leads to the new cationic systems, $[(OC)_3Fe(\mu\text{-}PPh_2)(\mu\text{-}H)_2PtL_2][BF_4]$, in which the phosphido bridge is augmented by bridging hydrides.[257] Reaction of PtClMe(cod) with $Na[W(CO)_3Cp]$ generates $[Cp(OC)_3WPtMe(cod)]$ which exhibits an unbridged Pt–W bond.[258] The analogous phenylplatinum derivative, $[Cp(OC)_3WPtPh(cod)]$, has also been characterized. X-ray diffractometric evaluation of the structure of $[Cp(OC)_3WPtPh(cod)]$ confirms the nominally unbridged metal–metal bond, with an essentially planar arrangement of Ph, the two C=C groups and tungsten around platinum, although there appears also to be weak interaction between platinum and two of the tungsten-bound carbonyl ligands. Differences in platinum–alkene distances indicate a stronger *trans*-influence for Ph than for the tungsten fragment.[129,259]

Different classes of binuclear complex are represented by the species which result from addition of terminal alkynes and allene to $[(OC)_3Fe(\mu\text{-}dppm)(\mu\text{-}CO)Pt(PPh_3)]$. With alkynes, HC≡CR (R = H, Me, Ph, CO_2Me, C_6H_4-4-Me, CH_2NHMe, CMe_2OH or CH_2CH_2OH), $[(OC)_2Fe(\mu\text{-}dppm)\{\mu,\eta^1,\eta^1,\eta^2\text{-}C(O)CH{=}CR\}Pt(PPh_3)]$ is formed via a carbonyl–alkyne coupling; the η^2-component of bridge bonding involves the iron centre, as confirmed by an x-ray diffractometric analysis. This product isomerizes slowly to $[(OC)_2Fe(\mu\text{-}dppm)\{\mu,\eta^1,\eta^1,\eta^2\text{-}C(O)CR{=}CH\}Pt(PPh_3)]$. With allene, $[(OC)_3Fe(\mu\text{-}dppm)\{\mu,\eta^1,\eta^1\text{-}CH_2(C{=}CH_2)\}Pt(PPh_3)]$ is obtained, the structure of which has also been confirmed diffractometrically. This, in turn, undergoes protonation, yielding the vinyl-bridged complex, $[(OC)_3Fe(\mu\text{-}dppm)(\mu,\eta^1,\eta^2\text{-}CMe{=}CH_2)Pt(PPh_3)]$. Reaction with BH_4^-, however, gives a dimetallacyclopentanol, whose exocyclic double bond prefers η^2-coordination to iron. Several analogues are also accessible using 1,1-(diphenylphosphino)ethene as an ancillary ligand instead of dppm.[260]

9.2.8 Polymetallic Alkene Complexes

9.2.8.1 Homometallic complexes: preparation, structure and reactivity

There are few examples of homometallic complexes. Interaction of ethyne with $[Pt_3(\mu_3\text{-}H)(\mu_2\text{-}dppm)_3]^+$ models its reaction with a [110] metal surface site and affords the triangulo cluster $[Pt_3(\eta^1\text{-}H)(\mu\text{-}dppm)_3(\mu,\eta\text{-}C{=}CH_2)]^+$ (Equation (24)). Deuterium-labelling studies suggest an equilibrium between $Pt_3(CH{=}CH_2)$, $Pt_3H(C{=}CH_2)$ and $Pt_3(C\text{–}Me)$ species.[261]

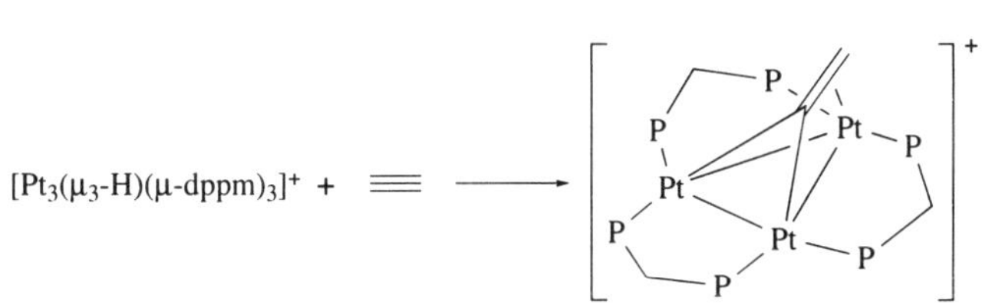

(24)

9.2.8.2 Heterometallic complexes: preparation, structure and reactivity

The majority of recent advances of heterometallic complexes have come about as a result of the interaction of other polymetallic species with $Pt(cod)_2$, and retain the Pt(cod) fragment in the product. Admixture of $Co_2(CO)_8$ and $Pt(cod)_2$ affords two new Pt–Co clusters—namely, $[PtCo_2(CO)_7(cod)]$ (3%) and $[Pt_2Co_2(CO)_6(cod)]$ (31%)—in addition to $Co_4(CO)_{12}$ (14%). $[Pt_2Co_2(CO)_8(cod)]$ reacts with EtC≡CEt to form $[Pt_2Co_2(CO)_8(cod)(EtC{\equiv}CEt)]$ and with PhC≡CPh to afford

[$Pt_3Co_2(CO)_6(cod)_2(PhC{\equiv}CPh)_2$]. X-ray diffraction studies reveal a structure for [$PtCo_2(CO)_7(cod)$] in which the Co–Co bond in the $Co_2(CO)_6$ unit is spanned by μ-Pt(cod) and μ-CO. In [$Pt_2Co_2(CO)_8(cod)$], a butterfly structure emerges with only weak (0.295 46(6) nm) Pt---Pt interaction. The cod ligand is η^2,η^2-bound to one platinum atom. X-ray diffraction studies on [$Pt_2Co_2(CO)_8(cod)(EtC{\equiv}CEt)$] confirm that the Pt–Pt bonding is augmented by η^2-parallel coordination of EtC≡CEt, which also coordinates to cobalt in an η^2-perpendicular fashion—that is, μ_3,η-face bridging of the metal triangle. There is no Co–Co bonding. Similar structure determination of [$Pt_3Co_2(CO)_6(cod)_2(PhC{\equiv}CPh)_2$] reveals that two platinum centres retain their diene ligand and that the Pt–Co framework is spanned by μ_3,η-PhC≡CPh groups.[262] Likewise, two different metal aggregates—[$PtRh_2(\mu_3\text{-}CO)(\mu_3,\eta\text{-}F_3CC{\equiv}CCF_3)Cp_2(cod)Cp_2$] (30%) and [$PtRh_4(\mu_2\text{-}CO)_2(\mu_3,\eta\text{-}F_3CC{\equiv}CCF_3)_2Cp_4$] (10%)—are recovered upon reaction of $Pt(cod)_2$ with two equivalents of [$Rh_2(CO)(\mu,\eta\text{-}F_3CC{\equiv}CCF_3)Cp_2$]. The molecular structure of [$PtRh_2(\mu_3\text{-}CO)(\mu_3,\eta\text{-}F_3CC{\equiv}CCF_3)Cp_2(cod)$] confirms the μ_3,η-mode of coordination for hexafluorobutyne and that platinum and rhodium both retain their cod and Cp ligands, respectively.[263] Similar reaction of $Pt(cod)_2$ with $Fe(CO)_5$ yields [$Pt_5Fe_2(CO)_{12}(cod)_2$] (40%), [$Pt_3Fe_3(CO)_{15}$] (9%) and the already known derivative [$PtFe_2(CO)_8(cod)$] (3%). Reaction of [$Pt_5Fe_2(CO)_{12}(cod)_2$] with PhC≡CPh yields [$Pt_5Fe_2(CO)_7(\mu,\eta\text{-}PhC{\equiv}CPh)(\mu_3,\eta\text{-}PhC{\equiv}CPh)(cod)$] (36%), while addition of PhC≡CH gives [$PtFe(CO)_3(\mu\text{-}PhC{=}CH{-}CH{=}CPh)(cod)$] (26%) via head-to-head dimerization of the alkyne. Crystallographic analysis of [$Pt_5Fe_2(CO)_{12}(cod)_2$] shows a tetrahedral Pt_4 array, with the fifth platinum atom bridging one edge and $Fe(CO)_4$ units spanning the Pt–Pt bonds to the unique edge-bridging platinum atom. Structural studies of [$Pt_5Fe_2(CO)_7(\mu,\eta\text{-}PhC{\equiv}CPh)(\mu_3,\eta\text{-}PhC{\equiv}CPh)(cod)$] and [$PtFe(CO)_3(\mu\text{-}PhC{=}CH{-}CH{=}CPh)(cod)$] indicate that each cod is bound to platinum, but that all η^2-interactions of other hydrocarbons are with iron.[264] [$PtRu_4(CO)_{13}(\mu\text{-}H)_2(cod)$] (37%), [$PtRu_3(CO)_9(\mu\text{-}CO)(\mu\text{-}H)_2(cod)$] (10%), [$Pt_2Ru_4(CO)_{11}(\mu_3\text{-}H)_2(cod)_2$] (2.5%) and [$Pt_5Ru_5(CO)_{18}(\mu_3\text{-}H)_2(cod)_2$] (1%) are obtained in similar fashion by reaction of [$Ru_4(CO)_{13}(\mu\text{-}H)_2$] with $Pt(cod)_2$. X-ray diffraction has been employed to deduce the structures of all the products of reaction of [$Ru_4(CO)_{13}(\mu\text{-}H)_2$] with $Pt(cod)_2$. [$PtRu_4(CO)_{13}(\mu\text{-}H)_2(cod)$] has a butterfly tetrahedral disposition for Ru_4, with one triangular face capped by the Pt(cod) unit. In [$PtRu_3(CO)_9(\mu\text{-}CO)(\mu\text{-}H)_2(cod)$], the metals form a tetrahedral array. [$Pt_2Ru_4(CO)_{11}(\mu_3\text{-}H)_2(cod)_2$] has a Ru_4 tetrahedron with two faces capped by platinum. In [$Pt_5Ru_5(CO)_{18}(\mu_3\text{-}H)_2(cod)_2$], the five ruthenium and four of the platinum units arrange as a face-shared bioctahedron, in which the shared face is Pt_3. The fifth platinum atom caps one Ru_2Pt triangle.[265] Unusual, carbyne-bridged clusters, [$PtNiMo_2(\mu_3\text{-}CC_6H_4\text{-}4\text{-}Me)(CO)_4(cod)Cp_2$] and [$PtMMo_2(\mu_2\text{-}CC_6H_4\text{-}4\text{-}Me)(\mu_3\text{-}CC_6H_4\text{-}4\text{-}Me)(CO)_4(cod)Cp_2$] (M = Ni or Pt), are also produced via reactions, initially, of $Pt(cod)_2$ with the mononuclear carbynemolybdenum precursor [$Mo({\equiv}CC_6H_4\text{-}4\text{-}Me)(CO)_2Cp$].[266] Relatedly, mixed molybdenum–tungsten–platinum clusters, [$Pt_4WMo_2(\mu_3\text{-}CMe)(\mu_3\text{-}CC_6H_4\text{-}4\text{-}Me)(CO)_6(cod)_2Cp_3$], [$Pt_5W_3Mo_2(\mu\text{-}CMe)(\mu\text{-}CC_6H_4\text{-}4\text{-}Me)(\mu_3\text{-}CC_6H_4\text{-}4\text{-}Me)(CO)_{10}(cod)Cp_5$] and [$Pt_6W_3Mo_2(\mu_3\text{-}CMe)(\mu_3\text{-}CC_6H_4\text{-}4\text{-}Me)_2(CO)_{10}(cod)_2Cp_5$] have been characterized, subsequent to addition of further $Pt(cod)_2$ to [$Pt_3WMo_2(\mu_3\text{-}CMe)(\mu_3\text{-}CC_6H_4\text{-}4\text{-}Me)_2(CO)_6(cod)_2Cp_3$], as well as congeners of higher nuclearity. An x-ray diffractometric determination of the structure of [$Pt_6W_3Mo_2(\mu_3\text{-}CMe)(\mu_3\text{-}CC_6H_4\text{-}4\text{-}Me)(CO)_{10}(PMe_2Ph)_4Cp_5$]—which is assumed to be isostructural with its precursor [$Pt_6W_3Mo_2(\mu_3\text{-}CMe)(\mu_3\text{-}CC_6H_4\text{-}4\text{-}Me)_2(CO)_{10}(cod)_2Cp_5$]—establishes that the molecular spine comprises 11 metal atoms in the sequence PtWPtMoPtMoPtWPtWPt and that each Pt_2M triangle is capped by a carbyne unit.[267]

Treatment of {[$(MeO)_3Si](OC)_3Fe(\mu\text{-}PR_2)Pt(cod)$} with $TlPF_6$ and two equivalents of $AuCl(PPh_3)$ leads to {$(OC)_3Fe(\mu\text{-}PR_2)[\mu_2\text{-}Au(PPh_3)]_2Pt(cod)$} (R = Ph or Cy). In this instance, phosphido bridging of the Fe–Pt bond in the cluster is reinforced by bridging $Au(PPh_3)$ units. The molecular structure of {$(OC)_3Fe(\mu\text{-}PR_2)[\mu_2\text{-}Au(PPh_3)]_2Pt(cod)$} has been confirmed by x-ray analysis and reveals no evidence of Au–Au interaction.[257] A novel polymetallic structure results from sequential treatment of [$Fe_2(CO)_6(\mu\text{-}Se_2)$] with HC≡CPh and $Pt(PPh_3)_2(\eta\text{-}C_2H_4)$ (Equation (25)). X-ray diffractometry shows that the coordination of the diselenoalkene in $(OC)_6Fe_2[\mu,\eta\text{-}SeC(Ph){=}C(H)Se]Pt(PPh_3)_2$ is essentially coplanar with the PtP_2 unit (Pt–C^1, 0.213 8(12) nm; Pt–C^2, 0.206 9(13) nm; C^1–C^2, 0.142 3(15) nm; C^1–Pt–C^2, 39.5(4)°).[268]

9.3 ALKYNE AND RELATED COMPLEXES

9.3.1 η^2-Monoalkyne Complexes

Comprehensive reviews of the chemistry of metal complexes of strained cyclic alkynes and arynes have appeared, which include synthetic, structural and reactive chemistry of η^2-cyclopentyne, η^2-cyclohexyne, η^2-cycloheptyne, η^2-cyclooctyne and η^2-aryne derivatives of zerovalent platinum, mainly

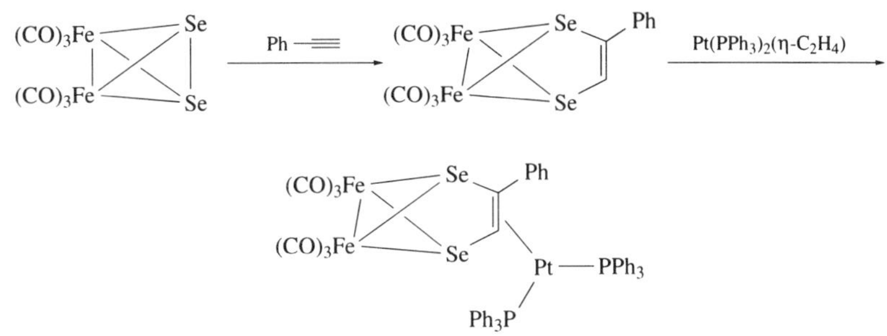

(25)

with ancillary phosphorus-donor ligands.[269,270] Two other related surveys (of cyclodiynes and their metal complexes and of group 4 metal–aryne derivatives) feature no platinum examples.[271]

9.3.1.1 Zerovalent derivatives

Like their monoalkene congeners (see Section 9.2.1.1) the ancillary ligands, if any, in (η^2-monoalkyne)platinum(0) complexes are usually other hydrocarbons or group 15 donor species. Three-coordinate (16-electron) structures are the most common, although dicoordination is not unknown. Some aspects have been reviewed.[25]

(i) Preparation

The preparation of zerovalent derivatives parallels the syntheses of the corresponding monoalkene complexes. Ligand displacements may accompany reduction of a platinum centre initially in a higher oxidation state, or may employ a zerovalent precursor.

(a) Preparations via reduction of higher valent platinum. A notable case is the high-yield synthesis of bis(η^2-alkyne)platinum complexes, $Pt(\eta\text{-}R^1C{\equiv}CR^2)_2$, by thermolysis of H_2PtCl_4 (Speier's catalyst; see Section 9.2.3.1(i)) in tetramethyldivinyldisiloxane, $O(SiMe_2CH{=}CH_2)_2$, followed by addition of alkyne ($R^1 = R^2 = Ph$; $R^1 = R^2 = CMe_2OH$; $R^1 = Ph$, $R^2 = CMePhOH$).[163] The nature of the reductive process—which is clearly implicated in catalysis of hydrosilation by these systems—is still unclear.

Intramolecular reductive processes may be used to afford (η^2-alkyne)platinum complexes. Thus, the reaction of $PtH_2(PMe_3)_2$ with $PhC{\equiv}CPh$ produces $Pt(PMe_3)_2(\eta\text{-}PhC{\equiv}CPh)$.[22] Thermolytic reductive elimination of neopentane from $Pt(H)(CH_2CMe_3)(dcpe)$ gives $[Pt^0(dcpe)]$. This highly reactive intermediate may be trapped by $PhC{\equiv}CPh$ to generate $Pt(dcpe)(\eta\text{-}PhC{\equiv}CPh)$. With $HC{\equiv}CPh$, a 1:1 mixture of $Pt(dcpe)(\eta\text{-}HC{\equiv}CPh)$ and the oxidative adduct $Pt(H)(C{\equiv}CPh)(dcpe)$ results (see Section 9.3.1.1(iii)).[21a] Photochemically induced reductive elimination of CO_2 from oxalatoplatinum(II) species has provided a versatile alternative route to η^2-alkene complexes (see Section 9.2.1.1(i)). Similarly, photolysis of $Pt(C_2O_4)L_2$ in presence of $PhC{\equiv}CPh$ yields $PtL_2(\eta\text{-}PhC{\equiv}CPh)$ (L_2 = dppe or dppm). There is an indication that, under some circumstances, $Pt(C_2O_4)L_2$ does not always generate CO_2 as the lone photochemical product, but that $Pt(CO_3)L_2$ and CO may also be products.[272] Electrochemical reduction may be exploited likewise. Controlled potential reduction of PtX_2L_2 generates nonlinear fragments, $[PtL_2]$, which can be trapped by $PhC{\equiv}CPh$ to form $PtL_2(\eta\text{-}PhC{\equiv}CPh)$ (X = Cl, Br or I; L_2 = dppe or dcpe).[273] Similarly, bulk reductive electrolysis of *cis*-$PtCl_2(PPh_3)_2$ in presence of $RC{\equiv}CR$ generates $Pt(PPh_3)_2(\eta\text{-}RC{\equiv}CR)$ (R = Ph or CO_2Me).[274]

In parallel with some alkene complexes, effective reduction of bromo- or dibromoalkenes may be employed. The 4-homoadamantyne complex $Pt(PPh_3)_2(\eta^2\text{-}C_{11}H_{14})$ is obtained via reduction of 4,5-dibromo-4-homoadamantene using sodium amalgam in the presence of $Pt(PPh_3)_3$ (Equation (26)).[275] Dehydrohalogenation of bromocycloheptene and bromocycloheptatriene induced by lithium diisopropylamide (LDA), again, in the presence of $Pt(PPh_3)_3$, leads to (η^2-cycloheptyne)- and (η^2-cycloheptadienyne)platinum complexes, respectively (Scheme 12). This may be contrasted with treatment of the same precursors with $KOBu^t$, which generates corresponding (η^2-cycloallene)platinum complexes.[230]

(b) Preparations via ligand displacements. Bis(η^2-alkyne)platinum complexes, $Pt(\eta\text{-}R^1C{\equiv}CR^2)_2$, are more generally accessible by treatment of $Pt(cod)_2$ or $Pt(\eta\text{-}C_2H_4)_3$ with excess $R^1C{\equiv}CR^2$. Reaction of $Pt(cod)_2$ is successful for a wide variety of alkynes ($R^1 = R^2$ = Ph, CH_2TMS, C_6H_4-4-OMe or CMe_2OH; $R^1 = Bu^t$, R^2 = TMS; R^1 = Ph, R^2 = TMS), but for many others ($R^1 = R^2$ = Me, Et, Bu^t, C_6H_4-4-Me or CMe_2OH; $R^1 = Bu^t$, R^2 = Me; R^1 = Ph, R^2 = Me), $Pt(\eta\text{-}C_2H_4)_3$ must be employed instead.[276]

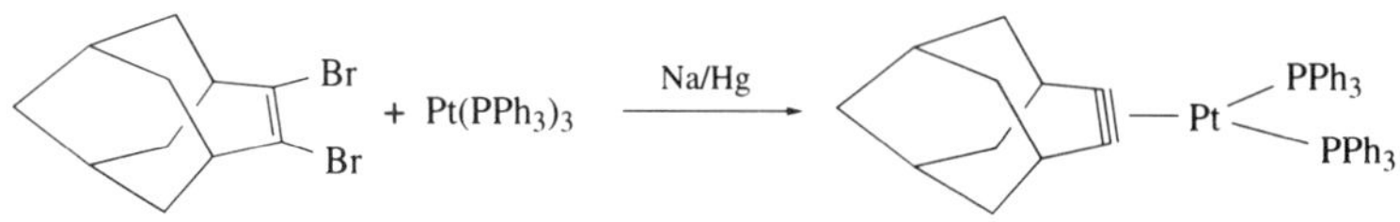

(26)

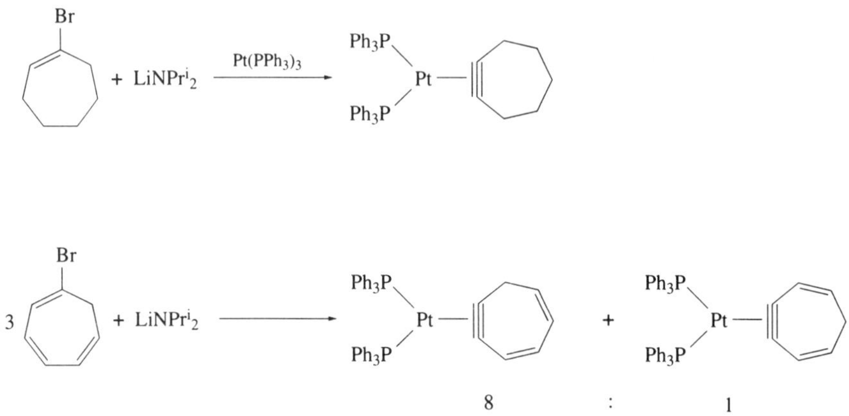

Scheme 12

Complexes of type $PtL_2(\eta\text{-}R^1C{\equiv}CR^2)$ may also be prepared from $Pt(cod)_2$. Equimolar amounts of $Pt(cod)_2$ and PhC≡CPh yield Pt(η-PhC≡CPh)(cod). In contrast, even excess of certain alkynes affords only $Pt(\eta\text{-}R^1C{\equiv}CR^2)(cod)$ ($R^1 = R^2 = C_6H_4$-4-Me, C_6F_5 or C_6F_4-4-OMe). Treatment of the latter ($R^1 = R^2 = C_6F_5$ or C_6F_4-4-OMe) with $CNBu^t$ affords $Pt(CNBu^t)_2(\eta\text{-}R^1C{\equiv}CR^2)$. Addition of $Pt(\eta\text{-}PhC{\equiv}CPh)_2$ to solutions of $CNBu^t$, PMe_3, PEt_3 and PPh_3 gives the corresponding species. $PtL_2(\eta\text{-}PhC{\equiv}CPh)$.[149]

The monofluoroethyne complex, $Pt(PPh_3)_2(\eta\text{-}FC{\equiv}CH)$, is obtained by treatment of $Pt(PPh_3)_2(\eta\text{-}C_2H_4)$ with FC≡CH. Similar reaction of ClC≡CCl does not afford a detectable (η^2-alkyne)platinum analogue. Instead, the product is the oxidative adduct, *trans*-$PtCl(C{\equiv}CCl)(PPh_3)_2$, whose formulation has been verified crystallographically.[277] Several new hexafluorobut-2-yne complexes, $PtL_2(\eta\text{-}F_3CC{\equiv}CCF_3)$, are accessible via sequential addition of phosphorus- or arsenic donor and alkyne to $Pt(cod)_2$ ($L = PCy_3$, $PPhCy_2$, PPh_2Cy or $AsPh_3$; $L_2 = Ph_2PN(CHMePh)PPh_2$). Reaction of $PhC{\equiv}CCO_2Et$ with $Pt(PPh_3)_4$ gives $Pt(PPh_3)_2(\eta\text{-}PhC{\equiv}CCO_2Et)$,[278] while displacement of ethene from $Pt[P(o\text{-}C_6H_4\text{-}2\text{-}OMe)_3]_2(\eta\text{-}C_2H_4)$ affords $Pt[P(o\text{-}C_6H_4\text{-}2\text{-}OMe)_3]_2(\eta\text{-}PhC{\equiv}CH)$.[13b]

(ii) Structural aspects

A crystallographic evaluation of the molecular structure of $Pt(\eta\text{-}PhC{\equiv}CPh)_2$ demonstrates an essentially tetrahedral PtC_4 arrangement. The angle between the PtC_2 planes is 82° and the bend-back angle is 26.8(5)°.[276] The structure of $Pt(PMe_3)_2(\eta\text{-}PhC{\equiv}CPh)$, as measured by x-ray diffraction, displays the normal tetragonal coplanar arrangement of ligating atoms (Table 4).[22] The molecular structure of the 4-homoadamantyne complex, $Pt(PPh_3)_2(\eta^2\text{-}C_{11}H_{14})$, also has an essentially coplanar arrangement of the $PtC^4C^5P^1P^2$ inner sphere (Table 4). The η^2-C–C bond shows slight elongation relative to the (η^2-cyclohexyne)- and (η^2-cycloheptyne)platinum congeners. The inherent molecular strain appears to be comparable with the η^2-cyclohexyne analogue.[275] The structure of the (η^2-cyclohepta-3,5-diene-1-yne)platinum complex, $Pt(PPh_3)_2(\eta^2\text{-}C_{17}H_6)$, displays an arrangement in which PtP_2 is coplanar with the $C^7C^1C^2C^3$ segment of the ring, which is folded along the C^7–C^3 vector. Selected distances and angles are included in Table 4.[230] The structures of several hexafluorobut-2-yne complexes, $PtL_2(\eta\text{-}F_3CC{\equiv}CCF_3)$ also show a similar, essentially coplanar, geometry, with F_3C–C–C angles which vary between 38° and 46°. Spectroscopic parameters ($L = PPh_3$, PPh_2Cy, $PPhCy_2$ or PCy_3; $L_2 = Ph_2PN(CHMePh)PPh_2$) are compared with those of $Pt(PPh_3)_2(\eta\text{-}R^1C{\equiv}CR^2)$ ($R^1 = R^2 = Ph$; $R^1 = Ph$, $R^2 = Me$ or CO_2Et) and correlated with diffractometric conclusions.[151,279] Platinum-195 NMR characteristics have been compiled for a related series of complexes, $Pt(PPh_3)_2(\eta\text{-}R^1C{\equiv}CR^2)$ ($R^1 = R^2 = H$, Me, Et, Ph, CF_3, CO_2Me, CO_2Et, $CO_2CH_2CF_3$ or CN; $R^1 = H$, $R^2 = Ph$ or CN; $R^1 = Me$, $R^2 = Ph$, CO_2Me or CN; $R^1 = Ph$, $R^2 = CO_2Me$). The platinum chemical shift is controlled predominantly by the energy of the π*-level of the alkyne. $^1J_{Pt\text{-}P}$, on the other hand, is highly sensitive

Table 4 Selected dimensions of the $PtC^1C^2E^1E^2$ skeleton in complexes of the type $Pt(ER_3)_2(\eta^2\text{-alkyne})$ (E = P or As).

Complex	Bond length (nm)					Bond angle (°)		R_w	Ref.
	Pt–C^1	*Pt–C^2*	*C^1–C^2*	*Pt–P^1*	*Pt–P^2*	*C^1–Pt–C^2*	*P^1–Pt–P^2*		
$Pt(PMe_3)_2(\eta\text{-}C_2Ph_2)$	0.206(1)	0.204(1)	0.129(2)	0.2284(4)	0.262(4)	36.7(5)	101.9(1)	0.056	22
$Pt(PPh_3)_2(\eta^2\text{-}C_7H_6)$[a]	0.2058(11)	0.200(2)	0.131(2)	0.2275(4)	0.2270(3)	37.5(6)	107.16(12)	0.0456	230
$Pt(PPh_3)_2(\eta^2\text{-}C_{11}H_{14})$[b]	0.2042(13)	0.2068(14)	0.1314(20)	0.2256(3)	0.2260(4)	37.3(5)	108.8(1)	0.081	275[c]
$Pt(AsPh_3)_2[\eta\text{-}C_2(CF_3)_2]$	0.2036(7)	0.2003(6)	0.1451(10)	0.23900(7)	0.23884(8)	37.0(3)	99.01(2)	0.0384	279
$Pt(Ph_2PN(CHMePh)PPh_2)[\eta\text{-}C_2(CF_3)_2]$	0.2019(18)	0.2014(15)	0.126(2)	0.2260(4)	0.2245(4)	36.4(6)	71.3(1)	0.069	278

[a] C_7H_6 = cyclohepta-3,5-dien-1-yne. [b] $C_{11}H_{14}$ = 4-homoadamantyne. [c] Also includes structural data for earlier analogues.

to the nature of R^1 and R^2 and correlation between $^1J_{Pt-P}$ and both the calculated polarizability and σ-bond populations (from charge-iterative extended Hückel calculations) suggest that substituent effects are transmitted primarily through the σ-bond framework.[280] In another, similar NMR study of $Pt(PPh_3)_2(\eta\text{-}R^1C{\equiv}CR^2)$ (variously, R^1 = H, Me, Et, Ph, TMS, $GeEt_3$, C_6H_4-4-Me or C≡CMe; R^2 = Me, Et, Ph, CMe_2OH, CHMeOH, CHPhOH, CPh(C≡CH)OH, TMS, $SiPh_3$, $GeEt_3$, $GePh_3$, $SnEt_3$, C_6H_4-4-Me, C_6H_4-4-F, C_6H_4-4-Br, C_6H_4-3-Br, C_6H_4-3-C≡C-TMS, C≡CMe, $(C{\equiv}C)_2Me$ or C≡C-TMS), $^1J_{Pt-P}$ also emerges as sensitive to the σ-bond infrastructure. In the triyne complex, coordination may involve either the internal or a terminal C≡C bond.[281]

(iii) Reactions

In an unusual intramolecular rearrangement, $Pt(PPh_3)_2(\eta\text{-}MeO_2CC{\equiv}CCO_2Me)$ undergoes photochemically induced oxidative addition of a C–C bond to yield *cis*-$Pt(CO_2Me)(C{\equiv}CCO_2Me)L_2$. Such photoisomerization does not occur, however, for analogues $Pt(PPh_3)_2(\eta\text{-}R^1C{\equiv}CR^2)$ ($R^1 = R^2$ = Ph, CH_2OH or CMe_2OH; R^1 = H, R^2 = CH_2OH, CO_2H or CH_2NH_2).[282] In a similar reaction, photolysis of Pt(dppe)(η-PhC≡CPh) affords Pt(Ph)(C≡CPh)(dppe).[272] This provides a contrast with the thermolytic behaviour of *cis*- and *trans*-$Pt(CH{=}CHMe)(C{\equiv}CBu^t)(PPh_3)_2$,[283a] and of $Pt(C{\equiv}CR)_2L_2$ (R = Ph or Bu^t; L = PPh_3; L_2 = dppe), which undergo reductive C–C elimination to generate—in the latter case—$Pt(PPh_2)_2(\eta\text{-}RC{\equiv}CR)$.[283b] Oxidative addition of haloalkynes, TMS-C≡CX, and the stannylalkyne, $PhC{\equiv}CSnR_3$ (R = Et or Ph) follows displacement of ethene from $Pt(PPh_2)_2(\eta\text{-}C_2H_4)$.[281] A range of alkynes, $RC{\equiv}CYPt(PPh_3)_2(\eta\text{-}RC{\equiv}CY)$ (R = Ph, Y = $SnEt_3$; R = TMS, Y = Cl or Br; R = $SiEt_3$ or Me, Y = I), also undergo oxidative addition of C–Y. The (η-alkyne)platinum precursor, $Pt(PPh_3)_2(\eta\text{-}RC{\equiv}CY)$, is observable in some cases (R = Ph, Y = $SnEt_3$; R = TMS, Y = Cl or Br). The triyne complex, $Pt(PPh_3)_2[\eta\text{-}Me(C{\equiv}C)_3Me]$, also yields $Pt(C{\equiv}CMe)[(C{\equiv}C)_2Me](PPh_3)_2$ via oxidative C–C addition.[281]

The highly reactive intermediate $[Pt^0(dcpe)]$ reacts with HC≡CPh to afford a 1:1 mixture of Pt(dcpe)(η-HC≡CPh) and the oxidative adduct Pt(H)(C≡CPh)(dcpe).[21a] An unusual example of C–H addition is stimulated by the reaction of $Pt(PPh_3)_2(\eta\text{-}HC{\equiv}CPh)$ with the triflato-rhodium derivative $Rh(OSO_2CF_3)_2(PMe_3)Cp^*$, to give a binuclear, cationic product, $[(Ph_3P)_2Pt(\mu\text{-}H)(\mu,\eta\text{-}PhC{\equiv}C)Rh(PMe_3)Cp^*](OSO_2CF_3)_2$. Here, the bridging ethynyl group is η^1-bound to platinum and η^2-bound to rhodium.[133]

9.3.1.2 Divalent derivatives

(i) Preparation

Mononuclear bis(η^2-alkyne)platinum(II) complexes remain relatively rare. Nevertheless, new examples of platinum(II) and palladium(II) derivatives of this type, *cis*-$M(C_6F_5)_2(\eta\text{-}PhC{\equiv}CPh)_2$, may be synthesized by treatment of *cis*-$M(C_6F_5)_2(THF)_2$ with PhC≡CPh.[284]

Pentacoordinate (η^2-alkyne)platinum complexes are also proving to be accessible. Like their η^2-alkene analogues (see Section 9.2.1.2), these seem most favourable for alkynes with electron-withdrawing substituents and feature bidentate nitrogen donors (N–N') as ancillary ligands. Such alkynes add readily to planar complexes, PtClR(N–N'), to give $PtClR^1(N\text{–}N')(\eta\text{-}R^2C{\equiv}CR^3)$ (R^1 = Me or C_6H_4-4-OMe; $R^2 = R^3$ = Me, CF_3, CO_2Me or CH_2Cl; R^2 = H, R^3 = CO_2Me; N–N' = 6-Me-C_5H_4N-2-CH=NMe (mimpy), 6-Me-C_5H_4N-2-CH=NPh (pimpy) or 2,9-Me_2-phen). In the presence of excess MeCN, the cationic complexes, $[PtMe(NCMe)(2,9\text{-}Me_2\text{-}phen)(\eta\text{-}R^2C{\equiv}CR^3)]^+$, may also be characterized ($R^2 = R^3$ = CF_3 or CO_2Me). Use of a 1:2 alkyne:platinum ratio (N–N' = 2,9-Me_2-phen) leads to isolation of the binuclear species $Pt_2Cl_2Me_2(2,9\text{-}Me_2\text{-}phen)(\eta\text{-}R^2C{\equiv}CR^3)$ ($R^2 = R^3$ = Me or CO_2Me; R^2 = H, R^3 = Ph or CO_2Me) (see Section 9.3.4.1).[285]

(ii) Structural aspects

The molecular structure of *cis*-$Pt(C_6F_5)_2(\eta\text{-}PhC{\equiv}CPh)_2$, determined by x-ray diffractometry, reveals η^2-alkyne (and η^1-aryl) coordination perpendicular to the tetragonal coordination plane.[284] As indicated by multinuclear NMR studies, the orientation of the alkyne in pentacoordinate derivatives, PtClR(N–N')(η^2-alkyne), is equatorially coplanar in a trigonal-bipyramidal geometry.[285]

(iii) Reactions

The mechanism of alkyne dissociation from PtXMe(bipy)(η-RC≡CR) is sensitive to the nature of both X and R. Where X = Cl and R = CO_2Me, dissociation occurs by a path involving preliminary ionization of Cl^-. If X = Cl or I and R = CF_3, however, simple dissociative loss of alkyne is proposed.[286]

Reaction of RC≡CR with *trans*-PtX(TMS)$(PEt_3)_2$ leads to migratory insertion, generating (ultimately) *trans*-PtX((*E*)-HRC=CR-TMS)$(PEt_3)_2$ (R = Ph or Pr; X = Br or I). Mechanistic studies suggest that this proceeds via dissociation of PEt_3 and association of RC≡CR to give *trans-X,Si*-PtX(TMS)(PEt_3)(η-RC≡CR). The reactivity of *trans*-PtX(TMS)$(PEt_3)_2$ towards PhC≡CPh is appreciably higher than that of *trans*-PtBrH$(PEt_3)_2$.[287]

9.3.1.3 *Theoretical aspects*

η^2-Monoalkyne complexes have received less attention than those of corresponding alkene derivatives (see Section 9.2.1.3), although there is a degree of overlap; the energy relationship between $M(CH)_2$ and M(η-C_2H_2) has been the subject of an extended Hückel study.[142] Extended Hückel calculations have also been carried out on the systems M(η-alkyne)L_3 and used to rationalize trends in structure and reactivity (M = Fe^0, Co^I, Ir^I or Pt^{II}), including the general energetic preference for perpendicular coordination in tetracoordinate planar platinum(II) complexes.[288] Charge-iterative extended Hückel calculations feature in a semiquantitative interpretation of the correlations between $^1J_{Pt-P}$ and both the calculated polarizability and σ-bond populations for a series of complexes: $Pt(PPh_3)_2$(η-$R^1C≡CR^2$) ($R^1 = R^2$ = H, Me, Et, Ph, CF_3, CO_2Me, CO_2Et, $CO_2CH_2CF_3$ or CN; R^1 = H, R^2 = Ph or CN; R^1 = Me, R^2 = Ph, CO_2Me or CN; R^1 = Ph, R^2 = CO_2Me).[280] The preference for transverse (perpendicular) or parallel μ-alkyne bridge bonding in $Pt_2(CO)_4$(μ-C_2H_2) has also been the subject of extended Hückel calculations.[289]

9.3.2 η^2-Aryne and Related Complexes

9.3.2.1 *Preparation, structure and reactivity*

The preparation, structure and reactivity of η^2-aryne and related complexes have been surveyed in two reviews. Organoplatinum examples are relatively few and all are zerovalent metal complexes with phosphorus-donor ancillary ligands, of general type $Pt(PR_3)_2(\eta^2$-aryne), typically prepared by reduction (Na/Hg) of the 1,2-dibromoarene in the presence of a suitable precursor to PtL_2, for example, PtL_2(η-C_2H_4) (L = PEt_3, PPr^i_3 or PCy_3; L_2 = dcpe).[269–71] Divalent examples also remain elusive but (η^2-aryne)platinum(II) derivatives are produced during thermolytic rearrangements of *cis*-Pt(C_6H_4-4-R)$_2py_2$ (R = H or Bu^t), which ultimately reductively eliminate mixtures of PhR, 4-R-C_6H_4–C_6H_4-4-R, 3-R-C_6H_4–C_6H_4-4-R and 3-R-C_6H_4–C_6H_4-3-R (Scheme 13). This represents a (to date) rare instance of (η^2-aryne)metal formation via reversible β-hydrogen transfer between ligand and metal. Labelling studies establish that β-hydrogen transfer from coordinated pyridine also occurs as a competitive process, to afford (η^2-pyridyne)platinum(II) intermediates.[193]

9.3.3 η^2-X≡CR Complexes

Developments in metal complexes of phosphaalkynes, P≡CR, have been the subject of a comprehensive survey.[238]

9.3.3.1 *Preparation, structure and reactivity*

P≡CBu^t reacts with various platinum(0) species, including $Pt(PPh_3)_2$(η-C_2H_4), $Pt(dppe)_2$, Pt(triphos)(PPh_3) and $Pt(cod)_2$ (in the presence of two equivalents of PMe_3 or PPh_2) to give $Pt(PR_3)_2(\eta^2$-P≡CBu^t).[238] The molecular structure of $Pt(PPh_3)_2(\eta^2$-P≡CBu^t), determined by x-ray diffractometry, confirms the π-coordination mode, with no significant interaction of the residual electron pair on

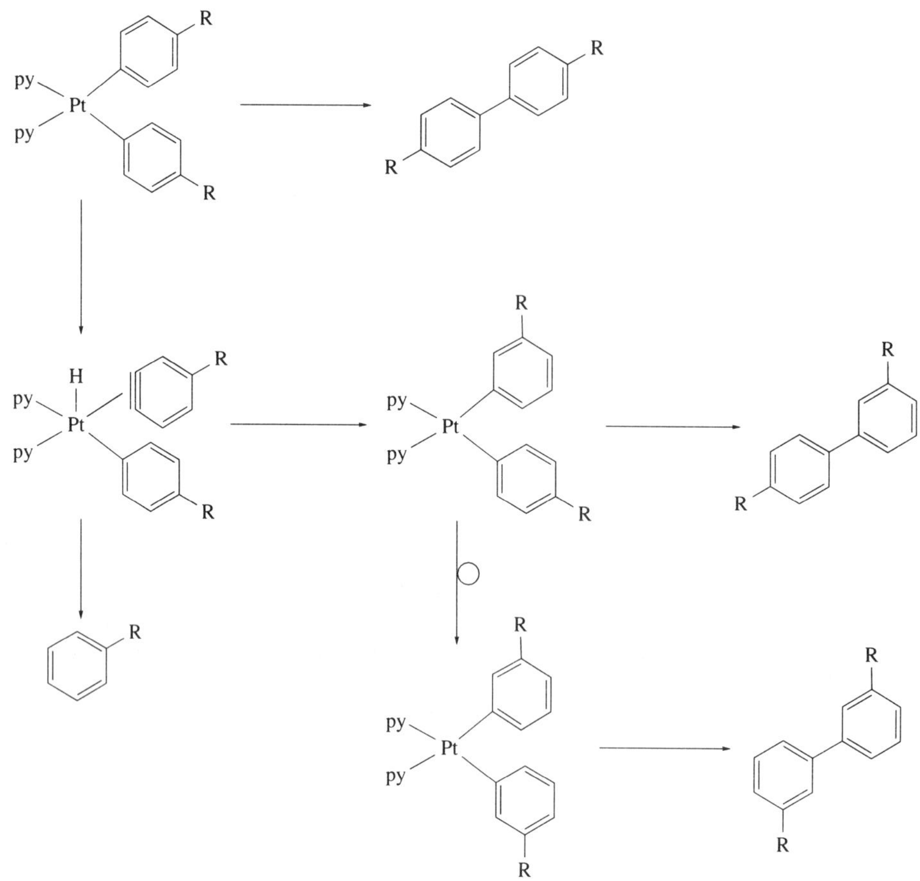

Scheme 13

phosphorus with platinum. The P–C bond shows considerable lengthening (relative to the free ligand) and the bend-back angle is 48(2)°. Spectroscopically, the phosphaalkyne ligand exhibits an extremely low value of $^1J_{P-Pt}$ (62 Hz). This is rationalized by low *s*-orbital character in the Pt–P bond, as a consequence of the noninteractive lone pair.[290] Reaction of $Pt(cod)_2$ with a 2:1 mixture of $MesP{=}CPh_2$ and $P{\equiv}CBu^t$ leads to isolation of $Pt(\eta^1\text{-}MesP{=}CPh_2)_2(\eta^2\text{-}P{\equiv}CBu^t)$, in which the phosphaalkene ligands show their established general preference for η^1-coordination (see Section 9.2.4.3). Only the phosphaalkyne is η^2-bound and, again, exhibits an unusually low value of $^1J_{Pt-P}$ (144 Hz; cf. 467 and 455 Hz for the phosphaalkenes).[291] These complexes have the potential to form derivatives of higher nuclearity, by use of either the unused terminal electron pair on phosphorus or the residual P≡C unsaturation. Examples of both have emerged. For example, $Pt(dppe)(\eta^2\text{-}P{\equiv}CBu^t)$ reacts with $M(CO)_5(THF)$, generating $PtM(dppe)(CO)_5(\mu,\eta^1,\eta^2\text{-}P{\equiv}CBu^t)$, in which the phosphaalkyne is η^1-phosphorus bound to M while retaining its η^2-coordination to platinum. Reaction of the same platinum precursor with $Fe(CO)_5$ or with $Fe_2(CO)_9$, on the other hand, affords $PtFe_2(dppe)(CO)_6(\mu,\eta^1,\eta^2\text{-}P{\equiv}CBu^t)$ which displays transverse bridge bonding across the Fe–Fe bond as well as η^1-phosphorus binding to platinum.[238] Interaction of $PtCl_4L_2$ and $(\mu,\eta\text{-}P{\equiv}CBu^t)[Mo(CO)_2Cp]_2$ leads to complexes where a *trans*-$PtCl_2L$ unit is simply η^1-coordinated via the terminal phosphorus in the cluster.[292]

9.3.4 Bimetallic Alkyne Complexes

9.3.4.1 Homobimetallic complexes: preparation, structure and reactivity

Treatment of $Pt(\eta\text{-}PhC{\equiv}CPh)_2$ with $Pt(PPh_3)_2(\eta\text{-}C_2H_4)$ or $Pt_2(PEt_3)_4(\mu\text{-}cod)$ produces the asymmetric dimeric platinum(0) species, $Pt_2(PR_3)_2(\mu\text{-}PhC{\equiv}CPh)(\eta\text{-}PhC{\equiv}CPh)$ (R = Et or Ph). $Pt_2(PPh_3)_2(\mu$-

$PhC{\equiv}CPh)(\eta\text{-}PhC{\equiv}CPh)$ is also accessible via reaction of $PhC{\equiv}CPh$ with $Pt(PPh_3)(\eta\text{-}C_2H_4)_2$. These examples feature transverse η-alkyne binding.[293]

Binuclear platinum(II) derivatives, $Pt_2Cl_2Me_2(2,9\text{-}Me_2\text{-phen})(\eta\text{-}R^1C{\equiv}CR^2)$, are formed when $R^1C{\equiv}CR^2$ and $PtClMe(2,9\text{-}Me_2\text{-phen})$ are reacted in a 1:2 ratio ($R^1 = R^2$ = Me or CO_2Me; R^1 = H, R^2 = Ph or CO_2Me). The molecular structure of $Pt_2Cl_2Me_2(2,9\text{-}Me_2\text{-phen})(\eta\text{-}MeC{\equiv}CMe)$ has been confirmed diffractometrically. The two platinum atoms have different coordination geometries and are bridged (transversely) by C_4H_6 and by one chlorine atom. Coordination around the pentacoordinate platinum atom is trigonal bipyramidal with μ-Cl and methyl axial and the alkyne equatorial coplanar with 2,9-Me_2-phen. There is essentially planar geometry at the tetracoordinate metal centre with C–C perpendicular to the coordination plane. An approximate mirror plane includes both platinum atoms, μ-Cl and Me (although the median of μ,η-MeC≡CMe is slightly displaced from this plane).[285]

The alternative, parallel mode of μ-alkyne binding is observed in the product of reacting the zerovalent precursor $Pt(PPh_3)_2(\eta\text{-}RC{\equiv}CR)$ with CO to afford $[Pt_2(CO)_2(PPh_3)_2(\mu\text{-}RC{\equiv}CR)]$ (R = CO_2Me). The same product is accessible by reaction of RC≡CR with $Pt(CO)_2(PPh_3)_2$. On the evidence of x-ray diffractometry, the alkyne lies coplanar with the Pt–Pt bond.[294] Parallel alkyne bridging is also encountered in A-frame diplatinum(II) complexes, most commonly with dppm as the ancillary bridging ligand(s). For example, $[Pt_2H_2(\mu\text{-}H)(\mu\text{-dppm})_2][PF_6]$ reacts with RC≡CR to yield the neutral species, $[Pt_2Cl(\eta^1\text{-}(E)\text{-}RC{=}CHR)(\mu\text{-dppm})_2(\mu\text{-}\eta^1,\eta^1\text{-}RC{=}CR)]$ (R = CO_2Me), with elimination of H_2 (Equation (27)); the chloro ligand is abstracted from the dichloromethane solvent).[295] The same coordinative preference is exhibited by $[Pt_2Cl(\eta^1(Z)RC{=}CHR)(\mu\text{-dppm})_2(\mu\text{-}\eta^1,\eta^1\text{-}RC{=}CR)]$, which is prepared from the reaction of $[Pt_2H_2(\mu\text{-}H)(\mu\text{-dppm})_2][PF_6]$ with RC≡CR (R = CF_3 or CO_2Me). Notably, here, the η^1-alkenyl group has *trans*-geometry. In donor solvents, for example, NCMe, NCPh and $OCMe_2$, the product is $[Pt_2H(solv)(\mu\text{-}RC{\equiv}CR)(\mu\text{-dppm})_2][PF_6]$, which may be converted to $Pt_2ClH(\mu\text{-}RC{\equiv}CR)(\mu\text{-dppm})_2$ or $Pt_2Cl_2(\mu\text{-}RC{\equiv}CR)(\mu\text{-dppm})_2$ with NaCl or HCl, respectively. If $HC{\equiv}CCF_3$ is used, $Pt_2(C{\equiv}CCF_3)_2(\mu\text{-}HC{\equiv}CCF_3)(\mu\text{-dppm})_2$ is recovered as the primary product.[296]

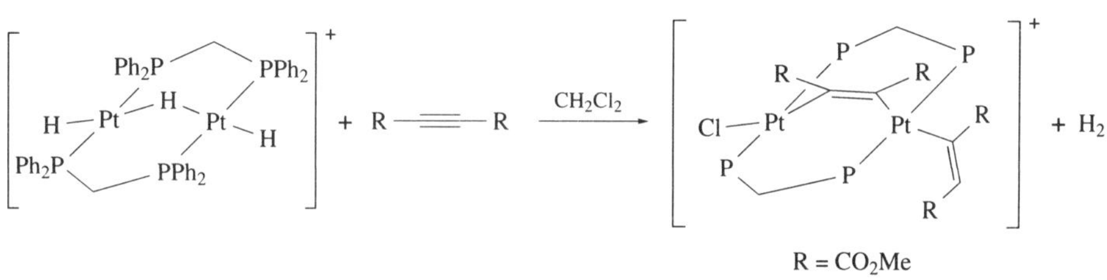

(27)

A somewhat different type of alkyne-bridging is exemplified in the A-frame derivatives, $[Pt_2(\eta^1\text{-}C{\equiv}CPh)_2(\mu\text{-dppm})_2(\mu,\eta^1,\eta^2\text{-}C{\equiv}CPh)](ClO_4)$ and $[Pt_2Cl(\eta^1\text{-}C{\equiv}CBu^t)(\mu\text{-dppm})_2(\mu,\eta^1,\eta^2\text{-}C{\equiv}CBu^t)](ClO_4)$, which are prepared by reaction of $[Pt(dppm)]Cl_2$ with $Hg(C{\equiv}CR)_2$ (X = Cl or ClO_4). Here, an η^1-alkynyl ligand on one platinum centre binds in an η^2-fashion to the neighbouring metal. These species have unusual photophysical properties, exhibiting long-lived photoluminescence ($\tau \approx 2\text{–}3$ μS; $\lambda_{em} \approx 620$ nm) in both the dissolved and the solid state.[297] The same binding accounts for the structures of $[Pt_2(\eta^1\text{-}C_6F_5)_2(PPh_3)_2(\mu,\eta^1,\eta^2\text{-}C{\equiv}CR)_2]$, which are obtained from admixtures of *trans*-$Pt(C{\equiv}CR)_2(PPh_3)_2$ and *cis*-$Pt(C_6F_5)_2(THF)_2$ (R = Ph, Bu^t or TMS). An x-ray diffractometric study (R = Ph) confirms the structure, which represents an exchange of substituents between metal centres. The bridging in these complexes is sufficiently substitution labile to allow isolation of the corresponding mononuclear complexes *trans*-$Pt(C_6F_5)(C{\equiv}CR)(PPh_3)L$ on addition of L (L = PPh_3 or py).[298]

9.3.4.2 Heterobimetallic complexes: preparation, structure and reactivity

In contrast with the μ,η^1,η^2-bridging mode favoured by related homobimetallic complexes,[298] $[Pt(\eta^1\text{-}C_6F_5)_2(\mu,\eta^1\text{-}C{\equiv}CR)_2TiCp_2]$—formed in an analogous way from $Ti(C{\equiv}CR)_2Cp_2$ and *cis*-$Pt(C_6F_5)_2(THF)_2$—displays no appreciable π-component in the bonding of the μ-alkynyl groups.[299] Interaction of *trans*-$PtCl_2(SMe_2)_2$ with one and two equivalents of $[Au(C{\equiv}CBu^t)]_n$ affords unstable species, $[PtCl(SMe_2)(C{\equiv}CBu^t)(AuCl)]$ and $[Pt(C{\equiv}CBu^t)_2(SMe_2)_2(AuCl)_2]$, respectively, in which AuCl units are proposedly linked by alkynylplatinum. These react further with dppm to produce *trans*-$[PtCl(C{\equiv}CBu^t)(\mu\text{-dppm})_2(Au)]Cl$, *trans*-$[Pt(C{\equiv}CBu^t)_2(\mu\text{-dppm})_2(Au)]Cl$ and *trans*-$[PtCl(C{\equiv}CBu^t)\{(\mu\text{-dppm})AuCl\}]$, in which, again, there is no evidence of η^2-binding to platinum.[300]

In contrast with the effect of adding terminal alkynes to $[(OC)_3Fe(\mu\text{-dppm})(\mu\text{-CO})Pt(PPh_3)]$ (see Section 9.2.7.2), reaction with the internal alkyne, $MeO_2CC{\equiv}CCO_2Me$, leads to $[(OC)_4Fe(\mu\text{-dppm})Pt(\eta\text{-}MeO_2CC{\equiv}CCO_2Me)]$, via carbonyl bridge cleavage and phosphine displacement. The Pt–Fe bond is spanned by dppm only, and the alkyne occupies a terminal coordination site on platinum.[260]

9.3.5 Polymetallic Alkyne Complexes

9.3.5.1 Homometallic complexes: preparation, structure and reactivity

Addition of $Pt(PPh_3)_2(\eta\text{-}C_2H_4)$ or $Pt_2(PEt_3)_4(\mu\text{-cod})$ to $Pt_2(PR_3)_2(\mu\text{-PhC}{\equiv}\text{CPh})(\eta\text{-PhC}{\equiv}\text{CPh})$ (R = Ph and Et, respectively) generates the triplatinum derivative, $Pt_3(PR_3)_4(\mu\text{-PhC}{\equiv}\text{CPh})_2$ (R = Et or Ph), also obtainable by direct reaction of $Pt(\eta\text{-PhC}{\equiv}\text{CPh})_2$ with two equivalents of $Pt(PR_3)_2(\eta\text{-}C_2H_4)$ (R = Ph). An x-ray structural analysis of $Pt_3(PPh_3)_4(\mu\text{-PhC}{\equiv}\text{CPh})_2$ confirms the open-chain structure of the trimer, with a substantial bend-back angle (41°) for the transversely bound μ-PhC≡CPh.[293]

Closed triplatinum clusters are more usual where μ-dppm is used as an ancillary ligand. One of the products of reaction of $[Pt_3(\mu_3\text{-H})(\mu\text{-dppm})_3]^+$ with excess HC≡CH is $[Pt_3(CH{=}CH_2)(\mu\text{-dppm})_3(\mu,\eta\text{-}C_2H_2)]^+$, whose structure, as determined by NMR, features a transversely bridging alkyne as well as a vinyl ligand.[301] The structure of $[Pt_3\{\mu,\eta\text{-}(Ph_2P)_2CHCH{=}CH_2\}(\mu\text{-dppm})_2(\mu,\eta\text{-}C_2H_2)]^+$ has also been analysed by x-ray diffraction.[302]

9.3.5.2 Heterometallic complexes: preparation, structure and reactivity

Many examples of heterometallic complexes also feature 1,5-cyclooctadiene as a ligand and arise via reactions of $Pt(cod)_2$ with established (alkene)metal complexes or by interaction of alkynes with metal aggregates derived from previous additions of $Pt(cod)_2$ to other metal complexes. Most employ the μ_3,η-face-bridging mode of alkyne coordination, in which the perpendicular η^2-coordination is to metals in the cluster other than platinum. As such, the preparations and structures of such clusters have been discussed in Section 9.2.8.2 and include $[Pt_2Co_2(CO)_8(cod)(\mu_3,\eta\text{-EtC}{\equiv}\text{CEt})]$, $[Pt_3Co_2(CO)_6(cod)_2(\mu_3,\eta\text{-PhC}{\equiv}\text{CPh})_2]$, $[Pt_2Co_2(CO)_8(cod)(\mu_3,\eta\text{-EtC}{\equiv}\text{CEt})]$, $[Pt_3Co_2(CO)_6(cod)_2(\mu_3,\eta\text{-PhC}{\equiv}\text{CPh})_2]$,[262] $[PtRh_2(\mu_3\text{-CO})(\mu_3,\eta\text{-}F_3CC{\equiv}CCF_3)(cod)Cp_2]$, $[PtRh_2(\mu_3\text{-CO})(\mu_3,\eta\text{-}F_3CC{\equiv}CCF_3)(cod)Cp_2]$,[263] $[Pt_5Fe_2(CO)_7(\mu,\eta\text{-PhC}{\equiv}\text{CPh})(\mu_3,\eta\text{-PhC}{\equiv}\text{CPh})(cod)]$ and $[Pt_2Fe(CO)_3(\mu\text{-PhC{=}CH{-}CH{=}CPh})(cod)_2]$.[264] In addition, a variety of related complexes without cod have been characterized structurally, all with μ,η-coordination of the alkyne, but none with perpendicular η^2-bonding to platinum. In $[PtRh_4(CO)(\mu_3,\eta\text{-}F_3CC{\equiv}CCF_3)_2Cp_4]$—one product of the reaction of $Pt(cod)_2$ and $[Rh_2(\mu_2\text{-CO})_2\text{-}(PhC{\equiv}CPh)Cp_2]$—platinum is a common apex for two linked $PtRh_2$ triangles, with the alkyne μ_3,η-face-bonded to each triangle. Each Rh–Rh bond is spanned by a semibridging CO.[263] A product from the interaction of PhC≡CPh and $Pt_2Ru(CO)_{18}$, $[Pt_2Ru_3(CO)_{18}(\mu_3,\eta\text{-PhC}{\equiv}\text{C Ph})(\mu_4,\eta\text{-PhC}{\equiv}\text{C Ph})]$ (6%), forms a distorted square-pyramidal core. For another product (24%) from the same preparative scheme, $[Pt_3Ru_6(CO)_{14}(\mu_3,\eta\text{-PhC}{\equiv}\text{C Ph})_3]$, a Pt_3Ru_3 octahedron is capped by three ruthenium units, with a Ru–Ru bond between each of the capping centres. Reaction of this latter species with additional CO gives two further clusters (16% and 20%, respectively). The first, $[Pt_2Ru_4(CO)_{14}(\mu_3,\eta\text{-PhC}{\equiv}\text{C Ph})(\mu_4,\eta\text{-PhC}{\equiv}\text{CPh})]$, is a square-pyramidal Pt_2Ru_3 cluster, which has a ruthenium bridge on the Ru–Pt axial–basal edge. The second, $[Pt_3Ru_6(CO)_{18}(\mu_3,\eta\text{-PhC}{\equiv}\text{C Ph})_3]$, has a central Pt_3 core with ruthenium bridging each Pt–Pt edge and an alkyne spanning each $PtRu_2$ triangle. Reaction of this complex with dppe affords the triangulo cluster $[PtRu_2(CO)_6(\mu_3,\eta\text{-PhC}{\equiv}\text{CPh})(dppe)]$.[303] In addition, $[PtRu_3(CO)_9(\mu\text{-H})(\mu_4,\eta\text{-C}{\equiv}\text{CBu}^t)(cod)]$ rearranges thermolytically to $[PtRu_6(CO)_{18}(\mu\text{-H})_2(\mu_4,\eta\text{-C}{\equiv}\text{CBu}^t)_2]$,[304] while addition of $Ph_2PC{\equiv}CPPh_2$ to $[PtRu_3(CO)_9(\mu\text{-H})(\mu_4,\eta\text{-C}{\equiv}\text{CBu}^t)(dppe)]$ gives $[PtRu_3(CO)_7(\mu\text{-PPh}_2)(\mu_4,\eta\text{-C}{\equiv}\text{CBu}^t)(dppe)]$.[305]

An unusual hexanuclear complex $[PtAg_2(C_6F_5)_2(C{\equiv}CR)_2(O{=}CMe_2)_2]_2$ is produced from the reaction of $(Bu^n{}_4N)_2[Pt(C_6F_5)_2(C{\equiv}CR)_2]$ with $AgClO_4$ in acetone (R = Ph or Bu^t). The structure features novel μ_3,η^2-binding to both platinum and silver, with no Pt–Ag bonding. The alkynyl is, however, bound only to platinum.[306] The structures of hexametallic $[Pt_2Ag_4(C{\equiv}CR)_8]$ (R = Ph or Bu^t) have also been determined.[307]

9.4 ALLYL AND RELATED COMPLEXES

9.4.1 η^3-Allyl Complexes

9.4.1.1 Preparation

The chemistry of (η^3-allyl)platinum derivatives is primarily that of the divalent state. These are often prepared from platinum(II) precursors, either with the allyl ligand already in place or by transmetallation, using an allyl derivative of another metal. Alternatively, oxidative addition of a suitable substrate to a platinum(0) parent may be used. Just as the Zeise anion and dimer are commonly employed in synthesis of η^2-alkene complexes, tetrameric [PtCl(η-C_3H_5)]$_4$ and dimeric species, [PtCl(η-C_3H_4R)]$_2$, are frequent starting points for other η^3-allyl compounds, by halogeno bridge cleavage or displacement reactions. The homoleptic species Pt(η-C_3H_4R)$_2$ also are useful precursors.

(i) Preparation from Pt^{II} precursors

Reaction of PtX(η^1-C_3H_5)(cod) (see Section 9.4.1.1(ii)) with pyridine results in diene displacement to give PtX(py)(η-C_3H_5). Likewise, treatment of PtX(η^1-C_3H_4-3-R)(cod) with PCy$_3$ or interaction of Pt(PCy$_3$)(η-C_2H_4)$_2$ with allyl halides leads to PtX(PCy$_3$)(η-C_3H_4-3-R) (R = H, Me or Ph; X = Cl or Br). Addition of L^1 (py or PPh$_3$) or L$^1_2$ (dppe) to [PtX(η-C_3H_4-3-R)(cod)]$^+$ (see Section 9.4.1.1(ii)) results in formation of [PtL$^1_2$(η-C_3H_4-3-R)]$^+$, while sequential introduction of one equivalent of PPh$_3$ and one of L^2 (CO, py or NCBu^t) affords [PtL^{1}L^2(η-C_3H_4-3-R)]$^+$.[308] Treatment of PtX(η^1-C_3H_4-3-R)(cod) with TlCp gives Pt(η-C_3H_4-3-R)Cp (R = H, Me or Ph; X = Cl or Br). When R is Ph, the σ-bound intermediate Pt(η^1-C_3H_4-3-Ph)(η^1-Cp)(cod) is isolable. Addition of excess CNBu^t to Pt(η-C_3H_4-3-R)Cp affords *trans*-Pt(η^1-C_3H_4-3-R)(η^1-Cp)(CNBu^t)$_2$, which rearranges to the thermodynamically more favourable *cis* isomer.[188]

Treatment of Pt(η-C_3H_5)$_2$ with PPh$_3$ generates Pt[η^1-H$_2$CCH=CH$_2$](PPh$_3$)(η-C_3H_5), with both η^1- and η^3-allyl groups.[309] Analogous reaction of Pt(η-C_3H_4-2-Me)$_2$ with PCy$_3$ results in the production of Pt[η^1-H$_2$CC(Me)=CH$_2$](PCy$_3$)(η-C_3H_4-2-Me).[310]

Similar reactivity attends the interaction of Pt(η-C_3H_4-2-Me)Cp and PCy$_3$, which affords the η^3-allyl complex, Pt(η^1-Cp)(PCy$_3$)(η-C_3H_4-2-Me), with the cyclopentadienyl group adopting the η^1-role.[311]

Addition of a range of diazabutadienes, R^{1}N=C(R^2)–C(R^2)=NR^1 to [PtCl(η-C_3H_5)]$_4$, in the presence of NaClO$_4$, yields mononuclear, cationic complexes, [Pt(η-C_3H_5){R^{1}N=C(R^2)–C(R^3)=NR^1}]ClO$_4$, in which the diimine is *N,N'*-bidentate (R^1 = C_6H_4-4-OMe, R^2 = R^3 = H; R^1 = C_6H_4-4-OMe, R^2 = H, R^3 = Me; R^1 = C_6H_4-4-OMe, R^2 = R^3 = Me; R^1 = Bu^t, R^2 = R^3 = H).[312] Related η^3-2-methallylplatinum complexes are accessible by reaction of pyridine-2-carbaldiimines, RN=CH-2-C_5H_5N, with [PtCl(η-C_3H_4-2-Me)]$_2$. The product in this case is [Pt(RN=CH-2-C_5H_5N)(η-C_3H_4-2-Me)]X, where X may be Cl$^-$ or [PtCl$_2$(η-C_3H_4-2-Me)]$^-$, depending on the stoichiometry of the reaction (R = C_6H_4-4-Me or Me).[313]

Reaction of [PtCl(η-C_3H_5)]$_4$ with Ph$_2$PCH$_2$CH$_2$SR affords neutral, tetracoordinate PtCl(η^1-C_3H_5)(Ph$_2$PCH$_3$CH$_2$SR), which has monodentate allyl and bidentate P,S ligands, respectively. This derivative is stable in nonpolar solvents, but in polar media (chlorocarbons or alcohols) chloride is displaced rapidly to give ionic [Pt(η^1-C_3H_5)(Ph$_2$PCH$_2$CH$_2$SR)(η-C_3H_5)]Cl.[314] Addition of the unusual triarylphosphine ligand L, P[C_6H_3-2,6-(OMe)$_2$], to [PtCl(η-C_3H_4-2-Me)]$_2$ furnishes PtCl(L)(η-C_3H_4-2-Me) in a simple bridge scission. This complex undergoes halide abstraction with AgBF$_4$, giving [Pt(L)(η-C_3H_4-2-Me)]BF$_4$, in which the arylphosphine is *P,O* bidentate, via one *o*-methoxy group. This group is replaced readily by CO to form [Pt(L)(CO)(η-C_3H_4-2-Me)]BF$_4$ which, upon treatment with an additional 0.5 equivalent of L, yields the allyl bridged, binuclear platinum(I) species, [Pt$_2$(L)$_2$(CO)$_2$(μ^2,η-C_3H_4-2-Me)]BF$_4$. If PtCl(L)(η-C_3H_4-2-Me) is reacted with Tl(acac), the product is [Pt{η^1-CH(COMe)$_2$}(L)(η-C_3H_4-2-Me)], whose structure contrasts with that of the corresponding triphenylphosphine complex which has η^1-allyl and η-pentanedionato ligands.[315] When Pt(η-acac)(η-C_3H_4-2-Me) is reacted with two equivalents of phosphine ligand, L, at low temperature, formation of the ion pair [PtL$_2$(η-C_3H_4-2-Me)][acac] is observed by NMR (L = PPh$_3$; L$_2$ = (Z)-Ph$_2$PCH = CHPPh$_2$ (dppen)).[316]

A range of allyl complexes PtCl(L^1)[η-H$_2$CH(R^1)CHR^2] react with MeLi to give corresponding methylplatinum derivatives, PtMe(L^1)[η-H$_2$CH(R^1)CHR^2] (R^1 = R^2 = H, L^1 = PCy$_3$, PPh$_3$, PMePh$_2$, PMe$_2$Ph or PBu$^n_3$; R^1 = Me, R^2 = H, L^1 = PPh$_3$; R^1 = H, R^2 = Me, L^1 = PPh$_3$). These complexes react with bidentate phosphines, L^2, to yield planar, tetracoordinate (η^1-allyl)platinum products, PtMe(L^2)[η-H$_2$CH(R^1)CHR^2] (L^2 = dppe or dppen).[317] Similar compounds, [PtMeL(η-C_3H_5)], may be prepared directly from $[PtCl(\eta\text{-}C_3H_5)]_4$ by addition of MeLi and L, either sequentially or simultaneously ($L = PPr_3$, $P(C_6H_4\text{-}2\text{-}Me)_3$, $PPh_2(C_6H_4\text{-}2\text{-}Me)$, $PPh_2(CH_2Ph)$ or $P(OPh)_3$). Thermolysis or treatment with further MeLi (in toluene) gives P-cyclometallated derivatives, with evolution of CH_4. Assistance by MeLi may be via transmetallation reactions, generating, first, a lithiated *ortho*-substituent, which then reacts with Pt–Me to regenerate MeLi.[318]

Neutral complexes $PtCl[\eta^3\text{-}H_2CC(Me)CH_2](\eta\text{-}H_2C{=}CHR)$ are accessible by reaction of $\{Pt[\eta^3\text{-}H_2CC(Me)CH_2](\mu\text{-}Cl)\}_2$ with $H_2C{=}CHR$ (R = Me, Et, CH_2Ph or $4\text{-}C_6H_4Y$; Y = H, Me, Cl, NO_2, NMe_2 or OMe).[66] Cationic homologues $[Pt(PPh_3)[\eta^3\text{-}H_2CC(Me)CH_2](\eta\text{-}H_2C{=}CHC_6H_4Y)]PF_6$ result via sequential treatment of $PtCl(PPh_3)[\eta^3\text{-}H_2CC(Me)CH_2]$ with $AgPF_6$ and $H_2C{=}CHC_6H_4\text{-}4\text{-}Y$ (Y = H, Me, Cl, OMe, NMe_2 or NO_2).[62] Other members of this family, $[Pt(PPh_3)(\eta^2\text{-alkene})\{\eta\text{-}H_2CC(Me)CH_2\}]^+$, are prepared in a similar fashion (alkene = C_2H_4, (*Z*)-C_4H_8, (*E*)-C_4H_8, $H_2C{=}CHPh$, $H_2C{=}CH\text{-}2\text{-}C_6H_4Cl$ or (*E*)-MeHC=CHPh).[67] The related complex, $Pt(C_6F_5)(\eta\text{-}H_2C{=}CHPh)\{\eta\text{-}H_2CC(Me)CH_2\}$, arises via treatment of $\{Pt[\eta^3\text{-}H_2CC(Me)CH_2](\mu\text{-}Cl)\}_2$ with LiC_6F_5.[68]

$Pt(SnCl_3)(\eta^2\text{-alkene})\{\eta\text{-}H_2CC(Me)CH_2\}$ is prepared by the addition of an alkene to a mixture of $SnCl_2$ and $[PtCl\{\eta\text{-}H_2CC(Me)CH_2\}]_2$ (alkene = ethene, styrene or norbornene).[69] The carbonyl complex, $[Pt(SnCl_3)(CO)(\eta\text{-}C_3H_4\text{-}2\text{-}Me)]_2$ is prepared by reaction of $[PtCl(\eta\text{-}C_3H_4\text{-}2\text{-}Me)]_2$ with $SnCl_2$ and CO.[319] $[PtCl(\eta\text{-}C_3H_4\text{-}2\text{-}Me)]_2$ reacts with 1,5-cyclooctadiene in the presence of $SnCl_2$ to generate ionic $[Pt(\eta\text{-}C_3H_4\text{-}2\text{-}Me)(cod)]_2[Pt(SnCl_3)_3(\eta\text{-}C_3H_4\text{-}2\text{-}Me)]$.[320]

An unusual, elegant synthesis of a cyclooctenylplatinum derivative also employs 1,5-cyclooctadiene, which is introduced to *trans*-$[PtH(HOMe)(PEt_3)_2](OSO_2CF_3)$ in the presence of $NaBPh_4$ to give $[Pt(PEt_3)_2(\eta^3\text{-}C_8H_{13})]BPh_4$.[321]

Cationic η^3-propargyl complexes, $[Pt(PPh_3)_2(\eta\text{-}C_3H_3)]^+$ (see Section 9.4.2.3) react with a variety of species, HY, via addition on the ligand to give corresponding η^3-allyl complexes $[Pt(PPh_3)_2(C_3H_4\text{-}2\text{-}Y)]^+$ (Y = OMe, OAc, NEt_2, C≡CPh, $CH(COMe)_2$ or SPh). Reaction with SiH_2Ph_2 gives $[Pt(PPh_3)_2(C_3H_5)]^+$.[322] Addition of HY to $[Pt(PPh_3)_2(\eta\text{-}C_3H_2\text{-}1\text{-}Ph)]^+$ leads, again, to 2-substituted η^3-allyl complexes $[Pt(PPh_3)_2\{\eta\text{-}H_2CC(Y)CHPh\}]^+$ (Y = OMe, OEt, NEt_2 or $CH(COMe)_2$). All have *endo*-orientations of the phenyl substituent, except the alkoxy derivatives; which show the *exo*-geometry.[323]

Reaction of the platinacyclopentadiene, $Pt\{CH{=}C(BEt_2)CH(Et){=}CH\}(Et_2PCH_2CH_2PEt_2)$ with excess BEt_3 leads to isolation of the unusual η^3-borallyl complex, $[Pt\{\eta^3\text{-}Me_2BC(Me)CHCHC(Me)\text{-}BMe_2\}(Et_2PCH_2CH_2PEt_2)]$.[330]

(ii) Preparation from Pt^0 precursors

Oxidative addition of allyl halide to $Pt(cod)_2$ yields $PtX(\eta^1\text{-}C_3H_4R)(cod)$ (X = Cl or Br; R = H, 2-Me, 3-Me or 3-Ph). Further reaction of these η^1-allyl derivatives with $AgBF_4$ affords $[Pt(\eta\text{-}C_3H_4R)(cod)]BF_4$.[187] Relatedly, oxidative addition of one C–Cl bond in 1,3-dichloro-2-methylpropane to $Pt(PPh_3)_2(\eta\text{-}C_2H_4)$ or $Pt(PPh_3)_4$ in the presence of KPF_6 affords the η-chloromethylallyl complex, $[Pt(PPh_3)_2\{\eta\text{-}H_2CC(CH_2Cl)CMeH\}]PF_6$. The remaining C–Cl bond is highly activated and will react with a range of nucleophiles to produce a further series of complexes, $[Pt(PPh_3)_2\{\eta\text{-}H_2CC(CH_2Y)CMeH\}](PF_6)_n$ (Scheme 14; Y = PPh_3, NEt_3 or py, $n = 2$; Y = Br^-, $n = 1$).[325] Treatment of $Pt(PPh_3)_4$ with $ClCH_2C({=}CH_2)OCH_2OMe$ results in oxidative C–Cl addition to form $[Pt(PPh_3)_2\{\eta\text{-}H_2CC(OCH_2OMe)CH_2\}]X$ (X = Cl or PF_6). Acid hydrolysis yields the 2-hydroxoallyl complex, $[Pt(PPh_3)_2\{\eta\text{-}H_2CC(OH)CH_2\}]X$. In contrast, base treatment affords the η^3-oxodimethylenemethane derivative, $Pt(PPh_3)_2\{\eta\text{-}H_2CC({=}O)CH_2\}]$ (see Section 9.4.3.1) which, on acid treatment, gives the same 2-hydroxoallyl complex (Scheme 15).[326] Oxidative addition of 5-(methoxycarbonyl)-1-chlorocyclohex-2-ene to $Pt(dba)_2$ and $Pt(\eta^3\text{-norbornene})_3$ leads to the chloro-bridged (η^3-cyclohexenyl)platinum dimer, $[PtCl(\eta^3\text{-}C_6H_8\text{-}5\text{-}CO_2Me)]_2$. The stereochemistry of formation is predominantly *anti* for this platinum system (although *syn* is favoured in the palladium analogues).[327] A more singular example of oxidative addition is provided by the interaction of $Pt(PPh_3)_4$ with triallylarsenite, $As(OCH_2CH{=}CH_2)_3$, which gives $[Pt(PPh_3)_2(\eta\text{-}C_3H_5)][As(OCH_2CH{=}CH_2)_2]$. Ion exchange with ClO_4 leads to isolation of $[Pt(PPh_3)_2(\eta\text{-}C_3H_5)]ClO_4$. Deuterium-labelling experiments suggest that the formation step—effectively the oxidative addition of a C–O bond—is readily reversible.[328] A superficially related addition results from the treatment of $[Pt(PPh_3)_2(\eta\text{-}C_2H_4)]$ with vinyl triflates, $R^1_2C{=}CR^2OSO_2CF_3$. The products—subsequent to migratory insertion of ethene in the transient σ-vinylplatinum intermediate—are $[Pt(PPh_3)_2\{\eta\text{-}R^1_2CC(R^2)CHMe\}](OSO_2CF_3)$ (R^1 = H, R^2 = Me or Bu^t; R^1 = Me, R^2 = H).[53]

In another unusual reaction, the η^3-allyl group is transferred from palladium(II) to platinum(0) in a redox transmetallation process. Interaction of $[Pd(\eta^3\text{-allyl})(PPh_3)_2]BF_4$ with $Pt(PPh_3)_4$ or $Pt(PPh_3)_2(\eta\text{-}$

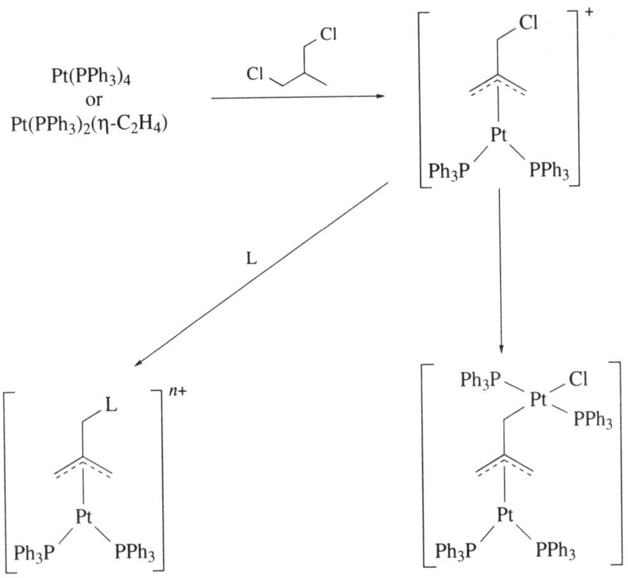

$L = PPh_3, NEt_3$, py, $n = 1$; $L = Br^-$, $n = 0$

Scheme 14

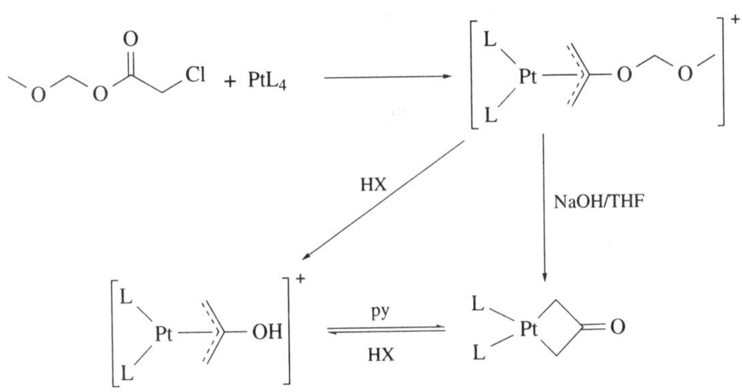

$L = PPh_3$; $X = Cl, PF_6$, OTf

Scheme 15

C_2H_4) generates $[Pt(\eta^3\text{-allyl})(PPh_3)_2]BF_4$ (η^3-allyl = η-C_3H_5, η-C_3H_4-2-Me or η^3-C_6H_8-5-CO_2Me). Neutral palladium(II) precursors may also be used, but the reaction is slower. Stereochemical analysis shows that platinum attacks at the η^3-allyl face opposite palladium.[329]

9.4.1.2 Structural aspects

The application of ^{13}C NMR to the characterization of allyl–metal complexes has been reviewed and the structure and dynamics of platinum derivatives described.[88] Platinum-195 NMR also features in a review which covers (η-allyl)platinum systems.[330] From an NMR study of allylmetals, which includes $Pt(\eta\text{-}C_3H_5)_2$, it emerges that the $^1J_{C-C}$ value reflects the relative donor–acceptor characteristics of the allyl ligand. $^1J_{C-H}$ and $^2J_{H-H}$, however, are sensitive to deviations of the hydrogen substituents from the C_3 plane.[331]

Helium(I) UV PES of $Pt(\eta\text{-}C_3H_5)_2$ and $Pt(\eta\text{-}C_3H_4\text{-2-Me})_2$—in addition to those of the nickel and palladium analogues—allow assignment of the first seven transitions to metal and ligand energy levels. These proposals are supported by INDO calculations (using ΔSCF and transition operator methods).[332] Some of these conclusions were revised in a more recent study.[333] Electron affinities in $M(\eta\text{-}C_3H_5)_2$ have been assessed by electron-transmission spectra and multiple scattering Xα calculations (the latter for M = nickel or palladium only). As emerges, attachment energies and ionization energies are rather insensitive to variations of the metal.[334]

In common with many other such organometallics, (η-allyl)platinum complexes frequently exhibit *syn–anti* hydrogen exchange in their ^1H NMR spectra. An overview of applications of two-dimensional (NOESY) NMR techniques includes an analysis of the fluxional behaviour of the η^3-methallyl complex $[Pt(SnCl_3)(\eta^2\text{-}C_2H_3Ph)\{\eta\text{-}H_2CC(Me)CH_2\}]$. This suggests both $SnCl_3$ and styrene exchange processes, but no η^1–η^3 isomerization of the allyl ligand.[109] The dynamic nature of $[Pt(SnCl_3)(PR_3)(\eta\text{-}C_3H_5)]$ has also been ascribed to Pt–Sn scission, either of $SnCl_3$ or $SnCl_2$ ($PR_3 = PPh_3$, PMe_2Ph or PPr^i_3). From $^1J_{Pt-C}$ measurements, $SnCl_3$ has a *trans* influence comparable with alkenes. Coordination of alkylphosphine results in improved Pt–Sn strength.[335] The gas-phase ion–molecule reaction between $[PtCl(PPh_3)(\eta\text{-}C_3H_5)]$ and $SnCl_2^+$ to form $[Pt(SnCl_3)(PPh_3)(\eta\text{-}C_3H_5)]^+$ has been investigated mass spectrometrically.[336]

In $[Pt(\eta\text{-}C_3H_4\text{-}2\text{-}Me)(cod)]_2[Pt(SnCl_3)_3(\eta\text{-}C_3H_4\text{-}2\text{-}Me)]$, the anion exhibits distorted square-pyramidal coordination, while the two cations have irregular square-planar geometry, as established by x-ray diffractometry. Comparisons with other Pt–Sn distances suggest that the *trans*-influence of the methallyl ligand is comparable with alkenes.[320] The molecular structure of $[Pt(SnCl_3)(CO)(\eta\text{-}C_3H_4\text{-}2\text{-}Me)]_2$ has also been the subject of an x-ray diffractometric determination, and reveals an essentially planar arrangement of ligands.[319] The molecular structure of $[Pt(PPh_3)_2\{\eta\text{-}Me_2CCHCHMe\}](OSO_2CF_3)$, confirmed by x-ray diffraction, favours an *exo*-disposition of the unique terminal methyl group.[53] In contrast with its fluxional palladium analogue, $Pt[\eta^1\text{-}H_2CC(Me){=}CH_2](PCy_3)(\eta\text{-}C_3H_4\text{-}2\text{-}Me)$ demonstrates rigidity in solution.[309] The choice between η^3- and η^1-coordination in cationic complexes $[Pt(PEt_3)\{\eta\text{-}C_3H_4\text{-}2,3\text{-}(CO_2Me)_2]\}cod)]^+$, where the allyl group is ambidentate, may be determined (in the solid state) by the nature of the counter ion. With PF_6^-, a conventional η^3-structure is favoured. With BPh_4^-, η^1-allyl coordination emerges, in which the ligand is overall *C,O*-bidentate.[337] A multinuclear NMR study of $PtY(L)(\eta\text{-}C_3H_5)$ demonstrates the asymmetric nature of the allyl coordination in these systems, and suggests an appreciable contribution from the η^1,η^2-bound form, depending on Y and L (Y = Cl or Ph; L = PMe_3, PMe_2Ph, $PMePh_2$, $P(C_6H_4\text{-}4\text{-}F)_3$, PCy_3 or $PR_2(CH_2CHMeEt)$).[338]

9.4.1.3 Reactions

Allowing $[PtL_2(\eta\text{-}C_3H_4\text{-}2\text{-}Me)][acac]$ to warm to ambient temperature leads to C–C coupling, yielding $CH_2CMeCH_2CH(COMe)_2$. A kinetic analysis suggests that ion-pair collapse is probably involved in the rate-determining process. Coupling is inhibited by solvent in the order: benzene > dichloromethane > acetonitrile/THF. Reaction of this platinum system is slower than that of its palladium analogue, but the difference is not as large as is generally observed for conventional reductive C–C elimination from comparable dialkylmetals.[316] Successive C–C couplings are the eventual outcome of sequential additions of TCNE and PPh_3 to $Pt(\eta\text{-}C_3H_5)_2$. Introduction of TCNE first gives $Pt(TCNE)(\eta\text{-}C_3H_5)_2$ as intermediate, while primary addition of PPh_3 affords $Pt(PPh_3)(\eta^1\text{-}C_3H_5)(\eta\text{-}C_3H_5)$.[339]

As noted in Section 9.4.1.1(i), addition of PR_3 to $Pt(\eta\text{-}C_3H_4\text{-}2\text{-}Me)Cp$ causes ring dislocation of the cyclopentadienyl ligand only, to yield $Pt(PR_3)(\eta^1\text{-}Cp)(\eta\text{-}C_3H_4\text{-}2\text{-}Me)$ ($PR_3 = PPr^i_3$, PBu^i_3, PBu^tPh_2, PCy_3 or $PPhCy_2$).[311] Introduction of $CNBu^t$ to $Pt(\eta\text{-}C_3H_4\text{-}3\text{-}R)Cp$, on the other hand, results in η^n–η^1 decoordination of both hydrocarbyl groups to give $Pt(CNBu^t)_2(\eta^1\text{-}Cp)(\eta^1\text{-}C_3H_4\text{-}3\text{-}R)$ (R = H, Me or Ph).[188] Another general exploration of $M(\eta\text{-}C_3H_4\text{-}2\text{-}R)Cp$ (R = H or Me; M = Ni, Pt or Pd) focuses on their reaction with H_2S. At −78 °C, dimeric species $[(\eta\text{-}C_3H_5)M](\mu\text{-}SH)_2[M(\eta\text{-}C_3H_5)]$ are formed. These eliminate propene at −60 °C to generate polymeric $\{[M(\eta\text{-}C_3H_5)]S\}_n$ of unknown structure.[340]

Pulsed laser photolysis of gaseous $Pt(\eta\text{-}C_3H_5)Cp$—and $PtMe_3(\eta\text{-}C_5H_4Me)$ (see Section 9.6.1.2)—initiates formation of metal aggregates (with diameters in the range 2–18 nm). These systems show some promise in the formation of metallic films by liquid chemical vapour deposition (LCVD), although there are spectroscopic indications of some incorporation of hydrocarbon in the films.[341]

The regioselectivity of nucleophilic addition at η^3-allyl ligands continues to stimulate interest. Reaction of ketene silylacetals $R^3MeC{=}C(OMe)(O\text{-}TMS)$ with $[PtL_2(\eta\text{-}C_3H_4R^1R^2)](O_2CMe)$ leads to platinacyclobutanes by attack of the derived enolate at the central carbon atom ($1\text{-}R^1 = 1\text{-}R^2$ = H or Me; $1\text{-}R^1$-H, $3 = R^2$ = Me; R^3 = H, Me; L = PPh_3 or PCy_3; L_2 = dppe, or dppb).[342]

9.4.1.4 Theoretical aspects

As noted in Section 9.4.1.2, MS Xα calculations for $M(\eta\text{-}C_3H_5)_2$ suggest that electron-attachment energies and ionization energies are rather insensitive to variations in the metal (M = Ni, Pd or Pt).[334] The interconversion between $[Pt(PH_3)_2(\eta^3\text{-}C_6H_7)]^+$ and $[Pt(PH_3)_2(\eta^5\text{-}C_6H_7)]^+$ has been the subject of *ab initio* calculations. The η^3-format is only 0.56 eV more stable, which is consistent with the low barrier to fluxionality in this type of system.[343] A new theoretical analysis, by extended Hückel methods, of $Pt(PH_3)_2(\eta\text{-}C_3H_5)$ suggests that the C_{int}-centred molecular orbital is not destabilized, as is usually assumed, but competes with another MO which is primarily located on the platinum and C_{term} atoms. As a consequence, nucleophilic attack at C_{int} and C_{term} are almost equally probable, disregarding other attractive–repulsive interactions that may occur during the approach of the nucleophile.[342]

9.4.2 Related η^3-Hydrocarbyl Complexes

9.4.2.1 η^3-Benzyl complexes: preparation, structure and reactivity

Protonation of $PtL_2(H_2C{=}CHC_6H_4Y)$ leads to complexes of the type $[Pt(\eta^3\text{-}anti\text{-}1\text{-}MeCHC_6H_4Y)\{Bu^t_2P(CH_2)_3PBu^t_2\}]BF_4$ (Y = H, 4-Me, 4-OMe, 4-Br or 3-Br). The structure has been confirmed (Y = 4-Br) by an x-ray diffraction study, which indicates that the ring aromaticity is substantially disrupted.[344] In similar fashion, several related η^2-styrene complexes, $Pt(P\text{–}P)(\eta^2\text{-}C_8H_8)$, undergo acid protonation (HBF_4), affording $[Pt(P\text{–}P)(\eta^3\text{-}MeCHPh)]^+$, which also exhibit η^3-coordination of the α-methylbenzyl ligand (P–P = dcpe, dcpp, dbpe or dbpp). These complexes are fluxional in solution at ambient temperature, via η^3–σ conversion coupled with β-hydrogen migration.[50]

9.4.2.2 η^3-Butadienyl complexes: preparation, structure and reactivity

Reaction of $Pt(PPh_3)_2(\eta\text{-}C_2H_4)$ with chloroprene or 4-chloro-3-methylbuta-1,2-diene generates η^1-2-butadienyl complexes. Treatment of these with $AgPF_6$, in turn, yields cationic η^3-butadienyl complexes $[Pt\{\eta\text{-}H_2CC(R)CCH_2\}(PPh_3)_2]PF_6$. These undergo attack at the central atom of the allylic fragment by $K[(CH(CO_2Me)_2]$ to produce 2-methyleneplatinacyclobutanes, which rearrange thermolytically to methylenecyclopropane complexes (Scheme 16; R = H or Me).[345] Reaction of *trans*-$[PtH(solvent)(PEt_3)_2]^+$ with $SiPh_2(C{\equiv}CBu^t)_2$ proceeds to give, first, the insertion product $[Pt\{C({=}CHBu^t)SiPh_2(C{\equiv}CBu^t)\}(solvent)(PEt_3)_2]^+$. This rearranges by alkynyl transfer from silicon to platinum, and then reacts with ROH to give $[Pt(C{=}C{=}CHBu^t)\{C({=}CHBu^t)SiPh_2OR\}(PEt_3)_2]^+$. The latter complex finally undergoes migratory insertion to yield $[Pt\{\eta^3\text{-}Bu^tHCC(SiPh_2OR)C{=}Bu^t\}(PEt_3)_2]^+$.[346]

9.4.2.3 η^3-Propargyl/allenyl complexes: preparation, structure and reactivity

Reaction of the (η^1-allenyl)platinum derivative, *trans*-$PtBr(CH{=}C{=}CH_2)(PPh_3)_2$, with Ag^+ leads to the cationic η^3-propargyl system, $[Pt(PPh_3)_2(\eta\text{-}C_3H_3)]^+$ (Equation (28)). This reacts with a range of species, HY, by addition on the C_3 ligand, to give η^3-allyl complexes $[Pt(PPh_3)_2(C_3H_4\text{-}2\text{-}Y)]^+$ (Y = OMe, OAc, NEt_2, C≡CPh, $CH(COMe)_2$ or SPh) (see Section 9.4.1.1(i)).[322] Oxidative addition of $BrCH_2C{\equiv}CPh$ to $Pt(PPh_3)_2(\eta\text{-}C_2H_4)$ leads—via the *cis* isomer—to *trans*-$PtBr(CH_2C{\equiv}CPh)(PPh_3)_2$. This, when treated with $Ag(OSO_2CF_3)$, generates the monosubstituted (η^3-propargyl)platinum complex, $[Pt(PPh_3)_2(\eta\text{-}C_3H_2\text{-}1\text{-}Ph)]^+$. The same cation is accessible by reaction of $Pt(PPh_3)_2(\eta\text{-}C_2H_4)$ with $MeOCH_2C{\equiv}CPh$. $[Pt(PPh_3)_2(\eta\text{-}C_3H_2\text{-}1\text{-}Ph)](OSO_2CF_3)$ reacts with PMe_3 to yield the (η^1-allenyl)platinum species, *trans*-$[Pt(CPh{=}C{=}CH_2)(PMe_3)(PPh_3)_2]^+$. Treatment with CO, on the other hand, results in a 1:1 mixture of corresponding *trans*-$[Pt(CPh{=}C{=}CH_2)(CO)(PPh_3)_2]^+$ and *trans*-$[Pt(CH_2C{\equiv}CPh)(CO)(PPh_3)_2]^+$. Addition of HY leads, again, to 2-substituted η^3-allyl complexes $[Pt(PPh_3)_2\{\eta\text{-}H_2CC(Y)CHPh\}]^+$ (Y = OMe, OEt, NEt_2 or $CH(COMe)_2$) (see Section 9.4.1.1(i)).[323] Likewise, oxidative addition of $[(RC{\equiv}C)IPh](OSO_2CF_3)$ to $Pt(PPh_3)_2(\eta\text{-}C_2H_4)$ affords a mixture of *trans*-$Pt(OSO_2CF_3)(C{\equiv}CR)(PPh_3)_2$ and the 1,3-disubstituted η^3-propargyl analogues, $[Pt(PPh_3)_2(\eta\text{-}MeCCCR)](OSO_2CF_3)$ (R = Me, Bu^n, Bu^t or TMS) (Equation (28)). The product distribution depends on R and on the reaction conditions. The molecular structure of $[Pt(PPh_3)_2(\eta\text{-}MeCCCBu^t)](OSO_2CF_3)$, solved diffractometrically, confirms the allenyl–propargyl nature of coordination, with differing C-2–C-3 (0.139 0(5) nm) and C-3–C-4 (0.126 6(5) nm) distances.[347]

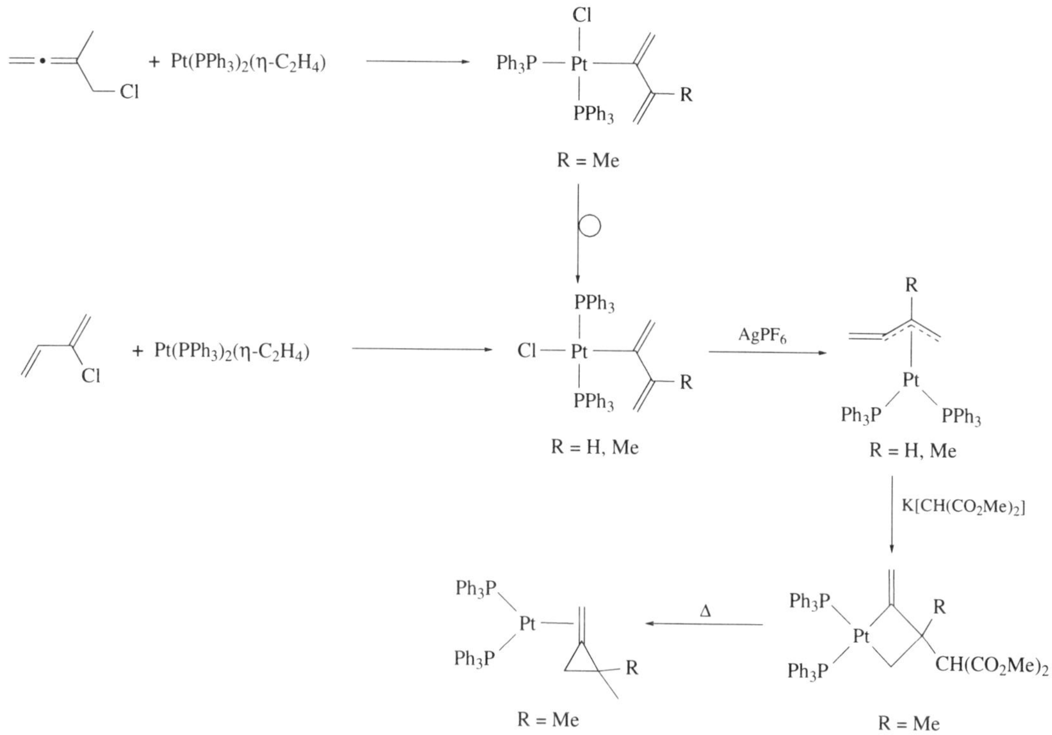

Scheme 16

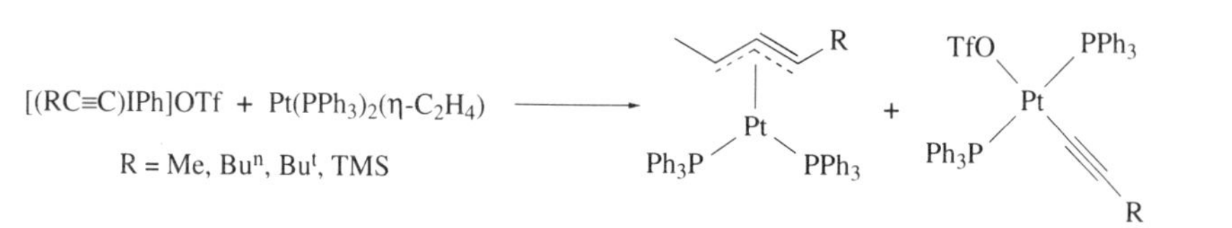

(28)

9.4.3 η-Trimethylenemethane and Related Complexes

9.4.3.1 Preparation, structure and reactivity

The general chemistry of transition metal complexes with η-trimethylenemethane ligands has been reviewed and reflects the lack of success, to date, with platinum derivatives of this type.[348] Like their (nominal) electronic analogues, η^4-butadiene complexes (see Section 9.2.5.1), mononuclear (η^4-trimethylenemethane)platinum derivatives remain a relative rarity. An asymmetric binuclear example, featuring trimethylenemethane in a bridging role, results from oxidative addition of the activated C–Cl bond in a 2-chloromethylallyl complex, $[Pt(PPh_3)_2\{\eta\text{-}H_2CC(CH_2Cl)CRH_2\}]PF_6$, to $Pt(PPh_3)_2(\eta\text{-}C_2H_4)$, giving $[(Ph_3P)_2Pt(\mu,\eta^4\text{-}C_4H_6)PtCl(PPh_3)_2]PF_6$ (Scheme 14).[325]

Complexes of the related η^3-oxodimethylenemethane fragment have begun to proliferate. Reactions of $Pt(CO_3)L_2$ with $RCH_2C(=O)CH_2R$ give $PtL_2\{RCHC(=O)CHR\}$ (R = CO_2Me, L = PPh_3, $PMePh_2$, PMe_2Ph or $AsPh_3$, L_2 = dppe; R = CO_2Et, L = PPh_3 or $AsPh_3$; R = CO_2Pr^n, L = PPh_3 or $AsPh_3$). The same compounds are, in general, accessible by treatment of PtL_4 with $RCH_2C(=O)CH_2R$ in air. An x-ray diffraction analysis of $Pt(PPh_3)_2\{MeO_2CCHC(=O)CHCO_2Me\}$ reveals an appreciable fold angle—between the CPtC and CC(O)C planes—of 50.4° and a weak Pt---C(O) transannular interaction (0.241 6(5) nm). The structure may be formulated as a highly puckered platinacyclobutan-3-one or as a slipped η^3-oxodimethylenemethane complex.[349] Likewise, heptane-2,4,6-trione reacts with $Pt(CO_3)L_2$ to yield $PtL_2\{RCHC(=O)CHR\}$ (R = COMe, L = PPh_3 or $AsPh_3$). Similar diffractometric analysis of $Pt(PPh_3)_2\{Me(O)CCHC(=O)CHC(O)Me\}$ shows a puckered PtC_4 unit, with a fold angle of 48.0(4)° and a transannular Pt---C(O) distance of 0.241 6(5) nm.[350] Synthesis of $PtL_2\{\eta\text{-}PhHCC(=O)CHPh\}$ is achieved by interaction of *cis*-$PtCl_2L_2$ with $K_2[PhHCC(=O)CHPh]$ (L = PPh_3 or $AsPh_3$; L_2 = cod). The structure of $Pt(AsPh_3)_2\{\eta\text{-}PhHCC(=O)CHPh\}$, measured diffractometrically, indicates a highly nonplanar metal–oxodimethylenemethane configuration. The fold angle is 56.7° and the Pt---C(O) transannular contact is relatively short (0.235 0(26) nm). The phenyl substituents adopt *exo*-orientations.[351] The corresponding unsubstituted analogue $Pt(PPh_3)_2\{\eta\text{-}H_2CC(=O)CH_2\}$ has been prepared by interaction of $Pt(PPh_3)_2(\eta\text{-}(E)\text{-}PhHC{=}CHPh)$ with $ClCH_2C(=CH_2)O$-TMS or by treatment of *cis*-$PtCl\{CH_2C(O)CH_2Cl\}(PPh_3)_2$ with Na/Hg. Structural characteristics obtained from x-ray diffraction are more consistent with the η^3-formulation than with a metallacyclobutane. In this instance, the fold angle is 51.0° and the Pt---C(O) transannular contact is 0.242 2(12) nm. These values are explained using extended Hückel calculation arguments based on the palladium analogue, $Pd(PH_3)_2\{\eta\text{-}H_2CC(=O)CH_2\}$.[352] Further analogues, $PtL(CNBu^t)\{\eta\text{-}(MeO_2C)HCC(=O)CH(CO_2Me)\}$, are obtained from $PtL_2\{\eta\text{-}(MeO_2C)HCC(=O)CH(CO_2Me)\}$ by treatment with one equivalent of $CNBu^t$ (L = PPh_3 or $AsPh_3$).[353] While acid hydrolysis of $[Pt(PPh_3)_2\{\eta\text{-}H_2CC(OCH_2OMe)CH_2\}]X$ (X = Cl or PF_6) yields the 2-hydroxoallyl complex, $[Pt(PPh_3)_2\{\eta\text{-}H_2CC(OH)CH_2\}]X$ (see Section 9.3.4.1), base treatment, by contrast, affords the η^3-oxodimethylenemethane derivative, $Pt(PPh_3)_2\{\eta\text{-}H_2CC(=O)CH_2\}]$. Acid treatment gives the same 2-hydroxoallyl complex in a reversible reaction (Scheme 15).[326]

9.5 CYCLOBUTADIENE AND RELATED COMPLEXES

9.5.1 η^4-Cyclobutadiene Complexes

9.5.1.1 Preparation, structure and reactivity

Activity in developing cyclobutadiene and related complexes appears to have been rather limited. Synthesis of new examples has featured established routes, either cyclodimerization of alkynes on a metal template, or ligand substitution on complexes with existing η-cyclobutadiene substituents.

Reaction of $PtCl_2(NCMe)_2$ with EtC≡CEt and $SnCl_2$ leads to the mixed-valence ion pair $[Pt(\eta\text{-}C_4Et_4)Cl_3]^+[Pt(\eta\text{-}C_4Et_4)(SnCl_3)_3]^-$. The neutral, tin-free dimeric species is obtained by treatment with an acetone solution of HCl.[354] In a similar fashion, interaction of $PtCl_2(CO)_2$ and PhC≡CMe yields $[(\eta\text{-}C_4Me_2Ph_2)Pt(\mu\text{-}Cl)_3Pt(\eta\text{-}C_4Me_2Ph_2)][PtCl_3(CO)]$, which features a dimeric cation. Treatment with Cl^- gives the neutral dimer $Pt_2Cl_4(\eta\text{-}C_4Me_2Ph_2)_2$, which, on subsequent addition of PPh_3, affords monomeric $PtCl_2(PPh_3)(\eta\text{-}C_4Me_2Ph_2)$. The molecular structure of the latter mononuclear complex, from x-ray diffractometric analysis, confirms the head-to-head nature of the original alkyne dimerization. Other relatives, $PtX_2L(\eta\text{-}C_4Me_2Ph_2)$, are available similarly (X = Cl, L = $AsPh_3$, py-3-Me or $H_2NC_6H_4$-4-Me; X = I, L = PPh_3).[355]

Highly reactive cations, $[Pt(\eta\text{-}C_4Me_4)(solvent)_n]^+$ (solvent = acetone or nitromethane), are obtained from $[Pt(\eta\text{-}C_4Me_4)Cl_2]_2$ by chlorine abstraction using $Ag(OSO_2CF_3)$. When the solvent is acetone, such species react with cyclopentadienes and with benzene to form novel mixed sandwich compounds: $[Pt(\eta\text{-}C_4Me_4)(Cp'')](OSO_2CF_3)$ (Cp'' = η-Cp, η-Cp*; or $\eta\text{-}C_5H_4Me$) and $[Pt(\eta\text{-}C_4Me_4)(\eta\text{-}C_6H_6)](OSO_2CF_3)$, respectively. These have been spectroscopically and electrochemically characterized. The structure of the cation $[Pt(\eta\text{-}C_4Me_4)(\eta\text{-}C_6H_6)]^+$ has been determined diffractometrically and reveals disparate mean Pt–C distances for the C_4 (0.2138 nm) and C_6 ligands (0.2351 nm), reflecting their differing structural *trans*-influences.[356]

9.6 CYCLOPENTADIENYL AND RELATED COMPLEXES

9.6.1 η^5-Cyclopentadienyl Complexes

Monocyclopentadienyl derivatives of *d*- and *f*-block transition metals, with halides as ancillary ligands, have been the focus of a review. There are no citations referring to platinum, however.[357] The same is true, to date, of pentadienylmetal complexes—the so-called 'open metallocene' derivatives—which have also been surveyed.[358] A recent review of chemistry at diplatinum centres includes complexes with $\mu,\eta^5\text{-}C_5H_4PPh_2$ ligands as well as other Pt_2 units with $\eta\text{-}C_5R_5$ substituents.[359]

9.6.1.1 Mono- and divalent derivatives: preparation, structure and reactivity

A review of the photochemistry of alkyltransition metals contains references of relevance to this section.[215]

Derivatives with η-allyl co-ligands are accessible via reaction of $PtX(\eta^1\text{-}C_3H_4\text{-}3\text{-}R)(cod)$ with TlCp to give $Pt(\eta\text{-}C_3H_4\text{-}3\text{-}R)Cp$ (R = H, Me or Ph; X = Cl or Br). When R = Ph, the σ-bound intermediate $Pt(\eta^1\text{-}C_3H_4\text{-}3\text{-}Ph)(\eta^1\text{-}Cp)(cod)$ is isolable. Additions of excess $CNBu^t$ to $Pt(\eta\text{-}C_3H_4\text{-}3\text{-}R)Cp$ affords *trans*-$Pt(\eta^1\text{-}C_3H_4\text{-}3\text{-}R)(\eta^1\text{-}Cp)(CNBu^t)_2$, which rearranges to the thermodynamically more favourable *cis*-isomer.[188] Cationic species $[PtL_2Cp](OSO_2CF_3)$ result from treating triflatoplatinum precursors, $Pt(OSO_2CF_3)_2L_2$, with TlCp ($L_2 = Ph_2P(CH_2)_nPPh_2$, n = 1–3). Structural features have been confirmed crystallographically for $[Pt(dppe)Cp](OSO_2CF_3)$. Addition of PEt_3 to the latter complex leads to $\eta^5\text{–}\eta^1$ ring decoordination and the formation of two σ-ring-linkage isomers of tetracoordinate $[Pt(dppe)(PEt_3)(\eta^1\text{-}Cp)](OSO_2CF_3)$.[360] Similar ring dislocation results when PEt_3 is introduced to the 2-metallated azobenzene derivative, $Pt(C_6H_4\text{-}2\text{-}N{=}NPh)Cp$, from which the ultimate product is *trans*-$Pt(\eta^1\text{-}C_6H_4\text{-}2\text{-}N{=}NPh)(\eta^1\text{-}Cp)(PEt_3)_2$.[361] In chloroform solution at low temperature, the η^0-cyclopentadienide derivative $[Pt(\eta^1\text{-}C_6H_4\text{-}2\text{-}N{=}NPh)(PEt_3)_3]Cp$ can also be detected.[362] Related platinum and palladium complexes with *ortho*-metallated amidine ligands are also accessible.[363]

Labile mononuclear, dicationic derivatives $[PtL_2(\eta\text{-}C_5H_4PPh_3)][PF_6]$ are formed on treatment of $[PtL_2(acetone)]^{2+}$ with the ylide, $C_5H_4PPh_3$ (L_2 = cod, nbd or cot).[364] The related ligand, $\eta\text{-}C_5H_4PPh_2$ more commonly features as a bridging ligand. $Tl(C_5H_4PPh_2)$ reacts with PtClR(cod) to form dimeric $Pt_2R_2(\mu,\eta\text{-}C_5H_4PPh_2)_2$, in which there is no Pt–Pt bond, as confirmed by a diffractometric study of $Pt_2Ph_2(\mu,\eta\text{-}C_5H_4PPh_2)_2$. Addition of CO causes $\eta^5\text{–}\eta^1$ ring dislocation, although the dimeric skeleton is maintained on the evidence of an x-ray diffraction analysis of $Pt_2Me_2(CO)_2(\mu,\eta^1\text{-}C_5H_4PPh_2)_2$. The latter complex eliminates acetone on thermolysis, but ethane on addition of $P(CH_2CH_2CN)_3$.[365] Dimeric products are also the outcome of reaction of the Zeise dimer, $Pt_2Cl_4(\eta\text{-}C_2H_4)_2$, with $MgCp_2$. The principal product is the fluxional $Pt(\eta\text{-}C_2H_4)(\eta^1\text{-}Cp)Cp$, with σ- and π-bound cyclopentadienyl groups. The other products of the reaction comprise a mixture of the *endo*- and *exo*-isomers of $[CpPt(\mu,\eta^4\text{-}Cp\text{–}Cp)PtCp]$, in which one ring of the coupled Cp_2 fragment spans the Pt–Pt bond in μ,η^4-fashion. Protonation of the mononuclear product yields the unusual cationic complex, $[Pt(\eta\text{-}C_2H_4)(\eta^2\text{-}C_5H_6)Cp]^+$, which contains both η^5-cyclopentadienyl and η^2-cyclopentadiene ligands (Equation (29)). Reaction with PhC≡CPh, on the other hand, leads to a η^1,η^2-7-norbornadienyl derivative, $Pt(\eta^1,\eta^2\text{-}C_7H_5\text{-}2,3\text{-}Ph_2)Cp$, via [4 + 2] *endo*-addition of the alkyne to the η^1-cyclopentadienyl ring. Reaction of $[Pt_2Cl_4(CO)_2]^{2-}$ with $MgCp_2$ yields another dimeric platinum(I) derivative, $Pt_2(CO)_2Cp_2$. This, in turn, reacts with PhC≡CPh, with CO coupling, to generate an unusual dimer, $Pt_2\{\mu,\eta\text{-}C(Ph)C(Ph)CO\}Cp_2$. With $Bu^tC{\equiv}CBu^t$, however, a transversely bridged alkyne complex, $Pt_2(\mu,\eta\text{-}Bu^tC{\equiv}CBu^t)Cp_2$, results.[366]

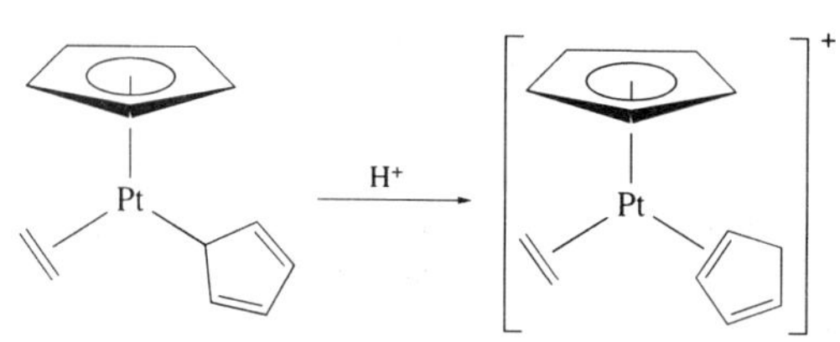

(29)

Compounds of general type PtRLCp are increasingly accessible. For example, reaction of the hydroxo-bridged dimer $Pt_2Ph_2(PPh_3)_2(\mu\text{-}OH)_2$ with two equivalents of C_5H_6 affords $PtPh(PPh_3)(OH)(\eta^2\text{-}C_5H_6)$, proposedly in equilibrium with its ionic isomer $[PtPh(PPh_3)(\eta^4\text{-}C_5H_6)](OH)$. Deprotonation of the cation (by its counter anion) yields $PtPh(PPh_3)Cp$.[367] Displacement of chloride from dimeric $Pt_2Cl_2Ph_2(PPh_3)_2$ using TlCp leads to $PtPh(PPh_3)Cp$, albeit in poor yield. Reaction of $PtPh(\eta^1\text{-}Cp)(cod)$ with PPh_3 in THF gives only $PtPh(PPh_3)\{\eta^1,\eta^2\text{-}C_8H_{12}Cp\}$—the result of cyclopentadienyl migration to the diene (see Sections 9.2.2.2(i) and 9.2.3.2(iii)(b))—while in diethyl ether a mixture of the latter complex and $PtPh(PPh_3)Cp$ is formed.[192]

As precursors have become more conveniently available, the chemistry of pentamethylcyclopentadienylplatinum systems has expanded. Investigations of the synthesis and reactivity of $Pt(CO)_2Cp^*$ have been reported in depth. Reaction of $[PtCl_3CO]^-$ with MgCp*Cl yields dimeric $Pt_2(CO)_2Cp^*_2$ via the observable, mononuclear intermediate PtCl(CO)Cp*. From x-ray diffractometric data, the molecular structure of $Pt_2(CO)_2Cp^*_2$ features staggered PtCOCp* units connected by an unbridged Pt–Pt bond. However, ^{13}C NMR studies indicate that the CO ligands readily exchange metal-binding sites.[368] Protonation of $Pt_2(CO)_2Cp^*_2$ leads to a tetrameric cluster $[Pt_4(\mu_3\text{-}CO)(CO)_2Cp^*_3]^+$, in which three PtCp* units, each defining one face of a tetrahedron, are triply bridged by a single CO ligand. The remaining carbonyl groups both bind terminally to the unique platinum atom.[369] Ultraviolet irradiation of

$Pt_2(CO)_2(\eta\text{-}C_5R_5)_2$ in argon, nitrogen and carbon monoxide matrices at 20 K—as monitored by time-resolved infrared spectroscopy—results in CO ejection and formation of symmetrically bridged dimers, $Pt_2(\mu\text{-}CO)(\eta\text{-}C_5R_5)_2$ (R = H or Me).[370]

Addition of NaCp* to $PtCl_2$(cod) generates a product, Pt{$\eta^1,\eta^2\text{-}C_8H_{12}$Cp*}Cp*, in which one C_5 group is η^5-bound to platinum, while the other has added *exo* to coordinated cyclooctadiene, as confirmed by an x-ray diffraction study. Introduction of PMe_3 leads to migration of the second C_5 ligand to the diene, yielding Pt{$\eta^1,\eta^1\text{-}C_8H_{12}Cp^*_2$}$(PMe_3)_2$, which has C_2 rotational symmetry as a result of the regio- and stereochemistry of ultimate addition to the C_8 ring.[205] A related migration of η^1-Cp has been noted in Section 9.2.3.2(iii)(b).[192]

A pentaphenylcyclopentadienylplatinum complex $Pt(\eta\text{-}C_3H_5)(\eta\text{-}C_5Ph_5)$ has been isolated from the reaction between $[PtCl(\eta\text{-}C_3H_5)]_2$ and $Na(C_5Ph_5)$.[371]

9.6.1.2 *Tetravalent derivatives: preparation, structure and reactivity*

The area of tetravalent derivatives continues to feature mainly $PtMe_3$Cp and its relatives and homologues. A review of the photochemistry of alkyltransition metals contains references of relevance to this section also.[215] Photolysis of $PtMe_3$Cp produces free radicals, detected by spin trapping and EPR analysis. The primary rearrangement route is Pt–C homolysis, generating methyl radicals. These abstract hydrogen with little selectivity from a neighbouring methyl group (45%), from the cyclopentadienyl ligand (5%) or from the cyclooctane solvent (50%).[372] Relatedly, perhaps, it emerges that $PtMe_3$Cp can furnish a photohydrosilation catalyst. $PtH(SiR_3)_2$Cp can be isolated, but does not appear to be a primary intermediate on the path to catalytically active species. The proposed mechanism involves $PtH_2(SiR_3)$Cp as a critical intermediate, leading to highly active colloidal platinum.[373]

A series of complexes $PtR^1R^2R^3$Cp has been subjected to spectroscopic analysis ($R^1 = R^2 = R^3$ = Me; $R^1 = R^2$ = Me, R^3 = Et; R^1 = Me, $R^2 = R^3$ = Et; R^1 = COMe, $R^2 = R^3$ = Me; R^1 = COMe, $R^2 = R^3$ = Et; R^1 = COMe, R^2 = Me, R^3 = Et). Multinuclear NMR indicates that Cp has a relatively weak *trans* influence. Introduction of a chiral europium(III) shift reagent allows spectroscopic resolution of enantiomers of Pt(COMe)MeEtCp, confirming the structure and establishing that it is stereochemically inert under the ambient conditions.[374] Estimates of ring rotation barriers for $PtR^1R^2R^3$Cp may be made from mechanical spectroscopic measurements, in which solid solutions are subjected to sinusoidal mechanical stress. Rotational barriers (ΔH_a) in the solid state fall within the range 20.5 ± 2.1 kJ mol^{-1} [$PtMe_3$Cp] to 25.9 ± 2.1 kJ mol^{-1} [$Pt(COMe)Et_2$Cp].[375] $PtMe_3$Cp* is accessible via reaction of LiCp* with $[PtMe_3I]_4$. Both UV photoelectron spectroscopic measurements and Fenske–Hall calculations on $PtMe_3$Cp and $PtMe_3$Cp* indicate an order of increasing binding energies Pt_{5d} < Pt–Me < Pt–Cp. Interaction of the e_1'' orbitals of the C_5 ring with metal orbitals is a major bonding contribution.[376]

$PtMe_3(\eta\text{-}C_5H_4Me)$ is obtained from the reaction of $[PtMe_3I]_4$ with $Na(C_5H_4Me)$. The structure has been confirmed by an x-ray diffraction study. NMR measurements, however, indicate rapid rotation of the $PtMe_3$ moiety in the solid state.[377] The relative volatility (and thermolytic or photolytic lability) of some derivatives in this class makes them attractive as precursors for vapour deposition (MOCVD) of metallic films, as has been noted for $Pt(\eta\text{-}C_3H_5)$Cp (see Section 9.4.1.3).[341] In a highly efficient low-temperature MOCVD process (120° in 4:1 Ar:H_2 carrier), $PtMe_3(\eta\text{-}C_5H_4Me)$ may be used to deposit a metallic platinum film incorporating less than 1 at.% carbon.[377]

9.6.2 η^5-Heterocyclopentadienyl Complexes

The majority of η-heterocyclopentadienyl compounds in this class feature boron as the incumbent heteroatom(s). Such boracyclopentadienylmetal complexes have been reviewed.[378] A recent review of di- and trinuclear metal complexes of diboraheterocyles notes no platinum derivatives of this type.[379] Multinuclear NMR studies of *trans*-$PtCl(P_3C_2Bu^t_2)(PR_3)_2$, *trans*-$Pt(P_3C_2Bu^t_2)_2(PR_3)_2$ and *cis*-$PtCl(P_3C_2Bu^t_2)(dppe)_2$ show the 1,2,4-triphospha-3,5-di-*t*-butylcyclopentadienyl ligand to be η^1-coordinated to platinum in these fluxional systems.[380]

9.6.2.1 *Preparation, structure and reactivity*

In contrast with the behaviour of its ethyl substituted analogue (see Section 9.4.1.1(i)), reaction of Pt{CH=C(BMe_2)CH(Me)=CH}(dppe) with BMe_3 leads to the η^5-borole derivative, Pt{η^5-

B(Me)CH$_2$CH(Me)CH(Me)CH$_2$}(dppe).[324] The complex Pt(cod)(η-BC$_4$Ph$_5$) is recovered from the reaction of pentaphenylborole with Pt(cod)$_2$, by the interaction of PtCl$_2$(cod) with pentaphenylborole in presence of cobalticene or by addition of K$_2$[BC$_4$Ph$_5$] to PtCl$_2$(cod).[381] The borole–ammonia adduct C$_4$H$_4$BPh·NH$_3$ reacts with Pt(cod)$_2$ to give Pt(cod)(η-C$_4$H$_4$BPh). The latter complex rearranges thermolytically to Pt(η-C$_4$H$_4$BPh)$_2$, with evolution of cyclooctadiene and metallic platinum. From NMR measurements, the rotational barriers in Pt(η-C$_4$H$_4$BPh)$_2$ and Pt(cod)(η-C$_4$H$_4$BPh) are 59 and 38 kJ mol^{-1}, respectively.[382]

9.7 ARENE AND RELATED COMPLEXES

Arene and related complexes have seen very little recent progress. No η-heteroarene derivatives appear to have emerged.

9.7.1 η^6-Arene Complexes: Preparation, Structure and Reactivity

The cation [Pt(η-C$_4$Me$_4$)(acetone)$_n$]$^+$ (see Section 9.5.1.1) reacts with benzene to form the mixed sandwich compound [Pt(η-C$_4$Me$_4$)(η-C$_6$H$_6$)](OSO$_2$CF$_3$), which has been characterized spectroscopically and electrochemically. From an x-ray diffraction study, the structure of [Pt(η-C$_4$Me$_4$)(η-C$_6$H$_6$)]$^+$ was found to show appreciably differing mean Pt–C distances for the C$_4$ (0.2138 nm) and C$_6$ ligands (0.2351 nm), reflecting their structural *trans*-influences.[356]

9.8 REFERENCES

1. (a) W. C. Zeise, *Pogg. Ann.*, 1827, **9**, 632; (b) W. C. Zeise, *Mag. Pharm.*, 1830, **35**, 105; (c) W. C. Zeise, *Pogg. Ann.*, 1831, **21**, 497.
2. F. R. Hartley, in 'COMC-I' vol. 6, p. 471.
3. G. K. Anderson, in 'Chemistry of the Platinum Group Metals: Recent Developments', ed. F. R. Hartley, Elsevier, Oxford, 1991, chap. 12, p. 338.
4. V. K. Jain, G. S. Rao and L. Jain, *Adv. Organomet. Chem.*, 1987, **27**, 113.
5. (a) P. A. Chaloner, *J. Organomet Chem.*, 1987, **324**, 283; (b) P. A. Chaloner, *ibid.*, 1987, **337**, 431; (c) P. A. Chaloner, *ibid.*, 1988, **357**, 51; (d) P. A. Chaloner, *ibid.*, 1989, **374**, 349; (e) P. A. Chaloner, *ibid.*, 1992, **432**, 387; (f) P. A. Chaloner, *ibid.*, 1992, **442**, 271.
6. (a) J. A. S. Howell, in 'Organometallic Chemistry', Specialist Periodical Report, eds. E. W. Abel and F. G. A. Stone, Royal Society of Chemistry, London, 1982, vol. **10**, p. 245; (b) J. A. S. Howell, *ibid.*, 1983, vol. **11**, p. 285; (c) J. A. S. Howell, *ibid.*, 1984, vol. **12**, p. 252; (d) J. A. S. Howell, *ibid.*, 1985, vol. **13**, p. 271; (e) J. A. S. Howell, *ibid.*, 1986, vol. **14**, p. 294; (f) J. A. S. Howell, *ibid.*, 1987, vol. **15**, p. 301; (g) M. H. Whitely, *ibid.*, 1987, vol. **16**, p. 302; (h) M. H. Whitely, *ibid.*, 1989, vol. **17**, p. 279; (i) M. H. Whitely, *ibid.*, 1989, vol. **18**, p. 272; (j) P. K. Baker, *ibid.*, 1990, vol. **19**, p. 280; (k) P. K. Baker, *ibid.*, 1991, vol. **20**, p. 279; (l) P. K. Baker, in 'Organometallic Chemistry', Specialist Periodical Report, ed. E. W. Abel, Royal Society of Chemistry, London, 1992, vol. **21**, p. 282; (m) D. G. Evans, *ibid.*, 1993, vol. **22**, p. 275.
7. (a) W. E. Watts, in 'Organometallic Chemistry', Specialist Periodical Report, eds. reps. E. W. Abel and F. G. A. Stone, Royal Society of Chemistry, London, 1982, vol. **10**, p. 283; (b) W. E. Watts, *ibid.*, 1983, vol. **11**, p. 323; (c) W. E. Watts, *ibid.*, 1984, vol. **12**, p. 288; (d) W. E. Watts, *ibid.*, 1985, vol. **13**, p. 317; (e) W. E. Watts, *ibid.*, 1986, vol. **14**, p. 342; (f) A. H. Wright, *ibid.*, 1987, vol. **15**, p. 354; (g) A. H. Wright, *ibid.*, 1987, vol. **16**, p. 351; (h) A. H. Wright, *ibid.*, 1989, vol. **17**, p. 326; (i) A. H. Wright, *ibid.*, 1989, vol. **18**, p. 315; (j) A. H. Wright, *ibid.*, 1990, vol. **19**, p. 323; (k) A. H. Wright, *ibid.*, 1991, vol. **20**, p. 324; (l) I. R. Butler, in 'Organometallic Chemistry', Specialist Periodical Report, ed. E. W. Abel, Royal Society of Chemistry, London, 1992, vol. **21**, p. 325; (m) I. R. Butler, *ibid.*, 1993, vol. **22**, p. 319.
8. A. Yamamoto, 'Organotransition Metal Chemistry, Fundamental Concepts and Applications', Wiley, New York, 1986.
9. R. H. Crabtree, 'The Organometallic Chemistry of the Transition Metals', Wiley, New York, 1988.
10. Ch. Elschenbroich and A. Salzer, 'Organometallics', VCH, Weinheim, 1989.
11. F. P. Pruchnik, 'Organometallic Chemistry of the Transition Elements', trans. S. A. Duraj, Plenum, New York, 1990.
12. L. E. Crascall and J. L. Spencer, *Inorg. Synth.*, 1990, **28**, 126.
13. (a) U. Nagel, *Chem. Ber.*, 1982, **115**, 1998; (b) M. J. Baker, K. N. Harrison, A. G. Orpen, P. G. Pringle and G. Shaw, *J. Chem. Soc., Dalton Trans.*, 1992, 2607.
14. (a) J. M. Brown, S. J. Cook and S. J. Kimber, *J. Organomet. Chem.* 1884, **269**, C58; (b) D. Parker and R. J. Taylor, *Tetrahedron*, 1988, **44**, 2241; (c) R. Fulwood, D. Parker, G. Ferguson and B. Kaitner, *J. Organomet. Chem.*, 1991, **419**, 269.
15. M. Camalli, F. Caruso, S. Chaloupka, E. M. Leber, H. Rimm and L. M. Venanzi, *Helv. Chim. Acta*, 1990, **73**, 2263.
16. (a) R. A. Head, *Inorg. Synth.*, 1985, **24**, 213; (b) R. A. Head, *ibid.*, 1990, **28**, 132.
17. (a) H. C. Clark, G. Ferguson, M. J. Hampden-Smith, B. Kaitner and H. Rügger, *Polyhedron*, 1988, **7**, 1349; (b) G. Ferguson, M. J. Hampden-Smith and B. L. Ruhl, *Acta Crystallogr., Sect. C*, 1987, **43**, 788.
18. P. G. Gassman and I. G. Cesa, *Organometallics*, 1984, **3**, 119.
19. (a) R. G. Nuzzo, T. J. McCarthy and G. M. Whitesides, *Inorg. Chem.*, 1981, **20**, 1312; (b) T. J. McCarthy, R. G. Nuzzo and G. M. Whitesides, *J. Am. Chem. Soc.*, 1981, **103**, 3396.
20. R. G. Nuzzo, T. J. McCarthy and G. M. Whitesides, *J. Am. Chem. Soc.*, 1981, **103**, 3404.

21. (a) M. Hackett, J. A. Ibers, P. Jernakoff and G. M. Whitesides, *J. Am. Chem. Soc.*, 1986, **108**, 8094; (b) M. Hackett, J. A. Ibers and G. M. Whitesides, *ibid.*, 1988, **110**, 1436; (c) M. Hackett and G. M. Whitesides, *ibid.*, 1988, **110**, 1449.
22. D. L. Packett, A. Syed and W. C. Trogler, *Organometallics*, 1988, **7**, 159.
23. (a) R. S. Paonessa and W. C. Trogler, *Organometallics*, 1982, **1**, 768; (b) R. S. Paonessa, A. L. Prignano and W. C. Trogler, *ibid.*, 1985, **4**, 647.
24. J. L. Spencer, *Inorg. Synth.*, 1990, **28**, 129.
25. F. G. A. Stone, *Acc. Chem. Res.*, 1981, **14**, 318.
26. A. Christofides, J. A. K. Howard, J. L. Spencer and F. G. A. Stone, *J. Organomet. Chem.*, 1982, **232**, 279.
27. J. L. Spencer, *Inorg. Synth.*, 1990, **28**, 130.
28. P. W. Frost, J. A. K. Howard, J. L. Spencer and J. E. Turner, *J. Chem. Soc., Chem. Commun.*, 1981, 1104.
29. M. J. Chetcuti, J. A. Herbert, J. A. K. Howard, M. Pfeffer, J. L. Spencer and F. G. A. Stone, *J. Chem. Soc., Dalton Trans.*, 1981, 284.
30. N. Carr, B. J. Dunne, L. Mole, A. G. Orpen and J. L. Spencer, *J. Chem. Soc., Dalton Trans.*, 1991, 863.
31. W. S. Hwang, *Polyhedron*, 1984, **3**, 513.
32. V. De Felice, M. Funicello, A. Panunzi and F. Ruffo, *J. Organomet. Chem.*, 1991, **403**, 243.
33. G. B. Stackhouse, IV, and L. L. Wright, *Inorg. Chim. Acta*, 1988, **150**, 5.
34. S. A. Godleski, K. B. Gundlach and R. S. Valpey, *Organometallics*, 1985, **4**, 296.
35. A. Kumar, J. D. Lichtenhan, S. C. Critchlow, B. Eichinger and W. T. Borden, *J. Am. Chem. Soc.*, 1990, **112**, 5633.
36. H. Choi, J. W. Hershberger, A. R. Pinhas and D. M. Ho, *Organometallics*, 1991, **10**, 2930.
37. F. Caruso, M. Camalli, G. Pellizer, F. Asaro and M. Lenarda, *Inorg. Chim. Acta*, 1991, **181**, 167.
38. M. Lenarda, R. Ganzerla, A. Lisini, M. Graziani and T. Boschi, *Transition Met. Chem.*, 1981, **6**, 199.
39. G. Pellizer, M. Graziani, M. Lenarda and B. T. Heaton, *Polyhedron*, 1983, **2**, 657.
40. A. L. Rheingold, C. J. Baldacchini, P. D. Macklin and G. L. Geoffroy, *Acta Crystallogr., Sect. C*, 1990, **46**, 496.
41. A. J. Fadiadi, *Synthesis*, 1987, 959.
42. A. Sen and J. Halpern, *Inorg. Chem.*, 1980, **19**, 1073.
43. P. A. Chaloner and G. T. L. Broadwood-Strong, *J. Organomet. Chem.*, 1989, **362**, C21.
44. J. A. K. Howard, J. L. Spencer and S. A. Mason, *Proc. R. Soc. London, Ser. A*, 1983, **386**, 145.
45. J. A. K. Howard, P. Mitrprachachon and A. Roy, *J. Organomet. Chem.*, 1982, **235**, 375.
46. P. Czaszar, P. L. Goggin, J. Mink and J. L. Spencer, *J. Organomet. Chem.*, 1989, **379**, 337.
47. M. Lenarda, R. Ganzerla, A. Lisini, M. Graziani and T. Boschi, *Transition Met. Chem.*, 1981, **6**, 199.
48. D. Parker and R. J. Taylor, *J. Chem. Soc., Chem. Commun.*, 1987, 1781.
49. P. Foley and G. M. Whitesides, *Inorg. Chem.*, 1980, **19**, 1402.
50. L. E. Crascall and J. L. Spencer, *J. Chem. Soc., Dalton Trans.*, 1992, 3445.
51. P. J. Stang, L. Song and B. Halton, *J. Organomet. Chem.*, 1990, **388**, 215.
52. J. Burgess, R. D. W. Kemmitt, N. Morton, C. T. Mortimer and M. P. Wilkinson, *J. Organomet. Chem.*, 1980, **191**, 477.
53. Z. D. Zhong, R. J. Hinkle, A. M. Anif and P. J. Stang, *J. Am. Chem. Soc.*, 1991, **113**, 6196.
54. D. Carmichael, P. B. Hitchcock, J. F. Nixon, F. Mathey and L. Ricard, *J. Chem. Soc., Dalton Trans.*, 1993, 1811.
55. W. Weigand, G. Bosl, C. Robl and W. Amrein, *Chem. Ber.*, 1992, **125**, 1047.
56. P. B. Chock, J. Halpern and F. E. Paulik, *Inorg. Synth.*, 1990, **28**, 349.
57. L. E. Erickson and D. C. Brower, *Inorg. Chem.*, 1982, **21**, 838.
58. L. E. Erickson, G. S. Jones, J. L. Blanchard and K. J. Ahmed, *Inorg. Chem.*, 1991, **30**, 3147.
59. P. Cavoli, R. Graziani, U. Casellato and P. Uguagliati, *Inorg. Chim. Acta*, 1986, **111**, L35.
60. V. Y. Kukushkin *et al.*, *Inorg. Chim. Acta*, 1991, **185**, 143.
61. M.-A. D. Stiakaki and A. Christofides, *Polyhedron*, 1993, **12**, 661.
62. H. Kurosawa and N. Asada, *J. Organomet. Chem.*, 1981, **217**, 259.
63. A. Albinati, W. R. Caseri and P. S. Pregosin, *Organometallics*, 1987, **6**, 788.
64. (a) J. R. Briggs, C. Crocker and B. L. Shaw, *Inorg. Chim. Acta*, 1980, **40**, 245; (b) J. R. Briggs, C. Crocker, W. S. McDonald and B. L. Shaw, *J. Chem. Soc., Dalton Trans.*, 1980, 64.
65. H. Alper, Y. J. Huang, D. Belli Dell'Amico, F. Calderazzo, N. Pasqualetti and C. A. Veracini, *Organometallics*, 1991, **10**, 1665.
66. H. Kurosawa and N. Asada, *Organometallics*, 1983, **2**, 251.
67. (a) K. Miki, K. Yamatoya, N. Kasai, H. Kurosawa, M. Emoto and A. Urabe, *J. Chem. Soc., Chem. Commun.*, 1984, 1520; (b) K. Miki *et al.*, *J. Am. Chem. Soc.*, 1988, **110**, 3191.
68. H. Kurosawa, H. Ohnishi, K. Miki, N. Kasai, K. Tatsumi and A. Nakamura, *Chem. Lett.*, 1987, 1623.
69. A. Musco *et al.*, *Organometallics*, 1988, **7**, 2130.
70. J. D. Scott and R. J. Puddephatt, *J. Chem. Soc., Chem. Commun.*, 1984, 193.
71. J. D. Scott and R. J. Puddephatt, *Organometallics*, 1986, **5**, 1253.
72. V. G. Albano, G. Natile and A. Panunzi, *Coord. Chem. Rev.*, 1994, **133**, 67.
73. G. van Koten and K. Vrieze. *Recl. Trav. Pays-Bas*, 1981, **100**, 129.
74. (a) H. van der Poel, G. van Koten, M. Kokker and C. H. Stam, *Inorg. Chem.*, 1981, **20**, 2941; (b) H. van der Poel and G. van Koten, *J. Organomet. Chem.*, 1980, **187**, C17.
75. A. de Renzi, B. Blasio, A. Saporito, M. Scalone and A. Vitagliano, *Inorg. Chem.*, 1980, **19**, 960.
76. V. G. Albano, D. Braga, V. De Felice, A. Panunzi and A. Vitagliano, *Organometallics*, 1987, **6**, 517.
77. S. Bartolucci, P. Carpinelli, V. De Felice, B. Giovanitti and A. De Renzi, *Inorg. Chim. Acta*, 1992, **197**, 51.
78. M. E. Cucciolito, V. De Felice, A. Panunzi and A. Vitagliano, *Organometallics*, 1989, **8**, 1180.
79. V. De Felice, M. L. Ferrara, A. Panunzi and F. Ruffo, *J. Organomet. Chem.*, 1992, **439**, C49.
80. M. E. Cucciolito, F. Giordano, A. Panunzi, F. Ruffo and V. De Felice, *J. Chem. Soc., Dalton Trans.*, 1993, 3421.
81. A. Panunzi, F. Ruffo, B. Åkermark and V. De Felice, *Acta Chem. Scand., Ser. A*, 1992, **46**, 499.
82. V. G. Albano, C. Castellari, V. De Felice, A. Panunzi and F. Ruffo, *J. Organomet. Chem.*, 1992, **425**, 177.
83. A. Sanchez, C. Castellari, A. Panunzi, A. Vitagliano and V. De Felice, *J. Organomet. Chem.*, 1990, **388**, 243.
84. V. G. Albano, C. Castellari, M. Monari, V. De Felice, A. Panunzi and F. Ruffo, *Organometallics*, 1992, **11**, 3665.
85. A. Nakamura, *Coord. Chem. Rev.*, 1991, **109**, 207.
86. K. Tatsumi and M. Tsutsui, *J. Mol. Catal.*, 1981, **13**, 117.

87. H. Kurosawa and I. Ikeda, *J. Organomet. Chem.*, 1992, **428**, 289.
88. P. W. Jolly and R. Mynott, *Adv. Organomet. Chem.*, 1981, **19**, 257.
89. T. H. Chang and J. I. Zink, *J. Am. Chem. Soc.*, 1984, **106**, 287.
90. Y. N. Huang, D. F. R. Gilson and I. S. Butler, *J. Chem. Soc., Dalton Trans.*, 1992, 2881.
91. S. Miya and K. Saito, *Inorg. Chem.*, 1981, **20**, 287.
92. M. Green and J. K. K. Sarhan, *Inorg. Chim. Acta*, 1981, **54**, L67.
93. (a) R. Lazzaroni, G. Uccello-Barretta, S. Bertozzi and P. Salvadori, *J. Organomet. Chem.*, 1984, **275**, 145; (b) R. Lazzaroni, G. Uccello-Barbetta, S. Bertozzi and P. Salvadori, *ibid.*, 1985, **297**, 117.
94. A. M. Al-Shalaan, S. S. Al-Showiman and I. M. Al-Najjar, *Inorg. Chim. Acta*, 1986, **121**, 127.
95. A. M. Al-Shalaan, S. S. Al-Showiman and I. M. Najjar, *J. Chem. Res. (S)*, 1986, 76.
96. I. M. Al-Najjar, *Spectrochim. Acta, Part A*, 1987, **44**, 57.
97. I. M. Al-Najjar and H. B. Amin, *J. Chem. Soc., Dalton Trans.*, 1990, 429.
98. D. A. Johnson, W. C. Deese and A. W. Cordes, *Acta. Crystallogr., Sect. B*, 1982, **38**, 2220.
99. D. A. Johnson, W. C. Deese and A. W. Cordes, *Inorg. Chem.*, 1981, **20**, 1519.
100. M. Camalli, F. Caruso and Z. Zambonelli, *Inorg. Chim. Acta*, 1980, **44**, L77.
101. H. Kurosawa, N. Asada, A. Urabe and M. Emoto, *J. Organomet. Chem.*, 1984, **272**, 321.
102. A. Olsson and P. Kofod, *Inorg. Chem.*, 1992, **31**, 183.
103. L. I. Elding and A. B. Groning, *Inorg. Chim. Acta*, 1980, **38**, 59.
104. S. Mija, K. Kashiwahara and K. Saito, *Inorg. Chem.*, 1980, **19**, 98.
105. G. Guillot-Edelheit and J. C. Chottard, *J. Chem. Soc., Dalton Trans.*, 1984, 169.
106. M. Öki, *Pure Appl. Chem.*, 1989, **61**, 699.
107. R. J. Cross and M. F. Davidson, *J. Chem. Soc., Dalton Trans.*, 1987, 139.
108. H. Kurosawa, A. Urabe and M. Emoto, *J. Chem. Soc., Dalton Trans.*, 1986, 891.
109. P. S. Pregosin and C. Ammann, *Pure Appl. Chem.*, 1989, **61**, 1771.
110. L. A. P. Kane-Maguire, *Mech. Inorg. Organomet. React.*, 1984, **2**, 301.
111. L. A. P. Kane-Maguire, E. D. Honig and D. A. Sweigart, *Chem. Rev.*, 1984, **84**, 525.
112. M. Green and J. K. K. Sarhan, *Inorg. Chim. Acta*, 1980, **45**, L31.
113. M. Green, J. K. K. Sarhan and I. M. Al-Najjar, *J. Chem. Soc., Dalton Trans.*, 1981, 1565.
114. M. Green, J. K. K. Sarhan and I. M. Al-Najjar, *J. Chem. Soc., Dalton Trans.*, 1984, 771.
115. L. Maresca, G. Natile, A. M. M. Lanfredi and A. Tiripicchio, *J. Am. Chem. Soc.*, 1982, **104**, 7661.
116. L. Maresca and G. Natile, *J. Chem. Soc., Chem. Commun.*, 1983, 40.
117. L. Maresca, G. Natile and G. Rizzardi, *Inorg. Chim. Acta*, 1980, **38**, 53.
118. L. Maresca, G. Natile and G. Rizzardi, *ibid.*, 1980, **38**, 137.
119. G. Annibale, L. Maresca, G. Natile, A. Tiripicchio and M. Tiripicchio-Camellini, *J. Chem. Soc., Dalton Trans.*, 1982, 1587.
120. F. Fanizzi, F. P. Intini, G. Annibale and L. Maresca, *J. Chem. Soc., Dalton Trans.*, 1982, 309.
121. D. Mansuy, M. Dreme, J. C. Chottard and J. P. Girault, *J. Am. Chem. Soc.*, 1980, **102**, 844.
122. R. J. Hinkle, P. J. Stang and A. M. Arif, *Organometallics*, 1993, **12**, 3510.
123. S. P. Ermer, G. E. Struck, S. P. Bitler, R. Richards, R. Bau and T. C. Flood, *Organometallics*, 1993, **12**, 2634.
124. R. J. Puddephatt, *Coord. Chem. Rev.*, 1980, **33**, 149.
125. (a) R. J. Al-Essa, R. J. Puddephatt, P. J. Thompson and C. F. H. Tipper, *J. Am. Chem. Soc.*, 1980, **102**, 7546; (b) R. J. Al-Essa, R. J. Puddephatt, D. C. L. Perkins, M. C. Rendle and C. F. H. Tipper, *J. Chem. Soc., Dalton Trans.*, 1981, 1738; (c) G. R. Riley, C. F. H. Tipper and R. J. Puddephatt, *J. Organomet. Chem.*, 1981, **208**, 429.
126. B. M. Cushman and D. B. Brown, *Inorg. Chem.*, 1981, **20**, 2490.
127. A. Miyashita, M. Takahashi and H. Takaya, *J. Am. Chem. Soc.*, 1981, **103**, 6257.
128. M. D. Waddington and P. W. Jennings, *Organometallics*, 1982, **1**, 385.
129. E. J. Parsons and P. W. Jennings, *J. Am. Chem. Soc.*, 1987, **109**, 3973.
130. J. O. Hoberg and P. W. Jennings, *Organometallics*, 1992, **11**, 3452.
131. J. O. Hoberg, R. D. Larsen and P. W. Jennings, *Organometallics*, 1990, **9**, 1334.
132. P. J. Stang, Y. H. Huang and A. M. Arif, *Organometallics*, 1992, **11**, 845.
133. P. J. Stang and D. Cao, *Organometallics*, 1993, **12**, 996.
134. T. Ziegler, *Inorg. Chem.*, 1985, **24**, 1547.
135. K. Morokuma and W. T. Borden, *J. Am. Chem. Soc.*, 1991, **113**, 1912.
136. P. J. Hay, *J. Am. Chem. Soc.*, 1981, **103**, 1390.
137. R. Ponec and R. Rericha, *J. Organomet. Chem.*, 1988, **341**, 549.
138. S. Sakaki, K. Maruta and K. Ohkubo, *Inorg. Chem.*, 1987, **26**, 2499.
139. (a) A. Stockis and R. Hoffmann, *J. Am. Chem. Soc.*, 1980, **102**, 2952; (b) O. Eisenstein and R. Hoffmann, *ibid.*, 1980, **102**, 6148; (c) O. Eisenstein and R. Hoffmann, *ibid.*, 1981, **103**, 4308.
140. A. D. Cameron, V. H. Smith, Jr. and M. C. Baird, *J. Chem. Soc., Dalton Trans.*, 1988, 1037.
141. T. A. Albright, *Tetrahedron*, 1982, **38**, 1339.
142. R. Hoffmann, C. N. Wilker and O. Eisenstein, *J. Am. Chem. Soc.*, 1982, **104**, 632.
143. I. Omae, *Angew. Chem., Int. Ed. Engl.*, 1982, **21**, 889.
144. M. A. Bennett, C. Chiraratvatana, G. B. Robertson and U. Tooptakong, *Organometallics*, 1988, **7**, 1394.
145. M. A. Bennett and C. Chiraratvatana, *J. Organomet. Chem.*, 1985, **296**, 255.
146. (a) R. D. Kelly and G. B. Young, *Polyhedron*, 1989, **8**, 433; (b) R. D. Kelly and G. B. Young, *J. Organomet. Chem.*, 1989, **361**, 123.
147. J. Browning, K. R. Dixon, N. J. Meanwell and F. Wang, *J. Organomet. Chem.*, 1993, **460**, 117.
148. M. K. Cooper and D. W. Yaniuk, *J. Organomet. Chem.*, 1981, **221**, 231.
149. M. K. Cooper, P. V. Stevens and M. McPartlin, *J. Chem. Soc., Dalton Trans.*, 1983, 553.
150. B. L. Simms and J. A. Ibers, *J. Organomet. Chem.*, 1987, **327**, 125.
151. M. K. Cooper, P. J. Guerney, H. J. Goodwin and M. McPartlin, *J. Chem. Soc., Dalton Trans.*, 1982, 757.
152. M. K. Cooper, P. J. Guerney and M. McPartlin, *J. Chem. Soc., Dalton Trans.*, 1980, 349.
153. L. Busetto, *J. Organomet. Chem.*, 1980, **186**, 411.

154. (a) E. W. Abel, D. G. Evans, J. R. Koe, V. Sik, P. A. Bates and M. B. Hursthouse, *J. Chem. Soc., Dalton Trans.*, 1989, 985; (b) E. W. Abel, D. G. Evans, J. R. Koe, V. Sik, M. B. Hursthouse and P. A. Bates, *ibid.*, 1989, 2315.
155. E. W. Abel *et al.*, *J. Chem. Soc., Dalton Trans.*, 1990, 1697.
156. E. W. Abel, D. G. Evans, J. R. Koe, M. B. Hursthouse and M. Mazid, *J. Chem. Soc., Dalton Trans.*, 1992, 663.
157. F. G. A. Stone, *Acc. Chem. Res.*, 1981, **14**, 318.
158. F. G. A. Stone, *Inorg. Chim. Acta*, 1981, **50**, 33.
159. I. Omae, *Coord. Chem. Rev.*, 1983, **51**, 1.
160. M. J. Chetcuti, J. A. K. Howard, M. Pfeffer, J. L. Spencer and F. G. A. Stone, *J. Chem. Soc., Dalton Trans.*, 1981, 276.
161. G. Benter, O. Heyke and I.-P. Lorenz, *Z. Naturforsch., Teil B*, 1991, **46**, 1694.
162. P. B. Hitchcock, M. F. Lappert and N. J. W. Warhurst, *Angew. Chem., Int. Ed. Engl.*, 1991, **30**, 438.
163. G. Chandra, P. Y. Lo, P. B. Hitchcock and M. F. Lappert, *Organometallics*, 1987, **6**, 191.
164. J. A. K. Howard, *Acta. Crystallogr., Sect. B*, 1982, **38**, 2896.
165. B. C. Ankianiec and G. B. Young, *J. Organomet. Chem.*, 1989, **361**, 123.
166. R. McCrindle, G. J. Arsenault and R. Farwaha, *J. Organomet. Chem.*, 1985, **296**, C51.
167. T. G. Appleton, J. R. Hall, C. H. L. Kennard, M. T. Mathieson, D. W. Neale, G. Smith and T. C. W. Mak, *J. Organomet. Chem.*, 1993, **453**, 299.
168. H. M. Buch, G. Schroth, R. Mynott and P. Binger, *J. Organomet. Chem.*, 1983, **247**, C63.
169. J. L. Haggitt and D. M. P. Mingos, *J. Organomet. Chem.*, 1993, **462**, 365.
170. D. Drew and J. R. Doyle, *Inorg. Synth.*, 1990, **28**, 346.
171. L. Mink, M. F. Rettig and R. M. Wing, *J. Am. Chem. Soc.*, 1991, **113**, 2065.
172. L. L. Wright, R. M. Wing, M. F. Rettig and G. R. Wiger, *J. Am. Chem. Soc.*, 1980, **102**, 5949.
173. M. H. Rakowsky, J. C. Woolcock, L. L. Wright, D. B. Green, M. F. Rettig and R. M. Wing, *Organometallics*, 1987, **6**, 1211.
174. M. H. Rakowsky, J. C. Woodcock, M. F. Rettig and R. M. Wing, *Organometallics*, 1988, **7**, 2149.
175. R. Bassan, K. H. Bryars, L. Judd, A. W. G. Platt and P. G. Pringle, *Inorg. Chem. Acta*, 1986, **121**, L41.
176. S. Komiya, Y. Mizuno and T. Shibuya, *Chem. Lett.*, 1986, 1065.
177. H. E. Bryndza, *J. Chem. Soc., Chem. Commun.*, 1985, 1696.
178. P. Foley, R. DiCosimo and G. M. Whitesides, *J. Am. Chem. Soc.*, 1980, **102**, 6713.
179. R. DiCosimo, S. S. Moore, A. F. Sowinski and G. M. Whitesides, *J. Am. Chem. Soc.*, 1982, **104**, 124.
180. M. Bochmann, G. Wilkinson and G. B. Young, *J. Chem. Soc., Dalton Trans.*, 1980, 1879.
181. D. C. Griffiths and G. B. Young, *Organometallics*, 1989, **8**, 875.
182. S. K. Thomson and G. B. Young, *Polyhedron*, 1988, **7**, 1953.
183. T. G. Appleton, J. R. Hall, M. T. Mathieson and D. W. Neale, *J. Organomet. Chem.*, 1993, **453**, 307.
184. J. F. Hoover and J. M. Stryker, *J. Am. Chem. Soc.*, 1989, **111**, 6466.
185. H. J. R. de Boer, O. S. Akkerman and F. Bickelhaupt, *J. Organomet. Chem.*, 1987, **336**, 447.
186. M. F. Lappert, T. R. Martin, C. L. Raston, B. W. Skelton and A. H. White, *J. Chem. Soc., Dalton Trans.*, 1982, 1959.
187. N. M. Boag, M. Green, J. L. Spencer and F. G. A. Stone, *J. Chem. Soc., Dalton Trans.*, 1980, 1200.
188. N. M. Boag, M. Green, J. L. Spencer and F. G. A. Stone, *J. Chem. Soc., Dalton Trans.*, 1980, 1220.
189. K. A. Fallis, G. K. Anderson and N. P. Rath, *Organometallics*, 1993, **12**, 2435.
190. C. S. Day, V. W. Day, A. Shaver and H. C. Clark, *Inorg. Chem.*, 1981, **20**, 2188.
191. D. G. Ibbott, N. C. Payne and A. Shaver, *Inorg. Chem.*, 1981, **20**, 2193.
192. G. K. Anderson, *Organometallics*, 1986, **5**, 1903.
193. S. E. Himmel and G. B. Young, *Organometallics*, 1988, **7**, 2440.
194. H.-A. Brune, J. Unsin, H. G. Alt, G. Schmidtberg and K.-H. Spohn, *Chem. Ber.*, 1984, **117**, 1606.
195. H.-A. Brune, J. Lerche, G. Schmidtberg and A. Baur, *J. Organomet. Chem.*, 1993, **450**, 269.
196. R. Baumgärtner, G. Schmidtberg and H.-A. Brune, *J. Organomet. Chem.*, 1988, **345**, 221.
197. G. B. Deacon and K. T. Nelson-Reed, *J. Organomet. Chem.*, 1987, **322**, 257.
198. R. Usón, J. Vicente, J. A. Cirac and M. Chicote, *J. Organomet. Chem.*, 1980, **198**, 105.
199. R. Hohenadel and H.-A. Brune, *J. Organomet. Chem.*, 1988, **350**, 101.
200. T. Debaerdemaeker, R. Hohenadel and H.-A. Brune, *ibid.*, 1991, **410**, 265.
201. (a) C. Weisemann, G. Schmidtberg and H.-A. Brune, *J. Organomet. Chem.*, 1989, **362**, 63. (b) H.-A. Brune, G. Schmidtberg and C. Weisemann, *ibid.*, 1989, **371**, 121.
202. T. Debaerdemaeker, B. Stapp and H.-A. Brune, *Acta Crystallogr., Sect. C*, 1987, **43**, 473.
203. K. A. Fallis, C. F. Xu and G. K. Anderson, *Organometallics*, 1993, **12**, 2243.
204. M. Hackett, J. A. Ibers, P. Jernakoff and G. M. Whitesides, *J. Am. Chem. Soc.*, 1986, **108**, 8094.
205. (a) D. O'Hare, *Organometallics*, 1987, **6**, 1766; (b) D. O'Hare, *J. Organomet. Chem.*, 1987, **323**, C13.
206. J. Vicente, J. A. Abad, F. Tereul and J. Garcia, *J. Organomet. Chem.*, 1988, **345**, 233.
207. R. Baumgärtner and H.-A. Brune, *J. Organomet. Chem.*, 1987, **332**, 379.
208. T. G. Appleton, J. R. Hall and M. A. Williams, *J. Organomet. Chem.*, 1986, **303**, 139.
209. V. Christou and G. B. Young, *J. Organomet. Chem.*, 1989, **368**, 391.
210. G. B. Deacon, B. M. Gatehouse and K. T. Nelson-Reed, *J. Organomet. Chem.*, 1989, **359**, 267.
211. T. Debaerdemaeker, C. Weisemann and H.-A. Brune, *J. Organomet. Chem.*, 1988, **350**, 91.
212. A. Sanchez, C. Castellari, A. Panunzi, A. Vitagliano and V. De Felice, *J. Organomet. Chem.*, 1990, **388**, 243.
213. (a) A. B. Goel, S. Goel and D. van der Veer, *Inorg. Chim. Acta*, 1982, **65**, L205; (b) A. Syed, E. D. Stevens and S. G. Cruz, *Inorg. Chem.*, 1984, **23**, 3673.
214. (a) D.-S. Yang, G. M. Bancroft, J. D. Bozek, R. J. Puddephatt and J. S. Tse, *Inorg. Chem.*, 1989, **28**, 1; (b) D.-S. Yang, G. M. Bancroft, L. Dignard-Bailey, R. J. Puddephatt and J. S. Tse, *ibid.*, 1990, **29**, 2487; (c) D.-S. Yang, G. M. Bancroft, R. J. Puddephatt and J. S. Tse, *ibid.*, 1990, **29**, 2496.
215. H. G. Alt, *Angew. Chem., Int. Ed. Engl.*, 1984, **23**, 766.
216. T. G. Appleton, J. R. Hall, D. W. Neale and M. A. Williams, *J. Organomet. Chem.*, 1984, **276**, C73.
217. G. K. Anderson, H. C. Clark and J. A. Davies, *Inorg. Chem.*, 1981, **20**, 3607.
218. G. K. Anderson, H. C. Clark and J. A. Davies, *Inorg. Chem.*, 1981, **20**, 1636.
219. (a) A. B. Goel and S. Goel, *J. Ind. Chem. Soc.*, 1982, **59**, 141; (b) A. B. Goel and S. Goel, *Inorg. Chim. Acta*, 1982, **65**, L77.

220. D. J. Evans and L. A. P. Kane-Maguire, *J. Organomet. Chem.*, 1986, **312**, C24.
221. (a) A. B. Goel, S. Goel and D. Van der Veer, *Inorg. Chim. Acta*, 1981, **54**, L169; (b) F. Giordano and A. Vitagliano, *Inorg. Chem.*, 1981, **20**, 633.
222. H. C. Clark, A. B. Goel and S. Goel, *J. Organomet. Chem.*, 1981, **216**, C25.
223. D. L. Oliver and G. K. Anderson, *Polyhedron*, 1992, **11**, 2415.
224. I. Zahn, K. Polborn and W. Beck, *J. Organomet. Chem.*, 1991, **412**, 397.
225. G. Ferguson, J. F. Gallagher, A. J. McAlees, R. McCrindle, J. Phillips and G. J. B. Williams, *J. Organomet. Chem.*, 1992, **430**, C23.
226. (a) T. R. Lee and G. M. Whitesides, *Acc. Chem. Res.*, 1992, **25**, 266. (b) T. R. Lee, P. E. Laibinis, J. P. Folkers and G. M. Whitesides, *Pure Appl. Chem.*, 1991, **63**, 821.
227. W. R. Winchester and W. M. Jones, *Organometallics*, 1985, **4**, 2228.
228. W. R. Winchester, M. Gawron, G. J. Palenik and W. M. Jones, *ibid.*, 1985, **4**, 1894.
229. Z. Lu, W. M. Jones and W. R. Winchester, *ibid.*, 1993, **12**, 1344.
230. Z. Lu, K. A. Abboud and W. M. Jones, *Organometallics*, 1993, **12**, 1471.
231. M. R. White and P. J. Stang, *Organometallics*, 1983, **2**, 1654.
232. P. Salvadori, G. Uccello-Barretta, R. Lazzaroni and A. M. Caporusso, *J. Chem. Soc., Chem. Commun.*, 1990, 1121.
233. A. de Renzi, P. Ganis, A. Panunzi, A. Vitagliano and G. Valle, *J. Am. Chem. Soc.*, 1980, **102**, 1722.
234. A. T. Hutton, D. M. McEwan, B. L. Shaw and S. W. Wilkinson, *J. Chem. Soc., Dalton Trans.*, 1983, 2011.
235. A. Miyashita, H. Shitara and H. Nohira, *Organometallics*, 1985, **4**, 1463.
236. L. D. Durfee and I. P. Rothwell, *Chem. Rev.*, 1988, **88**, 1059.
237. R. A. Head, *J. Chem. Soc., Dalton Trans.*, 1986, 132.
238. J. F. Nixon, *Chem. Rev.*, 1988, **88**, 1327.
239. (a) T. A. van der Knapp *et al.*, *Organometallics*, 1984, **3**, 1804; (b) H. W. Kroto *et al.*, *J. Organomet. Chem.*, 1985, **280**, 281.
240. J. Chatt, P. B. Hitchcock, A. Pidcock, C. P. Warrens and K. R. Dixon, *J. Chem. Soc., Chem. Commun.*, 1982, 932.
241. (a) E. K. Pham and R. West, *J. Am. Chem. Soc.*, 1989, **111**, 7667; (b) E. K. Pham and R. West, *Organometallics*, 1990, **9**, 1517.
242. P. D. Lickiss, *Chem. Soc. Rev.*, 1992, **21**, 271.
243. B. C. Ankianiec, V. Christou, D. T. Hardy, S. K. Thomson and G. B. Young, *J. Am. Chem. Soc.*, 1994, **116**, 9963.
244. R. A. Michelin *et al.*, *Organometallics*, 1993, **12**, 2372.
245. R. Benn, P. W. Jolly, T. Joswig, R. Mynott and K.-P. Schick, *Z. Naturforsch., Teil B*, 1986, **41**, 680.
246. D. G. Evans, G. R. Hughes, D. M. P. Mingos, J. M. Bassett and A. J. Welch, *J. Chem. Soc., Chem. Commun.*, 1980, 1255.
247. J. R. Briggs, C. Crocker, W. S. McDonald and B. L. Shaw, *J. Chem. Soc., Dalton Trans.*, 1982, 457.
248. A. L. Rheingold and K. A. Jacobson, *Acta Crystallogr., Sect. C*, 1992, **48**, 740.
249. P. J. Fagan, J. C. Calabrese and B. Malone, *Acc. Chem. Res.*, 1992, **25**, 134.
250. P. J. Fagan, J. C. Calabrese and B. Malone, *Science*, 1991, **252**, 1160.
251. P. J. Fagan, J. C. Calabrese and B. Malone, *J. Am. Chem. Soc.*, 1991, **113**, 9408.
252. S. A. Lerke, B. A. Parkinson, D. H. Evans and P. J. Fagan, *J. Am. Chem. Soc.*, 1992, **114**, 7807.
253. (a) J. D. Scott and R. J. Puddephatt, *J. Chem. Soc., Chem. Commun.*, 1984, 193; (b) J. D. Scott and R. J. Puddephatt, *Organometallics*, 1986, **5**, 1253.
254. K. Ikura, I. Ryu, A. Ogawa, N. Sonoda, S. Harada and N. Kasai, *Organometallics*, 1991, **10**, 528.
255. R. Usón, J. Forniés, M. Tomás, B. Menjón and A. J. Welch, *Organometallics*, 1988, **7**, 1318.
256. E. J. Parsons, R. D. Larsen and P. W. Jennings, *J. Am. Chem. Soc.*, 1985, **107**, 1793.
257. P. Braunstein, M. Knorr, T. Stährfeld, A. De Cian and J. Fischer *J. Organomet. Chem.*, 1993, **459**, C1.
258. S. Komiya and I. Endo, *Chem. Lett.*, 1988, 1709.
259. K. Miki, N. Kasai, I. Endo and S. Komiya, *Bull. Chem. Soc. Jpn.*, 1989, **62**, 4033.
260. (a) X. L. R. Fontaine, G. B. Jacobsen, B. L. Shaw and M. Thornton-Pett, *J. Chem. Soc., Chem. Commun.*, 1987, 662; (b) X. L. R. Fontaine, G. B. Jacobsen, B. L. Shaw and M. Thornton-Pett, *J. Chem. Soc., Dalton Trans.*, 1988, 741; (c) X. L. R. Fontaine, G. B. Jacobsen, B. L. Shaw and M. Thornton-Pett, *ibid.*, 1988, 1185; (d) X. L. R. Fontaine, G. B. Jacobsen and B. L. Shaw, *ibid.*, 1988, 2235.
261. M. Rashidi and R. J. Puddephatt, *J. Am. Chem. Soc.*, 1986, **108**, 7111.
262. R. D. Adams, G. Chenn, W. G. Wu and J. G. Yin, *Inorg. Chem.*, 1990, **29**, 4208.
263. (a) R. S. Dickson, G. D. Fallon, M. J. Liddell, B. W. Skelton and A. H. White, *J. Organomet. Chem.*, 1987, **327**, C51; (b) R. S. Dickson, G. D. Fallon, K. D. Heazle and M. J. Liddell, *ibid.*, 1992, **430**, 221.
264. (a) R. D. Adams, G. Chen and J.-G. Wang, *Polyhedron*, 1989, **8**, 2521; (b) R. D. Adams, I. Arafa, G. Chen, J.-C. Lii and J.-G. Wang, *Organometallics*, 1990, **9**, 2350.
265. R. D. Adams, Z.-Y. Li, J.-C. Lii and W.-G. Wu, *Inorg. Chem.*, 1992, **31**, 3445.
266. S. J. Davies and F. G. A. Stone, *J. Chem. Soc., Dalton Trans.*, 1989, 785.
267. S. J. Davies, J. A. K. Howard, R. J. Musgrove and F. G. A. Stone, *J. Chem. Soc., Dalton Trans.*, 1989, 2269.
268. P. Mathur, Md. M. Hossain, K. Das and U. C. Sinha, *J. Chem. Soc., Chem. Commun.*, 1993, 46.
269. M. A. Bennett and H. P. Schwemlein, *Angew. Chem., Int Ed. Engl.*, 1989, **28**, 1296.
270. M. A. Bennett, *Pure Appl. Chem.*, 1989, **61**, 1695.
271. S. L. Buchwald and R. B. Nielsen, *Chem. Rev.*, 1988, **88**, 1047.
272. G. K. Anderson, G. J. Lumetta and J. W. Siria, *J. Organomet. Chem.*, 1992, **434**, 253.
273. J. A. Davies and R. J. Staples, *Polyhedron*, 1991, **10**, 909.
274. J. A. Davies, C. T. Eagle, D. E. Otis and U. Venkataraman, *Organometallics*, 1986, **5**, 1264.
275. K. Komatsu, H. Kamo, R. Tsuji, H. Masuda and K. Takeuchi, *J. Chem. Soc., Chem. Commun.*, 1991, 71.
276. N. M. Boag, M. Green, D. M. Grove, J. A. K. Howard, J. L. Spencer and F. G. A. Stone, *J. Chem. Soc., Dalton Trans.*, 1980, 2170.
277. K. Sünkel, *J. Organomet. Chem.*, 1988, **348**, C12.
278. D. H. Farrar and N. C. Payne, *J. Organomet. Chem.*, 1981, **220**, 251.
279. D. H. Farrar and N. C. Payne, *J. Organomet. Chem.*, 1981, **220**, 239.
280. Y. Koie, S. Shimodo and Y. Saito, *J. Chem. Soc., Dalton Trans.*, 1981, 1082.

281. G. Butler, C. Eaborn and A. Pidcock, *J. Organomet. Chem.*, 1981, **210**, 403.
282. M. Kubota, W. G. Sly, B. D. Santarsiero, M. S. Clifton and L. Kuo, *Organometallics*, 1987, **6**, 1257.
283. (a) P. J. Stang and M. H. Kowalski, *J. Am. Chem. Soc.*, 1989, **111**, 3356; (b) K. H. Thompson (and G. B. Young), B.Sc. Dissertation, Imperial College, 1991.
284. R. Usón, J. Forniés, M. Tomás, B. Menjón and A. J. Welch, *J. Organomet. Chem.*, 1986, **304**, C24.
285. V. De Felice, A. De Renzi, F. Giordano and D. Tesauro, *J. Chem. Soc., Dalton Trans.*, 1993, 1927.
286. N. Chaudhury and R. J. Puddephatt, *Inorg. Chem.*, 1981, **20**, 467.
287. H. Yamashita, M. Tanaka and M. Goto, *Organometallics*, 1993, **12**, 988.
288. G. Marinelli *et al.*, *Polyhedron*, 1990, **9**, 1867.
289. D. M. Hoffman and R. Hoffmann, *J. Chem. Soc., Dalton Trans.*, 1982, 1471.
290. J. C. T. R. Burckett-St Laurent, P. B. Hitchcock, H. W. Kroto and J. F. Nixon, *J. Chem. Soc., Chem. Commun.*, 1981, 1141.
291. S. I. Al Resayes, S. I. Klein, H. W. Kroto, M. F. Meidine and J. F. Nixon, *J. Chem. Soc., Chem. Commun.*, 1983, 930.
292. M. F. Meidine, C. J. Meir, S. Morton and J. F. Nixon, *J. Organomet. Chem.*, 1985, **297**, 255.
293. N. M. Boag *et al.*, *J. Chem. Soc., Dalton Trans.*, 1980, 2182.
294. Y. Koie, S. Shinoda, Y. Saito, B. J. Fitzgerald and C. G. Pierpont, *Inorg. Chem.*, 1980, **19**, 770.
295. R. J. Puddephatt and M. A. Thomson, *Inorg. Chim. Acta*, 1980, **45**, L281.
296. R. J. Puddephatt and M. A. Thomson, *Inorg. Chem.*, 1982, **21**, 725.
297. V. W.-W. Yam, L.-P. Chan and T.-F. Lai, *Organometallics*, 1993, **12**, 2197.
298. J. R. Berenguer *et al.*, *Polyhedron*, 1993, **12**, 1797.
299. J. R. Berenguer, L. R. Falvello, J. Forniés, E. Lalinde and M. Tomás, *Organometallics*, 1993, **12**, 6.
300. L. Manojlovic-Muir, A. N. Henderson, I. Treurnicht and R. J. Puddephatt, *Organometallics*, 1989, **8**, 2055.
301. M. Rashidi and R. J. Puddephatt, *Organometallics*, 1988, **7**, 1636.
302. M. C. Jennings, L. Manojlovic-Muir and R. J. Puddephatt, *J. Am. Chem. Soc.*, 1989, **111**, 745.
303. R. D. Adams and W. G. Wu, *Organometallics*, 1993, **12**, 1248.
304. L. J. Farrugia, *Organometallics*, 1990, **9**, 105.
305. L. J. Farrugia, N. MacDonald and R. D. Peacock, *J. Chem. Soc., Chem. Commun.*, 1991, 163.
306. J. Forniés, M. A. Gomez-Saso, F. Martínez, E. Lalinde, M. T. Moreno and A. J. Welch, *New J. Chem.*, 1992, **16**, 483.
307. P. Espinet, *et al.*, *J. Chem. Soc., Dalton Trans.*, 1990, 791.
308. N. M. Boag, M. Green, J. L. Spencer and F. G. A. Stone, *J. Chem. Soc., Dalton Trans.*, 1980, 1208.
309. (a) T. Hung, P. W. Jolly and G. Wilke, *J. Organomet. Chem.*, 1980, **190**, C5; (b) A. Khun and H. Werner, *Chem. Ber.*, 1980, **113**, 2308.
310. R. Bertani, G. Carturan, A. Scrivanti and A. Wojcicki, *Organometallics*, 1985, **4**, 2139.
311. H. Werner, A. Kuhn and C. Burschka, *Chem. Ber.*, 1980, **113**, 2291.
312. B. Crociani, T. Boschi and E. M. Holt, *Inorg. Chim. Acta*, 1981, **48**, 9.
313. B. Crociani, F. Di Bianca, A. Giovenco and T. Boschi, *Inorg. Chim. Acta*, 1987, **127**, 169.
314. M. Bressan and A. Morvillo, *J. Organomet. Chem.*, 1986, **304**, 267.
315. H. Kurosawa, A. Tsuboi, Y. Kawasaki and M. Wada, *Bull. Chem. Soc. Jpn*, 1987, **60**, 3563.
316. H. Kurosawa, K. Ishii, Y. Kawasaki and S. Murai, *Organometallics*, 1989, **8**, 1756.
317. T. Suzuki, M. Ueda, R. Koumoto and Y. Nakamura, *Bull. Chem. Soc. Jpn.*; 1990, **63**, 804.
318. M. Sano and Y. Nakamura, *J. Chem. Soc., Dalton Trans.*, 1991, 417.
319. M. Grassi, S. V. Meille, A. Musco, R. Pontellini and A. Sironi, *J. Chem. Soc., Dalton Trans.*, 1989, 615.
320. M. Grassi, S. V. Meille, A. Musco, R. Pontellini and A. Sironi, *J. Chem. Soc., Dalton Trans.*, 1990, 251.
321. C. Sorato and L. M. Venanzi, *Inorg. Synth.*, 1989, **26**, 139.
322. T.-M. Huang, J.-T. Chen, G.-H. Lee and Y. Wang, *J. Am. Chem. Soc.*, 1993, **115**, 1170.
323. W. Blosser, D. G. Schimpff, J. G. Gallucci and A. Wojcicki, *Organometallics*, 1993, **12**, 1993.
324. A. Sebald and B. Wrackmeyer, *J. Organomet. Chem.*, 1986, **304**, 271.
325. R. D. W. Kemmitt and A. W. G. Platt, *J. Chem. Soc., Dalton Trans.*, 1986, 1603.
326. A. Ohsuka, T. Fujimori, T. Hirao, H. Kurosawa and I. Ikeda, *J. Chem. Soc., Chem. Commun.*, 1993, 1039.
327. H. Kurosawa *et al.*, *J. Am. Chem. Soc.*, 1992, **114**, 8417.
328. X. Y. Lu and L. Lu, *J. Organomet. Chem.*, 1986, **307**, 285.
329. (a) H. Kurosawa, S. Ogoshi, N. Chatani, Y. Kawasaki, S. Murai and I. Ikeda, *Chem. Lett.*, 1990, 1745; (b) S. Ogoshi and H. Kurosawa, *Organometallics*, 1993, **12**, 2869.
330. D. Rehder, *Coord. Chem. Rev.*, 1991, **110**, 161.
331. R. Benn and A. Rufinska, *Organometallics*, 1985, **4**, 209.
332. (a) M. C. Bohm, R. Gleiter and C. D. Batich, *Helv. Chem. Acta*, 1980, **63**, 990; (b) M. C. Bohm and R. Gleiter, *Angew. Chem., Int. Ed. Engl.*, 1983, **22**, 329.
333. X.-R. Li, G. M. Bancroft, R. J. Puddephatt, Y. F. Hu, Z. Liu and K. H. Tan, *Inorg. Chem.*, 1992, **31**, 5162.
334. M. Guerra, D. Jones, G. DiStefano, S. Torroni, A. Foffani and A. Modelli, *Organometallics*, 1993, **12**, 2203.
335. M. Gianotti, A. Musco, M. Sisti, M. Grassi and G. Gatti, *Inorg. Chim. Acta*, 1987, **133**, 255.
336. M. Grassi, A. Musco, B. Pelli and P. Traldi, *J. Organomet. Chem.*, 1988, **349**, 277.
337. C. P. Brock and T. G. Attig, *J. Am. Chem. Soc.*, 1980, **102**, 1319.
338. H. C. Clark, M. J. Hampden-Smith and H. Rügger, *Organometallics*, 1988, **7**, 2085.
339. R. Bertani, G. Carturan, A. Scrivanti and A. Wojcicki, *Organometallics*, 1985, **4**, 2139.
340. (a) B. Bogdanovic, M. Rubach and K. Seevogel, *Z. Naturforsch., Teil B*, 1983, **38**, 592; (b) B. Bogdanovic, P. Gottsch and M. Rubach, *ibid.*, 1983, **38**, 599; (c) R. Benn, B. Bogdanovic, P. Gottsch and M. Rubach, *ibid.*, 1983, **38**, 602.
341. J. Chaiken, M. J. Casey and M. Villarica, *J. Phys. Chem.*, 1992, **96**, 3183.
342. C. Carfagna, R. Galarini, K. Linn, J. A. López, C. Mealli and A. Musco, *Organometallics*, 1993, **12**, 3019.
343. D. M. P. Mingos and C. R. Nurse, *J. Organomet. Chem.*, 1980, **184**, 281.
344. L. E. Crascall, S. A. Litster, A. D. Redhouse and J. L. Spencer, *J. Organomet. Chem.*, 1990, **394**, C35.
345. S. A. Benyunes, L. Brandt, M. Green and A. W. Parkins, *Organometallics*, 1991, **10**, 57.
346. A. C. Dema and C. M. Lukehart, *Organometallics*, 1993, **12**, 583.
347. P. J. Stang, C. M. Crittell and A. M. Arif, *Organometallics*, 1993, **12**, 4799.
348. M. D. Jones and R. D. W. Kemmitt, *Adv. Organomet. Chem.*, 1987, **27**, 279.

349. D. A. Clarke *et al.*, *J. Chem. Soc., Dalton Trans.*, 1984, 1993.
350. A. Imram, R. D. W. Kemmitt, A. J. W. Markwick, P. McKenna, D. R. Russell and L. J. S. Sherry, *J. Chem. Soc., Dalton Trans.*, 1985, 549.
351. K. W. Chiu, W. Henderson, R. D. W. Kemmitt, L. J. S. Prouse and D. R. Russell, *J. Chem. Soc., Dalton Trans.*, 1988, 427.
352. J. Fawcett *et al.*, *Organometallics*, 1989, **8**, 1991.
353. R. D. W. Kemmitt, P. McKenna, D. R. Russell and L. J. S. Prouse, *J. Chem. Soc., Dalton Trans.*, 1989, 345.
354. H. Hoberg and A. Herrera, *Angew. Chem., Int. Ed. Engl.*, 1980, **19**, 927.
355. F. Canziani, C. Allevi, L. Garaschelli, M. C. Malatesta, A. Albinati and F. Ganazzoli, *J. Chem. Soc., Dalton Trans.*, 1984, 2637.
356. G. E. Herberich, U. Englert and F. Marken, *J. Chem. Soc., Dalton Trans.*, 1993, 1979.
357. R. Poli, *Chem. Rev.*, 1991, **91**, 509.
358. R. D. Ernst, *Chem. Rev.*, 1988, **88**, 1255.
359. G. K. Anderson, *Adv. Organomet. Chem.*, 1993, **35**, 1.
360. S. Fallis, L. Rodriguez, G. K. Anderson and N. P. Rath, *Organometallics*, 1993, **12**, 3851.
361. G. K. Anderson, D. M. Black, R. J. Cross, F. J. Robertson, D. S. Rycroft and R. K. M. Wat, *Organometallics*, 1990, **9**, 2568.
362. G. K. Anderson, R. J. Cross, S. Fallis and M. Rocamora, *Organometallics*, 1987, **6**, 1440.
363. J. Barker, N. D. Cameron, M. Kilner, M. M. Mahmoud and S. C. Wallwork, *J. Chem. Soc., Dalton Trans.*, 1991, 3435.
364. G. Tresoldi, F. Faraone, P. Piraino and F. A. Bottino, *J. Organomet. Chem.*, 1982, **231**, 265.
365. (a) G. K. Anderson, M. R. Lin and M. Y. Chiang, *Organometallics*, 1990, **9**, 288; (b) G. K. Anderson, M. R. Lin and N. P. Rath, *ibid.*, 1990, **9**, 2880.
366. N. M. Boag, R. J. Goodfellow, M. Green, B. Hessner, J. A. K. Howard and F. G. A. Stone, *J. Chem. Soc., Dalton Trans.*, 1983, 2585.
367. V. V. Grushin, C. Bensimon and H. Alper, *Organometallics*, 1993, **12**, 2737.
368. N. M. Boag, *Organometallics*, 1988, **7**, 1446.
369. N. M. Boag, *J. Chem. Soc., Chem. Commun.*, 1988, 617.
370. A. J. Dixon, S. Firth, A. Haynes, M. Poliakoff, J. J. Turner and N. M. Boag, *J. Chem. Soc., Dalton Trans.*, 1988, 1501.
371. J. Powell and N. I. Dowling, *Organometallics*, 1983, **2**, 1742.
372. O. Hackelberg and A. Wojcicki, *Inorg. Chim. Acta*, 1980, **44**, L63.
373. L. D. Boardman, *Organometallics*, 1992, **11**, 4194.
374. G. Hamer and A. Shaver, *Can. J. Chem.*, 1980, **58**, 2011.
375. A. Eisenberg, A. Shaver and T. Tsutsui, *J. Am. Chem. Soc.*, 1980, **102**, 1416.
376. D. S. Yang, G. M. Bancroft, R. J. Puddephatt, B. E. Bursten and S. D. McKee, *Inorg. Chem.*, 1989, **28**, 872.
377. Z. Xue *et al.*, *J. Am. Chem. Soc.*, 1989, **111**, 8779.
378. (a) W. Siebert, *Adv. Organomet. Chem.*, 1980, **18**, 301; (b) W. Siebert, *Angew. Chem., Int. Ed. Engl.*, 1985, **24**, 943.
379. W. Siebert, *Adv. Organomet. Chem.*, 1993, **35**, 187.
380. J. F. Nixon and G. J. D. Sillett, *J. Organomet. Chem.*, 1993, **461**, 237.
381. G. E. Herberich, B. Buller, B. Hessner and W. Oschmann, *J. Organomet. Chem.*, 1980, **195**, 253.
382. G. E. Herberich and W. M. Negele, *J. Organomet. Chem.*, 1988, **350**, 81.

Author Index

This Author Index comprises an alphabetical listing of the names of the authors cited in the text and the references listed at the end of each chapter in this volume.

Each entry consists of the author's name, followed by a list of numbers, for example

Templeton, J. L., 366, 385[233] (350, 366), 387[370] (363)

For each name, the page numbers for the citation in the reference list are given, followed by the reference number in superscript and the page number(s) in parentheses of where that reference is cited in the text. Where a name is referred to in text only, the page number of the citation appears with no superscript number. References cited both in the text and in the tables are included.

Although much effort has gone into eliminating inaccuracies resulting from the use of different combinations of initials by the same author, the use by some journals of only one initial, and different spellings of the same name as a result of the transliteration processes, the accuracy of some entries may have been affected by these factors.

Abad, J.-A., 278[107] (233, 262, 263), 280[224] (238, 262, 263), 519[73] (439), 585[206] (554, 555)
Abakumov, G. A., 289[914] (272), 289[915] (272)
Abakumova, L. G., 289[914] (272)
Abbott, E. H., 527[568] (497)
Abboud, K. A., 280[245] (239, 245, 262, 269, 271), 284[496] (245), 286[654] (252, 267), 586[230] (559, 565–7)
Abdul Malik, K. M., 282[417] (243), 285[569] (248)
Abe, J., 388[346] (360)
Abe, Y., 102[204] (49, 62)
Abel, E. W., 385[89] (305), 498, 500, 520[112] (441), 521[194] (450, 500), 527[580] (498), 527[581] (498), 527[582] (498, 504), 527[583] (498), 527[584] (498, 504), 527[585] (498, 504), 527[586] (498), 527[587] (498, 504), 528[588] (498, 504), 528[589] (498), 528[590] (498), 528[591] (498), 528[592] (498), 528[596] (499), 528[598] (499), 528[599] (499), 528[601] (499), 528[605] (499), 528[606] (499), 528[607] (499), 528[608] (499, 504), 528[609] (499, 500), 528[610] (500), 528[612] (500, 504), 528[613] (500, 504), 528[614] (500, 504), 528[621], 528[630] (500), 528[631] (504), 528[632] (504), 528[633] (504), 528[634] (504), 582[6a] (534), 582[6l] (534), 582[7a] (534, 535), 582[7l] (534, 535), 585[154a] (553), 585[154b] (553), 585[155] (553), 585[156] (553)
Abicht, H.-P., 101[148] (41, 47, 48, 63), 104[338] (81, 86), 221[77] (204), 279[174] (236), 281[294] (241), 281[295] (241, 264), 281[298] (241), 289[884] (268), 529[679] (509), 529[680] (509), 529[684] (509), 529[715], 530[745] (514), 530[746] (514)
Abis, L., 280[278] (240), 519[47] (437)
Absi-Halabi, M., 189[559] (172, 173), 221[79] (205)
Abu-Salah, O. M., 526[507] (490)
Achar, S., 523[307] (464)
Achatz, D., 223[175] (218)
Ackermann, K., 280[217] (238, 267, 270), 390[441] (377), 390[443] (377)
Adams, H., 288[827] (262–4), 289[848] (263), 289[908] (271)
Adams, R. D., 586[262] (564, 572), 586[264a] (564, 572), 586[264b] (564, 572), 586[265] (564), 587[303] (572)
Affolter, S., 277[73] (230, 231, 240), 283[432] (244, 248, 264), 283[489] (258, 263, 273), 284[565] (247, 263), 390[433] (375)
Afzal, D., 106[440] (96, 98), 526[486] (485, 494), 526[495]
Agapakis, S., 525[418] (475, 476)
Agawa, T., 25[27] (4), 25[28] (4)
Agbossou, S., 389[417] (372)
Agnès, G., 100[73] (37, 38)
Agnew, N. H., 528[654] (506), 529[656] (506), 529[657] (506)
Aguiló, M., 101[121] (42, 50, 58, 60)
Ahmar, M., 389[398] (370), 389[399] (370)
Ahmed, K. J., 583[58] (541, 545)
Ahmed, M., 288[838] (263)
Aime, S., 185[258] (130)
Aitchison, A. A., 27[193] (24)
Ajò, D., 185[258] (130), 189[542] (167)
Akai, Y., 24[20] (4), 101[90] (42, 58), 286[703] (255, 266, 268)
Akbarzadeh, M., 285[590] (249)
Åkermark, B., 285[584] (249), 285[589] (249), 386[170] (323), 386[180] (325, 326), 387[222] (340), 387[250] (346), 387[279] (352), 389[368] (363), 389[369] (364), 389[372] (365), 389[389] (368), 389[413] (371), 583[81] (543)
Akhmetov, N. S., 220[31a] (197), 220[32] (197)
Akimoto, M., 387[246] (345)
Akita, M., 105[428] (97), 105[429] (96, 97), 105[430] (96), 185[244] (131, 159, 172)
Akiyama, M., 287[775] (259)
Akiyoshi, K., 277[43] (230), 277[60] (230, 239, 266, 267, 270)
Akkerman, O. S., 525[426] (478), 525[434] (478), 585[185] (554, 555)
Akmedov, V. M., 186[283] (137, 139)
Al, E., 186[344] (139, 146)
Al Resayes, S. I., 587[291] (570)
Al Takhin, G., 521[183] (445)
Al-Ahmad, S., 519[21] (434)
Al-Allaf, T. A. K., 519[35] (436), 519[38] (436)
Albanese, J. A., 277[15] (227, 230), 277[17] (227, 230)
Albano, V. G., 25[34] (5), 280[256] (239, 262, 272), 286[706] (255, 256, 271), 385[95] (306), 385[96] (306), 520[82] (439), 583[72] (542, 544, 545, 548, 549), 583[76] (543), 583[82] (544), 583[84] (544)
Albéniz, A. C., 285[611] (249), 285[628] (250), 288[819] (262, 263), 288[821] (262), 288[830] (262), 289[864] (264), 385[106] (308), 385[107] (308), 387[220] (339)
Albert, J., 279[172] (236), 282[372], 282[373] (269), 282[375] (263, 264), 282[418] (243), 283[474] (245, 253), 283[477] (245, 263, 264), 283[479] (245), 385[93] (305)
Albert, M. R., 384[27] (293)
Albinati, A., 222[141] (214), 277[73] (230, 231, 240), 283[484] (245), 283[489] (258, 263, 273), 284[565] (247, 263), 285[618] (250, 252), 285[620] (250), 386[187] (328), 386[188] (328), 390[433] (375), 427[42b] (397), 429[152] (409), 429[154] (409, 422), 429[156d] (409), 520[141] (444), 520[145] (444), 525[394] (474), 525[395] (474), 525[396] (474), 529[702] (511), 583[63] (542), 588[355] (579)

Albright, T. A., 26[127] (16), 186[349] (139, 146), 384[54] (295), 390[464] (382), 584[141] (551)
Alcock, N. W., 283[445] (245), 428[98] (403, 406), 520[134] (443), 524[377] (471), 525[427] (478), 525[437] (479), 526[514] (490), 529[690] (509)
Aleksandrov, G. G., 430[207] (422)
Aleksanyan, V. T., 188[430] (152), 188[431] (152), 188[432] (152), 188[433] (152)
Al-Essa, R. J., 527[570] (497, 505), 527[571] (497), 528[637] (505), 528[642] (505), 584[125a] (549), 584[125b] (549)
Alexander, J. J., 522[263] (460)
Alfranca, G., 524[338] (467, 469)
Alibrandi, G., 428[121] (406), 520[83] (439), 521[167] (458, 459), 521[201] (453), 522[233] (457), 522[234] (457), 522[240] (458), 522[241] (458), 522[248] (458, 459), 522[258] (459), 522[259] (459), 522[260] (459)
Al-Juaid, S. S., 529[670] (508), 529[675] (508)
Allegra, G., 390[463] (381)
Allen, D. G., 283[440] (244)
Allen, F. H., 521[154] (445)
Allevi, C., 429[153] (409), 429[156a] (409), 429[156b] (409), 429[156c] (409), 429[156d] (409), 588[355] (579)
Allison, J., 99[7] (30), 100[30] (32), 187[400] (150, 174)
Allison, J. A. C., 284[512]
Allmann, R., 528[624]
Almeida, J. F., 519[66] (438)
Al-Najjar, I. M., 525[454] (480), 584[94] (545), 584[96] (545), 584[97] (545), 584[113] (548), 584[114] (548)
Al-Ohaly, A. R., 526[507] (490)
Alonso, G., 283[476] (245, 263, 264)
Alonso, M. T., 279[178] (236, 263, 264), 281[338] (242)
Alonso, M. Y., 288[825] (262, 272)
Alper, H., 26[106] (14), 290[931] (273), 390[439] (377), 428[108b] (403, 409), 583[65] (542), 588[367] (580)
Alphonse, P., 187[379] (148)
Al-Qahtani, H. A., 526[507] (490)
Al-Resayes, S. I., 223[182] (219)
Al-Salem, N. A., 282[396] (243, 245)
Al-Shalaan, A. M., 584[94] (545), 584[95] (545)
Al-Showiman, S. S., 584[94] (545), 584[95] (545)
Alsters, P. L., 278[101] (233, 262, 263, 268, 273), 278[102] (233, 250, 262–4), 279[176] (236, 240, 242, 243, 262, 269, 270, 272, 273, 275), 279[177] (236, 262), 289[892] (268), 289[893] (268), 530[752] (515)
Alt, H. G., 519[54] (437, 440), 521[185] (447), 585[194] (554, 555), 585[215] (556, 580, 581)
Altaba, M. F., 521[177]
Alvarez-Valdes, A., 283[462] (245)
Alyea, E. C., 282[410], 282[411], 282[412] (263), 285[607] (249), 385[90] (305), 529[664] (507), 530[718], 530[742] (514)
Amatore, C., 101[129] (42, 47, 48, 63), 102[207] (56), 278[93] (233), 278[94] (233), 278[95] (233)
Amin, H. B., 584[97] (545)
Ammann, C. J., 386[186] (328), 386[187] (328), 386[188] (328), 386[189] (328), 521[196] (450), 584[109] (547, 576)
Ammeter, J. H., 187[416] (151)
Amrein, W., 583[55] (540)
Anastasiou, D., 524[344] (468)
Andari, M. K., 386[176] (326), 386[177] (326)
Andell, O. S., 184[165] (123)
Andersen, L., 186[325] (139, 140)
Anderson, A. B., 100[82] (40)
Anderson, B. B., 100[35] (33), 100[38e] (31, 33), 278[86] (233)
Anderson, C. B., 285[590] (249)
Anderson, C. M., 428[88] (404), 523[299] (463), 523[317] (465), 528[617] (510), 528[618] (510)
Anderson, G. K., 103[249] (67), 189[514] (157), 280[220] (238, 255, 256, 261), 280[281] (240), 282[420] (243, 248, 263, 264), 282[421] (243), 287[717] (255), 288[794] (260), 289[860] (264), 289[910] (272), 384[16] (292), 390[454] (379), 390[455] (379), 390[456] (379), 428[100] (403, 405), 428[101] (403, 416), 428[111] (406, 417), 428[118] (406, 407), 428[119] (406), 428[120] (406), 428[125a] (407), 428[125b] (407), 428[125c] (407), 428[128] (407), 428[129] (407), 429[133] (408), 429[134] (408), 429[136] (408), 518[3] (432, 460), 519[63] (437, 458), 520[96] (440), 520[98] (440), 520[113] (441), 520[115] (441), 520[119] (442), 520[125] (442), 521[149] (444), 521[165], 521[191] (447), 522[251] (459), 522[264] (460), 522[267] (460), 522[268] (460), 522[270] (460), 523[273] (460), 523[274] (460), 526[524] (491), 530[771] (517), 582[3] (534), 585[189] (554, 555), 585[192] (554, 555, 557, 580, 581), 585[203] (554, 555), 585[217] (557), 585[218] (557), 586[223] (557), 586[272] (565, 568), 588[359] (579), 588[360] (580), 588[361] (580), 588[362] (580), 588[365a] (580), 588[365b] (580)
Anderson, L., 104[358] (81, 88, 90), 105[372] (81, 88, 89)
Anderson, L. G., 390[432] (375)
Anderson, O. P., 386[180] (325, 326)
Anderson, T. A., 279[179] (236)
Andersson, P. G., 389[384] (367), 389[410] (371), 389[411] (371)
Andrae, D., 220[5] (194)
Andreini, B. P., 220[18] (195, 196), 429[150b] (409–11), 429[150c] (409, 411)
Andrews, L. C., 384[47] (294)
Andrews, M. A., 285[612] (249), 384[49] (295)
Andrianov, Y. A., 187[404] (150), 187[405] (150, 155)
Andryukhova, N. P., 278[140] (235)
Angelici, R. J., 25[76] (10), 223[171] (217), 287[733] (256), 430[192] (418), 430[213] (423), 430[220] (425), 520[132] (443)
Angermund, K., 100[52] (31, 34, 35), 102[165a] (42, 59), 103[274] (73, 74), 182[40] (110), 183[70] (115), 185[210] (130, 132), 185[223] (130, 131, 134)
Anif, A. M., 583[53] (540, 548, 574, 576)
Ankianiec, B. C., 519[80] (439), 520[106] (440), 522[230] (456), 585[165] (553, 555), 586[243] (560)
Anklin, C. G., 283[483] (245, 263), 283[484] (245), 529[699] (510), 529[710] (512)
Annibale, G., 584[119] (548), 584[120] (548)
Antipin, M. Y., 279[194] (237)
Antón, M., 101[127] (42, 49), 101[128] (42, 47, 62)
Antonsson, T., 389[418] (373)
Antsyshkina, A. S., 222[93a] (207), 222[93b] (207), 222[93c] (207)
Aoki, K., 429[171] (412)
Aoki, T., 278[104] (233)
Aoyagui, S., 384[65] (300), 384[66] (300)
Aoyama, H., 279[209] (238, 269)
Aoyama, I., 287[725] (256, 257)
Apal'kova, G. M., 428[112] (406)
Apaolaza, J. A., 288[839] (263), 524[349] (468)
Apotecher, B., 183[93] (115, 130)
Appel, R., 24[4] (2), 184[175] (126, 127), 184[195] (126, 127), 184[205] (126), 184[206] (126, 145), 185[207] (126, 138, 145, 157), 186[319] (138, 145), 186[320] (138, 145), 387[283] (352)
Appleton, T. G., 286[667] (252), 519[75] (439, 464, 472, 497, 506), 521[189] (447), 524[356] (469), 525[422] (477, 497, 506), 528[627] (500), 528[654] (506), 529[655] (506), 529[656] (506), 529[657] (506), 529[658] (507), 529[659] (507), 585[167] (553–5), 585[183] (554, 555, 557), 585[208] (554, 555), 585[216] (557)
Appling, J. R., 187[403] (150)
Ara, I., 289[901] (270), 524[368] (470), 525[400] (474), 525[401] (474), 525[402] (474), 525[408] (475), 525[420] (477)
Aradi, A. A., 103[284] (73, 74), 103[285] (73, 74, 76), 183[96] (115, 148), 187[380] (148), 187[389] (148)
Arafa, I., 586[264b] (564, 572)
Arai, M., 286[638] (251)
Arbuckle, B. W., 520[110] (441)
Arcadi, A., 389[360] (363)
Arcas, A., 220[23] (196), 280[221] (238, 263), 280[222] (238, 263, 268), 280[223] (238), 286[689] (255, 271), 288[818] (262, 263)
Arena, G., 428[123b] (407)
Aresta, M., 25[34] (5), 184[182] (126, 128)
Aribi-Zouioueche, L., 387[256] (347)
Arif, A. M., 221[51] (201), 428[95b] (403), 429[140] (408), 526[472] (484, 487, 491, 493), 526[496] (487), 526[497] (487), 526[500]

(487), 584[122] (548), 584[132] (549), 587[347] (577)
Arimoto, Y., 287[775] (259)
Aristides, C., 429[167] (412)
Arlen, C., 100[67] (37, 38), 104[304] (81, 85), 281[292] (241, 253), 286[676] (253, 264)
Armentano, S., 282[360] (243), 282[423] (243), 282[424] (243), 284[506] (263)
Armentrout, P. B., 99[12] (30, 32), 99[13] (30, 32), 99[14] (30, 32), 99[15] (30, 32), 182[30] (109)
Arnek, R., 285[585] (249, 263)
Arnold, D. P., 521[187] (447), 523[284] (461), 523[286] (462), 529[676] (508, 512), 530[748] (515), 530[749] (515)
Arnup, P. A., 277[11] (227)
Arrieta, J. M., 524[390]
Arrigoni, A., 26[153] (20)
Arsenault, G. J., 280[219] (238, 259), 286[656] (252), 288[782] (260, 271), 523[316] (465), 523[317] (465), 524[371] (471, 495), 524[374] (471), 524[375] (471), 524[376] (471), 524[391] (472), 526[518] (490), 585[166] (553–5, 558)
Arsen'eva, T. I., 188[430] (152), 188[431] (152), 188[432] (152), 188[433] (152)
Arshady, R., 222[120] (211)
Arzoumanian, H., 25[80] (11), 25[81] (11), 525[417] (475)
Asada, N., 285[587] (249, 267), 384[61] (296, 297), 385[97] (306), 583[62] (541, 542, 547, 574), 583[66] (542, 547, 574), 584[101] (545, 547)
Asaro, F., 583[37] (536, 539)
Ascenso, J. R., 186[341] (139, 146)
Asfour, H. M., 384[53] (295)
Ashley-Smith, J., 287[766] (259)
Ashmawy, F. M., 184[139] (116, 122)
Ashworth, T. V., 105[423] (95, 96), 223[181] (219)
Ast, T., 187[401] (150)
Attig, T. G., 520[131] (443), 587[337] (576)
Atwood, J. D., 184[134] (116, 117, 157)
Atwood, J. L., 101[101] (41, 42, 48, 56, 58, 66, 67), 102[187] (48, 56, 62)
Auburn, P. R., 388[338] (360)
Aversa, M. C., 520[86] (439)
Avis, M. W., 287[713] (255), 526[499] (487)
Avshu, A., 283[445] (245), 428[98] (403, 406)
Awasthi, A. K., 277[35] (229), 389[408] (371)
Aye, K.-T., 290[948] (273, 274, 276), 523[312] (464), 523[313] (464), 523[324] (466)
Azam, K. A., 26[115] (16), 518[10] (432, 433, 453), 520[129] (442, 495), 521[193] (450), 522[210] (453), 522[223] (455)
Azar, M. C., 185[268] (136)
Azzabi, M., 278[94] (233), 278[95] (233)

Baba, S., 288[797] (260), 288[798] (260)
Bachechi, F., 390[440] (377), 520[141] (444), 520[142] (444), 529[710] (512)
Bachmann, J., 187[416] (151)
Bäckvall, J.-E., 184[165] (123), 285[582] (248), 385[122] (311), 385[124] (312), 385[125] (312), 387[224] (341), 387[225] (341), 387[226] (341), 389[369] (364), 389[384] (367), 389[390] (368), 389[391] (368), 389[392] (368), 389[394] (368), 389[395] (368), 389[403] (371), 389[404] (371), 389[405] (371), 389[406] (371), 389[407] (371), 389[408] (371), 389[409] (371), 389[410] (371), 389[411] (371), 389[412] (371)
Bader, M., 100[35] (33)
Baena, M. J., 283[464] (245, 264)
Baerends, E. J., 99[16] (30, 32)
Bagatur'yants, A. A., 384[50] (295)
Bahsoun, A., 286[675] (253)
Bailey, C. T., 385[103] (308)
Bailey, N. A., 288[827] (262–4), 289[848] (263), 289[908] (271)
Bailey, P. M., 385[111] (309), 385[139] (315)
Bailey, T. D., 25[61] (9)
Bain, A. C., 387[221] (340)
Baird, M. C., 277[11] (227), 285[598] (249), 286[695] (255, 263, 268, 270, 273, 275), 584[140] (551)
Bakač, A., 105[387] (91, 92), 105[388] (91, 92), 105[389] (92), 105[390] (91, 92), 105[391] (91, 92), 105[392] (91, 92), 105[393] (91, 92), 105[394] (91, 92), 105[395] (91, 92)
Bakakin, V. V., 427[22] (393)
Baker, G. R., 281[334] (242), 283[486] (245, 263), 385[92] (305), 529[691] (510)
Baker, M. J., 582[13b] (535, 537, 539, 566)
Baker, P. K., 582[6j] (534), 582[6k] (534), 582[6l] (534)
Bakhireva, S. I., 430[195] (419)
Bakirtzis, V., 189[527] (159, 170)
Balavoine, G., 279[175] (236), 281[337] (242, 263), 283[487] (245, 263), 289[898] (270), 388[312] (357)
Balazs, A. C., 102[208] (62), 521[203] (453)
Balch, A. L., 220[24] (196), 220[38] (197), 222[115] (210, 215), 222[135] (213), 223[149] (215), 223[151] (215), 278[119] (234, 236), 286[646], 286[702] (255, 256, 258, 269, 271), 289[911] (272), 428[69] (400), 429[174] (414), 518[7] (432, 441), 520[102] (440)
Baldacchini, C. J., 583[40] (536, 539)
Baldwin, J. E., 282[402]
Baldy, A., 25[80] (11), 182[23] (109)
Balegroune, F., 104[336] (81, 86), 104[337] (81, 86), 285[613] (249, 252), 286[665] (252), 286[666] (252), 526[491] (486, 509), 529[689] (509)
Ballester, L., 100[49] (31, 33), 189[500] (157), 277[42] (230, 231)
Ballesteros, A., 25[44] (7), 25[45] (7), 25[46] (7), 104[348] (81, 88), 104[363] (81, 88), 104[364] (81, 88)
Balme, G., 387[233] (343), 389[374] (365), 389[375] (365)
Baltensperger, U., 529[666] (507)
Baltzer, P. E. G., 187[413] (151), 187[416] (151)
Balzani, V., 530[728] (514), 530[729] (514), 530[730] (514), 530[732] (514)
Bancroft, D. P., 527[565] (497), 527[566] (497), 527[567] (497), 527[568] (497)
Bancroft, G. M., 186[277] (137), 386[172] (324), 521[200] (452), 524[392] (473), 528[625] (500), 585[214a] (556), 585[214b] (556), 585[214c] (556), 587[333] (575), 588[376] (581)
Bandini, A. L., 283[481] (245, 248), 430[224] (426)
Banditelli, G., 283[481] (245, 248), 430[224] (426)
Bandmann, H., 102[218] (65), 104[342] (81, 90), 186[303] (138, 140)
Bandoli, G., 287[744] (257), 387[270] (350)
Bandyopadhyay, D., 284[531] (263, 268), 284[535] (268), 289[891] (268)
Bandyopadhyay, P., 284[531] (263, 268), 289[857] (264), 289[891] (268)
Banister, A. J., 189[528] (159, 170)
Barabotti, P., 280[271] (240, 270)
Baralt, E., 526[503] (489, 494), 526[504] (489), 527[534] (493)
Baranova, T. V., 191[626] (180, 182)
Barborak, J. C., 285[634] (250, 255, 256, 267)
Barbosa, B. M., 25[73] (10)
Bard, A. J., 188[444] (153, 155)
Bardi, R., 287[726] (256, 272), 287[727] (256), 520[136] (443, 461), 520[137] (443), 521[155] (472, 493), 521[168], 524[383] (472)
Barefield, E. K., 101[105] (42, 47, 57), 187[403] (150), 188[420] (151, 153, 157, 163), 189[526] (159), 430[200] (420), 527[557] (495)
Bärhausen, D., 25[46] (7), 104[320] (80, 81), 104[348] (81, 88)
Barieux, J.-J., 389[374] (365), 389[398] (370)
Barigelletti, F., 530[733] (514)
Barker, J., 282[370] (263, 264), 529[697] (510), 588[363] (580)
Barnes, L. A., 220[13] (195)
Barofsky, D. F., 285[636] (250)
Barr, N., 283[429] (244), 285[617] (250)
Barro, J., 279[172] (236)
Barron, P. F., 528[628] (500)
Bars, O., 281[292] (241, 253), 286[676] (253, 264)
Bart, J. C. J., 100[73] (37, 38)
Bartczak, T. J., 289[909] (272, 276), 521[159] (493)
Bartel, K., 25[77] (11), 223[170] (217, 218), 430[222] (425)
Barth, R., 221[77] (204)
Bartik, T., 102[216] (65), 185[224] (130, 133), 185[260] (131, 132), 186[330] (139)
Bartlett, R. A., 184[172] (126)
Bartley, J. P., 285[617] (250)
Bartlone, A., 189[525] (159, 169)
Bartocci, C., 524[334] (467)
Bartolucci, S., 583[77] (543)
Barton, D. H. R., 386[199] (333)
Barton, L., 191[635] (180, 181)

Barton, R. J., 279[179] (236)
Bartsch, R., 184[179] (126, 127), 185[233] (131, 136, 159, 172), 186[321] (138, 155)
Basato, M., 222[120] (211)
Basch, H., 426[2a] (392), 426[2b] (392)
Base, K., 191[633] (180, 181)
Bashilov, V. V., 101[124] (41, 42, 47), 220[30] (197), 221[68d] (204), 222[95] (207), 222[96] (207), 222[97a] (207), 222[97b] (207), 279[199] (237), 279[202] (237), 279[203] (237)
Basoglu, R., 26[156] (20)
Basolo, F., 24[3] (2), 280[239] (239, 240)
Bassan, R., 519[44] (437), 585[175] (554, 555)
Basset, J.-M., 278[105] (233, 255, 256), 586[246] (560)
Basu, A., 427[57b] (399)
Basu, P., 284[532] (272), 284[534] (269), 530[773] (517)
Bates, P. A., 283[431] (244), 289[858] (264), 385[89] (305), 388[295] (354), 528[592] (498), 528[610] (500), 585[154a] (553), 585[154b] (553)
Batich, C. D., 386[171] (324), 587[332a] (575)
Batson, J. R., 521[192] (447)
Battiste, M. A., 103[257] (67, 73), 183[132] (116, 120, 157)
Bau, R., 285[614] (250), 384[47] (294), 524[386] (496, 497), 584[123] (548, 552, 554, 555)
Bauch, C. G., 25[70] (10)
Baum, G., 528[624]
Baumgärtner, R., 519[59] (437), 519[60] (437), 521[190] (447), 585[196] (554, 555), 585[207] (554, 555)
Baur, A., 522[213] (454), 585[195] (554, 555)
Bauschlicher, C. W. J., 220[13] (195)
Bauwe, E., 100[24b] (30)
Baxter, J. C., 519[64] (437), 522[262] (460)
Baxter, S. M., 521[151] (444), 521[152] (444)
Bayer, G. M., 100[62] (31, 36)
Bayer, U., 519[50] (437, 454)
Beach, D. L., 101[135] (42, 57)
Beak, P., 386[192] (331), 386[193] (331)
Beauchamp, J. L., 99[8] (30), 99[14] (30, 32), 182[29] (109), 182[30] (109), 290[952] (274)
Beauchamp, J. V., 187[397] (150)
Becalska, A., 189[494] (157, 163), 189[495] (157, 163)
Beck, W., 102[191] (48, 64), 222[121] (211), 284[537], 287[737] (256), 287[778] (260), 385[85] (304), 429[182] (416, 423), 430[190a] (418), 430[190b] (418), 519[30] (435), 526[477] (484), 527[532] (493), 528[623], 586[224] (558)
Becke, A., 99[17] (30, 32)
Becker, H. J., 190[584] (174, 176)
Becker, J., 279[162] (236), 388[333] (359)
Becker, Y., 278[111] (234), 288[791] (260, 270)
Beckett, M. A., 528[592] (498), 528[600] (499), 528[610] (500)
Beck-Hemetsberger, H., 101[122] (41, 42, 47–9, 51, 58, 60)
Behm, J., 25[31] (5), 188[476] (157, 167), 531[779] (517)
Behr, A., 101[133] (42, 62, 63), 105[373] (81, 88), 105[379] (80, 88), 288[788] (260), 387[214] (336)
Behrens, H., 25[74] (10)
Behrens, U., 278[83] (232)
Bein, T., 25[32] (5), 25[33] (5)
Belderraín, T. R., 24[16] (3), 101[104] (42, 47, 51, 57, 58, 66, 67), 186[346] (139, 146, 159), 188[482] (159)
Beletskaya, I. P., 278[140] (235), 290[927] (273)
Beley, M., 25[62] (9)
Bell, A., 385[115] (310)
Bell, A. T., 100[21] (30)
Bellenbaum, M., 103[236] (66, 67), 182[53] (111, 113, 115, 157)
Belli Dell'Amico, D., 583[65] (542)
Belluco, U., 223[161a] (216), 223[161b] (216), 223[169] (217, 218), 279[186] (237), 430[199] (419), 430[212] (423), 518[1] (432), 522[241] (458), 522[252] (459), 524[381] (472)
Belov, A. P., 386[174] (324), 386[176] (326), 386[177] (326)
Bénard, M., 26[135] (19), 105[409] (93, 94)
Bender, R., 220[29] (197), 221[75a] (204), 221[75b] (204), 221[76] (204), 221[78a] (205), 221[78b] (205), 221[79] (205), 221[80a] (205), 427[15] (393, 395), 427[31] (395), 427[65b] (400), 519[29] (435)
Benetollo, F., 223[163] (216, 218), 223[168] (217), 288[822] (262, 264, 267), 430[201b] (420, 426), 430[217] (424), 430[218] (424), 524[384]
Benn, F. R., 184[139] (116, 122)
Benn, R., 103[224] (66, 67), 103[229] (66, 67), 103[231] (67, 71), 103[232] (66, 67), 103[233] (66, 67), 103[251] (66, 67, 70), 103[252] (66, 67, 70), 103[260] (70), 103[281] (73, 74), 103[283] (73, 74), 104[297] (73, 74), 182[52] (111, 113, 155, 157), 182[57] (111, 113, 115, 119, 138, 157), 183[67] (113, 115, 117), 183[68] (113), 183[73] (114, 115, 117, 157), 183[84] (115, 123, 174), 183[116] (115, 117), 183[120] (114, 115), 184[135] (116, 117, 157), 184[136] (116, 117, 157), 184[143] (117, 139, 141), 184[145] (117, 118), 184[154] (117, 143), 186[306] (138, 143), 186[308] (138–40), 186[337] (139, 141), 187[364] (143), 187[365] (143), 187[366] (143), 285[623] (250), 285[624] (250), 287[779] (260), 288[787] (260, 267), 384[64] (299), 389[414] (371), 389[420] (373), 390[457] (380), 390[458] (380), 521[184] (447), 586[245] (560), 587[331] (575), 587[340c] (576)
Benneche, T., 279[154] (235), 385[134] (314)
Benner, L. S., 220[38] (197), 223[149] (215)
Bennett, M. A., 24[22] (4), 103[292] (73, 74, 81), 105[415] (42, 93, 95), 183[122] (115, 122), 183[128] (116, 122), 185[251] (131, 134), 185[252] (131), 185[253] (131, 134), 185[254] (131, 134), 278[143] (235), 279[197] (237), 280[264] (240, 261, 262, 266, 272, 275, 276), 282[394], 385[77] (303), 429[137a] (408), 429[137b] (408), 461, 521[173] (462), 521[187] (447), 523[284] (461), 523[285] (462), 523[286] (462), 524[382], 524[385], 526[487] (485), 526[488] (485), 526[498] (487), 529[676] (508, 512), 529[678] (509, 513, 515), 529[681] (509), 530[719] (513, 515), 530[748] (515), 530[749] (515), 584[144] (551), 584[145] (551), 586[269] (565, 569), 586[270] (565, 569)
Bensch, W., 277[70] (230, 245), 526[506] (490, 491)
Bensimon, C., 390[439] (377), 588[367] (580)
Benter, G., 585[161] (553)
Benyunes, S. A., 587[345] (577)
Bercaw, J. E., 521[162] (458, 462, 472), 521[182] (445)
Berenguer, J. R., 587[298] (571), 587[299] (571)
Bergamin, F., 278[127] (235, 264, 270)
Bergamini, P., 524[377] (471), 524[378] (471), 524[379] (471), 526[514] (490)
Bergens, S. H., 281[299] (241)
Berger, R., 102[217] (65), 102[218] (65), 102[219] (65), 186[301] (138, 142), 186[302] (138, 140), 186[303] (138, 140)
Berges, P., 528[611] (500)
Bergman, R. G., 100[71] (37, 38), 103[239] (66, 67), 188[488] (157), 281[290] (240), 531[776] (517)
Bergounhou, C., 26[166] (20, 21), 190[577] (174)
Berhalter, K., 521[176], 521[178]
Berke, H., 287[718] (255)
Berlin, A. J., 290[934] (273)
Bermudez, G., 188[445] (153)
Bernal, I., 522[266] (460)
Bernardinelli, G., 284[542] (272)
Bernhardt, W., 106[439] (96, 98)
Bernocchi, E., 389[360] (363)
Berry, A. D., 183[71] (115, 119)
Berry, D. E., 529[678] (509, 513, 515), 530[719] (513, 515)
Berry, M., 223[180] (219)
Berry, R. D., 524[356] (469), 525[422] (477, 497, 506), 529[659] (507)
Bertani, R., 187[367] (143), 222[123] (211), 222[126] (211, 213), 222[129] (211), 223[154] (215), 223[163] (216, 218), 223[168] (217), 223[173b] (218), 223[173c] (218), 277[26] (228, 230), 277[28] (228, 230), 278[129] (235, 257), 287[711] (255, 257, 264), 287[754] (257), 288[789] (260), 288[845] (263, 269), 289[862] (264), 289[880] (267), 387[266] (350), 387[267] (350), 387[270] (350), 430[214] (423, 425), 430[215] (423, 425), 430[216] (424, 426), 430[217] (424), 430[218] (424), 430[221] (425), 430[226] (426), 523[326] (466), 587[310]

(573), 587[339] (576)
Berthon, R. A., 284[525]
Bertleff, W., 278[124] (234), 286[700] (255), 289[905] (271)
Berton, A., 101[119] (42, 61), 222[129] (211), 287[711] (255, 257, 264), 288[789] (260), 289[862] (264), 289[880] (267), 385[75] (303), 523[277] (461), 523[278] (461)
Bertozzi, S., 584[93a] (544), 584[93b] (544)
Besenhard, J. O., 26[127] (16)
Besenyei, G., 26[117] (16)
Bespozvannyl, A. M., 428[108a] (403)
Bessler, E., 25[73] (10)
Betz, P., 25[48] (8), 101[97] (42, 47, 52), 104[319] (80, 81), 182[37] (110, 112), 183[85] (115), 184[146] (117), 284[541]
Beurich, H., 27[190] (23), 27[191] (23), 189[559] (172, 173)
Beveridge, K. A., 530[719] (513, 515)
Beydoun, N., 281[349] (242, 248), 283[434] (244, 253)
Beyer, A., 102[193] (48), 277[66] (230), 277[67] (230, 231)
Bezrukova, A. A., 519[40] (436)
Bhaduri, S., 427[57a] (399), 427[57b] (399)
Bharadwaj, P. K., 520[110] (441)
Bhargava, S. K., 527[584] (498, 504), 528[632] (504), 529[678] (509, 513, 515)
Bialy, S., 388[299] (355)
Bianchini, C., 184[189] (126, 129), 184[190] (126, 129), 184[191] (126, 129), 184[192] (126, 129), 184[196] (126, 129), 184[197] (126, 129), 185[208] (129)
Bickelhaupt, F., 525[426] (478), 525[434] (478), 585[185] (554, 555)
Bierschenk, T. R., 190[583] (174)
Bigorgne, M., 25[71] (10)
Billard, C., 428[125b] (407)
Billington, D. C., 185[273] (137)
Billups, W. E., 100[27] (32)
Bilton, M. S., 529[676] (508, 512)
Binger, P., 103[281] (73, 74), 103[283] (73, 74), 183[97] (115, 121, 138, 140), 184[159] (121), 187[384] (148, 149), 285[624] (250), 389[420] (373), 389[423] (373), 390[457] (380), 525[431] (478), 585[168] (554)
Bird, P. H., 26[125] (16)
Bitler, S. P., 523[293] (462), 584[123] (548, 552, 554, 555)
Bjarnason, A., 27[177] (21), 188[450] (154, 171), 189[552] (171), 190[604] (174)
Björkman, E. E., 385[122] (311), 385[124] (312), 385[125] (312), 387[224] (341), 387[225] (341), 389[391] (368)
Black, D. M., 521[191] (447), 588[361] (580)
Black, S. I., 101[98] (41, 42, 48, 66, 67)
Blackwell, C. M., 390[432] (375)
Bladon, P., 189[515] (157, 166)
Blagg, A., 527[552] (494), 527[553] (494)
Blake, A. J., 284[524] (245, 271, 272), 290[958] (275)
Blanchard, J. L., 583[58] (541, 545)
Blandamer, M. J., 522[256] (459), 522[257] (459)
Blankenship, C., 386[190] (328)
Blanton, C. B., 525[435] (478)
Bläser, D., 104[301] (73, 74), 186[284] (138, 145)
Blasio, B., 583[75] (543)
Blau, R. J., 518[14] (433), 527[554] (494)
Blau, W. J., 277[61] (230)
Bleasdale, C., 183[130] (116, 120, 157)
Bleeke, J. R., 390[464] (382)
Bliss, T., 223[166] (217, 218)
Blomberg, M. R. A., 100[24a] (30), 100[28] (32), 220[6] (194), 386[170] (323)
Blosser, W., 587[323] (574, 577)
Blum, O., 100[26] (32)
Blümel, J., 188[429] (152, 155, 156)
Blumhofer, R., 27[188] (23), 27[190] (23), 105[414] (93, 94, 96, 98), 106[438] (96, 98), 189[547] (171)
Boag, N. M., 188[478] (157, 159, 167), 220[26] (197, 200), 390[435] (375), 427[58] (399), 427[62] (400, 410), 427[63] (400, 410), 427[64] (400), 429[157] (410, 417), 429[158] (410, 416, 417), 429[159] (410), 525[448] (480), 585[187] (554, 555, 574), 585[188] (554, 555, 573, 576, 580), 586[276] (565, 566), 587[293] (571, 572), 587[308] (573), 588[366] (580), 588[368] (580), 588[369] (580), 588[370] (581)
Boardman, L. D., 588[373] (581)
Bochin, V. P., 429[147] (409)
Bochmann, A., 189[524] (159, 163)
Bochmann, M., 24[15] (3), 25[86] (11, 12), 101[117] (42, 58, 59, 61), 102[167] (50, 51, 58, 59), 102[174] (50, 56, 58, 59, 61, 62), 184[174] (126), 221[66] (204), 280[259] (239), 519[79] (439), 585[180] (554, 555)
Boegeman, S. C., 279[165] (236, 269, 273)
Boehm, M. C., 187[387] (147), 187[388] (147), 187[395] (150)
Boersma, J., 103[253] (67), 184[141] (117, 120, 159), 184[142] (117, 159), 190[571] (172), 190[572] (172), 190[573] (172, 173), 190[574] (172, 173), 277[57] (230, 232, 234, 239, 269, 270, 273), 278[98] (233, 239, 262), 278[99] (233, 255, 256, 262), 278[101] (233, 262, 263, 268, 273), 278[102] (233, 250, 262–4), 278[103] (233, 251, 262, 273), 278[114] (234), 278[142] (235), 280[260] (239, 255, 262, 272, 273), 280[261] (239, 240, 255, 262, 266, 267, 269–74), 280[262] (239, 262, 266, 271, 274, 276), 280[265] (240, 265, 269, 270, 272–4), 285[633] (250, 255, 262), 289[892] (268), 289[893] (268), 290[953] (274, 276)
Boese, R., 183[83] (115, 119), 185[260] (131, 132), 186[284] (138, 145), 187[381] (148, 149), 187[382] (148, 149), 187[383] (148, 149), 190[596] (175, 180)
Bogdanović, B., 186[327] (139, 141), 186[328] (139, 141), 186[337] (139, 141), 281[291] (240), 285[572] (248), 388[287] (352), 388[288] (352), 587[340a] (576), 587[340b] (576), 587[340c] (576)
Bögge, H., 184[144] (117, 121)
Böhm, M. C., 185[276] (137), 386[171] (324), 587[332a] (575), 587[332b] (575)
Bohm, T., 189[556] (172, 173)
Bohman, O., 386[173] (324)
Böhme, P., 100[33] (31, 33), 100[34] (31, 33), 183[133] (116, 124, 138), 186[310] (138), 186[311] (138), 187[363] (142)
Böhme, U., 103[244] (66, 67), 184[148] (117, 163)
Bohra, R., 520[111] (441)
Bois, C., 103[241] (66, 67), 189[490] (157, 158, 163)
Bökman, F., 386[173] (324)
Boleslawski, M. P., 103[285] (73, 74, 76), 183[96] (115, 148)
Bolletta, F., 530[729] (514)
Bolze-Kuhrt, D., 290[942] (273)
Bombieri, G., 223[163] (216, 218), 223[168] (217), 284[527], 287[745] (257), 288[822] (262, 264, 267), 430[201b] (420, 426), 430[217] (424), 430[218] (424), 524[384]
Bonaccorsi, P., 520[86] (439)
Bonacic-Koutecky, V., 220[9] (195)
Bonafede, S., 530[729] (514)
Bonati, F., 530[742] (514)
Boncella, J. M., 280[245] (239, 245, 262, 269, 271), 286[654] (252, 267)
Bondarenko, G. N., 384[42] (294)
Bonfa, M., 429[155] (409), 429[156b] (409)
Bonnet, J. J., 26[166] (20, 21), 190[577] (174)
Bonnet, M. C., 389[417] (372)
Bonrath, W., 103[273] (73, 74), 103[275] (73, 74, 80), 182[44] (110, 113), 182[46] (110, 111, 113, 115, 118, 126), 183[66] (113, 118), 184[150] (117, 118), 185[223] (130, 131, 134), 288[790] (260, 266), 385[82] (303, 318)
Bontempelli, G., 101[144] (42, 65), 101[145] (42, 47), 102[163] (42, 47, 65)
Booth, B. L., 100[63] (31, 36), 189[498] (157, 165), 189[499] (157, 165), 384[69] (302), 384[70] (302), 428[127] (407)
Booth, M., 528[612] (500, 504), 528[634] (504)
Borden, W. T., 583[35] (536, 538, 550), 584[135] (550)
Born, L., 100[59] (31, 36)
Borodko, Y. G., 280[237] (239)
Boron, P., 520[144] (444)
Borrachero, M. V., 220[23] (196), 280[221] (238, 263), 280[222] (238, 263, 268), 280[223] (238), 286[689] (255, 271), 288[818] (262, 263)
Boschi, T., 279[146] (235), 289[888] (268), 387[270] (350), 583[38] (536), 583[47] (538), 587[312] (573), 587[313] (573)
Bosl, G., 583[55] (540)
Bosnich, B., 281[299] (241), 388[338] (360), 388[339] (360)
Botnikov, M. Y., 223[179b] (219)

Bott, S. G., 427[41] (396, 413)
Botteghi, C., 430[211] (423), 523[275] (461), 523[276] (461), 523[277] (461), 523[278] (461)
Bottino, F. A., 390[437] (376), 588[364] (580)
Bottomley, A. R. H., 282[408] (263, 264), 529[665] (507)
Bouachir, F., 103[255] (67, 70), 189[497] (157, 159, 164), 189[521] (159, 163), 189[522] (159, 163), 529[685] (509)
Bouaoud, S.-E., 25[53] (8), 221[85a] (205), 221[85b] (205), 286[675] (253), 288[835] (263), 289[856] (264)
Boubekeur, K., 279[175] (236), 283[487] (245, 263)
Bouchareb, S. G. F., 182[25] (109)
Boucher, D., 220[26] (197, 200), 390[435] (375)
Boudjouk, P., 288[785] (260)
Bougeard, P., 105[434] (96, 98), 190[569] (172)
Bouquet, A., 25[71] (10)
Boutry, O., 102[188] (48, 56)
Bowie, C., 528[628] (500)
Bowmaker, G. A., 220[5] (194)
Boyter, H. A., 190[623] (180, 181)
Bozek, J. D., 521[200] (452), 585[214a] (556)
Bradford, A. M., 428[79] (402), 428[81] (403), 428[83] (403), 428[84] (404), 429[179] (415)
Bradshaw, J. D., 185[230] (131, 135)
Braga, D., 188[436] (152), 223[171] (217), 390[469] (383), 430[204b] (421), 430[205] (421), 430[213] (423), 430[220] (425), 520[82] (439), 583[76] (543)
Brainard, R. L., 522[232] (457), 522[235] (457), 531[783] (518)
Brammer, L., 521[154] (445)
Brandemark, U. B., 100[24a] (30), 182[19] (109)
Brandt, L., 587[345] (577)
Brant, P., 223[149] (215)
Braterman, P. S., 521[198] (450)
Braun, J., 185[227] (131, 136, 159), 185[271] (136, 172), 187[351] (139, 155, 159, 168, 171), 187[352] (139, 155, 159, 168, 171), 187[353] (139, 155, 159, 168, 172)
Braun, U., 104[357] (81, 88, 90)
Bräunlich, G., 25[42] (5), 25[47] (7), 104[342] (81, 90), 105[367] (81, 88, 90)
Braunstein, P., 25[53] (8), 27[167] (21), 104[336] (81, 86), 104[337] (81, 86), 105[377] (80), 105[380] (80, 88), 190[567] (172, 174), 204, 220[29] (197), 221[75a] (204), 221[75b] (204), 221[76] (204), 221[78a] (205), 221[78b] (205), 221[79] (205), 221[80a] (205), 221[80b] (205), 221[85a] (205), 221[85b] (205), 221[86] (205), 221[87a] (205), 221[87b] (205), 221[88a] (205), 221[88b] (205), 221[89a] (205), 221[89b] (205), 222[90] (205), 222[91] (205), 222[92] (206), 222[94] (207), 281[305] (241), 281[306] (241, 263), 286[665] (252), 286[666] (252), 288[835] (263), 288[836] (263), 289[856] (264), 289[902] (271), 387[275] (350), 427[15] (393, 395), 427[16] (393), 427[29] (395), 427[31] (395), 427[65b] (400), 430[193] (418), 519[29] (435), 523[330] (467), 526[491] (486, 509), 529[689] (509), 586[257] (563, 564)
Braye, E., 285[610] (249)
Brechot, P., 287[743] (257, 273)
Bressan, M., 587[314] (573)
Brevard, C., 530[765] (516)
Brezinski, M. M., 100[38e] (31, 33), 100[42] (33), 100[44] (31, 33), 190[605] (174), 190[606] (174), 190[607] (174)
Briant, C. E., 427[39] (396), 429[166] (411, 414, 419), 429[172] (412)
Bridge, M. E., 390[460] (381)
Briggs, J. R., 428[92] (403, 416), 429[180] (416, 423), 525[455] (480), 530[747] (514), 583[64a] (542), 583[64b] (542), 586[247] (561, 562)
Briggs, T. N., 288[827] (262–4), 289[848] (263)
Briguglio, J. J., 191[634] (180)
Brisdon, B. J., 281[356] (243, 244)
Broadley, K., 390[446] (378), 390[447] (378)
Broadwood-Strong, G. T. L., 583[43] (537)
Brock, C. P., 387[274] (350), 520[131] (443), 587[337] (576)
Brodt, G., 190[589] (177)
Broeker, G. K., 27[178] (21, 22)
Brookhart, M., 285[634] (250, 255, 256, 267)
Brower, D. C., 583[57] (541, 545)
Brown, C., 427[51] (398, 399)
Brown, D. B., 528[640] (505), 584[126] (549)
Brown, D. G., 290[956] (275, 276)
Brown, J. M., 278[117] (234, 250), 289[870] (266), 289[872] (266), 385[138] (315), 520[134] (443), 582[14a] (535, 558)
Brown, L. D., 286[660] (252)
Brown, M. P., 518[6] (432), 518[10] (432, 433, 453), 518[11] (432, 433), 519[43] (437), 520[117] (441, 442), 520[126] (442), 521[166] (453), 522[223] (455)
Brown, P. R., 429[148] (409)
Brown, T. L., 24[8] (2)
Browning, C. S., 427[33] (395, 400), 427[35] (395–7), 526[490] (486)
Browning, J., 220[21] (196), 280[277] (240), 519[65] (438), 584[147] (552)
Brownstein, S. K., 188[466] (155)
Bruce, M. I., 103[245] (66, 67), 103[246] (67, 72), 188[481] (157, 165), 189[492] (157, 161), 281[355] (243), 282[422] (243)
Bruder, G., 184[195] (126, 127)
Bruk, L. G., 220[31b] (197)
Brumbaugh, J. S., 279[164] (236, 250, 261, 262, 264, 265, 273), 285[629] (250, 261, 262, 264)
Brune, H.-A., 187[373] (148), 437, 454, 519[49] (437, 454), 519[50] (437, 454), 519[51] (437), 519[52] (437), 519[53] (437), 519[54] (437, 440), 519[55] (437), 519[56] (437), 519[57] (437), 519[58] (437), 519[59] (437), 519[60] (437), 519[61] (437, 454), 519[69] (438, 454), 519[70] (438), 519[71] (438), 520[107] (440, 454), 521[176], 521[178], 521[179], 521[180], 521[181], 521[185] (447), 521[190] (447), 521[199] (452), 522[211] (454), 522[212] (454), 522[213] (454), 525[436] (478, 481), 526[459], 526[462] (481), 585[194] (554, 555), 585[195] (554, 555), 585[196] (554, 555), 585[199] (554, 555), 585[200] (554, 555), 585[201a] (554, 555), 585[201b] (554, 555), 585[202] (554, 555), 585[207] (554, 555), 585[211] (554, 555)
Brunet, J. J., 187[362] (142)
Brunner, H., 105[411] (93, 94), 185[231] (131, 136), 522[266] (460)
Bruno, G., 223[156] (215), 521[155] (472, 493), 521[167] (458, 459), 521[169]
Bruno, M. C., 519[77] (439)
Bryars, K. H., 519[44] (437), 525[427] (478), 525[437] (479), 585[175] (554, 555)
Bryndza, H. E., 24[23] (4), 24[24] (4), 521[162] (458, 462, 472), 521[182] (445), 522[221] (455), 523[279] (461), 523[280] (461), 524[370] (471), 585[177] (554, 555, 558)
Büch, H. M., 103[283] (73, 74), 183[97] (115, 121, 138, 140), 184[159] (121), 285[624] (250), 285[625] (250), 389[420] (373), 390[457] (380), 525[431] (478), 585[168] (554)
Buchanan, R. M., 277[20] (227, 230)
Bucher, R., 187[416] (151)
Buchwald, S. L., 586[271] (565, 569)
Budzelaar, P. H. M., 190[571] (172), 190[572] (172), 190[573] (172, 173)
Bueno, C., 190[563] (172, 174), 190[576] (174)
Bulck, J., 279[191] (237)
Buller, B., 588[381] (582)
Bumagin, N. A., 278[140] (235), 290[927] (273)
Bumagina, I. G., 290[927] (273)
Bumgarner, R. E., 187[412] (151)
Bunel, E. E., 188[447] (153, 155), 188[465] (155, 156), 189[513] (157, 166)
Bunsenges, B., 187[395] (150)
Burckett-St Laurent, J. C. T. R., 587[290] (570)
Burgers, H., 185[255] (129)
Burgess, J., 277[54] (230), 522[256] (459), 522[257] (459), 525[450] (480), 583[52] (540)
Burgina, E. B., 427[22] (393)
Burkhardt, E. R., 103[239] (66, 67), 188[488] (157), 281[290] (240)
Burkhart, G., 26[99] (14), 104[344] (81, 88, 90), 105[369] (81, 88), 105[371] (81, 88), 184[162] (122)
Burkhart, J. P., 388[329] (358)
Burlakov, V. V., 185[212] (130, 131)
Burmeister, J. L., 277[15] (227, 230), 277[17] (227, 230), 277[19] (227, 229), 527[556] (495)
Burns, B., 388[316] (357), 388[317] (357), 388[319] (357), 388[320] (357)

Burns, G. T., 183[100] (115, 124), 183[102] (115, 124)
Burrow, R. A., 427[20] (393, 396)
Burrows, A. D., 221[62] (202, 209, 215), 222[110] (209, 215), 427[41] (396, 413)
Burschka, Ch., 288[792] (260), 387[223] (341), 388[285] (352), 587[311] (573, 576)
Bursten, B. E., 588[376] (581)
Burton, J. T., 527[572] (497), 528[649] (505)
Burtseva, O. Y., 283[446] (245)
Busetto, L., 584[153] (553)
Bushnell, G. W., 280[277] (240), 519[65] (438)
Busygina, I. N., 221[58] (202, 219), 222[93a] (207), 222[93b] (207), 222[93c] (207), 223[179a] (219), 223[179b] (219), 223[179c] (219)
Butler, G., 519[37] (436), 519[38] (436), 527[527] (491), 587[281] (568)
Butler, I. R., 283[456] (245), 582[71] (534, 535), 582[7m] (534, 535)
Butler, I. S., 584[90] (544)
Butters, T., 386[198] (333)
Bychkov, V. T., 189[538] (159)
Byers, L. R., 27[174] (21), 189[518] (159, 167), 189[543] (167, 171, 172)
Byers, P. K., 278[120] (234, 239, 240, 262, 266), 280[252] (239, 263), 280[253] (239, 261, 262), 280[254] (239, 262), 280[255] (239, 262), 280[263] (239, 242, 261–4, 272, 273, 275, 276), 288[816] (262, 274, 276), 289[896] (269, 273, 274), 290[916] (272, 273, 276), 290[943] (273–6), 290[945] (273), 290[946] (273), 290[947] (273, 274), 290[950] (273, 274, 276), 290[954] (275), 290[955] (275), 290[956] (275, 276), 520[91] (440), 523[305] (464), 523[318] (465), 528[615] (500)
Byrd, G. D., 187[396] (150)
Byrne, H. J., 277[61] (230)
Byrom, N. T., 389[415] (372)
Byström, S. E., 389[405] (371)

Cacchi, S., 389[360] (363)
Caglioti, L., 389[360] (363)
Cai, Z., 182[54] (111, 123), 182[55] (111, 123)
Calabrese, J. C., 26[136] (19), 183[127] (115, 117, 125), 184[168] (125), 518[20] (434), 520[90] (439), 521[162] (458, 462, 472), 524[370] (471), 586[249] (561), 586[250] (561), 586[251] (561)
Calcaterra, M., 100[73] (37, 38)
Calderazzo, F., 220[16] (195), 220[17] (195), 220[18] (195, 196), 220[19a] (196), 220[19b] (196), 428[108b] (403, 409), 428[116] (406), 429[146] (409), 429[149] (409), 429[150a] (409, 411), 429[150b] (409–11), 429[150c] (409, 411), 429[163] (411), 522[263] (460), 583[65] (542)
Calhorda, M. J., 289[872] (266)
Calligaris, M., 223[154] (215), 277[26] (228, 230), 277[28] (228, 230), 387[266] (350), 387[267] (350), 429[135] (408)
Calligaro, L., 222[130] (211), 287[750] (257), 522[241] (458)
Calvarin, G., 188[435] (152)
Calvin, G., 277[48] (230, 239, 240, 266)
Camalli, M., 582[15] (535), 583[37] (536, 539), 584[100] (545)
Camazon, M. J., 283[462] (245)
Camellini, M. T., 286[689] (255, 271)
Cameron, A. D., 285[598] (249), 584[140] (551)
Cameron, N. D., 282[370] (263, 264), 529[697] (510), 588[363] (580)
Cameron, T. S., 527[584] (498, 504), 528[588] (498, 504), 528[608] (499, 504), 528[612] (500, 504), 528[614] (500, 504)
Cammack, R., 105[383] (91)
Campagna, S., 428[121] (406), 521[201] (453)
Campagnaro, A., 222[127a] (211), 287[753] (257)
Campana, C. F., 527[577] (498)
Campbell, J. A., 527[577] (498)
Campbell, K. D., 522[266] (460)
Cámpora, J., 25[88] (12), 25[89] (12), 101[106] (42, 47, 50, 58, 81, 90), 101[107] (42, 47, 58, 62, 63), 101[108] (42, 47, 58, 62, 63), 104[294] (74, 78, 81), 186[343] (139), 186[345] (139)
Campos, P., 528[604] (499)
Campostrini, R., 278[87] (233), 288[789] (260), 289[880] (267), 429[135] (408), 430[204a] (421)
Camps, F., 25[26] (4)
Camus, A., 527[545] (494), 527[546] (494)
Cano, F. H., 288[819] (262, 263), 387[220] (339), 387[272] (350), 387[273] (350)
Cano, M., 100[49] (31, 33), 277[42] (230, 231)
Canovese, L., 277[27] (228), 385[75] (303)
Canty, A. J., 278[120] (234, 239, 240, 262, 266), 279[184] (236, 241), 280[252] (239, 263), 280[253] (239, 261, 262), 280[254] (239, 262), 280[255] (239, 262), 280[262] (239, 262, 266, 271, 274, 276), 280[263] (239, 242, 261–4, 272, 273, 275, 276), 280[264] (240, 261, 262, 266, 272, 275, 276), 282[385] (253, 263, 264), 282[389] (262), 282[390], 285[580] (248, 276), 288[816] (262, 274, 276), 289[896] (269, 273, 274), 290[916] (272, 273, 276), 290[917] (272, 273, 275, 276), 290[918] (272, 273, 275, 276), 290[919] (272, 273, 275), 290[943] (273–6), 290[945] (273), 290[946] (273), 290[947] (273, 274), 290[948] (273, 274, 276), 290[949] (273, 274, 276), 290[950] (273, 274, 276), 290[953] (274, 276), 290[954] (275), 290[955] (275), 290[956] (275, 276), 290[961] (276), 384[22] (292), 520[85] (439), 520[91] (440), 520[93] (440), 523[304] (464), 523[305] (464), 523[309] (464), 523[310] (464), 523[318] (465), 523[324] (466), 528[615] (500), 528[616] (510), 529[701] (510)
Canziani, F., 429[152] (409), 429[153] (409), 429[154] (409, 422), 588[355] (579)
Cao, D., 584[133] (549, 568)
Capat, C., 384[38] (294)
Capka, M., 187[392] (150)
Cápora, J., 184[193] (126, 129)
Caporusso, A. M., 586[232] (559)
Cardin, C. J., 280[276] (240), 429[139] (408), 526[480] (484, 487)
Cardin, D. J., 277[61] (230), 280[276] (240), 429[139] (408), 526[480] (484, 487)
Carfagna, C., 388[352] (361), 388[355] (361), 389[356] (361), 525[433] (478), 587[342] (576, 577)
Cargill Thompson, A. M. W., 280[229] (238)
Cariati, F., 283[480] (245)
Caridad, R., 186[343] (139)
Carl, R. T., 103[288] (74, 76, 80), 183[105] (115, 122), 183[106] (115, 122), 279[188] (237), 429[188] (417)
Carlton, L., 26[107] (15)
Carmichael, D., 279[192] (237), 279[193] (237), 529[675] (508), 583[54] (540)
Carmona, D., 520[142] (444)
Carmona, E., 24[16] (3), 24[17] (3), 25[87] (11), 25[88] (12), 25[89] (12), 25[92] (13), 100[78] (38, 39, 42, 56), 100[80] (38–40, 42), 101[99] (42, 56), 101[100] (42, 56), 101[101] (41, 42, 48, 56, 58, 66, 67), 101[102] (41, 42, 56, 62, 63, 73, 74, 78, 81, 90), 101[103] (41, 42, 47, 57, 58, 67), 101[104] (42, 47, 51, 57, 58, 66, 67), 101[106] (42, 47, 50, 58, 81, 90), 101[107] (42, 47, 58, 62, 63), 101[108] (42, 47, 58, 62, 63), 101[118] (42, 61), 101[139] (42, 56), 102[187] (48, 56, 62), 102[188] (48, 56), 102[220] (65), 103[268] (74, 78, 81), 104[293] (73–5, 78, 81, 90), 104[294] (74, 78, 81), 104[295] (74, 78), 104[296] (74, 78), 104[325] (81, 85), 183[108] (115, 124), 183[109] (115, 124), 184[134] (116, 117, 157), 184[173] (125, 126, 157), 184[193] (126, 129), 186[296] (138, 143), 186[340] (139, 146), 186[342] (139, 146, 157), 186[343] (139), 186[344] (139, 146), 186[345] (139), 187[371] (146), 188[482] (159), 288[843] (263)
Carneiro, T. M. G., 286[665] (252), 286[666] (252)
Carnicer, J., 288[820] (262), 289[865] (264), 524[358] (469, 470), 524[367] (470)
Carpenter, B. K., 99[4] (30, 31, 41, 42), 101[120] (42, 50, 55, 93, 94), 183[107] (115, 122), 184[152] (117, 118)
Carpinelli, P., 583[77] (543)
Carr, C. J., 427[44b] (397)
Carr, K., 282[401] (263, 270)
Carr, N., 385[81] (303), 430[194] (419), 519[24] (434), 522[222] (455), 583[30] (536, 538, 539)
Carr, S. W., 526[516] (490)
Carranza, M. M., 288[846] (263), 524[361] (469)
Carré, F., 183[99] (115, 124)
Carrié, M., 105[403] (92)
Carrier, A. M., 430[200] (420), 527[557]

(495)
Carroll, P. J., 191[637] (180)
Carter, E. A., 220[4] (194), 426[2d] (392)
Cartoon, M. E. K., 282[427] (244)
Carturan, G., 187[367] (143), 223[165] (216), 288[789] (260), 289[880] (267), 429[135] (408), 587[310] (573), 587[339] (576)
Carty, A. J., 185[241] (131, 172, 174), 185[242] (131, 172, 174), 185[243] (131, 172), 185[248] (131, 172, 174)
Carusi, P., 101[123] (42)
Caruso, F., 582[15] (535), 583[37] (536, 539), 584[100] (545)
Carvajal, J., 101[111] (42, 47, 48, 56, 61)
Casagrande, G. T., 390[463] (381)
Casarin, M., 185[258] (130), 189[542] (167)
Casas, J. M., 284[558] (245), 286[690] (255, 263, 264), 289[901] (270), 524[350] (468), 524[351] (468), 524[366] (470), 524[389], 525[400] (474), 525[401] (474), 525[403] (474), 525[405] (475), 525[406] (475)
Casellato, U., 520[138] (443, 461), 583[59] (541, 545)
Caseri, W. R., 583[63] (542)
Casey, C. P., 526[501] (487)
Casey, M. J., 587[341] (576, 581)
Casey, S., 384[69] (302), 384[70] (302)
Casser, C., 184[175] (126, 127), 184[205] (126)
Cassidy, D. P., 528[600] (499)
Castagnoli, C., 290[929] (273)
Castan, P., 284[517] (263), 288[840] (263)
Castanet, Y., 388[289] (353)
Caste, M. L., 190[563] (172, 174)
Castellani, C. B., 222[123] (211), 222[126] (211, 213), 278[129] (235, 257), 287[754] (257)
Castellani, M. P., 188[421] (151, 154, 155)
Castellari, C., 280[256] (239, 262, 272), 286[706] (255, 256, 271), 385[96] (306), 583[82] (544), 583[83] (544), 583[84] (544), 585[212] (554)
Castiglioni, M., 185[249] (131, 172, 174), 190[561] (172, 174), 190[562] (172, 174), 190[564] (172), 190[578] (174)
Castilho, T. J., 223[168] (217), 430[216] (424, 426)
Castineiras, A., 284[500]
Castle, M. L., 190[576] (174)
Catellani, M., 286[639] (251, 270, 272–4), 286[640] (251), 286[641] (251), 290[920] (272–4), 290[928] (273), 290[929] (273), 428[126] (407)
Cavalieri, A., 26[136] (19)
Cavell, K. J., 286[707] (255, 262), 286[708] (255, 262), 385[76] (303), 429[141] (408)
Cavigiolo, W., 100[73] (37, 38)
Cavinato, G., 287[726] (256, 272), 287[727] (256), 287[728] (256), 287[740] (257), 520[136] (443, 461), 520[137] (443), 520[138] (443, 461), 523[275] (461), 523[276] (461)
Cavoli, P., 287[726] (256, 272), 520[137] (443), 524[383] (472), 583[59] (541, 545)
Caygill, G. B., 284[507] (245, 246, 264), 284[514] (263, 264), 284[519], 284[553] (245, 246, 264), 285[579] (248)
Cazes, B., 389[398] (370), 389[399] (370), 389[400] (370), 389[401] (370), 389[402] (370)
Cebollada, J. I., 223[147] (214), 277[7] (226), 524[365] (469)
Cecconi, F., 27[182] (21), 187[355] (139, 147), 278[122] (234)
Ceder, R. M., 101[146] (42, 62, 63), 101[151] (41, 42, 50), 281[348] (242), 282[374], 282[418] (243), 284[557] (245, 246, 263, 264), 529[698] (510)
Cenini, S., 278[145] (235), 525[456] (480)
Ceriotti, A., 26[145] (19), 26[147] (20), 26[148] (20), 26[149] (20), 26[151] (20), 26[152] (20), 26[153] (20), 26[154] (20), 26[155] (20), 26[165] (20), 27[171] (21), 427[44a] (397), 427[52] (398)
Cesa, I. G., 582[18] (535)
Cesari, M., 287[731] (256)
Cesarotti, E., 282[425] (243, 264), 386[184] (327), 386[185] (328)
Cetini, G., 430[224] (426)
Ceulemans, A., 528[636] (505)
Chaiken, J., 587[341] (576, 581)
Chakladar, S., 284[548] (245, 263, 264), 284[549] (245, 263, 264), 284[551] (245, 263, 264)
Chakravorty, A., 284[531] (263, 268), 284[532] (272), 284[533] (268), 284[534] (269), 284[535] (268), 284[536] (268), 289[857] (264), 289[891] (268), 530[773] (517)
Chaloner, P. A., 182[7] (108), 182[8] (108), 182[9] (108), 182[10] (108), 384[17] (292), 384[18] (292), 384[19] (292), 384[20] (292), 384[21] (292), 582[5a] (534), 582[5b] (534), 582[5c] (534), 582[5d] (534), 582[5e] (534), 582[5f] (534), 583[43] (537)
Chaloupka, S., 582[15] (535)
Chan, D. M. T., 389[424] (373), 389[426] (374), 390[429] (375)
Chan, G. C., 428[110] (405)
Chan, L.-P., 587[297] (571)
Chan, M.-C., 188[470] (154)
Chandra, G., 585[163] (553, 565)
Chang, L. S., 521[174], 529[713] (512)
Chang, S.-C., 100[27] (32)
Chang, T. C.-T., 285[612] (249), 384[49] (295)
Chang, T. H., 584[89] (544)
Chao, C. H., 285[614] (250)
Chappell, S. D., 519[46] (437, 456)
Chaptal, N., 389[401] (370)
Chapuis, G., 430[201b] (420, 426), 524[384]
Charland, J.-P., 386[178] (326)
Chassot, L., 530[724] (514), 530[725] (514), 530[728] (514), 530[729] (514), 530[730] (514), 530[736] (514)
Chatani, N., 103[221] (65), 386[213] (336), 587[329a] (575)
Chatt, J., 280[239] (239, 240), 384[45] (294, 316), 586[240] (560)
Chattopadhyay, S., 284[532] (272), 284[533] (268), 284[534] (269), 284[536] (268), 530[773] (517)
Chaudhury, N., 286[670] (253), 587[286] (569)
Chaudhury, S., 528[593] (498)
Chaudret, B., 520[114] (441)
Chauvin, Y., 287[743] (257, 273), 289[861] (264, 268)
Chee, H.-K., 524[382]
Cheeseman, G. W. H., 282[427] (244)
Chelucci, G., 282[391], 529[693] (510), 530[775] (517)
Chen, G., 586[264a] (564, 572), 586[264b] (564, 572)
Chen, H., 184[172] (126)
Chen, J., 101[116] (42, 49, 55), 102[157] (42, 56)
Chen, J.-T., 279[167] (236, 259, 262), 429[142a] (408), 429[143] (408), 429[144] (408), 430[219] (425), 519[25] (434, 462), 519[27] (434), 523[288] (462), 523[289] (462), 523[291] (462), 523[292] (462), 587[322] (574, 577)
Chen, L., 185[266] (134, 138)
Chen, M. C., 277[38] (229, 230)
Chen, M. M. L., 25[94] (13)
Chen, R., 102[157] (42, 56)
Chen, Y., 187[406] (150, 151, 155), 187[407] (150, 151, 155)
Chen, Z., 182[54] (111, 123), 182[55] (111, 123)
Cheng, C.-H., 286[642] (251, 271), 286[643] (251), 288[807] (261, 273), 390[466] (382)
Cheng, C.-W. F., 285[612] (249), 384[49] (295)
Cheng, K.-M., 188[470] (154)
Cheng, M.-C., 277[32] (228–30), 277[39] (229, 230), 429[143] (408), 429[144] (408), 430[219] (425), 523[288] (462), 523[292] (462)
Cheng, W., 99[16] (30, 32)
Chenn, G., 586[262] (564, 572)
Cherkashin, G. M., 384[42] (294)
Cherkasov, V. K., 289[914] (272), 289[915] (272)
Cherkasova, K. L., 282[371]
Cheskin, B. K., 283[486] (245, 263)
Chetcuti, M. J., 26[108] (15), 26[109] (15, 19), 26[110] (15), 26[133] (19), 27[169] (21), 27[170] (21), 105[416] (95, 96), 105[417] (95, 96), 105[418] (95, 96), 105[419] (95, 96), 105[420] (95, 96), 105[421] (95, 96), 105[423] (95, 96), 184[160] (121, 144), 184[161] (121, 139, 144, 159), 185[236] (131, 136, 169), 185[239] (131, 169, 172), 185[268] (136), 185[269] (136), 185[270] (136), 188[483] (157, 159, 169), 189[525] (159, 169), 189[532] (159), 189[533] (169, 172, 173), 189[534] (159, 169), 189[545] (167, 169), 189[555] (172, 173), 190[570] (172, 173), 223[181] (219), 385[99] (307), 583[29] (536, 544), 585[160] (553)
Chia, L.-Y., 280[227] (238), 526[478] (484)
Chiang, M. Y., 428[111] (406, 417), 520[113] (441), 588[365a] (580)
Chicote, M.-T., 277[18] (227), 277[34] (228, 230), 529[706] (511), 585[198]

(554, 555)
Chierico, A., 390[465] (382)
Chiesi-Villa, A., 222[138] (213), 526[522] (490, 491, 493), 526[523] (490, 491)
Chin, K. W., 429[183] (416)
Chinakov, V. D., 220[36] (197)
Chini, P., 19, 26[136] (19), 26[137] (19), 26[141] (19), 26[145] (19), 26[165] (20), 427[51] (398, 399)
Chiraratvatana, C., 183[122] (115, 122), 183[128] (116, 122), 279[197] (237), 385[77] (303), 529[681] (509), 584[144] (551), 584[145] (551)
Chisholm, M. H., 527[554] (494)
Chiu, K. W., 103[266] (73, 74), 103[267] (73, 74), 280[285] (240), 525[443] (479), 525[445] (480), 588[351] (579)
Chiusoli, G. P., 186[326] (139, 140), 286[639] (251, 270, 272–4), 286[640] (251), 286[641] (251), 290[928] (273), 290[929] (273)
Chlor, K., 187[408] (150), 188[434] (152)
Chmielewski, P. J., 105[400] (92)
Chock, P. B., 583[56] (541)
Choe, S.-B., 190[608] (174), 100[38c] (31, 33), 100[38d] (31, 33), 100[39] (31, 33), 100[40] (31, 33), 100[41] (31, 33), 190[609] (174), 190[613] (175)
Choi, H., 183[103] (115), 183[104] (115), 583[36] (536, 539)
Chojnacki, S. S., 105[404] (92)
Chong, K. S., 186[286] (138, 140)
Chooi, S. Y. M., 283[443] (244, 263)
Chottard, J. C., 584[105] (547), 584[121] (548, 562)
Chottard, J.-C., 528[629] (500)
Choudhury, S. B., 284[536] (268)
Chowdhury, M. Z. A., 527[587] (498, 504)
Chrisope, D. R., 386[192] (331), 386[193] (331)
Christian, G., 430[223] (426)
Christidis, A., 390[451] (379), 390[452] (379)
Christofides, A., 222[108] (209), 583[26] (536, 537, 539, 543), 583[61] (541)
Christou, V., 519[78] (439), 585[209] (554, 555), 586[243] (560)
Chu, S.-Y., 186[278] (137)
Chujo, Y., 387[254] (347)
Chung, F. R. K., 26[144] (19)
Churchill, M. R., 190[563] (172, 174), 190[576] (174), 284[501]
Ciajolo, R., 388[292] (354)
Ciano, M., 530[729] (514)
Cinellu, M. A., 282[391], 283[480] (245), 283[481] (245, 248), 283[482] (245, 255), 286[694] (255, 256, 262), 529[693] (510), 530[774] (517), 530[775] (517)
Cinquantini, A., 278[126] (234), 429[187a] (417)
Cirac, J. A., 585[198] (554, 555)
Claire, P. P. K., 280[286] (240), 524[337] (467)
Clark, H. C., 286[653] (252), 286[667] (252), 286[669] (253), 286[671] (253), 290[959] (276), 426[1] (392), 428[101] (403, 416), 428[124] (407), 428[125a] (407), 428[125b] (407), 428[125c] (407), 429[134] (408), 429[136] (408), 430[206] (422), 519[41] (437), 519[68] (438), 520[92] (440), 520[98] (440), 520[119] (442), 521[189] (447), 522[267] (460), 523[274] (460), 523[308] (464), 526[481] (485, 487), 526[492] (486), 528[595] (498, 499), 529[660] (507, 514), 530[737] (514), 530[740] (514), 582[17a] (535, 539), 585[190] (554, 555, 557), 585[217] (557), 585[218] (557), 586[222] (557), 587[338] (576)
Clark, P. W., 279[171] (236), 279[173] (236, 263), 282[394], 282[428] (244, 270), 283[478] (245, 263), 285[617] (250)
Clarke, D. A., 525[441] (479, 481), 588[349] (578)
Cleary, D. G., 390[432] (375)
Clec'h, G., 188[435] (152)
Clegg, D. E., 528[628] (500)
Clegg, W., 189[528] (159, 170)
Clemens, J., 287[766] (259)
Clemens, P. R., 387[228] (342)
Clemente, D. A., 287[744] (257)
Clerici, M. G., 284[526]
Clifton, M. S., 527[529] (491, 493), 587[282] (568)
Clinet, J. C., 25[66] (10), 25[67] (10), 25[68] (10), 25[69] (10), 279[175] (236), 281[337] (242, 263), 283[487] (245, 263), 289[898] (270)
Cloke, F. G. N., 429[148] (409)
Clos, N., 101[127] (42, 49)
Coates, G. E., 277[48] (230, 239, 240, 266)
Coe, P. L., 280[286] (240), 524[337] (467)
Cohen, D., 426[2a] (392)
Cohen, H., 105[396] (91, 92), 105[397] (91, 92)
Colbran, S. B., 105[426] (95, 96), 189[536] (159), 284[525]
Cole-Hamilton, D. J., 519[46] (437, 456), 525[415] (475)
Coll, J., 25[26] (4)
Collin, J., 289[861] (264, 268)
Collin, J. P., 25[62] (9)
Collins, D. J., 386[196] (333), 386[197] (333), 388[300] (355)
Collum, D. B., 278[130] (235), 519[32] (436)
Colomer, E., 183[102] (115, 124)
Colovray-Gotteland, V., 389[400] (370), 389[401] (370)
Colpitts, D., 523[313] (464)
Colquhoun, H. M., 288[827] (262–4), 289[848] (263)
Colton, R., 285[580] (248, 276)
Commereuc, D., 287[743] (257, 273), 289[861] (264, 268)
Conder, H. L., 430[191] (418)
Cong, Q., 427[19] (393)
Connelly, N. G., 188[439] (152), 223[159] (215), 285[595] (249, 272), 385[101] (307), 390[446] (378), 390[447] (378), 390[448] (378)
Conroy-Lewis, F. M., 102[168] (47, 50, 54), 182[60] (111, 113), 279[198] (237)
Consiglio, G., 186[292] (138, 143), 388[349] (360)
Constable, A. G., 282[399], 283[488], 428[92] (403, 416), 523[303] (463), 529[696] (510), 530[747] (514)
Constable, E. C., 280[229] (238), 281[332] (242), 284[504] (263, 264), 284[511], 284[520] (261), 284[521], 284[522], 284[523], 529[692] (510), 529[694] (510)
Conzelmann, W., 521[170]
Cook, S. J., 582[14a] (535, 558)
Cooley, N. A., 278[117] (234, 250), 289[870] (266), 289[872] (266), 385[138] (315)
Cooney, R. P., 220[5] (194)
Cooper, G. R., 527[548] (494), 527[551] (494)
Cooper, M. K., 526[494] (487), 529[682] (509), 584[148] (552, 562), 584[149] (552, 566), 584[151] (552, 566), 584[152] (552)
Cooper, S. J., 520[117] (441, 442), 522[223] (455)
Cooper, S. R., 290[957] (275)
Cope, A. C., 285[603] (249)
Copijn, M., 279[176] (236, 240, 242, 243, 262, 269, 270, 272, 273, 275)
Corain, B., 222[120] (211), 223[173a] (218)
Corcoran, E. W., Jr., 103[288] (74, 76, 80), 183[106] (115, 122), 279[188] (237), 429[188] (417)
Cordes, A. W., 584[98] (545), 584[99] (545)
Cordier, G., 101[122] (41, 42, 47–9, 51, 58, 60)
Cornioley-Deuschel, C., 280[275] (240, 270, 273), 281[303] (241), 525[435] (478)
Coronas, J. M., 102[156] (42, 50, 55), 102[205] (41, 49, 56, 58), 102[206] (49, 63)
Corriu, R. J. P., 183[102] (115, 124)
Cosse-Mertens, C., 182[25] (109)
Costa, E., 524[378] (471), 524[379] (471)
Costa, M., 186[326] (139, 140), 290[928] (273)
Costanzo, L. L., 522[261] (460)
Coston, T. P. J., 521[194] (450, 500), 528[588] (498, 504), 528[589] (498)
Cotton, F. A., 430[191] (418), 524[350] (468), 524[389], 525[400] (474), 525[403] (474), 525[405] (475), 527[565] (497), 527[566] (497), 527[567] (497), 527[568] (497)
Countryman, R. M., 283[445] (245), 428[98] (403, 406)
Couture, C., 427[32] (395)
Coville, N. J., 187[394] (150)
Cowan, R. L., 278[76] (230), 520[133] (443)
Cowley, A. H., 184[176] (126, 127), 184[177] (126, 127)
Cox, D. N., 191[638] (180)
Coxon, J. M., 103[257] (67, 73), 183[132] (116, 120, 157)
Crabtree, R. H., 582[9] (534)
Craig, C. A., 284[498] (245), 530[723] (513)
Craig, D. C., 284[525], 528[624]
Crascall, L. E., 582[12] (535, 540, 553), 583[50] (538, 577), 587[344] (577)
Crespo, M., 290[947] (273, 274), 290[948] (273, 274, 276), 290[951] (273), 521[146] (444), 521[177], 523[299] (463), 523[316] (465), 523[321]

(466), 523[324] (466), 525[397] (474), 528[617] (510), 528[618] (510), 528[619] (510), 529[700] (510)
Crisan, D., 384[38] (294)
Crisp, G. T., 100[70] (37, 38), 288[786] (260), 385[78] (303), 385[80] (303)
Crispini, A., 282[423] (243), 282[424] (243), 283[465] (245), 283[490] (262)
Critchley, R. P., 384[70] (302)
Critchlow, S. C., 583[35] (536, 538, 550)
Crittell, C. M., 587[347] (577)
Crociani, B., 101[119] (42, 61), 222[123] (211), 222[125] (211, 213), 222[126] (211, 213), 222[128] (211), 222[129] (211), 222[142a] (214), 222[142b] (214), 223[161a] (216), 223[161b] (216), 223[165] (216), 278[129] (235, 257), 279[146] (235), 279[150] (235, 257), 279[153] (235, 262, 264, 267), 287[711] (255, 257, 264), 287[744] (257), 287[745] (257), 287[746] (257), 287[747] (257, 263), 287[751] (257), 287[754] (257), 288[822] (262, 264, 267), 288[845] (263, 269), 289[862] (264), 289[881] (267), 385[75] (303), 387[270] (350), 429[187a] (417), 429[187b] (417), 430[212] (423), 518[18] (434), 523[325] (466), 523[326] (466), 587[312] (573), 587[313] (573)
Crocker, C., 282[408] (263, 264), 429[180] (416, 423), 525[455] (480), 529[665] (507), 583[64a] (542), 583[64b] (542), 586[247] (561, 562)
Cronje, S., 103[247] (67, 70), 189[491] (157)
Cross, R. J., 103[249] (67), 189[514] (157), 280[218] (238), 280[281] (240), 282[420] (243, 248, 263, 264), 282[421] (243), 287[717] (255), 288[794] (260), 289[860] (264), 289[910] (272), 390[454] (379), 390[455] (379), 390[456] (379), 428[99] (403), 428[100] (403, 405), 428[103] (403), 428[104] (403), 428[105] (403), 428[109] (405, 416), 428[118] (406, 407), 428[119] (406), 428[120] (406), 428[122] (407), 428[128] (407), 428[129] (407), 429[130] (408), 429[133] (408), 430[203] (421), 521[191] (447), 522[244] (458), 522[251] (459), 522[264] (460), 522[268] (460), 522[269] (460), 523[273] (460), 526[508] (490), 526[509] (490), 526[520] (490, 493), 527[535] (493), 527[536] (493), 530[771] (517), 530[772] (517), 584[107] (547), 588[361] (580), 588[362] (580)
Crossley, J. G., 223[159] (215)
Crossley, T. E., 284[563] (246)
Crowe, W. E., 99[14] (30, 32)
Crutchley, R. J., 289[903] (271)
Cruz, S. G., 585[213b] (556)
Csaszar, P., 182[27] (109)
Csöregh, I., 385[119] (311)
Csuk, R., 25[30] (5)
Cubillo, J., 101[146] (42, 62, 63)
Cucciolito, M. E., 278[121] (234), 280[256] (239, 262, 272), 286[706] (255, 256, 271), 385[96] (306), 583[78] (543), 583[80] (543)
Cuevas, J. C., 288[821] (262)
Cui, Y. X., 283[442] (244)
Cunningham, P. N., 27[169] (21), 190[570] (172, 173)
Curtis, M. D., 189[531] (159, 169), 189[557] (172, 173)
Cushman, B. M., 528[640] (505), 584[126] (549)
Cusmano, F. P., 289[882] (267)
Cusumano, M., 520[86] (439), 522[258] (459)
Cutillas, N., 524[339] (468)
Cutler, A. R., 222[119b] (210)
Czaszar, P., 583[46] (537, 542, 550)

Dagnac, P., 220[41] (198), 429[162] (410)
D'Agostino, M. F., 105[433] (96, 98)
Dahl, L. F., 21, 26[136] (19), 26[137] (19), 26[140] (19, 20), 26[141] (19), 27[174] (21), 27[175] (21), 27[176] (21), 27[177] (21), 188[449] (154, 159), 188[450] (154, 171), 189[518] (159, 167), 189[543] (167, 171, 172), 189[549] (171), 189[552] (171), 189[554] (171, 172), 190[568] (171–3), 427[43] (397), 427[46] (398), 427[53] (398), 427[55] (398)
Dahlenburg, L., 531[782] (518)
Dahmen, K.-H., 427[9] (392), 427[42a] (397)
Dall'Antonia, P., 526[470] (483)
Dance, I. G., 528[624]
D'Angelo, J., 385[112] (309)
Daniele, S., 101[145] (42, 47)
D'Aniello, M. J., 427[44b] (397), 427[49] (398)
Daniels, L. M., 520[132] (443)
Dankosh, H. E., 101[120] (42, 50, 55, 93, 94), 184[152] (117, 118)
Danopoulos, A. A., 386[201] (333)
Danowski, F., 103[232] (66, 67), 103[233] (66, 67), 103[234] (66, 67, 72), 183[73] (114, 115, 117, 157), 183[120] (114, 115), 185[216] (130–2, 138, 157)
Daran, J. C.,
Daran, J.-C., 222[137a] (213), 282[365] (253, 254, 263), 286[692] (255, 258), 386[167] (321)
Darensbourg, D. J., 25[70] (10), 100[79] (38, 39, 42, 57)
Darensbourg, M. Y., 100[79] (38, 39, 42, 57), 100[84] (41, 42, 57), 105[404] (92)
Darlington, W. H., 388[351] (361)
Dartiguenave, M., 26[134] (19), 26[135] (19), 100[74] (37, 38), 100[75] (37–9, 42, 50, 56), 100[76] (37–9, 42, 56), 102[166] (50, 62), 105[408] (42, 93, 94), 105[409] (93, 94)
Dartiguenave, Y., 26[134] (19), 26[135] (19), 100[74] (37, 38), 100[75] (37–9, 42, 50, 56), 100[76] (37–9, 42, 56), 102[166] (50, 62), 105[408] (42, 93, 94), 105[409] (93, 94)
Das, K., 586[268] (564)
Das, M. K., 387[274] (350)
Datta, S., 388[299] (355)
Datz, K. H., 223[160] (216)
Dave, T., 187[352] (139, 155, 159, 168, 171)
Daves, G. D., Jr., 285[636] (250)
Davey, A. P., 277[61] (230)
Davidson, J. G., 101[105] (42, 47, 57), 189[526] (159)
Davidson, J. L., 27[181] (21, 22)
Davidson, J. M., 388[305] (356)
Davidson, M. F., 428[122] (407), 430[203] (421), 526[508] (490), 526[520] (490, 493), 530[756] (516), 584[107] (547)
Davies, B. W., 386[151] (317)
Davies, G. D., Jr., 386[205] (335)
Davies, J. A., 220[26] (197, 200), 288[781] (260), 384[23] (292), 386[158] (319), 386[161] (319), 390[435] (375), 390[462] (381), 428[101] (403, 416), 428[124] (407), 428[125a] (407), 428[125b] (407), 428[125c] (407), 429[134] (408), 429[136] (408), 519[22] (434), 520[98] (440), 520[119] (442), 522[267] (460), 523[274] (460), 531[784] (518), 531[785] (518), 585[217] (557), 585[218] (557), 586[273] (565), 586[274] (565)
Davies, P., 284[563] (246)
Davies, S. J., 27[185] (23), 277[71] (230), 526[519] (490), 586[266] (564), 586[267] (564)
Davis, J. H. J., 190[594] (175, 179)
Davis, R. E., 183[105] (115, 122)
Davis, W. M., 26[120] (16)
Day, C. S., 387[229] (342), 585[190] (554, 555, 557)
Day, R. O., 27[183] (22)
Day, V. W., 27[183] (22), 585[190] (554, 555, 557)
de Boer, H. J. R., 525[426] (478), 525[434] (478), 585[185] (554, 555)
de Boer, J. L., 103[253] (67), 184[142] (117, 159)
de Booys, J. L., 530[757] (516)
De Cian, A., 282[386], 282[387] (253), 283[434] (244, 253), 286[680] (253), 286[681] (253), 286[684] (253), 386[164] (320), 386[169] (323), 586[257] (563, 564)
De Crescenzi, M., 384[40] (294)
De Felice, V., 278[121] (234), 286[706] (255, 256, 271), 520[82] (439), 583[32] (536), 583[76] (543), 583[77] (543), 583[78] (543), 583[79] (543), 583[80] (543), 583[81] (543), 583[82] (544), 583[83] (544), 583[84] (544), 585[212] (554), 587[285] (568, 571)
de Goicoechea, M. L., 288[818] (262, 263)
de Graaf, W., 277[55] (230, 232), 277[57] (230, 232, 234, 239, 269, 270, 273), 278[98] (233, 239, 262), 278[142] (235), 280[260] (239, 255, 262, 272, 273), 280[261] (239, 240, 255, 262, 266, 267, 269–74), 280[265] (240, 265, 269, 270, 272–4)
de Haro, C., 288[823] (262, 263)
de Lu, Y., 387[221] (340)
de Meijere, A., 290[930] (273), 389[359] (363), 389[365] (363)
de Mendoza, J., 279[178] (236, 263, 264), 281[304] (241), 281[338]

(242), 288[821] (262)
De Meric de Bellefon, C., 221[85a] (205), 221[85b] (205), 221[87b] (205), 221[88a] (205), 221[88b] (205), 221[89a] (205), 221[89b] (205), 222[90] (205), 222[92] (206)
de Munno, G., 282[423] (243), 282[424] (243), 283[465] (245), 283[490] (262), 284[506] (263)
De Pieri, R., 222[120] (211)
De Renzi, A., 286[706] (255, 256, 271), 385[95] (306), 583[75] (543), 583[77] (543), 586[233] (559), 587[285] (568, 571)
De Simone, J. M., 285[634] (250, 255, 256, 267)
de Vaumas, R., 105[424] (95, 96), 184[153] (117, 159)
de Wit, M., 525[434] (478)
Deacon, G. B., 102[201] (41, 47, 48), 287[769] (259), 287[770] (259), 521[150] (444), 524[341] (468), 524[342] (468), 524[343] (468, 469), 524[344] (468), 524[346] (468), 524[387], 524[388], 585[197] (554, 555, 557), 585[210] (554, 555)
Deardorff, D. R., 389[380] (366), 389[381] (366)
Debad, J. D., 189[494] (157, 163), 189[495] (157, 163)
Debaerdemaeker, T., 519[51] (437), 519[55] (437), 519[57] (437), 519[59] (437), 521[176], 521[178], 521[179], 521[180], 521[181], 525[436] (478, 481), 526[459], 585[200] (554, 555), 585[202] (554, 555), 585[211] (554, 555)
DeCian, A., 222[91] (205)
Deck, K. J., 26[109] (15, 19), 188[483] (157, 159, 169)
Declercq, J. P., 524[390]
Decleva, P., 220[8] (194)
Dedieu, A., 222[102] (208), 286[687] (254)
Deeming, A. J., 281[330] (242), 281[351] (242, 243), 282[417] (243), 283[431] (244), 285[569] (248), 289[858] (264), 388[295] (354)
Deese, W. C., 584[98] (545), 584[99] (545)
DeGraff, B. A., 527[534] (493)
DeGray, J. A., 385[102] (307)
Dehand, J., 281[296] (241), 282[398] (243), 286[675] (253)
Deiss, E., 187[416] (151)
Dekker, G. P. C. M., 222[105] (208), 281[301] (241), 285[632] (250), 286[698] (255, 256, 262, 263), 286[699] (255, 256, 262), 287[738] (256), 429[138] (408), 522[271] (460), 529[677] (508, 513), 529[703] (511)
Dekker, M., 189[515] (157, 166)
Del Buttero, P., 278[145] (235)
Del Piero, G., 287[731] (256)
Del Pra, A., 287[727] (256), 520[136] (443, 461), 521[168], 524[381] (472)
Del Rosario, R., 26[104] (14), 26[105] (14), 26[119] (16, 18), 185[225] (130, 133)
DeLaet, D. L., 25[75] (10, 16, 17), 26[119] (16, 18), 26[128] (16, 17), 26[131] (17), 26[132] (18)
Delavaux, B., 520[114] (441)
Delgado, E., 27[186] (23)
DeLiberato, L. A., 185[239] (131, 169, 172)
Della Pergola, R., 26[145] (19), 26[151] (20), 26[153] (20)
Dell'Amico, D. B., 220[16] (195), 220[17] (195), 220[18] (195, 196), 220[19a] (196), 220[19b] (196), 428[108b] (403, 409), 428[116] (406), 429[149] (409), 429[150a] (409, 411), 429[150b] (409–11), 429[150c] (409, 411), 429[163] (411)
Dema, A. C., 526[485] (485), 587[346] (577)
Demartin, F., 282[391], 283[480] (245), 283[482] (245, 255), 385[95] (306), 386[184] (327), 386[185] (328), 525[396] (474), 525[456] (480)
Demas, J. N., 527[534] (493)
Denning, R. G., 384[58] (296)
Dérien, S., 25[65] (10), 25[66] (10), 25[67] (10), 25[68] (10), 25[69] (10), 105[366] (81, 88, 90)
Derome, A. E., 384[24] (292)
Derveer, D. G. V., 189[526] (159)
Deschler, U., 104[340] (81, 87, 90), 277[29] (228)
Desmartin, F., 529[693] (510)
Deuschel-Cornioley, C., 530[722] (513), 530[726] (514), 530[727] (514), 530[732] (514), 530[735] (514)
Deutsch, E., 290[962] (276)
Devlin, J., 283[470] (245)
Dewar, M. J. S., 384[44] (294, 316)
Dholakia, S., 529[695] (510)
Di Bianca, F., 101[119] (42, 61), 222[126] (211, 213), 222[128] (211), 222[129] (211), 279[150] (235, 257), 279[153] (235, 262, 264, 267), 287[711] (255, 257, 264), 287[754] (257), 288[822] (262, 264, 267), 288[845] (263, 269), 289[862] (264), 289[881] (267), 385[75] (303), 518[18] (434), 523[325] (466), 523[326] (466), 587[313] (573)
Dianoux, A. J., 187[410] (150)
Dias, A. A., 385[90] (305)
Dias, G. H. M., 525[448] (480)
Dias, S. A., 285[607] (249), 530[742] (514)
Dickson, R. S., 586[263a] (564, 572), 586[263b] (564, 572)
DiCosimo, R., 522[225] (456, 478, 481), 522[226] (456), 526[461] (481), 526[463] (481), 585[178] (554, 555), 585[179] (554, 555)
Diederich, F., 385[135] (314)
Dietrich, A., 521[150] (444)
Dietrich, H., 182[36] (110, 111, 113)
Dietrich-Buchecker, C. O., 284[524] (245, 271, 272)
Dignard-Bailey, L., 521[200] (452), 585[214b] (556)
Dinjus, E., 25[41] (5), 102[189] (41, 48), 102[214] (65), 102[215] (65), 104[356] (81, 88, 90), 104[357] (81, 88, 90), 104[358] (81, 88, 90), 104[359] (81, 88), 104[360] (81, 88), 104[361] (81, 88), 104[362] (81), 105[372] (81, 88, 89), 105[378] (80, 88), 183[81] (115), 183[90] (115), 184[184] (126, 127), 184[185] (126, 128), 185[240] (131, 135), 186[325] (139, 140), 187[356], 187[357] (137), 187[358] (137)
Dire, S., 429[135] (408)
DiSalvo, F. J., 427[52] (398)
Distefano, G., 185[275] (137), 386[175] (324), 587[334] (575, 576)
Ditzel, E. J., 529[678] (509, 513, 515)
Diversi, P., 280[257] (239, 240, 255, 269, 270, 273), 280[267] (240, 262, 267, 269–71), 280[268] (240, 262), 280[269] (240, 262, 267), 280[271] (240, 270), 289[906] (271)
Dixon, A. J., 427[64] (400), 588[370] (581)
Dixon, K. R., 280[277] (240), 519[65] (438), 584[147] (552), 586[240] (560)
Djebili, A., 185[266] (134, 138)
Dmitrieva, T. V., 280[237] (239)
Doddrell, D. M., 528[628] (500)
Doedens, R. J., 189[515] (157, 166)
Doh, M.-K., 279[189] (237)
Döhring, A., 25[35] (5), 184[180] (126, 128)
Dolg, M., 220[5] (194)
Dolgoplosk, B. A., 100[83] (41, 47), 102[182] (56)
Dolgopolova, N. A., 387[215] (337)
Doll, K.-H., 187[415] (151, 155), 188[424] (151, 155), 188[428] (151), 188[460] (154, 155)
Domachev, G. A., 189[538] (159)
Domaille, P. J., 101[95] (42, 54), 521[182] (445)
Domrachev, G. A., 388[297] (354)
Don, M. J., 27[192] (24)
Donaldson, W. A., 388[326] (358), 389[361] (363), 389[363] (363)
Donato, A., 285[591] (249), 289[882] (267)
Döring, M., 104[354] (81, 90)
Dorokhov, V. A., 282[371]
Douglas, G., 428[76] (402, 413), 428[81] (403), 428[83] (403), 428[88] (404), 428[89b] (404)
Douglas, W., 188[469] (154)
Dowben, P. A., 388[299] (355)
Dowling, N. I., 220[25] (197), 390[434] (375), 588[371] (581)
Downing, J. H., 521[161]
Doyle, J. R., 585[170] (554)
Doyle, M. J., 103[281] (73, 74)
Drage, J. S., 185[252] (131), 185[253] (131, 134)
Draux, M., 522[266] (460)
Dreissig, W., 182[36] (110, 111, 113)
Dreme, M., 584[121] (548, 562)
Drevs, H., 104[326] (81, 85)
Drew, D., 585[170] (554)
Drew, M. G. B., 280[286] (240), 524[337] (467)
Drewello, T., 187[401] (150)
Drickamer, H. G., 427[49] (398)
Drogunova, G. I., 101[126] (42)
Drouin, M., 220[45] (199), 223[158] (215)
Druliner, J. D., 101[95] (42, 54), 184[167] (123)
Druzhkov, O. N., 187[404] (150), 187[405] (150, 155)
Dryden, N. H., 522[238] (457)
du Plooy, K. E., 187[394] (150)

Dube, M. J., 288[834] (262)
Dubler, E., 190[586] (175, 177)
Dubrawski, J., 220[28] (197, 201)
Dudis, D. S., 103[265] (74)
Dufaud, V., 278[105] (233, 255, 256)
Duffin, A. J., 528[600] (499)
Duffy, D. N., 103[246] (67, 72), 189[492] (157, 161)
Duisenberg, A. J. M., 184[204] (125, 157), 190[571] (172), 190[573] (172, 173), 530[720] (513), 530[763] (516), 530[764] (516)
Duñach, E., 25[64] (10), 25[65] (10), 25[66] (10), 25[67] (10), 25[68] (10), 25[69] (10), 105[366] (81, 88, 90)
Duncanson, L. A., 384[45] (294, 316)
Dunina, V. V., 281[323] (242), 283[435] (244), 283[436] (244, 245), 283[446] (245)
Dunitz, J. D., 188[458] (154, 155)
Dunne, B. J., 385[81] (303), 519[24] (434), 583[30] (536, 538, 539)
Dunstan, P., 285[617] (250)
Dupont, J., 222[137a] (213), 222[137b] (213), 281[349] (242, 248), 282[365] (253, 254, 263), 282[386], 282[387] (253), 285[577] (248, 253), 286[692] (255, 258), 286[693] (255, 256, 258, 263), 386[164] (320), 386[167] (321)
Dupuis, M., 25[94] (13)
Duraj, S. A., 183[123] (115, 138, 145, 155), 582[11] (534)
Durfee, L. D., 586[236] (560)
Durocher, A., 182[23] (109)
Dusausoy, Y., 221[75a] (204), 221[75b] (204), 221[86] (205), 221[89a] (205), 281[306] (241, 263)
Dutremez, S., 288[781] (260)
Dutta, A. K., 526[475] (484)
Dyke, H. J., 282[428] (244, 270)
Dyke, S. F., 279[171] (236), 279[173] (236, 263), 281[356] (243, 244), 282[428] (244, 270), 283[429] (244), 283[469] (245), 283[478] (245, 263), 285[617] (250)
Dzhemilev, U. M., 102[169] (50)

Eaborn, C., 519[34] (436), 519[35] (436), 519[37] (436), 519[38] (436), 519[42] (437), 527[527] (491), 529[670] (508), 587[281] (568)
Eagle, C. T., 519[22] (434), 531[784] (518), 586[274] (565)
Eberspach, W., 189[523] (159, 163)
Ebina, S., 280[234] (238), 387[242] (344)
Eck, C., 220[28] (197, 201)
Edelstein, N., 187[419] (151, 155)
Edwards, A. J., 288[838] (263)
Edwin, J., 186[317] (138, 177), 189[520] (159, 175), 190[590] (175, 178), 190[591] (175, 178), 190[592] (175, 177), 190[598] (175, 179), 190[622] (179)
Efe, V. O., 100[86] (42, 55)
Efimov, O. N., 182[15] (108)
Egert, E., 188[448] (154, 155, 157)
Egolf, T., 385[110] (309)
Eguchi, K., 286[672] (253)
Ehara, K., 25[83] (11), 25[85] (11)
Eichinger, B., 583[35] (536, 538, 550)
Eigenbrot, C. W., 103[287] (75), 105[416] (95, 96), 183[125] (115, 121), 185[236] (131, 136, 169), 185[268] (136)
Eisch, J. J., 103[284] (73, 74), 103[285] (73, 74, 76), 104[299] (73, 74), 183[91] (115, 119), 183[96] (115, 148), 184[155] (119), 184[156] (119, 122, 131), 187[380] (148), 187[389] (148), 187[390] (148), 189[519] (159, 167), 190[610] (175), 190[611] (175), 190[612] (175), 190[613] (175)
Eisenberg, A., 588[375] (581)
Eisenstein, O., 182[22] (109), 285[597] (249), 285[599] (249), 385[116] (310), 385[117] (310), 584[139b] (551), 584[139c] (551), 584[142] (551, 569)
Ekeland, R. A., 527[562] (495), 527[578] (498), 528[647] (505)
El Khadem, H. S., 284[512]
Elbaze, G., 220[28] (197, 201)
Elding, L. I., 584[103] (547)
Eliseev, A. V., 281[345] (242), 290[925] (273), 386[195] (332)
Eller, K., 99[10] (30), 99[11] (30)
Ellern, A. M., 279[194] (237)
Elliott, G. P., 27[185] (23), 27[187] (23)
Ellis, J. E., 27[172] (21)
Ellis, J. W., 520[95] (440)
Elmurr, E., 188[441] (152, 155)
Elschenbroich, Ch., 582[10] (534)
Elsevier, C. J., 222[105] (208), 278[112] (234), 278[113] (234), 278[114] (234), 278[142] (235), 285[632] (250), 285[635] (250, 255, 256), 286[699] (255, 256, 262), 287[710] (255), 287[713] (255), 287[723] (256, 262, 263, 266), 287[738] (256), 287[761] (258), 290[944] (273, 274), 526[499] (487)
Emerson, K., 527[562] (495)
Emge, T. J., 285[612] (249), 384[49] (295)
Emoto, M., 103[221] (65), 103[222] (65), 186[294] (138), 289[877] (266), 289[899] (270, 273), 384[61] (296, 297), 389[366] (363), 583[67a] (542, 545, 574), 584[101] (545, 547), 584[108] (547)
Empsall, H. D., 282[396] (243, 245)
Endo, I., 521[148] (444), 586[258] (563), 586[259] (563)
Endoh, K., 283[454] (245, 263)
Engel, P. F., 279[176] (236, 240, 242, 243, 262, 269, 270, 272, 273, 275), 286[687] (254)
Engelhardt, L. M., 279[184] (236, 241), 280[252] (239, 263)
Engelter, C., 527[560] (495)
Englert, U., 588[356] (579, 582)
Enyo, M., 427[56] (399)
Erabi, T., 101[115] (42, 49, 55, 61)
Erdman, A. A., 191[626] (180, 182), 191[627] (180, 182), 191[628] (180, 182), 191[629] (180, 182)
Eremenko, N. K., 220[30] (197), 221[52] (201), 221[53] (201, 204), 221[64a] (203), 221[64b] (203), 221[67] (204), 221[68a] (204), 221[68b] (204), 221[68c] (204), 221[68e] (204), 221[68f] (204), 221[69] (204), 221[70] (204), 221[71] (204), 221[72] (204), 221[74a] (204), 221[74c] (204), 222[96] (207), 222[97a] (207), 222[97b] (207), 427[11] (393), 427[12] (393, 399), 427[22] (393), 427[25] (394), 427[28] (395)
Erickson, L. E., 583[57] (541, 545), 583[58] (541, 545)
Erker, G., 430[210] (423)
Ermakov, Yu. I., 220[36] (197), 388[306] (356)
Ermer, S. P., 584[123] (548, 552, 554, 555)
Ernst, R. D., 588[358] (579)
Errington, R. J., 282[397] (243), 284[543]
Ertl, J., 519[49] (437, 454), 519[51] (437)
Espenson, J. H., 105[387] (91, 92), 105[388] (91, 92), 105[389] (92), 105[390] (91, 92), 105[391] (91, 92), 105[392] (91, 92), 105[393] (91, 92), 105[394] (91, 92), 105[395] (91, 92), 518[14] (433), 520[127] (442), 520[128] (442)
Espinet, P., 220[1] (194–6, 198, 201, 205, 208–12, 215–18), 222[124] (211, 213), 222[132] (213), 222[133] (213), 222[134a] (213), 222[134b] (213), 222[144] (214), 223[145a] (214), 223[146] (214), 223[164] (216), 223[167] (217), 276[1] (226, 260), 277[5] (226, 227), 277[6] (226, 227), 277[10] (226, 227), 277[24] (228), 277[46] (230), 281[288] (240, 262, 264), 283[464] (245, 264), 283[492] (262, 263), 284[559] (246, 263), 284[560] (246, 263, 268), 285[581] (248, 251, 252), 285[611] (249), 285[628] (250), 286[644] (251, 252), 287[749] (257), 287[756] (258, 263), 287[757] (258), 287[758] (258, 259), 287[759] (258), 287[760] (258, 263), 288[819] (262, 263), 288[821] (262), 288[825] (262, 272), 288[830] (262), 289[864] (264), 289[867] (264), 383[1] (292), 383[2] (292), 383[3] (292, 294), 383[4] (292), 383[5] (292), 383[6] (292), 383[7] (292), 383[8] (292), 383[9] (292), 385[106] (308), 385[107] (308), 387[220] (339), 427[59] (399, 413), 429[181] (416, 417, 423), 518[5] (432), 518[9] (432, 433), 524[338] (467, 469), 524[359] (469), 524[364] (469), 525[398] (474, 490), 587[307] (572)
Estes, J. M., 389[382] (366)
Etxebarria, J., 284[559] (246, 263)
Evans, D. G., 26[142] (19), 385[89] (305), 427[8] (392, 396, 397), 427[39] (396), 428[78b] (402), 520[112] (441), 582[6m] (534), 585[154a] (553), 585[154b] (553), 585[156] (553), 586[246] (560)
Evans, D. H., 184[169] (125), 586[252] (561)
Evans, D. J., 285[593] (249), 385[123] (311), 586[220] (557)
Evans, D. W., 281[302] (241), 281[319] (241), 281[334] (242), 283[486] (245, 263), 385[92] (305), 529[691] (510)
Evdokimova, E. V., 221[61a] (202)
Even, U., 187[398] (150), 187[399] (150)
Eyring, M. W., 100[43] (31, 33), 183[92]

(115, 122)
Ezomo, O. J., 427[41] (396, 413)

Facchin, G., 223[154] (215), 277[26] (228, 230), 277[27] (228), 277[28] (228, 230), 278[128] (235), 279[146] (235), 387[266] (350), 387[267] (350), 429[186] (416), 430[201a] (420, 426), 430[204a] (421), 430[204b] (421), 430[205] (421), 525[421] (477)
Fackler, J. P., 100[82] (40), 103[265] (74), 105[375] (81, 91)
Fadiadi, A. J., 583[41] (536)
Fagan, P. J., 183[127] (115, 117, 125), 184[168] (125), 184[169] (125), 586[249] (561), 586[250] (561), 586[251] (561), 586[252] (561)
Faggiani, R., 289[903] (271)
Faglia, S., 522[208] (453)
Fahey, D. R., 279[148] (235, 237), 280[282] (240)
Fakhroeian, Z., 523[296] (463)
Falck, M., 519[54] (437, 440)
Faller, J. W., 386[190] (328)
Fallis, K. A., 519[63] (437, 458), 520[96] (440), 520[113] (441), 520[125] (442), 585[189] (554, 555), 585[203] (554, 555)
Fallis, S., 288[794] (260), 390[454] (379), 530[771] (517), 588[360] (580), 588[362] (580)
Fallon, G. D., 586[263a] (564, 572), 586[263b] (564, 572)
Falvello, L. R., 281[352] (243, 262, 271), 390[468] (382), 430[191] (418), 524[350] (468), 524[389], 525[400] (474), 525[403] (474), 525[404] (474), 525[405] (475), 527[565] (497), 587[299] (571)
Fañanás, F. J., 102[165a] (42, 59), 102[165b] (42, 59)
Fandos, R., 525[419] (477)
Fang, Y., 102[160] (42)
Fanizzi, F., 584[120] (548)
Fantucci, P., 220[9] (195), 220[12b] (195), 222[104] (208)
Fanwick, P. E., 25[84] (11, 23), 26[108] (15), 26[109] (15, 19), 26[119] (16, 18), 26[121] (16), 26[128] (16, 17), 26[130] (17), 26[131] (17), 26[132] (18), 26[133] (19), 27[178] (21, 22), 27[179] (21, 22), 27[180] (21), 104[341] (81, 87), 105[418] (95, 96), 105[420] (95, 96), 105[421] (95, 96), 184[161] (121, 139, 144, 159), 185[239] (131, 169, 172), 185[269] (136), 188[483] (157, 159, 169), 189[525] (159, 169), 189[532] (159), 189[533] (169, 172, 173), 189[534] (159, 169), 189[545] (167, 169)
Faraone, F., 223[156] (215), 280[280] (240), 390[437] (376), 588[364] (580)
Farina, V., 385[137] (315)
Farr, J. P., 220[24] (196), 278[119] (234, 236)
Farrar, D. H., 221[50a] (201), 221[50b] (201), 386[150] (317), 388[340] (360), 427[32] (395), 427[33] (395, 400), 427[35] (395–7), 430[227] (426), 430[228] (426), 526[490] (486), 586[278] (566, 567), 586[279] (566, 567)
Farrer, D. H., 427[20] (393, 396)
Farrugia, L. J., 26[161] (20), 27[185] (23), 27[193] (24), 190[566] (172), 587[304] (572), 587[305] (572)
Farwaha, R., 280[219] (238, 259), 524[371] (471, 495), 524[374] (471), 524[376] (471), 585[166] (553–5, 558)
Fasce, D., 280[257] (239, 240, 255, 269, 270, 273)
Faulks, S. J., 287[770] (259)
Fauvarque, J. F., 101[125] (42, 55, 63)
Favero, G., 101[132] (42, 47, 51), 186[338] (139, 145), 186[339] (139, 146), 187[370] (146), 522[237] (457)
Fawcett, J., 103[266] (73, 74), 103[267] (73, 74), 525[444] (479), 525[445] (480), 525[446] (480), 525[447] (480), 529[667] (507), 588[352] (579)
Fawzi, R., 103[243] (66, 67), 189[516] (157, 164, 177), 529[685] (509)
Fehlhammer, W. P., 25[77] (11), 25[78] (11, 13), 101[131] (42, 51, 81), 189[507] (157, 168), 223[166] (217, 218), 223[170] (217, 218), 223[175] (218), 430[222] (425), 430[223] (426)
Felixberger, J. K., 280[264] (240, 261, 262, 266, 272, 275, 276)
Felman, S. W., 389[396] (368)
Feltham, R. D., 25[61] (9), 220[28] (197, 201)
Fendesak, G. H. B., 103[276] (73, 74, 78, 80), 184[149] (117, 118), 186[334] (139, 144), 525[432] (478)
Fenske, D., 26[156] (20), 26[157] (20), 26[158] (20), 26[164] (20), 187[368] (143), 189[551] (171), 190[575] (171), 387[277] (351), 387[284] (352), 390[451] (379), 390[452] (379)
Fenske, R. F., 20, 186[280] (137), 220[20] (196), 390[428] (374)
Ferguson, G., 220[47] (200), 278[132] (235), 278[135] (235), 282[411], 282[412] (263), 285[605] (249), 285[607] (249), 286[656] (252), 287[768] (259), 288[782] (260, 271), 289[852] (263), 290[959] (276), 384[71] (302), 385[86] (305), 385[87] (305), 385[88] (305), 385[90] (305), 428[75] (401, 402), 428[77] (402, 404), 428[125c] (407), 523[308] (464), 523[312] (464), 523[313] (464), 523[319] (465), 524[371] (471, 495), 524[373] (471), 524[375] (471), 525[430] (478), 526[492] (486), 528[595] (498, 499), 528[617] (510), 528[618] (510), 530[718], 582[14c] (535, 539), 582[17a] (535, 539), 582[17b] (535, 539), 586[225] (558)
Ferguson, G. S., 521[151] (444)
Fernandez, J. J., 283[472] (245, 263), 283[476] (245, 263, 264), 289[854] (263)
Fernandez-Baeza, F., 277[18] (227)
Ferrara, J. D., 185[228] (131, 134), 185[229] (131, 134)
Ferrara, M. L., 583[79] (543)
Fessenbecker, A., 190[593] (175, 179)
Fiaud, J. C., 387[256] (347), 389[385] (367)
Ficini, J., 385[112] (309)
Fierro, C., 185[229] (131, 134)
Filatov, M. J., 384[48] (295)
Fink, M. J., 386[154] (317), 386[155] (318), 521[174], 529[713] (512)
Firsich, D. W., 102[190] (48, 56), 278[88] (233)
Firth, S., 427[64] (400), 588[370] (581)
Fischer, B., 103[253] (67), 184[141] (117, 120, 159), 184[142] (117, 159), 190[574] (172, 173)
Fischer, E. O., 21, 25[90] (12), 27[173] (21)
Fischer, H., 223[160] (216)
Fischer, J., 100[67] (37, 38), 104[304] (81, 85), 221[85a] (205), 221[85b] (205), 221[86] (205), 221[87a] (205), 221[87b] (205), 222[91] (205), 281[305] (241), 281[306] (241, 263), 282[386], 282[387] (253), 283[434] (244, 253), 286[679] (253), 286[680] (253), 286[681] (253), 286[684] (253), 286[687] (254), 288[836] (263), 289[902] (271), 386[164] (320), 386[169] (323), 427[15] (393, 395), 586[257] (563, 564)
Fischer, K., 26[114] (15), 100[54] (34), 105[414] (93, 94, 96, 98), 183[78] (115, 119), 189[547] (171)
Fischer, R., 25[47] (7), 104[342] (81, 90), 104[347] (81, 90), 105[365] (81, 90), 183[89] (115, 123, 174)
Fischer, R. A., 25[31] (5), 188[476] (157, 167), 531[779] (517)
Fischetti, W., 387[234] (343)
Fisher, D. S., 427[32] (395)
Fisher, E. R., 99[12] (30, 32), 99[13] (30, 32), 99[15] (30, 32)
Fisher, J. R., 428[71] (401)
Fisher, L. P., 290[934] (273)
Fittschen, C., 529[706] (511)
Fitzgerald, B. J., 427[60] (399), 587[294] (571)
Fladerer, E., 188[460] (154, 155)
Fleischer, H., 222[110] (209, 215)
Fleischer, R. S., 285[600] (249)
Fleming, I., 283[448] (245)
Floerke, U., 221[63] (203)
Flom, E. A., 27[172] (21)
Flood, T. C., 522[236] (457), 522[266] (460), 523[293] (462), 584[123] (548, 552, 554, 555)
Flores-Riveros, A., 521[204] (453)
Floriani, C., 102[178] (53), 222[138] (213)
Florke, U., 277[56] (230, 231)
Foces-Foces, C., 288[819] (262, 263), 387[220] (339), 387[272] (350), 387[273] (350)
Foffani, A., 185[275] (137), 386[175] (324), 587[334] (575, 576)
Folest, J. C., 101[125] (42, 55, 63)
Foley, P., 522[225] (456, 478, 481), 583[49] (538), 585[178] (554, 555)
Folkers, J. P., 586[226b] (558)
Folting, K., 527[554] (494)
Fong, L. K., 521[182] (445)
Fontaine, X. L. R., 26[122] (16), 102[192] (41, 47, 48), 278[75] (230, 236), 279[185] (236), 523[327] (466), 586[260a] (563, 572), 586[260b] (563, 572), 586[260c]

(563, 572), 586[260d] (563, 572)
Font-Altaba, M., 281[347] (242, 243, 271), 282[372], 284[557] (245, 246, 263, 264), 521[147] (444, 467)
Fontana, A., 288[845] (263, 269), 523[326] (466)
Font-Bardía, M., 105[427] (67, 96, 97), 283[493] (271), 528[619] (510)
Foong Murray, S.-W., 384[53] (295)
Forbes, E. J., 185[267] (135)
Forcolin, S., 525[449] (480)
Formica, M., 388[354] (361)
Forni, E., 25[34] (5)
Forniés, J., 220[22] (196), 222[124] (211, 213), 222[132] (213), 222[133] (213), 222[134a] (213), 222[134b] (213), 222[144] (214), 223[145a] (214), 223[146] (214), 223[147] (214), 223[148] (214), 223[164] (216), 223[167] (217), 277[4] (226, 237, 238, 240, 255, 257, 261, 262), 277[5] (226, 227), 277[6] (226, 227), 277[7] (226), 277[8] (226, 227, 252), 277[9] (226, 227, 252), 277[24] (228), 277[30] (228), 277[51] (230), 281[352] (243, 262, 271), 286[690] (255, 263, 264), 286[691] (255, 262), 287[749] (257), 287[756] (258, 263), 287[757] (258), 287[758] (258, 259), 287[759] (258), 287[760] (258, 263), 288[808] (261, 262), 288[809] (261), 288[813] (261), 288[814] (261), 288[815] (261), 288[820] (262), 288[828] (262, 263), 288[830] (262), 288[833] (262), 288[837] (262–4), 288[839] (263), 288[846] (263), 289[847] (263), 289[864] (264), 289[865] (264), 289[866] (264), 289[867] (264), 289[901] (270), 290[921] (273, 275), 290[922] (273, 275), 385[109] (309), 386[153] (317), 390[468] (382), 427[59] (399, 413), 428[96] (403), 428[113] (406), 429[151a] (409), 429[151b] (409), 429[181] (416, 417, 423), 518[5] (432), 518[9] (432, 433), 523[328a] (467), 524[338] (467, 469), 524[340] (468), 524[348] (468), 524[349] (468), 524[350] (468), 524[351] (468), 524[357] (469), 524[358] (469, 470), 524[359] (469), 524[360] (469), 524[361] (469), 524[362] (469), 524[363] (469), 524[364] (469), 524[365] (469), 524[366] (470), 524[367] (470), 524[368] (470), 524[369] (470), 524[386] (496, 497), 524[389], 525[398] (474, 490), 525[400] (474), 525[401] (474), 525[402] (474), 525[403] (474), 525[404] (474), 525[405] (475), 525[406] (475), 525[407] (475), 525[408] (475), 525[419] (477), 525[420] (477), 586[255] (562, 563), 587[284] (568), 587[299] (571), 587[306] (572)
Forsyth, C. M., 521[150] (444)
Fortuno, C., 223[145a] (214), 223[146] (214), 223[147] (214), 277[5] (226, 227), 277[6] (226, 227), 288[808] (261, 262), 427[59] (399, 413), 518[5] (432), 518[9] (432, 433), 524[348] (468), 524[351] (468), 524[364] (469)
Foulds, G. A., 428[117a] (406), 428[117b] (406)
Fournet, G., 387[233] (343), 389[374] (365), 389[375] (365)
Fox, G. A., 387[262] (348), 387[263] (348)
Francis, C. G., 222[107] (209, 215)
Francis, J. W., 385[131] (313), 385[132] (313)
Frank, A. J., 530[723] (513)
Frazier, C. C., 278[79] (231)
Freiser, B. S., 103[261] (73), 182[28] (109), 184[140] (116), 187[396] (150), 187[402] (150)
Freissl, F. R., 223[160] (216)
Frere, Y. A., 281[318] (241)
Frew, A. A., 26[115] (16), 427[24] (394, 402, 404), 519[43] (437), 520[120] (442), 520[129] (442, 495)
Frey, U., 522[249] (458, 459)
Frias, M. C., 277[9] (226, 227, 252), 518[9] (432, 433)
Friedrich, E. C., 285[603] (249)
Friess, B., 389[402] (370)
Frijns, J. H. G., 529[688] (509)
Fritz, M., 101[131] (42, 51, 81)
Fröhlich, C., 187[374] (148), 187[375] (148), 187[376] (148), 187[377] (148)
Fröhlich, H.-O., 103[271] (73, 74), 103[272] (73, 74)
Frolov, V. M., 384[42] (294)
Frolow, F., 220[34] (197), 386[179] (326)
Fronczek, F. R., 281[312] (241), 281[314] (241), 281[315] (241), 281[316] (241), 281[318] (241), 284[496] (245), 519[67] (438)
Frost, P. W., 583[28] (536)
Frunza, L., 384[38] (294)
Fryzuk, M. D., 101[93] (42, 58, 59), 101[94] (42), 101[138] (42, 58, 59), 280[279] (240)
Fu, Z., 182[54] (111, 123), 182[55] (111, 123)
Fuchikami, T., 279[157] (235)
Fuchita, Y., 223[162] (216), 223[172] (218), 277[68] (230), 278[144] (235), 281[336] (242), 282[358] (243, 248, 263), 282[378] (263, 264), 282[379] (263), 282[380], 282[382], 282[383] (263), 282[406] (264), 282[407] (248, 263), 282[416], 283[447] (245), 284[499], 284[502] (263), 284[503], 284[546] (245), 284[566] (248, 263), 285[578] (248), 286[650], 286[651], 287[773] (259), 287[774] (259), 287[775] (259), 287[776] (259)
Fuchs, J., 223[166] (217, 218)
Fujimori, M., 280[249] (239, 262), 280[272] (240, 267, 270)
Fujimori, T., 587[326] (574, 579)
Fujinami, T., 385[98] (307), 385[100] (307)
Fujisawa, F., 283[485] (245, 257)
Fujita, J., 285[575] (248)
Fujiwara, Y., 287[773] (259)
Fukui, H., 285[583] (249), 387[280] (352)
Fukuoka, A., 427[56] (399)
Fultz, W. C., 282[409], 284[538], 523[280] (461), 527[556] (495), 529[668] (507)
Fulwood, R., 582[14c] (535, 539)
Fumagalli, A., 26[165] (20), 427[51] (398, 399)
Funicello, M., 583[32] (536)
Furlani, A., 101[123] (42), 526[522] (490, 491, 493), 526[523] (490, 491), 527[526] (491), 527[545] (494), 527[546] (494)
Furrer, A., 187[413] (151)
Fürstner, A., 25[30] (5)
Fuss, B., 188[461] (155–7, 159, 166, 167)
Fyfe, C. A., 384[25] (292)

Gabe, E. J., 188[466] (155)
Gable, K. P., 99[4] (30, 31, 41, 42), 183[107] (115, 122)
Gabor, B., 25[91] (12), 103[275] (73, 74, 80), 182[42] (110–12), 182[43] (110, 112, 126), 184[150] (117, 118)
Gaffney, T. R., 427[67] (400)
Gaft, Yu. L., 388[293] (354)
Gal, M., 182[26] (109), 384[34] (293)
Galarini, R., 388[355] (361), 389[356] (361), 525[433] (478), 587[342] (576, 577)
Gale, R. J., 188[440] (152)
Galiano-Roth, A. S., 519[32] (436)
Galimberti, F., 429[152] (409), 429[154] (409, 422)
Gallagher, J. F., 524[373] (471), 586[225] (558)
Galle, J. E., 103[285] (73, 74, 76), 183[96] (115, 148)
Galli, B., 282[404], 282[405]
Gallicano, K. D., 26[126] (16)
Gallucci, J. G., 587[323] (574, 577)
Gambino, O., 105[431] (96, 98), 185[246] (131, 172, 174)
Ganadu, M. L., 283[480] (245)
Ganazzoli, F., 283[484] (245), 429[153] (409), 429[154] (409, 422), 429[156a] (409), 429[156c] (409), 429[156d] (409), 588[355] (579)
Ganis, P., 586[233] (559)
Ganzerla, R., 525[449] (480), 583[38] (536), 583[47] (538)
Gao, J., 26[128] (16, 17)
Garaschelli, L., 588[355] (579)
Garbuzova, I. A., 190[616] (177)
Garces, F. O., 530[723] (513)
Garcia, A., 223[164] (216)
Garcia, E., 188[444] (153, 155)
García, G., 100[46] (31, 33), 283[492] (262, 263), 288[810] (261), 288[811] (261), 288[823] (262, 263), 524[339] (468), 524[352] (469), 524[354] (469)
García, J., 100[46] (31, 33), 288[810] (261), 288[811] (261), 519[73] (439), 524[352] (469), 585[206] (554, 555)
Garcia, M. P., 387[273] (350)
Garcia-Cuesta, I., 386[147] (316)
Garcia-Herbosa, G., 288[825] (262, 272)
Garcia-Ruano, J. L., 283[463] (245)
Gardner, A. B., 187[409] (150)
Garkusha, O. G., 190[616] (177)
Garlaschelli, L., 429[152] (409), 429[153] (409), 429[154] (409, 422), 429[155] (409), 429[156a] (409), 429[156b] (409), 429[156c] (409), 429[156d] (409)
Garrou, P. E., 287[716] (255), 523[273]

(460)
Garty, N., 280[284] (240, 262)
Garza, D., 279[214] (238, 240, 250, 261, 265)
Garza, D. G., 280[251] (239), 285[626] (250, 252, 253, 255), 519[64] (437), 522[262] (460)
Garzon, G., 190[585] (175, 177)
Gasdaska, J. R., 289[897] (270), 290[926] (273)
Gasparrini, F., 282[405]
Gassman, P. G., 526[476] (484), 582[18] (535)
Gatehouse, B. M., 386[197] (333), 524[342] (468), 524[344] (468), 585[210] (554, 555)
Gates, A. M. R., 429[183] (416)
Gates, B. C., 427[45] (398)
Gates, J. A., 386[144] (316), 386[145] (316)
Gatti, G., 587[335] (576)
Gaube, W., 387[269] (350)
Gaunt, J. C., 283[453] (245)
Gavezzotti, A., 385[140] (316), 385[141] (316), 385[142] (316), 426[2c] (392)
Gawron, M., 525[438] (479), 586[228] (558)
Gayoso, E., 282[359] (243), 283[458] (245), 283[459] (245), 283[460] (245, 263, 264), 283[468] (245), 284[558] (245)
Gayoso, M., 282[359] (243), 283[458] (245), 283[459] (245), 283[460] (245, 263, 264), 283[468] (245), 283[472] (245, 263), 283[476] (245, 263, 264), 284[500], 284[554] (245, 263), 284[558] (245), 289[854] (263)
Gehrig, K., 282[376]
Gehrke, J.-P., 100[33] (31, 33), 100[34] (31, 33), 183[133] (116, 124, 138), 186[281] (137–9), 186[282] (137–9), 186[287] (138, 142), 186[288] (138, 142), 186[289] (138, 142), 186[293] (138, 142), 186[310] (138), 186[311] (138), 187[363] (142)
Geib, S. J., 188[421] (151, 154, 155), 283[438] (244), 283[461] (245)
Geiger, W. E., 183[94] (115, 124, 153), 188[439] (152), 285[595] (249, 272), 385[101] (307), 385[102] (307), 390[446] (378), 390[447] (378)
Geike, W. A., 100[64] (31, 36), 188[426] (151, 154, 155), 188[427] (151, 155), 188[460] (154, 155)
Geisenberger, J., 528[623]
Gelmini, L., 523[312] (464)
Gemmill, J., 428[105] (403), 429[130] (408), 522[269] (460), 526[509] (490), 527[535] (493)
Gennick Cesa, I., 526[476] (484)
Gentle, T. M., 386[146] (316)
Geoffroy, G. L., 521[197] (450), 583[40] (536, 539)
Geoffroy, M., 284[542] (272)
Georgiadis, R., 99[12] (30, 32)
Gerdes, I., 102[216] (65), 186[330] (139)
Gerfin, T., 427[42a] (397)
Germain, G., 524[390]
Germer, A., 389[423] (373)
Gersdorf, J., 188[461] (155–7, 159, 166, 167)
Geselbracht, M. J., 189[533] (169, 172, 173)
Ghedini, M., 282[360] (243), 282[419] (243), 282[423] (243), 282[424] (243), 282[425] (243, 264), 283[465] (245), 283[490] (262), 284[505] (263, 264), 284[506] (263), 428[73] (401), 519[77] (439)
Ghilardi, C. A., 27[182] (21), 184[191] (126, 129), 184[192] (126, 129), 184[196] (126, 129), 184[197] (126, 129), 187[354] (139, 147), 187[355] (139, 147), 187[372] (146), 278[122] (234), 278[123] (234), 524[335] (467), 529[686] (509)
Ghose, B. N., 100[86] (42, 55)
Giannetto, A., 523[306] (464)
Giannetto, P., 520[86] (439)
Giannoccaro, P., 24[18] (4)
Gianotti, M., 587[335] (576)
Gielens, E. E. C. G., 285[635] (250, 255, 256)
Giesbrecht, G. R., 284[544]
Gillard, R. D., 428[93] (403, 411), 528[638] (505), 529[695] (510)
Gillie, A., 278[118] (234, 237, 239, 266, 270, 273)
Gilmour, D. I., 429[166] (411, 414, 419), 429[172] (412)
Gilson, D. F. R., 584[90] (544)
Giordan, J. C., 187[411] (151)
Giordano, F., 583[80] (543), 586[221b] (557), 587[285] (568, 571)
Giordano, R., 185[249] (131, 172, 174), 190[561] (172, 174), 190[562] (172, 174), 190[578] (174)
Giordano, T. J., 284[545]
Giovanitti, B., 583[77] (543)
Giovenco, A., 101[119] (42, 61), 222[128] (211), 279[150] (235, 257), 279[153] (235, 262, 264, 267), 289[881] (267), 518[18] (434), 523[325] (466), 587[313] (573)
Giraudeau, A., 430[193] (418)
Girault, J. P., 584[121] (548, 562)
Girling, I. R., 283[475] (245)
Girolami, G. S., 388[296] (354)
Giuffrida, S., 522[261] (460)
Gladfelter, W. L., 27[184] (22)
Gladiali, S., 282[391], 283[482] (245, 255), 286[694] (255, 256, 262), 529[693] (510), 530[774] (517)
Glänzer, B. I., 25[30] (5)
Glavee, G. N., 520[132] (443)
Gleason, W. B., 220[35] (197), 519[28] (435), 526[489] (486, 487)
Gleiter, R., 185[276] (137), 386[171] (324), 587[332a] (575), 587[332b] (575)
Gleizes, A., 102[166] (50, 62)
Gliemann, G., 530[734] (514)
Glinskaya, L. A., 427[22] (393)
Gobetto, R., 184[182] (126, 128)
Goddard, J. D., 187[386] (147)
Goddard, R., 100[60] (36, 52), 103[277] (73, 74, 80), 182[38] (110, 111), 182[47] (110, 113, 117, 118), 183[93] (115, 130), 183[112] (115, 122), 184[145] (117, 118), 184[159] (121), 186[279] (137, 138), 186[316] (138, 144, 172), 189[553] (171, 172), 220[27b] (197, 210), 221[65] (203), 286[652] (252), 387[283] (352), 388[288] (352), 390[450] (378)
Goddard, W. A., III, 289[873] (266), 289[874] (266), 521[206] (453), 522[207] (453)
Godleski, S. A., 386[179] (326), 389[373] (365), 389[382] (366), 389[396] (368), 583[34] (536, 540, 542)
Goel, A. B., 282[361], 289[853] (263), 388[298] (355), 427[34] (395), 427[38] (396), 519[68] (438), 520[97] (440), 521[156], 526[492] (486), 529[660] (507, 514), 529[717], 530[740] (514), 530[741] (514), 530[743] (514), 530[744] (514), 585[213a] (556), 585[219a] (557), 585[219b] (557), 586[221a] (557), 586[222] (557)
Goel, R. G., 282[362], 427[13] (393–5, 411, 412), 427[14] (393, 403, 411), 428[102] (403), 529[660] (507, 514), 529[661] (507), 529[662] (507), 529[663] (507)
Goel, S., 282[361], 289[853] (263), 427[34] (395), 427[38] (396), 520[97] (440), 521[156], 529[660] (507, 514), 529[717], 530[740] (514), 530[741] (514), 530[743] (514), 530[744] (514), 585[213a] (556), 585[219a] (557), 585[219b] (557), 586[221a] (557), 586[222] (557)
Goggin, P. L., 182[26] (109), 182[27] (109), 220[14] (195), 220[15] (195), 220[21] (196), 384[34] (293), 427[58] (399), 518[4] (432), 583[46] (537, 542, 550)
Gogoll, A., 386[173] (324), 389[406] (371), 389[409] (371)
Goh, L. Y., 100[79] (38, 39, 42, 57)
Goldsworthy, D. H., 528[597] (499)
Goliaszewski, A., 387[259] (348), 387[260] (348)
Golič, L., 104[357] (81, 88, 90), 184[185] (126, 128), 187[361] (141, 142)
Golovchenko, L. S., 519[40] (436)
Gomes-Carneiro, T. M., 104[336] (81, 86), 104[337] (81, 86), 521[161], 526[491] (486, 509), 529[689] (509)
Gomez, M., 101[152] (41, 42, 47), 281[348] (242), 282[375] (263, 264), 282[418] (243), 520[140] (443)
Gomez, N., 428[96] (403)
Gomez-Saso, M. A., 277[51] (230), 524[363] (469), 587[306] (572)
Goncharov, V. S., 282[367]
Gong, J. K., 25[84] (11, 23), 26[118] (16), 26[121] (16), 26[130] (17), 104[341] (81, 87)
González, F., 101[99] (42, 56), 101[100] (42, 56), 102[187] (48, 56, 62), 184[134] (116, 117, 157)
Gonzalez, O., 284[500]
Goodall, B. L., 281[355] (243)
Goodby, A. J., 188[478] (157, 159, 167)
Goodfellow, R. J., 220[15] (195), 220[21] (196), 427[58] (399), 427[62] (400, 410), 427[65a] (400), 429[161] (410), 518[4] (432), 588[366] (580)
Gooding, R., 183[123] (115, 138, 145, 155)
Goodwin, H. J., 526[494] (487), 529[682] (509), 584[151] (552, 566)
Gordon, D. J., 390[428] (374)
Gordon, J. C., 26[109] (15, 19), 27[169] (21), 185[269] (136), 188[483] (157,

159, 169), 190[570] (172, 173)
Gore, J., 387[233] (343), 389[374] (365), 389[375] (365), 389[398] (370), 389[399] (370), 389[400] (370), 389[401] (370), 389[402] (370)
Görls, H., 103[272] (73, 74), 104[354] (81, 90), 185[209] (130, 132), 185[218] (130), 185[220] (130, 133), 186[325] (139, 140)
Gorman, E. H., 290[934] (273)
Gorrell, I. B., 189[528] (159, 170)
Gorzellik, M., 189[556] (172, 173)
Gosselink, J. W., 279[147] (235)
Gosser, L. W., 182[32] (110)
Goto, M., 587[287] (569)
Gottsch, P., 186[328] (139, 141), 186[337] (139, 141), 587[340b] (576), 587[340c] (576)
Göttsch, P., 388[287] (352)
Goursot, A., 103[263] (73)
Gouteron, J., 282[365] (253, 254, 263), 386[167] (321)
Gowda, N. M. N., 385[104] (308)
Gozum, J. E., 388[296] (354)
Grabowski, N. A., 525[451] (480)
Grafl, D., 519[49] (437, 454)
Graham, W. A. G., 531[776] (517)
Gramlich, V., 427[42a] (397)
Granberg, K. L., 389[394] (368), 389[395] (368)
Grandjean, C., 389[401] (370)
Grandjean, D., 25[53] (8), 104[336] (81, 86), 104[337] (81, 86), 221[85a] (205), 221[85b] (205), 281[292] (241, 253), 285[613] (249, 252), 285[619] (250), 286[665] (252), 286[666] (252), 288[835] (263), 289[856] (264), 526[491] (486, 509), 529[689] (509)
Granell, J., 279[172] (236), 281[347] (242, 243, 271), 282[372], 282[373] (269), 282[374], 282[375] (263, 264), 282[418] (243), 283[474] (245, 253), 283[477] (245, 263, 264), 283[479] (245), 283[493] (271), 284[555] (245, 262–4, 271), 284[556] (245, 262, 263), 385[93] (305), 521[186] (447)
Granozzi, G., 185[258] (130), 189[542] (167), 287[746] (257)
Grant, B. E., 26[109] (15, 19), 26[133] (19), 27[169] (21), 105[418] (95, 96), 105[420] (95, 96), 105[421] (95, 96), 185[239] (131, 169, 172), 188[483] (157, 159, 169), 189[532] (159), 189[533] (169, 172, 173), 189[534] (159, 169), 189[545] (167, 169), 190[570] (172, 173)
Grant, D. P., 388[290] (353)
Grassi, A., 428[123a] (407), 519[76] (439, 458, 459)
Grassi, M., 282[425] (243, 264), 386[184] (327), 386[185] (328), 388[334] (359), 428[115] (406), 587[319] (574, 576), 587[320] (574, 576), 587[335] (576), 587[336] (576)
Gray, H. B., 27[184] (22), 220[2b] (194), 280[239] (239, 240)
Grayson, I. L., 287[769] (259), 524[343] (468, 469), 524[346] (468)
Graziani, M., 279[186] (237), 285[610] (249), 525[449] (480), 526[470] (483), 583[38] (536), 583[39] (536, 538), 583[47] (538)
Graziani, R., 520[138] (443, 461), 583[59] (541, 545)
Grebenshchikova, E. A., 221[64b] (203)
Green, C. R., 287[733] (256)
Green, D. B., 585[173] (554, 556, 557, 561)
Green, K. A., 26[110] (15), 105[416] (95, 96), 105[417] (95, 96), 185[236] (131, 136, 169), 185[268] (136), 185[269] (136), 185[270] (136)
Green, M., 27[181] (21, 22), 287[766] (259), 384[53] (295), 385[105] (308), 427[62] (400, 410), 429[157] (410, 417), 429[158] (410, 416, 417), 525[448] (480), 525[454] (480), 584[92] (544), 584[112] (548), 584[113] (548), 584[114] (548), 585[187] (554, 555, 574), 585[188] (554, 555, 573, 576, 580), 586[276] (565, 566), 587[308] (573), 587[345] (577), 588[366] (580)
Green, M. L. H., 429[148] (409)
Greenberg, A., 390[462] (381)
Gregor, N., 385[129] (313)
Grennberg, H., 389[390] (368), 389[408] (371), 389[409] (371)
Grenouillet, P., 187[360] (140), 389[416] (372)
Grepioni, F., 188[436] (152)
Gretz, E., 288[829] (262, 264–6)
Greven, R., 388[333] (359)
Gridner, I. D., 185[265] (133)
Griffiths, D. C., 522[228] (456), 522[229] (456), 522[231] (456), 585[181] (554, 555)
Griffiths, K. D., 105[415] (42, 93, 95), 185[252] (131), 185[253] (131, 134), 185[254] (131, 134)
Grigg, R., 283[470] (245), 388[315] (357), 388[317] (357), 388[318] (357), 388[319] (357), 388[320] (357), 389[415] (372)
Griggs, B. G., Jr., 103[257] (67, 73), 183[132] (116, 120, 157)
Grimes, R. N., 190[593] (175, 179), 190[594] (175, 179), 190[601] (175, 177), 190[602] (175), 190[603] (175), 190[623] (180, 181)
Grinval'd, I. I., 188[430] (152), 188[431] (152), 188[432] (152), 188[433] (152)
Gritsenko, O. V., 384[48] (295), 384[50] (295)
Grobe, J., 185[226] (131, 136), 188[456] (154), 520[99] (440)
Groning, A. B., 584[103] (547)
Groshens, T. J., 24[10] (2), 24[11] (2), 100[36] (33, 41), 100[37] (33), 190[579] (174, 175), 190[580] (174, 175), 190[581] (174)
Gross, M., 221[78a] (205), 221[78b] (205), 430[193] (418)
Grossel, M. C., 521[192] (447)
Grove, D. M., 104[327] (81, 86), 104[328] (81, 86), 104[329] (81, 85), 104[332] (81, 86), 104[333] (81, 86), 104[334] (81, 86), 280[260] (239, 255, 262, 272, 273), 281[300] (241), 288[832] (262), 289[869] (264, 265), 385[105] (308), 530[721] (513), 530[750] (515), 530[751] (515), 530[754] (515), 530[756] (516), 530[761] (516), 530[762] (516), 530[770] (517), 586[276] (565, 566)
Grubbs, R. H., 183[64] (112), 280[270] (240, 267), 288[780] (260), 387[239] (344), 521[153] (444)
Gruetzmacher, J. A., 389[361] (363)
Grundy, K. R., 429[176] (414)
Gruppe, S., 186[319] (138, 145), 186[320] (138, 145)
Grushin, V. V., 390[439] (377), 588[367] (580)
Grutsch, P. A., 281[357] (243)
Guadalupi, G., 430[201b] (420, 426), 524[384]
Guarnieri, A., 285[571] (248)
Guastini, C., 222[138] (213), 526[522] (490, 491, 493), 526[523] (490, 491), 527[526] (491)
Gubin, S. P., 221[53] (201, 204), 221[68a] (204), 221[68c] (204), 221[69] (204), 221[70] (204), 221[72] (204), 427[11] (393), 427[12] (393, 399), 427[28] (395)
Gudat, D., 188[484] (157, 166)
Gudova, N. A., 428[108a] (403)
Guerney, P. J., 526[494] (487), 529[682] (509), 584[151] (552, 566), 584[152] (552)
Guerra, M., 185[275] (137), 386[175] (324), 587[334] (575, 576)
Guerra, M. A., 190[583] (174)
Guerts, P., 185[255] (129)
Guest, M. F., 102[178] (53)
Guggolz, E., 427[17] (393, 412)
Guha, S., 278[79] (231)
Guhl, D., 25[48] (8), 25[51] (8), 25[52] (8), 104[308] (80, 81), 104[316] (80, 81), 104[317] (80, 81), 104[318] (81), 104[319] (80, 81), 104[321] (80, 81), 184[146] (117)
Guibé, F., 387[253] (346), 388[312] (357)
Guibert, I., 290[941] (273)
Guillot-Edelheit, G., 584[105] (547)
Guimerani, R. R., 428[69] (400), 520[102] (440)
Gukathasan, R. R., 221[50a] (201), 221[50b] (201), 427[32] (395), 427[35] (395–7), 430[227] (426)
Guldi, D. M., 105[401] (92), 105[402] (92)
Gundlach, K. B., 386[179] (326), 583[34] (536, 540, 542)
Gunter, J. R., 529[666] (507)
Gupta, A., 524[391] (472)
Gupta, S. K., 519[39] (436)
Gupta, V. K., 281[312] (241), 281[314] (241), 281[316] (241), 281[318] (241), 281[320] (241, 242), 385[91] (305), 529[691] (510)
Gusev, A. I., 428[112] (406)
Gutierrez, M. A., 283[494] (245), 283[495] (245, 264), 284[496] (245), 284[497] (245)
Gutiérrez-Puebla, E., 24[17] (3), 25[88] (12), 25[89] (12), 101[106] (42, 47, 50, 58, 81, 90), 101[107] (42, 47, 58, 62, 63), 101[108] (42, 47, 58, 62, 63), 101[139] (42, 56), 102[188] (48, 56), 103[268] (74, 78, 81), 104[293] (73–5, 78, 81, 90), 104[294] (74, 78, 81), 104[295] (74,

78), 104[296] (74, 78), 183[108] (115, 124), 183[109] (115, 124), 184[193] (126, 129), 186[343] (139), 186[344] (139, 146), 186[345] (139), 189[500] (157), 288[843] (263)

Habazaki, H., 519[62] (437)
Habeeb, J. J., 278[91] (233)
Hachgenei, J., 26[156] (20), 26[157] (20), 26[158] (20), 26[164] (20)
Hackelberg, O., 588[372] (581)
Hackett, M., 529[674] (508), 531[778] (517), 583[21a] (535, 565, 568), 583[21b] (535), 583[21c] (535), 585[204] (554, 555)
Hacksell, U., 285[636] (250)
Hadj-Bagheri, N., 428[90] (404, 415), 520[124] (442, 463)
Haeffner, F., 386[170] (323)
Haenel, M. W., 284[541]
Haggerty, B. S., 189[531] (159, 169)
Haggitt, J. L., 585[169] (554)
Hagihara, N., 103[254] (67, 73), 189[493] (157, 161), 277[58] (230), 277[64] (230), 277[65] (230, 231), 277[69] (230, 231), 278[77] (231), 286[668] (252), 527[541] (493), 527[543] (493)
Hagihara, T., 387[244] (345), 388[342] (360)
Hall, J. R., 519[75] (439, 464, 472, 497, 506), 524[356] (469), 525[422] (477, 497, 506), 528[627] (500), 528[654] (506), 529[655] (506), 529[656] (506), 529[657] (506), 529[658] (507), 529[659] (507), 585[167] (553–5), 585[183] (554, 555, 557), 585[208] (554, 555), 585[216] (557)
Hall, M. B., 189[541]
Hall, P. S., 428[117a] (406), 428[117b] (406)
Hall, S. S., 387[222] (340)
Hall, T. L., 26[139] (19)
Hallam, M. F., 427[8] (392, 396, 397), 427[36] (396), 427[40] (396), 427[41] (396, 413), 427[68] (400)
Hallberg, A., 386[205] (335)
Halle, L. F., 99[14] (30, 32), 182[29] (109), 182[30] (109)
Hallenbeck, L. E., 184[155] (119)
Haller, K. A., 189[525] (159, 169)
Hallock, J. S., 519[32] (436)
Halpern, J., 280[278] (240), 427[4] (392), 428[126] (407), 519[47] (437), 583[42] (537, 560), 583[56] (541)
Halterman, R. L., 387[258] (347)
Halton, B., 279[187] (237), 525[439] (479), 525[440] (479), 583[51] (540)
Ham, N. S., 529[658] (507)
Hamamura, K., 283[452] (245), 285[576] (248)
Hamatani, Y., 285[583] (249), 387[280] (352)
Hambley, T. W., 103[292] (73, 74, 81), 185[251] (131, 134)
Hambright, P., 105[401] (92), 105[402] (92)
Hamer, G., 588[374] (581)
Haming, L., 278[113] (234)
Hammer, B., 522[266] (460)
Hampden-Smith, M. J., 524[371] (471, 495), 524[374] (471), 524[376] (471), 524[391] (472), 582[17a] (535, 539), 582[17b] (535, 539), 587[338] (576)
Hamrick, Y. M., 187[414] (151)
Han, K. I., 104[299] (73, 74), 183[91] (115, 119), 184[155] (119), 184[156] (119, 122, 131)
Han, S., 430[191] (418)
Hanan, G. S., 284[544], 529[712] (512)
Hancock, G. C., 186[280] (137)
Handschuh, D., 386[198] (333)
Hanes, R. M., 389[419] (373)
Hanko, R., 186[331] (139, 143)
Hanks, T. W., 527[562] (495), 528[647] (505)
Hanlan, A. J. L., 220[2a] (194), 220[2b] (194)
Hanson, A. W., 223[157] (215)
Hanson, B. E., 287[741] (257)
Hansson, S., 386[170] (323), 387[250] (346), 387[279] (352), 389[368] (363), 389[372] (365)
Hanzawa, Y., 388[328] (358)
Harada, S., 586[254] (562)
Hardcastle, K. I., 286[650]
Harder, S., 277[57] (230, 232, 234, 239, 269, 270, 273)
Hardy, D. T., 522[230] (456), 586[243] (560)
Harrison, K. N., 520[95] (440), 582[13b] (535, 537, 539, 566)
Harrison, L. W., 386[206] (335), 387[219] (339)
Hart, D. W., 285[614] (250)
Hart, W. P., 188[459] (154, 155)
Hartley, F. R., 277[12] (227, 228), 278[85] (233–5, 266, 273), 383[10] (292), 383[11] (292), 384[12] (292), 384[13] (292), 384[14] (292), 384[15] (292, 333), 384[16] (292), 384[23] (292), 384[58] (296), 518[1] (432), 518[2] (432, 434, 440, 442, 445, 453, 458, 460, 472, 477, 491, 496), 518[3] (432, 460), 582[2] (534, 560), 582[3] (534)
Hartshorn, R. M., 284[507] (245, 246, 264)
Harvey, P. D., 220[45] (199), 220[46] (199), 223[158] (215), 428[78c] (402)
Harvie, I. J., 289[890] (268)
Hase, T., 523[294] (462)
Hasegawa, N., 25[38] (5), 25[39] (5)
Hasegawa, S., 286[657] (252)
Hashimoto, H., 387[237] (343)
Hashmi, A. S. K., 290[939] (273)
Hassan, F. S. M., 520[103] (440, 466), 520[118] (441, 442, 445, 463), 523[327] (466)
Haszeldine, R. N., 384[69] (302), 384[70] (302), 428[127] (407)
Hatano, M., 387[240] (344)
Hauge, R. H., 99[6] (30, 31, 74), 100[27] (32), 182[24] (109), 185[257] (130)
Haupt, E. T. K., 290[942] (273)
Haupt, H.-J., 221[63] (203), 277[56] (230, 231)
Hausinger, R. P., 105[384] (91)
Hausmann, I., 190[587] (175, 179)
Havlas, Z., 519[31] (435)
Hawkins, I., 24[15] (3), 25[86] (11, 12), 101[117] (42, 58, 59, 61), 102[167] (50, 51, 58, 59), 102[174] (50, 56, 58, 59, 61, 62), 184[174] (126), 221[66] (204)
Hawthorne, M. F., 191[624] (180, 182)
Hay, P. J., 220[12a] (195), 384[46] (294), 427[3] (392), 584[136] (550)
Hayashi, R. K., 27[176] (21)
Hayashi, T., 285[631] (250), 286[705] (255, 256), 386[209] (335), 387[244] (345), 388[337] (360), 388[341] (360), 388[342] (360), 388[343] (360), 388[344] (360), 389[362] (363), 389[377] (365), 389[386] (367), 389[387] (368)
Hayashi, Y., 280[238] (239, 255, 256, 263), 281[289] (240, 267)
Haynes, A., 427[64] (400), 588[370] (581)
Head, R. A., 524[334] (467), 582[16a] (535, 539), 582[16b] (535, 539), 586[237] (560)
Heathcock, C. H., 103[239] (66, 67), 188[488] (157), 281[290] (240), 388[331] (359)
Heaton, B. T., 427[51] (398, 399), 583[39] (536, 538)
Heaton, D. E., 221[51] (201)
Heazle, K. D., 586[263b] (564, 572)
Hebendanz, N., 188[429] (152, 155, 156)
Heck, R. F., 278[139] (235, 244, 245, 255, 270), 283[438] (244), 283[461] (245), 285[614] (250), 286[674] (253, 255), 286[682] (253), 286[683] (253), 287[716] (255), 385[94] (305), 386[168] (322), 387[216] (337), 387[234] (343), 523[273] (460)
Heckmann, G., 185[227] (131, 136, 159), 185[271] (136, 172)
Hedden, D., 284[538], 385[114] (310), 529[668] (507)
Heeg, M. J., 223[153] (215), 429[175] (414)
Hegedus, L. S., 186[329] (139, 140), 285[589] (249), 287[724] (256), 386[180] (325, 326), 388[309] (357), 388[351] (361), 389[364] (363), 389[397] (369)
Hegemann, M., 185[226] (131, 136)
Heger, G., 186[300] (138, 143)
Heimbach, P., 102[216] (65), 185[260] (131, 132), 186[330] (139)
Hein, J., 27[186] (23)
Heiss, H., 531[780] (517)
Hejna, C. I., 220[40] (198)
Helm, L., 522[249] (458, 459)
Helms, M., 387[269] (350)
Hemmer, R., 187[373] (148), 519[54] (437, 440)
Hemond, R. C., 525[452] (480)
Henc, B., 288[784] (260)
Henderson, A. N., 526[517] (490), 587[300] (571)
Henderson, W., 103[266] (73, 74), 103[267] (73, 74), 222[131] (212), 429[185] (416), 525[443] (479), 525[445] (480), 525[446] (480), 526[469] (482), 588[351] (579)
Hendrickson, D. N., 27[184] (22)
Henling, L. M., 288[780] (260), 387[239] (344), 521[153] (444)
Henly, T. J., 387[276] (350)
Henney, R. P. G., 284[520] (261), 284[521], 284[522], 284[523], 529[692] (510), 529[694] (510)

Henry, P. M., 290[935] (273), 385[128] (313), 385[129] (313), 385[130] (313), 385[131] (313), 385[132] (313), 385[133] (314), 388[303] (356)
Herberich, G. E., 183[131] (116, 174, 176), 186[315] (138, 139, 148), 188[467] (155, 156, 175), 188[474] (156), 190[584] (174, 176), 190[587] (175, 179), 190[615] (176), 588[356] (579, 582), 588[381] (582), 588[382] (582)
Herbert, I. R., 220[15] (195), 427[58] (399), 427[65a] (400), 428[107] (403), 429[161] (410)
Herbert, J. A., 583[29] (536, 544)
Herbst-Irmer, R., 188[448] (154, 155, 157), 189[537] (159, 161)
Herdeis, C., 429[182] (416, 423)
Herdtweck, E., 25[31] (5), 188[476] (157, 167)
Hermanek, S., 191[633] (180, 181)
Hernandez, E., 25[49] (8), 25[50] (8), 102[175] (41, 42, 47, 62), 104[308] (80, 81), 104[311] (81), 104[313] (81), 104[314] (81), 104[315] (81), 104[321] (80, 81), 185[215] (130, 131, 133), 186[324] (139), 188[486] (157, 159, 164)
Hernández, J., 101[121] (42, 50, 58, 60)
Herrera, A., 103[279] (74, 76), 183[110] (115), 183[111] (115, 122), 185[213] (130), 588[354] (579)
Herrero, F. J., 283[492] (262, 263)
Herrero, S., 288[815] (261)
Herres, M., 185[234] (131)
Herring, A. M., 529[683] (509)
Hershberger, J. W., 100[32b] (31, 33), 183[103] (115), 583[36] (536, 539)
Hershberger, S. S., 286[637] (251), 386[203] (334), 388[329] (358)
Hervieu, J., 222[116] (210)
Herzig, J., 389[393] (368)
Herzog, V., 25[36] (5), 104[359] (81, 88), 104[362] (81)
Herzschuh, R., 186[281] (137–9), 186[282] (137–9)
Hess, F. W., 529[658] (507)
Hess, R., 519[50] (437, 454), 519[53] (437)
Hessner, B., 190[584] (174, 176), 190[587] (175, 179), 190[615] (176), 427[62] (400, 410), 588[366] (580), 588[381] (582)
Hettich, R. L., 182[28] (109)
Heumann, A., 285[594] (249), 385[119] (311), 385[120] (311), 389[394] (368), 389[413] (371)
Heuser, N., 188[475] (157, 161)
Heydenreich, F., 24[12] (3)
Heyke, O., 585[161] (553)
Hida, M., 388[345] (360)
Hidai, M., 278[104] (233), 287[720] (256), 287[721] (256)
Hidaka, K., 223[172] (218)
Hidaka, T., 280[249] (239, 262), 289[879] (267, 269, 270)
Hietkamp, S., 285[568] (248)
Higgins, K. M., 528[588] (498, 504)
Higgins, S. J., 26[122] (16), 102[192] (41, 47, 48), 278[75] (230, 236), 279[185] (236), 286[647] (252), 386[157] (319), 520[103] (440, 466), 523[327] (466), 526[510] (490)
Higuchi, T., 184[199] (125), 184[200] (125)
Hildebrandt, W., 520[99] (440)
Hill, R. H., 189[494] (157, 163), 189[495] (157, 163), 427[24] (394, 402, 404), 518[10] (432, 433, 453), 518[11] (432, 433), 518[12] (432, 453), 522[210] (453), 523[319] (465)
Hiller, W., 25[73] (10), 103[255] (67, 70), 189[497] (157, 159, 164), 189[521] (159, 163), 284[500]
Hillhouse, G. L., 25[54] (9), 25[60] (9), 101[96] (42, 54, 74, 80, 81, 90), 183[126] (115, 122, 125)
Himmel, S. E., 522[215] (454), 585[193] (554, 555, 569)
Hinkle, R. J., 526[500] (487), 583[53] (540, 548, 574, 576), 584[122] (548)
Hinrichs, W., 528[611] (500)
Hinton, P., 105[404] (92)
Hinze, S., 26[130] (17)
Hipler, B., 100[65] (37, 38, 41, 42, 47, 81), 103[271] (73, 74), 104[338] (81, 86)
Hirai, K., 286[660] (252), 286[661] (252)
Hiraki, H., 282[406] (264), 284[503]
Hiraki, K., 223[162] (216), 223[172] (218), 223[174a] (218), 223[174b] (218), 277[25] (228, 229), 277[68] (230), 278[125] (234, 262, 263), 278[144] (235), 279[161] (236, 262, 263), 279[181] (236, 264), 281[309] (241), 281[310] (241, 262, 263), 281[336] (242), 282[358] (243, 248, 263), 282[378] (263, 264), 282[379] (263), 282[380], 282[382], 282[383] (263), 282[406] (264), 282[407] (248, 263), 282[416], 284[499], 284[502] (263), 284[503], 284[546] (245), 284[566] (248, 263), 285[578] (248), 286[650], 286[651], 286[672] (253), 287[773] (259), 289[855] (263, 269), 289[859] (264), 385[84] (304), 390[436] (376)
Hiramatsu, M., 385[98] (307), 385[100] (307)
Hirao, T., 25[27] (4), 25[28] (4), 587[326] (574, 579)
Hiroe, Y., 287[721] (256)
Hiroi, K., 388[346] (360)
Hirotsu, K., 184[199] (125), 184[200] (125)
Hirsekorn, F. J., 27[183] (22)
Hishi, S., 387[254] (347)
Hitchcock, P. B., 184[179] (126, 127), 186[321] (138, 155), 223[182] (219), 279[192] (237), 279[193] (237), 529[670] (508), 529[675] (508), 583[54] (540), 585[162] (553), 585[163] (553, 565), 586[240] (560), 587[290] (570)
Ho, D. M., 183[103] (115), 583[36] (536, 539)
Ho, H. Y., 386[179] (326)
Ho, K.-C., 524[385]
Ho, N.-F., 188[451] (154, 170), 188[452] (154, 159, 170)
Hoberg, H., 25[43] (7), 25[44] (7), 25[45] (7), 25[46] (7), 25[48] (8), 25[49] (8), 25[50] (8), 25[51] (8), 25[52] (8), 25[55] (9), 26[99] (14), 102[165a] (42, 59), 102[165b] (42, 59), 103[279] (74, 76), 104[306] (81, 84), 104[307] (81), 104[308] (80, 81), 104[309] (80, 81), 104[310] (80, 81), 104[311] (81), 104[312] (81), 104[313] (81), 104[314] (81), 104[315] (81), 104[316] (80, 81), 104[317] (80, 81), 104[318] (81), 104[319] (80, 81), 104[320] (80, 81), 104[321] (80, 81), 104[322] (80, 81), 104[323] (81, 84), 104[324] (81), 104[344] (81, 88, 90), 104[346] (81, 88), 104[348] (81, 88), 104[349] (81, 88, 89), 104[350] (81, 88), 104[353] (81, 88), 104[355] (81, 88, 90), 104[363] (81, 88), 104[364] (81, 88), 105[369] (81, 88), 105[370] (81, 88), 105[371] (81, 88), 105[374] (81, 88), 105[376] (80), 183[93] (115, 130), 183[110] (115), 183[111] (115, 122), 184[146] (117), 184[162] (122), 185[213] (130), 186[300] (138, 143), 186[323] (137, 139), 186[324] (139), 187[359] (137), 187[374] (148), 187[375] (148), 187[376] (148), 187[377] (148), 187[378] (148, 174, 176), 190[614] (176), 385[113] (310), 588[354] (579)
Hoberg, J. O., 527[573] (497), 584[130] (549, 552), 584[131] (549, 552)
Höbold, W., 103[240] (66, 67), 189[496] (157, 159, 164)
Hodali, H. A., 280[243] (239, 267), 520[108] (441)
Hodgson, M., 278[135] (235), 384[71] (302)
Hoehne, S., 189[522] (159, 163)
Hoffman, D. M., 587[289] (569)
Hoffman, R., 289[871] (266)
Hoffmann, H. M. R., 384[30] (293), 388[353] (361)
Hoffmann, K., 278[83] (232)
Hoffmann, P., 223[160] (216)
Hoffmann, R., 100[23] (30), 102[209] (62), 103[262] (73), 182[22] (109), 191[638] (180), 285[597] (249), 285[630] (250, 253), 287[718] (255), 384[32] (293), 385[116] (310), 385[117] (310), 427[47] (398), 517, 519[31] (435), 521[202] (453), 528[635] (505), 531[777] (517), 584[139a] (551), 584[139b] (551), 584[139c] (551), 584[142] (551, 569), 587[289] (569)
Hofmann, B., 103[271] (73, 74)
Hofmann, P., 25[56] (9), 26[127] (16), 183[74] (115), 183[75] (115, 126, 128), 189[524] (159, 163), 189[530] (159, 163), 280[217] (238, 267, 270), 390[441] (377), 531[780] (517), 531[781] (517)
Hogerheide, M. P., 279[176] (236, 240, 242, 243, 262, 269, 270, 272, 273, 275)
Hohenadel, R., 520[107] (440, 454), 525[436] (478, 481), 526[459], 526[462] (481), 585[199] (554, 555), 585[200] (554, 555)
Hohner, U., 186[284] (138, 145)
Holland, K. D., 183[105] (115, 122)
Holle, S., 100[70] (37, 38), 385[80] (303)
Hollnagel, A., 189[551] (171), 190[575] (171), 387[277] (351)

Holm, R. H., 105[403] (92)
Holmes, R. G. G., 428[127] (407)
Hölscher, M., 190[601] (175, 177)
Holt, E. M., 386[202] (334), 587[312] (573)
Holt, S. L., Jr., 386[202] (334)
Holton, J., 277[23] (228), 527[559] (495)
Holton, R. A., 282[392], 282[393], 285[604] (249), 285[609] (249), 290[924] (273)
Holwerda, R. A., 289[900] (270)
Holy, N., 277[29] (228)
Holzmann, G., 223[166] (217, 218)
Honeyman, R. T., 280[253] (239, 261, 262), 280[263] (239, 242, 261–4, 272, 273, 275, 276), 282[390], 290[945] (273), 290[946] (273), 520[85] (439), 520[91] (440), 523[304] (464), 523[305] (464), 523[309] (464), 523[310] (464), 528[616] (510), 529[701] (510)
Hong, P., 103[254] (67, 73), 189[493] (157, 161)
Honig, E. D., 584[111] (548)
Honnick, W., 221[79] (205)
Hoover, J. F., 524[372] (471, 495), 525[424] (477), 526[467] (481), 527[561] (495), 585[184] (554, 555)
Hopkins, R. B., 389[408] (371)
Hopp, P., 183[83] (115, 119)
Hopton, F. J., 280[273] (240)
Hor, T. S. A., 283[443] (244, 263)
Hori, T., 223[150] (215)
Horino, H., 282[364] (250, 263, 269), 286[638] (251)
Horiuchi, C. A., 386[200] (333), 388[301] (355), 389[357] (362)
Hoskins, B. F., 284[564] (246)
Hosmane, N. S., 191[636] (180)
Hosokawa, T., 386[208] (335), 386[211] (336)
Hossain, Md. M., 586[268] (564)
Hostalek, M., 188[467] (155, 156, 175)
Houriet, R., 182[29] (109)
House, D. A., 284[513]
Howard, C. G., 429[183] (416)
Howard, J. A. K., 27[185] (23), 27[187] (23), 105[423] (95, 96), 187[409] (150), 190[566] (172), 223[180] (219), 223[181] (219), 427[62] (400, 410), 583[26] (536, 537, 539, 543), 583[28] (536), 583[29] (536, 544), 583[44] (537), 583[45] (537), 585[160] (553), 585[164] (553, 554), 586[267] (564), 586[276] (565, 566), 588[366] (580)
Howard, K. E., 187[385] (148, 149)
Howden, M. E., 277[54] (230)
Howell, J. A. S., 582[6a] (534), 582[6b] (534), 582[6c] (534), 582[6d] (534), 582[6e] (534), 582[6f] (534)
Howells, N. D., 427[36] (396)
Hoyano, J. K., 531[776] (517)
Hoye, P. A. T., 520[95] (440)
Hriljac, J. A., 26[150] (20)
Hsiao, Y.-M., 105[404] (92)
Hsu, C.-Y., 388[306] (356)
Hsu, M. H., 387[219] (339)
Hsu, W. S., 190[576] (174)
Hsu, W.-L., 190[563] (172, 174)
Hu, Y. F., 186[277] (137), 386[172] (324), 587[333] (575)
Huang, G.-L., 523[291] (462)
Huang, J., 25[84] (11, 23)
Huang, L., 222[98b] (207), 222[99a] (208), 222[99b] (208), 222[99c] (208), 286[688] (255), 287[734] (256), 287[735] (256, 257, 270), 429[145] (408), 523[290] (462)
Huang, N.-Z., 103[280] (74, 75), 103[287] (75), 183[124] (115, 121), 183[125] (115, 121)
Huang, T.-M., 429[142b] (408), 429[143] (408), 429[144] (408), 523[288] (462), 523[289] (462), 523[291] (462), 523[292] (462), 587[322] (574, 577)
Huang, Y., 184[140] (116)
Huang, Y. H., 584[132] (549)
Huang, Y. J., 428[108b] (403, 409), 583[65] (542)
Huang, Y. N., 584[90] (544)
Hübinger, E., 102[218] (65), 186[303] (138, 140)
Hubner, D., 529[685] (509)
Huch, V., 26[102] (14)
Huckerby, T. N., 284[563] (246)
Huckett, S. C., 281[291] (240)
Hudali, H., 519[21] (434)
Hudeczek, P., 188[425] (151, 154–6), 188[429] (152, 155, 156)
Hudson, S. A., 101[139] (42, 56), 288[843] (263)
Huffman, J. C., 27[170] (21), 189[555] (172, 173), 526[458], 528[648] (505)
Hugentobler, M., 282[376], 282[377] (244)
Huggins, J. M., 100[71] (37, 38)
Hughes, G. R., 586[246] (560)
Hughes, J. G., 284[561] (246)
Hughes, J. M., 427[53] (398)
Hughes, R. P., 103[288] (74, 76, 80), 103[289] (74, 76), 183[105] (115, 122), 183[106] (115, 122), 279[188] (237), 285[621] (250), 387[228] (342), 387[229] (342), 429[188] (417), 525[451] (480), 525[452] (480)
Huhn, M., 188[446] (153, 157), 188[448] (154, 155, 157), 189[537] (159, 161)
Hulliger, J., 187[413] (151)
Humphrey, M. G., 103[245] (66, 67)
Hung, A. Y. C., 522[250] (459)
Hung, T., 587[309a] (573, 576)
Hunt, C. T., 222[135] (213), 278[119] (234, 236), 286[646], 286[702] (255, 256, 258, 269, 271), 429[174] (414)
Hunter, A. D., 101[153] (41, 42, 47–9), 102[161] (41, 42, 47–9)
Hunter, B. K., 286[695] (255, 263, 268, 270, 273, 275)
Hupfer, E., 522[213] (454)
Hursthouse, M. B., 24[15] (3), 102[167] (50, 51, 58, 59), 102[174] (50, 56, 58, 59, 61, 62), 221[66] (204), 280[221] (238, 263), 282[417] (243), 283[431] (244), 285[569] (248), 287[772] (259), 289[858] (264), 385[89] (305), 388[295] (354), 429[183] (416), 520[112] (441), 528[592] (498), 528[606] (499), 528[610] (500), 528[622], 585[154a] (553), 585[154b] (553), 585[156] (553)
Huser, M., 278[108] (234, 255, 256)
Hütter, P., 386[198] (333)
Hutton, A. T., 288[817] (262, 265), 520[118] (441, 442, 445, 463), 523[297] (463), 527[548] (494), 527[549] (494), 527[551] (494), 527[552] (494), 586[234] (559)
Hutzinger, M. W., 388[313] (357)
Hux, J. E., 523[295] (463), 528[652] (506), 528[653] (506)
Hwan, L., 277[38] (229, 230)
Hwang, W. S., 583[31] (536, 537)
Hyde, T. I., 284[524] (245, 271, 272)
Hynes, R. C., 188[466] (155)

Ibbott, D. G., 585[191] (554, 555, 557)
Ibers, J. A., 184[194] (126), 286[658] (252), 286[660] (252), 289[909] (272, 276), 427[67] (400), 521[159] (493), 522[225] (456, 478, 481), 527[572] (497), 529[671] (508), 529[672] (508), 529[673] (508), 531[778] (517), 583[21a] (535, 565, 568), 583[21b] (535), 584[150] (552), 585[204] (554, 555)
Ibuka, T., 384[39] (294)
Ichikawa, M., 427[56] (399)
Igarashi, K., 105[382] (85)
Ihara, E., 287[763] (258), 287[764] (258)
Ihle, N. C., 388[331] (359)
Iitaka, Y., 25[82] (11), 184[163] (122), 525[453] (480)
Ikariya, T., 280[270] (240, 267)
Ikawa, T., 278[96] (233, 234), 280[244] (239)
Ikeda, I., 103[221] (65), 584[87] (544), 587[326] (574, 579), 587[329a] (575)
Ikura, K., 586[254] (562)
Illingsworth, M. L., 527[556] (495)
Ilsemann, G. v., 288[788] (260), 387[214] (336)
Imada, Y., 386[211] (336)
Imaizumi, S., 386[210] (335)
Imhof, W., 185[234] (131), 185[235] (131)
Immirzi, A., 390[463] (381), 520[144] (444)
Imran, A., 525[442] (479, 481), 588[350] (578)
Inaba, S., 101[154] (42, 47, 48)
Inanaga, J., 388[323] (358)
Ingram, G., 188[464] (155, 156, 177)
Ingrosso, G., 280[267] (240, 262, 267, 269–71), 280[268] (240, 262), 280[271] (240, 270)
Inoue, N., 282[364] (250, 263, 269), 286[638] (251)
Inoue, Y., 387[237] (343)
Intini, F. P., 584[120] (548)
Iranpoor, N., 185[267] (135)
Iraqi, A., 525[415] (475)
Irvine, D. J., 24[16] (3), 101[104] (42, 47, 51, 57, 58, 66, 67), 188[482] (159)
Irwin, J. J., 427[20] (393, 396)
Isaeva, L. S., 101[124] (41, 42, 47), 101[126] (42), 102[164] (41, 42, 47, 51, 58), 183[119] (115, 117, 119), 183[121] (115, 117, 119), 184[158] (119), 186[304] (138, 139, 143)
Isenberg, W., 285[570] (248)
Ishida, J., 100[85] (42, 53, 62)
Ishida, Y., 279[161] (236, 262, 263)
Ishii, K., 389[388] (368), 587[316] (573, 576)
Ishii, Y., 278[104] (233), 286[657] (252),

286[658] (252), 287[721] (256), 388[309] (357)
Ishikawa, M., 103[270] (73, 74), 184[171] (125, 126), 184[198] (125), 184[199] (125), 184[200] (125)
Ishimura, Y., 100[81] (40)
Ishio, N., 288[841] (263)
Ishizawa, S., 388[328] (358)
Ishizu, J., 24[21] (4), 186[285] (138, 140)
Ishizuka, K., 387[240] (344)
Isobe, K., 279[151] (235, 255, 262–4, 269), 279[152] (235, 255, 262), 279[180] (236), 279[182] (236), 279[183] (236, 255, 256, 264), 280[238] (239, 255, 256, 263), 281[289] (240, 267), 288[842] (263)
Isomura, K., 285[608] (249)
Issa, S., 186[339] (139, 146)
Issleib, K., 101[148] (41, 47, 48, 63), 104[338] (81, 86), 281[294] (241), 281[295] (241, 264), 281[298] (241), 529[679] (509), 529[680] (509), 530[745] (514), 530[746] (514)
Ito, N., 384[65] (300), 384[66] (300)
Ito, T., 278[125] (234, 262, 263), 280[230] (238, 239, 255, 273), 280[231] (238, 267), 280[232] (238, 239, 262, 266, 267), 286[657] (252), 289[904] (271)
Ito, Y., 103[270] (73, 74), 182[17] (109), 184[171] (125, 126), 279[209] (238, 269), 287[763] (258), 287[764] (258), 387[278] (351), 388[341] (360), 388[343] (360), 388[344] (360), 389[377] (365), 389[387] (368)
Itoh, K., 279[206] (238), 286[658] (252), 286[660] (252), 286[661] (252)
Itoh, T., 279[181] (236, 264), 286[672] (253), 289[859] (264)
Ittel, S. D., 101[136] (42, 47, 58, 59)
Ivleva, I. N., 280[237] (239)
Iwamoto, A., 289[855] (263, 269)
Iwasaki, M., 287[721] (256)
Iyoda, J., 103[270] (73, 74), 184[171] (125, 126)
Izod, K. J., 104[303] (81, 85)
Izumi, T., 283[454] (245, 263)

Jackson, R. D., 521[161]
Jackson, W. R., 386[196] (333), 386[197] (333), 388[300] (355)
Jacob, K., 102[189] (41, 48)
Jacobsen, G. B., 520[103] (440, 466), 520[121] (442), 523[327] (466), 586[260a] (563, 572), 586[260b] (563, 572), 586[260c] (563, 572), 586[260d] (563, 572)
Jacobson, D. B., 103[261] (73), 187[396] (150), 187[402] (150)
Jacobson, K. A., 586[248] (561)
Jacobson, R. A., 387[236] (343)
Jager, W. W., 279[211] (238)
Jahng, Y., 281[342] (242, 273)
Jain, L., 582[4] (534)
Jain, V. K., 290[959] (276), 426[1] (392), 430[206] (422), 519[68] (438), 520[111] (441), 523[308] (464), 528[593] (498), 528[594] (498), 528[595] (498, 499), 530[737] (514), 530[738] (514), 530[739] (514), 582[4] (534)
Jakkal, V. S., 520[111] (441), 528[593] (498)
Jakubik, D., 284[541]
Jama, M. A., 388[292] (354)
James, B. R., 26[117] (16), 286[649], 518[19] (434), 520[135] (443)
James, D. E., 285[588] (249)
James, D. I., 384[37] (293)
Jameson, G. B., 190[586] (175, 177)
Janecki, T., 281[297] (241, 255)
Janiak, C., 187[393] (150)
Janikowski, S. K., 24[11] (2), 100[42] (33), 190[607] (174)
Janousek, Z., 191[633] (180, 181)
Janowicz, A. H., 531[776] (517)
Jans, J., 520[141] (444)
Janssen, M. D., 280[262] (239, 262, 266, 271, 274, 276)
Jaouen, G., 105[434] (96, 98), 185[237] (131, 136), 185[238] (131, 136), 190[569] (172)
Jasffari, G. A., 189[515] (157, 166)
Jastrzebski, J. T. B. H., 280[226] (238, 240, 264, 265), 281[293] (241, 270, 273), 281[353] (243, 253, 255, 271), 390[467] (382)
Jaud, J., 278[105] (233, 255, 256), 284[517] (263)
Jaun, B., 105[406] (91, 92)
Jaw, H.-R. C., 427[30] (395)
Jawad, J. K., 522[243] (458)
Jaynes, B. S., 525[451] (480)
Jeannin, Y., 103[241] (66, 67), 189[490] (157, 158, 163), 222[137a] (213), 283[492] (262, 263), 285[611] (249), 286[692] (255, 258), 288[825] (262, 272), 385[106] (308)
Jeffery, J. C., 27[186] (23), 427[18] (393, 413), 524[385]
Jeffrey, J. G., 427[41] (396, 413)
Jeffreys, J. A. D., 281[297] (241, 255)
Jégat, C., 25[44] (7), 25[46] (7), 104[348] (81, 88), 184[181] (126, 128)
Jemmis, E. D., 186[347] (139, 146), 190[617] (177)
Jennings, M. C., 220[44] (199), 220[48] (200), 428[80] (403), 428[84] (404), 428[85] (404), 428[86] (404), 428[89b] (404), 528[617] (510), 587[302] (572)
Jennings, P. W., 498, 527[562] (495), 527[573] (497), 527[574] (497, 505), 527[575] (497), 527[576] (498), 527[577] (498), 527[578] (498), 527[579] (498, 505), 528[647] (505), 584[128] (549), 584[129] (549, 563), 584[130] (549, 552), 584[131] (549, 552), 586[256] (562)
Jensen, J. A., 388[296] (354)
Jeong, W.-Y., 289[900] (270)
Jernakoff, P., 531[778] (517), 583[21a] (535, 565, 568), 585[204] (554, 555)
Jiang, A., 190[596] (175, 180), 427[19] (393)
Jiang, X. D., 427[46] (398), 427[55] (398)
Jin, H., 286[707] (255, 262), 286[708] (255, 262), 429[141] (408)
Jitsukawa, K., 388[302] (355)
Job, R., 188[440] (152)
Jobe, I. R., 520[123] (442), 523[301] (463)
Jochem, K., 528[614] (500, 504)
Joh, T., 222[139a] (213), 222[139b] (213), 222[140] (213), 277[52] (230, 231), 277[53] (230, 231), 278[80] (231), 278[81] (231, 232), 278[82] (232), 527[531], 527[540] (493), 527[542] (493), 527[544] (494)
Johnson, B. F. G., 277[71] (230), 526[519] (490)
Johnson, D. A., 584[98] (545), 584[99] (545)
Johnson, J. R., 186[335] (139, 143)
Johnson, K. H., 102[208] (62), 521[203] (453)
Johnson, M. P., 529[713] (512)
Johnson, R. E., 26[138] (19)
Johnson, T. H., 528[643] (505)
Johnston, R. F., 289[900] (270)
Jolliet, P., 530[728] (514), 530[732] (514), 530[734] (514)
Jolly, P. W., 24[2] (1), 25[35] (5), 99[1] (29, 36, 41, 56, 94), 100[70] (37, 38), 183[67] (113, 115, 117), 184[180] (126, 128), 185[274] (137), 221[65] (203), 288[786] (260), 288[787] (260, 267), 288[793] (260), 289[907] (271), 385[78] (303), 385[80] (303), 386[191] (331), 387[281] (352, 359), 584[88] (544, 575), 586[245] (560), 587[309a] (573, 576)
Jonas, K., 100[51] (31, 34, 35), 100[54] (34), 100[55] (34), 182[12] (108), 182[16] (109), 183[62], 183[78] (115, 119)
Jonassen, H. B., 277[49] (230)
Jones, C. J., 280[286] (240), 288[838] (263), 289[908] (271), 524[337] (467)
Jones, D., 185[275] (137), 386[175] (324), 587[334] (575, 576)
Jones, D. W., 183[130] (116, 120, 157)
Jones, G. S., 583[58] (541, 545)
Jones, M. D., 389[422] (373), 389[425] (374), 525[444] (479), 587[348] (578)
Jones, N. L., 527[572] (497)
Jones, P. G., 102[212] (64), 222[132] (213), 222[134a] (213), 222[134b] (213), 277[34] (228, 230), 280[224] (238, 262, 263), 285[570] (248), 287[756] (258, 263), 287[757] (258), 287[759] (258), 289[847] (263), 428[113] (406), 524[360] (469), 529[706] (511)
Jones, R. A., 221[51] (201)
Jones, T. C., 283[430] (244)
Jones, W., 527[539] (493)
Jones, W. D., 390[462] (381)
Jones, W. M., 430[198] (419), 525[438] (479), 586[227] (558), 586[228] (558), 586[229] (558), 586[230] (559, 565–7)
Joó, F., 26[106] (14)
Jørgensen, K. A., 189[528] (159, 170)
Jortner, J., 187[398] (150), 187[399] (150)
Joswig, T., 183[67] (113, 115, 117), 586[245] (560)
Jou, D.-C., 286[643] (251)
Jouaiti, A., 284[542] (272)
Joy, L. G., 522[231] (456)
Juanes, O., 279[178] (236, 263, 264), 281[304] (241), 281[338] (242)
Jud, J. M., 221[75b] (204), 221[76] (204), 221[86] (205), 221[87a] (205)
Judd, L., 519[44] (437), 585[175] (554, 555)

Juhlke, T. J., 190[583] (174)
Julia, M., 103[241] (66, 67), 103[242] (67, 72), 189[489] (157, 163), 189[490] (157, 158, 163)
Jun, T. X., 389[379] (366)
Jund, R., 221[78a] (205), 221[78b] (205)
Jungmann, H., 524[334] (467), 527[563] (496), 527[564] (496)
Jutland, A., 101[125] (42, 55, 63), 101[129] (42, 47, 48, 63), 102[207] (56), 278[94] (233), 278[95] (233), 389[389] (368)
Jutzi, P., 184[144] (117, 121), 188[422] (151, 157, 159, 168), 188[477] (157)

Kacholdt, H., 528[611] (500)
Kaesz, H. D., 531[779] (517)
Kafafi, Z. H., 99[6] (30, 31, 74), 100[27] (32), 182[24] (109), 185[257] (130)
Kafuku, K., 279[182] (236)
Kage, Y., 282[358] (243, 248, 263), 282[380]
Kagi, S., 529[666] (507)
Kagotani, M., 282[414] (243, 263, 264), 388[327] (358)
Kaharu, T., 222[122] (211)
Kahr, G., 529[666] (507)
Kai, Y., 25[96] (13), 25[97] (13), 101[116] (42, 49, 55), 102[172] (49, 50, 64), 102[173b] (50, 64), 102[198] (48, 58), 102[199] (48, 58), 278[84] (232), 281[307] (241), 384[62] (299), 384[67] (300)
Kaiser, J., 103[223] (65, 66), 104[356] (81, 88, 90), 104[357] (81, 88, 90), 104[358] (81, 88, 90), 183[90] (115), 184[185] (126, 128), 187[361] (141, 142)
Kaitner, B., 582[14c] (535, 539), 582[17a] (535, 539)
Kaji, A., 389[379] (366)
Kajita, S., 284[510]
Kakkar, A. K., 188[462] (155, 156)
Kakos, G. A., 386[197] (333)
Kalinin, V. N., 102[164] (41, 42, 47, 51, 58), 279[156] (235), 281[345] (242)
Kalinkoski, H. T., 285[636] (250)
Kalyukova, E. N., 430[195] (419)
Kamalov, G. L., 384[50] (295)
Kambe, N., 388[309] (357), 389[397] (369)
Kamigata, N., 290[933] (273)
Kamitori, S., 184[199] (125), 184[200] (125)
Kamo, H., 586[275] (565–7)
Kamogawa, Y., 287[773] (259)
Kampmann, D., 186[284] (138, 145)
Kanai, H., 100[40] (31, 33), 100[41] (31, 33), 190[608] (174), 190[609] (174)
Kanatani, H., 384[67] (300)
Kanatomi, H., 281[311] (241)
Kanda, H., 279[180] (236)
Kaneda, K., 388[302] (355)
Kanehira, K., 388[337] (360), 388[342] (360)
Kane-Maguire, L. A. P., 285[593] (249), 385[123] (311), 584[110] (548), 584[111] (548), 586[220] (557)
Kang, K., 278[79] (231)
Kannan, S., 530[738] (514), 530[739] (514)
Kanters, J. A., 277[57] (230, 232, 234, 239, 269, 270, 273)
Kao, L. T. C., 277[22] (228), 527[558] (495)
Kapoor, P. N., 278[143] (235), 529[670] (508)
Kargol, J. A., 430[192] (418)
Karsch, H. H., 26[127] (16)
Kasahara, A., 283[454] (245, 263)
Kasai, N., 25[96] (13), 25[97] (13), 26[98] (13), 101[116] (42, 49, 55), 102[172] (49, 50, 64), 102[173a] (49, 50, 64), 102[173b] (50, 64), 102[196] (48, 64), 102[198] (48, 58), 102[199] (48, 58), 278[84] (232), 281[307] (241), 282[415], 285[627] (250, 260, 262, 267, 269), 288[802] (260), 384[62] (299), 384[67] (300), 386[208] (335), 389[366] (363), 583[67a] (542, 545, 574), 583[68] (542, 545, 547, 574), 586[254] (562), 586[259] (563)
Kaschube, W., 100[52] (31, 34, 35), 102[183] (41, 47, 48, 56, 63), 103[274] (73, 74), 182[40] (110), 182[44] (110, 113), 183[70] (115), 183[82] (115, 119)
Kashiwahara, K., 584[104] (547)
Kaska, W. C., 277[13] (227, 228)
Kataoka, S., 277[69] (230, 231), 278[77] (231)
Katser, S. B., 222[93a] (207), 222[93b] (207), 222[93c] (207)
Kavaliunas, A. V., 99[3] (30, 31, 47, 48), 278[90] (233, 262)
Kawada, M., 387[257] (347)
Kawaguchi, S., 279[151] (235, 255, 262–4, 269), 279[152] (235, 255, 262), 279[180] (236), 279[182] (236), 279[183] (236, 255, 256, 264), 281[307] (241), 281[308] (241), 288[795] (260), 288[796] (260), 288[797] (260), 288[798] (260), 288[800] (260, 261), 288[803] (260), 288[842] (263)
Kawahara, T., 105[428] (97)
Kawakami, M., 287[776] (259)
Kawanami, Y., 281[311] (241)
Kawasaki, N., 278[97] (233, 255, 262)
Kawasaki, Y., 25[95] (13), 101[149] (41, 42, 48, 57), 102[171] (41, 49, 50, 62), 103[221] (65), 103[222] (65), 186[294] (138), 289[877] (266), 289[899] (270, 273), 389[388] (368), 587[315] (573), 587[316] (573, 576), 587[329a] (575)
Kawatani, M., 282[407] (248, 263), 286[650], 286[651]
Kawato, T., 281[311] (241)
Kawazura, H., 384[63] (299)
Kayumov, F. F., 183[119] (115, 117, 119)
Kazankov, G. M., 281[343] (242), 283[451] (245)
Kazanskii, V. B., 384[50] (295)
Keil, R. N., 279[165] (236, 269, 273)
Keil, T., 103[237a] (66, 67, 71), 182[50] (111, 114, 116, 117, 157)
Keim, W., 99[2] (30, 36, 52), 100[60] (36, 52), 101[97] (42, 47, 52), 101[133] (42, 62, 63), 101[141] (42, 47, 58–60), 184[137] (116, 124, 140), 186[319] (138, 145), 186[320] (138, 145), 279[162] (236), 288[788] (260), 387[214] (336), 387[283] (352), 388[333] (359)
Keinan, E., 386[179] (326), 389[358] (362), 389[376] (365), 389[393] (368)
Kekre, M. G., 286[670] (253)
Kelley, D. G., 105[394] (91, 92), 105[395] (91, 92)
Kelley, E. A., 385[139] (315)
Kelly, J. D., 277[49] (230)
Kelly, K. P., 285[612] (249), 384[49] (295)
Kelly, R. D., 520[81] (439), 520[105] (440), 584[146a] (552, 555), 584[146b] (552)
Kelm, H., 522[253] (459), 522[254] (459)
Kemmitt, R. D. W., 103[266] (73, 74), 103[267] (73, 74), 222[131] (212), 277[54] (230), 287[777] (260), 389[422] (373), 389[425] (374), 429[184] (416, 417), 429[185] (416), 479, 525[441] (479, 481), 525[442] (479, 481), 525[443] (479), 525[444] (479), 525[445] (480), 525[446] (480), 525[447] (480), 525[450] (480), 525[457] (480), 526[460], 526[468] (482), 526[469] (482), 529[667] (507), 583[52] (540), 587[325] (574, 578), 587[348] (578), 588[350] (578), 588[351] (579), 588[353] (579)
Kemmitt, T., 527[585] (498, 504)
Kemp, T. J., 526[514] (490)
Kempe, R., 25[42] (5), 105[365] (81, 90), 105[367] (81, 88, 90), 105[368] (81, 88, 90), 186[298] (138, 142), 186[299] (138, 142), 186[309] (138, 142), 186[313] (138, 142)
Kennard, C. H. L., 283[429] (244), 283[478] (245, 263), 585[167] (553–5)
Kennard, O., 521[154] (445)
Kergoat, R., 105[410] (93, 94)
Kerkeni, A., 102[166] (50, 62)
Kermode, N. J., 277[23] (228), 527[559] (495)
Kervennal, J., 221[80a] (205)
Kesmodel, L. L., 386[144] (316), 386[145] (316)
Ketley, A. D., 290[934] (273)
Khac, T. N., 189[505] (157, 159, 161)
Khan, A. R., 527[582] (498, 504), 528[605] (499), 528[606] (499), 528[609] (499, 500), 528[631] (504)
Khan, M. A., 223[152] (215), 284[529], 285[605] (249), 385[87] (305)
Khan, M. S., 277[71] (230), 277[72] (230, 231), 526[519] (490), 527[525] (491)
Khan, S. I., 222[107] (209, 215), 531[779] (517)
Khandelwal, B. L., 519[39] (436)
Kharas, K. C. C., 427[43] (397)
Kharitonov, Yu. Ya., 428[112] (406)
Khivintseva, G. A., 221[67] (204)
Khodashova, T. S., 221[61b] (202)
Khouzami, F., 186[318] (138, 144, 155, 166), 188[461] (155–7, 159, 166, 167)
Khramov, A. S., 220[32] (197)
Khun, A., 587[309b] (573, 576)
Khusnutdinov, R. I., 102[169] (50)
Kickham, J. E., 284[544], 529[712] (512)

Kiefer, G. E., 281[318] (241), 281[319] (241), 281[320] (241, 242), 385[91] (305), 529[691] (510)
Kiji, Y., 279[155] (235)
Kikukawa, K., 279[196] (237), 287[771] (259)
Kildahl, N. K., 278[106] (233, 255)
Kilner, M., 282[370] (263, 264), 529[697] (510), 588[363] (580)
Kim, S., 387[236] (343)
Kim, Y.-G., 388[299] (355)
Kim, Y.-J., 25[25] (4), 101[88] (42, 53, 54), 101[89] (42, 47, 53, 58, 62, 63), 102[211] (63), 280[233] (238, 262, 263, 265, 268), 287[722] (256, 265, 268, 271), 521[157], 523[281] (461)
Kimber, S. J., 582[14a] (535, 558)
Kinder, J. D., 185[230] (131, 135)
King, G. D., 528[614] (500, 504), 528[634] (504)
King, R. B., 26[123] (16), 183[91] (115, 119), 187[389] (148), 187[390] (148), 189[519] (159, 167), 190[610] (175), 190[611] (175), 190[612] (175), 190[613] (175), 427[48] (398)
Kinoshita, K., 279[180] (236)
Kinsella, M. A., 390[432] (375)
Kinugawa, N., 388[324] (358)
Kirchner, R. M., 520[101] (440)
Kirmse, R., 102[214] (65), 102[215] (65), 183[89] (115, 123, 174)
Kirschbaum, K., 386[161] (319)
Kishi, K., 388[344] (360)
Kisin, A. V., 221[60c] (202)
Kissin, Y. V., 101[135] (42, 57)
Kita, M., 285[575] (248)
Kitaigorodskii, A. N., 531[786] (518), 531[787] (518), 531[790] (518)
Kitaike, Y., 25[40] (5)
Kitano, Y., 223[150] (215)
Kitaura, K., 182[20] (109), 287[719] (255), 429[131] (408), 521[205] (453), 522[265] (460)
Kitching, W., 282[403] (263), 283[448] (245)
Kite, K., 527[581] (498), 527[582] (498, 504), 528[590] (498), 528[597] (499), 528[598] (499), 528[599] (499), 528[605] (499), 528[606] (499), 528[609] (499, 500), 528[631] (504), 528[632] (504), 528[633] (504)
Kjonaas, R. A., 285[604] (249)
Klabunde, K. J., 24[10] (2), 24[11] (2), 100[35] (33), 100[36] (33, 41), 100[37] (33), 100[38a] (31, 33), 100[38b] (31, 33), 100[38c] (31, 33), 100[38d] (31, 33), 100[38e] (31, 33), 100[39] (31, 33), 100[40] (31, 33), 100[41] (31, 33), 100[42] (33), 100[44] (31, 33), 190[579] (174, 175), 190[580] (174, 175), 190[581] (174), 190[582] (174, 175), 190[605] (174), 190[606] (174), 190[607] (174), 190[608] (174), 190[609] (174), 190[610] (175), 190[611] (175), 190[612] (175), 190[613] (175), 278[86] (233), 278[87] (233), 278[89] (233), 386[156] (318), 429[160] (410), 523[333] (467)
Klabunde, U., 101[136] (42, 47, 58, 59)
Klaiss, J., 27[169] (21), 190[570] (172, 173)
Klar, G., 528[611] (500)
Klaüi, W., 188[448] (154, 155, 157), 188[471] (154, 155), 188[472] (155), 188[473], 189[523] (159, 163), 189[524] (159, 163), 189[537] (159, 161)
Klaus, A. J., 282[369], 282[376], 282[377] (244)
Kleijn, H., 190[574] (172, 173), 278[114] (234)
Kleimann, W., 183[63] (111, 115)
Klein, H.-F., 26[134] (19), 100[72] (37, 38, 42, 51, 58, 59), 100[74] (37, 38), 100[75] (37–9, 42, 50, 56), 100[76] (37–9, 42, 56), 100[77] (37–9, 42, 56), 101[122] (41, 42, 47–9, 51, 58, 60), 102[166] (50, 62), 105[408] (42, 93, 94), 221[63] (203), 277[56] (230, 231)
Klein, H.-P., 519[55] (437), 519[57] (437)
Klein, J., 103[260] (70), 184[154] (117, 143), 186[306] (138, 143)
Klein, R. A., 278[113] (234), 287[761] (258)
Klein, S. I., 587[291] (570)
Klein, W., 186[315] (138, 139, 148)
Klevtsova, R. F., 427[22] (393)
Kliegman, J. M., 285[603] (249)
Kline, E. S., 99[6] (30, 31, 74), 182[24] (109), 185[257] (130)
Klingler, R. J., 526[458], 528[648] (505)
Klotzbucher, R., 521[176], 522[212] (454)
Kluwe, C., 386[161] (319)
Knaap, C. T., 529[705] (511)
Knierzinger, A., 186[297] (138)
Knight, D. A., 24[16] (3), 101[104] (42, 47, 51, 57, 58, 66, 67), 188[482] (159)
Knobler, C. B., 191[624] (180, 182), 385[135] (314)
Knoch, F., 24[4] (2), 105[405] (92), 184[175] (126, 127), 184[205] (126), 184[206] (126, 145), 185[207] (126, 138, 145, 157), 186[319] (138, 145), 186[320] (138, 145), 223[177] (218), 430[202] (421)
Knorr, M., 387[275] (350), 586[257] (563, 564)
Knox, G. R., 103[250] (67, 70), 189[515] (157, 166)
Kobayashi, A., 277[40] (230)
Kobayashi, K., 182[17] (109), 222[143] (214), 223[155] (215), 287[748] (257), 390[445] (378)
Kobayashi, Y., 388[328] (358)
Kochi, J. K., 526[458], 528[648] (505)
Kochubei, D. I., 223[179a] (219), 223[179b] (219), 223[179c] (219)
Kocovský, P., 387[247] (345), 387[248] (345)
Kodama, G., 26[103] (14)
Koe, J. R., 385[89] (305), 520[112] (441), 585[154a] (553), 585[154b] (553), 585[156] (553)
Köelle, U., 183[95] (115, 124, 153), 188[461] (155–7, 159, 166, 167)
Koeppe, R., 220[3] (194)
Koestner, R. J., 386[143] (316)
Koetzle, T. F., 26[165] (20), 285[612] (249), 384[47] (294), 384[49] (295)
Köffer, D. P. J., 188[467] (155, 156, 175)
Kofod, P., 584[102] (547)
Koga, N., 222[100a] (208), 222[100b] (208), 222[101] (208), 222[103] (208), 287[715] (255), 385[126] (312), 429[132] (408), 522[265] (460)
Koh, J. J., 529[687] (509, 510)
Kohara, T., 24[9] (2), 24[21] (4), 100[68] (37, 38, 42, 53), 101[87] (42, 58), 101[92] (42, 53, 54, 62), 102[184] (48, 55, 61, 62)
Köhler, F. H., 100[64] (31, 36), 187[415] (151, 155), 188[424] (151, 155), 188[425] (151, 154–6), 188[426] (151, 154, 155), 188[427] (151, 155), 188[428] (151), 188[429] (152, 155, 156), 188[460] (154, 155), 190[592] (175, 177)
Kohli, D. K., 281[315] (241)
Koide, H., 285[631] (250), 286[705] (255, 256)
Koie, Y., 427[60] (399), 427[61] (399), 586[280] (568, 569), 587[294] (571)
Koike, Y., 386[200] (333)
Kojik-Prodic, B., 184[141] (117, 120, 159)
Kokker, M., 583[74a] (542)
Kolesov, V. S., 101[126] (42)
Kölle, U., 186[315] (138, 139, 148), 186[318] (138, 144, 155, 166), 186[333] (139, 148, 155, 159)
Kolomnikov, I. S., 428[112] (406)
Komatsu, K., 586[275] (565–7)
Komiya, S., 24[9] (2), 24[20] (4), 24[21] (4), 101[90] (42, 58), 101[91] (42, 53, 58, 62, 63), 102[179] (56, 62), 102[204] (49, 62), 102[210] (62), 104[305] (81, 85), 105[382] (85), 280[247] (239, 255, 266), 286[701] (255, 267), 286[703] (255, 266, 268), 518[16] (434, 455), 521[148] (444), 522[217] (455), 585[176] (554, 555), 586[258] (563), 586[259] (563)
Konda, H., 279[161] (236, 262, 263)
Kondo, T., 522[272] (460)
Kondo, Y., 277[36] (229)
Kondoh, A., 105[428] (97), 105[429] (96, 97)
Kong, K.-C., 288[807] (261, 273)
Kongkathip, B., 389[415] (372)
König, H., 26[134] (19), 105[408] (42, 93, 94)
Konishi, H., 279[155] (235)
Konishi, M., 386[209] (335), 387[244] (345), 389[362] (363), 389[386] (367)
Konrad, R., 427[17] (393, 412)
Kooijman, H., 278[103] (233, 251, 262, 273)
Koola, J. D., 519[36] (436), 519[48] (437, 453), 521[170]
Korff, J., 25[55] (9)
Kosemund, D., 104[354] (81, 90)
Köster, R., 187[381] (148, 149), 187[382] (148, 149), 187[383] (148, 149), 190[596] (175, 180)
Kostić, N. M., 186[280] (137), 220[20] (196), 390[442] (377)
Kostitsyna, N. N., 100[83] (41, 47)
Kotani, S., 102[202] (41, 48), 526[479] (484)

Kottnitz, J., 187[363] (142)
Koumoto, R., 587[317] (573)
Koutecky, J., 220[9] (195), 220[12b] (195), 222[104] (208)
Kovredov, A. I., 277[74] (230)
Kowaldt, F. H., 100[60] (36, 52)
Kowalski, M. H., 428[95a] (403), 526[471] (484), 526[472] (484, 487, 491, 493), 526[473] (484), 526[474] (484), 526[502] (487), 587[283a] (568)
Koyama, H., 281[311] (241)
Koyama, Y., 102[170] (41, 47, 49, 50, 62), 223[176a] (218), 223[176b] (218), 277[63] (230, 232), 430[208] (422), 430[209] (422)
Koyasu, Y., 287[721] (256)
Kozhevnikov, I. V., 390[461] (381)
Kozitsyna, N. Y., 279[194] (237)
Krafft, T. E., 220[40] (198)
Krakenberger, B., 389[368] (363), 389[372] (365)
Kramer, A. V., 278[115] (234)
Kramer, B., 187[369] (145, 159)
Kramer, P. A., 389[421] (373)
Kramer, R., 287[778] (260)
Kramolowsky, R., 279[191] (237)
Kraneburg, P., 388[333] (359)
Kraus, H.-J., 220[27b] (197, 210), 221[81] (205), 221[82] (205), 280[217] (238, 267, 270), 282[363] (264), 286[652] (252), 288[786] (260), 288[799] (260), 385[78] (303), 385[79] (303), 388[286] (352), 390[441] (377), 390[443] (377), 390[444] (377), 390[450] (378)
Krause, J., 288[790] (260, 266), 385[82] (303, 318)
Krause, L. J., 101[109] (41, 42, 47, 48), 279[210] (238), 524[336] (467)
Krause, M. E., 183[74] (115)
Kravtsov, D. N., 101[126] (42), 102[164] (41, 42, 47, 51, 58), 184[158] (119), 186[304] (138, 139, 143), 519[40] (436)
Krebs, B., 185[226] (131, 136)
Kretchmar, S. A., 24[23] (4), 523[279] (461)
Kriege-Simondsen, J., 25[61] (9)
Krishnan, B., 385[137] (315)
Kristof, E., 222[111] (209)
Kristoff, J. S., 27[183] (22)
Kronseder, C., 25[31] (5), 188[476] (157, 167)
Kroto, H. W., 586[239b] (560), 587[290] (570), 587[291] (570)
Krüger, C., 25[35] (5), 25[91] (12), 26[99] (14), 26[124] (16), 100[52] (31, 34, 35), 100[55] (34), 100[60] (36, 52), 101[97] (42, 47, 52), 102[165a] (42, 59), 103[252] (66, 67, 70), 103[274] (73, 74), 103[275] (73, 74, 80), 103[277] (73, 74, 80), 103[284] (73, 74), 103[291] (73, 74, 78), 104[298] (73, 74), 104[299] (73, 74), 104[302] (81), 104[344] (81, 88, 90), 104[349] (81, 88, 89), 105[413] (93, 94), 182[33] (110, 111, 117, 130–3), 182[34] (110, 111, 117, 130–2), 182[37] (110, 112), 182[38] (110, 111), 182[40] (110), 182[41] (110, 112), 182[42] (110–12), 182[43] (110, 112, 126), 182[44] (110, 113), 182[47] (110, 113, 117, 118), 182[48] (111, 128), 182[49] (111, 128), 183[63] (111, 115), 183[70] (115), 183[82] (115, 119), 183[85] (115), 183[117] (115), 184[143] (117, 139, 141), 184[145] (117, 118), 184[150] (117, 118), 184[151] (117, 118), 184[156] (119, 122, 131), 184[159] (121), 184[162] (122), 184[180] (126, 128), 185[210] (130, 132), 185[214] (130, 133), 185[223] (130, 131, 134), 186[279] (137, 138), 186[300] (138, 143), 186[316] (138, 144, 172), 186[317] (138, 177), 187[380] (148), 188[468] (155), 189[548] (171), 189[553] (171, 172), 190[596] (175, 180), 221[65] (203), 221[81] (205), 284[541], 285[572] (248), 285[625] (250), 288[786] (260), 288[788] (260), 385[78] (303), 387[214] (336), 387[281] (352, 359), 387[283] (352), 389[420] (373), 390[444] (377)
Kruschova, N. S., 283[457] (245, 250)
Krylov, A. V., 386[177] (326)
Krysan, D. J., 184[157] (119)
Kubiak, C. P., 25[75] (10, 16, 17), 25[84] (11, 23), 26[118] (16), 26[119] (16, 18), 26[121] (16), 26[128] (16, 17), 26[130] (17), 26[131] (17), 26[132] (18), 27[178] (21, 22), 27[179] (21, 22), 27[180] (21), 104[341] (81, 87), 220[39a] (198), 220[39b] (198), 222[118] (210), 222[119a] (210), 222[119b] (210), 223[145b] (214)
Kubota, M., 279[165] (236, 269, 273), 428[91] (403), 527[529] (491, 493), 587[282] (568)
Kucharczyk, D., 528[624]
Kudinov, A. R., 190[616] (177)
Kudo, K., 287[720] (256)
Kuhlmann, E. J., 522[263] (460)
Kuhlmann, T., 183[98] (115, 155, 157, 181), 187[391] (149, 175, 177), 190[588] (175, 177), 190[600] (175, 178), 190[621] (178), 191[639] (181), 191[640] (181)
Kühn, A., 288[783] (260), 288[792] (260), 387[223] (341), 387[282] (352), 388[285] (352), 587[311] (573, 576)
Kuhn, H., 183[118] (115, 156, 157)
Kuhn, M., 189[510] (157, 161)
Kuhn, N., 103[256] (67, 73), 183[129] (116, 121, 157, 161), 188[453] (154), 188[454] (154), 188[475] (157, 161), 188[480] (157, 160), 188[485] (157, 160), 188[487] (157, 161), 189[501] (157, 161), 189[502] (157, 160, 161), 189[503] (157, 160), 189[504] (157, 159, 161), 189[506] (157, 161), 189[508] (157, 161), 189[509] (157, 159, 160), 189[511] (157, 161), 189[539] (160), 189[540] (160), 190[599] (175, 177)
Kukharenko, S. V., 188[442] (153), 188[444] (153, 155)
Kukina, G. A., 388[293] (354)
Kukolich, S. G., 187[412] (151)
Kukushkin, V. Yu., 428[94] (403), 583[60] (541, 545)
Kukushkin, Yu. N., 428[108a] (403), 430[195] (419)
Kullberg, M. L., 220[39a] (198), 220[39b] (198), 222[119b] (210), 223[145b] (214)
Kumada, M., 386[209] (335), 387[244] (345), 388[337] (360), 388[342] (360), 389[362] (363), 389[386] (367)
Kumar, A., 583[35] (536, 538, 550)
Kumar, M., 105[402] (92)
Kumar, R., 430[197] (419)
Kumazoe, M., 101[114] (42, 49, 55, 61), 101[115] (42, 49, 55, 61), 101[116] (42, 49, 55), 102[195] (41, 48, 49, 56, 57), 102[200] (48, 56)
Kumberger, O., 183[74] (115)
Kundu, K., 519[34] (436), 519[35] (436), 519[39] (436), 519[42] (437)
Kunkely, H., 387[238] (344), 528[626] (500)
Kunz, R. W., 386[186] (328), 386[188] (328)
Kunze, U., 519[36] (436), 521[170]
Kuo, L., 527[529] (491, 493), 587[282] (568)
Kuo, Y.-M., 102[185] (48, 55, 63)
Kurasov, S. S., 220[30] (197), 427[25] (394)
Kurihara, K., 280[249] (239, 262), 280[250] (239), 289[876] (266)
Kurokawa, T., 288[802] (260)
Kurosaki, A., 390[436] (376)
Kurosawa, H., 103[221] (65), 103[222] (65), 186[290] (138, 143), 186[291] (138, 143), 186[294] (138), 279[215] (238, 267), 279[216] (238), 285[587] (249, 267), 285[627] (250, 260, 262, 267, 269), 289[877] (266), 289[899] (270, 273), 384[61] (296, 297), 384[67] (300), 385[97] (306), 386[213] (336), 387[249] (346), 389[366] (363), 389[367] (363), 389[388] (368), 583[62] (541, 542, 547, 574), 583[66] (542, 547, 574), 583[67a] (542, 545, 574), 583[68] (542, 545, 547, 574), 584[87] (544), 584[101] (545, 547), 584[108] (547), 587[315] (573), 587[316] (573, 576), 587[326] (574, 579), 587[327] (574), 587[329a] (575), 587[329b] (575)
Kurusu, Y., 388[321] (358), 388[322] (358), 388[324] (358)
Kushi, Y., 104[305] (81, 85), 184[186] (126, 128)
Kutal, C., 281[357] (243)
Kuz'mina, L. G., 221[55] (201), 283[435] (244), 283[446] (245), 387[215] (337)
Kyriakidis, C. E., 287[713] (255), 526[499] (487)

Laakmann, K., 103[291] (73, 74, 78)
Lachmann, J., 531[780] (517)
Läge, M., 185[226] (131, 136)
Lagow, R. J., 102[190] (48, 56), 190[583] (174), 278[88] (233)
Lagowski, J. J., 99[5] (30)
Lahiri, G. K., 105[399] (92)
Lahoz, F. J., 387[271] (350)
Lai, R., 25[81] (11)
Lai, T.-F., 587[297] (571)
Lai, T.-W., 289[885] (268)
Laibinis, P. E., 586[226b] (558)

Lakkis, D., 285[613] (249, 252)
Lalinde, E., 222[124] (211, 213), 222[132] (213), 222[133] (213), 222[134a] (213), 222[134b] (213), 222[144] (214), 223[167] (217), 277[51] (230), 284[560] (246, 263, 268), 287[749] (257), 287[756] (258, 263), 287[757] (258), 287[758] (258, 259), 287[759] (258), 287[760] (258, 263), 429[181] (416, 417, 423), 524[359] (469), 524[363] (469), 525[398] (474, 490), 587[299] (571), 587[306] (572)
Lallemand, J.-Y., 528[629] (500)
Lamberov, A. A., 220[31a] (197)
Lamberova, M. E., 220[32] (197)
Lambert, R. M., 384[30] (293), 384[31] (293)
Lancaster, J. R., 105[385] (91)
Landis, C. R., 280[258] (239, 268)
Lane, G. A., 183[94] (115, 124, 153), 285[595] (249, 272), 385[101] (307), 385[102] (307), 390[446] (378), 390[447] (378)
Lanfranchi, M., 105[431] (96, 98), 185[246] (131, 172, 174), 185[248] (131, 172, 174), 189[544] (172, 174), 190[564] (172), 222[92] (206), 287[739] (257), 523[282] (461)
Lanfredi, A. M. M., 185[245] (131, 172, 174), 282[419] (243), 584[115] (548)
Lanfredi, M., 288[818] (262, 263)
Lang, H., 185[234] (131), 185[235] (131)
Langer, G., 2, 24[1] (1, 2)
Langer, V., 389[390] (368)
Langrick, C. R., 102[192] (41, 47, 48), 277[47] (230), 429[189] (417), 518[8] (432, 494), 523[303] (463), 526[513] (490), 527[549] (494), 527[551] (494)
Lanza, S., 520[84] (439), 520[89] (439), 521[167] (458, 459), 521[169], 522[246] (458, 459), 522[247] (458, 459), 522[255] (459), 523[306] (464)
Lappert, M. F., 25[93] (13), 277[23] (228), 525[428] (478), 527[559] (495), 585[162] (553), 585[163] (553, 565), 585[186] (554, 555)
Lappin, A. G., 105[386] (91)
Larock, R. C., 286[637] (251), 386[203] (334), 386[204] (334), 386[206] (335), 386[207] (335), 387[217] (338), 387[218] (338), 387[219] (339), 387[221] (340), 387[231] (342), 387[232] (342), 387[236] (343), 388[329] (358)
Larsen, R. D., 527[562] (495), 527[579] (498, 505), 584[131] (549, 552), 586[256] (562)
Larson, S. B., 190[583] (174)
Lashanizadehgan, M., 523[295] (463)
Latos-Grazyński, L., 105[400] (92)
Latronico, M., 24[18] (4), 184[183] (126, 128)
Lau, J., 280[235] (239, 266), 280[236] (239, 266), 289[878] (266)
Lau, K. S. Y., 278[92] (233, 255, 256), 278[109] (234, 255), 278[116] (234, 273)
Laube, G., 519[59] (437)
Lauron, H., 103[241] (66, 67), 103[242] (67, 72), 189[489] (157, 163), 189[490] (157, 158, 163)
Lautens, M., 388[303] (356)
Laven, R., 188[467] (155, 156, 175)
Lavigne, G., 26[166] (20, 21), 190[577] (174)
Lawrie, C. J., 99[4] (30, 31, 41, 42), 101[120] (42, 50, 55, 93, 94), 183[107] (115, 122), 184[152] (117, 118)
Lazzaroni, R., 584[93a] (544), 584[93b] (544), 586[232] (559)
Le Borgne, G., 281[293] (241, 270, 273), 283[437] (244, 253, 254), 285[619] (250), 286[675] (253), 286[676] (253, 264), 286[677] (253, 262, 263), 286[678] (253, 254, 262, 263), 286[679] (253), 286[686] (254, 271), 386[165] (320), 386[166] (320)
Le Van, D., 185[226] (131, 136)
Leaman, S. A., 280[281] (240)
Leban, I., 187[358] (137)
Leber, E. M., 582[15] (535)
Lebioda, L., 279[214] (238, 240, 250, 261, 265), 280[251] (239)
Lebrilla, C. B., 187[401] (150), 220[6] (194)
Lechat, J. R., 281[296] (241)
Leconte, P., 287[742] (257)
Lee, C.-L., 26[117] (16), 222[135] (213), 223[151] (215), 278[119] (234, 236), 286[646], 286[649], 286[702] (255, 256, 258, 269, 271)
Lee, G.-H., 277[22] (228), 277[32] (228–30), 519[27] (434), 523[289] (462), 527[558] (495), 587[322] (574, 577)
Lee, K. S., 279[207] (238, 243)
Lee, T. R., 531[783] (518), 586[226a] (558), 586[226b] (558)
Lee, W.-H., 529[687] (509, 510)
Leese, T. A., 280[229] (238), 284[521], 284[522], 529[692] (510)
Lehmann, R., 390[440] (377)
Lehmkuhl, H., 103[224] (66, 67), 103[225] (67, 71), 103[226] (66, 67, 71), 103[227] (66, 67), 103[228] (67, 72), 103[229] (66, 67), 103[231] (67, 71), 103[232] (66, 67), 103[233] (66, 67), 103[234] (66, 67, 72), 103[236] (66, 67), 103[237a] (66, 67, 71), 103[237b] (66, 67, 71), 103[251] (66, 67, 70), 103[252] (66, 67, 70), 103[258] (70), 103[259] (70), 105[413] (93, 94), 182[50] (111, 114, 116, 117, 157), 182[51] (111, 113, 155, 157), 182[52] (111, 113, 155, 157), 182[53] (111, 113, 115, 157), 182[57] (111, 113, 115, 119, 138, 157), 182[58] (111, 114–17, 138, 140, 157), 182[61] (111, 115, 174, 175), 183[68] (113), 183[69] (113, 157), 183[72] (114, 115, 157, 171), 183[73] (114, 115, 117, 157), 183[120] (114, 115), 184[135] (116, 117, 157), 184[136] (116, 117, 157), 184[138] (116, 117, 157), 184[143] (117, 139, 141), 184[147] (114, 117, 138), 185[216] (130–2, 138, 157), 186[305] (138, 140, 157), 186[307] (138, 143), 186[322] (139), 189[548] (171)
Lehner, H., 525[394] (474), 525[395] (474)
Lehniger, P., 281[298] (241), 530[746] (514)
Leidl, E., 526[477] (484)
Lemenovskii, D. A., 188[457] (154)
Lemke, F. R., 220[39a] (198)
Lemoine, P., 221[78a] (205), 221[78b] (205), 430[193] (418)
Lenarda, M., 279[186] (237), 285[610] (249), 525[449] (480), 526[470] (483), 583[37] (536, 539), 583[38] (536), 583[39] (536, 538), 583[47] (538)
Lenhert, P. G., 526[486] (485, 494), 526[495]
Lentz, D., 25[72] (10), 189[529] (159, 168)
Leonard, S., 428[93] (403, 411), 528[638] (505)
Leoni, E., 101[123] (42)
Leoni, P., 390[469] (383)
L'Eplatteria, F. A., 284[530]
Lerche, J., 585[195] (554, 555)
Lerke, S. A., 184[169] (125), 586[252] (561)
Leschnitzer, D. H., 529[716]
Leuck, D. J., 386[206] (335)
Leung, P.-H., 281[299] (241), 283[443] (244, 263), 283[444] (244, 264)
Leutwyler, S., 187[398] (150), 187[399] (150)
Levason, W., 527[585] (498, 504)
Lever, A. B. P., 220[2b] (194)
Lewandos, G. S., 383[10] (292)
Lewis, G. J., 427[46] (398), 427[53] (398), 427[54] (398), 427[55] (398)
Lewis, J., 277[71] (230), 526[519] (490)
Ley, S. V., 283[448] (245)
Li, C.-S., 286[642] (251, 271), 286[643] (251), 390[466] (382)
Li, M. K., 188[470] (154)
Li, P., 189[557] (172, 173)
Li, S., 187[414] (151)
Li, X., 186[277] (137), 386[172] (324), 526[485] (485)
Li, X.-R., 587[333] (575)
Li, Y., 102[157] (42, 56)
Li, Z.-Y., 586[265] (564)
Lianza, F., 529[702] (511)
Liao, F.-L., 286[642] (251, 271), 286[643] (251), 390[466] (382)
Lichtenhan, J. D., 583[35] (536, 538, 550)
Lickiss, P. D., 586[242] (560)
Licoccia, S., 101[123] (42), 282[360] (243), 282[419] (243), 526[522] (490, 491, 493), 526[523] (490, 491), 527[526] (491), 527[545] (494)
Liddell, M. J., 103[246] (67, 72), 189[492] (157, 161), 282[422] (243), 586[263a] (564, 572), 586[263b] (564, 572)
Liebeskind, L. S., 289[897] (270), 290[926] (273)
Liebman, J. F., 390[462] (381)
Lii, J.-C., 586[264b] (564, 572), 586[265] (564)
Likholobov, V. A., 220[36] (197), 388[306] (356)
Lilga, M. A., 26[117] (16)
Limouzin-Marie, Y., 182[23] (109)
Lin, I. J. B., 277[22] (228), 277[31] (228, 230), 277[32] (228–30), 277[38] (229, 230), 277[39] (229, 230), 527[558] (495)

Lin, M., 520[113] (441), 521[149] (444), 521[165]
Lin, M. R., 428[111] (406, 417), 588[365a] (580), 588[365b] (580)
Lin, S., 288[785] (260)
Lin, S.-K., 105[406] (91, 92)
Lin, S.-T., 100[37] (33), 100[38a] (31, 33), 100[38b] (31, 33), 190[580] (174, 175), 190[582] (174, 175), 190[610] (175), 190[611] (175), 190[612] (175), 429[160] (410), 523[333] (467)
Linde, R. G., II, 389[381] (366)
Lindner, E., 103[243] (66, 67), 103[255] (67, 70), 189[497] (157, 159, 164), 189[516] (157, 164, 177), 189[521] (159, 163), 189[522] (159, 163), 529[685] (509)
Lindqvist, O., 104[358] (81, 88, 90), 105[372] (81, 88, 89), 186[325] (139, 140), 387[269] (350)
Lindsay, P. N., 27[168] (21)
Lindsell, W. E., 26[107] (15)
Lineham, J., 428[69] (400)
Linehan, J., 520[102] (440)
Linev, V. P., 220[31b] (197)
Ling, S. S. M., 428[90] (404, 415), 518[13] (433), 520[120] (442), 521[164], 523[295] (463), 523[300] (463), 523[301] (463), 523[302] (463), 528[641] (505), 528[642] (505)
Linn, K., 388[354] (361), 389[356] (361), 587[342] (576, 577)
Lippard, S. J., 285[622] (250), 385[115] (310)
Lisensky, G. C., 385[103] (308)
Lisini, A., 583[38] (536), 583[47] (538)
List, A. K., 25[60] (9), 183[126] (115, 122, 125)
Litster, S. A., 102[168] (47, 50, 54), 182[60] (111, 113), 279[198] (237), 587[344] (577)
Liu, A. T., 223[170] (217, 218), 430[222] (425)
Liu, C. H., 277[39] (229, 230)
Liu, C. W., 277[31] (228, 230)
Liu, L.-K., 277[31] (228, 230)
Liu, S., 386[178] (326)
Liu, Z., 386[172] (324), 587[333] (575)
Lloyd, B. R., 26[115] (16), 220[42a] (198, 199), 220[42b] (198, 199), 220[47] (200), 428[75] (401, 402), 428[77] (402, 404), 428[79] (402), 428[87] (404), 520[129] (442, 495)
Lloyd, D. R., 390[460] (381)
Lo, P. Y., 585[163] (553, 565)
Lo Schiavo, S., 223[156] (215)
Loar, M. K. S., 278[134] (235, 267, 270, 273)
Locatelli, P., 285[572] (248)
Lock, C. J. L., 289[903] (271)
Lockemeyer, J. R., 187[385] (148, 149)
Loeb, S. J., 284[544], 529[712] (512)
Loginova, I. V., 220[33] (197)
Lokshin, B. V., 190[616] (177)
Long, N. J., 527[586] (498), 528[596] (499)
Longeri, M., 519[77] (439)
Longford, D., 526[473] (484)
Longoni, G., 19, 26[136] (19), 26[137] (19), 26[141] (19), 26[144] (19), 26[145] (19), 26[147] (20), 26[148] (20), 26[151] (20), 26[153] (20), 26[165] (20), 427[44a] (397), 427[51] (398, 399), 427[52] (398)
Lopez, C., 283[455] (245, 263)
López, G., 100[45] (31, 33), 100[46] (31, 33), 100[47] (31, 33), 100[48] (31, 33), 281[287] (240), 288[810] (261), 288[811] (261), 288[812] (261), 288[823] (262, 263), 288[824] (262), 289[863] (264), 524[339] (468), 524[347] (468), 524[352] (469), 524[353] (469), 524[354] (469), 524[355] (469)
López, J. A., 389[356] (361), 587[342] (576, 577)
Lopez, J. C., 282[400]
Lopez, J. M., 102[158] (42, 56)
Lopez, M., 283[476] (245, 263, 264)
Lopez-Munoz, M.-J., 277[34] (228, 230)
Lopez-Solera, I., 283[463] (245)
Lorberth, J., 528[624]
Lorenz, I.-P., 585[161] (553)
Louati, A., 188[446] (153, 157)
Louey, M., 284[563] (246)
Lough, A. J., 523[312] (464), 528[617] (510), 528[618] (510)
Louw, W. J., 522[253] (459), 522[254] (459)
Louwen, J. N., 104[327] (81, 86), 281[300] (241), 530[721] (513), 530[750] (515)
Love, R. A., 384[47] (294)
Low, A. A., 189[541]
Low, J. J., 289[873] (266), 289[874] (266), 521[206] (453), 522[207] (453)
Lower, L. D., 26[137] (19), 26[141] (19)
Lozano, M. J., 288[830] (262), 289[864] (264)
Lozano, R., 283[462] (245)
Lu, J., 529[672] (508)
Lu, L., 587[328] (574)
Lu, Q., 525[440] (479)
Lu, X. Y., 587[328] (574)
Lu, Z., 419, 430[198] (419), 586[229] (558), 586[230] (559, 565–7)
Lucas, C. R., 386[178] (326)
Lucazeau, G., 187[408] (150), 187[410] (150), 188[434] (152)
Lucherini, A., 280[267] (240, 262, 267, 269–71), 280[268] (240, 262), 280[271] (240, 270)
Ludwig, M., 100[79] (38, 39, 42, 57), 100[84] (41, 42, 57)
Ludwig, W., 104[342] (81, 90)
Lueken, H., 186[318] (138, 144, 155, 166)
Luh, T.-Y., 188[451] (154, 170), 188[452] (154, 159, 170), 188[470] (154)
Lukacs, G., 282[400]
Lukas, J., 389[421] (373)
Luke, M. A., 427[68] (400)
Lukehart, C. M., 106[440] (96, 98), 526[484] (485), 526[485] (485), 526[486] (485, 494), 526[495], 526[504] (489), 527[534] (493), 587[346] (577)
Lumetta, G. J., 522[270] (460), 526[524] (491), 586[272] (565, 568)
Luond, R., 530[726] (514)
Lutsenko, A. I., 282[371]
Lutz, F., 183[117] (115), 184[151] (117, 118)
Lynch, M. W., 27[184] (22)
Lyons, J. E., 388[306] (356), 388[307] (356)
Lysyak, T. V., 428[112] (406)

Ma, E., 430[227] (426)
Maassarani, F., 104[304] (81, 85), 281[293] (241, 270, 273), 285[619] (250), 286[677] (253, 262, 263), 286[678] (253, 254, 262, 263), 286[685] (254, 264), 286[686] (254, 271), 289[868] (264), 386[165] (320), 386[166] (320), 387[261] (348)
McAlees, A. J., 223[152] (215), 223[157] (215), 278[132] (235), 279[169] (236), 279[170] (236, 267), 280[219] (238, 259), 285[605] (249), 285[606] (249), 285[607] (249), 286[656] (252), 288[782] (260, 271), 289[849] (263, 270), 289[852] (263), 385[86] (305), 385[87] (305), 385[90] (305), 524[371] (471, 495), 524[373] (471), 524[374] (471), 524[375] (471), 524[376] (471), 524[391] (472), 525[430] (478), 586[225] (558)
McAuley, A., 105[386] (91)
McAuliffe, C. A., 184[139] (116, 122)
McCallum, J. S., 289[897] (270), 290[926] (273)
McCarthy, T. J., 522[218] (455), 522[219] (455), 522[220] (455), 522[224] (455), 582[19a] (535), 582[19b] (535), 582[20] (535)
McCleverty, J. A., 280[286] (240), 288[827] (262–4), 288[838] (263), 289[848] (263), 289[908] (271), 390[448] (378), 524[337] (467)
McCrindle, R., 278[132] (235), 279[169] (236), 279[170] (236, 267), 280[219] (238, 259), 285[605] (249), 285[606] (249), 285[607] (249), 286[656] (252), 287[767] (259), 288[782] (260, 271), 289[849] (263, 270), 289[852] (263), 385[86] (305), 385[87] (305), 385[90] (305), 524[371] (471, 495), 524[373] (471), 524[374] (471), 524[375] (471), 524[376] (471), 524[391] (472), 525[430] (478), 585[166] (553–5, 558), 586[225] (558)
McCullough, K. J., 26[107] (15), 281[297] (241, 255)
McDonald, I. A. S., 284[564] (246)
MacDonald, N., 587[305] (572)
McDonald, R., 101[153] (41, 42, 47–9), 102[161] (41, 42, 47–9)
McDonald, S. R., 26[108] (15), 27[170] (21), 105[419] (95, 96), 184[160] (121, 144), 184[161] (121, 139, 144, 159), 189[555] (172, 173)
McDonald, W. S., 282[397] (243), 282[399], 283[488], 284[543], 428[92] (403, 416), 525[455] (480), 527[551] (494), 529[696] (510), 530[747] (514), 583[64b] (542), 586[247] (561, 562)
McEwan, D. M., 520[118] (441, 442, 445, 463), 526[513] (490), 526[515] (490), 527[547] (494), 527[548] (494), 527[549] (494), 527[550] (494), 527[551] (494), 586[234] (559)

McFeaters, J. S., 220[5] (194)
MacFerrin, K. D., 389[380] (366)
McGilligan, B. S., 525[399] (474)
McGinnety, J. A., 386[149] (317)
McGlinchey, M. J., 105[433] (96, 98), 105[434] (96, 98), 185[237] (131, 136), 190[569] (172)
Machell, J. C., 221[62] (202, 209, 215)
Machida, K., 427[56] (399)
McIntosh, D. F., 185[256] (129)
McIver, J. W., Jr., 384[55] (296)
Mack, J. U., 390[459] (381)
McKechnie, T. S., 26[140] (19, 20)
McKee, S. D., 588[376] (581)
McKee, V., 283[429] (244)
McKenna, P., 429[184] (416, 417), 525[441] (479, 481), 525[442] (479, 481), 526[460], 526[468] (482), 588[350] (578), 588[353] (579)
Mackenzie, P. B., 184[157] (119), 186[335] (139, 143), 288[780] (260), 387[239] (344), 388[338] (360), 388[339] (360), 521[153] (444)
MacKenzie, T. E., 528[613] (500, 504)
McKinney, R. J., 100[66] (37, 38), 103[262] (73), 182[56] (111), 184[164] (123), 184[166] (123), 184[167] (123), 188[463] (155, 156)
MacKinnon, P. I., 102[201] (41, 47, 48), 524[341] (468)
Macklin, P. D., 583[40] (536, 539)
McLain, S. J., 25[32] (5)
McLaughlin, G. M., 283[440] (244), 283[444] (244, 264), 524[385], 530[748] (515), 530[749] (515)
Maclaughlin, S. A., 185[248] (131, 172, 174)
MacLean, T. D., 520[134] (443)
McLennan, A. J., 428[72] (401), 428[103] (403), 428[104] (403), 520[122] (442), 522[244] (458), 526[508] (490), 526[521] (490), 527[536] (493)
McNeal, C. J., 427[53] (398)
MacNeil, P. A., 101[93] (42, 58, 59), 101[94] (42), 101[138] (42, 58, 59), 280[279] (240)
McPartlin, M., 526[494] (487), 529[682] (509), 584[149] (552, 566), 584[151] (552, 566), 584[152] (552)
McPhail, A. T., 101[142] (42, 47, 51), 526[484] (485)
McPhail, D. R., 526[484] (485)
McPherson, D. M., 220[5] (194)
McPherson, H. D., 278[133] (235, 263, 269, 273), 289[883] (267), 289[895] (269, 273), 518[15] (434)
McQuillin, F. J., 289[890] (268)
McWhinnie, W. R., 280[227] (238), 526[478] (484)
Madach, T., 189[550] (171)
Mader, M. M., 389[408] (371)
Madix, R. J., 384[33] (293)
Maeda, K., 279[181] (236, 264)
Maeda, M., 101[137] (42, 51, 61), 101[140] (42, 53, 62)
Maemura, M., 283[454] (245, 263)
Maestri, M., 530[722] (513), 530[728] (514), 530[730] (514), 530[732] (514), 530[735] (514)
Mager, M., 221[63] (203)
Magno, F., 101[145] (42, 47)
Mague, J. T., 286[648], 386[154] (317), 386[155] (318)
Maguire, J. A., 191[636] (180)
Mahan, J. E., 279[148] (235, 237)
Mahapatra, A. K., 284[531] (263, 268), 289[891] (268)
Mahé, C., 386[199] (333)
Mahmoud, M. M., 282[370] (263, 264), 529[697] (510), 588[363] (580)
Maichle, C., 103[243] (66, 67), 189[516] (157, 164, 177)
Maier, N. A., 191[626] (180, 182), 191[629] (180, 182)
Maier, W. B., 24[5] (2), 24[6] (2)
Main, P., 222[120] (211)
Maiorana, S., 278[145] (235)
Maisonnat, A., 220[24] (196)
Maitlis, P. M., 220[1] (194–6, 198, 201, 205, 208–12, 215–18), 222[113] (210), 276[1] (226, 260), 276[2] (226, 273), 285[581] (248, 251, 252), 286[644] (251, 252), 286[645] (251, 252), 383[1] (292), 383[2] (292), 383[3] (292, 294), 383[4] (292), 383[5] (292), 383[6] (292), 383[7] (292), 383[8] (292), 383[9] (292), 385[111] (309), 385[139] (315)
Maj, J. J., 188[449] (154, 159), 190[568] (171–3)
Majima, T., 285[587] (249, 267), 385[97] (306)
Mak, T. C. W., 188[451] (154, 170), 188[452] (154, 159, 170), 283[442] (244), 585[167] (553–5)
Malaba, D., 185[266] (134, 138)
Malatesta, M. C., 429[152] (409), 429[153] (409), 429[154] (409, 422), 429[155] (409), 429[156a] (409), 429[156b] (409), 429[156c] (409), 429[156d] (409), 588[355] (579)
Maldotti, A., 524[334] (467)
Malik, K. M. A., 528[606] (499)
Malito, J., 282[410], 282[411], 282[412] (263), 529[664] (507), 530[718]
Malleron, J. L., 389[385] (367)
Malone, B., 183[127] (115, 117, 125), 184[168] (125), 586[249] (561), 586[250] (561), 586[251] (561)
Manassero, M., 25[34] (5), 26[147] (20), 26[148] (20), 26[151] (20), 26[153] (20), 282[391], 283[480] (245), 529[693] (510)
Mandai, T., 387[257] (347)
Mangeot, J. P., 221[89a] (205)
Mani, D., 105[436] (96, 98)
Mann, B. E., 285[611] (249), 290[920] (272–4), 384[56] (296), 385[106] (308)
Manojlovic-Muir, L., 26[115] (16), 26[120] (16), 103[249] (67), 189[514] (157), 220[42a] (198, 199), 220[42b] (198, 199), 220[43] (199, 200), 220[44] (199), 220[47] (200), 282[420] (243, 248, 263, 264), 282[421] (243), 289[910] (272), 390[455] (379), 390[456] (379), 427[24] (394, 402, 404), 428[70] (400), 428[74] (401, 404), 428[77] (402, 404), 428[80] (403), 428[81] (403), 428[83] (403), 428[88] (404), 428[89b] (404), 428[90] (404, 415), 519[43] (437), 520[116] (441), 520[120] (442), 520[123] (442), 520[129] (442, 495), 521[164], 523[300] (463), 523[301] (463), 524[380], 526[517] (490), 526[518] (490), 530[771] (517), 587[300] (571), 587[302] (572)
Manotti-Lanfredi, A. M., 27[167] (21), 185[241] (131, 172, 174)
Manriquez, J. M., 189[513] (157, 166)
Mansuy, D., 584[121] (548, 562)
Mantovani, A., 222[125] (211, 213), 222[127a] (211), 222[127b] (211), 222[130] (211), 222[142a] (214), 222[142b] (214), 279[146] (235), 279[149] (235, 255, 262–4), 287[747] (257, 263), 287[750] (257), 287[751] (257), 287[752] (257), 287[753] (257), 518[17] (434)
Manzer, L. E., 519[41] (437), 521[189] (447)
Maqsudar Rahman, A. F. M., 429[175] (414)
Marchaj, A., 105[395] (91, 92)
Marchetti, F., 429[163] (411)
Marchionna, M., 427[44a] (397)
Marcos, M., 284[559] (246, 263), 284[560] (246, 263, 268)
Marczinke, B., 25[91] (12)
Mardanov, V. G., 186[283] (137, 139)
Marder, T. B., 188[462] (155, 156), 429[179] (415)
Maresca, L., 282[405], 584[115] (548), 584[116] (548), 584[117] (548), 584[118] (548), 584[119] (548), 584[120] (548)
Margerum, L. D., 387[228] (342)
Margrave, J. L., 99[6] (30, 31, 74), 100[27] (32), 182[24] (109), 185[257] (130)
Mari, M., 277[26] (228, 230), 387[266] (350)
Mariani, L., 388[352] (361)
Marín, J. M., 24[17] (3), 25[87] (11), 100[78] (38, 39, 42, 56), 100[80] (38–40, 42), 101[99] (42, 56), 101[100] (42, 56), 103[268] (74, 78, 81), 104[325] (81, 85), 183[108] (115, 124), 183[109] (115, 124), 186[342] (139, 146, 157), 187[371] (146)
Marinelli, F., 389[360] (363)
Marinelli, G., 587[288] (569)
Marinetti, A., 105[424] (95, 96), 184[153] (117, 159), 185[237] (131, 136), 529[675] (508)
Mark, F., 186[279] (137, 138)
Marken, F., 588[356] (579, 582)
Markham, D. P., 527[550] (494), 529[683] (509)
Markham, R., 282[396] (243, 245)
Markies, B. A., 278[99] (233, 255, 256, 262), 278[100] (233, 236, 239, 266, 271, 272), 278[103] (233, 251, 262, 273), 280[262] (239, 262, 266, 271, 274, 276), 285[633] (250, 255, 262), 290[953] (274, 276)
Markovich, L. M., 386[174] (324)
Markwick, A. J. W., 525[442] (479, 481), 588[350] (578)
Markzinke, B., 182[42] (110–12), 182[43] (110, 112, 126)
Marr, G., 182[1] (108, 150), 182[2] (108, 150), 182[3] (108, 150), 182[4] (108, 150), 384[12] (292), 384[13]

(292), 384[14] (292)
Marrs, C. M., 390[430] (375)
Marsella, A., 523[287] (462), 525[418] (475, 476)
Marsi, M., 521[162] (458, 462, 472)
Marsich, N., 527[545] (494), 527[546] (494)
Marti, J. M., 288[811] (261), 524[354] (469)
Martin, A., 289[901] (270), 525[401] (474)
Martin, A. M., 389[381] (366)
Martin, J., 529[706] (511)
Martin, J. W. L., 283[441] (244, 263), 283[444] (244, 264)
Martin, T. R., 525[428] (478), 585[186] (554, 555)
Martinengo, S., 26[136] (19)
Martinez, A., 101[99] (42, 56)
Martinez, F., 223[145a] (214), 223[148] (214), 277[5] (226, 227), 277[8] (226, 227, 252), 277[9] (226, 227, 252), 277[51] (230), 288[808] (261, 262), 288[813] (261), 288[837] (262–4), 518[9] (432, 433), 524[340] (468), 524[348] (468), 524[363] (469), 525[398] (474, 490), 587[306] (572)
Martinez, M., 101[143] (42, 56), 101[150] (42, 56), 528[619] (510), 529[700] (510)
Martinez-Carrera, S., 283[463] (245)
Martinez-Ilarduya, J. M., 281[288] (240, 262, 264)
Martinho Simões, J. A., 99[8] (30), 290[952] (274)
Martinon, S., 385[112] (309)
Martinov, Yu. N., 428[108a] (403)
Marty, W., 529[666] (507)
Maruta, K., 25[40] (5), 584[138] (551)
Maruta, T., 285[578] (248)
Maruya, K., 100[81] (40)
Maruyama, H., 286[651]
Mas, M. L., 289[866] (264)
Masaguer, J. R., 283[463] (245)
Masai, H., 277[58] (230)
Mascetti, J., 182[25] (109)
Mascharak, P. K., 289[857] (264)
Masciocchi, N., 26[148] (20), 525[456] (480)
Mashhood Ali, S., 388[308] (356), 388[310] (357)
Masi, D., 184[189] (126, 129)
Maskaeva, E. V., 279[203] (237)
Mason, M. G., 184[194] (126)
Mason, S., 287[777] (260), 525[447] (480), 525[457] (480), 529[667] (507)
Mason, S. A., 583[44] (537)
Mason, W. R., 427[30] (395), 521[197] (450)
Massa, M. A., 187[385] (148, 149)
Massa, W., 528[624]
Masters, C., 279[211] (238)
Mastrolli, G., 184[183] (126, 128)
Masuda, H., 586[275] (565–7)
Masumoto, T., 223[162] (216), 277[68] (230), 278[144] (235)
Masuyama, Y., 388[321] (358), 388[322] (358), 388[324] (358)
Mathey, F., 105[424] (95, 96), 184[153] (117, 159), 279[192] (237), 279[193] (237), 529[675] (508), 583[54] (540)
Mathieson, M. T., 585[167] (553–5), 585[183] (554, 555, 557)
Mathieu, R., 105[427] (67, 96, 97), 278[105] (233, 255, 256)
Mathur, P., 586[268] (564)
Matisons, J. G., 103[245] (66, 67)
Matson, G. B., 429[174] (414)
Matsubayashi, G.-E., 277[21] (228, 229), 277[36] (229)
Matsuda, K., 429[168] (412, 413), 429[169] (412)
Matsuda, T., 279[196] (237), 287[771] (259)
Matsuhiro, Y., 101[115] (42, 49, 55, 61)
Matsuhisa, T., 386[210] (335)
Matsukura, T., 279[206] (238)
Matsumoto, H., 101[154] (42, 47, 48)
Matsumoto, K., 281[289] (240, 267), 287[765] (259)
Matsumoto, M., 25[79] (11), 279[206] (238), 290[960] (276)
Matsumoto, N., 287[775] (259)
Matsumoto, Y., 281[336] (242), 282[416], 388[341] (360)
Matsumura, C., 279[155] (235)
Matsunaga, P. T., 25[54] (9), 101[96] (42, 54, 74, 80, 81, 90)
Matsuo, H., 281[309] (241)
Matsuoka, M., 388[341] (360)
Matsuoka, Y., 279[195] (237, 266), 290[932] (273)
Matsuzaka, H., 287[721] (256)
Matsuzawa, S., 184[199] (125), 184[200] (125)
Matt, D., 25[53] (8), 104[336] (81, 86), 104[337] (81, 86), 105[380] (80, 88), 281[305] (241), 281[306] (241, 263), 285[613] (249, 252), 286[665] (252), 286[666] (252), 288[835] (263), 288[836] (263), 289[856] (264), 289[902] (271), 525[394] (474), 526[491] (486, 509), 529[689] (509)
Mauret, P., 187[379] (148)
Mauro, A., 281[296] (241)
Maya, B. J., 520[123] (442)
Mayer, W., 184[145] (117, 118)
Mayr, A., 430[223] (426)
Mayr, U., 522[211] (454)
Mazany, A. M., 105[375] (81, 91)
Mazid, M. A., 520[112] (441), 528[606] (499), 585[156] (553)
Mazo, G. J., 221[55] (201)
Meah, M. N., 289[858] (264), 388[295] (354)
Mealli, C., 184[189] (126, 129), 184[190] (126, 129), 186[348] (139, 146), 186[349] (139, 146), 187[350] (139, 146, 171), 221[57] (201), 384[54] (295), 388[354] (361), 389[356] (361), 428[73] (401), 428[78a] (402), 587[342] (576, 577)
Meanwell, N. J., 584[147] (552)
Mednikov, E. G., 198, 220[30] (197), 221[52] (201), 221[53] (201, 204), 221[54a] (201), 221[54b] (201), 221[64a] (203), 221[64b] (203), 221[67] (204), 221[68a] (204), 221[68b] (204), 221[68c] (204), 221[68e] (204), 221[68f] (204), 221[69] (204), 221[70] (204), 221[71] (204), 221[72] (204), 221[73a] (204), 221[73b] (204), 221[74a] (204), 221[74b] (204), 221[74c] (204), 222[95] (207), 222[96] (207), 222[97a] (207), 222[97b] (207), 427[12] (393, 399), 427[28] (395)
Meegalla, S. K., 290[940] (273)
Meek, D. W., 520[101] (440), 524[380]
Mehler, G., 103[234] (66, 67, 72), 185[216] (130–2, 138, 157)
Mehner, T., 220[3] (194)
Meicheng, S., 101[134] (42, 47)
Meidine, M. F., 184[179] (126, 127), 587[291] (570), 587[292] (570)
Meille, S. V., 388[334] (359), 428[115] (406), 587[319] (574, 576), 587[320] (574, 576)
Meir, C. J., 587[292] (570)
Meli, A., 184[189] (126, 129), 184[190] (126, 129), 184[191] (126, 129), 184[192] (126, 129), 184[196] (126, 129), 184[197] (126, 129), 185[208] (129)
Mendivil, C., 277[24] (228)
Meng, B. H. H., 277[39] (229, 230)
Meng, X., 190[601] (175, 177)
Menjón, B., 220[22] (196), 223[145a] (214), 277[5] (226, 227), 286[691] (255, 262), 288[820] (262), 289[865] (264), 385[109] (309), 386[153] (317), 428[96] (403), 429[151a] (409), 429[151b] (409), 430[210] (423), 524[357] (469), 524[358] (469, 470), 524[362] (469), 524[367] (470), 524[369] (470), 524[386] (496, 497), 525[420] (477), 586[255] (562, 563), 587[284] (568)
Menu, M.-J., 26[134] (19), 100[74] (37, 38), 100[76] (37–9, 42, 56), 105[408] (42, 93, 94)
Merbach, A. E., 522[249] (458, 459)
Merchàn, M., 386[147] (316)
Merle-Mejean, T., 182[25] (109)
Merlino, S., 429[163] (411)
Merola, J. S., 287[741] (257)
Merwin, R. K., 519[48] (437, 453)
Merzweiler, K., 26[164] (20), 189[551] (171), 387[277] (351)
Meshulam, A., 105[397] (91, 92)
Messbauer, B., 223[153] (215), 429[175] (414)
Messmer, R. P., 185[256] (129)
Mesubi, M. A., 520[92] (440)
Metcalf, P. A., 222[119a] (210)
Metternich, H.-J., 184[179] (126, 127)
Mettler, H., 282[377] (244)
Metz, B., 189[553] (171, 172)
Metzger, J., 25[81] (11)
Metzner, P. J., 386[194] (331)
Meyer, H., 223[153] (215), 386[182] (327), 386[183] (327), 429[175] (414)
Meyer, K., 25[74] (10)
Meyer, W. R., 520[130] (443)
Meyer-Base, K., 289[847] (263), 428[113] (406), 524[360] (469)
Meyerstein, D., 105[396] (91, 92), 105[397] (91, 92)
Michaelis, S., 103[275] (73, 74, 80), 103[277] (73, 74, 80), 182[46] (110, 111, 113, 115, 118, 126), 182[47] (110, 113, 117, 118), 183[66] (113, 118), 184[150] (117, 118)
Michel, A., 220[45] (199), 223[158] (215)
Michelin, R. A., 223[161a] (216), 223[161b] (216), 223[163] (216, 218), 223[168] (217), 223[171] (217), 223[173a] (218), 223[173b] (218), 223[173c] (218), 278[128]

(235), 278[131] (235, 262, 264, 268), 289[888] (268), 429[186] (416), 430[201a] (420, 426), 430[201b] (420, 426), 430[204a] (421), 430[204b] (421), 430[205] (421), 430[212] (423), 430[213] (423), 430[214] (423, 425), 430[216] (424, 426), 430[217] (424), 430[218] (424), 430[220] (425), 430[221] (425), 430[225] (426), 430[226] (426), 521[172], 522[208] (453), 522[209] (453), 523[287] (462), 524[384], 524[393] (473), 525[409] (475), 525[410] (475), 525[411] (475), 525[412] (475), 525[413] (475), 525[414] (475), 525[416] (475), 525[421] (477), 586[244] (560)
Michman, M., 280[284] (240, 262)
Middleton, A. R., 528[622]
Midollini, S., 27[182] (21), 184[196] (126, 129), 186[348] (139, 146), 186[349] (139, 146), 187[350] (139, 146, 171), 187[354] (139, 147), 187[355] (139, 147), 187[372] (146), 278[122] (234), 278[123] (234), 384[54] (295), 524[335] (467), 529[686] (509)
Mija, S., 584[104] (547)
Mikaya, A. I., 100[83] (41, 47)
Mikhailov, V. A., 221[68c] (204), 221[69] (204)
Miki, K., 25[96] (13), 25[97] (13), 26[98] (13), 102[172] (49, 50, 64), 102[173a] (49, 50, 64), 102[173b] (50, 64), 102[196] (48, 64), 102[198] (48, 58), 102[199] (48, 58), 282[415], 285[627] (250, 260, 262, 267, 269), 288[802] (260), 384[62] (299), 384[67] (300), 386[208] (335), 389[366] (363), 583[67a] (542, 545, 574), 583[67b] (542, 545, 574), 583[68] (542, 545, 547, 574), 586[259] (563)
Milaev, A. G., 279[201] (237, 264, 272)
Milaeva, E. R., 279[201] (237, 264, 272), 289[913] (272), 387[264] (348)
Milchereit, A., 25[43] (7), 25[44] (7), 25[46] (7), 104[310] (80, 81), 104[312] (81), 104[346] (81, 88), 104[348] (81, 88), 104[350] (81, 88)
Milewski-Mahrla, B., 26[127] (16), 104[340] (81, 87, 90)
Miller, D. J., 389[382] (366)
Miller, G. A., 527[530] (493)
Miller, R. G., 280[282] (240)
Miller, R. W., 220[26] (197, 200), 388[299] (355), 390[435] (375)
Miller, T. D., 222[118] (210)
Miller, T. M., 522[235] (457), 526[465] (481)
Milne, C. R. C., 286[653] (252), 286[669] (253), 286[671] (253)
Milstein, D., 220[34] (197), 278[110] (234, 238, 270, 273), 279[158] (236), 280[228] (238, 255), 290[923] (273), 388[332] (359)
Min, D., 527[568] (497)
Min, S. F., 102[164] (41, 42, 47, 51, 58), 279[156] (235)
Minchin, N. J., 279[184] (236, 241), 282[385] (253, 263, 264), 282[389] (262), 520[93] (440)
Minghetti, G., 282[391], 283[480] (245), 283[481] (245, 248), 283[482] (245, 255), 286[694] (255, 256, 262), 427[66] (400, 414), 430[224] (426), 529[693] (510), 530[774] (517), 530[775] (517)
Mingos, D. M. P., 191[638] (180), 221[62] (202, 209, 215), 222[110] (209, 215), 427[7] (392), 427[8] (392, 396, 397), 427[36] (396), 427[39] (396), 427[40] (396), 427[41] (396, 413), 427[68] (400), 429[166] (411, 414, 419), 429[172] (412), 429[173] (413), 585[169] (554), 586[246] (560), 587[343] (577)
Mink, J., 182[26] (109), 182[27] (109), 220[14] (195), 220[21] (196), 384[34] (293), 583[46] (537, 542, 550)
Mink, L., 585[171] (554, 556)
Minniti, D., 519[72] (439), 520[83] (439), 520[86] (439), 521[167] (458, 459), 522[233] (457), 522[234] (457), 522[241] (458), 522[246] (458, 459), 522[247] (458, 459), 522[248] (458, 459), 522[255] (459), 522[257] (459), 522[258] (459), 522[260] (459)
Minton, R. P., 387[274] (350)
Minzhi, X., 101[134] (42, 47)
Miralles-Sabater, J., 386[147] (316)
Miravitlles, C., 101[111] (42, 47, 48, 56, 61), 102[205] (41, 49, 56, 58), 280[222] (238, 263, 268), 280[223] (238), 523[330] (467), 524[390]
Mirza, H. A., 26[120] (16)
Mise, T., 27[187] (23), 189[519] (159, 167)
Mishra, N. K., 277[35] (229)
Mitchell, M. A., 286[637] (251), 386[203] (334)
Mitchell, P. C., 388[305] (356)
Mitkov, M. I., 384[50] (295)
Mitrprachachon, P., 190[566] (172), 583[45] (537)
Mitschler, A., 100[67] (37, 38), 281[305] (241), 281[306] (241, 263), 427[15] (393, 395)
Mittal, P. K., 528[608] (499, 504)
Miura, H., 388[343] (360)
Miura, R., 279[206] (238)
Miwa, T., 279[152] (235, 255, 262)
Mix, A., 188[477] (157)
Miya, S., 584[91] (544)
Miyashita, A., 25[57] (9), 25[58] (9), 25[59] (9), 100[85] (42, 53, 62), 103[264] (74, 80), 103[269] (74, 75), 183[64] (112), 183[76] (115, 120, 128), 183[77] (115, 120), 184[187] (126, 128), 184[188] (126, 128), 280[241] (239, 266, 267), 523[294] (462), 584[127] (549), 586[235] (560)
Miyaura, N., 185[265] (133)
Miyoshi, K., 279[195] (237, 266), 290[932] (273)
Mizobe, Y., 278[104] (233)
Mizoroki, T., 100[81] (40)
Mizuno, H., 285[573] (248), 285[575] (248)
Mizuno, Y., 585[176] (554, 555)
Mlekuz, M., 105[433] (96, 98), 105[434] (96, 98), 190[569] (172)
Moberg, C., 285[594] (249), 385[119] (311), 385[120] (311), 385[121] (311), 389[391] (368), 389[418] (373)
Modelli, A., 185[275] (137), 386[175] (324), 587[334] (575, 576)
Mognaschi, E. R., 390[465] (382)
Mohamud, S. I., 187[370] (146)
Mohapatra, B. K., 288[804] (260)
Moiseev, A. I., 428[94] (403)
Moiseev, I. I., 221[55] (201), 221[56] (201), 221[58] (202, 219), 221[59] (202), 221[60c] (202), 221[61a] (202), 221[61b] (202), 222[93a] (207), 222[93b] (207), 222[93c] (207), 223[179a] (219), 223[179b] (219), 223[179c] (219), 279[194] (237)
Mok, C. Y., 428[110] (405)
Mok, K. F., 283[443] (244, 263)
Molander, G. A., 186[332] (139, 143)
Moldes, I., 105[427] (67, 96, 97)
Mole, L., 102[168] (47, 50, 54), 182[60] (111, 113), 279[198] (237), 385[81] (303), 522[222] (455), 583[30] (536, 538, 539)
Molins, E., 280[222] (238, 263, 268), 280[223] (238)
Moll, M., 105[405] (92), 223[177] (218), 430[202] (421)
Mollbach, A., 100[54] (34), 183[78] (115, 119)
Moller, K., 25[32] (5)
Monaghan, P. K., 523[311] (464), 523[314] (465), 523[319] (465), 523[320] (465), 523[322] (466)
Monari, M., 583[84] (544)
Mond, L., 2, 24[1] (1, 2)
Moneti, S., 186[348] (139, 146), 186[349] (139, 146), 187[350] (139, 146, 171), 187[355] (139, 147), 187[372] (146), 278[122] (234), 278[123] (234), 384[54] (295), 524[335] (467), 529[686] (509)
Monge, A., 24[17] (3), 25[89] (12), 101[106] (42, 47, 50, 58, 81, 90), 101[139] (42, 56), 103[268] (74, 78, 81), 104[293] (73–5, 78, 81, 90), 104[295] (74, 78), 104[296] (74, 78), 183[108] (115, 124), 183[109] (115, 124), 186[345] (139), 189[500] (157), 288[843] (263)
Montag, R. A., 26[140] (19, 20)
Moody, D. C., 427[37] (396)
Moor, A., 427[9] (392), 427[21] (393, 395), 427[26] (394, 395), 427[27] (395)
Moore, I., 27[187] (23), 427[18] (393, 413)
Moore, J. H., 187[411] (151)
Moore, J. J., 220[5] (194)
Moore, M. R., 525[447] (480)
Moore, P., 522[246] (458, 459), 522[247] (458, 459)
Moore, S. S., 520[100] (440), 522[226] (456), 526[461] (481), 585[179] (554, 555)
Moragas, R., 283[493] (271)
Moran, T. F., 187[403] (150)
Morandini, F., 186[292] (138, 143), 278[127] (235, 264, 270)
Moravskiy, A., 289[875] (266, 269, 273)
Morelli, G., 385[95] (306)
Moreno, M. T., 277[51] (230), 524[363] (469), 525[398] (474, 490), 587[306] (572)

Moreno-Mañas, M., 389[378] (366)
Moretó, J. M., 25[26] (4)
Moretti, G., 523[275] (461)
Morgan, C. R., 290[934] (273)
Morgenstern, D. A., 27[180] (21), 185[269] (136)
Mori, A., 388[309] (357)
Morikawa, S., 25[37] (5), 25[38] (5)
Morimoto, C. N., 520[101] (440)
Morimoto, H., 277[69] (230, 231), 278[78] (231), 527[543] (493)
Morimoto, Y., 522[217] (455)
Morinaga, S., 223[172] (218)
Moro, G., 184[183] (126, 128)
Morokuma, K., 182[20] (109), 222[100a] (208), 222[100b] (208), 222[101] (208), 222[103] (208), 287[715] (255), 287[719] (255), 385[126] (312), 427[6] (392), 429[131] (408), 429[132] (408), 521[205] (453), 522[265] (460), 584[135] (550)
Moronski, M., 183[123] (115, 138, 145, 155)
Moro-oka, Y., 105[428] (97), 105[429] (96, 97), 105[430] (96), 185[244] (131, 159, 172), 278[96] (233, 234), 280[244] (239), 285[583] (249), 387[280] (352), 525[429] (478)
Morozova, L. N., 101[124] (41, 42, 47), 101[126] (42), 102[164] (41, 42, 47, 51, 58)
Morris, M. J., 26[101] (14)
Morris, S. A., 221[50b] (201), 289[860] (264), 427[35] (395–7), 530[771] (517)
Morrison, J. A., 101[109] (41, 42, 47, 48), 279[210] (238), 523[328b] (467), 524[336] (467)
Morrone, S., 283[490] (262)
Mortimer, C. T., 525[450] (480), 583[52] (540)
Morton, N., 525[450] (480), 583[52] (540)
Morton, P. R., 222[107] (209, 215)
Morton, S., 587[292] (570)
Morvillo, A., 101[132] (42, 47, 51), 102[177] (49, 51), 186[338] (139, 145), 288[844] (263, 265), 522[237] (457), 587[314] (573)
Moseley, M. A., 384[59] (296), 384[60] (297)
Moser, W. R., 278[106] (233, 255)
Moss, I., 528[630] (500)
Moss, J. R., 527[560] (495)
Mossi, W., 282[369]
Motschi, H., 529[709] (512)
Moulding, R. P., 521[192] (447)
Moulton, C. J., 284[539]
Mozzon, M., 223[163] (216, 218), 223[168] (217), 223[173a] (218), 223[173b] (218), 223[173c] (218), 430[205] (421), 430[214] (423, 425), 430[216] (424, 426), 430[217] (424), 430[218] (424), 430[226] (426)
Muetterties, E. L., 22, 27[183] (22), 185[232] (131, 136), 386[146] (316), 390[464] (382)
Muetterties, M. C., 105[403] (92)
Muir, K. W., 26[115] (16), 26[120] (16), 103[249] (67), 189[514] (157), 220[42a] (198, 199), 220[42b] (198, 199), 220[43] (199, 200), 220[44] (199), 220[47] (200), 282[421] (243), 289[910] (272), 390[455] (379), 390[456] (379), 427[24] (394, 402, 404), 428[70] (400), 428[74] (401, 404), 428[77] (402, 404), 428[80] (403), 428[81] (403), 428[83] (403), 428[88] (404), 428[89b] (404), 428[90] (404, 415), 519[43] (437), 520[116] (441), 520[120] (442), 520[129] (442, 495), 523[300] (463), 523[301] (463), 524[380], 526[518] (490), 530[771] (517)
Mukai, K., 389[383] (366)
Mukhedkar, J. A., 278[136] (235, 237)
Mukhedkar, V. A., 278[136] (235, 237)
Mul, W. P., 104[331] (81, 85), 530[760] (516)
Müller, A., 184[144] (117, 121)
Müller, C., 183[113] (115, 117, 118), 183[114] (115, 117, 118), 280[242] (239, 271), 286[659] (252), 290[942] (273), 527[569] (497)
Muller, E. P., 519[23] (434), 530[724] (514)
Muller, F., 104[331] (81, 85), 530[758] (516), 530[760] (516)
Müller, G., 100[64] (31, 36), 101[111] (42, 47, 48, 56, 61), 101[121] (42, 50, 58, 60), 101[127] (42, 49), 101[128] (42, 47, 62), 101[143] (42, 56), 101[146] (42, 62, 63), 101[150] (42, 56), 101[151] (41, 42, 50), 102[205] (41, 49, 56, 58), 102[206] (49, 63), 183[74] (115), 190[596] (175, 180), 277[33] (228), 520[140] (443), 521[186] (447), 528[604] (499), 531[780] (517)
Muller, H.-J., 531[779] (517)
Muller, H.-P., 277[44] (230)
Müller, J., 187[415] (151, 155)
Müller, M., 26[163] (20), 190[560] (172, 173)
Müller, U., 101[97] (42, 47, 52)
Muller, W.-D., 519[52] (437), 519[71] (438), 521[185] (447)
Mullica, D. F., 529[716]
Munz, C., 286[659] (252), 286[662] (252), 286[663] (252), 525[432] (478)
Mura, P., 529[710] (512)
Murahashi, S.-I., 279[204] (237, 238, 240), 386[208] (335), 386[211] (336)
Murai, S., 103[221] (65), 103[222] (65), 186[294] (138), 289[877] (266), 386[213] (336), 389[388] (368), 587[316] (573, 576), 587[329a] (575)
Murakami, M., 287[763] (258), 287[764] (258)
Murali Mohan, K., 387[265] (350)
Muralidharan, S., 520[127] (442), 520[128] (442)
Murata, E., 277[65] (230, 231), 277[69] (230, 231)
Murphy, J. B., 386[202] (334)
Murphy, W. S., 282[393]
Murrall, N. W., 104[333] (81, 86), 386[181] (327), 388[290] (353)
Murray, M., 427[18] (393, 413)
Murray, T. F., 279[159] (236, 253)
Murtas, S., 280[267] (240, 262, 267, 269–71)
Murtaza, Z., 525[435] (478)
Musaev, A. A., 221[68d] (204), 279[200] (237), 279[202] (237)
Muschiol, M., 105[411] (93, 94), 185[231] (131, 136)
Musco, A., 387[268] (350), 388[334] (359), 388[352] (361), 388[354] (361), 388[355] (361), 389[356] (361), 428[115] (406), 520[144] (444), 525[433] (478), 583[69] (542, 545, 547, 574), 587[319] (574, 576), 587[320] (574, 576), 587[335] (576), 587[336] (576), 587[342] (576, 577)
Musgrove, R. J., 586[267] (564)
Musker, W. K., 520[110] (441)
Mutet, C., 282[398] (243)
Mwariri, B. N., 220[44] (199), 428[80] (403)
Myers, J. B., 526[484] (485)
Myles, D. C., 389[380] (366)
Mynott, R., 25[91] (12), 26[100] (14), 100[56] (31, 36), 103[232] (66, 67), 103[233] (66, 67), 103[235] (66, 67), 103[251] (66, 67, 70), 103[252] (66, 67, 70), 103[256] (67, 73), 103[275] (73, 74, 80), 103[277] (73, 74, 80), 104[300] (73, 74, 76), 182[35] (110, 112, 113, 115), 182[37] (110, 112), 182[42] (110–12), 182[43] (110, 112, 126), 182[47] (110, 113, 117, 118), 182[61] (111, 115, 174, 175), 183[67] (113, 115, 117), 183[73] (114, 115, 117, 157), 183[84] (115, 123, 174), 183[97] (115, 121, 138, 140), 183[120] (114, 115), 183[129] (116, 121, 157, 161), 184[135] (116, 117, 157), 184[143] (117, 139, 141), 184[150] (117, 118), 185[210] (130, 132), 185[261] (132, 138, 144), 185[262] (133), 186[307] (138, 143), 285[624] (250), 288[787] (260, 267), 288[793] (260), 385[80] (303), 390[457] (380), 525[431] (478), 584[88] (544, 575), 585[168] (554), 586[245] (560)
Myrvold, S., 430[228] (426)

Naegeli, R., 427[9] (392), 520[141] (444), 520[145] (444)
Nafees Meah, M., 283[431] (244)
Nag, K., 284[549] (245, 263, 264), 284[550] (245, 264), 284[551] (245, 263, 264)
Nagaki, D. A., 26[141] (19)
Nagasundara, K. R., 385[104] (308)
Nagata, H., 387[278] (351)
Nagata, S., 25[27] (4), 25[28] (4)
Nagel, U., 222[121] (211), 284[537], 385[85] (304), 430[190a] (418), 430[190b] (418), 528[623], 582[13a] (535, 537)
Nair, P., 281[356] (243, 244)
Najera, C., 282[402]
Najjar, I. M., 584[95] (545)
Naka, A., 184[198] (125)
Nakagawa, I., 279[195] (237, 266), 290[932] (273)
Nakamura, A., 102[210] (62), 427[47] (398), 583[68] (542, 545, 547, 574), 583[85] (544)
Nakamura, S., 427[6] (392)
Nakamura, Y., 100[81] (40), 101[137] (42, 51, 61), 101[140] (42, 53, 62), 279[151] (235, 255, 262–4, 269), 279[152] (235, 255, 262), 279[180] (236), 279[182] (236), 279[183]

(236, 255, 256, 264), 280[231] (238, 267), 280[232] (238, 239, 262, 266, 267), 280[238] (239, 255, 256, 263), 281[289] (240, 267), 286[661] (252), 288[796] (260), 288[797] (260), 288[800] (260, 261), 288[803] (260), 288[841] (263), 587[317] (573), 587[318] (574)
Nakano, M., 283[485] (245, 257), 287[725] (256, 257)
Nakashima, M., 282[378] (263, 264), 282[406] (264), 282[407] (248, 263), 284[503]
Nakatsu, K., 25[79] (11), 279[180] (236), 279[182] (236), 290[960] (276)
Nakaya, H., 284[499]
Nakazawa, H., 184[186] (126, 128), 279[195] (237, 266), 280[246] (239, 240), 290[932] (273)
Nalda, J. A., 288[830] (262), 289[864] (264)
Nanda, K. K., 284[550] (245, 264)
Nanjo, K., 279[151] (235, 255, 262–4, 269)
Nanninga, T. N., 389[427] (374), 390[428] (374), 390[429] (375)
Nardin, G., 223[154] (215), 277[26] (228, 230), 277[28] (228, 230), 285[610] (249), 387[266] (350), 387[267] (350), 429[135] (408)
Narske, R. N., 100[38a] (31, 33), 190[582] (174, 175)
Nassif, O. A., 430[228] (426), 526[490] (486)
Nassimbeni, L. R., 527[560] (495)
Nast, R., 102[193] (48), 102[194] (48), 277[41] (230), 277[44] (230), 277[66] (230), 277[67] (230, 231), 279[191] (237)
Natalie, K. J., 290[924] (273)
Natile, G., 282[405], 583[72] (542, 544, 545, 548, 549), 584[115] (548), 584[116] (548), 584[117] (548), 584[118] (548), 584[119] (548)
Navarro, R., 223[147] (214), 223[148] (214), 223[167] (217), 277[7] (226), 277[8] (226, 227, 252), 277[9] (226, 227, 252), 277[24] (228), 277[30] (228), 281[352] (243, 262, 271), 286[690] (255, 263, 264), 288[808] (261, 262), 288[809] (261), 288[813] (261), 288[814] (261), 288[820] (262), 288[828] (262, 263), 288[833] (262), 288[837] (262–4), 289[867] (264), 290[921] (273, 275), 290[922] (273, 275), 390[468] (382), 429[181] (416, 417, 423), 518[9] (432, 433), 524[348] (468), 524[358] (469, 470), 524[359] (469), 524[364] (469), 524[365] (469), 524[366] (470)
Navarro-Ranninger, M. C., 283[462] (245), 283[463] (245)
Naydowski, C., 103[224] (66, 67), 103[231] (67, 71), 103[236] (66, 67), 103[237b] (66, 67, 71), 103[252] (66, 67, 70), 182[52] (111, 113, 155, 157), 182[53] (111, 113, 115, 157), 182[57] (111, 113, 115, 119, 138, 157), 184[136] (116, 117, 157), 184[147] (114, 117, 138), 186[305] (138, 140, 157), 186[307] (138, 143)
Neale, D. W., 519[75] (439, 464, 472, 497, 506), 524[356] (469), 525[422] (477, 497, 506), 585[167] (553–5), 585[183] (554, 555, 557), 585[216] (557)
Nebot-Gil, I., 386[147] (316)
Negele, M., 188[474] (156), 190[587] (175, 179)
Negele, W. M., 588[382] (582)
Negishi, E.-I., 277[43] (230), 277[60] (230, 239, 266, 267, 270)
Nehls, I., 386[182] (327)
Neibecker, D., 187[360] (140), 290[941] (273), 389[416] (372)
Neidlein, R., 103[290] (73, 74, 78, 81), 104[298] (73, 74), 104[300] (73, 74, 76)
Neilsen, W. D., 527[579] (498, 505)
Neiteler, P., 531[780] (517)
Nekipelov, V. M., 220[36] (197), 531[787] (518)
Nelson, D. A., 26[117] (16)
Nelson, J. H., 277[49] (230)
Nelson, K. T., 524[342] (468), 524[344] (468), 524[387], 524[388], 585[197] (554, 555, 557), 585[210] (554, 555)
Nelson, P. R., 187[403] (150)
Nelson, R. V., 282[392]
Neri, G., 289[882] (267)
Nesterov, B. A., 388[297] (354)
Nestler, B., 24[14] (3), 25[47] (7), 104[347] (81, 90)
Neta, P., 105[401] (92), 105[402] (92)
Nettle, A., 528[603] (499)
Netzer, F. P., 390[459] (381)
Neuenschwander, K., 278[86] (233)
Neugebauer, D., 100[64] (31, 36)
Neumann, B., 188[422] (151, 157, 159, 168)
Neve, F., 282[360] (243), 282[423] (243), 282[424] (243), 282[425] (243, 264), 283[465] (245), 284[506] (263), 428[73] (401), 519[77] (439)
Newkome, G. R., 281[302] (241), 281[312] (241), 281[313] (241), 281[314] (241), 281[315] (241), 281[316] (241), 281[317] (241), 281[318] (241), 281[319] (241), 281[320] (241, 242), 281[321] (241), 281[334] (242), 283[486] (245, 263), 283[494] (245), 385[91] (305), 385[92] (305), 519[67] (438), 529[691] (510)
Newlands, M. J., 386[178] (326)
Newmark, R. A., 220[35] (197), 288[805] (261), 519[28] (435), 526[489] (486, 487)
Ng, F. T. T., 290[935] (273)
Nicasio, M. C., 101[139] (42, 56), 102[188] (48, 56), 288[843] (263)
Nicholas, K. M., 284[529], 285[599] (249)
Nichols, M., 288[834] (262)
Nicolini, M., 429[187b] (417)
Nicolo, F., 521[169]
Niebecker, D., 101[151] (41, 42, 50)
Niecke, E., 24[13] (3), 184[179] (126, 127), 187[369] (145, 159), 188[484] (157, 166)
Niedenzu, K., 387[274] (350)
Nieger, M., 184[195] (126, 127), 187[369] (145, 159)
Nielsen, R. B., 586[271] (565, 569)
Nielson, A. J., 283[430] (244), 283[450] (245, 255, 263)
Niemann, L. C., 288[832] (262), 530[761] (516)
Nikitaev, A. T., 531[787] (518), 531[788] (518), 531[789] (518)
Nilsson, Y. I. M., 389[384] (367)
Nimmons, H. L., 387[258] (347)
Nisar, M., 388[314] (357)
Nishi, S., 387[255] (347)
Nishii, S., 384[39] (294)
Nishijima, M., 384[35] (293), 384[36] (293)
Nishikawa, M., 388[350] (360)
Nishikimi, Y., 388[311] (357)
Nishioka, E., 388[343] (360)
Nishiwaki, K., 25[95] (13), 25[96] (13), 101[114] (42, 49, 55, 61), 101[149] (41, 42, 48, 57), 102[171] (41, 49, 50, 62), 102[172] (49, 50, 64), 102[198] (48, 58), 102[199] (48, 58)
Nishiyama, H., 279[206] (238), 286[661] (252)
Niven, M. L., 527[560] (495)
Nixon, B., 522[238] (457)
Nixon, J. F., 184[179] (126, 127), 185[233] (131, 136, 159, 172), 186[321] (138, 155), 223[182] (219), 279[192] (237), 279[193] (237), 529[675] (508), 583[54] (540), 586[238] (560, 569, 570), 587[290] (570), 587[291] (570), 587[292] (570), 588[380] (581)
Nizova, G. V., 531[788] (518), 531[789] (518), 531[790] (518)
Nobel, D., 25[53] (8), 105[377] (80), 105[380] (80, 88), 288[835] (263), 289[856] (264), 289[902] (271)
Nobile, C. F., 24[18] (4), 25[34] (5), 184[183] (126, 128), 287[736] (256, 257), 287[739] (257), 523[282] (461)
Nohira, H., 25[57] (9), 25[58] (9), 100[85] (42, 53, 62), 103[264] (74, 80), 103[269] (74, 75), 183[76] (115, 120, 128), 183[77] (115, 120), 184[188] (126, 128), 280[241] (239, 266, 267), 523[294] (462), 586[235] (560)
Nohlen, M., 104[322] (80, 81)
Nokami, J., 387[257] (347)
Noltes, J. G., 281[300] (241), 530[750] (515)
Nomura, T., 223[155] (215), 390[445] (378)
Nonoyama, K., 285[574] (248, 263)
Nonoyama, M., 282[381], 282[388], 283[452] (245), 284[508] (263, 264), 284[509] (263), 284[510], 284[515], 284[528], 285[573] (248), 285[574] (248, 263), 285[575] (248), 285[576] (248), 529[711] (512)
Nordberg, R. E., 387[226] (341), 389[369] (364), 389[391] (368), 389[392] (368), 389[403] (371), 389[404] (371), 389[405] (371)
Norin, T., 387[226] (341)
Norman, R. O. C., 388[304] (356)
Norrby, P.-O., 386[170] (323)
North, T. E., 27[177] (21), 188[450] (154, 171), 189[549] (171), 189[552] (171)
Norton, J. R., 279[159] (236, 253), 279[163] (236, 250, 253, 261)
Norton, M. G., 220[21] (196)

Norton, P. R., 522[238] (457)
Nosaka, H., 184[186] (126, 128)
Noskov, Yu. G., 388[335] (359)
Nothdurft, A., 103[243] (66, 67), 189[516] (157, 164, 177)
Novaro, O., 521[204] (453)
Novikov, N. A., 388[335] (359)
Nuber, B., 189[556] (172, 173)
Nudelman, A., 389[393] (368)
Nugent, W. A., 184[164] (123)
Numata, S., 279[216] (238)
Nunn, C. M., 27[185] (23), 27[187] (23), 221[51] (201)
Nurse, C. R., 587[343] (577)
Nussbaum, S., 520[94] (440)
Nuti, F., 280[271] (240, 270)
Nutt, W. R., 531[783] (518)
Nuzzo, R. G., 522[218] (455), 522[219] (455), 522[220] (455), 522[224] (455), 582[19a] (535), 582[19b] (535), 582[20] (535)
Nwokogu, G. C., 387[251] (346), 387[252] (346)
Nyberg, G. L., 384[31] (293)
Nyburg, S. C., 384[68] (302)
Nyström, J.-E., 389[404] (371)

O'Bannon, P., 100[26] (32)
Obara, S., 521[205] (453)
Obendorf, D., 188[464] (155, 156, 177)
Odaira, Y., 287[765] (259)
Oehlschlager, A. C., 388[313] (357)
Oehme, G., 185[212] (130, 131), 290[936] (273)
Ogawa, A., 586[254] (562)
Ogawa, H., 222[139a] (213), 222[139b] (213), 277[53] (230, 231), 278[80] (231), 278[81] (231, 232), 527[531], 527[540] (493)
Ogini, W. O., 282[362], 427[13] (393–5, 411, 412), 427[14] (393, 403, 411), 428[102] (403), 529[660] (507, 514), 529[661] (507), 529[662] (507), 529[663] (507)
Ogino, H., 519[62] (437)
Ogoshi, S., 386[213] (336), 587[329a] (575), 587[329b] (575)
Ogura, T., 288[798] (260)
Ohama, Y., 277[25] (228, 229), 279[161] (236, 262, 263), 289[859] (264), 390[436] (376)
O'Hare, D., 585[205a] (554, 555, 581), 585[205b] (554, 555, 581)
Ohe, K., 386[213] (336)
Ohga, K., 527[541] (493)
Ohizumi, O., 287[748] (257)
Ohizumi, T., 222[143] (214)
Ohkatsu, Y., 183[79] (115, 119), 183[80] (115, 119)
Ohkubo, K., 287[719] (255), 429[131] (408), 522[265] (460), 584[138] (551)
Ohmer, J., 26[158] (20), 26[164] (20)
Ohnishi, H., 103[221] (65), 103[222] (65), 186[294] (138), 289[877] (266), 583[68] (542, 545, 547, 574)
Ohno, M., 220[7a] (194), 220[7b] (194), 220[8] (194)
Ohnuma, K., 282[407] (248, 263)
Ohshiro, Y., 25[29] (4)
Ohshita, J., 103[270] (73, 74), 184[171] (125, 126)
Ohsuka, A., 587[326] (574, 579)
Ohta, T., 386[208] (335)
Ohyoshi, A., 388[350] (360)
Ohyoshi, M., 103[269] (74, 75), 280[241] (239, 266, 267)
Oikawa, T., 283[454] (245, 263)
Oka, H., 287[774] (259)
Okada, M., 387[255] (347)
Okada, N., 285[608] (249)
Okamoto, H., 278[97] (233, 255, 262)
Okamoto, T., 388[308] (356), 388[310] (357)
Okamura, M., 287[774] (259)
Okano, T., 105[415] (42, 93, 95), 185[254] (131, 134), 279[155] (235)
Okawa, H., 287[775] (259)
Okazaki, S., 184[198] (125)
Okeya, S., 280[238] (239, 255, 256, 263), 281[307] (241), 281[308] (241), 281[340] (242, 244, 245, 260, 264, 268), 288[800] (260, 261), 288[803] (260), 288[806] (261), 288[841] (263), 288[842] (263)
Okhlobystin, O. Yu., 279[201] (237, 264, 272), 289[913] (272), 387[264] (348)
Oki, A. R., 191[636] (180)
Öki, M., 584[106] (547)
Okuda, J., 188[479] (157, 159, 166)
Okunaka, M., 277[21] (228, 229)
Ol'dekop, Y. A., 191[626] (180, 182), 191[629] (180, 182)
Olesker, A., 282[400]
Oliver, D. L., 586[223] (557)
Oliver, J. P., 223[153] (215), 429[175] (414)
Ollivier, J., 389[365] (363)
Olmstead, M. M., 26[146] (20), 184[172] (126), 222[115] (210, 215), 223[151] (215), 278[119] (234, 236), 520[110] (441)
Olsen, D. J., 386[180] (325, 326)
Ol'shevskaya, V. A., 277[74] (230)
Olsson, A., 384[57] (296), 584[102] (547)
Olsson, L.-F., 285[589] (249), 384[57] (296)
Omae, I., 281[324] (242), 281[325] (242), 281[326] (242), 281[327] (242), 281[328] (242), 281[329] (242, 248), 584[143] (551), 585[159] (553)
O'Mahoney, C. A., 283[491]
Onchi, M., 384[35] (293), 384[36] (293)
Onishi, M., 223[174a] (218), 277[25] (228, 229), 278[125] (234, 262, 263), 279[161] (236, 262, 263), 279[181] (236, 264), 281[309] (241), 281[310] (241, 262, 263), 281[318] (241), 286[672] (253), 289[855] (263, 269), 289[859] (264), 385[84] (304), 390[436] (376)
Onitsuka, K., 222[139b] (213), 222[140] (213), 277[52] (230, 231), 277[53] (230, 231), 278[80] (231), 278[81] (231, 232), 278[82] (232), 527[531], 527[540] (493), 527[542] (493), 527[544] (494)
Oohama, T., 25[82] (11), 184[163] (122), 525[453] (480)
Oosawa, Y., 277[40] (230)
Operti, L., 430[224] (426)
Orama, O., 277[33] (228)
Orbell, J. D., 521[172], 525[410] (475)
Oreshkin, I. A., 100[83] (41, 47)
Orlandini, A., 184[191] (126, 129), 184[192] (126, 129), 184[196] (126, 129), 184[197] (126, 129), 187[354] (139, 147), 187[355] (139, 147), 187[372] (146), 278[122] (234), 278[123] (234), 524[335] (467), 529[686] (509)
Ormerod, R. M., 384[30] (293)
Oro, L. A., 387[271] (350), 387[272] (350), 387[273] (350)
Orpen, A. G., 220[15] (195), 223[159] (215), 385[81] (303), 427[65a] (400), 519[24] (434), 520[95] (440), 521[154] (445), 521[161], 522[222] (455), 524[378] (471), 524[379] (471), 529[688] (509), 582[13b] (535, 537, 539, 566), 583[30] (536, 538, 539)
Orrell, K. G., 521[194] (450, 500), 527[580] (498), 527[582] (498, 504), 527[583] (498), 527[584] (498, 504), 527[585] (498, 504), 527[586] (498), 527[587] (498, 504), 528[588] (498, 504), 528[589] (498), 528[596] (499), 528[605] (499), 528[608] (499, 504), 528[609] (499, 500), 528[612] (500, 504), 528[613] (500, 504), 528[614] (500, 504), 528[630] (500), 528[631] (504), 528[632] (504), 528[633] (504), 528[634] (504)
Ortega, A. M., 277[30] (228)
Ortega, R., 220[28] (197, 201)
Ortigueira, J. M., 284[554] (245, 263), 289[854] (263)
Ortiz, A., 102[155] (42, 56, 62, 63)
Ortiz, J. V., 519[31] (435)
Ortmann, I., 104[302] (81), 185[214] (130, 133)
Ortoleva, E., 385[140] (316), 385[141] (316)
Osa, T., 183[79] (115, 119), 183[80] (115, 119)
Osakada, K., 25[25] (4), 101[88] (42, 53, 54), 101[89] (42, 47, 53, 58, 62, 63), 101[130] (42, 62), 101[137] (42, 51, 61), 101[140] (42, 53, 62), 102[184] (48, 55, 61, 62), 102[211] (63), 104[305] (81, 85), 222[99a] (208), 222[99b] (208), 222[106] (208), 279[189] (237), 280[233] (238, 262, 263, 265, 268), 280[234] (238), 286[688] (255), 286[697] (255), 286[709] (255, 256), 287[722] (256, 265, 268, 271), 287[734] (256), 289[887] (268, 269, 271), 387[242] (344), 387[243] (344), 388[336] (359), 521[157], 523[281] (461)
Osborn, J. A., 26[125] (16), 278[108] (234, 255, 256), 278[115] (234)
Osborne, A. G., 527[586] (498), 528[596] (499)
Oschmann, W., 588[381] (582)
Osella, D., 189[542] (167)
Oshima, M., 387[241] (344), 388[314] (357)
Oshima, N., 285[583] (249), 387[280] (352)
Oskam, A., 104[327] (81, 86), 530[721] (513)
Ossain, M. B., 289[900] (270)
Ossor, H., 280[226] (238, 240, 264, 265), 281[296] (241), 281[353] (243, 253,

255, 271), 390[467] (382), 529[705] (511)
Oster, B. W., 104[306] (81, 84), 104[307] (81), 104[355] (81, 88, 90)
Ostoja Starzewski, K. A., 100[57] (31, 36), 100[58] (31, 36), 100[59] (31, 36), 100[61] (31, 36), 100[62] (31, 36)
O'Sullivan, R. D., 283[445] (245), 286[696] (255), 428[98] (403, 406)
Oswald, B., 222[90] (205), 222[91] (205), 222[92] (206), 222[94] (207), 427[16] (393)
Oswald, H. R., 190[586] (175, 177)
Otis, D. E., 519[22] (434), 531[784] (518), 586[274] (565)
Otsuka, K., 387[240] (344)
Otsuka, S., 221[50c] (201), 290[960] (276)
Otte, A. R., 388[353] (361)
Ou, E. C., 522[238] (457)
Ovalles, C., 25[70] (10)
Overbeek, O., 184[201] (125), 184[203] (125)
Overbosch, P., 184[201] (125), 184[202] (125), 184[203] (125), 184[204] (125, 157), 429[164] (411, 412)
Owen, M. D., 526[485] (485)
Ozaki, A., 100[81] (40)
Ozawa, F., 102[179] (56, 62), 102[180] (56), 222[98b] (207), 222[99a] (208), 222[99b] (208), 222[99c] (208), 222[106] (208), 278[97] (233, 255, 262), 278[138] (235, 255), 279[166] (236, 257, 259), 279[189] (237), 280[231] (238, 267), 280[232] (238, 239, 262, 266, 267), 280[234] (238), 280[246] (239, 240), 280[247] (239, 255, 266), 280[249] (239, 262), 280[250] (239), 280[270] (240, 267), 280[272] (240, 267, 270), 283[485] (245, 257), 285[631] (250), 286[688] (255), 286[697] (255), 286[705] (255, 256), 287[712] (255, 267, 268), 287[714] (255), 287[725] (256, 257), 287[734] (256), 287[735] (256, 257, 270), 288[780] (260), 289[876] (266), 289[879] (267, 269, 270), 289[904] (271), 387[239] (344), 387[241] (344), 387[242] (344), 388[336] (359), 429[145] (408), 521[153] (444), 523[290] (462)
Ozawa, Y., 286[709] (255, 256), 289[887] (268, 269, 271), 387[243] (344)
Ozin, G. A., 185[256] (129), 220[2a] (194), 220[2b] (194)

Paàl, Z., 384[28] (293)
Pacchioni, G., 220[12b] (195), 222[104] (208)
Paciello, R. A., 521[182] (445)
Packett, D. L., 583[22] (535, 565–7)
Paddock, N. L., 26[126] (16)
Paiaro, G., 184[170] (125)
Pain, G. N., 282[422] (243), 287[770] (259)
Pain, H. M., 528[596] (499)
Pakhomova, I. V., 428[108a] (403)
Pal, C. K., 284[533] (268)
Palenik, G. J., 284[545], 525[438] (479), 586[228] (558)
Palkovits, H., 521[199] (452)
Palm, C., 21, 27[173] (21)
Palma, P., 25[87] (11), 25[88] (12), 25[92] (13), 100[78] (38, 39, 42, 56), 101[108] (42, 47, 58, 62, 63), 101[118] (42, 61), 102[220] (65), 104[293] (73–5, 78, 81, 90), 104[325] (81, 85), 184[173] (125, 126, 157), 186[296] (138, 143), 186[343] (139), 187[371] (146)
Palmans, R., 530[723] (513)
Palmer, D. A., 522[253] (459)
Palmer, J. A. L., 283[441] (244, 263)
Palmieri, G., 282[405]
Pan, Y. H., 100[29] (32), 386[154] (317), 386[155] (318)
Pandey, K. K., 289[912] (272)
Pandolfo, L., 184[170] (125)
Paneque, M., 24[16] (3), 24[17] (3), 100[78] (38, 39, 42, 56), 100[80] (38–40, 42), 101[101] (41, 42, 48, 56, 58, 66, 67), 101[103] (41, 42, 47, 57, 58, 67), 101[104] (42, 47, 51, 57, 58, 66, 67), 101[118] (42, 61), 101[139] (42, 56), 102[188] (48, 56), 103[268] (74, 78, 81), 104[293] (73–5, 78, 81, 90), 104[295] (74, 78), 104[296] (74, 78), 183[108] (115, 124), 183[109] (115, 124), 184[173] (125, 126, 157), 186[340] (139, 146), 186[342] (139, 146, 157), 187[371] (146), 188[482] (159), 288[843] (263)
Pang, K., 182[54] (111, 123), 182[55] (111, 123)
Pank, V., 277[44] (230)
Pankratov, L. V., 189[512] (157, 168, 175, 177)
Panunzi, A., 278[121] (234), 280[256] (239, 262, 272), 385[96] (306), 520[82] (439), 583[32] (536), 583[72] (542, 544, 545, 548, 549), 583[76] (543), 583[78] (543), 583[79] (543), 583[80] (543), 583[81] (543), 583[82] (544), 583[83] (544), 583[84] (544), 585[212] (554), 586[233] (559)
Paolesse, R., 527[546] (494)
Paonessa, R. S., 384[72] (302), 427[5] (392), 583[23a] (535), 583[23b] (535)
Papageorgiou, F., 26[166] (20, 21), 190[577] (174)
Paquette, L. A., 390[432] (375)
Paquette, M. S., 189[554] (171, 172)
Paraskewas, S. M., 386[201] (333)
Pardy, R. B. A., 186[295] (138)
Parenago, O. P., 384[42] (294)
Parge, H. E., 280[276] (240), 429[139] (408), 526[480] (484, 487)
Parisi, M. F., 519[72] (439)
Park, J. W., 288[780] (260), 387[239] (344), 521[153] (444)
Park, S., 385[114] (310), 521[163], 522[239] (458)
Parker, D., 278[135] (235), 384[71] (302), 582[14b] (535), 582[14c] (535, 539), 583[48] (538)
Parkhurst, C. S., 389[396] (368)
Parkin, I. P., 283[491]
Parkins, A. W., 283[445] (245), 285[571] (248), 286[696] (255), 428[98] (403, 406), 587[345] (577)
Parkinson, B. A., 184[169] (125), 586[252] (561)
Parnell, D. R., 280[282] (240)
Parra-Hake, M., 285[601] (249), 285[602] (249), 384[53] (295), 385[108] (309), 387[227] (341), 387[230] (342)
Parshall, G. W., 289[886] (268, 270)
Parsons, E. J., 527[574] (497, 505), 527[575] (497), 584[129] (549, 563), 586[256] (562)
Parthasarathi, V., 105[415] (42, 93, 95), 185[254] (131, 134)
Parvez, M., 278[132] (235), 285[607] (249), 285[629] (250, 261, 262, 264), 286[656] (252), 290[959] (276), 385[90] (305), 428[125c] (407), 523[308] (464), 523[319] (465), 528[595] (498, 499)
Pasqualetti, N., 428[108b] (403, 409), 429[150c] (409, 411), 583[65] (542)
Pasquali, M., 390[469] (383)
Pasquetto, A., 222[130] (211), 287[750] (257)
Pasto, D. J., 103[280] (74, 75), 103[287] (75), 183[124] (115, 121), 183[125] (115, 121)
Pasynkiewicz, S., 103[228] (67, 72), 103[229] (66, 67), 103[230] (67, 72), 103[234] (66, 67, 72), 103[235] (66, 67), 103[238] (66, 67, 72), 103[244] (66, 67), 103[248] (67, 72), 105[413] (93, 94), 182[13] (108, 130–2, 138, 144, 157), 182[14] (108), 183[68] (113), 183[72] (114, 115, 157, 171), 184[148] (117, 163), 185[216] (130–2, 138, 157), 185[261] (132, 138, 144), 189[517] (157, 166), 189[548] (171)
Patai, S., 277[12] (227, 228), 278[85] (233–5, 266, 273), 383[10] (292), 383[11] (292), 384[12] (292), 384[13] (292), 384[14] (292), 384[15] (292, 333), 384[23] (292)
Patalinghug, W. C., 521[150] (444)
Patin, H., 386[199] (333)
Patrick, J. M., 520[93] (440)
Patrini, R., 427[45] (398)
Paul, J.-F., 384[29] (293)
Paul, P., 284[549] (245, 263, 264), 284[550] (245, 264), 284[551] (245, 263, 264)
Paul, R., 182[61] (111, 115, 174, 175), 184[143] (117, 139, 141)
Pauley, D. J., 187[412] (151)
Paulik, F. E., 583[56] (541)
Pauson, P. L., 103[250] (67, 70), 281[297] (241, 255)
Payne, J. D., 188[441] (152, 155)
Payne, N. C., 220[48] (200), 286[669] (253), 386[150] (317), 386[151] (317), 388[340] (360), 428[82] (403), 428[85] (404), 429[179] (415), 521[158], 521[160], 523[302] (463), 523[312] (464), 585[191] (554, 555, 557), 586[278] (566, 567), 586[279] (566, 567)
Peacock, R. D., 587[305] (572)
Pearson, R. G., 280[239] (239, 240)
Pebler, J., 100[64] (31, 36)
Peganova, T. A., 103[282] (73, 74), 183[119] (115, 117, 119), 184[158] (119), 186[304] (138, 139, 143)
Pelizzi, G., 220[18] (195, 196), 429[150b] (409–11)
Pelli, B., 283[481] (245, 248), 587[336]

(576)
Pellinghelli, M. A., 287[739] (257), 523[282] (461)
Pellizer, G., 285[610] (249), 525[449] (480), 583[37] (536, 539), 583[39] (536, 538)
Peluso, A., 103[263] (73)
Peng, S. M., 277[39] (229, 230)
Perego, M., 26[147] (20)
Peregudov, A. S., 519[40] (436)
Pereira, M. T., 279[208] (238, 243, 253, 254), 282[359] (243), 283[458] (245), 283[459] (245), 283[460] (245, 263, 264), 283[468] (245), 283[472] (245, 263), 283[476] (245, 263, 264), 284[554] (245, 263), 284[558] (245)
Perera, S. D., 277[62] (230), 520[109] (441)
Peres, Y., 25[43] (7), 104[346] (81, 88), 104[349] (81, 88, 89), 104[350] (81, 88)
Peretz, M., 389[376] (365)
Perevalova, E. G., 188[457] (154)
Perez, J., 284[559] (246, 263), 284[560] (246, 263, 268)
Perez, S., 189[500] (157)
Perez-Briso, C., 281[288] (240, 262, 264)
Perez-Moya, L. A., 25[56] (9), 183[74] (115), 183[75] (115, 126, 128)
Périchon, J., 25[64] (10), 25[65] (10), 25[66] (10), 25[67] (10), 25[68] (10), 25[69] (10), 101[125] (42, 55, 63), 105[366] (81, 88, 90)
Peringer, P., 519[23] (434)
Perkins, D. C. L., 527[570] (497, 505), 528[646] (505), 528[650] (506), 584[125b] (549)
Perkins, P. S., 528[598] (499), 528[599] (499)
Permin, A. B., 520[139] (443)
Perreault, D., 220[45] (199), 223[158] (215)
Perry, G., 282[428] (244, 270)
Peter, M., 519[23] (434)
Peters, E. M., 221[77] (204)
Peters, K., 221[77] (204)
Peters, T. B., 104[341] (81, 87)
Peterson, E. S., 527[568] (497)
Peterson, J. L., 524[380]
Petit, F., 388[289] (353)
Petrignani, J. F., 25[80] (11), 25[81] (11)
Petrosyan, V. S., 520[139] (443)
Petrov, E. S., 287[732] (256), 388[335] (359)
Petrov, G. B., 386[174] (324)
Petrova, N. E., 287[732] (256)
Petrovskii, P. V., 101[124] (41, 42, 47), 102[164] (41, 42, 47, 51, 58), 183[119] (115, 117, 119), 184[158] (119), 185[212] (130, 131), 186[304] (138, 139, 143), 220[30] (197), 279[156] (235), 279[203] (237)
Petter, W., 427[42a] (397)
Pettersson, L. G. M., 385[124] (312), 385[125] (312), 386[173] (324)
Peuckert, M., 184[137] (116, 124, 140)
Pfaltz, A., 388[348] (360)
Pfeffer, M., 100[67] (37, 38), 104[304] (81, 85), 222[137a] (213), 222[137b] (213), 279[208] (238, 243, 253, 254), 280[226] (238, 240, 264, 265), 281[292] (241, 253), 281[293] (241, 270, 273), 281[296] (241), 281[335] (242, 253), 281[339] (242, 243, 263), 281[349] (242, 248), 281[353] (243, 253, 255, 271), 282[365] (253, 254, 263), 282[386], 282[387] (253), 282[395] (243, 244, 263), 282[398] (243), 283[434] (244, 253), 283[437] (244, 253, 254), 285[577] (248, 253), 285[619] (250), 286[673] (253), 286[675] (253), 286[676] (253, 264), 286[677] (253, 262, 263), 286[678] (253, 254, 262, 263), 286[679] (253), 286[680] (253), 286[681] (253), 286[684] (253), 286[685] (254, 264), 286[686] (254, 271), 286[687] (254), 286[692] (255, 258), 286[693] (255, 256, 258, 263), 288[836] (263), 289[868] (264), 386[164] (320), 386[165] (320), 386[166] (320), 386[167] (321), 386[169] (323), 387[261] (348), 390[467] (382), 529[705] (511), 583[29] (536, 544), 585[160] (553)
Pfluger, F., 278[93] (233)
Pham, E. K., 586[241a] (560), 586[241b] (560)
Phillips, I. G., 284[552] (245, 246, 264), 428[109] (405, 416)
Phillips, J., 524[373] (471), 586[225] (558)
Phillips, J. R., 527[530] (493)
Philoche-Levisalles, M., 283[492] (262, 263), 285[611] (249), 288[825] (262, 272), 385[106] (308)
Piana, H., 387[275] (350)
Piazzesi, A. M., 287[726] (256, 272), 287[727] (256), 520[136] (443, 461), 520[137] (443), 521[155] (472, 493), 521[168], 524[381] (472), 524[383] (472)
Piazzi, O., 223[173a] (218)
Piccolo, O., 186[292] (138, 143)
Pickl, W., 187[394] (150)
Pidcock, A., 279[192] (237), 519[34] (436), 519[35] (436), 519[37] (436), 519[38] (436), 519[42] (437), 519[65] (438), 519[66] (438), 521[188] (447), 527[527] (491), 586[240] (560), 587[281] (568)
Piepgrass, K. W., 190[594] (175, 179), 190[601] (175, 177)
Piergiovanni, A., 287[739] (257), 523[282] (461)
Pierpont, C. G., 277[20] (227, 230), 387[262] (348), 387[263] (348), 427[60] (399), 587[294] (571)
Pierrot, M., 25[80] (11)
Pietropaolo, R., 280[280] (240), 285[591] (249), 289[882] (267)
Pietrzykowski, A., 103[238] (66, 67, 72), 103[248] (67, 72), 189[517] (157, 166), 281[297] (241, 255)
Pignolet, L. H., 288[801] (260), 288[805] (261), 518[7] (432, 441), 526[489] (486, 487)
Pilbrow, M. F., 428[93] (403, 411), 528[638] (505)
Pinhas, A. R., 100[32a] (31, 33), 100[32b] (31, 33), 183[103] (115), 183[104] (115), 583[36] (536, 539)
Pinillos, M. T., 387[271] (350), 387[272] (350)
Pinkerton, A. A., 220[26] (197, 200), 288[781] (260), 386[158] (319), 390[435] (375)
Pinna, F., 523[287] (462), 525[412] (475), 525[413] (475), 525[414] (475), 525[416] (475), 525[417] (475), 525[418] (475, 476)
Piotrowski, A. M., 103[284] (73, 74), 104[299] (73, 74), 183[91] (115, 119), 184[156] (119, 122, 131), 187[380] (148), 187[389] (148), 187[390] (148)
Pipal, J. R., 190[602] (175), 190[603] (175)
Piraino, P., 390[437] (376), 588[364] (580)
Piras, P. P., 389[365] (363)
Pittel, B., 102[219] (65), 186[302] (138, 140)
Pittman, C. U., Jr., 189[559] (172, 173), 221[79] (205), 389[419] (373)
Pitzer, R. M., 187[386] (147)
Pizzotti, M., 525[456] (480)
Plaia, U., 25[77] (11), 25[78] (11, 13), 189[507] (157, 168), 223[170] (217, 218), 223[175] (218), 430[222] (425)
Platt, A. W. G., 519[44] (437), 527[583] (498), 527[584] (498, 504), 529[690] (509), 585[175] (554, 555), 587[325] (574, 578)
Plesek, J., 191[633] (180, 181)
Pletcher, D., 188[445] (153)
Plumb, C. A., 191[637] (180)
Plummer, D. T., 25[76] (10)
Pluta, C., 104[302] (81), 182[37] (110, 112), 184[151] (117, 118), 185[214] (130, 133)
Poilblanc, R., 220[41] (198), 429[162] (410), 520[114] (441)
Poinsignon, C., 187[410] (150)
Poirier, J., 389[416] (372)
Polborn, K., 287[778] (260), 586[224] (558)
Pold, C., 102[156] (42, 50, 55)
Poli, R., 588[357] (579)
Poliakoff, M., 24[5] (2), 24[6] (2), 427[64] (400), 588[370] (581)
Poller, R. C., 286[667] (252)
Pollina, D. M., 388[296] (354)
Polo, A., 287[745] (257)
Polovnyak, V. K., 220[31a] (197), 220[31b] (197), 220[32] (197), 220[33] (197)
Polyachenok, L. D., 429[147] (409)
Polyachenok, O. G., 429[147] (409)
Polyakov, V. A., 282[384] (248, 262), 282[426] (244)
Pombeiro, A. J. L., 223[168] (217), 390[448] (378), 430[216] (424, 426)
Pombrik, S. I., 519[40] (436)
Pommier, C., 187[408] (150)
Ponec, R., 584[137] (551)
Ponec, V., 220[9] (195), 222[104] (208)
Pontellini, R., 388[334] (359), 388[354] (361), 428[115] (406), 587[319] (574, 576), 587[320] (574, 576)
Pontier-Johnson, M., 521[163], 521[174], 522[239] (458)
Poplawska, A., 185[261] (132, 138, 144)
Poplawska, J., 103[234] (66, 67, 72), 105[413] (93, 94), 185[216] (130–2, 138, 157), 189[548] (171)
Poplawska, M., 103[235] (66, 67), 103[238] (66, 67, 72)

Popod'ko, N. R., 102[169] (50)
Poppitz, W., 102[186] (41, 47, 48)
Porai-Koshits, M. A., 221[60c] (202), 221[61b] (202), 222[93a] (207), 222[93b] (207), 222[93c] (207), 283[446] (245), 388[293] (354)
Porri, L., 390[463] (381)
Pörschke, K.-R., 26[100] (14), 26[124] (16), 100[31] (31, 33), 100[50] (31, 34), 100[51] (31, 34, 35), 100[52] (31, 34, 35), 100[55] (34), 100[56] (31, 36), 102[183] (41, 47, 48, 56, 63), 103[273] (73, 74), 103[274] (73, 74), 103[275] (73, 74, 80), 103[277] (73, 74, 80), 103[278] (73, 74), 104[302] (81), 182[33] (110, 111, 117, 130–3), 182[34] (110, 111, 117, 130–2), 182[35] (110, 112, 113, 115), 182[37] (110, 112), 182[38] (110, 111), 182[39] (110), 182[40] (110), 182[41] (110, 112), 182[44] (110, 113), 182[45] (110), 182[46] (110, 111, 113, 115, 118, 126), 182[47] (110, 113, 117, 118), 182[48] (111, 128), 182[49] (111, 128), 182[59] (111, 117, 130–3), 183[62], 183[63] (111, 115), 183[65] (113, 117, 131, 133), 183[66] (113, 118), 183[70] (115), 183[82] (115, 119), 183[115] (115, 117, 118), 183[117] (115), 184[150] (117, 118), 184[151] (117, 118), 185[210] (130, 132), 185[214] (130, 133), 185[223] (130, 131, 134), 185[262] (133), 185[263] (133), 288[790] (260, 266), 385[82] (303, 318)
Porta, F., 278[145] (235), 525[456] (480)
Porter, P. L., 278[79] (231)
Portilla, A., 387[273] (350)
Portnoy, M., 220[34] (197)
Porzio, W., 529[714]
Potapov, V. M., 281[323] (242), 283[435] (244), 283[436] (244, 245), 283[446] (245)
Poveda, M. L., 24[16] (3), 24[17] (3), 25[87] (11), 25[88] (12), 25[89] (12), 25[92] (13), 100[78] (38, 39, 42, 56), 100[80] (38–40, 42), 101[99] (42, 56), 101[100] (42, 56), 101[101] (41, 42, 48, 56, 58, 66, 67), 101[103] (41, 42, 47, 57, 58, 67), 101[104] (42, 47, 51, 57, 58, 66, 67), 101[106] (42, 47, 50, 58, 81, 90), 101[107] (42, 47, 58, 62, 63), 101[108] (42, 47, 58, 62, 63), 101[118] (42, 61), 102[187] (48, 56, 62), 102[220] (65), 103[268] (74, 78, 81), 104[293] (73–5, 78, 81, 90), 104[294] (74, 78, 81), 104[295] (74, 78), 104[296] (74, 78), 104[325] (81, 85), 183[108] (115, 124), 183[109] (115, 124), 184[134] (116, 117, 157), 184[173] (125, 126, 157), 184[193] (126, 129), 186[296] (138, 143), 186[340] (139, 146), 186[342] (139, 146, 157), 186[343] (139), 186[345] (139), 187[371] (146), 188[482] (159)
Powell, D. R., 25[75] (10, 16, 17), 220[39a] (198)
Powell, H. H., 529[690] (509)
Powell, H. R., 221[62] (202, 209, 215)
Powell, J., 102[213] (62, 64), 220[25] (197), 285[621] (250), 289[903] (271), 390[434] (375), 588[371] (581)
Powell, P., 383[11] (292)
Power, P. P., 26[146] (20), 184[172] (126)
Power, W. J., 185[256] (129), 220[2b] (194)
Pozdeeva, A. A., 102[169] (50)
Prados, P., 288[821] (262)
Prasad, B. V., 186[347] (139, 146)
Prat, M., 389[378] (366)
Prati, L., 386[184] (327), 386[185] (328)
Prechtel, W., 223[177] (218), 430[202] (421)
Predieri, G., 185[241] (131, 172, 174), 185[250] (131, 172, 174), 190[567] (172, 174)
Pregosin, P. S., 222[141] (214), 277[73] (230, 231, 240), 283[432] (244, 248, 264), 283[433] (244, 245, 248), 283[467] (245), 283[483] (245, 263), 283[484] (245), 283[489] (258, 263, 273), 284[565] (247, 263), 285[618] (250, 252), 386[186] (328), 386[187] (328), 386[188] (328), 386[189] (328), 390[433] (375), 427[21] (393, 395), 427[26] (394, 395), 427[27] (395), 428[107] (403), 430[211] (423), 519[74] (439), 520[144] (444), 521[195] (450), 521[196] (450), 523[276] (461), 525[394] (474), 529[699] (510), 529[702] (511), 529[709] (512), 529[710] (512), 583[63] (542), 584[109] (547, 576)
Press, C. F. C., 384[25] (292)
Preston, P. N., 26[107] (15)
Pribytkova, I. M., 105[412] (94)
Prignano, A. L., 384[72] (302), 583[23b] (535)
Pring, G. M., 528[612] (500, 504), 528[634] (504)
Pringle, P. G., 26[116] (16), 277[47] (230), 288[817] (262, 265), 429[189] (417), 518[8] (432, 494), 519[44] (437), 520[95] (440), 520[118] (441, 442, 445, 463), 521[161], 524[377] (471), 524[378] (471), 524[379] (471), 525[425] (478), 525[427] (478), 525[437] (479), 526[513] (490), 526[514] (490), 526[515] (490), 526[516] (490), 527[547] (494), 527[548] (494), 527[549] (494), 527[550] (494), 527[551] (494), 527[552] (494), 529[690] (509), 582[13b] (535, 537, 539, 566), 585[175] (554, 555)
Priolo, F. C., 285[591] (249)
Pritzkow, H., 183[98] (115, 155, 157, 181), 183[101] (115, 124), 188[423] (151, 180, 181), 189[520] (159, 175), 190[590] (175, 178), 190[591] (175, 178), 190[593] (175, 179), 190[595] (175, 179), 190[597] (175, 180), 190[618] (178), 190[619] (178), 191[625] (180, 181), 191[639] (181), 191[640] (181)
Probst, M., 519[23] (434)
Proft, B., 183[117] (115), 184[151] (117, 118)
Prokoph'ev, A. I., 279[201] (237, 264, 272), 289[913] (272), 387[264] (348)
Propkopovich, V. P., 191[629] (180, 182)
Proserpio, D. M., 221[57] (201)
Prössdorf, W., 187[415] (151, 155), 188[424] (151, 155)
Protas, J., 221[75a] (204)
Prouse, L. J. S., 222[131] (212), 429[184] (416, 417), 429[185] (416), 525[443] (479), 525[446] (480), 526[460], 526[468] (482), 526[469] (482), 588[351] (579), 588[353] (579)
Provencher, R., 220[46] (199), 428[78c] (402)
Pruchnik, F. P., 582[11] (534)
Pucci, D., 282[425] (243, 264), 284[505] (263, 264), 284[506] (263)
Puckett, W. E., 281[316] (241), 281[320] (241, 242), 385[91] (305), 529[691] (510)
Puddephatt, R. J., 26[115] (16), 26[120] (16), 186[277] (137), 198, 220[37] (197, 214), 220[42a] (198, 199), 220[42b] (198, 199), 220[43] (199, 200), 220[44] (199), 220[47] (200), 220[48] (200), 221[49] (200), 222[111] (209), 222[112] (209), 279[212] (238), 286[667] (252), 286[670] (253), 290[947] (273, 274), 290[948] (273, 274, 276), 290[951] (273), 386[162] (320), 386[172] (324), 392, 401, 427[24] (394, 402, 404), 428[72] (401), 428[74] (401, 404), 428[75] (401, 402), 428[77] (402, 404), 428[79] (402), 428[80] (403), 428[81] (403), 428[82] (403), 428[83] (403), 428[84] (404), 428[85] (404), 428[86] (404), 428[87] (404), 428[88] (404), 428[89a] (404), 428[89b] (404), 428[90] (404, 415), 428[97] (403, 409–11), 429[178] (414, 415), 429[179] (415), 430[196] (419), 430[197] (419), 518[6] (432), 518[7] (432, 441), 518[10] (432, 433, 453), 518[11] (432, 433), 518[12] (432, 453), 518[13] (433), 519[43] (437), 519[45] (437), 520[88] (439), 520[116] (441), 520[117] (441, 442), 520[120] (442), 520[122] (442), 520[123] (442), 520[124] (442, 463), 520[129] (442, 495), 521[164], 521[193] (450), 521[200] (452), 522[210] (453), 522[223] (455), 522[238] (457), 522[242] (458, 462), 522[243] (458), 522[245] (458), 523[295] (463), 523[296] (463), 523[298] (463, 466), 523[299] (463), 523[300] (463), 523[301] (463), 523[302] (463), 523[307] (464), 523[311] (464), 523[312] (464), 523[313] (464), 523[314] (465), 523[315] (465), 523[316] (465), 523[317] (465), 523[319] (465), 523[320] (465), 523[321] (466), 523[322] (466), 523[323] (466), 523[324] (466), 524[392] (473), 526[466] (481), 526[482] (485), 526[483] (485), 526[512] (490), 526[517] (490), 526[518] (490), 526[521] (490), 527[570] (497, 505), 527[571] (497), 527[572] (497), 528[617] (510), 528[618] (510), 528[625] (500), 528[637] (505), 528[639] (505), 528[641] (505), 528[642]

(505), 528[644] (505), 528[646] (505), 528[649] (505), 528[650] (506), 528[651] (506), 528[652] (506), 528[653] (506), 583[70] (542), 583[71] (542), 584[124] (549), 584[125a] (549), 584[125b] (549), 584[125c] (549), 585[214a] (556), 585[214b] (556), 585[214c] (556), 586[253a] (562), 586[253b] (562), 586[261] (563), 587[286] (569), 587[295] (571), 587[296] (571), 587[300] (571), 587[301] (572), 587[302] (572), 587[333] (575), 588[376] (581)
Pueyo, L., 222[124] (211, 213), 287[760] (258, 263)
Puga, J., 427[45] (398)
Pukhnarevich, V. B., 430[207] (422)
Purucker, B., 287[737] (256)
Pye, P. L., 25[93] (13)

Qichen, H., 101[134] (42, 47)
Quaranta, E., 184[182] (126, 128)
Quessy, S. N., 283[469] (245)
Quett, C., 387[238] (344)
Quincke, F., 2, 24[1] (1, 2)
Quyoum, R., 188[478] (157, 159, 167), 429[159] (410)

Raabe, E., 183[131] (116, 174, 176)
Radecka-Paryzek, W., 520[122] (442)
Radeglia, R., 186[293] (138, 142)
Radonovich, L. J., 24[11] (2), 100[35] (33), 100[42] (33), 100[43] (31, 33), 100[44] (31, 33), 183[92] (115, 122), 190[606] (174), 190[607] (174)
Rae, A. D., 188[449] (154, 159), 190[568] (171–3)
Raghavan, N. S., 388[305] (356)
Raghu Veer, K. S., 26[123] (16)
Rahimi, R., 105[401] (92)
Rahman, A. F. M., 223[153] (215)
Rahman, H. U., 282[366] (270, 273)
Raida, V. S., 282[367]
Raithby, P. R., 284[520] (261), 529[694] (510)
Rakowsky, M. H., 585[173] (554, 556, 557, 561), 585[174] (554, 557, 558)
Ram, M. S., 105[389] (92), 105[390] (91, 92), 105[391] (91, 92), 105[392] (91, 92), 105[393] (91, 92)
Ramachandran, R., 428[82] (403)
Ramacher, L., 188[448] (154, 155, 157), 188[471] (154, 155), 188[472] (155), 188[473]
Ramamoorthy, V., 528[602] (499)
Ramirez, J. A., 278[123] (234), 524[335] (467)
Ramos, J. M., 288[825] (262, 272)
Randaccio, L., 521[172], 525[410] (475)
Rao, G. S., 430[206] (422), 530[737] (514), 582[4] (534)
Rao, K. M., 429[159] (410)
Rasch, G., 100[24b] (30)
Rashidi, M., 221[49] (200), 222[111] (209), 222[112] (209), 386[162] (320), 428[88] (404), 429[178] (414, 415), 430[197] (419), 518[6] (432), 523[295] (463), 523[296] (463), 526[466] (481), 526[483] (485), 586[261] (563), 587[301] (572)
Raspel, B., 288[793] (260)
Raston, C. L., 525[428] (478), 585[186] (554, 555)
Ratananukul, P., 388[320] (357)
Ratermann, A. L., 27[186] (23)
Rath, N. P., 26[108] (15), 105[419] (95, 96), 184[160] (121, 144), 184[161] (121, 139, 144, 159), 519[63] (437, 458), 520[96] (440), 520[113] (441), 521[165], 585[189] (554, 555), 588[360] (580), 588[365b] (580)
Ratliff, K. S., 26[128] (16, 17), 27[178] (21, 22), 27[179] (21, 22)
Ratovskii, G. V., 280[237] (239)
Rattray, A. J. M., 220[21] (196)
Raubenheimer, H. G., 103[247] (67, 70), 189[491] (157)
Rauchfuss, T. B., 187[385] (148, 149)
Raupach, L., 24[19] (4)
Ravenek, W., 99[16] (30, 32)
Rawle, S. C., 290[957] (275)
Raybuck, S. A., 105[407] (91)
Raymond, K. N., 188[463] (155, 156)
Razuvaev, G. A., 189[538] (159), 289[914] (272), 289[915] (272)
Re, N., 102[178] (53)
Reddy, A. C., 190[617] (177)
Reddy, G. K. N., 385[104] (308)
Redhouse, A. D., 102[168] (47, 50, 54), 182[60] (111, 113), 279[198] (237), 587[344] (577)
Redondo, A., 223[148] (214), 277[8] (226, 227, 252)
Reese, D. G. F., 280[229] (238)
Reger, D. L., 279[214] (238, 240, 250, 261, 265), 280[251] (239), 285[626] (250, 252, 253, 255), 519[64] (437), 522[262] (460)
Regitz, M., 187[384] (148, 149)
Rehder, D., 587[330] (574, 575)
Reibenspies, J. H., 105[404] (92)
Reid, G., 527[560] (495)
Reigel, H. J., 190[614] (176)
Reimer, G., 389[415] (372)
Reinecke, H. J., 427[29] (395)
Reinhardt, R-D., 521[184] (447)
Reinheckel, H., 102[189] (41, 48)
Reinhold, J., 105[368] (81, 88, 90)
Reinking, M. K., 222[118] (210), 222[119b] (210)
Reiser, O., 290[930] (273), 388[347] (360)
Reissova, A., 187[392] (150)
Reitzel, L., 100[72] (37, 38, 42, 51, 58, 59), 101[122] (41, 42, 47–9, 51, 58, 60)
Remon, A., 284[559] (246, 263)
Rendina, L. M., 280[264] (240, 261, 262, 266, 272, 275, 276)
Rendle, M. C., 428[93] (403, 411), 527[570] (497, 505), 528[638] (505), 528[639] (505), 528[645] (505), 528[646] (505), 584[125b] (549)
Reoyo, I., 524[340] (468)
Rericha, R., 584[137] (551)
Resconi, L., 26[148] (20)
Reshetova, M. D., 188[457] (154)
Rest, A. J., 280[273] (240)
Rettig, M. F., 222[113] (210), 285[592] (249), 285[596] (249), 285[600] (249), 285[601] (249), 285[602] (249), 384[51] (295), 384[52] (295), 385[108] (309), 385[118] (310), 387[227] (341), 387[230] (342), 522[250] (459), 585[171] (554, 556), 585[172] (554, 556), 585[173] (554, 556, 557, 561), 585[174] (554, 557, 558)
Rettig, S. J., 101[94] (42), 101[138] (42, 58, 59), 186[286] (138, 140), 280[279] (240), 286[649]
Reutov, O. A., 101[124] (41, 42, 47), 221[68d] (204), 279[199] (237), 279[200] (237), 279[202] (237)
Rheingold, A. L., 25[54] (9), 101[96] (42, 54, 74, 80, 81, 90), 188[421] (151, 154, 155), 189[531] (159, 169), 277[15] (227, 230), 277[17] (227, 230), 281[299] (241), 282[409], 283[438] (244), 283[461] (245), 283[471] (245), 284[538], 286[674] (253, 255), 286[682] (253), 286[683] (253), 385[94] (305), 385[114] (310), 386[168] (322), 525[451] (480), 525[452] (480), 527[556] (495), 529[668] (507), 583[40] (536, 539), 586[248] (561)
Ribas, J., 389[378] (366)
Ricard, L., 105[424] (95, 96), 184[153] (117, 159), 279[193] (237), 281[305] (241), 281[306] (241, 263), 427[15] (393, 395), 583[54] (540)
Rice, R. E., 524[376] (471), 524[391] (472)
Ricevuto, V., 428[121] (406), 521[201] (453)
Richards, R., 584[123] (548, 552, 554, 555)
Richards, R. E., 521[188] (447)
Richardson, R. M., 187[409] (150)
Riche, C., 385[112] (309)
Richmond, M. G., 27[192] (24), 189[559] (172, 173), 221[79] (205)
Richter, F., 189[559] (172, 173)
Richter, W., 187[374] (148)
Rickard, C. E., 283[430] (244)
Ridge, D. P., 100[29] (32)
Ridgwell, P. J., 385[111] (309), 385[139] (315)
Rieck, D. F., 26[140] (19, 20)
Riede, J., 25[56] (9), 183[75] (115, 126, 128)
Riediker, M., 389[371] (364)
Riegel, H. J., 187[359] (137), 187[378] (148, 174, 176), 385[113] (310)
Rieger, P. H., 385[102] (307)
Rieke, R. D., 99[3] (30, 31, 47, 48), 101[154] (42, 47, 48), 278[90] (233, 262)
Ries, M., 221[85a] (205), 221[85b] (205), 221[87b] (205), 221[88a] (205), 221[88b] (205), 221[89a] (205), 221[89b] (205), 222[92] (206)
Rietveld, M. H. P., 285[633] (250, 255, 262)
Rihs, G., 386[189] (328)
Rijnberg, E., 290[944] (273, 274)
Riley, G. E., 528[644] (505)
Riley, G. R., 584[125c] (549)
Rillema, D. P., 525[435] (478)
Rimml, H., 104[339] (81, 86), 284[540], 288[826] (262, 265), 529[669] (508), 582[15] (535)
Riordan, C. G., 100[84] (41, 42, 57)
Risen, W. M., Jr., 529[687] (509, 510)
Ritchey, J. M., 427[37] (396)
Rittner, W., 285[570] (248)
Rivetti, F., 287[730] (256)
Rix, F. C., 285[634] (250, 255, 256, 267)

Rizzardi, G., 584[117] (548), 584[118] (548)
Robbins, J. L., 187[417] (151), 187[418] (151), 187[419] (151, 155)
Roberts, D. A., 521[197] (450)
Roberts, J. S., 278[89] (233)
Roberts, N. K., 103[292] (73, 74, 81), 185[251] (131, 134), 185[252] (131), 185[253] (131, 134), 390[438] (376)
Roberts, P. J., 278[132] (235)
Robertson, B. E., 279[179] (236)
Robertson, F. J., 280[281] (240), 521[191] (447), 588[361] (580)
Robertson, G. B., 103[292] (73, 74, 81), 105[415] (42, 93, 95), 183[128] (116, 122), 185[251] (131, 134), 185[252] (131), 185[253] (131, 134), 185[254] (131, 134), 279[197] (237), 283[440] (244), 385[77] (303), 521[171], 521[173] (462), 524[382], 524[385], 529[676] (508, 512), 529[678] (509, 513, 515), 529[681] (509), 530[748] (515), 530[749] (515), 584[144] (551)
Robertson, K. N., 429[176] (414)
Robinson, B. H., 105[426] (95, 96), 189[536] (159)
Robinson, D. J., 525[452] (480)
Robl, C., 583[55] (540)
Robson, R., 284[561] (246), 284[562] (246), 284[563] (246)
Rocamora, M., 101[121] (42, 50, 58, 60), 101[146] (42, 62, 63), 102[205] (41, 49, 56, 58), 102[206] (49, 63), 280[281] (240), 282[420] (243, 248, 263, 264), 288[794] (260), 390[454] (379), 390[455] (379), 430[203] (421), 521[186] (447), 530[771] (517), 588[362] (580)
Rocherolle, V., 282[400]
Rockett, B. W., 182[1] (108, 150), 182[2] (108, 150), 182[3] (108, 150), 182[4] (108, 150), 384[12] (292), 384[13] (292), 384[14] (292)
Roddick, D. M., 519[48] (437, 453)
Rodenhäuser, B., 101[122] (41, 42, 47–9, 51, 58, 60)
Rodgers, J. R., 188[481] (157, 165)
Rodgers, R. D., 184[134] (116, 117, 157)
Rodrigo, R., 290[940] (273)
Rodriguez, J. H., 283[463] (245)
Rodriguez, L., 588[360] (580)
Rodriguez, M. C., 284[554] (245, 263)
Rodriguez, V., 524[354] (469)
Rodriguez-Ubis, J. C., 279[178] (236, 263, 264), 281[304] (241), 281[338] (242)
Roe, D. C., 100[66] (37, 38), 101[136] (42, 47, 58, 59), 182[56] (111), 184[166] (123), 521[162] (458, 462, 472)
Roelofsen, G., 184[204] (125, 157)
Rogel, F., 26[156] (20), 26[157] (20)
Rogers, R. D., 101[101] (41, 42, 48, 56, 58, 66, 67), 102[187] (48, 56, 62)
Roginski, R. T., 427[49] (398)
Rohlfing, C. M., 220[12a] (195), 427[3] (392)
Rokicki, A., 429[137a] (408), 429[137b] (408), 521[173] (462), 523[284] (461), 526[487] (485), 526[488] (485), 526[498] (487)
Roland, E., 106[437] (96, 98)
Roman, J., 283[462] (245)
Romano, U., 287[730] (256)
Romanov, G. V., 220[31b] (197)
Romão, M. J., 25[35] (5), 102[165a] (42, 59), 103[284] (73, 74), 104[344] (81, 88, 90), 184[162] (122), 184[180] (126, 128), 187[380] (148)
Romar, A., 283[472] (245, 263)
Romeo, R., 428[121] (406), 428[123a] (407), 428[123b] (407), 519[76] (439, 458, 459), 520[83] (439), 521[167] (458, 459), 521[201] (453), 522[233] (457), 522[234] (457), 522[241] (458), 522[246] (458, 459), 522[247] (458, 459), 522[248] (458, 459), 522[249] (458, 459), 522[252] (459), 522[255] (459), 522[256] (459), 522[257] (459), 522[258] (459), 522[259] (459), 522[260] (459), 522[261] (460)
Rommel, K., 105[368] (81, 88, 90)
Roncetti, L., 186[292] (138, 143)
Roobeek, C. F., 285[632] (250), 287[738] (256), 529[688] (509)
Roos, B. O., 182[18] (109), 386[148] (317)
Ros, J., 105[427] (67, 96, 97)
Ros, M. B., 283[464] (245, 264)
Ros, R., 278[126] (234), 278[131] (235, 262, 264, 268), 279[186] (237), 289[888] (268), 430[201a] (420, 426), 430[201b] (420, 426), 522[209] (453), 524[381] (472), 524[383] (472), 524[384], 524[393] (473), 525[409] (475), 525[421] (477)
Rosé, J., 27[167] (21)
Rosenberg, F. S., 428[91] (403)
Rosenthal, U., 185[209] (130, 132), 185[211] (130), 185[212] (130, 131), 185[217] (130), 185[218] (130), 185[219] (130), 185[220] (130, 133), 185[221] (130), 185[222] (130), 185[259] (130), 185[264] (133)
Rosevear, D. T., 280[273] (240)
Rosi, M., 102[178] (53), 220[13] (195)
Rossell, O., 521[146] (444), 521[147] (444, 467), 523[329] (467), 523[330] (467), 523[331] (467), 523[332] (467), 524[345] (468)
Roth, H., 525[436] (478, 481)
Roth, J. D., 427[46] (398), 427[55] (398)
Roth, S., 190[600] (175, 178), 190[621] (178), 528[602] (499)
Roth, Z., 389[393] (368)
Rothin, A. S., 288[838] (263), 289[908] (271)
Rothwell, I. P., 281[330] (242), 281[351] (242, 243), 282[417] (243), 285[569] (248), 586[236] (560)
Rotondo, E., 223[156] (215), 285[591] (249), 289[882] (267), 520[84] (439), 523[306] (464)
Rotteveel, M. A., 279[208] (238, 243, 253, 254), 282[386], 282[387] (253), 286[679] (253), 286[680] (253), 386[164] (320), 386[169] (323)
Röttinger, E., 26[113] (15)
Roulet, R., 289[888] (268)
Roundhill, D. M., 284[538], 385[114] (310), 521[163], 522[239] (458), 529[668] (507)
Roy, A., 583[45] (537)
Roy, S., 430[196] (419), 430[197] (419), 522[238] (457), 528[651] (506)
Roy, S. K., 103[245] (66, 67)
Royo, P., 101[152] (41, 42, 47), 102[175] (41, 42, 47, 62), 185[215] (130, 131, 133), 188[486] (157, 159, 164)
Rozière, J., 190[600] (175, 178), 190[621] (178)
Rubach, M., 186[327] (139, 141), 186[328] (139, 141), 186[337] (139, 141), 388[287] (352), 388[288] (352), 587[340a] (576), 587[340b] (576), 587[340c] (576)
Rubezhov, A. Z., 279[201] (237, 264, 272), 289[913] (272), 387[264] (348), 519[40] (436)
Rubin, Y., 385[135] (314)
Ruddick, J. D., 280[240] (239)
Rudisill, D. E., 290[937] (273)
Rudler, H., 279[213] (238)
Rudler-Chauvin, M., 279[213] (238)
Ruedi, R., 222[141] (214), 283[467] (245), 285[618] (250, 252)
Ruegg, H. J., 283[484] (245), 428[107] (403), 430[211] (423), 523[276] (461), 529[699] (510)
Rüegger, H., 386[186] (328), 386[187] (328), 521[195] (450), 529[709] (512), 582[17a] (535, 539), 587[338] (576)
Ruff, J. K., 26[139] (19), 27[175] (21)
Ruffo, F., 278[121] (234), 583[32] (536), 583[79] (543), 583[80] (543), 583[81] (543), 583[82] (544), 583[84] (544)
Rufińska, A., 103[224] (66, 67), 103[229] (66, 67), 103[231] (67, 71), 103[233] (66, 67), 103[251] (66, 67, 70), 103[252] (66, 67, 70), 103[260] (70), 103[283] (73, 74), 103[290] (73, 74, 78, 81), 182[52] (111, 113, 155, 157), 182[57] (111, 113, 115, 119, 138, 157), 183[68] (113), 183[120] (114, 115), 184[135] (116, 117, 157), 184[136] (116, 117, 157), 184[154] (117, 143), 186[306] (138, 143), 186[307] (138, 143), 187[365] (143), 187[366] (143), 281[291] (240), 389[420] (373), 521[184] (447), 587[331] (575)
Ruger, R., 285[570] (248)
Ruhl, B. L., 282[411], 282[412] (263), 285[605] (249), 287[768] (259), 289[852] (263), 385[86] (305), 385[87] (305), 385[88] (305), 524[371] (471, 495), 524[375] (471), 526[492] (486), 530[718], 582[17b] (535, 539)
Ruiz, A., 525[398] (474, 490)
Ruíz, C., 25[88] (12), 25[89] (12), 101[107] (42, 47, 58, 62, 63), 101[108] (42, 47, 58, 62, 63), 102[188] (48, 56), 104[294] (74, 78, 81), 184[193] (126, 129), 186[345] (139), 189[500] (157)
Ruiz, J., 288[810] (261), 288[811] (261), 524[339] (468), 524[352] (469), 524[354] (469)
Rulke, R. E., 285[635] (250, 255, 256), 286[704] (255, 256, 262)
Rund, J. V., 187[412] (151)
Ruppert, R., 25[62] (9)
Ruppin, C., 104[308] (80, 81)
Rusakov, S. L., 428[112] (406)

Rush, P. K., 191[635] (180, 181)
Russell, C. E., 387[221] (340), 388[351] (361)
Russell, D. R., 103[266] (73, 74), 103[267] (73, 74), 222[131] (212), 286[664] (252), 287[777] (260), 429[184] (416, 417), 429[185] (416), 525[441] (479, 481), 525[442] (479, 481), 525[443] (479), 525[444] (479), 525[445] (480), 525[446] (480), 525[447] (480), 525[457] (480), 526[460], 526[468] (482), 526[469] (482), 529[667] (507), 588[350] (578), 588[351] (579), 588[353] (579)
Russell, M. J. H., 220[1] (194–6, 198, 201, 205, 208–12, 215–18), 276[1] (226, 260), 285[581] (248, 251, 252), 286[644] (251, 252), 383[1] (292), 383[2] (292), 383[3] (292, 294), 383[4] (292), 383[5] (292), 383[6] (292), 383[7] (292), 383[8] (292), 383[9] (292)
Russo, M. V., 101[123] (42), 526[522] (490, 491, 493), 526[523] (490, 491), 527[526] (491), 527[545] (494), 527[546] (494)
Rutherford, N. M., 222[115] (210, 215)
Ryabov, A. D., 281[322] (242), 281[333] (242, 244, 268–70), 281[343] (242), 281[344] (242, 243), 281[345] (242), 281[346] (242, 245, 271), 281[350] (242), 282[368], 282[384] (248, 262), 282[426] (244), 283[437] (244, 253, 254), 283[451] (245), 285[615] (250), 285[616] (250), 289[850] (263, 268), 289[884] (268), 290[925] (273), 386[163] (320), 386[195] (332)
Ryabtsev, A. V., 386[174] (324)
Rybin, L. V., 278[137] (235), 385[74] (302)
Rybinskaya, M. I., 183[86] (115, 123, 174), 190[616] (177)
Rycroft, D. S., 280[281] (240), 428[119] (406), 428[120] (406), 521[191] (447), 588[361] (580)
Rys, P., 282[369], 282[376], 282[377] (244)
Ryu, I., 586[254] (562)
Rzehak, H., 182[23] (109)

Saare, A., 531[782] (518)
Sabat, M., 186[335] (139, 143), 190[594] (175, 179), 190[601] (175, 177)
Sabatino, P., 223[171] (217), 390[469] (383), 430[204b] (421), 430[205] (421), 430[213] (423), 430[220] (425)
Sacco, A., 287[736] (256, 257), 287[739] (257), 523[282] (461)
Sacconi, L., 186[348] (139, 146), 186[349] (139, 146), 187[350] (139, 146, 171), 384[54] (295)
Sachinidis, J., 522[246] (458, 459), 522[247] (458, 459)
Sackett, D., 103[257] (67, 73), 183[132] (116, 120, 157)
Sacksteder, L., 527[534] (493)
Sadler, N., 105[393] (91, 92)
Sadownick, J. A., 285[622] (250)
Saegusa, T., 25[37] (5), 25[38] (5), 25[39] (5), 279[209] (238, 269), 387[235] (343), 387[254] (347), 387[255] (347)
Saez, I., 102[175] (41, 42, 47, 62), 185[215] (130, 131, 133)
Safford, L. K., 427[46] (398)
Sahai, M., 389[358] (362), 389[393] (368)
Sahu, B. K., 288[804] (260)
Saillard, J.-Y., 100[23] (30), 185[237] (131, 136), 517, 531[777] (517)
Saint M'Leux, Y., 387[253] (346)
Sainz, D., 281[347] (242, 243, 271), 520[140] (443)
Saito, K., 584[91] (544), 584[104] (547)
Saito, O., 387[245] (345), 387[246] (345)
Saito, T., 277[40] (230)
Saito, Y., 427[60] (399), 427[61] (399), 586[280] (568, 569), 587[294] (571)
Saji, T., 384[65] (300), 384[66] (300)
Sakai, S., 385[98] (307), 385[100] (307)
Sakaki, S., 25[63] (9), 182[20] (109), 222[102] (208), 388[350] (360), 429[131] (408), 522[265] (460), 584[138] (551)
Sakamoto, H., 184[198] (125), 387[278] (351)
Sakodinskaya, I. K., 281[350] (242), 282[368], 285[615] (250), 285[616] (250)
Sala, M., 287[745] (257)
Salahub, D. R., 103[263] (73), 384[26] (292)
Salaün, J., 389[359] (363), 389[365] (363)
Saldana-Maldonado, M., 290[931] (273)
Saleem, A. M., 280[243] (239, 267), 520[108] (441)
Salem, G., 283[440] (244)
Sales, J., 101[111] (42, 47, 48, 56, 61), 101[128] (42, 47, 62), 101[146] (42, 62, 63), 101[151] (41, 42, 50), 102[155] (42, 56, 62, 63), 102[156] (42, 50, 55), 281[347] (242, 243, 271), 281[348] (242), 282[372], 282[373] (269), 282[374], 282[375] (263, 264), 282[418] (243), 283[455] (245, 263), 283[474] (245, 253), 283[477] (245, 263, 264), 283[479] (245), 283[493] (271), 284[556] (245, 262, 263), 284[557] (245, 246, 263, 264), 385[93] (305), 520[140] (443), 521[146] (444), 521[177], 523[331] (467), 523[332] (467), 524[345] (468), 525[397] (474), 528[619] (510), 529[698] (510), 529[700] (510)
Sallese, G., 388[352] (361)
Salvadori, P., 584[93a] (544), 584[93b] (544), 586[232] (559)
Salzer, A., 385[110] (309), 582[10] (534)
Sameshima, K., 25[95] (13), 101[112] (41, 42, 49, 50, 55, 64), 102[170] (41, 47, 49, 50, 62), 102[171] (41, 49, 50, 62), 223[176a] (218), 430[209] (422)
Samkoff, D. E., 103[288] (74, 76, 80), 183[105] (115, 122), 183[106] (115, 122), 279[188] (237), 429[188] (417)
Samsel, E. G., 279[163] (236, 250, 253, 261)
Sanati, H. K., 189[494] (157, 163), 189[495] (157, 163)
Sanau, A., 223[164] (216)
Sanchez, A., 583[83] (544), 585[212] (554)
Sanchez, G., 288[810] (261), 288[811] (261), 288[823] (262, 263), 524[352] (469)
Sánchez, G., 100[46] (31, 33)
Sanchez, J. A., 278[107] (233, 262, 263)
Sanchez, K. M., 427[45] (398)
Sànchez-de-Meràs, A. M., 386[147] (316)
Sandrini, D., 530[728] (514), 530[730] (514), 530[731] (514), 530[732] (514), 530[735] (514)
Sandrini, P. L., 222[142a] (214)
Sanmartín, A., 100[46] (31, 33)
Sano, K., 104[305] (81, 85), 104[335] (81), 104[343] (81, 90), 104[345] (81, 90), 104[351] (81, 90, 91), 104[352] (81, 90), 519[33] (436)
Sano, M., 587[318] (574)
Sansoni, M., 26[147] (20), 26[148] (20), 26[151] (20), 26[153] (20)
Santana, M. D., 100[46] (31, 33), 288[810] (261)
Santarsiero, B. D., 102[161] (41, 42, 47–9), 527[529] (491, 493), 587[282] (568)
Santhakumar, V., 388[319] (357)
Santhanam, M., 281[357] (243)
Santi, R., 280[278] (240), 388[352] (361), 388[355] (361), 519[47] (437), 525[433] (478)
Santini, C., 100[73] (37, 38)
Santini, R., 280[257] (239, 240, 255, 269, 270, 273)
Santos, A., 100[49] (31, 33), 277[42] (230, 231)
Santos, R. H. De A., 281[296] (241)
Santra, M., 287[712] (255, 267, 268)
Saporito, A., 385[95] (306), 583[75] (543)
Sappa, E., 26[160] (20), 27[167] (21), 105[431] (96, 98), 185[241] (131, 172, 174), 185[242] (131, 172, 174), 185[243] (131, 172), 185[245] (131, 172, 174), 185[246] (131, 172, 174), 185[247] (131, 172, 174), 185[248] (131, 172, 174), 185[249] (131, 172, 174), 185[250] (131, 172, 174), 189[544] (172, 174), 189[546] (171, 174), 190[561] (172, 174), 190[562] (172, 174), 190[564] (172), 190[565] (172), 190[567] (172, 174), 190[578] (174)
Sappenfield, E. L., 529[716]
Sarhan, J. K. K., 384[53] (295), 525[454] (480), 584[92] (544), 584[112] (548), 584[113] (548), 584[114] (548)
Sarjudeen, N., 185[233] (131, 136, 159, 172)
Saruwatari, M., 285[608] (249)
Sasaki, M., 286[661] (252)
Sasaki, S., 287[719] (255)
Sasaki, Y., 277[40] (230)
Sato, E., 183[79] (115, 119), 183[80] (115, 119)
Sato, M., 287[720] (256)
Satoh, J. Y., 386[200] (333), 388[301] (355), 389[357] (362)
Satoh, M., 290[933] (273)
Sauer, A., 105[396] (91, 92)
Sauer, D. R., 390[432] (375)
Saum, S. E., 289[860] (264), 530[771] (517)
Saunders, W. H., Jr., 386[193] (331)

Saura-Llamas, I., 277[34] (228, 230)
Saussine, L., 287[743] (257, 273)
Sautet, P., 285[599] (249), 384[29] (293)
Sauvage, J.-P., 25[62] (9), 284[524] (245, 271, 272)
Sawada, S., 387[235] (343)
Sawamura, M., 387[278] (351), 388[341] (360)
Sawkins, L. C., 282[399]
Saxena, A. K., 191[636] (180)
Saxton, H. M., 282[401] (263, 270)
Sayer, B. G., 185[237] (131, 136)
Sayers, D., 25[32] (5)
Scalone, M., 583[75] (543)
Scanlan, S. P., 527[585] (498, 504)
Scapacci, G., 184[190] (126, 129), 185[208] (129), 278[122] (234), 524[335] (467)
Schaad, D. R., 280[258] (239, 268)
Schaal, M., 430[190a] (418)
Schacht, H. T., 105[432] (96, 98)
Schade, W., 187[358] (137)
Schaefer, D., 104[344] (81, 88, 90), 104[353] (81, 88), 105[370] (81, 88), 105[371] (81, 88), 105[374] (81, 88), 184[162] (122), 186[323] (137, 139)
Schaefer, H. F., III, 25[94] (13), 187[386] (147)
Schaefer, W. P., 27[184] (22), 288[780] (260), 387[239] (344), 521[153] (444)
Schaefer-Stahl, H., 188[437] (152), 188[438] (152)
Schafer, W., 519[56] (437)
Schat, G., 525[434] (478)
Scheer, P., 428[114] (406)
Schenker, G., 288[787] (260, 267)
Schenkluhn, H., 102[217] (65), 102[218] (65), 102[219] (65), 185[260] (131, 132), 186[301] (138, 142), 186[302] (138, 140), 186[303] (138, 140)
Scherer, O. J., 185[227] (131, 136, 159), 185[271] (136, 172), 187[351] (139, 155, 159, 168, 171), 187[352] (139, 155, 159, 168, 171), 187[353] (139, 155, 159, 168, 172), 427[17] (393, 412), 524[334] (467), 527[563] (496), 527[564] (496)
Scherzer, K., 186[311] (138)
Schiavelli, M. D., 526[473] (484)
Schiavon, G., 101[144] (42, 65), 101[145] (42, 47), 102[163] (42, 47, 65)
Schick, K.-P., 183[67] (113, 115, 117), 221[65] (203), 288[793] (260), 387[281] (352, 359), 586[245] (560)
Schiebel, H.-M., 525[436] (478, 481)
Schillinger, H., 105[405] (92)
Schimpff, D. G., 587[323] (574, 577)
Schmid, G., 26[159] (20), 186[284] (138, 145)
Schmidbaur, H., 100[64] (31, 36), 104[340] (81, 87, 90), 277[14] (227, 228), 277[29] (228), 277[33] (228), 527[555] (495)
Schmidt, F. K., 280[237] (239)
Schmidt, H., 101[148] (41, 47, 48, 63)
Schmidt, H. R., 189[524] (159, 163)
Schmidt, K., 189[524] (159, 163)
Schmidt, T., 390[431] (375)
Schmidt, U., 186[287] (138, 142)
Schmidtberg, G., 519[49] (437, 454), 519[50] (437, 454), 519[52] (437), 519[53] (437), 519[57] (437), 519[58] (437), 519[59] (437), 519[61] (437, 454), 519[69] (438, 454), 519[70] (438), 520[107] (440, 454), 521[185] (447), 521[190] (447), 521[199] (452), 522[211] (454), 522[212] (454), 522[213] (454), 526[462] (481), 585[194] (554, 555), 585[195] (554, 555), 585[196] (554, 555), 585[201a] (554, 555), 585[201b] (554, 555)
Schnabel, R. C., 519[48] (437, 453)
Schneider, B. H., 188[456] (154)
Schneider, J., 100[44] (31, 33), 190[606] (174)
Schneider, J. J., 186[316] (138, 144, 172), 188[468] (155), 189[553] (171, 172)
Schneider, J. R., 25[90] (12)
Schnittger, J., 188[422] (151, 157, 159, 168)
Schnoeckel, H., 220[3] (194)
Schoettel, G., 221[49] (200), 386[162] (320), 428[89a] (404)
Schönberg, H., 25[41] (5), 185[240] (131, 135)
Schönecker, B., 105[365] (81, 90)
Schönholzer, P., 186[297] (138)
Schottenberger, H., 188[464] (155, 156, 177)
Schrauzer, G. N., 290[962] (276)
Schreurs, A. M. M., 289[868] (264)
Schröder, D., 100[26] (32)
Schroder, M., 284[524] (245, 271, 272), 290[958] (275)
Schröder, W., 103[273] (73, 74), 103[274] (73, 74), 182[48] (111, 128), 182[49] (111, 128), 183[70] (115), 183[115] (115, 117, 118)
Schroth, G., 103[224] (66, 67), 103[231] (67, 71), 103[232] (66, 67), 103[233] (66, 67), 103[251] (66, 67, 70), 103[252] (66, 67, 70), 103[260] (70), 103[291] (73, 74, 78), 182[52] (111, 113, 155, 157), 182[57] (111, 113, 115, 119, 138, 157), 183[73] (114, 115, 117, 157), 183[97] (115, 121, 138, 140), 183[120] (114, 115), 184[135] (116, 117, 157), 184[154] (117, 143), 186[306] (138, 143), 525[431] (478), 585[168] (554)
Schubert, D. M., 191[624] (180, 182)
Schubert, U., 189[530] (159, 163), 223[160] (216), 277[33] (228), 280[217] (238, 267, 270), 280[242] (239, 271), 387[275] (350), 390[441] (377), 390[443] (377), 527[569] (497)
Schuhn, W., 185[207] (126, 138, 145, 157)
Schüle, J., 100[19] (30), 100[28] (32)
Schulte, H. G., 186[330] (139)
Schulte, H.-G., 102[216] (65)
Schulz, B., 185[220] (130, 133)
Schulz, H., 183[101] (115, 124)
Schulz, W., 185[211] (130), 185[218] (130), 185[220] (130, 133), 185[264] (133)
Schumann, H., 187[393] (150), 188[455] (154, 159), 190[599] (175, 177), 521[150] (444)
Schurig, V., 428[114] (406)
Schussel, L. J., 105[399] (92)
Schütz, H., 104[347] (81, 90)
Schwager, H., 103[290] (73, 74, 78, 81), 103[291] (73, 74, 78), 104[297] (73, 74), 104[298] (73, 74), 104[300] (73, 74, 76), 287[779] (260), 390[458] (380)
Schwartz, H., 99[11] (30), 100[26] (32)
Schwartz, J., 387[259] (348), 387[260] (348), 389[370] (364), 389[371] (364)
Schwartz, M., 27[192] (24)
Schwartz, R., 530[734] (514)
Schwarz, H., 187[401] (150)
Schwarz, R., 189[523] (159, 163)
Schwemlein, H. P., 586[269] (565, 569)
Schwenk, U., 103[256] (67, 73), 183[129] (116, 121, 157, 161)
Schwerdtfeger, P., 220[5] (194)
Schwotzer, W., 527[565] (497)
Scolaro, L. M., 428[121] (406), 428[123a] (407), 428[123b] (407), 519[76] (439, 458, 459), 520[83] (439), 521[201] (453), 522[234] (457), 522[258] (459), 522[259] (459), 522[260] (459)
Scott, F., 103[247] (67, 70), 183[84] (115, 123, 174), 183[85] (115), 189[491] (157)
Scott, J. D., 290[947] (273, 274), 290[948] (273, 274, 276), 428[97] (403, 409–11), 430[196] (419), 519[45] (437), 520[88] (439), 522[245] (458), 523[298] (463, 466), 523[299] (463), 523[307] (464), 523[315] (465), 523[323] (466), 523[324] (466), 528[651] (506), 583[70] (542), 583[71] (542), 586[253a] (562), 586[253b] (562)
Scott, S. L., 105[393] (91, 92)
Scotti, M., 528[604] (499)
Scrivanti, A., 187[367] (143), 222[128] (211), 222[142b] (214), 223[165] (216), 279[150] (235, 257), 279[153] (235, 262, 264, 267), 287[747] (257, 263), 289[881] (267), 430[211] (423), 518[18] (434), 521[196] (450), 523[276] (461), 523[277] (461), 523[278] (461), 523[325] (466), 587[310] (573), 587[339] (576)
Sebald, A., 102[191] (48, 64), 277[70] (230, 245), 526[505] (490), 526[506] (490, 491), 526[511] (490), 527[532] (493), 527[537] (493), 527[538] (493), 527[539] (493), 528[623], 587[324] (581)
Secco, A. S., 101[94] (42), 280[279] (240)
Seco, M., 521[146] (444), 521[147] (444, 467), 523[329] (467), 523[330] (467), 524[345] (468)
Seddon, K. R., 518[6] (432), 521[192] (447)
Seeber, R., 278[126] (234), 429[187a] (417), 429[187b] (417)
Seevogel, K., 183[82] (115, 119), 186[327] (139, 141), 190[614] (176), 385[113] (310), 587[340a] (576)
Segre, A., 429[150b] (409–11)
Seidel, G., 187[381] (148, 149), 187[382] (148, 149), 187[383] (148, 149), 190[596] (175, 180)
Seidel, W. C., 24[19] (4), 101[95] (42, 54), 101[147] (41, 42, 47–9, 56), 182[32] (110), 184[167] (123)

Seiler, P., 188[458] (154, 155)
Sekitani, T., 384[35] (293), 384[36] (293)
Selbin, J., 283[494] (245), 283[495] (245, 264), 284[496] (245), 284[497] (245)
Seligson, A. L., 278[76] (230), 280[266] (240, 243, 245, 268, 270), 280[283] (240, 268, 271), 288[831] (262, 268, 271), 289[889] (268, 271)
Sellers, H., 384[43] (294)
Sellmann, D., 105[405] (92), 223[177] (218), 430[202] (421)
Semelhago, G., 430[227] (426), 430[228] (426)
Semikolenov, V. A., 384[41] (294)
Semmelhack, M. F., 101[113] (47, 51, 62)
Sen, A., 279[164] (236, 250, 261, 262, 264, 265, 273), 279[167] (236, 259, 262), 279[168] (236, 259), 285[629] (250, 261, 262, 264), 288[829] (262, 264–6), 289[885] (268), 427[4] (392), 429[142a] (408), 519[25] (434, 462), 519[26] (434, 462), 583[42] (537, 560)
Sen Gupta, P. K., 289[900] (270)
Sena, S., 386[190] (328)
Senda, V., 386[210] (335)
Seo, B.-S., 528[624]
Sepelak, D. J., 430[200] (420), 527[557] (495)
Seraglia, R., 430[226] (426)
Serdobov, M. V., 531[789] (518), 531[790] (518)
Sergeyenko, E. S., 281[345] (242)
Sergi, S., 280[280] (240)
Sergienko, V. S., 221[60c] (202), 388[293] (354)
Serrano, J. L., 283[464] (245, 264), 284[559] (246, 263), 284[560] (246, 263, 268)
Sevin, A., 385[112] (309)
Sexton, G. J., 386[148] (317)
Sgamellotti, A., 102[178] (53)
Shabanzadeh, B., 523[297] (463), 523[303] (463)
Shal'nova, K. G., 289[914] (272), 289[915] (272)
Shang, M., 529[672] (508)
Shapley, J. R., 387[276] (350), 427[49] (398)
Shaposhnikova, A. D., 105[422] (95, 96), 105[425] (95, 96), 189[558] (172, 173)
Sharapova, S. Ts., 390[461] (381)
Sharma, K. R., 427[57a] (399), 427[57b] (399)
Sharma, S., 390[431] (375)
Sharp, P. R., 188[463] (155, 156), 528[602] (499)
Shaver, A., 585[190] (554, 555, 557), 585[191] (554, 555, 557), 588[374] (581), 588[375] (581)
Shaver, R. J., 525[435] (478)
Shaw, B. L., 26[116] (16), 26[122] (16), 102[192] (41, 47, 48), 277[47] (230), 277[62] (230), 278[75] (230, 236), 279[185] (236), 280[239] (239, 240), 282[396] (243, 245), 282[397] (243), 282[399], 282[408] (263, 264), 283[449] (245), 283[453] (245), 283[488], 284[526], 284[539], 284[543], 285[567] (248), 286[647] (252), 288[817] (262, 265), 386[157] (319), 428[92] (403, 416), 429[180] (416, 423), 429[189] (417), 518[8] (432, 494), 520[103] (440, 466), 520[109] (441), 520[118] (441, 442, 445, 463), 520[121] (442), 523[297] (463), 523[303] (463), 523[327] (466), 525[425] (478), 525[455] (480), 526[510] (490), 526[513] (490), 526[515] (490), 526[516] (490), 527[547] (494), 527[548] (494), 527[549] (494), 527[550] (494), 527[551] (494), 527[552] (494), 527[553] (494), 529[665] (507), 529[683] (509), 529[696] (510), 530[747] (514), 583[64a] (542), 583[64b] (542), 586[234] (559), 586[247] (561, 562), 586[260a] (563, 572), 586[260b] (563, 572), 586[260c] (563, 572), 586[260d] (563, 572)
Shaw, G., 582[13b] (535, 537, 539, 566)
Sheldrick, G. M., 222[132] (213), 222[134a] (213), 222[134b] (213), 285[570] (248), 287[756] (258, 263), 287[757] (258), 287[759] (258), 529[706] (511)
Shemyakin, N. F., 102[203] (41, 48)
Shen, C., 102[160] (42)
Shepard, M. E., 101[142] (42, 47, 51)
Shepherd, I., 283[449] (245)
Sheppard, N., 384[37] (293)
Sherry, L. J. S., 525[441] (479, 481), 525[442] (479, 481), 588[350] (578)
Shiba, C. F., 523[283] (461)
Shibasaki, M., 388[311] (357)
Shibuya, T., 585[176] (554, 555)
Shifrina, R. R., 287[732] (256)
Shiina, K., 102[202] (41, 48), 526[479] (484)
Shiina, M., 25[82] (11), 184[163] (122), 525[453] (480)
Shilliday, L., 101[153] (41, 42, 47–9)
Shilov, A. E., 531[786] (518)
Shima, T., 388[345] (360)
Shimizu, I., 387[241] (344), 388[314] (357)
Shimodo, S., 586[280] (568, 569)
Shimoke, K., 287[776] (259)
Shindo, T., 518[16] (434, 455)
Shinoda, S., 427[60] (399), 427[61] (399), 587[294] (571)
Shintani, H., 277[25] (228, 229)
Shiotani, O., 384[67] (300)
Shiozaki, K., 385[98] (307)
Shiro, M., 287[763] (258)
Shitara, H., 25[58] (9), 103[264] (74, 80), 103[269] (74, 75), 183[76] (115, 120, 128), 183[77] (115, 120), 280[241] (239, 266, 267), 586[235] (560)
Shoner, S. C., 184[172] (126)
Shore, S. G., 190[563] (172, 174), 190[576] (174)
Short, R. L., 24[15] (3), 102[167] (50, 51, 58, 59), 102[174] (50, 56, 58, 59, 61, 62), 221[66] (204)
Shriver, D. F., 26[150] (20)
Shubert, D. C., 186[332] (139, 143)
Shuikina, L. P., 384[42] (294)
Shulman, M. J., 389[381] (366)
Shul'pin, G. B., 518, 531[786] (518), 531[787] (518), 531[788] (518), 531[789] (518), 531[790] (518)
Shur, V. B., 185[212] (130, 131)
Shustorovich, E., 100[20] (30), 100[21] (30)
Shy, H. C., 277[31] (228, 230), 277[38] (229, 230)
Sibi, M. P., 282[393]
Sicilia, V., 281[352] (243, 262, 271), 288[828] (262, 263), 288[833] (262), 288[837] (262–4), 390[468] (382)
Siebert, W., 183[98] (115, 155, 157, 181), 183[101] (115, 124), 186[317] (138, 177), 187[391] (149, 175, 177), 188[423] (151, 180, 181), 189[520] (159, 175), 190[588] (175, 177), 190[589] (177), 190[590] (175, 178), 190[591] (175, 178), 190[592] (175, 177), 190[593] (175, 179), 190[595] (175, 179), 190[597] (175, 180), 190[600] (175, 178), 190[618] (178), 190[619] (178), 190[620] (178, 181), 190[621] (178), 191[625] (180, 181), 191[639] (181), 191[640] (181), 191[641] (181), 588[378a] (581), 588[378b] (581), 588[379] (581)
Siedle, A. R., 220[35] (197), 281[341] (242, 243, 245, 248), 288[801] (260), 288[805] (261), 519[28] (435), 526[489] (486, 487)
Siegbahn, H. O. G., 386[173] (324)
Siegbahn, P. E. M., 100[19] (30), 100[24a] (30), 100[28] (32), 182[18] (109), 182[19] (109), 220[6] (194), 222[102] (208), 223[178] (219), 385[124] (312), 385[125] (312), 385[127] (312)
Siegmann, K., 519[74] (439)
Sieler, J., 25[41] (5), 25[42] (5), 103[223] (65, 66), 104[356] (81, 88, 90), 104[357] (81, 88, 90), 104[358] (81, 88, 90), 105[365] (81, 90), 105[367] (81, 88, 90), 105[368] (81, 88, 90), 105[372] (81, 88, 89), 105[378] (80, 88), 183[90] (115), 184[185] (126, 128), 185[240] (131, 135), 186[298] (138, 142), 186[299] (138, 142), 186[309] (138, 142), 186[325] (139, 140), 187[358] (137), 187[361] (141, 142), 387[269] (350)
Sieler, S., 186[313] (138, 142)
Siemeling, U., 184[144] (117, 121)
Siftar, J., 187[361] (141, 142)
Sigan, A., 25[44] (7), 25[45] (7), 25[46] (7), 104[348] (81, 88), 104[363] (81, 88)
Sik, V., 385[89] (305), 521[194] (450, 500), 527[582] (498, 504), 527[584] (498, 504), 527[586] (498), 527[587] (498, 504), 528[588] (498, 504), 528[589] (498), 528[596] (499), 528[605] (499), 528[608] (499, 504), 528[609] (499, 500), 528[613] (500, 504), 528[614] (500, 504), 528[630] (500), 528[631] (504), 528[632] (504), 528[633] (504), 528[634] (504), 585[154a] (553), 585[154b] (553)
Sillett, G. J. D., 588[380] (581)
Silverberg, L. J., 286[674] (253, 255), 386[168] (322)
Silvesto, L., 280[280] (240)
Silvestre, J., 186[349] (139, 146), 384[54] (295)
Simms, B. L., 529[672] (508), 529[673] (508), 584[150] (552)
Simon, K., 286[658] (252)

Simonetta, M., 385[140] (316), 385[141] (316), 385[142] (316), 426[2c] (392)
Simons, L. H., 99[5] (30)
Simpson, J., 105[426] (95, 96), 189[536] (159)
Simpson, K., 384[68] (302)
Simpson, M. B., 24[5] (2), 24[6] (2)
Simunic, J. L., 100[32a] (31, 33)
Singhal, R. K., 277[35] (229)
Sinha, C., 284[532] (272), 284[533] (268), 284[534] (269), 284[535] (268), 284[536] (268), 530[773] (517), 586[268] (564)
Sinigalia, R., 525[412] (475), 525[416] (475)
Sinkinson, J. A., 529[659] (507)
Sinn, E., 190[623] (180, 181)
Sinnema, J. C. M., 103[276] (73, 74, 78, 80), 184[149] (117, 118), 186[334] (139, 144)
Siria, J. W., 586[272] (565, 568)
Sironi, A., 388[334] (359), 428[115] (406), 587[319] (574, 576), 587[320] (574, 576)
Sirotkina, E. E., 282[367]
Sisti, M., 587[335] (576)
Sivade, A., 187[362] (142)
Skapski, A. C., 521[175], 522[214] (454), 522[231] (456)
Skelton, B. W., 277[23] (228), 279[184] (236, 241), 280[254] (239, 262), 280[255] (239, 262), 282[385] (253, 263, 264), 282[389] (262), 282[390], 286[707] (255, 262), 286[708] (255, 262), 288[816] (262, 274, 276), 290[916] (272, 273, 276), 290[943] (273–6), 290[946] (273), 290[949] (273, 274, 276), 290[950] (273, 274, 276), 290[954] (275), 290[961] (276), 390[438] (376), 429[141] (408), 523[304] (464), 523[305] (464), 523[309] (464), 523[310] (464), 523[318] (465), 525[428] (478), 527[559] (495), 528[615] (500), 528[616] (510), 529[701] (510), 585[186] (554, 555), 586[263a] (564, 572)
Skinner, H. A., 521[183] (445)
Skupińska, J., 102[176] (52, 63)
Slater, S., 185[232] (131, 136)
Sloan, M. P., 25[86] (11, 12), 101[117] (42, 58, 59, 61), 184[174] (126)
Sloan, O. D., 287[772] (259)
Slocum, D. W., 183[123] (115, 138, 145, 155)
Slovokhotov, Yu. L., 26[162] (20), 221[53] (201, 204), 221[64a] (203), 221[68c] (204), 221[69] (204), 221[70] (204), 221[72] (204), 221[73b] (204), 221[74a] (204), 221[74c] (204), 222[95] (207), 222[97a] (207), 222[97b] (207), 427[25] (394)
Sly, W. G., 527[529] (491, 493), 587[282] (568)
Smart, J. C., 187[419] (151, 155), 188[463] (155, 156)
Smeets, W. J. J., 278[102] (233, 250, 262–4), 280[261] (239, 240, 255, 262, 266, 267, 269–74), 528[620], 530[720] (513), 530[759] (516), 530[769] (517)
Smirnov, A. N., 188[430] (152), 188[431] (152)
Smirnov, A. S., 187[404] (150), 187[405] (150, 155)
Smith , D. E., 390[453] (379)
Smith, G., 283[429] (244), 283[478] (245, 263), 585[167] (553–5)
Smith, G. D., 287[741] (257)
Smith, G. W., 220[4] (194), 426[2d] (392)
Smith, J. D., 529[670] (508)
Smith, J. S., 220[40] (198)
Smith, K. G., 100[63] (31, 36), 189[498] (157, 165), 189[499] (157, 165)
Smith, M. B., 520[95] (440)
Smith, M. R., III, 25[60] (9), 183[126] (115, 122, 125)
Smith, V. H., Jr., 285[598] (249), 584[140] (551)
Smith, W. H., 102[185] (48, 55, 63)
Smolyakova, I. P., 283[436] (244, 245)
Sneddon, D. W., 280[219] (238, 259), 287[767] (259), 524[375] (471)
Sneddon, L. G., 191[634] (180), 191[637] (180)
Snow, M. R., 103[246] (67, 72), 188[481] (157, 165), 189[492] (157, 161)
Snow, S. A., 26[103] (14)
Snyder-Robinson, P. A., 189[527] (159, 170)
Sodeoka, M., 388[311] (357)
Sokolov, V. I., 101[124] (41, 42, 47), 220[30] (197), 221[68d] (204), 222[95] (207), 222[96] (207), 222[97a] (207), 222[97b] (207), 279[199] (237), 279[200] (237), 279[202] (237), 279[203] (237), 281[331] (242, 245), 283[457] (245, 250), 529[708] (511)
Solans, X., 101[111] (42, 47, 48, 56, 61), 101[121] (42, 50, 58, 60), 102[205] (41, 49, 56, 58), 105[427] (67, 96, 97), 281[347] (242, 243, 271), 282[372], 282[375] (263, 264), 283[455] (245, 263), 283[474] (245, 253), 283[493] (271), 284[557] (245, 246, 263, 264), 385[93] (305), 520[140] (443), 521[147] (444, 467), 521[177], 523[330] (467), 524[390], 528[619] (510)
Solntsev, K. A., 191[631] (180, 181), 191[632] (180, 181)
Solomun, T., 428[70] (400), 524[380]
Somers, J. S., 390[460] (381)
Sommovigo, M., 390[469] (383)
Somorjai, G. A., 386[143] (316)
Son, T.-I., 222[106] (208), 278[138] (235, 255), 280[234] (238), 286[697] (255), 387[242] (344), 388[336] (359)
Song, H., 387[236] (343)
Song, J.-I., 521[198] (450)
Song, L., 279[187] (237), 525[439] (479), 525[440] (479), 583[51] (540)
Soni, H. K., 526[484] (485)
Sonnek, G., 102[189] (41, 48)
Sonoda, N., 586[254] (562)
Sonogashira, K., 102[202] (41, 48), 103[254] (67, 73), 189[493] (157, 161), 277[58] (230), 277[64] (230), 277[65] (230, 231), 277[69] (230, 231), 278[77] (231), 278[78] (231), 286[668] (252), 526[479] (484), 527[540] (493), 527[541] (493), 527[543] (493)
Sorato, C., 520[143] (444), 587[321] (574)
Sostero, S., 522[209] (453), 524[334] (467), 524[377] (471), 524[378] (471), 524[379] (471)
Sotes, M., 525[398] (474, 490)
Soulie, J., 528[629] (500)
Sourisseau, C., 187[410] (150), 188[434] (152), 222[116] (210)
Sousa, L. R., 284[511], 284[520] (261), 529[694] (510)
Sowinski, A. F., 522[226] (456), 526[461] (481), 585[179] (554, 555)
Spangler, D., 25[94] (13)
Spek, A. L., 103[253] (67), 104[329] (81, 85), 184[141] (117, 120, 159), 184[142] (117, 159), 184[204] (125, 157), 190[571] (172), 190[572] (172), 190[573] (172, 173), 190[574] (172, 173), 277[55] (230, 232), 278[98] (233, 239, 262), 278[99] (233, 255, 256, 262), 278[102] (233, 250, 262–4), 278[103] (233, 251, 262, 273), 278[141] (235), 279[176] (236, 240, 242, 243, 262, 269, 270, 272, 273, 275), 280[260] (239, 255, 262, 272, 273), 280[261] (239, 240, 255, 262, 266, 267, 269–74), 280[262] (239, 262, 266, 271, 274, 276), 281[300] (241), 281[354] (243), 285[633] (250, 255, 262), 289[868] (264), 289[869] (264, 265), 528[620], 529[705] (511), 530[720] (513), 530[750] (515), 530[751] (515), 530[755] (516), 530[756] (516), 530[759] (516), 530[762] (516), 530[763] (516), 530[764] (516), 530[767] (517), 530[769] (517), 530[770] (517)
Spencer, B., 187[419] (151, 155), 189[549] (171), 189[552] (171)
Spencer, J. L., 102[168] (47, 50, 54), 182[26] (109), 182[27] (109), 182[60] (111, 113), 279[198] (237), 384[34] (293), 385[81] (303), 385[105] (308), 388[299] (355), 429[157] (410, 417), 429[158] (410, 416, 417), 519[24] (434), 522[222] (455), 525[448] (480), 582[12] (535, 540, 553), 583[24] (536), 583[26] (536, 537, 539, 543), 583[27] (536), 583[28] (536), 583[29] (536, 544), 583[30] (536, 538, 539), 583[44] (537), 583[46] (537, 542, 550), 583[50] (538, 577), 585[160] (553), 585[187] (554, 555, 574), 585[188] (554, 555, 573, 576, 580), 586[276] (565, 566), 587[308] (573), 587[344] (577)
Spevak, B. N., 102[158] (42, 56)
Spiers, J. C., 527[560] (495)
Spiliotis, D., 186[315] (138, 139, 148)
Spohn, K.-H., 519[56] (437), 521[185] (447), 585[194] (554, 555)
Sridhara, N. S., 277[54] (230)
Sridharan, V., 388[315] (357), 388[317] (357), 388[318] (357), 388[319] (357), 388[320] (357)
Srivastava, R. C., 427[13] (393–5, 411, 412), 427[14] (393, 403, 411), 428[102] (403), 428[106] (403), 529[661] (507), 529[662] (507), 529[663] (507)
Ssebuwufu, P. J. M., 518[8] (432, 494)

St. Clair, M. A., 222[118] (210)
Stach, J., 102[214] (65), 102[215] (65)
Stackhouse, G. B., 583[33] (536, 538)
Stacy, A. M., 189[533] (169, 172, 173)
Stader, C., 277[45] (230, 231), 277[70] (230, 245), 526[506] (490, 491)
Stahl, L., 26[102] (14)
Stährfeld, T., 586[257] (563, 564)
Stakem, F. G., 387[216] (337)
Staley, D. L., 277[17] (227, 230), 283[471] (245)
Stalteri, M. A., 522[242] (458, 462), 522[243] (458)
Stam, C. H., 104[331] (81, 85), 278[113] (234), 281[301] (241), 281[353] (243, 253, 255, 271), 287[713] (255), 287[761] (258), 288[832] (262), 525[434] (478), 526[499] (487), 529[703] (511), 530[757] (516), 530[758] (516), 530[760] (516), 530[761] (516), 583[74a] (542)
Stam, C. S., 390[467] (382)
Stammler, H.-G., 188[422] (151, 157, 159, 168)
Stamp, L., 183[113] (115, 117, 118), 183[114] (115, 117, 118), 385[83] (304)
Stang, P. J., 279[187] (237), 428[95a] (403), 428[95b] (403), 428[95c] (403, 408), 429[140] (408), 484, 525[439] (479), 525[440] (479), 526[471] (484), 526[472] (484, 487, 491, 493), 526[473] (484), 526[474] (484), 526[475] (484), 526[493], 526[496] (487), 526[497] (487), 526[500] (487), 526[502] (487), 583[51] (540), 583[53] (540, 548, 574, 576), 584[122] (548), 584[132] (549), 584[133] (549, 568), 586[231] (559), 587[283a] (568), 587[347] (577)
Stanger, A., 183[87] (115, 123, 174), 183[88] (115, 123, 174)
Stanley, G. G., 26[125] (16)
Stansfield, R., 186[279] (137, 138)
Stanton, R. E., 384[55] (296)
Stanton, S. A., 389[396] (368)
Staples, R. J., 531[785] (518), 586[273] (565)
Stapp, B., 519[58] (437), 519[61] (437, 454), 519[69] (438, 454), 521[179], 585[202] (554, 555)
Starý, I., 387[247] (345), 387[248] (345)
Statler, J. A., 522[236] (457)
Stauffer, R. D., 280[282] (240)
Stauffert, P., 26[127] (16), 189[524] (159, 163)
Stavropoulos, P., 105[403] (92)
Steadman, T. R., 290[934] (273)
Stebler, A., 187[413] (151)
Steel, P. J., 103[257] (67, 73), 183[132] (116, 120, 157), 284[507] (245, 246, 264), 284[513], 284[514] (263, 264), 284[519], 284[552] (245, 246, 264), 284[553] (245, 246, 264), 285[579] (248)
Steffen, W. L., 283[440] (244)
Stehling, L., 103[286] (74, 75)
Steigelmann, O., 25[56] (9), 183[75] (115, 126, 128)
Steimann, M., 526[477] (484)
Steinborn, D., 102[181] (56), 103[240] (66, 67), 189[496] (157, 159, 164)
Steiner, E., 284[530]
Steiner, N., 284[537], 385[85] (304)
Stephan, C., 286[662] (252), 286[663] (252)
Stephan, D. W., 521[158]
Stephan, M., 190[593] (175, 179)
Stephenson, D., 521[194] (450, 500), 527[580] (498), 527[585] (498, 504), 528[630] (500)
Stephenson, D. K., 279[170] (236, 267), 288[782] (260, 271)
Stepnowski, R., 100[30] (32), 187[400] (150, 174)
Stepuszek, D. J., 389[361] (363)
Stershic, M. T., 105[398] (92)
Sterzo, C. L., 385[136] (314)
Stevens, A. V., 187[397] (150)
Stevens, E. D., 585[213b] (556)
Stevens, P. V., 584[149] (552, 566)
Stevens, W. R., 184[167] (123)
Stevenson, P., 388[317] (357), 388[319] (357), 388[320] (357)
Stiakaki, M.-A. D., 583[61] (541)
Stibr, B., 191[630] (180, 181)
Stiegman, A. E., 26[129] (16)
Stille, J. K., 102[209] (62), 278[85] (233–5, 266, 273), 278[92] (233, 255, 256), 278[109] (234, 255), 278[110] (234, 238, 270, 273), 278[111] (234), 278[116] (234, 273), 278[118] (234, 237, 239, 266, 270, 273), 278[134] (235, 267, 270, 273), 279[158] (236), 285[588] (249), 288[791] (260, 270), 289[871] (266), 289[875] (266, 269, 273), 289[894] (269, 273), 290[923] (273), 290[937] (273), 385[136] (314), 521[202] (453)
Stoccoro, S., 283[482] (245, 255), 530[775] (517)
Stock, P., 187[368] (143), 387[284] (352)
Stockhammer, A., 427[42a] (397)
Stockis, A., 103[262] (73), 584[139a] (551)
Stoeckli-Evans, H., 530[727] (514)
Stolle, A., 389[359] (363), 389[365] (363)
Stolyarov, I. P., 221[60a] (202), 221[61a] (202)
Stolzenberg, A. M., 105[398] (92), 105[399] (92)
Stone, F. G. A., 21, 23, 27[181] (21, 22), 27[185] (23), 27[186] (23), 27[187] (23), 105[423] (95, 96), 190[566] (172), 223[180] (219), 223[181] (219), 280[273] (240), 281[355] (243), 287[766] (259), 385[105] (308), 427[18] (393, 413), 427[62] (400, 410), 429[157] (410, 417), 429[158] (410, 416, 417), 525[448] (480), 582[6a] (534), 582[7a] (534, 535), 583[25] (536, 565), 583[26] (536, 537, 539, 543), 583[29] (536, 544), 585[157] (553), 585[158] (553), 585[160] (553), 585[187] (554, 555, 574), 585[188] (554, 555, 573, 576, 580), 586[266] (564), 586[267] (564), 586[276] (565, 566), 587[308] (573), 588[366] (580)
Storeck, A., 387[283] (352)
Storr, A., 186[286] (138, 140), 520[94] (440)
Stovokhotov, Yu. L., 427[23] (393)
Strahle, J., 521[170]
Stransky, N., 186[281] (137–9), 186[282] (137–9)
Strauss, W., 188[429] (152, 155, 156)
Strelets, V. V., 182[15] (108), 188[442] (153), 188[443] (153), 188[444] (153, 155)
Strich, A., 222[102] (208)
Stringer, G., 188[462] (155, 156)
Stromnova, T. A., 221[55] (201), 221[56] (201), 221[58] (202, 219), 221[59] (202), 221[60b] (202), 221[60c] (202), 221[61b] (202), 222[93a] (207), 222[93b] (207), 222[93c] (207), 223[179a] (219), 223[179b] (219), 223[179c] (219)
Struchkov, Yu. T., 26[162] (20), 221[53] (201, 204), 221[55] (201), 221[64a] (203), 221[68c] (204), 221[69] (204), 221[70] (204), 221[73b] (204), 221[74a] (204), 221[74c] (204), 222[95] (207), 222[97a] (207), 222[97b] (207), 279[194] (237), 283[435] (244), 387[215] (337), 427[23] (393), 427[25] (394), 430[207] (422)
Struck, G. E., 584[123] (548, 552, 554, 555)
Strukul, G., 278[131] (235, 262, 264, 268), 521[172], 523[287] (462), 525[409] (475), 525[410] (475), 525[411] (475), 525[412] (475), 525[413] (475), 525[414] (475), 525[416] (475), 525[417] (475), 525[418] (475, 476)
Stryker, J. M., 524[372] (471, 495), 525[424] (477), 526[467] (481), 527[561] (495), 585[184] (554, 555)
Stucky, G. D., 25[32] (5)
Stufkens, D. J., 104[327] (81, 86), 104[331] (81, 85), 285[568] (248), 385[76] (303), 530[721] (513), 530[760] (516)
Stuhl, L. S., 26[104] (14), 26[105] (14), 185[225] (130, 133)
Sturge, K. C., 101[153] (41, 42, 47–9), 102[161] (41, 42, 47–9)
Stuve, E. M., 384[33] (293)
Su, M.-D., 186[278] (137)
Suarez, A., 282[359] (243), 283[459] (245), 283[460] (245, 263, 264), 283[468] (245), 284[518] (264), 289[854] (263)
Suenaga, M., 287[773] (259)
Sugai, R., 25[57] (9), 25[59] (9), 184[187] (126, 128), 184[188] (126, 128)
Suggs, J. W., 279[207] (238, 243), 288[834] (262)
Sugimoto, T., 279[166] (236, 257, 259), 287[712] (255, 267, 268)
Sugino, K., 223[174a] (218), 223[174b] (218), 385[84] (304)
Sugita, K., 25[25] (4), 101[89] (42, 47, 53, 58, 62, 63), 287[722] (256, 265, 268, 271)
Sukirthalingam, S., 388[315] (357), 388[317] (357), 388[318] (357)
Suld, G., 388[306] (356)
Suleimanov, G. Z., 279[200] (237)
Sullivan, A. C., 280[276] (240), 429[139] (408), 526[480] (484, 487)
Sumiya, R., 25[37] (5)
Sümmermann, K., 104[308] (80, 81),

104[309] (80, 81), 104[310] (80, 81), 104[311] (81), 104[312] (81), 104[323] (81, 84), 104[324] (81)
Sunderlin, L. S., 99[13] (30, 32)
Sünkel, K., 277[50] (230, 232), 524[386] (496, 497), 526[477] (484), 527[528] (491, 493), 586[277] (566)
Suschinskaya, S. P., 430[207] (422)
Suslick, K. S., 26[138] (19)
Sustmann, R., 183[83] (115, 119), 280[235] (239, 266), 280[236] (239, 266), 289[878] (266)
Sutcliffe, V. F., 520[87] (439), 521[175], 522[214] (454)
Sutherland, C., 280[264] (240, 261, 262, 266, 272, 275, 276)
Sutherland, J. K., 282[401] (263, 270)
Sutin, L., 285[594] (249), 385[119] (311), 385[120] (311), 385[121] (311)
Sutter, J.-P., 286[680] (253), 286[681] (253), 286[684] (253), 386[169] (323)
Suzuki, A., 185[265] (133)
Suzuki, H., 278[96] (233, 234), 280[244] (239), 285[583] (249), 286[658] (252), 286[660] (252), 387[280] (352), 525[429] (478)
Suzuki, M., 387[235] (343)
Suzuki, R., 388[327] (358)
Suzuki, T., 587[317] (573)
Svensson, A., 387[269] (350)
Sweigart, D. A., 584[111] (548)
Swepston, P. N., 26[150] (20), 184[194] (126)
Swincer, A. G., 103[245] (66, 67)
Swisher, R. G., 190[623] (180, 181)
Syed, A., 583[22] (535, 565–7), 585[213b] (556)
Syed, R., 220[26] (197, 200), 386[158] (319), 390[435] (375)
Szczendzina, G., 185[260] (131, 132)

Tabuchi, T., 388[323] (358), 388[328] (358)
Tada, T., 104[305] (81, 85)
Takagi, K., 286[637] (251), 386[203] (334), 386[204] (334), 387[217] (338), 387[218] (338), 388[329] (358)
Takagi, T., 105[428] (97)
Takahara, J. P., 388[321] (358), 388[322] (358)
Takahashi, H., 277[40] (230)
Takahashi, K., 25[83] (11), 25[85] (11), 222[114a] (210, 215), 429[168] (412, 413), 429[169] (412), 429[177a] (414)
Takahashi, M., 584[127] (549)
Takahashi, S., 222[122] (211), 222[139a] (213), 222[139b] (213), 222[140] (213), 277[52] (230, 231), 277[53] (230, 231), 277[64] (230), 277[65] (230, 231), 277[69] (230, 231), 278[77] (231), 278[78] (231), 278[80] (231), 278[81] (231, 232), 278[82] (232), 527[531], 527[540] (493), 527[541] (493), 527[542] (493), 527[543] (493), 527[544] (494)
Takahashi, T., 277[43] (230), 277[60] (230, 239, 266, 267, 270)
Takahashi, Y., 286[657] (252)
Takai, Y., 278[78] (231), 527[543] (493)
Takakura, S., 284[499], 284[502] (263)
Takamatsu, M., 100[69] (37, 38, 41, 48, 53)
Takaya, H., 584[127] (549)
Takayama, T., 386[200] (333)
Takechi, K., 282[382]
Takemoto, T., 388[311] (357)
Takenaka, A., 101[88] (42, 53, 54), 280[233] (238, 262, 263, 265, 268)
Takeuchi, K., 586[275] (565–7)
Takizawa, T., 25[82] (11), 184[163] (122), 525[453] (480)
Takusagawa, F., 26[165] (20)
Tam, W., 518[20] (434), 521[162] (458, 462, 472), 521[182] (445), 523[280] (461)
Tamao, K., 182[17] (109)
Tamaru, Y., 282[414] (243, 263, 264), 386[212] (336), 388[327] (358), 389[383] (366)
Tamura, R., 389[364] (363), 389[397] (369)
Tan, K. H., 186[277] (137), 386[172] (324), 587[333] (575)
Tan, S. G., 428[110] (405)
Tanaka, A. A., 185[229] (131, 134)
Tanaka, H., 288[842] (263), 384[63] (299)
Tanaka, K., 24[20] (4), 101[90] (42, 58), 286[703] (255, 266, 268)
Tanaka, M., 26[98] (13), 102[173a] (49, 50, 64), 102[196] (48, 64), 105[430] (96), 185[244] (131, 159, 172), 279[157] (235), 386[200] (333), 587[287] (569)
Tanaka, N., 282[415], 288[802] (260)
Tanaka, T., 277[21] (228, 229)
Tanase, T., 222[143] (214), 223[155] (215), 287[748] (257), 390[445] (378)
Tangestaninejad, S., 526[466] (481)
Tani, K., 283[439] (244)
Taniguchi, H., 25[96] (13), 102[172] (49, 50, 64), 285[608] (249)
Tanikaga, R., 389[379] (366)
Tanimoto, S., 388[308] (356), 388[310] (357)
Tantardini, G. F., 426[2c] (392)
Tao, W., 286[674] (253, 255), 386[168] (322)
Taqui Khan, B., 387[265] (350)
Tate, J. P., 288[838] (263)
Tatsumi, K., 102[209] (62), 102[210] (62), 289[871] (266), 427[47] (398), 521[202] (453), 583[68] (542, 545, 547, 574), 583[86] (544)
Tau, K. D., 520[101] (440)
Taube, R., 100[33] (31, 33), 100[34] (31, 33), 103[240] (66, 67), 183[133] (116, 124, 138), 186[281] (137–9), 186[282] (137–9), 186[287] (138, 142), 186[288] (138, 142), 186[289] (138, 142), 186[293] (138, 142), 186[298] (138, 142), 186[299] (138, 142), 186[310] (138), 186[311] (138), 186[312] (138, 142), 186[313] (138, 142), 186[314] (138, 142), 187[363] (142), 189[496] (157, 159, 164)
Tawara, K., 387[254] (347)
Taylor, A., 99[3] (30, 31, 47, 48), 223[157] (215), 278[90] (233, 262)
Taylor, B. F., 220[21] (196)
Taylor, H. C. R., 281[312] (241)
Taylor, N. J., 185[242] (131, 172, 174), 185[243] (131, 172), 188[462] (155, 156), 290[940] (273)
Taylor, R., 521[154] (445), 526[510] (490)
Taylor, R. J., 278[135] (235), 384[71] (302), 582[14b] (535), 583[48] (538)
Teagle, J. A., 277[19] (227, 229), 527[556] (495)
Tejel, C., 387[271] (350), 387[272] (350)
Temkin, O. N., 220[31b] (197)
Temple, J. S., 389[370] (364), 389[371] (364)
Tennent, N. H., 428[99] (403), 530[772] (517)
Teo, B. K., 19, 26[143] (19), 26[144] (19), 189[527] (159, 170), 427[52] (398)
Teplova, I. A., 289[914] (272), 289[915] (272)
Terada, M., 105[430] (96), 185[244] (131, 159, 172)
Teranishi, S., 388[302] (355)
Terekhova, M. I., 287[732] (256), 388[335] (359)
Terenghi, M. G., 186[326] (139, 140)
Tereul, F., 585[206] (554, 555)
Terheijden, J., 104[331] (81, 85), 277[59] (230), 280[274] (240), 281[354] (243), 289[869] (264, 265), 529[704] (511), 530[753] (515), 530[755] (516), 530[757] (516), 530[758] (516), 530[760] (516), 530[762] (516)
Terron, G., 284[542] (272)
Teruel, F., 519[73] (439)
Tesauro, D., 587[285] (568, 571)
Tessier-Youngs, C. A., 185[228] (131, 134), 185[229] (131, 134), 185[230] (131, 135), 185[266] (134, 138)
Tètènyi, P., 384[28] (293)
Teunissen, H. T., 289[892] (268)
Textor, M., 190[586] (175, 177)
Thanh, N. N., 183[90] (115), 187[358] (137)
Thayer, A. G., 521[160]
Thelen, G., 101[133] (42, 62, 63), 105[373] (81, 88)
Theocharis, C. R., 527[539] (493)
Theriot, K. J., 281[319] (241), 281[321] (241), 283[486] (245, 263), 519[67] (438)
Theurel, L., 282[387] (253)
Thewalt, U., 187[373] (148), 519[57] (437)
Thiele, K.-H., 102[189] (41, 48), 103[244] (66, 67), 184[148] (117, 163)
Thivolle-Cazat, J., 278[105] (233, 255, 256)
Thoden, J. B., 27[177] (21), 188[450] (154, 171), 189[549] (171), 189[552] (171)
Thomas, C. B., 388[304] (356)
Thomas, I. M., 285[580] (248, 276)
Thometzek, P., 220[27a] (197, 210), 220[27b] (197, 210), 221[81] (205), 221[82] (205), 221[83] (205), 221[84] (205), 286[652] (252), 390[449] (378), 390[450] (378)
Thompson, K. H., 587[283b] (568)
Thompson, P. J., 279[212] (238), 527[571] (497), 584[125a] (549)
Thomson, M. A., 519[43] (437), 520[120] (442), 526[482] (485), 526[512] (490), 587[295] (571), 587[296] (571)
Thomson, S. K., 520[104] (440, 447),

522[227] (456), 522[230] (456), 585[182] (554, 555), 586[243] (560)
Thorn, D. L., 103[262] (73), 285[630] (250, 253), 520[90] (439)
Thornton, D. A., 428[117a] (406), 428[117b] (406)
Thornton, P., 104[303] (81, 85), 287[772] (259)
Thornton-Pett, M., 26[122] (16), 279[185] (236), 283[472] (245, 263), 284[554] (245, 263), 520[103] (440, 466), 520[109] (441), 523[327] (466), 529[683] (509), 586[260a] (563, 572), 586[260b] (563, 572), 586[260c] (563, 572)
Thouvenot, R., 520[142] (444)
Thummel, R. P., 281[342] (242, 273)
Tieghi, G., 285[620] (250)
Tiekink, E. R. T., 103[246] (67, 72), 524[387], 524[388]
Tielink, E. R. J., 189[492] (157, 161)
Tikhonova, N. Y., 221[58] (202, 219)
Timms, R. N., 386[196] (333), 386[197] (333), 388[300] (355)
Tinyakova, E., 100[83] (41, 47)
Tipper, C. F. H., 428[93] (403, 411), 527[570] (497, 505), 527[571] (497), 528[638] (505), 528[639] (505), 528[644] (505), 528[645] (505), 528[646] (505), 528[650] (506), 584[125a] (549), 584[125b] (549), 584[125c] (549)

picchio, A., 26[160] (20), 27[167] (21), 105[431] (96, 98), 185[241] (131, 172, 174), 185[242] (131, 172, 174), 185[243] (131, 172), 185[245] (131, 172, 174), 185[246] (131, 172, 174), 185[247] (131, 172, 174), 185[248] (131, 172, 174), 185[250] (131, 172, 174), 189[544] (172, 174), 189[546] (171, 174), 190[564] (172), 190[565] (172), 190[567] (172, 174), 190[578] (174), 220[23] (196), 220[29] (197), 222[92] (206), 222[94] (207), 286[689] (255, 271), 288[818] (262, 263), 387[271] (350), 427[65b] (400), 428[73] (401), 519[29] (435), 584[115] (548), 584[119] (548)

Tiripicchio-Camellini, M., 105[431] (96, 98), 185[243] (131, 172), 185[246] (131, 172, 174), 185[247] (131, 172, 174), 185[250] (131, 172, 174), 189[544] (172, 174), 190[565] (172), 190[578] (174), 220[23] (196), 220[29] (197), 222[94] (207), 427[65b] (400), 519[29] (435), 584[119] (548)
Titova, S. N., 189[538] (159)
Tjandra, S., 100[25] (30)
Tkatchenko, I., 101[151] (41, 42, 50), 186[295] (138), 187[360] (140), 187[362] (142), 290[941] (273), 389[416] (372), 389[417] (372)
Tobe, M. L., 521[167] (458, 459), 522[246] (458, 459), 522[247] (458, 459), 522[248] (458, 459)
Tobita, H., 519[62] (437)
Tocher, D. A., 280[229] (238), 284[521], 284[522], 284[523], 529[692] (510)
Todd, L. J., 191[633] (180, 181)
Togni, A., 386[189] (328), 520[145] (444)
Tohda, Y., 277[64] (230), 286[668] (252)
Tohgoh, K., 287[773] (259)
Tolman, C. A., 24[7] (2), 101[95] (42, 54), 182[31] (110, 123), 182[32] (110), 184[167] (123)
Tolstikov, G. A., 102[169] (50)
tom Dieck, H., 103[276] (73, 74, 78, 80), 183[113] (115, 117, 118), 183[114] (115, 117, 118), 184[149] (117, 118), 186[334] (139, 144), 286[659] (252), 286[662] (252), 286[663] (252), 290[942] (273), 385[83] (304), 525[432] (478)
Tomás, M., 220[22] (196), 223[147] (214), 223[148] (214), 277[8] (226, 227, 252), 281[352] (243, 262, 271), 286[690] (255, 263, 264), 286[691] (255, 262), 288[814] (261), 288[820] (262), 288[833] (262), 289[865] (264), 289[901] (270), 385[109] (309), 386[153] (317), 390[468] (382), 427[59] (399, 413), 428[96] (403), 429[151a] (409), 429[151b] (409), 518[5] (432), 523[328a] (467), 524[340] (468), 524[350] (468), 524[351] (468), 524[357] (469), 524[358] (469, 470), 524[362] (469), 524[366] (470), 524[367] (470), 524[368] (470), 524[369] (470), 524[386] (496, 497), 524[389], 525[400] (474), 525[401] (474), 525[402] (474), 525[403] (474), 525[405] (475), 525[406] (475), 525[407] (475), 525[408] (475), 525[419] (477), 525[420] (477), 586[255] (562, 563), 587[284] (568), 587[299] (571)
Tometzki, G. B., 388[325] (358)
Tomita, S. S., 285[592] (249), 384[52] (295)
Tomkinson, J., 187[409] (150)
Tommasi, I., 184[182] (126, 128)
Tondello, E., 185[258] (130)
Toniolo, L., 287[726] (256, 272), 287[727] (256), 287[728] (256), 287[740] (257), 430[211] (423), 520[136] (443, 461), 520[137] (443), 520[138] (443, 461), 523[275] (461), 523[276] (461), 523[277] (461), 523[278] (461)
Tooptakong, U., 183[128] (116, 122), 279[197] (237), 385[77] (303), 529[681] (509), 584[144] (551)
Tooze, R., 280[285] (240)
Topalovic, I., 183[84] (115, 123, 174)
Torra, I., 521[147] (444, 467), 523[329] (467), 523[330] (467)
Torras, J., 25[26] (4)
Torresan, I., 523[287] (462)
Torroni, S., 185[275] (137), 386[175] (324), 587[334] (575, 576)
Tossell, J. A., 187[411] (151)
Toth, I., 287[723] (256, 262, 263, 266)
Toto, S. D., 520[110] (441)
Towl, A. D., 427[51] (398, 399)
Toyoshima, T., 280[249] (239, 262)
Traill, P. R., 285[580] (248, 276), 288[816] (262, 274, 276), 289[896] (269, 273, 274), 290[919] (272, 273, 275), 290[950] (273, 274, 276), 290[961] (276), 523[318] (465)
Traldi, P., 283[481] (245, 248), 430[226] (426), 587[336] (576)
Tranquille, M., 182[25] (109)
Traversi, A., 524[335] (467)
Traverso, O., 522[209] (453), 524[334] (467), 524[377] (471), 526[514] (490)
Tremont, S. J., 282[366] (270, 273), 289[897] (270), 290[926] (273)
Trenkle, A., 26[111] (15), 26[112] (15)
Tresoldi, G., 390[437] (376), 588[364] (580)
Treurnicht, I., 526[517] (490), 526[518] (490), 587[300] (571)
Trinquier, G., 26[135] (19), 105[409] (93, 94)
Trofimenko, S., 191[624] (180, 182), 284[547] (245)
Trogler, W. C., 188[421] (151, 154, 155), 278[76] (230), 280[266] (240, 243, 245, 268, 270), 280[283] (240, 268, 271), 288[831] (262, 268, 271), 289[889] (268, 271), 384[72] (302), 385[73] (302), 427[5] (392), 520[133] (443), 527[530] (493), 583[22] (535, 565–7), 583[23a] (535), 583[23b] (535)
Troitskaya, L. L., 283[457] (245, 250), 529[708] (511)
Trost, B. M., 283[448] (245), 290[938] (273), 290[939] (273), 386[194] (331), 388[325] (358), 389[424] (373), 389[426] (374), 389[427] (374), 390[428] (374), 390[429] (375), 390[430] (375), 390[431] (375)
Trotter, J., 101[94] (42), 186[286] (138, 140), 280[279] (240)
Trozzi, M., 521[168]
Trzcinska, B. M., 100[82] (40)
Trzeciak, A. M., 279[162] (236)
Tsai, F.-Y., 430[219] (425)
Tsay, Y.-H., 26[99] (14), 26[124] (16), 100[55] (34), 104[299] (73, 74), 104[349] (81, 88, 89), 182[33] (110, 111, 117, 130–3), 182[34] (110, 111, 117, 130–2), 182[48] (111, 128), 182[49] (111, 128), 183[63] (111, 115), 183[112] (115, 122), 184[143] (117, 139, 141), 184[156] (119, 122, 131), 186[300] (138, 143), 288[788] (260), 387[214] (336)
Tschinke, V., 99[17] (30, 32), 220[11] (195)
Tse, J. S., 521[200] (452), 524[392] (473), 585[214a] (556), 585[214b] (556), 585[214c] (556)
Tsetlina, E. O., 430[207] (422)
Tsuboi, A., 587[315] (573)
Tsuchiya, H., 280[230] (238, 239, 255, 273), 283[447] (245), 388[337] (360)
Tsuda, T., 25[37] (5), 25[38] (5), 25[39] (5), 25[40] (5), 387[254] (347), 387[255] (347)
Tsuji, J., 384[15] (292, 333), 387[257] (347), 388[314] (357)
Tsuji, R., 586[275] (565–7)
Tsuji, Y., 522[272] (460)
Tsukui, T., 525[429] (478)
Tsutsui, M., 583[86] (544)
Tsutsui, T., 588[375] (581)
Tsutsumi, S., 286[655] (252), 287[765] (259)
Tsutsumida, J., 284[546] (245)

Tuck, D. G., 278[91] (233)
Tucker, P. A., 286[664] (252), 521[171]
Tulip, T. H., 24[23] (4), 101[136] (42, 47, 58, 59), 523[279] (461)
Tully, P. S., 186[335] (139, 143)
Tuong, T. D., 102[201] (41, 47, 48), 524[341] (468)
Turco, A., 102[177] (49, 51), 186[338] (139, 145), 186[339] (139, 146), 187[370] (146), 278[127] (235, 264, 270), 288[844] (263, 265), 522[237] (457)
Turner, J. E., 583[28] (536)
Turner, J. J., 24[5] (2), 24[6] (2), 427[64] (400), 588[370] (581)
Turpin, R., 220[41] (198), 429[162] (410)
Tuzi, A., 388[292] (354)
Tyers, K. G., 519[68] (438)
Tyler, D. R., 26[129] (16)
Tysoe, W. T., 384[30] (293), 384[31] (293)
Tzeng, W.-H., 430[219] (425)

Ubbels, H. J. C., 104[327] (81, 86), 104[329] (81, 85), 281[300] (241), 288[832] (262), 530[721] (513), 530[750] (515), 530[751] (515), 530[754] (515), 530[757] (516), 530[761] (516), 530[762] (516)
Uccello-Barretta, G., 584[93a] (544), 584[93b] (544), 586[232] (559)
Uchibori, Y., 279[161] (236, 262, 263)
Uchida, Y., 287[720] (256)
Uchiyama, T., 282[378] (263, 264), 282[379] (263), 284[566] (248, 263)
Uchtman, V. A., 27[174] (21), 189[543] (167, 171, 172)
Udrea, M., 384[38] (294)
Uechi, T., 281[311] (241)
Ueda, M., 587[317] (573)
Ugozzoli, F., 282[419] (243), 428[73] (401)
Uguagliati, P., 222[127a] (211), 223[161a] (216), 223[161b] (216), 277[27] (228), 278[128] (235), 287[753] (257), 385[75] (303), 429[186] (416), 430[212] (423), 522[208] (453), 522[241] (458), 522[252] (459), 583[59] (541, 545)
Uhlig, E., 24[14] (3), 100[65] (37, 38, 41, 42, 47, 81), 102[186] (41, 47, 48), 104[338] (81, 86), 104[354] (81, 90), 183[89] (115, 123, 174)
Uhm, J.-K., 190[597] (175, 180)
Ukhin, L. Yu., 387[215] (337)
Ulrich, B., 189[530] (159, 163)
Underwood, D. J., 427[47] (398)
Undeutsch, B., 104[342] (81, 90)
Undheim, K., 385[134] (314)
Unfried, G., 531[781] (517)
Unsin, J., 521[185] (447), 585[194] (554, 555)
Upton, T. H., 100[22] (30)
Urabe, A., 285[627] (250, 260, 262, 267, 269), 384[61] (296, 297), 389[366] (363), 583[67a] (542, 545, 574), 584[101] (545, 547), 584[108] (547)
Urata, H., 278[96] (233, 234), 279[157] (235), 280[244] (239)
Urriolabeitia, E. P., 288[809] (261), 288[813] (261), 288[814] (261)
Ursenbach, C., 220[11] (195)
Usatov, A. V., 281[345] (242)
Uson, I., 525[404] (474), 525[407] (475)
Uson, M. A., 288[815] (261), 288[839] (263), 288[846] (263), 289[847] (263), 289[866] (264), 428[113] (406), 524[349] (468), 524[360] (469), 524[361] (469), 525[404] (474)
Usón, R., 220[22] (196), 222[124] (211, 213), 222[132] (213), 222[133] (213), 222[134a] (213), 222[134b] (213), 222[144] (214), 223[145a] (214), 223[146] (214), 223[147] (214), 223[164] (216), 223[167] (217), 277[4] (226, 237, 238, 240, 255, 257, 261, 262), 277[5] (226, 227), 277[6] (226, 227), 277[7] (226), 277[9] (226, 227, 252), 277[24] (228), 277[30] (228), 286[690] (255, 263, 264), 286[691] (255, 262), 287[749] (257), 287[755] (258), 287[756] (258, 263), 287[757] (258), 287[758] (258, 259), 287[759] (258), 287[760] (258, 263), 288[815] (261), 288[820] (262), 288[830] (262), 288[839] (263), 288[846] (263), 289[847] (263), 289[864] (264), 289[865] (264), 289[866] (264), 289[867] (264), 289[901] (270), 290[921] (273, 275), 290[922] (273, 275), 385[109] (309), 386[153] (317), 388[291] (353), 427[59] (399, 413), 428[113] (406), 429[151a] (409), 429[151b] (409), 429[181] (416, 417, 423), 518[5] (432), 518[9] (432, 433), 523[328a] (467), 524[338] (467, 469), 524[340] (468), 524[349] (468), 524[350] (468), 524[351] (468), 524[357] (469), 524[358] (469, 470), 524[359] (469), 524[360] (469), 524[361] (469), 524[362] (469), 524[364] (469), 524[365] (469), 524[366] (470), 524[367] (470), 524[368] (470), 524[369] (470), 524[386] (496, 497), 524[389], 525[400] (474), 525[401] (474), 525[402] (474), 525[403] (474), 525[404] (474), 525[405] (475), 525[406] (475), 525[407] (475), 525[408] (475), 525[419] (477), 525[420] (477) 585[198] (554, 555), 586[255] (562, 563), 587[284] (568)
Ustynyuk, Y. A., 105[412] (94)

Vågberg, J. O., 389[407] (371)
Vaglio, G. A., 430[224] (426)
Vahrenkamp, H., 20, 26[111] (15), 26[112] (15), 26[113] (15), 26[114] (15), 26[163] (20), 27[188] (23), 27[189] (23), 27[190] (23), 27[191] (23), 105[414] (93, 94, 96, 98), 105[432] (96, 98), 105[435] (96, 98), 105[436] (96, 98), 106[437] (96, 98), 106[438] (96, 98), 106[439] (96, 98), 189[547] (171), 189[550] (171), 189[559] (172, 173), 190[560] (172, 173)
Valderrama, M., 528[604] (499)
Valentini, G., 101[123] (42)
Valle, G., 586[233] (559)
Valle, L., 189[513] (157, 166)
Valle, M., 190[564] (172)
Valpey, R. S., 389[373] (365), 389[382] (366), 583[34] (536, 540, 542)
van Asselt, R., 285[635] (250, 255, 256), 287[710] (255), 290[944] (273, 274)
van Beek, J. A. M., 104[328] (81, 86), 104[330] (42, 81), 281[301] (241), 528[620], 529[703] (511), 530[720] (513), 530[758] (516), 530[759] (516), 530[769] (517)
Van der Avoird, A., 185[255] (129)
van der Helm, R., 289[900] (270)
van der Kerk, G. J. M., 190[571] (172), 190[572] (172), 190[573] (172, 173)
van der Knapp, T. A., 184[178] (126), 586[239a] (560)
van der Ploeg, A. F. M. J., 279[205] (237, 238, 264), 529[707] (511), 530[763] (516), 530[764] (516), 530[765] (516), 530[766] (516), 530[767] (517), 530[768] (517)
van der Poel, H., 289[851] (263), 583[74a] (542), 583[74b] (542)
van der Sluis, P., 278[141] (235), 530[769] (517), 530[770] (517)
van der Veer, D., 585[213a] (556), 586[221a] (557)
van der Voort, E., 277[55] (230, 232), 278[141] (235)
Van Derveer, D. G., 101[105] (42, 47, 57), 282[361], 430[200] (420), 521[156], 527[557] (495), 529[717], 530[741] (514)
van Eldik, R., 105[397] (91, 92), 283[437] (244, 253, 254), 522[254] (459)
Van Hove, M. A., 386[143] (316)
Van Hulle, M.-T., 386[199] (333)
van Koten, G., 104[328] (81, 86), 104[329] (81, 85), 104[331] (81, 85), 104[333] (81, 86), 184[201] (125), 184[202] (125), 184[203] (125), 184[204] (125, 157), 190[574] (172, 173), 277[57] (230, 232, 234, 239, 269, 270, 273), 278[98] (233, 239, 262), 278[99] (233, 255, 256, 262), 278[101] (233, 262, 263, 268, 273), 278[102] (233, 250, 262–4), 278[103] (233, 251, 262, 273), 278[142] (235), 279[147] (235), 279[176] (236, 240, 242, 243, 262, 269, 270, 272, 273, 275), 279[205] (237, 238, 264), 280[226] (238, 240, 264, 265), 280[260] (239, 255, 262, 272, 273), 280[261] (239, 240, 255, 262, 266, 267, 269–74), 280[262] (239, 262, 266, 271, 274, 276), 280[265] (240, 265, 269, 270, 272–4), 281[293] (241, 270, 273), 281[300] (241), 281[301] (241), 281[339] (242, 243, 263), 281[354] (243), 285[633] (250, 255, 262), 286[685] (254, 264), 286[686] (254, 271), 288[832] (262), 289[851] (263), 289[868] (264), 289[869] (264, 265), 289[892] (268), 289[893] (268), 290[953] (274, 276), 387[261] (348), 429[164] (411, 412), 511, 528[620], 529[703] (511), 529[705] (511), 529[707] (511), 530[720] (513), 530[750] (515), 530[751] (515), 530[754] (515), 530[755] (516), 530[756] (516), 530[757] (516), 530[758] (516), 530[759]

(516), 530[760] (516), 530[761] (516), 530[762] (516), 530[763] (516), 530[764] (516), 530[765] (516), 530[766] (516), 530[767] (517), 530[768] (517), 530[769] (517), 530[770] (517), 583[73] (542), 583[74a] (542), 583[74b] (542)
Van Leeuwen, P. W. N. M., 222[105] (208), 285[632] (250), 286[699] (255, 256, 262), 287[738] (256), 529[688] (509)
van Wegen, J., 278[98] (233, 239, 262)
Van Zee, R. J., 187[414] (151)
Vance, B., 528[624]
Varaprath, S., 186[329] (139, 140), 387[231] (342), 387[232] (342)
Vargaftik, M. N., 221[55] (201), 221[56] (201), 221[59] (202), 221[60b] (202), 221[61b] (202)
Varyukhin, V. A., 388[297] (354)
Vasapollo, G., 287[736] (256, 257), 287[739] (257), 523[282] (461)
Veillard, A., 385[126] (312)
Veith, M., 26[102] (14)
Venenzi, L. M., 104[339] (81, 86), 284[540], 288[826] (262, 265), 384[58] (296), 390[440] (377), 427[9] (392), 427[21] (393, 395), 427[26] (394, 395), 427[27] (395), 427[42a] (397), 519[74] (439), 520[130] (443), 520[141] (444), 520[142] (444), 520[143] (444), 520[144] (444), 520[145] (444), 521[188] (447), 525[394] (474), 525[395] (474), 525[396] (474), 525[399] (474), 529[669] (508), 582[15] (535), 587[321] (574)
Venkataraman, U., 519[22] (434), 531[784] (518), 586[274] (565)
Venkatsubramanian, K., 279[179] (236), 284[550] (245, 264), 284[551] (245, 263, 264), 528[593] (498)
Venturi, M. G., 220[18] (195, 196), 429[150b] (409–11)
Veracini, C. A., 220[19b] (196), 428[108b] (403, 409), 429[149] (409), 583[65] (542)
Vermeer, P., 278[112] (234), 278[114] (234)
Verpeaux, J.-N., 103[241] (66, 67), 103[242] (67, 72), 189[489] (157, 163), 189[490] (157, 158, 163)
Verschuuren, A. H. M., 104[328] (81, 86)
Verstuyft, A. W., 277[49] (230)
Vessey, J. D., 520[103] (440, 466)
Vetter, W. M., 279[167] (236, 259, 262), 279[168] (236, 259), 429[142a] (408), 519[25] (434, 462), 519[26] (434, 462)
Vettori, U., 186[339] (139, 146), 187[370] (146)
Veya, P., 222[138] (213)
Vicente, C., 524[352] (469), 524[354] (469)
Vicente, J., 220[23] (196), 277[18] (227), 277[34] (228, 230), 278[107] (233, 262, 263), 280[221] (238, 263), 280[222] (238, 263, 268), 280[223] (238), 280[224] (238, 262, 263), 280[225] (238, 263, 264, 269), 286[689] (255, 271), 288[818] (262, 263), 519[73] (439), 525[448] (480), 529[706] (511), 585[198] (554, 555), 585[206] (554, 555)
Vidal, J., 101[151] (41, 42, 50)
Vignali, C., 190[567] (172, 174)
Vila, J. M., 282[359] (243), 283[458] (245), 283[459] (245), 283[460] (245, 263, 264), 283[466] (245), 283[468] (245), 283[472] (245, 263), 283[473] (245, 263), 283[476] (245, 263, 264), 284[500], 284[554] (245, 263), 284[558] (245), 289[854] (263)
Vilarrasa, J., 284[556] (245, 262, 263), 521[186] (447)
Villanueva, L. A., 280[245] (239, 245, 262, 269, 271), 286[654] (252, 267)
Villar, B., 519[67] (438)
Villarica, M., 587[341] (576, 581)
Vilmer, M. P., 288[781] (260), 386[158] (319), 386[159] (319)
Vinay, I., 186[326] (139, 140)
Vinitskii, D. M., 191[642] (181)
Vinke, I. C., 281[354] (243), 530[755] (516)
Visalakshi, R., 528[594] (498)
Visser, J. P., 279[211] (238)
Vitagliano, A., 280[256] (239, 262, 272), 385[96] (306), 387[250] (346), 387[279] (352), 388[292] (354), 389[368] (363), 389[372] (365), 520[82] (439), 583[75] (543), 583[76] (543), 583[78] (543), 583[83] (544), 585[212] (554), 586[221b] (557), 586[233] (559)
Vitarelli, P., 522[233] (457)
Viterbo, D., 284[506] (263)
Vitorge, M. C., 222[116] (210)
Vittal, J. J., 221[49] (200), 222[111] (209), 222[112] (209), 386[162] (320), 428[89a] (404), 429[178] (414, 415), 523[312] (464)
Vitulli, G., 390[463] (381)
Voelkl, A., 223[170] (217, 218), 223[175] (218)
Voelter, W., 386[198] (333)
Vogel, J., 100[65] (37, 38, 41, 42, 47, 81)
Vogler, A., 387[238] (344), 528[626] (500)
Vogt, H., 522[266] (460)
Volkl, A., 430[222] (425)
Vollhardt, K. P. C., 183[88] (115, 123, 174)
Vol'pin, M. E., 185[212] (130, 131)
von Matt, P., 388[348] (360)
Von Niessen, W., 220[7a] (194), 220[7b] (194)
von Philipsborn, W., 385[110] (309)
von Schnering, H. G., 221[77] (204), 428[114] (406)
von Zelewsky, A., 280[275] (240, 270, 273), 281[303] (241), 513, 525[435] (478), 530[722] (513), 530[724] (514), 530[725] (514), 530[726] (514), 530[727] (514), 530[728] (514), 530[729] (514), 530[730] (514), 530[732] (514), 530[734] (514), 530[735] (514), 530[736] (514)
Voronkov, M. G., 430[207] (422)
Voyevodskaya, T. J., 105[412] (94)
Vreblevskaya, L. V., 430[195] (419)
Vrieze, K., 184[202] (125), 222[105] (208), 279[147] (235), 279[205] (237, 238, 264), 285[568] (248), 285[632] (250), 286[699] (255, 256, 262), 287[738] (256), 288[832] (262), 289[869] (264, 265), 385[76] (303), 529[707] (511), 530[761] (516), 530[762] (516), 530[763] (516), 530[764] (516), 530[766] (516), 530[767] (517), 530[768] (517), 583[73] (542)
Vyshinskii, N. N., 188[430] (152), 188[431] (152), 188[432] (152), 188[433] (152)

Wache, S., 186[298] (138, 142), 186[299] (138, 142), 186[309] (138, 142), 186[312] (138, 142), 186[313] (138, 142), 186[314] (138, 142)
Wada, F., 279[196] (237)
Wada, M., 25[95] (13), 25[96] (13), 25[97] (13), 26[98] (13), 101[112] (41, 42, 49, 50, 55, 64), 101[114] (42, 49, 55, 61), 101[115] (42, 49, 55, 61), 101[116] (42, 49, 55), 101[149] (41, 42, 48, 57), 102[170] (41, 47, 49, 50, 62), 102[171] (41, 49, 50, 62), 102[172] (49, 50, 64), 102[173a] (49, 50, 64), 102[173b] (50, 64), 102[195] (41, 48, 49, 56, 57), 102[196] (48, 64), 102[197] (41, 48, 57), 102[198] (48, 58), 102[199] (48, 58), 102[200] (48, 56), 223[176a] (218), 223[176b] (218), 277[63] (230, 232), 430[208] (422), 430[209] (422), 587[315] (573)
Waddell, W. H., 523[283] (461)
Waddill, G. D., 386[145] (316)
Waddington, M. D., 527[576] (498), 527[577] (498), 528[647] (505), 584[128] (549)
Waddington, T. C., 187[409] (150)
Wade, A. R., 389[415] (372)
Wadepohl, H., 189[520] (159, 175), 190[591] (175, 178), 190[595] (175, 179)
Wahlgren, U., 100[19] (30)
Waid, R., 520[101] (440)
Waissermann, J., 278[105] (233, 255, 256)
Wakabayashi, S., 101[130] (42, 62)
Wakatsuki, Y., 281[357] (243)
Walker, A., 430[227] (426), 430[228] (426), 526[490] (486)
Walker, N. P., 528[622]
Walker, N. P. C., 287[772] (259)
Waller, F. J., 287[741] (257)
Wallwork, S. C., 282[370] (263, 264), 529[697] (510), 588[363] (580)
Walther, B., 223[153] (215), 429[175] (414)
Walther, D., 25[36] (5), 25[41] (5), 25[42] (5), 25[47] (7), 102[214] (65), 102[215] (65), 103[223] (65, 66), 104[342] (81, 90), 104[356] (81, 88, 90), 104[357] (81, 88, 90), 104[358] (81, 88, 90), 104[359] (81, 88), 104[360] (81, 88), 104[361] (81, 88), 104[362] (81), 105[365] (81, 90), 105[367] (81, 88, 90), 105[368] (81, 88, 90), 105[372] (81, 88, 89), 105[378] (80, 88), 105[381] (80, 88), 183[90] (115), 184[184] (126, 127), 184[185] (126, 128), 185[240] (131, 135), 186[325] (139, 140), 186[336] (139, 141), 187[356], 187[357]

(137), 187[358] (137), 187[361] (141, 142)
Walther, P., 185[227] (131, 136, 159), 185[271] (136, 172), 187[353] (139, 155, 159, 168, 172)
Walton, R. A., 430[191] (418)
Walz, L., 428[114] (406)
Wan, W. K., 385[130] (313)
Wanat, R. A., 278[130] (235)
Wang, A. W., 278[106] (233, 255)
Wang, F., 584[147] (552)
Wang, H., 102[159] (41, 42, 56)
Wang, J., 389[363] (363)
Wang, J.-G., 586[264a] (564, 572), 586[264b] (564, 572)
Wang, L. P., 384[30] (293)
Wang, R. J., 527[554] (494)
Wang, S., 102[160] (42)
Wang, S.-L., 286[642] (251, 271), 286[643] (251), 390[466] (382)
Wang, X. C., 283[442] (244)
Wang, X. K., 102[157] (42, 56), 187[406] (150, 151, 155), 187[407] (150, 151, 155)
Wang, Y., 191[636] (180), 277[22] (228), 277[32] (228–30), 277[38] (229, 230), 429[143] (408), 429[144] (408), 430[219] (425), 519[27] (434), 523[288] (462), 523[289] (462), 523[292] (462), 526[501] (487), 527[558] (495), 587[322] (574, 577)
Ward, T., 281[303] (241)
Wardell, J. L., 278[133] (235, 263, 269, 273), 289[883] (267), 289[895] (269, 273), 518[15] (434)
Warden, G. M., 428[99] (403), 530[772] (517)
Wardle, R., 280[218] (238)
Wardle, R. W. M., 427[7] (392), 427[8] (392, 396, 397), 427[36] (396), 429[172] (412)
Warhurst, N. J. W., 585[162] (553)
Warner, S., 385[115] (310)
Warrens, C. P., 586[240] (560)
Wasserman, H. J., 284[501]
Waszczak, J. V., 427[52] (398)
Wat, R. K. M., 521[191] (447), 588[361] (580)
Watanabe, H., 387[257] (347)
Watanabe, Y., 522[272] (460)
Watkins, S. F., 284[496] (245)
Watkins, W. N., 522[230] (456)
Watson, A. A., 284[513], 288[816] (262, 274, 276), 289[896] (269, 273, 274), 290[945] (273), 290[948] (273, 274, 276), 290[949] (273, 274, 276), 290[950] (273, 274, 276), 523[318] (465), 523[324] (466)
Watson, D. G., 184[139] (116, 122), 521[154] (445)
Watson, G., 388[304] (356)
Watson, M. J., 429[173] (413)
Watson, P. L., 390[462] (381)
Watts, R. J., 284[498] (245), 530[723] (513)
Watts, W. E., 582[7a] (534, 535), 582[7b] (534, 535), 582[7c] (534, 535), 582[7d] (534, 535), 582[7e] (534, 535)
Waymouth, R. M., 388[349] (360)
Weatherell, C., 185[242] (131, 172, 174)
Weaver, M. J., 427[46] (398), 427[55] (398)
Webb, C. G., 279[165] (236, 269, 273)
Weber, J., 187[411] (151)
Weber, L., 277[12] (227, 228), 277[37] (229)
Weber, M., 290[930] (273)
Weckmann, E., 100[33] (31, 33), 183[133] (116, 124, 138)
Weeks, B., 282[396] (243, 245), 284[526]
Wehman, E., 280[226] (238, 240, 264, 265), 281[339] (242, 243, 263), 286[686] (254, 271), 529[705] (511)
Wehman-Ooyevaar, I. C. M., 530[758] (516), 530[769] (517), 530[770] (517)
Wehrle, G., 282[377] (244)
Wei, D., 187[406] (150, 151, 155), 187[407] (150, 151, 155)
Wei, L., 385[115] (310)
Weidlein, J., 25[73] (10)
Weigand, W., 222[121] (211), 430[190a] (418), 430[190b] (418), 583[55] (540)
Weihua, X., 101[134] (42, 47)
Weimann, B., 102[217] (65), 186[301] (138, 142)
Weimer, T., 100[74] (37, 38), 100[75] (37–9, 42, 50, 56), 100[76] (37–9, 42, 56), 100[77] (37–9, 42, 56)
Weinberg, E. L., 286[695] (255, 263, 268, 270, 273, 275)
Weinberger, B., 25[77] (11)
Weisemann, C., 519[56] (437), 519[70] (438), 521[178], 521[180], 521[181], 585[201a] (554, 555), 585[201b] (554, 555), 585[211] (554, 555)
Weisenberger, C. R., 190[563] (172, 174), 190[576] (174)
Weiss, J., 189[535] (159, 163)
Weiss, K., 223[160] (216)
Weisshaar, J. C., 99[9] (30)
Welch, A. J., 27[181] (21, 22), 104[333] (81, 86), 223[147] (214), 223[148] (214), 277[8] (226, 227, 252), 288[808] (261, 262), 288[837] (262–4), 289[865] (264), 385[109] (309), 386[153] (317), 386[181] (327), 388[290] (353), 390[453] (379), 427[21] (393, 395), 427[59] (399, 413), 518[5] (432), 524[348] (468), 524[362] (469), 524[367] (470), 524[369] (470), 525[398] (474, 490), 586[246] (560), 586[255] (562, 563), 587[284] (568), 587[306] (572)
Wells, A. P., 282[403] (263)
Weltner, J. W., 187[414] (151)
Wendoloski, J. J., 25[94] (13)
Wenkert, E., 101[142] (42, 47, 51)
Werner, H., 183[95] (115, 124, 153), 189[505] (157, 159, 161), 189[530] (159, 163), 205, 220[27a] (197, 210), 220[27b] (197, 210), 221[81] (205), 221[82] (205), 221[83] (205), 221[84] (205), 278[124] (234), 280[217] (238, 267, 270), 282[363] (264), 286[652] (252), 286[700] (255), 288[783] (260), 288[786] (260), 288[792] (260), 288[799] (260), 289[905] (271), 385[78] (303), 385[79] (303), 385[80] (303), 387[223] (341), 387[282] (352), 388[285] (352), 388[286] (352), 390[441] (377), 390[443] (377), 390[444] (377), 390[449] (378), 390[450] (378), 528[604] (499), 587[309b] (573, 576), 587[311] (573, 576)
Werner, S., 189[553] (171, 172)
Wertz, D. W., 384[59] (296), 384[60] (297)
West, R., 586[241a] (560), 586[241b] (560)
Westcott, S. A., 188[462] (155, 156)
Wetherell, S. N., 385[139] (315)
Wettling, T., 187[384] (148, 149)
Whelan, J., 388[339] (360)
White, A. H., 277[23] (228), 279[184] (236, 241), 280[252] (239, 263), 280[254] (239, 262), 280[255] (239, 262), 282[385] (253, 263, 264), 282[389] (262), 282[390], 286[707] (255, 262), 286[708] (255, 262), 288[816] (262, 274, 276), 290[916] (272, 273, 276), 290[943] (273–6), 290[946] (273), 290[949] (273, 274, 276), 290[950] (273, 274, 276), 290[954] (275), 290[961] (276), 390[438] (376), 429[141] (408), 520[93] (440), 521[150] (444), 523[304] (464), 523[305] (464), 523[318] (465), 525[428] (478), 527[559] (495), 528[615] (500), 528[616] (510), 529[701] (510), 585[186] (554, 555), 586[263a] (564, 572)
White, A. J., 523[309] (464), 523[310] (464)
White, D. A., 289[886] (268, 270)
White, M. R., 586[231] (559)
White, R. P., 27[175] (21)
Whiteley, M. W., 190[618] (178), 190[619] (178)
Whitely, M. H., 582[6g] (534), 582[6h] (534), 582[6i] (534)
Whitesides, G. M., 102[208] (62), 517, 520[100] (440), 521[203] (453), 522[216] (454), 522[218] (455), 522[219] (455), 522[220] (455), 522[224] (455), 522[225] (456, 478, 481), 522[226] (456), 522[232] (457), 522[235] (457), 526[461] (481), 526[463] (481), 526[464] (481), 526[465] (481), 529[674] (508), 531[778] (517), 531[783] (518), 582[19a] (535), 582[19b] (535), 582[20] (535), 583[21a] (535, 565, 568), 583[21b] (535), 583[21c] (535), 583[49] (538), 585[178] (554, 555), 585[179] (554, 555), 585[204] (554, 555), 586[226a] (558), 586[226b] (558)
Whitmore, B., 386[190] (328)
Whittaker, M. J., 530[748] (515)
Whitten, J. L., 100[18] (30)
Whittle, R. R., 279[167] (236, 259, 262), 285[629] (250, 261, 262, 264), 429[142a] (408), 519[25] (434, 462)
Wickramasinghe, W. A., 185[252] (131), 185[253] (131, 134), 521[173] (462)
Widdowson, D. A., 283[475] (245)
Widmark, P.-O., 182[18] (109), 386[148] (317)
Wiege, M., 519[55] (437), 519[57] (437), 519[59] (437)
Wiegreffe, P., 100[79] (38, 39, 42, 57)

Wieland, W., 188[422] (151, 157, 159, 168)
Wiger, G. R., 285[592] (249), 384[51] (295), 384[52] (295), 585[172] (554, 556)
Wijkens, P., 278[99] (233, 255, 256, 262), 278[103] (233, 251, 262, 273)
Wilcox, D. E., 285[600] (249)
Wilczok, U., 281[291] (240)
Wild, S. B., 283[440] (244), 283[441] (244, 263), 283[444] (244, 264), 390[438] (376)
Wilde, A., 388[353] (361)
Wilhelm, D., 387[226] (341), 389[392] (368)
Wilke, G., 24[2] (1), 25[91] (12), 26[100] (14), 100[51] (31, 34, 35), 100[52] (31, 34, 35), 100[53] (34), 100[54] (34), 100[56] (31, 36), 102[183] (41, 47, 48, 56, 63), 103[286] (74, 75), 103[290] (73, 74, 78, 81), 103[291] (73, 74, 78), 104[297] (73, 74), 104[298] (73, 74), 104[300] (73, 74, 76), 182[5] (108, 138), 182[6] (108, 138), 182[35] (110, 112, 113, 115), 182[38] (110, 111), 182[40] (110), 182[41] (110, 112), 182[42] (110–12), 182[43] (110, 112, 126), 182[44] (110, 113), 182[45] (110), 183[62], 183[63] (111, 115), 183[78] (115, 119), 184[145] (117, 118), 185[223] (130, 131, 134), 185[272] (137), 221[65] (203), 287[779] (260), 387[281] (352, 359), 390[458] (380), 587[309a] (573, 576)
Wilker, C. N., 528[635] (505), 584[142] (551, 569)
Wilkes, D. J., 522[231] (456)
Wilkinson, G., 280[240] (239), 280[259] (239), 280[285] (240), 429[183] (416), 519[79] (439), 528[622], 585[180] (554, 555)
Wilkinson, M. P., 525[450] (480), 583[52] (540)
Wilkinson, S. W., 586[234] (559)
Williams, B. L., 527[581] (498), 528[590] (498), 528[632] (504), 528[633] (504)
Williams, D. J., 283[491], 289[848] (263)
Williams, G. J. B., 384[47] (294), 524[373] (471), 586[225] (558)
Williams, I. D., 385[115] (310), 427[68] (400), 429[173] (413)
Williams, J. L., 285[602] (249), 387[230] (342)
Williams, M. A., 519[75] (439, 464, 472, 497, 506), 529[655] (506), 529[658] (507), 585[208] (554, 555), 585[216] (557)
Williard, P. G., 529[687] (509, 510)
Willis, A. C., 280[264] (240, 261, 262, 266, 272, 275, 276), 529[678] (509, 513, 515)
Willison, D., 103[250] (67, 70), 189[515] (157, 166)
Willson, J. M., 279[170] (236, 267)
Wilson, C. A., 284[512]
Wilson, S. R., 187[385] (148, 149), 387[276] (350)
Wimmer, F. L., 284[516], 284[517] (263), 288[840] (263), 529[695] (510)
Wimmer, S., 284[516], 284[517] (263), 288[840] (263)
Winchester, W. R., 430[198] (419), 525[438] (479), 586[227] (558), 586[228] (558), 586[229] (558)
Wing, R. M., 285[592] (249), 285[596] (249), 285[601] (249), 285[602] (249), 384[51] (295), 384[52] (295), 384[53] (295), 385[108] (309), 385[118] (310), 387[227] (341), 387[230] (342), 522[250] (459), 585[171] (554, 556), 585[172] (554, 556), 585[173] (554, 556, 557, 561), 585[174] (554, 557, 558)
Winkhaus, V., 24[4] (2), 184[206] (126, 145)
Winter, M., 103[256] (67, 73), 183[118] (115, 156, 157), 183[129] (116, 121, 157, 161), 188[475] (157, 161), 188[480] (157, 160), 188[485] (157, 160), 188[487] (157, 161), 189[501] (157, 161), 189[502] (157, 160, 161), 189[503] (157, 160), 189[504] (157, 159, 161), 189[506] (157, 161), 189[508] (157, 161), 189[509] (157, 159, 160), 189[510] (157, 161), 189[511] (157, 161), 189[539] (160), 189[540] (160)
Winter, W., 386[198] (333)
Winzenburg, M. L., 430[192] (418)
Wisbey, S. J., 105[423] (95, 96), 223[181] (219)
Wisner, J. M., 289[909] (272, 276), 521[159] (493)
Wissing, E., 281[301] (241), 529[703] (511)
Witte, J., 100[57] (31, 36), 100[58] (31, 36), 100[61] (31, 36)
Wittrig, R. E., 27[180] (21)
Wojcicki, A., 587[310] (573), 587[323] (574, 577), 587[339] (576), 588[372] (581)
Wolczanski, P. T., 521[151] (444), 521[152] (444)
Wolfer, M., 525[395] (474), 525[399] (474)
Wolfer, M. K., 525[396] (474)
Wolmershäuser, G., 185[227] (131, 136, 159), 185[271] (136, 172), 187[351] (139, 155, 159, 168, 171), 187[352] (139, 155, 159, 168, 171), 187[353] (139, 155, 159, 168, 172), 187[384] (148, 149)
Wombacher, F. J., 283[433] (244, 245, 248), 529[699] (510), 529[702] (511)
Won, K., 221[50a] (201)
Wong, C. S., 286[671] (253), 428[125b] (407), 519[68] (438)
Wong, F. S., 188[481] (157, 165)
Wong, H. N. C., 283[442] (244)
Wong, P. K., 278[109] (234, 255), 289[894] (269, 273)
Wong, Y.-T., 384[32] (293)
Wong-Ng, W., 384[68] (302)
Wood, F. E., 428[69] (400)
Woodgate, P. D., 389[397] (369)
Woodward, P., 105[423] (95, 96), 190[566] (172), 223[181] (219)
Woodward, S., 189[531] (159, 169)
Woolcock, J. C., 285[601] (249), 387[227] (341), 522[250] (459), 585[173] (554, 556, 557, 561), 585[174] (554, 557, 558)
Woolery, G., 25[32] (5)
Woolley, R. G., 427[50] (398)
Woollins, J. D., 283[491]
Worakun, T., 388[317] (357), 388[318] (357), 388[319] (357), 388[320] (357)
Wörner, K.-F., 190[597] (175, 180)
Wouters, J. M. A., 278[113] (234), 287[713] (255), 287[761] (258), 526[499] (487)
Wozniak, B., 280[240] (239)
Wrackmeyer, B., 102[191] (48, 64), 187[381] (148, 149), 187[382] (148, 149), 187[383] (148, 149), 277[45] (230, 231), 277[70] (230, 245), 526[505] (490), 526[506] (490, 491), 526[511] (490), 527[532] (493), 527[533] (493), 527[537] (493), 527[538] (493), 527[539] (493), 587[324] (581)
Wreford, S. S., 521[162] (458, 462, 472), 524[370] (471)
Wright, A. H., 582[7f] (534, 535), 582[7g] (534, 535), 582[7h] (534, 535), 582[7i] (534, 535), 582[7j] (534, 535), 582[7k] (534, 535)
Wright, L. L., 285[596] (249), 385[118] (310), 583[33] (536, 538), 585[172] (554, 556), 585[173] (554, 556, 557, 561)
Wright, W. F., 191[633] (180, 181)
Wu, F. J., 277[22] (228), 527[558] (495)
Wu, G. Z., 283[438] (244), 283[461] (245), 286[682] (253), 286[683] (253), 385[94] (305)
Wu, R. F., 277[32] (228–30)
Wu, W.-G., 586[262] (564, 572), 586[265] (564), 587[303] (572)
Wu, Y. C., 26[122] (16), 279[185] (236)
Wucherer, E. J., 390[464] (382)
Wurminghausen, Th., 427[29] (395)
Wyrwa, R., 103[272] (73, 74)

Xie, Y., 286[649], 518[19] (434)
Xu, C., 520[125] (442)
Xu, C. F., 585[203] (554, 555)
Xu, D., 102[173a] (49, 50, 64), 102[173b] (50, 64), 102[198] (48, 58), 102[199] (48, 58)
Xu, D. J., 25[97] (13), 26[98] (13)
Xue, Z., 588[377] (581)

Yague, J. F., 289[847] (263), 428[113] (406), 524[360] (469)
Yam, V. W.-W., 587[297] (571)
Yama, M., 384[62] (299)
Yamada, K., 288[797] (260)
Yamada, Y., 389[383] (366)
Yamagami, I., 283[485] (245, 257)
Yamagishi, T., 388[345] (360)
Yamaguchi, M., 388[323] (358), 388[345] (360)
Yamaguchi, T., 285[578] (248)
Yamamoto, A., 24[9] (2), 24[20] (4), 25[25] (4), 100[68] (37, 38, 42, 53), 100[69] (37, 38, 41, 48, 53), 101[87] (42, 58), 101[88] (42, 53, 54), 101[89] (42, 47, 53, 58, 62, 63), 101[90] (42, 58), 101[91] (42, 53, 58, 62, 63), 101[92] (42, 53, 54, 62), 101[137] (42, 51, 61), 101[140] (42, 53, 62), 102[179] (56, 62), 102[180] (56), 102[184] (48, 55, 61, 62), 102[204] (49, 62), 102[209] (62), 102[210] (62), 102[211] (63),

104[305] (81, 85), 104[335] (81), 104[343] (81, 90), 104[345] (81, 90), 104[351] (81, 90, 91), 104[352] (81, 90), 105[382] (85), 186[285] (138, 140), 222[98a] (207), 222[98b] (207), 222[99a] (208), 222[99b] (208), 222[99c] (208), 222[106] (208), 278[97] (233, 255, 262), 278[138] (235, 255), 279[166] (236, 257, 259), 279[189] (237), 280[230] (238, 239, 255, 273), 280[231] (238, 267), 280[232] (238, 239, 262, 266, 267), 280[233] (238, 262, 263, 265, 268), 280[234] (238), 280[246] (239, 240), 280[247] (239, 255, 266), 280[248] (239, 255, 266), 280[249] (239, 262), 280[250] (239), 280[270] (240, 267), 280[272] (240, 267, 270), 283[485] (245, 257), 285[631] (250), 286[688] (255), 286[697] (255), 286[701] (255, 267), 286[703] (255, 266, 268), 286[705] (255, 256), 286[709] (255, 256), 287[712] (255, 267, 268), 287[714] (255), 287[722] (256, 265, 268, 271), 287[725] (256, 257), 287[729] (256), 287[734] (256), 287[735] (256, 257, 270), 289[871] (266), 289[876] (266), 289[879] (267, 269, 270), 289[887] (268, 269, 271), 289[904] (271), 387[241] (344), 387[242] (344), 387[243] (344), 387[245] (345), 387[246] (345), 388[336] (359), 388[343] (360), 388[344] (360), 389[377] (365), 389[387] (368), 429[145] (408), 519[33] (436), 521[157], 521[202] (453), 522[217] (455), 523[281] (461), 523[290] (462), 582[8] (534)
Yamamoto, H., 281[310] (241, 262, 263)
Yamamoto, J., 25[59] (9), 184[187] (126, 128)
Yamamoto, K., 388[330] (359)
Yamamoto, T., 24[9] (2), 24[20] (4), 24[21] (4), 25[25] (4), 100[68] (37, 38, 42, 53), 100[69] (37, 38, 41, 48, 53), 101[87] (42, 58), 101[89] (42, 47, 53, 58, 62, 63), 101[90] (42, 58), 101[91] (42, 53, 58, 62, 63), 101[92] (42, 53, 54, 62), 101[110] (42, 47, 51, 58), 101[130] (42, 62), 101[137] (42, 51, 61), 101[140] (42, 53, 62), 102[162] (42), 102[179] (56, 62), 102[180] (56), 102[184] (48, 55, 61, 62), 102[204] (49, 62), 102[210] (62), 104[305] (81, 85), 104[335] (81), 104[343] (81, 90), 104[345] (81, 90), 104[351] (81, 90, 91), 104[352] (81, 90), 105[382] (85), 186[285] (138, 140), 278[97] (233, 255, 262), 279[166] (236, 257, 259), 279[190] (237), 280[247] (239, 255, 266), 280[250] (239), 280[272] (240, 267, 270), 286[701] (255, 267), 286[703] (255, 266, 268), 287[712] (255, 267, 268), 287[722] (256, 265, 268, 271), 287[725] (256, 257), 289[876] (266), 289[879] (267, 269, 270), 387[245] (345), 387[246] (345), 519[33] (436), 522[217] (455)
Yamamoto, Y., 25[82] (11), 25[83] (11), 25[85] (11), 184[163] (122), 222[109] (209, 212), 222[114a] (210, 215), 222[114b] (210, 215), 222[117] (210, 215), 222[136] (213), 222[139b] (213), 222[143] (214), 223[155] (215), 277[16] (227), 277[53] (230, 231), 278[81] (231, 232), 279[160] (236), 287[748] (257), 287[762] (258), 384[39] (294), 390[445] (378), 427[47] (398), 429[165] (411), 429[168] (412, 413), 429[169] (412), 429[170] (412), 429[171] (412), 429[177a] (414), 429[177b] (414), 525[453] (480), 527[531], 527[540] (493)
Yamana, Y., 25[28] (4)
Yamanaka, I., 387[240] (344)
Yamaoka, H., 279[182] (236)
Yamasaki, H., 285[608] (249)
Yamashita, H., 587[287] (569)
Yamashita, J., 387[237] (343)
Yamashita, R., 279[196] (237)
Yamatoya, K., 583[67a] (542, 545, 574)
Yamazaki, H., 25[82] (11), 25[83] (11), 184[163] (122), 189[519] (159, 167), 222[109] (209, 212), 222[114a] (210, 215), 222[114b] (210, 215), 222[117] (210, 215), 222[136] (213), 222[139b] (213), 277[53] (230, 231), 278[81] (231, 232), 281[357] (243), 287[762] (258), 388[314] (357), 429[165] (411), 429[168] (412, 413), 429[169] (412), 429[170] (412), 429[171] (412), 429[177a] (414), 429[177b] (414), 525[453] (480), 527[531], 527[540] (493)
Yanagi, A., 279[160] (236)
Yanagi, K., 388[343] (360)
Yanagihara, H., 278[138] (235, 255)
Yanase, N., 288[796] (260)
Yáñez, R., 105[427] (67, 96, 97)
Yang, D.-S., 428[82] (403), 429[179] (415), 521[200] (452), 524[392] (473), 528[625] (500), 585[214a] (556), 585[214b] (556), 585[214c] (556), 588[376] (581)
Yang, G., 102[160] (42)
Yang, H., 100[18] (30), 284[529]
Yang, J. J., 389[419] (373)
Yang, X. G., 187[385] (148, 149)
Yang, Y. P., 286[649]
Yang, Z.-Y., 525[423] (477)
Yaniuk, D. W., 584[148] (552, 562)
Yanlong, Q., 101[134] (42, 47)
Yasinetskaya, S. V., 429[147] (409)
Yasuda, T., 278[84] (232)
Yasuoka, N., 26[98] (13), 102[173a] (49, 50, 64), 278[84] (232), 281[307] (241)
Yatake, T., 277[64] (230)
Yates, J. T., Jr., 384[27] (293)
Yatsimirsky, A. K., 281[350] (242), 282[368], 282[426] (244), 283[451] (245), 285[615] (250), 285[616] (250), 289[884] (268), 290[925] (273), 386[195] (332)
Yavari, A., 518[10] (432, 433, 453), 518[11] (432, 433)
Yeh, S.-K., 277[31] (228, 230)
Yeh, Y.-S., 519[27] (434)
Yin, J. G., 586[262] (564, 572)
Yokota, K.-I., 389[362] (363)
Yoneda, A., 281[321] (241)
Yoneda, H., 184[186] (126, 128)
Yoshida, M., 290[933] (273)
Yoshida, T., 24[22] (4), 221[50c] (201), 290[960] (276)
Yoshida, Z., 388[327] (358), 389[383] (366)
Yoshida, Z.-I., 282[414] (243, 263, 264)
Yoshimatsu, H., 288[803] (260)
Yoshinobu, J., 384[35] (293), 384[36] (293)
Yoshioka, H., 389[383] (366)
Youinou, M.-T., 278[108] (234, 255, 256)
Young, G. B., 101[98] (41, 42, 48, 66, 67), 280[259] (239), 519[78] (439), 519[79] (439), 519[80] (439), 520[81] (439), 520[87] (439), 520[104] (440, 447), 520[105] (440), 520[106] (440), 521[175], 522[214] (454), 522[215] (454), 522[227] (456), 522[228] (456), 522[229] (456), 522[230] (456), 522[231] (456), 525[423] (477), 584[146a] (552, 555), 584[146b] (552), 585[165] (553, 555), 585[180] (554, 555), 585[181] (554, 555), 585[182] (554, 555), 585[193] (554, 555, 569), 585[209] (554, 555), 586[243] (560), 587[283b] (568)
Young, G. J., 284[501]
Young, P. A., 522[238] (457)
Young, S. J., 276[3] (226, 255, 256, 266, 272)
Youngs, W. J., 185[228] (131, 134), 185[229] (131, 134), 185[230] (131, 135), 185[266] (134, 138), 282[413], 529[671] (508), 529[672] (508)
Youqi, T., 101[134] (42, 47)
Yuan, P., 27[192] (24)
Yuasa, Y., 287[712] (255, 267, 268)
Yudanov, V. F., 427[28] (395)
Yukawa, T., 286[655] (252)
Yurchenko, E. N., 427[22] (393)
Yus, M., 282[402]
Yusupova, F. G., 183[119] (115, 117, 119)

Zaera, F., 100[25] (30), 384[30] (293)
Zaghal, M. H., 519[21] (434)
Zahn, I., 586[224] (558)
Zähres, M., 102[219] (65), 186[302] (138, 140)
Zajíček, J., 387[248] (345)
Zakeeruddin, S. M., 387[265] (350)
Zakharkin, L. I., 102[203] (41, 48), 277[74] (230), 281[345] (242)
Zakharov, L. N., 189[538] (159)
Zakharova, I. A., 388[293] (354), 388[294] (354)
Zaki, A. A., 521[183] (445)
Zakosi, S., 283[454] (245, 263)
Zalevskaya, O. A., 281[323] (242), 283[435] (244), 283[436] (244, 245), 283[446] (245)
Zambonelli, L., 520[142] (444), 529[710] (512)
Zambonelli, Z., 584[100] (545)
Zammit, M. G., 427[44b] (397)
Zanardo, A., 523[287] (462), 525[412] (475), 525[413] (475), 525[414] (475)
Zandona, N., 220[17] (195), 220[19b] (196), 429[149] (409)

Zanello, P., 278[126] (234), 429[187a] (417), 429[187b] (417)
Zanotti, G., 524[381] (472)
Zanotto, G., 223[173a] (218)
Zanotto, L., 223[154] (215), 223[171] (217), 223[173c] (218), 277[27] (228), 277[28] (228, 230), 387[267] (350), 430[213] (423), 430[218] (424), 430[220] (425), 430[221] (425), 529[670] (508)
Zarati, E. A., 185[266] (134, 138)
Zaw, K., 385[129] (313), 385[130] (313), 385[133] (314), 388[303] (356)
Zebrowski, J. P., 27[176] (21)
Zecca, M., 222[120] (211), 223[173a] (218)
Zegi, Y., 277[68] (230)
Zeigan, D., 186[336] (139, 141)
Zeigler, T., 99[16] (30, 32), 99[17] (30, 32)
Zeile Krevor, J. V., 531[786] (518)
Zeise, W. C., 582[1a] (534), 582[1b] (534), 582[1c] (534)
Zelenka, L., 190[584] (174, 176)
Zenkert, K., 220[27a] (197, 210), 220[27b] (197, 210), 286[652] (252), 390[449] (378), 390[450] (378)
Zennaro, R., 525[417] (475)
Zenneck, U., 182[11] (108), 183[98] (115, 155, 157, 181), 188[423] (151, 180, 181), 190[592] (175, 177), 190[593] (175, 179), 190[618] (178), 190[619] (178), 191[625] (180, 181), 191[639] (181), 191[640] (181)
Zeppenfeld, E., 185[260] (131, 132)
Zerbib, P., 279[175] (236), 283[487] (245, 263)
Zerner, M. C., 384[26] (292)
Zettel, B., 277[56] (230, 231)
Zetterberg, K., 285[584] (249), 285[585] (249, 263), 285[586] (249, 263, 265), 285[589] (249), 386[180] (325, 326), 389[368] (363), 389[369] (364)
Zettlitzer, M., 385[83] (304)
Zhai, W., 427[10] (392, 393)
Zhang, G., 182[54] (111, 123), 182[55] (111, 123)
Zhang, H., 191[636] (180)
Zhang, L., 285[586] (249, 263, 265)
Zhang, Q., 102[157] (42, 56)
Zhang, X., 186[279] (137, 138)
Zhang, Z. Z., 102[157] (42, 56), 102[159] (41, 42, 56), 102[160] (42), 187[406] (150, 151, 155), 187[407] (150, 151, 155)
Zhao, Z., 427[10] (392, 393)
Zhdanov, S. I., 102[169] (50)
Zheng, S., 386[160] (319)
Zhidomirov, G. M., 384[48] (295)
Zhong, Z. D., 428[95a] (403), 428[95b] (403), 428[95c] (403, 408), 429[140] (408), 526[474] (484), 526[493], 526[496] (487), 526[497] (487), 526[500] (487), 583[53] (540, 548, 574, 576)
Zhou, Z.-H., 102[162] (42)
Zhu, H., 182[54] (111, 123), 182[55] (111, 123)
Zhu, L., 390[442] (377)
Ziegler, M. L., 189[556] (172, 173), 427[17] (393, 412)
Ziegler, R. J., 520[131] (443)
Ziegler, T., 182[21] (109), 220[10] (195), 220[11] (195), 386[152] (317), 584[134] (550)
Ziegler, U., 521[199] (452), 526[462] (481)
Zigna, A.-M., 388[312] (357)
Zimmer-Gasser, B., 189[530] (159, 163)
Zimmermann, H., 188[456] (154)
Zink, J. I., 584[89] (544)
Zinsius, M., 286[675] (253)
Zipp, M., 280[235] (239, 266), 289[878] (266)
Zocchi, M., 285[620] (250)
Zoeller, J. R., 285[609] (249)
Zoellner, R. W., 386[156] (318)
Zoet, R., 104[329] (81, 85), 104[333] (81, 86), 530[751] (515)
Zoller, L., 187[416] (151)
Zotti, G., 101[144] (42, 65), 102[163] (42, 47, 65)
Zoutberg, M. C., 281[301] (241), 287[761] (258), 529[703] (511)
Zquiak, R., 283[455] (245, 263)
Zschunke, V. A., 386[182] (327), 386[183] (327)
Zsolnai, L., 185[234] (131)
Zubreichuk, Z. P., 191[626] (180, 182), 191[629] (180, 182)
Zucca, A., 530[775] (517)
Zudin, V. N., 220[36] (197)
Zwecker, J., 183[98] (115, 155, 157, 181), 188[423] (151, 180, 181), 191[625] (180, 181)

Subject Index

JOHN NEWTON
David John (Services), Slough, UK

This Subject Index contains individual entries to the text pages of Volume 9. The index covers general types of organometallic compound, specific organometallic compounds, general and specific organic compounds where their synthesis or use involves organometallic compounds, types of reaction (insertion, oxidative addition, etc.), spectroscopic techniques (NMR, IR, etc.), and topics involving organometallic compounds.

Because authors may have approached similar topics from different viewpoints, index entries to those topics may not always appear under the same headings. Both synonyms and alternatives should therefore be considered to obtain all the entries on a particular topic. Commonly used synonyms include alkyne/acetylene, compound/complex, preparation/synthesis, etc. Entries where the oxidative state of a metal has been specified occur after all the entries for the unspecified oxidation state, and the same or similar compounds may occur under both types of heading. Thus $Cr(C_6H_6)_2$ occurs under Chromium, bis(η-benzene) and again under Chromium(0), bis(η-benzene). Similar ligands may also occur in different entries. Thus a carbene–metal complex may occur under Carbene complexes, Carbene ligands, or Carbenes, as well as under the specific metal. Individual organometallic compounds may also be listed in the Cumulative Formula Index in Volume 14.

Acetyl coenzyme A synthase
  nickel complexes, 91
Alkenes
  hydroformylation
    platinum complex catalysts, 443
  hydrogenation
    platinum complex catalysts, 443
  palladium complexes
    bonding, 294–302
    structure, 294–302
    theory, 294–6
  reactions with metallic palladium, 293, 294
  reactions with organonickel complexes, 62
Alkynes
  cyclodimerization
    nickel catalysts, 66
  insertion into organonickel complexes, 58
  reactions with metallic palladium, 316
  reactions with palladium complexes, 320–3
  Reppe cyclotetramerization, 122
  terminal
    carboxylation, nickel complexes, 10
Alkynes, thio-
  hydroboration
    nickel catalysts, 133
Allenes
  reactions with palladium salts, 341
Allyl alcohol
  oxidation
    palladium catalysis, 314
Allylic compounds
  oxidative addition
    palladium(0) complexes, 343–7
Arsenic ylides
  palladium(II) complexes, 229
5-Azanickelacyclopentan-4-one
  synthesis, 8
Azanickelacyclopentanones
  synthesis, 80

Borate anions, bis(brenzcatechnicato)-
  in organonickel synthesis, 33
Butadiene
  oligomerization
    nickel crotyl complex catalysts, 142
  polymerization
    mechanism, 142
  1,4-*trans*-polymerization
    nickel catalyst, 33
  reaction with phenyl isocyanate
    nickel complexes, 8

Carbene ligands
  platinum complexes, 391–426
Carbenes
  alkyl(alkoxy)-
    nickel complexes, 13
Carbon–hydrogen bonds
  activation
    platinum complexes, 517, 518
Carbon dioxide
  cocyclization with alkenes
    nickel complexes
  cocyclization with alkynes
    nickel complexes
  copolymerization with diynes, 5
Carbon monoxide
  insertion into organonickel complexes, 58
  reaction with organonickel complexes, 62
Carbon suboxide
  nickel complexes, 9
Carbonyl ligands
  platinum complexes, 391–426
Carbynes
  nickel complexes, 12
Chromium complexes
  palladium carbonyls, 204
Cinnamanilide
  reductive elimination, 8
Cobalt complexes
  palladium carbonyls, 205
Cobalt dehydrogenase
  nickel complexes, 91
Cyclam
  nickel complexes, 91
Cycloaddition reactions
  trimethylenemethanepalladium, 373–5
Cycloalkanes
  ring opening
    palladium salts, 341–3
Cyclobut-3-ene-1,2-dione
  reaction with nickel isocyanide complexes, 11
Cyclohexane, methylene-

reaction with palladium dichloride, 331–40
Cyclooctene
reaction with carbon dioxide
nickel complexes, 7
Cyclooctenylcarboxylic acid
β,γ-unsaturated
synthesis, 7
Cyclopalladation
palladium(II) complexes, 242–8
Cyclopent-4-ene-1,2,3-trione, 4,5-diphenyl-
reaction with nickel isocyanide complexes, 11
Cyclopropane, *gem*-dibromo-
ring opening carbonylation
nickel carbonyl induced, 4
Cyclopropenone, diphenyl-
addition to isothiocyanates
nickel carbonyl promoted, 4

Decarbonylation
acylplatinum(II) complexes, 408
Diazepam
synthesis, 245
Dienes
oligomerization
palladium catalysts, 371–3
reactions with palladium salts, 331–40
telomerization
palladium catalysts, 371–3
1,2-Dienes
reactions with palladium salts, 341
1,3-Dienes
sulfonylpalladation, 336
Dipalladation
tetraethylxylenediamines, 245

Ethene
adsorption onto palladium, 293
dimerization
nickel catalysts, 40
Ethene, 1,1-difluoro-
reaction with phenyl isocyanate, 8
Ethyne
polymerization
nickel ylide catalysts, 36
Ethyne, phenyl-
oligomerization
nickel catalyst, 30

Fischer-type carbene complexes
nickel, 12
1-Hexene
amination by diethylamine
palladium catalyst, 311

Hydroboration
thioalkynes
nickel catalysts, 133
Hydrocarbons
unsaturated
electrocatalytic carboxylation, 9
Hydrocyanation
nickel catalysts, 122
Hydroformylation
platinum complex catalysts, 461

Iron complexes
palladium carbonyls, 205
Isocyanates
alkene functionalization, 7
coupling with alkenes
nickel complexes, 8
Isocyanide ligands
platinum complexes, 391–426
Isocyanides
insertion into organonickel complexes, 61

Manganese complexes
palladium carbonyls, 205
Mercury complexes
palladium carbonyls, 207
Methyl coenzyme M reductase
nickel complexes, 91
Methylenomycin B
synthesis, 4
Molybdenum complexes
palladium carbonyls, 204

Nickel alkoxides
carbonylation, 3
Nickel alkyls
carbonylation, 3
Nickel, bipyridyl(cyclooctadienyl)-
synthesis, 119
Nickel carbonyl
reaction with diboroles, 177
Nickel cluster complexes, 19–24
carbenes, 23, 24
carbonyl, 15, 19–22
carbynes, 23, 24
cyclopentadienyl, 170–4
homonuclear, 171, 172
heterometallic, 95–8
heteronuclear
cyclopentadienyl, 172–4
homometallic, 93–5
homonuclear
synthesis, metal vapor techniques, 171
hydrido, 19, 22
iodo-bridged, 19, 22
isocyanide, 22, 23
methylene-bridged, 15
Ni–C σ-bonds, 93–8
trinuclear
paramagnetic, 171
sulfide-bridged, 171
Nickel complexes, 107–82
acyl, 124
aldehydes, 127
alkaneimidoyl, 125
alkenes, 113–25
thermal decomposition, 114
alkoxyvinyl, 47
alkyl, 37–40
alkyl aryl, 47
alkyl-methylallyl, 143
alkyl phenyl ketene
preparation, 9
alkynes, 129–36
physical studies, 129–31
theory, 129–31
allenes, 121
allyl, 136
mass spectroscopy, 137
molecular orbital calculations, 137
physical studies, 137
pyrazolyl gallates, 140
reactions, 137–44
theory, 137
thermolysis, 140
allylic, 132
anionic alkynes, 133
arene, 33
π-arenes, 174–6
reactions, 174–6
synthesis, 174–6
aryl, 37–40
aryl azides, 125
benzyl, 145, 146
benzyne, 133, 134
binuclear, 15–19
carbonyl, 15, 16
methylenediphosphine bridged, 16
tetramethyleneethane bridged, 15
biological significance, 91, 92
π-bonded ligands, 30–6
boracyclopentadienyl, 156
boranes, 180–2
bridging alkynes, 15
butadiene, 114
carbenes, 1–24, 47
carbon σ-bonded, 29–98
carbon dioxide, 5–10, 128
carbon disulfide, 129
carbon suboxide, 122
carbonyl, 1–24
carboranes, 180–2
carboxymethyl butadienes, 121
crotyl, 142
cyclam, 9
cycloalkenyl, 142
cyclobutadiene
cationic, reduction, 148
cyclooctenyl, 124
cyclopentadienyl, 149
cyclic voltammetry, 152
electrochemistry, 152
photolysis, 163
cyclopropylium, 146, 147
DAD-Ar_2, 118
decamethylbimetallic cyclopentadienyl, 154
dialkynes, 134, 135
diazabutadiene, 113, 118
diethyl-1,4-dimethyl-1,4-diboranaphthalene, 124
dimetallic
alkynes, 135, 136
1,1-dimethylsilole, 124
dinuclear
aminocarbyne, 17, 18
carbenes, 18, 19
diazoalkanes, 19
isocyanide, 17, 18
nitrosyl, 17, 18
diphosphaallyl, 144, 145
diphosphines, 113
diphosphorus, 136
19-electron, 162
electron pair donor ligands, 14
ethene, 110–13
ethoxycarbonyl
preparation, 4

Fischer-type carbenes, 111
fullerene, 125
fulvalene, 151
heteroalkenes, 125–9
heteroallyl, 144, 145
heterocyclobutadiene, 149
indenyl, 156
isocyanides, 1–24
isocyanoethanol, 11
isoquinolone, 12
ketenes, 120, 128
ketones, 127
ketophosphonates, 128
ligand-free, 30–6
η^2-ligands, 108–36
ab initio calculations, 109
physical studies, 109, 110
theory, 109, 110
XPES studies, 109
η^3-ligands, 136–47
η^4-ligands, 147–9
reactions, 148, 149
synthesis, 148, 149
theory, 147
η^5-ligands, 149–74
dinuclear, 167–70
gas-phase studies, 150
heterodinuclear, 169, 170
ion cyclotron resonance spectroscopy, 150
magnetic studies, 150–2
mass spectroscopy, 150
physical studies, 150–4
spectroscopy, 150–2
theory, 150–4
thiolato-bridged, 170
mercury, 11
methylenecyclopropane, 119
monoalkynes, 132, 133
monoazadienes, 118
naphthalene, 123
norbornadiene, 119
pentafluorophenyl, 122
pentafulvalenes, 168
phosphaalkene, 126
phosphaalkynes, 136
phosphine cyclopentadienyl, 160
phosphinoenolates, 8
polyalkynes, 134, 135
polydentate arsenic ligands, 122
polymetallic
alkynes, 135, 136
reactions with carbon dioxide and alkynes, 121
silylene, 125
tethered alkenes, 117
trimethylsiloxyallyl, 143
vinyl halides, 119
xylyl, 146
Nickel(0) complexes
dinuclear
dmpm-bridged, 16
dppm-bridged, 16
ketenes, 9
polyalkene, 119
reactions with alkenes, 80
reactions with isocyanates, 80
trimethoxysilylalkene, 119
trimethylsilylalkene, 119
Nickel(II) complexes
carbonyl
structure, 2
methyl(aryloxide)
reactions, 4
Nickel, dicarbonyldicyclopentadienyldi-
chemistry, 167, 168
Nickel effect
ethene dimerization, 34
Nickel polydecker complexes
π-bonded, 176, 177, 180
physical properties, 177
synthesis, 177–80
theory, 177
Nickel powder
preparation, 30
Nickel sandwich complexes
diborolene, 176, 177, 180
fulvalene, 155, 156
polysubstituted cyclopentadienyl, 155, 156
Nickel tetracarbonyl
photolysis, 2
preparation, 1
reactions with alkyl isocyanides, 10
substitution reactions, 2
Nickel triple decker complexes
slipped, 177
Nickel, tris(norbornene)-
structure, 119
Nickel ylides, 30–6
catalyst
ethene polymerization, 36
Nickelacarbocyclic complexes
chemistry, 73–80
properties, 78–80
reactions, 78–80
synthesis, 73–8
Nickelacycles
formation from carbon dioxide, 6
Nickelacyclic complexes
diazabutadiene, 80
Nickelacyclobutane
synthesis, 30
Nickelacyclobutenones
formation, 15
Nickelaheterocyclic complexes
chemistry, 80–91
Nickel–carbon bonds
enthalpies, 30
Nickel–cobalt polydecker complexes, 177
Nickelocene
chemistry, 149, 154–6
electrochemistry, 152–4
pentamethyl, 156
peralkylation, 151
reactions with dialkylphosphine sulfides, 163
reactions with ylides, 165
Nickel–platinum complexes
synthesis, 98
Nickel–zinc clusters, 173
Nitrogen ylides
palladium(II) complexes, 229
Nitrous oxide
nickel complexes, 9

Organoaluminum compounds
in organonickel complex synthesis, 41
Organocadmium compounds
in organonickel complex synthesis, 41
Organolithium compounds
in cyclopentadienyl alkyl nickel complex synthesis, 66
in organonickel complex synthesis, 41
Organomagnesium complexes
in cyclopentadienyl alkyl nickel complex synthesis, 66
in organonickel complex synthesis, 41
Organomercury compounds
in organonickel complex synthesis, 41
Organonickel complexes
allyl alkyl, 65, 66
allyl aryl, 65, 66
cyclopentadienyl alkyl, 66–73
properties, 70–3
reactions, 70–3
synthesis, 66–70
cyclopentadienyl aryl, 66–73
properties, 70–3
reactions, 70–3
synthesis, 66–70
properties, 64, 65
reactions, 56–63
alkyl group displacement, 62, 63
aryl group displacement, 62, 63
Ni–C bond generated, 58–62
Ni–C bond not involved, 56, 57
structures, 64, 65
synthesis, 41–56
Organopalladium carbonyls
properties, 196
Organopalladium(II) complexes
decomposition, 265–7
isomerization, 265–7
metathesis, 261–5
reactions, 261–71
Organoplatinum(IV) polymers
preparation, 464
Organothallium compounds
in organonickel complex synthesis, 41
Organoytterbium compounds
in organonickel complex synthesis, 41
Oxanickelacyclopentanones
synthesis, 90
1-Oxa-3-platinacyclobutane
reactions, 481
Oxidative addition
allylic compounds
to palladium(0) complexes, 343–7
nickel complex synthesis, 47–53

Palladacycles
formation, 250
Palladacycloalkane complexes
formation, 240
Palladacyclobutane
formation, 237
Palladacyclobutarene

structure, 237
Palladacyclopentane
decomposition, 274
Pallada(IV)cyclopentane
complexes
decomposition, 273
Palladium
carbon σ-bonds
formation, 233–61
metallic
reactions with alkenes, 293, 294
reactions with alkynes, 316
reactions with arenes, 381
Palladium clusters
alkyne
synthesis, 320
asymmetric carbonyl caps, 202
Palladium complexes
alkenes, 293–316
electrochemistry, 300
electronic spectroscopy, 296
NMR, 297–300
nucleophilic attack on alkene, 310
reactions, 310–16
substitution reactions, 310
substitution reactions, alkene, 310
synthesis, 302–10
thermodynamics, 300–2
vibrational spectroscopy, 296, 297
alkynes, 316–23
bonding, 316, 317
structure, 316, 317
synthesis, 317–20
theory, 316, 317
alkynyl, 230–2
dinuclear, 231
reactions, 231, 232
structures, 231, 232
synthesis, 230, 231
allyl
polynuclear, 337
allylic, 320–3
allyl transfer, 349
bonding, 323–8
charge reversal, 357, 358
conjugated dienes, addition reactions, 370, 371
halide exchange, 349, 353, 354
halogenation, 349, 356, 357
NMR, 325–8
nucleophilic attack, 349
nucleophilic attack at carbon, 349, 360–71
nucleophilic attack at carbon, regiochemistry, 361–7
nucleophilic attack at carbon, stereochemistry, 367, 368
nucleophilic attack at palladium, 350–2
oxidation, 349, 355, 356
photochemical decomposition, 354, 355
photoelectron spectroscopy, 324, 325
reactions, 349–71
reactions with acids, 349
reactions with carbon monoxide, 349, 358, 359
reactions with dienes, 349
reduction, 349, 357, 358
structure, 323–75
synthesis, 328–48
theory, 323, 324
thermal decomposition, 349, 354, 355
arene, 381–3
butadiene
NMR, 299
carbene ligands, 193–219
carbenes, 216–19
heterocyclic, synthesis, 217
novel, 219
structures, 218, 219
synthesis, 216–18
carbon σ-bonded, 225–76
carbon π-bonded, 291–383
carbonyl halides
carbonyl-bridged, 196
chloride-bridged, 196
synthesis, 195, 196
carbonyl ligands, 193–219
carbonyls
carbonylation, allylic compounds, 208
catalysis, 207, 208
catalysis, reactions of the carbonyl ligands, 207, 208
metallic derivatives, 204–7
cyclobutadiene
synthesis, 308
1,5-cyclooctadiene
vibrational spectroscopy, 296
cyclopentadienyl, 375–80
bridged, synthesis, 377, 378
reactions, 378–80
synthesis, 375–7
decanuclear carbonyls, 204
dibenzylideneacetone
NMR, 299
dicyclopentadiene
nucleophilic attack, 310
dinuclear carbonyls, 196–8
A-frame, 197
carbonyl-bridged, 197
water soluble, 198
4,7-diphenyl-1,10-phenanthroline
electrochemistry, 300
divinylcyclohexane
structure, 311
dodecanuclear carbonyls, 204
hexadecanuclear carbonyls, 204
hexanuclear carbonyls, 202, 203
high nuclearity carbonyls, 203, 204
homoleptic carbonyls, 194
catalysis, theoretical studies, 194, 195
stability, theoretical studies, 195
theoretical studies, 194, 195
isocyanide ligands, 193–219
isocyanides, 208–15
trinuclear, 209
multinuclear carbonyls, 198–204
norbornadiene
vibrational spectroscopy, 297
octanuclear carbonyls, 203
pentanuclear carbonyls, 202
radicals, 272
reaction with alkynes, 251–4, 320–3
reaction with benzene, 259
reaction with carbon monoxide, 255–7
reaction with diazoacetonitrile, 259
reaction with diisocyanoarenes
oligomers, 258
reaction with hexafluorodiazopropane, 259
reaction with isocyanides, 257, 258
tetranuclear carbonyls, 201, 202
degradation, 201
synthesis, mechanism, 201
trieicosanuclear carbonyls, 204
trinuclear carbonyls, 198–201
coordination chemistry, 199
oxidative addition reactions, 200
reactions with alkynes, 200
structure, 198, 199
synthesis, 198, 199
vinyl ether
structure, 305
Palladium(0) complexes
alkyne
synthesis, 317, 318
diene
synthesis, 307
isocyanides, 208, 209
monoene
synthesis, 302, 303
Palladium(I) complexes
alkyne
synthesis, 319
carbon σ-bonded, 226, 227
diene
synthesis, 307
hydrido-bridged, 226
isocyanides, 210
bridging isocyanides, 210
Palladium(II) complexes
alkyl, 233–72
alkyne
synthesis, 320
aryl, 233–72
cyclobutadiene
synthesis, 309, 310
diene
synthesis, 308, 309
diorgano
reactions, 266, 267
formation
from palladium(0) complexes, 233–7
from palladium metal, 233–7
organoaluminum reagents, 238

organolithium reagents, 239
organomagnesium reagents, 240
organomercury reagents, 238
organometal reagents, 238–41
organosilicon reagents, 238
organotin reagents, 238
isocyanides, 211–15
multiple insertions, 213, 214
structures, 215
monoalkyl
stability, 265
monoene
cationic, 5-coordinate, synthesis, 306
neutral, 4-coordinate, synthesis, 304, 305
neutral, 5-coordinate, synthesis, 306
synthesis, 304–6
reaction with acids, 267, 268
reaction with alcohols, 267, 268
reaction with alkenes, 248–51
reaction with amines, 267, 268
reaction with carbon dioxide, 271
reaction with carbon disulfide, 271
reaction with halogens, 268–70
reaction with molybdenum peroxide, 268
reaction with nitric oxide, 271
reaction with organohalides, 268–70
reaction with oxygen, 271
reaction with peracids, 267, 268
reaction with peroxides, 267, 268
reaction with sulfur dioxide, 271
reaction with thiols, 267, 268
reduction, 270
structures, 271, 272
vinyl alcohol
synthesis, 305
Palladium(IV) complexes
chemistry, 272–6
hydrocarbyl
isolation, 272
intramolecular coordination, 273–5
phenyl
formation, 274
reactions, 276
structures, 276
sulfur donor ligands, 275, 276
tripodal nitrogen ligands, 275, 276
Palladium(IV), pentafluorophenyl-
synthesis, 272
Palladium salts
reactions with dienes, 329
reactions with monoenes, 328, 331–5
Palladium surfaces
carbon monoxide binding, 194
Palladium trifluoroacetate
reactions with alkenes, 331, 336
Palladium, trimethylenemethane-
cycloaddition reactions, 373–5
Palladium(II) ylides
carbon σ-bonded, 227–30
structures, 230
Phosphines
reductive elimination
organonickel complexes, 62
Phosphorus ylides
palladium(II) complexes, 227, 228
3-Platinaazetidine
preparation, 478
Platinacycloalkanes, 477–83
preparation, 478–80
properties, 481
reactions, 481–3
structures, 481, 500
Platinacyclobutane sulfones
synthesis, 480
Platinacyclobutane sulfoxides
synthesis, 480
Platinacyclobutanes
photolysis, 505
preparation, 497, 498
reactions, 504
structure, 481
thermal decomposition, 505
thermolysis, 505
2-Platinacyclobutanone
preparation, 479
Platinacyclobutanones
structure, 481
Platinacyclohexane
preparation, 478
Platinacyclopentadiene
preparation, 480, 493
Platinacyclopentane
preparation, 478, 505
Platinacyclopentanediones
preparation, 480
Platinum complexes
acyl, 560
alkenes, 534–64
alkynes, 564–72
allenes, 558–60
allenyl, 577
preparation, 577
structure, 577
allyls, 573–9
preparation, 573–5
reactions, 576
structure, 575, 576
theory, 576
arenes, 582
preparation, 582
reactivity, 582
structure, 582
arynes, 569
preparation, 569
reactivity, 569
structure, 569
benzyl
preparation, 577
reactivity, 577
structure, 577
bimetallic alkenes, 561–3
bimetallic alkynes, 570–2
butadienyl
preparation, 577
reactivity, 577
structure, 577
carbenes, 419–26
preparation from electrophilic attack on coordinated DF_3, 420
preparation from nitrogen ylides, 420
preparation from thiolate amines, 402, 421
carbon σ-bonded, 431–518
carbon π-bonded, 533–82
carbonyl anions
polynuclear, 397–9
polynuclear, magnetic moments, 398
polynuclear, mass spectrometry, 398
polynuclear, NMR, 398
polynuclear, reactions, 399
polynuclear, spectroscopy, 398
polynuclear, structures, 398
carbonyl, isocyanide and carbene ligands, 391–426
carbonyl ligands, 392–411
C,N-carbenes, 402, 421
C,O-carbenes, 402–21, 423
cumulenes
preparation, 558, 559
reactivity, 558, 559
structure, 558, 559
cyclic 5-membered N,O-carbenes, 425
cyclobutadiene, 579
preparation, 579
reactivity, 579
structure, 579
cyclopentadienyl, 579–82
diselenes, 560
fullerenes, 561
preparation, 561
reactivity, 561
structure, 561
heterobimetallic alkenes
preparation, 563
reactivity, 563
structure, 563
heterobimetallic alkynes
preparation, 571, 572
reactivity, 571, 572
structure, 571, 572
heterocyclopentadienyl, 581, 582
preparation, 581, 582
structure, 581, 582
heterometallic alkenes
preparation, 563, 564
reactivity, 563, 564
structure, 563, 564
heterometallic alkynes
preparation, 572
reactivity, 572
structure, 572
homobimetallic alkenes
preparation, 561, 562
reactivity, 561, 562
structure, 561, 562
homobimetallic alkynes
preparation, 570, 571
reactivity, 570, 571
structure, 570, 571
homometallic alkenes

preparation, 563
reactivity, 563
structure, 563
homometallic alkynes
preparation, 572
reactivity, 572
structure, 572
hydrocarbyls, 577
isocyanides, 411–19
ketenes
preparation, 559, 560
reactivity, 559, 560
structure, 559, 560
metallated arsenic ligands
preparation, 507–9
reactions, 514, 515
metallated ligands, 507–17
preparation, 507–12
properties, 512–14
reactions, 514–17
structures, 512–14
metallated nitrogen ligands
preparation, 510, 511
reactions, 515–17
metallated phosphorus ligands
preparation, 507–9
reactions, 514, 515
monoalkenes
monodentate, 534–51
polydentate, 551–3
monoalkynes, 564–9
N,N-carbenes, 423, 424
N,O-carbenes, 425, 426
O,O-carbenes, 426
palladium carbonyls, 205
phosphaalkenes, 560
phosphaalkynes, 569, 570
preparation, 569, 570
reactivity, 569, 570
structure, 569, 570
polyalkenes
conjugated, 560, 561
conjugated, preparation, 560, 561
conjugated, reactivity, 560, 561
conjugated, structure, 560, 561
polyenes
unconjugated, 553–8
polymetallic alkenes, 563, 564
polymetallic alkynes, 572
propargyl
preparation, 577
reactivity, 577
structure, 577
S,S-carbenes, 426
trimethylenemethane, 578, 579
preparation, 578, 579
reactivity, 578, 579
structure, 578, 579
Platinum(0) complexes
carbonyl clusters
NMR, 395
reactions, 395–7
replacement of carbonyl by sulfur dioxide, 396
spectroscopy, 394, 395
tetranuclear, NMR, 395
carbonyl ligands
monomeric, 392
polynuclear, 392–7
polynuclear, structures, 393, 394
polynuclear, synthesis, 392, 393
isocyanides
mononuclear, 411
polynuclear, 411–13
monoalkenes
monodentate, 535–40
monodentate, displacement reactions, 538
monodentate, dissociation, 538
monodentate, oxidative addition, 540
monodentate, preparation, 535–7
monodentate, protonation, 538
monodentate, reactions, 538–40
monodentate, structure, 537, 538
polydentate, 551
monoalkynes, 565–8
preparation, 565, 566
reactions, 568
structure, 566–8
polyenes
unconjugated, 553, 554
unconjugated, preparation, 553
unconjugated, reactions, 554
unconjugated, structure, 553
Platinum(I) complexes
carbon σ-bonded, 432, 433
carbonyls, 399–401
ligand substitution, 401
triangular dppm-bridged clusters, 401–4
cyclopentadienyl
preparation, 580, 581
reactivity, 580, 581
structure, 580, 581
dppm-bridged dimers, 432
isocyanides, 413–15
IR spectroscopy, 414
NMR, 414
metal–metal bonds, 432
Platinum(II) complexes
alkenyls, 483–9
preparation, 483–6
properties, 487
reactions, 487–9
structure, 487
alkyls, 433–66
C bonds, stability, 445
coordinated ligands, reactions, 466
dppm, 441, 442
electrophilic cleavage, 458
homolytic cleavage, 457
insertion reactions, 460–2
isomerization, 458–60
luminescence spectra, 453
NMR, 445
oxidative addition, 454, 463–6
photoelectron spectroscopy, 452
photoisomerization, 460
preparation, 433–44
preparation, bimetallic compounds, 444
preparation, displacement of weakly bound ligands, 439–41
preparation, insertion reactions, 442, 443
preparation, main group organometallics, 436–9
preparation, oxidative addition, 434–6
properties, 445–53
reactions, 453–66
reductive elimination, 453
spectroscopy, 445–53
structure, 445–53
substitution, 458–60
thermal decomposition, 453–8
x-ray diffraction, 445
alkynyls, 490–4
preparation, 490, 491
properties, 491–3
reactions, 493, 494
structure, 491–3
aryl, 433–66
C bonds, stability, 445
coordinated ligands, reactions, 466
dppm, 441, 442
electrophilic cleavage, 458
homolytic cleavage, 457
insertion reactions, 460–2
isomerization, 458–60
luminescence spectra, 453
NMR, 445
oxidative addition, 454, 463–6
photoelectron spectroscopy, 452
preparation, 433–44
preparation, bimetallic compounds, 444
preparation, displacement of weakly bound ligands, 439–41
preparation, insertion reactions, 442, 443
preparation, main group organometallics, 436–9
preparation, oxidative addition, 434–6
properties, 445–53
reactions, 453–66
reductive elimination, 453
spectroscopy, 445–53
structure, 445–53
substitution, 458–60
thermal decomposition, 453–8
x-ray diffraction, 445
bis(alkynyl)
preparation, 490
bis(isocyanide), 417, 418
electrochemistry, 418
carbon σ-bonded, 433–96
carbonylation, 460
carbonyls, 404–11
18-electron, 410
chloro
hydroformylation catalysts, 460

hydrogenation catalysts, 460
cyclopentadienyl
preparation, 580, 581
reactivity, 580, 581
structure, 580, 581
dicarbonyls, 409
binuclear, 410, 411
dithiocarbamate
alkyl group isomerization, 460
double carbonylation, 462
haloalkyls, 467–77
preparation, 467–71
preparation, displacement of weakly bound ligands, 468–70
preparation, insertion reactions, 471
preparation, main group organometallics, 467, 468
preparation, oxidative addition, 467
properties, 472, 473, 490
Pt–C bond stability, 472
reactions, 473–7
reactions, other metal complexes, 473–5
spectroscopy, 473, 494
structure, 472, 473
x-ray diffraction, 472
haloaryls, 467–77
preparation, 467–81
preparation, displacement of weakly bound ligands, 468–70
preparation, insertion reactions, 471
preparation, main group organometallics, 467, 468
preparation, oxidative addition, 467
properties, 472, 473
Pt–C bond stability, 472, 494
reactions, 473–7
reactions, other metal complexes, 473–5
spectroscopy, 473
structure, 472, 473
x-ray diffraction, 472
Hg bonds, 444
hydride-bridged
preparation, 444
hydroxy
reactions, 461
isocyanides, 415–19
α-ketoacyl, 434
methyl
chemical vapor deposition, 457
monoalkenes
monodentate, 541–9
monodentate, displacement reactions, 547
monodentate, dissociation, 547
monodentate, migratory alkene insertion, 548
monodentate, NMR, 544
monodentate, nucleophilic addition, 548
monodentate, pentacoordinate, 542–4
monodentate, preparation, 541–4
monodentate, reactions, 547–9
monodentate, structure, 544, 545
monodentate, theory, 550, 551
polydentate, 551–3
monoalkynes, 568, 569
preparation, 568
reactions, 569
structure, 568
theory, 569
monocarbonyl, 405–8
cis–trans isomerization, 405
electronic spectroscopy, 406
ligand replacement reactions, 407
migratory insertion reactions, 407, 408
spectroscopy, 406
structures, 406
synthesis, 405, 406
monoisocyanides, 416, 417
pentafluorophenyl
preparation, 468
polyenes
unconjugated, 554–8
unconjugated, diene displacement, 557
unconjugated, insertion reactions, 558
unconjugated, migration reactions, 558
unconjugated, nucleophilic addition, 557
unconjugated, pentacoordination, 554
unconjugated, preparation, 554
unconjugated, reactions, 556–8
unconjugated, redox reactions, 558
unconjugated, structure, 554–6
preparation
organotin compounds, 438
tetrakis(isocyanide), 419
thienyl
preparation, 484
trifluoromethyl
catalysts, 475, 476
preparation, 467
tris(isocyanide), 419
ylides, 495, 496
preparation, 495
properties, 495
reactions, 496
structure, 495
ytterbium bonds, 444
zirconium bonds, 444
Platinum(III) complexes, 496, 497
Platinum(IV) complexes, 497–507
carbonyls, 411
cyclopentadienyl
preparation, 581
reactivity, 581
structure, 581
glycinato, 506
isocyanides, 419
monoalkenes
polydentate, 553
preparation, 497–500
oxidative addition to platinum(II) complexes, 497
properties, 500–4
structures, 500–4
Platinum, halotrimethyl-
reactions, 498–500
Platinum(0), dichlorocarbonyl(trialkylphosphine)-
reduction, 392
Platinum(0), tricarbonyltetrakis(triphenylphosphine)-
reactions, 397
Platinum(0), tricarbonyltris(ditertiarybutylphenylphosphine)-
reactions, 397
Platinum(II), dichlorodicarbonyl-
reactions with alkynes, 409
Prazepam
synthesis, 245
Protonolysis
organonickel complexes, 62
α-Pyrone
synthesis
nickel complexes, 5
2-Pyrone, tetraethyl-
synthesis, 5

Reductive elimination
organonickel complexes
phosphines, 62

Shell Higher Olefin Process
nickel catalysts, 52
nickel ylide catalysts, 36
Silver(I) reagents
reactions
with platinum(II) haloaryl complexes, 473
Sodium carbonylplatinates
$Na_2[Pt_{30}(CO)_{60}]$
synthesis, 397
Styrene
hydrogenation
palladium catalysts, 294
Sulfonylpalladation
1,3-dienes, 331, 336
Sulfur ylides
palladium(II) complexes, 229
1,4,8,11-Tetraazacyclotetradecane
nickel complexes, 91

Tetralladia cations
formation, 272
Thermolysis
organonickel complexes, 62
Tin complexes
palladium carbonyls, 207
Tungsten complexes
palladium carbonyls, 204

Vanadium complexes

palladium carbonyls, 205

Water–gas shift catalyst
platinum complexes, 393

Ylides
nickel complexes, 14